10 8B	11 1B	12 2B		13 3A	14 4A	15 5A	16 6A	17 7A	18 8A

(1 – 18 is IUPAC system, A,B designation is older U.S. System)

2 He 4.0026 $1s^2$

Main Group (p block)

5 B 10.811 $2s^2 2p^1$	6 C 12.011 $2s^2 2p^2$	7 N 14.007 $2s^2 2p^3$	8 O 15.999 $2s^2 2p^4$	9 F 18.998 $2s^2 2p^5$	10 Ne 20.180 $2s^2 2p^6$
13 Al 26.982 $3s^2 3p^1$	14 Si 28.086 $3s^2 3p^2$	15 P 30.974 $3s^2 3p^3$	16 S 32.066 $3s^2 3p^4$	17 Cl 35.453 $3s^2 3p^5$	18 Ar 39.948 $3s^2 3p^6$

28 Ni 58.693 $4s^2 3d^8$	29 Cu 63.546 $4s^1 3d^{10}$	30 Zn 65.39 $4s^2 3d^{10}$	31 Ga 69.723 $4s^2 4p^1$	32 Ge 72.61 $4s^2 4p^2$	33 As 74.922 $4s^2 4p^3$	34 Se 78.96 $4s^2 4p^4$	35 Br 79.904 $4s^2 4p^5$	36 Kr 83.80 $4s^2 4p^6$
46 Pd 106.42 $4d^{10}$	47 Ag 107.87 $5s^1 4d^{10}$	48 Cd 112.41 $5s^2 4d^{10}$	49 In 114.82 $5s^2 5p^1$	50 Sn 118.71 $5s^2 5p^2$	51 Sb 121.76 $5s^2 5p^3$	52 Te 127.60 $5s^2 5p^4$	53 I 126.90 $5s^2 5p^5$	54 Xe 131.29 $5s^2 5p^6$
78 Pt 195.08 $6s^1 5d^9$	79 Au 196.97 $6s^1 5d^{10}$	80 Hg 200.59 $6s^2 5d^{10}$	81 Tl 204.38 $6s^2 6p^1$	82 Pb 207.2 $6s^2 6p^2$	83 Bi 208.98 $6s^2 6p^3$	84 Po (209) $6s^2 6p^4$	85 At (210) $6s^2 6p^5$	86 Rn (222) $6s^2 6p^6$
110 Ds (271) $7s^2 6d^8$	111	112		114		116		

A few very unstable nuclei each of elements # 111, 112, 114, and 116 have been created using high-energy nuclear accelerators

64 Gd 157.25 $6s^2 4f^7 5d^1$	65 Tb 158.93 $6s^2 4f^9$	66 Dy 162.50 $6s^2 4f^{10}$	67 Ho 164.93 $6s^2 4f^{11}$	68 Er 167.26 $6s^2 4f^{12}$	69 Tm 168.93 $6s^2 4f^{13}$	70 Yb 173.04 $6s^2 4f^{14}$
96 Cm (247) $7s^2 5f^7 6d^1$	97 Bk (247) $7s^2 5f^9$	98 Cf (251) $7s^2 5f^{10}$	99 Es (252) $7s^2 5f^{11}$	100 Fm (257) $7s^2 5f^{12}$	101 Md (258) $7s^2 5f^{13}$	102 No (259) $7s^2 5f^{14}$

eGrade Plus

www.wiley.com/college/olmsted
Based on the Activities You Do Every Day

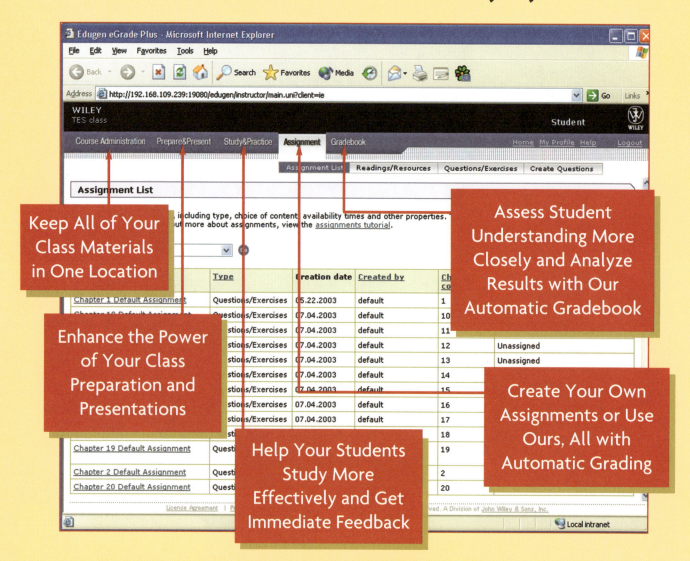

Keep All of Your Class Materials in One Location

Enhance the Power of Your Class Preparation and Presentations

Help Your Students Study More Effectively and Get Immediate Feedback

Assess Student Understanding More Closely and Analyze Results with Our Automatic Gradebook

Create Your Own Assignments or Use Ours, All with Automatic Grading

All the content and tools you need, all in one location, in an easy-to-use browser format.

Choose the resources you need, or rely on the arrangement supplied by us.

Now, many of Wiley's textbooks are available with eGrade Plus, a powerful online tool that provides a completely integrated suite of teaching and learning resources in one easy-to-use website. eGrade Plus integrates Wiley's world-renowned content with media, including a multimedia version of the text, PowerPoint slides, and more. Upon adoption of eGrade Plus, you can begin to customize your course with the resources shown here.

See for yourself!

Go to www.wiley.com/college/egradeplus for an online demonstration of this powerful new software.

Students,
eGrade Plus Allows You to:

Study More Effectively

Get Immediate Feedback When You Practice on Your Own

Our website links directly to **electronic book content,** so that you can review the text while you study and complete homework online. Additional resources include **self-assessment quizzing** with detailed feedback, **Interactive Learningware** with step by step problem solving tutorials, and **interactive dialogs** to help you review key topics.

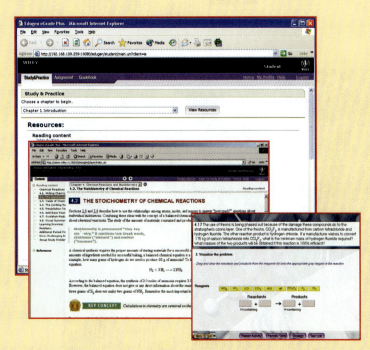

Complete Assignments / Get Help with Problem Solving

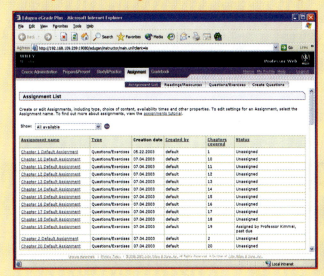

An **"Assignment"** area keeps all your assigned work in one location, making it easy for you to stay on task. In addition, many chapter problems are **linked** to the relevant section of the **electronic book,** providing you with a text explanation to help you conquer problem-solving obstacles as they arise.

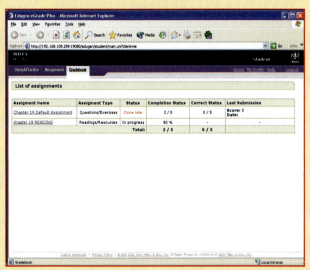

Keep Track of How You're Doing

A **Personal Gradebook** allows you to view your results from past assignments at any time.

About the Authors

(Photo courtesy of the Friends of the Fullerton Arboretum.)

John Olmsted III is Professor Emeritus of Chemistry at California State University, Fullerton, from which he retired in 2003 after nearly 40 years of teaching and research in general and experimental physical chemistry. John was honored as the CSUF Outstanding Professor in 1997-98 and served as department chair from 1998 to 2001. In addition to 25 years at CSUF, he taught for 12 years at the American University of Beirut. He had visiting teaching/research appointments at UCLA and the University of North Carolina at Chapel Hill and did research at the Max-Planck-Institut für Biophysikalische Chemie (Göttingen, Germany), the University of California at San Diego, and Sandia National Laboratory in Albuquerque, NM.

John received his BS degree in chemistry from Carnegie Institute of Technology (now Carnegie-Mellon University) and his PhD in physical chemistry from UC Berkeley, where he also did postdoctoral work at the Lawrence Berkeley Laboratory. He has more than 30 refereed research publications and has also published regularly on chemical education topics in the *Journal of Chemical Education*.

In his retirement, besides continuing to write chemistry textbooks, John keeps busy with his interests in gardening, photography, and the philosophy of chemistry. He and his wife Eileen enjoy traveling, dancing, and visiting with their three married children and two grandchildren.

Greg Williams is an Adjunct Professor of Chemistry at the University of Oregon. He earned an undergraduate degree in chemistry at UCLA and a Ph.D. in inorganic chemistry at Princeton University. He has taught and conducted research at the University of Oregon, California State University, Fullerton, UCLA, and the University of California, Irvine. Outside the classroom, Greg's professional work is concentrated on developing graphics, digital animation, and interactive multimedia for teaching chemistry.

When he is not teaching or writing about chemistry, Greg can be found somewhere in the western United States, backpacking, climbing, skiing, fly fishing, or kayaking. He also sings low bass with the Eugene Vocal Arts Ensemble. Greg lives in Eugene, Oregon, with his wife Trudy Cameron, a Professor of Economics at the University of Oregon, and their daughters, Casey and Perry. He absolutely insists on enjoying life.

(Photo courtesy of Fred E. Skinner.)

CHEMISTRY

FOURTH EDITION

John Olmsted III

California State University, Fullerton

Gregory M. Williams

University of Oregon

WILEY

John Wiley & Sons, Inc.

ACQUISITIONS EDITOR	Kevin Molloy
SENIOR DEVELOPMENT EDITOR	Ellen Ford
PROJECT EDITOR	Jennifer Yee
SENIOR MEDIA EDITOR	Martin Batey
PRODUCTION EDITOR	Barbara Russiello
MARKETING MANAGER	Amanda Wygal
DESIGN DIRECTOR	Harry Nolan
TEXT DESIGNER	Delgado and Company
COVER DESIGN	Harry Nolan
ILLUSTRATION EDITOR	Sigmund Malinowski
ART MANAGEMENT	Edward T. Starr & Associates
ELECTRONIC ILLUSTRATIONS	Precision Graphics
MOLECULAR ART	Gregory Williams, Davi Erickson
VISUAL SUMMARIES	Gregory Williams, John Olmsted III
SENIOR PHOTO EDITOR	Jennifer MacMillan
PHOTO RESEARCHER	Elyse Rieder
FRONT AND BACK COVER PHOTOGRAPHS	*Red coffee beans on plant:* Fernando Bueno/ The Image Bank/Getty Images *Micrograph of caffeine:* Michael W. Davidson at Florida State University *Roasted coffee beans:* Nick Gunderson/Stone/ Getty Images *Decaffeination processing:* Maximilian Stock LTD/Phototake
BACK COVER PHOTOGRAPH	*Coffee mug:* Andy Washnik

This book was set in 10.5/12 Bembo by Progressive Information Technologies.

This book is printed on acid free paper. ∞

Library of Congress Cataloging-in-Publication Data

Olmsted, John.
  Chemistry/John Olmsted III, Gregory M. Williams. — 4th ed.
    p. cm.
  ISBN 0-471-47811-3 (CLOTH/CD-ROM)
  1. Chemistry. I. Williams, Gregory M. II. Title.

QD33.2.O38 2006
540—dc22

2004059800

Printed in the United States of America
10 9 8 7 6 5 4 3 2

Preface

According to our dictionary, a preface serves to tell the subject, purpose, and plan of what follows. The subject of this book is obvious from its title, and the general plan is clear from the table of contents. What remains defines our goal in this preface: to explain our purpose and clarify the plan. We want to help our potential readers discover how the fourth edition of *Chemistry* can guide students to a confident mastery of the fundamentals of chemistry.

THE MOLECULAR PERSPECTIVE: INFORMING QUANTITATIVE REASONING

The purpose of our book is to describe the fundamentals of chemistry as a chemist thinks about the subject. Matter is composed of atoms and molecules, and chemists explain the behavior of matter by describing how atoms and molecules behave and interact. We believe that students will master chemistry more quickly and thoroughly when they grasp this molecular perspective; hence our general theme: "Think molecules."

When we published our first edition, no major general chemistry textbook emphasized the molecular approach. Now almost every textbook shows some molecular pictures. *Our text remains unique, however, in reinforcing each chemical topic with molecular-level explanations.* Our molecular approach to concepts appears throughout the book, but it can be seen most clearly in the following chapters:

- Chapter 4, where we emphasize the molecular underpinnings of stoichiometry;
- Chapter 5, where we introduce the ideal gas equation by describing how molecular kinetic behavior varies with macroscopic variables;
- Chapter 15, where we develop chemical kinetics starting with reaction mechanisms;
- Chapters 17 and 18, where we emphasize the importance of identifying the species in solution as the first step in understanding aqueous chemical equilibria;
- Chapter 19, where we present molecular descriptions of electrode surfaces to support the chemical basis of electrochemical cells.

Because chemists work quantitatively just as frequently as they think molecularly, we use the molecular approach to facilitate successful quantitative reasoning, not to replace it. To see one example of this approach, explore our presentation (Section 16.6) of the species in aqueous solution as an essential prologue to stoichiometric and equilibrium calculations involving solution phase reactions. Look, too, at how we introduce the ideal gas equation: We use a qualitative description of how gaseous molecules behave to develop the quantitative statement (Section 5.2).

EMPHASIS ON PROBLEM SOLVING

Students usually attempt to solve problems using easy algorithms. Instructors want their students to reason through problems rather than relying mindlessly on algorithms. We believe there is validity to both views, because whereas reasoning is at the heart of any science, every experienced scientist regularly employs algorithms as "short-cuts." Reflecting this view, we have developed step-wise approaches to problem-solving that blend sound reasoning with the efficient application of algorithms. We use a seven-step problem-solving template consistently for quantitative problems, including a version designed specifically for attacking equilibrium problems. We also provide step-by-step procedures for the construction of Lewis structures and for the balancing of redox reactions. Beyond this, for each Example Problem we provide a "strategy" of attack, designed to encourage students to reason their way to solutions rather than search aimlessly for algorithms.

NEW IN THE FOURTH EDITION

Although chemistry is, in many ways, a mature science, our understanding of how best to teach chemistry continues to undergo significant changes. The challenge to authors of general chemistry textbooks is to incorporate contemporary teaching and learning strategies while maintaining the established core of subject matter. As was true in the third edition, our electronic materials—web-based—are integrated with the textbook. We have participated closely in the development of the media components, to ensure that these supplements present and develop concepts in the same manner that we use in the textbook.

We have made several organizational changes in this edition:

- What was Chapter 12, *Chemical Energetics,* is now Chapter 6, *Energy and Its Conservation,* providing early, concentrated coverage of this essential topic.
- The chapter on *Spontaneity of Chemical Processes* has been revised both to reflect the earlier placement of energy and to emphasize energy dispersal as an important driving force.
- The content of the molecular structure chapters has been reorganized in the sequence, *Fundamentals of Molecular Structure and Theories of Chemical Bonding.*
- We have given more emphasis to solutions properties by adding a chapter, *Properties of Solutions.*

Coverage of Energy

In previous editions, we introduced energy concepts in "just-in-time" fashion, followed by tandem chapters on energetics and spontaneity. While this approach provided integrated coverage of thermodynamic principles, it also de-emphasized energy in the first part of the text. Chapter 6 of this edition, *Energy and Its Conservation,* emphasizes the importance of energy to chemistry. This chapter has sections on Properties of Energy, Energy Transfers, Energy Changes in Chemical Processes, Measuring Energy Changes: Calorimetry, Enthalpy, and Energy Sources. Much of this material was in Chapter 12 of the Third Edition.

Moving this material forward has two advantages. First, an early chapter that focuses explicitly on energy emphasizes to students the centrality of energy to chemistry. Second, an early description of calorimetry makes it possible to include calorimetry experiments early in the laboratory portion of the course. To accommodate those who prefer a more integrated approach, we have organized Chapter 6 so that it is possible to defer Calorimetry and Enthalpy until later in the course.

Although entropy has historically been linked with disorder, it can also be explained in terms of the dispersal of energy from concentrated to diffuse forms. We believe that both linkages are valid and useful, and we have revised our coverage of spontaneity to describe energy dispersal in detail, while reducing emphasis on entropy-disorder connections.

Molecular Structure

We have reorganized our coverage of molecular structure into two separate parts—facts and theories. Chapter 9, *Fundamentals of Molecular Structure,* concentrates on more concrete, readily grasped, features of molecules. We use Lewis structures to organize the facts of molecular properties: molecular shapes (VSEPR) and bond lengths, strengths, and polarities. Chapter 10, *Theories of Chemical Bonding,* introduces the more abstract concepts of quantum chemistry: orbital overlap, hybridization, molecular orbitals, delocalization, and band theory. The presentation of facts before theory allows students to become comfortable with tangible features of Lewis structures and molecular shapes before confronting the abstractions inherent in bonding theory.

Solutions

Aqueous solutions are central to chemistry, and they also play important roles in biology and geology. In recognition of this, we have expanded our coverage of solution properties, removing the coverage of solutions from our chapter on *Effects of Intermolecular Forces* and creating a new chapter devoted exclusively to *Solution Properties*.

Problems

No matter how extensive and varied the problems in a general chemistry textbook are, there is always room for improvement. This fourth edition incorporates the following improvements:

- An "Extra Practice Exercise" example accompanies and reinforces every Example done in the text.
- New Examples illustrate how to calculate yields in limiting reactant problems (Chapter 4), how to determine reaction orders using the initial rate method (Chapter 15), how to treat polyprotic acid titrations (Chapter 17), and how to analyze the common-ion effect on solubility (Chapter 18).
- Our supporting materials include a greatly expanded electronic homework feature and additional Interactive Learning Ware problem-solving tutorials.
- Chapter problems have been reviewed, and we have replaced those that appear ambiguous or do not emphasize important concepts.
- More elaborate Group Study Problems provide platforms for group learning.
- Additional qualitative and visualization problems encourage the development of conceptual and visual skills.

OTHER UNIQUE FEATURES

The first edition of *Chemistry, the Molecular Science,* introduced features that were not present in other general chemistry textbooks but that we felt reflected how contemporary chemistry should be taught. These resonated well with our users, so we have retained them: molecular approaches to traditional topics (described above), use of organic and biochemical examples equally with inorganic ones, a "rational" approach to Lewis structures, and a separate chapter on macromolecules.

Organic and Biochemical Examples

Whereas most other general chemistry textbooks restrict their use of organic and biochemical molecules to a separate chapter, we feel that general chemistry is just that—a general introduction to principles that apply equally well to all categories of chemical compounds. Moreover, students encounter many organic and biochemical substances in their daily lives, so the use of examples in these areas can help excite student interest. We introduce organic line structures in Chapter 3, allowing us to use interesting molecules as we introduce basic chemical concepts. As one reviewer of our second edition wrote, "Olmsted and Williams cleverly take advantage of this [line structures] by immediately showing students interesting molecules including cholesterol, malathion, various amino acids and hundreds of other examples. Seeing these molecules piques students' interests and, furthermore, makes them feel like scientists because they are learning the language used by practicing chemists."

Macromolecules

Macromolecules, both natural and synthetic, are everywhere in the contemporary world. In recognition of their importance, we devote Chapter 13 to describing macromolecule formation and structure. The chapter builds on the structure and bonding information of earlier chapters, but it can be omitted without loss of continuity. We place this chapter at the natural break between semesters for a two-semester sequence, allowing instructors to cover it at the end of semester one, beginning of semester two, or not at all.

PROVIDING WHAT STUDENTS NEED

Our discussions with students and colleagues, over the many years that we have taught general chemistry, have convinced us that relevance, timeliness, readability, and problem-solving strategies all are important for student success.

Most students who take general chemistry do not intend to major in chemistry. They are pursuing careers in the health sciences, biology, agriculture, engineering, geology, or physics. Many of these students need to be convinced that a sound understanding of chemical principles is essential to their field of interest. Our chapter introductions offer short descriptions of topics that have "real-world" importance and are related to material that appears in the chapter. Boxed material provides further examples under five general themes: *Chemistry and Life, Chemistry and the Environment, Chemistry and Technology, Chemical Milestones,* and *Tools for Discovery.* In addition, some chapters end with sections that may interest a particular group of students but can be omitted without loss of continuity. For examples, see sections 6.6, Energy Sources; 14.6, Bioenergetics; 18.5, Industrial Equilibria; and 20.6, Transition Metals in Biology.

LEARNING RESOURCES

Learners have a variety of styles, which are best served by supplements to textual and oral explanations. The supplementary resources that accompany our fourth edition can be divided into three categories: *features in the textbook, tools designed for the student, and ancillaries addressed to the instructor.*

Text Features

Copious examples: To provide students with early reinforcement, each worked example in the text ends with an *Extra Practice Exercise.* In addition, each section of the text ends with three *Section Exercises* designed to test comprehension of concepts introduced in that section. Answers to the extra practice exercises and the section exercises appear at the end of each chapter.

 Does the Result Make Sense? Every instructor has anecdotes about student answers that are completely nonsensical. To help students learn how to recognize whether or not a result is reasonable, we make a check for sensibility the last step in our quantitative problem-solving strategy. To reinforce this habit, every worked example ends with a brief analysis showing that the result makes sense. We identify this feature using an icon of Auguste Rodin's famous sculpture, *The Thinker.* This provides a visual reminder that thinking is essential to successful problem solving.

 Key Concepts: Certain ideas in chemistry, such as the laws of conservation, are both essential and powerful. We identify these as Key Concepts and emphasize them using green color screens and an icon showing Leonardo da Vinci's best-known drawing, *Vitruvian Man.* This drawing is renowned for illustrating the key concept of ideal proportions as conveyed in the human figure.

 Skills to Master: In addition to learning to think critically and assimilating key concepts, students of chemistry must master many skills. When we describe one of these skills, we flag it with an icon showing a weathervane depicting Diana, the Greek goddess of the hunt who was a skilled archer. This graceful image, created by Augustus St. Gaudens, visually conveys skills mastery.

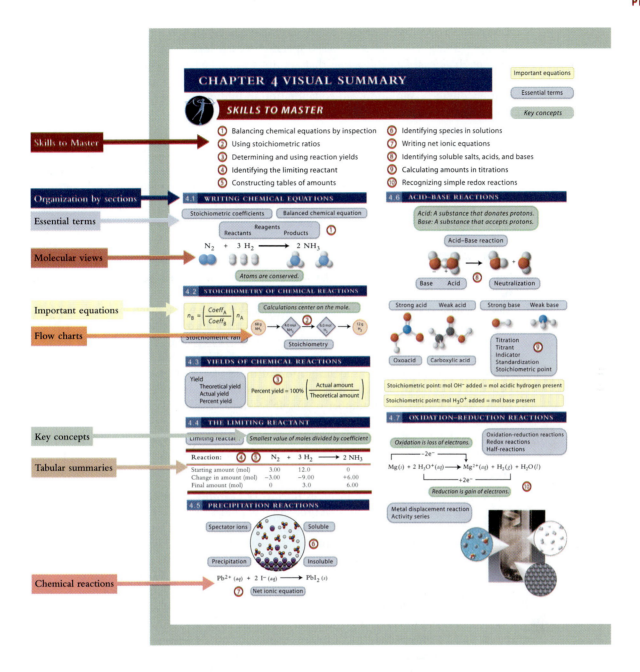

Visual Summaries: Each chapter features a full-page Visual Summary of its most important elements, presented in pictorial fashion. These concept maps provide students with what we believe will be more powerful learning tools than standard Chapter Summaries can provide. The Visual Summaries include the following visual reminders of important ideas developed in the chapter:

- Sequential **Organization by section** that shows the logical flow of the chapter
- A list of the **Skills to Master,** whose numbers appear in the chapter as well as in the Visual Summary
- All the **Important equations,** color highlighted in yellow as they are in the text
- **Molecular views** of chemical processes, emphasizing our molecular approach to chemistry
- **Flow charts** that summarize the logic of analytical reasoning
- All the **Key Concepts,** which also appear in boldface in the chapter
- **Tabular summaries,** providing visual representations of how data can be organized
- **Chemical reactions,** because reactions are ultimately what chemistry is about

Chapter Problems: Our chapter *Problems* have been praised for their variety and quality. After a set of paired *Problems* specific to each section of the chapter, we present *Additional Paired Problems* that are not section-specific. These are followed by *More Challenging Problems* that are unpaired. These include *Group Study Problems* that encourage exploration and collaboration in finding solutions to more complex problems.

Student Tools

Technology-based resources have become essential parts of every general chemistry toolkit, but all too often these have not been well integrated with the textbook. Our technology components, found at *www.wiley.com/college/olmsted*, fall into two distinct categories: *Study & Practice* and *Assessment*.

Study & Practice

To enhance visualization, molecular understanding, and problem solving effectively, we believe that computer-based materials for student use should closely mirror the molecular and problem-solving approach of the textbook. We have worked actively with the software developers to ensure products that truly complement our text.

The **website** that accompanies the text (*www.wiley.com/college/olmsted*) features animations, 3-D molecules, video clips, and tutorials that we designed and edited. These address *key concepts* including identification of molecular/ionic species in solution, construction of Lewis structures, properties of atomic orbitals, and other molecular visualizations. Also included are a number of simple *molecular visualization* clips to aid students in learning how to "think molecules." The content is linked to the on-line problem tutorials. A lightbulb icon in the text margin calls the reader's attention to the existence of a web-based element.

Assessment

- The text's website also includes **Interactive Learning Ware (ILW),** a step-by-step problem solving tutorial program that guides students through selected problems from the book. The program includes text excerpts relevant to each ILW problem, which students can access while on-line if they have difficulty understanding the underlying concepts. When appropriate, students can also link to molecular representations of the problems. The ILW problems are representative of those that students frequently find most difficult and they reinforce students' critical thinking and problem solving skills. An ILW icon identifies each ILW problem in the textbook.
- eGrade Plus includes all of the companion website's assets and more. In classes using eGrade Plus, students are able to complete homework and receive immediate feedback on their progress, as well as have their worked tracked by their instructor.
- The **Student Study Guide** includes chapter overviews, learning objectives, sample exercises with worked out solutions, and self-test questions for each chapter.
- The **Student Solutions Manual** provides worked-out solutions for the odd-numbered problems in the text.

Instructor Ancillaries

These ancillaries provide resources both for lecture and course management.

- The **Instructor's Solutions Manual** is available on the Instructors' website and permits instructors to select, collate, and print solutions to the specific set of chapter problems that have been assigned as homework.
- The **Instructor Resource CD** contains PowerPoint presentations, including selected illustrations from the text. The PowerPoint slides are editable so instructors can modify the content to suit their own preferences. Also, the art has been optimized for large-scale projection in the lecture hall.

- The *Color Transparencies* are colorful and informative acetates suitable for overhead presentation. These reproduce important art from the textbook, rendered for projection clarity.
- The *Instructor's Manual,* available both in print and on the web, provides chapter overviews and lecture outlines.
- The *Test Bank,* extensively expanded and revised, includes a significant number of molecular reasoning problems.
- The *Web Resources for Classroom Management* include Web CT, Blackboard, Web Assign, and eGrade Plus. eGrade Plus is a powerful online tool that provides instructors with an integrated suite of teaching and learning resources, including an online version of the text, in one easy-to-use website. Organized around the essential activities you perform in class, eGrade Plus allows you to create class presentations, assign homework and quizzes that will be automatically graded, and track your students' progress. The system links homework problems to the relevant section of the online text, providing students with context-sensitive help. View a demo and learn more about eGrade Plus by visiting www.wiley.com/college/egradeplus.

ACKNOWLEDGMENTS

No textbook can be created without dedicated effort on the part of an editorial staff. This edition has benefited immensely from a new and imaginative editor, Kevin Molloy, as well as a seasoned yet equally enthusiastic Ellen Ford. As with the third edition, the sharp eyes and minds of Barbara Russiello and Gloria Hamilton resulted in a significantly cleaner and more accurate final product. The supplementary materials have been efficiently coordinated by Jennifer Yee. We are grateful to Harry Nolan for an extraordinary cover design.

We are forever indebted to Jim Smith, the editor who discovered us and had the courage and skills to mold our first edition into a cohesive and successful product. We also owe huge debts to our wives, Eileen Olmsted and Trudy Cameron, who have now tolerated and usually understood more than a decade of single-minded dedication to this project on the part of their author husbands.

Significant amounts of new art have been created for this edition. The art program was ably coordinated by Ed Starr and Sigmund Malinowski. For the many fine photographs that help convey the beauty of chemistry, we are indebted to Wiley's photo research team, Hilary Newman, Jennifer MacMillan, and Elyse Rieder, and to two excellent professional photographers, Stephen Frisch and Andy Washnik (John Olmsted also contributed his amateur skills). For this edition, many new molecular views were created by Davi Erickson. The line art was elegantly rendered by Precision Graphics.

The electronic ancillary materials for this project were the work of a diverse and talented group, coordinated by Martin Batey and Jennifer Yee at Wiley.

Peer reviewers perform an essential service, not only identifying weaknesses and strengths but also assisting authors to find the right balance of coverage and assuring accuracy. The following provided their insights for the fourth edition:

Shylaja Akkaraju
College of DuPage

Patricia Amateis
Virginia Tech

Tim Anstine
Northwest Nazarene University

Neil Baker
The Ohio State University

Robert Balahura
University of Guelph

Sharmistha Basu-Dutt
State University of West Georgia

Gary Beall
Texas State University

David Belt
Johnson County Community College and Penn Valley Community College

Donna Bivans
Pitt Community College

Lesley Blair
Oregon State University

Robert Boyd
Auburn State University

Dana Brown-Haine
Central Piedmont Community College

Brian Buffin
Western Michigan University

David Burgess
Rivier College

David Byres
Florida Community College Jacksonville

Harvey Carroll
Kingsborough Community College

Todd Carter
Seward County Community College

Jocelyn Cash
Central Piedmont Community College

Pam Clarke
Red Deer College

Jan Coles
Kansas State University

Jerry Cook,
Sam Houston State University

Laura Crews
Mesa Community College

Dean Cuebas
Southwest Missouri State University

Todd Deal
Georgia Southern University

Gary DeBoer
LeTourneau University

Allison Dobson
Georgia Southern University

Anne Donnelly
State University of New York, Cobleskill

William Donovan
The University of Akron

Marly Eidsness
University of Georgia

Joe Free
Art Institutes International, Phoenix

Margaret Geselbracht
Reed College

Stephen Goldberg
Adelphi University

Jack Goldsmith
University of South Carolina, Aiken

Eric Goll
Brookdale Community College

Amy Gottfried
University of Michigan, Ann Arbor

Paul Hanson
University of New Orleans

Daniel Haworth
Marquette University

Mike Jezercak
University of Oklahoma

Mitrick Johns
Northern Illinois University

Lori Jones
University of Guelph

Matthew Johnston
Lewis & Clark College

Booker Juma
Fayetteville State University

Arne Lekven
Texas A&M University, College Station

Maureen Leupold
Genesee Community College

David Lippman
Texas State University, San Marcos

Patrick Lloyd
Kingsborough Community College

Ann Lumsden
Florida State University

Edward Lyons
Albany State University

Yong Ma
Oxnard College

Suzanne Martin
Maricopa Community Colleges

David Mascotti
John Carroll University

Amy Massengill
Middle Tennessee State University

Clyde Metz
College of Charleston

Nancy Miller
Diablo Valley College

Randy Moore
University of Minnesota

Michael Mueller
Rose-Hulman Institute of Technology

Wyatt Murphy
Seton Hall University

Mark Ott
Jackson Community College

Jason Overby
College of Charleston

Melissa Overcash
Frederick Community College

Kenneth Overly
Providence College

Maria Parr
Trinity College, Hartford

Matthew Partin
Bowling Green State University

Jack Pennington
Forest Park University

Thomas Pentecost
Aims Community College

David Peyton
Morehead State University

Gregory Phelan
Seattle Pacific University

Jay Phelan
University of California, Los Angeles

Paula Piehl
Potomac State College

Roz Potter
University of Chicago

James Reho
East Carolina University

Donald Reinhardt
Georgia State University

Thomas Richardson
North Georgia College & State University

Kim Sadler
Middle Tennessee State University

Shamili Sandiford
College of DuPage

Fred Schaefer
University of the Sciences in Philadelphia

Tarak Sharma
College of Staten Island

Susan Schelble
University of Colorado, Denver

Jeff Schoonover
St. Mary's University

Richard Schwenz
University of Northern Colorado

James Seizler
Allan Hancock College

David Shaw
Madison Area Technical College

Marilyn Shopper
Johnson County Community College

Suzanne Simoneau
Augusta State University

Joseph Sinski
Bellarmine University

Dianne Snyder
Augusta State University

Mary Sohn
Florida Institute of Technology

Thomas Sorenson
University of Wisconsin, Milwaukee

Joan Stover
South Seattle Community College

Eric Strauss
Boston College

Douglas Strout
Alabama State University

Stephen Summers
Seminole Community College

Kathy Tehrani
*Cincinnati State Technical and
Community College*

Donald Thompson
Oklahoma State University

Russ Tice
California Polytechnic State University

Ramaiyer Venkatraman
Jackson State University

Carol Wake
South Dakota State University

Jerry Waldvogel
Clemson University

Randall Wanke
Augustana College

Thomas Webb
Auburn University

Dale Wheeler
Appalachian State University

Art Weiner
University of Alaska, Anchorage

Christine Whitlock
Georgia Southern University

Ken Wunch
Sam Houston State University

Sidney Young
University of South Alabama

To those reviewers who tested and provided feedback on the media resources, we also send our thanks. Their feedback was practical, insightful, and helpful. These media reviewers include Paul Hanson, *University of New Orleans*; Mufeed Basti, *North Carolina A&T University*; Pamela Brown, *New York City Technical College*; John J. Dolhun, *Norwalk Community College*; Richard H. Langley, *Stephen F. Austin State University*; Jack McKenna, *St. Cloud State University*; Michael Louis Norton, *Marshall University*; Thomas Pentecost, *Aims Community College*; Michael Russell, *Mt. Hood Community College*; Richard S. Treptow, *Chicago State University*; In addition, thanks go to reviewers on the previous three editions who gave us the foundation on which to build. They include:

**Reviewers of
the First Edition**

Bruce Ault
University of Cincinnati

Caroline Ayers
East Carolina University

George Baldwin
University of Manitoba

Jon M. Bellama
University of Maryland

Allan R. Burkett
Dillard University

Donald Campbell
University of Wisconsin, Eau Claire

John F. Cannon
Brigham Young University

Grover W. Everett
University of Kansas

Michael D. Fryzuk
University of British Columbia

Steven D. Gammon
University of Idaho

Michael F. Golde
University of Pittsburgh

Paul Hunter
Michigan State University

Richard F. Jordan
University of Iowa

Paul J. Karol
Carnegie Mellon University

Robert Kiser
University of Kentucky

Joseph W. Kolis
Clemson University

David F. Koster
Southern Illinois University at Carbondale

Glenn D. Kuehn
New Mexico State University

Richard S. Lumpkin
University of Alabama

John Luoma
Cleveland State University

Bruce E. Norcross
SUNY, Binghamton

Henry W. Offen
University of California, Santa Barbara

M. Larry Peck
Texas A & M University

John V. Rund
University of Arizona

Martha E. Russell
Iowa State University

Sanford A. Safron
Florida State University

Caesar V. Senoff
University of Guelph

Joanne Stewart
Hope College

Dwight A. Sweigart
Brown University

Wayne Tikkanen
*California State University,
Los Angeles*

Charles A. Trapp
University of Louisville

D. Rodney Truax
University of Calgary

**Reviewers of
the Second Edition**

Lavoirs Banks
Elgin Community College

Paul Braterman
University of North Texas

Larry Brown
Texas A & M University

John F. Cannon
Brigham Young University

Terry S. Carlton
Oberlin College

T.A. George
University of Nebraska, Lincoln

John M. Halpin
New York University

Thomas A. Holme
University of Wisconsin, Milwaukee

James A. Ibers
Northwestern University

Virginia Indivero
Swarthmore College

Milton Johnston, Jr.
University of South Florida

Glenn D. Kuehn
New Mexico State University

Glenn Millhauser
University of California, Santa Cruz

Gholam Mirafzal
Drake University

Gary Mort
Dixie Community College

Melinda E. Oliver
Louisiana State University

John V. Rund
University of Arizona

Venkatesh M. Shanbhag
Mississippi State University

William Zoller
University of Washington

**Reviewers of
the Third Edition**
Robert Balahura
University of Guelph

David Ball
Cleveland State University

Gary Buckley
Cameron University

Larry Brown
Texas A&M University

Jim Byrd
California State University-Stanislaus

Michael Chetcuti
Universite Louis Pasteur, France

Stephen R. Daniel
Colorado School of Mines

James Falcone
West Chester University

Debbie Finocchio
University of San Diego

J. S. Francisco
Purdue University

L. Peter Gold
Pennsylvania State University

Thomas Greenbowe
Iowa State University

Hans Gunderson
Northern Arizona University

James Harrison
Michigan State University

James Hovick
University of North Carolina, Charlotte

Ronald C. Johnson
Emory University

Philip Keller
University of Arizona

Pamela Kerrigan
*Manhattan College/College of Mt.
St Vincent*

Robley Light
Florida State University

Jeffry Madura
Duquesne University

Gary Mort
Lane Community College

Wyatt Murphy
Seton Hall University

George Reilly
University of Delaware

Dale Russell
Boise State University

Brian Sanctuary
McGill University

John R. Sowa
Union College

Stephen Summers
Seminole Community College

Larry Thompson
University of Minnesota, Duluth

Worth Vaughan
University of Wisconsin, Madison

John S. Winn
Dartmouth College

Contents in Brief

Contents

CHEMISTRY

Matter Is Molecular

A view of the Earth from space shows that our planet is an integrated whole. At the same time, we know that the Earth is mind-boggling in its diversity. Nevertheless, the stunning complexity of our world can be described using a small set of chemical principles. These fundamental aspects of chemistry are the subject of this book.

The entire universe is made up of matter, from the vast reaches of the galaxies to a simple glass of water. As we describe in the coming chapters, matter is composed of tiny particles called "atoms." On Earth there are around 100 different kinds of atoms, each kind with its own unique combination of prop-

erties. The complexity of our world arises from the unlimited number of ways that atoms can combine to form different molecules. The principles of modern chemistry are organized around the molecular nature of matter. Our book presents this perspective while at the same time emphasizing the quantitative aspects of chemistry.

A drop of water contains an unimaginable number of molecules, as our molecular inset shows. Water is essential to life as we know it. The simple yet unusual fact that solid water (ice) floats atop liquid water allows life to exist on our planet. Just as important is the fact that water dissolves an immense range of chemical compounds: Water is the solvent of life. In fact, water is so important to our perspective of life that the search for water

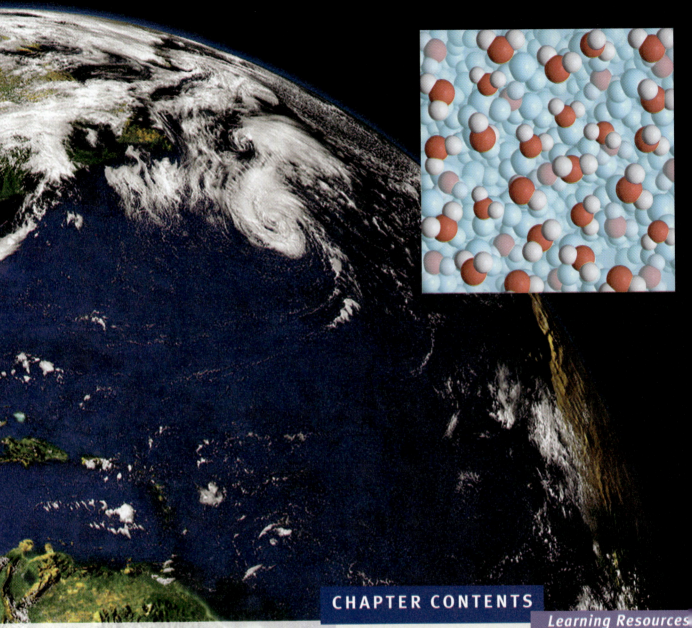

CHAPTER CONTENTS

is a key feature of our quest to discover life in other quarters of the galaxy. The inset photo of the surface of Mars, for example, shows no sign of water at present, but some erosional features appear to have been caused by flowing water in the past.

Does the same chemistry that takes place on the Earth occur within the galaxies and nebulae in the far reaches of the universe? We have no way to know for certain, but observations made by astronomers are consistent with chemistry being the same throughout the universe. Moreover, from research on stars, chemists have learned that the various kinds of atoms probably form during stellar evolution and are dispersed throughout the universe by supernova explosions.

Chemists are interested in a huge range of problems, extending from the galactic scale to what

1.1 **What Is Chemistry?**
1.2 **Atoms, Molecules, and Compounds**
1.3 **The Periodic Table of the Elements**
1.4 **Characteristics of Matter**
1.5 **Measurements in Chemistry**
1.6 **Calculations in Chemistry**
1.7 **Chemical Problem Solving**

Learning Resources

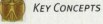 KEY CONCEPTS

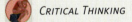 CRITICAL THINKING

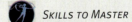 SKILLS TO MASTER

 ADDITIONAL HELP
www.wiley.com/college/olms
- TUTORIALS
- ANIMATIONS

takes place between individual atoms and molecules. Here are some of the practical problems identified by a 2003 report from the National Research Council, "Beyond the Molecular Frontier: Challenges for Chemistry and Chemical Engineering."

- Develop new materials that will protect citizens against terrorism.
- Develop medicines and therapies that can cure currently untreatable diseases.
- Develop unlimited and inexpensive energy to pave the way to a truly sustainable future.

- Revolutionize the design of chemical processes to make them safe and environmentally benign.
- Understand the complex chemistry of the Earth, to design policies that will prevent environmental degradation.
- Learn how to design substances with predictable properties, to streamline the search for new and useful substances.

As this list suggests, chemistry is important in many respects. We hope you enjoy your study of chemistry!

1.1 WHAT IS CHEMISTRY?

Science, in the broadest sense, can be viewed as an attempt to organize and understand our observations of nature. Because this is a vast undertaking, science is subdivided into various disciplines, including chemistry, biology, geology, and physics. Chemistry is the science that studies the properties and interactions of matter. Chemists seek to understand how chemical transformations occur by studying the properties of matter. Because of the broad scope of chemistry, the interests of chemists intertwine with those of physicists, biologists, engineers, and geologists.

Matter is anything that possesses mass and occupies space.

How Chemistry Advances

Chemists learn about chemical properties by performing experiments. They organize information about chemical properties using general principles and theories. The periodic table, for example, organizes the elements according to chemical properties. Chemists use general principles and theories to make predictions about yet-unknown substances. These predictions generate experiments whose results may extend the scope of the principles and theories.

Chemical research is driven by many goals, and it progresses in many different ways. The essential traits of a good researcher are curiosity, creativity, flexibility, and dedication. Some chemical advances come from a direct assault on a known problem. A classic example is the development of the Hall process for refining aluminum from its ores, which we describe in Chapter 21. As a contemporary example, many scientists around the world are working at a feverish pace to develop a vaccine against the AIDS virus.

Chemistry also advances when an imaginative researcher recognizes the potential of a lucky accident. Synthetic dye-making, the first major chemical industry, arose from one such lucky accident. While searching for a way to synthesize quinine, a drug for the prevention of malaria, English chemist William Perkin accidentally made a beautiful reddish-violet dye, which he called mauveine. Perkin had the imagination to realize the commercial potential of the new substance and switched his research interests from drugs to dyes. His insight made him rich and famous.

Chemical advances frequently are driven by technology. The discovery that atoms have inner structure was an outgrowth of the technology for working with radioactive materials. In Chapter 2 we describe a famous experiment in which the structure of atoms was studied by bombarding a thin gold foil with subatomic particles. A contemporary example is the use of lasers to study the details of chemical reactions. We introduce these ideas in Chapters 7 and 8.

Mauveine is a vivid red-violet dye

Methods of Science

However a new chemical discovery arises, an essential component of science is to explain that discovery on the basis of general principles. When a new general principle is posed, it is termed a **hypothesis.** A hypothesis is tentative until it can be confirmed in two ways. First, additional observations must be consistent with the hypothesis; second, the hypothesis must predict new results that can be confirmed by experiments. If a hypothesis meets these tests, it is promoted to the status of a **theory.** A theory is a unifying principle that explains a collection of facts.

Experimental observations are at the heart of chemical research. Many experiments are designed specifically to answer some particular chemical question. Often, the results of these experiments are unexpected and lead to new hypotheses. New hypotheses, in turn, suggest additional experiments. The Chemistry and Life Box describes how the hypothesis of extraterrestrial life can be tested.

A typical example of the interaction between hypothesis and experiment is the story of the work that resulted in worldwide concern over the depletion of the ozone layer in the stratosphere. These studies led to the awarding of the 1995 Nobel Prize for Chemistry to Paul Crutzen, Mario Molina, and F. Sherwood Rowland. Figure 1-1 provides a schematic view of how this prize-winning research advanced. It began in 1971 when experiments revealed that chlorofluorocarbons, or CFCs, had appeared in the Earth's atmosphere. At the time, these CFCs were widely used as refrigerants and as aerosol propellants. Rowland wondered what eventually would happen to these gaseous compounds. He carried out a theoretical analysis, from which he concluded that CFCs are very durable and could persist in the atmosphere for many years.

Meanwhile, Crutzen had done experiments showing that ozone in the upper atmosphere can be destroyed easily by reactions with nitrogen oxides. This work demonstrated that the ozone layer is in a delicate balance that could be disturbed significantly by changes in atmospheric composition. In 1974, Molina and Rowland combined Crutzen's experimental work with their own theoretical analysis and published a prediction (hypothesis) that CFCs pose a serious threat to the ozone layer.

Following this interplay between observations and theory, many atmospheric scientists began studying chemical reactions of ozone in the upper atmosphere. Chemists duplicated atmospheric conditions in the laboratory and measured how fast various chemical reactions occur. The results of these experiments were used to create theoretical models of the upper atmosphere and predict how the ozone concentration would change as CFCs were introduced. Meanwhile, atmospheric scientists carried out measurements showing that ozone was being depleted in the upper atmosphere at a rate even faster than had been predicted.

Today, scientists realize that the chemistry of the upper atmosphere is quite complex. In addition to gaseous molecules, solid particles such as tiny ice crystals play important roles in the chemistry that affects ozone. The original hypothesis of Rowland and Molina, that CFCs reach the upper atmosphere and deplete the ozone layer, has been fully confirmed. Exactly how this occurs, what other chemicals are involved, and how this process might be controlled, are still under intense study by chemists and other scientists, leading to yet more hypotheses and experiments.

The story of research into the depletion of atmospheric ozone is just one example of how scientific understanding and theories develop. The fundamental theories and laws of chemistry that we present in this text all went through similar intense scrutiny and study.

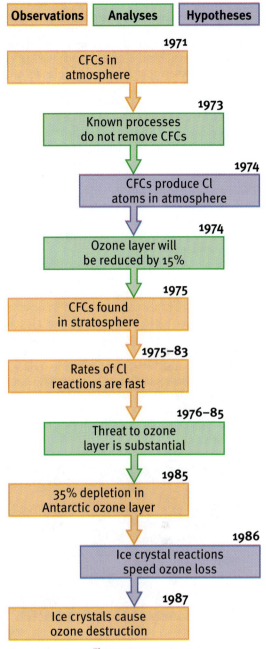

Observations | Analyses | Hypotheses

1971 CFCs in atmosphere

1973 Known processes do not remove CFCs

1974 CFCs produce Cl atoms in atmosphere

1974 Ozone layer will be reduced by 15%

1975 CFCs found in stratosphere

1975–83 Rates of Cl reactions are fast

1976–85 Threat to ozone layer is substantial

1985 35% depletion in Antarctic ozone layer

1986 Ice crystal reactions speed ozone loss

1987 Ice crystals cause ozone destruction

Figure 1-1

This flow chart illustrates how the scientific process led to worldwide concern over the effect of chlorofluorocarbons on the ozone layer.

Box 1-1   Chemistry and Life: Is There Life on Other Planets?

Speculation about life on other planets probably began when humans discovered that the Earth is not unique. We know that several other planets of the solar system bear at least some resemblance to our own. Why, then, should there not be life on Mars, or Venus, or perhaps on undiscovered Earthlike planets orbiting some other star?

How can scientists collect experimental evidence about possible life on another planet? Sending astronauts to see for themselves is impractical at our current level of technology. Nevertheless, it is possible to search for life on other worlds without sending humans into space. In the late 1970s, NASA's *Viking* spacecraft lander collected a sample of dirt from Mars, the planet in our solar system most like Earth. The sample showed no signs of life. Nevertheless, speculation continues about Martian life.

The photo below, taken by the *Viking* spacecraft, shows that the surface of Mars has been eroded, apparently by liquid water. More recent photos transmitted by *Spirit* and *Opportunity* convince scientists that this was the case. Apparently, Mars was once much warmer than it is today. Planetary scientists speculate that at one time the atmosphere of Mars may have contained large amounts of carbon dioxide, setting up a "greenhouse" effect that made the surface of that planet warmer and wetter. Might there, then, have been life on Mars at some earlier time? Molecular structures found in meteorites thought to come from Mars have been interpreted to show that there was once life there, but these results are controversial.

require the presence of water, this observation indicates that there could be life on Europa.

Conditions on other planets seem too hostile for life as we know it, but recent discoveries on our own planet indicate that life is much more robust than was once thought. Deep-sea explorers have discovered flourishing life around hydrothermal vents. Whereas life on the surface of the Earth relies on sunlight and photosynthesis for energy, these deep-sea life forms exploit energy-rich compounds spewed forth by volcanic vents. The warm waters around hydrothermal vents (photo below) teem with bacteria, which in turn support higher life forms such as worms and crustaceans. This terrestrial life thrives in an environment similar to one that might exist on Europa, reinforcing speculation that this moon of Jupiter could support some forms of life.

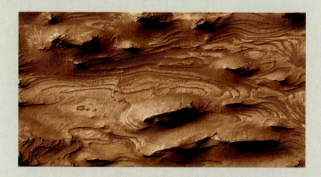

Indirect evidence can be collected without actually visiting a planet. Recent photographs taken from flyby spacecraft offer tantalizing hints. NASA's *Galileo* took photographs, shown above right, of Europa, one of Jupiter's moons. The close-up photo of the surface of Europa shows what appear to be huge broken chunks of ice, which suggests that there may be liquid water under the ice, warmed by tidal forces generated by Jupiter's huge mass and strong gravity. Because life seems to

Outside our own solar system, might there be planetary environments where life flourishes? In recent years, astronomers have discovered planets orbiting stars other than our own. Whether or not these planets support life is still impossible to say. Nevertheless, the more we discover about the variety of the universe, the more likely it becomes that we are not alone.

Section Exercises

1.1.1 List four ways that chemistry applies to cooking.

1.1.2 Describe how chemistry applies to the automobile industry.

1.1.3 A planetary scientist announces a theory predicting that substances on Venus react differently than they do on Earth. Write a paragraph that describes ways to test this theory.

1.2 ATOMS, MOLECULES, AND COMPOUNDS

Every substance has physical properties that we can measure and describe, including shape, color, and texture. For example, the iron girder shown in Figure 1-2 has a lustrous silvery color and a smooth texture. Substances also have chemical properties, such as their flammability. Physical and chemical properties that can be observed with the eye are called **macroscopic.** The underlying structure of a chemical substance, which is called **microscopic,** can be explored using magnifying devices. The magnified view of iron shown in Figure 1-2 reveals fissures and pits that are not visible in the macroscopic view. Still further magnification eventually reveals the building blocks of matter. This, the **molecular** or **atomic view,** is an essential part of every chemist's thinking.

Atoms

The fundamental unit of a chemical substance is called an **atom.** The word is derived from the Greek *atomos,* meaning "uncuttable." An atom is the smallest possible particle of a substance.

Atoms are extremely small. Measurements show that the diameter of a single carbon atom is approximately 0.000 000 0003 meters (about 0.000 000 001 feet). To give you some idea of just how small that is, a sample of carbon the size of the period at the end of this sentence contains more atoms than the number of stars in the Milky Way. Any sample of matter large enough for us to see or feel contains an unfathomable number of atoms.

Molecules

Atoms combine to make all the substances in the world around us, but they do so in very orderly ways. Most substances that we encounter in day-to-day life are made up of small units called **molecules.**

Figure 1-2
Iron appears different at the macroscopic, microscopic, and atomic levels.

> **KEY CONCEPT**
> *A molecule is a combination of two or more atoms held together in a specific shape by attractive forces.*

The simplest molecules contain just two atoms. For example, a molecule of hydrogen is made up of two hydrogen atoms. A molecule that contains two atoms is classified as a *diatomic* molecule. Figure 1-3 represents a diatomic hydrogen molecule as two spheres connected together.

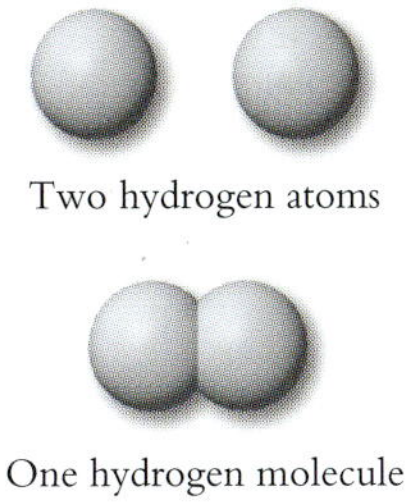

Because chemistry deals mostly with the behavior of molecules, this book emphasizes chemistry's molecular foundation. Throughout this book you will see many figures that represent molecules, with each atom represented by a sphere. Although there are more than 100 different types of atoms, only about 20 are encountered frequently in our world. Many of the molecules described in this book are made up of just 10 different types of atoms: hydrogen, carbon, nitrogen, oxygen, phosphorus, sulfur, fluorine, chlorine, bromine, and iodine. Figure 1-4 shows the color scheme that we use to represent these atoms. We introduce other types of atoms as the need arises. Although many substances are colored, individual atoms do not have the colors shown in Figure 1-4.

Two hydrogen atoms

One hydrogen molecule

Figure 1-3
A hydrogen molecule can be represented by connecting two spheres together, with each sphere representing one hydrogen atom.

The name *hydrogen* refers to both atoms and molecules. To minimize confusion, we refer to *atomic hydrogen* when we mean hydrogen atoms and *molecular hydrogen* when we mean hydrogen molecules.

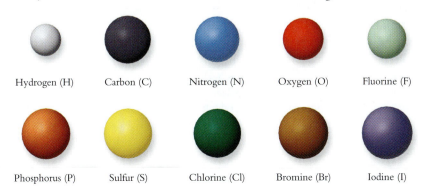

Hydrogen (H) Carbon (C) Nitrogen (N) Oxygen (O) Fluorine (F)

Phosphorus (P) Sulfur (S) Chlorine (Cl) Bromine (Br) Iodine (I)

Figure 1-4
Color-coded scale models of 10 types of atoms that appear frequently in this book.

Figure 1-5
Models of seven relatively simple molecules

Molecular oxygen

Water

Molecular chlorine

Carbon dioxide

Animation

Ammonia

Methanol

Acetylene

With practice, you can recognize simple molecules, such as carbon dioxide and water, just by looking at their models. Figure 1-5 shows scale models of a few molecules whose names may be familiar to you. The structures of larger molecules, however, are too complex to be recognized at a glance. Consequently, chemists have created a shorthand language of symbols, formulas, and equations that convey information about atoms and molecules in a simple manner. A specific letter or pair of letters designates each type of atom. These symbols, in turn, are combined into formulas that describe the compositions of more complicated chemical substances. Formulas can then be used to write chemical equations that describe how molecules change in chemical reactions.

The Elements

A substance that contains only one type of atom is called a chemical **element,** and each different element contains its own specific type of atom. Each chemical element has a unique name and number (Z), such as hydrogen (Z = 1), carbon (Z = 6), oxygen (Z = 8), iron (Z = 26), and uranium (Z = 92). The name of an element may refer to its history or to one of its properties. The Romans gave copper its name, *cuprum,* after the island of Cyprus, where copper was mined as early as 5000 BC. Bromine received its name from the Greek word *bromos,* meaning "stench." If you ever work with bromine, you will understand the reason for its name. Other examples of evocative names include xenon (from the Greek *xenos,* "stranger"), rubidium (from the Latin "dark red"), and neptunium (named for the planet Neptune).

Each element is represented by a unique one- or two-letter symbol. For example, the symbol for hydrogen is H, oxygen's symbol is O, and nitrogen's symbol is N. When two or more elements have names that begin with the same English letter, all but one of the elemental symbols has a second letter. The second letter is always lower case. For example, carbon is C, chlorine is Cl, cobalt is Co, and chromium is Cr. Chemists understand that the symbol for an element represents more than one or two letters. Instead, a chemist sees the symbol Ni and immediately thinks of nickel *atoms.*

The elements listed in Table 1-1 have symbols derived from their names in other languages. Most of these elements were known in ancient times, so their symbols reflect the Latin language that was dominant when they were named.

Copper, one of the first elements to be purified by humans, is used for sculpture worldwide.

Tutorial

Table 1-1   Elemental Symbols with Non-English Roots

Name	Symbol	Root	Language	Name	Symbol	Root	Language
Antimony	Sb	stibium	Latin	Potassium	K	kalium	Latin
Copper	Cu	cuprum	Latin	Silver	Ag	argentum	Latin
Gold	Au	aurum	Latin	Sodium	Na	natrium	Latin
Iron	Fe	ferrum	Latin	Tin	Sn	stannum	Latin
Lead	Pb	plumbum	Latin	Tungsten	W	wolfram	German
Mercury	Hg	hydrargyrum	Latin				

Chemical Formulas

A chemical **compound** is a substance that contains more than one element. The relative amounts of the elements in a particular compound do not change: Every molecule of a particular chemical substance contains a characteristic number of atoms of its constituent elements. For example, every water molecule contains two hydrogen atoms and one oxygen atom. To describe this atomic composition, chemists write the **chemical formula** for water as H_2O.

The chemical formula for water shows how formulas are constructed. The formula lists the symbols of all elements found in the compound, in this case H (hydrogen) and O (oxygen). A subscript number after an element's symbol denotes how many atoms of that element are present in the molecule. The subscript 2 in the formula for water indicates that each molecule contains two hydrogen atoms. No subscript is used when only one atom is present, as is the case for the oxygen atom in a water molecule. Atoms are indivisible, so molecules always contain whole numbers of atoms. Consequently, the subscripts in chemical formulas of molecular substances are always integers. We explore chemical formulas in greater detail in Chapter 3.

Molecules vary considerably in complexity. Molecular oxygen is made up of two oxygen atoms, so its chemical formula is O_2. A carbon monoxide molecule contains one atom of carbon and one atom of oxygen, so its chemical formula is CO. Each molecule of methane, the major constituent of natural gas, contains one carbon atom and four hydrogen atoms, so its formula is CH_4. You will encounter still more complicated structures, such as methanol (three different elements, CH_4O) as you progress through this book.

Molecular oxygen
O_2

Water
H_2O

Carbon monoxide
CO

Methane
CH_4

Section Exercises

■ **1.2.1** What are the elemental symbols for cerium, cesium, copper, calcium, and carbon?

■ **1.2.2** What are the names of the elements represented by the symbols Zr, Ni, Sn, W, Se, Be, and Au?

■ **1.2.3** Molecular pictures of some common molecules are shown here. What are their chemical formulas?

Hydrogen peroxide
(a common disinfectant
and bleaching agent)

Sulfur dioxide
(a common air
pollutant)

Dinitrogen oxide
(laughing gas)

Acetic acid
(vinegar)

Ethylene
(used to make
polyethylene plastic)

1.3 THE PERIODIC TABLE OF THE ELEMENTS

More chemical reactions exist than anyone can imagine. Nevertheless, certain patterns of chemical reactivity have been recognized for more than 100 years. These patterns remain valid even though new reactions are always being discovered. Each chemical element has characteristic chemical properties. Moreover, certain groups of elements display similar chemical properties. In 1869, the Russian chemist Dmitri Mendeleev and the German chemist Julius Lothar Meyer independently discovered how to arrange the chemical elements in a table so that the elements in each column have similar chemical properties. This arrangement, the **periodic table,** contains all the known chemical elements.

Arrangement

The periodic table lists all the known elements in numerical order, starting with the lightest (hydrogen) and proceeding to the heaviest (uranium, among naturally occurring elements). The list is broken into seven rows. Each row is placed below the previous row in a way that places elements with similar chemical properties in the same column of

We show in Chapter 2 that the periodic table is based on the structure of atoms rather than on their masses. Elemental masses correlate closely with atomic structure, however, so ordering by mass is almost the same as ordering by structure. There are only three exceptions among more than 100 elements.

Period																																
1	1 H																															2 He
2	3 Li	4 Be																									5 B	6 C	7 N	8 O	9 F	10 Ne
3	11 Na	12 Mg																									13 Al	14 Si	15 P	16 S	17 Cl	18 Ar
4	19 K	20 Ca														21 Sc	22 Ti	23 V	24 Cr	25 Mn	26 Fe	27 Co	28 Ni	29 Cu	30 Zn	31 Ga	32 Ge	33 As	34 Se	35 Br	36 Kr
5	37 Rb	38 Sr														39 Y	40 Zr	41 Nb	42 Mo	43 Tc	44 Ru	45 Rh	46 Pd	47 Ag	48 Cd	49 In	50 Sn	51 Sb	52 Te	53 I	54 Xe
6	55 Cs	56 Ba	57 La	58 Ce	59 Pr	60 Nd	61 Pm	62 Sm	63 Eu	64 Gd	65 Tb	66 Dy	67 Ho	68 Er	69 Tm	70 Yb	71 Lu	72 Hf	73 Ta	74 W	75 Re	76 Os	77 Ir	78 Pt	79 Au	80 Hg	81 Tl	82 Pb	83 Bi	84 Po	85 At	86 Rn
7	87 Fr	88 Ra	89 Ac	90 Th	91 Pa	92 U	93 Np	94 Pu	95 Am	96 Cm	97 Bk	98 Cf	99 Es	100 Fm	101 Md	102 No	103 Lr	104 Rf	105 Db	106 Sg	107 Bh	108 Hs	109 Mt	110 Ds	111	112		114				

Figure 1-6
Periodic table of the elements with all the elements included in their proper rows and columns.

the table. Moving *across a row* of the periodic table, the elements generally increase in mass and change dramatically in their chemical properties. Moving *down a column,* mass also increases, but the elements have similar chemical properties.

Figure 1-6 shows the periodic table. Notice that rows 6 and 7 are quite long, which makes the table rather cumbersome. For convenience, 14 elements in the sixth row and 14 in the seventh row usually are separated from the rest of the table and placed beneath the main portion, as shown in Figure 1-7 and on the inside front cover of the book. This is the most common format for the periodic table.

Metals, Nonmetals, and Metalloids

The elements can be divided into categories: metals, nonmetals, and metalloids. Examples of each appear in Figure 1-7. Except for hydrogen, all the elements in the left and central regions of the periodic table are **metals.** Metals display several characteristic properties. For example, they are good conductors of heat and electricity and usually appear shiny. Metals are malleable, meaning that they can be hammered into thin sheets, and ductile, meaning that they can be drawn into wires. Except for mercury, which is a liquid, all metals are solids at room temperature.

As Figure 1-8 shows, the **nonmetals** are found in the upper right corner of the periodic table. The properties of nonmetals are highly variable, but most nonmetals are poor conductors of electricity and heat. Running diagonally across the table between the metals and the nonmetals are six elements that are categorized as **metalloids** (B, Si, Ge, As, Sb, and Te). These dull-appearing, brittle solids are sometimes called semiconductors because they conduct electricity better than nonmetals but not as well as metals. Silicon and germanium are used in the manufacture of semiconductor chips in the electronics industry.

Periodic Properties

The first column of the periodic table, Group 1, contains elements that are soft, shiny solids. These **alkali metals** include lithium, sodium, potassium, rubidium, and cesium. At the other end of the table, fluorine, chlorine, bromine, iodine, and astatine appear in the next-to-last column. These are the **halogens,** or Group 17 elements. These four elements exist as diatomic molecules, so their formulas have the form X_2. A sample of chlorine appears in Figure 1-7. Each alkali metal combines with any of the halogens in a 1:1 ratio to form a white crystalline solid. The general formula of these compounds is AX, where A represents the alkali metal and X represents the halogen (AX = NaCl, LiBr, CsBr, KI, etc.).

The elements in the second column of the table (Group 2) are the **alkaline earth metals.** These resemble the alkali metals in their appearance, but they have different chemical properties. For example, each of these metals combines with the halogens in a 1:2 ratio (AX_2 = $CaCl_2$, $MgBr_2$, BaF_2, etc.). Each also reacts with atmospheric oxygen to form a solid with the formula AO (BaO, CaO, etc.).

The last column of the periodic table contains the **noble gases,** or Group 18 elements, all of which occur in nature as gases. With a few exceptions, these elements do not undergo chemical reactions.

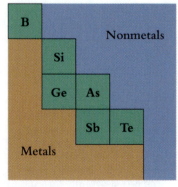

Figure 1-8
The six metalloids occupy a diagonal region of the periodic table between the metals and the nonmetals.

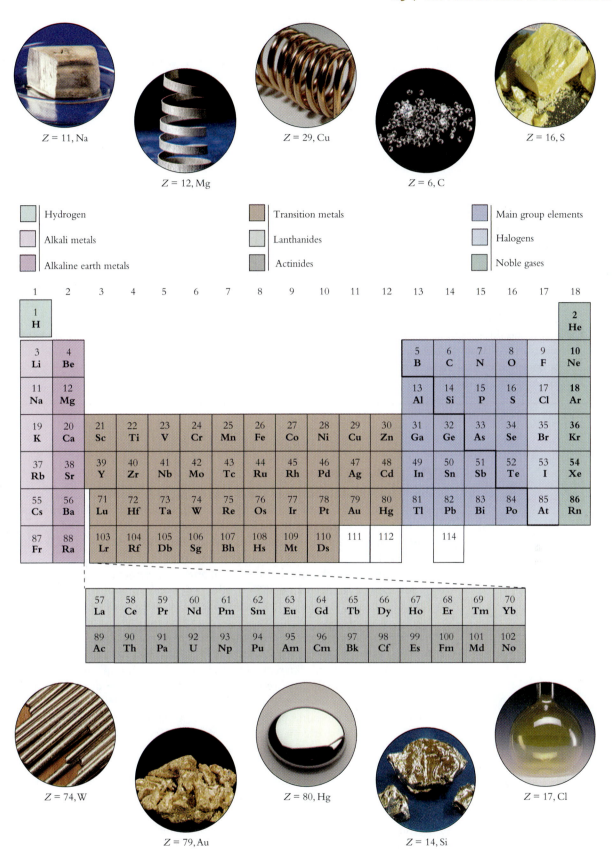

Figure 1-7
The periodic table of the elements as used in common practice. The photos show ten pure elements, including six metals (Na, Mg, Cu, W, Au, Hg), one metalloid (Si), and three nonmetals (C, S, Cl).

The elements in Groups 3 through 12 are known as **transition metals.** The elements in rows 6 and 7 that are normally shown below the rest of the table are the inner transition metals, subdivided into **lanthanides** (row 6) and **actinides** (row 7). All other elements are **main group elements.**

The periodic table is a useful way to organize chemical properties. To help you see the patterns, the periodic table on the inside front cover of this book highlights the various groups of elements. As you learn more about chemical structure and behavior, you will discover the principles that account for similarities and differences in the chemical behavior of the elements.

Section Exercises

■ **1.3.1** Predict the formulas of the compounds formed in the reactions between (a) calcium and chlorine; (b) cesium and iodine; (c) barium and oxygen; and (d) magnesium and fluorine.

■ **1.3.2** Boron and fluorine form a compound with the formula BF_3. Based on this, suggest formulas for compounds of aluminum with bromine and gallium with chlorine.

■ **1.3.3** Classify each of the following elements as alkali metal, alkaline earth metal, halogen, noble gas, main group, transition metal, lanthanide, or actinide: aluminum, fluorine, cobalt, phosphorus, krypton, europium, thorium, barium, and sodium.

1.4 CHARACTERISTICS OF MATTER

Matter is anything that has mass and occupies space. A sample of matter can contain a single substance or any number of different substances. As already described, the building blocks of most substances are molecules, which in turn are composed of atoms. It is convenient to classify samples of matter according to the complexity of their composition, both at the atomic level and at the macroscopic level.

The elements are the simplest form of matter. An element contains only one type of atom and cannot be decomposed into other chemical components. Of the more than 100 known chemical elements, only a few are found in nature in their pure form. Figure 1-7 shows three of these: Diamonds are pure carbon, nuggets of pure gold can be found by panning in the right stream bed, and sulfur is found in abundance in its elemental form.

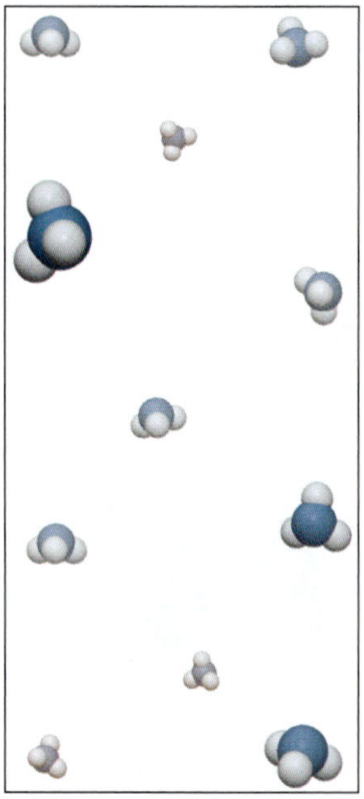

When two or more different chemical elements combine, they form a chemical compound. Even though they are made of more than one type of element, pure chemical compounds are uniform in composition; that is, all samples of a particular chemical compound contain the same proportions of each element. For example, ammonia is a chemical compound that contains the elements nitrogen and hydrogen in a 1:3 atomic ratio. A sample of pure NH_3 always contains nothing but ammonia molecules, each one containing three hydrogen atoms bonded to one nitrogen atom (Figure 1-9).

Although pure elements and pure compounds occur often, both in nature and in the laboratory, matter is usually a **mixture** of substances. A mixture contains two or more chemical substances. Unlike pure compounds, mixtures vary in composition because the proportions of the substances in a mixture can change. For example, dissolving sucrose, table sugar, in water forms a mixture that contains water molecules and sucrose molecules. A wide range of mixtures can be prepared by varying the relative amounts of sucrose and water.

A sample is **homogeneous** if it always has the same composition, no matter what part of the sample is examined. Pure elements and pure chemical compounds are homogeneous. Mixtures can be homogeneous, too; a homogeneous mixture usually is called a

Figure 1-9
The composition of a pure substance is homogeneous and invariant. A sample of pure ammonia contains nothing but molecules made of nitrogen atoms and hydrogen atoms in a 1:3 ratio.

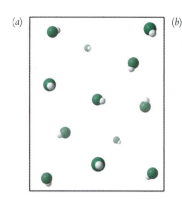
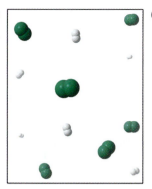
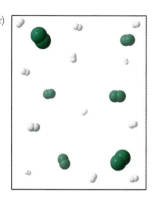

Figure 1-10
A sample of hydrogen chloride gas (*a*) is homogeneous and has constant composition. A mixture of hydrogen gas and chlorine gas is homogeneous but can have different compostions (*b* and *c*).

solution. As shown in Figure 1-10, the difference between a pure substance and a homogeneous mixture can be illustrated using hydrogen and chlorine. Under the right conditions, these two elements react in a 1:1 atomic ratio to give diatomic molecules of hydrogen chloride. Hydrogen chloride gas is a homogeneous *pure substance* and always contains equal numbers of hydrogen atoms and chlorine atoms linked in HCl molecules. Under other conditions, molecular hydrogen and molecular chlorine do not react with each other. The two gases form a homogeneous *solution* whose composition can be changed by adding more of either substance.

When different portions of a mixture have different compositions, the mixture is said to be **heterogeneous.** For example, quartz is a pure chemical compound made from silicon and oxygen, and gold is a pure element, but the lump of quartz containing a vein of gold that appears in Figure 1-11*a* is a heterogeneous mixture because different parts of the lump have different compositions.

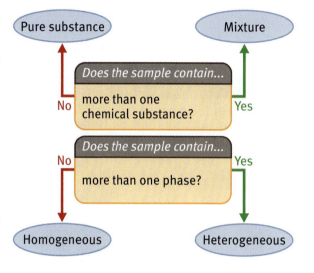

Phases of Matter

Matter can also be categorized into three distinct phases: solid, liquid, and gas. An object that is **solid** has a definite shape and volume that cannot be changed easily. Trees, automobiles, ice, and coffee mugs are all in the solid phase. Matter that is **liquid** has a definite volume but changes shape quite easily. A liquid flows to take on the shape of its container. Gasoline, water, and cooking oil are examples of common liquids. Solids and liquids are termed *condensed phases* because of their well-defined volumes. A **gas** has neither specific shape nor constant volume. A gas expands or contracts as its container expands or contracts. Helium balloons are filled with helium gas, and the Earth's atmosphere is made up of gas that flows continually from place to place. Molecular pictures that illustrate the three phases of matter appear in Figure 1-12.

The gas of the atmosphere is held in place by gravity, not by the walls of a container.

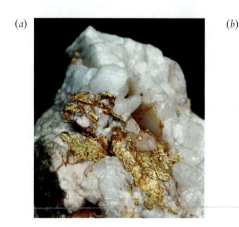

Figure 1-11
(*a*) Quartz containing a vein of gold is a heterogeneous mixture of two different solid substances. (*b*) A glass of ice water is a heterogeneous mixture of two different phases of a pure substance.

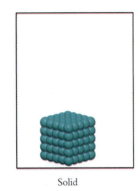  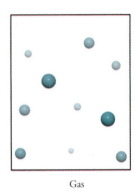

Solid   Liquid   Gas

Figure 1-12
Atomic views of the three different phases of matter.

Different phases can be mixed together, even when a substance is pure. A glass of water containing ice cubes (Figure 1-11*b*) is a heterogeneous mixture containing two separate phases of a homogeneous pure substance.

Example 1-1
Characteristics of Matter

Decide whether each of the following molecular pictures represents a pure substance, a homogeneous mixture, or a heterogeneous mixture. Tell whether the sample is a solid, a liquid, or a gas.

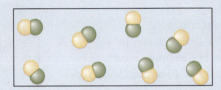 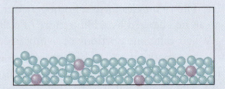

Strategy: Apply the characteristics described in the preceding paragraphs. Each circle in the figure represents an atom. Different colors distinguish one type of atom from another.

Solution: The sample on the left contains a collection of eight diatomic molecules. All the molecules have the same composition, so this is a pure substance. The molecules in the sample are distributed evenly through the entire volume of the container. This is the defining characteristic of a gas.

In the sample on the right, four atoms of one type are distributed evenly through a larger collection of atoms of a second type. This is a homogeneous mixture. Notice that the sample is spread across the bottom of the container, but it is also confined to a specific volume. These features identify the sample as a liquid.

 Do the Results Make Sense? Compare the pictures in the example with those in Figure 1-12 to see that the first sample has the distribution characteristic of a gas, while the second sample has the distribution characteristic of a liquid.

Extra Practice
Exercise 1.1

Draw a molecular picture of dry air, which is a gaseous mixture of 80% molecular nitrogen and 20% molecular oxygen.

Transformations of Matter

Although every substance exists in one particular phase under ordinary circumstances, under the right conditions most substances can be converted from one phase to another. Figure 1-13 shows water in all three phases. Under normal conditions, water changes from the liquid phase into a solid (ice) when the temperature drops below 0 °C. When the temperature rises above 100 °C, water changes from a liquid into a gas (steam). Similarly, lava is rock that has been heated sufficiently to convert it to the liquid phase, and "dry ice" is carbon dioxide that has been cooled enough to change it from the gas phase to the solid phase.

Figure 1-13
Water can exist as a solid (ice and snow), as a liquid, and as a gas (steam).

A process that changes the properties of a substance is a transformation. Transformations of matter are either physical or chemical. In a **physical transformation,** physical properties change but the substance's chemical nature remains the same. For example, when water freezes, it undergoes a physical transformation because the chemical makeup of ice is the same as that of liquid water: Both ice and liquid water contain molecules made up of two atoms of hydrogen and one atom of oxygen. Another physical transformation is the dissolving of a sugar cube in a cup of hot coffee. Although the sugar mixes uniformly with the coffee in this process, the chemical nature of each remains unchanged. The Tools for Discovery Box describes some atomic-level transformations.

A **chemical transformation,** on the other hand, produces new substances. For example, when magnesium metal burns in air, elemental magnesium metal and molecular oxygen combine chemically to form magnesium oxide, a white solid containing Mg and O atoms in 1:1 ratio. This is a chemical transformation that rearranges the atoms in Mg and O_2 to yield MgO.

Section Exercises

■ **1.4.1** Decide whether each of the following molecular pictures represents a single substance or a mixture. Identify any mixture as homogeneous or heterogeneous. If the picture represents a single substance, tell whether it is an element or a chemical compound.

 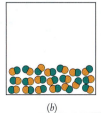 

(a)   (b)   (c)

■ **1.4.2** Classify each of the following as a pure substance, a solution, or a heterogeneous mixture: a cup of coffee, a lump of sugar, seasoned salt, a silver coin, and seawater.

■ **1.4.3** Classify each of the following as a physical or a chemical transformation: melting snow, burning coal, chopping wood, and digesting food.

Box 1-2 Tools for Discovery: Atomic-Level Microscopy

Atoms and molecules are much too tiny to see, even with the most powerful light microscopes. In recent years, however, scientists have developed a set of immensely powerful magnifying techniques that make it possible to visualize how individual atoms are arranged in solids. These techniques are scanning tunneling microscopy (STM) and atomic force microscopy (AFM).

Scanning tunneling microscopy and atomic force microscopy generate images of atoms and molecules using highly miniaturized lever arms with atomically sharp tips—like phonograph needles scaled down to the atomic level. These probes respond to the contours of individual atoms. In atomic force microscopy, the tip can be moved extremely precisely across a surface. The tip responds to individual atoms on that surface, moving up and down by tiny amounts as it passes over each atom. A laser beam (like the scanner at a supermarket checkout stand) focused on the edge of the lever arm detects this tiny motion, and an optical detector creates an image of the surface.

At their highest sensitivities, STM and AFM generate images that show how atoms are arranged on the surfaces they probe. At first, scientists used these tools to explore how atoms are arranged on surfaces. The example below shows individual atoms on the surface of nickel metal.

Increasingly, scientists use STM and AFM to explore chemical reactions. For example, molecules often undergo chemical reactions when they "stick" to a metal surface. Scanning tunneling microscopy can be used to monitor the atomic changes that take place during such chemical reactions. It is even possible to use STM to manipulate individual atoms. The figures at the bottom of this box show how iron atoms can be arranged on the surface of copper metal to make an "atomic corral."

Whereas STM is best suited for imaging atoms, atomic force microscopy is more appropriate for larger structures. The following image shows a strand of DNA. The two blue regions of the figure are protein molecules bound to the DNA.

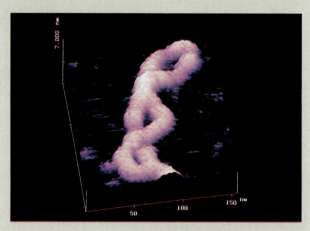

A strand of DNA

The next generation of molecular probes may be able to look within atoms to image their underlying structure. These probes combine magnetic resonance imaging (MRI) with the atomic force microscope. Physicians use MRI to do brain scans to pinpoint tumors or blood clots. By coupling MRI with an atomic force scanner and cooling the sample to extremely low temperatures, it may be possible to create images of the interiors of individual atoms.

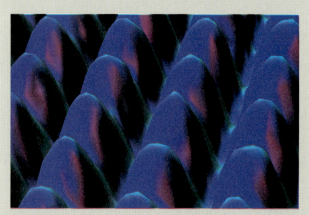

Ni metal surface

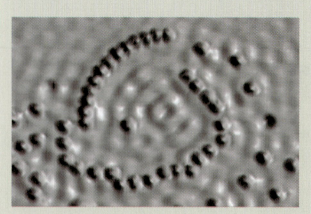

Iron atoms forming a ring

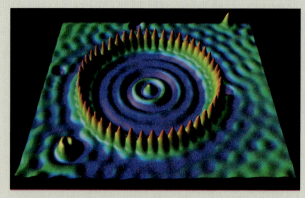

A complete "corral" of iron atoms

1.5 MEASUREMENTS IN CHEMISTRY

The knowledge that allows chemists to describe, interpret, and predict the behavior of chemical substances is gained by making careful experimental measurements. The properties of a sample can be divided into *physical properties,* which can be measured without observing a chemical reaction, and *chemical properties,* which are displayed only during a chemical transformation. Physical properties include familiar attributes such as size, color, and mass. Some chemical properties also are familiar to us. As examples, bleach reacts chemically with many colored substances to destroy their colors, and molecular oxygen reacts chemically with many fuels to generate heat.

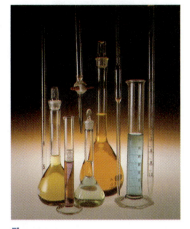

Figure 1-14
Some common laboratory glassware for measuring volume.

Physical Properties

Length, area, and volume measure the size of an object. **Length** refers to one dimension, **area** refers to two dimensions, and **volume** refers to three dimensions of space. Size measurements require standard measuring devices, such as rulers or measuring cups. Figure 1-14 shows some standard laboratory equipment for measuring volume.

In addition to its volume, every object possesses a certain quantity of matter, called its **mass (*m*).** Mass measurements are particularly important in chemistry. Consequently, highly accurate mass-measuring machines, called analytical balances, are essential instruments in chemistry laboratories. Analytical balances work by comparing forces acting on masses. A modern balance, whether in a delicatessen or in a chemistry laboratory (Figure 1-15), compares forces quickly and automatically, providing a digital readout of an object's mass.

Chemists measure **time (*t*)** because they want to know how long it takes for chemical transformations to occur. Some chemical reactions, such as the conversion of green plants into petroleum, may take millions of years. Other chemical processes, such as an explosion of dynamite, are incredibly fast. Whereas wristwatches typically measure time only to the nearest second, chemists have developed instruments that make it possible to study processes that occur in less than 0.000 000 000 000 01 second.

Most of us associate **temperature (*T*)** with the concepts of hot and cold. More accurately, however, temperature is the property of an object that determines the direction of heat flow. Heat flows naturally from a warm object to a cool object, from higher temperature to lower temperature. Heat is a form of energy, and because energy changes in chemical systems have important consequences, chemists are interested in temperature changes that occur during chemical transformations.

All experimental sciences rely on quantitative measurements of properties. Every measurement gives a numerical result that has three aspects: a numerical *magnitude;* an indicator of scale, called a *unit;* and a *precision.* Each aspect is essential, and all three must be reported to make a measurement scientifically valuable.

The process of determining mass is called *weighing.* Mass and weight are related, but they are not the same property. Mass is a fundamental characteristic of an object, whereas weight results from gravitational force acting on an object's mass. In outer space, objects have mass but no weight because there is no gravitational force.

(*a*)

(*b*)

Figure 1-15
Masses are determined using balances that compare two forces. (*a*) Delicatessen balance. (*b*) Students wearing protective glasses while using analytical balance.

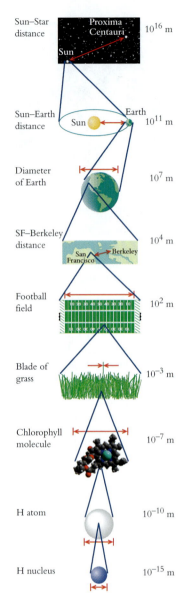

Figure 1-16
Dimensions of objects known to humans span more than 30 orders of magnitude, from interstellar distances (>10^{16} m) to the diameter of a hydrogen nucleus (~10^{-15} m).

Table 1-2   Frequently Used Scientific Prefixes for Magnitudes

Prefix	Symbol	Number	Exponential Notation
tera	T	1,000,000,000,000	10^{12}
giga	G	1,000,000,000	10^{9}
mega	M	1,000,000	10^{6}
kilo	k	1000	10^{3}
—	—	1	10^{0}
centi	c	0.01	10^{-2}
milli	m	0.001	10^{-3}
micro	μ	0.000 001	10^{-6}
nano	n	0.000 000 001	10^{-9}
pico	p	0.000 000 000 001	10^{-12}

Magnitude

The magnitudes of experimental values in science range from vanishingly small to astronomically large, as summarized in Figure 1-16. To simplify manipulating very large and very small numbers, we use powers of ten, also called **scientific notation** (see Appendix A for a review). For example, the diameter of a carbon atom is 0.000 000 0003 meters (symbol: m). This cumbersome number can be simplified by the use of scientific notation: 0.000 000 0003 m = 3×10^{-10} m.

Scientists routinely study objects whose sizes extend far beyond the narrow range encountered in daily life. Physicists, for example, study atomic nuclei measuring 10^{-15} m across, and astronomers study our universe, which spans about 10^{25} m. Chemists are most often interested in matter on the smaller side of this range. Length measurements in the laboratory vary from meters to subatomic sizes, 10^{-12} m. To further simplify the use of very large and very small numbers, scientists use prefixes that change the unit sizes by multiples of 10. For instance, the prefix pico means "10^{-12}." The symbol for picometer is pm. The diameter of a carbon atom is 3×10^{-10} m, which is 300 pm:

$$\text{Carbon diameter} = 3 \times 10^{-10} \text{ m} = 300 \times 10^{-12} \text{ m}$$
$$= 300 \text{ pm} \qquad \uparrow \text{pico}$$

The most common magnitude prefixes are listed in Table 1-2.

Units

The units associated with a numerical value are just as important as the value itself. If a recipe instruction reads, "Add 1 of sugar," that instruction is useless. The unit of measure, such as "teaspoon," must also be included. It is *essential* to include a unit with every experimental value.

A unit of measurement is an agreed-upon standard with which other values are compared. Scientists use the meter as the standard unit of length. The meter was originally chosen to be 10^{-7} times the length of a line from the North Pole to the equator. Volume can be measured in pints, quarts, and gallons, but the scientific units are the cubic meter and the liter. Temperature can be measured in degrees Fahrenheit (°F), degrees Celsius (°C), or kelvins (K).

The international scientific community prefers to work exclusively with a single set of units, the Système International (SI), which expresses each fundamental physical quantity in decimally (power of 10) related units. The seven base units of the SI are listed in Table 1-3. The SI unit for volume is obtained from the base unit for length: A cube that measures 1 meter on a side has a volume of 1 cubic meter.

Table 1-3   Base SI Units

Quantity	Unit	Symbol
Mass	Kilogram	kg
Length	Meter	m
Time	Second	s
Temperature	Kelvin	K
Amount	Mole	mol
Electric current	Ampere	A
Luminous intensity	Candela	cd

Unit Conversions

It is often necessary to convert measurements from one set of units to another. As an everyday example, travelers between the United States and Canada need to be able to convert between miles and kilometers. Chemists frequently need to convert volumes from one unit to another. The SI unit of volume is the cubic meter, but chemists usually work with much smaller volumes. Hence chemists often express volume using the liter (L), which is defined to be exactly 10^{-3} m³. Another volume unit in common use is the milliliter (mL), or 10^{-3} L. The milliliter is the same as the cubic centimeter (cm³). Figure 1-17 illustrates these volume measurements.

$$1 \text{ L} = 10^{-3} \text{ m}^3 \qquad 1 \text{ mL} = 10^{-3} \text{ L} = 1 \text{ cm}^3$$

A unit conversion starts with an equality between different units. For example, the equivalence between quarts and liters is

$$1 \text{ quart} = 0.946\ 353 \text{ L}$$

Each equality gives two conversion ratios. The equality between quarts and liters can be rearranged to give two ratios:

$$1 = \frac{1 \text{ quart}}{0.946\ 353 \text{ L}} \quad and \quad 1 = \frac{0.946\ 353 \text{ L}}{1 \text{ quart}}$$

To convert from one unit to another, we multiply by the ratio that leads to an appropriate cancellation of units. For example, a volume of exactly two liters is expressed in quarts as follows:

$$2 \text{ L} \left(\frac{1 \text{ quart}}{0.946\ 353 \text{ L}} \right) = 2.11338 \text{ quarts}$$

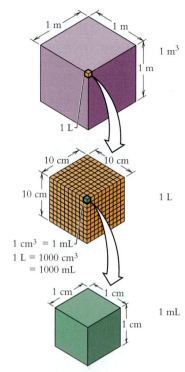

Figure 1-17
The defined unit of volume in SI is the cubic meter (m³). Chemists more commonly use the liter (L) or the milliliter (mL).

The correct conversion ratio leads to cancellation of unwanted units.

KEY CONCEPT

②

A unit equality may link SI units and non-SI units (1 quart = 0.946 353 L), decimally related units (10^6 cm³ = 1 m³), or base units and derived units (1 L = 10^{-3} m³). Some of the more common unit equalities are given on the inside back cover of this text. Examples 1-2 and 1-3 treat unit conversions.

Example 1-2

Unit Conversions

Two rock climbers stand at the bottom of a rock face they estimate to be 155 feet high. Their rope is 65 m long. Is the rope long enough to reach the top of the cliff?

Strategy: The team must compare the English measure for the height of the cliff and the SI measure for the length of their rope. Is 65 m more or less than 155 feet? The equivalence between feet and meters can be found on the inside back cover of the book: 1 ft = 0.3048 m.

Solution: The length of the rope is given in meters. To set up a ratio that converts meters to feet, we need to have feet in the numerator and meters in the denominator:

$$0.3048 \text{ m} = 1 \text{ ft} \qquad so \qquad 1 = \frac{1 \text{ ft}}{0.3048 \text{ m}}$$

Now multiply the length of the rope by the conversion ratio:

$$(65 \text{ m}) \left(\frac{1 \text{ ft}}{0.3048 \text{ m}} \right) = 2.1 \times 10^2 \text{ ft}$$
$$\uparrow \text{ conversion ratio}$$

| **Example 1-2** | The two climbers have plenty of rope to climb the rock face. |

Unit Conversion
(continued)

Does the Result Make Sense?   Notice that we multiply the length of the rope in meters by feet/meter so that meters cancel. Multiplying the height of the cliff by the same conversion ratio would have given nonsensical units: (ft)(ft/m) = ft²/m. For a quick estimate, realize that there are a bit over three feet in one meter, so 65 m is a bit more than 195 ft.

Extra Practice Exercise 1.2

A bottle of lemon juice concentrate contains 24 fluid ounces. What is this volume in liters? (There are 32 fluid ounces in 1 quart.)

Example 1-3

Multiple Unit Conversions

The speed limit on many highways in the United States is 55 miles/hr. What is this speed limit in SI units?

Strategy: We are asked to make a unit conversion. The SI base unit of length is the meter, and the SI base unit of time is the second. It is necessary to convert from miles to meters and from hours to seconds. The appropriate unit equivalences are

$$1 \text{ mile} = 1.6093 \text{ km}, \quad 1 \text{ km} = 10^3 \text{ m}, \quad 1 \text{ hr} = 60 \text{ min}, \quad \text{and} \quad 1 \text{ min} = 60 \text{ s}$$

Solution: The speed limit in miles per hour is given. To cancel miles, multiply by a ratio that has miles in the denominator. To obtain meters, use a ratio that has meters in the numerator. We convert miles to kilometers and then convert kilometers to meters:

$$\left(\frac{55 \text{ miles}}{\text{hr}}\right)\left(\frac{1.6093 \text{ km}}{1 \text{ mile}}\right)\left(\frac{10^3 \text{ m}}{\text{km}}\right) = 8.9 \times 10^4 \text{ m/hr}$$

conversion ratio ↑   ↑ conversion ratio

Notice that the conversion ratios are set up to cancel the unwanted units.

This completes the conversion into SI units of length. Now convert from hours to seconds, following a similar procedure. Because hours appear in the denominator, multiply by a ratio that has hours in the numerator:

$$\left(\frac{8.9 \times 10^4 \text{ m}}{\text{hr}}\right)\left(\frac{1 \text{ hr}}{60 \text{ min}}\right)\left(\frac{1 \text{ min}}{60 \text{ s}}\right) = 25 \text{ m/s}$$

Does the Result Make Sense?   The units are correct, and 25 m/s is a reasonable speed for an automobile.

Extra Practice Exercise 1.3

A hybrid automobile gets 45 miles per gallon of gasoline in city traffic. What is this mileage in km/L?

The Fahrenheit scale, in which water freezes at 32 °F and boils at 212 °F, is still in common use in the United States, but scientists rarely use the Fahrenheit scale. The formula for converting temperature from Fahrenheit to Celsius is

$$T_C = \frac{5 \text{ °C}}{9 \text{ °F}} (T_F - 32 \text{ °F})$$

Scientists use two units for temperature, the Celsius (°C) scale and the Kelvin (K) scale. These scales are shown schematically in Figure 1–18. Unlike other scientific units, the unit size of the Celsius and Kelvin scales is the same, but their zero points differ. For both scales, the difference in temperature between the freezing and boiling points of water is defined to be 100 units. However, the temperature at which ice melts to liquid water is 0 °C and 273.15 K.

The conversion between kelvins and degrees Celsius is straightforward because their temperature (T) units are the same size. A temperature change of 1 °C is the same as a temperature change of 1 K. To convert from one scale to the other, add or subtract 273.15:

$$T(K) = T(°C) + 273.15 \qquad T(°C) = T(K) - 273.15 \qquad \textbf{(1-1)}$$

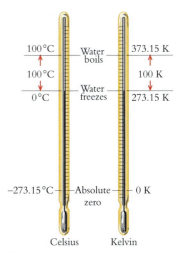

Figure 1-18
Celsius and Kelvin temperature scales. Kelvins and degrees Celsius have the same unit size but different zero points. "Room temperature" is typically about 295 K (22 °C).

Precision and Accuracy

The exactness of a measurement is expressed by its **precision.** This concept can be explained with an example. Suppose three swimmers are discussing the temperature of a swimming pool. The first dips a finger in the water and says that the temperature is "about 24 °C." The second examines an immersed pool thermometer and reports the temperature to be 26 °C. The third swimmer, who has been monitoring daily variations in the pool's temperature, uses a portable precision digital thermometer and reports, "According to my precision thermometer, the pool temperature is 25.8 °C."

The swimmers have measured the water temperature using different measuring devices with different levels of precision. The first swimmer's "finger test" is precise to about 3 °C: $T = 24 \pm 3$ °C (read "twenty-four plus or minus three degrees"). The pool thermometer gives a reading that is precise to the nearest degree: $T = 26 \pm 1$ °C. The precision thermometer measures to the nearest tenth of a degree: $T = 25.8 \pm 0.1$ °C. An actual temperature between 25.7 and 25.9 °C falls within the precision ranges of all three measurements, so all three are correct to within their stated limits of precision.

Whereas precision describes the exactness of a measurement, **accuracy** describes how close a measurement is to the true value. Figure 1-19 illustrates the difference between precision and accuracy. An expert archer shoots arrows with high precision, all the arrows hitting the target near the same spot (*a* and *b*). Under calm conditions, the expert is also quite accurate, shooting all the arrows into the bull's-eye (*a*). In a strong wind, however, the archer may be precise but not accurate, regularly missing the bull's-eye (*b*). In contrast, a novice shooter is not very precise (*c* and *d*): The arrows scatter over a large area. The novice usually is not accurate (*c*), but an occasional arrow may accurately find the bull's-eye (*d*).

The goal of any measurement in science is to be as accurate as possible. However, determining the accuracy of a measurement is much harder than determining precision. An archer can examine the target to see if the arrows found their mark, but a scientific researcher studying a new phenomenon does not know the correct value. A scientist can assess precision by making additional measurements to find out how closely repeated measurements agree with one another. In contrast, assessing accuracy requires careful attention to the design of an experiment and the instruments used in that experiment.

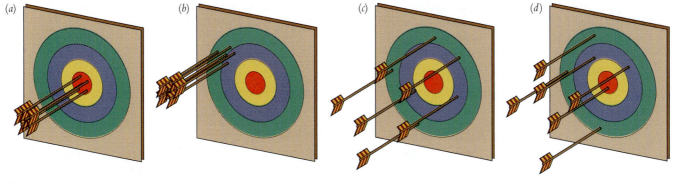

(*a*) (*b*) (*c*) (*d*)

Figure 1-19
Patterns of arrows striking a target illustrate the notions of precision and accuracy. (*a*) is precise and accurate. (*b*) is precise but inaccurate. (*c*) is neither precise nor accurate. (*d*) is an imprecise pattern with one accurate shot.

Significant Figures

Scientific measurements should always include both magnitude and precision. Indicating the precision limits with a plus/minus ($\pm$) statement is cumbersome, particularly when many numerical values are reported at one time. Scientists have agreed to simplify the reporting of precision. Experimental measurements are written so that there is an uncertainty of up to one unit in the last reported digit. For example, a temperature reported as 26 °C is greater than 25 °C but less than 27 °C, or 26 $\pm$ 1 °C. A temperature reported as 25.8 °C is 25.8 $\pm$ 0.1 °C. The number of digits expressed in a numerical value is called the number of **significant figures**. The value 26 has two significant figures, whereas 25.8 has three significant figures.

Zeros can present a problem when determining the precision of a numerical value. This is because zeros are needed both to locate the decimal point and to express precision. From the statement that the sun is 93,000,000 miles from the Earth, we cannot tell whether the measurement is precise to eight significant figures or whether the zeros are there only to put the decimal point in the right place. Scientific notation eliminates this ambiguity because a power of 10 locates the decimal point, leaving us free to indicate the precision by the number of digits. For example, a distance of 9.3×10^7 miles means that the number is precise to $\pm 0.1 \times 10^7$ miles; the number has two significant figures. If this distance is known to a precision of $\pm 0.01 \times 10^7$ miles, it is written as 9.30×10^7 miles, with three significant figures. Zeros at the end of a number with a decimal point always indicate increased precision. Writing "0.010" in scientific notation clarifies that this number has two significant figures rather than three or four: $0.010 = 1.0 \times 10^{-2}$.

To determine how many significant figures there are in a particular numerical value, read the number from left to right and count all the digits, starting with the first digit that is not zero. To avoid ambiguities about trailing zeros, we place a decimal point after a value when its trailing zeros are significant. For example, "110" has only two significant figures (110 $\pm$ 10), whereas "110." has three significant figures (110 $\pm$ 1). Here are some examples of significant figures:

500	1 significant figure	505	3 significant figures
0.05	1 significant figure	5.00×10^3	3 significant figures
55	2 significant figures	5.000	4 significant figures
50.	2 significant figures	505.0	4 significant figures

Section Exercises

1.5.1 Convert the following measurements to scientific notation and express in base SI units: 0.000 463 L, 17,935 km, and 260,000 hr (precise to three significant figures).

1.5.2 One light-year is the distance light travels in exactly one year. The speed of light is 6.7×10^8 miles/hr. Express the speed of light and the length of one light-year in SI units.

1.5.3 Convert each of these measurements to SI units: 155 pounds (mass of a typical person), 120.0 yards (full length of a football field), 39 °C (body temperature of someone with a slight fever), and 365.2422 days (length of 1 year).

1.6 CALCULATIONS IN CHEMISTRY

During an experiment, a chemist may measure physical quantities such as mass, volume, and temperature. Usually the chemist seeks information that is related to the measured quantities but must be found by doing calculations. In later chapters we develop and use equations that relate measured physical quantities to important chemical properties. Calculations are an essential part of all of chemistry; therefore, they play important roles in much of general chemistry. The physical property of density illustrates how to apply an equation to calculations.

Table 1-4 Densities of Some Solids and Liquids

Element	Density (g/cm³)	Substance	Density (g/cm³)
Aluminum (Al, solid)	2.70	Acetic acid ($C_2H_4O_2$, liquid)	1.05
Bromine (Br_2, liquid)	3.10	Acetone (C_3H_6O, liquid)	0.791
Copper (Cu, solid)	8.96	Benzene (C_6H_6, liquid)	0.885
Gold (Au, solid)	19.3	Chloroform ($CHCl_3$, liquid)	1.49
Iodine (I_2, solid)	4.93	Cork (solid)	0.24
Iron (Fe, solid)	7.87	Diethyl ether ($C_4H_{10}O$, liquid)	0.714
Lithium (Li, solid)	0.532	Ethanol (C_2H_6O, liquid)	0.785
Lead (Pb, solid)	11.34	Octane (C_8H_{18}, liquid)	0.703
Mercury (Hg, liquid)	13.55	Quartz (SiO_2, solid)	2.65
Silver (Ag, solid)	10.50	Sodium chloride (NaCl, solid)	2.165
Sodium (Na, solid)	0.968	Water (H_2O, liquid)	1.00
Titanium (Ti, solid)	4.54	Wood (balsa, solid)	0.12
Zinc, (Zn, solid)	7.14	Wood (pine, solid)	0.35 – 0.50

Density

The **density** (ρ) of an object is its mass (m) divided by its volume (V):

$$\text{Density} = \frac{\text{Mass}}{\text{Volume}} \quad \text{or in symbols} \quad \rho = \frac{m}{V} \qquad (1\text{-}2)$$

The symbol for density is the Greek lower-case letter rho, ρ.

Every pure liquid or solid has a characteristic density that helps distinguish it from other substances. To give one example, the density of pure gold is 19.3 g/cm³, whether the sample is a nugget in a miner's pan or an ingot in a bank vault. Pyrite, an iron compound that resembles gold, has a much lower density of 5.0 g/cm³. Table 1-4 lists the densities of several common solids and liquids. The densities of liquids and solids range from 0.1 to 20 g/cm³. This 200-fold variation is readily apparent to us and dictates how different materials are used in applications in which density is important. Fishermen use floats and sinkers made of low- and high-density materials, respectively, as Figure 1-20 illustrates. Low-density cork is used for floats, because low-density materials float in water. High-density lead is used for sinkers, because high-density materials sink.

Unlike mass and volume, density does not vary with the amount of a substance. Notice in Figure 1-20 that all the corks float, regardless of their sizes. Notice also that all the pieces of lead sink, regardless of their sizes. Dividing a sample into portions changes the mass and volume of each portion but leaves the density unchanged. A property that depends on amount is called **extensive.** Mass and volume are two examples of extensive properties. A property that is independent of amount is called **intensive.** Density and temperature are intensive properties.

Mass and volume often can be measured easily, and density is then calculated using Equation 1-2. The equation can also be rearranged to find an object's volume or mass, as Example 1-4 illustrates.

Figure 1-20
Cork has a lower density than water, so corks of all sizes float on water. Lead has a higher density than water, so lead pieces of all sizes sink in water.

A diamond is a pure sample of the element carbon. The tabulated density of diamond is 3.51 g/cm³. Jewelers use a unit called a carat to describe the mass of a diamond: 1 carat = 0.200 g. What is the volume of the stone in a 2.00-carat diamond engagement ring?

Strategy: Given the diamond's mass and density, we are asked to find its volume. Rearranging the density equation makes this possible:

$$\text{Density} = \frac{\text{Mass}}{\text{Volume}} \quad so \quad \text{Volume} = \frac{\text{Mass}}{\text{Density}}$$

Example 1-4

Using Density

Example 1-4

Using Density
(continued)

Solution: A list of the information given in the problem allows us to determine what to substitute into the equation:

$$Density_{diamond} = 3.51 \text{ g/cm}^3$$

$$Mass = 2.00 \text{ carat}$$

$$1 \text{ carat} = 0.200 \text{ g}$$

In the volume equation, density and mass must have consistent units. Thus mass must be converted from carats to grams using a conversion ratio that eliminates the unwanted unit:

$$2.00 \text{ carat} \left(\frac{0.200 \text{ g}}{1 \text{ carat}} \right) = 0.400 \text{ g}$$

↑ conversion ratio

Now substitute into the equation for volume:

$$Volume = \frac{Mass}{Density} = \frac{0.400 \text{ g}}{3.51 \text{ g/cm}^3} = 0.114 \text{ cm}^3$$

Does the Result Make Sense?   Think of a cube that is 1 cm on each side. The calculated volume is about $^1/_{10}$ this size, a reasonable value for a diamond.

**Extra Practice
Exercise 1.4**

Jewelers measure masses of precious metals in troy ounces (1 troy oz. = 31.103 g). A jeweler has a silver ingot weighing 6.44 troy oz. What is the volume of this ingot?

Precision of Calculations

The precision of a measuring instrument determines the precision of a single measurement. However, many scientific investigations require several measurements, often involving more than one instrument. Data obtained from multiple measurements are then used to calculate a quantity of interest. How precise is a value obtained by calculations? As a general rule, the least precise measurement determines the precision of a result. The following example, again using density, illustrates this guideline.

A chemical manufacturer prepares a silicone fluid for use as a lubricant. To determine the density of the fluid, a technician measures 25.0 cm³ into a graduated cylinder and determines the mass of this amount of fluid to be 39.086 g. Dividing mass by volume on a calculator, the technician reads the following result: 1.56344. What value does the technician report for the fluid's density? Are all six of these digits significant? A set of measurements is limited by the precision of the least-sensitive instrument used in the experiments. The mass is known to five significant figures, but there are only three significant figures in the volume. Thus only three significant figures are used to report the density of the fluid:

$$\rho = \frac{39.0876 \text{ g}}{25.0 \text{ cm}^3} = 1.56 \text{ g/cm}^3$$

Three guidelines can be used to determine the precision of a sequence of mathematical operations:

1. **When *adding* or *subtracting*, the number of decimal places in the result is the *number of decimal places* in the number with the fewest places.**

0.0120	**4 decimal places**	3 significant figures
+1.6	**1 decimal place**	2 significant figures
+8.49026	**5 decimal places**	6 significant figures
10.1	**1 decimal place**	3 significant figures

The value with the *fewest decimal places* determines the number of decimal places in a sum.

2. When *multiplying* or *dividing*, the number of significant figures in the result is the same as in the quantity with the *fewest significant figures*.

	(0.0120)	×	(1.6)	×	(8.49026)	=	(0.16)
Significant figures	3		2		6		2
Decimal place(s)	4		1		5		2

The value with the *fewest significant figures* determines the number of significant figures in a product.

3. **Postpone adjusting results to the correct number of significant figures until a calculation is complete.**

In a sequence of computations, adjusting the number of significant figures in intermediate results can lead to errors in the final value. Instead, wait until the computations are complete, and then express the final value with the appropriate number of significant figures.

When exact numbers are used, their presence has no effect on the significant figures in the result. For example, we can convert 1.855 hours into seconds.

$$1.855 \ \text{hr} \left(\frac{60 \ \text{min}}{1 \ \text{hr}} \right) \left(\frac{60 \ \text{s}}{1 \ \text{min}} \right) = 6678 \ \text{s}$$

Calculators often display more significant figures than the data justify.

How many significant figures should be used? Although the conversion ratios contain only two digits, they are exact numbers. By definition, there are *exactly* 60 minutes in an hour and *exactly* 60 seconds in a minute. Thus the precision of this product is determined by the precision of the time in hours, which is four significant figures. We report the time in seconds to four significant figures.

When a number is exact (such as 60 min/hr), precision need not be specified.

Calculators usually give more significant figures than are justified. For example:

$$0.0120 + 1.6 + 8.49026 = \boxed{10.1}0226 \ \text{(calculator result)}$$

$$(0.0120)(1.6)(8.49026) = 0.\boxed{163} \ 012 \ 992 \ \text{(calculator result)}$$

In each of these calculations, the calculator displays extra digits that are not significant. The significant digits are highlighted in brown. A calculator cannot distinguish the number of significant figures that is appropriate for a calculation. Many calculators can be set so that they display a predetermined number of digits; however, these also do not necessarily represent the correct number of significant figures.

KEY CONCEPT

Always adjust calculator results to the appropriate number of significant figures using the three guidelines given earlier.

In the process of dropping extra digits beyond the precision of the result, a conventional guideline is used for rounding off results. If the first digit that will be removed is 5 or greater, round the last remaining digit upward by one unit. If the first digit that will be removed is smaller than 5, leave the last remaining digit unchanged.

Example 1-5 shows how to apply these guidelines.

Example 1-5

Significant Figures

A farmer owns a rectangular field that fronts along a road. State highway engineers surveyed the frontage and found that it measures 138.3 m in length. The farmer, who wants to build a fence around the field, paces off the field's width and estimates it to be 52 m. How many meters of fence will the farmer have to build? What mass of fertilizer will the farmer need to fertilize the field with 0.0050 kg of fertilizer for each square meter of field?

Example 1-5

Significant Figures
(continued)

Strategy: First, calculate the field's perimeter to determine the amount of fencing needed. Then calculate the area of the field to determine how much fertilizer is needed. The precision of the results must also be determined.

Solution:

$$\text{Perimeter} = \text{Length} + \text{Width} + \text{Length} + \text{Width}$$

$$\text{Perimeter} = 138.3 \text{ m} + 52 \text{ m} + 138.3 \text{ m} + 52 \text{ m} = 380.6 \text{ m}$$

According to the guideline for adding or subtracting, this result must have the same number of decimal places as the least precise measurement. In this case, pacing off the width of the field limits the precision of the calculation to the nearest meter. The value should be rounded up because the "6" in "380.6" is larger than 5. The farmer requires 381 m of fence.

Calculate the area of the field to determine how much fertilizer is needed:

$$\text{Area} = (\text{Length})(\text{Width}) = (138.3 \text{ m})(52 \text{ m}) = 7191.6 \text{ m}^2$$

This result has too many significant figures, but we carry the extra digits until the calculation is complete. Each square meter of field requires 0.0050 kg of fertilizer. Find the required amount of fertilizer by multiplying:

$$\text{Mass}_{\text{fertilizer}} = (7191.6 \text{ m}^2)(0.0050 \text{ kg/m}^2) = 35.958 \text{ kg}$$

Now we are ready to round off. The multiplication steps include two numbers that have only two significant figures. The guideline for multiplying or dividing indicates that the result should also have two significant figures. The farmer requires 36 kg of fertilizer for this field.

Does the Result Make Sense?   Notice that the units work out correctly in this calculation: area multiplied by mass per unit area gives mass. There are 2.2 pounds in 1 kg, so a mass of 36 kg is about 100 lb. This is a reasonable mass of fertilizer for a field.

Extra Practice Exercise 1.5

A rectangular glass tabletop measures 60.0 in. by 36.0 in. and is 0.66 in. thick. What are the perimeter and volume of the tabletop, in metric units (1 in. = 2.54 cm)?

Section Exercises

1.6.1 Calculate the mass of a cylindrical cork of radius 1.00 cm and length 4.00 cm. (Volume of a cylinder is $V = \pi r^2 h$).

1.6.2 Carry out the following calculations and state each result with the correct number of significant figures: (a) 14.55 cm − 135.5 mm;

(b) $\dfrac{(175.5 \text{ g} - 143.5 \text{ g})}{(187.5 \text{ mL} - 165.3 \text{ mL})}$;

(c) 37.7 mi divided by 1.5 hr.

1.6.3 Which of the following quantities can be determined exactly, and which must be measured with some degree of uncertainty? (a) mass of a gold nugget, (b) number of seconds in 1 day, (c) time it takes a sprinter to run the 100-m dash, (d) speed of light, and (e) number of centimeters in exactly 1 mile.

1.7 CHEMICAL PROBLEM SOLVING

Many students have difficulty solving chemistry problems, but problem solving is a skill that you can learn and master. To help you, we devote extra attention to setting up and solving problems. Throughout this book, the Examples include information about methods for finding solutions.

Each Example includes a stategy that describes how to approach the problem. After describing the solution, we offer an Extra Practice Exercise designed to help you master the particular type of problem while the example is fresh in your mind. The answers to these Extra Practice Exercises appear at the end of each chapter, along with the answers to the Section Exercises.

Chemical problems that call for calculations often involve several tasks. It can be hard to recognize where to start and how to negotiate these tasks. A stepwise procedure can help you master theses types of calculations. We recommend the sequence of steps shown in the box. When you see a problem that appears complicated, apply the stepwise procedure.

The first step in any problem is to identify what you are asked to find. Next, try to visualize what is going on. Think about the chemistry and draw pictures to help you see what is taking place. Once you have a good idea about the problem, determine which equations apply to the calculation and organize the data that are provided. We identify the most important equations by giving them numbers and screening them in yellow, and we recommend that you memorize these equations. Once you have organized the equations and data, you can perform appropriate manipulations and substitutions to carry out the calculation correctly. After you have finished the problem, check to see that your answer makes sense. Checking your answer for its reasonableness will keep you from making careless errors.

Example 1-6 provides a first illustration of this stepwise procedure.

SOLVING QUANTITATIVE PROBLEMS

Step 1 Determine what is asked for.
What type of problem is it?
What are you asked to find?

Step 2 Visualize the problem.
Draw pictures that illustrate what takes place.
If chemical changes occur, draw molecular pictures.

Step 3 Organize the data.
What data are available?
How are the data related to what is asked for?

Step 4 Identify a process to solve the problem.
What concepts are required?
What equations apply?

Step 5 Manipulate the equations.
If necessary, do calculations in steps.
Solve for what is asked for.

Step 6 Substitute and calculate.
Keep track of units.
Use the correct number of significant figures.

Step 7 Does the result make sense?
Are the units consistent?
Is the result sensible?

Example 1-6

Problem-Solving Strategy

Electroplating is a process by which a metal such as copper is coated with another metal, such as silver or chromium. The transfer of metal atoms is driven by an electrical current. In an electroplating process, a spoon is coated with silver from a silver rod. In the process, 1.0×10^{21} atoms are transferred from the rod to the spoon, and the rod loses 0.179 g of its mass. Use this information and the density of silver to estimate the volume occupied by one silver atom.

Strategy: Apply the seven-step approach. The first two steps are part of the strategy.

1. Determine what is asked for. The problem asks for the volume of a silver atom.

2. Visualize the problem. Draw a picture that shows the setup and the information provided in the problem. A picture often helps to interpret and summarize a problem. In this case, the figure should show the spoon, the silver wire, and some indication that atoms are transferred from the wire to the spoon.

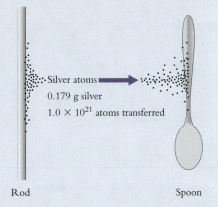

Silver atoms
0.179 g silver
1.0×10^{21} atoms transferred

Rod   Spoon

Spoons are electroplated with silver to give them an attractive finish.

Solution:

3. Organize the data. First, what data are given in the problem?

$$\text{Mass of silver} = 0.179 \text{ g}$$

$$\text{Number of atoms transferred} = 1.0 \times 10^{21} \text{ atoms}$$

Second, what information can be found in tables?

$$\text{Density of silver} = \rho = 10.50 \text{ g/cm}^3 \text{ (see Table 1-4)}$$

Example 1-6

Problem–Solving Strategy *(continued)*

4. Identify a process to solve the problem. The question asks about the volume of one silver atom. Mass and volume are related through density: $\rho = m/V$. From this equation, we can calculate the total volume of the silver atoms. The problem also gives the total number of silver atoms transferred from the wire to the spoon. The volume of a single atom is the total volume divided by the number of atoms. Oftentimes, a flow chart helps to summarize the process:

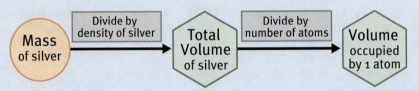

5. Manipulate the equations as needed. The flow chart shows two division steps.

   To begin the calculations, we use the mass and density to calculate the total volume:

   $$\rho = m/V \quad \text{so} \quad V = m/\rho$$

   This is V_{total}, which we divide by the number of atoms to find the volume of a single atom:

   $$V_{\text{atom}} = \frac{V_{\text{total}}}{\text{Number of atoms}}$$

6. Substitute and calculate.

   $$V_{\text{total}} = \frac{m}{\rho} = \frac{0.179 \text{ g}}{10.50 \text{ g/cm}^3} = 1.70 \times 10^{-2} \text{cm}^3$$

   $$V_{\text{atom}} = \frac{V_{\text{total}}}{\text{\# of atoms}} = \frac{1.70 \times 10^{-2} \text{ cm}^3}{1.0 \times 10^{21} \text{ atoms}} = 1.7 \times 10^{-23} \text{ cm}^3/\text{atom}$$

   Notice that units and significant figures are included in each step.

7. Does the Result Make Sense? Is this a "reasonable" value for the volume of an atom? At this stage of your study of chemistry, you cannot easily answer this question. Atoms are unimaginably small, however, and a volume of 10^{-23} cm^3 certainly is unimaginably small. Also, the units are right: cm^3 is a measure of volume, and the calculation yields units of volume per atom, as the question asked.

Extra Practice Exercise 1.6

A cylindrical jar has an inside diameter of 3.00 cm and a height of 8.00 cm. The empty jar weighs 185.65 g. Filled with gasoline, the same jar weighs 225.40 g. Find the density of gasoline.

Section Exercises

1.7.1 The standard unit of length in the land of Ferdovia is the frud (2.000 frud = 10^{-2} m). Fast-food restaurants in Ferdovia sell fruit drinks in rectangular cartons measuring 50.00 by 40.00 by 80.0 fruds. The Frod family left a freshly opened carton on their kitchen table, and their pet ferret (Fred) drank some of the fruit drink. Mrs. Frod found the height of the remaining liquid to be 61 fruds. How many cubic meters of fruit drink did Fred drink?

1.7.2 A technician used a section of glass tubing to measure the density of a liquid that the laboratory needed to identify. The inside diameter of the tubing was known to be 0.87 mm. An empty piece of the tubing weighed 0.785 g. When liquid was drawn into the tubing to a height of 4.0 cm, the tubing and liquid weighed 0.816 g. Find the density of the liquid.

CHAPTER 1 VISUAL SUMMARY

Important equations

Essential terms

Key concepts

SKILLS TO MASTER

① Using the periodic table of the elements ③ Keeping track of significant figures

② Applying unit conversions ④ Analyzing and solving problems

1.1 WHAT IS CHEMISTRY?

Studies: properties and interactions of matter

Advances via: technology and imagination

Method: observations, hypotheses, experiments, and theories

1.2 ATOMS, MOLECULES AND COMPOUNDS

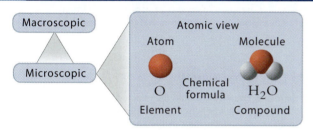

Macroscopic

Microscopic

Atomic view

Atom

Molecule

O

Chemical formula

H_2O

Element

Compound

A molecule is two or more atoms bound in a specific shape.

1.3 THE PERIODIC TABLE OF THE ELEMENTS

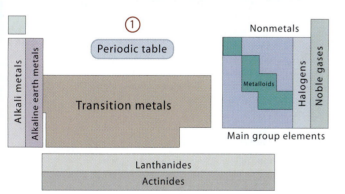

①

Periodic table

Alkali metals

Alkaline earth metals

Transition metals

Nonmetals

Metalloids

Halogens

Noble gases

Main group elements

Lanthanides

Actinides

1.4 CHARACTERISTICS OF MATTER

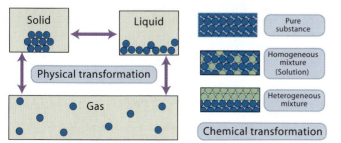

Solid

Liquid

Physical transformation

Gas

Pure substance

Homogeneous mixture (Solution)

Heterogeneous mixture

Chemical transformation

1.5 MEASUREMENTS IN CHEMISTRY

Precision   Accuracy

Significant figures

Time (t)   ②   ③   Volume (V)

Temperature (T)   Length

Scientific notation   Mass (m)

$$T \text{ (K)} = T \text{ (°C)} + 273.15$$
$$T \text{ (°C)} = T \text{ (K)} - 273.15$$

1.6 CALCULATIONS IN CHEMISTRY

Adjust calculator results to the appropriate number of significant figures.

Extensive   Intensive

Density $\rho = \dfrac{m}{V}$ Mass

Volume

1.7 CHEMICAL PROBLEM SOLVING

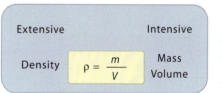

SOLVING QUANTITATIVE PROBLEMS

Step 1: Determine what is asked for.   ④

Step 2: Visualize the problem.

Step 3: Organize the data.

Step 4: Identify a process to solve the probem.

Step 5: Manipulate the equations.

Step 6: Substitute and calculate.

Step 7: Does the result make sense?

Learning Exercises

1.1 Make a list of all terms new to you in Chapter 1. Give a one-sentence definition for each. Consult the glossary if you need help.
1.2 The following skills are introduced in Chapter 1. Write down a "plan of attack" for accomplishing each: (a) classifying samples of matter; (b) interpreting and drawing simple molecular pictures; (c) working with the density equation; (d) using scientific notation; (e) determining significant figures; (f) converting between different units; and (g) solving chemistry problems.

1.3 Go through the end-of-chapter problems and identify those that require the skills listed in Learning Exercise 1.2. Some problems may require more than one skill, and some require skills that are not listed.
1.4 "Memory bank" equations are those important enough for you to memorize. Start a list of these equations and learn them. Chapter 1 has only two memory bank equations.

 Problems   <u>ilw</u> = interactive learning ware problem. Visit the website at www.wiley.com/college/olmsted

What Is Chemistry?

1.1 Describe three political problems for which knowledge of chemistry would be helpful.
1.2 Describe three environmental problems for which knowledge of chemistry would be helpful.
1.3 List three reasons why it is important for a pharmacy student to learn about chemistry.
1.4 List three reasons why it is important for an engineering student to learn about chemistry.
1.5 A chemist says, "My results must be wrong because they don't agree with the theory." Criticize this statement, and make recommendations for what the chemist should do next.
1.6 A leading manufacturer announces a new detergent ingredient that is "completely biodegradable and nontoxic." If you were a chemist working for a consumer products testing company, what recommendations would you make for testing this claim?

Atoms, Molecules, and Compounds

1.7 What is the elemental symbol for each of the following elements? (a) hydrogen; (b) helium; (c) hafnium; (d) nitrogen; (e) neon; and (f) niobium.
1.8 What is the elemental symbol for each of the following elements? (a) potassium; (b) platinum; (c) plutonium; (d) lead; (e) palladium; and (f) phosphorus.
1.9 What is the name of each of the following elements? (a) As; (b) Ar; (c) Al; (d) Am; (e) Ag; (f) Au; (g) At; and (h) Ac.
1.10 What is the name of each of the following elements? (a) Br; (b) Be; (c) B; (d) Bk; (e) Ba; and (f) Bi.
1.11 Examine the following molecular pictures and determine the corresponding molecular formulas.

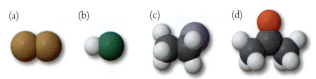

(a)   (b)   (c)   (d)

1.12 Examine the following molecular pictures and determine the corresponding molecular formulas.

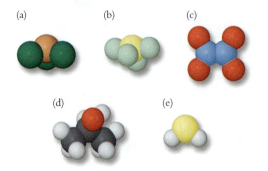

(a)   (b)   (c)

(d)   (e)

1.13 Write the chemical formula of the compound whose molecules contain: (a) four atoms of chlorine and one atom of carbon; (b) two atoms of oxygen and two atoms of hydrogen; (c) four phosphorus atoms and ten oxygen atoms; (d) two atoms of iron and three atoms of sulfur.
1.14 Write the chemical formula of the compound whose molecules contain: (a) five carbon atoms and twelve hydrogen atoms; (b) four atoms of fluorine and one atom of silicon; (c) two atoms of nitrogen and five atoms of oxygen; (d) one iron atom and three chlorine atoms.

The Periodic Table of the Elements

1.15 What element is just after xenon in the periodic table?
1.16 What element is just before sodium in the periodic table?
1.17 What are the names and chemical symbols of the elements that are vertical and horizontal neighbors of sulfur in the periodic table? Which of these have chemical properties similar to those of sulfur?
1.18 What are the names and chemical symbols of the elements that are vertical and horizontal neighbors of tin in the periodic table? Which of these have chemical properties similar to those of tin?
1.19 Name two metals that react with bromine to give compounds with the chemical formula MBr.
1.20 Name two elements that react with oxygen to give compounds with the chemical formula MO.
1.21 Write the names and symbols of all elements that occupy the same row of the periodic table as nitrogen.
1.22 Write the names and symbols of all elements that occupy the same column of the periodic table as nitrogen.

Characteristics of Matter

1.23 Classify each of the following as a pure substance, solution, or heterogeneous mixture: (a) block of iron; (b) cup of coffee; (c) glass of milk; (d) atmosphere, when free of dust; (e) atmosphere, when dusty; and (f) block of wood.
1.24 Classify each of the following as a pure substance, solution, or heterogeneous mixture: (a) blood; (b) dry ice; (c) krypton gas; (d) a rusty nail; (e) table salt; and (f) glass of lemonade.
1.25 Classify each of the following as a solid, liquid, or gas: (a) gasoline; (b) Teflon tape; (c) snow; and (d) water vapor.
1.26 Classify each of the following as a solid, liquid, or gas: (a) tree sap; (b) ozone; (c) dry ice; and (d) motor oil.
1.27 Classify each of the following as a chemical or a physical transformation: (a) formation of frost; (b) drying of clothes; and (c) burning of leaves.
1.28 Classify each of the following as a chemical or a physical transformation: (a) water boiling; (b) coffee brewing; and (c) photographic film being developed.
1.29 Classify each of the following as an element, a compound, or a mixture: (a) lake water; (b) distilled water; (c) mud; (d) helium inside a balloon; (e) rubbing alcohol; and (f) paint.

1.30 Classify each of the following as an element, a compound, or a mixture: (a) Earth's atmosphere; (b) beer; (c) iron magnet; (d) ice; (e) liquid bromine; and (f) mercury in a barometer.

Measurements in Chemistry

1.31 Convert to scientific notation: (a) 100,000, precise to ± 1; (b) ten thousand, precise to ± 1000; (c) 0.000 400; (d) 0.0003; and (e) 275.3.

1.32 Convert to scientific notation: (a) 175,906; (b) 0.000 0605; (c) two and a half million, precise to ± 100; and (d) two and a half billion, precise to $\pm$ one million.

1.33 Express each of the following in SI base units using scientific notation (example: 1.45 mm = 1.45×10^{-3} m): (a) 432 kg; (b) 624 ps; (c) 1024 ng; (d) 93,000 km, precise to ± 10; (e) 1 day; and (f) 0.0426 in.

1.34 Express each of the following in SI base units using scientific notation: (a) 1 week; (b) 1.35 mm; (c) 15 miles; (d) 4.567 μs; (e) 6.45 mL; and (f) 47 kg.

1.35 The mass unit most commonly used for precious stones is the carat. 1 carat = 3.168 grains, and 1 gram = 15.4 grains. Find the total mass in kilograms (kg) of a ring that contains a 5.0×10^{-1} carat diamond and 7.00 grams of gold.

1.36 What is the total mass in grams, expressed in scientific notation with the correct number of significant figures, of a solution containing 2.000 kg of water, 6.5 g of sodium chloride, and 47.546 g of sugar?

Calculations in Chemistry

1.37 What is the mass of 1 quart of water (1 L = 1.057 quarts)?

1.38 What is the mass of 1 quart of mercury (1 L = 1.057quarts)?

ilw **1.39** A plastic block measures 15.5 cm by 4.6 cm by 1.75 cm, and its mass is 98.456 g. Compute the density of the plastic.

1.40 A penny has a diameter of 1.9 cm and a thickness of 0.12 cm, and its mass is 2.51 g. Compute the density of the penny (cylinder volume, $V = \pi r^2 h$).

1.41 Calculate the volume of an aluminum spoon whose mass is 15.4 g.

1.42 Calculate the volume of a quartz crystal of mass 0.246 g.

Chemical Problem Solving

1.43 A chemist who wished to verify the density of water constructed a cylindrical container of aluminum 4.500 inches high, whose inside radius measured 0.875 inch. The empty cylinder had a mass of 93.054 g. When filled with water, its mass was 270.064 g. Find the density of water from these data, expressing your result in SI units with the correct precision (cylinder volume, $V = \pi r^2 h$).

1.44 A chemist who prepared a new organic liquid wanted to determine its density. Having only a small sample to work with, the chemist had to use a small container. A tube whose volume was 8.00×10^{-3} cm^3 weighed 0.4763 g when empty and 0.4827 g when filled with the liquid. Compute the density of the liquid in g/cm^3.

ilw **1.45** Determine which possesses more mass, a sphere of gold with a diameter of 2.00 cm or a cube of silver measuring 2.00 cm on each side (sphere volume, $V = 4\pi r^3/3$).

1.46 Determine which possesses more mass, a sphere of zinc with a diameter of 2.00 cm or a cube of aluminum measuring 2.00 cm on each side (sphere volume, $V = 4\pi r^3/3$).

1.47 Bromine is one of the two elements that is a liquid at room temperature (mercury is the other). The density of bromine at room temperature is 3.12 g/mL. What volume of bromine is required if a chemist needs 36.5 g for an experiment?

1.48 The density of gasoline at room temperature is 0.70 g/mL. If the gas tank of a car holds 12.0 gallons (45.4 L), what is the mass of a tankful of gasoline?

Additional Paired Problems

1.49 How many elements have symbols beginning with T? Give the name and symbol of each.

1.50 How many elements have symbols beginning with P? Give the name and symbol of each.

1.51 Draw a molecular picture that represents a homogeneous mixture of molecular fluorine and molecular chlorine in a 3:1 ratio of fluorine to chlorine. Both are gases, and both are diatomic molecules. The drawing should contain at least 20 molecules. Use Figure 1-4 as a guide.

1.52 Gasoline does not dissolve in water. Instead, molecules of gasoline float on the surface of water. Draw a molecular picture that shows a heterogeneous mixture of gasoline and water. Draw enough molecules of each to show the structure of the mixture clearly. Use circles for water molecules and ovals for gasoline molecules.

1.53 What are the name and symbol of the element of highest mass whose symbol and English name do not match?

1.54 What are the name and symbol of the element of lowest mass whose symbol and English name do not match?

1.55 An athlete runs the 100-yard dash in 10.17 s. At the same speed, how many seconds does it take the same athlete to run 100 meters?

1.56 An athlete runs the mile in 3 min 57 s. At the same speed, how many seconds does it take the same athlete to run 1500 meters?

1.57 Which elements would you expect to have chemical behavior similar to that of gold?

1.58 Which elements would you expect to have chemical behavior similar to that of sodium?

1.59 Draw a molecular picture of SO_2, in which each S atom is flanked by two O atoms in a bent arrangement. (SO_2 is a major contributor to acid rain.)

1.60 Draw a molecular picture of O_3, in which a central O atom is connected to the other two O atoms in a bent arrangement. (O_3 is a major contributor to smog.)

1.61 For each of the following statements, determine whether the property in italics is extensive or intensive: (a) The *boiling point* of ammonia (NH_3) is 239.8 K. (b) The *mass* of a diamond is 2.34 carats. (c) The *density* of nickel is 8.90 g/cm^3. (d) A copper wire is 3.2 cm in *length*. (e) Bromine is a *dark red* liquid.

1.62 For each of the following statements, determine whether the property in italics is extensive or intensive: (a) The *density* of iron is 7.86 g/cm^3. (b) Liquid oxygen has a *pale blue color*. (c) The crankcase of an automobile holds 5 *quarts* of oil. (d) The *melting point* of gallium metal is 30 °C. (e) A recipe calls for 100 *grams* of sugar.

1.63 Ethylene glycol is used as antifreeze in car radiators. The freezing temperature of ethylene glycol is -11.5 °C. Convert this freezing temperature to kelvins (K).

1.64 The melting point of silver metal is 1235 K. Convert this melting point to degrees Celsius (°C).

1.65 Perform the following calculations and report your answers with the correct number of significant figures:

(a) $\dfrac{(6.531 \times 10^{13})(6.02 \times 10^{23})}{(435)(2.000)}$

(b) $\dfrac{4.476 + (3.44)(5.6223) + 5.666}{(4.3)(7 \times 10^4)}$

1.66 Perform the following calculations, and report your answers with the correct number of significant figures:

(a) $\dfrac{(3.14159)(4.599 \times 10^6) - (1.12 \times 10^7)}{(4.756 \times 10^8) + (3.67 \times 10^4)}$

(b) $\dfrac{(6.577 \times 10^{-6}) + 0.00369 + (8.234 \times 10^{-4})}{(0.0002567) + (6.9377 \times 10^{-5})}$

1.67 Match each photograph with the appropriate molecular picture. Write a short justification of your choices.

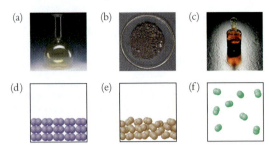

1.68 Decide whether each of the following molecular pictures represents a homogeneous solution, a single substance, or a heterogeneous mixture. If the picture represents a single substance, tell whether it is an element or a chemical compound. Explain your choices.

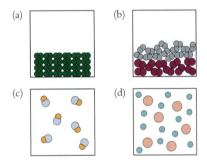

1.69 Examine the following molecular pictures and determine the chemical formulas of the compounds they represent:

(a) Freon 21
(a refrigerant suspected of damaging the ozone layer in the stratosphere)

(b) Formic acid
(the stinging compound in an ant bite)

(c) Bromine trifluoride
(used in the production of fuel for nuclear reactors)

(d) Butane
(the fluid in disposable lighters)

1.70 Examine the following molecular pictures and determine the chemical formulas of the compounds they represent:

(a) **Phosphoric acid**
(the 7th-ranked industrial chemical)

(b) **Acetic acid**
(the component that gives vinegar an acid taste)

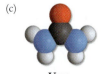

(c) **Urea**
(an important fertilizer and component of human urine)

(d) **Dichloroethylene**
(used to manufacture plastics)

1.71 Convert each of the following into SI units: (a) engine displacement of 454 cubic inches; (b) car speed of 35 mph; (c) height of 6 feet 9 inches; and (d) boulder mass of 227 pounds.

1.72 Convert each of the following into SI units: (a) gold nugget mass of 1.5 ounces; (b) light speed of 6.71×10^8 mph; (c) hike length of 11 miles; and (d) car mileage of 32 miles/gallon.

1.73 From the elements Ne, Cs, Sr, Br, Co, Pu, In, and O, choose one that fits each of the following descriptions:
(a) alkaline earth metal;
(b) element whose properties are similar to those of aluminum;
(c) element that reacts with potassium; (d) transition metal; (e) noble gas; (f) actinide.

1.74 From the elements Ar, K, Ca, Cl, Cu, U, P, and S, choose one that fits each of the following descriptions: (a) alkaline earth metal;
(b) element whose properties are similar to those of nitrogen;
(c) element that reacts with potassium; (d) transition metal; (e) noble gas; (f) actinide.

1.75 Draw appropriately scaled and colored molecular pictures of each of the following molecules: HCN (linear); H_2O (bent, O in the middle); CO; and NNO (linear). Use Figure 1-4 as a guideline.

1.76 Draw appropriately scaled and colored molecular pictures of each of the following molecules: CO_2 (linear, C in the middle); ClO_2 (bent, Cl in the middle); ClBr; and HOCl (linear). (Use Figure 1-4 as a guide.)

1.77 The distance from New York to Los Angeles is 2786 miles. How many minutes would it take an airplane flying at a constant speed of 685 km/hr to fly between the two cities?

1.78 The distance between Seattle, Washington and Portland, Oregon is 279 km. How many minutes would it take a traveler moving at a constant speed of 65 miles per hour to drive from one city to the other?

1.79 Potassium metal has a lustrous silvery appearance. It melts at 336.8 K. It reacts with chlorine gas to give a compound whose formula is KCl. The density is 0.862 g/mL. Potassium is soft enough to cut with a knife. When added to water, it often bursts into flame. Which of these properties are physical properties, and which are chemical properties?

1.80 Elemental oxygen is a colorless gas. It exists as diatomic molecules. It reacts readily with metals to form oxides, such as FeO. When cooled below 90 K, it condenses to form a pale blue liquid. Categorize each of these properties as physical or chemical.

1.81 Write the names and chemical symbols for three examples of each of the following: (a) halogens; (b) alkaline earth metals; (c) actinides; and (d) noble gases.

1.82 Write the names and chemical symbols for three examples of each of the following: (a) transition metals; (b) lanthanides; (c) main group elements; and (d) alkali metals.

1.83 Determine how many kilometers through outer space a beam of light travels in one year (365.24 days).

1.84 The distance from the Earth to the sun is 9.3×10^7 miles. How long does it take for light from the sun to reach our planet?

1.85 The photo below shows blue copper sulfate between samples of elemental copper metal and elemental sulfur. The third element present in copper sulfate is oxygen, which exists as a colorless gas under terrestrial conditions. (a) What are the elemental symbols of the three elements present in copper sulfate? (b) In copper sulfate, copper and sulfur combine in a 1:1 atomic ratio, and there are four oxygen atoms for every copper atom in the compound. What is the formula of copper sulfate?

1.86 Aluminum phosphate is a compound that contains four oxygen atoms and one phosphorus atom for every atom of aluminum. (a) What are the elemental symbols for the elements in this compound? (b) What is its chemical formula?

More Challenging Problems

1.87 The diameter of a chlorine atom is 2.00×10^2 pm. How many chlorine atoms lined up end to end would form a line 1.0 inch long?

1.88 Some chemists refer to chemistry as the "central" science because of its importance to other sciences. Do you agree? Give your reasons.

1.89 One of Jules Verne's most famous novels is *20,000 Leagues Under the Sea*. The league is a nautical unit of distance. Here are some conversion factors: 1 league = 3 nautical miles; 1 fathom = 6 ft; 1 nautical mile = 10 cable lengths; and 1 cable length = 100 fathoms. How deep is 20,000 leagues in feet and in kilometers?

1.90 A square of aluminum foil measuring 3.00 inches on each side weighs 255 mg. Find the thickness in micrometers (μm) of the aluminum foil.

1.91 We began this chapter by listing several questions that chemists are interested in answering. List five more.

1.92 Almost all the elements whose symbols come from Latin names are metals. Suggest a chemical reason why these elements were named many years ago.

1.93 Using everyday observations, decide which one in the following pairs of substances has the greater density. Explain your reasoning. (a) oil or vinegar; (b) table salt or water; and (c) iron or aluminum.

1.94 Carbon monoxide is a common pollutant in urban environments. On one particular day, the air contains 5.5 mg of carbon monoxide per 1.000 cubic meter of air. How many grams of carbon monoxide are present in a room whose dimensions are 12 feet × 9.5 feet × 10.5 feet?

1.95 The day is defined to contain exactly 24 hours of 60 minutes each, and each minute contains exactly 60 seconds. The year is 365.24 days long. How many seconds are there in a century? Express your answer in scientific notation with the correct number of significant figures.

Group Study Problems

1.96 Water for irrigation is usually expressed in units of acre-feet. One acre-foot is enough water to cover one acre of land to a depth of 1 foot (1 acre = 43,560 ft^2). A water reservoir has a maximum capacity of 8.97×10^5 acre-feet. What is the lake's volume in (a) liters; (b) cubic feet; and (c) cubic meters?

1.97 The diameter of metal wire is given by its wire gauge number. For example, 16-gauge wire has a diameter of 0.0508 in. Calculate the length in meters of a 5.00-pound spool of 16-gauge copper wire. The density of copper is 8.92 g/cm^3.

1.98 An object will float on water, even though it is made of a material that is denser than water, if its net density is less than 1. Thus, a hollow sphere with a large enough interior volume will float, because air has a very low density (1.189 g/L at room temperature and pressure). Determine the volume percentage that the interior of a sphere must have to float if it is made of aluminum. If the sphere has an interior volume of 1.50 L, how thin must its metal shell be? (Hint: for a sphere, $V = \frac{4}{3}\pi r^3$ and $A = 4\pi r^2$. Repeat your calculations for copper.

1.99 The Greek scientist Archimedes was given the task of determining whether a crown was pure gold or an alloy of gold and silver. No part of the crown could be destroyed. To do this, Archimedes invented volume determination by displacement. First, he measured the crown's mass to be 2.65 kg. Next, he put the crown in a full basin of water and found the

amount of overflow to be 145 cm^3. Was the crown pure gold or a gold–silver alloy? Do a calculation that supports your answer.

1.100 On average, a sample of human blood whose dimensions are 0.1 mm × 0.1 mm × 0.1 mm contains 6.0×10^3 red blood cells. The volume of blood in a typical adult is about 5 L. How many red blood cells are there in an adult?

1.101 U.S. pennies are composed of copper and zinc. The table gives the results of mass determinations on a set of pennies minted in different years. The size of the penny remained the same during this time period.

Year	Mass (g)	Year	Mass (g)	Year	Mass (g)
1979	3.075	1981	3.066	1983	2.518
1980	3.095	1982	2.545	1984	2.526

(a) What can you conclude about the composition of pennies during this time period? (b) In 1979, pennies were 95% copper and 5% zinc by mass. Calculate the volume of a penny (see Table 1-4). (c) Measure the diameter of a penny, then determine its thickness to as many significant figures as you can. (d) Determine the mass percentages of copper and zinc in post-1981 pennies. (e) Post-1981 pennies are essentially pure zinc with a thin copper overlayer. Determine the thickness of the copper overlayer (this will be easier if you neglect the overlayer around the circumference).

Answers to Section Exercises

1.1.1 Chemistry is involved in sugar dissolving, yeast causing bread to rise, caramelizing sugar, and marinating meat.

1.1.2 Chemistry is involved in gasoline additives, plastics for interior fittings, ceramics for engine parts, and paint that protects finishes.

1.1.3 Develop a set of chemical reactions that can be performed on Earth and on Venus. Observe the results carefully on both planets. If the results obtained on Venus differ from the results obtained on Earth, the theory may be valid. However, if the results are the same, the theory cannot be valid.

1.2.1 Ce, Cs, Cu, Ca, and C

1.2.2 Zirconium, nickel, tin, tungsten, selenium, beryllium, and gold

1.2.3 (a) H_2O_2; (b) SO_2; (c) N_2O; (d) $C_2H_4O_2$; and (e) C_2H_4

1.3.1 (a) $CaCl_2$; (b) CsI; (c) BaO; and (d) MgF_2

1.3.2 $AlBr_3$ and $GaCl_3$

1.3.3 Al, main group; F, halogen and main group; Co, transition metal; P, main group; Kr, noble gas; Eu, lanthanide; Th, actinide; Ba, alkaline earth; and Na, alkali metal

1.4.1 (a) Single substance, an element; (b) single substance, a compound; and (c) heterogeneous mixture

1.4.2 Solution, pure substance, heterogeneous mixture, pure substance, and solution

1.4.3 Physical, chemical, physical, and chemical

1.5.1 4.63×10^{-7} m^3, 1.7935×10^7 m, and 9.36×10^8 s

1.5.2 3.0×10^8 m/s and 9.5×10^{15} m

1.5.3 70.3 kg, 110. m, 312 K, and $3.155\,693 \times 10^7$ s

1.6.1 3.0 g

1.6.2 (a) 1.00 cm; (b) 1.44 g/mL; (c) 25 mi/hr

1.6.3 (a), (c), and (d), uncertain; (b) and (e), exact

1.7.1 4.8×10^{-3} m^3

1.7.2 1.3 g/cm^3

Answers to Extra Practice Exercises

1.1 Your picture should show diatomic molecules in 4:1 ratio of N_2 to O_2, scattered randomly about.

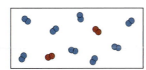

1.2 0.71 L
1.3 19 km/L
1.4 19.1 cm^3
1.5 Perimeter = 487.7 cm; volume = 2.3×10^4 cm^3
1.6 0.703 g/cm^3

The Atomic Nature of Matter

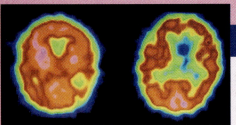

Imaging Using Nuclear Radiation

One of the great challenges facing medical science is how to diagnose and treat afflictions of the brain such as Alzheimer's disease. In part this is because we cannot "see" directly inside an active human brain. Recently, however, medical researchers have developed a powerful diagnostic tool called positron emission tomography (PET).

The PET technique relies on radioactive unstable atoms that disintegrate spontaneously, giving off particles called positrons. As soon as an atom emits a positron, the positron combines with an electron. Both particles are annihilated, producing a brief flash of gamma-ray radiation that is easily detected by radiation monitors.

A PET scan requires a substance called a *tracer*. A suitable tracer must accumulate in the target organ, and it must be modified to contain unstable radioactive atoms that emit positrons. Glucose is used for brain imaging, because the brain processes glucose as the fuel for mental and neural activities. A common tracer for PET brain scans is glucose modified to contain radioactive fluorine atoms. Our molecular inset shows a simplified model of this modified glucose molecule.

In preparation for a PET scan, a patient is injected with a dose of the tracer, which quickly accumulates in the brain. The patient is placed inside the PET scanner, and the instrument detects

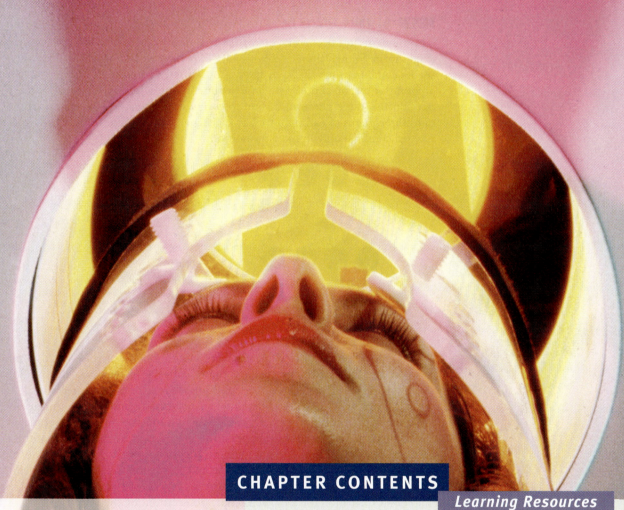

CHAPTER CONTENTS

Learning Resources

 KEY CONCEPTS

CRITICAL THINKING

SKILLS TO MASTER

 ADDITIONAL HELP
www.wiley.com/college/olmsted

● TUTORIALS
● ANIMATIONS

the gamma rays emitted by the tracer. The result is an image showing the distribution of glucose in the brain, which indicates where brain activity is greatest.

Our inset photo shows a PET brain scan of a 9-year-old girl suffering from seizures. Bright regions of the image reveal high concentrations of tracer, indicating active areas of the brain. Dark regions of the scan show inactive areas. Notice that the lower portion of the image is bright on one side but dark on the other. The dark side is inactive, indicating a brain malfunction responsible for the girl's seizures.

Brain scans are used to study epilepsy, brain tumors, strokes, Alzheimer's disease, and mental illness. Each of these disorders generates a unique brain activity pattern that differs from the pattern seen in normal brains. Physicians interpret these patterns both for diagnosis and to indicate appropriate treatment.

In this chapter, we present the atomic perspective of matter, as expressed by atomic theory and the principles of atomic structure. We describe the building blocks of atoms: electrons, protons, and neutrons. Then we show how these interact to form all the chemical elements and explain which combinations are stable. Next we describe how atomic masses are related to these building blocks. We end the chapter by introducing ions, atoms that have either lost or gained electrons. Further applications of radioactive atoms in medicine are found within the chapter.

2.1 ATOMIC THEORY

The modern atomic theory of matter was developed 200 years ago, primarily by the Englishman John Dalton (1766–1844). Experiments had revealed that when two elements combine, they do so in fixed proportions. For example, no matter what starting amounts of carbon and oxygen an experimenter uses, just two products form. One product contains three grams of carbon for every four grams of oxygen, and the other contains three grams of carbon for every eight grams of oxygen. In other words, carbon and oxygen combine only in fixed proportions.

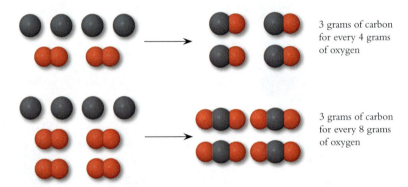

3 grams of carbon for every 4 grams of oxygen

3 grams of carbon for every 8 grams of oxygen

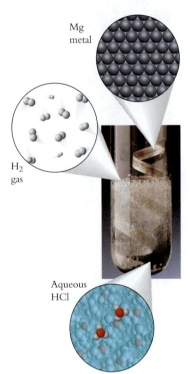

Mg metal

H₂ gas

Aqueous HCl

Figure 2-1
Magnesium metal reacts with aqueous acid solution to generate hydrogen gas. The expanded views are molecular pictures of each phase.

Dalton reasoned that if matter were continuous, elements would combine in all proportions, much as we can make a cup of coffee as sweet as we like by adding various amounts of sugar. He realized that fixed proportions result if matter is made up of indivisible atoms, with atoms of different elements (carbon and oxygen, for instance) having different characteristic masses. Dalton's theory stimulated further experiments to test its predictions. All experiments supported the theory, and by the middle of the nineteenth century, most scientists accepted atomic theory as correct. Still, no one had actually observed atoms! More than 100 years would pass before scientists succeeded in producing actual images of atoms. Atomic theory was accepted because it provided the simplest and best explanation for the macroscopic observations of chemistry.

Dalton's observations remind us that chemists have always observed changes in large-scale appearance and properties as a chemical reaction occurs. For example, Figure 2-1 shows that when solid magnesium metal reacts with a liquid solution of an acid, the metal disappears and gas bubbles out of the solution. These macroscopic changes result from interactions at the atomic level, as shown by the inset views. An understanding of chemistry requires knowledge of how matter looks at the atomic level and how atomic-sized objects interact. We can summarize the modern view of atomic theory in four general statements:

Features of Atomic Theory

1. All matter is composed of tiny particles called atoms.
2. All atoms of a given element have identical chemical properties that are characteristic of that element.
3. Atoms form chemical compounds by combining in whole-number ratios.
4. Atoms can change how they are combined, but they are neither created nor destroyed in chemical reactions.

Figure 2-2 shows the vigorous reaction that takes place between molecular hydrogen and molecular oxygen when they mix and burn. The only product is water molecules that contain hydrogen atoms and oxygen atoms bound to one another. If the reaction is performed with varying amounts of hydrogen and oxygen, the *amount* of water changes from one experiment to the next, but the *composition* of water does not. Water always contains 1.0 g of hydrogen for every 8.0 g of oxygen, and two atoms of hydrogen for every atom of oxygen.

It is possible to reverse this reaction between hydrogen and oxygen. An electrical current causes water molecules to decompose into hydrogen gas and oxygen gas. These gases can be captured and their amounts measured. Repeated observations show that the two substances are always produced in the same mass ratio: Every 9.0 g of decomposed water produces 1.0 g of hydrogen and 8.0 g of oxygen.

Chemists think of these processes in terms of atoms, the building blocks of all matter. The molecular insets in Figure 2-2 show how chemists visualize hydrogen gas, oxygen gas, and gaseous water. Notice that hydrogen gas contains hydrogen atoms (Feature 1 of the atomic theory). Experiments on hydrogen reveal that all its atoms behave identically by combining into diatomic molecules (Feature 2). Oxygen gas consists of diatomic molecules containing oxygen atoms, but these molecules act differently than molecules made from hydrogen atoms (Feature 2). When molecules of hydrogen and oxygen react to give water, atoms of hydrogen combine with atoms of oxygen in a 2:1 ratio (Feature 3). Atoms rearrange during this chemical process, but the total number of each type of atom remains the same (Feature 4).

Conservation of Atoms and Mass

One of the cornerstones of the atomic theory is that atoms are neither created nor destroyed in chemical or physical processes. In other words, the number of atoms of each type is constant and unchanging. When a quantity does not change, we say that the quantity is *conserved*. A statement that some quantity is conserved is a **conservation law.**

Figure 2-2
When hydrogen and oxygen burn to form water, atoms change the manner in which they are combined together.

Atoms are conserved during physical and chemical transformations

KEY CONCEPT

The numbers of atoms of each element are conserved in any chemical reaction. When methane burns to produce heat, for example, the carbon atom from each molecule of methane ends up in a molecule of carbon dioxide. The four hydrogen atoms of methane end up in two molecules of water. The atoms rearrange and recombine during the course of the reaction, but the numbers of each type of atom do not change. The molecular picture shown in Figure 2-3 represents the chemical reaction of methane with molecular oxygen. Counting the atoms, we see that there are four oxygen atoms (red), four hydrogen atoms (white), and one carbon atom (black) before and after the reaction.

Chemists keep track of individual atoms and electrons at the atomic level, but in the laboratory, chemists measure mass. Neither the numbers nor the masses of atoms and electrons change during chemical transformations, so mass is also conserved. For example, the burning of 1 g of methane and 2 g of oxygen produces 3 g of carbon dioxide and water.

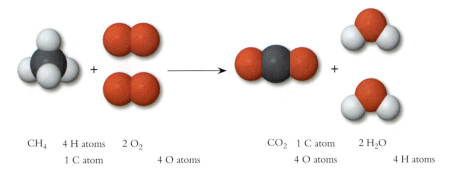

CH_4 4 H atoms   2 O_2

1 C atom   4 O atoms

CO_2 1 C atom   2 H_2O

4 O atoms   4 H atoms

Figure 2-3
Molecular picture of the reaction of methane (CH_4) with oxygen (O_2) to produce carbon dioxide (CO_2) and water (H_2O). Atoms of each element are conserved.

Figure 2-4
It appears that mass is lost when wood burns, but the mass lost by the solids is transferred to the atmosphere as molecules of carbon dioxide and water.

 KEY CONCEPT *Mass is conserved during physical and chemical transformations.*

As Albert Einstein theorized in 1905, mass can be converted into energy and energy into mass. However, the amount of mass lost or gained in chemical transformations is always too small to measure.

In some chemical transformations, it appears that mass is not conserved. When wood burns, for example (Figure 2-4), the mass of the ash is much less than the mass of the original wood. This is because the carbon, hydrogen, and oxygen atoms that make up most of the wood are incorporated into carbon dioxide and water vapor, both of which are gases that escape into the atmosphere. Careful experiments that capture and weigh all the products show that the mass of ash plus gases equals the mass of wood plus oxygen. The total mass remains unchanged.

Atoms Combine to Make Molecules

Another essential feature of the atomic theory is that atoms combine in whole-number ratios to make molecules. In the vigorous reaction of hydrogen with oxygen shown in Figure 2-2, every two molecules of hydrogen combine with one molecule of oxygen to form two molecules of water. Even when this reaction occurs in the presence of a large amount of oxygen, the product molecules contain hydrogen atoms and oxygen atoms in fixed 2:1 ratios, as Figure 2-5 illustrates.

Many elements can combine with one another in more than one way. Hydrogen and oxygen provide a simple example. When these elements react directly, they form water molecules containing two hydrogen atoms and one atom of oxygen. Under other conditions the same two elements form a different compound, hydrogen peroxide. Each molecule of this substance contains two atoms of hydrogen and two atoms of oxygen.

Although water and hydrogen peroxide contain the same elements, their chemical properties are very different. Water is stable under most conditions. It is easy to form and difficult to destroy. Hydrogen peroxide, on the other hand, is a very reactive molecule. It is not difficult to make, but it can be destroyed in many ways. Water is life-sustaining; in fact, life depends completely on water. Hydrogen peroxide can be life-destroying. Because it kills microorganisms, hydrogen peroxide is a good disinfectant.

From an understanding of how atoms join to make molecules, chemists can explain why two compounds that seem so similar have profoundly different reactivity patterns. We describe how atoms link together in Chapters 9 and 10. Meanwhile, remember that chemists try to visualize chemical reactions at the molecular level. Contemporary chemists also manipulate individual atoms to make elaborate structures, as described in our Chemistry and Technology Box.

H_2O H_2O_2

Figure 2-5
Even in the presence of excess amounts of oxygen, hydrogen and oxygen combine in 2:1 atomic ratios.

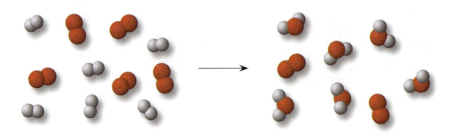

Box 2-1   Chemistry and Technology: Molecular Machines?

A student using a molecular graphics computer program can take images of individual atoms and connect them together to make any desired molecular structure. Imagine this program interfaced to a molecular assembly line that could build actual molecules to match those structures. Science fiction? Perhaps not. Theoretical and computational models indicate that molecular manufacturing doesn't violate any known physical or chemical laws. Nanotechnology, including the construction of molecular structures, one at a time, appears to be within the realm of possibility.

What uses might "molecular machines" have? Imaginative researchers have come up with numerous possibilities. Current medical technology cannot get inside the human brain to remove the blood clots responsible for strokes. What if we could manufacture "nanosubmarines" that could propel themselves through the arteries, navigate to the proper site, and either delicately excise the clot or deliver a nanodose of a clot-dissolving drug? The arena of biotechnology poses many similar challenges that nanotechnology might meet.

Alternatively, consider the field of computer technology. The computer chip, the heart of modern computing, has been miniaturized nearly to the limits of current technology, resulting in amazingly fast machines with prodigious memories. Computer designers dream of further miniaturization of computer circuitry, down to the "molecular wire" level. This would increase the speed and capacity of computers by several orders of magnitude. To construct computer chips with molecular dimensions will require nanotechnology.

Building things at the molecular level will require tools that are about the sizes of atoms and molecules. Designing and building such tools is a major challenge, but the research is already under way. The figure shows a molecular gear based on tubes of carbon atoms studded with atomic appendages that mesh as the gear turns.

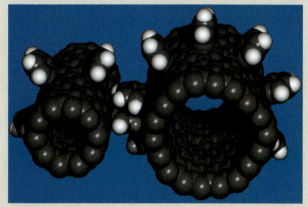

A molecular gear

How might we build these fantastic molecular machines? One approach uses the scanning tunneling microscope (STM) as a molecular "tweezers." This instrument, which we describe in Chapter 1, has been used to place single atoms onto a germanium surface and to place organic molecules precisely onto graphite surfaces. More spectacularly, as the photo shows, iron atoms have been arranged on a copper surface in the form of the Japanese word for *atom*.

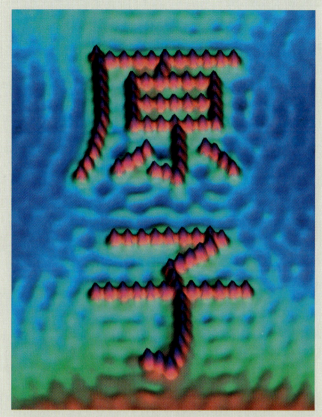

Iron atoms arranged to form the Japanese character for *atom*.

We are still years away from building molecular machines. In fact, using STM as a tool for molecular engineering has been compared to trying to build a wristwatch with a sharpened stick.

Supposing that scientists succeed in constructing molecular tools, they must overcome another obstacle for nanotechnology to be effective. A medical nanosubmarine is likely to contain about a billion (10^9) atoms. At an assembly speed of one atom per second, it would take 10^9 seconds to construct one such device. That's almost 32 years! If the assembly rate can be increased to one atom per microsecond, the construction time for a 1-billion-atom machine drops to 1000 seconds, or just under 17 minutes. That's not bad if only a few machines are needed, but molecular machines are tiny, so large numbers of machines will be required for any practical application. Consequently, scientists will have to discover ways to mass-produce nanodevices.

To be practical, then, nanotechnology must be precise, extremely fast, and amenable to mass production. Perhaps this strikes you as definitely in the realm of science fiction rather than science fact, and perhaps it is. Nevertheless, scientists at many universities are vigorously tackling the challenges of this field, and major technology companies have active research groups as well.

The atomic–molecular view is so ingrained that chemists often take it for granted. As you study chemistry, strive to attain this point of view. Example 2-1 provides some practice.

<table>
<tr><td>

Example 2-1

Molecular Pictures

</td><td>

When charcoal burns in air, carbon atoms combine with oxygen atoms from molecular oxygen to form carbon dioxide. One molecule of carbon dioxide contains one carbon atom and two oxygen atoms. Experiments on carbon dioxide show that each molecule is linear, with a carbon atom in the middle. Draw a molecular picture that illustrates this reaction.

Strategy: This is a qualitative problem requiring you to visualize and represent molecules. In molecular pictures, circles represent atoms, and different colors or shadings identify different elements. The problem does not state how many atoms and molecules to draw, so we can start with any convenient amounts. However, the numbers of atoms of each element must not change during the reaction.

Solution: Every carbon atom requires two oxygen atoms (one molecule of O_2) and generates one molecule of carbon dioxide. Here is how the molecular picture looks if we choose to start with six atoms of carbon and six molecules of oxygen:

6 O_2 molecules   6 C atoms   6 O_2 molecules

In our representation, six carbon atoms have combined with six oxygen molecules to form six molecules of carbon dioxide.

</td></tr>
<tr><td>

</td><td>

Does the Result Make Sense? Atoms of each element must be conserved in any chemical or physical process. Count the atoms: 6 C atoms before reaction, 6 C atoms (in CO_2 molecules) after reaction; 12 O atoms before reaction, 12 O atoms after reaction. Thus, the picture makes sense.

</td></tr>
<tr><td>

Extra Practice Exercise 2.1

</td><td>

In lightning strikes, ozone molecules (3 O atoms in a bent row) can form from diatomic oxygen molecules. Draw a molecular picture showing the ozone molecules formed from six oxygen molecules.

</td></tr>
</table>

Atoms and Molecules Are Continually in Motion

In all matter, even substances as solid as steel, individual atoms or molecules are in never-ending motion. Monatomic substances, such as argon, contain individual atoms that are not bound to any other atoms. Other substances, including O_2, H_2, and H_2O, contain groups of atoms bound into molecules. Continuous motion at the atomic–molecular level involves atoms for monatomic species and molecules for species composed of atomic groups.

In solids, atoms "rattle around"—vibrate—in the "cages" formed by the surrounding atoms. In liquids, atoms or molecules move past one another continuously, like minnows in a stream endlessly changing positions. In gases, atoms or molecules are free to move over large distances. Figure 2-6 is a schematic illustration of motion of a monatomic substance in these three phases.

As we describe in Section 2.5, the atoms in solids such as sodium chloride (NaCl) exist as positive and negative ions that vibrate back and forth.

Animation

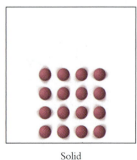  

Solid   Liquid   Gas

Figure 2-6
Atomic pictures of a monatomic solid (*left*), liquid (*center*), and gas (*right*), showing how atoms move about in each phase.

The pressures exerted by gases demonstrate molecular motion. Gases are collections of molecules, so the pressure exerted by a gas must come from these molecules. Just as the basketball in Figure 2-7 exerts a force when it collides with a backboard, moving gas molecules exert forces when they collide with the walls of their container. The collective effect of many molecular collisions generates pressure.

The **diffusion** of one liquid into another also demonstrates molecular motion. Figure 2-8 shows that if a drop of ink is added to a beaker of still water, the color slowly but surely spreads throughout the water. The water molecules and the molecules that give ink its color move continuously. As they slide by one another, the ink molecules eventually become distributed uniformly throughout the volume of liquid.

Figure 2-7
When a basketball strikes a backboard, it exerts a force on the backboard. The gas molecules inside the basketball (*inset*) also exert forces when they strike the walls of the basketball. The net result of many collisions is gas pressure, which we describe in Chapter 5.

Dynamic Molecular Equilibrium

Atoms and molecules are always moving, even when no visible changes take place. In our ink example, ink molecules move randomly in all directions. As the molecular view in Figure 2-8*c* indicates, however, the total number of ink molecules and water molecules in any region of the liquid does not change once the molecules are evenly distributed. As a result, there is no further change in color.

This condition of balanced motion is called **dynamic equilibrium.** Although a dynamic system contains objects that move continuously, a system at equilibrium shows no change in its observable properties. Our example of ink in water is dynamic because the water and ink molecules continually move about. The mixture is at equilibrium when the color is uniform and unchanging. In any part of the solution, ink molecules continue to move, but the number of ink molecules in each region does not change.

Dynamic equilibria occur frequently in chemical systems. Chemical processes reach a state of equilibrium if allowed to continue for a sufficient time. Nevertheless, molecular activity always goes on after equilibrium has been reached. The following example, illustrated schematically in Figure 2-9, should help you grasp this important idea.

Wet towels hung on a clothesline eventually dry, because the continual motion of molecules in liquid water allows some molecules to escape from the liquid phase (Figure 2-9*a*). A wet towel left in a closed washing machine, however, stays wet for a long time. This is because water molecules that escape from the surface of the towel remain within the washing chamber (Figure 2-9*b*). The number of water molecules in the gas phase increases, and the towel recaptures some of these molecules when they collide with its surface. The system soon reaches a condition of dynamic equilibrium in which, for every water molecule that leaves the surface of the towel, one water molecule returns from the gas phase to the towel (Figure 2-9*c*). Under these conditions, the towel remains wet indefinitely.

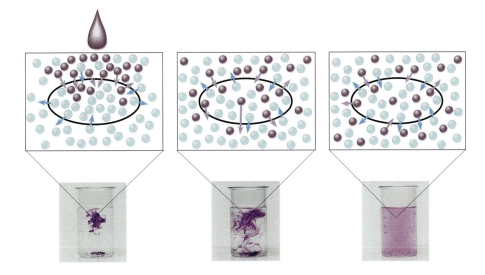

Figure 2-8
After a drop of ink is added to a beaker of water (*left*), the ink diffuses slowly through the liquid (*center*) until eventually the ink is distributed uniformly (*right*). The molecular views indicate that the motion of ink molecules and water molecules is responsible for this diffusion. Ink molecules (*red-violet circles*) and water molecules (*blue circles*) move about continually, even after they are well mixed.

Figure 2-9
(*a*) When water evaporates from an open container, molecules escape but are not recaptured. (*b*) In a closed container, the number of water molecules in the gas above the liquid increases, and some are recaptured when they collide with the liquid surface. (*c*) The container reaches dynamic equilibrium when enough water molecules are present in the gas to make the rate of recapture equal to the rate of escape.

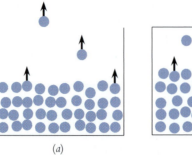
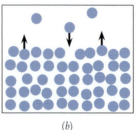
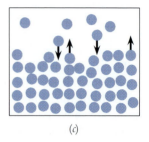

(*a*)   (*b*)   (*c*)

Animation

Summarizing, once this system has reached dynamic equilibrium, molecules continue to leave the liquid phase for the gas phase, but the liquid captures equal numbers of molecules from the gas. The amount of water in each phase remains the same (equilibrium) even though molecules continue to move back and forth between the gas and the liquid (dynamic). As with dye dispersed in water, no net change occurs after equilibrium is established.

Section Exercises

2.1.1 Elemental bromine, chlorine, and iodine exist as diatomic molecules. Chlorine is a gas at room temperature, bromine is a liquid, and iodine is a solid. Draw molecular pictures that show the molecular distributions in samples of chlorine, bromine, and iodine.

2.1.2 Chlorine and hydrogen molecules can react with each other to form molecules of hydrogen chloride, which contain one atom of each element. Draw a molecular picture showing three molecules of chlorine reacting with enough molecules of hydrogen to convert all the chlorine into hydrogen chloride.

2.1.3 After a summer shower, rain puddles on the road quickly disappear. Describe what happens in molecular terms.

2.2 ATOMIC ARCHITECTURE: ELECTRONS AND NUCLEI

Atoms are the fundamental building blocks of chemistry, but are atoms made of other, still smaller particles? It turns out that atoms do have internal structures. Furthermore, the internal structure of atoms of one particular element differs from that of every other element. These differences in structure are what make the chemistry of one element different from that of any other. The rich diversity of chemical behavior results from the different internal structures of atoms of different elements.

A series of elegant experiments, designed and performed in the first three decades of the twentieth century, revealed the essential structure of the atom. In this section, we outline the main features of the most important experiments to give you an idea of how scientists probe matter at the atomic level.

Forces

To understand the atom, we have to know something about the nature of forces. Intuition tells us that forces either hold things together or push them apart. Physicists have identified four types of forces: gravitational, electromagnetic, strong nuclear, and weak nuclear. Of these, the most familiar is **gravitational force.**

The force of gravity pulls all objects toward the center of the Earth. Rain falls from the clouds, skydivers plunge toward the Earth, and balls thrown into the air return to the ground. The sun's gravitational force holds the planets in their regular orbits. In fact, every mass exerts a gravitational attraction on all other masses. The existence of gravitational force between any two bodies is a fundamental law of the uni-

Gravitational force pulls all objects toward the Earth.

verse. Gravitation is obviously an important force for large objects such as airplanes, baseballs, and humans.

For tiny objects such as atoms and molecules, electrical and magnetic forces are most important. **Electrical force** can either attract or repel. Studies show that when the charges of electrically charged objects have opposite signs, electrical force pulls the objects together (Figure 2-10). When both charges have the same sign, however, the objects repel one another. Electrical force between two objects increases with the amount of charge on each, and it decreases as the objects move farther apart.

A charged object in motion is also subject to **magnetic force.** Objects that generate magnetic force are called magnets (Figure 2-11a). A magnet has two ends, a north (N) pole and a south (S) pole. As with electrical force, magnetic force may attract or repel. Opposite poles (N and S) attract, but like poles (N–N or S–S) repel.

To understand some of the early experiments that probed atomic structure, you need to know that a moving charged object is deflected along a curved path as it passes between the poles of a magnet (Figure 2-11b). The amount of curvature reveals information about the properties of moving charges, so magnets are very useful in studies of atomic structure.

Electrons

Important clues about the structure of atoms came from experiments that used electrical force. One of these experiments, shown schematically in Figure 2-12, used two perforated metal plates sealed inside a glass tube along with a sample of a gas. One plate was given a large positive electrical charge, and the other was given a large negative charge. These charges attracted each other, resulting in a high voltage between the plates. At high enough voltage and low enough gas pressure, a steady electrical current passed through the space between the plates.

This electrical discharge broke apart some of the atoms of the gas into charged particles. Particles with positive charges moved toward the negative plate, and particles with negative charges moved toward the positive plate. Because the plates had holes in them, some of these charged particles passed through the plates and reached collectors at the ends of the tube. This experiment showed that atoms contain smaller parts that possess positive and negative charges.

Changing the gas in the tube changed the behavior of the positively charged particles, but the negatively charged particles always acted the same. These negatively charged fragments, which are common to all atoms, are called **electrons.** The atomic or molecular charged particles created by removing electrons are called *positive ions*. The gas discharge experiment showed that atoms can be decomposed to negatively charged electrons and positively charged ions.

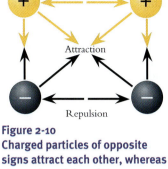

Figure 2-10
Charged particles of opposite signs attract each other, whereas charged particles of the same sign repel each other.

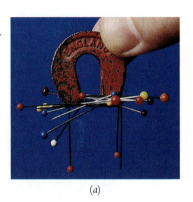

(a)

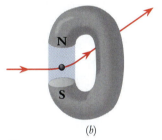

(b)

Figure 2-11
(*a*) Magnetic force attracts iron-containing objects. (*b*) Magnetic force also causes a moving charge to change direction when it passes between the poles of a magnet. The amount of bending can be related quantitatively to the charge, mass, and speed of the particle.

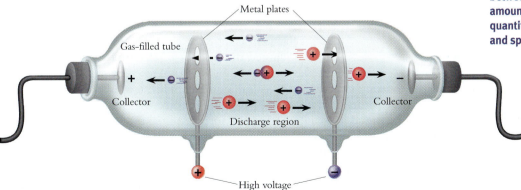

Figure 2-12
Schematic drawing of a gas discharge tube in operation. When a high voltage is applied to the two perforated plates, an electrical discharge occurs between them. The positively charged and negatively charged particles that form in the gas move to collectors at the ends of the tube.

Figure 2-13
Schematic drawing of a cathode-ray tube. A beam of electrons is deflected by a pair of charged plates (*bent line*), but magnetic force can be adjusted to exactly counterbalance the effect of the electrical force (*straight line*).

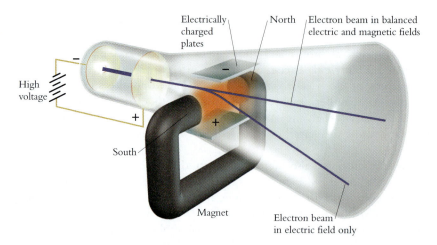

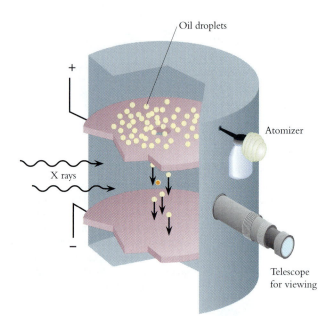

Figure 2-14
Schematic view of Millikan's oil drop experiment. An atomizer generated a fine mist of oil droplets (*yellow circles*). Bombarding the droplets with X rays gave some of them extra negative charge (*orange circle*). In the presence of sufficient electrical force, these negatively charged droplets could be suspended in space.

The fundamental unit of electrical charge is the coulomb (C), defined as the quantity of electricity transferred by a current of 1 ampere in 1 second.

The discovery of the electron prompted a series of more sophisticated experiments. J. J. Thomson experimented with a device called a cathode-ray tube, illustrated in Figure 2-13. A cathode ray is a beam of electrons. Because an electron beam is a collection of moving electrical charges, electrical and magnetic forces affect the beam. Application of either type of force at right angles to the direction of electron motion causes the beam to bend. The amount of bending depends on the speed, charge, and mass of the electrons. Thomson subjected cathode rays simultaneously to electrical and magnetic forces. By measuring the amount of magnetic force required to exactly counterbalance the deflection of the beam by a known electrical force, Thomson was able to calculate the ratio of the electron's charge to its mass:

$$\frac{\text{Charge}}{\text{Mass}} = \frac{e}{m} = -1.76 \times 10^{11} \text{ C/kg}$$

This experiment showed that one kilogram of electrons has a charge of -1.76×10^{11} C, but Thomson was unable to find out how much charge resides on a single electron.

American physicist Robert A. Millikan and his student Harvey Fletcher designed an experiment to determine the charge on the electron. As shown in Figure 2-14, the apparatus was a chamber containing two electrical plates. An atomizer sprayed a mist of oil droplets into the chamber, where the droplets drifted through a hole in the top plate. A telescope allowed the experimenters to measure how fast the droplets moved downward under the force of gravity. The mass of each droplet could then be calculated from its rate of downward motion.

When the chamber was irradiated with X rays, some of the droplets gained electrical charges. The charge on the metal plates generated an upward electrical force on these droplets. Fletcher adjusted the amount of electrical charge on the plates until the upward–acting electrical force exactly counterbalanced the downward–acting gravitational force on a droplet. This stopped the downward drift of these negative particles. Millikan and Fletcher calculated the charge on a droplet from its previously measured mass and the electrical force required to suspend its motion.

The measurements gave several different values, but the charge was always equal to $n(-1.6 \times 10^{-19}$ C), where n was an integer (1, 2, 3, . . .). Millikan concluded that n was the number of extra electrons carried by an oil droplet. Thus, the charge of an individual electron is -1.6×10^{-19} C. Combining this value with Thomson's measurement of charge/mass ratio, Millikan computed the mass of a single electron:

$$m_{\text{electron}} = \frac{e}{e/m} = \frac{-1.6 \times 10^{-19} \text{ C}}{-1.76 \times 10^{11} \text{ C/kg}} = 9.1 \times 10^{-19} \text{ kg}$$

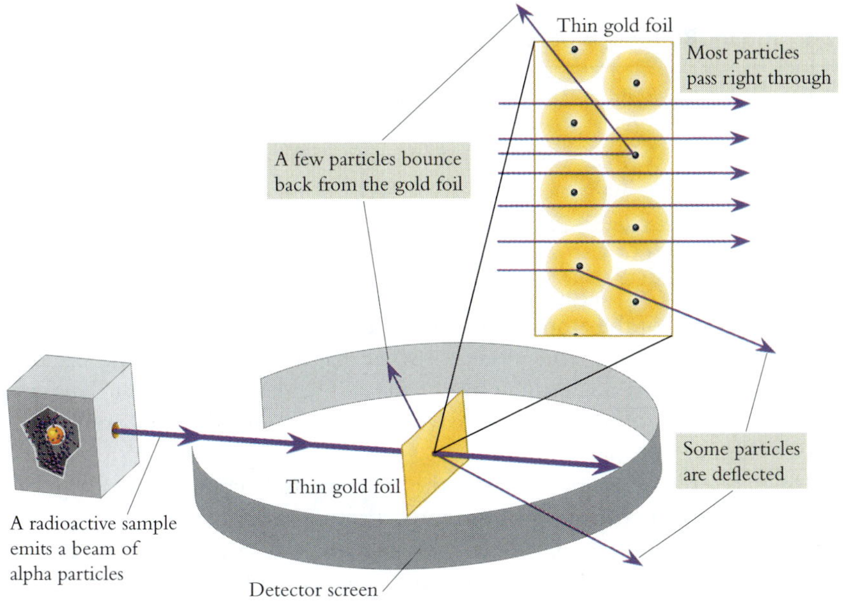

Thin gold foil

Most particles pass right through

A few particles bounce back from the gold foil

Some particles are deflected

A radioactive sample emits a beam of alpha particles

Thin gold foil

Detector screen

Figure 2-15
Schematic view of Rutherford's scattering experiment. When a beam of positively charged helium particles was "shot" at a thin gold foil, most of them passed through without much effect. Some, however, were reflected backward.

The Nucleus

By the early twentieth century, scientists had discovered that atoms contain electrons and positively charged particles. The experiments of Thomson and Millikan revealed the nature of electrons, but the nature of the positive particles was not yet known. Also, it was not known how the particles fit together to make an atom.

In 1911, Ernest Rutherford performed an experiment that showed how charges and masses are distributed in an atom. J. J. Thomson had hypothesized that the atom was similar to a chocolate chip cookie. Negative electrons (the "chips") nestled in the positive "dough" of the atom in a way that balanced repulsion between like charges and attraction between unlike charges. Rutherford tested this model of the atom by using subatomic projectiles to bombard a target of atoms. The projectiles, called *alpha particles,* had been discovered during research on radioactivity. Alpha particles are high-energy, positively charged fragments of helium atoms emitted during radioactive decay of unstable elements such as uranium.

Figure 2-15 diagrams Rutherford's experiment and its results. Alpha particles bombarded a thin film of gold metal. According to the "cookie model," the mass of each gold atom in the foil should have been spread evenly over the entire atom. Rutherford knew that alpha particles had enough energy to pass directly through such a uniform distribution of mass. He expected the particles to slow and change direction slightly as they passed through the foil.

The results were quite unexpected. Although most alpha particles passed directly through the gold film and some showed slight deflections, a few particles bounced back in the direction from which they came. Rutherford was astonished. He said it was similar to shooting an artillery shell at a piece of tissue paper and having it bounce back at him. Somewhere within the atom there had to be a positively charged mass capable of blocking the path of high-energy, positively charged alpha particles. The cookie model of the atom crumbled.

To explain his observations, Rutherford proposed a new hypothesis for atomic structure. He suggested that every atom has a tiny central core, called the **nucleus,** within which all the positive charge and most of the mass is concentrated. Electrons surround this central core, as shown schematically in Figure 2-16. Electrons occupy a volume that is huge compared with the size of the nucleus, but each electron has such a small mass that alpha particles are not deflected by the electrons. Consequently, an alpha particle is deflected only when it passes very near a nucleus, and it bounces back only when it collides head-on with a nucleus. Because most of the volume of an atom contains only electrons, most projectiles pass through the foil without being affected.

Thomson, who was from England, used plum pudding with raisins as his analogy.

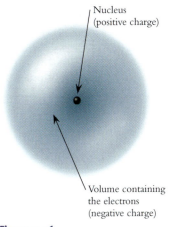

Nucleus (positive charge)

Volume containing the electrons (negative charge)

Figure 2-16
Schematic drawing of an atom, showing a central, positive nucleus surrounded by a cloud of electrons. This model of the atom is consistent with the results of Rutherford's scattering experiments.

Table 2-1   Atomic Building Blocks

Name	Symbol	Charge	Mass
Electron	e	-1.6022×10^{-19} C	9.1094×10^{-31} kg
Proton	p	$+1.6022 \times 10^{-19}$ C	1.6726×10^{-27} kg
Neutron	n	0	1.6749×10^{-27} kg

From the number of particles deflected and the pattern of deflection, Rutherford calculated the fraction of the atomic volume occupied by the positive nucleus. That fraction is 1 part in 10^{14}. To give you an idea of what that means, an atom the size of a baseball stadium would have a nucleus the size of a pea. The density of the nucleus is so great that a nucleus the size of a pea would have a mass of more than 250 million tons, as much as 33 million elephants!

Experiments on nuclei showed that the nucleus itself contains two types of subatomic particles called **protons** and **neutrons.** Protons account for the positive charges of nuclei, whereas neutrons contribute mass but are electrically neutral. The positive charge of a proton is equal in magnitude to the negative charge of an electron. The mass of a proton, on the other hand, is almost 2000 times greater than the mass of an electron. The mass of a neutron is almost the same as the mass of a proton. The experiments leading to the discovery of protons and neutrons were as important to our understanding of matter as were the experiments leading to the discovery of the electron. Our interest is in chemistry, however, and we show in later chapters that the chemistry of atoms and molecules depends mainly on their electrons.

Our picture of atomic architecture is now complete. Three kinds of particles—electrons, protons, and neutrons—combine in various numbers to make the different atoms of all the elements of the periodic table. Table 2-1 summarizes the characteristics of these three atomic building blocks.

> Electrical repulsion between protons should cause a nucleus that contains more than one proton to fly apart. In Section 2.3 we describe how the third type of fundamental force, called the strong nuclear force, acts within nuclei and generates enough attraction among nuclear particles to hold nuclei together.

Section Exercises

2.2.1 Draw a sketch (including appropriate signs for the electrical plates) for a tube in which positive ions are accelerated into a deflection region. Use Figure 2-13 as a guide.

2.2.2 In an experiment such as Millikan's (refer to Figure 2-14), an oil droplet with a mass of 1.2 μg and carrying three extra electrons is held motionless by an electrical force. (a) In which direction is a droplet of mass 1.2 μg and carrying four electrons moving? (b) In which direction is a droplet of mass 2.0 μg and carrying three extra electrons moving? Explain your answers.

2.2.3 (a) How many electrons are required to give a mass of 1.0 μg? (1 kg = 10^3 g; 1 μg = 10^{-6} g) (b) What is the mass of the same number of neutrons?

2.3   ATOMIC DIVERSITY: THE ELEMENTS

According to the atomic theory, each element has unique properties. The differences among elements are caused by differences in their atoms. Each element is unique because its atoms contain characteristic numbers of protons and electrons.

Early experiments showed that strong electrical forces can strip electrons from atoms. Atoms can also gain electrons under the influence of electrical force. In fact, much of the chemistry that takes place in the world around us involves electrons shifting from one chemical substance to another. Chemical reactions have no effect, however, on the structures of nuclei. All atoms of a particular element have the same number of protons in the nucleus, and these do not change during chemical processes. The defining feature of an element, therefore, is the charge carried by the protons in its nucleus.

KEY CONCEPTS   *An element is identified by the charge of its nucleus.*

Every element has a unique nuclear charge and a specific and unchanging number of protons. The number of protons in the nucleus is called the **atomic number** and is symbolized Z. All atoms with the same value of Z belong to the same element. Each element has its own name, symbol, and atomic number. As examples, the element with $Z = 1$ is hydrogen, symbol H; the element with $Z = 2$ is helium, symbol He; and the element with $Z = 92$ is uranium, symbol U. The periodic table lists all the elements in order of increasing atomic number.

The amount of charge on a nucleus is determined by its number of protons. The charge on an electron is opposite in sign but equal in magnitude to the charge on a proton. Thus, a neutral atom contains the same number of electrons as it does protons.

> As stated in Chapter 1, the periodic table lists elements in order of increasing masses, with a few exceptions. In fact, nuclear charge is the organizing feature of the periodic table. As nuclear charge increases, so also does nuclear mass (with few exceptions), so these two characteristics of elements are closely related.

Isotopes

The *identity* of an atom is determined by its nuclear charge, but its *mass* is the sum of the contributions from all its atomic building blocks. Recall from Table 2-1 that the mass of an electron is almost 2000 times smaller than the mass of a proton or a neutron. Consequently, the mass of an atom is determined almost entirely by the mass of its nucleus. Thus, the mass of an atom depends on the number of protons and neutrons in its nucleus.

Two atoms with the same number of protons but different numbers of neutrons are called **isotopes.** For example, every chlorine atom has 17 protons in its nucleus, but whereas most chlorine nuclei contain 18 neutrons, others contain 20 neutrons. Naturally occurring chlorine contains both isotopes, so any sample that contains chlorine includes atoms with two different masses.

An isotope is usually specified by its **mass number.** The mass number is the total number of protons and neutrons contained in a nucleus. We can represent any isotope of a chemical element completely by writing its chemical symbol (X) preceded by a superscript giving its mass number (A) and a subscript giving its atomic number (Z):

> The subscript Z is redundant because each chemical symbol already defines a unique atomic number. The subscripts are nevertheless useful for isotopic bookkeeping.

$$\text{Mass number} \longrightarrow {}^{A}_{Z}X \longleftarrow \text{Elemental symbol}$$
$$\text{Atomic number} \longrightarrow$$

②

For example, the two isotopes of chlorine are ${}^{35}_{17}\text{Cl}$ and ${}^{37}_{17}\text{Cl}$.

Another way to describe an isotope is to state its elemental name and mass number. The stable isotopes of chlorine are ${}^{35}\text{Cl}$, or chlorine-35, and ${}^{37}\text{Cl}$, or chlorine-37. Example 2-2 provides practice in determining the composition of atoms.

Example 2-2

Determining the Composition of Atoms

The following isotopes have medical applications. Determine the number of protons, neutrons, and electrons in each isotope: (a) ${}^{24}_{11}\text{Na}$ (used to study electrolytes within the body); (b) ${}^{51}\text{Cr}$ (used to label and monitor red blood cells); and (c) iridium-192 (internal radiotherapy source for cancer treatment).

Strategy: The number of protons is determined from the atomic number (Z), the number of neutrons is found from Z and the mass number (A), and the number of electrons in a neutral atom must equal the number of protons.

Solution:

a. Na is the elemental symbol for sodium. The subscript 11 is Z, which is the number of protons in the nucleus. The superscript 24 is A. Find the number of neutrons by subtracting Z from A: $A - Z = 24 - 11 = 13$ neutrons. Because this is a neutral atom, the number of electrons must equal the number of protons. Sodium-24 has 11 protons, 13 neutrons, and 11 electrons.

b. Cr is the symbol for chromium. A is 51. Find Z by consulting the periodic table. Chromium has a Z value of 24, which tells us the nucleus contains 24 protons.

Example 2-2

Determining the Composition of Atoms (*continued*)

Subtracting Z from A, we find that there are 27 neutrons in this isotope. Finally, the atom is neutral, so 24 electrons are present.

c. Iridium has Z of 77. A neutral atom of iridium-192 has 77 protons, 77 electrons, and $192 - 77 = 115$ neutrons.

Do the Results Make Sense? An isotopic symbol makes sense if the elemental symbol matches the value of Z. Consult a periodic table to verify that all three of these isotopic symbols are sensible.

Extra Practice Exercise 2.2

The following radioactive isotopes have medical applications. Determine the number of protons and neutrons for each. (a) ^{59}Fe (used in studies of iron metabolism); (b) samarium-153 (very effective in relieving the pain of bone cancer); (c) $^{42}_{19}$K (used in studies of coronary blood flow)

A few elements, among them fluorine and phosphorus, occur naturally with just one isotope, but most elements are isotopic mixtures. For example, element number 22 is titanium (Ti), a light and strong metal used in jet engines and in artificial human joints. There are five naturally occurring isotopes of Ti. Each one has 22 protons in its nuclei, but the number of neutrons varies from 24 to 28. In a chemical reaction, all isotopes of an element behave nearly identically. This means that the isotopic composition of an element remains essentially constant. The isotopic composition of Ti (number percentages) is

^{46}Ti, 8.25% ^{47}Ti, 7.44% ^{48}Ti, 73.72% ^{49}Ti, 5.41% ^{50}Ti, 5.18%

Wherever titanium is found, for example in TiO_2, a compound used as a white pigment in paint, or pure Ti metal in an artificial knee joint, it has this isotopic composition.

Figure 2-17 illustrates the range of isotopic composition. One element shown in the figure is tin, the element with the largest number of stable isotopes.

Hydrogen, the simplest of all chemical elements, has two naturally occurring isotopes. Its most common isotope, ^{1}H, contains a single electron and a single proton, and 99.98% of the hydrogen atoms on the Earth have this composition. The second isotope of hydrogen, ^{2}H, has a nucleus made up of both a proton and a neutron. The addition of this neutron doubles the mass of the atom. Doubling the mass is enough to make the chemical behavior of ^{2}H differ somewhat from that of ^{1}H. Thus, the element hydrogen is an exception to our generalization about the common chemical properties of isotopes. Chemists give the "heavy" isotope its own name and symbol: deuterium (D). In naturally occurring substances, the abundance of deuterium is so low that the subtle differences in its chemistry are rarely observed. It is possible to prepare molecules that are enriched in deuterium, however, and this allows chemists to take advantage of its special properties. The most common compound of deuterium is "heavy" water, D_2O, which is available from chemical supply houses at a price of about $50 for 100 g.

Figure 2-17
The natural abundances of the isotopes of four elements (Cl, Cr, Ge, and Sn) illustrate the diversity of isotopic distributions. The mass number and percent abundance of each isotope are indicated.

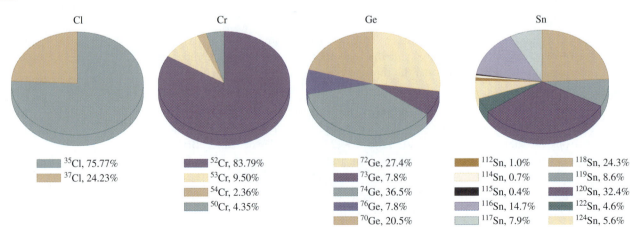

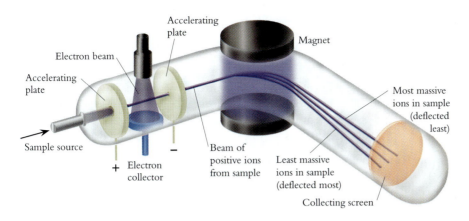

Mass Spectrometry

The existence of isotopes is demonstrated dramatically by research done with a **mass spectrometer,** one of which is represented in Figure 2-18. In this instrument, a sample of matter passes through an electrical discharge. The discharge removes electrons from the atoms or molecules, changing them into positively charged ions. A molecular ion may remain intact or may lose one or more of its atoms. Electrical force shapes these ions into a beam that passes between the poles of a magnet. Recall from Section 2.2 that charged particles moving through a magnetic field are deflected (see Figure 2-11). The curvature of the path depends on the mass of the ion: Ions with large masses are deflected less than ions with small masses. The mass spectrometer separates positively charged ions according to their masses and produces a graph of their abundance as a function of mass.

Figure 2-19 shows the mass spectrum of the element neon. The three peaks in the mass spectrum come from three different isotopes of neon, and the peak heights are proportional to the natural abundances of these isotopes. The most abundant isotope of neon has a mass number of 20, with 10 protons and 10 neutrons in its nucleus, whereas its two minor isotopes have 11 and 12 neutrons. Example 2-3 illustrates how to read and interpret a mass spectrum.

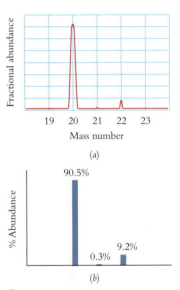

Figure 2-19
Mass spectrum of neon: (a) actual appearance, (b) bar graph representation.

Example 2-3

Identifying Isotopes

A sample of lead atoms is analyzed by mass spectrometry. The bar graph in the margin shows the results. Use information from the graph to write the elemental symbol that represents each Pb isotope and estimate the natural abundance of each. List the number of protons and neutrons for each isotope.

Strategy: Each peak in a mass spectrum corresponds to an ion with a different mass. From the mass numbers of these peaks, we can determine the isotopic symbols and obtain a count of the protons and the neutrons. The height of each peak is proportional to the abundance of fragments of that particular mass.

Solution: The four peaks in the mass spectrum represent four Pb isotopes. Their A values are 204, 206, 207, and 208. Consulting the periodic table, we find that Z of lead is 82. Thus, the elemental symbols are as follows:

$$^{204}_{82}\text{Pb} \qquad ^{206}_{82}\text{Pb} \qquad ^{207}_{82}\text{Pb} \qquad ^{208}_{82}\text{Pb}$$

The atomic number (Z) gives the number of protons in the nucleus, so all four isotopes have 82 protons. On the other hand, A gives the sum of the number of protons

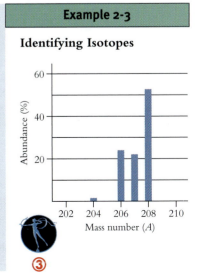

Example 2-3

Identifying Isotopes
(*continued*)

and neutrons. Subtracting 82 from each *A* value yields the number of neutrons in the nucleus of each isotope. Read the abundances from the heights of the peaks in the mass spectrum. To summarize:

^{204}Pb	82 protons	122 neutrons	1% abundant
^{206}Pb	82 protons	124 neutrons	24% abundant
^{207}Pb	82 protons	125 neutrons	22% abundant
^{208}Pb	82 protons	126 neutrons	52% abundant

Does the Result Make Sense? Percentages should sum to 100%. In this case, the sum is 1 + 24 + 22 + 52 = 99%, but we can read the graph only to about 1% accuracy, so a sum that is within 1% of 100% makes sense.

Extra Practice Exercise 2.3

Determine the isotopic information for antimony, whose mass spectrum appears in the margin below.

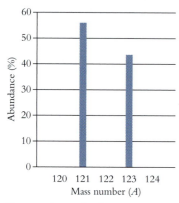

Mass spectrum of antimony

Nuclear Stability

For each different element, there are a few specific values of *A* that result in stable nuclei. Figure 2-20 shows all the stable nuclei on a plot of the number of neutrons (*N*) versus the number of protons (*Z*). These data show a striking pattern: All stable nuclei fall within a "belt of stability." Any nucleus whose ratio of neutrons to protons falls outside the belt of stability is unstable and decomposes spontaneously. Lighter nuclei lie along the $N = Z$ line, but the N/Z ratio of stable nuclei rises slowly until it reaches 1.54. The trend is illustrated by the N/Z ratios of the following stable nuclei: $^{4}_{2}$He (1.00), $^{31}_{15}$P (1.07), $^{56}_{26}$Fe (1.15), $^{81}_{35}$Br (1.31), $^{197}_{79}$Au (1.49), and $^{208}_{82}$Pb (1.54).

Electrical forces cause positively charged protons to repel each other, so an additional attractive force is needed to stabilize nuclei containing more than one proton. This attractive force, called the *strong nuclear force*, acts between protons and neutrons. The strong nuclear force operates only at very small distances within the nucleus, but at those distances, it is about 100 times stronger than proton−proton repulsion. This strong force is what binds neutrons and protons tightly in the nucleus. As *Z* increases, the effect of electrical repulsion increases. This requires more neutrons to generate enough nuclear force to stabilize the nucleus.

The main features of nuclear stability can be summarized qualitatively. The balance between electrical repulsion and strong nuclear attraction changes with the number of protons and neutrons in the nucleus. Compressing more than one positively charged proton into a volume as small as the nucleus leads to strongly repulsive electrical forces that must be offset by the presence of specific numbers of neutrons. Neutrons provide strong nuclear binding forces without adding electrical repulsive forces. The extra binding force of the neutrons holds the protons within the nucleus.

When *Z* gets big enough, no number of neutrons is enough to stabilize the nucleus. Notice in Figure 2-20 that there are no stable nuclei above bismuth, $Z = 83$. Some elements with higher *Z* are found on Earth, notably radium ($Z = 88$), thorium ($Z = 90$), and uranium ($Z = 92$), but all such elements are unstable and eventually disintegrate into nuclei with $Z < 83$. Consequently, the set of stable nuclei, those that make up the world of "normal" chemistry and provide the material for all terrestrial chemical reactions, is a small subset of all possible nuclei.

Figure 2-20
Plot of the *Z* and *N* values of stable nuclei. All stable nuclei fall within a belt that lies between $N = Z$ and $N = 1.54 Z$.

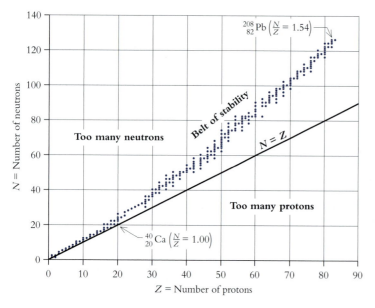

Unstable Nuclei

Some isotopes that occur in nature are unstable and are said to be **radioactive.** A few radioactive isotopes, such as uranium-238 and carbon-14, are found on Earth, and many others can be synthesized in nuclear chemistry laboratories, as we describe in Chapter 22. Over time, radioactive isotopes decompose into other stable isotopes. Unstable isotopes decompose in several ways. Most nuclei that have $Z > 83$ decompose by giving off a helium nucleus. Helium-4 nuclei ejected from an unstable nucleus are called **alpha particles.** For example, uranium-238 decomposes to thorium-234 by this pathway:

$$^{238}_{92}U \longrightarrow \, ^{234}_{90}Th + \, ^{4}_{2}He$$

Thorium is also unstable ($Z = 90$), so it will undergo its own decomposition.

Other radioactive atoms decompose by ejecting an electron from the nucleus. These electrons have high kinetic energies, so they are given a special designation and symbol, **beta particles,** $_{-1}^{0}\beta$. Radioactive carbon-14 decomposes by emission of a beta particle:

$$^{14}_{6}C \longrightarrow \, ^{14}_{7}N + \, ^{0}_{-1}\beta$$

A third nuclear decomposition pathway begins with the emission of a **positron** (symbolized $_{+1}^{0}\beta^{+}$), a particle with the same mass as an electron but a positive charge. The isotope of fluorine used for PET scans decays by positron emission, as described in our chapter introduction:

$$^{18}_{9}F \longrightarrow \, ^{18}_{8}O + \, ^{0}_{+1}\beta^{+}$$

As soon as a positron encounters an electron, the two particles destroy each other and generate a pulse of gamma radiation. Notice that in each of our examples of nuclear decomposition, total charge (the subscripts) is conserved, and so is the mass number (the superscripts). This is true for all nuclear decay processes.

Each different unstable isotope has its own characteristic rate of decomposition. Some isotopes survive for only a fraction of a second, but others decompose slowly, sometimes over thousands of years. Most of chemistry involves the stable isotopes, so we defer further consideration of nuclear decomposition until Chapter 22, which covers nuclear processes in detail.

Isotopes in Medicine

Medical scientists have learned how to exploit the products of nuclear decay in the diagnosis and treatment of disease. As a diagnostic tool, nuclear decomposition allows physicians to examine internal organs by noninvasive procedures. Although X rays are good for forming images of bones and teeth, most soft tissue organs are transparent to X rays. For these organs, γ rays resulting from nuclear decay can be used to produce images. Like an X-ray film, a γ-ray image reveals abnormalities, allowing physicians to diagnose ailments of the heart, spleen, liver, brain and other vital organs.

Radioactive tracers for medical imaging are designed to concentrate in specific organs. Our chapter introduction describes one such tracer, modified glucose, which is used to image the brain during a PET scan. Other examples include iron-52 for bone marrow scans, xenon-133 for studying lung functions, and iodine-131 for the thyroid gland. Our Chemistry and Life Box describes the medical applications of the technetium isotope ^{99m}Tc.

Because exposure to radiation is a health risk, the administration of radioactive isotopes must be monitored and controlled carefully. Isotopes that emit alpha or beta particles are not used for imaging, because these radiations cause substantial tissue damage. Specificity for a target organ is essential so that the amount of radioactive material can be kept as low as possible. In addition, an isotope for medical imaging must have a decay rate that is slow enough to allow time to make and administer the tracer compound, yet fast enough rid the body of radioactivity in as short a time as possible.

Radioactivity also is used to treat certain diseases. Some cancers respond particularly well to radiation therapy. Radioactivity must be used with care, because exposure to

BOX 2-2	Chemistry and Life: Medical Applications of Technetium

Technetium compounds are the most important radio-pharmaceuticals in use today. Over 80% of all medical applications of radioactive atoms involve a single isotope, ^{99m}Tc (the "m" stands for "metastable," because this technetium isotope converts to ^{99}Tc by emitting γ rays). This form of technetium has near-perfect properties for medical imaging. First, ^{99m}Tc emits γ rays of moderate energy that are easy to detect but are not highly damaging to living tissue. Second, although technetium is not normally found in the body, the element can be bound to many chemical tracers that the body recognizes and processes. Third, the isotope decomposes at a rate that is long enough to deliver ^{99m}Tc-containing tracers to the body and generate an image, but short enough to minimize long-term effects of γ radiation. About 95% of the ^{99m}Tc administered for a medical scan decomposes within the first 24 hours after introduction into the body.

Technetium-99m can be used to collect images of the thyroid gland, brain, lungs, heart, liver, stomach, kidneys, and bones, making it a very versatile imaging agent. The figure below shows how ^{99m}Tc imaging reveals thyroid tumors. The uniform distribution of ^{99m}Tc in the figure on the left indicates a normal thyroid gland. The asymmetrical distribution shown in the figure on the right reveals a thyroid tumor on the right side of the organ.

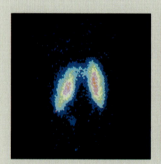

^{99m}Tc images of thyroid glands

Technetium isotopes also help tremendously in the diagnosis of breast cancer. A technetium complex preferentially binds to cancer cells, so if a patient has cancer, radioactivity imaging will reveal high levels of radioactivity from the cancerous tissues. The red spot in the image below marks the location of cancerous cells.

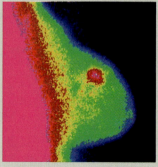

^{99m}Tc image of a cancerous breast

All isotopes of technetium ($Z = 43$) are unstable, so the element is not found anywhere in the Earth's crust. Its absence left a gap in the periodic table below manganese. The search for this "missing element" occupied researchers for many years. It was not until 1937 that the first samples of technetium were prepared in a nuclear reactor. In fact, technetium was the first element to be made artificially in the laboratory. To date, 21 radioactive isotopes of technetium have been identified, some of them requiring millions of years to decompose.

One of the key features that accounts for the widespread use of ^{99m}Tc as an imaging agent is the convenience with which the isotope can be obtained in a clinical setting. A nuclear facility provides a "technetium generator," shown below in schematic form. The starting material for the generator is a radioactive molybdenum compound, ^{99}MoO$_4^{2-}$, generated as a byproduct of the nuclear power industry. Molybdenum-99 decomposes to technetium-99m:

$$^{99}_{42}\text{Mo} \longrightarrow {}^{99m}_{43}\text{Tc} + {}^{0}_{-1}\beta$$

The molybdenum compound in the generator is bound to a column of solid alumina, Al_2O_3. To collect ^{99m}TcO$_4^-$ formed by radioactive decay of starting material, a salt solution is passed periodically through the column. The molybdenum starting material decomposes at a convenient rate: About half of the available ^{99}Mo decomposes to ^{99m}Tc every 65 hours. The product is released into the saline solution, washed out of the reactor, and collected for medical imaging.

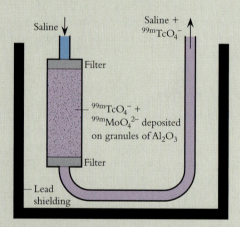

Schematic view of the separation process for ^{99m}TC

In some cases, ^{99m}TcO$_4^-$ can be used directly as an imaging agent. For example, the TcO$_4^-$ anion shares several physical properties with the iodide ion I$^-$, including a tendency to pool in the thyroid gland. For most applications, however, the TcO$_4^-$ anion is used to make other tracer molecules that tend to collect in various and specific organs of the body. One of the most promising uses of ^{99m}Tc and other radioactive nuclei relies on the body's own defense mechanisms. Researchers are learning to attach radioactive atoms to antibodies, natural disease-fighting proteins of the body. Antibodies can be extremely specific in targeting proteins contained in tumors. Attaching ^{99m}Tc to an antibody is an exciting strategy for delivering a "magic bullet" of radioactivity to a specific organ or tumor.

radiation damages healthy cells and eventually causes cancer. The key to radiation therapy is that cancer cells reproduce more rapidly than normal cells, and rapidly reproducing cells are more sensitive to radiation. If concentrated doses of radiation are focused on the malignant cells, a cancer may be destroyed with minimal damage to healthy tissue. Nevertheless, radiation therapy always has unpleasant side effects, including nausea and hair loss.

Thyroid cancers are often treated with radioactive iodine because the thyroid gland preferentially absorbs iodine. If a patient is treated with ^{131}I, the β emissions from this isotope are concentrated in the gland, destroying cancerous cells more rapidly than normal cells. At the correct dosage, the cancer may be eliminated without destroying the healthy part of the thyroid gland. In an alternative approach, a metal wire of radioactive iridium-197 is implanted within the target area through a catheter. The wire implant is removed after the correct dose of radiation has been administered.

The radioactive source need not always be introduced into the body. Inoperable brain tumors can be treated with γ rays from an external source, usually a sample of cobalt-60. The patient is placed in a position where the γ-ray beam passes through the tumor. The patient is moved so that the γ rays irradiate the tumor from several angles. In this manner the tumor receives a much higher dose of radiation than the dose received by any surrounding tissues.

Section Exercises

2.3.1 Use the periodic table (inside front cover) to fill in the missing information:

Name	Symbol	Z
Carbon	C	6
_____	Au	_____
_____	_____	33
Iron	_____	_____

2.3.2 Write isotopic symbols for (a) a cobalt atom with 30 neutrons; (b) the element with $Z = 3$ and $A = 7$; and (c) potassium with one more neutron than protons.

2.3.3 Figure 2-17 gives the isotopic composition of naturally occurring chromium (Cr). Draw a bar graph that shows the mass spectrum of Cr.

2.4 COUNTING ATOMS: THE MOLE

Chemistry is a quantitative science, and chemists frequently measure amounts of matter. As the atomic theory states, matter consists of atoms, so measuring amounts means measuring numbers of atoms. Counting atoms is difficult, but we can easily measure the mass of a sample of matter. To convert a mass measurement into a statement about the number of atoms in a sample, we must know the mass of an individual atom.

Mass spectrometry allows us to measure masses of individual atoms. Still, there is an enormous difference between the mass of one atom and the masses of samples measured in the laboratory. For example, a good laboratory balance measures mass values from about 10^{-6} to 10^3 g. An atom, on the other hand, has a mass between 10^{-24} and 10^{-21} g. Even a tiny mass of 10^{-6} g contains approximately 10^{16} atoms.

The Mole and Avogadro's Number

To avoid having to work with unimaginably large numbers, chemists use a convenient unit called the **mole (mol):**

One mole is the number of atoms in exactly 12 g of the pure isotope carbon-12. **KEY CONCEPT**

Using mass spectrometers, scientists have determined that the mass of a ^{12}C atom is $1.992\,646 \times 10^{-23}$ g. This experimental mass combined with the definition of the mole gives the number of atoms in one mole:

$$\left(\frac{12 \text{ g}^{-12}\text{C}}{1 \text{ mol}}\right)\left(\frac{1 \text{ atom}}{1.992\,646 \times 10^{-23} \text{ g}^{-12}\text{C}}\right) = 6.022\,142 \times 10^{23} \text{ atoms/mol}$$

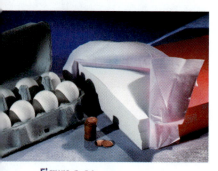

Figure 2-21
Different units are useful for measuring different-sized items. Eggs are sold by the dozen, paper is sold by the ream, and atoms are measured by the mole. Twenty-four pre-1982 pennies contain 1 mol of copper atoms.

The number of items in one mole is important enough to be given its own name and symbol. It is known as **Avogadro's number,** commonly symbolized N_A. Although Avogadro's number is known with seven-figure accuracy, four figures are enough for most calculations: $N_A = 6.022 \times 10^{23}$ items/mol.

The mole is a convenient unit for chemical amounts because the mass of most typical laboratory samples contains between 0.01 and 10 mol of atoms or molecules. For example, 1.20 g of ^{12}C contains 6.02×10^{22} atoms of carbon, which is 0.100 mol. The convenience of using a larger unit to describe many small items is not limited to atoms and molecules. Eggs are sold by the dozen, and paper is sold by the ream; chemists "sell" atoms and molecules by the mole. There are 12 eggs in a dozen eggs, 500 sheets of paper in a ream of paper and 6.022×10^{23} atoms in a mole of copper (Figure 2-21).

How large is a mole? If a computer were to count items at one million per second, it would take about 19 billion years to reach Avogadro's number. If every person on the planet (population = 6 billion) were to spend 1 million dollars per second, it would take about 300 years to spend one mole of dollars.

Molar Mass

The *molar mass (MM)* of any substance is the mass of one mole of that substance. As described in Section 2.3, each isotope of a particular element has a different mass. Therefore, the mass of one mole of any isotope has a unique value, its *isotopic molar mass*. This characteristic molar mass can be found by multiplying the mass of one atom of that isotope by Avogadro's number. For example, mass spectrometry experiments reveal that one atom of carbon-13 has a mass of 2.15928×10^{-23} g, from which we can calculate the isotopic molar mass of ^{13}C atoms:

$$\left(\frac{2.15928 \times 10^{-23}\text{ g}}{\text{atom of }^{13}\text{C}}\right)\left(\frac{6.022142 \times 10^{23}\text{ atoms of }^{13}\text{C}}{1\text{ mol}}\right) = 13.0035\text{ g/mol}$$

A sample of a chemical element is usually a mixture of all of its stable isotopes. The carbon in a sample of graphite is 99.892% ^{12}C and 1.108% ^{13}C; elemental titanium is a mixture of five stable isotopes, and tin has ten isotopes, each with its own particular mass. Every element has a characteristic distribution of isotopes, and each isotope of the element contributes proportionally to the molar mass of the element.

KEY CONCEPT *The molar mass of any naturally occurring element is the sum of the contributions from its isotopes.*

The mass contributed by each isotope is proportional to its fractional abundance, as given by Equation 2-1:

The symbol Σ is the Greek letter sigma. It means "Take the sum of the following quantities."

Elemental molar mass = Σ(Fractional abundance)(Isotopic molar mass)   (2-1)

Equation 2-1 combines multiplication and addition steps. A good way to carry out such calculations is by making a table. Here is such a table for determining the elemental molar mass of titanium:

⑤

Isotope	Fractional Abundance	Isotopic Molar Mass (g/mol)	Contribution to Sum (g/mol)
^{46}Ti	0.0825	45.95263	3.79
^{47}Ti	0.0744	46.95176	3.49
^{48}Ti	0.7372	47.94795	35.35
^{49}Ti	0.0541	48.94787	2.65
^{50}Ti	0.0518	49.94479	2.59

The molar mass of naturally occurring titanium is the sum of all five contributions, 47.87 g/mol.

Example 2-4 illustrates the use of isotopic molar masses and natural abundances to calculate the molar mass of elemental iron.

Example 2-4

Calculating a Molar Mass

The mass spectral analysis of an iron sample gives the following data.

Isotope	Isotopic Molar Mass	Abundance
^{54}Fe	53.940 g/mol	5.82%
^{56}Fe	55.935 g/mol	91.66%
^{57}Fe	56.935 g/mol	2.19%
^{58}Fe	57.933 g/mol	0.33%

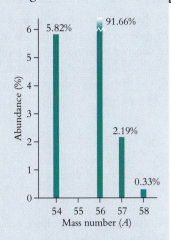

Calculate the molar mass of iron (Fe).

Strategy: The molar mass of a naturally occurring mixture of isotopes is the weighted average of the isotopic molar masses. Each isotope contributes to the total in proportion to its percentage (fractional) abundance, and the average is calculated using Equation 2-1.

Solution: First, the percentages must be converted to fractions by dividing by 100%. Then we multiply these fractional abundances by the isotopic masses and add the results. A table helps in organizing these manipulations:

Isotope	Fractional Abundance	Isotopic Molar Mass (g/mol)	Contribution to Sum (g/mol)
^{54}Fe	0.0582	53.940	3.14
^{56}Fe	0.9166	55.935	51.27
^{57}Fe	0.0219	56.935	1.25
^{58}Fe	0.0033	57.933	0.19

Sum the contributions to get the final result: molar mass of natural Fe = 55.85 g/mol.

Does the Result Make Sense? Look again at the percentage distribution of the isotopes of iron. Just over 90% of the composition of this element is ^{56}Fe, so the molar mass of the mixture of isotopes should be close to the molar mass of this isotope, 55.935 g/mol. Moreover, there is more of the lighter isotope (5.82% ^{54}Fe) than of the two heavier isotopes, so it makes sense that the weighted average is slightly smaller than the molar mass of ^{56}Fe. Why does the result have four significant figures, when the isotopic molar masses are known to five figures, whereas some of the percentages are known to only three? This calculation involves both multiplication and addition, so we must apply both rules for determining significant figures:(0.0582)(53.940 g/mol) = 3.14 g/mol, a result that has three significant figures because when we multiply, the number of significant figures in the result is the same as the least number of significant figures among the numbers being multiplied. On the other hand, 3.14 g/mol + 51.27 g/mol + 1.25 g/mol + 0.19 g/mol = 55.85 g/mol, a result that has two decimal places because when we add, the number of decimal places in the result is the same as the least number of decimal places among the numbers being added.

Naturally occurring magnesium contains three isotopes, with atomic masses 23.985 04 (78.99%), 24.985 84 (10.00%), and 25.982 59 (11.01%) g/mol. Determine the molar mass of magnesium.

For purposes of chemical bookkeeping, it is unnecessary to know the isotopic molar masses and isotopic distributions of the elements. All we need to know is the mass of one mole of an element containing its natural composition of isotopes. These molar masses usually are included in the periodic table, and they appear on the inside front and back covers of this textbook.

Mass–Mole–Atom Conversions

Molar mass can be thought of as a conversion factor between mass in grams and number of moles. These conversions are essential in chemistry, because chemists count amounts of substances in moles but routinely measure masses in grams. To determine the amount (number of moles) in a sample, we measure the mass of the sample and then divide by the molar mass of the substance:

$$n = \frac{m}{MM} \tag{2-2}$$

If we know moles and wish to convert to mass, we rearrange Equation 2-2 and multiply that number of moles by the molar mass:

$$m = n\,MM$$

Suppose, for example, that a bracelet contains 168 g of silver (Ag). To determine the number of moles of silver in the bracelet, divide the mass by the molar mass of Ag, which is 107.87 g/mol:

$$n = \frac{m}{MM} \quad so \quad n = \frac{168 \text{ g Ag}}{107.87 \text{ g/mol}} = 1.56 \text{ mol Ag}$$

If a chemical reaction calls for 0.250 mol of silver, we multiply moles by molar mass to determine the mass of Ag that should be used:

$$m = n\,MM \quad so \quad m = (0.250 \text{ mol Ag})(107.87 \text{ g/mol}) = 27.0 \text{ g Ag}$$

Avogadro's number is the conversion factor that links the number of moles with the number of individual particles. To determine the number of atoms in a sample of an element, we multiply the number of moles by Avogadro's number:

$$\# = nN_A$$

Likewise, if a sample contains a certain number of atoms, the number of moles in the sample can be calculated by dividing by Avogadro's number:

$$n = \frac{\#}{N_A} \tag{2-3}$$

The number of silver atoms in the bracelet that contains 1.56 mol of Ag is found by multiplying the amount in moles by Avogadro's number:

$$\# = nN_A \quad so \quad \# = (1.56 \text{ mol Ag})(6.022 \times 10^{23} \text{ atoms/mol}) = 9.39 \times 10^{23} \text{ atoms Ag}$$

Radioactivity detectors, which "count" the number of nuclei that decay, provide an example of a conversion from number of particles to moles. If a radioactivity measurement on a blood sample from a patient receiving radiation treatment indicates that 1.75×10^4 nuclei of ^{99m}Tc decay in one hour, dividing by Avogadro's number converts to moles:

$$n = \frac{\#}{N_A} \quad so \quad n = \frac{1.75 \times 10^4 \text{ nuclei } ^{99m}\text{Tc}}{6.022 \times 10^{23} \text{ nuclei/mol}} = 2.91 \times 10^{-20} \text{ mol } ^{99m}\text{Tc}$$

Avogadro's number and molar mass make it possible to convert readily among the mass of a pure element, the number of moles, and the number of atoms in the sample. These conversions are represented schematically in the flowchart shown in Figure 2-22.

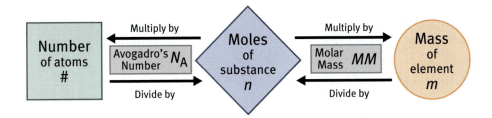

Figure 2-22
This flowchart shows the conversion processes among the number of moles, the mass, and the number of atoms for a sample of an element.

Moles occupy the central position of this flowchart because the mole is the unit that chemists use in almost all chemical calculations. When you set out to solve a chemical problem, first interpret the question on the atomic/molecular level. The second part of chemical problem solving often involves quantitative calculations, which usually require working with *moles*.

Chemical calculations are built around the mole.

KEY CONCEPT

Example 2-5 provides a practical application of these ideas.

Example 2-5

Mass–Mole–Atom Conversions

⑥

The eruption of Mount St. Helens on May 18, 1980, provided geologists with a unique opportunity to study the action of volcanos. Gas samples from the plume were collected and analyzed for toxic heavy metals. To collect mercury (Hg), gas samples were passed over a piece of gold metal, which binds Hg atoms very tightly. The mass of the metal increased as it absorbed Hg from the plume. From a plume-gas sample containing 200 g of ash, 3.60 μg of Hg was deposited on the gold. How many moles of mercury were present in the gas sample? How many atoms is this?

Strategy: This problem contains several pieces of information, not all of which are needed to answer the questions. The seven-step approach described in Chapter 1 is particularly useful for extracting the essential features from such a problem.

1. *Determine what is asked for.* The problem requires quantitative calculation, because it asks for the number of moles and the number of atoms of Hg present in the plume sample.
2. *Visualize the problem.* Mercury atoms from the gas plume stick to the gold metal, causing an increase in mass. Thus, the increase in mass of the piece of gold equals the mass of Hg in the plume sample.

Solution:
3. *Organize the data.* The data available include what is given in the problem and what can be looked up in tables:

Given in the problem: 3.60 μg of Hg deposited

From tables: 1 μg = 10^{-6} g

MM of Hg = 200.6 g/mol

N_A = 6.022 × 10^{23} atoms/mol

Notice that the mass of the ash is not included in the organized data. The problem asks about the amount of mercury in the gas sample. The mass of ash in that sample is an extra piece of information that is not needed to solve the problem.

4. *Identify a process to solve the problem.* The mass of mercury is given. Conversion from mass to number of moles and number of atoms requires Equations 2-2 and 2-3:

$$n = \frac{m}{MM} \quad and \quad n = \frac{\#}{N_A}$$

Example 2-5
Mass–Mole–Atom Conversions *(continued)*

5. *If necessary, manipulate the equations.* Moles and atoms of mercury are the goal of the calculation. We need to solve Equation 2–3 for the number of atoms:

$$\text{\# atoms} = nN_A$$

6. *Substitute and calculate.* Pay careful attention to units:

$$\text{Moles Hg} = \frac{1.00 \, \mu\text{g Hg}}{(200.6 \, \text{g/mol})(10^6 \, \mu\text{g/g})} = 1.795 \times 10^{-8} \, \text{mol Hg}$$

$$\text{Atoms Hg} = (1.795 \times 10^{-8} \, \text{mol Hg})(6.022 \times 10^{23} \, \text{atoms/mol}) = 1.08 \times 10^{16} \, \text{atoms}$$

7. Do the Results Make Sense? First of all, the units cancel to give moles and atoms, which are the units that the problem asked for. Further, the mass of Hg is quite small, so we expect the number of moles to be small also. The number of atoms, 1.08×10^{16}, is large but much smaller than Avogadro's number.

Extra Practice Exercise 2.5

Although 10^{-8} mol is not very much mercury, the initial eruption blanketed the Yakima Valley with 35 metric tons of volcanic ash per acre. Given that a metric ton is 1000 kg, determine how much Hg the volcano deposited on every acre of the Yakima Valley, in grams, moles, and number of atoms.

Section Exercises

■ **2.4.1** Use the data in the following table to calculate the molar mass of naturally occurring sulfur.

Isotope	Isotopic Molar Mass	Abundance
^{32}S	31.97207 g/mol	95.02%
^{33}S	32.97146 g/mol	0.75%
^{34}S	33.96786 g/mol	4.21%
^{36}S	35.96709 g/mol	0.02%

■ **2.4.2** Calculate the number of atoms present in 2.55 g of each of these elements: (a) lithium; (b) titanium; and (c) gold.

■ **2.4.3** Recently, there has been concern about pollution in the home from radon, a radioactive gas whose elemental molar mass is 222 g/mol. The Environmental Protection Agency believes that a level of radon of 3.6×10^{-17} g/L of air is unhealthy. At this level, how many moles of radon would there be in a living room whose volume is 2455 L? How many atoms is this?

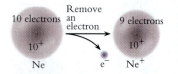

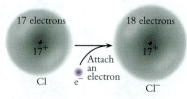

Figure 2-23
Cations form by electron removal, and anions form by electron attachment.

Ion is pronounced "eye'-un," cation "cat'-eye-un," and anion "an'-eye-un."

2.5 CHARGED ATOMS: IONS

An atom is neutral when it has equal numbers of protons and electrons. In a neutral atom, the negative charge of the electrons exactly balances the positive charge of the nucleus. Experiments using mass spectrometers and electrical discharge tubes show that electrons can be removed from neutral atoms. The product species have positive electrical charges. To distinguish these species from their neutral counterparts, scientists add superscript plus signs to the chemical symbols:

$$\text{Ne} \longrightarrow \text{Ne}^+ + e^-$$

Neutral atoms or molecules can also capture electrons, producing negatively charged species, which scientists distinguish from their neutral counterparts by adding superscript minus signs to the chemical symbols:

$$\text{Cl} + e^- \longrightarrow \text{Cl}^-$$

Figure 2-23 illustrates these processes.

Electrically charged atomic or molecular particles are called **ions.** When the charge is positive, the particles are **cations.** We do not classify electrons as ions, but when neutral atoms or molecules capture electrons, the negative ions that form are **anions.**

Ion formation always involves the movement of electrons from one location to another, yet no matter how ions form, electrons are neither created nor destroyed. In other words, electrons are conserved. The equation for an ionization process makes it appear that an electron has been created: $Ne \longrightarrow Ne^+ + e^-$. Notice from Figure 2-23, however, that a neutral neon atom has 10 electrons, but a neon cation has only 9 electrons. Thus, 10 electrons are present both before and after the formation of the cation.

The key to keeping track of electrons is net charge. Moving electrons around may generate positive charges in one location and negative charges in another location, but the net charge of the entire system remains unchanged. The ionization of neon atoms provides an illustration. The net charge of a neon cation plus an electron is zero, the same as the net charge of a neon atom.

Net electrical charge is always conserved　　　　　　　　**KEY CONCEPT**

Ionic Compounds

Electrical charges of the same type repel each other, so a collection containing only cations is highly unstable. Likewise, a collection containing only anions is unstable. Even the cations in a mass spectrometer capture electrons and become neutral as soon as they strike the collector. In contrast, opposite charges attract each other. This makes it possible for a collection containing both cations and anions to be stable, even though collections of either type alone are not. To be stable, a collection of cations and anions must be electrically neutral overall. That is, the amount of positive charge carried by the cations must exactly balance the amount of negative charge carried by the anions.

A solid that contains cations and anions in balanced whole-number ratios is called an **ionic compound.** Sodium chloride, commonly known as table salt, is a simple example. Sodium chloride can form through the vigorous chemical reaction of elemental sodium and elemental chlorine. The appearance and composition of these substances are very different, as Figure 2-24 shows. Sodium is a soft, silver-colored metal that is an array of Na atoms packed closely together. Chlorine is a faintly yellow-green toxic gas made up of diatomic, neutral Cl_2 molecules. When these two elements react, they form colorless crystals of NaCl that contain Na^+ and Cl^- ions in a 1:1 ratio.

Ionic compounds have chemical properties very different from those of the neutral atoms from which they form. Sodium metal reacts very violently with water, and chlorine gas is poisonous and highly corrosive. In contrast, sodium chloride simply dissolves in water and is a substance that most people use to season their food.

Ions form during the reaction between sodium and chlorine. Each sodium atom loses one electron, leaving one less electron than the number of protons in the nucleus:

$$Na \longrightarrow Na^+ + e^-$$

Chlorine molecules decompose into atoms, and the electron lost by a sodium atom becomes attached to a chlorine atom to produce an anion:

$$Cl + e^- \longrightarrow Cl^-$$

Figure 2-24
Photographs and molecular pictures of (*a*) sodium metal, (*b*) chlorine gas, and (*c*) crystalline sodium chloride.

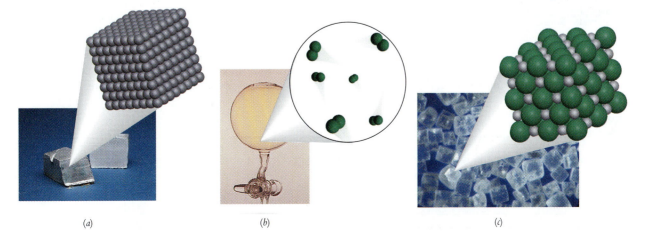

(a)　　　　　　　　(b)　　　　　　　　(c)

A sodium chloride crystal contains equal numbers of Na^+ cations and Cl^- anions packed together in an alternating cubic array. Figure 2-24 illustrates a portion of the sodium chloride array. Electrical forces hold the cations and anions in place. Each Na^+ cation attracts all the nearby Cl^- anions. Likewise, each Cl^- anion attracts all its Na^+ neighbors. Positive cations and negative anions group together in equal numbers to make the entire collection neutral.

Several elements have a strong tendency to form **salts** which are composed of cations and anions. The formulas of salts are dictated by the fact that the numbers of cations and anions must lead to overall charge neutrality. Sodium and the other Group 1 elements (alkali metals) form salts containing atomic cations with $+1$ charges. Chlorine and the other Group 17 elements (halogens) form salts containing atomic anions with -1 charges. The metals in Group 2 (alkaline earth metals) form salts containing atomic cations that have lost two electrons. Magnesium chloride, for example, contains Mg^{2+} cations and Cl^- anions. To maintain electrical neutrality, this compound has a 1:2 ratio of cations to anions, giving it the chemical formula $MgCl_2$. Two elements from Group 16, oxygen and sulfur, form ionic compounds containing atomic anions that have gained two electrons. Examples are calcium oxide (CaO), which contains Ca^{2+} cations and O^{2-} anions in a 1:1 ratio, and potassium sulfide (K_2S), which contains K^+ cations and S^{2-} anions in a 2:1 ratio. Example 2-6 examines some common ionic compounds.

Example 2-6 **Formulas of Ionic Compounds**	Identify the ionic compounds formed from the following elements: (a) calcium and fluorine; (b) lithium and oxygen; (c) magnesium and sulfur; (d) potassium and iodine. **Strategy:** To identify these salts, we must specify the cation or anion formed by each element. The ions combine in a ratio that gives a neutral salt. In other words, the net positive charge of the cation must balance the overall negative charge of the anion. We need to remember that the Group 1 and Group 2 elements form $+1$ and $+2$ cations, respectively. The halogens, from Group 17, form -1 anions, and oxygen forms -2 anions. **Solution:** a. Calcium (Group 2) forms Ca^{2+}, and fluorine forms F^-. These combine in a 1:2 ratio: CaF_2. b. Lithium (Group 1) forms Li^+, and oxygen forms O^{2-}. The neutral salt is Li_2O. c. Magnesium $= Mg^{2+}$ and sulfur $= S^{2-}$. The salt is MgS. d. Potassium $= K^+$ and iodine $= I^-$. The salt is KI.

Do the Results Make Sense? Verify that each element has been assigned to the correct Group and that the ionic charges correspond to the values that characterize the Group. If the ions are combined in proportions that give a neutral compound, the results are reasonable.

Extra Practice Exercise 2.6	Identify the ionic compounds formed from the following elements: (a) magnesium and oxygen; (b) cesium and bromine; (c) strontium and chlorine; (d) sodium and sulfur.

Salts are compounds that contain cations and anions. Many other compounds are molecular rather than ionic. We turn to these compounds in the next chapter.

Section Exercises

2.5.1 Write the symbols for the ions that form from magnesium, lithium, fluorine, and oxygen.

2.5.2 How many protons and electrons are there in a Mg^{2+} ion?

2.5.3 Draw a molecular picture that shows six atoms of magnesium reacting with three molecules of O_2 to form ionic MgO.

CHAPTER 2 VISUAL SUMMARY

Important equations

Essential terms

Key concepts

SKILLS TO MASTER

① Drawing atomic and molecular pictures
② Writing and interpreting atomic symbols
③ Interpreting mass spectra

④ Identifying stable and unstable isotopes
⑤ Determining elemental molar masses
⑥ Doing mole-mass-number conversions

2.1 ATOMIC THEORY

  ①

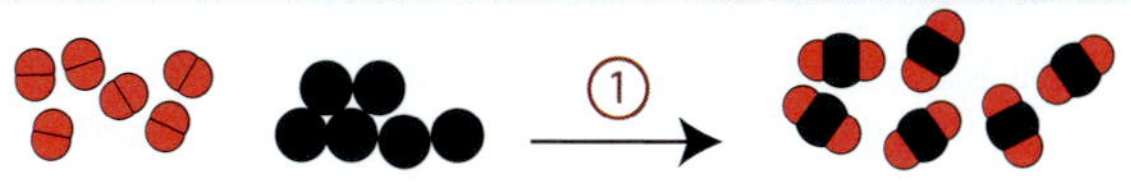

6 O_2 molecules   6 C atoms   6 CO_2 molecules

Mass is conserved.
Atoms are conserved.

Features of Atomic Theory
All matter is composed of atoms.
Each element is unique.
Atoms combine in whole-number ratios.
Atoms rearrange in chemical reactions.

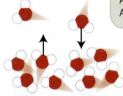

Atoms are always moving:
Diffusion
Dynamic equilibrium

2.2 ATOMIC ARCHITECTURE: ELECTRONS & NUCLEI

Gravitational force
Electrical force
Magnetic force

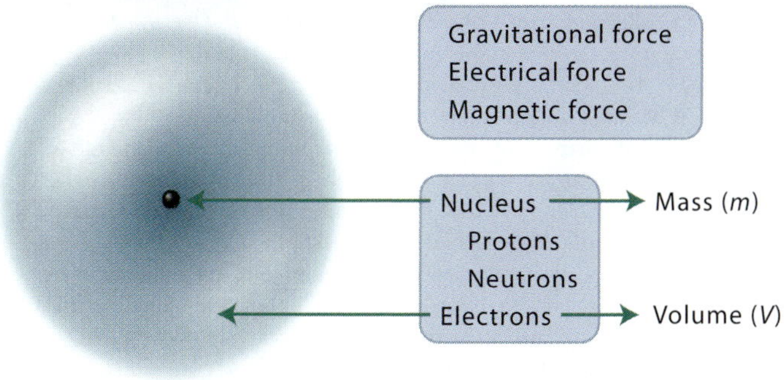

Nucleus ──→ Mass (m)
Protons
Neutrons
Electrons ──→ Volume (V)

2.3 ATOMIC DIVERSITY: THE ELEMENTS

Mass number ←
Elemental symbol ← $^A_Z X$ ②
Atomic number ←

Nuclear charge
identifies an element.

Radioactive
Alpha particle
Positron

Mass spectrometer

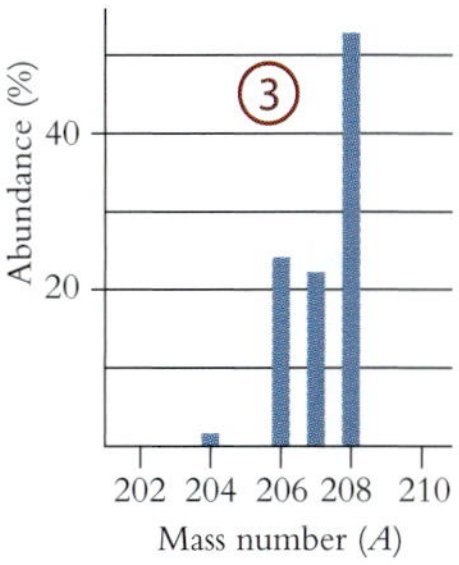

Belt of stability

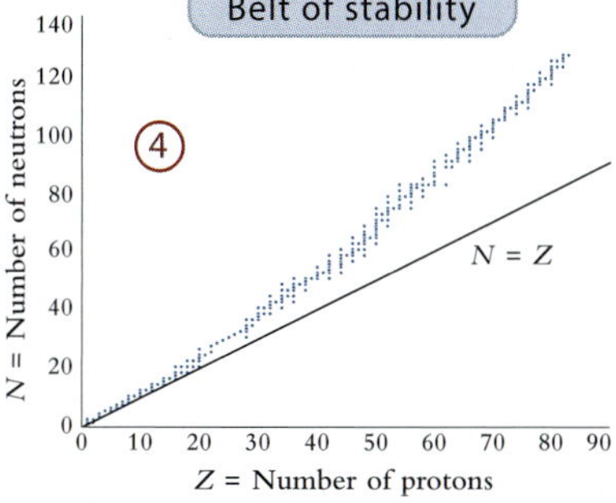

③
④ $N = Z$

Abundance (%)
40
20
202 204 206 208 210
Mass number (A)

N = Number of neutrons
140
120
100
80
60
40
20
0   10 20 30 40 50 60 70 80 90
Z = Number of protons

2.4 COUNTING ATOMS: THE MOLE

One mole is the number of atoms in exactly 12 g of pure carbon-12.

Molar mass (MM)
Avogadro's number (N_A)

Elemental molar mass is the sum of the contribution of its isotopes.

$$MM = \Sigma(\text{Fractional abundance})(\text{Isotopic } MM)$$ ⑤

Chemical calculations are built around the mole.

⑥ $n = \dfrac{\#}{N_A}$ $n = \dfrac{m}{MM}$

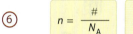

| Number of atoms # | Multiply by Avogadro's Number N_A / Divide by | Moles of substance, n | Multiply by Molar mass MM / Divide by | Mass of element, m |

2.5 CHARGED ATOMS: IONS

Salts contain cations and anions.

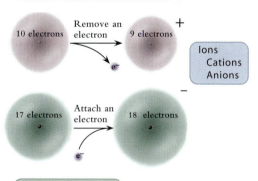

10 electrons → Remove an electron → 9 electrons +

Ions
Cations
Anions

17 electrons → Attach an electron → 18 electrons

−

Charge is conserved.

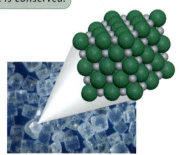

Learning Exercises

2.1 List the conservation laws that appear in this chapter. Describe each one in your own words.

2.2 Describe what atoms "look like," what they are composed of, and how they behave.

2.3 Prepare a table listing all the different types of forces that you are aware of. For each type of force, give an example of a situation where this particular force is important.

2.4 Write a paragraph summarizing nuclear stability and instability.

2.5 List all terms new to you that appear in this chapter. Use your own words to write a one-sentence definition of each. Consult the glossary if you need help.

 Problems <u>ilw</u> = interactive learning ware problem. Visit the website at www.wiley.com/college/olmsted

Atomic Theory

2.1 Draw molecular pictures that show part of a sample of each of the following: (a) helium, a monatomic gas; (b) tungsten, an atomic solid; and (c) gallium, an atomic liquid at human body temperature.

2.2 Draw molecular pictures that show part of a sample of each of the following: (a) mercury, an atomic liquid; (b) iron, an atomic solid; and (c) neon, a monatomic gas.

2.3 The following is a molecular picture of carbon reacting with oxygen to form carbon monoxide, a deadly poisonous gas. Describe how this picture illustrates the features of atomic theory.

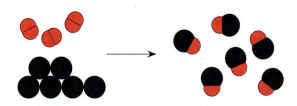

2.4 When lightning strikes, some N_2 reacts with O_2 to generate NO, as shown in the following molecular picture. Describe how this picture illustrates the features of atomic theory.

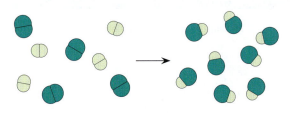

2.5 When a strip of magnesium metal burns in air, the mass of the resulting residue is greater than the mass of the original strip of metal. Explain this observation in terms of conservation of mass.

2.6 Liquid mercury metal can be obtained from a mercury ore called cinnabar simply by heating the ore in air. There is a significant mass loss when this process occurs. How can this mass loss be reconciled with conservation of mass?

2.7 Bromine exists as diatomic molecules. This element is a liquid at room temperature but gaseous above 332 K and solid below 266 K. Draw molecular pictures that represent bromine in each of its three phases.

2.8 Carbon dioxide can exist as a liquid, solid (dry ice), or gas. Each carbon dioxide molecule is linear, with a carbon atom in the middle and an oxygen atom on either side. Draw molecular pictures that represent carbon dioxide in each phase.

2.9 When we smell the odor of a rose, our olfactory nerves are sensing molecules of the scent. Explain how smelling a rose demonstrates that molecules are always moving.

2.10 If a drop of liquid bromine is placed inside a sealed flask, the entire volume above the liquid soon takes on a deep red-brown color. Explain how this demonstrates that molecules are always moving.

2.11 Iodine is an element whose molecules can move directly from the solid to the gas phase. A sample of solid iodine in a stoppered flask stood undisturbed for several years. As the photo shows, crystals of solid iodine grew on the sides of the flask. Use the principle of dynamic equilibrium to explain at the molecular level what happened. Include an observation about the color of the atmosphere inside the flask.

2.12 The photo shows a stoppered flask containing a highly concentrated salt solution with many salt crystals on the bottom. If the flask stands for a long time, some crystals become smaller while others grow in size, but the total mass of crystals remains constant. Explain what is happening at the molecular level in terms of a dynamic equilibrium.

Atomic Architecture: Electrons and Nuclei

2.13 The collection screen of a cathode-ray tube registers the total charge of a beam of electrons to be -1.00×10^{-6} C. (a) How many electrons is this? (b) What mass does this number of electrons have?

2.14 In a gas discharge tube containing hydrogen gas, a total charge of 2.44×10^{-12} C is measured at the collector. Assume that this charge is entirely due to protons. (a) How many protons is this? (b) What mass does this number of protons have?

2.15 In Millikan's oil drop experiment, some droplets have negative charges, so others must have positive charges. Suppose the upper electrical plate in Figure 2-14 is made negative and the lower plate positive. Describe the results that would be obtained under these conditions.

2.16 Describe the scattering pattern that would have been observed in Rutherford's experiment if atoms were like chocolate chip cookies.

<u>ilw</u> **2.17** How many protons does it take to give a mass of 1.5 g? What charge does this sample carry?

2.18 How many electrons does it take to give a mass of 1.5 g? What charge does this sample carry?

Atomic Diversity: The Elements

2.19 How many protons, neutrons, and electrons are contained in each of the following atoms or ions?
(a) $^{16}_{8}O^{2-}$; (b) $^{11}_{5}B$; (c) $^{55}_{25}Mn^{3+}$; (d) $^{35}_{17}Cl^{-}$; and (e) $^{37}_{17}Cl^{+}$

2.20 How many protons, neutrons, and electrons are contained in each of the following atoms or ions?
(a) $^{66}_{30}Zn^{2+}$; (b) $^{15}_{7}N$; (c) $^{81}_{35}Br^{+}$; (d) $^{79}_{35}Br^{-}$; and (e) $^{238}_{92}U$

2.21 Write the symbols of the following isotopes: (a) $Z = 26$ with 30 neutrons; (b) ^{236}U; (c) argon with two more neutrons than protons; and (d) an atom with 9 protons, 10 neutrons, and 9 electrons.

2.22 Write the symbols of the following isotopes: (a) helium with 1 neutron; (b) zinc with $A = 66$; (c) element number 54 with 78 neutrons; and (d) nitrogen with the same number of protons and neutrons.

2.23 Elemental boron is 80.0% boron-11, the rest being boron-10. Sketch the mass spectrum of this element.

2.24 Elemental copper is 69.17% ^{63}Cu, the rest being ^{65}Cu. Sketch the mass spectrum of this element.

2.25 Platinum has four stable isotopes whose mass numbers and percentages are 194, 32.9%; 195, 33.8%; 196, 25.3%; and 198, 7.2%. Construct a pie chart illustrating these isotopic abundances.

2.26 Zinc consists of five isotopes whose mass numbers and percentages are 64, 48.6%; 66, 27.9%; 67, 4.1%; 68, 18.8%; and 70, 0.6%. Construct a pie chart illustrating these isotopic abundances.

2.27 Predict whether each of the following nuclei is stable or unstable and give the reason for each prediction: (a) carbon-10; (b) neptunium-239; and (c) cadmium with 60 neutrons.

2.28 Predict whether each of the following nuclei is stable or unstable, and give the reason for each prediction: (a) carbon-15; (b) ruthenium with 58 neutrons; and (c) ^{211}At.

2.29 Use the periodic table (inside front cover) to fill in the missing information:

Name	Symbol	Z
——	W	——
——	——	27
Mercury	——	

2.30 Use the periodic table (inside front cover) to fill in the missing information:

Name	Symbol	Z
——	U	——
——	——	15
Copper	——	——

Counting Atoms: The Mole

ilw 2.31 Use the data in the following table to calculate the molar mass of naturally occurring argon:

Isotope	Isotopic Molar Mass	Abundance
^{36}Ar	35.96755 g/mol	0.337%
^{38}Ar	37.96272 g/mol	0.063%
^{40}Ar	39.96238 g/mol	99.600%

2.32 Use the data in the following table to calculate the molar mass of naturally occurring silicon:

Isotope	Isotopic Molar Mass	Abundance
^{28}Si	27.97693 g/mol	92.23%
^{29}Si	28.97649 g/mol	4.67%
^{30}Si	29.97376 g/mol	3.10%

2.33 Calculate the number of moles in the following: (a) 7.85 g of Fe; (b) 65.5 μg of carbon; (c) 4.68 mg of Si; and (d) 1.46 metric tons of Al (1 metric ton = 10^3 kg).

2.34 Calculate the number of moles in the following: (a) 3.67 kg of titanium; (b) 7.9 mg of calcium; (c) 1.56 g of ruthenium; and (d) 9.63 pg of technetium.

2.35 Determine the number of atoms present in 5.86 mg of each of the following elements: (a) beryllium; (b) phosphorus; (c) zirconium; and (d) uranium.

2.36 Determine the number of atoms present in 2.44 kg of each of the following elements: (a) argon; (b) silver; (c) iodine; and (d) potassium.

Charged Atoms: Ions

2.37 Identify each of the following as a cation, anion, or neutral atom. For each ion, write the reaction by which it forms from a neutral atom: C, Cl$^-$, and Cr^{3+}.

2.38 Identify each of the following as a cation, anion, or neutral atom. For each ion, write the reaction by which it forms from a neutral atom: O^{2-}, Os^{2+}, and Os.

2.39 Write the chemical formula for the species resulting from each of the following processes: (a) a chloride anion loses an electron; (b) a sodium atom loses an electron; and (c) an oxygen atom gains two electrons.

2.40 Write the chemical formula for the species resulting from each of the following processes: (a) an argon atom loses an electron; (b) a doubly-positive magnesium cation gains two electrons; and (c) a bromine atom gains an electron.

2.41 Based on their positions in the periodic table, decide what ion is likely to form from each of the following elements: (a) rubidium; (b) fluorine; and (c) barium.

2.42 Based on their positions in the periodic table, decide what ion is likely to form from each of the following elements: (a) cesium; (b) strontium; and (c) iodine.

2.43 Write the chemical formulas of all ionic compounds that can form between the elements listed in Problem 2.41.

2.44 Write the chemical formulas of all ionic compounds that can form between the elements listed in Problem 2.42.

2.45 Aluminum is one of the few elements that forms cations with + 3 charge. What are the chemical formulas of the ionic compounds that form between aluminum and (a) oxygen; (b) fluorine?

2.46 Nitrogen forms a few compounds in which it exists as an anion with -3 charge. What are the chemical formulas of ionic compounds that form between nitrogen and (a) sodium; (b) magnesium?

Additional Paired Problems

2.47 The density of gold is 19.32 g/cm^3. An ingot of gold measures $15 \times 7.5 \times 2.0$ cm. How many gold atoms are in the ingot?

2.48 Mercury is an unusual metal in that it is a liquid at room temperature. The density of liquid mercury is 13.55 g/mL. The photo shows a sample of mercury in a graduated cylinder. How many atoms of mercury are in the sample?

2.49 Determine Z, A, and N for each of the following nuclei: (a) ^{6}Li; (b) ^{43}Ca; (c) ^{238}U; (d) ^{130}Te; (e) the nucleus of neon that contains the same number of protons and neutrons; and (f) the nucleus of lead that contains 1.5 times as many neutrons as protons.

2.50 Determine Z, A, and N for each of the following nuclei: (a) ^{22}Ne; (b) ^{202}Pb; (c) ^{41}K; (d) ^{109}Ag; (e) the helium nucleus with one less neutron than proton; and (f) the nucleus of barium whose neutron–proton ratio is 1.25.

2.51 Write the correct elemental symbols for the following nuclei (include the atomic number subscript): (a) helium with the same number of neutrons and protons; (b) tungsten with 110 neutrons; (c) the nucleus with $Z = 28$ and $N = 32$; and (d) the nucleus with 12 protons and 14 neutrons.

2.52 Write the correct elemental symbols for the following nuclei (include the atomic number subscript): (a) the nucleus with $Z = 50$ and $N = 64$; (b) atomic number 86 possessing 133 neutrons; (c) sulfur with two more neutrons than protons; and (d) antimony with 70 neutrons.

2.53 A copper atom has a diameter of 127.8 pm. What is the length in meters of a line containing 1.000 mol of copper atoms?

2.54 A dollar bill is 0.10 mm thick. How high would a stack of 1.00 mol of dollar bills be?

2.55 Air is mostly diatomic molecules of nitrogen and oxygen, in a molecular ratio of 4:1. Draw a molecular picture of a sample of air containing a total of 10 molecules.

2.56 One breathing mixture for deep-sea divers contains 25% molecular oxygen and 75% helium. Draw a molecular picture of a sample of this mixture that contains three molecules of oxygen.

2.57 The pie chart in Figure 2-17 shows the isotopic abundances for tin. Sketch the mass spectrum of this element.

2.58 The pie chart in Figure 2-17 shows the isotopic abundances for chromium. Sketch the mass spectrum of this element.

2.59 The following isotopes are useful for medical imaging. Determine the number of protons, neutrons, and electrons of each of them: (a) ^{99m}Tc, used to study head function (shown in the figure); (b) iron-52, used in bone marrow scans; (c) ^{133}Xe, for studying lung function; and (d) ^{131}I, for examination of the thyroid gland.

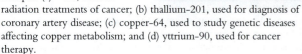

2.60 Determine the number of protons, neutrons, and electrons of each of the following isotopes: (a) ^{60}Co, used in radiation treatments of cancer; (b) thallium-201, used for diagnosis of coronary artery disease; (c) copper-64, used to study genetic diseases affecting copper metabolism; and (d) yttrium-90, used for cancer therapy.

2.61 Calculate the mass of lithium that has the same number of atoms as 5.75 g of platinum.

2.62 Calculate the mass of neon that has the same number of atoms as 37.5 mg of xenon.

2.63 When calcium reacts with chlorine, each calcium atom loses two electrons, and each chlorine atom gains one electron. (a) What are the symbols of the resulting ions? (b) What is the chemical formula of the ionic compound formed from calcium and chlorine? (c) Draw a molecular picture that shows the reaction between five atoms of Ca metal and enough molecules of Cl_2 gas to react completely.

2.64 Approximately 20% of the iron and steel produced in the United States each year is used to replace rusted metal. Rust forms when iron reacts with oxygen. Each iron atom loses three electrons, and each oxygen atom gains two electrons. (a) What are the symbols of the resulting ions? (b) What is the chemical formula of the compound formed from iron and oxygen? (c) Draw a molecular picture that shows the reaction between four atoms of Fe metal and enough O_2 gas to react completely.

2.65 Each of the elements from chlorine to scandium has a nucleus with $A = 40$. Write correct atomic symbols for all these nuclei.

2.66 Except for beryllium, each of the elements with Z values from 1 to 8 has a stable isotope with the same number of protons as neutrons. Write the correct atomic symbols for each of these isotopes.

2.67 Naturally occurring magnesium has three isotopes whose mass numbers and percent abundances are 24, 78.99%; 25, 10.00%; and 26, 11.01%. Sketch the mass spectrum of Mg.

2.68 Naturally occurring nickel has five stable isotopes whose mass numbers and percent abundances are 58, 68.27%; 60, 26.10%; 61, 1.13%; 62, 3.59%; and 64, 0.91%. Sketch the mass spectrum of Ni.

2.69 How many protons and electrons are present in the following ions? (a) Na$^+$; (b) N^{3-}; (c) Ti^{4+}; and (d) I$^-$.

2.70 How many protons and electrons are present in the following ions? (a) Al^{3+}; (b) Se^{2-}; (c) K$^+$; and (d) Ca^{2+}.

More Challenging Problems

2.71 The element bromine exists as diatomic molecules and is a liquid under normal conditions. Bromine evaporates easily, however, giving a red-brown color to the gas phase above liquid bromine, as shown in the photo. Draw molecular pictures showing liquid bromine, the gas above it, and the dynamic equilibrium between the phases.

2.72 The ratio of neutrons to protons in stable nuclei varies from 1:1 to about 1.5:1. Write the isotopic symbols for all isotopes that contain 14 nuclear particles and have ratios in this range.

2.73 Give the atomic symbols for all the isotopes of elements between $Z = 20$ and $Z = 40$ that have exactly 1.25 times as many neutrons as protons.

2.74 A scientist claims to have prepared a sample of chlorine gas that is isotopically pure. Which scientific apparatus described in this chapter could be used to test this claim? Explain.

2.75 Table salt, which has the formula NaCl, contains Na$^+$ and Cl$^-$ ions arranged in a cubic crystal. When heated to a sufficiently high temperature, salt crystals melt to give a liquid that still contains these ions. Draw molecular pictures of table salt crystals and of liquid NaCl.

2.76 How many atoms of chromium-53 are there in a 0.123 g sample of chromium metal? (Consult Figure 2-17.)

2.77 Is the following molecular picture of a reaction correct? If so, explain why. If not, redraw the right-hand portion so the picture is correct.

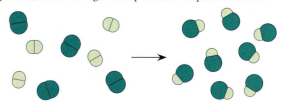

2.78 A student drew the following molecular picture to show what happens when solid I_2 is partially vaporized to give gaseous I_2 molecules. Is this molecular picture correct? If so, explain why. If not, redraw the right-hand view so that it correctly represents this process.

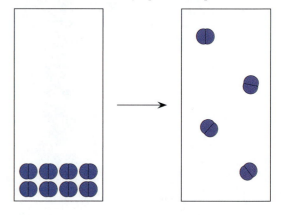

Group Study Problems

2.79 People with little knowledge of science often are cautious about "synthetic" versions of naturally occurring molecules that are used by humans. Vitamins, for example, can be isolated from plants or manufactured in the laboratory. Is pure vitamin C isolated from oranges better for you than pure vitamin C made in the laboratory? Use the atomic theory to support your answer.

2.80 Consult the *CRC Handbook of Chemistry and Physics* and compile two lists, one of the *Z* values of the elements that have just one stable isotope and the other of the *Z* values of the elements that have four or more stable isotopes. Determine if there are any patterns in the atomic number values of the elements occurring in each of these lists.

2.81 In discovering the electron, Thomson was able to measure the ratio of the electron's mass to its charge, but he was unable to determine the mass of a single electron. How did Millikan's experiment allow the mass of the electron to be determined?

2.82 The process in which a solid is converted directly to a gas is called sublimation. Solid carbon dioxide (CO_2), which is known as dry ice, is one common substance that sublimes. Draw a series of molecular pictures that show how a piece of dry ice in a closed container at low temperature illustrates the principle of dynamic equilibrium.

Answers to Section Exercises

2.1.1

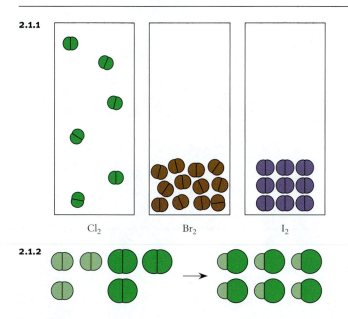

Cl$_2$ Br$_2$ I$_2$

2.1.2

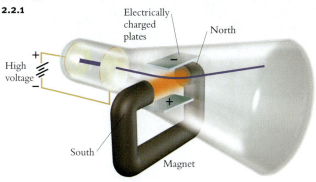

2.1.3 Water molecules in the warm rain puddle escape from the liquid phase to the gas phase, where they diffuse away or are swept away by the wind.

2.2.1

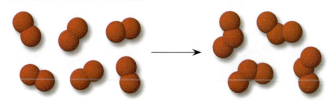

(You are not expected to draw pictures as elaborate as this one, but your picture should include the essential features: an enclosed tube, two sets of charged plates, a magnet, and a beam that deflects away from the positively charged plate.)

2.2.2 (a) The larger charge increases the electric force, so this droplet moves upward. (b) The larger mass increases the gravitational force, so this droplet moves downward.

2.2.3 (a) 1.1×10^{21} electrons; (b) 1.8 mg

2.3.1

Element	Symbol	Z
Carbon	C	6
Gold	Au	79
Arsenic	As	33
Iron	Fe	26

2.3.2 (a) ^{57}Co; (b) ^{7}Li; and (c) ^{39}K

2.3.3

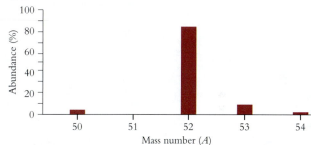

2.4.1 32.06 g/mol

2.4.2 (a) 2.21×10^{23} atoms; (b) 3.21×10^{22} atoms; and (c) 7.80×10^{21} atoms

2.4.3 4.0×10^{-16} mol and 2.4×10^{8} atoms

2.5.1 Mg^{2+}, Li^{+}, F^{-}, and O^{2-}

2.5.2 12 protons and 10 electrons

2.5.3

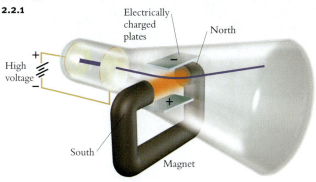

Answers To Extra Practice Exercises

2.1 The number of O atoms must be conserved, so your picture should show four molecules of ozone:

2.2 ^{59}Fe, 26 protons and 33 neutrons; samarium-153, 62 protons and 91 neutrons; and ^{42}K, 19 protons and 23 neutrons

2.3 $Z = 51$; two isotopes, one with 70 neutrons, 57% abundance, the other with 72 neutrons, 43% abundance

2.4 $MM = 24.305$ g/mol

2.5 0.63 g, 3.1×10^{-3} mol, 1.9×10^{21} atoms

2.6 (a) MgO; (b) CsBr; (c) $SrCl_2$; and (d) Na_2S

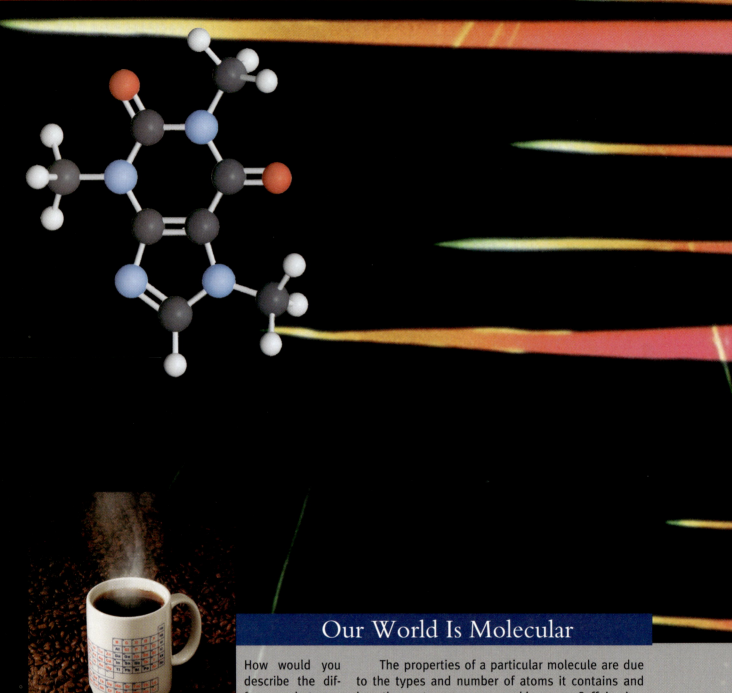

Our World Is Molecular

How would you describe the differences between a cup of coffee and a cup of hot water? What probably come to mind are the aroma, the dark color, and the taste of a good cup of coffee. Coffee's action as a stimulant is another obvious difference. These properties come from the chemical compounds that hot water dissolves from ground coffee beans. These compounds are molecules constructed from different atoms bound together in very specific arrangements. The molecule that makes coffee a stimulant is caffeine. Our background photo is a magnification of crystals of pure caffeine, and the inset is a ball-and-stick model of this molecule.

The properties of a particular molecule are due to the types and number of atoms it contains and how those atoms are arranged in space. Caffeine is a stimulant because it has the same shape as one part of cyclic adenosine monophosphate (cyclic AMP), a molecule that helps to regulate the supply of energy in the brain. When caffeine is absorbed into the blood and carried to the brain, it binds to an enzyme that normally controls the supply of cyclic AMP. As a result, the enzyme can no longer bind cyclic AMP, the brain's supply of this energy-regulating molecule is increased, and we feel stimulated.

Molecular sizes and shapes play key roles in determining chemical and physical properties. The immense variety of chemical and physical properties displayed by substances in the natural world mirrors

The Composition of Molecules

Learning Resources

 KEY CONCEPTS

 CRITICAL THINKING

 SKILLS TO MASTER

 ADDITIONAL HELP
www.wiley.com/college/olmsted
● TUTORIALS
● ANIMATIONS

an equally immense variety of different types of molecules. However, variety need not come from a large number of different elements. The molecules that make up a cup of coffee are made up almost entirely of atoms of just five elements: hydrogen, carbon, oxygen, nitrogen, and sulfur. Carbon, in particular, is capable of combining in many different ways, generating molecules with elaborate structures.

Even simple molecules often have strikingly different properties. For example, carbon and oxygen form two different simple compounds. Whereas a molecule of carbon monoxide contains one oxygen atom and one carbon atom, carbon dioxide contains two atoms of oxygen and one atom of carbon. Although these molecules have some common properties (both are colorless, odorless gases), the difference in chemistry caused by a change of one atom is profound. We produce and exhale carbon dioxide as a natural by-product of metabolism. This compound is relatively harmless to humans. In contrast, carbon monoxide is a deadly poison, even at very low concentrations.

Chapter 3 examines molecules from two perspectives: What is it? and How much is there? As we describe in this chapter, chemists answer these questions in terms of *names* and *numbers* of molecules.

3.1 REPRESENTING MOLECULES

Chemists use both chemical names and molecular pictures to describe molecules. Molecular pictures take several forms, including structural formulas, ball-and-stick models, space-filling models, and line structures. These molecular representations can help you improve your ability to "think molecules."

Chemical Formulas

A **chemical formula** describes the composition of a substance by giving the relative numbers of atoms of each element. When a substance contains discrete molecules, a chemical formula is also a molecular formula. A chemical formula contains elemental symbols to represent atoms and subscripted numbers to indicate the number of atoms of each type. The simplest chemical formulas describe pure elements. The chemical formulas of most elements are their elemental symbols: helium is He, silicon is Si, copper is Cu. However, seven elements occur naturally as diatomic molecules (Figure 3–1), so their chemical formulas take the form X_2. A few other elements occur as atomic clusters, notably P_4 and S_8.

A chemical compound is a substance that contains a combination of atoms of different elements. Because a compound contains more than one element, there is more than one way to write its formula. For example, hydrogen chloride is a diatomic molecule with one atom each of hydrogen and chlorine. Its chemical formula might be written as HCl or ClH. To avoid possible confusion, chemists have standardized the writing of chemical formulas.

For **binary compounds**—those containing only two elements—the following guidelines apply:

Guidelines for Formulas of Binary Compounds

1. Except for hydrogen, the element farther to the left in the periodic table appears first: KCl, PCl_3, Al_2S_3, and Fe_3O_4.

2. If hydrogen is present, it appears last except when the other element is from Group 16 or 17 of the periodic table: LiH, NH_3, B_2H_6, and CH_4 but H_2O_2, H_2S, HCl, and HI.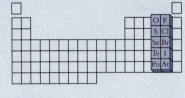

3. If both elements are from the same group of the periodic table, the lower one appears first: SiC and BrF_3.

Example 3-1 shows how to use these guidelines.

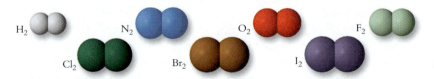

Figure 3-1
The seven chemical elements that normally exist as diatomic molecules. A mnemonic for these is Hair Needs Oil For Clean Bright Images.

Write the correct chemical formulas for the molecules containing sulfur whose molecular pictures follow:

Example 3-1

Writing
Chemical Formulas

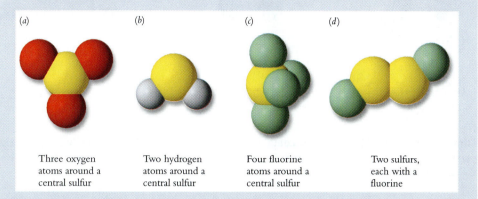

(a) Three oxygen atoms around a central sulfur

(b) Two hydrogen atoms around a central sulfur

(c) Four fluorine atoms around a central sulfur

(d) Two sulfurs, each with a fluorine

Strategy: To determine a chemical formula from a molecular picture, count the atoms of each type and arrange the symbols in the correct order using the guidelines.

Solution:

a. Oxygen and sulfur are in the same group of the periodic table. According to the third guideline, the lower element appears first. Thus, the correct formula is SO_3.

b. This compound contains hydrogen and sulfur, an element from Group 16 of the periodic table. According to the second guideline, hydrogen is placed before elements from this group. Thus, the correct formula is H_2S.

c. The elements in this compound fall in two different groups of the periodic table. According to the first guideline, the element farther to the left appears first. Thus, the correct formula is SF_4.

d. Once again, S is listed before F. The formula is S_2F_2. It is important to list all the atoms in the compound. The formula SF would be ambiguous because it could mean any combination of sulfur and fluorine in a 1:1 atomic ratio: SF, S_2F_2, S_3F_3, etc.

Do the Results Make Sense? Compare each chemical formula with the corresponding molecular model. Each formula and model represents the same compound, so they should contain the same number of atoms of each element.

Write the correct chemical formulas for the compounds whose molecules contain the following: (a) two atoms each of hydrogen and carbon; (b) three atoms of oxygen for every two atoms of aluminum; (c) two atoms of oxygen for every atom of sulfur.

Extra Practice
Exercise 3.1

When three or more different elements occur in a compound, the order depends on whether or not the compound contains ions. We describe ionic compounds in Section 3.3. Many multiple-element compounds that do not contain ions contain carbon. The formulas of carbon-containing compounds start with carbon, followed by hydrogen. After that, any other elements appear in alphabetical order, as illustrated by the following examples: C_2H_6O, C_4H_9BrO, CH_3Cl, and $C_8H_{10}N_4O_2$.

Structural Formulas

The chemical formula of a substance gives only the number of atoms of each element present in one of its molecules. A **structural formula,** on the other hand, not only gives the number of atoms but also shows how the atoms are connected to one another. The atoms in molecules have specific arrangements because they are held together by attractive forces called bonds. In brief, a bond is the result of the electrical force of attraction between positively charged nuclei and negatively charged electrons. We discuss chemical bonding in Chapters 9 and 10. For now, it is enough to know that a pair of electrons shared between two atoms generates a chemical bond.

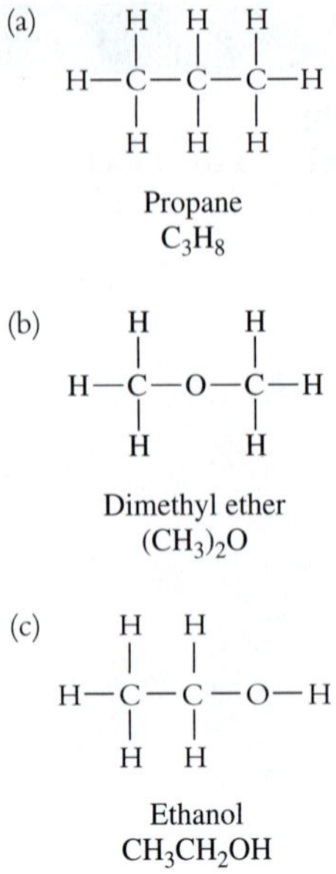

(a)

$$H—\underset{\underset{H}{|}}{\overset{\overset{H}{|}}{C}}—\underset{\underset{H}{|}}{\overset{\overset{H}{|}}{C}}—\underset{\underset{H}{|}}{\overset{\overset{H}{|}}{C}}—H$$

Propane
C_3H_8

(b)

$$H—\underset{\underset{H}{|}}{\overset{\overset{H}{|}}{C}}—O—\underset{\underset{H}{|}}{\overset{\overset{H}{|}}{C}}—H$$

Dimethyl ether
$(CH_3)_2O$

(c)

$$H—\underset{\underset{H}{|}}{\overset{\overset{H}{|}}{C}}—\underset{\underset{H}{|}}{\overset{\overset{H}{|}}{C}}—O—H$$

Ethanol
CH_3CH_2OH

Figure 3-2
Structural formulas of three organic compounds.

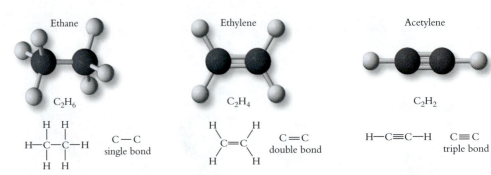

Ethane
C_2H_6

$$H—\underset{\underset{H}{|}}{\overset{\overset{H}{|}}{C}}—\underset{\underset{H}{|}}{\overset{\overset{H}{|}}{C}}—H \qquad \begin{matrix} C—C \\ \text{single bond} \end{matrix}$$

Ethylene
C_2H_4

$$\overset{H}{\underset{H}{}}C=C\overset{H}{\underset{H}{}} \qquad \begin{matrix} C=C \\ \text{double bond} \end{matrix}$$

Acetylene
C_2H_2

$$H—C\equiv C—H \qquad \begin{matrix} C\equiv C \\ \text{triple bond} \end{matrix}$$

Figure 3-3
Ethane, a component of natural gas, contains a C—C single bond. Ethylene, widely used to make plastics, contains a C=C double bond. Acetylene, used as fuel for welding torches, contains a C≡C triple bond.

In a structural formula, lines that connect the atoms represent bonds. For example, the major component of bottled cooking gas is propane. Each molecule of propane contains three carbon atoms and eight hydrogen atoms, so its chemical formula is C_3H_8. The three carbon atoms of a propane molecule link to form a chain. Each outer carbon is bonded to three hydrogens, and the inner carbon is bonded to two hydrogens. This arrangement results in the structural formula of propane that is shown in Figure 3-2a. Notice that the structural formula of propane contains more information than the chemical formula. Both formulas identify the number of atoms (three C atoms and eight H atoms), but the structural formula also shows how the atoms are connected.

Quite often, molecules have the same chemical formula but differ in their structures. For example, two compounds have the chemical formula C_2H_6O (Figure 3-2b and c). Dimethyl ether has a C—O—C linkage, while ethanol contains C—O—H. To distinguish between these two different substances, chemists write the formula of dimethyl ether as CH_3OCH_3 and the formula of ethanol as CH_3CH_2OH. The first formula indicates that the molecule has two CH_3 units linked to an oxygen atom. The second formula tells us that there is a C_2H_5 unit attached to an OH group. Molecules that have the same molecular formula but different arrangements of atoms are called *isomers*.

Each line in a structural formula represents one pair of shared electrons, but atoms can share more than one pair of electrons. When two atoms share one pair of electrons, the bond is called a *single bond*, and the structural formula shows a single line. When two atoms share four electrons, the bond is called a *double bond*, and the structural formula shows two lines between the atoms. Similarly, when two atoms share six electrons, the bond is called a *triple bond*, and the structural formula shows three lines between the atoms. Two carbon atoms can bond to each other through any of these three kinds of bonds, as the compounds in Figure 3-3 illustrate.

Animation

Three-Dimensional Models

A molecule is a three-dimensional array of atoms. In fact, many of a molecule's properties, such as its odor and chemical reactivity, depend on its three-dimensional shape. Although molecular and structural formulas describe the composition of a molecule, they do not represent the molecule's shape. To provide information about shapes, chemists frequently use ball-and-stick models or space-filling models.

In a **ball-and-stick model,** balls represent atoms, and sticks represent chemical bonds. The balls are labeled with elemental symbols or with different colors to distinguish among different elements. Figure 3-3 includes ball-and-stick models, and Figure 3-4 shows a ball-and-stick model of propane.

A **space-filling model** recognizes that a molecule is defined by the space occupied by its electrons. Recall from Rutherford's experiment that electrons make up nearly the entire volume of any atom. Therefore, each atom in a space-filling model is shown as a distorted sphere representing its electrons. These spheres merge into one another to build up the entire molecule. Examples of space-filling models appear in Figures 3-1 and 3-4. Figure 3-5 shows ball-and-stick and space-filling models for several chemical compounds common in everyday life.

Ball-and-stick model

Space-filling model

Figure 3-4
Three-dimensional models of propane, C_3H_8.

Molecule	Water	Ammonia	Methane	Ethanol
Chemical formula	H_2O	NH_3	CH_4	C_2H_5OH

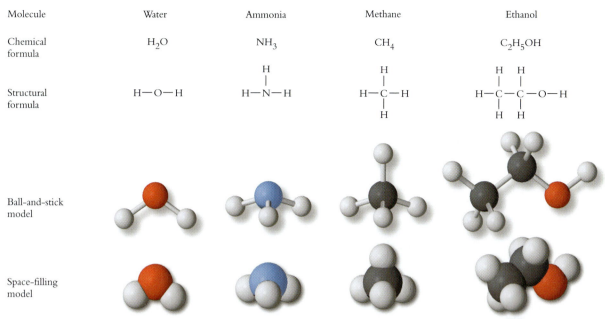

Structural formula	H—O—H	H—N—H (with H above)	H—C—H (with H above and below)	H—C—C—O—H (with H's)
Ball-and-stick model				
Space-filling model				

Figure 3-5
Different ways of representing some common chemical substances. See Figure 1-4 for the color codes for the various elements.

Line Structures

The chemistry of living processes is complex, and many carbon-based molecules found in living organisms have extremely complicated structures. Because of this complexity, chemists have developed **line structures,** which are compact representations of the structural formulas of carbon compounds. Line structures are constructed according to the following guidelines:

Guidelines for Line Structures

1. All bonds except C—H bonds are shown as lines.
2. C—H bonds are not shown in the line structure.
3. Single bonds are shown as single lines; double bonds are shown as two lines; triple bonds are shown as three lines.
4. Carbon atoms are not labeled.
5. All atoms except carbon and hydrogen are labeled with their elemental symbols.
6. Hydrogen atoms are labeled when they are attached to any atom other than carbon.

Example 3-2 illustrates these guidelines for three relatively simple molecules.

Construct line structures for the compounds with the following structural formulas:

$C_5H_{12}O$
Methyl tert-butyl ether
(a key antiknock ingredient
in gasoline)

2-propanol
(rubbing alcohol)

C_5H_8
Isoprene
(the building block of
natural rubber)

Example 3-2

Drawing Line Structures

Example 3-2	**Strategy:** A line structure is built by using information about how atoms are connected in the molecule. The structural formula is the basis for the line structure. Apply the guidelines to convert the structural formula representation into a line-structure representation.
Drawing Line Structures (*continued*)	

Solution: The first two guidelines state that bonds are represented by lines and that all C—H bonds are ignored. Thus we remove the C—H bonds from the structural formula, leaving a bond framework. According to the third guideline, the double bonds remain as two lines:

$$\underset{\overset{\displaystyle C}{|}}{\overset{\overset{\displaystyle C}{|}}{C-C-O-C}} \qquad \underset{C \qquad C}{\overset{\overset{H}{O}}{\underset{|}{C}}} \qquad \underset{C}{\overset{C=C}{C=C}}$$

According to the fourth guideline, carbon atoms are not labeled, so we remove the C's:
These are the line structures for the three molecules:

Do the Results Make Sense? To check that the line structures represent the correct substances, count the number of intersections and line ends, which should equal the number of C atoms in the compound. The first line structure has five, the second has three, and the third has five, matching the chemical formulas.

Extra Practice Exercise 3.2	Draw the line structures for the compounds that appear in Figures 3-2 and 3-3.

C≡O is the only molecule in which a C atom has fewer than four bonds.

Although line structures are convenient, a chemist must be able to convert them back into chemical formulas. The reconstruction of a complete formula from a line structure relies on the most important general feature of carbon chemistry: In all neutral molecules containing carbon, each carbon atom has *four* chemical bonds. Look again at Figure 3-3, and count the bonds around each carbon atom. The total is always four. Each carbon atom in acetylene has three bonds to the other carbon atom and one C—H bond. Each carbon atom of ethylene has two bonds to its neighboring carbon atom and two bonds to hydrogen atoms. Finally, each carbon atom of ethane has one bond to the other carbon atom and three bonds to hydrogen atoms.

Keep in mind that carbon atoms are not shown in a line structure, so the first step in constructing a structural formula from a line structure is to place a C at every line intersection and at the end of every line. Then add singly bonded hydrogen atoms (—H) until every carbon atom has four bonds. Example 3-3 illustrates this procedure.

Example 3-3	Construct the structural formulas and determine the chemical formulas from the following line drawings:
Converting Line Structures	

Strategy: Line drawings show all structural features except carbon atoms and C—H bonds. Convert a line drawing into a structural formula in two steps. First, place a C at any unlabeled line end and at each line intersection. Second, add C—H bonds until each carbon atom has four bonds. The chemical formula is then obtained by counting the atoms of each element.

Solution: Place a C at each intersection and line end:

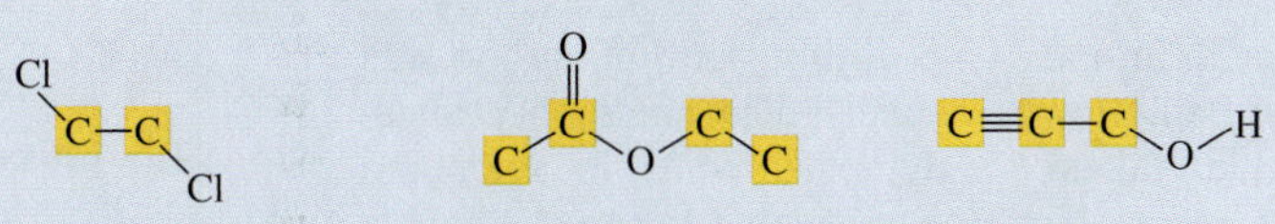

Now add C—H bonds until each carbon atom has a total of four bonds. Each carbon atom in the first structure has two bonds, so each carbon needs two C—H bonds to complete the structural formula:

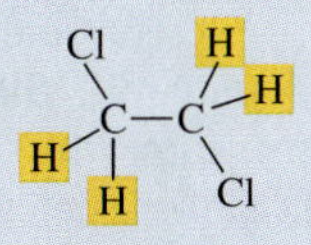

The second structure has four carbon atoms. The two end carbons have just one bond, so each needs three C—H bonds. The carbon with the double bond to oxygen already has its complete set of four bonds. The other inner carbon atom has a bond to C and a bond to O, so it needs two C—H bonds:

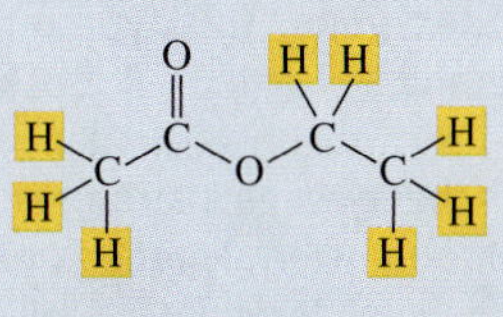

The third structure contains a triple bond. The end carbon needs one C—H bond, but the other carbon atom of the triple bond already has four bonds. The next carbon has two bonds, one to its carbon neighbor and one to the oxygen. Two C—H bonds are needed to give this atom its usual set of four bonds:

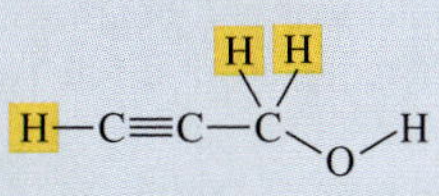

These are the correct structural formulas. Now it is a simple matter to count the number of atoms of each element and write the chemical formulas of the compounds: $C_2H_4Cl_2$, $C_4H_8O_2$, and C_3H_4O.

Do the Results Make Sense? Check the consistencies of the structural formulas by counting the number of bonds associated with each carbon atom. If you have converted the line structure correctly, each carbon atom will have four bonds.

Convert the following line structure into structural and chemical formulas:

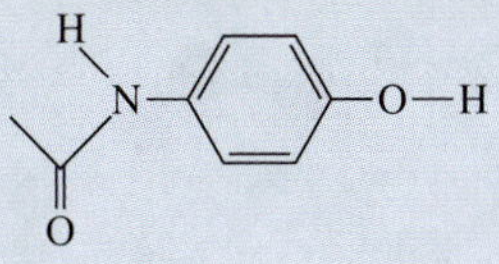

Example 3-3

Converting Line Structures (*continued*)

Extra Practice Exercise 3.3

The structural formulas of the compounds in these examples are not very complicated, yet their line structures are clearer than other representations. In ever-larger carbon-containing compounds, the simplification provided by line drawings is a great help to chemists. Line structures are not cluttered by the C—H bonds in the molecule, making them easier both to draw and to interpret. Figure 3-6 shows the line drawings of two biologically important molecules that may be familiar to you, and the Chemical Milestones Box describes unusual molecules made up entirely of carbon.

Box 3-1	Chemical Milestones: Forms of Elemental Carbon

C an you imagine atoms connected together to form a molecule shaped like a minuscule soccer ball? How about connections that result in molecular tubes? Remarkably, the element carbon can form these molecular shapes. Perhaps even more remarkably, chemists did not discover this until late in the twentieth century.

Two elemental forms of carbon with different appearances and properties are well known. These are graphite and diamond. The most common form of carbon is graphite, a black, soft, brittle solid. When a candle burns in a limited supply of air, it generates black powdery soot that is mostly graphite. In graphite, carbon atoms are connected in the planar honeycomb arrangement shown below. Each carbon atom is connected to *three* neighbors in the same plane. The atoms of any plane are connected in a network, but there are no connections between adjacent planes. This allows layers to slide past each other, giving graphite useful lubricating properties. Graphite is brittle because, lacking connections between atoms in separate layers, it fractures easily along its layer planes.

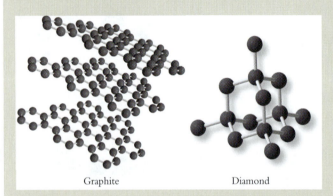

Graphite   Diamond

The carbon atoms in a diamond are connected in a three-dimensional network, each atom connected to *four* others. Each atom is at the center of a regular tetrahedron, as shown above. We describe this geometry, which occurs in many compounds of carbon, in Chapter 9. The three-dimensional connections result in a solid that is transparent, hard, and durable. The diamond structure forms naturally only at extremely high temperature and pressure, deep within the Earth. That's why diamonds are rare and precious.

A third form of carbon was observed in mass spectrometry experiments in 1985 and was isolated in 1990. Whereas transparent diamond and black graphite are extended lattices of interconnected carbon atoms, this red-violet form contains discrete molecules with the formula C_{60}. Sixty carbon atoms form the regular structure, shown below, that resembles a soccer ball. Five carbon atoms bond together in a pentagonal ring. Each side of

A solution of C_{60}

the pentagonal ring is also a side of a hexagonal ring. These clusters of rings join to form a ball of 60 carbon atoms in which each carbon atom is at a corner of one pentagon and two hexagons. This type of carbon was named buckminsterfullerene, in honor of R. Buckminster Fuller. Fuller invented the geodesic dome, a similar structure of pentagons and hexagons that creates a strong and spacious building design.

Buckminsterfullerene is one of a group of molecules having the general formula C_n, with *n* having several possible integral values. This group is collectively called "fullerenes." Fullerenes can be made by heating graphite intensely in a helium atmosphere. Under these conditions, graphite vaporizes as molecular fragments, some of which recondense as fullerenes. Chemists propose a variety of applications for fullerenes, including molecular ball bearings, lubricants, and plastics. It has also been suggested that fullerenes could lead to new cancer drugs by encapsulating a radioactive atom inside the fullerene cage.

The newest addition to the forms of elemental carbon is the nanotube. A carbon nanotube is a long cylinder of carbon atoms, connected together in much the same way as in a fullerene. Both the diameter and the length of carbon nano-tubes can vary. Properties of nanotubes, such as their ability to conduct electrical charge, change dramatically with the dimensions of the tube. Carbon nanotubes are under intensive study. For example, a carbon nanotube laid down on a silicon chip forms a molecular transistor. Such devices may eventually lead to further miniaturization of the chips that are at the heart of modern computers.

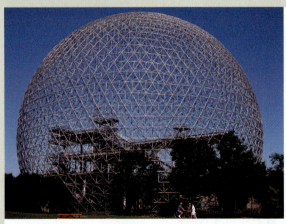

Geodesic dome

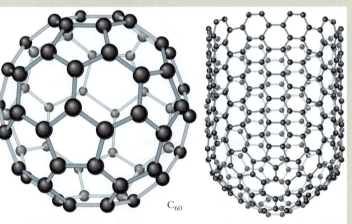

C_{60}

Carbon nanotube

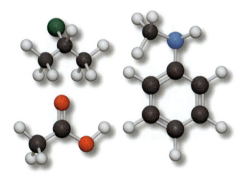

Figure 3-6
Structural formula and line drawing of caffeine, and line drawing of cholesterol.

Caffeine
$C_8H_{10}N_4O_2$

Cholesterol
$C_{27}H_{46}O$

Section Exercises

■ **3.1.1** Write the chemical formula for each of the following substances: (a) stearic acid, whose molecules contain 36 hydrogen atoms, 18 carbon atoms, and 2 oxygen atoms; (b) silicon tetrachloride, whose molecules contain one silicon atom and four chlorine atoms; and (c) Freon-113, whose molecules contain three atoms each of fluorine and chlorine and two atoms of carbon.

■ **3.1.2** Determine the molecular formula, structural formula, and line structure for each compound whose ball-and-stick model follows (see Figure 1-4 for atom colors):

■ **3.1.3** The line structures that follow represent starting materials for making plastics. For each of them, draw the structural formula and give the chemical formula.

Isoprene
(used in nature to produce rubber)

Styrene
(used to make styrofoam)

Methyl methacrylate
(used to make plexiglass)

3.2 NAMING CHEMICAL COMPOUNDS

In the early days of chemistry, the list of known compounds was short, so chemists could memorize the names of all of them. New compounds were often named for their place of origin, physical appearance, or properties. As the science of chemistry grew, the number of known compounds increased quickly. Soon, nobody could keep track of all of the common names. Today, more than 20 million compounds are known, and thousands of new ones are discovered or created each year. Consequently, chemists need systematic procedures for naming chemical compounds. The International Union of Pure and Applied Chemistry (IUPAC) has established uniform guidelines for naming various types of chemical substances, and chemists increasingly use IUPAC-approved names rather than their common counterparts. Systematic names are less colorful than common names, but they make chemistry less hectic because it is much easier to learn a few systematic guidelines than to memorize the names of thousands of individual compounds.

We focus on naming the substances that are most commonly encountered in general chemistry: binary compounds and compounds containing ions. Because many useful and

A system for naming is called *nomenclature,* from two Latin words: *nomen* means "name" and *calare* means "call."

Table 3-1   Common Roots for Naming Compounds

Element	Root		Element	Root
As	Arsen-		I	Iod-
Br	Brom-		Mn	Mangan-
C	Carb-		N	Nitr-
Cl	Chlor-		O	Ox-
Cr	Chrom-		P	Phosph-
F	Fluor-		S	Sulf-
H	Hydr-			

interesting chemical compounds contain carbon, we also describe a few principles for naming this rich array of substances. This section presents guidelines for compounds that do not contain metallic elements. We describe the naming of compounds containing metals in Section 3.3.

Naming Nonmetallic Binary Compounds

The written name of a compound includes the names of the elements it contains and information about the numbers of atoms of each element. The elements have to occur in some order, and this order is set by the same guidelines as for the chemical formula (see Section 3.1). Names can contain element names, roots derived from element names, and prefixes indicating the number of atoms of each element. Tables 3-1 and 3-2 list the more important roots and prefixes that appear in the names of binary compounds. We can summarize the rules for naming binary compounds in three guidelines:

②

Guidelines for Naming Binary Compounds

1. The element that appears first retains its elemental name.
2. The second element begins with a root derived from its elemental name and ends with the suffix-*ide*. Some common roots are listed in Table 3-1.
3. When there is more than one atom of a given element in the formula, the name of the element usually contains a prefix that specifies the number of atoms present. Common prefixes are given in Table 3-2.

Table 3-2   Number Prefixes For Chemical Names

Number	Prefix*	Example	Name
1	Mon(o)-	CO	Carbon monoxide
2	Di-	SiO_2	Silicon dioxide
3	Tri-	NI_3	Nitrogen triiodide
4	Tetr(a)-	CCl_4	Carbon tetrachloride
5	Pent(a)-	PCl_5	Phosphorus pentachloride
6	Hex(a)-	SF_6	Sulfur hexafluoride
7	Hept(a)-	IF_7	Iodine heptafluoride

*If the numerical prefix ends with the letter "o" or "a" and the name of the element begins with a vowel, drop the last letter of the prefix.

Numerical prefixes are essential in naming similar binary compounds. For example, nitrogen and oxygen form six different molecules: NO, nitrogen oxide; NO_2, nitrogen dioxide; N_2O, dinitrogen oxide; N_2O_3, dinitrogen trioxide; N_2O_4, dinitrogen tetroxide; and N_2O_5, dinitrogen pentoxide. Example 3-4 provides additional practice.

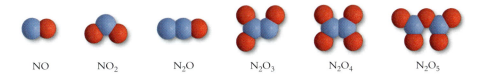

NO NO_2 N_2O N_2O_3 N_2O_4 N_2O_5

Name the following binary compounds: SO_2, CS_2, BCl_3, and BrF_5.

Strategy: None of these compounds contains a metallic element, so we apply the guidelines for binary compound nomenclature.

Solution: Name the first element, use a root plus *-ide* for the second element, and indicate the number of atoms with prefixes.
Here are the correct names: sulfur dioxide, carbon disulfide, boron trichloride, and bromine pentafluoride.

Do the Results Make Sense? Names make sense if they conform to the rules for nomenclature. The element further to the left in the periodic table appears first, the second element has an –ide suffix, and prefixes denote the number of atoms.

Name the compounds that contain (a) four atoms of bromine and one atom of carbon; (b) three atoms of oxygen and one atom of sulfur; (c) two atoms of fluorine and one atom of xenon.

Example 3-4

Naming
Binary Compounds

**Extra Practice
Exercise 3.4**

Binary Compounds of Hydrogen

Hydrogen requires special consideration because it may appear first or second in the formula and name of a compound. With elements from Groups 1 and 17, hydrogen forms diatomic molecules named according to our guidelines. For example, LiH is lithium hydride, and HF is hydrogen fluoride. With elements from Groups 2 and 16, hydrogen forms compounds containing two atoms of hydrogen. Except for oxygen, there is only one commonly occurring binary compound for each element, so the prefix *di-* is omitted. Examples are H_2S, hydrogen sulfide, and CaH_2, calcium hydride. Oxygen forms two binary compounds with hydrogen. These have unsystematic names: one is water, H_2O, and the other is hydrogen peroxide, H_2O_2. The binary compounds of hydrogen with elements from Groups 13 through 15 have unsystematic names, some of which are listed in Table 3-3. Carbon, boron, and silicon form many different binary compounds with hydrogen; only the simplest is listed in the table.

Compounds That Contain Carbon

The chemistry of carbon and its compounds is called *organic chemistry*. Biology and biochemistry build on the foundations of organic chemistry because the world of living matter is composed largely of carbon-based compounds. Hence, we present examples of organic molecules throughout this book, and here we list a few of the guidelines for naming organic compounds.

Table 3-3   Names of Some Hydrogen Compounds

Group 13	Group 14	Group 15
B_2H_6 Diborane	CH_4 Methane	NH_3 Ammonia
	SiH_4 Silane	PH_3 Phosphine

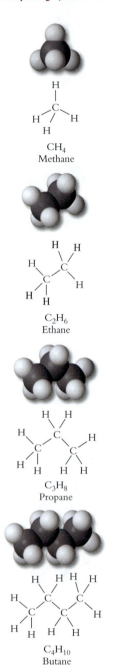

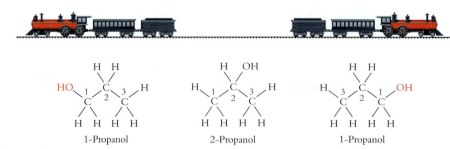

Figure 3-8

Structural formulas of 1-propanol and 2-propanol. Just as a train has its engine as the first car whether it is moving to the left or to the right, an —OH group on either end of propane makes the same molecule, 1-propanol.

Carbon forms a huge number of binary compounds with hydrogen. Three major categories of these compounds are *alkanes, alkenes,* and *alkynes.* An alkane has only single bonds between carbon atoms. The four simplest alkanes, which are shown in Figure 3-7, are methane, ethane, propane, and butane. An alkene, on the other hand, contains one or more double bonds between carbons, and an alkyne has one or more triple bonds between carbon atoms. Figure 3–3 shows the structures of ethylene, the simplest alkene, and acetylene, the simplest alkyne.

One of the major organizing principles in organic chemistry is the presence of special arrangements of atoms. These so-called *functional groups* convey particular chemical properties. For example, a substance that contains an —OH group is called an *alcohol.* The systematic name of an alcohol is obtained by adding the suffix *-ol* to the name of the alkane with the same carbon framework. Thus, CH_3OH has the carbon framework of methane and is called *methanol,* whereas C_2H_5OH has the carbon framework of ethane and is called *ethanol.*

The two different alcohols shown in Figure 3-8 are derivatives of propane. If one hydrogen atom on either of the *end* carbons of propane is replaced with OH, the alcohol is called *1-propanol.* If OH replaces a hydrogen atom on the central carbon, the alcohol is *2-propanol.* The numerical prefix specifies the carbon atom that bears the functional group. Even though there are three carbon atoms in propane, there is no compound named 3-propanol. That is because the two end carbons of propane are identical, so replacing a hydrogen atom on either of them gives the same molecule. This situation is analogous to the train shown in Figure 3-8. We identify the engine as the first car of the train whether the train travels from left to right or from right to left.

The halogens make up another important organic functional group. The functional group is named by adding the *-o* suffix to the end of the halogen's root name: fluoro-, chloro-, bromo-, and iodo-. For these compounds the halogen is identified at the beginning of the name rather than at the end: 1-bromopropane or 2-chloropropane, for example. We introduce additional families of organic molecules and their names as we proceed through the chapters of this book.

Figure 3-7
The four simplest alkanes are methane, ethane, propane, and butane.

Section Exercises

■ **3.2.1** Write chemical formulas for the following compounds: chlorine monofluoride, xenon trioxide, hydrogen bromide, silicon tetrachloride, sulfur dioxide, and hydrogen peroxide.

■ **3.2.2** Name the following compounds: ClF_3, H_2Se, ClO_2, $SbCl_3$, PCl_5, N_2O_5, N_2Cl_4, and NH_3.

■ **3.2.3** Draw structural formulas of the following molecules: butane, 2–butanol, 1–butanol, and 1–bromobutane.

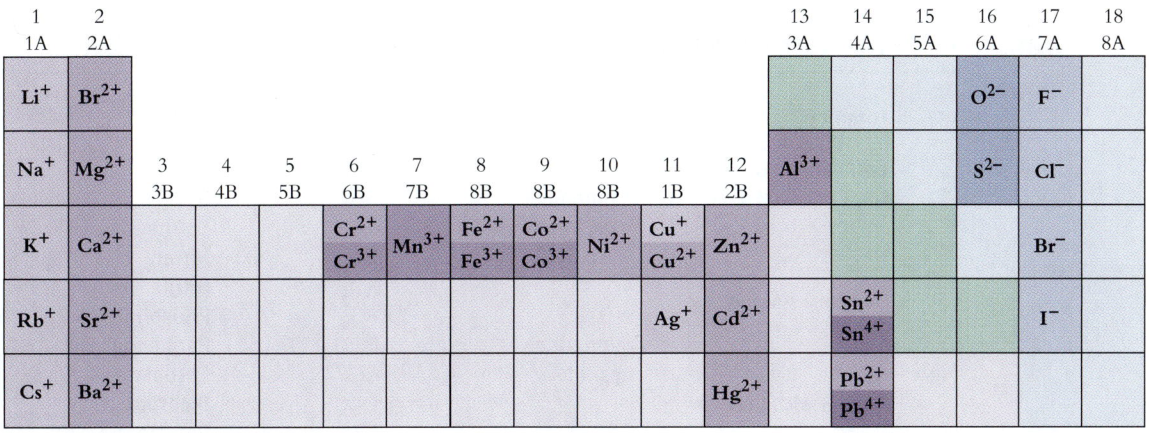

1 (1A)	2 (2A)	3 (3B)	4 (4B)	5 (5B)	6 (6B)	7 (7B)	8 (8B)	9 (8B)	10 (8B)	11 (1B)	12 (2B)	13 (3A)	14 (4A)	15 (5A)	16 (6A)	17 (7A)	18 (8A)
Li^+	Be^{2+}														O^{2-}	F^-	
Na^+	Mg^{2+}											Al^{3+}			S^{2-}	Cl^-	
K^+	Ca^{2+}				Cr^{2+} Cr^{3+}	Mn^{3+}	Fe^{2+} Fe^{3+}	Co^{2+} Co^{3+}	Ni^{2+}	Cu^+ Cu^{2+}	Zn^{2+}					Br^-	
Rb^+	Sr^{2+}									Ag^+	Cd^{2+}		Sn^{2+} Sn^{4+}			I^-	
Cs^+	Ba^{2+}										Hg^{2+}		Pb^{2+} Pb^{4+}				

Figure 3-9
Whereas many metals form monatomic cations, only six nonmetallic elements commonly form anions.

3.3 FORMULAS AND NAMES OF IONIC COMPOUNDS

Recall from Chapter 2 that ionic compounds contain positively charged cations and negatively charged anions. Some ionic compounds contain atomic cations and anions such as Na^+ and Cl^-. In many other cases, ionic compounds contain groups of atoms that have net electrical charges. Any stable sample of matter must be electrically neutral, and this requires that ionic compounds have the same amounts of positive and negative charges. This requirement for electrical neutrality, together with knowledge of the charges on various ions, allows us to determine the chemical formulas and names of ionic compounds.

Atomic Cations and Anions

The elements classified as metals have a strong tendency to lose electrons and form atomic cations. Almost every compound whose formula contains a metallic element from Group 1 or Group 2 is ionic. Other metals not only form ionic compounds in which they exist as cations but also commonly form compounds in which they share electrons.

Cations formed from metals in Group 1 always have +1 charges, and cations formed from Group 2 metals always have + 2 charges. Four other common metals have only one stable cation: Ag^+, Zn^{2+} Cd^{2+}, and Al^{3+}. Most transition metals, on the other hand, can exist in more than one cationic form. Figure 3-9 shows the important cations formed by the common elements. Figure 3-9 also indicates that, while many elements form stable atomic cations, only six form common stable atomic anions. Four elements in Group 17 form anions with -1 charge, and oxygen and sulfur form anions with -2 charge.

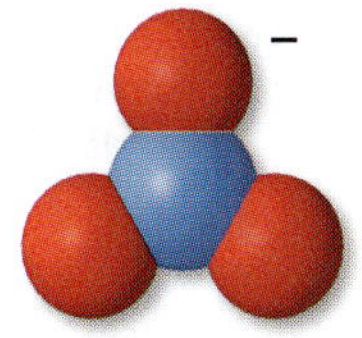

Nitrate anion
NO_3^-

Figure 3-10
The nitrate anion is a group of four atoms held together by chemical bonds. The entire unit bears a −1 electrical charge.

Polyatomic Ions

Sodium nitrate, $NaNO_3$, is an example of an ionic substance that contains a group of atoms with a net charge. Sodium is present as Na^+ atomic cations. The other atoms of sodium nitrate are grouped together in one structure, NO_3^-, which carries a -1 charge (Figure 3-10). This anion is a molecular ion, the nitrate ion. The nitrate ion displays some properties of ions and some properties of molecules: It has a negative electrical charge, but it also contains chemical bonds. Molecular ions are called **polyatomic ions** to distinguish them from neutral molecules and atomic ions.

Two important polyatomic *cations* appear in introductory chemistry. These are the ammonium ion, NH_4^+, and the hydronium ion, H_3O^+, both of which are shown in Figure 3-11. These cations always have +1 charges. The diatomic cation Hg_2^{2+} is the only relatively common polyatomic metal cation.

There are many different polyatomic *anions,* including several that are abundant in nature. Each is a stable chemical species that maintains its structure in the solid state and in aqueous solution. Polyatomic anions are treated as distinct units when writing chemical formulas, naming compounds, or drawing molecular pictures. The names,

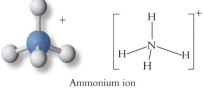

Ammonium ion

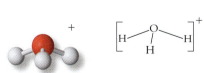

Hydronium ion

Figure 3-11
Structures of the ammonium and hydronium cations. Each always bears a + 1 charge.

Table 3-4   Common Polyatomic Ions

Formula	Name	Formula	Name
Cations		**Oxoanions**	
NH_4^+	Ammonium	SO_4^{2-}	Sulfate
H_3O^+	Hydronium	SO_3^{2-}	Sulfite
Hg_2^{2+}	Mercury(I)	NO_3^-	Nitrate
		NO_2^-	Nitrite
Diatomic Anions		PO_4^{3-}	Phosphate
OH^-	Hydroxide	MnO_4^-	Permanganate
CN^-	Cyanide	CrO_4^{2-}	Chromate
Anions with Carbon		$Cr_2O_7^{2-}$	Dichromate
CO_3^{2-}	Carbonate	ClO_4^-	Perchlorate
$CH_3CO_2^-$	Acetate	ClO_3^-	Chlorate
$C_2O_4^{2-}$	Oxalate	ClO_2^-	Chlorite
		ClO^-	Hypochlorite

Species **means "a distinct kind."** *Chemical species* **refers to a distinct kind of structure at the molecular level. Chemical species, which may be atoms, molecules, or ions, are represented by chemical formulas.**

formulas, and charges of the more common polyatomic anions are listed in Table 3-4. You should memorize the common polyatomic ions because they appear regularly throughout this textbook.

Most polyatomic anions contain a central atom surrounded by one to four oxygen atoms. These species are called **oxoanions,** and they are named according to the following guidelines:

Guidelines for Naming Oxoanions

1. The name has a root taken from the name of the central atom (for example, <u>carbo</u>nate, $\underline{C}O_3^{2-}$, and <u>nitr</u>ite, $\underline{N}O_2^-$).

2. When an element forms two different oxoanions, the one with fewer oxygen atoms ends in *-ite,* and the other ends in *-ate* (for example, SO_3^{2-}, sulf<u>ite</u>, and SO_4^{2-}, sulf<u>ate</u>).

3. Chlorine, bromine, and iodine each form four different oxoanions that are distinguished by prefixes and suffixes. The nomenclature of these ions is illustrated for bromine, but it applies to chlorine and iodine as well: BrO^-, <u>hypo</u>brom<u>ite</u>; BrO_2^-, brom<u>ite</u>; BrO_3^-, brom<u>ate</u>; and BrO_4^-, <u>per</u>brom<u>ate</u>.

4. A polyatomic anion with a charge more negative than -1 may add a hydrogen cation (H^+) to give another anion. These anions are named from the parent anion by adding the word *hydrogen.* For example, HCO_3^- is hydrogen carbonate, HPO_4^{2-} is hydrogen phosphate, and $H_2PO_4^-$ is dihydrogen phosphate.

Carbonate
CO_3^{2-}

Hydrogen carbonate
HCO_3^-

Recognizing Ionic Compounds

Ionic compounds show distinctive chemical behavior, because they contain positively charged cations and negatively charged anions. Because every substance must be electrically neutral, every ionic compound contains both a cationic part and an anionic part. A compound is ionic if it contains

1. The ammonium (NH_4^+) cation, or

2. Any metal and a polyatomic anion, or

3. Any Group 1 or 2 metal and a halide (X^-), oxide (O^{2-}), or sulfide (S^{2-}) ion.

To identify an ionic compound, follow the process shown in Figure 3-12. Look first for the ammonium ion, then for a metal. If the compound contains neither ammonium nor a metal, it is not ionic. For example, $SiCl_4$, which contains a halogen (Cl) but no metal, is not ionic. Neither is nitric acid, HNO_3. Although this compound contains the

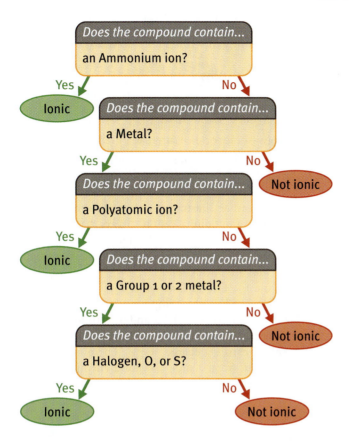

Figure 3-12
Use this decision tree to determine whether or not a compound is ionic.

nitrate group, hydrogen does not exist as a separate cation. If the compound contains a metal, look next for a polyatomic ion. If one is present, the compound is ionic. If there is no polyatomic ion, determine if the metal is from Group 1 or 2. If it is not, the compound is not ionic. If it is, look for one of the nonmetals that form anions: F, Cl, Br, I, O, or S. Example 3-5 provides practice in identifying ionic compounds.

Determine which of the following substances are ionic: CCl_4, $CaCl_2$, Li, $Co(NO_3)_2$.

Strategy: To identify an ionic compound, follow the steps described in Figure 3-12.

Solution:
CCl_4: There is neither ammonium nor a metal, so the compound is not ionic.

$CaCl_2$: Ammonium is not present, but there is a metal (Ca). There is no polyatomic anion, but Ca is from Group 2, and there is a halide (Cl), so this compound is ionic.

Li: Ammonium is not present, but there is a metal (Li). However, there is nothing that could be an anion, so lithium is not ionic.

$Co(NO_3)_2$: Ammonium is not present, but there is a metal (Co^{2+}), and there is a polyatomic anion (NO_{3-}), so this compound is ionic.

Do the Results Make Sense? To verify that two of the compounds are ionic, we can identify the cations and anions that they contain. $CaCl_2$ is made up of Ca^{2+} cations and Cl^- anions. $Co(NO_3)_2$ contains nitrate (NO_3^-) anions and Co^{2+} cations.

Determine which of the following substances are ionic: KCN, $TiCl_4$, $MgHPO_4$, and H_2SO_4.

Example 3-5

Recognizing
Ionic Compounds

**Extra Practice
Exercise 3.5**

Ionic Formulas

Every ionic compound contains discrete ionic units with specific charges. In addition, ionic compounds must always contain equal amounts of positive and negative charge.

These requirements dictate the ratio of cations to anions in an ionic substance. The following guidelines ensure uniformity in writing ionic formulas:

Guidelines for Ionic Formulas

1. The cation–anion ratio must give a net charge of zero.
2. The cation is always listed before the anion.
3. The formula of any polyatomic ion is written as a unit.
4. Polyatomic ions are placed in parentheses with a following subscript to indicate ratios different from 1:1.

Here are some specific examples illustrating chemical formulas of ionic compounds.

Ammonium nitrate: The ammonium cation always has $+1$ charge, and nitrate anions are always -1, so these ions combine in 1:1 ratio. The cation is listed before the anion: NH_4NO_3. Notice that we do not lump together the two nitrogen atoms because the ammonium and nitrate ions are distinct entities.

Sodium carbonate: Sodium, a Group 1 metal, always forms $+1$ atomic cations, and carbonate anions are always -2. Thus, the compound formed from sodium and carbonate must contain two Na^+ ions for every CO_3^{2-}, and its formula is Na_2CO_3.

Calcium phosphate: Calcium ions, from Group 2 of the periodic table, always carry a $+2$ charge, whereas phosphate carries a charge of -3. There must be three cations (total charge $+6$) for every two anions (total charge -6) for a chemical formula of $Ca_3(PO_4)_2$.

Ionic compounds are named using the same guidelines used for naming binary molecules, except that the cation name always precedes the anion name. Thus, NH_4NO_3 is ammonium nitrate, Na_2CO_3 is sodium carbonate, and $Ca_3(PO_4)_2$ is calcium phosphate. The subscripts are not specified in these names because the fixed ionic charges determine the cation–anion ratios unambiguously. Example 3-6 reinforces these guidelines by showing how to construct chemical formulas from chemical names.

Example 3-6 **Formulas of Ionic Compounds**	Determine the chemical formulas of calcium chloride, magnesium nitrate, and potassium dihydrogen phosphate. **Strategy:** The names of the cations and anions identify the chemical formulas, including the charges of each ion. The charges, in turn, dictate the ion ratio that must appear in the chemical formula. **Solution:** Calcium chloride contains Ca^{2+} (Group 2) and Cl^- (Group 17), so there must be two anions for every cation: $CaCl_2$. Magnesium nitrate contains Mg^{2+} (Group 2) and the polyatomic anion NO_3^-. Again, there must be two anions for every cation: $Mg(NO_3)_2$. Potassium dihydrogen phosphate requires an extra step in reasoning. Potassium (Group 1) is K^+, and phosphate is PO_4^{3-}. However, dihydrogen means that two hydrogen $+1$ cations are attached to phosphate, reducing its charge to -1. The dihydrogen phosphate ion, $H_2PO_4^-$, is a single structural unit, so the two ions are present in a 1:1 ratio: KH_2PO_4.
	Do the Results Make Sense? You can check that a chemical formula is reasonable by adding up the charges on its various components and verifying that they sum to zero. For $CaCl_2$, $(+2) + 2(-1) = 0$; for $Mg(NO_3)_2$, $(+2) + 2(-1) = 0$; and for KH_2PO_4, $(+1) + (-1) = 0$.
Extra Practice Exercise 3.6	Determine the chemical formulas of potassium sulfide, barium perchlorate, and calcium phosphate.

Cations of Variable Charge

By convention, the chemical formulas of many ionic compounds do not explicitly state the charges of the ions. It is not necessary to do so when the species involved form ions with only one possible charge. However, many metals form more than one type of stable cation. For example, copper forms two different oxides, black CuO and red Cu_2O. The oxide anion has a -2 charge, so for the first compound to be neutral the copper cation must bear a $+2$ charge. In Cu_2O, each copper ion must have $+1$ charge.

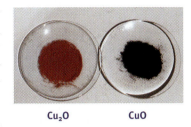

Cu₂O   CuO

Without an additional guideline for nomenclature, each of these ionic compounds would be called *copper oxide*. These are different compounds, however, so we must introduce a way to distinguish between them. For any metal that forms *more than one* stable cation, the charge is specified by placing a Roman numeral in parentheses after the name of the metal. According to this guideline, CuO is copper(II) oxide, and Cu_2O is copper(I) oxide. Notice, however, that we add Roman numerals *only* when a metal forms more than one stable cation. Figure 3-9 identifies the metals that form more than one cation: Group 1 $= M^+$; Group 2 $= M^{2+}$; Ag^+, Cd^{2+}, Zn^{2+}, and Al^{3+}. Example 3-7 shows how to apply the guidelines to a set of binary compounds.

Name the chlorine-containing compounds that have the following chemical formulas: $MgCl_2$, $CoCl_3$, PCl_3, $SnCl_2$, $SnCl_4$, and $GeCl_4$.

Strategy: Name compounds by applying the guidelines. The guidelines for naming binary compounds that contain metals differ from those for compounds containing no metal. Unless a metal forms only one stable atomic cation, its charge must be specified with a Roman numeral in parentheses.

Solution:

$MgCl_2$: Magnesium is a metal that always forms a $+2$ ion. This ionic compound is named without Roman numerals or prefixes: magnesium chloride.

$CoCl_3$: Cobalt, a transition metal, forms more than one stable cation: cobalt(III) chloride.

PCl_3: Phosphorus is not a metal, so use the guidelines for binary compounds: phosphorus trichloride.

$SnCl_2$: Tin is not one of the metals that forms a single cation. With two -1 chloride ions, this compound is tin(II) chloride.

$SnCl_4$: By the same reasoning used for tin(II) chloride, this is tin(IV) chloride.

$GeCl_4$: Germanium is not a metal, so the standard binary guidelines apply: germanium tetrachloride.

Do the Results Make Sense? From the chemical formulas, we might expect similar names for these compounds, but notice that they fall into three classes with different rules for each class: PCl_3 and $GeCl_4$ do not contain metals, $MgCl_2$ contains a metal that forms a unique cation, while $CoCl_3$, $SnCl_2$, and $SnCl_4$ contain metals with more than one stable cation.

Example 3-7

Binary Nomenclature

Name the compounds whose chemical formulas are $Al(NO_3)_3$, $FeBr_3$, and Cu_2CO_3.

Extra Practice Exercise 3.7

Hydrates

Many ionic compounds can have water molecules incorporated into their solid structures. Such compounds are called **hydrates.** To emphasize the presence of discrete water molecules in the chemical structure, the formula of any hydrate shows the waters of hydration separated from the rest of the chemical formula by

Anhydrous $CuSO_4$ is white, but $CuSO_4 \cdot 5H_2O$ is blue.

a dot. A coefficient before H_2O indicates the number of water molecules in the formula. Copper(II) sulfate pentahydrate is a good example. The formula of this beautiful deep blue solid is $CuSO_4 \cdot 5H_2O$, indicating that five water molecules are associated with each $CuSO_4$ unit. Upon prolonged heating, $CuSO_4 \cdot 5H_2O$ loses its waters of hydration along with its color. Other examples of hydrates include aluminum nitrate nonahydrate, $Al(NO_3)_3 \cdot 9H_2O$, with nine water molecules for every one Al^{3+} cation and three NO_3^- anions, and nickel(II) sulfate hexahydrate, $NiSO_4 \cdot 6H_2O$.

Whereas we can deduce the ratios of anions to cations in an ionic compound from the electrical charges on the individual ions, we cannot determine the number of water molecules in a hydrate from the nature of the anions and cations in the ionic compound. The number of water molecules, which can range from 0 to as high as 18, must be determined by doing experiments. In fact, some ionic substances exist in several different forms with different numbers of water molecules.

Examples 3-8 and 3-9 integrate the procedures used to convert between names and chemical formulas.

Example 3-8	Name the following compounds: CrO_3, ClF_3, Ag_2SO_4, and NH_4HSO_4.
Naming Chemical Compounds	**Strategy:** Apply the guidelines for naming compounds.

Solution:

CrO_3: As a transition metal, chromium forms more than one stable cation. Name the metal first, using a Roman numeral to designate chromium's charge. Each of the three oxide anions has a -2 charge. To maintain net charge neutrality, Cr must be $+6$, so the name of the compound is chromium(VI) oxide.

ClF_3: This compound contains two elements from Group 17 of the periodic table. Chlorine is named first because it is lower in the group, and we add a prefix that specifies the number of fluorine atoms: chlorine trifluoride.

Ag_2SO_4: The polyatomic sulfate ion indicates that this is an ionic compound. Silver is always $+1$, so no Roman numeral is needed: silver sulfate.

NH_4HSO_4: The polyatomic ammonium cation is combined with a hydrogen-containing anion: ammonium hydrogen sulfate.

Do the Results Make Sense? Verify that the components of the name correspond to the components in the chemical formula, and that compounds that lack a metal have been named following different guidelines than those used for metal-containing compounds.

Extra Practice Exercise 3.8	Name the compounds whose chemical formulas are NO_2, FeO, and $NiSO_4 \cdot 6H_2O$.

Example 3-9	Determine the correct chemical formulas of potassium permanganate, dinitrogen tetroxide, nickel(II) chloride hexahydrate, sodium hydrogen phosphate, and iron(III) oxide.
Chemical Formulas	**Strategy:** Work from name to formula, using information about polyatomic ions and being careful to build a formula that is electrically neutral.

Solution:

Potassium permanganate: Permanganate is MnO_4^- (Table 3-4), and potassium always has a $+1$ charge (Group 1). The formula is $KMnO_4$.

Dinitrogen tetroxide: The prefixes identify the formula: N_2O_4.

Example 3-9

Chemical Formulas
(*continued*)

Nickel(II) chloride hexahydrate: Nickel(II) is Ni^{2+}, and chloride is Cl^-. To achieve electrical neutrality, there must be two Cl^- and one Ni^{2+}. "Hexa" indicates six water molecules; $NiCl_2 \cdot 6H_2O$.

Sodium hydrogen phosphate: One hydrogen cation (H^+) attached to phosphate (PO_4^{3-}) leaves two negative charges, which requires two sodium ions for neutrality: Na_2HPO_4.

Iron(III) oxide: Iron(III) is Fe^{3+}, and oxide is O^{2-}. To be a neutral compound, there must be two Fe^{3+} ions (total charge = +6) for every three O^{2-} ions (total charge = −6): Fe_2O_3.

Do the Results Make Sense? Verify that the components of the name correspond to the components in the chemical formula, and that compounds that lack a metal have been named following different guidelines than those used for metal-containing compounds.

Determine the chemical formulas of barium chloride dihydrate, chromium(III) hydroxide, and sulfur trioxide.

**Extra Practice
Exercise 3.9**

Section Exercises

3.3.1 Name the following compounds: SCl_2, $CaCl_2$, $PbCl_2$, $NaNO_3$, MnO_2, $ZrCl_4$, NaH, and $NaIO_3$.

3.3.2 Write correct molecular formulas for aluminum oxide, potassium dichromate, lead(II) nitrate, nitrogen dioxide, sodium sulfate, iodine pentafluoride, manganese(II) acetate, and sodium hypochlorite.

3.3.3 Name the following compounds (refer to Figure 1-4 for atom color codes):

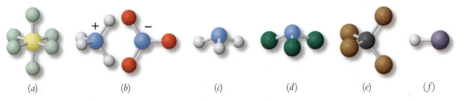

(a)   (b)   (c)   (d)   (e)   (f)

3.4 AMOUNTS OF COMPOUNDS

Sections 3.1–3.3 address how chemists answer the question, "What is it?" It is just as important to ask, "How much is there?" As described in Section 2.4, chemists use the mole to describe amounts of substances. The procedures introduced in that section work equally well for compounds as for elements. To apply these procedures, we need to be able to determine the molar masses of chemical compounds.

Molar Masses of Chemical Compounds

Just as each element has a characteristic molar mass, so does every chemical compound. Chemical compounds are composed of atoms bound together into molecules or ions clustered together in electrically neutral aggregates. In either case a chemical formula describes the atomic composition of a compound.

One mole of any chemical compound is one mole of its chemical formula unit. Here are some examples:

- One mole of O_2 is one mole of O_2 molecules. Each molecule contains two oxygen atoms, so one mole of O_2 molecules contains two moles of oxygen atoms.
- One mole of methane (CH_4) is one mole of CH_4 molecules, so it contains one mole of carbon atoms and four moles of hydrogen atoms.
- One mole of sodium chloride $(NaCl)$ contains one mole each of Na^+ cations and Cl^- anions.
- One mole of magnesium chloride $(MgCl_2)$ contains one mole of Mg^{2+} and two moles of Cl^-.

When atoms combine to form molecules, the atoms retain their atomic identities and characteristic molar masses. Thus, we can add elemental molar masses to obtain the molar mass of any compound.

> **KEY CONCEPT**   *The molar mass of a compound is found by adding together the molar masses of all of its elements, taking into account the number of moles of each element present.*

One mole of O_2 molecules contains two moles of O atoms, so the molar mass of molecular O_2 is calculated as follows:

$$\left(\frac{2 \text{ mol O}}{1 \text{ mol O}_2}\right)\left(\frac{15.999 \text{ g}}{1 \text{ mol O}}\right) = 31.998 \text{ g/mol O}_2$$

$$MM \text{ of } O_2 = 31.998 \text{ g/mol}$$

You can verify that the molar masses of CH_4, $NaNO_3$, and $MgCl_2$ are 16.043 g/mol, 84.994 g/mol, and 95.211 g/mol, respectively.

As chemical formulas become more complicated, the calculation of molar mass requires more steps, as shown in Examples 3-10 and 3-11.

Example 3-10

Molar Mass of a Molecule

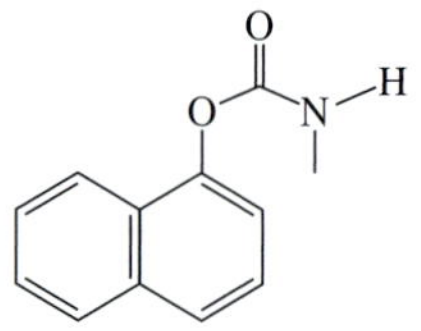

Sevin
$C_{12}H_{11}NO_2$

The line drawing and chemical formula of Sevin, a common insecticide, appear in the margin. Determine the molar mass of Sevin.

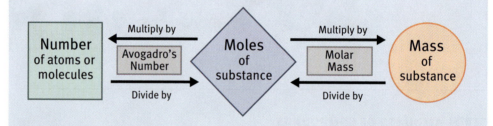

Strategy: To determine the molar mass of a substance, we need its chemical formula and elemental molar masses. From the chemical formula, determine the number of moles of each element contained in one mole of the substance. Multiply each elemental molar mass by the number of moles of that element, and add.

Solution: The line structure demonstrates how the atoms are connected in the molecule, but we do not need this information because the chemical formula is provided. We use the chemical formula to calculate the mass of each element in one mole of the compound:

$$\left(\frac{12 \text{ mol C}}{1 \text{ mol C}_{12}H_{11}NO_2}\right)\left(\frac{12.011 \text{ g C}}{\text{mol C}}\right) = \left(\frac{144.13 \text{ g C}}{1 \text{ mol C}_{12}H_{11}NO_2}\right)$$

$$\left(\frac{11 \text{ mol H}}{1 \text{ mol C}_{12}H_{11}NO_2}\right)\left(\frac{1.0079 \text{ g H}}{\text{mol H}}\right) = \left(\frac{11.087 \text{ g H}}{1 \text{ mol C}_{12}H_{11}NO_2}\right)$$

$$\left(\frac{1 \text{ mol N}}{1 \text{ mol C}_{12}H_{11}NO_2}\right)\left(\frac{14.007 \text{ g N}}{\text{mol N}}\right) = \left(\frac{14.007 \text{ g N}}{1 \text{ mol C}_{12}H_{11}NO_2}\right)$$

$$\left(\frac{2 \text{ mol O}}{1 \text{ mol C}_{12}H_{11}NO_2}\right)\left(\frac{15.999 \text{ g O}}{\text{mol O}}\right) = \left(\frac{31.998 \text{ g O}}{1 \text{ mol C}_{12}H_{11}NO_2}\right)$$

Now sum the individual components to find the molar mass of the compound:

$$MM = \left(\frac{144.13 \text{ g C}}{1 \text{ mol } C_{12}H_{11}NO_2}\right) + \left(\frac{11.087 \text{ g H}}{1 \text{ mol } C_{12}H_{11}NO_2}\right)$$
$$+ \left(\frac{14.007 \text{ g N}}{1 \text{ mol } C_{12}H_{11}NO_2}\right) + \left(\frac{31.998 \text{ g O}}{1 \text{ mol } C_{12}H_{11}NO_2}\right)$$
$$= 201.22 \text{ g/mol}$$

Does the Result Make Sense? We can do a quick check to see that the magnitude of the molar mass is about right by limiting the use of significant figures. Sevin contains 12 C and 11 H atoms, which contribute $(12 \times 12) + (11 \times 1) = 155$ g/mol to its molar mass. Add the contributions of two O atoms and one N atom, $(2 \times 16) + (1 \times 14) = 46$, giving 201 g/mol.

Determine the molar mass of malathion, an insecticide whose line structure and chemical formula appear in the margin.

Example 3-10

Molar Mass
of a Molecule (*continued*)

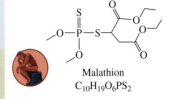

Malathion
$C_{10}H_{19}O_6PS_2$

**Extra Practice
Exercise 3.10**

Metal salts often exist as hydrates. One example is iron(II) nitrate hexahydrate. Determine the molar mass of this compound.

Strategy: Before we can calculate a molar mass, we need a chemical formula. Then we can calculate the masses of each of the elements in one mole of the compound.

Solution: Iron(II) has a charge of $+2$, and the nitrate anion carries -1 charge, so there must be two nitrates for every one iron ion, and *hexa-* stands for 6, so the chemical formula is $Fe(NO_3)_2 \cdot 6H_2O$.

In Example 3-10 we included a complete analysis of units. Chemists, however, usually abbreviate the designations for the units, as we show here.

It is convenient to decompose the formula into a list of elements, removing the parentheses and including the atoms in the waters of hydration.

One mole of $Fe(NO_3)_2 \cdot 6H_2O$ contains 1 mol Fe, 12 mol H, 2 mol N, and 12 mol O:

$$(1 \text{ mol Fe})(55.845 \text{ g/mol}) = 55.845 \text{ g Fe}$$
$$(12 \text{ mol H})(1.0079 \text{ g/mol}) = 12.095 \text{ g H}$$
$$(2 \text{ mol N})(14.007 \text{ g/mol}) = 28.014 \text{ g N}$$
$$(12 \text{ mol O})(15.999 \text{ g/mol}) = 191.99 \text{ g O}$$

$$MM \text{ } Fe(NO_3)_2 \cdot 6H_2O = 287.94 \text{ g/mol}$$

Does the Result Make Sense? Another way to calculate this molar mass is by working with the individual components of the formula. Each formula unit of the compound contains one iron(II) cation, two nitrate anions, and six waters of hydration: $MM \text{ } Fe(NO_3)_2 \cdot 6H_2O = MM \text{ } Fe^{2+} + 2 \text{ } MM \text{ } NO_3^- + 6 \text{ } MM \text{ } H_2O$. Verify that this procedure gives the same result.

Determine the molar mass of $Al(NO_3)_3 \cdot 9H_2O$.

Example 3-11

Molar Mass
of an Ionic Compound

**Extra Practice
Exercise 3.11**

Mass–Mole–Number Conversions for Compounds

As emphasized in Section 2.4, many of the calculations in chemistry involve converting back and forth among the mass of a substance, the number of moles, and the number of atoms and/or molecules. These calculations are all centered on the mole. The connections shown in Figure 2-22 apply to chemical compounds as well as to atoms of pure elements. Molar mass and Avogadro's number provide links between mass of a sample, the number of moles, and the number of molecules.

$$n = \frac{m}{MM} \qquad \text{(2-2)}$$

$$n = \frac{\#}{N_A} \qquad \text{(2-3)}$$

For example, the following calculation shows how many moles and molecules of Sevin are in a 5.0-g sample (the molar mass of the insecticide Sevin was calculated in Exercise 3-10):

$$n = \frac{m}{MM} = \frac{5.0 \text{ g}}{201.22 \text{ g/mol}} = 0.0248 \text{ mol Sevin}$$

$$\# = nN_A = (0.0248 \text{ mol Sevin})(6.022 \times 10^{23} \text{ molecules/mol}) = 1.5 \times 10^{22} \text{ molecules of Sevin}$$

Notice that once again moles are at the center of the scheme. To repeat, *calculations of chemical amounts center on the mole.*

Mass–mole–number calculations often involve atoms *within* a compound as well as the compound itself. The chemical formula provides the link between moles of a compound and the number of moles of the compound's individual elements:

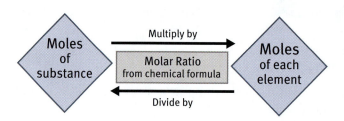

Example 3-12 illustrates mass–mole–number conversions involving both elements and compounds.

| **Example 3-12** | Ammonium nitrate (NH_4NO_3) is used as fertilizer because it is a good source of nitrogen atoms. It is such a good source, in fact, that it ranks among the top 15 industrial chemicals produced yearly in the United States. How many moles of nitrogen atoms are present in a 1.00-pound bag of NH_4NO_3 fertilizer? How many atoms is this? (1 pound = 453.592 g.) |

Elemental Content

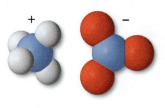

NH_4^+ NO_3^-
Ammonium   Nitrate

Strategy: First think about the chemistry of the problem, and then construct a mathematical solution. Use the seven-step procedure.

1. The problem asks for the number of moles and atoms of nitrogen in a 1.00-pound sample of NH_4NO_3.

2. Visualize NH_4NO_3. (Always think atoms and molecules.) Ammonium and nitrate are common polyatomic ions whose chemical formulas you should remember.

Solution:

3. The problem gives the mass of the sample:

$$\text{Mass of sample} = (1.00 \text{ pound})(453.592 \text{ g/pound}) = 453.6 \text{ g } NH_4NO_3$$

4. Equations 2-2 and 2-3 link mass, moles, and atoms:

$$n = \frac{m}{MM} \qquad and \qquad n = \frac{\#}{N_A}$$

5. Often, it helps to draw a flowchart that organizes the steps necessary to analyze and solve a problem. A flowchart for this problem appears in the margin.

6. Now we are ready to work through the calculations. Determine the molar mass of NH_4NO_3:

$$MM = (2 \text{ mol N})(14.007 \text{ g/mol}) + (4 \text{ mol H})(1.0079 \text{ g/mol})$$
$$+ (3 \text{ mol O})(15.999 \text{ g/mol})$$
$$= 80.043 \text{ g/mol}$$

Divide the sample mass by the molar mass to determine moles of NH_4NO_3:

$$n = \frac{m}{MM} = \left(\frac{453.6 \text{ g NH}_4\text{NO}_3}{80.043 \text{ g/mol}} \right) = 5.667 \text{ mol NH}_4\text{NO}_3$$

The chemical formula reveals that every mole of NH_4NO_3 contains 2 mol of nitrogen atoms:

$$(5.667 \text{ mol NH}_4\text{NO}_3)\left(\frac{2 \text{ mol N atoms}}{1 \text{ mol NH}_4\text{NO}_3} \right) = 11.33 \text{ mol N atoms}$$

As a final answer, this value should be rounded to three significant figures to match the data (1.00 pound): 11.3 mol N atoms. For the next step of the overall calculation, however, the value should retain a fourth digit to avoid possible rounding errors. Finish the problem by rearranging Equation 2-3 and use Avogadro's number to convert from the number of moles to the number of atoms:

$$\# = nN_A = (11.33 \text{ mol N})(6.022 \times 10^{23} \text{ atoms/mol}) = 6.82 \times 10^{24} \text{ N atoms}$$

The final result is rounded to three significant figures to match the three significant figures given for the mass of the bag of fertilizer.

7. **Do the Results Make Sense?** Although 10^{24} is an immense number of atoms, we know that one mole contains 6×10^{23} atoms. Numbers of the order of 10^{24} are reasonable when calculating the number of atoms in a sample of everyday size.

How many moles and atoms of hydrogen are present in 125 g of water (H_2O)?

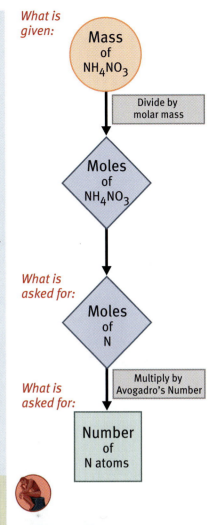

What is given:

Mass of NH_4NO_3

Divide by molar mass

Moles of NH_4NO_3

What is asked for:

Moles of N

Multiply by Avogadro's Number

What is asked for:

Number of N atoms

Extra Practice Exercise 3.12

Section Exercises

■ **3.4.1** Calculate the molar masses of the following compounds:

Compound	Name	Uses
(a) $NaAsO_2$	Sodium arsenite	Insecticide, herbicide, fungicide
(b) Na_4SiO_4	Sodium silicate	Soaps, adhesives, flame retardant
(c) $Ca(ClO)_2$	Calcium hypochlorite	Bactericide, bleach, pool cleaner
(d) $Ba(NO_3)_2$	Barium nitrate	Green fireworks
(e) Tl_2SO_4	Thallium sulfate	Rat and ant poison

■ **3.4.2** Twenty different amino acids are the essential building blocks of proteins. Calculate the molar masses of these three.

Histidine
$C_6H_9N_3O_2$

Cysteine
$C_3H_7NO_2S$

Asparagine
$C_4H_8N_2O_3$

■ **3.4.3** Adenosine triphosphate (ATP) is the body's principal energy storage molecule. The formula of ATP is $C_{10}H_{16}N_5O_{13}P_3$. (a) How many moles of ATP are in a 50.0-mg sample? (b) How many molecules of ATP is this? (c) How many atoms of phosphorus are in the sample? (d) How many atoms of nitrogen? (e) What is the mass of oxygen in 50.0 mg of ATP? (f) How many total atoms are there in 50.0 mg of ATP?

3.5 DETERMINING CHEMICAL FORMULAS

The chemical formula of a compound contains essential information about its composition. The formula identifies which elements are present, and it states the number of atoms of each kind present in one unit of the compound. We need the chemical formula of a substance to calculate its molar mass. In fact, almost all chemical calculations require the correct chemical formula. How are chemical formulas determined in the first place?

Every time a chemist prepares a "new compound" (one that has never been reported in the scientific literature), its chemical formula must be established beyond reasonable doubt. Usually, the molecule is analyzed by several methods, and each technique reveals information about the compound's identity.

To determine a chemical formula, we would like to count the atoms of each element in one molecule of the compound. Atoms are too small to count, but we might hope to measure the number of moles of each element present in one mole of the compound. Unfortunately, there is no direct experimental method for measuring moles. Instead, laboratory experiments give the *masses* of the various elements contained in some *total mass* of the compound.

Mass Percent Composition

The **mass percent composition** is a listing of the mass of each element present in 100 g of a compound. This percentage by mass listing is also called the compound's **elemental analysis.**

We illustrate how the mass percent composition of a compound is related to its chemical formula using ammonium nitrate (NH_4NO_3). The molar masses of NH_4NO_3 and its constituent elements can be used to convert the chemical formula into mass percentages.

The formula describes how many moles of each element are present in one mole of the substance. One mole of NH_4NO_3 contains two moles of nitrogen, four moles of hydrogen, and three moles of oxygen. Multiplying these numbers of moles by elemental molar masses gives the mass of each element contained in one mole of ammonium nitrate:

$$(2 \text{ mol N})(14.007 \text{ g/mol}) = 28.014 \text{ g N}$$

$$(4 \text{ mol H})(1.0079 \text{ g/mol}) = 4.0316 \text{ g H}$$

$$(3 \text{ mol O})(15.999 \text{ g/mol}) = 47.997 \text{ g O}$$

Summing, we see that one mole of NH_4NO_3 has a total mass of 80.043 g. Of this total, 28.014 g is nitrogen. The ratio of these masses is the mass fraction of nitrogen, and the percent nitrogen by mass is the mass fraction multiplied by 100%:

$$\frac{28.014 \text{ g}}{80.043 \text{ g}} = 0.34999 \text{ (mass fraction of nitrogen in } NH_4NO_3)$$

$$(0.34999)(100\%) = 34.999\% \text{ (mass percentage of nitrogen in } NH_4NO_3)$$

In other words, 34.999% of the mass of NH_4NO_3 comes from its nitrogen atoms. This means that every 100.000 g of NH_4NO_3 contains 34.999 g of nitrogen. The mass percentages of hydrogen and oxygen can be found in the same way:

$$\left(\frac{4.032 \text{ g}}{80.043 \text{ g}}\right)(100\%) = 5.037\% \text{ H} \qquad \left(\frac{47.997 \text{ g}}{80.043 \text{ g}}\right)(100\%) = 59.964\% \text{ O}$$

If the calculations have been done properly, the sum of the percent compositions of the individual elements will be 100%:

$$34.999\% \text{ N} + 5.037\% \text{ H} + 59.964\% \text{ O} = 100.000\%$$

This example shows how to compute mass percentages from a chemical formula. When the formula of a compound is unknown, chemists must work in the opposite direction. First, they do experiments to find the mass percentage of each element, and then they deduce what chemical formula matches those percentages.

The elemental analysis of a compound is usually determined by a laboratory that specializes in this technique. A chemist who has prepared a new compound sends a sample to the laboratory for analysis. The laboratory charges a fee that depends on the type and number of elements analyzed. The results are returned to the chemist as a listing of mass percent composition. The chemist must then figure out which chemical formula matches this composition. If a chemist has reason to expect a particular chemical formula, the observed percentages can be matched against the calculated percentages for the expected formula. This process is illustrated in Example 3-13.

Example 3-13

Formulas and Mass Percentages

A sample thought to be caffeine, the stimulant found in coffee, tea, and cola beverages, gave the following elemental analysis:

$$49.5\% \text{ C} \qquad 5.2\% \text{ H} \qquad 28.8\% \text{ N} \qquad 16.5\% \text{ O}$$

Does this elemental analysis agree with the chemical formula of caffeine, which is $C_8H_{10}N_4O_2$?

Strategy: We are asked to compare an elemental analysis with a chemical formula. To do so, we can either convert mass percentages to mole amounts or convert the formula to mass percentages. It is easier to compute mass percentages and compare them with the measured values.

Solution: Begin by calculating the mass of one mole of caffeine (because the percentage analysis is reported to only three significant figures, we need the molar mass to no more than two decimals):

$$
\begin{aligned}
(8 \text{ mol C})(12.01 \text{ g/mol}) &= 96.08 \text{ g C} \\
(10 \text{ mol H})(1.008 \text{ g/mol}) &= 10.08 \text{ g H} \\
(4 \text{ mol N})(14.01 \text{ g/mol}) &= 56.04 \text{ g N} \\
(2 \text{ mol O})(16.00 \text{ g/mol}) &= 32.00 \text{ g O} \\
\text{TOTAL} &= 194.20 \text{ g}
\end{aligned}
$$

Caffeine
$C_8H_{10}N_2O_4$
MM = 194.2 g/mol

Divide the mass of each element by 194.20 g to obtain its mass fraction. Next multiply by 100% to convert mass fractions to mass percentages:

Element	Expected % for $C_8H_{10}N_4O_2$	% Found
C	49.47	49.5
H	5.19	5.2
N	28.86	28.8
O	16.48	16.5

Does the Result Make Sense? Within the precision of the measured elemental analysis, the experimental percentages are the same as those expected for $C_8H_{10}N_4O_2$. Thus, the data are consistent with caffeine.

A toxic substance thought to be Sevin (see Example 3-10) is analyzed for its nitrogen content, which is found to be 8.5%. Might the substance be Sevin?

Extra Practice Exercise 3.13

Empirical Formula

Elemental analysis is a powerful tool for confirming molecular formulas, but it has its limitations. Consider caffeine again. Its chemical formula is $C_8H_{10}N_4O_2$, which means that the four elements are present in molar ratios 8:10:4:2. Dividing this set of numbers by 2 does not change the relative molar amounts, so a compound whose formula is $C_4H_5N_2O$ has exactly the same elemental analysis as caffeine. A compound whose formula is $C_{12}H_{15}N_6O_3$ also has the same elemental analysis, and so do numerous other compounds. Elemental analysis cannot distinguish among these possibilities. For that, we need to know the molar mass of the substance. Mass spectrometry, described in Chapter 2, is one way of measuring molar mass. If the mass spectrum of the substance shows that the molar mass is 194.2 g/mol, the formula is $C_8H_{10}N_4O_2$. On the other hand, a molar mass of 291.3 g/mol would correspond to $C_{12}H_{15}N_6O_3$. We describe other techniques for determining molar mass in later chapters.

If the molar mass of the compound is not known, the best we can do is to find the simplest formula that agrees with the elemental analysis. This simplest formula, or **empirical formula,** contains the smallest set of whole–number subscripts that match the elemental analysis. The empirical formula of caffeine is $C_4H_5N_2O$.

Sometimes chemists have to analyze substances about which they know very little. A chemist may isolate an interesting molecule from a natural source, such as a plant or an insect. Under these conditions the chemical formula must be deduced from mass percentage data, without the help of an "expected" formula. A four-step procedure accomplishes this by using mass–mole conversions, the molar masses of the elements, and the fact that a chemical formula must contain integral numbers of atoms of each element.

Guidelines for Elemental Analysis

1. Divide each mass percentage by the molar mass of the element. This gives the number of moles of each element in a 100-g sample.

2. Divide the results of Step 1 by whichever number of moles is the smallest. This maintains the mole ratios from Step 1 but bases them on one mole of the least abundant element.

3. If some results from Step 2 are far from whole numbers, multiply through by a common factor that makes all molar amounts close to whole numbers.

4. Round off each molar number to the nearest whole number.

These guidelines for solving elemental analysis problems are applied in Example 3-14.

Example 3-14 **Chemical Formula from Composition** 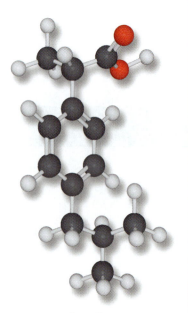 Ibuprofen	Analysis of ibuprofen, the active ingredient in several over-the-counter pain relievers, shows that it contains 75.7% carbon, 8.8% hydrogen, and 15.5% oxygen. The mass spectrum of ibuprofen shows that its molar mass is less than 210 g/mol. Determine the chemical formula of this compound. **Strategy:** First, follow the four-step process for finding the empirical formula. Then compare the empirical formula with the molar mass information to find the true formula.

Solution:

1. A 100-g sample of this compound would contain the following:

$$\left(\frac{75.7 \ \cancel{g \ C}}{100 \ g \ ibuprofen}\right)\left(\frac{1 \ mol \ C}{12.01 \ \cancel{g \ C}}\right) = 6.30 \ mol \ C/100 \ g \ ibuprofen$$

The calculation of carbon is shown in detail to highlight the unit cancellation. Here are the condensed calculations for hydrogen and oxygen:

$$\left(\frac{8.8 \ g \ H/100 \ g}{1.008 \ g \ H/mol}\right) = 8.7 \ mol \ H/100 \ g$$

$$\left(\frac{15.5 \ g \ O/100 \ g}{16.00 \ g \ O/mol}\right) = 0.969 \ mol \ O/100 \ g$$

2. Dividing each molar number by the smallest gives the following:

$$\frac{6.30 \ mol \ C}{0.969 \ mol \ O} = 6.50 \ mol \ C/mol \ O \qquad \frac{8.7 \ mol \ H}{0.969 \ mol \ O} = 9.0 \ mol \ H/mol \ O$$

$$\frac{0.969 \ mol \ O}{0.969 \ mol \ O} = 1 \ mol \ O/mol \ O$$

3. The value for carbon is not an integer, but if we multiply all values by 2, we obtain integers: 13 mol C, 18 mol H, and 2 mol O.

4. The empirical formula of ibuprofen is $C_{13}H_{18}O_2$.

Example 3-14

Chemical Formula from Composition (*continued*)

The mass spectrum of the compound indicates that its molar mass is less than 210 g/mol. The molar mass calculated from the empirical formula is 206.27 g/mol. This tells us that the chemical formula of ibuprofen is the same as its empirical formula.

Does the Result Make Sense? The chemical formula of ibuprofen shows that this biologically active compound contains similar numbers of C, H, and O atoms as does caffeine, another biologically active compound. A ball-and-stick model of ibuprofen appears in the margin opposite.

A liquid extracted from petroleum contains 92.3% carbon, the rest being hydrogen. Its molar mass is around 80 g/mol. Determine the empirical formula and molecular formula of the liquid.

Extra Practice Exercise 3.14

These examples illustrate the important role that mass spectrometry plays in the analysis of molecules. Our Tools for Discovery Box describes additional applications of this powerful technique.

We have shown how to convert mass percentages to formulas, but we have not yet shown how mass percentages are determined. Somehow a compound of unknown formula must be analyzed for the masses of each of its elements.

Analysis by Decomposition

Elemental analysis is relatively easy to accomplish when compounds can be decomposed into pure elements. Figure 3-13, for example, shows that an orange-red solid compound of mercury decomposes on heating to yield elemental mercury, a silver-colored liquid. A colorless gas is also produced. To exploit this decomposition reaction for elemental analysis, the masses of these products must be measured carefully. Suppose we start with 5.00 g of the compound and collect the gas formed during the decomposition. Weighing the gas gives its mass as 0.37 g, and mass spectroscopy identifies the gas as molecular oxygen. The liquid mercury has a mass of 4.63 g. The sum of the masses of oxygen and mercury equals the original mass of the compound: 4.63 g + 0.37 g = 5.00 g. Because mass is always conserved and the entire 5.00 g has been accounted for, this tells us that the compound contains only mercury and oxygen.

Because each chemical element is conserved, the masses of the products are equal to the masses of the elements contained in the original compound. This lets us complete the elemental analysis of the compound. The 5.00-g sample contained 4.63 g mercury and 0.37 g oxygen. The percent composition is found by dividing each elemental mass by the total mass and multiplying by 100:

$$\left(\frac{4.63 \text{ g Hg}}{5.00 \text{ g sample}}\right)(100\%) = 92.6\% \text{ Hg} \qquad \left(\frac{0.37 \text{ g O}}{5.00 \text{ g sample}}\right)(100\%) = 7.4\% \text{ O}$$

A 100.0-g sample would contain 92.6 g Hg and 7.4 g O.

To get from elemental analysis to a chemical formula, we begin by dividing each mass by the appropriate molar mass:

$$\frac{92.6 \text{ g Hg}}{200.6 \text{ g/mol}} = 0.462 \text{ mol Hg} \qquad \frac{7.4 \text{ g O}}{16.0 \text{ g/mol}} = 0.46 \text{ mol O}$$

These amounts are the same to within the accuracy of the measurements, so we conclude that the compound contains mercury and oxygen in 1:1 molar ratio. The empirical formula is HgO.

Figure 3-13
When mercury(II) oxide is heated, it decomposes into liquid mercury, driving off oxygen gas in the process.

Box 3-2   Tools for Discovery: Applications of Mass Spectrometry

Mass spectrometers can measure the masses of isotopes and molecules with high accuracy. Scientists use this capability to solve a variety of problems. Mass spectrometers are essential instruments in laboratories that study topics as varied as drug identification and climatic change.

Chemical analysis. The most widespread modern use of mass spectrometers is to identify chemical substances. When a molecule is placed in a mass spectrometer, the electrical discharge strips away one of its electrons. This so-called "parent ion" has virtually the same mass as the neutral molecule. If the mass of the parent ion is measured with high enough accuracy, the data can provide the molecular formula of the substance.

The parent ion usually breaks apart into a collection of smaller pieces, many of which are also positively charged. The masses of these fragments provide a chemical "fingerprint" that indicates how the atoms in a molecule are connected together. A relatively simple example is the mass spectrum of methane, shown in the figure. There are peaks at mass numbers 16, 15, 14, 13, 12, and 1. Nearly all hydrogen atoms have $A = 1$, and nearly all carbon atoms have $A = 12$, so these peaks can be identified as CH_4^+, CH_3^+, CH_2^+, CH^+, C^+, and H^+. From this fingerprint, we can infer the formula and structure of methane.

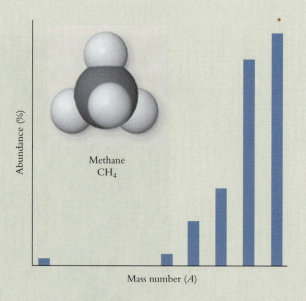

Methane
CH_4

Abundance (%)

Mass number (A)

The number of peaks in a mass spectrum grows rapidly with the complexity of the molecule, so each chemical substance has a unique mass spectral pattern. These patterns help verify the presence of a particular compound in a mixture. In crime investigation laboratories, forensic chemists use the mass spectrometer to identify illegal drugs by comparing the mass spectrum of a sample with the known fragmentation pattern of a substance. Mass spectrometry is also used in analyzing urine samples for evidence of substance abuse. The figure at the top of the next column shows the characteristic mass spectral fragmentation patterns for cocaine and heroin.

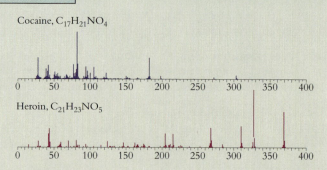

Cocaine, $C_{17}H_{21}NO_4$

Heroin, $C_{21}H_{23}NO_5$

Isotopic ratios. Different isotopes differ in their atomic masses. The intensities of the signals from different isotopic ions allow isotopic abundances to be determined with high accuracy. Mass spectrometry reveals that the isotopic abundances in elemental samples from different sources have slightly different values. Isotopic ratios vary because isotopes with different masses have slightly different properties; for example, they move at slightly different speeds. These differences have tiny effects at the level of parts per ten thousand (0.0001). The effects are too small to appear as variations in the elemental molar masses. Nevertheless, high-precision mass spectrometry can measure relative abundances of isotopes to around 1 part in 100,000.

For example, water molecules that contain ^{18}O evaporate from the oceans slightly more slowly than water molecules that contain ^{16}O. This rate difference is larger at low temperature than at high temperature. The difference is enough to monitor global warming and cooling. Marine carbonate sediments from different depths have different ages, and they also have slightly different values for $^{18}O/^{16}O$. The figure below shows that global warming and cooling has undergone regular cyclic variations over the past 500,000 years.

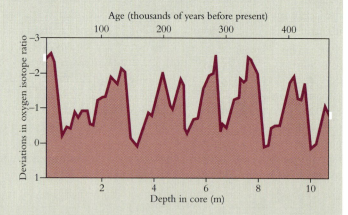

Age (thousands of years before present)

Deviations in oxygen isotope ratio

Depth in core (m)

Sulfur also shows variations in the abundance ratio of ^{34}S and ^{32}S. Tiny differences in the rates of chemical reactions of the two isotopes cause these variations. The abundance ratio can identify the source of sulfur contaminants in the atmosphere. For example, mining and smelting operations release harmful SO_2 into the atmosphere. Iron ores such as pyrites (FeS_2) have different values of $^{34}S/^{32}S$ than ores of zinc (ZnS) have. Thus, the value of $^{34}S/^{32}S$ in atmospheric SO_2 can identify whether the source of pollutant was an iron or zinc smelting operation.

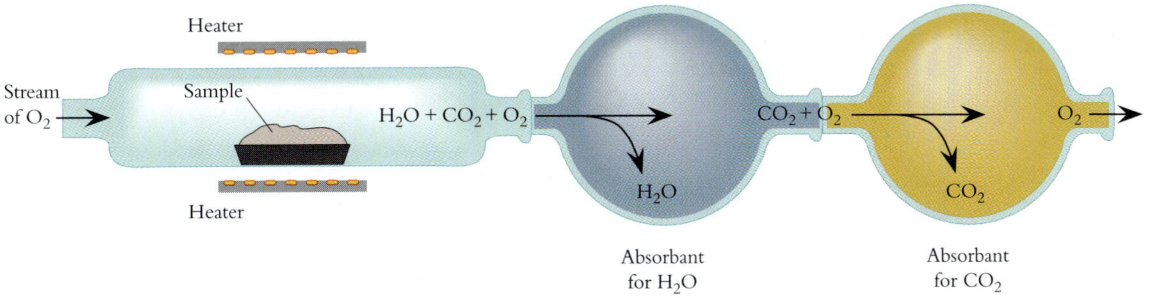

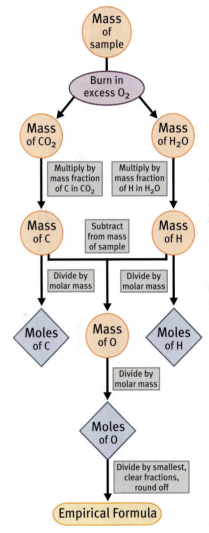

Figure 3-14
In combustion analysis, a hot stream of O₂ gas reacts with a compound to form CO₂ and H₂O, which are trapped and weighed.

Combustion Analysis

Compounds that do not decompose cleanly into their elements must be analyzed by other means. **Combustion analysis** is particularly useful for determining the empirical formulas of carbon-containing compounds. In combustion analysis, an accurately known mass of a compound is burned in a stream of oxygen gas. The conditions are carefully controlled so that all of the carbon in the sample is converted to carbon dioxide, and all of the hydrogen is converted to water. Certain other elements present in the sample are also converted to their oxides.

Figure 3-14 shows a schematic view of an apparatus for combustion analysis. The stream of oxygen used to burn the substance carries the combustion products out of the reaction chamber and through a series of traps. Each trap is designed to collect just one combustion product. The mass of each trap is measured before and after combustion, and the difference is the mass of that particular product.

The flowchart in Figure 3-15 shows one way to analyze the data from a combustion analysis. The most important feature in this analysis is that atoms of each element are conserved. Every *carbon* atom in the original sample ends up in a CO_2 molecule, and every *hydrogen* atom in the original sample ends up in a molecule of H_2O. Mass is also conserved in a combustion reaction, so the mass of carbon contained in the sample of CO_2 is the same as the mass of carbon in the original sample, and the mass of hydrogen in the sample of H_2O is the same as the mass of hydrogen in the original sample. After the masses of CO_2 and H_2O produced in the combustion reaction have been determined, the mass percentages of carbon and hydrogen in these products can be used to calculate the masses of carbon and hydrogen present in the original sample. This leads to an elemental analysis of the unknown compound from which we can deduce the empirical formula.

The products of a combustion reaction contain oxygen, carbon, and hydrogen. Did the oxygen in these products come from the original sample or from the oxygen used in the combustion process? To answer this question, we note that *all the mass of the original sample must be accounted for,* because mass is never destroyed in a chemical reaction. The masses of CO_2 and H_2O tell us how much carbon and hydrogen were present in the original sample. Subtracting the combined masses of carbon and hydrogen from the mass of the original sample gives the total mass of any other elements that were present in the original sample. If the unknown compound contained only carbon and hydrogen, this difference in mass will be zero. If the mass difference is not zero, then other elements must have been present. If a compound is known to contain only carbon, hydrogen, and oxygen, then the mass difference must be the mass of oxygen in the original sample.

Example 3-15 demonstrates how to use combustion analysis to determine the formula of a compound containing only C and H, and Example 3-16 shows how combustion analysis is conducted when the compound contains O.

The central feature of any combustion analysis is to keep track of all the atoms involved in the decomposition. Figure 3-16 illustrates this point with a molecular picture for Example 3-16.

Figure 3-15
This flowchart summarizes the method for using combustion analysis to determine the empirical formula of a compound that contains no elements other than C, H, and O.

Example 3-15

Combustion Analysis

A petroleum chemist isolated a component of gasoline and found its molar mass to be 114 g/mol. When a 1.55-g sample of this compound was burned in excess oxygen, 2.21 g of H_2O and 4.80 g of CO_2 were produced. Find the empirical and molecular formulas of the compound.

Strategy: The flowchart in Figure 3-15 outlines the process. From masses of products, determine masses of elements. Then convert masses of elements to moles of elements. From moles of the elements, find the empirical formula. Finally, use information about the molar mass to obtain the molecular formula.

Solution: Begin by computing masses of carbon and hydrogen in the combustion products. This requires the mass fractions of C in CO_2 and H in H_2O:

$$C: (4.80 \text{ g } CO_2)\left[\frac{(1 \text{ mol C/mol } CO_2)(12.01 \text{ g C/mol C})}{(44.01 \text{ g } CO_2/\text{mol } CO_2)}\right] = 1.31 \text{ g C}$$

$$H: (2.21 \text{ g } H_2O)\left[\frac{(2 \text{ mol H/mol } H_2O)(1.008 \text{ g H/mol H})}{(18.02 \text{ g } H_2O/\text{mol } H_2O)}\right] = 0.247 \text{ g H}$$

(Because the final calculations will give integers, it is not necessary to carry an extra significant figure in this problem.)

Now we convert mass to moles:

$$\frac{1.31 \text{ g C}}{12.01 \text{ g/mol}} = 0.109 \text{ mol C} \quad and \quad \frac{0.247 \text{ g H}}{1.008 \text{ g/mol}} = 0.245 \text{ mol H}$$

Next, compare the masses of C and H with the mass of the original sample to see whether the sample contained oxygen:

$$1.31 \text{ g C} + 0.247 \text{ g H} = 1.56 \text{ g} \qquad \text{Sample mass} = 1.55 \text{ g}$$

These are virtually the same, so C and H account for all of the mass in the original sample. Thus, there are no other elements in the unknown compound.

We now have a molar relationship between hydrogen and carbon in the original compound: 0.245 mol H to 0.109 mol C. Divide each by the smaller value:

$$\frac{0.245 \text{ mol H}}{0.109 \text{ mol C}} = 2.25 \text{ mol H/mol C} \qquad \frac{0.109 \text{ mol C}}{0.109 \text{ mol C}} = 1.00 \text{ mol C/mol C}$$

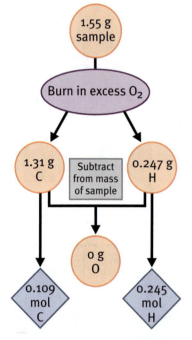

Several decimals are readily converted to integers by multiplying by a small whole number. When you see any of these, think about multiplying rather than rounding:

$(0.20)(5) = 1$	$(0.25)(4) = 1$
$(0.33)(3) = 1$	$(0.40)(5) = 2$
$(0.50)(2) = 1$	$(0.60)(5) = 3$
$(0.67)(3) = 2$	$(0.75)(4) = 3$
	$(0.80)(5) = 4$

The mole ratio for hydrogen is not close to a whole number, so we must multiply each ratio by a common integer that will make the value for hydrogen a whole or near-whole number. In this case, 4 is the smallest integer that accomplishes this goal, with $(2.25)(4) = 9$ and $(1.00)(4) = 4$. This means that the empirical formula is C_4H_9. Now, what is the molecular formula of the unknown? We must compare the empirical molar mass with the known molar mass, which is 114 g/mol. Compute the empirical mass from the empirical formula and elemental molar masses:

$$\text{Empirical mass} = 4 \text{ (mol C)}(12.01 \text{ g/mol}) + 9 \text{ (mol H)}(1.008 \text{ g/mol})$$
$$= 48.04 \text{ g/mol} + 9.072 \text{ g/mol} = 57.11 \text{ g/mol}$$

$$\frac{MM}{\text{Empirical mass}} = \frac{114 \text{ g/mol}}{57.11 \text{ g/mol}} = 2$$

Because the molar mass is twice the empirical mass, we know that the molecular formula of this hydrocarbon is twice its empirical formula: C_8H_{18}.

Does the Result Make Sense? You can verify that C_8H_{18} has a molar mass of 114 g/mol. This compound, containing only carbon and hydrogen, is an example of a hydrocarbon. Its name is *octane*, from the eight carbon atoms in the molecule. Gasoline is known to contain hydrocarbons, so this is a reasonable result.

Extra Practice Exercise 3.15

When 0.405 g of a gaseous substance was burned in excess oxygen, the products were 1.11 g of CO_2 and 0.91 g of H_2O. What is the empirical formula of this gas?

Butyric acid, a component of rancid butter, has a vile stench. Burning 0.440 g of butyric acid in excess oxygen yields 0.882 g of CO_2 and 0.360 g of H_2O as the only products. The molar mass of butyric acid is 88 g/mol. What are its empirical formula and molecular formula?

Example 3-16

Empirical Formula

Strategy: Proceed exactly as in Example 3-15. The flowchart in Figure 3-15 outlines the process. From masses of products, determine masses of elements. Then convert masses of elements to moles of elements. From moles of the elements, find the empirical formula. Finally, use information about the molar mass to obtain the molecular formula.

Solution: Start by determining how much carbon and hydrogen were present in the sample of butyric acid:

$$C: (0.882 \text{ g } CO_2)\left[\frac{(1 \text{ mol C/mol } CO_2)(12.01 \text{ g C/mol C})}{(44.01 \text{ g } CO_2/\text{mol } CO_2)}\right] = 0.241 \text{ g C}$$

$$H: (0.360 \text{ g } H_2O)\left[\frac{(2 \text{ mol H/mol } H_2O)(1.008 \text{ g H/mol H})}{(18.02 \text{ g } H_2O/\text{mol } H_2O)}\right] = 0.0403 \text{ g H}$$

$$\frac{0.241 \text{ g C}}{12.01 \text{ g/mol}} = 0.0201 \text{ mol C} \quad and \quad \frac{0.0403 \text{ g H}}{1.008 \text{ g/mol}} = 0.0400 \text{ mol H}$$

Next, compute the total mass of carbon and hydrogen in the original sample:

$$0.241 \text{ g C} + 0.0403 \text{ g H} = 0.281 \text{ g C and H}$$

The combustion of butyric acid gives CO_2 and H_2O as the only products, so the only other element that might be present is oxygen. The difference between the mass of the sample and the mass of C + H is the mass of O in the original sample:

$$\text{Mass C} + \text{Mass H} + \text{Mass O} = \text{Mass of sample} = 0.440 \text{ g}$$

$$0.440 \text{ g sample} - 0.281 \text{ g (C + H)} = 0.159 \text{ g O in the sample}$$

Divide this mass by the elemental molar mass of oxygen:

$$\frac{0.159 \text{ g O}}{16.00 \text{ g/mol}} = 0.00994 \text{ mol O}$$

Divide each number of moles by the smallest among them, 0.00994 mol O, to obtain moles of each element per mole of oxygen:

$$\frac{0.0201 \text{ mol C}}{0.00994 \text{ mol O}} = 2.02 \qquad \frac{0.0400 \text{ mol H}}{0.00994 \text{ mol O}} = 4.02 \qquad \frac{0.00994 \text{ mol O}}{0.00994 \text{ mol O}} = 1.00$$

Round each value to the nearest integer: 4.02 becomes 4, and 2.02 becomes 2. The empirical formula for butyric acid is C_2H_4O, giving an empirical mass of 44 g/mol. We are told that the molar mass of butyric acid is 88 g/mol. This is twice as large as the empirical mass. You should be able to use this information to show that the molecular formula of butyric acid is $C_4H_8O_2$.

Does the Result Make Sense? The fact that the ratios of the three elements all came out very close to integers gives us confidence that our analysis is correct. As we describe in Chapter 4, organic acids contain the $-CO_2H$ group. Thus, a chemical formula containing two O atoms is consistent with an organic acid.

When a 1.000-g sample of caproic acid (contains, C, H, and O, *MM* between 100 and 120) is burned in excess oxygen, 2.275 g of CO_2 and 0.929 g of H_2O are collected. Determine the empirical formula and molecular formula of caproic acid.

Figure 3-16
A molecular picture of the combustion of butyric acid to give carbon dioxide and water. Atoms are conserved in the reaction.

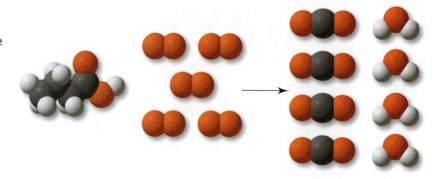

If an unknown sample contains elements in addition to C, H, and O, more elaborate measurements are required to determine the empirical formula. These techniques are beyond the scope of our coverage.

Section Exercises

3.5.1 Determine the percentage composition of acetic acid, $C_2H_4O_2$, the ingredient that gives vinegar its tart taste.

3.5.2 One of the major iron ores is an oxide with the following percentage composition: Fe, 72.36% and O, 27.64%. What is the chemical formula of this ore?

3.5.3 Our bodies can neither make nor store much ascorbic acid, also known as vitamin C, so this essential compound must be supplied in our diets. Combustion of 7.75 mg of vitamin C gives 11.62 mg CO_2 and 3.17 mg H_2O. Vitamin C contains only C, H, and O, and its molar mass is between 150 and 200 g/mol. Determine its molecular formula.

3.6 AQUEOUS SOLUTIONS

We developed the concept of the mole in terms of pure chemical substances, but many chemical reactions take place in solution. To treat solution reactions quantitatively, we need ways to apply the mole concept to solutions. A substance used to dissolve solutes is a **solvent,** and a pure substance dissolved in solution is a **solute.** Most of the time, the solvent is a liquid and is present in much larger quantities than any solutes.

Chemists use many different liquid solvents, but we focus most of our attention on water. When water is the solvent, the solution is said to be **aqueous.** A rich array of chemistry occurs in aqueous solution, including many geological and biochemical processes. Aqueous solutions dominate the chemistry of the Earth and the biosphere. The oceans, for instance, are rich broths of various cations and anions, sodium and chloride being the most abundant. The oceans can be thought of as huge aqueous solvent vessels for the remarkably complex chemistry of our world. Blood is an aqueous system that contains many ionic species, most notably carbonate, sodium, and potassium. In fact, the human body is mostly water, and much of the biochemistry of life takes place in aqueous solution.

Molarity

Any solution contains at least two chemical species, the solvent and one or more solutes. The mass of a solution is the sum of the masses of the solvent and all dissolved solutes. To answer questions such as "How much is there?" about solutions, we need to know the amount of each solute present in a specified volume of solution. The amount of a solute in a solution is given by the **concentration,** which is the ratio of the amount of solute to the amount of solution. In chemistry the most common measure of concentration is **molarity (M).** Molarity is the number of *moles of solute (n)* divided by the total *volume of the solution (V)* in liters:

$$\text{Molarity (mol/L)} = \frac{\text{Moles of solute}}{\text{Total volume of solution}} \quad or \quad M = \frac{n}{V} \quad \text{(3-1)}$$

Equation 3-1 defines molarity. When we need to calculate the number of moles of a substance in a solution of known molarity, we use a rearranged form of this equation: $n = MV$.

An aqueous solution is prepared by dissolving a measured quantity of solid or liquid in enough water to give a desired final volume. For example, we might dissolve 2.05 g of table sugar (sucrose) in just enough water to make the final volume 25.0 mL. What is the molarity of this solution? The units of molarity are moles per liter (mol/L). To determine the molarity of the sugar solution, we must convert the mass of sucrose into moles and convert the solution volume into liters. These conversions involve no new ideas. To begin, use the formula of sucrose, which is $C_{12}H_{22}O_{11}$, to calculate the molar mass, 342 g/mol. Next, convert mass to moles in the usual manner:

$$n = \frac{m}{MM} = \frac{2.05 \text{ g sugar}}{342 \text{ g/mol}} = 5.994 \times 10^{-3} \text{ mol sugar}$$

Molarity always is expressed in units of mol/L, so convert the volume of the solution from milliliters to liters:

$$(25.0 \text{ mL})(10^{-3} \text{ L/mL}) = 2.50 \times 10^{-2} \text{ L}$$

Finally, divide moles of sugar by volume in liters to obtain molarity:

$$\frac{5.994 \times 10^{-3} \text{ mol}}{2.50 \times 10^{-2} \text{ L}} = 0.240 \text{ M}$$

Concentrations are used so frequently in chemistry that a shorthand notation for concentration is almost essential. Chemists represent the molar concentration of a species by enclosing its formula in brackets: $[C_{12}H_{22}O_{11}] = 0.240$ M for the sugar solution.

In the laboratory, a solution of known concentration is often prepared in a glass vessel called a *volumetric flask*. Volumetric flasks typically allow volumes to be measured with an accuracy of four significant figures. Figure 3-17 summarizes the general procedure for making a solution in a volumetric flask. This procedure is further described in Example 3-17.

We use an italicized *M* when molarity appears as a quantity in an equation, as in $M = n/V$. We use a Roman M when molarity is the unit associated with a numerical value, as in 0.240 M.

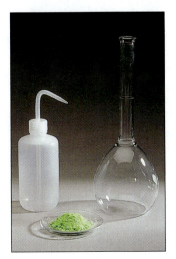

Figure 3-17
A solution of known concentration is prepared using a volumetric flask. When the container is filled with liquid to the etched line, the volume of solution in the container is the amount specified for that volumetric flask.

Example 3-17

Preparing a Solution

What mass of nickel(II) chloride hexahydrate is required to prepare 250. mL of aqueous solution whose concentration is 0.255 M?

Strategy: As with all calculations of chemical amounts, we must work with moles. Because grams are asked for, we must do a mole–mass conversion; this requires the molar mass of the substance, which in turn requires that we know the chemical formula.

Solution: Begin by finding the chemical formula from the name, nickel(II) chloride hexahydrate. Chloride ion carries −1 charge, and the (II) indicates that nickel is a +2 cation. Electrical neutrality requires two chlorides for every nickel. The name also tells us that the salt contains six water molecules for each unit of nickel(II) chloride. Thus, each formula unit of the salt contains one Ni^{2+} cation, two Cl^- anions, and six water molecules: $NiCl_2 \cdot 6H_2O$.

To determine the mass of $NiCl_2 \cdot 6H_2O$ required to prepare the solution, first calculate the number of moles of the salt required, and then use the molar mass to determine the mass in grams:

$$n = MV$$

$$n = \left(\frac{0.255 \text{ mol}}{L}\right)(250. \text{ mL})\left(\frac{10^{-3} \text{ L}}{1 \text{ mL}}\right) = 6.375 \times 10^{-2} \text{ mol } NiCl_2 \cdot 6H_2O$$

We need the molar mass of $NiCl_2 \cdot 6H_2O$ to convert moles to mass. By now, molar mass calculations should be routine. The molar mass of $NiCl_2 \cdot 6H_2O$ is 237.69 g/mol:

$$(6.375 \times 10^{-2} \text{ mol})(237.69 \text{ g/mol}) = 15.15 \text{ g}$$

To prepare the desired solution, we would first weigh 15.2 g of solid using a balance. (Rounding to three significant figures is consistent with the target molarity stated in the problem.) We would transfer this solid into a 250-mL volumetric flask, add enough water to dissolve the solid, then continue adding water and mixing until the solution level matched the mark on the flask.

Does the Result Make Sense? The mass is substantially less than the molar mass of the compound. Given that the volume is considerably less than 1 L and the target concentration is considerably less than 1 M, this is a reasonable result.

Extra Practice Exercise 3.17

Calculate the mass of ammonium nitrate needed to prepare 100. mL of aqueous solution whose concentration is 0.400 M.

Species in Solution

To understand the chemical behavior of solutions, we must "think molecules." Before working any problem involving aqueous solutions, begin with the question, "What chemical species are present in the solution?" There will always be an abundance of water molecules in an aqueous solution. In addition, there will be solute species, which may be molecules or ions.

An aqueous solution of a molecular substance such as sugar ($C_{12}H_{22}O_{11}$) or ethanol (C_2H_5OH) contains individual molecules in a sea of water molecules (Figure 3-18). We know that these solutes dissolve as neutral molecules from measurements of electrical conductivity. Figure 3-19 shows that pure water does not conduct electricity, and neither does a solution of sugar in water. This result shows that these solutions contain no mobile charged particles. Sugar and ethanol dissolve as neutral molecules.

In contrast to sugar, solid sodium chloride dissolves in water to give a liquid that conducts electricity. Figure 3-19 shows that a solution of NaCl is a good conductor. When an ionic compound dissolves in water, its component cations and anions are free to move about in the solution. Mixing leads to a uniform distribution of Na^+ and Cl^- ions through the entire solution, with each ion surrounded by a sheath of water molecules as shown in Figure 3-20.

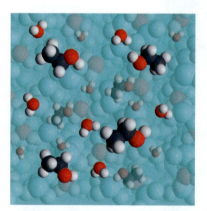

Figure 3-18
Molecular picture of an aqueous solution of ethanol, which contains ethanol molecules and water molecules.

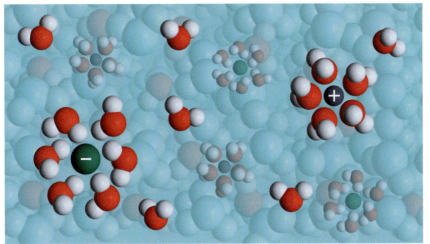

Figure 3-20
Molecular picture of a solution of sodium chloride in water. All the molecules and ions move freely about, but overall electrical neutrality is maintained because the total amount of anionic charge equals the total amount of cationic charge.

Figure 3-19
Pure water *(left)* and a solution of sugar *(right)* do not conduct electricity because they contain virtually no ions. A solution of salt *(center)* conducts electricity well because it contains mobile cations and anions.

 Tutorial

The presence of ions in solution is what gives a sodium chloride solution the ability to conduct electricity. If positively and negatively charged wires are dipped into the solution, the ions in the solution respond to the charges on the wires. Chloride anions move toward the positive wire, and sodium cations move toward the negative wire. This directed movement of ions in solution is a flow of electrical current. Pure water, which has virtually no dissolved ions, does not conduct electricity. Any solution formed by dissolving an ionic solid in water conducts electricity. Ordinary tap water, for example, contains ionic impurities that make it an electrical conductor.

Notice that the molecular view in Figure 3-20 includes ions as well as neutral molecules. "Molecular view" means a view of how matter looks at the atomic/molecular level. Thus, such illustrations may contain ions and individual atoms as well as molecules.

When a salt containing polyatomic ions dissolves in water, the cations separate from the anions, but each polyatomic ion remains intact. An example is ammonium nitrate, composed of NH_4^+ polyatomic cations and NO_3^- polyatomic anions. Ammonium nitrate dissolves in water to give a solution containing NH_4^+ cations and NO_3^- anions, as Figure 3-21 illustrates.

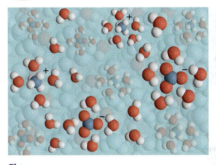

Figure 3-21
Molecular view of an aqueous solution of ammonium nitrate. Ammonium ions separate from nitrate ions, but both these species remain intact as polyatomic clusters.

Polyatomic ions remain intact when a salt dissolves in water.   **KEY CONCEPT**

Because aqueous solutions containing ions play a central role in the world around us, the chemistry of such solutions is discussed in depth in several chapters of this book. To understand the chemistry of aqueous solutions containing ions, *it is essential that you learn to recognize the common ions at a glance.* This is especially true of the polyatomic ions.

Concentrations of Ionic Solutions

The ratio of cations and anions in a solution of a salt is determined by its chemical formula. For example, a solution of NaCl contains Na^+ cations and Cl^- anions in 1:1 ratio: a 1.0 M solution of sodium chloride is 1.0 M in each of Na^+ and Cl^-. In contrast, $NiCl_2$ contains two moles of Cl^- anions for every one mole of Ni^{2+} cations. This ratio is maintained when nickel(II) chloride dissolves in water; a 1.0 M solution of $NiCl_2$ is 1.0 M in Ni^{2+} and 2.0 M in Cl^-:

$$NiCl_2(s) \xrightarrow{H_2O} Ni^{2+}(aq) + Cl^-(aq) + Cl^-(aq)$$

1 mol ⇑   1 mol ⇑   2 mol ⇑

The (*s*) and (*aq*) in the equation indicate phases: (*s*) designates solid and (*aq*) designates aqueous solution.

Notice that preserving the cation–anion ratio ensures that the solution is electrically neutral overall. The total positive charge carried by one mole of Ni^{2+} equals the total negative charge carried by two moles of Cl^-.

Example 3-18 treats concentrations of polyatomic ions.

Example 3-18

Molarity of Ions in Solution

Find the molarities of the ionic species present in 250. mL of an aqueous solution containing 1.75 g of ammonium sulfate, $(NH_4)_2SO_4$.

Strategy: The chemical formula identifies the ions that are present in the final solution. The formula also tells us how many moles of each ion are present in one mole of the salt. Use mass, molar mass, and volume to calculate molarity.

Solution: Looking at the formula, we recognize the ammonium cation, NH_4^+, and the sulfate anion, SO_4^{2-}. When ammonium sulfate dissolves in water, it dissociates into its component polyatomic ions. Each mole of salt will produce 2 mol of NH_4^+ and 1 mol of SO_4^{2-}.

After identifying what species are present in solution, we can calculate molarities. Begin with the molar mass of the salt:

$$\begin{aligned}
(2 \text{ mol N})(14.01 \text{ g/mol}) &= 28.02 \text{ g N} \\
(8 \text{ mol H})(1.008 \text{ g/mol}) &= 8.064 \text{ g H} \\
(1 \text{ mol S})(32.06 \text{ g/mol}) &= 32.06 \text{ g S} \\
(4 \text{ mol O})(16.00 \text{ g/mol}) &= 64.00 \text{ g O} \\
\hline
1 \text{ mol } (NH_4)_2SO_4 &= 132.14 \text{ g}
\end{aligned}$$

Use the molar mass to determine the number of moles of salt and, in turn, the number of moles of each ion. Notice that even though we are aiming for molarities of ionic species, we determine moles of the salt. This is because the mass that is provided is the *mass of the salt*. To find moles of salt (*n*), this mass must be divided by the *molar mass of the salt*:

$$n = \frac{m}{MM} = \frac{1.75 \text{ g}}{132.14 \text{ g/mol}} = 1.324 \times 10^{-2} \text{ mol } (NH_4)_2SO_4$$

From the number of moles of salt, calculate the number of moles of each ion:

$$(1.324 \times 10^{-2} \text{ mol } (NH_4)_2SO_4)\left[\frac{2 \text{ mol } NH_4^+}{1 \text{ mol } (NH_4)_2SO_4}\right] = 2.648 \times 10^{-2} \text{ mol } NH_4^+$$

$$(1.324 \times 10^{-2} \text{ mol } (NH_4)_2SO_4)\left[\frac{1 \text{ mol } SO_4^{2-}}{1 \text{ mol } (NH_4)_2SO_4}\right] = 1.324 \times 10^{-2} \text{ mol } SO_4^{2-}$$

> Remember to carry an extra significant figure until the final calculation.

Now find the molarities of the individual ions:

$$(250. \text{ mL})(10^{-3} \text{ L/mL}) = 0.250 \text{ L}$$

> Use dimensional analysis to obtain correct conversion factors. Rearrange 1 mL = 10^{-3} L to give 1 = (10^{-3} L/mL), with mL in the denominator so it cancels mL in the numerator.

$$[NH_4^+] = \frac{2.648 \times 10^{-2} \text{ mol}}{0.250 \text{ L}} = 1.06 \times 10^{-1} \text{ M}$$

$$[SO_4^{2-}] = \frac{1.324 \times 10^{-2} \text{ mol}}{0.250 \text{ L}} = 5.30 \times 10^{-2} \text{ M}$$

Does the Result Make Sense? The concentration of ammonium ions is twice that of sulfate ions. This is consistent with the 2:1 ratio of ammonium to sulfate in the chemical formula. You should understand that when we describe this solution as "0.0530 M $(NH_4)_2SO_4$," stating moles of salt per liter of solution, we are describing a solution that actually contains a *mixture of ions, each with its own molarity*.

Extra Practice Exercise 3.18

Calculate the molarities of the ions present in the solution of nickel(II) chloride hexahydrate prepared in Example 3-17.

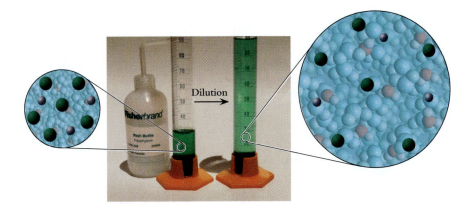

Figure 3-22
When a concentrated solution of NiCl$_2$ *(left)* is diluted by adding more solvent, the resulting solution *(right)* contains the same numbers of Ni^{2+} and Cl$^-$ ions. The solution is more dilute, however, because the ions are spread around in a larger volume of solvent.

Dilutions

One of the common methods for preparing a solution of known concentration is to dilute an existing solution of higher concentration by adding more solvent. In a **dilution** the *amount of solute* remains the same, but the *volume of the solution* increases. Thus, a dilution results in a solution of lower molarity.

The most important feature of a dilution is that *the number of moles of solute does not change during the dilution.* Consider the dilution of an aqueous solution of nickel(II) chloride, as shown in Figure 3-22. The photo shows that the green color, which is due to the Ni^{2+} ions, fades when the solution is diluted. This fading is visible evidence that the concentration decreases. At the molecular level, there are the same *numbers* of Ni^{2+} and Cl$^-$ ions in the solution before the dilution and after the dilution. However, because the volume of the solution increases, the *concentration* of the new solution is lower than the concentration of the original solution.

The quantitative aspects of dilutions can be found from the fact that the number of moles of solute does not change during a dilution:

$$n_{\text{solute, initial}} = n_{\text{solute, final}}$$

Because we can arrange Equation 3-1 to give $n = MV$, this leads to a simple equation:

$$M_i V_i = M_f V_f \qquad (3\text{-}2)$$

Equation 3-2 is convenient for dilution calculations. If any three of the quantities are known, we can calculate the fourth, as Example 3-19 shows.

Aqueous hydrochloric acid, HCl, is usually sold as a 12.0-M solution, commonly referred to as *concentrated HCl*. A chemist needs to prepare 2.00 L of 1.50 M HCl for a number of different applications. What volume of concentrated HCl solution should the chemist use in the dilution?

Strategy: The seven-step problem-solving approach is appropriate.
1. The question asks for the volume of concentrated HCl required to prepare a dilute solution.
2. Visualization: A concentrated solution is diluted to a larger volume. The chemist will remove a sample from the bottle of concentrated HCl solution and mix it with water.

Solution:
3. The information available is as follows:

$$\text{Concentrated HCl} = 12.0 \text{ M} \qquad \text{Diluted HCl} = 1.50 \text{ M}$$

$$\text{Final volume} = 2.00 \text{ L}$$

Example 3-19

Dilution of a Solution

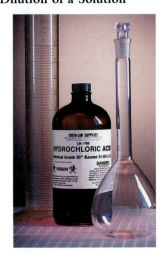

Example 9-4

Dilution of a Solution (*continued*)

4. This is a dilution process, so we use Equation 3-2.

5. Rearrange Equation 3-2 to solve for V_i, the volume required: $V_i = M_f V_f / M_i$.

6. Do the calculation. In Step 3 we identified the appropriate values:

$$M_f = 1.50 \text{ M} \qquad V_f = 2.00 \text{ L} \qquad M_i = 12.0 \text{ M} \qquad V_i = ?$$

$$V_i = \frac{(1.50 \text{ M})(2.00 \text{ L})}{(12.0 \text{ M})} = 0.250 \text{ L}$$

7. Does the Result Make Sense? The volume has the correct units, and the amount required is smaller than the total volume. This is reasonable.

Extra Practice Exercise 3.19

What volume (in mL) of concentrated ammonia (14.8 M) should be used to prepare 0.500 L of 0.350 M solution?

When preparing acid solutions, water should never be added directly to concentrated acid. Addition of water often causes rapid local heating that may lead to a dangerous acid splash. Instead, concentrated acid is always added to water. Additional water can then be added safely to the less concentrated solution that results. The solution in Example 3-19 is prepared by using a volume-measuring device, such as a graduated cylinder, to measure 0.250 L (250 mL) of concentrated HCl. About 0.5 L of water is added to a 2.00-L volumetric flask, and then the concentrated acid from the graduated cylinder is poured slowly into the water. Filling the volumetric flask to the mark with additional water completes the dilution.

Chemists often need to work with solutions with a variety of concentrations. Often, a relatively concentrated "stock" solution is prepared by weighing. Then, a portion of this solution is diluted volumetrically. In procedures of this kind, it is important to keep track of the volumes, as illustrated in Example 3-20 and shown in the flowchart in Figure 3-23.

Example 3-20

Preparing Solutions

An agricultural chemist wished to study the effect of varying fertilizer applications on the growth of tomato plants. The chemist prepared a stock aqueous solution of urea, $(NH_2)_2CO$, by dissolving 1.75 g of this compound in water to make 1.00 L of solution. Then she prepared a series of more dilute solutions to apply to her tomato plants. One of these solutions contained 5.00 mL of stock solution diluted to give a final volume of 25.00 mL. What was the concentration of urea in this diluted solution?

Strategy: Use the seven-step method.
1. We are asked to determine the concentration of the dilute solution. This requires us first to determine the concentration of the stock solution, so we need to do two separate calculations.

2 and 3. A flowchart, shown below, helps us to visualize the problem and organizes the data. The flowchart shows that the calculation has two parts. First we treat the stock solution, then we use the result to treat the diluted solution.

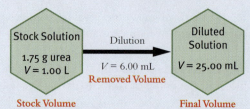

Solution:
4. We have mass and volume information for the stock solution, so we use Equations 2-2 and 3-1:

$$n = \frac{m}{MM} \quad and \quad M = \frac{n}{V}$$

For the diluted solution, we use Equation 3-2:

$$M_i V_i = M_f V_f$$

5. Manipulate the equations to obtain equations for the molarity of the stock and diluted solutions:

$$M_{stock} = \frac{m}{(MM)(V_{stock})} \quad and \quad M_{final} = \frac{(M_{stock})(V_{removed})}{V_{final}}$$

We need the molar mass of urea:

$$MM = (4\ mol\ H)(1.0079\ g/mol) + (2\ mol\ N)(14.007\ g/mol)$$
$$+ (1\ mol\ C)(12.011\ g/mol) + (1\ mol\ O)(15.999\ g/mol)$$
$$= 60.056\ g/mol$$

6. Substitute and calculate:

$$M_{stock} = \frac{m}{(MM)(V_{stock})} = \frac{1.75\ g}{(60.056\ g/mol)(1.00\ L)} = 2.914 \times 10^{-2}\ M$$

$$M_{final} = \frac{(M_{stock})(V_{removed})}{V_{final}} = \frac{(2.914 \times 10^{-2}\ M)(5.00\ mL)}{25.00\ mL} = 5.82 \times 10^{-3}\ M$$

7. Do the Results Make Sense? To check for reasonableness, make sure you have used the correct volumes. The *stock solution* was prepared in a volume of 1.00 L. The *dilution* was from 5.00 mL to 25.00 mL, a factor of 5. We can see by inspection that the final concentration is lower than the stock concentration by a factor of about 5.

Calculate the concentration of ammonium nitrate in a solution prepared by dissolving 3.20 g of the salt in enough water to make 100. mL of solution, then diluting 5.00 mL of this solution to a volume of 25.00 mL.

Example 3-20

Preparing Solutions
(*continued*)

**Extra Practice
Exercise 3.20**

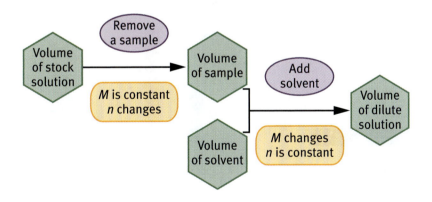

Figure 3-23
Flowchart showing the connections among measured quantities in dilution procedures.

Remove a sample

Volume of stock solution

M is constant
n changes

Volume of sample

Add solvent

Volume of solvent

M changes
n is constant

Volume of dilute solution

Section Exercises

3.6.1 An urban gardener buys a 1.0-kg bag of ammonium nitrate to fertilize house plants. Compute the concentrations of the ionic species present in a stock solution prepared by dissolving 7.5 g of the salt in 0.300 L of water.

3.6.2 Plants can be badly damaged by solutions that are highly concentrated. A friend of the urban gardener recommends against using fertilizer solutions that are greater than 1.0 mM

$(1.0 \times 10^{-3}\ M)$ in total nitrogen. What volume (in mL) of the stock solution from Section Exercise 3.6.1 will be needed to prepare 750 mL of 1.0 mM fertilizer solution?

3.6.3 Draw a molecular picture of a portion of the solution described in Section Exercise 3.6.2. Make sure the solution is electrically neutral (omit water molecules to simplify your drawing).

CHAPTER 3 VISUAL SUMMARY

Important equations

Essential terms

Key concepts

SKILLS TO MASTER

① Drawing various types of chemical structures
② Naming binary and ionic compounds
③ Recognizing ionic compounds
④ Calculating molar masses
⑤ Doing mole-mass-number conversions

⑥ Converting mass % and chemical formulas
⑦ Determining empirical and molecular formulas
⑧ Visualizing ionic solutions
⑨ Calculating solution molarities

3.1 REPRESENTING MOLECULES

Binary compound Fe_2O_3, $CaCl_2$, CH_4

Chemical formula C_2H_6O or C_2H_5OH or CH_3CH_2OH

Structural formula

Line structure   ①

Ball-and-stick model

Space-filling model

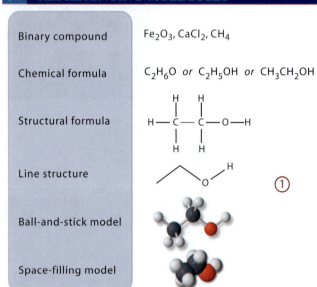

3.2 NAMING CHEMICAL COMPOUNDS

Nonmetallic binary compounds: Sulfur dioxide, SO_2 ②

Binary compounds of hydrogen: Ammonia, NH_3

Carbon-based compounds: Propane, C_3H_8

3.3 FORMULAS AND NAMES OF IONIC COMPOUNDS

Polyatomic ions   ③

Oxoanion

$CuSO_4 \cdot 5\,H_2O$
$NiCl_2 \cdot 6\,H_2O$
$Ca(NO_3)_2 \cdot 4\,H_2O$

Hydrates

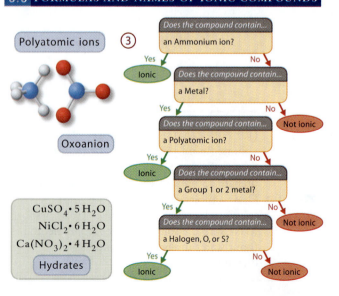

3.4 AMOUNTS OF COMPOUNDS

$$MM\ Compound = \Sigma(\#\ Atoms)(Element\ MM)$$

Element	#	MM (g/mol)	Sum (g/mol)
N	2	14.007	28.014
H ④	4	1.008	4.032
O	3	15.999	47.997

NH_4NO_3 MM = 80.043 g/mol

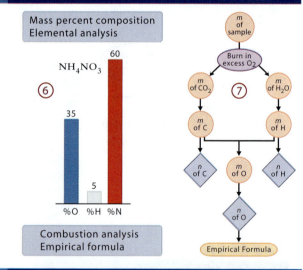

⑤ Moles of substance —— Multiply by —→ Molar Ratio —→ Moles of each element
←—— Divide by ——

3.5 DETERMINING CHEMICAL FORMULAS

Mass percent composition
Elemental analysis

⑥ NH_4NO_3

Combustion analysis
Empirical formula

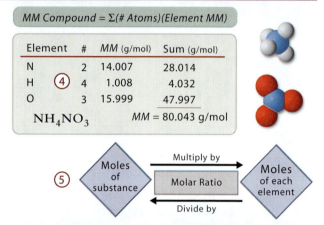

⑦

Empirical Formula

3.6 AQUEOUS SOLUTIONS

Solvent
Solute
Aqueous

⑧   Polyatomic ions remain intact.

⑨ $$M = \frac{n}{V} \qquad M_i V_i = M_f V_f$$

Concentration
Molarity (M)
Dilution

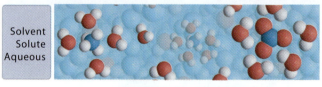

Learning Exercises

3.1 List the various ways of representing molecules. Describe how they differ from one another.

3.2 Explain how to determine whether a chemical compound is ionic or shares electrons.

3.3 Write a paragraph that defines molar mass, mole, and molarity and explains the differences among them.

3.4 Diagram the process for converting from the mass of a compound of a known chemical formula to the number of atoms of one of its constituent elements. Include all necessary equations and conversion factors.

3.5 Explain in words the reasoning used to deduce an empirical formula from combustion analysis of a compound containing C, H, and O.

3.6 Describe what a solution of magnesium nitrate looks like to a molecular-sized observer.

3.7 Update your list of "memory bank" equations.

3.8 List all the terms in Chapter 3 that are new to you. Using your own words, write a one-sentence definition of each. Consult the glossary if you need help.

 Problems   <u>ilw</u> = interactive learning ware problem. Visit the website at www.wiley.com/college/olmsted

Representing Molecules

3.1 Write chemical formulas for the molecules whose ball-and-stick models follow:

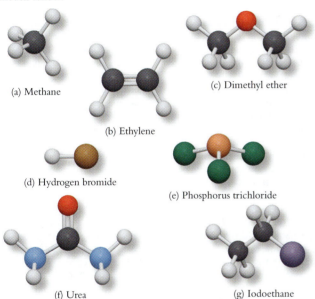

(a) Methane

(b) Ethylene

(c) Dimethyl ether

(d) Hydrogen bromide

(e) Phosphorus trichloride

(f) Urea

(g) Iodoethane

3.2 Write chemical formulas for the molecules whose ball-and-stick models follow:

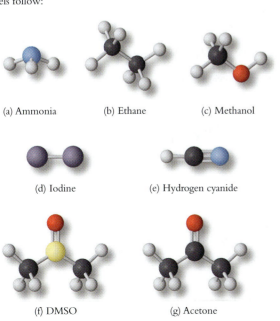

(a) Ammonia   (b) Ethane   (c) Methanol

(d) Iodine   (e) Hydrogen cyanide

(f) DMSO   (g) Acetone

3.3 Write structural formulas for the molecules in Problem 3.1.

3.4 Write structural formulas for the molecules in Problem 3.2.

3.5 Convert the following structural formulas into line structures:

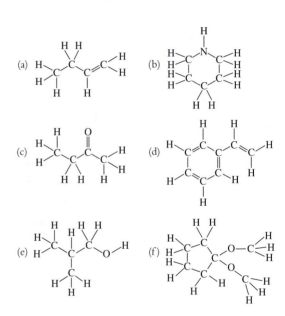

3.6 Convert the following structural formulas into line structures:

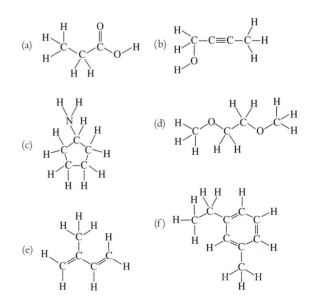

3.7 Convert the following line structures into structural formulas:

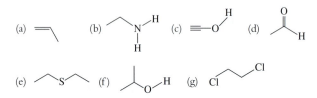

(a)   (b)   (c)

(d)   (e)   (f)

(g)   (h)   (i)

3.8 Convert the following line structures into structural formulas:

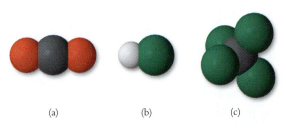

(a)   (b)   (c)   (d)

(e)   (f)   (g)

Naming Chemical Compounds

3.9 Write chemical formulas and names for the compounds whose space-filling models follow:

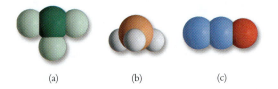

(a)   (b)   (c)

3.10 Write chemical formulas and names for the compounds whose space-filling models follow:

(a)   (b)   (c)

3.11 Write chemical formulas and names for the compounds whose ball-and-stick models follow:

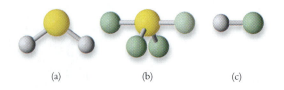

(a)   (b)   (c)

3.12 Write chemical formulas and names for the compounds whose ball-and-stick models follow:

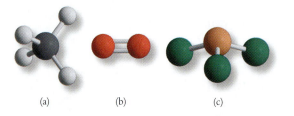

(a)   (b)   (c)

3.13 Write chemical formulas for these compounds: (a) methane; (b) hydrogen iodide; (c) calcium hydride; (d) phosphorus trichloride; (e) dinitrogen pentoxide; (f) sulfur hexafluoride; and (g) boron trifluoride.

3.14 Write chemical formulas for these compounds: (a) ammonia; (b) hydrogen sulfide; (c) 2-chloropropane; (d) silicon dioxide; (e) molecular nitrogen; (f) xenon tetrafluoride; and (g) bromine pentafluoride.

3.15 Name the following compounds: (a) S_2Cl_2; (b) IF_7; (c) HBr; (d) N_2O_3; (e) SiC; and (f) CH_3OH.

3.16 Name the following compounds: (a) XeF_2; (b) $GeCl_4$; (c) N_2F_4; (d) LiH; (e) SeO_2; and (f) CH_3CH_2OH.

Formulas and Names of Ionic Compounds

3.17 Which of the following compounds are ionic? Write the formula of each compound. (a) hydrogen fluoride; (b) calcium fluoride; (c) aluminum sulfate; (d) ammonium sulfide; (e) sulfur dioxide; and (f) carbon tetrachloride.

3.18 Which of the following compounds are ionic? Write the formula of each compound. (a) manganese(II) acetate; (b) sodium hypochlorite; (c) silicon tetrachloride; (d) lithium periodate; (e) magnesium bromide; and (f) hydrogen selenide.

3.19 Which of the following compounds are ionic? Name each compound. (a) CH_2Cl_2; (b) CO_2; (c) CaO; (d) K_2CO_3; (e) PBr_3; (f) HBr; and (g) Na_2HPO_4.

3.20 Which of the following compounds are ionic? Name each compound. (a) $(NH_4)_2SO_4$; (b) KBr; (c) SF_6; (d) H_2S; (e) Na_2S; (f) NH_3; and (g) C_2H_6.

3.21 Write chemical formulas for these compounds: (a) sodium sulfate; (b) potassium sulfide; (c) potassium dihydrogen phosphate; (d) cobalt(II) fluoride tetrahydrate; (e) lead(IV) oxide; (f) sodium hydrogen carbonate; and (g) lithium perbromate.

3.22 Write chemical formulas for these compounds: (a) potassium chlorate; (b) ammonium hydrogen carbonate; (c) iron(II) phosphate; (d) copper(II) nitrate hexahydrate; (e) aluminum chloride; (f) cadmium(II) chloride; and (g) potassium oxide.

3.23 Name the following compounds: (a) $CaCl_2 \cdot 6H_2O$; (b) $Fe(NH_4)_2(SO_4)_2$; (c) K_2CO_3; (d) $SnCl_2 \cdot 2H_2O$; (e) NaClO; (f) Ag_2SO_4; (g) $CuSO_4$; (h) KH_2PO_4; (i) $NaNO_3$; (j) $CaSO_3$; and (k) $KMnO_4$.

3.24 Name the following compounds: (a) $K_2Cr_2O_7$; (b) $NaNO_2$; (c) $Mg_3(PO_4)_2$; (d) $CrCl_3$; (e) V_2O_3; (f) $KHSO_4$; (g) CsBr; (h) $In(NO_3)_3 \cdot 5H_2O$; (i) $Al(ClO_4)_3$; (j) $SnCl_4$; and (k) $TaCl_5$.

Amounts of Compounds

3.25 Calculate the molar mass of each of the following substances: (a) carbon tetrachloride; (b) potassium sulfide; (c) O_3 (ozone); (d) lithium bromide; (e) GaAs (a semiconductor); and (f) silver nitrate.

3.26 Calculate the molar mass of each of the following compounds: (a) ammonium carbonate; (b) N_2O (laughing gas); (c) calcium carbonate; (d) NH_3 (ammonia); (e) sodium sulfate; and (f) C_4H_{10} (butane).

3.27 Determine the molecular formula and calculate the molar mass of each of the following essential amino acids:

(a) Tyrosine

(b) Tryptophan

(c) Glutamic acid

(d) Lysine

3.28 Determine the molecular formula and calculate the molar mass of each of the following B vitamins:

(a) Biotin

(b) Nicotinamide

(c) Pyridoxamine

(d) Pantothenic acid

3.29 Calculate the number of molecules present in 5.86 mg of each of the following compounds: (a) methane (CH_4); (b) phosphorus trichloride; (c) dimethyl ether (C_2H_6O); and (d) uranium hexafluoride.

3.30 The U.S. Recommended Daily Allowances (RDAs) of several vitamins follow. In each case, calculate how many molecules are in the U.S. RDA. (a) 60. mg vitamin C, $C_6H_8O_6$; (b) 400. mg folic acid, $C_{19}H_{19}N_7O_6$; (c) 1.5 mg vitamin A, $C_{20}H_{30}O$; and (d) 1.70 mg vitamin B_2 (riboflavin), $C_{17}H_{20}N_4O_6$.

3.31 Calculate the mass of the following: (a) 3.75×10^5 molecules of methane; (b) 2.5×10^9 molecules of adrenaline, $C_9H_{13}NO_3$; and (c) one molecule of chlorophyll, $C_{55}H_{72}MgN_4O_5$.

3.32 Calculate the mass of the following: (a) 1.0×10^{15} molecules of ozone; (b) 5.000×10^3 molecules of cholesterol, $C_{27}H_{46}O$; and (c) one molecule of vitamin B_{12}, $C_{63}H_{88}CoN_{14}O_{14}P$.

ilw 3.33 One carrot may contain 0.75 mg of vitamin A ($C_{20}H_{30}O$). How many moles of vitamin A is this? How many molecules? How many hydrogen atoms are in 0.75 mg of vitamin A? What mass of hydrogen is this?

3.34 A particular oral contraceptive contains 0.035 mg ethynyl estradiol in each pill. The formula of this compound is $C_{20}H_{24}O_2$. How many moles of ethynyl estradiol are there in one pill? How many molecules is this? How many carbon atoms are in a 0.035-mg sample of ethynyl estradiol? What mass of carbon is this?

Determining Chemical Formulas

3.35 Portland cement contains CaO, SiO_2, Al_2O_3, and Fe_2O_3. Calculate the mass percent composition of each compound.

3.36 Small amounts of the following compounds are added to glass to give it color: CaF_2 (milky white), MnO_2 (violet), CoO (blue), Cu_2O (green). Calculate the mass percent composition of each additive.

3.37 Nicotine is an addictive compound found in tobacco leaves. Elemental analysis of nicotine gives these data: C: 74.0%, H: 8.65%, N: 17.35%. What is the empirical formula of nicotine? The molar mass of nicotine is 162 g/mol. What is the molecular formula of nicotine?

3.38 Tooth enamel is composed largely of hydroxyapatite, which has the following mass percent composition: O: 41.41%, P: 18.50%, H: 0.20%, Ca: 39.89%. What is the empirical formula of hydroxyapatite? The molar mass of hydroxyapatite is 1004 g/mol. What is the molecular formula of hydroxyapatite?

3.39 Cinnabar is an ore of mercury known to contain only Hg and S. When a 0.350-g sample of cinnabar is heated in oxygen, the ore decomposes completely, giving 0.302 g of pure Hg metal. Find the empirical formula of cinnabar.

3.40 When galena, an ore of lead that contains only Pb and S, is heated in oxygen, the ore decomposes to produce pure lead. In one such process, 7.85 g of galena gave 6.80 g of lead. Find the empirical formula of galena.

3.41 A chemical compound was found to contain only Fe, C, and H. When 5.00 g of this compound was completely burned in O_2, 11.8 g of CO_2 and 2.42 g of H_2O were produced. Find the percentage by mass of each element in the original compound and its empirical formula.

3.42 A 1.45-g sample of a compound containing only C, H, and O was burned completely in excess oxygen. It yielded 1.56 g of H_2O and 3.83 g of CO_2. Find the percentage by mass of each element in the original compound and its empirical formula.

ilw 3.43 Combustion analysis of 0.60 g of an unknown organic compound that contained only C, H, and O gave 1.466 g of carbon dioxide and 0.60 g of water in a combustion analysis. Mass spectral analysis showed that the compound had a molar mass around 220 g/mol. Determine the empirical formula and molecular formula.

3.44 Combustion analysis of 0.326 g of a compound containing only C, H, and O produces 0.7160 g CO_2 and 0.3909 g H_2O. Mass spectral analysis showed that the compound had a molar mass around 120 g/mol. Determine the empirical formula and molecular formula.

Aqueous Solutions

3.45 (a) Calculate the molarities of the ionic species in 1.50×10^2 mL of aqueous solution that contains 4.68 g of magnesium chloride. (b) Draw a molecular picture that shows a portion of this solution, showing the relative proportions of the ions present.

3.46 A chemist places 3.25 g of sodium carbonate in a 250.-mL volumetric flask and fills it to the mark with water. (a) Calculate the molarities of the major ionic species. (b) Draw a molecular picture that shows a portion of this solution, making sure the portion is electrically neutral.

ilw 3.47 A student prepares a solution by dissolving 4.75 g of solid KOH in enough water to make 275 mL of solution. (a) Calculate the molarities of the major ionic species present. (b) Calculate the molarities of the major ionic species present if 25.00 mL of this solution is added to a 100-mL volumetric flask, and water is added to the mark. (c) Draw molecular pictures of portions of the solutions in (a) and (b), showing how they differ.

3.48 A mildly antiseptic mouthwash can be prepared by dissolving sodium chloride in water. (a) Calculate the molarities of the ionic species present in 0.150 L of solution containing 27.0 g of sodium chloride. (b) Calculate the new molarities if 50.0 mL of this solution is diluted with water to give 450. mL of a new solution. (c) Draw molecular pictures of portions of the solutions described in (a) and (b), showing how they differ.

3.49 Concentrated hydrochloric acid is 12.1 M. What volume of this solution should be used to prepare 0.500 L of 0.125 M HCl?

3.50 Concentrated ammonia is 14.8 M. What volume of this solution should be used to prepare 1.25 L of 0.500 M NH_3?

3.51 Calculate the concentrations of the major ionic species present in each of the following solutions: (a) 4.55 g Na_2CO_3 in 245 mL of solution; (b) 27.45 mg of NH_4Cl in 1.55×10^{-2} L of solution; and (c) 1.85 kg potassium sulfate in 5.75×10^3 L of solution.

3.52 Calculate the concentrations of the major ionic species present in each of the following solutions: (a) 1.54 g sodium hydrogen carbonate in 75.0 mL of solution; (b) 1.44 mg $FeCl_3$ in 2.75 mL of solution; and (c) 8.75 kg KNO_3 in 235 L of solution.

Additional Paired Problems

3.53 In everyday life, we encounter chemicals with unsystematic names. What is the name a chemist would use for each of the following substances? (a) dry ice (CO_2); (b) saltpeter (KNO_3); (c) salt (NaCl); (d) baking soda ($NaHCO_3$); (e) soda ash (Na_2CO_3); (f) lye (NaOH); (g) lime (CaO); and (h) milk of magnesia ($Mg(OH)_2$).

3.54 A mineral is a chemical compound found in the Earth's crust. What are the chemical names of the following minerals? (a) TiO_2 (rutile); (b) PbS (galena); (c) Al_2O_3 (bauxite); (d) $CaCO_3$ (limestone); (e) $BaSO_4$ (barite); (f) $Mg(OH)_2$ (brucite); (g) HgS (cinnabar); and (h) Sb_2S_3 (stibnite).

3.55 Calculate the mass percentages of all of the elements in each of the following minerals that are found in meteorites: (a) fayalite: Fe_2SiO_4; (b) albite: $NaAlSi_3O_8$; (c) kaolinite: $Al_2Si_2O_5(OH)_4$; and (d) serpentine silicate: $MgSi_4O_{10}(OH)_8$.

3.56 Calculate the mass percentages of all of the elements in the following semiprecious minerals: (a) lapis lazuli, $Na_4Al_3Si_3O_{12}Cl$; (b) garnet, $Mg_3Al_2(SiO_4)_3$; (c) turquoise, $CuAl_6(PO_4)_4(OH)_8 \cdot 4H_2O$; and (d) jade, $Ca_2Mg_5Si_8O_{22}F_2$.

3.57 Four commonly used fertilizers are NH_4NO_3 (ammonium nitrate), $(NH_4)_2SO_4$ (ammonium sulfate), $(NH_2)_2CO$ (urea), and $(NH_4)_2HPO_4$ (ammonium hydrogen phosphate). How many kilograms of each of these would be required to provide 1.00 kg of nitrogen?

3.58 What mass percentage of carbon is contained in each of the following fuels? (a) propane (C_3H_8); (b) octane (C_8H_{18}); (c) ethanol (C_2H_5OH); and (d) methane (CH_4).

3.59 The line structures of three different plant growth hormones are given below. For each one, write the chemical formula for the compound and calculate its molar mass.

(a)

Abscisic acid
(inhibits germination)

(b)

Indole acetic acid
(promotes growing shoots)

(c)

Zeatin
(promotes root growth)

3.60 The following molecules are known for their characteristic fragrances. For each one, convert the line structure into a complete structural formula and calculate its molar mass.

(a)

Benzaldehyde
(cherry)

(b)

Methylbutyl acetate
(banana)

(c)

Jasmone
(jasmine)

(d)

Limonene
(lemon)

(e)

Vanillin
(vanilla)

3.61 What species are present in solution when the following compounds dissolve in water? (a) ammonium sulfate; (b) carbon dioxide; (c) sodium fluoride; (d) potassium carbonate; (e) sodium hydrogen sulfate; and (f) chlorine.

3.62 What species are present in solution when the following compounds dissolve in water? (a) sodium dichromate; (b) copper(II) chloride; (c) barium hydroxide; (d) methanol; (e) sodium hydrogen carbonate; and (f) iron(III) nitrate.

3.63 The following pairs of substances are quite different despite having similar names. Write correct formulas for each. (a) sodium nitrite and sodium nitrate; (b) potassium carbonate and potassium hydrogen carbonate; (c) iron(II) oxide and iron(III) oxide; and (d) iodine and iodide ion.

3.64 The following pairs of substances are quite different despite having similar names. Write correct formulas for each. (a) sodium chloride and sodium hypochlorite; (b) nitrogen oxide and nitrogen dioxide; (c) potassium chlorate and potassium perchlorate; and (d) ammonia and ammonium ion.

3.65 Aluminum sulfate is $Al_2(SO_4)_3$. (a) Compute its molar mass. (b) Compute the number of moles contained in 25.0 g of this compound. (c) Determine its percent composition. (d) Determine the mass of this compound that contains 1.00 mol of O.

3.66 Nickel sulfate hexahydrate is $NiSO_4 \cdot 6H_2O$. (a) Compute its molar mass. (b) Compute the number of moles contained in 25.0 g of this compound. (c) Determine its percent composition. (d) Determine the mass of this compound that contains 1.00 mol of O.

3.67 Sea water at 25 °C contains 8.3 mg/L of oxygen. What molarity is this?

3.68 The oceans contain 1.3 mg of fluoride ions per liter. What molarity is this?

3.69 Name the following compounds: (a) NH_4Cl; (b) XeF_4; (c) Fe_2O_3; (d) SO_2; and (e) $KClO_4$.

3.70 Name the following compounds: (a) $KClO_3$; (b) $KClO_2$; (c) KClO; (d) KCl; and (e) Na_2HPO_4.

3.71 Heart disease causes 37% of the deaths in the United States. However, the death rate from heart disease has dropped significantly in recent years, partly because of the development of new drugs for heart therapy by chemists working in the pharmaceutical industry. One of these drugs is verapamil, used for the treatment of arrhythmia, angina, and hypertension. A tablet contains 120.0 mg of verapamil. Determine the following quantities: (a) the molar mass of verapamil; (b) the number of moles of verapamil in one tablet; and (c) the number of nitrogen atoms in one tablet.

Verapamil
$C_{27}H_{38}O_4N_2$

3.72 Penicillin was discovered in 1928 by Alexander Fleming, who was a bacteriologist at the University of London. This molecule was originally isolated from a mold that contaminated some of Fleming's experiments. Penicillin destroys bacterial cells without harming animal cells, so it has been used as an antibiotic and has saved countless lives. The molecular formula of penicillin G is $C_{16}H_{18}N_2O_4S$. (a) Calculate the molar mass of penicillin G. (b) How many moles of penicillin are in a 50.0-mg sample? (c) How many carbon atoms are in a 50.0-mg sample? (d) What is the mass of sulfur in a 75.0-mg sample?

Penicillin G
$C_{16}H_{18}N_2O_4S$

3.73 Concentrated aqueous sulfuric acid is 80.% by weight H_2SO_4 and has a density of 1.75 g/mL. (a) Calculate the molarity of the concentrated solution. (b) How many mL of the concentrated solution are needed to make 2.50 L of 0.65 M sulfuric acid?

3.74 Aqueous solutions of potassium permanganate are often used in general chemistry laboratories. A laboratory instructor weighed out 474.1 g of $KMnO_4$, added the solid to a 2.00-L volumetric flask, and filled to the mark with water. A 50.00-mL sample was removed from the volumetric flask and mixed with enough water to make 1.500 L of solution. What was the final molarity of the $KMnO_4$ solution?

3.75 For each of the following salt solutions, calculate the amounts in moles and the numbers of each type of ion: (a) 55.6 mL of 1.25 M magnesium nitrate; (b) 25.7 mL of a solution that contains 9.03×10^{21} formula units of potassium sulfate; and (c) 3.55 mL of a solution that contains 13.5 g/L of sodium phosphate.

3.76 For each of the following salt solutions, calculate the amounts in moles and the numbers of each type of ion: (a) 2.87 L of 0.0550 M lithium carbonate; (b) 325 mL of a solution that contains 1.02×10^{19} formula units of sodium hydrogen sulfate; and (c) 2.55 mL of a solution that contains 263 mg/L of sodium oxalate.

3.77 The density of water is 1.0 g/mL. Calculate the molarity of water molecules in pure water.

3.78 The density of pure liquid sulfuric acid (H_2SO_4) is 1.84 g/mL. Calculate the molarity of this liquid.

3.79 A 3.75-g sample of compound that contains sulfur and fluorine contains 2.93 g of fluorine. The molar mass is less than 200 g/mol. Calculate the percent composition of the compound and determine its molecular formula.

3.80 A 6.82-g sample of a compound that contains silicon and chlorine is 16.53% by mass silicon. The molar mass is less than 200 g/mol. Determine the molecular formula of the compound.

3.81 For each of the following pairs, identify which has the larger of the indicated amounts: (a) mass: 0.35 mol water or 0.25 mol hydrogen peroxide; (b) atoms: 0.88 mol carbon monoxide or 0.68 mol carbon dioxide; and (c) moles: 2.5 mL of methanol ($\rho = 0.791$ g/mL) or 3.5 mL of ethanol ($\rho = 0.789$ g/mL).

3.82 Consider 35 mL of carbon tetrachloride ($\rho = 1.594$ g/mL) and 32 mL of sulfur dichloride ($\rho = 1.62$ g/mL). (a) Which sample has the larger mass? (b) Which sample contains more moles of molecules? (c) Which sample contains more atoms?

More Challenging Problems

3.83 Answer the following questions about Sevin, whose chemical formula is $C_{12}H_{11}NO_2$: (a) How many moles of carbon atoms are there in 8.3 g of Sevin? (b) How many grams of oxygen are there in 4.5 g of Sevin? (c) The label on a 75-mL bottle of garden insecticide states that the solution contains 0.010% Sevin and 99.99% inert ingredients. How many moles and how many molecules of Sevin are in the bottle? (Assume that the density of the solution is 1.00 g/mL.) (d) The instructions for the bottle of insecticide from part (c) say to dilute the insecticide by adding 1.0 mL of the solution to 1.0 gallon of water. If you spray 15 gallons of the diluted insecticide mixture on your rose garden, how many moles of Sevin are dispersed?

3.84 Solution A is prepared by dissolving 90.0 g of Na_3PO_4 in enough water to make 1.5 L of solution. Solution B is 2.5 L of 0.705 M Na_2SO_4. (a) What is the molar concentration of Na_3PO_4 in Solution A? (b) How many milliliters of Solution A will give 2.50 g of Na_3PO_4? (c) A 50.0-mL sample of Solution B is mixed with a 75.00 mL sample of Solution A. Calculate the concentration of Na^+ ions in the final solution.

3.85 The seventh-ranked industrial chemical in U.S. production is phosphoric acid, whose chemical formula is H_3PO_4. One method of manufacture starts with elemental phosphorus, which is burned in air; the product of this reaction then reacts with water to give the final product. In 1995, the United States manufactured 2.619×10^8 pounds of phosphoric acid. How many moles is this? If 15% of this material was made by burning elemental phosphorus, how many moles and how many kilograms of phosphorus were consumed?

3.86 Police officers confiscate a packet of white powder that they believe contains heroin. Purification by a forensic chemist yields a 38.70-mg sample for combustion analysis. This sample gives 97.46 mg CO_2 and 20.81 mg H_2O. A second sample is analyzed for its nitrogen content, which is 3.8%. Show by calculations whether these data are consistent with the formula for heroin, $C_{21}H_{22}NO_5$.

3.87 When an unknown compound is burned completely in O_2, 1.23 g of CO_2 and 97.02 g of H_2O are recovered. What additional information is needed before the molecular formula of the unknown compound can be determined?

3.88 A sulfur-containing ore of copper releases sulfur dioxide when heated in air. A 5.26-g sample of the ore releases 2.12 g of SO_2 on heating. Assuming that the ore contains only copper and sulfur, what is the empirical formula?

3.89 Turquoise has the formula $CuAl_6(PO_4)_4(OH)_8 \cdot 4H_2O$. Answer the following questions about turquoise: (a) How many grams of aluminum are in a 7.25-g sample of turquoise? (b) How many phosphate ions are in a sample of turquoise that contains 5.50×10^{-3} g of oxygen? (c) What is the charge of the copper ion in turquoise?

3.90 A chemist needs a solution that contains aluminum ions, sodium ions, and sulfate ions. Around the lab she finds a large volume of 0.355 M sodium sulfate solution and a bottle of solid $Al_2(SO_4)_3 \cdot 18H_2O$. The chemist places 250. mL of the sodium sulfate solution and 5.13 g of aluminum sulfate in a 500-mL volumetric flask. The flask is filled to the mark with water. Determine the molarity of aluminum ions, sodium ions, and sulfate ions in the solution.

3.91 Adenosine triphosphate (ATP) is used to generate chemical energy in plant and animal cells. The molecular formula of ATP is $C_{10}H_{16}N_5O_{13}P_3$. Answer the following about ATP: (a) What is its percent composition? (b) How many P atoms are in 1.75 μg of ATP? (c) If a cell consumes 3.0 pmol of ATP, what mass has it consumed? (d) What mass of ATP contains the same number of atoms of H as the number of N atoms in 37.5 mg of ATP?

3.92 The waters of the oceans contain many elements in trace amounts. Rubidium, for example, is present at the level of 2.2 nM. How many ions of rubidium are present in 1.00 L of seawater? How many liters would have to be processed to recover 1.00 kg of rubidium, assuming the recovery process was 100% efficient?

3.93 Vitamin B_{12} is a large molecule called cobalamin. There is one atom of cobalt in each molecule of vitamin B_{12}, and the mass percentage of cobalt is 4.34%. Calculate the molar mass of cobalamin.

3.94 An element E forms a compound whose formula is ECl_5. Elemental analysis shows that the compound is 85.13% by mass chlorine. Identify the element E.

3.95 Answer the following questions about Malathion, whose chemical formula is $C_{10}H_{19}O_6PS_2$: (a) How many sulfur atoms are there in 6.5 g of Malathion? (b) How many grams of oxygen are there in 17.8 g of Malathion? (c) A 200.0-mL container of Malathion is 50% by mass Malathion and 50% inert ingredients (mostly water). What is the molarity of Malathion in this solution? (Assume a density of 1.00 g/mL.) (d) When dispensed using a hose-end sprayer, this solution of Malathion becomes diluted by a factor of 10^4. If the sprayer delivers 25 mL of this diluted solution onto the leaves of a fruit tree, how many moles of Malathion have been applied?

3.96 Pure acetic acid, CH_3CO_2H, is 17.4 M. A laboratory worker measured out 100.0 mL of pure acetic acid and added enough water to make 500.0 mL of solution. A 75.0-mL portion of the acetic acid solution was then mixed with enough water to make 1.50 L of dilute solution. What was the final molarity of acetic acid in the dilute solution?

3.97 A worker in a biological laboratory needed a solution that was 0.30 M in sodium acetate ($NaCH_3CO_2$) and 0.15 M in acetic acid (CH_3CO_2H). On hand were stock solutions of 5.0 M sodium acetate and 5.0 M acetic acid. Describe how the worker prepared 1.5 L of the desired solution.

3.98 The thyroid gland produces hormones that help regulate body temperature, metabolic rate, reproduction, the synthesis of red blod cells, and more. Iodine must be present in the diet for these thyroid hormones to be produced. Iodine deficiency leads to sluggishness and weight gain, and can cause severe problems in the development of a fetus. One thyroid hormone is thyroxine, whose chemical

Thyroxine
$C_{15}H_{11}I_4NO_4$

formula is $C_{15}H_{11}I_4NO_4$. How many milligrams of thyroxine can be produced from 210 mg of iodine atoms, the amount a typical 180-lb adult consumes per day? How many molecules is this?

3.99 Anhydrous copper(II) sulfate is $CuSO_4$, a pale blue solid. Hydrated copper(II) sulfate is a deep blue solid containing five water molecules for every $CuSO_4$ unit: $CuSO_4 \cdot 5H_2O$. How many moles of $CuSO_4$ are there in 125 g of anhydrous copper(II) sulfate? What mass of hydrated copper(II) sulfate contains the same number of moles?

3.100 A solution is prepared by combining 755 mL of 0.725 M calcium chloride with 325 mL of 1.15 M lithium chloride. Assuming volume is conserved upon mixing, calculate the concentrations of all species in the final solution.

Group Study Problems

3.101 In the United States, the federal government regulates the amounts of pollutants that industry emits. Emissions in excess of allotted levels are taxed. As an incentive to clean up emissions, companies that emit less than their allotted amounts are allowed to sell their unused portions (so-called emission credits) to other companies. Several years ago, the Tennessee Valley Authority (TVA), which operates 11 coal-fired electricity plants, purchased emission credits from Wisconsin Power and Light. The TVA bought "pollution rights" for the emission of 1.00×10^4 tons of sulfur dioxide per year at a price of $275 per ton. How much will it cost to emit 1 mol of SO_2? How many molecules can be emitted for $1.00?

3.102 Disodium aurothiomalate ($Na_2C_4H_3O_4SAu$) has the trade name Myochrysine and is used to treat arthritis. (a) What is its molar mass? (b) A solution is prepared by dissolving 0.25 g in 10.0 mL of water. The drug is diluted by adding 2.00 mL of the stock solution to a 10.00-mL volumetric flask and filling to the mark with water. Patients receive 0.40 mL of the dilute Myochrysine solution by intramuscular injection. Calculate the mass of Myochrysine administered in each injection. (c) How many moles of Myochrysine is this? (d) During treatment, the concentration of Myochrysine is 50.0 mg per 1.00 mL of serum. How many atoms of gold are in a 5.0-mL sample of serum?

3.103 A biologist needed to prepare a solution for growing a cell culture. For the culture, he needed to make 1.5 L of solution with these concentrations: $[KH_2PO_4] = 0.55$ M and $[K_2HPO_4] = 0.85$ M. He made two stock solutions. Solution A: 545 g of KH_2PO_4 was dissolved in enough water to give a volume of 2.0 L. Solution B: 1045 g of K_2HPO_4 was dissolved in enough water to give a total volume of

3.0 L. Describe how to make the solution needed for the cell culture using stock solutions A and B.

3.104 A sample of a component of petroleum was subjected to combustion analysis. An empty vial of mass 2.7534 g was filled with the sample, after which vial plus sample had a mass of 2.8954 g. The sample was burned in a combustion train whose CO_2 trap had a mass of 54.4375 g and whose H_2O trap had a mass of 47.8845 g. At the end of the analysis, the CO_2 trap had a new mass of 54.9140 g and the H_2O trap had a new mass of 47.9961 g. Determine the empirical formula of this component of petroleum.

3.105 Describe how the results of the analysis in Problem 3.104 would be affected by the following errors in operation of the apparatus. Which measurements would change, and in what direction? How would that affect the calculation of the empirical formula? (a) A limited supply of oxygen was provided, rather than excess oxygen; (b) some of the H_2O leaked through the H_2O trap but was collected in the CO_2 trap; and (c) the H_2O trap functioned properly, but some of the CO_2 leaked through the CO_2 trap.

3.106 Aluminum sulfate is used in the manufacture of paper and in the water purification industry. In the solid state, aluminum sulfate is a hydrate. The formula is $Al_2(SO_4)_3 \cdot 18H_2O$. (a) How many grams of sulfur are there in 0.570 moles of solid aluminum sulfate? (b) How many water molecules are there in a 5.1-g sample of solid aluminum sulfate? (c) How many moles of sulfate ions are there in a sample of solid aluminum sulfate that contains 12.5 moles of oxygen atoms? (d) An aqueous solution of aluminum sulfate contains 1.25% by mass aluminum and has a density of 1.05 g/mL. What is the molarity of aluminum ions in the solution?

Answers to Section Exercises

3.1.1 (a) $C_{18}H_{36}O_2$; (b) $SiCl_4$; and (c) $C_2Cl_3F_3$

3.1.2

Molecular formula	Structural formula	Line structure
C_3H_7Cl		
C_7H_9N		
$C_2H_4O_2$		

3.1.3

Isoprene		C_5H_8
Styrene		C_8H_8
Methyl methacrylate		$C_5H_8O_2$

3.2.1 ClF, XeO_3, HBr, $SiCl_4$, SO_2, and H_2O_2

3.2.2 Chlorine trifluoride, hydrogen selenide, chlorine dioxide, antimony trichloride, phosphorus pentachloride, dinitrogen pentoxide, dinitrogen tetrachloride, and ammonia

3.2.3

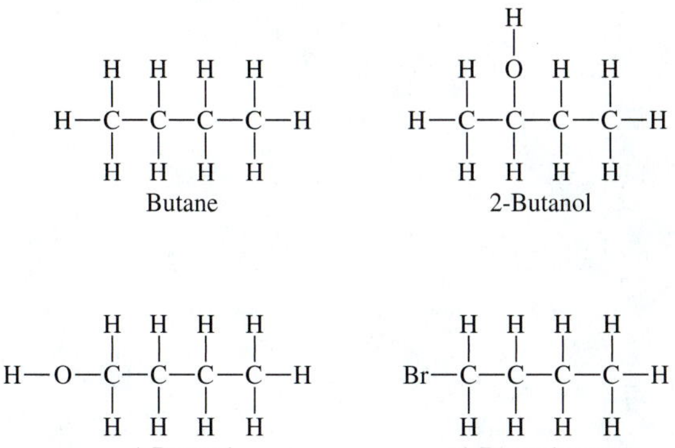

H H H H
| | | |
H—C—C—C—C—H
| | | |
H H H H

Butane

H O H H
| | | |
H—C—C—C—C—H
| | | |
H H H H

2-Butanol

H H H H
| | | |
H—O—C—C—C—C—H
| | | |
H H H H

1-Butanol

H H H H
| | | |
Br—C—C—C—C—H
| | | |
H H H H

1-Bromobutane

3.3.1 Sulfur dichloride, calcium chloride, lead(II) chloride, sodium nitrate, manganese(IV) oxide, zirconium(IV) chloride, sodium hydride, and sodium iodate

3.3.2 Al_2O_3, $K_2Cr_2O_7$, $Pb(NO_3)_2$, NO_2, Na_2SO_4, IF_5, $Mn(CH_3CO_2)_2$, and NaClO

3.3.3 (a) Sulfur hexafluoride; (b) ammonium nitrate; (c) ammonia; (d) nitrogen trichloride; (e) carbon tetrabromide; and (f) hydrogen iodide

3.4.1 (a) 129.91 g/mol; (b) 184.04 g/mol; (c) 142.98 g/mol; (d) 261.34 g/mol; and (e) 504.82 g/mol

3.4.2 (a) 155.16 g/mol; (b) 121.16 g/mol; (c) 132.12 g/mol

3.4.3 (a) 9.86×10^{-5} mol; (b) 5.94×10^{19} molecules; (c) 1.78×10^{20} atoms; (d) 2.97×10^{20} atoms; (e) 20.5 mg; and (f) 2.79×10^{21} atoms

3.5.1 C, 40.00%; H, 6.71%; O, 53.29%

3.5.2 Fe_3O_4

3.5.3 $C_6H_8O_6$

3.6.1 $[NH_4^+] = 0.31$ M, $[NO_3^-] = 0.31$ M

3.6.2 1.2 mL

3.6.3

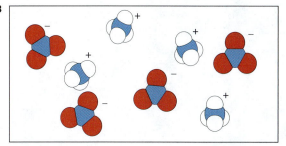

Answers to Extra Practice Exercises

3.1 (a) C_2H_2; (b) Al_2O_3; and (c) SO_2

3.2

C_3H_8 (CH$_3$)$_2$O C_2H_5OH C_2H_6 C_2H_4 C_2H_2

3.3 Structural formula:

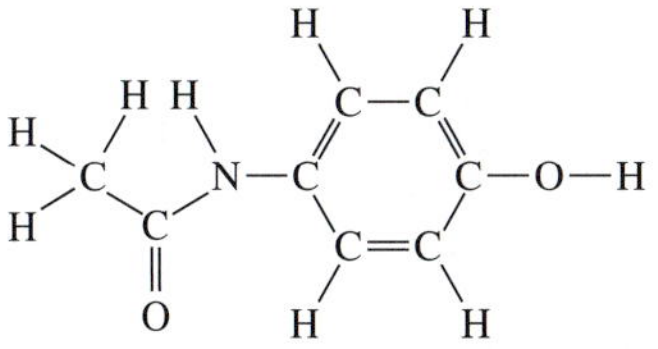

Chemical formula: $C_8H_9NO_2$

3.4 (a) carbon tetrabromide; (b) sulfur trioxide; and (c) xenon difluoride

3.5 KCN and $MgHPO_4$ are ionic; $TiCl_4$ and H_2SO_4 are not ionic.

3.6 K_2S, $Ba(ClO_4)_2$, and $Ca_3(PO_4)_2$

3.7 Aluminum nitrate, iron(III) bromide, and copper(I) carbonate

3.8 Nitrogen dioxide, iron(II) oxide, and nickel(II) sulfate hexahydrate

3.9 $BaCl_2\cdot 2H_2O$, $Cr(OH)_3$, and SO_3

3.10 330.36 g/mol

3.11 375.13 g/mol

3.12 13.9 moles and 8.36×10^{24} atoms

3.13 No. Sevin contains 7.0% by mass nitrogen, not 8.5%.

3.14 Empirical formula is CH, molecular formula is C_6H_6.

3.15 CH_4

3.16 Empirical formula is C_3H_6O, molecular formula is $C_6H_{12}O_2$.

3.17 3.20 g

3.18 $[Ni^{2+}] = 0.255$ M; $[Cl^-] = 0.510$ M

3.19 11.8 mL

3.20 8.00×10^{-2} M

Chemical Reactions and Stoichiometry

The Dye Industry

Humans have used dyes to create color since the dawn of history. Until the mid-nineteenth century, all dyes were of natural origin. Many came from plants, such as indigo, a dark blue dye that was extracted from the leaves of a native East Indian plant. In 1856, the young English chemist William Perkin stumbled upon the first synthetic dye. Perkin was trying to synthesize quinine, a valuable antimalaria drug. None of his experiments met with success. As he was about to discard the residue from yet another failed reaction, Perkin noticed that it was colored with a purple tinge. He washed the residue with hot alcohol and obtained a purple solution from which strikingly beautiful purple crystals precipitated. Perkin had no idea what the substance was or what reactions had created it, but he immediately saw its potential as a new dye.

Perkin turned his full attention to the purple substance, and within six months he was producing the new dye commercially. The dye was named mauveine, and the color is known as mauve. Today we know the structure of mauveine, which appears in the inset. We also know that Perkin was very lucky, because mauveine was produced from an impurity in the chemicals that he used.

Mauveine was an immediate hit with French textile dyers, allowing Perkin to retire wealthy at the age of 35, having founded the first industry based on a synthetic chemical. Within 20 years of

Learning Resources

 KEY CONCEPTS

 CRITICAL THINKING

 SKILLS TO MASTER

 ADDITIONAL HELP
www.wiley.com/college/olmsted
- TUTORIALS
- ANIMATIONS

Perkin's discovery, the German chemist Johann Baeyer developed a method to synthesize indigo in the laboratory. Soon thereafter the German dye industry blossomed. Thus, both the English and the German chemical industries began with syntheses of dyes.

Contemporary synthetic chemists know detailed information about molecular structures and use sophisticated computer programs to simulate a synthesis before trying it in the laboratory. Nevertheless, designing a chemical synthesis requires creativity and a thorough understanding of molecular structure and reactivity. No matter how complex, every chemical synthesis is built on the principles and concepts of general chemistry. One such principle is that quantitative relationships connect the amounts of materials consumed and the amounts of products formed in a chemical reaction. We can use these relationships to calculate the amounts of materials needed to make a desired amount of product and to analyze the efficiency of a chemical synthesis. The quantitative description of chemical reactions is the focus of Chapter 4.

4.1 WRITING CHEMICAL EQUATIONS

Balanced Equations

In chemical reactions, *the amount of each element is always conserved.* This is consistent with the statements of Dalton's atomic theory. In addition, *the total amount of electrical charge is always conserved.* This is the law of conservation of charge. A **balanced chemical equation** describes a chemical reaction in which the amounts of all elements and of electrical charge are conserved. We refer to the species that take part in a chemical reaction as **reagents.** **Reactants** are reagents that are consumed in a reaction, while **products** are reagents that are produced in a reaction. In addition to providing a list of reactants and products of the reaction, a balanced chemical equation tells us the relative amounts of all the reagents.

As an example, consider the industrial synthesis of ammonia (NH_3). Ammonia is made by the Haber process, a single chemical reaction between molecules of hydrogen (H_2) and nitrogen (N_2). Although it is simple, this synthesis has immense industrial importance. The United States produces more than 16 billion kilograms of ammonia annually.

In this synthesis, the reactants are N_2 and H_2, which react to produce NH_3. Here is the balanced chemical equation:

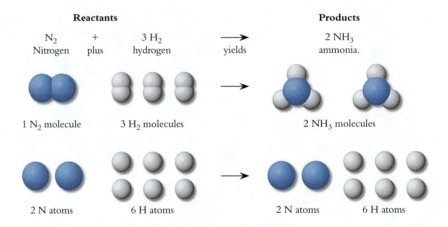

All balanced chemical equations have the following features:

1. The reactants appear on the left and the products appear on the right. The arrow joining them indicates the direction of reaction.

2. An integer precedes the formula of each substance. These numbers are the **stoichiometric coefficients.** When no number appears (as for N_2 in this equation), the stoichiometric coefficient is 1.

3. The stoichiometric coefficients in a chemical equation are the smallest integers that give a balanced equation.

4. Charge is conserved. In this equation, all participants are neutral species, so charge is conserved regardless of the stoichiometric coefficients.

In Section 4.5 we show how to apply charge conservation to reactions that include ions.

Stoichiometric coefficients describe the relative numbers of molecules involved in the reaction. In any actual reaction, immense numbers of molecules are involved, but the relative numbers are always related through the stoichiometric coefficients. Further, these coefficients describe both the relative numbers of *molecules* and the relative numbers of *moles* involved in the reaction. For example, the Haber reaction always involves immense numbers of molecules, but the equation describing the synthesis of ammonia tells us the following:

a. Each molecule of nitrogen reacts with three molecules of hydrogen to give two molecules of ammonia.

b. Each mole of molecular nitrogen reacts with three moles of molecular hydrogen to give two moles of ammonia.

Keep in mind that the key feature of balanced chemical equations is the conservation law:

The number of atoms of each element is conserved in any chemical reaction.   **KEY CONCEPT**

Balancing Equations

The stoichiometric coefficients in a balanced chemical equation must be chosen so that the atoms of each element are conserved. Many chemical equations can be balanced by inspection. Balancing by inspection means changing stoichiometric coefficients until the number of atoms of each element is the same on each side of the arrow. Usually, we can tell what changes need to be made by looking closely at the reaction and matching the numbers of atoms of each element on both sides of the equation. Consider the following example.

Propane, which is used as a fuel for gas barbecues, reacts with molecular oxygen to form carbon dioxide and water:

$$C_3H_8 + O_2 \longrightarrow CO_2 + H_2O$$

To determine if this equation is balanced, make a list of the elements and numbers of atoms on each side:

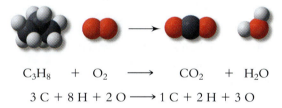

$$C_3H_8 \quad + \quad O_2 \quad \longrightarrow \quad CO_2 \quad + \quad H_2O$$

$$3\,C + 8\,H + 2\,O \longrightarrow 1\,C + 2\,H + 3\,O$$

The equation is not balanced, because there are too many carbon and hydrogen atoms on the left and too many oxygen atoms on the right. We need to change the numbers of molecules by changing stoichiometric coefficients until the numbers of atoms of each element are equal.

It is easiest to balance a chemical equation one element at a time, starting with the elements that appear in only one substance on each side. Notice that all of the carbon atoms in propane end up in carbon dioxide molecules, and all of propane's hydrogen atoms appear in water molecules. This feature allows us to balance carbon and hydrogen easily.

①

To take care of the three carbon atoms per propane molecule, we need three molecules of CO_2. Thus, the carbon atoms are balanced by changing the stoichiometric coefficient of CO_2 from 1 to 3. In this reaction the ratio of CO_2 to propane is 3:1. Similarly, we need four molecules of water for the eight hydrogen atoms in one molecule of propane. Using this information, we modify the equation as follows:

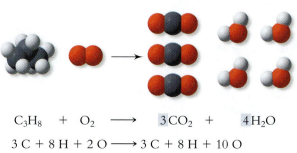

$$C_3H_8 \quad + \quad O_2 \quad \longrightarrow \quad 3\,CO_2 \quad + \quad 4\,H_2O$$

$$3\,C + 8\,H + 2\,O \longrightarrow 3\,C + 8\,H + 10\,O$$

The situation is looking better because atoms of carbon and hydrogen are now conserved. However, the equation still is not balanced because there are too few oxygen atoms on the left side. To balance oxygen, we must give O_2 a stoichiometric coefficient of 5:

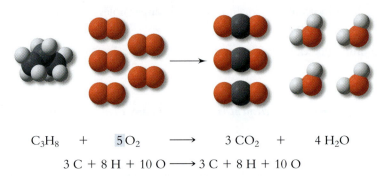

$$C_3H_8 \ + \ 5\,O_2 \ \longrightarrow \ 3\,CO_2 \ + \ 4\,H_2O$$
$$3\,C + 8\,H + 10\,O \longrightarrow 3\,C + 8\,H + 10\,O$$

Now the equation is balanced. Combustion of one molecule of propane gas produces three molecules of CO_2 and four molecules of H_2O and consumes five molecules of O_2. Also, combustion of one mole of propane produces three moles of CO_2 and four moles of H_2O and consumes five moles of O_2. Example 4-1 shows how to balance another reaction.

Example 4-1	Ammonium nitrate, a colorless ionic solid used as a fertilizer, explodes when it is

Balancing Chemical Reactions

Ammonium nitrate, a colorless ionic solid used as a fertilizer, explodes when it is heated above 300 °C. The products are three gases: molecular nitrogen, molecular oxygen, and steam (water vapor). Write a balanced equation for the explosion of ammonium nitrate.

Strategy: The description in the problem tells us what happens to the starting material: NH_4NO_3 breaks apart into molecules of N_2, O_2, and H_2O. An unbalanced form of the equation can be written from this description. Then we must balance each element in turn by inspection.

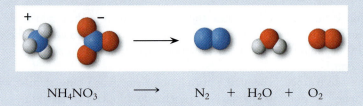

$$NH_4NO_3 \ \longrightarrow \ N_2 \ + \ H_2O \ + \ O_2$$

Solution: Count atoms of each element to see if the elements are in balance:

$$2\,N + 4\,H + 3\,O \longrightarrow 2\,N + 2\,H + 3\,O$$

Focus first on the elements that appear in only one reactant and one product, nitrogen and hydrogen in this case. Nitrogen is already balanced. To balance the H atoms, change the stoichiometric coefficient of water from 1 to 2:

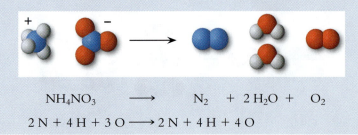

$$NH_4NO_3 \ \longrightarrow \ N_2 \ + \ 2\,H_2O + \ O_2$$
$$2\,N + 4\,H + 3\,O \longrightarrow 2\,N + 4\,H + 4\,O$$

Now atoms of nitrogen and hydrogen are conserved, but the equation is not balanced with respect to oxygen. We cannot change the 1:1 ratio of NH_4NO_3 to N_2 and the 1:2 ratio of NH_4NO_3 to H_2O, because these ratios must be retained to keep hydrogen and nitrogen atoms conserved. Thus, we must adjust the coefficient of O_2 to balance with respect to oxygen. Numerically, we can take care of oxygen by changing the coefficient of O_2 to $\frac{1}{2}$:

$$NH_4NO_3 \longrightarrow N_2 + 2\,H_2O + \tfrac{1}{2}O_2$$

$$2\,N + 4\,H \longrightarrow 2\,N + 4\,H + 3\,O$$

This balances the equation, but it is unrealistic from a molecular perspective because there is no such thing as half a molecule. To get rid of the $\frac{1}{2}$ without unbalancing the reaction, we multiply *all* the coefficients by 2:

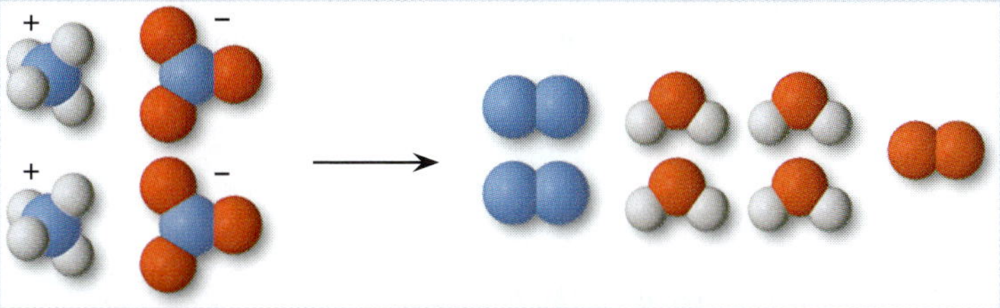

$$2\,NH_4NO_3 \quad\longrightarrow\quad 2\,N_2 + \quad 4\,H_2O \quad + \quad O_2$$

$$4\,N + 8\,H + 6\,O \longrightarrow 4\,N + 8\,H + 6\,O$$

Does the Result Make Sense?   Count the atoms of each element to be sure that the reaction is balanced and reasonable from a molecular perspective.

Some racing cars use methanol, CH_3OH, as their fuel. Balance the combustion reaction of methanol.

Example 4-1

Balancing Chemical Reactions (*continued*)

Later, we show that it can be convenient to use fractional coefficients such as $\frac{1}{2}$. When this is the case, the coefficient refers to $\frac{1}{2}$ mol, not $\frac{1}{2}$ molecule.

Extra Practice Exercise 4.1

Section Exercises

4.1.1 Although gasoline is a complex mixture of molecules, the chemical reaction that takes place in an automobile engine can be represented by combustion of one of its components, octane (C_8H_{18}). Such burning of fossil fuels releases millions of tons of carbon dioxide into the Earth's atmosphere each year. Write a balanced equation for the combustion of octane.

4.1.2 Additional reactions accompany the combustion of octane. For example, molecular nitrogen reacts with molecular oxygen in an automobile cylinder to give nitrogen oxide. After leaving the engine, nitrogen oxide reacts with atmospheric oxygen to give nitrogen dioxide, the red-brown gas seen in the air over many urban areas. Write balanced equations for these two reactions.

4.1.3 Acrylonitrile is used to make synthetic fibers such as Orlon. About 1.5 billion kg of acrylonitrile are produced each year. Balance the following chemical equation, which shows how acrylonitrile is made from propene, ammonia, and oxygen:

$$C_3H_6 + NH_3 + O_2 \longrightarrow C_3H_3N + H_2O$$

Propene

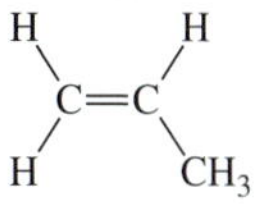

Acrylonitrile

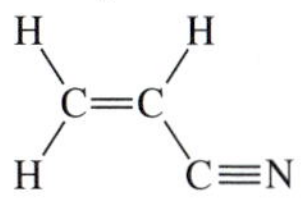

4.2 THE STOICHIOMETRY OF CHEMICAL REACTIONS

Sections 2.4 and 3.4 describe how to use the relationships among atoms, moles, and masses to answer "how much?" questions about *individual substances*. Combining these ideas with the concept of a balanced chemical equation lets us answer "how much?" questions about *chemical reactions*. The study of the amounts of materials consumed and produced in chemical reactions is called **stoichiometry.**

Stoichiometry is pronounced "stoy-key-om'-etry." It combines two Greek words, *stoicheion* ("element") and *metron* ("measure").

A chemical synthesis requires the proper amounts of starting materials for a successful outcome. Just as a cake recipe provides the amounts of ingredients needed for successful baking, a balanced chemical equation is a chemical recipe for successful synthesis. For example, how many grams of hydrogen do we need to produce 68 g of ammonia? To find out, we begin with the balanced chemical equation:

$$N_2 + 3\,H_2 \longrightarrow 2\,NH_3$$

According to the balanced equation, the synthesis of 2.0 moles of ammonia requires 3.0 moles of hydrogen and 1.0 mole of nitrogen. However, the balanced equation does not give us any direct information about the *masses* involved in the synthesis. One gram of N_2 plus three grams of H_2 *does not* make two grams of NH_3. Remember the most important lesson from Chapter 3:

 KEY CONCEPT *Calculations in chemistry are centered on the mole.*

In other words, a chemical recipe for making ammonia requires amounts in moles, not masses. Thus, to prepare 68 g of ammonia, we need to know the number of moles of ammonia in 68 g. To do that, we must divide the mass of ammonia by the molar mass of ammonia, 17.0 g/mol, to convert grams to moles:

$$n = \frac{m}{MM} = \frac{68\ \text{g NH}_3}{17.0\ \text{g/mol}} = 4.0\ \text{mol NH}_3$$

How many moles of H_2 are needed to make 4.0 moles of NH_3? According to the balanced equation, 2 mol of NH_3 requires 3 mol of H_2. The ratio of stoichiometric coefficients from a balanced equation is called the **stoichiometric ratio.** We now apply this ratio to the desired amount of NH_3:

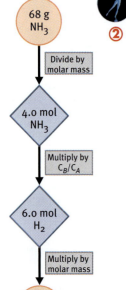

$$(4.0\ \overline{\text{mol NH}_3})\left(\frac{3\ \text{mol H}_2}{2\ \overline{\text{mol NH}_3}}\right) = 6.0\ \text{mol H}_2$$

↑ Stoichiometric ratio

All that remains is to convert from moles of H_2 to mass:

$$m = n\,(MM) = (6.0\ \text{mol H}_2)\,(2.02\ \text{g/mol}) = 12\ \text{g H}_2$$

The same procedure can be used to show that the synthesis of 68 g of NH_3 requires 56 g of N_2.

To summarize, the amounts of different reagents that participate in a chemical reaction are related through the stoichiometric coefficients in the balanced chemical equation. To convert from moles of one reagent to moles of any other reagent, multiply by the stoichiometric ratio that leads to proper cancellation of units:

$$\text{Moles}_B = \left(\frac{\text{Coefficient}_B}{\text{Coefficient}_A}\right)\text{Moles}_A \tag{4-1}$$

A flowchart illustrates the logic of the stoichiometric ratio.

Examples 4-2 and 4-3 illustrate this procedure.

Geranyl formate is used as a synthetic rose essence in cosmetics. The compound is prepared from formic acid and geraniol:

$$HCO_2H + C_{10}H_{18}O \longrightarrow C_{11}H_{18}O_2 + H_2O$$

Formic acid   Geraniol   Geranyl formate Water

A perfumery needs some geranyl formate for a batch of perfume. How many grams of geranyl formate can a chemist make from 375 g of geraniol?

Strategy: There are some exotic chemical names here, but they should not distract you from the basic principles of reaction stoichiometry. The stoichiometric coefficients state that one mole of each reactant will produce one mole of each product. A flowchart summarizes the steps used to convert the mass of geraniol into the mass of geranyl formate.

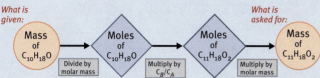

Solution: First, convert the mass of geraniol into moles by dividing by the molar mass:

$$\frac{375 \text{ g geraniol}}{154.2 \text{ g/mol}} = 2.432 \text{ mol geraniol}$$

A 1:1 stoichiometric ratio links moles of geraniol to moles of geranyl formate:

$$\text{Moles geranyl formate} = \left(\frac{1 \text{ mol geranyl formate}}{1 \text{ mol geraniol}}\right)(2.432 \text{ mol geraniol})$$

If the chemist starts the reaction with 2.432 mol of geraniol, the synthesis can produce 2.432 mol of geranyl formate. To finish the problem, multiply by the molar mass of geranyl formate to determine the mass of geranyl formate produced in the reaction. The mole units are eliminated, leaving gram units for the answer:

$$(2.432 \text{ mol geranyl formate})\left(\frac{182.3 \text{ g geranyl formate}}{1 \text{ mol geranyl formate}}\right) = 443 \text{ g geranyl formate}$$

Does the Result Make Sense? The result has three significant figures, just like the initial mass of geraniol. The answer makes sense. Geranyl formate has a slightly larger molar mass than geraniol, and we can see that the mass of the product is slightly more than the mass of the starting material.

What mass of formic acid is required to produce 443 g of geranyl formate?

Example 4-2

How Much Product Can Be Made?

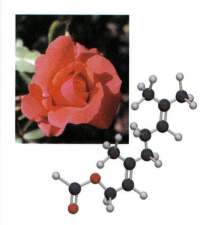

Extra Practice Exercise 4.2

Example 4-3

Amounts of Reactants and Products

Poisonous hydrogen cyanide (HCN) is an important industrial chemical. It is produced from methane (CH_4), ammonia, and molecular oxygen. The reaction also produces water. An industrial manufacturer wants to convert 175 kg of methane into HCN. How much molecular oxygen will be required for this synthesis?

Strategy: This problem looks complicated, so it is a good idea to apply the seven-step problem-solving method.

1. This is a stoichiometry problem (how much?), in which we are asked to find the mass of a reactant.

2. To visualize this kind of problem, we need a balanced chemical equation:

   Write the unbalanced form of the equation from the information given, namely methane, ammonia, and molecular oxygen react to form hydrogen cyanide and water:

$$CH_4 + NH_3 + O_2 \longrightarrow HCN + H_2O$$

Example 4-3

Amounts of Reactants and Products (*continued*)

Balance the equation by inspection, one element at a time. Carbon and nitrogen are in balance when there are equal numbers of molecules of CH_4, NH_3, and HCN. Hydrogen and oxygen, however, are not in balance. There are seven hydrogen atoms on the left but only three on the right. The coefficient for HCN cannot be changed without unbalancing C and N. Thus, to balance H, we must multiply the coefficient for H_2O by 3:

$$CH_4 + NH_3 + O_2 \longrightarrow HCN + 3\,H_2O$$

This gives three oxygen atoms on the right, which corresponds to $\frac{3}{2}$ O_2 molecules among the starting materials:

$$CH_4 + NH_3 + \tfrac{3}{2}\,O_2 \longrightarrow HCN + 3\,H_2O$$

To avoid the impossible $\frac{1}{2}$ molecule, multiply all of the stoichiometric coefficients by 2. This gives the balanced equation:

$$2\,CH_4 + 2\,NH_3 + 3\,O_2 \longrightarrow 2\,HCN + 6\,H_2O$$

Solution

3. The problem gives only one piece of data: 175 kg of CH_4 will be converted into HCN. The problem asks about another starting material, so we will need two molar masses:

$$MM_{(CH_4)} = 16.04 \text{ g/mol}, \quad MM_{(O_2)} = 32.00 \text{ g/mol}$$

4. Knowing the mass of one reagent, we are asked to find the mass of another. We need to determine molar amounts and then convert to kilograms. The calculations require two equations for interconversion of moles and mass:

$$n = \frac{m}{MM} \quad \text{and} \quad n_B = \left(\frac{\text{Coefficient}_B}{\text{Coefficient}_A}\right)n_A$$

5. After converting mass to moles and n_A to n_B, we need to convert moles back to mass: $m = nMM$

6. Begin by converting mass of methane to moles:

$$(175 \text{ kg } CH_4)\left(\frac{10^3 \text{ g}}{1 \text{ kg } CH_4}\right) = 1.75 \times 10^5 \text{ g } CH_4$$

$$\frac{1.75 \times 10^5 \text{ g } CH_4}{16.04 \text{ g/mol}} = 1.091 \times 10^4 \text{ mol } CH_4$$

Next use the ratio of coefficients to find the number of moles of O_2 that are required for the synthesis:

$$(1.091 \times 10^4 \text{ mol } CH_4)\left(\frac{3 \text{ mol } O_2}{2 \text{ mol } CH_4}\right) = 1.637 \times 10^4 \text{ mol } O_2$$

Notice that including the formula in the unit, "mol O_2" instead of "mol", for example, clarifies the cancellation of units. Finally, use the molar mass of O_2 to convert from moles back to mass:

$$(1.637 \times 10^4 \text{ mol } O_2)\left(\frac{32.00 \text{ g } O_2}{1 \text{ mol } O_2}\right)\left(\frac{1 \text{ kg}}{10^3 \text{ g}}\right) = 524 \text{ kg } O_2$$

7. Does the Result Make Sense?   The units of the result, kilograms, indicate that the calculations are in order, because the question asked for the mass of oxygen. The mass of O_2 required for the synthesis, 524 kg, is a large quantity, but it is reasonable because the reaction also involves a large quantity of methane (175 kg).

Extra Practice Exercise 4.3

In this synthesis, what mass of ammonia will be required, and what mass of HCN could be prepared?

Section Exercises

■ **4.2.1** Phosphorus trichloride is produced from the reaction of solid phosphorus and chlorine gas:

$$P_4 + 6\ Cl_2 \longrightarrow 4\ PCl_3$$

(a) What mass of phosphorus trichloride can be prepared from 75.0 g of phosphorus?
(b) What mass of chlorine will be consumed in the reaction?

■ **4.2.2** Tooth enamel consists, in part, of $Ca_5(PO_4)_3(OH)$. Tin(II) fluoride (toothpaste labels call it *stannous fluoride*) is added to some toothpastes because it exchanges F^- for OH^- to give a more decay-resistant compound, $Ca_5(PO_4)_3F$. In addition to fluoroapatite, this reaction produces tin(II) oxide and water. What mass of $Ca_5(PO_4)_3(OH)$ can be converted to $Ca_5(PO_4)_3F$ by reaction with 0.115 g of tin(II) fluoride?

■ **4.2.3** Combustion reactions require molecular oxygen. In an automobile the fuel-injection system must be adjusted to provide the right "mix" of gasoline and air. Compute the number of grams of oxygen required to react completely with 1.00 L of octane (C_8H_{18}, $\rho = 0.80$ g/mL). What masses of water and carbon dioxide are produced in this reaction? (*Hint:* Recall the equation for density in Chapter 1: $\rho = m/V$.)

4.3 YIELDS OF CHEMICAL REACTIONS

Under practical conditions, chemical reactions almost always produce smaller amounts of products than the amounts predicted by stoichiometric analysis. There are three major reasons for this.

1. Many reactions stop before reaching completion. For example, solid sodium metal reacts with oxygen gas to give sodium peroxide (Na_2O_2). As the reaction proceeds, a solid crust of the product builds up on the surface of the metal, as Figure 4-1 shows. The crust prevents oxygen from reaching the sodium metal, and the reaction stops, even though both starting materials are still present.

   Other reactions do not go to completion because they reach dynamic equilibrium. While reactant molecules continue to form product molecules, product molecules also interact to re-form reactant molecules. The Haber reaction and many precipitation reactions, described later in this chapter, are examples of reactions that reach dynamic equilibrium rather than going to completion. We treat chemical equilibria in detail in Chapters 16–18.

2. Competing reactions often consume some of the starting materials. For example, sodium metal reacts with water to produce sodium hydroxide. If a sample of oxygen is contaminated with water vapor, both O_2 and H_2O will react with the sodium metal. The more water present in the gas mixture, the less Na_2O_2 will be formed.

Figure 4-1
The reaction of sodium metal with O_2 gas produces a crust of sodium peroxide, which blocks additional reactant molecules from reaching the unreacted metal.

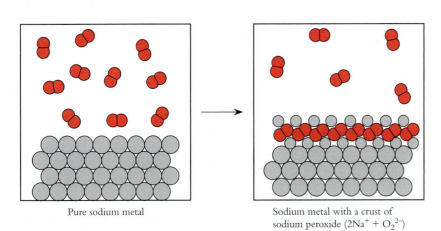

Pure sodium metal   Sodium metal with a crust of sodium peroxide ($2Na^+ + O_2^{2-}$)

3. When the product of a reaction is purified and isolated, some of it is inevitably lost during the collection process. Gases may escape while being pumped out of a reactor. Liquids adhere to glass surfaces, making it impossible to transfer every drop of a liquid product. Likewise, it is impossible to scrape every trace of a solid material from a reaction vessel.

We can compare the incomplete formation of a chemical product with the process of harvesting. When a farmer harvests a fruit crop, some of the fruit is left on the trees. Some may spill off the truck on the way to market. Some of the fruit may be unsuitable for sale. The result is that less than 100% of the potential crop eventually finds its way to market. Likewise, a synthetic chemist loses some of the potential product of a reaction at each step in the preparation.

The amount of a product obtained from a reaction is often reported as a **yield.** The amount of product predicted by stoichiometry is the **theoretical yield,** whereas the amount actually obtained is the **actual yield.** The **percent yield** is the percentage of the theoretical amount that is actually obtained:

$$\text{Percent yield} = 100\%\left(\frac{\text{Actual amount}}{\text{Theoretical amount}}\right) \tag{4-2}$$

When we calculate a percent yield, the amounts can be expressed in either moles or mass, provided both the actual and theoretical amounts are in the same units. Example 4-4 shows how to use Equation 4-2.

Example 4-4

Calculating Percent Yield

Extra Practice Exercise 4.4

According to Example 4-2, it is possible to make 443 g of geranyl formate from 375 g of geraniol. A chemist making geranyl formate for the preparation of a perfume uses 375 g of starting material and collects 417 g of purified product. What is the percent yield of this synthesis?

Strategy: To calculate a percent yield, we need to compare the actual amount obtained in the synthesis with the theoretical amount that could be produced, using Equation 4-2.

Solution: The theoretical yield of geranyl formate is 443 g. This is the amount of product that would result from complete conversion of geraniol into geranyl formate. The actual yield, 417 g, is the quantity of the desired product that the chemist collects. The percent yield is their ratio multiplied by 100%:

$$\text{Percent yield} = (100\%)\left(\frac{417\text{ g}}{443\text{ g}}\right) = 94.1\%$$

Does the Result Make Sense? Yields can never exceed the theoretical amount, so a percent yield always must be less than 100%. A percent yield between 90% and 100% is quite reasonable.

If the reaction described in Example 4-3 produces 195 kg of HCN from 175 kg of methane reactant, what is the percent yield?

If the percent yield of a reaction is already known, we can calculate how much of a product to expect from a synthesis that uses a known amount of starting material. For example, the Haber synthesis of ammonia stops when 13% of the starting materials have formed products. Knowing this, how much ammonia could an industrial producer expect to make from 2.0 metric tons of molecular hydrogen? First, calculate the theoretical yield:

$$\text{Theoretical yield} = \frac{(2.0\text{ ton H}_2)(10^6\text{ g/ton})}{(2.016\text{ g/mol H}_2)}\left(\frac{2\text{ mol NH}_3}{3\text{ mol H}_2}\right) = 6.61 \times 10^5\text{ mol NH}_3$$

Rearrange Equation 4-2 and substitute known values:

$$\text{Actual yield} = (\text{Theoretical yield})\left(\frac{\%\ \text{yield}}{100\%}\right)$$

$$\text{Actual yield} = (6.61 \times 10^5\ \text{mol})\left(\frac{13\%}{100\%}\right) = 8.60 \times 10^4\ \text{mol NH}_3$$

Finally, convert back to mass:

$$\text{Actual mass} = (8.60 \times 10^4\ \text{mol})(17.03\ \text{g/mol})\ (10^{-6}\ \text{ton/g}) = 1.5\ \text{metric ton}$$

Chemists often do yield calculations in reverse to determine the masses of reactants needed to obtain a desired amount of product. If the percent yield is known from previous experiments, we can calculate the amount of starting material needed to make a specific amount of product. Example 4-5 shows how to do this.

Example 4-5

Calculating Reactant Mass from Yield

The industrial production of hydrogen cyanide is described in Example 4-3. If the yield of this synthesis is 97.5%, how many kilograms of methane should be used to produce 1.50×10^5 kg of HCN?

Strategy: This is a two-stage problem that requires a yield calculation and a conversion among molar amounts of products and reactants. This is typical of yield calculations. Use the percent yield to determine the theoretical yield, and then use the stoichiometric ratios to calculate masses of starting materials. A flowchart summarizes the stepwise strategy.

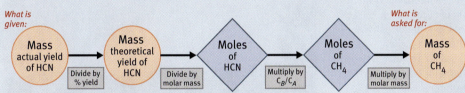

Solution: First, rearrange the equation for percent yield to find the theoretical yield that will give an actual yield of 1.50×10^5 kg:

$$\text{Theoretical yield} = (100\%)\left(\frac{\text{Actual yield}}{\text{Percent yield}}\right)$$

$$\text{Theoretical yield} = (100\%)\left(\frac{1.50 \times 10^5\ \text{kg}}{97.5\%}\right) = 1.538 \times 10^5\ \text{kg HCN}$$

$$(1.538 \times 10^5\ \text{kg HCN})(10^3\ \text{g/kg}) = 1.538 \times 10^8\ \text{g HCN}$$

$$\frac{1.538 \times 10^8\ \text{g HCN}}{27.03\ \text{g/mol}} = 5.690 \times 10^6\ \text{mol HCN}$$

Next, to convert this mass to moles apply the stoichiometric ratio, and multiply by the molar mass of methane:

$$(5.690 \times 10^6\ \text{mol HCN})\left(\frac{2\ \text{mol CH}_4}{2\ \text{mol HCN}}\right) = 5.690 \times 10^6\ \text{mol CH}_4$$

$$(5.690 \times 10^6\ \text{mol CH}_4)(16.04\ \text{g/mol})(10^{-3}\ \text{kg/g}) = 9.13 \times 10^4\ \text{kg CH}_4$$

Does the Result Make Sense?   Compare the mass of the reactant with the mass of the product. Notice that both masses are the same order of magnitude, 10^5 kg. This makes sense when the yield is close to 100%, the stoichiometric ratio is 1:1, and the molar masses are comparable.

What mass of NH_3 is required for this synthesis?

Extra Practice Exercise 4.5

Box 4-1   Chemistry and Life: Feeding the World

Fertilizers are immensely important to humanity. Agriculture requires fertilizers because growing plants remove various chemical elements from the soil. In a fully contained ecosystem, decaying organic matter replenishes the soil, but the elements contained in crops that are harvested and shipped elsewhere are not replenished. Thus, intensive agriculture inevitably depletes the soil of essential elements, which must be replaced by fertilization.

Lacking means of fertilization, primitive people farmed a plot of land until the nutrients in the soil were exhausted. Then they moved to new ground, where they burned the natural vegetation and began farming again. This slash-and-burn method of growing food is still used extensively in South America, where this human activity is destroying vast tracts of rain forest.

Fertilization using animal products has been practiced since ancient times. Animal manure returns nutrients to the soil, replenishing elements that are depleted as crops are grown and harvested. It is likely that the use of animal fertilizers quickly followed the domestication of goats, sheep, and cattle.

Growing plants require a variety of chemical elements. Nitrogen, phosphorus, and potassium are required in the greatest amounts, but plants also need trace amounts of calcium, copper, iron, zinc, and other elements. By far the most substantial need is for nitrogen. The Earth's atmosphere is 80% molecular nitrogen, but plants cannot use N_2. Instead, most plants absorb nitrogen from the soil in the form of nitrate ions.

The production of nitrogen fertilizers is a major activity of the chemical industry. Every year, the top 15 chemicals in industrial production in the United States include several nitrogen-containing compounds whose major use is in fertilizers. Molecular nitrogen serves as the primary source of nitrogen for chemical production. Gaseous ammonia (NH_3), which is synthesized from N_2 and H_2, can be injected directly into the ground, where it dissolves in moisture in the soil and serves as a fertilizer. Ammonia is more widely used in reactions with acids to produce other fertilizers: Ammonia and nitric acid produce ammonium nitrate (NH_4NO_3), while ammonia and sulfuric acid produce ammonium sulfate. These chemicals and urea, $(H_2N)_2CO$, are among the most important fertilizers.

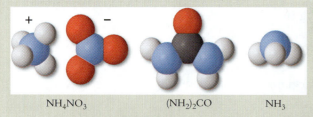

NH$_4$NO$_3$ (NH$_2$)$_2$CO   NH$_3$

Crop yields can rise dramatically with the use of commercial fertilizers. For example, in 1800 an acre of land in the United States produced about 25 bushels of corn. In the 1980s the same acre of land produced 110 bushels. Worldwide, approximately 4 billion acres of land are used to grow food crops. This would probably be enough land to feed the world's population if the entire acreage could be fertilized commercially. It has been estimated that world crop production would increase by about 50% if about $40 per acre were spent to apply modern chemical fertilizers. However, it would cost about $160 trillion

to produce this additional food. Furthermore, the use of chemical fertilizers can lead to the contamination of streams, lakes, and bays with phosphates and nitrates.

Contamination of the world's oceans by fertilizer runoff is an emerging environmental problem. Ironically, the very fertilizers that make the land bloom can cause the oceans to wither. When water containing excess nitrogen drains into the seas, the nitrogen triggers rapid growth of phytoplankton. These microscopic algae eventually die, and as they rot, they consume the oxygen dissolved in seawater. Without oxygen, animal life of all types, including shellfish, crustaceans, and fish, can no longer survive. Oxygen-starved areas have existed in the Gulf of Mexico and Chesapeake Bay for some time, but in 2003 the United Nations Environmental Program reported that the known dead zones doubled in the last decade of the twentieth century, to nearly 150 dead zones around the world. Among these zones are portions of the Baltic and Adriatic seas.

In addition, the production of fertilizers from molecular nitrogen is very energy-intensive. In the United States alone, hundreds of millions of barrels of oil are used every year to produce fertilizers. Finding ways to produce noncontaminating and inexpensive fertilizers without huge energy investment is a major challenge to today's chemical industry.

The left side of this field has been fertilized.

As world supplies of petroleum are depleted, and as the Earth's population steadily increases, society will be forced to develop more efficient ways to make fertilizer. Genetic engineering offers a promising solution. There is a remarkable bacterium that lives in the roots of leguminous plants such as soybeans, peas, and peanuts. This organism can convert molecular nitrogen into ammonia. The plant and the bacterium have a symbiotic relationship. Ammonia produced by the bacterium nourishes the plant, and the plant provides other nutrients to the bacterium. Exactly how the bacterium converts nitrogen to ammonia is the subject of vigorous research. Scientists hope eventually to transfer the bacterial gene responsible for the conversion of nitrogen to ammonia into the cells of nonleguminous plants.

Suppose a chemical reaction gives a 50% yield. Is that a good yield or a poor yield? The answer depends on the circumstances. Many valuable chemicals are manufactured in processes that require several steps. The overall yield of a multistep synthesis is the product of the yields of the individual steps. For example, suppose that a reaction requires two steps. If the first reaction has a yield of 85% and the second a yield of 65%, the overall yield is $(100\%)(0.85)(0.65) = 55\%$. Notice that the overall yield is lower than the yield of the least efficient step. For this reason, the more steps there are in a synthesis, the more important the yields of the individual reactions become. For example, the synthesis of a growth hormone might require a sequence of 16 different chemical reactions. At an average yield per step of 95%, the overall yield of this synthesis is 44%, but at an average yield of 75% the overall yield of the synthesis falls to 1.0%!

Many-step reactions that have only moderate yields at each step are wasteful and expensive. For this reason, chemists devote much time, effort, and ingenuity in devising reaction sequences and conditions that improve the yields of chemical syntheses. Our most important industrial chemicals are produced in billion-pound quantities on an annual basis. Here, improving the synthesis yield by even a few tenths of a percent can save a company millions of dollars each year. A good example is fertilizer production, which we describe in our Chemistry and Life Box.

Section Exercises

4.3.1 When heated, potassium chlorate decomposes to potassium chloride and gaseous molecular oxygen:

$$2 \, KClO_3 \longrightarrow 2 \, KCl + 3 \, O_2$$

What is the theoretical yield of oxygen when 5.00 g of potassium chlorate decomposes? Calculate the percent yield if a 5.00-g sample gives 1.84 g O_2 on decomposition. Give possible reasons why the actual yield is less than the theoretical yield.

4.3.2 Sulfuric acid is made from elemental sulfur in a three-step industrial process. How many tons of sulfur are required to produce 15 tons of H_2SO_4 if the three steps have yields of 94%, 92.5%, and 97%?

4.3.3 Aspirin is produced by treating salicylic acid with acetic anhydride:

$$2 \, C_7H_6O_3 + C_4H_6O_3 \longrightarrow 2 \, C_9H_8O_4 + H_2O$$

Salicylic acid   Acetic   Aspirin
             anhydride

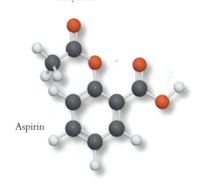

Aspirin

If the synthesis has an 87% yield, what mass of salicylic acid should be used to produce 75 g of aspirin?

4.4 THE LIMITING REACTANT

The reactions encountered so far in this chapter were set up so that all the starting materials were used up at the same time. In other words, there were no leftovers at the end of the reaction. Often, chemical reactions are run with an excess of one or more starting materials. This means that one reactant will "run out" before the others. The reactant that runs out is called the **limiting reactant** because it limits the amount of product that can be made. The other starting materials are said to be *in excess*.

We illustrate the limiting reactant concept with the bicycle analogy shown in Figure 4-2. It takes two wheels, a frame, and a chain to build a bicycle. Imagine that a bicycle shop has the following parts on hand: 8 wheels, 5 frames, and 5 chains. How many bicy-

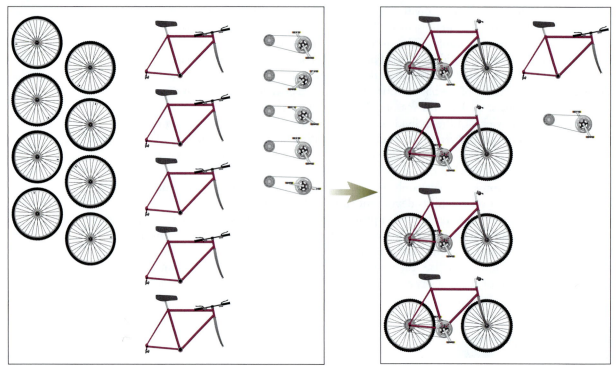

Figure 4-2
The number of bicycles that can be assembled is limited by whichever part runs out first. In the inventory shown in this figure, wheels are that part.

cles can be built with these parts? There are enough frames and chains to make five bicycles. However, because each bicycle needs two wheels, eight wheels are enough to make only four bicycles. Even though there are more wheels around the shop than any of the other parts, the shop will run out of wheels first. After four bicycles are built, the shop inventory will be 4 bicycles, 0 wheels, 1 frame, and 1 chain. There are not enough parts left to make a fifth bicycle because there is a shortage of two wheels. In this bicycle shop, wheels are the limiting reactant.

Chemical reactions must be analyzed in this same way. Instead of comparing numbers of bicycle parts, we compare the number of moles of each starting material on hand with the number of moles required to make the desired product. Consider the Haber synthesis again. Figure 4–3 shows a limiting reactant situation for this synthesis. If we start with six molecules of H_2 and four molecules of N_2, the six molecules of H_2 will combine with two molecules of N_2 to make four molecules of NH_3. When all the H_2 molecules are consumed, two molecules of N_2 will be left over. Here, H_2 is the limiting reactant. This molecular analysis applies equally well to numbers of moles. In molar terms, if six moles of H_2 and four moles of N_2 react completely, all the H_2 is consumed, four moles of NH_3 are produced, and two moles of N_2 are left over.

Figure 4-3
A molecular view of a limiting reactant situation for the ammonia synthesis. To make 4 molecules of NH_3 requires 2 molecules of N_2 and 6 molecules of H_2. If we start with 4 molecules of N_2 and 6 molecules of H_2, H_2 is the limiting reactant.

To solve a quantitative limiting reactant problem, we identify the limiting reactant by working with amounts in moles and the stoichiometric coefficients from the balanced equation. For the ammonia synthesis, if we start with 84.0 g of molecular nitrogen and 24.2 g of molecular hydrogen, what mass of ammonia can be prepared? First, convert from masses to moles:

$$\frac{84.0 \text{ g } N_2}{28.01 \text{ g/mol}} = 3.00 \text{ mol } N_2 \qquad \frac{24.2 \text{ g } H_2}{2.016 \text{ g/mol}} = 12.0 \text{ mol } H_2$$

Animation

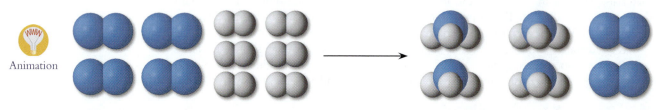

According to the balanced equation, the stoichiometric ratio of nitrogen to ammonia is 1:2.

$$N_2 + 3\,H_2 \longrightarrow 2\,NH_3$$

There are 3.00 moles of N_2, which is enough to make 6.00 moles of NH_3.

$$(3.00 \text{ mol } N_2)\left(\frac{2 \text{ mol } NH_3}{1 \text{ mol } N_2}\right) = 6.00 \text{ mol } NH_3$$

The stoichiometric ratio of hydrogen to ammonia is 3:2. There are 12.0 moles of H_2, which is enough to make 8.00 moles of NH_3.

$$(12.0 \text{ mol } H_2)\left(\frac{2 \text{ mol } NH_3}{3 \text{ mol } H_2}\right) = 8.00 \text{ mol } NH_3$$

There is enough H_2 on hand to make 8.00 mol of NH_3, but only enough N_2 to make 6.00 mol of NH_3. Nitrogen is the limiting reactant, and hydrogen is present in excess. Once six moles of ammonia have been produced, there will still be three moles of hydrogen left over. However, no more ammonia can be produced because there is no nitrogen left for the reaction.

We can identify the limiting reactant by dividing each amount in moles by the stoichiometric coefficient for that reactant. To see how this works, rearrange the ratios for H_2 and N_2:

$$\left(\frac{12.0 \text{ mol } H_2}{3 \text{ mol } H_2}\right)(2 \text{ mol } NH_3) = 4.00\,(2 \text{ mol } NH_3) = 8.00 \text{ mol } NH_3$$

$$\left(\frac{6.00 \text{ mol } N_2}{2 \text{ mol } N_2}\right)(2 \text{ mol } NH_3) = 3.00\,(2 \text{ mol } NH_3) = 6.00 \text{ mol } NH_3$$

The reactant with the smaller ratio of amount to coefficient can produce a smaller amount of product. That reactant (N_2 in this case) is limiting because it runs out first. This leads to the following generalization:

The limiting reactant is the one whose number of moles divided by its stoichiometric coefficient has the smallest value.

KEY CONCEPT

Tables of Amounts

A table of amounts is a convenient way to organize the data and summarize the calculations of a stoichiometry problem. Such a table helps to identify the limiting reactant, shows how much product will form during the reaction, and indicates how much of the excess reactant will be left over. A table of amounts has the balanced chemical equation at the top. The table has one column for each substance involved in the reaction and three rows listing amounts. The first row lists the starting amounts for all the substances. The second row shows the changes that occur during the reaction, and the last row lists the amounts present at the end of the reaction. Here is a table of amounts for the ammonia example:

Reaction	N_2	+	$3\,H_2$	$\longrightarrow$	$2\,NH_3$
Starting amount (mol)	3.00		12.0		0
Change in amount (mol)	−3.00		−9.00		+6.00
Final amount (mol)	0		3.0		6.00

A table of amounts has three important features, all involving the *changes in amounts:*

1. *Changes are negative for reactants and positive for products.* This is because the amounts of reactants decrease during the reaction and the amounts of products increase during the reaction.

2. *All changes are related by stoichiometry.* Each ratio of changes in amount equals the ratio of stoichiometric coefficients in the balanced equation. In the example above, the changes in amounts for H_2 and N_2 are in the ratio 3:1, the same as the ratio for the coefficients of H_2 and N_2 in the balanced equation.

3. In each column, *the starting amount plus the change in amount gives the final amount.* Each entry in the final amount row of the table is the sum of the previous entries in the same column.

The table of amounts for the ammonia example illustrates these features:

1. The changes are negative for N_2 and H_2, which are consumed, and positive for NH_3, which is produced.

2. The changes in amount for H_2 and N_2 are in the ratio of 3:1, the same as the stoichiometric ratio for the coefficients of H_2 and N_2 in the balanced equation.

3. The change of -9.00 mol for H_2 leaves $(12.0 - 9.00) = 3.0$ mol as the final amount.

You can use these three key features to construct a table of amounts for any reaction.

When a reaction goes to completion, the limiting reactant is consumed completely, so its final amount must be zero. Two facts help identify the limiting reactant. First, final amounts can never be negative. If a negative amount appears in the bottom row of the table, a mistake has been made in identifying the limiting reactant. Second, the limiting reactant is always the one whose ratio of moles to stoichiometric coefficient is smallest. The smallest ratio indicates the limiting reactant, in this case N_2. Thus, dividing starting amounts by coefficients is a quick way to identify the limiting reactant. Example 4-6 applies the features of limiting reactants and amounts tables.

Example 4-6

Limiting Reactant Calculation

Nitric acid, a leading industrial chemical, is used in the production of fertilizers and explosives. One step in the industrial production of nitric acid is the reaction of ammonia with molecular oxygen to form nitrogen oxide:

$$4\,NH_3 + 5\,O_2 \longrightarrow 4\,NO + 6\,H_2O$$

In a study of this reaction, a chemist mixed 125 g of ammonia with 256 g of oxygen and allowed them to react to completion. What masses of NO and H_2O were produced, and what mass of which reactant was left over?

Strategy: We apply the seven-step problem-solving method, but without numbering the steps. We are asked about masses of reagents in a chemical reaction. From the fact that starting amounts of both reactants are given, we identify this as a limiting reactant problem (one reactant will be consumed before the other is consumed). A table of amounts will help organize the information, and we need the balanced chemical equation, which is given in the statement of the problem.

Solution: The data available are starting masses of the reactants, 125 g of NH_3 and 256 g of O_2. We also need the molar masses, which are easily calculated:

$$MM_{NH_3} = 17.03 \text{ g/mol}; MM_{O_2} = 32.00 \text{ g/mol}$$

$$MM_{NO} = 30.01 \text{ g/mol}; MM_{H_2O} = 18.02 \text{ g/mol}$$

Example 4-6

Limiting Reactant Calculation (*continued*)

As always in stoichiometric calculations, we need to work with moles:

$$n_{NH_3} = \frac{125 \text{ g}}{17.03 \text{ g/mol}} = 7.34 \text{ mol} \qquad n_{O_2} = \frac{256 \text{ g}}{32.00 \text{ g/mol}} = 8.00 \text{ mol}$$

Determine the limiting reactant by dividing each starting amount by the stoichiometric coefficient for that reactant:

$$\frac{7.34 \text{ mol NH}_3}{4 \text{ mol NH}_3} = 1.84 \qquad \frac{8.00 \text{ mol O}_2}{5 \text{ mol O}_2} = 1.60$$

The ratio for O_2 is smaller, so oxygen is the limiting reactant.

Now set up and complete a table of amounts. Because O_2 is limiting, it will be completely consumed, so its change is -8.00 mol. Knowing this, we calculate all other changes in amounts using ratios of stoichiometric coefficients:

$$\text{Change in NH}_3 = (8.00 \text{ mol O}_2)\left(\frac{4 \text{ mol NH}_3}{5 \text{ mol O}_2}\right) = 6.40 \text{ mol NH}_3 \text{ used}$$

$$\text{Change in NO} = (8.00 \text{ mol O}_2)\left(\frac{4 \text{ mol NO}}{5 \text{ mol O}_2}\right) = 6.40 \text{ mol NO produced}$$

$$\text{Change in H}_2\text{O} = (8.00 \text{ mol O}_2)\left(\frac{6 \text{ mol H}_2\text{O}}{5 \text{ mol O}_2}\right) = 9.60 \text{ mol H}_2\text{O produced}$$

Remember that changes in amounts are negative for reactants and positive for products. Here is the complete amounts table:

Reaction	4 NH$_3$ +	5 O$_2$ $\longrightarrow$	4 NO +	6 H$_2$O
Starting amount (mol)	7.34	8.00	0	0
Change in amount (mol)	-6.40	-8.00	$+6.40$	$+9.60$
Final amount (mol)	0.94	0	6.40	9.60

All that remains is to convert back to masses using $m = n\,MM$. We leave it to you to verify that the masses present at the end of the reaction are 16 g NH_3, 192 g NO, and 173 g H_2O.

Does the Result Make Sense? Are these masses reasonable? First, it is reasonable that none of the final amounts is negative. Second, we can check to see if total mass is conserved. We started with 125 g ammonia and 256 g oxygen, a total of 381 g. The masses present at the end total 381 g, too, indicating that the calculations are consistent, and the results are reasonable.

Construct an amounts table for this reaction if the starting amounts are 85.15 g of NH_3 and 224 g of O_2.

You should verify that if ammonia is selected as the limiting reactant, the amounts table will have a negative amount of oxygen at the end of the reaction. A negative final amount is impossible.

A reaction that is carried out under limiting reactant conditions nevertheless has a yield that generally will be less than 100%. The reasons why reactions yield less than the theoretical amounts, given in Section 4.3, apply to all reactions. When a reaction operates under limiting reactant conditions, we calculate the theoretical yield assuming that the limiting reactant will be completely consumed. We then determine the percent yield as described in Section 4.3. Example 4-7 shows how to do this.

Example 4-7

Yield Under Limiting Reactant Conditions

The synthesis of aspirin appears in Section Exercise 4.3.3:

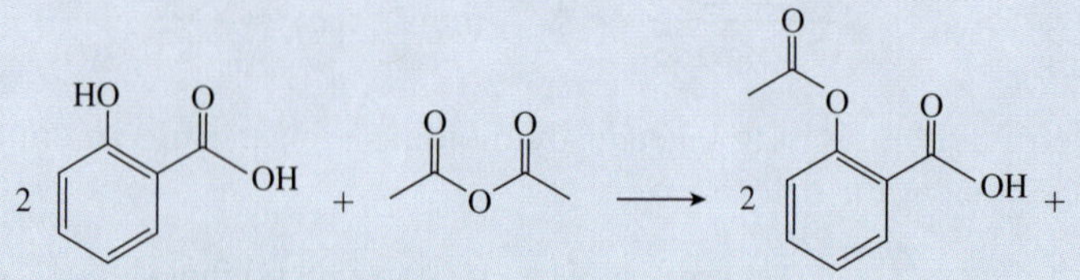

Chemical formula:	$C_7H_6O_3$	$C_4H_6O_3$	$C_9H_8O_4$
Name:	Salicylic acid	Acetic anhydride	Aspirin
MM:	138.1 g/mol	102.1 g/mol	180.2 g/mol

Suppose a chemist started with 152 g of salicylic acid and 86.8 g of acetic anhydride and produced 133 g of aspirin. What is the yield of this reaction?

Strategy: The problem asks for a yield, so we identify this as a yield problem. In addition, we recognize this as a limiting reactant situation because we are given the masses of both starting materials. First, identify the limiting reactant by working with moles and stoichiometric coefficients; then carry out standard stoichiometry calculations to determine the theoretical amount that could form. A table of amounts helps organize these calculations. Calculate the percent yield from the theoretical amount and the actual amount formed.

Solution: Begin by converting mass to moles. The problem gives the molar masses.

$$\frac{152 \text{ g } C_7H_6O_3}{138.1 \text{ g/mol}} = 1.10 \text{ mol } C_7H_6O_3 \qquad \frac{86.8 \text{ g } C_4H_6O_3}{102.1 \text{ g/mol}} = 0.850 \text{ mol } C_4H_6O_3$$

Now divide moles by stoichiometric coefficients to identify the limiting reactant.

$$\frac{1.10 \text{ mol } C_7H_6O_3}{2 \text{ mol } C_7H_6O_3} = 0.555 \qquad \frac{0.850 \text{ mol } C_4H_6O_3}{1 \text{ mol } C_4H_6O_3} = 0.850$$

Because it has the smaller ratio, $C_7H_6O_3$ is the limiting reactant.
Next we organize the data by using a table of amounts.

Reaction	$2 \text{ } C_7H_6O_3$	$+$	$C_4H_6O_3$	$\longrightarrow$	$2 \text{ } C_9H_8O_4$	$+$	H_2O
Starting amount (mol)	1.10		0.850		0		0
Change in amount (mol)							
Final amount (mol)							

We determine the entries in the change row using the amount of the limiting reactant and stoichiometric ratios.

$$(1.10 \text{ mol } C_7H_6O_3)\left(\frac{1 \text{ mol } C_4H_6O_3}{2 \text{ mol } C_7H_6O_3}\right) = 0.550 \text{ mol } C_4H_6O_3 \text{ consumed}$$

$$(1.10 \text{ mol } C_7H_6O_3)\left(\frac{2 \text{ mol } C_9H_8O_4}{2 \text{ mol } C_7H_6O_3}\right) = 1.10 \text{ mol } C_9H_8O_4 \text{ produced}$$

$$(1.10 \text{ mol } C_7H_6O_3)\left(\frac{1 \text{ mol } H_2O}{2 \text{ mol } C_7H_6O_3}\right) = 0.550 \text{ mol } H_2O \text{ produced}$$

Example 4-7

Yield Under Limiting Reactant Conditions (*continued*)

We can now complete the table of amounts:

Reaction	$2\ C_7H_6O_3$	$+\ C_4H_6O_3$	$\longrightarrow\ 2\ C_9H_8O_4$	$+\ H_2O$
Starting amount (mol)	1.10	0.850	0	0
Change in amount (mol)	−1.10	−0.550	+1.10	+0.550
Final amount (mol)	0	0.300	1.10	0.550

The final amount in this table represents the theoretical yield. To determine the percent yield, we need to have actual and theoretical amounts in the same units. In this example, we work with moles:

$$\text{Actual yield} = \frac{133\text{ g C}_9\text{H}_8\text{O}_4}{180.2\text{ g C}_9\text{H}_8\text{O}_4/\text{mol}} = 0.738\text{ mol}$$

$$\text{Percent yield} = (100\%)\left(\frac{0.738\text{ mol}}{1.10\text{ mol}}\right) = 67.1\%$$

Does the Result Make Sense? The entries in the amounts table all are reasonable, and notice that the yield turns out to be less than 100%, as we know must be the case.

If 152 g of salicylic acid and 41.2 g of acetic anhydride react to form 114 g of aspirin, what is the percent yield?

Extra Practice Exercise 4.7

Notice that neither the masses of the reactants nor the number of moles indicate directly which reactant is limiting. In this example, salicylic acid ($C_7H_6O_3$) is the limiting reactant even though the mass of salicylic acid and the number of moles of salicylic acid are both larger than the corresponding amounts of acetic anhydride ($C_4H_6O_3$).

It is often desirable to run practical syntheses under limiting reactant conditions, even though this may appear to be wasteful. For example, the aspirin synthesis typically is run with salicylic acid as the limiting reactant. Making salicylic acid the limiting reactant means that the product mixture will contain acetic anhydride and aspirin but a negligible amount of salicylic acid. This is advantageous for two reasons. First, salicylic acid and aspirin have similar molecular structures, making it difficult to separate these two substances. In contrast, aspirin is easy to separate from acetic anhydride. Second, salicylic acid is more expensive than acetic anhydride, and it is more economical to consume the more expensive reactant completely than to leave some of it unreacted.

> Masses and number ratios cannot be used to determine the "limiting reactant" in bicycle manufacture either. Five chains are fewer in number than eight wheels and have less mass than eight wheels. Nevertheless, the wheels are used up before the chains are.

Section Exercises

4.4.1 Suppose that the industrial synthesis of HCN described in Example 4-3 is carried out using 5.00×10^2 kg each of ammonia and methane in excess oxygen. What is the maximum mass of HCN that could be produced, and what mass of which reactant would be left over?

4.4.2 Lithium metal is one of the few substances that reacts directly with molecular nitrogen:

$$6\text{ Li} + \text{N}_2 \longrightarrow 2\text{ Li}_3\text{N}$$

What mass of the product, lithium nitride, can be prepared from 4.5 g of lithium metal and 9.5 g of molecular nitrogen?

4.4.3 Solid carbon reacts with chlorine gas to give carbon tetrachloride. If 150 g of carbon reacts with 250 g of Cl_2 with an 85% yield, what mass of CCl_4 is produced?

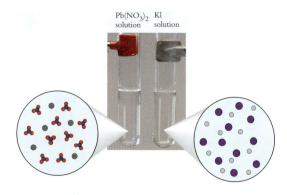

Pb(NO₃)₂   KI
solution   solution

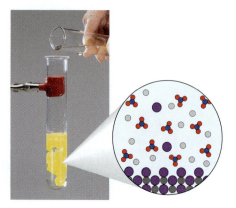

Figure 4-4
Mixing colorless aqueous solutions of potassium iodide and lead(II) nitrate results in the formation of a yellow precipitate of lead(II) iodide.

4.5 PRECIPITATION REACTIONS

The diversity of chemical reactions is immense. To make sense of this vast expanse of chemistry, we need a system for grouping chemical reactions into categories. The reactions within each category should share some characteristics or follow a common theme. One relatively simple category is **precipitation** reactions, in which cations and anions in aqueous solution combine to form neutral insoluble solids.

Species in Solution

To understand precipitation reactions, it is essential to work with the ionic species that exist in aqueous solution. For instance, mixing colorless solutions of lead(II) nitrate and potassium iodide causes a brilliant yellow solid to precipitate from the mixture (Figure 4-4). To identify this yellow solid, we must examine the chemical species present in the solutions.

When a salt dissolves in water, it produces cations and anions. Lead(II) nitrate and potassium iodide are **soluble** salts. Lead(II) nitrate dissolves in water to generate Pb^{2+} cations and NO_3^- anions. Potassium iodide dissolves in water to generate K^+ and I^- ions. Mixing the solutions combines all four types of ions. A precipitate forms if any of the new combinations of the ions forms a salt that is **insoluble** in water. The new combinations when these two solutions mix are K^+ combining with NO_3^- or Pb^{2+} combining with I^-:

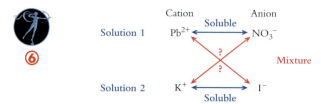

If we were to conduct a second solubility experiment in which solutions of KI and $NaNO_3$ were mixed, we would find that no precipitate forms. This demonstrates that K^+ and NO_3^- ions do not form a solid precipitate, so the bright yellow precipitate must be lead(II) iodide, PbI_2. As the two salt solutions mix, Pb^{2+} cations and I^- anions combine to produce lead(II) iodide, which precipitates from the solution. On standing, the yellow precipitate settles, leaving a colorless solution that contains potassium cations and nitrate anions. The molecular blowups in Figure 4-4 depict these solutions at the molecular level.

Actually, PbI_2 is not completely insoluble. Tiny amounts of Pb^{2+} and I^- ions remain in the aqueous solution after precipitation, but these amounts are so small that we consider PbI_2 to be an insoluble salt. Only 4.1×10^{-6} mol of PbI_2 dissolve in one liter of water at 25 °C, whereas 3.5 mol of KNO_3 dissolve in one liter of water at 25 °C. Figure 4-5 highlights this difference in solubility.

Figure 4-5
One liter of water can dissolve 354 g of KNO₃, a soluble salt. In contrast, PbI₂ is considered to be an insoluble salt because only 2 mg dissolves in 1 L of water.

These ions in solution exist in dynamic equilibrium with the precipitate. The system is dynamic because lead ions and iodide ions continually move in and out of solution; it is at equilibrium because the net amount of dissolved ions remains constant. We treat solubility equilibria quantitatively in Chapter 18.

Net Ionic Equations

The simplest balanced chemical equation for a precipitation reaction is a **net ionic equation** that has ions as the reactants and a neutral solid as the product. In a precipitation reaction, reactant ions combine to form a *neutral ionic solid*. One reactant carries positive charge and the other carries negative charge, but the product is electrically neutral. Because electrical charge always is conserved, the total positive charge of the reacting cations must be the same as the total negative charge of the reacting anions. For example, the precipitate that forms from Pb^{2+} and I^- must contain two iodide ions for every lead ion, to make the precipitate neutral:

$$Pb^{2+}(aq) + 2\,I^-(aq) \longrightarrow PbI_2(s)$$

⑦

In reactions such as this one, where reagents are in different phases, we use parentheses to convey that information: (s) for solid, (l) for liquid, (g) for gas, and (aq) for aqueous, meaning dissolved in water.

Another example of a precipitation reaction is the process that occurs when we mix aqueous solutions of potassium hydroxide (KOH) and iron(III) chloride ($FeCl_3$). A precipitate forms. A list of species present helps us to determine the net ionic equation describing this process:

Species present: K^+, Fe^{3+}, Cl^-, OH^-, H_2O

Other experiments show that KCl is a soluble salt, so the only possible precipitate contains Fe^{3+} cations and OH^- anions. We describe this precipitation by a balanced net ionic equation that contains only those species involved in the reaction. The substance that forms contains three OH^- anions combined with every Fe^{3+} cation, resulting in a neutral product. Thus, the reactant species are Fe^{3+} cations and OH^- anions with stoichiometric coefficients of 1 and 3. Here is the net ionic equation:

$$Fe^{3+}(aq) + 3\,OH^-(aq) \longrightarrow Fe(OH)_3(s)$$

A net ionic equation contains only those species that participate in a chemical reaction. Notice that neither K^+ nor Cl^- appears in the equation for the precipitation of $Fe(OH)_3$. Similarly, neither K^+ nor NO_3^- appears in the equation for the precipitation of PbI_2. Although these other ions are present in the solution, they undergo no change during the precipitation reaction. Ions that are not involved in the chemical change are referred to as **spectator ions.** Spectator ions are omitted from net ionic equations.

Solubility Guidelines

Through many years of experience and research, chemists have discovered patterns in the solubilities of ionic substances. Most salts are insoluble. The soluble salts are summarized in Table 4-1, and the flowchart in Figure 4-6 shows how to determine if a salt is soluble or insoluble.

The flowchart lets us predict whether a salt is soluble or insoluble. For example, will $Cu(OH)_2$ dissolve in water? The compound does not contain NH_4^+ or a Group 1 cation, nor does it contain any anion that confers solubility. Because $Cu(OH)_2$ does not appear among the exceptions (Table 4-1), it is insoluble. Example 4-8 further illustrates the solubility flowchart and net ionic equations.

Figure 4-6
A flowchart can be used to predict whether a salt is soluble or insoluble in water.

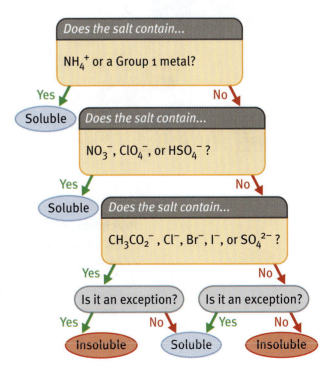

Table 4-1   Soluble Salts

Cations whose salts are soluble: NH_4^+ and Group 1 (Li^+, Na^+, K^+, etc.)
Anions whose salts are soluble:

Anion:	NO_3^-	ClO_4^-	HSO_4^-	$CH_3CO_2^-$	Cl^-	Br^-	I^-	SO_4^{2-}
Exceptions:	None	None	None	None	$AgCl$	$AgBr$	AgI	Hg_2SO_4
					Hg_2Cl_2	Hg_2Br_2	Hg_2I_2	$PbSO_4$
					$PbCl_2$	$PbBr_2$	PbI_2	$CaSO_4$
								$SrSO_4$
								$BaSO_4$

Other soluble compounds: $Ca(OH)_2$, $Sr(OH)_2$, $Ba(OH)_2$.

Example 4-8

Salt Solubility

Will a precipitate form when solutions of magnesium sulfate ($MgSO_4$) and barium chloride ($BaCl_2$) are mixed? If so, write the net ionic equation for the reaction.

Strategy: The solution that results from mixing contains all the ions of the original solutions. If any cation–anion combination results in an insoluble salt, that salt will precipitate from solution. List the ions and then apply the flowchart to find out whether any new combination of cations and anions gives an insoluble salt. If there is an insoluble salt, write the net ionic equation for its formation.

Solution: The ions present are Mg^{2+} and SO_4^{2-} from one solution, Ba^{2+} and Cl^- from the other, so the possible new combinations are $MgCl_2$ and $BaSO_4$.

According to the flowchart, salts of chloride and sulfate are soluble. Therefore, no precipitate will form unless one of these compounds is included among the exceptions listed in Table 4-1. Barium sulfate is one of these exceptions, so it is an insoluble salt.

Solid barium sulfate will form when the two solutions are combined. Each ion carries two units of charge, so they combine in a 1:1 stoichiometric ratio:

$$Ba^{2+}(aq) + SO_4^{2-}(aq) \longrightarrow BaSO_4(s)$$

Does the Result Make Sense?   The photo in the margin shows a cloudy precipitate of barium sulfate forming as solutions of barium chloride and magnesium sulfate mix.

Extra Practice Exercise 4.8

Will a precipitate form when a solution of iron(II) acetate is added to a solution of magnesium chloride?

Precipitation Stoichiometry

Tables of amounts are useful in stoichiometry calculations for precipitation reactions. For example, a precipitate of $Fe(OH)_3$ forms when 50.0 mL of 1.50 M NaOH is mixed with 35.0 mL of 1.00 M $FeCl_3$ solution. We need a balanced chemical equation and amounts in moles to calculate how much precipitate forms. The balanced chemical equation is the net reaction for formation of $Fe(OH)_3$:

$$Fe^{3+}(aq) + 3\ OH^-(aq) \longrightarrow Fe(OH)_3(s)$$

Notice that the net ionic reaction focuses our attention on the ions that actually participate in the reaction. Now calculate the number of moles of Fe^{3+} and OH^-, using $n = MV$:

$$(1.00 \text{ mol } FeCl_3/L)\left(\frac{1 \text{ mol } Fe^{3+}}{1 \text{ mol } FeCl_3}\right)(35.0 \text{ mL})\left(\frac{10^{-3} \text{ L}}{1 \text{ mL}}\right) = 3.50 \times 10^{-2} \text{ mol } Fe^{3+}$$

$$(1.50 \text{ mol } NaOH/L)\left(\frac{1 \text{ mol } OH^-}{1 \text{ mol } NaOH}\right)(50.0 \text{ mL})\left(\frac{10^{-3} \text{ L}}{1 \text{ mL}}\right) = 7.50 \times 10^{-2} \text{ mol } OH^-$$

We know the amounts of both Fe^{3+} and OH^-, so this is a limiting reactant situation. The ion that is consumed first determines how much $Fe(OH)_3$ precipitates. Dividing the numbers of moles by the stoichiometric coefficients identifies the limiting reactant:

$$\frac{3.50 \times 10^{-2} \text{ mol } Fe^{3+}}{1 \text{ mol } Fe^{3+}} = 3.50 \times 10^{-2}$$

$$\frac{7.50 \times 10^{-2} \text{ mol } OH^-}{3 \text{ mol } OH^-} = 2.50 \times 10^{-2}$$

Iron(III) hydroxide precipitate

The hydroxide anion has the smaller ratio, so it is the limiting reactant. A table of amounts helps determine the amount of $Fe(OH)_3$ that precipitates from solution:

Reaction	Fe^{3+}	+	$3 OH^-$	$\longrightarrow$	$Fe(OH)_3$
Starting amount (mol)	3.50×10^{-2}		7.50×10^{-2}		0
Change in amount (mol)	-2.50×10^{-2}		-7.50×10^{-2}		$+2.50 \times 10^{-2}$
Final amount (mol)	1.00×10^{-2}		0		2.50×10^{-2}

Mixing the two solutions will produce 2.50×10^{-2} mol of $Fe(OH)_3$ precipitate, which is 2.67 g. The mixed solution contains Na^+ cations and Cl^- anions, too, but we can ignore these spectator ions in our calculations. Notice that this precipitation reaction is treated just like other limiting reactant problems. Examples 4-9 and 4-10 further illustrate the application of general stoichiometric principles to precipitation reactions.

Silver bromide is a major component of photographic films and paper. A film manufacturer mixed 75.0 L of a 1.25 M solution of silver nitrate with 90.0 L of a 1.50 M potassium bromide solution and obtained 17.0 kg of silver bromide precipitate. What was the percent yield of this precipitation reaction?

Strategy: Apply the seven-step strategy to this quantitative problem. The first two steps are part of the strategy; the rest are part of the solution.

1. The problem asks for a percent yield, so we need to compare the actual yield and the theoretical yield. Information is provided about amounts of both starting materials, so this is a limiting reactant situation.

2. We need a balanced net ionic equation. The precipitate, AgBr, forms from two ions, Ag^+ and Br^-:

$$Ag^+ (aq) + Br^- (aq) \longrightarrow AgBr (s)$$

Example 4-9

Percent Yield of Precipitation

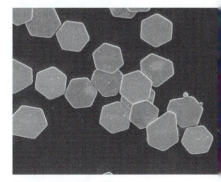

Hexagonal crystals of AgBr

Example 4-9

Percent Yield of Precipitation (*continued*)

Solution:

3. The data available are 75.0 L of 1.25 M silver nitrate, 90.0 L of 1.50 M potassium bromide, and 17.0 kg of silver bromide product. The molar mass of AgBr is 187.8 g/mol.

4. Use $n = MV$ to calculate amounts of each ion present in the two solutions before mixing, and determine which is limiting. Then solve for the theoretical yield, and apply Equation 4-2 to calculate percent yield.

5 and 6.

$$\text{mol Ag}^+ = MV = (1.25 \text{ mol/L})(75.0 \text{ L}) = 93.8 \text{ mol}$$

$$\text{mol Br}^- = MV = (1.50 \text{ mol/L})(90.0 \text{ L}) = 135 \text{ mol}$$

The 1:1 stoichiometric ratio between Ag^+ and Br^- makes it easy to identify the limiting reactant: Ag^+ is limiting, so the precipitation reaction can produce no more than 93.8 mol of AgBr. This is the theoretical yield.

Now convert the mass of precipitate into moles:

$$n = \frac{m}{MM} = \frac{(17.0 \text{ kg})(10^3 \text{ g/kg})}{(187.8 \text{ g/mol})} = 90.5 \text{ mol (actual yield)}$$

Use the theoretical and actual yields in Equation 4-2 to determine the percent yield:

$$\%\text{Yield} = (100\%)\left(\frac{\text{Actual yield}}{\text{Theoretical yield}}\right) = (100\%)\left(\frac{90.5 \text{ mol}}{93.8 \text{ mol}}\right) = 96.5\%$$

7. Does the Result Make Sense? A yield slightly less than 100% seems reasonable. Silver is an expensive substance, so manufacturers seek to maximize the yield of AgBr based on silver. Consequently, silver ion is always made the limiting reactant in this preparation. In other words, the precipitation is always carried out with an excess of bromide ions to ensure that as much Ag^+ precipitates as possible.

Extra Practice Exercise 4.9

In a test to see if silver precipitates chloride ion from solution quantitatively, 100 mL of 0.125 M NaCl solution was treated with 25.0 mL of 1.0 M aqueous AgNO_3. After precipitation was complete, 1.78 g of AgCl solid was recovered. What was the percent yield of this process?

Example 4-10

Precipitation Stoichiometry

A white precipitate forms when 2.00×10^2 mL of 0.200 M potassium phosphate solution is mixed with 3.00×10^2 mL of 0.250 M calcium chloride solution. Write the net ionic equation that describes this process. Calculate the mass of the precipitate that forms, and identify the ions remaining in solution.

Strategy: We could again apply the seven-step process in detail. Instead, we take a more compact approach. Begin by determining what species are present in the reaction mixture. Next, use the solubility guidelines to identify the precipitate. After writing the balanced net ionic reaction, use solution stoichiometry and a table of amounts to find the required quantities.

Solution:: The combined solutions contain four different ions: Ca^{2+}, K^+, Cl^-, and PO_4^{3-}. According to the solubility guidelines, potassium salts and chloride salts are

Example 4-10

Precipitation Stoichiometry (*continued*)

soluble. The white precipitate must be calcium phosphate, the only combination that does not involve these ions. To balance the charges, three Ca^{2+} ions must combine with two PO_4^{3-} ions. Potassium and chloride are spectator ions that do not appear in the net ionic reaction:

$$3\ Ca^{2+}(aq) + 2\ PO_4^{3-}(aq) \longrightarrow Ca_3(PO_4)_2(s)$$

To calculate the mass of the precipitate, we need to know the numbers of moles of Ca^{2+} and PO_4^{3-}:

$$(0.200\ mol\ K_3PO_4/L)\left(\frac{1\ mol\ PO_4^{3-}}{1\ mol\ K_3PO_4}\right)(2.00 \times 10^2\ mL)\left(\frac{10^{-3}\ L}{1\ mL}\right)$$
$$= 4.00 \times 10^{-2}\ mol\ PO_4^{3-}$$

$$(0.250\ mol\ CaCl_2/L)\left(\frac{1\ mol\ Ca^{2+}}{1\ mol\ CaCl_2}\right)(3.00 \times 10^2\ mL)\left(\frac{10^{-3}\ L}{1\ mL}\right)$$
$$= 7.50 \times 10^{-2}\ mol\ Ca^{2+}$$

Identify the limiting reactant by dividing the numbers of moles by the stoichiometric coefficients:

$$\frac{7.50 \times 10^{-2}\ mol\ Ca^{2+}}{3\ mol\ Ca^{2+}} = 2.50 \times 10^{-2}$$

$$\frac{4.00 \times 10^{-2}\ mol\ PO_4^{3-}}{2\ mol\ PO_4^{3-}} = 2.00 \times 10^{-2}$$

The smaller value identifies phosphate as the limiting reactant, so its final amount will be zero. Using this information, construct a table of amounts such that PO_4^{3-} is completely consumed:

Reaction	3 Ca^{2+}	+ 2 PO_4^{3-} $\longrightarrow$	$Ca_3(PO_4)_2$
Starting amount (10^{-2} mol)	7.50	4.00	0
Change in amount (10^{-2} mol)	−6.00	−4.00	+2.00
Final amount (10^{-2} mol)	1.50	0	2.00

The table of amounts shows that 2.00×10^{-2} mol of calcium phosphate will precipitate. Use this amount to compute the mass of product:

$$(2.00 \times 10^{-2}\ mol\ Ca_3(PO_4)_2)\ (310.2\ g/mol) = 6.20\ g\ Ca_3(PO_4)_2$$

The second part of the problem asks us to identify the ions remaining in the solution. Two spectator ions are present: K^+ and Cl^-. In addition, the table of amounts reveals that some of the Ca^{2+} remains in solution after the precipitation reaction. The ions remaining in solution are K^+, Cl^-, and excess Ca^{2+}.

Does the Result Make Sense? We started with less than 1 L each of two solutions whose concentrations are in the 0.2 M range, so an amount that is less than 0.1 mol makes sense.

What mass of precipitate forms when 125 mL of 0.222 M aqueous magnesium chloride is mixed with 225 mL of 0.105 M aqueous sodium phosphate?

Extra Practice Exercise 4.10

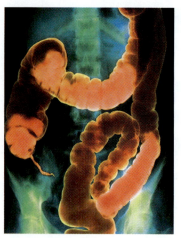

A barium sulfate X-ray image

Synthesis via Precipitation

The solubility guidelines can be used to design ways of making salts. Suppose that we want to prepare barium sulfate, $BaSO_4$. This substance is opaque to X rays, so it is often used to visualize the intestinal tract. Patients are given a "barium cocktail," and then the areas of interest are irradiated. Barium sulfate absorbs the X rays to give a picture of the intestines. Soluble barium salts are poisonous, but $BaSO_4$ is insoluble in water, so it can be administered safely. (Only 1.0×10^{-5} mol dissolves in 1 L of water at 25 °C.)

A source of soluble Ba^{2+} cations is required for the synthesis of barium sulfate. We can use any barium salt that is soluble, such as barium nitrate, barium acetate, or barium chloride. A soluble source of sulfate ions is also needed, such as Na_2SO_4 or $(NH_4)_2SO_4$. Because Ba^{2+} ions are toxic, it is important to precipitate all of the barium, so the two salts are measured out in quantities that will ensure that barium is the limiting reactant. Each salt is dissolved separately in water, and then the two solutions are mixed. After precipitation, the $BaSO_4$ is collected by filtration, the mass of the product is measured, and the yield of the reaction is calculated. It is important to know the yield of the reaction because any toxic barium ions left in the solution will have to be disposed of properly.

Example 4-11 shows how to apply these principles to the synthesis of CaF_2.

Example 4-11 **Synthesis via Precipitation** 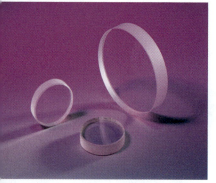 Calcium fluoride windows	Calcium fluoride is insoluble in water and is transparent to light. Because of these properties, calcium fluoride is sometimes used to make windows for optical devices. Design a synthesis of 1.0 kg of CaF_2.

Strategy: The synthesis of calcium fluoride requires soluble sources of Ca^{2+} ions and F^- ions. We can choose any pair of appropriate salts that happen to be available in the laboratory. Once we have selected suitable salts, we use stoichiometry to calculate the masses of each one.

Solution: Common cations that confer solubility are Na^+ and K^+, so the fluoride of either would be an appropriate choice. As a source of Ca^{2+}, choose a salt containing an anion that confers solubility, such as Cl^- or NO_3^-. Two compounds that are inexpensive and found in many laboratories are NaF and $CaCl_2$.

To calculate the masses that should be used, begin by writing the net ionic equation for the synthesis:

$$Ca^{2+}(aq) + 2\,F^-(aq) \longrightarrow CaF_2(s)$$

You should be able to do the mass–mole conversion showing that 1.0 kg of CaF_2 is 13 mol. Thus, we need 13 mol of Ca^{2+} and 26 mol of F^-. These ions can be provided by 13 mol of $CaCl_2$ and 26 mol of NaF. Use molar masses to determine the masses of the two starting materials:

$$(13 \text{ mol } CaCl_2)(111.0 \text{ g/mol}) = 1.4 \times 10^3 \text{ g } CaCl_2$$

$$(26 \text{ mol NaF})(41.90 \text{ g/mol}) = 1.1 \times 10^3 \text{ g NaF}$$

To carry out the synthesis, weigh the proper amounts of calcium chloride and sodium fluoride and dissolve each in water. Then mix the two solutions to generate a precipitate, and filter the solution to isolate the product.

Does the Result Make Sense? Our target was 1.0 kg of product, and we start with slightly more than 1.0 kg of each reactant. That makes sense, because some of the mass of the starting materials is due to spectator ions, Na^+ and Cl^-. An attractive feature of this synthesis is that these ions, which remain in solution, are nontoxic.

Extra Practice **Exercise 4.11**	Devise a synthesis that will produce 250 g of lead(II) iodide.

Section Exercises

■ **4.5.1** Determine whether the following salts are soluble or insoluble: (a) sodium acetate; (b) AgNO$_3$; (c) barium hydroxide; (d) CaO; (e) lead(II) sulfate; (f) ZnCl$_2$; and (g) manganese(II) sulfide.

■ **4.5.2** Design a synthesis of 1.5 kg of calcium phosphate that starts with soluble salts.

■ **4.5.3** Cadmium ions are environmental pollutants found in mining waste, metal plating, water pipes, and industrial discharge. Cadmium ions replace zinc ions in biochemistry and cause kidney damage, high blood pressure, and brittle bones. Dissolved Cd^{2+} impurities can be removed from a water sample by precipitation with sulfide ions. What is the minimum mass of ammonium sulfide required to precipitate all the Cd^{2+} from 5.0×10^3 L of water contaminated with 0.077 M cadmium ions?

4.6 ACID–BASE REACTIONS

One of the most fundamental chemical reactions is the combination of a hydroxide ion (OH$^-$) and a hydronium ion (H$_3$O$^+$) to produce two molecules of water:

$$\text{OH}^-(aq) + \text{H}_3\text{O}^+(aq) \longrightarrow 2\,\text{H}_2\text{O}\,(l)$$

A molecular view of this reaction (Figure 4-7) shows that the hydroxide anion accepts one hydrogen atom from the hydronium cation. Taking account of charges, it is a hydrogen *cation* (H$^+$) that is transferred. The reaction occurs rapidly when H$_3$O$^+$ and OH$^-$ ions collide. The hydroxide anion accepts a hydrogen cation from the hydronium cation, forming two neutral water molecules.

> Many books abbreviate the hydronium ion as H$^+$(*aq*) or just H$^+$. We prefer H$_3$O$^+$ because it serves as a reminder of the molecular structure of the hydronium ion and of the proton-transfer nature of acid–base reactions.

Proton Transfer

A hydrogen cation is a hydrogen atom that has lost its single electron, leaving a bare hydrogen nucleus. A bare hydrogen nucleus is a proton. Thus, any reaction in which H$^+$ moves from one species to another is called a *proton-transfer reaction*. Protons readily form chemical bonds. In aqueous solution, they associate with water molecules to form hydronium ions.

Any reaction in which a proton is transferred from one substance to another is an **acid–base reaction.** More specifically, the proton-transfer view is known as the Brønsted-Lowry definition of acids and bases. In an acid–base reaction, an acid donates a proton, and a base accepts that proton. Any species that can give up a proton to another substance is an **acid,** and any substance that can accept a proton from another substance is a **base.** The production of two water molecules from a hydroxide anion (a base) and a hydronium ion (an acid) is just one example of an acid–base reaction; acids and bases are abundant in chemistry.

> In Chapter 21, we introduce a second definition of acids and bases, the Lewis definition, which focuses attention on *electron* movement rather than *proton* movement. Until then, acid–base always means "proton transfer."

KEY CONCEPT

Acid: A substance that donates protons.
Base: A substance that accepts protons.

In Figure 4-7, the hydronium ion acts as an acid because it donates a proton to a base. The hydroxide anion acts as a base because it accepts a proton from an acid. When a hydronium ion with charge +1 transfers a proton to a hydroxide ion with charge −1,

⑧

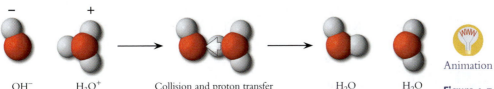

| OH$^-$ | H$_3$O$^+$ | Collision and proton transfer | H$_2$O | H$_2$O |
| (Base: a hydrogen ion acceptor) | (Acid: a hydrogen ion donor) | | | |

Animation

Figure 4-7
Proton transfer between H$_3$O$^+$ and OH$^-$.

Figure 4-8
Proton transfer between HCl and H₂O.

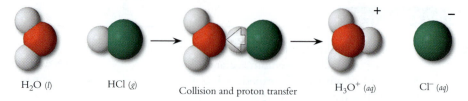

H₂O (l)   HCl (g)   Collision and proton transfer   H₃O⁺ (aq)   Cl⁻ (aq)

Animation

the two resulting water molecules have zero charges. The pair of charges becomes a neutral pair. A proton transfer reaction such as this one, in which water is one product and a pair of charges has been neutralized, is called a **neutralization reaction.**

Strong and Weak Acids

Hydrogen chloride, a colorless gas with a pungent, irritating odor, dissolves in water to give a solution of hydrochloric acid. In water, hydrogen chloride acts as an acid because it donates a proton to a water molecule, giving a hydronium ion and a chloride ion. As shown in Figure 4–8, water acts as the base, accepting a proton from the acid, hydrogen chloride. Remember that proton transfer always involves an acid and a base.

Hydrogen chloride produces hydronium ions and chloride ions *quantitatively* when it dissolves in water. This means that virtually every molecule of HCl transfers its proton to a water molecule. Therefore, the concentration of hydronium ions equals the concentration of the acid. The species present in an aqueous solution of HCl are Cl^-, H_3O^+, and, of course, H_2O.

Any acid that undergoes quantitative reaction with water to produce hydronium ions and the appropriate anion is called a **strong acid.** Table 4–2 gives the structures and formulas of six common strong acids, all of which are supplied commercially as concentrated aqueous solutions. These solutions are corrosive and normally are diluted for routine use in acid−base chemistry. At the concentrations normally used in the laboratory, a solution of any strong acid in water contains H_3O^+ and anions that result from the loss of a proton. Example 4-12 shows a molecular view of the proton transfer reaction of a strong acid.

Example 4-12
Proton Transfer

In nitric acid, three oxygen atoms are bonded directly to a nitrogen atom, and a hydrogen is bonded to one of the oxygens. Write the net ionic equation and draw a molecular picture that illustrates the reaction between nitric acid and water.

Strategy: Nitric acid is one of the six common strong acids. This means that when nitric acid dissolves in water, each acid molecule transfers a proton to a water molecule, generating a hydronium ion and the appropriate anion. Both the reaction and its molecular representation must show this proton transfer.

Solution: In this example the anion is nitrate:

$$H_2O + HNO_3 \longrightarrow H_3O^+ + NO_3^-$$

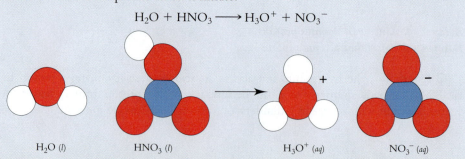

H₂O (l)   HNO₃ (l)   H₃O⁺ (aq)   NO₃⁻ (aq)

Does the Result Make Sense?   The picture shows that H^+ has moved from nitric acid to water, forming a cation and leaving behind an anion. The process is proton transfer, and electrical charge is conserved.

Extra Practice
Exercise 4.12

Draw a molecular picture showing what happens when hydrogen bromide dissolves in water.

Table 4-2   Common Strong Acids

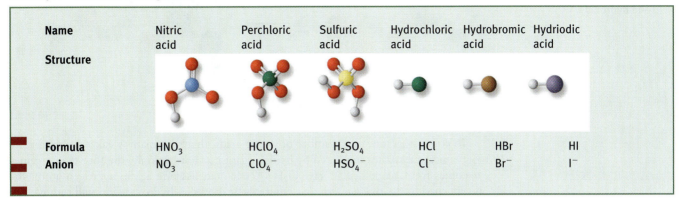

Name	Nitric acid	Perchloric acid	Sulfuric acid	Hydrochloric acid	Hydrobromic acid	Hydriodic acid
Formula	HNO_3	$HClO_4$	H_2SO_4	HCl	HBr	HI
Anion	NO_3^-	ClO_4^-	HSO_4^-	Cl^-	Br^-	I^-

Among the strong acids, three are significant industrial chemicals. More than 40 billion kilograms of sulfuric acid are produced in the United States each year at a cost of about 8¢ per kilogram. About 60% is used in the production of fertilizers. Another 30% is used to manufacture detergents, drugs, explosives, dyes, paint, paper, and other chemicals. Most of the remaining 10% is used in petroleum refining and metallurgy. Nitric acid is also a valuable industrial chemical. About 90% of the nearly 8 billion kilograms produced each year is used to make NH_4NO_3, which is used as a fertilizer. Ammonium nitrate is also used to make explosives and other nitrogen-containing chemicals.

About 3 billion kilograms of hydrochloric acid are produced each year, mostly as a by-product of the plastics industry. The largest single use of hydrochloric acid is the "pickling" of steel. The pickling process removes iron(III) oxide (Fe_2O_3, rust) from the surface of the metal. About a third of all hydrochloric acid is used to produce other chemicals, mostly ionic compounds. Other strong acids have specialized applications in industry and research laboratories, but none approaches the importance of sulfuric, nitric, and hydrochloric acids.

Weak acids are compounds that donate protons quantitatively to hydroxide ions but not to water. All acids are proton donors, but whereas a strong acid quantitatively donates protons to water, a weak acid does not. Aqueous solutions of weak acids contain small concentrations of hydronium ions, making the solutions acidic, but nearly all the weak acid molecules remain intact. Representative of strong acids, HCl generates H_3O^+ and Cl^- quantitatively when dissolved in water. Representative of weak acids, HF remains predominantly as HF molecules when dissolved in water. However, HF donates protons quantitatively to OH^- ions to give H_2O molecules and F^- ions, as shown in Figure 4-9.

Sulfuric acid, H_2SO_4, is both a strong acid and a weak acid. The compound has two acidic hydrogen atoms. In aqueous solution, one hydrogen atom undergoes quantitative proton transfer to water:

$$H_2SO_4\,(aq) + H_2O\,(l) \longrightarrow HSO_4^-\,(aq) + H_3O^+(aq)$$

However, the second hydrogen atom remains attached to HSO_4^-. Thus, an aqueous solution of sulfuric acid contains H_3O^+ and HSO_4^- ions. Like other weak acids, the hydrogen sulfate ion reacts quantitatively with hydroxide to give sulfate ion and water.

$$HSO_4^-\,(aq) + OH^-\,(aq) \longrightarrow SO_4^{2-}\,(aq) + H_2O\,(l)$$

As we describe in Chapter 17, any weak acid transfers protons to water molecules to a small extent.

Animation

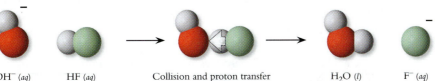

$OH^-\,(aq)$ $HF\,(aq)$ Collision and proton transfer $H_2O\,(l)$ $F^-\,(aq)$

Figure 4-9
Proton transfer between HF and OH^-.

Figure 4-10
Proton transfer between acetic acid and hydroxide.

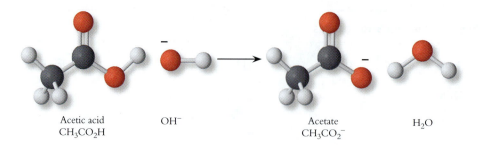

Acetic acid   OH⁻   Acetate   H₂O
CH_3CO_2H $CH_3CO_2^-$

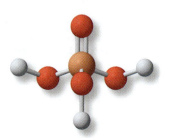

Formic acid
HCO_2H

Benzoic acid
$C_6H_5CO_2H$

In contrast to the small number of strong acids, there are many weak acids. Some are biochemical substances produced by living organisms. For example, the sour tang of vinegar comes from acetic acid, CH_3CO_2H. Acetic acid has one acidic hydrogen atom, the one attached to the oxygen atom, as Figure 4-10 shows. A hydroxide ion will remove this proton, converting acetic acid into the acetate anion. The hydrogen atoms bound to the carbon atom of acetic acid are not acidic. Neither water nor hydroxide will remove these hydrogen atoms. Acetic acid is a simple example of a very large family of organic compounds that contain the CO_2H grouping of atoms. Acids that contain the CO_2H group are called **carboxylic acids.** Two other common carboxylic acids are formic acid, HCO_2H, and benzoic acid, $C_6H_5CO_2H$. All carboxylic acids are weak acids.

Another example of a weak acid is phosphoric acid, H_3PO_4, an important industrial compound used primarily to make fertilizer. All three of the hydrogen atoms of phosphoric acid will react with hydroxide ions. Example 4-13 illustrates the acid–base properties of phosphoric acid.

Example 4-13	Write the balanced net ionic equation for the reaction of phosphoric acid with an excess of aqueous potassium hydroxide.

Weak Acid Reaction with OH⁻ Ions

Phosphoric acid
H_3PO_4

Strategy: When an acid and a base react, the acid transfers a proton to the base. We must identify the species in solution and then evaluate the proton transfer chemistry.

Solution: An aqueous solution of potassium hydroxide contains K^+, OH^-, and H_2O. Phosphoric acid is a weak acid, so most of its molecules remain as H_3PO_4 in aqueous solution. The species present at the beginning of the reaction are K^+ and OH^- ions and molecules of H_3PO_4 and H_2O. The hydroxide ion is a powerful base that removes all of the acidic hydrogen atoms from both strong and weak acids. An excess of potassium hydroxide means that there are enough OH^- ions to remove all three acidic hydrogens from phosphoric acid. This reaction produces three molecules of water and one phosphate anion:

$$H_3PO_4\,(aq) + 3\,OH^-\,(aq) \longrightarrow PO_4^{3-}\,(aq) + 3\,H_2O\,(l)$$

Does the Result Make Sense? Check the reaction to be sure that everything is in balance. There are 9 H atoms, 7 O atoms, and 1 P atom on each side, and the net charge on each side is −3, so this is a balanced proton-transfer reaction.

Extra Practice Exercise 4.13

Write the net ionic reaction and draw a picture of the resulting anion when two hydroxide ions react with a phosphoric acid molecule.

Phosphoric acid, sulfuric acid, and the hydrogen sulfate ion are members of a group of acids known as **oxoacids.** An oxoacid has a central atom bonded to a variable number of oxygen atoms and OH groups. Except for the three oxoacids shown in Table 4-2 (sulfuric acid, nitric acid, and perchloric acid), all of the oxoacids described in this textbook are weak acids. Chapter 17 describes in detail the chemistry of strong and weak acids, including carboxylic acids and oxoacids.

Strong and Weak Bases

The hydroxide ion is a powerful proton acceptor. Thus, any compound that generates stoichiometric quantities of hydroxide ions when it dissolves in water is called a **strong base.** All Group 1 hydroxides, AOH (where A is any of the Group 1 cations), are strong bases. Sodium hydroxide is one example:

$$NaOH(s) \longrightarrow Na^+(aq) + OH^-(aq)$$

Three of the Group 2 hydroxides, $Ca(OH)_2$, $Sr(OH)_2$, and $Ba(OH)_2$, are soluble strong bases.

Just as there are weak acids, there are also weak bases. A **weak base** does not readily accept protons from water molecules but does quantitatively accept protons from hydronium ions. Ammonia is the most common weak base. Ammonia exists predominantly as NH_3 molecules in aqueous solution, but it undergoes quantitative proton transfer with hydronium ions to generate ammonium ions:

$$NH_3(aq) + H_3O^+(aq) \longrightarrow NH_4^+(aq) + H_2O(l)$$

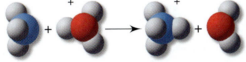

We describe other examples of weak bases in Chapter 17.

Acid–Base Stoichiometry

The quantitative aspects of acid–base chemistry obey the principles introduced earlier in this chapter. The common acid–base reactions that are important in general chemistry take place in aqueous solution, so acid–base stoichiometry uses molarities and volumes extensively. Example 4-14 illustrates the essential features of aqueous acid–base stoichiometry.

Example 4-14

Acid–Base Stoichiometry

What volume of 0.050 M nitric acid solution is needed to neutralize 0.075 L of 0.065 M $Ba(OH)_2$?

Strategy: Apply the seven-step method: Determine what's asked for, visualize the chemistry, organize the data, set up the appropriate molar equalities, and work toward a quantitative solution. The problem asks for the volume of acid that reacts stoichiometrically with a given amount of base. To visualize and identify the reaction, start with species in solution. Nitric acid is one of the six strong acids, so an aqueous solution of nitric acid contains H_3O^+ cations and NO_3^- anions. A solution of $Ba(OH)_2$ contains Ba^{2+} and OH^- ions. (Of course, both solutions contain large quantities of H_2O molecules.) When the solutions are mixed, there might be precipitate formation and/or proton transfer. According to the solubility guidelines, all nitrate salts are soluble, so no precipitate forms. Hydronium ion is a strong acid, and hydroxide ion is a strong base. When a strong acid is mixed with a strong base, acid–base neutralization occurs rapidly. By the time the solutions are mixed thoroughly, the proton transfer reaction is complete:

$$H_3O^+(aq) + OH^-(aq) \longrightarrow 2 H_2O(l)$$

Solution: The data provided are the molarities of the acid and base and the volume of base:

Base	Acid
$[Ba(OH)_2] = 0.065$ M	$[HNO_3] = 0.050$ M
$V = 0.075$ L	$V =$ asked for

Example 4-14

Acid−Base Stoichiometry
(*continued*)

The process to reach a quantitative solution to the problem requires working with moles. Thus, we need the relationship linking moles to molarity and volume:

$$n = MV$$

We must use the equation in two ways:

a. Calculate moles of OH^- using $n = MV$ and stoichiometric reasoning.

b. Calculate volume of nitric acid solution using $V = n/M$.

First, determine how many moles of hydroxide must be neutralized:

$$(0.065 \text{ mol/L Ba(OH)}_2)\left(\frac{2 \text{ mol OH}^-}{1 \text{ mol Ba(OH)}_2}\right)(0.075 \text{ L}) = 9.75 \times 10^{-3} \text{ mol OH}^-$$

From the balanced net ionic equation, we see that neutralization requires one hydronium ion for every hydroxide ion, so we need 9.75×10^{-3} mol of hydronium ions. Knowing the concentration of the nitric acid, we can calculate the volume of solution that contains this number of moles:

$$V_{\text{acid}} = \frac{9.75 \times 10^{-3} \text{ mol H}_3\text{O}^+}{0.050 \text{ mol H}_3\text{O}^+/\text{L}} = 0.20 \text{ L}$$

Does the Result Make Sense? The hydroxide ions are neutralized by 0.20 L of the acid solution. This is a bit more than twice the volume of the base solution. The answer is reasonable because each mole of $Ba(OH)_2$ contains two moles of hydroxide ions, but each mole of HNO_3 supplies just one mole of H_3O^+. Furthermore, the molarity of the base solution is larger than the molarity of the acid solution.

Extra Practice Exercise 4.14

Calculate the volume of 0.108 M NaOH solution required to react completely with 145 mL of phosphoric acid solution that is 1.15×10^{-2} M.

Titration

The amount of an acid in a sample can be determined by adding a solution of base of known concentration until the amount of base added exactly matches the amount of acid in the sample. Similarly, the amount of a base in a sample can be determined by adding a solution of acid of known concentration until the amount of acid added exactly matches the amount of base in the sample. This procedure is called **titration.** The solution to be analyzed is placed in a beaker or flask. A second solution of known concentration, called the **titrant,** is added slowly by means of a calibrated measuring vessel called a buret.

In one type of titration, a solution of a strong base such as sodium hydroxide is added slowly to a solution that contains an unknown amount of an acid. Each hydroxide ion added to the acid solution accepts one proton from a molecule of acid. As the titration proceeds, fewer and fewer acid molecules remain in the acid solution, but the solution is still acidic. At the **stoichiometric point,** just enough hydroxide ions have been added to react with every acidic proton present in the acid solution before the titration was started. The hydroxide ions in the next drop of titrant do not react because acid molecules are no longer present in the solution. Before the stoichiometric point, the solution contains excess acid. After the stoichiometric point, the solution contains excess OH^-. Figure 4-11 shows a titration setup and molecular views illustrating titration of a strong acid by a strong base.

At the stoichiometric point, the amount of base added exactly matches the amount of acid originally present.

Stoichiometric point: mol OH^- added = mol acidic hydrogen present   (4-3)

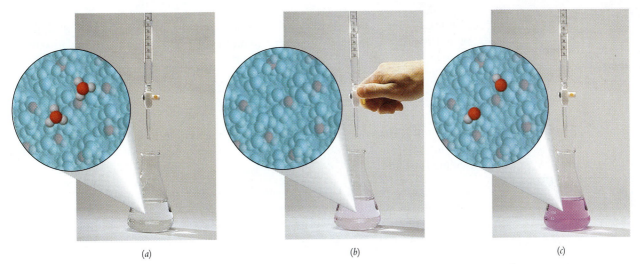

(a) (b) (c)

Figure 4-11
Titration of HCl by NaOH.
Spectator ions have been omitted for clarity. (*a*) Initially, the HCl solution contains many H_3O^+ ions. (*b*) At the stoichiometric point, the solution contains only H_2O molecules and spectator ions. (*c*) Past the stoichiometric point, the solution contains many OH^- ions.

If we know the molarity of the titrant and measure the volume of titrant required to reach the stoichiometric point, we can calculate the number of moles of hydroxide required to react with all the acid. This allows us to determine the concentration of the unknown acid solution.

How do we know when the stoichiometric point of a titration has been reached? At the stoichiometric point, the nature of the acid–base species in the solution changes. We cannot "see" ions directly, but we can obtain a visual measure by placing a tiny amount of a substance called an **indicator** in the solution to be titrated. An indicator is a molecule whose color depends on the concentration of hydronium ions. A properly chosen indicator shows a distinct color change when the titration reaches the stoichiometric point and the acid in the solution has all reacted. In Figure 4-11, the indicator is phenolphthalein, a molecule that is colorless when hydronium ions are present but pink when hydroxide ions are present. This color change signals an end to the titration. As soon as we see the color change, we stop adding base and measure the volume that has been delivered from the buret. Example 4-15 illustrates how to calculate concentration from a titration volume.

	Example 4-15
	Acid–Base Titration

Industrial wastewater often is contaminated with strong acids. Environmental regulations require that such wastewater be neutralized before it is returned to the environment. A 1.50×10^2 mL sample of wastewater was titrated with 0.1250 M sodium hydroxide, and 38.65 mL of the base was required to reach the stoichiometric point. What was the molarity of hydronium ions in this sample of wastewater?

Strategy: We are asked for the molarity of an acid. The analysis is a titration. Knowing that the wastewater contains strong acid, we can write the general acid–base neutralization reaction:

$$H_3O^+ + OH^- \longrightarrow 2\,H_2O$$

The key feature of a titration is that at the stoichiometric point, the number of moles of OH^- added equals the number of moles of acid originally present. This lets us use the data for the base to calculate the number of moles of H_3O^+ in the sample.

Solution: Use the titration to calculate the number of moles of OH^- required to reach the stoichiometric point:

$$\text{mol } OH^- \text{ added} = MV$$

$$\text{mol } OH^- \text{ added} = (0.1250 \text{ mol/L})(38.65 \text{ mL})(10^{-3} \text{ L/mL}) = 4.831 \times 10^{-3} \text{ mol}$$

<table>
<tr><td>

Example 4-15

Acid–Base Titration
(*continued*)

</td><td>

According to Equation 4-3, this is also the number of moles of acid in the original sample. Because the original sample had a volume of 1.50×10^2 mL, its molarity is as follows:

$$M = \frac{n}{V} = \frac{4.831 \times 10^{-3} \text{ mol}}{(1.50 \times 10^2 \text{ mL})(10^{-3} \text{ L/mL})} = 3.22 \times 10^{-2} \text{ mol/L}$$

</td></tr>
</table>

Does the Result Make Sense? An acid concentration around 10^{-2} M may not appear high, but it is more than 1000 times the concentration of hydronium ion in unpolluted streams.

<table>
<tr><td>

Extra Practice Exercise 4.15

</td><td>

The acidic component of vinegar is acetic acid. When 5.00 mL of vinegar was added to water and the solution titrated with the 0.1250 M NaOH, it took 33.8 mL to reach the stoichiometric point. Calculate the molarity of acetic acid in vinegar.

</td></tr>
</table>

A titration requires a solution whose concentration is known. In Example 4–15 the NaOH solution used as the titrant was known to be 0.1250 M. A titrant of known concentration is known as a standard solution, and the concentration of such a solution is determined by a **standardization** titration. In a standardization titration, the solution being titrated contains a known amount of acid or base. An excellent acid for standardization is potassium hydrogen phthalate, $KHC_8H_4O_4$. This substance, a carboxylic acid that contains one weakly acidic hydrogen atom per molecule, is easily obtained as a highly pure solid. A known number of moles can be weighed on an analytical balance, dissolved in pure water, and then titrated with the base solution to be standardized. Example 4-16 illustrates the standardization procedure.

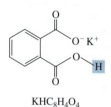

$KHC_8H_4O_4$

<table>
<tr><td>

Example 4-16

Standardization

</td><td>

A biochemist needed to standardize a solution of KOH. A sample of potassium hydrogen phthalate weighing 0.6745 g was dissolved in 100.0 mL of water and a drop of indicator was added. The solution was then titrated with the KOH solution. The titration required 41.75 mL of base to reach the stoichiometric point. Find the molarity of the KOH solution.

Strategy: First, we must identify the chemistry. This is an acid–base titration in which hydrogen phthalate anions (the acid) react with OH^- (the base). We use the molar equality of acid and base at the stoichiometric point together with the equations that link moles with mass and volume.

Solution: Hydrogen phthalate has one acidic hydrogen atom. As the net reaction shows, there is a 1:1 molar ratio between hydrogen phthalate and hydroxide:

$$HC_8H_4O_4^- + OH^- \longrightarrow C_8H_4O_4^{2-} + H_2O$$

The number of moles of hydrogen phthalate anions is the same as the number of moles of $KHC_8H_4O_4$, which is calculated from its mass and molar mass:

$$n_{HC_8H_4O_4^-} = n_{KHC_8H_4O_4} = \frac{0.6745 \text{ g}}{204.2 \text{ g/mol}} = 3.303 \times 10^{-3} \text{ mol}$$

At the stoichiometric point of the titration, the number of moles of hydroxide added from the buret equals the number of moles of hydrogen phthalate:

$$n_{OH^-} \text{ added} = 3.303 \times 10^{-3} \text{ mol}$$

$$\text{Volume added} = 41.75 \text{ mL} = 41.75 \times 10^{-3} \text{ L}$$

$$M = \frac{n}{V} = \frac{3.303 \times 10^{-3} \text{ mol}}{41.75 \times 10^{-3} \text{ L}} = 7.911 \times 10^{-2} \text{ M}$$

</td></tr>
</table>

Example 4-16

Standardization (*continued*)

**Extra Practice
Exercise 4.16**

Does the Result Make Sense? As we have already seen, base concentrations around 10^{-1} M are common. This concentration is a bit lower but still in the range that we would expect for laboratory solutions.

To check the results of the first standardization, the biochemist repeated the titration. This time, a mass of 0.7015 g of potassium hydrogen phthalate dissolved in 135 mL of water required 43.36 mL of the base. Find the molarity and decide if the two results are consistent.

The technique of titration is equally useful for the titration of an unknown base by a solution of strong acid. The calculations proceed exactly as described previously. For the titration of a base, the stoichiometric point is reached when the number of moles of added acid in the titrant equals the number of moles of base in the unknown solution.

Stoichiometric point: Moles H_3O^+ added $=$ Moles base present $\qquad$ (4-4)

Section Exercises

4.6.1 Carbonic acid, H_2CO_3 (molecular model shown below), is a weak oxoacid that forms when carbon dioxide dissolves in water. Carbonic acid contains two acidic hydrogen atoms. Write the net ionic reaction that occurs when carbonic acid reacts with an excess of hydroxide ions. Draw a molecular picture of the process.

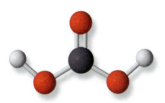

Carbonic acid

4.6.2 The wastewater in Example 4-15 has $[H_3O^+] = 3.22 \times 10^{-2}$ M. What mass of NaOH would have to be added to 1.00×10^3 L of this wastewater to make it neutral?

4.6.3 While cleaning a laboratory, a technician discovers a large bottle containing a colorless solution. The bottle is labeled "$Ba(OH)_2$," but the molarity of the solution is not given. Concerned because of the toxicity of Ba^{2+} ions, the technician titrates with a solution of hydrochloric acid standardized at 0.1374 M. A 25.00-mL sample of the barium hydroxide solution requires 36.72 mL of the HCl solution to reach the stoichiometric point. What is the concentration of Ba^{2+} in the solution?

4.7 OXIDATION–REDUCTION REACTIONS

As described in Section 4.6, one important class of chemical reactions involves transfers of protons between chemical species. An equally important class of chemical reactions involves transfers of electrons between chemical species. These are **oxidation–reduction reactions.** Commonplace examples of oxidation–reduction reactions include the rusting of iron, the digestion of food, and the burning of gasoline. Paper manufacture, the subject of our Chemistry and the Environment Box, employs oxidation–reduction chemistry to bleach wood pulp. All metals used in the chemical industry and manufacturing are extracted and purified through oxidation–reduction chemistry, and many biochemical pathways involve the transfer of electrons from one substance to another.

The reaction of magnesium metal with aqueous strong acid, which appears in Figure 4-12, illustrates the fundamental principles of oxidation–reduction. When a piece of magnesium is dropped into a solution of hydrochloric acid, a reaction starts almost

Figure 4-12
Magnesium metal reacts with H_3O^+, generating H_2 gas and a solution containing Mg^{2+} cations. The spectator ions are not shown in the aqueous solution.

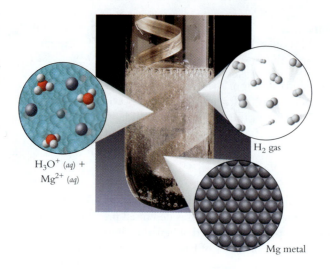

$H_3O^+ (aq) +$
$Mg^{2+} (aq)$

H_2 gas

Mg metal

immediately. The metal dissolves, and gas bubbles from the solution. The gas is H_2, and analysis of the solution reveals the presence of Mg^{2+} ions. A list of chemical species before and after the reaction indicates what has taken place:

Before Reaction		After Reaction	
$Mg (s)$	(reactant)	$Mg^{2+}(aq)$	(product)
$H_3O^+(aq)$	(reactant)	$H_2 (g)$	(product)
$Cl^- (aq)$	(spectator)	$Cl^- (aq)$	(spectator)
$H_2O (l)$	(solvent)	$H_2O (l)$	(solvent)
		$H_3O^+(aq)$	(excess reactant)

Solid magnesium has been transformed into Mg^{2+} ions, and hydronium ions have decomposed to give H_2 gas and water molecules. Quantitative measurements reveal that for every mole of Mg consumed, the reaction also consumes two moles of H_3O^+, and it produces one mole of H_2 and two moles of water. The reaction can be summed up in the following balanced chemical equation:

$$Mg(s) + 2\,H_3O^+(aq) \longrightarrow Mg^{2+}(aq) + H_2(g) + 2\,H_2O(l)$$

By examining the changes occurring for magnesium and hydronium ions, we discover that Mg atoms lose electrons during this reaction and hydronium ions gain electrons:

Each side of this balanced equation has a net charge of +2. Remember, however, that the net ionic equation does not show all of the species present in solution. In this example, Cl^- ions balance out the positive charges of Mg^{2+} and H_3O^+. All solutions have a net charge of zero, so whenever a net ionic equation has an overall charge, there are spectator ions present that neutralize the charge.

The loss of electrons by magnesium atoms to form Mg^{2+} cations indicates that this reaction between magnesium metal and hydronium ions involves oxidation and reduction. An atom of magnesium is oxidized, losing two electrons to form a Mg^{2+} cation.

Box 4-2	Chemistry and the Environment: Paper Without Pollution

E xamine the page before you: Not the words, but the material itself, paper. We often take this product for granted, but paper-making is one of the most important developments in the advance of civilization. According to legend, the first sheets of paper were made from mulberry leaves in China in AD 105. For many centuries paper was made in individual sheets, so it was a rare and expensive commodity. Paper-making machines were first developed in the early years of the nineteenth century. The development of machinery that allowed high-speed paper production was partially responsible for the increase in literacy and education of people around the world.

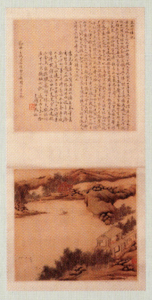

Early Chinese artists worked on paper

Modern paper-making is chemically intensive. Strong bases are used to convert wood fibers into pliable pulp, and oxidizing agents are used to bleach the pulp. Inorganic materials are added as fillers that increase opacity, and polymer binders are added to increase strength. Finally, paper is treated with organic agents called "sizers" that improve its printing qualities.

A modern paper mill

Paper-making carries with it considerable potential for environmental pollution because of the high need for chemicals in the various processing steps. Nevertheless, this industry has made it a priority to eliminate polluting wastes.

Making paper without pollution requires that each part of the process be nonpolluting. The chemicals most commonly used in the production of pulp are NaOH and Na_2S. In modern paper mills, sulfur-containing by-products are scrubbed from the plant exhaust, and the aqueous sodium hydroxide is reclaimed and recycled. The fillers used to make paper opaque—titanium dioxide, calcium carbonate, and kaolin (a clay)—are natural, nonpolluting minerals. The polymer binders and sizers are relatively easy to recapture from the aqueous waste stream.

The bleaching process, in contrast, poses major difficulties. Traditional paper bleaching uses chlorine gas, which is reduced to chloride anions, Cl^-, as it oxidizes the colored pigments in wood pulp. The chloride anion is not a pollutant, as it is a major species in the oceans. Unfortunately, chlorine processing also generates small quantities of chlorine-containing dioxins such as 2,3,7,8-tetrachloro-dibenzo-*p*-dioxin, whose structure (below) appears less formidable than its name:

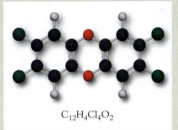

$C_{12}H_4Cl_4O_2$

Dioxins are of concern because they accumulate in the biosphere, where they have highly deleterious effects. Tests have shown that when the concentration of dioxins in the blood of laboratory animals reaches a critical level, reproductive and immune system defects result. Moreover, recent data indicate that the concentration of dioxins in the blood of the average U.S. resident has nearly reached that level. A major reason is that dioxins are not very water-soluble, so they accumulate in the body rather than being readily processed and excreted. Consequently, several groups, including the American Public Health Association, have issued calls for phasing out the use of industrial chlorine.

The paper industry has responded by replacing chlorine gas with other oxidizing agents. Among the alternatives are sodium chlorate ($NaClO_3$) and hydrogen peroxide (H_2O_2). By 1997 it was estimated that half the U.S. paper industry had converted its bleaching process to chlorine dioxide generated from sodium chlorate. Although advocates of sodium chlorate claim that no detectable dioxins are produced by this bleaching agent, others contend that only completely chlorine-free bleaches such as hydrogen peroxide will be completely safe. More recently, the use of enzymes to pretreat wood pulp has allowed the amount of bleach to be reduced significantly. It may be only a matter of time before all chlorine-based bleaches in paper-making are discontinued. If that occurs, the paper industry may be the first major chemical-intensive industry to succeed at sustainable development.

Because electrons must be conserved in every chemical process, electrons lost by Mg must be gained by some other species. In this example the electrons lost by Mg are gained by H_3O^+, which is reduced to form H_2 and H_2O.

> ### KEY CONCEPT
> *Oxidation is the loss of electrons from a substance.*
> *Reduction is the gain of electrons by a substance.*

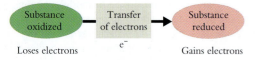

Substance oxidized — Transfer of electrons e^- → Substance reduced

Loses electrons — Gains electrons

Electrons are conserved. That means that oxidation (electron loss) and reduction (electron gain) always go together. Because of this necessary connection of reduction with oxidation, the process is often referred to simply as a **redox reaction** (*reduction–oxidation*).

Redox reactions are more complicated than precipitation or proton transfer reactions because the electrons transferred in redox chemistry do not appear in the balanced chemical equation. Instead, they are "hidden" among the starting materials and products. However, we can keep track of electrons by writing two **half-reactions** that describe the oxidation and the reduction separately. A half-reaction is a balanced chemical equation that includes electrons and describes either the oxidation or reduction but not both. Thus, a half-reaction describes half of a redox reaction. Here are the half-reactions for the redox reaction of magnesium and hydronium ions:

$$\begin{array}{rcll} \text{Mg} & \longrightarrow & \text{Mg}^{2+} + \quad 2e^- & \text{Oxidation} \\ 2\,\text{H}_3\text{O}^+ + 2e^- & \longrightarrow & \text{H}_2 + 2\,\text{H}_2\text{O} & \text{Reduction} \\ \hline \text{Net:} \quad 2\,\text{H}_3\text{O}^+ + \text{Mg} & \longrightarrow & \text{Mg}^{2+} + \text{H}_2 + 2\,\text{H}_2\text{O} & \end{array}$$

Separating oxidation from reduction makes it possible to verify that electrons are conserved in a redox reaction. Note that the electrons produced in the oxidation of magnesium are consumed in the reduction of hydronium ions. The electrons required for a reduction must come from an oxidation.

A detailed discussion of redox reactions must wait until Chapter 19, after we explore the nature of the atom, periodic properties of the elements, and thermodynamics. For now, we focus on only a few types of redox reactions that are common and relatively simple.

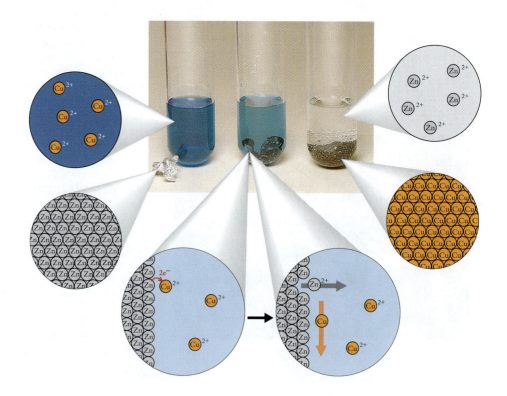

Figure 4-13
Zinc metal displaces copper ions from aqueous solution. The blue color signals the presence of copper ions. The color fades and copper metal appears as the reaction proceeds.

Table 4-3 Activity Series of the Metals

Increasing ease of reduction ↓		Ion	Metal		Increasing ease of oxidation ↑
	Cations difficult to displace	$K^+ +\ e^- \rightleftharpoons K$		Metals that react with both H_2O and H_3O^+	
		$Ca^{2+} + 2\,e^- \rightleftharpoons Ca$			
		$Na^+ +\ e^- \rightleftharpoons Na$			
		$Mg^{2+} + 2\,e^- \rightleftharpoons Mg$			
		$Al^{3+} + 3\,e^- \rightleftharpoons Al$			
		$Zn^{2+} + 2\,e^- \rightleftharpoons Zn$		Metals that react with H_3O^+	
		$Fe^{2+} + 2\,e^- \rightleftharpoons Fe$			
		$Ni^{2+} + 2\,e^- \rightleftharpoons Ni$			
		$Pb^{2+} + 2\,e^- \rightleftharpoons Pb$			
	Hydronium ion reduction	$H_3O^+ +\ e^- \rightleftharpoons \frac{1}{2}H_2 + H_2O$			
		$Cu^{2+} + 2\,e^- \rightleftharpoons Cu$			
	Easily displaced cations	$Ag^+ +\ e^- \rightleftharpoons Ag$		Unreactive metals	
		$Au^{3+} + 3\,e^- \rightleftharpoons Au$			

Metal Displacement

When a strip of zinc metal is added to a solution of copper(II) sulfate, the blue color slowly fades, and the zinc metal is replaced by copper metal (Figure 4–13). As copper ions in the solution are reduced to copper metal, zinc atoms are oxidized to Zn^{2+} cations. This is an example of a **metal displacement reaction,** in which a metal ion in solution (Cu^{2+}) is displaced by another metal (Zn) by means of a redox reaction. Figure 4–13 also shows molecular views of this displacement reaction.

$$\text{Oxidation: } Zn\,(s) \longrightarrow Zn^{2+}(aq) + 2\ e^- \qquad \text{Reduction: } Cu^{2+}(aq) + 2\ e^- \longrightarrow Cu\,(s)$$

$$\text{Redox: } Zn\,(s) + Cu^{2+}(aq) \longrightarrow Zn^{2+}(aq) + Cu\,(s)$$

Many other metal displacement reactions can be visualized, but not all of them occur. Some metals are oxidized readily, but others are highly resistant to oxidation. Likewise, some metal cations are highly susceptible to reduction, but others resist reduction. Zinc displaces copper ions from aqueous solutions, but copper will not replace zinc ions, because Cu^{2+} is easier to reduce than Zn^{2+}. Zinc will not displace Mg^{2+} ions, because magnesium cations are harder to reduce than zinc cations. Copper ions are displaced by aluminum but not by silver. After many pairs of elements are compared, it becomes possible to arrange the metals in order of their reactivity. Such a list, called an **activity series,** is shown in Table 4–3. Metals at the top of the list are easier to oxidize than metals at the bottom. Conversely, metal cations at the bottom of the list are easier to reduce than those at the top.

A metal will transfer electrons to any cation that is lower on the list. Furthermore, the list is in order of reactivity, so the greater the separation between the species, the more vigorous the reaction. Example 4–17 shows how to use the activity series to predict the outcome of a metal displacement reaction.

Predict whether a displacement reaction will occur in the following instances. Explain your conclusions.
a. A small piece of calcium is added to a solution of nickel(II) chloride.
b. Iron pellets are added to a solution of magnesium chloride.
c. A strip of aluminum foil is dipped in aqueous silver nitrate.

Strategy: A displacement reaction occurs when a metal high in the activity series is added to a solution containing a cation lower on the list. For each case, we need to identify the species in the mixture and evaluate reactivity using the activity series.

Example 4-17

The Activity Series

Example 4-17	**Solution:**

The Activity Series (*continued*)

a. Solid Ca metal is added to a solution that contains Ni^{2+} and Cl^- ions. Calcium, near the top of the activity series, is one of the most reactive metals. Calcium will displace most other metal ions from solution, including Ni^{2+}:

$$\text{Oxidation: } Ca\,(s) \longrightarrow Ca^{2+}(aq) + 2\,e^- \qquad \text{Reduction: } Ni^{2+}(aq) + 2\,e^- \longrightarrow Ni\,(s)$$

$$\text{Redox: } Ca\,(s) + Ni^{2+}(aq) \longrightarrow Ca^{2+}(aq) + Ni\,(s)$$

b. Solid iron metal is added to a solution that contains Mg^{2+} and Cl^-. Iron is below Mg^{2+} in the activity series, so no reaction will occur.

c. Solid aluminum metal is dipped in a solution that contains Ag^+ cations and NO_3^- anions. Aluminum is in the middle of the metals, and Ag^+ is near the bottom of the ions. Thus, aluminum displaces Ag^+:

$$\text{Oxidation: } Al\,(s) \longrightarrow Al^{3+}(aq) + 3\,e^- \qquad \text{Reduction: } 3\,[Ag^+(aq) + e^- \longrightarrow Ag\,(s)]$$

$$\text{Redox: } 3\,Ag^+(aq) + Al\,(s) \longrightarrow 3\,Ag\,(s) + Al^{3+}(aq)$$

 Do the Results Make Sense? In everyday life, we encounter silver as an unreactive solid that is used for jewelry, and we know that iron, while it rusts, is relatively stable. Calcium and magnesium, in contrast, are not normally encountered as pure metals but instead as salts. Thus, the directions of the reactions predicted by the activity series are in accord with everyday observations.

Extra Practice Exercise 4.17

Predict the reaction, if any, that occurs when a lead sinker is immersed in a solution that contains Fe^{2+} and Cu^{2+} cations.

Oxidation of Metals by H_3O^+ and H_2O

The redox reaction opening this section is the oxidation of magnesium metal by hydronium ions. It turns out that most metals are oxidized by H_3O^+, as can be seen by the location of the hydronium ion reduction reaction in the activity series of Table 4-3. All the metals above hydrogen will displace hydronium ions from solution. In each case the reduction half-reaction is the same one involved in the magnesium reaction: Hydronium ions are reduced to water and hydrogen gas. The electrons required to reduce H_3O^+ are provided by the metal, which is oxidized to its most stable cation. Remember that Group 1 metals form cations with $+1$ charge, and Group 2 metals form cations with $+2$ charge. Aluminum forms Al^{3+}, zinc forms Zn^{2+}, cadmium forms Cd^{2+}, and silver forms Ag^+. Other metals, such as iron and copper, form more than one stable cation. In these cases the cationic charge must be specified before a balanced redox reaction can be written. Table 4-3 lists reduction half-reactions of several metals. Here are some examples of metals reacting with H_3O^+:

$$2\,Li\,(s) + 2\,H_3O^+(aq) \longrightarrow 2\,Li^+(aq) + H_2\,(g) + 2\,H_2O\,(l)$$

$$Fe\,(s) + 2\,H_3O^+(aq) \longrightarrow Fe^{2+}(aq) + H_2\,(g) + 2\,H_2O\,(l)$$

$$2\,Al\,(s) + 6\,H_3O^+(aq) \longrightarrow 2\,Al^{3+}(aq) + 3\,H_2\,(g) + 6\,H_2O\,(l)$$

A few metals are so easy to oxidize that they react vigorously with water. Sodium, for example, reduces water to H_2 and OH^-. Here are the half-reactions:

$$2\,Na\,(s) \longrightarrow 2\,Na^+(aq) + 2\,e^-$$

$$2\,H_2O\,(l) + 2\,e^- \longrightarrow H_2\,(g) + 2\,OH^-(aq)$$

$$\text{Net: } 2\,Na\,(s) + 2\,H_2O\,(l) \longrightarrow 2\,Na^+(aq) + H_2\,(g) + 2\,OH^-(aq)$$

(a)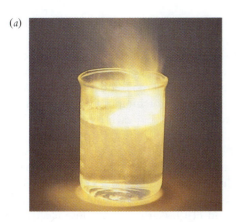

(b)

Figure 4-14
Sodium and other Group 1 metals react vigorously with water. The energy released in this redox reaction can ignite the hydrogen gas with spectacular and potentially dangerous results (a). Consequently, Group 1 metals must be protected from exposure to the environment. Shown here (b) is a sample of cesium metal sealed in a glass ampule.

As Figure 4-14 shows, this redox reaction releases enough energy to ignite the highly flammable hydrogen gas. All the Group 1 metals react in a manner similar to sodium, so these elements must be protected from exposure to water and oxygen. The Group 2 metals also react with water, but much more slowly.

Oxidation by Molecular Oxygen

Almost all elements combine with molecular oxygen to form binary oxides. In fact, loss of electrons is called *oxidation* because elements lose electrons when they combine with oxygen. The metals of Groups 1 and 2 form oxides that are ionic solids composed of metal cations and oxide anions. The oxygen atoms of O_2 capture two electrons each to form a pair of oxide anions. Every O^{2-} anion requires a counterbalancing cationic charge of +2. Group 1 metals form A^+ cations, so their oxides have the chemical formula A_2O, whereas Group 2 metals form A^{2+} cations and oxides with the chemical formula AO.

The stoichiometry of transition metal oxides is more variable. Iron, for example, forms three binary oxides. In FeO the iron atoms have lost two electrons each (Fe^{2+}, O^{2-}), and in Fe_2O_3 they have lost three electrons each (Fe^{3+}, O^{2-}). The third oxide, Fe_3O_4, contains one Fe^{2+} cation for every two Fe^{3+} cations. Other transition metals lose anywhere from one to eight electrons in forming oxides. There is no simple pattern that determines which oxide is most stable, but we can tell how many electrons the metal has lost by assigning a gain of two electrons to each oxygen atom in the chemical formula. Thus, the metal atoms in Cu_2O have lost one electron each, those in CrO_3 have lost six electrons each, and those in TiO_2 have lost four electrons each.

As we might expect, the more easily a metal is oxidized, the more likely it is to react with molecular oxygen. The least reactive metals—those at the bottom of the activity series—are among the few elements found in their elemental forms in the Earth's crust. These metals occupy Group 11 of the periodic table. They are used for durable items such as coins and jewelry. In contrast, the metals at the top of the activity series (those in Groups 1 and 2 of the periodic table) react vigorously with oxygen. Magnesium, for example, can be set ablaze with a match, and cesium bursts into flame spontaneously when exposed to air. The resulting oxides react readily with water to generate aqueous solutions of the metal cations. Eventually, these cations precipitate as metal salts, so Group 1 and 2 metals are found in the Earth's crust as salts rather than oxides. Examples are rock salt (sodium chloride, NaCl), limestone (calcium carbonate, $CaCO_3$), and epsomite (magnesium sulfate heptahydrate, $MgSO_4 \cdot 7H_2O$).

Almost all other metals react relatively rapidly with oxygen to form insoluble oxides. These metals are found in nature as minerals. Metallurgical processes combine chemistry and engineering to reduce metals in minerals to their elemental forms, as we describe in Chapter 20. In the presence of air, these metals tend to return to their oxidized states: Objects made from iron must be protected from oxygen and water, or they react to form

Copper, silver, and gold are used for jewelry.

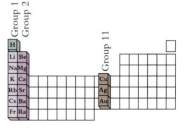

hydrated iron(III) oxide, better known as rust. Objects made of aluminum also react with O_2, but the product of this redox reaction, Al_2O_3, forms an impervious film on the metal surface that prevents O_2 from reaching the underlying metal. Moving farther up the activity series, the reactions with oxygen become more vigorous.

Compounds and nonmetallic elements also undergo redox reactions with oxygen. The nonmetals form binary oxides of varying stoichiometry. Once again, we can determine electron losses for oxygen's partner by viewing each oxygen atom as gaining two electrons during the redox reaction. Examples include CO_2 (C loses 4 e^-), SO_2 (S loses 4 e^-), SO_3 (S loses 6 e^-), and N_2O_5 (each N loses 5 e^-). These products do not contain ions, but the reactions by which they form are redox reactions nonetheless. The combustion reactions described in Chapter 3 are additional examples of redox reactions of chemical compounds. In a combustion reaction, molecular oxygen combines with carbon and hydrogen to form CO_2 and H_2O. Carbon atoms lose electrons in this process, and oxygen atoms gain electrons.

Example 4-18 shows how to analyze the various reaction types discussed in Sections 4.5 to 4.7.

Example 4-18 **Types of Chemical Reactions**	Predict what will happen when the following pairs of substances are allowed to react. Write a balanced chemical equation for each reaction. When the reaction involves ions, write a net ionic equation. Identify each reaction as precipitation, as acid–base, or as redox. (a) $AgNO_3(aq)$ and $NaCl(aq)$; (b) $HCl(aq)$ and $Zn(s)$; (c) $NaOH(s)$ and $CH_3CO_2H(aq)$; (d) $Ca(s)$ and $H_2O(l)$; and (e) $K(s)$ and $O_2(g)$.

Strategy: Begin by identifying what chemical species are present in the mixture. Then classify the species and identify which category of reaction is involved. Recognizing a few key species can give important clues about chemical reactivity. For example, the presence of H_3O^+ signals either an acid–base reaction or the oxidation of a metal. Hydroxide ions suggest either an acid–base reaction or a precipitation. A solid metal indicates a redox reaction with either hydronium ion or molecular oxygen.

Solution:

a. Silver nitrate and sodium chloride are both salts. They dissolve in water to generate the ionic species Ag^+, NO_3^-, Na^+, and Cl^-. Neither solution contains H_3O^+ or OH^-, so this is not an acid–base reaction. There is no solid metal present, so a redox reaction is unlikely. A precipitation reaction is possible, so we apply the solubility guidelines. These indicate that $NaNO_3$ is soluble in water, but $AgCl$ is not. We conclude that this is a precipitation reaction:

$$Ag^+(aq) + Cl^-(aq) \longrightarrow AgCl(s)$$

b. Hydrochloric acid contains H_3O^+ and Cl^-. The presence of hydronium ions and a solid metal suggests a redox reaction. According to the activity series, zinc is one of the metals that reacts with H_3O^+. We can write the appropriate half-reactions and then combine them to give the overall redox reaction:

$$\text{Reduction: } 2\,H_3O^+(aq) + 2\,e^- \longrightarrow H_2(g) + 2\,H_2O(l)$$

$$\text{Oxidation: } Zn(s) \longrightarrow Zn^{2+}(aq) + 2\,e^-$$

$$\text{Redox: } 2\,H_3O^+(aq) + Zn(s) \longrightarrow H_2(g) + 2\,H_2O(l) + Zn^{2+}(aq)$$

c. Acetic acid is a weak acid, so its solution contains mostly CH_3CO_2H and H_2O molecules. Sodium hydroxide dissolves to produce Na^+ ions and OH^- ions. Hydroxide ion, a strong base, removes the acidic hydrogen from acetic acid in an acid–base reaction:

$$CH_3CO_2H(aq) + OH^-(aq) \longrightarrow CH_3CO_2^-(aq) + H_2O(l)$$

d. The presence of solid calcium metal suggests a redox reaction. This Group 2 metal is oxidized by water according to the following half-reactions:

Example 4-18

Types of Chemical Reactions
(*continued*)

$$\text{Oxidation: Ca}(s) \longrightarrow \text{Ca}^{2+}(aq) + 2\ e^-$$

$$\text{Reduction: } 2\ \text{H}_2\text{O}(l) + 2\ e^- \longrightarrow \text{H}_2(g) + 2\ \text{OH}^-(aq)$$

$$\text{Redox: Ca}(s) + 2\ \text{H}_2\text{O}(l) \longrightarrow \text{Ca}^{2+}(aq) + 2\ \text{OH}^-(aq) + \text{H}_2(g)$$

e. Potassium, a Group 1 metal, is readily oxidized by molecular oxygen. The redox chemistry is best seen through the individual half-reactions:

$$\text{Oxidation: K}(s) \longrightarrow \text{K}^+ + e^- \qquad \text{Reduction: O}_2(g) + 4\ e^- \longrightarrow 2\ \text{O}^{2-}$$

$$\text{Redox: } 4\ \text{K}(s) + \text{O}_2(g) \longrightarrow 2\ \text{K}_2\text{O}(s)$$

Do the Results Make Sense? Another way of analyzing these combinations is by type of substance. Parts a and c involve mixing of two aqueous solution, parts b and d involve adding a metal to an aqueous system, and part e is the interaction of a metal with O$_2$ gas. When solutions mix, we look first for acid–base reactions (part c), then for formation of a precipitate (part b). When a metal contacts an aqueous system, the most likely reaction, if any, is oxidation of the metal. Any time molecular oxygen is present, we can expect oxidation to be one possibility.

Predict the process that occurs, if any, when each of the following solids is placed in aqueous hydrochloric acid: Fe , FeCl$_3$, and Fe(OH)$_3$.

**Extra Practice
Exercise 4.18**

Section Exercises

4.7.1 Although aluminum cans are not attacked by water, strong acid oxidizes Al to Al^{3+} cations, liberating hydrogen gas in the process. How many moles of H$_2$ gas will be liberated if a 2.43-g sample of pure Al metal reacts completely with an excess of 3.00 M sulfuric acid solution?

4.7.2 Metallic titanium is produced from rutile ore (TiO$_2$) by a direct replacement reaction with magnesium that frees titanium metal and produces MgO. (a) Write the balanced redox reaction for this process and identify the species that are oxidized and reduced.

(b) How many kilograms of magnesium are required to manufacture 1.00×10^2 kg of titanium metal, assuming that the reaction has a 100% yield?

4.7.3 Predict whether or not a reaction will occur, and if a reaction does take place, write the half-reactions and the balanced redox reaction: (a) a strip of nickel wire is dipped in 6.0 M HCl; (b) aluminum foil is dipped in aqueous CaCl$_2$; (c) a lead rod is dipped in a beaker of water; (d) an iron wire is immersed in a solution of silver nitrate.

CHAPTER 4 VISUAL SUMMARY

Important equations

Essential terms

Key concepts

SKILLS TO MASTER

① Balancing chemical equations by inspection
② Using stoichiometric ratios
③ Determining and using reaction yields
④ Identifying the limiting reactant
⑤ Constructing tables of amounts

⑥ Identifying species in solutions
⑦ Writing net ionic equations
⑧ Identifying soluble salts, acids, and bases
⑨ Calculating amounts in titrations
⑩ Recognizing simple redox reactions

4.1 WRITING CHEMICAL EQUATIONS

Stoichiometric coefficients   Balanced chemical equation

Reagents
Reactants   Products   ①

$$N_2 + 3 H_2 \longrightarrow 2 NH_3$$

Atoms are conserved.

4.2 STOICHIOMETRY OF CHEMICAL REACTIONS

Calculations center on the mole.

$$n_B = \left(\frac{Coeff_A}{Coeff_B} \right) n_A$$

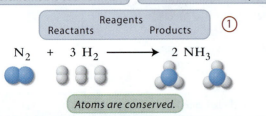

Stoichiometric ratio

Stoichiometry

4.3 YIELDS OF CHEMICAL REACTIONS

Yield
  Theoretical yield
  Actual yield
  Percent yield

③

$$\text{Percent yield} = 100\% \left(\frac{\text{Actual amount}}{\text{Theoretical amount}} \right)$$

4.4 THE LIMITING REACTANT

Limiting reactant   Smallest value of moles divided by coefficient

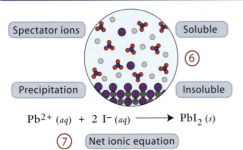

Reaction: ④ ⑤	N_2	$+$	$3 H_2$	$\longrightarrow$	$2 NH_3$
Starting amount (mol)	3.00		12.0		0
Change in amount (mol)	−3.00		−9.00		+6.00
Final amount (mol)	0		3.0		6.00

4.5 PRECIPITATION REACTIONS

Spectator ions   Soluble

⑥

Precipitation   Insoluble

$$Pb^{2+} (aq) + 2 I^- (aq) \longrightarrow PbI_2 (s)$$

⑦   Net ionic equation

4.6 ACID–BASE REACTIONS

Acid: A substance that donates protons.
Base: A substance that accepts protons.

Acid–Base reaction

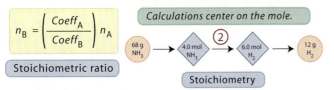

−   +   ⑧

Base   Acid   Neutralization

Strong acid   Weak acid   Strong base   Weak base

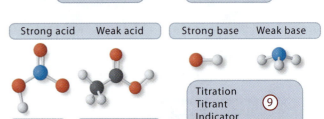   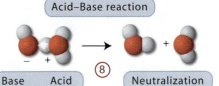

Oxoacid   Carboxylic acid

Titration
Titrant
Indicator   ⑨
Standardization
Stoichiometric point

Stoichiometric point: mol OH^- added = mol acidic hydrogen present

Stoichiometric point: mol H_3O^+ added = mol base present

4.7 OXIDATION–REDUCTION REACTIONS

Oxidation is loss of electrons.

Oxidation-reduction reactions
Redox reactions
Half-reactions

$$\overset{\displaystyle \overset{-2e^-}{\overbrace{\hspace{3cm}}}}{Mg(s) + 2 H_3O^+(aq) \longrightarrow Mg^{2+}(aq) + H_2(g) + H_2O(l)}$$

$+2e^-$

Reduction is gain of electrons.   ⑩

Metal displacement reaction
Activity series

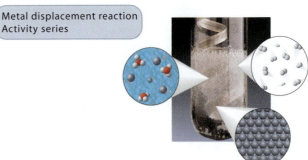

Learning Exercises

4.1 Several examples of chemical reasoning are introduced in this chapter. Write out the reasoning steps that you will follow in (a) balancing a chemical equation; (b) identifying the limiting reactant; (c) determining whether a precipitate forms; and (d) computing a reaction yield.

4.2 List features that can be used to distinguish each of the three major types of reactions introduced in this chapter.

4.3 Draw one specific molecular picture that illustrates each of the reaction types introduced in this chapter.

4.4 Construct a flowchart that summarizes the problem-solving strategy used for Example 4-18.

4.5 Define each of these terms: (a) percent yield; (b) limiting reactant; (c) spectator ion; (d) precipitate; (e) titration; and (f) oxidation.

4.6 Update your list of memory bank equations. For each equation, add a phrase that describes the kinds of chemical problems for which the equation is used.

4.7 List all terms new to you that appear in Chapter 4. In your own words, give a one-sentence definition of each. Consult the glossary if you need help.

 Problems   **ilw = interactive learning ware problem. Visit the website at www.wiley.com/college/olmsted**

Writing Chemical Equations

4.1 Balance the following chemical equations:
(a) $NH_4NO_3 \rightarrow N_2O + H_2O$
(b) $P_4O_{10} + H_2O \rightarrow H_3PO_4$
(c) $HIO_3 \rightarrow I_2O_5 + H_2O$
(d) $As + Cl_2 \rightarrow AsCl_5$

4.2 Balance the following chemical equations:
(a) $N_2O_5 + H_2O \rightarrow HNO_3$
(b) $KClO_3 \rightarrow KCl + O_2$
(c) $Fe + O_2 + H_2O \rightarrow Fe(OH)_2$
(d) $P_4 + Cl_2 \rightarrow PCl_3$

4.3 Draw a molecular picture that illustrates reaction 4.1 (d).

4.4 Draw a molecular picture that illustrates reaction 4.2 (d).

4.5 Balance the chemical equations for the following important industrial processes:
(a) Molecular hydrogen and carbon monoxide react to form methanol (CH_3OH).
(b) $CaO + C \rightarrow CO + CaC_2$
(c) $C_2H_4 + O_2 + HCl \rightarrow C_2H_4Cl_2 + H_2O$

4.6 The following reactions play roles in the manufacture of nitric acid. Write a balanced equation for each of them.
(a) $NH_3 + O_2 \rightarrow NO + H_2O$
(b) Nitrogen oxide and molecular oxygen combine to give nitrogen dioxide.
(c) Nitrogen dioxide and water react to form nitric acid and nitrogen oxide.
(d) $NH_3 + O_2 \rightarrow N_2 + H_2O$
(e) $NH_3 + NO \rightarrow N_2 + H_2O$

4.7 Draw a molecular picture that illustrates reaction 4.5 (a).

4.8 Draw a molecular picture that illustrates reaction 4.6 (b).

4.9 Here are some reactions involving water. Balance the equations:
(a) $Ca(OH)_2 + H_3PO_4 \rightarrow H_2O + Ca_3(PO_4)_2$
(b) $Na_2O_2 + H_2O \rightarrow NaOH + H_2O_2$
(c) $BF_3 + H_2O \rightarrow HF + H_3BO_3$
(d) $NH_3 + CuO \rightarrow Cu + N_2 + H_2O$

4.10 Here are some reactions of phosphorus and its compounds. Balance the equations:
(a) $P_4 + Na \rightarrow Na_3P$
(b) $Na_3P + H_2O \rightarrow PH_3 + NaOH$
(c) $PH_3 + O_2 \rightarrow P_4O_{10} + H_2O$
(d) $P_4O_{10} + H_2O \rightarrow H_3PO_4$

The Stoichiometry of Chemical Reactions

4.11 For each reaction in Problem 4.5, calculate the mass of the second reactant that is required to react completely with 5.00 g of the first reactant.

4.12 For each reaction in Problem 4.6, calculate the mass (in kg) of the first reactant that is required to react completely with 875 kg of the second reactant.

4.13 Iodine can be prepared by bubbling chlorine gas through an aqueous solution of sodium iodide:

$$2\ NaI\,(aq) + Cl_2\,(g) \longrightarrow I_2\,(s) + 2\ NaCl\,(aq)$$

What mass (in g) of NaI is required to produce 1.50 kg of I_2?

4.14 The fertilizer ammonium sulfate is prepared by the reaction between ammonia and sulfuric acid:

$$2\ NH_3\,(g) + H_2SO_4\,(aq) \longrightarrow (NH_4)_2SO_4\,(aq)$$

What mass (in kg) of NH_3 is needed to make 3.50 metric tons (1 metric ton = 1000 kg) of ammonium sulfate?

4.15 The fermentation of sugar to produce ethyl alcohol occurs by the following reaction:

$$C_6H_{12}O_6\,(s) \xrightarrow{\text{yeast}} 2\ C_2H_5OH\,(l) + 2\ CO_2\,(g)$$

What mass of ethyl alcohol can be made from 1.00 kg of sugar?

4.16 One starting material for the preparation of nylon is adipic acid. Adipic acid is produced from the oxidation of cyclohexane:

$$2 \bigcirc + 5\ O_2 \longrightarrow 2\ \text{adipic acid} + 2\ H_2O$$

Cyclohexane
C_6H_{12} Adipic acid
$C_6H_{10}O_4$

If 375 kg of cyclohexane reacts with an unlimited supply of oxygen, how much adipic acid (in kg) can be formed?

ilw 4.17 The use of Freons is being phased out because of the damage these compounds do to the stratospheric ozone layer. One of the Freons, CCl_2F_2, is manufactured from carbon tetrachloride and hydrogen fluoride. The other reaction product is hydrogen chloride. If a manufacturer wishes to convert 175 kg of carbon tetrachloride into CCl_2F_2, what is the minimum mass of hydrogen fluoride required? What masses of the two products will be obtained if this reaction is 100% efficient?

CCl_2F_2

4.18 One way to clean up emissions of sulfur dioxide formed during the burning of coal is to pass the exhaust gases through a water slurry of powdered limestone. In the reaction, sulfur dioxide reacts with calcium carbonate and oxygen gas to produce calcium sulfate and carbon dioxide. A coal-burning plant that uses coal with a typical sulfur content produces 2.00×10^4 metric tons (1 metric ton = 1000 kg) of sulfur dioxide in a month. How many metric tons of calcium carbonate will be consumed during the scrubbing process, assuming 100% efficiency? How many metric tons of calcium sulfate waste will the plant have to dispose of? (The process is actually about 90% efficient.)

Yields of Chemical Reactions

ilw 4.19 Approximately 12 billion kilograms of phosphoric acid are produced annually for fertilizers, detergents, and agents for water treatment. Phosphoric acid can be prepared by heating the mineral fluoroapatite with sulfuric acid in the presence of water:

$$Ca_5(PO_4)_3F + 5 H_2SO_4 + 10 H_2O \longrightarrow$$

Fluoroapatite

$$3 H_3PO_4 + 5 CaSO_4 \cdot 2 H_2O + HF$$

Phosphoric acid

If every kilogram of fluoroapatite yields 4.00×10^2 g of phosphoric acid, what is the percent yield?

4.20 When HgO is heated, it decomposes into elemental mercury and molecular oxygen gas. If 60.0 g of Hg is obtained from 80.0 g of the oxide, what is the percent yield of the reaction?

4.21 Phenobarbital is a sleep-inducing drug whose chemical formula is $C_{12}H_{12}N_2O_3$. It is manufactured in an eight-step process starting from toluene, C_7H_8. Theoretically, each molecule of toluene yields one molecule of phenobarbital. If each of the steps has a yield of 88%, what mass of toluene is needed to manufacture 25 kg of phenobarbital?

4.22 Most of the ammonia produced by the Haber process is used as fertilizer. A second important use of NH_3 is in the production of nitric acid, a top-15 industrial chemical. Nitric acid is produced by a three-step synthesis called the *Ostwald process*:

$$4 NH_3 + 5 O_2 \longrightarrow 4 NO + 6 H_2O$$

$$2 NO + O_2 \longrightarrow 2 NO_2$$

$$3 NO_2 + H_2O \longrightarrow 2 HNO_3 + NO$$

The NO is recycled so that every mole of ammonia theoretically yields one mole of nitric acid. Starting with 7.50×10^2 kg of ammonia, what mass of nitric acid can be produced if the average yield of the three steps is 94.5% efficient?

4.23 What is the yield of the reaction described in Problem 4.17 if the manufacturer obtains 105 kg of CCl_2F_2 from 175 kg of CCl_4? Given this reaction yield, what masses of CCl_4 and hydrogen fluoride should be used in order to make 155 kg of CCl_2F_2?

4.24 If the oxidation reaction described in Problem 4.16 is 76.5% efficient, what mass of cyclohexane is required to produce 3.50 kg of adipic acid?

The Limiting Reactant

4.25 A fast-food restaurant makes double cheeseburgers using one hamburger roll, two quarter-pound patties of beef, one slice of cheese, a fourth of a tomato, and 15 g of shredded lettuce. At the start of the day, the restaurant manager has 12 dozen rolls, 40 lb of beef patties, 2 packages of sliced cheese containing 65 slices each, 40 tomatoes, and 1 kg of lettuce. Which ingredient will run out first, and how many cheeseburgers can be made?

4.26 A hardware store sells a "handyman assortment" of wood screws containing the following:

Type	Number of screws	Mass per screw (g)
#8, $\frac{1}{2}''$	10	12.0
#8, 1''	8	21.5
#10, $\frac{1}{2}''$	8	14.5
#10, $1\frac{1}{2}''$	5	31.0
#12, 2''	4	45.0

The store buys these screws in 1-pound boxes (1 lb = 454 g). From two boxes of each type of screws, how many handyman assortments can the store prepare?

ilw 4.27 Ammonia is produced industrially using the Haber process:

$$N_2 + 3 H_2 \longrightarrow 2 NH_3$$

Suppose that an industrial reactor is charged with 75.0 kg each of N_2 and H_2. Use a table of amounts to determine what mass of ammonia could be produced if the reaction went to completion.

4.28 Acrylonitrile is an important building block for synthetic fibers and plastics. Over 1.4 billion kilograms of acrylonitrile are produced in the United States each year. The compound is synthesized from propene in the following reaction:

$$2 C_3H_6(g) + 2 NH_3(g) + 3 O_2(g) \longrightarrow 2 C_3H_3N(l) + 6 H_2O(g)$$

Propene   Acrylonitrile

Use a table of amounts to determine how many kilograms of acrylonitrile can be prepared from 1.50×10^3 kg of propene, 6.80×10^2 kg of ammonia, and 1.92×10^3 kg of oxygen.

4.29 For each of the reactions given in Problem 4.5, consider an industrial process that starts with 1.00 metric ton (1000 kg) of each reactant. Construct a table of amounts, identify the limiting reactant, and determine the maximum mass of each product that could be produced.

4.30 For each of the reactions given in Problem 4.6, consider an industrial process that starts with 7.50×10^3 kg of each reactant. Construct a table of amounts, identify the limiting reactant, and determine the maximum mass of each product that could be produced.

4.31 Elemental phosphorus, P_4, reacts vigorously with oxygen to give P_4O_{10}. Use a table of amounts to determine how much P_4O_{10} can be prepared from 3.75 g of P_4 and 6.55 g O_2, and how much of the excess reactant will remain at the end of the reaction.

4.32 The following unbalanced reaction is called the *thermite reaction*. It releases tremendous amounts of energy and is sometimes used to generate heat for welding:

$$Al + Fe_3O_4 \longrightarrow Fe + Al_2O_3$$

Use a table of amounts to determine the masses of all substances present after the reaction if 2.00×10^2 g of Al and 7.00×10^2 g of Fe_3O_4 react to completion.

Precipitation Reactions

4.33 What are the major species present in aqueous solutions of each of the following? (a) NH_4Cl; (b) $Fe(ClO_4)_2$; (c) Na_2SO_4; (d) Br_2; and (e) KBr

4.34 What are the major species present in aqueous solutions of each of the following? (a) $(NH_4)_2CO_3$; (b) $NaHSO_4$; (c) $CoCl_2$; (d) $Mg(NO_3)_2$; and (e) CH_3OH (methanol)

4.35 What are the major species present in aqueous solutions of each of the following? (a) potassium hydrogenphosphate; (b) acetic acid; (c) sodium acetate; (d) ammonia; and (e) ammonium chloride

4.36 What are the major species present in aqueous solutions of each of the following? (a) carbonated water (water saturated with carbon dioxide); (b) lithium carbonate; (c) potassium sulfite; (d) hydrogen sulfide; (e) sodium hydrogen sulfate

4.37 Consider the addition of each of the following aqueous solutions to each solution described in Problem 4.33. In each case, decide whether a precipitate will form. If so, identify the precipitate. (a) $AgNO_3$; (b) Na_2CO_3; (c) $Ba(OH)_2$

4.38 Consider the addition of each of the following aqueous solutions to each solution described in Problem 4.34. In each case, decide whether a precipitate will form. If so, identify the precipitate. (a) $Pb(CH_3CO_2)_2$; (b) $Ca(OH)_2$; (c) KOH

4.39 Write the balanced net ionic equation for each of these precipitation reactions. Also, identify the spectator ions.

(a) $AgNO_3(aq) + KOH(aq) \rightarrow$?
(b) $Fe(ClO_4)_3(aq) + (NH_4)_2C_2O_4(aq) \rightarrow$?
(c) $Pb(NO_3)_2(aq) + NaBr(aq) \rightarrow$?
(d) $KOH(aq) + NiSO_4(aq) \rightarrow$?

4.40 Write the balanced net ionic equation for each of these precipitation reactions. Also, identify the spectator ions.

(a) $LiCl(aq) + AgNO_3(aq) \rightarrow$?
(b) $MgSO_4(aq) + Na_3PO_4(aq) \rightarrow$?
(c) $Ba(OH)_2(aq) + Na_2SO_4(aq) \rightarrow$?
(d) $AlCl_3(aq) + KOH(aq) \rightarrow$?

ilw 4.41 If 55.0 mL of a 5.00×10^{-2} M solution of silver nitrate is mixed with 95.0 mL of 3.50×10^{-2} M potassium carbonate, what mass of solid forms, and what ions remain in solution?

4.42 If 75.0 mL of a 0.750 M solution of lead(II) nitrate is mixed with 125 mL of 0.855 M ammonium chloride, what mass of solid forms, and what ions remain in solution?

Acid – Base Reactions

4.43 Write the balanced net ionic equation for each of these acid–base reactions. Also, identify the spectator ions.

(a) $HCl(aq) + Ca(OH)_2(aq) \rightarrow$?
(b) $H_3PO_4(aq) + $ excess $LiOH(aq) \rightarrow$?
(c) $NH_3(aq) + HNO_3(aq) \rightarrow$?
(d) $CH_3CO_2H(aq) + KOH(aq) \rightarrow$?

4.44 Write the balanced net ionic equation for each of these acid–base reactions. In each case, write the formulas of the spectator ions.

(a) $NH_3(aq) + HBr(aq) \rightarrow$?
(b) $HClO(aq)$ (a weak acid) $+ NaOH(aq) \rightarrow$?
(c) $HClO_4(aq) + Ca(OH)_2(aq) \rightarrow$?
(d) $H_2SO_4(aq) + $ excess $KOH(aq) \rightarrow$?

4.45 One common component of antacids is $Al(OH)_3$. If an upset stomach contains 155 mL of 0.175 M HCl, what mass of $Al(OH)_3$ is required to completely neutralize the acid?

4.46 One common component of antacids is $Mg(OH)_2$. If an upset stomach contains 125 mL of 0.115 M HCl, what mass of $Mg(OH)_2$ is required to completely neutralize the acid?

4.47 Calculate the molarities of all ions present in a solution made by mixing 150.0 mL of 2.00×10^{-2} M $Ba(OH)_2$ solution with 100.0 mL of 5.00×10^{-2} M HCl solution.

4.48 Calculate the molarities of all ions present in a solution made by adding 1.53 g of solid NaOH to 215 mL of 0.150 M $HClO_4$.

ilw 4.49 A student wishes to determine the concentration of a solution of KOH. The student adds 5.00 mL of the KOH solution to 150 mL of water, adds an indicator, and titrates with 0.1206 M HCl. If the titration requires 27.35 mL of acid, what is the concentration of the KOH solution?

4.50 A student wishes to determine the concentration of a solution of HCl. The student adds 10.00 mL of the HCl solution to 150 mL of water, adds an indicator, and titrates with 0.0965 M NaOH. If the titration requires 32.45 mL of base, what is the concentration of the HCl solution?

4.51 A technician standardizes a solution of NaOH by titrating 0.6634 g of potassium hydrogen phthalate dissolved in 200 mL of water. It takes 36.55 mL of base solution to reach the stoichiometric point. Determine the concentration of the NaOH solution.

4.52 A technician standardizes a solution of KOH by titrating 0.7455 g of potassium hydrogen phthalate dissolved in 150 mL of water. It takes 39.20 mL of base solution to reach the stoichiometric point. Determine the concentration of the KOH solution.

Oxidation – Reduction Reactions

4.53 Some of the following react when added together, but others do not. For those that do, write a balanced net ionic redox equation. (a) $Cu + HCl(aq)$; (b) $Cu + MgCl_2(aq)$; (c) $Cu + AgNO_3(aq)$; and (d) $K + H_2O$

4.54 Some of the following react when added together, but others do not. For those that do, write a balanced net ionic redox equation. (a) $Al + HCl(aq)$; (b) $Zn + Au^{3+}(aq)$; (c) $Ni + MgCl_2(aq)$; and (d) $Na + H_2O$

4.55 Write the balanced redox reactions for the formation of each of the following oxides from the reaction of molecular oxygen with pure metal: (a) strontium oxide; (b) chromium(III) oxide; (c) tin(IV) oxide.

4.56 Write the balanced redox reactions for the formation of each of the following oxides from the reaction of molecular oxygen with pure metal: (a) chromium(VI) oxide; (b) zinc oxide; (c) copper(I) oxide.

ilw 4.57 Aluminum metal generates H_2 gas when dropped into 6 M HCl. Calculate the mass of H_2 that will form from the complete reaction of 0.355 g Al with 8.00 mL of 6.00 M HCl.

4.58 Iron metal reacts with hydrochloric acid to give H_2 gas and Fe^{2+} ions. If 5.8 g of iron is to be dissolved in 1.5 M HCl, what is the minimum volume of the acid solution required to react with all of the iron?

Additional Paired Problems

4.59 The Solvay process is a commercial method for producing sodium carbonate. In one step of this process, sodium bicarbonate is precipitated by mixing highly concentrated aqueous solutions of sodium chloride and ammonium bicarbonate:

$$NH_4HCO_3(aq) + NaCl(aq) \longrightarrow NaHCO_3(s) + NH_4Cl(aq)$$

(a) Write the net ionic reaction for this step of the Solvay process. (b) What are the spectator ions? (c) If 5.00×10^2 L of 1.50 M NH_4HCO_3 are treated with 5.00×10^2 L of 6.00 M NaCl, and 35.0 kg of $NaHCO_3$ are produced, what is the percent yield of the process? (d) Calculate the concentrations of all dissolved ions at the end of the reaction.

4.60 A white precipitate forms when aqueous calcium nitrate is mixed with aqueous ammonium sulfate. (a) Identify the precipitate and write the net ionic equation for the reaction. (b) What are the spectator ions? (c) Calculate the mass of the precipitate that forms when 100.0 mL of 1.50 M aqueous calcium nitrate is mixed with 75.0 mL of 3.00 M ammonium sulfate. (d) Calculate the concentrations of all dissolved ions at the end of the reaction.

4.61 Although xenon is a "noble gas," it reacts with fluorine and oxygen. When xenon gas reacts with F_2 gas, XeF_4 is one of the products. If 5.00 g of xenon reacts with excess fluorine and generates 4.00 g of XeF_4, what is the percent yield and how much xenon remains unreacted, assuming no other product forms?

4.62 Hydrogen fluoride is produced industrially by the action of sulfuric acid on CaF_2. If 365 kg of CaF_2 is treated with excess sulfuric acid and 155 kg of HF is produced, what is the percent yield and how much CaF_2 remains unreacted, assuming no other fluorine-containing product forms?

4.63 Pyrite (FeS_2) reacts with excess molecular oxygen to give iron(III) oxide and SO_2. If the extraction process is 85% efficient, what mass of iron(III) oxide can be formed in the processing of 175 metric tons of ore that contains 55% iron pyrite by mass?

4.64 The element titanium is commonly found as the ore ilmenite, $FeTiO_3$. Much of the world reserves of titanium is found in Canada. At a particular mine, a sample of earth was found to contain 15% ilmenite by mass. What mass of pure titanium metal can be isolated from 1.00×10^3 metric tons of earth, if the extraction process is 95% efficient?

4.65 Predict the product(s) of the following reactions by writing balanced equations. When the reaction involves ions, write a net ionic equation. Identify each reaction as precipitation, acid–base, or redox.

    (a) $Al(s) + O_2(g) \longrightarrow$

    (b) $C_3H_8(g) + O_2(g) \longrightarrow$

    (c) $Mg(s) + HBr(aq) \longrightarrow$

    (d) $NaOH(aq) + HCl(aq) \longrightarrow$

    (e) $Pb(NO_3)_2(aq) + (NH_4)_2CO_3(aq) \longrightarrow$

    (f) $Ca(OH)_2(aq) + H_2SO_4(aq) \longrightarrow$

4.66 Predict the product(s) of the following reactions by writing balanced equations. When the reaction involves ions, write a net ionic equation. Identify each reaction as precipitation, acid–base, or redox.

    (a) $Ca(OH)_2(s) + HCl(aq) \longrightarrow$

    (b) $Ni(s) + HCl(aq) \longrightarrow$

    (c) $AgNO_3(aq) + KOH(aq) \longrightarrow$

    (d) $Ba(OH)_2(aq) + HClO_4(aq) \longrightarrow$

    (e) $Ca(s) + H_2O(l) \longrightarrow$

    (f) $HNO_3(aq) + NH_3(aq) \longrightarrow$

4.67 Ruthenium forms four different binary oxides: Ru_2O_3, RuO_2, RuO_3, and RuO_4. Write a balanced chemical equation for molecular oxygen reacting with metallic ruthenium to form each of these oxides, and determine how many electrons a ruthenium atom loses in each case.

4.68 Nitrogen forms the binary oxides NO, N_2O, NO_2, N_2O_4, and N_2O_5. Write a balanced chemical equation for O_2 reacting with N_2 to form each of these oxides, and determine how many electrons a nitrogen atom loses in each case.

4.69 The reaction of atom X with atom Y is represented in the following diagram. Write the balanced equation for this reaction.

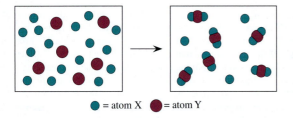

    ● = atom X ● = atom Y

4.70 The reaction of element U with element V is represented in the diagram shown below. Write the balanced equation for this reaction.

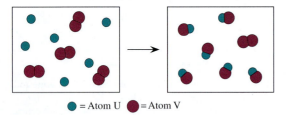

    ● = Atom U ● = Atom V

4.71 The gases leaving an automobile cylinder after combustion include CO_2, CO (from incomplete combustion), NO, H_2O, H_2, N_2, O_2, and unburned hydrocarbons. After leaving the engine, these exhaust gases are passed through a catalytic converter whose purpose is to change pollutants into less harmful substances. Many reactions occur in a catalytic converter, including those shown below. Write a balanced chemical equation for each.

    (a) $H_2 + NO \longrightarrow NH_3 + H_2O$

    (b) $CO + NO \longrightarrow N_2 + CO_2$

    (c) $NH_3 + O_2 \longrightarrow N_2O + H_2O$

    (d) Nitrogen oxide and ammonia react to give nitrogen and water.

    (e) Water and nitrogen oxide react to give molecular oxygen and ammonia.

    (f) Molecular hydrogen and oxygen react to give water.

4.72 Recall that combustion is a reaction with O_2. Write balanced chemical equations for the combustion reactions of the following substances. Assume that the only products are H_2O and CO_2.

    (a) C_4H_{10} (butane, the substance used in lighter fluid)

    (b) C_6H_6 (benzene, a solvent that is carcinogenic)

    (c) C_2H_5OH (ethanol, the intoxicant in alcoholic drinks)

    (d) C_5H_{12} (pentane, a highly volatile liquid)

    (e) $C_6H_{11}OH$ (cyclohexanol, a useful solvent)

4.73 Devise a synthesis, write the net ionic reaction, and compute masses of each starting material needed to make 2.50 kg of each of the following solid ionic compounds: (a) $FePO_4$; (b) $Zn(OH)_2$; (c) $NiCO_3$.

4.74 Lithopone, a brilliant white pigment used in paints, paper, and white rubber products, is a mixture of two insoluble ionic solids, ZnS and $BaSO_4$. Suggest how 1.0 kg of lithopone could be prepared by a precipitation reaction.

4.75 The following diagram represents a small portion of a reaction vessel that contains the starting materials for the Haber synthesis of ammonia:

$$N_2 + H_2 \longrightarrow NH_3 \text{ (unbalanced)}$$

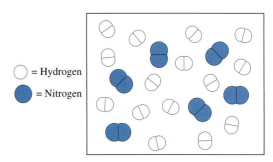

    ○ = Hydrogen

    ● = Nitrogen

(a) What is the limiting reactant in this reaction? (b) Draw a picture that shows how this portion of the vessel will look when the reaction is complete. (c) If each molecule in the picture represents one mole of compound, what mass of ammonia will the system produce, and what mass of the excess reactant will be left behind?

4.76 Carbon dioxide, which is used to carbonate beverages and as a coolant (dry ice), is produced from methane and water vapor:

$$CH_4(g) + H_2O(g) \longrightarrow CO_2(g) + H_2(g) \text{ (unbalanced)}$$

The diagram shown below represents a small portion of a vessel that contains starting materials for this reaction.

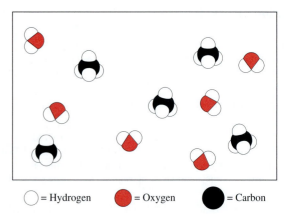

    ○ = Hydrogen ● = Oxygen ● = Carbon

(a) What is the limiting reactant in the reaction? (b) Draw a picture that shows what the vessel will look like when the reaction is complete. (c) If each molecule in the picture represents one mole of compound, what mass of each product will the system produce and what mass of the excess reactant will be left behind?

4.77 A 4.6-g sample of ethanol (C_2H_5OH) is burned completely in air. (a) How many moles of H_2O are formed? (b) How many molecules? (c) How many grams?

4.78 An automobile engine completely burns 3.5 g of octane (C_8H_{18}). (a) How many moles of CO_2 are formed? (b) How many molecules? (c) How many grams?

4.79 Identify the species present in the following aqueous solutions. In each case, write a net ionic equation that describes what reaction occurs upon mixing equimolar portions of the two solutions. (a) NH_3 and HCl; (b) $CaCl_2$ and Na_2SO_4; (c) KOH and HBr; and (d) HNO_2 (a weak acid) and KOH.

4.80 Identify the species present in the following aqueous solutions. In each case, write a net ionic equation that describes what reaction occurs upon mixing equimolar portions of the two solutions. (a) $HgCl_2$ and K_2S; (b) HNO_3 and $Ba(OH)_2$; (c) $NaOH$ and CH_3CO_2H; and (d) KOH and $FeCl_3$.

4.81 Propylene oxide is used primarily in the synthesis of poly(propylene glycol), a polymer used in the manufacture of automobile seats, bedding, and carpets. Around 2 billion kilograms of this compound are produced annually in the United States. Propylene oxide is produced by the following reaction:

t-Butyl hydroperoxide $C_4H_{10}O_2$ Propene C_3H_6 Propylene oxide C_3H_6O *t*-Butanol $C_4H_{10}O$

(a) How many kilograms of propylene oxide can be prepared from 75 kg of *t*-butyl hydroperoxide? (b) What mass of propene will be required for the synthesis?

4.82 Vinyl chloride, one of the top 20 industrial compounds, is used primarily to make the polymer poly(vinyl chloride), better known as PVC. This versatile polymer is used to make piping, siding, gutters, floor tile, clothing, and toys. Vinyl chloride is made by oxychlorination of ethylene (C_2H_4). The overall balanced equation follows:

Ethylene   Vinyl chloride

Annual production of vinyl chloride is about 6.8 billion kilograms. What minimum masses (in billions of kilograms) of C_2H_4 and HCl are required to produce this much vinyl chloride?

4.83 Write the net ionic reaction and draw a molecular picture that shows what happens when 50.0 mL of 0.010 M HCl solution is mixed with 50.0 mL of 0.0050 M $Ba(OH)_2$ solution.

4.84 Write the net ionic reaction and draw a molecular picture that shows what happens when 20.0 mL of 0.200 M NaOH solution is mixed with 40.0 mL of 0.100 M HCl solution.

4.85 Magnesium metal burns with a bright flame in oxygen gas, and the product is solid white magnesium oxide. Draw a molecular picture showing six magnesium atoms and four oxygen molecules. Then draw another molecular picture of this same system after reaction occurs.

4.86 Sodium metal reacts vigorously with chlorine gas to form solid white sodium chloride. Draw a molecular picture showing ten sodium atoms and three chlorine molecules. Then draw another molecular picture of this same system after reaction occurs.

4.87 Calcium carbonate (limestone) reacts with hydrochloric acid to generate water and carbon dioxide gas. In a certain experiment, 5.0 g of $CaCO_3$ is added to 0.50 L of 0.10 M HCl. (a) Write a balanced net equation for this reaction. (b) What mass of carbon dioxide is formed? (c) Calculate the concentrations of all species present in solution at the end of the reaction.

4.88 Magnesium metal reacts with HCl solution, liberating H_2 gas and generating Mg^{2+} cations in solution. A 1.215-g sample of Mg metal is added to 50.0 mL of a 4.0 M HCl solution, and the reaction goes to completion. (a) Write a balanced net equation for this reaction. (b) What mass of H_2 is formed? (c) Calculate the concentrations of all ions present in the solution at the end of the reaction.

4.89 Predict the products of the following reactions by writing balanced equations. When the reaction involves ions, write a net ionic equation. Identify each reaction as precipitation, acid–base, or redox.
(a) $HCl(aq) + Ca(s) \longrightarrow$
(b) $Li(s) + O_2(g) \longrightarrow$
(c) $HBr(aq) + NH_3(aq) \longrightarrow$
(d) $C_3H_8(g) + O_2(g) \longrightarrow$

4.90 Predict the products of the following reactions by writing balanced equations. When the reaction involves ions, write a net ionic equation. Identify each reaction as precipitation, acid–base, or redox.
(a) $KOH(aq) + Co(NO_3)_2(aq) \longrightarrow$
(b) $HCl(aq) + Fe(s) \longrightarrow Fe^{3+}(aq)$
(c) $KI(aq) + Pb(CH_3CO_2)_2(aq) \longrightarrow$
(d) $HClO_4(aq) + NH_3(aq) \longrightarrow$
(e) $C_6H_5CO_2H(aq) + Ba(OH)_2(aq) \longrightarrow$

4.91 Compounds that undergo explosions typically produce large quantities of hot gases from much smaller volumes of highly reactive solids or liquids. Balance the reaction for the decomposition of nitroglycerine, a violent explosive:

$$C_3H_5N_3O_9(l) \longrightarrow N_2(g) + CO_2(g) + H_2O(g) + O_2(g)$$

4.92 Compounds that undergo explosions typically produce large quantities of hot gases from much smaller volumes of highly reactive solids or liquids. Balance the reaction for the decomposition of trinitrotoluene (TNT), a violent explosive:

$$C_7H_5N_3O_6(l) \longrightarrow N_2(g) + CO_2(g) + H_2O(g) + C(s)$$

More Challenging Problems

4.93 In the Haber synthesis of ammonia, N_2 and H_2 react at high temperature, but they never react completely. In a typical reaction, 24.0 kg of H_2 and 84.0 kg of N_2 react to produce 68 kg of NH_3. Using a table of amounts, find the theoretical yield, the percent yield, and the masses of H_2 and N_2 that remain unreacted, assuming that no other products form.

4.94 A student prepared 1.00 L of a solution of NaOH for use in titrations. The solution was standardized by titrating a sample of potassium hydrogen phthalate whose mass was 0.7996 g. Before titration, the buret reading was 0.15 mL. When the indicator changed color, the buret reading was 43.75 mL. Calculate the molarity of the NaOH solution.

4.95 One set of reactants for rocket fuel is hydrazine and hydrogen peroxide, which react vigorously when mixed:

$$N_2H_4(l) + H_2O_2(l) \longrightarrow N_2(g) + H_2O(g) \text{ (unbalanced)}$$

The density of liquid hydrazine is 1.44 g/mL, and that of hydrogen peroxide is 1.01 g/mL. What volume ratio of these two liquids should be used if both fuels are to be used up at the same time?

4.96 Sulfur dioxide is an atmospheric pollutant that is converted to sulfuric acid by O_2 and water vapor. This is one source of acid rain, a serious environmental problem. The sulfur dioxide content of an air sample can be determined. A sample of air is bubbled through an aqueous solution of hydrogen peroxide to convert all of the SO_2 to H_2SO_4.

$$H_2O_2 + SO_2 \longrightarrow H_2SO_4$$

Titration of the resulting solution completes the analysis (both H atoms of H_2SO_4 are titrated). In one such case, the analysis of 1.55×10^3 L of Los Angeles air gave a solution that required 5.70 mL of 5.96×10^{-3} M NaOH to complete the titration. Determine the number of grams of SO_2 present in the air sample.

4.97 Vitamin C (also called ascorbic acid) is an acid whose formula is $HC_6H_7O_6$. When treated with strong base, it undergoes the following reaction:

$$HC_6H_7O_6 + OH^- \longrightarrow C_6H_7O_6^- + H_2O$$

A pharmacist suspects that the vitamin C tablets received in a recent shipment are not pure. When a single 500.0-mg tablet is dissolved in 200.0 mL of water and titrated with a standard base that is 0.1045 M, it takes 24.45 mL to reach the stoichiometric point. Are the tablets pure? If not, what is the mass percentage of impurities? (Assume no impurity is either an acid or a base.)

4.98 Silicon tetrachloride is used in the electronics industry to make elemental silicon for computer chips. Silicon tetrachloride is prepared from silicon dioxide, carbon graphite, and chlorine gas.

$$SiO_2(s) + 2\,C(s) + 2\,Cl_2(g) \longrightarrow SiCl_4(l) + 2\,CO(g)$$

If the reaction goes in 95.7% yield, how much silicon tetrachloride can be prepared from 75.0 g of each starting material, and how much of each reactant remains unreacted?

4.99 The largest single use of sulfuric acid is for the production of phosphate fertilizers. The acid reacts with calcium phosphate in a 2:1 mole ratio to give calcium sulfate and calcium dihydrogen phosphate. The mixture is crushed and spread on fields, where the salts dissolve in rain water. (Calcium phosphate, commonly found in phosphate rock, is too insoluble to be a direct source of phosphate for plants.) (a) Write a balanced equation for the reaction of sulfuric acid with calcium phosphate. (b) How many kilograms each of sulfuric acid and calcium phosphate are required to produce 50.0 kg of the calcium sulfate–dihydrogen phosphate mixture? (c) How many moles of phosphate ion will this mixture provide?

4.100 Write the balanced equation and determine the number of moles of water produced when 2.95 mL of pyridine (C_5H_5N, $\rho = 0.982$ g/mL) reacts with excess O_2 to give water, carbon dioxide, and molecular nitrogen.

4.101 A former antiknock ingredient in gasoline is a colorless liquid whose formula is $C_5H_{12}O$. Write the balanced equation, and determine the number of moles of carbon dioxide produced when 3.15 mL of the ingredient reacts with excess oxygen to give carbon dioxide and water (density of $C_5H_{12}O$ = 0.740 g/mL).

4.102 Surface deposits of elemental sulfur around hot springs and volcanoes are believed to come from a two-step redox process. Combustion of hydrogen sulfide (H_2S) produces sulfur dioxide and water. The sulfur dioxide reacts with more hydrogen sulfide to give elemental sulfur and water. Write balanced chemical equations for these two reactions, and determine the minimum mass of hydrogen sulfide that a volcano must emit in order to deposit 1.25 kg of sulfur.

4.103 Silver jewelry is usually made from silver and copper alloys. The amount of copper in an alloy can vary considerably. The finest-quality alloy is sterling silver, which is 92.5% by mass silver. To determine the composition of a silver–copper alloy, a jeweler dissolved 0.135 g of metal shavings in 50 mL of concentrated nitric acid and then added 1.00 M KCl solution until no more precipitate formed. Filtration and drying yielded 0.156 g of AgCl precipitate. What was the mass composition of the silver alloy?

4.104 The CO_2 exhaled by astronauts must be removed from the spacecraft atmosphere. One way to do this is with solid LiOH:

$$CO_2(g) + 2\,LiOH(s) \longrightarrow Li_2CO_3(s) + H_2O(l)$$

The CO_2 output of an astronaut is about 1.0 kg/day. What is the minimum mass of LiOH required for a six-day space shuttle flight involving five astronauts?

4.105 Potassium hydroxide is considerably cheaper than lithium hydroxide and undergoes the analogous reaction with CO_2. Repeat the calculation of Problem 4.104 for KOH. Why is the more expensive substance used on the space shuttle?

4.106 Phosphorus is essential for plant growth, and it is often the limiting nutrient in aqueous ecosystems. However, too much phosphorus can cause algae to grow at an explosive rate. This process, known as *eutrophication,* robs the rest of the ecosystem of essential oxygen, often destroying all other aquatic life. One source of aquatic phosphorus pollution is the HPO_4^{2-} used in detergents in sewage plants. The simplest way to remove HPO_4^{2-} is to treat the contaminated water with lime, CaO, which generates Ca^{2+} and OH^- ions in water. The phosphorus precipitates as $Ca_5(PO_4)_3OH$. (a) Write the balanced equation for CaO dissolving in water. (b) Write the balanced equation for the precipitation reaction. (*Hint:* Proton transfer occurs as well as solid formation.) (c) How many kilograms of lime are required to remove all the phosphorus from a 1.00×10^4 L holding tank filled with contaminated water that is 0.0156 M in HPO_4^{2-}?

Group Study Problems

4.107 Analysis of a sample of natural gas gives the following results: 74% (by mass) methane (CH_4), 18% ethane (C_2H_6), and 8% propane (C_3H_8). How many moles of CO_2 would be produced in the combustion of 750 g of this gas?

4.108 Although H_2SO_4 has two acidic hydrogen atoms, only one of them transfers readily to water. Thus, an aqueous solution of sulfuric acid contains water molecules, hydronium cations, and hydrogen sulfate anions. The following drawing represents one part of an aqueous solution of sulfuric acid (the solvent water molecules have been omitted for clarity):

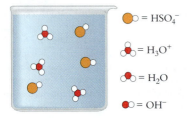

= HSO_4^-

= H_3O^+

= H_2O

= OH^-

For each of the following situations, draw a new picture that shows what this portion of the solution looks like after the reaction is complete. Include any water molecules produced in the reaction. (*Hint:* Hydroxide reacts with hydronium ions in preference to HSO_4^-.) (a) Three OH^- ions are added to the solution. (b) Three additional OH^- ions are added to solution (a). (c) Two additional OH^- ions are added to solution (b).

4.109 Decaborane, $B_{10}H_{14}$, was used as fuel for the Redstone rockets of the 1950s. Decaborane reacts violently with molecular oxygen according to the following equation:

$$B_{10}H_{14} + O_2 \longrightarrow B_2O_3 + H_2O \quad \text{(unbalanced)}$$

The two starting materials are stored in separate containers. When mixed, they ignite spontaneously, releasing large amounts of energy. Both fuel materials should run out at the same time because this minimizes the excess mass that the rocket must carry. If the total mass of both components is to be 12.0×10^4 kg, what mass of liquid oxygen and what mass of decaborane should be used?

4.110 As a final examination in the general chemistry laboratory, a student was asked to determine the mass of $Ca(OH)_2$ that dissolves in 1.000 L water. Using a published procedure, the student did the following: (1) About 1.5 mL of concentrated HCl (12 M) was added to 750 mL of distilled water. (2) A solution of KOH was prepared by adding approximately 1.37 g KOH to 1.0 L distilled water. (3) A sample of potassium hydrogen phthalate (185.9 mg) was dissolved in 100 mL of distilled water. Titration with the KOH solution required 25.67 mL to reach the stoichiometric point. (4) A 50.00-mL sample of the HCl solution prepared in step 1 was titrated with the KOH solution. The titration required 34.02 mL of titrant to reach the stoichiometric point. (5) The student was given a 25.00-mL sample of a saturated solution of $Ca(OH)_2$ for analysis. Titration with the HCl solution required 29.28 mL to reach the stoichiometric point. How many grams of calcium hydroxide dissolve in 1.00 L of water?

4.111 CF$_3$CH$_2$F (HFC-134a) has replaced chlorofluorocarbon compounds for use as refrigerants. HFC-134a is produced from trichloroethylene by the following reactions:

$$CCl_2=CHCl + 3\ HF \xrightarrow{catalyst} CF_3CH_2Cl + 2\ HCl$$

$$CF_3CH_2Cl + HF \xrightarrow{catalyst} CF_3CH_2F + HCl$$

In 1999, over 1.3 million metric tons of HFC-134a were produced, and production is increasing at a rate of nearly 20% annually. Manufacturers sell HFC-134a for $5.50/kg. Assume that you are a chemist working for a company that sold 65 million kilograms of HFC-134a last year. (a) If the synthesis had a reaction yield of 47%, how much trichloroethylene was used to produce the HFC-134a? (b) You have developed a new catalyst that improves the yield of the synthesis of HFC-134a from 47% to 53%. If the company uses the same mass of trichloroethylene for next year's production, how many kilograms of HFC-134a can the company produce if they use your new process? (c) Assuming that the selling price of HFC-134a does not change, how much more money will the company make next year because of your work?

4.112 Bronze is an alloy of copper and zinc. When a 5.73-g sample of bronze was treated with excess aqueous HCl, 21.3 mg of H$_2$ was produced. What was the percentage by mass composition of the bronze? (*Hint:* Consult the activity series shown in Table 4-3.)

Answers to Section Exercises

4.1.1 $2\ C_8H_{18} + 25\ O_2 \rightarrow 16\ CO_2 + 18\ H_2O$
4.1.2 $N_2 + O_2 \rightarrow 2\ NO$ $2\ NO + O_2 \rightarrow 2\ NO_2$
4.1.3 $2\ C_3H_6 + 2\ NH_3 + 3\ O_2 \rightarrow 2\ C_3H_3N + 6\ H_2O$
4.2.1 (a) 333 g PCl$_3$ can be prepared; (b) 258 g Cl$_2$ is consumed.
4.2.2 0.737 g
4.2.3 2.80×10^3 g O$_2$ is required; 1.14×10^3 g H$_2$O and 2.47×10^3 g CO$_2$ are produced.
4.3.1 Theoretical yield is 1.96 g, and percent yield is 94.0%. Possible reasons are that not all the KClO$_3$ decomposes or that some O$_2$ escapes and is not collected.
4.3.2 Start with 5.8 tons of sulfur.
4.3.3 66 g salicylic acid is required.
4.4.1 The maximum mass is 793 kg HCN, and 29 kg CH$_4$ is left.
4.4.2 7.5 g
4.4.3 2.3×10^2 g
4.5.1 d, e, and g are insoluble; a, b, c, and f are soluble.
4.5.2 Mix solutions of a soluble calcium salt and a soluble phosphate salt; for example, Ca(NO$_3$)$_2$ and Na$_3$PO$_4$. To make 1.5 kg of product requires 2.4 kg Ca(NO$_3$)$_2$ and 1.6 kg Na$_3$PO$_4$.

4.5.3 26 kg
4.6.1 Net ionic reaction:
$$H_2CO_3(aq) + 2\ OH^-(aq) \rightarrow CO_3^{2-}(aq) + 2\ H_2O(l)$$
Molecular picture:

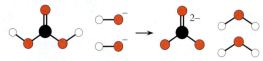

4.6.2 1.29×10^3 g
4.6.3 0.1009 M
4.7.1 0.135 mol
4.7.2 (a) $2\ Mg + TiO_2 \rightarrow Ti + 2\ MgO$. Mg is oxidized and TiO$_2$ is reduced. (b) 1.02×10^2 kg
4.7.3 (a) Half-reactions: Ni$(s) \rightarrow$ Ni$^{2+}(aq) + 2$ e$^-$ and $2\ H_3O^+(aq) + 2$ e$^- \rightarrow H_2(g) + 2\ H_2O(l)$; balanced redox equation: Ni$(s) + 2\ H_3O^+(aq) \rightarrow$ Ni$^{2+}(aq) + H_2(g) + 2\ H_2O(l)$. (b) and (c) No reaction; (d) Half-reactions: Fe$(s) \rightarrow$ Fe$^{2+}(aq) + 2$ e$^-$ and Ag$^+(aq) + $ e$^- \rightarrow$ Ag(s); balanced redox equation: Fe$(s) + 2\ Ag^+(aq) \rightarrow$ Fe$^{2+}(aq) + 2\ Ag(s)$

Answers to Extra Practice Exercises

4.1 $2\ CH_3OH + 3\ O_2 \rightarrow 2\ CO_2 \rightarrow 4\ H_2O$
4.2 112 g formic acid is required.
4.3 186 kg of NH$_3$ are required as a starting material, and the reaction produces 295 kg of HCN.
4.4 66.1% yield
4.5 9.69×10^4 kg of NH$_3$ are required.
4.6

Reaction	4 NH$_3$	+	5 O$_2$	$\longrightarrow$	4 NO	+	6 H$_2$O
Start (mol)	5.00		7.00		0		0
Change (mol)	−5.00		−6.25		+5.00		+7.50
Final (mol)	0		0.75		5.00		7.50

4.7 Yield = 78.4%
4.8 No precipitate
4.9 The yield is 99.4%.
4.10 The mass of magnesium phosphate precipitate is 2.43 g.
4.11 Suitable salts: NaI (162.6 g) and Pb(NO$_3$)$_2$ (180 g)
4.12

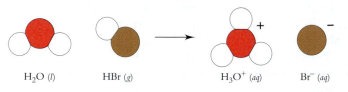

H$_2$O (*l*)   HBr (*g*)   H$_3$O$^+$ (*aq*)   Br$^-$ (*aq*)

4.13 $H_3PO_4(aq) + 2\ OH^-(aq) \longrightarrow HPO_4^{2-}(aq) + 2\ H_2O(l)$.

HPO$_4^{2-}$

4.14 46.3 mL of base is required.
4.15 The concentration of acetic acid is 0.845 M.
4.16 7.923×10^{-2} M, which is consistent with the previous result to within 1 part in 800.
4.17 Pb is above Cu^{2+} in the activity series, so it transfers electrons: Pb$(s) + $ Cu$^{2+}(aq) \longrightarrow$ Pb$^{2+}(aq) + $ Cu(s)
4.18 Fe undergoes redox: $2\ H_3O^+(aq) + $ Fe$(s) \longrightarrow H_2(g) + 2\ H_2O(l) + $ Fe$^{2+}(aq)$
FeCl$_3$ dissolves: FeCl$_3(s) \longrightarrow$ Fe$^{3+}(aq) + 3$ Cl$^-(aq)$
Fe(OH)$_3$ undergoes proton transfer: Fe(OH)$_3(s) + 3\ H_3O^+(aq) \longrightarrow$ Fe$^{3+}(aq) + 6\ H_2O(l)$

The Behavior of Gases

Earth's Atmosphere

The atmosphere of the Earth is a shroud of gas that surrounds our planet. Most of the atmosphere lies within 100 km of the surface. The atmosphere contains two major components: a bit less than 80% N_2 and a bit more than 20% O_2. In addition to these two species, the atmosphere contains small amounts of water and traces of other species. Our molecular inset is a representation of the atmosphere.

In recent years scientists have discovered that some of the trace components of the atmosphere are changing in ways that may cause dramatic changes in the environment. All of us have heard the alarms: Acid rain damages ecosystems. Burning fossil fuels may lead to global warming. Human activities have caused a "hole" in the ozone layer. More immediately visible are the spectacular sunsets — such as the one in our background photo — that result when the atmosphere contains a significant concentration of fine particles.

How could molecules present in mere trace amounts have such extensive effects on the environment? Acid rain provides a good example. Burning gasoline and coal produces some gaseous NO_2 and SO_2. These oxides dissolve in water droplets and are oxidized to produce nitric acid (HNO_3) and sulfuric acid (H_2SO_4). Rain that forms from such acid-containing water vapor is highly acidic and can damage forests and aquatic life. For example, rain falling in Europe now regularly registers 10 times "normal"

CHAPTER CONTENTS

Learning Resources

 KEY CONCEPTS

 CRITICAL THINKING

 SKILLS TO MASTER

 ADDITIONAL HELP
www.wiley.com/college/olmsted
● *TUTORIALS*
● *ANIMATIONS*

acidity. In 1994 it was reported that 25% of the trees in Europe suffer from leaf loss or leaf damage, most likely as a result of acid rain.

Sunlight acting on trace components in automobile exhaust gases causes the formation of smog. Cities worldwide, including Los Angeles, Mexico City, and Shanghai (shown in our inset photo), suffer from smog. Smog contains nitrogen oxides, ozone, and larger molecules. The chemistry of smog is complex and not fully understood. Atmospheric scientists are studying how smog forms and how it can be prevented.

Acid rain and smog show that changes in the amounts of trace components of the atmosphere can have dramatic consequences on the environment. Ozone depletion in the upper atmosphere (treated in Chapter 15) and gradual global warming (treated in Chapter 7) are long-term effects that may be equally or even more damaging. Atmospheric research scientists seek for ways to mitigate these effects.

We begin this chapter with a discussion of the variables that characterize gases. Then we develop a molecular description that explains gas behavior.

Next, we explore additional gas properties and show how to do stoichiometric calculations for reactions involving gas-phase species. Finally, we return to the Earth's atmosphere and describe some aspects of its composition and chemical reactions.

Figure 5-1
When a basketball strikes a backboard, it exerts a force on the backboard. The gas molecules inside the basketball (*inset*) also exert forces when they strike the walls of the basketball. The net result of many collisions is gas pressure.

5.1 PRESSURE

Like a liquid or a solid, a gas can be characterized by its volume (V), temperature (T), and amount (n). Unlike liquids and solids, however, the properties of gases are highly sensitive to pressure (P). We can understand this sensitivity using the molecular perspective.

An object that strikes a surface exerts a force against that surface. For example, Figure 5-1 shows that a basketball exerts force when it strikes a backboard. Gas molecules also exert forces. Recall from Chapter 2 that atoms and molecules are always in motion. At the molecular level, atoms and molecules exert forces through never-ending collisions, among themselves and with the walls of their container.

Although pressure is caused by molecular collisions, it is a macroscopic property, the collective result of countless collisions. We can get a feel for the macroscopic characteristics of gas pressure by examining Earth's atmosphere.

The air around us is a huge reservoir of gas that exerts pressure on the Earth's surface. This pressure of the atmosphere can be measured with an instrument called a **barometer.** Figure 5-2 shows a schematic view of a simple mercury barometer. A long glass tube, closed at one end, is filled with liquid mercury. The filled tube is inverted carefully into a dish that is partially filled with more mercury. The force of gravity pulls downward on the mercury in the tube. With no opposing force, the mercury would all run out of the tube and mix with the mercury in the dish. The mercury does fall, but the flow stops at a fixed height. The column of mercury stops falling because the atmosphere exerts pressure on the mercury in the dish, pushing the column up the tube. The column is in balance when the height of the mercury column generates a downward force on the inside of the tube that exactly balances the force exerted by the atmosphere on the outside of the tube.

At sea level, atmospheric pressure supports a mercury column approximately 760 mm in height. Changes in altitude and weather cause fluctuations in atmospheric pressure. Nevertheless, at sea level the height of the mercury column seldom varies by more than 10 mm, except under extreme conditions, such as in the eye of a hurricane, when the mercury in a barometer may fall below 740 mm.

A **manometer** is similar to a barometer, but in a manometer gases exert pressure on *both* liquid surfaces. Consequently, a manometer measures the *difference* in pressures exerted by two gases. A simple manometer, shown in Figure 5-3, is a U-shaped glass tube containing mercury. One side of the tube is exposed to the atmosphere and the other to a gas whose pressure we want to measure. In Figure 5-3, the pressure exerted by the atmosphere is less than the pressure exerted by the gas in the bulb. The difference in heights of mercury (Δh, in mm) between the two sides of the manometer depends on the difference in the pressures.

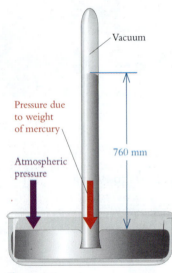

Vacuum

Pressure due to weight of mercury

Atmospheric pressure

760 mm

Figure 5-2
Diagram of a mercury barometer.

Units of Pressure

Traditionally, chemists define the units of pressure in terms of the Earth's atmosphere and the mercury barometer. The **standard atmosphere (atm)** is the pressure that will support a column of mercury 760 mm in height.

A second common pressure unit, the **torr,** also is based on the mercury barometer. One torr is the pressure exerted by a column of mercury 1 mm in height. Because the standard atmosphere supports a 760-mm column of mercury, the relationship between the atmosphere and the torr is 1 atm = 760 torr = 760 mm Hg.

The accepted SI unit for pressure is the **pascal (Pa).** Pressure is defined as force per unit area, F/A, so the pascal can be expressed by combining the SI units for these two variables. The SI unit of force is the newton (N), and area is measured in square meters (m^2). Thus, the pascal is 1 N/m^2. Expressed in pascals, the numerical value of atmospheric pressure is quite large. By international agreement, 1 atm is defined exactly in terms of pascals: 1 atm = 1.01325×10^5 Pa. A more convenient SI unit for pressures in the vicinity of 1 atm is the **bar,** defined to be exactly 10^5 pascals. As we describe in Chapters 6 and 14, chemists consider 1 bar to be the standard pressure for tabulating various properties of substances. Here are the equivalences between bar, atm, and torr:

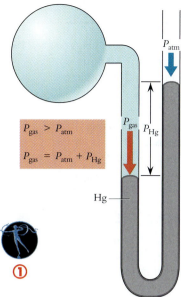

$$P_{gas} > P_{atm}$$
$$P_{gas} = P_{atm} + P_{Hg}$$

Hg

1 atm = 1.01325 bar   1 bar = 770.07 torr

In this chapter, we work primarily with pressures in atmospheres or torr, but we use pressures in bars in later chapters when we work with standard chemical conditions. Example 5-1 illustrates pressure measurement and unit conversions.

Figure 5-3
The difference in heights, Δh, of liquid on the two sides of a manometer, is a measure of the difference in gas pressures (ΔP) applied to the two sides.

| Δ (Delta) stands for "difference in."

A scientist collected an atmospheric gas sample. A manometer attached to the gas sample gave the reading shown in the figure (see also Figure 5-3), and the barometric pressure in the laboratory was 752 mm Hg. Calculate the pressure of the sample in atmospheres and in bars.

Example 5-1

Measuring Pressure

Strategy: We must combine the reading from the manometer with the barometric pressure to find the pressure of the gas sample. The manometer displays the pressure difference in millimeters of mercury, so conversion factors are needed to express the pressure in atmospheres and bars.

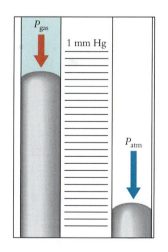

Solution: The difference in mercury levels in the manometer is 19 mm. This is the pressure difference in torr between the gas sample and atmospheric pressure. The latter, as measured with a barometer, is 752 torr. Shall we add or subtract the 19 torr pressure difference? Notice that the mercury level in the manometer is lower on the side exposed to the atmosphere. Thus, the atmosphere pushes on the mercury harder than does the gas sample, meaning that the pressure of the gas sample is lower than the pressure of the atmosphere. Subtract the pressure difference:

$$P_{gas} = P_{atm} - \Delta P = 752 \text{ torr} - 19 \text{ torr} = 733 \text{ torr}$$

Use the appropriate conversion factors to give the pressure in atm and bars.

1 atm = 760 torr   *and*   1 atm = 1.01325 bar

P (atm) = 733 torr (1 atm/760 torr) = 0.964 atm

P (bar) = 0.964 atm (1.01325 bar/atm) = 0.977 bar

Do the Results Make Sense? The manometer shows that the pressure is lower on the inside than on the outside, so pressures that are smaller than 1 atm and 1 bar are reasonable.

In Figure 5-3, suppose that $P_{Hg} = 34.5$ mm Hg on a day when the barometer reads 765.3 torr. Determine the gas pressure in the bulb in torr, atm, and bar.

Extra Practice
Exercise 5.1

Section Exercises

■ **5.1.1** In the eye of a severe hurricane, the height of a mercury barometer may fall to 710 mm. Express this pressure in torr and atmospheres. What percentage change from standard atmospheric pressure is this?

■ **5.1.2** Oxygen gas is sold in pressurized tanks. The gas in a tank exerts a pressure of 15.75 atm. Convert this pressure to pascals, kPa, and bars.

■ **5.1.3** What pressure is exerted on the left side of the manometer shown in Figure 5-3 when the difference in column heights is 4.75 cm, the right side is open to the atmosphere, and barometric pressure is 752.8 torr? Express your result in torr, in atmospheres, and in kPa.

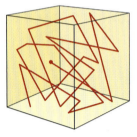

Figure 5-4
Molecules in a gas move freely throughout the entire volume of a container, changing direction whenever they collide with other molecules or with a wall. The line traces a possible path of a single molecule.

5.2 **DESCRIBING GASES**

The volume occupied by a gas changes dramatically in response to changes in conditions. This variability in volume is an obvious difference between gases and the other two phases, liquids and solids. Gas volumes change because the atoms or molecules of a gas move freely about, as shown schematically in Figure 5-4. Gas molecules move about to fill whatever volume is available to them.

Variations in Gas Volume

Early scientists carried out experiments that determined how the volume of a gas changes with conditions. Starting around 1660, the Englishman Robert Boyle investigated the properties of gases confined in J-shaped glass tubes. Figure 5-5a illustrates Boyle's experiments. He added liquid mercury to the open end of a tube, trapping a fixed amount of air on the closed side. When Boyle added additional mercury to the open end of the tube, the pressure exerted by the additional mercury compressed the trapped gas into a smaller volume. Boyle found that doubling the total pressure on the trapped air reduced its volume by a factor of two. In more general terms, the volume of the trapped air was inversely proportional to the total pressure applied by the mercury plus the atmosphere, as the graph in Figure 5-5b shows:

$$V_{gas} \propto \frac{1}{P_{gas}} \quad \text{(fixed temperature and amount)}$$

Boyle also observed that heating a gas causes it to expand in volume, but more than a century passed before Jacques-Alexandre-César Charles reported the first quantitative studies of gas volume as a function of temperature. Charles found that for a fixed amount

Figure 5-5
(*a*) Schematic illustration of Boyle's experiments on air trapped in J-shaped tubes.
(*b*) Graph showing the linear variation of *V* vs. 1/*P*.

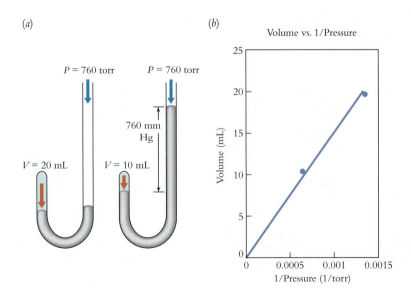

of gas, a graph of gas volume vs. temperature gives a straight line, as shown in Figure 5-6. In other words, the volume of a gas is directly proportional to its temperature:

$$V_{gas} \propto T_{gas} \qquad \text{(fixed pressure and amount)}$$

The volume of a gas also changes when the amount of the gas changes. In this respect, gases are like liquids or solids. If we double the amount of any gas while keeping the temperature and pressure fixed, the gas volume doubles. In other words, gas volume is proportional to the amount of gas.

$$V_{gas} \propto n_{gas} \qquad \text{(fixed pressure and temperature)}$$

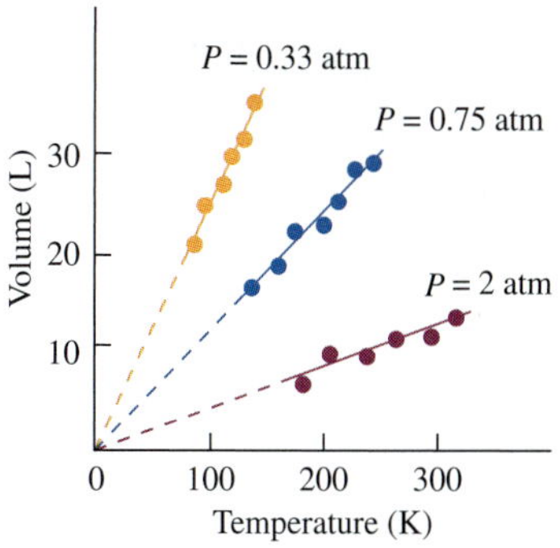

The Ideal Gas Equation

The work of Boyle, Charles, and other early scientists showed that the volume of any gas is proportional to the amount of gas, proportional to the temperature, and inversely proportional to the pressure of the gas. As the experiments show, *P, T,* and *n* can vary independently. This independence lets us link the four variables in a single proportionality:

$$V \propto \frac{nT}{P}$$

We can convert this proportionality into an equality by introducing a constant. This constant is known as the **universal gas constant** and is represented by the symbol **R.** Introducing *R* and multiplying through by *P* gives the master equation for describing gas behavior:

$$PV = nRT \qquad \qquad \text{(5-1)}$$

We call Equation 5-1 the **ideal gas equation,** because it refers to the behavior of a so-called ideal gas. As we describe in the following section, no gas is truly ideal, but under conditions close to 1 atm and room temperature, the ideal gas equation is adequate to describe most real gases.

> *Solving quantitative problems about gases requires only one equation, the ideal gas equation.*

KEY CONCEPT

To use the ideal gas equation for quantitative calculations, we must express each quantity in appropriate units. The ideal gas equation holds only when temperature is expressed using an absolute scale. We will always use the Kelvin scale, applying the conversion introduced in Chapter 1: T (K) $= T$ (°C) $+ 273.15$. Typical laboratory pressures are expressed in atmospheres, and typical laboratory volumes are expressed in liters. For this choice of units and *n* in moles, $R = 0.08206$ L atm mol^{-1} K^{-1}. Examples 5-2 and 5-3 show how to apply the ideal gas equation.

Figure 5-6
Plots of volume vs. temperature for 1 mol of air at three constant pressures.

A 265-gallon steel storage tank contains 88.5 kg of methane (CH_4). If the temperature is 25 °C, what is the pressure inside the tank?

Example 5-2

Calculation
of Gas Pressure

Strategy: We the seven-step strategy for problem solving.

1. The problem asks for the pressure exerted by a gas.

2. The gas is stored in a steel tank, so the volume of the gas cannot change. No chemical changes are described in the problem.

Solution:

3. We summarize the data given in the problem:

$$V = 265 \text{ gallons}, \; T = 25 \text{ °C}, \; m = 88.5 \text{ kg of methane}, \; P = ?$$

Example 5-2

**Calculation
of Gas Pressure** (*continued*)

4. We can calculate the pressure of the gas using the ideal gas equation, but we need to make sure all the variables are expressed in consistent units. Temperature must be in kelvins, amount of methane in moles. For $R = 0.08206$ L atm mol^{-1} K^{-1}, we need the volume in liters, and the units of the calculated pressure will be atmospheres.

5. We rearrange the ideal gas equation so that pressure is isolated on the left:

$$PV = nRT \qquad so \qquad P = \frac{nRT}{V}$$

Use the standard procedure for unit conversions.
The gallon–liter equivalence is given on the inside back cover of this book:

$$V = (265 \text{ gallons})(3.7854 \text{ L/gallon}) = 1.003 \times 10^3 \text{ L}$$

Add 273.15 to the temperature in °C to convert to kelvins:

$$T = 25 + 273.15 = 298 \text{ K}$$

Note that when we add 25 + 273.15, the result is known to the nearest degree, giving three significant figures.
Use the molar mass of methane to calculate the number of moles:

$$n = \frac{(88.5 \text{ kg})(10^3 \text{ g/kg})}{(16.04 \text{ g/mol})} = 5.517 \times 10^3 \text{ mol CH}_4$$

6. Now substitute and calculate the pressure:

$$P = \frac{(5.517 \times 10^3 \text{ mol})(0.08206 \text{ L atm mol}^{-1} \text{ K}^{-1})(298 \text{ K})}{(1.003 \times 10^3 \text{ L})} = 135 \text{ atm}$$

7. **Does the Result Make Sense?** You should verify that the units cancel properly, giving a result in pressure units. The problem describes a large amount of methane in a relatively small volume, so a high pressure is reasonable. This high value indicates why gases such as methane must be stored in tanks made of materials such as steel that can withstand high pressures.

**Extra Practice
Exercise 5.2**

Determine the new pressure if the tank is stored in a hot basement where the temperature reaches 42 °C.

Example 5-3

**Calculation
of Gas Amount**

A pressure gauge on a tank of molecular oxygen gas reads 5.67 atm. If the tank holds 75.0 L and the temperature in the lab is 27.6 °C, how many grams of molecular oxygen are in the tank?

Strategy: The question asks for the mass of oxygen. We can use the ideal gas equation to calculate the number of moles of oxygen, and then molar mass leads us from moles to grams.

Solution: Begin by summarizing the data:

$$P = 5.67 \text{ atm} \qquad V = 75.0 \text{ L} \qquad T = 27.6 + 273.15 = 300.8 \text{ K}$$

Rearrange the ideal gas equation to solve for the number of moles:

$$PV = nRT \qquad so \qquad n = \frac{PV}{RT}$$

Now substitute and do the calculation:

$$n = \frac{(5.67 \text{ atm})(75.0 \text{ L})}{(0.08206 \text{ L atm/mol K})(300.8 \text{ K})} = 17.23 \text{ mol}$$

Complete the problem by converting moles to grams:

$$(17.23 \text{ mol O}_2)(32.00 \text{ g/mol}) = 551 \text{ g O}_2$$

Does the Result Make Sense? The units are correct and the numbers are all within the appropriate magnitudes, so the result is reasonable.

A 1.23-g sample of neon in a 1.50-L glass bulb has a pressure of 745 torr. What is the temperature in the lab?

Example 5-3

Calculation of Gas Amount (*continued*)

Extra Practice Exercise 5.3

Variations on the Gas Equation

During chemical and physical transformations, any of the four variables in the ideal gas equation *(P, V, n, T)* may change, and any of them may remain constant. The experiments carried out by Robert Boyle are a good example. Boyle worked with a fixed amount of air trapped in a glass tube, so the number of moles of gas remained the same during his experiments. In other words, *n* was held constant. Boyle also worked at only one temperature, so *T* remained constant. Example 5-4 applies the ideal gas equation to this situation.

A sample of helium gas is held at constant temperature inside a cylinder whose volume is 0.80 L when a piston exerts a pressure of 1.5 atm. If the external pressure on the piston is increased to 2.1 atm, what will be the new volume?

Example 5-4

Pressure–Volume Variations

Strategy: We follow the seven-step strategy for problem solving.
1. The question asks for the new volume.
2 and 3. Visualize the conditions by drawing a schematic diagram of the initial and final conditions:

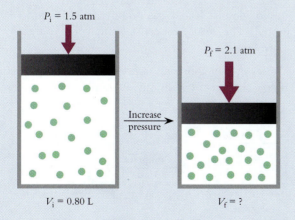

Solution:
4. Gas behavior is involved, so the equation that applies is the ideal gas equation, $PV = nRT$.

5. Rearranging the gas equation to solve for *V* will not help because we do not know *n*, the number of moles of He present in the system, nor do we know *T*, the temperature of the gas. We do know that *n* and *T* remain unchanged as the pressure increases.

Example 5-4

Pressure–Volume Variations (continued)

To determine the final volume of the helium gas, apply the ideal gas equation to the initial (i) and final (f) conditions:

$$P_iV_i = n_iRT_i \quad and \quad P_fV_f = n_fRT_f$$

In this problem, the quantity of He inside the cylinder and the temperature of the gas are constant:

$$n_i = n_f \quad and \quad T_i = T_f \quad so \quad n_iRT_i = n_fRT_f$$

Therefore;

$$P_iV_i = P_fV_f \text{ (constant } n \text{ and } T)$$

Notice that this equality can be solved for V_f without knowing the values for n and T.

6. Now substitute and calculate the final volume:

$$V_f = \frac{P_iV_i}{P_f} = \frac{(1.5 \text{ atm})(0.80 \text{ L})}{(2.1 \text{ atm})} = 0.57 \text{ L}$$

7. Does the Result Make Sense? This answer is reasonable because a pressure increase has caused a volume decrease.

Extra Practice Exercise 5.4

The piston in Example 5-4 is withdrawn until the gas volume is 2.55 L. Calculate the final pressure.

In Example 5-4, the quantities on the right side of the ideal gas equation are fixed, whereas the quantities on the left are changing. A good strategy for organizing gas calculations is to determine which variables do not change, and rearrange the gas equation to group them all on the right.

Examples 5-5 and 5-6 further demonstrate the analysis of gas problems.

Example 5-5

Changing Gas Conditions

A sample of carbon dioxide in a 10.0-L gas cylinder at 25 °C and 1.00 atm pressure is compressed and heated. The final temperature and volume are 55 °C and 5.00 L. Compute the final pressure.

Strategy: We use the seven-step approach in compact form. The problem asks for the pressure of a gas after a change in conditions. A simple diagram helps us organize the information:

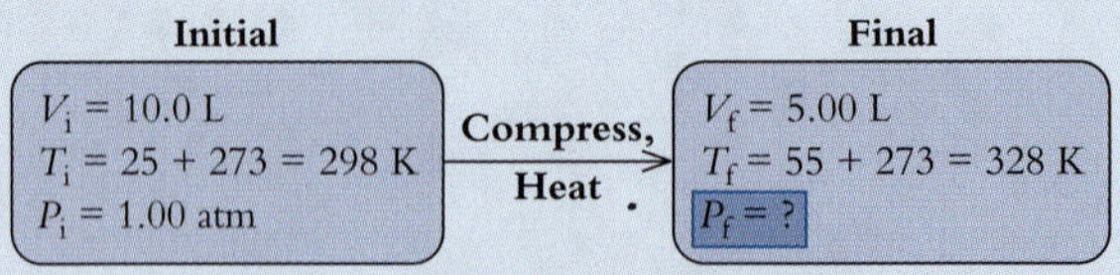

Initial		Final
$V_i = 10.0$ L	Compress,	$V_f = 5.00$ L
$T_i = 25 + 273 = 298$ K	Heat	$T_f = 55 + 273 = 328$ K
$P_i = 1.00$ atm		$P_f = ?$

The data tell us that P, V, and T all change. The number of moles of CO_2 is not mentioned directly, but the physical transformation performed on the gas does not affect the number of molecules, so we conclude that $n_i = n_f$.

Solution: We rearrange $PV = nRT$, dividing both sides by T to group the constants n and R on one side:

$$\frac{P_iV_i}{T_i} = nR = \frac{P_fV_f}{T_f}$$

Rearrange this expression to solve for the final pressure, insert the known information, and perform the calculations:

$$P_f = \left(\frac{P_iV_i}{T_i}\right)\left(\frac{T_f}{V_f}\right) = \frac{(1.00\ \text{atm})(10.0\ \text{L})(328\ \text{K})}{(298\ \text{K})(5.00\ \text{L})} = 2.20\ \text{atm}$$

Example 5-5

Changing Gas Conditions (*continued*)

Does the Result Make Sense? The result has the correct units and number of significant figures. The pressure has increased, which is what we expect, because compression increases the pressure and a temperature increase also increases the pressure.

The sample of gas in Example 5-5 is placed in an ice bath and the pressure is increased to 1.35 atm. What is the new volume?

Extra Practice Exercise 5.5

Example 5-6

Gas Calculations

Two natural gas storage tanks, with volumes of 1.5×10^4 and 2.2×10^4 L, are at the same temperature. The tanks are connected by pipes that equalize their pressures. What fraction of the stored natural gas is in the larger tank?

Strategy: The problem asks us to determine how the gas is distributed between the two tanks. This is a gas problem, so we use the ideal gas equation.

Solution: There are two pieces of data:

$$V_1 = 1.5 \times 10^4\ \text{L} \quad and \quad V_2 = 2.2 \times 10^4\ \text{L}$$

Examine the conditions to determine which variables are the same for both tanks. The problem states that the tanks have the same pressure and temperature. Rearrange the gas equation to group the constants and find an equation that applies:

$$PV = nRT \quad so \quad \frac{n}{V} = \frac{P}{RT} = \text{constant}$$

The problem asks about the fraction of gas in the larger tank. This fraction is the ratio of the number of moles in the larger tank to the total number of moles in both tanks. Because n/V is constant, we can write

$$\frac{n_{\text{large tank}}}{V_{\text{large tank}}} = \frac{n_{\text{total}}}{V_{\text{total}}}$$

Now solve for the fraction of gas in the larger container:

$$\frac{n_{\text{large tank}}}{n_{\text{total}}} = \frac{V_{\text{large tank}}}{V_{\text{total}}} = \frac{2.2 \times 10^4\ \text{L}}{(2.2 \times 10^4\ \text{L}) + (1.5 \times 10^4\ \text{L})} = 0.59$$

Does the Result Make Sense? A fraction of 0.59 means that 59% of the gas is in the larger tank. This seems reasonable because more than half the gas has to be in the larger tank.

Natural gas storage tanks have movable roofs that rise and fall as the volume of gas changes, keeping the pressure equal to that of the atmosphere. By what percentage does the volume of a tank change if the pressure changes from 768 torr to 749 torr?

Extra Practice Exercise 5.6

Section Exercises

■ **5.2.1** According to the ideal gas equation, what will happen to the pressure of a gas if we make each of the following changes, holding all other conditions constant? (a) Double T (in kelvins); (b) reduce the container's volume by a factor of two; (c) triple the amount of gas in the container; and (d) replace the gas with an equal number of moles of another gas whose molar mass is twice as great.

■ **5.2.2** The molecular picture represents a small portion of an ideal gas exerting a pressure P. Redraw the picture in two ways, each of

which gives a new pressure that is half as great.

■ **5.2.3** A 1.54-L gas bulb in a chemistry laboratory contains oxygen gas at 21 °C and 758 torr. The air conditioning in the laboratory breaks down, and the temperature rises to 31 °C. What pressure does the gauge show now?

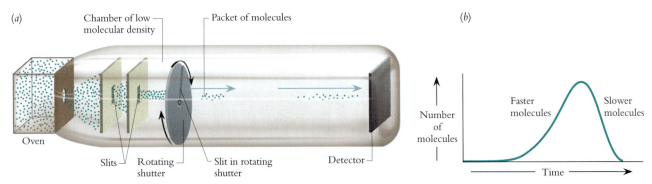

(a)

Chamber of low molecular density — Packet of molecules

Oven

Slits — Rotating shutter — Slit in rotating shutter — Detector

(b)

Number of molecules

Faster molecules   Slower molecules

Time

Figure 5-7
(a) Diagram of a molecular beam apparatus designed to measure the speeds of gas molecules. (b) Distribution of molecules observed by the detector as a function of time after opening the shutter.

5.3 MOLECULAR VIEW OF GASES

To understand why all gases can be described by a single equation, we need to explore how gases behave at the molecular level. In this section we examine the molecular properties of gases and how they result in the ideal gas equation.

Molecular Speeds

How fast do gas molecules move? Molecular speeds can be measured using a molecular beam apparatus, shown schematically in Figure 5-7a. Gas molecules escape from an oven through a small hole into a chamber in which the molecular density is very low. A set of slits blocks the passage of all molecules except those moving in the forward direction. The result is a beam of molecules, all moving in the same direction. A rotating shutter blocks the beam path except for a small slit that allows a packet—"pulse"—of molecules to pass through. Each molecule moves down the beam axis at its own speed. The faster a molecule moves, the less time it takes to travel the length of the chamber. A detector at the end of the chamber measures the number of molecules arriving as a function of time, giving a profile of speeds.

When the speed profile of a gas is measured in this way, the results give a distribution like the one shown in Figure 5-7b. If all the molecules traveled at the same speed, they would reach the detector at the same time, in a single clump. Instead, faster molecules move ahead of the main packet, and slower molecules fall behind. This experiment shows that molecules in a gas have a distribution of speeds.

A pattern emerges when this molecular beam experiment is repeated for various gases at a common temperature: Molecules with small masses move faster than those with large masses. Figure 5-8 shows this for H_2, CH_4, and CO_2. Of these molecules, H_2 has the smallest mass and CO_2 the largest. The vertical line drawn for each gas shows the

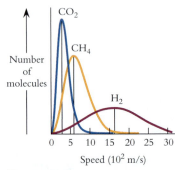

Number of molecules

CO_2

CH_4

H_2

0   5   10   15   20   25   30
Speed (10^2 m/s)

Figure 5-8
Molecular speed distributions for CO_2, CH_4, and H_2 at a temperature of 300 K.

speed at which the distribution reaches its maximum height. More molecules have this speed than any other, so this is the most probable speed for molecules of that gas. The most probable speed for a molecule of hydrogen at 300 K is 1.57×10^3 m/s, which is 3.51×10^3 mi/hr.

Example 5-7 describes an experiment with a molecular beam apparatus.

Example 5-7

A Molecular Beam Experiment

The figures below represent mixtures of neon atoms and hydrogen molecules.

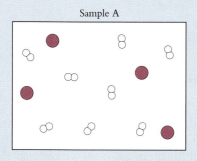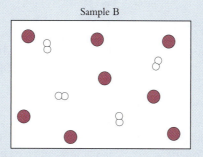

One of the gas mixtures was used in a pulsed molecular beam experiment. The result of the experiment is shown below. Which of the two gas samples, A or B, was used for this experiment?

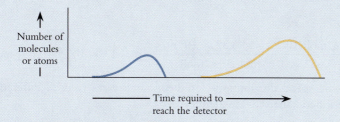

Strategy: Begin by taking an inventory of the two samples, then decide how each of them would behave in a beam experiment.

Solution: Sample A contains eight H_2 molecules and four Ne atoms, and Sample B contains four H_2 molecules and eight Ne atoms. In a beam experiment, both samples would give two peaks in relative areas of 2:1. Which sample was used for this experiment? Particles with small mass move faster than particles with large mass, so we expect H_2 ($MM = 2.02$ g/mol) to reach the detector before Ne ($MM = 20.2$ g/mol). The data show that the first substance to reach the detector is present in the smaller amount. Consequently, the sample used in the beam experiment is the one with the smaller amount of the hydrogen molecules, Sample B.

Does the Result Make Sense? Figure 5-8 shows that H_2 moves much faster than other substances, so it makes sense that the H_2 molecules get to the detector faster.

Suppose that Sample A is used and that the second gas is helium instead of neon. Sketch a graph showing the number of molecules/atoms vs. time.

Extra Practice Exercise 5.7

Speed and Energy

The energy of a molecule is related to its speed. Any moving object has kinetic energy ($E_{kinetic}$) whose magnitude is given by Equation 5-2:

$$E_{kinetic} = \tfrac{1}{2} mu^2 \qquad (5\text{-}2)$$

In this equation, m is the object's mass and u its speed. The most probable speed of hydrogen molecules at 300 K is 1.57×10^3 m/s. We need the mass of one H_2 molecule to calculate its most probable kinetic energy. The SI unit of energy is the joule (J), which equals 1 kg m^2 s^{-2}. Thus, we need the mass (m) in kilograms. The molar mass gives the mass of one mole of molecules, so dividing molar mass by Avogadro's number (N_A) gives the mass per molecule:

$$m = \frac{MM\ (1\ \text{kg}/10^3\ \text{g})}{N_A} = \frac{(2.016\ \text{g/mol})(1\ \text{kg}/10^3\ \text{g})}{6.0223 \times 10^{23}\ \text{molecules/mol}}$$

$$m = 3.348 \times 10^{-27}\ \text{kg}/H_2\ \text{molecule}$$

Now apply Equation 5-2:

$$E_{\text{kinetic}}\ (\text{most probable}) = \tfrac{1}{2}(3.348 \times 10^{-27}\ \text{kg molecule}^{-1})(1.57 \times 10^3\ \text{m s}^{-1})^2$$

$$= 4.13 \times 10^{-21}\ \text{kg m}^2\ \text{s}^{-2}\ \text{molecule}^{-1}$$

$$E_{\text{kinetic}}\ (\text{most probable}) = 4.13 \times 10^{-21}\ \text{J/molecule}$$

The most probable speeds of methane and carbon dioxide are slower than the most probable speed of hydrogen, but CH_4 and CO_2 molecules have larger masses than H_2. When kinetic energy calculations are repeated for these gases, they show that the most probable kinetic energy is the same for all three gases.

For CH_4:

$$\frac{(16.04\ \text{g/mol})(10^{-3}\ \text{kg/g})(5.57 \times 10^2\ \text{m/s})^2}{2(6.022 \times 10^{23}\ \text{molecules/mol})} = 4.13 \times 10^{-21}\ \text{J}/CH_4\ \text{molecule}$$

For CO_2:

$$\frac{(44.01\ \text{g/mol})(10^{-3}\ \text{kg/g})(3.37 \times 10^2\ \text{m/s})^2}{2(6.022 \times 10^{23}\ \text{molecules/mol})} = 4.13 \times 10^{-21}\ \text{J}/CO_2\ \text{molecule}$$

Even though the speed distributions for these three gases peak at different values, the most probable kinetic energies are identical.

KEY CONCEPT   *At a given temperature, all gases have the same molecular kinetic energy distribution.*

Molecular beam experiments show that molecules move faster as temperature increases. Molecules escaping from the oven at 900 K take less time to reach the detector than molecules escaping at 300 K. As molecular speed increases, so does kinetic energy (Equation 5-2). Figure 5-9 shows the molecular kinetic energy distributions at 300 and 900 K. Comparing this figure to Figure 5-8, we see that molecular energies show a wide distribution that is similar to the distribution in molecular speeds. Unlike speeds, however, molecular energy distributions are the same for all gases at any particular temperature. The distributions shown in Figure 5-9 apply to any gas.

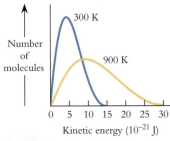

Figure 5-9
Distribution in molecular energies for a gas at 300 K and 900 K.

Average Kinetic Energy

The most probable kinetic energy is not the same as the average kinetic energy. Figure 5-9 shows that the distribution of kinetic energies is not symmetrical. To find the average kinetic energy per molecule ($\overline{E}_{\text{kinetic}}$), we must add all the individual molecular energies and divide by the total number of molecules. The result is Equation 5-3, which describes how the average kinetic energy of gas molecules depends on the temperature of the gas:

A bar over a symbol means "average value."

$$\overline{E}_{\text{kinetic}} = \frac{3RT}{2N_A} \tag{5-3}$$

In this equation, T is the temperature in kelvins, N_A is Avogadro's number (units of molecules/mol), and R is the gas constant. For energy calculations, we express R in SI units, which gives kinetic energy in joules per molecule. The value of R in SI units is

$$R = 8.314\ \text{J mol}^{-1}\ \text{K}^{-1}$$

The average kinetic energy expressed by Equation 5-3 is kinetic energy *per molecule*. We find the total kinetic energy ($E_{\text{kinetic, molar}}$) of one mole of gas molecules by multiplying Equation 5-3 by Avogadro's number:

$$E_{\text{kinetic, molar}} = N_A \frac{3RT}{2N_A} = \tfrac{3}{2}RT$$

Thus, 1 mole of any gas has a total molecular kinetic energy of $\tfrac{3}{2}RT$. Example 5-8 applies these equations to sulfur hexafluoride.

Example 5-8

Molecular Kinetic Energies

Determine the average molecular kinetic energy and molar kinetic energy of gaseous sulfur hexafluoride, SF_6, at 150 °C.

Strategy: The problem asks for a calculation of kinetic energies and provides a temperature. We have two equations for the kinetic energy of a gas:

$$\overline{E}_{\text{kinetic}} = \frac{3RT}{2N_A} \quad and \quad E_{\text{kinetic, molar}} = \tfrac{3}{2}RT$$

To carry out the calculations, we need the constants R and N_A as well as T in kelvins. The constants are found in standard reference tables, but both are used so frequently in chemical calculations that it is wise to memorize them:

$$R = 8.314 \, \text{J mol}^{-1} \, \text{K}^{-1} \quad and \quad N_A = 6.022 \times 10^{23} \, \text{molecules/mol}$$

Solution: Convert the temperature from °C to K by adding 273.15:

$$T = 150 \, °C + 273.15 = 423 \, \text{K}$$

Now substitute and calculate:

$$\overline{E}_{\text{kinetic}} = \frac{3(8.314 \, \text{J mol}^{-1} \, \text{K}^{-1})(423 \, \text{K})}{2(6.022 \times 10^{23} \, \text{molecules/mol})} = 8.76 \times 10^{-21} \, \text{J/molecule}$$

$$E_{\text{kinetic, molar}} = \tfrac{3}{2}(8.314 \, \text{J mol}^{-1} \, \text{K}^{-1})(423 \, \text{K}) = 5.28 \times 10^3 \, \text{J/mol}$$

Do the Results Make Sense? Does it bother you to find that neither the chemical formula nor the molar mass is needed for these calculations? Remember that not all data are necessarily required for any particular calculation. Because average kinetic energy depends on temperature but not on molar mass, we do not need mass information to do this problem.

Extra Practice Exercise 5.8

Atmospheric temperature at the altitudes where jetliners fly is around −35 °C. Calculate the average and molar kinetic energies of molecular nitrogen at this temperature.

Ideal Gases

The behavior of gases suggests that their molecules have little effect on one another. Gases are easy to compress, showing that there is lots of empty space between molecules. Gases also escape easily through any opening, indicating that their molecules are not strongly attracted to one another. An **ideal gas** is defined as one for which both the volume of molecules and the forces among the molecules are so small that they have no effect on the behavior of the gas.

KEY CONCEPTS

The volume occupied by the molecules of an ideal gas is negligible compared with the volume of its container.
The energies generated by forces among ideal gas molecules are negligible compared with molecular kinetic energies.

According to this model, a substance will be a gas when its molecules are small and do not attract one another strongly. As we describe in Chapter 11, small molecules usu-

Table 5-1   Some Gaseous Substances*

Elemental Gases			Binary Gases		
Substance	Formula	*MM* (g/mol)	Substance	Formula	*MM* (g/mol)
Hydrogen	H_2	2.0	Methane	CH_4	16.0
Helium	He	4.0	Ammonia	NH_3	17.0
Neon	Ne	20.2	Carbon monoxide	CO	28.0
Nitrogen	N_2	28.0	Ethane	C_2H_6	30.0
Oxygen	O_2	32.0	Nitrogen oxide	NO	30.0
Fluorine	F_2	38.0	Hydrogen sulfide	H_2S	34.1
Argon	Ar	39.9	Hydrogen chloride	HCl	36.5
Ozone	O_3	48.0	Carbon dioxide	CO_2	44.0
Chlorine	Cl_2	71.0	Nitrogen dioxide	NO_2	46.0

*Under "normal" conditions: 298 K, 1 atm pressure.

ally have small attractive forces. Table 5-1 lists the substances that exist as gases under normal conditions. In accord with the model, they mostly are small molecules with molar masses less than 50 g/mol.

How does an ideal gas behave? We can answer this question by considering how changes in the other conditions affect the pressure, *P*. Each time a molecule strikes a wall, it exerts a force on the wall. During each second, many collisions exert many such forces. Pressure is the sum of all these forces per unit area. In an ideal gas each molecule is independent of all others. This independence means that the total pressure is the sum of the pressure created by each individual molecule.

To see how pressure depends on the other properties, we consider the effect of changing one property at a time, holding the other properties constant. We analyze what happens to the molecular collisions as each property changes.

First, consider increasing the amount of the gas while keeping the temperature and volume fixed. Figure 5-10 shows that doubling the amount of gas in a fixed volume doubles the number of collisions with the walls. Thus, pressure is directly proportional to the amount of gas. This agrees with the ideal gas equation.

Next, consider keeping the temperature and amount fixed and changing the volume of the gas. Figure 5-11 shows that compressing a gas into a smaller volume has the same effect as adding more molecules. The result is more collisions with the walls. If the molecules act independently, cutting the volume in half will double the pressure. In other words, pressure is inversely proportional to volume, again in agreement with the ideal gas equation.

To complete our analysis, we must determine the effect of a change in temperature. According to Equation 5-3, kinetic energy is proportional to temperature, and according to Equation 5-2, kinetic energy is proportional to the square of the molecular speed. Thus, the square of the molecular speed is proportional to temperature.

Figure 5-10
Schematic view of the effect of doubling the number of gas molecules in a fixed volume. Container (*b*) has twice as many molecules as (*a*). Consequently, the molecular density is twice as large in (*b*), with twice as many collisions per second with the walls.

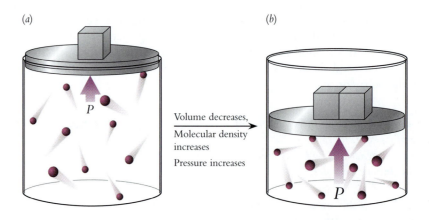

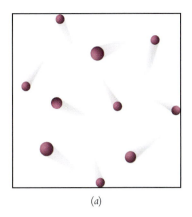
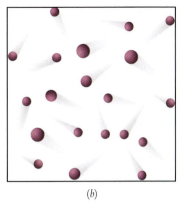

(a) (b)

Figure 5-11
Schematic view of the effect of compressing a fixed quantity of gas into a smaller volume at constant *T*. Decreasing the volume increases the molecular density, which increases the number of collisions per second with the walls.

Molecular speed affects pressure in two ways that are illustrated in Figure 5-12. First, faster-moving molecules hit the walls more often than slower-moving molecules. The *number of collisions* each molecule makes with the wall is proportional to the molecule's speed. Second, the *force* exerted when a molecule strikes the wall depends on the molecule's speed. A fast-moving molecule exerts a larger force than the same molecule moving slower. Force per collision increases with speed, and number of collisions increase with speed, so the total effect of a single molecule on the pressure of a gas is proportional to the square of its speed.

Pressure is proportional to the square of molecular speed, which in turn is proportional to temperature. For an ideal gas, then, the pressure is directly proportional to temperature, and a plot of *P* vs. *T* yields a straight line. Again, this agrees with the ideal gas equation.

A gas will obey the ideal gas equation whenever it meets the conditions that define the ideal gas. Molecular sizes must be negligible compared to the volume of the container, and the energies generated by forces between molecules must be negligible compared to molecular kinetic energies. The behavior of any real gas departs somewhat from ideality because real molecules occupy volume and exert forces on one another. Nevertheless, departures from ideality are small enough to neglect under many circumstances. We consider departures from ideal gas behavior in Chapter 11.

(a)   (b)

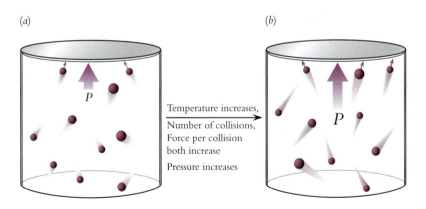

Temperature increases,
Number of collisions,
Force per collision
both increase

Pressure increases

Figure 5-12
Schematic view of the effect of increasing the temperature of a gas. Molecular speeds increase, resulting in more wall collisions and more force per collision, so the pressure increases.

Section Exercises

■ **5.3.1** A molecular beam experiment of the type illustrated in Figure 5-7 is performed with an equimolar mixture of He and CO_2. Sketch the appearance of a graph of the number of molecules reaching the detector as a function of time.

■ **5.3.2** Calculate each of the following energies: (a) total kinetic energy of 1.00 mol of He atoms at -100 °C; (b) total kinetic energy of 1.00 g of N_2 molecules at 0 °C; and (c) kinetic energy of a single molecule of SF_6 that has twice as much energy as the average molecular energy at 200 °C.

■ **5.3.3** Is a gas more likely to behave ideally at high pressure or at low pressure? Explain your answer in terms of the two assumptions made for ideal gases.

5.4 ADDITIONAL GAS PROPERTIES

The ideal gas equation and the molecular view of gases lead to several useful applications. We have already described how to carry out calculations involving $P-V-n-T$ relationships. In this section, we examine the use of the gas equation to determine molar masses, gas density, and rates of gas movement.

Determination of Molar Mass

The ideal gas equation can be combined with the mole–mass relation to find the molar mass of an unknown gas:

$$PV = nRT \text{ (ideal gas equation)} \qquad and \qquad n = \frac{m}{MM} \text{ (mole–mass relation)}$$

If we know the pressure, volume, and temperature of a gas sample, we can use this information to calculate how many moles are in the sample:

$$n = \frac{PV}{RT}$$

If we also know the mass of the gas sample, we can use that information to determine the molar mass of the gas:

$$MM = \frac{m}{n}$$

Example 5-9 illustrates a molar mass calculation using gas data.

Example 5-9	Calcium carbide (CaC_2) is a hard, gray-black solid that has a melting point of 2000 °C. This compound reacts strongly with water to produce a gas and a solution containing OH^- ions. A 12.8-g sample of CaC_2 was treated with excess water. The resulting gas was collected in an evacuated 5.00-L glass bulb with a mass of 1254.49 g. The filled bulb had a mass of 1259.70 g and a pressure of 0.988 atm when its temperature was 26.8 °C. Calculate the molar mass and determine the formula of the gas.
Molar Mass Determination	

Strategy: We can use the ideal gas equation to calculate the molar mass. Then we can use the molar mass to identify the correct molecular formula among a group of possible candidates, knowing that the products must contain the same elements as the reactants. The problem involves a chemical reaction, so we must make a connection between the gas measurements and the chemistry that takes place. Because the reactants and one product are known, we can write a partial equation that describes the chemical reaction:

$$CaC_2\,(s) + H_2O\,(l) \longrightarrow Gas + OH^-(aq)$$

In any chemical reaction, atoms must be conserved, so the gas molecules can contain only H, O, C, and/or Ca atoms. To determine the chemical formula of the gas, we must find the combination of these elements that gives the observed molar mass.

Solution: Use the gas data to determine the molar mass. The problem gives the following data about the unknown gas:

$$V_{bulb} = V_{gas} = 5.00 \text{ L} \qquad T = 26.8 \text{ °C} = 300.0 \text{ K} \qquad P = 0.988 \text{ atm}$$

$$m_{bulb\,+\,gas} = 1259.70 \text{ g} \qquad m_{bulb} = 1254.49 \text{ g}$$

$$m_{gas} = m_{bulb\,+\,gas} - m_{bulb} = 5.21 \text{ g}$$

Example 5-9

Molar Mass
Determination (*continued*)

We use V, T, P, and the ideal gas equation to find the number of moles of gas. Then, with the mass of the gas sample, we can determine the molar mass:

$$\frac{PV}{RT} = n_{gas} = \frac{m_{gas}}{MM}$$

We could determine n and then solve for MM, but we can also solve the equality for MM and then substitute the data directly:

$$MM = \frac{RTm_{gas}}{PV}$$

$$MM = \frac{(0.08206 \text{ L atm mol}^{-1}\text{K}^{-1})(300.0 \text{ K})(5.21 \text{ g})}{(0.988 \text{ atm})(5.00 \text{ L})} = 26.0 \text{ g/mol}$$

To identify the gas, we examine the formulas and molar masses of known compounds that contain H, O, C, and Ca:

Formula	*MM*	Comment
Ca	40	A gas with $MM = 26.0$ g/mol cannot contain Ca
CO	28	This is close but too high
O_2	32	This is also too high, and H_2O_2 is even higher
H_2O	18	$H_{10}O$, $MM = 26.0$ g/mol, does not exist
CH_4	16	CH_{14}, $MM = 26.0$ g/mol, does not exist
C_2H_2	26	This substance has the observed molar mass

A little trial and error leads to the molecular formula: C_2H_2, commonly known as acetylene.

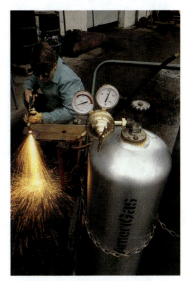

Acetylene is a high-energy molecule used as a fuel for oxyacetylene welding.

Does the Result Make Sense? Knowing the formula of the gaseous product and that hydroxide is another product, we can write a balanced equation that generates acetylene and a basic solution:

$$CaC_2(s) + 2\,H_2O(l) \longrightarrow Ca^{2+}(aq) + 2\,OH^-(aq) + C_2H_2(g)$$

The balanced equation suggests that the result is reasonable.

The glass bulb of Example 5-9 is filled with an unknown gas until the pressure is 774 torr at a temperature of 24.5 °C. The mass of bulb plus contents is 1260.33 g. Determine the molar mass of the unknown gas.

Gas Density

Whereas liquids and solids have well-defined densities, the density of a gas varies strongly with the conditions. To see this, we combine the ideal gas equation and the mole–mass relation and rearrange to obtain an equation for density ($\rho = m/V$):

$$n = \frac{PV}{RT} \quad and \quad n = \frac{m}{MM}$$

Set the two expressions for n equal to each other:

$$\frac{m}{MM} = \frac{PV}{RT}$$

Now multiply both sides of the equality by MM and divide both sides by V:

$$\rho_{gas} = \frac{m}{V} = \frac{P\,MM}{RT}$$

Hot-air balloons rise because hot air is less dense than cooler air.

This equation reveals three features of gas density:

1. For any given gas at fixed temperature, the density increases linearly with pressure. The reason is that increasing the pressure compresses the gas into a smaller volume without changing its mass.

2. For any given gas at fixed pressure, the density decreases linearly with temperature. The reason is that increasing the temperature causes the gas to expand without changing its mass.

3. For different gases at the same temperature and pressure, the density increases linearly with molar mass. The reason is that equal numbers of *moles* of different gases occupy equal volumes at a given temperature and pressure.

There are practical applications of Features 2 and 3. Balloons inflated with helium rise in the atmosphere because the molar mass of helium is substantially lower than that of air. Consequently, the density of a helium-filled balloon is less than the density of air, and the balloon rises, just as a cork released underwater rises to the surface. Hot-air balloons exploit Feature 3. When the air beneath a hot-air balloon is heated, its density decreases, becoming smaller than the density of the outside air. With sufficient heating, the balloon rises and floats over the landscape. In contrast, cold air is less dense than warm air, so cold air sinks. For this reason, valleys often are colder than the surrounding hillsides during winter.

When a gas is released into the atmosphere, whether it rises or sinks depends on its molar mass. If the molar mass of the gas is greater than the average molar mass of air, the gas remains near the ground. Carbon dioxide fire extinguishers are effective because of this feature. The molar mass of CO_2 is greater than that of N_2 or O_2, so a CO_2 fire extinguisher lays down a blanket of this gas that excludes oxygen from the fire, snuffing it out. Example 5-10 treats another practical example of gas density.

Example 5-10	A hot-air balloon will rise when the density of its air is 15% lower than that of the atmospheric air. Calculate the density of air at 295 K, 1.0 atm (assume that dry air is 78% N_2 and 22% O_2) and determine the minimum temperature of air that will cause a balloon to rise.
Gas Density	

Strategy: The problem has two parts. First we must calculate the density of atmospheric air. To do this, we need to determine the molar mass of dry air, which is the weighted average of the molar masses of its components. Then we must calculate the temperature needed to reduce the density by 15%. For both calculations, we use the ideal gas equation as rearranged to give gas density:

$$\rho_{gas} = \frac{P\,MM}{RT}$$

Solution: Begin by calculating the molar mass of dry air. Multiply the fraction of each component by its molar mass:

$$MM_{air} = \left(\frac{78\%}{100\%}\right)(28.02 \text{ g/mol}) + \left(\frac{22\%}{100\%}\right)(32.00 \text{ g/mol}) = 28.9 \text{ g/mol}$$

The other conditions are stated in the problem:

$$T = 295 \text{ K} \qquad P = 1.0 \text{ atm}$$

Substitute into the equation for gas density and calculate:

$$\rho_{gas} = \frac{P\,MM}{RT} = \frac{(1.0 \text{ atm})(28.9 \text{ g/mol})}{(0.08206 \text{ L atm/mol K})(295 \text{ K})} = 1.19 \text{ g/L}$$

As a final result, we would round to two significant figures, but for purposes of the next calculation we retain the extra significant figure until the calculations are complete.

The balloon will rise when the density of its contents is 15% less than the density of the exterior air:

$$\rho_{gas} = 1.19 \text{ g/L} - \left(\frac{15\%}{100\%}\right)(1.19 \text{ g/L}) = 1.01 \text{ g/L}$$

Example 5-10

Gas Density (*continued*)

Rearrange the density equation to solve for temperature, then substitute and calculate:

$$\rho_{gas} = \frac{P\,MM}{RT} \qquad so \qquad T = \frac{P\,MM}{R\rho}$$

$$T = \frac{(1.0 \text{ atm})(28.9 \text{ g/mol})}{(0.08206 \text{ L atm/mol K})(1.01 \text{ g/L})} = 349 \text{ K}$$

Because some of the data are stated only to two significant figures, round both results to two significant figures: $\rho_{gas} = 1.2$ g/L and $T = 350$ K.

Do the Results Make Sense? A temperature of 350 K is 77 °C, a reasonable result. Notice that the molar mass and the gas constant appear in both calculations, so we can find the temperature requirement by simple proportions:

$$T_{inside} = T_{outside}\frac{\rho_{outside}}{\rho_{inside}} = \frac{(295 \text{ K})(1.19 \text{ g/L})}{(1.01 \text{ g/L})} = 348 \text{ K, rounds to } 350 \text{ K}$$

Helium is used for lighter-than-air blimps, whereas argon is used to exclude air from flasks in which air-sensitive syntheses are performed. Calculate the densities of these two noble gases at 295 K and 1.0 atm, and explain why the two gases have different uses.

Extra Practice Exercise 5.10

Gas density has a significant effect on the interactions among molecules of a gas. As molecules move about, they collide regularly with one another and with the walls of their container. Figure 5-13 shows that the frequency of collisions depends on the density of the gas. At low density, a molecule may move all the way across a container before it encounters another molecule. At high density, a molecule travels only a short distance before it collides with another molecule. As our Tools for Discovery Box describes, many scientific experiments require gas densities low enough to provide collision-free environments.

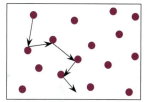
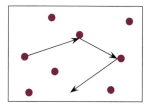
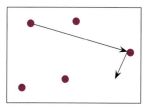

Rates of Gas Movement

The average speed of a gas molecule depends on the temperature of the gas. To state this dependence quantitatively, we start by applying Equation 5-2 to the average speed and energy:

$$\overline{E}_{kinetic} = \tfrac{1}{2}m\overline{u^2}$$

Equation 5-3 provides a second expression for the average kinetic energy:

$$\overline{E}_{kinetic} = \frac{3RT}{2N_A}$$

We set the two expressions for kinetic energy equal to each other:

$$\tfrac{1}{2}m\overline{u^2} = \frac{3RT}{2N_A}$$

Now solve this equality for u, noting that the product mN_A is the mass of one mole of gas molecules:

$$\overline{u^2} = \frac{3RT}{mN_A} = \frac{3RT}{MM} \qquad so$$

$$\overline{u} = \left(\frac{3RT}{MM}\right)^{1/2}$$

Figure 5-13
As molecular density decreases, the average distance traveled between molecular collisions increases.

(5-4)

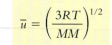

Box 5-1   Tools for Discovery: High Vacuum

"Nature abhors a vacuum." True? On Earth, it is certainly difficult to remove the gas from a container, thereby generating and maintaining a vacuum—the absence of gas. In outer space, on the other hand, a vacuum is the rule rather than the exception. Most of the volume of the universe is nearly empty space, close to a perfect vacuum. If a spacecraft sprang a leak, its gaseous atmosphere would quickly escape into that vacuum. Perhaps it would be better to say that conditions on Earth are unfavorable for vacuums.

Over the years, scientists have spent immense efforts to achieve high vacuums. Powerful and sophisticated pumps are required to remove the gas from a chamber. Once a high vacuum is established, the chamber must be leak-free at the molecular level to keep atmospheric gas from seeping back into an evacuated system. Moreover, the materials used to make a vacuum chamber cannot contain substances that give off gases. A modern high-vacuum instrument is an elaborate array of pumps, valves, seals, and windows, as the photograph at right shows.

Why bother to remove the gases from a small volume of space? High vacuum is used in chemistry and physics research to achieve any of the following conditions: the absence of molecular collisions, the maintenance of an ultraclean environment, or a simulation of conditions in outer space.

Mass spectrometers, workhorse instruments described in Chapter 2, require a vacuum to function. A mass spectrometer generates a beam of ions that is sorted according to specifications of the particular instrument. Usually, the sorting depends on differences in speed, trajectory, and mass. For instance, one type of mass spectrometer measures how long it takes ions to travel from one end of a tube to another. Residual gas must be removed from the tube to eliminate collisions between gas molecules and the ions that are being analyzed. As the diagram shows, collisions with unwanted gas molecules deflect the ions from their paths and change the expected mass spectral pattern.

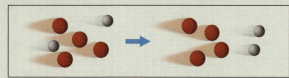

In a high-vacuum, collision-free tube, a beam of ions is sorted according to ion speed (mass).

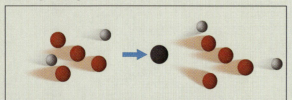

If the vacuum is not high enough, residual gas molecules (black circle) collide with ions, deflecting them from their paths and destroying the sorting process.

A mass spectrometer provides an example of a *molecular beam,* in this case a beam of molecular ions. Molecular beams are used in many studies of fundamental chemical interactions. In a high vacuum, a molecular beam allows chemists to study the reactions that take place through specifically designed types of collisions. For example, a crossed-beam experiment involves the intersection of two molecular beams of two different substances. The types of substances, molecular speeds, and orientations of the beams can be changed systematically to give detailed information about how chemical reactions occur at the molecular level. Chemists also have learned how to create molecular beams in which the molecules have very little energy of motion. These isolated, low-energy molecules are ideal for studies of fundamental molecular properties.

An ultraclean environment is another major reason for generating high vacuum. At atmospheric pressure, every atom on a solid surface is bombarded with gas molecules at a rate of trillions per second. Even under a reasonably high vacuum, 10^{-9} atm, a gas molecule strikes every atom on a solid surface about once per second. If the surface is reactive, these collisions result in chemical reactions that contaminate the surface. The study of pure surfaces of metals or semiconductors requires ultrahigh vacuum, with pressures on the order of 10^{-12} atm.

High vacuum on a large scale has long played a role in studies of nuclear reactions. To study fundamental nuclear processes, physicists have constructed very long tubes through which they can accelerate beams containing nuclei of various elements. The photo shows one of the longest of these, the high-energy linear accelerator at Stanford University. The interior of the accelerator is a one-mile-long pipe that is an obvious feature in this photo. The pipe must be maintained at high vacuum. Otherwise, the particles in the beam would be deflected by molecular collisions, ruining the experiments. This long pipe has massive vacuum pumps located at regular intervals along its entire length, pumping continually to keep the pressure inside the pipe at an acceptably low level.

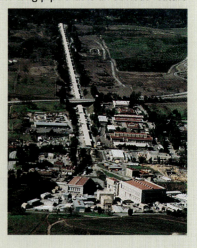

This speed is called the **root–mean–square** speed, because it is found by taking the square root of $\overline{u^2}$. According to Equation 5-4, the average speed of gas molecules is directly proportional to the square root of the temperature and is inversely proportional to the square root of the molar mass.

Equation 5-4 can be applied directly to the movement of molecules escaping from a container into a vacuum. This process is **effusion.** Effusion is exemplified by the escape of molecules from the oven of Figure 5-7.

A second type of gas movement is **diffusion,** the movement of one type of molecule through molecules of another type. Diffusion is exemplified by air escaping from a punctured tire. The escaping molecules must diffuse among the molecules already present in the atmosphere. Diffusing molecules undergo frequent collisions, so their paths are similar to that shown in Figure 5-4. Nevertheless, their average rate of movement depends on temperature and molar mass according to Equation 5-4. Figure 5-14 shows a molecular-level comparison of effusion and diffusion.

An example of diffusion is shown in Figure 5-15, in which aqueous solutions of hydrochloric acid (HCl) and ammonia (NH_3) are placed at opposite ends of a glass tube. Molecules of HCl gas and NH_3 gas escape from the solutions and diffuse through the air in the tube. When the two gases meet, they undergo an acid–base reaction to make ammonium chloride, a white solid salt:

$$HCl(g) + NH_3(g) \longrightarrow NH_4Cl(s)$$

$$\text{Strong acid} \quad \text{Weak base} \qquad \text{White solid}$$

The lighter NH_3 molecules diffuse more rapidly than the heavier HCl molecules, so the white band of salt forms closer to the HCl end of the tube, as can be seen in Figure 5-15.

Rates of molecular motion are directly proportional to molecular speeds, so Equation 5-4 predicts that for any gas, rates of effusion and diffusion increase with the square root of the temperature in kelvins. Also, at any particular temperature, effusion and diffusion are faster for molecules with small molar masses.

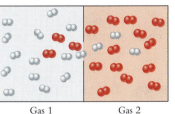

Effusion

Gas 1 Vacuum

Diffusion

Gas 1 Gas 2

Figure 5-14
Effusion is the movement of a gas into a vacuum. Diffusion is the gradual mixing of two or more gases.

Animation

The movement of gases through the atmosphere, such as the fragrance of a rose moving from the flower to our noses, generally involves air currents as well as diffusion. Convection, the flow of gas in a current, moves molecules much more rapidly than diffusion.

HCl (aq) NH₃ (aq)

NH_4Cl (s)

Figure 5-15
NH_3 diffuses through a glass tube faster than HCl. When the two gases meet, they form ammonium chloride (NH_4Cl), which appears as a white band closer to the end of the tube that contains HCl. (An aqueous solution of ammonia is also known as ammonium hydroxide.)

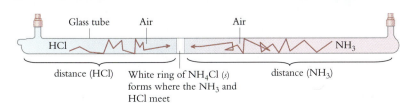

Glass tube Air Air

HCl NH₃

distance (HCl) White ring of NH_4Cl (s) forms where the NH_3 and HCl meet distance (NH₃)

Section Exercises

■ **5.4.1** On a summer day the temperature inside a house rises to 37 °C (99 °F). In the evening the house may cool to 21 °C (68 °F). Does air enter or leave the house as it cools? If the volume of the house is 9.50×10^5 L and the barometric pressure has held steady at 768 mm Hg, what mass of air is transferred? (Average molar mass of air is 28.8 g/mol.)

■ **5.4.2** A 2.96-g sample of a compound of mercury and chlorine is vaporized in a 1.000-L bulb at 307 °C, and the final pressure is found to be 394 torr. What are the molar mass and chemical formula of the compound?

■ **5.4.3** Calculate the average speeds of H_2O and D_2O molecules when $T = 500$ °C. (D represents deuterium, the isotope of hydrogen that contains one neutron.) Use these speeds to calculate the ratio of effusion rates for these two gases from an oven at 500 °C.

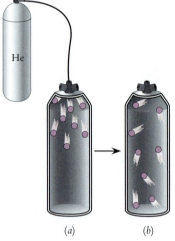
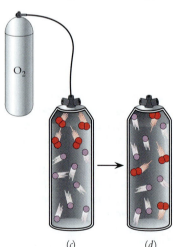

Figure 5-16
When 8.0 mol of He is added to a diver's tank (*a*) the atoms quickly become distributed uniformly throughout the tank (*b*). When 4.0 mol of O₂ is added to this tank (*c*) the molecules move about independently of the He atoms, causing the gases to mix uniformly (*d*).

Animation

5.5 GAS MIXTURES

Many gases are mixtures of two or more species. The atmosphere, with its mixture of nitrogen, oxygen, and various trace gases, is an obvious example. Another example is the gas used by deep-sea divers, which contains a mixture of helium and oxygen. The ideal gas model provides guidance as to how we describe mixtures of gases.

In an ideal gas, all molecules act independently. This notion applies to gas mixtures as well as to pure substances. Gas behavior depends on the *number* of gas molecules but not on the *identity* of the gas molecules. The ideal gas equation applies to each gas in the mixture, as well as to the entire collection of molecules.

Suppose we pump 4.0 mol of helium into a deep-sea diver's tank. If we pump in another 4.0 mol of He, the container now contains 8.0 mol of gas. The pressure can be calculated using the ideal gas equation, with $n = 4.0 + 4.0 = 8.0$ mol. Now suppose that we pump in 4.0 mol of molecular oxygen. Now the container holds a total of 12.0 mol of gas. According to the ideal gas model, it does not matter whether we add the same gas or a different gas. Because all molecules in a sample of an ideal gas behave independently, the pressure increases in proportion to the increase in the total number of moles of gas. Thus, we can calculate the total pressure from the ideal gas equation, using $n = 8.0 + 4.0 = 12.0$ mol.

How does a 1:2 mixture of O_2 and He appear on the molecular level? As O_2 is added to the container, its molecules move throughout the volume and become distributed uniformly. Diffusion causes gas mixtures to become homogeneous. Figure 5-16 shows this.

Dalton's Law of Partial Pressures

The pressure exerted by an ideal gas mixture is determined by the total number of moles:

$$P = \frac{n_{\text{total}}RT}{V}$$

We can express the total number of moles of gas as the sum of the amounts of the individual gases. For the diver's He and O_2 mixture,

$$n_{\text{total}} = n_{\text{He}} + n_{O_2}$$

Substitution gives a two-term equation for the total pressure:

$$P = \frac{(n_{\text{He}} + n_{O_2})RT}{V} = \frac{n_{\text{He}}RT}{V} + \frac{n_{O_2}RT}{V}$$

Notice that each term on the right resembles the ideal gas equation rearranged to express pressure. Each term therefore represents the **partial pressure (*p*)** of one of the

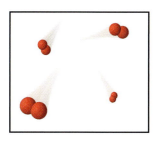 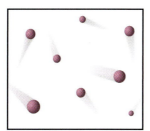 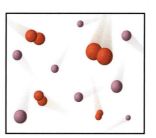

3.0 L at 273 K
4.0 mol O_2
P_{O_2} = 30 atm

3.0 L at 273 K
8.0 mol He
P_{He} = 60 atm

3.0 L at 273 K
12.0 mol gas
p_{O_2} = 30 atm
p_{He} = 60 atm
P_{total} = 90 atm

Figure 5-17
Molecular pictures of a sample of O_2, a sample of He, and a mixture of the two gases. Both components are distributed uniformly throughout the gas volume. Each gas behaves the same, whether it is pure or part of a mixture.

gases. As Figure 5-17 illustrates, partial pressure is the pressure that would be present in a gas container if one gas were present by itself:

$$p_{He} = \frac{n_{He}RT}{V} \quad and \quad p_{O_2} = \frac{n_{O_2}RT}{V}$$

The total pressure in the container is the sum of the partial pressures:

$$P_{total} = p_{He} + p_{O_2}$$

We have used He and O_2 to illustrate the behavior of a mixture of ideal gases, but the same result is obtained regardless of the number and identity of the gases.

The partial pressure of each gas is designated with a lower case *p* to distinguish it from the total pressure of the mixture, *P*.

⑦

In a mixture of gases, each gas contributes to the total pressure the amount that it would exert if the gas were present in the container by itself.

KEY CONCEPT

This is **Dalton's law of partial pressures.** To obtain a total pressure, simply add the contributions from all gases present:

$$P_{total} = p_1 + p_2 + p_3 + \cdots + p_i$$

When doing calculations on a mixture of gases, we can apply the ideal gas equation to each component to find its partial pressure (p_i). Alternatively, we can treat the entire gas as a unit, using the total number of moles to determine the total pressure of the mixture (*P*).

John Dalton was the first to describe gas mixtures in this way as an application of his atomic theory.

Describing Gas Mixtures

There are several ways to describe the chemical composition of a mixture of gases. The simplest method is merely to list each component with its partial pressure or number of moles. Two other descriptions, mole fractions and parts per million, also are used frequently.

Chemists often express chemical composition in fractional terms, stating the number of moles of a substance as a fraction of the number of moles of all substances in the mixture. This way of stating composition is the **mole fraction (*X*):**

$$\text{Mole fraction of A} = X_A = \frac{n_A}{n_{total}}$$

Mole fractions provide a simple way to relate the partial pressure of one component to the total pressure of the gas mixture:

$$p_A = \frac{n_A RT}{V} \quad and \quad P_{total} = \frac{n_{total}RT}{V}$$

Dividing p_A by P_{total} gives

$$\frac{p_A}{P_{total}} = \frac{\left(\frac{n_A RT}{V}\right)}{\left(\frac{n_{total}RT}{V}\right)} = \frac{n_A}{n_{total}} = X_A$$

Rearranging gives Equation 5-5:

$$p_A = X_A P_{total}$$

(5-5)

The partial pressure of a component in a gas mixture is its mole fraction times the total pressure. Example 5-11 illustrates calculations with gas mixtures.

Example 5-11

Gas Mixtures

Deep-sea divers use mixtures of helium and molecular oxygen in their breathing tanks.

The amount of gas introduced into a diving tank can be determined by weighing the tank before and after charging the tank with gas. A diving shop placed 80.0 g of O_2 and 20.0 g of He in a 5.00-L tank at 298 K. Determine the total pressure of the mixture, and find the partial pressures and mole fractions of the two gases.

Strategy: We have a mixture of two gases in a container whose volume and temperature are known. The problem asks for pressures and mole fractions. Because molecular interactions are negligible, each gas can be described independently by the ideal gas equation. As usual, we need molar amounts for the calculations.

Solution: Begin with the data provided:

$$V = 5.00 \text{ L} \qquad T = 298 \text{ K} \qquad m_{He} = 20.0 \text{ g} \qquad m_{O_2} = 80.0 \text{ g}$$

Convert the mass of each gas into an amount in moles:

$$n = m/MM \qquad n_{He} = 5.00 \text{ mol} \qquad n_{O_2} = 2.50 \text{ mol}$$

Next, use the ideal gas equation to compute the partial pressure of each gas:

$$p_{He} = \frac{(5.00 \text{ mol})(0.08206 \text{ L atm mol}^{-1} \text{ K}^{-1})(298 \text{ K})}{5.00 \text{ L}} = 24.5 \text{ atm}$$

The same calculation for O_2 gives $p_{O_2} = 12.2$ atm.

The total pressure is the sum of the partial pressures:

$$P_{total} = p_{He} + p_{O_2} = 24.5 \text{ atm} + 12.2 \text{ atm} = 36.7 \text{ atm}$$

Mole fractions can be calculated from moles or partial pressures:

$$X_{He} = \frac{n_{He}}{n_{total}} = \frac{5.00 \text{ mol}}{7.50 \text{ mol}} = 0.667 \qquad and \qquad X_{O_2} = \frac{p_{O_2}}{P_{total}} = \frac{12.2 \text{ atm}}{36.7 \text{ atm}} = 0.333$$

Do the Results Make Sense? The partial pressures are large, but diving tanks are built to stand high pressures. To check for consistency, add the mole fractions, which must sum to 1.00.

Extra Practice Exercise 5.11

Another 5.00-L diving tank is charged with 75.0 g of O_2 and 25.0 g of He. Calculate its mole fractions and the partial pressures at 25 °C.

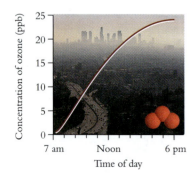

Air quality in Los Angeles, California, on Sept. 11, 2000

When referring to sparse components of gas mixtures, scientists typically use **parts per million (ppm)** to designate the relative number of molecules of a substance present in a sample. Parts per million measures how many molecules of a substance are present in one million molecules of sample. As an example, concentrations of atmospheric pollutants often are given in ppm. If a pollutant is present at a concentration of 1 ppm, there is one molecule of the pollutant in every one million molecules of the atmosphere. In molar terms, 1 ppm means that there is one mole of pollutant for every one million moles of gas. Or, to put it another way, there is 10^{-6} mol of pollutant in every mole of air.

The city of Los Angeles issues a smog alert when the ozone in its atmosphere reaches 0.5 ppm. This means that in every mole of air, there is 0.5×10^{-6} mol of ozone. This may not seem like much, but ozone is a very toxic substance that is particularly damaging to soft tissue such as the lungs.

Even lower concentrations in a gas mixture are expressed in **parts per billion (ppb).** Parts per billion measures how many molecules of substance are present in one billion molecules of sample.

$$1 \text{ ppm} = 1 \text{ molecule out of every } 10^6 \text{ molecules}$$

$$1 \text{ ppb} = 1 \text{ molecule out of every } 10^9 \text{ molecules}$$

Mole fractions, parts per million, and parts per billion all are ratios of moles of a particular substance to total moles of sample. Mole fraction is moles per mole, ppm is moles per million moles, and ppb is moles per billion moles. These measures are related by scale factors: ppm $= 10^6 X$, and ppb $= 10^9 X$. In other words, a concentration of 1 ppm is a mole fraction of 10^{-6}, and a concentration of 1 ppb is a mole fraction of 10^{-9}. When the ozone concentration in the atmosphere reaches 0.5 ppm, the mole fraction of ozone is 0.5×10^{-6}, or 5×10^{-7}. Example 5-12 shows how to work with parts per million.

Example 5-12
Working with ppm

The exhaust gas from an average automobile contains 206 ppm of the pollutant nitrogen oxide, NO. If an automobile emits 125 L of exhaust gas at 1.00 atm and 350 K, what mass of NO has been added to the atmosphere?

Strategy: The question asks for mass of NO. Information about ppm tells us how many moles of NO are present in one mole of exhaust gas. We can use the ideal gas equation to determine the total number of moles of gas emitted, use the ppm information to find moles of NO, and do a mole–mass conversion to get the required mass.

Solution:

Data: Concentration of NO = 206 ppm, $V = 125$ L, $P = 1.00$ atm, $T = 350$ K

$$n_{\text{gas}} = \frac{PV}{RT} = \frac{(1.00 \text{ atm})(125 \text{ L})}{(0.08206 \text{ L atm mol}^{-1} \text{ K}^{-1})(350 \text{ K})} = 4.35 \text{ mol gas}$$

$$X_{\text{NO}} = 10^{-6}(206 \text{ ppm}) = 2.06 \times 10^{-4}$$

$$n_{\text{NO}} = X_{\text{NO}} \, n_{\text{gas}} = (2.06 \times 10^{-4})(4.35 \text{ mol}) = 9.00 \times 10^{-4} \text{ mol}$$

$$m = n \, MM = (9.00 \times 10^{-4} \text{ mol})(30.0 \text{ g/mol}) = 2.7 \times 10^{-2} \text{ g}$$

Does the Result Make Sense? The mass of NO is rather small, only 27 mg, but the mole fraction also is small, so this is a reasonable result.

The maximum NO emission allowed by the state of California is 762 ppm. What mass of NO is this per liter of exhaust emitted at a temperature of 50 °C?

Extra Practice Exercise 5.12

The description presented in this section applies to a gas mixture that is not undergoing chemical reactions. As long as reactions do not occur, the number of moles of each gas is determined by the amount of that substance initially present. When reactions occur, the numbers of moles of reactants and products change as predicted by the principles of stoichiometry. Changes in composition must be taken into account before the properties of the gas mixture can be computed. Gas stoichiometry is described in the next section.

Section Exercises

5.5.1 Find the partial pressures and mole fractions of a gas mixture that contains 1.00 g each of H_2 and N_2, if the total pressure of this mixture is 2.30 atm.

5.5.2 Draw a molecular picture that shows how a sample of the gas mixture in Section Exercise 5.5.1 appears at the molecular level.

5.5.3 An atmospheric chemist reported that the air in an urban area contained CO at a concentration of 35 ppm. The air temperature was 29 °C, and the atmospheric pressure was 745 torr. (a) What was the mole fraction of CO? (b) What was the partial pressure of CO? (c) How many molecules of CO were in 1.0×10^3 L of this air?

5.6 GAS STOICHIOMETRY

The principles of stoichiometry apply equally to solids, liquids, and gases. That is, no matter what phase substances are in, their chemical behavior can be described in molecular terms, and their transformations must be visualized and balanced using molecules and moles.

The ideal gas equation relates the number of moles of gas to the physical properties of that gas. When a chemical reaction involves a gas, the ideal gas equation provides the link between $P-V-T$ data and molar amounts.

$$n_i = \frac{p_i V}{RT}$$

Stoichiometric calculations always require amounts in moles. For gases, amounts in moles are usually calculated from the ideal gas equation. Example 5-13 shows how to do this.

Example 5-13
Gas Stoichiometry

Example 5-9 describes the synthesis of acetylene (C_2H_2) from calcium carbide (CaC_2). Modern industrial production of acetylene is based on a reaction of methane (CH_4) under carefully controlled conditions. At temperatures greater than 1600 K, two methane molecules rearrange to give three molecules of hydrogen and one molecule of acetylene:

$$2\ CH_4\,(g) \xrightarrow{\ 1600\ K\ } C_2H_2\,(g) + 3\ H_2\,(g)$$

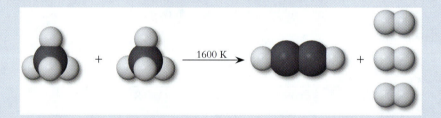

A 50.0-L steel vessel, filled with CH_4 to a pressure of 10.0 atm at 298 K, is heated to 1600 K to convert CH_4 into C_2H_2. What mass of C_2H_2 can be produced? What pressure does the reactor reach at 1600 K?

Strategy: We apply the seven-step approach to this problem.

1. This is a stoichiometry problem involving gases. The mass of a product and the final pressure must be calculated.

2 and 3. Draw a diagram showing the process and listing the data:

CH_4 (g)		C_2H_2 (g) + H_2 (g)
$V = 50.0$ L	Heat	$V = 50.0$ L
$T = 298$ K	$\longrightarrow$	$T = 1600$ K
$P = 10.0$ atm		$P = ?$ atm

Solution:

4. In any stoichiometry problem, work with moles. This problem involves gases, so use the ideal gas equation to convert $P-V-T$ information into moles.

5. Use the ideal gas equation to calculate the amount of CH_4 present initially:

$$n = \frac{P_i V}{RT_i} = \frac{(10.0\ \text{atm})(50.0\ \text{L})}{(0.08206\ \text{L atm mol}^{-1}\,\text{K}^{-1})(298\ \text{K})} = 20.45\ \text{mol}\ CH_4$$

(Carry one additional significant figure to avoid errors caused by premature rounding.)

Now construct an amounts table to determine how much of each product forms:

Example 9-13

Gas Stoichiometry (*continued*)

Reaction	2 CH$_4$ $\longrightarrow$	C$_2$H$_2$ +	3 H$_2$
Initial amount (mol)	20.45	0	0
Change in amount (mol)	-20.45	$+10.22$	$+30.67$
Final amount (mol)	0	10.22	30.67

Do a mole–mass conversion to determine the mass of acetylene formed:

$$m = n\, MM = (10.22\text{ mol})(26.04\text{ g/mol}) = 266\text{ g acetylene}$$

6. Use the ideal gas equation and the total number of moles present at the end of the reaction to calculate the final pressure:

$$n_{\text{total}} = 10.22 + 30.67 = 40.89\text{ mol}$$

$$\text{Rearrange } PV = nRT \quad \text{to} \quad P = \frac{nRT}{V}$$

$$P = \frac{(40.89\text{ mol gas})(0.08206\text{ L atm mol}^{-1}\text{ K}^{-1})(1600\text{ K})}{50.0\text{ L}} = 107\text{ atm}$$

7. Do the Results Make Sense? From the initial amount of methane, we can calculate the initial mass of methane, which is 327 g. The mass of acetylene, 266 g, is less than this, a reasonable result given that H$_2$ also is produced. The final pressure, 107 atm, seems high, but the temperature of the reactor has increased substantially and the amount of products is twice the amount of reactants, so a large pressure increase is to be expected in this reaction.

Calculate the volume of hydrogen gas generated when 1.52 g of Mg reacts with an excess of aqueous HCl, if the gas is collected at 755 torr and 22.5 °C (1 mol of Mg produces 1 mol of H$_2$).

Extra Practice Exercise 5.13

Any of the types of problems discussed in Chapters 3 and 4 can involve gases. The strategy for doing stoichiometric calculations is the same whether the species involved are solids, liquids, or gases. In this chapter, we add the ideal gas equation to our equations for converting measured quantities into moles. Example 5-14 is a limiting reactant problem that involves a gas.

Example 5-14

Limiting Reactants in a Gas Mixture

Margarine can be made from natural oils such as coconut oil by hydrogenation:

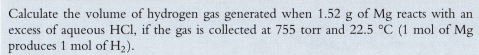

$$\underset{\text{Oil}}{C_{57}H_{104}O_6(l)} + 3\,H_2(g) \xrightarrow{\text{200 °C, 7atm, Ni catalyst}} \underset{\text{Margarine}}{C_{57}H_{110}O_6(s)}$$

An industrial hydrogenator with a volume of 2.50×10^2 L is charged with 12.0 kg of oil and 7.00 atm of hydrogen (H$_2$) at 473 K (200 °C), and the reaction goes to completion. What is the final pressure of H$_2$ and how many kilograms of margarine will be produced?

Strategy: We have data for the amounts of both starting materials, so this is a limiting reactant problem. Given the chemical equation, the first step in a limiting reactant problem is to determine the number of moles of each starting material present at the beginning of the reaction. Next compute ratios of moles to coefficients to identify the limiting reactant. After that, a table of amounts summarizes the stoichiometry.

Solution: The ideal gas equation is used for the gaseous starting material:

$$\text{Moles H}_2 = \frac{PV}{RT} = \frac{(7.00\text{ atm})(2.50 \times 10^2\text{ L})}{(0.08206\text{ L atm mol}^{-1}\text{ K}^{-1})(473\text{ K})} = 45.09\text{ mol}$$

Example 5-14

**Limiting Reactants
in a Gas Mixture**
(*continued*)

A commercial hydrogenator

For the other starting material, do a mole–mass conversion:

$$\text{Moles oil} = \frac{m}{MM} = \frac{1.20 \times 10^4 \text{ g}}{885.4 \text{ g/mol}} = 13.55 \text{ mol}$$

Now divide each number of moles by the stoichiometric coefficient to identify the limiting reactant.

$$\frac{45.09 \text{ mol H}_2}{3 \text{ mol H}_2} = 15.03 \qquad \frac{13.55 \text{ mol oil}}{1 \text{ mol oil}} = 13.55$$

The reactant with the smaller ratio, oil, is limiting.

The table of amounts follows:

Reaction	3 H$_2$	+	C$_{57}$H$_{104}$O$_6$	$\longrightarrow$	C$_{57}$H$_{110}$O$_6$
Initial amount (mol)	45.09		13.55		0
Change in amount (mol)	−40.65		−13.55		+ 13.55
Final amount (mol)	4.4		0		13.6

(The final amounts have been rounded off correctly.)

Use the ideal gas equation to calculate the pressure of hydrogen at the end of the reaction:

$$p = \frac{nRT}{V} = \frac{(4.4 \text{ mol})(0.08206 \text{ L atm mol}^{-1} \text{ K}^{-1})(473 \text{ K})}{2.50 \times 10^2 \text{ L}} = 0.68 \text{ atm}$$

Now calculate the mass of margarine:

$$m = n \, MM = (13.6 \text{ mol})(891.5 \text{ g/mol}) = 1.21 \times 10^4 \text{ g} = 12.1 \text{ kg}$$

Do the Results Make Sense? In this example, oil is the limiting reactant. Excess hydrogen is easily recovered from a gas-phase reactor, so margarine manufacturers make the oil the limiting reactant to ensure complete conversion of oil into margarine. The excess H$_2$ gas is recovered and used again in a subsequent reaction.

**Extra Practice
Exercise 5.14**

In the industrial production of nitric acid, one step is oxidation of nitrogen oxide: $2 \text{ NO} + \text{O}_2 \rightarrow 2 \text{ NO}_2$ (all gases). A reaction chamber is charged with 5.00 atm each of NO and O$_2$. If the reaction goes to completion without a change in temperature or volume, calculate the final pressures of each reagent.

Summary of Mole Conversions

Because moles are the currency of chemistry, all stoichiometric computations require amounts in moles. In the real world, we measure mass, volume, temperature, and pressure. With the ideal gas equation, our catalog of relationships for mole conversion is complete. Table 5-2 lists three equations, each of which applies to a particular category of chemical substances.

All three of these equations should be firmly embedded in your memory, *along with the substances to which they apply*. Using $PV = nRT$ on an aqueous solution gives impossible results. Example 5-15 uses all three relationships. Viewed as a whole, the example may seem complicated. As the solution illustrates, however, breaking the problem into separate parts allows each part to be solved using simple chemical and stoichiometric principles. Complicated problems are often simplified considerably by looking at them one piece at a time.

Table 5-2   Summary of Mole Relationships

⑧

Substance	Relationship	Equation
Pure liquid or solid	Moles = Mass/Molar mass	$n = m/MM$
Liquid solution	Moles = (Molarity)(Volume)	$n = MV$
Gas	Moles $= \dfrac{\text{(Pressure)(Volume)}}{\text{(Constant)(Temperature)}}$	$n = PV/RT$

Example 5-15

General Stoichiometry

Redox reactions of metals with acids are described in Chapter 4. Oxidation of the metal generates hydrogen gas and an aqueous solution of ions. Suppose that 3.50 g of magnesium metal is dropped into 0.150 L of 6.00 M HCl in a 5.00-L cylinder at 25.0 °C whose initial gas pressure is 1.00 atm, and the cylinder is immediately sealed. Find the final partial pressure of hydrogen, the total pressure in the container, and the concentrations of all ions in solution.

Strategy: Data are given for all reactants, so this is a limiting reactant problem. We must balance the chemical equation and then work with a table of molar amounts.

Solution: Begin by analyzing the chemistry. First, list all major species present in the system before reaction: $Mg\,(s)$, $H_3O^+\,(aq)$, $Cl^-\,(aq)$, and $H_2O\,(l)$.

Magnesium, a Group 2 metal, reacts with acids to generate +2 cations:

$$Mg\,(s) + H_3O^+(aq) \longrightarrow Mg^{2+}(aq) + H_2\,(g) \qquad \text{(unbalanced)}$$

This reaction can be balanced by inspection. Charge balance requires that the +2 charge on the Mg^{2+} be matched by two hydronium ions among the starting materials. Two water molecules on the right side must then balance two oxygen atoms on the left:

$$Mg\,(s) + 2\,H_3O^+(aq) \longrightarrow Mg^{2+}(aq) + H_2\,(g) + 2\,H_2O\,(l)$$

The problem asks for pressures and ion concentrations. The final pressure can be determined from $P-V-T$ data and n_{H_2}. Moles of hydrogen can be found from the mass of magnesium, the stoichiometric ratio, and a table of amounts. Here is a summary of the data:

$$V_{\text{container}} = 5.00\,\text{L} \qquad T = 298\,\text{K} \qquad m_{Mg} = 3.50\,\text{g}$$

$$V_{\text{solution}} = 0.150\,L \qquad P_{\text{air}} = 1.00\,\text{atm} \qquad [H_3O^+] = [Cl^-] = 6.00\,\text{M}$$

Now we analyze the stoichiometry of the reaction. The starting amounts of H_3O^+ and Mg are given, but the data must be converted to moles before we can construct a table of amounts:

$$\text{Moles Mg} = n_{Mg} = \frac{m}{MM} = \frac{3.50\,\text{g Mg}}{24.31\,\text{g/mol}} = 0.1439\,\text{mol Mg}$$

$$\text{Moles } H_3O^+ = n_{H_3O^+} = MV = (6.00\,\text{mol/L})(0.150\,\text{L}) = 0.900\,\text{mol } H_3O^+$$

You should be able to use the numbers of moles and the stoichiometric ratio to show that magnesium is the limiting reactant. The table of amounts follows. We omit the amounts of water because as the solvent, it is present in excess:

Mg reacting with aqueous HCl

Reaction	Mg	+ 2 H₃O⁺ ⟶	Mg²⁺ +	H₂	+ 2 H₂O
Initial amount (mol)	0.1439	0.900	0	0	
Change in amount (mol)	−0.1439	−0.2878	+ 0.1439	+ 0.1439	
Final amount (mol)	0	0.6122	0.1439	0.1439	

| Example 5-15 | From the final numbers of moles, we can calculate the final concentrations: |

General Stoichiometry
(continued)

$$[Mg^{2+}] = \frac{0.1439 \text{ mol}}{0.150 \text{ L}} = 0.959 \text{ M} \quad [H_3O^+] = \frac{0.6122 \text{ mol}}{0.150 \text{ L}} = 4.08 \text{ M} \quad [Cl^-] = 6.00 \text{ M}$$

Remember, chloride is a spectator ion, so its concentration does not change.

Before calculating the pressures, we must visualize the reaction vessel. The container's total volume is 5.00 L, but 0.150 L is occupied by the aqueous solution. This leaves 4.85 L for the gas mixture. The partial pressure of hydrogen is calculated using the ideal gas equation and assuming that no H_2 remains in solution; this is a good assumption because hydrogen gas is not very soluble in water:

$$p_{H_2} = \frac{nRT}{V} = \frac{(0.1439 \text{ mol})(0.08206 \text{ L atm mol}^{-1} \text{ K}^{-1})(298 \text{ K})}{4.85 \text{ L}} = 0.726 \text{ atm}$$

The amount of air originally present does not change in the reaction, so the pressure exerted by the air remains constant at 1.00 atm. The final total pressure is the sum of the partial pressures:

$$P_{total} = p_{H_2} + p_{initial} = 0.726 \text{ atm } H_2 + 1.00 \text{ atm air} = 1.73 \text{ atm}$$

Do the Results Make Sense? The values of all the quantities calculated in this example are in reasonable ranges for a reaction that consumes somewhat less than 1 mol of reactants.

| Extra Practice Exercise 5.15 | Repeat the calculations if all conditions are the same except that 14.0 g of Mg is added to the HCl solution. |

Section Exercises

■ **5.6.1** In an oxyacetylene torch, acetylene (C_2H_2) burns in a stream of O_2. This combustion reaction produces a flame with a temperature exceeding 3000 K. A robotic welding machine welds the inside of a tank whose air volume is 1.50×10^2 L, at a temperature of 30 °C. What mass of C_2H_2 must be burned to raise the CO_2 partial pressure to 0.10 atm? Assume that a negligible amount of the CO_2 escapes from the tank before all the C_2H_2 has been burned. What mass of water will be produced during the burn?

■ **5.6.2** A reaction vessel contains H_2 gas at 2.40 atm. If just enough O_2 is added to react completely with the H_2 to form H_2O, what will the total pressure in the vessel be before any reaction occurs? (H_2 and O_2 can also be used as a torch. The flame produced in an oxyhydrogen torch is about 4000 K.)

■ **5.6.3** When ammonium nitrate (NH_4NO_3) explodes, all the products are gases:

$$2\ NH_4NO_3\ (s) \longrightarrow 4\ H_2O\ (g) + O_2\ (g) + 2\ N_2\ (g)$$

If 5.00 g of NH_4NO_3 explodes in a closed 2.00-L container that originally contains air at $P = 1.00$ atm and $T = 25$ °C, and the temperature rises to 205 °C, what total pressure develops?

5.7 CHEMISTRY OF THE EARTH'S ATMOSPHERE

The blanket of air that cloaks our planet behaves as an ideal gas, but the atmosphere is bound to the Earth by gravitational attraction, not by confining walls. The pressure exerted by the atmosphere can be thought of as the pressure of a column of air, just as the pressure exerted by mercury in a barometer is the pressure of the column of mercury. The higher we rise into the atmosphere, the less air there is above us. Less air above us means that the pressure exerted by the column of air is lower.

Lower pressure, in turn, means lower molecular density, as indicated by the ideal gas equation:

$$\frac{n}{V} = \frac{P}{RT}$$

Figure 5-18 shows a molecular profile of a column of atmospheric air.

Pressure also varies with atmospheric conditions. For example, molecules in the atmosphere move faster when they are heated by the sun. This in turn causes an increase in pressure that is described by the ideal gas equation:

$$P = \frac{nRT}{V}$$

Molecules in high-pressure areas flow into regions of lower pressure, generating wind. This continual interplay of temperature and pressure plays a major role in determining our planet's weather.

Composition of the Lower Atmosphere

The atmosphere is a complex, dynamic mixture of gases. The composition and chemical reactivity patterns of the atmosphere change significantly as altitude above the surface of the Earth increases, as we explore in Chapter 7. Here, we focus on the layer closest to the surface, called the **troposphere.** Nitrogen and oxygen make up more than 99% of the troposphere. Only three other substances, H_2O, Ar, and CO_2, are present in amounts greater than 0.01%. Of these five principal constituents, all of them but water are present in nearly constant amounts. Because water can condense and evaporate readily, the water content of the atmosphere changes from day to day, depending on temperature and geography. The composition of dry air at sea level, expressed in mole fractions, is shown in Table 5-3.

Vapor Pressure

After a rainfall, puddles of water slowly disappear. The higher the temperature, the faster the puddles vanish. Puddles disappear because water molecules move from the liquid phase to the gas phase through **evaporation.** Evaporation is common to all substances in condensed phases, not just to water. We use the term **vapor** to describe a gaseous substance that forms by evaporation. Evaporation from bodies of water guarantees that the Earth's atmosphere always contains water vapor.

Evaporation can be understood by examining the kinetic energies of molecules. All molecules—solid, liquid, and gas—have some kinetic energy of motion, and any sample of molecules has a distribution of kinetic energies determined by its temperature (see Fig-

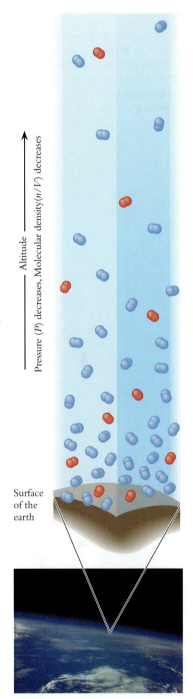

Figure 5-18

Molecular profile of the Earth's atmosphere, showing a column above some point on the Earth's surface. As altitude increases, both pressure (P) and molecular density (n/V) decrease.

Altitude ⟶

Pressure (P) decreases, Molecular density(n/V) decreases

Surface of the earth

Table 5-3 Composition of Dry Air at Sea Level*

Major Components	Mole Fraction	Trace Components	Abundance (ppm)
N_2	0.7808	Ne	18.2
O_2	0.2095	He	5.24
Minor Components		CH_4	1.4
		Kr	1.14
Ar	9.34×10^{-3}	H_2	0.50
CO_2	3.25×10^{-4}	NO	0.50
		N_2O	0.25

*Note: Figures are for water-free air. The H_2O content of air varies between $X = 0$ and $X = 0.04$.

Figure 5-19
Schematic view of evaporation and condensation in a closed system. (*a*) Initial conditions. (*b*) Intermediate conditions. (*c*) Equilibrium conditions: partial pressure = vapor pressure.

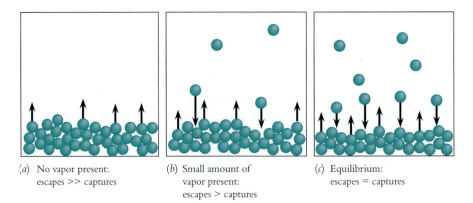

(*a*) No vapor present: escapes >> captures

(*b*) Small amount of vapor present: escapes > captures

(*c*) Equilibrium: escapes = captures

ure 5-9). Attractive forces between molecules confine liquids and solids to a fixed volume, but some molecules at the surface of that volume have enough kinetic energy to overcome the forces holding them in the condensed phase. These molecules evaporate from a liquid or solid and move into the vapor phase. The rate of evaporation increases with temperature because the fraction of molecules with sufficient energy to escape into the vapor phase increases as the sample warms up.

The water vapor that evaporates from a puddle moves away into the atmosphere. This process continues until the puddle has vaporized completely. The situation is quite different when a substance is confined to an enclosed space. If the vaporized molecules cannot escape, their numbers increase, and they exert pressure on the walls of the closed container. As evaporation proceeds, more and more molecules enter the vapor phase, and the partial pressure of the vapor rises accordingly. Figure 5-19 shows that as this happens, some gas molecules are recaptured when they collide with the surface of the liquid. Eventually, the partial pressure of the substance in the vapor phase reaches a level at which the number of molecules being recaptured (condensing) during any time interval equals the number of molecules escaping (evaporating) from the surface during that same time interval.

When the rate of evaporation equals the rate of condensation, the system is in a state of **dynamic equilibrium.** The system is dynamic because molecular transfers continue, and it has reached equilibrium because no further net change occurs. The pressure of the vapor at dynamic equilibrium is called the **vapor pressure (*vp*)** of the substance. The vapor pressure of any substance increases rapidly with temperature because the kinetic energies of the molecules increase as the temperature rises. Table 5-4 lists the vapor pressures for water at various temperatures. We describe intermolecular forces and vapor pressure in more detail in Chapter 11.

A vapor pressure is the pressure exerted by a gas in equilibrium with its condensed phase. When this equilibrium has been reached, the gas is saturated with that particular vapor. Notice in Table 5-4 that at 25 °C the atmosphere is saturated with water vapor

Table 5-4   Vapor Pressures (*vp*) of Water at Various Temperatures (*T*)

T (°C)	*vp* (torr)	*T* (°C)	*vp* (torr)	*T* (°C)	*vp* (torr)
0	4.579	35	42.175	70	233.7
5	6.543	40	55.324	75	289.1
10	9.209	45	71.88	80	355.1
15	12.788	50	92.51	85	433.6
20	17.535	55	118.04	90	525.76
25	23.756	60	149.38	95	633.90
30	31.824	65	187.54	100	760.00

when the partial pressure of H_2O is 23.756 torr. At this pressure, the molecular density of H_2O in the gas phase is sufficient to make the rate of condensation equal to the rate of evaporation. Any attempt to add more water molecules to the gas phase results in condensation to hold the partial pressure of H_2O fixed at 23.756 torr.

In most cases the atmosphere contains less water vapor than the maximum amount it can hold; that is, $p_{H_2O} < vp_{H_2O}$. The amount of water vapor actually present in the atmosphere is described by the relative humidity, which is the partial pressure of water present in the atmosphere, divided by the vapor pressure of water at that temperature and multiplied by 100% to convert to percent:

$$\text{Relative humidity} = (100\%)\left(\frac{p_{H_2O}}{vp_{H_2O}}\right) \tag{5-6}$$

Relative humidity describes how close the atmosphere is to being saturated with water vapor.

Because the vapor pressure of water varies with temperature, a given amount of water in the atmosphere represents a higher relative humidity as the temperature falls. For example, a partial pressure of H_2O of 6.54 torr at 15 °C corresponds to a relative humidity of about 50% (vp = 12.788 torr from Table 5-4):

$$\text{Relative humidity} = (100\%)\left(\frac{6.54 \text{ torr}}{12.788 \text{ torr}}\right) = 51.1\%$$

At 5 °C, however, this same partial pressure is 100% relative humidity.

The formation of dew and fog are consequences of this variation in relative humidity. Warm air at high relative humidity may cool below the temperature at which its partial pressure of H_2O equals the vapor pressure. When air temperature falls below this temperature, called the *dew point,* some H_2O must condense from the atmosphere. Example 5-16 shows how to work with vapor pressure variations with temperature, and our Chemistry and the Environment Box explores how variations in other trace gases affect climate.

Dew and fog form when water vapor condenses from the atmosphere.

	Example 5-16

Water Vapor Pressure

Fog forms when humid warm air from above a body of water moves inland and cools. What is the highest temperature at which fog could form from air that is at 65% relative humidity when its temperature is 27.5 °C?

Strategy: This problem asks about the partial pressure of water vapor in the atmosphere. Fog forms when that partial pressure exceeds the vapor pressure. Partial pressures are not given among the data, but relative humidity describes how close the partial pressure of water vapor is to its vapor pressure at the given temperature. Because vapor pressure varies strongly with temperature, we must use the information given in Table 5-4.

Solution: To convert relative humidity at 27.5 °C into a partial pressure, we need the vapor pressure of water at that temperature. Table 5-4 lists vapor pressures in 5 °C increments, but we can interpolate to find the correct value. Because 27.5 °C is halfway between 25 °C and 30 °C, the vapor pressure is about halfway between its values at 25 °C and 30 °C:

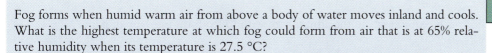

T (°C)	25	30	27.5
vp (torr)	23.756	31.824	$\frac{1}{2}(23.756 + 31.824) = 27.790$

Next, we find the partial pressure from a rearranged version of Equation 5-6:

$$p_{H_2O} = \frac{(\text{Relative humidity})(vp_{H_2O})}{100\%} = \frac{(65\%)(27.790 \text{ torr})}{100\%} = 18 \text{ torr}$$

<table>
<tr><td>

Example 5-16

Water Vapor Pressure (*continued*)

Extra Practice Exercise 5.16

</td><td>

This partial pressure remains constant as the temperature drops, meaning that fog can form below the temperature at which the vapor pressure of water is 18 torr. Examining Table 5-4, we find that the first listed temperature at which $vp_{H_2O} < 18$ torr is $T = 20\ °C$, where $vp_{H_2O} = 17.535$ torr. Thus, under these conditions, fog can form if the temperature drops below 20 °C.

Does the Result Make Sense? 20 °C is 68 °F. If you live near a large body of water, you probably have encountered fog at temperatures similar to this.

If the relative humidity is 65% at 20 °C, at what temperature will dew form?

</td></tr>
</table>

Chemistry in the Troposphere

Every year, humanity releases millions of tons of gaseous and particulate pollutants into the Earth's atmosphere. The main source of atmospheric pollution is the burning of fossil fuels. For example, diesel fuel and gasoline provide energy for automobiles, trucks, and buses, but burning these mixtures also produces vast quantities of CO_2 and lesser amounts of CO. Carbon monoxide is highly toxic, and carbon dioxide is a principal agent in global warming, as we discuss in Chapter 7. In addition, internal combustion engines are rather inefficient, so significant amounts of unburned hydrocarbons are emitted in exhaust gases. Combustion also generates nitrogen oxides and sulfur oxides, both of which undergo reactions in the atmosphere that contribute to pollution, including acid rain.

Oxides of Nitrogen

Normally, N_2 is a stable, unreactive molecule, as noted in Chapter 4. However, under the extreme conditions found in an automobile cylinder, nitrogen molecules may react with oxygen molecules to produce nitrogen oxide:

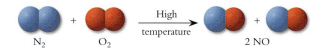

Nitrogen oxide reacts in the atmosphere with O_2 to form nitrogen dioxide:

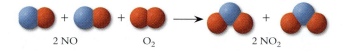

This compound is a red-brown gas that can be seen in the atmosphere over many large cities, where levels of NO_2 often reach 0.9 ppm. Tolerable limits for this toxic, irritating gas are around 3 to 5 ppm.

Nitrogen dioxide absorbs energy from sunlight and decomposes into NO molecules and oxygen atoms. Remember that oxygen is one of the elements that exists as a diatomic molecule in its natural state, so oxygen *atoms* are very reactive. Molecular nitrogen is too stable to react with oxygen atoms, but molecules of oxygen react with oxygen atoms to produce ozone (O_3), a toxic and highly reactive molecule:

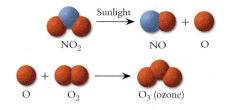

Box 5-2	Chemistry and the Environment: Does Human Activity Change the Weather?

I n 1997–98, meteorologists predicted successfully that an observed increase in temperature of Pacific Ocean waters, known as El Niño, would lead to unusual and devastating global weather patterns. As predicted, rain and snowfall in California were more than double the normal values, and spring tornadoes in the South and Midwest were more devastating than usual. Weather elsewhere was also significantly affected. A relatively small change in the temperature of one portion of the Pacific Ocean resulted in substantial changes in global weather patterns. The El Niño phenomenon highlights the fact that the Earth's weather patterns can be changed by variations in the environment.

Can human activities modify the environment in ways that cause changes in weather patterns? Growing evidence suggests that they can. The activities of modern society appear to be changing the average temperature of the atmosphere. That temperature is determined by the balance between the amount of sunlight reaching the Earth's surface and the amount of heat radiated back into space. Water, CO_2, and CH_4 in the atmosphere act as "greenhouse" gases that reduce the planet's heat loss. Increased amounts of these trace components thus are expected to increase the average temperature of our world.

Atmospheric concentrations of both CO_2 and CH_4 are increasing. The graph shows data for CO_2, whose amount has increased by 30% as humans consume more and more fossil fuel. The graph also shows that the global mean temperature has risen by nearly 1 °C.

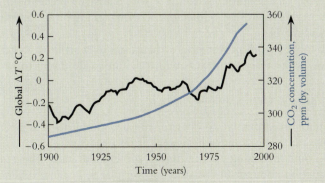

The three main variables that most influence the weather are air temperature, pressure, and water content. Their intricate interplay, energized by sunlight, creates weather patterns. Given this complexity, how can we determine if the increase in global temperature is due to the higher levels of greenhouse gases, and how can we predict the effects of further increases? One approach is computer modeling to simulate the atmosphere and project how the weather might change as a result of changes in conditions. These models predict that if the current output of greenhouse gases continues, average global temperature will rise by at least 1 °C over the next 50 to 100 years. The models also predict the consequences of global warming, including increases in extreme weather events such as heat waves and tornadoes.

A rise in average temperature of 1 °C may not seem enough to cause dramatic shifts in the weather, but the historical record shows otherwise. The period from 1500 to 1850 is called the Little Ice Age, because there were extensive increases in the sizes of the glaciers in all alpine regions. During that period, the average global temperature was just 0.5 °C lower than in 1900.

While current trends in climate lead experts to predict further global warming, natural events could change the balance. The geological record shows a series of Ice Ages alternating with eons of global warming. The cataclysmic eruption of the Indonesian volcano Krakatau in 1883 spewed huge quantities of volcanic dust and ash into the atmosphere. This caused enough global cooling that the following year was known in Europe, half a globe away, as "the year without a summer."

In the light of such uncertainties, calls to curb the pace of greenhouse emissions are controversial. Proponents point to the potentially devastating consequences of predicted global warming. Opponents cite the high current costs of reducing the use of fossil fuels, coupled with the uncertainties of how natural events affect the weather. Meanwhile, climate experts continue to collect data on current weather trends and refine their computer models, hoping to make more definitive projections about what lies ahead for our planet's weather.

Both ozone and oxygen atoms react with unburned hydrocarbons to produce many compounds that are harmful to the respiratory system. The mixture of all these pollutants is sometimes called **photochemical smog.** To control photochemical smog, governments have set limits on the levels of NO and unburned hydrocarbons allowed in automobile exhaust. Today, chemists and engineers are researching cleaner-burning fuels, more efficient engines, and fuel cells as alternatives to combustion engines.

Catalytic converters convert much of the NO and NO_2 from exhaust gases into N_2 and O_2 before they are released into the atmosphere. These have helped alleviate pollution from nitrogen oxides at only a small additional cost. We could reduce pollution emissions even further, but consumers and manufacturers are reluctant to pay the higher costs required to develop and produce cleaner fuels and engines.

Oxides of Sulfur

About half the electricity produced in the United States is generated by burning coal. Like petroleum, coal was formed millions of years ago from decaying plants under conditions of high temperature and pressure. Coal is a network of carbon atoms that also contains small amounts of other elements, including hydrogen, oxygen, nitrogen, and sulfur. When coal burns, its sulfur combines with O_2 to produce sulfur dioxide:

$$S\,(s,\ from\ coal) + O_2\,(g) \longrightarrow SO_2\,(g)$$

In the presence of dust particles or ultraviolet (UV) light, atmospheric SO_2 reacts further with O_2 to form SO_3:

$$2\,SO_2 \quad + \quad O_2 \quad \xrightarrow[\text{or UV light}]{\text{Dust particles}} \quad 2\,SO_3$$

The combustion products from sulfur impurities in coal are particularly damaging to the environment. In humans, prolonged exposure to sulfur dioxide diminishes lung capacity and aggravates respiratory problems such as asthma, bronchitis, and emphysema. Concentrations as low as 0.15 ppm can incapacitate persons with these diseases, and at about 5 ppm everyone experiences breathing difficulties. In 1952 a particularly serious episode of SO_2 pollution in London caused approximately 4000 deaths over several days.

One particularly troubling aspect of SO_2 and SO_3 pollution is acid rain, which occurs when these gases combine with water to produce acid mists:

$$SO_2\,(g) + H_2O\,(g) \longrightarrow H_2SO_3\ (mist)$$

$$SO_3\,(g) + H_2O\,(g) \longrightarrow H_2SO_4\ (mist)$$

Raindrops passing through such mists become acidic, increasing the acidity of rainwater as much as 1000-fold. Acid rain is common in the heavily industrialized areas of the United States, Canada, and Europe. In areas where acid rain is particularly severe, lakes are becoming so acidic that fish can no longer survive. In addition to the devastating effect on aquatic life, acid rain leaches nutrients from the soil, alters the metabolism of organisms in the soil, accelerates the corrosion of metals, and damages important building materials such as limestone and marble.

Sulfur dioxide can be removed from power plant exhaust gas by a scrubber system. One common method involves the reaction of SO_2 with calcium oxide (lime) to form calcium sulfite:

$$SO_2\,(g) + CaO\,(s) \longrightarrow CaSO_3\,(s)$$

Unfortunately, scrubber systems are expensive to operate, and the solid $CaSO_3$ is generated in large enough quantities to create significant disposal problems.

Sulfur pollution may become increasingly significant in the United States if increasing amounts of energy production come from coal-fired power plants. Because different deposits of coal have different sulfur concentrations, the amount of pollution will depend

Statues are slowly decomposed by acid rain.

on the type of coal used. The oldest coal, anthracite, is 90% carbon and has a sulfur content lower than 1%, whereas bituminous coal can contain as much as 5% sulfur by mass. Unfortunately, vast North American coal reserves are mostly high-sulfur bituminous coal. Thus, increased dependence on coal as a source of energy will come with a high cost to society. The choices are to find other energy sources; to spend more money to develop efficient scrubber systems; to reduce consumption through conservation; or to face potentially disastrous consequences for our environment.

Section Exercises

5.7.1 What are the mole fractions of the three most abundant trace atmospheric constituents listed in Table 5-3?

5.7.2 Argon gas can be recovered from the atmosphere by appropriate cooling and liquefaction processes. How many liters of dry air at 1.00 atm and 27 °C must be processed to collect 10.0 mol of Ar?

5.7.3 A weather report gives the current temperature as 18 °C and sets the dew point at 10 °C. Using data from Table 5-4, determine the partial pressure of water vapor in the atmosphere and calculate the relative humidity.

CHAPTER 5 VISUAL SUMMARY

SKILLS TO MASTER

1. Converting among pressure units
2. Using the ideal gas equation
3. Relating final conditions to initial conditions
4. Relating molecular properties to gas properties
5. Using gas properties to determine molar mass
6. Describing molecular motion in gases
7. Determining partial pressures
8. Solving stoichiometry problems involving gases
9. Working with vapor pressure

5.1 PRESSURE

Standard atmosphere (atm)
Torr
Pascal (Pa)
Bar
(1)

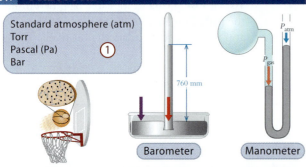

P_{atm}

P_{gas}

760 mm

Barometer

Manometer

5.2 DESCRIBING GASES

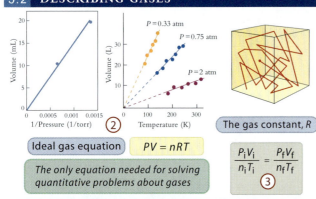

$P = 0.33$ atm
$P = 0.75$ atm
$P = 2$ atm

Volume (mL)

1/Pressure (1/torr) (2)

Volume (L)

Temperature (K)

The gas constant, R

Ideal gas equation $PV = nRT$

$$\frac{P_i V_i}{n_i T_i} = \frac{P_f V_f}{n_f T_f}$$

(3)

The only equation needed for solving quantitative problems about gases

5.3 MOLECULAR VIEW OF GASES

$$E_{kinetic} = \frac{1}{2}mu^2 \qquad \overline{E}_{kinetic} = \frac{3RT}{2N_A}$$

300 K

900 K

Number of molecules

Kinetic energy

At a given T, all gases have the same kinetic energy distribution.

Ideal gas   Negligible molecular sizes
Negligible intermolecular forces   (4)

5.4 ADDITIONAL GAS PROPERTIES

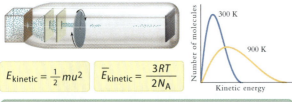

Diffusion

Effusion   (6)

Molar mass (5) Gas density

$$\overline{u} = \left(\frac{3RT}{MM}\right)^{\frac{1}{2}}$$

Root-mean-square speed

5.5 GAS MIXTURES

Dalton's law of partial pressures

$$P_{total} = p_1 + p_2 + p_3 \cdots$$

(7)   Partial pressure

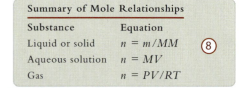

Mole fraction (X)
Parts/million (ppm)
Parts/billion (ppb)

$$p_A = X_A P_{total}$$

5.6 GAS STOICHIOMETRY

Reaction:	N_2	$+$	$3\ H_2$	$\longrightarrow$	$2\ NH_3$
Initial p (atm)	1		4		0
Change in p (atm)	-1		-3		$+2$
Final p (atm)	0		1		2

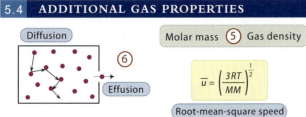

Summary of Mole Relationships

Substance	Equation	
Liquid or solid	$n = m/MM$	(8)
Aqueous solution	$n = MV$	
Gas	$n = PV/RT$	

5.7 CHEMISTRY OF EARTH'S ATMOSPHERE

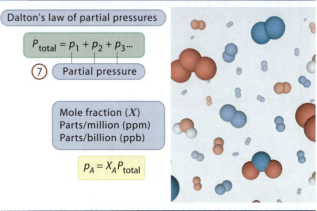

$H_2O\ (l) \rightleftharpoons H_2O\ (g)$

Dynamic equilibrium

Troposphere

NO SO_2
O_3
NO_2 SO_3

Photochemical smog

(9)

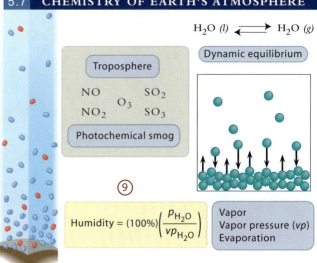

$$Humidity = (100\%)\left(\frac{p_{H_2O}}{vp_{H_2O}}\right)$$

Vapor
Vapor pressure (vp)
Evaporation

Learning Exercises

5.1 List the macroscopic properties and the microscopic properties of gases.

5.2 Update and reorganize your list of memory bank equations so that all equations that involve moles are grouped together.

5.3 Summarize the strategy for calculating final pressures of gases in a limiting reactant problem.

5.4 Define partial pressure, vapor pressure, and relative humidity. Explain how they are related.

5.5 List all terms new to you that appear in Chapter 5. Write a one-sentence definition of each in your own words. Consult the glossary if you need help.

 Problems <u>ilw</u> = **interactive learning ware problem. Visit the website at www.wiley.com/college/olmsted**

Pressure

5.1 Describe what would happen to the barometer in Figure 5-2 if the tube holding the mercury had a pinhole at its top.

5.2 Redraw the U-tube shown in Figure 5-3 to show how the manometer appears if the pressure on the side exposed to the atmosphere is greater than the pressure on the side exposed to the gas by an amount $2\Delta h$.

5.3 Describe how the difference between an inflated and a flat automobile tire shows that a gas exerts pressure.

5.4 A sailboat moves across the water making use of the wind. How does the motion of a sailboat demonstrate that gas molecules exert pressure?

5.5 Express the following in units of pascals and bars: (a) 455 torr; (b) 2.45 atm; (c) 0.46 torr; and (d) 1.33×10^{-3} atm.

5.6 Convert the following to torr: (a) 1.00 Pa; (b) 125.6 bar; (c) 75.0 atm; and (d) 4.55×10^{-10} atm.

Describing Gases

5.7 A sample of air was compressed to a volume of 20.0 L. The temperature was 298 K and the pressure was 5.00 atm. How many moles of gas were in the sample? If the sample had been collected from air at $P = 1.00$ atm, $T = 298$ K, what was the original volume of the gas?

5.8 A bicycle pump inflates a tire whose volume is 565 mL until the internal pressure is 6.47 atm at a temperature of 21.7 °C. How many moles of air does the tire contain? What volume of air at 1.01 atm and 21.7 °C did the pump transfer?

5.9 Rearrange the ideal gas equation to give the following expressions: (a) an equation that relates P_i, T_i, P_f, and T_f when n and V are constant; (b) $V = ?$; and (c) an equation that relates P_i, V_i, P_f, and V_f when n and T are constant.

5.10 Rearrange the ideal gas equation to give the following expressions: (a) $n = ?$; (b) an equation that relates V_i, n_i, V_f, and n_f when P and T are constant; and (c) $n/V = ?$

ilw 5.11 It requires 0.255 L of air to fill a metal foil balloon to 1.000 atm pressure at 25 °C. The balloon is tied off and placed in a freezer at −15 °C. What is the new volume of air in the balloon?

5.12 A pressurized can of whipping cream has an internal pressure of 1.075 atm at 25 °C. If it is placed in a freezer at −15 °C, what is the new value for its internal pressure?

5.13 Under which of the following conditions could you use the equation $P_f V_f = P_i V_i$? (a) A gas is compressed at constant T. (b) A gas-phase chemical reaction occurs. (c) A container of gas is heated. (d) A container of liquid is compressed at constant T.

5.14 Under which of the following conditions could you *not* use the equation, $P_f V_f / T_f = P_i V_i / T_i$? (a) P is expressed in torr. (b) T is expressed in °C. (c) V is changing. (d) n is changing.

Molecular View of Gases

5.15 Redraw Figure 5-7*b* to show the distribution of molecules if the temperature of the oven is doubled.

5.16 Redraw Figure 5-8 to show a temperature of 200 K.

5.17 Draw a single graph that shows the *speed* distributions of N_2 at 200 K, N_2 at 300 K, and He at 300 K.

5.18 Draw a single graph that shows the *speed* distributions of Cl_2 at 500 K, Br_2 at 500 K, and Cl_2 at 300 K.

5.19 Redraw the graph in Problem 5.17 as a graph of *energy* distributions for the three gases.

5.20 Redraw the graph in Problem 5.18 as a graph of *energy* distributions for the three gases.

5.21 Calculate the most probable speed and average kinetic energy per mole for each of the following: (a) He at 627 °C; (b) O_2 at 27 °C; and (c) SF_6 at 627 °C.

5.22 Calculate the most probable speed and average kinetic energy per mole for each of the following: (a) Ar at 127 °C; (b) Xe at 127 °C; and (c) C_3H_8 at 327 °C.

5.23 Explain in molecular terms why each of these statements is true: (a) At very high pressure, no gas behaves ideally. (b) At very low temperature, no gas behaves ideally.

5.24 A cylinder with a movable piston contains a sample of gas. Describe in molecular terms the effect on pressure exerted by the gas for each of the following changes: (a) The piston is pushed in. (b) Some gas is removed while the piston is held in place. (c) The gas is heated while the piston is held in place.

5.25 The figure represents an ideal gas in a container with a movable friction-free piston.

(a) The external pressure on the piston exerted by the atmosphere is 1 atm. If the piston is not moving, what is the pressure inside the container? Explain in terms of molecular collisions.

(b) Redraw the sketch to show what would happen if the temperature of the gas in the container is doubled. Explain in terms of molecular collisions.

1 atm

5.26 Consider the figure appearing in Problem 5.25. (a) The monatomic gas shown in the figure is replaced by an equal number of molecules of a diatomic gas, all other conditions remaining the same. What is the pressure inside the container? Explain in terms of molecular collisions. (b) The external pressure is reduced to 0.75 atm, and the piston moves as a result. Redraw the sketch to show the new situation.

5.27 Describe the molecular changes that account for the result in Problem 5.11.

5.28 Describe the molecular changes that account for the result in Problem 5.12.

Additional Gas Properties

ilw 5.29 Freons (CFCs) are compounds that contain carbon, chlorine, and fluorine in various proportions. They are used as foaming agents, propellants, and refrigeration fluids. Freons are controversial because of the damage they do to the ozone layer in the stratosphere. A 2.55-g sample of a particular Freon in a 1.50-L bulb at 25.0 °C has a pressure of 262 torr. What is the molar mass and formula of the compound?

5.30 Gaseous hydrocarbons, which contain only carbon and hydrogen, are good fuels because they burn in air to generate large amounts of heat. A sample of hydrocarbon with $m = 1.65$ g exerts a pressure of 1.50 atm in a 945-mL bulb at 21.5 °C. Determine the molar mass and chemical formula of this hydrocarbon.

5.31 What is the density (g/L) of SF_6 gas at 755 torr and 27 °C?

5.32 What is the density (g/L) of H_2 gas at 380 torr and −23 °C?

5.33 Determine the root-mean-square speed of SF_6 molecules under the conditions of Problem 5.31.

5.34 Determine the root-mean-square speed of H_2 molecules under the conditions of Problem 5.32.

5.35 If a gas line springs a leak, which will diffuse faster through the atmosphere and why, CH_4 or C_2H_6?

5.36 A student proposes to separate CO from N_2 by an effusion process. Is this likely to work? Why or why not?

Gas Mixtures

ilw 5.37 The amount of NO_2 in a smoggy atmosphere was measured to be 0.78 ppm. The barometric pressure was 758.4 torr. Compute the partial pressure of NO_2 in atmospheres.

5.38 California's automobile emission standards require that exhaust gases contain less than 220 parts per million hydrocarbons and less than 1.2% CO (both of these values are in moles per mole of air). At standard atmospheric pressure, what are the partial pressures, in torr and in atmospheres, that correspond to these values?

5.39 In dry atmospheric air, the four most abundant components are N_2, $X = 0.7808$; O_2, $X = 0.2095$; Ar, $X = 9.34 \times 10^{-3}$; and CO_2, $X = 3.25 \times 10^{-4}$. Calculate the partial pressures of these four gases, in torr, under standard atmospheric conditions.

5.40 Find the partial pressures, total pressure, and mole fractions of a gas mixture in a 4.00-L container at 375 °C if it contains 1.25 g each of Ar, CO, and CH_4.

5.41 The following questions refer to Figure 5.16*d*: (a) Which gas exerts a higher pressure, O_2 or He? (b) Which has a higher mole fraction? (c) Compute the mole fraction of He.

5.42 The figures shown below represent very small portions of three gas mixtures, all at the same volume and temperature.

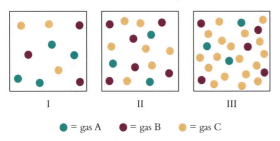

| I | II | III |

$\bullet$ = gas A $\bullet$ = gas B $\bullet$ = gas C

(a) Which sample has the highest partial pressure of gas A ? (b) Which sample has the highest mole fraction of gas B? (c) In sample III, what is the concentration of gas A in ppm?

5.43 The natural gas in a storage tank is analyzed. A 2.00-g sample of the gas contains 1.57 g CH_4, 0.41 g C_2H_6, and 0.020 g C_3H_8. Calculate the partial pressures of each gas in the storage tank if the total pressure in the tank is 2.35 atm.

5.44 A sample of automobile exhaust gas is analyzed at a smog station. The gas contains 487.4 ppm CO_2, 10.3 ppb NO, and 4.2 ppb CO. Calculate the partial pressures of these gases if the exhaust is emitted at 113.1 kPa pressure.

Gas Stoichiometry

ilw 5.45 Humans consume glucose to produce energy. The products of glucose consumption are CO_2 and H_2O:

$$C_6H_{12}O_6\,(s) + O_2\,(g) \longrightarrow CO_2\,(g) + H_2O\,(l) \qquad \text{(unbalanced)}$$

What volume of CO_2 is produced under body conditions (37 °C, 1.00 atm) during the consumption of 4.65 g of glucose?

5.46 Oxygen gas can be generated by heating $KClO_3$ in the presence of a catalyst:

$$KClO_3 \longrightarrow KCl + O_2 \qquad \text{(unbalanced)}$$

What volume of O_2 gas will be generated at $T = 25$ °C and $P = 765.1$ torr from 1.57 g of $KClO_3$?

5.47 Sodium metal reacts with molecular chlorine gas to form sodium chloride. A closed container of volume 3.00×10^3 mL contains chlorine gas at 27 °C and 1.25×10^3 torr. Then 6.90 g of solid sodium is introduced, and the reaction goes to completion. What is the final pressure if the temperature rises to 47 °C?

5.48 Carbon monoxide and molecular oxygen react to form carbon dioxide. A 50.0-L reactor at 25.0 °C is charged with 1.00 atm of CO. The gas is then pressurized with O_2 to give a total pressure of 3.56 atm. The reactor is sealed, heated to 350 °C to drive the reaction to completion, and cooled back to 25.0 °C. Compute the final partial pressure of each gas.

ilw 5.49 Ammonia is produced industrially by reacting N_2 with H_2 at elevated pressure and temperature in the presence of a catalyst:

$$N_2 + H_2 \longrightarrow NH_3 \qquad \text{(unbalanced)}$$

Assuming 100% yield, what mass of ammonia would be produced from a 1:1 mole ratio mixture in a reactor that has a volume of 8.75×10^3 L, under a total pressure of 275 atm at $T = 455$ °C?

5.50 Ethylene oxide is produced industrially from the reaction of ethylene with oxygen at atmospheric pressure and 280 °C, in the presence of a silver catalyst:

$$C_2H_4\,(g) + O_2\,(g) \longrightarrow C_2H_4O\,(g) \qquad \text{(unbalanced)}$$

Assuming a 100% yield, how many kilograms of ethylene oxide would be produced from 5.00×10^4 L of a mixture containing ethylene and oxygen in 1:1 mole ratio?

5.51 In actual practice, the reaction of Problem 5.49 gives a yield of only 13%. Repeat the calculation of Problem 5.49 taking this into account.

5.52 In actual practice, the ethylene oxide reaction of Problem 5.50 gives only a 65% yield under optimal conditions. Repeat the calculation of Problem 5.50 assuming a 65% yield.

Chemistry of the Earth's Atmosphere

5.53 How much does the mole fraction of O_2 in the atmosphere change at 20 °C as the relative humidity varies from 100% to 0%?

5.54 How much does the partial pressure of N_2 gas in the atmosphere change at 30 °C and 1.00 atm as the relative humidity varies from zero to 100%?

5.55 Find the dew point for relative humidity 78% at 25 °C.

5.56 In the tropics, water will condense in human lungs when the temperature and relative humidity are too high. Using Table 5-4, estimate the vapor pressure of water at body temperature of 37 °C. If atmospheric temperature is 40 °C, at what relative humidity does this life-threatening process occur?

ilw 5.57 Neon was discovered by cooling dry air until it is liquefied and then boiling off the components one at a time. What volume of dry air at 298 K and 1.00 atm must be treated in this way to obtain 10 mg of neon? (See Table 5-3.)

5.58 Compute the total mass of krypton contained in 1.00 cubic kilometer of dry air, assuming constant atmospheric pressure of 1.00 atm (see Table 5-3).

Additional Paired Problems

5.59 What mass of krypton is present in a 725-mL container at 925 °C in which the pressure of krypton is 10.0 atm? How many atoms is this?

5.60 Nitrogen gas is available commercially in pressurized 9.50-L steel cylinders. If a tank has a pressure of 145 atm at 298 K, how many moles of N_2 are in the tank? What is the mass of N_2?

5.61 The figure shows three chambers with equal volumes, all at the same temperature, connected by closed valves. Each chamber contains helium gas, with amounts proportional to the number of atoms shown.

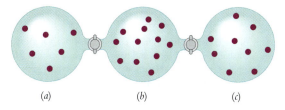

(a)   (b)   (c)

Answer each of the following questions, briefly stating your reasoning: (a) Which of the three has the highest pressure? (b) If the pressure of B is 1.0 atm, what is the pressure of A? (c) If the pressure of A starts at 1.0 atm, and then all of the atoms from B and C are added to A's container, what will be the new pressure? (d) If the pressure in B is 0.50 atm, what will the pressure be after the valves are opened?

5.62 The following figure shows two tanks connected by a valve. Each tank contains a different gas, both at 0.0 °C.

(a) Redraw the system to show how it appears after the valve has been opened. (b) If each molecule in the diagram represents 1.0 mol of gas, what is the system's total pressure after the valve is opened? (Assume that the valve has negligible volume.) (c) If all the gas is pumped into the smaller tank, what are the partial pressures of the two gases?

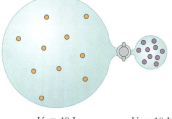

$V_1 = 40$ L $V_2 = 10$ L

$T_1 = 0$ °C $T_2 = 0$ °C

5.63 At an altitude of 40 km above the Earth's surface, the temperature is about −25 °C, and the pressure is about 3.0 torr. Calculate the average molecular speed of ozone (O_3) at this altitude.

5.64 The average kinetic energy of a 1.55-g sample of argon gas in a 5.00-L bulb is 1.02×10^{-20} J/atom. What is the pressure of the gas? What is the average speed of the argon atoms under these conditions?

5.65 At an altitude of 150 km above the Earth's surface, atmospheric pressure is 10^{-10} atm and the temperature is 310 K. What is the molecular density at this altitude?

5.66 Molecular clouds composed mostly of hydrogen molecules have been detected in interstellar space. The molecular density in these clouds is about 10^{10} molecules m^{-3}, and their temperature is around 25 K. What is the pressure in such a cloud?

5.67 Use data from Table 5-4 to estimate the dew point for air that has (a) relative humidity of 80% at 35 °C; (b) relative humidity of 50% at 15 °C; and (c) relative humidity of 30% at 25 °C.

5.68 A sample of warm moist air (100% relative humidity) is collected in a container at $P = 756$ torr and $T = 30.0$ °C. Find (a) the new pressure after a drying agent is added to remove the water vapor and (b) the mole fraction and concentration in grams per liter of water vapor in the original air sample.

5.69 Describe a gas experiment that would show that the element oxygen exists naturally as diatomic molecules.

5.70 A student proposes that SO_2 gas contains a significant fraction of dimer with formula S_2O_4. Describe a gas experiment that would verify or disprove this proposal.

5.71 A mixture of cyclopropane gas (C_3H_6) and oxygen (O_2) in 1.00:4.00 mole ratio is used as an anesthetic. What mass of each of these gases is present in a 2.00-L bulb at 23.5 °C if the total pressure is 1.00 atm?

5.72 A gas cylinder of volume 5.00 L contains 1.00 g of Ar and 0.0500 g of Ne. The temperature is 275 K. Find the partial pressures, total pressure, and mole fractions.

5.73 Consider two gas bulbs of equal volume, one filled with H_2 gas at 0 °C and 2 atm, the other containing O_2 gas at 25 °C and 1 atm. Which bulb has (a) more molecules; (b) more mass; (c) higher average kinetic energy of molecules; and (d) higher average molecular speed?

5.74 Consider two gas bulbs of equal volume, one filled with Cl_2 gas at 100 °C and 2 atm, the other containing N_2 gas at 25 °C and 1 atm. Which bulb has (a) more molecules; (b) more mass; (c) higher average kinetic energy of molecules; and (d) higher average molecular speed?

5.75 Refer to Figure 5-9 to determine the following for O_2 gas at 300 K: (a) What is the most probable kinetic energy? (b) What is the most probable speed?

5.76 Molecular beam experiments on ammonia at 425 K give the speed distribution shown in the figure:

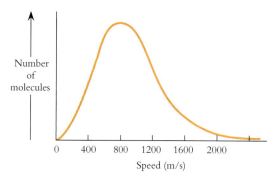

Number of molecules

0   400   800   1200   1600   2000

Speed (m/s)

(a) What is the most probable speed? (b) What is the most probable kinetic energy?

5.77 Explain in your own words how sulfur impurities in coal can lead to acid rain. Use balanced equations when appropriate.

5.78 Explain in your own words how the reactions occurring in the cylinders of automobile engines contribute to the production of photochemical smog. Use balanced equations when appropriate.

5.79 A sample of gas is found to exert a pressure of 525 torr when it is in a 3.00-L flask at 0.00 °C. Compute (a) the new volume if P becomes 755 torr and T is unchanged; (b) the new pressure if V becomes 2.00 L and T is unchanged; and (c) the new pressure if the temperature is raised to 50.0 °C and V is unchanged.

5.80 A sample of gas is found to exert a pressure of 322 torr when it is in a 2.00-L flask at 100.00 °C. Compute (a) the new volume if P becomes 525 torr and T is unchanged; (b) the new pressure if the volume is reduced to 1.50 L and T becomes 50.0 °C; and (c) the new pressure if half the gas is removed but V and T remain the same.

5.81 Redraw the central figure in Figure 5-17 to show each of the following changes in conditions: (a) The temperature increases by 50%. (b) 25% of the helium atoms are removed. (c) Hydrogen molecules replace half the helium atoms.

5.82 Redraw the central figure in Figure 5-17 to show each of the following changes in conditions: (a) Half of the helium atoms are removed. (b) The helium atoms are replaced with an equal number of nitrogen molecules. (c) The temperature is lowered by 50%.

5.83 Determine whether each of the following statements is true or false. If false, rewrite the statement so that it is true. (a) At constant T and V, P is inversely proportional to the number of moles of gas. (b) At constant V, the pressure of a fixed amount of gas is directly proportional to T. (c) At fixed n and V, the product of P and T is constant.

5.84 Determine whether each of the following statements is true or false. If false, rewrite the statement so that it is true. (a) At fixed n and P, V is independent of T. (b) At fixed n and T, the product of P and V is constant. (c) When gas is added to a chamber at fixed V and T, the pressure increases as n^2.

5.85 A 3.00-g sample of an ideal gas occupies 0.963 L at 22 °C and 0.969 atm. What will be its volume at 15 °C and 1.00 atm?

5.86 A 96.0-g sample of O_2 gas at 0.0 °C and 380 torr is compressed and heated until the volume is 3.00 L and the temperature is 27 °C. What is the final pressure in torr?

5.87 The gas SF_6 is used to trace air flows because it is nontoxic and can be detected selectively in air at a concentration of 1.0 ppb. What partial pressure is this? At this concentration, how many molecules of SF_6 are contained in 1.0 cm^3 of air at $T = 21$ °C?

5.88 In 1990, carbon dioxide levels at the South Pole reached 351.5 parts per million by volume. (The 1958 reading was 314.6 ppm by volume.) Convert this reading to a partial pressure in atmospheres. At this level, how many CO_2 molecules are there in 1.0 L of dry air at −45 °C?

More Challenging Problems

5.89 A 15.00-g piece of dry ice (solid CO_2) is dropped into a bottle with a volume of 0.750 L, and the air is pumped out. The bottle is sealed and the CO_2 is allowed to evaporate. What will be the final pressure in the bottle if the final temperature is 0.0 °C?

5.90 Two chambers are connected by a valve. One chamber has a volume of 15 L and contains N_2 gas at a pressure of 2.0 atm. The other has a volume of 1.5 L and contains O_2 gas at 3.0 atm. The valve is opened, and the two gases are allowed to mix thoroughly. The temperature is constant at 300 K throughout this process. (a) How many moles each of N_2 and O_2 are present? (b) What are the final pressures of N_2 and O_2, and what is the total pressure? (c) What fraction of the O_2 is in the smaller chamber after mixing?

5.91 Calculate the density of (a) dry air at 1 atm and 298 K and (b) air of 100% relative humidity at 1 atm and 298 K.

5.92 On a smoggy day, the ozone content of air over Los Angeles reaches 0.50 ppm. Compute the partial pressure of ozone and the number of molecules of ozone per cubic centimeter if atmospheric pressure is 762 torr and the temperature is 28 °C.

5.93 Liquid oxygen, used in some large rockets, is produced by cooling dry air to −183 °C. How many liters of dry air at 25 °C and 750 torr would have to be processed to produce 150 L of liquid oxygen (density = 1.14 g/mL)?

5.94 Benzaldehyde is a fragrant molecule used in artificial cherry flavoring. Combustion of 125 mg of benzaldehyde gives 363 mg of carbon dioxide and 63.7 mg of water. In another experiment, a 110-mg sample is vaporized at 150 °C in a 0.100-L bulb. The vapor gives a pressure of 274 torr. Determine the molecular formula of benzaldehyde.

5.95 The methanation reaction, $3 H_2 + CO \rightarrow CH_4 + H_2O$, is used commercially to prepare methane. A gas reactor with a volume of 1.00×10^2 L is pressurized at 575 K with 20.0 atm of H_2 gas and 10.0 atm of CO gas; 145 g of CH_4 is produced. What is the percent yield of the synthesis?

5.96 In an explosion, a compound that is a solid or a liquid decomposes very rapidly, producing large volumes of gas. The force of the explosion results from the rapid expansion of the hot gases. For example, TNT (trinitrotoluene) explodes according to the following balanced equation:

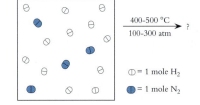

TNT
$C_7H_5(NO_2)_3$

$$2 C_7H_5(NO_2)_3\,(s) \longrightarrow 12\ CO\,(g) + 2\ C\,(s) + 5\ H_2\,(g) + 3\ N_2\,(g)$$

(a) How many moles of gas are produced in the explosion of 1.0 kg of TNT? (b) What volume will these gases occupy if they expand to a total pressure of 1.0 atm at 25 °C? (c) At 1.0 atm total pressure, what would be the partial pressure of each gas?

5.97 The Haber synthesis of ammonia occurs in the gas phase at high temperature (400 to 500 °C) and pressure (100 to 300 atm). The starting materials for the Haber synthesis are placed inside a container, in proportions shown in the figure. Assuming 100% yield, draw a sketch that illustrates the system at the end of the reaction.

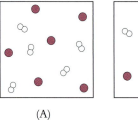

400–500 °C
100–300 atm → ?

① = 1 mole H_2
① = 1 mole N_2

5.98 When a sealed bulb containing a gas is immersed in an ice bath, it has a gas pressure of 345 torr. When the same bulb is placed in an oven, the pressure of the gas rises to 745 torr. What is the temperature of the oven in Celsius?

5.99 Many of the transition metals form complexes with CO; these complexes are called metal carbonyls and have the general formula $M(CO)_x$. A 0.500-g sample of gaseous nickel carbonyl in a 0.100-L bulb generates a pressure of 552 torr at 30 °C. What is the formula of nickel carbonyl?

5.100 At low temperature nitrogen dioxide molecules join together to form dinitrogen tetroxide.

$$2\ NO_2 \longrightarrow N_2O_4 \qquad \text{(low temperature)}$$

A sample of NO_2 sealed inside a glass bulb at 23 °C gave a pressure of 691 torr. Lowering the temperature to −5 °C converted the NO_2 to N_2O_4. What was the final pressure inside the bulb?

5.101 People often remark that "the air is thin" at higher elevation. Explain this comment in molecular terms using the fact that the atmospheric pressure atop Mount Everest is only about 250 torr.

5.102 The figures shown below represent mixtures of argon atoms and hydrogen molecules. The volume of container B is twice the volume of container A.

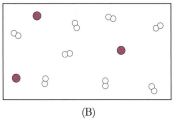

(A)   (B)

(a) Which container has a higher total gas pressure? Explain. (b) Which container has a higher partial pressure of molecular hydrogen? Explain. (c) One of the two gas mixtures shown above was used in a pulsed molecular beam experiment. The result of the experiment is shown below. Which of the two gas samples, A or B, was used for this experiment? Explain.

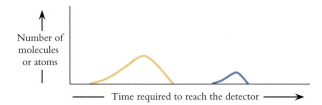

Number of molecules or atoms

Time required to reach the detector

5.103 Find the partial pressures in atmospheres of the eight most abundant atmospheric components listed in Table 5-3 at 25 °C, 50% relative humidity, and $P = 765$ torr.

5.104 A 0.1054-g mixture of $KClO_3$ and a catalyst was placed in a quartz tube and heated vigorously to drive off all the oxygen as O_2. The O_2 was collected at 25.17 °C and a pressure of 759.2 torr. The volume of gas collected was 22.96 mL. (a) How many moles of O_2 were produced? (b) How many moles of $KClO_3$ were in the original mixture? (c) What was the mass percent of $KClO_3$ in the original mixture?

5.105 Elemental analysis of an organic liquid with a fishy odor gives the following elemental mass percentages: H, 14.94%; C, 71.22%; and N, 13.84%. Vaporization of 250 mg of the compound in a 150-mL bulb at 150 °C gives a pressure of 435 torr. What is the molecular formula of the compound?

5.106 When heated to 150 °C, $CuSO_4 \cdot 5H_2O$ loses its water of hydration as gaseous H_2O. A 2.50-g sample of the compound is placed in a sealed 4.00-L steel vessel containing dry air at 1.00 atm and 27 °C and the vessel is then heated to 227 °C. What are the final partial pressure of H_2O and the final total pressure?

Group Study Problems

5.107 A balloon filled with helium gas at 1.0 atm and 25 °C is to lift a 350-kg payload. What is the minimum volume required for the balloon? Assume that dry air at the same pressure and temperature makes up the atmosphere.

5.108 What temperature must the air be if the balloon described in Problem 5.107 is going to use hot air instead of helium?

5.109 A mouse is placed in a sealed chamber filled with air at 765 torr and equipped with enough solid KOH to absorb any CO_2 and H_2O produced. The gas volume in the chamber is 2.05 L, and its temperature is held at 298 K. After 2 hours, the pressure inside the chamber has fallen to 725 torr. What mass of oxygen has the mouse consumed?

5.110 Liquid helium at 4.2 K has a density of 0.147 g/mL. A 2.00-L metal bottle at 95 K contains air at 1.0 atm pressure. We introduce 0.100 L of liquid helium, seal the bottle, and allow the entire system to warm to room temperature (25 °C). What is the pressure inside the bottle?

5.111 You are on vacation in Hawaii, where the temperature is 27 °C. You wish to go for a scuba dive. Your tank, with a volume of 12.5 L, is filled with dry air to a pressure of 2425 psi (165 atm). If your body requires you to consume 14.0 g O_2 per minute, how long can you dive at a depth of 70 feet if you must allow 6.0 minutes to return to the surface?

5.112 If 1.00 metric ton of soft coal containing 4.55% (by mass) sulfur is burned in an electric power plant, what volume of SO_2 is produced at 55 °C and 1.00 atm? What mass of CaO is required to "scrub" this SO_2? What mass of $CaSO_3$ must be disposed of?

5.113 Construct the following graphs for an ideal gas when $T = 298$ K: (a) V vs. P; (b) PV vs. P; and (c) $1/V$ vs. P. Describe in words the shape of each of these graphs.

Answers to Section Exercises

5.1.1 710 torr, 0.93 atm, and 6.6% change
5.1.2 1.596×10^6 Pa, 1.596×10^3 kPa, and 15.96 bar
5.1.3 800.3 torr, 1.053 atm, and 1.067×10^2 kPa
5.2.1 (a) and (b) pressure doubles; (c) pressure triples; and (d) no change
5.2.2

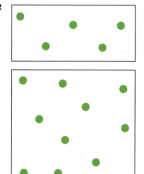

5.2.3 784 torr
5.3.1

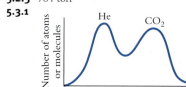

5.3.2 (a) 2.16×10^3 J; (b) 122 J; and (c) 1.96×10^{-20} J
5.3.3 Low pressure, because fewer molecular interactions and more space between the molecules
5.4.1 59.1 kg of air enters as the house cools.
5.4.2 272 g/mol, $HgCl_2$
5.4.3 Speeds are 1.04×10^3 m/s for H_2O and 9.82×10^2 m/s for D_2O. Ratio of rates is 1.05.
5.5.1 H_2, $X = 0.933$ and $p = 2.15$ atm; N_2, $X = 0.067$ and $p = 0.15$ atm
5.5.2

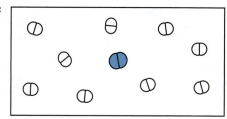

5.5.3 (a) 3.5×10^{-5}; (b) 2.6×10^{-2} torr; and (c) 8.3×10^{20} molecules
5.6.1 7.8 g C_2H_2 and 5.4 g H_2O
5.6.2 3.60 atm
5.6.3 5.89 atm (Remember to calculate the new partial pressure of air at the higher temperature.)
5.7.1 Ne, 1.82×10^{-5}; He, 5.24×10^{-6}; and CH_4, 1.4×10^{-6}
5.7.2 2.64×10^4 L
5.7.3 $vp = 9.209$ torr, 59% relative humidity

Answers to Extra Practice Exercises

5.1 $P_{gas} = 799.8$ torr $= 1.052$ atm $= 1.066$ bar
5.2 $P = 143$ atm
5.3 $T = 294$ K $= 21$ °C
5.4 $P = 0.47$ atm
5.5 $V = 6.79$ L
5.6 Increase of 2.5%
5.7 He atoms move slower than H_2 molecules.

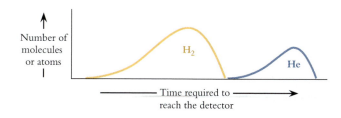

5.8 $\bar{E}_{kinetic} = 4.93 \times 10^{-21}$ J/molecule; $E_{kinetic, molar} = 2.97 \times 10^3$ J/mol
5.9 $MM = 28.0$ g/mol
5.10 $\rho_{He} = 0.17$ g/L and $\rho_{Ar} = 1.7$ g/L. Ar is more dense than air, making it a good gas blanket, whereas He is considerably less dense than air, giving it good lifting ability.
5.11 $p_{O_2} = 11.5$ atm; $p_{He} = 30.6$ atm; $X_{He} = 0.727$
5.12 $m = 8.6 \times 10^{-4}$ g
5.13 $V = 1.53$ L
5.14 $p_{NO} = 0$ (LR), $p_{O_2} = 2.50$ atm, $p_{NO_2} = 5.00$ atm
5.15 $[H_3O^+] = 0$ M (LR), $m_{Mg} = 3.06$ g, $[Mg^{2+}] = 3.00$ M, $[Cl^-] = 6.00$ M, $P_{H_2} = 2.27$ atm, $P = 3.27$ atm
5.16 Dew forms at around 13 °C.

Energy and Its Conservation

Our Energy Future

A plentiful supply of energy is essential for life. Human societies make a vast enterprise out of obtaining energy in useful forms. The sun provides nearly all of the Earth's energy. Green plants use sunlight for photosynthesis, which produces energy-rich chemicals such as glucose, $C_6H_{12}O_6$, shown in our molecular inset. Living organisms extract energy from glucose through a complex web of chemical reactions, the net result of which is the combustion of glucose with oxygen:

$$C_6H_{12}O_6\,(s) + 6\,O_2\,(g) \longrightarrow 6\,CO_2\,(g)$$
$$+ \,6\,H_2O\,(l) + Energy$$

The fuels of modern society — natural gas, petroleum products, and coal — are called fossil fuels because they are formed from the decomposition of green plants over many millions of years. Like glucose, fossil fuels release energy when they react with oxygen to form carbon dioxide and water.

Fossil fuels are nonrenewable, and combustion products contaminate the atmosphere. Consequently, scientists are searching for new sources of energy. One possibility is molecular hydrogen, which releases energy when it reacts with oxygen:

$$2\,H_2\,(g) + O_2\,(g) \longrightarrow 2\,H_2O\,(l) + Energy$$

Hydrogen powers the rockets of the space shuttle, but it is not a major fuel source at this time, for several reasons. There is no economical source of molec-

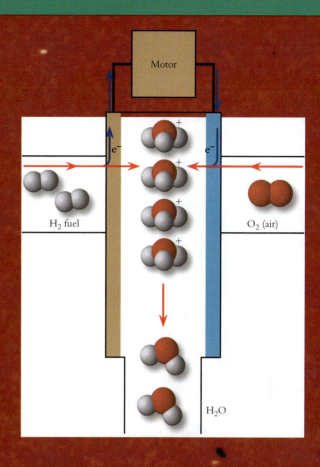

H₂ fuel

O₂ (air)

Motor

H₂O

CHAPTER CONTENTS

ular hydrogen, transporting this gas is difficult, and small hydrogen-powered engines have not been developed.

Water could become a source of molecular hydrogen, but the decomposition of H_2O into H_2 and O_2 requires the input of just as much energy as is released during the combustion of molecular hydrogen:

$$2\,H_2O\,(l) + \text{Energy} \longrightarrow 2\,H_2\,(g) + O_2\,(g)$$

Sunlight may be the future source of energy for "splitting water," but this remains a major technological challenge.

Small engines may incorporate the fuel cell as a power source. Fuel cells combine hydrogen and oxygen in a way that generates electricity, which drives a

Learning Resources

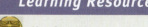 KEY CONCEPTS

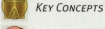 CRITICAL THINKING

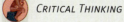 SKILLS TO MASTER

 ADDITIONAL HELP
www.wiley.com/college/olmsted
● MOLECULAR MODELS
● ANIMATIONS

motor. Leading automobile manufacturers have already developed fuel-cell vehicles.

Iceland, an island nation with abundant hydroelectric and geothermal power sources close to where people live, is using natural power sources to convert water into hydrogen and oxygen. In April 2003, Iceland opened its first hydrogen filling station for fuel-cell vehicles. If this initiative proves successful, perhaps the rest of the industrialized world will eventually convert to a hydrogen-fueled economy.

This chapter describes the properties and behavior of energy. We discuss the flow of energy from place to place and the conversions of energy from one form to another. We describe the law of conservation of energy and how to measure energy changes. Next, we introduce enthalpy, a useful energy-related quantity. To end the chapter, we describe current and future energy sources.

6.1 TYPES OF ENERGY

Personal experiences provide an intuitive sense of what energy is about. Backpackers climbing toward a mountain pass use energy to raise themselves and their packs up the mountain. When they reach the top, the hikers are tired after spending energy to make the climb. Scientifically, however, we need more than an intuitive sense of a concept as important as energy.

In science, energy is defined as the ability to do work, and work is defined as the displacement of an object against an opposing force. Here are some examples:

- Backpackers do work by climbing against the force of gravity.
- A baseball player does work by hitting the ball over the fence.
- Water does work by turning a turbine that generates electricity.
- A gasoline engine does work by moving an automobile along the highway.

These examples involve two categories of energy. One is **kinetic energy,** the energy of moving objects. The other is **potential energy,** which is energy that is stored. In addition to kinetic and potential energy, **thermal energy** refers to the energy content of a hot object, and **radiant energy** is the energy content of electromagnetic radiation (light). Kinetic, potential, thermal, and radiant energies all play important roles in chemistry.

Kinetic and Potential Energy

Every moving object has kinetic energy. A speeding jet plane has a large amount of kinetic energy, and so do atoms and molecules. Whether we describe an airplane or a molecule, the amount of kinetic energy always is given by the kinetic energy equation, introduced in Chapter 5:

$$E_{\text{kinetic}} = \tfrac{1}{2}mu^2 \qquad \text{(5-2)}$$

The SI units for kinetic energy are kilogram-meters^2/second^2. Because energy is so fundamental, this combination of units has its own name and symbol, the *joule* (J):

$$1\,\text{J} = 1\,\text{kg m}^2/\text{s}^2$$

The kinetic energy of an object changes with its conditions. For example, on accelerating from a stoplight, an automobile increases its kinetic energy. Conversely, when that automobile brakes to a stop, its kinetic energy decreases.

There are several types of potential energy, but the most familiar form is **gravitational energy.** A rock teetering high on a ledge has gravitational energy that will change to kinetic energy as the rock falls to lower ground. Water behind a dam has stored gravitational energy that is converted into electrical energy when the water drives

Figure 6-1
Charged objects attract one another if the charges have opposite signs (*a*). Charged objects with the same sign, either positive (*b*) or negative (*c*), repel one another.

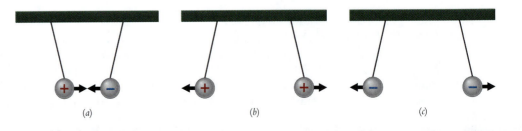

(a)   *(b)*   *(c)*

a turbine. Although gravitational energy is most familiar in everyday life, **electrical energy, chemical energy,** and **mass** are the forms of potential energy that are particularly important in chemistry.

Electrical Energy

Matter is electrical by its very nature. As we describe in Chapter 2, atoms are composed of nuclei and electrons that are subject to electrical force. The nucleus bears a positive electrical charge, while the electrons are negatively charged. Countless studies have shown that when two objects have opposite charges, electrical forces pull the objects together. Figure 6-1 demonstrates the interactions of charged objects by showing two small metal balls that bear electrical charges. The balls are pulled together when they have opposite charges. This attractive force between opposite charges is what holds electrons and nuclei together in atoms. It is also the reason that atoms combine into molecules, and it is why gases condense to form liquids and solids. Figure 6-1 also shows that objects possessing like charges push one another apart. Electrons repel one another, and so do nuclei.

Electrical forces lead to electrical potential energy. Energy is released when oppositely charged objects move close together, whereas energy must be supplied to pull oppositely charged objects apart. For example, separating an electron from its nucleus requires an energy input, just as lifting a backpack requires an energy input. Figure 6-2 shows this in diagrammatic form.

Just as kinetic energy can be calculated from an equation, the electrical potential energy of two charges, q_1 and q_2, separated by distance r can be calculated using Equation 6-1:

$$E_{\text{electrical}} = k \frac{q_1 q_2}{r} \qquad (6\text{-}1)$$

When distance is expressed in picometers (1 pm = 10^{-12} m) and charges are in electronic units, the constant in the equation is $k = 2.31 \times 10^{-16}$ J pm. To remove electrons from atoms or molecules, energy must be supplied to overcome this electrical energy. We use Equation 6-1 in later chapters to analyze the energetics of electrons, ions, and nuclei.

Chemical Energy

Atoms bind together to form molecules through the influence of electrical forces. As we describe in Chapters 9 and 10, chemical bonds result from negative electrons shared between positive nuclei. Electrons and nuclei in a molecule are arranged in a way that minimizes electrical potential energy. Chemists give this form of potential energy a specific name, chemical energy or bond energy. Every different molecule has a specific stability that is related to its chemical energy. Consequently, energy is either absorbed or released during any particular chemical reaction. Figure 6-3 shows the chemical energy changes involved in the formation of molecular hydrogen, oxygen, and water from atoms of hydrogen and oxygen. The formation of chemical bonds makes H_2 and O_2 more stable than collections of separated atoms, but molecules of H_2O are even more stable than these diatomic molecules, so energy is released when H_2 and O_2 react to form H_2O.

Mass

We think of mass (m) and energy (E) as separate quantities, but, as Albert Einstein was the first to show, mass actually represents a form of potential energy. Einstein's most famous equation, $E = mc^2$, where c is the speed of light, describes the mass–energy equivalence. Conversion of mass into other forms of energy is not important in standard chemical processes, but, as we describe in Chapter 22, nuclear energy is the result of transformations of nuclear mass into useable energy.

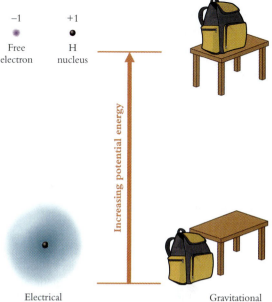

Figure 6-2
Removing an electron from a nucleus increases the electrical potential energy. Lifting a backpack from the floor to a tabletop increases the gravitational potential energy.

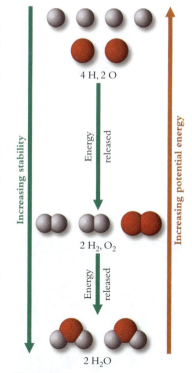

Figure 6-3
A collection of 2 H_2 and 1 O_2 molecules is more stable than a collection of 4 H and 2 O atoms, but a pair of H_2O molecules is more stable than either of these.

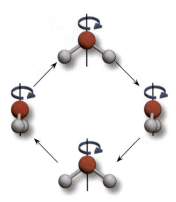

Figure 6-4
A water molecule has rotational energy.

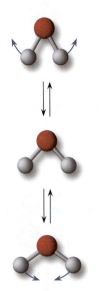

Figure 6-5
A water molecule can vibrate by internal motion of its atoms.

Thermal Energy

Consider a monatomic gas such as helium or argon. Recall from Chapter 5 that the atoms of this gas move continuously with a distribution of kinetic energies. The total energy of these random movements is called thermal energy. Equation 5-3 expresses the average kinetic energy of a gas:

$$\overline{E}_{\text{kinetic}} = \frac{3RT}{2N_A} \tag{5-3}$$

As the equation shows, average molecular kinetic energy increases as temperature increases. For a monatomic gas such as He or Ar, the only contribution to thermal energy is the kinetic energy of the atoms.

Equation 5-3 also describes the kinetic energy of gaseous molecules, but molecules possess additional thermal energy. Figure 6-4 shows that molecules rotate in space, giving them rotational kinetic energy, and Figure 6-5 shows that a molecule also vibrates by internal movement of its atoms. The energies of molecular vibrations are combinations of kinetic and potential energies. Like translational kinetic energy, rotational and vibrational energies of molecules increase with temperature and are part of the total thermal energy of the gas.

In the liquid phase, atoms and molecules still have energies of translation, rotation, and vibration, but the nearby presence of many other molecules significantly restricts their freedom of motion. In the solid phase, in contrast, atoms and molecules no longer translate freely. Each atom or molecule in a solid can vibrate back and forth but cannot easily change places with other atoms and molecules.

Whatever factors contribute to thermal energy, heating an object causes all its atoms and molecules to move faster, leading to an increase in thermal energy. A higher temperature indicates greater thermal energy, and a lower temperature indicates lesser thermal energy.

Radiant Energy

When you bake in the sun, your body absorbs energy from sunlight. Infrared radiation from a heat lamp in a restaurant keeps food warm until the server delivers the meal to the customer. When a microwave oven cooks food, the food absorbs energy from microwave radiation. Sunlight, infrared light, and microwaves are examples of electromagnetic radiation, which possesses radiant energy, as we discuss in Chapter 7.

Energy Transfers and Transformations

Energy can be transferred from one object to another. Anyone who has played a game of pool has firsthand experience with energy transfers. When the cue ball strikes the pack, the balls carom off in all directions. The cue ball loses most of its speed, and it may even come to a complete stop. The collision transfers kinetic energy from the cue ball to the other balls. Whereas transfers of energy among pool balls occur when initiated by a pool cue, transfers of energy are never-ending at the molecular level. When atoms or molecules of substance collide with one another, energy transfers causes some molecules to speed up and others to slow down.

Thermal energy also can be transferred between objects. When a hot object is placed in contact with a cold object, thermal energy flows from the hot object to the cold object until the two reach the same temperature. At the molecular level, there is a decrease in the average energy of the molecules in the hotter object, so its temperature decreases. There is a corresponding increase in the average energy of the molecules in the cooler object, so its temperature increases.

If you stir a cup of hot coffee with a cool spoon, energy flows from the coffee to the spoon. If the spoon is left in the coffee long enough, the two will reach the same temperature. You can detect the energy transfer easily by touching the spoon after it has been removed from the hot liquid.

Energy can also be transformed from one type to another. For example, when a rock falls from a ledge, gravitational potential energy is transformed into kinetic energy. When the rock hits the ground, its kinetic energy is transformed into thermal energy. As backpackers hike up a trail, their bodies consume glucose as they climb higher and higher. In this process, chemical energy is transformed into gravitational potential energy.

Energy transformations accompany chemical reactions. If a reaction releases energy, chemical energy is converted into other forms of energy. A reaction that releases energy almost always produces thermal energy, but chemical energy can be transformed into a variety of other forms, depending on the conditions under which the reaction occurs. An example is the combustion of gasoline. When a gasoline engine powers an automobile, chemical energy is transformed into kinetic energy. When a gasoline engine operates a water pump, chemical energy is transformed into gravitational potential energy, whereas if the engine operates an electrical generator, chemical energy is transformed into electrical potential energy.

Energy transformations and energy transfers often occur together. For instance, when a sidewalk absorbs sunlight, radiant energy of the sunlight is transferred to the sidewalk and transformed into thermal energy. When water courses through a turbine, gravitational potential energy of the water is transferred to the turbine blades and eventually transformed into electrical energy.

In addition to its many different types, energy comes in amounts that span a huge range. Figure 6-6 gives some indication of the magnitudes of energies encountered in chemistry.

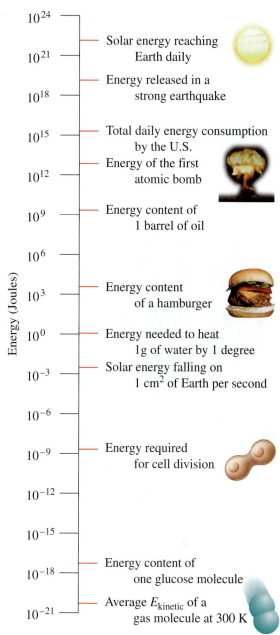

Figure 6-6
Chemists must deal with energies that span over 40 orders of magnitude, from 10^{-21} J (the kinetic energy of a typical gas molecule at room temperature) to 10^{22} J (the energy reaching the Earth from the sun daily).

Section Exercises

■ **6.1.1** Calculate the kinetic energy (in J) of an electron whose speed is 75 km/hr (see inside back cover for electron properties).

■ **6.1.2** What type of energy is each of the following? (a) a single molecule of methane moves at 100 km/s; (b) 1 mol of methane has an average speed of 100 km/s; (c) the flame from burning methane emits blue light; (d) methane is a fuel that releases energy when it burns.

■ **6.1.3** What transformations of energy take place during the following processes? (a) water is heated in a microwave oven; (b) snow breaks loose in an avalanche; (c) a speeding automobile brakes to a stop.

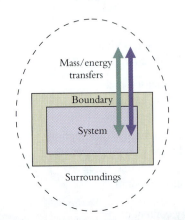

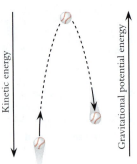

An aquarium is an open system.

6.2 THERMODYNAMICS

In this section, we develop the basic principles of **thermodynamics,** which is the quantitative study of energy transfers and transformations. Simply put, thermodynamics is the study of "how much energy goes where."

Terms of Thermodynamics

A thermodynamic **system** is whatever we want to describe and study by itself. Once we select something as a system, everything else is the **surroundings.** A system is separated from its surroundings by a **boundary,** across which matter and/or energy can move. An aquarium is a good example of a system. The aquarium and its contents are separated from the surroundings by glass walls and by the interface between air and water. Matter and energy move between the aquarium system and the surroundings as water evaporates, as food is added, and as the aquarium warms or cools during the course of a day.

Conservation of Energy

Energy can be transferred from one object to another, and energy can be transformed from one type to another. Nevertheless, all studies of thermodynamics indicate that the total amount of energy remains the same during transfers and transformations. That is, *energy is conserved,* as summarized in the **law of conservation of energy:**

KEY CONCEPT *Energy is neither created nor destroyed in any process, although it may be transferred from one body to another or transformed from one form into another.*

②

Innumerable experiments confirm this fundamental law of science. Whenever the energy of one body increases, a compensating decrease occurs in the energy of some other body. Whenever the amount of one form of energy increases, the amount of some other form decreases by an equal amount.

A baseball helps to illustrate features of conservation of energy. A baseball that has been popped up has kinetic energy. As it rises in the air, the ball slows until it reaches its highest point. At that instant, the ball is not moving, so it has no kinetic energy. What has happened to the initial kinetic energy of the baseball? During its climb, the baseball slows because gravitational force acts to convert the ball's kinetic energy into potential energy. When the ball reaches its highest point, all the kinetic energy of its initial upward motion has been stored as gravitational potential energy. If the ball were to be trapped at this point (for example, if it landed on top of the screen behind home plate), the energy would remain stored until the ball fell back to the playing field. During that fall, the stored energy would be released again as kinetic energy.

When a catcher catches a fastball, the ball loses the kinetic energy it had while it was in flight. This energy has not been transformed into gravitational potential energy because the ball is no higher above the Earth than before the pitch. It seems that energy has not been conserved. However, careful measurements of the temperature of the ball and of the glove would reveal that as the ball came to rest, the temperature of both objects increased slightly. When a catcher catches a fastball, the kinetic energy lost by the ball shows up as thermal energy of the glove and the ball. This increase in thermal energy results in a higher temperature.

Kinetic energy

Extending the baseball analogy, a baseball pitcher is very tired after throwing fastballs for nine innings. That tired feeling is the body's signal that stored chemical energy was consumed in throwing all those pitches. Each time your body moves, it does so by transforming chemical energy. The body releases this stored energy by breaking down sugars and fats into carbon dioxide and water. To replenish our supply of stored chemical energy, we must eat.

Thermal energy

Heat

A system can exchange thermal energy with its surroundings. When this happens, the amount of energy that is transferred is referred to as **heat (q).** Heat flows are changes in energy, so they are measured in joules (J). Chemical reactions frequently result in heat flows. When a reaction releases energy, some or all of the released energy may flow from the chemical system to its surroundings. When a reaction absorbs energy, that energy is likely to be supplied by a heat flow from the surroundings. Figure 6-7 illustrates heat flows in schematic fashion.

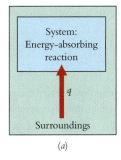
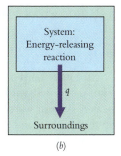

System:
Energy-absorbing
reaction

q

Surroundings

(a)

System:
Energy-releasing
reaction

q

Surroundings

(b)

Figure 6-7
(*a*) **In a chemical reaction that absorbs energy, heat flows from the surroundings to the system.** (*b*) **In a chemical reaction that releases energy, heat flows from the system to the surroundings.**

Chemists use *q* to stand for electrical charge (Eq 6-1) and to stand for heat.

Some heat flows result only in changes in temperature. Under these conditions, the amount of heat can be determined from the temperature change ($\Delta T = T_{final} - T_{initial}$). Experiments show that changes in temperature depend on four factors:

1. ΔT depends on q, the *amount of heat transferred*. As an example, the transfer of 50 J of heat to an object causes an increase in temperature that is twice as large as the increase caused by 25 J of heat.

2. ΔT depends on the *direction of the heat flow*. If a substance absorbs heat, ΔT will be positive; but if a substance releases heat, ΔT will be negative.

3. ΔT depends inversely on the *amount of material*. For example, ΔT for the transfer of 50 J of heat to 1 mol of a substance is twice as large as ΔT for the transfer of 50 J of heat to 2 mol of the same substance.

4. ΔT depends on the identity of the material. For instance, 50 J of heat increases the temperature of 1 mol of gold more than it increases the temperature of 1 mol of water. The dependence of ΔT on the identity of the material is expressed by the **molar heat capacity (C,** units J mol^{-1} °C^{-1}). The molar heat capacity is the amount of heat needed to raise the temperature of 1 mol of substance by 1 °C. Every substance has a different value for C. The molar heat capacities of several chemical substances are listed in Table 6-1.

Heat capacities can also be expressed in terms of *mass* rather than *moles*. When expressed in terms of mass, heat capacity is called the *specific heat* of a substance, and it has the units J g^{-1} °C^{-1}. We do not use specific heats in this book, but you may encounter them in other sources.

One equation incorporates all four factors and describes the temperature change resulting from a heat flow:

$$\Delta T = \frac{q}{nC}$$

In this equation, q is the amount of heat transferred, n is the number of moles of material involved, and C is the molar heat capacity of the substance. Solving for q gives Equation

Table 6-1 Molar Heat Capacities (*C*) of Representative Substances at 25 °C

Substance	C (J mol^{-1}°C^{-1})	Substance	C (J mol^{-1}°C^{-1})	Substance	C (J mol^{-1}°C^{-1})
Ag (s)	25.351	C$_2$H$_5$OH (l)	111.46	KCl (s)	51.30
Al (s)	24.35	CO (g)	29.142	K$_2$Cr$_2$O$_7$ (s)	219.24
Al$_2$O$_3$ (s)	79.04	CO$_2$ (g)	37.11	MgSO$_4$ (s)	96.48
Ar (g)	20.786	Cl$_2$ (g)	33.907	N$_2$ (g)	29.125
BaCl$_2$ (s)	75.14	Cu (s)	24.435	NH$_3$ (g)	35.06
Br$_2$ (l)	75.689	Fe (s)	25.10	NH$_4$Cl (s)	84.1
C (s, graphite)	8.527	H$_2$ (g)	28.824	NO (g)	29.844
CaCO$_3$ (s)	81.88	HCl (g)	29.12	Na (s)	28.24
CCl$_4$ (l)	131.75	H$_2$S (g)	34.23	NaOH (s)	59.54
CH$_3$CN (l)	91.5	He (g)	20.786	O$_2$ (g)	29.355
CH$_3$CO$_2$H (l)	124.3	Hg (l)	27.983	S (s)	22.64
CH$_3$OH (l)	81.6	H$_2$O (l)	75.291	Si (s)	20.00

6-2, the essential equation for calculating heat flows from temperature measurements:

$$q = nC\Delta T \qquad (6\text{-}2)$$

Energy transfer is directional, and for this reason it is essential to keep track of the signs associated with heat flows. For example, if we drop a heated block of metal into a beaker of cool water, we know that heat will flow from the metal to the water: q is negative for metal and positive for the water. The *amount* of heat flow is the same, however, for both objects. In other words, $q_{water} = -q_{metal}$. In more general terms:

$$q_{surroundings} = -q_{system}$$

Note that neither temperature nor a change in temperature equates directly with the heat flow, q. We can relate q to ΔT only if we know the identity and amount of the material that undergoes a change of state. Example 6-1 provides some practice.

Example 6-1 **Heat Capacity and Temperature Change**	Calculate the temperature change that results from adding 250 J of thermal energy to each of the following: (a) 0.75 mol of Hg; (b) 0.35 mol of Hg; and (c) 0.35 mol of H_2O.

Strategy: This is a simple quantitative calculation, so we apply the seven-step method in condensed form. We are asked to determine the change in temperature, ΔT, that accompanies a heat flow. Thermal energy is added to each substance, so we expect an increase in temperature for each case. A diagram similar to Figure 6-7 summarizes the process:

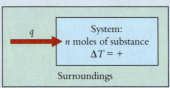

Solution: The data provided are q and n. These are related to ΔT through Equation 6-2, which requires values of C from Table 6-1:

$$q = nC\Delta T \qquad \text{from which} \qquad \Delta T = \frac{q}{nC}$$

Substitute values of q, n, and C and calculate:
a. $q = 25$ J, $n = 0.75$ mol, $C = 27.983$ J mol^{-1} °C^{-1}

$$\Delta T = \frac{q}{nC} = \frac{250\text{ J}}{(0.75\text{ mol})(27.983\text{ J mol}^{-1}\text{ °C}^{-1})} = 12\text{ °C}$$

b. $q = 25$ J, $n = 0.35$ mol, $C = 27.983$ J mol^{-1} °C^{-1}

$$\Delta T = \frac{q}{nC} = \frac{250\text{ J}}{(0.35\text{ mol})(27.983\text{ J mol}^{-1}\text{ °C}^{-1})} = 26\text{ °C}$$

c. $q = 25$ J, $n = 0.35$ mol, $C = 75.291$ J mol^{-1} °C^{-1}

$$\Delta T = \frac{q}{nC} = \frac{250\text{ J}}{(0.35\text{ mol})(75.291\text{ J mol}^{-1}\text{ °C}^{-1})} = 9.5\text{ °C}$$

Do the Results Make Sense? In each case, the sign of ΔT is positive because the substance absorbs heat from its surroundings. Notice that the magnitudes of ΔT make sense. The smaller quantity of Hg experiences a larger temperature change than the larger quantity, and the substance with the larger heat capacity, water, experiences a smaller temperature change than the substance with the smaller heat capacity, Hg.

Extra Practice Exercise 6.1	Determine ΔT for 27.5 g of methanol (CH_3OH) that transfers 76.5 J of energy to its surroundings.

Equation 6-2 demonstrates that a change of temperature accompanies a heat flow. The equation is also used to compute the amount of heat transferred in a specific temperature change, as Example 6-2 shows.

Example 6-2

Heat Transfer and Temperature Change

An aluminum frying pan that weighs 745 g is heated on a stove from 25 °C to 205 °C. What is q for the frying pan?

Strategy: Again, we use a condensed form of the seven-step strategy. A temperature change signals a heat flow. In this example, an increase in the temperature of the aluminum pan means that heat flows from the surroundings (including the stove) to the pan, which represents the system. As in Example 6-1, a diagram summarizes the process.

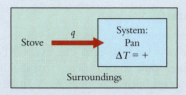

Solution: Equation 6-2 is used to calculate how much heat flows, provided we know n, C, and ΔT. Use the mass of the frying pan and the molar mass of aluminum to calculate n:

$$n = \frac{m}{MM} = \frac{745 \text{ g Al}}{26.98 \text{ g/mol}} = 27.61 \text{ mol Al}$$

Consult Table 6-1 for the molar heat capacity of Al:

$$C_{Al} = 24.35 \text{ J mol}^{-1} \text{ °C}^{-1}$$

Temperatures are given in the problem:

$$T_i = 25 \text{ °C} \qquad T_f = 205 \text{ °C}$$

$$\Delta T = T_f - T_i = 205 \text{ °C} - 25 \text{ °C} = 180. \text{ °C}$$

Now substitute and calculate, converting to kilojoules to avoid large numbers:

$$q = nC\Delta T = (27.61 \text{ mol})(24.35 \text{ J/mol °C})(180. \text{ °C})(10^{-3} \text{ kJ/J}) = 121 \text{ kJ}$$

As a frying pan is heated, it loses some energy as heat to the surrounding air at the same time as it gains energy as heat from the burner. The calculation in Example 6-2 gives the *net* heat gain for the pan.

Does the Result Make Sense? The result has the right units, and its significant figures are consistent with the data. The quantity is positive, which is what we expect for a temperature increase. The magnitude seems rather large, but so are the temperature change and the amount of Al being heated.

A bottle that contains 235 g of drinking water cools in a refrigerator from 26.5 °C to 4.9 °C. Calculate q for the water.

Work

Backpackers climbing a mountain trail convert chemical energy into gravitational potential energy. Their bodies undergo displacements (increases in altitude) against an external force (force of gravity). Energy used to move an object against an opposing force is called **work** (w). Like heat, work is a flow of energy between objects or between a chemical system and its surroundings. The amount of work depends on the magnitude of the force that must be overcome and the amount of movement or displacement. Work is the product of force (F) and displacement (d): $w = Fd$.

It is just as important to keep track of the signs associated with work as it is to keep track of the signs associated with heat flows. We view work from the perspective of the

system. When a system does work on the surroundings, the sign of w is *negative* because the system *loses* energy. When the surroundings do work on the system, w is *positive* for the system because the system *gains* energy. Work always is energy transferred between system and surroundings, so the sign of $w_{surroundings}$ always is opposite to the sign of w_{system}. Moreover, the *amount* of work is the same for both system and surroundings. In other words,

$$w_{surroundings} = -w_{system}$$

Chemical energy fuels all human activity, whether it be work or play (Figure 6-8). In any physical activity, a person does work to accomplish movement, and the energy for this work comes from the chemical energy stored in food. Our Chemistry and Life Box explores human energy requirements.

First Law of Thermodynamics

After carrying out countless observations and experiments, scientists concluded that energy is exchanged in only two ways, as heat or as work. For example, when gasoline burns in an automobile engine, some of the energy that is released does work in moving a piston, and some of the energy heats the engine. A simple equation summarizes the relationship linking work and heat to the energy change of a system:

$$\Delta E = q + w \qquad (6\text{-}3)$$

Equation 6-3 applies to any process. This equation is one version of the **first law of thermodynamics.**

Equation 6-3 is a restatement of the law of conservation of energy. We can show this by substituting into Equation 6-3 the equalities for heat and work:

$$q_{sys} = -q_{surr} \qquad and \qquad w_{sys} = -w_{surr}$$

$$\Delta E_{sys} = q_{sys} + w_{sys} = (-q_{surr}) + (-w_{surr}) = -(q_{surr} + w_{surr}) = -\Delta E_{surr}$$

$$\Delta E_{sys} = -\Delta E_{surr} \qquad (6\text{-}4)$$

Our primary interest is in a chemical system rather than its surroundings. For convenience, the subscript denoting the *system* usually is omitted in thermodynamic equations. Thus, an energy term without a subscript always refers to a system. To avoid ambiguity, we *never* omit the subscript indicating the surroundings.

Equation 6-4 is a quantitative statement of the law of conservation of energy. Whenever a system undergoes a change in energy, its surroundings undergo an equal and opposite change in energy, ensuring that the total energy remains unchanged. This leads to still another expression for the first law of thermodynamics:

$$\Delta E_{total} = \Delta E_{sys} + \Delta E_{surr} = 0$$

This relationship is profound, because it means that regardless of the conditions and circumstances, we can always completely account for energy changes. This powerful thermodynamic tool helps us analyze many different types of chemical processes.

Figure 6-8
All physical human activities involve doing thermodynamic work, which the body supplies from the chemical energy stored in foods.

State and Path Functions

The properties that describe a system and its transformations can be grouped in two broad categories. Some properties depend only on the conditions that describe the system. These properties are called **state functions.** Other properties depend on *how* the change occurs. Properties that depend on how a change takes place are called **path functions.**

An everyday example illustrates the difference between state functions and path functions. The Daltons, who live in San Francisco, decide to visit relatives in Denver.

Box 6-1   Chemistry and Life: Human Energy Requirements

A person's weight depends on how much energy is taken in and how much energy is expended. Food brings stored chemical energy into the body, and metabolism releases this chemical energy. Some energy is transferred out of the body as work and heat, but when energy intake exceeds the immediate needs of the body, the extra energy is stored as fat.

To lose weight, people must take in less energy than their bodies require. When this happens, the body "burns" fat to meet its energy requirements. People can modify the energy balances of their bodies by eating less and by exercising more.

"Eat less and exercise more" is overly simplistic for several reasons. Foods of various types have different energy contents. Exercise of various types requires different amounts of energy. Moreover, foods that are readily converted to sugar can stress the body's metabolic processes, and individuals of various metabolic types process foods with different efficiencies.

Foods can be grouped into types, each with a characteristic energy content. The table below lists average energy contents in kilojoules per gram (1 kJ/g = 6.78 Calorie/ounce) for some foods. Fats (energy content = 39 kJ/g) are most energy-rich. Carbohydrates (16 kJ/g) and proteins (17 kJ/g) are the other main sources of energy. Vegetables and fruits contain much water but very little fat, so their energy content per gram is low. Margarine is mostly fat, so its energy content per gram is extremely high.

Type of Food	Energy (kJ/g)	Type of Food	Energy (kJ/g)
Green vegetables	1.2	Regular yogurt	10
Beer	2.0	Bread, cheese	12
Fruit	2.5	Ground beef	16
Low-fat yogurt	4.5	Sugar	16
Broiled chicken	6.0	Margarine	30

The table shows that *what* we eat is as important as *how much* we eat. For example, 10 g of margarine provides the same energy content as 250 g of green vegetables. A 100-g serving of low-fat yogurt contains less than half the energy of a 100-g serving of regular yogurt. The easiest way to reduce energy intake is by eliminating fats and greasy foods such as hamburgers and pizza.

Energy intake is only part of the equation. We can also adjust our energy balance by exercising. Various forms of exercise require different average energy outputs. Exercise involves doing thermodynamic work. The table in the next column indicates that the amount of work depends on the type of exercise and the amount of mass being displaced.

Energy Consumed (kJ/hr)		
Body Weight:		
55 kg (120 lb)	70 kg (155 lb)	85 kg (185 lb)

Activity	55 kg (120 lb)	70 kg (155 lb)	85 kg (185 lb)
Resting	290	335	380
Doing housework	650	750	850
Walking at 4 km/hr	770	880	990
Walking at 6 km/hr	1090	1250	1410
Bicycling at 9 km/hr	775	880	985
Bicycling at 20 km/hr	2400	2760	3120
Playing volleyball	1380	1590	1800
Skiing at 16 km/hr	2175	2510	2825
Running at 16 km/hr	3285	3770	4245

A single example illustrates the relative importance of diet and exercise in weight control. A typical can of a soft drink contains about 31 g of sugar. Sugar provides 16 kJ/g, so the total energy content of a can of soft drink is 490 kJ. To expend 490 kJ of energy, a 70-kg person needs to walk at 6 km/hr for 23 minutes or play volleyball for 18 minutes. A 100-g (about a 1/4-pound) hamburger on a bun contains 1600 kJ of energy. Thus, a 70-kg person has to play volleyball continuously for over an hour to burn off the energy contained in a hamburger and soft drink.

This comparison demonstrates that it is much easier to control energy balance through diet than by exercise. The human body stores excess energy in fatty tissues. When half a kilogram of this fatty tissue is consumed, it releases 20,000 kJ of energy, which is enough to fuel about 9 hours of tennis, 6 hours of skiing, or 4 hours of running.

A proper diet is the key to weight control, but what is a proper diet? Nutritional experts have not always agreed, in part because good health depends on other factors besides weight. Nevertheless, recent scientific studies strongly indicate that we should avoid foods that rapidly generate blood sugar. Foremost among these are sweets and refined flour. In other words, bran muffins are much healthier than sugar donuts.

Figure 6-9
Distance between two cities is a state function, but distance traveled is not a state function. Each of the Daltons travels a different distance, but the distance from San Francisco to Denver is a fixed amount.

They take different routes, as Figure 6-9 shows. Mr. Dalton takes a train directly from San Francisco to Denver, but Ms. Dalton goes to Dallas for a business meeting and then flies from Dallas to Denver. The Daltons' daughter drives to Los Angeles, where she catches a flight to Denver. On arrival, each of the Daltons is asked two questions by their relatives: "How far is Denver from San Francisco?" and "How far did you travel to get here?" Each answers "950 miles" to the first question, because the distance between the two cities is a difference in values of a *state function*. Each answers differently to the second question, however, because each Dalton traveled a different distance to reach Denver. Distance traveled depends on the path and is a change in a *path function*.

The travels of the Dalton family illustrate two features of state functions:

1. The change in a state function can be determined without knowing the details of a process, because that change is independent of path. Denver is 950 miles from San Francisco, no matter how the Daltons travel between the two cities. This means that state function values can be tabulated and looked up when needed. The distance from San Francisco to Denver can be found in an atlas of the United States. Energy is a state function, and changes in energy that accompany standard chemical reactions are tabulated in reference books such as the *CRC Handbook of Chemistry and Physics*.

2. The change in a state function can be measured using the most convenient method available. To determine the distance between two cities, modern geographers make use of satellite mapping. We describe convenient methods for determining energy changes of chemical reactions later in this chapter.

These two features of state functions are more important for scientists than for cross-country travelers like the Daltons. Chemists determine values of state function changes by doing careful experiments using convenient paths. These values are collected in tables, just as distances between cities are collected in an atlas. Energy is one such state function, and chemists use tabulated energy values to analyze chemical processes from a thermodynamic point of view.

How do we know that energy is a state function? Consider the combustion of methane from a molecular perspective. Every molecule of methane that burns reacts with two molecules of molecular oxygen to produce one molecule of carbon dioxide and two molecules of water. Figure 6-10 shows that the value of ΔE for this process is the difference between the energy required to break all the bonds in the reactant molecules and the energy released in forming all the bonds in the product molecules. This energy difference depends only on the

Figure 6-10
Regardless of the conditions, the reaction of methane with molecular oxygen to form water and carbon dioxide involves breakage of two O=O bonds and four C—H bonds and subsequent formation of four O—H bonds and two C=O bonds for every molecule of methane that reacts.

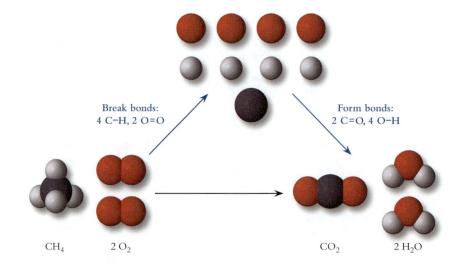

Break bonds:
4 C–H, 2 O=O

Form bonds:
2 C=O, 4 O–H

CH_4 2 O_2 CO_2 2 H_2O

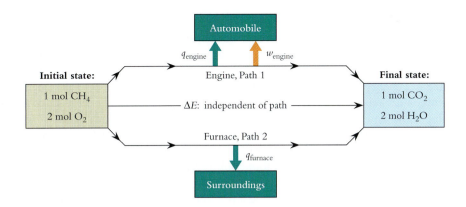

Figure 6-11
A block diagram of two different paths for the reaction of methane and oxygen. Occurring inside an engine (Path 1), the process transfers heat and work. Occurring in a furnace (Path 2), the process does no work and transfers only heat.

strengths of the various chemical bonds, not on how the bonds are broken and formed. Whether this process occurs in a car fueled by natural gas or in a home furnace, the combustion reaction has the same overall energy change. In summary, ΔE depends on the net change in the conditions of the system but not on *how* that change occurs.

Thermodynamic Path Functions

Energy is a state function, but heat and work are path functions. To illustrate this, Figure 6-11 describes two different paths for the combustion of 1 mol of methane. Path 1 represents what happens in an automobile fueled by natural gas: As methane burns, the system does work on its surroundings by driving back the piston. At the same time, some heat is transferred from the chemicals to the engine block:

$$\Delta E_{engine} = q_{engine} + w_{engine}$$

Path 2 represents what happens when natural gas burns in a furnace: The heat released from burning methane is transferred completely to the surroundings:

$$\Delta E_{furnace} = q_{furnace}$$

No work is done along Path 2, so $w_{furnace} = 0$. For Path 1, $w_{engine} > 0$, showing that the work done is different for these two paths. Work is a path function, not a state function.

Knowing that energy is a state function, we can state that ΔE is the same for both paths:

$$\Delta E_{engine} = \Delta E_{furnace}$$

In the engine, some of this energy accomplishes work and the rest is transferred as heat. In the furnace, all of this energy is transferred as heat. In other words, $q_{engine} < q_{furnace}$. The heat transferred is different for the two paths, so q is a path function, not a state function.

As with the distance each Dalton traveled in going from San Francisco to Denver (see Figure 6-9), q and w are path functions. As with the distance between San Francisco and Denver, ΔE is a state function. The fact that heat transfer depends on the path while energy change is independent of path has important consequences in chemistry, as we describe later in this chapter.

Section Exercises

■ **6.2.1** How much heat is required to raise the temperature of 25.0 g of water from 25 °C to 65 °C on a methane-burning stove (see Table 6-1)?

■ **6.2.2** Bicyclists in the Tour de France are on the road an average of 4.0 hr/day, and their average speed is at least 20 km/hr. How much ground beef must a 55-kg bicyclist consume daily to maintain his weight? If the cyclist ate fruit rather than ground beef, how much fruit would he require? (Consult the tables in the Chemistry and Life Box.)

■ **6.2.3** Which of the following are state functions? (a) height of a mountain; (b) distance traveled in climbing that mountain; (c) energy consumed in climbing the mountain; and (d) gravitational potential energy of a climber on top of the mountain

6.3 ENERGY CHANGES IN CHEMICAL REACTIONS

Chemical reactions involve rearrangements of atoms: Some chemical bonds break, and others form. Bond breakage always requires an input of energy, and bond formation always results in a release of energy. The balance between these opposing trends determines the net energy change for the reaction.

Figure 6-12 summarizes the balance between gains and losses of energy in chemical reactions. If the bonds of the products are more stable than the bonds of the reactants, energy is released during the reaction. For example, the reactions of fuels with oxygen gas release large amounts of energy that can heat a home or drive an engine. For chemical fuels, the energy released by bond formation is greater than the energy consumed by bond breakage (Figure 6-12a).

Conversely, energy is absorbed during a reaction if the bonds of the products are less stable than the bonds of the reactants. The surroundings must transfer energy to the chemical system for the reaction to occur (Figure 6-12b). An example is the photosynthesis reaction, in which plants use radiant energy from sunlight to form energy-rich products from water and carbon dioxide. In photosynthesis, the energy consumed in bond breakage exceeds the energy released by bond formation, so energy must be provided by the surroundings.

If a reaction releases energy when going in one direction, it must absorb an equal amount of energy to go in the opposite direction. Reversing the *direction* of a reaction changes the *sign* of the energy change, but it does not affect the *amount* of energy change. This applies to all energy transfers, not only those occurring in chemical reactions. If a change is "uphill" going in one direction, it must be "downhill" by an equal amount in the other direction.

Uphill: energy absorbed

Downhill: energy released

Features of Reaction Energies

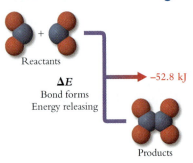

Reactants

ΔE
Bond forms
Energy releasing

−52.8 kJ

Products

The thermodynamic conventions for amounts and sign allow us to include the energy change when we write the equation for a balanced chemical reaction. For example, when two NO_2 molecules collide, they may form a chemical bond in a reaction that releases energy. The energy flow for NO_2 reacting to produce N_2O_4 corresponds to Figure 6-12a:

$$2\,NO_2 \longrightarrow N_2O_4 \qquad \Delta E = -52.8 \text{ kJ}$$

The balanced equation indicates that the reaction of two moles of NO_2 to form one mole of N_2O_4 releases 52.8 kJ of energy to the surroundings.

When a chemical reaction releases energy, ΔE has a negative sign. When a reaction absorbs energy, ΔE has a positive sign. If we reverse the direction of a chemical reaction, we also reverse the direction of energy flow. A reaction that releases energy in one direction must absorb energy to run in the opposite direction. For example, upon heating,

Figure 6-12
Schematic diagrams showing energy flows that accompany reactions. (a) When a reaction releases energy, the change in energy (ΔE) is negative for the chemical reaction and positive for the surroundings. (b) When a reaction absorbs energy, ΔE is positive for the chemical reaction and negative for the surroundings.

(a) Energy-releasing Reaction

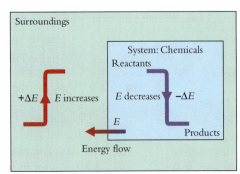

(b) Energy-absorbing Reaction

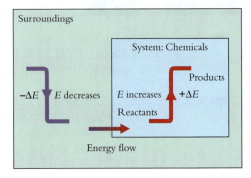

N_2O_4 decomposes into a pair of NO_2 molecules:

$$N_2O_4 \longrightarrow 2\ NO_2 \qquad \Delta E = +52.8 \text{ kJ}$$

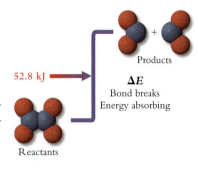

Products

52.8 kJ $\rightarrow$

ΔE

Bond breaks
Energy absorbing

Reactants

The decomposition reaction absorbs energy, with an energy profile similar to the one in Figure 6-12b.

The amount of energy released or absorbed in a chemical reaction is proportional to the amounts of chemicals that react. For instance, when 10 moles of NO_2 react to form 5 moles of N_2O_4, five times as much energy is released as when 2 moles of NO_2 form 1 mole of N_2O_4:

$$2\ NO_2 \longrightarrow N_2O_4 \qquad \Delta E = -52.8 \text{ kJ}$$

$$10\ NO_2 \longrightarrow 5\ N_2O_4 \qquad \Delta E = -264 \text{ kJ}$$

We must take stoichiometric coefficients into account if we are interested in the energy change per mole of any specific reagent. The molar energy change for any component of a reaction can be found by dividing the overall energy change by the stoichiometric coefficient of the specific reagent. Here is the calculation for the energy change per mole of NO_2 consumed to produce N_2O_4:

$$2\ NO_2 \longrightarrow N_2O_4 \qquad \Delta E = -52.8 \text{ kJ}$$

$$\Delta E_{\text{molar, } NO_2} = \frac{-52.8 \text{ kJ}}{2 \text{ mol } NO_2} = -26.4 \text{ kJ/mol } NO_2$$

Knowing the overall energy change for a reaction, we can use stoichiometric principles to determine ΔE for a given amount of starting materials. Example 6-3 shows how to do this.

Example 6-3

Energy Change in a Reaction

The Haber reaction for the formation of ammonia releases energy:

$$N_2 + 3\ H_2 \longrightarrow 2\ NH_3 \qquad \Delta E = -40.9 \text{ kJ}$$

How much energy is released in the production of 1.00 kg of ammonia?

Strategy: The problem asks for the amount of energy released in a chemical reaction that forms a specified mass of product. The balanced equation is as much visualization as is needed for this problem. The data provided are ΔE for the balanced equation and the mass of product formed, 1.00 kg.

Our goal is to link the mass of the product with the energy released. We must determine the number of moles of ammonia and take account of the stoichiometric coefficients in the balanced equation. A flowchart summarizes the calculations:

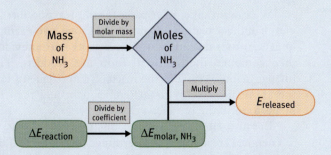

Solution: Use the mass and the molar mass of ammonia to calculate n. You should verify that the molar mass of ammonia is 17.03 g/mol.

$$n = \frac{m}{MM} = \frac{1.00 \text{ kg } (1000 \text{ g/kg})}{17.03 \text{ g/mol}} = 58.72 \text{ mol } NH_3$$

Example 6-3
Energy Change in a Reaction (*continued*)

Use the stoichiometric coefficient for NH_3 and ΔE for the balanced equation to determine ΔE_{molar} for NH_3:

$$\frac{-40.9 \text{ kJ}}{2 \text{ mol } NH_3} = -20.45 \text{ kJ/mol } NH_3$$

Now multiply ΔE_{molar} by the number of moles of NH_3 produced:

$$\Delta E = (-20.45 \text{ kJ/mol } NH_3)(58.72 \text{ mol } NH_3) = -1.20 \times 10^3 \text{ kJ}$$

Does the Result Make Sense? The units and number of significant figures are correct. The sign is negative, as it should be for an energy-releasing reaction. A magnitude of 1200 kJ is reasonable for the amount produced, about 60 mol, with an energy change of about -20 kJ/mol.

Extra Practice Exercise 6.3

The acid–base neutralization reaction releases energy:

$$H_3O^+ + OH^- \longrightarrow 2 H_2O \qquad \Delta E = -55.8 \text{ kJ}$$

Calculate the energy change when 25.0 mL of 0.100 M HCl reacts with excess aqueous NaOH.

Figure 6-13
An energy-level diagram demonstrates that the net energy change for a chemical reaction (ΔE) has the same value regardless of the path used to accomplish the reaction.

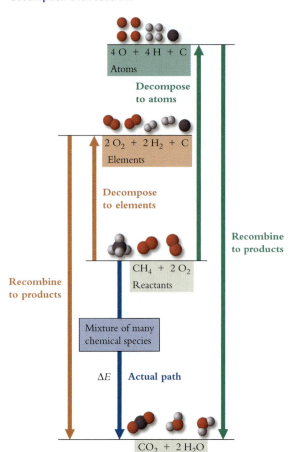

4 O + 4 H + C
Atoms

Decompose to atoms

2 O_2 + 2 H_2 + C
Elements

Decompose to elements

Recombine to products

CH_4 + 2 O_2
Reactants

Recombine to products

Mixture of many chemical species

ΔE Actual path

CO_2 + 2 H_2O
Products

Path Independence

A change in any thermodynamic state function is independent of the path used to accomplish that change. This feature of state functions tells us that the energy change in a chemical reaction is independent of the manner in which the reaction takes place. In the real world, chemical reactions often follow very complicated paths. Even a relatively simple overall reaction such as the combustion of CH_4 and O_2 can be very complicated at the molecular level:

$$CH_4 (g) + 2 O_2 (g) \longrightarrow CO_2 (g) + 2 H_2O (l)$$

Nevertheless, path independence tells us that we can calculate the energy change for any reaction by choosing the path that makes the calculation easiest.

Figure 6-13 shows three different paths for the combustion reaction of methane. One path, indicated with the blue arrow, is the path that might occur when natural gas burns on a stove burner. As CH_4 and O_2 combine in a flame, all sorts of chemical species can form, including OH, CH_3O, and so on. This is not a convenient path for calculating the energy change for the net reaction, because the process involves many steps and several unstable chemical species.

We can imagine other paths for this reaction that are much simpler than the actual one. Figure 6-13 shows two examples. One path (green arrows) decomposes the reactants into individual atoms, and then combines the atoms to form the products. The other path (orange arrows) converts the reactants into elements in the form that they normally take at room temperature, and then reacts the elements to form the products.

Because energy is a state function, the change of energy in a chemical reaction is determined by the conditions of the reactants and products, as the figure shows. Thus, we can calculate this change by following whichever path we choose. Just as geographers can determine the distance between two cities using the most convenient method, chemists can select a simple path to determine the change in energy for a chemical process.

Bond Energies

If we wish to use the path that breaks molecules apart into atoms, we must imagine breaking chemical bonds. A chemical bond is a stable arrangement of electrons shared between the bonded atoms. Therefore, a bond cannot be broken without an input of the **bond energy (BE).** Bond energies are positive quantities because a molecule must absorb energy to break its chemical bonds. Imagine grasping a molecule and "pulling it apart" at a junction between two particular atoms. The energy that you must supply to accomplish this separation is the bond energy.

Bond energies are expressed most often in units of kilojoules of energy per mole of bonds. To separate one mole of hydrogen molecules into hydrogen atoms, for example, requires an input of 435 kJ to break the chemical bonds:

$$H_2(g) \longrightarrow H(g) + H(g) \qquad \Delta E_{\text{bond breaking}} = BE = +435 \text{ kJ/mol}$$

In the opposite direction, energy is released when a bond forms. The formation of one mole of H_2 from two moles of hydrogen atoms releases 435 kJ of energy:

$$H(g) + H(g) \longrightarrow H_2(g) \qquad \Delta E_{\text{bond making}} = -BE = -435 \text{ kJ/mol}$$

Bond energies depend on the types of atoms that are bonded together. The bond energies of representative diatomic molecules show that bond energies span a wide range of values from substance to substance:

$$Br_2(g) \longrightarrow Br(g) + Br(g) \qquad \Delta E = BE = 190 \text{ kJ/mol}$$

$$HCl(g) \longrightarrow H(g) + Cl(g) \qquad \Delta E = BE = 430 \text{ kJ/mol}$$

$$N_2(g) \longrightarrow N(g) + N(g) \qquad \Delta E = BE = 945 \text{ kJ/mol}$$

In polyatomic molecules, the bond energies depend not only on the specific atoms in the bonds but also on the rest of the molecular structure. Figure 6-14 shows that an ethanol molecule has five C—H bonds in two different bonding environments. The C—H bonds shown in red are part of a CH_3 group that is bonded to another carbon atom. The C—H bonds in blue belong to a CH_2 group that is bonded to both another carbon atom and to an oxygen atom. The different bonding patterns result in different C—H bond energies. To break a red C—H bond requires 410 kJ/mol, whereas to break a blue C—H bond requires slightly less energy, 393 kJ/mol. These variations in the C—H bond energies show that individual bond energies change with the structure of the entire molecule.

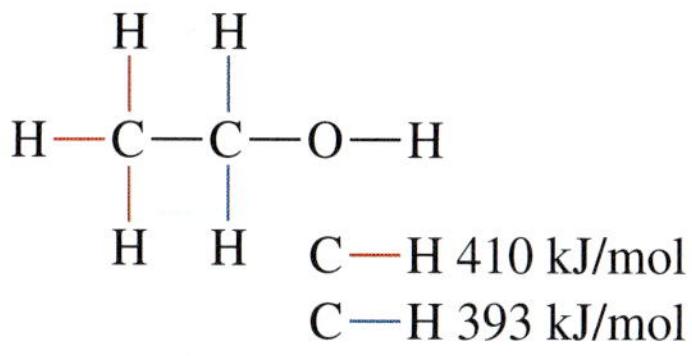

Despite the fact that any particular bond energy depends on the entire molecule, many bonds between specific atoms cluster around an average value. For example, C—H bond energies have an average value of 415 kJ/mol. The two C—H bond energies for ethanol are close to the average value, although neither matches the average exactly.

Table 6-2 gives average bond energies for a variety of chemical bonds. The table has the following features:

1. All values (other than those for diatomic molecules) are averages over different compounds. A bond in any given compound may have a value that differs from the average by between 5 and 20 kJ/mol.

2. The values are given to the nearest 5 kJ/mol.

3. Each value is the energy per mole required to break a gaseous molecule into a pair of neutral gaseous fragments. As we describe in Chapter 11, molecules in liquids and solids experience attractive forces between the molecules that modify the energies of individual bonds. Consequently, average bond energies are less reliable when they are applied to substances in the liquid phase and the solid phase.

Table 6-2 reveals that a few elements, most notably carbon, nitrogen, and oxygen, form more than one type of bond. The average energies of the different types of bonds show striking differences. As we describe in detail in Chapter 10, these differences arise from differences in the number of electrons involved in bond formation. Most of the bonds listed in the table are single bonds, which involve the sharing of two electrons. Double bonds, which involve the sharing of four electrons, are substantially stronger than

Figure 6-14
The two different types of C—H bonds in ethanol have slightly different bond energies, because the bonding environment about one C atom differs slightly from the bonding environment about the other C atom.

C—H 410 kJ/mol
C—H 393 kJ/mol

Table 6-2 Average Bond Energies*

Diatomic Molecules (Dissociation Energy of Gaseous Molecules)

H—H	435	F—F	155	O=O	495
H—F	565	Cl—Cl	240	N=O	605
H—Cl	430	Br—Br	190	N≡N	945
H—Br	365	I—I	150	C≡O	1070
H—I	295				

Single Covalent Bonds (Average Values)

H—C	415	C—C	345	Si—Si	225	N—N	160	O—O	145
H—N	390	C—N	305	Si—F	565	N—O	200	O—Si	450
H—O	460	C—O	360	Si—Cl	390	N—F	285	O—P	335
H—Si	320	C—Si	300	Si—Br	310	N—Cl	200	O—F	190
H—P	320	C—P	265	Si—N	320	P—P	210	O—Cl	220
H—S	365	C—S	270	Sn—Sn	145	P—F	490	O—Br	200
H—Te	240	C—F	485	Sn—Cl	315	P—Cl	320	S—O	265
		C—Cl	330			P—Br	270	S—F	285
		C—Br	275			P—I	185	S—S	240
		C—I	215			As—Cl	295	S—Cl	255

Multiple Covalent Bonds (Average Values)

C=C	615	N=N	420	C≡C	835
C=N	615	N=O	605	C≡N	890
C=O	750 800[†]	O=P	545		
C=S	575	O=S	515		

* All values are in kilojoules per mole (kJ/mol) and are rounded to the nearest 5 kJ/mol.
† C=O bond energy in CO_2

single bonds. Triple bonds, which involve the sharing of six electrons, are stronger yet. The bond energies of carbon–carbon bonds illustrate this trend:

C—C	Single bond	BE = 345 kJ/mol
C=C	Double bond	BE = 615 kJ/mol
C≡C	Triple bond	BE = 835 kJ/mol

Reaction Energy

A chemist who runs a chemical reaction needs to know if that reaction releases or absorbs energy. A reaction that releases energy may go out of control or even cause an explosion unless provisions are made to transfer energy efficiently to the surroundings. Conversely, a reaction that absorbs energy is unlikely to proceed efficiently unless provisions are made to provide energy from the surroundings. Fortunately, tabulated data can be used to calculate the energy change for any chemical reaction, including one that has not been previously carried out.

Bond energies can be used to estimate the energy change that occurs in a chemical reaction. The reaction of molecular hydrogen with molecular oxygen to form gaseous water provides a simple example:

$$2\,H_2\,(g) + O_2\,(g) \longrightarrow 2\,H_2O\,(g)$$

Imagine that this reaction occurs along a two-step path. In the first step, the bonds in the reactants break, yielding four hydrogen atoms and two oxygen atoms. In the second step, the hydrogen and oxygen atoms form two water molecules that contain four O—H bonds.

We can calculate the energy change for this reaction, on a molar basis, using bond energies from Table 6-2. The energy required for the bond-breaking step ($\Delta E_{\text{bond breaking}}$) is the sum of the bond energies for all bonds in the reactant molecules:

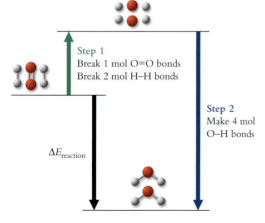

Step 1
Break 1 mol O=O bonds
Break 2 mol H–H bonds

Step 2
Make 4 mol O–H bonds

$\Delta E_{\text{reaction}}$

$$\Delta E_{\text{bond breaking}} = (2 \text{ mol } H_2)(BE \text{ } H_2) + (1 \text{ mol } O_2)(BE \text{ } O_2)$$

$$\Delta E_{\text{bond breaking}} = (2 \text{ mol})(435 \text{ kJ/mol}) + (1 \text{ mol})(495 \text{ kJ/mol}) = 1365 \text{ kJ}$$

This energy change is *positive* because the chemical system *gains* the energy required to break the chemical bonds.

The energy change in the second step, $\Delta E_{\text{bond forming}}$, is the sum of all the energies released during bond formation of the products. The reaction produces two moles of water containing four moles of O—H bonds. Bond formation always releases energy, so the chemical system loses energy to the surroundings. Consequently, the energy change is negative:

$$\Delta E_{\text{bond forming}} = -(4 \text{ mol O—H bonds})(460 \text{ kJ/mol}) = -1840 \text{ kJ}$$

The overall energy change for the reaction is the sum of these two terms:

$$\Delta E_{\text{reaction}} = \Delta E_{\text{bond breaking}} + \Delta E_{\text{bond forming}} = (1365 \text{ kJ}) + (-1840 \text{ kJ}) = -475 \text{ kJ}$$

The large negative ΔE for this reaction is one of the reasons why molecular hydrogen is a candidate to replace fossil fuels, as described in our chapter opener.

We can apply the analysis of this reaction to any chemical reaction. The energy change for a chemical reaction can be estimated as the sum of all bond energies of the reactants minus the sum of all bond energies of the products, as summarized in Equation 6-5:

$$\Delta E_{\text{reaction}} = \Sigma \text{ } BE_{\text{bonds broken}} - \Sigma \text{ } BE_{\text{bonds formed}} \qquad (6\text{-}5)$$

The reaction of H_2 with O_2 fuels the space shuttle.

The Greek letter Σ means "the sum of." Thus, Σ *BE* means "the sum of all the bond energies."

To apply Equation 6-5, we need complete inventories of the bonds present in all the reagents that participate in the reaction. We build such inventories from structural formulas, from which we count the number of bonds of each type. The procedures for generating structural formulas are not introduced until Chapter 9, so we provide the correct structures for examples in this chapter.

As an example that uses structural formulas and Equation 6-5, consider the energy change that takes place during the combustion reaction of propane (C_3H_8). Recall from Chapter 3 that combustion is a reaction with molecular oxygen. The products of propane combustion are carbon dioxide and water:

⑤

$$C_3H_8 \text{ } (g) + 5 \text{ } O_2 \text{ } (g) \longrightarrow 3 \text{ } CO_2 \text{ } (g) + 4 \text{ } H_2O \text{ } (g)$$

To estimate the amount of energy absorbed or released in this reaction, we must compile an inventory of all the bonds that break and all the bonds that form. A ball-and-stick model shows that propane contains 8 C—H bonds and 2 C—C bonds. These bonds break in each propane molecule, and one O=O bond breaks in each oxygen molecule. Two C=O bonds form in each CO_2 molecule, and two O—H bonds form in each H_2O molecule. In summary:

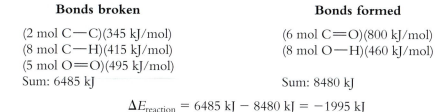

Now we apply Equation 6-5, using average bond energy values from Table 6-2:

$$\Delta E_{\text{reaction}} = \Sigma \text{ } BE_{\text{bonds broken}} - \Sigma \text{ } BE_{\text{bonds formed}}$$

A table helps us carry out the sums:

Bonds broken	**Bonds formed**
(2 mol C—C)(345 kJ/mol)	(6 mol C=O)(800 kJ/mol)
(8 mol C—H)(415 kJ/mol)	(8 mol O—H)(460 kJ/mol)
(5 mol O=O)(495 kJ/mol)	
Sum: 6485 kJ	Sum: 8480 kJ

$$\Delta E_{\text{reaction}} = 6485 \text{ kJ} - 8480 \text{ kJ} = -1995 \text{ kJ}$$

Table 6-3 Comparison of Combustion Reaction Bond Energies

Bond Type	Bond Energy (kJ/mol)	Electron Pairs per Bond	Energy per Electron Pair (kJ)
Reactant Bonds			
O=O	495	2	248
C—H	415	1	415
C—C	345	1	345
Product Bonds			
C=O	800	2	400
O—H	460	1	460

Equation 6-5 predicts that burning one mole of propane releases 1995 kJ of energy. The energy change accompanying this reaction has been measured accurately to be −2044 kJ/mol of propane. Thus, average bond energies predict a value that is off by about 50 kJ/mol, an error of about 2.5%. This error occurs because the bond energies in any given molecule usually differ somewhat from average bond energies. These variations are typically only a few percent, so calculations using average bond energies provide reliable estimates of reaction energies.

The energy released by combusting one mole of propane is large compared to the energy changes in typical chemical reactions. Propane is typical of compounds that come from fossil fuels: large negative molar energy values characterize the burning of all these compounds. Why do combustion reactions release so much energy? Table 6-3 summarizes the properties of the bonds involved in the combustion of fossil fuels. The most important features of the data are the bond energies *per pair of electrons*. Notice that the bond of molecular oxygen has a significantly smaller bond energy per electron pair than any of the other bonds in the table. In addition, the O—H and C=O bonds that make up the products of combustion are more stable than the C—H and C—C bonds found in fossil fuels. Thus, burning a compound that contains carbon and hydrogen releases a great amount of energy that can be used to heat homes, cook food, power vehicles, or produce electricity. In fact, most of the energy requirements of our society are provided by the energy released during the combustion of hydrocarbons derived from three fossil fuels: petroleum, natural gas, and coal.

Energy analyses can be extended to reactions other than combustion, as shown in Example 6-4.

Example 6-4
Reaction Energies

The United States produces more than 7 billion kilograms of vinyl chloride annually. Most is converted to the polymer poly(vinyl chloride) (PVC), which is used to make piping, siding, gutters, floor tiles, clothing, and toys. Vinyl chloride is made in a two-step process. The balanced overall equation is as follows:

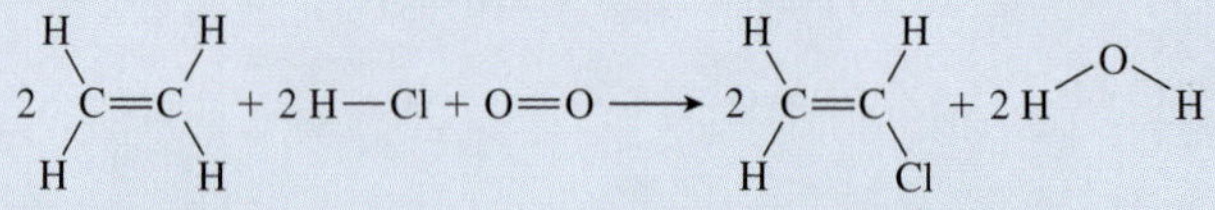

$$2\ \underset{\text{Ethylene}}{\overset{\text{H}\quad\text{H}}{\text{C}=\text{C}}} + 2\,\text{H}-\text{Cl} + \text{O}=\text{O} \longrightarrow 2\ \underset{\text{Vinyl chloride}}{\overset{\text{H}\quad\text{H}}{\text{C}=\text{C}}} + 2\,\text{H}-\text{O}-\text{H}$$

Based on average bond energies, what energy change accompanies the formation of one mole of vinyl chloride? Does the synthesis require an input of energy?

Strategy: This is a quantitative problem, so apply the seven-step strategy. As a review of the process, we present the steps as a list.

1. The question asks for an energy change.

2. The balanced equation presented in the problem is the visual aid we need to identify the bonds that break and form.

Example 6-4

Reaction Energies (*continued*)

Solution:

3. The structures are given in the equation. Here is the inventory:

Bonds broken	Atom Inventory	Bonds formed
8 C—H bonds	10 H atoms	6 C—H bonds
2 C=C bonds	4 C atoms	2 C=C bonds
2 H—Cl bonds	2 Cl atoms	2 C—Cl bonds
1 O=O bond	2 O atoms	4 O—H bonds

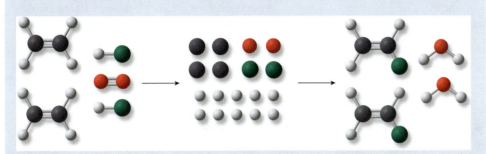

Look up bond energies in Table 6-2:

C—H: 415 kJ/mol   C=C: 615 kJ/mol   H—Cl: 430 kJ/mol
O=O: 495 kJ/mol   C—Cl: 330 kJ/mol   O—H: 460 kJ/mol

4 and 5. We use Equation 6-5, which does not need to be manipulated:

$$\Delta E_{\text{reaction}} = \Sigma\ BE_{\text{bonds broken}} - \Sigma\ BE_{\text{bonds formed}}$$

6. Sum the bond energies for bonds broken and bonds formed and subtract the latter from the former:

Bonds broken	Bonds formed
(8 mol C—H)(415 kJ/mol)	(6 mol C—H)(415 kJ/mol)
(2 mol C=C)(615 kJ/mol)	(2 mol C=C)(615 kJ/mol)
(2 mol H—Cl)(430 kJ/mol)	(2 mol C—Cl)(330 kJ/mol)
(1 mol O=O)(495 kJ/mol)	(4 mol O—H)(460 kJ/mol)

$$\Delta E_{\text{reaction}} = (5905\ \text{kJ}) - (6220\ \text{kJ}) = -315\ \text{kJ}$$

Our calculation predicts that 315 kJ of energy is released for every two moles of vinyl chloride. However, the problem asks for the energy change associated with one mole of product, so divide by two to obtain an estimate of 157.5 kJ of energy released per mole of vinyl chloride. Remember, though, that this is an estimate and that tabulated values are given to the nearest 5 kJ, so round this value to 160 kJ. The negative sign for $\Delta E_{\text{reaction}}$ indicates that the chemicals release energy to the surroundings.

7. Does the Result Make Sense? Units and significant figures are correct. You should be developing a sense of the energies involved in chemical reactions. Although the sum of reactant bond energies and the sum of product bond energies are very large, the result is a typical reaction energy.

Vinyl chloride can also be formed by reacting ethylene with Cl_2, which produces 1,2-dichloroethane ($ClCH_2$—CH_2Cl). This compound is heated to eliminate HCl. Estimate ΔE for each of these reactions.

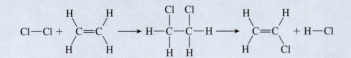

Section Exercises

■ **6.3.1** Methanol (CH₃OH) is used in the engines of some high-powered racing cars. Estimate how much energy is released when one mole of methanol is burned in air.

$$H-\underset{\underset{H}{\vert}}{\overset{\overset{H}{\vert}}{C}}-O-H$$

■ **6.3.2** Draw a diagram like the one in Figure 6-13 to illustrate the different paths for the reaction in Section Exercise 6.3.1.

$$H-\underset{\underset{H}{\vert}}{\overset{\overset{Cl}{\vert}}{C}}-H$$

■ **6.3.3** Methyl chloride (CH₃Cl) can be produced from methane in either of the following reactions:

$$CH_4(g) + HCl(g) \longrightarrow CH_3Cl(g) + H_2(g)$$
$$CH_4(g) + Cl_2(g) \longrightarrow CH_3Cl(g) + HCl(g)$$

Create a bond inventory and estimate the energy change for each reaction. Based on the values that you calculate, which reaction is more suitable for industrial production of methyl chloride? Explain.

6.4 MEASURING ENERGY CHANGES: CALORIMETRY

The bond energies that appear in Table 6-2 were computed from the results of thousands of measurements of the energy changes accompanying chemical reactions. How are such energy changes measured? Recall that any energy change is the sum of two transfers: heat (q) and work (w). To determine the energy change that accompanies a chemical process, chemists run that process under controlled conditions that allow them to determine heat and work. This section describes those controlled conditions.

Calorimeters

A **calorimeter** is a device used to measure heat flows that accompany chemical processes. The basic features of a calorimeter include an insulated container and a thermometer that monitors the temperature of the calorimeter. A block diagram of a calorimeter appears in Figure 6-15. In a calorimetry experiment, a chemical reaction takes place within the calorimeter, resulting in a heat flow between the chemicals and the calorimeter. The temperature of the calorimeter rises or falls in response to this heat flow.

Calorimeters are insulated to ensure that negligible heat is transferred between the calorimeter and its surroundings during the experiment. As a result, the only heat exchange is between the chemicals and the calorimeter:

Exo is Greek for "out," endo is Greek for "within," and therm is Greek for "heat."

$$q_{\text{chemicals}} = -q_{\text{calorimeter}}$$

Figure 6-15
Block diagram of a typical calorimeter.

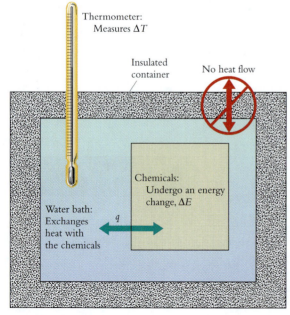

If the chemicals release heat, $q_{\text{chemicals}}$ is negative and the reaction is termed **exothermic.** The calorimeter gains this heat ($q_{\text{calorimeter}}$ is positive), and the temperature of the calorimeter rises. Conversely, if the chemicals absorb heat, $q_{\text{chemicals}}$ is positive and the reaction is termed **endothermic.** The calorimeter loses this heat ($q_{\text{calorimeter}}$ is negative), and the temperature of the calorimeter falls.

We can determine $q_{\text{calorimeter}}$ from its temperature change using Equation 6-6, which is similar to Equation 6-2:

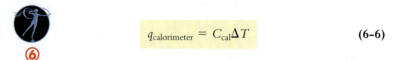

$$q_{\text{calorimeter}} = C_{\text{cal}}\Delta T \qquad \text{(6-6)}$$

Here, C_{cal} is the total heat capacity of the calorimeter. That is, C_{cal} is the amount of heat required to raise the temperature of the entire calorimeter (water bath, container, and thermometer) by 1 °C.

The total heat capacity, C_{cal}, can be found by determining the temperature change resulting from a known amount of heat.

One way of doing this uses an electrical heater to supply heat (q_{heater}) that can be determined accurately by measuring current, voltage, and time. Then, q_{heater} and the measured temperature increase can be used in Equation 6-6 to calculate C_{cal}. Example 6-5 illustrates this technique.

Example 6-5

**Determining
Total Heat Capacity**

A calorimeter is calibrated with an electrical heater. Before the heater is turned on, the calorimeter temperature is 23.6 °C. The addition of 2.02×10^3 J of electrical energy from the heater raises the temperature to 27.6 °C. Determine the total heat capacity of this calorimeter.

Strategy: Follow the standard seven-step problem-solving procedure. We are asked to determine the total heat capacity of a calorimeter.

A sketch of the process helps to identify what takes place. The electrical heater converts electrical energy into heat that flows into the calorimeter and raises the temperature of the water bath.

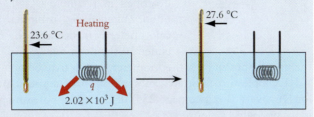

Solution: The problem statement gives us the temperature data and the heat flow. The heater releases energy to the calorimeter, so q_{heater} has a negative value:

$$q_{heater} = -2.02 \times 10^3 \text{ J} \qquad \Delta T = T_f - T_i = (27.6\,°\text{C} - 23.6\,°\text{C}) = 4.0\,°\text{C}$$

Calorimetry experiments are designed so that the heat transfer is confined to the calorimeter. Equation 6-6 relates heat flow and temperature change:

$$q_{calorimeter} = -q_{heater} \quad and \quad q_{calorimeter} = C_{cal}\Delta T$$

Combine these equations and rearrange and solve for C_{cal}:

$$C_{cal} = \frac{-q_{heater}}{\Delta T}$$

A *total* heat capacity has units of J/°C, whereas a *molar* heat capacity has units of J/mol °C. If we have n moles of a pure substance, its total heat capacity is $C_{total} = nC_{molar}$.

Now, substitute and calculate:

$$C_{cal} = \frac{2.02 \times 10^3 \text{ J}}{4.0\,°\text{C}} = 5.1 \times 10^2 \text{ J/°C}$$

Does the Result Make Sense? The result has the correct units, the value is positive, and is rather large; this is what we expect for a calorimeter, so this is a reasonable result. We round to two significant figures because, although each temperature is precise to three significant figures, the temperature *difference* is only known to two significant figures.

What will be the final temperature of this calorimeter if the initial temperature is 23.7 °C and the heater supplies 4.65 kJ of electrical energy?

**Extra Practice
Exercise 6.5**

Types of Calorimeters

Volume, pressure, temperature, and amounts of substances may change during a chemical reaction. When scientists make experimental measurements, however, they prefer to control as many variables as possible, to simplify the interpretation of their results. In general, it is possible to hold volume or pressure constant, but not both. In *constant-volume calorimetry,* the volume of the system is fixed, whereas in *constant-pressure calorimetry,* the pressure of the system is fixed. Constant-volume calorimetry is most often used to study reactions that involve gases, while constant-pressure calorimetry is particularly convenient for studying reactions in liquid solutions. Whichever type of calorimetry is used, temperature changes are used to calculate q.

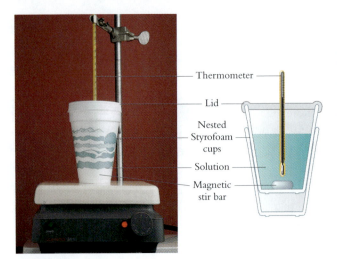

Figure 6-16
A constant-pressure calorimeter can be constructed from two Styrofoam cups, a cover, a stirrer, and a thermometer.

Constant-pressure calorimetry requires only a thermally insulated container and a thermometer. A simple, inexpensive constant-pressure calorimeter can be made using two nested Styrofoam cups. Figure 6-16 shows an example. The inner cup holds the water bath, a magnetic stir bar, and the reactants. The thermometer is inserted through the cover. The outer cup provides extra thermal insulation.

In one use of this calorimeter, aqueous solutions containing the reactants are mixed in the cup, and the thermometer registers the resulting temperature change. The heat capacity of the calorimeter must be determined independently. One way to do this would be through electrical heating, as described in Example 6-5. However, satisfactory accuracy is often obtained by assuming that the heat capacity of the calorimeter is the same as the heat capacity of the water contained in the calorimeter. This neglects the contribution from the nested cups and the thermometer, which may be only about 1% of C_{cal}.

Example 6-6 illustrates an application of constant-pressure calorimetry. Our Tools for Discovery Box (see page 234) describes uses of constant-pressure calorimetry in studies of biological systems.

Example 6-6

Constant-Pressure Calorimetry

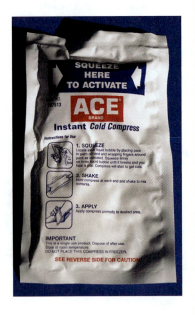

Ammonium nitrate (NH_4NO_3, $MM = 80.05$ g/mol) is used in cold packs to "ice" injuries. When 20.0 g of this compound dissolves in 125 g of water in a coffee-cup calorimeter, the temperature falls from 23.5 °C to 13.4 °C. Determine q for the dissolving of the compound. Is the process exothermic or endothermic?

Strategy: Follow the seven-step procedure. For this Example, we review the procedure by specifying the steps in numerical order. The first two steps are part of the strategy.

1. We are asked to find q and determine its sign for ammonium nitrate dissolving in water.

2. A sketch helps to identify the process. The temperature of the calorimeter decreases as NH_4NO_3 dissolves.

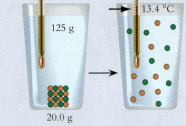

Solution:
3. We are given temperature data and information about the calorimeter's contents:

$$\Delta T = T_f - T_i = 13.4 \text{ °C} - 23.5 \text{ °C} = -10.1 \text{ °C}$$

$$NH_4NO_3: 20.0 \text{ g} \qquad H_2O: 125 \text{ g}$$

4. Equation 6-6 lets us determine q after heat capacity and temperature change are known.

$$q_{calorimeter} = C_{cal}\Delta T$$

5. Given no additional information, we make the approximation that C_{cal} is the heat capacity of its water content:

$$C_{cal} \cong n_{water}C_{water}$$

$$n = \frac{m}{MM} = \frac{125 \text{ g}}{18.016 \text{ g/mol}} = 6.938 \text{ mol}$$

$$C_{water} = (6.938 \text{ mol})(75.291 \text{ J/mol °C}) = 522.4 \text{ J/°C}$$

6. Now we are ready to solve for the quantities asked for. Begin by calculating the heat flow for the calorimeter.

$$q_{calorimeter} = C_{cal}\Delta T$$

$$q_{calorimeter} = (522.4 \text{ J/°C})(-10.1 \text{ °C}) = -5276 \text{ J} = -5.28 \text{ kJ}$$

The temperature of the calorimeter falls during the process, which means the calorimeter loses heat as the ammonium nitrate dissolves. The ions gain energy, so the dissolving process is endothermic and $q_{salt} = +5.28$ kJ.

7. Does the Result Make Sense? The units are kJ, appropriate for an energy calculation. The drop in temperature of the calorimeter indicates a heat-absorbing process for the calorimeter, so the chemical system absorbs heat from the calorimeter as the compound dissolves. Thus, a positive q is reasonable. Does the magnitude seem reasonable to you?

When 10.00 mL of 1.00 M HCl solution is mixed with 115 mL of 0.100 M NaOH solution in a constant-pressure calorimeter, the temperature rises from 22.45 °C to 23.25 °C. Assuming that the heat capacity of the calorimeter is the same as that of 125 g of water, calculate q for this reaction.

Example 6-6

Constant–Pressure Calorimetry (*continued*)

Extra Practice Exercise 6.6

Figure 6-17 illustrates a constant-volume calorimeter of a type that is often used to measure q for combustion reactions. A sample of the substance to be burned is placed inside the sealed calorimeter in the presence of excess oxygen gas. When the sample burns, energy flows from the chemicals to the calorimeter. As in a constant-pressure calorimeter, the calorimeter is well insulated from its surroundings, so all the heat released by the chemicals is absorbed by the calorimeter. The temperature change of the calorimeter, with the calorimeter's heat capacity, gives the amount of heat released in the reaction.

Calculating Energy Changes

In a calorimetry experiment, the heat flow resulting from a process is determined by measuring the temperature change of the calorimeter. Then q can be related to energy change through the first law of thermodynamics (Equation 6-3):

$$\Delta E = q + w$$

Figure 6-17

A commercial instrument for constant-volume calorimetry is called a *bomb calorimeter*, because the container in which the reaction occurs resembles a bomb.

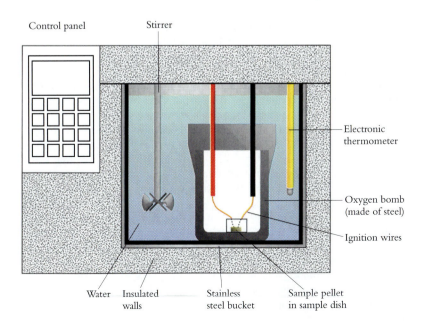

Control panel   Stirrer

Electronic thermometer

Oxygen bomb (made of steel)

Ignition wires

Water   Insulated walls   Stainless steel bucket   Sample pellet in sample dish

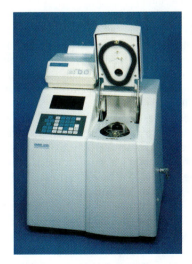

Box 6-2	Tools for Discovery: Calorimetry in Biology

Calorimetry is an important technique in biology as well as in chemistry. The inventor of the calorimeter was Antoine Lavoisier, who is shown in the illustration. Lavoisier was a founder of modern chemistry, but he also carried out calorimetric measurements on biological materials. Lavoisier and Pierre Laplace reported in 1783 that respiration is a very slow form of combustion. Thus, calorimetry has been applied to biology virtually from its invention.

Calorimetry is valuable in biological studies because every living thing is an energy-processing system. An analysis of energy flows in biology can provide information about how organisms use energy for growth, reproduction, and many other processes.

Despite Lavoisier's early work on the link between energy and life, calorimetric measurements played a relatively minor role in biology until recent years, primarily because of practical obstacles. Every organism must take in and give off matter as part of its normal function, and it is very difficult to make accurate heat-flow measurements when matter is transferred. Moreover, the sizes of many organisms are poorly matched to the sizes of calorimeters. Although a chemist can adjust the amount of a substance on which to carry out calorimetry, a biologist often cannot.

Improvements in calorimeters have led to a blossoming of biology-related studies within the last two decades. One development is the differential scanning calorimeter, in which the temperature difference is measured between two matched chambers, only one of which contains the material being studied. A second development is improved sensitivity of temperature detection using differential amplifiers and microscopic-sized sensors. These and other advances in instrumentation make it possible to detect temperature changes of less than 0.001 °C and heat flows as small as 0.001 J. The ability to measure very small amounts of heat has been exploited by biologists, as the following examples illustrate.

Calorimetry shows that the rates of metabolism of plant tissues vary widely with species, with cell types, and with environmental conditions. This provides a means of exploring the mechanisms by which various agents influence the health of a plant community. Studies are being done on beneficial agents such as growth promoters and detrimental ones such as atmospheric pollutants. For example, a correlation has been found between the metabolic heat rates and the extent of damage to pine needles by ozone.

Muscle activity is accompanied by cellular pumping of sodium ions. The energy requirements of the sodium pump have been studied on an individual cardiac muscle mounted inside a tiny differential calorimeter and stimulated by electrical impulses. The heat evolved was different in the presence and absence of a known inhibitor of the sodium pump.

The metabolism of microbes is relatively easy to study by calorimetry because microbial growth is accompanied by energy release in the form of heat. Pharmaceutical calorimetry uses the heat profiles of microbial cultures to assess how various drugs and drug combinations affect the growth of the culture. A graph of heat profiles shows that a drug acting as a growth inhibitor reduces the heat output of a culture, whereas a lethal drug kills the culture and eliminates its heat production. A flow calorimeter, in which a bathing solution is slowly passed through the cell culture, is particularly well suited for these studies, because it mimics a living system and provides information about how quickly a given drug acts.

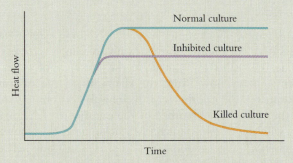

The pharmaceutical industry has found that microcalorimetry can be a powerful tool both for quality control and in improving drug action. A drug generally releases energy when it degrades. Because of this, measurements of the heat released when a drug is subjected to temperature or humidity changes allow manufacturers to determine how rapidly a drug will lose its effectiveness under different conditions. Bacteria such as those that cause tooth decay continually release energy. The heat flow from a bacterial colony is significantly reduced when a drug reduces bacterial action, and this feature makes microcalorimetry useful in probing the effectiveness of antibacterial agents.

Microcalorimetry also provides insight into how drugs act at the molecular level. Drug action generally requires the drug to form a chemical interaction with some specific biomolecule in a cell, and these interactions involve energy changes. Using computational techniques like those described in Section 6.3, pharmaceutical chemists can predict the heat profile that will accompany various pharmaceutical interactions. Microcalorimetric measurements can then verify whether a drug is acting effectively.

These are just some of the ways in which calorimetry is used in contemporary biological research. Our examples highlight studies at the cellular level, but ecologists also use calorimetry to explore the energy balances in ecosystems, and whole-organism biologists have found ways to carry out calorimetric measurements on fish, birds, reptiles, and mammals, including humans.

To determine ΔE using measured values of q, we also must know w. Because heat and work are path functions, however, we proceed differently for constant volume than for constant pressure. To distinguish between these different paths, we use a subscript v for constant-volume calorimetry and a subscript p for constant-pressure calorimetry. This gives different expressions for the two types of calorimeters:

$$\Delta E \text{ (constant volume)} = q_v + w_v \qquad and \qquad \Delta E \text{ (constant pressure)} = q_p + w_p$$

As described in Section 6-2, work is the product of force and displacement. In a constant-volume calorimeter, the chemical reaction is contained within the sealed calorimeter, so there is no displacement and $w_v = 0$. Thus:

$$\Delta E = q_v \text{ (Constant-volume process)}$$

For a constant-pressure calorimeter, the volume of the reacting chemicals may change, so $w_p \neq 0$ and must be evaluated. We do this in Section 6.5.

Molar Energy Change

Energy change is an extensive quantity, which means that the amount of energy released or absorbed depends on the amounts of substances that react. For example, as more methane burns, more energy is released. When we report an energy change, we must also report the amounts of the chemical substances that generate the energy change. For tabulation purposes, all changes in thermodynamic functions such as ΔE are given *per mole,* just as heat capacities are conveniently tabulated as molar values. We use ΔE_{molar} for the energy change that accompanies the reaction of one mole of a particular substance.

Experimental measurements by calorimetry usually involve amounts different from one mole. The molar energy change can be found from an experimental energy change by dividing by the number of moles that reacted, as expressed by Equation 6-7:

$$\Delta E_{molar} = \frac{\Delta E}{n} \qquad\qquad \textbf{(6-7)}$$

Example 6-7 shows how to evaluate ΔE_{molar} for the combustion of octane.

Example 6-7

Molar Energy of Combustion

A 0.1250-g sample of octane (C_8H_{18}, $MM = 114.2$ g/mol) is burned in excess O_2 in the constant-volume calorimeter described in Example 6-5. The temperature of the calorimeter rises from 21.1 to 32.9 °C. What is ΔE_{molar} for the combustion of octane?

Strategy: Again, the seven-step process serves us well.

1. We are asked to find the molar energy change for combustion of octane.

2. A sketch helps to identify the process. The heat given off in the combustion reaction is absorbed by the calorimeter.

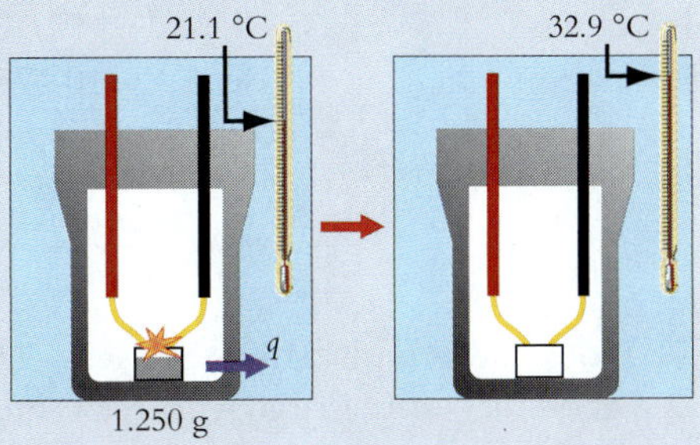

Solution:

3. The value of C_{cal} is calculated in Example 6-5, and ΔT can be found from the data:

$$C_{cal} = 5.1 \times 10^2 \text{ J/°C} \qquad \Delta T_{reaction} = 32.9 \text{ °C} - 21.1 \text{ °C} = 11.8 \text{ °C}$$

We also know the mass of octane (0.1250 g) and its molar mass (114.2 g/mol).

Example 6-7

**Molar Energy
of Combustion** (*continued*)

4. The calculation takes more than one step, so we need to identify a process. Use Equation 6-6 to find $q_{\text{calorimeter}}$. The heat gained by the calorimeter is supplied by the chemical reaction, so $q_{\text{reaction}} = -q_{\text{calorimeter}}$. Because the calorimeter operates at constant volume, $w = 0$, so $\Delta E = q_{\text{reaction}}$. This energy change is for 0.1250 g of octane. Use $n = m/MM$ to determine n, then use Equation 6-7 to convert to the molar energy change: $\Delta E_{\text{molar}} = \Delta E/n$. A flowchart summarizes the process:

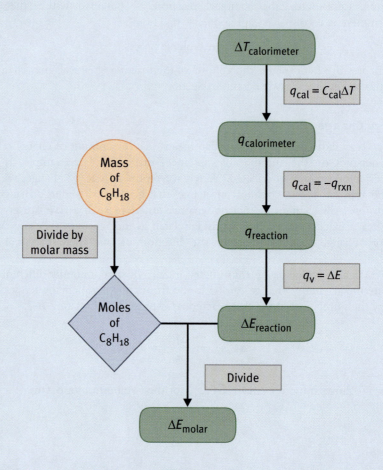

5. None of the equations needs to be manipulated.

6. Now substitute and calculate, step by step:

$$q_{\text{calorimeter}} = C_{\text{cal}}\Delta T = (5.1 \times 10^2 \text{ J/°C})(11.8 \text{ °C})(10^{-3} \text{ kJ/J}) = 6.0 \text{ kJ}$$

$$\Delta E = q_{\text{reaction}} = -q_{\text{calorimeter}} = -6.0 \text{ kJ}$$

$$n = \frac{m}{MM} = \frac{0.1250 \text{ g}}{114.2 \text{ g/mol}} = 1.09 \times 10^{-3} \text{ mol}$$

$$\Delta E_{\text{molar}} = \frac{\Delta E}{n} = \frac{-6.0 \text{ kJ}}{1.1 \times 10^{-3} \text{ mol}} = -5.5 \times 10^3 \text{ kJ/mol octane}$$

7. Does the Result Make Sense?   The units are correct, and we know that octane is a good fuel, so the large negative value is reasonable.

When 0.100 g of graphite (elemental carbon) burns in the calorimeter of Example 6-7, the temperature rises from 23.5 °C to 29.9 °C. Determine the molar energy of combustion of graphite.

Section Exercises

6.4.1 A coffee-cup calorimeter is calibrated using a small electrical heater. The addition of 3.45 kJ of electrical energy raises the calorimeter temperature from 21.65 °C to 28.25 °C. Calculate the heat capacity of the calorimeter.

6.4.2 Adding 1.530×10^3 J of electrical energy to a constant-pressure calorimeter changes the water temperature from 20.50 °C to 21.85 °C. When 1.75 g of a solid salt is dissolved in the water, the temperature falls from 21.85 °C to 21.44 °C. Find the value of q_p for the solution process.

6.4.3 In a constant-volume calorimeter, 3.56 g of solid sulfur is burned in excess oxygen gas:

$$S\,(s) + O_2\,(g) \longrightarrow SO_2\,(g)$$

The calorimeter has a total heat capacity of 4.32 kJ/°C. The combustion reaction causes the temperature of the calorimeter to increase from 25.93 °C to 33.56 °C. Calculate the molar energy for combustion of sulfur.

6.5 ENTHALPY

In our world, most chemical processes occur in contact with the Earth's atmosphere at a virtually constant pressure. For example, plants convert carbon dioxide and water into complex molecules; animals digest food; water heaters and stoves burn fuel; and running water dissolves minerals from the soil. All these processes involve energy changes at constant pressure. Nearly all aqueous-solution chemistry also occurs at constant pressure. Thus, the heat flow measured using constant-pressure calorimetry, q_p, closely approximates heat flows in many real-world processes. As we saw in the previous section, we cannot equate this heat flow to ΔE, because work may be involved. We can, however, identify a new thermodynamic function that we can use without having to calculate work. Before doing this, we need to describe one type of work involved in constant-pressure processes.

Expansion Work

Whenever a chemical process occurs at constant pressure, the volume can change, particularly when gases are involved. In a constant-pressure calorimeter, for instance, the chemical system may expand or contract. In this change of volume, the system moves against the force exerted by the constant pressure. Because work is force times displacement, $w = Fd$, this means that work is done whenever a volume change occurs at constant pressure.

The cylinder and piston in Figure 6-18 illustrate expansion work. If the chemical system expands, it pushes the piston through a displacement (d). The opposing force is related to the pressure (P) on the piston. As described in Chapter 5, pressure is force per unit area:

$$\text{Pressure} = \frac{\text{Force}}{\text{Area}} = \frac{F}{A} \quad or \quad F = PA$$

Substituting this into the equation for work, we see how the work done by the expanding system (w) is related to the constant pressure (P):

$$w = Fd = PAd$$

As Figure 6-18 shows, the volume change of the chemical system is the product of area multiplied by displacement:

$$\Delta V = (\text{Area})(\text{Displacement}) = Ad$$

This lets us write an expression giving the work done on the surroundings in terms of the opposing constant pressure and the change of volume:

$$w_{\text{surr}} = P\Delta V$$

 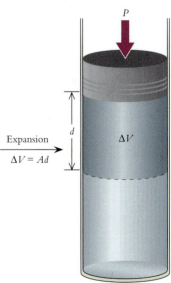

Figure 6-18
When a reaction occurs in a constant-pressure calorimeter, the external pressure is fixed but the volume of the chemical system can change, so work is done.

Every energy transfer has a *direction*. When a system expands, it does work on the surroundings, transferring energy to the surroundings. This means that w is positive for the surroundings and negative for the system in an expansion:

$$w_{sys} = -w_{surr}$$

$$w_{sys} = -P\Delta V_{sys} \qquad \text{(constant pressure)} \qquad \textbf{(6-8)}$$

Equation 6-8 is used to calculate the amount of work done during any process that occurs at constant pressure. When the process is an expansion, ΔV_{sys} is positive and w_{sys} is negative. In a contraction, on the other hand, ΔV_{sys} is negative and w_{sys} is positive.

Because of this expansion work, a process at constant pressure involves both heat and work:

$$\Delta E = q_p + w_p = q_p - P\Delta V_{sys}$$

Rearranging this equation gives an equation for q_p:

$$q_p = \Delta E + P\Delta V$$

We omit the subscript for V because all quantities refer to the chemical system.

Definition of Enthalpy

Equation 6-9 defines a new state function, called **enthalpy (H),** that we can relate to q_p:

$$H = E + PV \qquad \textbf{(6-9)}$$

We can use Equation 6-9 to relate a change in enthalpy to changes in energy, pressure, and volume:

$$\Delta H = \Delta(E + PV) = \Delta E + \Delta(PV)$$

For a process occurring at constant pressure, $\Delta(PV) = P\Delta V$;

$$\Delta H = \Delta E + P\Delta V \qquad \text{(constant pressure)}$$

As shown above, the quantity on the right is q_p, so a simple equality links ΔH and the heat flow in a constant-pressure process:

$$\Delta H = q_p \qquad \textbf{(6-10)}$$

Equation 6-10 lets us describe enthalpy in words:

KEY CONCEPT | *Enthalpy is a thermodynamic state function that describes heat flow at constant pressure.*

Just as the heat flow in a constant-volume process gives the energy change, the heat flow in a constant-pressure process gives the enthalpy change:

$$q_v = \Delta E \qquad and \qquad q_p = \Delta H$$

Whereas a constant-volume calorimeter provides direct measurements of energy changes, a constant-pressure calorimeter provides direct measurements of enthalpy changes. Table 6-4 summarizes the properties of the two different types of calorimetry.

The units of enthalpy are the same as the units of energy: joules or kilojoules. Tables give enthalpy changes per mole of substance, usually expressed in kilojoules per mole.

Table 6-4   Properties of Calorimeters

Type	Reaction Medium	Measurement	Example
Constant pressure	Solution	$q_p = \Delta H$	Coffee cup
Constant volume	Gas	$q_v = \Delta E$	Bomb

Energy and Enthalpy

Energy is a *fundamental* thermodynamic property. In contrast, enthalpy is a *defined* thermodynamic property that is convenient when working at constant pressure. Enthalpy changes and energy changes are related through the PV product:

$$\Delta H = \Delta E + \Delta(PV)$$

For solids and liquids, volume changes during chemical reactions are small enough to neglect:

$$\Delta(PV)_{\text{(condensed phases)}} \cong 0$$

Thus, enthalpy changes and energy changes are essentially equal for processes that involve only liquids and solids.

An example of such a process is the dissolving of ammonium nitrate in water to produce aqueous ions:

$$NH_4NO_3\,(s) \longrightarrow NH_4^+(aq) + NO_3^-(aq)$$

As Example 6-6 shows, 20.0 g of NH_4NO_3 absorbs 5.28 kJ of energy when the compound dissolves in a constant-pressure calorimeter. This information lets us calculate the molar enthalpy change for the dissolving process:

$$\Delta H = q_p = 5.28 \text{ kJ} \qquad and \qquad n = \frac{20.0 \text{ g}}{80.05 \text{ g/mol}} = 0.2498 \text{ mol}$$

$$\Delta H_{\text{molar}} = \frac{\Delta H}{n} = \frac{5.28 \text{ kJ}}{0.2498 \text{ mol}} = 21.1 \text{ kJ/mol}$$

Since none of the reactants or products is a gas, the energy and enthalpy changes for dissolving are essentially equal:

$$\Delta E_{\text{molar}} \cong \Delta H_{\text{molar}} = 21.1 \text{ kJ/mol}$$

Both pressure and volume may change significantly when a reaction produces or consumes gases. Consequently, the enthalpy change is likely to differ from the energy change for any chemical reaction involving gaseous reagents. To determine the difference, we can use the ideal gas equation to relate the change in the pressure–volume product to the change in number of moles:

$$PV = nRT \qquad so \qquad \Delta(PV)_{\text{gases}} = \Delta(nRT)_{\text{gases}}$$

Combining the equations for enthalpy and $\Delta(PV)$, we find that the enthalpy and energy changes are related through Equation 6-11:

$$\Delta H_{\text{reaction}} = \Delta E_{\text{reaction}} + \Delta(PV)_{\text{reaction}}$$

$$\Delta(PV)_{\text{reaction}} \cong \Delta(PV)_{\text{gases}} = \Delta(nRT)_{\text{gases}}$$

$$\Delta H_{\text{reaction}} \cong \Delta E_{\text{reaction}} + \Delta(nRT)_{\text{gases}}$$

Because R and T are constant, RT can be brought outside the parentheses.

$$\Delta H_{\text{reaction}} \cong \Delta E_{\text{reaction}} + RT\Delta n_{\text{gases}} \qquad \text{(at constant } T) \qquad \textbf{(6-11)}$$

In this equation Δn is the difference between the numbers of moles of gaseous products and gaseous reactants in the balanced chemical equation:

$$\Delta n_{\text{gases}} = n_{\text{gas, products}} - n_{\text{gas, reactants}}$$

Example 6-8 demonstrates a typical magnitude for the difference between $\Delta E_{\text{reaction}}$ and $\Delta H_{\text{reaction}}$.

Example 6-8

ΔE and ΔH for Combustion

Find the difference between the molar ΔH and ΔE for the combustion of octane at 298 K.

Strategy: As always, we use the seven-step method. In this example, we need to pay particular attention to relationships between different quantities.

1. The problem asks about the difference between a molar ΔH and ΔE.

2 and 3. Because we are given no specific data, visualizing includes figuring out what information we need. Equation 6-11 links ΔH and ΔE:

$$\Delta H_{reaction} \cong \Delta E_{reaction} + RT\Delta n_{gases}$$

To calculate the difference between $\Delta H_{reaction}$ and $\Delta E_{reaction}$, we need information about the $RT\Delta n$ term as it applies to the combustion reaction.

Solution:

4. The balanced chemical equation for combustion applies to two moles of octane:

$$2\, C_8H_{18}(l) + 25\, O_2(g) \longrightarrow 16\, CO_2(g) + 18\, H_2O(l)$$

The question asks about a molar amount of octane, so divide each coefficient by 2 to obtain a reaction in which one mole of octane burns at 298 K:

$$C_8H_{18}(l) + \tfrac{25}{2} O_2(g) \longrightarrow 8\, CO_2(g) + 9\, H_2O(l)$$

5. To find the difference between ΔH and ΔE, rearrange Equation 6-11:

$$\Delta H - \Delta E = RT\Delta n_{gases} \quad \text{(at constant T)}$$

6. Here are the values needed to complete the calculation:

$$n_f = \text{mol gas (products)} = \text{mol } CO_2 = 8$$

$$n_i = \text{mol gas (reactants)} = \text{mol } O_2 = \tfrac{25}{2}$$

$$\Delta n_{gases} = 8 \text{ mol } CO_2 - \tfrac{25}{2} \text{ mol } O_2 = -\tfrac{9}{2} \text{ mol gas}$$

$$T = 298 \text{ K (Temperature must be in kelvins to match the units of } R)$$

$$\Delta H - \Delta E = (8.314 \text{ J/mol K})(298 \text{ K})(-\tfrac{9}{2} \text{ mol gas/mol octane})$$

$$\Delta H - \Delta E = -1.11 \times 10^4 \text{ J/mol octane} = -11.1 \text{ kJ/mol octane}$$

7. Does the Result Make Sense? The units are kJ/mol, as required for molar energy and enthalpy, and the number of significant figures is appropriate. The difference is negative because the number of moles of gas decreases during the reaction. This means the volume change is negative, $\Delta(PV)$ is negative, so $\Delta H = \Delta E + \Delta(PV)$ is more negative than ΔE. The magnitude of the difference is in the range of typical molar energy values.

Extra Practice Exercise 6.8

Determine the difference between molar ΔH and ΔE for the combustion of graphite (C).

In Example 6-7, we used calorimetry data to determine that the energy change when one mole of octane burns is -5.5×10^3 kJ/mol. The result of Example 6-8 shows that the enthalpy change for 1 mol of octane differs from this by -11.1 kJ. To two significant figures, ΔH and ΔE are the same, even though the value of Δn_{gases} is significant. The fractional difference between $\Delta H_{reaction}$ and $\Delta E_{reaction}$ is 11.1/5500, or about 0.2%.

This calculation shows that reaction energies and reaction enthalpies are usually about the same, even when reactions involve gases. For this reason, chemists often use $\Delta E_{reaction}$ and $\Delta H_{reaction}$ interchangeably. Because many everyday processes occur at constant pressure, thermodynamic tables usually give values for enthalpy changes. Nevertheless, bear in mind that these are different thermodynamic quantities. For processes with modest ΔE values and significant volume changes, ΔE and ΔH can differ substantially.

Energy and Enthalpy of Vaporization

One process for which ΔH and ΔE differ significantly is vaporization, the conversion of a liquid or solid into a gas. Here are two such conversions:

$$H_2O\,(l) \longrightarrow H_2O\,(g) \qquad \Delta H_{vap}(373\text{ K}) = 40.79\text{ kJ/mol}$$

$$I_2\,(s) \longrightarrow I_2\,(g) \qquad \Delta H_{vap}(298\text{ K}) = 62.4\text{ kJ/mol}$$

A substance vaporizes from a condensed phase (liquid or solid) into the gas phase, so Δn (gas) = 1 mol. The change in volume for this type of transformation is almost equal to the volume of the resulting gas. For example, according to the ideal gas equation, one mole of water vapor at 373 K has a volume of 30.6 L:

$$\frac{V}{n} = \frac{RT}{P} = \frac{(0.08206\text{ L atm mol}^{-1}\text{K}^{-1})(373\text{ K})}{(1.00\text{ atm})} = 30.6\text{ L/mol}$$

The same amount of liquid water at 373 K has a volume of only 18 mL, or 0.018 L. This small volume can be neglected compared with the larger volume of the gas. Thus, we can simplify Equation 6-11 for vaporizations:

$$\Delta H_{vap} \cong \Delta E_{vap} + RT_{vap} \qquad and \qquad \Delta E_{vap} \cong \Delta H_{vap} - RT_{vap}$$

Applying this equation, we calculate ΔE_{vap} for the two typical processes listed above:

$$H_2O\,(l) \longrightarrow H_2O\,(g)$$

$$\Delta E_{vap}(373\text{ K}) = 40.79\text{ kJ/mol} - (8.314\text{ J mol}^{-1}\text{ K}^{-1})(373\text{ K})(10^{-3}\text{ kJ/J})$$
$$= 40.79\text{ kJ/mol} - 3.10\text{ kJ/mol} = 37.69\text{ kJ/mol}$$

$$I_2\,(s) \longrightarrow I_2\,(g)$$

$$\Delta E_{vap}(298\text{ K}) = 62.4\text{ kJ/mol} - (8.314\text{ J mol}^{-1}\text{ K}^{-1})(298\text{ K})(10^{-3}\text{ kJ/J})$$
$$= 62.4\text{ kJ/mol} - 2.48\text{ kJ/mol} = 59.9\text{ kJ/mol}$$

Notice that the difference between ΔE_{vap} and ΔH_{vap} depends only on temperature. At 373 K (100 °C) ΔH_{vap} always is 3.10 kJ/mol greater than ΔE_{vap}. At 298 K (25 °C, room temperature) ΔH_{vap} always is 2.48 kJ/mol greater than ΔE_{vap}. The difference between ΔE_{vap} and ΔH_{vap} arises because, in addition to overcoming intermolecular forces in the condensed phase (ΔE), the escaping vapor must do work, $w = \Delta(PV) = RT$ as it expands against the constant external pressure of the atmosphere.

Enthalpies of Formation

How do we determine the energy and enthalpy changes for a chemical reaction? We could perform calorimetry experiments and analyze the results, but to do this for every chemical reaction would be an insurmountable task. Furthermore, it turns out to be unnecessary. Using the first law of thermodynamics and the idea of a state function, we can calculate enthalpy changes for almost any reaction using experimental values for one set of reactions, the **formation reactions.**

A formation reaction produces 1 mol of a chemical substance from the elements in their most stable forms.

KEY CONCEPT

Formation reactions have the following features:

1. There is a *single product* with a stoichiometric coefficient of *1*.

2. All the starting materials are *elements*, and each is in its most stable form.

3. Enthalpies of reactions involving gases vary with pressure (because of the PV term in the definition of H), so pressures must be specified.

4. Enthalpies of reactions occurring in solution vary with concentration, so concentrations must be specified.

Form that is most stable refers to both phase and chemical form. For example, bromine is most stable as Br_2, not Br, and as a liquid, not as a gas or a solid.

$$\underset{\substack{\uparrow\\ \text{Change} \quad \text{Enthalpy}}}{\Delta H}\underset{\substack{\uparrow\\ }}{{}^{\circ}_{f}}\ \substack{\longleftarrow \text{Standard}\\ \longleftarrow \text{Formation}}$$

When the partial pressure of each gaseous reagent is 1 bar and the concentration of each species in solution is 1 M, the conditions are defined to be "standard." Under these conditions, the enthalpy change in a formation reaction is the **standard enthalpy of formation (ΔH°_{f}).**

Reaction enthalpies may also vary with temperature. Most tabulations of thermodynamic functions are for $T = 298$ K, so we assume that is the temperature unless otherwise specified.

Here are several examples of formation reactions:

$$Fe\,(s) + \tfrac{1}{2}\,O_2\,(g) \longrightarrow FeO\,(s) \qquad\qquad \Delta H^{\circ}_{f} = -272.0 \text{ kJ/mol}$$

$$Br_2\,(l) \longrightarrow Br_2\,(g) \qquad\qquad \Delta H^{\circ}_{f} = 30.9 \text{ kJ/mol}$$

$$\tfrac{1}{2}\,N_2\,(g) + \tfrac{1}{2}\,O_2\,(g) \longrightarrow NO\,(g) \qquad\qquad \Delta H^{\circ}_{f} = 91.3 \text{ kJ/mol}$$

$$C\,(graphite) + 2\,H_2\,(g) + \tfrac{1}{2}\,O_2\,(g) \longrightarrow CH_3OH\,(l) \qquad \Delta H^{\circ}_{f} = -239.2 \text{ kJ/mol}$$

$$C\,(graphite) \longrightarrow C\,(graphite) \qquad\qquad \Delta H^{\circ}_{f} = 0 \text{ kJ/mol}$$

These examples illustrate additional features of formation reactions. First, formation reactions can be either exothermic or endothermic. Second, an element in a state that is different from the standard state has a positive standard formation enthalpy, as the formation reaction for $Br_2\,(g)$ shows. Third, fractional stoichiometric coefficients are common in formation reactions because the reaction must generate exactly one mole of product. Finally, an element that is already in its standard state has $\Delta H^{\circ}_{f} = 0$, because the formation "reaction" involves no change.

Some formation reactions, such as the oxidation of metals, proceed readily and directly. Others do not. Methanol, for example, cannot form by direct combination of elemental carbon, hydrogen, and oxygen. Whether or not a formation reaction occurs in nature, however, the enthalpy of formation is a well-defined quantity.

Over the years, scientists have measured standard enthalpies of formation of many chemical substances. Reference books such as the *CRC Handbook of Chemistry and Physics* contain tables of these values, and Appendix D of this text lists values for many common substances.

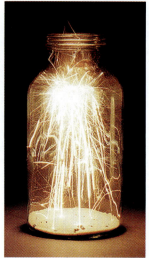

The formation reaction of FeO proceeds vigorously.

Enthalpy Changes for Chemical Reactions

Animation

Figure 6-19
The enthalpy change for the reaction of NO₂ to produce N₂O₄ can be calculated using a two-step path. In the first step, NO₂ decomposes to N₂ and O₂. In the second step, the elements react to form N₂O₄. Enthalpy is a state function, so ΔH for the overall reaction is the sum of the enthalpy changes for these two steps.

Reactants

2 NO$_2$ (g)

ΔH° reaction

ΔH° decomposition

N$_2$O$_4$ (g)

Products

ΔH° formation

Elements

2 O$_2$ (g) + N$_2$ (g)

Standard enthalpies of formation are particularly useful because they can be used to calculate the enthalpy change for any reaction. As a first example, consider nitrogen dioxide reacting to form N_2O_4:

$$2\,NO_2\,(g) \longrightarrow N_2O_4\,(g)$$

Imagine this reaction following the two-step pathway that appears in Figure 6-19. In the first step, 2 mol of NO_2 molecules decompose into nitrogen molecules and oxygen molecules:

$$2\,NO_2\,(g) \longrightarrow N_2\,(g) + 2\,O_2\,(g)$$

In the second step, these nitrogen and oxygen molecules react to produce N_2O_4:

$$N_2\,(g) + 2\,O_2\,(g) \longrightarrow N_2O_4\,(g)$$

The pathway shown in Figure 6-19 is not how the reaction actually occurs, but enthalpy is a state function. Because the change of any state function is independent of the path of the reaction, we can use any convenient path for calculating the enthalpy change. **Hess' law** summarizes this feature:

> *The enthalpy change for any overall process is equal to the sum of enthalpy changes for any set of steps that leads from the starting materials to the products.*

KEY CONCEPT

In this example, the enthalpy change for the overall reaction can be determined by adding the enthalpy changes of the two steps of Figure 6-19:

$$\Delta H^\circ_{\text{reaction}} = \Delta H^\circ_{\text{decomposition}} + \Delta H^\circ_{\text{formation}}$$

$$2\,NO_2(g) \longrightarrow \cancel{N_2}(g) + 2\,\cancel{O_2}(g) \quad \Delta H^\circ_{\text{decomposition}}$$

$$\underline{\cancel{N_2}(g) + 2\,\cancel{O_2}(g) \longrightarrow N_2O_4(g) \qquad\qquad \Delta H^\circ_{\text{formation}}}$$

$$2\,NO_2(g) \longrightarrow N_2O_4(g) \qquad\qquad \Delta H^\circ_{\text{reaction}}$$

The decomposition step is related to the formation reaction of NO_2 in two ways. First, decomposition is the opposite of formation. Second, this decomposition step involves two moles of NO_2, whereas the formation reaction involves one mole of NO_2. Using these features, we can express the enthalpy change of the decomposition reaction for NO_2 in terms of the standard enthalpy of formation of NO_2. The sign for the enthalpy change of the decomposition reaction is opposite that of the corresponding formation reaction, and the total enthalpy of decomposition requires that we multiply by the stoichiometric coefficient:

$$\Delta H^\circ_{\text{decomposition}} = -2\,\Delta H^\circ_{\text{f}}(NO_2)$$

The enthalpy change of the formation reaction is just $\Delta H^\circ_{\text{f}}$ of N_2O_4:

$$\Delta H^\circ_{\text{formation}} = \Delta H^\circ_{\text{f}}(N_2O_4)$$

Thus, the molar enthalpy change of the overall reaction can be expressed entirely in terms of standard formation reactions:

$$\Delta H^\circ_{\text{reaction}} = \Delta H^\circ_{\text{f}}(N_2O_4) - 2\,\Delta H^\circ_{\text{f}}(NO_2)$$

The enthalpy change of the overall reaction is the sum of the formation enthalpies of the products minus the sum of the formation enthalpies of the reactants.

We can calculate the enthalpy change for the overall reaction using standard enthalpies of formation from Appendix D:

$$\Delta H^\circ_{\text{reaction}} = (1\text{ mol }N_2O_4)(11.1\text{ kJ/mol }N_2O_4) - (2\text{ mol }NO_2)(33.2\text{ kJ/mol }NO_2)$$

$$\Delta H^\circ_{\text{reaction}} = -55.3\text{ kJ}$$

The negative value for the enthalpy change indicates that this reaction is exothermic. Qualitatively, this is logical because during the reaction, a chemical bond forms between the nitrogen atoms, but none of the bonds of the reactants breaks.

Our analysis of the reaction of nitrogen dioxide molecules is not unique. The same type of path can be visualized for any chemical reaction, as Figure 6-20 shows. The reaction enthalpy for any chemical reaction can be found from the standard enthalpies of

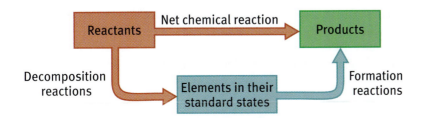

Figure 6-20
Any chemical reaction can be imagined to proceed in two stages. First, reactants decompose into their constituent elements. Second, these elements recombine to form products.

formation for all the reactants and products. Multiply each standard enthalpy of formation by the appropriate stoichiometric coefficient, add the values for the products, add the values for the reactants, and subtract the sum for reactants from the sum for products. Equation 6-12 summarizes this procedure:

$$\Delta H^{\circ}_{reaction} = \Sigma \, coeff_{products} \, \Delta H^{\circ}_{f \, products} - \Sigma \, coeff_{reactants} \, \Delta H^{\circ}_{f \, reactants} \qquad \text{(6-12)}$$

Remember that the Greek symbol Σ means "the sum of," and "coeff" refers to the stoichiometric coefficients in the balanced equation.

As can be seen from Equation 6-12, the value of $\Delta H^{\circ}_{reaction}$ depends on the coefficients used for the reagents. This means that the enthalpy change for one mole of any particular reactant may differ from $\Delta H^{\circ}_{reaction}$. Returning to our earlier example, the enthalpy change of -55.3 kJ is for two moles of NO_2 reacting to form one mole of N_2O_4. The enthalpy change per mole of NO_2 is $(-55.3 \text{ kJ}/2 \text{ mol}) = -28.6$ kJ/mol. We can generalize this calculation: To find the molar enthalpy change for any reagent, divide $\Delta H^{\circ}_{reaction}$ by the appropriate stoichiometric coefficient.

As another illustration of the use of Equation 6-12, we calculate the enthalpy of combustion of methane, the principal component of natural gas. We begin with the balanced chemical equation:

$$CH_4 \, (g) + 2 \, O_2 \, (g) \longrightarrow CO_2 \, (g) + 2 \, H_2O \, (l)$$

We could evaluate the enthalpy for this reaction step by step, first decomposing methane into its elemental constituents and then recombining the elements into carbon dioxide and liquid water. There is no need to do this, however, because Equation 6-12 summarizes the step-by-step processes. The products are one mole of CO_2 and two moles of H_2O, and the reactants are one mole of CH_4 and two moles of O_2:

$$\Delta H^{\circ}_{reaction} = [\Delta H^{\circ}_f \, (CO_2) + 2 \, \Delta H^{\circ}_f \, (H_2O)] - [\Delta H^{\circ}_f \, (CH_4) + 2 \, \Delta H^{\circ}_f \, (O_2)]$$

The data required for the calculation come from Appendix D. Notice that the value for O_2 is zero, because this is an element in its standard state:

$$\Delta H^{\circ}_{reaction} = \left[(1 \text{ mol } CO_2)\left(\frac{-393.5 \text{ kJ}}{\text{mol } CO_2}\right) + (2 \text{ mol } H_2O)\left(\frac{-285.8 \text{ kJ}}{\text{mol } H_2O}\right) \right]$$

$$- \left[(1 \text{ mol } CH_4)\left(\frac{-74.81 \text{ kJ}}{\text{mol } CH_4}\right) + (2 \text{ mol } O_2)\left(\frac{0 \text{ kJ}}{\text{mol } O_2}\right) \right]$$

$$\Delta H^{\circ}_{reaction} = [-965.1 \text{ kJ}] - [-74.81 \text{ kJ}] = -890.3 \text{ kJ}$$

Enthalpy Changes Under Nonstandard Conditions

Standard formation enthalpies refer to 1 bar pressure, 1 M concentration, and, unless the temperature is specified otherwise, $T = 298$ K. What if a reaction occurs under other conditions? Energies and enthalpies of substances change as temperature, concentration, and pressure change, so we must expect that ΔH for a reaction depends on these variables, too. Calculations of ΔH under nonstandard conditions are beyond the scope of our coverage. Fortunately, changing conditions have only a small effect on ΔH, so calculations of reaction enthalpies using standard enthalpies of formation provide reasonable results, even when the temperature, pressure, and concentration depart somewhat from 298 K, 1 bar, and 1 M.

Example 6-9 applies tabulated enthalpies to a reaction occurring at elevated temperature and pressure.

Nitric acid (HNO_3), which is produced in the gas phase at elevated temperature and pressure, is an important chemical in the fertilizer industry because it can be converted into ammonium nitrate. The following overall chemical equation summarizes the industrial reaction:

$$12 \, NH_3\,(g) + 21 \, O_2\,(g) \longrightarrow 8 \, HNO_3\,(g) + 4 \, NO\,(g) + 14 \, H_2O\,(g)$$

Determine whether this overall process is exothermic or endothermic and estimate the energy change per mole of HNO_3 formed.

Strategy: Again, we follow the seven-step procedure. The first question asks whether this reaction is exothermic (releases energy) or endothermic (absorbs energy). The second question asks us to estimate the energy change for the reaction per mole of HNO_3. Although the reaction conditions differ from standard conditions, ΔH varies slowly with temperature and pressure, so we can use Equation 6-12 and tabulated standard enthalpies of formation to estimate the reaction enthalpy. Because reaction enthalpy is almost the same as reaction energy, this calculation will give a satisfactory estimate of the reaction energy.

Solution: Appendix D lists the following standard heats of formation, all in kJ/mol:

$$NH_3\,(g): -45.9; \quad HNO_3\,(g): -133.9; \quad H_2O\,(g): -241.8; \quad NO\,(g): +91.3$$

Substitute these values into Equation 6-12 and use the stoichiometric coefficients from the balanced chemical equation:

$$\begin{aligned}
\Delta H_{reaction} \cong & \; [(8 \text{ mol})(-133.9 \text{ kJ/mol}) + (4 \text{ mol})(91.3 \text{ kJ/mol}) \\
& + (14 \text{ mol})(-241.8 \text{ kJ/mol})] \\
& - [(12 \text{ mol})(-45.9 \text{ kJ/mol}) + (21 \text{ mol})(0 \text{ kJ/mol})] \\
= & \; [-1071.2 \text{ kJ} + 365.2 \text{ kJ} - 3385.2 \text{ kJ}] - [-550.8 \text{ kJ} + 0 \text{ kJ}] \\
= & \; [-4091.2 \text{ kJ}] - [-550.8 \text{ kJ}] \\
\Delta H_{reaction} \cong & \; -3540.4 \text{ kJ} = -3.5404 \times 10^3 \text{ kJ}
\end{aligned}$$

It is important to pay careful attention to the signs, which we have done by carrying out the calculations in sequence. First, multiply each standard enthalpy, including its sign, by the appropriate coefficient. Then combine values for reactants and values for products, adding or subtracting according to the sign of each enthalpy. Finally, subtract the sum for reactants from the sum for products to obtain the result.

The negative sign for $\Delta H_{reaction}$ indicates that this reaction is exothermic. This is the value for a reaction that generates 8 moles of $HNO_3\,(g)$. Therefore, the value *per mole* of nitric acid produced is

$$\Delta H_{molar} = \frac{-3.5404 \times 10^3 \text{ kJ}}{8 \text{ mol } HNO_3} = -442.55 \text{ kJ/mol } HNO_3$$

Does the Result Make Sense? The units are correct and the numerical value is typical of reaction energies and enthalpies. Moreover, it makes sense for the chemical industry to develop exothermic reactions for large-scale industrial production.

Determine the energy change per mole of NH_3 consumed for the first step in nitric acid formation, reaction of ammonia with oxygen:

$$4 \, NH_3\,(g) + 7 \, O_2\,(g) \longrightarrow 4 \, NO_2\,(g) + 6 \, H_2O\,(l)$$

Example 6-9

Molar Enthalpy Change

The heat of formation of a substance depends on its phase, because heat is required to melt or boil a substance. ΔH_f° for liquid H_2O is -285.8 kJ/mol, and ΔH_f° for gaseous H_2O is -242 kJ/mol. This difference illustrates the importance of keeping track of the phase of each species.

Extra Practice Exercise 6.9

Equation 6-12 can also be used to calculate the standard enthalpy of formation of a substance whose formation reaction does not proceed cleanly and rapidly. The enthalpy change for some other chemical reaction involving the substance can be determined by calorimetric measurements. Then Equation 6-12 can be used to calculate the unknown standard enthalpy of formation. Example 6-10 shows how to do this using experimental data from a constant-volume calorimetry experiment combined with standard heats of formation.

Example 6-10 **Enthalpy of Formation of Octane**	The enthalpy of combustion of octane (C_8H_{18}) is -5.5×10^3 kJ/mol. Using tabulated standard enthalpies of formation in Appendix D, determine the standard enthalpy of formation of octane. **Strategy:** Again, it is convenient to follow the seven-step procedure to solve this problem. We are asked to find an enthalpy of formation. Because enthalpy is a state function, we can visualize the reaction as occurring through decomposition and formation reactions. Appendix D lists enthalpies of formation, and the experimental heat of combustion is provided. We can use Equation 6-12 to relate the enthalpy of combustion to the standard enthalpy of formation for octane. **Solution:** Begin with the balanced chemical equation for the combustion of one mole of octane:

$$C_8H_{18}\,(l) + \tfrac{25}{2}\,O_2\,(g) \longrightarrow 8\,CO_2\,(g) + 9\,H_2O\,(l)$$

Use Equation 6-12 to set the heat of combustion, ΔH_{molar}, equal to the sum of the enthalpies of formation and solve for the value for octane:

$$\Delta H_{molar} = \Sigma\,coeff_{products}\,\Delta H^\circ_{f\,products} - \Sigma\,coeff_{reactants}\,\Delta H^\circ_{f\,reactants}$$

$$\Delta H_{molar} = 8\,\Delta H^\circ_f(CO_2) + 9\,\Delta H^\circ_f(H_2O) - \Delta H^\circ_f(octane) - \left(\tfrac{25}{2}\right)\Delta H^\circ_f(O_2)$$

$$\Delta H^\circ_f(octane) = 8\,\Delta H^\circ_f(CO_2) + 9\,\Delta H^\circ_f(H_2O) - \left(\tfrac{25}{2}\right)\Delta H^\circ_f(O_2) + \Delta H_{molar}$$

Substitute the heat of combustion and experimental values for the heats of formation from Appendix D:

$$\Delta H^\circ_f(octane) = (8\text{ mol }CO_2)(-393.5\text{ kJ/mol}) + (9\text{ mol }H_2O)(-285.8\text{ kJ/mol})$$

$$- \left(\tfrac{25}{2}\text{ mol }O_2\right)(0\text{ kJ/mol}) + 5.5 \times 10^3\text{ kJ/mol}$$

$$\Delta H^\circ_f(octane) = -220\text{ kJ/mol}$$

 Does the Result Make Sense? The units are correct, and the negative value indicates that octane is somewhat more stable than the elements from which it forms. From Appendix D, we see that methane (CH_4) and ethane (C_2H_6), similar substances to octane, have ΔH°_f values that are negative, so the result is reasonable.

Extra Practice Exercise 6.10	Determine the enthalpy of formation of phosphoric acid from data in Appendix D and the following information:

$$P_4O_{10}\,(s) + 6\,H_2O\,(l) \longrightarrow 4\,H_3PO_4\,(s) \qquad \Delta H^\circ_{reaction} = -417\text{ kJ}$$

Bond Energies and Enthalpies of Formation

As described earlier, we can estimate the energies of chemical reactions from bond energies rather than calculating them from formation reactions. Equation 6-5, like Equation 6-12, contains two sums:

$$\Delta E_{reaction} = \Sigma\,BE_{bonds\ broken} - \Sigma\,BE_{bonds\ formed} \qquad (6\text{-}5)$$

The similarity arises because both equations describe similar processes and are consequences of Hess' law. For Equation 6-12, we imagine a process in which reactants decompose into elements in their standard states, which then recombine to give products. For Equation 6-5, we imagine a process in which reactants break up entirely into gaseous atoms, which then recombine to give products.

It is preferable to use Equation 6-12 when standard enthalpies of formation are available for the reagents taking part in a chemical reaction, because this equation uses actual experimental values for standard enthalpies of formation. Consequently, Equation 6-12 is exact at $T = 298$ K, pressure = 1 bar, and concentration = 1 M, and provides very good estimates even under other conditions. Equation 6-5, on the other hand, uses tabulated bond energies, which are average values for each chemical bond rather than exact values for the bonds in any specific substance. Consequently, Equation 6-5 provides *estimates* of reaction energies rather than *exact* values. The advantage of average values is that we can make a reasonable estimate of the sign and magnitude of ΔE and ΔH for reactions that involve substances for which standard formation enthalpies are not available.

Section Exercises

6.5.1 In a steam engine, water vapor drives a piston to accomplish work. What amount of work, in joules, is done when 25.0 g of water at 100.0 °C vaporizes to steam, pushing a piston against 1.00 bar of external pressure? (*Hint:* Use the ideal gas equation to find the volume change when liquid water vaporizes.)

6.5.2 Which of the following substances have zero values for their standard enthalpy of formation? (a) ozone, $O_3(g)$; (b) solid mercury,

$Hg(s)$; (c) liquid bromine, $Br_2(l)$; (d) graphite, $C(s)$; (e) atomic fluorine, $F(g)$; and (f) solid sulfur, $S_8(s)$

6.5.3 Hydrogen gas is prepared industrially from methane and steam:

$$CH_4(g) + H_2O(g) \longrightarrow CO(g) + 3\,H_2(g)$$

Use standard enthalpies of formation to determine $\Delta H_{reaction}$ and $\Delta E_{reaction}$ for the balanced reaction at 298 K.

6.6 ENERGY SOURCES

Our chapter introduction emphasizes the importance of energy to human civilizations. Civilization may have begun when humans learned how to use wood fires as energy sources. The mastery of fire was just the first step in the use of a variety of energy sources to accomplish increasingly energy-intense goals, such as the launching of spacecraft that escape from the gravitational pull of the Earth.

Energy and Civilization

The advances of human civilization can be viewed as the results of people figuring out how to increase the availability of energy. As Figure 6-21 shows, the amount of energy used per person per day has multiplied by more than a factor of 100 as humans moved beyond the application of their own muscles, by making use of animals, water power, fossil fuels, and nuclear energy.

The taming of fire was undoubtedly the first advance in energy usage by humans. Burning wood provided heat, light, cooking, and eventually the mastery of chemical arts, such as the firing of pottery and the smelting of metals. While these applications did not directly increase the amount of energy that humans could use for locomotion or construction, they made life considerably easier and more comfortable.

The domestication of animals occurred at least 6000 years ago. Besides providing a reliable food source, domestic animals supplied energy for many purposes. Hitched to a plow, oxen or water buffalo allowed humans to cultivate larger parcels of land. Land travel over long distances became much easier with horses, camels, and even elephants.

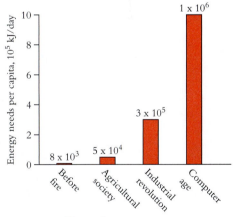

Figure 6-21
As civilization has progressed, the energy needs per capita have multiplied over 100-fold.

Figure 6-22
Since 1800, the main sources of energy have been fossil fuels, with other sources contributing less than 20% of the total.

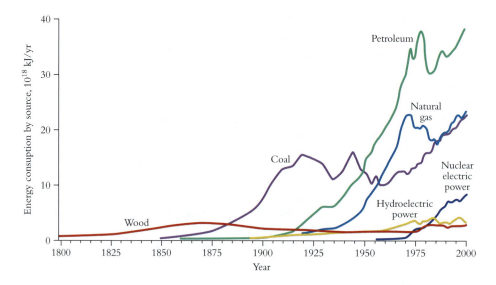

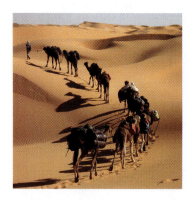

The next stage in exploitation of energy sources was the harnessing of moving air or water. This was the first method of creating mechanical energy to replace work by humans or animals. The Greek poet Antipater of Thessalonica, writing around 80 BC, describes how young women were freed from work by watermills. Wind energy was used to move sailboats from very early times: Egyptian drawings from around 3200 BC show sailboats. It took a long time, however, for humans to learn how to use wind energy for stationary applications. The first windmills, used for grinding grain and pumping water, appear to have been developed in Persia some time around 600 AD.

The Industrial Revolution came hand-in-hand with the use of fossil fuels. Although coal had been used for heating and in metallurgy since at least the thirteenth century, it was not until the invention and refinement of the steam engine that coal consumption increased greatly. By the middle of the nineteenth century, work done by machines exceeded the work done by animal power. While steam engines were mainly fueled by coal, the advent of the internal combustion engine required a volatile fuel, and petroleum distillates are perfectly suited for this purpose.

Figure 6-22 illustrates how important fossil fuel sources have been since the dawn of the Industrial Revolution. From about 1900 until the present, these nonrenewable sources have supplied more than 80% of the energy needs of our society. In the year 2001, total world energy production was 4.25×10^{20} J, of which 85% came from petroleum, coal, and natural gas.

Ultimate Energy Sources

The vast majority of our energy sources originate in solar energy. The sun bathes the Earth in a flood of electromagnetic radiation from the sun, about 8×10^{18} J/min. About half of this energy reaches the Earth's surface, the rest being reflected by the atmosphere or radiated from the earth. Almost none of this incoming energy is used directly, however, because solar energy is too diffuse to accomplish energy-intensive tasks like running an automobile.

Fortunately, natural processes convert some solar energy into more concentrated forms. The most obvious process is photosynthesis, by which green plants use the energy of visible light to drive energy-storing chemical reactions. Photosynthesis combines carbon dioxide and water to form compounds that store substantial amounts of chemical energy. One example is glucose:

$$6 \; CO_2 \, (g) + 6 \; H_2O \, (l) + 2880 \; kJ \longrightarrow C_6H_{12}O_6 \, (s) + 6 \; O_2 \, (g)$$

Plants use photosynthesis to produce energy-rich compounds. Cattle and other herbivores obtain energy-rich compounds by eating plants. Cougars and other carnivores eat herbivores, but the food chain ultimately leads back to energy storage via photosynthesis. When a compound such as glucose reacts with oxygen, the energy that was stored during photosynthesis is released and can be used to drive the various processes of life:

$$C_6H_{12}O_6\,(s) + 6\ O_2\,(g) \longrightarrow 6\ CO_2\,(g) + 6\ H_2O\,(l) + 2880\ kJ$$

From an energy perspective, forests are bank accounts that store the results of photosynthesis. When we cut and burn wood, we make withdrawals from these energy accounts. So long as we do not cut more rapidly than trees can replace themselves, forests represent a renewable energy resource. In the past, as well as in some parts of the world today, forests were harvested much more rapidly than they reproduced. Fortunately, attitudes about forest use are changing, and modern forestry practices make it possible to sustain this energy resource.

Fossil fuels also can be thought of as bank accounts, because these fuels originally were forests. Geological processes acting over millions of years converted the carbon-containing molecules of these ancient forests into coal, petroleum, and natural gas that remained trapped underground until humans discovered them. Unlike today's living forests, however, fossil fuel accounts cannot be replenished. Once consumed, they will be gone forever. Moreover, the combustion of fossil fuels releases carbon dioxide into the atmosphere. As we describe in Chapter 7, CO_2 is a greenhouse gas that contributes significantly to global climate change.

Photosynthesis is the most important conversion process by which solar energy is stored, but sunlight also drives two other energy sources. The sun heats and evaporates water from the oceans and lakes, and this evaporated water eventually returns to earth as precipitation. In addition to providing the fresh water that is essential to terrestrial life, the precipitation that falls at high altitudes has stored gravitational energy that can be harnessed through watermills, dams, and hydroelectric turbines. The sun also heats air masses at different rates. This leads to pressure differences between different locations, as we describe in Chapter 5. Wind is air moving from high-pressure zones to low-pressure zones, and sails and windmills can harness this air movement.

The most prominent nonsolar energy source is nuclear energy. Worldwide, nuclear energy supplies about 6.5% of human energy production. Nuclear power plants are used exclusively to produce electricity. Some countries, most notably Belgium (76%) and France (58%), rely heavily on nuclear-generated electricity. The United States currently produces the largest quantity of nuclear energy, but the future of this energy source is uncertain. Many people have strong concerns over the safety of nuclear power plants and the long-term storage of the radioactive waste from these plants, as we describe in Chapter 22.

The Earth's hot interior is an additional nonsolar energy source that has not been extensively exploited. Beneath the Earth's crust is a layer called the mantle, where the temperature reaches 4000 °C. For the most part, this energy source is too deep underground to be exploited. In certain areas, however, geological formations allow water to act as a carrier of energy that comes to the surface as superheated water or steam. Power plants have been installed in some of these geothermal areas, most notably in Iceland (look again at the photo inset on the opening pages of this chapter), but these represent only a tiny fraction of the overall world energy usage.

Future Resources

Industrial civilization was built by the consumption of fossil fuels. Currently, well over 80% of world energy comes from the three main fossil fuels: petroleum, natural gas, and coal. Because these are nonrenewable resources, there will come a time when these energy bank accounts are exhausted. Experts disagree on how soon the depletion of fossil fuels will occur, but eventually it will be necessary to find other energy sources.

Even if fossil fuel supplies are sufficient for many years to come, there are environmental reasons to reduce our reliance on them. If world consumption of fossil fuels continues at the current pace, there are likely to be significant undesirable climate effects.

From an economic perspective, ideal energy sources have high intensity and are readily extracted and transported. From an environmental perspective, the ideal energy source is both renewable and environmentally benign. Fossil fuels have been nearly ideal from the economic standpoint, which is why they dominate world usage even though they have environmentally undesirable characteristics. It is a major challenge to find and develop other energy sources that are both economically feasible and environmentally friendly.

The solutions to world energy problems most likely will differ by sector of use. Currently, nearly 40% of U.S. energy consumption goes into production of electrical power, and another 28% is for transportation. In addition to relying on electrical power, industrial, commercial, and residential users consume other types of energy. Industry, which accounts for 33% of total energy consumption, gets about a third of its total energy needs from electricity. Commercial and residential users account for 39% of total energy needs, of which nearly two-thirds comes from electricity.

The production of electricity traditionally requires massive installations such as nuclear power plants, coal-fired generators, or dams with huge turbines. Each of these carries environmental liabilities. Advocates of clean power propose a dual approach to reduce the environmental impacts of electricity generation. One part is large-scale power plants using renewable energy sources: biomass, wind, and geothermal energy. The economics of such power generators generally has not been favorable, but several states have mandated increased use of so-called green power sources, and power companies increasingly are buying electricity from green power producers. The second part is a web of lower-intensity power generators such as photovoltaic cells and small windmills. Although such generators cannot provide reliable electricity continuously, they can augment central power sources, thereby reducing demand and the consumption of fossil fuels.

Industries that require large amounts of energy have the option of establishing dedicated power plants. Often these can reduce both the cost and the environmental impact of power generation. One example is cogeneration, in which electrical power generation is coupled with heating and cooling. The heat that would be dumped into the environment at a large central power plant is instead used to operate the heating and cooling system for the facility. Another example is the use of biomass, such as the waste from paper mills, wood products manufacturing, orchard pruning, and agricultural byproducts. In the United States, more than 350 biomass power plants generate over 7500 megawatts of electricity—enough power to meet the energy needs of several million homes.

The transportation sector relies nearly totally on fossil fuels. This is because the best energy sources for transportation are fluids that deliver the largest possible amount of energy per unit of volume. Gasoline is nearly ideal in this regard, but because it is nonrenewable and contributes to the buildup of carbon dioxide in the atmosphere, there is growing interest in replacing gasoline with a clean, renewable fluid fuel.

As described in our chapter introduction, hydrogen may be such a fuel. It can be produced by electrolysis of water and consumed in a fuel cell that generates electricity, with the only product being nonpolluting water. As an energy source for transportation, however, hydrogen is far from ideal. One reason is that as a gas, the energy content per unit volume of hydrogen is much smaller than that of liquid fuels like gasoline or methanol. A second reason is that hydrogen must be produced by electrolysis, which requires substantial energy, or by chemical treatment of a fossil fuel, which releases carbon dioxide to the atmosphere. Hydrogen may be the clean transportation fuel of the future, but only if efficient means of production, distribution, and storage can be developed.

Energy Conservation

Our Box on Human Energy Requirements notes that weight depends on fuel intake and expenditure. Similarly, an economy needs to balance its energy demands with its energy supplies. In addition to finding clean, renewable sources of energy, society can reduce the environmental impact of its energy uses by using energy more efficiently.

One good example of improved energy efficiency is cogeneration, described above. Generating electricity at a central power plant and dumping the heat by-products of that generation into the environment, then using that electricity at another site to heat or cool a facility, is hugely wasteful. Cogeneration is much more energy-efficient. Electricity is generated for the other needs of the facility, such as running motors, lighting, and computers. The waste heat from the generation facility is used to operate heaters and refrigerators.

Improved energy efficiency in heating and cooling also can be accomplished through proper design. Throughout history, humans have designed their homes to maximize comfort. To stay cool in hot climates, they have used contact with the ground (cliff dwellings throughout the world) or air (wind towers in Iran), the evaporation of water (pools and fountains at Alhambra in Spain) or the reflection of the sun's rays (whitewash in Southern Europe and North Africa). To stay warm in cold climates, humans have used southern exposures that capture as much solar energy as possible and thick walls that absorb heat during the day and release it slowly at night.

Contemporary architects and builders have rediscovered and improved on these passive heating and cooling techniques. Passive heating relies on capturing incoming solar radiation and storing it in walls of concrete, brick, or rock. Use of high-quality insulation, including double-glazing of windows, reduces the need for additional heating. Passive cooling relies on shading, good ventilation, evaporators, and contact with cool sources such as the ground.

There is a major potential for energy conservation in transportation, by increasing the energy efficiency of automobiles. The recent commercialization of hybrid vehicles, which combine electric and gasoline motors, demonstrates how much more efficient automobile transport can be. Hybrid power systems deliver double the fuel efficiency of conventional engines. Moreover, as fuel cells are perfected, even greater energy efficiencies may be achieved.

No matter how efficient we become at our uses of energy, however, it will remain true that technological modern societies require copious energy supplies. Meeting those needs without overwhelming our planet with pollution and greenhouse gases is an ongoing challenge that many chemists are working to meet.

Section Exercises

6.6.1 What was the most important energy source in each of the following years: 1850, 1900, 1950, and 2000?

6.6.2 In the nineteenth century, coal-fueled steam engines were widespread. Today, petroleum-fueled internal combustion engines have replaced them. Suggest reasons for this replacement.

CHAPTER 6 VISUAL SUMMARY

Important equations

Essential terms

Key concepts

SKILLS TO MASTER

① Identifying forms of energy
② Applying the law of conservation of energy
③ Carrying out heat flow calculations
④ Using the first law of thermodynamics

⑤ Using bond energies to estimate reaction energies
⑥ Carrying out calorimetry calculations
⑦ Relating energy and enthalpy calculations
⑧ Working with standard enthalpies of formation

6.1 TYPES OF ENERGY

Energy transfer

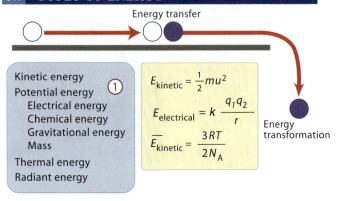

Energy transformation

Kinetic energy
Potential energy
  Electrical energy
  Chemical energy
  Gravitational energy
  Mass
Thermal energy
Radiant energy

①

$$E_{kinetic} = \frac{1}{2}mu^2$$

$$E_{electrical} = k\frac{q_1 q_2}{r}$$

$$\overline{E}_{kinetic} = \frac{3RT}{2N_A}$$

6.2 THERMODYNAMICS

Law of conservation of energy ②

Energy is neither created nor destroyed.

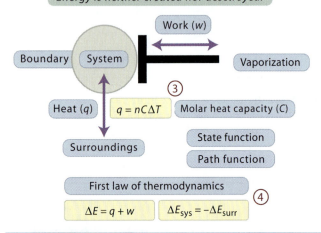

Work (w)

Boundary   System

Vaporization

③

Heat (q) $q = nC\Delta T$ Molar heat capacity (C)

State function
Path function

Surroundings

First law of thermodynamics
④
$$\Delta E = q + w \qquad \Delta E_{sys} = -\Delta E_{surr}$$

6.3 ENERGY CHANGES IN CHEMICAL REACTIONS

Bond energy (BE)

$$\Delta E_{reaction} = \Sigma BE_{bonds\ broken} - \Sigma BE_{bonds\ formed}$$ ⑤

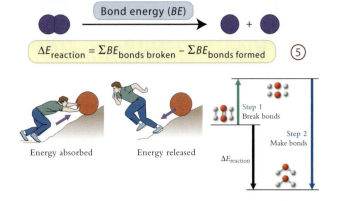

Energy absorbed   Energy released

Step 1
Break bonds

Step 2
Make bonds

$\Delta E_{reaction}$

6.4 MEASURING ENERGY CHANGES: CALORIMETRY

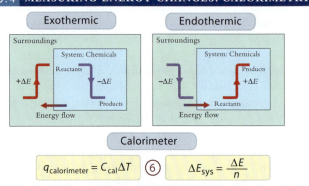

Exothermic   Endothermic

Surroundings   Surroundings

System: Chemicals   System: Chemicals

Reactants   Products
$+\Delta E$ $-\Delta E$ $-\Delta E$ $+\Delta E$
           Products   Reactants

Energy flow   Energy flow

Calorimeter

$$q_{calorimeter} = C_{cal}\Delta T$$ ⑥ $$\Delta E_{sys} = \frac{\Delta E}{n}$$

6.5 ENTHALPY

$$w_{sys} = -P\Delta V_{sys} \qquad \Delta H = q_p$$
⑦
$$H = E + PV$$

Enthalpy (H) = Heat flow at constant pressure

Hess's Law

Overall ΔH = Sum of ΔH for individual steps

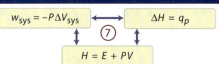

Net reaction

Reactants   Products

Decomposition   Formation
reactions   reactions

Elements

Standard enthalpy of formation (ΔH_f^o)   Formation reactions

Formation reaction = 1 mol of a substance produced from elements

$$\frac{1}{2}N_2(g) + \frac{1}{2}O_2(g) \longrightarrow NO(g)$$ ⑧

$$\Delta H_{reaction}^o = \Sigma coeff_{products}\Delta H_{f\ products}^o - \Sigma coeff_{reactants}\Delta H_{f\ reactants}^o$$

$$\Delta H_{reaction} \cong \Delta E_{reaction} + RT\Delta n_{gases}$$

6.6 ENERGY SOURCES

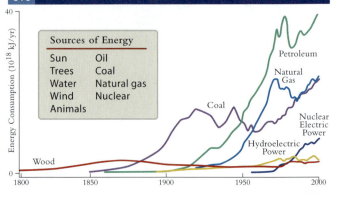

Sources of Energy

Sun   Oil
Trees   Coal
Water   Natural gas
Wind   Nuclear
Animals

Energy Consumption (10^{18} kJ/yr)

Petroleum
Natural Gas
Coal
Nuclear Electric Power
Hydroelectric Power
Wood

1800   1850   1900   1950   2000

Learning Exercises

6.1 List as many different types of energy as you can, and provide a specific example illustrating each energy type.

6.2 Write a paragraph that explains the relationships among energy, heat, work, and temperature.

6.3 Prepare a table listing all the bond energies involving halogen atoms (F, Cl, Br, I), organized so it matches the periodic table. What regularities can you find in this listing, and what exceptions are there?

6.4 Write a paragraph describing what happens to the energy released during a chemical reaction that occurs in a constant-pressure calorimeter.

6.5 Describe in your own words what enthalpy is and why it is preferred to energy in describing the thermochemistry of many chemical processes.

6.6 Make a list of energy sources that you personally use. For each item on your list, describe a way that you might be able to reduce your energy consumption.

6.7 Update your list of memory bank equations. Write a sentence that describes the restrictions on each equation. (For example, $H = E + PV$ is a definition that has no restrictions.)

6.8 Prepare a list of the terms in Chapter 6 that are new to you. Write a one-sentence definition for each, using your own words. If you need help, consult the glossary.

 Problems   **ilw = interactive learning ware problem. Visit the website at www.wiley.com/college/olmsted**

Types of Energy

6.1 Explain how each of the following observations is consistent with conservation of energy: (a) An apple gains kinetic energy as it falls from a tree. (b) When that apple hits the ground, it loses its kinetic energy.

6.2 Explain how each of the following observations is consistent with conservation of energy: (a) Water running into a pond loses kinetic energy. (b) A rock lying in the sun increases in temperature.

6.3 Calculate the kinetic energy of an electron moving at a speed of 4.55×10^5 m/s (see inside back cover for electron properties).

6.4 Calculate the kinetic energy of a proton moving at a speed of 2.32×10^3 m/s (see inside back cover for proton properties).

6.5 In KCl solid, the distance between each K^+ cation and the nearest Cl^- anion is 320 pm. Calculate the attractive electrical energy for this pair of ions.

6.6 In CaO solid, the distance between each Ca^{2+} cation and the nearest O^{2-} anion is 240 pm. Calculate the attractive electrical energy for this pair of ions.

6.7 In each of the following processes, energy is transformed from one type to another. Identify what type of energy is consumed and what type of energy is produced. (a) Sunlight heats the roof of a house. (b) A baseball pitcher throws a fastball. (c) Wax burns in a candle flame.

6.8 In each of the following processes, energy is transformed from one type to another. Identify what type of energy is consumed and what type is produced. (a) Methane burns on the element of a stove. (b) An elevator carries passengers from the ground floor to the fourth floor. (c) A firefly produces light on a summer evening.

6.9 Compute the speed of a neutron that has a kinetic energy of 3.75×10^{-23} J (see inside back cover for neutron properties).

6.10 Compute the speed of a proton that has a kinetic energy of 2.75×10^{-25} J (see inside back cover for proton properties).

Thermodynamics

6.11 An electric heater supplies 25.0 joules of energy to each of the following samples. Compute the final temperature in each case: (a) 10.0-g block of Al originally at 15.0 °C; (b) 25.0-g block of Al originally at 29.5 °C; (c) 25.0-g block of Ag originally at 29.5 °C; and (d) 25.0-g sample of H_2O originally at 22.0 °C.

6.12 Each of the following is placed in an ice bath until it has lost 65.0 J of energy. Compute the final temperature in each case: (a) 35.0-g block of Al originally at 65.0 °C; (b) 50.0-g block of Al originally at 65.0 °C; (c) 50.0-g block of Ag originally at 65.0 °C; and (d) 50.0-g sample of H_2O originally at 32.5 °C.

6.13 An iron kettle weighing 1.35 kg contains 2.75 kg of water at 23.0 °C. The kettle and water are heated to 95.0 °C. How many joules of energy are absorbed by the water and by the kettle?

6.14 A piece of silver whose mass is 15.0 g is immersed in 25.0 g of water. This system is heated electrically from 24.0 to 37.6 °C. How many joules of energy are absorbed by the silver and how many by the water?

ilw 6.15 A silver coin weighing 27.4 g is heated to 100.0 °C in boiling water. It is then dropped into 37.5 g of water initially at 20.5 °C. Find the final temperature of coin + water.

6.16 A stainless steel spoon weighs 24.7 g and is at a temperature of 18.5 °C. It is immersed in 85.0 mL of hot coffee ($T = 84.0$ °C) in a Thermos flask. What is the final temperature of spoon + coffee? (Assume that the heat capacity of the spoon is the same as that of pure iron and that the heat capacity of the coffee is the same as that of pure water.)

6.17 A pot containing 475 mL of water at 21.5 °C is heated on a stove until its temperature is 87.6 °C. What is q for the water?

6.18 A dish containing 145 g of water at 54.0 °C is put in a refrigerator to cool. It is removed when its temperature is 5.50 °C. What is q for the water?

6.19 Refer to the data in the Chemistry and Life Box and calculate how far a person weighing 55 kg must walk at a rate of 6.0 km/hr to consume the additional energy contained in 250 g of ground beef relative to 250 g of broiled chicken.

6.20 Refer to the data in the Chemistry and Life Box and calculate how far a person weighing 85 kg must run at 16 km/hr to consume the energy contained in 1 lb (0.455 kg) of sugar.

Energy Changes in Chemical Reactions

6.21 When 1.350 g of benzoic acid ($C_7H_6O_2$) burns completely in excess O_2 gas at constant volume and 298 K, it releases 35.61 kJ of energy. (a) What is the balanced chemical equation for this reaction? (b) What is the molar energy of combustion of benzoic acid? (c) How much energy is released per mole of O_2 consumed?

6.22 Acetylene (C_2H_2) is used in welding torches because it has a high heat of combustion. When 1.00 g of acetylene burns completely in excess O_2 gas at constant volume, it releases 48.2 kJ of energy. (a) What is the balanced chemical equation for this reaction? (b) What is the molar energy of combustion of acetylene? (c) How much energy is released per mole of O_2 consumed?

6.23 Consult Table 6-2 to find the bond energies for bonds between hydrogen and Group 16 elements. Is there a trend? If so, use the trend to predict the bond energies of H—Se and H—Po.

6.24 Consult Table 6-2 to find the single-bond energies for bonds between identical row 3 elements (for example, P—P). Is there a trend? If so, use the trend to predict the bond energy of an Al—Al bond.

ilw 6.25 Use average bond energies to estimate the energy change for the reaction of N_2 and O_2 to give N_2O_4.

$$N \equiv N + 2\, O = O \longrightarrow \begin{matrix} O & & O \\ \diagdown & & \diagup \\ N - N \\ \diagup & & \diagdown \\ O & & O \end{matrix}$$

6.26 Acrylonitrile is an important starting material for the manufacture of plastics and synthetic rubber. The compound is made from propene in a gas-phase reaction at elevated temperature:

$$2\ C_3H_6 + 2\ NH_3 + 3\ O_2 \longrightarrow 2\ H_2C{=}CH{-}C{\equiv}N + 6\ H_2O$$

Use average bond energies to estimate the energy change for the acrylonitrile synthesis.

6.27 Draw a diagram like the one in Figure 6-13 to illustrate the different paths for the reaction, $N_2O_4 \rightarrow 2\ NO_2$, which absorbs energy.

6.28 Draw a diagram like the one in Figure 6-13 to illustrate the different paths for the reaction in Problem 6.26.

6.29 Estimate the reaction energy for each of the following reactions (all bonds are single bonds except as noted):

$$2\ H_2 + O_2 \text{ (double bond)} \longrightarrow 2\ H_2O$$
$$3\ H_2 + N_2 \text{ (triple bond)} \longrightarrow 2\ NH_3$$
$$3\ H_2 + CO \text{ (triple bond)} \longrightarrow CH_4 + H_2O$$

6.30 Estimate the energy change for the following reaction when X is (a) F; (b) Cl; (c) Br; and (d) I (all bonds are single bonds):

$$2\ HX(g) \longrightarrow H_2(g) + X_2(g)$$

Measuring Energy Changes: Calorimetry

6.31 When 5.34 g of a salt dissolves in 155 mL of water (density = 1.00 g/mL) in a coffee-cup calorimeter, the temperature rises from 21.6 °C to 23.8 °C. Determine q for the solution process, assuming that $C_{cal} \cong C_{water}$.

6.32 When 10.00 mL of a solution of a strong acid is mixed with 100.0 mL of a solution of a weak base in a coffee-cup calorimeter, the temperature falls from 24.6 °C to 22.7 °C. Determine q for the acid–base reaction, assuming that the liquids have densities of 1.00 g/mL and the same heat capacity as pure water.

6.33 Constant-volume calorimeters are sometimes calibrated by running a combustion reaction of known ΔE and measuring the change in temperature. For example, the combustion energy of glucose is 15.57 kJ/g. When a 1.7500-g sample of glucose burns in a constant-volume calorimeter, the calorimeter temperature increases from 21.45 °C to 23.34 °C. Find the total heat capacity of the calorimeter.

6.34 Constant-pressure calorimeters can be calibrated by electrical heating. When a calorimeter containing 125 mL of water is supplied with 1150. J of electrical energy, its temperature rises from 23.45 °C to 25.25 °C. What is the total heat capacity of the calorimeter, and what percentage of this is due to the water?

ilw **6.35** A 1.35-g sample of caffeine ($C_8H_{10}N_4O_2$) is burned in a constant-volume calorimeter that has a heat capacity of 7.85 kJ/°C. The

temperature increases from 24.65 °C to 29.04 °C. Determine the amount of heat released and the molar energy of combustion of caffeine.

6.36 An electric heater adds 19.75 kJ of heat to a constant-volume calorimeter. The temperature of the calorimeter increases by 4.22 °C. When 1.75 g of methanol is burned in the same calorimeter, the temperature increases by 8.47 °C. Calculate the molar energy of combustion of methanol.

Enthalpy

6.37 How much work is done in blowing up a balloon from zero volume to a volume of 2.5 L, assuming that $P = 1.00$ atm and no work is required to stretch the rubber? (In reality, the work that goes into stretching the rubber is substantial.)

6.38 A typical hot-air balloon has a volume of 19.5 m^3 when inflated. How much work must be done to inflate such a balloon when atmospheric pressure is 755 torr?

6.39 Determine the standard enthalpy change for each of the following reactions:

(a) $C_2H_4(g) + 3\ O_2(g) \longrightarrow 2\ CO_2(g) + 2\ H_2O(l)$
(b) $2\ NH_3(g) \longrightarrow N_2(g) + 3\ H_2(g)$
(c) $5\ PbO_2(s) + 4\ P\ (s,\ white) \longrightarrow P_4O_{10}(s) + 5\ Pb(s)$
(d) $SiCl_4(l) + 2\ H_2O(l) \rightarrow SiO_2(s) + 4\ HCl(aq)$

6.40 Determine the standard enthalpy change for each of the following reactions:

(a) $2\ NH_3(g) + 3\ O_2(g) + 2\ CH_4(g) \longrightarrow 2\ HCN(g) + 6\ H_2O(g)$
(b) $2\ Al(s) + 3\ Cl_2(g) \longrightarrow 2\ AlCl_3(s)$
(c) $3\ NO_2(g) + H_2O(l) \longrightarrow 2\ HNO_3(g) + NO(g)$
(d) $2\ C_2H_2(g) + 5\ O_2(g) \longrightarrow 4\ CO_2(g) + 2\ H_2O(l)$

6.41 Find $\Delta E^{\circ}_{reaction}$ for each of the reactions in Problem 6.39.

6.42 Find $\Delta E^{\circ}_{reaction}$ for each of the reactions in Problem 6.40.

6.43 Write a balanced equation for the formation reaction of each of the following substances: (a) $K_3PO_4(s)$; (b) acetic acid, $CH_3CO_2H(l)$; (c) trimethylamine, $(CH_3)_3N(g)$; and (d) bauxite, $Al_2O_3(s)$.

6.44 Write the balanced equation for the formation reaction of each of the following substances: (a) butanol, $C_4H_9OH(l)$; (b) sodium carbonate, $Na_2CO_3(s)$; (c) ozone, $O_3(g)$; and (d) rust, $Fe_3O_4(s)$.

ilw **6.45** Using standard heats of formation, determine ΔH for the following reactions:

(a) $4\ NH_3(g) + 5\ O_2(g) \longrightarrow 4\ NO(g) + 6\ H_2O(l)$
(b) $4\ NH_3(g) + 3\ O_2(g) \longrightarrow 2\ N_2(g) + 6\ H_2O(l)$

6.46 Using standard heats of formation, determine ΔH for the following reactions:

(a) $Fe_2O_3(s) + 3\ H_2O(l) \longrightarrow 2\ Fe(OH)_3(s)$
(b) $B_2O_3(s) + 3\ H_2O(l) \longrightarrow 2\ H_3BO_3(s)$

Energy Sources

6.47 The total population of the United States, according to the 2000 Census, was 281,421,906. From Figure 6-21, calculate the annual total energy usage of the United States.

6.48 The total population of the United States in 1860 was 31,443,321. Using data in Figure 6-21 and assuming that the energy usage at that time was typical of the Industrial Revolution, calculate the annual total energy usage of the United States in 1860.

6.49 Using information in Figure 6-22, calculate the total world energy consumption in 1975. Then construct a pie chart showing the percentages of that total contributed by each major source.

6.50 Repeat the calculation described in Problem 6.49 for the years 1923 and 2000.

Additional Paired Problems

6.51 Does more heat have to be removed from an automobile engine when it burns one gram of gasoline while idling in a traffic jam or when it burns one gram of gasoline while accelerating? Explain in terms of ΔE, q, and w.

6.52 One way to vaporize a liquid is to inject a droplet into a high vacuum. If this is done, is the heat absorbed by the droplet equal to ΔE or ΔH for the phase change? Explain in terms of q and w.

6.53 Estimate the difference between $\Delta H_{reaction}$ and $\Delta E_{reaction}$ for the reaction

$$C\ (graphite) + H_2O(l) \longrightarrow CO(g) + H_2(g),\ \text{at}\ T = 298\ K$$

6.54 Estimate the difference between $\Delta H_{reaction}$ and $\Delta E_{reaction}$ for the combustion of liquid butanol (C_4H_9OH) in excess $O_2(g)$, at $T = 298$ K.

6.55 The amount of heat produced in an "ice calorimeter" is determined from the quantity of ice that melts, knowing that it takes 6.01 kJ of heat to melt exactly 1 mol of ice. Suppose that a 12.7-g copper block at 200.0 °C is dropped into an ice calorimeter. How many grams of ice will melt?

6.56 A gold coin whose mass is 7.65 g is heated to 100.0 °C in a boiling water bath and then quickly dropped into an ice calorimeter. What mass of ice melts? (See Problem 6.55 for useful information.)

6.57 The human body "burns" glucose ($C_6H_{12}O_6$) for energy. Burning 1.00 g of glucose produces 15.7 kJ of heat. (a) Write the balanced equation for the combustion (burning) of glucose. (b) Determine the molar heat of combustion of glucose. (c) Using appropriate thermodynamic data, determine the heat of formation of glucose.

6.58 Solid urea, $(NH_2)_2CO$, burns to give CO_2, N_2, and liquid H_2O. Its heat of combustion is -632.2 kJ/mol. (a) Write the balanced combustion equation. (b) Calculate the heat generated per mole of H_2O formed. (c) Using this heat of combustion and the appropriate thermodynamic data, determine the heat of formation of urea.

6.59 Use average bond energies (Table 6-2) to compare the stabilities of ethanol, C_2H_5OH, and dimethyl ether, $(CH_3)_2O$, which have the same empirical formula, C_2H_6O (all the bonds are single bonds).

6.60 Use average bond energies (Table 6-2) to compare the stabilities of allyl alcohol, $H_2C=CHCH_2OH$, and acetone, $(CH_3)_2C=O$, which have the same empirical formula, C_3H_6O (each compound has one double bond, as indicated in the formulas).

6.61 What energy transformations take place when an airplane accelerates down a runway and takes off into the air?

6.62 What energy transformations take place when a moving automobile brakes and skids to a stop?

6.63 It takes 100.0 J of heat to raise the temperature of 52.5 g of lead from 28.0 °C to 47.6 °C. What is the molar heat capacity of lead?

6.64 It takes 87.7 J of heat to raise the temperature of 27.0 g of tin from 0.0 °C to 15.0 °C. What is the molar heat capacity of lead?

6.65 Use average bond energies (Table 6-2) to compare the combustion energies of ethane, ethylene, and acetylene. Calculate which of these hydrocarbons releases the most energy per gram.

6.66 Isooctane, C_8H_{18}, is the basis of the octane rating system for gasoline because it burns smoothly, with minimal engine "knocking." Use average bond energies to estimate the energy released during the combustion of 1 mol of pure isooctane.

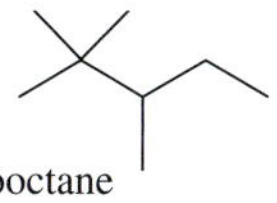

Isooctane
C_8H_{18}

6.67 What is the kinetic energy (in J) of a 2250-pound automobile traveling at 57.5 miles per hour?

6.68 What is the speed (in miles per hour) of a 60-kg runner whose kinetic energy is 345 J?

6.69 Use standard enthalpies of formation to determine $\Delta H°_{reaction}$ for the following reactions:
(a) $2 SO_2 (g) + O_2 (g) \longrightarrow 2 SO_3 (g)$
(b) $2 NO_2 (g) \longrightarrow N_2O_4 (g)$
(c) $Fe_2O_3 (s) + 2 Al (s) \longrightarrow Al_2O_3 (s) + 2 Fe (s)$

6.70 Using standard enthalpies of formation, calculate $\Delta H°_{reaction}$ for the following reactions:
(a) $2 NH_3 (g) + 3 O_2 (g) + 2 CH_4 (g) \rightarrow 2 HCN (g) + 6 H_2O (g)$
(b) $2 C_2H_2 (g) + 5 O_2 (g) \longrightarrow 4 CO_2 (g) + 2 H_2O (g)$
(c) $C_2H_4 (g) + O_3 (g) \longrightarrow CH_3CHO (g) + O_2 (g)$

6.71 Use tabulated bond energies in Table 6-2 to estimate the energy change when HCl adds to ethylene ($H_2C=CH_2$) to produce CH_3CH_2Cl and when Cl_2 adds to ethylene to produce $ClCH_2CH_2Cl$.

6.72 Phosgene ($Cl_2C=O$) is a highly toxic gas that was used for chemical warfare during World War I. Use the bond energies in Table 6-2 to estimate the energy change that occurs when carbon monoxide and chlorine combine to make phosgene.

$$C\equiv O (g) + Cl_2 (g) \longrightarrow Cl_2C=O (g)$$

6.73 "Strike anywhere" matches contain P_4S_3, a compound that ignites when heated by friction. It reacts vigorously with oxygen, as follows:

$$P_4S_3 (s) + 8 O_2 (g) \longrightarrow P_4O_{10} (s) + 3 SO_2 (g) \quad \Delta H°_{reaction} = -3677 \text{ kJ}$$

Use data from Appendix D to determine $\Delta H°_f$ for $P_4S_3 (s)$.

6.74 When a corpse decomposes, much of the phosphorus in the body is converted to phosphine, PH_3, a colorless gas with the odor of rotting fish. Phosphine is a highly reactive molecule that ignites spontaneously in air. In the graveyard, phosphine that escapes from the ground ignites in air, giving small flashes of flame. These flashes are sometimes attributed to supernatural causes, such as a will-o'-the-wisp. Determine $\Delta H°$ for the combustion of phosphine.

$$PH_3 (g) + O_2 (g) \longrightarrow P_4O_{10} (s) + H_2O (g) \text{ (unbalanced)}$$

6.75 A 44.0-g sample of an unknown metal at 100.0 °C is placed in a constant-pressure calorimeter containing 80.0 g of water at 24.8 °C. Assume that the heat capacity of the calorimeter equals the heat capacity of the water it contains. The final temperature is 28.4 °C. Calculate the heat capacity of the metal and use the result to identify the metal: Al = 0.903 J/g K; Cr = 0.616 J/g K; Co = 0.421 J/g K; or Cu = 0.385 J/g K.

6.76 A coin dealer, offered a rare silver coin, suspected that it might be a counterfeit nickel copy. The dealer heated the coin, which weighed 15.5 g, to 100.0 °C in boiling water and then dropped the hot coin into 21.5 g of water at $T = 15.5$ °C in a coffee-cup calorimeter. The temperature of the water rose to 21.5 °C. Was the coin made of silver or nickel?

6.77 In some liquid-fuel rockets, such as the lunar lander module of the Apollo moon missions, the fuels are liquid hydrazine (N_2H_4) and dinitrogen tetroxide gas (N_2O_4). The two chemicals ignite on contact to release very large amounts of energy:

$$2 N_2H_4 (l) + N_2O_4 (g) \longrightarrow 3 N_2 (g) + 4 H_2O (g)$$

(a) Calculate the enthalpy change that takes place when one mole of hydrazine burns in a lunar lander. (b) Calculate the enthalpy change if O_2 was used in the lander instead of N_2O_4.

6.78 For spacecraft fuels, the energy content *per gram* of fuel should be as large as possible. Which of the following has the largest energy content per gram and which has the smallest?
(a) dimethylhydrazine, $(CH_3)_2NNH_2$, $\Delta H°_{combustion} = -1694$ kJ/mol;
(b) methanol, CH_3OH, $\Delta H°_{combustion} = -726$ kJ/mol; or
(c) octane, C_8H_{18}, $\Delta H°_{combustion} = -5590$ kJ/mol

6.79 Refer to Figure 6-22 to answer the following questions: (a) Before 1850, how many main sources of energy were there, and what were they? (b) Before 1980, what was the highest value for natural gas production, and in what year? (c) What percentage of energy was provided by fossil fuels (petroleum, natural gas, and coal) in 1950?

6.80 Refer to Figure 6-22 to answer the following questions: (a) Before 1985, what was the highest value for petroleum production, and in what year? (b) What percentage of energy was provided by coal in 1945? (c) When was the first year in which coal provided less energy than either natural gas or petroleum?

6.81 For the constant-temperature process that follows, give the sign ($+$, $-$, or 0) for each of the specified thermodynamic functions. In each case give a brief account of your reasoning: (a) ΔH_{sys}; (b) ΔE_{surr}; and (c) ΔE_{univ}.

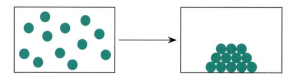

6.82 For the constant-temperature process that follows, give the sign ($+$, $-$, or 0) for each of the specified thermodynamic functions. In each case give a brief account of your reasoning: (a) w_{sys}; (b) q_{sys}; and (c) ΔE_{surr}.

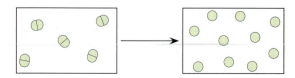

More Challenging Problems

6.83 A piece of rhodium metal whose mass is 4.35 g is heated to 100.0 °C and then dropped into an ice calorimeter. As the Rh metal cools to 0.0 °C, 0.316 g of ice melts (it takes 6.01 kJ of heat to melt exactly 1 mol of ice). What is the molar heat capacity of rhodium?

6.84 Gases are sold and shipped in metal tanks under high pressure. A typical tank of compressed air has a volume of 30.0 L and is pressurized to 15.0 bar at $T = 298$ K. What work had to be done in filling this tank if the compressor operates at 15.0 bar? (*Hint*: What volume did the air occupy before it was compressed?)

6.85 A 70-kg person uses 220 kJ of energy to walk 1.0 km. This energy comes from "burning" glucose (see Problem 6.57), but only about 30.% of the heat of combustion of glucose can be used for propulsion. The rest is used for other bodily functions or is "wasted" as heat. Assuming that a "sugar-coated" breakfast cereal contains 35% sugar (which can be considered as glucose) and no other energy source, calculate how many grams of cereal provide enough energy to walk 1.0 km.

6.86 A home swimming pool contains 155 m³ of water. At the beginning of swimming season, the water must be heated from 20 °C to 30 °C. (a) How much heat energy must be supplied? (b) If a natural gas heater supplies this energy with an 80% heat transfer efficiency, how many grams of methane must be burned? (The heat of combustion of methane is −803 kJ/mol.)

6.87 One way to cool a hot beverage is with a cold spoon. A silver spoon weighing 99 g is placed in a Styrofoam cup containing 205 mL of hot coffee at 83.2 °C. Find the final temperature of the coffee, if the initial temperature of the spoon is 22.0 °C and assuming that coffee has the same heat capacity as water. Would an aluminum spoon of the same mass cool the coffee more or less effectively?

6.88 A room in a home measures 3.0 m by 5.0 m by 4.0 m. Assuming no heat or material losses, how many grams of natural gas (methane, CH_4) must be burned to heat the air in this room from 15 °C to 25 °C? Assume that air is 78% N_2 and 22% O_2, and use data from Table 6-1 and Appendix D.

6.89 Only one of the following expressions describes the heat of a chemical reaction under all possible conditions. Which is it? For each of the others, give an example for which the expression gives the wrong value for the heat. (a) ΔE; (b) ΔH; (c) q_v; (d) q_p; and (e) $\Delta E - w$.

6.90 For the following constant-temperature process, give the sign (+, −, or 0) for each of the specified thermodynamic functions. In each case, give a brief account of your reasoning: (a) w_{sys}; (b) ΔE_{surr}; and (c) q_{sys}.

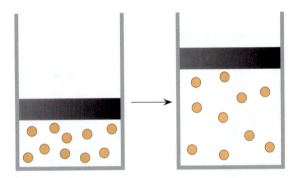

6.91 The power ratings of microwave ovens are listed in watts (1 watt = 1 J/s). How many seconds will it take a microwave oven rated at 750 watts to warm 575 mL of soup from 25 °C to 85 °C? Assume that the heat capacity and the density of the soup are the same as that of pure water.

6.92 In metric terms, a typical automobile averages 6.0 km/L of gasoline burned. Gasoline has a heat of combustion of 48 kJ/g and a density of 0.68 g/mL. How much energy is consumed in driving an automobile 1.0 km?

6.93 A 9.50-g copper block, initially at 200.0 °C, is dropped into a Thermos flask containing 200 mL of water initially at 5.00 °C. What is the final temperature of the Thermos flask contents?

6.94 Use average bond energies (see Table 6-2) to estimate the net energy change per mole of silicon for the conversion of a silicon chain into an Si—O—Si chain. Repeat this calculation to estimate the net energy change per mole of carbon for the conversion of a carbon chain into a C—O—C chain.

6.95 The following figure represents a piston and cylinder containing a collection of gas molecules. The piston can move in either direction. Assume that the gas molecules are the system. (a) Redraw the figure to show what happens when some work is done on the system. (b) Redraw the figure to show what happens when some heat is added to the system.

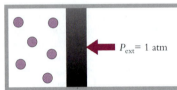

6.96 A sample of carbon monoxide gas whose mass is 7.75 g is heated from 25 °C to 175 °C at constant pressure of 5.00 atm. Calculate q, w, and ΔE for the process.

6.97 Determine the values of ΔH_1° and ΔH_2° and the identity and coefficient of the missing species in the diagram below. (*Hint*: Use Appendix D).

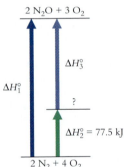

6.98 The most common form of elemental sulfur is S_8, in which eight sulfur atoms link in a ring of single bonds. At high temperature, in the gas phase, S_8 can break apart to give S_2, the sulfur analog of molecular oxygen:

$$S_8\,(g) \longrightarrow 4\,S_2\,(g) \qquad \Delta H = +239 \text{ kJ} \qquad \text{at } T = 800.\text{ K}$$

Using the appropriate data in Table 6-2, calculate the S=S double-bond energy in $S_2\,(g)$.

6.99 The reaction that forms calcium carbonate deposits in caves is as follows:

$$Ca^{2+}(aq) + 2\,HCO_3^{-}(aq) \longrightarrow CaCO_3\,(s) + CO_2\,(g) + H_2O\,(l)$$
$$\Delta H^\circ = +39 \text{ kJ}$$

(a) Assuming that the inside of the cave is at $T = 15$ °C, $P = 1.00$ atm, determine ΔE for this reaction.
(b) Using data from Appendix D, determine the heat of formation of $HCO_3^{-}(aq)$.

6.100 The five stable oxides of nitrogen are NO, NO_2, N_2O, N_2O_4, and N_2O_5. Balance each of the following oxidation reactions, and then use standard formation enthalpies to calculate the heats of reaction per mole of *atomic* nitrogen for each reaction:

(a) $N_2 + O_2 \rightarrow NO$
(b) $N_2O + O_2 \rightarrow NO$
(c) $NO + O_2 \rightarrow NO_2$
(d) $NO_2 + O_2 \rightarrow N_2O_5$

Group Study Problems

6.101 The heat required to sustain animals that hibernate comes from the biochemical combustion of fatty acids, one of which is arachidonic acid. For this acid, (a) determine its structural formula; (b) write its balanced combustion reaction; (c) use average bond energies to estimate the energy released in the combustion reaction; and (d) calculate the mass of arachidonic acid needed to warm a 500-kg bear from 5 to 25 °C. (Assume that the average heat capacity of bear flesh is 4.18 J/g K.)

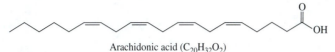

Arachidonic acid ($C_{20}H_{32}O_2$)

6.102 Photosynthesis on the Earth's land surfaces converts approximately 5.5×10^{16} g/yr of CO_2 into energy-rich compounds. Assume that all of the carbon is incorporated into glucose, and use information found in this chapter to determine the following: (a) How much energy is stored per day? (b) The land mass of the Earth covers 29% of its surface. What fraction of the solar energy that reaches the land surface is stored via photosynthesis? (c) If the biosphere is at steady state, consuming all the energy that it stores, at what rate is the biosphere using energy? Express your result in MW. (1 W = 1 J/s) (d) If a typical nuclear power plant is rated at 10^3 MW, how many nuclear power plants would it take to generate energy at the same rate as the biosphere does?

6.103 According to Table 6-1, molar heat capacities of monatomic gases (He, Ar) are significantly smaller than those of diatomic gases (N_2, O_2, H_2). Explain in molecular terms why more heat must be supplied to raise the temperature of 1 mol of diatomic gas by 1 K than to raise the temperature of 1 mol of monatomic gas by 1 K.

6.104 Suppose 100.0 mL of 1.00 M HCl and 100.0 mL of 1.00 M NaOH, both initially at 25.0 °C, are mixed in a Thermos flask. When the reaction is complete, the temperature is 31.8 °C. Assuming that the solutions have the same heat capacity as pure water, compute the heat released. Use this value to evaluate the molar heat of the neutralization reaction:

$$H_3O^+(aq) + OH^-(aq) \longrightarrow 2\,H_2O\,(l)$$

6.105 Ethanol, CH_3CH_2OH, is used as a gasoline additive because it boosts octane ratings. Gasoline that contains ethanol is known as gasohol. Calculate the amount of energy released by burning one gallon of ethanol. The density of ethanol is 0.787 g/mL. Use the data in Appendix D and average bond energies from Table 6-2.

6.106 Using the information in Box 6-1, estimate the daily energy balance for each member of your group. For any member of the group whose energy expenditure differs from his or her energy intake, propose appropriate changes that would bring the energy flow into balance.

Answers to Section Exercises

6.1.1 2.0×10^{-28} J
6.1.2 (a) kinetic energy; (b) thermal energy; (c) radiant energy; and (d) chemical energy
6.1.3 (a) Radiant energy is transformed into thermal energy; (b) gravitational (potential) energy is transformed into kinetic energy; and (c) kinetic energy is transformed into thermal energy.
6.2.1 $q_{sys} = 4.2 \times 10^3$ J
6.2.2 0.60 kg of ground beef or 3.8 kg of fruit
6.2.3 (a) and (d) are state functions
6.3.1 -630 kJ
6.3.2 See figure in right column.
6.3.3 CH_4, four C—H bonds; CH_3Cl, three C—H bonds, one C—Cl bond; others have one bond each. The first reaction absorbs 80 kJ, and the second reaction releases 105 kJ. The second reaction is used in industry because it does not require an energy input.
6.4.1 $C = 523$ J/K
6.4.2 $q_p = +4.6 \times 10^2$ J
6.4.3 $\Delta E_{molar} = -297$ kJ/mol
6.5.1 $w_{sys} = -4.31 \times 10^3$ J
6.5.2 c, d, and f
6.5.3 $\Delta H_{reaction} = +206.1$ kJ and $\Delta E_{reaction} = +201.1$ kJ
6.6.1 1850, wood; 1900, coal; 1950 and 2000, petroleum
6.6.2 Petroleum is easier to transport and store; internal combustion engines are more powerful and compact than steam engines.

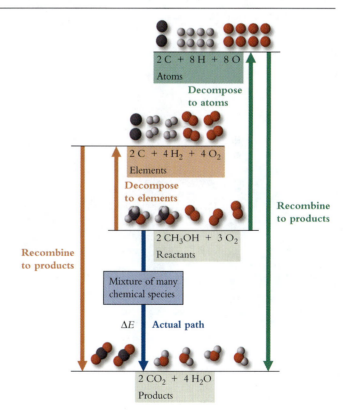

Answers to Extra Practice Exercises

6.1 $\Delta T = -1.09$ K
6.2 $q = -21.2$ kJ
6.3 $\Delta E = -0.140$ kJ
6.4 $\Delta E_1 = -150$ kJ; $\Delta E_2 = +45$ kJ
6.5 Final temperature would be 32.8 °C.

6.6 $q = -4.2 \times 10^2$ J
6.7 $\Delta E_{molar} = -3.9 \times 10^2$ kJ/mol of graphite
6.8 $\Delta n = 0$, so $\Delta H - \Delta E = 0$
6.9 $\Delta E = -349$ kJ/mol NH_3
6.10 ΔH_f° (H_3PO_4) $= -1279$ kJ/mol

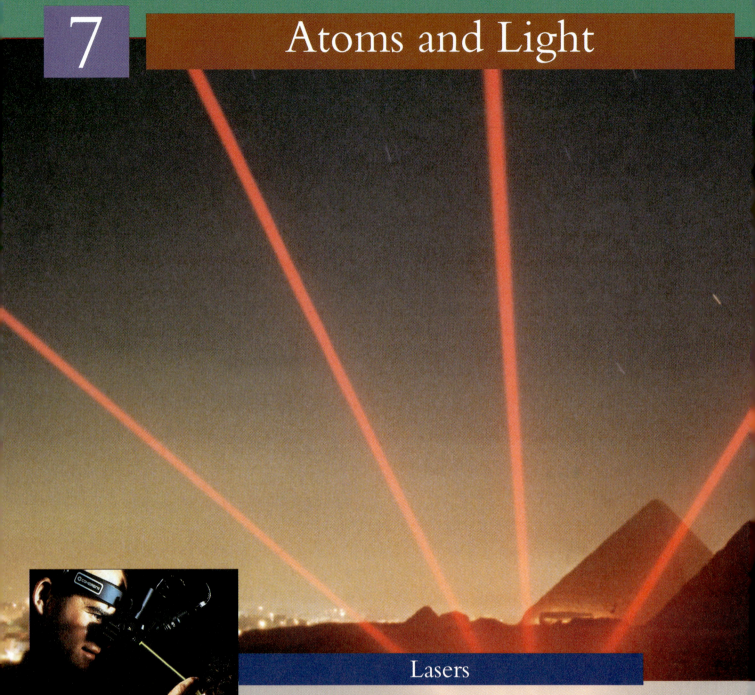

Atoms and Light

Lasers

In the right combination, atoms and light combine to create one of the most remarkable tools of modern technology, the laser. Lasers have many commercial applications, ranging from laser light shows to CD scanners, eye surgery, and fiber optics communications. Lasers have also become versatile tools for advanced scientific research. To give just one example, lasers can be finely tuned to deposit vapor-phase atoms in regular patterns. Our inset shows chromium atoms deposited in a dot array using laser focusing. Each peak is 13 nm high.

The word *laser* stands for *light amplification by stimulated emission of radiation*. In a laser, atoms and light interact to generate a beam of light that has precisely defined properties. In simple terms, when a sample of atoms (or molecules) is "pumped" uphill in energy, light of a specific wavelength (color) may interact with the atoms or molecules to produce more light of that wavelength. Mirrors reflect the light back into the sample, so the process can repeat itself many times, amplifying the intensity of the light immensely.

Light from a laser differs substantially from conventional light. Laser light is monochromatic — has a single color — whereas conventional light sources typically produce light of many colors. A laser is highly directional, whereas conventional sources send light in all directions. Consequently, laser light is more highly organized than light from normal sources. Many

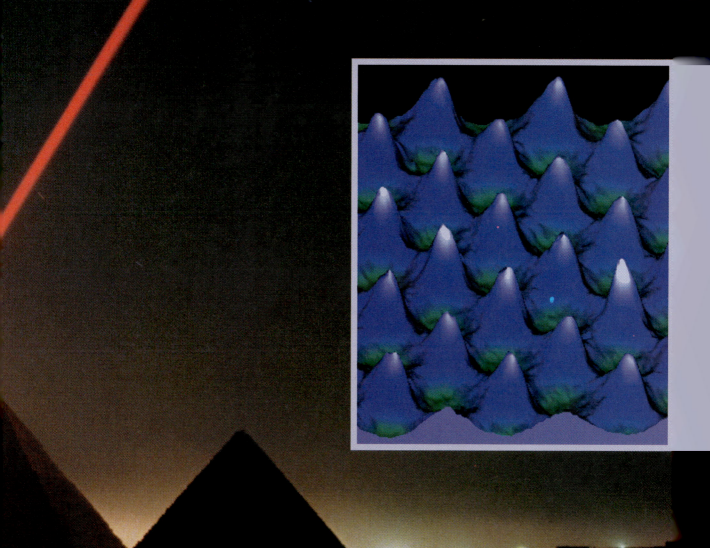

of the applications of lasers take advantage of this high degree of organization.

Bar code scanners, for example, exploit the directionality of a laser beam. The laser light reflects more strongly from white stripes than from black. Because of the directionality of the laser beam, the reflected laser beam faithfully mirrors the pattern of the bar code. A sensor reads these variations and converts the light pattern into an electronic representation of the bar code. The sensor transmits the electronic representation to a computer.

The combination of high intensity and directionality is the basis for laser surgery. Essentially, a laser acts as a "light knife." In eye surgery, a laser beam can be focused to a tiny spot at the back of retina. This allows the surgeon to carry out delic operations such as repairing a detached retina w out physically invading the eyeball.

CHAPTER CONTENTS

Learning Resource

 KEY CONCEPTS

 CRITICAL THINKING

 SKILLS TO MASTER

 ADDITIONAL HELP
www.wiley.com/college/or
● TUTORIALS
● ANIMATIONS

Regardless of the application, every laser is based on the interactions between light and atoms or molecules. We describe these interactions in this chapter. First, we summarize the properties of atoms and light. Next, we discuss the energy changes that accompany the interactions between electrons and light. Finally, we describe the properties of electrons bound to atoms and construct a picture of atomic structure. Differences in atomic structure distinguish the elements from one another and generate the rich array of chemical behavior exhibited by different types of atoms.

7.1 CHARACTERISTICS OF ATOMS

We begin with a review of the fundamental characteristics of atoms, many of which we introduced in Chapter 2:

Atoms possess mass. Matter possesses mass, and matter is made up of atoms, so atoms possess mass. This property was already recognized in the time of John Dalton, who made it one of the postulates of his atomic theory.

Atoms contain positive nuclei. Recall from Chapter 2 what Rutherford's scattering experiment demonstrated. Every atom contains a tiny central core where all the positive charge and most of the mass are concentrated. Subsequent experiments showed that while the masses of nuclei can have various values, the mass of every nucleus is at least 1800 times larger than the mass of the electron. Thus, more than 99.9% of the mass of an atom is contained in its nucleus.

Atoms contain electrons. For an atom to be neutral, the number of electrons that it contains must equal the total positive charge on its nucleus. Because each element has a characteristic positive charge associated with its nucleus, ranging from +1 for hydrogen to greater than +100 for the heaviest elements, atoms of different elements have different numbers of electrons.

Atoms occupy volume. Matter occupies space, and matter is made up of atoms, so atoms occupy space. It is extremely difficult to compress a solid such as copper or a liquid such as mercury, because the electron cloud of each atom occupies some volume that no other atom is able to penetrate because of electron–electron repulsion. Example 7-1 shows how to estimate the volume of an atom from the density of a sample, the molar mass of the substance, and Avogadro's number.

Example 7-1

Atomic Volumes

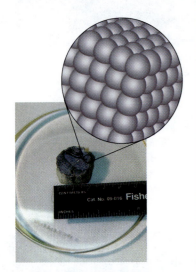

A cylindrical block of lithium with diameter and height of 2.4 cm contains approximately 1 mol of Li atoms.

The density of lithium, the lightest metal, is 0.534 g/cm^3. Estimate the volume occupied by a single atom in solid lithium.

Strategy: Our task is to estimate the volume occupied by one atom of lithium. As usual, the mole is a convenient place to begin the calculations. Visualize a piece of lithium containing one mole of atoms. The molar mass, taken from the periodic table, tells us the number of grams of Li in one mole. The density equation can be used to convert from mass to volume. Once we have the volume of one mole of lithium, we divide by the number of atoms per mole to find the volume of a single atom.

Solution: First, assemble the data:

$$\rho = 0.534 \text{ g/cm}^3 \text{ (stated in the problem)}$$

$$MM = 6.941 \text{ g/mol (from periodic table)}$$

$$N_A = 6.022 \times 10^{23} \text{ atom/mol}$$

Now rearrange the density equation and solve for the molar volume of lithium.

$$\rho = \frac{m}{V} \quad so \quad V_{molar} = \frac{MM}{\rho} = \frac{6.941 \text{ g/mol}}{0.534 \text{ g/cm}^3} = 13.00 \text{ cm}^3/\text{mol}$$

Finally, divide by Avogadro's number to find volume per atom:

$$V_{atom} = \frac{V_{molar}}{N_A} = \frac{13.00 \text{ cm}^3/\text{mol}}{6.022 \times 10^{23} \text{ atom/mol}} = 2.16 \times 10^{-23} \text{ cm}^3/\text{atom}$$

Does the Result Make Sense? The units are correct. This is a very tiny volume, as we would expect, knowing that atoms are very tiny. A cube with this volume has sides measuring 2.78×10^{-10} m in length (0.278 nm).

Estimate the volume occupied by a single atom of iron, whose density is 7.874 g/cm³.

**Extra Practice
Exercise 7.1**

The volume of an atom is determined by the size of its electron cloud. Example 7-1 demonstrates that atomic dimensions are a little over 10^{-10} m, whereas Rutherford's experiments showed that nuclear dimensions are only about 10^{-15} m. This is 100,000 times smaller than atomic dimensions, so the nucleus is buried deep within the electron cloud. If an atom were the size of a sports stadium, its nucleus would be the size of a pea. Figure 7-1 shows a schematic view of two atoms with their electron clouds in contact with each other.

Atoms have various properties. The periodic table is a catalog of the elements, each with its own unique set of physical and chemical properties. Each element has a unique value for *Z*, the positive charge on its nucleus. The number of electrons possessed by a neutral atom of that element is also equal to *Z*. The different properties of elements arise from these variations in nuclear charges and numbers of electrons.

Atoms attract one another. Every gas changes into a liquid if the pressure is high enough and the temperature is low enough. The atoms or molecules of a liquid or solid stick together in a finite volume rather than expanding, as a gas does, to fill all available space. This cohesiveness comes from electrical forces of attraction between the negative electron cloud of each atom and the positive nuclei of other atoms. We describe inter-molecular forces in Chapter 11.

Atoms can combine with one another. Atoms can form chemical bonds with one another to construct molecules. As we point out in Chapter 2, this is one of the fundamental features of the atomic theory. The details of chemical bonding appear in Chapters 9 and 10.

What does an atom experience in an encounter with another atom? The nucleus, which contains most of the atom's mass, is confined to a tiny volume. Electrons, on the other hand, are spread out through space. Therefore, a collision between two atoms is a collision of their electron clouds. The electron clouds repel each other but are attracted by the nuclei. Chemists describe molecular structure, properties of materials, and chemical reactions in terms of how electrons respond to these electrical forces.

Our catalog of atomic characteristics emphasizes electrons, because electrons determine the chemical properties of atoms. For the same reason, the next several chapters examine electrons and the way they influence chemical properties. First, however, we describe light and its interaction with atoms, because light is an essential tool for probing properties of electrons.

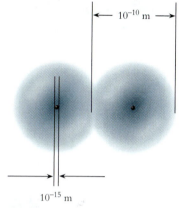

← 10^{-10} m →

10^{-15} m

**Figure 7-1
Depiction of two atoms with their electron clouds in contact.**

Section Exercises

■ **7.1.1** The density of gold metal is 18.9 g/cm³. (a) What is the volume occupied by one gold atom? (b) If a gold nucleus is 1/100,000 times as large as a gold atom, what is the volume of one gold nucleus?

■ **7.1.2** (a) Use the volume occupied by a gold atom to estimate the thickness of one atomic layer of gold. (b) Estimate how many layers of gold atoms there are in a strip of gold foil that is 1.0 μm thick (1 μm = 10^{-6} m).

7.2 CHARACTERISTICS OF LIGHT

The most useful tool for studying the structure of atoms is **electromagnetic radiation.** What we call **light** is one form of this radiation. We need to know about the properties of light in order to understand what electromagnetic radiation reveals about atomic structure.

Light Has Wave Aspects

Light has wave-like properties. A wave is a regular oscillation in some particular property, such as the up-and-down variation in position of water waves. Water waves vary with time. A surfer waiting for a "big one" bobs up and down as "small ones" pass by. Light waves vary with time, too. This variation is characterized by the wave's **frequency** (v), which is the number of wave crests passing a point in space in one second. The frequency unit is s^{-1}, also designated Hertz (Hz). Water waves also vary in space; that is, wave height differs from one place to another. Light waves vary in space in a manner illustrated in Figure 7-2. This variation in space is characterized by the **wavelength (λ),** which is the distance between successive wave crests. Wavelengths are measured in units of distance, such as meters or nanometers.

The height of a wave is called its **amplitude.** The amplitude of a light wave measures the **intensity** of the light. As Figure 7-3 shows, a bright light is more intense than a dim one.

Light waves always move through empty space at the same speed. The speed of light is a fundamental constant, denoted by the symbol c: $c = 2.99792458 \times 10^8$ m/s. For any wave, its wavelength (in units of m) multiplied by its frequency (in units of s^{-1}) equals its speed (m/s). Thus, light obeys Equation 7-1:

$$\lambda v = c \tag{7-1}$$

Example 7-2 applies Equation 7-1.

Like electromagnetic radiation, water waves vary in time (frequency) and space (wavelength).

λ is the Greek letter *lambda,* and v is the Greek letter *nu.*

The value of c can be rounded to 2.998×10^8 m/s for most calculations.

①

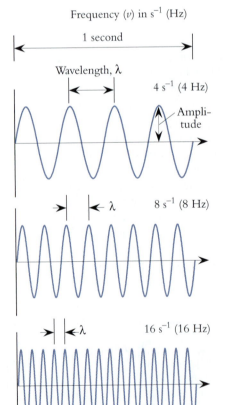

Figure 7-2
A light wave can be described by its wavelength and frequency, which are connected. As wavelength increases, frequency decreases, and vice versa.

Frequency (v) in s^{-1} (Hz)

1 second

Wavelength, λ

4 s^{-1} (4 Hz)

Amplitude

λ 8 s^{-1} (8 Hz)

λ 16 s^{-1} (16 Hz)

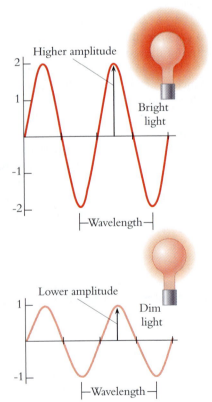

Figure 7-3
The amplitude of a light wave represents the intensity of the light. A bright light is more intense and has greater amplitude than a dim light.

Higher amplitude

Bright light

Wavelength

Lower amplitude

Dim light

Wavelength

An FM radio station transmits its signal at 88.1 MHz. What is the wavelength of the radio signal?

Example 7-2

Wavelength–Frequency Conversion

Strategy: This is a simple conversion problem. The link between wavelength (λ) and frequency (ν) is given by Equation 7-1.

Solution: First, summarize the data:

$$c = 2.998 \times 10^8 \text{ m/s} \qquad \nu = 88.1 \text{ MHz}$$

Next, rearrange Equation 7-1 to solve for wavelength.

$$\nu\lambda = c \qquad so \qquad \lambda = \frac{c}{\nu}$$

To obtain equivalent units, convert the frequency units from MHz to Hz. The prefix "M" stands for "mega," which is a factor of 10^6. Remember that Hz is equivalent to s^{-1}.

$$\lambda = \frac{2.998 \times 10^8 \text{ m/s}}{88.1 \times 10^6/\text{s}} = 3.40 \text{ m}$$

Does the Result Make Sense? The wavelength is rather long—3.40 m—but radio waves are known as long wavelength radiation. See Figure 7-4 for a sense of the wavelengths of electromagnetic radiation.

What is the frequency of light that has a wavelength of 1.40 cm?

The wavelengths and frequencies of electromagnetic radiation cover an immense range. What we call light is the tiny part of the spectrum that the human eye can detect. Figure 7-4 shows that the visible spectrum of light covers the wavelength range from about 400 nm (violet) to 700 nm (red). The center of this range is yellow light, with a wavelength around 580 nm and a frequency around $5.2 \times 10^{14} \text{ s}^{-1}$. Although visible light is extremely important to living creatures for seeing, the gamma-ray, X-ray, ultraviolet, infrared, microwave, and radio frequency portions of the electromagnetic spectrum have diverse effects and applications in our lives.

Figure 7-4
The electromagnetic spectrum, showing its various regions and the wavelengths and frequencies associated with each.

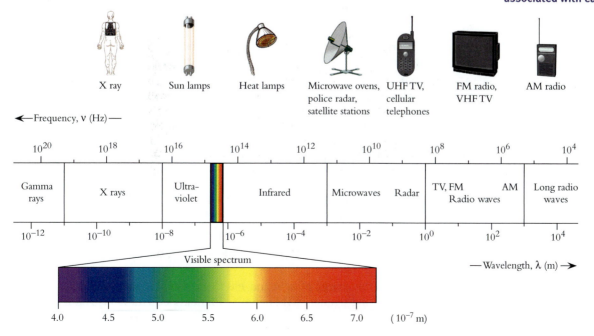

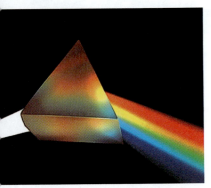

Light rays bend as they pass through a prism, causing white light to separate into its rainbow of colors.

Radiation with short wavelengths, in the X-ray and gamma-ray regions, can generate ions by removing electrons from atoms and molecules. These ions are highly reactive and can cause serious damage to the material that absorbs the light. However, under closely controlled conditions, X rays are used in medical imaging, and gamma rays are used to treat cancer. Ultraviolet (UV) radiation falls in between X rays and visible light. Ultraviolet radiation can also damage materials, especially in high doses.

Radiation with long wavelengths falls in the infrared, microwave, or radio frequency regions. Heat lamps make use of infrared radiation, microwave ovens cook with microwave radiation, and radio and television signals are transmitted by radio waves.

What we perceive as white light actually contains a range of wavelengths. These wavelengths are revealed when sunlight passes through a prism or through raindrops. These objects bend different wavelengths of light through different angles, so the light that passes through spreads out in space, with each wavelength appearing at its own characteristic angle.

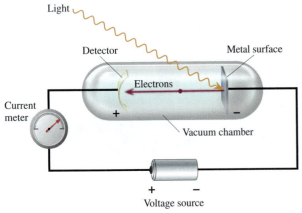

The Photoelectric Effect

Light carries energy. When our bodies absorb sunlight, for example, we feel warm because the energy of the sunlight has been transferred to our skin. Moreover, the total energy of a beam of light depends on its intensity (how bright the light is). Sunlight at midday is more intense than the rays of the setting sun, because light from the setting sun is reduced in intensity as it passes through a larger thickness of the atmosphere.

A phenomenon known as the **photoelectric effect** shows how the energy of light depends on its frequency and intensity.

As Figure 7-5 illustrates, the photoelectric effect involves light and electrons. In a photoelectric experiment, a beam of light strikes the surface of a metal. Under the right conditions, the light causes electrons to be ejected from the metal surface. These electrons strike a detector that measures the number of electrons. The kinetic energies of the electrons can be determined by changing the voltage applied to the detector.

Figure 7-5
A diagrammatic view of the photoelectric effect. When light of high enough frequency strikes a metal surface, electrons are ejected from the surface.

A detailed study of the photoelectric effect reveals how the behavior of the electrons is related to the characteristics of the light:

The photoelectric effect is the basis for many light-sensing devices, such as automatic door openers and camera exposure meters.

1. Below a characteristic threshold frequency, ν_0, no electrons are observed, regardless of the light's intensity.

2. Above the threshold frequency, the maximum kinetic energy of ejected electrons increases linearly with the frequency of the light, as shown in Figure 7-6.

3. Above the threshold frequency, the *number* of emitted electrons increases with the light's intensity, but the *kinetic energy* per electron does not depend on the light's intensity.

4. All metals exhibit the same pattern, but as Figure 7-6 shows, each metal has a different threshold frequency.

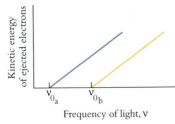

Figure 7-6
Variation in the maximum kinetic energy of electrons ejected from two different metal surfaces (a and b) by light of various frequencies.

In 1905, Albert Einstein provided an elegant explanation of the photoelectric effect. Einstein postulated that light comes in packets or bundles, called **photons.** Each photon has an energy that is directly proportional to the frequency:

Planck's constant (h) is named in honor of Max Planck, who first introduced this constant in 1900 to explain another phenomenon of light.

$$E_{photon} = h\nu_{photon} \qquad (7\text{-}2)$$

In this equation, E is the energy of light and ν is its frequency. The proportionality constant between energy and frequency is known as **Planck's constant (h)** and has a value of $6.626\,068\,76 \times 10^{-34}$ J s. Example 7-3 illustrates the use of Equation 7-2.

Example 7-3

The Energy of Light

What is the energy of a photon of red light of wavelength 655 nm?

Strategy: This conversion problem requires two steps. Equations 7-1 and 7-2 relate the energy of a photon to its frequency and wavelength.

Solution: Summarize the data:

$$h = 6.626 \times 10^{-34} \, \text{J s} \qquad \lambda = 655 \, \text{nm} \qquad c = 2.998 \times 10^8 \, \text{m/s}$$

Combine the equations into an equation that relates energy to wavelength:

$$E = h\nu \qquad and \qquad \nu = \frac{c}{\lambda} \qquad so \qquad E = \frac{hc}{\lambda}$$

Substitute and evaluate:

$$E_{\text{photon}} = \frac{(6.626 \times 10^{-34} \, \text{J s})(2.998 \times 10^8 \, \text{m/s})}{(655 \, \text{nm})(10^{-9} \, \text{m/nm})} = 3.03 \times 10^{-19} \, \text{J}$$

Does the Result Make Sense? An energy of 10^{-19} J seems very small, but as we describe later in this section, a calculation of the energy of 1 mol of these photons gives $(3.03 \times 10^{-19} \, \text{J})(6.022 \times 10^{23}/\text{mol})(10^{-3} \, \text{kJ/J}) = 182 \, \text{kJ/mol}$. This energy is comparable to the energies of chemical bonds.

What is the energy of an UV photon whose wavelength is 254 nm?

Einstein applied the law of conservation of energy to the photoelectric effect, as shown schematically in Figure 7-7. When a metal surface absorbs a photon, the energy of the photon is transferred to an electron:

$$\Delta E_{\text{electron}} = E_{\text{photon}}$$

Some of this energy is used to overcome the forces that bind the electron to the metal, and the remainder shows up as kinetic energy of the ejected electron.

The energy of a photon that has the threshold frequency (ν_0) corresponds to the binding energy of the electron. In other words, the energy of a photon at the threshold frequency equals the minimum energy needed to overcome the forces that bind the electron to the metal. Putting these ideas together, Einstein obtained a simple linear equation that matches the graph in Figure 7-6:

Electron kinetic energy = Photon energy − Binding energy

$$E_{\text{kinetic}} \, (\text{electron}) = h\nu - h\nu_0 \qquad (7\text{-}3)$$

②

Einstein's explanation accounts for the observed properties of the photoelectric effect. First, when the energy of the photon is less than $h\nu_0$ (low-frequency light), there is not enough energy per photon to overcome the electron's binding energy. Under these conditions, no electrons can escape from the metal surface, no matter how intense the light. Second, after the energy of the photon exceeds the threshold value ($h\nu > h\nu_0$), electrons are ejected. The extra energy of the photon is transferred to the ejected electron as kinetic energy; this extra kinetic energy increases linearly with ν. Third, the intensity of a light beam is a measure of the number of photons: Light with higher amplitude carries more photons than light of lower amplitude. The intensity of the light *does not* determine the amount of energy per photon. "Higher intensity" means more photons but not more energy per photon. More photons striking the metal result in more electrons being ejected, but the energy of each photon and each electron is unchanged. Finally, each metal has its own characteristic threshold frequency because electrons are bound more tightly to some metals than to others.

Example 7-4 shows how to apply Einstein's analysis of the photoelectric effect.

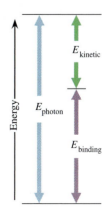

Figure 7-7
Diagram of the energy balance for the photoelectric effect.

<table>
<tr><td>

Example 7-4

The Photoelectric Effect

</td><td>

The minimum energy needed to remove an electron from a potassium metal surface is 3.7×10^{-19} J. Will photons of frequencies 4.3×10^{14} s^{-1} (red light) and of 7.5×10^{14} s^{-1} (blue light) trigger the photoelectric effect? If so, what is the maximum kinetic energy of the ejected electrons?

Strategy: This problem asks if red and blue photons can cause potassium metal to lose electrons. We must analyze the energy requirements for ejection of an electron. No electrons will be ejected unless the energy of the photons exceeds some threshold value characteristic of the metal. If the photon energy exceeds this threshold value, electrons will be ejected with kinetic energy given by Equation 7-3. An important part of this problem is the conversion of photon frequency to photon energy.

Solution: The threshold energy of potassium metal, 3.7×10^{-19} J, is given in the problem. To determine whether the red and blue photons will eject electrons, we convert the frequencies of these photons to their corresponding energies:

$$E_{\text{red photon}} = h\nu = (6.626 \times 10^{-34} \text{ J s})(4.3 \times 10^{14} \text{ s}^{-1}) = 2.8 \times 10^{-19} \text{ J}$$

$$E_{\text{blue photon}} = h\nu = (6.626 \times 10^{-34} \text{ J s})(7.5 \times 10^{14} \text{ s}^{-1}) = 5.0 \times 10^{-19} \text{ J}$$

According to these calculations, a photon of red light does not have enough energy to overcome the forces that bind electrons to the metal. A blue photon, however, will eject an electron from the surface of potassium metal because its energy exceeds the threshold value. What happens to the extra energy of a blue photon? Equation 7-3 indicates that it is transferred to the electron as kinetic energy:

$$E_{\text{kinetic}} \text{ (electron)} = h\nu - h\nu_0$$

$$E_{\text{kinetic}} \text{ (electron)} = (5.0 \times 10^{-19} \text{ J}) - (3.7 \times 10^{-19} \text{ J}) = 1.3 \times 10^{-19} \text{ J}$$

</td></tr>
</table>

Do the Results Make Sense? Figure 7-4 shows that red light has a lower frequency than blue light, and Equation 7-2 shows that photon energy is directly proportional to frequency. It makes sense that the photon of higher energy can eject an electron, but the photon of lower energy cannot.

Extra Practice Exercise 7.4

Will a green photon ($\lambda = 515$ nm) eject an electron from a potassium surface? If so, what is the maximum kinetic energy of ejected electrons?

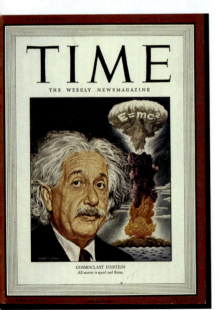

TIME

THE WEEKLY NEWSMAGAZINE

$E = mc^2$

COSMOCLAST EINSTEIN
All matter is speed and flame.

Albert Einstein is considered the greatest modern physicist because his theories influenced topics in physics ranging from the basis for lasers to the expanding universe.

Light Has Particle Aspects

Einstein's explanation of the photoelectric effect showed that light has some properties of particles. Light consists of photons, each of which is like a "bullet of energy" with the discrete energy $E = h\nu$. Although simple, this particle-like explanation of light was revolutionary, because before 1905, the properties of light had been explained with a wave picture. A complete description of light includes wave-like and particle-like properties.

Neither the particle nor the wave view of light is wrong or right. Light has some properties of waves and some properties of particles. When light interacts with a relatively large body such as a raindrop or a prism, its wave properties dominate the interaction. On the other hand, when light interacts with a small body such as an atom or an electron, particle properties dominate the interaction. Each view provides different information about the properties of light, and when we think about light, we must think of *wave-particles* that combine both types of features.

A photon of a particular frequency has a specific energy. As the photoelectric effect shows, when something absorbs that photon, the energy is transferred to whatever absorbs the photon. Sunlight striking your skin is absorbed and warms you up. When an individual atom absorbs a photon, the atom's energy increases by an amount equal to the energy of the photon. We analyze these energy exchanges in the next section.

Section Exercises

■ **7.2.1** A compact disc player uses light of frequency 3.85×10^{14} s^{-1} to read the information on the disc. (a) What is this light's wavelength? (b) In what portion of the electromagnetic spectrum (visible, ultraviolet, and so on) does this wavelength fall? (c) What is the energy of one mole of photons at this frequency?

■ **7.2.2** The light reaching us from distant stars is extremely dim, so astronomers use instruments capable of detecting small numbers of photons. An infrared photon detector registered a signal at 1250 nm from Alpha Centauri with an energy of 1.20×10^{-16} J. How many photons were detected?

■ **7.2.3** In a photoelectric effect experiment, a metal absorbs photons with $E = 6.00 \times 10^{-19}$ J. The maximum kinetic energy of the ejected electrons is 2.00×10^{-19} J. Calculate the frequency of the light and the binding energy in joules per electron, and convert this energy into kilojoules per mole.

7.3 ABSORPTION AND EMISSION SPECTRA

In the photoelectric effect, energy absorbed from photons provides information about the binding energies of electrons to metal surfaces. When light interacts with free atoms, the interaction reveals information about electrons bound to individual atoms.

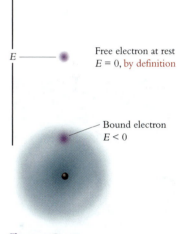

Figure 7-8
By convention, a free stationary electron has zero energy, so bound electrons have negative energies.

Light and Atoms

The absorption of photons by free atoms has two different results, depending on the energy of the photon. When an atom absorbs a photon of sufficiently high energy, an electron is ejected. We describe this process, photoionization, in Chapter 8. Here, we focus on the second type of result, in which the atom gains energy but does not ionize. Instead, the atom is transformed to a higher energy state called an **excited state.** Atoms in excited states subsequently give up their excess energy to return to lower energy states. The lowest energy state of an atom, which is its most stable state, is called the **ground state.**

Attractive electrical force holds a bound electron within an atom, and energy must be supplied to remove a bound electron from an atom. The lower an energy state is, the more energy must be supplied to remove its electron. These are energy changes that are measured relative to the energy of a free electron. By convention, scientists define the energy of a free stationary electron to be zero. As Figure 7-8 indicates, the kinetic energy of a freely moving electron is positive relative to this conventional zero point. In contrast, bound electrons are lower in energy than free stationary electrons, so they have negative energy values relative to the zero point. Therefore, the term "most stable" in this context corresponds to "lowest" or "most negative." A bound electron has "negative energy" only in relation to a free electron, because energy is released when a free stationary electron binds to an atom.

Atomic energy transformations can be represented using an **energy level diagram** such as the one shown in Figure 7-9. These diagrams show energy along the vertical axis. Each energy state of the atom is represented by a horizontal line. Absorption of a photon, shown by an upward arrow, causes a transition from the ground state to one of the excited states. Excited states are not stable. They lose their excess energy in collisions with other atoms or by emitting photons, shown by downward arrows.

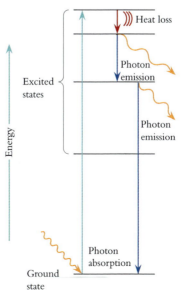

Figure 7-9
An atomic energy level diagram showing the relationships among atomic energy levels and photon absorption and emission.

Energy level diagrams are compact ways to summarize large amounts of information about atomic energies.

The key feature in the exchange of energy between atoms and light is that energy is conserved. This requires that the change in energy of the atom exactly equals the energy of the photon. Equation 7-4 states this equality:

$$\Delta E_{\text{atom}} = \pm h\nu_{\text{photon}} \tag{7-4}$$

When an atom *absorbs* a photon, the atom gains the photon's energy, so ΔE_{atom} is positive. When an atom *emits* a photon, the atom loses the photon's energy, so ΔE_{atom} is negative. As an atom returns from an excited state to the ground state, it must lose exactly the amount of energy that it originally gained. However, excited atoms usually lose excess energy in several steps involving small energy changes, so the frequencies of emitted photons often are lower than those of absorbed photons. Example 7-5 analyzes energy changes associated with the emission of light.

Example 7-5 **Emission Energies**	A sodium-vapor street lamp emits yellow light at wavelength $\lambda = 589$ nm. What is the energy change for a sodium atom involved in this emission? How much energy is released per mole of sodium atoms?

Strategy: This problem relates energies of photons to energy changes of atoms. The solution requires a conversion involving wavelength and energy.

Solution: Sodium atoms lose energy by emitting photons of light ($\lambda = 589$ nm). We can use Equations 7-1 and 7-2 to relate the wavelength of one of these photons to its energy:

$$E_{\text{photon}} = h\nu = \frac{hc}{\lambda}$$

Energy is conserved, so the energy of the emitted photon must exactly equal the energy lost by the atom:

$$\Delta E_{\text{atom}} = -E_{\text{photon}} = -\frac{hc}{\lambda}$$

The equality includes a negative sign because the atom *loses* energy.

To calculate the energy of a photon, we need the following data:

$$h = 6.626 \times 10^{-34}\,\text{J s} \qquad c = 2.998 \times 10^{8}\,\text{m/s} \qquad \lambda = 589\,\text{nm} = 589 \times 10^{-9}\,\text{m}$$

$$\Delta E_{\text{atom}} = -E_{\text{photon}} = \frac{(6.626 \times 10^{-34}\,\text{J s})(2.998 \times 10^{8}\,\text{m/s})}{(589 \times 10^{-9}\,\text{m})}$$

$$= -3.37 \times 10^{-19}\,\text{J}$$

This calculation gives the energy change for one sodium atom emitting one photon. We use Avogadro's number to convert from energy per atom to energy per mole of atoms:

$$\Delta E_{\text{mol}} = (\Delta E_{\text{atom}})(N_{\text{A}}) = (-3.37 \times 10^{-19}\,\text{J})(6.022 \times 10^{23}\,\text{mol}^{-1})$$

$$= -2.03 \times 10^{5}\,\text{J/mol} = -203\,\text{kJ/mol}$$

Sodium vapor lamps emit yellow light.

Do the Results Make Sense? The energy change is negative, as required by the fact that the emitted photon carries energy. The magnitude is reasonable, being the same order of magnitude as bond energies.

Extra Practice **Exercise 7.5**	Mercury lamps emit photons with wavelength of 436 nm. Calculate the energy change of the mercury atoms in J/atom and kJ/mol.

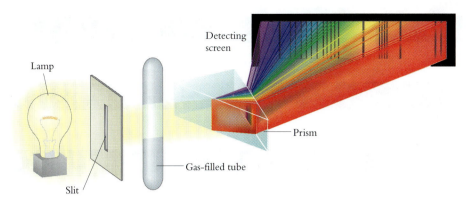

Figure 7-10
Schematic representation of an apparatus that measures the absorption spectrum of a gaseous element. The gas in the tube absorbs light at specific wavelengths, called lines, so the intensity of transmitted light is low at these particular wavelengths.

Atomic Spectra

When a light beam passes through a tube containing a monatomic gas, the atoms absorb specific and characteristic frequencies of the light. As a result, the beam emerging from the sample tube contains fewer photons at these specific frequencies. We can see where the frequencies are missing in the visible portion of the spectrum by passing the emerging light through a prism. The prism deflects the light, with different frequencies deflecting through different angles. After leaving the prism, the beam strikes a screen, where the missing frequencies appear as gaps or dark bands. These are the frequencies of light absorbed by the atoms in the sample tube. The resulting pattern, shown schematically in Figure 7-10, is called an **absorption spectrum.**

An absorption spectrum measures the frequencies of the photons that an atom *absorbs*. A similar experiment can be performed to measure the energies of the photons *emitted* by atoms in excited states. Figure 7-11 outlines the features of an apparatus that measures these emitted photons. An electrical discharge excites a collection of atoms from their ground state into higher-energy states. These excited atoms lose all or part of their excess energy by emitting photons. This emitted light can be analyzed by passing it through a prism to give an **emission spectrum.** This is a plot of the intensity of light emitted as a function of frequency. The emission spectrum for hydrogen, shown in Figure 7-11, shows several sharp emission lines of high intensity. The frequencies of these lines correspond to photons emitted by the hydrogen atoms as they return to their ground state.

Each element has unique absorption and emission spectra. That is, each element has its own set of characteristic frequencies of light that it can absorb or emit. Also, Figures 7-10 and 7-11 show only the visible portions of absorption and emission spectra. Electron transitions also take place in regions of the electromagnetic spectrum that the human eye cannot detect. Instruments allow scientists to "see" into these regions.

Each frequency absorbed or emitted by an atom corresponds to a particular energy change for the atom. These characteristic patterns of energy gains and losses provide information about atomic structure.

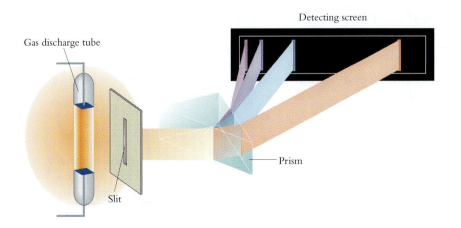

Figure 7-11
Schematic representation of an apparatus that measures the emission spectrum of a gaseous element. Emission lines appear bright against a dark background. The spectrum shown is the emission spectrum for hydrogen atoms.

Each element has a unique emission pattern that provides valuable clues about atomic structure.

Sodium (Na)

Neon (Ne)

Mercury (Hg)

Quantization of Energy

When an atom absorbs light of frequency ν, the light beam loses energy $h\nu$, and the atom gains that amount of energy. What happens to the energy that the atom gains? A clue is that when the frequency of the bombarding light is high enough, it produces cations and free electrons. In other words, a photon with high enough energy can cause an atom to lose one of its electrons. This implies that absorption of a photon results in an energy gain for an electron in the atom. Consequently, the energy change for the atom equals the energy change for an atomic electron:

$$\Delta E_{\text{atom}} = \Delta E_{\text{electron}} = h\nu$$

The atomic spectra of most elements are complex and show little regularity. However, the emission spectrum of the hydrogen atom is sufficiently simple to be described by a single formula:

$$\nu_{\text{emission}} = (3.29 \times 10^{15} \text{ s}^{-1})\left(\frac{1}{n_1^2} - \frac{1}{n_2^2}\right)$$

Niels Bohr (1885–1962) was a Danish physicist whose discovery of the quantization of atomic energy levels won him a Nobel Prize in 1922. Bohr headed a world-renowned institute for atomic studies in Copenhagen in the 1920s and 1930s.

The Swiss mathematician and physicist Johann Balmer proposed a form of this equation in 1885. At that time, the link between frequency and energy was not known. Balmer did not understand the significance of n_1 and n_2, both of which are integers (1, 2, 3, and so on). Then, in 1913, Niels Bohr used the discovery that $E = h\nu$ to interpret Balmer's observations. Bohr realized that the emission frequencies have specific values because the electron in a hydrogen atom is restricted to specific energies described by an equation containing an integer n:

$$E_n = -\frac{2.18 \times 10^{-18} \text{ J}}{n^2} \tag{7-5}$$

The constants in Bohr's equation and Balmer's equation are related through $E = h\nu$.

The negative sign in Equation 7-5 reflects the fact that bound electrons have negative energies relative to a stationary free electron.

Bohr's idea of restricted energy levels was revolutionary, because scientists at that time thought that the electron in a hydrogen atom could have any energy, not just the ones described by Equation 7-5. In contrast, Bohr interpreted the hydrogen emission spectrum to mean that electrons bound to atoms can have only certain specific energy values. A property that is restricted to specific values is said to be **quantized.** The atomic energy levels of hydrogen (and other elements) are quantized. In Equation 7-5, each integral value of n describes one of the allowed energy levels of the hydrogen atom. For example, the energy of an electron in hydrogen's fourth level is

$$E_4 = -\frac{2.18 \times 10^{-18} \text{ J}}{4^2} = -1.36 \times 10^{-19} \text{ J}$$

When an electron changes energy levels, the change is an electronic transition between quantum levels. When a hydrogen atom absorbs or emits a photon, its electron changes from one energy level to another. Thus, the change in energy of the atom is the difference between the two levels:

$$\Delta E_{\text{atom}} = E_{\text{final}} - E_{\text{initial}}$$

Photons always have positive energies, but energy changes (ΔE) can be positive or negative. When absorption occurs, an atom gains energy, ΔE for the atom is positive, and a photon disappears:

$$E_{\text{absorbed photon}} = \Delta E_{\text{atom}}$$

When emission occurs, an atom loses energy, ΔE for the atom is negative, and a photon appears:

$$E_{\text{emitted photon}} = -\Delta E_{\text{atom}}$$

We can combine these two equations by using absolute values:

$$E_{\text{photon}} = |\Delta E_{\text{atom}}| \qquad\qquad (7\text{-}6)$$

Example 7-6 applies Equations 7-5 and 7-6 to the hydrogen atom.

Example 7-6

Hydrogen Energy Levels

What is the energy change when the electron in a hydrogen atom changes from the fourth energy state to the second energy state? What is the wavelength of the photon emitted?

Strategy: The problem asks about energy and the wavelength of a photon emitted by a hydrogen atom. Wavelength is related to energy, and photon energy is determined by the difference in energy between the two levels involved in the transition. In this case, the electron moves from the fourth to the second energy level. The energy difference between these two states is given by

$$\Delta E_{\text{atom}} = E_{\text{final}} - E_{\text{initial}} = (E_2 - E_4)$$

Solution: As shown in the text, $E_4 = -1.36 \times 10^{-19}$ J. A similar calculation using Equation 7-5 gives the energy of the second level:

$$E_2 = -\frac{2.18 \times 10^{-18} \text{ J}}{2^2} = -5.45 \times 10^{-19} \text{ J}$$

This lower energy level has a more negative energy. The energy difference is

$$\Delta E_{\text{atom}} = E_{\text{final}} - E_{\text{initial}} = (-5.45 \times 10^{-19} \text{ J}) - (-1.36 \times 10^{-19} \text{ J})$$

$$= -4.09 \times 10^{-19} \text{ J}$$

This energy change is *negative* because the atom *loses* energy. This lost energy appears as a photon whose energy is given by Equation 7-6:

$$E_{\text{photon}} = |\Delta E_{\text{atom}}| = |-4.09 \times 10^{-19} \text{ J}| = 4.09 \times 10^{-19} \text{ J}$$

To determine the wavelength of the photon, we use Equations 7-1 and 7-2:

$$E_{\text{photon}} = h\nu = \frac{hc}{\lambda}$$

Solving for λ gives

$$\lambda_{\text{photon}} = \frac{hc}{E_{\text{photon}}}$$

$$= \frac{(6.626 \times 10^{-34} \text{ J s})(2.998 \times 10^8 \text{ m/s})(10^9 \text{ nm/m})}{(4.09 \times 10^{-19} \text{ J})} = 486 \text{ nm}$$

Do the Results Make Sense? The signs of the energies are consistent—the atom loses energy, the photon has positive energy. The wavelength of light falls in the visible region. According to Figure 7-4, a photon with a wavelength of 486 nm has a blue color.

What energy and wavelength must a photon have to excite a hydrogen atom from its ground state to the $n = 4$ level?

Extra Practice
Exercise 7.6

Energy Level Diagrams

The quantum levels of an electron bound to an atom are crudely analogous to the gravitational potential energies available to a ball on a staircase. As illustrated in Figure 7-12, a ball may sit on any of the steps. If we define the top of the steps to be $E = 0$, the ball has a negative potential energy when it is on any of the lower steps. To move a ball from the bottom of the staircase to step 5 requires the addition of a specific amount of energy, $\Delta E = E_5 - E_1$. If too little energy is supplied, the ball cannot reach this step. Conversely, if a ball moves down the staircase, it releases specific amounts of energy. If a ball moves from step 5 to step 3, it loses energy, $\Delta E = E_3 - E_5$. Although a ball may rest squarely on any step, it cannot be suspended at some position between the stair steps. Electrons in atoms, like balls on steps, cannot exist "between steps" but must occupy one of the specific, quantized energy levels. (Remember that this is an analogy; atomic energy levels are not at all similar to staircases except in being quantized.)

According to Equation 7-5, a hydrogen atom has a regular progression of quantized energy levels. Figure 7-13 shows the energy level diagram for hydrogen atoms; arrows represent some of the absorption and emission transitions. Notice that the energies of absorption from the lowest energy level are identical to the energies of emission to the lowest energy level. This means that the wavelengths of light absorbed in these upward transitions are identical to the wavelengths of light emitted in the corresponding downward transitions.

Elements other than hydrogen also have quantized energy levels, but they lack the regular spacing described by Equation 7-5. Scientists use experimental values for observed absorption and emission lines to calculate the allowed energy levels for each different element. As an example, Figure 7-14 shows the energy level diagram for mercury. The absorption spectrum of mercury shows two dominant lines (upward arrows). Its emission spectrum, generated by an electrical discharge, contains light of many different wavelengths (downward arrows). Energy level diagrams such as Figures 7-13 and 7-14 summarize the quantized energy levels of an atom as deduced from light measurements and energy conservation. Example 7-7 provides more practice in this type of reasoning, and our Tools for Discovery Box (see page 275) discusses how spectroscopy—the observation of spectra—can be used to study otherwise inaccessible objects.

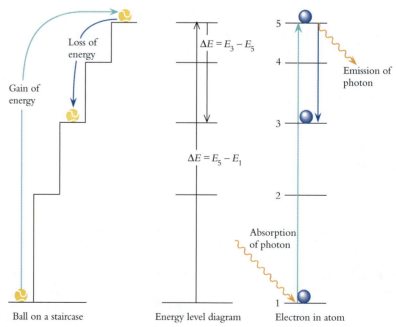

Figure 7-12
A ball on a staircase shows some properties of quantized energy states.

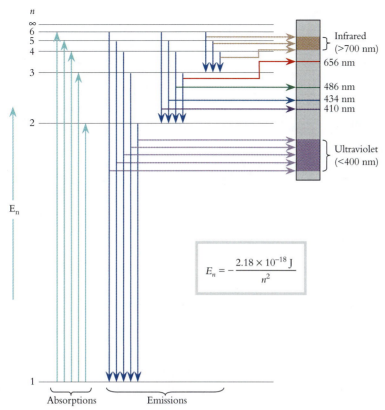

Figure 7-13
Energy levels for the hydrogen atom and some of the transitions that occur between levels.
Upward arrows represent absorption transitions, and downward arrows represent emissions.

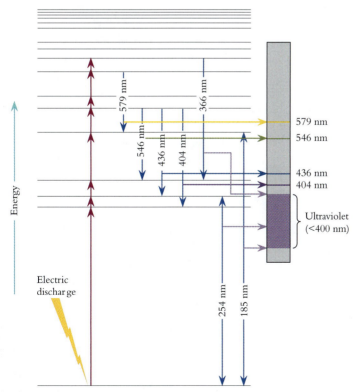

Figure 7-14
Energy level diagram for mercury (Hg) atoms showing the most prominent absorption and
emission lines. The numbers accompanying the arrows are the wavelengths in nanometers (nm)
of the photons associated with these transitions.

Energy Level Diagrams

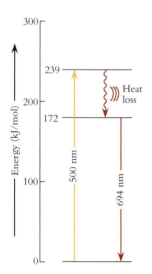

Light from a ruby laser is generated from Cr^{3+} ions.

④

Ruby lasers use crystals of Al_2O_3. The crystals contain small amounts of Cr^{3+} ions, which absorb light between 400 and 560 nm. These excited-state ions lose some energy as heat. After losing heat, the Cr^{3+} ions return to the ground state by emitting red light of wavelength 694 nm. Calculate (a) the molar energy of the 500-nm radiation used to excite the Cr^{3+} ions, (b) the molar energy of the emitted light, and (c) the fraction of the excitation energy emitted as red photons and the fraction lost as heat. (d) Draw an energy level diagram, in kilojoules per mole, that summarizes these processes.

Strategy: This problem asks about energies, light, and atoms. With multipart problems, the best strategy is to work through the parts one at a time. Parts (a) and (b) concern the link between light and energy, discussed in Section 7.2. Once the transition energies have been determined, we can calculate the fraction of the excited-state energy lost as heat, part (c), and we can draw an energy level diagram that shows how the levels are related, part (d).

Solution: We solve part (a) using Equations 7-1 and 7-2 and Avogadro's number:

$$E_{photon} = h\nu \quad and \quad \lambda\nu = c \quad so \quad E_{photon} = \frac{hc}{\lambda}$$

$$E_{photon\ absorbed} = \frac{(6.626 \times 10^{-34}\ J\ s)(2.998 \times 10^8\ m/s)}{(500\ nm)(10^{-9}\ m/nm)} = 3.97 \times 10^{-19}\ J$$

This is the energy change per atom. To calculate the change per mole, we multiply by Avogadro's number. We also need to convert from J to kJ:

$$E_{photon\ absorbed} = (3.97 \times 10^{-19}\ J)(6.022 \times 10^{23}\ mol^{-1})(10^{-3}\ kJ/J) = 239\ kJ/mol$$

Apply the same reasoning to part (b). The result gives the energy of the red photon.

$$E_{photon\ emitted} = 172\ kJ/mol$$

To solve part (c), remember that energy is conserved. The sum of the emitted heat and the emitted photon must equal the energy absorbed by the ion:

$$E_{photon\ absorbed} = E_{photon\ emitted} + E_{heat\ emitted}$$

Because 239 kJ/mol is absorbed and 172 kJ/mol is emitted, the fraction of the excitation energy re-emitted is 172/239 = 0.720. The fraction converted to heat is the difference between this value and 1.000, or 0.280. In other words, 72.0% of the energy absorbed by the chromium ion is emitted as red light, and the other 28.0% is lost as heat.

Part (d) asks for an energy level diagram for this process. The electron starts in the ground state. On absorption of a photon, the electron moves to an energy level that is higher by 239 kJ/mol. The chromium ion loses 28.0% of its excited-state energy as heat as the electron moves to a different level that is 172 kJ/mol above the ground state. Finally, emission of the red photon returns the Cr^{3+} ion to the ground state. The numerical values allow us to construct an accurate diagram.

Do the Results Make Sense? When Cr^{3+} ions absorb light, they are pumped to an energy level higher than the energy that they later emit. If calculations had shown that the emitted light had a larger energy than the absorbed light, the result would have been unreasonable because the system would violate the law of conservation of energy.

After excitation in an electric discharge, an atom of Hg returns to the ground state by emitting two photons with wavelengths of 436 nm and 254 nm. Identify the excited state on Figure 7-14 and calculate its excitation energy in kJ/mol.

Box 7-1 **Tools for Discovery: Spectroscopy, Observing from Afar**

How do we know the composition of the sun and other stars? How can we measure the temperature inside a flame so hot that any thermometer would melt? How can we explore chemical reactions among molecules that are much too tiny to see directly? Light allows us to do all these things. The study of matter with electromagnetic radiation is called *spectroscopy.*

Astronomers use spectroscopy to identify the composition of the sun and other stars. A striking example is the discovery of the element helium. In 1868, astronomers viewing a solar eclipse observed emission lines that did not match any known element. The English astronomer Joseph Lockyer attributed these lines to a new element that he named *helium,* from *helios,* the Greek word for the sun. For 25 years the only evidence for the existence of helium was these solar spectral lines.

In 1894, the Scottish chemist William Ramsay removed nitrogen and oxygen from air through chemical reactions. From the residue, Ramsay isolated argon, the first noble gas to be discovered. A year after discovering argon, Ramsay obtained an unreactive gas from uranium-containing mineral samples. The gas exhibited the same spectral lines that had been observed in the solar eclipse of 1868. After helium was shown to exist on Earth, this new element was studied and characterized.

The tip of the flame of a gas stove is much too hot to measure using physical probes, but its temperature can be measured spectroscopically. The outer portion of a gas flame glows bright blue. Other hot objects emit different colors: A candle flame is yellow, and the heating element of a hairdryer glows red. All hot objects lose energy by giving off light, and the hotter the object, the higher the average energy of the emitted light. An object at around 1000 K gives off red light. At around 2000 K, an object emits predominantly yellow light, and blue light indicates a temperature of around 3000 K.

Just as hot objects on Earth display different colors, the colors of stars range from distinctly red through yellow to quite

blue, indicating different temperatures at their surfaces. The stars making up the Orion constellation illustrate the differences in star colors. Betelgeuse, at the upper left, is a relatively cool red star whose surface temperature is around 3000 K. The distinctly blue color of Rigel, at lower right, results from its surface temperature of about 11,000 K. The surface temperature of our own sun is around 6000 K, resulting in its lovely yellow hue. The photo of Orion is a time exposure using different focal lengths, which results in the color trails that highlight different star hues.

The previous examples illustrate the use of light given off—emission spectra—to explore inaccessible objects. Absorption spectroscopy is a powerful complementary technique. In this application, light shines on the system to be studied. The researcher then examines how that light changes during the interaction with the system. Absorption spectroscopy has been enhanced greatly by the development of powerful lasers. For example, by using a combination of a pulsed laser (laser #1 in the illustration below) and a continuous laser (laser #2), chemists can observe chemical bonds in the process of breaking. The light from laser #1 is absorbed by a specific chemical bond, which stretches and breaks. The molecule absorbs light from laser #2 only while it is in the process of breaking. The graph shows the observed intensity of the second laser beam as a function of time.

If the pulse from laser #1 is ultrafast, the bond-energizing step occurs in a very short time, about 10 fs (1 fs = 10^{-15} s). As the bond stretches through the specific length at which it absorbs photons from laser #2, the molecule can absorb a photon from the second laser beam. This absorption causes the transmitted intensity of laser #2 to fall rapidly as the bond stretches. When the bond breaks, photons from laser #2 are no longer absorbed and the transmitted intensity returns to its original value. By measuring the time it takes for this to occur, chemists have determined that it takes about 200 fs for a chemical bond to break.

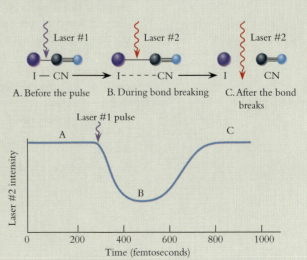

Laser #1   Laser #2   Laser #2

I — CN   I - - - - CN   I   CN

A. Before the pulse   B. During bond breaking   C. After the bond breaks

Section Exercises

7.3.1 When minerals absorb invisible ultraviolet light from a mercury lamp, they emit visible light of a longer wavelength, converting the remaining energy into heat. How much energy per mole is converted to heat by a mineral that absorbs ultraviolet light at 366 nm and emits green light at 545 nm?

7.3.2 Calculate the wavelengths that hydrogen atoms in the fifth energy level can emit.

7.3.3 Gaseous helium atoms absorb UV light of wavelength 53.7 nm. After absorbing a photon of this wavelength, a helium atom may emit light of wavelength 501.6 nm. What is the net energy change for a helium atom that has gone through this absorption–emission sequence? Draw an energy level diagram that shows the sequence.

7.4 PROPERTIES OF ELECTRONS

The energies of electrons in atoms play a central role in determining chemical behavior. Several other properties of electrons also influence the physical and chemical characteristics of atoms and molecules. Some properties are characteristic of all electrons, but others arise only when electrons are bound to atoms or molecules. In this section, we describe the properties possessed by all electrons.

Properties Shared by All Electrons

Each electron has the same mass and charge. Every electron has a mass of 9.109×10^{-31} kg and a charge of 1.602×10^{-19} C, as the experiments described in Chapter 2 demonstrate (see Figures 2-13 and 2-14 on page 42).

Electrons behave like magnets. Some types of atoms behave like tiny magnets. The best-known example is iron, the material used to make many permanent magnets. Experiments have shown that the magnetic behavior of atoms is caused by magnetic properties of their component parts, especially electrons.

The magnetic properties of electrons arise from a property called **spin,** which we describe in more detail in Chapter 8. All electrons have spin of the same magnitude, but electron spin can respond to a magnet in two different ways. Most magnetic effects associated with atoms are caused by the spins of their electrons. Iron and nickel are permanent magnets because of the cooperative effect of many electrons.

Electrons have wave properties. We are used to thinking of electrons as particles. As it turns out, electrons display both particle properties and **wave properties.** The French physicist Louis de Broglie first suggested that electrons display wave–particle duality like that exhibited by photons. De Broglie reasoned from nature's tendency toward symmetry: If things that behave like waves (light) have particle characteristics, then things that behave like particles (electrons) should also have wave characteristics.

Experiments had shown that a beam of light shining on an object exerts a pressure, and this, in turn, implies that a photon has momentum. Quantitative measurements of the pressure exerted by light showed that a simple equation relates the momentum of light (p) to its energy:

$$E = pc$$

As we have already described, light energy also is related to its wavelength:

$$E = h\nu = \frac{hc}{\lambda}$$

Setting these two energy expressions equal to each other gives an expression relating p to λ:

$$pc = \frac{hc}{\lambda}$$

 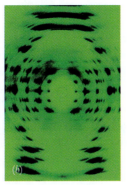 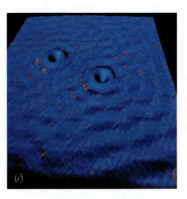

Figure 7-15
Examples of wave patterns.
(a) Floats produce standing water waves. (b) X rays generate wave interference patterns. (c) Protruding atoms on a metal surface generate standing electron waves.

The speed of light cancels, leaving an equation that de Broglie suggested should apply to electrons and other particles as well as to photons:

$$p = \frac{h}{\lambda}$$

The momentum of a particle is the product of its mass and speed, $p = mu$. Making this substitution and solving for λ gives a form of the de Broglie equation that links the wavelength of a particle with its mass and speed:

$$\lambda_{\text{particle}} = \frac{h}{mu} \qquad (7\text{-}7)$$

⑤

De Broglie's theory predicted that electrons are wave-like. How might this be confirmed? Figure 7-15 shows examples of the characteristic intensity patterns displayed by waves. In Figure 7-15a, water waves radiate away from two bobbing floats and form a standing pattern. In Figure 7-15b, X rays form a similar wave pattern. Here, high-energy photons have passed through the regular array of atoms in a crystal, whose nuclei scatter the photon waves. If electrons have wave properties, they should display regular wave patterns like these.

In 1927, American physicists Clinton Davisson and Lester Germer and British physicist George Thomson carried out experiments in which they exposed metal films to electron beams with well-defined kinetic energies. Both experiments generated patterns like those shown in Figure 7-15b, confirming the validity of the de Broglie equation for electron wavelengths. This established the wave nature of electrons.

In recent years, scanning tunneling electron microscopes have produced pictures of electron waves, an example of which appears in Figure 7-15c. Here, two atoms on an otherwise smooth metal surface act like the floats in Figure 7-15a, and cause the electrons in the metal to set up a standing wave pattern.

Both photons and electrons are particle-waves, but different equations describe their properties. Table 7-1 summarizes the properties of photons and free electrons, and Example 7-8 shows how to use these equations.

De Broglie received the Nobel Prize in physics in 1929, only two years after experiments confirmed his theory. Davisson, a student of Nobel laureate Robert Millikan, and Thomson, the son and student of J. J. Thomson (who won the Nobel prize for discovering the electron), shared the Nobel Prize in physics in 1937.

Table 7-1   Equations for Photons and Free Electrons

Property	Photon Equation	Electron Equation
Energy	$E = h\nu$	$E_{\text{kinetic}} = \dfrac{mu^2}{2}$
Wavelength	$\lambda = \dfrac{hc}{E}$	$\lambda = \dfrac{h}{mu}$
Speed	$c = 3 \times 10^8 \text{ m/s}$	$u = \sqrt{\dfrac{2E_{\text{kinetic}}}{m}}$

h, Planck's constant; ν, frequency; m, mass; u, speed.

Example 7-8
Wavelengths

The structure of a crystal can be studied by observing the wave interference patterns that result from passing particle-waves through the crystal lattice. To generate well-defined patterns, the wavelength of the particle-wave must be about the same as the distance between atomic nuclei in the crystal. In a typical crystal, this distance is 0.25 nm. Determine the energy of a photon particle-wave beam with this wavelength and of an electron particle-wave beam with this wavelength.

Strategy: This problem has two parts, one dealing with photons and the other with electrons. We are asked to relate the wavelengths of the particle-waves to their corresponding energies. Table 7-1 emphasizes that photons and electrons have different relationships between energy and wavelength. Thus, we use different equations for the two calculations.

Solution:

Photon energy is $E = h\nu = hc/\lambda$. We substitute and evaluate, being careful with units:

$$E_{photon} = \frac{(6.626 \times 10^{-34}\,\text{J s})(2.998 \times 10^{8}\,\text{m/s})}{(0.25\,\text{nm})(10^{-9}\,\text{m/nm})} = 7.9 \times 10^{-16}\,\text{J}$$

For an electron, we need to work with two equations. The de Broglie equation links the speed of an electron with its wavelength:

$$\lambda_{particle} = \frac{h}{mu}$$

The kinetic energy equation links the speed of an electron with its kinetic energy:

$$E_{kinetic} = \tfrac{1}{2}\,mu^2$$

Begin by determining the speed of the electron:

$$u_{electron} = \frac{h}{m\lambda} = \frac{(6.626 \times 10^{-34}\,\text{kg m}^2/\text{s})}{(9.109 \times 10^{-31}\,\text{kg})(0.25 \times 10^{-9}\,\text{m})} = 2.91 \times 10^{6}\,\text{m/s}$$

Next, use the speed to find the kinetic energy of the electron:

$$E_{kinetic} = \tfrac{1}{2}\,mu^2 = \frac{(9.109 \times 10^{-31}\,\text{kg})(2.91 \times 10^{6}\,\text{m/s})^2}{2}$$

$$E_{kinetic} = 3.9 \times 10^{-18}\,\text{kg m}^2/\text{s}^2 = 3.9 \times 10^{-18}\,\text{J}$$

Do the Results Make Sense? To see if these results are reasonable, we need to compare the values to other values of these quantities. The photon has a much higher energy than those associated with visible light, which is reasonable for X rays. The speed of the electron is about 1% of the speed of light—very fast but not out of the realm of possibility.

Extra Practice
Exercise 7.8

In a photoelectric-effect experiment, a photon with energy of 1.25×10^{-18} J is absorbed, causing ejection of an electron with kinetic energy of 2.5×10^{-19} J. Calculate the wavelengths (in nm) associated with each.

The de Broglie equation predicts that every particle has wave characteristics. The wave properties of subatomic particles such as electrons and neutrons play important roles in their behavior, but larger particles such as Ping-Pong balls or automobiles do not behave like waves. The reason is the scale of the waves. For all except subatomic particles, the wavelengths involved are so short that we are unable to detect the wave properties. Example 7-9 illustrates this.

Example 7-9

Matter Waves

Compare the wavelengths of an electron traveling at 1.00×10^5 m/s and a Ping-Pong ball of mass 11 g traveling at 2.5 m/s.

Strategy: This problem deals with particle-waves that have mass. Equation 7-7, the de Broglie equation, relates the mass and speed of an object to its wavelength.

Solution: For the electron: $m_e = 9.109 \times 10^{-31}$ kg, $u = 1.00 \times 10^5$ m/s

$$\lambda_{electron} = \frac{h}{mu} = \frac{6.626 \times 10^{-34} \text{ kg m}^2/\text{s}}{(9.109 \times 10^{-31} \text{ kg})(1.00 \times 10^5 \text{ m/s})} = 7.27 \times 10^{-9} \text{ m}$$

For the Ping-Pong ball: $m_{ball} = 11$ g, $u = 2.5$ m/s

$$\lambda_{ball} = \frac{h}{mu} = \frac{6.626 \times 10^{-34} \text{ kg m}^2/\text{s}}{(11 \text{ g})(1 \text{ kg}/10^3 \text{ g})(2.5 \text{ m/s})} = 2.4 \times 10^{-32} \text{ m}$$

Do the Results Make Sense? The wavelength of the electron is significant because it is about the same size as the radius of an atom, but the wavelength of the Ping-Pong ball is inconsequential compared to its size.

Calculate the wavelength associated with a proton that is moving at a speed of 2.85×10^5 m/s.

Extra Practice Exercise 7.9

Heisenberg's Uncertainty Principle

A particle occupies a particular location, but a wave has no exact position. A wave extends over some region of space. Because of their wave properties, electrons are always spread out rather than located in one particular place. As a result, *the position of a moving electron cannot be precisely defined*. We describe electrons as *delocalized* because their waves are spread out rather than pinpointed.

Mathematically, the position and momentum of a wave-particle are linked. Werner Heisenberg, a German physicist, found in the 1920s that the momentum and position of a particle-wave cannot be simultaneously pinned down. If a particular particle-wave can be pinpointed in a specific location, its momentum cannot be known. Conversely, if the momentum of a particle-wave is known precisely, its location cannot be known. Heisenberg summarized this uncertainty in what has become known as the **uncertainty principle:** The more accurately we know position, the more uncertain we are about momentum, and vice versa. Uncertainty is a feature of all objects, but it becomes noticeable only for very tiny objects like electrons.

> Momentum is the product of mass times velocity.

Heisenberg's uncertainty principle forced a change in thinking about how to describe the universe. In a universe subject to uncertainty, many things cannot be measured exactly, and it is never possible to predict with certainty exactly what will occur next. This uncertainty has become accepted as a fundamental feature of the universe at the scale of electrons, protons, and neutrons.

Section Exercises

7.4.1 Calculate the wavelength associated with a photon whose energy is 1.00×10^{-19} J and the wavelength associated with an electron having a kinetic energy of 1.00×10^{-19} J.

7.4.2 Describe the differences between the properties of a free electron and those of a Ping-Pong ball.

7.4.3 The smallest distance that can be resolved by a microscope is typically 1.25 λ, where λ is the wavelength of light or electrons used. The microscope can be greatly improved if a beam of electrons is used instead of a beam of light. If an electron microscope is to successfully resolve objects 0.600 nm apart, what kinetic energy must the electrons have?

7.5 QUANTIZATION AND QUANTUM NUMBERS

The properties of electrons described so far (mass, charge, spin, and wave nature) apply to all electrons. Electrons traveling freely in space, electrons moving in a copper wire, and electrons bound to atoms all have these characteristics. Bound electrons, those held in a specific region in space by electrical forces, have additional important properties relating to their energies and the shapes of their waves. These additional properties can have only certain specific values, so they are said to be quantized.

Energies of bound electrons are quantized. As described in Section 7.3, atoms of each element have unique, quantized electronic energy levels (see Figures 7-12 and 7-13). This quantization of energy is a property of *bound* electrons. The absorption and emission spectra of atoms consist of discrete energies because electrons undergo transitions from one bound state to another. In contrast, if an atom absorbs enough energy to remove an electron completely, the electron is no longer bound and can take on any amount of kinetic energy. *Bound* electrons have quantized energies; *free* electrons can have any amount of energy.

Absorption and emission spectroscopies provide experimental values for the quantized energies of atomic electrons. The theory of quantum mechanics provides a mathematical explanation that links quantized energies to the wave characteristics of electrons. These wave properties of atomic electrons are described by the Schrödinger equation, a complicated mathematical equation with numerous terms describing the kinetic and potential energies of the atom.

The Schrödinger equation has solutions only for specific energy values. In other words, the energy of an atom is quantized, restricted to certain values. For each quantized energy value, the Schrödinger equation generates a wave function that describes how the electrons are distributed in space.

To picture the spatial distribution of an electron around a nucleus, we must try to visualize a three-dimensional wave. Scientists have coined a name for these three-dimensional waves that characterize electrons: they are called **orbitals.** The word comes from *orbit,* which describes the path that a planet follows when it moves about the sun. An orbit, however, consists of a specific path, typically a circle or an ellipse. In contrast, an orbital is a three-dimensional volume; for example, a sphere or an hourglass. The shape of a particular orbital shows how an atomic or a molecular electron fills three-dimensional space. Just as energy is quantized, orbitals have specific shapes and orientations. We describe the details of orbitals in Section 7.6.

Each quantized property can be identified, or indexed, using a **quantum number.** These are integers that specify the values of the electron's quantized properties. Each electron in an atom has three quantum numbers that specify its three variable properties. A set of three quantum numbers is a shorthand notation that describes a particular energy, orbital shape, and orbital orientation. A fourth quantum number, with a value of $+\frac{1}{2}$ or $-\frac{1}{2}$, describes spin orientation. To describe an atomic electron completely, chemists specify a value for each of its four quantum numbers.

Principal Quantum Number

The most important quantized property of an atomic electron is its energy. The quantum number that indexes energy is the **principal quantum number (n).** For the simplest atom, hydrogen, we can use Equation 7-5 to calculate the energy of the electron if we know n. However, that equation applies only to the hydrogen atom.

No known equation provides the exact energies of an atom that has more than one electron. Nevertheless, each electron in an atom can be assigned a value of n that is a positive integer and that correlates with the energy of the electron. The most stable energy for an atomic electron corresponds to $n = 1$, and each successively higher value of n describes a less stable energy state.

KEY CONCEPT *The principal quantum number must be a positive integer: n = 1, 2, 3, . . . , 8, 9, etc.*

Values such as $n = 0$, $n = -3$, and $n = \frac{5}{2}$ are unacceptable because they do not represent solutions to the Schrödinger equation, meaning that they do not correspond to reality.

The principal quantum number also tells us something about the size of an atomic orbital, because the energy of an electron is correlated with its distribution in space. That is, the more energy the electron has, the more it is spread out in space. As illustrated by Figure 7-16, this situation is roughly analogous to a bouncing tennis ball: The more kinetic energy the ball has, the higher it bounces and the greater its average distance above the court. Similarly, the higher the principal quantum number, the more energy the electron has and the greater its average distance from the nucleus.

Summarizing, the principal quantum number (**n**) can have any positive integral value. It indexes the energy of the electron and is correlated with orbital size. As **n** increases, the energy of the electron increases, its orbital gets bigger, and the electron is less tightly bound to the atom.

$n = 1$ $n = 2$

Sizes of atomic orbitals

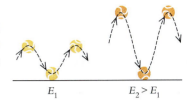

E_1 $E_2 > E_1$

Average heights of bouncing balls

Figure 7-16
The orbital size of an atomic electron (top) is correlated with its orbital energy, just as the bounce height of a tennis ball is correlated with its kinetic energy (bottom).

Azimuthal Quantum Number

In addition to size, an atomic orbital also has a specific shape. The solutions for the Schrödinger equation and experimental evidence show that orbitals have a variety of shapes. A second quantum number indexes the shapes of atomic orbitals. This quantum number is the **azimuthal quantum number (l).**

We can categorize the shapes of objects—such as the basketball, football, and tire iron shown in Figure 7-17—according to their preferred axes. A basketball has no preferred axis because its mass is distributed equally in all directions about its center. A football has one preferred axis, with more mass along this axis than in any other direction. A tire iron has two preferred axes, at right angles to each other. In analogous fashion, electron density in an orbital can be concentrated along preferred axes.

The value of **l** correlates with the number of preferred axes in a particular orbital and thereby identifies the orbital shape. According to quantum theory, orbital shapes are highly restricted. These restrictions are linked to energy, so the value of the principal quantum number (**n**) limits the possible values of **l**. The smaller **n** is, the more compact the orbital and the more restricted its possible shapes:

> When the Schrödinger equation for a one-electron atom is solved mathematically, the restrictions on **n** and **l** emerge as quantization conditions that correlate with energy and the shape of the wave function.

> **The azimuthal quantum number (l) can be zero or any positive integer smaller than n:**
> $$l = 0, 1, 2, \ldots, (n - 1).$$

KEY CONCEPT

Historically, orbital shapes have been identified with letters rather than numbers. These letter designations correspond to the values of **l** as follows:

Value of **l**	0	1	2	3	4
Orbital letter	s	p	d	f	g

An orbital is named by listing the numerical value for **n**, followed by the letter that corresponds to the numerical value for **l**. Thus, a 3s orbital has quantum numbers **n** = 3, **l** = 0. A 5f orbital has **n** = 5, **l** = 3. Notice that the restrictions on **l** mean that many combinations of **n** and **l** do not correspond to orbitals that exist. For example, when **n** = 1, **l** can only be zero. In other words, 1s orbitals exist, but there are no 1p, 1d, 1f, or 1g orbitals. Similarly, there are 2s and 2p orbitals but no 2d, 2f, or 2g orbitals. Remember that **n** restricts **l**, but **l** does not restrict **n**. Thus, a 10d orbital (**n** = 10, **l** = 2) is rather high in energy but perfectly legitimate, but there is no orbital with **n** = 2, **l** = 10.

Figure 7-17
Some everyday objects have their masses concentrated along preferred axes. A basketball has no such axes, but a football has one, and a tire iron has two.

Magnetic and Spin Orientation Quantum Numbers

A sphere has no preferred axis, so it has no directionality in space. When there is a preferred axis, as for a football, Figure 7-18 shows that the axis can point in many different directions relative to an *xyz* coordinate system. Thus, objects with preferred axes have directionality as well as shape.

Among atomic orbitals, s orbitals are spherical and have no directionality. Other orbitals are nonspherical, so, in addition to having shape, every orbital points in some direction. Like energy and orbital shape, orbital direction is quantized. Unlike footballs,

> The magnetic quantum number derives its name from the fact that different orbital orientations generate different behaviors in the presence of magnetic fields.

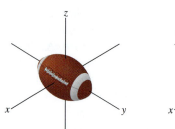

Figure 7-18
A football has directionality and shape. The figure shows four of the many ways in which a football can be oriented relative to a set of x-, y-, and z-axes.

p, d, and f orbitals have restricted numbers of possible orientations. The **magnetic quantum number (m_l)** indexes these restrictions.

Just as orbital size (n) limits the number of preferred axes (l), the number of preferred axes (l) limits the orientations of the preferred axes (m_l). When $l = 0$, there is no preferred axis and there is no orientation, so $m_l = 0$. One preferred axis ($l = 1$) can orient in any of three directions, giving three possible values for m_l: $+1$, 0, and -1. Two preferred axes ($l = 2$) can orient in any of five directions, giving five possible values for m_l: $+2$, $+1$, 0, -1, and -2. Each time l increases in value by one unit, two additional values of m_l become possible, and the number of possible orientations increases by two:

KEY CONCEPT | *The magnetic quantum number (m_l) can have any positive or negative integral value between 0 and l: $m_l = 0, \pm1, \pm2, \ldots, \pm l$*

As mentioned in Section 7-4, an electron has magnetism associated with a property called spin. Magnetism is directional, so the spin of an electron is directional, too. Like orbital orientation, spin orientation is quantized: Electron spin has only two possible orientations, "up" or "down." The **spin orientation quantum number (m_s)** indexes this behavior. The two possible values of m_s are $+\frac{1}{2}$ (up) and $-\frac{1}{2}$ (down).

Spin "up"   Spin "down"

Sets of Quantum Numbers

A complete description of an atomic electron requires a set of four quantum numbers, n, l, m_l, and m_s, which must meet all the restrictions summarized in Table 7-2. Any set of quantum numbers that does not obey these restrictions does not correspond to an orbital and cannot describe an electron.

An atomic orbital is designated by its n and l values, such as $1s$, $4p$, $3d$, and so on. When $l > 0$, there is more than one orbital of each designation: three n p orbitals, five n d orbitals, and so on. When an electron occupies any orbital, its spin quantum number, m_s, can be either $+\frac{1}{2}$ or $-\frac{1}{2}$. Thus, there are many sets of valid quantum numbers. An electron in a $3p$ orbital, for example, has six valid sets of quantum numbers:

$n = 3, l = 1, m_l = +1, m_s = +\frac{1}{2}$ $n = 3, l = 1, m_l = +1, m_s = -\frac{1}{2}$

$n = 3, l = 1, m_l = 0, m_s = +\frac{1}{2}$ $n = 3, l = 1, m_l = 0, m_s = -\frac{1}{2}$

$n = 3, l = 1, m_l = -1, m_s = +\frac{1}{2}$ $n = 3, l = 1, m_l = -1, m_s = -\frac{1}{2}$

Table 7-2   Restrictions on Quantum Numbers for Atoms

Quantum Number	Restrictions	Range
n	Positive integers	$1, 2 \ldots, \infty$
l	Positive integers less than n	$0, 1 \ldots, (n-1)$
m_l	Integers between l and $-l$	$-l \ldots, -1, 0, +1, \ldots, +l$
m_s	Half-integers, $+\frac{1}{2}$ or $-\frac{1}{2}$	$-\frac{1}{2}, +\frac{1}{2}$

As **n** increases, so does the number of valid sets of quantum numbers. Example 7-10 shows this.

Example 7-10

Valid Quantum Numbers

Determine how many valid sets of quantum numbers exist for $4d$ orbitals, and give two examples.

Strategy: The question asks for the sets of quantum numbers that have $n = 4$ and $l = 2$. Each set must meet all the restrictions listed in Table 7-2. The easiest way to see how many valid sets there are is to list all the valid quantum numbers.

Solution: Because this is a $4d$ orbital, **n** and **l** are specified and cannot vary. The other two quantum numbers, however, have several acceptable values. For each value of m_l, either value of m_s is acceptable, so the total number of possibilities is the product of the number of possible values for each quantum number:

Quantum number	n	l	m_l	m_s
Possible values	4	2	2, 1, 0, −1, −2	$+\frac{1}{2}, -\frac{1}{2}$
Number of possible values	1	1	5	2

Possible sets of values for a $4d$ electron: $(1)(1)(5)(2) = 10$

There are ten sets of quantum numbers that describe a $4d$ orbital. Here are two of them, chosen randomly:

$$n = 4, l = 2, m_l = 1, m_s = +\frac{1}{2}$$
$$n = 4, l = 2, m_l = -2, m_s = -\frac{1}{2}$$

You should be able to list the other eight sets.

Do the Results Make Sense? If you write all the possible sets, you will find there are ten of them. Notice that there are always $2l + 1$ possible values of m_l and 2 possible values of m_s, so every nd set of orbitals has ten possible sets of quantum numbers.

Write all the valid sets of quantum numbers of the $5p$ orbitals.

Section Exercises

■ **7.5.1** List all the valid sets of quantum numbers for $n = 3$. Give each its proper orbital name (for example, the orbital with $n = 1$, $l = 0$ is $1s$).

■ **7.5.2** Which of the following sets of quantum numbers describe actual orbitals, and which are nonexistent? For each one that is nonexistent, list the restriction that makes it forbidden.

	n	l	m_l	m_s
(a)	4	1	1	0
(b)	4	4	1	$+\frac{1}{2}$
(c)	4	0	1	$+\frac{1}{2}$
(d)	4	2	2	$-\frac{1}{2}$

■ **7.5.3** Determine the number of different allowable sets of quantum numbers that have $n = 4$.

7.6 SHAPES OF ATOMIC ORBITALS

Wave-like properties cause electrons to be smeared out rather than localized at an exact position. This smeared-out distribution can be described using the notion of electron density: Where electrons are most likely to be found, there is high electron density. Low electron density correlates with regions where electrons are least likely to be found. Each electron, rather than being a point charge, is a three-dimensional particle-wave that is distributed over space in an orbital. Orbitals describe the delocalization of electrons. Moreover, when the energy of an electron changes, the size and shape of its distribution in space change as well.

KEY CONCEPT   *Each atomic energy level is associated with a specific three-dimensional atomic orbital.*

An atom that contains many electrons can be described by superimposing (adding together) the orbitals for all of its electrons to obtain the overall size and shape of the atom.

The chemical properties of atoms are determined by the behavior of their electrons. Because atomic electrons are described by orbitals, the interactions of electrons can be described in terms of orbital interactions. The two characteristics of orbitals that determine how electrons interact are their shapes and their energies. Orbital shapes, the subject of this section, describe the distribution of electrons in three-dimensional space. Orbital energies, which we describe in Chapter 8, determine how easily electrons can be moved.

The quantum numbers n and l determine the size and shape of an orbital. As n increases, the size of the orbital increases, and as l increases, the shape of the orbital becomes more elaborate.

Orbital Depictions

We need ways to visualize electrons as particle-waves delocalized in three-dimensional space. Orbital pictures provide maps of how an electron wave is distributed in space. There are several ways to represent these three-dimensional maps. Each one shows some important orbital features, but none shows all of them. We use three different representations: plots of electron density, pictures of electron density, and pictures of electron contour surfaces.

An **electron density plot** is useful because it represents the electron distribution in an orbital as a two-dimensional plot. These graphs show electron density along the y-axis and distance from the nucleus, r, along the x-axis. Figure 7-19*a* shows an electron density plot for the 2*s* orbital.

Electron density plots are useful because those for several orbitals can be superimposed to indicate the relative sizes of various orbitals. The simplicity of such a graph is also a drawback, however, because it does not show the three-dimensionality of an orbital.

Orbital pictures have an advantage over electron density plots in that they can indicate the three-dimensional nature of orbitals. One type of orbital picture is a two-dimensional color pattern in which the density of color represents electron density. Figure 7-19*b* shows such an **orbital density picture** of the 2*s* orbital. This two-dimensional pattern of color density shows a cross-sectional slice through the middle of the orbital.

Orbital density pictures are probably the most comprehensive views we can draw, but they require much time and care. An **electron contour drawing** provides a simplified orbital picture. In this representation, we draw a contour surface that encloses almost all the electron density. Commonly, "almost all" means 90%. Thus, the electron density is high inside the contour surface but very low outside the surface. Figure 7-19*c* shows a contour drawing of the 2*s* orbital.

A useful analogy for understanding the value of contour surfaces is a swarm of bees around a hive. At any one time, some bees will be off foraging for nectar, so a contour surface drawn around *all* the bees might cover several acres. This would not be a very useful map of bee density. A contour surface containing 90% of the bees, on the other hand, would be just a bit bigger than the hive itself. This would be a very useful map of bee density, because anyone inside that contour surface would surely interact with bees.

The drawback of contour drawings is that all details of electron density *inside* the surface are lost. Thus, if we want to convey the maximum information about orbitals, we must use combinations of the various types of depictions.

The advantages and disadvantages of the three types of plots are highlighted by how they show one characteristic feature of orbitals. Figure 7-19*a* shows clearly that there is a value for r where the electron density falls to zero. A place where electron density is zero is called a **node.** Figure 7-19*b* shows the node for the 2*s* orbital as a white ring. In three

(*a*)

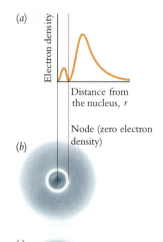

(*b*)

(*c*)

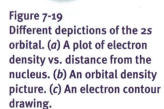

Figure 7-19
Different depictions of the 2s orbital. (*a*) A plot of electron density vs. distance from the nucleus. (*b*) An orbital density picture. (*c*) An electron contour drawing.

dimensions, this node is a spherical surface. Figure 7-19c does not show the node, because the spherical nodal surface is hidden inside the 90% contour surface. The graph shows the location of the node most clearly, the orbital density picture gives the best sense of the shape of the node, while the contour drawing fails to show it at all.

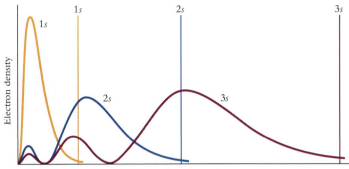

Figure 7-20
Electron density plots for the 1s, 2s, and 3s atomic orbitals of the hydrogen atom. The vertical lines indicate the values of r where the 90% contour surface would be located.

Orbital Size

How large are orbitals? Experiments that measure atomic radii provide information about the size of an orbital. In addition, theoretical models of the atom predict how the electron density of a particular orbital changes with distance from the nucleus, r. When these sources of information are combined, they reveal several regular features about orbital size.

*In any particular atom, orbitals get larger as the value of **n** increases.* For any particular atom, the **n** = 2 orbitals are larger than the 1s orbital, the **n** = 3 orbitals are larger than the **n** = 2 orbitals, and so on. The electron density plots in Figure 7-20 show this trend for the first three s orbitals of the hydrogen atom. This plot also shows that the number of nodes increases as **n** increases.

In any particular atom, all orbitals with the same principal quantum number are similar in size. As an example, Figure 7-21 shows that the **n** = 3 orbitals of the copper atom have their maximum electron densities at similar distances from the nucleus. The same regularity holds for all other atoms. The quantum numbers other than **n** affect orbital size only slightly. We describe these small effects in the context of orbital energies in Chapter 8.

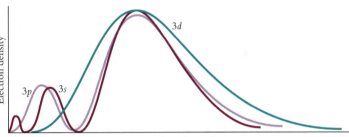

Figure 7-21
Electron density plots for the 3s (*maroon line*), 3p (*lavender line*), and 3d (*teal line*) orbitals for the copper atom. All three orbitals are nearly the same size.

Each orbital becomes smaller as nuclear charge increases. As the positive charge of the nucleus increases, the electrical force exerted by the nucleus on the negatively charged electrons increases, too, and electrons become more tightly bound. This in turn reduces the radius of the orbital. As a result, each orbital shrinks in size as atomic number increases. For example, the 2s orbital steadily decreases in size across the second row of the periodic table from Li (Z = 3) to Ne (Z = 10). The atomic number, Z, is equal to the number of protons in the nucleus, so increasing Z means increasing nuclear charge.

Details of Orbital Shapes

The shapes of orbitals strongly influence chemical interactions. Hence, we need to have detailed pictures of orbital shapes to understand the chemistry of the elements.

The quantum number $l = 0$ corresponds to an s orbital. According to the restrictions on quantum numbers, there is only one s orbital for each value of the principal quantum number. All s orbitals are spherical, with radii and number of nodes that increase as **n** increases.

The quantum number $l = 1$ corresponds to a p orbital. A p electron can have any of three values for m_l, so for each value of **n** there are three different p orbitals. The p orbitals, which are not spherical, can be shown in various ways. The most convenient representation shows the three orbitals with identical shapes but pointing in three different directions. Figure 7-22 shows electron contour drawings of the 2p orbitals. Each p orbital has high electron density in one particular direction, perpendicular to the other two orbitals, with the nucleus at the center of the system. The three different orbitals can be represented so that each has its electron density concentrated on both sides of the nucleus along a preferred axis. We can write subscripts on the orbitals to distinguish the three distinct orientations: p_x, p_y, and p_z. Each p orbital also has a nodal plane that passes through the nucleus. The nodal plane for the p_x orbital is the yz plane, for the p_y orbital the nodal plane is the xz plane, and for the p_z orbital it is the xy plane.

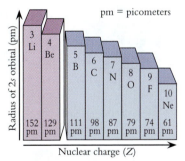

The radius of the 2s orbital decreases as Z increases.

Figure 7-22
Contour drawings of the three 2*p* orbitals. The three orbitals have the same shape, but each is oriented perpendicular to the other two.

Tutorial

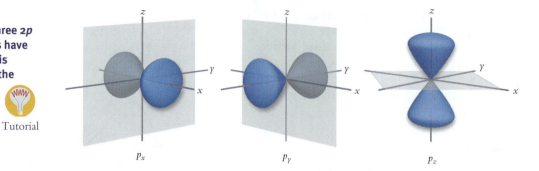

p_x p_y p_z

As *n* increases, the detailed shapes of the *p* orbitals become more complicated (the number of nodes increases, just as for *s* orbitals). Nevertheless, the *directionality* of the orbitals does not change. Each *p* orbital is perpendicular to the other two in its set, and each *p* orbital has its lobes along its preferred axis, where electron density is high. To an approaching atom, therefore, an electron in a 3*p* orbital presents the same characteristics as one in a 2*p* orbital, except that the 3*p* orbital is bigger. Consequently, the shapes and relative orientations of the 2*p* orbitals in Figure 7-22 represent the prominent spatial features of all *p* orbitals.

The quantum number *l* = 2 corresponds to a *d* orbital. A *d* electron can have any of five values for m_l (−2, −1, 0, +1, and +2), so there are five different orbitals in each set. Each *d* orbital has two nodal planes. Consequently, the shapes of the *d* orbitals are more complicated than their *s* and *p* counterparts. The contour drawings in Figure 7-23 show these orbitals in the most convenient way. In these drawings, three orbitals look like three-dimensional cloverleaves, each lying in a plane with the lobes pointed between the axes. A subscript identifies the plane in which each lies: d_{xy}, d_{xz}, and d_{yz}. A fourth orbital is also a cloverleaf in the *xy* plane, but its lobes point along the *x* and *y* axes. This orbital is designated $d_{x^2-y^2}$. The fifth orbital looks quite different. Its major lobes point along the *z* axis, but there is also a "doughnut" of electron density in the *xy* plane. This orbital is designated d_{z^2}.

The chemistry of all the common elements can be described completely using *s*, *p*, and *d* orbitals, so we need not extend our catalog of orbital shapes to the *f* orbitals and beyond.

Tutorial

Figure 7-23
Contour drawings of the *d* orbitals. Four of the five have two preferred axes and two nodal planes at right angles to each other. The d_{z^2} orbital has the *z*-axis as its only preferred axis. It has two nodes that are shaped like cones, one above and one below the *xy* plane.

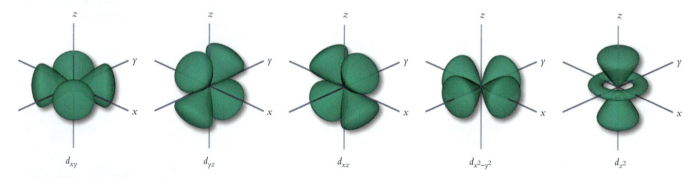

d_{xy} d_{yz} d_{xz} $d_{x^2-y^2}$ d_{z^2}

Section Exercises

7.6.1 Construct contour drawings of *s*, *p*, and *d* orbitals. Label the coordinate axes.

7.6.2 Draw orbital pictures that show a 2*p* orbital on one atom interacting with a 2*p* orbital on a different atom: (a) end-on (preferred axes pointing toward each other) and (b) side by side (preferred axes parallel to each other).

7.6.3 Construct an accurately scaled composite contour drawing on which you superimpose a 2*s* orbital, a $2p_x$ orbital, and the outermost portion of a $3p_x$ orbital of the same atom. (Use different colors to distinguish the different orbitals.) What does your contour drawing tell you about the importance of the *n* = 2 orbitals for chemical interactions when the $3p_x$ orbital contains electrons?

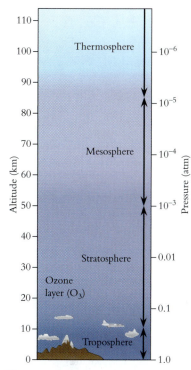

Figure 7-24
The Earth's atmosphere is divided into several regions. Note that the scale for pressure is logarithmic.

7.7 SUNLIGHT AND THE EARTH

The sun bathes the Earth in a never-ending flux of electromagnetic radiation. This light interacts with atoms and molecules in the atmosphere according to the principles developed in this chapter. We can divide the atmosphere into the regions designated in Figure 7-24. Each region occupies a particular altitude range, and each region absorbs different wavelengths of incoming light. Although these regions blend into one another without distinct boundaries, there are distinctive interactions of light with the atoms and molecules of each region. We describe these interactions from the top down, starting with the first layer encountered by incoming light from the sun.

Reactions in the Thermosphere

The **thermosphere** is the outermost part of the atmosphere, above an altitude of about 85 km. Here, molecules of nitrogen and oxygen absorb X-ray radiation coming from the sun. These photons have enough energy to ionize molecules and to break chemical bonds. Here are two examples:

$$N_2 + h\nu \longrightarrow N_2^+ + e^- \qquad O_2 + h\nu \longrightarrow O + O$$

The products of these reactions are unstable, so they eventually recombine, releasing energy in the form of heat:

$$N_2^+ + e^- \longrightarrow N_2 + Heat \qquad O + O \longrightarrow O_2 + Heat$$

The thermosphere, which gets its name from the heat released in these reactions, is a complex mixture of atoms, ions, and molecules, including a high mole fraction of oxygen atoms. At the same time, however, the total atmospheric pressure in the thermosphere is less than 10^{-7} atm, which means that the density of atoms and molecules in the thermosphere is quite low.

The aurora borealis, a spectacular atmospheric light show shown in Figure 7-25, originates in the thermosphere. In addition to electromagnetic radiation, the sun emits a steady stream of protons and electrons. The Earth's magnetic field deflects most of these particles, but some reach the thermosphere above the north and south poles of the planet, particularly during times of solar storms, when their emission from the sun is at its highest. Upon entering the atmosphere, these high-energy particles collide with the atoms and molecules in the thermosphere, transferring energy and generating highly excited states. As these excited atoms and molecules relax back to their ground states, they emit visible light. The color of the emission depends on the species involved. For example, nitrogen molecules emit red light during an aurora event. Oxygen atoms can emit either red or green light, and nitrogen atoms emit blue light. In other words, the aurora is based on the same principles as emission spectra: Atoms and molecules that reach excited states by gaining energy from some external source return to their ground states by emitting photons.

As solar radiation passes through the thermosphere, absorption by atoms and molecules removes its highest-energy photons. The result is that the intensity of high-energy light decreases as sunlight moves down through the atmosphere toward the Earth's surface. The boundary between the thermosphere and the mesosphere is the altitude where almost all the ionizing radiation has been removed. This altitude is about 85 km above the Earth's surface.

Sunlight not absorbed in the thermosphere passes through the mesosphere, which is about 35 km thick, with little absorption. Consequently, there are no important light-induced chemical reactions in the mesosphere.

Reactions in the Ozone layer

Near the top of the **stratosphere**, solar radiation generates an abundance of ozone (O_3) molecules. The resulting **ozone layer** has important consequences for life on Earth. Ozone forms in two steps. First, a photon with a wavelength between 180 and 240 nm breaks an O_2 molecule into two atoms of oxygen:

$$O_2 + h\nu_{(\lambda\,=\,180\text{-}240\ nm)} \longrightarrow O + O$$

Figure 7-25
The aurora borealis is due to the emission of photons by excited-state atoms and molecules in the thermosphere. In the Northern Hemisphere, the aurora is called the Northern Lights.

Second, an oxygen molecule captures one of these oxygen atoms to form an ozone molecule:

$$O_2 + O \longrightarrow O_3 + \text{Heat}$$

The second step occurs twice for each O_2 fragmentation, giving the overall balanced process for ozone formation:

$$3\,O_2 + h\nu_{(\lambda\,=\,180\text{-}240\,\text{nm})} \longrightarrow 2\,O_3 + \text{Heat}$$

Why is the ozone layer confined to one region of the atmosphere? The production of ozone requires both a source of oxygen atoms and frequent collisions between the atoms and the molecules that make up the atmosphere. At altitudes lower than 20 km, all the light energetic enough to split oxygen molecules into oxygen atoms has already been absorbed. Consequently, below this altitude, the dissociation of O_2 does not occur and ozone does not form. At altitudes higher than 35 km, on the other hand, there is plenty of light to dissociate O_2, but the molecular density and the rate of molecular collisions are too low. Above 35 km, there are not enough atoms and molecules for the combination reaction to occur.

Another light-induced reaction limits the ozone concentration in the ozone layer. Ozone strongly absorbs UV light in the region between 200 and 340 nm. The energies of these photons are high enough to break ozone apart into O_2 molecules and oxygen atoms:

$$O_3 + h\nu_{(\lambda\,=\,200\text{-}340\,\text{nm})} \longrightarrow O + O_2$$

The interactions of molecules and light in the ozone layer result in a delicate balance that holds ozone concentration at a relatively constant value. Three reactions maintain this balance:

1. Photons with wavelengths of 180 to 240 nm break apart O_2 molecules.
2. Photons with wavelengths of 200 to 340 nm break apart O_3 molecules.
3. Oxygen atoms combine with O_2 molecules to produce O_3 molecules and heat.

The absorption of UV light by ozone and oxygen molecules is critical for life on Earth. If this radiation reached the Earth's surface, it would cause severe biological damage because light of wavelengths around 300 nm has enough energy to break biological molecules apart. If we were continually bathed in such light, death rates would increase dramatically for almost all living species. Our Chemistry and the Environment Box (see page 290) explores the effects of ultraviolet light, and in Chapter 15 we examine the ozone layer in more detail and discuss how this delicate balance has been endangered by human activities.

The Greenhouse Effect

The energy balance between light absorbed and emitted determines the Earth's temperature. The sunlight that reaches the Earth's surface is in the ultraviolet, visible, and infrared (IR) regions. The Earth's surface absorbs much of this sunlight, which heats the Earth's surface to comfortable temperatures. This sunlight also heats and evaporates water from the oceans. This water later condenses and falls as rain or snow. Furthermore, sunlight drives photosynthesis, the chemical engine of life.

Warm bodies emit radiation, and the Earth is no exception. The Earth's temperature is determined by the balance between solar energy absorbed by the planet and the energy lost through infrared radiation. Most of the radiation emitted by the Earth lies in the infrared region of the spectrum. Nitrogen and oxygen, the major components of the atmosphere, are transparent to infrared radiation. However, some trace gases in the **troposphere,** the layer of atmosphere closest to the surface of the Earth, are strong IR absorbers. Foremost among these are carbon dioxide, water vapor, and methane. These gases are known as **greenhouse gases.** Figure 7-26 shows how this works. Much as the glass panes of a greenhouse keep the greenhouse warm, greenhouse gases moderate temperature changes from day to night by absorbing some of the infrared photons emitted

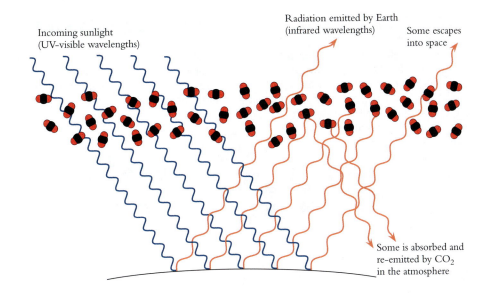

Incoming sunlight
(UV–visible wavelengths)

Radiation emitted by Earth
(infrared wavelengths)

Some escapes
into space

Some is absorbed and
re-emitted by CO_2
in the atmosphere

Figure 7-26
The role of atmospheric CO_2 in the greenhouse effect. Carbon dioxide is transparent to incoming sunlight, but it absorbs and re-emits a significant amount of the infrared radiation emitted by the Earth. This alters Earth's energy balance, raising its average temperature.

⑧

by the Earth. Following absorption, the gases re-emit still longer wavelength photons, some of which return to the surface where they are reabsorbed by the planet. This process keeps the temperature from falling dramatically at night when the ground is no longer absorbing energy from the sun.

The degree of temperature control depends on the amount of water and carbon dioxide in the atmosphere. The amount of H_2O in the atmosphere is very low in desert regions, and the atmosphere is less dense at high altitude. Consequently, deserts and mountains experience more severe day-to-night temperature variations than regions that have higher humidity or are at lower altitude.

Over the past century, the atmospheric concentration of one greenhouse gas, CO_2, has increased dramatically. Figure 7-27 shows these concentrations over the last 1100 years. The data up to 1960 are from air trapped in ice cores collected in Antarctica, while data from 1958 to the present are from an air sampling station atop Mauna Loa volcano in Hawaii. Many atmospheric chemists have warned for years that this increase is causing a small but dramatic rise in the Earth's average temperature. In 1995, the report of the United Nations Intergovernmental Panel on Climate Change documented an increase in average global air temperature of between 0.3 and 0.6 °C over the past century. More dramatically, an increase of between 1.0 and 3.5 °C is predicted for the current century.

A temperature change of 3 °C may seem trivial, but data indicate that the average global temperature during the last major ice age was only a few degrees lower than it is today. Moreover, computer models predict that higher average temperatures lead to increased heat waves and more severe weather patterns such as hurricanes. There is evidence that such effects may already be occurring. The extreme heat suffered by much of Europe in the summer of 2003 is a recent example. Additional predicted consequences include a rise in sea levels that could inundate coastal urban centers and disturbances of the ecological balance in many parts of the globe.

Among the components of our atmosphere, the concentration of carbon dioxide is a mere 325 parts per million (ppm). In other words, 999,675 of every million molecules in the air are *not* CO_2. (Almost all the molecules are N_2 or O_2.) At such a low concentration, how could CO_2 possibly cause a measurable change in the Earth's surface temperature? The answer lies in the role that minor atmospheric species play in the global energy balance.

Carbon dioxide does not affect the energy *input* to the planet because CO_2 is transparent to most of the incoming solar radiation. In contrast, CO_2 is extremely effective at absorbing infrared radiation, so the energy *output* from the planet decreases when the amount of carbon dioxide in the atmosphere increases by even a small amount. Thus, a little CO_2 goes a long way.

At present, the atmospheric CO_2 content is increasing by about 1.5 ppm/year. Moreover, concentrations of two other greenhouse gases, methane and N_2O, have also

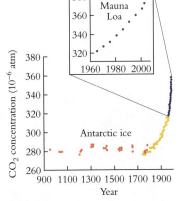

Figure 7-27
The concentration of CO_2 in the atmosphere has increased dramatically in the course of the last century, as humans have increased their consumption of fossil fuels.

Box 7-2	Chemistry and the Environment: Effects of Ultraviolet Light

U ltraviolet photons have enough energy to break chemical bonds. When UV light breaks bonds in biochemical molecules, the products can undergo chemical reactions that lead to cell damage.

The sunlight that normally reaches the Earth's surface has very little UV light with wavelengths shorter than about 360 nm. As the diagram shows, the thermosphere and stratosphere absorb photons with shorter wavelengths.

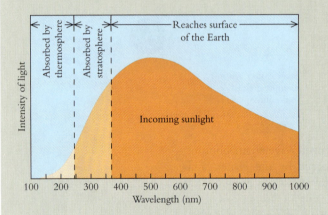

Ultraviolet light with wavelengths between 360 and 400 nm, which is designated UVA, reaches the Earth's surface. This long-wavelength UV light, while not as damaging as shorter-wavelength radiation, nevertheless can damage skin cells. Our bodies have developed a mechanism to protect us from such damage. When exposed to sunlight, skin produces melanin molecules, which absorb UV light and convert its energy into heat. Every photon that is absorbed by a melanin molecule is prevented from being absorbed by some other component of skin.

While melanin provides some protection from the damaging effects of ultraviolet light, studies have shown that prolonged exposure to sunlight increases the risk that skin cells will become cancerous. Skin cancer is much more prevalent among people who have a long history of exposure to the sun. Fair-skinned people are more prone to skin cancer, because their bodies produce less melanin and therefore less protection from the damage caused by UV light.

Sun worshipers and those who work outside can protect themselves from ultraviolet exposure with skin creams that contain UV-absorbing molecules. The UV-absorbing molecule in the first sunscreens was PABA, para-aminobenzoic acid. However, this compound may have toxic effects of its own. The UV-absorbing components of current sunscreens are derivatives of cinnamic acid and benzophenone.

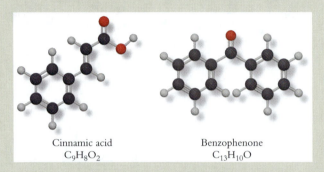

Cinnamic acid
$C_9H_8O_2$

Benzophenone
$C_{13}H_{10}O$

Ultraviolet light is also harmful to the eyes. Although our eyes do not detect UV photons, the lens of the eye is particularly susceptible to damage by UV light. The lens focuses visible photons, but it does not absorb them. In contrast, the lens does absorb ultraviolet photons, causing cell damage that eventually can result in the formation of cataracts. Optometrists and ophthalmologists recommend that eyeglasses contain UV absorbers to protect the eyes from exposure to ultraviolet light.

In recent years, the amount of ozone in the stratosphere has diminished through reactions with human-generated pollutants, as we describe in Chapter 15. As the concentration of ozone in the stratosphere fell, the amount of ultraviolet light that reached the Earth's surface increased. There is great concern about the biological effects of this higher-energy UV light, which includes light in the UVB range (290-320 nm). In the lower Southern Hemisphere, where ozone depletion has been greatest, the incidence of eye damage among sheep has shown a sharp increase in recent years, perhaps because of increased UV exposure.

Sunlight is not the only source of UV light. Ultraviolet light is emitted by mercury vapor black lights. These high-energy UV photons (UVC, between 200 and 290 nm) can be absorbed by nearby atoms and molecules. The electrons excited by this UV absorption often lose energy by emitting visible light. For example, metal ions in certain minerals absorb UV photons and emit visible light, as shown in the photos below.

increased during the industrial age. Methane has more than doubled, CO_2 has increased by 30%, and N_2O has increased by 15%. The increased presence of each of these gases in the atmosphere means that less energy escapes from the Earth's surface, causing the surface to heat up until the amount of radiant energy that escapes once again balances the incoming amount of solar energy.

The increase in greenhouse gases over the past century is a documented reality, but several uncertainties make it difficult to say how much the greenhouse effect will change Earth's climate. For example, the CO_2 buildup may not continue at the accelerating pace of recent times, because the biosphere and the oceans may be able to absorb much of the additional CO_2 that results from human activities. Although increased greenhouse gases inevitably result in global warming, some argue that these effects may be smaller than changes in the Earth's temperature caused by natural causes. The ice ages demonstrate, for example, that the Earth undergoes significant warming and cooling cycles without human intervention. Nevertheless, the human species is running a huge meteorological experiment on our planet, the long-term outcome of which might be very damaging. By the time we know the results of this experiment, it may be too late to do anything about it.

Section Exercises

■ **7.7.1** Volcanic eruptions change the Earth's energy balance by adding large amounts of smoke particles to the troposphere and stratosphere. Spectacular red sunsets result, because these particles scatter shorter-wavelength light so that less of it reaches the Earth's surface. What effect does this have on the Earth's average temperature? Draw a graph similar to the one in the Box to support your answer. On this graph, show the effect on incoming sunlight and the resulting shift, if any, in outgoing radiation.

■ **7.7.2** The ozone cycle is a delicate balance described by the three equations mentioned in the text. Combine these three equations in appropriate numbers with cancellation of common species to show that the overall ozone cycle converts UV light into heat:

$$h\nu_{(\lambda\ =\ 180\text{-}340\ nm)} \longrightarrow \text{Heat}$$

(*Hint:* You need to show one of the three reactions running backward.)

CHAPTER 7 VISUAL SUMMARY

Important equations

Essential terms

Key concepts

SKILLS TO MASTER

① Calculating properties of photons
② Analyzing the photoelectric effect
③ Calculating energies of hydrogen atoms
④ Working with energy-level diagrams
⑤ Using the de Broglie equation
⑥ Determining sets of quantum numbers
⑦ Drawing shapes of s, p, and d orbitals
⑧ Understanding how sunlight interacts with the Earth

7.1 CHARACTERISTICS OF ATOMS

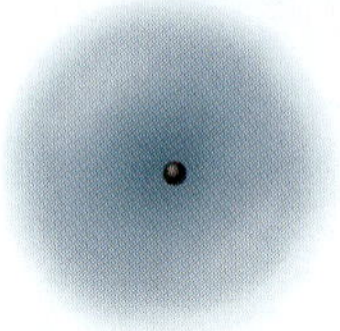

Mass
Volume
Nuclei
Electrons

7.2 CHARACTERISTICS OF LIGHT

$E_{photon} = h\nu$ $\lambda\nu = c$

①

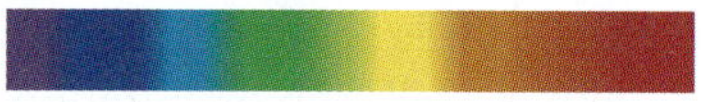

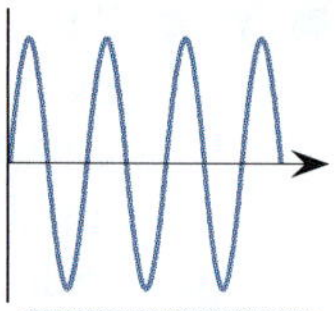

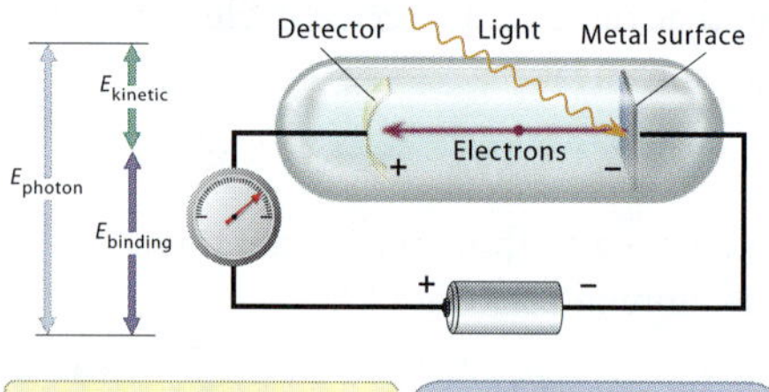

Detector   Light   Metal surface

$E_{kinetic}$

E_{photon}

Electrons

$E_{binding}$

Frequency (ν)
Wavelength (λ)
Amplitude
Intensity
Photons

$E_{electron} = h\nu - h\nu_o$ Photoelectric effect ②

7.3 ABSORPTION AND EMISSION SPECTRA

$$E_{H\ atom} = \frac{2.18 \times 10^{-18}\ J}{n^2}$$

$\Delta E_{atom} = h\nu$ ③

$E_{photon} = |\Delta E_{atom}|$

Energy level diagram

Heat loss
Photon emission

Quantized

Excited state

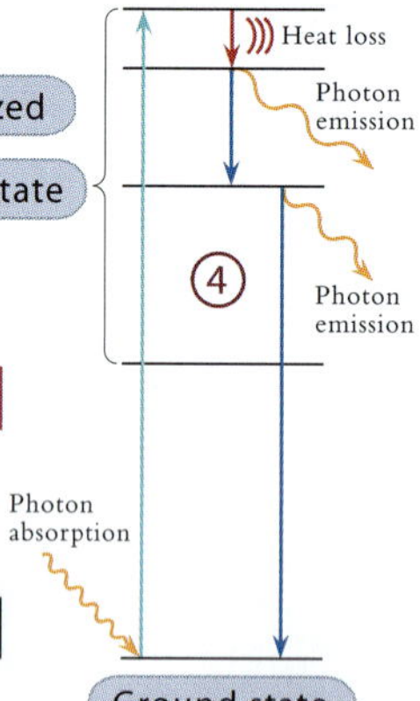

④

Photon emission

Absorption spectrum

Emission spectrum

Photon absorption

Ground state

7.4 PROPERTIES OF ELECTRONS

Wave properties   Spin

$\lambda = \dfrac{h}{mu}$ Mass Charge   ⑤

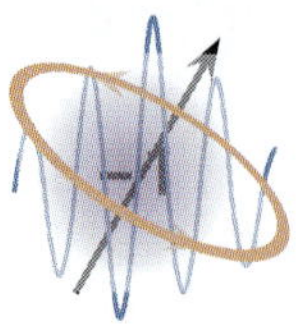

Uncertainty principle

7.5 QUANTIZATION AND QUANTUM NUMBERS

Quantum numbers:		Property:
Principal	n	Energy
Azimuthal	l	Shape
Magnetic	m_l	Orientation
Spin orientation	m_s	Spin direction

⑥

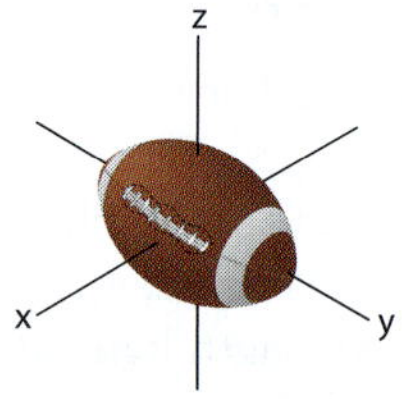

z

n is a positive integer.
l ranges from 0 to $n-1$.
m_l ranges from $-l$ to $+l$.

x   y

Planck's constant (h)   Orbitals

7.6 SHAPES OF ATOMIC ORBITALS

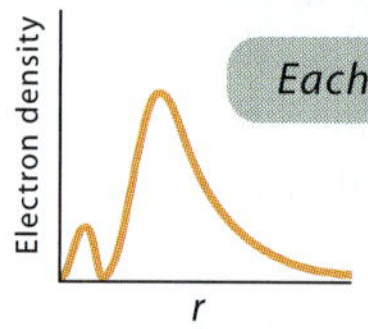

Electron density

Each energy level matches a particular orbital.

r

Electron density plot

Orbital density picture

Node

Electron contour drawing

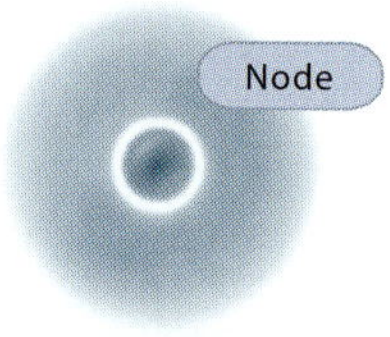

⑦

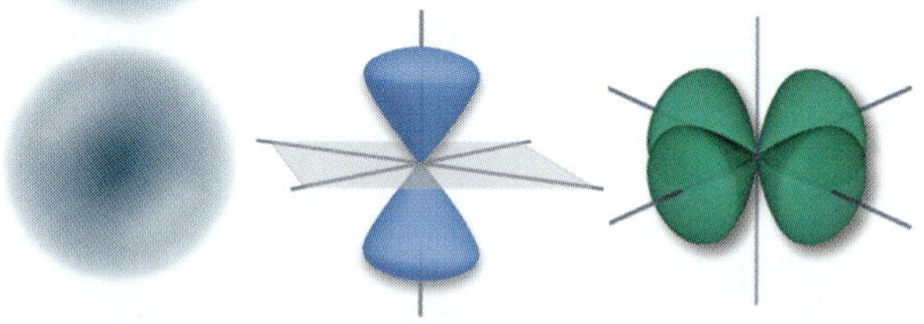

s orbital   p orbital   d orbital

7.7 SUNLIGHT AND THE EARTH

Greenhouse gases   ⑧

Incoming sunlight   Infrared radiation emitted by the Earth

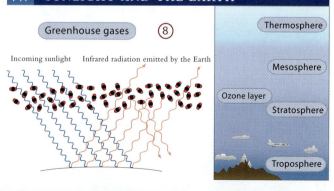

Thermosphere

Mesosphere

Ozone layer

Stratosphere

Troposphere

Learning Exercises

7.1 List the properties of electrons and of photons, including the equations used to describe each.

7.2 Write a short description of (a) the photoelectric effect, (b) wave-particle duality, (c) electron spin, and (d) the uncertainty principle.

7.3 Describe an atomic energy level diagram and the information it incorporates.

7.4 Update your list of memory bank equations. For each new equation, specify the conditions under which it can be used.

7.5 Make a list of all terms in this chapter that are new to you. Write a one-sentence definition of each in your own words. Consult the glossary if you need help.

 Problems   ilw = interactive learning ware problem. Visit the website at www.wiley.com/college/olmsted

Characteristics of Atoms

7.1 The density of silver is 1.050×10^4 kg/m^3, and the density of lead is 1.134×10^4 kg/m^3. For each metal, (a) calculate the volume occupied per atom; (b) estimate the atomic diameter; and (c) using this estimate, calculate the thickness of a metal foil containing 6.5×10^6 atomic layers of the metal.

7.2 The density of aluminum metal is 2.700×10^3 kg/m^3, and the density of copper is 8.960×10^3 kg/m^3. For each metal, (a) calculate the volume occupied per atom; (b) estimate the atomic diameter; and (c) using this estimate, calculate the thickness of a metal foil containing 6.5×10^6 atomic layers of the metal.

7.3 Describe evidence that indicates that atoms have mass.

7.4 Describe evidence that indicates that atoms have volume.

7.5 Draw an atomic picture of a layer of aluminum metal atoms.

7.6 Steel is iron containing a small amount of carbon, with the relatively small carbon atoms occupying "holes" between larger iron atoms. Draw a picture that shows what a layer of stainless steel looks like on the atomic level.

Characteristics of Light

7.7 Convert the following wavelengths into frequencies (Hz), using power-of-ten notation: (a) 4.33 nm; (b) 2.35×10^{-10} m; (c) 735 mm; (d) 4.57 μm.

7.8 Convert the following wavelengths into frequencies (Hz), using power-of-ten notation: (a) 2.76 km; (b) 1.44 cm; (c) 3.77×10^{-7} m; (d) 348 nm.

7.9 Convert the following frequencies into wavelengths, expressing the result in the indicated units: (a) 4.77 GHz (m); (b) 28.9 kHz (cm); (c) 60. Hz (mm); (d) 2.88 MHz (μm).

7.10 Convert the following frequencies into wavelengths, expressing the result in the indicated units: (a) 2.77 MHz (mm); (b) 90.1 kHz (m); (c) 50. Hz (km); (d) 8.88 GHz (μm).

7.11 Calculate the energy in joules per photon and in kilojoules per mole of the following: (a) blue-green light with a wavelength of 490.6 nm; (b) X rays with a wavelength of 25.5 nm; and (c) microwaves with a frequency of 2.5437×10^{10} Hz.

7.12 Calculate the energy in joules per photon and in kilojoules per mole of the following: (a) red light with a wavelength of 665.7 nm; (b) infrared radiation whose wavelength is 1255 nm; and (c) ultraviolet light with a frequency of 4.5528×10^{15} Hz.

ilw **7.13** A nitrogen laser puts out a pulse containing 10.0 mJ of energy at a wavelength of 337.1 nm. How many photons is this?

7.14 A dye laser emits a pulse at 450 nm that contains 2.75×10^{15} photons. What is the energy content of this pulse?

7.15 What are the wavelength and frequency of photons with the following energies: (a) 745 kJ/mol; (b) 3.55×10^{-19} J/photon?

7.16 What are the wavelength and frequency of photons with the following energies: (a) 355 J/mol; (b) 2.50×10^{-18} J/photon?

ilw **7.17** When light of frequency 1.30×10^{15} s^{-1} shines on the surface of cesium metal, electrons are ejected with a maximum kinetic energy of 5.2×10^{-19} J. Calculate (a) the wavelength of this light; (b) the binding energy of electrons to cesium metal; and (c) the longest wavelength of light that will eject electrons.

7.18 The binding energy of electrons to a chromium metal surface is 7.21×10^{-19} J. Calculate (a) the longest wavelength of light that will eject electrons from chromium metal; (b) the frequency required to give electrons with kinetic energy of 2.5×10^{-19} J; and (c) the wavelength of the light in part (b).

7.19 Draw energy level diagrams that illustrate the difference in electron binding energy between cesium metal and chromium metal. Refer to Problems 7.17 and 7.18.

7.20 A phototube delivers an electrical current when a beam of light strikes a metal surface inside the tube. Phototubes do not respond to infrared photons. Draw an energy level diagram for electrons in the metal of a phototube and use it to explain why phototubes do not respond to infrared light.

7.21 Refer to Figure 7-4 to answer the following questions: (a) What is the wavelength range for radio waves? (b) What color is light whose wavelength is 5.8×10^{-7} m? (c) In what region does radiation with frequency of 4.5×10^8 Hz lie?

7.22 Refer to Figure 7-4 to answer the following questions: (a) What is the wavelength range for infrared radiation? (b) What color is light whose wavelength is 4.85×10^{-7} m? (c) In what region does radiation with frequency of 4.5×10^{18} Hz lie?

Absorption and Emission Spectra

ilw **7.23** From Figure 7-14, calculate the energy difference in kilojoules per mole between the excited state of mercury that emits 404-nm light and the ground state.

7.24 From Figure 7-14, determine the wavelength of light needed to excite an electron from the ground state to the lowest excited state of the mercury atom.

7.25 Determine the wavelengths that hydrogen atoms absorb to reach the $n = 8$ and $n = 9$ states from the ground state. In what region of the electromagnetic spectrum do these photons lie?

7.26 Determine the frequencies that hydrogen atoms emit in transitions from the $n = 6$ and $n = 5$ levels to the $n = 3$ level. In what region of the electromagnetic spectrum do these photons lie?

7.27 Using Figure 7-13, explain why more lines appear in emission spectra than in absorption spectra.

7.28 Using Figure 7-14, explain why mercury emits photons at longer wavelengths than the wavelengths that mercury absorbs.

Properties of Electrons

7.29 What is the mass in grams of one mole of electrons?

7.30 What is the charge in coulombs of one mole of electrons?

7.31 Determine the wavelengths of electrons with the following kinetic energies: (a) 1.15×10^{-19} J; (b) 3.55 kJ/mol; (c) 7.45×10^{-3} J/mol.

7.32 Determine the wavelengths of electrons with the following kinetic energies: (a) 76.5 J/mol; (b) 4.77×10^{-18} J; (c) 3.21×10^{-11} J.

7.33 Determine the kinetic energies in joules of electrons with the following wavelengths: (a) 3.75 nm; (b) 4.66 m; (c) 8.85 mm.

7.34 Determine the kinetic energies in joules of electrons with the following wavelengths: (a) 3.75 m; (b) 4.66 μm; (c) 2.85 mm.

Quantization and Quantum Numbers

7.35 List all the valid sets of quantum numbers for a $6p$ electron.

7.36 List all the valid sets of quantum numbers for a $4f$ electron.

ilw 7.37 If you know that an electron has $n = 3$, what are the possible values for its other quantum numbers?

7.38 If you know that an electron has $m_l = -2$, what are the possible values for its other quantum numbers?

7.39 For the following sets of quantum numbers, determine which describe actual orbitals and which are nonexistent. For each one that is nonexistent, list the restriction that forbids it:

	n	l	m_l	m_s
(a)	5	3	-2	-1
(b)	5	3	-3	$\frac{1}{2}$
(c)	3	3	-3	$\frac{1}{2}$
(d)	3	0	0	$-\frac{1}{2}$

7.40 For the following sets of quantum numbers, determine which describe actual orbitals and which are nonexistent. For each one that is nonexistent, list the restriction that forbids it:

	n	l	m_l	m_s
(a)	3	-1	-1	$\frac{1}{2}$
(b)	3	1	-1	$-\frac{1}{2}$
(c)	3	1	2	$\frac{1}{2}$
(d)	3	2	2	$\frac{1}{2}$

7.41 List the values for the quantum numbers for a $3d$ electron that has spin up and the largest possible value for its magnetic quantum number.

7.42 List the values for the quantum numbers for a $4p$ electron that has spin down and the most negative possible value for its magnetic quantum number.

Shapes of Atomic Orbitals

7.43 Refer to Figures 7-19 and 7-22. Draw the analogous set of three depictions for an orbital that has $n = 2$, $l = 1$.

7.44 Refer to Figures 7-19 and 7-21. Draw the analogous set of three depictions for an orbital that has $n = 3$, $l = 0$.

7.45 Draw pictures showing how the p_y orbital looks when viewed along the z-axis, the y-axis, and the x-axis (some views may look the same).

7.46 Draw pictures showing how the d_{xz} orbital looks when viewed along the z-axis, the y-axis, and the x-axis (some views may look the same).

7.47 Identify each of the following orbitals, and state the values for n and l for each. (*Hint:* Use the size of the $3s$ orbital to identify the other three.)

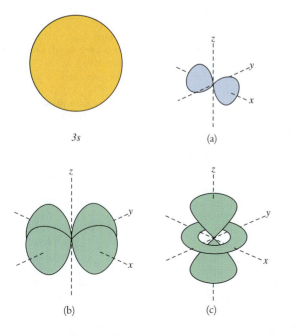

7.48 Shown below are electron density pictures and electron density plots for the $1s$, $2s$, $2p$, and $3p$ orbitals. Assign the various depictions to their respective orbitals.

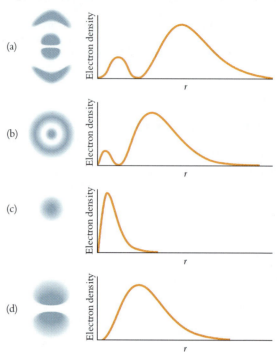

7.49 What are the limitations of plots of electron density vs. r?

7.50 The conventional method of showing the three-dimensional shape of an orbital is an electron contour surface. What are the limitations of this representation?

7.51 Construct contour drawings for the orbitals graphed in Figure 7-20, appropriately scaled to illustrate the size differences among these orbitals.

7.52 Draw contour drawings for the orbitals graphed in Figure 7-21, illustrating the shape differences among these orbitals.

Sunlight and the Earth

7.53 When molecules of nitrogen and oxygen in the thermosphere absorb short-wavelength light, N_2 molecules ionize and O_2 molecules break into atoms. Draw molecular pictures that illustrate these processes.

7.54 When ozone molecules in the mesosphere absorb UV light, they fragment into oxygen atoms and oxygen molecules. Draw a molecular picture that illustrates this process.

ilw 7.55 Light of wavelength 340 nm or shorter is required to fragment ozone molecules. What is the minimum energy in kJ/mol for this process? If an ozone molecule absorbs a 250-nm photon, how much excess kinetic energy will the fragments possess?

7.56 It requires 496 kJ/mol to break O_2 molecules into atoms and 945 kJ/mol to break N_2 molecules into atoms. Calculate the maximum wavelengths of light that can break these molecules apart. What part of the electromagnetic spectrum contains these photons?

7.57 List the region of the atmosphere and the atmospheric gases that absorb light in each of the following spectral regions: (a) less than 200 nm; (b) 240 to 310 nm.

7.58 List the region of the atmosphere and the atmospheric gases that absorb light in each of the following spectral regions: (a) 200 to 240 nm; (b) 700 to 2000 nm.

7.59 A high-altitude balloon equipped with a transmitter and pressure sensor reports a pressure of 10^{-3} atm (see Figure 7-24). (a) What altitude has the balloon reached? (b) In what region of the atmosphere is the balloon? (c) What chemical processes take place in this region?

7.60 At one stage of the return to Earth of a space shuttle flight, its instruments report that the atmospheric pressure is 10^{-6} atm (see Figure 7-24). (a) What is the altitude? (b) In what region of the atmosphere is the shuttle? (c) What chemical processes take place in this region?

Additional Paired Problems

7.61 Calculate the following for a photon of frequency 4.5×10^{13} Hz that is reflected off the moon: (a) its wavelength; (b) its energy; and (c) how long it takes to reach the Earth, which is 2.86×10^5 miles from the moon.

7.62 Calculate the following for a photon ($\lambda = 525$ nm) emitted by the sun: (a) its frequency; (b) its energy; and (c) the time it takes to reach the Earth, which is 93 million miles from the sun.

7.63 It requires 243 kJ/mol to fragment Cl_2 molecules into Cl atoms. What is the longest wavelength (in nm) of sunlight that could accomplish this? Will Cl_2 molecules in the troposphere fragment?

7.64 It requires 364 kJ/mol to break the chemical bond in HBr molecules. What is the longest wavelength (in nm) of light that has enough energy to cause this bond to break? Will HBr molecules in the troposphere be fragmented into atoms?

7.65 How many sets of quantum number values are there for a $4p$ electron? List them.

7.66 How many sets of quantum number values are there for a $3d$ electron? List them.

7.67 Redraw the first light wave in Figure 7-3 to show a wave whose frequency is twice as large as that shown in the figure, but whose amplitude is the same.

7.68 Redraw the first light wave in Figure 7-3 to show a wave whose wavelength and amplitude are each twice as large as that shown in the figure.

7.69 A minimum of 216.4 kJ/mol is required to remove an electron from a potassium metal surface. What is the longest wavelength of light that can do this?

7.70 A minimum of 216.4 kJ/mol is required to remove an electron from a potassium metal surface. If UV light at 255 nm strikes this surface, what is the maximum speed of the ejected electrons?

7.71 In a photoelectric effect experiment, photons whose energy is 6.00×10^{-19} J are absorbed by a metal, and the maximum kinetic energy of the resulting electrons is $E_{kinetic} = 2.70 \times 10^{-19}$ J. Calculate (a) the binding energy of electrons in the metal; (b) the wavelength of the light; and (c) the wavelength of the electrons.

7.72 In a photoelectric effect experiment, the minimum frequency needed to eject electrons from a metal is 7.5×10^{14} s^{-1}. Suppose that a 366-nm photon from a mercury discharge lamp strikes the metal. Calculate (a) the binding energy of the electrons in the metal; (b) the maximum kinetic energy of the ejected electrons; and (c) the wavelength associated with those electrons.

7.73 One frequency of a CB radio is 27.3 MHz. Calculate the wavelength and energy of photons at this frequency.

7.74 Microwave ovens use radiation whose wavelength is 12.5 cm. What is the frequency and energy in kJ/mol of this radiation?

7.75 Refer to Figure 7-24 to answer the following questions: (a) What is the pressure at an altitude of 60 km? (b) What atomic and molecular species are present at that altitude? (c) At what altitude is the pressure 8 torr? (d) What region of the atmosphere is this?

7.76 Refer to Figure 7-24 to answer the following questions: (a) What is the pressure at an altitude of 100 km? (b) Describe the chemistry that takes place at that altitude. (c) What are the atomic and molecular species present at that altitude?

7.77 Barium salts in fireworks generate a yellow-green color. Ba^{2+} ions emit light with $\lambda = 487, 514, 543, 553,$ and 578 nm. Convert these wavelengths into frequencies and into energies in kJ/mol.

7.78 The bright red color of highway safety flares comes from strontium ions in salts such as $Sr(NO_3)_2$ and $SrCO_3$. Burning a flare produces strontium ions in excited states, which emit red photons at 606 nm and at several wavelengths between 636 and 688 nm. Calculate the frequency and energy (kJ/mol) of emissions at 606, 636, and 688 nm.

7.79 A hydrogen atom emits a photon as its electron changes from $n = 5$ to $n = 1$. What is the wavelength of the photon? In what region of the electromagnetic spectrum is this photon found?

7.80 A hydrogen atom emits a photon as its electron changes from $n = 7$ to $n = 3$. What is the wavelength of the photon? In what region of the electromagnetic spectrum is this photon found?

More Challenging Problems

7.81 The photoelectric effect for magnesium metal has a threshold frequency of 8.95×10^{14} s^{-1}. Can Mg be used in photoelectric devices that sense visible light? Do a calculation in support of your answer.

7.82 Energetic free electrons can transfer their energy to bound electrons in atoms. In 1913, James Franck and Gustav Hertz passed electrons through mercury vapor at low pressure to determine the minimum kinetic energy required to produce the excited state that emits ultraviolet light at 253.7 nm. What is that minimum kinetic energy? What wavelength is associated with electrons of this energy?

7.83 The radius of a typical atom is 10^{-10} m, and the radius of a typical nucleus is 10^{-15} m. Compute typical atomic and nuclear volumes and determine what fraction of the volume of a typical atom is occupied by its nucleus.

7.84 Neutrons, like electrons and photons, are particle-waves whose diffraction patterns can be used to determine the structures of molecules. Calculate the kinetic energy of a neutron with a wavelength of 75 pm.

7.85 The human eye can detect as little as 2.35×10^{-18} J of green light of wavelength 510 nm. Calculate the minimum number of photons that can be detected by the human eye.

7.86 Gaseous lithium atoms absorb light of wavelength 323 nm. The resulting excited lithium atoms lose some energy through collisions with other atoms. The atoms then return to their ground state by emitting two photons with $\lambda = 812.7$ and 670.8 nm. Draw an energy level diagram that shows this process. What fraction of the energy of the absorbed photon is lost in collisions?

7.87 Calculate the wavelengths associated with an electron and a proton, each traveling at 5.000% of the speed of light.

7.88 One hydrogen emission line has a wavelength of 486 nm. Identify the values for n_{final} and $n_{initial}$ for the transition giving rise to this line.

7.89 An atomic energy level diagram, shown to scale, follows:

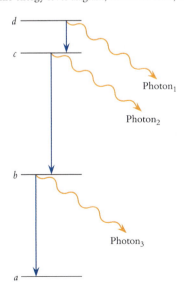

An excited-state atom emits photons when the electron moves in succession from level d to level c, from level c to level b, and from level b to the ground state (level a). The wavelengths of the emitted photons are 565 nm, 152 nm, and 121 nm (not necessarily in the proper sequence). Match each emission with the appropriate wavelength and calculate the energies of levels $b, c,$ and d relative to level a.

7.90 Small helium–neon lasers emit 1.0 mJ/s of light at 634 nm. How many photons does such a laser emit in one minute?

7.91 The argon-ion laser has two major emission lines, at 488 and 514 nm. Each of these emissions leaves the Ar^+ ion in an energy level that is 2.76×10^{-18} J above the ground state. (a) Calculate the energies of the two emission wavelengths in joules. (b) Draw an energy level diagram (in joules per atom) that illustrates these facts. (c) What frequency and wavelength radiation is emitted when the Ar^+ ion returns to its lowest energy level?

7.92 The series of emission lines that results from excited hydrogen atoms undergoing transitions to the $n = 3$ level is called the "Paschen series." Calculate the energies of the first five lines in this series of transitions, and draw an energy level diagram that shows them to scale.

7.93 It takes 486 kJ/mol to remove electrons completely from sodium atoms. Sodium atoms absorb and emit light of wavelengths 589.6 and 590.0 nm. (a) Calculate the energies of these two wavelengths in kJ/mol. (b) Draw an energy level diagram for sodium atoms that shows the levels involved in these transitions and the ionization energy. (c) If a sodium atom has already absorbed a 590.0-nm photon, what is the wavelength of the second photon a sodium atom must absorb in order to remove an electron completely?

7.94 As a general rule, the temperature drops more on clear nights than on cloudy nights. What feature of clouds accounts for this?

7.95 The graph below shows the results of photoelectron experiments on two metals, using light of the same energy:

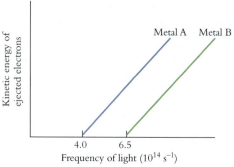

(a) Calculate the binding energy of each metal. Which has the higher binding energy? Explain. (b) Calculate the kinetic energies of electrons ejected from each metal by photons with wavelength of 125 nm. (c) Calculate the wavelength range over which photons can eject electrons from one metal but not from the other.

Group Study Problems

7.96 For the $3d_{yz}$ orbital, draw graphs of electron density vs. r: (a) for r lying along the z-axis; (b) for r lying along the x-axis; (c) for r pointing halfway between the y- and z-axes in the yz plane. (*Hint:* Consult Figure 7-23.)

7.97 A hydrogen atom undergoes an electronic transition from the $n = 4$ to the $n = 2$ state. In the process the H atom emits a photon, which then strikes a cesium metal surface and ejects an electron. It takes 3.23×10^{-19} J to remove an electron from Cs metal. Calculate (a) the energy of the $n = 4$ state of the H atom; (b) the wavelength of the emitted photon; (c) the energy of the ejected electron; and (d) the wavelength of the ejected electron.

7.98 The sun's atmosphere contains vast quantities of He^+ cations. These ions absorb some of the sun's thermal energy, promoting electrons from the He^+ ground state to various excited states. A He^+ ion in the fifth energy level may return to the ground state by emitting three successive photons: an infrared photon ($\lambda = 1014$ nm), a green photon ($\lambda = 469$ nm), and an X ray ($\lambda = 26$ nm). (a) Calculate the excitation energies of each of the levels occupied by the He^+ ion as it returns to the ground state. (b) Draw an energy level diagram for He^+ cations that illustrates these processes.

7.99 Gaseous Ca atoms absorb light at 422.7, 272.2, and 239.9 nm. After the 272.2-nm absorption, an emission at 671.8 nm occurs. Absorption of light at 239.9 nm is followed by emission at 504.2 nm. Construct an energy level diagram for Ca atoms and answer the following questions: (a) After the 504.2-nm emission, what wavelength would have to be emitted to return to the lowest energy state? (b) What wavelength of light corresponds to the energy difference between the state reached using 422.7-nm light and that reached using 272.2-nm light? (c) Do any sequences described by the data lead to a common energy level? If so, which ones?

7.100 Design a figure that summarizes the chemistry described in Section 7.7. Begin your figure with the regions of the atmosphere. Add the key atomic, molecular, and ionic species found in each region. Include three downward arrows that represent the penetration of sunlight into the atmosphere. Label two of these arrows "Ionizing radiation" and "High-energy UV light." Label the third arrow with the components of sunlight that reach the Earth's surface. Show on your figure the ozone layer, the region associated with the greenhouse effect, and the location of the aurora borealis. The text and figures of Section 7.7 contain all of the information you need.

Answers to Section Exercises

7.1.1 (a) 1.73×10^{-29} m³; and (b) 1.73×10^{-34} m³
7.1.2 (a) 2.59×10^{-10} m; and (b) 3.9×10^3 layers
7.2.1 (a) 779 nm; (b) infrared; and (c) 154 kJ
7.2.2 755 photons
7.2.3 $\nu = 9.06 \times 10^{14}$ s^{-1}; $E_{binding} = 4.00 \times 10^{-19}$ J = 241 kJ/mol
7.3.1 107 kJ/mol
7.3.2 4.05×10^3 nm, 1.28×10^3 nm, 434 nm, and 95.0 nm
7.3.3 3.30×10^{-18} J
See energy level diagram at right.
7.4.1 Photon: 1990 nm; electron: 1.55 nm
7.4.2 Their masses are very different. A free electron carries electrical charge, is a small magnet, and sometimes displays wave behavior; a Ping-Pong ball is electrically neutral and nonmagnetic and displays particle behavior.
7.4.3 1.05×10^{-18} J

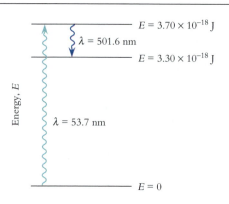

7.5.1

$n = 3$	$l = 0$	$m_l = \ \ \ 0$	$m_s = +\frac{1}{2}$	$3s$
$n = 3$	$l = 0$	$m_l = \ \ \ 0$	$m_s = -\frac{1}{2}$	$3s$
$n = 3$	$l = 1$	$m_l = \ \ \ 0$	$m_s = +\frac{1}{2}$	$3p$
$n = 3$	$l = 1$	$m_l = \ \ \ 0$	$m_s = -\frac{1}{2}$	$3p$
$n = 3$	$l = 1$	$m_l = +1$	$m_s = +\frac{1}{2}$	$3p$
$n = 3$	$l = 1$	$m_l = +1$	$m_s = -\frac{1}{2}$	$3p$
$n = 3$	$l = 1$	$m_l = -1$	$m_s = +\frac{1}{2}$	$3p$
$n = 3$	$l = 1$	$m_l = -1$	$m_s = -\frac{1}{2}$	$3p$
$n = 3$	$l = 2$	$m_l = \ \ \ 0$	$m_s = +\frac{1}{2}$	$3d$
$n = 3$	$l = 2$	$m_l = \ \ \ 0$	$m_s = -\frac{1}{2}$	$3d$
$n = 3$	$l = 2$	$m_l = +2$	$m_s = +\frac{1}{2}$	$3d$
$n = 3$	$l = 2$	$m_l = +2$	$m_s = -\frac{1}{2}$	$3d$
$n = 3$	$l = 2$	$m_l = -2$	$m_s = +\frac{1}{2}$	$3d$
$n = 3$	$l = 2$	$m_l = -2$	$m_s = -\frac{1}{2}$	$3d$
$n = 3$	$l = 2$	$m_l = +1$	$m_s = +\frac{1}{2}$	$3d$
$n = 3$	$l = 2$	$m_l = +1$	$m_s = -\frac{1}{2}$	$3d$
$n = 3$	$l = 2$	$m_l = -1$	$m_s = +\frac{1}{2}$	$3d$
$n = 3$	$l = 2$	$m_l = -1$	$m_s = -\frac{1}{2}$	$3d$

7.5.2 (a) Nonexistent, m_s must be $+\frac{1}{2}$ or $-\frac{1}{2}$; (b) nonexistent, l must be less than n; (c) nonexistent, m_l cannot exceed l; and (d) actual

7.5.3 32

7.6.1 Refer to Figures 7-19, 7-22, and 7-23.

7.6.2 (a)   (b)

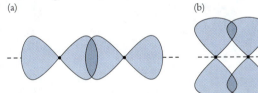

7.6.3

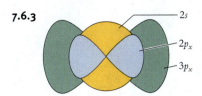

The $n = 2$ orbitals are not important when $3p_x$ is occupied.

7.7.1 Decreases the temperature, because amount of incoming blue light is reduced.

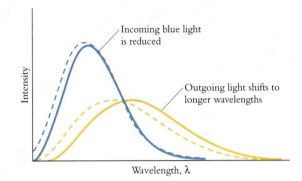

7.7.2

$$h\nu + O_2 \longrightarrow 2O$$
$$O + O_2 \longrightarrow O_3 + \text{Heat}$$
$$O + O_2 \longrightarrow O_3 + \text{Heat}$$
$$h\nu + O_3 \longrightarrow O_2 + O$$
$$h\nu + O_3 \longrightarrow O_2 + O$$
$$\underline{2O \longrightarrow O_2 + \text{Heat}}$$
$$h\nu \longrightarrow \text{Heat}$$

Answers to Extra Practice Exercises

7.1 $V_{atom} = 1.178 \times 10^{-23}$ cm^3/atom

7.2 $\nu = 2.14 \times 10^{10}$ Hz

7.3 $E_{photon} = 7.82 \times 10^{-19}$ J

7.4   Yes; $E_{kinetic}$ (electron) $= 2 \times 10^{-20}$ J

7.5 $\Delta E = 4.56 \times 10^{-19}$ J $= 274$ kJ/mol

7.6 $E = 2.04 \times 10^{-18}$ J, $\lambda = 97.2$ nm

7.7   The level is the sixth from the bottom of Figure 7-14; $E = 745$ kJ/mol

7.8   Photon has $\lambda = 159$ nm, electron has $\lambda = 0.98$ nm.

7.9   Proton wavelength is 1.39×10^{-12} m.

7.10

$n = 5, l = 1, m_l = +1, m_s = +\frac{1}{2}$	$n = 5, l = 1, m_l = +1, m_s = -\frac{1}{2}$
$n = 5, l = 1, m_l = \ \ \ 0, m_s = +\frac{1}{2}$	$n = 5, l = 1, m_l = \ \ \ 0, m_s = -\frac{1}{2}$
$n = 5, l = 1, m_l = -1, m_s = +\frac{1}{2}$	$n = 5, l = 1, m_l = -1, m_s = -\frac{1}{2}$

The Chemistry of Fireworks

Quick-burning fuse

Time-delay fuses

Red pellets (Sr) and bursting charge

Blue pellets (Cu) and bursting charge

Golden pellets (Fe) and bursting charge

Black powder propellant

People love the spectacle of fireworks. From Bastille Day in France to Guy Fawkes Day in Britain, from Chinese New Year to Canada Day, fireworks bring joy to celebrations all around the world. In the United States, about $100 million worth of fireworks are discharged every year in honor of Independence Day. Fireworks date back more than 1000 years to the discovery of black powder in China. This first gunpowder was brought to Europe during the Middle Ages and was used widely in weapons, in construction, and for fireworks.

Aerial fireworks are launched using fast-burning fuses that ignite propellants of black powder. After launching, a delayed fuse sets off a "bursting charge" that contains pellets containing an oxidizing agent and a fuel. An exothermic redox reaction within the pellets causes the components of the fireworks to emit light. A random distribution of pellets gives an irregular spray of light, while arranging the pellets on a sphere gives a highly symmetrical explosion.

To produce white light in fireworks, a reactive metal such as magnesium produces particles of solid magnesium oxide that reach temperatures in excess of 3000 °C during the explosion of the fireworks. These hot particles appear white by emitting light across the visible spectrum. Iron particles burn at a lower temperature and generate golden sparks. Larger particles stay hot longer than smaller ones and continue to burn using O_2 in the atmosphere,

Atomic Energies and Periodicity

CHAPTER CONTENTS

Learning Resources

 Key Concepts

 Critical Thinking

 Skills to Master

 Additional Help
www.wiley.com/college/olmsted

● *Tutorials*

● *Animations*

emitting long-lived sparks rather than short flashes.

The colors of fireworks displays are produced by emission from atomic ions as described in Chapter 7. The explosions of fireworks promote electrons to excited states. The energy level scheme of every element is different, so fireworks manufacturers can change colors by incorporating different elements. Sodium ions emit yellow light, strontium ions produce red light, and green comes from barium ions. Excited copper ions emit blue light, but good blue fireworks require precisely optimized conditions, so a fireworks show can be judged by the quality of its blue explosions.

The colors of fireworks depend on the energies of the atomic orbitals of the various atomic ions, but orbital energy levels have consequences that are much more far-reaching. Orbital energies determine the stabilities of atoms and how atoms react. The structure of the periodic table is based on orbital energy levels. In this chapter we explore the details of orbital energies and relate them to the form and structure of the periodic table. This provides the foundation for interpreting chemical behavior patterns.

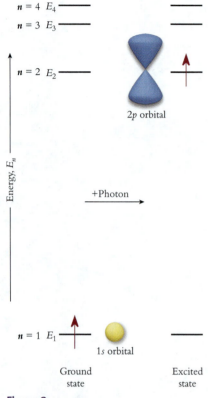

Energy, E_n

$n = 4$ E_4 —

$n = 3$ E_3 —

$n = 2$ E_2 —

2p orbital

+Photon

$n = 1$ E_1 —

1s orbital

Ground state   Excited state

Figure 8-1
When a hydrogen atom absorbs light, the energy of the photon converts it from the ground state to an excited state. In this process, its electron transfers to an orbital that has a higher energy and a larger principal quantum number.

8.1 ORBITAL ENERGIES

A hydrogen atom can absorb a photon and change from its most stable state (ground state) to a less stable state (excited state), as described in Section 7.3. We can account for this process in terms of atomic orbitals. When a hydrogen atom absorbs a photon, its electron transfers to an orbital that is higher in energy and has a larger principal quantum number. Figure 8-1 illustrates this process.

The atomic orbital model explains perfectly the spectra and the energy levels of the hydrogen atom. Does this model apply to other atoms? Experiments show that although the details are different for each kind of atom, the underlying principles are the same. Variations in nuclear charge and in the number of electrons change the magnitudes of the electrical forces that hold electrons in their orbitals. Differences in forces cause changes in orbital energies that can be understood qualitatively using forces of electrical attraction and repulsion, as we describe later in this chapter.

The Effect of Nuclear Charge

A helium +1 cation, like a hydrogen atom, has just one electron. Absorption and emission spectra show that He^+ has energy levels that depend on n, just like the hydrogen atom. Nevertheless, Figure 8-2 shows that the emission spectra of He^+ and H differ, which means that these two species must have different energy levels. We conclude that something besides n influences orbital energy.

The difference between He^+ and H can be found in their nuclei. A hydrogen nucleus is a single proton with a +1 charge, whereas a helium nucleus contains two protons and two neutrons and has a charge of +2. The larger nuclear charge of He^+ attracts the single electron more strongly than does the smaller charge of H. As a result, He^+ binds the electron with stronger force. Thus, any given energy level in the helium ion is more stable (lower in energy) than the corresponding level in the hydrogen atom.

The stability of an orbital can be determined by measuring the amount of energy required to remove an electron completely. This is the **ionization energy (IE):**

$$H \longrightarrow H^+ + e^- \qquad IE_H = 2.18 \times 10^{-18} \, J$$

$$He^+ \longrightarrow He^{2+} + e^- \qquad IE_{He^+} = 8.72 \times 10^{-18} \, J$$

The ionization energy of He^+ is four times as large as the IE of H. Thus, the ground state orbital for He^+ must be four times as stable as the ground state orbital for H. Spectral analysis shows that each orbital of a helium cation is four times as stable as its coun-

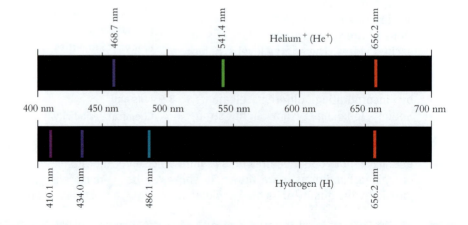

Figure 8-2
The emission spectra of He^+ and H reveal transitions at characteristic energies. The emitted photons have different wavelengths and energies because He^+ has quantized energy levels that are different from those of H.

468.7 nm   541.4 nm   Helium$^+$ (He$^+$)   656.2 nm

400 nm   450 nm   500 nm   550 nm   600 nm   650 nm   700 nm

410.1 nm   434.0 nm   486.1 nm   Hydrogen (H)   656.2 nm

Table 8-1   Comparative Ionization Energies

Orbital	H Atom	He Atom	He$^+$ Ion
1s	2.18×10^{-18} J	3.94×10^{-18} J	8.72×10^{-18} J
2p	0.545×10^{-18} J	0.585×10^{-18} J	2.18×10^{-18} J

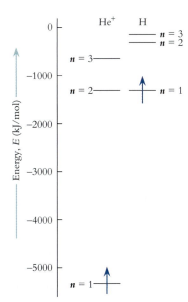

Figure 8-3
An energy level diagram for He$^+$ and H. Each species has just one electron, but the different nuclear charges make each He$^+$ orbital four times more stable than the corresponding H orbital.

terpart orbital in a hydrogen atom, showing that orbital stability increases with Z^2. Figure 8-3 shows the relationship among the energy levels of He$^+$ and H. The diagram is in exact agreement with calculations based on the Schrödinger equation.

Effect of Other Electrons

A hydrogen atom or a helium cation contains just one electron, but nearly all other atoms and ions contain *collections* of electrons. In a multielectron atom, each electron affects the properties of all the other electrons. These electron–electron interactions make the orbital energies of every element unique.

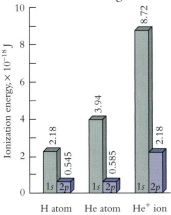

A given orbital is less stable in a multielectron atom than it is in the single-electron ion with the same nuclear charge. For instance, Table 8-1 shows that it takes more than twice as much energy to remove the electron from He$^+$ (one electron) as it does to remove one of the electrons from a neutral He atom (two electrons). This demonstrates that the 1s orbital in He$^+$ is more than twice as stable as the 1s orbital in neutral He. The nuclear charge of both species is +2, so the smaller ionization energy for He must result from the presence of the second electron. A negatively charged electron in a multielectron atom is attracted to the positively charged nucleus, but it is repelled by the other negatively charged electrons. This electron–electron repulsion accounts for the lower ionization energy of the helium atom.

Screening

Figure 8-4 shows a free electron approaching a helium cation. The +2 charge of the nucleus attracts the incoming electron, but the negative charge on the He$^+$ 1s electron repels the incoming electron. This electron–electron repulsion cancels a portion of the attraction between the nucleus and the incoming electron. Chemists call this partial cancellation **screening.**

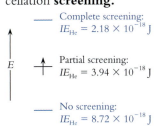

With its −1 charge, a bound electron could reduce the total charge by a maximum of one charge unit. Indeed, when an approaching electron is far enough away from a He$^+$ ion, it feels an attraction due to the net charge on the ion, +1. However, the 1s orbital is spread out all around the nucleus. This means that as an approaching electron gets close enough, the 1s electron screens only part of the total nuclear charge. Consequently, an approaching electron feels a net attraction resulting from some **effective nuclear charge (Z_{eff})** less than +2 but greater than +1.

Incomplete screening can be seen in the ionization energies of hydrogen atoms, helium atoms, and helium ions (Table 8-1). Without *any* screening, the ionization energy of a helium atom would be the same as that of a helium ion; both would be 8.72×10^{-18} J. With *complete* screening, one helium electron would compensate for one of the protons in the nucleus, making $Z_{eff} = +1$. The energy required to remove an

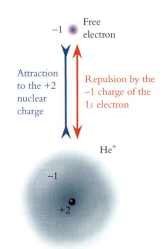

Figure 8-4
As a free electron approaches a He$^+$ cation, it is attracted to the +2 charge on the nucleus but repelled by the −1 charge on the 1s electron. When it is far from the cation, the electron experiences a net charge of +1.

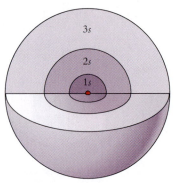

Figure 8-5
Cutaway view of the first three *s* orbitals. The 1*s* orbital screens the 2*s* and 3*s* orbitals. The 2*s* orbital screens the 3*s* orbital but not the 1*s* orbital. The 3*s* orbital is ineffective in screening the 1*s* and 2*s* orbitals.

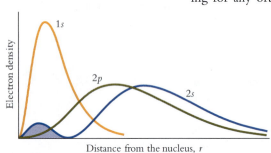

Figure 8-6
Electron density plots for the 1*s*, 2*s*, and 2*p* orbitals. Unlike the 2*p* orbital, the 2*s* orbital has significant electron density very near the nucleus (shaded region).

electron from a helium atom would then be the same as the energy required to remove an electron from a hydrogen atom, 2.18×10^{-18} J. The actual ionization energy of a helium atom is 3.94×10^{-18} J, about twice the fully screened value and about half the totally unscreened value. Screening is incomplete because both helium electrons occupy an extended region of space, so neither is completely effective at shielding the other from the +2 charge of the nucleus.

Electrons in compact orbitals pack around the nucleus more tightly than do electrons in large, diffuse orbitals. As a result, the effectiveness in screening nuclear charge decreases as orbital size increases. Because the size of an orbital increases with *n*, an electron's ability to screen decreases as *n* increases. In a multielectron atom, lower-*n* electrons are concentrated between the nucleus and higher-*n* electrons. The negative charges of these inner electrons counteract most of the positive charge of the nucleus.

The efficient screening by electrons with small values of *n* can be appreciated by comparing the ionization energies of the 2*p* orbitals listed in Table 8-1. Consider an excited-state helium atom that has one of its electrons in the 1*s* orbital and its other electron in a 2*p* orbital. It takes 0.585×10^{-18} J to remove the 2*p* electron from this excited-state helium atom. This value is almost the same as that of an excited hydrogen atom with its lone electron in a 2*p* orbital, 0.545×10^{-18} J. It is much less than the 2*p* orbital ionization energy of an excited He$^+$ ion, 2.18×10^{-18} J. These data show that Z_{eff} is quite close to +1 for the 2*p* orbital of an excited atom. In the excited He atom, the electron in the 1*s* orbital is very effective at screening the electron in the 2*p* orbital from the full +2 charge of the nucleus.

In multielectron atoms, electrons with any given value of *n* provide effective screening for any orbital with a larger value of *n*. That is, *n* = 1 electrons screen the *n* = 2, *n* = 3, and larger orbitals, whereas *n* = 2 electrons provide effective screening for the *n* = 3, *n* = 4, and larger orbitals but provide little screening for the *n* = 1 orbital. Figure 8-5 illustrates this for the 1*s*, 2*s*, and 3*s* orbitals.

The amount of screening also depends on the shape of the orbital. The shaded area of the electron density plot shown in Figure 8-6 emphasizes that the 2*s* orbital has a region of significant electron density near the nucleus. A 2*p* orbital lacks this inner layer, so virtually all of its electron density lies outside the region occupied by the 1*s* orbital. Consequently, a 1*s* orbital screens the 2*p* orbital more effectively than the 2*s* orbital, even though both *n* = 2 orbitals are about the same size. Thus, a 2*s* electron feels a larger effective nuclear charge than a 2*p* electron. This results in stronger electrical attraction to the nucleus, which makes the 2*s* orbital more stable than the 2*p* orbitals. The 2*p* orbitals of any multielectron atom are always less stable (higher in energy) than the 2*s* orbital.

The screening differences experienced by the 2*s* and 2*p* orbitals also extend to larger values of *n*. The 3*s* orbital is more stable than the 3*p* orbital, the 4*s* is more stable than the 4*p*, and so on. Orbitals with higher *l* values show similar effects. The 3*d* orbitals are always less stable than the 3*p* orbitals, and the 4*d* orbitals are less stable than the 4*p* orbitals. These effects can be summarized in a single general statement:

KEY CONCEPT   *The higher the value of the l quantum number, the more that orbital is screened by electrons in smaller, more stable orbitals.*

In a one-electron system (H, He$^+$, Li^{2+}, and so on) the stability of the orbitals depends only on *Z* and *n*. In multielectron systems, orbital stability depends primarily on *Z* and *n*, but it also depends significantly on *l*. In a sense, *l* fine-tunes orbital energies.

Electrons with the same *l* value but different values of m_l do not screen one another effectively. For example, when electrons occupy different *p* orbitals, the amount of mutual screening is slight. This is because screening is effective only when much of the electron density of one orbital lies between the nucleus and the electron density of another. Recall from Chapter 7 (Figure 7-22) that the *p* orbitals are perpendicular to one another, with high electron densities in different regions of space. The electron density of the $2p_x$ orbital does not lie between the $2p_y$ orbital and the nucleus, so there is little screening. The d orbitals also occupy different regions of space from one another, so mutual screening among electrons in these orbitals is also small. Example 8-1 provides another look at screening.

Example 8-1

Screening

Make an electron density plot showing the $1s$, $2p$, and $3d$ orbitals to scale. Label the plot in a way that summarizes the screening properties of these orbitals.

Strategy: This is a qualitative problem that asks us to combine information about three different orbitals on a single plot. We need to find electron density information and draw a single graph to scale.

Solution: Figures 7-20, 7-21, and 8-6 show electron density plots of the $n = 1$, $n = 2$, and $n = 3$ orbitals. We extract the shapes of the $1s$, $2p$, and $3d$ orbitals from these graphs. Then we add labels that summarize the screening properties of these orbitals. Screening is provided by small orbitals whose electron density is concentrated inside larger orbitals. In this case, $1s$ screens both $2p$ and $3d$; $2p$ screens $3d$, but not $1s$; and $3d$ screens neither $1s$ nor $2p$. The screening patterns can be labeled as shown.

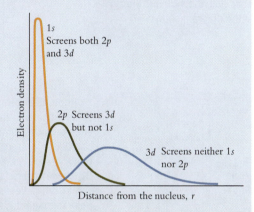

Does the Result Make Sense? We know that the most important factor for orbital size is the value of n and that small orbitals screen better than large ones, so the screening sequence makes sense.

Construct and label a graph illustrating that the $2s$ orbital screens the $3s$ orbital more effectively than the $3s$ screens the $2s$.

Quantitative information about energies of atomic orbitals is obtained using **photoelectron spectroscopy,** which applies the principles of the photoelectric effect to gaseous atoms. Our Tools for Discovery Box (on the next page) explores this powerful spectroscopic technique.

There are many different atomic orbitals, and each has a characteristic energy and shape. How the electrons of an atom distribute themselves among the atomic orbitals is the subject of the next two sections.

Section Exercises

■ **8.1.1** Figures 8-6, 7-20, and 7-21 show electron density plots of $n = 1$, $n = 2$, and $n = 3$ orbitals. Draw a plot that shows the $n = 1$ and $n = 3$ orbitals to scale. Use different colors to keep the figure as clear as possible. Shade the regions of the $3s$ and $3p$ plots where screening by $1s$ electrons is relatively ineffective.

■ **8.1.2** The outer layers of the sun contain He atoms in various excited states. One excited state contains one $1s$ electron and one $3p$ electron. Based on the effectiveness of screening, estimate the ionization energy of the $3p$ electron in this excited atom.

■ **8.1.3** Redraw Figure 8-3 so that it includes the energy levels for Li^{2+}.

Box 8-1	**Tools for Discovery: Photoelectron Spectroscopy**

Absorption and emission spectra of atoms and ions yield information about energy differences *between* orbitals, but they do not give an orbital's *absolute* energy. The most direct measurements of orbital energies come from a technique called *photoelectron spectroscopy*.

Photoelectron spectroscopy works like the photoelectric effect described in Chapter 7, except that the sample often is a gas. Light shines on the sample, and when the sample absorbs a photon, the photon's energy is transferred to an electron. If the photon energy is high enough, the electron is ejected from the sample. The kinetic energy of the ejected electron is equal to the difference between the photon energy and the binding energy (ionization energy) of the electron, as the diagram shows.

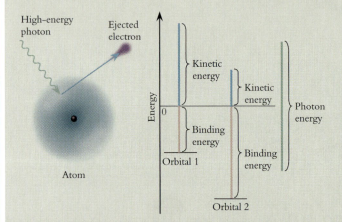

A photoelectron spectrometer uses high-energy photons, typically 11,900 kJ/mol (emitted by excited Mg atoms). These photons have more than enough energy to eject an electron from an atom or molecule of the sample. The kinetic energy of the departing electron can be determined by measuring its speed. Knowing the photon energy and the electron's kinetic energy, we can find the binding energy of the orbital from which the electron came:

Orbital binding energy = $E_{photon} - E_{kinetic}$ (electron)

The photoelectron spectrum of a monatomic gas is a set of peaks representing the energies of orbitals. The figure shows the spectrum of neon. Two peaks correspond to orbital energies of 2080 and 4680 kJ/mol. The smaller value is the ionization energy of the 2p orbital and the larger value is the ionization

energy of the 2s orbital. The ionization energy of neon's 1s electrons is so great that they cannot be ejected by the photons used in photoelectron experiments. The ionization energy of an electron measures its stability when the electron is bound to the atom. In other words, a 2p electron of a neon atom is 2080 kJ/mol more stable than a free electron in a vacuum.

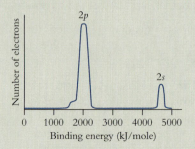

Photoelectron spectra of the first five elements, shown schematically in the graph, illustrate how atomic energy levels change with atomic number and with quantum numbers. The peaks in black are due to 1s electrons. As *Z* increases, the stability of the 1s orbital increases dramatically (for Be and B, these orbitals have ionization energies greater than 10^4 kJ/mol). Peaks in blue are due to 2s electrons, and the peak in magenta is attributed to a 2p electron. The table below gives values for orbital ionization energies (all in kJ/mol) for elements 11-21 obtained from photoelectron spectra.

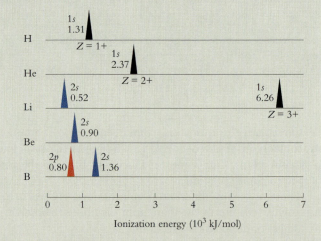

Ionization energy (10^3 kJ/mol)

						Element					
Orbital	**Na**	**Mg**	**Al**	**Si**	**P**	**S**	**Cl**	**Ar**	**K**	**Ca**	**Sc**
2s	6840	9070	*								
2p	3670	5310	7790	*							
3s	500	740	1090	1460	1950	2050	2440	2820	3930	4650	5440
3p		580	790	1010	1000	1250	1520	2380	2900	3240	
3d											770
4s									420	590	630

All values in kJ/mol.
*For this and all other elements with higher *Z* values, *IE* > 11,900 kJ/mol.

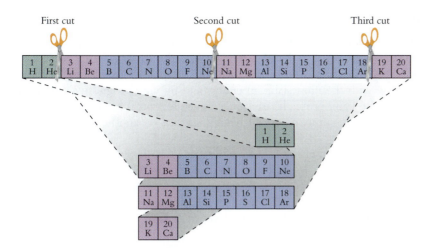

First cut   Second cut   Third cut

Figure 8-7
To generate the periodic table, the ribbon of elements is cut, and the segments are placed in rows so that each column contains elements with similar chemical properties.

8.2 STRUCTURE OF THE PERIODIC TABLE

The periodicity of chemical properties, which is summarized in the periodic table, is one of the most useful organizing principles in chemistry. Periodic patterns also provide information about electron arrangements in atoms.

The periodic table lists the elements in order of increasing atomic number. Because the number of electrons in a neutral atom is the same as its atomic number, this list is also in order of increasing number of atomic electrons. Hydrogen, with $Z = 1$ and one electron, appears first, followed sequentially by helium (two electrons), lithium (three electrons), and so on.

A list of the elements in one long row is not a periodic table, because it does not reveal periodic patterns. To convert this single, long row into a periodic table, the "ribbon" of elements is cut at appropriate points to generate strips that are placed in rows. The rows are positioned so that each column contains elements with similar chemical properties. Figure 8-7 illustrates the positioning for the first 20 elements. The first cut comes between helium (two electrons) and lithium (three electrons); the next comes between neon (10 electrons) and sodium; and the third is between argon (18 electrons) and potassium. With this arrangement, the first column contains Li, Na, and K—three highly reactive metals—while the last column contains He, Ne, and Ar—three unreactive gases.

Before establishing the connection between atomic orbitals and the periodic table, we must first describe two additional features of atomic structure: the Pauli exclusion principle and the aufbau principle.

The Pauli Exclusion Principle

It is a general principle of nature that any system tends to find its most stable arrangement. Atoms obey this principle, so hydrogen atoms are normally in their most stable state, the **ground state.** A hydrogen atom in its ground state has its one electron in the 1s orbital. We might expect a multielectron atom in its ground state to have all its electrons in the most stable, 1s orbital. However, studies show that this is not the case. Instead, bound electrons obey a fundamental law of quantum mechanics called the **Pauli exclusion principle:**

Wolfgang Pauli won the Nobel Prize for Physics in 1945 for discovering the exclusion principle.

Each electron in an atom has a unique set of quantum numbers.   **KEY CONCEPT**

Named for the Austrian physicist Wolfgang Pauli (1900–1958), this principle can be derived from the mathematics of quantum mechanics, but it cannot be rationalized in a simple way. Nevertheless, all experimental evidence upholds the idea. When one electron in an atom has a particular set of quantum numbers, no other electron in the atom is described by that same set. There are no exceptions to the Pauli exclusion principle.

The Aufbau Principle

The ground state of an atom is, by definition, the most stable arrangement of its electrons. *Most stable* means that the electrons occupy the lowest-energy orbitals available. We construct the **ground-state configuration** of an atom by placing electrons in the orbitals starting with the most stable in energy and moving progressively upward. In accordance with the Pauli principle, each successive electron is placed in the *most stable orbital* whose quantum numbers *are not already assigned* to another electron. This is the **aufbau principle.**

Aufbau **is a German word meaning "construction."**

In applying the aufbau principle, remember that a full description of an electron requires four quantum numbers: n, l, m_l, and m_s. Each combination of the quantum numbers n and l describes one quantized energy level. Moreover, each level includes one or more orbitals, each with a different value of m_l. Within a set, all orbitals with the same value of l have the same energy. For example, the 2p energy level ($n = 2$, $l = 1$) consists of three distinct p orbitals ($m_l = -1$, 0, and $+1$), all with the same energy. In addition, different values of the spin orientation quantum number, m_s, describe the different spin orientations of an electron. When the two possible values of m_s are taken into account, we find that six different sets of quantum numbers can be used to describe an electron in a 2p energy level:

$$n = 2 \qquad l = 1 \qquad m_l = +1 \qquad m_s = +\tfrac{1}{2}$$

$$n = 2 \qquad l = 1 \qquad m_l = +1 \qquad m_s = -\tfrac{1}{2}$$

$$n = 2 \qquad l = 1 \qquad m_l = \;\;\;0 \qquad m_s = +\tfrac{1}{2}$$

$$n = 2 \qquad l = 1 \qquad m_l = \;\;\;0 \qquad m_s = -\tfrac{1}{2}$$

$$n = 2 \qquad l = 1 \qquad m_l = -1 \qquad m_s = +\tfrac{1}{2}$$

$$n = 2 \qquad l = 1 \qquad m_l = -1 \qquad m_s = -\tfrac{1}{2}$$

In other words, the 2p energy level consists of three orbitals with different m_l values that can hold as many as six electrons without violating the Pauli exclusion principle. The same is true of every set of p orbitals (3p, 4p, etc.). Each set has six equivalent descriptions and can hold six electrons. A similar analysis for other values of l shows that each s energy level contains a single orbital and can hold up to two electrons, each d energy level consists of five different orbitals than can hold up to 10 electrons, and each f energy level consists of seven different orbitals that can hold up to 14 electrons.

The Pauli and aufbau principles dictate where the cuts occur in the ribbon of elements. After two electrons have been placed in the 1s orbital (He), the next electron must go in a less stable, $n = 2$ orbital (Li). After eight additional electrons have been placed in the 2s and 2p orbitals (Ne), the next electron must go in a less stable, $n = 3$ orbital (Na). The ends of the rows in the periodic table are the points at which the next electron occupies an orbital of next higher principal quantum number.

Which $n = 2$ orbital does the third electron in a lithium atom occupy? Screening causes the orbitals with the same principal quantum number to decrease in stability as l increases. Consequently, the 2s orbital, being more stable than the 2p orbital, fills first. Similarly, 3s fills before 3p, which fills before 3d, and so on.

Here is a summary of the conditions for atomic ground states:

1. Each electron in an atom occupies the most stable available orbital.
2. No two electrons can have identical descriptions.
3. Orbital capacities are as follows: s, 2 electrons; p set, 6 electrons; d set, 10 electrons; f set, 14 electrons.
4. The higher the value of n, the less stable the orbital.
5. For equal n, the higher the value of l, the less stable the orbital.

Armed with these conditions, we can correlate the rows and columns of the periodic table with values of the quantum numbers n and l. This correlation appears in the periodic table shown in Figure 8-8. Remember that the elements are arranged so that Z

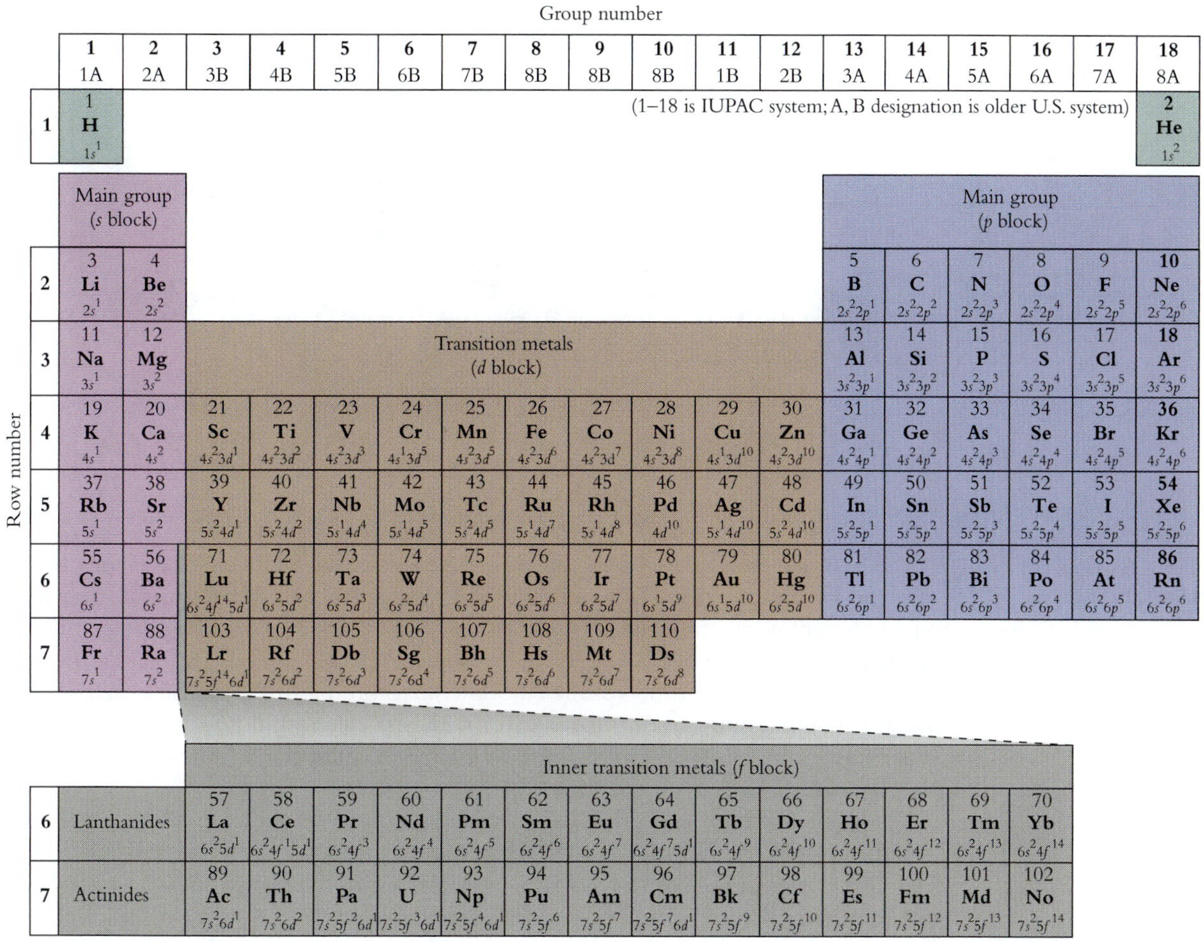

Figure 8-8 (Periodic Table)

Group number

1 1A	2 2A	3 3B	4 4B	5 5B	6 6B	7 7B	8 8B	9 8B	10 8B	11 1B	12 2B	13 3A	14 4A	15 5A	16 6A	17 7A	18 8A

(1–18 is IUPAC system; A, B designation is older U.S. system)

Main group (s block) — Main group (p block) — Transition metals (d block)

Row 1:
- 1 H $1s^1$
- 2 He $1s^2$

Row 2:
- 3 Li $2s^1$ | 4 Be $2s^2$ | 5 B $2s^22p^1$ | 6 C $2s^22p^2$ | 7 N $2s^22p^3$ | 8 O $2s^22p^4$ | 9 F $2s^22p^5$ | 10 Ne $2s^22p^6$

Row 3:
- 11 Na $3s^1$ | 12 Mg $3s^2$ | 13 Al $3s^23p^1$ | 14 Si $3s^23p^2$ | 15 P $3s^23p^3$ | 16 S $3s^23p^4$ | 17 Cl $3s^23p^5$ | 18 Ar $3s^23p^6$

Row 4:
- 19 K $4s^1$ | 20 Ca $4s^2$ | 21 Sc $4s^23d^1$ | 22 Ti $4s^23d^2$ | 23 V $4s^23d^3$ | 24 Cr $4s^13d^5$ | 25 Mn $4s^23d^5$ | 26 Fe $4s^23d^6$ | 27 Co $4s^23d^7$ | 28 Ni $4s^23d^8$ | 29 Cu $4s^13d^{10}$ | 30 Zn $4s^23d^{10}$ | 31 Ga $4s^24p^1$ | 32 Ge $4s^24p^2$ | 33 As $4s^24p^3$ | 34 Se $4s^24p^4$ | 35 Br $4s^24p^5$ | 36 Kr $4s^24p^6$

Row 5:
- 37 Rb $5s^1$ | 38 Sr $5s^2$ | 39 Y $5s^24d^1$ | 40 Zr $5s^24d^2$ | 41 Nb $5s^14d^4$ | 42 Mo $5s^14d^5$ | 43 Tc $5s^24d^5$ | 44 Ru $5s^14d^7$ | 45 Rh $5s^14d^8$ | 46 Pd $4d^{10}$ | 47 Ag $5s^14d^{10}$ | 48 Cd $5s^24d^{10}$ | 49 In $5s^25p^1$ | 50 Sn $5s^25p^2$ | 51 Sb $5s^25p^3$ | 52 Te $5s^25p^4$ | 53 I $5s^25p^5$ | 54 Xe $5s^25p^6$

Row 6:
- 55 Cs $6s^1$ | 56 Ba $6s^2$ | 71 Lu $6s^24f^{14}5d^1$ | 72 Hf $6s^25d^2$ | 73 Ta $6s^25d^3$ | 74 W $6s^25d^4$ | 75 Re $6s^25d^5$ | 76 Os $6s^25d^6$ | 77 Ir $6s^25d^7$ | 78 Pt $6s^15d^9$ | 79 Au $6s^15d^{10}$ | 80 Hg $6s^25d^{10}$ | 81 Tl $6s^26p^1$ | 82 Pb $6s^26p^2$ | 83 Bi $6s^26p^3$ | 84 Po $6s^26p^4$ | 85 At $6s^26p^5$ | 86 Rn $6s^26p^6$

Row 7:
- 87 Fr $7s^1$ | 88 Ra $7s^2$ | 103 Lr $7s^25f^{14}6d^1$ | 104 Rf $7s^26d^2$ | 105 Db $7s^26d^3$ | 106 Sg $7s^26d^4$ | 107 Bh $7s^26d^5$ | 108 Hs $7s^26d^6$ | 109 Mt $7s^26d^7$ | 110 Ds $7s^26d^8$

Inner transition metals (f block)

Lanthanides (Row 6):
- 57 La $6s^25d^1$ | 58 Ce $6s^24f^15d^1$ | 59 Pr $6s^24f^3$ | 60 Nd $6s^24f^4$ | 61 Pm $6s^24f^5$ | 62 Sm $6s^24f^6$ | 63 Eu $6s^24f^7$ | 64 Gd $6s^24f^75d^1$ | 65 Tb $6s^24f^9$ | 66 Dy $6s^24f^{10}$ | 67 Ho $6s^24f^{11}$ | 68 Er $6s^24f^{12}$ | 69 Tm $6s^24f^{13}$ | 70 Yb $6s^24f^{14}$

Actinides (Row 7):
- 89 Ac $7s^26d^1$ | 90 Th $7s^26d^2$ | 91 Pa $7s^25f^26d^1$ | 92 U $7s^25f^36d^1$ | 93 Np $7s^25f^46d^1$ | 94 Pu $7s^25f^6$ | 95 Am $7s^25f^7$ | 96 Cm $7s^25f^76d^1$ | 97 Bk $7s^25f^9$ | 98 Cf $7s^25f^{10}$ | 99 Es $7s^25f^{11}$ | 100 Fm $7s^25f^{12}$ | 101 Md $7s^25f^{13}$ | 102 No $7s^25f^{14}$

Figure 8-8
The periodic table with its rows and blocks labeled to show the relationship between sectors of the table and ground-state configurations. Rows are labeled with the highest principal quantum number of the occupied orbitals, and each block is labeled with the letter (s, p, d, f) indicating the orbital set that is filling.

---

increases one unit at a time from left to right across a row. At the end of each row, we move down one row, to the next higher value of n, and return to the left side to the next higher Z value. Inspection of Figure 8-8 reveals that the ribbon of elements is cut after elements 2, 10, 18, 36, 54, and 86.

As the atomic number increases, the length of ribbon between cuts increases, too. The first segment contains only hydrogen and helium. Then there are two 8-element segments, followed by two 18-element segments, and finally two 32-element pieces. The last segment stops before reaching its full length, however, because these elements have not been discovered in nature, nor have they been prepared in the laboratory. As we describe in Chapter 22, nuclei of elements with very high Z are unstable.

In Figure 8-8, each *row* is labeled with the highest principal quantum number of its occupied orbitals. For example, elements of the third row (Na → Ar) have electrons in orbitals with $n = 3$ (in addition to electrons with $n = 1$ and 2). Each *column* is labeled with its group number, starting with Group 1 on the left and proceeding to Group 18 on the right (the f block does not have group numbers). In general, elements in the same group of the periodic table have the same arrangement of electrons in their least stable occupied orbitals.

Order of Orbital Filling

The variations in orbital stability with n and l ensure that $1s$ fills before $2s$ and $2s$ fills before $2p$. After $2p$, the next orbitals to fill are $3s$ and $3p$, but then what? Both $3d$ and $4s$ are less stable than $3p$, so either might be the next orbital to fill. Calculations of screening show that the $4s$ and $3d$ orbital energies for the elements from $Z = 19$ to $Z = 30$ are nearly the same, as shown schematically in Figure 8-9. We cannot predict which orbital fills next from this stability ladder.

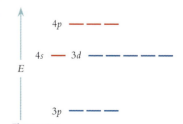

Figure 8-9
The calculated energy level diagram for neutral atoms with Z between 19 and 30 shows that the $3d$ and $4s$ atomic orbitals have nearly the same energy.

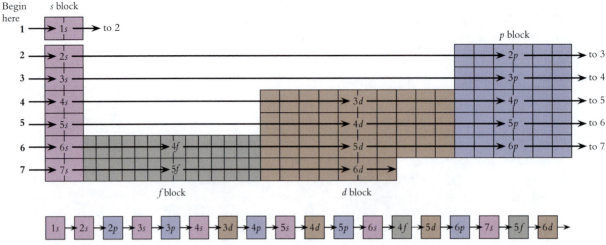

Figure 8-10
The periodic table in block form, showing the filling sequence of the atomic orbitals. Filling proceeds from left to right across each row and from the right end of each row to the left end of the succeeding row.

The periodic table provides the answer. Each cut in the ribbon of the elements falls at the end of the p block. This indicates that when the np orbitals are full, the next orbital to accept electrons is the $(n + 1)s$ orbital. For example, after filling the $3p$ orbitals from Al ($Z = 13$) to Ar ($Z = 18$), the next element, potassium, has its final electron in the $4s$ orbital rather than in one of the $3d$ orbitals. According to the aufbau principle, this shows that the potassium atom is more stable with one electron in its $4s$ orbital than with one electron in one of its $3d$ orbitals. The $3d$ orbitals fill after the $4s$ orbital is full, starting with scandium ($Z = 21$).

A similar situation exists at the end of the next row. When the $4p$ orbital is full (Kr, $Z = 36$), the next element (Rb, $Z = 37$) has an electron in the $5s$ orbital rather than either a $4d$ orbital or a $4f$ orbital. In fact, electrons are not added to the $4f$ orbitals until element 58, after the $5s$, $5p$, and $6s$ orbitals have filled.

The arrangement of the periodic table provides a simple way to determine the filling order of the elements, as shown in Figure 8-10 and applied in Example 8-2.

Example 8-2

Orbital Filling Sequence

Which orbitals are filled, and which set of orbitals is partially filled, in a germanium atom?

Strategy: For this qualitative problem, use the periodic table to determine the order of orbital filling. Locate the element in a block and identify its row and column. Move along the ribbon of elements to establish the sequence of filled orbitals.

Solution: Germanium is element 32. Consult Figure 8-8 to determine that Ge is in Group 14, row 4 of the p block:

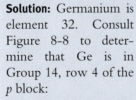

Starting from the top left of the periodic table and working left to right across the rows until we reach Ge, we identify the filled orbitals: $1s$, $2s$, $2p$, $3s$, $3p$, $4s$, $3d$. Germanium is in row 4 of the p block, so the $4p$ set of orbitals is partially filled.

Does the Result Make Sense? Ge has $Z = 32$, meaning its neutral atoms contain 32 electrons. We can count how many each orbital can hold, 2 for each s orbital, 6 for each p orbital set, and 10 for each d orbital set. 2 ($1s$) + 2 ($2s$) + 6 ($2p$) + 2 ($3s$) + 6 ($3p$) + 2 ($4s$) + 10 ($3d$) = 30 electrons, leaving 2 in the partially filled $4p$ orbital set.

Extra Practice
Exercise 8.2

Determine which orbitals are filled and which is partially filled for the element Zr.

Valence Electrons

The chemical behavior of an atom is determined by the electrons that are accessible to an approaching chemical reagent. Accessibility, in turn, has a spatial component and an energetic component. An electron is accessible *spatially* when it occupies one of the largest orbitals of the atom. Electrons on the perimeter of the atom, farthest from the nucleus, are the first ones encountered by an incoming chemical reagent. An electron is accessible *energetically* when it occupies one of the least stable occupied orbitals of the atom. Electrons in less stable orbitals are more chemically active than electrons in more stable orbitals.

Similar electron accessibility generates similar chemical behavior. For example, iodine has many more electrons than chlorine, but these two elements display similar chemical behavior, as reflected by their placement in the same group of the periodic table. This is because the chemistry of chlorine and iodine is determined by the number of electrons in their largest and least stable occupied orbitals: $3s$ and $3p$ for chlorine and $5s$ and $5p$ for iodine. Each of these elements has seven accessible electrons, and this accounts for the chemical similarities.

Accessible electrons are called **valence electrons,** and inaccessible electrons are called **core electrons.** Valence electrons participate in chemical reactions, but core electrons do not. Orbital size increases and orbital stability decreases as the principal quantum number n gets larger. Therefore, the valence electrons for most atoms are the ones in orbitals with the largest value of n. Electrons in orbitals with lower n values are core electrons. In chlorine, valence electrons have $n = 3$, and core electrons have $n = 1$ and $n = 2$. In iodine, valence electrons have $n = 5$, and all others are core electrons.

The nearly equal energies of ns and $(n - 1)d$ orbitals creates some ambiguity about valence and core electrons for elements in the d and f blocks. For example, titanium forms a chloride and an oxide whose chemical formulas are consistent with four valence electrons: $TiCl_4$ and TiO_2. This shows that the two $3d$ electrons of titanium participate in its chemistry, even though they have a lower principal quantum number than its two $4s$ electrons. On the other hand, zinc forms compounds such as $ZnCl_2$ and ZnO, indicating that only two electrons are involved in its chemistry. This chemical behavior indicates that zinc's ten $3d$ electrons, which completely fill the $3d$ orbitals, are inaccessible for chemical reactions. When the d orbitals are partially filled, the d electrons participate in chemical reactions, but when the d orbitals are completely filled, the d electrons do not participate in reactions. Analogous behavior is observed for the f orbitals, leading to a general rule for identifying valence electrons:

Valence electrons are all those of highest principal quantum number plus those in partially filled d and f orbitals.

KEY CONCEPT

Applying this general rule, we find that the number of valence electrons can be determined easily from group numbers. For Groups 1–8, the number of valence electrons equals the group number. As examples, potassium and rubidium, members of Group 1, have just one valence electron each. Tungsten, in Group 6, has six valence electrons: two $6s$ electrons and four $5d$ electrons. For Groups 12–18, the number of valence electrons equals the group number minus 10 (the number of electrons it takes to fill the d orbitals). Thus, antimony and nitrogen, in Group 15, have $15 - 10 = 5$ valence electrons each (two s and three p). For Groups 9–11, the number of valence electrons cannot be stated unambiguously, because the d electrons may or may not participate in bonding.

We have described the layout of the periodic table in terms of the orbital descriptions of the various elements. As our Chemical Milestones Box describes, the periodic table was first proposed well before quantum theory was developed, when the only guidelines available were patterns of chemical and physical behavior.

Box 8-2   Chemical Milestones: History of the Periodic Table

Today we work confidently with the rows and columns of the periodic table. Yet less than 150 years ago, only about half of all elements known today had been discovered, and these presented a bewildering collection of chemical and physical properties. The discovery of the patterns that underlie this apparent randomness is a tale of inspired chemical detective work.

One early attempt to organize the elements clustered them into groups of three, called *triads,* whose members display similar chemical properties. Lithium, sodium, and potassium, for example, have many common properties and were considered to be a triad. This model was severely limited, for many elements could not be grouped into triads. The triad model is just one of nearly 150 different periodic arrangements of the elements that have been proposed.

Our modern periodic table was developed independently in the late 1860s by Dimitri Mendeleev (Russian) and Julius Lothar Meyer (German). At that time, about 60 elements had been discovered, but nothing was known about atomic structure. Lothar Meyer and Mendeleev had to work with elemental molar masses and other known elemental properties.

Lothar Meyer, a physicist, examined atomic volumes. He plotted atomic volume against molar mass and observed the pattern shown in the figure. There is a clear pattern of "waves," cresting successively at Li, Na, K, Rb, and Cs.

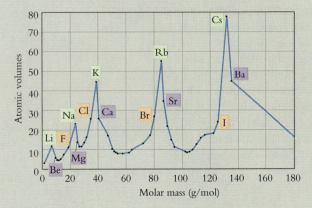

Mendeleev, a chemist, examined the relative numbers of the atoms of different elements that combine in chemical compounds. In $MgCl_2$, for example, each magnesium atom combines with two chlorine atoms. When he matched combining ability against molar mass, Mendeleev found the same sort of pattern as Lothar Meyer, with Li, Na, K, Rb, and Cs all combining 1:1 with Cl. Thus, each scientist was led to propose a table in which elements are arranged in rows of increasing mass, with breaks so that these five elements fall in the same column.

Mendeleev was bolder in his interpretation than Lothar Meyer, and for this reason we honor him as the primary discoverer of the modern periodic table. A few elements did not fit the pattern of variation in combining numbers with molar mass. Mendeleev proposed that these irregularities meant that the element's molar mass had been measured incorrectly. For example, Mendeleev predicted that the correct molar mass of indium is 113 g/mol, not 75 g/mol, the value assigned at that time on the assumption that the formula for indium oxide is InO. Later experiments showed that the correct formula is In_2O_3, and indium's true molar mass is 114.8 g/mol.

Mendeleev also predicted the existence of elements that had not yet been discovered. His arrangement of the then-known elements left some obvious holes in the periodic table. For instance, between zinc (combines with 2 Cl) and arsenic (combines with 5 Cl) were holes for one element that would combine with three chlorine atoms and another that would combine with four. Mendeleev assigned these holes to two new elements. He predicted that one element would have a molar mass of 68 g/mol and chemical properties like those of aluminum, while the other would have a molar mass of 72 g/mol and chemical properties similar to silicon. These elements, gallium ($Z = 31$, $MM = 69.7$ g/mol) and germanium ($Z = 32$, $MM = 72.6$ g/mol), were discovered within 15 years. Chemists soon verified that gallium resembles aluminum in its chemistry, while germanium resembles silicon, just as Mendeleev had predicted.

Ge (*left*) has many similarities with Si (*right*).

The predictions made by Mendeleev provide an excellent example of how a scientific theory allows far-reaching predictions of as-yet-undiscovered phenomena. Today's chemists still use the periodic table as a predictive tool. For example, modern semiconductor materials such as gallium arsenide were developed in part by predicting that elements in the appropriate rows and columns of the periodic table should have the desired properties. At present, scientists seeking to develop new superconducting materials rely on the periodic table to identify elements that are most likely to confer superconductivity.

Section Exercises

■■ **8.2.1** What is the atomic number of the element that would occupy the position in Row 7, Column 17 of the periodic table?

■■ **8.2.2** Determine which orbitals are filled and which is partially filled for Br.

■■ **8.2.3** List all elements that have two valence electrons and specify the orbitals to which the valence electrons belong.

8.3 ELECTRON CONFIGURATIONS

A complete specification of how an atom's electrons are distributed in its orbitals is called an **electron configuration.** There are three common ways to represent electron configurations. One is a complete specification of quantum numbers. The second is a shorthand notation from which the quantum numbers can be inferred. The third is a diagrammatic representation of orbital energy levels and their occupancy.

A list of the values of all quantum numbers is easy for the single electron in a hydrogen atom: $n = 1$, $l = 0$, $m_l = 0$, and $m_s = +\frac{1}{2}$ or $n = 1$, $l = 0$, $m_l = 0$, and $m_s = -\frac{1}{2}$. Either designation is equally valid, because under normal conditions these two states are equal in energy. In a large collection of hydrogen atoms, half the atoms have one designation and the other half have the other designation.

As the number of electrons in an atom increases, a listing of all quantum numbers quickly becomes tedious. For example, iron, with 26 electrons, would require the specification of 26 sets of 4 quantum numbers. To save time and space, chemists have devised a shorthand notation to write electron configurations. The orbital symbols ($1s$, $2p$, $4d$, etc.) are followed by superscripts designating how many electrons are in each set of orbitals. The compact configuration for a hydrogen atom is $1s^1$, indicating one electron in the $1s$ orbital.

The third way to represent an atomic configuration uses an energy level diagram similar to the one shown in Figure 8-9 to designate orbitals. Each electron is represented by an arrow and is placed in the appropriate orbital. The direction of the arrow indicates the value of m_s. The arrow points upward for $m_s = +\frac{1}{2}$ and downward for $m_s = -\frac{1}{2}$. The configuration of hydrogen can be represented by a single arrow in a $1s$ orbital.

$1s$

A neutral helium atom has two electrons. To write the ground-state electron configuration of He, we apply the aufbau principle. One unique set of quantum numbers is assigned to each electron, moving from the most stable orbital upward until all electrons have been assigned. The most stable orbital is always $1s$ ($n = 1$, $l = 0$, $m_l = 0$). Both helium electrons can occupy the $1s$ orbital, provided one of them has $m_s = +\frac{1}{2}$ and the other has $m_s = -\frac{1}{2}$. Here are the three representations of helium's ground-state electron configuration:

$n = 1, l = 0, m_l = 0, m_s = +\frac{1}{2}$
$\qquad\qquad\qquad\qquad\qquad 1s^2 \qquad\qquad 1s$
$n = 1, l = 0, m_l = 0, m_s = -\frac{1}{2}$

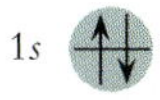

The two electrons in this configuration are said to be *paired* electrons, meaning that they are in the same energy level, with opposing spins. Opposing spins cancel, so paired electrons have zero net spin.

A lithium atom has three electrons. The first two electrons fill lithium's lowest possible energy level, the $1s$ orbital, and the third electron occupies the $2s$ orbital. The three representations for the ground-state electron configuration of a lithium atom are as follows:

$n = 1, l = 0, m_l = 0, m_s = +\frac{1}{2}$
$n = 1, l = 0, m_l = 0, m_s = -\frac{1}{2} \qquad 1s^2 2s^1$
$n = 2, l = 0, m_l = 0, m_s = +\frac{1}{2}$

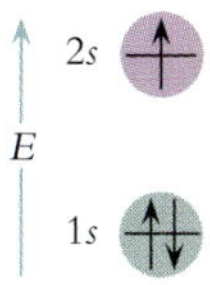

The set $n = 2$, $l = 0$, $m_l = 0$, $m_s = -\frac{1}{2}$ is equally valid for the third electron.

The next atoms of the periodic table are beryllium and boron. You should be able to write the three different representations for the ground-state configurations of these elements. The filling principles are the same as we move to higher atomic numbers. Example 8-3 shows how to apply these principles to aluminum.

Example 8-3

**An Electron
Configuration**

Construct an energy level diagram and the shorthand representation of the ground-state configuration of aluminum. Provide one set of valid quantum numbers for the highest-energy electron.

Strategy: First consult the periodic table to locate aluminum and determine how many electrons are present in a neutral atom. Then construct the electron configuration using the patterns of the periodic table.

Solution: Aluminum has $Z = 13$, so a neutral atom of Al has 13 electrons. Place the 13 electrons sequentially, using arrows, into the most stable orbitals available. Two electrons fill the $n = 1$ orbital, eight electrons fill the $n = 2$ orbitals, two electrons fill the $3s$ orbital, and one electron goes in a $3p$ orbital.

The last electron could be placed in any of the $3p$ orbitals, because these three orbitals are equal in energy. The final electron also could be given either spin orientation. By convention, we place electrons in unfilled orbitals starting with the left-hand side, with spins pointed up.

The shorthand configuration is $1s^2\, 2s^2\, 2p^6\, 3s^2\, 3p^1$.

The least stable electron is in a $3p$ orbital, meaning $n = 3$ and $l = 1$. The value of m_l can be any of three values: $+1$, -1, or 0. The spin quantum number, m_s, can be $+\frac{1}{2}$ or $-\frac{1}{2}$. One valid set of quantum numbers is

$$n = 3,\ l = 1,\ m_l = 1,\ \text{and}\ m_s = +\tfrac{1}{2}$$

You should be able to write the other five possible sets.

Do the Results Make Sense?   Aluminum, in Group 13, has three valence electrons. The configurations show three electrons with $n = 3$, so the configuration is consistent with the valence electron count.

Determine the energy level diagram and shorthand notation for the electron configuration of the fluorine atom.

Electron configurations become longer as the number of electrons increases. To make the writing of a configuration even more compact, chemists make use of the regular pattern for the electrons with lower principal quantum numbers. Compare the configurations of neon and aluminum:

Ne (10 electrons) $1s^2\, 2s^2\, 2p^6$

Al (13 electrons) $1s^2\, 2s^2\, 2p^6\ 3s^2\, 3p^1$

The description of the first 10 electrons in the configuration of aluminum is identical to that of neon, so we can represent that portion as [Ne]. With this notation, the configuration of Al becomes [Ne] $3s^2\, 3p^1$. The element at the end of each row of the periodic table has a **noble gas configuration.** These configurations can be written in the following shorthand notation:

Notation		Configuration	Element
[He]	=	$1s^2$	He (2 electrons)
[Ne]	=	[He] $2s^2\, 2p^6$	Ne (10 electrons)
[Ar]	=	[Ne] $3s^2\, 3p^6$	Ar (18 electrons)
[Kr]	=	[Ar] $4s^2\, 3d^{10}\, 4p^6$	Kr (36 electrons)
[Xe]	=	[Kr] $5s^2\, 4d^{10}\, 5p^6$	Xe (54 electrons)
[Rn]	=	[Xe] $6s^2\, 5d^{10}\, 4f^{14}\, 6p^6$	Rn (86 electrons)

To write the configuration of any other element, we first consult the periodic table to find its location relative to the noble gases. Then we specify the noble gas configuration and build the remaining portion of the configuration according to the aufbau principle. Example 8-4 applies this procedure to indium.

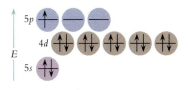

Example 8-4

A Shorthand Electron Configuration

Determine the configuration of indium, first in shorthand form and then in full form.

Strategy: Locate the element in the periodic table, and find the nearest noble gas with smaller atomic number. Start with the configuration of that noble gas, and add enough additional electrons to the next filling orbitals to give the neutral atom.

Solution: Indium (In, $Z = 49$) is in Row 5, Group 13. The nearest noble gas of smaller Z is Kr ($Z = 36$). Thus, the configuration of In has 36 electrons in the Kr configuration and 13 additional electrons. The last orbital to fill in Kr is $4p$, and the periodic table shows that the next orbitals to fill are the $5s$, $4d$, and $5p$ orbitals:

Configuration of indium: $[Kr]\, 5s^2\, 4d^{10}\, 5p^1$

To write the full configuration, decompose the krypton configuration:

$[Kr] = [Ar]\, 4s^2\, 3d^{10}\, 4p^6 = 1s^2\, 2s^2\, 2p^6\, 3s^2\, 3p^6\, 4s^2\, 3d^{10}\, 4p^6$

Full configuration of In (49 electrons):

$1s^2\, 2s^2\, 2p^6\, 3s^2\, 3p^6\, 4s^2\, 3d^{10}\, 4p^6\, 5s^2\, 4d^{10}\, 5p^1$.

Does the Result Make Sense? Indium, in Group 13, has three valence electrons. The configurations show three electrons with $n = 5$, so the configuration is consistent with the valence electron count.

Determine the compact notation for the electron configuration of the cadmium atom.

Extra Practice Exercise 8.4

Electron–Electron Repulsion

The aufbau principle allows us to assign quantum numbers to aluminum's 13 electrons without ambiguity. The first 12 electrons fill the $1s$, $2s$, $2p$, and $3s$ energy levels, and the last electron can occupy any $3p$ orbital with either spin orientation. But what happens when more than one electron must be placed in a p energy level? Carbon atoms, for example, have six electrons, two of which occupy $2p$ orbitals. How should these two electrons be arranged in the $2p$ orbitals? As Figure 8-11 shows, three different arrangements of these electrons obey the Pauli principle and appear to be consistent with the aufbau principle:

1. The electrons could be paired in the same $2p$ orbital (same m_l value but different m_s values).
2. The electrons could occupy different $2p$ orbitals with the same spin orientation (different m_l values but the same m_s value).
3. The electrons could occupy different $2p$ orbitals with opposite spin orientations (different m_l values and different m_s values).

These three arrangements have different energies, because electrons that are close together repel each other more than electrons that are far apart. As a result, for two or more orbitals having nearly equal energies, greatest stability results when electrons occupy the orbitals that keep them farthest apart. Placing two electrons in different p orbitals keeps them relatively far apart, so an atom is more stable with the two electrons in different p orbitals. Thus, arrangements 2 and 3 are more stable than arrangement 1.

Arrangements 2 and 3 look spatially equivalent, but experiments show that a configuration that gives unpaired electrons the same spin orientation is always more stable than one that gives them opposite orientations. **Hund's rule** summarizes the way in which electrons occupy orbitals of equal energies.

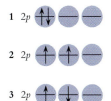

Figure 8-11
Three different arrangements of two 2p electrons obey the Pauli and the aufbau principles.

KEY CONCEPT *The most stable configuration involving orbitals of equal energies is the one with the maximum number of electrons with the same spin orientation.*

According to Hund's rule, the ground-state configuration for carbon atoms is arrangement 2. Example 8-5 provides practice in the application of Hund's rule.

Example 8-5

Applying Hund's Rule

Write the shorthand electron configuration and draw the ground-state orbital energy level diagram for the valence electrons in a sulfur atom.

Strategy: From the periodic table, we see that sulfur has 16 electrons and is in the p block, Group 16. To build the ground-state configuration, apply the normal filling rules and then apply Hund's rule if needed.

Solution: The first 12 electrons fill the four lowest-energy orbitals:

$$1s^2\, 2s^2\, 2p^6\, 3s^2$$

Sulfur's remaining 4 electrons occupy the three $3p$ orbitals. The complete configuration is $1s^2\, 2s^2\, 2p^6\, 3s^2\, 3p^4$, or [Ne] $3s^2\, 3p^4$.

To minimize electron–electron repulsion, put three of the $3p$ electrons in different orbitals, all with the same spin, and then place the fourth electron, with opposite spin, in the first orbital. In accord with Hund's rule, this gives the same value of m_s to all electrons that are not paired. Here is the energy level diagram for the valence electrons:

Do the Results Make Sense? Sulfur, in Group 16, has six valence electrons. The configurations show six electrons with $n = 3$, so the configuration is consistent with the valence electron count. The electrons are spread among the three $3p$ orbitals, which minimizes electron–electron repulsion.

Extra Practice Exercise 8.5

Determine the compact configuration for the nitrogen atom and write a valid set of quantum numbers for its valence electrons.

Orbitals with Nearly Equal Energies

The filling order embodied in the periodic table predicts a regular progression of ground-state configurations. Experiments show, however, that some elements have ground-state configurations different from the predictions of the regular progression. Among the first 40 elements, there are only two exceptions: copper and chromium. Chromium ($Z = 24$) is in Group 6, four elements into the d block. We would predict that chromium's valence configuration should be $4s^2\, 3d^4$. Instead, experiments show that the ground-state configuration of this element is $4s^1\, 3d^5$. Likewise, the configuration of copper ($Z = 29$) is $4s^1\, 3d^{10}$ rather than the predicted $4s^2\, 3d^9$.

Look again at Figure 8-9, which shows that these two sets of orbitals are nearly the same in energy. Each $(n - 1)d$ orbital has nearly the same energy as its ns counterpart. In addition, each $(n - 2)f$ orbital has nearly the same energy as its $(n - 1)d$ counterpart. Table 8-2 lists these orbitals and the atomic numbers for which the filling sequence differs from the expected pattern. These configurations are also indicated in Figure 8-8.

Often, an s orbital contains only one electron rather than two. Five of the exceptional ground-state configurations have a common pattern and are easy to remember: Cr and Mo are $s^1\, d^5$, and Cu, Ag, and Au are $s^1\, d^{10}$. The other exceptional cases follow no recognizable patterns, because they are generated by subtle interactions among all the electrons. Among elements whose valence electrons are filling orbitals with nearly equal energies, several factors help to determine the ground-state configuration. The details are beyond the scope of general chemistry, except that you should recognize that even a subtle change can cause variations in the filling pattern predicted by the periodic table.

Table 8-2 Atomic Orbitals with Nearly Equal Energies

Orbitals	Atomic Numbers Affected	Example
$4s, 3d$	24, 29	Cr: [Ar] $4s^1 3d^5$
$5s, 4d$	41–47	Ru: [Kr] $5s^1 4d^7$
$6s, 5d, 4f$	57, 58, 64, 78, 79	Au: [Xe] $6s^1 4f^{14} 5d^{10}$
$6d, 5f$	89, 91–93, 96	U: [Rn] $7s^2 5f^3 6d^1$

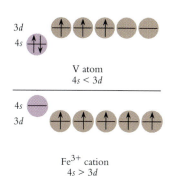

Figure 8-12
Transition metal atoms and cations have different valence configurations, even when the species contain the same number of electrons.

Configurations of Ions

The electron configurations of atomic ions are written using the same procedure as for neutral atoms, taking into account the proper number of electrons. An anion has one *additional* electron for each unit of negative charge. A cation has one *fewer* electron for each unit of positive charge.

For most atomic ions, the filling order of orbitals is the same as that of neutral atoms. For example, Na^+, Ne, and F^- all contain 10 electrons, and each has the configuration $1s^2 2s^2 2p^6$. Atoms and ions that have the same number of electrons are said to be **isoelectronic.**

The nearly-equal energies of ns and $(n-1)d$ orbitals causes the configurations of some cations to differ from the configurations predicted by the filling pattern of the periodic table. This feature is particularly important for the transition metals. Experiments show that in transition metal cations the $(n-1)d$ orbitals are *always* more stable than the ns orbitals. For example, an Fe^{3+} cation contains 23 electrons. The first 18 electrons fill the $1s$, $2s$, $2p$, $3s$, and $3p$ orbitals, as predicted by the periodic table. However, the five remaining electrons populate the $3d$ set, leaving the $4s$ orbital empty. Thus the configuration of the Fe^{3+} cation is [Ar] $3d^5$.

Vanadium atoms ([Ar] $4s^2 3d^3$) and Fe^{3+} cations ([Ar] $3d^5$) have different configurations, even though each has 23 electrons. Remember that the energy ranking of orbitals such as $4s$ and $3d$ depends on a balance of several factors, and that even a small variation in that balance can change the filling order of the orbitals. That is the case for transition metal atoms and cations, as shown in Figure 8-12. In neutral transition metal atoms, a configuration that fills the ns orbital is slightly more stable than one that places all valence electrons in the $(n-1)d$ orbital. In transition metal cations, however, the stable configuration always places all valence electrons in the $(n-1)d$ orbitals.

Example 8-6 shows how to write the configuration of a transition metal cation.

What is the ground-state electron configuration of a Cr^{3+} cation?

Strategy: Use the aufbau approach, remembering that because Cr^{3+} is a transition metal cation, its $3d$ orbital is more stable than is its $4s$ orbital.

Solution: A neutral chromium atom has 24 electrons, so the corresponding Cr^{3+} cation has 21 electrons. The first 18 electrons follow the usual filling order to give the argon core configuration: $1s^2 2s^2 2p^6 3s^2 3p^6$, or [Ar]. Place the remaining three electrons in the $3d$ set of orbitals, following Hund's rule: [Ar] $3d^3$

Example 8-6

Configuration of a Cation

Does the Result Make Sense? For any cation, the empty $4s$ orbital is slightly higher in energy than the partially filled $3d$ orbital. Thus, the isoelectronic V^{2+} and Cr^{3+} cations both have the [Ar] $3d^3$ configuration. On the other hand, the isoelectronic neutral atom scandium has the configuration [Ar] $4s^2 3d^1$.

Determine the ground-state electron configuration (compact form) of a Ru^{3+} cation.

Extra Practice Exercise 8.6

Here are the guidelines for building atomic or ionic electron configurations:

1. Count the total number of electrons.
   1a. Add electrons for anions.
   1b. Subtract electrons for cations.
2. Fill orbitals to match the nearest noble gas of smaller atomic number.
3. Add remaining electrons to the next filling orbitals according to Hund's rule.
   3a. For neutral atoms and anions, place electrons in ns before $(n-1)d$.
   3b. For cations, place electrons in $(n-1)d$ before ns.
4. Look for exceptions and correct the configuration, if necessary.

Magnetic Properties of Atoms

How do we know that an Fe^{3+} ion in its ground state has the configuration [Ar] $3d^5$ rather than the [Ar] $4s^2\ 3d^3$ configuration predicted by the periodic table? Remember from Chapter 7 that electron spin gives rise to magnetic properties. Consequently, any atom or ion with unpaired electrons has nonzero net spin and is attracted by a strong magnet. We can divide the electrons of an atom or ion into two categories with different spin characteristics. In the filled orbitals, all the electrons are paired. Each electron with spin orientation $+\frac{1}{2}$ has a partner with spin orientation $-\frac{1}{2}$. The spins of these electrons cancel each other, giving a net spin of zero. An atom or ion with all electrons paired is not attracted by strong magnets and is termed **diamagnetic.** In contrast, spins do not cancel when unpaired electrons are present. An atom or ion with unpaired electrons is attracted to strong magnets and is termed **paramagnetic.** Moreover, the spins of all the unpaired electrons are additive, so the amount of paramagnetism shown by an atom or ion is proportional to the number of unpaired spins.

The magnetic properties of chemical species can be measured with an instrument known as a Gouy balance, as shown schematically in Figure 8-13. The paramagnetic sample is suspended from one pan of the balance and placed just above the poles of a magnet. The paramagnetic substance is pulled into the magnetic field, creating a downward force on the sample. Weights are added to the second pan until the force is balanced. The paramagnetism of the sample is proportional to the mass required to balance the pans.

In Fe^{3+}, Hund's rule dictates that the five d electrons all have the same spin orientation. For these five electrons, the spins all act together, giving a net spin of $5 \times \frac{1}{2} = \frac{5}{2}$. The alternative configuration for Fe^{3+}, [Ar] $4s^2\ 3d^3$, is paramagnetic, too, but its net spin is $3 \times \frac{1}{2} = \frac{3}{2}$. Experiments show that Fe^{3+} has a net spin of $\frac{5}{2}$. In fact, magnetic measurements on a wide range of transition metal cations all are consistent with the $(n-1)d$ orbitals being occupied rather than the ns orbital.

Example 8-7 shows how to use configurations to predict whether an atom or ion is diamagnetic or paramagnetic.

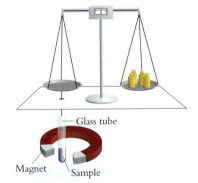

Figure 8-13
The number of unpaired electrons in a paramagnetic substance can be measured with a Gouy balance.

Example 8-7	Which of these species is paramagnetic: F^-, Zn^{2+}, and Ti?

Unpaired Electrons

Strategy: Paramagnetism results from unpaired spins, which exist only in partially filled sets of orbitals. We need to build the configurations and then look for any orbitals that are partially filled.

Solution:
F^-: A fluorine atom has 9 electrons, so F^- has 10 electrons. The configuration is $1s^2\ 2s^2\ 2p^6$. There are no partially filled orbitals, so fluoride ion is diamagnetic.

Zn^{2+}: The parent zinc atom has 30 electrons, and the cation has 28, so the configuration for Zn^{2+} is [Ar] $3d^{10}$. Again, there are no partially filled orbitals, so this ion is also diamagnetic.

Example 8-7

Unpaired Electrons
(*continued*)

Ti: A neutral titanium atom has 22 electrons. The ground-state configuration is [Ar] $4s^2 3d^2$. The spins of the $4s$ electrons cancel, but the two electrons in $3d$ orbitals have the same spin orientation, so their effect is additive. This ion is paramagnetic, with net spin of $(\frac{1}{2}) + (\frac{1}{2}) = 1$.

[Ar]

Does the Result Make Sense? Filled orbitals always have all electrons paired, and two of these three species have completely filled orbitals. Only Ti has a partially filled orbital set.

Most transition metal cations are paramagnetic. Which cations in the first transition metal series have net charges less than +4 and are exceptions to this generalization?

The ground-state configurations of most neutral atoms and many ions contain unpaired electrons, so we might expect most materials to be paramagnetic. On the contrary, most substances are diamagnetic. This is because stable substances seldom contain free atoms. Instead, atoms are bonded together in molecules, and as we show in Chapter 9, bonding results in the pairing of electrons and the cancellation of spin. As a result, paramagnetism is observed primarily among salts of the transition and rare-earth metals, whose cations have partially filled d and f orbitals.

Excited States

A ground-state configuration is the most stable arrangement of electrons, so an atom or ion will usually have this configuration. When an atom absorbs energy, however, it can reach an excited state with a new electron configuration. For example, sodium atoms normally have the ground-state configuration [Ne] $3s^1$, but when sodium atoms are in the gas phase, an electrical discharge can induce transitions that transfer the $3s$ electron to a higher-energy orbital, such as $3p$ or $4s$. Excited atoms are unstable and spontaneously return to the ground-state configuration, giving up their excess energy in the process. This is the principle behind sodium vapor lamps, which are used for street lighting. The light from a sodium vapor lamp comes from photons emitted as excited sodium atoms return to their ground states.

Excited-state configurations are perfectly valid as long as they meet the restrictions given in Table 7-2. In the electrical discharge of a sodium vapor lamp, for instance, we find some sodium atoms in excited states with configurations such as $1s^2 2s^2 2p^6 3p^1$ or $1s^2 2s^2 2p^5 3s^2$. These configurations use valid orbitals and are in accord with the Pauli principle, but they describe atoms that are less stable than those in the ground state. Each atom or ion has only one state that is most stable, and that is the ground state.

Excited states play important roles in chemistry. Recall from Chapter 7 that the properties of atoms can be studied by observing excited states. In fact, chemists and physicists use the characteristics of excited states extensively to probe the structure and reactivity of atoms, ions, and molecules. Excited states also have practical applications. For example, street lamps use the emissions from excited sodium atoms, the dazzling colors of a fireworks display come from photons emitted by metal ions in excited states, and the red light in highway flares often comes from excited Sr^{2+} ions.

The red light in highway flares comes from excited Sr^{2+} ions.

Section Exercises

8.3.1 Determine the ground-state electron configurations and predict the net spin of (a) Au^+; (b) the neutral element with $Z = 118$; and (c) S^{2-}.

8.3.2 The ground-state configuration of Np is [Rn] $7s^2 5f^4 6d^1$. Draw an energy level diagram that shows how the $7s$, $5f$, and $6d$ orbitals are related. Include arrows representing the seven electrons that occupy these orbitals.

8.3.3 Determine the ground-state configurations of Mo, I, and Hg^+. Use shorthand notation.

Figure 8-14
Underlying features of the periodic table.

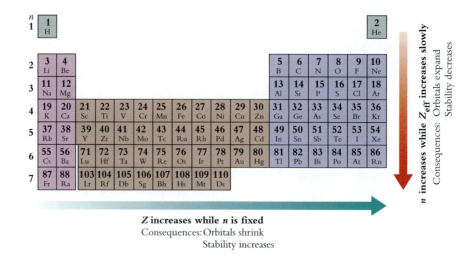

Z increases while *n* is fixed
Consequences: Orbitals shrink
Stability increases

8.4 PERIODICITY OF ATOMIC PROPERTIES

The physical and chemical properties of the elements show regular periodic trends that can be explained using electron configurations and nuclear charges. We focus on the physical properties of the elements in this section. A preliminary discussion of the chemical properties of some of the elements appears in Section 8.6. Other chemical properties are discussed after we introduce the principles of chemical bonding in Chapters 9 and 10.

Underlying Patterns

Two fundamental features of orbitals form the basis of periodicity and are summarized on the periodic table in Figure 8-14.

KEY CONCEPTS

As principal quantum number n increases, atomic orbitals become larger and less stable.

As atomic number Z increases, any given atomic orbital becomes smaller and more stable.

Moving from left to right across a row of the periodic table, the ***n*** value of the least stable occupied orbital remains the same while *Z* increases. A larger nuclear charge exerts a stronger electrical attraction on the electron cloud, and this stronger attraction results in smaller orbitals. Furthermore, electrons closer to the nucleus are energetically more stable than those farther from the nucleus. Moving from *left to right* across a row, orbitals become *smaller and more stable*.

Proceeding down a column of the periodic table, ***n*** and *Z* both increase. As ***n*** increases, orbitals become larger and less stable, but as *Z* increases, orbitals become smaller and more stable. Which trend dominates here? Recall that the number of core electrons increases as we move down any column. For example, sodium ($Z = 11$) has 10 core electrons and 1 valence electron. In the next lower row, potassium ($Z = 19$) has 18 core electrons and 1 valence electron. The screening provided by potassium's additional eight core electrons largely cancels the effect of the additional eight protons in its nucleus. Consequently, increased screening largely offsets the increase in *Z* value from one row to the next. For this reason, ***n*** is the most important factor in determining orbital size and stability within a column. From *top to bottom* of a column, valence orbitals become *larger and less stable*.

Decreases

Increases

Atomic radii

Atomic Radii

Because most of the volume of an atom is occupied by its electron cloud, the size of an atom is determined by the sizes of its orbitals. Atomic size follows these periodic trends:

Atomic size decreases from left to right and increases from top to bottom of the periodic table.

KEY CONCEPT

A convenient measure of atomic size is the radius of the atom. Figure 8-15 shows the trends in **atomic radii.** For example, the atomic radius decreases smoothly across row 3, from 186 pm for sodium to 100 pm for chlorine. The atomic radius increases smoothly down Group 1, from 152 pm for lithium to 265 pm for cesium. Notice, however, that the atomic radius changes very little across the d and f blocks of the table. This is due to screening. For these elements, the largest orbital is the filled *ns* orbital. Moving from left to right across a row, Z increases, but electrons add to the smaller $(n-1)d$ or $(n-2)f$ orbitals. An increase in Z by one unit is matched by the addition of one screening electron. From the perspective of the outlying s orbital, the increases in Z are balanced by increased screening from the added d or f electrons. Thus, the electron in the outermost occupied orbital, *ns*, feels an effective nuclear charge that changes very little across these blocks. As a consequence, atomic size remains nearly constant across each row of the d and f blocks.

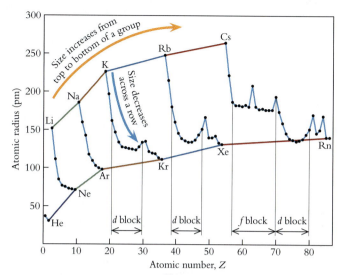

Figure 8-15
The radii of gaseous atoms vary in periodic fashion. Atomic radius decreases from left to right within any row (*blue lines*) and increases from top to bottom within any group (*red lines*).

It is important to be familiar with periodic trends in physical and chemical properties, but it is just as important to understand the principles that give rise to these trends. Example 8-8 shows how to analyze trends in terms of the underlying principles.

For each of the following pairs, predict which atom is larger and why: Si or Cl, S or Se, and Mo or Ag.

Example 8-8

Trends in Atomic Radii

Strategy: Qualitative predictions about atomic size can be made on the basis of electron configurations and the effects of Z and *n* on size.

Solution: Silicon and chlorine are in the third row of the periodic table:

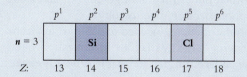

Chlorine's nuclear charge ($+17$) is larger than silicon's ($+14$), so chlorine's nucleus exerts a stronger pull on its electron cloud. Chlorine also has three more electrons than silicon, which raises the possibility that screening effects could counter the extra nuclear charge. Remember, however, that electrons in the same type of orbital do a poor job of screening one another from the nuclear charge. For Si and Cl, screening comes mainly from the core electrons, not from the electrons in the 3p orbitals. Because screening effects are similar for these elements, nuclear charge determines which of the two atoms is larger. Therefore, we conclude that chlorine, with its greater nuclear attraction for the electron cloud, is the smaller atom.

Sulfur and selenium are in Group 16 of the periodic table.

Although they both have the s^2p^2 valence configurations, selenium's least stable electrons are in orbitals with a larger *n* value. Orbital size increases with *n*. Selenium also has a greater nuclear charge than sulfur, which raises the possibility that nuclear attraction could offset increased *n*.

	Group 16
n	
2	
3	S
4	Se

Example 8-8

Trends in Atomic Radii
(*continued*)

Remember, however, that much of this extra nuclear charge is offset by the screening influence of the core electrons. Selenium has 18 core electrons, and sulfur has 10. Thus, we conclude that selenium, with its larger *n* value, is larger than sulfur.

Molybdenum and silver are in the same row of the *d* block:

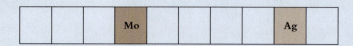

They have the following configurations:

$$Mo = [Kr]\ 5s^1\ 4d^5 \qquad Ag = [Kr]\ 5s^1\ 4d^{10}$$

In each case, 5*s* is the largest occupied orbital. The 4*d* orbitals are smaller, with their electron density located mostly inside the 5*s* orbital. Consequently, 4*d* is effective at screening 5*s*. The nuclear charge of silver is five units larger than that of molybdenum, but silver also has five extra screening electrons. These offset the extra nuclear charge, making Mo and Ag nearly the same size.

Do the Results Make Sense? The trends in Figure 8-15 confirm the results. Chlorine lies to the right of silicon in the same row of the periodic table. Size decreases from left to right in any row; thus, chlorine is smaller than silicon. Selenium is immediately below sulfur in the same column of the periodic table. Size increases down a column; thus, selenium is larger than sulfur. Molybdenum and silver occupy the same row of the *d* block of the periodic table, across which size changes very little; thus, molybdenum and silver are nearly the same size.

**Extra Practice
Exercise 8.8**

Use periodic trends to determine which of the following are smaller than As and which are larger than As: P, Ge, Se, and Sb.

Ionization Energy

When an atom absorbs a photon, the gain in energy promotes an electron to a less stable orbital. As electrons move into less stable orbitals, they have less electrical attraction for the nucleus. If the absorbed photon has enough energy, an electron can be ejected from the atom, as occurs in photoelectron spectroscopy.

The minimum amount of energy needed to remove an electron from a neutral atom is the first ionization energy (IE_1). Variations in ionization energy mirror variations in orbital stability, because an electron in a less stable orbital is easier to remove than one in a more stable orbital.

KEY CONCEPT *First ionization energy increases from left to right across each row and decreases from top to bottom of each column of the periodic table.*

Increases

Decreases

1st *IE*

Figure 8-16 shows how the first ionization energies of gaseous atoms vary with atomic number. Notice the trends in ionization energy. Ionization energy increases regularly from left to right across each row (Row 3: 496 kJ/mol for Na to 1520 kJ/mol for Ar) and decreases regularly from top to bottom of each column (Group 18: 2372 kJ/mol for He to 1037 kJ/mol for Rn). As with atomic radius, ionization energy does not change much for elements in the *d* and *f* blocks, because increased screening from the *d* and *f* orbitals offsets increases in *Z*.

Higher Ionizations

A multielectron atom can lose more than one electron, but ionization becomes more difficult as cationic charge increases. The first three ionization energies for a magnesium

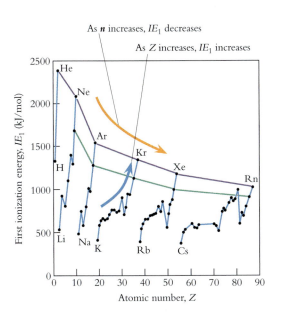

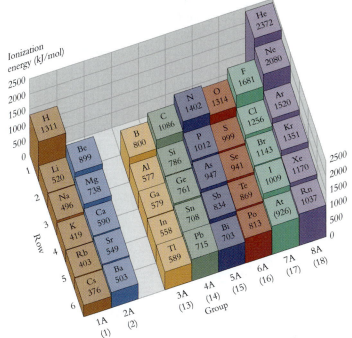

Figure 8-16
The first ionization energy increases from left to right (*orange arrow*) and decreases from top to bottom (*blue arrow*) of the periodic table.

Appendix C gives the first three ionization energies for the first 36 elements.

atom in the gas phase provide an illustration. (Ionization energies are measured on gaseous elements to ensure that the atoms are isolated from one another.)

Process	Configurations	IE
$Mg(g) \longrightarrow Mg^+(g) + e^-$	$[Ne]\, 3s^2 \longrightarrow [Ne]\, 3s^1$	738 kJ/mol
$Mg^+(g) \longrightarrow Mg^{2+}(g) + e^-$	$[Ne]\, 3s^1 \longrightarrow [Ne]$	1450 kJ/mol
$Mg^{2+}(g) \longrightarrow Mg^{3+}(g) + e^-$	$[Ne] \longrightarrow [He]\, 2s^2\, 2p^5$	7730 kJ/mol

Notice that the second ionization energy of magnesium is almost twice as large as the first, even though each electron is removed from a $3s$ orbital. This is because Z_{eff} increases as the number of electrons decreases. That is, the positive charge on the magnesium nucleus remains the same throughout the ionization process, but the net charge of the electron cloud decreases with each successive ionization. As the number of electrons decreases, each electron feels a greater electrical attraction to the nucleus, resulting in a larger ionization energy.

The third ionization energy of magnesium is more than ten times the first ionization energy. This large increase occurs because the third ionization removes a *core* electron ($2p$) rather than a *valence* electron ($3s$). Removing core electrons from any atom requires much more energy than removing valence electrons. The second ionization energy of any Group 1 metal is substantially larger than the first ionization energy; the third ionization energy of any Group 2 metal is substantially larger than the first or second ionization energy; and so on.

Electron Affinity

A neutral atom can add an electron to form an anion. The energy change when an electron is added to an atom is called the **electron affinity (EA).** Both ionization energy (*IE*) and electron affinity measure the stability of a bound electron, but for different species. Here, for example, are the values for fluorine:

$$F \longrightarrow F^+ + e^- \qquad IE_1 = 1681 \text{ kJ/mol}$$

$$F + e^- \longrightarrow F^- \qquad EA = -322 \text{ kJ/mol}$$

Energy is released when an electron is added to a fluorine atom to form a fluoride anion. In other words, a fluoride anion is more stable than a fluorine atom plus a free electron. Another way of saying this is that fluorine atoms have an affinity for electrons.

Tables in reference sources often give electron affinities as positive values when the negative ion is more stable than the neutral atom. This convention is contrary to the sign convention for other energetic processes, which uses negative values when energy is released.

The energy associated with removing an electron to convert an anion to a neutral atom (that is, the reverse of electron attachment) has the same magnitude as the electron affinity, but the opposite sign. Removing an electron from F^-, for example, requires energy, giving a positive energy change:

$$F^- \longrightarrow F + e^- \qquad \Delta E = 322 \text{ kJ/mol}$$

The aufbau principle must be obeyed when an electron is added to a neutral atom, so the electron goes into the most stable orbital available. Hence, we expect trends in electron affinity to parallel trends in orbital stability. However, electron–electron repulsion and screening are more important for negative ions than for neutral atoms, so there is no clear trend in electron affinities as n increases. Thus, there is only one general pattern:

KEY CONCEPT   *Electron affinity tends to become more negative from left to right across a row of the periodic table.*

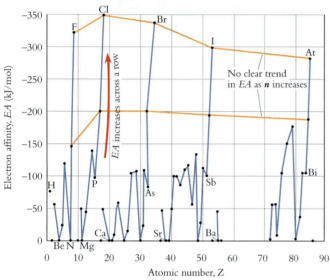

Figure 8-17

The electron affinity of atoms varies with atomic number. In moving across any main group (blue lines), the electron affinity becomes more negative, but this is the only clear trend.

The plot in Figure 8-17 shows how electron affinity changes with atomic number. The blue lines reveal the trend: Electron affinity increases in magnitude across each row of the periodic table. This trend is due to increasing effective nuclear charge, which binds the added electron more tightly to the nucleus. Notice that in contrast to the pattern for ionization energies (Figure 8-16), values for electron affinities remain nearly constant among elements occupying the same column of the periodic table.

The electron affinity values for many of the elements shown in Figure 8-17 appear to lie on the x axis. Actually, these elements have positive electron affinities, meaning the resulting anion is less stable than the neutral atom. Moreover, the second electron affinity of every element is large and positive. Positive electron affinities cannot be measured directly. Instead, these values are estimated by other methods, as we show in Section 8.5.

Although electron affinity values show only one clear trend, there is a recognizable pattern in the values that are positive. When the electron that is added must occupy a new orbital, the resulting anion is unstable. Thus, all the elements of Group 2 have positive electron affinities, because their valence ns orbitals are filled. Similarly, all the noble gases have positive electron affinities, because their valence np orbitals are filled. Elements with half-filled orbitals also have lower electron affinities than their neighbors. As examples, N (half-filled $2p$ orbital set) has a positive electron affinity, and so does Mn (half-filled $3d$ orbital set).

Irregularities in Ionization Energies

Ionization energies deviate somewhat from smooth periodic behavior. These deviations can be attributed to screening effects and electron–electron repulsion. Aluminum, for example, has a smaller ionization energy than either of its neighbors in Row 3:

Element	Z	Atom Configuration	IE_1	Cation Configuration
Mg	12	$[Ne]\,3s^2$	738 kJ/mol	$[Ne]\,3s^1$
Al	13	$[Ne]\,3s^2\,3p^1$	577 kJ/mol	$[Ne]\,3s^2$
Si	14	$[Ne]\,3s^2\,3p^2$	786 kJ/mol	$[Ne]\,3s^2\,3p^1$

The configurations of these elements show that a $3s$ electron is removed to ionize magnesium, whereas a $3p$ electron is removed to ionize aluminum or silicon. Screening makes the $3s$ orbital significantly more stable than a $3p$ orbital, and this difference in stability more than offsets the increase in nuclear charge in going from magnesium to aluminum.

As another example, oxygen has a smaller ionization energy than either of its neighbors in Row 2:

Element	Z	Atom Configuration	IE$_1$	Cation Configuration
N	7	$1s^2\,2s^2\,2p^3$	1402 kJ/mol	$1s^2\,2s^2\,2p^2$
O	8	$1s^2\,2s^2\,2p^4$	1314 kJ/mol	$1s^2\,2s^2\,2p^3$
F	9	$1s^2\,2s^2\,2p^5$	1681 kJ/mol	$1s^2\,2s^2\,2p^4$

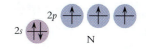

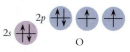

Remember that electron–electron repulsion has a destabilizing effect. The ionization energy of oxygen is less than that of nitrogen, despite the increased nuclear charge, because the p^4 configuration in the O atom has significantly greater electron-electron repulsion than the p^3 configuration in the N atom.

Sizes of Ions

An atomic cation is always smaller than the corresponding neutral atom. Conversely, an atomic anion is always larger than the neutral atom. Figure 8-18 illustrates these trends, and electron–electron repulsion explains them. A cation has fewer electrons than its parent neutral atom. This reduction in the number of electrons means that the cation's remaining electrons experience less electron–electron repulsion. An anion has more electrons than its parent neutral atom. This increase in the number of electrons means that there is greater electron–electron repulsion in the anion than in the parent neutral atom.

Figure 8-18 also highlights the relationships among isoelectronic species, those possessing equal numbers of electrons. As noted in Section 8.3, the F$^-$ anion and the Na$^+$ cation are isoelectronic, each having 10 electrons and the configuration [He] $2s^2\,2p^6$. For isoelectronic species, properties change regularly with Z. For example, Table 8-3 shows two properties of the 10-electron isoelectronic sequence. A progressive increase in nuclear charge results in a corresponding decrease in ionic radius, a result of stronger electrical force between the nucleus and the electron cloud. For the same reason, as Z increases, it becomes progressively more difficult to remove an electron.

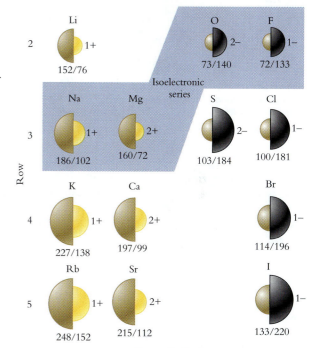

Figure 8-18
Comparison of the sizes (r, in pm) of neutral atoms and their ions for some representative elements.

Table 8-3   Trends in an Isoelectronic Sequence

Property	O^{2-}	F$^-$	Ne	Na$^+$	Mg^{2+}
Z	8	9	10	11	12
Radius (pm)	140	133	—*	102	72
Ionization energy (kJ/mol)	<0 ($-EA_2$)	322 ($-EA$)	2100 (IE_1)	4560 (IE_2)	7730 (IE_3)

*Radii of ions are determined from dimensions of ionic crystals, and neutral atoms cannot be measured by this method.

Section Exercises

8.4.1 Explain the following observations, using variations in Z, Z_{eff}, and n: (a) Cl has a higher ionization energy than Al. (b) Cl has a higher ionization energy than Br. (c) Sr is larger than Mg. (d) Cl$^-$ is larger than K$^+$. (e) The first electron affinity of O is negative, but the second is positive.

8.4.2 Use periodic properties to explain the following observations: (a) Niobium and indium have nearly the same atomic size. (b) The first ionization energy of gallium is significantly lower than that of zinc. (c) Manganese has a positive electron affinity.

8.4.3 Explain why ionization energy decreases substantially as the configuration changes from np^6 to $np^6\,(n+1)s^1$.

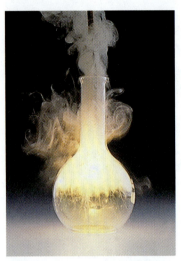

Sodium and chlorine react rapidly to form sodium chloride, the white cloud rising from the flask.

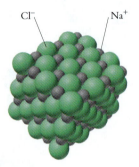

Cl^- Na^+

Figure 8-19
Sodium chloride contains Na^+ cations and Cl^- anions organized in a three-dimensional cubic array.

8.5 ENERGETICS OF IONIC COMPOUNDS

The electronic structures of atoms developed in the preceding sections allow us to determine why some elements form ionic compounds whereas others do not. Soft, lustrous sodium metal reacts vigorously with yellow-green chlorine gas to form sodium chloride, a white crystalline solid that we know as common table salt. Sodium chloride contains sodium and chlorine in a 1:1 elemental ratio. As first described in Section 2.5 and shown in Figure 8-19, a crystal of sodium chloride is an ionic compound made up of equal numbers of Na^+ and Cl^- ions. This contrasts starkly with the result of the reaction of elemental carbon with oxygen gas. Under appropriate conditions, these elements form CO, containing carbon and oxygen in a 1:1 elemental ratio, but CO does not contain ions. In fact, carbon and oxygen never react directly to form an ionic compound.

A limited number of elements form ionic compounds. As we describe in the next two chapters, most substances contain neutral molecules rather than charged ions. The trends in ionization energies and electron affinities indicate which elements tend to form ions. Ionic compounds form when the stabilization gained through ionic attraction exceeds the energy required to create ions from neutral atoms. In this section, we use ionization energy, electron affinity, and electrical forces to analyze the energetics of ionic compounds.

Recall from Chapter 6 that energy changes for reactions do not depend on the path by which starting materials are converted to products. Thus, we can determine the energy change accompanying a reaction using any convenient path. To evaluate the energy change for formation of ionic NaCl, we make use of a reaction path that includes an ionization step and an electron attachment step. Here is one such path:

$$Na(s) \longrightarrow Na(g)$$
$$\tfrac{1}{2} Cl_2(g) \longrightarrow Cl(g)$$
$$Na(g) \longrightarrow Na^+(g) + e^- \qquad \text{Ionization energy}$$
$$Cl(g) + e^- \longrightarrow Cl^-(g) \qquad \text{Electron affinity}$$
$$\underline{Na^+(g) + Cl^-(g) \longrightarrow NaCl(s)}$$
$$Na(s) + \tfrac{1}{2} Cl_2(g) \longrightarrow NaCl(s)$$

Notice that adding all the individual steps of this path gives a net reaction that matches the overall stoichiometry of the reaction. Consequently, the net energy change for the reaction is the sum of the energies of the individual steps.

Step 1: Vaporization. Sodium atoms must be removed from the solid to form sodium gas. Energy must be supplied to do this because, as we describe in Chapter 11, interatomic forces hold the atoms together in the solid metal. The tabulated value for the enthalpy of vaporization of Na is 107.5 kJ/mol. As described in Chapter 6, at 298 K the energy of vaporization is 2.5 kJ/mol less than this:

$$Na(s) \longrightarrow Na(g) \qquad \Delta E_{\text{vaporization}} = 105 \text{ kJ/mol}$$

Step 2: Bond Breakage. Chlorine molecules must be broken apart into chlorine atoms. Table 6-2 gives the bond energy (BE) of molecular chlorine, 240 kJ/mol. We need $\tfrac{1}{2}$ mole of Cl_2 to form 1 mole of NaCl, so the energy requirement is half this amount:

$$\tfrac{1}{2} Cl_2(g) \longrightarrow Cl(g) \qquad \Delta E = \tfrac{1}{2}BE = 120 \text{ kJ/mol}$$

Step 3: Ionization of Na. Ionizing sodium atoms requires that energy be supplied, the amount being the first ionization energy for sodium:

$$Na(g) \longrightarrow Na^+(g) + e^- \qquad \Delta E = IE = 495.5 \text{ kJ/mol}$$

Step 4: Electron Attachment to Cl. The electron affinity of chlorine is negative, which means that energy is released when a chlorine atom gains an electron:

$$Cl(g) + e^- \longrightarrow Cl^-(g) \qquad \Delta E = EA = -348.5 \text{ kJ/mol}$$

Step 5: Condensation. Individual Na^+ and Cl^- ions must condense into a three-dimensional array of ions. Energy is released in this condensation, because cations and anions attract each other. We need to examine this attraction in detail to determine how much energy is released in the condensation step.

Figure 8-20 reminds us that an ionic crystal contains many ions. All the ions exert electrical forces on one another. To calculate the energy resulting from these forces, we start with Equation 6-1, which describes the energy arising from the electrical forces between one pair of charged particles:

$$E_{electrical} = k \frac{q_1 q_2}{r} \tag{6-1}$$

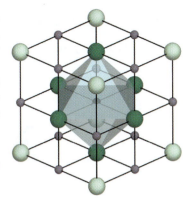

Figure 8-20
In a sodium chloride crystal, each sodium cation has six chloride anions closest to it in an octahedral array, but all the other sodium cations and chloride anions also exert electrical forces.

Recall that q_1 and q_2 are the electrical charges of the objects and r is the distance between them. In this set of calculations, all the energies are in kJ/mol. When charges are expressed in units of electron charge, the value of k is 1.389×10^5 kJ pm/mol.

For sodium cations and chloride anions, $q_1 = +1$ and $q_2 = -1$. To complete the calculation, we need to know how closely the ions approach each other before their mutual attraction is balanced by electron cloud repulsion. In the sodium chloride crystal this distance is 313 pm. Using this value for r, we can calculate the energy released in forming one mole of ion pairs:

$$Na^+(g) + Cl^-(g) \longrightarrow NaCl \, (g, \text{ ion pair})$$

$$E_{ion\ pair} = \frac{(1.389 \times 10^5 \text{ kJ pm/mol})(+1)(-1)}{(313 \text{ pm})} = -444 \text{ kJ/mol}$$

This is only part of the energy released in forming the crystal, because electrical interactions do not stop at individual ion pairs. In solid sodium chloride, each sodium cation is attracted to all the surrounding chloride anions. As Figure 8-20 shows, one Na^+ ion has six nearby Cl^- ions. Moreover, there are many other chloride ions farther away, and all of them contribute to the total energy of the crystal in proportion to their distance from a particular sodium ion. At the same time, there are repulsive interactions from ions of the same charge that are close to one another. Equation 6-1 can be applied to all these ion–ion interactions in all three dimensions. When all these energies are added up for sodium chloride, the calculated value of the energy released is -769 kJ/mol:

$$Na^+(g) + Cl^-(g) \longrightarrow NaCl \, (s) \qquad \Delta E_{calculated} = -769 \text{ kJ/mol}$$

This is the energy released when the solid forms from separated gaseous ions. The reverse process, in which an ionic solid decomposes into gaseous ions, is termed the **lattice energy (*LE*)** and is a positive quantity:

Figure 8-21
The reaction of sodium metal with molecular chlorine gas to produce solid sodium chloride can be analyzed by breaking the overall process into a series of steps involving ions in the gas phase.

$$NaCl \, (s) \longrightarrow Na^+(g) + Cl^-(g) \qquad \Delta E = LE = 769 \text{ kJ/mol}$$

The overall energy change for the sodium chloride reaction is obtained by summing the energies of the five steps. Figure 8-21 summarizes the process.

$$\Delta E_{calculated} = \Delta E_{vaporization} + \tfrac{1}{2} BE + IE + EA - LE$$

$$\Delta E_{calculated} = 105 + 120 + 495.5 + (-348.5) - 769 = -397 \text{ kJ/mol}$$

This calculated energy is within 5% of the experimental value for the energy released in the actual reaction:

$$Na \, (s) + \tfrac{1}{2} Cl_2 \, (g) \longrightarrow NaCl \, (s) \qquad \Delta E_{experimental} = -411 \text{ kJ/mol}$$

This close agreement between the actual energy released in the reaction and the energy calculated by assuming that the final product is composed of ions shows that solid sodium chloride is best described as made up of sodium cations and chloride anions.

An energy cycle like the one in Figure 8-21 is called a *Born–Haber cycle*. Example 8-9 uses a Born–Haber cycle to estimate an electron affinity.

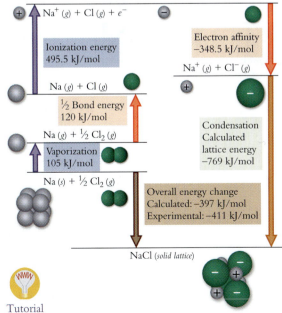

Na$^+$ (g) + Cl (g) + e$^-$

Ionization energy
495.5 kJ/mol

Electron affinity
-348.5 kJ/mol

Na$^+$ (g) + Cl$^-$ (g)

Na (g) + Cl (g)

½ Bond energy
120 kJ/mol

Na (g) + ½ Cl$_2$ (g)

Vaporization
105 kJ/mol

Condensation
Calculated
lattice energy
-769 kJ/mol

Na (s) + ½ Cl$_2$ (g)

Overall energy change
Calculated: -397 kJ/mol
Experimental: -411 kJ/mol

NaCl (*solid lattice*)

Tutorial

Example 8-9

Born–Haber Cycle

⑦

Magnesium metal burns in air to produce magnesium oxide, a white solid that contains Mg^{2+} and O^{2-}. The oxide anion, O^{2-}, is not stable except in a crystalline solid such as MgO. This makes it impossible to measure directly the second electron affinity of oxygen. Use a Born–Haber cycle and the following data to calculate oxygen's second electron affinity.

$$\text{Vaporization enthalpy of Mg} = \Delta H_{vap} = 147.1 \text{ kJ/mol}$$
$$1^{st} \text{ ionization energy of Mg} = IE_1 = 738 \text{ kJ/mol}$$
$$2^{nd} \text{ ionization energy of Mg} = IE_2 = 1451 \text{ kJ/mol}$$
$$\text{Bond energy of } O_2 = BE = 495 \text{ kJ/mol}$$
$$1^{st} \text{ electron affinity of O} = EA_1 = -141 \text{ kJ/mol}$$
$$\text{Calculated lattice energy of MgO} = LE = 3795 \text{ kJ/mol}$$
$$\text{Experimental reaction energy} = \Delta E_{reaction} = -602 \text{ kJ/mol}$$

Strategy: We are asked to find the second electron affinity of oxygen:

$$O^-(g) + e^- \longrightarrow O^{2-}(g) \qquad EA_2 = ?$$

As noted in the previous section, second electron affinities are all large and positive.

The different steps of a Born–Haber cycle can be connected together as shown in Figure 8-21. The overall energy change for the reaction described by the cycle is equal to the sum of the energy changes for the individual steps. A diagram of the steps, similar to Figure 8-21, helps sort out the calculation. Remember that the overall reaction consumes $\frac{1}{2}$ mole of O_2 for each mole of Mg. Therefore, our calculation requires $\frac{1}{2}$ the bond energy of molecular oxygen.

Solution: All the energy changes for the cycle are found among the data, except for the electron affinity of O^-, which is EA_2. We need to subtract 2.48 kJ/mol from the enthalpy of vaporization to convert it into an energy of vaporization. Then we set up the sum of the individual energy changes and solve for EA_2.

$$\Delta E_{vap} = 147.1 \text{ kJ/mol} - 2.48 \text{ kJ/mol} = 144.6 \text{ kJ/mol, rounds to } 145 \text{ kJ/mol}$$

$$\Delta E_{reaction} = \Delta E_{vap} + IE_1 + IE_2 + \tfrac{1}{2}BE + EA_1 + EA_2 - LE$$

$$EA_2 = \Delta E_{reaction} - \Delta E_{vap} - IE_1 - IE_2 - \tfrac{1}{2}BE - EA_1 + LE$$

$$EA_2 = [-602 - 145 - 738 - 1451 - 248 - (-141) + 3795] \text{ kJ/mol}$$

$$EA_2 = 752 \text{ kJ/mol}$$

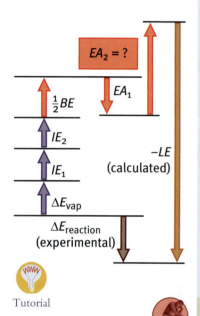

Tutorial

Does the Result Make Sense? The result is a large positive value, reflecting the fact that an isolated O^{2-} anion is very unstable.

Extra Practice Exercise 8.9

The formation reaction of Na_2O has $\Delta E = -413$ kJ/mol. Use this value and energy values in Example 8-9 and the text to calculate the lattice energy of this compound.

Why Not Na²⁺ Cl²⁻?

The electrical attraction between doubly charged ions is significantly greater than the attraction between singly charged ions, because q_1 and q_2 are twice as large. This suggests that the transfer of a second electron might give an even more stable crystal composed of Na^{2+} and Cl^{2-} ions. We can extend our calculations to see if this is true.

The energy of attraction between Na^{2+} and Cl^{2-} ions in one $Na^{2+}Cl^{2-}$ ion pair would be about four times larger than that of one Na^+Cl^- pair. If we assume that the dimensions and arrangement of the $Na^{2+}Cl^{2-}$ crystal are the same as those for the Na^+Cl^- crystal, this gives a calculated lattice energy for $Na^{2+}Cl^{2-}$ of around 3075 kJ/mol. To collect this energy, however, the ion pair must pay the price of forming the doubly charged ions. Appendix C provides the first two ionization energies for sodium:

$$Na \longrightarrow Na^+ + e^- \qquad 1s^2\,2s^2\,2p^6\,3s^1 \longrightarrow 1s^2\,2s^2\,2p^6 \qquad IE_1 = 495.5 \text{ kJ/mol}$$

$$Na^+ \longrightarrow Na^{2+} + e^- \qquad 1s^2\,2s^2\,2p^6 \longrightarrow 1s^2\,2s^2\,2p^5 \qquad IE_2 = 4562 \text{ kJ/mol}$$

We do not know the second electron affinity for Cl, but the result of Example 8-9 suggests that the value will be large and positive, perhaps around 700 kJ/mol.

These energies demonstrate why sodium and chlorine do not form $Na^{2+}Cl^{2-}$. Forming Na^{2+} and Cl^{2-} from neutral gaseous atoms requires more than 5000 kJ/mol, much more energy than the 3075 kJ/mol that would be released in the formation of the $Na^{2+}Cl^{2-}$ lattice.

The second ionization energy of sodium is much larger than its first ionization energy because a *core 2p* electron must be removed to create Na^{2+} from Na^{+}. Removal of a core electron always requires a great deal of energy, so it is a general feature of ionic systems that ions formed by removing core electrons are not found in stable ionic compounds.

Cation Stability

Knowing that the energy cost of removing core electrons is always excessive, we can predict that the ionization process will stop when all valence electrons have been removed. Thus, a knowledge of ground-state configurations is all that we need to make qualitative predictions about cation stability.

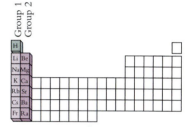

Each element in Group 1 of the periodic table has one valence electron. These elements form ionic compounds containing A^{+} cations. Examples are KCl and Na_2CO_3. Each element in Group 2 of the periodic table has two valence electrons and forms ionic compounds containing A^{2+} cations. Examples are $CaCO_3$ and $MgCl_2$.

Beyond these two columns, the removal of all valence electrons is usually not energetically possible. For example, iron has eight valence electrons but forms only two stable cations, Fe^{2+} and Fe^{3+}. Compounds of iron containing these ions are abundant in the Earth's crust. Pyrite (FeS_2) and iron(II) carbonate ($FeCO_3$, or siderite) are examples of Fe^{2+} salts. Iron(III) oxide (Fe_2O_3, or hematite) can be viewed as a network of Fe^{3+} cations and O^{2-} anions. One of the most abundant iron ores, magnetite, has the chemical formula Fe_3O_4 and contains a 2:1 ratio of Fe^{3+} and Fe^{2+} cations. The formula of magnetite can also be written as $FeO \cdot Fe_2O_3$ to emphasize the presence of two different cations.

Other metallic elements form ionic compounds with cation charges ranging from +1 to +3. Aluminum nitrate nonahydrate, $Al(NO_3)_3 \cdot 9H_2O$, is composed of Al^{3+} cations, NO_3^{-} anions, and water molecules. Silver nitrate ($AgNO_3$), which contains Ag^{+} cations, is a soluble silver salt that is used in silver plating.

Anion Stability

Halogens, the elements in Group 17 of the periodic table, have the largest electron affinities of all the elements, so halogen atoms ($ns^2\,np^5$) readily accept electrons to produce halide anions ($ns^2\,np^6$). This allows halogens to react with many metals to form binary compounds, called *halides,* which contain metal cations and halide anions. Examples include NaCl (chloride anion), CaF_2 (fluoride anion), AgBr (bromide anion), and KI (iodide anion).

Isolated atomic anions with charges more negative than −1 are always unstable, but oxide (O^{2-}, $1s^2\,2s^2\,2p^6$) and sulfide (S^{2-}, [Ne] $3s^2\,3p^6$) are found in many ionic solids, such as CaO and Na_2S. The lattice energies of these solids are large enough to make the overall reaction energy-releasing despite the large positive second electron affinity of the anions. In addition, three-dimensional arrays of surrounding cations stabilize the −2 anions in these solids.

Trends in Lattice Energies

The lattice energy is the sum of all ion interactions, each of which is described by Equation 6-1. Looking at this equation, we can predict that lattice energy will increase as ionic charge increases and that it will decrease as ionic size increases. A third trend occurs in the summing of all the ion contributions: Lattice energy increases with the number of ions in the chemical formula of the salt.

Table 8-4   Lattice Energies of Halides and Oxides

Cation	Anion				
	F^-	Cl^-	Br^-	I^-	O^{2-}
Li^+	1030	834	788	730	2799
Na^+	910	769	732	682	2481
K^+	808	701	671	632	2238
Rb^+	774	680	651	617	2163
Cs^+	744	657	632	600	*
Mg^{2+}	2913	2326	2097	1944	3795
Ca^{2+}	2609	2223	2132	1905	3414
Sr^{2+}	2476	2127	2008	1937	3217
Ba^{2+}	2341	2033	1950	1831	3029
Fe^{2+}	2769	2525	2464	2382	3795
Fe^{3+}	5870	5364	5268	5117	14309

All values are in kJ/mol.
*Exists as superoxide rather than oxide.

These trends are apparent in the values of lattice energy that appear in Table 8-4. Notice, for example, that the lattice energies of the alkali metal chlorides decrease as the size of the cation increases, and the lattice energies of the sodium halides decrease as the size of the anion increases. Notice also that the lattice energy of MgO is almost four times the lattice energy of LiF. Finally, notice that the lattice energy of Fe_2O_3, which contains five ions in its chemical formula, is four times as large as that of FeO, which contains only two ions in its chemical formula.

The ionic model describes a number of metal halides, oxides, and sulfides, but it does not describe most other chemical substances adequately. Whereas substances such as CaO, NaCl, and MgF_2 behave like simple cations and anions held together by electrical attraction, substances such as CO, Cl_2, and HF do not. In a crystal of MgF_2, electrons have been *transferred* from magnesium atoms to fluorine atoms, but the stability of HF molecules arises from the *sharing* of electrons between hydrogen atoms and fluorine atoms. We describe electron sharing, which is central to molecular stability, in Chapters 9 and 10.

Section Exercises

8.5.1 Magnesium fluoride forms from the elements as follows:

$$Mg\,(s) + F_2\,(g) \longrightarrow MgF_2\,(s)$$

$$\Delta E_{reaction} = -1123 \text{ kJ/mol}$$

The energy of vaporization of Mg is 145 kJ/mol, and the bond energy of F_2 is 155 kJ/mol. Use this information and data from Appendix C to calculate the lattice energy of MgF_2. Compare your value with the value in Table 8-4.

8.5.2 Iron and cobalt form compounds that can be viewed as containing A^{3+} cations, but nickel does not. Use the ionization energies in Appendix C to predict which other transition metal elements are unlikely to form stable cations with charges greater than +2.

8.5.3 From the location of each element in the periodic table, predict which ion of each of the following elements will be found in ionic compounds: Ca, Cs, Al, and Br.

8.6   IONS AND CHEMICAL PERIODICITY

The elements that form ionic compounds are found in specific places in the periodic table. Atomic anions are restricted to elements on the right side of the table: the halogens, oxygen and sulfur. All the elements in the *s, d,* and *f* blocks, on the other hand, form compounds containing atomic cations.

Ion formation is only one pattern of chemical behavior. Many other chemical trends can be traced ultimately to valence electron configurations, but we need the description of chemical bonding that appears in Chapters 9 and 10 to explain such periodic properties. Nevertheless, we can relate important patterns in chemical behavior to the ability of some elements to form ions. One example is the subdivision of the periodic table into metals, nonmetals, and metalloids, first introduced in Chapter 1.

The elements that can form cations relatively easily are metals. All metals have similar properties, in part because their outermost s electrons are relatively easy to remove. All elements in the s block have ns^1 or ns^2 valence configurations. The d-block elements have one or two ns electrons and various numbers of $(n-1)d$ electrons. Examples are titanium ($4s^2 3d^2$) and silver ($5s^1 4d^{10}$). Elements in the f block have two ns electrons and a number of $(n-2)f$ electrons. Samarium, for example, has the valence configuration $6s^2 4f^6$. As we describe in Chapter 10, the metallic behavior of these elements occurs partly because the s electrons are shared readily among all atoms. Metals form ionic salts because s electrons (and some d, p, and f electrons) can be readily removed from the metal atoms to form cations.

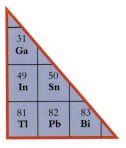

Whereas the other blocks contain only metals, elemental properties vary widely within the p block. We have already noted that aluminum ($3s^2 3p^1$) can lose its three valence electrons to form Al^{3+} cations. Draw a right triangle in the lower left portion of this block whose hypotenuse runs from Ga to Bi. The six elements within this triangle also lose p electrons easily and therefore have metallic properties. Examples are tin ($5s^2 5p^2$) and bismuth ($6s^2 6p^3$). In contrast, the halogens and noble gases on the right of this block are distinctly nonmetallic. The noble gases, Group 18 of the periodic table, are monatomic gases that resist chemical attack because their electron configurations contain completely filled s and p orbitals.

Elements in any intermediate column of the p block display a range of chemical properties even though they have the same valence configurations. Carbon, silicon, germanium, and tin all have $ns^2 np^2$ valence configurations; yet carbon is a nonmetal, silicon and germanium are metalloids, and tin is a metal.

Qualitatively, we can understand this variation by recalling that as the principal quantum number increases, the valence orbitals become less stable. In tin, the four $n = 5$ valence electrons are bound relatively loosely to the atom, resulting in the metallic properties associated with electrons that are easily removed. In carbon, the four $n = 2$ valence electrons are bound relatively tightly to the atom, resulting in nonmetallic behavior. Silicon ($n = 3$) and germanium ($n = 4$) fall in between these two extremes. Example 8-10 describes the elements with five valence electrons.

Example 8-10 **Classifying Elements** 	Nitrogen is a colorless diatomic gas. Phosphorus has several elemental forms, but the most common is a red solid that is used for match tips. Arsenic and antimony are gray solids, and bismuth is a silvery solid. Classify these elements of Group 15 as metals, nonmetals, or metalloids. **Strategy:** All elements except those in the p block are metals. Group 15, however, is part of the p block, within which elements display all forms of elemental behavior. To decide the classifications of these elements, we must examine this group relative to the diagonal arrangement of the metalloids: **Solution:** We see that Group 15 passes through all three classes of elements. The elements with the lowest Z values, nitrogen and phosphorus, are nonmetals. The element with highest Z value, bismuth, is a metal, and the two elements with intermediate Z values, arsenic and antimony, are metalloids.
	Do the Results Make Sense? As the principal quantum number increases, valence electrons become progressively easier to remove, and metals are those elements with valence electrons that are easily removed.
Extra Practice **Exercise 8.10**	Classify the $4p$ set of elements, from Ga to Kr.

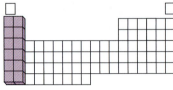

s block

s-Block Elements

The electron configuration of any element in Groups 1 and 2 of the periodic table contains a core of tightly bound electrons and one or two *s* electrons that are loosely bound. The **alkali metals** (Group 1, ns^1 configuration) and the **alkaline earth metals** (Group 2, ns^2 configuration) form stable ionic salts because their valence electrons are easily removed. Nearly all salts of alkali metals and many salts of alkaline earth metals dissolve readily in water, so naturally occurring sources of water frequently contain these ions.

The four most abundant *s*-block elements in the Earth's crust are sodium, potassium, magnesium, and calcium; their occurrence is summarized in Table 8-5. These elements are found in nature in salts such as $NaCl$, KNO_3, $MgCl_2$, $MgCO_3$, and $CaCO_3$. Portions of these solid salts dissolve in rainwater as it percolates through the Earth's crust. The resulting solution of anions and cations eventually finds its way to the oceans. When water evaporates from the oceans, the ions are left behind. Over many eons the continual influx of river water containing these ions has built up the substantial salt concentrations found in the Earth's oceans.

Table 8-5 shows that each of the four common *s*-block ions is abundant not only in seawater but also in body fluids, where these ions play essential biochemical roles. Sodium is the most abundant cation in fluids that are outside of cells, and proper functioning of body cells requires that sodium concentrations be maintained within a narrow range. One of the main functions of the kidneys is to control the excretion of sodium. Whereas sodium cations are abundant in the fluids outside of cells, potassium cations are the most abundant ions in the fluids inside cells. The difference in ion concentration across cell walls is responsible for the generation of nerve impulses that drive muscle contraction. If the difference in potassium ion concentration across cell walls deteriorates, muscular activity, including the regular muscle contractions of the heart, can be seriously disrupted.

The cations Mg^{2+} and Ca^{2+} are major components of bones. Calcium occurs as hydroxyapatite, a complicated substance whose chemical formula is $Ca_5(PO_4)_3(OH)$. The structural form of magnesium in bones is not fully understood. In addition to being essential ingredients of bone, these two cations also play key roles in various biochemical reactions, including photosynthesis, the transmission of nerve impulses, and the formation of blood clots.

Beryllium behaves differently from the other *s*-block elements because the $n = 2$ orbitals are more compact than orbitals with higher principal quantum number. The first ionization energy of beryllium, 899 kJ/mol, is comparable with those of nonmetals, so beryllium does not form compounds that are clearly ionic.

Some compounds of the *s*-block elements are important industrial chemicals, too. For example, more than 1.4 billion kilograms of K_2CO_3 (potassium carbonate, whose common name is potash), is produced in the United States each year. This compound, which is obtained from mineral deposits, is the most common source of potassium for fertilizers. Fertilization with potassium is necessary because this element is essential for healthy plant growth. Moreover, potassium salts are highly soluble in water, so potassium quickly becomes depleted from the soil. Consequently, agricultural land requires frequent addition of potassium fertilizers.

The calcium compound that makes up bones is very durable.

The first U.S. patent was for a method of making potash, issued in 1790 to Samuel Hopkins of Pittsford, Vermont. The patent examiner was Thomas Jefferson and the signator was George Washington.

Table 8-5 Abundance of *s*-Block Elements

Element	Abundance in Crust (% by Mass)	Abundance in Seawater (mol/L)	Abundance in Plasma (mol/L)
Na	2.27	0.462	0.142
K	1.84	0.097	0.005
Mg	2.76	0.053	0.003
Ca	4.66	0.100	0.005

Three other compounds of *s*-block elements—calcium oxide (CaO, known as "lime"), sodium hydroxide (NaOH), and sodium carbonate (Na_2CO_3)—are among the top 15 industrial chemicals in annual production. Lime is perennially in the top 10 because it is the key ingredient in construction materials such as concrete, cement, mortar, and plaster. Two other compounds, calcium chloride ($CaCl_2$) and sodium sulfate (Na_2SO_4), rank just below the top 50 in industrial importance.

Many industrial processes make use of chemically useful anions such as hydroxide (OH^-), carbonate (CO_3^{2-}), and chlorate (ClO_3^-). These anions must be supplied as chemical compounds that include cations. Sodium is most frequently used as this spectator cation because it is abundant, inexpensive, and nontoxic. Hydroxide ion is industrially important because it is a strong base. Sodium hydroxide is used to manufacture other chemicals, textiles, paper, soaps, and detergents. Sodium carbonate and sand are the major starting materials in the manufacture of glass. Glass contains sodium and other cations embedded in a matrix of silicate (SiO_3^{2-}) anions. About half the sodium carbonate produced in the world is used in glass making.

p-Block Elements

The properties of elements in the *p* block vary across the entire spectrum of chemical possibilities. The elements in Group 13, with a single electron in a *p* orbital as well as two valence *s* electrons, display chemical reactivity characteristic of three valence electrons. Except for boron, the elements of this group are metals that form stable cations with $+3$ charge. Examples are $Al(OH)_3$ and GaF_3. Metallic character diminishes rapidly as additional *p* electrons are added. This change culminates in the elements in Group 18. With filled *p* orbitals, these elements are so unreactive that for many years they were thought to be completely inert. Xenon is now known to form compounds with the most reactive nonmetals, oxygen, fluorine, and chlorine; krypton forms a few highly unstable compounds with these elements.

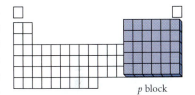

p block

Although the nonmetals do not readily form cations, many of them combine with oxygen to form polyatomic oxoanions. These anions have various stoichiometries, but there are some common patterns. Two second-row elements form oxoanions with three oxygen atoms: carbon (four valence electrons) forms carbonate, CO_3^{2-}, and nitrogen (five valence electrons) forms nitrate, NO_3^-. In the third row, the most stable oxoanions contain four oxygen atoms: SiO_4^{4-}, PO_4^{3-}, SO_4^{2-}, and ClO_4^-.

Many of the minerals that form the Earth's crust contain oxoanions. Examples of carbonates are $CaCO_3$ (limestone) and $MgCa(CO_3)_2$ (dolomite). Barite ($BaSO_4$) is a sulfate mineral. An important phosphate is $Ca_5(PO_4)_3F$ (apatite). Two silicates are zircon ($ZrSiO_4$) and olivine (a mixture of $MgSiO_4$ and $FeSiO_4$).

Several leading industrial chemicals contain these anions or are acids resulting from addition of H^+ to the anions. Sulfuric acid, H_2SO_4, is the #1 industrial chemical in the United States. Two other industrially important acids are nitric acid, HNO_3, and phosphoric acid, H_3PO_4. Heating $CaCO_3$ drives off CO_2 and forms CaO, the essential ingredient of building materials mentioned earlier. Industrially important salts include ammonium sulfate, $(NH_4)_2SO_4$; aluminum sulfate, $Al_2(SO_4)_3$; sodium carbonate, Na_2CO_3; and ammonium nitrate, NH_4NO_3.

Section Exercises

■ **8.6.1** Consult the table of first ionization energies in Appendix C and calculate the average values for the nonmetals, metalloids, and *s*-block elements. How does the trend in these averages relate to the ionic chemistry of these elements?

■ **8.6.2** Classify each of the following elements as a metal, nonmetal, or metalloid: S, Si, Sr, Se, Sc, Sg, and Sn.

■ **8.6.3** Some metals form oxoanions with the same relationship among charge, valence electrons, and number of oxygen atoms as described in this section. Consult Table 3-4 and identify the oxoanions of metals that satisfy this relationship.

CHAPTER 8 VISUAL SUMMARY

SKILLS TO MASTER

1. Arranging orbitals in order of energy
2. Counting valence electrons
3. Writing electron configurations
4. Recognizing near-equivalent orbitals

5. Correlating configurations with periodicity
6. Predicting periodic variations in properties
7. Calculating energy changes for ion formation

8.1 ORBITAL ENERGIES

Photoelectron spectroscopy

Attraction | Repulsion

IE

Screening

Ionization energy (IE)

Screening increases as l increases.

Effective nuclear charge (Z_{eff})

8.2 STRUCTURE OF THE PERIODIC TABLE

1. Aufbau principle
   Pauli exclusion principle
   Ground state configuration
2.

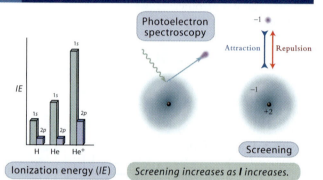

s block   f block   d block   p block

Valence electrons = highest n + partially filled d and f

Each electron has a unique set of quantum numbers.

Valence electrons $2p$
                       $2s$

Core electrons $1s$

8.3 ELECTRON CONFIGURATIONS

3. Electron configuration

4.

Paramagnetic   Diamagnetic

Hund's rule   Noble gas configuration

As many electrons as possible have the same spin orientation.

Isoelectronic

8.4 PERIODICITY OF ATOMIC PROPERTIES

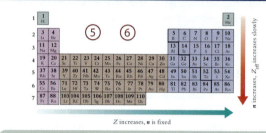

n increases, Z_{eff} increases slowly

Z increases, n is fixed

Orbital energy and size increase with n and decrease with Z.

Electron affinity increases across each row of the periodic table.

Atomic radii
Ionization energy
Electron affinity

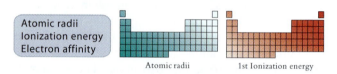

Atomic radii   1st Ionization energy

8.5 ENERGETICS OF IONIC COMPOUNDS

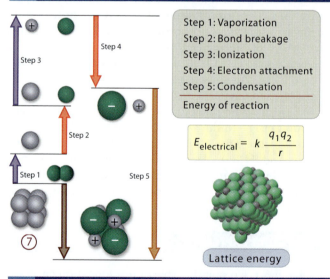

Step 3

Step 4

Step 2

Step 1

Step 5

7.

Step 1: Vaporization
Step 2: Bond breakage
Step 3: Ionization
Step 4: Electron attachment
Step 5: Condensation
Energy of reaction

$$E_{electrical} = k\frac{q_1 q_2}{r}$$

Lattice energy

8.6 IONS AND CHEMICAL PERIODICITY

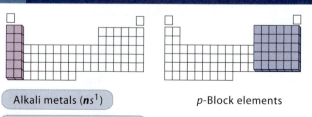

Alkali metals (ns^1)

Alkaline earth metals (ns^2)

p-Block elements

Learning Exercises

8.1 Draw appropriately scaled pictures of all the occupied orbitals in a krypton atom.

8.2 Write brief explanations of (a) screening; (b) the Pauli exclusion principle; (c) the aufbau principle; (d) Hund's rule; and (e) valence electrons.

8.3 Construct an orbital energy level diagram for all orbitals with $n < 8$ and $l < 4$. Use the periodic table to help determine the correct order of the energy levels.

8.4 Describe periodic variations in electron configurations; explain how they affect ionization energy and electron affinity.

8.5 List the 18 elements classified as nonmetals. Give the name, symbol, and atomic number for each.

8.6 Update your list of memory bank equations by adding principles for atomic configurations.

8.7 List all terms new to you that appear in Chapter 8. Write a one-sentence definition of each, using your own words. Consult the glossary if you need help.

 Problems   **ilw** = interactive learning ware problem. Visit the website at www.wiley.com/college/olmsted

Orbital Energies

8.1 For each pair of orbitals, determine which is more stable and explain why: (a) He 1s and He 2s; (b) Kr 5p and Kr 5s; and (c) He 2s and He$^+$ 2s.

8.2 For each pair of orbitals, determine which is more stable and explain why: (a) C 2s and C 2p; (b) Ar 5p and Ar$^+$ 5p; and (c) Ar 4s and Ar 5s.

8.3 In a hydrogen atom the 3s, 3p, and 3d orbitals all have the same energy. In a helium atom, however, the 3s orbital is lower in energy than the 3p orbital, which in turn is lower in energy than the 3d orbital. Explain why the energy rankings of hydrogen and helium are different.

8.4 The energy of the $n = 2$ orbital of the He$^+$ ion is the same as the energy of the $n = 1$ orbital of the H atom. Explain this fact.

8.5 Refer to Table 8-1 to answer the following questions. In each case, provide a brief explanation of your choice. (a) Which ionization energies show that an electron in a 1s orbital provides nearly complete screening of an electron in a 2p orbital? (b) Which ionization energies show that the stability of $n = 2$ orbitals increases with Z^2?

8.6 Refer to Table 8-1 to answer the following questions. In each case, provide a brief explanation of your choice. (a) Which ionization energies show that stability of $n = 1$ orbitals increase with Z? (b) Which ionization energies show that an electron in a 1s orbital incompletely screens another electron in the same orbital?

Structure of the Periodic Table

8.7 Draw the periodic table in block form, and outline and label each of the following sets: (a) Elements that are one electron short of filled p orbitals; (b) elements for which $n = 3$ orbitals are filling; (c) elements with half-filled d orbitals; and (d) the first element that contains a 5s electron.

8.8 Draw the periodic table in block form, and outline and label each of the following sets: (a) Elements for which $n = 1$ orbitals are filling; (b) elements for which the 5f orbitals are filling; (c) elements with $s^2 p^4$ configurations; and (d) elements with filled valence s orbitals but empty valence p orbitals.

8.9 Determine the atomic number and valence orbital that is filling for the element that would appear below lead in the periodic table.

8.10 Determine the atomic number and valence orbital that is filling for the element that would appear below francium in the periodic table.

8.11 Predict the location in the periodic table (row and column) of element 111.

8.12 Predict the location in the periodic table (row and column) of element 113.

8.13 How many valence electrons does each of the following atoms have? O, V, Rb, Sn, and Cd.

8.14 How many valence electrons does each of the following elements have? P, Cr, Kr, I, and Ba.

Electron Configurations

ilw **8.15** Write the complete electron configuration, and list a correct set of values of the quantum numbers for each of the valence electrons in the ground-state configurations of (a) Be, (b) O, (c) Ne, and (d) P.

8.16 Write the complete electron configuration, and list a correct set of values of the quantum numbers for each of the valence electrons in the ground-state configurations of (a) Li, (b) C, (c) F, and (d) Mg.

8.17 Which of the atoms of Problem 8.15 are paramagnetic? Draw orbital energy level diagrams to support your answer.

8.18 Which of the atoms of Problem 8.16 are paramagnetic? Draw orbital energy level diagrams to support your answer.

8.19 The following are hypothetical configurations for a beryllium atom. Which use nonexistent orbitals, which are forbidden by the Pauli principle, which are excited states, and which is the ground-state configuration? (a) $1s^3 2s^1$; (b) $1s^1 2s^3$; (c) $1s^1 2p^3$; (d) $1s^2 2s^1 2p^1$; (e) $1s^2 2s^2$; (f) $1s^2 1p^2$; and (g) $1s^2 2s^1 2d^1$.

8.20 None of the following hypothetical configurations describes the ground state of a fluorine atom. For each, state the reason why it is not correct: (a) $1s^2 2s^2 2p^4$; (b) $1s^2 2s^1 2p^6$; (c) $1s^3 2s^2 2p^4$; and (d) $1s^2 2s^2 1p^5$.

8.21 The ground state of Mo has higher spin than that of Tc. Construct energy level diagrams for the valence electrons that show how electron configurations account for this difference.

8.22 The ground state of V has lower spin than that of Cr. Construct energy level diagrams for the valence electrons that show how electron configurations account for this difference.

8.23 Write the correct ground-state electron configuration, in shorthand notation, for C, Cr, Sb, and Br.

8.24 Write the correct ground-state electron configuration, in shorthand notation, for N, Ti, As, and Xe.

8.25 For nitrogen, how many different excited-state configurations are there in which no electron has $n > 2$? Write all of them.

8.26 For fluorine, how many different excited-state configurations are there in which no electron has $n > 2$? Write all of them.

Periodicity of Atomic Properties

8.27 Arrange the following atoms in order of decreasing first ionization energy (smallest last): Ar, Cl, Cs, and K.

8.28 Arrange the following atoms in order of increasing size (largest last): Cl, F, P, and S.

8.29 One element has these ionization energies and electron affinity (all in kJ/mol): $IE_1 = 376$, $IE_2 = 2420$, $IE_3 = 3400$, $EA = -45.5$. In what column of the periodic table is this element found? Give your reasoning. Refer to Appendix C if necessary.

8.30 One element has these ionization energies and electron affinity (all in kJ/mol): $IE_1 = 503$, $IE_2 = 965$, $IE_3 = 3600$, $EA = 46$. In what group of the periodic table is this element found? Give your reasoning. Refer to Appendix C if necessary.

8.31 According to Appendix C, each of the following elements has a positive electron affinity. For each one, construct its valence orbital energy level diagram and use it to explain why the anion is unstable: N, Mg, and Zn.

8.32 According to Appendix C, each of the following elements has a positive electron affinity. For each one, construct its valence orbital energy level diagram and use it to explain why the anion is unstable: Be, Ar, and Mn.

8.33 List the atomic ions that are isoelectronic with Br^- and have net charges (absolute values) that are less than 4 units. Arrange these in order of increasing size.

8.34 List the ionic species that are isoelectronic with Ar and have net charges (absolute values) that are less than 4 units. Arrange these in order of increasing size.

Energetics of Ionic Compounds

8.35 From the following list, select the elements that form ionic compounds: Ca, C, Cu, Cs, Cl, and Cr. Indicate whether each forms a stable cation or a stable anion.

8.36 From the following list, select the elements that form ionic compounds: B, Ba, Be, Bi, and Br. Indicate whether each forms a stable cation or a stable anion.

8.37 Given the following data, estimate the energy released when gaseous K and I atoms form gaseous $[K^+ I^-]$.

Element	EA (kJ/mol)	IE_1 (kJ/mol)	IE_2 (kJ/mol)	Ion Radius (pm)
K	−48.4	418.8	3051	133 (cation)
I	−295.3	1008.4	1845.9	220 (anion)

8.38 Repeat the calculation of Problem 8.37 for K^{2+} and I^{2-}, using 500 kJ/mol as the estimated second electron affinity of iodine and assuming no change in distance of closest approach.

8.39 Consider three possible ionic compounds formed by barium and oxygen: $Ba^+ O^-$, $Ba^{2+} O^{2-}$, and $Ba^{3+} O^{3-}$. (a) Which would have the greatest lattice energy? (b) Which would require the least energy to form the ions? (c) Which compound actually exists, and why?

8.40 Consider three possible ionic compounds formed by calcium and the chloride anion: $CaCl$, $CaCl_2$, and $CaCl_3$. (a) Which would have the greatest lattice energy? (b) Which would require the least energy to form the ions? (c) Which compound actually exists, and why?

ilw **8.41** Calculate the overall energy change for the formation of lithium fluoride from lithium metal and fluorine gas. In addition to data found in Appendix C and Table 8-4, the following information is needed: The bond energy of F_2 is 155 kJ/mol, and lithium's enthalpy of vaporization is 159.3 kJ/mol.

8.42 Calculate the overall energy change for the formation of calcium bromide from calcium metal and liquid bromine. In addition to data found in Appendix C and Table 8-4, the following information is needed: The bond energy of Br_2 is 224 kJ/mol; bromine's enthalpy of vaporization is 30.9 kJ/mol; calcium's enthalpy of vaporization is 177.8 kJ/mol.

8.43 Draw a Born−Haber diagram similar to Figure 8-21 for the reaction described in Problem 8.41.

8.44 Draw a Born−Haber diagram similar to Figure 8-21 for the reaction described in Problem 8.42.

8.45 Refer to the lattice energy values in Table 8-4. If Cs_2O existed, what lattice energy would you predict it would have? Explain your prediction.

8.46 Refer to the lattice energy values in Table 8-4. Predict the lattice energy of MgS, and explain your prediction.

Ions and Chemical Periodicity

8.47 Classify each of the elements from Group 16 of the periodic table as a metal, a nonmetal, or a metalloid.

8.48 Classify each of the elements from Group 14 of the periodic table as a metal, a nonmetal, or a metalloid.

8.49 Classify each of the elements listed in Problem 8.35 as a metal, a nonmetal, or a metalloid.

8.50 Classify each of the elements listed in Problem 8.36 as a metal, a nonmetal, or a metalloid.

8.51 We list polonium as a metal, but some chemists classify it as a metalloid. List other metals that might be expected to show properties in between those of metals and metalloids.

8.52 What is the maximum number of valence p electrons possessed by a metallic element? Which metal(s) have this configuration?

Additional Paired Problems

8.53 Use the periodic table to find and list (a) all elements whose ground-state configurations indicate that the $4s$ and $3d$ orbitals are nearly equal in energy; (b) the elements in the column that has two elements with one valence configuration and two with another valence configuration; and (c) a set of elements whose valence configurations indicate that the $6d$ and $5f$ orbitals are nearly equal in energy.

8.54 Use the periodic table to find and list (a) all elements whose ground-state configurations indicate that the $5s$ and $4d$ orbitals are nearly equal in energy; (b) two elements whose configurations indicate that the $5d$ and $4f$ orbitals are nearly equal in energy; and (c) three elements in the same group that have different valence configurations.

8.55 Write the correct electron configuration for the Mn^{2+} ground state, and give a correct set of quantum numbers for all electrons in the least stable occupied orbital.

8.56 Write the correct electron configuration for the Co^{3+} ground state, and give a correct set of quantum numbers for all electrons in the least stable occupied orbital.

8.57 Predict the total electron spin for P, Br^-, and Cu^+.

8.58 Predict the total electron spin for Gd, Sr, and Ag^+.

For Problems 8.59−8.62, explain your rankings in terms of quantum numbers and electrical interactions.

8.59 Arrange the following in order of increasing ionization energy: N, O, Ne, Na, and Na^+.

8.60 Arrange the following in order of decreasing ionization energy: Br, Ar, Ar^+, and Cl.

8.61 Arrange the following in order of decreasing size (radius): Cl^-, K^+, Cl, and Br^-.

8.62 Arrange the following in order of increasing size: K, K^+, Ar, and Ca.

8.63 Write correct ground-state electron configurations for the neutral atoms with atomic numbers 9, 20, and 33.

8.64 Write correct ground-state electron configurations for the neutral atoms and doubly positive ions with atomic numbers 14, 34, and 92.

8.65 Using information in Table 8-4, determine the average difference in lattice energy between alkali fluorides and alkali chlorides, and the average difference in lattice energy between alkali bromides and alkali iodides. Is there a trend in the individual values? If so, describe and explain it.

8.66 Using information in Table 8-4, determine the average difference in lattice energy between lithium halides and sodium halides, and the average difference in lattice energy between rubidium halides and cesium halides. Is there a trend in the individual values? If so, describe and explain it.

8.67 Draw energy level diagrams that show the ground-state valence electron configurations for Cu^+, Mn^{2+}, and Au^{3+}.

8.68 Draw energy level diagrams that show the ground-state valence electron configurations for Ir^+, Cd^{2+}, and V^{2+}.

8.69 Write a brief explanation for each of the following: (a) In a hydrogen atom, the $2s$ and $2p$ orbitals have identical energy. (b) In a helium atom, the $2s$ and $2p$ orbitals have different energies.

8.70 Write a brief explanation for each of the following: (a) All three $2p$ orbitals of a helium atom have identical energy. (b) The IE of the He $1s^1 2p^1$ excited state is nearly the same as the IE of the H $2p^1$ excited state.

8.71 Refer to Figure 8-16 to answer the following questions about first ionization energies: (a) Which element shows the greatest decrease from its neighbor of next lower Z? (b) What is the atomic number of the element with the lowest value? (c) Identify three ranges of Z across which the value changes the least. (d) List the atomic numbers of all elements whose values are between 925 and 1050 kJ/mol.

8.72 Refer to Figure 8-17 to answer the following questions about electron affinities:
(a) Which column of the periodic table has the second-most negative set of values?
(b) Which element has the highest value?
(c) Apart from the end of a row, which element shows the greatest decrease compared with its neighbor of immediately lower Z?
(d) What do elements 6, 14, and 32 have in common?
(e) List the atomic numbers of all elements whose values are within 10% of -50 kJ/mol.

8.73 Which has the most unpaired electrons, S^+, S, or S^-? Use electron configurations to support your answer.

8.74 Which has the most unpaired electrons, Si, P, or S? Use electron configurations to support your answer.

8.75 The figure below shows four proposed electron energy diagrams for a phosphorus atom. Which are forbidden by the Pauli principle, which are excited states, and which is the ground-state configuration?

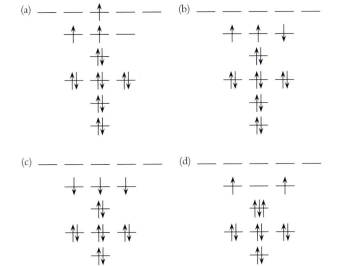

8.76 None of the four proposed electron energy diagrams shown below describes the ground state of a sulfur atom. For each, state the reason why it is not correct:

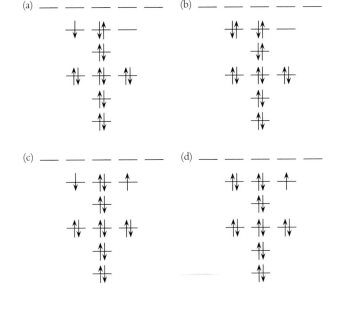

8.77 Identify the ionic compounds that best fit the following descriptions:
(a) the alkaline earth cation with the second-smallest radius and a Group 16 anion that is isoelectronic with the noble gas from Row 3;
(b) a $+1$ ion that is isoelectronic with the noble gas from Row 3, combined with the anion formed from the Row 2 element with the highest electron affinity; and
(c) the alkaline earth metal with the highest second ionization energy that combines in a 1:2 ratio with an element from Row 3.

8.78 Identify the ionic compounds that best fit the following descriptions:
(a) the naturally occurring alkali metal with the lowest ionization energy combined with the Group 17 element with the most negative electron affinity;
(b) the anion from Group 16 that has the smallest radius and is isoelectronic with a noble gas, combined with an alkali cation that is isoelectronic with the same noble gas; and
(c) the compound formed from the elements of Row 4 that have the lowest ionization energy and highest electron affinity.

8.79 Consider the three atomic orbitals that follow. They are drawn to scale, and orbital c has $n = 3$.

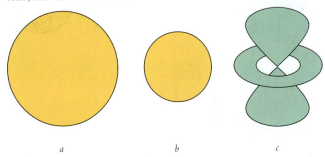

(a) Rank the orbitals in order of stability for a multielectron atom.
(b) Provide two sets of quantum numbers for an electron in orbital a.
(c) Give the atomic number of an element that has two electrons in orbital a but no electrons in orbital c.
(d) Give the name of a cation that has one electron in orbital c.
(e) How many other orbitals have the same principal quantum number as orbital c?
(f) An element has its two least stable electrons in orbital b. If an atom of that element loses one electron, will orbital b become larger or smaller or remain the same size?

8.80 Consider the following three atomic orbitals. They are drawn to scale, and orbital a has $n = 2$.

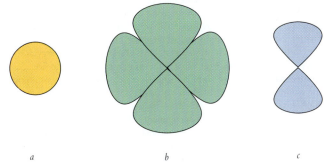

(a) How many electrons can be placed in orbital b?
(b) Provide three sets of quantum numbers that describe an electron in orbital c.
(c) Arrange these orbitals in order of increasing stability for a multielectron atom.
(d) In a sodium atom, how many other orbitals have exactly the same energy as orbital b?
(e) Give the name of the element that has orbital a filled but orbitals b and c empty.
(f) Give the name of an element that has both an empty and a partially filled orbital c.
(g) Give the name of a common anion in which all orbitals of type c are filled.

More Challenging Problems

8.81 Write the electron configuration for the lowest-energy *excited* state of each of the following: Be, O^{2-}, Br^-, Ca^{2+}, and Sb^{3+}.

8.82 Write the ground-state configurations for the isoelectronic species Ce^{2+}, La^+, and Ba. Are they the same? What features of orbital energies account for this?

8.83 The ionization energy of lithium atoms in the gas phase is about half as large as the ionization energy of beryllium atoms in the gas phase. In contrast, the ionization energy of Li^+ is about four times larger than the ionization energy of Be^+. Explain the difference between the atoms and the ions.

8.84 The figures show Cl^-, Ar, and K^+ drawn to scale. Decide which figure corresponds to which species and explain your reasoning.

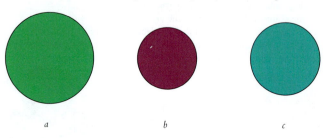

a   *b*   *c*

8.85 Make an electron density plot that shows how the 3s and 3p orbitals are screened effectively by the 2p orbitals. Provide a brief explanation of your plot.

8.86 Use the data in Appendix C to explain why the noble gases seldom take part in chemical reactions.

8.87 From its location in the periodic table, predict some of the physical and chemical properties of francium. What element does it most closely resemble?

8.88 What would be the next two orbitals to fill after the 7p orbital?

8.89 Which has the more stable 2s orbital, a lithium atom or a Li^{2+} cation? Explain your reasoning.

8.90 The first four ionization energies of aluminum are as follows: $IE_1 = 579$ kJ/mol. $IE_2 = 1817$ kJ/mol, $IE_3 = 2745$ kJ/mol, and $IE_4 = 11578$ kJ/mol. (a) Explain the trend in ionization energies. (b) Rank ions of aluminum in order of ionic radius, from largest to smallest. (c) Which ion of aluminum has the largest electron affinity?

8.91 The idea of an atomic radius is inherently ambiguous. Explain why this is so.

8.92 Use electron–electron repulsion and orbital energies to explain the following irregularities in first ionization energies: (a) Boron has a lower ionization energy than beryllium. (b) Sulfur has a lower ionization energy than phosphorus.

Group Study Problems

8.93 Suppose that the effects of screening were always less than the effects of increasing principal quantum number, making (for example) the 3d orbital always more stable than the 4s orbital. Construct the periodic table for the resulting filling order of the orbitals. What is the lowest value of Z that would have a different configuration, and how does it differ? Use this periodic table to identify the atomic numbers of elements that would be alkali metals. Which elements do you think would be halogens (adding one electron easily)? Which would be noble gases (chemically unreactive)? What similarities do you see, and what differences are there? Do you think the periodic table would have been easier to discover, harder to discover, or about the same had the filling order been like this? Why?

8.94 No elements have ground-state configurations with electrons in g ($l = 4$) orbitals, but excited states can have such electrons. (a) How many different g orbitals are there? (b) What are the possible values of m_l? (c) What is the lowest principal quantum number for which there are g orbitals? (d) Which orbitals may have nearly the same energy as the lowest-energy g orbitals? (e) What is the atomic number of the first element that has a g electron in its ground-state configuration, assuming the g orbital begins to fill after an s orbital?

8.95 Use data in Appendix C, Table 8-4, and the following information to calculate the overall energy changes for the formation of CaCl and $CaCl_2$. Use your results to determine which compound is more likely to form in the reaction of solid calcium with chlorine gas.
Vaporization enthalpy of Ca = 178 kJ/mol
Cl_2 bond energy = 240 kJ/mol
Lattice energy for CaCl (estimated) = 720 kJ/mol

8.96 In the imaginary galaxy Topsumturvum, all of the rules for atomic configurations are the same as ours, except that orbital stability increases slightly as l increases. As a result, the 3s orbital is more stable than the 4d orbitals, but the 4f orbitals are slightly more stable than the 3s orbital. (a) Draw an orbital energy ladder for Topsumturvum. (b) Determine the ground-state configuration for the Topsumturvum atom containing 27 electrons. (c) Draw a block diagram that shows the first 52 elements in the periodic table for Topsumturvum.

Questions 8.97–8.100 refer to a hypothetical universe named Morspin, where electrons have three possible spin orientations ($m_s = +\frac{1}{2}, -\frac{1}{2}, 0$) rather than the two ($m_s = +\frac{1}{2}, -\frac{1}{2}$) in our own universe. All other physical laws in the two universes are the same. The periodic table of Morspin follows:

1	2													3
4	5	6			7	8	9	10	11	12	13	14	15
16	17	18			19	20	21	22	23	24	25	26	27
28	29	30	31–45	46	47	48	49	50	51	52	53	54
55	56	57	58–72	73	74	75	76	77	78	79	80	81

8.97 (a) Which would have the larger first ionization energy, Morspin element 18 or Morspin element 30? Explain.
(b) Which would have the larger radius, Morspin element 15 or cation 17^{2+}? Explain.
(c) Which would have the larger electron affinity, Morspin element 47 or 48? Explain.

8.98 (a) What would be the electron configuration of Morspin cation 38^{3+}?
(b) Sketch the d_{xy} orbital as it would appear in this alternative universe. Be sure to include the coordinate axes.
(c) What would be the correct ground-state configurations for Morspin atoms of atomic numbers 3 and 14?

8.99 (a) How many electrons would the first three noble gases in Morspin have?
(b) Write acceptable sets of quantum numbers for the valence electrons of Morspin elements 4, 7, and 32.
(c) Which would screen a 4s electron more effectively, the $n = 3$ electrons of Morspin or the $n = 3$ electrons of our own universe? Explain.

8.100 (a) What would be the electron configuration of Morspin element 19?
(b) How many unpaired electrons would there be in element 28? Explain.
(c) What would be the highest positive charge of cations found in ionic salts that could form from Morspin element 30? Explain.

Answers to Section Exercises

8.1.1

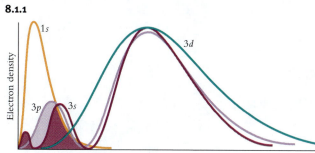

8.1.2 2.4×10^{-19} J
8.1.3

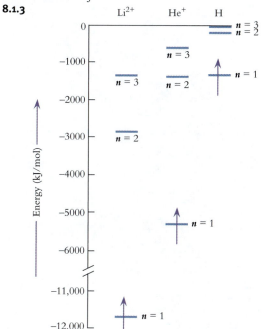

8.2.1 117
8.2.2 Br, with 35 electrons, is in the *p* block. Filled orbitals are 1*s*, 2*s*, 2*p*, 3*s*, 3*p*, 4*s*, and 3*d*. The 4*p* orbital is partially filled.
8.2.3 Be, Mg, Ca, Sr, Ba, Ra, Zn, Cd, and Hg; all have s^2.
8.3.1 (a) [Xe] $4f^{14} 5d^{10}$, zero spin; (b) [Rn] $7s^2 5f^{14} 6d^{10} 7p^6$, zero spin; and (c) [Ar], zero spin
8.3.2

8.3.3 Mo, [Kr] $5s^1 4d^5$; I, [Kr] $5s^2 4d^{10} 5p^5$; and Hg^+, [Xe] $6s^1 5d^{10} 4f^{14}$
8.4.1 (a) Cl has larger *Z*; (b) Cl has lower *n*; (c) Sr has higher *n*; (d) Cl^- has smaller *Z*; and (e) O^- has smaller Z_{eff}.
8.4.2 (a) Between Nb and In, a *d* block fills, so Z_{eff} remains nearly constant; (b) Ga loses a 4*p* electron, which is easier to remove than the 4*s* electrons of Zn; and (c) Mn has d^5 valence configuration, so the next electron must pair with an existing electron.
8.4.3 Stability decreases markedly with *n*.
8.5.1 $LE = 2956$ kJ/mol, compares closely with the value in Table 8-4, 2913 kJ/mol
8.5.2 Cu and Zn
8.5.3 Ca^{2+}, Cs^+, Al^{3+}, and Br^-
8.6.1 Nonmetals, 1390; metalloids, 824; *s* block, 610 (all in kJ/mol); low ionization energy leads to formation of ions
8.6.2 S, nonmetal; Si, metalloid; Sr, metal; Se, nonmetal; Sc, metal; Sg, metal; and Sn, metal
8.6.3 MnO_4^- (7 valence electrons, charge −1, total 8); CrO_4^{2-} (6 valence electrons, charge −2, total 8); and $Cr_2O_7^{2-}$ (12 valence electrons, charge −2, total 14)

Answers to Extra Practice Exercises

8.1

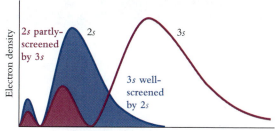

8.2 Zr, with 40 electrons, is in the *d* block. Filled orbitals are 1*s*, 2*s*, 2*p*, 3*s*, 3*p*, 4*s*, 3*d*, 4*p*, and 5*s*. The 4*d* orbital is partially filled.
8.3 Configuration is $1s^2 2s^2 2p^5$.

8.4 Configuration is [Kr] $5s^2 4d^{10}$.
8.5 Compact configuration is [He] $2s^2 2p^3$. Valid quantum numbers for the valence electrons are as follows:

$n = 2, l = 0, m_l = 0, \quad m_s = +\frac{1}{2}$ $n = 2, l = 0, m_l = 0, m_s = -\frac{1}{2}$
$n = 2, l = 1, m_l = +1, m_s = +\frac{1}{2}$ $n = 2, l = 1, m_l = 0, m_s = +\frac{1}{2}$
$n = 2, l = 1, m_l = -1, m_s = +\frac{1}{2}$

8.6 Ru^{3+} cation has $44 - 3 = 41$ electrons. Configuration is [Kr] $4d^5$.
8.7 Exceptions will be the ions with the 3*d* orbital set completely filled or completely empty: Zn^{2+}, Cu^+, and Sc^{3+}.
8.8 P and Se are smaller than As, Ge and Sb are larger than As.
8.9 Lattice energy = 2473 kJ/mol.
8.10 Ga is a metal, Ge and As are metalloids, and Se, Br, and Kr are nonmetals.

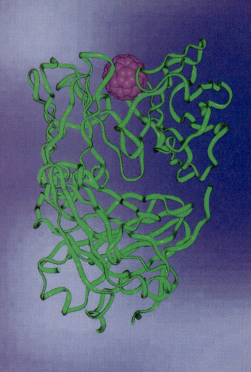

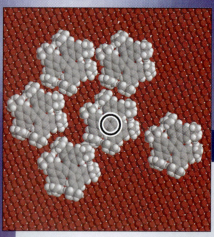

Nanotechnology

A gear-shaped molecule spins freely on a copper metal surface. A molecular "soccer ball" nestles snugly in a cavity on a protein. It sounds like science fiction, but these creations are real developments in the blossoming field of nanotechnology. The term "nanotechnology" refers to science carried out at the nanometer scale ($1 \text{ nm} = 10^{-9}$ m).

Our background image, taken by a scanning tunneling microscope, shows a single layer of gear-shaped organic molecules lying on a copper metal surface. Most of the molecules are locked in place by nesting close to their neighbors. One of them, however, appears distinctly "fuzzy", which suggests that the molecule may be spinning rapidly. This particular molecule was "unlocked" by moving it a fraction of a nanometer with the tip of a scanning tunneling microscope. (See Box 1-2 for a description of scanning tunneling microscopy.)

The inset image is a computer simulation of this system. The molecule on the right has been moved slightly. Computer-based calculations show that this small change in position allows the molecule in the center to rotate freely, just as is observed experimentally.

Another example of nanotechnology research is an attempt to develop biological molecules that can interact with fullerene, C_{60}. By themselves, C_{60} molecules are difficult to manipulate because they are "greasy" and inert. Scientists envision using

Fundamentals of Chemical Bonding

proteins bound to C_{60}, like the one illustrated here, as molecular machines that can deliver C_{60} units to build larger carbon structures.

A spinning molecule on a copper surface and a soccer-ball molecule tethered to a protein may seem no more useful than a spinning ice-skater or a tetherball. Nonetheless, advocates of nanotechnology cite a wealth of potential applications for this new field, including tailored synthetic membranes that can collect specific toxins from industrial waste and computers that process data much faster than today's best models. The list of possible benefits from nanotechnology is limited only by our imaginations.

CHAPTER CONTENTS

Learning Resources

 KEY CONCEPTS

 CRITICAL THINKING

 SKILLS TO MASTER

 ADDITIONAL HELP
www.wiley.com/college/olmsted
- TUTORIALS
- ANIMATIONS

Nanotechnology involves designing and constructing specific molecular structures, one atom at a time. This requires a thorough understanding of chemical bonding, because every molecule is held together by chemical bonds in which electrons are shared between atoms. The basic principles of bond formation are always the same, but different elements have different bond-forming abilities, and chemical bonds show a variety of forms and strengths.

This chapter and the next describe chemical bonding. First, we explore the interactions among electrons and nuclei that account for bond forma-tion. Then we show how atoms are connected together in simple molecules such as water (H_2O). We show how these connections lead to a number of characteristic molecular geometries. In Chapter 10, we discuss more elaborate aspects of bond-ing that account for the properties of materials as diverse as deoxyribonucleic acid (DNA) and transistors.

9.1 OVERVIEW OF BONDING

A collection of electrons and nuclei responds to electrical forces, just as a body of water responds to the force of gravity. Without a barrier to block its path, water runs "down-hill," finding its most stable position. In a similar way, any group of charged particles arranges itself to maximize electrical attraction and give the most stable arrangement of charges.

Equation 6-1 describes electrical potential energy as a function of charges (q_1 and q_2) and their separation distance (r):

$$E_{\text{electrical}} = k\frac{q_1 q_2}{r} \tag{6-1}$$

The equation states that electrical energy between charged species is proportional to the magnitudes of the charges and inversely proportional to the distance between them. Also, charges of opposite sign attract one another, but like charges repel.

Equation 6-1 describes the potential energy of one pair of charges. Molecules, however, contain two or more nuclei and two or more electrons. To obtain the total electrical energy of a molecule, Equation 6-1 must be applied to every possible pair of charged species. These pair-wise interactions are of three types. First, electrons and nuclei attract one another. Attractive interactions release energy, so an electron attracted to a nucleus is at lower energy, and therefore more stable, than a free elec-tron. Second, electrons repel each other, raising the energy and reducing the stability of a molecule. Third, nuclei repel each other, so these interactions also reduce the sta-bility of a molecule.

The electrons and nuclei in a molecule balance these three interactions in a way that gives the molecule its greatest possible stability. This balance is achieved when the elec-trons are concentrated between the nuclei. We view the electrons as shared between the nuclei and call this sharing a **covalent bond.** In any covalent bond, the attractive energy between nuclei and electrons exceeds the repulsive energy arising from nuclear−nuclear and electron−electron interactions.

Figure 9-1
When electrons are in the region between two nuclei, attractive electrical forces exceed repulsive electrical forces, leading to the stable arrangement of a chemical bond. Remember that electrons are not point charges but are spread out over a relatively large volume.

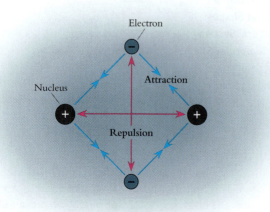

The Hydrogen Molecule

We demonstrate these ideas for the simplest stable neutral molecule, molecular hydrogen. A hydrogen molecule contains just two nuclei and two electrons.

Consider what happens when two hydrogen atoms come together and form a covalent bond. As the atoms approach, each nucleus attracts the opposite electron, pulling the two atoms closer together. At the same time, the two nuclei repel each other, and so do the two electrons. These repulsive interactions drive the atoms apart.

For H_2 to be a stable molecule, the sum of the attractive energies must exceed the sum of the repulsive energies. Figure 9-1 shows a static

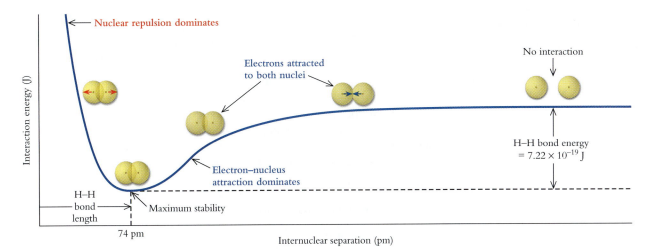

Figure 9-2
The interaction energy of two hydrogen atoms depends on the distance between the nuclei.

arrangement of electrons and nuclei in which the electron−nucleus distances are shorter than the electron−electron and nucleus−nucleus distances. In this arrangement, attractive interactions exceed repulsive interactions, leading to a stable molecule. Notice that the two electrons occupy the region between the two nuclei, where they can interact with both nuclei at once. In other words, the atoms share the electrons in a covalent bond.

An actual molecule is dynamic, not static. Electrons move continuously and can be thought of as being spread over the entire molecule. In a covalent bond, nevertheless, the distribution of electrons has the general characteristics shown by the static view in the figure. The most probable electron locations are between the nuclei, where they are best viewed as being shared between the bonded atoms.

Bond Length and Bond Energy

As two hydrogen atoms come together to form a molecule, attractive forces between nuclei and electrons make the hydrogen molecule more stable than the individual hydrogen atoms. The amount of increased stability depends on the distance between the nuclei, as shown in Figure 9-2. At distances greater than 300 pm, there is almost no interaction between the atoms, so the total energy is the energy of the two individual atoms. At closer distances, the attraction between the electron of one atom and the nucleus of the other atom increases, and the combined atoms become more stable. Moving the atoms closer together generates greater stability until the nuclei are 74 pm apart. At distances closer than 74 pm, however, the nucleus−nucleus repulsion increases more rapidly than the electron−nucleus attraction. Thus, at a separation distance of 74 pm, the hydrogen molecule is at the bottom of a potential "energy well." The atoms are combined in a molecule, sharing two electrons in a covalent bond.

Experimental studies of molecular motion reveal that nuclei vibrate continuously, oscillating about their optimum separation distance like two balls attached to opposite ends of a spring. Figure 9-3 shows this in schematic fashion for a hydrogen molecule vibrating about its optimum separation distance of 74 pm.

Figure 9-2 shows two characteristic features of chemical bonds. The separation distance where the molecule is most stable (74 pm for H_2) is known as the **bond length.** As described in Chapter 6, the amount of stability at this separation distance is the **bond energy,** or strength of the bond. Bond lengths and strengths are important properties of bonds that describe the characteristics of chemical bonding. The bond strength of the hydrogen molecule is 435 kJ/mol. Dividing this value by N_A (6.022×10^{23} mol^{-1}) and converting to joules gives a bond energy of 7.22×10^{-19} J/molecule.

Every bond has a characteristic length and strength. Example 9-1 compares molecular hydrogen with molecular fluorine.

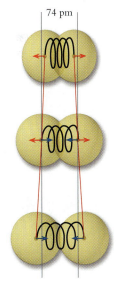

Figure 9-3
Molecules vibrate continually about their bond length, like two balls attached to a spring.

Example 9-1 **Bond Length and Energy**	The bond length of molecular fluorine is 142 pm, and the bond energy is 155 kJ/mol. Draw a figure similar to Figure 9-2 that includes both F_2 and H_2. Write a caption for the figure that summarizes the comparison of these two diatomic molecules. **Strategy:** We are asked to make a graph that compares the bond properties of F_2 and H_2. The bond energy is the lowest point on the interaction energy curve, representing the energy minimum for the pair of atoms. Energy in joules is plotted on the *y*-axis. The bond distance represents optimal separation of the nuclei, plotted along the *x*-axis in units of pm. Scale the axes so that both sets of data can be seen clearly. The caption should be a succinct statement that summarizes the main point of the figure. **Solution:** Begin by assembling the data:

$$\text{Bond length (}x\text{-axis): } H_2 = 74 \text{ pm, } F_2 = 142 \text{ pm}$$

$$\text{Interaction energy (}y\text{-axis): } H_2 = 7.22 \times 10^{-19} \text{ J/molecule}$$

$$F_2 = \frac{(155 \text{ kJ/mol})(10^3 \text{ J/kJ})}{6.022 \times 10^{23} \text{ molecules/mol}} = 2.57 \times 10^{-19} \text{ J/molecule}$$

The curves for the two molecules should have the same general appearance, with zero interaction energy at long distance, and with interaction energy rising very sharply at distances shorter than the bond length. The caption should summarize the content of the figure: Interaction energy plots show that the bond in F_2 is both longer and weaker than the bond in H_2.

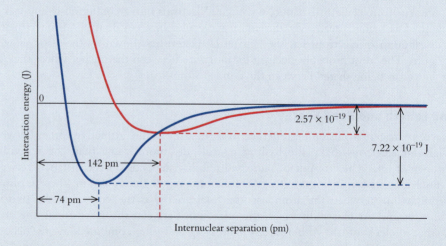

 Do the Results Make Sense? Fluorine is a second-row atom, so its atoms are much bigger than first-row hydrogen atoms. Thus, it makes sense that F atoms cannot get as close to one another as can H atoms.

Extra Practice **Exercise 9.1**	Make an interaction energy plot similar to Figure 9-2 for H—F (92 pm, 565 kJ/mol) and H—Br (141 pm, 365 kJ/mol). Write a brief caption for the figure that summarizes the data.

Other Diatomic Molecules: F_2

Bond formation in H_2 is relatively easy to describe, because we need to place just two electrons. Bond formation in other molecules, however, requires us to consider the fates of many electrons. This is a complicated task, but one simplified way to describe many bonds requires the sharing of only two electrons, as we illustrate using F_2.

As two fluorine atoms approach each other, the electrons of each atom feel the attraction of the nucleus of the other atom, just as happens in H_2. Remember from

Chapter 8 that atomic electrons are described by an electron configuration, which assigns each electron to an atomic orbital. The electron configuration of the F atom is $1s^2 \, 2s^2 \, 2p^5$. Orbitals have different sizes and shapes, so when one fluorine atom approaches another, the electrons in each orbital experience different forces of electrical attraction. The largest of these orbitals is the $2p$ set, so $2p$ electrons are closest to the neighboring nucleus and experience the strongest attraction. We can arrange the $2p$ orbitals so that one orbital of each atom points directly at the approaching atom, as shown in Figure 9-4. An electron in this orbital gets closer to the opposite nucleus than any of the other electrons do. Consequently, we can describe the F—F bond as the sharing of two electrons from the fluorine $2p$ orbitals that point directly at the approaching atom.

Unequal Electron Sharing

Each nucleus of a hydrogen molecule has a charge of $+1$. Consequently, both nuclei attract electrons equally. The result is a symmetrical distribution of the electron density between the atoms. Each nucleus of a fluorine molecule has a charge of $+9$, and again the electrons experience the same net attraction toward both nuclei. In a chemical bond between identical atoms, the two nuclei share the bonding electrons equally.

In contrast to the symmetrical forces in H_2 and F_2, the bonding electrons in HF experience *unsymmetrical* attractive forces. The bonding electrons are attracted in one direction by the $+1$ charge of the hydrogen nucleus and in the other direction by the $+9$ charge of the fluorine atom. Recall from Chapter 8 that the electrons close to the nucleus effectively screen the nuclear charge of a multielectron atom. This means that the electron in an approaching H atom will not experience a charge of $+9$. Still, the effective nuclear charge of a fluorine atom is significantly greater than the $+1$ nuclear charge of a hydrogen atom, so the electrons shared between H and F feel stronger attraction to the F atom than to the H atom.

Unequal attractive forces lead to an unsymmetrical distribution of the bonding electrons. The HF molecule is most stable when its electron density is concentrated closer to the fluorine atom and away from the hydrogen atom. This unequal distribution of electron density gives the fluorine end of the molecule a partial negative charge and the hydrogen end a partial positive charge. These partial charges are less than one charge unit and are equal in magnitude. To indicate such partial charges, we use the symbols $\delta+$ and $\delta-$ or an arrow pointing from the positive end toward the negative end, as shown for HF in Figure 9-5. This unequal electron sharing results in a **polar covalent bond.** We describe details of bond polarity and its consequences in Section 9.5.

Electronegativity and Polar Covalent Bonds

Any covalent bond between atoms of different elements is polar to some extent, because each element has a different effective nuclear charge. Each element has a characteristic ability to attract bonding electrons. This ability is called **electronegativity** and is symbolized by the Greek letter chi (χ). When two elements have different electronegativity values, a bond between their atoms is polar, and the greater the difference ($\Delta\chi$), the more polar the bond.

Electronegativity measures how strongly an atom attracts the electrons in a chemical bond. This property of an atom involved in a bond is related to but distinct from ionization energy and electron affinity. As described in Chapter 8, ionization energy measures how strongly an atom attracts one of its own electrons. Electron affinity specifies how strongly an atom attracts a free electron. Figure 9-6 provides a visual summary of these three different types of attraction.

Electronegativities, which have no units, are estimated by using combinations of atomic and molecular properties. The American chemist Linus Pauling developed one commonly used set of electronegativities. The periodic table shown in Figure 9-7 presents these values. Modern X-ray techniques can measure the electron density distributions of chemical bonds. The distributions obtained in this way agree with those predicted from estimated electronegativities.

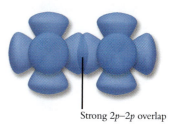

Strong 2p–2p overlap

Figure 9-4
The chemical bond in F$_2$ forms from strong electrical attraction of the electron in the fluorine 2p orbital that points directly at the approaching atom.

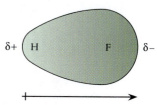

$\delta+$ H   F $\delta-$

Figure 9-5
Unequal sharing of electron density in HF results in a polar covalent bond. The color gradient represents the variation in electron density shared between the atoms.

Chi is pronounced *kai* (rhymes with *eye*).

Electronegativity

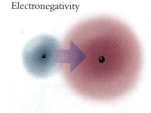

Ionization energy

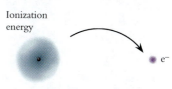

e^-

Electron affinity

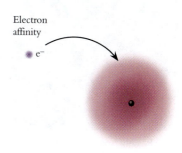

e^-

Figure 9-6
Electronegativity, ionization energy, and electron affinity are distinct properties.

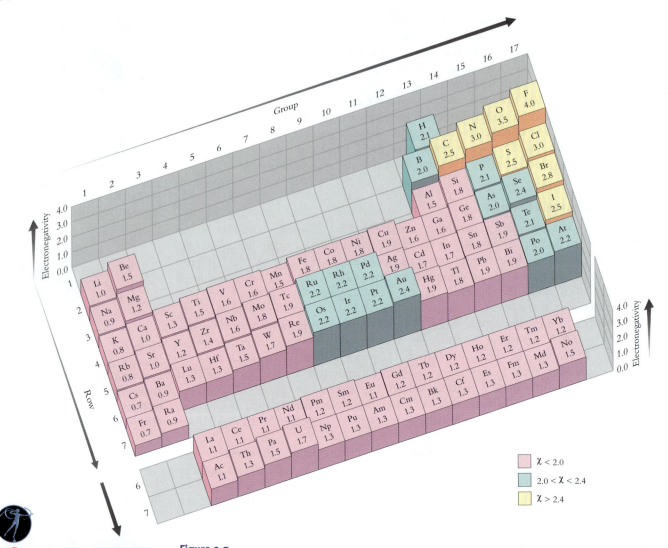

Figure 9-7
Electronegativities of the elements vary periodically, increasing from left to right and decreasing from top to bottom of the periodic table. Values for Group 18 have not been determined.

①

Linus Pauling (1901–1994) won the 1954 Nobel Prize in chemistry for his ideas about chemical bonds. Pauling was a leader in the movement to limit nuclear weapons, for which he was awarded the Nobel Peace Prize in 1963.

Figure 9-7 shows that electronegativity is a periodic property. Electronegativities increase from the lower left to the upper right of the periodic table. Cesium ($\chi = 0.70$) has the lowest value, and fluorine ($\chi = 4.0$) has the highest value. Notice also that electronegativities decrease down most columns and increase from left to right across the s and p blocks. As with ionization energies and electron affinities, variations in nuclear charge and principal quantum number explain electronegativity trends. Metals generally have low electronegativities ($\chi = 0.7$ to 2.4) and nonmetals have high electronegativities, ranging from $\chi = 2.1$ to 4.0. As we would expect from their intermediate character between metals and nonmetals, metalloids have electronegativities that are larger than the values for most metals and smaller than those for most nonmetals.

Electronegativity differences ($\Delta\chi$) between bonded atoms provide a measure of where any particular bond lies on the continuum of bond polarities. Three fluorine-containing substances, F_2, HF, and CsF, represent the range of variation. At one end of the continuum, the bonding electrons in F_2 are shared equally between the two fluorine atoms ($\Delta\chi = 4.0 - 4.0 = 0$). At the other limit, CsF ($\Delta\chi = 4.0 - 0.7 = 3.3$) is an ionic compound in which electrons have been fully transferred to give Cs^+ cations and F^- anions. Most bonds, including the bond in HF ($\Delta\chi = 4.0 - 2.1 = 1.9$), fall between these extremes. These are polar covalent bonds, in which two atoms share electrons unequally but do not fully transfer the electrons. Example 9-2 illustrates the periodicity of electronegativity and bond polarity variations.

Use the periodic table, without looking up electronegativity values, to rank each set of three bonds from least polar to most polar: (a) S—Cl, Te—Cl, Se—Cl; and (b) C—S, C—O, C—F.

Example 9-2

Electronegativities

Strategy: The larger the difference in electronegativity, the more polar the bond. Therefore, we can use periodic trends in electronegativities to arrange these bonds in order of polarity. Electronegativities decrease down most columns and increase from left to right across the s and p blocks. Use the periodic table to compare electronegativity values and rank the bond polarities.

Solution: (a) All these bonds contain chlorine, so the trend in bond polarity matches the trend in electronegativity of the bonding partners, S, Te, and Se. These elements are from Group 16 of the periodic table, so their electronegativities decrease going down the column: S > Se > Te. The trends also reveal that chlorine is more electronegative than its three partners are, so $\Delta\chi$ and bond polarity increases as χ of the bonding partner decreases:

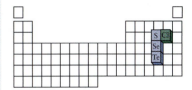

$$S\text{—}Cl < Se\text{—}Cl < Te\text{—}Cl$$

Least polar   Most polar

(b) These three bonds contain carbon, so the trend in bond polarities depends on the trend in electronegativities of S, O, and F. Sulfur is below oxygen in Group 16 of the periodic table, so sulfur is less electronegative than oxygen. Fluorine is to the right of oxygen in the same row, so fluorine is more electronegative than oxygen, because electronegativity increases across a row. All three have larger electronegativities than carbon, so the electronegativity difference and bond polarity increases as electronegativity increases:

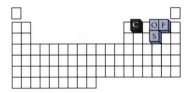

$$C\text{—}S < C\text{—}O < C\text{—}F$$

Least polar   Most polar

Do the Results Make Sense? You can use actual electronegativity values to verify that these rankings are correct.

List in order of increasing polarity all the possible bonds formed from the following elements: H, O, F. Consult Figure 9-7 for electronegativity values.

Section Exercises

▪ **9.1.1** Describe bond formation between hydrogen atoms and chlorine atoms to form HCl molecules.

▪ **9.1.2** For each of the following pairs, identify which element tends to attract electron density from the other in a covalent bond: (a) Si and O; (b) C and H; (c) As and Cl; and (d) Cl and Sn.

▪ **9.1.3** List the bonds of Exercise 9.1.2 from least polar to most polar.

9.2 LEWIS STRUCTURES

In this section, we develop a process for making schematic drawings of molecules called **Lewis structures.** A Lewis structure shows how the atoms in a molecule are bonded together. A Lewis structure also reveals the distribution of bonding and nonbonding valence electrons in a molecule. In a sense, a Lewis structure is a "molecular blueprint" that shows how a molecule is laid out. From this perspective, writing a Lewis structure is the first step in developing a bonding description of a molecule. Lewis structures are named for their inventor, G. N. Lewis, an American chemist who was a professor of chemistry at the University of California, Berkeley during the first half of the twentieth century.

The Conventions

Lewis structures are drawn according to the following conventions, which we illustrate in Figure 9-8 for the HF molecule:

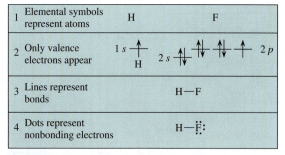

Figure 9-8
The Lewis structure conventions for hydrogen fluoride.

1	Elemental symbols represent atoms	H	F
2	Only valence electrons appear		
3	Lines represent bonds	H—F	
4	Dots represent nonbonding electrons	H—F̈:	

1. *Each atom is represented by its elemental symbol.* In this respect, a Lewis structure is an extension of the chemical formula.

2. *Only the valence electrons appear in a Lewis structure.* Recall from Chapter 8 that valence electrons are accessible for bonding. Core electrons are too close to the nucleus and too tightly bound to be shared with other atoms.

3. *A line joining two elemental symbols represents one pair of electrons shared between two atoms.* Two atoms may share up to three pairs of electrons: in **single bonds** (two shared electrons, one line), **double bonds** (four shared electrons, two lines), or **triple bonds** (six shared electrons, three lines).

4. *Dots placed next to an elemental symbol represent nonbonding electrons on that atom.* Nonbonding electrons usually occur in pairs with opposing spins.

These conventions divide molecular electrons into three groups. Core electrons are purely atomic in nature and do not appear in Lewis structures. Bonding valence electrons are shared between atoms and appear as lines. Nonbonding valence electrons are localized on atoms and appear as dots.

The Bonding Framework

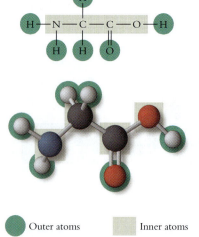

Figure 9-9
Outer atoms bond to only one other atom. Inner atoms bond to two or more other atoms.

A central feature of a Lewis structure is the bonding framework, which shows all the atoms connected as they are in the actual molecule. There is no foolproof method for putting the atoms together in the correct arrangement, but the following guidelines frequently lead from the chemical formula to the correct arrangement of atoms.

a. An *outer atom* bonds to only one other atom. An *inner atom* bonds to more than one other atom. Figure 9-9 shows a ball-and-stick model of glycine, an amino acid, using colored highlights to show the difference between inner atoms and outer atoms.

b. Hydrogen atoms are always outer atoms in the compounds discussed in this text. Notice that all five of glycine's hydrogen atoms are outer atoms. (Hydrogen does adopt an inner position in some molecules, but these examples are relatively uncommon and are not described in this textbook.)

c. In inorganic compounds, outer atoms other than hydrogen usually are the ones with the highest electronegativities. Here are some examples. Sulfur dioxide (SO_2) has two outer oxygen atoms ($\chi = 3.5$) bonded to an inner sulfur atom ($\chi = 2.5$). Silicon tetrachloride ($SiCl_4$) contains an inner silicon atom ($\chi = 1.8$) bonded to four outer chlorine atoms ($\chi = 3.0$). Chlorine trifluoride (ClF_3) has an inner chlorine atom that bonds to three outer fluorine atoms ($\chi = 4.0$):

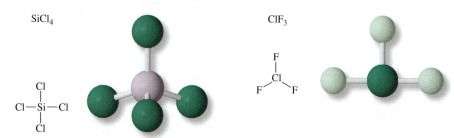

These three frameworks and the framework for glycine in Figure 9-9 illustrate an important point about Lewis structures. Although Lewis structures show how atoms are connected to one another, a Lewis structure is not intended to show the actual shape of a molecule. Silicon tetrachloride is not flat and square, SO_2 is not linear, and the fluorine atoms in ClF_3 are not all equivalent. We describe how to use Lewis structures to determine the shapes of molecules later in this chapter.

d. The order in which atoms appear in the formula often indicates the bonding pattern. For example, in HCN the framework is H—C—N, and the atoms in the OCN⁻ anion are in the order O—C—N. This guideline is particularly useful for organic substances. As an example, CH_3NH_2 has the framework shown in the margin:

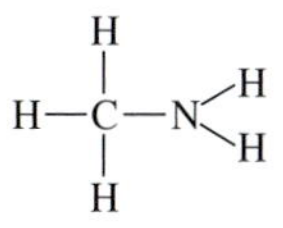

Notice in this case that the hydrogen atoms are grouped in the formula with their bonding partners, three hydrogen atoms with carbon and two more with nitrogen.

e. The hydrogen atoms appear first in the formula of an oxoacid. Nevertheless, in almost all cases these acidic hydrogen atoms bond to oxygen atoms, not to the central atom of the oxoacid. In phosphoric acid, H_3PO_4, for example, there are three O—H bonds but *no* H—P bond. This exception to Guideline *d* arises because of the convention of listing acidic hydrogen atoms first in the formula of an oxoacid, as we describe in Chapter 17.

These five guidelines allow us to figure out the frameworks of many compounds from their chemical formulas. When the guidelines do not provide an unambiguous framework, however, we need additional structural information. For example, the chemical formula of benzene is C_6H_6. The guidelines tell us that the six H atoms are in outer positions but provide no help in deducing the rest of the connections. In fact, there are several possible frameworks for the chemical formula C_6H_6. We must be told that the six C atoms of benzene form a ring, with one H bonded to each C, before we can write the framework for this molecule.

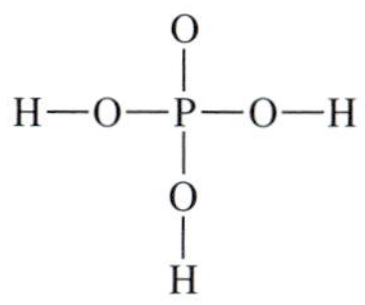

Building Lewis Structures

We build Lewis structures according to the six-step procedure described in the box, Building Lewis Structures. Learning how to apply these steps is best done through examples. We begin with phosphorus trichloride, PCl_3.

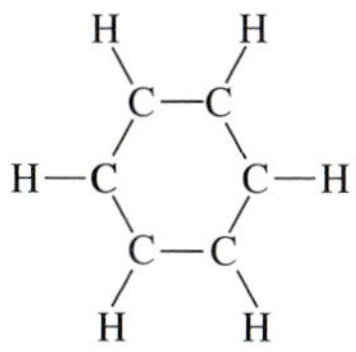

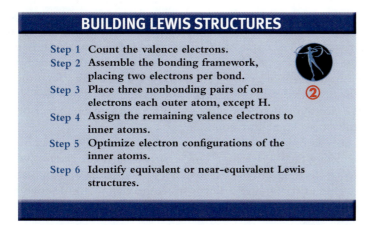

H_3PO_4

The bonding framework of phosphoric acid

C_6H_6

The bonding framework of benzene

> ## BUILDING LEWIS STRUCTURES
>
> **Step 1** Count the valence electrons.
> **Step 2** Assemble the bonding framework, placing two electrons per bond.
> **Step 3** Place three nonbonding pairs of on electrons each outer atom, except H.
> **Step 4** Assign the remaining valence electrons to inner atoms.
> **Step 5** Optimize electron configurations of the inner atoms.
> **Step 6** Identify equivalent or near-equivalent Lewis structures.

Step 1. *Count the valence electrons.* A Lewis structure shows *all* valence electrons and *only* valence electrons, so a correct count of valence electrons is essential. Recall from Section 8.2 that the number of valence electrons of an atom can be found from its position in the periodic table. Add the contributions from all atoms to obtain a total count of valence electrons. If the species is an anion, *add* one electron for each negative charge; if the species is a cation, *subtract* one electron for each positive charge.

In PCl_3, phosphorus contributes five valence electrons (Group 15, s^2p^3). Each chlorine atom contributes seven electrons (Group 17, s^2p^5), so the three chlorine atoms of phosphorous trichloride contribute 21 electrons, giving a total valence electron count of $5 + 21 = 26$.

Step 2. *Assemble the bonding framework.* To draw a Lewis structure correctly, we must know how the atoms are connected. The guidelines help us do this. The Lewis structure does not have to match the actual three-dimensional molecular shape, so we can arrange the framework in any convenient way.

The three chlorine atoms of PCl_3, being more electronegative than phosphorus, are outer atoms (Guideline *c*). Each Cl atom forms a bond to the inner P atom. To build the bonding framework, draw lines to indicate the bonds between the atoms. In the following structures, we show the most recent modification in red:

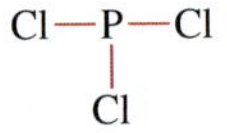

$$Cl-P-Cl$$
$$|$$
$$Cl$$

It is convenient to show the framework of phosphorus trichloride as a T shape. Later in this chapter, we will see that this is not the actual geometry of the molecule.

Step 3. *Place three nonbonding pairs of electrons on each outer atom, except H.* Studies of the electron distributions around outer atoms consistently show that hydrogen is always associated with two electrons (one pair). All other outer atoms always have eight electrons (four pairs). The reason for this regularity is that each atom in a molecule is most stable when its valence shell of electrons is complete. For hydrogen, this requires a single pair of electrons, enough to make full use of the hydrogen $1s$ orbital. Any other atom needs four pairs of electrons, the maximum number that can be accommodated by an *ns np* valence shell. Details of these features can be traced to the properties of atoms (Chapter 8) and are discussed further in Chapter 10.

One bonding electron pair plus three nonbonding pairs, also called **lone pairs,** give each outer atom four electron pairs. Such a set of four pairs of electrons associated with an atom is often called an **octet.**

KEY CONCEPT *An outer atom other than hydrogen is most stable when it is associated with an octet of electrons.*

Working from the bonding framework for PCl_3, we add three lone pairs to each Cl atom in phosphorus trichloride, completing the octets for all three.

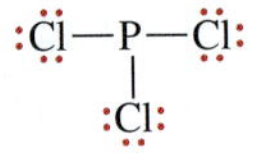

$$:\ddot{C}l-P-\ddot{C}l:$$
$$|$$
$$:\ddot{C}l:$$

Step 4. *Assign the remaining valence electrons to inner atoms.* This step is straightforward when there is only one inner atom. If the molecule has more than one inner atom, place nonbonding pairs around the most electronegative atom until it has an octet of electrons. If there are still unassigned electrons, do the same for the next most electronegative atom. Continue in this manner until all the electrons have been assigned.

Continuing with our example of PCl_3, we need to take an inventory of the valence electrons before completing Step 4. The framework contains three bonds that require a total of six electrons. Each of three Cl atoms has three lone pairs, for a total of 18 electrons. Thus, the first three steps of the procedure account for all but two of the 26 valence electrons. Step 4 directs us to place the remaining pair of electrons on the inner phosphorus atom.

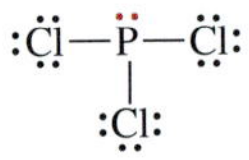

$$:\ddot{C}l-\ddot{P}-\ddot{C}l:$$
$$|$$
$$:\ddot{C}l:$$

We have now accounted for all the valence electrons of PCl_3.

After completing Step 4, we have built a *provisional Lewis structure,* which accounts for all the valence electrons but may or may not represent the optimum arrangement of electrons. Steps 5 and 6 optimize a Lewis structure. Before we describe these steps in detail, Examples 9-3 and 9-4 reinforce Steps 1–4.

Example 9-3

Lewis Structure of an Anion

Determine the provisional Lewis structure of the BF_4^- anion.

Strategy: The strategy for a Lewis structure is always the same. Follow the six-step procedure.

Solution:
1. Use the periodic table to count valence electrons. Boron, in Group 13, has three valence electrons ($s^2 p^1$). Each fluorine atom contributes seven electrons ($s^2 p^5$). The species is an anion, so we *add* one extra electron to account for the negative charge:

$$1\ B = (1)(3\ e^-) = 3\ e^-$$
$$4\ F = (4)(7\ e^-) = 28\ e^-$$
$$-1\ ion = 1\ e^-$$
$$\overline{}$$
$$Total\ valence\ e^- = 32\ e^-$$

2. Fluorine, having the highest electronegativity among the elements, is always an outer atom. Four F atoms bond to the inner B atom, accounting for eight valence electrons:

Example 9-3

Lewis Structure of an Anion
(*continued*)

$$
\begin{array}{c}
\text{F} \\
| \\
\text{F—B—F} \\
| \\
\text{F}
\end{array}
\qquad
\begin{array}{l}
\text{32 valence e}^- \\
\underline{-\ 8\ \text{framework e}^-} \\
\text{24 e}^-\ \text{left to place}
\end{array}
$$

3. Next, complete the octets of the four fluorine atoms:

$$
\begin{array}{c}
:\!\ddot{\text{F}}\!: \\
| \\
:\!\ddot{\text{F}}\text{—B—}\ddot{\text{F}}\!: \\
| \\
:\!\ddot{\text{F}}\!:
\end{array}
\qquad
\begin{array}{l}
\text{32 valence e}^- \\
-\ 8\ \text{framework e}^- \\
\underline{-24\ \text{lone pair e}^-} \\
\text{0 e}^-\ \text{left to place}
\end{array}
$$

4. At the end of Step 3, we have placed all 32 valence electrons, so Step 4 is not needed. The result of Step 3 is the provisional Lewis structure of the BF_4^- ion.

Does the Result Make Sense? Check a provisional Lewis structure by determining that all the valence electrons are assigned. Also check that identical atoms—the four F atoms in this example—all have the same electron distribution.

Determine the provisional Lewis structure of the hydronium ion, H_3O^+.

Determine the provisional Lewis structure of diethylamine, $(CH_3CH_2)_2NH$.

Example 9-4

**Parentheses and
Lewis Structures**

Strategy: Follow the six-step procedure, using the chemical formula to determine the bonding framework.

Solution:

1. The molecule has one nitrogen atom (s^2p^3). Interpreting the parentheses in the formula reveals that there are four carbon atoms (s^2p^2) and eleven hydrogen atoms (s^1). You should be able to verify that the valence electron count is 32 e$^-$.

2. The parentheses in the formula tell us that diethylamine has two CH_3CH_2 groups bonded to the nitrogen atom:

$$
\begin{array}{ccccccccc}
 & \text{H} & & \text{H} & \text{H} & & \text{H} & \text{H} \\
 & | & & | & | & & | & | \\
\text{H—} & \text{C} & \text{—} & \text{C} & \text{—N—} & \text{C} & \text{—} & \text{C} & \text{—H} \\
 & | & & | & | & & | & | \\
 & \text{H} & & \text{H} & \text{H} & & \text{H} &
\end{array}
$$

The framework contains 15 bonds, accounting for 30 of the 32 valence electrons.

3. All of the outer atoms are hydrogen atoms, so step 3 is not required for diethylamine.

4. Nitrogen is more electronegative than carbon. Place the last two valence electrons on this atom, completing its octet and giving the provisional Lewis structure.

$$
\begin{array}{ccccccccc}
 & \text{H} & & \text{H} & \text{H} & \text{H} & & \text{H} & \text{H} \\
 & | & & | & | & | & & | & | \\
\text{H—} & \text{C} & \text{—} & \text{C} & \text{—N—} & \text{C} & \text{—} & \text{C} & \text{—H} \\
 & | & & | & | & & | & | \\
 & \text{H} & & \text{H} & & \text{H} & & \text{H} &
\end{array}
$$

Does the Result Make Sense? Remember from line structure rules that carbon always has four bonds. Do all the carbon atoms in this structure meet that requirement?

Build the provisional Lewis structure of $(CH_3)_3COH$, whose proper name is 2-methyl-2-propanol.

Optimizing the Structure

The first four steps in our procedure lead to a provisional Lewis structure that contains the correct bonding framework and the correct number of valence electrons. Although the provisional structure is the correct structure in some cases, many other molecules require additional reasoning to reach the optimum Lewis structure. This is because the *distribution* of electrons in the provisional structure may not be the one that makes the molecule most stable. Step 3 of the procedure places electrons preferentially on outer atoms, ensuring that each *outer* atom has its full complement of electrons. However, this step does not always give the optimal configuration for the *inner* atoms. Step 5 of the procedure addresses this need.

Step 5. *Optimize electron configurations of the inner atoms.* To optimize the electron distributions about inner atoms, we first check to see if any inner atom lacks an octet. Examine the provisional structures of PCl_3, BF_4^-, and $(CH_3CH_2)_2NH$. All the inner atoms have octets in these structures, indicating that the provisional structures are the optimum Lewis structures. No optimization is required for these three examples.

As the next examples show, the provisional structure may contain one or more inner atoms with less than octets of valence electrons. These provisional structures must be optimized in order to reach the most stable molecular configuration. To optimize the electron distribution about an inner atom, move electrons from adjacent outer atoms to make double or triple bonds until the octet is complete. Examples 9-5 and 9-6 illustrate this procedure.

Example 9-5 **Optimizing a Provisional Structure**	Aqueous solutions of formaldehyde, H_2CO, are used to preserve biological specimens. Determine the Lewis structure of formaldehyde. **Strategy:** Apply the six-step procedure. **Solution:** 1−4. You should be able to follow the first four steps of the procedure to obtain the following provisional structure:

$$\ddot{O}:$$
$$|$$
$$\underset{H \quad \quad H}{C}$$

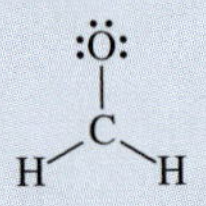

5. In the provisional structure, there are only six electrons (three single bonds) associated with the inner carbon atom. The carbon atom is more stable when it is associated with eight electrons. All the valence electrons have been placed, so the only source of additional electrons is the nonbonding electrons on the outer oxygen atom. We move one pair of electrons from the outer oxygen to make a second bond between carbon and oxygen (the three pairs are equivalent, so it doesn't matter which pair we move). In the completed Lewis structure, the double bond signifies four electrons shared between carbon and oxygen:

Sharing four electrons allows both carbon and oxygen to have octets of electrons. The right-hand figure above is the correct Lewis structure for formaldehyde.

 Does the Result Make Sense? Carbon has four bonds and oxygen has an octet, as expected.

Extra Practice Exercise 9.5 — Determine the Lewis structure of ethylene, H_2CCH_2.

Acrylonitrile, H_2CCHCN, is used to manufacture polymers for synthetic fibers. Draw the Lewis structure of acrylonitrile.

Example 9-6

Working with Multiple Inner Atoms

Strategy: Apply the six-step procedure.

Solution:

1. You should be able to show that the molecule has 20 valence electrons.

2. To assemble the framework, remember that hydrogen atoms are always outer atoms and that the atoms are usually bonded in the order listed in the formula:

$$H-\overset{\overset{\displaystyle H}{|}}{C}-\overset{\overset{\displaystyle H}{|}}{C}-C-N$$

20 valence e^-
−12 framework e^-
8 e^- left to place

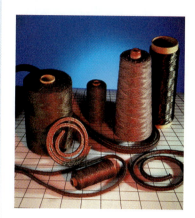

The framework has six bonds, which use 12 of the 20 valence electrons, leaving 8 left to place.

3. Use six of the remaining electrons to complete the octet of the outer nitrogen atom:

$$H-\overset{\overset{\displaystyle H}{|}}{C}-\overset{\overset{\displaystyle H}{|}}{C}-C-\ddot{\ddot{N}}\colon$$

20 valence e^-
−12 framework e^-
−6 outer e^-
2 e^- left to place

4. There are two electrons left to place in the molecule, and none of the carbon atoms has an octet. Which carbon atom gets that last pair of electrons? Anticipating step 5, we see that we will be able to complete the octet on the carbon adjacent to the nitrogen atom by shifting lone pairs. This suggests that we place the electrons on either the left-most or the middle carbon atom:

$$H-\overset{\overset{\displaystyle H}{|}}{\ddot{C}}-\overset{\overset{\displaystyle H}{|}}{C}-C-\ddot{\ddot{N}}\colon$$

5. In the provisional structure, two of the carbon atoms lack octets of electrons. Complete the Lewis structure by shifting lone pairs and making a double bond and a triple bond, as shown at right:

$$H-\overset{\overset{\displaystyle H}{|}}{\ddot{C}}-\overset{\overset{\displaystyle H}{|}}{C}-C-\ddot{\ddot{N}}\colon \longrightarrow H-\overset{\overset{\displaystyle H}{|}}{C}=\overset{\overset{\displaystyle H}{|}}{C}-C\equiv N\colon$$

Provisional structure   Final Lewis structure

Does the Result Make Sense? Because the inner atoms are electron-deficient in the provisional structure, we should pay particular attention to their electron count in the final structure. Each of the carbon atoms has an octet composed of four bonds, in accordance with line structure rules.

Acrolein, $H_2CCHCHO$, is used to manufacture polymers for plastics. Draw the Lewis structure of acrolein.

Beyond the Octet

The concept of an octet of electrons is one of the foundations of chemical bonding. In fact, C, N, and O, the three elements that occur most frequently in organic and biological molecules, rarely stray from the pattern of octets. Nevertheless, an octet of electrons does not guarantee that an inner atom is in its most stable configuration. In particular, elements that occupy the third and higher rows of the periodic table and have more than four valence electrons may be most stable with more than an octet of electrons. Atoms of these elements have valence *d* orbitals, which allow them to accommodate more than eight electrons. In the third row, phosphorus, with five valence electrons, can form as many as five bonds. Sulfur, with six valence electrons, can form six bonds, and chlorine, with seven valence electrons, can form as many as seven bonds.

A molecule that contains an inner atom from the third or higher row of the periodic table may have electrons left over after each inner atom has an octet of electrons. In such cases, we place any remaining electrons on those inner atoms. Example 9-7 describes one such molecule.

Example 9-7

Lewis Structure of ClF$_3$

Chlorine trifluoride is used to recover uranium from nuclear fuel rods in a high-temperature reaction that produces gaseous uranium hexafluoride:

$$2 \, ClF_3 \, (g) + U \, (s) \longrightarrow UF_6 \, (g) + Cl_2 \, (g)$$

Determine the Lewis structure of ClF$_3$.

Strategy: "Lewis structure" means "six-step procedure."

Solution:

1. All four atoms are halogens (Group 17, s^2p^5), so ClF$_3$ has 28 valence electrons.

2. Build the framework. Chlorine, with lower electronegativity than fluorine, is the inner atom. Make a single bond to each of the three fluorine atoms:

$$\begin{array}{c} F \\ | \\ F-Cl-F \end{array}$$

3. Add three nonbonding pairs to each fluorine outer atom.

$$\begin{array}{c} :\ddot{F}: \\ | \\ :\ddot{F}-Cl-\ddot{F}: \end{array}$$

4. Four electrons remain to be assigned. Place the last two electron pairs on the inner chlorine atom.

$$\begin{array}{c} :\ddot{F}: \\ | \\ :\ddot{F}-\ddot{Cl}-\ddot{F}: \end{array}$$

Does the Result Make Sense? Step 4 results in 10 electrons around the chlorine atom. However, chlorine, from row 3 of the periodic table, has d orbitals available to participate in bonding, so this is a legitimate Lewis structure. Notice that shifting electrons to make double bonds would not reduce the electron count around chlorine. Instead, making a Cl—F double bond would give fluorine more than an octet, which is not possible because fluorine is a second-row element.

Determine the Lewis structure of sulfur tetrafluoride, SF$_4$.

When do we entertain the possibility of valence electron counts greater that eight? For some molecules, such as ClF$_3$, the first four steps of the procedure lead to an inner atom with an electron count that exceeds an octet. In other cases, the optimization carried out in Step 5 may result in electron counts higher than eight. Optimization leads to an inner atom with more than eight electrons when the provisional Lewis structure leaves an inner atom with a deficiency of electrons. To determine if there is a deficiency of electrons, we make use of the concept of **formal charge (FC).** The formal charge of an atom is the difference between the number of valence electrons in the free atom and the number of electrons assigned to that atom in the Lewis structure:

$$\text{Formal charge} = \text{(Valence electrons of free atom)} - \text{(Valence electrons assigned in Lewis structure)} \tag{9-1}$$

When an inner atom has a positive formal charge, a better Lewis structure may be obtained by shifting electrons to make multiple bonds.

We determine the number of electrons assigned to an atom in the Lewis structure by making two assumptions. First, lone pairs are localized on the atoms to which they are assigned. Second, bonding electrons are shared equally between bonded atoms. In this way, an atom is assigned all of its lone-pair electrons and one-half of its bonding electrons. The other half of the bonding electrons are assigned to the atom's bonding partners. Thus, a quick way to determine electrons assigned to an atom is by counting dots and lines around that atom. Each dot represents a nonbonding electron, and each line represents a bonding pair, one of whose electrons is assigned to the atom.

For the molecule treated in Example 9-7, the formal charge on the inner chlorine atom is calculated as follows:

$$FC_{Cl} = \text{Valence electrons} - \text{Assigned electrons}$$

$$= (7 \text{ valence electrons}) - [\tfrac{1}{2}(6 \text{ bonding electrons}) + (4 \text{ lone-pair electrons})] = 0$$

Each of the three fluorine atoms also has a formal charge of zero:

$$FC_F = (7 \text{ valence electrons}) - [\tfrac{1}{2}(2 \text{ bonding electrons}) + (6 \text{ lone-pair electrons})] = 0$$

A useful feature of formal charges is that the sum of the formal charges on all atoms equals the charge of the species. For a neutral molecule, the sum of the formal charges must be zero. For a cation or anion, the sum of the formal charges equals the charge on the ion.

In general, the best Lewis structure is the one that results in minimum formal charges. Thus, the formal charge of each atom is zero in the Lewis structure of ClF_3. This leads us to a general strategy for optimizing electron configurations of inner atoms: *If Step 4 of the procedure leaves a positive formal charge on an inner atom beyond the second row, shift electrons to make double or triple bonds to minimize formal charge, even if this gives an inner atom with more than an octet of electrons.* Example 9-8 illustrates the use of formal charge in optimizing the configuration of an inner atom.

As described in Chapter 5, sulfur dioxide, a by-product of burning fossil fuels, is the primary contributor to acid rain. Determine the Lewis structure of SO_2.

Strategy: Use the six-step procedure.

Solution:

1. All three atoms are from Group 16 of the periodic table, so each has six valence electrons, giving 18 for the molecule.

2. The more electronegative oxygen atoms are outer atoms: O—S—O

3. The two bonds use 4 valence electrons, leaving 14 to place. Six go on each of the outer oxygen atoms, using 12:   :Ö—S—Ö:

4. Place the last two valence electrons on the sulfur atom: :Ö—S̈—Ö:

5. Because sulfur is from row 3, we determine how to optimize the structure by evaluating formal charge. Sulfur has six valence electrons (Group 16) and four assigned electrons (2 bonds + 2 lone-pair electrons):

$$FC_S = (6 \text{ valence e}^-) - [\tfrac{1}{2}(4 \text{ bonding e}^-) + (2 \text{ lone-pair e}^-)] = +2$$

It is helpful also to evaluate the formal charges on the O atoms:

$$FC_O = (6 \text{ valence e}^-) - [\tfrac{1}{2}(2 \text{ bonding e}^-) + (6 \text{ lone-pair e}^-)] = -1$$

All three formal charges can be reduced to zero if we move two electrons from each of the outer oxygen atoms to form two S=O double bonds.

FC = +2

:Ö—S̈—Ö: ⟹   :O=S̈=O:

FC = −1   FC = −1   All FC = 0

Example 9-8

Lewis Structure of Sulfur Dioxide

Does the Result Make Sense? Sulfur has more than an octet, but that is permissible because of the d orbitals accessible to elements whose valence orbitals have $n \geq 3$. The two oxygen atoms have identical electron distributions. We describe additional evidence that this is a reasonable structure in Section 9.5.

Determine the Lewis structure of phosphoric acid, H_3PO_4.

Extra Practice Exercise 9.8

To summarize, the provisional Lewis structure reached after Step 4 may not allocate an optimum number of electrons to one or more of the inner atoms. The electron distribution must be optimized when any inner atom does not have at least eight electrons or when an inner atom from beyond the second row has a positive formal charge. In either of these situations, a more stable structure results from transferring nonbonding electrons from outer atoms to inner atoms to create double bonds (four shared electrons) or triple bonds (six shared electrons).

We do not need to calculate the formal charges of an inner atom from row 2, because these species are limited to octets. Nevertheless, formal charges can indicate the best way to optimize a structure, as Example 9-9 shows.

Example 9-9 **Lewis Structure** **of an Organic Molecule**	Acetic acid (CH_3CO_2H, a carboxylic acid) is an important industrial chemical and is the sour ingredient in vinegar. Build its Lewis structure. **Strategy:** Once again, follow the stepwise procedure. **Solution:** 1. The valence electron count is 24 e^-.

2. The way the chemical formula is written helps us construct the correct framework. Three of the hydrogen atoms are connected to one of the carbon atoms, the two oxygen atoms are connected to the other carbon atom, and the fourth hydrogen atom is connected to one of the oxygen atoms:

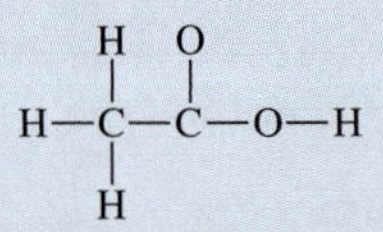

$$\begin{array}{c} \text{H} \quad \text{O} \\ | \quad\quad | \\ \text{H}-\text{C}-\text{C}-\text{O}-\text{H} \\ | \\ \text{H} \end{array}$$

3. Only one outer atom is not a hydrogen atom. This outer oxygen atom requires three pairs of nonbonding electrons:

$$\begin{array}{c} \text{H} \quad :\ddot{\text{O}}: \\ | \quad\quad | \\ \text{H}-\text{C}-\text{C}-\text{O}-\text{H} \\ | \\ \text{H} \end{array}$$

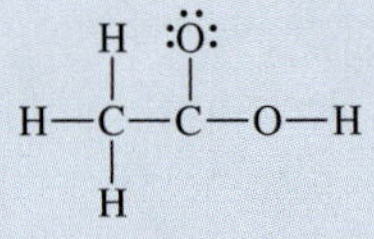

4. Four valence electrons must be placed around the three inner atoms. The most electronegative inner atom is oxygen, so place two pairs of nonbonding electrons on that atom, completing its octet. All the valence electrons have now been placed:

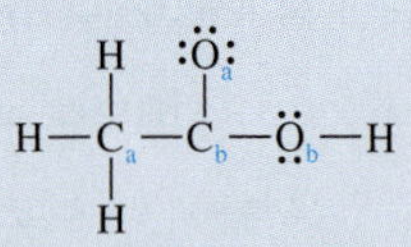

$$\begin{array}{c} \text{H} \quad :\ddot{\text{O}}:_a \\ | \quad\quad | \\ \text{H}-\text{C}_a-\text{C}_b-\ddot{\text{O}}_b-\text{H} \\ | \\ \text{H} \end{array}$$

(We label the different atoms of each element to help identify them in Step 5.)

5. Among the inner atoms, C_a and O_b have eight electrons each, but C_b has only six. Both oxygen atoms have lone pairs that could be transferred, but O_a has a formal charge of -1, so using one of its pairs of electrons to form a double bond gives C_b its octet and reduces all formal charges in the molecule to zero:

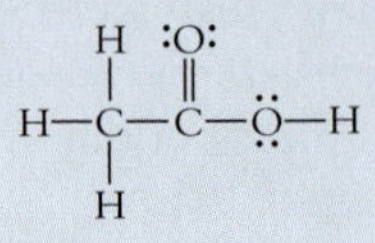

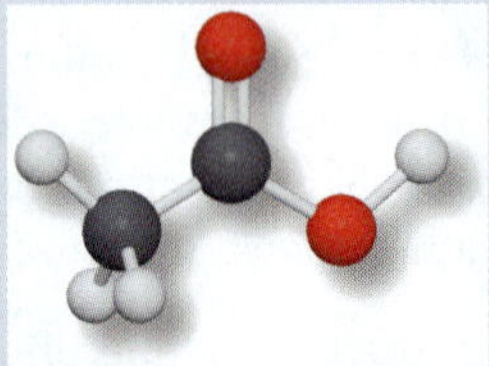

$$\begin{array}{c} \text{H} \quad :\text{O}: \\ | \quad\quad || \\ \text{H}-\text{C}-\text{C}-\ddot{\text{O}}-\text{H} \\ | \\ \text{H} \end{array}$$

Another way to complete the octet of C_b would be to make a double bond to the inner oxygen, O_b. Notice, however, that doing so creates formal charge on both of the oxygen atoms:

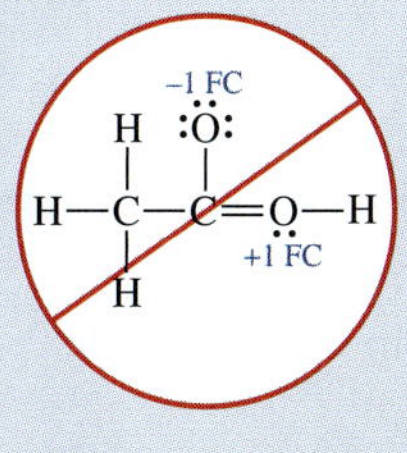

$$
\begin{array}{c}
\quad\quad -1\text{ FC} \\
H \quad\; :\ddot{O}: \\
\mid \quad\quad\; \mid \\
H-C-C=O-H \\
\mid \quad\quad\; {\scriptstyle +1\text{ FC}} \\
H
\end{array}
$$

This is not a realistic depiction, particularly when compared with a structure with no formal charges.

Does the Result Make Sense? Each carbon atom has four bonds, and each oxygen atom has zero formal charge. The $-CO_2H$ group, with one $C=O$ double bond and an acidic $C-O-H$ linkage, is characteristic of carboxylic acids.

Build the Lewis structure of carbonic acid, H_2CO_3.

Example 9-9

Lewis Structure of an Organic Molecule (*continued*)

Extra Practice Exercise 9.9

Resonance Structures

In completing Step 5 of the Lewis structure procedure, we often find that there is more than one way to optimize the configurations of inner atoms. This leads to the sixth and final step of the Lewis structure procedure.

Step 6. *Identify equivalent or near-equivalent Lewis structures.* The nitrate ion, NO_3^-, is a good example for this step. The anion has 24 valence electrons, and nitrogen is the inner atom. Three $N-O$ bonds use 6 electrons, and completing the octets of the three outer oxygen atoms places the remaining 18 electrons, giving the provisional Lewis structure shown in Figure 9-10. The nitrogen atom has only six electrons in the provisional structure, so optimization is required. We need to make one $N=O$ double bond to complete nitrogen's octet. As Figure 9-10 shows, a lone pair from any of the three oxygen atoms can be moved to form this double bond.

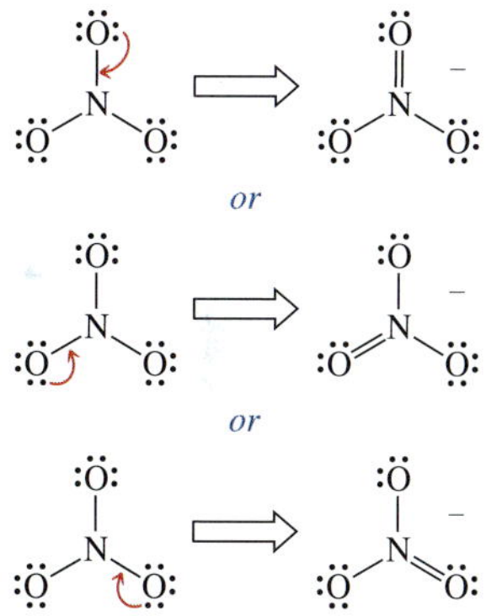

Which of these options is the "best" Lewis structure? Actually, no single Lewis structure by itself is an accurate representation of NO_3^-. Any single structure of the anion shows nitrate with one $N=O$ double bond and two $N-O$ single bonds. In Section 9-5, we show that single and double bonds between the same types of atoms have different lengths and different energies. In contrast, experiments show that the three nitrate $N-O$ bonds are identical. To show that the nitrate $N-O$ bonds are all alike, we use a composite of the three equivalent Lewis structures. These are traditionally called **resonance structures.** Resonance structures are connected by double-headed arrows to emphasize that a complete depiction requires all of them.

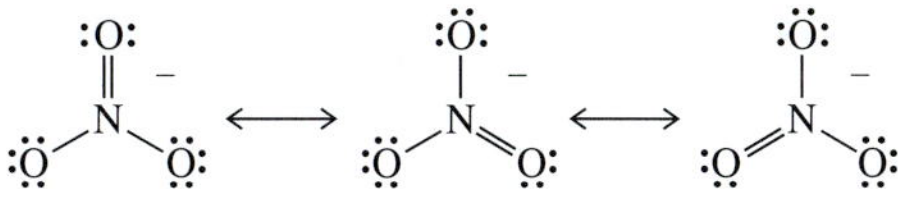

Figure 9-10
To convert from the provisional structure of the nitrate anion to complete the nitrogen octet, any of the three oxygen atoms can supply a pair of electrons.

It is essential to realize that electrons in the nitrate anion *do not* "flip" back and forth among the three bonds, as implied by separate structures. The true character of the anion is a blend of the three, in which all three nitrogen−oxygen bonds are equivalent. The need to show several equivalent structures for such species reflects the fact that Lewis structures are approximate representations. They reveal much about how electrons are distributed in a molecule or ion, but they are imperfect instruments that cannot describe the entire story of chemical bonding. In Chapter 10, we show how to interpret these structures from a more detailed bonding perspective.

Example 9-10 examines another anion that displays resonance structures.

Example 9-10

Lewis Structure of an Oxoanion

Determine the Lewis structure of dihydrogen phosphate, $H_2PO_4^-$.

Strategy: Follow the six-step procedure.

Solution:

1. The phosphorus atom contributes five valence electrons, the four oxygen atoms contribute six valence electrons each, and the hydrogen atoms each contribute one electron. The negative charge indicates that there is an extra electron, for a total of 32 valence electrons.

2. Guidelines for the framework indicate that the phosphorus atom is in the middle, bonded to the four oxygen atoms. By recognizing the link between dihydrogen phosphate and phosphoric acid, we know that the hydrogen atoms bond to two of the oxygen atoms.

3 and 4. Twelve electrons are required for the framework, leaving 20 to complete the octets of the four oxygen atoms. Here is the provisional Lewis structure:

$$\begin{array}{c} \ddot{\underset{..}{O}} \\ | \\ H - \ddot{\underset{..}{O}} - P - \ddot{\underset{..}{O}} - H \\ | \\ \ddot{\underset{..}{O}} \end{array}$$

5. Although the inner phosphorus atom has its octet of electrons, a formal charge calculation indicates that we can shift electrons to optimize the provisional structure:

$$FC_P = (5 \text{ valence e}^-) - [\tfrac{1}{2}(8 \text{ bonding e}^-) + (0 \text{ lone-pair e}^-)] = +1$$

The +1 formal charge can be eliminated by making one P=O double bond. As in acetic acid, we do not shift electrons from either inner oxygen atom, because that would create +1 formal charge on that atom. We can shift electrons from either outer O atom, so there are two options for the double bond, and this signals the possibility of resonance structures.

6. The two candidates are equivalent Lewis structures, so the best depiction of $H_2PO_4^-$ shows two resonance structures connected by a double-headed arrow.

$$\begin{array}{c} \underset{}{\overset{:O:}{\underset{||}{}}} \\ H - \ddot{\underset{..}{O}} - P - \ddot{\underset{..}{O}} - H \\ | \\ \underset{FC = -1}{\ddot{\underset{..}{O}}} \end{array} \longleftrightarrow \begin{array}{c} \overset{.. \; FC = -1}{\underset{}{\overset{:O:}{\underset{|}{}}}} \\ H - \ddot{\underset{..}{O}} - P - \ddot{\underset{..}{O}} - H \\ || \\ :O: \end{array}$$

Does the Result Make Sense? Notice that the two outer O atoms have identical descriptions and so do the two inner O atoms. Evaluating formal charges on all the atoms gives a net −1 charge, matching the overall charge of the anion.

Extra Practice Exercise 9.10

Draw the Lewis structure of the acetate anion, $CH_3CO_2^-$.

In the examples presented so far, all the resonance structures are equivalent, but resonance structures are not always equivalent. Resonance structures that are not equivalent occur when Step 5 requires shifting electrons from atoms of different elements. In such cases, different possible structures may have different formal charge distributions, and the optimal set of resonance structures includes those forms with the least amount of formal charge. Example 9-11 treats a molecule that has near-equivalent resonance structures.

Determine the Lewis structure of dinitrogen oxide (NNO), a gas used as an anesthetic, a foaming agent, and a propellant for whipped cream.

Strategy: Once again, follow the six-step procedure. We condense the steps.

Solution:

1–4. By now the first four steps should be familiar to you. Here is the provisional Lewis structure: :N̈—N—Ö:

5 and 6. In the provisional structure, the inner nitrogen atom has only four electrons, so we complete its octet by transferring two pairs of electrons from the outer atoms. We can choose any two pairs, which leads to three possible structures:

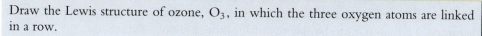

Evaluate the three potential resonance structures of dinitrogen oxide using their formal charges. The third structure shows more accumulation of formal charge than the first two. Thus, the optimal Lewis structure of the NNO molecule is a composite of the first two structures, but not the third: :N̈=N=Ö: ⟷ :N≡N—Ö̈:

Does the Result Make Sense? Experimental studies support our conclusion about the resonance structures of NNO. As we describe in Section 9.5, double bonds are shorter than single bonds. The length of the nitrogen–oxygen bond in NNO is less than typical N—O single bonds but greater than typical N=O double bonds.

Draw the Lewis structure of ozone, O₃, in which the three oxygen atoms are linked in a row.

Example 9-11

Lewis Structure of N₂O

Extra Practice Exercise 9.11

The Lewis structure of a molecule shows how its valence electrons are distributed. These structures present simple, yet information-filled views of the bonding in chemical species. In the remaining sections of this chapter, we build on Lewis structures to predict the shapes and some of the properties of molecules. In Chapter 10, we use Lewis structures as the starting point to develop orbital overlap models of chemical bonding.

Section Exercises

■ **9.2.1** Determine Lewis structures for the following molecules: (a) Br₂; (b) NCl₃; and (c) BrF₅.

■ **9.2.2** Determine Lewis structures for the following molecules: (a) H₂NCH₂CH₂NH₂; (b) HCCH (acetylene, used in welding torches); and (c) (H₃C)₂CO (acetone, an organic solvent).

■ **9.2.3** Determine Lewis structures for the following ions: (a) OH⁻; (b) CO₃²⁻; and (c) ClO₃⁻.

9.3 MOLECULAR SHAPES: TETRAHEDRAL SYSTEMS

The Lewis structure of a molecule shows how the valence electrons are distributed among the atoms. This gives a useful qualitative picture, but a more thorough understanding of chemistry requires more detailed descriptions of molecular bonding and molecular shapes. In particular, the *three-dimensional* structure of a molecule, which plays an essential role in determining chemical reactivity, is not shown directly by a Lewis structure.

The Shape of Methane

The Lewis structure of methane (CH₄) shows that the molecule contains four C—H single covalent bonds. How are these bonds arranged in three-dimensional space? Based on its chemical behavior, chemists concluded more than a century ago that the methane molecule has a highly symmetrical shape. Modern experiments show that the carbon atom is at the center of the molecule, with the four hydrogen atoms located at the four

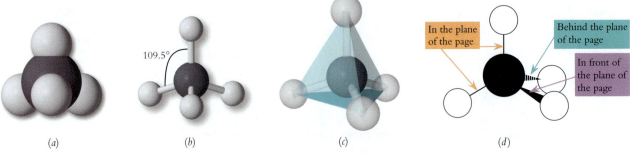

(a)   (b)   (c)   (d)

Figure 9-11
Representations of tetrahedral methane: (*a*) space-filling model; (*b*) ball-and-stick model; (*c*) ball-and-stick model, highlighting the tetrahedral faces; (*d*) ball-and-stick drawing using wedge representations for the out-of-plane bonds.

The Lewis structure of methane

corners of a regular **tetrahedron.** Figure 9-11 shows four representations of the tetrahedral methane molecule. Each H—C—H set makes a bond angle of 109.5° (Figure 9-11*b*), and all C—H bond lengths are 109 pm. A regular tetrahedron has the shape of a pyramid with four identical faces and four identical corners (Figure 9-11*c*). Each face is an equilateral triangle.

Chemists use a variation on the ball-and-stick model to depict more clearly the three-dimensional character of molecules, as shown for methane in Figure 9-11*d*. The central carbon atom is placed in the plane of the paper. In these models, solid lines represent bonds lying in the plane of the paper, solid wedges represent bonds that protrude outward from the plane of the paper, and dashed wedges represent bonds extending backward, behind the plane.

Methane is the simplest molecule with a tetrahedral shape, but many molecules contain atoms with tetrahedral geometry. Because tetrahedral geometry is so prevalent in chemistry, it is important to be able to visualize the shape of a tetrahedron.

Why a Tetrahedron?

Animation

The most stable shape for any molecule maximizes electron–nuclear attractive interactions while minimizing nuclear–nuclear and electron–electron repulsions. The distribution of electron density in each chemical bond is the result of attractions between the electrons and the nuclei. The distribution of chemical bonds relative to one another, on the other hand, is dictated by electrical repulsion between electrons in different bonds. The spatial arrangement of bonds must minimize electron–electron repulsion. This is accomplished by keeping chemical bonds as far apart as possible. The principle of minimizing electron–electron repulsion is called **valence shell electron pair repulsion,** usually abbreviated **VSEPR.**

Methane has four pairs of valence electrons, each shared in a chemical bond between the carbon atom and one of the hydrogen atoms. The electron density in each C—H bond is concentrated between the two nuclei. At the same time, methane's four pairs of bonding electrons repel one another. Electron–electron repulsion in methane is minimized by keeping the four C—H bonds as far apart as possible.

Consider building methane by sequential addition of H atoms to a C atom, as shown schematically in Figure 9-12. The first hydrogen atom can

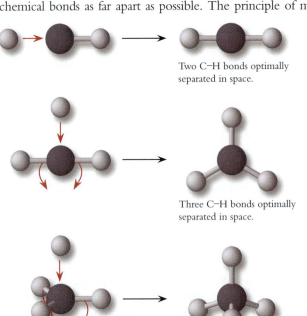

Two C–H bonds optimally separated in space.

Three C–H bonds optimally separated in space.

Four C–H bonds optimally separated in space.

Figure 9-12
The tetrahedron can be visualized as built by the sequential addition of hydrogen atoms to carbon, always keeping the C—H bonds as far apart as possible.

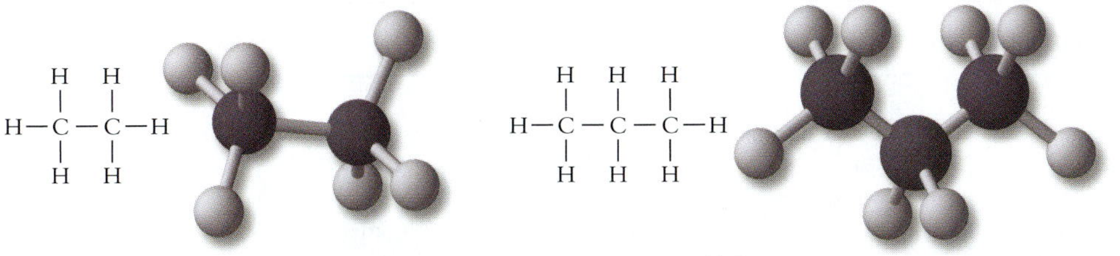

(a) Ethane

(a) Propane

Figure 9-13
Lewis structure and ball-and-stick models of ethane (a) and propane (b). All the carbon atoms have tetrahedral shapes, because each has four pairs of electrons to separate in three-dimensional space.

Animation

approach from any direction. To stay as far away from the first as possible, the second atom approaches from the opposite side of the carbon atom, generating a linear array. The third hydrogen atom approaches the CH_2 fragment from one side and repels the two existing C—H bonds to form a triangular shape, also called *trigonal planar*. The fourth atom approaches from above or below the plane of the existing bonds and repels the three existing C—H bonds. This converts the triangular shape into the tetrahedral geometry of methane.

Carbon and the Tetrahedron

Methane is the smallest member of a huge class of compounds called *hydrocarbons,* whose molecules contain only carbon and hydrogen. Hydrocarbons in which each carbon atom forms bonds to four other atoms are called *alkanes*. The three smallest alkanes are methane (CH_4), ethane (C_2H_6), and propane (C_3H_8). The structures of ethane and propane, shown in Figure 9-13, are unambiguous, because the carbon atoms can be placed only in a row. We can think of ethane as a methane molecule with one hydrogen atom replaced by a CH_3 (methyl) group. Similarly, propane is a methane molecule with two hydrogen atoms replaced by methyl groups. The Lewis structures of both these alkanes show each carbon atom surrounded by four pairs of bonding electrons. The bonds around each carbon are arranged in a tetrahedron to keep the four bonding pairs as far apart as possible.

Any hydrocarbon with four or more carbon atoms has more than one possible backbone. Figure 9-14 shows the possibilities for hydrocarbons with four carbon atoms.

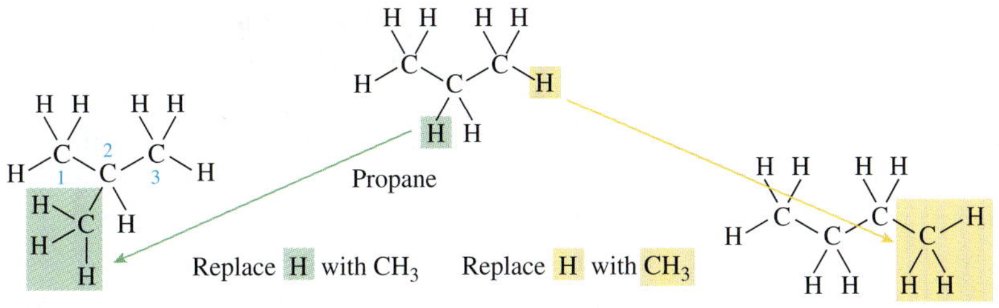

Propane

Replace H with CH_3 Replace H with CH_3

2-Methylpropane

Butane

Figure 9-14
The two structural isomers of C_4H_{10} at left have the same chemical formula but different bonding arrangements and different properties.

Building on propane, we can replace a hydrogen atom on a terminal carbon with a methyl group to form butane, an alkane with four carbon atoms in a row. Alternatively, we can replace either hydrogen atom on the inner carbon of propane to give a different compound, 2-methylpropane. In this molecule, three carbon atoms are in a row, but the fourth carbon atom is off to one side.

Two or more compounds that have the same molecular formula but different arrangements of atoms are called *structural isomers*. Butane and 2-methylpropane are structural isomers with the formula C_4H_{10}.

As the number of carbon atoms in the alkane increases, so does the number of possible structural isomers. Thousands of different alkanes exist, because there are no limits on the length of the carbon chain. Regardless of the number of the chain length, alkanes have tetrahedral geometry around all of their carbon atoms. The structure of decane, $C_{10}H_{22}$, shown in Figure 9-15, illustrates this feature. Notice that the carbon backbone of decane has a zigzag pattern because of the 109.5° bond angles that characterize the tetrahedron.

Petroleum, a complex mixture of many different hydrocarbons, is the main source of alkanes. Petroleum can be processed into various "fractions" by boiling a mixture in

Figure 9-15
Decane and other hydrocarbons consist of chains of tetrahedral carbon groups, emphasized here by highlighting one of the groups.

Table 9-1 Major Fractions of Petroleum and Their Uses

Fraction	Formulas	Boiling Point Range (°C)	Uses
Natural gas	CH_4 to C_4H_{10}	−160 to + 20	Fuel, cooking gas
Petroleum ether	C_5H_{12} to C_6H_{14}	30 to 60	Solvent for organic compounds
Gasoline	C_6H_{14} to $C_{12}H_{26}$	60 to 180	Fuel, solvent
Kerosene	$C_{12}H_{26}$ to $C_{16}H_{34}$	170 to 275	Rocket and jet engine fuel, domestic heating
Heating oil	$C_{15}H_{32}$ to $C_{18}H_{38}$	250 to 350	Industrial heating, fuel for electricity production
Lubricating oil	$C_{16}H_{34}$ to $C_{24}H_{50}$	300 to 370	Lubricants for automobiles and machines
Residue	$C_{20}H_{42}$ and up	Over 350	Asphalt, paraffin

Oil refineries use immense distillation towers to separate crude petroleum into useful fractions.

The word *ligand* is derived from the Latin verb *ligare,* which means "to bind."

huge distilling towers. Each fraction contains alkanes with a relatively narrow range of molar masses. Table 9-1 lists these major fractions of petroleum and some of their uses. As the table shows, these compounds are the principal sources of energy in our society.

The VSEPR Model

Having introduced methane and the tetrahedron, we now begin a systematic coverage of the VSEPR model and molecular shapes. The valence shell electron pair repulsion model assumes that electron−electron repulsion determines the arrangement of valence electrons around each inner atom. This is accomplished by positioning electron pairs as far apart as possible. Figure 9-12 shows the optimal arrangements for two electron pairs (linear), three electron pairs (trigonal planar), and four electron pairs (tetrahedral). Using Lewis structures, we can explore these and other shapes in detail.

We begin by introducing a few terms that are useful in describing molecular shapes. An **electron group** is a set of electrons that occupies a particular region around an atom. A group can be two electrons in a single bond, four electrons in a double bond, or six electrons in a triple bond. Nonbonding electrons almost always occur in groups of two electrons (pairs) with opposite spins, but when a species has an odd number of electrons, a single electron may be a "group." We use **ligand** to refer to an atom or a group of atoms bonded to an inner atom. A ligand can be as simple as a hydrogen atom or as complex as dozens of atoms held together as a unit by covalent bonds of their own. The **steric number (SN)** of an inner atom is the sum of the number of ligands plus the number of lone pairs; in other words, the total number of groups associated with that atom.

Figure 9-16 shows the molecular shapes of methane, ammonia, and water, all of which have hydrogen ligands bonded to an inner atom. These molecules have different numbers of ligands, but they all have the same steric number.

The steric number identifies how many groups of electrons must be widely separated in three-dimensional space. In ammonia, for example, the nitrogen atom bonds to three

Steric number = 4
Electron group geometry = Tetrahedral

Tetrahedral

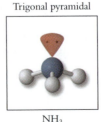

CH_4
Ligands = 4
Lone pairs = 0

Trigonal pyramidal

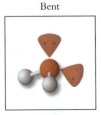

NH_3
Ligands = 3
Lone pairs = 1

Bent

H_2O
Ligands = 2
Lone pairs = 2

Figure 9-16
Electron group geometries and molecular shapes for steric number of 4.

hydrogen atoms, and it has one lone pair of electrons. How are the three hydrogen atoms and the lone pair oriented in space? Just as in methane, the four groups of electrons are positioned as far apart as possible, thus minimizing electron–electron repulsion.

The **electron group geometry** of an atom refers to the three-dimensional arrangement of the valence shell electron groups. Since electrons repel one another regardless of whether they are bonding pairs or lone pairs, it is the steric number that identifies the electron group geometry of an atom. The tetrahedral arrangement leads to greatest stability for steric number 4.

> *An inner atom with a steric number of 4 has tetrahedral electron group geometry.*   **KEY CONCEPT**

The inner atoms of methane, ammonia, and water, shown in Figure 9-16, all have tetrahedral electron group geometry. Nevertheless, each molecule has a different shape. The steric number dictates electron group geometry, but the **molecular shape** describes how the *ligands,* not the *electron groups,* are arranged in space. This is because atoms, not electrons, define molecular shape. For example, the shape of ammonia is found by ignoring the tetrahedral arm occupied by the lone pair. What remains is a nitrogen atom atop three hydrogen ligands. This shape is called a **trigonal pyramid,** consistent with the arrangement of the four atoms that make up the molecule. Nitrogen is at the apex of the pyramid, and the three hydrogen atoms make up the triangular base of the pyramid. To find the shape of a water molecule, we ignore the two tetrahedral arms that contain lone pairs. This leaves a planar H—O—H atomic system with a **bent shape.**

Our approach to these molecules illustrates the general strategy for determining the electron group geometry and the molecular shape of each inner atom in a molecule. The process has four steps, beginning with the Lewis structure and ending with the molecular shape.

Examples 9-12 and 9-13 illustrate this strategy.

DETERMINING MOLECULAR SHAPES

Step 1 Determine the Lewis structure.

Step 2 Use the Lewis structure to find steric numbers for inner atoms.

Step 3 Determine electron group geometries from steric numbers.

Step 4 Use the ligand count to derive molecular shapes from electron group geometries.

Describe the shape of the hydronium ion (H_3O^+). Make a sketch of the ion that shows the three-dimensional shape, including any lone pairs that may be present.

Example 9-12

Shape of the Hydronium Ion

Strategy: Follow the four-step process described in the flowchart. Begin with the Lewis structure. Use this structure to determine the steric number, which indicates the electron group geometry. Then take into account any lone pairs to deduce the molecular shape.

Solution:

1. Determine the Lewis structure. A hydronium ion has eight valence electrons. Six are used to make three O—H single bonds, and two are placed as a lone pair on the oxygen atom.

$$H—\underset{\underset{\displaystyle H}{|}}{\overset{\displaystyle H}{\underset{..}{O}}}—H \quad +$$

Eight electrons surround the oxygen atom, a Row 2 element, so this is the correct Lewis structure.

2. From the Lewis structure, we see that oxygen bonds to three hydrogen ligands and has one lone pair. The sum of the lone pairs and the ligands yields a steric number of 4.

3. A steric number of 4 identifies four electron groups that must be separated in three-dimensional space. Four groups are as far apart as possible in tetrahedral electron group geometry.

Example 9-12

Shape of the Hydronium Ion (*continued*)

Extra Practice Exercise 9.12

4. The number of ligands determines the molecular shape. With three ligands, the hydronium ion has trigonal pyramidal geometry, just like ammonia. Sketch the structure using solid and dashed wedges to indicate hydrogen atoms in front of and behind the plane of the page. The lone pair can be at any of the four tetrahedral positions, but the structure is easiest to visualize and to draw if the lone pair occupies the top position of the tetrahedron. A ball-and-stick model helps us to visualize the molecular shape.

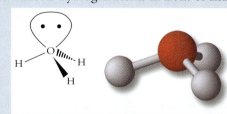

Does the Result Make Sense? The hydronium ion has the same number of electrons and atoms as the ammonia molecule, so it is reasonable for these two species to have the same shape.

Describe the shape of chloromethane (ClCH$_3$). Make a sketch of the molecule that shows its three-dimensional shape.

Example 9-13

Shapes Around Inner Atoms

Extra Practice Exercise 9.13

Describe the shape of hydroxylamine, HONH$_2$.

Strategy: Apply the four-step process to each of the inner atoms.

Solution:
1. There are 14 valence electrons, and the atoms are bonded in the order listed. In the final Lewis structure, oxygen has two lone pairs and nitrogen has one.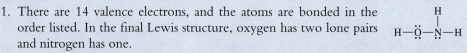
2. The oxygen atom has two ligands; one is a hydrogen atom and the other is the NH$_2$ group. Combined with two lone pairs, this gives oxygen a steric number of 4. The nitrogen atom also has a steric number of 4, with one lone pair and three ligands (H, H, and OH).
3. Both atoms have tetrahedral electron group geometry.
4. Oxygen has two ligands, so the geometry around the oxygen atom is bent. The nitrogen atom has three ligands, giving it trigonal pyramidal geometry. A ball-and-stick model shows the overall shape of the molecule. We show the lone pairs in this model to emphasize the electron group geometries of the two inner atoms.

Does the Result Make Sense? You can visualize hydroxylamine as ammonia (trigonal pyramid) with one H replaced by —OH, or as water (bent) with one H replaced by —NH$_2$.

Describe the shape of ethanol, CH$_3$CH$_2$OH.

Silicon

Many elements of the periodic table, from titanium and tin to carbon and chlorine, exhibit tetrahedral electron group geometry and tetrahedral molecular shapes. In particular, silicon displays tetrahedral shapes in virtually all of its stable compounds.

Compounds of silicon with oxygen are prevalent in the Earth's crust. About 95% of crustal rock and its various decomposition products (sand, clay, soil) are composed of silicon oxides. In fact, oxygen is the most abundant element in the Earth's crust (45% by mass) and silicon is second (27%). In the Earth's surface layer, four of every five atoms are silicon or oxygen.

The principal oxide of silicon is *silica*. The empirical formula of silica is SiO$_2$, but the compound consists of a continuous network of Si—O bonds rather than individual SiO$_2$

molecules. Figure 9-17 shows part of this net-work. Each silicon atom is at the center of a regular tetrahedron, bonded to four oxygen atoms. As in water molecules, each oxygen atom displays bent shape, with two lone pairs and two O—Si bonds generating a steric number of 4.

More than 20 different forms of silica exist, because the bonds and lone pairs around the oxygen atoms can be arranged in various ways. Each arrangement creates a different structural form for the silica network. Quartz, the most common form of silica, is found in granite, sandstone, and beach sand.

Closely related to silica are the silicate minerals, all of which contain polyatomic anions made of silicon and oxygen. The sim-plest silicates, called *orthosilicates,* contain SiO_4^{4-} anions. The SiO_4^{4-} anion is tetrahedral, with a central silicon atom bonded to four outer oxygen atoms.

Different minerals contain different metal cations to balance the -4 charge on the orthosilicate ion. Examples include calcium silicate (Ca_2SiO_4), an important ingredient in cement, and zircon ($ZrSiO_4$), which is often sold as "artificial diamond." One of the most prevalent minerals in the Earth's mantle is olivine, $M_2(SiO_4)$, in which M is one or two of the abundant metal cations, Fe^{2+}, Mg^{2+}, and Mn^{2+}.

The orthosilicates have entirely different structures than silica. In contrast to the fully covalent bonding network in silica, orthosilicates are ionic, containing metal cations and discrete SiO_4^{4-} anions. Orthosilicate minerals have 4:1 ratios of oxygen atoms to silicon atoms, and every oxygen atom is a negatively charged outer atom. In contrast, every oxygen atom in silica is inner, bonded to two silicon atoms. Silica has no ionic units and a 2:1 ratio of oxygen atoms to silicon atoms. Nevertheless, the silicon atoms in both structures have four Si—O bonds and tetrahedral shapes.

Many other silicate minerals contain inner *and* outer oxygen atoms. In *metasilicates,* for example, two oxygen atoms bonded to each silicon atom are inner and two are outer, with negative formal charges. Metasilicate networks can be linear chains or rings, examples of which appear in Figure 9-18. As the examples show, the ratio of silicon to oxygen in a metasilicate is 3:1. The models also show that metasilicates consist of large networks of Si—O—Si linkages bearing negatively charged outer O atoms. These charges must be counterbalanced by metal cations. In jade ($NaAlSi_2O_6$), which has a lin-ear chain structure, the metal cations are Na^+ and Al^{3+}. Beryl ($Be_3Al_2Si_6O_{18}$), the princi-

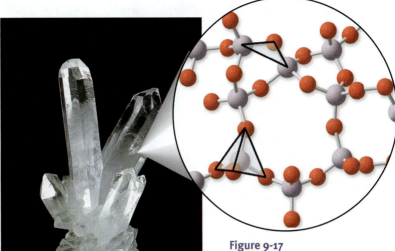

Figure 9-17
Quartz, a common form of silica, is a network of Si—O bonds. Silicon and oxygen both have tetrahedral electron group geometry. All the silicon atoms have tetrahedral shapes and all the oxygen atoms have bent shapes.

Lewis structure

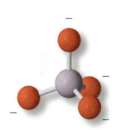

Ball-and stick model

Lewis structure and ball-and-stick model of the orthosilicate ion

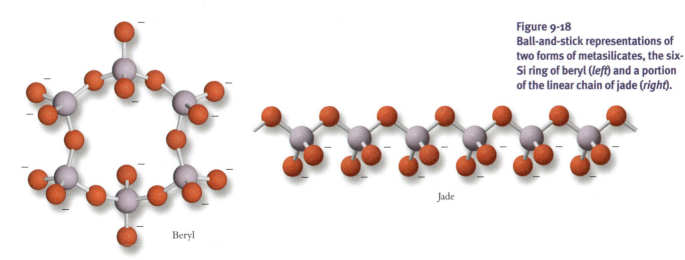

Beryl

Jade

Figure 9-18
Ball-and-stick representations of two forms of metasilicates, the six-Si ring of beryl (*left*) and a portion of the linear chain of jade (*right*).

| Box 9-1 | Chemistry and Life: The Importance of Shape |

T he regular tetrahedron is a simple yet elegant geometric form. The ancient Greeks identified it as one of only five regular solids that can be placed inside a sphere so that every vertex touches the surface of the sphere. The Greeks had no idea, however, of the importance that tetrahedral shapes have for the chemical processes of life.

Start with water, which is essential for life as we know it. If the water molecule were linear rather than bent, it would lack the properties that life-forms require. Linear water would not be polar and would be a gas like carbon dioxide. Why is water bent? Its four electron pairs adopt tetrahedral geometry, putting lone pairs at two vertices of a tetrahedron and hydrogen atoms at the other two vertices.

Biological molecules contain carbon, nitrogen, and oxygen atoms in inner positions. These second-row atoms are frequently associated with four electron pairs, giving them steric number 4 and tetrahedral electron group geometry. When many tetrahedra are attached to one another, an immense variety of shapes become possible. One relatively simple example is glucose, $C_6H_{12}O_6$. This sugar molecule forms a puckered ring of tetrahedral

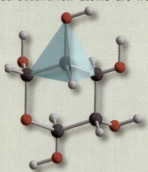

atoms. Polar O—H groups extend off the ring, and this is why glucose can dissolve in water or blood. Glucose can fuel life processes because its shape makes it easy to transport in and out of biological cells.

Glucose molecules can link together into chains, with each ring tethered to the next by a bridging oxygen atom. In one form, this is cellulose, the stiff material that gives the stalks of plants and the trunks of trees their structural strength. Chitin, a variation on cellulose, is an even stiffer material that forms the exoskeletons of crustaceans such as crabs and lobsters.

The variety of molecular shapes becomes even larger for proteins, molecules constructed from amino acids. An amino acid contains a tetrahedral nitrogen atom bound to a tetrahedral carbon atom, followed by a planar carbon atom and a tetrahedral oxygen atom. Any number of amino acids can be linked together. Moreover, the tetrahedral geometry allows bending and folding that leads to an infinite variety of protein sizes and shapes, including the helix and the pleated sheet:

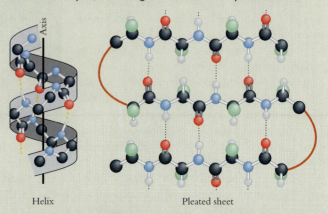

Helix   Pleated sheet

From water to sugars to proteins, tetrahedral shapes are at the heart of the molecules of biology. Our sex drives are controlled by the release of testosterone, a steroid molecule with many tetrahedral carbon atoms. Soothing odors and salves such as those made with menthol (oil of mint) are the result of molecules with tetrahedral carbon atoms. The disgusting smell of rotting flesh is due to putrescine, another molecule whose atoms have tetrahedral shapes. As long as we are alive, we are awash in that elegant shape that owes its name to the Greeks.

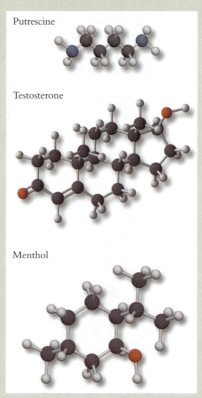

Putrescine

Testosterone

Menthol

Cellulose

Chitin

Figure 9-19
Photos of four silicate minerals, red beryl, emerald, mica, and asbestos. Silicates take many forms, depending on the detailed structure of the Si—O bonding network.

pal commercial source of beryllium metal, has six silicon tetrahedra linked in a ring and contains Be^{2+} and Al^{3+} cations. Pure beryl is colorless, but most beryl contains small amounts of transition metal ions such as Mn^{2+} (red) or Cr^{3+} (green). Emerald gemstones are beryl with 2% of its Al^{3+} ions replaced by Cr^{3+} ions. Figure 9-19 shows some of these colorful silicate minerals.

Silicates also exist in which each silicon atom bonds to one outer oxygen and to three inner oxygen atoms. The result is a linked network in which every silicon atom forms three Si—O—Si links, giving a planar, sheet-like structure. The empirical formula of this silicate is $Si_2O_5^{2-}$. In many minerals, aluminum atoms replace some of the silicon atoms to give aluminosilicates. The *micas*—one has the chemical formula $KMg_3(AlSi_3O_{10})(OH)_2$—contain sheets in which every fourth silicon atom is replaced by an aluminum atom. Because of the planar arrangement of its aluminosilicate network, mica is easily broken into flakes. Figure 9-19 includes a photograph of mica.

Even more complicated chemical formulas result when some silicon atoms have one outer oxygen atom, whereas others in the same mineral have two outer oxygen atoms. The asbestos minerals—one is crocidolite, $Na_2Fe_5(OH)_2(Si_4O_{11})_2$—have this type of structure. The photograph in Figure 9-19 shows that this silicate contains long fibrous chains. At the molecular level, the chains can be visualized as linked $Si_4O_{11}^{6-}$ units.

In summary, despite the structural diversity exhibited by the silicates, their silicon atoms always have tetrahedral geometry. In addition, every outer oxygen atom contributes a net charge of −1 to the structure, while every inner oxygen atom is electrically neutral and has an Si—O—Si bond angle close to 109.5°.

Tetrahedral shapes dominate the structures of biological molecules, as our Chemistry and Life Box describes.

When asbestos is handled, microscopic fibers become suspended in the atmosphere and are breathed into the lungs. There, they lodge in lung tissue, where they remain for many years, causing irritation that eventually leads to loss of lung function. Asbestos, which was once used extensively as insulation, is now recognized as a significant health hazard.

Section Exercises

■ **9.3.1** Tin tetrachloride ($SnCl_4$) is an important starting material for the preparation of a variety of tin compounds. Build the Lewis structure for $SnCl_4$ and determine its shape.

■ **9.3.2** Hydrazine, occasionally used as a rocket fuel, has the chemical formula N_2H_4. Determine the Lewis structure of hydrazine and draw a picture that shows the shape of the molecule.

■ **9.3.3** Draw a ball-and-stick model of the $Si_2O_5^{2-}$ unit found in mica. Your sketch should have one Si—O—Si linkage and should show that it connects to four other $Si_2O_5^{2-}$ units to form an interlocking sheet.

9.4 OTHER MOLECULAR SHAPES

Tetrahedral geometry may be the most common shape in chemistry, but several other shapes also occur frequently. This section applies the VSEPR model to four additional electron group geometries and their associated molecular shapes.

Steric Number 2: Linear geometry

A molecule with two ligands and no lone pairs has a steric number of 2, meaning there are two electron groups that must be separated in three-dimensional space. This is accomplished by placing the groups on opposite sides of the inner atom, separated by 180°. We illustrate this arrangement using a compound of zinc.

Zinc forms both ionic and covalent compounds. One important covalent example is dimethylzinc, $Zn(CH_3)_2$, a substance that has two Zn—C bonds. Used in synthesis reactions since the mid-1800s, dimethylzinc finds uses today in the preparation of catalysts and semiconductors. Zinc is in Group 12 of the periodic table (configuration $[Ar]3d^{10}4s^2$), so it has only two valence electrons. Each CH_3 group contributes seven electrons, giving the molecule 16 valence electrons. The bonding framework uses all 16:

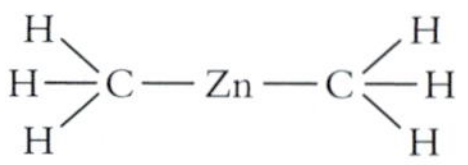

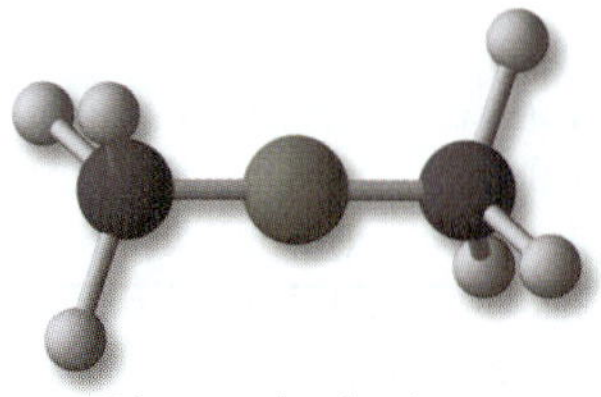

Coordination number = 2
Steric number = 2

Linear molecular shape
Bond angle = 180°

The other elements in Group 12, Cd and Hg, also form dimethyl compounds. Dimethylmercury, which can be synthesized by bacteria from industrial wastes, is fatal in very small quantities.

Notice that the zinc atom is associated with only four valence electrons. Although this is less than an octet, the adjacent carbon atoms have no lone pairs available to form multiple bonds. In addition, the formal charge on the zinc atom is zero. Thus, Zn has only four electrons in the optimal Lewis structure of dimethylzinc. This Lewis structure shows two pairs of bonding electrons and no lone pairs on the inner atom, so Zn has a steric number of 2. Two pairs of electrons are kept farthest apart when they are arranged along a line. Thus, the C—Zn—C bond angle is 180°, and **linear geometry** exists around the zinc atom.

KEY CONCEPT   *An inner atom with a steric number of 2 has linear electron group geometry.*

A second example of linear geometry comes from the familiar gas, carbon dioxide:

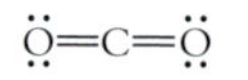

$$\ddot{O}{=}C{=}\ddot{O}$$

Coordination number = 2
Steric number = 2

Linear molecular shape
Bond angle = 180°

The carbon atom in CO_2 has two groups of electrons. Recall from our definition of a group that a double bond counts as one group of four electrons. Although each double bond includes four electrons, all four are concentrated between the nuclei. Remember also that the VSEPR model applies to electron *groups,* not specifically to electron pairs (despite the name of the model). It is the number of ligands and lone pairs, not the number of shared electrons, that determines the steric number and hence the molecular shape of an inner atom.

Steric Number 3: The trigonal plane

The structure and bonding of alkenes appear in Chapter 10.

Triethylaluminum, $Al(CH_2CH_3)_3$, has long been used in the chemical industry in the production of alkenes—hydrocarbons that have C=C double bonds. In the presence of triethylaluminum, two or more ethylene molecules link together to form straight-chain hydrocarbons that contain an even number of carbon atoms and one double bond. For example, four ethylene molecules form octene:

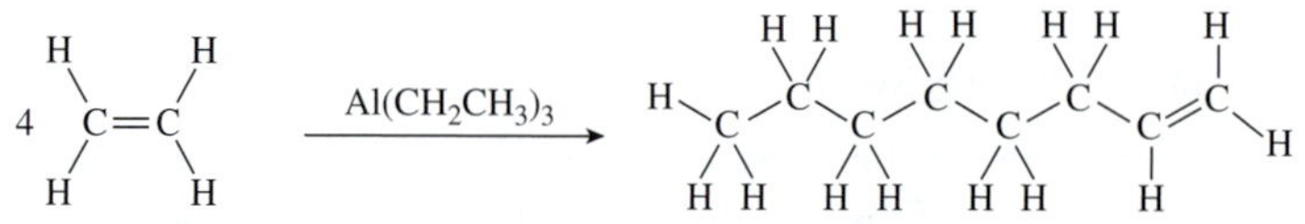

Ethylene

Octene

A description of the bonding in triethyl-aluminum begins with the Lewis structure. The chemical formula, $Al(CH_2CH_3)_3$, indicates that aluminum bonds to three CH_2CH_3 fragments. There are 42 valence electrons, all of which are used to complete the bonding framework. Each of the six carbon atoms in triethylaluminum has an octet of electrons and a steric number of 4. Thus, each ethyl group of $Al(CH_2CH_3)_3$ can be described exactly as in ethane (see Figure 9-13), except that one C—H bond is replaced by a C—Al bond. Although aluminum has less than an octet of electrons, the formal charge on the Al atom is zero. Thus, the Al atom has only six electrons in the optimal Lewis structure of triethyl-aluminum.

(a) Ball-and-stick model of $Al(C_2H_5)_3$

(b) Top view of simplified ball-and-stick model

Figure 9-20
Ball-and-stick model of triethylaluminum (*a*) and a stylized model (*b*) that emphasizes the trigonal planar geometry around the aluminum atom.

With three ligands and no lone pairs, the aluminum atom has a steric number of 3. The three pairs of bonding electrons must be as far apart as possible to minimize electron–electron repulsion. This is accomplished by placing the ligands in a triangular array called a **trigonal plane,** with bond angles of 120°. Figure 9-20*a* shows the molecule in a ball-and-stick model, and Figure 9-20*b* is another view, with the ethyl groups shown schematically to emphasize the trigonal planar geometry around the aluminum atom.

> **An inner atom with a steric number of 3 has trigonal planar electron group geometry.**   KEY CONCEPT

Trigonal planar geometry is common in carbon compounds that have double bonds. For example, ethylene has two trigonal planar carbon atoms. Each carbon atom has three ligands: two hydrogen atoms and one CH_2 group.

Each C atom has
Steric number = 3

Each C atom has
Trigonal planar geometry

④

Steric Number 5: The Trigonal Bipyramid

The elements beyond Row 2 of the periodic table can accommodate more than four groups of electrons, and this results in steric numbers greater than 4.

Phosphorus pentachloride exemplifies molecules with a steric number of 5. The Lewis structure of PCl_5 shows the molecule with five P—Cl bonds. The five groups of bonding electrons must be arranged to keep the chlorine atoms as far apart as possible. The resulting geometry, shown in Figure 9-21, is called **trigonal bipyramidal** because it can be viewed as two pyramids that share a triangular base. As Figure 9-21*b* shows, a trigonal bipyramid has six triangular faces.

Lewis structure of PCl_5

> **An inner atom with a steric number of 5 has trigonal bipyramidal electron group geometry.**   KEY CONCEPT

Unlike the geometries for other steric numbers, the five positions in a trigonal bipyramid are not all equivalent, as shown in Figure 9-21*a*. Three positions lie at the corners of an equilateral triangle around the phosphorus atom, separated by 120° bond angles. Atoms in the trigonal plane are in *equatorial positions*. The other two positions lie

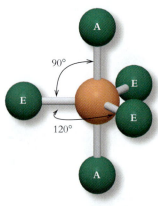

90°

120°

A = Axial position
E = Equatorial position

(a)

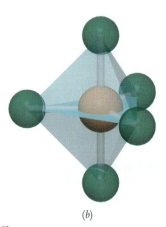

(b)

Figure 9-21
(a) Ball-and-stick model of PCl₅ with axial and equatorial sites labeled. (b) Model showing the equatorial plane and opposing pyramids that make up the trigonal bipyramid.

Lewis structure

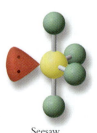

Trigonal pyramid

Seesaw

Figure 9-22
Sulfur tetrafluoride has two possible molecular geometries. The stable isomer is the seesaw form.

along an axis above and below the trigonal plane, separated from equatorial positions by 90° bond angles. Atoms in these sites are in *axial positions*.

The difference between equatorial and axial positions determines the arrangement of bonding pairs and lone pairs around an atom with a steric number of 5. An example is provided by sulfur tetrafluoride, a colorless gas that has industrial uses as a potent fluorinating agent. The Lewis structure of SF_4 shows four S—F bonds and one lone pair of electrons on the sulfur atom. These five pairs of electrons are distributed in a trigonal bipyramid around the sulfur atom.

Because equatorial and axial positions differ, two molecular geometries are possible for SF_4. As Figure 9-22 shows, placing the lone pair in an axial position gives a trigonal pyramid, whereas placing the lone pair in an equatorial position gives a **seesaw** shape.

Experiments show that SF_4 has the seesaw geometry, which means that this shape is more stable than the trigonal pyramid. This is explained by the fact that lone pairs are attracted to just one nucleus while bonding pairs are attracted to two nuclei. As a result, lone pairs are more spread out in space than are electrons in bonds. Consequently, a structure is more stable when lone pairs are placed as far as possible from other lone pairs and from bonding pairs. The greater stability resulting from placing the lone pair in an equatorial position indicates that less net electron–electron repulsion is associated with this arrangement. Studies of the geometries about other atoms with steric number 5 show that lone pairs *always* occupy equatorial positions.

The trigonal bipyramid (PCl_5) and the seesaw (SF_4) are two of the four geometries for an atom with steric number 5. Example 9-14 introduces the other two, **T-shaped** and linear.

Example 9-14	The Lewis structure of chlorine trifluoride is treated in Example 9-7. Determine the molecular geometry and draw a three-dimensional picture of the molecule.
Geometry of ClF₃	

Strategy: Use the Lewis structure of ClF_3 to determine the steric number of the chlorine atom. Obtain the molecular shape from the orbital geometry after placing lone pairs in appropriate positions.

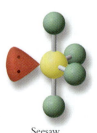

Solution: The steric number for chlorine is 5, leading to a trigonal bipyramidal electron group geometry.

When the steric number is 5, there are two distinct positions, equatorial and axial. Placing lone pairs in equatorial positions always leads to the greatest stability. Thus, ClF_3 is T-shaped with two equatorial lone pairs.

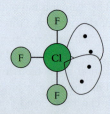

Does the Result Make Sense? Think of making ClF_3 from SF_4 by replacing one F atom with a lone pair. An axial lone pair would be at 90° to the existing equatorial lone pair, whereas a second equatorial lone pair is at 120° to the existing one. The larger angle means less repulsion between lone pairs.

Extra Practice **Exercise 9.14**	The fourth molecular shape arising from a steric number of 5 is represented by the triiodide anion I_3^-. Determine the molecular geometry and draw a three-dimensional picture of the triiodide ion.

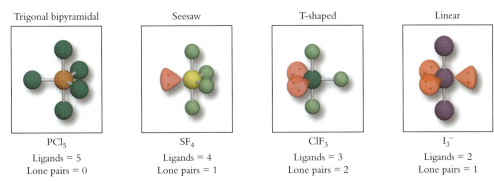

Steric number = 5
Electron group geometry = Trigonal bipyramidal

Trigonal bipyramidal	Seesaw	T-shaped	Linear
PCl_5	SF_4	ClF_3	I_3^-
Ligands = 5	Ligands = 4	Ligands = 3	Ligands = 2
Lone pairs = 0	Lone pairs = 1	Lone pairs = 2	Lone pairs = 1

Figure 9-23
Electron group geometries and molecular shapes for steric number of 5.

Figure 9-23 summarizes the characteristics of atoms with steric number 5.

Steric Number 6: The Octahedron

Sulfur hexafluoride is a colorless, odorless, nontoxic, unreactive gas. It is prepared commercially by reacting sulfur with excess fluorine. Because of its unusual stability, sulfur hexafluoride is used as an insulating gas for high-voltage electrical devices.

The Lewis structure of SF_6, shown in Figure 9-24a, indicates that sulfur has six S—F bonds and no lone pairs. The molecular geometry that keeps the six fluorine atoms as far apart as possible is **octahedral** in shape, as Figure 9-24b shows. Figure 9-24c shows that an octahedron has eight triangular faces.

> **KEY CONCEPT** *An inner atom with a steric number of 6 has octahedral electron group geometry.*

The six positions around an octahedron are equivalent, as Figure 9-25 demonstrates. Replacing one fluorine atom in SF_6 with a chlorine atom gives SF_5Cl. No matter which fluorine is replaced, the SF_5Cl molecule has four fluorine atoms in a square, with the fifth fluorine and the chlorine atom on opposite sides, at right angles to the plane of the square.

Three common molecular shapes are associated with octahedral electron group geometry. Most often, an inner atom with a steric number of 6 has octahedral molecular shape with no lone pairs. Example 9-15 uses a compound of xenon, whose chemical

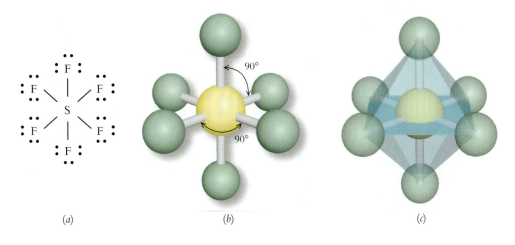

Figure 9-24
Views of sulfur hexafluoride: (a) Lewis structure; (b) ball-and-stick model; (c) ball-and-stick model showing the triangular faces of the octahedron.

(a)   (b)   (c)

| Box 9-2 | Chemical Milestones: Do Noble Gases React? |

T he noble gases are the only elements that exist naturally as individual atoms. Electron configurations make it clear why noble gas atoms prefer to remain as single atoms. Each noble gas has a filled shell configuration: $1s^2$ for He and s^2p^6 for the others. All electrons are paired, and there are no vacant valence orbitals. Moreover, the ionization energies of these elements are extremely high—over 2000 kJ/mol for He and Ne, and over 1000 kJ/mol for the other noble gases.

Nevertheless, as early as 1933 it was suggested that xenon might form stable compounds with the most electronegative elements, fluorine and oxygen. Early attempts to react xenon directly with fluorine were unsuccessful, and chemists ignored this possibility for the next 30 years.

In 1962, the English chemist Neil Bartlett overturned the conventional wisdom. Bartlett was exploring the reactions of platinum hexafluoride, an extremely reactive molecule. He found that PtF_6 reacted cleanly and rapidly with molecular oxygen:

$$O_2 + PtF_6 \longrightarrow O_2^+ PtF_6^-$$

Bartlett knew that the ionization energies of O_2 and Xe are nearly identical (1180 and 1170 kJ/mol, respectively). He reasoned that if PtF_6 reacted with molecular oxygen, it might also react with xenon. Sure enough, mixing Xe gas with PtF_6 resulted in an immediate reaction that formed a yellow solid. With this simple experiment, the field of noble gas chemistry was inaugurated. Within a year of Bartlett's first experiment, eight different compounds of xenon had been made and studied. For example, when heated or illuminated, xenon reacts with fluorine to form a mixture of three different xenon fluorides: XeF_2, XeF_4, and XeF_6. The hexafluoride reacts with one mole of water to give an oxofluoride:

$$XeF_6 + H_2O \longrightarrow XeOF_4 + 2\,HF$$

Reaction with excess water produces xenon trioxide:

$$XeF_6 + 3\,H_2O \longrightarrow XeO_3 + 6\,HF$$

Stable noble gas compounds are restricted to those of xenon. Most of these compounds involve bonds between xenon and the most electronegative elements, fluorine and oxygen. More exotic compounds containing Xe—S, Xe—H, and Xe—C bonds can be formed under carefully controlled conditions, for example in solid matrices at liquid nitrogen temperature. The three Lewis structures below are examples of these compounds in which the xenon atom has a steric munber of 5 and trigonal bipyramidal electron group geometry.

Chlorine, the next most electronegative element, reacts with xenon to form transient species that decompose at room temperature. Krypton forms KrF_2 and a few other compounds, but the chemistry of krypton is much more restricted than that of xenon.

The ability of xenon to react is easily explained. Although it has a closed-shell configuration, $[Kr]\,5s^2\,4d^{10}\,5p^6$, xenon's outermost electrons can be promoted relatively easily into $5d$ orbitals, $[Kr]\,5s^2\,4d^{10}\,5p^{6-n}\,5d^n$. This produces incompletely filled orbital sets that can form stable chemical bonds.

At first, noble gas chemistry had almost no practical applications. Recently, however, lasers have been developed that are based on the chemical reactions of noble gases.

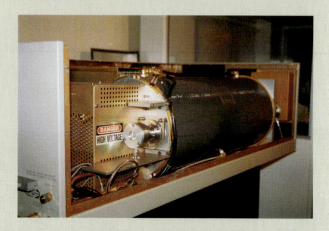

The gas in an excimer laser is a mixture containing about 2% of one noble gas and 0.2% halogen in second noble gas, typically neon. The mixtures include Ar—F_2, Kr—F_2, and Xe—Cl_2. In a strong electric discharge, the noble gas atoms become excited and ionized and the halogen molecules fragment into atoms. Excited states of the halogen atoms and rare gas atoms are created. These species collide and form excited diatomic molecules: ArF*, KrF*, XeCl*. These excited molecules emit ultraviolet light and break up into atoms in the process:

$$KrF^* \longrightarrow Kr + F + h\nu$$

Excimer lasers emit ultraviolet light that can vaporize solid substances like semiconductor surfaces and polymers, making it possible for such a laser to serve as a cutting device. Laser light can be focused to very small areas, so an excimer laser can perform very delicate cutting tasks.

A recent application of the excimer laser is eye surgery to correct for nearsightedness or astigmatism. In this technique, known as PRK or photorefractive keratectomy, a surgeon wields a computer-controlled excimer laser to selectively vaporize a portion of the corneal lens. Thinning the lens in just the right way can improve the ability of the eye to focus light correctly without the need for glasses or contact lenses. This procedure can be performed in just a few minutes with only a local anesthetic.

The capability to control an excimer laser beam also is exploited in the semiconductor industry, where these lasers are used to etch elaborate features during the fabrication of semiconductor chips. Neil Bartlett probably never dreamed that his explorations of the chemistry of xenon would lead to such exotic applications.

xenon difluorodioxide

hydrogen xenon
hydrogen sulfide

bis (2,6-difluorophenyl) xenon

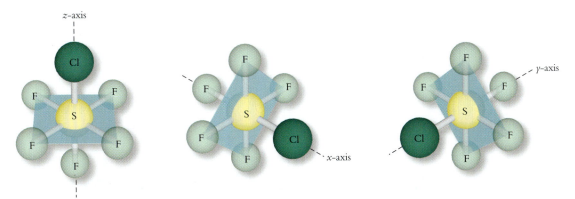

Figure 9-25
Replacing any of the fluorine atoms of SF₆ gives a molecule with a square of fluorine ligands capped by one fluorine and one chlorine. This shows that all six positions of the octahedron are equivalent.

behavior is described in the Chemical Milestones Box, to show a second common molecular shape, **square planar.**

Example 9-15

Structure and Bonding of XeF₄

Describe the geometry and draw a ball-and-stick sketch of xenon tetrafluoride.

Strategy: Follow the usual procedure. Determine the Lewis structure, then use it to find the steric number for xenon and to deduce electron group geometry. Next, use the number of ligands to identify the molecular shape.

Solution:
1. The xenon atom contributes eight valence electrons to the molecule ($5s^2\ 5p^6$). Four fluorine atoms add 28 more for a total of 36 valence electrons. Eight electrons are used to make the four Xe—F bonds, and 24 more fill the valence shells of the fluorine atoms. We are left with four electrons that must be placed as lone pairs on the xenon atom:

8 e⁻ used to make Xe—F bonds
24 e⁻ used to fill the F valence shells
4 e⁻ used in two Xe lone pairs
_____
36 total valence electrons

2. The steric number is the sum of ligands (four) and lone pairs (two): 6.

3. The ligands must be located at the corners of an octahedron to minimize electron–electron repulsion between the electron pairs. To give the greatest stability, the two lone pairs must be as far apart as possible, because lone pairs take up more space than bonding pairs. Placing the lone pairs at opposite ends of one axis, 180° apart, minimizes their mutual repulsion. This leaves the four fluorine atoms in a square plane around xenon:

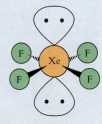

Does the Result Make Sense? Where we place the first lone pair makes no difference, inasmuch as all six positions are identical. If we place the second lone pair anywhere except opposite the first, there is a 90° angle between the two lone pairs, generating large lone-pair repulsion.

The third molecular shape arising from an octahedron is exemplified by chlorine pentafluoride, ClF₅. Describe the geometry and draw a ball-and-stick sketch of chlorine pentafluoride.

The **square pyramidal** geometry of ClF_5 completes our inventory of molecular shapes. Figure 9-26 summarizes the characteristics of atoms with steric number 6.

Figure 9-26
Electron group geometries and molecular shapes for steric number of 6.

Steric number = 6
Electron group geometry = Octahedral

Octahedral

SF_6
Ligands = 6
Lone pairs = 0

Square pyramidal

ClF_5
Ligands = 5
Lone pairs = 1

Square planar

XeF_4
Ligands = 4
Lone pairs = 2

Section Exercises

■ **9.4.1** Describe the geometry and draw a ball-and-stick sketch of gallium triiodide.

■ **9.4.2** Describe the geometry and draw a ball-and-stick sketch of the $AsCl_4^-$ anion.

■ **9.4.3** Determine the Lewis structure and describe the geometry of $XeOF_4$.

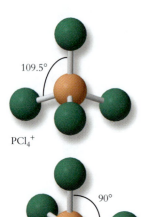

109.5°

PCl_4^+

90°

120°

PCl_5

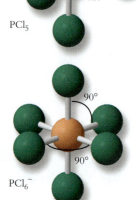

90°

90°

PCl_6^-

9.5 **PROPERTIES OF COVALENT BONDS**

Having developed ideas about Lewis structures and shapes of molecules, we are now in a position to explore some of the important properties of covalent bonds. These properties provide revealing evidence about molecular shapes.

Bond Angles

Each of the steric numbers described in Sections 9.3 and 9.4 results in electron groups separated by well-defined **bond angles.** If the VSEPR model is accurate, the actual bond angles found by experimental measurements on real molecules should match the optimal angles predicted by applying the model.

Experimental results agree with the predictions of the model. For instance, measurements show bond angles of 109.5° in CH_4, 120° in $Al(C_2H_5)_3$, and 90° in SF_6. Moreover, when the steric number of an atom changes, bond angles change exactly as the model predicts. Phosphorus pentachloride provides an example. Experiments show that gaseous PCl_5 is a trigonal bipyramid with two sets of bond angles, 120° and 90°. When this compound solidifies, however, it forms ionic crystals that contain PCl_4^+ cations and PCl_6^- anions. Molecular structure determination reveals that the bond angles in the cation are 109.5°, whereas those in the anion are 90°. These are exactly what the VSEPR model predicts for tetrahedral and octahedral geometries.

Symmetrical molecules without lone pairs on their inner atoms, such as CH_4, PCl_5, and SF_6, have exactly the bond angles of regular geometric shapes. However, lone pairs alter bond angles by small but important amounts. In ammonia, for example, the experimental H—N—H angles are 107°, and in water the H—O—H bond angle is only 104.5°. These distortions from the ideal 109.5° angle of tetrahedral geometry are caused by greater electron–electron repulsion for lone pairs than for bonding pairs. A pair of electrons in a bonding orbital is attracted to both nuclei involved in the bond, but the lone-pair electrons in a nonbonding orbital are attracted to just one nucleus. As a result, lone-pair electron density is spread more widely over space than the electron density in a bonding orbital. This means that the electron–electron repulsion generated by nonbonding pairs is always greater than that generated by bonding pairs.

Distorting the tetrahedral geometry to increase the angles around the lone pair minimizes the overall electron–electron repulsion in an ammonia molecule. Thus, the three hydrogen atoms move slightly closer together to minimize the repulsion between their bonding pairs and the larger lone pair. Similarly, the angle between the two lone pairs in a water molecule increases by a few degrees, pushing the two hydrogen atoms closer together. The magnitude of distortion depends on the details of electron distribution in the molecule, but the qualitative effect is always the same.

> **KEY CONCEPT**
>
> *Lone pairs in a molecule cause bond angles to be a few degrees smaller than predicted for symmetrical geometry.*

Example 9-16 examines bond angles in another molecule.

⑤

Example 9-16

Bond Angles

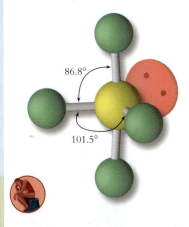

Experiments show that sulfur tetrafluoride has bond angles of 86.8° and 101.5°. Give an interpretation of these bond angles.

Strategy: To interpret bond angles, we must construct a model of the molecule by using Lewis structures and steric numbers.

Solution: Refer to the previous section, where we describe sulfur tetrafluoride. The sulfur atom has a steric number of 5, with one lone pair in the equatorial plane. The model predicts that SF_4 is a seesaw-shaped molecule, with F—S—F bond angles of 90° and 120°. However, the equatorial lone pair extends over more space than a bonding pair. To reduce overall electron–electron repulsion, the two axial fluorine atoms move slightly away from the lone pair and toward the equatorial fluorine atoms to give a bond angle slightly smaller than the expected 90°. Likewise, the two equatorial fluorine atoms move closer together, resulting in a bond angle of 101.5°.

Does the Result Make Sense? Notice that the change of the equatorial–equatorial bond angle is much greater than the change in the angle involving the axial fluorine atoms. Starting from an ideal angle of 120°, the equatorial ligands have much more space for angle reduction than do the 90° axial ligands.

Experiments show that chlorine trifluoride has bond angles of 87.5°. Explain these bond angles, and make a sketch of the molecule.

Extra Practice Exercise 9.16

Three triatomic molecules, water, ozone, and carbon dioxide, provide particularly strong experimental evidence to support the VSEPR:

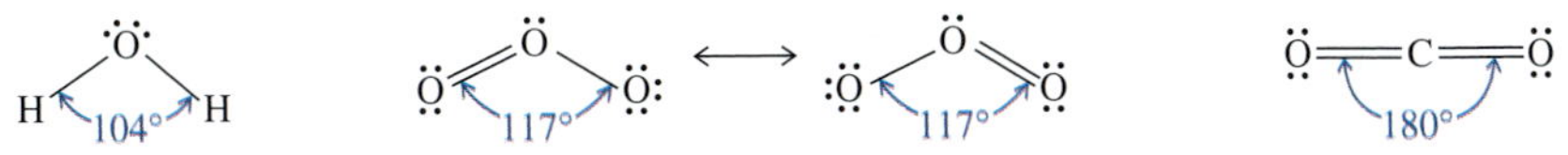

The inner O atom in water has a steric number of 4, for which VSEPR predicts tetrahedral geometry and a bond angle of 109.5°. The H—O—H bond angle in water is reduced by 5.5°, to 104°, to minimize repulsion between the lone pairs on the oxygen atom. The inner oxygen atom of ozone has just one lone pair and a steric number of 3, giving a predicted bond angle of 120°. Reducing the bond angle by 3° minimizes electron–electron repulsion between the lone pair and the two bonding pairs. In con-

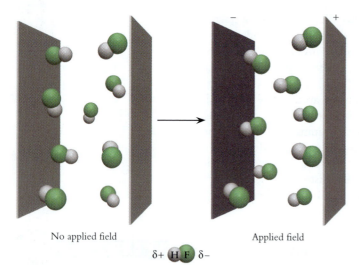

No applied field   Applied field

δ+ H F δ−

Figure 9-27
An applied electrical field causes an alignment of polar HF molecules. The extent of alignment depends on the magnitude of the dipole moment.

trast, the carbon atom in CO_2 has no lone pairs and a steric number of 2. In agreement with the observed bond angle of 180°, the model predicts that the two $C=O$ bonds point in opposite directions along a line.

Dipole Moments

As described in Section 9.1, most chemical bonds are polar, meaning that one end is slightly negative, and the other is slightly positive. Bond polarities, in turn, tend to give a molecule a negative end and a positive end. A molecule with this type of asymmetrical distribution of electron density is said to have a **dipole moment,** symbolized by the Greek letter mu, μ.

The dipole moment of a polar molecule can be measured by placing a sample in an electrical field. For example, Figure 9-27 shows a sample of hydrogen fluoride molecules between a pair of metal plates. In the absence of an applied field, the molecules are oriented randomly throughout the volume of the device. When an electrical potential is applied across the plates, the HF molecules align spontaneously according to the principles of electrical force. The positive ends of the molecules point toward the negative plate, and the negative ends point toward the positive plate. The extent of alignment depends on the magnitude of the dipole moment. The SI unit for the dipole moment is the coulomb meter (C m), but experimental values are usually recorded in units of debyes, D (1 D = 3.34×10^{-30} C m). Dipole moments span a range of about 2 debyes. Hydrogen fluoride has $\mu = 1.78$ D. A nonpolar molecule like F_2 has no dipole moment, $\mu = 0$ D.

The large dipole moment of the HF molecule indicates that electron density in the chemical bond of this molecule is strongly concentrated at the fluorine end. This simple molecule demonstrates that formal charge calculations, done to determine if Lewis structures are correct, do not indicate how charge is *actually* distributed in a molecule. The formal charges of the H atom and the F atom in HF are zero, yet HF has a large dipole moment. We determine formal charges by *assuming* that all bonding electrons are shared equally. In reality, most bonds, the H—F bond being just one example, involve unequal electron sharing. Consequently, formal charge calculations are extremely useful for assessing whether a valence electron distribution is reasonable, but they do not reliably predict bond polarity or the distribution of actual charge.

Dipole moments depend on bond polarities. For example, the trend in dipole moments for the hydrogen halides follows the trend in electronegativity differences; the more polar the bond (indicated by $\Delta\chi$), the larger the molecular polarity (indicated by the dipole moment, μ:

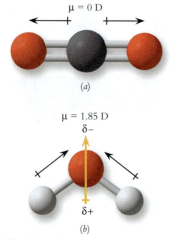

μ = 0 D

(a)

μ = 1.85 D
δ−

δ+

(b)

Figure 9-28
(a) When identical polar bonds point in opposite directions, their polarity effects cancel, giving zero net dipole moment. (b) When identical polar bonds do not point in opposite directions there is a net dipole moment, symbolized by μ.

HCl
$\Delta\chi = 1.9$
$\mu = 1.07$ D

HBr
$\Delta\chi = 0.7$
$\mu = 0.79$ D

HI
$\Delta\chi = 0.4$
$\mu = 0.38$ D

Dipole moments also depend on molecular shape. Any diatomic molecule with different atoms has a dipole moment. For more complex molecules, we must evaluate dipole moments using both bond polarity and molecular shape. A molecule with polar bonds has no dipole moment if a symmetrical shape causes polar bonds to cancel one another.

Figure 9-28 illustrates the dramatic effect of shape on the dipole moments of triatomic molecules. Recall from Section 9.1 that an arrow with a crossbar at one end can be used to represent a polar bond. The head of the arrow points to the partial negative end ($\delta-$) of the polar bond, and the "+" end of the arrow points toward the partial positive end ($\delta+$). Figure 9-28a shows that although both bonds in linear CO_2 are polar ($\Delta\chi = 1.0$), the two arrows indicating individual bond polarities point in opposite directions along a line. Thus, the effect of one polar bond exactly cancels the effect of the other. In bent H_2O, in contrast, Figure 9-28b shows that the effects of the two polar bonds do not cancel. Water has a partial negative charge on its oxygen atom and partial positive charges on its hydrogen atoms. The resulting dipole moment of the molecule is $\mu = 1.85$ D, shown by the colored arrow in Figure 9-28b.

No fully symmetrical molecule has a dipole moment. Phosphorus pentachloride, for example, has $\mu = 0$ D. The two axial P—Cl bonds point in opposite directions, and although three P—Cl bonds arranged in a trigonal plane have no counterparts pointing in opposite directions, trigonometric analysis shows that the polar effects of three identical bonds in a trigonal plane cancel exactly. Likewise, the bonds in a tetrahedron are arranged so that their polarities cancel exactly, so neither CH_4 nor CCl_4 has a dipole moment.

The perfect symmetry of these geometric forms is disrupted when a lone pair replaces a bond, giving a molecule with a dipole moment. Examples include SF_4 (see-saw), ClF_3 (T shape), NH_3 (trigonal pyramid), and H_2O (bent), all of which have dipole moments. Replacing one or more bonds with a bond to a different kind of atom also introduces a dipole moment. Thus, chloroform ($CHCl_3$) has a dipole moment but CCl_4 does not. The carbon atom of chloroform has four bonds in a near-regular tetrahedron, but the four bonds are not identical. The C—Cl bonds are more polar than the C—H bond, so the polarities of the four bonds do not cancel. Figure 9-29 shows that chloroform has $\mu = 1.04$ D.

In a symmetrical octahedral system such as SF_6, each polar S—F bond has a counterpart pointing in the opposite direction. The bond polarities cancel in pairs, leaving this molecule without a dipole moment. Example 9-17 examines molecular variations on octahedral geometry.

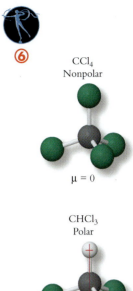

⑥ CCl₄
Nonpolar

$\mu = 0$

CHCl₃
Polar

$\mu = 1.04$ D

Figure 9-29
Carbon tetrachloride is a symmetrical tetrahedral molecule, so the individual bond polarities cancel. Chloroform is also a tetrahedral molecule, but the four bonds are not identical, so the bond polarities do not cancel.

Example 9-17

Predicting Dipole Moments

Does either ClF_5 or XeF_4 have a dipole moment?

Strategy: Molecules have dipole moments unless their symmetries are sufficient to cancel their bond polarities. Therefore, we must examine the structure and layout of bonds in each molecule.

Solution: Chlorine pentafluoride and xenon tetrafluoride appear in Figure 9-26. Each has an inner atom with a steric number of 6, but their electron group arrangements include lone pairs. As a result, ClF_5 has a square pyramidal shape, whereas XeF_4 has a square planar shape. Pictures can help us determine whether or not the bond polarities cancel:

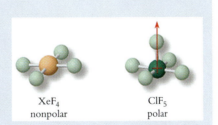

XeF₄
nonpolar

ClF₅
polar

Each molecule has four fluorine atoms at the corners of a square. The Xe—F bond polarities cancel in pairs, leaving XeF_4 with no dipole moment. Four bond polarities also cancel in ClF_5, but the fifth Cl—F bond has no counterpart in the opposing direction, so ClF_5 has a dipole moment that points along the axis containing the lone pair and the fifth Cl—F bond.

Do the Results Make Sense? XeF_4, like I_3^-, is a species that has lone pairs but zero dipole moment. Symmetry explains this. We can see why by examining how the lone pairs are placed. In I_3^-, the three lone pairs form a symmetrical trigonal plane. In XeF_4, the two lone pairs oppose each other.

Use molecular symmetry to determine if ethane (C_2H_6) and ethanol (C_2H_5OH) have dipole moments.

Extra Practice Exercise 9.17

Bond Length

As we describe in Section 9.1, the **bond length** of a covalent bond is the nuclear separation distance where the molecule is most stable. The H—H bond length in molecular hydrogen is 74 pm (picometers). At this distance, attractive interactions are maximized relative to repulsive interactions (see Figure 9-2). Having developed ideas about Lewis structures and molecular shapes, we can now examine bond lengths in more detail.

Table 9-2 lists average bond lengths for the most common chemical bonds. The table displays several trends. One trend is that bonds become longer as the radii of the atoms become larger. Compare the bond lengths of the diatomic halogens:

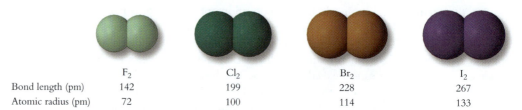

	F_2	Cl_2	Br_2	I_2
Bond length (pm)	142	199	228	267
Atomic radius (pm)	72	100	114	133

The trend is consistent with our introduction to orbital overlap. Recall that bonding involves valence orbitals, and it is the filled valence orbitals that determine the size of an atom. Because atomic size increases with the principal quantum number, bond lengths vary predictably with the **n**-value of the valence orbitals.

Nuclear charge also affects bond length. Recall from Section 8.4 that atomic size decreases from left to right across a row of the periodic table because of the progressive rise in nuclear charge. Smaller orbitals form shorter bonds, so Cl—Cl bonds are shorter than S—S bonds, and S—S bonds are shorter than P—P bonds.

Table 9-2   Average Bond Lengths*

H—X Bonds

n_a	n_b							
1	1	H—H	74					
1	2	H—C	109	H—N	101	H—O	96	H—F	92
1	3	H—Si	148	H—P	144	H—S	134	H—Cl	127
1	4							H—Br	141

Second-Row Elements

n_a	n_b							
2	2	C—C	154	C—N	147	C—O	143	C—F	135
2	2			N—N	145	O—O	148	F—F	142
2	3	C—Si	185	C—P	184	C—S	182	C—Cl	177
2	3	O—Si	166	O—P	163	O—S	158	N—Cl	175
2	3	F—Si	157	F—P	157	F—S	156		
2	4, 5			F—Xe	190	C—Br	194	C—I	214

Larger Elements

n_a	n_b							
3	3	Si—Si	235	P—P	221	S—S	205	Cl—Cl	199
3	3	Si—Cl	202	P—Cl	203	S—Cl	207		
4	4							Br—Br	228
5	5							I—I	267

Multiple Bonds

C=C	133	C=N	138	C=O	120	O=O	121
P=O	150	S=O	143				
C≡C	120	C≡N	116	C≡O	113	N≡N	110

*All values are in picometers; 1 pm = 10^{-12} m.

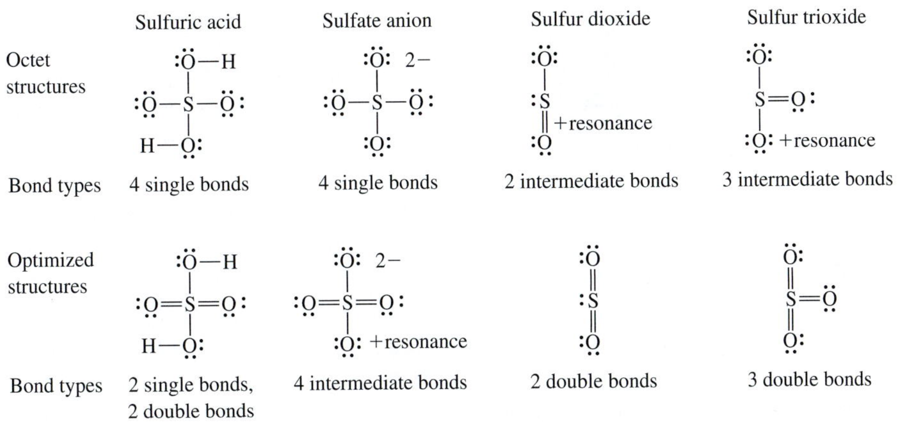

Sulfuric acid Sulfate anion Sulfur dioxide Sulfur trioxide

Figure 9-30
Octet structures and optimized structures predict different bond types for sulfur–oxygen bonds.

Bond polarity also contributes to bond length because partial charges generate electrical attraction that pulls the atoms closer together. For example, notice in Table 9-2 that C—O bonds are slightly shorter than either C—C or O—O bonds. This is a result of the polarity of the C—O bond.

The bond lengths in Table 9-2 show one final feature: A multiple bond is shorter than the corresponding single bond between the same two atoms. This is because placing additional electrons between atoms increases net attraction and pulls the atoms closer together. Thus, triple bonds are the shortest of all bonds among second-row elements.

Because bond lengths depend on the number of electrons involved in the bond, we can use bond lengths to decide which Lewis structure best represents the actual electron distribution in a molecule. Sulfur–oxygen bonds provide a good example. Figure 9-30 shows two possible sets of Lewis structures for species that contain sulfur–oxygen bonds. One set places octets of electrons on the inner sulfur atoms, while the other set reduces the formal charge on sulfur to zero at the expense of exceeding the octet on the S atom. As the figure shows, the two sets of structures predict different bond types.

Experimental bond length values clearly support the optimized Lewis structures. In sulfuric acid, there are two distinctly different bond types: the S—OH bond lengths are 157 pm, while the S—O bond lengths are 142 pm. This indicates that an S—O single bond has a length of 157 pm, and an S=O double bond has a length of 142 pm. In the sulfate anion, the bond lengths are 147 pm, indicating intermediate bond character consistent with the optimized Lewis structure. The bond lengths in SO_2 and SO_3 are 143 pm and 142 pm, respectively. These values also support the optimized Lewis structures, which predict double bonds rather than intermediate bonds.

To summarize, the following factors influence bond lengths:

1. The smaller the atoms, the shorter the bond.
2. The higher the bond multiplicity, the shorter the bond.
3. The higher the effective nuclear charges of the bonded atoms, the shorter the bond.
4. The larger the electronegativity difference of the bonded atoms, the shorter the bond.

Example 9-18 provides practice in the use of these factors.

Example 9-18

Bond Lengths

What factor accounts for each of the following differences in bond length?

a. I_2 has a longer bond than Br_2.

b. C—N bonds are shorter than C—C bonds.

c. H—C bonds are shorter than the C≡O bond.

d. The carbon–oxygen bond in formaldehyde, $H_2C=O$, is longer than the bond in carbon monoxide, C≡O.

Strategy: Bond lengths are controlled by four factors, some of which are more influential than others. To explain a difference in bond length, we need to determine the way that the factors are balanced.

Example 9-18

Bond Lengths (*continued*)

Solution:

a. I—I > Br—Br. Iodine is just below bromine in the periodic table, so the valence orbitals of iodine are larger than the valence orbitals of bromine. Thus, the I_2 bond is longer than the Br_2 bond because iodine has larger atomic radius.

b. C—N < C—C. Carbon and nitrogen are second-row elements. Nitrogen has a higher nuclear charge than carbon, however, so nitrogen has the smaller radius. This makes C—N bonds shorter than C—C bonds. In addition, a C—N bond is polar, which contributes to the shortening of the C—N bond.

c. H—C < C≡O. Here, we are comparing bonds in which the atomic radii and the amount of multiple bonding influence bond length. The experimental fact that H—C bonds are shorter than the triple C≡O bond indicates that the size of the hydrogen orbital is a more important factor than the presence of multiple bonding.

d. C=O > C≡O. Both bonds are between carbon and oxygen, so *n*, atomic number (*Z*), and electronegativity difference ($\Delta\chi$) are the same. However, carbon monoxide contains a triple bond, whereas formaldehyde has a double bond. The triple bond in CO is shorter than the double bond in H_2CO because more shared electrons means a shorter bond.

Do the Results Make Sense? When factors work in the same direction (effective nuclear charge and bond polarity in the case of C— N < C—C), we can make confident predictions. When factors work in the opposite direction (*n* value and bond multiplicity in the case of H—C < C≡O), we can explain the observed values but could not predict them.

Extra Practice Exercise 9.18

What factor accounts for each of the following differences in bond length? (a) The C=C bond is longer than the C≡C bond. (b) The C—Cl bond is shorter than the Si—Cl bond. (c) The C—C bond is longer than the O—O bond.

Bond Energy

As described in Section 6.3, energy must be supplied to break any chemical bond. Bond energies, like bond lengths, vary in ways that can be traced to atomic properties. There are three consistent trends in bond strengths:

1. *Bond strength increases as more electrons are shared between the atoms.* We described this trend in Section 6.3. Shared electrons are the "glue" of chemical bonding, so sharing more electrons strengthens the bond.

2. *Bond strength increases as the electronegativity difference ($\Delta\chi$) between bonded atoms increases.* Polar bonds gain stability from the electrical attraction between the negative and positive fractional charges around the bonded atoms. Bonds between oxygen and other second-row elements exemplify this trend:

O—O	$\Delta\chi = 0.0$	$BE = 145$ kJ/mol
O—N	$\Delta\chi = 0.5$	$BE = 200$ kJ/mol
O—C	$\Delta\chi = 1.0$	$BE = 360$ kJ/mol

3. *Bond strength decreases as bonds become longer.* As atoms become larger, the electron density of a bond is spread over a wider region. This decreases the net attraction between the electrons and the nuclei. The following bond energies illustrate this effect:

H—F	92 pm	$BE = 565$ kJ/mol
H—Cl	127 pm	$BE = 430$ kJ/mol
H—Br	141 pm	$BE = 360$ kJ/mol
H—I	161 pm	$BE = 295$ kJ/mol

Like bond lengths, bond energies result from the interplay of several factors, including nuclear charge, principal quantum number, electrical forces, and electronegativity. Thus, it should not be surprising that there are numerous exceptions to these three trends in bond energies. Although it is possible to *explain* many differences in bond energy, it frequently is not possible to *predict* differences with confidence.

Table 9-3 Features of Molecular Geometries

Steric Number	Ligands	Lone Pairs	Electron Group Geometry	Molecular Shape	Bond Angles	Dipole Moment*	Picture
2	2	0	Linear	Linear	180°	No	
3	3	0	Trigonal planar	Trigonal planar	120°	No	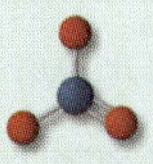
	2	1	Trigonal planar	Bent	<120°	Yes	
4	4	0	Tetrahedral	Tetrahedral	109.5°	No	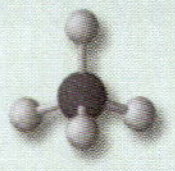
	3	1	Tetrahedral	Trigonal pyramidal	<109.5°	Yes	
	2	2	Tetrahedral	Bent	<109.5°	Yes	
5	5	0	Trigonal bipyramidal	Trigonal bipyramidal	90°, 120°	No	
	4	1	Trigonal bipyramidal	Seesaw	<90°, <120°	Yes	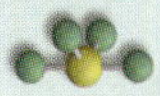
	3	2	Trigonal bipyramidal	T-shaped	<90°, <120°	Yes	
	2	3	Trigonal bipyramidal	Linear	180°	No	
6	6	0	Octahedral	Octahedral	90°	No	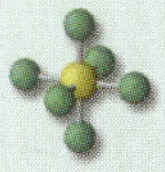
	5	1	Octahedral	Square pyramidal	90°, <90°	Yes	
	4	2	Octahedral	Square planar	90°	No	

*Applies only to molecules with identical ligands.

Summary of Molecular Shapes

Table 9-3 summarizes the relationships among steric number, electron group geometry, and molecular shape. If you remember the electron group geometry associated with each steric number, you can deduce molecular shapes, bond angles, and existence of dipole moments.

This relatively small catalog of molecular shapes accounts for a remarkable number of molecules. Even complicated molecules such as proteins and other polymers have shapes that can be traced back to these relatively simple templates. The overall shape of a large molecule is a composite of the shapes associated with its inner atoms. The shape around each inner atom is determined by steric numbers and the number of lone pairs.

Section Exercises

9.5.1 Predict the bond angles for BCl_3, SF_4, and $SnCl_4$. Which of these molecules, if any, has a dipole moment?

9.5.2 Experimental evidence shows that PF_3Cl_2 has a dipole moment, whereas PCl_3F_2 does not. Determine the structures of these two compounds and explain how the structures minimize total repulsion.

9.5.3 Arrange the following bonds in order of increasing length (shortest first) and increasing strength (weakest first). State the factors responsible for the position of each bond in your sequences: $C-C$, $C=O$, $C=C$, $C-H$, and $C-Cl$.

CHAPTER 9 VISUAL SUMMARY

SKILLS TO MASTER

① Using electronegativity to assess bond polarity
② Drawing Lewis structures
③ Determining electron group geometry
④ Predicting and sketching molecular shapes
⑤ Predicting and explaining bond angles
⑥ Identifying molecular polarity

9.1 OVERVIEW OF BONDING

Attraction
Repulsion

①

Polar covalent bond
Electronegativity

$$E_{\text{electrical}} = k\,\frac{q_1 q_2}{r}$$

Covalent bond
Bond length
Bond energy

Interaction energy (J)

Bond energy

Internuclear separation (pm)

9.2 LEWIS STRUCTURES

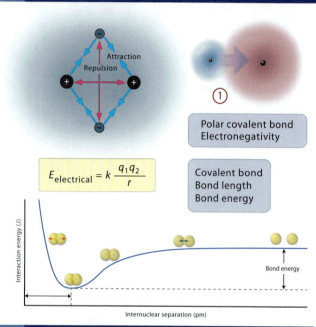

Count valence e^- — $O_3 = 18$ valence e^-

Build framework

②

Outer octets

Inner atoms

Optimize structure

Eqiuvalent structures

Single bonds   Lewis structures   Octet
Double bonds   Lone pairs
Triple bonds   Resonance structures

Outer atoms other than hydrogen have octets.

$$FC = \text{Valence } e^- \text{ (free atom)} - \text{Valence } e^- \text{ (Lewis)}$$

9.3 MOLECULAR SHAPES: TETRAHEDRAL SYSTEMS

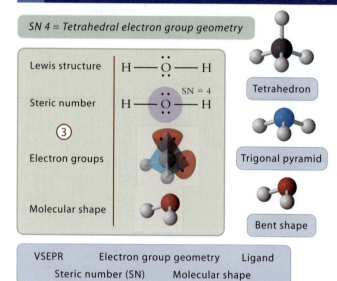

SN 4 = Tetrahedral electron group geometry

Lewis structure   H — O — H

Steric number   H — O — H   SN = 4

③

Electron groups

Molecular shape

Tetrahedron

Trigonal pyramid

Bent shape

VSEPR   Electron group geometry   Ligand
Steric number (SN)   Molecular shape

9.4 OTHER MOLECULAR SHAPES

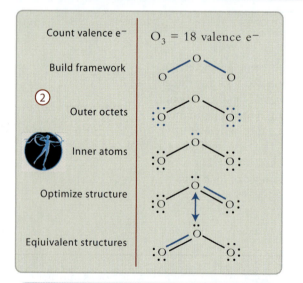

Linear   Linear   Bent shaped

Trigonal planar

T-shaped

SN	Electron group geometry
2	Linear
3	Trigonal planar
5	Trigonal bipyramidal
6	Octahedral

④

Square planar

Trigonal bipyramidal   Seesaw   Square pyramidal   Octahedral

9.5 PROPERTIES OF COVALENT BONDS

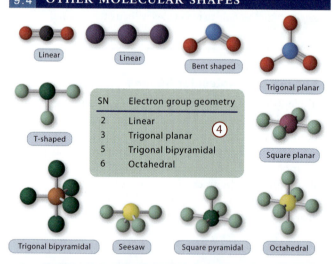

Bond length $\delta-$ Dipole moment (μ)

r

⑥

⑤ $\delta+$

Bond angle

Bond strength increases as:
More electrons are shared
$\Delta\chi$ increases
Bonds become shorter

Lone pairs increase bond angles.

Learning Exercises

9.1 Design a flowchart that shows how to determine the Lewis structure of a molecule.

9.2 Design a flowchart that shows how to determine the shape of a molecule.

9.3 Describe the role that electrical forces play in determining each of the following properties: (a) bond length; (b) bond polarity; (c) bond angle; and (d) molecular shape.

9.4 Write a paragraph that explains why formal charges do not match actual charges.

9.5 Draw as many different pictures as you can that illustrate geometric shapes of molecules that have no dipole moments.

9.6 Explain in your own words the meaning of each of the following terms: (a) electronegativity; (b) bonding framework; (c) inner atom; (d) ligand; (e) steric number; (f) multiple bond; (g) silicate; (h) alkane; and (i) dipole moment.

9.7 In your own words, write a one-sentence definition of each term in Chapter 9 that is new to you. Consult the glossary if you need help.

 Problems **ilw** = interactive learning ware problem. Visit the website at www.wiley.com/college/olmsted

Overview of Bonding

9.1 For the following elements, write the complete electron configuration and identify which of the electrons will be involved in bond formation: (a) O; (b) P; (c) B; and (d) Br.

9.2 For the following elements, write the complete electron configuration and identify which of the electrons will be involved in bond formation: (a) Si; (b) N; (c) Cl; and (d) S.

9.3 Describe bond formation between a hydrogen atom and an iodine atom to form a molecule of HI, and include a picture of the overlapping orbitals.

9.4 Describe bond formation between two bromine atoms to form a molecule of Br_2, and include a picture of the overlapping orbitals.

9.5 Give the group number and the number of valence electrons for the following elements: (a) aluminum; (b) arsenic; (c) fluorine; and (d) tin.

9.6 Give the group number and the number of valence electrons for the following elements: (a) selenium; (b) iodine; (c) germanium; and (d) nitrogen.

9.7 Hydrogen forms diatomic molecules with elements from Group 1 of the periodic table. Describe the bonding in LiH and include a picture of the overlapping orbitals.

9.8 Molecules of dilithium can form in the gas phase at low pressure. Describe the bonding in Li_2 and include a picture of the overlapping orbitals.

9.9 For each of the following pairs, identify which element tends to attract electron density from the other in a covalent bond: (a) C and N; (b) S and H; (c) Zn and I; and (d) S and As.

9.10 For each of the following pairs, identify which element tends to attract electron density from the other in a covalent bond: (a) C and O; (b) O and H; (c) Hg and C; and (d) Si and Cl.

9.11 Show the direction of bond polarity for the following bonds using $\delta+/\delta-$ notation: (a) Si—O; (b) N—C; (c) Cl—F; and (d) Br—C.

9.12 Show the direction of bond polarity for the following bonds using $\delta+/\delta-$ notation: (a) Cl—P; (b) B—O; (c) C—Sn; and (d) N—H.

9.13 Arrange the following molecules in order of increasing bond polarity: H_2O, NH_3, PH_3, and H_2S.

9.14 Arrange the following molecules in order of increasing bond polarity: $SiCl_4$, SCl_2, NCl_3, and PCl_3.

Lewis Structures

9.15 Count the total number of valence electrons in the following species: (a) H_3PO_4; (b) $(C_6H_5)_3C^+$; (c) $(NH_2)_2CO$; and (d) SO_4^{2-}.

9.16 Count the total number of valence electrons in the following species: (a) $(CH_3)_4N^+$; (b) H_2SO_4; (c) PO_4^{3-}; and (d) $(CH_3)_3SiCl$.

9.17 Convert the following formulas into molecular frameworks. For each molecule, tell how many valence electrons are required to construct the framework. (a) $(CH_3)_3CBr$; (b) $(CH_3CH_2CH_2)_2NH$; (c) $HClO_3$; and (d) $OP(OCH_3)_3$.

9.18 Convert the following formulas into molecular frameworks. For each molecule, tell how many valence electrons are required to construct the framework. (a) $(CH_3)_3SiCl$; (b) $(CH_3CH_2)_3N$; (c) H_2CO_3; and (d) $CH_3CH_2CH_2OH$.

9.19 Determine the Lewis structure of (a) NH_3; (b) NH_4^+; and (c) H_2N^-.

9.20 Determine the Lewis structure of (a) H_3O^+; (b) H_2O; and (c) OH^-.

ilw 9.21 Determine the Lewis structure of (a) PBr_3; (b) SiF_4; and (c) BF_4^-.

9.22 Determine the Lewis structure of (a) CBr_4; (b) $AlCl_4^-$; and (c) H_2S.

9.23 Use the standard procedures to determine the Lewis structure of (a) H_3CNH_2; (b) CF_2Cl_2; and (c) OF_2.

9.24 Use the standard procedures to determine the Lewis structure of (a) $HCCl_3$; (b) H_2NNH_2; and (c) H_2SiCl_2.

9.25 Use the standard procedures to determine the Lewis structure of (a) $(CH_3)_2CO$; (b) CH_3CN; (c) CH_2CHCH_3; and (d) CH_3CHNH.

9.26 Use the standard procedures to determine the Lewis structure of (a) CH_2CHCHO; (b) $HCCCH_2OH$; (c) $(CH_3)_2CCH_2$; and (d) H_2CCCH_2.

9.27 Determine the Lewis structure of (a) IF_5; (b) SO_3; (c) $OPCl_3$; and (d) XeF_2.

9.28 Determine the Lewis structure of (a) PBr_5; (b) O_2SCl_2; (c) $AsCl_3$ and (d) KrF_4.

9.29 Determine the Lewis structure of each of the following polyatomic ions. Include all resonance structures and formal charges, where appropriate: (a) NO_3^-; (b) HSO_4^-; (c) CO_3^{2-}; and (d) ClO_2^-.

9.30 Determine the Lewis structure of each of the following polyatomic ions. Include all resonance structures and formal charges, where appropriate: (a) SO_3^{2-}; (b) HPO_4^{2-}; (c) BrO_3^-; and (d) HCO_3^-.

Molecular Shapes: Tetrahedral Systems

9.31 Sketch and name the shapes of the following molecules: (a) CF_2Cl_2; (b) SiF_4; and (c) PBr_3.

9.32 Sketch and name the shapes of the following molecules: (a) OF_2; (b) $HCCl_3$; and (c) NH_2Cl.

9.33 Draw a ball-and-stick model that shows the geometry of 1,2-dichloroethane, ClH_2CCH_2Cl.

9.34 Draw a ball-and-stick model that shows the geometry of 1-bromopropane, $BrCH_2CH_2CH_3$.

9.35 Draw Lewis structures of all possible structural isomers of the alkanes with formula C_6H_{14}.

9.36 Draw Lewis structures of all possible structural isomers of the bromoalkanes with formula C_4H_9Br.

9.37 Write the Lewis structure of dimethylamine, $(CH_3)_2NH$. Determine its geometry, and draw a ball-and-stick model of the molecule, showing it as an ammonia molecule with two hydrogen atoms replaced by CH_3 groups.

9.38 Write the Lewis structure of dimethyl ether, $(CH_3)_2O$. Draw a ball-and-stick model of this molecule, showing it as a water molecule with each hydrogen atom replaced by a CH_3 group.

9.39 Silicon forms a tetramethyl compound with the formula $(CH_3)_4Si$. Determine the Lewis structure, determine the shape about each inner atom, and draw a ball-and-stick model of the compound.

9.40 The second simplest silicate is the disilicate anion, $Si_2O_7^{6-}$, in which one oxygen bridges between the two silicon atoms. Determine the Lewis structure, determine the shape about each inner atom, and draw a ball-and-stick model of the anion.

Other Molecular Shapes

9.41 Name the molecular shape of an inner atom that has the following characteristics: (a) two lone pairs and three ligands; (b) steric number of 5 and one lone pair; (c) steric number of 3 and no lone pairs; and (d) five ligands and steric number of 6.

9.42 Name the molecular shape of an inner atom that has the following characteristics: (a) two lone pairs and steric number of 6; (b) two ligands and three lone pairs; (c) one lone pair and five ligands; and (d) steric number of 5 and two lone pairs.

9.43 Fluorine forms compounds whose chemical formula is XF_4 with elements from groups 14, 16, and 18. Determine the Lewis structure, describe the shape, and draw a ball-and-stick model of GeF_4, SeF_4, and XeF_4.

9.44 Iodine forms three compounds with chlorine: ICl, ICl_3, and ICl_5. Determine the Lewis structure, describe the shape, and draw a ball-and-stick model of each compound.

9.45 Determine the molecular shape and the ideal bond angles of each of the following: (a) SO_2; (b) SbF_5; (c) ClF_4^+; and (d) ICl_4^-.

9.46 Determine the molecular shape and the ideal bond angles of each of the following: (a) SO_3; (b) I_3^+; (c) $SbCl_5^{2-}$; and (d) XeO_4.

Properties of Covalent Bonds

9.47 Determine the Lewis structures of the following compounds, and determine which ones have dipole moments. For each molecule that has a dipole moment, draw a ball-and-stick model and include an arrow to indicate the direction of the dipole moment: (a) SiF_4; (b) H_2S; (c) XeF_2; (d) $GaCl_3$; and (e) NF_3.

9.48 Determine the Lewis structures of the following compounds, and determine which ones have dipole moments. For each molecule that has a dipole moment, draw a ball-and-stick model and include an arrow to indicate the direction of the dipole moment: (a) CH_4; (b) $CHCl_3$; (c) CH_2Cl_2; (d) CH_3Cl; and (e) CCl_4.

ilw 9.49 Carbon dioxide has no dipole moment, but sulfur dioxide has $\mu = 1.63$ D. Use Lewis structures to account for this difference in dipole moments.

9.50 Sulfur trioxide has no dipole moment, but arsenic trifluoride has $\mu = 2.59$ D. Use Lewis structures to account for this difference in dipole moments.

9.51 Which of the following molecules would you expect to have bond angles that deviate from the ideal VSEPR values? For the molecules that do, make sketches that illustrate the deviations. (a) PF_5; (b) CH_3I; and (c) BrF_5.

9.52 Which of the following molecules would you expect to have bond angles that deviate from the ideal VSEPR values? For the molecules that do, make sketches that illustrate the deviations. (a) $TeCl_4$; (b) XeF_4; and (c) $SbCl_3$.

9.53 Arrange the following bonds in order of increasing length (shortest first). List the factors responsible for each placement: $H—N$, $N—N$, $Cl—N$, $N\equiv N$, and $C\equiv N$.

9.54 Arrange the following bonds in order of increasing bond length (shortest first). List the factors responsible for each placement: $Cl—Cl$, $Br—Br$, $O—Cl$, $P=O$, and $H—F$.

9.55 Use Table 6-2 to arrange the following bonds in order of increasing bond strength (weakest first). List the single most important factor for each successive increase in strength: $C=C$, $H—N$, $C=O$, $N\equiv N$, and $C—C$.

9.56 Use Table 6-2 to arrange the following bonds in order of increasing bond strength (weakest first). List the single most important factor for each successive increase in strength: $Si—Si$, $C—C$, $C—Si$, $H—C$, and $Sn—Sn$.

Additional Paired Problems

9.57 Draw Lewis structures and ball-and-stick structures showing the correct geometries for molecules of the following substances: (a) Cl_2O, dichlorine oxide (used for bleaching wood pulp and water treatment, about 10^5 tons produced each year); (b) C_6H_6, benzene, which contains a ring of six carbon atoms, each bonded to one hydrogen atom (one of the top 20 industrial chemicals, used in production of polymers); and (c) C_2H_4O, ethylene oxide, which contains a $C—C—O$ triangular ring (one of the top 50 industrial chemicals, used in polymer production).

9.58 Draw Lewis structures and ball-and-stick structures showing the geometries of molecules of the following substances: (a) CH_3NCO, methyl isocyanate (toxic compound responsible for thousands of deaths in Bhopal, India, in 1984); and (b) N_2F_4, tetrafluorohydrazine (colorless liquid used as rocket fuel).

9.59 Determine the molecular geometries of the following molecules: $SiCl_4$, SeF_4, and CI_4.

9.60 Determine the molecular geometries of the following ions: ClF_2^-, BF_4^-, and PF_4^+.

9.61 Shorter bonds are usually stronger bonds, but this is not always the case. Using Tables 9-2 and 6-2, find and list any $X—Y$ bonds, for X and Y both Row 2 elements, that are (a) shorter but weaker and (b) longer but stronger than the corresponding $X—X$ bond.

9.62 Using Tables 9-2 and 6-2, prepare a list of bonds to fluorine that are both longer and stronger than F—F bonds.

9.63 How many different structural isomers are there for octahedral molecules with the general formula AX_3Y_3? Draw three-dimensional structures of each.

9.64 Identify which of the four octahedral molecules shown here are equivalent:

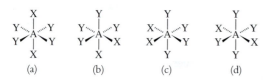

(a)   (b)   (c)   (d)

9.65 Describe the bonding and determine the empirical chemical formula of the silicon–oxygen network of zircon.

9.66 Describe the bonding and determine the empirical chemical formula of the silicon–oxygen network of asbestos.

9.67 Write Lewis structures and calculate formal charges for the following polyatomic ions: (a) bromate; (b) nitrite; (c) phosphate; and (d) hydrogen carbonate.

9.68 Carbon, nitrogen, and oxygen form two different polyatomic ions: cyanate ion, NCO^-, and isocyanate ion, CNO^-. Write Lewis structures for each anion, including near-equivalent resonance structures and indicating formal charges.

9.69 Species with chemical formula XY_4 can have the following shapes. For each, name the molecular geometry, identify the ideal VSEPR bond angles, tell how many lone pairs are present in the structure, and give a specific example.

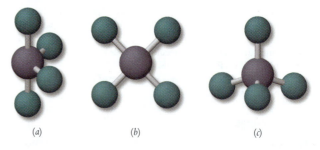

(a)   (b)   (c)

9.70 Species with chemical formula XY_3 can have the following shapes. For each, name the molecular geometry, identify the ideal VSEPR bond angles, tell how many lone pairs are present in the structure, and give a specific example.

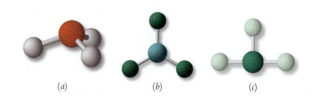

(a)   (b)   (c)

9.71 Four different bonds are found in H_2O, CO_2, and HCN. Arrange the four different bonds in order of increasing bond length. Give the reasons for your placements.

9.72 Four different bonds are found in NH_3, CO, and HOCl. Arrange the four different bonds in order of increasing bond length. Give the reasons for your placements.

9.73 The following partial structure is that of azodicarbonamide:

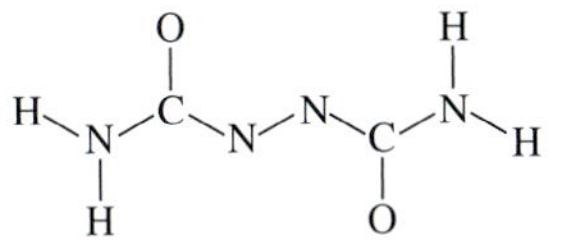

When heated, azodicarbonamide breaks apart into gaseous carbon monoxide, nitrogen, and ammonia. Azodicarbonamide is used as a foaming agent in the polymer industry. (a) Add nonbonding electron pairs and multiple bonds as required to complete the Lewis structure of this molecule. (b) Determine the geometry around each inner atom.

9.74 In the lower atmosphere, NO_2 participates in a series of reactions in air that is also contaminated with unburned hydrocarbons. One product of these reactions is peroxyacetyl nitrate (PAN). The skeletal arrangement of the atoms in PAN appears above. (a) Complete the Lewis structure of this compound. (b) Determine the shape around each atom marked with an asterisk. (c) Give the approximate values of the bond angles indicated with arrows.

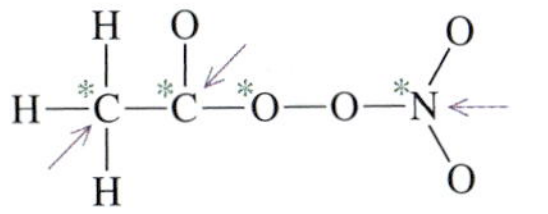

9.75 The H—O—H bond angle in a water molecule is 104.5°. The H—S—H bond angle in hydrogen sulfide is only 92.2°. Explain these variations in bond angles, using orbital sizes and electron-electron repulsion arguments. Draw space-filling models to illustrate your explanation.

9.76 The bond angles are 107.3° in NH_3, 100.3° in PCl_3, and 93.3° in PH_3. Explain these variations in bond angles, using orbital sizes and electron-electron repulsion arguments. Draw space-filling models to illustrate your explanation.

9.77 List the following X—H bonds from smallest bond polarity to largest bond polarity: C—H, F—H, N—H, O—H, and Si—H.

9.78 List the following X—O bonds from smallest bond polarity to largest bond polarity: C—O, N—O, S—O, O—H, and Br—O.

9.79 Do the following structures represent the same compound or are they different compounds? Explain.

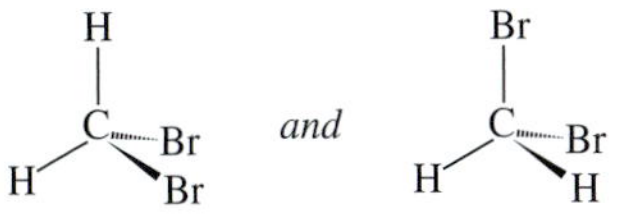

9.80 Do the following structures represent the same compound or are they different compounds? Explain.

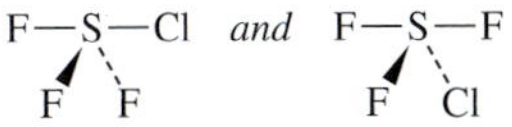

9.81 Benzyne is an unstable molecule that can be generated as a short-lived species in solution. Suggest a reason why benzyne is very reactive. (*Hint:* Examine the bond angles.)

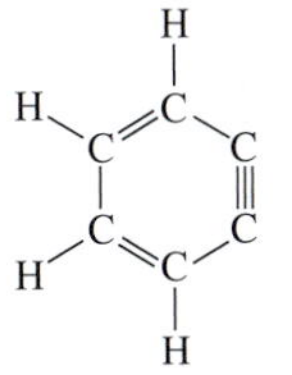

9.82 Cyclopropane, C_3H_6, which has three carbon atoms in a ring, is far more reactive than other alkanes. Determine the Lewis structure and the molecular shapes of the carbon atoms of this molecule. Suggest a reason for the reactivity of C_3H_6. (*Hint:* Examine the bond angles.)

Benzyne

More Challenging Problems

9.83 Determine the Lewis structures for the two possible arrangements of the N_2O molecule, N—N—O and N—O—N. Experiments show that the molecule is linear and has a dipole moment. What is the arrangement of atoms? Justify your choice.

9.84 The inner atom of a triatomic molecule can have any of four different electron group geometries. Identify the four, describe the shape associated with each, and give a specific example of each.

9.85 The molecule dichlorobenzene, $C_6H_4Cl_2$, which contains a ring of six carbon atoms, has three structural isomers. These differ in the relative positions of the Cl atoms around the ring. Determine the Lewis structures and geometries of these three isomers. Draw line structures of each. Which isomers have dipole moments? Which isomer has the largest dipole moment?

9.86 Determine the Lewis structure of borazine, $B_3N_3H_6$. The molecule contains a planar ring of alternating boron and nitrogen atoms, with a hydrogen atom attached to each ring atom.

9.87 Both PF_3 and PF_5 are known compounds. NF_3 also exists, but NF_5 does not. Why is there no molecule with the formula NF_5?

9.88 Phosphorous acid (H_3PO_3) is an exception to the rule that hydrogen always bonds to oxygen in oxoacids. In this compound, one of the hydrogen atoms bonds to phosphorus. Determine the Lewis structure for phosphorous acid and determine the geometry around the phosphorus atom. Draw a ball-and-stick sketch of the molecule.

9.89 Tellurium compounds, which are toxic and have a hideous stench, must be handled with extreme care. Predict the formula of the tellurium–fluorine molecule or ion that has the following molecular geometry: (a) bent; (b) T-shaped; (c) square pyramid; (d) trigonal bipyramid; (e) octahedron; and (f) seesaw.

9.90 Indium triiodide exists in the gas phase as individual InI_3 molecules. In the liquid phase, however, two InI_3 molecules combine to give In_2I_6, in which two iodine atoms bridge between indium atoms, and there are four outer iodine atoms. Describe the bonding in InI_3 and In_2I_6. Draw a ball-and-stick model of In_2I_6.

9.91 Imagine a square planar molecule, XY_2Z_2, in which X is the central atom and Z is more electronegative than Y. Two structural isomers are possible. (a) Draw a ball-and-stick model of each isomer. (b) Suppose you have made one isomer, but do not know which one. Explain how measuring the dipole moment of the substance would identify the isomer.

9.92 Sulfur and fluorine form seven different molecules: SF_2, SSF_2, FSSF, F_3SSF, SF_4, F_5SSF_5, and SF_6. Draw the Lewis structure of each molecule, and identify the geometry around each inner sulfur atom.

9.93 Among the halogens, only one known molecule has the formula XY_7. It has pentagonal bipyramidal geometry, with five Y atoms in a pentagon around the central atom X. The other two Y atoms are in axial positions. Draw a ball-and-stick model of this compound. Based on electron–electron repulsion and atomic size, determine the identities of atoms X and Y. Explain your reasoning. (Astatine is not involved. This element is radioactive and highly unstable.)

9.94 Spiroalkanes are hydrocarbons in which two carbon-containing rings share one carbon atom. One of the simplest spiroalkanes is spiroheptane. Describe the bonding in spiroheptane. Draw a ball-and-stick model that shows the geometry of the molecule.

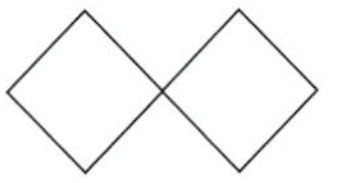

Spiroheptane
C_7H_{12}

Group Study Problems

9.95 Many transition metal cations form metal complexes with geometries that match those described in this chapter. In developing Lewis structures for these complexes, we assume that the metal cation has no valence electrons. Each ligand forms a metal-ligand bond using one of its lone electron pairs. Determine the Lewis structure and describe the geometry around each inner atom of (a) $[Sc(H_2O)_6]^{3+}$; (b) $[Ni(CN)_4]^{2-}$ (not tetrahedral); and (c) $[CuCl_5]^{3-}$.

9.96 Sulfur forms neutral compounds with oxygen, fluorine, and chlorine that display a variety of steric numbers and molecular shapes. Find as many examples as you can, describe the geometry of each, and identify the steric numbers and shapes that are not found.

9.97 Convert each of the following line structures into a complete Lewis structure. For each inner atom, identify the electron group geometry and shape. Draw ball-and-stick pictures that show the molecular geometries accurately.

9.98 In the following reactions, phosphorus forms a bond to a Row 2 element. In one reaction, phosphorus donates two electrons to make the fourth bond, but in the other reaction, phosphorus accepts two electrons to make the fourth bond. Use Lewis structures of starting materials and products to determine in which reaction phosphorus is a donor and in which it acts as an acceptor.

$$PCl_3 + N(CH_3)_3 \longrightarrow Cl_3PN(CH_3)_3$$

$$PCl_3 + BBr_3 \longrightarrow Cl_3PBBr_3$$

9.99 Describe the bonding differences among silica, silicate minerals, and metasilicate minerals.

9.100 Determine the Lewis structures, electron group geometries, and molecular shapes of the following compounds, which contain odd numbers of valence electrons. (*Hint:* Treat the odd electron like a lone pair.) (a) NO_2, nitrogen dioxide (red-brown gas that pollutes the air over many cities); and (b) ClO_2, chlorine dioxide (highly explosive gas used as an industrial bleach).

Answers to Section Exercises

9.1.1 The Cl nucleus attracts the electron in the H atom, and the H nucleus most strongly attracts an electron in the $3p$ orbital of Cl that points along the bond axis. Two electrons are shared in the resulting bond, which is polar with greater electron density at the Cl end of the molecule.

9.1.2 (a) O; (b) C; (c) Cl; and (d) Cl

9.1.3 C—H < As—Cl < Sn—Cl < Si—O

9.2.1

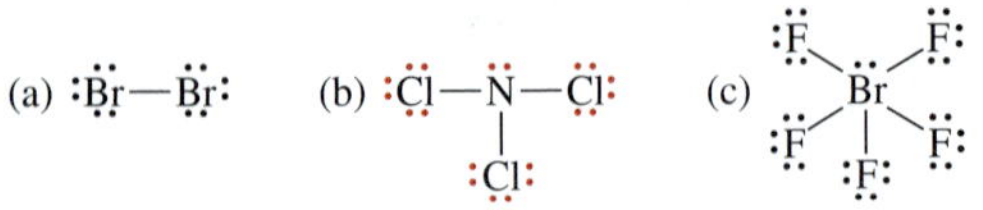

(a) :B̈r—B̈r: (b) :C̈l—N—C̈l: (c) [Br with F's]
                      :C̈l:

9.2.2

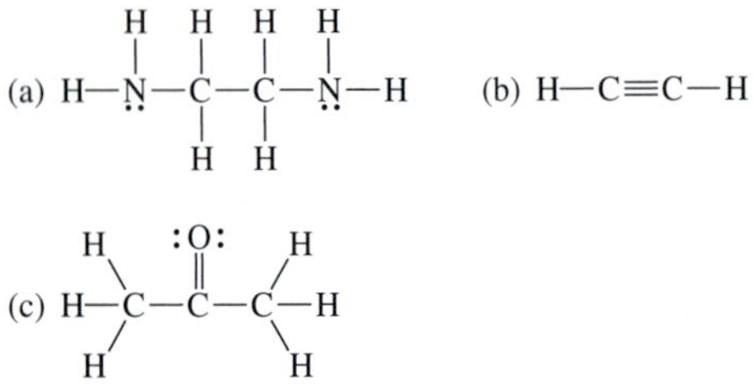

(a) H—N—C—C—N—H (with H atoms)   (b) H—C≡C—H

(c) H—C—C—C—H (with :O: double bond and H atoms)

9.2.3

(a) H—Ö:⁻   (b) [carbonate resonance structures, 2−]

(c) [chlorate resonance structures]

9.3.1 Sn is in the same group of the periodic table as C. It forms tetrahedral compounds, of which $SnCl_4$ is one.

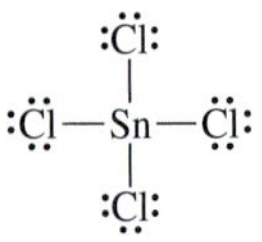

:C̈l—Sn—C̈l:
(with Cl above and below)

9.3.2 Each N atom has electron group number 4 and one lone pair, and the shape around each N atom is a trigonal pyramid.

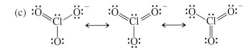

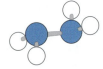

9.3.3 The atoms inside the circle make up one $Si_2O_5^{2-}$ unit:

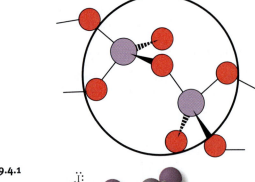

9.4.1

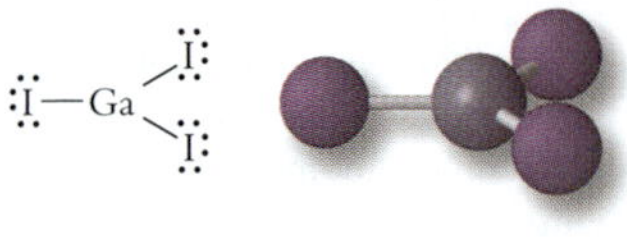

:I—Ga (with I's)

The bonding about Ga is similar to that around Al in $Al(C_2H_5)_3$. Gallium forms three bonds and has no lone pairs. The molecule is trigonal planar in shape.

9.4.2

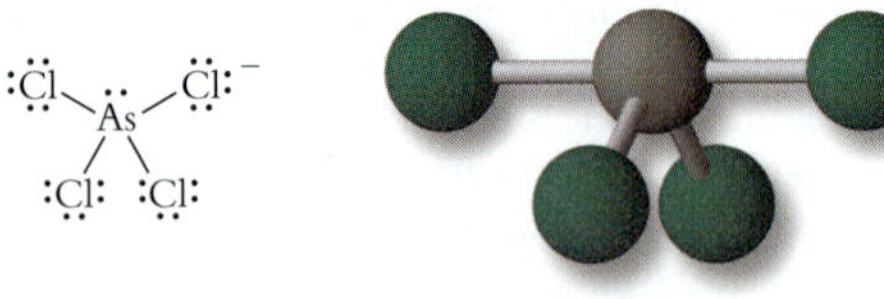

:C̈l···As···C̈l:⁻ (with Cl's)

The bonding is similar to that in SF_4. The ion has a seesaw shape.

9.4.3 Xe is the inner atom. It has steric number 6 and one lone pair, giving square pyramidal geometry:

:F̈—Xe—F̈: (with :O: and F's)

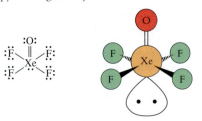

9.5.1 BCl_3, 120°; SF_4, < 90° and < 120°; and $SnCl_4$, 109.5°; only SF_4 has a dipole moment.

9.5.2

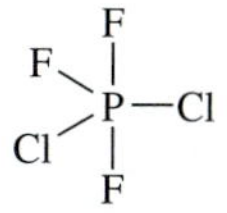

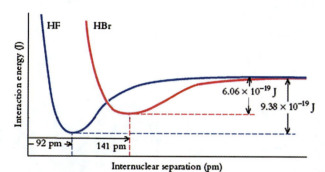

Asymmetric 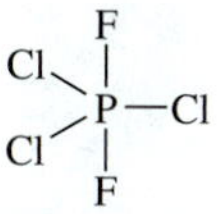   Symmetric

The chlorine atoms, being larger than fluorine atoms, occupy the equatorial positions because this minimizes atom–atom repulsion.

Answers to Extra Practice Exercises

9.1 Your plots should show that the bond in HBr is both longer and weaker than the bond in HF.

9.2 H—H, O—O, F—F < O—F < O—H < H—F

9.3

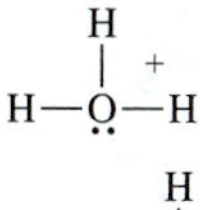

9.4

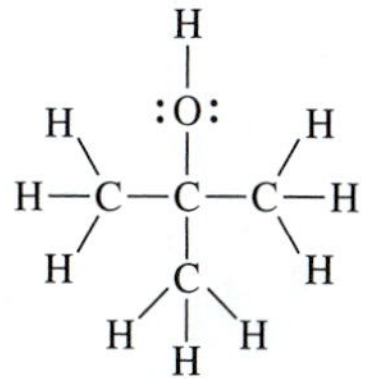

9.5

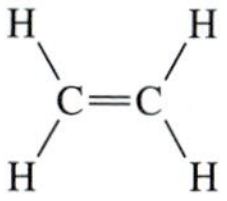

9.6

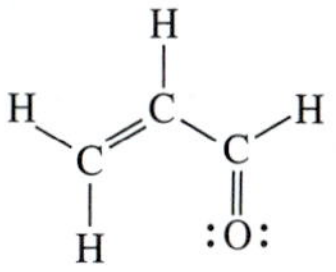

9.7

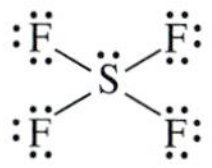

9.8

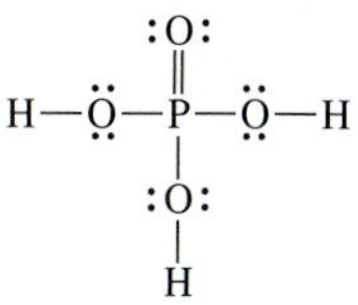

9.9 H—Ö—C—Ö—H

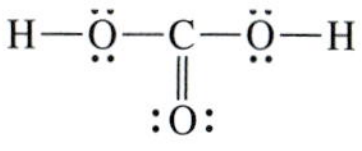

9.10

9.5.3 Increasing length: C—H ($n = 1$) < C=O (polar double) < C=C (double) < C—C (single) < C—Cl ($n = 3$); increasing strength: C—Cl ($n = 3$) < C—C < C—H ($n = 1$, polar) < C=C (double) < C=O (polar double)

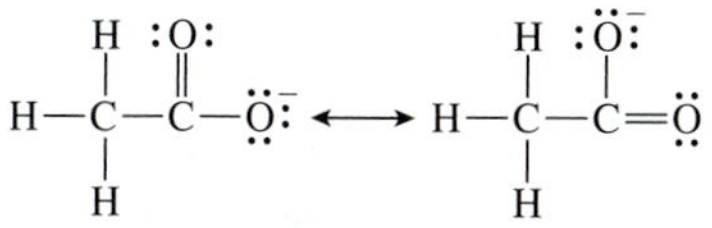

9.11

9.12 Chloromethane is tetrahedral like methane, with a Cl atom replacing one H atom.

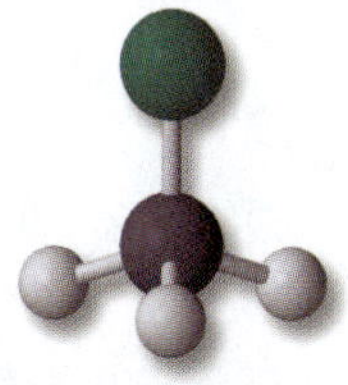

Chloromethane

9.13 The three inner atoms in ethanol all have tetrahedral group geometry. Each of the carbon atoms has tetrahedral geometry, and so does the oxygen atom. With two lone pairs, the geometry about the O atom is bent.

9.14 The inner I atom has a steric number of 5 and trigonal bipyramidal electron group geometry. Three lone pairs occupy equatorial positions, so the ion is linear.

9.15 The inner Cl atom has a steric number of 6 and octahedral electron group geometry. One lone pair occupies one octahedral position. The resulting shape is a pyramid built on a square, or square pyramid.

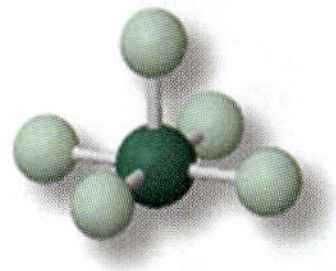

9.16 The electron pair geometry of ClF_3 is trigonal bipyramidal, with predicted bond angles of 90° and 120°. Lone pairs occupy two of the equatorial positions, however, so there is only one equatorial F atom and no 120° bond angles. Lone-pair repulsion distorts the axial positions, pushing them away and reducing the bond angle below 90°.

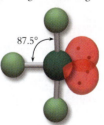

9.17 Ethane is symmetrical, so it has no dipole moment. Ethanol has a polar O—H group at one end, so it has a dipole moment.

9.18 (a) Bond lengths decrease as the number of electron pairs in the bond increases. (b) Bond lengths increase with the principal quantum number of the valence electrons. (c) Both bonds are between identical second-row atoms. The O—O bond is shorter because orbital size decreases across a row.

Photosynthesis

Photosynthesis is the process used by green plants to harvest light energy from the sun and store it in the form of chemical energy. Through a complex sequence of chemical reactions, plants use solar energy to convert water and carbon dioxide into energy-rich glucose and molecular oxygen. Plants then convert glucose into more complex sugars, starch, or cellulose. These compounds, in turn, are available to plant-eating animals as sources of energy and building blocks for biochemical synthesis.

Photosynthesis begins with chlorophyll, whose ball-and-stick structure appears in our inset figure. This molecule is a pigment that gives plants their green colors. Chlorophyll molecules absorb photons of visible light, and the photon energy creates a molecular excited state. This process is analogous to the atomic absorption leading to excited states, as described in Chapter 7. The excited electron generates a cascade of reactions in which an electron moves from chlorophyll to plastoquinone to cytochrome. The line structure of plastoquinone appears later in this chapter, and we discuss cytochromes briefly in Chapter 20. These three molecules are packed together in chloroplasts, subcellular components in the cells of higher plants and animals. Chloroplasts can be seen in our inset photo of a plant cell.

Chlorophyll, plastoquinone, and cytochrome are complicated molecules, but each has an extended pattern of single bonds alternating with double

Theories of Chemical Bonding

CHAPTER CONTENTS

Learning Resources

 KEY CONCEPTS

 CRITICAL THINKING

 SKILLS TO MASTER

 ADDITIONAL HELP
www.wiley.com/college/olmsted
- TUTORIALS
- ANIMATIONS

bonds. Molecules that contain such networks are particularly good at absorbing light and at undergoing reversible oxidation–reduction reactions. These properties are at the heart of photosynthesis.

In this chapter, we develop a model of bonding that can be applied to molecules as simple as H_2 or as complex as chlorophyll. We begin with a description of bonding based on the idea of overlapping atomic orbitals. We then extend the model to include the molecular shapes described in Chapter 9. Next we apply the model to molecules with double and triple bonds. Then we present variations on the orbital overlap model that encompass electrons distributed across three, four, or more atoms, including the extended systems of molecules such as chlorophyll. Finally, we show how to generalize the model to describe the electronic structures of metals and semiconductors.

10.1 LOCALIZED BONDS

Lewis structures are blueprints that show the distribution of valence electrons in molecules. However, the dots and lines of a Lewis structure do not show any details of how bonds form, how molecules react, or the shape of a molecule. In this respect, a Lewis structure is like the electron configuration of an atom: both tell us about electron distributions, but neither provides detailed descriptions. Just as we need atomic orbitals to understand how electrons are distributed in an atom, we need an orbital view to understand how electrons are distributed in a molecule.

The localized bond model is often called *valence bond theory*.

In this chapter, we describe two ways to think about bonding: **localized bonds** and **delocalized bonds.** The localized bond approach to molecules takes its cue from Lewis structures. According to Lewis structures, electrons are either localized in bonds between two atoms or localized on a single atom, usually in pairs. Accordingly, the localized bonding model develops orbitals of two types. One type is a bonding orbital that has high electron density between two atoms. The other type is an orbital located on a single atom. In other words, any electron is restricted to the region around a single atom, either in a bond to another atom or in a nonbonding orbital.

Localized bonds are easy to apply, even to very complex molecules, and they do an excellent job of explaining much chemical behavior. In many instances, however, localized bonds are insufficient to explain molecular properties and chemical reactivity. In the second half of this chapter, we show how to construct delocalized bonds, which spread over several atoms. Delocalization requires a more complicated analysis, but it explains chemical properties that localized bonds cannot.

Orbital Overlap

As described in the opening pages of Chapter 9, the electrons in a hydrogen molecule are "smeared out" between the two nuclei in a way that maximizes electron/nucleus attraction. To understand chemical bonding, we must develop a new orbital model that accounts for shared electrons. In other words, we need to develop a set of **bonding orbitals.**

The model of bonding used by most chemists is an extension of the atomic theory that incorporates a number of important ideas from Chapters 7 and 8 of this textbook. Bonding orbitals are described by combining atomic orbitals. Remember that orbitals have wave properties. Waves interact by addition of their amplitudes. Two waves that occupy the same region of space become superimposed, generating a new wave that is a composite of the original waves. In regions where the amplitudes of the superimposed waves have the same sign, the waves add. This gives a new wave amplitude that is larger than either original wave, as Figure 10-1*a* shows. In regions where the amplitudes have opposite signs, the waves subtract, giving a new wave amplitude that is smaller than either of the original waves, as illustrated in Figure 10-1*b*.

Because electrons have wave-like properties, orbital interactions involve similar addition or subtraction of wave functions. When two orbitals are superimposed, one result is a new orbital that is a composite of the originals, as shown for molecular hydro-

Figure 10-1
(*a*) When two wave amplitudes (*dashed lines*) add, the resulting new wave (*solid line*) has a large amplitude in the overlap region. (*b*) When two wave amplitudes (*dashed lines*) subtract, the resulting new wave (*solid line*) has a small amplitude in the overlap region.

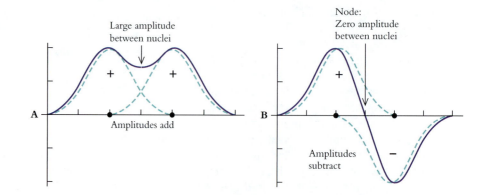

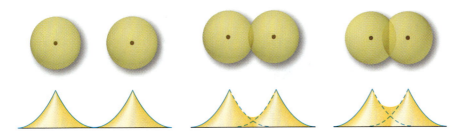

Figure 10-2
As two hydrogen atoms approach each other, the overlap of their 1s atomic orbitals increases. The wave amplitudes add, generating a new orbital with high electron density between the nuclei.

gen in Figure 10-2. This interaction is called **orbital overlap,** and it is the foundation of the bonding models described in this chapter.

The simplest bonding model requires only the addition of wave functions. Later in this chapter, we show that a complete description of bonding often requires both addition and subtraction of wave functions.

Bonding orbitals are constructed by combining atomic orbitals from adjacent atoms.   **KEY CONCEPT**

Conventions of the Orbital Overlap Model

Orbital overlap models proceed from the following assumptions:

1. Each electron in a molecule is assigned to a specific orbital.
2. No two electrons in a molecule have identical descriptions, because the Pauli exclusion principle applies to electrons in molecules as well as in atoms.
3. The electrons in molecules obey the aufbau principle, meaning that they occupy the most stable orbitals available to them.
4. Even though every atom has an unlimited number of atomic orbitals, the valence orbitals are all that are needed to describe bonding.

Bonding involves the valence orbitals almost exclusively, because these orbitals have the appropriate sizes and energies to interact strongly. Examine the electron energy level diagram of a fluorine atom, shown in Figure 10-3. Recall from Section 7.6 that the sizes of the atomic orbitals increase substantially as the principal quantum number (n) increases. Thus, the core 1s orbitals on a fluorine atom are much smaller than any of the $n = 2$ orbitals and do not participate effectively in orbital overlap. On the other hand, orbitals with $n > 2$ lie at considerably higher energy than the $n = 2$ orbitals. These orbitals can interact, but the resulting orbitals are too unstable to form strong covalent bonds. The only orbitals of fluorine that form chemical bonds are the valence orbitals, those with $n = 2$.

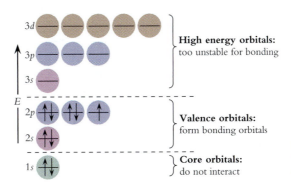

Figure 10-3
The energy level diagram for a fluorine atom ($1s^2 \ 2s^2 \ 2p^5$).

Diatomic Molecules: HF and F_2

Consider the orbital interactions of a hydrogen atom and a fluorine atom as they combine to form a molecule of hydrogen fluoride. The electron in the hydrogen atom occupies the 1s orbital. According to the orbital overlap model, this orbital interacts with the valence orbitals from the fluorine atom. The 2s orbital of fluorine is substantially more stable than the set of 2p orbitals. Moreover, each p orbital has a strong directional component that is not present in the spherical 2s orbital. Both factors make a fluorine 2p orbital a better choice for orbital interaction than the 2s orbital. Consequently, the bond-

H—F̈:
Lewis structure of hydrogen fluoride

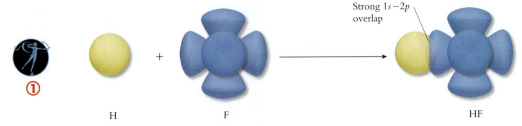

Figure 10-4
The bond in H—F forms from orbital overlap between the hydrogen 1s orbital and the fluorine 2p orbital that points along the bond axis.

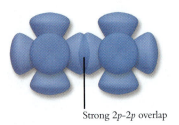

Strong 2p-2p overlap

Figure 10-5
The bond in F₂ forms from orbital overlap between a pair of fluorine 2p orbitals that point along the bond axis.

ing in HF can be described as shown in Figure 10-4. One covalent bond forms from the overlap between the hydrogen 1s orbital and the fluorine 2p orbital that points along the bond axis.

The Lewis structure of hydrogen fluoride shows three lone pairs on the fluorine atom. These nonbonding electrons are localized in atomic orbitals that belong solely to fluorine. Remembering that one of the fluorine 2p orbitals is used to form the H—F bond, we conclude that the three lone pairs must occupy the remaining pair of 2p orbitals and the 2s orbital of the fluorine atom.

A similar situation arises when two fluorine atoms approach each other. The first valence orbitals to overlap are the 2p orbitals pointing along the axis that joins the atoms. Figure 10-5 shows that bond formation in molecular fluorine results from the strong directional overlap of these two atomic 2p orbitals.

Bonding in Hydrogen Sulfide

Hydrogen sulfide is a toxic gas with the foul odor of rotten eggs. The Lewis structure of H_2S shows two bonds and two lone pairs on the S atom. Experiments show that hydrogen sulfide has a bond angle of 92.1°. We can describe the bonding of H_2S by applying the orbital overlap model.

Sulfur has the valence configuration $3s^2 3p^4$, with two of the valence p orbitals occupied by single electrons. Recall from Section 7.6 that atomic p orbitals are separated by 90° and thus are in just the right orientation to form a pair of H—S bonds at the appropriate bond angle. As Figure 10-6 shows, the 1s orbitals of the two hydrogen atoms overlap with two of the sulfur 3p orbitals to form a pair of bonding orbitals at right angles to one another, consistent with the experi-mental data for H_2S. The slight increase of the H—S—H bond angle from the ideal value of 90° is consistent with repulsion of the two hydrogen atoms. This minor deviation is not large enough to justify alteration of the model. The Lewis structure of H_2S shows the molecule with two lone pairs localized on the sulfur atom. These nonbonding electrons occupy the 3s and the remaining 3p orbital of the sulfur atom.

Example 10-1 applies the orbital overlap model to phosphine, PH_3.

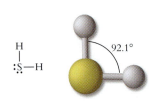

Lewis structure and ball-and-stick model of hydrogen sulfide

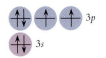

Valence electron configuration of sulfur

Figure 10-6
The bonding in hydrogen sulfide can be described as the overlap of two hydrogen 1s orbitals with a pair of sulfur 3p orbitals.

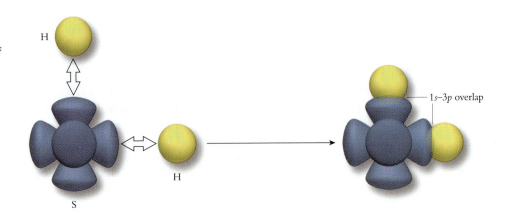

Example 10-1

Phosphine is a colorless, highly toxic gas with bond angles of 93.6°. Describe the bonding in PH_3.

Orbital Overlap in Phosphine

Strategy: We begin with the Lewis structure of PH_3, and apply the principles of orbital overlap. The bonding description of the molecule must be consistent with the bond angles, so we must pay particular attention to the orientation of the valence orbitals.

Solution: The Lewis structure of phosphine shows three P—H bonds and one lone pair on the phosphorus atom.

$$H—\overset{\displaystyle\cdot\cdot}{\underset{\displaystyle |}{P}}—H$$
$$H$$

The valence electron configuration of phosphorus is $3s^2 3p^3$, with a single electron in each of the three $3p$ orbitals. The angles between these p orbitals are 90°, so the phosphorus atom is in the proper configuration to form three P—H bonds by direct orbital overlap with three hydrogen $1s$ orbitals. The angles increase by 3.6° because of repulsion among the hydrogen atoms. Finally, the lone pair of electrons on the phosphorus atom occupies the $3s$ atomic orbital. A ball-and-stick model shows the shape of the molecule.

Valence configuration of phosphorus

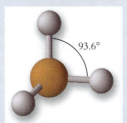

Ball-and-stick model of phosphine. The bond angles are 93.6°.

Does the Result Make Sense? When bond angles are close to 90°, as in phosphine, it is sensible to describe the bonding using atomic orbitals that point at right angles to one another; in other words, the valence p orbitals.

Hydrogen selenide is yet another offensive-smelling toxic gas. The bond angle in H_2Se is 92.1°. Describe the bonding of hydrogen selenide and draw a ball-and-stick model that shows the shape of the molecule.

Methane: The Need for an Expanded Model

As we describe in Chapter 9, methane adopts tetrahedral molecular geometry because of electron–electron repulsions among the four pairs of bonding electrons. Figure 10-7 shows 109.5° angles between the four C—H bonds, as predicted by the VSEPR model. How can the carbon atom bond to four hydrogen atoms in this geometry? The valence $2p$ orbitals of the carbon atom are separated by 90° angles, and the $2s$ orbital has a spherical shape. Clearly, the four valence orbitals of carbon cannot overlap individually with four hydrogen $1s$ orbitals and give the known tetrahedral geometry of methane. Nevertheless, experiments verify the bond angles predicted by the VSEPR model, 109.5°. When theory and experiment disagree, the theory must be modified. In the next section we present additional features of the orbital overlap model that allow for bond angles that are impossible to describe using traditional atomic orbitals.

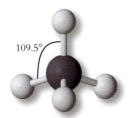

Figure 10-7
Tetrahedral molecular geometry, with 109.5° bond angles, minimizes repulsion among the bonding electron pairs of methane.

Section Exercises

■ **10.1.1** Draw a figure similar to Figure 10-2 that shows changes in the wave functions as a hydrogen atom approaches a fluorine atom to form an H—F bond.

■ **10.1.2** Describe the bond between a hydrogen atom and a bromine atom in a molecule of hydrogen bromide, HBr.

■ **10.1.3** The gaseous compound stibine, SbH$_3$, has experimental bond angles of 91.3°. Give a complete description of the bonding in stibine, including a ball-and-stick sketch that shows the shape of the molecule.

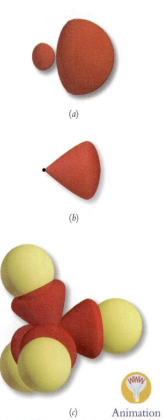

(a)

(b)

(c)   Animation

Figure 10-8
(*a*) The shape of an *sp*³ hybrid orbital. (*b*) For clarity and convenience, hybrid orbitals are represented in a stylized form. (*c*) Orbital depiction of methane, showing four *sp*³ hybrid orbitals interacting with 1s orbitals from four H atoms.

10.2 HYBRIDIZATION OF ATOMIC ORBITALS

We cannot generate a tetrahedron by simple overlap of atomic orbitals, because atomic orbitals do not point toward the corners of a tetrahedron. In this section, we present a modification of the localized bond model that accounts for tetrahedral geometry and several other common molecular shapes.

Methane: *sp*³ Hybrid Orbitals

The most convenient way to visualize tetrahedral bonding in methane is to imagine the four carbon valence orbitals mixing together to generate a set of four new orbitals. This can be done in such a way that each new orbital points toward a different corner of a tetrahedron. Combining atomic orbitals on a particular atom to form a special set of directional orbitals is referred to as **hybridization.** Every hybrid orbital is directional, with a lobe of high electron density pointing in one specific direction.

Any hybrid orbital is named from the atomic valence orbitals from which it is constructed. To match the geometry of methane, we need four orbitals that point at the corners of a tetrahedron. We construct this set from one *s* orbital and three *p* orbitals, so the hybrids are called ***sp*³ hybrid orbitals.** Figure 10-8*a* shows the detailed shape of an *sp*³ hybrid orbital. For the sake of convenience and to keep our figures as uncluttered as possible, we use the stylized view of hybrid orbitals shown in Figure 10-8*b*. In this representation, we omit the small backside lobe, and we slim down the orbital in order to show several orbitals around an atom. Figure 10-8*c* shows a stylized view of an *sp*³ hybridized atom. This part of the figure shows that all four *sp*³ hybrids have the same shape, but each points to a different corner of a regular tetrahedron.

Orbital mixing always generates changes in orbital energies as well as in orbital shapes. Hybrid orbitals always have different energies than do the pure atomic orbitals from which they form. The energy level diagram shown in Figure 10-9 summarizes the hybridization process.

A complete orbital overlap view of methane appears in Figure 10-10. Hybridization gives each carbon orbital a strongly favored direction for overlap with an atomic 1s orbital from an approaching hydrogen atom. Four such interactions generate four localized bonds that use all the valence electrons of the five atoms involved.

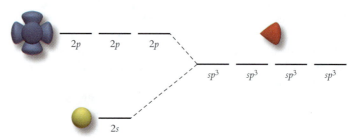

2p	2p	2p

sp³	sp³	sp³	sp³

2s

Figure 10-9
Hybrid orbitals form from combinations of atomic valence orbitals.

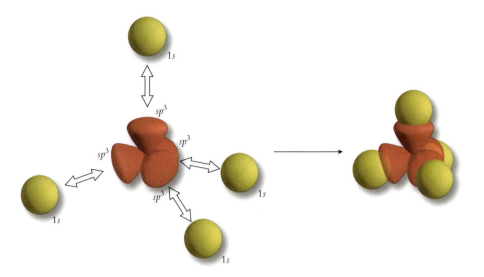

Figure 10-10
Methane forms from orbital overlap between the hydrogen 1s orbitals and the sp³ hybrid orbitals of the carbon atom.

Other Tetrahedral Systems

According to the VSEPR model developed in Chapter 9, an inner atom with a steric number of 4 adopts tetrahedral electron group geometry. This tetrahedral arrangement of four electron groups is very common, the only important exceptions being the hydrides of elements beyond the second row, such as H_2S and PH_3. Thus, we can make a general statement:

> *An inner atom with a steric number of 4 has tetrahedral electron group geometry and can be described using sp³ hybrid orbitals.*

KEY CONCEPT

Example 10-2 applies the orbital overlap molecule to the hydronium ion, and Example 10-3 treats a molecule with two inner atoms.

Describe the bonding of the hydronium ion, H_3O^+.

Strategy: Use the strategies from Chapter 9 to determine the Lewis structure, steric number of the inner atom, and electron group geometry. The steric number also determines the hybridization.

$$H-\overset{\displaystyle H}{\underset{\displaystyle \cdot\cdot}{O}}-H \quad +$$

Solution: A hydronium ion has eight valence electrons. Six are used to make three O—H single bonds, and two are placed as a lone pair on the oxygen atom.

The Lewis structure shows the inner oxygen atom with three ligands and one lone pair, which gives a steric number of 4. A steric number of 4 indicates that the electron group geometry is tetrahedral and that the bonding can be described with sp^3 hybrids for the oxygen atom. Three hybrid orbitals overlap with hydrogen 1s atomic orbitals to form bonds. The fourth sp^3 hybrid contains a lone pair of electrons.

	Example 10-2
	Bonding in the Hydronium Ion

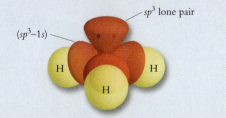

Does the Result Make Sense? Remember that the molecular shape ignores the lone pair. The hydronium ion has a trigonal pyramidal shape described by the three sp^3 hybrid orbitals that form bonds to hydrogen atoms.

Describe the bonding of a water molecule.

Extra Practice
Exercise 10.2

Example 10-3

A Molecule with Two Inner Atoms

Describe the bonding of methanol, CH_3OH. Sketch an orbital overlap picture of the molecule.

Strategy: Begin with the Lewis structure, and then survey each inner atom for its steric number and hybridization. Here is a summary of the information:

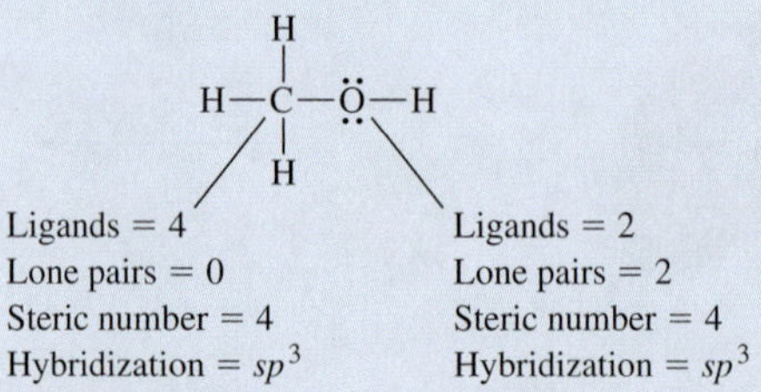

$$H-C-\ddot{O}-H$$ (with H above and below C)

Ligands = 4
Lone pairs = 0
Steric number = 4
Hybridization = sp^3

Ligands = 2
Lone pairs = 2
Steric number = 4
Hybridization = sp^3

Solution: Both inner atoms have steric numbers of 4 and tetrahedral electron group geometry, so both can be described using sp^3 hybrid orbitals. All four hydrogen atoms occupy outer positions, and these form bonds to the inner atoms through $1s-sp^3$ overlap. The oxygen atom has two lone pairs, one in each of the two hybrid orbitals not used to form O—H bonds.

Sketching an orbital overlap picture for a molecule like methanol takes practice. Because drawing is an important skill in chemistry, time spent practicing is a good investment. One way to begin is to draw the bond between the inner atoms, showing direct end-on overlap of the hybrid orbitals. Next, fill in the tetrahedral geometry around each inner atom, and then add the outer hydrogen atoms. Finish the sketch by adding the two lone pairs. Remember that the four positions of a tetrahedron are equivalent, so it does not matter which orbitals hold the lone pairs and which ones form bonds. The use of color helps to clarify the sketch. At the right is one possible sketch of the molecule, but other orientations would be equally valid.

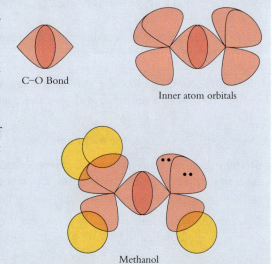

C–O Bond

Inner atom orbitals

Methanol

Does the Result Make Sense? You can visualize methanol either as methane with one —H replaced by —OH or as water with one —H replaced by —CH₃. Either way of visualizing the molecule reveals a tetrahedral inner atom, for which sp^3 hybridization is appropriate.

Extra Practice Exercise 10.3

Describe the bonding of methylamine, CH_3NH_2. Sketch an orbital overlap picture of the molecule.

General Features of Hybridization

We generate hybrid orbitals on inner atoms whose bond angles are not readily reproduced using direct orbital overlap with standard atomic orbitals. Consequently, each of the electron group geometries described in Chapter 9 is associated with its own specific set of hybrid orbitals. Each type of hybrid orbital scheme shares the characteristics described in our discussion of methane:

1. The number of valence orbitals generated by the hybridization process equals the number of valence atomic orbitals participating in hybridization.

2. The steric number of an inner atom uniquely determines the number and type of hybrid orbitals.

3. Hybrid orbitals form localized bonds by overlap with atomic orbitals or with other hybrid orbitals.

4. There is no need to hybridize orbitals on outer atoms, because atoms do not have limiting geometries. Hydrogen always forms localized bonds with its $1s$ orbital. The bonds formed by all other outer atoms can be described using valence p orbitals.

To visualize bond formation by an outer atom other than hydrogen, recall the bond formation in HF. One valence p orbital from the fluorine atom overlaps strongly with the hydrogen $1s$ orbital to form the bond. We can describe bond formation for any outer atom except H through overlap of one of its valence p orbitals with the appropriate hybrid orbital of the inner atom. An example is dichloromethane, CH_2Cl_2, which appears in Figure 10-11. We describe the C—H bonds by sp^3–$1s$ overlap, and we describe the C—Cl bonds by sp^3–$3p$ overlap. Each C—Cl bond can be visualized as resulting from overlap between a carbon sp^3 hybrid and a chlorine $3p$ atomic orbital.

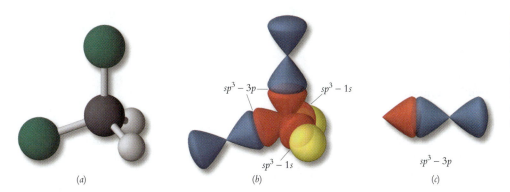

$sp^3 - 3p$ $sp^3 - 1s$

$sp^3 - 1s$

$sp^3 - 3p$

(a) (b) (c)

Figure 10-11
(*a*) Ball-and-stick model of CH_2Cl_2. (*b*) The bonding in CH_2Cl_2 involves two sp^3–$1s$ C—H bonds and two sp^3–$3p$ C—Cl bonds; (*c*) The C—Cl bond forms by overlap of a carbon sp^3 hybrid and a chlorine $3p$ atomic orbital.

It is possible to describe outer atoms as hybridized, and some computer programs for predicting molecular structures describe bonding by hybridizing every atom except hydrogen. We prefer to use unhybridized orbitals to describe outer atoms, for two reasons. First, hybridization is invoked in order to produce sets of orbitals that point in directions that match observed bond angles. There are no bond angles centered on outer atoms, so there is no need for such directional orbitals. Second, hybridization of an outer atom places all non-bonding electrons in hybrid orbitals that have the same energy. In contradiction to this prediction, photoelectron spectroscopy reveals that the nonbonding electrons on outer atoms have two distinct energies that better correspond to the energies of s and p atomic orbitals. Remember, however, that all bonding descriptions are models that oversimplify the electron distribution of any actual molecule. All models of bonding describe some features well but fall short of describing all features accurately.

sp^2 Hybrid Orbitals

Triethylaluminum appears in Chapter 9 as an example of a molecule with a trigonal planar inner atom, with 120° angles separating the three Al—C bonds (Figure 9-20). The valence orbitals of the Al atom have the wrong geometry to form a trigonal plane. Furthermore, a set of sp^3 hybrids is not appropriate, because these would give C—Al—C bond angles of 109.5° rather than 120°. We need a different set of hybrid orbitals to represent an atom with trigonal planar electron group geometry. Proper mixing of the s orbital and two p orbitals gives a set of three **sp^2 hybrid orbitals**. A set of sp^2 hybrid orbitals makes use of two of the valence p orbitals, but not the third. An individual sp^2

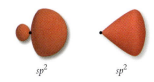

sp^2 sp^2

Actual and stylized views of an sp^2 hybrid orbital

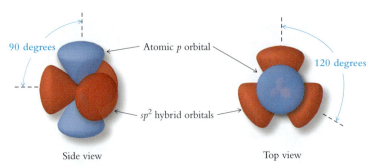

Tutorial

Side view   Top view

Figure 10-12
An *sp²* hybridized atom has three coplanar hybrid orbitals separated by 120° angles. One unchanged *p* orbital is perpendicular to the plane of the hybrids.

Side view

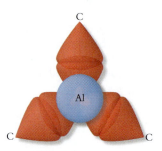

Top view

Figure 10-13
An orbital overlap picture for the Al atom in triethylaluminum, showing its three equivalent bonds and the remaining 3*p* valence orbital. For clarity, we show only one hybrid orbital from each *sp³*-hybridized carbon atom.

Dimethylzinc

hybrid orbital looks very much like its *sp³* counterpart, with high electron density along one axis. As Figure 10-12 shows, however, these hybrid orbitals differ from the *sp³* set in their orientations. The three *sp²* hybrid orbitals point to the three corners of an equilateral triangle. Figure 10-12 also shows that the unused *p* orbital is perpendicular to the plane of the three hybrids.

In triethylaluminum, each Al—C bond can be visualized as an *sp²* hybrid on aluminum overlapping with an *sp³* hybrid on a carbon atom. Figure 10-13 shows this bonding representation, with three equivalent Al—C bonds and the unused 3*p* orbital on the aluminum atom.

Remember from Chapter 6 that energy is released when a bond forms. Consequently, atoms that form covalent bonds tend to use all their valence *s* and *p* orbitals to make as many bonds as possible. We might expect the *sp²*-hybridized aluminum atom to form a fourth bond with its unused 3*p* orbital. A fourth bond does not form in $Al(C_2H_5)_3$ because the carbon atoms bonded to aluminum have neither orbitals nor electrons available for additional bond formation. The potential to form a fourth bond makes triethylaluminum a very reactive molecule.

sp Hybrid Orbitals

Dimethylzinc appears in Chapter 9 as an example of linear geometry. There are two bonds to the zinc atom and a C—Zn—C bond angle of 180°. To describe linear orbital geometry, we need a hybridization scheme that generates two orbitals pointing in opposite directions. This new scheme is a pair of **sp hybrid orbitals,** formed from the zinc 4*s* orbital and one of its 4*p* orbitals. Figure 10-14 shows the shapes and orientations of the two *sp* hybrid orbitals, and Figure 10-15 presents the orbital overlap picture of dimethylzinc. Each C—Zn bond in dimethylzinc can be described as a localized orbital resulting from the overlap of an *sp³* hybrid on the carbon atom and an *sp* hybrid on the zinc atom. For clarity, the figure shows only one of the *sp³* hybrid orbitals for each methyl group.

s + p → sp sp

Figure 10-14
A pair of *sp* hybrid orbitals forms by interacting an *s* atomic orbital and one *p* atomic orbital. The two hybrids point in opposite directions.

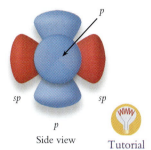

Figure 10-15
The Zn—C bonds of dimethylzinc are formed by overlap of an *sp* hybrid orbital from the metal with an *sp*³ hybrid orbital from carbon.

As with the aluminum atom in triethylaluminum, the bonding description of the zinc atom in dimethylzinc includes vacant *p* orbitals as well as hybrid orbitals for bonding. As Figure 10-16 shows, a set of *sp* hybrid orbitals makes use of only one valence *p* orbital. The remaining two *p* orbitals are perpendicular to each other and perpendicular to the pair of hybrids.

Compounds like triethylaluminum and dimethylzinc that have metal–carbon bonds are rather uncommon. Nevertheless, trigonal planar geometry (*sp*²) and linear geometry (*sp*) occur frequently in nature. As we show in Section 10.3, these geometries and their corresponding hybridizations occur in molecules with double bonds and triple bonds.

Figure 10-16
An *sp*-hybridized atom has two hybrid orbitals separated by 180°. The remaining two atomic *p* orbitals are perpendicular to the hybrids and perpendicular to each other.

> *An inner atom with a steric number of 3 has trigonal planar electron group geometry and can be described using sp² hybrid orbitals.*
>
> *An inner atom with a steric number of 2 has linear electron group geometry and can be described using sp hybrid orbitals.*

KEY CONCEPTS

The Participation of *d* Orbitals

Elements beyond the second row of the periodic table can form bonds to more than four ligands and can be associated with more than an octet of electrons. These features are possible for two reasons. First, elements with $n > 2$ have atomic radii that are large enough to bond to 5, 6, or even more ligands. Second, elements with $n > 2$ have *d* orbitals whose energies are close to the energies of the valence *p* orbitals. An orbital overlap description of the bonding in these species relies on the participation of *d* orbitals of the inner atom.

Phosphorus pentachloride (Figure 9-21) exemplifies the electron group geometry associated with a steric number of 5 and trigonal bipyramidal geometry. Four valence 3*s* and 3*p* orbitals can form no more than four covalent bonds, but phosphorus can form additional bonds using the 3*d* orbitals in its valence shell. One way to involve a 3*d* orbital in bond formation is by combining the 3*s* orbital, the set of 3*p* orbitals, and one 3*d* orbital to form a set of five ***sp*³*d* hybrid orbitals.** In PCl_5, each P—Cl bond can be visualized as resulting from overlap between a chlorine 3*p* orbital and an *sp*³*d* hybrid on the phosphorus atom.

A molecule with a steric number of 6 requires six hybrid orbitals arranged in octahedral geometry. In Chapter 9, sulfur hexafluoride appears as the primary example of a molecule with a steric number of 6 (Figure 9-24). Six equivalent orbitals for sulfur can be constructed for the inner sulfur atom by combining the 3*s*, the three 3*p*, and two of the 3*d* orbitals to form a set of ***sp*³*d*² hybrid orbitals.** Each S—F bond can be visualized as resulting from overlap of an *sp*³*d*² hybrid orbital from sulfur with a 2*p* orbital from fluorine. Figure 10-17 shows ball-and-stick models and orbital overlap sketches of PCl_5 and SF_6.

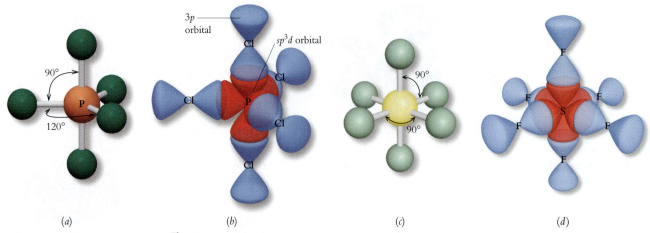

Figure 10-17
Ball-and-stick models and orbital overlap pictures of PCl₅ (*a* and *b*) and SF₆ (*c* and *d*)

KEY CONCEPTS

An inner atom with a steric number of 5 has trigonal bipyramidal electron group geometry and can be described using sp³d hybrid orbitals.

An inner atom with a steric number of 6 has octahedral electron group geometry and can be described using sp³d² hybrid orbitals.

Example 10-4 examines the role of lone pairs in *sp³d* and *sp³d²* hybrid systems.

Example 10-4

sp³d and *sp³d²*
Hybrid Systems

Describe the bonding of chlorine trifluoride and xenon tetrafluoride.

Strategy: To describe bonding, start with a Lewis structure, determine the steric number, and use the steric number to assign the electron group geometry and hybridization of the inner atom. Treat each molecule separately.

Solution: These molecules appear in Examples 9-14 and 9-15. We reproduce the results here, but you should review those examples if the procedure is not clear to you:

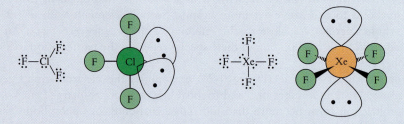

With a steric number of 5, chlorine has trigonal bipyramidal electron group geometry. This means the inner atom requires five directional orbitals, which are provided by an *sp³d* hybrid set. Fluorine uses its valence 2*p* orbitals to form bonds by overlapping with the hybrid orbitals on the chlorine atom. Remember that the trigonal bipyramid has nonequivalent axial and equatorial sites. As we describe in Chapter 9, lone pairs always occupy equatorial positions. See the orbital overlap view on the next page.

With a steric number of 6, xenon has octahedral electron group geometry. This means the inner atom requires six directional orbitals, which are provided by an *sp³d²* hybrid set. Fluorine uses its valence 2*p* orbitals to form bonds by overlapping with the hybrid orbitals on the xenon atom. The two lone pairs are on opposite sides of a square plane, to minimize electron–electron repulsion. See the orbital overlap view on the next page.

Example 10-4

sp^3d and sp^3d^2
Hybrid Systems (*continued*)

ClF$_3$

XeF$_4$

Do the Results Make Sense?   The hybridization schemes match the shapes that we describe in Chapter 9. Notice that the steric numbers of inner atoms uniquely determine both the electron group geometry and the hybridization. This makes sense, because the steric number describes how many electron groups must be accommodated around an inner atom.

Describe the bonding and draw orbital overlap pictures of SF$_4$ and ClF$_5$.

**Extra Practice
Exercise 10.4**

The sp^3d and sp^3d^2 hybrid orbital sets complete our survey of the common chemical geometries. Table 10-1 summarizes the results. Example 10-5 completes this section of the Chapter.

Table 10-1   A Summary of Valence Orbital Hybridization

Steric Number	Electron Group Geometry	Hybridization	Number of Hybrid Orbitals	Number of Unused p Orbitals	Picture
2	Linear	sp	2	2	
3	Trigonal planar	sp^2	3	1	
4	Tetrahedral	sp^3	4	0	
5	Trigonal bipyramidal	sp^3d	5	0	
6	Octahedral	sp^3d^2	6	0	

Example 10-5	For each of the following Lewis structures, name the electron group geometry and the hybrid orbitals used by the inner atoms.
Assigning Hybrid Orbitals	

$$\text{(a) } :\ddot{\text{Cl}}-\underset{\overset{|}{\text{N}}}{\text{N}}-\text{H} \qquad \text{(b) } :\ddot{\text{I}}-\ddot{\text{I}}-\ddot{\text{I}}:^- \qquad \text{(c) } \text{H}_3\text{C}-\underset{\overset{|}{\text{Ga}}}{\text{Ga}}-\text{CH}_3$$

with $:\ddot{\text{Cl}}:$ above the N in (a) and CH_3 above the Ga in (c)

Strategy: The steric number of an inner atom determines the electron group geometry, each of which is associated with one specific type of hybrid orbital.

Solution:

a. The inner nitrogen atom bonds to three ligands and has one lone pair. The steric number is 4, and the electron group geometry is tetrahedral. Four groups of electrons require four hybrid orbitals, so the nitrogen atom uses sp^3 hybrid orbitals.

b. The inner iodine atom bonds to two ligands and has three lone pairs. The steric number is 5, and the electron group geometry is trigonal bipyramidal. The inner atom adopts sp^3d hybridization,

c. Gallium bonds to three methyl groups and has no lone pairs, thus giving a steric number of 3. The electron group geometry is trigonal planar, and the gallium atom uses sp^2 hybrid orbitals. (The carbon atoms of the methyl group ligands are tetrahedral and utilize sp^3 hybridization.)

Do the Results Make Sense? The hybridization choices for inner atoms are correct when they match the steric numbers as indicated by the Lewis structures.

Extra Practice Exercise 10.5	For each of the following Lewis structures, name the electron group geometry and the hybrid orbitals used by the inner atoms.

$$\text{(a) } \quad \text{(b) } :\ddot{\text{F}}-\ddot{\text{O}}-\ddot{\text{F}}: \quad \text{(c) } :\ddot{\text{Cl}}-\ddot{\text{Sn}}-\ddot{\text{Cl}}:$$

with (a) showing a PF$_6$-type structure with P center surrounded by six F atoms, labeled $:\ddot{\text{F}}:^-$

Section Exercises

■ **10.2.1** Draw an energy level diagram that summarizes the sp^2 hybridization process. Use Figure 10-9 as a guide.

■ **10.2.2** Give a complete bonding description of SCl_2. Your answer should include an orbital overlap sketch in which you label all of the orbitals with their names. Name the electron group geometry and the molecular shape.

■ **10.2.3** The following species all have four ligands, but each has a different geometry and different hybridization: SiF_4, TeF_4, and ClF_4^-. Draw a complete Lewis structure for each of them, name each electron group geometry, and identify the hybrid orbitals used by each inner atom.

10.3 MULTIPLE BONDS

Many of the Lewis structures in Chapter 9 and elsewhere in this book represent molecules that contain double bonds and triple bonds. From simple molecules such as ethylene and acetylene to complex biochemical compounds such as chlorophyll and plastoquinone, multiple bonds are abundant in chemistry. Double bonds and triple bonds can be described by extending the orbital overlap model of bonding. We begin with ethylene, a simple hydrocarbon with the formula C_2H_4.

Ethylene is a colorless, flammable gas with a boiling point of -104 °C. More than 22 billion kilograms of ethylene are produced annually in the United States, making it one of the top five industrial chemicals. The manufacture of plastics (polyethylene is the most common example) consumes 75% of this output, and much of the rest is used to make antifreeze. Because ethylene stimulates the breakdown of cell walls, it is used commercially to hasten the ripening of fruit, particularly bananas.

Bonding in Ethylene

Every description of bonding starts with a Lewis structure. Ethylene has twelve valence electrons. The bond framework of the molecule has one C—C bond and four C—H bonds, requiring ten of these electrons. We place the final two electrons as a lone pair on one of the carbon atoms, leaving the second carbon atom with only six electrons. Making a double bond between the carbon

Lewis structure   Ball-and-stick Model

atoms gives both carbon atoms octets and completes the Lewis structure.

What orbital overlap picture best describes ethylene? Both carbon atoms have three ligands, and neither has a lone pair. Thus, each carbon atom has a steric number of 3, which predicts trigonal planar electron group geometry. Experiments show that C_2H_4 is a planar molecule with bond angles close to 120°. These features are consistent with sp^2 hybridization, so a bonding framework can be constructed using carbon sp^2 hybrid orbitals and hydrogen $1s$ orbitals. According to this model, each C—H bond involves overlap of a carbon sp^2 hybrid with a hydrogen $1s$ atomic orbital, and the C—C bond forms from the overlap of two sp^2 hybrid orbitals. Figure 10-18 shows the bonding framework for ethylene.

The orbital overlap view in Figure 10-18 shows one bond joining the carbon atoms in ethylene, but the Lewis structure indicates that a *double* bond exists between the carbon atoms. A look at the full set of valence orbitals lets us see how to construct a second bond between the carbon atoms of ethylene. The sp^2 hybrid orbitals require three of the valence orbitals, leaving one valence p orbital that is not part of the hybrid set. The electron density of this p orbital is concentrated above and below the sp^2 plane. As Figure 10-19 shows, these "leftover" p orbitals on adjacent carbon atoms overlap in a side-by-side fashion. This is *one* bond that has two regions of high electron density, one on either side of the plane defined by the bonding framework. Figure 10-19 shows a π bond from four perspectives.

σ Bonds and π Bonds

In the bond framework in Figure 10-18, all the bonds form from end-on overlap of orbitals directed toward each other. As illustrated by the three examples in Figure 10-20, this type of overlap gives high electron density distributed symmetrically *along* the internuclear axis. A bond of this type is called a **sigma (σ)** bond, and a bonding orbital that describes a σ bond is a σ orbital.

As Figure 10-19 shows, bonds that form from the side-by-side overlap of atomic p orbitals have different electron density profiles than σ bonds. A p orbital has zero electron density—a node—in a plane passing through the nucleus, so bonds that form from side-by-side overlap have no electron density directly on the bond axis. High electron density exists between the bonded atoms, but it is concentrated *above and below* the bond axis. A bond of this type is called a **pi (π)** bond, and a bonding orbital that describes a π bond is a π orbital.

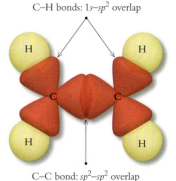

C–H bonds: $1s$–sp^2 overlap

C–C bond: sp^2–sp^2 overlap

Figure 10-18
A top view shows that the bonding framework of ethylene involves sp^2 hybrids on the carbon atoms and $1s$ orbitals on the hydrogen atoms.

The symbols σ and π are Greek letters that correspond to the English letters *s* and *p*.

Figure 10-19
Four views of a π bond.

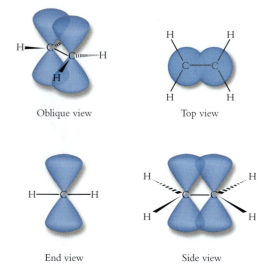

Oblique view   Top view

End view   Side view

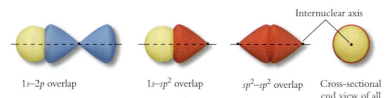

$1s$–$2p$ overlap $1s$–sp^2 overlap sp^2–sp^2 overlap   Cross-sectional end view of all three orbitals

Internuclear axis

Figure 10-20
All σ bonds have high electron density concentrated along the internuclear axis and axial symmetry, so their end-on profiles are circles.

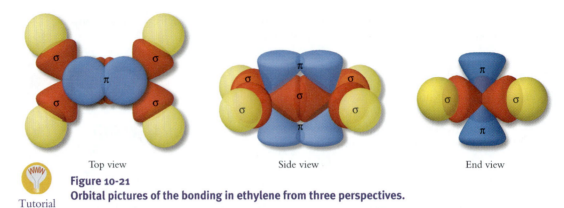

Top view   Side view   End view

Tutorial

Figure 10-21
Orbital pictures of the bonding in ethylene from three perspectives.

KEY CONCEPTS   *A σ bond has high electron density distributed symmetrically along the bond axis.*
A π bond has high electron density concentrated above and below the bond axis.

The double bond in ethylene contains one σ bond and one π bond. The σ bond forms from the end-on overlap of two hybrid orbitals, and the π bond forms from the side-by-side overlap of two atomic *p* orbitals. Figure 10-21 shows the complete orbital picture of the bonding in ethylene. Ethylene is the simplest of a class of molecules, the alkenes, all of which contain C=C double bonds. The alkenes are the subject of our Chemistry and Life Box on page 404.

The availability of a leftover valence *p* orbital to form π bonds is characteristic of sp^2 hybridization and is not restricted to carbon atoms. Example 10-6 shows how the same bonding picture applies to a compound that has a C=N double bond. We construct this bonding picture using the following procedure:

1. Determine the Lewis structure.
2. Use the Lewis structure to determine steric numbers and hybridizations.
3. Construct the σ bond framework.
4. Add the π bonds.

Example 10-6

**Orbital Overlap
in Double Bonds**

An imine is a molecule that contains a carbon—nitrogen double bond. Describe the bonding of the simplest possible imine, H_2CNH, by sketching the σ and π bonding systems.

Strategy: Follow the four-step procedure. Use the Lewis structure to identify the appropriate hybridization. Then draw sketches of the various orbitals involved in the bonding.

Solution: Each molecule of H_2CNH has twelve valence electrons. The bonding framework requires eight electrons, leaving four electrons. Placing these on nitrogen, the more electronegative inner atom, leaves the carbon atom with only three pairs of electrons. Give the carbon atom an octet by transferring one lone pair from the nitrogen atom to make a C=N bond:

The carbon and nitrogen atoms have steric numbers of 3, so their bonding frameworks can be represented using sp^2 hybrid orbitals. Now we can sketch the σ bonding system of imine:

sp^2 lone pair

sp^2–1s σ bond

sp^2–1s σ bond

sp^2–sp^2
σ bond

Example 10-6

**Orbital Overlap
in Double Bonds**
(*continued*)

The carbon and nitrogen atoms have a p orbital left over after construction of the sp^2 hybrids. These two p orbitals are perpendicular to the plane that contains the five nuclei. Side-by-side overlap of the p orbitals gives a π bond that completes the bonding description of this molecule:

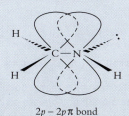

$2p - 2p\ \pi$ bond

Does the Result Make Sense? Imine and ethylene have the same number of valence electrons and similar bonding frameworks. Compare these bonding pictures with those for ethylene. Notice that they are identical, except that a lone pair on the nitrogen atom replaces one of the C—H bonds.

Chloroethylene, H_2CCHCl (also known as vinyl chloride), is the precursor to the common polymer "PVC". Describe the bonding of chloroethylene, by sketching the σ and π bonding systems.

**Extra Practice
Exercise 10.6**

π Bonds Involving Oxygen Atoms

In addition to carbon and nitrogen, oxygen is a common participant in the formation of π bonds. The bonding patterns for oxygen are illustrated by acetic acid (CH_3CO_2H), whose Lewis structure appears in Example 9-9. The Lewis structure shows that acetic acid has a carbon atom and an inner oxygen whose steric numbers are 4, indicating tetrahedral orbital geometry and bonding that can be described using sp^3 hybrids. The second carbon atom has a steric number of 3, indicating trigonal planar orbital geometry and bonding that can be described using sp^2 hybrids. Remember that outer atoms do not require hybrid orbitals, so the outer oxygen atom in acetic acid uses a $2p$ orbital to form a σ bond to carbon. The π bond framework appears in Figure 10-22a.

Once the σ bonding framework is completed, the only atoms with unused valence orbitals are the sp^2-hybridized carbon atom and the outer oxygen atom. As shown in Figure 10-22b and c, each of these atoms has a $2p$ atomic orbital that is positioned perfectly for side-by-side overlap, just as in ethylene and imine.

We complete the description of acetic acid by identifying the orbitals that contain the two lone pairs on the outer oxygen atom. The σ bond and the π bond account for two valence $2p$ orbitals of the oxygen atom. This leaves the third $2p$ orbital and the $2s$ orbital for the lone pairs.

Lewis structure of acetic acid

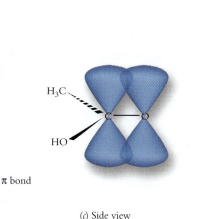

(a)   (b) Top view   (c) Side view

Figure 10-22
(a) The bonding framework of acetone is composed of σ bonds. (b) A top view shows how the π bond forms by side-by-side overlap of $2p$ orbitals from oxygen and carbon. (c) Side view of the π bond.

Box 10-1	Chemistry and Life: The Alkenes

Alkenes are hydrocarbons that contain one or more carbon–carbon double bonds. The simplest alkene, ethylene, is produced by plants to stimulate the ripening of fruit. Larger alkenes are abundant in the biological world and play a wide variety of roles.

Animals communicate by excreting minute amounts of chemical signaling agents known as *pheromones*. Some pheromones repel predators, others mark trails, still others signal alarm. Pheromones have a wide variety of chemical structures, but many of them are alkenes. For example, the trail marker used by termites is neocembrene, an alkene with four carbon–carbon double bonds. Animals produce pheromones to attract members of the opposite sex. One such pheromone is released by female moths to signal that they are ready to mate. A male moth that senses this pheromone immediately seeks out the female. Remarkably, the same molecule has been isolated from the urine of female elephants that are about to ovulate. The presence of this pheromone sends a signal to bull elephants that the mating season has arrived. The similar function of this same compound in two different species is one of the surprising coincidences of evolution.

Neocembrene ($C_{20}H_{32}$)
(a termite trail marker)

Moth/elephant sex attractant

Plants produce a vast array of *terpenes,* alkenes built in multiples of five carbon atoms. Many terpenes have characteristic fragrances. For example, the fresh odor of a pine forest is due to pinene, a ten-carbon molecule with a ring structure and one double bond. The fragrances of terpenes make them important in the flavor and fragrance industry. Limonene, another ten-carbon molecule with a ring and two double bonds, is the principal component of lemon oil. Geraniol, a chainlike molecule with two double bonds, is one of the molecules that is responsible for the fragrance of roses and is used in many perfumes. Many other terpenes have important medicinal properties.

Pinene ($C_{10}H_{16}$)   Limonene ($C_{10}H_{16}$)   Geraniol ($C_{10}H_{18}O$)

Alkenes with many double bonds in a row are colored. Some plant pigments are alkenes of this kind. One example is β-carotene, which gives carrots their distinctive orange color. Animals break down β-carotene into vitamin A, which is essential for vision. Xanthin molecules, relatives of β-carotene that contain oxygen atoms, occur in corn, orange juice, and shellfish. The xanthin below makes the flamingo pink.

Xanthin ($C_{38}H_{48}O_2$)

Isoprene may be the naturally occurring alkene with the greatest economic impact. This compound, a major component of the sap of the rubber tree, is used to make the long-chain molecules of natural rubber (polyisoprene). As we describe in Chapter 13, the synthetic rubbers that make up most of today's tires are made from other alkenes.

Long chain   Long chain

Polyisoprene

Used tires create serious environmental problems, because they do not degrade easily and they release noxious contaminants when they burn. Recently, a process has been developed that breaks tires down into a polyisoprene oil that can then be further decomposed into limonene. The yield of limonene is only a few percent, but if this process can be made more efficient it may become possible to turn ugly, smelly old tires into fragrant oil of lemon.

The bonding pattern of the oxygen atoms in acetic acid repeats in most other oxygen-containing compounds. Outer oxygen atoms frequently have some π character in their bonding. Inner oxygen atoms, on the other hand, do not participate in π bonding. Any inner oxygen atom has a steric number of 4 and can be described appropriately using sp^3 hybrid orbitals.

> Ozone, O_3, is the only exception to this generalization about inner oxygen atoms.

To π Bond or Not To π Bond: Carbon vs. Silicon

Carbon forms π bonds readily, but silicon, its Group 14 neighbor, forms σ bonds with tetrahedral geometry in almost all its compounds. Compare the most stable oxide of carbon, CO_2, with the most stable oxide of silicon, SiO_2. Despite their similar chemical formulas, these two oxides could hardly be different. Carbon dioxide exists as discrete linear triatomic molecules whose atoms are strongly bonded together by a combination of σ bonds and π bonds. Silicon dioxide, on the other hand, exists as a network of silicon and oxygen atoms containing tetrahedral bond angles and no π bonds. In fact, the most abundant silicon-containing materials are silica and the silicate minerals, in which silicon and oxygen form extensive networks of σ bonds formed by overlap of sp^3 hybrid orbitals (see Figures 9-17 and 9-18). These differences in bond formation indicate that the $3p$ valence orbitals of silicon atoms form much more stable σ bonds than π bonds. In contrast, the $2p$ valence orbitals of carbon form stable π and σ bonds, and carbon atoms can have linear, trigonal planar, and tetrahedral geometries. The versatility in bonding of carbon atoms gives rise to the vast, intricate worlds of organic chemistry and biochemistry.

Why does silicon not form π bonds as readily as carbon, its Group 14 neighbor? The answer can be found by examining the sizes of the valence orbitals of silicon and carbon. Figure 10-23 compares side-by-side p-orbital overlap for carbon atoms and silicon atoms. Notice that for the larger silicon atom, the $3p$ orbitals are too far apart for strong side-by-side overlap. The result is a very weak π bond. Consequently, silicon makes more effective use of its valence orbitals to form four σ bonds, which can be described using sp^3 hybrid orbitals.

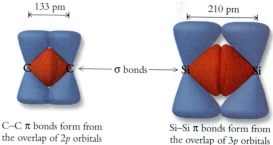

C–C π bonds form from the overlap of $2p$ orbitals

Si–Si π bonds form from the overlap of $3p$ orbitals

Figure 10-23
Comparison of π bonding between two carbon atoms and between two silicon atoms.

In general, π bonding requires the presence of a second-row element. Carbon, nitrogen, and oxygen readily form π bonds with one another. Oxygen also readily forms π bonds with third- and fourth-row elements. Phosphate (PO_4^{3-}), sulfate (SO_4^{2-}), perchlorate (ClO_4^{-}), and other similar anions contain $O{-}X$ π bonds, which we describe in Section 10.6. Many other examples of π bonding exist, but we confine our descriptions in this textbook to bonds between second-row elements and in oxoanions.

Acetylene: Formation of a Triple Bond

The Lewis structure of acetylene (C_2H_2) shows a triple bond between the carbon atoms:

$$H{-}C{\equiv}C{-}H$$

Each carbon atom has a steric number of 2, indicating that acetylene is a linear molecule and that sp hybrid orbitals can be used to construct the bonding orbital framework. Figure 10-24a shows the σ bonding system of acetylene.

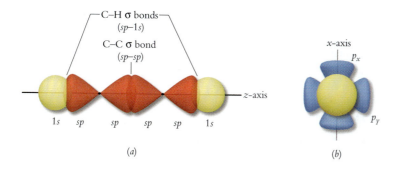

(a)

(b)

Figure 10-24
(a) A side view of the σ bond framework of acetylene. The unhybridized p orbitals of carbon are not shown in this view. (b) In an end view, the hydrogen atom obscures the bonding orbitals. This view shows the p orbitals of carbon, perpendicular to the σ bond axis.

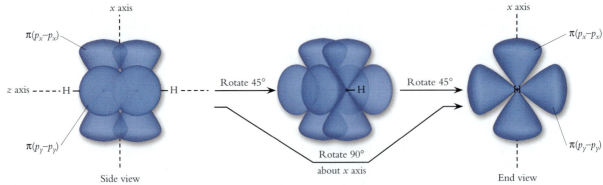

x axis

$\pi(p_x-p_x)$

z axis ----- H

$\pi(p_y-p_y)$

Side view

Rotate 45°

Rotate 90°
about x axis

Rotate 45°

x axis

$\pi(p_x-p_x)$

$\pi(p_y-p_y)$

End view

Figure 10-25
Three views of the π bonding in acetylene. The two π bonds are perpendicular to each other and have electron density above and below the internuclear axis.

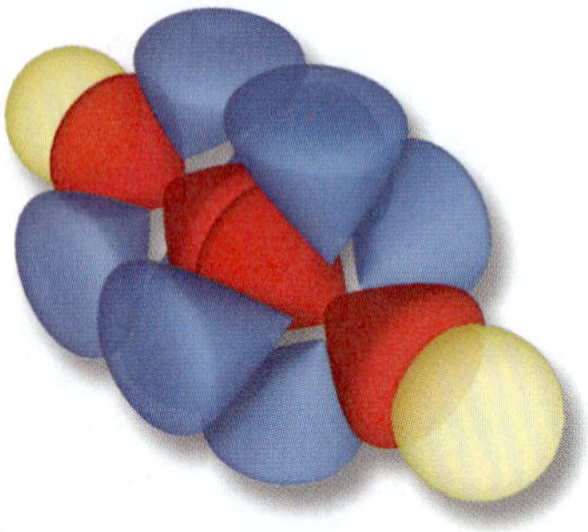

Figure 10-26
Composite orbital overlap view of the σ and π bonds of acetylene. The π bonds are shown in an "exploded" view that makes the C—C σ bond visible.

The sp hybrid set requires just one of the valence p orbitals. Figure 10-24b shows that the remaining two p orbitals are oriented at right angles to the axis of the acetylene molecule and at right angles to each other. Each p orbital can overlap with its counterpart on the other carbon atom, just as in ethylene. Thus, acetylene has two π bonds. Each one forms from the side-by-side overlap of p orbitals on each carbon atom. Figure 10-25 illustrates this. According to this description, a triple bond is made up of one σ bond and two π bonds. Figure 10-26, which superposes the views in Figures 10-24 and 10-25, depicts all the bonds in the acetylene molecule. To make all the bonding orbitals visible, the π orbitals have been moved outward, displaced slightly from their true locations.

Example 10-7 treats another molecule with bonding that can be described by sp hybrid orbitals.

Example 10-7	Hydrogen cyanide (HCN) is an extremely poisonous gas with an odor resembling that of almonds. Approximately one billion pounds of HCN are produced each year, most of which are used to prepare starting materials for polymers. Construct a complete bonding picture for HCN and sketch the various orbitals.
Orbital Overlap in Triple Bonds	**Strategy:** Use the four-step procedure described earlier. Begin with the Lewis structure for the molecule, and then identify the appropriate hybrid orbitals. Construct a σ bond framework, and complete the bonding picture by assembling the π bonds from the unhybridized p orbitals.

Solution: The six steps of our procedure for writing Lewis structures lead to a triple bond between carbon and nitrogen: H—C≡N:

Hydrogen and nitrogen are outer atoms, so they use atomic orbitals to form covalent bonds. The carbon atom has a steric number of 2, so it can be described using sp hybrids. With this information, we can construct the σ bonding network for HCN:

$\sigma(1s{-}sp)$ $\sigma(sp{-}2p)$

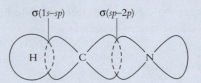

H   C   N

The triple bond includes two π bonds formed by the pair-wise overlap of the p orbitals that remain on the carbon and nitrogen atoms. Two figures from different perspectives help illustrate the triple bond:

C   N

N

Side view   End view

Example 10-7

Orbital Overlap in Triple Bonds (*continued*)

The nonbonding pair on the nitrogen atom occupies the $2s$ orbital, which is the only valence orbital of this atom not used for bonding.

The σ and π bonding networks are shown separately for clarity. Keep in mind, however, that the molecule is a *composite* of both networks, so visualize the HCN molecule with the two bonding networks superimposed.

Does the Result Make Sense? Just as imine resembles ethylene, hydrogen cyanide is similar to acetylene, with the same number of valence electrons. A lone pair on the N atom of HCN replaces one C—H bond of acetylene.

Describe and draw the bonding system of 2-butyne: CH_3CCCH_3

Extra Practice Exercise 10.7

Section Exercises

10.3.1 Determine the Lewis structure and describe the bonding completely (including an orbital sketch) for the formaldehyde molecule (H_2CO). Formaldehyde is a pungent, colorless gas that is highly toxic and is a suspected carcinogen.

10.3.2 Propene is the three-carbon alkene, and propyne is the three-carbon hydrocarbon that contains a triple bond:

Propene   Propyne

Determine the chemical formulas and Lewis structures of these two substances. Describe their bonding completely, including the geometry and hybridization for each carbon atom.

10.3.3 Convert the line structure of limonene (see Chemistry and Life Box) into a Lewis structure. Determine how many carbon atoms participate in π bonding and draw orbital sketches for a π bond, a C—C σ bond, and a C—H σ bond in this molecule.

10.4 MOLECULAR ORBITAL THEORY: DIATOMIC MOLECULES

Hybrid orbitals and localized bonds provide a model of bonding that can be applied easily to a wide range of molecules. This model does an excellent job of rationalizing and predicting chemical structures, but localized bonds cannot predict or interpret other aspects of bonding and reactivity. For example, localized bonds cannot explain either the green color of chlorophyll or the ability of this molecule to collect energy from sunlight. Another example is the electrical properties of metals and semiconductors, which cannot be accounted for by a localized bonding model. In the remaining sections of this chapter we explore the properties of electrons that are said to be *delocalized* over three, four, or even dozens of atoms. Delocalized electrons occupy **molecular orbitals (MOs),** so called because they may span entire molecules.

Molecular orbital theory is more complex than the hybrid orbital approach, but the foundations of the model are readily accessible. Though complex, molecular orbital theory opens the door to many fascinating aspects of modern chemistry. In this section, we introduce the molecular orbital approach through diatomic molecules.

Molecular Orbitals of H₂

The opening pages of Chapter 10 introduce the principles of orbital overlap by describing molecular hydrogen as a composite of overlapping spherical $1s$ orbitals. This interaction is the starting point for our discussion of molecular orbital theory.

Because electrons have wave-like properties, orbital interactions involve addition or subtraction of amplitudes, as we describe in Section 10-1. So far, we have described only additive orbital interactions. The wave amplitudes add in the overlap region, generating a new bonding orbital with larger amplitude between the nuclei. However, a complete mathematical treatment of orbital overlap requires that orbitals be conserved. In other words, whenever several orbitals interact, they must generate an equal number of new orbitals.

The total number of molecular orbitals produced by a set of interacting atomic orbitals is equal to the number of interacting orbitals.

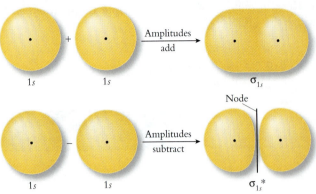

Figure 10-27
When two hydrogen 1s orbitals interact, they generate two molecular orbitals, one bonding (σ_{1s}) and one antibonding (σ_{1s}*).

When two hydrogen atoms interact, their 1s orbitals generate two new molecular orbitals, as shown in Figure 10-27. One of these interactions is additive, leading to a molecular orbital with high electron density between the nuclei. High electron density between the nuclei results in more attraction than repulsion (Figure 9-1), so this combination is a **bonding molecular orbital.** Electrons in a bonding orbital are more stable than they would be in individual atoms. The bonding MO of molecular hydrogen is labeled σ_{1s}. It is σ because the electron distribution is concentrated along the bond axis and 1s because the MO is constructed from 1s atomic orbitals.

The second orbital interaction is subtractive, generating an orbital with low electron density between the nuclei. Low electron density between the nuclei results in more nucleus–nucleus repulsion than electron–nucleus attraction. This combination is destabilizing compared to individual atoms and is referred to as an **antibonding molecular orbital.** The antibonding combination is labeled as σ_{1s}*. The MO is a σ orbital because the electron density lies along the bond axis, although the electron density in this orbital is concentrated outside the nuclei rather than between the nuclei. The asterisk in the symbol signifies an antibonding orbital. Figure 10-27 shows that the σ_{1s}* orbital has a node between the nuclei. Nodes that pass between nuclei are one of the characteristics of antibonding molecular orbitals. Recall from Section 7.6 that nodes represent regions within an orbital where there is no electron density. It is important to realize that σ_{1s}* is one orbital with lobes of electron density on either side of the node. This is analogous to atomic p orbitals, as described in Chapter 7.

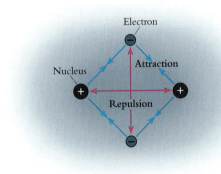

Figure 9-1: a bonding interaction

Figure 10-28 is an energy level diagram, called a **molecular orbital diagram,** summarizing the MO treatment of H_2. It shows that the bonding orbital is more stable than the orbitals from which it forms, whereas the antibonding orbital is less stable. When a hydrogen molecule forms, its two electrons obey the same principles for distributing electrons as we describe in Chapter 8. Electrons must obey the aufbau process, the Pauli exclusion principle, and Hund's rule, no matter what types of orbitals they occupy. The two valence electrons of molecular hydrogen fill the most stable orbital, the σ_{1s} orbital, with their spins oriented in opposite directions. The 1s orbitals of the hydrogen atoms have been used to form the MO pair, and the σ_{1s}* orbital, being higher in energy, is empty.

As their names suggest, molecular orbitals can span an entire molecule, while localized bonds cover just two nuclei. Because diatomic molecules contain just two nuclei, the localized view gives the same general result as molecular orbital theory. The importance of molecular orbitals and delocalized electrons becomes apparent as we move beyond diatomic molecules in the follow-ing sections of this chapter. Meanwhile, diatomic molecules offer the simplest way to develop the ideas of molecular orbital theory.

The MO diagram shown in Figure 10-28 applies to all diatomic species from the first row of the periodic table. We can, for instance, use molecular orbital theory to rationalize why helium atoms do not combine to form molecules of He_2. This hypothetical diatomic molecule would have four electrons. Two of these would populate the bonding σ_{1s} orbital, imparting stability to the molecule. The remaining two electrons would have to occupy the antibonding σ_{1s}* orbital, and their destabilizing

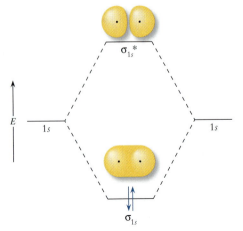

Figure 10-28
Molecular orbital diagram for molecular hydrogen.

effect would cancel the influence of the bonding electrons, as shown by the MO diagram in Figure 10-29. A convenient way to summarize this argument is by calculating the **bond order,** which represents the net amount of bonding between two atoms:

Bond order = $\frac{1}{2}$(number of electrons in bonding MOs − number of electrons in antibonding MOs)　　　　**(10-1)**

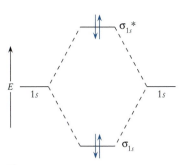

Figure 10-29
Diatomic He$_2$ does not exist because the stability imparted by the bonding electrons is offset by the destabiliztion due to the antibonding electrons.

For molecular hydrogen, with two bonding electrons and zero electrons in the antibonding orbitals, the bond order is 1. A bond order of 1 represents a single bond, consistent with the Lewis structure of H$_2$. For the hypothetical He$_2$ molecule, equal numbers of bonding and antibonding electrons give a bond order of 0. Helium atoms do not bond together into stable diatomic molecules. Example 10-8 looks at an intermediate case.

Example 10-8

Does He$_2{}^+$ Exist?

Use a molecular orbital diagram to predict if it is possible to form the He$_2{}^+$ cation.

Strategy: The MO diagram shown in Figure 10-28 can be applied to any of the possible diatomic molecules or ions formed from the first-row elements, hydrogen and helium. Count the electrons of He$_2{}^+$, place the electrons in the MO diagram, and calculate the bond order. If the bond order is greater than zero, the species can form, under the right conditions.

Solution: One He atom has two electrons, so a He$_2{}^+$ cation has three electrons. Following the aufbau process, two electrons fill the lower-energy σ_{1s} orbital, so the third must be placed in the antibonding $\sigma_{1s}{}^*$ orbital in either spin orientation. A shorthand form of the MO diagram appears at right. The bond order of the cation is $\frac{1}{2}$:

$$\text{Bond order} = \tfrac{1}{2}(2 - 1) = \tfrac{1}{2}$$

We predict that He$_2{}^+$, with a net bond order of $\frac{1}{2}$, can be prepared in the laboratory.

Does the Result Make Sense? Electrical discharges through samples of helium gas generate He$^+$ cations, some of which bond with He atoms to form He$_2{}^+$ cations. These fall apart as soon as they capture electrons, but they last long enough to be studied spectroscopically. The bond dissociation energy is 250 kJ/mol, approximately 60% as strong as the bond in the H$_2$ molecule, whose bond order is 1.

Which species has the stronger bond, H$_2$ or H$_2{}^-$? Use a molecular orbital diagram to support your answer.

**Extra Practice
Exercise 10.8**

Second-Row Diatomic Molecules

The next step in the development of molecular orbital theory is to consider the MOs formed by additive and subtractive overlap of atomic p orbitals. We use molecular oxygen as a case study.

The Lewis structure of molecular oxygen shows the two atoms connected by a double bond, with two nonbonding electron pairs on each oxygen atom. Both atoms in O$_2$ are outer atoms, so there are no constraining bond angles and no need for hybridization. Atomic valence $2s$ and $2p$ orbitals can be used to describe the bonding in this molecule.

The valence orbitals on the two oxygen atoms interact to form a set of molecular orbitals. Each oxygen atom has four valence orbitals, so the MO diagram for O$_2$ contains eight molecular orbitals. The first interaction involves the $2s$ atomic orbitals. This pair interacts in exactly the same way as the $1s$ orbitals described for H$_2$. The additive combination generates σ_{2s} and the subtractive combination forms $\sigma_{2s}{}^*$. Because of the larger

$\overset{..}{O}=\overset{..}{O}$

Lewis structure of O$_2$

Figure 10-30
Additive and subtractive combinations of p orbitals lead to bonding and antibonding orbitals. (a) End-on overlap gives σ orbitals. (b) Side-by-side overlap gives π orbitals.

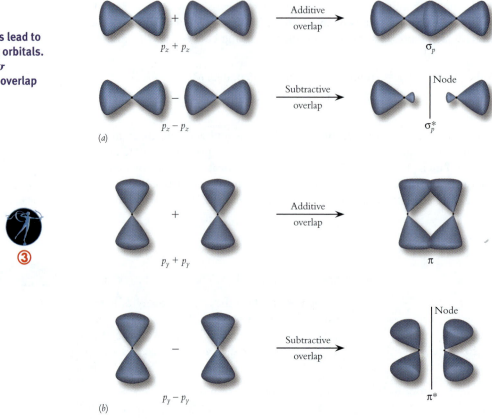

size of the atomic orbitals, these two MOs are larger than their hydrogen σ_{1s} and $\sigma_{1s}*$ counterparts, but their overall appearances are similar.

Figure 10-30 shows the construction of the 2p-based molecular orbitals. One pair of MOs forms from the p orbitals that point toward each other along the bond axis. By convention, we label this as the z-axis. This end-on overlap gives σ_p and σ_p* orbitals that concentrate electron density between the two oxygen nuclei, as shown in Figure 10-30a. The remaining four p orbitals form pairs of π and $\pi*$ MOs through side-by-side overlap. One of these pairs comes from the p_y orbitals, and the other pair comes from the p_x. Figure 10-30b shows only the p_y pair of π orbitals. The p_x pair has the same appearance but is perpendicular to the one shown in the figure. Figure 10-31 shows complete sets of the π and $\pi*$ orbitals from three perspectives. Notice that the π molecular orbitals closely resemble π bonds of acetylene (Figure 10-25).

Remember that electrons preferentially fill the lowest-energy (most stable) orbitals that are available. Before placing the valence electrons in these MOs, therefore, we need an orbital energy level diagram. The orbital overlap model and the properties of electrons allow us to draw some general conclusions about the relative energies of the eight molecular orbitals generated by $n = 2$ atomic orbitals:

1. The bonding and antibonding σ_s orbitals are more stable than any of the six molecular orbitals derived from the 2p orbitals. This is because the 2s orbitals that generate the σ_s and σ_s* orbitals are more stable than the 2p atomic orbitals.

2. The two π bonding orbitals have identical energies, because the corresponding atomic p orbitals have identical energies. Likewise, the two $\pi*$ orbitals have identical energies.

3. The antibonding orbitals formed from the atomic 2p orbitals are at the top of the energy level diagram, with the σ_p* orbital higher than $\pi*$.

After applying these features, we are left with the bonding σ_p and π molecular orbitals. These must be placed between σ_s* and $\pi*$, but which of them is lower in energy? Orbital overlap arguments suggest that σ_p should be more stable than π, because end-on overlap is more effective than side-by-side overlap. Molecular oxygen displays

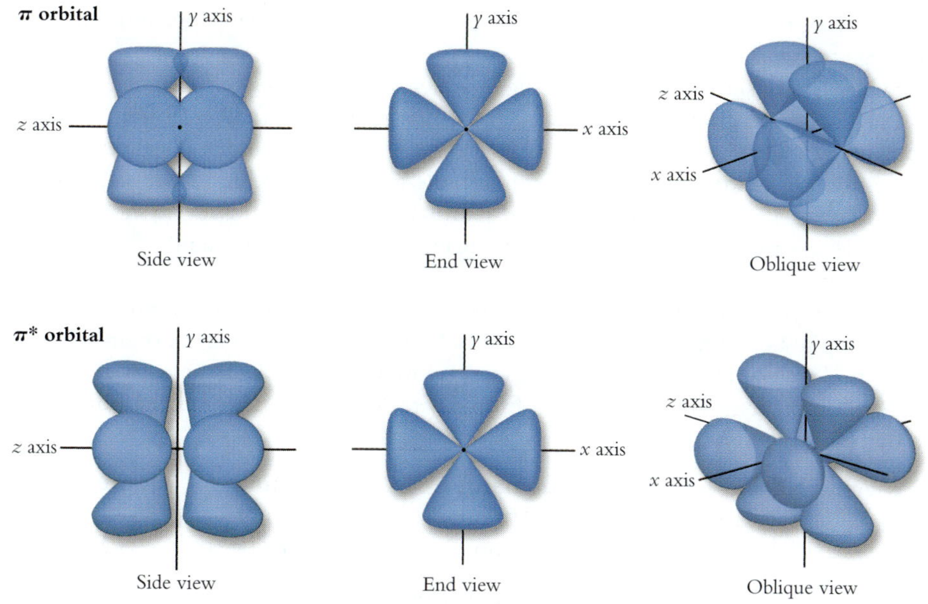

π orbital

Side view   End view   Oblique view

π^* orbital

Side view   End view   Oblique view

Figure 10-31
Three views of the π and π^* sets of molecular orbitals.

this ranking. Figure 10-32 shows the complete diagram and places the valence electrons of O_2 in accordance with the Pauli and aufbau principles.

We can write configurations for diatomic molecules by naming the orbitals (σ_s, π_x, etc.), using superscripts to show how many electrons each orbital contains. Here is the configuration for the ground state of O_2:

$$(\sigma_s)^2 \, (\sigma_s{}^*)^2 \, (\sigma_p)^2 \, (\pi_x)^2 \, (\pi_y)^2 \, (\pi_x{}^*)^1 \, (\pi_y{}^*)^1$$

Evidence for Antibonding Orbitals

The molecular orbital model developed in this section is more elaborate than the localized bonds described earlier in this chapter. Is this more complicated model necessary to

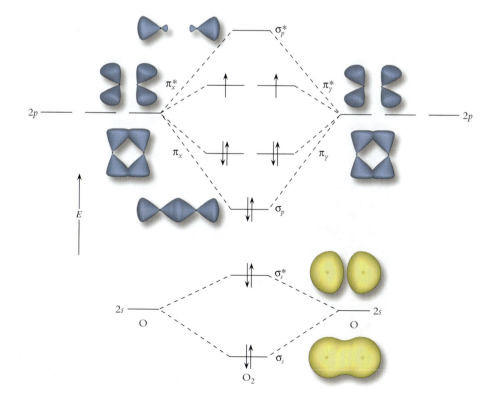

Figure 10-32
The molecular orbital diagram for O_2. The energy ranking of the MOs applies to second-row diatomic molecules with $Z > 7$.

Figure 10-33
Molecular oxygen is paramagnetic, so it clings to the poles of a magnet.

give a thorough picture of chemical bonding? Experimental evidence for molecular oxygen suggests that the answer is yes.

Figure 10-33 shows that liquid oxygen adheres to the poles of a magnet. Attraction to a magnetic field shows that molecular oxygen is paramagnetic. Recall from Chapter 8 that paramagnetism arises when the electron configuration includes unpaired electrons. Neither the Lewis structure of O_2 nor the simple localized bond model reveals the presence of unpaired electrons. The molecular orbital description, however, shows that the least stable electrons of the oxygen atom occupy the two equivalent π^* antibonding orbitals (Figure 10-32).

Bond length and bond energy measurements on oxygen and its cation also indicate that the highest-energy occupied orbital is antibonding. Removing a bonding electron from a molecule gives a cation with a weaker, longer bond than the neutral molecule has. In contrast, removing an electron from O_2 to form O_2^+ increases the bond energy and decreases the bond length:

Species	Bond length	Bond energy	Configuration	Bond order
O_2	121 pm	496 kJ/mol	. . . $(\sigma_p)^2 (\pi)^4 (\pi^*)^2$	2
O_2^+	112 pm	643 kJ/mol	. . . $(\sigma_p)^2 (\pi)^4 (\pi^*)^1$	2.5

These data show that the least stable occupied orbital of O_2 is antibonding in character: removing an electron reduces the amount of antibonding and strengthens the bond.

Our treatment of O_2 shows that the extra complexity of the molecular orbital approach explains features that a simpler description of bonding cannot explain. The Lewis structure of O_2 does not reveal its two unpaired electrons, but an MO approach does. The simple $\sigma-\pi$ description of the double bond in O_2 does not predict that the bond in O_2^+ is stronger than that in O_2, but an MO approach does. As we show in the following sections, the molecular orbital model has even greater advantages in explaining bonding when Lewis structures show the presence of resonance.

Homonuclear Diatomic Molecules

The MO diagram for molecular oxygen can be generalized for any of the second-row diatomic molecules. For instance, we can apply it to B_2. Although elemental boron is a solid under normal conditions, gaseous B_2 molecules form when the solid is strongly heated. The MO diagram for O_2 predicts that B_2, with six valence electrons, should have the configuration $(\sigma_s)^2 (\sigma_s^*)^2 (\sigma_p)^2$. With all its electrons paired, the species should be diamagnetic. In contradiction to this prediction, experiments show that B_2 molecules are paramagnetic, with two unpaired electrons. When theory and experiment conflict, theory must be revised. The diatomic molecular orbital diagram must be revised to account for the paramagnetism of B_2.

The view of molecular orbital theory that leads to the MO diagram in Figure 10-32 assumes that the $2s$ and $2p$ orbitals act independently. A more refined treatment considers interactions among the $2s$ and $2p$ sets of orbitals. Remember that the $2s$ and $2p$ orbitals have similar radii. Consequently, when two atoms approach each other, the $2s$ and $2p_z$ orbitals on one atom overlap with *both* the $2s$ and the $2p_z$ orbitals of the other atom. This mixed interaction of $2s$ and $2p_z$ stabilizes σ_s and destabilizes σ_p. In other words, **orbital mixing** causes σ_s and σ_p to move farther apart in energy. The amount of mixing depends on the difference in energy between the $2s$ and $2p$ atomic orbitals. Mixing is largest when the energies of the orbitals are nearly the same. As Figure 10-34 shows, the energies of atomic

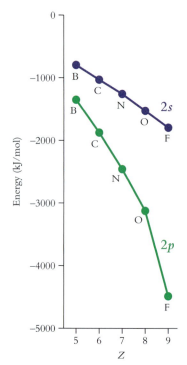

Figure 10-34
Energies of the $n = 2$ valence orbitals as a function of Z.

Figure 10-35
Mixing of the $2s$ and $2p_z$ orbitals causes the σ_s MOs to become more stable and the σ_p MOs to become less stable. The amount of mixing decreases moving across Row 2. For B_2, C_2, and N_2 the π MOs are more stable than σ_p. For O_2 and F_2, σ_p is more stable than π.

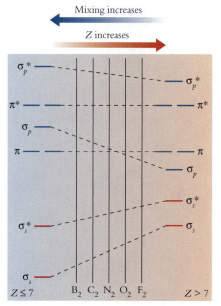

orbitals diverge as Z increases across the second row, so mixing is large for B_2 but small for F_2. Figure 10-35 shows how mixing affects the stability of the σ_s and σ_p orbitals. Notice that crossover of the energy levels takes place between molecular nitrogen and molecular oxygen. Because of orbital mixing, we require two generalized molecular orbital diagrams for diatomic molecules, one for B_2, C_2, and N_2, and the other for O_2 and F_2. Figure 10-35 shows both diagrams.

These molecular orbital diagrams help to rationalize experimental observations about molecules, as Example 10-9 shows.

Example 10-9

Trends in Bond Energy

Use molecular orbital diagrams to explain the trend in the following bond energies: $B_2 = 290$ kJ/mol, $C_2 = 600$ kJ/mol, and $N_2 = 942$ kJ/mol.

Strategy: The data show that bond energies for these three diatomic molecules increase moving across the second row of the periodic table. We must construct molecular orbital diagrams for the three molecules and use the results to interpret the trend.

Solution: The crossover point for the σ_p/π energy levels takes place just after N_2, so the general diagram for $Z \leq 7$ in Figure 10-35 applies to all three molecules. The valence electron counts are $B_2 = 6$ e⁻, $C_2 = 8$ e⁻, and $N_2 = 10$ e⁻. Place these electrons in the MO diagram following the aufbau process. Here are the results:

B_2	C_2	N_2
σ_p^*	σ_p^*	σ_p^*
π_y^*, π_y^*	π_y^*, π_y^*	π_y^*, π_y^*
σ_p	σ_p	⇅ σ_p
↑ ↑ π_x, π_y	⇅ ⇅ π_x, π_y	⇅ ⇅ π_x, π_y
⇅ σ_s^*	⇅ σ_s^*	⇅ σ_s^*
⇅ σ_s	⇅ σ_s	⇅ σ_s

The diagrams show that the increasing electron counts as we progress from boron to carbon to nitrogen result in the filling of bonding molecular orbitals. This is revealed most clearly by calculating the bond orders for the three molecules: $B_2 = 1$, $C_2 = 2$, and $N_2 = 3$. Increasing bond order corresponds to greater electron sharing between the nuclei, resulting in stronger bonding between the atoms.

Does the Result Make Sense? Of these three diatomic molecules, only N_2 exists under normal conditions. Boron and carbon form solid networks rather than isolated diatomic molecules. However, molecular orbital theory predicts that B_2 and C_2 are stable molecules under the right conditions, and in fact both molecules can be generated in the gas phase by vaporizing solid boron or solid carbon in the form of graphite.

Extra Practice Exercise 10.9

Use molecular orbital diagrams to explain the trend in bond energies for the following diatomic molecules: $N_2 = 942$ kJ/mol, $O_2 = 495$ kJ/mol, and $F_2 = 155$ kJ/mol.

Heteronuclear Diatomic Molecules

Nitrogen oxide (NO) is an example of heteronuclear diatomic molecules, those composed of different atoms. This interesting molecule has been in the news several times in recent years, because of important discoveries about the role of NO as a biological messenger, as we describe in our introduction to Chapter 21.

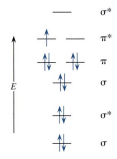

Molecular orbital diagram of NO

Because the qualitative features of orbital overlap do not depend on the identity of the atoms, the bonding in NO can be described by the same sets of orbitals that describe homonuclear diatomic molecules. Which of the two general MO diagrams for diatomic molecules applies to nitrogen oxide? The crossover point for the energy rankings of the σ_p and π molecular orbitals falls between N and O, so we expect the orbital energies to be nearly equal. Photoelectron spectroscopy confirms this but also shows that the σ_p MO is a bit more stable than the π MO. Consequently, the MO configuration for the eleven valence electrons of NO mirrors that for the 12 valence electrons of O_2, except that there is a single electron in the π^* orbital:

$$(\sigma_s)^2 \, (\sigma_s{}^*)^2 \, (\sigma_p)^2 \, (\pi_x)^2 \, (\pi_y)^2 \, (\pi_x{}^*)^1$$

There are eight bonding electrons and three antibonding electrons, so the bond order of NO is 2.5. This bond is stronger and shorter than the double bond of O_2 but weaker and longer than the triple bond of N_2.

Species	Bond length	Bond energy	Configuration	Bond Order
O_2	121 pm	495 kJ/mol	$\ldots (\sigma_p)^2 (\pi)^4 (\pi^*)^2$	2
NO	115 pm	605 kJ/mol	$\ldots (\sigma_p)^2 (\pi)^4 (\pi^*)^1$	2.5
N_2	110 pm	945 kJ/mol	$\ldots (\pi)^4 (\sigma_p)^2$	3

Electron configurations for these molecules lead to a guideline for molecular orbital configurations based on average nuclear charge, $Z_{average}$:

KEY CONCEPT *For second-row diatomic molecules and ions:*

σ_p *is lower in energy than* π *when* $Z_{average} > 7$, *and*

π *is lower in energy than* σ_p *when* $Z_{average} \leq 7$.

To conclude this section, Example 10-10 compares three diatomic molecules. Section 10.5 explores molecular orbitals in triatomic molecules.

Example 10-10

Comparing Ionization Energies

The first ionization energy of NO is 891 kJ/mol, that of N_2 is 1500 kJ/mol, and that of CO is 1350 kJ/mol. Use electron configurations to explain why NO ionizes more easily than either N_2 or CO.

Strategy: Recall from Chapter 8 that ionization energy refers to the removal of an electron from an atom, or, in this case, from a molecule. We must count the valence electrons, choose the correct MO diagram, follow the aufbau process in placing the electrons, and then use the configurations to explain the ionization energy data.

Solution: Begin with a summary of the important information for each molecule: NO has 11 valence electrons and $Z_{average} = 7.5$; N_2 has 10 valence electrons and $Z_{average} = 7$; CO has 10 valence electrons and $Z_{average} = 7$. Applying the general diatomic MO diagrams of Figure 10-35 gives the following configurations:

$$NO = (\sigma_s)^2 \, (\sigma_s{}^*)^2 \, (\sigma_p)^2 \, (\pi_x)^2 \, (\pi_y)^2 \, (\pi_x{}^*)^1$$

$$N_2 = (\sigma_s)^2 \, (\sigma_s{}^*)^2 \, (\sigma_p)^2 \, (\pi_x)^2 \, (\pi_y)^2$$

$$CO = (\sigma_s)^2 \, (\sigma_s{}^*)^2 \, (\sigma_p)^2 \, (\pi_x)^2 \, (\pi_y)^2$$

CO and N_2 are isoelectronic species, meaning they have the same electron configurations (see Section 8.3). Each has a bond order of 3, and the least stable electrons are in π bonding orbitals. It makes sense that these two molecules have comparable ionization energies. NO has one electron more than N_2 and CO, and this electron occupies one of the antibonding π^* orbitals. It is much easier to remove this antibonding electron from NO than it is to remove a bonding electron from either of the other two species.

Does the Result Make Sense? The difference in ionization energies between NO and the other two molecules is around 500 kJ/mol. We can say that an electron in a π orbital strengthens a bond by about 250 kJ/mol, and an electron in a π^* orbital weakens a bond by about 250 kJ/mol. Look at the bond energies for NO, O_2, and N_2 and decide if this conclusion is valid.

Use electron configurations to predict which of the following is the most stable diatomic combination of carbon and nitrogen: CN^-, CN, or CN^+.

Extra Practice Exercise 10.10

Section Exercises

■ **10.4.1** Dilithium molecules can be generated from lithium metal at very low pressure. Construct a molecular orbital diagram of dilithium. What is the bond order of this molecule?

■ **10.4.2** Draw an orbital sketch of each of the occupied valence orbitals for dilithium.

■ **10.4.3** Draw a complete molecular orbital diagram for the diatomic molecule Ne_2. Use the diagram to explain why Ne_2 is not a stable molecule.

10.5 **THREE-CENTER π ORBITALS**

The π molecular orbitals described so far involve two atoms, so the orbital pictures look the same for the localized bonding model applied to ethylene and the MO approach applied to molecular oxygen. In the organic molecules described in the introduction to this chapter, however, π orbitals spread over three or more atoms. Such **delocalized π orbitals** can form when *more than two p* orbitals overlap in the appropriate geometry. In this section, we develop a molecular orbital description for three-atom π systems. In the following sections, we apply the results to larger molecules.

Ozone and the Nature of Resonance

We begin our exploration of delocalized π bonds with ozone, O_3. As described in Chapter 7, ozone in the upper stratosphere protects plants and animals from hazardous ultraviolet radiation. Ozone has 18 valence electrons and a Lewis structure that appears in Figure 10-36a. Experimental measurements show that ozone is a bent molecule with a bond angle of 118°.

The bond framework of ozone can be represented by the localized σ bonds shown in Figure 10-36b. The inner oxygen atom, with a steric number of 3, is trigonal planar and can be described using sp^2 hybrid orbitals. Two of these sp^2 orbitals form σ bonds by overlap with $2p$ orbitals from the outer oxygen atoms. The third sp^2 hybrid contains a lone pair of electrons. The Lewis structure also indicates the presence of a π bond and five nonbonding electron pairs on the two outer oxygen atoms.

The sp^2 hybridized inner oxygen atom has one unused $2p$ orbital, which is oriented perpendicular to the plane defined by the hybrid orbitals (review Figure 10-12). This p orbital is in the perfect position to form a π bond through side-by-side overlap with a p orbital on an outer oxygen atom. The resonance structures in the Lewis structure of ozone indicate that either outer oxygen is an equally good choice for this overlap. However, the two distinct localized π bonds shown in the ozone resonance structures are misleading, because all three of the perpendicular p orbitals overlap simultaneously. Neither Lewis structures nor the localized bonding model gives an adequate description of the π bonding in ozone. Molecular orbital theory, on the other hand, does an excellent job of treating the interactions of multiple p orbitals.

Molecular orbital interactions are possible only for atomic orbitals with overlapping electron densities. In particular, π molecular orbitals form when p atomic orbitals overlap

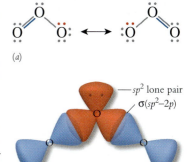

(a)

(b)

Figure 10-36
The structure of ozone. (a) The Lewis structure shows resonance that involves the π bond (blue lines) and one lone pair (red dots). (b) The σ bonding framework.

Figure 10-37
The three π MOs of ozone. (*a*) The π orbital spreads over all three atoms. (*b*) The π_{nb} orbital has a node on the inner O atom. (*c*) The π^* orbital has nodes in each of the bonding regions.

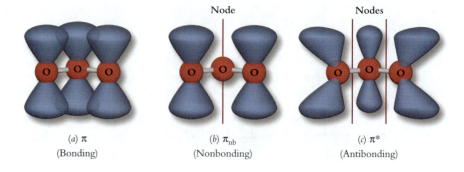

(*a*) π
(Bonding)

(*b*) π_{nb}
(Nonbonding)

(*c*) π^*
(Antibonding)

side by side, on adjacent atoms. As always, the resulting number of molecular orbitals must equal the number of contributing atomic orbitals. Another general feature of orbital interactions is that there are equal numbers of bonding MOs and antibonding MOs. Thus, ozone, with three overlapping atomic p orbitals, has three π MOs, one of which is bonding and one of which is antibonding. What is the third molecular orbital?

A clue to the nature of the third π MO can be found in the placement of electrons in the two resonance structures for ozone, which are shown with color highlights in Figure 10-36*a*. Notice that in one resonance structure, the left outer atom has three lone pairs and a single bond, while the right outer atom has two lone pairs and a double bond. In the other resonance structure, the third lone pair is on the *right* outer atom, with the double bond to the *left* outer atom. The double bond appears in different positions in the two structures, and one of the lone pairs also appears in different positions. These variations signal delocalized π orbitals.

The double bond shown in different locations in the two resonance structures represents a pair of electrons in a delocalized π bonding molecular orbital that spans all three of the oxygen atoms, as shown in Figure 10-37*a*. One lone pair also appears in different positions in the two resonance structures, again signaling a delocalized orbital. This lone pair is spread over both *outer* atoms but not across the *inner* atom, as shown in Figure 10-37*b*. This is a **nonbonding molecular orbital,** π_{nb}. The lone pair shown in different positions in the resonance structures occupies the delocalized π_{nb} orbital.

To summarize, the interactions among the three p orbitals of ozone generate three molecular orbitals, of which the π and π_{nb} are filled. The third orbital, a π^* antibonding orbital (Figure 10-37*c*), is empty. Except in diatomic molecules, ground-state configurations seldom include occupied antibonding π orbitals.

KEY CONCEPT *A delocalized π system is present whenever p orbitals on more than two adjacent atoms are in position for side-by-side overlap.*

The bond lengths in ozone provide evidence that the π electrons extend over the entire molecule rather than being isolated between two atoms. Both bond lengths are 128 pm, intermediate between the length of the double bond in O_2 (121 pm) and the length of the single O—O bond in H_2O_2 (148 pm).

A Composite Model of Bonding

Notice that our description of ozone mixes the two approaches to bonding described in this chapter. We describe the framework of the molecule using localized bonds made from hybrid and atomic orbitals. We describe the π system of ozone using molecular orbitals. This mixed approach is useful, because although the π systems of many molecules require delocalized molecular orbitals, the bonding framework of any molecule is easy to describe using hybridization and localized bonds. A combination of the two models is a straightforward and powerful way to describe many molecules. The composite approach to bonding begins with the Lewis structure and relies on the following guidelines:

1. Construct the σ bonding framework using hybrid orbitals for inner atoms and atomic orbitals for outer atoms, as described in Section 10.2. Any hybrid orbitals not used to form σ bonds contain lone pairs of electrons.

2. If the molecule contains multiple bonds, construct the π bonding system using molecular orbital theory, as described in this section and in the remaining pages of Chapter 10. Watch for resonance structures, which signal the presence of delocalized electrons.

3. Place one pair of valence electrons in any atomic orbital that is not used in hybridization or in the π system.

4. Sum the electrons allocated in Steps 1–3. The result must match the total number of valence electrons used in the Lewis structure. In addition, a complete bonding description must account for all the valence orbitals.

A review of ozone illustrates the four-step procedure. The Lewis structure shows nine pairs of valence electrons. As Figure 10-36a highlights, two pairs form the bond framework, two pairs are in different locations in the two resonance structures, and there are five localized lone pairs. The σ bonding system and the sp^2 lone pair account for 6 of the 18 valence electrons (Step 1, Figure 10-36b). The π system is constructed using MO theory, paying attention to the orbital overlap characteristics of the atomic p orbitals not used for hybridization. There are three π molecular orbitals, two of which are filled. As is often the case, the antibonding π* orbital is empty. The π bonds account for four valence electrons, leaving eight left to be placed (Step 2, Figure 10-37).

The remaining eight valence electrons are the four pairs shown in gray on the outer atoms of Figure 10-36a. Which orbitals hold these lone pairs? We answer this question by taking an inventory of the valence orbitals. For the inner atom, the sp^2 hybrid orbitals account for the 2s orbital and the two 2p orbitals that lie in the plane of the molecule, and the third p orbital, perpendicular to the molecular plane, is part of π system. All of the valence orbitals of the inner oxygen atom are accounted for. For each outer atom, one p orbital in the plane of the page contributes to the σ bonds, and the p orbital perpendicular to the plane of the page is part of the π system. The remaining valence orbitals on each outer atom are the 2s orbital and the 2p orbital that lies in the molecular plane and is perpendicular to the σ bonds. These four atomic orbitals hold the final eight electrons (Step 3). Figure 10-38 shows the orientation of these four nonbonding orbitals. Our inventory accounts for all 18 valence electrons and the full set of the 12 valence orbitals (Step 4).

Example 10-11 applies the composite model of bonding to the acetate anion.

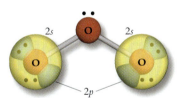

Figure 10-38
The 2s orbital and one 2p orbital on each outer atom account for four lone pairs in ozone.

	Example 10-11

Bonding in Acetate

The acetate anion ($CH_3CO_2^-$) forms when acetic acid, the acid present in vinegar, is treated with hydroxide ion:

$$CH_3CO_2H + OH^- \longrightarrow CH_3CO_2^- + H_2O$$

Use the four guidelines of the composite model of bonding to describe the bonding of this anion. Sketch the σ bonding system and the occupied π orbitals.

Strategy: As with any description of bonding, the procedure begins with the Lewis structure, from which the steric numbers and hybridizations can be determined. Then the σ and π bonds in the species can be described using localized bonds for the framework and molecular orbitals for delocalized π systems.

Solution: The acetate anion has 24 valence electrons, leading to a Lewis structure with two resonance forms:

Example 10-11

Bonding in Acetate
(*continued*)

From the Lewis structure, we see that the CH_3 carbon has a steric number of 4, so its bonding can be described by sp^3 hybrid orbitals. Three of the sp^3 hybrid orbitals form σ bonds by overlap with $1s$ orbitals on the hydrogen atoms. The fourth sp^3 hybrid of the CH_3 group forms a σ bond with the neighboring carbon atom. This second carbon has a steric number of 3, indicating trigonal planar geometry and sp^2 hybrid orbitals, one of which bonds to the CH_3 group. The two remaining sp^2 hybrids bond to $2p$ orbitals from the outer oxygen atoms. This gives a total of six σ bonds and uses 12 of the 24 valence electrons, as the sketch at right shows:

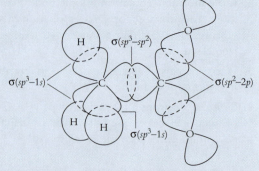

The resonance structures indicate that acetate has a delocalized π system. The sp^2 hybridized carbon atom and the two outer oxygen atoms all have $2p$ orbitals available for side-by-side overlap. From the perspective of the orbital overlap drawing shown above, these three p orbitals are perpendicular to the plane of the page. As in the ozone molecule, these three orbitals form a set of three delocalized π orbitals. The $C=O$ double bond is in different positions in the two resonance structures, which indicates two delocalized π electrons. One of the oxygen lone pairs also exchanges position in the resonance structures, accounting for two more delocalized π electrons. The π system for acetate has the same orbital assignments as the π system for ozone. At right are sketches of the occupied orbitals, π and π_{nb}:

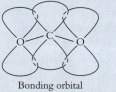 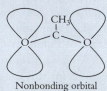

Bonding orbital   Nonbonding orbital

Does the Result Make Sense? In the acetate anion, the two oxygen atoms are equivalent, so we expect that the orbital descriptions about each O atom should be identical. You can verify that they are, and experiments further verify this: the two carbon–oxygen bond lengths in acetate are identical.

Extra Practice
Exercise 10.11

Describe the bonding in the nitrite anion, NO_2^-.

Carbon Dioxide

Carbon dioxide is essential to photosynthesis because plants use the carbon atoms of CO_2 to make glucose. Furthermore, animals and plants use fats and carbohydrates for energy production and eliminate the carbon atoms from these molecules as CO_2.

The molecule has 16 valence electrons. Its Lewis structure shows that the molecule has two double bonds, with a steric number of 2 for the carbon atom. Consistent with this, the molecule is linear. Figure 10-39 shows the two σ bonds formed by end-on overlap between sp hybrid orbitals on the carbon atom and $2p_z$ atomic orbitals of oxygen.

The $2p_x$ and $2p_y$ orbitals of the carbon atom are not used in forming the hybrid set. Each of these orbitals is oriented perfectly to overlap side by side with p orbitals of each oxygen neighbor. As in ozone, the three $2p_y$ orbitals interact to generate three delocalized π orbitals. One of these orbitals is bonding, one is nonbonding, and the third is antibonding, as shown in Figure 10-40. These MOs have the same forms as their counterparts in ozone but follow the linear shape of the bond framework. The $2p_x$ atomic orbitals combine in exactly the same way, except that the resulting three orbitals point at right angles to the p_y set. These two sets of delocalized π orbitals have exactly the same energies, but they are independent of one another because they point in different directions.

Figure 10-39
The σ bonding system of carbon dioxide. The bonds result from sp–p_z overlap.

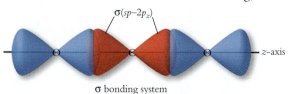

σ bonding system

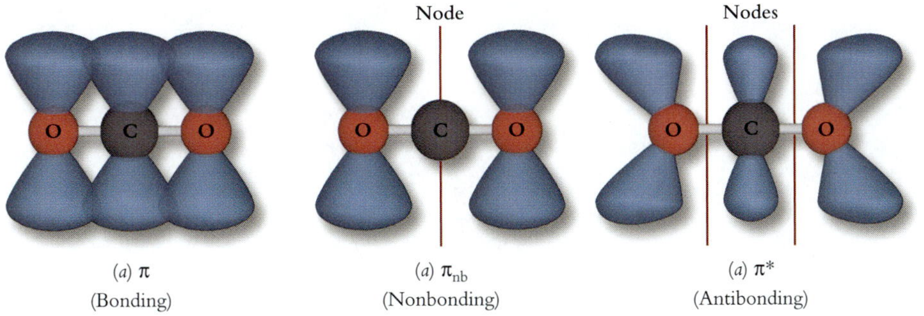

Figure 10-40
One set of delocalized π orbitals of carbon dioxide. Notice the similarity to the π orbitals of ozone.

(a) π
(Bonding)

(a) π_{nb}
(Nonbonding)

(a) π^*
(Antibonding)

Node

Nodes

The eight pairs of valence electrons in the CO_2 molecule occupy the eight most stable valence orbitals. Two pairs form the σ bonds shown in Figure 10-39. Another two pairs fill the two delocalized π bonding orbitals, and two more pairs fill the two delocalized π nonbonding MOs, leaving the π^* antibonding orbitals empty. The remaining two pairs are in the $2s$ atomic orbitals on the outer oxygen atoms. This completes the valence configuration and accounts for all the valence orbitals.

Other Second-Row Triatomics

Triatomic species can be linear, like CO_2, or bent, like O_3. The principles of orbital overlap do not depend on the identity of the atoms involved, so all second-row triatomic species with 16 valence electrons have the same bonding scheme as CO_2 and are linear. For example, dinitrogen oxide (N_2O) has 16 valence electrons, so it has an orbital configuration identical to that of CO_2. Each molecule is linear with an inner atom whose steric number is 2. As in CO_2, the bonding framework of N_2O can be represented with sp hybrid orbitals. Both molecules have two perpendicular sets of three π molecular orbitals. The resonance structures of N_2O, described in Example 9-11, reveal four pairs of delocalized π electrons, enough to fill the bonding and nonbonding MOs. Second-row triatomic ions with 16 valence electrons include azide (N_3^-), cyanate (NCO^-), and isocyanate (CNO^-). These are also linear, with sp hybridized inner atoms and two sets of delocalized π orbitals.

Ozone, which has 18 valence electrons, exemplifies bent molecules. Another example is the nitrite anion, the subject of Extra Practice Exercise 10-11. The bonding of NO_2^- can be represented using sp^2 hybrid orbitals for the inner nitrogen atom and one set of delocalized π orbitals.

$\ddot{N}{=}N{=}\ddot{O}$
Dinitrogen oxide
16 electrons

$\ddot{O}{=}c{=}\ddot{O}$
Carbon dioxide
16 electrons

Section Exercises

10.5.1 Construct the bonding description for the formate anion (HCO_2^-) and compare its π system with that of the acetate anion.

10.5.2 Explain the difference between the nitride ion, N^{3-}, and the azide ion, N_3^-. Describe in detail the distribution of electrons in both species.

10.5.3 Describe the bonding in NO_2^+ and contrast it with the bonding in NO_2^-. Your answer should include separate sketches of the σ and π bonding systems. Tell how many delocalized π electrons, if any, are present in each species.

10.6 EXTENDED π SYSTEMS

Ozone and carbon dioxide demonstrate that p orbitals can overlap side by side with more than one neighbor. This feature can lead to a system of π bonds that can extend over many atoms. Extended systems can be long chains, or they can be more compact clusters or rings. We begin by describing two examples of four-atom π systems, butadiene and the carbonate anion.

Figure 10-41
Lewis structure and orbital overlap view of the bonding framework of butadiene. The framework contains three C—C bonds (sp^2–sp^2) and six C—H bonds (sp^2–1s).

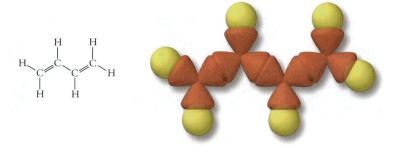

Butadiene

Butadiene, used in the chemical industry as a precursor of synthetic rubber, is a hydrocarbon with the formula C_4H_6. (See Box 13-1 to learn more about the rubber industry.) The Lewis structure of butadiene contains two double bonds on sequential pairs of carbon atoms. The chemistry of butadiene, including its ability to form rubber, can be traced to the delocalized electrons in the π system of the molecule.

The four carbon atoms of butadiene have steric number 3 and can be described using sp^2 hybridization. There are no lone pairs in the molecule, so all of the hybrid orbitals are used to form σ bonds. There are three C—C bonds formed by sp^2–sp^2 overlap and six C—H bonds formed by sp^2–1s overlap. The σ bonding system, which Figure 10-41 shows, has nine bonds that account for 18 of the 22 valence electrons, leaving 4 for the π system.

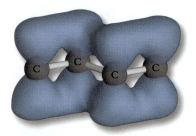

Figure 10-42
Each of the carbon atoms in butadiene has a p orbital oriented to overlap with those on neighboring carbon atoms, generating a set of four π MOs, of which one is shown.

All the atoms of butadiene lie in a plane defined by the sp^2 hybrid orbitals. Each carbon atom has one remaining p orbital that points perpendicular to the plane, in perfect position for side-by-side overlap. Figure 10-42 shows that all four p orbitals interact to form four delocalized π molecular orbitals; two are bonding MOs and two are antibonding. The four remaining valence electrons fill the π orbitals, leaving the two π^* orbitals empty.

Whenever the Lewis structure of a molecule shows double or triple bonds alternating with single bonds, the bonding description includes delocalized π orbitals. Molecules that have alternating single and double bonds, like butadiene, are said to have **conjugated π systems.** Each double bond in a conjugated system adds two electrons to the corresponding set of delocalized MOs. Example 10-12 looks at another molecule with conjugated double bonds.

Example 10-12	Describe the bonding of methyl methacrylate. This compound is an important industrial chemical, used mainly to make plastics such as poly(methyl methacrylate) (PMMA).
Conjugated Double Bonds	

Strategy: Follow the four-step procedure for the composite model of bonding. Use localized bonds and hybrid orbitals to describe the bonding framework and the inner atom lone pairs. Next, analyze the π system, paying particular attention to resonance structures or conjugated double bonds. Finally, make sure the bonding inventory accounts for all the valence electrons and all the valence orbitals.

PMMA is a common polymer used to make plastics. See Chapter 13 for more information.

Solution: The Lewis structure shows that methyl methacrylate has the formula $C_5H_8O_2$, with 40 valence electrons. You should be able to verify that the two CH_3 groups have sp^3-hybridized carbons, the inner oxygen atom is sp^3 hybridized, the outer oxygen atom uses $2p$ atomic orbitals, and the three double-bonded carbons are sp^2 hybridized. These assignments lead to the following inventory of σ bonds and inner-atom lone pairs:

6 sp^3–1s C—H σ bonds	2 sp^2–1s C—H σ bonds	2 sp^2–sp^2 C—C σ bonds
1 sp^3–sp^2 C—C σ bond	1 sp^3–sp^3 O—C σ bond	1 sp^2–sp^3 C—O σ bond
1 sp^3–$2p$ C—O σ bond	2 sp^3 O lone pairs	

Summing these contributions, the 14 σ bonds and 2 sp^3 lone pairs account for 32 of the 40 valence electrons.

As the Lewis structure shows, one single bond separates the two double bonds, which signals conjugation and delocalized π orbitals. The π system of methyl methacrylate forms from the leftover p orbitals on the three sp^2 hybridized carbon atoms, plus one p orbital from the outer oxygen atom. These four p orbitals are located on adjacent atoms, positioned for side-by-side overlap. They form a set of four delocalized π molecular orbitals, two bonding and two antibonding. Four electrons occupy the π bonding orbitals. The π^* antibonding MOs are empty. Finally, the two lone pairs on the outer oxygen atom complete the valence electron count, filling the only remaining valence orbitals, the $2s$ orbital and the final $2p$ orbital on the outer oxygen atom.

A ball-and-stick model of methyl methacrylate shows the planar part of the molecule covered by the delocalized π orbitals. The atoms that contribute p orbitals to the delocalized π system are marked with stars, and the other atoms that lie in the same plane are marked with dots.

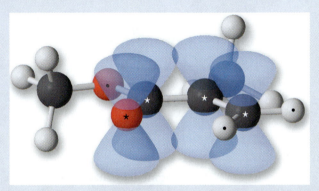

Does the Result Make Sense? Notice that the π system in methyl methacrylate is nearly the same as that in butadiene. The only difference is the replacement of one C atom by an O atom. Both systems have four overlapping p orbitals, giving rise to two bonding MOs and two antibonding MOs, so it is reasonable that the π systems are quite similar.

Describe the bonding in acryloyl chloride, C_3H_3OCl.

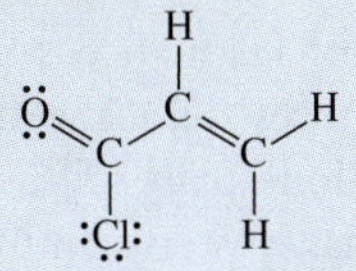

Example 10-12

Conjugated Double Bonds
(*continued*)

**Extra Practice
Exercise 10.12**

The effect of conjugation on bond stability is revealed by comparing 1,3-pentadiene and 1,4-pentadiene. Figure 10-43 shows that both have eight C—H bonds, two C—C bonds, and two C=C bonds. The only significant difference between these molecules is that the π system of 1,3-pentadiene is conjugated, generating delocalized π orbitals. In contrast, 1,4-pentadiene has an sp^3-hybridized CH_2 group acting as a "spacer" to separate two localized π bonds.

Does the delocalized π system of 1,3-pentadiene stabilize the molecule relative to 1,4-pentadiene? Each molecule reacts with two molecules of H_2 to give a common product, pentane. Thus, a comparison of their energies of hydrogenation allows us to determine the relative stabilities of two delocalized π bonds compared with two isolated π bonds. The hydrogenation of 1,4-pentadiene releases 252 kJ/mol, but the hydrogenation of 1,3-pentadiene releases only 224 kJ/mol. As Figure 10-43 demonstrates, these energies of hydrogenation show that 1,3-pentadiene is 28 kJ/mol *more stable* than is 1,4-pentadiene. Because the bonding in these two molecules is otherwise the same, we can conclude that the extra stability comes from the delocalized π system.

Figure 10-43
Energy level diagram for the hydrogenation of 1,3- and 1,4-pentadiene. The difference in these energies, 28 kJ/mol, represents the additional stability that delocalization provides to 1,3-pentadiene.

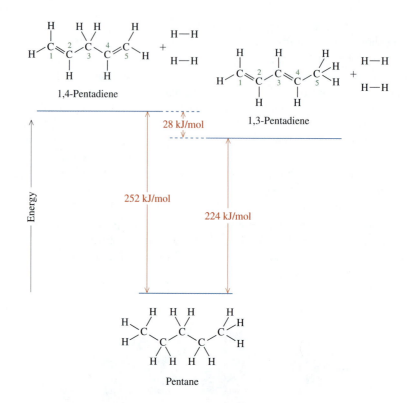

Carbonate Ion

In addition to four atoms in a row, a four-atom delocalized framework can have a central atom surrounded by three other atoms with which it interacts. An example is the carbonate anion, whose Lewis structure displays resonance:

$$\text{Carbonate anion (CO}_3{}^{2-})$$

The existence of resonance structures indicates that a complete description of the carbonate ion requires delocalized π orbitals. The changing positions of the double bond and one of the lone pairs tell us there are four delocalized electrons in the π system. The carbon atom has a steric number of 3, meaning that the σ bonding framework for $CO_3{}^{2-}$ can be described using sp^2 hybrid orbitals on the carbon atom. These hybrid orbitals overlap end-on with $2p$ orbitals of the three oxygen atoms, as shown in Figure 10-44a.

The carbon atom has one unused $2p$ orbital perpendicular to the molecular plane. As Figure 10-44b shows, all three oxygen atoms also have perpendicular p orbitals, oriented

Figure 10-44
(a) The framework of the carbonate anion consists of three sp^2–$2p$ σ bonds. (b) The π system forms from four overlapping p orbitals on adjacent atoms. (c) Viewed from above, the bonding π MO resembles a propeller. The other three π MOs are not shown.

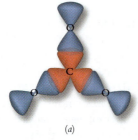

(a)

σ bonding framework in $CO_3{}^{2-}$

(b) Side view

(c) Top view

Delocalized π bond

properly for side-by-side overlap. These four p orbitals give four π molecular orbitals delocalized over the entire anion. Figure 10-44c shows that the bonding π molecular orbital resembles a propeller when viewed from above. The remaining π orbitals, which consist of two nonbonding MOs and one π^* antibonding MO, are not shown in the figure. Depictions of molecular orbitals become complicated as the number of contributing atomic orbitals increases. These details are beyond the scope of an introductory course. In agreement with the Lewis structure, a bonding orbital and one nonbonding orbital are filled in the carbonate π system.

Larger Delocalized π Systems

The dienes are the first members of a set of chain compounds with conjugated double bonds. These chains can be quite long, as shown by the delocalized π systems of some biological molecules. Figure 10-45 shows line structures of three examples: vitamin A, retinal, and β-carotene. β-Carotene is an orange compound that gives carrots their color. Our bodies convert carotene into vitamin A. This compound, in turn, is used to make retinal, a molecule that plays an essential role in the chemistry of vision. Notice that each of these molecules contains a sequence of conjugated double bonds. Vitamin A has a row of ten carbon atoms with sp^2 hybridization. The leftover p orbitals on these ten atoms generate ten π-type molecular orbitals, five bonding and five antibonding. The line structure shows five double bonds, consistent with the five bonding π orbitals being filled and the five π^* orbitals being empty. β-Carotene has 22 atoms that participate in the delocalized π system. Retinal has 11 sequential sp^2 carbon atoms and an outer oxygen atom, resulting in six conjugated double bonds.

Plastoquinone, whose line structure appears in the margin, has ten double bonds. Unlike the π bonds in the tail of retinal, however, the π bonds in the long tail of plastoquinone are not delocalized because sp^3-hybridized carbon atoms separate them. The delocalized π system of plastoquinone is its planar ring of six carbon atoms, with two of the carbon atoms double-bonded to outer oxygen atoms.

The orange color of β-carotene and the pink of xanthin (See Box 10-1) demonstrate that compounds with extended π systems often are colored. A molecule is colored when it absorbs photons in the visible region of the electromagnetic spectrum. Recall from Chapter 7 that substances absorb only photons whose energies match differences among energy levels. Visible photons have wavelengths between 700 and 400 nm and energies between 180 and 330 kJ/mol. Thus, in a colored substance, the energy needed to promote an electron from the highest filled orbital to the lowest unoccupied orbital is between 180 and 330 kJ/mol.

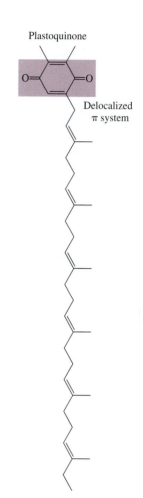

Plastoquinone

Delocalized π system

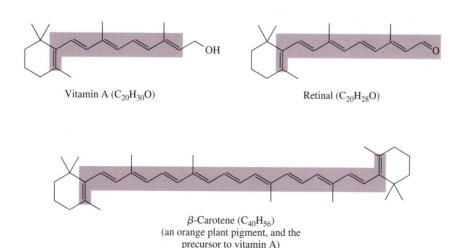

Vitamin A (C$_{20}$H$_{30}$O)

Retinal (C$_{20}$H$_{28}$O)

β-Carotene (C$_{40}$H$_{56}$)
(an orange plant pigment, and the
precursor to vitamin A)

Figure 10-45
The line structures of three biologically important molecules that contain extended conjugated π systems. The conjugated systems are highlighted with color screens.

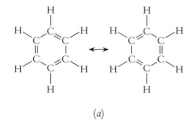

(a)

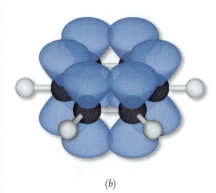

(b)

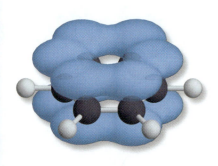

(c)

Figure 10-46
(a) The Lewis structure of benzene. (b) The p orbitals that interact to form delocalized MOs. (c) The most stable bonding π molecular orbital of benzene has two rings of electron density above and below the molecular plane.

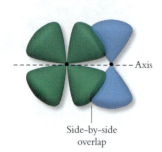

Side-by-side overlap

Figure 10-47
A π bond can form through side-by-side overlap between a d orbital and a p orbital.

Molecules with isolated π bonds, such as ethylene or 1,4-pentadiene, have energy gaps between the filled π and empty π* that are greater than 330 kJ/mol. These molecules absorb UV radiation but are colorless. However, it is a general feature of π orbital energies that the larger the number of atoms over which the orbital extends, the smaller the energy difference between orbitals. The long conjugated chains of β-carotene and xanthin reduce their energy gaps into the visible region and make these compounds colored.

In addition to long conjugated chains, planar rings with delocalized π bonding are common in biological systems. Look at the structure of chlorophyll in the introduction to this chapter. Chlorophyll contains four five-atom rings (four C and one N) connected in a larger ring that forms a plane. The alternating double and single bonds in the ring portion of chlorophyll indicates a delocalized π system.

Benzene (C_6H_6) is the simplest example of a stable ring molecule with a delocalized π system. This compound has a ring of six carbon atoms, and the Lewis structure in Figure 10-46a shows alternating double bonds and single bonds. The figure also shows that benzene has two resonance structures. The molecule has six adjacent sp^2-hybridized atoms, each with a p orbital perpendicular to the plane of the ring (Figure 10-46b). These six p orbitals are in perfect position for side-by-side overlap. Together they form a set of six molecular orbitals: three bonding π orbitals and three antibonding π* orbitals. There are six π electrons in benzene, just enough to fill the bonding molecular orbitals. The antibonding orbitals remain empty. The most stable π orbital of benzene consists of circular orbital rings above and below the plane of the carbon atoms (Figure 10-46c). Two other bonding π orbitals have more complicated shapes and are not shown. These delocalized molecular orbitals are highly resistant to chemical attack, making the benzene ring remarkably stable.

Benzene is highly stable to chemical attack. β-Carotene absorbs visible light and is orange. Both properties are direct consequences of delocalized π systems. Delocalized π systems also affect the oxidation–reduction characteristics of organic substances. Our chapter introduction noted that plastoquinone and cytochrome play important redox roles in photosynthesis. As another example, β-carotene, with its delocalized π system, is thought to act as an antioxidant, protecting the body from cancer-causing oxidation processes. Oxidation and reduction reactions involve the transfer of electrons from occupied orbitals in one substance to empty orbitals in another substance. Whether or not a redox reaction takes place depends on orbital energies, so the stabilization of π systems by delocalization has a direct effect on redox behavior.

π Bonding Beyond the Second Row

Several common polyatomic oxoanions, including sulfate, perchlorate, and phosphate, have inner atoms from the third row of the periodic table. In these anions, valence d orbitals are available to participate in π bonding. Figure 10-47 shows how a π orbital can form through side-by-side overlap of a d orbital on one atom with a 2p orbital on another atom. As with other π bonds, electron density is concentrated above and below the bond axis.

Several elements in row 3 of the periodic table form π bonds to oxygen through side-by-side overlap of 3d and 2p orbitals. An example is the sulfate anion, whose Lewis structure appears in Figure 10-48. The steric number of the sulfur atom in SO_4^{2-} is 4, indicating that the anion has tetrahedral geometry. We can describe the bonding framework using sp^3 hybrid orbitals. In addition, however, the six resonance structures show that there are delocalized π bonds that involve all five atoms. These bonds are the result of delocalized π orbitals created from a 2p orbital on each oxygen atom and valence d orbitals on the sulfur atom. The details of these interactions, which also occur for other third-row oxoanions such as phosphate, are beyond the scope of this text. You need only recognize that they include π orbitals formed through the side-by-side overlap of d orbitals and p orbitals.

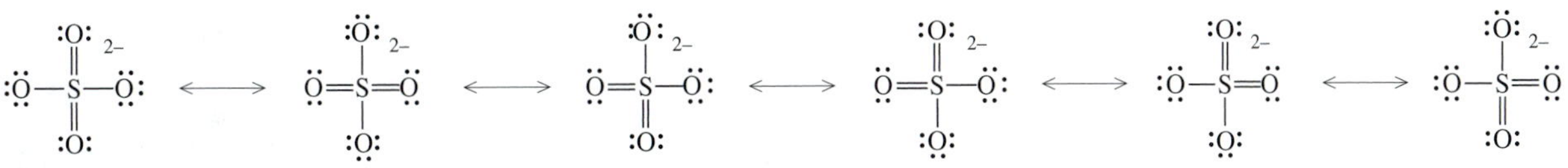

Figure 10-48
The six resonance structures of the sulfate ion indicate that delocalized orbitals span the entire anion.

Section Exercises

■ **10.6.1** Describe the π bonding in the hydrogen carbonate ion, HCO_3^-.

■ **10.6.2** Shown at right is the line structure of azulene, a deep purple solid, which forms a blue solution (see photo in margin). How many delocalized orbitals are in an azulene molecule? How many of these orbitals are occupied?

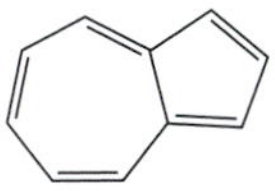

Azulene
$C_{10}H_8$

■ **10.6.3** The perchlorate ion (ClO_4^-) has the same number of valence electrons as the sulfate anion. Describe the bonding in ClO_4^-.

10.7 BAND THEORY OF SOLIDS

Copper, iron, and aluminum are three common metals in modern society. Copper wires carry the electricity that powers most appliances, including the lamp by which you may be reading. The chair in which you are sitting may have an iron frame, and you may be sipping a soft drink from an aluminum can. The properties that allow metals to be used for such a wide range of products can be traced to the principles of bonding and electronic structure.

Iron and other metals have tremendous mechanical strength, which suggests that the bonds between their atoms must be strong. At the same time, most metals are malleable, which means they can be shaped into thin sheets to make objects such as aluminum cans. Metals are also ductile, which means they can be drawn into wires. The properties of malleability and ductility suggest that atoms in metals can be moved about without weakening the bonding. Finally, metals conduct electricity, which shows that some of the electrons in a metal are free to move throughout the solid.

All these properties of metals are consistent with a bonding description that places the valence electrons in delocalized orbitals. This section describes the **band theory of solids,** an extension of the delocalized orbital ideas that are the theme of this chapter. Band theory accounts for the properties of metals, and it also explains the properties of metalloids such as silicon.

A solution of azulene has a blue color.

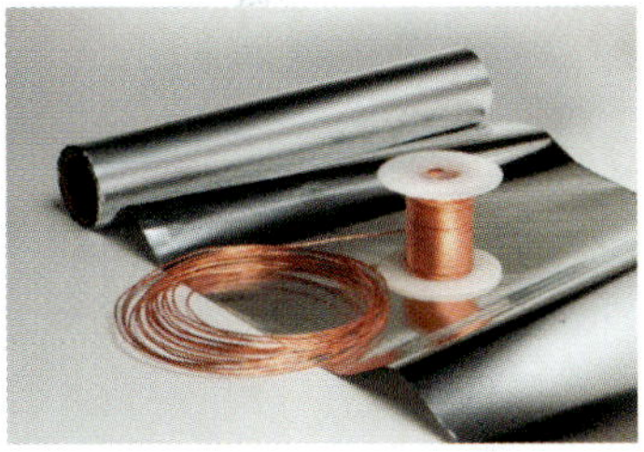

Metals are malleable and ductile.

Delocalized Orbitals in Lithium Metal

We use lithium, the lightest metal, to demonstrate the principles of band theory. Solid lithium contains atoms held together in a three-dimensional crystal lattice. Bonding interactions among these atoms can be described by orbital overlap. To see how this occurs, consider building an array of lithium atoms one at a time.

Figure 10-49 indicates the orbital energies resulting from such sequential construction of ever-larger atom clusters. We describe the bonding using overlapping valence orbitals. In Li_2, the two valence $2s$ orbitals interact to form a σ bonding orbital and a σ^* antibonding orbital. The bonding orbital is filled, and the antibonding orbital is empty. The single bond in Li_2 is relatively weak, 105 kJ/mol, because compact $2s$ orbitals do not overlap strongly.

Clusters of lithium atoms such as Li_4, Li_6, and Li_{18} can be generated in the gas phase by bombarding a piece of lithium metal with gaseous metal atoms in a chamber at very low pressure. The Li_4 molecule is bonded together by four $2s$ orbitals that interact to give two filled delocalized bonding orbitals and two empty delocalized antibonding orbitals. Likewise, Li_6 has 6 delocalized orbitals; Li_{18} has 18 delocalized orbitals, and so on. As with all molecular orbitals, these interactions generate equal numbers of bonding and antibond-

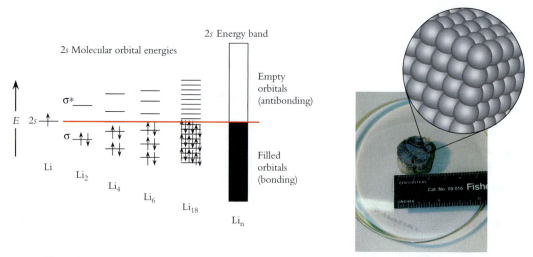

Figure 10-49
The energy band describing the bonding in lithium metal can be constructed by adding atoms one at a time, allowing all the 2s valence orbitals to interact to form delocalized orbitals.

ing orbitals. In 1 cm^3 of Li metal there are roughly 4.6×10^{22} atoms whose 2s atomic orbitals generate approximately 4.6×10^{22} delocalized orbitals. This is a huge number of bonding orbitals and antibonding orbitals; yet they all obey the principles developed in this chapter. Most important, each orbital extends over the entire lattice of atoms.

Figure 10-49 shows that the energy spacing between orbitals decreases as the number of delocalized orbitals increases. For the huge number of atoms in a piece of metal, the orbitals are spaced so closely that they behave as if they were merged into an energy band. This bonding model is called band theory because energy bands are one of its main features.

To keep our picture as simple as possible, we include only the 2s orbitals in Figure 10-49. For most metals, however, a comprehensive picture of the bonding includes interactions among all valence s, p, d, and even f orbitals. Each group of orbitals interacts to generate an energy band. Interactions among bands are complex, and often the bands overlap. Despite this complexity, however, the fundamental features illustrated in Figure 10-49 remain valid.

Electrical Conductivity

The most important feature of metallic bonding is the energy bands that arise because of the close spacing between orbital energies. The countless orbitals within the band are separated by infinitesimally small energy gaps. The valence electrons of the metal atoms occupy these orbitals according to the aufbau and Pauli principles. In a metal, the occupied orbitals of highest energy are so close in energy to the unoccupied orbitals of lowest energy that it takes little energy to transfer an electron from an occupied to an unoccupied orbital.

When an electrical potential is applied to a metal, the negative pole repels electrons and the positive pole attracts them. In energy terms, the occupied orbitals near the negative pole are pushed higher in energy than unoccupied orbitals near the positive pole. This "tilts" the energy levels, as shown in Figure 10-50, and electrons can move downhill out of filled orbitals into empty orbitals at lower energy. As a result, electrons flow through the metal from the negative end to the positive end, generating an electrical current.

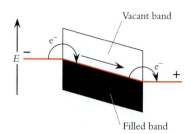

Figure 10-50
An electrical potential shifts the energy levels in a metal. The arrows show the direction of electron flow.

Insulators and Conductors: Carbon vs. Lead

Metallic lead is dark in color and is an electrical conductor. Diamond, the most valuable form of carbon, is transparent and is an electrical insulator. These properties are very different; yet both lead and carbon are in Group 14 of the periodic table and have the same valence configuration, s^2p^2. Why, then, are diamonds transparent insulators, whereas lead is a dark-colored conductor?

Both diamond and lead can be described using delocalized orbitals and energy bands, but, as Figure 10-51 shows, the energy distributions of the bands are quite different. Lead is metallic because the energy separation between its filled and vacant orbitals is infinitesimally small. One continuous band of orbitals results in highly mobile electrons and electrical conductivity. The valence orbitals of carbon, on the other hand, form two distinct bands. The filled bonding orbitals are well separated in energy from the empty, antibonding orbitals. The energy difference between these two bands is called the **band gap (E_g).** The band gap for diamond is 580 kJ/mol.

Lead and diamond have distinctly different band structures for two reasons. First, the valence orbitals of carbon are smaller than those of lead, resulting in stronger overlap for diamond. This results in a large stabilization energy and a large band gap. Second, lead has valence d and f orbitals that are involved in band formation. These d and f bands overlap the s and p bands, contributing to the continuous distribution of orbital energies that gives lead its metallic properties.

Carbon in the form of diamond is an electrical insulator because of its huge band gap. In fact, its band gap of 580 kJ/mol substantially exceeds the C—C bond energy of 345 kJ/mol. In other words, it requires more energy to promote an electron from band to band in diamond than to break a covalent bond. Lead, in contrast, is a metallic conductor because it has $E_g = 0$.

The differences in their band structures also explain why diamond is transparent and lead is dark. Remember that substances can absorb only photons whose energies match differences between energy levels, and a colored substance absorbs visible photons, whose energies are between 180 and 330 kJ/mol. For diamond, the energy difference between the top of the filled band and the bottom of the empty band is 580 kJ/mol. Diamond cannot absorb visible light because there are no energy levels between 180 and 330 kJ/mol above the filled band. Visible light passes straight through a diamond. Lead, on the other hand, has many vacant orbitals into which visible photons can promote electrons. Thus, lead can absorb all visible wavelengths, making it gray.

Figure 10-51
Lead and carbon have different properties because the energy differences between their filled bonding orbitals and their empty antibonding orbitals are different.

Metalloids

Orbital energies are determined largely by the amount of spatial overlap. Because atomic orbitals become increasingly diffuse (spread out in space) as their principal quantum number (n) increases, the spatial overlap of the valence orbitals decreases as the valence shell n value increases (see Figure 10-23). Less overlap means less difference between the energy of the bonding and antibonding levels, so the band gap shrinks from carbon to lead. Carbon ($n = 2$) has a large band gap and is a nonmetal, but tin ($n = 5$) and lead ($n = 6$) have $E_g = 0$ and are metals. Silicon and germanium ($n = 3$ and $n = 4$, respectively) have intermediate band gap values, which cause them to behave as **metalloids.**

Like carbon, silicon and germanium have low-energy bands of bonding orbitals that are completely filled with electrons and higher-energy bands that are empty. For silicon, $E_g = 105$ kJ/mol, whereas $E_g = 64$ kJ/mol for germanium. Electrons cannot make the transition between these bands unless sufficient energy is supplied. Thus, silicon and germanium are nonconductors in the absence of an energy source. However, electrons can move between the bands at higher temperatures or in the presence of photons with energy that matches E_g. When electrons are transferred into the high-energy band, they can move freely among the many vacant orbitals of the band. In addition, electron transfer creates vacancies (holes) among the bonding orbitals in the low-energy band. The holes and high-energy electrons allow electricity to flow. Thus, silicon and germanium become conductors in the presence of an appropriate energy source. They are called **semiconductors** to indicate their ability to conduct electricity under special conditions. Figure 10-52 uses band gap diagrams to summarize the behavior of a semiconductor.

Transfer of electrons from one band to another requires energy equal to the band gap. This energy can be provided by heating (exploited in thermistors) or by absorption of light (used in photoconductors). Example 10-13 deals with photoconductors.

Figure 10-52
When a semiconductor absorbs heat or light, electrons are excited from the filled band to the empty band, and the semiconductor conducts electricity.

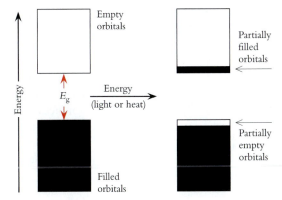

Example 10-13	
Properties of Photoconductors	"Electric eye" door openers use photoconductors that respond to infrared light with a wavelength (λ) of 1.5 μm. Which is suitable for photoconductors operating at this wavelength, germanium (E_g = 64 kJ/mol) or silicon (E_g = 105 kJ/mol)?

Strategy: To conduct electricity, a semiconductor must be provided with energy that is *at least* equal to its band gap. In a photoconductor, this energy must come from photons. Thus, we need to compare the band gaps with the energy of the infrared photons.

Solution: First, calculate the energy of 1.5-μm light, recalling from Chapter 7 that photon energy is related to wavelength:

$$E_{photon} = h\nu = \frac{hc}{\lambda}$$

$$E_{photon} = \frac{(6.626 \times 10^{-34}\,\text{J s})(2.998 \times 10^8\,\text{m/s})}{(1.5\,\mu\text{m})(10^{-6}\,\mu\text{m/nm})} = 1.32 \times 10^{-19}\,\text{J}$$

Multiply by Avodagro's number to convert the energy of one photon into the energy of one mole of photons in kilojoules per mole:

$$(1.32 \times 10^{-19}\,\text{J/photon})(6.022 \times 10^{23}\,\text{photons/mol})(10^{-3}\,\text{kJ/J}) = 79\,\text{kJ/mol}$$

This energy is insufficient to overcome the band gap of silicon, so silicon could not be used for an infrared photoconductor. Germanium, however, with its smaller band gap (E_g = 64 kJ/mol), becomes an electrical conductor when illuminated by infrared light with λ = 1.5 μm.

Does the Result Make Sense? Silicon is closer to carbon, and germanium is closer to lead, in the periodic table. We expect the band gap in silicon to be larger than that in germanium, so it makes sense that lower-energy radiation can generate electrical conductivity in germanium than in silicon.

Extra Practice Exercise 10.13	The band gap of silicon is 105 kJ/mol. What is the minimum wavelength of light that can promote an electron from the valence band to the conduction band? What region of the electromagnetic spectrum matches this wavelength?

Doped Semiconductors

The relatively large band gaps of silicon and germanium limit their usefulness in electrical devices. Fortunately, adding tiny amounts of other elements that have different numbers of valence electrons alters the conductive properties of these solid elements. When a specific impurity is added deliberately to a pure substance, the resulting material is said to be *doped*. A **doped semiconductor** has almost the same band structure as the pure material, but it has different electron populations in its bands.

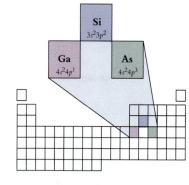

In pure silicon, all orbitals of the low-energy band are filled, and all orbitals of the high-energy band are empty. When pure silicon is doped with arsenic atoms, As atoms replace some Si atoms in the crystal lattice. Each arsenic atom contributes five valence electrons instead of the four contributed by a silicon atom. The fifth valence electron cannot enter the low-energy band of the solid because that band is already filled. The fifth valence electron from each arsenic atom must therefore occupy the higher-lying energy band. Silicon or germanium doped with atoms from Group 15 is called an **n-type semiconductor** because extra *negative* charges exist in its high-energy band. These few electrons in a partially occupied band make the doped material an electrical conductor.

Silicon can also be doped with gallium, a Group 13 element. Gallium has three valence electrons, so each Ga atom in the Si lattice has one less electron than the atom that the Ga atom replaces. In gallium-doped silicon, there are not quite enough electrons

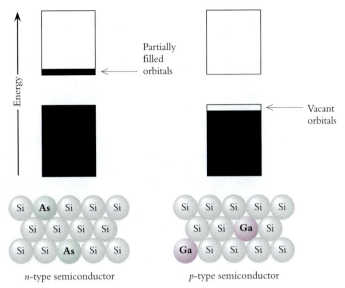

Figure 10-53
Schematic views of an *n*-type and a *p*-type semiconductor.

to fill all the bonding orbitals. Silicon and germanium doped with atoms from Group 13 are known as ***p*-type semiconductors** because their low-energy bands have *positive* vacancies. Electrons move through the crystal by flowing from filled orbitals into these vacant ones.

Semiconductors such as silicon are extremely sensitive to impurities. Replacing just 0.00001% of the Si atoms with a dopant can cause as much as a 100,000-fold increase in electrical conductivity.

Figure 10-53 shows band-gap diagrams of *n*-type and *p*-type semiconductors. Electrical current flows in a doped semiconductor in the same way as current flows in a metal (see Figure 10-50). Only a small energy difference exists between the top of the filled band and the next available orbital, so the slightest applied potential tilts the bands enough to allow electrons to move and current to flow.

Not all semiconductors are made from silicon or germanium. Compounds made from equimolar amounts of Group 13 and Group 15 elements also are semiconductors. Gallium arsenide (GaAs) is a typical example. One Ga atom and one As atom have a total of eight valence electrons, so gallium arsenide is isoelectronic with germanium. Gallium arsenide can be doped with zinc atoms to make a *p*-type semiconductor or with tellurium atoms to make an *n*-type semiconductor. Other semiconductors contain equimolar compositions of an element such as zinc that has two valence electrons and a Group 16 element (S, Se, or Te) with six valence electrons. Zinc sulfide is an example. Our Chemistry and Technology Box on the next page describes the light-emitting characteristics of this type of semiconductor.

Section Exercises

■ **10.7.1** CdS, a semiconductor studied for use in solar cells, absorbs blue light at 470 nm. Calculate its band gap in kJ/mol.

■ **10.7.2** Draw a band-gap diagram that illustrates a semiconductor made of gallium arsenide doped with zinc atoms.

■ **10.7.3** You have been asked to develop a *p*-type semiconductor using Al, Si, and/or P. Which two elements would you choose, and what role would each element play in the semiconductor?

Box 10-2 — **Chemistry and Technology: Light-Emitting Diodes**

One application of modern solid-state electronic devices is semiconductor materials that convert electrical energy into light. These *light-emitting diodes* (LEDs) are used for visual displays and solid-state lasers. Many indicator lights are LEDs, and diode lasers read compact discs in a CD player. The field of diode lasers is expanding particularly rapidly, driven by such applications as fiber optic telephone transmission.

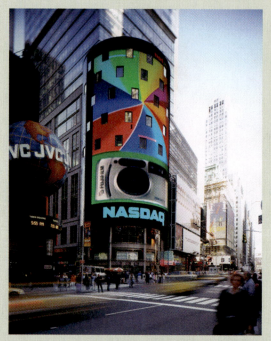

Smartvision Display, Courtesy of Saco Technologies, Inc.

The heart of a light-emitting diode is a junction between a *p*-type semiconductor and an *n*-type semiconductor. The different semiconductor types have different electron populations in their bands. The lower-energy band of a *p* semiconductor is deficient in electrons, while the upper-energy band of an *n* semiconductor has a small population of electrons. The band structure in the junction region is shown schematically in the figure below.

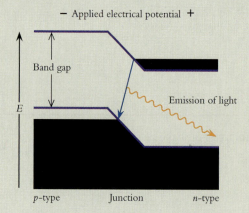

When an electrical potential is applied to an LED device, the *p* region is made positive relative to the *n* region. The positive potential attracts electrons, which can jump from the higher-energy band of the *n* region to the lower-energy band of the *p* region. In doing so, these electrons emit their excess energy as photons of light.

The color of the light depends on the energy loss for the electrons and increases with the size of the band gap of the semiconductor. Thus, red LEDs use semiconductors with relatively small band gaps, whereas green LEDs have larger band gaps.

Semiconductors containing arsenic are particularly versatile. A GaAs semiconductor can be made a *p* type by doping with a small excess of Ga or an *n* type by doping with a small excess of As. The band gap of these semiconductors can be varied by replacing some Ga with Al or some As with P. Since GaAs forms solid solutions with both AlAs and GaP, the composition can be varied continuously, from pure AlAs to pure GaAs and from pure GaAs to pure GaP. These materials have nonstoichiometric chemical formulas, $Al_xGa_{1-x}As$ and GaP_xAs_{1-x}.

The band gap of the LED varies with composition for both these solid solutions, as shown in the figure below. The cause of the variation is different for the two substances. Semiconductor band gaps increase when orbital overlap decreases. A decrease in orbital overlap can arise from increased spacing between atoms or increased ionic character of the bonds.

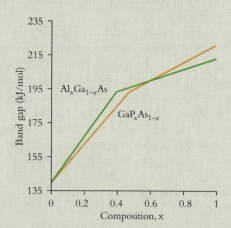

Phosphorus and arsenic have nearly identical electronegativities, so in GaP_xAs_{1-x}, the dominant effect is the smaller atomic radius of P relative to As. Substituting P atoms for As atoms shrinks the dimensions of the semiconductor lattice. This leads to greater overlap of the valence orbitals, increased stability of the bonding orbitals (valence band), and an increased band gap.

Substituting Al atoms for Ga atoms does not change the lattice dimensions, which are determined by the size of As, the largest atom of this set. Here, the lower electronegativity of Al relative to Ga leads to more ionic character and a smaller band gap.

CHAPTER 10 VISUAL SUMMARY

Important equations

Essential terms

Key concepts

SKILLS TO MASTER

1. Drawing sketches that show orbital overlap
2. Using hybrid orbitals to describe bonding
3. Drawing molecular orbitals
4. Drawing and interpreting MO diagrams
5. Recognizing delocalized orbitals
6. Using the composite model of bonding
7. Describing energy bands in solids

10.1 LOCALIZED BONDS

Bonding orbitals
Orbital overlap
Localized bonds
Delocalized bonds

Attraction
Repulsion

Bonding orbitals result from combining atomic orbitals.

(1)

10.2 HYBRIDIZATION OF ATOMIC ORBITALS

sp hybrid orbitals | sp^2 hybrid orbitals | sp^3 hybrid orbitals

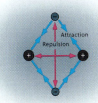

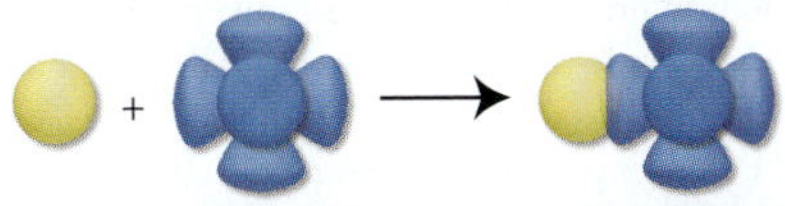

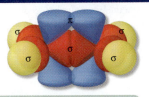

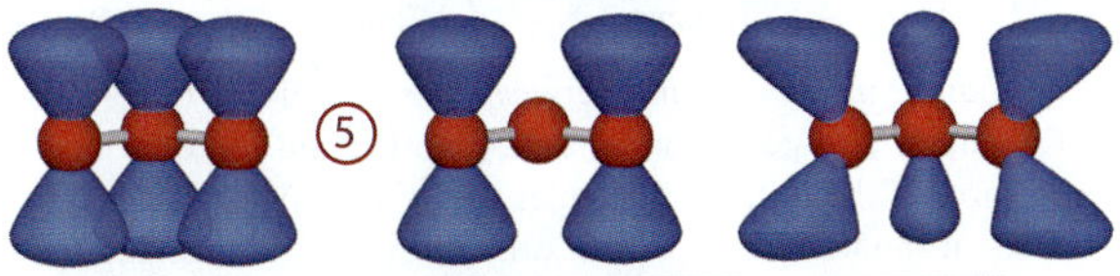

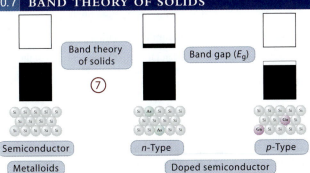

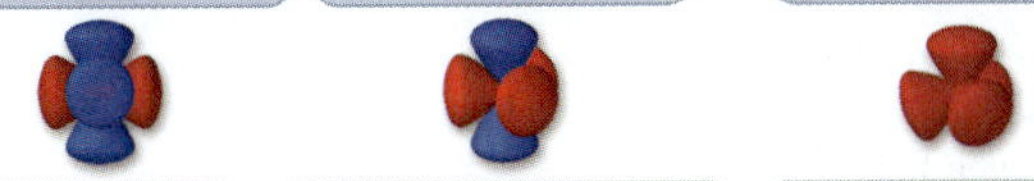

SN = 2, linear | *SN = 3, trigonal planar* | *SN = 4, tetrahedral*

sp^3d hybrid orbitals | sp^3d^2 hybrid orbitals

(2)

SN = 5, trigonal bipyramidal | *SN = 6, octahedral*

10.3 MULTIPLE BONDS

Sigma bond (σ)
Pi bond (π)

π
σ
σ
σ
σ
σ

σ bond: high electron density on the bond axis
π bond: high electron density off the bond axis

10.4 MOLECULAR ORBITAL THEORY

# of MOs = # of AOs | MO diagram | Molecular orbitals (MO)

Orbital mixing

$Z_{average} > 7$, $\sigma_p < \pi$
$Z_{average} \leq 7$, $\pi < \sigma_p$

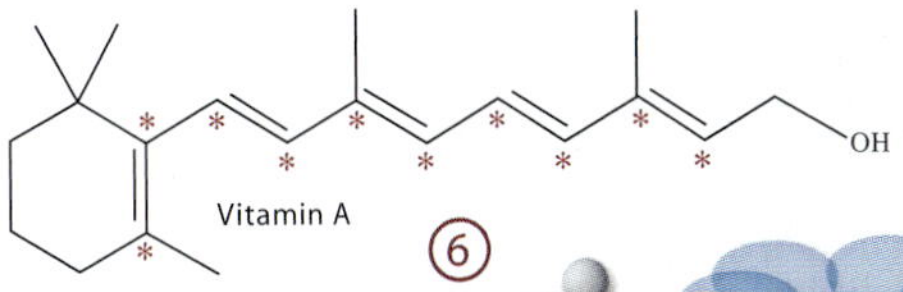

Antibonding MO

Bonding MO

$\sigma_p{}^*$
(4)
π^*
σ_p
π
(3)
$\sigma_s{}^*$
σ_s

Bond order (BO)

:N≡N:

$BO = \frac{1}{2}$(# e⁻ in bonding MOs − # e⁻ in antibonding MOs)

10.5 THREE-CENTER ORBITALS

Delocalized π orbitals

(5)

π orbital | Nonbonding MO (π_{nb}) | π* orbital

Delocalized π orbitals result from side-by-side overlap of atomic p orbitals on more than two adjacent atoms.

10.6 EXTENDED π SYSTEMS

OH
Vitamin A

(6)

Conjugated π systems

10.7 BAND THEORY OF SOLIDS

Band theory of solids

Band gap (E_g)

(7)

Si
As
Ga

Semiconductor | *n*-Type | *p*-Type

Metalloids

Doped semiconductor

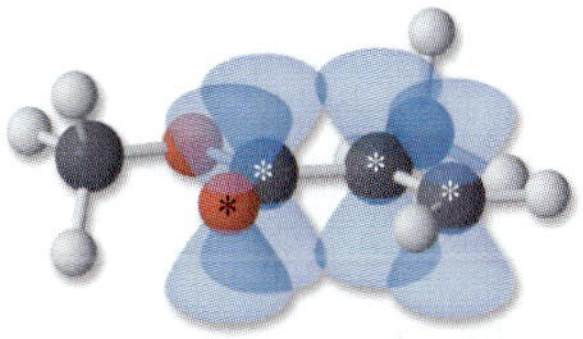

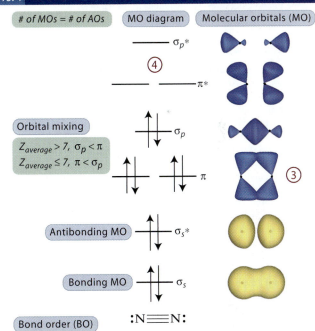

Learning Exercises

10.1 Design a detailed flowchart that shows how to describe the bonding of a second-row diatomic molecule.

10.2 Design a detailed flowchart that shows how to describe the bonding of a polyatomic species that has multiple bonds.

10.3 Prepare a table of hybridizations and their various shapes.

10.4 Prepare a list of molecules and ions mentioned in this chapter that contain electrons in delocalized orbitals. Organize your list by the number of atoms over which delocalization occurs.

10.5 Design a flowchart that summarizes the composite model of bonding.

10.6 Describe in your own words your understanding of each of the following terms: (a) σ bond; (b) π bond; (c) antibonding molecular orbital; (d) nonbonding electrons; (e) bond order; and (f) delocalized orbital.

10.7 Summarize the types of experimental evidence that support the idea of delocalized orbitals.

10.8 Create a single figure that explains the difference between a conductor, an insulator, and a semiconductor.

10.9 Prepare a list of all terms new to you that appear in Chapter 10. Write a one-sentence definition of each. Consult the glossary if you need help.

 Problems **ilw = interactive learning ware problem. Visit the website at www.wiley.com/college/olmsted**

Localized Bonds

10.1 Describe the bonding between two bromine atoms in Br_2.

10.2 Describe the bonding between fluorine and chlorine atoms in the mixed halogen FCl.

10.3 Hydrogen forms polar diatomic molecules with elements of Group 1. Describe the bonding in LiH and include a picture of the overlapping orbitals.

10.4 Describe the bonding between hydrogen and chlorine atoms in HCl and include a picture of the overlapping orbitals.

10.5 The bond angles in antimony trifluoride are 87°. Describe the bonding in SbF_3, including a picture of the orbital overlap interaction that creates the Sb—F bonds.

10.6 Hydrogen telluride, H_2Te, is stable only at low temperature. The compound forms a colorless gas, with bond angles of 90°. Describe the bonding of H_2Te, including an orbital overlap sketch.

Hybridization of Atomic Orbitals

10.7 Name the hybridization of an inner atom that has the following characteristics: (a) two lone pairs and two ligands; (b) three ligands and one lone pair; (c) three ligands and no lone pairs; and (d) five ligands and one lone pair.

10.8 Name the hybridization of an inner atom that has the following characteristics: (a) two ligands and no lone pairs; (b) two ligands and three lone pairs; (c) two lone pairs and four ligands; and (d) one lone pair and two ligands.

10.9 Name the hybrid orbitals formed by combining the following sets of atomic orbitals: (a) $3s$ and three $3p$ orbitals; (b) $2s$ and one $2p$ orbital; and (c) $3s$, three $3p$, and two $3d$ orbitals.

10.10 Name the hybrid orbitals formed by combining the following sets of atomic orbitals: (a) $2s$ and two $2p$ orbitals; (b) $3s$, three $3p$, and one $3d$ orbital; and (c) $4s$ and three $4p$ orbitals.

10.11 Identify the hybridization of the underlined atom in the following species: (a) $(CH_3)_2\underline{N}H$; (b) $\underline{S}bF_5$; (c) $\underline{S}O_2$; and (d) $\underline{C}S_2$.

10.12 Identify the hybridization of the underlined atom in the following species: (a) $\underline{S}F_6$, (b) $(CH_3)_2\underline{C}O$; (c) $\underline{B}F_4{}^-$, and (d) $HC\underline{C}CH_3$.

10.13 Identify the hybrid orbitals used by the inner atoms.

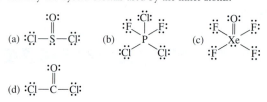

10.14 Identify the hybrid orbitals used by the inner atoms.

Multiple Bonds

10.19 Describe the bonding in the common solvent acetone, $(CH_3)_2CO$. Include sketches of all the bonding orbitals.

10.20 Describe the bonding in the solvent used in nail polish, ethyl acetate, $C_2H_5CO_2CH_3$. Include sketches of all the bonding orbitals.

ilw 10.21 Describe the steric number, geometry, and hybridization of each of the atoms of neocembrene, whose line structure appears in the Chemistry and Life Box.

10.22 Describe the steric number, geometry, and hybridization of each of the atoms of geraniol, whose line structure appears in the Chemistry and Life Box.

10.23 The five-carbon compounds 1,4-pentadiene, 1-pentyne, and cyclopentene all have molecular formula C_5H_8. Use steric numbers and hybridization to develop bonding pictures of these three molecules.

1,4-Pentadiene   1-Pentyne   Cyclopentene

10.24 The four-carbon compounds butane, 2-butene, and 2-butyne differ in the bonding between the second and third carbon atoms in the chain. Use steric numbers and hybridization to develop bonding pictures of these three molecules.

Butane   2-Butene   2-Butyne

10.25 Decide if the following pairs of orbitals overlap to form a σ bond, π bond, or no bond at all. Explain your reasoning in each case, including a sketch of the orbitals. Assume the bond lies along the z-axis. (a) $2p_z$ and $2p_z$; (b) $2p_y$ and $2p_x$; (c) sp^3 and $2p_z$; and (d) $2p_y$ and $2p_y$.

10.26 Decide if the following pairs of orbitals overlap to form a σ bond, π bond, or no bond at all. Explain your reasoning in each case, including a sketch of the orbitals. Assume the bond lies along the z-axis. (a) sp^2 and $2p_x$; (b) $2p_x$ and $2p_x$; (c) sp and $2p_z$; and (d) $2p_y$ and $2p_z$.

Molecular Orbital Theory: Diatomic Molecules

10.27 Use molecular orbital diagrams to rank the bond energies of the following diatomic species, from weakest to strongest: H_2, $H_2{}^-$, and $H_2{}^{2-}$.

10.28 Use molecular orbital diagrams to rank the bond energies of the following diatomic species, from weakest to strongest: He_2, $He_2{}^+$, and $He_2{}^{2+}$.

10.29 In each of the following pairs, which has the stronger bond? Use orbital configurations to justify your selections: (a) CO or CO^+; (b) N_2 or $N_2{}^+$; and (c) CN^- or CN.

10.30 In each of the following pairs, which has the stronger bond? Use orbital configurations to justify your selections: (a) O_2 or $O_2{}^+$; (b) CO or CO^-; and (c) F_2 or $F_2{}^+$.

10.15 Describe the bonding of chloroform, $CHCl_3$. Sketch an orbital overlap picture of the molecule.

10.16 Describe the bonding of dichlorosilane, SiH_2Cl_2. Sketch an orbital overlap picture of the molecule.

10.17 Describe the bonding of hydrazine, H_2NNH_2. Sketch an orbital overlap picture of the molecule.

10.18 Describe the bonding of hydrogen peroxide, HOOH. Sketch an orbital overlap picture of the molecule.

10.31 For each of the following interactions between orbitals of two different atoms, sketch the resulting molecular orbitals. Assume that the nuclei lie along the z-axis, and include at least two coordinate axes in your drawing. Label each MO as bonding or antibonding and σ or π. (a) $2s$ and $2p_z$; and (b) $2p_x$ and $2p_x$.

10.32 For each of the following interactions between orbitals of two different atoms, sketch the resulting molecular orbitals. Assume that the nuclei lie along the z-axis, and include at least two coordinate axes in your drawing. Label each MO as bonding or antibonding and σ or π. (a) $2p_z$ and $2p_z$; and (b) $2p_y$ and $2p_y$.

10.33 The active ingredient in commercial laundry bleach is sodium hypochlorite (NaClO). Describe the bonding in the hypochlorite anion, ClO^-. (*Hint:* In this ion, the valence orbitals of chlorine behave like $n = 2$ orbitals.)

10.34 The superoxide ion, O_2^-, is a reactive species that may play a role in the chemistry of aging. Describe the bonding in this ion.

Three-center π Orbitals

10.35 Describe the bonding of carbon disulfide, CS_2, in terms of orbital overlap and delocalized electrons. How many valence electrons occupy delocalized π orbitals in this molecule?

10.36 Describe the bonding of sulfur dioxide, SO_2, in terms of orbital overlap and delocalized electrons. How many valence electrons occupy delocalized π orbitals in this molecule?

10.37 Calcium cyanamide (CaNCN) is used as a pesticide, herbicide, and plant growth regulator. Write a Lewis structure for the NCN^{2-} anion and describe its bonding, including any delocalized orbitals.

10.38 Detonators for explosive charges of trinitrotoluene (TNT) contain mercury fulminate, $Hg(CNO)_2$, which explodes when struck. Write a Lewis structure for the CNO^- anion and describe its bonding, including any delocalized orbitals.

10.39 Hydrazoic acid has the formula HN_3, with the three nitrogen atoms in a row. Determine the Lewis structure, hybridization, and bond angles of this compound. Describe its π bonding network.

10.40 Two molecules that have been detected in interstellar space are HNCO and OCS. Describe the bonding in each molecule, identify the hybrid orbitals used to make σ bonds, and identify their delocalized π systems.

Extended π Systems

10.41 Which of the molecules whose line structures appear in the Chemistry and Life Box are stabilized by delocalized π electrons? For those that are, identify the atoms that contribute to the delocalized orbitals.

10.42 Which of the following molecules are stabilized by delocalized π orbitals? For those that are, identify the atoms over which delocalization extends.

10.43 Vitamin C has the structure shown at right:
(a) Identify the hybridization of each carbon and oxygen atom. (b) How many π bonds are there? (c) How many electrons are in delocalized π orbitals? (d) Redraw the structure and circle the largest continuous plane of atoms.

Vitamin C
($C_6H_8O_6$)

10.44 Carvone, whose line structure appears at right, is the principal flavor and fragrance ingredient in spearmint.
(a) Identify the hybridization of each carbon atom. (b) How many π bonds are there? (c) How many electrons are in delocalized π orbitals? (d) Redraw the structure and circle the largest continuous plane of atoms.

Carvone
($C_{10}H_{14}O$)

10.45 The two common oxides of carbon are CO and CO_2, which are linear. A third oxide of carbon, which is rare, is carbon suboxide, OCCCO. Draw a Lewis structure for carbon suboxide. Determine its geometry and identify all orbitals involved in delocalized π orbitals. Make separate sketches of the σ and π bonding systems.

10.46 The oxalate anion ($C_2O_4^{2-}$) has a C—C bond and two outer oxygen atoms bonded to each carbon atom. Determine the Lewis structure of this anion, including all resonance structures. What is the geometry about the carbon atoms, and how many atoms contribute p orbitals to the delocalized π system?

10.47 Determine the Lewis structure and describe the bonding of the perchlorate anion (ClO_4^-).

10.48 The permanganate ion (MnO_4^-) gives potassium permanganate solutions a rich burgundy color. Determine the Lewis structure, including all resonance structures, and describe the bonding in this ion.

Band Theory of Solids

10.49 Identify the following as p-type, n-type, or undoped semiconductors: (a) GaP; (b) InSb doped with Te; and (c) CdSe.

10.50 Identify the following as p-type, n-type, or undoped semiconductors: (a) Ge; (b) Ge doped with P; and (c) AlAs doped with Zn.

10.51 Use band theory to explain why iron is harder and melts at a higher temperature than potassium. (*Hint:* Consider the number of valence electrons in each element.)

10.52 The following energies measure the strength of bonding in sodium, magnesium, and aluminum metals:

$$Na\,(s) \longrightarrow Na\,(g) \qquad \Delta E = 99 \text{ kJ/mol}$$
$$Mg\,(s) \longrightarrow Mg\,(g) \qquad \Delta E = 127 \text{ kJ/mol}$$
$$Al\,(s) \longrightarrow Al\,(g) \qquad \Delta E = 291 \text{ kJ/mol}$$

Use band theory to explain these data. (*Hint:* Consider the number of valence electrons in each element.)

10.53 Draw a band-gap diagram for gallium arsenide doped with zinc.

10.54 Draw a band-gap diagram for germanium doped with phosphorus.

Additional Paired Problems

10.55 Sulfur forms two stable oxides, SO_2 and SO_3. Describe the bonding and geometry of these compounds.

10.56 Nitrogen forms two stable oxoanions, NO_2^- and NO_3^-. Describe the bonding and geometry of these anions.

10.57 Indium arsenide is a semiconductor with a band gap of 34.7 kJ/mol. What is the minimum frequency of light that will cause InAs to conduct electricity? What region of the electromagnetic spectrum corresponds to this frequency?

10.58 Gallium arsenide is a common semiconductor. This material becomes a conductor when it is irradiated with infrared light whose frequency is 3.43×10^{14} s^{-1}. Calculate the band gap of GaAs in kJ/mol.

10.59 Acrolein has the Lewis structure shown at right. Give a complete description of its bonding, including orbital overlap sketches of the σ and π bonding systems, and an accounting of its delocalized electrons.

10.60 The allyl cation has the Lewis structure shown at right. Give a complete description of its bonding, including orbital overlap sketches of the σ and π bonding systems, and an accounting of its delocalized electrons.

10.61 Cinnamic acid is one of the natural components of cinnamon oil.
(a) Redraw the structure of cinnamic acid, adding the lone pairs. (b) What orbitals are used for bonding by atoms *A, B, C, D,* and *E*? (c) What are the bond angles about atoms *A, C,* and *D*? (d) How many π bonds does cinnamic acid have?

Cinnamic acid ($C_9H_8O_2$)

10.62 Capsaicin is the molecule responsible for the hot spiciness of chili peppers:

Capsaicin ($C_{18}H_{27}O_3N$)

(a) How many π bonds does capsaicin have? (b) What orbitals are used for bonding by each of the labeled atoms? (c) What are the bond angles about each of the labeled atoms? (d) Redraw the structure of capsaicin, adding the lone pairs.

10.63 Use electron configurations to decide if the following species are paramagnetic or diamagnetic: (a) CO; (b) N_2^+; (c) O_2^+; and (d) CN^-.

10.64 Use electron configurations to decide if the following species are paramagnetic or diamagnetic: (a) NO; (b) C_2; (c) O_2^-; and (d) Cl_2

10.65 Determine the type of orbitals (atomic, sp^3, or sp^2) used by each atom in the molecules shown at right:

H_3CNH_2

10.66 Determine the type of orbitals (atomic, sp^3, or sp^2) used by each atom in the molecules shown at right:

N_2H_4

10.67 Draw band-gap diagrams for a metallic conductor and an *n*-type semiconductor.

10.68 Draw band-gap diagrams for an insulator and a *p*-type semiconductor.

10.69 Carbocations are unstable high-energy compounds that contain positively charged carbon atoms. Carbocations form during the course of a reaction but are usually consumed rapidly. Describe the bonding and geometry of the carbocation $(CH_3)_3C^+$.

10.70 The methylene fragment CH_2 has been identified as a reactive species in some gas-phase chemical reactions. Determine its Lewis structure, describe its bonding, and predict the H—C—H bond angle.

More Challenging Problems

10.71 Combine the features of Figures 10-50 and 10-52 in a single drawing that shows how silicon and germanium conduct electricity.

10.72 Write the electron configuration and determine the bond order of NO, NO^+, and NO^-. Which ones display paramagnetism?

10.73 Oxygen forms three different ionic compounds with potassium: potassium oxide (K_2O), potassium superoxide (KO_2), and potassium peroxide (K_2O_2). Write electron configurations for the superoxide and peroxide anions. Compare these configurations with the configuration of the oxygen molecule. Rank the three species in order of increasing bond order, bond energy, and bond length. Which of the three are magnetic? Which has the largest paramagnetism?

10.74 Nitrogen molecules can absorb photons to generate excited-state molecules. Construct an energy level diagram and place the valence electrons so that it describes the most stable *excited* state of an N_2 molecule. Is the N—N bond in this excited-state N_2 molecule stronger or weaker than the N—N bond in ground-state nitrogen? Explain your answer.

10.75 Two oxoanions containing chlorine are ClO_3^- (chlorate) and ClO_2^- (chlorite). Determine Lewis structures and describe the bonding for each of these anions, including delocalized π bonds.

10.76 The ionization energy of molecular oxygen is smaller than that of atomic oxygen (1314 vs. 1503 kJ/mol). In contrast, the ionization energy for molecular nitrogen is larger than that of atomic nitrogen (1503 vs. 1402 kJ/mol). Explain these data, using the electron configurations of the diatomic molecules.

10.77 What is the relationship among bond length, bond energy (see Chapter 6), and bond order? Which of them can be measured?

10.78 Buckminsterfullerene contains 60 carbon atoms, each of which uses sp^2 hybrid orbitals and contributes one $2p$ orbital to make delocalized molecular orbitals. How many π orbitals and how many π bonding electrons does buckminsterfullerene have?

10.79 Acrylonitrile is used to manufacture polymers for synthetic fibers. Describe the bonding of acrylonitrile. (a) Identify the hybrid and/or atomic orbitals involved in bonding for each atom. (b) Draw sketches that show the σ and π bonding systems. (c) If there are delocalized π orbitals, identify and describe them.

10.80 Use orbital energy diagrams to decide whether NF would be stabilized or destabilized by adding one electron to make the corresponding anion and by removing an electron to form the corresponding cation. Sketch the orbital involved in these changes.

10.81 The molecule CN is not stable in our atmosphere, but it has been detected in the interstellar regions. Describe its bonding, and sketch its least stable occupied molecular orbital.

10.82 The concept of threshold energy appears in Chapter 7 as part of the discussion of the photoelectric effect. What feature of band theory would you associate with the threshold energy? Consider electrons ejected by photons with a frequency significantly higher than the threshold frequency. Would these electrons all have the same kinetic energy? Explain.

10.83 Chlorine forms one neutral oxide, ClO_2. Describe the bonding in this unusual compound. Explain why it is considered unusual.

10.84 Use orbital sketches to illustrate the bonding of allene (H_2CCCH_2). Identify all orbitals in the bonding scheme. Make sure your sketches show the three-dimensional arrangement of the hydrogen atoms. Make separate sketches of the σ and π bonding systems.

10.85 Dilithium molecules can be generated by vaporizing lithium metal at very low pressure. Do you think it is possible to prepare diberyllium? Explain your reasoning using MO diagrams for Li_2 and Be_2.

10.86 Silicon carbide (carborundum) is a hard crystalline material similar to diamond. It is widely used for cutting tools. Describe the bonding in silicon carbide, whose empirical formula is SiC. Draw an energy band picture showing the band gap for SiC compared with those of diamond and silicon.

10.87 The bond strengths in the H—X molecules (X = F, Cl, Br, I) show that bonds become weaker as orbitals become more diffuse. Make a series of orbital overlap sketches for the H—X molecules to illustrate this. Shade your figures in a way that illustrates the changes in electron density.

10.88 When an oxalate anion, $C_2O_4^{2-}$, adds two protons to form oxalic acid, two C—O bonds become longer and two become shorter than the bonds in oxalate anions. Which bonds get longer and which shorter? Use bonding principles to explain these changes.

10.89 Elements in the same group of the periodic table usually have similar chemical properties. Nitrogen and oxygen are unique among elements in their groups, however, because they are stable as diatomic molecules. Phosphorus and sulfur, the next two members of these groups, are stable as long-chain molecules. Use the properties of orbitals to account for these differences in chemical behavior.

10.90 Explain how the following bond lengths (in picometers) provide evidence for the existence of delocalized π orbitals in the carbonate ion: C—O (average), 143; C=O (average), 122; and CO_3^{2-}, 129.

10.91 Describe the bonding in ketene (H_2CCO). Make separate sketches of the σ and π bonding systems.

10.92 Consider the bond lengths of the following diatomic molecules: N_2, 110 pm; O_2, 121 pm; and F_2, 143 pm. Explain the variation in length in terms of the molecular orbital descriptions of these molecules.

Group Study Problems

10.93 Pyrrole, C₄H₅N, is a common organic molecule. Pyrrole rings are common substructures of many biological molecules. According to the localized bonding model presented in this chapter, what geometry and hybridization are predicted for the nitrogen atom of pyrrole? In fact, pyrrole is a planar molecule with a trigonal planar nitrogen atom. Apply the composite model of bonding developed in this chapter to explain why the bonding of pyrrole differs from the predictions of the localized bond model. Your answer should include a sketch of the most stable π orbital of pyrrole.

Pyrrole

10.94 The pentadienyl cation (C₅H₇⁺) is a high-energy species that can be generated in the laboratory under carefully controlled conditions. (a) Describe the bonding orbitals used by each of the carbon atoms. (b) How many p orbitals contribute to the π bonding system? How many molecular orbitals do they form? (c) How many delocalized π electrons are there in C₅H₇⁺? (d) Would you expect this ion to have a more or less stable π bonding system than 1,3-pentadiene or 1,4-pentadiene? (Give reasons.)

10.95 One of the first compounds discovered for the treatment of AIDS was azidothymidine (AZT). Here is the line structure of AZT: (a) Identify the type of orbitals used for σ bonding by each of the nitrogen atoms. (b) What is the bond angle around the oxygen atom in the five-membered ring? (c) How many carbon atoms can be described as using sp³ hybrid orbitals? (d) How many π bonds are there? What orbitals are used to form these bonds?

10.96 Metal carbonyls are compounds in which molecules of carbon monoxide form covalent bonds to transition metals. Examples include Cr(CO)₆, Fe(CO)₅, and Ni(CO)₄. The principal bonding interaction in a metal carbonyl is a σ bond formed between the carbon atom of CO

and one of the valence d orbitals of the metal. (a) What is the hybridization of the C atoms in metal carbonyls? (b) Use the d_{z²} orbital to make an orbital overlap sketch of the M—CO σ bond. (c) Draw structures showing the geometries of these carbonyls. (d) Metal carbonyls also have π bonds that result from overlap of the π* orbital of CO with one of the metal d orbitals. Use the d_{xz} orbital to make an orbital overlap sketch of the π bond in a metal carbonyl.

10.97 Use the principles of orbital overlap to explain why two different isomers of 1,2-dichloroethylene exist but only one isomer of 1,2-dichloroethane exists.

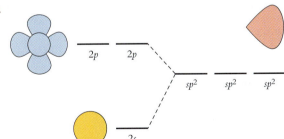

trans-1,2-Dichloro-ethylene *cis*-1,2-Dichloro-ethylene 1,2-Dichloro-ethane

Use bond strength arguments to estimate the amount of energy required to interconvert the cis and trans isomers.

10.98 Titanium(IV) oxide, TiO₂, is a white solid that absorbs light in the ultraviolet at 350 nm. Heating this solid in the presence of molecular hydrogen reduces some of the Ti(IV) ions to Ti(III), which act as dopants. The product shows semiconducting properties and is blue, indicating that it absorbs red light. Develop appropriate band-gap diagrams for TiO₂ and the doped product. Use the diagrams to explain what type of semiconductor the product is and illustrate the changes in energy gaps indicated by the light-absorbing properties of the materials. Suggest possible chemical formulas for the product. (*Hints:* There is more than one possibility; they are nonstoichiometric but must be electrically neutral.)

10.99 Imagine making an O₄ molecule by attaching an oxygen atom to the nonbonding pair of electrons on the central oxygen of ozone (O₃). Describe the bonding in O₄, which is isoelectronic with NO₃⁻ and CO₃²⁻. Compare the expected stability of the O₄ molecule with that of two O₂ molecules. Does the simple picture of bonding explain why O₄ does not exist?

Answers to Section Exercises

10.1.1

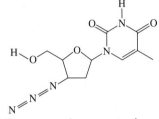

10.1.2 The 1s orbital of the H atom overlaps with the 4p orbital of Br that points along the bond axis, giving a bond that looks similar to that in HF.

10.1.3 Sb uses its 5p orbitals, which overlap with 1s orbitals of the H atoms, forming three bonds at right angles to one another.

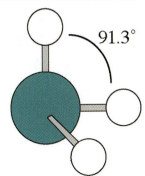

91.3°

10.2.1

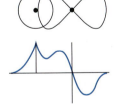

10.2.2 S has SN = 4, uses sp³ hybridization, has tetrahedral electron group geometry, and is bent, with bond angles somewhat less than 109.5°.

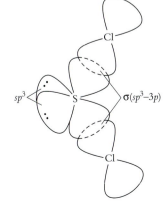

10.2.3

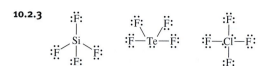

SiF$_4$ has tetrahedral electron group geometry and uses sp^3 hybridization; TeF$_4$ has trigonal bipyramidal electron group geometry and uses sp^3d hybridization; ClF$_4^-$ has octahedral electron group geometry and uses sp^3d^2 hybridization.

10.3.1

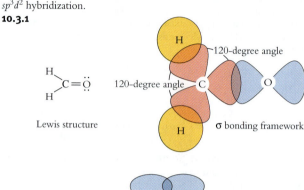

Lewis structure   120-degree angle   σ bonding framework

π bonding system

The carbon atom uses sp^2 hybrids to form three σ bonds. Its unused p orbital overlaps with an oxygen p orbital to form a π bond.

10.3.2

Propene (C$_3$H$_6$)   Propyne (C$_3$H$_4$)

The CH$_3$ carbon in each compound uses sp^3 hybrids and has tetrahedral geometry. In propene, the other two carbon atoms have trigonal planar geometry, use sp^2 hybrids for three σ bonds, and have a π bond between them. In propyne, the other two carbon atoms have linear geometry, use sp hybrids for two σ bonds, and have two π bonds between them.

10.3.3

Four carbon atoms participate in π bonding. For appropriate orbital sketches, see Figures 10-20 and 10-21.

10.4.1

—— σ*

⇅ σ

Li$_2$

With two bonding electrons, Li$_2$ has a bond order of 1.

10.4.2 There is only one occupied valence orbital, σ_s.

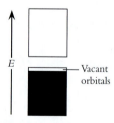

10.4.3 Ne$_2$ has 16 valence electrons, which fill all the MOs formed from its valence orbitals. Your diagram should look like the right side of Figure 10-35, with a pair of electrons in each level. The calculated bond order is zero, which explains why this molecule cannot be prepared.

10.5.1 The formate anion is just like acetate, except it has a hydrogen atom in place of the methyl (CH$_3$) group of acetate. The bonding description is similar to that given for acetate in Example 10-11. In the bonding framework, replace the sp^3 hybrid–sp^2 hybrid σ bond with a $1s$–sp^2 hybrid σ bond. The π bonding system is identical to that of acetate.

10.5.2 With 16 valence electrons, N$_3^-$ is isoelectronic with CO$_2$. Its bonding is identical with that of CO$_2$. It is linear, with the inner atom having a steric number of 2. There are two delocalized π systems. The nitride ion, N^{3-}, contains a single atom with an atomic configuration identical to that of Ne. Electrons fill a spherical shell in this atomic anion.

10.5.3 NO$_2^+$ contains 16 valence electrons, making it isoelectronic with CO$_2$. It is linear, with the inner atom having a steric number of 2. There are two delocalized π systems, each containing four electrons, for a total of eight delocalized electrons. NO$_2^-$ is treated in Extra Practice Exercise 10-11. With 18 valence electrons, this anion is isoelectronic with ozone. Its inner atom has a steric number of 3, so it is bent. There is one delocalized π system containing four electrons.

10.6.1 The Lewis structure shows two resonance forms, indicating three-atom π delocalization that can be described exactly as in the acetate or formate anions.

$$\overset{-}{:}\ddot{O}-\overset{\overset{\displaystyle :O:}{\|}}{C}-\ddot{O}-H \longleftrightarrow \overset{-}{:}\ddot{O}=\overset{\overset{\displaystyle :\ddot{O}:}{|}}{C}-\ddot{O}-H$$

10.6.2 Azulene has ten delocalized π orbitals; five of the orbitals are occupied.

10.6.3 The Lewis structure and bonding are the same as in the sulfate anion. The ion is tetrahedral. The Cl atom uses sp^3 hybrids to form four σ bonds and uses d orbitals to form delocalized π bonds with the O atoms.

10.7.1 250 kJ/mol

10.7.2 This is a p-type semiconductor because Zn has one fewer valance electron than Ga. The band-gap diagram appears as follows:

10.7.3 Si would be the semiconductor and Al would be the dopant. Aluminum has one less valence electron than silicon.

Answers to Extra Practice Exercises

10.1 Two bonds form by overlap of H 1s orbitals with Se 4p orbitals. Your ball-and-stick model should look like the model of H_2S.

10.2 The oxygen atom has SN = 4, so it uses four sp^3 hybrid orbitals. Two overlap with 1s orbitals from the hydrogen atoms to form the bonds, and the other two hold the two lone pairs.

10.3 Each inner atom has SN = 4, so each uses four sp^3 hybrid orbitals. The carbon atom forms three bonds with H atoms, each described by sp^3–1s overlap. The fourth orbital bonds to an sp^3 orbital of the nitrogen atom to form an sp^3–sp^3 bond. Two sp^3 hybrids on N overlap with 1s orbitals on H to form bonds, and the fourth hybrid holds the lone pair.

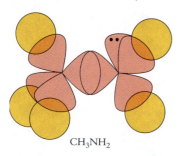

CH$_3$NH$_2$

10.4 Bonds are between F 2p orbitals and hybrids on the inner atoms.

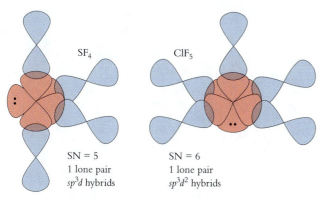

SF$_4$

SN = 5
1 lone pair
sp^3d hybrids

ClF$_5$

SN = 6
1 lone pair
sp^3d^2 hybrids

10.5 (a) The P atom in the PF_6^- anion has SN = 6, octahedral electron group geometry and sp^3d^2 hybridization. (b) The O atom in OF_2 has SN = 4, tetrahedral electron group geometry and sp^3 hybridization. (c) The Sn atom in $SnCl_2$ has SN = 3, trigonal planar electron group geometry and sp^2 hybridization.

10.6 View chloroethylene as ethylene with one H atom replaced by a Cl atom. The bonding is identical to that of ethylene, except for a σ bond formed by overlap of a Cl 3p orbital with an sp^2 hybrid.

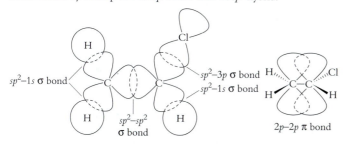

10.7 View 2-butyne as acetylene with both H atoms replaced by methyl groups. The two inner C atoms have sp hybridization and a triple bond between them. The two outer C atoms use sp^3 hybrids to form three C—H bonds and one C—C bond.

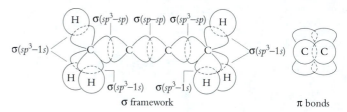

σ framework π bonds

10.8 The third electron of H_2^- must occupy an antibonding orbital, so H_2 has the stronger bond.

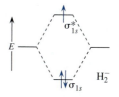

10.9 The MO diagrams show that whereas N_2 has bond order of $(8 - 2)/2 = 3$, O_2 has bond order of $(8 - 4)/2 = 2$ and F_2 has bond order of $(8 - 6)/2 = 1$. After N_2, the extra electrons occupy antibonding orbitals, causing bond energy to decrease in this set.

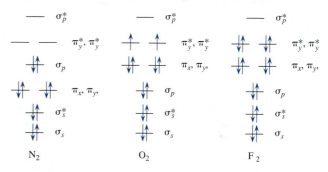

N$_2$ O$_2$ F$_2$

10.10 The CN species have the same MO diagrams as CO, leading to these configurations: $CN^- = (\sigma_s)^2 (\sigma_s^*)^2 (\sigma_p)^2 (\pi_x)^2 (\pi_y)^2$; $CN = (\sigma_s)^2 (\sigma_s^*)^2 (\sigma_p)^2 (\pi_x)^2 (\pi_y)^1$; and $CN^+ = (\sigma_s)^2 (\sigma_s^*)^2 (\sigma_p)^2 (\pi_x)^2$. CN^- has the strongest bond (bond order = 3).

10.11 The nitrite anion is isoelectronic with the ozone molecule. Describe its bonding exactly like that of ozone, but replace the inner O atom with N and include the -1 charge on the anion.

10.12 The Lewis structure of acryloyl chloride shows a four-atom chain with alternating single and double bonds, so the π system of this molecule is identical to that of methacrylate. Here is an inventory of the bond framework:

3 sp^2–1s C—H σ bonds	1 sp^2–2p C—O σ bond	2 sp^2–sp^2 C—C σ bonds
1 sp^2–3p C—Cl σ bond	2 3p Cl lone pairs	1 3s Cl lone pair
1 2s O lone pair	1 2p O lone pair	

10.13 The wavelength is 1.14 μm, in the near infrared spectral region.

Opals and Butterfly Wings

Opals and butterfly wings seem very different, yet both shimmer with beautiful colors. Although they are made of very different chemical substances, both have highly regular structures containing hexagonal patterns.

Opals contain microscopic spheres of silica (SiO_2), packed into planar hexagonal layers. The shimmering colors of opals result from the way this layered structure interacts with light. Diffraction and interference of light passing through these layers depends on the wavelength, with the result that some colors are enhanced while other colors are diminished.

The surface of a butterfly wing has an "inverse opal" structure. Butterfly wings are composed of chitin, the same cellulose-like polymer that forms the shells of crabs. In a butterfly wing, however, the chitin forms a hexagonal network. Instead of tightly packed spheres, the result is a solid structure with a regular array of spherical holes. The shimmering colors of butterfly wings arise from interference of light scattered and refracted by this regular array of chitin at the wing surface. As with opals, color variations arise from the way different wavelengths interact with molecular-level structure.

Materials scientists have learned how to make synthetic opals. The procedure requires solid spheres of uniform diameter around 250 nm. Silica spheres are made by reacting an organosilicon compound, $Si(OCH_3)_4$, with water. Hydrogen and methyl groups trade places, forming $Si(OH)_4$ and CH_3OH. The silicon-

Effects of Intermolecular Forces

11

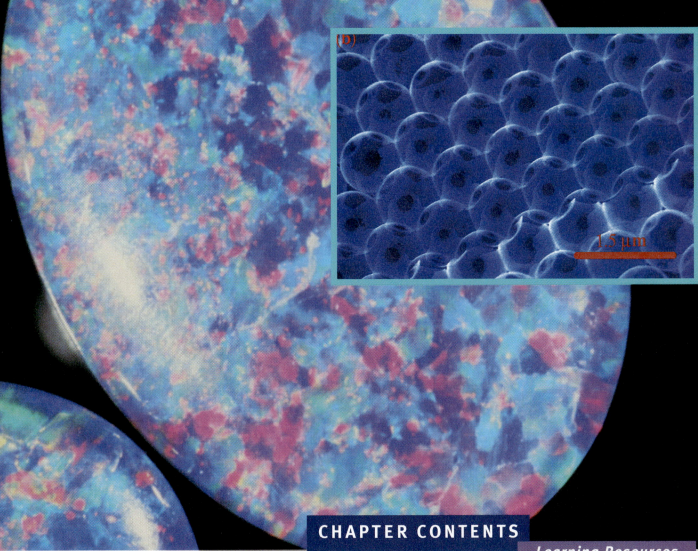

(D)

1.5 µm

CHAPTER CONTENTS

containing product molecules eliminate water and combine to form Si—O—Si bonds, and if the reaction conditions are right, the network of bonds closes on itself to form spheres of uniform size. When these spheres are suspended in water, they slowly solidify into the regular array of opals.

Scientists also have learned how to mimic the surface of a butterfly wing. Polystyrene beads and smaller silica nanoparticles are suspended in water and mixed thoroughly using ultrasound. When a glass slide is dipped into the suspension and slowly withdrawn, a thin film forms on the glass surface. This film is a regular array of beads encased in a matrix of nanoparticles. Heating the film destroys the poly-

Learning Resources

KEY CONCEPTS

CRITICAL THINKING

SKILLS TO MASTER

ADDITIONAL HELP
www.wiley.com/college/olmsted

- TUTORIALS
- ANIMATIONS

styrene beads but leaves the silica web intact. The result is a silica "inverse opal" film.

The iridescent properties of opals and butterfly wings are due to their highly regular molecular structures. To understand how solid structures form ordered patterns, we need information about the forces of attraction that exist between molecules. This chapter begins with descriptions of intermolecular forces and their effects on gases. Then we show how these forces, even though considerably weaker than the chemical bonds described in Chapters 9 and 10, are important in determining the behavior of solids and liquids. We end the chapter with a discussion of phase changes.

11.1 REAL GASES AND INTERMOLECULAR FORCES

Forces of attraction between molecules are responsible for the existence of liquids and solids. In the absence of these **intermolecular forces,** all molecules would move independently, and all substances would be gases. The natural phases of the elements indicate the importance of intermolecular forces. At room temperature and pressure, only 11 elements are gases. Mercury and bromine are liquids, and all the rest of the elements are solids. For all but the 11 gaseous elements, intermolecular forces are too large to ignore under normal conditions.

The Halogens

The halogens, the elements from Group 17 of the periodic table, provide an introduction to intermolecular forces. These elements exist as diatomic molecules: F_2, Cl_2, Br_2, and I_2. The bonding patterns of the four halogens are identical. Each molecule contains two atoms held together by a single covalent bond that can be described by end-on overlap of valence p orbitals.

Although they have the same bonding patterns, bromine and iodine differ from chlorine and fluorine in their macroscopic physical appearance and in their molecular behavior. As Figure 11-1 shows, at room temperature and pressure, fluorine and chlorine are gases, bromine is a liquid, and iodine is a solid.

Gases and condensed phases look very different at the molecular level. Molecules of F_2 or Cl_2 move freely throughout their gaseous volume, traveling many molecular diameters before colliding with one another or with the walls of their container. Because much of the volume of a gas is empty space, samples of gaseous F_2 and Cl_2 readily expand or contract in response to changes in pressure. This freedom of motion exists because the intermolecular forces between these molecules are small.

Molecules of liquid bromine also move about relatively freely, but there is not much empty space between molecules. A liquid cannot be compressed significantly by increasing the pressure, because its molecules are already in close contact with one another. Neither does a liquid expand significantly if the pressure is reduced, because the intermolecular forces in a liquid are large enough to prevent the molecules from breaking away from one another.

Solid iodine, like liquid bromine, has little empty space between molecules. Like liquids, solids have sufficiently strong intermolecular forces that they neither expand nor contract significantly when pressure changes. In solids, the intermolecular forces are strong enough to prevent molecules from moving freely past one another. Instead, the I_2 molecules in a sample of solid iodine are arranged in ordered arrays. Each molecule vibrates back and forth about a single most-stable position, but it cannot slide easily past its neighbors.

The interplay between molecular kinetic energies of motion and intermolecular attractive forces accounts for these striking differences, as illustrated in Figure 11-2. Recall from Section 5.3 that molecules are always moving. Intermolecular attractive forces tend to hold molecules together in a condensed phase, but molecules that are moving fast enough can overcome these forces and move freely in the gas phase. When the average kinetic energy of motion is large enough, molecules remain separated from one another and the substance is a gas. Conversely, when intermolecular attractive forces are large enough, molecules remain close to one another and the substance is a liquid or solid.

The bars in Figure 11-3 compare the attractive energies arising from intermolecular forces acting on the halogens. The figure also shows the average molecular kinetic energy

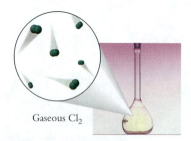

Gaseous Cl_2

Liquid Br_2

Solid I_2

Figure 11-1
Under normal conditions, chlorine is a pale yellow-green gas, bromine is a dark red liquid, and iodine is a dark crystalline solid.

at room temperature. For fluorine and chlorine, the attractive energy generated by intermolecular forces is smaller than the average molecular kinetic energy at room temperature. Thus, these elements are gases under normal conditions. In contrast, bromine molecules have enough kinetic energy to move freely about, but their energy of motion is insufficient to overcome the intermolecular forces of the liquid phase. Finally, attractive forces are strong enough in iodine to lock the I_2 molecules in position in the solid state.

Real Gases

As we describe in Chapter 5, the ideal gas model makes two assumptions: A gas has negligible forces between its molecules, and gas molecules have negligible volumes. The behavior of bromine shows that neither of these assumptions is completely correct for a real gas. Although bromine is a liquid at room temperature, the red color above the liquid shown in Figure 11-4 indicates that the gas above a sample of liquid Br_2 contains a substantial number of Br_2 molecules in the gas phase. Under these conditions, intermolecular forces are strong enough to condense most but not all of the molecules to the liquid phase. Moreover, exerting pressure on liquid bromine hardly changes its volume. The molecules in liquid bromine are packed as close together as their molecular volumes allow. Regardless of their phase, bromine molecules have the same physical properties, including intermolecular forces and molecular volumes. In other words, both these properties have finite values, so bromine gas cannot be ideal.

How close do real gases come to ideal behavior? To answer this question, we rearrange the ideal gas equation to examine the ratio PV/nRT. Figure 11-5 shows how PV/nRT varies with pressure for chlorine gas at room temperature. If chlorine were ideal, the PV/nRT ratio would always be 1, as shown by the red line on the graph. Instead, chlorine shows deviations from 1 as the pressure increases.

Notice in the inset of Figure 11-5 that chlorine is nearly ideal at pressures around one atmosphere. In fact, PV/nRT deviates from 1.0 by less than 4% at pressures below 4 atm. As the pressure increases, however, the deviations become increasingly significant. At first, the PV/nRT ratio for chlorine drops below 1. This is because the chlorine molecules are close enough together for attractive forces to play a significant role. These intermolecular attractions hold molecules together and reduce the forces exerted when the molecules strike the walls of the container. Intermolecular attractions tend to make the pressure of a real gas lower than the ideal value.

Figure 11-5 also shows that at pressures greater than 375 atm, PV/nRT becomes *larger* than 1. This is the effect of molecular size. At high enough pressure, the molecules

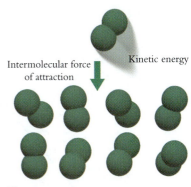

Figure 11-2
A substance exists in a condensed phase when its molecules have too little average kinetic energy to overcome intermolecular forces of attraction.

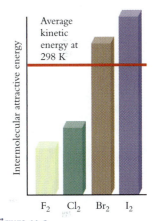

Figure 11-3
The balance of average kinetic energy (*red line*) and intermolecular forces of attraction favors the gas phase for F_2 and Cl_2 and the condensed phases for Br_2 and I_2.

Figure 11-4
Although bromine is a liquid at room temperature and pressure, enough molecules escape into the gas phase to give the gas above a liquid sample a distinct red color.

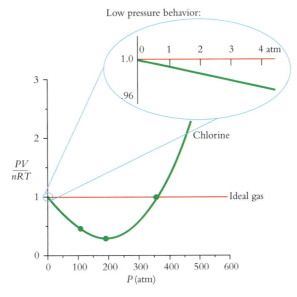

Figure 11-5
The variation in *PV/nRT* with pressure shows that chlorine is not an ideal gas.

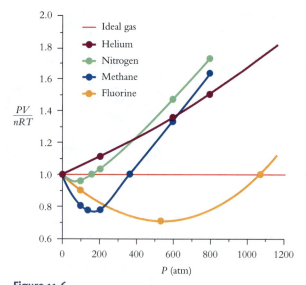

$\frac{PV}{nRT}$

Legend: Ideal gas, Helium, Nitrogen, Methane, Fluorine

P (atm)

Figure 11-6
Variations in *PV/nRT* for He, F₂, CH₄, and N₂ at 300 K.

are packed so close together that the total volume of the molecules is no longer negligible compared to the overall volume of the gas. The added volume of the molecule means that the volume of the real gas is greater than the volume of the ideal gas.

Every gas shows deviations from ideal behavior at high pressure. Figure 11-6 shows PV/nRT for He, F_2, CH_4, and N_2, all of which are gases at room temperature. Notice that PV/nRT for helium increases steadily as pressure increases. Interatomic forces for helium are too small to reduce the ratio below 1, but the finite size of the helium atom generates deviations from ideality that become significant at pressures above 100 atm.

Given that every gas deviates from ideal behavior, can we use the ideal gas model to discuss the properties of real gases? The answer is yes, as long as conditions do not become too extreme. The gases with which chemists usually work, such as chlorine, helium, and nitrogen, are nearly ideal at room temperature at pressures below about 10 atm.

The van der Waals Equation

It would be useful to have an equation that describes the relationship between pressure and volume for a real gas, just as $PV = nRT$ describes an ideal gas. One way to approach real gas behavior is to modify the ideal gas equation to account for attractive forces and molecular volumes. The result is the **van der Waals equation,** named for the scientist who first proposed it in 1873, Johannes van der Waals:

$$\left(P + \frac{n^2 a}{V^2}\right)(V - nb) = nRT \qquad \textbf{(11-1)}$$

The van der Waals equation adds two correction terms to the ideal gas equation. Each correction term includes a constant that has a specific value for every gas. The first correction term, $n^2 a/V^2$, adjusts for attractive intermolecular forces. The van der Waals constant a measures the strength of intermolecular forces for the gas; the stronger the forces, the larger the value of a. The second correction term, nb, adjusts for molecular sizes. The van der Waals constant b measures the size of molecules of the gas; the larger the molecules, the larger the value of b.

The van der Waals constants for a number of gases appear in Table 11-1, and the magnitude of van der Waals corrections is explored in Example 11-1.

Table 11-1   van der Waals Constants

Substance	a (L² atm /mol²)	b (L/mol)	Substance	a (L² atm /mol²)	b (L/mol)
He	0.0341	0.0237	Cl_2	6.493	0.0562
Ne	0.211	0.0171	CO_2	3.592	0.0427
Ar	1.345	0.0322	H_2O	5.464	0.0305
H_2	0.244	0.0266	NH_3	4.170	0.0371
N_2	1.390	0.0391	CH_4	2.253	0.0428
O_2	1.360	0.0318	C_2H_6	5.489	0.0638
CO	1.453	0.0395	C_6H_6	18.00	0.0115
F_2	1.156	0.0290			

Gases such as methane are sold and shipped in compressed gas cylinders. A typical cylinder has a volume of 15.0 L and, when full, contains 62.0 mol of CH_4. After prolonged use, 0.620 mol of CH_4 remains in the cylinder. Use the van der Waals equation to calculate the pressures in the cylinder when full and after use, and compare the values to those obtained from the ideal gas equation. Assume a temperature of 27 °C.

Strategy: We are asked to calculate and compare the pressures of methane gas using the van der Waals equation (Equation 11-1) and the ideal gas equation ($PV = nRT$). Van der Waals constants a and b must be looked up in a data table such as Table 11-1. To calculate pressures, we rearrange the van der Waals equation and the ideal gas equation.

$$P_{real} = \frac{nRT}{V - nb} - \frac{n^2 a}{V^2} \qquad and \qquad P_{ideal} = \frac{nRT}{V}$$

Solution: The van der Waals constants for CH_4 are

$$a = 2.253 \text{ L}^2 \text{ atm/mol}^2 \qquad and \qquad b = 0.0428 \text{ L/mol}$$
$$V = 15.0 \text{ L and } T = 300 \text{ K}$$

When the tank is full, $n = 62.0$ mol:

$$P_{real} = \frac{(62.0 \text{ mol})(0.08206 \text{ L atm/mol K})(300 \text{ K})}{15.0 \text{ L} - (62.0 \text{ mol})(0.0428 \text{ L/mol})} - \frac{(62.0 \text{ mol})^2 (2.253 \text{ L}^2 \text{ atm/mol})^2}{(15.0 \text{ L})^2}$$

$$P_{real} = \frac{1526 \text{ L atm}}{15.0 \text{ L} - 2.654 \text{ L}} - 38.49 \text{ atm} = 85.1 \text{ atm}$$

$$P_{ideal} = \frac{nRT}{V} = \frac{(62.0 \text{ mol})(0.08206 \text{ L atm/mol K})(300 \text{ K})}{15.0 \text{ L}} = 102 \text{ atm}$$

After use, $n = 0.620$ mol:

$$P = \frac{(0.620 \text{ mol})(0.08206 \text{ L atm/mol K})(300 \text{ K})}{15.0 \text{ L} - (0.620 \text{ mol})(0.0428 \text{ L/mol})} - \frac{(0.620 \text{ mol})^2 (2.253 \text{ L}^2 \text{ atm/mol})^2}{(15.0 \text{ L})^2}$$

$$P = \frac{15.26 \text{ L atm}}{15.0 \text{ L} - 0.0265 \text{ L}} - 0.00385 \text{ atm} = 1.02 \text{ atm}$$

$$P_{ideal} = \frac{nRT}{V} = \frac{(0.620 \text{ mol})(0.08206 \text{ L atm/mol K})(300 \text{ K})}{15.0 \text{ L}} = 1.02 \text{ atm}$$

Organize the results to make a comparison.

$$P_{full, real} = 85.1 \text{ atm} \qquad and \qquad P_{full, ideal} = 102 \text{ atm}$$
$$P_{used, real} = 1.02 \text{ atm} \qquad and \qquad P_{used, ideal} = 1.02 \text{ atm}$$

Notice that the van der Waals correction is appreciable (16.6 %) at high pressure but is negligible for a pressure near 1 atm.

Do the Results Make Sense? We know that methane normally is a gas at 298 K and atmospheric pressure, so it makes sense that the gas behaves ideally under these conditions. A pressure above 100 atm, on the other hand, is substantially higher than normal conditions, so we expect to see deviations from ideality.

Methane boils at −164 °C. Use the ideal gas equation to calculate the molar volume of methane gas at this temperature at 1 atm pressure. Then use the van der Waals equation to calculate the pressure exerted by one mole of the gas if it occupies this volume.

Example 11-1

Magnitudes
of van der Waals Corrections

**Extra Practice
Exercise 11.1**

Melting and Boiling Points

Melting points and boiling points can be used as indicators of the strengths of intermolecular forces. Remember from Chapter 5 that the average kinetic energy of molecular motion increases with temperature in kelvins. The boiling point of a substance is the temperature at which the average kinetic energy of molecular motion balances the attractive energy of intermolecular attractions. When the pressure is 1 atm, that temperature is the **normal boiling point.** For example, the normal boiling point of bromine is 332 K (59 °C). Above this temperature, the average kinetic energy of motion exceeds the attractive energies created by intermolecular forces of attraction, and bromine exists as a gas. Under atmospheric pressure, bromine is a liquid at temperatures below 332 K.

The conversion of a liquid into a gas is called **vaporization.** A liquid vaporizes when molecules leave the liquid phase faster than they are captured from the gas. **Condensation** is the reverse process. A gas condenses when molecules leave the gas phase more rapidly than they escape from the liquid. We explore these processes in more detail in Section 11.6.

The molecules in a liquid are able to move about freely, even though they do not escape from the liquid. When a liquid is cooled, however, its molecular energies decrease. At temperatures below the freezing point, the molecules become locked in place and the liquid solidifies. When the pressure is 1 atm, that temperature is the **normal freezing point.** A liquid freezes when liquid molecules have too little energy of motion to slide past one another. Conversely, a solid melts when its molecules have enough energy of motion to move freely past one another. Just as intermolecular forces determine the normal boiling point, these forces determine the freezing point. The larger the intermolecular forces, the higher the freezing point: bromine freezes at 266 K, chlorine at 172 K, and fluorine at 53.5 K.

Boiling points and melting points depend on the strengths of intermolecular forces. This is because the rates of escape and capture depend on the balance between molecular kinetic energies and intermolecular forces of attraction. A substance with large intermolecular forces must be raised to a high temperature before its molecules have sufficient kinetic energies to overcome those forces. A substance with small intermolecular forces must be cooled to a low temperature before its molecules have small enough kinetic energies to coalesce into a condensed phase. Table 11-2 lists the boiling and melting points of some representative elemental substances.

Table 11-2 Melting and Boiling Points

Substance	mp (K)	bp (K)	Substance	mp (K)	bp (K)
He	0.95*	4.2	Br$_2$	266	332
H$_2$	14.0	20.3	I$_2$	387	458
N$_2$	63.3	77.4	P$_4$	317	553
F$_2$	53.5	85.0	Na	371	1156
Ar	83.8	87.3	Mg	922	1363
O$_2$	54.8	90.2	Si	1683	2628
Cl$_2$	172	239	Fe	1808	3023

*Under high pressure.
He remains liquid at 0 K under normal pressure.

Section Exercises

■ **11.1.1** Based on the behavior of the other elements of Group 17, predict whether At$_2$ will be a gas, liquid, or solid at room temperature. Redraw the bar graph of Figure 11-3 to include At$_2$.

■ **11.1.2** From Figure 11-6, determine which of the four gases has the largest intermolecular forces and which has the smallest. State your reasoning.

■ **11.1.3** Use the data in Table 11-2 to answer the following: (a) Does Xe boil at a higher or lower temperature than Ar? (b) What is the order of increasing intermolecular forces for F$_2$, N$_2$, H$_2$, and Cl$_2$? (c) Which group in the periodic table has the smallest intermolecular forces?

11.2 TYPES OF INTERMOLECULAR FORCES

There are three general types of intermolecular forces. **Dispersion forces** describe the attractions between the negatively charged electron cloud of one molecule and the positively charged nuclei of neighboring molecules. All substances have dispersion forces. **Dipolar forces** describe the attractions between the negatively charged end of a polar molecule and the positively charged ends of neighboring polar molecules. Dipolar forces exist only for compounds that possess permanent dipole moments (see Section 9.5). **Hydrogen bonding forces** involve lone pairs of electrons on an electronegative atom of one molecule and a polar bond to hydrogen in another molecule. Hydrogen bonds occur primarily in molecules that contain O—H, N—H, and F—H covalent bonds.

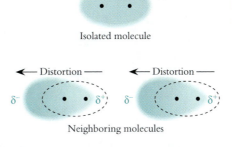

Figure 11-7
Exaggerated view of how dispersion forces arise.

Isolated molecule

←— Distortion ——   ←— Distortion ——

Neighboring molecules

Dispersion Forces

Dispersion forces exist because the electron cloud of any molecule distorts easily. For example, consider what happens when two halogen molecules approach each other. Each molecule contains positive nuclei surrounded by a cloud of negative electrons. As two molecules approach, the nuclei of one molecule attract the electron cloud of the other. Electrons are highly mobile, so their orbitals change shape in response to this attraction. At the same time, the two electron clouds repel each other, which leads to further distortion to minimize electron–electron repulsion. As Figure 11-7 indicates, this distortion of the electron cloud creates a charge imbalance, giving the molecule a slight positive charge at one end and a slight negative charge at the other. Dispersion forces are the net attractive forces among molecules generated by all these induced charge imbalances.

The magnitude of dispersion forces depends on how easy it is to distort the electron cloud of a molecule. This ease of distortion is called the **polarizability** because distortion of an electron cloud generates a temporary polarity within the molecule. We can explore how polarizability varies by examining boiling points of the halogens, as shown by the graph in Figure 11-8. The data reveal that boiling points increase with the total number of electrons. Fluorine, with 18 total electrons, has the lowest boiling point (85 K). Iodine, with 106 electrons, has the highest boiling point (458 K). The large electron cloud of I_2 distorts more readily than the small electron cloud of F_2. Figure 11-9 shows schematically how a large electron cloud distorts more than a small one, generating larger dispersion forces and leading to a higher boiling point.

Molecular size increases with chain length as well as with the size of individual atoms. Figure 11-10 shows how the boiling points of alkanes increase as the carbon chain gets longer. As alkanes get longer, their electron clouds become larger and more polarizable, making dispersion forces larger and raising the boiling point. As examples, methane (CH_4, 10 electrons) is a gas at 298 K, pentane (C_5H_{12}, 45 electrons) is a low-boiling liquid, decane ($C_{10}H_{22}$, 115 electrons) is a high-boiling liquid, and eicosane ($C_{20}H_{42}$, 225 electrons) is a waxy solid.

The boiling point increases progressively as molecules become larger and more polarizable. We apply this reasoning to the elemental noble gases in Example 11-2.

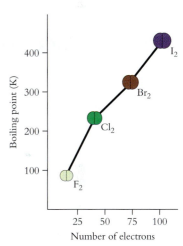

Figure 11-8
Boiling points of the halogens increase with the number of electrons.

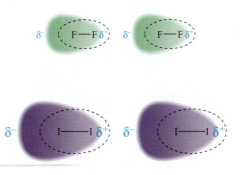

Figure 11-9
With many more electrons, the electron cloud of iodine is much larger and more polarizable than that of fluorine.

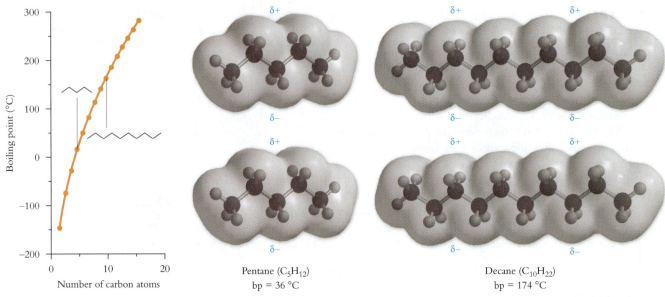

Pentane (C_5H_{12})
bp = 36 °C

Decane ($C_{10}H_{22}$)
bp = 174 °C

Figure 11-10
The boiling points of alkanes increase with the length of the carbon chain, because a long electron cloud is more polarizable than a short one.

Example 11-2	Neon and xenon are gases at room temperature, but both become liquids if the temperature is low enough. Draw a molecular picture showing the relative sizes and polarizabilities of atoms of neon and xenon, and use the picture to determine which substance has the lower boiling point.
Boiling Point Trends	

Strategy: The boiling point of a substance depends on the magnitude of its intermolecular forces, which in turn depends on the polarizability of its electron cloud. Monatomic gases contain atoms rather than molecules, so we must assess interatomic forces for these substances.

Solution: The dispersion forces that act between atoms of the noble gases depend on the polarizabilities of their electron clouds. The total electron counts for these atoms are 10 for neon and 54 for xenon. When two atoms approach each other, the smaller electron cloud of neon distorts less than the larger electron cloud of xenon, as a molecular picture illustrates:

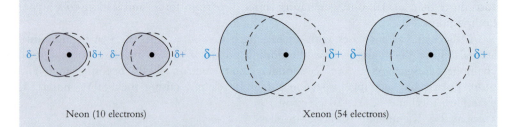

Neon (10 electrons)   Xenon (54 electrons)

Less polarizability means smaller partial charges and weaker intermolecular forces. Thus, neon has the lower boiling point.

Does the Result Make Sense? The experimental values are 166.1 K for Xe and 27.1 K for Ne, in agreement with our analysis.

Extra Practice Exercise 11.2	Among H_2S, H_2Se, and H_2Te, use dispersion arguments to determine which has the highest boiling point and which has the lowest.

Dispersion forces increase in strength with the number of electrons, because larger electron clouds are more polarizable than smaller electron clouds. For molecules with comparable numbers of electrons, the shape of the molecule makes an important secondary contribution to the magnitude of dispersion forces. For example, Figure 11-11 shows the shapes of pentane and 2,2-dimethylpropane. Both of these molecules have the formula C_5H_{12}, with 72 total electrons. Notice that 2,2-dimethylpropane has a more compact structure than pentane. This compactness results in a less polarizable electron cloud and smaller dispersion forces. Accordingly, pentane has a boiling point of 36 °C, while 2,2-dimethylpropane boils at 10 °C.

Pentane (C_5H_{12})
bp = 36 °C

2,2-Dimethylpentane (C_5H_{12})
bp = 10 °C

Figure 11-11
The boiling point of *n*-pentane is higher than the boiling point of 2,2-dimethylpropane, because an extended electron cloud is more polarizable than a compact one.

Dipolar Forces

Dispersion forces exist among all molecules, but some substances remain liquid at much higher temperatures than can be accounted for by dispersion forces alone. Consider 2-methylpropane and acetone, whose structures are shown in Figure 11-12. These molecules have similar shapes and nearly the same number of electrons (34 vs. 32). They are so similar that we might expect the two to have nearly equal boiling points, but acetone is a liquid at room temperature, whereas 2-methylpropane is a gas.

Why does acetone remain a liquid at temperatures well above the boiling point of 2-methylpropane? The reason is that acetone has a large dipole moment. Remember from Chapter 9 that chemical bonds are polarized toward the more electronegative atom. Thus, the C=O bond of acetone is highly polarized, with a partial negative charge on the O atom ($\chi = 3.5$) and a partial positive charge on the C atom ($\chi = 2.5$). The C—H bonds in these molecules, in contrast, are only slightly polar, because the electronegativity of hydrogen ($\chi = 2.1$) is only slightly smaller than that of carbon.

When two polar acetone molecules approach each other, they align so that the positive end of one molecule is close to the negative end of the other (Figure 11-12). In a liquid array, this repeating pattern of head-to-tail alignment gives rise to significant net attractive dipolar forces among the molecules.

The dispersion forces in acetone are nearly the same as those in 2-methylpropane, but the addition of dipolar forces makes the total amount of intermolecular attraction between acetone molecules substantially greater than the attraction between molecules of 2-methylpropane. Consequently, acetone boils at a considerably higher temperature than 2-methylpropane. Example 11-3 provides some additional comparisons of dispersion forces and dipole forces.

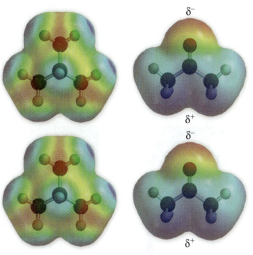

δ^-

δ^+

δ^-

δ^+

Figure 11-12
Acetone and 2-methylpropane have similar molecular shapes, but acetone has a large dipole moment resulting from its polar C=O bond.

The line structures of butane, methyl ethyl ether, and acetone follow. Explain the trend in boiling points: butane (0 °C), methyl ethyl ether (8 °C), and acetone (56 °C).

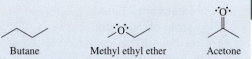

Butane   Methyl ethyl ether   Acetone

Example 11-3

Boiling Points and Structure

Strategy: These boiling points can be explained in terms of dispersion forces and dipolar forces. First, assess the magnitudes of dispersion forces, which are present in all substances, and then look for molecular polarity.

Solution: A table helps to organize the available information:

Substance	Boiling point	Total electrons
Butane	0 °C	34
Methyl ethyl ether	8 °C	34
Acetone	56 °C	32

Example 11-3

Boiling Points and Structure
(*continued*)

The table shows that dispersion forces alone cannot account for the range in boiling temperatures. Methyl ethyl ether and butane have the same number of electrons and similar shapes; yet their boiling points are different. Acetone, which has fewer electrons than the other compounds, has slightly smaller dispersion forces; yet it boils at a higher temperature. The order of boiling points indicates that acetone is a more polar molecule than methyl ethyl ether, which in turn is more polar than butane.

We expect butane to be nonpolar because of the small electronegativity difference between carbon and hydrogen. Acetone and methyl ethyl ether, on the other hand, contain polar carbon–oxygen bonds. Molecular geometry reveals why acetone is more polar than methyl ethyl ether. The full Lewis structures of these molecules show that the oxygen atom in the ether has a steric number of 4 and bent geometry. The two C—O bond dipoles in methyl ethyl ether partially cancel each other, leaving a relatively small molecular dipole moment. On the other hand, the polar C=O bond in acetone is unopposed, so acetone has a larger dipole moment and is more polar than methyl ethyl ether.

Does the Result Make Sense?
Drawings using arrows help us see why acetone has a larger dipole moment than dimethyl ether. The arrows show the charge displacement for each polar bond. Experimental values for the dipole moments are butane, 0 D; methyl ethyl ether, 1.12 D; and acetone, 2.88 D.

Methyl ethyl ether

Acetone

Acetaldehyde, CH_3COH, has a structure like that of acetone but with one CH_3 group replaced by H. This substance boils at 21 °C. Explain its boiling point relative to the three compounds described in this Example.

Hydrogen-Bonding Forces

Methyl ethyl ether is a gas at room temperature (boiling point = 8 °C), but 1-propanol, shown in Figure 11-13, is a liquid (boiling point = 97 °C). The compounds have the same molecular formula, C_3H_8O, and each has a chain of four inner atoms, C—O—C—C and O—C—C—C. Consequently, the electron clouds of these two molecules are about the same size, and their dispersion forces are comparable. Each molecule has an sp^3-hybridized oxygen atom with two polar single bonds, so their dipolar forces should be similar. The very different boiling points of 1-propanol and methyl ethyl ether make it clear that dispersion and dipolar forces do not reveal the entire story of intermolecular attractions.

The forces of attraction between 1-propanol molecules are stronger than those between methyl ethyl ether molecules because of an intermolecular interaction called a **hydrogen bond.** A hydrogen bond occurs when a highly electronegative atom with a lone pair of electrons shares its nonbonding electrons with a positively polarized hydrogen atom. Hydrogen bonds are intermolecular interactions, not intramolecular chemical bonds. These interactions are comparable to and sometimes stronger than dipolar and dispersion interactions, but they are only 5 to 10% as strong as covalent bonds.

There are two requirements for hydrogen bond formation. First, there must be an electron-deficient hydrogen atom that can accept an electron pair. Hydrogen atoms in O—H, F—H, and N—H bonds meet this requirement. Second, there must be a highly electronegative atom that can donate an electron pair. Three second-row elements, O, N, and F, meet this requirement, and there is recent evidence that S atoms in biological compounds also can donate electron pairs to form hydrogen bonds. Figure 11-14 shows representative examples of hydrogen bonding. The hydrogen bonds, which

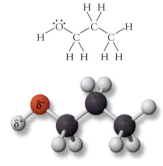

Figure 11-13
The Lewis structure and ball-and-stick model of 1-propanol.

Hydrogen fluoride

Water–ethanol

Salicylic acid

Formic acid

Ammonia–water

Glycine
(an amino acid)

Figure 11-14
Examples of hydrogen bonding.

are highlighted, are shown as dashed lines to indicate the weakly bonding nature of these interactions.

Notice from the examples shown in Figure 11-14 that hydrogen bonds can form between different molecules (for example, $H_3N \cdots H_2O$) or between identical molecules (for example, $HF \cdots HF$). Also notice that molecules can form more than one hydrogen bond (glycine, for example) and that hydrogen bonds can form within a molecule (salicylic acid, for example) as well as between molecules. Example 11-4 explores the possibilities for hydrogen bond formation.

In which of the following systems will hydrogen bonding play an important role: CH_3F, $(CH_3)_2CO$ (acetone), CH_3OH, and NH_3 dissolved in $(CH_3)_2CO$?

Example 11-4

Formation of Hydrogen Bonds

Strategy: Hydrogen bonds require electron-deficient hydrogen atoms in polar H—X bonds and highly electronegative atoms with nonbonding pairs of electrons are present. Use Lewis structures to determine whether these requirements are met.

Solution: Here are the Lewis structures of the four molecules:

Acetone and CH_3F contain electronegative atoms with nonbonding pairs, but neither has any highly polar H—X bonds. Thus, there is no hydrogen bonding between molecules of these substances.

The O—H bond in CH_3OH meets both of the requirements for hydrogen bonding. The O—H hydrogen atom on one molecule interacts with the oxygen atom of a neighboring molecule:

For a solution of ammonia in acetone, we must examine both components. Acetone has an electronegative oxygen atom with nonbonding pairs, whereas NH_3 has a polar N—H bond. Consequently, a mixture of these two compounds displays hydrogen bonding between the hydrogen atoms of ammonia and oxygen atoms of acetone:

Do the Results Make Sense? Methanol has a considerably higher boiling point than its alkane relatives, methane and ethane, consistent with significant intermolecular forces. Ammonia dissolves readily in acetone, also consistent with significant intermolecular forces.

Draw a picture that shows the hydrogen-bonding interactions of an acetone molecule dissolved in water.

**Extra Practice
Exercise 11.4**

Hydrogen bonding is particularly important in biochemical systems, because bio-molecules contain many oxygen and nitrogen atoms that participate in hydrogen bonding. For example, the amino acids from which proteins are made contain NH_2 (amino) and CO_2H (carboxylic acid) groups. Four different types of hydrogen bonds exist in these systems: O----H—N, N----H—O, O----H—O, and N----H—N. When biological molecules contain S atoms, S----H—O and S----H—N hydrogen bonds also can form. Figure 11-14 includes a view of hydrogen bonding between glycine molecules, and we examine more details of hydrogen bonding in biomolecules in Chapter 13.

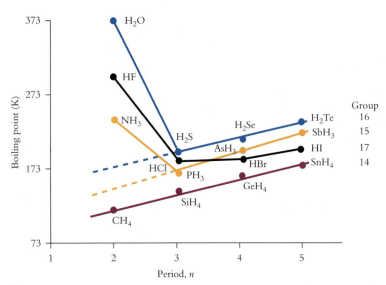

Figure 11-15
Periodic trends in the boiling points of binary hydrogen compounds. Notice that H_2O, HF, and NH_3 are exceptions to the trends.

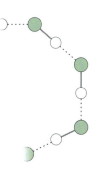

Binary Hydrogen Compounds

Boiling points of the binary hydrogen compounds illustrate the interplay among different types of intermolecular forces. The graph in Figure 11-15 shows that there are periodic trends in the boiling points of these compounds. In general, the boiling points of the binary hydrogen compounds increase from top to bottom of each column of the periodic table. This trend is due to increasing dispersion forces: The more electrons the molecule has, the stronger the dispersion forces and the higher the boiling point. In Group 16, for example, H_2S (18 e^-) boils at −60 °C, H_2Se (36 e^-) at −41 °C, and H_2Te (54 e^-) at −4 °C.

As Figure 11-15 shows, ammonia, water, and hydrogen fluoride depart dramatically from this periodic behavior. This is because their molecules experience particularly large intermolecular forces resulting from hydrogen bonding. In hydrogen fluoride, for instance, partial donation of an electron pair from the highly electronegative fluorine atom of one HF molecule to the electron-deficient hydrogen atom of another HF molecule creates a hydrogen bond. Similar interactions among many HF molecules result in a network of hydrogen bonds that gives HF a boiling point much higher than those of HCl, HBr, and HI.

Fluorine has the highest electronegativity, so the strongest *individual* hydrogen bonds are those in HF. Every hydrogen atom in liquid HF is involved in a hydrogen bond, but there is only one polar hydrogen atom per molecule. Thus each HF molecule participates in two hydrogen bonds with two other HF partners. There is one hydrogen bond involving the partially positive hydrogen atom and a second involving the partially negative fluorine atom.

Water has a substantially higher boiling point than hydrogen fluoride has, which indicates that the total amount of hydrogen bonding in H_2O is greater than that in HF. The higher boiling point of water reflects the fact that it forms more hydrogen bonds *per molecule* than hydrogen fluoride. A water molecule has two hydrogen atoms that can form hydrogen bonds and two nonbonding electron pairs on each oxygen atom. This permits every water molecule to be involved in four hydrogen bonds to four other H_2O partners, as shown in Figure 11-16.

Hydrogen bonding in solid ice creates a three-dimensional network that puts each oxygen atom at the center of a distorted tetrahedron. Figure 11-16 shows that two arms of the tetrahedron are regular covalent O—H bonds, whereas the other two arms of the tetrahedron are hydrogen bonds to two different water molecules.

Dispersion forces, dipole interactions, and hydrogen bonds all are significantly weaker than covalent intramolecular bonds. For example, the average C—C bond energy is 345 kJ/mol, whereas dispersion forces are just 0.1 to 5 kJ/mol for small alkanes such as propane. Dipolar interactions between polar molecules such as acetone range between 5 and 20 kJ/mol, and hydrogen bonds range between 5 and 50 kJ/mol.

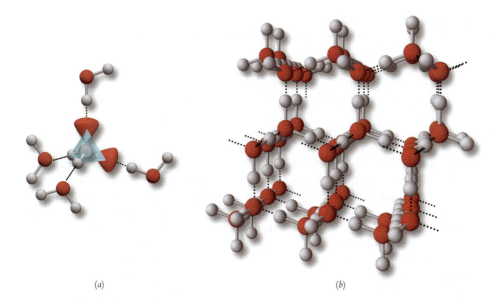

(a)   (b)

Figure 11-16
The structure of ice. (*a*) Each oxygen atom is at the center of a distorted tetrahedron of hydrogen atoms. The tetrahedron is composed of two short covalent O—H bonds and two long H----O hydrogen bonds. (*b*) Water molecules are held in a network of these tetrahedra.

Section Exercises

■ **11.2.1** Explain the following differences in boiling points: (a) Kr boils at −152 °C, and propane boils at −42 °C; (b) C(CH$_3$)$_4$ boils at 10 °C, and CCl$_4$ boils at 77 °C; and (c) N$_2$ boils at −196 °C, and CO boils at −191.5 °C.

■ **11.2.2** Acetone and methanol have nearly equal boiling points. What types of intermolecular forces do each exhibit? What does the similarity in boiling points tell you about the relative magnitudes of each type of force in these two compounds?

■ **11.2.3** There are nine important hydrogen-bonding interactions. One of them is O----HO. Draw the other eight. For each of the nine, draw a Lewis structure of a specific example using real molecules.

11.3 LIQUIDS

A gas condenses to a liquid if it is cooled sufficiently. Condensation occurs when the average kinetic energy of motion of molecules falls below the value needed for the molecules to move about independently. Thus, the molecules in a liquid are confined to a specific volume by intermolecular forces of attraction. Although they cannot readily escape, liquid molecules remain free to move about within the liquid phase. In this behavior, liquid molecules behave like the molecules of a gas. The large-scale consequences of the molecular-level properties are apparent. Like gases, liquids are fluid, so they flow easily from place to place. Unlike gases, however, liquids are compact, so they cannot expand or contract significantly.

Properties of Liquids

In a liquid, intermolecular forces are strong enough to confine the molecules to a specific volume, but they are not strong enough to keep molecules from moving from place to place within the liquid. The relative freedom of motion of liquid molecules leads to three liquid properties arising from intermolecular forces: **surface tension, capillary action,** and **viscosity.**

A small amount of a liquid tends to take a spherical shape: For example, mercury drops are nearly spherical and water drips from a faucet in nearly spherical liquid droplets. Surface tension, which measures the resistance of a liquid to an increase in its surface area, is the physical property responsible for this behavior.

Figure 11-17 illustrates at the molecular level why liquids exhibit surface tension. A molecule in the interior of a liquid is completely surrounded by other molecules. A molecule at a liquid surface, on the other hand, has other molecules beside it and beneath it but very few above it in the gas phase. This difference means that there is a net intermolecular force on molecules at the surface that pulls them toward the interior of the liquid.

Mercury forms near-spherical drops.

Molecule is pulled away from surface

No net "pull" in any direction

④

Figure 11-17
In the interior of a liquid (*bottom*), each molecule experiences equal forces in all directions. A molecule at the surface of a liquid (*top*) is pulled back into the liquid by intermolecular forces.

The net attraction of surface molecules to the interior of the liquid indicates that molecules are most stable when attractive forces are maximized by as many neighbor molecules as possible. Consequently, a liquid is most stable when the fewest molecules are at its surface. This occurs when the liquid has minimal surface area. Spheres have less surface area per unit volume than any other shape, so small drops of a liquid tend to be spheres. Large drops are distorted from ideal spheres by the force of gravity.

Molecules in contact with the surface of their container experience two sets of intermolecular forces. *Cohesive forces* attract molecules in the liquid to one another. In addition, *adhesive forces* attract molecules in the liquid to the molecules of the container walls.

One result of adhesive forces is the curved surface of a liquid, called a **meniscus.** As Figure 11-18 shows, water in a glass tube forms a concave meniscus that increases the number of water molecules in contact with the walls of the tube. This is because adhesive forces of water to glass are stronger than the cohesive forces among water molecules.

Figure 11-18 shows another result of adhesive forces. In a tube of small enough diameter, water actually climbs the walls, pulled upward by strong adhesive forces. This upward movement of water against the force of gravity is capillary action. Capillary action between sap and the cellulose walls of wood fiber plays a role in how trees transport sap from their roots to their highest branches.

Water can be poured very quickly from one container to another, salad oil pours more slowly, and honey sometimes seems to take forever. A liquid's resistance to flow is called its viscosity; the greater its viscosity, the more slowly the liquid pours. Viscosity measures how easily molecules slide by one another, and this depends strongly on molecular shapes. Liquids such as water, acetone, and benzene, whose molecules are small and compact, have low viscosity. In contrast, large molecules such as the sugars in honey and the hydrocarbons found in oils tend to be tangled up with each other. Tangling inhibits the flow of molecules and leads to high viscosity. In addition, strong intermolecular cohesive forces make it harder for molecules to move about.

Molecules move faster as temperature increases, and this allows them to slide by one another more easily. Thus, viscosity decreases as temperature increases. This dependence is quite noticeable for highly viscous substances such as honey and syrup, which are much easier to pour when hot than when cold.

Vapor Pressure

Whenever a liquid has an exposed surface, some of its molecules will escape into the gas phase. Remember from Section 5.3 that any collection of molecules has a distribution of kinetic energies. A liquid has a fixed volume because, although its molecules have sufficient energy of motion to move about within the liquid, on average they do not have sufficient energy of motion to escape into the gas phase. Nevertheless, the distribution of molecular energies guarantees that some of the molecules in any liquid have enough kinetic energy to overcome the intermolecular forces that confine the liquid. These molecules escape into the gas phase if they reach the surface of the liquid.

Our senses tell us that molecules escape from a liquid. The red gas phase above the liquid bromine in Figure 11-4 shows that bromine molecules are in both phases. The evaporation of a rain puddle in the sunshine shows that water molecules escape from the liquid into the gas phase. The smell of gasoline around an open tank informs our noses that gasoline molecules escape from the liquid phase into the gas phase.

How many of the molecules of a liquid have enough energy to escape into the vapor phase? That depends on the temperature and the strength of intermolecular forces for the liquid. As Figure 11-19 shows, at any particular temperature more molecules can escape from liquid bromine than from liquid water. This is because the intermolecular forces among water molecules are larger than those between bromine molecules. A water molecule needs more energy of motion to escape the liquid phase. The figure also illustrates that when the temperature rises, the fraction of molecules having enough energy to escape increases.

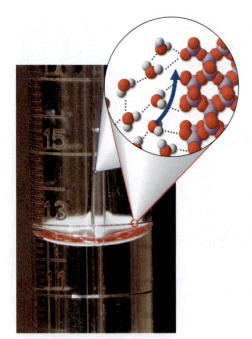

Figure 11-18
Water in contact with glass takes on a concave meniscus, and water rises inside a small-diameter tube because of capillary action.

A liquid in an open container continually loses molecules until eventually it has evaporated completely. In a closed container such as the one shown in Figure 11-4, however, the partial pressure of the vapor increases as more and more molecules enter the gas phase. As this partial pressure builds, increasing numbers of molecules from the gas strike the liquid surface and are recaptured. Eventually, as Figure 11-20 shows, the number of molecules escaping from the liquid exactly matches the number of molecules being captured by the liquid. There is no net change, so this is a dynamic equilibrium. The pressure at which this equilibrium exists is the **vapor pressure** of the liquid. We can conclude from Figure 11-19 that the vapor pressure of any liquid increases with increasing temperature.

When molecular energies are nearly sufficient to overcome intermolecular forces, molecules of a substance move relatively freely between the liquid phase and the vapor phase. We describe these phase changes in Section 11.6.

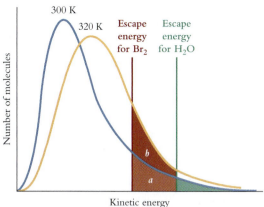

Figure 11-19
The fraction of molecules with enough kinetic energy to escape a liquid depends on the strength of intermolecular forces and temperature. (*a*) At 300 K, more bromine molecules can escape than water molecules. (*b*) More bromine molecules can escape at 320 K than at 300 K.

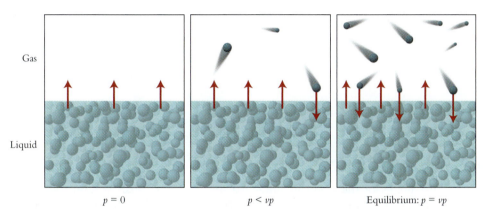

$p = 0$ $p < vp$ Equilibrium: $p = vp$

Figure 11-20
As the partial pressure of a substance in the gas above a liquid increases, the difference between escapes and recaptures decreases until, at the vapor pressure of the substance, equilibrium exists.

Section Exercises

■ **11.3.1** The alcohol *n*-pentanol ($C_5H_{11}OH$) is more than ten times as viscous as *n*-hexane (C_6H_{14}) at room temperature. Why is this?

■ **11.3.2** The mercury inside the glass column of a barometer displays a convex meniscus rather than the concave meniscus shown by water inside a glass column. Explain in terms of intermolecular forces.

■ **11.3.3** If the equilibrium view in Figure 11-20 represents bromine at 298 K, redraw the figure to show, qualitatively, how the liquid−vapor equilibrium for water at 298 K differs.

11.4 FORCES IN SOLIDS

Like liquids, solids are compact, with molecules nestled close to one another. In solids, though, intermolecular forces are so strong that the molecules cannot move past one another. Instead, they are locked in position. This rigidity gives solids stable shapes that we see in structures ranging from bones to airplane wings. One of the most active areas of research in modern chemistry, physics, and engineering is the development of new solid materials. Solids play an ever-larger role in society, from high-temperature superconductors, to heat-resistant tiles for the outer "skin" of the space shuttle, to new tissue-compatible solids for surgical implants. In this section, we describe the various forces that exist in solids.

Table 11-3 Forces in Different Types of Solids

Solid type	Molecular	Molecular	Molecular	Metallic	Network	Ionic
Attractive forces	Dispersion	Dispersion + dipolar	Dispersion + dipolar + H bonding	Delocalized bonding	Covalent	Electrical
Energy (kJ/mol)	0.05 – 40	5 – 25	10 – 40	75 – 1000	150 – 500	400 – 4000
Example	Ar	HCl	H_2O	Cu	SiO_2	NaCl
Melting point (K)	84	158	273	1357	1983	1074
Molecular picture						

⑤

Magnitudes of Forces

The melting points of the elements presented in Table 11-2 span an immense range, from <15 K (H_2) to >1500 K (Si and Fe). These values indicate that the forces holding solids together range from very small to extremely large. This is because in addition to the intermolecular forces described in Section 11.2, solids can be bound together by covalent bonds, metallic bonding (with delocalized electrons), and ionic interactions. There are four distinct types of solids—molecular, metallic, network, and ionic—each characterized by a different type of force. Table 11-3 contrasts the forces and energies in these four types of solids.

The molecules (or atoms, for noble gases) of a **molecular solid** are held in place by the types of forces already discussed in this chapter: dispersion forces, dipolar interactions, and/or hydrogen bonds. The atoms of a **metallic solid** are held in place by the delocalized bonding described in Section 10.7. A **network solid** contains an array of covalent bonds linking every atom to its neighbors. An **ionic solid** contains cations and anions, attracted to one another by electrical forces as described in Section 8.5.

Naphthalene
$C_{10}H_8$

Molecular Solids

Molecular solids are aggregates of molecules bound together by intermolecular forces. Substances that are gases under normal conditions form molecular solids when they condense at low temperature. Many larger molecules have sufficient dispersion forces to exist as solids at room temperature. One example is naphthalene ($C_{10}H_8$), a white solid that melts at 80 °C. Naphthalene has a planar structure like that of benzene (see Section 10.6), with a cloud of ten delocalized π electrons that lie above and below the molecular plane. Naphthalene molecules are held in the solid state by strong dispersion forces among these highly polarizable π electrons. The molecules in crystalline naphthalene are stacked face-to-face to maximize dispersion forces. This intermolecular molecular structure leads to plate-like crystals in the macroscopic solid state.

In addition to dispersion forces, molecular solids often involve dipolar interactions and hydrogen bonding. Benzoic acid ($C_6H_5CO_2H$) provides a good example, as illustrated in Figure 11-21. Molecules of benzoic acid are held in place by a combination of dispersion forces among the π electrons and hydrogen bonding among the $-CO_2H$

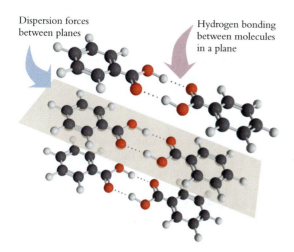

Dispersion forces between planes

Hydrogen bonding between molecules in a plane

Figure 11-21
Crystals of benzoic acid contain pairs of molecules held together head to head by hydrogen bonds. These pairs then stack in planes that are held together by dispersion forces.

groups. With fewer π electrons, benzoic acid has weaker dispersion forces than naphthalene, but its hydrogen bonding gives benzoic acid a higher melting point, 122 °C.

The effect of extensive hydrogen bonding is revealed by the relatively high melting point of glucose ($C_6H_{12}O_6$), the sugar found in blood and human tissue. Glucose melts at 155 °C, because each of its molecules has five —OH groups that form hydrogen bonds to neighboring molecules. Although glucose lacks the highly polarizable π electrons found in naphthalene and benzoic acid, its extensive hydrogen bonding gives this sugar the highest melting point of the three compounds.

The molecules of a molecular solid retain their individual properties. Solid I_2, for example, is dark in color because individual I_2 molecules absorb visible light, as shown by the fact that I_2 vapor also is colored. Glucose dissolved in coffee tastes just as sweet as solid glucose because the sweet taste of glucose comes from its shape, which is the same in the solid phase as it is in solution.

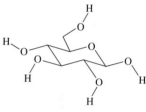

Glucose $C_6H_{12}O_6$

Network Solids

In sharp contrast to molecular solids, network solids have very high melting points. Compare the behavior of phosphorus and silicon, third-row neighbors in the periodic table. As listed in Table 11-2, phosphorus melts at 317 K, but silicon melts at 1683 K. Phosphorus is a molecular solid that contains individual P_4 molecules, but silicon is a network solid in which covalent bonds among Si atoms connect all the atoms. The vast array of covalent bonds in a network solid makes the entire structure behave as one giant "molecule."

The cause of the great difference in the melting points of these two elements is evident from Table 11-3: Bonding forces are much stronger than intermolecular forces. For solid silicon to melt, a significant fraction of its Si—Si covalent bonds must break. The average Si—Si bond energy is 225 kJ/mol, whereas attractive energies due to intermolecular forces are less than 40 kJ/mol, so it takes a much higher temperature to melt silicon than phosphorus.

Bonding patterns determine the properties of network solids. Diamond and graphite, the two forms of elemental carbon that occur naturally on Earth, are both network solids, but they have very different physical and chemical properties. Diamond contains a three-dimensional array of σ bonds, with each tetrahedral (sp^3) carbon atom linked to all the others through a network of covalent bonds. This three-dimensional network of strong covalent bonds makes diamond extremely strong and abrasive. Graphite, in contrast, has trigonal planar (sp^2) carbon atoms in a two-dimensional array of σ bonds. The structure is supplemented by delocalized π bonding above and below the plane of the σ bonds. Each two-dimensional layer is attracted to its

Phosphorus is a molecular solid.

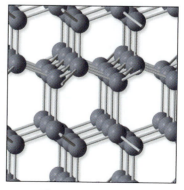

Silicon is a network solid.

Diamond has three-dimensional bonding.

Graphite has a planar bond network.

neighboring layers only by dispersion forces among the π electrons. As a result, planes of carbon atoms easily slide past one another, making graphite a brittle lubricant.

Several oxides and sulfides display the characteristics of network solids. The bond network of silica appears in Section 9.3. Other examples are titania (TiO_2) and alumina (Al_2O_3). These two substances have extremely high melting points because their atoms are held together by networks of strong σ covalent bonds. Like graphite, MoS_2 is a two-dimensional network solid that serves as a solid lubricant.

Covalent bonds make network solids extremely durable. Geological examples include the "everlasting sands" and granite formations such as the Rock of Gibraltar. Like diamonds, other valuable gemstones are network solids, too. Rubies and sapphires are covalent crystals of aluminum oxide with small amounts of colored transition metal ion impurities. Carborundum is a 1:1 network solid of silicon and carbon that has the same lattice structure as diamond. Carborundum is much less expensive than diamond but almost as strong and wear-resistant, so it is used for the edges of cutting tools. Example 11-5 compares the structures and properties of a network solid and a molecular solid.

Example 11-5 **Network and Molecular Solids**	Whereas SiO_2 melts at 1710 °C, other nonmetal oxides melt at much lower temperatures. For example, P_4O_6 melts at 25 °C. Referring to the accompanying bonding pictures, describe the forces that hold these solids together.

P_4O_6

SiO_2 network

Strategy: Identify the type of solid. Solids may be network, ionic, metallic, or molecular. Different forces account for the stability of each type.

Solution: Because these are nonmetal oxides, they cannot be described as metallic. Neither oxide contains ions, so they must be network or molecular. The melting points provide the information needed to categorize the oxides, and the molecular views support the identifications.

SiO_2: The high melting point of 1710 °C indicates that this is a network solid, in which strong covalent bonds must break to liquefy the substance. The bonding picture shows atoms linked together, each O atom bonded to two Si atoms and each Si atom bonded to four O atoms. This gives a three-dimensional array of strong covalent bonds, many of which must break for silica to melt.

P_4O_6: The relatively low melting point of 25 °C indicates a molecular solid. The molecular structure shows that P_4O_6 is a discrete molecule. Strong covalent bonding holds the atoms in each molecule together, but each molecule is attracted to the others only by dispersion forces. In this molecular solid, little energy is required to overcome dispersion forces and allow P_4O_6 solid to melt.

Do the Results Make Sense? Much of the Earth's silicon is found in durable rock formations based on silica. Granite is a common example. This is consistent with a network solid. Phosphorus, in contrast, occurs naturally in ionic solids containing the phosphate anion, indicating that phosphorus oxides readily undergo chemical reactions. The latter observation is consistent with a molecular solid.

One neighbor of arsenic, germanium, melts at 1210 °C. Its other neighbor, selenium, melts at 490 °C. Based on these melting points, what type of solid is each of these elements?

**Extra Practice
Exercise 11.5**

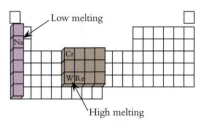

Sodium can be cut with a knife.

Metallic Solids

As described in Section 10.7, the bonding in solid metals comes from electrons in highly delocalized valence orbitals. There are so many such orbitals that they form energy bands, giving the valence electrons high mobility. Consequently, each metal atom can be viewed as a cation embedded in a "sea" of mobile valence electrons. The properties of metals can be explained on the basis of this picture. Section 10.7 describes the most obvious of these properties, electrical conductivity.

Metals display a range of melting points, indicating that the strength of metallic bonding is variable. The alkali metals of Group 1 are quite soft and melt near room temperature. Sodium, for example, melts at 98 °C and cesium at 28.5 °C. Bonding is weak in these metals because each atom of a Group 1 metal contributes only one valence electron to the bond-forming energy band. Metals near the middle of the *d* block, on the other hand, are very hard and have some of the highest known melting points. Tungsten melts at 3407 °C, rhenium at 3180 °C, and chromium at 1857 °C. Atoms of these metals contribute two *s* electrons and several *d* electrons to the bond-forming energy band, leading to extremely strong metallic bonding.

Metals are ductile and malleable, meaning they can be drawn into wires or hammered into thin sheets. When a piece of metal forms a new shape, its atoms change position. Because the bonding electrons are fully delocalized, however, changing the positions of the atoms does not cause corresponding changes in the energy levels of the electrons. The "sea" of electrons is largely unaffected by the pattern of metal cations, as Figure 11-22 illustrates. Thus, metals can be forced into many shapes, including sheets and wires, without destroying the bonding nature of their filled bands of delocalized orbitals.

The transition metals display a range of properties. Copper and silver are much better electrical conductors than chromium. Tungsten has very low ductility. Mercury is a liquid at room temperature. All these differences result from variations in numbers of

Tungsten can be heated to incandescence without melting.

**Figure 11-22
When a metal changes shape, its atoms shift position. However, because the valence electrons are fully delocalized, the energy of these electrons is unaffected.**

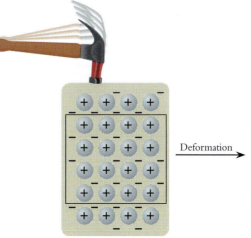

Deformation

valence electrons. A transition metal such as vanadium or chromium has five or six valence electrons per atom, all of which occupy bonding orbitals in the metal lattice. As a result, there are strong attractive forces among the metal atoms, and vanadium and chromium are strong and hard. Beyond the middle of the transition metal series, the additional valence electrons occupy antibonding orbitals, which reduces the net bonding. This effect is most pronounced at the end of the d block, where the number of antibonding electrons nearly matches the number of bonding electrons. Zinc, cadmium, and mercury, with d^{10} configurations, have melting temperatures that are more than 600 °C lower than those of their immediate neighbors.

Ionic Solids

As described in Chapter 8, ionic solids contain cations and anions strongly attracted to each other by electrical forces. These forces act between *ions* rather than between *molecules*. Ionic solids must be electrically neutral, so their stoichiometries are determined by the charges carried by the positive and negative ions.

Many ionic solids contain metal cations and polyatomic anions. Here again, the stoichiometry of the solid is dictated by the charges on the ions and the need for the solid to maintain electrical neutrality. Some examples of 1:1 ionic solids containing polyatomic anions are $NaOH$, KNO_3, $CuSO_4$, $BaCO_3$, and $NaClO_3$. Some metallic ores have 1:1 stoichiometry, such as scheelite, $CaWO_4$ (contains WO_4^{2-}), zircon, $ZrSiO_4$ (contains SiO_4^{4-}), and ilmenite, $FeTiO_3$ (contains TiO_3^{2-}).

Minerals often contain more than one cation or anion. For example, apatite, $Ca_5(PO_4)_3F$, contains both phosphate and fluoride anions. Beryl, $Be_3Al_2Si_6O_{18}$, contains beryllium and aluminum cations as well as the $Si_6O_{18}^{12-}$ polyatomic anion. An even more complicated example is garnierite, $(Ni,Mg)_6Si_4O_{10}(OH)_2$, which has a variable composition of Ni^{2+} and Mg^{2+} cations. Although the relative proportions of Mg^{2+} and Ni^{2+} vary, garnierite always has six cations and two anions of OH^- for every anion of $Si_4O_{10}^{10-}$.

One of the most exciting developments in materials science in recent years involves mixed oxides containing rare earth metals. Some of these compounds are *superconductors*, as described in our Chemistry and Technology Box. Below a certain temperature, a superconductor can carry an immense electrical current without losses from resistance. Before 1986, it was thought that this property was limited to a few metals at temperatures below 25 K. Then it was found that a mixed oxide of lanthanum, barium, and copper showed superconductivity at around 30 K, and since then the temperature threshold for superconductivity has been advanced to 135 K.

Ceramic superconductor materials display a characteristic that is unexpected in traditional chemistry but relatively common in solids: nonstoichiometric composition. When $YBa_2Cu_3O_{7-x}$ contains a stoichiometric amount of oxygen ($x = 0$ or 1), it has inferior superconducting properties. Optimal superconducting behavior requires a slight deviation from stoichiometric composition. To maintain overall electrical neutrality, the average charge on the copper cations must deviate from integral values. Assigning characteristic charges to the other elements (+3 for Y, +2 for Ba, and −2 for O), we find that Cu must have an average charge of $+6.8/3 = 2.27$ when $x = 0.1$. Because fractional charges are impossible, this means that most of the Cu atoms in the substance bear +2 charges, but a few have +3 charges. We return to nonstoichiometric compositions of solid materials in Section 11.5.

Section Exercises

11.4.1 Describe the forces that exist in (a) solid CO_2 (dry ice); (b) crystalline yellow elemental sulfur, S_8; (c) tin (soft, malleable, melts at 232 °C); and (d) Li_2O.

11.4.2 Arrange the following in order of increasing melting point and explain the reasons for your rankings: C (diamond), F_2, K, Co.

11.4.3 Determine the chemical formulas of the following ionic substances: (a) magnesium phosphate; (b) spodumene, a mineral containing lithium and aluminum cations and $Si_2O_6^{4-}$; and (c) the superconducting oxide that contains Bi^{3+}, Sr^{2+}, and Cu^{2+} in 2:2:1 atomic ratio.

Box 11-1   Chemistry and Technology: Superconductors

T rains that run on frictionless tracks and computer chips smaller than those of the present generation yet faster and with much larger capacities—these are potential applications of room-temperature superconductors. Research groups around the world are developing new materials in hopes of reaching this spectacular goal.

Metals conduct electricity because their valence electrons are highly mobile. Although charge flows easily through a metal wire, normal electrical conductors always have significant resistivity. Qualitatively, electrons are slowed by bumping into atoms along their paths. In contrast to normal conductivity, superconductivity is characterized by zero electrical resistance. Electrons move freely through a superconductor without any friction. A superconductor can carry immense amounts of current without losses resulting from heating.

A material that displays superconductivity does so only below a critical temperature, T_c. Above T_c, the material has normal resistivity, but as the temperature drops below T_c, its resistance abruptly disappears, as the graph shows.

O   Cu
La, Sr, or Ba

The transition between normal conductivity and superconductivity occurs at different temperatures for different materials. In general, T_c is near 4.2 K, the boiling point of liquid helium. For this reason, any device that makes use of superconductivity must be immersed in a bath of liquid helium. Little wonder that superconductivity was not discovered until early in this century and remained a laboratory curiosity until the mid-1980s.

Early work on superconductors concentrated on metals or metal mixtures (alloys). Niobium alloys are particularly good superconductors, and in 1973 a niobium alloy, Nb_3Ge, was found to have $T_c = 23$ K, the highest known value for a metal superconductor. In 1986, a ceramic oxide with formula $La_{2-x}Ba_xCuO_4$ was found to show superconductivity at 30 K. Through intense research efforts on ceramic oxides, $YBa_2Cu_3O_{7-x}$, with $T_c = 93$ K, was discovered in 1987.

A temperature above 77 K brought superconductivity to a temperature that can be achieved using inexpensive liquid nitrogen as a coolant. Moreover, the compound is easy to make: Three oxides, Y_2O_3, BaO, and CuO, are ground together in the correct stoichiometric ratio and heated to 950 °C. The mixture is cooled and pressed into pellets, after which the pellets are heated to just below their melting point to bind the grains in the pellets tightly together. Finally, the pellets are heated in oxygen at around 550 °C.

Ceramic oxide superconductors have distinct atomic layers. The Cu-containing superconductors contain planes of copper and oxygen atoms, as the molecular view shows. These planes alternate with layers containing oxygen and the other metals that make up the superconductor. Superconductivity takes place in the Cu—O planes.

The record for the highest superconducting temperature in the year 2004 is 138 K, held by a nonstoichiometric ceramic oxide, $Hg_{0.8}Tl_{0.2}Ba_2Ca_2Cu_3O_{8.33}$. This is still far below room temperature, but research continues.

Superconductivity has also been discovered in rather exotic materials, including the following: Buckminsterfullerene (C_{60}) doped with ICl; Carbon nanotubes (superconductivity in just one direction); Nickel borocarbides, which contain Ni_2B_2 layers alternating with RC sheets, where R is a rare earth element such as Er; and organic superconductors that contain planar organic cations and oxoanions. Chemists and physicists continue to study these and other families of superconductors.

For many applications, a superconductor must first be drawn into a wire. This has recently been accomplished. The photo shows a superconductor ribbon wrapped around the copper wires that it could replace.

Flexible superconducting tapes provide promise of uses for superconductors in motors, generators, and even electric transmission lines. Meanwhile, superconducting magnets cooled to the temperature of liquid helium already are in use. High-field nuclear magnetic resonance (NMR) spectrometers have become standard instruments in chemical research laboratories, and the same type of machine (called an MRI spectrometer) is used for medical diagnosis in hospitals worldwide.

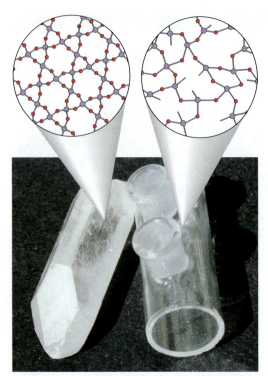

Figure 11-23
Quartz (*left*) is a crystalline form of silicon dioxide containing a regular three-dimensional array of SiO$_2$ units. Silica glass (*right*) is an amorphous form containing irregular arrays of SiO$_2$ units, but in this case the bond arrangement is irregular.

11.5 ORDER IN SOLIDS

In a solid, the atoms, molecules, or ions are in fixed positions. Their motions are restricted to vibrations about preferred locations. This contrasts with gases and liquids, whose molecules change their positions continuously. A solid that displays highly regular repeating patterns of organization is classified as a **crystalline solid.** Examples include diamonds, sugar crystals, quartz, and table salt. Precious gemstones like rubies, sapphires, and emeralds are highly valued crystals of rare and richly colored minerals. A solid whose arrangement of particles is irregular, with no observable pattern, is classified as an **amorphous solid.** Amorphous solids include cotton candy, glass, and wax.

Whether a particular substance is crystalline or amorphous can depend on how the solid is prepared. Silicon dioxide exists in both forms. Quartz is a crystalline form of silicon dioxide found in minerals all over the world. Each tetrahedral silicon atom bonds to four oxygens in a highly symmetrical three-dimensional network. This strong bonding network gives silicon dioxide its high melting point of 1710 °C. Slow cooling of molten SiO$_2$ gives crystalline quartz, but rapid cooling gives an amorphous glass. Figure 11-23 shows molecular representations of quartz and glass. Both structures are three-dimensional networks, but the two-dimensional representations shown in the figure are sufficient to show the differences between crystalline and amorphous solids.

The photographs in Figure 11-24 illustrate the following characteristics of crystalline solids:

1. Parallel faces and edges characterize the shape of a crystal. The edges of a crystal usually intersect at characteristic angles.

2. When a crystal breaks into smaller pieces, fragmentation occurs along crystal edges. Small crystals have the same characteristic angles as larger crystals.

3. A crystal has a high degree of symmetry.

Figure 11-24
Minerals can form colorful regular crystals. Shown here are vanadinite, Pb$_5$(VO$_4$)$_3$Cl (*left*); quartz, SiO$_2$ (*center*); fluorapatite, Ca$_5$(PO$_4$)$_3$F (*top right*); and stibnite, Sb$_2$S$_3$ (*bottom right*).

The Crystal Lattice and the Unit Cell

Any crystal is a vast array of atoms, molecules, or ions arranged in some regular repeating pattern. In Section 8.5 we present the regular arrangement of ions in an ionic crystal, known as the crystal lattice. In the crystal lattice, every ion of a given type has an identical environment. The notion of a crystal lattice extends to any crystalline substance. We place a point at the center of each species (atom, molecule, or ion) contained in the crystal. The crystal lattice is the pattern formed throughout the crystal by these points. This pattern is regular and symmetrical, and the symmetry of a crystal reflects the symmetry of its lattice.

Because the pattern of the lattice repeats exactly, every crystal has one smallest unit from which the entire pattern can be assembled. This minimum unit is called a **unit cell.** The idea of a unit cell is illustrated in two dimensions by the art of M. C. Escher. Escher often used symmetrical patterns aligned together to create an overall design, one example of which appears in Figure 11-25. The repeating units can be visualized as tiles placed edge to edge. The faces of unit cells can be squares, rectangles, or parallelograms. In Figure 11-25 each tile is a parallelogram.

A unit cell is a three-dimensional block that can be stacked in orderly fashion to create an entire crystal. We show several different unit cells in the figures that follow.

Cubic Crystals

The easiest crystal lattice to visualize is the **simple cubic** structure. In a simple cubic crystal, layers of atoms stack one directly above another, so that all atoms lie along straight lines at right angles, as Figure 11-26 shows. Each atom in this structure touches six other atoms: four within the same plane, one above the plane, and one below the plane. Within one layer of the crystal, any set of four atoms forms a square. Adding four atoms directly above or below the first four forms a cube, for which the lattice is named. The unit cell of the simple cubic lattice, shown in the cutaway portion of Figure 11-26, contains $\frac{1}{8}$ of an atom at each of the eight corners of a cube. The total number of atoms in the unit cell is $8(\frac{1}{8}) = 1$.

When eight atoms form a cube, there is a cavity at the center of the cube. The cavity is not large enough to hold an additional atom, but moving the eight corner atoms slightly away from each other does allow the structure to accommodate another atom. The result is the **body-centered cubic** structure. Each atom in a body-centered cubic lattice is at the center of a cube and contacts eight neighboring corner atoms. In other

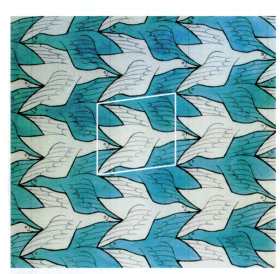

Figure 11-25
A drawing by M. C. Escher that contains a repeating pattern. One two-dimensional unit cell is highlighted.

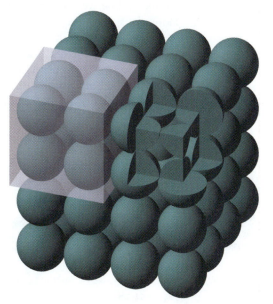

Figure 11-26
The simple cubic lattice is built from layers of spheres stacked one directly above another. The cutaway view shows one unit cell.

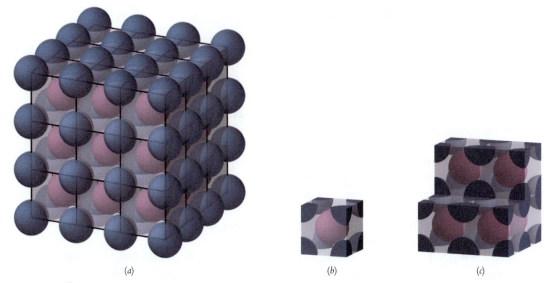

Figure 11-27
Views of the iron crystal: (a) A molecular view, with the lattice representation; (b) the unit cell; (c) six unit cells stacked to form part of the crystal.

words, every atom in the structure can be viewed as the center atom of one body-centered cube or as a corner atom in an alternate cube. The unit cell of a body-centered cube contains a total of $[1 + 8(\frac{1}{8})] = 2$ atoms. The views of the iron crystal lattice in Figure 11-27 illustrate this arrangement.

Although the corner atoms must move apart to convert a simple cube into a body-centered cube, the extra atom in the center of the structure makes the body-centered cubic lattice more compact than the simple cubic structure. All the alkali metals, as well as iron and the transition metals from Groups 5 and 6, form crystals with body-centered cubic structures.

Yet another common crystal lattice based on the simple cubic arrangement is known as the **face-centered cubic** structure. When four atoms form a square, there is open space at the center of the square. A fifth atom can fit into this space by moving the other four atoms away from one another. Stacking together two of these five-atom sets creates a cube. When we do this, additional atoms can be placed in the centers of the four faces along the sides of the cube, as Figure 11-28 shows.

The unit cell of the face-centered cube consists of six $\frac{1}{2}$-atoms embedded in the faces of the cube and eight $\frac{1}{8}$-atoms at the corners of the cube. This unit cell contains $[6(\frac{1}{2}) + 8(\frac{1}{8})] = 4$ atoms. As we describe below, each atom in a face-centered cube contacts 12 neighboring atoms. Consequently, the face-centered cube is more compact than either of the other cubic structures.

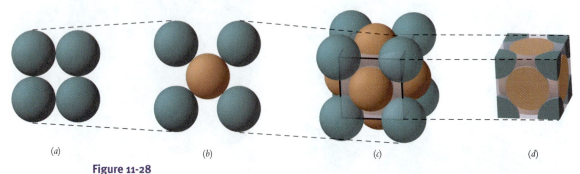

Figure 11-28
The face-centered cubic lattice can be viewed as a simple cube with each face (a) expanded just enough to fit an additional atom in the center of each face (b), giving eight atoms at the corners and six atoms embedded in the faces (c). The unit cell (d) has $\frac{1}{2}$ of an atom at the center of each face and $\frac{1}{8}$ of an atom at each corner.

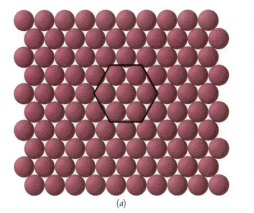
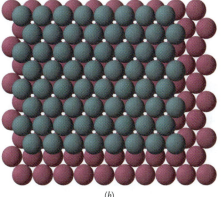

(a)   (b)

Figure 11-29
(*a*) Spheres close-packed in a layer generate a hexagonal pattern. (*b*) When a second layer is packed on top of the first, each sphere in the second layer nestles in the dimple created by three adjacent spheres in the lower layer. The second layer has two different sets of dimples, one directly above the spheres in the first layer (*maroon*) and the other offset from the spheres in the layer below it (*unshaded*).

Close-packed Crystals

A solid is most stable when each atom, molecule, or ion has as many close neighbors as possible, thus maximizing intermolecular attractions. An arrangement that accomplishes this is described as a **close-packed** structure. Close-packed structures are arranged so that the empty space around the atoms or molecules is minimized.

To visualize a close-packed atomic solid, think of the atoms as spheres placed as compactly as possible. Instead of square planar layers, a close-packed structure has hexagonal planar layers as shown in Figure 11-29*a*. Notice that this is a more compact planar arrangement than the square, because it places each sphere within a regular hexagon formed by six others. Each sphere has six nearest neighbors in the same plane. Now add a second layer of spheres. To achieve a most compact arrangement, each sphere sits in one of the "dimples" between a trio of spheres in the first layer, as shown in Figure 11-29*b*. This adds three more nearest neighbors for each sphere in the first layer, one in each of three neighboring dimples. As additional spheres are added, the second layer eventually looks identical to the lower layer, except that it is offset slightly to allow the spheres to nestle in the dimples formed by the layer below.

Now consider adding a third layer of close-packed spheres. This new layer can be placed in two different ways because there are two sets of dimples in the second layer. Notice in Figure 11-29*b* that the view through one set of dimples reveals the maroon spheres of the first layer. If spheres in the third layer lie in these dimples, the third layer is directly above the first, and the structure is called **hexagonal close-packed.** The view through the second set of dimples passes unobstructed through both layers. If the spheres of the third layer lie in the unobstructed set of dimples, the third layer is offset from both of the lower layers. It turns out that this arrangement is the face-centered cubic lattice, also called **cubic close-packed.** Figure 11-30 compares the two close-packed structures.

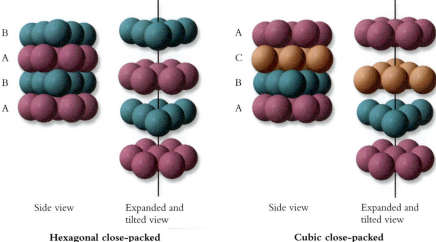

Side view   Expanded and   Side view   Expanded and
                tilted view   tilted view

Hexagonal close-packed   **Cubic close-packed**

Animation

Figure 11-30
Side and expanded views of hexagonal and cubic close-packed crystal types. In the hexagonal close-packed structure, spheres on both sides of any plane are in the same positions, and the *third* layer is directly above the first. In the cubic close-packed structure, layers take up three different positions, and the *fourth* layer is directly above the first.

Figure 11-31
Three views of the hexagonal layers contained within a face-centered cubic array. (*a*) A view perpendicular to the hexagonal layers, with all but one atom removed from the top layer. (*b*) A side view showing an outline of the cube and atoms from three successive layers of the cubic array. (*c*) The same side view, with two hexagonal layers screened for emphasis.

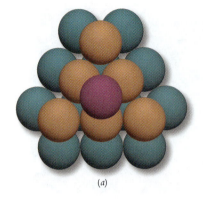

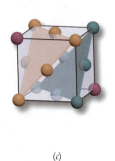

(*a*) (*b*) (*c*)

Animation

In either of these close-packed structures, each sphere has 12 nearest neighbors: 6 in the same plane, 3 in the dimples above, and 3 in the dimples below. The expanded views in Figure 11-30 show the different arrangements of the hexagonal and cubic close-packed crystalline types. In the hexagonal close-packed structure, notice that the third layer lies directly above the first, the fourth above the second, and so on. The layers can be labeled ABAB. . . . In the cubic close-packed structure, the third layer is offset from the other two, but the fourth layer is directly above the first. This arrangement can be labeled ABCABC. . . .

To see how a set of hexagonal layers can generate a cubic close-packed structure, we need to remove all but one of the atoms from the top layer, as shown in Figure 11-31*a*. The remaining atom from this top layer (shown in maroon in the figure) defines one corner of a cube. Each face of the cube contains an atom at each corner and one atom in the center, as shown by Figure 11-31*b*. Three atoms from the second layer (shown in gold in the figure) lie at corners of the cube, and three more fall in the centers of three faces of the cube. Atoms from the third row (shown in teal in the figure) occupy the remaining three corners and three faces of the cube. Figure 11-31*c* shows that the atoms from two hexagonal layers fall along a diagonal plane through the cube. The perspectives of Figures 11-31*b* and *c* reveal the face-centered cubic arrangement.

Atoms and molecules with spherical symmetry often form crystals with hexagonal or cubic close-packed geometry. For instance, magnesium and zinc crystallize with their atoms in a hexagonal close-packed array, while silver, aluminum, and gold crystallize in the cubic close-packed arrangement. Argon solidifies at low temperature as a cubic close-packed crystal, and neon can solidify in either form.

Up to now, we have described the crystalline arrays favored by spherical objects such as atoms, but most molecules are far from spherical. Stacks of produce illustrate that non-spherical objects require more elaborate arrays to achieve maximal stability. Compare a stack of bananas with a stack of oranges. Just as the stacking pattern for bananas is less symmetrical than that for oranges, the structural patterns for most molecular crystals are less symmetrical than those for crystals of spherical atoms, reflecting the lower symmetry of the molecules that make up molecular crystals.

Ionic Solids

The packing in ionic crystals requires that ions of opposite charges alternate with one another to maximize attractions among ions. A second important feature of ionic crystals is that the cations and anions usually are of different sizes. Usually the cations are smaller than the anions. Consequently, ionic compounds adopt a variety of structures that depend on the charges and sizes of the ions. One way to discuss ionic structures is to identify a crystal lattice for one set of ions, and then describe how the other ions pack within the lattice of the first set.

For many 1:1 ionic crystals such as NaCl, the most stable arrangement is a face-centered cubic array of anions with the cations packed into the holes between the anions. This structure appears in Figure 11-32. In addition to the face-centered cubic

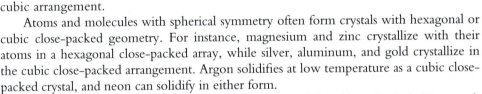

Nonspherical bananas require more elaborate packing schemes than do spherical oranges.

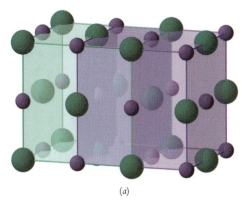

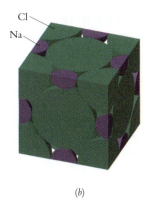

(a) (b)

Figure 11-32
(a) A sodium chloride crystal can be viewed as two overlapping face-centered cubic lattices, one of Na⁺ cations (*purple*) and the other of Cl⁻ anions (*green*). (b) Because Cl⁻ anions are considerably larger than Na⁺ cations, the most convenient unit cell places anions at the corners. There is a cation at the center of this unit cell.

Figure 11-33
The cesium chloride lattice consists of a simple cubic array of chloride anions with a cesium cation at the center of each unit cell.

arrangement of chloride ions, the unit cell shows $\frac{1}{4}$ of a sodium cation along each edge of the cell, and there is an additional sodium cation in the center of the unit cell. Summing the components reveals that each unit cell contains four complete NaCl formula units and is electrically neutral:

Chloride anions: 6 faces ($\frac{1}{2}$ anion/face) + 8 corners ($\frac{1}{8}$ anion/corner) = 4 Cl⁻ anions

Sodium cations: 12 edges ($\frac{1}{4}$ cation/edge) + 1 in center = 4 Na⁺ cations

Many ionic solids have the same unit cell structure as the sodium chloride lattice. Examples are all the halides of Li^+, Na^+, K^+, and Rb^+, as well as the oxides of Mg^{2+}, Ca^{2+}, Sr^{2+}, and Ba^{2+}.

The cations are significantly smaller than the anions in most 1:1 ionic salts, but cesium, which forms the largest monatomic cation, is an exception. Because its cations and anions are close to the same size, cesium chloride is most stable in a body-centered cubic lattice. There are Cl⁻ anions at the corners of the cube, with a Cs⁺ in the central body position, as Figure 11-33 shows. The unit cell contains $8(\frac{1}{8}) = 1$ chloride ion and 1 cesium ion. The cesium chloride lattice is found for CsCl, CsBr, CsI, and several other 1:1 ionic crystals.

Many ionic compounds have stoichiometries that differ from 1:1. To give just a single example, the fluorite lattice is named for the naturally occurring form of CaF_2 and is common for salts of the general formula MX_2. This lattice is easiest to describe by thinking of the Ca^{2+} cations arranged in a face-centered cubic structure. The F⁻ anions form a simple cube of eight anions in the interior of the face-centered cube. Figure 11-34*a* shows the unit cell of the fluorite lattice, which contains a total of four Ca^{2+} cations and eight F⁻ anions:

Calcium anions: 6 faces ($\frac{1}{2}$ cation/face) + 8 corners ($\frac{1}{8}$ cation/corner) = 4 Ca^{2+} cations

Fluoride anions: 8 complete anions within the face-centered cube = 8 F⁻ anions

(a)

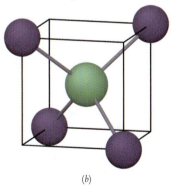

(b)

This gives the correct 1:2 stoichiometry, resulting in electrical neutrality.

From the perspective of the fluoride ions, the fluorite structure can be seen as a body-centered cube with an F⁻ anion in the center. Instead of eight corner ions, however, the alternating corners are vacant, placing calcium ions at four of the corners. This structure appears in Figure 11-34*b*. If you look closely at this structure, you will see that the calcium ions are arranged in a tetrahedron about the fluoride ion.

We have touched briefly on three simple ionic lattices, but there are many others. Moreover, the structures of many crystalline network solids can also be described by the methods we have introduced here for NaCl, CsCl, and CaF_2.

Figure 11-34
(a) The unit cell of the fluorite lattice has cations in a face-centered cubic arrangement, with a simple cube of anions within the unit cell. (b) The fluorite lattice can also be visualized as a body-centered cube with calcium ions at four of the corners, the other four corners vacant, and a fluoride ion in the body position.

Cotton candy is amorphous. Rock candy is crystalline.

Amorphous Solids

Solid materials are most stable as crystals. When a liquid is cooled slowly, it generally solidifies as crystals. When solids form rapidly, on the other hand, their atoms or molecules may become locked into positions other than those of a regular crystal, giving materials that are amorphous, meaning "without form." Ordinary cane sugar is crystalline, and crystalline rock candy forms when melted sugar is cooled slowly. In contrast, rapid cooling of melted sugar gives an amorphous solid: Cotton candy contains long threads of amorphous sugar.

What we call **glass** is an entire family of amorphous solids based on silicon dioxide, whose common name is silica (SiO_2). Pure silica is usually found as crystals containing the regular array of covalent bonds shown in Figure 11-23. Quartz is crystalline silica. When quartz is melted and then quickly cooled, however, it forms fused silica, an amorphous solid glass. Silica glass has many desirable properties. It resists corrosion, transmits light well, and withstands wide variations in temperature. Unfortunately, pure silica is very difficult to work with because of its high melting point (1710 °C). Consequently, silica glass is used only for special applications.

Mixing sodium oxide (Na_2O) with silica results in a glass that can be shaped at a lower temperature. Sodium oxide is ionic, so it breaks the Si—O—Si chain of covalent bonds, as shown in Figure 11-35. This weakens the lattice strength of the glass, lowers its melting point, and reduces the viscosity of the resulting liquid. However, the weakened lattice also means that glass made from mixed sodium and silicon oxides is vulnerable to chemical attack.

A desirable glass melts at a reasonable temperature, is easy to work with, and yet is chemically inert. Such a glass can be prepared by adding a third component that has bonding characteristics intermediate between those of purely ionic sodium oxide and those of purely covalent silicon dioxide. Several different components are used, depending on the properties desired in the glass.

The glass used for windowpanes and bottles is soda–lime–silica glass, a mixture of sodium oxide, calcium oxide, and silicon dioxide. The addition of CaO strengthens the lattice enough to make the glass chemically inert to most common substances. (Strong bases and HF, however, attack this glass.) Pyrex, the glass used in coffeepots and laboratory glassware, is a composite of B_2O_3, CaO, and SiO_2. This glass can withstand rapid temperature changes that would crack soda–lime–silica glass. Lenses and other optical components are made from glass that contains PbO. Light rays are strongly bent as they pass through lenses made of this glass. Colored glasses contain small amounts of colored metal oxides such as Cr_2O_3 (amber), NiO (green), or CoO (brown).

Many contemporary materials are amorphous solids composed of extremely large molecules called *polymers*. Polymeric solids are intermediate between molecular and covalent solids. They have discrete but extremely large molecules held together by dispersion forces and/or hydrogen bonds. Because polymer molecules are so large, their covalent bonding plays an important role in determining the properties of the solid. Plastic polymers can be shaped and molded because the intermolecular forces between polymer molecules are weaker than chemical bonds, but a plastic polymer also has relatively high strength because its long-chain molecules are held together by strong covalent bonds. We discuss polymers in Chapter 13.

Amorphous solids resemble liquids in that their molecules are not organized in regular arrays. In fact, there is no clear distinction between amorphous solids and very viscous liquids. When asphalt gets hot, it becomes sticky and eventually melts, but whether sticky asphalt is a soft solid or a highly viscous liquid is a matter of perspective. Our Chemistry and Life Box explores another type of structure, the liquid crystal, that shares characteristics of liquids and solids.

Figure 11-35
Adding sodium oxide to quartz glass breaks some of the Si—O covalent bonds (*left view*, red outlines) to create terminal O⁻ ions associated with Na⁺ cations (*right view*). This softens the glass, reducing its melting point.

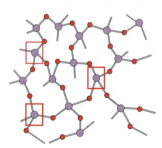

Quartz glass

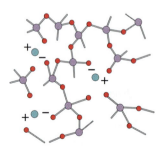

Sodium-containing glass

Box 11-2   Chemistry and Life: Liquid Crystals

"L iquid crystal" sounds like a contradiction. Liquids are fluid, their molecules continually changing places in a manner that is not particularly well organized. Crystals are immobile, their molecules locked into fixed positions that form regular patterns. Yet, not only does this unusual combination of fluidity and regular patterns exist, it plays important roles in biological organisms.

We can visualize a liquid crystal as not quite liquid and not quite solid. The molecules in a liquid crystal can move about, but they do so in patterned ways. Molecules that form liquid crystals have cylindrical shapes. In the liquid phase, these "molecular rods" have random orientations, whereas in the solid phase, they are all aligned like boards of a fence. In the liquid crystal phase, there is partial alignment of molecules in layers. These crystalline layers rearrange easily, making it possible for small changes in forces to change the properties of the liquid crystal. The illustration below shows the most common arrangements of liquid crystals. The nematic arrangement has the molecules aligned along their long axes, but the ends do not align. Smectic arrangements have greater alignment, with the ends of the molecules closely aligned. The molecules in a smectic arrangement can be upright (smectic A) or tilted (smectic C).

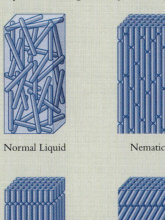

Normal Liquid

Nematic

Smectic A

Smectic C

Liquid crystals were discovered by an Austrian biologist, Frederich Reinitzer, in 1888. Reinitzer found that cholesteryl benzoate, a biological chemical, melts to form a hazy liquid. At a higher temperature, the haziness disappears. This clear state is what we know now as a liquid crystal.

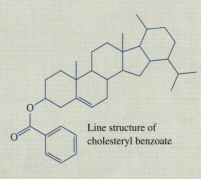

Line structure of cholesteryl benzoate

The development of high-magnification microscopy made it possible to create images of biological materials at the molecular level. Many of these images show structures that have liquid crystalline aspects. Shown here are aligned mosaic virus molecules and protein molecules in voluntary muscles. In addition, all cell walls are "picket fences" of rod-shaped molecules in regular yet fluid arrays.

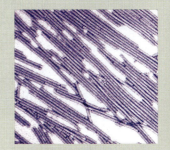

Chemists learned how to design molecules that display liquid crystal behavior, and physicists discovered how to use electrical force to make rapid changes in liquid crystal orientations. The result is LCD (liquid crystal display) technology, which brought us digital watches, laptop computer screens, and flat-screen television sets. This field continues to expand, with some predicting that "electronic paper" based on LCD technology will eventually make it possible to carry computer images in our pockets.

During the late twentieth century, most of the excitement about liquid crystals was in chemistry, physics, and technology. Now, however, the knowledge that made flat TV screens possible is being applied to problems in biology. These applications take a variety of forms. For example, DNA packs inside chromosomes in liquid-crystal-like structures. The physics of liquid crystals will give molecular biologists insights about chromosome structure. Furthermore, drugs and antibiotics that interact with DNA often have liquid crystal properties. Rather than being cylindrical, these liquid crystal molecules are planar, allowing them to slide into a DNA coil without destroying the organized pattern of the DNA. Do liquid crystal structures also exist in living cells, and, if so, what roles do liquid crystal properties play in the regulation of life processes?

Another direction in liquid crystal research is the fabrication of new molecules that mimic natural materials. The molecule shown here forms two different liquid crystals because of the cylindrical properties of the color-shaded portions of the molecule. These two trios are tethered together by long hydrocarbon chains. In addition to forming liquid crystals, this molecule self-assembles into organized larger units, as do proteins and DNA. The techniques used to synthesize this particular molecule can be adapted to incorporate different cylindrical components that give a variety of properties, paving the way for the creation of new materials with potential applications in biotechnology and nanotechnology.

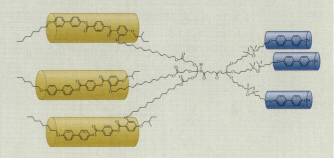

Sapphires are blue; rubies are red.

Crystal Imperfections

The two extremes of ordering in solids are perfect crystals with complete regularity and amorphous solids that have little symmetry. Most solid materials are crystalline but contain defects. **Crystalline defects** can profoundly alter the properties of a solid material, often in ways that have useful applications. Doped semiconductors, described in Section 10.7, are solids into which impurity "defects" are introduced deliberately in order to modify electrical conductivity. Gemstones are crystals containing impurities that give them their color. Sapphires and rubies are imperfect crystals of colorless Al_2O_3, Ti^{3+} or Fe^{3+} makes sapphires blue, and Cr^{3+} makes rubies red.

As noted earlier, yttrium–barium–copper superconductors have optimal properties when they have a slight deficiency of oxygen. This departure from stoichiometric composition is created by defects in the crystal structure. The superconductor crystal has oxygen anions missing from some positions in the crystal lattice, and the number of missing anions can vary sufficiently to give the material a nonstoichiometric overall composition. The solid remains electrically neutral when its anion content varies, because some of the cations take on different charges. The copper cations in the superconductor can have a +2 or +3 charge. The relative number of Cu^{3+} ions decreases as oxygen anions are removed from the structure.

Substitutional impurities replace one metal atom with another, while interstitial impurities occupy the spaces between metal atoms. Interstitial impurities create imperfections that play important roles in the properties of metals. For example, small amounts of impurities are deliberately added to iron to improve its mechanical properties. Pure iron is relatively soft and easily deformed, but the addition of a small amount of carbon creates steel, a much harder material. Carbon atoms fill some of the open spaces between iron atoms in the crystalline structure of steel. Although carbon atoms fit easily into these spaces, their presence reduces the ability of adjacent layers of iron atoms to slide past each other, as Figure 11-36 illustrates.

This section describes only a few of the organizing features of solid materials. The rapidly growing field of materials science addresses these and many other atomic and molecular aspects of solids that determine the technologically useful properties of materials.

Figure 11-36
The presence of even a few carbon atoms (*gray spheres*) in the interstitial holes in an iron lattice prevents adjacent layers of iron atoms from sliding past one another and hardens the iron into steel.

Section Exercises

11.5.1 Predict the angles found in the crystalline fragments broken from a hexagonal close-packed crystal. What additional angles would be found in fragments from cubic close-packed crystals? (*Hint:* Think about the angles found in hexagons and cubes.)

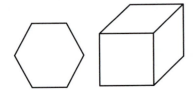

11.5.2 Is the view of fluorite shown in Figure 11-34*b* a unit cell? If so, show how to assemble it to create the fluorite crystal. If not, explain why it is not a unit cell.

11.5.3 Carbon atoms in steel occupy interstitial positions, as shown in Figure 11-36. Manganese atoms, which are also added to steel to increase its hardness, occupy substitutional positions, replacing iron atoms in the crystalline structure. Draw an atomic picture of a portion of manganese-containing steel.

11.6 PHASE CHANGES

Phase changes are characteristic of all substances. The normal phases displayed by the halogens appear in Section 11.1, where we also show that a gas liquefies or a liquid freezes at low enough temperatures. Vapor pressure, which results from molecules escaping from a condensed phase into the gas phase, is one of the liquid properties described in Section 11.3. Phase changes depends on temperature, pressure, and the magnitudes of intermolecular forces.

Heats of Phase Changes

Any phase change that results in increased molecular mobility requires that intermolecular forces be overcome. This, in turn, requires an energy input. Consider a teakettle on a hot stove. The burner supplies energy that heats the water until the temperature reaches the boiling point of 100 °C. As Figure 11-37 shows, the temperature then stops rising even though heat is still transferred to the water. The added energy now causes molecules of water to move from the liquid phase into the gas phase, and the temperature remains constant as the water boils away. Steam escapes into the atmosphere as the burner transfers heat to the water molecules.

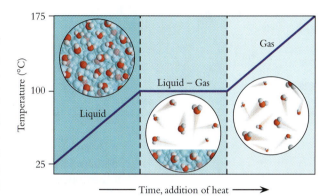

Figure 11-37
When heat is supplied at a constant rate to a sample of liquid water, the temperature rises to 100 °C and then remains constant until all the water has evaporated.

A molecular perspective reveals why energy must be supplied to boil water. A molecule of water cannot escape the liquid phase unless it has enough energy of motion to overcome the hydrogen bonding forces that hold liquid water together. About 40 kJ of heat must be supplied to transfer 1 mol of water molecules from the liquid phase into the vapor phase.

Phase changes require that energy must be either supplied or removed from the substance undergoing the phase change. This energy transfer process normally occurs in the form of heat transfers. Moreover, phase changes generally occur at constant pressure. Recall from Chapter 6 that for constant pressure changes, heat transfer equals the enthalpy change: $q_p = \Delta H$. A change of phase always involves changes in both the energy and the enthalpy of the substance. The magnitude of the changes depends on the strength of intermolecular forces in the substance undergoing the phase change.

The amount of heat required to vaporize a substance also depends on the size of the sample. Twice as much energy is required to vaporize two moles of water than one mole. The heat needed to vaporize one mole of a substance at its normal boiling point is called the **molar heat of vaporization, ΔH_{vap}**.

Energy must also be provided to melt a solid substance. This energy is used to overcome the intermolecular forces that hold molecules or ions in fixed positions in the solid phase. Thus, the melting of a solid also has characteristic energy and enthalpy changes. The heat needed to melt one mole of a substance at its normal melting point is called the **molar heat of fusion, ΔH_{fus}**.

Fusion and vaporization are the most familiar phase changes, but **sublimation** is also common. Sublimation is a phase change in which a solid converts directly to a vapor without passing through the liquid phase. Dry ice (solid CO_2) sublimes at 195 K with $\Delta H_{subl} = 25.2$ kJ/mol. Mothballs contain naphthalene ($C_{10}H_8$, $\Delta H_{subl} = 73$ kJ/mol), a crystalline white solid that sublimes to produce a vapor that repels moths. The purple color of the gas above iodine crystals in a closed container provides visible evidence that this solid also sublimes at room temperature ($\Delta H_{subl} = 62.4$ kJ/mol).

Phase changes can go in either direction: Steam condenses upon cooling, and liquid water freezes at low temperature. Each of these is *exothermic* because each is the reverse of an endothermic phase change. That is, heat is released as a gas condenses to a liquid and as a liquid freezes to a solid. To make ice cubes, for instance, water is placed in a freezer that absorbs the heat released during the formation of ice. A phase change that is exothermic has a negative enthalpy change:

The purple color of the vapor above solid iodine is due to I₂ molecules in the gas phase.

$$\Delta H_{solidification} = -\Delta H_{fus} \quad and \quad \Delta H_{condensation} = -\Delta H_{vap}$$

Table 11-4   Molar Heats of Phase Change

Substance	Formula	T_{fus} (K)	ΔH_{fus} (kJ/mol)	T_{vap} (K)	ΔH_{vap} (kJ/mol)
Argon	Ar	83	1.3	87	6.3
Oxygen	O_2	54	0.45	90	9.8
Methane	CH_4	90	0.84	112	9.2
Ethane	C_2H_6	90	2.85	184	15.5
Diethyl ether	$(C_2H_5)_2O$	157	6.90	308	26.0
Bromine	Br_2	266	10.8	332	30.5
Ethanol	C_2H_5OH	156	7.61	351	39.3
Benzene	C_6H_6	278.5	10.9	353	31.0
Water	H_2O	273	6.01	373	40.79
Mercury	Hg	234	23.4	630	59.0

By convention, tabulated values of heats of phase changes are always specified in the endothermic direction. Consequently, the reverse processes have the same magnitude but the opposite sign:

$$\text{Solid} \longrightarrow \text{Liquid} \quad \Delta H_{fus} \qquad \text{Liquid} \longrightarrow \text{Solid} \quad -\Delta H_{fus}$$

$$\text{Liquid} \longrightarrow \text{Vapor} \quad \Delta H_{vap} \qquad \text{Vapor} \longrightarrow \text{Liquid} \quad -\Delta H_{vap}$$

$$\text{Solid} \longrightarrow \text{Vapor} \quad \Delta H_{subl} \qquad \text{Vapor} \longrightarrow \text{Solid} \quad -\Delta H_{subl}$$

Table 11-4 lists values of ΔH_{fus}, ΔH_{vap}, melting points, and boiling points for different chemical substances. Example 11-6 provides practice in using enthalpies of phase changes.

Example 11-6

Heat of Phase Change

A swimmer emerging from a pool is covered with a film containing about 75 g of water. How much heat must be supplied to evaporate this water?

Strategy: Energy in the form of heat is required to evaporate the water from the swimmer's skin. The energy needed to vaporize the water can be found using the molar heat of vaporization and the number of moles of water. The process can be shown with a simple block diagram:

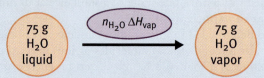

Solution: The ΔH_{vap} of water is 40.79 kJ/mol (see Table 11-4). The molar mass of water is 18.02 g/mol, so 75 g of water is 4.16 mol. Therefore, the heat that must be supplied is $q_p = n\,\Delta H_{vap} = (4.16\text{ mol})(40.79\text{ kJ/mol}) = 1.7 \times 10^2$ kJ.

 Does the Result Make Sense? If the swimmer's body must supply all this heat, a substantial chilling effect occurs. Thus, swimmers usually towel off (to reduce the amount of water that must be evaporated) or lie in the sun (to let the sun provide most of the heat required).

Extra Practice Exercise 11.6

Determine how much heat is involved in freezing 125 g of water in an ice cube tray. What is the direction of heat flow in this process?

The graph in Figure 11-37 shows that adding heat to boiling water does not cause the temperature of the water to increase. Instead, the added energy is used to overcome inter-molecular attractions as molecules leave the liquid phase and enter the gas phase. Other two-phase systems, such as an ice–water mixture, show similar behavior. Heat added to an ice–water mixture melts some of the ice, but the mixture remains at 0 °C. Similarly, when an ice–water mixture in a freezer loses heat to the surroundings, the energy comes from some liquid water freezing, but the mixture remains at 0 °C until all the water has frozen. This behavior can be used to hold a chemical system at a fixed temperature. A temperature of 100 °C can be maintained by a boiling water bath, and an ice bath holds a system at 0 °C. Lower temperatures can be achieved with other substances. Dry ice maintains a temperature of −78 °C; a bath of liquid nitrogen has a constant temperature of −196 °C (77 K); and liquid helium, which boils at 4.2 K, is used for research requiring ultracold temperatures.

Phase Diagrams

The stable phase of a substance depends on temperature and pressure. A **phase diagram** is a map of the pressure–temperature world showing the phase behavior of a substance. As Figure 11-38 shows, a phase diagram is a *P-T* graph that shows the ranges of temperature and pressure over which each phase is stable. Pressure is plotted along the *y*-axis, and temperature is plotted along the *x*-axis. In the upper left-hand region (low *T*, high *P*), the substance is stable as a solid. In the lower right-hand region (high *T*, low *P*), the substance is stable as a gas. In some intermediate range of *T* and *P*, the substance is stable as a liquid.

As illustrated by Figure 11-38, phase diagrams have many characteristic features:

1. Boundary lines between phases separate the regions where each phase is stable.

2. Movement across a boundary line corresponds to a phase change. The arrows on the figure show six different phase changes: sublimation and its reverse, deposition; melting and its reverse, freezing; and vaporization and its reverse, condensation.

3. At any point along a boundary line, the two phases on either side of the line coexist in a state of dynamic equilibrium. The normal freezing point and normal boiling point of a substance (shown by red dots) are the points where the phase boundary lines intersect the horizontal line that represents *P* = 1 atm.

4. Three boundary lines meet in a single point (shown by a red dot), called a **triple point.** All three phases are stable simultaneously at this unique combination of temperature and pressure. Notice that, although *two* phases are stable under any of the conditions specified by the boundary lines, *three* phases can be simultaneously stable only at a triple point.

5. Above a sufficiently high temperature and pressure, called the **critical point** (shown by a red dot), the distinction between the gas phase and the liquid phase disappears. Instead, the substance is a supercritical fluid, with viscosity typical of a liquid but able to expand or contract like a gas.

6. What happens to a substance as temperature changes at constant pressure can be determined by drawing a horizontal line at the appropriate pressure on the phase diagram.

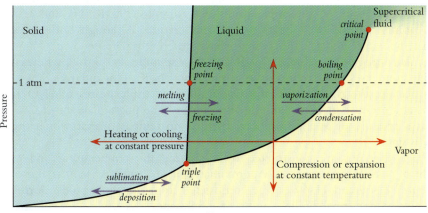

Figure 11-38
The general form of a phase diagram. Any point on the diagram corresponds to a specific temperature and pressure. Lines trace conditions under which phase changes occur, and the blue arrows show six types of phase transitions.

7. What happens to a substance as pressure changes at constant temperature can be determined by drawing a vertical line at the appropriate temperature on the phase diagram.

8. The temperature for conversion between the vapor and a condensed phase depends strongly on pressure. Qualitatively, this is because compressing a gas increases the collision rate and makes condensation more favorable. We describe the quantitative details of this variation in Section 14.5.

Figure 11-39
The phase diagram for water. The critical point is at 647 K, 218 atm, well off-scale on this graph.

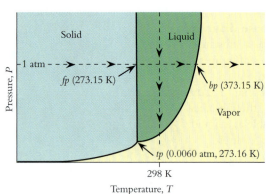

9. The melting temperature is almost independent of pressure, making the boundary line between solid and liquid nearly vertical. Qualitatively, this is because pressure has hardly any effect on the compact liquid and solid phases.

10. The solid–vapor boundary line extrapolates to $P = 0$ and $T = 0$. This is a consequence of the direct link between temperature and molecular energy. At $T = 0$ K, molecules have minimum energy, so they cannot escape from the solid lattice. At 0 K, the vapor pressure of every substance would be 0 atm.

The phase diagram for water, shown in Figure 11-39, illustrates these features for a familiar substance. The figure shows that liquid water and solid ice coexist at the normal freezing point, $T = 273.15$ K and $P = 1.00$ atm. Liquid water and water vapor coexist at the normal boiling point, $T = 373.15$ K and $P = 1.00$ atm. The triple point of water occurs at $T = 273.16$ K and $P = 0.0060$ atm. The figure shows that when P is lower than 0.0060 atm, there is no temperature at which water is stable as a liquid. At sufficiently low pressure, ice sublimes but does not melt.

The dashed lines on Figure 11-39 show two paths that involve phase changes for water. The *horizontal dashed line* shows what happens as the temperature increases at a constant pressure of 1 atm. As ice warms from a low temperature, it remains in the solid phase until the temperature reaches 273.15 K. At that temperature, solid ice melts to liquid water, and water remains liquid as the temperature increases until the temperature reaches 373.15 K. At 373.15 K, liquid water changes to water vapor. When the pressure is 1 atm, water is most stable in the gas phase at all higher temperatures. The *vertical dashed line* shows what happens as the pressure on water is reduced at a constant temperature of 298 K (approximately room temperature). Water remains in the liquid phase until the pressure drops to 0.03 atm. At 298 K, water is most stable in the gas phase at any pressure lower than 0.03 atm. Notice that at 298 K there is no pressure at which water is most stable in the solid state.

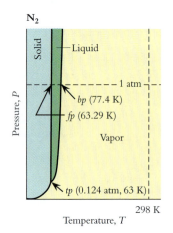

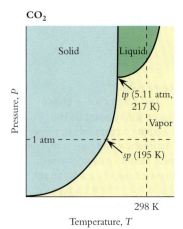

Figure 11-40
Phase diagrams for molecular nitrogen and carbon dioxide, two substances that are gases at room temperature and pressure.

All phase diagrams share the ten common features listed above. However, the detailed appearance of a phase diagram is different for each substance, as determined by the strength of the intermolecular interactions for that substance. Figure 11-40 shows two examples, the phase diagrams for molecular nitrogen and for carbon dioxide. Both these substances are gases under normal conditions. Unlike H_2O, whose triple point lies close to 298 K, N_2 and CO_2 have triple points that are well below room temperature. Although both are gases at room temperature and pressure, they behave differently when cooled at $P = 1$ atm. Molecular nitrogen liquefies at 77.4 K and then solidifies at 63.3 K, whereas carbon dioxide condenses directly to the solid phase at 195 K. This difference in behavior arises because the triple point of CO_2, unlike the triple points of H_2O and N_2, occurs at a pressure greater than one atmosphere. The phase diagram of CO_2 shows that at a pressure of one atmosphere, there is no temperature at which the liquid phase is stable.

Phase diagrams are constructed by measuring the temperatures and pressures at which phase changes occur. Approximate phase diagrams such as those shown in Figures 11-39 and 11-40 can be constructed from the triple point, normal melting point, and normal boiling point of a substance. Example 11-7 illustrates this procedure.

Example 11-7

Constructing
a Phase Diagram

Ammonia is a gas at room temperature and pressure. Its normal boiling point is 239.8 K, and it freezes at 195.5 K. The triple point for NH_3 is $P = 0.0604$ atm and $T = 195.4$ K. Use this information to construct an approximate phase diagram for NH_3.

Strategy: The normal melting, boiling, and triple points give three points on the phase boundary curves. To construct the curves from knowledge of these three points, use the common features of phase diagrams: the vapor–liquid and vapor–solid boundaries of phase diagrams slope upward, the liquid–solid line is nearly vertical, and the vapor–solid line begins at $T = 0$ K and $P = 0$ atm.

Solution: Begin by choosing appropriate scales, drawing the $P = 1$ atm line and locating the given data points. An upper temperature limit of 300 K encompasses all the data:

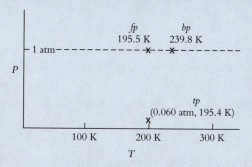

Next, connect the points and label the domains:

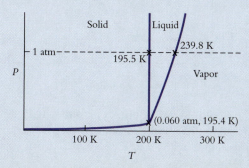

Does the Result Make Sense? We know that ammonia is a gas under normal pressure and temperature, and the diagram shows that 298 K, 1 atm lies within the vapor region of the phase diagram.

Molecular chlorine melts at 172 K, boils at 239 K, and has a triple point at 172 K, 0.014 atm. Sketch the phase diagram for this compound.

**Extra Practice
Exercise 11.7**

Phase diagrams can be used to determine what phase of a substance is stable at any particular pressure and temperature. They also summarize how phase changes occur as either condition is varied. Example 11-8 provides an illustration.

A chemist wants to perform a synthesis in a vessel at $P = 0.50$ atm using liquid NH_3 as the solvent. What temperature range would be suitable? When the synthesis is complete, the chemist wants to boil off the solvent without raising T above 220 K. Is this possible?

Example 11-8

Interpreting
a Phase Diagram

Strategy: The phase diagram for NH_3 shows the boundary lines for the liquid domain. These boundary lines can be used to determine the conditions under which phase changes occur.

Example 11-8

Interpreting a Phase Diagram (*continued*)

Solution: Because the chemist wants to work at $P = 0.50$ atm, draw a horizontal line across the phase diagram at $P = 0.50$ atm. Here is an expanded view of the phase diagram between 150 and 300 K:

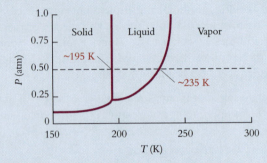

The horizontal line intersects the boundary lines at about 235 and 195 K. Liquid NH_3 is stable between these temperatures at this pressure.

At the completion of the synthesis, the chemist wants to remove the solvent without raising T above 220 K. A vertical line at 220 K on the phase diagram represents this condition:

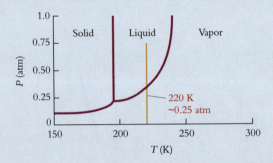

The line intersects the liquid–vapor boundary at about 0.25 atm. A vacuum pump capable of reducing P below 0.25 atm can be used to vaporize and remove the NH_3 while keeping the temperature below 220 K.

Do the Results Make Sense?   We know that ammonia is a gas at normal temperature and pressure, so it is sensible that the temperature must be lowered substantially to run a reaction in liquid ammonia. Knowing that ammonia is liquid at 0.50 atm and 220 K, it makes sense that a still lower pressure is required to boil off the liquid at this temperature.

Extra Practice Exercise 11.8

A sample of N_2 gas is at $P = 0.1$ atm and $T = 63.1$ K. Use Figure 11-40 to determine what will happen to this sample if the temperature remains fixed while the pressure slowly increases to 1 atm.

Variations on Phase Diagrams

We tend to think of three phases of matter: solid, liquid, and vapor. However, many substances have more than one solid phase, while others may display liquid crystal phases that are distinct from either the standard liquid phase or the solid phase. Phase diagrams of solids are particularly useful in geology, because many minerals undergo solid–solid phase transitions at the high temperatures and pressures found deep in the Earth's crust. Figure 11-41 is the phase diagram for silica, an important geological substance. Notice that the pressure scale ranges from 0 to 130 atm, and the temperature scale ranges from 0 to 2000 °C. These are geological pressures and temperatures. Notice also that there are six different forms of crystalline silica, each stable in a different temperature–pressure region. A geologist who encounters a sample of stishovite can be confident that it solidified under extremely high pressure conditions.

Another complicating feature of phase diagrams at high pressure is the critical point, above which the liquid–vapor transition no longer occurs. As the pressure on a gas increases, the gas is compressed into an ever-smaller volume. If the temperature is low enough, this compression eventually results in liquefaction: the volume decreases dramatically as intermolecular forces hold the molecules close together in the condensed phase. At high enough temperature, however, no amount of compression is enough to cause liquefaction. The substance is a supercritical fluid, able to expand to fill the available space, but resistant to further compression.

The phase diagrams in Figures 11-39 and 11-40 do not show critical points, because the critical points of water, carbon dioxide and nitrogen occur at higher pressures than those shown on these diagrams. The critical point of water is $P = 218$ atm, $T = 647$ K; that of CO_2 is $P = 72.9$ atm, $T = 304$ K; and that of N_2 is $P = 33.5$ atm, $T = 126$ K.

Although critical pressures are many times greater than atmospheric pressure, supercritical fluids have important commercial applications. The most important of these is the use of supercritical carbon dioxide as a solvent. Supercritical CO_2 diffuses through a solid matrix rapidly, and it transports materials well because it has a lower viscosity than ordinary liquids. Because CO_2 is a natural product of biological processes, it is nontoxic, making it well suited for "Green Chemistry" applications. Supercritical CO_2 currently is used as a solvent for dry cleaning, for petroleum extraction, for decaffeination, and for polymer synthesis.

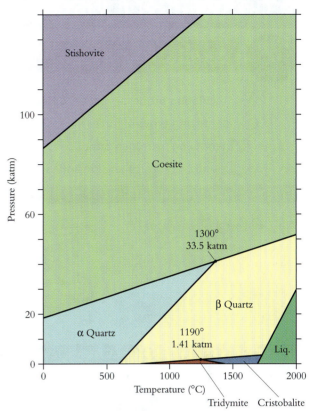

Figure 11-41
The phase diagram for silica shows six different forms of the solid, each stable under different temperature and pressure conditions.

Section Exercises

■ 11.6.1 Use information in Table 11-4 to determine how much heat must be supplied to vaporize 15 g of the following substances: (a) methane, (b) ethane, and (c) ethanol.

■ 11.6.2 If dry ice is heated under a pressure greater than 10 atm, it melts instead of subliming. If the pressure is then reduced to 1 atm, the liquid boils to give gaseous CO_2. Show these processes on a phase diagram for CO_2.

■ 11.6.3 Refer to the phase diagrams in this chapter to answer the following: (a) What is the maximum pressure at which solid N_2 can sublime? (b) What happens if the pressure above ice at -10 °C is reduced to 0.5 atm, to 0.05 atm, and finally to 0.005 atm? (c) What occurs if dry ice (solid CO_2) is heated from 180 K to room temperature in a container held at constant pressure of 7 atm?

SKILLS TO MASTER

① Calculating properties of real gases
② Identifying types of forces
③ Identifying hydrogen bonds
④ Describing surface tension and viscocity

⑤ Recognizing types of solids
⑥ Depicting simple crystal types
⑦ Drawing and interpreting phase diagrams

Important equations

Essential terms

Key concepts

11.1 REAL GASES AND INTERMOLECULAR FORCES

van der Waals equation

① $\left(P + \dfrac{n^2 a}{V^2}\right)(V - nb) = nRT$

Normal boiling point   Vaporization
Normal freezing pont   Condensation

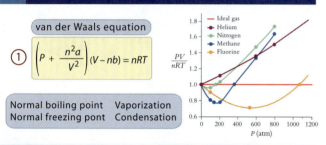

— Ideal gas
— Helium
— Nitrogen
— Methane
— Fluorine

$\dfrac{PV}{nRT}$

P (atm)

11.2 TYPES OF INTERMOLECULAR FORCES

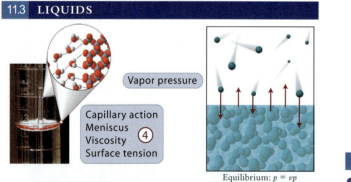

Dispersion forces

Dipolar forces

③ Hydrogen bonding

②

11.3 LIQUIDS

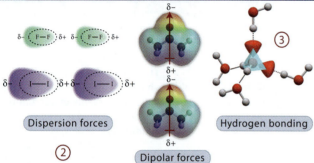

Vapor pressure

Capillary action
Meniscus
Viscosity   ④
Surface tension

Equilibrium: $p = vp$

11.4 FORCES IN SOLIDS

Ionic solid

Network solid

⑤

Molecular solid

Metallic solid

11.5 ORDER IN SOLIDS

Amorphous solid
Glass

Crystalline solid

Simple cubic

⑥

Body-centered cubic

Face-centered cubic

Unit cell

Close-packed structure

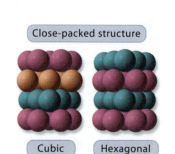

Cubic

Hexagonal

Crystalline defects

11.6 PHASE CHANGES

ΔH_{sub} Sublimation

ΔH_{vap}

ΔH_{fus}

Molar heat of fusion, ΔH_{fus}

Molar heat of vaporization, ΔH_{vap}

Phase diagram

Triple point   Critical point

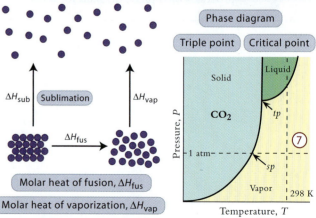

Solid

Liquid

CO_2

tp

1 atm

sp

Vapor

298 K

Pressure, P

Temperature, T

⑦

Learning Exercises

11.1 List all the types of interactions that can act to hold a solid together. Organize the list from strongest to weakest.

11.2 Draw molecular pictures that show every type of hydrogen bond that exists in a solution containing methanol, water, and ammonia.

11.3 Write a paragraph that describes the phenomenon of superconductivity.

11.4 Define and give an example of each of the following: (a) close-packed structure; (b) unit cell; (c) molecular solid; (d) covalent solid; (e) amorphous solid; and (f) lattice.

11.5 Write a paragraph describing what happens at the molecular level during each of the six phase changes shown in Figure 11-38.

11.6 Prepare a list of the terms in Chapter 11 that are new to you. Write a one-sentence definition for each, using your own words. If you need help, consult the glossary.

 Problems **ilw** = interactive learning ware problem. Visit the website at www.wiley.com/college/olmsted

Real Gases and Intermolecular Forces

11.1 Xenon condenses at 166 K, krypton condenses at 121 K, and argon condenses at 87 K. Draw a bar graph similar to Figure 11-3 showing the relative magnitudes of interatomic attractive energies for these three noble gases. Include two horizontal lines representing the average kinetic energy at 140 and 100 K.

11.2 Methane condenses at 121 K, but carbon tetrachloride boils at 350 K. Draw a bar graph similar to that in Figure 11-3 showing the relative attractive energies between molecules for these two substances. Include a horizontal line representing the average kinetic energy at room temperature (298 K).

11.3 Predict whether the effects of molecular volume and of intermolecular attractions become more or less significant when the following changes are imposed: (a) A gas is expanded to a larger volume at constant temperature. (b) More gas is introduced into a container of constant volume at constant temperature. (c) The temperature of a gas increases at constant pressure.

11.4 Predict whether the effects of molecular volume and of intermolecular attractions become more or less significant when the following changes are imposed: (a) The temperature of a gas is lowered at constant volume. (b) A gas is compressed into a smaller volume at constant temperature. (c) Half the gas is removed from a sample at constant volume.

11.5 Draw pictures showing the atomic arrangements of samples of $Ag(s)$, $Ar(g)$, and $Hg(l)$.

11.6 Draw pictures showing the atomic arrangements of samples of $Zn(s)$, $Kr(g)$, and $Ga(l)$.

ilw 11.7 From the following experimental data, calculate the percent deviation from ideal behavior: 1.00 mol CO_2 in a 1.20-L container at 40.0 °C exerts 19.7 atm pressure.

11.8 From the following experimental data, calculate the percent deviation from ideal behavior: 3.000 g H_2 at 0.00 °C and 193.5 atm occupies a volume of 189.18 cm^3.

11.9 Chlorine gas is commercially produced by electrolysis of seawater and then stored under pressure in metal tanks. A typical tank has a volume of 15.0 L and contains 1.25 kg of Cl_2. Use the van der Waals equation to calculate the pressure in this tank if the temperature is 295 K, and compare the result with the ideal gas value.

11.10 Chlorine gas used in chemical syntheses is transferred from tanks under high pressure to reactor flasks at considerably lower pressure. If 10.5 g of Cl_2 is introduced into a 5.00-L reaction flask at 145 °C, calculate the pressure of Cl_2 gas using the van der Waals equation and using the ideal gas equation.

Types of Intermolecular Forces

11.11 Arrange the following in order of ease of liquefaction: CCl_4, CH_4, and CF_4. Explain your ranking.

11.12 Arrange the following in order of increasing boiling point: Ar, He, Ne, and Xe. Explain your ranking.

11.13 Draw molecular pictures of CCl_4 and CH_4 showing the relative amount of polarizability of each.

11.14 Draw molecular pictures of Ne and Xe showing the relative amount of polarizability of each.

11.15 List ethanol (CH_3CH_2OH), propane ($CH_3CH_2CH_3$), and *n*-pentane ($CH_3CH_2CH_2CH_2CH_3$) in order of increasing boiling point, and explain what features determine this order.

11.16 List *n*-propanol ($CH_3CH_2CH_2OH$), dimethyl ether (CH_3OCH_3) and diethyl ether ($CH_3CH_2OCH_2CH_3$) in order of increasing boiling point and explain what features determine this order.

11.17 Which of the following will form hydrogen bonds with another molecule of the same substance? Draw molecular pictures illustrating these hydrogen bonds. (a) CH_2Cl_2; (b) H_2SO_4; (c) H_3COCH_3; and (d) $H_2NCH_2CO_2H$.

11.18 Which of the following form hydrogen bonds with water? Draw molecular pictures illustrating these hydrogen bonds. (a) CH_4; (b) I_2; (c) HF; (d) H_3COCH_3; and (e) $(CH_3)_3COH$.

11.19 Draw Lewis structures showing all possible hydrogen-bonding interactions for (a) two NH_3 molecules; and (b) one NH_3 molecule and one H_2O molecule.

11.20 Draw Lewis structures that show the hydrogen-bonding interactions for (a) two CH_3OH molecules; and (b) one HF molecule and one acetone molecule [$(CH_3)_2C{=}O$].

Liquids

11.21 Pentane is a C_5 hydrocarbon, gasoline contains mostly C_8 hydrocarbons, and fuel oil contains hydrocarbons in the C_{12} range. List these three hydrocarbons in order of increasing viscosity, and explain what molecular feature accounts for the variation.

11.22 Given that a lubricant must flow easily to perform its function, which grade of motor oil is preferred for winter use, high- or low-viscosity? Why?

11.23 Aluminum tubing has a thin surface layer of aluminum oxide. What shape meniscus would you expect to find for water and for mercury inside aluminum tubing? Explain your answers in terms of intermolecular forces.

11.24 Platinum tubing has a clean metal surface. What shape meniscus would you expect to find for water and for mercury inside platinum tubing? Explain your answers in terms of intermolecular forces.

11.25 Paper towels are better at absorbing spilled wine than at absorbing spilled salad oil. Why is this?

11.26 Water forms "beads" on the surface of a freshly waxed automobile, but it forms a film on the clean windshield. Why is this?

11.27 For each of the following pairs of liquids, choose which has the higher vapor pressure at room temperature and state the reason why: (a) benzene (C_6H_6) or chlorobenzene (C_6H_5Cl); (b) hexane (C_6H_{14}) or 1-hexanol ($C_6H_{13}OH$); and (c) octane (C_8H_{18}) or heptane (C_7H_{16}).

11.28 For each of the following pairs of liquids, choose which has the lower vapor pressure at room temperature and state the reason why: (a) water (H_2O) or methanol (CH_3OH); (b) 1-pentanol ($C_5H_{12}OH$) or 1-hexanol ($C_6H_{13}OH$); and (c) methyl chloride (CH_3Cl) or chloroform ($CHCl_3$).

Forces in Solids

11.29 Classify each of the following as an ionic, network, molecular, or metallic solid: Sn, S_8, Se, SiO_2, and Na_2SO_4.

11.30 Classify each of the following as an ionic, network, molecular, or metallic solid: Pt, P_4, Ge, As_2O_3, and $(NH_4)_3PO_4$.

11.31 Describe the differences in (a) bonding characteristics and (b) macroscopic properties between metals and network solids.

11.32 Describe the differences in (a) interparticle forces and (b) macroscopic properties between molecular and ionic solids.

11.33 Indicate what type of solid (ionic, network, metallic, or molecular/atomic) each of the following forms on solidification: (a) Br_2; (b) KBr; (c) Ba; (d) SiO_2; and (e) CO_2.

11.34 Indicate what type of solid (ionic, covalent, metallic, or molecular/atomic) each of the following forms on solidification: (a) HCl; (b) KCl; (c) NH_4NO_3; (d) Mn; and (e) Si.

Order in Solids

11.35 The density of graphite is 2260 kg m^{-3}, whereas that of diamond is 3513 kg m^{-3}. Describe the bonding features that cause these two forms of carbon to have different densities.

11.36 Amorphous silica has a density of around 2.3 g/cm^3, and crystalline quartz has a density of 2.65 g/cm^3. Describe the bonding features that cause these two forms of the same substance to have different densities.

ilw 11.37 The unit cell of the mineral perovskite follows. What is the formula of perovskite?

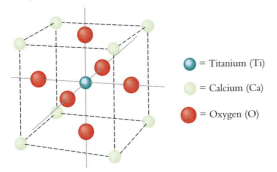

= Titanium (Ti)

= Calcium (Ca)

= Oxygen (O)

11.38 The unit cell of a compound of sodium (purple) and oxygen (red) appears at right. Determine the chemical formula of the compound.

11.39 Construct part of the Lewis structure of carborundum, the diamond-like compound of empirical formula SiC.

11.40 Construct part of the Lewis structure of quartz, the network solid of empirical formula SiO_2.

11.41 For the pattern shown below, draw a tile that represents a unit cell for the pattern and contains only complete fish. Draw a second tile that is a unit cell but contains no complete fish.

11.42 For the pattern shown at right, draw a tile that represents a unit cell for the pattern, but does not contain any complete stars. Draw a different tile that contains only complete stars.

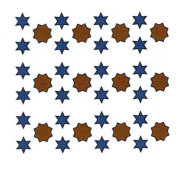

11.43 In the unit cell of lithium sulfide, the sulfide ions adopt a face-centered cubic lattice. The lithium ions are in the holes within the face-centered cube. If the unit cell is charge-neutral, how many complete lithium ions must be packed within each face-centered cube?

11.44 In the unit cell of calcium oxide, the oxide ions adopt a face-centered cubic lattice. The calcium ions are in the holes within the face-centered cube. If the unit cell is charge-neutral, how many complete calcium ions must be packed within each face-centered cube?

Phase Changes

11.45 From Table 11-4, determine which of each pair has the higher heat value and use intermolecular forces to explain why: (a) heats of vaporization of methane and ethane; (b) heats of vaporization of ethanol and diethyl ether; and (c) heats of fusion of argon and methane.

11.46 From Table 11-4, determine which of each pair has the higher heat value and use intermolecular forces to explain why: (a) heats of vaporization of water and methane; (b) heats of fusion of benzene and ethane; and (c) heats of vaporization of molecular oxygen and argon.

11.47 Sketch the approximate phase diagram for Br_2 from the following information: normal melting point is 265.9 K, normal boiling point is 331.9 K, triple point at $P = 5.79 \times 10^{-2}$ atm and $T = 265.7$ K. Label the axes and the area where each phase is stable.

11.48 Oxygen has a normal melting point of 55 K and a normal boiling point of 90.2 K, and its triple point occurs at $P = 0.0015$ atm and $T = 54$ K. Sketch the approximate phase diagram for oxygen, and label the axes and the area where each phase is stable.

11.49 Using the phase diagram of Problem 11.47, describe what happens to a sample of Br_2 as the following processes take place. Draw lines on the phase diagram showing each process. (a) A sample at $T = 400$ K is cooled to 250 K at constant $P = 1.00$ atm. (b) A sample is compressed at constant $T = 265.8$ K from $P = 1.00 \times 10^{-3}$ atm to $P = 1.00 \times 10^3$ atm. (c) A sample is heated from 250 to 400 K at constant $P = 2.00 \times 10^{-2}$ atm.

11.50 Using the phase diagram of Problem 11.48, describe what happens to a sample of O_2 as the following processes take place. Draw lines on the phase diagram showing each process. (a) A sample at $T = 125$ K is cooled to 25 K at constant $P = 1.00$ atm. (b) A sample is compressed at constant $T = 54.5$ K, starting at $P = 1.00 \times 10^{-2}$ atm, to $P = 1.00 \times 10^2$ atm. (c) A sample is heated from 25 to 100 K at constant $P = 1.00 \times 10^{-3}$ atm.

ilw 11.51 The heat capacity of ethanol is 111 J °C^{-1} mol^{-1}. Combine this information with the data in Table 11-4 to calculate the amount of heat required to convert 1.00 g of ethanol at 25 °C to ethanol vapor.

11.52 White phosphorus is a molecular solid made of molecules of P_4. A 1.00-g sample of white phosphorus starts at 25 °C. Calculate the amount of heat required to melt the sample. Use the following data: mp = 44 °C; C = 94 J °C^{-1} mol^{-1}); ΔH_{fusion} = 2.5 kJ/mol.

Additional Paired Problems

11.53 List the different kinds of forces that must be overcome to convert each of the following from a liquid to a gas: (a) NH_3; (b) $CHCl_3$; (c) CCl_4; and (d) CO_2.

11.54 List all the intermolecular forces that stabilize the liquid phase of each of the following: (a) Xe; (b) SF_4; (c) CF_4; and (d) CH_3CO_2H (acetic acid).

11.55 Manganese forms crystals in either body-centered cubic or face-centered cubic geometry. Which phase has the higher density? Explain your choice.

11.56 Beryllium forms crystals in either body-centered cubic or hexagonal close-packed geometry. Which phase has the higher density? Explain your choice.

11.57 For each of the following pairs, identify which has the higher boiling point, and identify the type of force that is responsible: (a) H_3COCH_3 and CH_3OH; (b) SO_2 and SiO_2; (c) HF and HCl; and (d) Br_2 and I_2.

11.58 The boiling points of the Group 16 binary hydrides are as follows: H_2O, 100 °C; H_2S, −60 °C; H_2Se, −41 °C. Explain in terms of intermolecular forces why H_2S has a lower boiling point than either H_2O or H_2Se.

11.59 Refer to Figure 11-40. Describe in detail what occurs when each of the following is carried out: (a) The pressure on a sample of liquid N_2 is reduced from 1.00 to 0.010 atm at a constant temperature of 70 K. (b) A sample of N_2 gas is compressed from 1.00 to 50.0 atm at a constant temperature of 298 K. (c) A sample of N_2 gas is cooled at a constant pressure of 1.00 atm from 298 to 50 K.

11.60 Refer to Figure 11-40. Describe in detail what occurs when each of the following is carried out: (a) A sample of CO_2 gas is compressed from 1.00 to 50.0 atm at $T = 298$ K. (b) Dry ice at 195 K is heated to 350 K at $P = 6.00$ atm. (c) A sample of CO_2 gas at $P = 1.00$ atm is cooled from 298 to 50 K.

11.61 The compound 1,2-dichloroethylene (ClCHCHCl) exists as *cis* and *trans* isomers. One isomer boils at 47 °C, the other at 60 °C. Draw Lewis structures of the two isomers. Use dipole moments and symmetry arguments to assign the boiling points to the isomers.

11.62 Why do the two isomers shown have very different melting points?

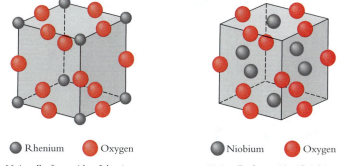

C₈H₈O₃
Methyl 2-hydroxybenzoate
(Oil of wintergreen)
mp = −8 °C

C₈H₈O₃
Methyl 4-hydroxybenzoate
mp = 127 °C

11.63 Which gas deviates more from ideal PV/nRT behavior, F_2 or Cl_2? Explain your choice.

11.64 Which gas deviates more from ideal PV/nRT behavior, CH_4 or SnH_4? Explain your choice.

11.65 Refer to Figure 11-41, and describe what happens to silica as it is slowly heated from room temperature to 2000 °C at atmospheric pressure.

11.66 Refer to Figure 11-41. How many triple points appear on this phase diagram? For each one, describe the conditions and state the three phases that coexist under these conditions.

11.67 Solid sodium metal can exist in two different crystalline forms, body-centered cubic and hexagonal. One is stable only below 5 K at relatively low pressure; the other is stable under all other conditions. Draw atomic pictures of both phases, identify the phase that is stable at low temperature and pressure, and explain your choice.

11.68 When neon solidifies, it can form either face-centered cubic or hexagonal close-packed crystals of virtually identical stability. Draw atomic pictures of both phases and explain why they are equally stable.

11.69 What type of solid is each of the following? Give reasons for your assignments. (a) ZrO_2 (mp = 2677 °C); (b) Zr; and (c) $ZrSiO_4$.

11.70 What type of solid is each of the following? Give reasons for your assignments. (a) SO_2; (b) $ZnSO_4$; and (c) Zn.

11.71 The figure below shows the phase diagram for elemental sulfur. Use it to answer the following questions: (a) Under what conditions does rhombic sulfur melt? (b) Under what conditions does rhombic sulfur convert to monoclinic sulfur? (c) Under what conditions does rhombic sulfur sublime?

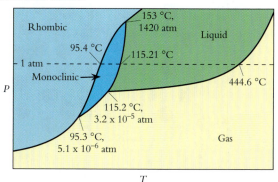

11.72 Use the figure in Problem 11.71 to answer the following questions: (a) How many triple points are there in the diagram? (b) What phases coexist at each triple point? (c) If sulfur is at the triple point at 153 °C, what happens if the pressure is reduced?

11.73 The unit cell of rhenium oxide appears in the figure below. (a) Name the lattice occupied by the rhenium atoms. (b) Calculate the number of rhenium atoms and the number of oxygen atoms in the unit cell and identify the formula of rhenium oxide.

🔵 Rhenium 🔴 Oxygen   🔵 Niobium 🔴 Oxygen

Unit cell of an oxide of rhenium   Unit cell of an oxide of niobium

11.74 One form of niobium oxide crystallizes in the unit cell shown above. Calculate the number of niobium and oxygen atoms in the unit cell and identify the formula of niobium oxide.

11.75 Mercury is a liquid at room temperature, and both cesium and gallium melt at body temperature. How do interatomic forces in these three metals compare with those in more typical metals such as copper or iron?

11.76 Tungston does not melt even when heated to incandescence. How do interatomic forces in tungsten compare with those in more typical metals such as gold or nickel?

More Challenging Problems

11.77 Use the phase diagram of Problem 11.71 to answer the following questions: (a) Are there constant pressures at which each of the four phases becomes stable as the temperature is raised? If so, identify those pressures and describe what happens as the temperature increases. (b) Are there constant temperatures at which each of the four phases becomes stable as the pressure is raised? If so, identify those temperatures and describe what happens as the pressure increases.

11.78 Does the boiling point of HCl (see the graph in Figure 11-15) suggest that it may form hydrogen bonds? Explain your answer and draw a molecular picture that shows the possible hydrogen bonds between HCl molecules.

11.79 Molecular hydrogen and atomic helium have two electrons, but He boils at 4.2 K, whereas H_2 boils at 20 K. Neon boils at 27.1 K, whereas methane, which has the same number of electrons, boils at 114 K. Explain why molecular substances boil at a higher temperature than atomic substances with the same number of electrons.

11.80 Quartz and glass are both forms of silicon dioxide. A piece of quartz breaks into a collection of smaller regular crystals with smooth faces. A piece of glass breaks into irregular shards. Use molecular structures to explain why the two solids break so differently.

11.81 Draw the unit cell of the NaCl crystal and determine the number of nearest neighbors of opposite charge for each ion in this unit cell.

11.82 In the two figures shown below, the green molecule is about to strike the wall of its container. Assuming all other conditions are identical, which collision will exert greater pressure on the wall? Explain in terms of intermolecular interactions.

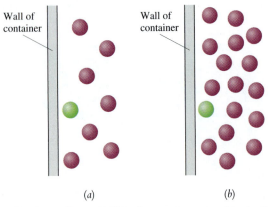

Wall of container

Wall of container

(a)   (b)

11.83 Aluminum silicate, Al_2SiO_5, is a prevalent mineral in the Earth's crust. The figure below shows a portion of the phase diagram for this mineral, which has three different crystalline forms. Use the phase diagram to answer the following questions: (a) What can you say about

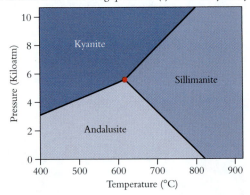

the melting point of Al_2SiO_5? (b) When liquid Al_2SiO_5 solidifies, what crystalline form results? (c) Might there be a pressure above which the liquid solidifies to give a different crystalline form? If so, identify the form and estimate the minimum pressure needed. (d) Among the three forms of solid, which is the densest, and which is the least dense? State your reasoning.

11.84 Arrange the following liquids in order of increasing viscosity, and state the factors that determine the ranking: *n*-butanol, $CH_3CH_2CH_2CH_2OH$; *n*-pentane, $CH_3CH_2CH_2CH_2CH_3$; propane-1,3-diol, $HOCH_2CH_2CH_2OH$; and 2,2-dimethylpropane, $(CH_3)_4C$.

11.85 Construct a qualitative graph similar to the one in Figure 11-37 that summarizes the energy changes that accompany the following process. A sample of water at 35 °C is placed in a freezer until the temperature reaches −25 °C. Plot temperature along the *y*-axis and kilojoules of heat removed along the *x*-axis.

11.86 Recently, a new group of solids was prepared that can act as superconductors at temperatures near the boiling point of liquid nitrogen. The unit cell of one of these new superconductors follows. Identify the formula of the compound.

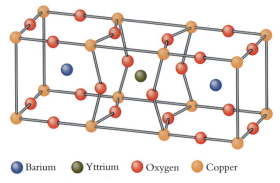

● Barium ● Yttrium ● Oxygen ● Copper

11.87 As a vapor, acetic acid (CH_3CO_2H) exists as a mixture of individual molecules and pairs of molecules held together by hydrogen bonds. (a) Draw a Lewis structure that shows how two acetic molecules can form a pair by hydrogen bonding. (b) Does the fraction of paired acetic acid molecules increase or decrease as the temperature rises? Explain.

Group Study Problems

11.88 Classify each of the following solids as network, metallic, ionic, or molecular: (a) a solid that conducts electricity; (b) a solid that does not conduct electricity but dissolves in water to give a conducting solution; and (c) a solid that does not conduct electricity and melts below 100 °C to give a nonconducting liquid.

11.89 Describe the similarities and differences between hexagonal close-packed and body-centered cubic structures.

11.90 The packing efficiency of spheres in a crystal is the percentage of the space occupied by the spheres. Calculate the packing efficiencies of the simple cubic, body-centered cubic, and face-centered cubic cells. *Hints:* The volume of a sphere is $4\pi r^3/3$. Use the number of spheres contained in each unit cell and determine the lengths of the sides of the unit cells from the spheres that are in contact with each other (edge of cube, body diagonal of body-centered, face diagonal of face-centered).

11.91 How many grams of methane must be burned to provide the heat needed to melt a 75.0-g piece of ice and bring the resulting water to a boil? Assume a constant pressure of 1 atm. Use the data in Tables 6-1 and 11-4 to develop your solution.

11.92 Solid silver adopts a face-centered cubic lattice. The metallic radius of a silver atom is 144 pm. (a) How many silver atoms occupy one unit cell of solid silver? (b) Calculate the length of one side of a unit cell. (c) Calculate the volume of the unit cell. (d) What percentage of the volume of the unit cell is empty? Use the diagram at right to develop your answers.

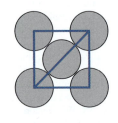

11.93 A chemist has stored pure bromine in a freezer at 265.9 K, the melting point of Br_2. A sample with a mass of 50.0 g is removed from the freezer and placed in a distillation flask, where it is heated until all the bromine has vaporized. The heater provides a constant flow of heat at 25.0 J/s. Prepare a heating curve for this process, showing temperature vs. time. Draw it as quantitatively as you can, using data from Table 11-4 and the following heat capacity information: $C_{liquid} = 76$ J °C^{-1} mol^{-1}; $C_{gas} = 20.8$ J °C^{-1} mol^{-1}.

Answers to Section Exercises

11.1.1 At_2 is a solid, with stronger intermolecular forces than I_2. Continuing the trends shown in Figure 11-3, the bar for At_2 should be higher than that for I_2.

11.1.2 Intermolecular forces cause PV/nRT to be smaller than 1.0. Fluorine reaches the smallest value, so it has the largest intermolecular

forces. Helium never dips below 1.0, indicating very small intermolecular forces.

11.1.3 (a) Xe boils at a higher temperature than Ar. (b) The order is H_2, N_2, F_2, Cl_2. (c) Group 18 has the smallest intermolecular forces.

11.2.1 (a) Propane has larger polarizability because its electrons are spread over a larger volume than those in krypton. (b) CCl_4 has a larger polarizability because it has more electrons. (c) CO has a dipole moment, generating dipole–dipole attractions not present in N_2.

11.2.2 Acetone has dispersion forces and dipole–dipole forces. Methanol has dispersion forces, dipole–dipole forces, and hydrogen bonding forces. Their boiling points are similar because the dipolar and dispersion forces in acetone are larger than the analogous forces in methanol, counterbalancing the hydrogen bonding in methanol.

11.2.3

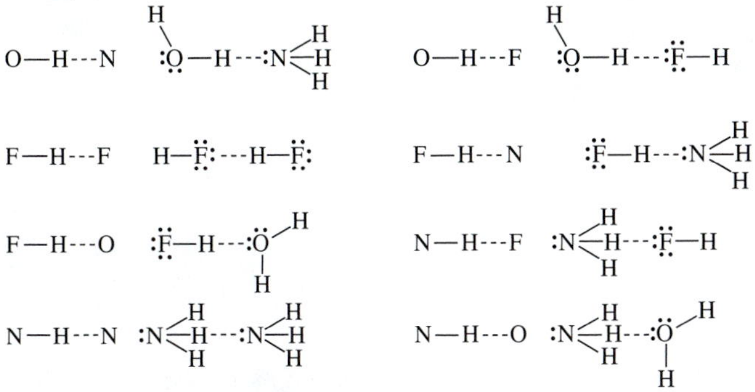

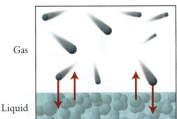

11.3.1 Each *n*-pentanol molecule, with its highly polar O—H bond, forms hydrogen bonds with neighboring *n*-pentanol molecules. These hydrogen bonds must break for *n*-pentanol molecules to move past one another, increasing the viscosity of this liquid relative to that of *n*-hexane.

11.3.2 A meniscus forms because of differences between the forces holding molecules in the liquid and those attracting them to the walls of the container. For water in glass, water–wall forces are greater than water–water forces, so water is attracted to the wall and the meniscus is concave. The convex meniscus for mercury in glass signals that mercury–mercury forces are greater than mercury–wall forces, so the wall repels mercury.

11.3.3 Figure 11-19 shows that fewer water molecules than bromine molecules have enough energy to escape the liquid at 298 K, so your figure should show fewer molecules in the gas phase and fewer arrows:

Equilibrium: $p = vp$

11.4.1 (a) CO_2 is a molecular solid, with dispersion forces between the molecules. (b) S_8 is another molecular solid, with dispersion forces between the molecules. (c) Tin is a metal, with delocalized metallic bonding holding the atoms together. (d) Li_2O is ionic, with electrical attractions between Li^+ cations and O^{2-} anions.

11.4.2 The order of melting points is F_2, K, Co, C (diamond). F_2 is molecular, with dispersion forces only. K and Co are metals, but Co has more valence electrons, resulting in stronger bonding. Diamond is a network solid with strong covalent bonds.

11.4.3 (a) $Mg_3(PO_4)_2$; (b) $LiAlSi_2O_6$; and (c) $Bi_2Sr_2CuO_6$

11.5.1 All angles in a hexagonal crystal are multiples of 60°, so we expect angles of 60°, 120°, and so on. A cubic close-packed crystal also has right angles, so we expect 90° angles as well.

11.5.2 The view is not a unit cell, because if two of these views are placed side by side, the ion from one view occupies a vacancy from the other view.

11.5.3

11.6.1 (a) methane, 8.6 kJ; (b) ethane, 7.8 kJ; and (c) ethanol, 13 kJ

11.6.2 CO_2

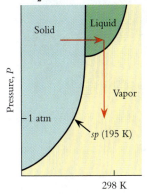

11.6.3 (a) 0.124 atm; (b) ice remains solid until the pressure drops to about 0.005 atm, when it sublimes; and (c) at 7 atm, dry ice melts at around 217 K and is liquid at room temperature.

Answers to Extra Practice Exercises

11.1 $V_{in} = 8.96$ L/mol; $P = 0.977$ atm

11.2 H_2Te, with the largest inner atom, has largest dispersion forces and highest boiling point. H_2S, with the smallest inner atom, has smallest dispersion forces and lowest boiling point.

11.3 Acetaldehyde is more compact than acetone and has fewer electrons, so it boils at a lower temperature. Like acetone, acetaldehyde has a polar C=O bond, giving it a higher boiling point than either butane or methyl ethyl ether.

11.4 The O atom in an acetone molecule can form two H bonds to two different water molecules as shown at right:

11.5 Selenium is a molecular solid; germanium is a network solid.

11.6 41.7 kJ flows from the tray to the freezer compartment.

11.7

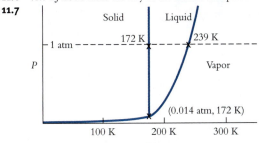

11.8 The gas liquefies at $P = 0.124$ atm, and the liquid freezes when the pressure reaches about 0.5 atm. At 1 atm, 63.1 K, N_2 is a solid.

The Chemistry of Shampoo

Products for personal care are big business: Americans spend close to $12 billion annually on shampoos, conditioners, and styling gels. These products rely on intermolecular forces for their effectiveness. Shampoos, in particular, interact with water in ways that we can understand by knowing the properties of solutions.

The active ingredients in a shampoo play three fundamental roles. Some allow water to wash away the substances that make hair "dirty." Others adhere to hair to impart a desirable feel and texture. The rest are emulsifiers that keep the mixture from separating into its components. To accomplish these effects, ingredients combine two types of interactions: a strong attraction to water (hydrophilic) and an aversion to water (hydrophobic). It may seem that these properties are incompatible, but shampoos contain molecules that are designed to be simultaneously hydrophilic and hydrophobic. One example is sodium lauryl sulfate, our inset molecule. The ionic "head" of the molecule is hydrophilic, so it interacts attractively with water. The hydrocarbon "tail" is hydrophobic, so it interacts attractively with grease and dirt. Molecules of the shampoo associate with hydrophobic dirt particles to form hydrophilic clumps that dissolve in water and wash away.

Washing removes the natural coating of protective oil that gives hair its body and shine. To counter this, shampoos and conditioners contain hydrophobic oils that cling to the surface of hair and remain in

Properties of Solutions

place upon rinsing. Dimethicone, shown in our inset, is an artificial oil that contains hydrophobic silicon–oxygen chains and methyl groups.

To keep their various ingredients mixed together uniformly, shampoos also contain emulsifiers whose structures include hydrophobic and hydrophilic components. Our inset shows the structure of panthenol, one such emulsifier. Panthenol has hydrophobic hydrocarbon regions interspersed with hydrophilic polar groups such as N—H and O—H.

In addition to cleaning and conditioning hair, a shampoo must be nontoxic, biodegradable, and pleasant-smelling. Therefore, chemists who formu-

late hair-care products face challenging chemical design problems. That is why a typical list of ingredients contains a multitude of exotic-sounding chemicals.

CHAPTER CONTENTS

Learning Resources

 KEY CONCEPTS

 CRITICAL THINKING

 SKILLS TO MASTER

 ADDITIONAL HELP
www.wiley.com/college/olmsted
● TUTORIALS
● ANIMATIONS

This chapter explores solutions. Its main emphasis is on aqueous solutions, because water is the most important solvent for life. We first describe the nature of solutions. Then we examine the molecular features that determine whether one substance dissolves in another. We introduce some characteristic properties of solutions, including effects on phase behavior. The chapter concludes by looking more closely at dual-nature molecules such as those present in shampoos.

12.1 THE NATURE OF SOLUTIONS

A solution is a homogeneous mixture of two or more substances. As described in Chapter 3, a solution contains a **solvent** and one or more **solutes.** The solvent determines the state of the solution, and normally the solvent is the component present in the greatest quantity. The most common solutions are liquids with water as solvent, but solutions exist in all three states of matter. The atmosphere of our planet, air, is a gaseous solution with molecular nitrogen as the solvent. Steel is a solid solution containing solutes such as chromium and carbon that add strength to the solvent, iron.

Components of Solutions

The solvent determines the phase of a solution, but the solutes may be substances that would normally exist in different phases. Table 12-1 lists an example of each possibility.

Gaseous solutions are easy to prepare and easy to describe. The atoms or molecules of a gas move about freely. When additional gases are added to a gaseous solvent, each component behaves independently of the others. Unless a chemical reaction occurs, the ideal gas equation and Dalton's law of partial pressures describe the behavior of gaseous solutions at and below atmospheric pressure (see Chapter 5).

At the other extreme, solid solutions are difficult to prepare. To form a solid solution, individual atoms or molecules must mix homogeneously within the solid network. Given that the atoms or molecules in a solid cannot move freely about, such mixing is difficult to achieve. Even if we grind two solids to fine powders and then mix them thoroughly, the resulting mixture will not be a solution, because the mixture will remain heterogeneous at the microscopic and molecular levels.

Solid solutions generally form in one of two ways, both of which involve forming the solid from the liquid phase. One way is to heat the solid solvent until it melts, add the solutes into the molten material, and then cool the melt until it solidifies. Solid solutions of one metal in another, such as brass and steel, are prepared in this way. A second method is to dissolve the solid solvent and solutes in an appropriate liquid, then cool or evaporate the liquid until a solid precipitates. Solid solutions of organic substances can form in this manner.

Usually we think of solutions as liquids, and aqueous solutions are the most common liquid solutions in everyday life. Blood, sweat, and tears all are aqueous solutions.

Ocean water is an aqueous solution containing many solutes.

Table 12-1 Phases of Solutions

Solvent Type	Solute Type		
	Gas	**Liquid**	**Solid**
Gas	Diving gas (He, O_2)	Humid air (N_2, O_2, H_2O)	Air above I_2 (N_2, O_2, I_2)
Liquid	Carbonated water (H_2O, CO_2)	Vodka (H_2O, C_2H_5OH)	Saline solution (H_2O, NaCl)
Solid	H_2 storage alloy (La, Ni, H_2)	Plastic (PVC, dioctylphthalate)	Steel (Fe, C, Mn)

So are the beverages that we drink. The oceans are aqueous solutions of sodium chloride and other mineral salts. Even the fresh waters of streams and lakes are aqueous solutions containing salts and organic solutes. Most of this chapter discusses aqueous solutions.

Solution Concentration

The composition of a solution can vary, so we must specify the concentrations of solutes as well as their identities. There are several ways to express concentration, each having advantages as well as limitations. Any concentration value is a ratio of amounts. The amount of one component, usually a solute, appears in the numerator, and some other amount, describing either the solvent or the total solution, appears in the denominator.

A convenient concentration measure for gaseous solutions is the mole fraction, introduced in Section 5.5. The mole fraction (X, dimensionless) is the number of moles of one component divided by the total moles of all components.

$$\text{Mole fraction of } A = \frac{\text{Moles of } A}{\text{Total number of moles}} \quad or \quad X_A = \frac{n_A}{n_{\text{total}}}$$

Mole fractions are convenient for expressing concentrations of gas mixtures, because we often measure gas pressures, and partial pressures are related to total pressure through the mole fraction.

Parts per million (ppm) and parts per billion (ppb) are additional concentration measures related to mole fraction. As described in Section 5.5, parts per million is the number of items of one kind present in a sample containing a million (10^6) total items. Parts per billion is the number of items of one kind present in a sample containing a billion (10^9) items. When the items are molecules,

$$1 \text{ ppm} = 1 \text{ molecule out of every } 10^6 \text{ molecules}$$

$$1 \text{ ppb} = 1 \text{ molecule out of every } 10^9 \text{ molecules}$$

These measures are useful when the concentration of a solute is quite small. Krypton, for example, is present in the Earth's atmosphere at a mole fraction of 1.14×10^{-6}. This is more conveniently expressed as 1.14 ppm.

As described in Section 3.6, molarity is the most common concentration measure for aqueous solutions. Molarity (M, units of mol/L) is the number of moles of a solute divided by the volume of solution:

$$\text{Molarity} = \frac{\text{Moles of solute}}{\text{Total volume of solution}} \quad or \quad M = \frac{n_{\text{solute}}}{V_{\text{solution}}} \qquad \textbf{(3-1)}$$

Molarity is convenient for expressing concentrations of liquid solutions, because we often measure liquid solutions volumetrically.

Molarity is not a convenient concentration measure when the temperature of the solution changes. This is because a solution expands as it is heated. Consequently, the total volume of solution changes with temperature, causing molarity to vary even though the amounts of solutes and solvent remain fixed.

For applications where the solution temperature changes, chemists prefer to use the **molality** (c_m, units of mol/kg). Molality is defined to be the number of moles of solute divided by the mass of solvent in kilograms.

$$\text{Molality} = \frac{\text{Moles of solute}}{\text{Kilograms of solvent}} \quad or \quad c_m = \frac{n_{\text{solute}}}{m_{\text{solvent}}} \qquad \textbf{(12-1)}$$

Chemists traditionally use m to designate molality, but this symbol easily leads to confusion because m also represents mass. For this reason, we use the symbol c_m (concentration, molal) for molality.

Neither moles of solute nor mass of solvent changes with temperature, so solution molality is independent of temperature. Example 12-1 illustrates molality calculations.

Example 12-1

Molality

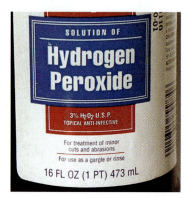

Hydrogen peroxide disinfectant typically contains 3.0% H_2O_2 by mass. Assuming that the rest of the contents is water, what is the molality of this disinfectant?

Strategy: To calculate molality, we use Equation 12-1, for which we need moles of solute and mass of solvent in kilograms. To find these quantities, it is convenient to consider 100 g of solution, for which the mass percentage is the mass of one component present.

Solution: In 100 g of disinfectant, there are 3.0 g of hydrogen peroxide and 97 g of water. We need to convert the mass of solute into moles and the mass of solvent into kilograms:

$$MM_{H_2O_2} = 2(16.00 \text{ g/mol}) + 2(1.008 \text{ g/mol}) = 34.02 \text{ g/mol}$$

$$n_{H_2O_2} = \frac{m}{MM} = \frac{3.0 \text{ g}}{97 \text{ g/mol}} = 0.0309 \text{ mol}$$

$$m_{water} = (97 \text{ g})\left(\frac{1 \text{ kg}}{10^3 \text{ g}}\right) = 0.0970 \text{ kg}$$

$$c_m = \frac{n_{solute}}{m_{solvent}} = \frac{0.0309 \text{ mol}}{0.0970 \text{ kg}} = 0.32 \text{ mol/kg}$$

Does the Result Make Sense?   The units match those of molality, and the result is rounded to two significant figures to match the precision of the mass percentage information.

Determine the molality of a sugar solution that contains 3.94 g of sucrose ($C_{12}H_{22}O_{11}$) dissolved in 285 g of water.

Each concentration measure is convenient for some types of calculations but inconvenient for others. Consequently, a chemist may need to convert a concentration from one measure to another. Example 12-1 illustrates the conversion from a mass-based concentration, percent by mass, to a mole-based concentration, molality. Mole fraction, molarity, and molality all are mole-based for the solute, but the denominators of these concentration measures are quite different. Mole fraction uses total number of moles, molarity uses total volume, and molality uses mass of *solvent*.

To convert between molarity and the other mole-based concentration measures, we must relate volume to mass and number of moles. Whereas molar mass lets us convert between mass and moles, we need density to convert between volume and mass:

$$\text{Density} = \frac{\text{Mass}}{\text{Volume}} \quad or \quad \rho = \frac{m}{V} \tag{1-2}$$

Conversions between molarity and other concentration measures require two steps. First, we use density to convert volume of solution to mass of solution. Then we find the mass of solvent by subtracting the mass of solute from the mass of solution. Example 12-2 provides practice in conversions among mole-based concentration measures.

Example 12-2

Concentration Conversions

Concentrated aqueous ammonia (also known as ammonium hydroxide) is 14.8 M and has a density of 0.898 g/mL. Determine the molality and mole fraction of ammonia in this solution.

Strategy: We have information about molarity (mol/L) and density (g/mL) and are asked to find molality (mol/kg) and mole fraction (mol/mol). A good way to approach conversions from molarity to another measure is to choose a convenient volume for the solution, determine its mass and the mass of solute, and find the mass of water by difference. Then convert mass of water to kilograms and to moles to complete the calculations.

Solution: Because we know molarity (mol/L), it is most convenient to work with 1.000 L of solution, which has a volume of 1000. mL and contains 14.8 mol NH_3. Use the density to calculate the mass of this solution:

$$\rho = \frac{m}{V} \qquad \text{so} \qquad m = \rho V$$

$$m_{solution} = (0.898 \text{ g/mL}) (1000. \text{ mL}) = 898 \text{ g}$$

Calculate the mass of NH_3 from moles of NH_3:

$$m = n \, MM = (14.8 \text{ mol})(17.0 \text{ g/mol}) = 252 \text{ g } NH_3$$

Subtract this mass from the total mass to find mass of H_2O:

$$m_{solvent} = 898 \text{ g} - 252 \text{ g} = 646 \text{ g}$$

Convert solvent mass to kilograms and to moles:

$$m_{solvent} = (646 \text{ g})\left(\frac{1 \text{ kg}}{10^3 \text{ g}}\right) = 0.646 \text{ kg}$$

$$n_{solvent} = \frac{m}{MM} = \frac{646 \text{ g}}{18.0 \text{ g/mol}} = 35.9 \text{ mol}$$

Now calculate molality and mole fraction:

$$c_m = \frac{n_{solute}}{m_{solvent}} = \frac{14.8 \text{ mol}}{0.646 \text{ kg}} = 22.9 \text{ mol/kg}$$

$$X = \frac{n_{solute}}{n_{solute} + n_{solvent}} = \frac{14.8 \text{ mol}}{14.8 \text{ mol} + 35.9 \text{ mol}} = 0.292$$

Do the Results Make Sense? Concentrated ammonia has a high molarity, so we expect that it will also have a high molality and mole fraction.

Determine the molality and mole fraction of concentrated aqueous HCl, which is 12.0 M and has a density of 1.19 g/mL.

Example 12-2

Concentration Conversions
(*continued*)

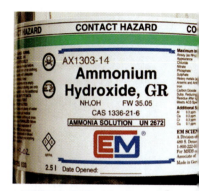

Concentrated aqueous ammonia is known commercially as ammonium hydroxide (NH_4OH), because H_2O molecules can transfer protons to NH_3 molecules to generate NH_4^+ cations and OH^- anions.

**Extra Practice
Exercise 12.2**

Section Exercises

12.1.1 Brass contains 63% copper and 37% zinc. Identify the solvent and solute, and describe what type of solution this is.

12.1.2 Tonic water contains 31 ppm *by mass* of quinine ($C_{20}H_{24}O_2N_2$). Calculate the mole fraction of quinine in tonic water.

12.1.3 Determine the molarity and molality of quinine in the tonic water described in Section Exercise 12.1.2, assuming that the density of tonic water is 1.00 g/mL.

12.2 DETERMINANTS OF SOLUBILITY

Solubilities vary tremendously. At one extreme, some substances form solutions in all proportions and are said to be **miscible.** For example, acetone and water can be mixed in any proportion, from pure water to pure acetone. At the other extreme, a substance may be **insoluble** in another. One example is common salt, NaCl, whose solubility in gasoline is virtually zero.

Many combinations display solubility that is between the two extremes of misci-ble and insoluble. In other words, the substance dissolves, but there is a limit to the amount of solute that will dissolve in a given amount of solvent. When that limit has been reached, the solution is **saturated.** The concentration of a saturated solution is the **solubility** of the substance in that particular solvent at a specified temperature. Solubilities vary over a wide range. For instance, 350 g of NaCl will dissolve in 1 L of water at room temperature, but only 0.0015 g of AgCl will dissolve under the same conditions.

Like Dissolves Like

Solubility is a complex phenomenon that depends on the balance of several prop-erties. A complete understanding of solubility is beyond an introductory course, but the general features of solubility are summarized by the expression *like dis-solves like.*

KEY CONCEPT *Substances that dissolve in each other usually have similar types of intermolecular interactions.*

One substance dissolves in another if the forces of attraction between the solute and the solvent are similar to the solvent–solvent and solute–solute interactions. This gener-alization can be applied to a variety of solution types.

When two liquids are mixed, three sets of intermolecular interactions must be assessed. Water (H_2O) and methanol (CH_3OH) provide a first example. When water and methanol mix, $H_2O \cdots H_2O$ attractive forces and $CH_3OH \cdots CH_3OH$ attractive forces must be overcome, but they are replaced by $H_2O \cdots CH_3OH$ attractive forces. The two liquids are alike in that both substances contain O—H groups that form hydrogen bonds readily. When these liquids are mixed, $H_2O \cdots H_2O$ hydrogen bonds and $CH_3OH \cdots CH_3OH$ hydrogen bonds break, but $H_2O \cdots CH_3OH$ hydrogen bonds form. The net result is that the degree of hydrogen bonding in the solution is about the same as in either of the pure liquids, making these two liquids miscible.

Octane and cyclohexane are another liquid pair whose intermolecular interactions are alike. Both have low polarities, so these molecules in the pure liquids are held together by the dispersion forces caused by their polarizable electron clouds. Dispersion forces in solutions of octane and cyclohexane are about the same as in the pure liquids. Again, these two liquids are miscible.

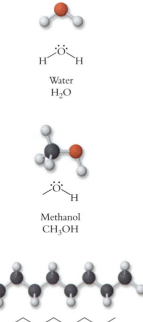

Water
H_2O

Methanol
CH_3OH

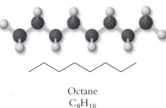

Octane
C_8H_{18}

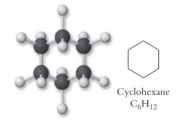

Cyclohexane
C_6H_{12}

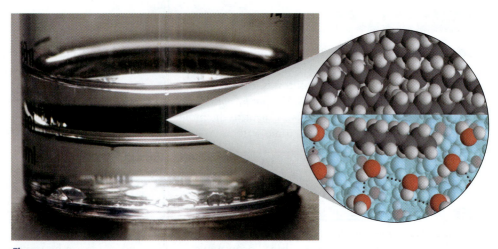

Figure 12-1
Water and octane are not alike. Rather than dissolving, octane floats on water. The presence of an octane molecule in the water layer disrupts a portion of the hydrogen-bonding network in water.

In contrast, water and octane are not alike. As Figure 12-1 shows, an octane molecule disrupts the hydrogen-bonding network of water. Octane does not form hydrogen bonds, so the only forces of attraction between water molecules and octane molecules are dispersion forces. Because hydrogen bonds are stronger than dispersion forces, the cost of disrupting the hydrogen-bonding network in water is far greater than the stability gained from octane–water dispersion forces. Thus, octane and water are nearly insoluble in each other. When mixed, octane and water partition into distinct layers, one of nearly pure water and the other of nearly pure octane. For the same reason, water and cyclohexane are also insoluble in each other and form two layers when mixed.

Some liquids can interact with other substances in multiple ways. Acetone, for instance, has a polar $C{=}O$ bond and a three-carbon bonding framework (Figure 12-2). The bonding framework is similar to that of a hydrocarbon, so acetone mixes with cyclohexane and octane. The polar $C{=}O$ group makes acetone miscible with other polar molecules such as acetonitrile (CH_3CN). Finally, the polar oxygen atom in acetone has lone pairs of electrons that can form hydrogen bonds with hydrogen atoms of ammonia or water. Because of its versatility, acetone is an important industrial solvent. It is also used in laboratories to clean and rinse glassware. Example 12-3 treats several alcohols that also display multiple types of interactions.

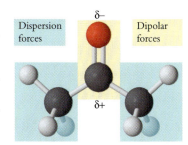

Figure 12-2
Acetone is a versatile solvent that dissolves both polar and nonpolar substances. The three-carbon chain is compatible with nonpolar molecules such as octane and cyclohexane, whereas the polar $C{=}O$ bond is compatible with polar molecules such as methanol and water.

Example 12-3

Solubility Trends

Give a molecular explanation for the following trend in alcohol solubilities in water:

n-Propanol	$CH_3CH_2CH_2OH$	Completely miscible
n-Butanol	$CH_3CH_2CH_2CH_2OH$	1.1 M
n-Pentanol	$CH_3CH_2CH_2CH_2CH_2OH$	0.30 M
n-Hexanol	$CH_3CH_2CH_2CH_2CH_2CH_2OH$	0.056 M

Strategy: Solubility limits depend on the stabilization generated by solute–solvent interactions balanced against the destabilization that occurs when solvent–solvent interactions are disrupted by solute. Thus, we must examine intermolecular interactions involving water and alcohol molecules.

Solution: When an alcohol dissolves in water, the nonpolar hydrocarbon part of the alcohol disrupts the hydrogen-bonding network of water. Counterbalancing this solvent disruption, the —OH groups of the alcohol form hydrogen-bonding interactions with water molecules.

As the nonpolar region of an alcohol grows longer, each solute molecule disrupts more and more hydrogen bonds in the solvent. At the same time, each of these alcohols has only one —OH group, so the amount of compensating solute–solvent hydrogen bonding is the same for all the alcohols.

Longer-chain alcohols are progressively less soluble in water because as the hydrocarbon chain gets longer, more destabilization is involved in inserting the alcohol into the water matrix.

Does the Result Make Sense? Another way to look at this solubility trend is to note that *n*-propanol, $CH_3CH_2CH_2OH$, is not much different from ethanol, which mixes completely with water. In contrast, *n*-hexanol, $CH_3CH_2CH_2CH_2CH_2CH_2OH$, is not much different from octane, which is insoluble in water.

Here are the room-temperature solubilities (mass percentages) of some liquids in cyclohexane (C_6H_{12}): water, 0.01%; methanol, 5%; ethanol, completely miscible. Give a molecular explanation for the trend.

**Extra Practice
Exercise 12.3**

Solubility of Solids

As described in Chapter 11, there are four different kinds of solids: network, molecular, metallic, and ionic. Each is held together by a different kind of interaction, so each has its own solubility characteristics.

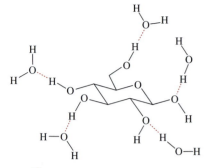

Figure 12-3
All five polar O—H groups of glucose can form hydrogen bonds with water molecules (*red dotted lines*).

Network solids such as diamond, graphite, or silica cannot dissolve without breaking covalent chemical bonds. Because intermolecular forces of attraction are always much weaker than covalent bonds, solvent–solute interactions are never strong enough to offset the energy cost of breaking bonds. Covalent solids are insoluble in all solvents. Although they may react with specific liquids or vapors, covalent solids will not dissolve in solvents.

At the opposite extreme, *molecular solids* contain individual molecules bound together by various combinations of dispersion forces, dipole forces, and hydrogen bonds. Conforming to "like dissolves like," molecular solids dissolve readily in solvents with similar types of intermolecular forces. Nonpolar I_2, for instance, is soluble in nonpolar liquids such as carbon tetrachloride (CCl_4). Many organic compounds are molecular solids that dissolve in organic liquids such as cyclohexane and acetone.

Hydrogen bonding allows water to dissolve materials that form hydrogen bonds. Hydrogen bonding makes sugars such as sucrose ($C_{12}H_{22}O_{11}$) and glucose ($C_6H_{12}O_6$) highly soluble in water. Figure 12-3 shows that glucose is an organic molecule with five polar O—H groups, each of which forms hydrogen bonds with water molecules. When glucose dissolves, hydrogen bonds between water and glucose replace the hydrogen bonds lost by the water molecules of the solvent. This balance means that the energy requirements for solution formation are small, and glucose is quite soluble in water. On the other hand, naphthalene, a similarly sized solid hydrocarbon limited to dispersion forces, is nearly insoluble in water.

The best solvent for a molecular solid is one whose intermolecular forces match the forces holding the molecules in the crystal. For a solid held together by dispersion forces, good solvents are nonpolar liquids such as carbon tetrachloride (CCl_4) and cyclohexane (C_6H_{12}). For polar solids, a polar solvent such as acetone works well. Example 12-4 provides some practice in recognizing solubility types.

Example 12-4	Vitamins are organic molecules that are required for proper function but not synthesized by the human body. Thus, vitamins must be present in the foods people eat. Vitamins fall into two categories: fat-soluble, which dissolve in fatty hydrocarbon-like tissues, and water-soluble. The structures of several vitamins follow. Assign each one to the appropriate category.
Solubilities of Vitamins	

Vitamin E

Vitamin A

Pantothenic acid
(Vitamin B$_5$)

Vitamin C

Pyridoxamine
(Vitamin B$_6$)

Vitamin D

Strategy: At first glance it may seem that "like dissolves like" does not apply here. Certainly, none of these complex molecules looks like water, and the resemblance to simple hydrocarbons such as cyclohexane also is remote. Keep in mind, however, that the basis for the principle is that similar compounds dissolve in each other because they have common patterns of intermolecular interactions. Example 12-3 indicates that alcohols containing large nonpolar segments do not dissolve well in water. We can categorize vitamins similarly by the amounts of their structures that can be stabilized by hydrogen bonding to water molecules.

Example 12-4

Solubilities of Vitamins
(*continued*)

Solution: A hydrogen-bond donor must have a hydrogen atom bonded to F, O, or N, and a hydrogen-bond acceptor is an electronegative atom with a lone pair of electrons. By these criteria, all the vitamins shown are capable of some hydrogen bonding. However, vitamins A, D, and E have large regions containing only nonpolar C—C and C—H bonds. Like the longer alcohols in Example 12-3, the hydrogen-bonding sites in these molecules are insufficient to make them soluble in water. Their large hydrocarbon-like regions make these vitamins fat-soluble.

The remaining molecules, vitamin C, pantothenic acid, and pyridoxamine, have comparatively large numbers of O—H and N—H groups. These groups allow each of these vitamin molecules to form many hydrogen bonds, so they are all water-soluble (all the B vitamins are soluble in water).

Do the Results Make Sense? The different solubilities of these two kinds of vitamins have important metabolic consequences. Aqueous body fluids do not dissolve fat-soluble vitamins, so these molecules can be stored in fatty body tissue for a long time. As a result, too much of a fat-soluble vitamin can overload the body's storage capabilities and lead to a toxic reaction. In contrast, the body cannot store water-soluble vitamins; instead, it excretes anything more than the amount it can use immediately. People must therefore have a steady supply of water-soluble vitamins in their diets to remain healthy.

Classify each of the following as water-soluble or fat-soluble:

Uric acid   Testosterone

Metals do not dissolve in water, because they contain extensive delocalized bonding networks that must be disrupted before the metal can dissolve. A few metals *react* with water, and several *react* with aqueous acids, but no metal will simply *dissolve* in water. Likewise, metals do not dissolve in nonpolar liquid solvents.

When an alkali metal contacts water, metal atoms donate electrons to water molecules, producing hydrogen gas and a solution of the metal cation (for example, Na^+). When a metal such as Ca, Zn, or Fe is treated with a strong aqueous acid, hydronium ions in the acid solution accept electrons from metal atoms, creating cations that then dissolve. We describe these redox reactions in Chapter 4. Zinc metal, for example, reacts with hydrochloric acid to generate H_2 gas and Zn^{2+} cations in solution:

$$Zn\,(s) + 2\,H_3O^+(aq) \longrightarrow Zn^{2+}(aq) + H_2\,(g) + 2\,H_2O\,(l)$$

These processes differ from other solution processes in that the solute undergoes a chemical transformation. The aqueous medium dissolves the metal by a chemical reaction that converts the insoluble metal into soluble cations. The solution produced when zinc reacts with aqueous HCl is an aqueous solution of Zn^{2+} ions, not a solution of Zn metal in water. If this solution is boiled to dryness, the remaining solid is $ZnCl_2$, not Zn metal.

Alloys

Metals are insoluble in common liquid solvents but can dissolve in each other (like dissolves like). A mixture of substances with metallic properties is called an **alloy.** Some

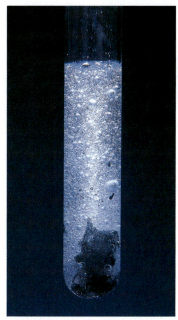

Zinc metal reacts with aqueous acids to produce hydrogen gas and Zn^{2+} ions.

Figure 12-4
Metal alloys are solid solutions that can be either substitutional (copper in silver) or interstitial (carbon in iron).

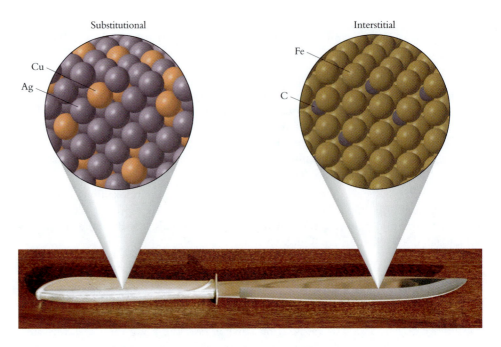

alloys are true solutions, but microscopic views show that others are heterogeneous mixtures. Brass, for instance, is a homogeneous solution of copper (20 to 97%) and zinc (80 to 3%), but common plumber's solder is a heterogeneous alloy of lead (67%) and tin (33%). When solder is examined under a microscope, separate regions of solid lead and solid tin can be seen. When brass is examined, no such regions can be detected.

One way that a solid metal can accommodate another is by substitution. For example, sterling silver is a solid solution containing 92.5% silver and 7.5% copper. Copper and silver occupy the same column of the periodic table, so they share many properties, but copper atoms (radius of 128 pm) are smaller than silver atoms (radius of 144 pm). Consequently, copper atoms can readily replace silver atoms in the solid crystalline state, as shown schematically in Figure 12-4.

A second way for a solid to accommodate a solute is interstitially, with solute atoms fitting in between solute atoms in the crystal structure. An important alloy of this type is carbon steel, a solid solution of carbon in iron, also shown in Figure 12-4. Steels actually are both substitutional and interstitial alloys. Iron is the solvent and carbon is present as an interstitial solute, but varying amounts of manganese, chromium, and nickel are also present and can be in substitutional positions.

Mercury, the only metal that is a liquid at room temperature, dissolves a number of metals to give liquid solutions. Any solution of another metal dissolved in mercury is called an *amalgam*. Metals close to mercury in the periodic table, such as silver, gold, zinc, and tin, are particularly soluble in mercury. An amalgam of silver, tin, and mercury is used to make dental fillings. When the intermetallic compound Ag_3Sn is ground with mercury, it forms a semisolid amalgam that can be shaped to fill a cavity. On standing, mercury binds with the other metals to form a hard solid mixture of Ag_5Hg_8 and Sn_7Hg_8. The mixture expands slightly during reaction, forming a tight fit within the cavity. The mercury atoms in dental fillings are chemically bound and do not dissolve, so they have been considered to be safe for the wearer, despite the fact that free mercury metal is highly toxic.

Because of concerns about possible leaching of mercury from fillings, new resin materials have replaced mercury amalgams as the material of choice for dental fillings.

Solubility of Salts

Ionic solids, also known as salts, contain cations and anions held in three-dimensional lattices by strong electrical attractions, as described in Section 8.5. We expect ionic liquids to be good solvents for ionic solids. However, typical ionic salts have such strong attractive forces that these substances are solids at room temperature. Thus, we might think that there are no solvents that dissolve ionic solids at room temperature. Experience tells us differently, however, because many ionic solids dissolve in water. Moreover, chemists have discovered how to make ionic substances that combine inorganic ions with organic

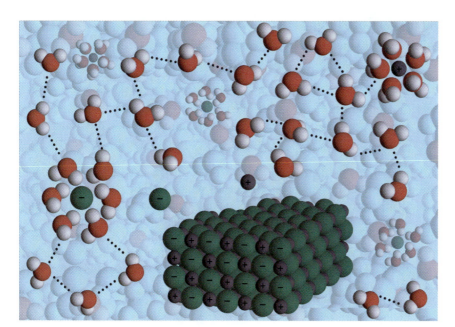

Figure 12-5
An ionic solid dissolves in water through the formation of ion–dipole interactions that overcome the forces of the crystal lattice. The arrows indicate ion–dipole interactions, and the dotted lines represent hydrogen bonds.

groups. These substances, although ionic, can be liquids at room temperature, and their potential as solvents is being extensively studied.

Water is highly polar, but it is not ionic. How, then, can water act as a solvent for ionic solids? A salt dissolves only if the interactions between the ions and the solvent are strong enough to overcome the attractive forces that hold ions in the crystal lattice. When an ionic solid forms an aqueous solution, the cations and anions are *solvated* by strong ion–dipole interactions with water molecules.

Figure 12-5 illustrates the solvation of Na^+ and Cl^- ions as NaCl dissolves in water. A cluster of water molecules surrounds each ion in solution. Notice how the water molecules are oriented so that their dipole moments align with charges of the ions. The partially negative oxygen atoms of water molecules point toward Na^+ cations, whereas the partially positive hydrogen atoms of water molecules point toward Cl^- anions.

Remember that any liquid is dynamic, with molecules and ions moving about continuously. The solvent molecules do not remain fixed in their positions. In fact, the water molecules solvating the ions change places many millions of times per second with water molecules in the bulk solvent. Nevertheless, the solvation of ions is the primary driving force in the formation of a salt solution. We describe the energy changes that accompany the formation of aqueous solutions of salts in more detail in the following section.

Section Exercises

12.2.1 List the types of intermolecular interactions that stabilize a solution of acetone in methanol, and draw molecular pictures that illustrate any dipole–dipole and hydrogen-bonding interactions that exist between molecules of these substances.

12.2.2 On the basis of their molecular structures, predict which of the following silicon-containing materials is soluble in water: elemental Si, SiO_2, Na_4SiO_4, and $Si(CH_3)_4$.

12.2.3 Explain in terms of attractive forces why some salts containing Zn^{2+} are soluble in water, but Zn metal is insoluble.

12.3 CHARACTERISTICS OF AQUEOUS SOLUTIONS

Although there are many different solvents, water is by far the most important. Water is the liquid responsible for much of the Earth's characteristics. Furthermore, water is the medium in which life begins and is sustained. In this section, we focus specifically on the characteristics of water and its solutions.

The Uniqueness of Water

Why does water play central roles in geology and biology? One important reason is that water is the only small molecule that is not a gas under terrestrial conditions. Compare the boiling point of water, 100 °C, with the boiling points of methane (−91 °C) and ammonia (−34 °C) or the sublimation temperature of carbon dioxide (−80 °C). Moreover, water is a liquid over a uniquely large temperature range, 0–100 °C. Compare this 100-degree range with the 44-degree range for ammonia. Because water is a liquid under the most prevalent terrestrial conditions, it is uniquely available to play the role of solvent.

Life as we know it depends on this existence of water as a liquid. Biochemical processes require free movement of chemicals, which cannot occur in the solid phase. Biochemical structures contain many interlocking parts that would not be stable in the gas phase. Thus, the liquid phase is best suited for life. Moreover, water is an excellent solvent, particularly for molecules that can form hydrogen bonds. As we describe in Chapter 13, the molecular building blocks of living matter are rich in groups that form hydrogen bonds. This allows biological molecules to be synthesized, move about, and assemble into complex structures, all in aqueous solution.

The structure of the water molecule accounts for its unique properties. The O—H bonds are highly polar, and the geometry of the molecule is bent. These features give H_2O a large dipole moment, generating strong dipole–dipole attractive forces. Equally important are the hydrogen bonds that water forms, both with other water molecules and with other substances. Water can form hydrogen bonds because of the two lone pairs of valence electrons on its oxygen atom and its polar O—H bonds.

The combination of a large dipole moment and hydrogen-bonding ability allows water molecules to establish strong attractive interactions of several types. Ionic salts that dissolve in water do so because of the strongly attractive forces between the dipoles of water molecules and the ions of the solute. Water-soluble organic and biological molecules contain —OH and —NH_2 groups that form strongly attractive hydrogen-bonding interactions with solvent water molecules. Even small nonpolar molecules such as O_2 dissolve in water to a certain extent because of dispersion forces.

The biochemical structures of even the simplest life forms require a liquid medium.

Solubility Equilibrium

Nearly every substance that dissolves in water has an upper limit to its solubility. Solids, liquids, and gases all display this characteristic. The room-temperature solubility of solid NaCl in water is about 6 M. Liquid *n*-hexanol forms a saturated aqueous solution at a concentration of 5.6×10^{-2} M. Gaseous O_2 in the Earth's atmosphere dissolves in natural waters to give concentrations of about 2.5×10^{-4} M.

If the concentration of a solute is lower than its solubility, additional solute can dissolve, but once the concentration of solute reaches the solubility of that substance, no further net changes occur. Individual solute molecules still enter the solution, but the solubility process is balanced by precipitation, as Figure 12-6 illustrates. A saturated solution in contact with excess solute is in a state of dynamic equilibrium. For every molecule or ion that enters the solution, another returns to the solid state. We represent dynamic equilibria by writing the equations using double arrows, showing that both processes occur simultaneously:

$$\text{Solute } (\textit{pure state}) \rightleftharpoons \text{Solute } (aq)$$

$$NaCl\,(s) \rightleftharpoons Na^+\,(aq) + Cl^-\,(aq)$$

$$CH_3CH_2CH_2CH_2CH_2CH_2OH\,(l) \rightleftharpoons CH_3CH_2CH_2CH_2CH_2CH_2OH\,(aq)$$

$$O_2\,(g) \rightleftharpoons O_2\,(aq)$$

Several observations show that saturated solutions are at dynamic equilibrium. For example, if O_2 gas enriched in the oxygen-18 isotope is introduced into the gas phase above water that is saturated with oxygen gas, the gas in the solution eventually also becomes enriched in the heavier isotope. As another example, if finely divided crystalline

Solution

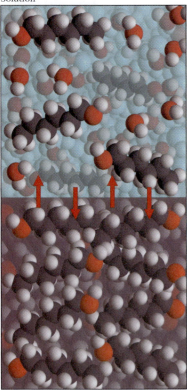

Solute

Figure 12-6
When a solution is saturated, there is dynamic equilibrium. Solute molecules move back and forth between the solute phase and the solution phase at equal rates.

salt is in contact with a saturated solution of the salt, the small crystals slowly disappear and are replaced by larger crystals. Each of these observations shows that molecules are moving between the two phases, yet the concentrations of the saturated solutions remain constant.

Energetics of Salt Solubility

When a salt dissolves, energy must be supplied to separate the ions and to break hydrogen bonds in the solvent. At the same time, energy is released when solvent molecules form ion–dipole interactions with the ions. The **molar heat of solution (ΔH_{soln})** measures the net energy flow resulting from these three factors. In any actual dissolving process, all these energy changes occur simultaneously, but we can visualize a path that treats each step separately. Figure 12-7 illustrates this path for NaCl. The first step is separating the ions, for which energy equal to the lattice energy (see Section 8.5) must be supplied. Because ionic attractions extend throughout the crystal, lattice energies are quite large. The lattice energy of NaCl is 787 kJ/mol, and that of MgO is 3791 kJ/mol.

Additional energy is needed to move solvent molecules apart to make room for the dissolving ions. In this step, hydrogen bonds break. Hydrogen bonds are substantially weaker than ion–ion interactions, so this step requires much less energy than breaking apart the lattice.

The third step is solvation of the ions by solvent molecules. Water molecules cluster around each ion, oriented to give attractive ion–dipole interactions. This step releases energy. Although each individual ion–dipole interaction is weak, each ion forms from four to eight such interactions, depending on the size of the ion and the concentration and temperature of the solution. Taken together, the vast number of ion–dipole interactions results in a substantial release of energy.

Large positive energy changes accompany the breaking apart of a lattice, and large negative energy changes accompany ion solvation. Nevertheless, molar heats of solution (ΔH_{soln}) are typically less than 100 kJ/mol. As Figure 12-7 suggests, the reason for this is that the energy absorbed in separating ions of opposite charge is largely, if not entirely, balanced by the energy released when water molecules form ion–dipole interactions with the ions.

The dissolving process can be either exothermic or endothermic. If the energy released through ion–dipole attractions is larger than the crystal lattice energy, energy is released as the salt dissolves, giving an exothermic process with negative ΔH_{soln}. Some salts, including calcium chloride (CaCl$_2$, $\Delta H_{soln} = -83$ kJ/mol) and magnesium sulfate (MgSO$_4$, $\Delta H_{soln} = -91.2$ kJ/mol), release relatively large amounts of energy when they dissolve in water. Conversely, if the crystal lattice energy is larger than the ion–dipole attractive energy, the salt must absorb energy from the surroundings as it dissolves, giving an endothermic solution process with positive enthalpy change. One soluble salt with a relatively large positive ΔH_{soln} is ammonium nitrate (NH$_4$NO$_3$, $\Delta H_{soln} = 21.1$ kJ/mol). This balance between the lattice energy and the energies of ion–dipole interactions is summarized in Figure 12-8.

Heats of solution are the basis for "instant cold packs" and "instant hot packs" used for the first-aid treatment of minor sprains and pulled muscles. These packs have two separate compartments. One contains water, and the other contains a salt: NH$_4$NO$_3$ for cold packs and MgSO$_4$ or CaCl$_2$ for hot packs. Kneading the pack breaks the wall between the compartments, allowing the salt to mix with water. As the salt dissolves to form an aqueous solution, the temperature of the pack changes. Heat is

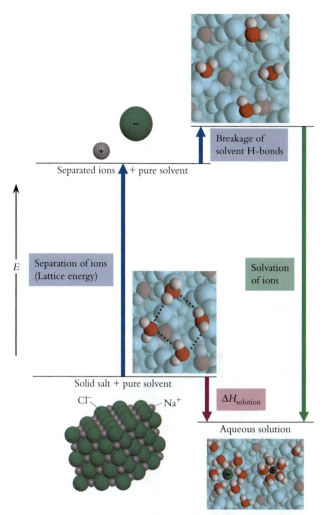

Figure 12-7
The dissolving of an ionic salt to produce an aqueous solution can be analyzed by breaking the overall process into three steps.

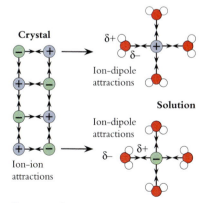

Figure 12-8
Whether the dissolving of a salt is exothermic or endothermic depends on the balance between the attractive forces of the crystal lattice and the ion–dipole forces that stabilize the ions in solution.

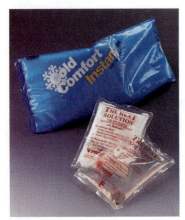

Instant hot and cold packs rely on heats of solution to generate instant heating or cooling.

When diluting a strong acid, the acid always should be added slowly to the water, with vigorous stirring.

absorbed or released only as the salt dissolves, however, so after all of the salt has dissolved, the pack gradually returns to room temperature. Further manipulation of the pack has no effect.

As mentioned above, complete solvation of ions involves from four to eight water molecules per ion. In highly concentrated aqueous solutions, there are not enough water molecules to fully solvate all the ions. Because of this, ion–solvent interactions may increase as water is added to a concentrated solution. When a concentrated solution is diluted by adding water, energy is released and the temperature of the solution increases. The molar energy change accompanying this process is called the **heat of dilution.**

Strong acids often have large heats of dilution, because proton transfer from a strong acid to water is highly exothermic. This can lead to problems when a concentrated acid is diluted with water. When water contacts the solution, the dilution process releases a large amount of energy. This energy heats the solution, and the heating can occur rapidly enough to form a "hot spot" in which the temperature is greater than the boiling temperature of water. If this occurs, droplets of acid will be ejected from the solution. Hot acid is a caustic material that rapidly burns the skin and attacks many other materials, including clothing. Sulfuric acid is particularly likely to splatter because concentrated H_2SO_4 has a high viscosity and does not mix readily with added water. Therefore, acids should always be diluted by slowly adding the concentrated acid to a larger volume of water. In this way the transformation of concentrated acid into a more dilute solution occurs relatively slowly, and no local "hot spots" of high temperature can develop.

Effect of Temperature

In general, solids are more soluble in hot water than in cold water. This is true for nearly all organic solids and for most ionic salts. Figure 12-9 shows how the solubilities of several salts vary with temperature. The figure indicates that some ionic salts behave contrary to this general rule. Whereas the solubility of potassium nitrate increases dramatically with temperature, that of sodium chloride changes hardly at all, while the solubility of lithium sulfate decreases. Solubility in water decreases with increasing temperature for a number of other sulfate salts, including sodium sulfate and calcium sulfate.

A compound whose solubility increases with temperature can be purified by recrystallization. The impure solid is dissolved in a minimum volume of hot water. The hot solution is filtered to remove insoluble impurities, and then the solution is cooled in an ice bath. The solubility of the compound decreases as the temperature drops, causing the substance to precipitate from solution. Soluble impurities usually remain in solution. Purification by recrystallization is not restricted to aqueous solutions. An organic solid can be purified by recrystallization from an appropriate organic solvent.

Gas–Solution Equilibria

Nonpolar gases are only slightly soluble in water. For example, water in contact with the Earth's atmosphere contains O_2 at a concentration of only about 2.5×10^{-4} M and CO_2 at about 1×10^{-5} M. Nevertheless, these small concentrations are essential for aquatic life. Fish and other aquatic animals use their gills to extract O_2 dissolved in water, and unless that oxygen is replenished, these species die. Submerged green plants carry out photosynthesis using dissolved carbon dioxide, which also must be replenished for these plants to survive.

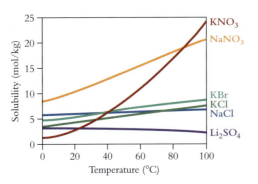

**Figure 12-9
Solubilities of salts in water vary with temperature.**

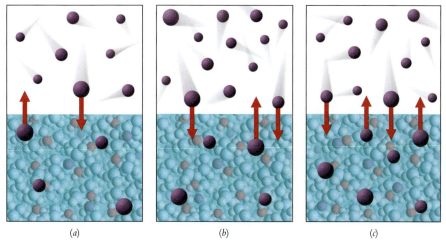

(a)   (b)   (c)

Figure 12-10
Molecular view of a gas–solution equilibrium. (*a*) At equilibrium, the rate of escape of gas molecules from the solution equals the rate of capture of gas molecules by the solution. (*b*) An increase in gas pressure causes more gas molecules to dissolve, throwing the system out of equilibrium. (*c*) The concentration of solute increases until the rates of escape and capture once again balance.

Figure 12-10 is a molecular view showing that the equilibrium concentration of a dissolved gas varies with the partial pressure of that gas. An increase in the partial pressure of gas results in an increase in the rate at which gas molecules enter the solution. This increases the concentration of gas in solution. The increased concentration in solution, in turn, results in an increase in the rate at which gas molecules escape from the solution. Equilibrium is reestablished when the solute concentration is high enough that the rate of escape equals the rate of capture.

Experiments show that the equilibrium concentration of dissolved gas increases linearly with the partial pressure of the gas. The equation describing this linear behavior is **Henry's law:**

$$[gas\,(aq)]_{eq} = K_H(p_{gas})_{eq} \tag{12-2}$$

The proportionality constant for a gas-solution equilibrium is designated K_H and is called the Henry's law constant. This constant has a different value for each combination of gas and solvent, and it also varies with temperature. Table 12-2 lists values for K_H for representative gases dissolving in water.

Fish and submerged plants depend on the equilibria between O_2 and CO_2 in the atmosphere and dissolved in the water. Home aquariums use air bubblers to maintain equilibrium concentrations by replenishing these gases.

Table 12-2   Henry's Law Constants* (K_H)

Gas	0.0 °C	25 °C	30 °C
N_2	1.1×10^{-3}	6.7×10^{-4}	4.0×10^{-4}
O_2	2.5×10^{-3}	1.3×10^{-3}	8.9×10^{-4}
CO	1.6×10^{-3}	9.6×10^{-4}	4.4×10^{-4}
Ar	2.5×10^{-3}	1.5×10^{-3}	1.0×10^{-3}
He	4.1×10^{-4}	3.8×10^{-4}	3.8×10^{-4}
CO_2	7.8×10^{-2}	3.4×10^{-2}	1.6×10^{-2}

*Aqueous solutions, with solution concentrations in mol/L and gas pressures in atm.

Example 12-5 uses Henry's law to determine the concentrations of atmospheric gases that dissolve in water, and our Chemistry and Life Box addresses an important consequence of Henry's law.

Example 12-5

Solubilities of Atmospheric Gases

The Earth's atmosphere contains 78% N_2, 21% O_2, and minor amounts of other gases, including CO_2 (0.0325%). Determine the concentrations of N_2, O_2, and CO_2 in water at equilibrium with the Earth's atmosphere at 25 °C.

Strategy: Each gas establishes its own dynamic equilibrium with water. The concentration depends on the partial pressure of the gas in the atmosphere and on the value of its Henry's law constant at 25 °C. Recall from Chapter 5 that the partial pressure of any gas in a mixture is given by the mole fraction (X_i) multiplied by total pressure.

Solution: Use 1.00 atm for the total pressure:

$$p_{O_2} = X_{O_2}P = \left(\frac{21\% \; O_2}{100\%}\right)(1.00 \text{ atm}) = 0.21 \text{ atm } O_2$$

Analogous calculations show that the partial pressure of N_2 is 0.78 atm, and that of CO_2 is 3.25×10^{-4} atm.

Now use Equation 12-2 and values of K_H from Table 12-2 to calculate the concentration of each dissolved gas:

$$[\text{gas}(aq)] = K_H(p_{\text{gas}})$$

$$[N_2(aq)] = (6.7 \times 10^{-4} \text{ M/atm})(0.78 \text{ atm}) = 5.2 \times 10^{-4} \text{ M } N_2$$

$$[O_2(aq)] = (1.3 \times 10^{-3} \text{ M/atm})(0.21 \text{ atm}) = 2.7 \times 10^{-4} \text{ M } O_2$$

$$[CO_2(aq)] = (3.4 \times 10^{-2} \text{ M/atm})(3.25 \times 10^{-4} \text{ atm}) = 1.1 \times 10^{-5} \text{ M } CO_2$$

Do the Results Make Sense? All the molarities are rather small, in the 10^{-5} to 10^{-4} M range. Given that these are nonpolar gases, it does make sense that their solubilities are small in a polar solvent such as water.

Extra Practice Exercise 12.5

What are the concentrations of these gases in water at equilibrium with the atmosphere at 30 °C?

A few gases interact with water to form concentrated aqueous solutions. For example, a 12 M solution can be prepared by bubbling HCl gas through water, because proton transfer occurs to generate H_3O^+ and Cl^- ions:

$$HCl(g) + H_2O(l) \longrightarrow Cl^-(aq) + H_3O^+(aq)$$

Ammonia is another gas that is very soluble in water, giving solutions as concentrated as 14.8 M. Ammonia dissolves because it forms hydrogen bonds with water molecules. When an ammonia molecule displaces a water molecule, one hydrogen-bonding interaction is exchanged for another.

Section Exercises

12.3.1 Draw a molecular picture illustrating the hydrogen bonds that form between a methanol (CH_3OH) molecule and water molecules in a dilute aqueous solution of methanol.

12.3.2 List the salts in Figure 12-9 in order of increasing solubility at 0 °C.

12.3.3 Some gases can be collected by liquid displacement. The gas is bubbled through water into an inverted container. If 0.18 mol of CO_2 at $P = 0.98$ atm is bubbled through 450 mL of water at 298 K, what fraction of the gas dissolves in the water?

Box 12-1	Chemistry and Life: Scuba Diving and Henry's Law

According to Henry's law, gases become more soluble as pressure increases. This solubility property has minimum effects on everyday life, because changes in altitude or weather cause only modest variations in atmospheric pressure. Scuba divers, however, must pay careful attention to the solubility equilibria of gases. The pressure exerted on a diver increases by 1 atm for every 30 feet of descent. Increasing pressure causes more and more gas to dissolve in a diver's blood. When the diver returns to the surface, the amount of dissolved gas far exceeds its solubility. Consequently, ascending too rapidly can be disastrous.

Carbonated beverages illustrate what happens when a dissolved gas undergoes a rapid drop in pressure. Soft drinks, soda water, and champagne are bottled under several atmospheres pressure of carbon dioxide. When a bottle is opened, the total pressure quickly falls to 1 atm. At this lower pressure, the concentration of CO_2 in the solution is much higher than its solubility, so the excess CO_2 forms gas bubbles and escapes from the liquid. As the photo shows, this process can be dramatic.

Scuba divers experience similar pressure changes. The amount of air dissolved in the blood increases significantly as the diver descends. If a diver returns to the surface too quickly, nitrogen gas dissolved in the blood forms bubbles in the same way as the CO_2 in a freshly opened carbonated drink. These bubbles interfere with the transmission of nerve impulses and restrict the flow of blood. The effect is extremely painful and can cause paralysis or death. The bubbles tend to collect in the joints, where they cause severe contractions. This is the source of the name of this dangerous condition — the "bends".

Divers avoid the bends by returning to the surface slowly, taking short "decompression stops" at intermediate depths to allow excess gas to escape from their blood without forming bubbles. Another way divers reduce the risk of the bends is by using helium – oxygen gas mixtures instead of compressed air. Helium is only half as soluble in water as nitrogen is, so less gas dissolves in blood.

Scuba divers face other hazards from the effects of Henry's law. At depths between 80 and 130 feet, divers experience an intoxicating feeling from the high concentration of N_2 in the blood. This dissolved N_2 is thought to interfere with nerve transmission, giving rise to feelings of euphoria known as "rapture of the deep." Like other forms of intoxication, this causes a loss of judgment that can be deadly. The famous oceanographer Jacques-Yves Cousteau wrote: "I am personally quite receptive to nitrogen rapture. I like it and fear its doom. It destroys the instinct of life."

Still another hazard associated with Henry's law is oxygen toxicity. In high enough concentration, molecular oxygen is poisonous. Breathing pure oxygen at 1 atm pressure for just a few hours causes pain in the chest and coughing. Lung congestion and permanent tissue damage result after 24 hours of breathing pure O2, and longer exposure causes death. Breathing compressed air (21% O_2) at a dive depth of 125 feet is equivalent to breathing pure oxygen at sea level. At 200 feet ($p_{O_2} = 1.3 - 1.5$ atm) oyxgen toxicity causes muscle twitching, vomiting, and dizziness, and can result in deadly seizures. Although recreational divers seldom reach depths of 200 feet, those who do venture this far into the sea must breathe mixtures of gas that contain less than 10% oxygen.

Despite these hazards, humans continue to venture beneath the surface of the seas. Adventurers seek to salvage materials from sunken ships. Archaeologists wish to explore ancient seaport sites that have since sunk beneath the waves. Navy Seals practice warfare underwater. Biologists and lovers of nature study marine life in its natural habitat. In addition, hordes of amateur divers pursue this sport for its sheer enjoyment.

Scuba divers experience an undersea world filled with mystery and beauty, but diving can be a perilous hobby. Little wonder that novice divers must undergo rigorous training courses before they are free to explore the depths.

12.4 COLLIGATIVE PROPERTIES

Solute molecules alter many properties of a liquid. For instance, adding salt to water gives a solution that boils at a higher temperature than pure water, and adding ethylene glycol to the water in an automobile radiator gives a solution that protects against freezing. The presence of solute molecules causes changes in four common properties of solutions: vapor pressure, freezing point, boiling point, and osmotic pressure. These four are known as the **colligative properties.** Changes in each of them can be understood from a molecular perspective by examining how added solute molecules affect the dynamics of phase changes. These examinations reveal that the colligative properties generally are independent of the nature of the solute.

KEY CONCEPT | *For a broad range of solutes, colligative properties depend on the amount of solute but not on the nature of that solute.*

Vapor Pressure Reduction

The easiest of the colligative properties to visualize is the effect of solute molecules on the vapor pressure exerted by a liquid. In a closed system, the solvent and its vapor reach dynamic equilibrium at a partial pressure of solvent equal to the vapor pressure. At this pressure, the rate of condensation of solvent vapor equals the rate of evaporation from the liquid.

Figure 12-11 is a molecular view of how a solute changes this liquid–vapor equilibrium of the solvent. The presence of a solute means that there are fewer solvent molecules at the surface of the solution. As a result, the rate of solvent evaporation from a solution is slower than the rate of evaporation of pure solvent. At equilibrium, the rate of condensation must be correspondingly slower than the rate of condensation for the pure solvent at equilibrium with its vapor. In other words, the vapor pressure drops when a solute is added to a liquid. A solute decreases the concentration of solvent molecules in the gas phase by reducing the rates of both evaporation and condensation.

This molecular view of Figure 12-11 suggests that the extent of vapor pressure lowering will depend on the fraction of solvent molecules that has been replaced. In other words, the vapor pressure should be proportional to the mole fraction of the solvent. The molecular view also suggests that this effect does not depend on the nature of the

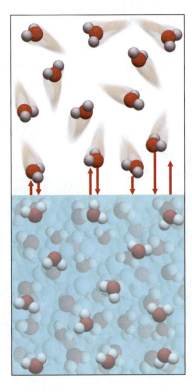
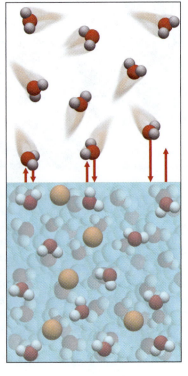

Figure 12-11
Addition of solute molecules (*right view, orange balls*) reduces the rate of escape of solvent molecules compared to pure solvent (*left view*). As a result, the vapor pressure of a solution is lower than the vapor pressure of a pure solvent.

solute, but only on its mole fraction. Experiments show that this is often the case, particularly for dilute solutions. A simple equation, **Raoult's law,** expresses this proportionality between vapor pressure and mole fraction:

$$vp_{solution} = X_{solvent} \, vp_{pure\ solvent} \qquad\qquad (12\text{-}3)$$

Raoult's law states that the vapor pressure of a solution is the vapor pressure of pure solvent times the mole fraction of solvent in the solution.

Equation 12-3 describes the total vapor pressure above a solution when the solute does not have a significant vapor pressure of its own. In other words, Raoult's law applies only to nonvolatile solutes. When the solute is volatile, such as for a solution of acetone in water, the total vapor pressure above the solution is a sum of contributions from both solvent and solute.

④

Example 12-6 shows how to apply Raoult's law.

Example 12-6

Vapor Pressure of Vinegar

A closed bottle contains vinegar, which is 5.0% by mass acetic acid in water. Calculate the vapor pressure of water above the vinegar at 25 °C. The vapor pressure of pure water at 25 °C is 23.76 torr.

Strategy: To calculate vapor pressure above a solution, we need to know the mole fraction of the solvent as well as the vapor pressure of the pure solvent. The information about amounts of acetic acid and water must be converted into moles.

Solution: First, use the mass percentage to find the masses of acetic acid and water in a convenient amount of vinegar, exactly 100 g:

$$m_{acetic\ acid} = \left(\frac{5.0\%}{100\%}\right)(100.\ g) = 5.00\ g \qquad m_{water} = 100.\ g - 5.0\ g = 95.0\ g$$

Convert from mass to moles using molar mass in the usual manner:

$$MM_{acetic\ acid} = 60.0\ g/mol \quad so \quad n_{acetic\ acid} = \frac{m}{MM} = \frac{5.00\ g}{60.0\ g/mol} = 0.120\ mol$$

$$MM_{water} = 18.0\ g/mol \quad so \quad n_{water} = \frac{m}{MM} = \frac{95.0\ g}{18.0\ g/mol} = 5.00\ mol$$

Use these amounts to determine the mole fraction of the solvent, water:

$$X_{water} = \frac{n_{water}}{n_{water} + n_{acetic\ acid}} = \frac{5.00\ mol}{5.00\ mol + 0.120\ mol} = 0.977$$

Now use Equation 12-3, Raoult's law, to calculate the vapor pressure:

$$vp_{solution} = X_{solvent} \, vp_{pure\ solvent} = (0.977)(23.76\ torr) = 23\ torr$$

The result is rounded to two significant figures to match the data.

Does the Result Make Sense? The vapor pressure of pure water under these conditions is 23.76 torr. Any nonvolatile solute reduces the vapor pressure, so we expect a result that is smaller than this value.

Vinegar is a dilute solution of acetic acid (CH₃CO₂H) in water.

How much is the vapor pressure reduced below 760. torr by dissolving 12.5 g of sucrose ($C_{12}H_{22}O_{11}$) in 225 g of boiling coffee?

Boiling and Freezing Points

The lower vapor pressure of a solution compared to pure solvent results in changes in the boiling point and freezing point of the solution. The molecular perspective is similar to that for vapor pressure reduction. In a solution, solute molecules replace some of the

Figure 12-12
Molecular views of the rates of solid–liquid phase transfer of a pure liquid and a solution at the normal freezing point. The addition of solute does not change the rate of escape from the solid, but it decreases the rate at which the solid captures solvent molecules from the solution. This disrupts the dynamic equilibrium between escape and capture.

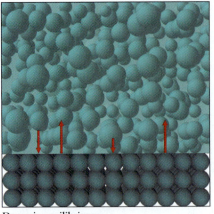

Dynamic equilibrium:
Two solid molecules escape,
two liquid molecules are captured

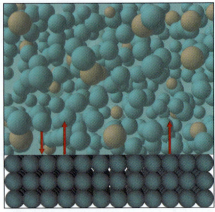

Solute disrupts equilibrium:
Two solid molecules escape,
one liquid molecule is captured

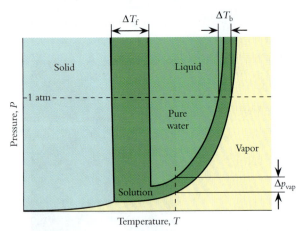

Figure 12-13
The phase diagram for water, showing how the phase boundaries change when a nonvolatile solute dissolves in the liquid.

solvent molecules, so a given volume of a solution contains a smaller number of solvent molecules than does the same volume of pure solvent. Consequently, the presence of solute molecules reduces the rate at which solvent molecules leave the liquid phase. Figure 12-11 shows the effect on liquid–vapor equilibrium, and Figure 12-12 shows how this affects a liquid–solid equilibrium: Changing one rate without changing the other rate throws the dynamic equilibrium out of balance.

The addition of solutes *decreases* the freezing point of a solution. In the solution, solvent molecules collide with crystals of solid solvent less frequently than they do in the pure solvent. Consequently, fewer molecules are captured by the solid phase than escape from the solid to the liquid. Cooling the solution restores dynamic equilibrium because it simultaneously reduces the number of molecules that have sufficient energy to break away from the surface of the solid and increases the number of molecules in the liquid with small enough kinetic energy to be captured by the solid.

The effect of a solute on the boiling point of a solution is opposite to its effect on the freezing point. A nonvolatile solute *increases* the boiling point of a solution. This is because the solute blocks some of the solvent molecules from reaching the surface of the solution and thus decreases the rate of escape into the gas phase. To get back to dynamic equilibrium, the solution must be heated so that more molecules acquire sufficient energy to escape from the liquid phase.

These effects can be displayed on a phase diagram that compares the behavior of pure solvent and a solution. Figure 12-13 shows such a phase diagram for water and aqueous solutions.

Experiments show that at low solute concentration, the changes in freezing point and boiling point of a solution, ΔT_f and ΔT_b, depend on the concentration of the solution, expressed as molality (c_m):

⑤

Equations 12-4 and 12-5 can be derived from our simple molecular picture and kinetic molecular theory. The derivation is independent of the nature of solute and solvent, so the equations are valid for other solvents besides water, except that K_f and K_b have different values for each solvent.

$$\Delta T_f = K_f\, c_m \qquad\qquad (12\text{-}4)$$

$$\Delta T_b = K_b\, c_m \qquad\qquad (12\text{-}5)$$

We use molality in these equations because they describe temperature changes. The constant K_f is called the **freezing point depression constant**, and K_b is called the **boiling point elevation constant**. These constants are different for different solvents but do not depend on the identity of the solutes. For water, K_f is 1.858 °C kg/mol and K_b is 0.512 °C kg/mol.

Equations 12-4 and 12-5 are both used in the same manner. Example 12-7 shows the procedure.

Ethylene glycol (1,2-ethanediol) is added to automobile radiators to prevent cooling water from freezing. Estimate the freezing point of coolant that contains 2.00 kg of ethylene glycol and 5.00 L of water.

Example 12-7

Freezing Point Depression

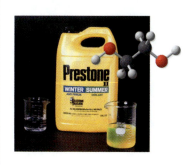

Strategy: The question asks for the freezing point of a solution. The phrase *to prevent the water from freezing* reveals that we are dealing with depression of the freezing point of water. Equation 12-4 describes this process for a dilute solution: $\Delta T_f = K_f\, c_m$. A coolant solution is quite concentrated, so this equation will not apply exactly, but we can use it to obtain an estimate of the freezing point.

Solution: The freezing point depression constant for water is known from experiments and can be found in tables: $K_f = 1.858\ °C\ kg/mol$. To calculate the freezing point, we must first determine the molality of the solute in this solution, using Equation 12-1:

$$c_m = \frac{n_{\text{solute}}}{m_{\text{solvent}}}$$

Determine the number of moles of solute from the mole−mass relationship. The molar mass of ethylene glycol is obtained from its chemical formula, $C_2H_6O_2$:

$$MM = 62.07\ g/mol$$

$$n\ C_2H_6O_2 = \frac{(2.00\ kg)(10^3 g/kg)}{62.07\ g/mol} = 32.22\ mol$$

Find the mass of water from its density, $\rho = 1.00\ g/mL$, then convert to kilograms:

$$m_{\text{water}} = \rho V = (1.00\ g/mL)(5.00\ L)(10^3\ mL/L)(10^{-3}\ kg/g) = 5.00\ kg$$

Now calculate the molality and substitute into Equation 12-4 to estimate the difference between the freezing point of the solution and that of pure water:

$$c_m = \frac{32.22\ mol}{5.00\ kg} = 6.444\ mol/kg$$

$$\Delta T_f = K_f\, c_m = (1.858\ °C\ kg/mol)(6.444\ mol/kg) = 12\ °C$$

We round to two significant figures because we do not expect the equation to be accurate at this high concentration.

The freezing point of the solution differs from that of pure water by this amount. Because the freezing point of water is 0 °C and freezing points are depressed by adding solutes, the new freezing point is below 0 °C: $T_f = -12\ °C$.

Does the Result Make Sense? Antifreeze is designed to prevent engine coolant from freezing at the low temperatures encountered in winter and at high altitudes. These temperatures often fall several degrees below freezing, so the value calculated in this example is what we would expect for antifreeze.

Calculate the boiling point of the sugar−coffee solution described in Extra Practice Exercise 12-6.

The *number of solute particles* present in a given amount of solvent causes the temperature changes described by Equations 12-4 and 12-5. When an ionic salt dissolves in water, each mole of salt produces two or more moles of ions. A dilute solution of sodium chloride (NaCl), for example, contains two moles of ions for every mole of NaCl—one mole of Na^+ cations and one mole of Cl^- anions. The proper application of Equations 12-4 and 12-5 takes this into account by viewing c_m as the *total* number of moles of particles per kilogram of solvent. When this is done, however, the predicted values of ΔT_f or ΔT_b are higher than experimental values. This is because some of the cations and anions in aqueous solutions form ion pairs, reducing the total number of solute particles.

Osmosis

Osmosis is the movement of solvent molecules through a semipermeable membrane. The membrane is a thin, pliable sheet of material, perforated with molecular-scale holes. The holes are large enough to allow water molecules to pass back and forth through the membrane, but too small to allow the passage of solute molecules or hydrated ions; hence, the membrane is semipermeable.

If a semipermeable membrane separates two identical solutions, solvent molecules move in both directions at the same rate, and there is no net osmosis. The two sides of the membrane are at dynamic equilibrium. The situation changes when the solutions on the two sides of the membrane are different. Consider the membrane in Figure 12-14a, which has pure water on one side and a solution of sugar in water on the other. The sugar molecules reduce the concentration of solvent molecules in the solution. Consequently, fewer solvent molecules pass through the membrane from the solution side than from the pure solvent side. Water flows from the side containing pure solvent to the side containing solution, so there is a net rate of osmosis.

In the absence of other forces, osmosis continues until the concentration of solvent is the same on both sides of the membrane. However, pressure can be used to stop this process. An increase in pressure on the solution side pushes solvent molecules against the membrane and thereby increases the rate of transfer of water molecules from the solution side to the solvent side. Figure 12-14b shows that dynamic equilibrium can be established by increasing the pressure on the solution until the rate of solvent transfer is equal in both directions.

The pressure increase needed to equalize the transfer rates is called the **osmotic pressure** (Π). Osmotic pressure is a pressure *difference*. Pressure is exerted on both sides of a semipermeable membrane, but Π is the extra pressure that must be exerted on the solution to maintain dynamic equilibrium.

Like freezing point depression and boiling point elevation, osmotic pressure is proportional to the concentration of solute molecules. Experiments show that osmotic pressure is proportional to both concentration (expressed as molarity, M) and temperature:

$$\Pi = MRT \qquad\qquad (12\text{-}6)$$

In Equation 12-6, M is the total molarity of all solutes, T is the temperature in kelvins, and R is the gas constant. If osmotic pressure is expressed in atmospheres, the fact that molarity is in moles per liter requires us to use $R = 0.08206$ L atm/mol K.

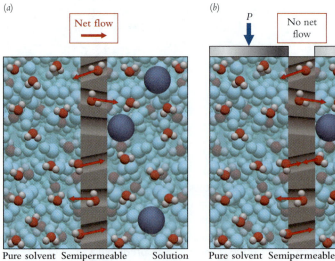
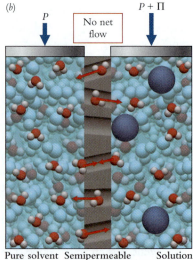

Figure 12-14
(a) The solvent concentration in a solution is lower than the solvent concentration in pure liquid, leading to an imbalance in flow through a semipermeable membrane. (b) By increasing the pressure on the solution, we can increase the rate of solvent flow out of the solution until it matches the flow out of the pure solvent.

Osmotic pressure effects can be substantial. For example, the waters of the oceans contain dissolved salts at a total ionic molarity of about 1.13 M. The osmotic pressure of ocean water can be calculated:

$$\Pi = MRT = (1.13 \text{ mol/L})(0.08206 \text{ L atm/mol K})(298 \text{ K}) = 27.6 \text{ atm}$$

This means that if samples of pure water and ocean water were placed on opposite sides of a semipermeable membrane, an external pressure of 27.6 atm would have to be applied to the ocean water side to prevent osmosis. The osmotic pressure of ocean water is more than 25 times atmospheric pressure. By comparison, the freezing point of ocean water is depressed by only about 1% from the freezing point of pure water, from 273 K to about 271 K (-2 °C).

Osmotic pressure plays an important role in biological chemistry because the cells of the human body are encased in semipermeable membranes and bathed in body fluids. Under normal physiological conditions, the body fluid outside the cells has the same total solute molarity as the fluid inside the cells, and there is no net osmosis across cell membranes. Solutions with the same solute molarity are called *isotonic* solutions.

The situation changes when there is a concentration imbalance. Figure 12-15 shows red blood cells immersed in solutions of different concentrations. When the fluid outside the cell has a higher solute concentration, the result is slower movement of water through the membrane into the cell. The net result is that water leaves the cell, causing it to shrink. When the fluid outside the cell has a lower concentration, movement of water into the cell increases. The extra water in the cell causes an increase in internal pressure. Eventually, the internal pressure of the cell matches the osmotic pressure, and water transport reaches dynamic equilibrium. Unfortunately, osmotic pressures are so large that cells can burst under the increased pressure before they reach equilibrium.

Red blood cells are particularly susceptible to these potentially damaging concentration changes because they are suspended in the aqueous medium of the blood. Consequently, solutions used for intravenous feeding must be isotonic. Example 12-8 deals with isotonic solutions.

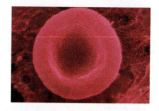

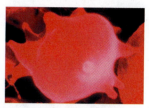

Figure 12-15
In isotonic solution, red blood cells are spherical (*top*). At higher ion concentration (*center*), osmotic flow removes water from the cell interior, shrinking the cell. At lower ion concentration, osmotic flow pumps water into the cell and may cause it to burst (*bottom*).

Isotonic intravenous solutions contain 49 g/L of glucose ($C_6H_{12}O_6$). What is the osmotic pressure of blood?

Example 12-8

Isotonic Solutions

Strategy: Isotonic solutions, by definition, exert equal osmotic pressure. Therefore, Π for blood is the same as Π for the glucose solution. We can calculate Π from Equation 12-6 after converting the concentration into moles per liter:

$$\Pi = MRT \quad and \quad M = \frac{n}{V} = \frac{m/MM}{V}$$

Solution: From the formula of glucose, $MM = 180$ g/mol. Substitute to find the molarity of the glucose solution:

$$M = \frac{(49 \text{ g})}{(180 \text{ g/mol})(1 \text{ L})} = 0.272 \text{ mol/L}$$

Because we are working with blood in the human body, T is human body temperature, 37 °C.

$$T = 37 \text{ °C} = 37 + 273 = 310 \text{ K} \qquad R = 0.08206 \text{ L atm/mol K}$$

$$\Pi = (0.272 \text{ mol/L})(0.08206 \text{ L atm/mol K})(310 \text{ K}) = 6.9 \text{ atm}$$

We round this result to two significant figures to match the data (49 g/L).

If a sample of red blood cells is added to pure water, osmosis carries water into the cells. This process would continue until the internal pressure of the cell was

Example 12-8
Isotonic Solutions
(*continued*)

Extra Practice
Exercise 12.8

6.9 atm higher than the pressure on the outside of the cell. However, 6.9 atm is much more than the cell membrane can tolerate. Consequently, red blood cells burst when immersed in pure water.

Does the Result Make Sense? The high osmotic pressure of seawater, 27.6 atm, makes it clear that osmotic pressures can be quite large. Thus, an osmotic pressure of 6.9 atm is a reasonable value for a solution such as blood.

Calculate the osmotic pressure of the coffee solution described in Extra Practice Exercise 12-6, assuming that it has cooled to 50 °C and the density of the solution is 1.00 g/mL.

Reverse osmosis is carried out on both large and small scales.

When the additional pressure applied to a solution is less than the osmotic pressure, solvent molecules flow from pure solvent into the solution. When the additional pressure equals the osmotic pressure, equilibrium is established, and there is no net flow of solvent molecules. What if we apply an additional pressure that is greater than the osmotic pressure? Now the pressure on the solution is sufficient to drive solvent molecules from the solution into the pure solvent. Osmosis now transfers solvent in the opposite direction; **reverse osmosis** occurs.

Reverse osmosis can be used to purify water, because the liquid passing through the semipermeable membrane is pure solvent. A water purifier that uses reverse osmosis requires semipermeable membranes that do not rupture under the high pressures required for reverse osmosis. Recall that seawater has an osmotic pressure of nearly 28 atm and that red blood cells rupture at 7 atm. Nevertheless, membranes have been developed that make it feasible to purify water using this technique. Reverse osmosis currently supplies pure drinking water to individual households as well as entire municipalities.

Determination of Molar Mass

The magnitude of osmotic pressure is large enough that measurements of Π provide a convenient way to determine the molar mass of a compound. The osmotic pressure equation (Equation 12-6) can be solved for molar mass after molarity is expressed in terms of mass and molar mass:

$$\Pi = MRT \qquad so \qquad \Pi = \frac{mRT}{V(MM)}$$

Rearrangement gives an equation for molar mass:

$$MM = \frac{mRT}{\Pi V} \tag{12-7}$$

For determination of the molar mass of an unknown compound, a measured mass of material is dissolved to give a measured volume of solution. The system is held at constant temperature, and the osmotic pressure is determined. Osmotic pressure measurements are particularly useful for determining the molar masses of large molecules such as polymers and biological materials, as Example 12-9 illustrates.

Example 12-9
Determining Molar Mass

A 25.00-mL aqueous solution containing 0.420 g of hemoglobin has an osmotic pressure of 4.6 torr at 27 °C. What is the molar mass of hemoglobin?

Strategy: This is a straightforward calculation. We have mass, volume, temperature, and osmotic pressure. We use Equation 12-7 to calculate the molar mass.

Solution: The problem contains all of the necessary data:

Example 12-9

Determining Molar Mass
(*continued*)

$$m_{solute} = 0.420 \text{ g} \qquad R = 0.08206 \text{ L atm/mol K} \qquad T = 27 \text{ °C} + 273 = 300. \text{ K}$$

$$\Pi = (4.6 \text{ torr})(1 \text{ atm/760 torr}) = 6.05 \times 10^{-3} \text{ atm}$$

$$V_{solution} = (25.00 \text{ mL})(1 \text{ L/1000 mL}) = 2.500 \times 10^{-2} \text{ L}$$

$$MM = \frac{mRT}{\Pi V} = \frac{(0.420 \text{ g})(8.206 \times 10^{-2} \text{ L atm/mol K})(300. \text{ K})}{(6.05 \times 10^{-3} \text{ atm})(2.500 \times 10^{-2} \text{ L})}$$

$$MM = 6.8 \times 10^4 \text{ g/mol}$$

Does the Result Make Sense? This is a large value for a molar mass, but it is reasonable because biological molecules often are quite large. As in any calculation, be careful to express all data in appropriate units. The osmotic pressure was measured to two significant figures, so the result has two significant figures.

When 7.50 mg of a protein is dissolved in water to give 10.00 mL of solution, the osmotic pressure is found to be 1.66 torr at 21°C. Determine the molar mass of the protein.

Section Exercises

12.4.1 A water-soluble protein molecule has a molar mass of 985 g/mol. Calculate the freezing point depression, boiling point elevation, and osmotic pressure at 27 °C of an aqueous solution containing 0.750 g/L of this protein. (Assume that the solution has a density of 1.000 g/mL.)

12.4.2 Redraw Figure 12-14*b* using arrows to represent water movement across the membrane during reverse osmosis.

12.4.3 In your own words, write a detailed, molecular-level description of how reverse osmosis can be used to desalinate seawater.

12.5 BETWEEN SOLUTIONS AND MIXTURES

Sugar dissolves in water to give a solution that contains individual sugar molecules distributed uniformly among the water molecules. The aqueous sugar solution is stable and remains uniform indefinitely. Recall from Chapter 1 that a solution is a homogeneous mixture. On the microscopic scale, one microscopic portion of a solution looks the same as every other microscopic portion.

On the other hand, when cornstarch and water are stirred together and allowed to stand, the resulting mixture soon separates into its components. Because cornstarch is insoluble in water, a combination of cornstarch and water forms a heterogeneous mixture. That is, one microscopic portion of this mixture has a different composition than another microscopic portion does. An insoluble granule of cornstarch has an entirely different composition than a microscopic portion of water.

One of the characteristic differences between a solution and a heterogeneous mixture is the sizes of the particles composing each of them. Molecule-sized particles such as sugar molecules, with dimensions in the nanometer range, tend to form solutions. Particles such as cornstarch granules, with dimensions larger than micrometers, tend to form heterogeneous mixtures.

Cornstarch granules are too large to dissolve or to remain suspended in water.

Colloidal Suspensions

Particles whose dimensions are between 1 nanometer and 1 micrometer, called **colloids,** are larger than the typical molecule but smaller than can be seen under an optical microscope. When a colloid is mixed with a second substance, the colloid can become

Table 12-3   Types of Colloidal Suspensions

Colloid Phase	Phase of Medium	Name	Natural Example	Technical Example
Liquid	Gas	Aerosol	Fog	Hair spray
Solid	Gas	Aerosol	Pollen	Fumigant
Gas	Liquid	Foam	Whipped cream	Fire extinguisher
Liquid	Liquid	Emulsion	Milk	Shampoo
Solid	Liquid	Sol	River water	Ink
Gas	Solid	Solid foam	Pumice	Styrofoam
Liquid	Solid	Solid emulsion	Opal	Butter
Solid	Solid	Solid sol	Bone	Plumber's solder

uniformly spread out, or dispersed, throughout the dispersing medium. Such a dispersion is a **colloidal suspension** that has properties intermediate between those of a true solution and those of a heterogeneous mixture. As Table 12-3 demonstrates, colloidal suspensions can involve nearly any combination of the three phases of matter. Gas–gas mixtures are the exception, because any gas mixes uniformly with any other gas to form a true solution.

The examples in Table 12-3 illustrate the diversity of colloidal suspensions. Colloidal suspensions are commonly found in nature, and humans have learned how to make useful suspensions. Some of these are foods such as butter, ice cream, and whipped cream. Others are personal care items like shampoos, shaving creams, and hair sprays. Ink, paint, and solder also fall in this category. Our Chemistry and the Environment Box explores aerosols in detail.

The stable condition for a liquid or solid is the bulk phase, so we would expect the colloidal particles making up aerosols and emulsions to coagulate into larger particles, destroying the suspension. Although this frequently happens, for example when rain condenses from clouds, many colloidal suspensions last indefinitely. Why is this? Intermolecular forces pull colloidal particles together, so a colloidal suspension must be stabilized by opposing repulsive forces. Electrical charges provide these repulsive forces. The particles of some colloids preferentially attract positive ions, and the particles of other colloids preferentially attract negative ions. When particles possess charges of the same sign, the particles repel one another, and this electrical repulsion can be sufficient to prevent condensation into larger particles.

Surfactants

As described in our chapter introduction, substances that do not dissolve in water, such as organic fats and oils, are called **hydrophobic.** Substances that are miscible with water, including hydrogen-bonding molecules such as methanol and acetone, are called **hydrophilic.** Some molecules contain both hydrophilic and hydrophobic regions. Such a dual-nature molecule, which is known as a **surfactant,** may have a polar or ionic "head" that is compatible with water and a long hydrocarbon "tail" that is incompatible with water. Figure 12-16 shows the structure of one dual-nature molecule, sodium stearate. The head of the stearate anion resembles the water-soluble acetate anion, and the tail is a hydrocarbon chain containing 17 carbon atoms.

Dual-nature molecules such as sodium stearate can form three different types of structures in water. They may form a molecular *monolayer* on the surface. The polar head groups immerse themselves in the water while the nonpolar tails aggregate together on the surface. Agitating the solution may cause the molecules to arrange into spherical aggregates called *micelles,* in which the hydrophobic tails point inward. The polar heads lie on the outside of the micelle, where they interact with the aqueous solvent. Dual-nature molecules may also form enclosed bilayers, called *vesicles.* Vesicles

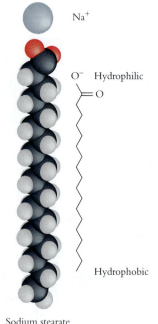

Na$^+$

O$^-$ Hydrophilic

O

Hydrophobic

Sodium stearate
NaC$_{17}$H$_{35}$CO$_2$

**Figure 12-16
Sodium stearate is a typical surfactant molecule. It has an ionic, hydrophilic head and a nonpolar, hydrophobic tail.**

Box 12-2   Chemistry and the Environment: Aerosols

You have undoubtedly used one of the common products of modern society, the aerosol spray can. We use these handy containers to deliver fine mists of many useful products. Do you need to keep your hair in place? Use an aerosol spray can to apply mousse. Do you want to get rid of unpleasant room odors? Spray the air with an aerosol air freshener. Has the paint worn off your favorite outdoor chair? Touch it up using an aerosol paint can. These are just three of the many products that can be delivered using aerosol sprays.

Aerosols, like solutions, are mixtures. Unlike solutions, however, they are not single phases. Instead, an aerosol is a suspension in a gas of tiny particles of a condensed phase, either liquid or solid. The particles that make up an aerosol can have a range of diameters between 10 nm and 10 μm.

Individual aerosol particles are too small to be seen with the naked eye, but we can nevertheless see an aerosol. Aerosol particles scatter light, and we can observe this scattered light from the side. Our atmosphere always contains some aerosol particles, and spectacular sky views sometimes result from such scattering, especially just before sunset. Closer to the ground, we can "see" a searchlight beam from the side because of light scattered by aerosol particles. Light scattering also reduces visibility, so aerosols generate haziness. The Great Smoky Mountains are named for the nearperpetual haziness caused by aerosols emitted from their forested slopes.

People invented the aerosol spray can, but naturally occurring aerosols play critical roles in regulating the Earth's weather. Erupting volcanos spew immense aerosol plumes into the stratosphere. These aerosols persist for years, scattering incoming sunlight and reducing the Earth's average temperature. The water in droplets of sea spray evaporates, leaving behind an aerosol suspension of minuscule salt crystals. These solid particles provide sites where water vapor in the atmosphere can condense, resulting in fog. Indeed, it is believed that the formation of clouds and fog always requires the presence of aerosol particles.

Even though aerosols occur naturally in the atmosphere, anthropogenic aerosols can have substantial effects on atmospheric processes. The burning of tropical forests releases significant aerosol smoke clouds that can change surface temperatures far from their point of origin. Desertification caused by human activities has been followed by the generation of huge dust clouds, whipped up by prevailing winds and carried around the world. Probably the best-known atmospheric effect caused by humans, the ozone hole, was contributed to by freons that were used as propellants in aerosol cans. This usage has now been banned, and recent measurements show that the ozone layer may be on its way to recovery.

Aerosol spray cans were invented in 1929, and perfection of a reliable valve and development of disposable cans took place in the 1940s. Shortly thereafter, "aerosol" became a household word. Like many other modern conveniences, however, the aerosol spray can has drawbacks as well as advantages. Because the particles in an aerosol are extremely tiny, they are quite mobile. They last for a long time in the atmosphere and can affect the climate, as already described. They can penetrate deep into our lungs and cause adverse health effects. Thus, anthropogenic aerosols have both global and local side effects. Despite increasing scientific studies, these effects are not yet fully understood.

Figure 12-17
Cross-sectional molecular views of the structures that can form when surfactant molecules are placed in water.

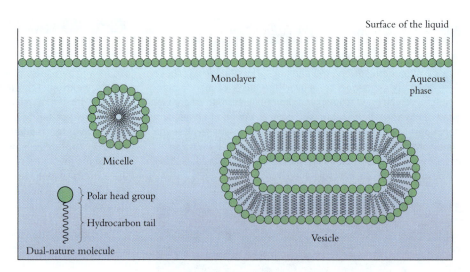

have two parallel rows of molecules oriented so that their hydrocarbon tails cluster together. Figure 12-17 illustrates these three types of structures.

All these arrangements obey the principle of like dissolving like. The hydrocarbon tails aggregate through dispersion forces because they are incompatible with the aqueous medium. Incorporating these tails into the solution would disrupt the hydrogen-bonding network of the solvent. The ionic heads, on the other hand, interact strongly with water to maximize ion–dipole and hydrogen-bonding interactions.

Surfactants are widely used in industrial chemistry to modify the behavior of aqueous solutions. Common surfactant head groups include carboxylate ($-CO_2^-$), sulfonate ($-SO_3^-$), sulfate ($-OSO_3^-$), and ammonium ($-NH_3^+$). The negative charge of anionic head groups usually is neutralized by Na^+, and the positive charge of ammonium usually is neutralized by Cl^-. These ions are used because they are nontoxic and their salts are highly soluble.

The most familiar surfactants are soaps and detergents. Many types of substances soil clothing; some are water-soluble, but others are not. In a washing machine, a soap or detergent removes water-insoluble grease (for example, butter, fat, and oil) from the clothing surface. Dispersion forces stabilize the attachment of grease particles to the hydrocarbon tails of surfactant molecules, which form aggregates. Agitation removes these aggregates from the fabric, suspending them in solution as tiny micelles with grease particles trapped inside. The micelles do not redeposit on the fabric because the hydrophilic heads of the surfactants hold them in solution. As water drains from the washing machine, it carries away the grease-containing micelles, leaving clean clothes behind.

Soaps are derived from natural sources such as animal fats. Soap made by boiling animal fat in an alkaline solution obtained from ashes has been known since the time of the ancient Sumerians, 2500 BC. Such soaps contain stearic acid and other long-chain organic acids. These carboxylate surfactants form insoluble salts with Ca^{2+} and Mg^{2+}. In regions where water is hard, these soaps precipitate calcium and magnesium stearate scum that inhibits cleansing action and is responsible for bathtub rings. Detergents such as sodium lauryl sulfate, on the other hand, are "synthetic" compounds, originally prepared in the laboratory, that contain sulfonate and sulfate head groups. These groups do not form precipitates with +2 cations. The cleaning action of soaps and detergents is similar, but detergents have largely replaced soaps because of their superior behavior in hard water.

Some surfactants are used as emulsifiers in processed foods such as bottled salad dressing. An emulsifier causes normally incompatible liquids such as the oil and water in salad dressing to disperse in each other, by forming molecular connections between the liquids. The hydrophobic tails of emulsifier molecules interact with oil molecules, while the hydrophilic heads on the emulsifier molecules interact with water molecules.

Surfactants are used in such a wide variety of ways that billions of dollars are spent on them every year. They appear in many household products, including cleansing agents and shampoos (recall the chapter introduction). Half of the surfactants produced in the United States are used in household and industrial cleaning products, but the remaining half are used in a wide range of industries.

In agriculture, surfactants are used as wetting agents that assist in the uniform application of sprayed pesticides. They also are used to prevent caking of fertilizers. Additives to agricultural products must not interfere with the active agents and must be biodegradable and environmentally benign. In the food industry, various surfactants are used as emulsifiers, cleaners, foaming agents, and antifoaming agents. Paints are dispersions of dyes, binding agents, and fillers. Most paints contain surfactants that improve their flow and mixing properties. Surfactants are used widely in the plastics industry as foaming agents to assist in the production of plastic foams and to improve moldability and extrudability of specially shaped products. In the manufacture of textiles, surfactants are used to clean natural fibers, as lubricants that reduce friction during the spinning and weaving processes, as emulsifiers that improve the application of dyes and finishes, and as antistatic agents.

Cell Membranes

It may seem like a huge conceptual leap from industrial surfactants to biological cell membranes, but the same principles apply to both sets of substances. Every biological cell is surrounded by a membrane only a few molecules thick. Among the major components of membranes are *phospholipids,* which are dual-nature molecules. Although their chemical structures are more complex than simple surfactants such as sodium stearate, phospholipids nevertheless have hydrophilic heads and hydrophobic tails. One phospholipid, lecithin, has a cationic $N(CH_3)_3^+$ group and eight oxygen atoms with nonbonding pairs of electrons, all of which form hydrogen bonds with water molecules. The hydrophobic portion of lecithin consists of two hydrocarbon tails.

Phospholipids form bilayers in aqueous media. The molecules form two approximately parallel rows with tails on the inside and heads toward the outside, in contact with the solution. This arrangement, shown in Figure 12-18, is analogous to the vesi-

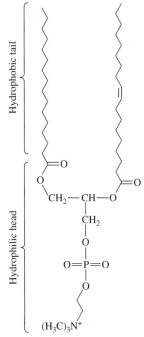

Lecithin, a common phospholipid, has a hydrophobic tail and a hydrophilic head.

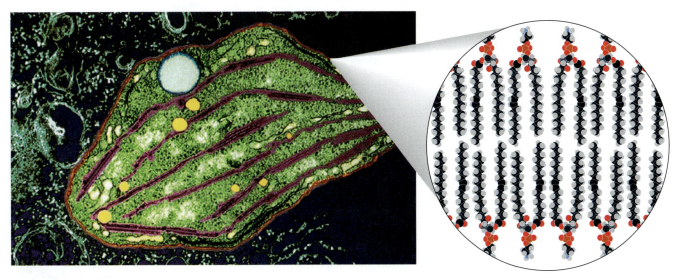

Figure 12-18
The lipid bilayer of a cell membrane contains two layers of a phospholipid such as lecithin, arranged tail-to-tail.

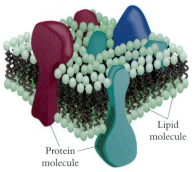

Figure 12-19
In addition to phospholipids, cell membranes contain protein molecules that carry out special functions such as transporting ions and molecules through the membrane.

Lipid molecule

Protein molecule

cles in Figure 12-17. The bilayer forms a closed sac that contains the aqueous cytoplasm and all the cellular components. Thus, a cell can be viewed as a large and complex vesicle.

One purpose of a cellular lipid bilayer is to control the passage of molecules into and out of the cell. Uncharged small molecules such as water, ammonia, and oxygen diffuse readily through the membrane. Hydrophobic molecules such as hydrocarbons can also pass through, because they are soluble in the overlapping tails that make up the interior of the bilayer. Ions and water-soluble polar molecules such as glucose and urea, on the other hand, cannot pass through the membrane.

For cells to carry out their functions, glucose and other nutrients must be brought in, and urea and other waste products must be expelled. This would be an impossible task if cell membranes were composed only of phospholipids. Large protein molecules act as molecular "gates" through the membranes (see Chapter 13 for the structures of proteins). These proteins are embedded in the bilayers but protrude into the surrounding water and/or into the cell interiors, as Figure 12-19 indicates.

Section Exercises

■ **12.5.1** Line drawings of some molecules follow. Identify the hydrophilic and hydrophobic regions of each and determine which are surfactants.

$H_3C-N^{+}-CH_3$ CH_3 Cl^- Tetramethylammonium chloride

Benzalkonium chloride
(used as a disinfectant)

Dipentyl ether

■ **12.5.2** Explain why glucose and other large, water-soluble molecules cannot pass through a lipid bilayer.

Glucose, hydrogen-bonded to water molecules

CHAPTER 12 VISUAL SUMMARY

SKILLS TO MASTER

① Calculating solution concentrations
② Predicting solubility patterns
③ Applying Henry's law

④ Applying Raoult's Law
⑤ Calculating colligative properties
⑥ Describing surfactant properties

12.1 THE NATURE OF SOLUTIONS

Solution concentration
  Molarity
  Mole fraction
  ppm
  ppb ①

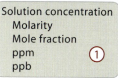

Solute
Solvent

$$c_m = \frac{n_{solute}}{m_{solvent}}$$

Molality

12.2 DETERMINANTS OF SOLUBILITY

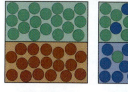

Insoluble   Saturated   Miscible

Solubility ②

Like dissolves like.

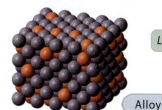

Alloy

12.3 CHARACTERISTICS OF AQUEOUS SOLUTIONS

Molar heat of solution (ΔH_{soln})

$$[\text{gas } (aq)]_{eq} = K_H (p_{gas})_{eq}$$

Heat of dilution

Henry's Law ③

12.4 COLLIGATIVE PROPERTIES

Colligative properties
  Raoult's Law
  Boiling point elevation
  Freezing point depression ④

$$\Delta T_f = K_f c_m$$

$$\Delta T_b = K_b c_m$$

$$vp_{solution} = X_{solvent} \, vp_{pure\ solvent}$$

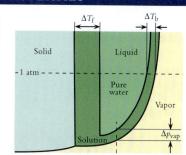

Colligative properties depend on the amount of solute, but not on the nature of the solute.

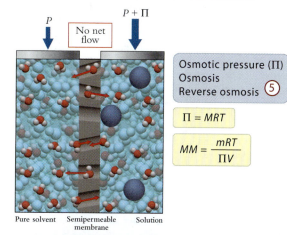

Osmotic pressure (Π)
Osmosis
Reverse osmosis ⑤

$$\Pi = MRT$$

$$MM = \frac{mRT}{\Pi V}$$

Pure solvent   Semipermeable membrane   Solution

12.5 BETWEEN SOLUTIONS AND MIXTURES

Hydrophilic   Sodium stearate $NaC_{17}H_{35}CO_2$ Hydrophobic

Colloid   Colloidal suspension

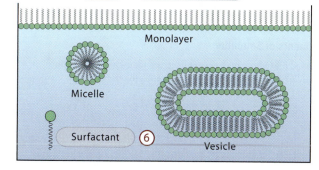

Monolayer

Micelle

Surfactant ⑥   Vesicle

Learning Exercises

12.1 Prepare a table of the various solution concentration measures. Describe how to convert from each of them to each of the others.

12.2 Draw molecular pictures that show every type of hydrogen bond that exists in a solution containing methanol, water, and ammonia.

12.3 Write a paragraph that explains the reasons why "like dissolves like" and includes several examples involving different types of solvents and solutes.

12.4 Define and give an example of each of the following: (a) alloy; (b) amalgam; (c) aerosol; (d) colligative property; and (e) surfactant.

12.5 Update your list of memory-bank equations. Be sure to mention how the equations in this chapter are used.

12.6 Write a paragraph that describes the types of substances that form monolayers, micelles, and vesicles in water. Explain the differences among these structures.

12.7 Prepare a list of the terms in Chapter 12 that are new to you. Write a one-sentence definition for each, using your own words. If you need help, consult the glossary.

 Problems   ilw = interactive learning ware problem. Visit the website at www.wiley.com/college/olmsted

The Nature of Solutions

12.1 Identify the solvent, the primary solutes, and the normal phase of each in each of the following solutions: (a) carbonated water; (b) white wine; and (c) humid air.

12.2 Identify the solvent, the primary solutes, and the normal phase of each in each of the following solutions: (a) seawater; (b) steel; and (c) vinegar.

12.3 Calculate the mole fractions of a solution made by mixing 14.5 g of methanol (CH_3OH) with 101 g of water.

12.4 Calculate the mole fractions of a solution made by mixing 1.755 g of p-xylene (C_8H_{10}) with 125 g of cyclohexane (C_6H_{12}).

12.5 A solution contains 1.521 g of maleic acid, $HO_2CCH=CHCO_2H$, dissolved in 85.0 mL of acetone ($\rho = 0.818$ g/mL). Calculate the molality, mole fraction, and mass percent of maleic acid in the solution.

12.6 Camphor is a natural compound from a tree found in Java, Brazil, China and other countries. A solution of camphor in ethanol is a mild topical antiseptic. One such solution contains 1.47 g of camphor dissolved in 125 mL of ethanol ($\rho = 0.785$ g/mL). Calculate the molality, mole fraction, and mass percent of camphor in the solution.

Camphor ($C_{10}H_{16}O$)

ilw **12.7** A saturated solution of hydrogen peroxide in water contains 30.% by mass H_2O_2 and has a density of 1.11 g/mL. Calculate the mole fractions, molarity, and molality of this solution.

12.8 Vodka is 35% by mass ethanol (C_2H_5OH) and has a density of 0.94 g/mL. Assuming no other components other than water, calculate the mole fractions, molarity, and molality of vodka.

12.9 Aqueous ammonia at a concentration of 2.30 M has a density of 0.9811 g/mL. Calculate the mole fractions, mass fraction, and molality of this solution.

12.10 Aqueous ammonia at a concentration of 6.69 M has a density of 0.9501 g/mL. Calculate the mole fractions, mass fraction, and molality of this solution.

12.11 You are asked to prepare an aqueous solution of methanol (CH_3OH) with a mole fraction of 0.105. (a) If you use 25.0 g of methanol, what mass of water should you use? (b) What is the molality of the resulting solution?

12.12 You are asked to prepare an aqueous solution of ethylene glycol ($HOCH_2CH_2OH$) with a mole fraction of 0.095. (a) If you use 545 g of water, what mass of ethylene glycol should you use? (b) What is the molality of the resulting solution?

Determinants of Solubility

12.13 Salad oil (whose major ingredients are long-chain fatty acids such as $C_{17}H_{33}CO_2H$) and vinegar (a dilute aqueous solution of acetic acid, CH_3CO_2H) are not miscible. Explain.

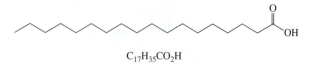

$C_{17}H_{35}CO_2H$

12.14 Acetonitrile, CH_3CN, is miscible with water and miscible with cyclohexane (C_6H_{12}), but water and cyclohexane are nearly insoluble in each other. Explain.

$H_3C-C{\equiv}N\colon$ Acetonitrile   Cyclohexane

12.15 Ammonia condenses to a liquid at low temperature. What kinds of solids would you expect to be soluble in liquid ammonia?

12.16 Mercury is the only metallic element that is a liquid at room temperature. What kinds of solids would you expect to be soluble in liquid mercury?

12.17 Select all the miscible pairs from among the following liquids, and explain your choices: H_2O, CH_3OH, C_8H_{18}, and CCl_4.

12.18 Select all the miscible pairs from among the following liquids, and explain your choices: H_2O, CCl_4, C_6H_{12}, and Br_2.

12.19 Match each solute to its most appropriate solvent, and explain your choices:

Solute:	I_2	NaCl	Au	paraffin
Solvent:	Hg	CCl_4	n-octane	water

12.20 Match each solute to its most appropriate solvent, and explain your choices:

Solute:	benzene	Br_2	glucose	Sn
Solvent:	Hg	CCl_4	gasoline	water

12.21 Draw a molecular picture of an alloy that contains 90% copper and 10% nickel.

12.22 Draw a molecular picture of an alloy that contains 95% iron and 5% boron.

Characteristics of Aqueous Solutions

12.23 Titan, one of the moons of Jupiter, appears to have oceans composed of liquid methane. Describe how this liquid differs from liquid water, and predict whether methane-based life forms are likely.

12.24 At high pressures, carbon dioxide can be liquefied at room temperature. Describe how this liquid differs from liquid water, and predict whether carbon-dioxide-based life forms are likely.

12.25 Draw a molecular picture illustrating the solubility equilibrium between KCl (s) and KCl (aq).

12.26 Draw a molecular picture illustrating the solubility equilibrium between NH_3 (g) and NH_3 (aq).

12.27 A drop in temperature accompanies the escape of CO_2 from a carbonated beverage. Explain this phenomenon in terms of intermolecular forces.

12.28 A rise in temperature accompanies the addition of water to a concentrated solution of HCl. Explain this phenomenon in terms of intermolecular forces.

12.29 One of the reasons that different aquatic life-forms thrive in water of different temperatures is the variation with temperature in the concentration of dissolved oxygen. Using data in Table 12-2, calculate the percentage change in the equilibrium oxygen concentration when water warms from 0.0 °C to 25.0 °C.

12.30 One of the detrimental effects of the "thermal pollution" of water supplies is that a rise in temperature reduces the amount of dissolved oxygen available for fish. Using the information in Table 12-2, calculate how many liters of water a fish requires at 30.0 °C to obtain the same amount of oxygen that it could obtain from 1.00 L of water at 25.0 °C.

ilw 12.31 At 25 °C, the equilibrium pressure of ammonia vapor above a 0.500 M aqueous ammonia solution is 6.8 torr. Calculate the Henry's law constant and determine the equilibrium pressure of ammonia vapor above a 2.5 M solution.

12.32 At 25 °C, the equilibrium pressure of HCl vapor above a 0.100 M aqueous solution of HCl is 4.0 torr. Calculate the Henry's law constant and determine the equilibrium pressure of HCl vapor above a 6.0 M solution.

Colligative Properties

12.33 Urea, a fertilizer, has the chemical formula $(NH_2)_2CO$. Calculate the vapor pressure of water above a fertilizer solution containing 7.50 g of urea in 15.0 mL of water (density = 1.00 g/mL), at a temperature for which the vapor pressure of pure water is 33.00 torr.

12.34 Ethylene glycol, an automobile coolant, has the chemical formula $HOCH_2CH_2OH$. Calculate the vapor pressure of water above a coolant solution containing 65.0 g of ethylene glycol dissolved in 0.500 L of water (density = 1.00 g/mL), at 100 °C, the boiling point of pure water.

ilw 12.35 Estimate the freezing point of a wine that is 12% by mass ethanol. (Ignore all other solutes.)

12.36 Calculate the freezing point of a solution that contains 12.50 g sucrose $(C_{12}H_{22}O_{11})$ in 155 mL of water.

12.37 Do you have enough information to calculate the boiling point of the wine in Problem 12.35? If so, calculate it. If not, explain what feature of wine prevents you from doing this calculation.

12.38 Do you have enough information to calculate the boiling point of the solution in Problem 12.36? If so, calculate it. If not, explain what feature of the solution prevents you from doing this calculation.

12.39 List the following aqueous solutions in order of decreasing freezing point, and explain your listing: 0.30 M NaCl, 0.75 M NH_3, 0.25 M MgCl_2, 0.25 M NH_3.

12.40 List the following aqueous solutions in order of increasing boiling point and explain your listing: 0.20 M NaCl, 0.20 M sucrose, 0.25 M MgCl_2, 0.40 M sucrose.

12.41 An aqueous solution contains 1.00 g/L of a derivative of the detergent lauryl alcohol. The osmotic pressure of this solution at 25.0 °C is measured to be 64.8 torr. (a) What is the molar mass of the detergent? (b) The hydrocarbon portion of the molecule is an 11-carbon chain. What is the molar mass of the polar portion?

12.42 An aqueous solution containing 1.00 g of a sugar in 100. mL of solution has an osmotic pressure of 1.36 atm at 25 °C. (a) What is the molar mass of this sugar? (b) Sugars have the empirical formula $C_x(H_2O)_y$, where $y = x$, $x - 1$, or $x - 2$. What is the molecular formula of this sugar?

ilw 12.43 The boiling point of an aqueous solution containing sucrose $(C_{12}H_{22}O_{11})$ is 101.45 °C. Calculate the osmotic pressure of this solution at 35 °C, at which the solution density is 1.036 g/mL.

12.44 The freezing point of an aqueous solution containing pantothenic acid $(MM = 205.3$ g/mol) is -0.65 °C. Calculate the osmotic pressure of this solution at 18 °C, at which the density of the solution is 1.015 g/mL.

Between Solutions and Mixtures

12.45 Describe the similarities and differences between fog and smoke.

12.46 Describe the similarities and differences between milk and whipped cream.

12.47 Of the following compounds, which will be the best and which will be the worst surfactant? Support your choices with molecular pictures. (a) propanoic acid, $H_3CCH_2CO_2H$; (b) sodium lauryl sulfate, $H_3C(CH_2)_{11}OSO_3^- Na^+$; and (c) lauryl alcohol, $H_3C(CH_2)_{11}OH$.

12.48 Of the following compounds, which will be the best and which will be the worst surfactant? Support your choices with molecular pictures. (a) sodium alkylbenzenesulfonate, $C_{18}H_{29}SO_3Na$; (b) octadecane, $C_{18}H_{38}$; and (c) decanoic acid, $H_3C(CH_2)_8CO_2H$.

12.49 The figure below represents a monolayer of the surfactant sodium stearate on a liquid surface. Decide if the solvent is polar or nonpolar. Explain your reasoning.

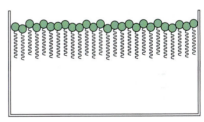

Sodium stearate | Simplified depiction of the stearate anion

12.50 Stearic acid forms a monolayer on the surface of water. Draw a molecular picture that shows how stearic acid molecules are arranged in this monolayer.

Additional Paired Problems

12.51 Commercial phosphoric acid is 85% by mass H_3PO_4 and 15% water. The density of the acid solution is 1.685 g/mL. Calculate the molality, the mole fraction, and the molarity of this solution.

12.52 A commercial solution contains 2.0 M ammonia in methanol. The density of the solution is 0.787 g/mL. Calculate the molality, the mole fraction, and the percent by mass of the ammonia solution.

12.53 Dichlorodiphenyltrichloroethane (DDT) has the following line structure:

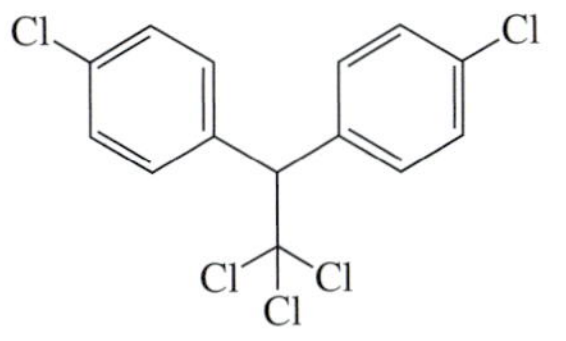

Is this compound hydrophilic or hydrophobic? Do animals excrete DDT readily, or will the compound concentrate in fatty tissues? Does your answer explain why DDT has been banned as a pesticide?

12.54 Butylated hydroxytoluene (BHT) is used as a food preservative. It has the following line structure:

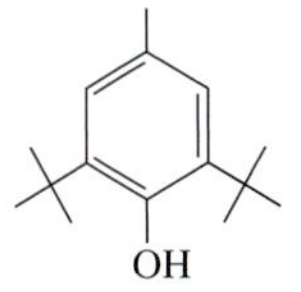

Would you expect to find this compound to be excreted in urine or stored in body fat? Explain your reasoning. BHT is nontoxic to humans.

12.55 What molality of ethylene glycol is required to protect the water in an automobile cooling system from freezing at -20 °C?

12.56 Compute the molar mass of vitamin C, if a solution containing 22.0 g in 1.00×10^2 g of water freezes at -2.33 °C.

12.57 Rank the following substances in order of increasing solubility in water, and state the reasons for your rankings: C_6H_6 (benzene), $HOCH_2CH(OH)CH(OH)CH_2OH$ (erythritol), and $C_5H_{11}OH$ (pentanol).

12.58 Rank the following substances in order of increasing solubility in cyclohexane (C_6H_{12}) and state the reasons for your rankings: KCl, C_2H_5OH, and C_3H_8.

12.59 Water and carbon tetrachloride are not miscible. When mixed together, they form two layers, like water and oil. If an aqueous solution of I_2 is shaken with CCl_4, the iodine moves into the CCl_4 layer. Explain this behavior based on your knowledge of intermolecular forces.

12.60 If some benzene is shaken with a mixture of water and carbon tetrachloride, the resulting mixture contains two layers (see Problem 12.59). Which layer contains the benzene? Explain this behavior based on your knowledge of intermolecular forces.

12.61 To make a good solder joint, the liquid metal solder must adhere well to the metal surfaces being joined. "Flux" is used to clean the metal surfaces. What types of substances must flux remove?

12.62 A pipet is "dirty" when water forms beads on its walls rather than forming a thin film that drains well. Which of the following on the surface of a pipet wall will make it "dirty"? In each case, explain the intermolecular forces underlying your classification. (a) grease; (b) Mg^{2+} ions; (c) acetone; and (d) SiO_2.

12.63 Carbon disulfide (CS_2) has a density of 1.261 g/mL and boils at 46.30 °C. When 0.125 mol of a solute is dissolved in 200.0 mL of CS_2, the solution boils at 47.46 °C. (a) Determine the molal boiling point constant for this solvent. (b) When 2.7 g of an unknown compound is dissolved in 25.0 mL of CS_2, the solution boils at 47.08 °C. Determine the molar mass of the unknown.

12.64 Benzene (C_6H_6) has a normal freezing point of 5.50 °C and a density of 0.88 g/mL. When 1.28 g of naphthalene ($C_{10}H_8$) is dissolved in 125 mL of benzene, the freezing point of the solution is 5.03 °C. (a) Determine the molal freezing point constant for this solvent. (b) When 0.125 g of an unknown compound is dissolved in 25.0 mL of benzene, the solution freezes at 5.24 °C. Determine the molar mass of the unknown.

More Challenging Problems

12.65 One of the earliest methods of preserving fish was by salting. Explain what happens when fish is placed in concentrated salt solution.

12.66 Would water dissolve salts as well as it does if it had a linear structure (such as CO_2) instead of a bent one? Explain.

12.67 When an aqueous solution cools to low temperature, part of the water freezes as pure ice. What happens to the freezing point of the remaining solution when this occurs? A glass of wine placed in a freezer at −10 °C for a very long time forms some ice crystals but does not com-pletely freeze. Compute the molality of ethanol in the remaining liquid phase.

12.68 Some surfactants form membranes that span small holes between two aqueous solutions. These membranes are liquid bilayers two molecules thick. Draw a molecular picture of one of these membranes.

12.69 Explain using intermolecular forces whether water can dissolve gasoline. Knowing that gasoline is less dense than water, would you use water to fight a gasoline fire? Explain.

12.70 Saltwater fish have blood that is isotonic with seawater, which freezes at −1.96 °C. What is the osmotic pressure of fish blood at 15 °C?

12.71 Brackish water, with a salt content around 0.5% by mass, is found in semiarid regions such as the American southwest. Assuming that brackish water contains only sodium chloride and that the ions form no ion pairs, estimate the osmotic pressure of brackish water at 298 K.

12.72 Hard candy is made from very hot solutions of sugar and water with small amounts of added flavorings. In a typical preparation, the boiling point of the sugar/water mixture reaches 145 °C. What mass ratio of sugar ($C_{12}H_{22}O_{11}$) to water is required to raise the boiling point of a sugar/water solution to 145 °C?

12.73 Using the information in Table 12-2, calculate the following: (a) the mass of CO_2 that dissolves in 225 mL of a carbonated beverage at 1.10 atm pressure and 25 °C; and (b) the partial pressure of CO_2 in the gas space above the liquid if a bottle of this carbonated beverage is stored in an ice chest at 0.0 °C.

Group Study Problems

12.74 Drinking seawater can be dangerous because it is more concentrated than a person's body fluids. Discuss the consequences, at both the cellular and the molecular levels, of drinking too much seawater.

12.75 Polystyrene is a plastic that dissolves in benzene. When a benzene solution (density = 0.88 g/mL) containing 8.5 g/L of polystyrene is placed in an osmometer at 25 °C, the solution reaches equilibrium when the height of liquid inside the osmometer is 12.9 cm higher than outside. Calculate the molar mass of the polystyrene.

12.76 Colloidal suspensions of gold are used to label biological molecules for imaging by electron microscopy. A typical colloidal suspension contains 0.10 g of gold per liter of suspension, and the particle size is 15 nm diameter. The density of gold is 19.3 g/cm³. The volume of a sphere is $4\pi r^3/3$, and the surface area of a sphere is $4\pi r^2$. (a) What is the volume in liters of one colloidal particle? (b) How many gold atoms are there in one colloidal particle? (c) How many colloidal particles are there in 1 L of this solution? (d) What is the total surface area of gold in part (c)? (e) What would be the surface area of a cube of gold of mass 0.10 g?

12.77 The table lists the parts per million by mass of the principal ions in seawater:

Ion	Cl^-	Na^+	Mg^{2+}	SO_4^{2-}	Ca^{2+}	K	Br^-
10^2 ppm	194	108	12.9	9.04	4.11	3.92	0.67

Assuming that each ion acts independently of all the others and that seawater has a density of 1.026 g/mL, calculate the freezing point and osmotic pressure of seawater. The actual freezing point of seawater is −1.96 °C. What conclusion can you reach about your assumptions? Take this into account and recalculate the osmotic pressure of seawater.

12.78 At body temperature (37 °C), the Henry's law constant for N_2 dissolving in water is 3.8×10^{-4}. Calculate the solubility of dinitrogen in blood that is in contact with air (78% N_2) at atmospheric pressure and at 4.0 atm, the pressure at a sea depth of 100 feet. Determine the volume of N_2 that will be released from 1 L of blood if a deep-sea diver surfaces quickly from this depth. If this escape of N_2 occurs in the form of gas bubbles that are 1 mm in diameter, how many bubbles per liter is this?

12.79 Consider Raoult's law, as given in Equation 12-3. Raoult's law applies to the total vapor pressure only when the vapor pressure of the solute is negligible compared to the vapor pressure of the solvent. Explain this limitation in molecular terms. What equation from this chapter might you use to correct for the limitation? Will a solution containing a volatile solute have a higher or lower vapor pressure and boiling point than that predicted using Raoult's law? Use Figure 12-11 as a starting point.

Answers to Section Exercises

12.1.1 Copper is the solvent, zinc is the solute. This is a substitutional solid solution.
12.1.2 $X = 1.7 \times 10^{-6}$
12.1.3 $M = 9.6 \times 10^{-5}$ mol/L; $c_m = 9.6 \times 10^{-5}$ mol/kg
12.2.1 Intermolecular interactions are dispersion, dipole–dipole, and hydrogen bonding.

Dipole-dipole   Hydrogen bonding

12.2.2 Only Na_4SiO_4, which is an ionic solid, is water-soluble.
12.2.3 Zn^{2+} is soluble because the solution is stabilized by ion–dipole interactions between partially negative O atoms in water molecules and the Zn^{2+} cations, which replace ion–ion interactions in a zinc salt. Zn metal is insoluble because the "sea" of electrons in delocalized orbitals strongly binds the cations in the metal.
12.3.1 The O atom and the H atom of the methanol —OH group can form hydrogen bonds to water molecules:

12.3.2 $KNO_3 < Li_2SO_4 < KCl < KBr < NaCl < NaNO_3$

Answers to Extra Practice Exercises

12.1 $c_m = 4.04 \times 10^{-2}$ mol/kg
12.2 $c_m = 15.9$ mol/kg; $X_{HCl} = 0.223$
12.3 Each liquid has extensive hydrogen bonding that must be overcome for it to dissolve in cyclohexane. Solubility increases as the hydrocarbon portion gets longer because the extent of hydrogen bonding decreases and the magnitude of cyclohexane–solute dispersion forces increases.
12.4 Uric acid forms several hydrogen bonds and is water-soluble.

12.3.3 The fraction that dissolves is 8.3×10^{-2} (8.3%). This is not an efficient way to collect CO_2 because its solubility in water is significant.
12.4.1 Freezing point depression $= 1.42 \times 10^{-3}$ °C; boiling point elevation $= 3.90 \times 10^{-4}$ °C; and osmotic pressure $= 1.87 \times 10^{-2}$ atm $= 14.2$ torr
12.4.2 The figure should look the same, except there should be more arrows moving from right to left than from left to right.
12.4.3 Your description should include the features shown in your figure for 12.4.2 and the prevention of ion passage through the pores of the semipermeable membrane.
12.5.1

12.5.2 Because they are large, they cannot slip easily between the molecules of a lipid bilayer membrane. Because they are hydrophilic, they do not dissolve in the bilayer.

Testosterone contains mostly nonpolar C—C and C—H bonds and is fat-soluble.
12.5 3.1×10^{-4} M N_2, 1.9×10^{-4} M O_2, 5.2×10^{-6} M CO_2
12.6 $\Delta vp = 2.22$ torr
12.7 $T_b = 100.08$ °C
12.8 $\Pi = 4.3$ atm
12.9 $MM = 8.28 \times 10^3$ g/mol

Macromolecules

Silk, A Natural Macromolecule

Stumble into a spider web and you will experience one of the strongest substances known. Spiders spin their webs from silk, a huge molecule made up of a vast chain of atoms, as shown in our molecular inset. A spider's web is at least five times stronger than steel. What's more, spider silk is waterproof and highly flexible. These properties come from the chemical bonding of the silk structure, as well as the intermolecular forces between the chains.

The feeling of a spider web may be unsettling, but a similar natural material has been used for centuries to make silk fabric that is prized for its smooth texture. Silkworms produce the silk fibers used to make clothing. They feast on mulberry leaves and convert the molecules from these leaves into silk, from which they spin cocoons.

The ancient Chinese discovered how to harvest silkworm cocoons, boil them to loosen the tangle, and unravel the silk into a fiber from which elegant clothing could be produced. A single silkworm cocoon can yield nearly a mile-long filament of silk, but the filament is so fine that it takes around 30 mulberry trees to yield enough cocoons to make one kilogram of silk.

The discovery of silk in China occurred many centuries BC, and by the time of the Roman empire, silk fabric was a prized trade commodity. The caravan routes across Asia became known as the Silk Roads. It is estimated that nearly 90% of the imports into the Roman Empire consisted of silk goods.

CHAPTER CONTENTS

As we describe in this chapter, silk is a macromolecule, an extremely long molecule built from relatively simple molecular building blocks. Silk is made up of proteins, which are long sequences of amino acids linked together into near-infinite chains. Spiders and silkworms secrete liquid proteins that solidify into strands of silk.

Chemists learned the basic composition of silk many years ago, but the reasons why this macromolecule is so strong, yet flexible, are still not fully understood. Recent studies indicate that the secret lies in the way the chains of this protein nestle together. Current research efforts focus on using techniques of

Learning Resources

 KEY CONCEPTS

 CRITICAL THINKING

 SKILLS TO MASTER

 ADDITIONAL HELP
www.wiley.com/college/olmsted
● TUTORIALS
● ANIMATIONS

genetic engineering to replicate natural spider silk on a useful scale.

Silk is just one example of macromolecules, also known as polymers. Macromolecules are the subject of this chapter. The principles introduced in Chapters 9–11 help to explain the properties of these molecules, many of which are carbon-based. In this chapter, we outline the principles of the structure and synthesis of the major classes of macromolecules and describe the properties that give these chemical substances central roles in industrial chemistry and biochemistry. We describe the components from which macromolecules are constructed, some important industrial polymers, and the macromolecules found in living systems.

13.1   STARTING MATERIALS FOR POLYMERS

The prefix "poly" means "many," and **polymers** are very large molecules constructed by linking together many copies of much smaller molecules called **monomers.** You can think of a polymer as the molecular equivalent of a very long chain of paper clips, of which a single paper clip is the monomer. Polymers are one class of **macromolecules,** which are very large molecules. Whereas all polymers are macromolecules, not all macromolecules are polymers. Some macromolecules that are not polymers appear in earlier chapters. A diamond, for example, is a macromolecule made entirely of carbon atoms. A grain of sand is a macromolecule made of silicon and oxygen atoms, in 1:2 ratio, bonded together in a three-dimensional array.

In this chapter we focus on polymers, from the plastics that are essential in our everyday lives to the proteins and DNA that are the materials of life. To understand the chemistry of polymers, we need to understand how monomers are able to link together into long chains. We begin with a description of the various monomers that are the most important starting materials for industrial and natural polymers.

For a polymer to form, monomers must react with one another to form links. Most monomers are organic molecules containing particular **functional groups.** These are groups of atoms that impart specific chemical functions. In this section, we introduce functional groups that can form the linkages that are characteristic of polymers.

Any functional group is only part of an organic molecule. When chemists wish to emphasize a functional group, they commonly use the symbol R to indicate some other portion of the molecule that plays an unimportant role in the molecule's chemical behavior. An R group in these structures can be hydrogen (H) or any organic fragment containing a carbon atom that bonds directly to the functional group. An organic fragment can be as simple as a methyl group (CH_3) or a phenyl group (C_6H_5), or it can represent an elaborate structure containing dozens of atoms.

The most important functional groups that participate in polymerization reactions are listed in Table 13-1. All these groups have electron pairs that can be incorporated into new chemical bonds relatively easily. The π electrons in a double bond are more reactive than electrons in σ bonds, and the lone pairs of electrons on O, N, and S atoms are available for bond formation.

Figure 13-1
The π electrons are readily available for chemical reactions that form bonds to other molecules.

C═C Double Bonds

A carbon–carbon double bond is a reactive functional group because of its π electrons. Remember from Chapter 10 that ethylene has a C═C bond made up of one σ bond plus one π bond. As shown in Figure 13-1, the electrons in the π bond are located off the bond axis, making them more readily available for chemical reactions. Moreover, π electrons are less tightly bound than σ electrons. Consequently, the reactivity patterns of ethylene are dominated by the chemistry of its π electrons. Polyethylene is one familiar polymer whose monomer is ethylene. We describe the polymerization reaction of ethylene and other monomers containing C═C bonds in Section 13.2.

—OH and —SH

An **alcohol** is a molecule that contains a hydroxyl group (—OH) covalently bonded to carbon. Methanol and ethanol are the simplest alcohols. Sugar

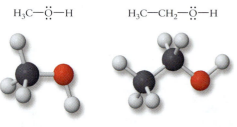

H_3C—$\ddot{O}$—H H_3C—CH_2—$\ddot{O}$—H

CH$_3$OH   CH$_3$CH$_2$OH
Methanol   Ethanol

Table 13-1   Polymerizable Functional Groups

Formula	Name	Class	Example		Reaction Partners
$C{=}C$	Double bond	Alkene	Ethylene	═	Alkenes
$-\overset{..}{\underset{..}{O}}-H$	Hydroxyl	Alcohol	Ethanol		Acids, amines, aldehydes, alcohols
Aldehyde	Aldehyde	Aldehyde	Formaldehyde		Alcohols, amines
Carboxyl	Carboxyl	Acid	Acetic acid		Alcohols, amines
Amine	Amine	Amine	Ethylenediamine		Acids, alcohols
Phosphate	Phosphate	Phosphate	Phosphoric acid		Phosphates, alcohols
$-\overset{..}{\underset{..}{S}}-H$	Sulfhydryl	Thiol	Cysteine		Thiols

molecules, which we describe in Section 13.5, contain several hydroxyl groups bonded to carbon. Starch and cellulose are biological polymers containing sugar monomer units. Some chemical reactions break the O—H bond, while others break the C—O bond of an alcohol or sugar. In addition to forming linkages, the O—H bond is highly polar, so alcohols and sugars form hydrogen bonds readily. We describe later how hydrogen bonds play important roles in the properties of polymers, especially those that have biochemical importance.

A **thiol** contains an —SH group covalently bonded to carbon. Sulfur is just below oxygen in the periodic table, so a thiol is somewhat similar to an alcohol. Still, the chemical and physical properties of thiols differ significantly from those of alcohols. For example, whereas alcohols have inoffensive odors, thiols smell bad. The stench of skunk scent is due to thiols, including 3-methylbutanethiol. Thiols are important in proteins because of their abilities to form S—S linkages, which we describe in Section 13-7.

> The hydroxyl group of an alcohol, which is covalently bonded to a carbon atom, has completely different properties than the hydroxide ion, OH^-, which is present in salts such as NaOH and $Mg(OH)_2$.

Skunk scent contains 3-methylbutanethiol.

—NH₂

An **amine** is a derivative of ammonia in which hydrogen atoms are replaced by carbon atoms:

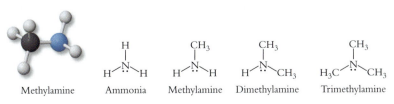

Methylamine	Ammonia	Methylamine	Dimethylamine	Trimethylamine

The —NH_2 group is polar and forms hydrogen bonds. Moreover, the lone pair of electrons on the nitrogen atom can accept a proton from a hydronium ion:

$$CH_3 \quad\quad\quad CH_3$$

$$H-\overset{..}{\underset{H}{N}} \quad + \quad H_3O^+ \quad\longrightarrow\quad H-\overset{+}{\underset{H}{N}}-H \quad + \quad H_2O$$

When they are protonated, amines are compatible with water, a property critical in biochemical processes because it helps biological macromolecules dissolve in water. The N—H bond in an amine is fairly easy to break, so amines participate in the formation of several important classes of polymers, including nylon and proteins.

The Carbonyl and Carboxyl Groups

An oxygen atom doubly bonded to carbon, known as the *carbonyl group,* occurs frequently in organic molecules. The C=O linkage occurs in several types of compounds.

When a carbonyl group bonds to two carbon atoms, the compound is a **ketone.** An **aldehyde** has a hydrogen atom bonded to the carbonyl group.

Aldehyde   Ketone

The chemistry of aldehydes and ketones is dominated by the polarity of the carbonyl group. The difference in electronegativity between carbon and oxygen polarizes the double bond, creating a partial positive charge on the carbon atom and a partial negative charge on the oxygen atom. Aldehydes are more important than ketones in polymerization reactions, but both groups are important in all aspects of chemistry.

A **carboxylic acid** can be represented as R—CO_2H. Many different carboxylic acids participate in organic chemistry and biochemistry. Although carboxylic acids react in many different ways, breaking the C—OH bond is the only reaction that is important in polymer formation. A carboxylic acid is highly polar and can give up H^+ to form a carboxylate anion, R—CO_2^-. The carboxyl group (—CO_2H) also forms hydrogen bonds readily. These properties enhance the solubility of carboxylic acids in water, a particularly important property for biochemical macromolecules.

Phosphates

The **phosphate** group is derived from phosphoric acid (H_3PO_4) by replacing an O—H bond by an O—C or O—P bond. Phosphate is an important functional group in biochemistry, being involved in cellular energy production as well as acting as an important monomer in biopolymers, particularly in DNA. Bonds to phosphate groups form or break in the course of a number of important biochemical reactions.

To summarize from the perspective of polymer formation, the most important role of functional groups in polymerization is to provide bonds that are relatively easy to break. Because C—H and C—C σ bonds are relatively strong and do not break easily, polymerization requires monomers that contain reactive functional groups. To form polymers, bonds in these groups must break, and new bonds that link monomers into macromolecules must form.

Linkage Groups

When two monomers react by combining their functional groups, the result is a new functional group called a *linkage group.* The three linkage groups listed in Table 13-2 are particularly important in polymerization reactions. As a linkage forms, a small molecule is produced, often but not always water. A reaction that produces a linkage group and eliminates a small molecule is called a **condensation reaction.**

An **ester** is an organic compound formed from the condensation reaction between a carboxylic acid and an alcohol. A water molecule is eliminated as the oxygen atom of the

Formaldehyde

$EN_O = 3.5$
$EN_C = 2.5$

δ^-
δ^+

Carboxylic acid

Phosphate

Table 13-2 Important Polymer Linkage Groups

Linkage	Name	Precursors	Polymer Type
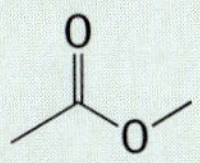 (ester structure)	Ester	Acid + alcohol	Polyesters
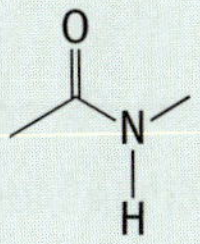 (amide structure)	Amide	Acid + amine	Polyamides, proteins
C—O—C	Ether	Alcohol + alcohol	Cellulose, starch

Tutorial

alcohol links to the carbon atom of the carbonyl group. For example, ethanol reacts with acetic acid to give ethyl acetate:

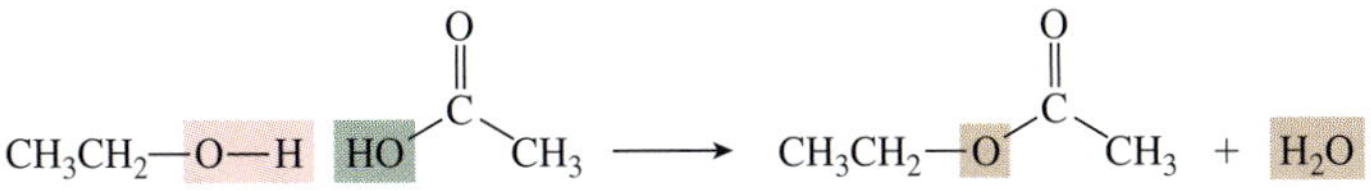

$$CH_3CH_2{-}O{-}H \;\; HO{-}\underset{\underset{\text{CH}_3}{|}}{\overset{\overset{O}{\|}}{C}} \longrightarrow CH_3CH_2{-}O{-}\underset{\underset{\text{CH}_3}{|}}{\overset{\overset{O}{\|}}{C}} \; + \; H_2O$$

Ethanol   Acetic acid   Ethyl acetate
(an alcohol) (a carboxylic acid)   (an ester)

A carboxylic acid can condense with an amine to form an **amide.** Water is eliminated as the new N—C bond forms:

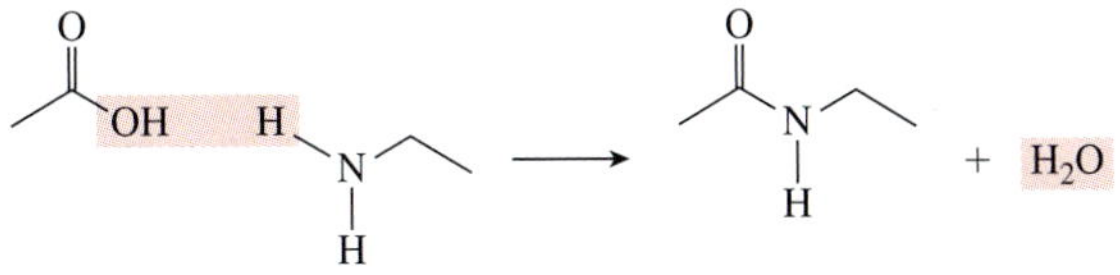

$$CH_3CO_2H \quad H_2NCH_2CH_3 \longrightarrow CH_3C(O)NHCH_2CH_3 \; + \; H_2O$$

(a carboxylic acid) (an amine)   (an amide)

The amide linkage occurs in nature in proteins, biological macromolecules with amazing structural and functional diversity. Silk, described in the introduction to this chapter, is one such molecule. The monomers from which proteins form are *amino acids,* each of which contains both an amine group and a carboxyl group. The amino acids are strung together through a series of condensation reactions that give amide linkages.

The condensation reaction of two alcohol molecules to eliminate water and form a C—O—C bond sequence is yet another linkage reaction. A molecule that contains a C—O—C linkage is called an **ether:**

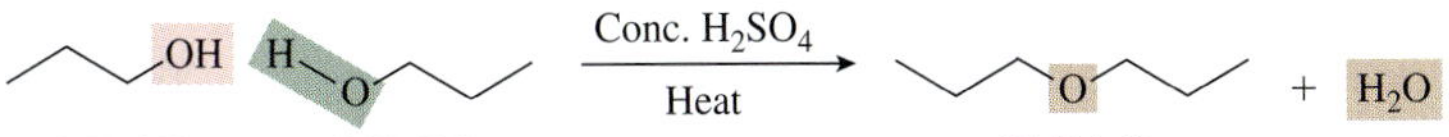

$$C_3H_7OH \quad C_3H_7OH \xrightarrow[\text{Heat}]{\text{Conc. } H_2SO_4} (C_3H_7)_2O \; + \; H_2O$$

C₃H₇OH   C₃H₇OH   (C₃H₇)₂O
(an alcohol) (an alcohol)   (an ether)

This direct condensation of alcohols has limited use as a route to polymers because many alcohols undergo other reactions under the conditions required for condensation. Nature forges ether linkages in a more roundabout way than by eliminating water directly. The most important polymers containing ether linkages are two biochemical macromolecules, starch and cellulose.

Two molecules of phosphoric acid can undergo a condensation reaction, eliminating a water molecule and forming a P—O—P phosphate linkage:

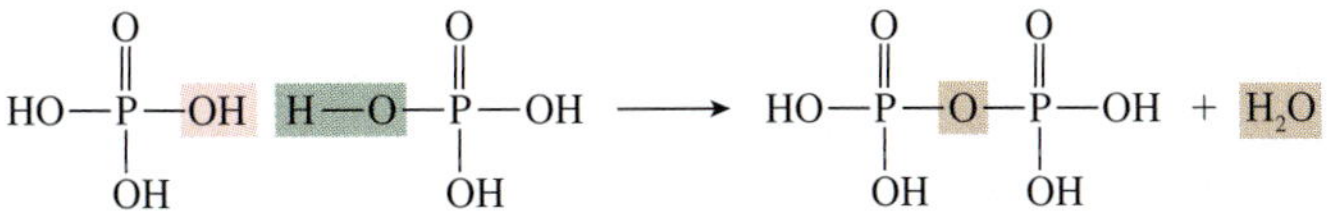

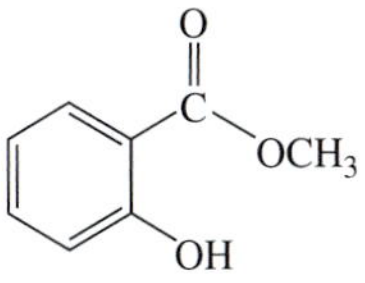

$$HO{-}\underset{\underset{OH}{|}}{\overset{\overset{O}{\|}}{P}}{-}OH \;\; H{-}O{-}\underset{\underset{OH}{|}}{\overset{\overset{O}{\|}}{P}}{-}OH \longrightarrow HO{-}\underset{\underset{OH}{|}}{\overset{\overset{O}{\|}}{P}}{-}O{-}\underset{\underset{OH}{|}}{\overset{\overset{O}{\|}}{P}}{-}OH \; + \; H_2O$$

Methyl salicylate

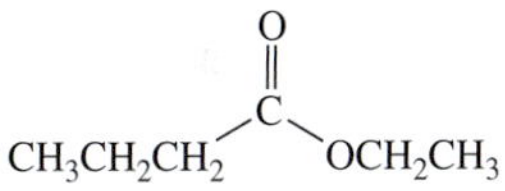

Ethyl butyrate

Many esters are important components of flavorings and fragrances. Methyl salicylate has the fragrance of wintergreen, and ethyl butyrate occurs in pineapples.

Amine   Carboxyl
group   group

Phosphoric acid or phosphate can also condense with an alcohol to generate a phosphate linkage. As described later, such linkages are found in the structure of DNA.

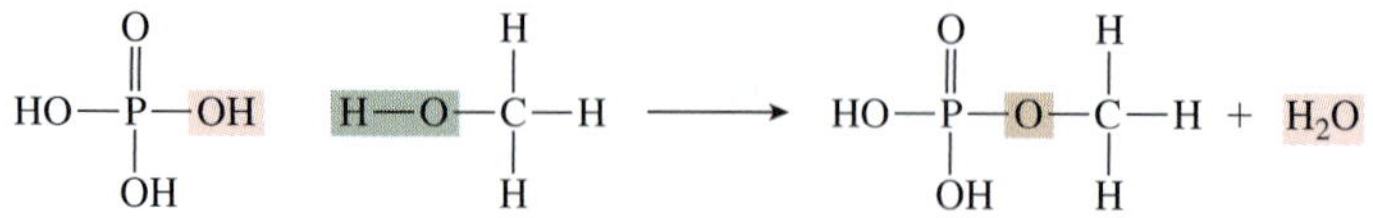

$$HO-\overset{\overset{\displaystyle O}{\|}}{\underset{\underset{\displaystyle OH}{|}}{P}}-OH \quad H-O-\overset{\overset{\displaystyle H}{|}}{\underset{\underset{\displaystyle H}{|}}{C}}-H \longrightarrow HO-\overset{\overset{\displaystyle O}{\|}}{\underset{\underset{\displaystyle OH}{|}}{P}}-O-\overset{\overset{\displaystyle H}{|}}{\underset{\underset{\displaystyle H}{|}}{C}}-H + H_2O$$

Section Exercises

■ **13.1.1** Molecular pictures of several molecules appear below. Identify the functional groups or linkage groups present in each. (Some contain more than one functional group.)

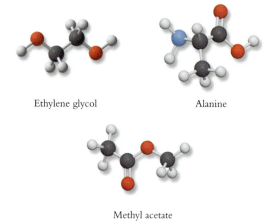

Ethylene glycol   Alanine

Methyl acetate

■ **13.1.2** Amino acids are the molecular building blocks of proteins. Any two amino acids can condense in two ways; each creates an amide linkage. Draw the structures of the two condensation products that link these two amino acids:

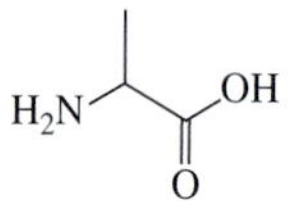

Alanine
($C_3H_7NO_2$)

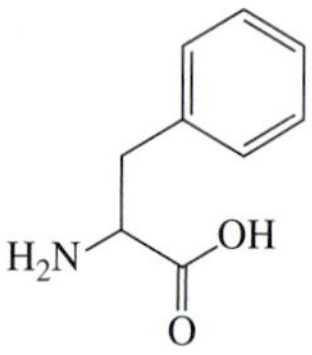

Phenylalanine
($C_9H_{11}NO_2$)

■ **13.1.3** Draw one complete specific line structure for each of the following: (a) an amine with the formula $C_6H_{15}N$; (b) an ester with at least seven carbon atoms; (c) an alcohol with a molar mass of 74 g/mol; and (d) a molecule with at least ten carbons that contains an aldehyde group and a phenyl ring (C_6H_5).

13.2 FREE RADICAL POLYMERIZATION

The simplest of all polymers is **polyethylene,** formed by bonding many ethylene molecules together in long chains containing 500 or more repeating CH_2 groups.

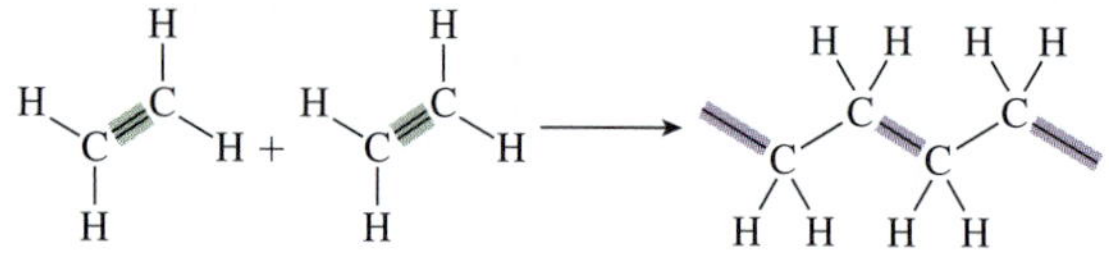

The linkage between two ethylene molecules involves reactions of their π electrons. An electron from the π bond of one ethylene molecule pairs with an electron from the π bond of another ethylene molecule to form a new σ bond between the two molecules. The second electron from each π bond pairs with a π electron from another ethylene molecule to continue the chain.

Writing the structural formula of a macromolecule such as polyethylene with thousands of atoms would be very time-consuming and tedious. Fortunately, the entire structure of a polyethylene molecule can be represented by simply specifying its repeat unit, as shown in Figure 13-2.

Synthesis of Polyethylene

Ethylene forms a polymer through a three-step sequence. First is an **initiation** step, in which a reactive chemical substance attacks the π bond of a single ethylene molecule. The product of the initiation step is able to react readily with the π bond of another ethylene molecule in a **propagation** step. Many propagation steps occur, each one lengthening the growing polymer chain by adding one monomer unit. Finally, chain growth comes to an end through a **termination** step. Initiation, propagation, and termination are important steps in the synthesis of many polymers.

The polymerization of ethylene starts with the thermal decomposition of an initiator molecule, whose general formula is $R-O-O-R$. Heating breaks the weak $O-O$ single bond to form a pair of $R-O\cdot$ **free radicals.** Free radicals are highly reactive molecules that contain unpaired electrons. A free radical will attack any bond that has exposed electron density. In this case, a free radical attacks the π bond of an ethylene molecule:

In the initiation reaction, one π electron of the ethylene molecule pairs with the single electron on the oxygen atom of the free radical. Together, these two electrons create a new $C-O$ σ bond (shown highlighted in purple). The second electron from the ethylene π bond remains as an unpaired electron on the outermost carbon atom of the product.

The carbon atom with the unpaired electron is another free radical, so the stage is set for propagation. The unpaired electron on the carbon atom attacks the π bond of another molecule of ethylene, making a new $C-C$ σ bond and leaving yet another carbon free radical:

This addition reaction can be repeated over and over in the propagation process, and each step adds two CH_2 groups to the growing polymer chain. The propagation can be written in the following general form:

Because chain growth involves free radicals, this type of polymerization is called **free radical polymerization.**

Free radical polymerization does not go on indefinitely, because sometimes two free radicals collide and react. This destroys two free radicals and terminates the chain. For example, another initiator fragment can react with the carbon atom at the end of the chain:

The product of the last reaction may not look like polyethylene, but remember that n is a large number, around 250 or more. Thus, the OR ends of the macromolecule are insignificant compared with the many CH_2-CH_2 repeat units that make up the bulk of the polymer.

Other Polyalkenes

A variety of polymers can be made using derivatives of ethylene in which one or more of the hydrogen atoms is replaced with other groups of atoms. Changing the structure of

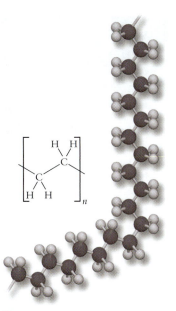

Figure 13-2
Polyethylene has a repeating pattern of $[-CH_2-CH_2-]$ units, each one derived from an ethylene molecule. Shown here are the Lewis structure of the repeat unit and a ball-and-stick model of a segment of a polyethylene chain.

Polyalkenes are used to make many of the items used in everyday life.

Table 13-3   Important Polymers Made from Alkenes

Polymer	Uses	Production (10^9 kg)*
Polyethylene	Piping, bottles, toys	15.7
Polypropylene	Carpets, labware, toys	8.0
Polyvinylchloride (PVC)	Piping, floor tile, clothing	6.6
Polystyrene (Styrofoam)	Containers, heat insulators	2.9
Styrene–butadiene	Tires, fibers	2.0

*2003 figures.

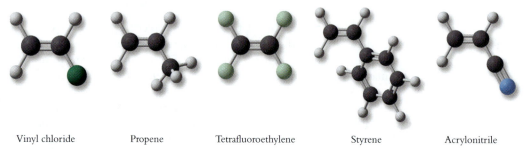

| Vinyl chloride | Propene | Tetrafluoroethylene | Styrene | Acrylonitrile |

Figure 13-3
Structures of alkene monomers used in the manufacture of important polymers.

the monomer makes it possible to adjust the properties of the polymer. Table 13-3 lists important polymers that are made from alkene monomers. You may recognize some of their common names. Figure 13-3 shows the molecular structures of five of these monomers. Notice the variety of groups that can replace hydrogen atoms: CH_3, Cl, F, CN, and C_6H_5. Teflon is tetrafluoroethylene, and Example 13-1 shows how to write the structure of another polyethylene derivative.

Example 13-1	Polyacrylonitrile, known commercially as Orlon, is made by polymerizing acryloni-trile (see Figure 13-3). Orlon is used to make fibers for carpeting and clothing. Draw the Lewis structure of polyacrylonitrile, showing at least three repeat units.
Drawing the Structure of a Polymer	

Strategy: Polyalkenes form by linking carbon atoms in a free radical polymerization. The polymer structure is constructed by connecting monomer units. The polymerization process converts the π bonds of the monomers to σ bonds between polymer repeat units.

Solution: Draw three monomers connected in a symmetrical way that shows the poly-mer chain. A CN group dangles from every other carbon:

Does the Result Make Sense?   Acrylonitrile polymerizes in the same way as eth–ylene. Notice that this polymer has the same structure as polyethylene, except that a CN group is attached to every second carbon atom, so the structure is a reasonable one. A line structure of polyacrylonitrile eliminates the clutter caused by the H atoms. A ball-and-stick model of the same polymer segment is included for comparison.

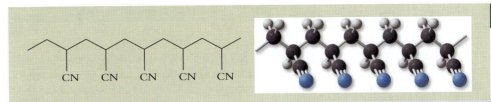

Example 13-1

Drawing the Structure of a Polymer (*continued*)

The polymer known as PVC is used to make plastic pipes. The monomer used to make PVC is vinyl chloride. Draw a segment of the PVC polymer containing four repeat units. See Figure 13-3 for the structure of vinyl chloride.

Extra Practice
Exercise 13.1

Rubber

The first alkene polymer to be used in society was polyisoprene, a natural product extracted from the sap of rubber trees. See our Chemical Milestones Box for a description of the history of rubber. The monomer from which this polymer is constructed is isoprene, an alkene with two double bonds. The line structure of isoprene and the repeat unit of polyisoprene are shown in Figure 13-4. The figure also shows polybutadiene, a synthetic rubber, and its monomer, butadiene. When isoprene or butadiene polymerize, attack on one π bond causes the second π bond to shift, giving a macromolecule containing a four-carbon repeat unit with a double bond in its center.

Natural rubber and polybutadiene each contains just one monomer, but the versatility of rubber materials is greatly increased by polymerizing mixtures of two different monomers to give **copolymers.** Because of their diverse properties, copolymers account for over 60% of the 2.1×10^9 kg of rubber produced in the United States each year. The largest production, accounting for 8×10^8 kg of total rubber production, is a polymer of three parts butadiene to one part styrene. During polymerization of butadiene and styrene, the chain grows by adding whichever monomer happens to collide with the chain end. The propagation process is random, so the exact sequence varies, but the resulting polymer contains three butadiene units for every styrene unit. Here is the line structure of a representative portion of such a polymer, with styrene units highlighted in light brown and butadiene units in gold:

The structures here and on subsequent pages of this chapter are line drawings. See Section 3.1 for a review.

Butadiene–styrene copolymer

At any point in the chain, there is a 75% probability of finding a butadiene unit and a 25% probability of finding a styrene unit.

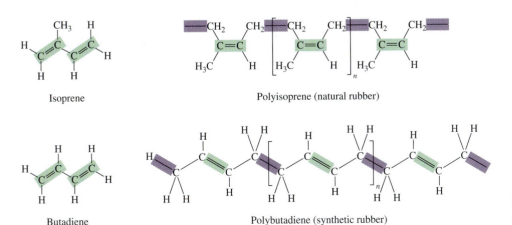

Isoprene

Polyisoprene (natural rubber)

Butadiene

Polybutadiene (synthetic rubber)

Figure 13-4
Structures of isoprene, butadiene, and their polymers.

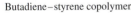

Box 13-1 Chemical Milestones: The Natural Rubber Industry

The rubber industry has a long and colorful history. Natural rubber is produced from latex, a milky fluid found in cells that lie between the bark and the wood of many plants. You may have seen latex flow from the broken stalks of milkweed plants, but the source of commercial rubber is the Hevea tree, a native of Brazil. When the bark of this tree is slashed, its milky white sap oozes out and can be collected in cups mounted on the tree's trunk. The people of the Amazon jungle made bouncing balls, shoes, and water jars out of rubber, and Portuguese explorers sent waterproof boots and a rubber-coated coat back to their king. The first commercial exports included some rubber shoes shipped to Boston in 1823.

Entrepreneurs in Europe and the United States experimented with rubber hats, coats, erasers, life preservers, and a number of other products. Through the first half of the nineteenth century, however, rubber items were impractical because natural rubber becomes sticky in hot weather and brittle in cold weather. In 1830 the world consumed just 156 tons of rubber, all from Brazil.

Nine years later, Charles Goodyear discovered how to stabilize rubber through vulcanization, and rubber quickly became an essential part of the world economy. Goodyear's discovery set off a rubber boom that rivaled the California gold rush. Adventurers swarmed to the jungles to amass fortunes as the price of rubber soared. The Amazon people were captured and

forced to work at harvesting the latex fluid. Many of them died under the harsh conditions. New towns sprang to life along the river as paddle-wheel steamers carried luxury goods to rubber barons, who built grand mansions up the tributaries of the Amazon. The pinnacle of the rubber boom was the town of Manaus, which is located a thousand miles up the Amazon system on the Negro river. At its height, Manaus teemed with mansions, cafes, and fine hotels. It had a streetcar system and even an enormous ornate opera house.

The Manaus opera house

The Amazon rubber industry collapsed almost overnight. In 1876 the English botanist Henry Wickham shipped 70,000 Hevea seeds to the Royal Botanic Gardens in London. New strains of Hevea were developed that produced three to four times as much rubber and were more disease-resistant than their wild Amazonian cousins. Soon, seedlings were sent to Malaya, Java, and other islands of the East Indies. Thirty-five years later, rubber plantations on these islands took control of the industry.

In 1910 the price of rubber was $2.88 a pound, world production was 94,000 tons, and wild rubber from Brazil accounted for 83,000 tons. In 1912 the price collapsed, and by 1932 a pound of rubber cost 2.5¢. In 1937 the world bought more than 1.1 million tons of rubber, less than 2% of which came from Brazil. The rubber cities of the Amazon became near ghost towns as the market for wild rubber vanished. The opulent opera house in Manaus had barely opened its doors when the curtain came down on the Brazilian rubber economy.

The rubber industry changed again when the Japanese captured the East Indian rubber plantations during World War II. The resulting shortage of rubber prompted an intensive research program to produce synthetic rubber. Today, more than 2 million tons of synthetic rubber is produced each year in the United States. Natural rubber is still produced in the tropics, but its importance pales compared to the glory days of the Brazilian rubber plantations.

Cross-Linking

The long-chain molecules described so far do not have the durability and strength associated with rubber products such as rubber bands and automobile tires. To achieve these properties, rubber must be treated chemically to create chemical bonds between long-chain molecules. This process is called **cross-linking** because links are formed *across* the chains in addition to bonds *along* the chains. Vulcanization, the first way to form cross-links in rubber, was discovered in 1839 by Charles Goodyear, founder of the first U.S. rubber company.

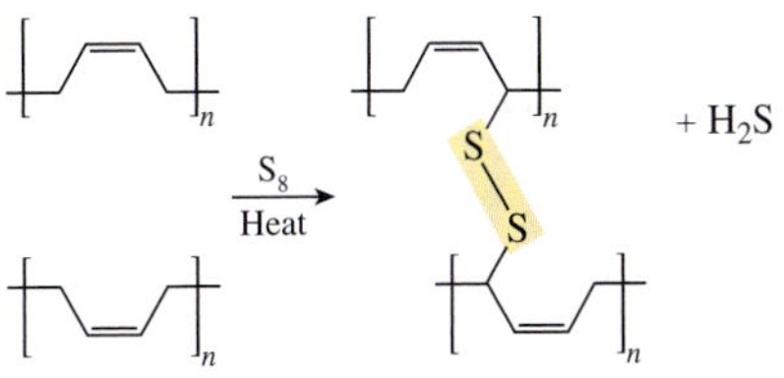

The basic chemistry of vulcanization is simple. Sulfur is added to the polymer, and the mixture is heated under controlled conditions. Some of the polymer C—H bonds break and are replaced by C—S bonds. The cross-links consist of C—(S_n)—C chains, with n being 2 or more. Figure 13-5 represents the overall scheme as it applies to polybutadiene.

Cross-linking increases the tensile strength of rubber while also increasing its rigidity. A small amount of cross-linking gives rubber bands the ability to stretch and rebound, while a larger amount gives truck tires their toughness. Too much cross-linking is detrimental, however. If the sulfur content exceeds 10% by mass, vulcanized rubber becomes hard and brittle. Thus, the amount of cross-linking must be controlled to optimize flexibility and tensile strength without generating a brittle product.

Figure 13-5
When heated in the presence of sulfur, rubber forms cross-links between polymer chains.

Sulfur cross-links also play an important role in determining the structures of proteins, as discussed in Section 13.7.

Section Exercises

13.2.1 Draw the structure of the polymer formed from each of these monomers:

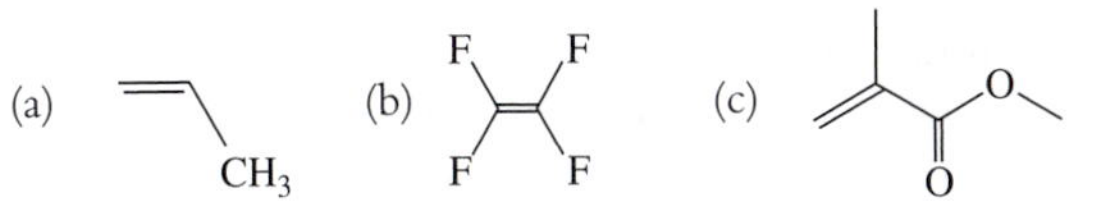

(a) Propene (C_3H_6)

(b) Tetrafluoroethylene (C_2F_4)

(c) Methyl methacrylate ($C_5H_8O_2$)

13.2.2 In a 3:1 copolymer of butadiene and styrene, the placement of butadiene and styrene fragments along the chain is random. Draw a line structure for a portion of the copolymer that has this sequence: –butadiene–styrene–styrene–butadiene–.

13.2.3 Neoprene is a synthetic rubber used to make gaskets. A section of neoprene follows. Draw the structure of the monomer used to make neoprene.

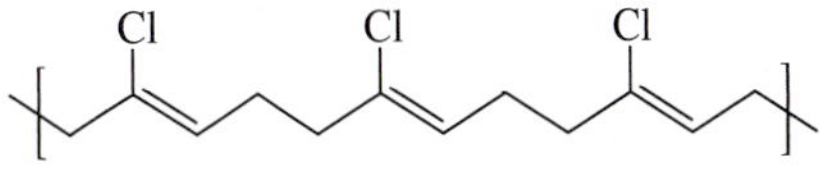

13.3 CONDENSATION POLYMERIZATION

In a condensation reaction, two molecules form bonds between their functional groups by eliminating water or some other small molecule. If no additional functional groups are present, that is as far as the linkage process goes. Polymer formation via condensation requires that each monomer contain *two* linkage-forming functional groups. After the two molecules link together, there are still two functional groups that can link to two more monomers, allowing the chain to grow into a polymer.

Figure 13-6 shows that there are two arrangements of functional groups that make it possible for condensation polymers to form. A single monomer may contain two *different* functional groups capable of linking, or one monomer may contain two of one particular functional group while a second monomer contains two of another functional group that can link with the first.

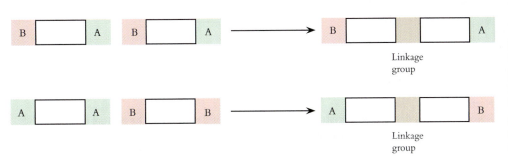

Figure 13-6
Two different arrangements of functional groups can give a condensation polymer. Notice that both condensation reactions generate a product that has an A end and a B end, allowing condensation to continue.

Amino acids belong to the first category. As its name implies, an amino acid contains an amine and a carboxyl group. Amino acids join through amide linkages:

The product is a new amino acid with an amine group at one end and a carboxyl group at the other. Consequently, each end can form another amide linkage, extending the chain.

The condensation of a diacid with a diol (di–alcohol) belongs to the second category. Here, a monomer with two carboxyl groups links to a monomer with two hydroxyl groups to give an ester:

The product is a bifunctional molecule with a carboxyl group at one end and a hydroxyl group at the other. Chain growth can continue through repeated condensation reactions at both ends of the molecule.

Most polymers in production today are made from two different monomers because this synthetic route offers advantages over the use of bifunctional monomers. First, monomers with two identical functional groups are easier and less expensive to produce than monomers with two different groups. Second, the properties of the polymer can be varied easily by changing the structure of one of the monomers.

Polyamides

Polymers that contain amide linkage groups are called **polyamides.** Proteins, which are biological polyamides, are described in Section 13.7. Here we focus on two commercially important polyamides: Nylon 66 and Kevlar.

Nylon 66 was the first polyamide to be produced commercially. Developed by Wallace Carothers at the DuPont Chemical Company in 1935, it still leads the polymer industry in annual production. Figure 13-7 illustrates that Nylon 66, made from

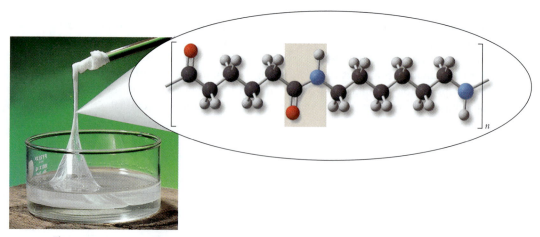

Figure 13-7
Nylon 66 can be drawn from a dish containing two starting materials. The bottom layer is an aqueous solution of 1,6-hexanediamine, and the top layer is adipoyl chloride (a more reactive derivative of adipic acid). Nylon 66 forms at the interface between the two liquids.

adipic acid and hexamethylenediamine, is so easy to make that it is often used for a classroom demonstration:

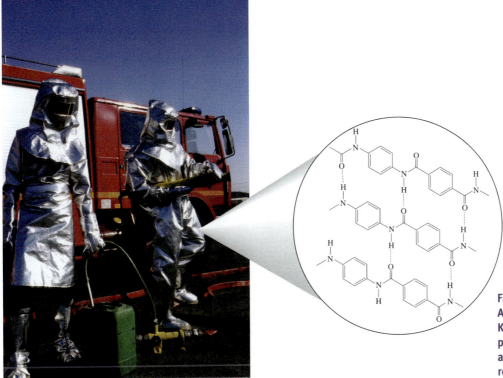

Many other useful polymers can be made by combining different diamines and diacids. One example is Kevlar, a polyamide made by condensing terephthalic acid with phenylenediamine:

Figure 13-8 shows a portion of the structure of Kevlar, which is so strong that it is used to make bulletproof vests. The tremendous strength of Kevlar comes from a combination of intramolecular and intermolecular forces. The stiff phenyl rings make each polymer chain strong and rigid, while the chains are held together by hydrogen bonds. Interchain hydrogen bonds enhance the strength and durability of all polyamides, but Kevlar has a particularly large number of hydrogen bonds per unit volume because of its compact, highly regular molecular structure.

We can identify the monomers from which a condensation polymer is made by visualizing the condensation reaction operating in reverse. The linkage bond breaks, and a small molecule is inserted. Example 13-2 illustrates this type of reasoning for a polyamide that was developed for use in fabrics.

Figure 13-8
A portion of the structure of Kevlar. Pound for pound, this polymer is stronger than steel, and it has excellent flame resistance.

Example 13-2

The Structure of a Polyamide

Qiana cloth looks and feels much like silk.

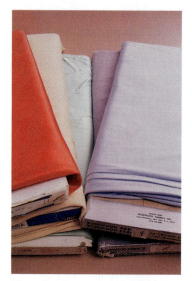

Qiana, a polyamide that feels much like silk, has the following structure:

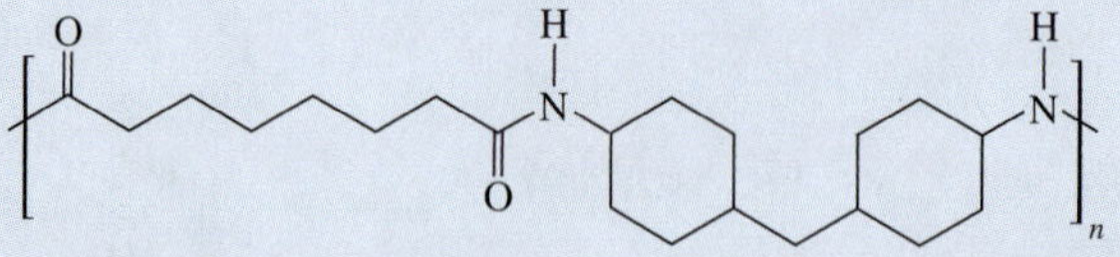

Identify the monomers used to make Qiana.

Strategy: Visualize the condensation reaction that forms a polymer. Two monomers with appropriate functional groups combine, forming a new bond and eliminating water. Reverse this process to see what monomers make Qiana.

Solution: A polyamide is made from the condensation reaction of a diamine and a dicarboxylic acid. To identify the monomers, separate the amide linkage group and add water across the C—N bond:

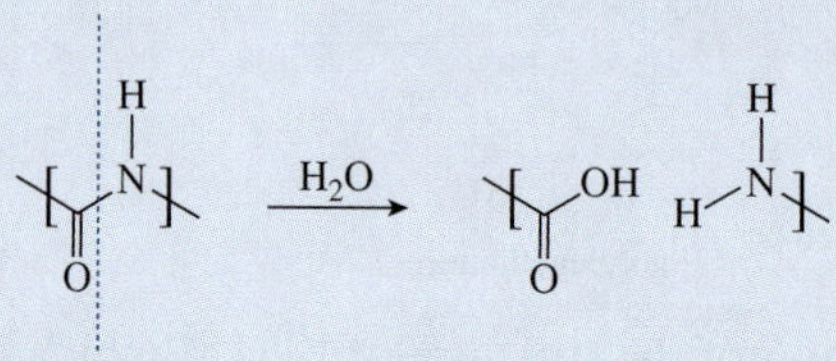

Now construct the monomers:

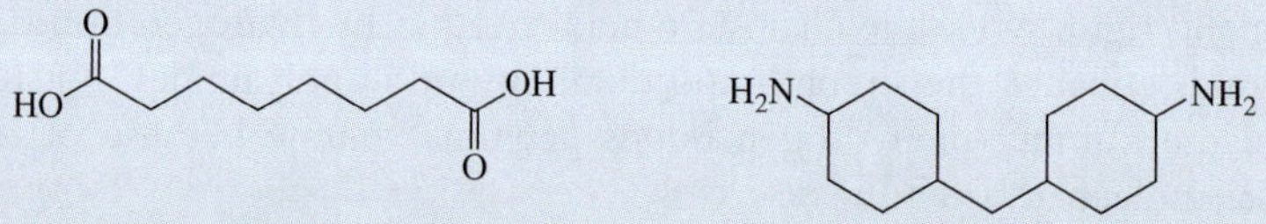

Does the Result Make Sense? Deconstruction of the polyamide gives a pair of monomers, each one containing two of a different functional group. One monomer contains a pair of carboxyl groups, and the other contains a pair of amine groups. These are the components used to build amide linkages, so this monomer pair is reasonable.

Extra Practice Exercise 13.2

Nylon 612 forms stiff fibers that are used to make bristles for brushes. The repeat unit of Nylon 612 follows. Identify the monomers used to make this polyamide.

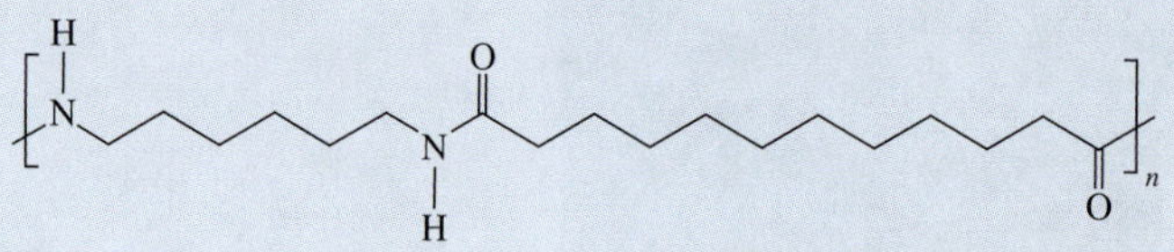

Polyesters

Nylon was the first commercial polymer to make a substantial impact on the textile industry, but **polyesters** now comprise the largest segment of the market for synthetic fibers. In fact, polyesters account for 40% of the more than 4 billion kilograms of synthetic fibers produced in the United States each year. The leading polyester, by far, is poly(ethylene terephthalate), or PET. This polymer is made from terephthalic acid and ethylene glycol in an acid–alcohol condensation reaction:

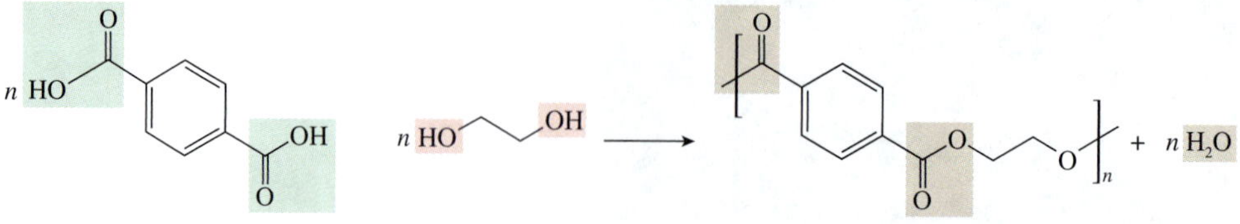

Terephthalic acid — Ethylene glycol — PET

A typical polyester molecule has 50 to 100 repeat units and a molar mass between 10,000 and 20,000 g/mol. PET can be formed into fibers (such as Dacron) or films (such as Mylar). Mylar films, which can be rolled into sheets 30 times thinner than a human hair, are used to make magnetic recording tape and packaging for frozen food. Dacron is best known for its use in clothing, but it has many other applications. For example, tubes of Dacron are used as synthetic blood vessels in heart bypass operations because Dacron is inert, nonallergenic, and noninflammatory.

More than 2 billion kilograms of PET are produced in the United States each year. This polymer is used to make tire cord, beverage bottles, home furnishings, small appliances, and many other common items.

Dacron can be used for synthetic blood vessels.

Section Exercises

■ **13.3.1** The following monomers can be used to synthesize polymers:

Draw structures that show the repeat units of the polyamide and the polyester arising from these monomers.

■ **13.3.2** Polycarbonates are colorless polymers nearly as tough as steel. One of the most common polycarbonates, Lexan, is used in bulletproof windows and as face plates in the helmets worn by astronauts. Lexan is made by condensing phosgene and a compound that contains two —OH functional groups, with the elimination of HCl:

Write the structure of Lexan. Show at least two complete repeat units of the polymer.

■ **13.3.3** The following polymer is used to make carpets because its fibers have a single three-dimensional structure. Footprints quickly disappear from a carpet made of these fibers because compressed fibers quickly return to their most stable orientation. Identify the monomer used to make the polymer and write a balanced equation that shows the condensation of two of these monomers.

13.4 TYPES OF POLYMERS

To a chemist, a polymer is described by its chemical structure. To a manufacturer or a consumer, the important features of a polymer are its macroscopic properties. A tire manufacturer needs a flexible, tough, and durable polymer. A shirtmaker wants to spin a polymer into fibers that wear well, hold dyes, and are free of wrinkles. Makers of lenses look for transparent polymers that can be molded into precise shapes. Each of these polymer applications has somewhat different requirements.

Although they have an endless variety of properties, polymers can be divided into three general categories, based on their form and resistance to stretching. These are **plastics, fibers,** and **elastomers.** Plastics differ in form from fibers: whereas plastics exist as blocks or sheets, fibers have been drawn into long threads. Unlike plastics or fibers, elastomers can be stretched without breaking. Polyethylene packaging films and polyvinylchloride (PVC) pipe are examples of plastics. Orlon carpets are made from polymer fibers, and rubber bands are elastomers. Some polymers, such as Nylon, can be formed into both plastics and fibers.

Plastics

To the general public, the term *plastic* has become synonymous with *polymer*. More precisely, a plastic is the type of polymer that hardens on cooling or on evaporation

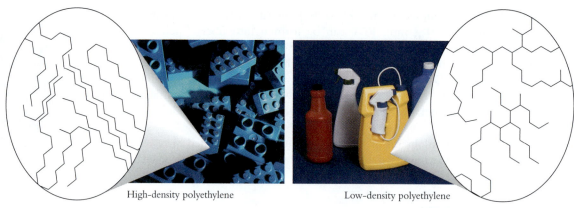

High-density polyethylene   Low-density polyethylene

Figure 13-9
High-density polyethylene has aligned chains of CH₂ units, giving a tough, rigid polymer with a high melting point. Low-density polyethylene has branching chains that give a soft polymer with a low melting point.

of solvent, allowing it to be molded or extruded into specific shapes or spread into thin films.

Plastics fall into two groups based on their response to heating. Those that melt or deform on heating are classed as *thermoplastic* materials, whereas plastics that retain their structural integrity are said to be *thermosetting*. Cross-linking makes it harder for a polymer to deform, so highly cross-linked polymers are thermosetting. Thermoplastic polymers have small amounts of cross-linking. For example, polyethylene consists of huge alkane molecules without cross-linking, held together only by dispersion forces between the chains. Polyethylene is a thermoplastic polymer that melts when heated because individual molecules acquire enough kinetic energy to overcome dispersion forces and slide past one another. In contrast, a rubber truck tire may be so extensively cross-linked that it can be viewed as an immense single molecule. To melt this structure would require the breakage of covalent bonds. This is a thermosetting polymer, which may decompose irreversibly if it is heated to a high enough temperature but will not reversibly melt or deform.

Polyethylene, the best known thermoplastic material, exists in two general forms with different properties. High-density polyethylene is a rigid, strong polymer used to make bottlecaps, toys, pipes, and cabinets for electronic devices such as computers and televisions. Low-density polyethylene is a soft, semirigid polymer used to make plastic bags, squeeze bottles, food packaging films, and other common items. Both kinds of polyethylene have the same repeat unit, CH₂, but different structures of individual polyethylene molecules, as shown in Figure 13-9.

High-density polyethylene forms under conditions that produce polymers made of straight chains of CH₂ units. These linear molecules maximize attraction resulting from dispersion forces by lining up in rows that create crystalline regions within the polymer. Maximizing attraction between the chains imparts strength and rigidity to the polymer. In contrast, low-density polyethylene has chains of CH₂ groups that branch off the main backbone of the polymer. These branches prevent the polymer molecules from packing closely together, thus decreasing dispersion forces and weakening the attraction between the chains. The result is an amorphous polymer that is flexible and melts at a relatively low temperature.

The flexibility of some plastics can be improved by the addition of small molecules called plasticizers. For example, pure PVC turns brittle and cracks too easily to make useful flexible plastic products. With an added plasticizer, however, PVC can be used to make seat covers for automobiles, raincoats, garden hoses, and other flexible plastic objects. Plasticizers must be liquids that mix readily with the polymer. In addition, they must have low volatility so that they do not escape rapidly from the plastic. Dioctylphthalate is a liquid plasticizer that is formed by condensing two alcohol molecules with one molecule of phthalic acid, as illustrated in Figure 13-10.

Fibers

For centuries, humans have woven and spun cotton, wool, and silk into fabrics for clothing and other purposes. Like silk (see page 518), cotton and wool are composed of polymer molecules. Today, synthetic fibers are equally important to the fabrics industry.

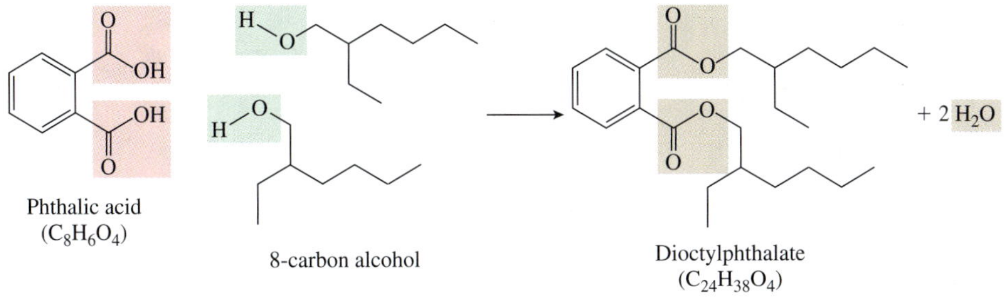

Phthalic acid
$(C_8H_6O_4)$

8-carbon alcohol

Dioctylphthalate
$(C_{24}H_{38}O_4)$

$+ 2\ H_2O$

Figure 13-10
The leading plasticizer, dioctylphthalate, is made by condensing two molecules of an eight-carbon alcohol with phthalic acid.

Nylon 6, a polyamide with six carbon atoms in its repeat unit, makes resilient fibers that are used in carpeting. Part of the reason for the strength of these fibers is hydrogen bonding between neighboring strands of the nylon polymer, as Figure 13-11 illustrates.

Synthetic fibers are thin threads of polymer made by forcing a fluid thermoplastic material through a set of tiny pores, as shown in Figure 13-12. This process requires that the polymer be in the liquid phase, so a fiber-forming polymer must melt at low temperature or dissolve in a convenient volatile solvent. A good fiber has to have high tensile strength, so most synthetic fibers are polyesters, polyamides, or polyacrylonitrile, all of which contain polar functional groups. These groups produce strong intermolecular forces that add significant tensile strength to the material.

Elastomers

An elastomer is a polymer that is flexible, allowing it to be distorted from one shape to another. Polyisoprene (natural rubber), polybutadiene, and butadiene–styrene copolymer are the most important commercial elastomers. All contain some C=C bonds, and their bulk properties are affected by the varying geometries about the carbon atoms that make up the polymer backbone.

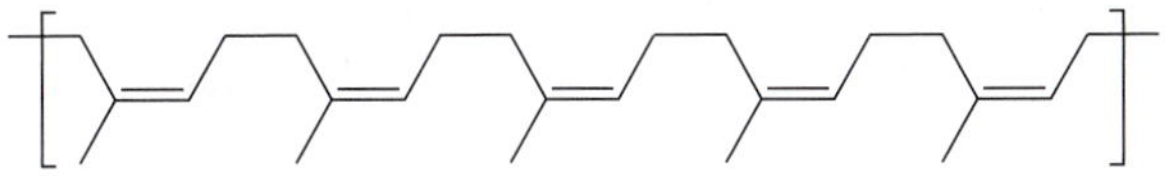

The geometry about the carbon atoms of the CH_2 groups is tetrahedral, whereas that about the carbon atoms involved in double bonds is trigonal. Because of these varied shapes, the molecules lack the structural regularity required to form a polymer with crystalline properties. Consequently, elastomers such as polyisoprene are amorphous solids with individual polymer strands tangled together. This lack of structural regularity keeps the molecules from approaching too closely, thus minimizing attraction resulting from dispersion forces. This in turn allows individual polymer molecules to slide past one another and makes the polymer flexible.

Table 13-4 shows that the flexibility and strength of an elastomer are highly dependent on its amount of cross-linking. The linking of chains increases the strength of the

Figure 13-11
One of the polymers used for carpet fibers is Nylon 6, in which neighboring polymer strands are held together by hydrogen bonds.

Figure 13-12
Fibers form when a fluid thermoplastic polymer is forced through tiny pores into long, thin threads, which solidify as they cool or as solvent evaporates.

Table 13-4   Effects of Cross-Linking on Rubber

Monomer Units between Cross-Links	Degree of Flexibility	Strength	Product
5 – 10	Small	Very high	Casing for calculators
10 – 20	Restricted	High	Tires
20 – 30	Moderate	Moderate	Tire tubes
30 – 40	Moderate	Moderate	Artificial heart membrane
50 – 80	Moderate	Low	Kitchen gloves
100 – 150	High	Low	Surgical gloves

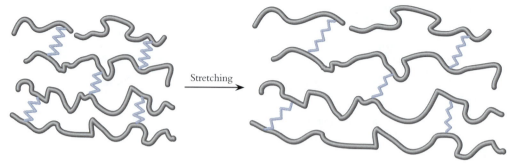

Stretching

Figure 13-13
The amount of cross-linking in an elastomer determines its strength and flexibility. A modest number of cross-links allows the material to stretch but then pulls it back to its original shape.

elastomer but reduces its flexibility. Cross-linking also increases the ability of the elastomer to return to its original shape after distortion, as Figure 13-13 illustrates. For uses such as automobile tires that require both strength and flexibility, the amount of cross-linking must be carefully controlled.

Polymer Stability

Any polymer that has commercial value must be stable under a variety of conditions. This means that it must not degrade when exposed to light, heat, or a variety of chemicals, including acids, bases, and oxidizing agents. Over time, polymers with superior stability have replaced less stable materials. For example, the first synthetic plastic, celluloid, is so highly flammable that it is no longer an important commercial polymer.

Among elastomers, artificial rubbers have replaced natural rubber for many uses because of their high resistance to chemical attack by ozone, an atmospheric pollutant. When ozone reacts with polymer chains, it breaks $C=C$ π bonds and introduces additional cross-linking. Breaking π bonds causes the rubber to soften, and cross-linking makes it more brittle. Both changes eventually lead to rupture of the polymer structure.

Among fibers, nylon and polyester have captured large segments of the clothing and floor-covering markets, in part because they last longer than natural fibers such as silk and wool. Moth holes in a wool sweater demonstrate that living creatures degrade natural fibers. Artificial fibers, by contrast, have no such natural enemies.

The stability of polymers is advantageous while they are being used, but it becomes a liability when these materials need to be discarded. Up to 15% of the volume of municipal waste dumps is polymeric material. Unlike natural materials, polymeric trash does not readily decompose after it is discarded. Thus, polymer disposal has become a cumulative problem.

Recycling polymers is one way to minimize the disposal problem, but not much recycling occurs at present. Only about 25% of the plastic made in the United States is recycled each year, compared with 55% of the aluminum and 40% of the paper. A major obstacle to recycling plastics is the great variation in the composition of polymeric material. Polyethylene and polystyrene have different properties, and a mixture of the two is inferior to either. Recyclers must either separate different types of plastics or process the recycled material for less specialized uses. Manufacturers label plastic containers with numbers that indicate their polymer type and make it easier to recycle these materials. Table 13-5 shows the recycling number scheme.

Thermoplastics are more suitable for recycling than elastomers or thermosetting polymers. Thermoplastics can be heated above their melting temperatures and then recast into new shapes. Elastomers and thermosets, on the other hand, have extensive cross-linking networks that must be destroyed and then reformed in the process of recycling. Processes that destroy cross-linking, however, generally break down the polymer beyond the point at which it can be easily reconstituted.

Table 13-5   Recycling Categories for Plastics

Number	Polymer	Abbreviation
1	Poly(ethylene terephthalate)	PET
2	High-density polyethylene	HDPE
3	Polyvinylchloride (PVC)	P
4	Low-density polyethylene	LDPE
5	Polypropylene	PP
6	Polystyrene	PS
7	Other	—

One way to reduce disposal problems is to make polymers degradable. This poses a substantial challenge to polymer chemists: to fine-tune polymer properties so that the materials are stable while in use but degrade readily to innocuous materials when their useful life is over. One promising approach is to use polymers structurally similar to those found in nature. The goal is to develop polymers that can be degraded by biological organisms. Starch, which is described in the next section, is a natural polymer whose properties can be tailored to various commercial requirements.

Section Exercises

13.4.1 Both polystyrene (see Figure 13-3) and Kevlar (Figure 13-8) contain phenyl rings; yet polystyrene is a thermoplastic material like polyethylene, whereas Kevlar is tough and rigid. Explain why the presence of phenyl rings in polystyrene does not lead to the rigidity characteristic of Kevlar.

13.4.2 Describe what category of polymer is used to make the following items: (a) toothbrush bristles; (b) toothbrush handles; (c) chewing gum.

13.4.3 Make a drawing similar to Figure 13-13 that illustrates how cross-linking reduces the stretchability of a polymer.

13.5 CARBOHYDRATES

Cellulose and starch are macromolecules with empirical formulas that resemble "hydrated carbon," $C_x(H_2O)_y$, where x and y are integers. The monomers from which these macromolecules are constructed are sugars such as glucose and fructose. These monomers and macromolecules are the **carbohydrates.** Structurally, carbohydrates are very different from simple combinations of carbon and water. Even the smallest carbohydrates contain carbon chains with hydrogen atoms, OH groups, and occasional ether linkages.

Carbohydrates are an important food source for most organisms. Glucose, fructose, and sucrose are small carbohydrate molecules that can be broken down rapidly to provide quick energy for cells. Large amounts of energy are stored in carbohydrate macromolecules called **polysaccharides.** For example, glycogen is a polysaccharide used by humans for long-term energy storage. Other polysaccharides, such as cellulose and chitin, are building materials for plants and animals.

Monosaccharides, the molecular units of the saccharides, are carbohydrate molecules containing between three and six carbon atoms. Oligosaccharides contain small chains of two to ten monosaccharide units, and polysaccharides contain long-chain polymers of monosaccharides.

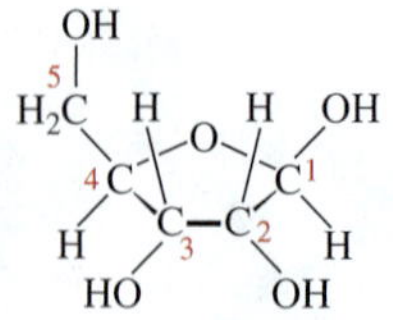

Ribose
($C_5H_{10}O_5$)

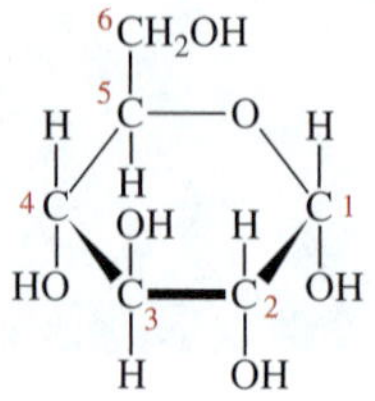

α-Glucose
($C_6H_{12}O_6$)

Figure 13-14
Ribose and α-glucose are monosaccharides that occur in nature. The carbon atoms are numbered for identification purposes.

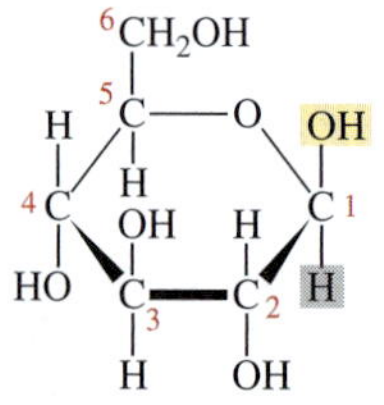

β-Glucose

Figure 13-15
The structure of β-glucose differs from that of α-glucose in the placement of the H atom and the —OH group on carbon atom number 1.

Monosaccharides

Monosaccharides have the formula $(CH_2O)_n$, where n is between 3 and 6. Of the 70 or so monosaccharides that are known, 20 occur in nature. The most important naturally occurring monosaccharides contain five carbons (pentoses) or six carbons (hexoses). Structures of ribose, an important pentose, and α-glucose, a hexose that is the most common monosaccharide, are shown in Figure 13-14. As shown in the figure, it is customary to number the carbon atoms in a monosaccharide, beginning with the HCOH group adjacent to the ether linkage.

The structures of ribose and α-glucose exemplify the characteristics of most monosaccharides. Each is a cyclic compound with an oxygen atom forming an ether linkage in one of the ring positions. Monosaccharides exist in several structural forms, but their ring forms are the basic building blocks of polysaccharides. Ribose has four hydroxyl groups bonded to the ring, and α-glucose has five. Monosaccharides are soluble in water because their several hydroxyl groups form hydrogen bonds with water.

Because the six-membered ring of α-glucose is an important structural component in many polysaccharides, we examine its structure in more detail. As Figure 13-14 shows, each carbon atom in the glucose ring is bonded to one hydrogen atom. Four of the ring carbons are bonded to hydroxyl groups (—OH), and the fifth ring carbon atom is bonded to a —CH_2OH fragment.

Glucose exists as two different isomers, α- and β-glucose. The β-glucose isomer, which is slightly more stable, is shown in Figure 13-15. The two isomers differ only in the placement of —H and —OH on carbon atom number 1. Whereas the —OH on carbon 1 of α-glucose lies below the ring in the view shown in Figure 13-14, that —OH lies above the ring in β-glucose. This difference may not appear significant, but it has profound importance, especially for glucose polymers. The shape of a polysaccharide formed from α-glucose is quite different from the shape of a polysaccharide formed from β-glucose.

Every atom in the glucose ring has a steric number of 4 and approximately tetrahedral geometry, so the glucose ring is puckered rather than planar. This gives glucose a "lawn chair" shape. Although the ring is not planar, we can define a molecular plane that passes through the midpoints of the ring bonds, as shown in Figure 13-16. Each carbon atom of the ring is bonded to two non-ring atoms that lie in different positions relative to this molecular plane. The bond to one of the two atoms is perpendicular to the plane, whereas the other lies roughly parallel to the plane.

Hexose monosaccharides can form both five- and six-membered rings. In most cases, the six-membered ring structure is more stable, but fructose is an important example of a hexose that is more stable as a five-membered ring. The structure of β-fructose is shown in Figure 13-17. Notice that there are —CH_2OH fragments bonded to two positions of this five-membered ring. Examples 13-3 and 13-4 explore the structures of monosaccharides in more detail.

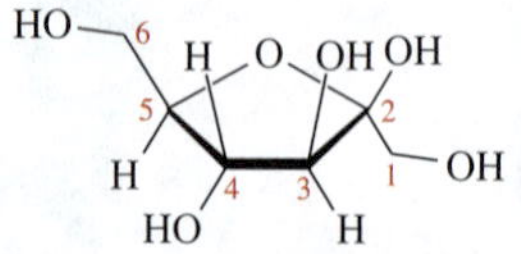

Figure 13-17
Fructose is a hexose monosaccharide that forms a five-membered ring structure. In the ball-and-stick model, the ring is highlighted in blue.

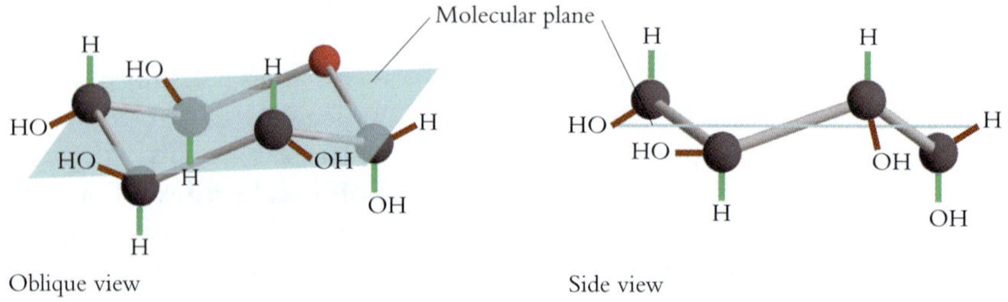

Oblique view

Side view

Figure 13-16
The six-membered ring of a hexose such as α-glucose takes on a puckered structure. A molecular plane can be defined that passes through the midpoints of all the bonds of the ring. Bonds to non-ring atoms are either perpendicular to this plane (shown in green) or roughly parallel to the plane (shown in brown).

Describe the differences in the structures of ribose and fructose.

Example 13-3

Monosaccharide Structures

Strategy: Monosaccharides can differ in their formulas, their ring sizes, and the spatial orientations of their hydroxyl groups. To analyze the differences between two monosaccharides, begin with structural drawings of the molecules, oriented so the ether linkages are in comparable positions. Then examine the structures to locate differences in constituents and bond orientations.

Solution: Here are the structures of ribose (Figure 13-14) and β-fructose (Figure 13-17):

Although the chemical formulas indicate that ribose is a pentose and fructose is a hexose, the ring portions of the structures are the same size. Proceeding clockwise around the rings from the oxygen atom, we see that the structures differ at the first two positions. In the first position, ribose has a carbon atom bonded to —H and —OH, while β-fructose has a carbon bonded to —OH and —CH$_2$OH. In the second position the molecules have the same two bonds but in different orientations. The OH group points "up" in β-fructose and "down" in ribose. The two molecules have the same structures at the other positions.

Ribose
($C_5H_{10}O_5$)

β-Fructose
($C_6H_{12}O_6$)

Does the Result Make Sense? The answer makes sense if the molecules have been described clearly, and if all the structural features have been accounted for.

Describe the differences in the structures of α-glucose and α-idose, a rare naturally occurring monosaccharide.

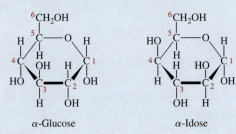

α-Glucose

α-Idose

The six-carbon sugar α-galactose is identical to α-glucose except at carbon atom number 4, where the orientations are different. Draw the molecular structure of α-galactose. Simplify the structure by using flat rings rather than the true three-dimensional forms.

Example 13-4

Drawing Monosaccharides

Strategy: When a carbohydrate with a six-membered ring is drawn, it is best to start with the ring itself. Next, use α-glucose (see Figure 13-14) as a convenient template to obtain the proper orientations for the groups that galactose and glucose have in common. Finally, switch positions of groups as needed to obtain the correct final structure for galactose.

Solution: Here is the flat-ring structure of α-glucose:

6-membered ring

α-Glucose

Example 13-4
Drawing Monosaccharides *(continued)*

The problem states that α-galactose differs from α-glucose at carbon atom number 4. This position is highlighted in the structure. Starting from α-glucose, exchange the hydroxyl group with the hydrogen on this carbon atom:

α-Galactose

Does the Result Make Sense? The two structures differ in only one place, the orientation at position 4. The numbering system is the same for both structures, so the proposed structure of galactose meets the specifications of the problem.

Extra Practice Exercise 13.4

The six-carbon sugar β-mannose is identical to α-glucose except at carbon atoms in positions 1 and 2, where the orientations are different. Draw the molecular structure of β-mannose.

Disaccharides

Two monosaccharides can combine by a condensation reaction between two hydroxyl groups. This reaction forms a linkage in which an oxygen atom connects two saccharide rings:

A C—O—C linkage between two sugar molecules is termed a *glycosidic bond*. Monosaccharides contain several hydroxyl groups, so many glycosidic bonds are possible. Nevertheless, in all natural glycosidic bonds the linkage uses a hydroxyl group on a carbon atom next to a ring oxygen atom of one sugar. The second sugar can link to the first through any of its hydroxyl groups. For example, α-maltose forms from two molecules of α-glucose. The glycosidic bond occurs at the position adjacent to the ring oxygen for one glucose molecule and a different position on the second glucose, as shown in Figure 13-18. Maltose is formed from starch and decomposes in the presence of yeast, first to give glucose and then ethanol and water.

Some disaccharides serve as soluble energy sources for animals and plants, whereas others are important because they are intermediates in the decomposition of polysaccharides. A major energy source for humans is sucrose, which is common table sugar. Sucrose contains α-glucose linked to β-fructose. About 80 million tons of sucrose are produced each year. Of that, 60% comes from sugar cane and 40% comes from sugar beets. Example 13-5 treats a disaccharide that is an energy source for insects.

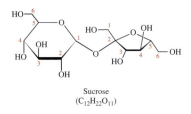

Sucrose, a disaccharide containing α-glucose and β-fructose, is extracted from sugar cane.

Sucrose
($C_{12}H_{22}O_{11}$)

Figure 13-18
The structure of maltose (*far right*). A glycosidic bond links different ring carbon atoms of two glucose molecules.

α-Glucose
$C_6H_{12}O_6$

α-Glucose
$C_6H_{12}O_6$

Maltose
$C_{12}H_{22}O_{11}$

Whereas humans can obtain energy from sucrose, insects obtain energy from trehalose, whose line structure follows. Identify the monosaccharides from which trehalose is constructed.

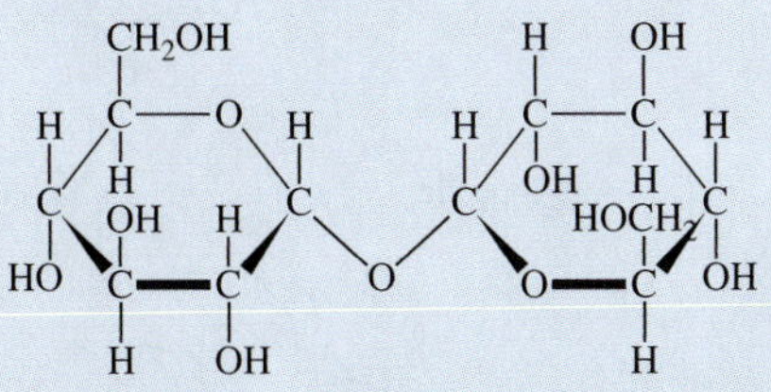

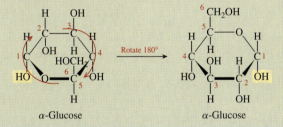

Trehalose
$(C_{12}H_{22}O_{11})$

Example 13-5

Decomposing a Sugar

Insects extract energy from trehalose.

Strategy: Identifying monomer building blocks requires us to visualize a condensation reaction operating in reverse. The linkage bond breaks, and water is added.

Solution: Break the glycosidic bond of trehalose and add water to generate these two monosaccharides:

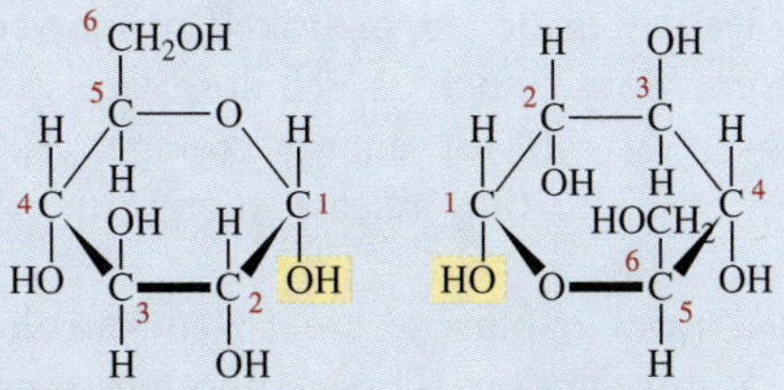

α-Glucose
$(C_6H_{12}O_6)$

α-Glucose
$(C_6H_{12}O_6)$

The monosaccharide on the left is oriented with its ring oxygen atom in the back right position. Compare the locations of its —OH groups with those in the structures in Figure 13-15 to see that this sugar is α-glucose. The sugar on the right is also α-glucose.

Do the Results Make Sense? The monosaccharide on the right must be rotated clockwise about the plane of the ring to bring its ring oxygen atom to the back right position. When that is done, you should recognize that this sugar is identical to the other. Both the monosaccharides in trehalose are α-glucose.

Rotate 180°

α-Glucose   α-Glucose

The structure of the disaccharide melibiose is shown below. Identify the monosaccharides from which melibiose is constructed.

Extra Practice Exercise 13.5

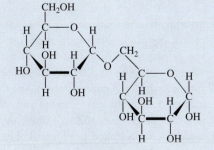

Figure 13-19
Stylized view of amylopectin. The polymer is made up entirely of α-glucose units.

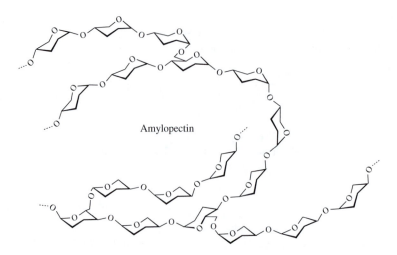

Amylopectin

Polysaccharides

Polysaccharides, macromolecules made up of linked monosaccharides, play two major roles in biological organisms. Some, such as cellulose, are structural materials. Others, including glycogen, act as reservoirs for energy storage. All carbohydrates are good sources of chemical energy, because they release energy upon reacting with oxygen to produce water and carbon dioxide.

Polysaccharides formed from α-glucose are called **starches.** A starch stores sugar until it is needed for energy production. Three important starches are glycogen, which animals produce in their livers, and amylose and amylopectin, produced by plants through photosynthesis. On average, plant starch is about 20% amylose and 80% amylopectin. Each of these polysaccharides contains glucose as its monomer, but they differ in how the monosaccharide units are linked.

Starch collects in granules in plant cells and in animal liver cells.

The simplest starch is amylose, which consists of long chains of α-glucose linked end to end by glycosidic bonds. One molecule of amylose contains about 200 glucose units. Amylopectin and glycogen are predominantly linear chains, too, but in these polysaccharides, some of the rings form a second glycosidic linkage, generating branches along the main chain. Figure 13-19 shows a portion of the structure of amylopectin. Amylopectin contains about 1000 glucose monomers. Branches that are about 30 units long occur at intervals of 20 to 25 glucose units.

Starch is the principal carbohydrate reservoir in plants and is the major component of rice, grains, corn, and potatoes. The long chains are sparingly soluble in water, so they collect in cells as granules. Starch is digested by a series of enzymes into glucose molecules, which are metabolized for energy production. Humans, other animals, and plants have the enzymes necessary to digest plant starch.

Animals store glucose in the form of glycogen. This form of starch is structured like amylopectin but with more frequent, shorter branches. Glycogen is synthesized and stored as granules in the liver and in muscle tissue, where the sugar is readily available for rapid energy production. The liver has a limited capacity to store glycogen. When that capacity is exceeded, the body activates alternative biochemical pathways that convert sugar into fat.

The most abundant organic molecule in the biosphere is **cellulose,** a polysaccharide that is the principal building material for plants. Like amylose, cellulose is a linear polymer of glucose. Unlike amylose, however, the glucose monomers in cellulose are in the β configuration (see Figure 13-15).

Figure 13-20 compares the glycosidic linkages of starch and cellulose. These polysaccharides are made from different forms of glucose, and as a result they have distinctly different shapes. Starch is a chain of α-glucose units linked head to tail. The glycosidic linkage of two α-glucose units imparts a distinct kink to the structure of the polysaccharide, resulting in a

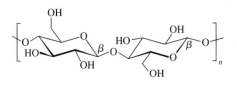

Cellulose

Starch

Figure 13-20
Starch and cellulose have different shapes, even though both polysaccharides are made from glucose monomers.

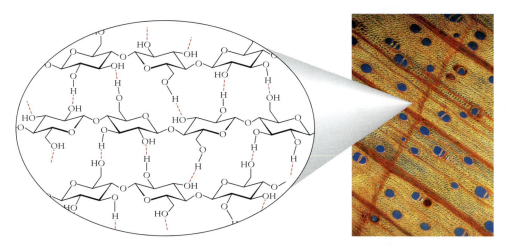

Figure 13-21
Plant cell walls are made of bundles of cellulose chains laid down in a cross-hatched pattern that gives cellulose strength in all directions. Hydrogen bonding between the chains gives cellulose a sheetlike structure.

chain that coils on itself to produce the granular shape of starch deposits. In contrast, cellulose is made entirely from β-glucose units, whose glycosidic linkages lead to a flat structure.

Cellulose is a long, ribbon-like chain of β-glucose units. As shown in Figure 13-21, the planar arrangement of cellulose makes it possible for hydrogen bonds to form between polysaccharide chains, generating extended packages of cellulose ribbons. A typical cellulose molecule contains 2000 to 3000 glucose units in an unbranched linear chain. Individual chains form hydrogen bonds to other chains, giving organized bundles. The cell walls of plants are made up of these bundles laid down in a cross-hatched pattern that gives cellulose strength in all directions.

The enzymes that cleave α-glucose units from starches cannot attack the β-glucose linkages of cellulose because the geometry of the glycosidic linkage is different. As a result, cows and other ruminants must rely on bacteria in their digestive tracts to break down cellulose. These microorganisms, which are also present in termites, use a different group of enzymes to cleave β-glucose units.

Cellulose is the most abundant structural polysaccharide, but it is by no means the only one. Plants and animals use a variety of polysaccharides for a wide range of structural applications. Some polysaccharides contain sugars other than glucose, and others are derivatives of cellulose in which some of the hydroxyl groups on the ring have been converted into other functional groups. One important example is *chitin*, a derivative of cellulose present in many animals. Chitin makes up the exoskeletons of crustaceans, spiders, insects, and other arthropods. The structure of chitin is the same as that of cellulose, except that one hydroxyl group is replaced by an amide group.

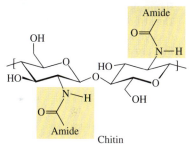

Chitin, found in the exoskeletons of arthropods, is a derivative of cellulose in which an amide replaces one hydroxyl group.

Section Exercises

■ **13.5.1** Gulose is a six-carbon sugar that differs from glucose in two positions. In gulose the hydroxyl orientations at positions 2 and 3 are reversed from their orientations in glucose. Draw the structures of α- and β-gulose.

■ **13.5.2** Approximately 5% of milk is lactose, which is made from β-galactose and β-glucose. The hydroxyl group adjacent to the ring oxygen atom in galactose links to the hydroxyl group in glucose that is in position 4. Draw the structure of lactose using the flattened views of the monosaccharide rings.

(The structure of galactose appears in Example 13-4.)

■ **13.5.3** Most animals cannot digest cellulose. Even ruminant animals such as cattle and deer rely on bacteria in the gut for the enzyme to decompose cellulose. Enzymes that break down a polysaccharide recognize the specific shapes of the various pieces formed in the decomposition process; thus, the enzyme that breaks down starch cannot break down cellulose because the pieces of the two polysaccharides have different shapes. Examine the structures of starch and cellulose (Figure 13-20) and identify the differences in their shapes that might account for this.

Figure 13-22
The organic bases found in DNA are of two types. Guanine and adenine are pyrimidines, with two rings. Uracil, thymine, and cytosine are purines, with single rings. The hydrogen atoms that are eliminated during condensation are highlighted.

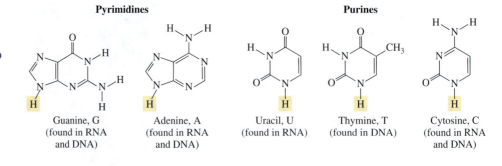

Pyrimidines

Purines

Guanine, G
(found in RNA
and DNA)

Adenine, A
(found in RNA
and DNA)

Uracil, U
(found in RNA)

Thymine, T
(found in DNA)

Cytosine, C
(found in RNA
and DNA)

Ribose
$C_5H_{10}O_5$

Deoxyribose
$C_5H_{10}O_4$

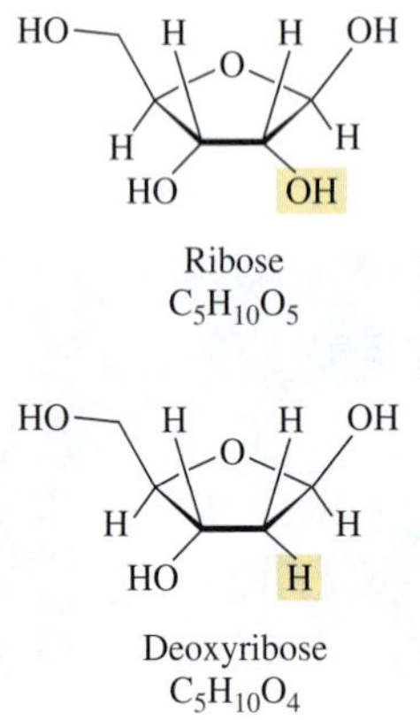

Figure 13-23
The structures of ribose and deoxyribose. The hydrogen atoms that are eliminated during condensation are highlighted.

13.6 NUCLEIC ACIDS

All biological organisms have the ability to reproduce themselves. The instructions for self-replication are stored and transmitted by macromolecules called **nucleic acids.** There are two types of nucleic acids, one that stores genetic information and one that transmits the information. Genetic information is *stored* in molecules of **deoxyribonucleic acid (DNA),** which are located in cell nuclei. These huge molecules have molar masses as large as 10^9 g/mol. The information stored in DNA is *transmitted* by **ribonucleic acid (RNA).** There are several kinds of RNA; each has its own role in the operation of a cell. Molecules of RNA have molar masses of 20,000 to 40,000 g/mol, so they are much smaller than their DNA counterparts. In this section, we examine the structures of nucleic acids and survey their biochemical functions.

The Building Blocks

Nucleic acids are macromolecules made of three component parts:

1. A nitrogen-containing organic base. There are five such bases, whose line structures appear in Figure 13-22. **Adenine** and **guanine** are *purines,* two-ring structures. **Thymine, cytosine,** and **uracil** are *pyrimidines,* one-ring structures. Thymine occurs only in DNA, and uracil occurs only in RNA.

2. A pentose sugar. In RNA the sugar is ribose, and in DNA the sugar is deoxyribose, a ribose in which one OH group has been replaced with one H atom (Figure 13-23).

3. A phosphate linkage derived from phosphoric acid.

These three components are linked through condensation reactions. As Figure 13-24 shows, the —OH group on the 1-position in the ribose ring condenses with an N—H from the base, thus connecting the sugar to the base with a C—N bond. The combination of a base and a sugar is named for its base: cytidine, uridine, thymidine, guanosine, and adenosine. Phosphoric acid condenses with the CH_2OH hydroxyl group of the sugar to complete the linkage and form a nucleotide. Adenosine monophosphate (AMP), the nucleotide that contains adenine, is shown in Figure 13-25. Example 13-6 shows the formation of another nucleotide.

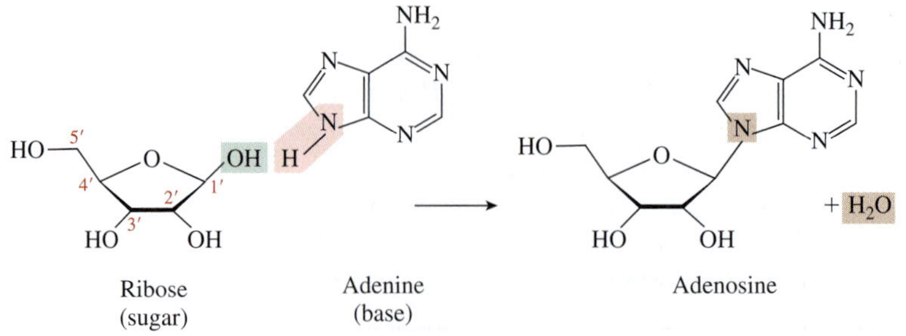

The carbon atoms of DNA and RNA are numbered with primes to distinguish these numbers from the numbers used for positions around the rings of the nitrogen bases.

Ribose
(sugar)

Adenine
(base)

Adenosine

$+ H_2O$

Figure 13-24
Adenosine is formed by a condensation reaction between adenine and ribose.

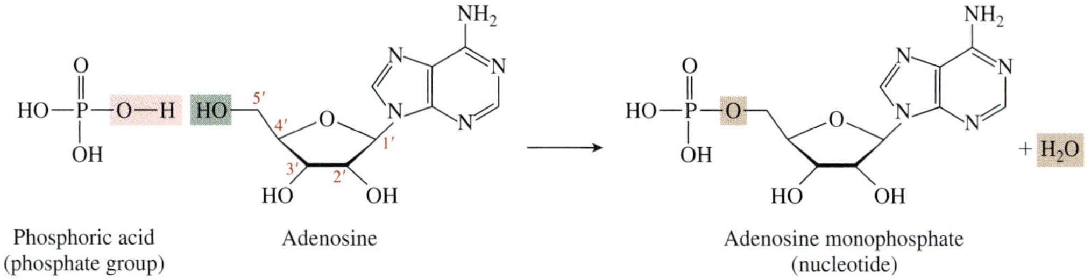

Figure 13-25
The formation of adenosine monophosphate (AMP) by condensation of adenosine and phosphoric acid. The three linked units form the nucleotide building blocks required for nucleic acid synthesis.

Phosphoric acid
(phosphate group)   Adenosine

Adenosine monophosphate
(nucleotide) $+ H_2O$

Example 13-6

Drawing Nucleotides

Draw the structure of uridine monophosphate (UMP).

Strategy: To construct a molecule from component parts, a good approach is to draw each piece separately and then combine them in the proper order. The pieces are linked by condensation reactions.

Solution: The UMP nucleotide contains uracil, ribose, and one phosphate group. The structure of uracil and the hydrogen eliminated during the condensation appear in Figure 13-22. Here are the component parts, drawn in position to eliminate water molecules and link:

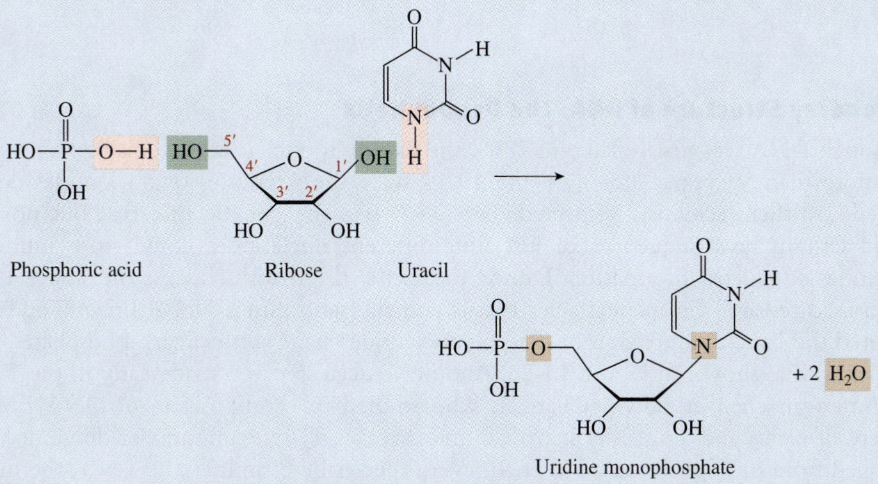

Phosphoric acid   Ribose   Uracil

Uridine monophosphate $+ 2 \ H_2O$

After the components have been placed next to one another, drawing the final structure requires removal of the H_2O units. UMP has three components. From right to left in the drawing, they are uracil, a single-ring base; ribose, the sugar; and one phosphate group.

Does the Result Make Sense? The best way to verify that this structure is reasonable is to compare the parts of the molecule with the components shown in Figures 13-22 and 13-23.

Draw the structure of cytosine monophosphate.

**Extra Practice
Exercise 13.6**

The Primary Structure of Nucleic Acids

A nucleic acid polymer contains nucleotide chains in which the phosphate group of one nucleotide links to the sugar ring of a second. The resulting backbone is an alternating sequence of sugars and phosphates, as shown in Figure 13-26. As the figure indicates, the backbone has directionality. Each phosphate unit forms a bridge from the 5′ position of one ribose to the 3′ position of another ribose.

Each position along a nucleic acid sequence is identical except for the identity of its base. The sequence of bases is called the **primary structure** of a polynucleotide chain. By convention, the listing always begins with the nucleotide that has the terminal

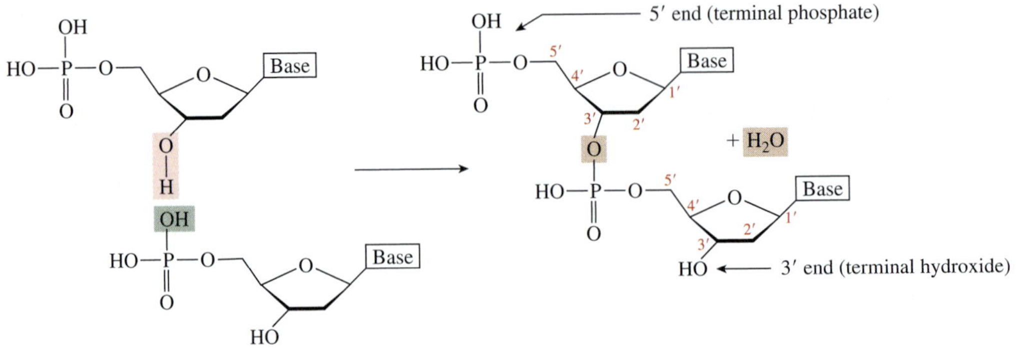

Figure 13-26
The backbone of a nucleic acid is formed by condensation reactions between nucleotides.

5′ end (terminal phosphate)

+ H_2O

3′ end (terminal hydroxide)

phosphate group (5′ end), and it continues to the opposite end of the chain, where the sugar has an unreacted hydroxyl group in the linkage position (3′ end). The nucleotides are listed as their one-letter abbreviations, A, C, G, and T or U. For example, ACGT stands for the following sequence:

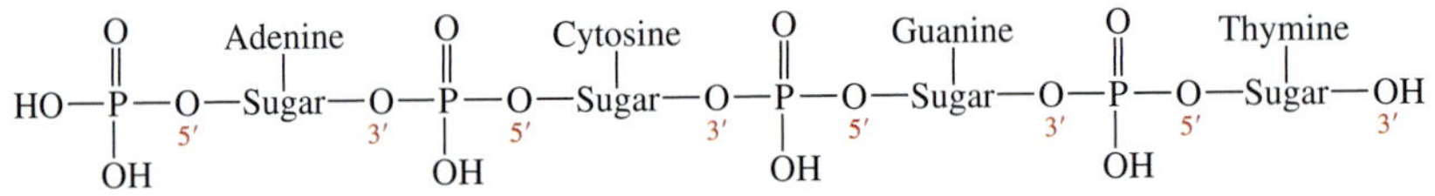

Secondary Structure of DNA: The Double Helix

Although DNA was first isolated in 1868, the nature of nucleic acids remained a mystery for more than 50 years. Not until the 1920s were the structures of nucleotides determined. By then, scientists suspected that DNA was the genetic material, but no one could fathom how sequences of just four different nucleotides could store immense quantities of genetic information. During the 1940s, the British chemist Alexander Todd performed research on nucleic acids that eventually won him a Nobel Prize. Todd discovered the basic composition of the polynucleotide chain, with sugars, phosphates, and bases linked as shown in Figure 13-26. Another crucial step was made early in the 1950s by American chemist Edwin Chargaff, who studied the composition of DNA from a variety of plants and animals. Chargaff found that the relative amounts of different bases changed from one species to another. In every species he examined, however, the molar ratios of guanine to cytosine and of adenine to thymine were always very close to 1.0. Chargaff concluded that these constant ratios could not be coincidence. Somehow, adenine and thymine are paired in DNA, and so are guanine and cytosine.

By now the stage was set for the discovery of the three-dimensional structure of DNA, the so-called **secondary structure** of the molecule. Some of the best minds in science were working on the problem. In 1953, James D. Watson and Francis Crick of Cambridge University announced that they had discovered the structure of DNA. Watson and Crick relied on X-ray diffraction patterns of crystalline DNA. The diffraction photographs were taken by Rosalind Franklin, a researcher in the laboratory of Maurice Wilkins at King's College in London. Combining Franklin's data with the earlier insights of Chargaff, Todd, and many others, Watson and Crick concluded that DNA must consist of two helices wound around one another in a **double helix.** In the double helix, shown schematically in Figure 13-27, the hydrophilic sugars and the phosphate groups lie on the outside of the molecule, with the hydrophobic bases tucked inside the structure. For their brilliant insight, Watson, Crick, and Wilkins shared the 1963 Nobel Prize in medicine and physiology.

According to the Watson–Crick model, hydrogen bonding holds the double helix together. The bases of one strand of DNA form hydrogen bonds to the bases of the second strand. Because of their matching structures, adenine pairs with thymine through two hydrogen bonds, and guanine pairs with cytosine through three hydrogen bonds. These sets are said to be **complementary base pairs.** As shown in Figure 13-28, complementary base pairs fit together like matching gears. The matching of bases in this way keeps the distance between the two strands constant through the entire length of the DNA molecule.

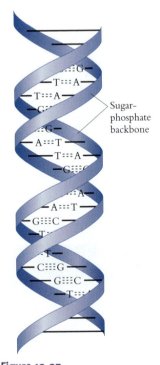

Sugar-phosphate backbone

Figure 13-27
DNA consists of two strands of sugar–phosphate backbones wound around each other in a double helix. The two helices are connected by hydrogen bonds between the bases.

The structure of DNA was discovered by (shown left to right) Francis Crick, James Watson, Maurice Wilson, and Rosalind Franklin.

The nucleotide bases are flat molecules. Each base pair is parallel to the one below it, with 340 picometers separating the two. There is a rotation of 36° between pairs, giving ten base pairs per complete turn of the helix. The two sugar–phosphate backbone strands wind around these stacked pairs, as shown in Figure 13–29. The two strands of DNA run in opposite directions, with the terminal phosphate end of one polynucleotide matched with the free hydroxyl end of the other.

Figure 13-28
The complementary base pairs of DNA. Thymine fits with adenine, and cytosine fits with guanine.

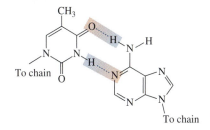

Thymine · · · · · Adenine

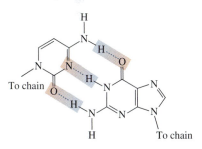

Cytosine · · · · · Guanine

Figure 13-29
The structure of DNA. (*a*) A ball-and-stick model, with the sugar–phosphate backbone colored blue and the bases colored red. (*b*) A space-filling model, showing C atoms in blue, N atoms in dark blue, H atoms in white, O atoms in red, and P atoms in yellow.

| Box 13-2 | Chemistry and Technology: DNA "Chips" |

Every gene contains DNA with a unique sequence of bases forming a genetic code containing the information an organism uses to live and replicate itself. Many years of research have resulted in an understanding of how the information content of DNA is translated into particular biochemical substances and how DNA replicates. The processes include unwinding of the DNA double helix so its code can be read or duplicated.

In an unwound helix, the bases are exposed and can form new hydrogen-bonding interactions. In protein synthesis, the unwound DNA sequence serves as a template to build a molecule of RNA whose base sequence is complementary to that of the DNA sequence. The RNA molecule, in turn, serves as a blueprint for protein manufacture. In replication, as the illustration shows, unwinding and duplication generates two identical DNA helices from a single helix.

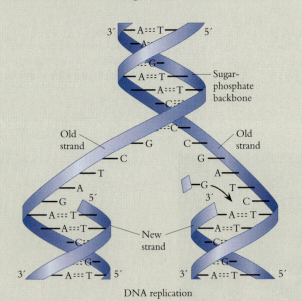

DNA replication

In both RNA synthesis and DNA replication, the bases of the original DNA strand are matched to a set of complementary bases through strong hydrogen-bonding interactions. This process is highly specific because each base interacts only with its complementary base. In DNA replication, the growing chain incorporates the correct complementary base by binding that base, then forming the backbone linkages.

The promise of being able to predict and modify the genetic characteristics of an organism fuelled massive efforts to determine the base sequences of human genes. The human genome project has now reached the goal of sequencing all the important DNA carried by humans.

A developing application of DNA technology uses a DNA "chip" that contains many small segments of bases of known sequence. Such DNA chips are being used to attack cancers. Cancers occur when defective genes cause cells to divide uncontrollably, but usually the process of protein synthesis also is modified. Different types of cancer result in different modifications to protein synthesis, depending on how the DNA sequence of the cancerous cell differs from that of a normal cell. The DNA chips allow clinicians to identify the proteins modified by a particular cancer.

This technology combines microfabrication techniques with the ability of DNA to recognize and bind to its complementary sequence. Light-sensitive chemical reactions are used to construct a DNA chip containing up to 10,000 distinct short DNA sequences, each one occupying a specific position on the chip. In one use of such chips, DNA is extracted from a cancerous cell, fragmented into segments, and chemically reacted with small fluorescent molecules that bind preferentially to specific sites along the DNA sequence. A solution of these fluorescent DNA segments is then applied to the chip.

As the schematic illustration shows, a segment containing a particular base sequence molecule will bind to an area of the chip that contains its complementary sequence, but it will not bind to any area that has a mismatched sequence.

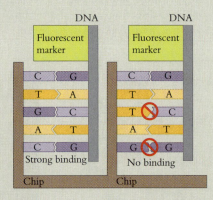

When the chip is rinsed, fluorescent DNA remains bound to all those areas of the chip that have complementary sequences. Areas of the chip whose sequences are not complementary do not bind to the DNA. When the chip is rinsed and subsequently exposed to ultraviolet light, the areas with bound DNA fluoresce, generating light from each area to which the DNA has bound. In contrast, areas of the chip that contain noncomplementary sequences remain dark. The fluorescence pattern reveals which protein synthesis processes have been modified in the cancerous cell, and this in turn helps physicians determine what treatment is most likely to be effective against the cancer.

The power of this new technique comes from the large number of DNA sequences that can be placed on a single chip, several tens of thousands on a single 1.3 × 1.3-cm chip. This gives a richly detailed fluorescence pattern that can be interpreted using a high-resolution optical scanner and computer analysis. Such detailed information makes it possible to distinguish among a large number of different types of cancerous tissues.

In recent years, our detailed understanding of the structure of DNA and how it functions has led to many new methods of genetic manipulation. Our Chemistry and Technology Box explores one aspect of this subject.

The Structure of RNA

The structures of DNA and RNA are similar in that each has a sugar–phosphate backbone with one organic base bound to each sugar. However, there are four distinct differences between RNA and DNA:

1. The sugar in RNA is ribose, not deoxyribose.
2. RNA uses uracil instead of thymine. The component bases in RNA are adenine, uracil, guanine, and cytosine.
3. RNA is much smaller than DNA. The molecules of RNA range in molar mass between 20,000 and 40,000 g/mol instead of as much as 10^9 g/mol.
4. RNA is usually single-stranded, not double-stranded.

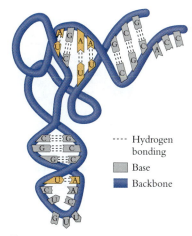

Figure 13-30
The structure of an RNA molecule. Notice the folding caused by the intrastrand base pairing.

Although most of an RNA molecule is single-stranded, there often are some double-stranded regions. Intramolecular base pairing between guanine and cytosine and between adenine and uracil creates loops and kinks in the RNA molecule. The structure of one kind of RNA molecule is shown in Figure 13-30.

Whereas DNA has a single role as the storehouse of genetic information, RNA plays many roles in the operation of a cell. There are several different types of RNA, each having its own function. The principal job of RNA is to provide the information needed to synthesize proteins. Protein synthesis requires several steps, each assisted by RNA. One type of RNA copies the genetic information from DNA and carries this blueprint out of the nucleus and into the cytoplasm, where construction of the protein takes place. The protein is assembled on the surface of a ribosome, a cell component that contains a second type of RNA. The protein is constructed by sequential addition of amino acids in the order specified by the DNA. The individual amino acids are carried to the growing protein chain by yet a third type of RNA. The details of protein synthesis are well understood, but the process is much too complex to be described in an introductory course in chemistry.

Section Exercises

- **13.6.1** Draw the structure of the RNA nucleotide that contains guanine.
- **13.6.2** Part of a DNA sequence is G-C-C-A-T-A-G-G-T. What is its complementary sequence?
- **13.6.3** Nucleotides can contain more than one phosphate group. One example is energy-storing adenosine triphosphate (ATP), which is discussed in Chapter 14. An ATP molecule is formed in two sequential condensation reactions between phosphoric acid and the phosphate group of AMP:

$$AMP + 2\ H_3PO_4 \longrightarrow ATP + 2\ H_2O$$

Draw the structure of ATP.

13.7 PROTEINS

Life is organized around the cell, the smallest functioning unit of an organism. The most important biochemicals in cells are proteins, including enzymes, antibodies, hormones, transport molecules, and the structural materials that make up the cell itself. Proteins protect organisms from disease, extract energy from food, move essential cellular components from place to place, and are responsible for vision, taste, and smell. Proteins even synthesize the genetic material contained in all cells. In other words, proteins are the molecular machinery of the cell. Structural materials in most animals, including skin, hooves, and feathers, are also made of proteins. How can this single group of molecules play so many different roles in the chemistry of life? To answer that question, we need to look at the structure of proteins. Remarkably, all proteins have amino acids as their common structural components.

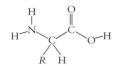

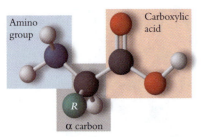

Figure 13-31
Ball-and-stick representation and structural formula common to all of the naturally occurring amino acids.

Amino Acids

As the name implies, an **amino acid** is a bifunctional molecule with a carboxylic acid group at one end and an amine group at the other. All proteins are polyamides made from condensation reactions of amino acids. Every amino acid in proteins has a central carbon atom bonded to one hydrogen atom and to a second group, symbolized in Figure 13-31 as *R*.

Twenty different amino acids are used to build proteins in living cells. The simplest is glycine, where $R = H$. In each of the 19 amino acids other than glycine, the side chain begins with a carbon atom. The side chains in naturally occurring amino acids are shown in Figure 13-32.

Among the common amino acids, eleven have side chains that contain polar functional groups that can form hydrogen bonds, such as $-OH$, $-NH_2$, and $-CO_2H$. These hydrophilic amino acids are commonly found on the outside of a protein, where their interactions with water molecules increase the solubility of the protein. The other nine amino acids have nonpolar hydrophobic side chains containing mostly carbon and hydrogen atoms. These amino acids are often tucked into the inside of a protein, away from the aqueous environment of the cell.

Biochemists represent each amino acid with a three-letter abbreviation. These abbreviations appear under the names of the amino acids in Figure 13-32. For example, the abbreviation for glycine is Gly.

Polypeptides

Proteins form in a sequence of condensation reactions in which the amine end of one amino acid combines with the carboxyl end of another, eliminating a water molecule to create an amide linkage. The amide group that connects two amino acids is called a **peptide linkage,** and the resulting molecule is known as a *peptide*. When two amino acids are linked, the product is a *dipeptide*. A dipeptide formed from alanine and glycine is shown in Figure 13-33.

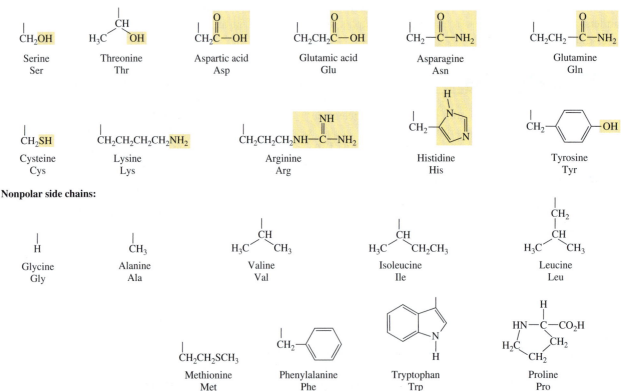

Figure 13-32
Of the 20 amino acids, 11 have polar side chains (color screened), and 9 have nonpolar side chains. One, proline, has a unique ring structure. Under the name of each amino acid is its three-letter abbreviation.

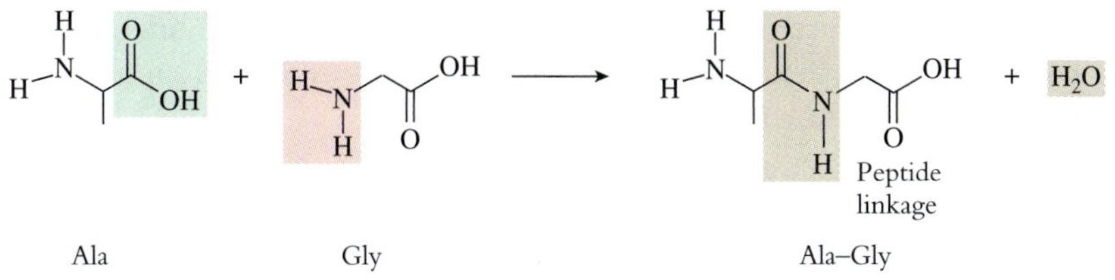

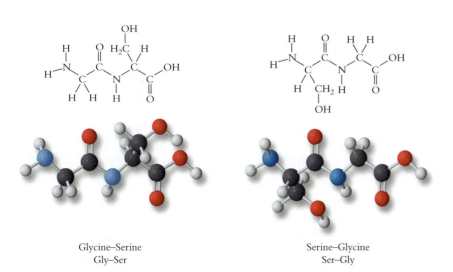

Ala   Gly   Ala–Gly

Figure 13-33
Alanine and glycine form a dipeptide by a condensation reaction.

Notice that this dipeptide is also an amino acid because the molecule retains an amine group at one end and a carboxyl group at the other end. Consequently, an additional amino acid can add to either end of a dipeptide to form a new peptide that also has an amine terminal group at one end and a carboxyl terminal group at its other end. Figure 13-34 shows the peptide that results from addition of another alanine molecule at the amine end and a cysteine molecule at the carboxyl end of the Ala-Gly dipeptide.

Ala   Ala-Gly dipeptide   Cys

Ala-Ala-Gly-Cys tetrapeptide

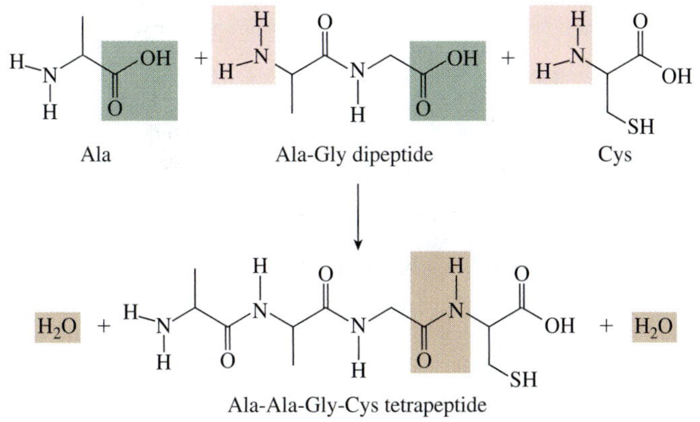

Figure 13-34
Condensation reactions can take place at both ends of a dipeptide. The product can continue to grow into a polypeptide or a protein. The figure shows alanine and cysteine (Cys) adding to opposite ends of the Ala–Gly dipeptide.

Protein synthesis in cells occurs by sequential condensations, always at the carboxyl end of the growing chain, eventually leading to a macromolecule called a **polypeptide.** All proteins are macromolecular polypeptides, but the lengths of proteins vary tremendously. Some small hormone proteins contain as few as eight or nine amino acids. Myosin, a very large muscle protein, has approximately 1750 amino acid units. Proteins exist with amino acid chain lengths that range between these extremes.

The **primary structure** of a polypeptide is its sequence of amino acids. It is customary to write primary structures of polypeptides using the three-letter abbreviation for each amino acid. By convention, the structure is written so that the amino acid on the left bears the terminal amino group of the polypeptide and the amino acid on the right bears the terminal carboxyl group. Figure 13-35 shows the two dipeptides that can

Glycine–Serine
Gly–Ser

Serine–Glycine
Ser–Gly

Figure 13-35
Glycine and serine form two dipeptides, Gly–Ser and Ser–Gly. As the ball-and-stick models show, even dipeptides have distinctive shapes that depend on their primary structure.

be made from glycine and serine. Although they contain the same amino acids, they are different molecules whose chemical and physical properties differ. Example 13-7 shows how to draw the primary structure of a peptide.

| Example 13-7 | Draw the line structure of the peptide Asp–Met–Val–Tyr. |

The Primary Structure of a Peptide

Strategy: We are asked to translate a shorthand designation into a line drawing. First, construct a backbone containing three amide linkages, putting the terminal NH_2 group on the left end and the terminal CO_2H group on the right end of the peptide. Then attach the appropriate side groups, as determined from the molecular structures of the amino acids in Figure 13-32.

Solution: The four amino acids are joined with peptide linkages in the order given. Put the backbone in place as a line structure. Remember, however, that carbon atoms are not shown in line structures and hydrogens are included only for atoms other than carbon.

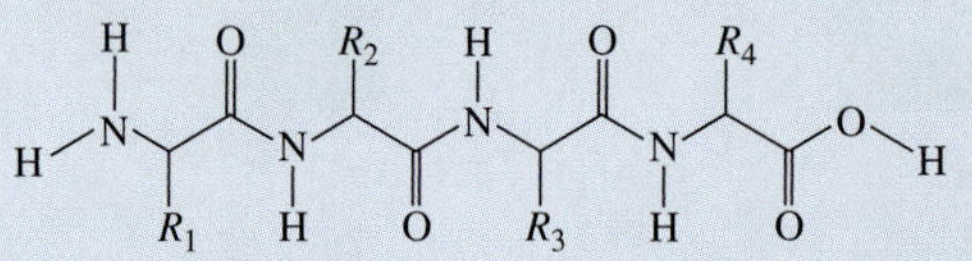

Next, add the four side chains in the positions marked R_1, R_2, R_3, and R_4 to give the final structure of the peptide. The leftmost amino acid, aspartic acid, is at the amino end. Aspartic acid is followed by methionine, valine, and tyrosine. Tyrosine is at the carboxyl end. Here are the individual amino acids, with their R groups highlighted:

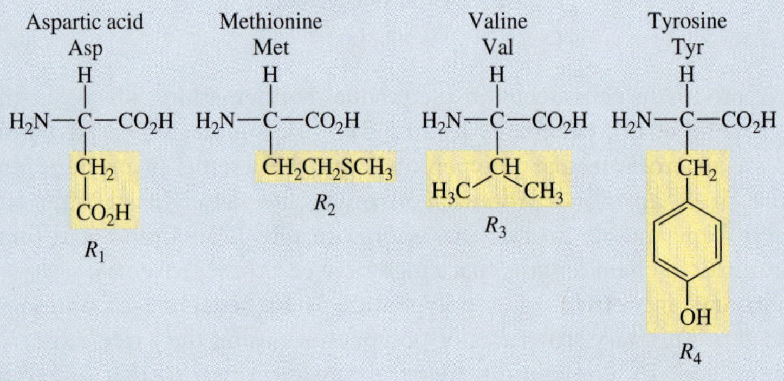

To finish the structure, we replace the R groups on the peptide backbone with their appropriate line structures:

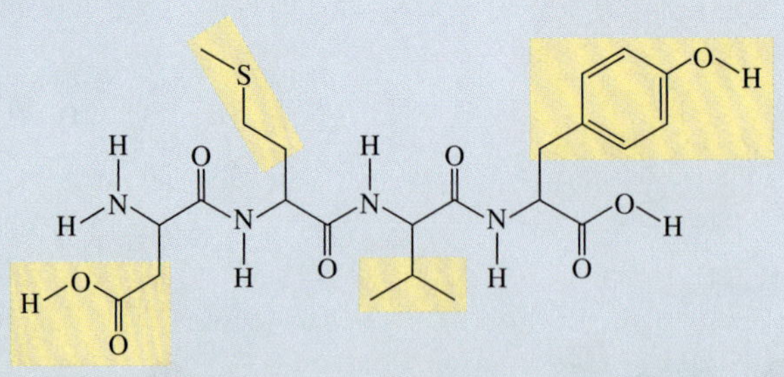

Does the Result Make Sense? The structure is reasonable if the peptide backbone is correct and if the four amino acid side chains match the structures shown in Figure 13-32.

| Extra Practice Exercise 13.7 | Draw the line structure of the peptide Cys–Lys–Leu–Asn. |

Proteins with similar primary structures can serve very different functions. For example, Figure 13-36 shows the primary structures of two pituitary hormones, vasopressin and oxytocin. These structures differ by just two amino acids (those highlighted in color), but changing these two amino acids has a profound effect on the biochemistry of these molecules. Vasopressin regulates the rate at which water is reabsorbed by the kidneys and intestine. Alcohol suppresses the release of vasopressin, which is why consumption of alcoholic beverages leads to excessive urine production and dehydration. Oxytocin, on the other hand, induces contractions of the uterus during childbirth, and it triggers the release of milk by contracting muscles around the ducts that come from the mammary glands.

The enormous diversity of protein structure and function comes from the many ways in which 20 amino acids can combine into polypeptide chains. Consider how many tetrapeptide chains can be made using just two amino acids, cysteine and aspartic acid:

Cys–Cys–Cys–Cys	Cys–Cys–Cys–Asp	Cys–Cys–Asp–Cys	Cys–Cys–Asp–Asp
Cys–Asp–Cys–Cys	Cys–Asp–Cys–Asp	Cys–Asp–Asp–Cys	Cys–Asp–Asp–Asp
Asp–Cys–Cys–Cys	Asp–Cys–Cys–Asp	Asp–Cys–Asp–Cys	Asp–Cys–Asp–Asp
Asp–Asp–Cys–Cys	Asp–Asp–Cys–Asp	Asp–Asp–Asp–Cys	Asp–Asp–Asp–Asp

In all, there are $2^4 = 16$ different ways to combine two amino acids in chains of four units. The 20 common amino acids combine to give 20^4 tetrapeptides, which is 1.6×10^5 different molecules. Each time the chain adds an amino acid, the number of possibilities is multiplied by 20, so a set of n amino acids can form 20^n different polypeptides. Because proteins can contain hundreds of amino acids, each of which can be any of the 20 possibilities, the number of possible structures becomes immense. In fact, 20^{100}, the number of possible ways to construct a protein of 100 units, is greater than the estimated number of atoms in the universe.

Secondary Protein Structure

A primary structure represents a polypeptide as a simple linear string of amino acids. Actually, within long polypeptides, certain sections fold into sheets or twist into coils. These regions with specific structural characteristics constitute the **secondary structure** of the protein. Figures 13-37 and 13-38 show the two most common secondary structures.

The secondary structure of a protein is determined by hydrogen bonding between C=O and N—H groups of the peptide linkages that make up the backbone of the protein. Hydrogen bonds can exist within the same protein chain or between different chains. Hydrogen bonding within a single protein chain gives a **helix** (Figure 13-37). Notice that the helix is held together by hydrogen bonds between the O atom of the carbonyl group in one amino acid and the H atom from an amide nitrogen three amino acids away. One complete turn of the helix contains 3.6 amino acids.

Another common form of secondary protein structure is a **pleated sheet.** (Figure 13-38). The chains in a sheet are fully extended rather than coiled, and hydrogen bonds exist between different portions of the protein chains, with one part of the chain bonded to another. In pleated regions, the oxygen and the amide hydrogen protrude at 90° angles

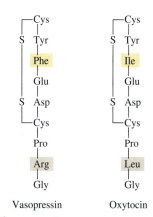

Vasopressin   Oxytocin

Figure 13-36
The primary structures of the pituitary hormones vasopressin and oxytocin differ by just two amino acids, but they have remarkably different properties. (The sulfur cross-links between Cys residues are described later in this section.)

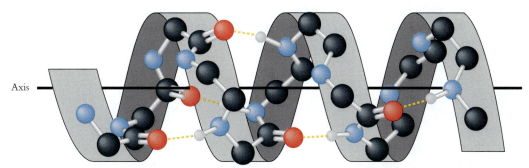

Axis

Figure 13-37
A helical peptide backbone. The side chains are omitted to emphasize the shape of the helix. Notice the hydrogen bonding between N—H and C=O groups.

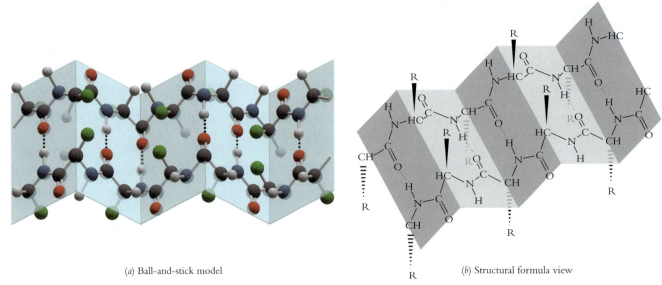

(a) Ball-and-stick model   (b) Structural formula view

Figure 13-38
Two views of a pleated sheet. (*a*) The ball-and-stick view emphasizes the hydrogen bonding between the strands of the sheet. (*b*) The structural formula view emphasizes the folds of the sheet.

to the axis of the extended protein backbone. This allows row on row of polypeptides to form hydrogen bonds and make a sheet of protein. The pleats in the sheet are caused by the bond angles of the peptide linkages.

Although other secondary protein structures play roles in determining the shapes of proteins, the helix and pleated sheet occur most frequently. A discussion of less common secondary structures is beyond the scope of this text.

Tertiary Protein Structure

Figure 13-39
The cord that connects this computer to its keyboard illustrates the primary, secondary, and tertiary levels of protein structure.

Each protein has a unique three-dimensional shape called its **tertiary structure.** The tertiary structure is the result of the bends and folds that a polypeptide chain adopts to achieve the most stable structure for the protein. As an analogy, consider the cord in Figure 13-39 that connects a computer to its keyboard. The cord can be pulled out so that it is long and straight; this corresponds to its primary structure. The cord has a helical region in its center; this is its secondary structure. In addition, the helix may be twisted and folded on top of itself. This three-dimensional character of the cord is its tertiary structure.

The tertiary structure of a protein is determined primarily by the way in which water interacts with the side chains on the amino acids in the polypeptide chain. When the polypeptide folds into a three-dimensional shape that arranges its hydrophobic regions inside the overall structure, polar interactions with water molecules are maximized, making the system most stable. The folding of the protein is further directed and strengthened by a large number of hydrogen bonds. Hydrophilic side chains form hydrogen bonds with other side chains and with the protein backbone. In addition, water forms hydrogen bonds with the protein backbone and with hydrophilic side chains such as those in arginine, aspartic acid, and tyrosine, as Example 13–8 shows.

Example 13-8	Draw a line structure that shows the various ways in which water molecules form hydrogen bonds with a protein backbone.
Hydrogen Bonding in Proteins	**Strategy:** Remember from Chapter 11 that the hydrogen atoms of water molecules form hydrogen bonds with electronegative O and N atoms, whereas the oxygen atom of a water molecule forms hydrogen bonds with hydrogen atoms in highly polar N—H and O—H bonds.

Example 13-8

Hydrogen Bonding
in Proteins (*continued*)

Solution: Begin by drawing a section of protein backbone. Because the problem asks only about hydrogen bonds to the backbone, the side chains are not involved in this problem and may be designated simply as *R*.

Now add water molecules to illustrate the hydrogen-bonding interactions. Each hydrogen bond is shown as a dotted line. Because the problem does not ask that all hydrogen bonds be shown, it is sufficient to show one or two examples of each type. Lone pairs are shown only for atoms involved in hydrogen bonding.

Does the Result Make Sense? The structure involves water molecules acting as hydrogen bond acceptors to N—H and O—H groups or as hydrogen bond acceptors to electronegative oxygen atoms. These are all legitimate hydrogen-bonding interactions, so the result is reasonable.

Amino acid side chains on the outside of a protein can form hydrogen bonds to water molecules. Draw a line structure that shows how water molecules can form hydrogen bonds to the side chains in the protein segment shown below.

The amino acid cysteine plays a unique role in tertiary protein structure. The —SH groups of two cysteine side chains can cross-link through an S—S bond called a *disulfide bridge,* as shown below.

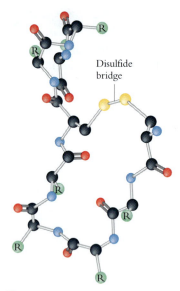

Figure 13-40
The three-dimensional structure of oxytocin. The sulfur atoms of the disulfide bridge are in yellow. The side chains have been simplified as green spheres to emphasize the shape of the polypeptide chain.

The primary structures of oxytocin and vasopressin (Figure 13-36) have two cysteine residues connected by a disulfide bridge. The three-dimensional structure of the oxytocin backbone is shown in Figure 13-40. Notice that the disulfide bridge locks the peptide chain into a compact cyclic structure.

Globular Proteins

A huge and diverse group of molecules called **globular proteins** carries out most of the work done by cells, including synthesis, transport, and energy production. Globular proteins have compact, roughly spherical tertiary structures containing folds and grooves. *Enzymes* are globular proteins that speed up biochemical reactions. The reactions of metabolism proceed too slowly to be of use to living organisms unless enzymes make them go faster. Antibodies, the agents that protect humans from disease, are also globular proteins. Other globular proteins transport smaller molecules through the blood. Hemoglobin, the macromolecule that carries oxygen in the blood, is a transport protein. Globular proteins, including oxytocin and vasopressin, also act as hormones. Others are bound in cell membranes, facilitating passage of nutrients and ions into and out of the cell. Many globular proteins have hydrophilic side chains over the outer surfaces of their tertiary structures, making them soluble in the aqueous environment of the cell.

The secondary structures of globular proteins include helices and pleated sheets in varying proportions. The unique primary structure of each globular protein leads to a unique distribution of secondary structures and to a specific tertiary structure. The stylized ribbon views of globular proteins in Figure 13-41 illustrate the diversity and intricate structures of these molecules. Regions that are helical are shown in blue, regions that are pleated are magenta, and sections that have no specific secondary structure are orange.

Fibrous Proteins

Structural components of cells and tissues are made of proteins that form fibers. These **fibrous proteins** are the cables, girders, bricks, and mortar of organisms.

The helix is the prevalent secondary structure in keratins, which include wool, hair, skin, fingernails, and fur. These molecules are long strands of helical protein that lie with their axes parallel to the axis of the fiber. In hair, individual keratin molecules are wound like the strands of a rope. Because of its compact helical chains, keratin has a certain amount of elasticity. Stretching the fibers stretches and breaks the relatively weak hydrogen bonds of the helix, but as long as none of the stronger covalent bonds is broken, the

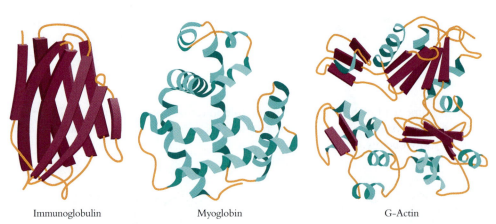

Immunoglobulin   Myoglobin   G–Actin

Figure 13-41
Ribbon views of proteins with varying amounts of helices and pleated sheets. Immunoglobulin, an antibody, is made up almost entirely of pleated sheets (magenta). Myoglobin, which stores oxygen in muscle tissue, is composed of about 70% helix (blue). G-Actin, a component of muscle protein fibers, is a complex mixture of helices and pleated sheets. Regions with no specific secondary structure are shown in orange.

protein relaxes to its original length when released. Hydrogen bonds can easily break and form again, but covalent bonds cannot.

A "permanent wave" changes the shape of a person's hair by changing the tertiary structure of hair protein. A solution that breaks the disulfide bridges between the protein chains is applied to the hair. The hair is then set in curlers to give the desired shape. A second solution that recreates S—S bonds is applied to the hair. When the curlers are removed, the new disulfide bridges hold the hair in its new configuration.

Other fibrous proteins contain extensive regions of pleated sheets. The fibers spun by a silkworm, for example, are made almost entirely of fibroin, a protein composed primarily of just three amino acids: glycine (45%), alanine (30%), and serine (12%). Each chain of fibroin contains extensive regions where a sequence of six amino acids occurs repeatedly: . . . –Gly–Ser–Gly–Ala–Gly–Ala– . . . Notice that every other amino acid is glycine, which is the smallest amino acid. This alternating arrangement is an important feature in the packing of the strands that make up the pleated sheet.

Section Exercises

13.7.1 Write the three-letter shorthand form of this peptide sequence:

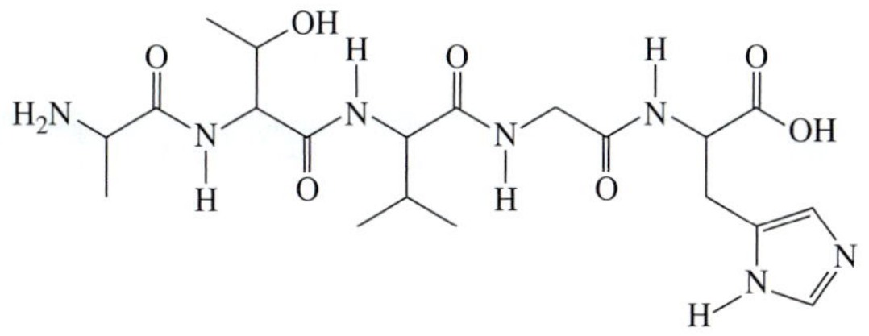

13.7.2 Draw a line structure of the peptide sequence Ser–Gly–Lys–Asp and show how the side chains form hydrogen bonds to water molecules.

13.7.3 Write a few sentences that explain the difference among primary, secondary, and tertiary protein structures. Your description should identify which types of intramolecular and intermolecular forces are responsible for each.

SKILLS TO MASTER

1. Recognizing functional groups
2. Recognizing linkage groups
3. Describing free radical polymerization
4. Describing condensation polymerization
5. Drawing structures of polymers
6. Drawing structures of carbohydrates
7. Drawing DNA and RNA structures
8. Describing protein structures
9. Recognizing hydrophilic and hydrophobic amino acids

13.1 STARTING MATERIALS FOR POLYMERS

Monomers   Macromolecules   Polymers

1 Functional groups

Thiol   Alcohol   Amine

Aldehyde   Ketone   Carboxylic acid

Phosphate   Alkene

2 Linkage groups

Amide

Ether

Ester

CH_3CH_2-O-H HO C CH_3 CH_3CH_2-O C CH_3 + H_2O

Condensation reaction

13.2 FREE RADICAL POLYMERIZATION

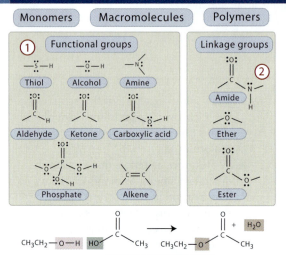

$RO\bullet$ + ⟶ RO • 3 Initiation

RO • + n ⟶ RO [ ]$_n$ • Propagation

RO [ ]$_n$ • + $RO\bullet$ ⟶ RO [ ]$_n$ OR Termination

Free radical polymerization   Polyethylene   Free radicals

Copolymer

Cross-Linking

13.3 CONDENSATION POLYMERIZATION

4

5

Polyamide

Polyester

13.4 TYPES OF POLYMERS

Plastics
Fibers
Elastomers

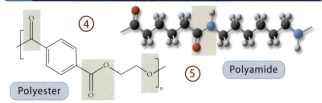

Thermoplastic   Thermosetting   Plasticizer

13.5 CARBOHYDRATES

$C_x(H_2O)_y$

Carbohydrate

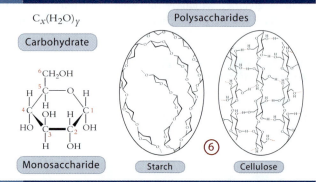

Monosaccharide

Polysaccharides

Starch   Cellulose   6

13.6 NUCLEIC ACIDS

Primary structure

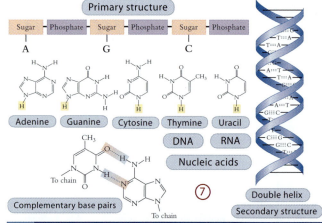

Sugar — Phosphate — Sugar — Phosphate — Sugar — Phosphate

A   G   C

Adenine   Guanine   Cytosine   Thymine   Uracil

DNA   RNA

Nucleic acids

Complementary base pairs

To chain   To chain

7

Double helix

Secondary structure

13.7 PROTEINS

Amino acid

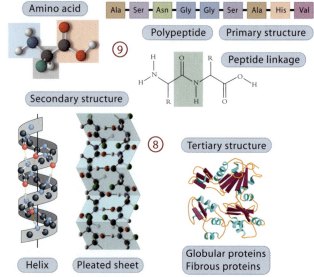

Ala — Ser — Asn — Gly — Gly — Ser — Ala — His — Val

Polypeptide   Primary structure

9   Peptide linkage

Secondary structure

8   Tertiary structure

Helix   Pleated sheet

Globular proteins
Fibrous proteins

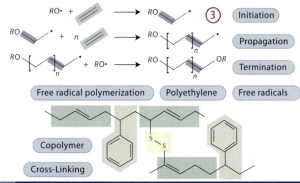

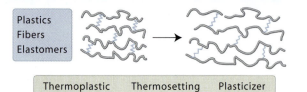

Learning Exercises

13.1 Prepare a list of the types of polymer linkage reactions described in this chapter. List at least one important polymer that forms from each type of reaction.

13.2 Define primary, secondary, and tertiary structure. Give examples of each type of structure for a protein and for DNA.

13.3 Draw molecular pictures that illustrate the linkages in each of the following polymer types: polyethylene, polyester, polyamide, polyether, and silicone.

13.4 Draw pictures that illustrate the essential features of the helix, the pleated structure, and the double helix.

13.5 Write a paragraph that describes the role of hydrogen bonding in the structures of biological macromolecules.

13.6 List all terms new to you that appear in Chapter 13. Write a one-sentence definition of each, using your own words. Consult the index and glossary if you need help.

 Problems   **ilw = interactive learning ware problem. Visit the website at www.wiley.com/college/olmsted**

Starting Materials for Polymers

13.1 Draw structural formulas and circle and identify the functional groups in the following molecules:

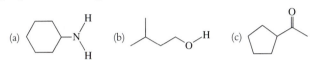

(a)   (b)   (c)

(d)

13.2 Draw structural formulas and circle and identify the functional groups in the following molecules:

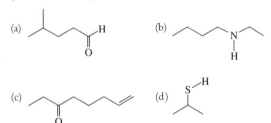

(a)   (b)

(c)   (d)

13.3 Convert the ball-and-stick structures to line drawings and identify the functional groups in the following molecules:

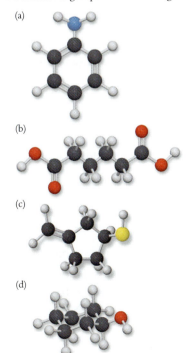

(a)

(b)

(c)

(d)

13.4 Convert the ball-and-stick structures to line drawings and identify the functional groups in the following molecules:

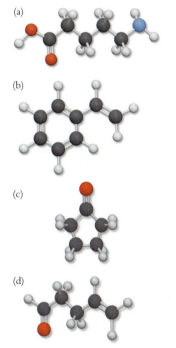

(a)

(b)

(c)

(d)

13.5 Draw structural formulas and circle and identify the linkage groups in the following molecules:

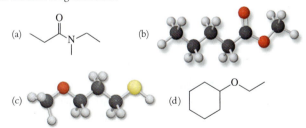

(a)   (b)

(c)   (d)

13.6 Draw structural formulas and circle and identify the linkage groups in the following molecules:

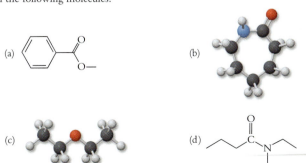

(a)   (b)

(c)   (d)

13.7 Draw one example of (a) an amine with the formula $C_5H_{13}N$; (b) an ester with eight carbon atoms; (c) an aldehyde whose molar mass is at least 80 g/mol; and (d) an ether that contains a phenyl ring (C_6H_5).

13.8 Draw one example of (a) an alcohol with the formula $C_4H_{10}O$; (b) a carboxylic acid with a molar mass of at least 60 g/mol; (c) a thiol with 6 C atoms; and (d) an amide that contains at least 10 carbons, including a phenyl ring (C_6H_5).

13.9 Write balanced equations that show how the following molecules are produced in condensation reactions:

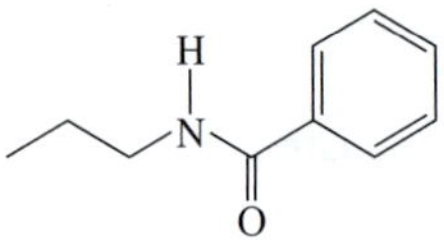

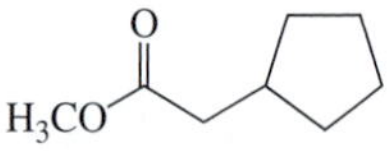

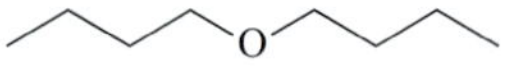

13.10 Write balanced equations that show how the following molecules are produced in condensation reactions:

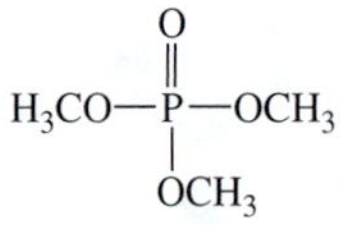

13.11 Draw the structures of all possible products resulting from condensation reactions between aspartic acid and isoleucine:

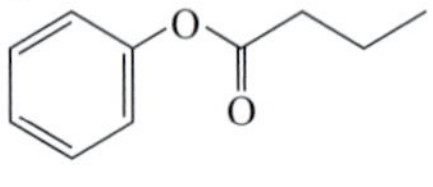

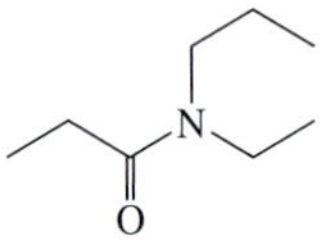

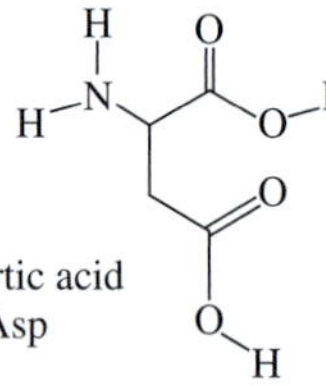

Aspartic acid
Asp

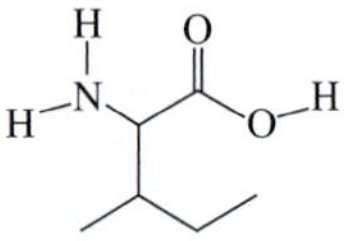

Isoleucine
Ile

13.12 Draw the structures of the condensation products of the following reactions:

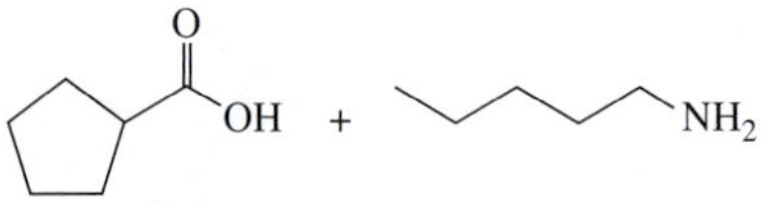

Free Radical Polymerization

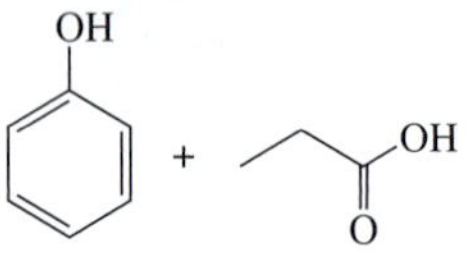

ilw **13.13** Saran is a copolymer made from vinyl chloride and vinylidene chloride $(H_2C=CCl_2)$. Draw the structure of this polymer, showing at least four repeat units in the polymer.

13.14 Draw the structure of polypropylene, showing at least five repeat units.

13.15 Identify the monomers used to make the following polymers:

13.16 Nitrile rubber, whose structure follows, is a copolymer made from two monomers. The polymer is used to make automotive hoses and gaskets. Draw the structures of the monomers and name them.

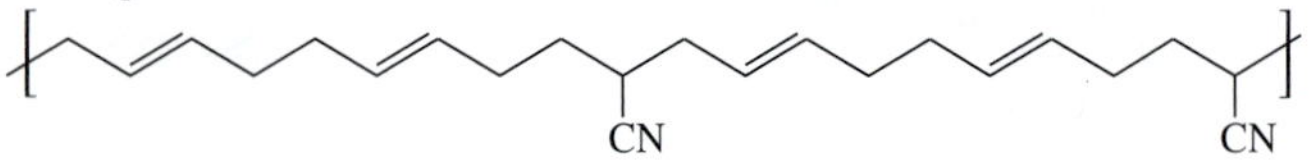

13.17 Draw a section of the polymer chain for polybutadiene and describe how it differs from polyethylene.

13.18 Draw a section of the polymer chain showing at least four repeat units for each of the following polymers: (a) Teflon; (b) PVC; and (c) Styrofoam.

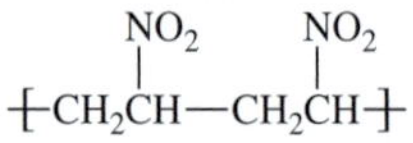

Condensation Polymerization

13.19 The structure of Nylon 11 follows. Draw a line structure of each monomer used to make this polymer.

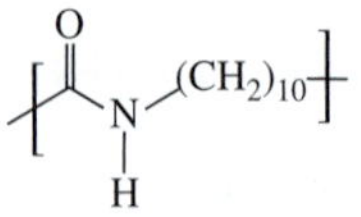

13.20 Polybutylene terephthalate, used to make countertops and sinks, has the following structure. Draw the structural formulas of the monomers from which this polymer is made.

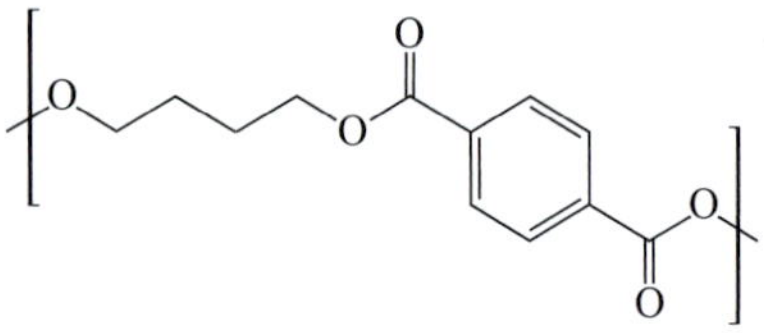

13.21 Draw the structure of the repeat unit of the polymer that forms from ethylene glycol and p-phenylene diamine.

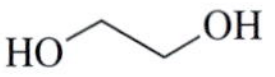

Ethylene glycol

p-Phenylene diamine

13.22 Kodel is a polyester fiber made from terephthalic acid and cyclohexanedimethanol. Draw a segment of Kodel that contains at least four repeat units.

Cyclohexanedimethanol $HOCH_2$—⬡—CH_2OH

Types of Polymers

13.23 Which is more extensively cross-linked, the rubber in an automobile tire or the rubber in a pair of surgical gloves? Explain.

13.24 Plastic wrap can be stretched slightly to fit snugly over a food container. Is plastic wrap a thermoplastic or a thermosetting polymer? Explain.

13.25 Describe what category of polymer to use to make each of the following items: (a) a balloon; (b) rope; and (c) a camera case.

13.26 Describe what category of polymer to use to make each of the following items: (a) a countertop; (b) artificial turf; and (c) a bungee cord.

13.27 Describe the changes in polymer properties that occur on adding dioctylphthalate to the reaction mixture.

13.28 Describe the changes in polymer properties that occur on cross-linking.

Carbohydrates

13.29 The six-carbon sugar talose differs from glucose in the orientations of the hydroxyl groups at the 2 and 4 positions. Draw the structure of α-talose.

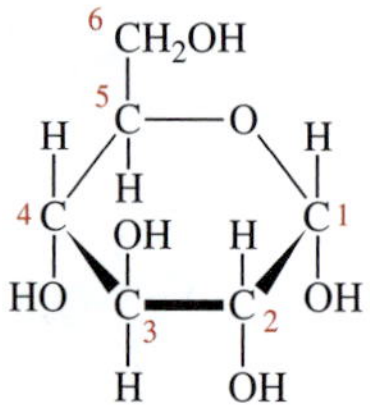

α-Glucose

13.30 The five-carbon sugar ribose can form a six-membered ring, ribopyranose, that differs from glucose in the orientation of the —OH at position 3 and in having —H instead of —CH$_2$OH in position 6. Draw the ring structure of α-ribopyranose.

13.31 Draw the structure of β-talose (refer to Problem 13.29).

13.32 Draw the structure of β-ribopyranose (refer to Problem 13.30).

13.33 Glycogen is a glucose polymer that collects in granules in the liver. Cellulose is also a polymer of glucose, but cellulose forms sheetlike arrangements that are used in the cell walls of plants. What are the distinguishing structural features that allow glycogen to form granules whereas cellulose forms sheets? Illustrate your answer with structural drawings.

13.34 Glycogen and cellulose are both polymers of glucose. Explain why humans can use glycogen but not cellulose as an energy source. Why can cows digest cellulose, but humans cannot?

Nucleic Acids

13.35 Part of a DNA sequence is A-A-T-G-C-A-C-T-G. What is its complementary sequence?

13.36 Part of a DNA sequence is C-G-T-A-G-G-A-A. What is its complementary sequence?

13.37 Draw the complete structure of the following segment of DNA: A-T-C-G.

13.38 Draw the complete structure of the following segment of RNA: G-U-A-C.

13.39 One strand of DNA contains the base sequence A-T-C. What is the base sequence of the complementary strand? Draw a structure of this section of DNA that shows the hydrogen bonding between the base pairs.

13.40 Although RNA is a single-stranded molecule, it can have extensive regions of double-stranded structure resulting from intramolecular hydrogen bonding between guanine and cytosine and between adenine and uracil. Draw the hydrogen-bonding interaction between adenine and uracil.

Proteins

13.41 Draw the structures of the following amino acids: (a) Tyr; (b) Phe; (c) Glu; and (d) Met.

13.42 Draw the structures of the following amino acids: (a) Cys; (b) His; (c) Leu; and (d) Pro.

13.43 Assign each amino acid in Problem 13.41 as possessing hydrophobic or hydrophilic side chains. Explain each assignment.

13.44 Assign each amino acid in Problem 13.42 as possessing hydrophobic or hydrophilic side chains. Explain each assignment.

13.45 Identify the following amino acids and characterize them as possessing hydrophilic or hydrophobic side chains:

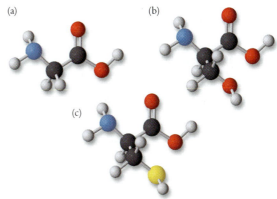

(a) (b) (c)

13.46 Identify the following amino acids and characterize them as possessing hydrophilic or hydrophobic side chains:

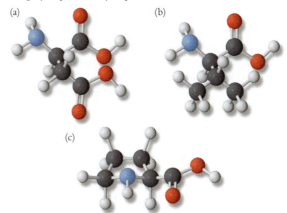

(a) (b) (c)

13.47 Draw the line structures of all possible dipeptides that can form in condensation reactions between alanine, glutamic acid, and methionine.

13.48 Draw the line structures of all possible dipeptides that can form from Cys, Gly, and Asn.

Additional Paired Problems

13.49 Compute the molar mass of a polyethylene molecule that has 744 monomer units.

13.50 Compute the molar mass of a polystyrene molecule that has 452 monomer units.

13.51 There are four different bases in DNA strands. How many different 12-base combinations are there?

13.52 The smallest proteins contain about 50 amino acids. How many different proteins containing 50 amino acids can be formed from the 20 common amino acids? Express your answer in power of 10 notation.

13.53 Hair spray is a solution of a polymer in a volatile solvent. When the solvent evaporates, a thin film of polymer that holds the hair in place is left behind. Many hair sprays contain a copolymer made of the following monomers:

(a) Assume that the polymer is made from a 1:1 mixture of the two monomers. Will the two monomers alternate in the polymer or will the

arrangement be random? Explain. (b) Draw a section of the copolymer containing at least six repeat units. (c) What structural features of the copolymer allow it to hold hair in place?

13.54 Chewing gum is mostly polyvinylacetate. The monomer used to make chewing gum is vinyl acetate:

Vinyl acetate
(C$_4$H$_6$O$_2$)

(a) Draw a portion of the polyvinylacetate polymer. Show at least three repeat units. (b) Is polyvinylacetate a plastic, a fiber, or an elastomer? Use the behavior of chewing gum to justify your choice. (c) Chewing gum can be removed from clothing by cooling the polymer with a piece of ice. At low temperature, the gum crumbles easily and can be removed from the fabric. Explain this procedure at the molecular level.

13.55 One nucleotide and its hydrogen-bonded partner in double-stranded DNA is called a *duplex*. Identify the duplex formed by guanine and draw its structure.

13.56 Identify the duplex (see Problem 13.55) formed by thymine and draw its structure.

13.57 Use the polymerization of acrylonitrile to describe each of the three steps of free radical polymerization. Write structures that illustrate the steps.

13.58 Condensation polymers are usually made from two different monomers, each of which has two of the same functional group. Give three reasons that this strategy is used so often in polymer synthesis.

13.59 Almost 1000 g of α-glucose will dissolve in 1 L of water. Draw the structure of α-glucose and include enough hydrogen-bonded water molecules to account for its tremendous solubility.

13.60 Nylon 6 has the following repeat structure:

Draw two strands of Nylon 6, each with four repeat units. Show the two strands connected by hydrogen bonds.

13.61 What features do nylon and proteins have in common? In what ways are they distinctly different?

13.62 Complete the following table:

	RNA	DNA
Sugar		
One-ring bases		
Two-ring bases		

13.63 Identify the monomers used to make the following polymers: (a) Kevlar; (b) PET; and (c) Styrofoam.

13.64 Identify the monomers used to make the following polymers: (a) Dacron; (b) PVC; and (c) Teflon.

13.65 High-density polyethylene has more CH_2 groups per unit volume than low-density polyethylene. Explain why this is so in terms of the structures of the two forms of the polymer.

13.66 Explain the major differences between fibrous proteins and globular proteins. What role does each type of protein play in biological organisms?

13.67 In the 1950s, Edwin Chargaff of Columbia University studied the composition of DNA from a variety of plants and animals. He found that the relative amounts of different bases changed from one species to another. However, in every species studied, the molar ratios of guanine to cytosine and of adenine to thymine were found to be very close to 1.0. Explain Chargaff's observations in terms of the Watson–Crick model of DNA structure.

13.68 The melting point of DNA, which is the temperature at which the double helix unwinds, increases as the amount of guanine and cytosine increases relative to the amount of adenine and thymine. Explain this observation.

13.69 For a cell to synthesize a particular protein, genetic information must be transmitted from the cell nucleus to the cytoplasm, where protein synthesis takes place. This shuttling of genetic information is accomplished by a type of RNA, messenger RNA (mRNA). The bases of mRNA are complementary to those in DNA. Assuming that mRNA is made from strand A of the DNA that follows, identify bases 1, 2, 3, and 4 in the mRNA.

13.70 Transfer-RNA molecules (tRNA) have a set of three bases at their tip that are exposed and can bind their complementary bases. What sequence of bases will be recognized by the following tRNA sequences: (a) GAU, (b) AGG, and (c) CCU?

13.71 Describe how interactions with solvents affect the tertiary structure of a protein. Include explanations of the roles of regions containing mostly hydrophobic side groups and of regions containing mostly hydrophilic side groups. State where hydrophobic and hydrophilic side groups would be located on a water-soluble globular protein.

13.72 Globular proteins adopt three-dimensional structures that place some of the amino acids on the inside of the molecule, out of contact with the aqueous environment, and others on the outer surface of the molecule. Which of the following amino acids would be most likely to be found on the inside of a globular protein? (a) Arg; (b) Val; (c) Met; (d) Thr; and (e) Asp. Explain your choices.

More Challenging Problems

13.73 The first step of glucose metabolism is an enzyme-catalyzed condensation reaction between phosphoric acid and the CH_2OH hydroxyl on glucose. Draw the structure of this glucose phosphate.

13.74 Proteins are synthesized in the cell by adding one amino acid at a time to the growing polypeptide chain. Each amino acid is carried to the protein in a form in which the amino acid is linked to adenosine monophosphate. The amino acid is joined to AMP by a condensation of its carboxylic acid with the phosphate group. Draw the structure of Ala–AMP.

13.75 Copolymerization of styrene with a small amount of divinylbenzene gives a cross-linked polymer that is hard and insoluble. Draw a picture of this polymer that shows at least two cross-links. (*Hint:* The cross-linking starts when divinylbenzene is incorporated into the growing polystyrene chain.)

Divinylbenzene

13.76 Ethylene oxide forms a polyether by ring opening followed by chain formation.

Ethylene oxide

(a) Draw the Lewis structure of the intermediate that results from ring opening.
(b) Draw the Lewis structure of the adduct that forms between two of these intermediates.
(c) Draw a line structure showing four repeat units of the polymer.
(d) What type of polymerization reaction is this?

13.77 Fungal laccase is an enzyme found in fungi that live on rotting wood. The enzyme is blue and contains 0.40% by mass copper. The molar mass of the enzyme is approximately 64,000 g/mol. How many copper atoms are there in one molecule of fungal laccase?

13.78 One of the problems encountered in the polymerization of monomers that contain two different functional groups is that the molecules tend to cyclize rather than polymerize. Draw the structure of the cyclized product that would be produced from the following monomer:

13.79 In the synthesis of glycogen, an enzyme catalyzes the transfer of a glucose molecule from glucose–UDP to the growing end of a glycogen polysaccharide. Glucose–UDP is a uridine diphosphate molecule linked to the hydroxyl group on the carbon adjacent to the ring oxygen atom of α-glucose. The glucose is at the end of UDP's phosphate chain. Draw the structure of glucose–UDP.

13.80 The artificial sweetener aspartame (NutraSweet) is the methyl ester of the following dipeptide:

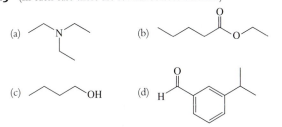

What two amino acids are used to make aspartame?

13.81 The nitrogen atom of a peptide linkage has trigonal planar geometry. What is the hybridization of the nitrogen atom in a peptide linkage? Explain why nitrogen adopts this form of hybridization.

13.82 The development of artificial substances compatible with human tissue is an important area of research. One example is a polymer of lactic acid:

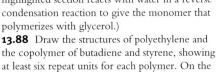

This polymer is used to make body implants needed for only a short time. Eventually the polymer is converted back to lactic acid, which is metabolized to CO_2 and water in the same manner as natural lactic acid. Thus, the body absorbs the polymer without leaving any permanent residue. Draw at least four repeat units of the structure of the polymer made from lactic acid.

13.83 Gentobiose is a disaccharide found bonded to a number of biological molecules. Gentobiose contains two linked β-glucose molecules. One molecule links through the hydroxyl group in position 1, and the other links through the CH_2OH hydroxyl group. Draw the structure of gentobiose.

13.84 Suppose that a polypeptide is constructed with alanine as the only monomer. What is the empirical formula of this polypeptide? If the polypeptide has a molar mass of 1.20×10^3 g/mol, how many repeat units of alanine does it contain?

13.85 Draw the structure of the nucleotide formed from cytosine, ribose, and a phosphate.

Group Study Problems

13.86 Automobiles and major appliances such as refrigerators and washing machines require very tough, long-lasting paints that are baked onto the surface of the object. One group of such paints is known as *alkyds,* which stands for *alc*ohol and ac*id*. These polyesters have extensive cross-linking that characterizes a tough coating material. One of the simplest alkyds is formed from glycerol and phthalic acid. Heating at 130 °C for about an hour maximizes the amount of cross-linking. Draw the structure of the condensation polymer that forms from these two monomers:

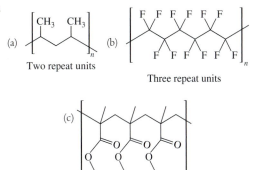

13.87 Glyptal is a highly cross-linked polymer made by heating glycerol and phthalic anhydride. Show the structure of glyptal. (*Hint:* The highlighted section reacts with water in a reverse condensation reaction to give the monomer that polymerizes with glycerol.)

13.88 Draw the structures of polyethylene and the copolymer of butadiene and styrene, showing at least six repeat units for each polymer. On the basis of their molecular structures, explain why polyethylene is more rigid than butadiene–styrene copolymer.

13.89 Design a protein containing ten amino acids whose tertiary structure would be roughly spherical with a hydrophobic interior and a hydrophilic exterior. Include one S—S bridge that would help stabilize the structure.

Answers to Section Exercises

13.1.1 (a) hydroxyl; (b) amine and carboxyl; and (c) ester
13.1.2

13.1.3 (In each case there are several correct answers.)

13.2.2

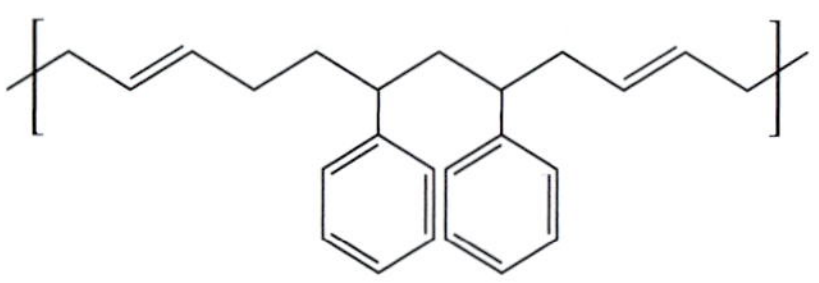

13.2.3

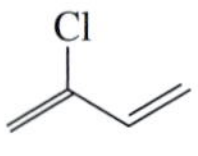

13.3.1

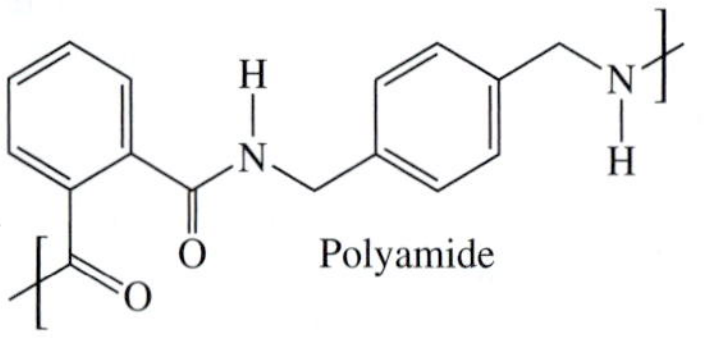

Polyamide

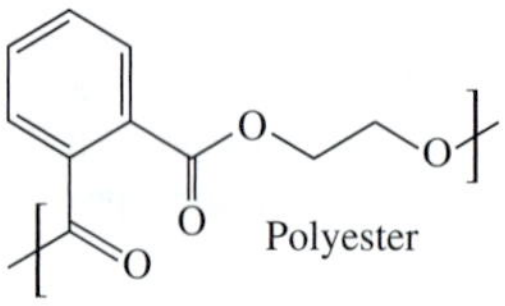

Polyester

13.3.2

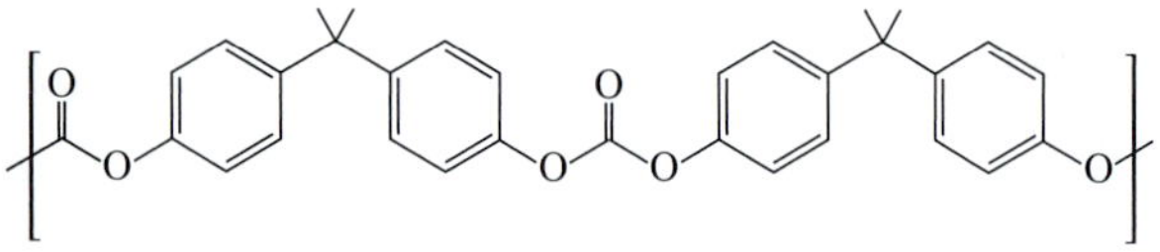

13.3.3

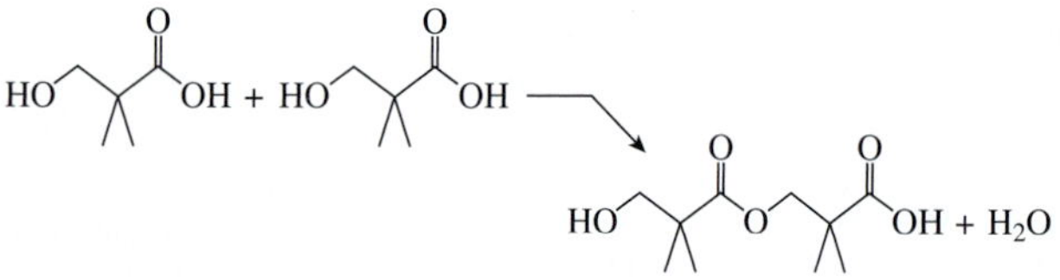

13.4.1 In Kevlar, hydrogen bonding between the amide linkage groups holds the polymer chains in a sheetlike arrangement. This gives the phenyl rings a specific ordered orientation and makes the polymer highly crystalline. On the other hand, polystyrene chains are held together by weaker dispersion forces. The polymer is less ordered and more flexible than Kevlar.

13.4.2 (a) fiber; (b) plastic; and (c) elastomer

13.4.3

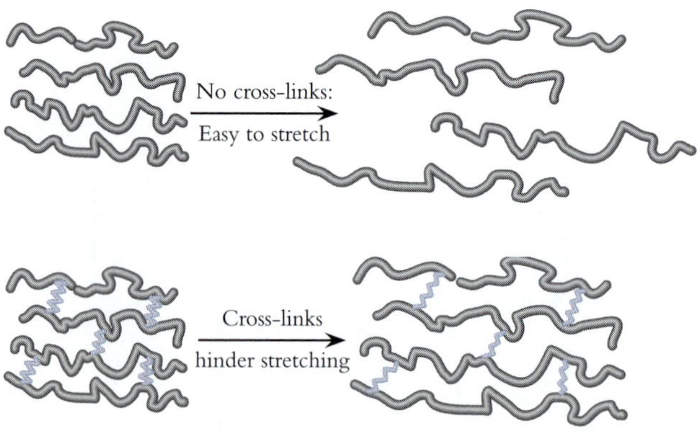

No cross-links: Easy to stretch

Cross-links hinder stretching

13.5.1

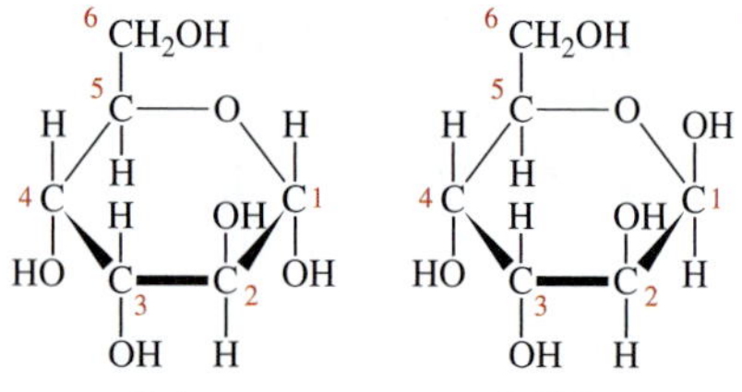

α-Gulose   β-Gulose

13.5.2

β-Galactose

β-Glucose

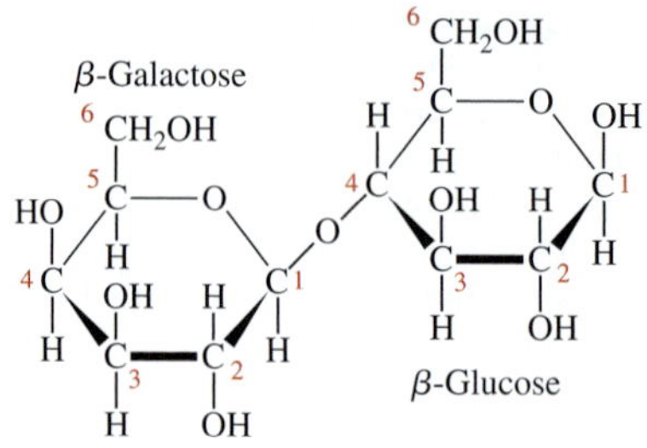

13.5.3 Refer to Figure 13.20. The β linkages in cellulose result in a flat structrue, whereas the α linkages in starch result in a curved structure. Thus, an enzyme shaped to bind starch cannot bind cellulose.

13.6.1

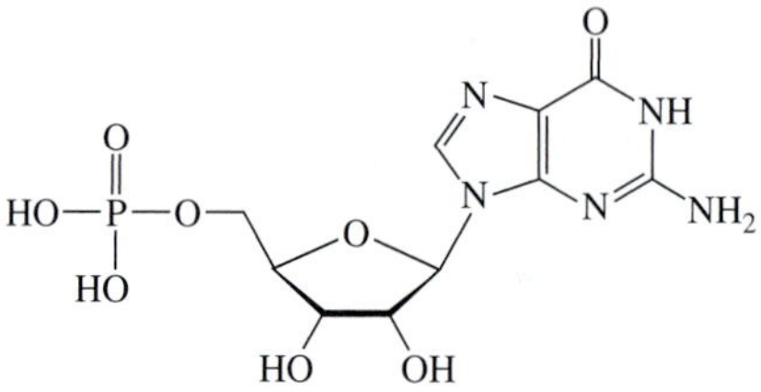

13.6.2 C–G–G–T–A–T–C–C–A

13.6.3

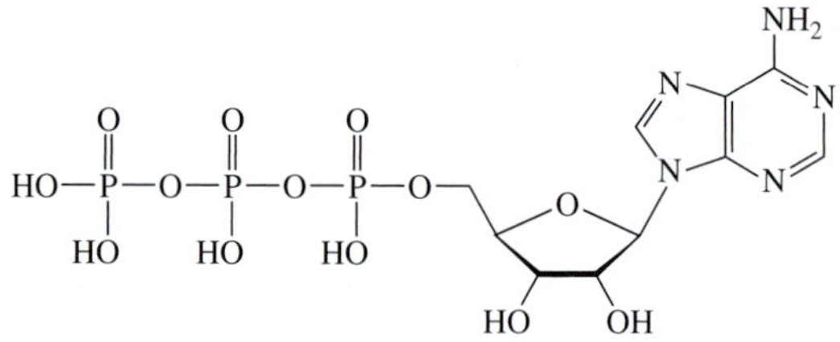

13.7.1 Ala–Thr–Val–Gly–His

13.7.2 (Lone pairs are shown only on hydrogen-bonded atoms.)

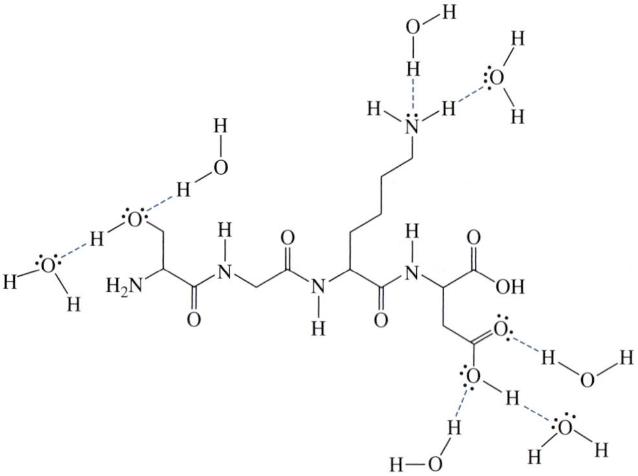

13.7.3 The primary structure of a protein is the linear sequence of amino acids, which are held together by covalent bonds. Secondary structure involves regions of the protein chain that twist into coils or fold into sheets. Hydrogen bonding between the amide residues of the protein backbone accounts for secondary structure. The overall shape of a protein is its tertiary structure. Dispersion forces, hydrogen bonding, and covalent cross-links between cysteine side chains contribute to the tertiary structure.

Answers to Extra Practice Exercises

13.1

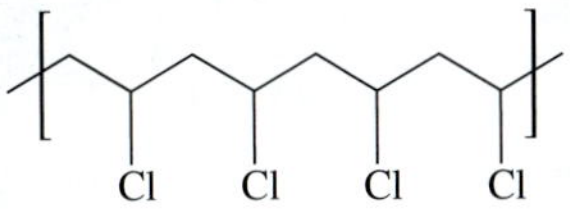

13.2

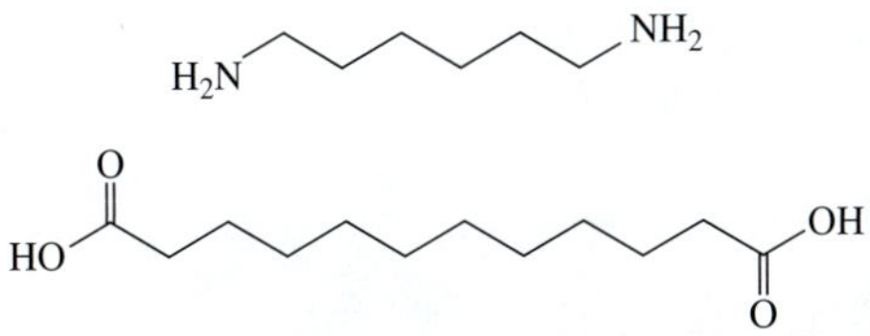

13.3 The structures differ at positions 2, 3, and 4. At each of these, the molecules have the same same types of bonds but in different orientations. The OH group points "down−up−down" in α-glucose and "up−down−up" in α-idose. The two molecules have the same structures at the other positions.

13.4

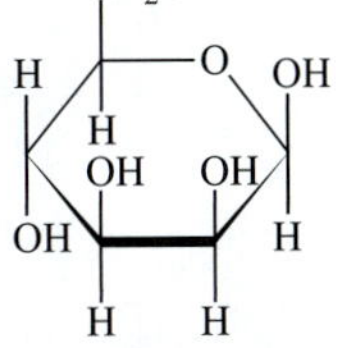

β-Mannose

13.5 Both monosaccharides of melibiose are β-glucose. The structure is shown in Figure 13–15.

13.6

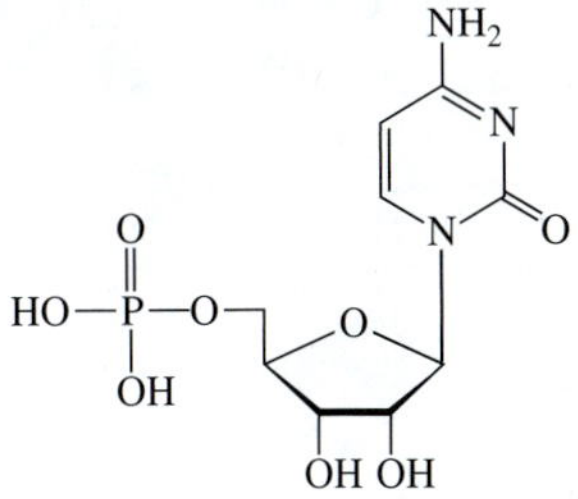

13.7

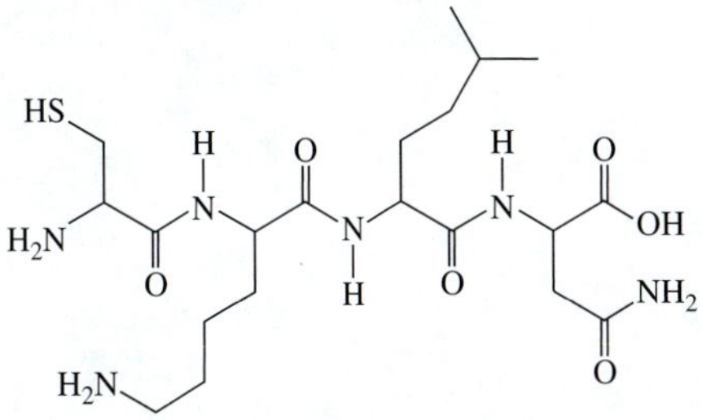

13.8

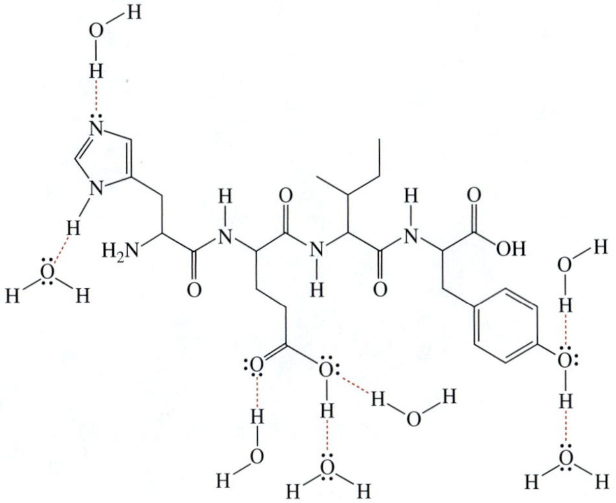

Things Become Disorganized

Perhaps you have noticed that there is a natural tendency for structures to become disorganized. Death is followed by decay. The organized biological structures of the once-living creature decompose. The inscription on a gravestone erodes and becomes illegible. The gravestone tilts and eventually falls over.

Despite this natural tendency, living organisms are highly structured. At the molecular level, DNA is a regular spiral of repeating units carrying specific genetic messages. The order of a DNA sequence is very precise, and the slightest deviation can have disastrous consequences. Macroscopically, animals create regular structures, from the simple hexagonal pattern of a honeycomb to the elaborate network of a highway bridge. Living organisms expend large quantities of energy to produce and maintain organized structures such as these.

Chemists would like to invent molecules that assemble themselves into specific, well-organized arrays. Recently there have been some exciting successes. Our inset is a molecular view of a self-assembled "molecular wreath." This organized structure forms when four molecular chains weave themselves together in the presence of 12 copper cations. Self-assembling molecular systems could lead to new materials with useful properties. They may also shed light on how life originated.

Why do structures tend to disintegrate, and why must living organisms spend much time and effort

Spontaneity of Chemical Processes

CHAPTER CONTENTS

Learning Resources

 KEY CONCEPTS

 CRITICAL THINKING

 SKILLS TO MASTER

 ADDITIONAL HELP
www.wiley.com/college/olmsted
- TUTORIALS
- ANIMATIONS

creating organized structures? The first law of thermodynamics, presented in Chapter 6, requires that energy must be conserved but places no restrictions on what direction a process must take. The explanation for why processes have preferred directions lies in the second law of thermodynamics, which we describe in this chapter. The second law lets us determine whether or not a chemical process can occur. It also lets us determine whether a change of conditions will change the direction of a chemical process. These predictive features make the second law as important as the first law. Chemists need to know in advance if a particular product can form when a set of reactants is mixed, and if a different set of conditions would favor the production of products.

Two quantitative thermodynamic functions, energy and enthalpy, appear in Chapter 6. In the first section of this chapter, we show that we cannot predict the directions of processes using only energy and enthalpy changes. To extend thermodynamics to cover directionality, we need new thermodynamic functions. Sections 14.2 and 14.3 introduce and describe one of these functions, entropy. Section 14.4 describes a combination of energy and entropy, free energy, which is useful for treating chemical systems at constant temperature and pressure. We complete this chapter with two sections that address useful applications of thermodynamics.

14.1   SPONTANEITY

Every process has a preferred direction, which is referred to in thermodynamics as the **spontaneous** direction. Left to itself, a process follows its spontaneous direction. For example, the spontaneous direction for water movement is downhill, from higher altitude to lower altitude. A spontaneous process can be reversed only by the action of some outside force. Water runs uphill only if an external agent, such as a pump, forces it to do so.

A process may be spontaneous, and yet the process may not occur. Extending our water example, water can be stored behind a dam for a very long time unless a spillway is opened. A chemical example is the reaction of methane and oxygen to form carbon dioxide and water:

$$CH_4(g) + 2\,O_2(g) \longrightarrow CO_2(g) + 2\,H_2O(l) \qquad \text{Spontaneous direction}$$

Methane–oxygen mixtures can be stored indefinitely, but a spark will cause the mixture to burst into flames.

A process that does not appear to occur may be spontaneous but very slow, or it may be nonspontaneous. Water does not run uphill, even when a spillway is present, without the action of a pump, because this is the nonspontaneous direction. Water and carbon dioxide do not rearrange into methane and molecular oxygen, even in the presence of a spark, because this is the nonspontaneous direction:

$$CO_2(g) + 2\,H_2O(l) \longrightarrow CH_4(g) + 2\,O_2(g) \qquad \text{Nonspontaneous direction}$$

Notice that the word *spontaneous* has a different meaning in thermodynamics than it does in everyday speech. Ordinarily, spontaneous refers to an event that takes place without any effort or premeditation. For example, a crowd cheers spontaneously for an outstanding performance. In thermodynamics, spontaneous refers only to the natural direction of a process, without regard to whether it occurs rapidly and easily. Chemical kinetics, which we introduce in Chapter 15, describes the factors that determine the speeds of chemical reactions. Thermodynamic spontaneity refers to the direction that a process will take if left alone and given enough time.

Dispersal of Matter

To gain some insight into why processes have preferred directions, consider some large-scale events that go in one particular direction:

1. A bag of marbles dropped on the floor will scatter, but the marbles will not roll spontaneously back into the bag.

2. An untended wooden fence eventually falls apart, but piles of wood will not assemble spontaneously into fences.

3. A completed jigsaw puzzle can be disassembled by a sweep of the hand, but shaking a box containing jigsaw puzzle pieces never generates an assembled puzzle.

Each of these processes has a spontaneous direction, in which matter becomes more spread out or dispersed. Another way to describe these processes states that each reduces the *constraints* on the objects. The marbles in a bag, the boards of a fence, and the interlocking pieces of a jigsaw puzzle are restricted, or constrained, in their positions. We can summarize these observations in a common-sense law:

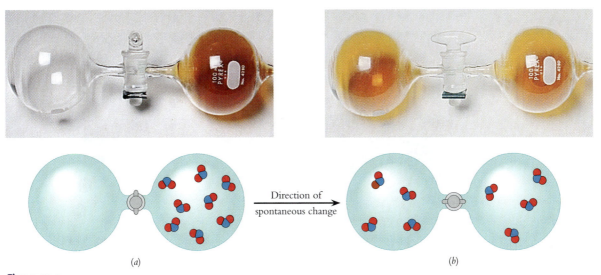

Figure 14-1
When the valve is opened between two bulbs, one of which is empty and the other filled with a gas, gas molecules move spontaneously from the filled bulb until the gas molecules are dispersed equally between the bulbs. The reverse process never occurs spontaneously.

Things tend to become dispersed.

 KEY CONCEPT

Does this common-sense law also apply to events at the molecular level? Consider the two glass bulbs shown in Figure 14-1. One bulb contains nitrogen dioxide, a red-brown gas, and the second bulb is empty. When the valve that connects the bulbs is opened, the red-brown gas expands to fill both bulbs. The opposite process never occurs spontaneously. That is, if both bulbs contain NO_2 at the same pressure, opening the valve never causes the pressure to rise in one bulb and fall in the other.

The second part of Figure 14-1 shows a molecular view of what happens in the two bulbs. Recall from Chapter 5 that the molecules of a gas are in continual motion. The NO_2 molecules in the filled bulb are always moving, undergoing countless collisions with one another and with the walls of their container. When the valve between the two bulbs is opened, some molecules move into the empty bulb, and eventually the concentration of molecules in each bulb is the same. At this point, the gas molecules are in a state of dynamic equilibrium. Molecules still move back and forth between the two bulbs, but the concentration of molecules in each bulb remains the same.

The gas molecules escaping from a bulb behave similarly to marbles rolling out of a bag. Molecules, like marbles, tend to become more dispersed, filling a larger volume. The opposite process, in which gas molecules become more constrained by all moving into one bulb, never occurs spontaneously. Thus, our common-sense law applies to this molecular example.

For a second example, consider Figure 14-2, which shows sugar dissolving in coffee from a molecular perspective. Before the solid sugar dissolves, the sugar

Figure 14-2
Sugar dissolving in coffee involves dispersal of molecules. Sugar molecules are distributed more widely in solution than they are in the solid crystal.

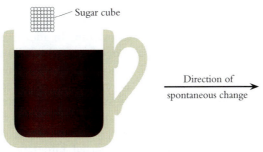

Mug of coffee

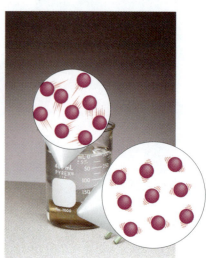

Figure 14-3
When a hot object contacts a cold object, heat always flows from higher to lower temperature.

molecules are organized in a crystal. As the molecules dissolve, they become dispersed uniformly throughout the liquid coffee. The opposite process, in which the molecules become more constrained by forming sugar cubes from sweet coffee, does not occur. Once again, the dir-ection of spontaneous change at the molecular level conforms to the common-sense law. Sugar molecules are more dispersed when they are dissolved in coffee than when they are part of a solid crystal. A sugar cube is the molecular equivalent of an assembled jig-saw puzzle. Sugar molecules disperse, leaving the compact molecular organization of a sugar cube as the cube dissolves in coffee.

Energy Dispersal

Processes that do not involve obvious dispersal of matter may nevertheless have preferred directions. For example, Figure 14–3 shows that when a hot block of metal is placed in a cold glass of water, the metal block cools and the water warms. This process continues until the two are at the same temperature. Whenever two objects at different temperatures contact each other, the object at higher temperature transfers energy to the object at lower temperature.

The molecular views shown in Figure 14-3 show that this is an example of energy dispersal. Recall from Chapter 6 that higher temperature means larger average thermal energy. The metal atoms in the hot block have higher thermal energies than the water molecules in the glass. When the two objects contact each other, the hotter atoms transfer energy to the colder molecules until the average thermal energy of each object is the same.

Figure 14-4 shows another example of energy dispersal. When a baseball pitcher throws a fastball, all the molecules of the ball move together through space. This represents an energy constraint. In addition to their random thermal energies of motion, all the molecules making up the ball have a portion of their kinetic energies that point in the same direction. As described in Chapter 6, when the ball comes to rest in the catcher's glove, the conversion of kinetic energy to thermal energy causes an increase in the temperature of the glove. The molecules in the glove move with larger average speeds. Moreover, these increased molecular speeds point randomly in all directions. In other words, the constrained, directional character of the kinetic energy of the ball has been dispersed into randomly directed energy of molecules in the glove.

Figure 14-4
As a thrown baseball is caught, its kinetic energy is dispersed from the organized kinetic energy of the ball to randomly oriented kinetic energies of many individual molecules in the catcher's glove.

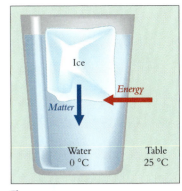

Figure 14-5
Schematic view of the spontaneous process for a water-and-ice mixture on a table. The energy-absorbing process, melting, is spontaneous under these conditions.

Energy and Spontaneity

Spontaneous processes result in the dispersal of matter and energy. In many cases, however, the spontaneous direction of a process may not be obvious. Can we use energy changes to predict spontaneity? To answer that question, consider two everyday events, the melting of ice at room temperature and the formation of ice in a freezer.

As diagrammed in Figure 14-5, ice melts spontaneously if a mixture of water and ice is placed on a table at 25 °C:

$$H_2O \ (s, 0\ °C) \xrightarrow{\text{Table top, 25 °C}} H_2O \ (l, 0\ °C)$$

Recall from Section 11.6 that a phase change from solid to liquid requires energy to overcome the intermolecular forces holding molecules in place in the solid. In this spontaneous process, then, the ice−water system gains energy as ice melts. The enthalpy change of the system also is positive.

Now consider what happens if the same mixture is placed in a freezer at −15 °C. Water freezes spontaneously, as shown schematically in Figure 14-6:

$$H_2O \ (l, 0\ °C) \xrightarrow{\text{Freezer, -15 °C}} H_2O \ (s, 0\ °C)$$

This process is the reverse of melting. To form a solid from a liquid, we must remove enough energy to allow the molecules to stick together in the solid phase. In this spontaneous process, the ice−water system releases energy as water solidifies, and the enthalpy of the system also decreases.

Both these everyday processes are spontaneous, but whereas one process is endothermic, the other is exothermic. The energy and enthalpy of the system increase in one process, but these quantities decrease in the other process. This simple example demonstrates that analyzing energy changes and enthalpy changes is not enough to predict whether a process will occur spontaneously. We need a property other than energy and enthalpy if we hope to use thermodynamics to determine when a process will be spontaneous.

Opposing Dispersal Trends

Liquid water contains molecules that are free to move about through their entire volume, whereas the water molecules in ice are held in a highly structured three-dimensional lattice. Thus, when water freezes, the spontaneous conversion leads to more constraint within the system. This seems like a violation of the common-sense law, but to complete our analysis we need to consider the energy change that accompanies freezing. As Figure 14-7 illustrates, water releases energy as it freezes. Energy is dispersed in the freezer as a result of the energy transfer. Freezing constrains matter but disperses energy.

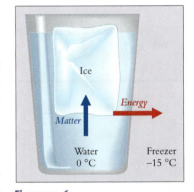

Figure 14-6
Schematic view of the spontaneous process for a water-and-ice mixture in a freezer. The energy-releasing process, freezing, is spontaneous under these conditions.

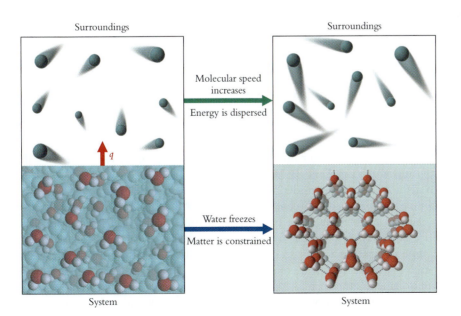

Figure 14-7
When water freezes, its molecules becomes more constrained, but the energy released in this process becomes more dispersed in the surroundings.

The melting of ice is the reverse of the freezing of water. Energy becomes more constrained as it is transferred from the air in the room to the melting ice. At the same time, the molecules in the ice cube become less constrained, because they are free to move about in the liquid phase. Melting disperses matter but constrains energy.

If the common-sense law of dispersal is correct, then the freezing and melting processes both must have a net increase in dispersion. In freezing, the energy dispersal caused by the heat flow must be larger than the molecular constraints created by the fixed positions of the ice molecules. In melting, the dispersal of matter that results from releasing molecules from their fixed positions must be larger than the energy constraints created by absorbing energy from the surroundings. The spontaneity of any process must be evaluated in terms of the dispersal of both energy and matter. In Section 14.2, we define a new thermodynamic function that allows us to do such evaluations.

Section Exercises

■ **14.1.1** Explain the following observations in terms of dispersal and constraints: (a) untended fences eventually fall down; (b) a plate is easy to break but difficult to mend; and (c) a wine cooler beverage does not spontaneously separate into alcohol, water, and fruit juice.

■ **14.1.2** Solid sugar can be recovered from coffee by boiling off the water. (Coffee candy can be made in this way.) Sugar molecules become more constrained in this process. What else must occur for total dispersal to increase?

■ **14.1.3** Draw molecular pictures of liquid water and water vapor that show what happens to the constraints among H_2O molecules when water evaporates.

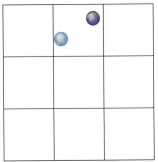

Two marbles confined
to one compartment
$W = 9 \times 1 = 9$

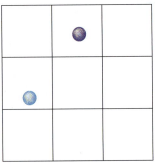

Two marbles occupying
different compartments
$W = 9 \times 8 = 72$

Figure 14-8
Two marbles distributed among nine compartments provides an example of *W* and its relationship to the dispersal of matter.

14.2 ENTROPY: THE MEASURE OF DISPERSAL

The spontaneous direction of any process is toward greater dispersal of matter plus energy. If we are to apply this criterion in a quantitative way, we need ways to measure amounts of dispersal. Scientists analyze the constraints on a system to measure the dispersal of matter. The more the system is constrained, the less dispersed it is. Scientists do calculations on the flow of heat to measure changes in the dispersal of energy.

Entropy and Dispersal of Matter

The function that provides a quantitative measure of dispersal is called **entropy** and is symbolized S. In 1877, the Austrian physicist Ludwig Boltzmann derived Equation 14-1, which defines the entropy of a substance in terms of W, the number of ways of describing the system.

$$S = k \ln W \qquad (14\text{-}1)$$

The details of the derivation are complicated, but the essence of this equation is that the more possible descriptions the system has, the greater is its entropy. The equation states that entropy increases in proportion to the natural logarithm of W, the proportionality being given by the Boltzmann constant, $k = 1.3806 \times 10^{-23}$ J/K. Equation 14-1 also establishes a starting point for entropy. If there is only one way to describe the system, it is fully constrained and $W = 1$. Because ln (1) = 0, $S = 0$ when $W = 1$.

To get a feel for W, consider placing two marbles of different colors in a box containing nine compartments of equal size. Figure 14-8 shows two different ways to distribute the two marbles. We could place both marbles in any of the nine compartments. There are nine ways to place the first marble, but then we must place the second marble in the same compartment, so this distribution has $W = 9$. Another type of distribution places each marble in a *different* compartment. After placing the first marble in one of the nine compartments, the second can go into any of the other eight. The nine possible placements of the first marble multiplied by the eight possibilities for the second gives a total of 72 possible arrangements, so $W = 72$. The fact that W is larger for the second distribution than for the first means that the second distribution has a larger value of S.

How does this relate to dispersal and constraint? When both marbles are confined to the same box, the marbles are more constrained (less dispersed) than when they occupy two different boxes.

A second example using marbles lets us tie this way of measuring dispersal to molecular structures. Consider placing nine identical marbles in the nine compartments, as shown in Figure 14-9. There is only one distribution that places each marble in a different compartment, so this arrangement has $W = 1$ and $S = 0$. Now consider removing one marble from a compartment and adding it to another one. We can remove the marble from any of the nine compartments and add it to any of the other eight, giving $W = 72$. There are fewer constraints on the second distribution, which has higher entropy. Compare this with ice melting to form water. In an ice crystal, all the molecules are constrained to fixed positions, analogous to one marble per compartment. In liquid water, molecules are free to move around, analogous to moving marbles between compartments.

Although the Boltzmann equation may appear simple, applying it to a molecular system always is challenging. The reason is that there are immense numbers of molecules in any realistic molecular system, so it is necessary to count huge numbers of possibilities to determine the value of W. Instead, scientists have found ways to measure entropy by analyzing energy dispersal.

Entropy and Dispersal of Energy

Any energy transfer results in either dispersal or constraint of energy, and thus generates a *change* in entropy (ΔS). When a flow of heat occurs at constant temperature, Equation 14-2 provides a quantitative measure of the entropy change:

$$\Delta S = \frac{q_T}{T} \qquad (14\text{-}2)$$

To put this in words, when heat flows at constant temperature, the entropy change is equal to the heat transferred (q_T) divided by the temperature in kelvins (T). The units of ΔS are energy/temperature, or J/K. The subscript T in Equation 14-2 is a reminder that the quantity of heat transferred depends on the conditions. This equation is restricted to processes that occur at constant temperature.

The presence of q in this equation is consistent with the earlier observation that heat flows change the dispersal of energy. The amount of change depends on the magnitude of the heat flow, but it depends on temperature, too. Temperature tracks the thermal energy of motion of molecules in an object. At low temperature, the average molecular energy of motion is small, so the dispersal of a given amount of heat has a greater impact at low temperature than dispersing the same amount of heat at higher temperature. Dividing by T takes this effect into account.

Heat flow at constant temperature may seem paradoxical, because we usually associate a flow of heat with a change in temperature. Nevertheless, three important types of chemical processes occur in which heat flows take place without a temperature change:

1. Phase changes take place at constant temperature, absorbing or releasing heat in the process.

2. A chemical reaction may occur under conditions in which temperature is held constant, such as in the human body or a thermostatted automobile engine. The exothermicity or endothermicity of the reaction generates a heat flow, even though the temperature is constant.

3. The surroundings may be so large that they can absorb or release significant amounts of heat before the temperature changes by a measurable amount. An experiment performed in a constant-temperature water bath is a common example of this category.

Example 14-1 applies the entropy equation to ice forming in a freezer, the process diagrammed in Figure 14-6.

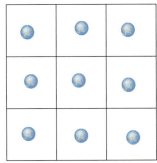

Ice-like distribution of marbles
$W = 1$

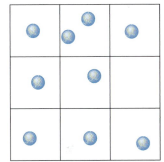

Water-like distribution of marbles
$W = 9 \times 8 = 72$

Figure 14-9
Nine identical marbles distributed among nine compartments have distributions analogous to the distribution of water molecules in ice and water.

Equation 14-2 can be generalized to cover processes that are not at constant temperature, but this requires the use of calculus and is beyond the scope of this text.

Example 14-1

Entropy Change During Freezing

What is the total entropy change when 2.00 mol of water ($\Delta H_{fus} = 6.01$ kJ/mol) freezes at 0.0 °C in a freezer compartment whose temperature is held at −15 °C?

Strategy: The seven-step problem-solving approach served us well in doing the thermodynamics examples of Chapter 6 and is equally valuable for quantitative entropy calculations.

The problem asks for the *total* entropy change, which includes ΔS for the water and ΔS for the freezer compartment. When water freezes, heat flows from the water to its surroundings, the freezer compartment (see Figure 14-6). Thus, q is negative for the water, whose entropy decreases. At the same time, q is positive for the freezer compartment, whose entropy increases.

Solution: We need a way of determining q, and we must identify T for the water and for the freezer compartment. The water/ice mixture remains at a constant temperature of 0.0 °C (273.15 K) while the phase change occurs. (Eventually, the ice will cool to −15 °C, but the question asks only about the freezing process.) The freezer is at a constant temperature of −15 °C = 258 K. The problem states that $n = 2.00$ mol and $\Delta H_{fus} = 6.01 \times 10^3$ J/mol. Remember that ΔH for a phase change is the amount of heat required for the phase conversion of one mole of the substance. This gives us an equation to calculate q for the phase change:

$$q = n \, \Delta H_{fus}$$

We calculate the entropy changes for these constant-temperature processes using Equation 14-2.

$$\Delta S = \frac{q_T}{T}$$

Remember that ΔH_{fus}, which is positive, refers to ice melting. Remember also that the temperature must be in kelvins. The ice–water mixture releases heat as the water freezes, so q for the water has a negative value:

$$q_{H_2O} = -(2.00 \text{ mol})(6.01 \times 10^3 \text{ J/mol}) = -1.202 \times 10^4 \text{ J}$$

$$\Delta S_{H_2O} = \frac{q_{H_2O}}{T_{H_2O}} = \frac{-1.202 \times 10^4 \text{ J}}{273.15 \text{ K}} = -44.0 \text{ J/K}$$

The freezer compartment absorbs the heat released by the water, but it does so at *its* temperature, which is −15 °C + 273 = 258 K. Because the freezer absorbs heat, $q_{freezer}$ has a positive sign. Thus, for the freezer:

$$q_{freezer} = -q_{H_2O} = +1.202 \times 10^4 \text{ J}$$

$$\Delta S_{freezer} = \frac{q_{freezer}}{T_{freezer}} = \frac{1.202 \times 10^4 \text{ J}}{258 \text{ K}} = 46.6 \text{ J/K}$$

The total entropy change is the sum of these changes:

$$\Delta S_{total} = \Delta S_{H_2O} + \Delta S_{freezer} = (-44.0 \text{ J/K}) + (46.6 \text{ J/K}) = +2.6 \text{ J/K}$$

Do the Results Make Sense? The negative sign for the entropy change of the ice/water mixture is consistent with our qualitative view that matter is more dispersed in a liquid than in a solid. The positive sign for the entropy change of the freezer is consistent with heat being absorbed by the freezer, which increases the dispersal of energy.

Extra Practice Exercise 14.1

The enthalpy of vaporization of water is 40.79 kJ/mol, and $T_b = 100$ °C. Calculate the entropy changes for water and for the hot plate, and the overall entropy change when a teakettle on a hot plate at 250 °C boils off 3.55 mol of water.

Applying Equation 14–2 to both the ice–water mixture and the freezer, we find that the total entropy change is positive for the spontaneous freezing of water in a freezer. We can analyze the total entropy change for 2.00 mol of ice melting in a room at 25 °C (see Figure 14–5) in the same way. The heat required to melt the ice has the same magnitude but the opposite sign as the heat released in freezing the water in Example 14–1: 1.202×10^4 J/mol. Ice melts at the same temperature as water freezes, 273.15 K, so the entropy change for the water is also equal in magnitude but is opposite in sign: $\Delta S = 44.0$ J/K. To calculate the entropy change of the room, we must use its temperature, 25 °C + 273 = 298 K. The heat required to melt the ice comes from the air in the room, but the amount of heat is so small and the volume of the room is so large that the temperature remains essentially constant:

$$\Delta S_{room} = \frac{-1.202 \times 10^4 \, J}{298 \, K} = -40.3 \, J/K$$

$$\Delta S_{total} = (44.0 \, J/K) + (-40.3 \, J/K) = +3.7 \, J/K$$

We see that the total change in entropy is a positive quantity for both these spontaneous processes, even though one process is exothermic and the other is endothermic. When this type of calculation is carried out for other processes, the same result is always obtained. For any spontaneous process, the total change of entropy is a positive quantity. Thus, this new state function of entropy provides a thermodynamic criterion for spontaneity, which is summarized in the **second law of thermodynamics:**

KEY CONCEPT

In any spontaneous process, the total entropy increases: $\Delta S_{total} > 0.$

Figure 14–10 summarizes the ice and water example and illustrates potential pitfalls in evaluating entropy changes. First, be careful about the sign of q. Heat flows between a system (the ice–water mixture in this example) and its immediate surroundings (the freezer or air in the room). The sign for the system depends on whether the process is endothermic or exothermic, and the sign for the surroundings always is the opposite of the sign for the system. Table 14-1 summarizes the signs. Second, always use the absolute temperature of the component whose entropy change is being evaluated. In our examples, the freezer, the room, and the water–ice mixture all have different temperatures. Third, find the *total* entropy change by summing the entropy changes of the system and its surroundings, as described by Equation 14-3:

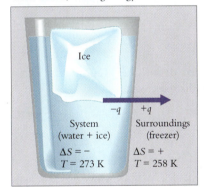

Water freezes, releasing energy to the freezer

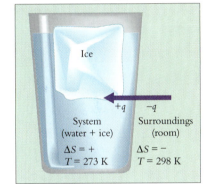

Ice melts, absorbing energy from the room

In both cases, $\Delta S_{total} = \Delta S_{system} + \Delta S_{surroundings} > 0$

Figure 14-10
When calculating entropy changes, be careful about the sign of q, use the appropriate temperatures, and sum the changes for system and surroundings.

$$\Delta S_{total} = \Delta S_{system} + \Delta S_{surroundings} \qquad (14\text{-}3)$$

Example 14-2 does another entropy calculation.

Table 14-1   Signs Associated with Heat Transferred (q)

Process	System	Surroundings	Example
Exothermic	−	+	Water freezing
Endothermic	+	−	Ice melting

Example 14-2

**Entropy Change
of a Refrigerant**

In a refrigerator, a liquid refrigerant absorbs heat from the contents of the refrigerator. This heat vaporizes the refrigerant. Later, a mechanical pump compresses the vapor and recondenses it to a liquid. One common refrigerant is HFC-134a, CH_2FCF_3 ($MM = 102.0$ g/mol). HFC-134a boils at -27 °C with $\Delta H_{vap} = 22.0$ kJ/mol. Calculate the total entropy change when 150. g of HFC-134a vaporizes at its boiling point, exchanging heat with contents of the refrigerator at 4 °C.

Strategy: The problem describes a relatively complicated process, so again we apply the seven-step problem-solving method. The problem asks for the total entropy change (ΔS_{total}) caused by the refrigeration process. We must determine the entropy change of the liquid refrigerant, HFC-134a, and the entropy change of the contents of the refrigerator.

A block diagram helps us to visualize the thermodynamic processes. The liquid refrigerant evaporates at constant temperature as it absorbs heat from the contents of the refrigerator, which are at a different constant temperature.

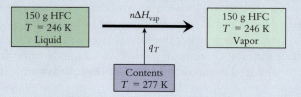

Solution: Here are the data:
For CH_2FCF_3, $m = 150.$ g, $MM = 102.0$ g/mol, $T_b = -27$ °C $+ 273 = 246$ K, $\Delta H_{vap} = 22.0$ kJ/mol.
For the contents of the refrigerator, $T_{refrig} = 4$ °C $+ 273 = 277$ K.

Equation 14-3 gives the total entropy change:

$$\Delta S_{total} = \Delta S_{system} + \Delta S_{surroundings}$$

We can define the HFC as the system and the contents of the refrigerator as the immediate surroundings:

$$\Delta S_{system} = \Delta S_{HFC}$$

$$\Delta S_{surroundings} = \Delta S_{contents}$$

$$\Delta S_{total} = \Delta S_{HFC} + \Delta S_{contents}$$

Use the data to find q for the HFC, then use Equation 14-2 to calculate ΔS for the HFC and for the contents:

$$q = n\,\Delta H_{vap} \qquad and \qquad \Delta S = \frac{q_T}{T}$$

To evaluate q, we need the number of moles of HFC-134a:

$$n_{HFC} = \frac{150.\ g}{102.0\ g/mol} = 1.471\ mol$$

$$q_{HFC} = (1.471\ mol)(22.0\ kJ/mol)(10^3\ J/kJ) = 3.235 \times 10^4\ J$$

$$\Delta S_{HFC} = \frac{q_{HFC}}{T_{HFC}} = \frac{3.235 \times 10^4\ J}{246\ K} = 132\ J/K$$

This is the entropy change of the refrigerant fluid. The contents of the refrigerator give up heat equal to the heat absorbed by the fluid:

$$q_{contents} = -q_{HFC}$$

$$\Delta S_{contents} = \frac{-q_{HFC}}{T_{contents}} = \frac{-3.235 \times 10^4\ J}{277\ K} = -117\ J/K$$

Sum, being careful about signs, to get the total entropy change:

$$\Delta S_{\text{total}} = \Delta S_{\text{HFC}} + \Delta S_{\text{contents}} = 132 + (-117) = 15 \text{ J/K}$$

$\underset{\text{System}}{\uparrow} \qquad \underset{\text{Surroundings}}{\uparrow}$

Example 14-2

Entropy Change of a Refrigerant (*continued*)

Do the Results Make Sense? The refrigerant vaporizes, dispersing its matter, so a positive ΔS is reasonable. The contents release energy to the refrigerant, constraining their energy, so a negative ΔS is reasonable. The total entropy change is positive, as it must be for any spontaneous process.

Calculate the total entropy change when 175 g of water at 0 °C freezes, transferring heat that causes HFC-134a to boil at −27 °C.

Extra Practice Exercise 14.2

Direction of Heat Flow

Equation 14-2 states that an entropy change at constant temperature is heat flow divided by temperature, and according to the second law of thermodynamics, the total entropy change is always positive. Taken together, these two requirements dictate that spontaneous heat flow between two bodies at different temperatures always goes from the warmer body to the colder body. For example, when a kettle of boiling water sits on the heating element of a stove, heat always flows from the hot burner to the relatively cooler boiling water.

To understand why this must occur, consider the entropy changes that would accompany heat transfer in the *opposite* direction. Suppose the burner is at 455 K and the water is at 373 K. We can calculate the entropy change that would occur if 100. J of heat flowed from the water to the burner. In this scenario, q for the burner is positive, so the burner gains entropy. For the water, q is negative, so the water loses entropy:

$$\Delta S = \frac{100. \text{ J}}{455 \text{ K}} = 0.220 \text{ J/K}; \qquad \Delta S_{\text{water}} = \frac{-100. \text{ J}}{373 \text{ K}} = -0.268 \text{ J/K}$$

$$\Delta S_{\text{total}} = \Delta S_{\text{burner}} + \Delta S_{\text{water}} = 0.220 \text{ J/K} + (-0.268 \text{ J/K}) = -0.048 \text{ J/K}$$

A spontaneous "uphill" flow of heat, from the cooler water to the hotter burner, would result in a *decrease* in total entropy, which the second law of thermodynamics forbids. According to all observations, which the second law summarizes, heat never flows spontaneously from a cold to a hot body.

We can summarize the first and second laws of thermodynamics in two simple equations:

$$\text{First law:} \qquad \Delta E_{\text{total}} = 0 \text{ (always)}$$

$$\text{Second law:} \qquad \Delta S_{\text{total}} > 0 \text{ (always)}$$

Section Exercises

14.2.1 Benzene (C_6H_6) has $\Delta H_{\text{fus}} = 10.9$ kJ/mol and a freezing point of 5.5 °C, whereas water has $\Delta H_{\text{fus}} = 6.01$ kJ/mol and freezes at 0.0 °C. Suppose that a sealed jar containing a mixture of solid and liquid benzene at 5.5 °C is immersed in a mixture of ice and water at 0.0 °C. What will happen?

14.2.2 Compute ΔS_{total} for 10.0 g of benzene changing phase as described in Section Exercise 14.2.1.

14.2.3 Suppose that a mixture of ice and water is placed in a refrigerator that is held at the freezing point of water, exactly 0.0 °C. Calculate ΔS for the ice–water mixture, ΔS for the refrigerator, and ΔS_{total} when 5.00 g of ice forms. Is this process spontaneous? Is the reverse process spontaneous?

Figure 14-11
(*a*) When there is only one possible distribution, *W* = 1 and *S* = 0. (*b*) When items are not identical, there is more than one possible distribution, making *W* > 1 and *S* > 0. (*c*) When there are more places to put the items than there are items, *W* > 1 and *S* > 0.

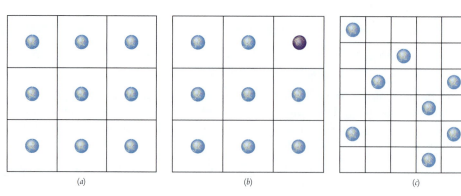

(a) (b) (c)

14.3   ENTROPIES OF PURE SUBSTANCES

The Boltzmann equation, $S = k \ln W$, establishes the zero point for entropies. Because $\ln(1) = 0$, this equation predicts that the entropy will be zero for a system with only one possible description. Figure 14-11*a* shows a system with $W = 1$, nine identical marbles placed in nine separate compartments. The figure also shows two types of conditions where $W > 0$. Figure 14-11*b* shows a condition when the marbles are not identical. If one marble is a different color than the other eight, there are nine different places to place the different marble. Figure 14-11*c* shows a condition where there are more compartments than marbles. If there are more places to put the marbles than there are marbles, there are multiple ways to place the marbles in the compartments.

Minimization of Entropy

We can apply the same reasoning to atoms or molecules as to marbles. First, any substance that is impure is like the depiction in Figure 14-11*b*. The impurity atoms or molecules can occupy any locations in the sample, increasing W and giving the sample additional entropy. Second, if there is any extra space available for the atoms or molecules of a pure substance, the situation is like the depiction in Figure 14-11*c*. A gas has much empty space between its atoms or molecules, so gases have very large values of W and correspondingly high entropies. Liquids have much less available space, but the atoms and molecules still have considerable freedom of motion, meaning there are multiple locations available, so $W > 1$ and $S > 0$.

A pure crystalline solid comes closest to the depiction in Figure 14-11*a*. Nevertheless, each atom or molecule in a pure crystalline solid vibrates back and forth in its compartment, and this vibration can be thought of as similar to the depiction in 14-11*c*. The vibrations move the atoms or molecules randomly about over the space available to them, making $W > 1$ and $S > 0$.

To reach $W = 1$ and $S = 0$, we must remove as much of this vibrational motion as possible. Recall that temperature is a measure of the amount of thermal energy in a sample, which for a solid is the energy of the atoms or molecules vibrating in their cages. Thermal energy reaches a minimum when $T = 0$ K. At this temperature, there is only one way to describe the system, so $W = 1$ and $S = 0$. This is formulated as the **third law of thermodynamics,** which states that a pure, perfect crystal at 0 K has zero entropy. We can state the third law as an equation, Equation 14-4:

$$S_{(\text{pure, perfect crystal; } T=0\,\text{K})} = 0 \qquad (14\text{-}4)$$

Each of the three specifications is essential in order to make $W = 1$. If the sample is not *pure,* the impurities can have multiple locations. If the sample is not a *perfect crystal,* the imperfections can have multiple locations. If the sample is not at $T = 0$ K, the vibrational energy gives the atoms or molecules multiple locations.

Absolute Entropies

The third law of thermodynamics makes it possible to measure the absolute entropy of any substance at any temperature. Figure 14-12 shows the stages of a pure substance, in

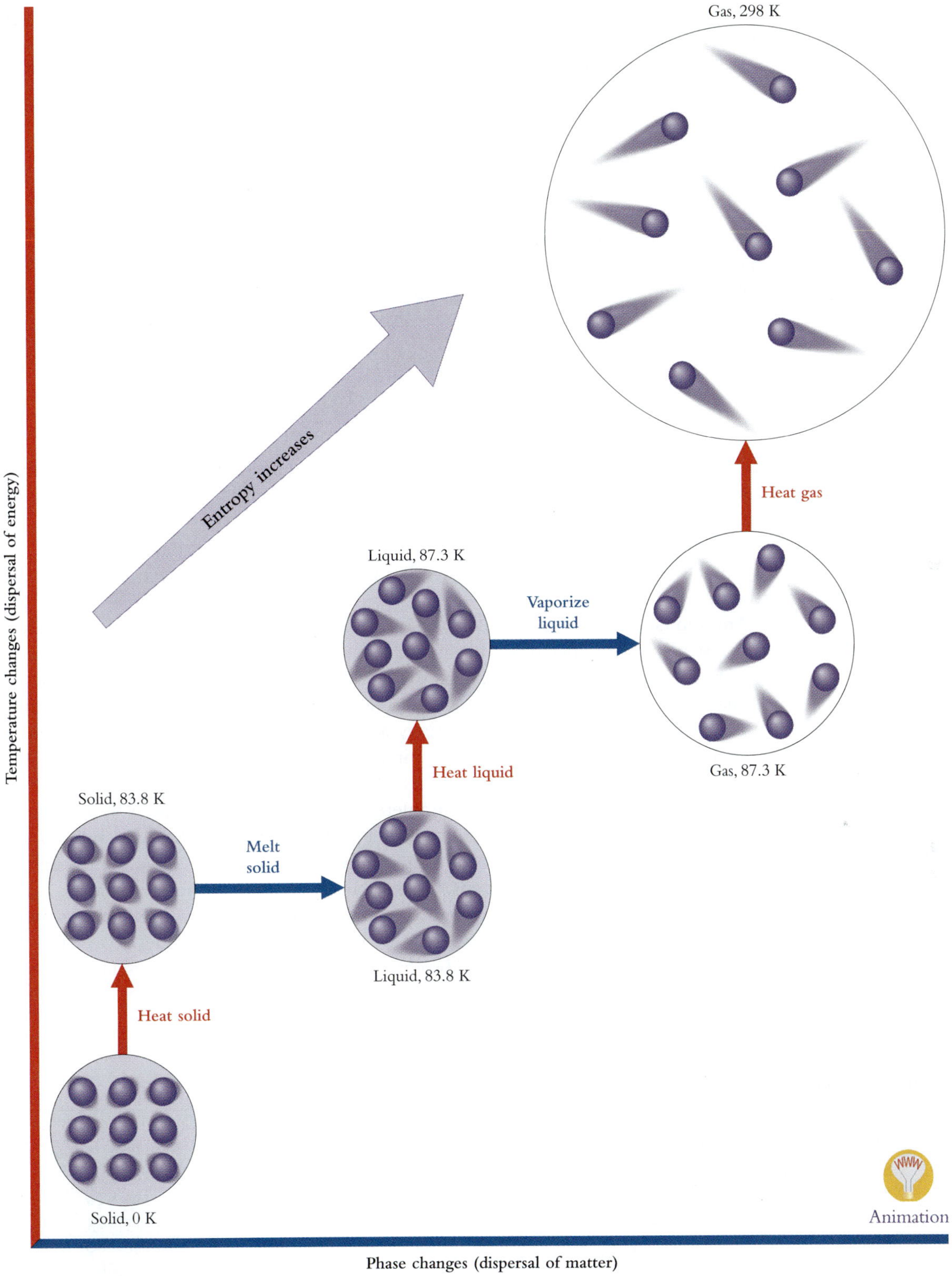

Figure 14-12
At $T = 0$ K, a pure perfect solid crystal has $S = 0$ J/K. The absolute entropy of the substance increases as the temperature increases and the phase changes. The temperatures shown here are for pure argon.

this case argon, as the temperature is raised from $T = 0$, $S = 0$ to $T = 298$ K. As temperature increases, entropy increases steadily.

When solid argon is heated, its atoms "rattle around" in their crystalline cages with increasing energy of motion. Equation 14-2 does not apply to this process because the temperature is changing, but the techniques of calculus can be used to determine the entropy change from the energy required for the heating.

When the solid melts at 83.8 K, entropy increases by an amount given by Equation 14-2:

$$q_{Ar} = n\,\Delta H_{fus} = (1.00 \text{ mol})(1.21 \text{ kJ/mol})(10^3 \text{ J/kJ}) = 1.21 \times 10^3 \text{ J}$$

$$\Delta S_{fus} = \frac{q_{Ar}}{T_{fus}} = \frac{1.21 \times 10^3 \text{ J}}{83.8 \text{ K}} = 14.4 \text{ J/K}$$

Raising the temperature of liquid argon from 83.8 K to its boiling point, 87.3 K, results in an additional increase in entropy as the argon atoms gain thermal energy. When argon vaporizes, there is another large increase in entropy that can be calculated using Equation 14-2:

$$q_{Ar} = n\,\Delta H_{vap} = (1.00 \text{ mol})(6.53 \text{ kJ/mol})(10^3 \text{ J/kJ}) = 6.53 \times 10^3 \text{ J}$$

$$\Delta S_{vap} = \frac{q_{Ar}}{T_{vap}} = \frac{6.53 \times 10^3 \text{ J}}{87.3 \text{ K}} = 74.8 \text{ J/K}$$

This large increase in entropy comes mainly from the increase in volume that accompanies vaporization. At 87.3 K, the molar volume of liquid argon is only 0.29 L, whereas the molar volume of argon gas is 7.17 L.

More energy is required to heat one mole of argon gas from 87.3 to 298 K, and this energy gets dispersed among the argon atoms, increasing the entropy still further.

Any substance can be subjected to the path outlined in Figure 14-12. Starting from the lowest possible temperature, the substance is heated to 298 K at a constant pressure of 1 bar. The amount of heat required to increase the temperature and the energies of any phase changes are measured using calorimetry. These calorimetry experiments may sound commonplace, but as our Chemistry and Technology Box describes, working at temperatures near 0 K is both difficult and intriguing.

The third law of thermodynamics establishes a "starting point" for entropies. At 0 K, any pure perfect crystal is completely constrained and has $S = 0$ J/K. At any higher temperature, the substance has a positive entropy that depends on the conditions. The molar entropies of many pure substances have been measured at standard thermodynamic conditions, $P° = 1$ bar. The same thermodynamic tables that list standard enthalpies of formation usually also list standard molar entropies, designated $S°$, for $T = 298$ K. Table 14-2 lists representative values of $S°$ to give you an idea of the magnitudes of absolute entropies. Appendix D contains a more extensive list.

Table 14-2 Standard Molar Entropies ($S°$) of Selected Substances at 298 K

Substance	Phase	$S°$ ($J\,mol^{-1}K^{-1}$)	Substance	Phase	$S°$ ($J\,mol^{-1}K^{-1}$)
C	Solid (diamond)	2.4	He	Gas	126.153
C	Solid (graphite)	5.74	Ar	Gas	154.846
Si	Solid	18.8	Xe	Gas	169.685
Al	Solid	28.3	H_2	Gas	130.680
Cu	Solid	33.2	CO	Gas	197.660
Ag	Solid	42.6	F_2	Gas	202.79
SiO_2	Solid	41.5	Cl_2	Gas	223.08
NaCl	Solid	72.1	CO_2	Gas	213.78
I_2	Solid	116.1	CH_4	Gas	184.3
H_2O	Liquid	69.95	C_2H_2	Gas	200.9
Hg	Liquid	75.9	C_2H_4	Gas	219.3
Br_2	Liquid	152.2	C_2H_6	Gas	229.2

Box 14-1 Chemistry and Technology: Seeking Absolute Zero

Atoms and molecules at absolute zero temperature truly would be fully constrained. Because all excess energy would be removed, the inherent properties of atoms and molecules would be sharply defined. For this reason, chemists and physicists have worked extensively to develop techniques to reduce the temperature of a sample as close as possible to absolute zero. Along the way, a number of astonishing discoveries have been made, including superconductivity and superfluidity.

Reducing the temperature becomes ever more difficult, the nearer we approach absolute zero. If we wish to reduce the temperature of a sample below the temperature of its surroundings, we must find some way other than a heat flow to remove energy of motion, because "uphill" heat flows are never spontaneous. To cool a refrigerator below the temperature of its surroundings, we use a refrigerating fluid that condenses when compressed and absorbs energy from its surroundings as it vaporizes. Molecular nitrogen is liquefied by rapid expansion. As the gas expands, the molecules must overcome intermolecular forces of attraction. Their potential energy increases at the expense of their kinetic energy of motion. This slows them down, so the temperature drops. Nitrogen condenses at 77 K. Still lower temperatures require even more elaborate procedures.

Low-temperature research requires hard work and imagination, but successful advances are richly rewarded. Seven Nobel Prizes in physics and chemistry have been awarded for low-temperature research. The first, in 1913, went to the Dutch physicist Heike Kamerlingh Onnes, who discovered how to cool He gas to 4.2 K and convert it into a liquid. The American William Giauque received the 1949 prize in chemistry and the Russian Pyotr Kapitsa won the 1978 prize in physics. Each was honored for a variety of discoveries resulting from low-temperature research, and each developed a new technique for achieving low temperature.

Giauque's technique still is used in contemporary low-temperature research. A sample containing paramagnetic ions (Fe^{3+}, for example) is bathed in liquid helium under reduced pressure to chill it to a temperature below 4.2 K. The sample is held in a strong magnetic field, which aligns the magnetic directions of the paramagnetic ions. Then the liquid helium is removed and the magnet turned off. The paramagnetic ions absorb energy in order to randomize their magnetic alignments. This absorption of energy causes the temperature to fall. Temperatures in the range of 10^{-3} to 10^{-6} K can be achieved using this technique.

A superfluid fountain of liquid helium

The 1996 Nobel Prize in physics went to three researchers who studied liquid helium at a temperature of 0.002 K, discovering superfluid helium. A superfluid behaves completely unlike conventional liquids. Liquids normally are viscous because their molecules interact with one another to reduce fluid motion. Superfluid helium has zero viscosity, because all of its atoms move together like a single "superatom." This collective behavior also causes superfluid liquid helium to conduct heat perfectly, so heating a sample at one particular spot results in an immediate and equal increase in temperature throughout the entire volume. A superfluid also flows extremely easily, so it can form a fountain, shown in the photo, in apparent defiance of gravity.

Superfluidity is just one of the surprising new properties discovered through low-temperature research. Another example is superconductivity, described in our Chemistry and Technology Box in Chapter 11. The 2003 Nobel Prize in physics went to three theoreticians who developed theories explaining these phenomena.

In 1997 and 2001, the Nobel Prizes in physics were awarded for low-temperature research on gases. When an atomic gas expands through a small opening, it forms a beam of atoms travelling in one direction (recall the molecular beams described in Section 5.3). Normally, these atoms have speeds around 1 km/s. When the atoms are bombarded head-on with photons from an intense laser beam with just the right frequency, however, the interaction between photons and atoms slows down the atoms, much as a stiff headwind slows down a runner. A sophisticated combination of laser beams and magnetic fields makes it possible to cool a beam of helium atoms to temperatures around 2×10^{-7} K. At this temperature, the atoms move at a leisurely 2 cm/s. The winners of the 2001 Nobel Prize studied beams of alkali metal atoms at a temperature around 1×10^{-7} K. At this temperature, the atoms would instantaneously condense to form a metallic solid if they collided with one another, but the use of atomic beams ensures that the atoms do not collide, so they keep their atomic identities. These clusters of gaseous alkali metal atoms act like single "superatoms," just like the atoms in superfluid helium.

The existence of "superatoms" was first predicted in 1924 by an Indian physicist, S. N. Bose, and elaborated further by Albert Einstein. Over 70 years later, studies at ultralow temperature confirmed the predictions. Physicists and chemists continue to work at the limits of low temperature to test some of the most fundamental predictions of quantum mechanics. Undoubtedly, additional Nobel Prizes will reward such research.

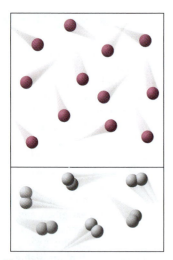

Figure 14-13
At the same temperature and pressure, the atoms in a 1-mol sample of H₂ (*bottom*) are more constrained than the atoms in a 2-mol sample of gaseous He (*top*).

One important feature not revealed directly by Table 14-2 is that binding atoms into molecules always constrains the atoms and reduces total entropy. As illustrated in Figure 14-13, a sample containing 2 mol He has considerably more entropy (252 J/K) than 1 mol H₂ (131 J/K), even though both samples contain the same total number of atoms.

The representative values of $S°$ listed in Table 14-2 reveal the following additional general trends in absolute entropies:

1. Unlike enthalpies of formation, $S°$ values are *never* zero at any temperature above 0 K. Absolute entropies can be zero only at 0 K.

2. The absolute entropy of a substance depends on its phase. For example, compare three substances that are quite similar except for their phase at 298 K, 1 bar (Figure 14-14):

   $I_2 (s), S° = 116 \, J \, mol^{-1} K^{-1}$ $Br_2 (l), S° = 152 \, J \, mol^{-1} K^{-1}$ $Cl_2 (g), S° = 223 \, J \, mol^{-1} K^{-1}$

   The substance in the more constrained phase has the lower entropy. For substances that are otherwise similar, $S_{solid} < S_{liquid} < S_{gas}$.

3. Among substances that share a phase and similar structure, $S°$ values increase with molar mass. For example:

Substance	MM (g/mol)	S° (J mol⁻¹K⁻¹)	Substance	MM (g/mol)	S° (J mol⁻¹K⁻¹)
He (g)	4.00	126	Cu (s)	63.6	33
Ar (g)	39.9	155	Ag (s)	108	43

   The entropies of gases increase with molar mass because translational energy levels, which are quantized just like other forms of energy, are spaced more closely as mass increases, increasing the number of ways to distribute energy among the levels. The entropy of metallic solids increases with molar mass because atomic number also increases, meaning there are more electrons among which to disperse the available energy.

4. Molar entropies increase as the size and complexity of the molecule increases. Compare, for example, the standard molar entropies of the three two-carbon hydrocarbons:

C₂H₂
200.9 J/(mol K)

C₂H₄
219.3 J/(mol K)

C₂H₆
229.2 J/(mol K)

Figure 14-14
The absolute entropies of otherwise similar substances depend on their phases:
$S° \, I_2 (s) < S° \, Br_2 (l) < S° \, Cl_2 (g)$

Iodine (solid)   Bromine (liquid)   Chlorine (gas)

The reason behind this trend is that a molecule with many atoms has more variations in how it can vibrate than a molecule with few atoms. This means more ways to distribute energy—in other words, greater energy dispersal at any given temperature—and larger entropy.

Example 14-3 provides practice in identifying how absolute entropies vary with molecular properties.

For each of the following pairs of substances under standard conditions, predict which has the larger standard entropy and give a reason why: (a) 1 mol each of Hg and Au; (b) 1 mol each of NO and NO$_2$; (c) 2 mol of NO$_2$ and 1 mol of N$_2$O$_4$; and (d) 1 mol each of Xe and Kr.	**Example 14-3** **Absolute Entropies**

Strategy: Analyze each pair, keeping in mind that several features affect absolute entropy, including phase, number of atoms or molecules, amount of bonding, and molar mass.

Solution:
a. At 298 K, mercury is a liquid metal, but gold is a solid metal. In general, a liquid is less constrained than a solid. The two elements have almost the same molar mass, so liquid Hg should have a larger $S°$ value than solid Au. Values in Appendix D confirm the prediction: Hg, 76 J mol^{-1}K^{-1}; Au, 47 J mol^{-1}K^{-1}.
b. Both NO and NO$_2$ are gases under standard conditions. Each molecule of NO$_2$ has three atoms, and each molecule of NO has two atoms. Thus, NO$_2$ should have a larger standard molar entropy than NO, and experimental values confirm this: NO$_2$, 240 J mol^{-1}K^{-1}; NO, 211 J mol^{-1}K^{-1}.
c. Two moles of NO$_2$ contain the same number of atoms as 1 mol of N$_2$O$_4$. In N$_2$O$_4$, however, pairs of NO$_2$ units are tied together by a bond that constrains them. Thus, 2 mol of NO$_2$ should have more entropy than 1 mol of N$_2$O$_4$, as confirmed by experimental values: NO$_2$, (240 J mol^{-1}K^{-1})(2 mol) = 480 J/K; N$_2$O$_4$, (304 J mol^{-1}K^{-1})(1 mol) = 304 J/K.
d. Both Xe and Kr are monatomic gases from Group 18 of the periodic table. Because Xe has a higher molar mass than Kr, Xe should have greater entropy: Xe, 170 J mol^{-1}K^{-1}; Kr, 164 J mol^{-1}K^{-1}.

Do the Results Make Sense? In these examples, we apply atomic and molecular reasoning to predict trends in the absolute entropies. We verify that our predictions make sense by consulting tables of quantitative $S°$ values.

For each of the following pairs, select the one with smaller $S°$ at 298 K and give the reason: (a) CH$_4$ or SiH$_4$; (b) CH$_4$ or H$_2$; (c) CH$_4$ or H$_2$O.	**Extra Practice Exercise 14.3**

Entropy and Concentration

Entropy increases as a substance spreads out. The molar entropy of a gas is larger than the molar entropy of the same substance as a liquid or a solid, because molecules are more dispersed in a gas than in a condensed phase. As shown in Figure 14-15a, as a gas expands into a larger volume, it also becomes more dispersed, so the entropy of a gas increases when a gas expands. There is a similar increase in dispersion when a solution is diluted. Figure 14-15b illustrates that the dilution of a *solution* has the same dispersing effect as the

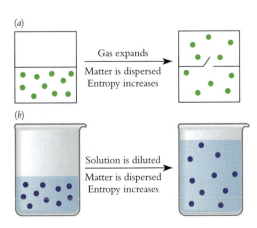

Figure 14-15
(a) When a gas expands, it becomes dispersed and entropy increases. (b) When a solution is diluted by a factor of two, the increase in dispersion of the solute is the same as when the pressure of a gas is reduced by a factor of two.

expansion of a *gas* into a larger volume. Entropy increases when molecules are dispersed throughout a larger volume. This means that entropy increases as concentration decreases.

The mathematical machinery of thermodynamics allows this qualitative statement to be expressed quantitatively. Experiments and theory show that the molar entropy of a gas or solute varies logarithmically with concentration (c):

$$S = S° - R \ln\left(\frac{c}{c°}\right)$$

The ideal gas equation (Equation 5-1) can be rearranged to show that the concentration of a gas, which is moles of gas per unit volume, is conveniently measured by its partial pressure (p):

$$c = \frac{n}{V} = \frac{p}{RT}$$

Applying this equality to both c and $c°$, we find that for a gas, the concentration ratio equals the pressure ratio:

$$\frac{c}{c°} = \frac{p}{p°}$$

Since $c° = 1$ M and $p° = 1$ bar, we can simplify the equations for entropies of gases and solutes when their concentrations differ from standard conditions, obtaining Equation 14-5:

$$S_{(p \neq 1\ \text{bar})} = S° - R \ln p \qquad S_{(c \neq 1\ \text{M})} = S° - R \ln c \qquad \text{(14-5)}$$

Remember, however, that concentration always must be expressed using the same units as the standard state of the substance: bar for gases and molarity for solutes.

Equation 14-5 refers to molar quantities. To obtain the total entropy of a sample, we must multiply its molar entropy by n, the number of moles. Example 14-4 illustrates the calculation of ΔS for a change in concentration.

| **Example 14-4** | Oxygen gas has many applications, from welders' torches to respirators. The gas is sold commercially in pressurized steel tanks. One such tank contains O_2 at $p = 6.50$ bar and $T = 298$ K. Using standard thermodynamic data, compute the molar entropy of the gas in the tank at 6.50 bar and the change in entropy of a 0.155–mol sample of gas withdrawn from the tank at 1.10 bar and constant temperature. |

Entropy Change on Expansion

Strategy: This is a quantitative thermodynamics problem that asks us to determine entropy under nonstandard conditions using standard thermodynamic data, so we apply the seven-step approach. The problem asks for two quantities: the molar entropy of O_2 gas in the pressurized tank and the entropy change when a sample of that gas undergoes a pressure change.

Begin with a diagram of the process that includes a summary of all the information given in the problem:

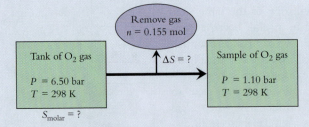

We also need the absolute entropy for O_2, which we find in Appendix D:

$$O_2\,(g),\ S° = 205\ \text{J mol}^{-1}\text{K}^{-1}$$

Solution: The gas stored in the tank is not at standard pressure, so apply Equation 14-5 to calculate its molar entropy. As the gas leaves the tank, it expands and its entropy increases. The final pressure is not standard pressure, so again use Equation 14-5 to calculate its molar entropy at the final pressure. Then calculate the entropy change for the expansion, taking the difference in molar entropies between initial and final conditions and multiplying by the number of moles undergoing the expansion.

$$S_{(p\,=\,6.50\,\text{bar})} = S° - R \ln p = 205\,\text{J mol}^{-1}\text{K}^{-1} - (8.314\,\text{J mol}^{-1}\text{K}^{-1}) \ln 6.50$$

$$= 205\,\text{J mol}^{-1}\text{K}^{-1} - 15.562\,\text{J mol}^{-1}\text{K}^{-1} = 189\,\text{J mol}^{-1}\text{K}^{-1}$$

$$S_{(p\,=\,1.10\,\text{bar})} = S° - R \ln p = 205\,\text{J mol}^{-1}\text{K}^{-1} - (8.314\,\text{J mol}^{-1}\text{K}^{-1}) \ln 1.10$$

$$= 205\,\text{J mol}^{-1}\text{K}^{-1} - 0.7924\,\text{J mol}^{-1}\text{K}^{-1} = 204\,\text{J mol}^{-1}\text{K}^{-1}$$

$$\Delta S = n(S_{\text{final}} - S_{\text{initial}}) = (0.155\,\text{mol})(204\,\text{J mol}^{-1}\text{K}^{-1} - 189\,\text{J mol}^{-1}\text{K}^{-1}) = 2.3\,\text{J/K}$$

Do the Results Make Sense? The positive value for this entropy change reflects the fact that the expansion is a spontaneous process.

Calculate the entropy change when 125 g of He gas initially at 298 K, 1.00 bar is compressed to a new pressure of 15.7 bar.

Example 14-4

Entropy Change on Expansion (*continued*)

Extra Practice Exercise 14.4

Standard Reaction Entropies

Entropy changes are important in every process, but chemists are particularly interested in the effects of entropy on chemical reactions. If a reaction occurs under standard conditions, its entropy change can be calculated from absolute entropies using the same reasoning used to calculate reaction enthalpies from standard enthalpies of formation. The products of the reaction have molar entropies, and so do the reactants. The total entropy of the products is the sum of the molar entropies of the products multiplied by their stoichiometric coefficients in the balanced chemical equation. The total entropy of the reactants is a similar sum for the reactants. Equation 14-6 expresses the entropy change accompanying a reaction as the difference between these two quantities:

$$\Delta S°_{\text{reaction}} = \Sigma\,\text{coeff}_{\text{products}}\,S°_{\text{products}} - \Sigma\,\text{coeff}_{\text{reactants}}\,S°_{\text{reactants}} \qquad (14\text{-}6)$$

Notice that Equation 14-6 is similar to Equation 6-12, which is used to calculate reaction *enthalpies*.

$$\Delta H°_{\text{reaction}} = \Sigma\,\text{coeff}_{\text{products}}\,\Delta H°_{\text{f products}} - \Sigma\,\text{coeff}_{\text{reactants}}\,\Delta H°_{\text{f reactants}} \qquad (6\text{-}12)$$

As Example 14-5 shows, the applications of these equations follow parallel paths.

Acrylonitrile is an essential monomer in the polymer industry because it is used to make polyacrylonitrile for synthetic fibers.
Acrylonitrile is made from propene:

$$2\,\text{C}_3\text{H}_6\,(g) + 2\,\text{NH}_3\,(g) + 3\,\text{O}_2\,(g) \longrightarrow 2\,\text{C}_3\text{H}_3\text{N}\,(l) + 6\,\text{H}_2\text{O}\,(l)$$

| Propene | | Acrylonitrile |

Calculate $\Delta H°_{\text{reaction}}$ and $\Delta S°_{\text{reaction}}$ for the synthesis of acrylonitrile using standard thermodynamic data.

Example 14-5

Entropy and Enthalpy Changes

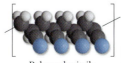

Polyacrylonitrile — Many repeat units

Example 14-5	**Strategy:** We use a short version of the seven-step method. The problem asks for the entropy and enthalpy changes accompanying a chemical reaction, so we focus on the balanced chemical equation and the thermodynamic properties of the reactants and products.

Entropy and Enthalpy Changes (*continued*)

Solution: Appendix D lists standard enthalpies of formation and standard entropies. Here are the values for the substances involved in this reaction:

Substance	$C_3H_6(g)$	$NH_3(g)$	$O_2(g)$	$C_3H_3N(l)$	$H_2O(l)$
ΔH_f° (kJ/mol)	20.0	−45.9	0	172.9	−285.83
S° (J/mol^{-1}K^{-1})	226.9	192.8	205.15	188	69.95
Coefficient	2	2	3	2	6

Standard thermodynamic changes for a reaction can be calculated using Equations 6-12 and 14-6:

$$\Delta H_{reaction}^\circ = \Sigma\, coeff_{products}\, \Delta H_{f\,products}^\circ - \Sigma\, coeff_{reactants}\, \Delta H_{f\,reactants}^\circ$$

$$\Delta S_{reaction}^\circ = \Sigma\, coeff_{products}\, S^\circ_{products} - \Sigma\, coeff_{reactants}\, S^\circ_{reactants}$$

We leave it to you to substitute the values and calculate the answers. The results are

$$\Delta H_{reaction}^\circ = -1317 \text{ kJ} \qquad \Delta S_{reaction}^\circ = -659 \text{ J/K}$$

Do the Results Make Sense? According to these values, the reaction is exothermic, but the products are less dispersed than the reactants. Because the reaction produces liquids from gases, this decrease in dispersion is a reasonable result.

Extra Practice Exercise 14.5	Calculate the entropy change accompanying the combustion of methane if the water is produced in the liquid phase.

Section Exercises

14.3.1 Of the following pairs, which has the greater entropy? Explain each choice. (a) 1 g of dew or 1 g of frost; (b) 1 mol of gaseous hydrogen atoms or 0.5 mol of gaseous hydrogen molecules; (c) "perfect" diamond or flawed diamond, each $\frac{1}{4}$ carat; and (d) 5 mL of liquid ethanol at 0 °C or 5 mL of liquid ethanol at 50 °C.

14.3.2 Explain the following differences in entropies in molecular terms (substances are at standard conditions unless otherwise noted): (a) 1 mol of O_2 has less entropy than 1 mol of O_3; (b) 3 mol of O_2 has more entropy than 2 mol of O_3; (c) 1 mol of I_2 has less entropy than 1 mol of O_2; and (d) 1 mol of HCl (*aq*) in concentrated solution (12 M) has less entropy than 1 mol of HCl (*aq*) in dilute solution (0.100 M).

14.3.3 Draw molecular pictures to illustrate your answers to part (b) of Section Exercise 14.3.1 and part (b) of Section Exercise 14.3.2.

14.3.4 Compute the standard entropy change for the following reaction:

$$12\, NH_3(g) + 21\, O_2(g) \longrightarrow$$
$$8\, HNO_3(g) + 4\, NO(g) + 12\, H_2O(l)$$

14.4 SPONTANEITY AND FREE ENERGY

The second law of thermodynamics states that total entropy always increases during a spontaneous process. That is, the sign of ΔS_{total} is always positive for every actual process. A calculation of ΔS_{total} determines whether a particular chemical process is spontaneous. Unfortunately, this is not practical for most processes. It is usually possible to calculate the entropy change for a *system*. However, the surroundings often are much more complicated than the system, making it difficult to calculate ΔS of the *surroundings*. It would be much more convenient to have a way to determine the direction of spontaneous change using *just the system,* not the surroundings.

There is no single criterion for the system alone that applies to *all* processes. However, if we restrict the conditions to constant temperature and pressure, there is a state function whose change for the system predicts spontaneity. This new state function is the **free energy (G),** which was introduced by the American J. Willard Gibbs and is defined by Equation 14-7:

$$G = H - TS \qquad\qquad (14\text{-}7)$$

As usual, H is enthalpy, T is absolute temperature, and S is entropy.

How does free energy allow us to predict spontaneity while ignoring the surroundings? To answer this question, we examine the change in free energy of a system,

$$\Delta G_{sys} = \Delta H_{sys} - \Delta (TS)_{sys}$$

If we hope to use ΔG_{sys} to predict spontaneity, we must relate ΔG_{sys} to ΔS_{total}, which must be positive if a process is spontaneous. To do that, we must restrict the conditions.

First, the process must occur at *constant temperature*. This lets us relate $\Delta (TS)_{sys}$ to ΔS_{sys}, resulting in Equation 14-8:

$$\Delta (TS)_{sys} = (TS)_{final} - (TS)_{initial} = T(S_{final} - S_{initial}) = T \Delta S_{sys}$$

$$\Delta G_{sys} = \Delta H_{sys} - T \Delta S_{sys} \qquad (\text{constant } T) \qquad\qquad (14\text{-}8)$$

Second, the process must occur at *constant pressure*. Then we can relate the *enthalpy* change for the system to the *entropy* change for the surroundings. Recall that ΔH equals q when P is constant:

$$\Delta H_{sys} = q_{sys} \qquad (\text{constant } P)$$

Remember also that the heat flow for a system is always equal in magnitude but opposite in sign to the heat flow of the surroundings:

$$q_{sys} = -q_{surr} \qquad so \qquad \Delta H_{sys} = -q_{surr}$$

Furthermore, we have already required constant temperature, so the heat flow of the surroundings measures the entropy change of the surroundings. That is,

$$\frac{q_{surr}}{T} = \Delta S_{surr} \qquad and \qquad q_{surr} = T \Delta S_{surr}$$

We combine these equalities into an equation that relates ΔH_{sys} to ΔS_{surr}:

$$\Delta H_{sys} = -T \Delta S_{surr} \qquad (\text{constant } P \text{ and } T)$$

Substituting this result into Equation 14-8 gives

$$\Delta G_{sys} = -T \Delta S_{surr} - T \Delta S_{sys} = -T(\Delta S_{surr} + \Delta S_{sys}) \quad (\text{constant } P \text{ and } T)$$

Finally, because $\Delta S_{surr} + \Delta S_{sys} = \Delta S_{total}$:

$$\Delta G_{sys} = -T \Delta S_{total} \quad (\text{constant } P \text{ and } T)$$

In words, in any process that occurs at constant T and P, the free energy change for the *system* is negative whenever the *total* entropy change is positive; that is, whenever the overall process is spontaneous. Defining a new function and imposing some restrictions provides a way to use properties of a system to determine whether a process is spontaneous.

ΔG_{sys} is negative for any spontaneous process taking place at constant temperature and pressure.

KEY CONCEPT

Although the restrictions of constant T and P are stringent, they apply to many important chemical processes, including reactions that occur in the human body, which has a nearly constant temperature of 37 °C and nearly constant pressure of 1 bar. Any

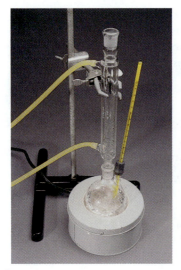

Figure 14-16
Many real-life situations operate under nearly constant temperature and pressure and thus are subject to $\Delta G_{sys} < 0$ as the condition for spontaneity. Examples shown here are a laboratory synthesis, a biological system, and an industrial process.

biochemical reaction that occurs in the body occurs under conditions in which the immediate surroundings are at constant T and P. Figure 14-16 shows three examples of systems at constant T and P.

Because free energy is a state function, its values can be tabulated for use in chemical calculations. As with standard enthalpies of formation, the **standard molar free energy of formation (ΔG_f°)** for any substance is defined to be the change of free energy when one mole of that substance forms from its elements in their standard states. The same reasoning that is used for enthalpy changes leads to Equation 14-9, from which we can calculate the free energy change for any chemical reaction:

$$\Delta G_{reaction}^\circ = \Sigma \, coeff_{products} \, \Delta G_{f \, products}^\circ - \Sigma \, coeff_{reactants} \, \Delta G_{f \, reactants}^\circ \qquad (14\text{-}9)$$

The form of Equation 14-9 should look familiar to you, because it is analogous to Equation 6-12 for reaction enthalpies and Equation 14-6 for reaction entropies.

As Equation 14-10 describes, the standard free energy change for a reaction can also be calculated from ΔH° and ΔS° for the reaction by making use of Equation 14-8 under standard conditions:

$$\Delta G^\circ = \Delta H^\circ - T \, \Delta S^\circ \qquad (14\text{-}10)$$

Either of these equations can be used to find standard free energy changes. Which equation to use depends on the available data. Example 14-6 illustrates both types of calculations.

Example 14-6	Find the standard free energy change for the acrylonitrile synthesis discussed in Example 14-5.
Free Energy of Reaction	**Strategy:** There are two ways to calculate $\Delta G_{reaction}^\circ$. The first method uses standard free energies of formation and Equation 14-9. The second method uses Equation 14-10 and the values of ΔH° and ΔS° calculated earlier. Either method requires the balanced chemical equation: $$2 \, C_3H_6 \, (g) + 2 \, NH_3 \, (g) + 3 \, O_2 \, (g) \longrightarrow 2 \, C_3H_3N \, (l) + 6 \, H_2O \, (l)$$ We perform both calculations to show that they give the same result. **Solution:** a. Using Equation 14-9 and ΔG_f° values from Appendix D: $$\Delta G_{reaction}^\circ = \Sigma \, coeff_{products} \, \Delta G_{f \, products}^\circ - \Sigma \, coeff_{reactants} \, \Delta G_{f \, reactants}^\circ$$

Substitute values from Appendix D:

$$\Delta G^\circ_{reaction} = [(6\ mol)(-237.1\ kJ/mol) + (2\ mol)(208.6\ kJ/mol)]$$
$$-[(3\ mol)(0\ kJ/mol) + (2\ mol)(-16.4\ kJ/mol) + (2\ mol)(74.62\ kJ/mol)]$$
$$\Delta G^\circ_{reaction} = -1122\ kJ$$

b. Using Equation 14-10 and the enthalpy and entropy results of Example 14-5:

$$\Delta H^\circ_{reaction} = -1317\ kJ \quad and \quad \Delta S^\circ_{reaction} = -659\ J/K$$
$$\Delta G^\circ_{reaction} = \Delta H^\circ_{reaction} - T\Delta S^\circ_{reaction}$$
$$\Delta G^\circ_{reaction} = [-1317\ kJ] - [(298\ K)(-659\ J/K)(10^{-3}\ kJ/J)] = -1121\ kJ$$

Does the Result Make Sense? In the second calculation, the entropy term $T\Delta S$ is positive, making ΔG° less negative than ΔH°. This is due to the formation of liquid water, which makes the products of this reaction more constrained than the reactants. Nevertheless, the large negative enthalpy change accompanying the reaction means that much heat becomes dispersed in the surroundings, making the overall process highly favorable. The large negative value for $\Delta G^\circ_{reaction}$ indicates that the production of acrylonitrile is highly spontaneous under standard conditions.

Calculate $\Delta G^\circ_{reaction}$ for the combustion of methane, producing liquid water, under standard conditions.

Example 14-6

Free Energy of Reaction *(continued)*

Pay close attention to units when using Equation 14-10. The values of ΔH° and ΔG° are usually given in kJ or kJ/mol, but S° is usually expressed in J/K or J mol^{-1} K^{-1}. Thus, entropies must be multiplied by 10^{-3} kJ/J before adding the two terms.

Extra Practice Exercise 14.6

The tabulated values for thermodynamic functions refer to unit concentrations and 298 K, but chemical reactions occur at many different concentrations and temperatures. To use ΔG as a measure of spontaneity under all conditions, we must know how ΔG depends on temperature and concentration.

Free Energy and Temperature

To see how free energy varies with temperature, we turn to Equation 14-10. This equation is valid at *any* temperature as long as the temperature is *constant*. However, application of the equation at a temperature different from 298 K requires knowledge of $\Delta H_{reaction}$ and $\Delta S_{reaction}$ at the new temperature.

The entropy of any chemical substance increases as temperature increases. These changes in entropy as a function of temperature can be calculated, but the techniques require calculus. Fortunately, temperature affects the entropies of reactants and products similarly. The absolute entropy of every substance increases with temperature, but the entropy of the reactants often changes with temperature by almost the same amount as the entropy of the products. This means that the temperature effect on the entropy *change* for a reaction is usually small enough that we can consider $\Delta S_{reaction}$ to be independent of temperature.

As mentioned in Section 6.5, $\Delta H_{reaction}$ also does not change rapidly with temperature. This is because the enthalpy change is closely related to the energy change, and the energy change is determined by the strengths of the chemical bonds in reactants and products. These, in turn, are nearly temperature-independent. As a result, changes in free energy at temperatures reasonably close to 298 K can be estimated by assuming that standard enthalpies and entropies at 298 K also apply at any other temperature. This is described by Equation 14-11:

$$\Delta G^\circ_{reaction,\ T} \cong \Delta H^\circ_{reaction,\ 298\ K} - T\Delta S^\circ_{reaction,\ 298\ K} \qquad (14\text{-}11)$$

An immediate consequence of Equation 14-11 is that the spontaneity of any reaction with a large ΔS° is very sensitive to temperature. Example 14-7 shows this.

Example 14-7

Temperature and Spontaneity

Dinitrogen tetroxide can decompose into two molecules of nitrogen dioxide:

$$N_2O_4(g) \longrightarrow 2\,NO_2(g)$$

a. Show that this reaction is not spontaneous at 298 K under standard conditions.

b. Find the temperature at which the reaction becomes spontaneous at standard pressure.

Strategy: The key to this problem is the first step in the seven-step strategy: What is asked for? The word *spontaneous* suggests that free energies are involved in this problem. The criterion for spontaneity at constant temperature and pressure is $\Delta G_{reaction} < 0$, so the problem asks us to show that $\Delta G^{\circ}_{reaction} > 0$ at $T = 298$ K. Then we can use Equation 14-11 to calculate the temperature at which $\Delta G^{\circ}_{reaction} < 0$. As temperature changes, ΔG° must become zero before it becomes negative. The reaction will be spontaneous at any temperature farther from 298 K than the temperature at which $\Delta G^{\circ} = 0$.

Solution:

a. Because standard free energies of formation at 298 K are available in Appendix D, we can use Equation 14-9 to determine whether the decomposition reaction is spontaneous at 298 K, 1 bar:

$$\Delta G^{\circ} = (2\ mol)(51.3\ kJ/mol) - (1\ mol)(99.8\ kJ/mol) = 2.8\ kJ$$

The positive value for ΔG° indicates that this reaction is *not* spontaneous at 298 K and standard pressures. In fact, the calculation tells us that the reaction will be spontaneous in the opposite direction. Under standard conditions at 298 K, NO_2 reacts to form N_2O_4:

$$2\,NO_2(g) \longrightarrow N_2O_4(g) \qquad \Delta G^{\circ} = -2.8\ kJ$$

b. For temperatures different from 298 K, we use Equation 14-11 to estimate $\Delta G^{\circ}_{reaction}$. This equation lets us determine the temperature at which N_2O_4 decomposition becomes spontaneous when the partial pressure of each gas is 1 bar:

$$\Delta G^{\circ}_{reaction,\ T} \cong \Delta H^{\circ}_{reaction,\ 298\ K} - T\,\Delta S^{\circ}_{reaction,\ 298\ K}$$

To find the temperature at which $\Delta G^{\circ} = 0$, begin by calculating ΔH° and ΔS° using tabulated data:

$$\Delta H^{\circ} = (2\ mol)(33.2\ kJ/mol) - (1\ mol)(11.1\ kJ/mol) = 55.3\ kJ$$

$$\Delta S^{\circ} = (2\ mol)(240.1\ J\ mol^{-1}\ K^{-1}) - (1\ mol)(304.4\ J\ mol^{-1}\ K^{-1}) = 175.8\ J/K$$

Now set ΔG° equal to zero and rearrange the equation to solve for temperature:

$$\Delta G^{\circ} = 0 = \Delta H^{\circ} - T\,\Delta S^{\circ} \qquad or \qquad T\,\Delta S^{\circ} = \Delta H^{\circ}$$

$$T = \frac{\Delta H^{\circ}}{\Delta S^{\circ}} = \frac{55.3\ kJ}{(175.8\ J/K)(10^{-3}\ kJ/J)} = 315\ K$$

At 315 K, $\Delta G^{\circ} = 0$, so all temperatures greater than 315 K, ΔG° is negative, and decomposition of N_2O_4 is spontaneous at partial pressures of 1 bar.

Do the Results Make Sense? The decomposition of N_2O_4 requires a bond to break. This is the reason why the decomposition has a positive ΔH°. At the same time, the number of molecules doubles during decomposition, which is the reason ΔS° has a positive value. The positive enthalpy change means that energy is removed from the surroundings and constrained, whereas the positive entropy change means that matter is dispersed. At temperatures below 315 K, the enthalpy term dominates and decomposition is not spontaneous, but at temperatures above 315 K, the entropy term dominates and decomposition is spontaneous.

Extra Practice Exercise 14.7

Estimate ΔG° for the combustion reaction of methane at 1920 °C, the temperature of a gas flame (*Reminder:* H_2O is a gas at this temperature!)

Free Energy and Concentration

Gases that participate in chemical reactions typically are at pressures different from one bar. Substances in solution are likely to be at concentrations different from one molar. For example, a biochemist who wants to know what processes are spontaneous under physiological conditions will find that the substances dissolved in biological fluids are rarely at one molar concentration. How does ΔG vary with changes in molarity and pressure? Recall that enthalpy is virtually independent of concentration but that entropy obeys Equation 14-5:

$$S_{(p \neq 1\,\text{bar})} = S° - R \ln p \qquad\qquad S_{(c \neq 1\,\text{M})} = S° - R \ln c$$

Equation 14-5 applies to individual substances. To see how these equations affect ΔS for a chemical reaction, consider the decomposition of limestone ($CaCO_3$) to produce lime (CaO), carried out industrially in kilns in which the partial pressure of CO_2 is different from 1 bar:

$$CaCO_3\,(s) \longrightarrow CaO\,(s) + CO_2\,(g)$$

The entropy change for the reaction is the difference in entropy between products and reactants:

$$\Delta S_{\text{reaction}} = S\,(CaO) + S\,(CO_2) - S\,(CaCO_3)$$

Older limestone kilns have air vents through which CO₂ can escape.

The entropy of each solid is its standard entropy, $S°$, but the entropy of the gas must be corrected for the deviation of pressure from standard conditions. Equation 14-5 gives the molar entropy of carbon dioxide as a function of its partial pressure:

$$S\,(CO_2) = S°\,(CO_2) - R \ln(p_{CO_2})$$

Now substitute the corrected entropy of CO_2 into the equation for $\Delta S_{\text{reaction}}$:

$$\Delta S_{\text{reaction}} = S°\,(CaO) + [S°\,(CO_2) - R \ln(p_{CO_2})] - S°\,(CaCO_3)$$

Next, rearrange the equation so that all the standard entropies are together:

$$\Delta S_{\text{reaction}} = S°\,(CaO) + S°\,(CO_2) - S°\,(CaCO_3) - R \ln(p_{CO_2})$$

The first three terms are the standard entropy change for the reaction, allowing simplification:

$$\Delta S_{\text{reaction}} = \Delta S°_{\text{reaction}} - R \ln(p_{CO_2})$$

Modern limestone kilns use blowers as well as vents and are computer-controlled.

In this example, only one of the reagents has a concentration that can vary, and each stoichiometric coefficient is one. What happens for a more complicated reaction? Consider the synthesis of ammonia carried out in a pressurized reactor containing N_2, H_2, and NH_3 at partial pressures different from 1 bar:

$$N_2\,(g) + 3\,H_2\,(g) \longrightarrow 2\,NH_3\,(g)$$

The entropy change for the reaction is the difference in entropy between products and reactants, obtained by multiplying each corrected entropy by the appropriate stoichiometric coefficient:

$$\Delta S_{\text{reaction}} = 2\,S\,(NH_3) - 3\,S\,(H_2) - S\,(N_2)$$

Applying Equation 14-5 gives the molar entropy of each gas:

$$S\,(N_2) = S°\,(N_2) - R \ln(p_{N_2})$$

$$S\,(H_2) = S°\,(H_2) - R \ln(p_{H_2})$$

$$S\,(NH_3) = S°\,(NH_3) - R \ln(p_{NH_3})$$

Now substitute the corrected entropies:

$$\Delta S_{\text{reaction}} = 2\,[S°\,(NH_3) - R \ln(p_{NH_3})] - 3\,[S°\,(H_2) - R \ln(p_{H_2})] - [S°\,(N_2) - R \ln(p_{N_2})]$$

Next, rearrange the equation so that all the logarithmic terms are together:

$$\Delta S_{\text{reaction}} = 2\,S°\,(NH_3) - 3\,S°\,(H_2) - S°\,(N_2) - 2\,R \ln(p_{NH_3}) + 3\,R \ln(p_{H_2}) + R \ln(p_{N_2})$$

As before, the first three terms are the standard entropy change for the reaction:

$$\Delta S_{\text{reaction}} = \Delta S^\circ_{\text{reaction}} - 2\,R\ln(p_{\text{NH}_3}) + 3\,R\ln(p_{\text{H}_2}) + R\ln(p_{\text{N}_2})$$

We use the properties of logarithms to combine the logarithmic terms. Because $a\ln x = \ln x^a$,

$$2\,R\ln(p_{\text{NH}_3}) = R\ln(p_{\text{NH}_3})^2 \quad and \quad 3\,R\ln(p_{\text{H}_2}) = R\ln(p_{\text{H}_2})^3$$

With these changes, the equation becomes

$$\Delta S_{\text{reaction}} = \Delta S^\circ_{\text{reaction}} - R\,[\ln(p_{\text{NH}_3})^2 - \ln(p_{\text{H}_2})^3 - \ln(p_{\text{N}_2})]$$

Additional logarithmic properties, $\ln u + \ln v = \ln(uv)$ and $\ln u - \ln v = \ln(u/v)$, allow us to put all the logarithmic terms into a single ratio:

$$\Delta S_{\text{reaction}} = \Delta S^\circ_{\text{reaction}} - R\ln\left[\frac{(p_{\text{NH}_3})^2}{(p_{\text{N}_2})(p_{\text{H}_2})^3}\right]$$

Note that the pressure ratio has product concentrations raised to their stoichiometric coefficients in the numerator and reactant concentrations raised to their stoichiometric coefficients in the denominator. The form of this equation applies to all reactions, not just the synthesis of ammonia. A general reaction can be written as

$$a\text{A} + b\text{B} \longrightarrow d\text{D} + e\text{E}$$

Here, A and B represent any reactants, D and E represent any products, and the lower-case letters represent the stoichiometric coefficients. The expression for the entropy change accompanying the reaction is

$$\Delta S_{\text{reaction}} = \Delta S^\circ_{\text{reaction}} - R\ln\left[\frac{(c_{\text{D}})^d(c_{\text{E}})^e}{(c_{\text{A}})^a(c_{\text{B}})^b}\right]$$

The concentrations must, as always, be expressed in the units consistent with the standard state: pressure in bar for any gas, molarity for any solute. The ratio in the logarithmic term is called the **reaction quotient (Q)** (Equation 14-12).

$$\left[\frac{(c_{\text{D}})^d(c_{\text{E}})^e}{(c_{\text{A}})^a(c_{\text{B}})^b}\right] = Q \tag{14-12}$$

The entropy change for a reaction under nonstandard concentrations can be expressed in terms of the standard entropy change and Q:

$$\Delta S_{\text{reaction}} = \Delta S^\circ_{\text{reaction}} - R\ln Q$$

This general expression describing how $\Delta S_{\text{reaction}}$ varies with concentrations leads to an equation for the free energy change when concentrations are nonstandard:

$$\Delta G_{\text{reaction}} = \Delta H^\circ_{\text{reaction}} - T\,(\Delta S^\circ_{\text{reaction}} - R\ln Q)$$

Because $\Delta H^\circ - T\,\Delta S^\circ = \Delta G^\circ$, this equation reduces to

$$\Delta G_{\text{reaction}} = \Delta G^\circ_{\text{reaction}} + RT\ln Q \tag{14-13}$$

An immediate consequence of Equation 14-13 is that the direction of spontaneity of a reaction depends on the concentrations of reactants and products. Product concentrations appear in the numerator of Q, so when the concentration of a product increases, $\ln Q$ increases as well. Increasing the $\ln Q$ term makes ΔG less negative, so a reaction becomes less spontaneous as product concentrations increase. Conversely, reactant concentrations appear in the denominator of Q, so increasing the concentration of a reactant decreases $\ln Q$, which in turn makes ΔG more negative. Thus, a reaction becomes more spontaneous as reactant concentrations increase.

If a reaction is not spontaneous at unit concentrations, we may be able to make it spontaneous by increasing the concentrations of reactants or by removing products as they form. If a reaction has a very positive $\Delta G^\circ_{\text{reaction}}$, however, it may not be possible to change the value of Q sufficiently to make $\Delta G_{\text{reaction}} < 0$. Remember that $\ln Q$ changes more slowly than Q itself. A tenfold change in Q, for instance, changes $\ln Q$ only by a factor of 2.3. Nonetheless, reactions that are almost spontaneous under standard conditions can be driven forward by making appropriate changes in concentration. Example 14-8 provides an illustration.

Example 14-8

Effect of Concentration on Spontaneity

The decomposition of dinitrogen tetroxide produces nitrogen dioxide:

$$N_2O_4\,(g) \longrightarrow 2\,NO_2\,(g)$$

Find the minimum partial pressure of N_2O_4 at which the reaction is spontaneous if $p_{NO_2} = 1$ bar and $T = 298$ K.

Strategy: We apply the condensed form of the seven-step strategy. We need to calculate a pressure that is just sufficient to make the reaction spontaneous. That means solving for the pressure at which $\Delta G = 0$, using Equation 14-13:

$$\Delta G = \Delta G^\circ + RT \ln Q = \Delta G^\circ + RT \ln \left[\frac{(p_{NO_2})^2}{p_{N_2O_4}} \right]$$

Solution: The standard free energy change was determined in Example 14-7:

$$\Delta G^\circ = 2.8 \text{ kJ/mol}$$

A visual diagram summarizes the information:

N_2O_4 $\longrightarrow$ 2 NO_2	$\Delta G^\circ = 2.8$ kJ	
$p = ?$ $\qquad$ $p = 1$ bar	$\Delta G^\circ \leq 0$	

The problem asks for the partial pressure of N_2O_4 that will make the decomposition spontaneous when $T = 298$ K and p of $NO_2 = 1$ bar. The value of ΔG must be zero before it can become negative. Therefore, to find the threshold pressure of N_2O_4 that makes the decomposition spontaneous, set $\Delta G = 0$ and $p_{NO_2} = 1$ bar and then rearrange to solve for the partial pressure of N_2O_4:

$$0 = \Delta G^\circ + RT \ln \left[\frac{(1 \text{ bar})^2}{p_{N_2O_4}} \right]$$

$$\ln \left[\frac{1 \text{ bar}}{p_{N_2O_4}} \right] = -\frac{\Delta G^\circ}{RT} = -\frac{(2.8 \text{ kJ/mol})(10^3 \text{ J/kJ})}{(8.314 \text{ J mol}^{-1} \text{ K}^{-1})(298 \text{ K})} = -1.13$$

$$\frac{1 \text{ bar}}{p_{N_2O_4}} = e^{-1.13} = 0.323 \quad and \quad p_{N_2O_4} = \frac{1 \text{ bar}}{0.323} = 3.1 \text{ bar}$$

This decomposition is spontaneous as long as the pressure of N_2O_4 is greater than 3.1 bar. As always, we have to be careful about units. ΔG° is converted to joules to use the appropriate value of the gas constant (R) in J mol^{-1} K^{-1}.

> The mol unit in the value of ΔG° refers to "per mole of reaction" and is included in the calculation to cancel the mole unit in R.

> Recall that $y = \ln x$ implies that $x = e^y$, where e is the base for natural logarithms.

Does the Result Make Sense? Under standard conditions (both pressures 1 bar), the reaction is not spontaneous. To force the reaction to go, we need more reactants, so a reactant pressure that is greater than 1 bar makes sense.

Find the maximum partial pressure of NO_2 at which the reaction is spontaneous if $p_{N_2O_4} = 1$ bar and $T = 298$ K.

This example shows that a reaction with a small positive ΔG° can be made spontaneous by relatively small changes in concentrations.

If neither temperature nor concentrations are at standard values, free energy calculations must be done in two steps. First, correct for temperature to obtain ΔG_T° using Equation 14-11:

$$\Delta G^\circ_{\text{reaction},T} \cong \Delta H^\circ_{\text{reaction, 298 K}} - T \Delta S^\circ_{\text{reaction, 298 K}}$$

Second, use this value of ΔG_T° in Equation 14-13 to complete the calculation of ΔG:

$$\Delta G_{\text{reaction}} = \Delta G_{\text{reaction},T}^\circ + RT \ln Q$$

Influencing Spontaneity

Suppose a chemist wants to carry out a particular chemical synthesis, but the reaction has a positive value for ΔG°. The thermodynamic calculation indicates that the reaction is spontaneous in the wrong direction at 298 K and *standard* concentrations, but this does not prevent it from occurring under *all* conditions. What can be done to make the reaction go in the desired direction?

Example 14-8 illustrates that changing the reaction quotient changes ΔG for a reaction. Changing Q alters the entropy change of the chemical system. In particular, increasing the pressure of N_2O_4 above 3.1 bar would cause spontaneous decomposition of N_2O_4, even though this reaction is not spontaneous under standard conditions. Reactions in liquid solutions likewise may proceed spontaneously if the concentrations of reactants are high enough or the concentrations of products are low enough.

Changing the temperature of the system is another way to influence the spontaneity of a reaction. The equation for ΔG has two parts, ΔH and $T\,\Delta S$, which can work together or in opposition:

$$\Delta G_T^\circ = \Delta H^\circ - T\,\Delta S^\circ$$

A *positive* ΔS° promotes spontaneity because it makes ΔG° more negative. This reflects the fact that a positive ΔS° means the system becomes more dispersed during the reaction. A *negative* ΔH° promotes spontaneity as well because it also makes ΔG° more negative. This reflects the fact that energy becomes more dispersed in the surroundings when a reaction releases energy. Thus, a reaction that has a positive ΔS° and a negative ΔH° is spontaneous at all temperatures.

The combustion of propane to form gaseous products is an example of a reaction that is spontaneous at all temperatures:

$$C_3H_8\,(g) + 5\,O_2\,(g) \longrightarrow 3\,CO_2\,(g) + 4\,H_2O\,(g)$$

$$\Delta H^\circ = -897\text{ kJ} \qquad \Delta S^\circ = +145\text{ J/K}$$

At 298 K, 1 bar, the water product would be in its condensed phase, making the value of ΔS° negative. Nevertheless, the highly negative ΔH° value keeps ΔG° highly negative.

The products of this reaction are more dispersed than the reactants, and the reaction releases energy. Consequently, ΔG° is negative at all temperatures, and the reverse reaction cannot be made spontaneous by changing the temperature.

By the same reasoning, a *negative* ΔS° and a *positive* ΔH° oppose spontaneity, so a reaction in which the system becomes constrained and energy is absorbed is nonspontaneous regardless of temperature. The system and its surroundings both would experience decreases in entropy if such a process were to occur, and this would violate the second law of thermodynamics.

A reaction that has the same sign for ΔS° and ΔH° will be spontaneous at some temperatures but nonspontaneous at others. At low temperature, ΔS° is multiplied by a small value for T, so at sufficiently low temperature, ΔH° contributes more to ΔG° than $T\,\Delta S^\circ$ does. At high temperature, ΔS° is multiplied by a large value for T, so at sufficiently high temperature, ΔS° contributes more to ΔG° than does ΔH°.

Reactions that have positive ΔH° and positive ΔS° are favored by entropy but disfavored by enthalpy. Such reactions are spontaneous at high temperature, where the $T\,\Delta S^\circ$ term dominates ΔG°, because matter becomes dispersed during the reaction. A reaction is entropy-driven under these conditions. These reactions are nonspontaneous at low temperature, where the ΔH° term dominates ΔG°.

The opposite situation holds for reactions that have negative values for ΔH° and ΔS°. These reactions are spontaneous at low temperature because their release of heat disperses energy into the surroundings. The favorable ΔH° dominates ΔG° as long as T does not become too large, and the reaction is enthalpy-driven. At high temperature, however, the unfavorable ΔS° dominates ΔG°, and the reaction is no longer spontaneous. The effects of temperature on spontaneity are summarized in Table 14-3.

Table 14-3 Influence of Temperature on Spontaneity

$\Delta H°$	$\Delta S°$	$\Delta G°$ (high T)	$\Delta G°$ (low T)	Spontaneous
−	+	−	−	All T
+	−	+	+	No T
+	+	−	+	High T (entropy driven)
−	−	+	−	Low T (enthalpy driven)

Calcium sulfate, the substance used to absorb water in desiccators, provides an example of this temperature sensitivity. Anhydrous calcium sulfate absorbs water vapor from the atmosphere to give the hydrated salt. The reaction has a negative $\Delta S°$ because water molecules become more constrained when gaseous water molecules move into the solid state. The reaction also has a negative $\Delta H°$ because of the electrical forces of attraction between the ions of the salt and the polar water molecules:

$$CaSO_4\,(s) + 2\,H_2O\,(g) \longrightarrow CaSO_4{\cdot}2H_2O\,(s)$$

$$\Delta H° = -104.9\ \text{kJ} \qquad \Delta S° = -290.2\ \text{J/K}$$

At 298 K, the favorable $\Delta H°$ contributes more to $\Delta G°$ than the unfavorable $\Delta S°$:

$$\Delta G°_{298\ \text{K}} = (-104.9\ \text{kJ}) - (298\ \text{K})(-290.2\ \text{J/K})(10^{-3}\ \text{kJ/J}) = -18.4\ \text{kJ}$$

Thus, anhydrous calcium sulfate acts as a "chemical sponge" at room temperature, trapping water vapor spontaneously to form calcium sulfate dihydrate.

The calcium sulfate in a desiccator is effective at removing water vapor only as long as some anhydrous salt remains. When all the anhydrous salt has been converted to the dihydrate, the desiccator can no longer maintain a dry atmosphere. Fortunately, the thermodynamics of this reaction makes it possible to regenerate the drying agent. At 100 °C (373 K), $\Delta S°$ contributes more to $\Delta G°$ than does $\Delta H°$:

$$\Delta G°_{373\ \text{K}} = (-104.9\ \text{kJ}) - (373\ \text{K})(-290.2\ \text{J/K})(10^{-3}\ \text{kJ/J}) = +3.3\ \text{kJ}$$

At this temperature, the reverse reaction is spontaneous. Calcium sulfate dihydrate can be converted to anhydrous calcium sulfate in a drying oven at 100 °C. Then it can be cooled and returned to a desiccator, ready once more to act as a chemical sponge for water.

Desiccators are used to store chemicals that react slowly with water. Calcium sulfate chips in the bottom of the desiccator absorb water vapor from the atmosphere. The blue chips contain an indicator that turns pink when the calcium sulfate is saturated with water.

Section Exercises

■ **14.4.1** Estimate $\Delta G°$ for the formation of gaseous water at $T = 373$ K.

■ **14.4.2** Using Examples 14-7 and 14-8, find the minimum partial pressure of N_2O_4 at which decomposition of N_2O_4 occurs, if T is 127 °C and p of $NO_2 = 0.50$ bar.

■ **14.4.3** Does a temperature exist at which the water formation reaction is nonspontaneous under standard pressure? If so, compute this temperature. If not, explain why in molecular terms.

14.5 SOME APPLICATIONS OF THERMODYNAMICS

Nitrogen Fixation

Nitrogen is unevenly distributed between the Earth's crust and the atmosphere. In the crust, nitrogen is present at the level of 19 parts per million (ppm) by mass, four orders of magnitude less than oxygen (4.55×10^5 ppm) and silicon (2.72×10^5 ppm). In contrast, 80% of the atmosphere is molecular nitrogen. Paradoxically, nitrogen is essential for all life, but the sea of atmospheric nitrogen is virtually inaccessible to higher life forms. Most biochemical systems lack the ability to break the strong triple bond between the nitrogen atoms in N_2.

$$:N \equiv N:$$

Bond energy = 945 kJ/mol

Molecular nitrogen must be converted to some other species, usually ammonia (NH_3) or nitrate anions (NO_3^-), before most life forms can incorporate nitrogen atoms into their biochemical molecules. This process is known as **nitrogen fixation.**

In nature, nitrogen fixation is accomplished by nitrogenase, an enzyme that binds N_2 and weakens its bonding sufficiently to break the triple bond. Only a few algae and bacteria contain nitrogenase. Our Chemistry and Life Box describes what is known about this enzyme.

The thermodynamics of nitrogen chemistry helps explain why N_2 is so abundant in the atmosphere, and yet the element remains inaccessible to most life forms. Table 14-4 shows that most of the abundant elements react with O_2 spontaneously under standard conditions. This is why many of the elements occur in the Earth's crust as their oxides. However, N_2 is resistant to oxidation, as shown by the positive ΔG_f° for NO_2.

According to Table 14-4, chlorine is also resistant to oxidation. Unlike nitrogen, however, chlorine reacts spontaneously with metals to generate salts such as NaCl and $MgCl_2$. Thus, among abundant elements on Earth, nitrogen is uniquely stable in its elemental form.

Because of their resistance to chemical attack, nitrogen atoms are not "locked up" in solid or liquid substances as are other elements such as Si, Al, Fe, and H. The most stable form of the element nitrogen is a gaseous diatomic molecule with a very strong triple bond. Therefore, the element nitrogen is concentrated in the Earth's gaseous atmosphere even though it is only a trace element in overall abundance.

Every breath of air we take is 80% nitrogen, but our bodies must obtain the nitrogen required for biosynthesis from nitrogen contained in the proteins we eat. In the plant kingdom, the most important sources of nitrogen are NH_3 and the ammonium cation (NH_4^+).

Why is the nitrogen atom in NH_3 accessible to living organisms? Consider an organism synthesizing the amino acid glycine from its nitrogen-deficient precursor, acetic acid. In elemental terms, the net reaction is

$$CH_3CO_2H + N + H \longrightarrow H_2NCH_2CO_2H$$

Acetic acid

Glycine

Neither nitrogen nor hydrogen exists as free atoms, so the synthesis of glycine from acetic acid requires molecular sources of nitrogen and hydrogen. Atmospheric nitrogen and water are the most abundant sources of these two elements, but the production of glycine from N_2 and H_2O is significantly nonspontaneous under standard conditions:

$$4\ CH_3CO_2H + 2\ N_2 + 2\ H_2O \longrightarrow 4\ H_2NCH_2CO_2H + O_2$$

$$\Delta G_{reaction}^\circ = (1\ mol)(0\ kJ/mol) + (4\ mol)(-367\ kJ/mol) - (4\ mol)(-389\ kJ/mol)$$

$$-(2\ mol)(-237\ kJ/mol) - (2\ mol)(0\ kJ/mol) = +564\ kJ$$

Per mole of glycine, this reaction is "uphill" by 141 kJ.

Table 14-4   Surface-Abundant Elements and Their Oxides

Element	% by Mass	Oxide	ΔG_f° (kJ/mol)
O	49.1	O_2	0
Si	26.1	SiO_2	−856
Al	7.5	Al_2O_3	−1582
Fe	4.7	Fe_3O_4	−1015
Ca	3.4	CaO	−603
Na	2.6	Na_2O	−376
K	2.4	KO_2	−239
Mg	1.9	MgO	−569
H	0.88	H_2O	−237
Ti	0.58	TiO_2	−885
Cl	0.19	Cl_2O	+98
C	0.09	CO_2	−394
N	<0.1	NO_2*	+51

*Several other oxides of nitrogen exist. All have even more positive free energies of formation than NO_2 does.

Box 14-2	Chemistry and Life: How Does Nitrogenase Work?

I n the epic poem, *Rime of the Ancient Mariner,* a becalmed sailor laments "Water, water, everywhere, nor any drop to drink." With regard to the element nitrogen, we might say "N_2, N_2, everywhere, nor any N to eat." Even though the atmosphere of our planet is 80% molecular nitrogen, plants are unable to break the extremely strong $N \equiv N$ triple bond. Instead, plants must rely on other sources for the nitrogen atoms that are essential for the synthesis of biomolecules such as amino acids and DNA.

For nitrogen to be available to life-forms, it must be either oxidized to nitrate (NO_3^-) or reduced to ammonia (NH_3). Either process is known as *nitrogen fixation.* Lightning oxidizes atmospheric nitrogen to nitrate, but lightning accounts for only about 1% of the fixed nitrogen required by the biosphere. The fertilizer industry reduces N_2 to NH_3 by reaction with H_2, but fertilizers supply only about 30% of fixed nitrogen. The bulk—about 70%—comes from the activity of one group of bacteria that can convert N_2 to NH_3. These bacteria contain nitrogenase, an enzyme that carries out this difficult chemical transformation.

Researchers are working to understand how this enzyme works. Research on nitrogenase takes two main forms. One is an examination of the structure and operation of the enzyme to determine the details of the reactions by which N_2 is converted to ammonia. The other form is the synthesis of artificial catalysts that mimic the operation of nitrogenase.

Nitrogenase is a complex protein with two metal-containing parts that "dock" together during the fixation of nitrogen. Part 1 contains an interconnected group of seven iron atoms, nine sulfur atoms, and a single molybdenum atom. Molecular nitrogen interacts with this cluster, where it gains the six electrons and six protons needed to become two ammonia molecules. However, exactly how N_2 binds to the [7Fe–9S–Mo] group and how it is reduced to NH_3 remain a mystery. Recent structural determinations indicate that a relatively light atom, perhaps nitrogen, sits in the center of the cluster, interacting simultaneously with six of the iron atoms.

Part 1 of the nitrogenase protein contains another interconnected group of Fe–S atoms, this one with eight iron atoms and seven sulfur atoms. This [8Fe–7S] group collects electrons and transmits them to the binding center. Part 2 of nitrogenase contains a third Fe–S group, this one made up of four iron atoms and four sulfur atoms. This part of the enzyme also binds two molecules of ATP.

When the two parts of the enzyme join, the [4Fe–4S] group of Part 2 ends up close to the [8Fe–7S] group of Part 1. It is thought that when the parts of the enzyme join, two electrons are pumped from the ATP molecules to the [4Fe–4S] site, then to the [8Fe–7S] site, and finally to the [7Fe–9S–Mo] binding site. Next, the parts separate, and two new ATP molecules recharge Part 2. This process must occur four times to provide the eight electrons needed for the fixation reaction. The entire fixation process requires eight electrons and eight protons, six of each to form the ammonia molecule and two more that are "wasted" by forming an H_2 molecule.

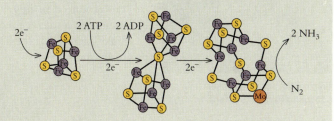

Although the mechanistic details of N_2 reduction have not been determined, chemists have been able to make a molybdenum-containing complex, shown below, that can bind N_2 and convert it to NH_3. In this complex, N_2 binds to Mo and is locked in place by three bulky groups. The complex enables N_2 to gain 6 electrons and 6 H^+, in steps, from other reagents. Further exploration of Mo–N_2 chemistry could lead to success in mimicking nitrogenase, or at least to a better understanding of how this remarkable enzyme works.

Perhaps chemists will be able to mimic nature without duplicating the iron-sulfur-molybdenum structure. For example, a zirconium complex with tetramethyl cyclopentadiene can bind dinitrogen in a manner that breaks the $N \equiv N$ bond, as shown below. Treatment with hydrogen gas results in formation of small amounts of ammonia. Although the yields are too low to make this a viable commercial process, researchers hope to make the process more efficient through chemical modifications and changes in conditions.

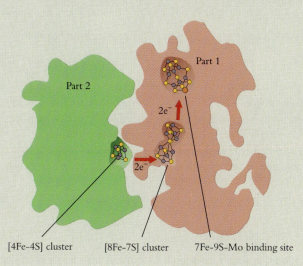

[4Fe-4S] cluster   [8Fe-7S] cluster   7Fe-9S-Mo binding site

On the other hand, NH_3 can provide both nitrogen and hydrogen atoms. Each NH_3 molecule also contains two extra hydrogen atoms that can be "burned" to produce water:

$$CH_3CO_2H + NH_3 + \tfrac{1}{2}O_2 \longrightarrow H_2NCH_2CO_2H + H_2O$$

$$\Delta G^\circ_{reaction} = [(1\ mol)(-237\ kJ/mol) + (1\ mol)(-367\ kJ/mol)]$$

$$-[(1\ mol)(-389\ kJ/mol) + (1\ mol)(-17\ kJ/mol) + (0.5\ mol)(0\ kJ/mol)]$$

$$\Delta G^\circ_{reaction} = -198\ kJ$$

This reaction is significantly spontaneous under standard conditions because it couples the production of water with the formation of the N—C bond in glycine.

The intensive agriculture that is characteristic of industrialized countries requires much more fixed nitrogen than is readily available from natural sources. Consequently, one of the major products of the chemical industry is nitrogen-containing fertilizers. Among the top 50 industrial chemicals, nitrogen gas (separated from air by cooling and liquefaction) perennially ranks number 2, with NH_3 (ammonia), HNO_3 (nitric acid), $(NH_2)_2CO$ (urea), and NH_4NO_3 (ammonium nitrate) all in the top 15. All these chemicals are produced in huge amounts because of their roles as fertilizers (urea and ammonium nitrate), fertilizer precursors (nitrogen gas and nitric acid), or both (ammonia).

Free energies of formation suggest that ammonia and ammonium nitrate could be produced spontaneously in nature:

$$N_2(g) + 3\ H_2(g) \longrightarrow 2\ NH_3(g) \qquad \Delta G^\circ_f = -32.8\ kJ$$

$$2\ N_2(g) + 3\ O_2(g) + 4\ H_2(g) \longrightarrow 2\ NH_4NO_3(s) \qquad \Delta G^\circ_f = -367.8\ kJ$$

Commercial agriculture uses substantial amounts of fertilizer.

These reactions might occur in an atmosphere containing significant amounts of molecular hydrogen, but on Earth, water is the only readily available source of hydrogen. The free energies of the reactions that yield NH_3 and NH_4NO_3 from gaseous nitrogen and water are highly unfavorable:

$$2\ N_2 + 6\ H_2O \longrightarrow 4\ NH_3 + 3\ O_2 \qquad \Delta G^\circ_{reaction} = +1358\ kJ$$

$$2\ N_2 + 4\ H_2O + O_2 \longrightarrow 2\ NH_4NO_3 \qquad \Delta G^\circ_{reaction} = +580\ kJ$$

Thus, large energy costs are involved in generating fixed nitrogen.

The strategy used in fertilizer production is to synthesize molecular hydrogen from methane:

$$CH_4 + H_2O \longrightarrow CO + 3\ H_2$$

$$\Delta G^\circ = +142\ kJ \qquad \Delta H^\circ = +205.9\ kJ \qquad \Delta S^\circ = +214.6\ J/K$$

Although this reaction is not spontaneous at room temperature, it becomes thermodynamically favorable at a temperature of 1000 K:

$$\Delta G^\circ_{1000\ K} = \Delta H^\circ - T\Delta S^\circ = +205.9\ kJ - (1000\ K)(214.6\ J/K)(10^{-3}\ kJ/J)$$

$$= -8.7\ kJ$$

Heat energy must be supplied continuously to keep the temperature from falling while the endothermic reaction proceeds. Otherwise, the temperature would quickly fall below the minimum value at which H_2 synthesis is spontaneous. For this reason, the production of H_2 (and ultimately of fertilizer) requires considerable amounts of energy. The energy cost is even greater if we consider that methane is obtained from nonrenewable reservoirs of natural gas, created over countless eons through the photosynthetic storage of solar energy.

With an ample supply of hydrogen, production of ammonia from N_2 becomes feasible:

$$N_2 + 3\ H_2 \longrightarrow 2\ NH_3 \qquad \Delta G^\circ_{reaction} = -32.8\ kJ$$

Figure 14-17 shows other spontaneous reactions that lead from ammonia to nitric acid, urea, and ammonium nitrate.

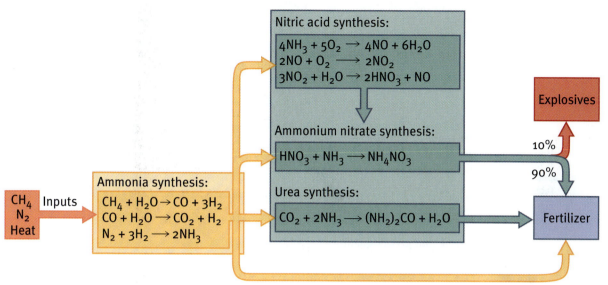

Nitric acid synthesis:

$4NH_3 + 5O_2 \longrightarrow 4NO + 6H_2O$
$2NO + O_2 \longrightarrow 2NO_2$
$3NO_2 + H_2O \longrightarrow 2HNO_3 + NO$

Ammonium nitrate synthesis:

$HNO_3 + NH_3 \longrightarrow NH_4NO_3$

Urea synthesis:

$CO_2 + 2NH_3 \longrightarrow (NH_2)_2CO + H_2O$

Ammonia synthesis:

$CH_4 + H_2O \rightarrow CO + 3H_2$
$CO + H_2O \rightarrow CO_2 + H_2$
$N_2 + 3H_2 \longrightarrow 2NH_3$

CH_4
N_2 Inputs
Heat

Explosives

10%

90%

Fertilizer

Figure 14-17
Block diagram of the industrial routes from nitrogen gas to fertilizers. Methane and energy in the form of heat are key ingredients in the first step.

Phase Changes

Phase changes, which convert a substance from one phase to another, have characteristic thermodynamic properties: Any change from a more constrained phase to a less constrained phase increases both the enthalpy and the entropy of the substance. Recall from our description of phase changes in Chapter 11 that enthalpy increases because energy must be provided to overcome the intermolecular forces that hold the molecules in the more constrained phase. Entropy increases because the molecules are more dispersed in the less constrained phase. Thus, when a solid melts or sublimes or a liquid vaporizes, both ΔH and ΔS are positive. Figure 14-18 summarizes these features.

As described in Section 14.4, when ΔH and ΔS have the same sign, the spontaneous direction of a process depends on T. For a phase change, enthalpy dominates ΔG at low temperature, and the formation of the more constrained phase is spontaneous. In contrast, entropy dominates ΔG at high temperature, and the formation of the less constrained phase is spontaneous. At one characteristic temperature, $\Delta G = 0$, and the phase

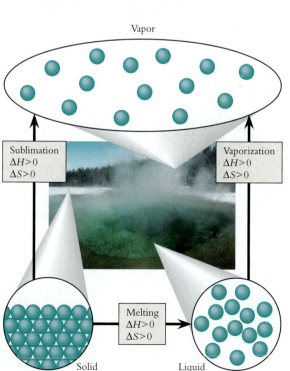

Vapor

Sublimation
$\Delta H > 0$
$\Delta S > 0$

Vaporization
$\Delta H > 0$
$\Delta S > 0$

Melting
$\Delta H > 0$
$\Delta S > 0$

Solid

Liquid

Figure 14-18
Schematic view of the three phase changes leading from more constrained to less constrained phases, illustrated by the phase changes for water. Each is accompanied by positive enthalpy and entropy changes for the substance.

change proceeds in both directions at the same rate. The two phases coexist, and the system is in a state of dynamic equilibrium.

The spontaneous direction of a phase change also depends on pressure, primarily because the molar entropy of a gas exhibits strong pressure-dependence:

$$S \text{ (gas)} = S° - R \ln p_{\text{gas}}$$

$$\Delta G_{\text{vap}} = \Delta H_{\text{vap}} - T \Delta S_{\text{vap}} = \Delta H_{\text{vap}} - T \Delta S°_{\text{vap}} + RT \ln p_{\text{gas}}$$

See Chapter 5 for information about vapor pressure. The vapor pressure of water at various temperatures appears in Table 5-4.

The result of this pressure dependence is that at any temperature, there is some pressure at which $\Delta G_{\text{vap}} = 0$. This is the vapor pressure (*vp*) of the substance at that temperature. We can relate *vp* to thermodynamic properties:

$$\Delta G_{\text{vap}} = 0 = \Delta H_{\text{vap}} - T \Delta S°_{\text{vap}} + RT \ln vp$$

$$RT \ln vp = -\Delta H_{\text{vap}} + T \Delta S°_{\text{vap}}$$

Rearranging gives Equation 14-14:

$$\ln vp = -\frac{\Delta H_{\text{vap}}}{RT} + \frac{\Delta S°_{\text{vap}}}{R} \qquad \text{(14-14)}$$

⑧

Equation 14-14 is one form of the Clausius–Clapeyron equation, which was first developed by Rudolf Clausius and Emile Clapeyron in the nineteenth century. Example 14-9 shows how to apply this equation.

Example 14-9
Vapor Pressure

Liquid bromine has a significant vapor pressure.

Although molecular bromine is a liquid at 298 K, 1 bar, it has a significant vapor pressure at this temperature. Calculate this vapor pressure in torr.

Strategy: We apply the condensed form of the seven-step strategy. To calculate a vapor pressure, we need the enthalpy and entropy of vaporization. Then we can apply Equation 14-14.

Solution: Appendix D contains thermodynamic information for molecular bromine:

$$Br_2 (l): \qquad \Delta H°_f = 0 \text{ kJ/mol} \qquad S° = 152.2 \text{ J mol}^{-1} \text{ K}^{-1}$$

$$Br_2 (g): \qquad \Delta H°_f = 30.9 \text{ kJ/mol} \qquad S° = 245.5 \text{ J mol}^{-1} \text{ K}^{-1}$$

We apply Equations 6-12 and 14-6 to the vaporization reaction:

$$Br_2 (l) \longrightarrow Br_2 (g)$$

$$\Delta H°_{\text{reaction}} = \Sigma \text{ coeff}_{\text{products}} \Delta H°_{f \text{ products}} - \Sigma \text{ coeff}_{\text{reactants}} \Delta H°_{f \text{ reactants}}$$

$$\Delta H°_{\text{reaction}} = 30.9 \text{ kJ} - 0 \text{ kJ} = 30.9 \text{ kJ}$$

$$\Delta S°_{\text{reaction}} = \Sigma \text{ coeff}_{\text{products}} S°_{\text{products}} - \Sigma \text{ coeff}_{\text{reactants}} S°_{\text{reactants}}$$

$$\Delta S°_{\text{reaction}} = 245.5 \text{ J/K} - 152.2 \text{ J/K} = 93.3 \text{ J/K}$$

Now substitute these values into Equation 14-14 and calculate the result:

$$\ln vp = -\frac{\Delta H_{\text{vap}}}{RT} + \frac{\Delta S°_{\text{vap}}}{R}$$

$$= -\frac{(30.9 \text{ kJ/mol})(10^3 \text{ J/kJ})}{(8.314 \text{ J mol}^{-1} \text{ K}^{-1})(298 \text{ K})} + \frac{(93.3 \text{ J mol}^{-1} \text{ K}^{-1})}{(8.314 \text{ J mol}^{-1} \text{ K}^{-1})}$$

$$\ln vp = -12.47 + 11.22 = -1.25 \qquad so \qquad vp = e^{-1.25} = 0.287 \text{ bar}$$

Remember that the standard unit of pressure is the bar. To complete the calculation, we need to convert from bar to torr (1 bar = 770 torr):

$$vp = 0.287 \text{ bar} \left(\frac{770 \text{ torr}}{1 \text{ bar}} \right) = 221 \text{ torr}$$

Does the Result Make Sense? We know that Br_2 is a liquid under standard conditions, but the margin photograph shows a distinct red color in the vapor phase, indicating a significant vapor pressure. The calculated result is significant but well below atmospheric pressure of 760 torr.

Use the values of standard thermodynamic functions to determine the vapor pressure of water, in torr, at 50 °C.

**Extra Practice
Exercise 14.9**

Thermal Pollution

One of the most important practical consequences of the second law of thermodynamics is that heat must be transferred from the system to its surroundings whenever a process is driven in the direction that increases constraints on the system. This heat raises the temperature and increases the entropy of the surroundings, thus offsetting the entropy decrease of the system. Refrigerators, computers, automobile engines, nuclear power plants, and humans are subject to this requirement. All these objects increase the temperature of their surroundings as they operate. For industrial-scale operations such as nuclear power plants, the amount of heat generated can have serious consequences. This type of heating is called **thermal pollution.**

Any thermally powered electrical power plant uses heat flows to convert stored energy into work. The stored energy may be chemical (coal, oil, or natural gas) or nuclear (uranium). When this potential energy is released, it creates a heat flow that is used to generate steam, which drives turbines to generate electricity. Electricity is the organized motion of electrons, all going in the same direction in a wire. The electrons become constrained, so there is a decrease in entropy that must be paid for by dispersing energy somewhere else. Consequently, a significant portion of the energy used for thermal generation of electrical power has to be "wasted" by dumping it into the surroundings as heat.

Power plants require large cooling towers to dissipate heat.

The increase in temperature that results from thermal power generation creates biological stresses on local ecosystems. Fish, being cold-blooded, are particularly sensitive to changes in water temperature. As water temperature increases, the blood temperature of the fish increases, and this in turn increases its rate of metabolism. At the same time, however, the amount of oxygen available to supply the metabolic demands of the fish decreases, because oxygen is less soluble in warm water than in cold water. Species that thrive in cold water, such as trout, die if the water temperature rises by more than a few degrees. Simpler forms of aquatic life also are highly sensitive to temperature. Plankton, which provide the food base for many aquatic ecosystems, thrive at temperatures between 14 °C and 24 °C. When water temperatures exceed 24 °C, blue-green algae crowd out these essential plankton, leading to serious disruption of the ecological balance. Furthermore, algal blooms consume the oxygen dissolved in water and can cause fish to suffocate.

Hydroelectric generation of electricity need not lead to thermal pollution, because flowing water can drive turbines without a large increase in the energy constraints. Unfortunately, there are other detrimental ecological consequences of hydroelectric power plants, such as the flooding of vast natural habitats, disruption of rivers and streams, and changes in local weather patterns caused by the creation of huge bodies of water. Although these effects may be less destructive than thermal pollution, no truly "clean" power source seems to exist. The second law of thermodynamics states that a price must always be paid for concentrating energy. Inevitably, human existence does this, so we pay this price continually. Conservation measures minimize the thermodynamic costs, but these costs can never be eliminated completely.

Section Exercises

14.5.1 Atmospheric nitrogen can be transformed to ammonia by various bacteria that fix nitrogen. The overall process is complicated, but the net reaction can be written as

$$2 N_2 + 6 H_2O \longrightarrow 4 NH_3 + 3 O_2$$

Assuming standard conditions (not quite true but close enough for approximations), how much free energy must the bacteria consume to fix one mole of N atoms?

14.5.2 All automobile engines operate at relatively high temperature and require continual cooling. Using our description of thermal pollution, explain how the second law of thermodynamics requires automobile engines to "dump" heat to their surroundings.

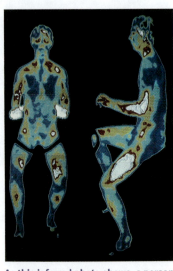

As this infrared photo shows, a person exercising emits heat to his/her surroundings.

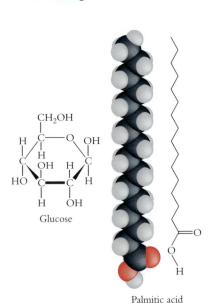

Glucose

Palmitic acid

14.6 BIOENERGETICS

Life creates order out of chaos, but thermodynamics shows that any spontaneous process must increase the total amount of dispersion in the universe. To constrain matter and energy in a living organism, it is necessary to release stored energy as heat, which, in turn, disperses energy in the surroundings. Consequently, living things must use large amounts of energy to survive. In this section, we describe representative energetic processes that operate in living organisms.

Biochemical Energy Production

Living organisms use carbohydrates as their source of energy. Plants make their own carbohydrates through photosynthesis. Animals, on the other hand, obtain carbohydrates by eating plants or other animals. Plants and animals use the oxidation of carbohydrates by O_2 as a source of free energy. They also transform carbohydrates into fats, which react spontaneously with O_2. The extraction of chemical energy from these reactions is an important component of metabolism. The highly spontaneous nature of these metabolic oxidation reactions is illustrated by glucose (a carbohydrate) and palmitic acid (a fatty acid, one component of fat):

$$\text{Glucose: } C_6H_{12}O_6 + 6 O_2 \longrightarrow 6 CO_2 + 6 H_2O \qquad \Delta G° = -2870 \text{ kJ}$$

$$\text{Palmitic acid: } C_{15}H_{31}CO_2H + 23 O_2 \longrightarrow 16 CO_2 + 16 H_2O \qquad \Delta G° = -9790 \text{ kJ}$$

The negative standard free energy changes of these reactions are due to two features. There is a large negative enthalpy change because the relatively weak $O{=}O$ bond in molecular oxygen is converted into stronger $O{-}H$ and $C{=}O$ bonds in H_2O and CO_2. Entropy also favors these reactions because there are many more molecules on the product side of the equation than on the reactant side (recall from Section 14.3 that entropy always increases when large molecules break apart into smaller ones).

The large negative free energy change that accompanies the oxidation of glucose and palmitic acid means that a little fat or carbohydrate goes a long way as a fuel for life processes. However, a living cell would be destroyed quickly if all this change in free energy occurred in a single burst. To harness the free energy released in these reactions without being destroyed, cells use elaborate chains of sequential reactions that harvest free energy a little at a time.

The free energy changes in metabolic oxidation reactions serve several purposes. Part of this energy appears as heat flows that maintain body temperature and disperse energy to the surroundings. Another portion of the energy is used to synthesize high-energy molecules that the body uses to drive the many reactions that occur within cells. In addition to storing part of the energy produced in the oxidation reactions, these high-energy species serve as energy transport molecules, moving to different regions of the cell where energy is required for various metabolic processes.

Adenosine triphosphate (ATP) is the most important of these energy transport molecules. Some of the energy released during the oxidation of glucose is used to drive a condensation reaction in which adenosine diphosphate (ADP) and phosphoric acid link

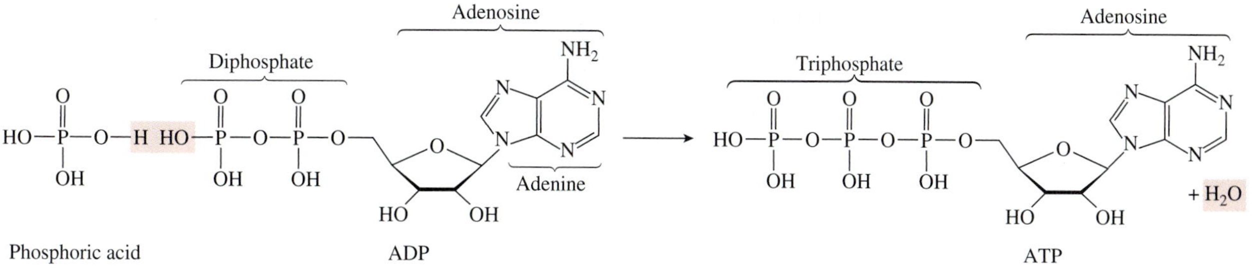

Figure 14-19
Phosphoric acid reacts with ADP to produce water and ATP.

together and eliminate water. This reaction stores chemical energy, as indicated by its positive standard free energy change:

$$ADP + H_3PO_4 \longrightarrow ATP + H_2O \qquad \Delta G° = +30.6 \text{ kJ}$$

Figure 14-19 illustrates the molecular details of this reaction. Although ATP is a complex molecule, the adenosine portion does not change during the condensation reaction. The condensation reaction adds a third phosphate group to the end of an existing chain. This reaction absorbs energy because the P—O—P and H—O—H bonds that form are somewhat weaker than the two polarized P—O—H linkages that break.

The exact processes by which carbohydrates and fats are converted to CO_2 and H_2O depend on the conditions and the particular needs of the cell. Each possible route involves a complex series of chemical reactions, many of which are accompanied by the conversion of ADP to ATP. One molecule of glucose, for example, is oxidized to CO_2 and H_2O in a sequence of many individual reactions that can convert as many as 36 ADP molecules into ATP molecules:

$$C_6H_{12}O_6 + 6 \, O_2 + 36 \, ADP + 36 \, H_3PO_4 \longrightarrow 6 \, CO_2 + 36 \, ATP + 42 \, H_2O$$

Coupled Reactions

Cells obtain energy by reversing the ADP–ATP reaction, converting ATP back to ADP and phosphoric acid:

$$ATP + H_2O \longrightarrow ADP + H_3PO_4 \qquad \Delta G° = -30.6 \text{ kJ}$$

The energy released in converting ATP to ADP is used to drive reactions that are non-spontaneous under physiological conditions. This is accomplished by coupling the two reactions in a single system.

For example, the amino acid glutamine is synthesized in cells by the reaction of ammonia with another amino acid, glutamic acid:

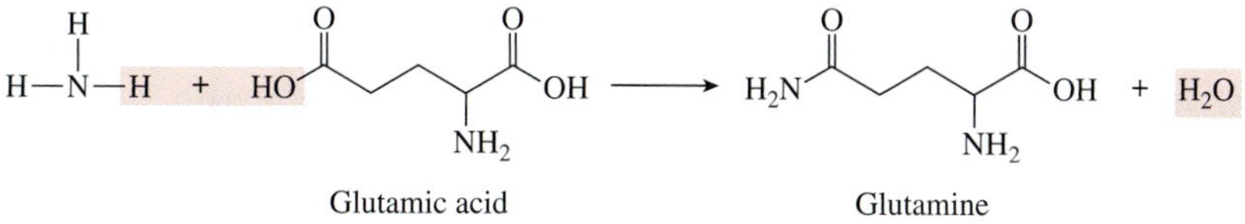

This reaction is thermodynamically unfavorable, $\Delta G° = +14$ kJ, but it can be driven by coupling it with the conversion of ATP into ADP. The net free energy change for the coupled process is the sum of the $\Delta G°$ values for the individual reactions:

$$\text{Glutamic acid} + NH_3 \longrightarrow \text{Glutamine} + H_2O \qquad \Delta G°_{glutamine} = +14 \text{ kJ}$$

$$ATP + H_2O \longrightarrow ADP + H_3PO_4 \qquad \Delta G°_{ATP} = -30.6 \text{ kJ}$$

Net: Glutamic acid + ATP + NH_3 ⟶ Glutamine + ADP + H_3PO_4

$$\Delta G°_{reaction} = \Delta G°_{glutamine} + \Delta G°_{ATP} = 14 \text{ kJ} + (-30.6 \text{ kJ}) = -17 \text{ kJ}$$

The negative value of $\Delta G°_{reaction}$ indicates that the free energy released in the ATP reaction is more than enough to drive the conversion of glutamic acid into glutamine.

For a spontaneous reaction to drive one that is nonspontaneous, the two sets of reactants must interact chemically. A coupled biochemical reaction occurs on the surface of a protein. The protein has a particular shape that accommodates the molecule that participates in the coupled reaction. In this way, the protein acts as an enabler for a particular biochemical reaction.

The net reaction between glutamic acid and ATP actually requires two steps, each of which is spontaneous. In the first step, an ATP molecule transfers a phosphate group to glutamic acid:

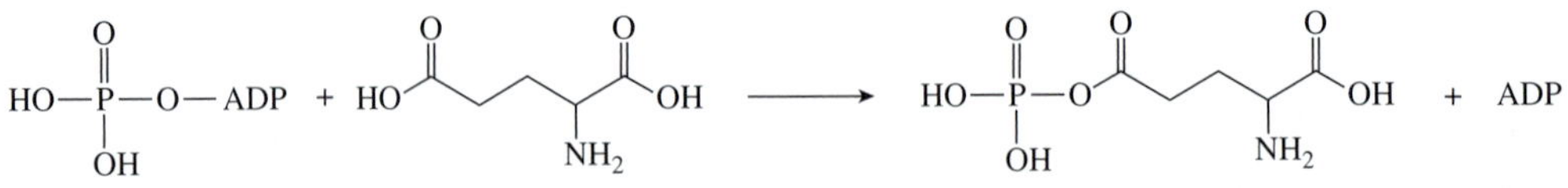

Glutamic acid

Next, an ammonia molecule reacts with the phosphorylated form of glutamic acid, producing phosphoric acid and glutamine:

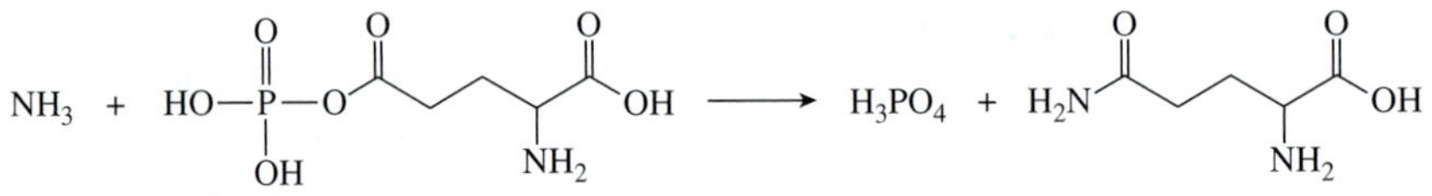

Phosphorylated glutamic acid

Overall, one molecule of ATP is converted to ADP and phosphoric acid for each molecule of glutamine produced from glutamic acid.

The synthesis of ATP from ADP and phosphoric acid is nonspontaneous. Consequently, ATP synthesis must be coupled to some more spontaneous reaction. Example 14-10 describes one of these reactions.

Example 14-10

ATP-Forming Reactions

One biochemical reaction that produces ATP involves the conversion of acetyl phosphate to acetic acid and phosphoric acid:

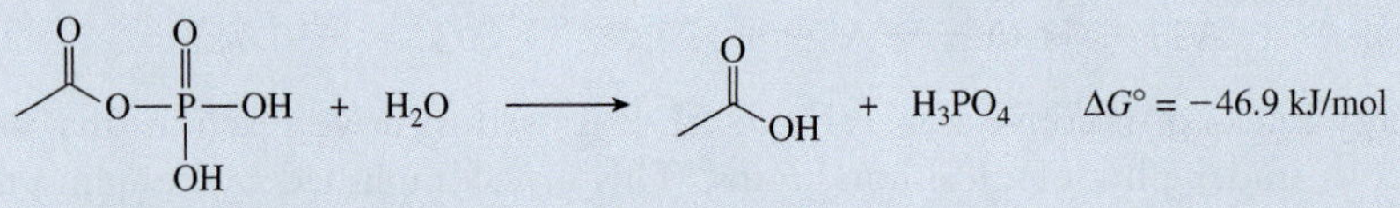

Acetyl phosphate ... Acetic acid ... $\Delta G° = -46.9$ kJ/mol

Couple this reaction with the ADP–ATP reaction, balance the net equation, and show that the coupled reaction is spontaneous.

Strategy: A coupled process links a spontaneous reaction with a nonspontaneous one. In this case, the negative free energy change of the acetyl phosphate reaction drives the conversion of ADP to ATP.

Solution: Combining the two reactions gives the overall balanced equation:

$$CH_3CO_2PO_3H_2 + H_2O \longrightarrow CH_3CO_2H + H_3PO_4$$

$$ADP + H_3PO_4 \longrightarrow ATP + H_2O$$

$$\text{Net: } CH_3CO_2PO_3H_2 + ADP \longrightarrow CH_3CO_2H + ATP$$

The net energy change for the coupled process is the sum of the $\Delta G°$ values for the individual reactions:

$$\Delta G°_{net} = \Delta G°_{acetyl\ phosphate} + \Delta G°_{ATP} = -46.9 \text{ kJ} + 30.6 \text{ kJ} = -16.3 \text{ kJ}$$

Does the Result Make Sense? The negative value of $\Delta G°_{net}$ shows that the free energy released in the acetyl phosphate reaction is more than enough to drive the conversion of ADP to ATP.

The coupled reaction that adds phosphate to glucose is spontaneous:

**Extra Practice
Exercise 14.10**

$$\text{Glucose} + \text{ATP} \longrightarrow \text{Glucose phosphate} + \text{ADP}, \Delta G_{net}^{\circ} = -16.7 \text{ kJ}$$

Write the separate reactions involved in this process and determine ΔG° for each.

Energy Efficiency

Cells store the energy released during the oxidation of glucose by converting ADP into ATP. The storage process cannot be perfectly efficient, however, because each step in the reaction sequence must have a negative free energy change. In practical terms, this requires that some energy be released to the surroundings as heat.

The oxidation of one mole of glucose can produce as many as 36 molecules of ATP under normal physiological conditions:

$$C_6H_{12}O_6 + 6\ O_2 + 36\ ADP + 36\ H_3PO_4 \longrightarrow 6\ CO_2 + 36\ ATP + 42\ H_2O$$

The overall free energy change for this process can be determined from the values for its uncoupled parts:

$$C_6H_{12}O_6 + 6\ O_2 \longrightarrow 6\ CO_2 + 6\ H_2O \qquad \Delta G^{\circ} = -2870 \text{ kJ}$$

$$36\ (ADP + H_3PO_4 \longrightarrow ATP + H_2O) \quad \Delta G^{\circ} = (36 \text{ mol})(+30.6 \text{ kJ/mol}) = +1100 \text{ kJ}$$

$$\Delta G_{overall}^{\circ} = -2870 \text{ kJ} + 1100 \text{ kJ} = -1770 \text{ kJ}$$

Although this coupled process stores 1100 kJ of free energy under standard conditions, 1770 kJ of free energy is "wasted." Thus, cells convert less than 40% of the chemical energy stored in glucose into chemical energy stored in ATP and later used to drive the biochemical machinery of metabolism. Well over half of the energy is dispersed as heat, increasing the entropy of the surroundings.

A functioning cell does not operate under standard conditions, so the above calculation is only an estimate. The standard free energy change for conversion of ADP to ATP is 30.6 kJ/mol. Because the concentrations of phosphate species in cells are far from 1 M, however, the free energy change under cellular conditions can be as low as about 28 kJ/mol or as high as about 50 kJ/mol, depending on the exact concentrations of the various species involved in the reaction. Nevertheless, the qualitative result holds under cellular conditions: over half the free energy released in the oxidation of glucose is dispersed as heat.

Fats such as palmitic acid are metabolized through pathways similar to the ones for the oxidation of glucose. The complete oxidation of palmitic acid has a standard free energy change of −9790 kJ/mol and produces 130 ATP molecules per molecule of palmitic acid consumed. You should be able to verify that this metabolic process has about the same efficiency as the oxidation of glucose.

Section Exercises

14.6.1 Nitrogen-fixing bacteria react N_2 with H_2O to produce NH_3 and O_2 using ATP as their energy source. About 24 molecules of ATP are consumed per molecule of N_2 fixed. What percentage of the free energy derived from ATP is stored in NH_3?

14.6.2 For the conversion of ATP to ADP, $\Delta H^{\circ} = -21.0$ kJ/mol and $\Delta G^{\circ} = -30.6$ kJ/mol at 298 K. Calculate

ΔS° for this reaction. What happens to the spontaneity of this reaction as the temperature increases to 37 °C?

14.6.3 In running a mile, an average person consumes about 500 kJ of free energy. (a) How many moles of ATP does this represent? (b) Assuming 38% conversion efficiency, how many grams of glucose must be "burned"?

SKILLS TO MASTER

① Assessing dispersal of matter and energy
② Calculating entropy changes at constant T
③ Calculating entropy changes with concentration
④ Calculating reaction entropies and free energies
⑤ Determining spontaneity from change in free energy
⑥ Estimating changes in free energy with T
⑦ Calculating changes in free energy with concentration
⑧ Applying thermodynamics to phase changes
⑨ Estimating energy efficiencies of biological reactions

14.1   SPONTANEITY

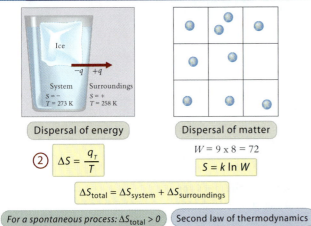

Spontaneous

①

Dispersal of energy   Dispersal of matter

Things tend to become dispersed.

14.2   ENTROPY: THE MEASURE OF DISPERSAL

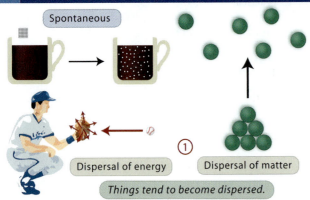

Ice
$-q$ $+q$
System   Surroundings
$S = -$ $S = +$
$T = 273$ K $T = 258$ K

Dispersal of energy   Dispersal of matter

$W = 9 \times 8 = 72$

② $\Delta S = \dfrac{q_T}{T}$

$S = k \ln W$

$\Delta S_{total} = \Delta S_{system} + \Delta S_{surroundings}$

For a spontaneous process: $\Delta S_{total} > 0$ Second law of thermodynamics

14.3   ENTROPIES OF PURE SUBSTANCES

Third law of thermodynamics $S_{pure,\ perfect\ crystal;\ 0\ K} = 0$

$S_{solid} < S_{liquid} < S_{gas}$

$S_{(p \neq 1\ bar)} = S^o - R \ln p$ $S_{(c \neq 1\ M)} = S^o - R \ln c$

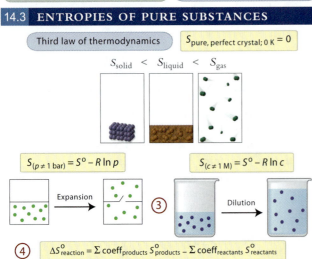

Expansion   ③   Dilution

④ $\Delta S^o_{reaction} = \Sigma\ coeff_{products}\ S^o_{products} - \Sigma\ coeff_{reactants}\ S^o_{reactants}$

14.4   SPONTANEITY AND FREE ENERGY

$\Delta G_{sys} < 0$ for any spontaneous process at constant T and P.

Free energy (G) $G = H - TS$

$\Delta G_{sys} = \Delta H_{sys} - T\Delta S_{sys}$ ⑤

$\Delta G^o_{sys} = \Delta H^o_{sys} - T\Delta S^o_{sys}$

⑥ $\Delta G^o_{reaction,\ T} = \Delta H^o_{reaction,\ 298\ K} - T\Delta S^o_{reaction,\ 298\ K}$

Standard molar free energy of formation (ΔG^o_f)

$\Delta G^o_{reaction} = \Sigma\ coeff_{products}\ \Delta G^o_{products} - \Sigma\ coeff_{reactants}\ \Delta G^o_{reactants}$

Reaction quotient (Q) $\dfrac{(c_D)^d (c_E)^e}{(c_A)^a (c_B)^b} = Q$

⑦ $\Delta G_{reaction} = \Delta G^o_{reaction} + RT \ln Q$

14.5   SOME APPLICATIONS OF THERMODYNAMICS

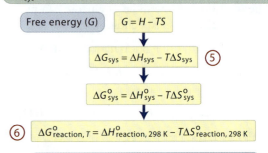

or   Phase changes

Nitrogen fixation

⑧

$\ln vp = \dfrac{\Delta H_{vap}}{RT} + \dfrac{\Delta S^o_{vap}}{R}$

Thermal pollution

Sublimation
$\Delta H > 0, \Delta S > 0$ Vapor   Vaporization
$\Delta H > 0, \Delta S > 0$

Melting
$\Delta H > 0$
$\Delta S > 0$

Solid   Liquid

14.6   BIOENERGETICS

Adenosine triphosphate (ATP)

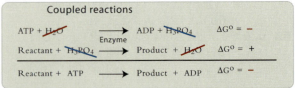

⑨

Coupled reactions

| | | $\Delta G^o = -$ |
ATP + H₂O $\xrightarrow{}$ ADP + H₃PO₄ $\Delta G^o = -$
Reactant + H₃PO₄ $\xrightarrow{Enzyme}$ Product + H₂O $\Delta G^o = +$

Reactant + ATP $\xrightarrow{}$ Product + ADP $\Delta G^o = -$

Learning Exercises

14.1 Describe the thermodynamic criteria for spontaneity. Describe in your own words what entropy is and why it is not a conserved quantity.
14.2 Make a list of everyday processes that illustrate that matter becomes less constrained.
14.3 Write a paragraph that explains why living creatures must convert some stored energy into heat in the course of their activities.

14.4 Update your list of memory bank equations. Describe the restrictions on each equation.
14.5 Make a list of all terms new to you that appear in Chapter 14. Using your own words, write a one-sentence definition for each. Consult the Glossary if you need help.

 Problems <u>ilw</u> = interactive learning ware problem. Visit the website at www.wiley.com/college/olmsted

Spontaneity

14.1 Describe the dispersals and constraints that accompany the following processes: (a) Ocean waves wash away a sand castle. (b) Water and acetone, two liquids, mix to form a homogeneous liquid solution. (c) In the child's game "pick up sticks," a bundle of sticks is dropped to the floor. (d) Water evaporates from a puddle after a summer shower.
14.2 Describe the dispersals and constraints accompanying the following processes: (a) A secretary "straightens up" the boss' desk. (b) Wood burns, producing CO_2 and H_2O vapors. (c) I_2 crystals form as a hot solution of I_2 in CCl_4 cools. (d) A skilled mechanic reassembles a torn-down engine.
14.3 The figure at right shows what happens when a drop of ink is added to a beaker of water. Using ideas of dispersal and constraints, explain what is occurring. 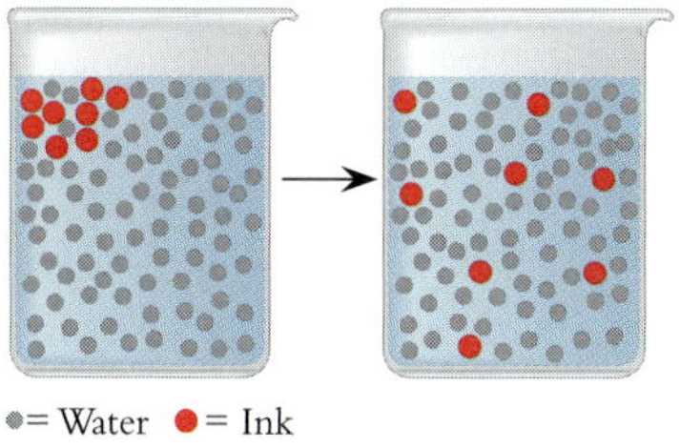
14.4 "All the king's horses and all the king's men couldn't put Humpty together again." Describe Humpty-Dumpty's fate in terms of constraints and dispersals.

●= Water ●= Ink

14.5 Explain each of the following observations using constraints and dispersals: (a) A puncture causes a tire to deflate. (b) An open bottle of perfume on a table eventually fills the room with the fragrance of the perfume.
14.6 Explain each of the following observations using constraints and dispersals: (a) A glass dropped on the floor shatters into many pieces. (b) The wind scatters raked leaves.
14.7 Use dispersal notions to explain why the following do not occur without the action of some outside force: (a) The handle of a spoon placed in ice water becomes too hot to handle. (b) Sand on a beach forms a sand castle.
14.8 Use dispersal notions to explain why the following do not occur without the action of some outside force: (a) Water runs uphill from the bottom of a dam to its top. (b) Air inflates a flat tire.

Entropy: The Measure of Dispersal

14.9 A tic-tac-toe game contains nine compartments. What is W for all possible first moves by each player (one X and one O placed in two different compartments)?
14.10 Consider a box of 16 equal-sized compartments into which two different colored marbles are to be placed. Calculate W for placing each in a different compartment and for placing both in the same compartment.
14.11 Ice melts at 273.15 K with $\Delta H_{fus} = 6.01$ kJ/mol. An ice cube whose mass is 13.8 g is dropped into a swimming pool whose temperature is held at 27.5 °C. (a) What is the entropy change (ΔS) for melting the ice? (b) What is ΔS of the pool? (c) What is the overall ΔS?
14.12 Solid CO_2 (dry ice) sublimes with $\Delta H_{subl} = 25.2$ kJ/mol at 195 K. A block of dry ice whose mass is 27.5 g sublimes in a room whose temperature is 26.5 °C. (a) What is the entropy change (ΔS) of the CO_2? (b) What is ΔS of the room? (c) What is the overall ΔS?

14.13 Without doing any calculations, determine the signs of ΔS for the system, for the surroundings, and for the overall process in each of the following: (a) Water boils in a teakettle on a hot stove. (b) Ice in an ice cube tray, left on a counter top, melts. (c) A cup of coffee is reheated in a microwave oven.
14.14 Without doing any calculations, determine the signs of ΔS for the system, for the surroundings, and for the overall process in each of the following: (a) Ice forms on the surface of a birdbath in winter. (b) A hot cup of coffee cools when left to stand. (c) A Popsicle melts when left on a table.
<u>ilw</u> **14.15** Calculate the entropy change of 15.5 g of steam that condenses to liquid water at 373.15 K. Without doing additional calculations, what can you say about the entropy change of the surroundings?
14.16 Calculate the entropy change of 75.4 g of water that freezes to ice at 273.15 K. Without doing additional calculations, what can you say about the entropy change of the surroundings?
14.17 Table 11-4 lists molar enthalpies of fusion of several substances. Calculate the molar entropy of fusion at its normal melting point for each of the following: (a) argon; (b) methane; (c) ethanol; and (d) mercury.
14.18 Table 11-4 lists molar enthalpies of vaporization of several substances. Calculate the molar entropy of vaporization at its normal boiling point for each of the following: (a) molecular oxygen; (b) ethane; (c) benzene; and (d) mercury.

Entropies of Pure Substances

14.19 For each of the following pairs of substances, determine which has the larger molar entropy at 298 K and state the main reason for the difference: (a) NaCl (aq) and $MgCl_2$ (aq); (b) HgO (s) and HgS (s); and (c) Br_2 (l) and I_2 (s).
14.20 For each of the following pairs of substances, determine which has the larger molar entropy at 298 K and state the main reason for the difference: (a) Br_2 and Cl_2; (b) Ni and Pt; (c) C_5H_{12} (pentane, liquid) and C_8H_{18} (octane, liquid); and (d) SiF_4 and CH_4 (both gases).
14.21 Oxygen, ozone, and hydrogen are gases at $T = 298$ K, $P = 1$ bar. Their molar entropies are in the sequence $H_2 < O_2 < O_3$. Using molecular properties, explain why ozone has more entropy than oxygen but hydrogen has less entropy than either.
14.22 Mercury, water, and bromine are liquids at $T = 298$ K, $P = 1$ bar. Their molar entropies are in the sequence $H_2O < Hg < Br_2$. Using molecular properties, explain why bromine has more entropy than mercury but water has the least entropy of these three.
14.23 Using tabulated values of $S°$, calculate the standard entropy per mole of atoms for He, H_2, CH_4, and C_3H_6. Explain the trend that you find in terms of what bond formation does to atomic constraints.
14.24 Using tabulated values of $S°$, calculate the standard entropy per mole of atoms for Ar, O_2, O_3, and C_2H_6. Explain the trend that you find in terms of what bond formation does to atomic constraints.
14.25 Compute the absolute entropy of the following: (a) 2.50 mol of Ar gas at $p = 0.25$ bar; (b) 0.75 mol of O_3 gas at $p = 2.75$ bar; and (c) 0.45 mol of a mixture of N_2 ($X = 0.78$) and O_2 ($X = 0.22$) at $P_{tot} = 1.00$ bar.
14.26 Compute the absolute entropy of the following: (a) 1.00 mol of molecular hydrogen gas at $p = 5.0$ bar; (b) 0.25 mol of ethane gas at $p = 0.10$ bar; and (c) 1.00 mol of a mixture of N_2 ($p = 125$ bar) and H_2 ($p = 375$ bar).

14.27 Compute the standard entropy change for the following reactions:
(a) $N_2(g) + 3 H_2(g) \longrightarrow 2 NH_3(g)$
(b) $3 O_2(g) \longrightarrow 2 O_3(g)$
(c) $PbO_2(s) + 2 Ni(s) \longrightarrow Pb(s) + 2 NiO(s)$
(d) $C_2H_4(g) + 3 O_2(g) \longrightarrow 2 CO_2(g) + 2 H_2O(l)$

ilw

14.28 Compute the standard entropy change for the following reactions:
(a) $2 H_2(g) + O_2(g) \longrightarrow 2 H_2O(l)$
(b) $C(s) + 2 H_2(g) \longrightarrow CH_4(g)$
(c) $CH_3CH_2OH(l) + 3 O_2(g) \longrightarrow 2 CO_2(g) + 3 H_2O(l)$
(d) $Fe_2O_3(s) + 2 Al(s) \longrightarrow Al_2O_3(s) + 2 Fe(s)$

14.29 For each reaction in Problem 14.27, explain what features of the atomic constraints for the reactants and products account for the magnitude and sign of $\Delta S°$.

14.30 For each reaction in Problem 14.28, explain what features of the atomic constraints for the reactants and products account for the magnitude and sign of $\Delta S°$.

14.31 Without doing calculations or looking up absolute entropy values, determine the sign of $\Delta S°$ for the following processes:
(a) $CO_2(s) \longrightarrow CO_2(g)$
(b) $2 Na(s) + F_2(g) \longrightarrow 2 NaF(s)$
(c) $2 CO(g) + O_2(g) \longrightarrow 2 CO_2(g)$

14.32 Without doing calculations or looking up absolute entropy values, determine the sign of $\Delta S°$ for the following processes:
(a) $MgCl_2(s) \longrightarrow Mg^{2+}(aq) + 2 Cl^-(aq)$
(b) $SF_6(g) \longrightarrow SF_6(l)$
(c) $2 CH_3OH(l) + 3 O_2(g) \longrightarrow 2 CO_2(g) + 4 H_2O(g)$

Spontaneity and Free Energy

14.33 Each of the following statements is false. Rewrite each so that it makes a correct statement about free energy: (a) $\Delta G_{total} > 0$ for any spontaneous process. (b) ΔG_{system} increases in any process at constant T and P. (c) $\Delta H = \Delta G - T \Delta S$.

14.34 Each of the following statements is false. Rewrite each so that it makes a correct statement about free energy: (a) In any process at constant T and P, the overall free energy decreases. (b) $\Delta G_{sys} = 0$ for any spontaneous process. (c) $\Delta G = \Delta H + \Delta S$.

14.35 Compute the standard free energy change for each reaction in Problem 14.27.

14.36 Compute the standard free energy change for each reaction in Problem 14.28.

14.37 Estimate $\Delta G°$ at -85 °C for the reactions in Problem 14.27 (b) and (c).

14.38 Estimate $\Delta G°$ at 825 °C for the reactions in Problem 14.28 (b) and (d).

14.39 Calculate $\Delta G_{reaction}$ at 298 K for the reactions in Problem 14.27 (a) and (b) if $p = 15$ bar for each gaseous substance.

14.40 Calculate $\Delta G_{reaction}$ at 298 K for the reactions in Problem 14.28 (a) and (b) if $p = 0.25$ bar for each gaseous substance.

Some Applications of Thermodynamics

ilw **14.41** Compute $\Delta H°$, $\Delta S°$, and $\Delta G°$ for the production of NH_4NO_3 from ammonia and oxygen:

$$2 NH_3(g) + 2 O_2(g) \longrightarrow NH_4NO_3(s) + H_2O(l)$$

(This reaction is not feasible industrially because NH_3 combustion cannot be controlled to give NH_4NO_3 as a product.)

14.42 Compute $\Delta H°$, $\Delta S°$, and $\Delta G°$ for the production of urea from ammonia (see Figure 14-17).

14.43 Without doing any calculations, answer the following for the formation of NO_2 from N_2 and O_2, and state your reasons: (a) Is there any temperature at which the reaction is spontaneous when all reagents are at 1 bar pressure? (b) At 298 K, is there any combination of partial pressures of the reagents that would make the reaction spontaneous?

14.44 Without doing any calculations, answer the following for the formation of ClO_2 from Cl_2 and O_2, and state your reasons: (a) Is there any temperature at which the reaction is spontaneous when all reagents are at 1 bar pressure? (b) At 298 K, is there any combination of partial pressures of the reagents that would make the reaction spontaneous?

ilw **14.45** Using data from Appendix D, calculate the vapor pressure of $P_4(g)$ over solid white phosphorus at 298 K.

14.46 Using data from Appendix D, calculate the vapor pressure of $Hg(g)$ over liquid mercury at 298 K.

Bioenergetics

14.47 Two children on opposite ends of a seesaw can be used as an analogy for a coupled reaction. Describe the coupling of spontaneous and nonspontaneous processes during the actions of a seesaw.

14.48 A pair of weights and a pulley can be used as an analogy for a coupled reaction. Use the figure to describe a coupled process. Identify spontaneous and nonspontaneous portions of the process.

14.49 Glucose and fructose combine to make sucrose in a condensation reaction for which $\Delta G° = 23.0$ kJ/mol:

$$C_6H_{12}O_6 + C_6H_{12}O_6 \longrightarrow C_{12}H_{22}O_{11} + H_2O$$

Glucose   Fructose   Sucrose

Write the balanced equation for this reaction coupled with the ADP–ATP reaction and operating in the direction of overall spontaneity. Calculate $\Delta G°$ for the overall process.

14.50 Although the ATP–ADP reaction is the principal energy shuttle in metabolic pathways, many other examples of coupled reactions exist. For example, the glutamic acid–glutamine reaction discussed in the text can couple with the acetyl phosphate reaction shown in Example 14-10. Write the balanced equation for the coupled reaction operating in the direction of overall spontaneity and calculate $\Delta G°$ for the overall process.

Additional Paired Problems

14.51 For the chemical reaction

$$Al_2O_3(s) + 3 H_2(g, p = 1 \text{ bar}) \longrightarrow 2 Al(s) + 3 H_2O(l)$$

at $T = 298$ K, answer each of the following by doing a quantitative calculation of the appropriate thermodynamic function: (a) Is this a spontaneous reaction? (b) Does the reaction absorb or release heat? (c) Are the products more or less dispersed than the reactants?

14.52 For the chemical reaction

$$3 Fe(s) + 4 H_2O(l) \longrightarrow Fe_3O_4(s) + 4 H_2(g, p = 1 \text{ bar})$$

at $T = 298$ K, answer each of the following by doing a quantitative calculation of the appropriate thermodynamic function: (a) Is this a spontaneous reaction? (b) Does the reaction absorb or release heat? (c) Are the products more or less dispersed than the reactants?

14.53 Use data from Appendix D to compute $\Delta S°$ for each of the following reactions:
(a) $2 ClO_2^-(aq) + O_2(g) \longrightarrow 2 ClO_3^-(aq)$
(b) $4 FeCl_3(s) + 3 O_2(g) \longrightarrow 2 Fe_2O_3(s) + 6 Cl_2(g)$
(c) $3 N_2H_4(l) + 4 O_3(g) \longrightarrow 6 NO(g) + 6 H_2O(l)$

14.54 Calculate the standard entropy change at 298 K of each of the following reactions, which are important in the chemistry of coal. Assume that coal has the same thermodynamic properties as graphite.

(a) C $(s, coal)$ + H_2O (g) ⟶ CO (g) + H_2 (g)
(b) C $(s, coal)$ + O_2 (g) ⟶ CO_2 (g)
(c) C $(s, coal)$ + $\frac{1}{2}$ O_2 (g) ⟶ CO (g)
(d) CO (g) + H_2O (g) ⟶ CO_2 (g) + H_2 (g)

14.55 Use data from Appendix D to compute $\Delta G°$ for each reaction in Problem 14.53.

14.56 Use data from Appendix D to compute $\Delta G°$ for each reaction in Problem 14.54.

14.57 Use data from Appendix D to compute $\Delta G°$ at 85.0 °C for each reaction in Problem 14.53.

14.58 Use data from Appendix D to compute $\Delta G°$ at 200.0 °C for each reaction in Problem 14.54.

14.59 For the following constant-temperature process, give the sign (+, −, or 0) for each of the specified thermodynamic functions. In each case give a brief account of your reasoning: (a) w_{sys}; (b) q_{sys}; and (c) ΔS_{surr}. (*Hint:* Review Chapter 6.)

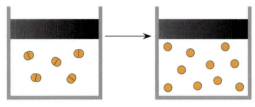

14.60 For the following constant-temperature process, give the sign (+, −, or 0) for each of the specified thermodynamic functions. In each case give a brief account of your reasoning: (a) q_{sys}; (b) ΔS_{sys}; and (c) ΔE_{total}. (*Hint:* Review Chapter 6.)

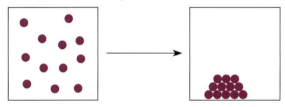

14.61 One possible source of acid rain is the reaction between NO_2, a pollutant from automobile exhausts, and water:

$$3 NO_2 (g) + H_2O (l) \longrightarrow 2 HNO_3 (g) + NO (g)$$

Determine whether this is thermodynamically feasible (a) under standard conditions at 298 K; (b) under standard conditions at 277 K; and (c) at 298 K with each product gas present at $p = 1.00 \times 10^{-6}$ bar.

14.62 Filaments of elemental boron might be made by reducing boron trichloride with molecular hydrogen:

$$2 BCl_3 (g) + 3 H_2 (g) \longrightarrow 2 B (s) + 6 HCl (g)$$

Determine whether this is thermodynamically feasible (a) under standard conditions at 298 K; (b) under standard conditions at 995 K; and (c) at 995 K if each reactant gas is present at $p = 5.5$ bar and HCl pressure is 0.1 bar.

14.63 ATP and ADP have been referred to as the "energy currency" of the cell. Explain this analogy.

14.64 Explain why a nuclear power plant must contribute thermal pollution to the environment.

14.65 Use data from Appendix D to answer quantitatively the following questions about the dissolving of table salt in water under standard conditions:

$$NaCl (s) \longrightarrow Na^+ (aq) + Cl^- (aq)$$

(a) Is this a spontaneous reaction? (b) Does it release energy? (c) Does the chemical system become more constrained?

14.66 Repeat the calculations of Problem 14.65 for silver chloride, AgCl, dissolving in water. What do the results reveal about the difference in solubility of these two salts and the reasons for the difference?

14.67 An ice cube tray containing 155 g of liquid H_2O at 0.0 °C is placed in a freezer whose temperature is −20.0 °C. As soon as all the H_2O has frozen, the tray is removed from the freezer. (a) Find ΔS for the H_2O. (b) Find ΔS for the overall process. (c) If the tray were left in the freezer until its temperature reached −20 °C, would there be an additional net entropy change? Explain.

14.68 Find ΔS for the system, surroundings, and overall when 25.0 g of liquid H_2O is evaporated at 100. °C, if the heat required is provided by a hot plate whose temperature is 315 °C.

14.69 Although ammonia can be used directly as a fertilizer, much of it is converted to urea or ammonium nitrate before being applied in the field. Considering the physical properties of these substances, suggest why ammonia is not preferred as a fertilizer.

14.70 What are the four most important nitrogen-containing chemicals, and for what are they used?

14.71 Determine $\Delta G_{combustion}$ per gram of palmitic acid and of glucose. On a per-gram basis, which is the better energy source?

14.72 The most important commercial process for generating hydrogen gas is the water–gas shift reaction:

$$CH_4 (g) + H_2O (g) \longrightarrow CO (g) + 3 H_2 (g)$$

Use tabulated thermodynamic data to find $\Delta G°$ and $\Delta G°_{1300}$.

14.73 Without doing any calculations, determine the sign of ΔS for the system for each of the following changes: (a) A soft drink is chilled in an ice chest. (b) The air in a bicycle pump is compressed. (c) A carton of juice concentrate is mixed with water.

14.74 Without doing any calculations, determine the sign of ΔS for the system for each of the following changes: (a) A cold soft drink warms as it sits in the sun. (b) Air leaks out of a punctured automobile tire. (c) Solid AgCl precipitates from solution when silver nitrate solution is mixed with sodium chloride solution.

14.75 A teaspoon from a freezer is placed in a glass of water, and the two equilibrate to the same temperature. In this process, what can you deduce about each of the following? (a) ΔE_{total}; (b) $\Delta E_{teaspoon}$; (c) ΔS_{total}; (d) ΔS_{water}; and (e) $q_{teaspoon}$.

14.76 A pie is removed from a hot oven and allowed to cool to room temperature. In this process, what can you deduce about the changes in each of the following? (a) ΔE_{total}; (b) ΔE_{pie}; (c) ΔS_{total}; (d) ΔS_{pie}; and (e) q_{pie}.

14.77 Arrange the following in order of increasing entropy, from smallest to largest value: 1.0 mol H_2O (liquid, 373 K); 0.50 mol H_2O (liquid, 298 K); 1.0 mol H_2O (liquid, 298 K); 1.0 mol H_2O (gas, 373 K, 1.0 bar); and 1.0 mol H_2O (gas, 373 K, 0.1 bar).

14.78 Arrange the following in order of increasing entropy, from smallest to largest value: 1.00 g Br_2 (gas, 331.9 K, 0.10 bar); 1.00 g Br_2 (gas, 331.9 K, 1.00 bar); 1.00 g Br_2 (liquid, 331.9 K, 1.00 bar); and 1.00 g Br atoms (gas, 331.9 K, 0.10 bar).

14.79 One reaction that generates pollutants in the internal combustion engine is the oxidation of nitrogen by oxygen:

$$N_2 (g) + O_2 (g) \longrightarrow 2 NO (g)$$

Determine $\Delta H°$, $\Delta S°$, and $\Delta G°$ for this reaction at room temperature, and estimate the temperature above which it becomes spontaneous at standard pressure. Given this result, explain how NO can form in automobile engines.

14.80 Advertisements that state "diamonds are forever" assume that the reaction, C (*diamond*) ⟶ C (*graphite*) is not spontaneous. Determine $\Delta H°$, $\Delta S°$, and $\Delta G°$ for this reaction at room temperature. Is there a temperature at which this reaction is not spontaneous? If so, determine that temperature.

14.81 The normal boiling point of benzene is 80.1 °C, at which $\Delta H°_{vap} = 30.77$ kJ/mol. (a) Determine $\Delta S°_{vap}$ and $\Delta G°_{vap}$ for benzene at its normal boiling point. (b) Estimate $\Delta G°_{vap}$ for benzene at 21 °C. (c) Calculate the vapor pressure of benzene at 25 °C.

14.82 The normal boiling point of ethane is 184 K, at which $\Delta H°_{vap} = 15.5$ kJ/mol. (a) Determine $\Delta S°_{vap}$ and $\Delta G°_{vap}$ for ethane at its normal boiling point. (b) Estimate $\Delta G°_{vap}$ for ethane at 21 °C. (c) Calculate the vapor pressure of ethane at −65 °C.

More Challenging Problems

14.83 A sample of liquid Br_2 (0.080 mol) at 332 K is placed in a cylinder equipped with a piston at $P_{ext} =$ 1.0 bar. The cylinder is immersed in a constant temperature bath at 332 K, the boiling point of Br_2 ($\Delta H_{vap} =$ 30.9 kJ/mol). The system undergoes a constant-temperature, constant-pressure expansion to a final volume of 2.2 L. The sketch shows the apparatus immediately after the cylinder is placed in the bath. (a) Draw a sketch that shows how the system looks when it reaches its final state. (b) Determine the total ΔS for this process.

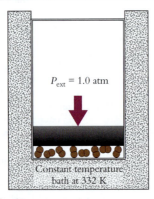

$P_{ext} = 1.0$ atm

Constant temperature bath at 332 K

14.84 The notion of thermodynamic coupling of a nonspontaneous process with a spontaneous process is not restricted to chemical reactions. Identify the spontaneous and nonspontaneous portions of the following coupled processes: (a) Water behind a dam passes through a turbine and generates electricity. (b) A gasoline engine pumps water from a valley to the top of a hill.

14.85 Crystalline KCl has $S° = 83$ J mol^{-1} K^{-1}, and crystalline CaO has $S° = 55$ J mol^{-1} K^{-1}. What accounts for the larger entropy of KCl crystals?

14.86 Both CCl_4 (carbon tetrachloride) and CS_2 (carbon disulfide) are liquids used as solvents in special industrial applications. (a) Using data from Appendix D, calculate $\Delta H°$ and $\Delta G°$ for combustion of these liquids:

$$CCl_4(l) + 5\,O_2(g) \longrightarrow CO_2(g) + 4\,ClO_2(g)$$

$$CS_2(l) + 3\,O_2(g) \longrightarrow CO_2(g) + 2\,SO_2(g)$$

(b) What features of the reactants and products account for the differences? (c) Based on your results, would you recommend special precautions against fires for industrial plants using either solvent? Explain your recommendations.

14.87 In a system operating without any restrictions, heat is not a state function. However, heat flow describes a state function change under certain restricted conditions. For each of the following conditions, identify the state function that corresponds to heat flow: (a) q_V; (b) q_P; and (c) q_T.

14.88 A piece of dry ice at $T = 195$ K is dropped into a beaker containing water at $T = 273.15$ K. You may assume that all of the resulting CO_2 vapor escapes into the atmosphere. (a) Describe the process that takes place at the molecular level. (b) Calculate the overall entropy change if 12.5 g of dry ice undergoes this process.

14.89 At its triple point, a dynamic equilibrium can exist among all three phases of matter. Draw a molecular picture of argon that shows what happens at the triple point. What is ΔG for each of the processes under these conditions? Describe the matter and energy dispersal taking place for each of the processes.

14.90 The following diagram represents two flasks connected by a valve. Each flask contains a different gas. (a) If the valve is opened, the two gases can move back and forth between the two flasks. Redraw the figure to show the system at maximum entropy. (b) Redraw the figure in a way that shows the state of the system at its lowest possible entropy. (The substances need not remain gaseous.)

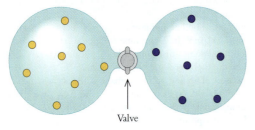

Valve

14.91 Without doing any calculations, predict the signs (+, −, or 0) of ΔH and ΔS for the following processes, all occurring at constant T and P. Explain your predictions.
(a) $2\,NO_2(g) \longrightarrow N_2O_4(g)$
(b) A block of gold melts in a jeweler's crucible
(c) $CH_4(g) + 2\,O_2(g) \longrightarrow CO_2(g) + 2\,H_2O(g)$

14.92 Heat pumps extract heat from a cold reservoir (usually the ground) and use it to maintain the temperature of a warmer reservoir (for example, a home). Explain why heat pumps cannot operate without converting some additional energy into heat.

14.93 For the ATP → ADP reaction, $\Delta G° = -30.6$ kJ/mol. Under cellular conditions, this reaction can release more than 50 kJ/mol. List the ways in which cellular conditions differ from standard conditions and describe the effect that each difference has on the free energy change.

14.94 On a hot day, one of your friends suggests opening the door of your refrigerator to cool your kitchen. Will this strategy work? Explain your answer using principles of thermodynamics.

14.95 What is the efficiency of the metabolic conversion of palmitic acid to ATP? Compute the number of grams of palmitic acid that would have to be metabolized to provide the heat to warm a swimmer from whose skin 75 g of water evaporates.

14.96 Few of the elements exist on the Earth in their pure form. Most elements exist as oxides or sulfides. Explain why this is so, and explain why nitrogen exists as a pure element.

14.97 A firefly produces light by converting the compound luciferin to oxyluciferin:

COOH

+ ATP ⟶

HO

Luciferin

COAMP

+ Diphosphate + H_2O

HO

Luciferyl adenylate

COAMP

+ O_2 ⟶

HO

O

+ AMP + CO_2 + light

O

Oxyluciferin

Answer the following for each of these processes: (a) Is energy absorbed or released? (b) Does system entropy increase, decrease, or remain nearly the same? (c) Is ΔG negative or positive? Explain your reasoning.

14.98 Humans perspire as a way of keeping their bodies from overheating during strenuous exercise. The evaporation of perspiration transfers heat from the body to the surrounding atmosphere. Calculate the total ΔS for evaporation of 1.0 g of water if the skin is at 37.5 °C and air temperature is 23.5 °C.

14.99 Two moles of HCl gas have a larger standard entropy, 374 J mol^{-1} K^{-1}, than 1 mol each of H_2 gas and Cl_2 gas, 354 J mol^{-1} K^{-1}. What feature of the molecular structures of these substances accounts for this difference?

Group Study Problems

14.100 The enthalpy of sublimation of ice at 273.15 K is not the simple sum of the enthalpies of fusion and vaporization of water, but it can be calculated using Hess' law and an appropriate path that includes fusion and vaporization. Devise such a path, show it on a phase diagram for water, and carry out the calculation, making reasonable assumptions if necessary (C(liquid water) = 75.3 J mol^{-1} K^{-1}, and C(water vapor) = 33.6 J mol^{-1} K^{-1}).

14.101 In the upper atmosphere, ozone is produced from oxygen:

$$3\ O_2\ (g) \longrightarrow 2\ O_3\ (g)$$

(a) Compute $\Delta H°$, $\Delta S°$, and $\Delta G°$ for this reaction. (b) Is there a temperature at which this reaction becomes spontaneous at 1 bar pressure? If so, find it. If not, explain why one does not exist. (c) Assume an atmosphere with $p_{O_2} = 0.20$ bar and $T = 298$ K. Below what pressure of O_3 is O_3 production spontaneous? (d) In view of your answers to parts (a) through (c), how can the ozone layer form? (You may need to review Section 7.7.)

14.102 $S°$ of graphite is 3 times larger than $S°$ of diamond. Explain why this is so. (You may need to review the structures and properties of graphite and diamond in Section 11.4.) Buckminsterfullerene is a solid that consists of individual molecules with formula C_{60}. Is the molar entropy of buckminsterfullerene larger or smaller than that of graphite? How about the entropy per gram? Explain.

14.103 Here are thermodynamic data for the fusion of NH_3:

$$NH_3\ (s) \longrightarrow NH_3\ (l) \qquad \Delta H° = 5.65\ kJ/mol,\ \Delta S° = 28.9\ J\ mol^{-1}\ K^{-1}$$

(a) Calculate $\Delta G°$ for the melting of 1.00 mol of NH_3 at 298 K; (b) Calculate the freezing point of NH_3.

14.104 Construct a table of the absolute entropies of all the gaseous substances listed in Appendix D, arranged in order of increasing $S°$. Use the general trends described in the text to explain the order of entries in this table. Are there substances that do not appear to fit the general trends? If so, suggest reasons why their $S°$ values are higher or lower than expected.

14.105 Phosphorus forms white crystals made up of P_4 molecules. There are two forms of white crystalline phosphorus, called α and β. The difference between $P_{4\alpha}$ and $P_{4\beta}$ is determined by the way the P_4 molecules pack together in the crystal lattice. The α form is always obtained when liquid phosphorus freezes. However, at temperatures below −77 °C, the $P_{4\alpha}$ crystals change spontaneously to $P_{4\beta}$:

$$P_{4\alpha} \xrightarrow{-77\ °C} P_{4\beta}$$

Answer the following, and explain your reasoning: (a) Which form of phosphorus has a more constrained crystalline structure? (b) What are the signs of ΔH and ΔS for this process? (c) Would the entropy of $P_{4\alpha}$ crystals be 0 at 0 K?

Answers to Section Exercises

14.1.1 (a) A fence is constrained; when it falls down, it becomes more dispersed; (b) a plate has a relatively constrained structure, which becomes dispersed when the plate breaks; and (c) forming a solution from pure substances involves dispersal of the components throughout the solution.

14.1.2 Constraining one substance (formation of sugar crystals) must be accompanied by even greater dispersal elsewhere. In this example, the water molecules that boil away are much more dispersed than water molecules in the liquid.

14.1.3

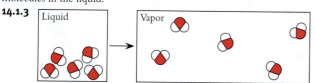

14.2.1 Heat will flow from the warmer mixture to the cooler mixture, so benzene will freeze and ice will melt.

14.2.2 0.10 J/K

14.2.3 $\Delta S_{mixture} = -6.11$ J/K, $\Delta S_{refrigerator} = 6.11$ J/K, $\Delta S_{total} = 0$; neither process is spontaneous.

14.3.1 (a) 1 g of dew because liquids are less constrained than solids of the same material; (b) 1 mol of gaseous hydrogen atoms, because although both samples contain the same number of atoms, binding atoms into molecules constrains them; (c) the flawed diamond, because a perfect diamond is more constrained; and (d) the sample at 50 °C, because increasing the temperature of a sample increases its molecular motion and makes it less constrained.

14.3.2 (a) There are more atoms in the sample of O_3, so there is more dispersal of matter in the molecular arrangement; (b) both samples have the same number of atoms, but there are more molecules in 3 mol of O_2, so there is more dispersal of matter in its molecular arrangement; (c) I_2 is a solid under standard conditions, so it is more constrained than gaseous O_2; and (d) diluting a solution spreads its solute particles (ions in this case) over a larger region, giving it greater dispersal.

14.3.3

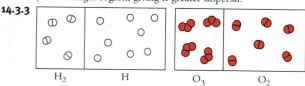

14.3.4 −2804.0 J/K
14.4.1 −225 kJ/mol
14.4.2 2.7 × 10^{-3} bar
14.4.3 Yes, $T = 5.4 × 10^3$ K
14.5.1 340 kJ
14.5.2 When an automobile engine moves an automobile, it does work and constrains energy (all parts of the automobile move together in the same direction). To do this without violating the second law of thermodynamics, enough heat must be transferred to the surroundings to disperse a greater amount of energy.
14.6.1 92%
14.6.2 +32.2 J mol^{-1} K^{-1}; the spontaneity increases as temperature increases.
14.6.3 (a) about 16 mol of ATP; and (b) about 80 g of glucose

Answers to Extra Practice Exercises

14.1 $\Delta S_{water} = 388$ J/K, $\Delta S_{hot\ plate} = -277$ J/K, and $\Delta S_{total} = 111$ J/K
14.2 $\Delta S_{total} = 23$ J/K
14.3 (a) CH_4, lower MM; (b) H_2, fewer atoms; (c) H_2O, liquid
14.4 −715 J/K
14.5 −242.9 J/K
14.6 − 818 kJ

14.7 −791 kJ (This estimate is only approximate, inasmuch as the temperature is much higher than 298 K.)
14.8 0.57 bar
14.9 96.4 torr
14.10 $ATP + H_2O \longrightarrow ADP + H_3PO_4 \qquad \Delta G° = -30.6$ kJ
$Glucose + H_3PO_4 \longrightarrow Glucose\ phosphate + H_2O \qquad \Delta G° = 13.9$ kJ

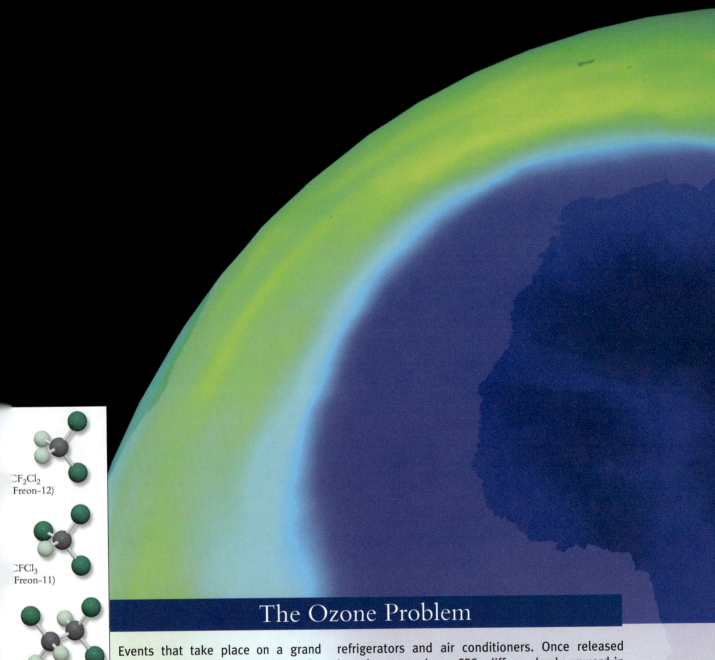

CF₂Cl₂ (Freon-12)

CFCl₃ (Freon-11)

C₂F₃Cl₃ (Freon-113)

The Ozone Problem

Events that take place on a grand scale often can be traced to the molecular level. An excellent example is the depletion of the ozone layer in the Earth's stratosphere. The so-called ozone hole was first observed above the Antarctic in the 1980s and is now being observed above both the Arctic and Antarctic poles. The destruction of ozone in the stratosphere is caused primarily by reactions between chlorine atoms and ozone molecules, as depicted in our molecular inset view.

The chlorine atoms in the upper atmosphere come from the breakdown of CF_2Cl_2 and other similar chlorofluorocarbons (CFCs), known commercially as Freons. Production of these compounds was more than one million tons in 1988, largely for use in refrigerators and air conditioners. Once released into the atmosphere, CFCs diffuse slowly upward in the atmosphere until they reach the ozone layer. There, ultraviolet light from the sun splits off chlorine atoms. These react with ozone, with dramatic results. Annual ozone decreases have exceeded 50% above Antarctica. The background photo shows the Antarctic hole (red-violet) on September 24, 2003.

Ozone in the stratosphere removes high-energy ultraviolet radiation coming from the sun. Because this radiation is lethal, loss of ozone in the stratosphere imperils life on Earth. This danger led world nations to agree to ban CFC production by 1996. Despite dramatic reductions in the use of CFCs, these and other ozone-depleting pollutants remain

Kinetics: Mechanisms and Rates of Reactions

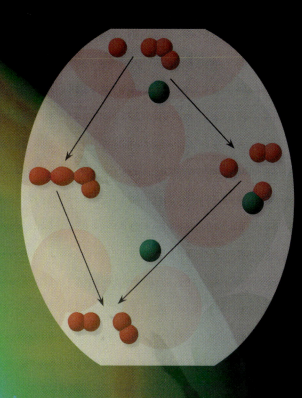

CHAPTER CONTENTS

in the atmosphere. The Antarctic ozone hole was so large in the year 2000 that populated regions of Chile and Australia were exposed to increased amounts of ultraviolet radiation.

The ozone hole would almost certainly be much worse if chemists had not studied the reactions of CFCs with atmospheric gases before ozone depletion was discovered. The 1995 Nobel Prize in chemistry was awarded to the three pioneers in this effort. A German chemist, Paul Crutzen, discovered how ozone concentration is regulated in a normal stratosphere, while two Americans, F. Sherwood Rowland and Mario Molina, showed that CFCs can destroy ozone.

Learning Resources

KEY CONCEPTS

CRITICAL THINKING

SKILLS TO MASTER

ADDITIONAL HELP
www.wiley.com/college/olmsted
- TUTORIALS
- ANIMATIONS

These studies of molecular reactions allowed quick determination that CFCs are a likely cause of ozone depletion and led to the international restrictions described above.

The story of the ozone hole illustrates how important it is to learn the molecular details of chemical reactions. Some chemists use information about how reactions occur to design and synthesize useful new compounds. Others explore how to modify reaction conditions to minimize the cost of producing industrial chemicals. This chapter explores how chemical reactions occur at the molecular level. We show how to describe a reaction from the molecular perspective, introduce the basic principles that govern these processes, and describe some experimental methods used to study chemical reactions.

15.1   WHAT IS A REACTION MECHANISM?

Atoms, ions, and molecules rearrange and recombine during chemical reactions. These processes usually do not occur all at once. Instead, each reaction consists of a sequence of molecular events called a **reaction mechanism.**

> **KEY CONCEPT**   *A reaction mechanism is the* **exact molecular pathway** *that starting materials follow on their way to becoming products.*

We introduce the principles of mechanisms using simple chemical reactions. After presenting the principles, we will be able to return to the more complicated mechanisms involved in the ozone problem.

Example of a Mechanism: Formation of N_2O_4

Nitrogen dioxide, a red-brown gas, is stable at room temperature. However, Figure 15-1 shows that reducing the temperature causes a sample of NO_2 to become colorless. The color change that happens at low temperature occurs because two NO_2 molecules combine to form one molecule of a colorless gas, N_2O_4.

The mechanism of this reaction describes what happens at the molecular level. Two NO_2 molecules collide, forming a collision complex. In this collision complex, a bond may form between the two nitrogen atoms, producing an N_2O_4 molecule.

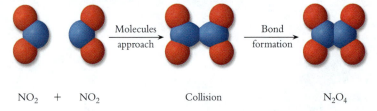

$$NO_2 \quad + \quad NO_2 \qquad\qquad \text{Collision} \qquad\qquad N_2O_4$$

NO_2

N_2O_4

The Lewis structure of NO_2 indicates that the nitrogen atom has an unpaired electron. Two NO_2 molecules combine by using their unpaired electrons to form an N—N bond.

The formation of N_2O_4 requires more than a simple collision between two NO_2 molecules. The product contains a bond between the nitrogen atoms, so the collision must form a collision complex that brings the two nitrogen atoms into contact. A collision in which the nitrogen atom of one NO_2 molecule strikes an oxygen atom of a second will not lead to N_2O_4.

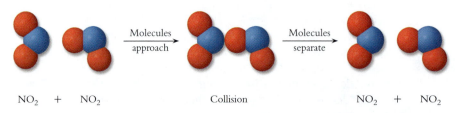

$$NO_2 \quad + \quad NO_2 \qquad\qquad \text{Collision} \qquad\qquad NO_2 \quad + \quad NO_2$$

Elementary Reactions

A mechanism is a description of the actual molecular events that occur during a chemical reaction. Each such event is an **elementary reaction.** Elementary reactions involve one, two, or occasionally three reactant molecules or atoms. In other words, elementary reactions can be unimolecular, bimolecular, or termolecular. A typical mechanism consists of a sequence of elementary reactions. Although an *overall* reaction describes the starting materials and final products, it usually is not *elementary* because it does not represent the individual steps by which the reaction occurs.

The most common type of elementary process is a **bimolecular reaction** that results from the collision of two molecules, atoms, or ions. The collision of two NO_2 molecules to give N_2O_4 is a bimolecular reaction. Here is another example:

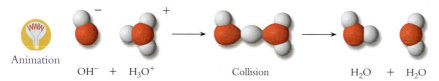

Animation

$$OH^- + H_3O^+ \qquad \text{Collision} \qquad H_2O + H_2O$$

Notice that the characteristic feature of a bimolecular elementary reaction is a collision between two species, giving a collision complex that results in a rearrangement of chemical bonds. The two reaction partners stick together by forming a new bond (N_2O_4 forming from two NO_2 molecules), or they form two new species by transferring one or more atoms or ions from one partner to another (2 H_2O forming from OH^- and H_3O^+).

In a **unimolecular reaction,** a single molecule fragments into two pieces or rearranges to a new isomer. A simple example of a unimolecular reaction is the decomposition of N_2O_4. An N_2O_4 molecule continually vibrates, and these vibrations cause its N—N bond to stretch. If the molecule has sufficient energy, the bond breaks and the molecule separates into two molecules of NO_2, much like a spring that breaks if it is stretched too far.

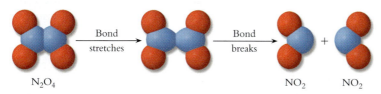

$$N_2O_4 \qquad \text{Bond stretches} \qquad \text{Bond breaks} \qquad NO_2 \quad NO_2$$

A stable molecule such as N_2O_4 does not decompose unless it first acquires the energy needed to break one of its bonds. Recall from Section 6-1 that any collection of molecules has a distribution of kinetic energies. Molecular collisions can convert kinetic energy of one molecule into vibrational energy of another molecule, giving some molecules enough energy to break apart.

In a **termolecular reaction**, three chemical species collide simultaneously. Termolecular reactions are rare because they require a collision of three species at the *same time* and in exactly the right orientation to form products. The odds against such a simultaneous three-body collision are high. Instead, processes involving three species usually occur in two-step sequences. In the first step, two molecules collide and form a collision complex. In a second step, a third molecule collides with the complex before it breaks apart. Most chemical reactions, including all those introduced in this book, can be described at the molecular level as sequences of bimolecular and unimolecular elementary reactions.

To summarize, the mechanism of a reaction converts starting materials to products through a specific sequence of bimolecular collisions and unimolecular rearrangements.

Alternative Mechanisms

Very few chemical reactions are as simple as the low-temperature reaction of NO_2 to form N_2O_4. For example, at high temperature, NO_2 molecules have enough energy to decompose into nitrogen oxide and molecular oxygen rather than forming N_2O_4:

$$2 NO_2 \longrightarrow 2 NO + O_2$$

In this reaction, one N—O bond breaks and a new O=O bond forms. We can visualize these changes occurring in more than one sequence.

One possible sequence that describes the decomposition of NO_2 molecules starts with a unimolecular reaction. At high temperature, collisions may transfer enough

Figure 15-1
When a flask containing orange NO_2 gas at room temperature (*top*) is cooled in a low-temperature bath (*bottom*), NO_2 molecules combine to form colorless N_2O_4 molecules.

Animation

energy to allow some NO_2 molecules to break apart. That is, an N—O bond in NO_2 may break to produce a molecule of NO and an oxygen atom. Because oxygen atoms are highly reactive, a subsequent bimolecular collision between an oxygen atom and an NO_2 molecule would then generate NO and O_2. This reaction sequence can be summarized as Mechanism I for NO_2 decomposition:

Mechanism I

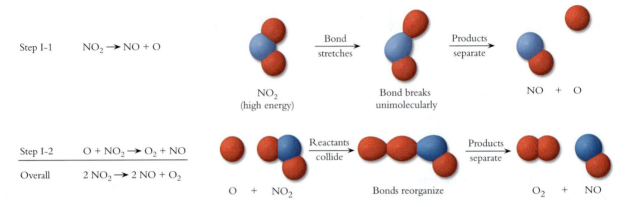

Step I-1 $NO_2 \longrightarrow NO + O$

Step I-2 $O + NO_2 \longrightarrow O_2 + NO$

Overall $2\,NO_2 \longrightarrow 2\,NO + O_2$

Mechanism I illustrates an important requirement for reaction mechanisms. Because a mechanism is a summary of events at the molecular level, a mechanism *must* lead to the correct stoichiometry to be an accurate description of the chemical reaction. The sum of the steps of a mechanism must give the balanced stoichiometric equation for the overall chemical reaction. If it does not, the proposed mechanism must be discarded. In Mechanism I, the net result of two sequential elementary reactions is the observed reaction stoichiometry.

A second possible mechanism for NO_2 decomposition starts with a bimolecular reaction. When two fast-moving NO_2 molecules collide, an oxygen atom may be transferred between them to form molecules of NO_3 and NO. Molecules of NO_3 are unstable and readily break apart into NO and O_2. This reaction sequence can be summarized as Mechanism II for NO_2 decomposition:

Mechanism II

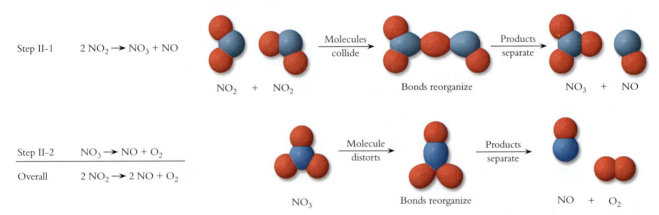

Step II-1 $2\,NO_2 \longrightarrow NO_3 + NO$

Step II-2 $NO_3 \longrightarrow NO + O_2$

Overall $2\,NO_2 \longrightarrow 2\,NO + O_2$

Mechanism II gives the observed overall stoichiometry because the NO_3 molecule produced in the first step is consumed in the second step.

Either Mechanism I or Mechanism II can account for the decomposition of NO_2. Each consists of elementary steps, each generates a highly reactive species that undergoes further chemistry, and each accounts for the observed reaction stoichiometry. Does either mechanism represent what really happens at the molecular level? Before describing how to test whether a mechanism is realistic, we must introduce some additional features of mechanisms.

Intermediates

Both proposed mechanisms for NO_2 decomposition contain chemical species produced in the first step and consumed in the second step. This is the defining characteristic of an **intermediate.** An intermediate is a chemical species *produced* in an early step of a mechanism and *consumed* in a later step. Intermediates never appear in the overall chemical equation. Notice that neither the O atoms of Mechanism I nor the NO_3 molecules of Mechanism II appear in the balanced chemical equation for NO_2 decomposition.

Intermediates are reactive chemical species that usually exist only briefly. They are consumed rapidly by bimolecular collisions with other chemical species or by unimolecular decomposition. The intermediate in Mechanism I is an oxygen atom that reacts rapidly with NO_2 molecules. The intermediate in Mechanism II is an unstable NO_3 molecule that rapidly decomposes.

The most direct way to test the validity of a mechanism is to determine what intermediates are present during the reaction. If oxygen atoms were detected, we would know that Mechanism I is a reasonable description of NO_2 decomposition. Likewise, the observation of NO_3 molecules would suggest that Mechanism II is reasonable. In practice, the detection of intermediates is quite difficult because they are usually reactive enough to be consumed as rapidly as they are produced. As a result, the concentration of an intermediate in a reaction mixture is very low. Highly sensitive measuring techniques are required for the direct detection of chemical intermediates.

The Rate-Determining Step

How fast do chemical reactions occur? The speed of a reaction is described by its **rate.** Rate is the number of events per unit time, such as the number of molecules reacting per second. Every elementary reaction has a characteristic rate. Some reactions are so fast that they are complete in the smallest measurable fraction of a second, whereas others are so slow that they require almost an eternity to reach completion. The observed rate of an *overall* chemical reaction is determined by the rates of the elementary reactions that make up the mechanism.

Each elementary reaction in a mechanism proceeds at its own unique rate. Consequently, every mechanism has one step that proceeds more slowly than any of the other steps. The slowest elementary step in a mechanism is called the **rate-determining step.** The rate-determining step governs the rate of the overall chemical reaction because no net chemical reaction can go faster than its slowest step. The idea of the rate-determining step is central to the study of reaction mechanisms.

Anyone who has flown on a commercial airline is familiar with the consequences of rate-determining steps, as illustrated by Figure 15-2. At a busy airport, for example, airplanes queue up while waiting for a runway to become available. After the pilot receives permission to take off, the plane becomes airborne very quickly. No matter how quickly an airplane is loaded and ready to leave the gate, the rate of takeoffs is determined by availability of runways. That is, the rate-determining step for airplane takeoff is the rate of takeoffs from the runways. Baggage claim is the rate-determining step at the end of a flight. No matter how quickly the passengers leap from their seats, push through the aisle, and race through the terminal, passengers who checked their baggage cannot leave the airport any faster than the baggage is delivered.

Chemical knowledge often can be used to assess which step of a mechanism is likely to be rate-determining. For example, in both proposed mechanisms for the decomposition of NO_2, chemical knowledge suggests that the first step is much slower than the second step. In Mechanism I an O atom is produced in the first step, $NO_2 \rightarrow NO + O$. Oxygen atoms are known to be highly reactive, so it is reasonable to predict that this intermediate reacts rapidly with NO_2 molecules. Compared with the fast second step of this mechanism, the step that forms the oxygen atoms is expected to be slow and rate-determining. Similarly, the first step of Mechanism II, $2\,NO_2 \rightarrow NO_3 + NO$, produces NO_3. This species is known to be unstable, so it will decompose in the second step of Mechanism II almost as soon as it forms. Again, the second step of the mechanism is expected to be fast, so the step that forms the reactive intermediate is slow and rate-determining. Later in this chapter, we discuss experiments that make it possible to distinguish between Mechanisms I and II.

Figure 15-2
The slowest step in any sequential process is rate-determining. Airplanes take off at the rate that runways become available, and arriving passengers leave the terminal at the rate their baggage is delivered.

Section Exercises

■ **15.1.1** Nitrogen dioxide (NO_2) in polluted air forms from NO and CO_2 by a single bimolecular reaction. Identify the second molecular product of this reaction, and draw a molecular picture similar to the ones in this section that shows how this chemical reaction occurs.

■ **15.1.2** One common analogy for the rate-determining step involves automobiles on a highway passing through a tollbooth. In your own words, explain rate-determining steps using the tollbooth analogy.

■ **15.1.3** The following mechanism has been proposed for the reaction of NO with H_2:

$$2\,NO \longrightarrow N_2O_2$$

$$N_2O_2 + H_2 \longrightarrow H_2O + N_2O$$

$$N_2O + H_2 \longrightarrow N_2 + H_2O$$

Determine the balanced chemical equation for the reaction that occurs by this mechanism and identify any chemical intermediates.

15.2 RATES OF CHEMICAL REACTIONS

Some of the most important information about a mechanism comes from experiments that determine how fast a chemical reaction occurs under various conditions. In chemical reactions, amounts of reactants and products change, so reaction rates are given in units of amount per unit time; for example, molecules per second. Amounts also can be expressed as concentrations, so rates can be measured in units of concentration per unit time; for example, molar per minute.

Recall that M is the symbol for molarity, expressed as mol/L.

The study of the rates of chemical reactions is called **kinetics.** Chemists study reaction rates for many reasons. To give just one example, Rowland and Molina used kinetic studies to show the destructive potential of CFCs. Kinetic studies are essential to the explorations of reaction mechanisms, because a mechanism can never be determined by calculations alone. Kinetic studies are important in many areas of science, including biochemistry, synthetic chemistry, biology, environmental science, engineering, and geology. The usefulness of chemical kinetics in elucidating mechanisms can be understood by examining the differences in rate behavior of unimolecular and bimolecular elementary reactions.

A Molecular View

To obtain a molecular perspective of reaction rates, consider the unimolecular reaction shown in Figure 15-3. At elevated temperature, the compound *cis*-2-butene can rearrange to form its isomer, *trans*-2-butene. The reaction occurs after collisions transfer enough energy to a *cis*-2-butene molecule to break the C—C π bond. Once the bond breaks, rotation around the C—C σ bond takes place rapidly until a π bond forms again.

For a molecular view of rates, consider a sample containing 12 molecules of *cis*-2-butene, which represents a tiny portion of an immense reaction mixture. Figure 15-4 illustrates schematically how this reaction progresses with time. During the first minute, six *cis*-2-butene molecules isomerize to *trans*-2-butene. The average rate for this time period is 6 molecules/min. As the reaction proceeds, however, the rate of reaction decreases. During the second minute, only three molecules isomerize, giving an average rate for that period of 3 molecules/min.

Figure 15-3

***cis*-2-Butene can rearrange to form *trans*-2-butene in a unimolecular reaction if collisions give the molecule enough extra energy to break the C—C π bond.**

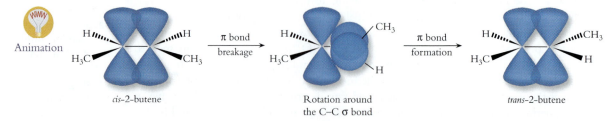

Animation

cis-2-butene π bond breakage   Rotation around the C–C σ bond π bond formation   *trans*-2-butene

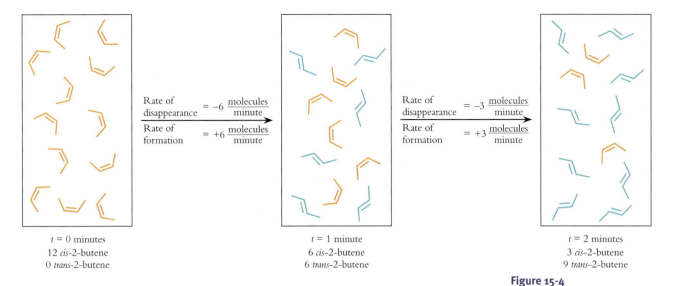

t = 0 minutes
12 *cis*-2-butene
0 *trans*-2-butene

Rate of disappearance $= -6 \dfrac{\text{molecules}}{\text{minute}}$

Rate of formation $= +6 \dfrac{\text{molecules}}{\text{minute}}$

t = 1 minute
6 *cis*-2-butene
6 *trans*-2-butene

Rate of disappearance $= -3 \dfrac{\text{molecules}}{\text{minute}}$

Rate of formation $= +3 \dfrac{\text{molecules}}{\text{minute}}$

t = 2 minutes
3 *cis*-2-butene
9 *trans*-2-butene

Figure 15-4
A molecular view of the isomerization of *cis*-2-butene illustrates that the rate of isomerization decreases as the number of *cis*-2-butene molecules decreases.

Why does the rate become slower as this reaction proceeds? Dividing the number of *cis*-2-butene molecules that react by the total number of *cis*-2-butene molecules present at the beginning of the interval reveals the answer:

First minute: Fraction reacting $= \dfrac{6 \; cis\text{-2-butene molecules react}}{12 \; cis\text{-2-butene molecules present}} = 0.5$

Second minute: Fraction reacting $= \dfrac{3 \; cis\text{-2-butene molecules react}}{6 \; cis\text{-2-butene molecules present}} = 0.5$

The fraction of the *cis*-2-butene molecules present that react during each one-minute interval is the same. That is, the rate of reaction is constant on a *per molecule* basis. As the reaction proceeds, however, fewer *cis*-2-butene molecules remain, causing the overall rate of reaction to decrease. This is characteristic of unimolecular elementary reactions. The rate per molecule is constant, but if the number of reactant molecules is cut in half, the rate of reaction is cut in half, as well.

Bimolecular elementary reactions also slow down as the number of reactant molecules decreases, as can be seen for the formation of N_2O_4 from NO_2. Figure 15-5 shows 12 molecules of starting material, again representing a tiny fraction of a much larger sample. Four NO_2 molecules combine during the first minute, so the average rate during this period is 4 molecules/min. As the reaction proceeds, however, the rate decreases. During the second minute, only two molecules react, giving an average rate for that interval of 2 molecules/min.

Figure 15-5
A molecular view of the formation of N_2O_4 from NO_2. The rate of reaction decreases as the number of molecules decreases because fewer NO_2 molecules are present to undergo collisions.

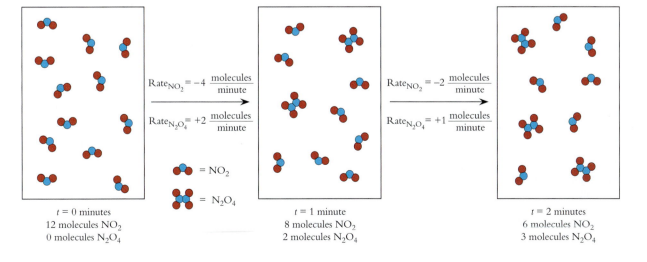

t = 0 minutes
12 molecules NO_2
0 molecules N_2O_4

$\text{Rate}_{NO_2} = -4 \dfrac{\text{molecules}}{\text{minute}}$

$\text{Rate}_{N_2O_4} = +2 \dfrac{\text{molecules}}{\text{minute}}$

= NO_2

= N_2O_4

t = 1 minute
8 molecules NO_2
2 molecules N_2O_4

$\text{Rate}_{NO_2} = -2 \dfrac{\text{molecules}}{\text{minute}}$

$\text{Rate}_{N_2O_4} = +1 \dfrac{\text{molecules}}{\text{minute}}$

t = 2 minutes
6 molecules NO_2
3 molecules N_2O_4

Dividing the number of molecules reacting during each time interval by the number of reactant molecules present shows that this reaction behaves differently from a unimolecular reaction:

First minute: $\text{Fraction reacting} = \dfrac{4 \text{ NO}_2 \text{ molecules react}}{12 \text{ NO}_2 \text{ molecules present}} = 0.33$

Second minute: $\text{Fraction reacting} = \dfrac{2 \text{ NO}_2 \text{ molecules react}}{8 \text{ NO}_2 \text{ molecules present}} = 0.25$

In this case, not only do fewer molecules react, but also the rate of reaction per molecule is slower. If it were possible to count the number of collisions, we would find that the reaction slows down not only because fewer reactant molecules are present, but also because fewer collisions occur. For all bimolecular reactions, the rate *per collision* is constant, but as the reaction proceeds, fewer reactant molecules remain to collide with one another, so the rate per molecule decreases.

Figure 15-5 also shows that each species has its own rate, but the individual rates are linked by the stoichiometric coefficients. The reaction generates one molecule of N_2O_4 for every two molecules of NO_2 that are destroyed. That is, the 1:2 stoichiometry of this reaction results in a 1:2 relationship between the rate of disappearance of NO_2 and the rate of appearance of N_2O_4. The ratio of rates for different species is always equal to the ratio of their stoichiometric coefficients.

A Macroscopic View: Concentration Changes

Molecules are too small and much too numerous to follow on an individual basis. Therefore, a chemist interested in measuring the rate of a reaction monitors the *concentration* of a particular compound as a function of time. The concentrations of reactants, products, or both may be monitored. For example, Figure 15-6 shows some experimental data obtained from a series of concentration measurements on the decomposition of NO_2:

$$2 \text{ NO}_2 \longrightarrow 2 \text{ NO} + \text{O}_2$$

The rate of the reaction at any specific time is given by how fast the concentration changes. The plots in Figure 15-6 show the same features that the molecular pictures of Figures 15-4 and 15-5 show: Rates of reaction decrease as starting materials are consumed, and rates for different species are linked by stoichiometry.

The average rate of a reaction can be expressed as the change in concentration (Δc) over some time interval (Δt). For oxygen formation during the decomposition of NO_2:

$$\text{Rate of production of } O_2 = \frac{\Delta [O_2]}{\Delta t}$$

Because rates change continuously with time, accurate rate determinations must use small time intervals. At the outset of the NO_2 decomposition experiment shown in Figure

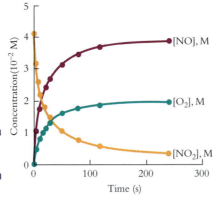

Figure 15-6
A plot of [NO₂], [O₂], and [NO] as a function of time (seconds) for the decomposition reaction of NO₂. The concentration data are shown in the table.

Time	(s)	0	5.0	10	15	20	30	50	80	120	240
[NO₂]	(10⁻² M)	4.1	3.1	2.5	2.1	1.8	1.4	1.0	0.70	0.50	0.30
[NO]	(10⁻² M)	0	1.0	1.6	2.0	2.3	2.7	3.1	3.4	3.6	3.8
[O₂]	(10⁻² M)	0	0.50	0.80	1.0	1.1	1.3	1.6	1.7	1.8	1.9

15-6, the time interval between measurements is 5 seconds. For the first 5 seconds, the average rate of O_2 production is as follows:

$$\text{Rate of production of } O_2 = \frac{\Delta[O_2]}{\Delta t} = \frac{0.50 \times 10^{-2} \text{ M} - 0 \text{ M}}{5.0 \text{ s} - 0 \text{ s}} = 1.0 \times 10^{-3} \text{ M/s}$$

At later times the rate is different. Verify this for yourself by calculating the average rates for the second and third 5-second intervals.

We can also write a rate expression for the consumption of NO_2:

$$\text{Rate of consumption of } NO_2 = -\frac{\Delta[NO_2]}{\Delta t}$$

The negative sign appears in this equation to make the rate positive, even though NO_2 concentration decreases with time. By convention, rate expressions are always written in a way that gives positive rates. Over the first 5 seconds of this reaction, the average rate of NO_2 consumption is as follows:

$$\text{Rate of consumption of } NO_2 = -\frac{\Delta[NO_2]}{\Delta t} = -\frac{3.1 \times 10^{-2} \text{ M} - 4.1 \times 10^{-2} \text{ M}}{5.0 \text{ s} - 0 \text{ s}}$$
$$= 2.0 \times 10^{-3} \text{ M/s}$$

Notice that over the same period the rate of O_2 formation is only half the rate of NO_2 consumption. This follows from the molecular view of the mechanism and from the stoichiometry of the reaction. The rate relationship among the three species involved in NO_2 decomposition is given by the following expression:

$$\text{Reaction rate} = \frac{\Delta[O_2]}{\Delta t} = \left(\frac{1}{2}\right)\left(\frac{\Delta[NO]}{\Delta t}\right) = \left(-\frac{1}{2}\right)\left(\frac{\Delta[NO_2]}{\Delta t}\right)$$

The relationship among reaction rates and stoichiometric coefficients can be applied to any reaction. Equation 15-1 gives the relationships among the rate expressions for a reaction of the form $a\,A + b\,B \longrightarrow d\,D + e\,E$:

$$\text{Reaction rate} = \left(-\frac{1}{a}\right)\left(\frac{\Delta[A]}{\Delta t}\right) = \left(-\frac{1}{b}\right)\left(\frac{\Delta[B]}{\Delta t}\right) = \left(\frac{1}{d}\right)\left(\frac{\Delta[D]}{\Delta t}\right) = \left(\frac{1}{e}\right)\left(\frac{\Delta[E]}{\Delta t}\right)$$

$$(15\text{-}1)$$

Here, lowercase letters represent stoichiometric coefficients and uppercase letters represent chemical substances.

Equation 15-1 is one of the few rate statements that follow from knowledge of the reaction stoichiometry. Almost all other rate information must be determined by carrying out experiments on how concentrations change with time. Example 15-1 illustrates the application of Equation 15-1.

Example 15-1

Relative Rates of Reaction

Acrylonitrile is produced from propene, ammonia, and oxygen by the following balanced equation (see Example 14-5):

$$2\,C_3H_6 + 2\,NH_3 + 3\,O_2 \longrightarrow 2\,CH_2CHCN + 6\,H_2O$$

Relate the rates of reaction of starting materials and products.

Strategy: The rates for different species participating in a chemical reaction are related by the stoichiometric coefficients of the balanced chemical equation. Equation 15-1 provides the exact relationship.

Solution: The balanced equation shows that three molecules of oxygen are consumed for every two molecules of propene and two molecules of ammonia. Thus, the rate of C_3H_6 and NH_3 consumption is only two-thirds the rate of O_2 consumption. Those seven molecules of starting materials produce two molecules of CH_2CHCN and six

Example 15-1

Relative Rates of Reaction
(*continued*)

molecules of H_2O. Thus, CH_2CHCN is produced at the same rate as C_3H_6 is consumed, whereas H_2O is produced three times as fast as CH_2CHCN is. The link between relative reaction rates and reaction stoichiometry is Equation 15-1. Therefore,

$$\text{Reaction rate} = \left(-\frac{1}{2}\right)\left(\frac{\Delta[C_3H_6]}{\Delta t}\right) = \left(-\frac{1}{3}\right)\left(\frac{\Delta[O_2]}{\Delta t}\right) = \left(-\frac{1}{2}\right)\left(\frac{\Delta[NH_3]}{\Delta t}\right)$$

$$= \left(\frac{1}{2}\right)\left(\frac{\Delta[CH_2CHCN]}{\Delta t}\right) = \left(\frac{1}{6}\right)\left(\frac{\Delta[H_2O]}{\Delta t}\right)$$

Does the Result Make Sense? Each of the reactants disappears as the reaction proceeds, so negative signs are appropriate for these three species. Three molecules of O_2 are consumed for every six molecules of H_2O formed, so it makes sense that the rate of disappearance of O_2 is half as large as the rate of disappearance of H_2O; in other words

$$\left(-\frac{1}{3}\right)\left(\frac{\Delta[O_2]}{\Delta t}\right) = \left(\frac{1}{6}\right)\left(\frac{\Delta[H_2O]}{\Delta t}\right) \qquad or$$

$$-\left(\frac{\Delta[O_2]}{\Delta t}\right) = \left(\frac{3}{6}\right)\left(\frac{\Delta[H_2O]}{\Delta t}\right) = \left(\frac{1}{2}\right)\left(\frac{\Delta[H_2O]}{\Delta t}\right)$$

Extra Practice
Exercise 15.1

Relate the rate of formation of N_2O_4 to the rate of disappearance of NO_2 for the reaction described in Section 15-1.

Section Exercises

■ **15.2.1** Ammonia is produced by the following reaction:

$$N_2 + 3\,H_2 \longrightarrow 2\,NH_3$$

Consider a portion of a flask that contains five molecules of N_2 and nine molecules of H_2. Draw a molecular picture showing the contents of this portion of the flask after two molecules of NH_3 have formed. Rep-

resent nitrogen atoms as dark circles and hydrogen atoms as light circles.

■ **15.2.2** In the ammonia synthesis from the previous problem, which of the starting materials is consumed at a faster rate?

■ **15.2.3** State the relative rates for the consumption of starting materials and the formation of products for the ammonia synthesis.

15.3 **CONCENTRATION AND REACTION RATES**

Both concentration and temperature affect the rate of a chemical reaction. This section examines how changes in the concentrations of starting materials and products affect the rate of a chemical reaction. We describe temperature effects in Section 15.6.

Concentration Effects

A reaction of ozone provides an example of concentration effects. Ozone in the atmosphere near the Earth's surface is a serious pollutant that damages soft tissues such as the lungs. In major urban areas, smog alerts are issued whenever there are elevated concentrations of ozone in the lower atmosphere. Nitrogen oxide, another component of photochemical smog, is a colorless gas produced in a side reaction in automobile engines. One reaction that links these species is the reaction of NO and O_3 to produce O_2 and NO_2:

$$NO + O_3 \longrightarrow NO_2 + O_2$$

Experiments indicate that this reaction occurs through a single bimolecular collision, as shown in Figure 15-7.

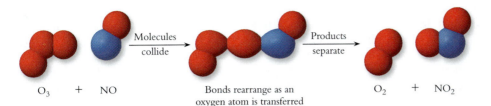

Figure 15-7
The reaction between O_3 and NO is believed to occur by a mechanism that consists of the single bimolecular step shown here in a molecular view.

O_3 +   NO   Molecules collide   Bonds rearrange as an oxygen atom is transferred   Products separate O_2 + NO_2

For a better understanding of the effect of changing concentrations on the rate of a chemical reaction, it helps to visualize the reaction at the molecular level. In this one-step bimolecular reaction, a collision between molecules that are in the proper orientation leads to the transfer of an oxygen atom from O_3 to NO. As with the formation of N_2O_4, the rate of this bimolecular reaction is proportional to the number of collisions between O_3 and NO. The more such collisions there are, the faster the reaction occurs.

To relate collision rates with concentrations, consider two containers of equal volume, each containing NO and O_3. Figure 15-8 shows molecular views of portions of two such containers. The container in Figure 15-8b has the same number of O_3 molecules as the container in Figure 15-8a, but the container in Figure 15-8b holds twice as many NO molecules as the container in Figure 15-8a. The larger amount of NO in the container in Figure 15-8b gives an O_3 molecule in this container twice as many opportunities to collide with an NO molecule. Thus, the rate of reaction in the container in Figure 15-8b is twice the rate in the container in Figure 15-8a. To generalize, collision rates and reaction rates of bimolecular reactions increase linearly with the concentration of each participant. Example 15-2 extends this reasoning to O_3 molecules.

Figure 15-8
A schematic illustration of two containers of equal volume containing different amounts of O_3 and NO. Container b has the same amount of O_3 but twice as much NO as container a. With twice as many NO molecules, twice as many NO–O_3 collisions occur, so the reaction in container b proceeds twice as fast as the reaction in container a.

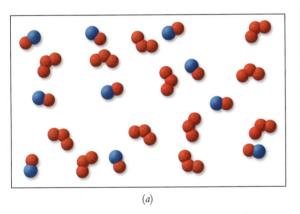

(a)

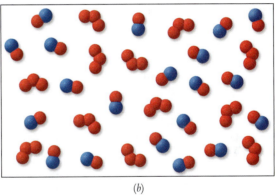

(b)

The container that follows contains O_3 and NO. Compared with the container in Figure 15-8a, how fast will the reaction proceed?

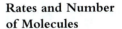

Example 15-2

Rates and Number of Molecules

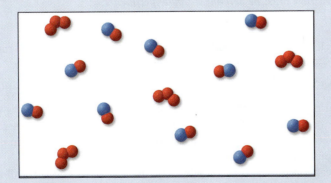

Strategy: The mechanism of this reaction requires a collision between an NO molecule and an O_3 molecule, as shown in Figure 15-7. The rate of the reaction therefore depends on the number of collisions that occur, and the collision frequency is proportional to the number of molecules.

Example 15-2

Rates and Number of Molecules (continued)

Solution: In Figure 15-8a there are 12 molecules of O_3 and 10 of NO. The container shown in this example contains 4 molecules of O_3 and 10 molecules of NO. The concentration of O_3 is only one-third as great, so any one molecule of NO will encounter an O_3 molecule three times less often than it would in the container in Figure 15-8a. This new mixture will react one-third as fast as the one in Figure 15-8a.

Does the Result Make Sense? If we reduce the amount of one reactant in a reaction flask, it makes sense that there are fewer collisions, and if there are fewer collisions, it makes sense that the rate of reaction will decrease.

Extra Practice Exercise 15.2

Suppose the container contains 10 molecules of O_3 and 12 molecules of NO. Compare the rate of reaction with the rates of the two views in Figure 15-8 and the view in this example.

Rate Laws

The effect of concentration on the rate of a particular chemical reaction can be summarized in an algebraic expression known as a **rate law.** A rate law links the rate of a reaction with the concentrations of the reactants through a **rate constant (k).** In addition, as we show later in this chapter, the rate law may contain concentrations of chemical species that are not part of the balanced overall reaction.

Although every reaction has its own unique rate law, many rate laws have a general form:

$$\text{Rate} = k[A]^y[B]^z \tag{15-2}$$

In general, each concentration has some exponent (here, y and z). Each exponent is called the *order* of the reaction with respect to that particular species. In Equation 15-2, y is the order of reaction with respect to species A, and z is the order with respect to species B. When the value of y is 1, the reaction is called **first order** in A; when the value of z is 2, the reaction is called **second order** in B, and so on. Orders of reaction are small integers or simple fractions. The most common orders are 1 and 2. The sum of the exponents is known as the *overall order* of the reaction.

The exponents in a rate law depend on the reaction *mechanism* rather than on the *stoichiometry* of the overall reaction. *The order of reaction often differs from the stoichiometric coefficient.* Consequently, a rate law must always be determined by conducting experiments; it can never be derived from the stoichiometry of the overall chemical reaction.

KEY CONCEPT *The order of a reaction must be determined by experiments.*

For example, experimental studies show that the rate law for the reaction of O_3 with NO_2 to give N_2O_5 and O_2 is first order in each reactant:

$$2\,NO_2 + O_3 \longrightarrow N_2O_5 + O_2 \qquad \text{Experimental rate} = k[NO_2][O_3]$$

Notice that for this reaction, the order of reaction with respect to NO_2 is 1, whereas the stoichiometric coefficient is 2. This shows that the order of a reaction for a particular species cannot be predicted by looking at the overall balanced equation. We describe additional examples in Section 15.5.

When a reaction proceeds in a single elementary step, its rate law will mirror its stoichiometry. An example is the rate law for O_3 reacting with NO. Experiments show that this reaction is first order in each of the starting materials and second order overall:

$$NO + O_3 \longrightarrow NO_2 + O_2 \qquad \text{Experimental rate} = k[NO][O_3]$$

This rate law is fully consistent with the molecular view of the mechanism shown in Figure 15-7. If the concentration of either O_3 or NO is doubled, the number of collisions between starting material molecules doubles too, and so does the rate of reaction. If the concentrations of *both* starting materials are doubled, the collision rate and the reaction rate increase by a factor of four.

Mechanisms and Rate Laws

The relationship between a mechanism and its rate law can be illustrated for the decomposition of NO_2. Experiments that we describe in Section 15.4 reveal that the rate of this reaction is proportional to the square of NO_2 concentration but is independent of the concentration of O_2:

$$2\,NO_2 \longrightarrow 2\,NO + O_2 \qquad \text{Experimental rate} = k[NO_2]^2$$

As described earlier, either Mechanism I or Mechanism II might describe this reaction:

Mechanism I
$$NO_2 \longrightarrow NO + O \qquad \text{Step I-1 (rate-determining)}$$
$$O + NO_2 \longrightarrow O_2 + NO \qquad \text{Step I-2 (fast)}$$

Mechanism II
$$2\,NO_2 \longrightarrow NO_3 + NO \qquad \text{Step II-1 (rate-determining)}$$
$$NO_3 \longrightarrow NO + O_2 \qquad \text{Step II-2 (fast)}$$

Each mechanism *predicts* a rate behavior that can be compared with the experimental rate law. If the prediction differs from the experimental observation, the mechanism is incorrect.

Remember that the overall rate of a reaction is determined by the rate of the slowest step. In other words, no reaction can proceed faster than the rate-determining step. Any step that comes *after* the rate-determining step cannot influence the overall rate of reaction. In both proposed mechanisms, the first step of each mechanism is rate-determining. That is, each proposed mechanism predicts an overall rate of NO_2 decomposition that is the same as the rate of the first step in the mechanism.

The first step in Mechanism I is the unimolecular decomposition of NO_2. Our molecular analysis shows that the rate of a unimolecular reaction is constant on a *per molecule* basis. Thus, if the concentration of NO_2 is doubled, twice as many molecules decompose in any given time. In quantitative terms, if NO_2 decomposes by Mechanism I, the rate law will be

$$\text{Predicted rate (Mechanism I)} = k[NO_2]$$

Once an NO_2 molecule decomposes, the O atom that results from decomposition very quickly reacts with another NO_2 molecule.

The rate-determining step of Mechanism II is a bimolecular collision between two identical molecules. A bimolecular reaction has a constant rate on a *per collision* basis. Thus, if the number of collisions between NO_2 molecules increases, the rate of decomposition increases accordingly. Doubling the concentration of NO_2 doubles the number of molecules present, and it also doubles the number of collisions for *each* molecule. Each of these factors doubles the rate of reaction, so doubling the concentration of NO_2 increases the rate for this mechanism by a factor of *four*. Consequently, if NO_2 decomposes by Mechanism II, the rate law will be

$$\text{Predicted rate (Mechanism II)} = k[NO_2][NO_2] = k[NO_2]^2$$

The two proposed mechanisms for this reaction predict different rate laws. Whereas Mechanism I predicts that the rate is proportional to NO_2 concentration, Mechanism II predicts that the rate is proportional to the *square* of NO_2 concentration. Experiments agree with the prediction of Mechanism II, so Mechanism II is consistent with the experimental behavior of the NO_2 decomposition reaction. Mechanism I predicts rate behavior contrary to what is observed experimentally, so Mechanism I cannot be correct.

We return to the relationship between rate laws and mechanisms in Section 15.5, after discussing experimental methods for determining the rate law of a reaction.

Rate Constants

Every reaction has its own characteristic rate constant that depends on the intrinsic speed of that particular reaction. For example, the value of k in the rate law for NO_2 decomposition is different from the value of k for the reaction of O_3 with NO. Rate constants are independent of concentration and time, but as we discuss in Section 15.6, rate constants are sensitive to temperature.

Rates of reaction have units of (concentration)(time)$^{-1}$. Because time does not appear in any other term on the right-hand side of the rate law, the units of k must always include time in the denominator. The concentration units of the rate constant depend on the overall order of the rate law, however, because the units of the rate constant must cancel concentration units to give the proper units for the rate of reaction. As an example, consider the rate law for the reaction of NO with O_3:

$$\text{Rate} = k[\text{NO}][\text{O}_3]$$

To agree with the units of (concentration)(time)$^{-1}$ for the rate of reaction, the rate constant must have units of (concentration)$^{-1}$(time)$^{-1}$ (such as $M^{-1} s^{-1}$). These units fit into the rate law as follows:

Rate law:	Rate	=	k	[NO]	[O_3]
Units:	$(M\ s^{-1})$	=	$(M^{-1}\ s^{-1})$	(M)	(M)

Example 15-3 provides practice in deducing the units of a rate constant.

<table>
<tr><td>

Example 15-3

Units of the Rate Constant

</td><td>

Reactions in aqueous solution can have complicated kinetics. An example is the reaction between arsenic acid and iodide ions:

$$\text{H}_3\text{AsO}_4\,(aq) + 3\,\text{I}^-\,(aq) + 2\,\text{H}_3\text{O}^+\,(aq) \longrightarrow \text{H}_3\text{AsO}_3\,(aq) + \text{I}_3^-\,(aq) + 3\,\text{H}_2\text{O}\,(l)$$

The rate law for this reaction has been found experimentally to be as follows:

$$\text{Rate} = k[\text{H}_3\text{AsO}_4][\text{I}^-][\text{H}_3\text{O}^+]$$

What are the units of the rate constant when the time is expressed in minutes?

Strategy: Units on the left must be the same as units on the right. The rate constant must have units that achieve this. Analyze the units for each component of the rate equation to find the units for k.

Solution: On the left side of the rate law, the rate of reaction is expressed in terms of changes in concentration and time:

$$\text{Rate of reaction} = \frac{\Delta c}{\Delta t}$$

For species in solution, concentration (c) is in molarity (M). According to the conditions of the problem, time (t) is in minutes. Therefore, the units of the rate must be

$$M\ \text{min}^{-1}$$

The right-hand side of the rate law contains a product of three concentrations, so it shows third-order behavior overall. This gives concentration units of M^3. This requires units of $M^{-2}\ \text{min}^{-1}$ for the rate constant:

Term in equation:	Rate	k	[H_3AsO_4]	[I^-]	[H_3O^+]
Units:	$(M\ \text{min}^{-1})$	$(M^{-2}\ \text{min}^{-1})$	(M)	(M)	(M)

</td></tr>
</table>

 Does the Result Make Sense? Units of $M^{-2}\ \text{min}^{-1}$ for a rate constant may look strange, but the test of reasonableness is that the units are the same on both sides of the equality. A rate always is in concentration/time, so we have to adjust the units of the rate constant according to the overall order of the reaction.

<table>
<tr><td>

**Extra Practice
Exercise 15.3**

</td><td>

What are the units of the rate constant for the reaction of hydroxide anions with hydronium cations, which proceeds by the single collision shown in Section 15.1?

</td></tr>
</table>

Once again, we emphasize that the order of reaction and the value of the rate constant must be determined by doing experiments. Knowing the order of reaction then makes it possible to write the specific rate law for the chemical process. In the next three sections, we discuss how chemists determine orders of reactions and further explore how rate laws are related to chemical mechanisms.

Section Exercises

■ **15.3.1** Draw a container with numbers of NO and O_3 molecules that would react at half the rate of the one in Figure 15-8*a*.

■ **15.3.2** The isomerization reaction of *cis*-2-butene, shown in Figure 15-3, is found experimentally to follow first-order kinetics. What is the rate law, and what are the units of the rate constant?

■ **15.3.3** The reaction between hydrogen and bromine is as follows: $H_2(g) + Br_2(g) \rightarrow 2\,HBr(g)$. Experiments show that this reaction is first order in H_2 and one-half order in Br_2. What is the rate law, what is the overall order of the reaction, and what are the units of the rate constant?

15.4 EXPERIMENTAL KINETICS

Every rate law must be determined experimentally. A chemist may imagine a reasonable mechanism for a reaction, but that mechanism must be tested by comparing the *actual* rate law for the reaction with the rate law *predicted* by the mechanism. To determine a rate law, chemists observe how the rate of a reaction changes with concentration. The graph of the data for the NO_2 decomposition reaction shown in Figure 15-6 is an example of such observations.

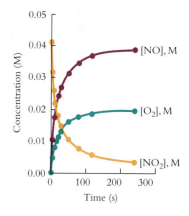

Regardless of the rate law, the rate of a reaction generally decreases with time because the concentrations of reactants decrease. The form of a rate law is determined by exploring the details of how the rate decreases with time. Because rate laws describe how rates vary with *concentration,* it is necessary to do mathematical analysis to convert rate laws into equations that describe how concentrations vary with *time.* These equations are different for different reaction orders.

First-Order Reactions

Recall that Mechanism I for the decomposition of NO_2 and the mechanism for the isomerization reaction of *cis*-2-butene both predict a simple first-order rate law:

$$\text{Rate} = k[NO_2] \quad \text{and} \quad \text{Rate} = k[\textit{cis}\text{-2-butene}]$$

The general form of a first-order rate law is rate = $k[A]$, where A is a reactant in the overall reaction. Mathematical treatment converts this general form into an equation relating concentration and time. For a first-order reaction, a logarithmic relationship links concentration and time:

$$\ln\left(\frac{[A]_0}{[A]}\right) = kt \tag{15-3}$$

> Students who have taken calculus will recognize that Equation 15-3 results from integration of the first-order rate law.

In this expression, [A] is the concentration of reactant A at time t, and $[A]_0$ is the concentration of reactant A at the beginning of the reaction ($t = 0$).

The most important feature of Equation 15-3 is that a graph of $\ln\left(\dfrac{[A]_0}{[A]}\right)$ vs. t gives a straight line whose slope is k. Consequently, experimental data can be tested for first-order behavior by preparing such a graph and observing whether or not it is linear.

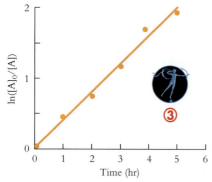

Time (hr)	[C$_5$H$_{11}$Br] (M)	[C$_5$H$_{11}$Br]$_0$/[C$_5$H$_{11}$Br]	ln([A]$_0$/[A])
0	0.150	1.00	0
1.00	0.099	1.52	0.42
2.00	0.073	2.05	0.72
3.00	0.047	3.19	1.16
4.00	0.028	5.36	1.68
5.00	0.022	6.82	1.92

Figure 15-9
A plot of ln ([A]$_0$/[A]) vs. time in hours for the conversion of an alkyl bromide to an alkene. The experimental data are shown in the table. The linearity of this plot verifies that this reaction obeys a first-order rate law.

Figure 15-9 illustrates this procedure for measurements done on the conversion of an alkyl bromide to an alkene:

Alkyl bromide (C$_5$H$_{11}$Br) + H$_2$O $\longrightarrow$ Alkene (C$_5$H$_{10}$) + H$_3$O$^+$ + Br$^-$

The linear appearance of the plot shows that this reaction obeys a first-order rate law. Additional mechanistic studies suggest that alkene formation proceeds in a two-step sequence. In the first step, which is rate-determining, the C—Br bond breaks to generate a bromide anion and an unstable cationic intermediate. In the second step, the intermediate transfers a proton (H$^+$) to a water molecule, forming the alkene and H$_3$O$^+$:

C$_5$H$_{11}$Br $\xrightarrow{\text{Slow}}$ Intermediate (C$_5$H$_{11}^+$) + Br$^-$

Intermediate (C$_5$H$_{11}^+$) + H$_2$O $\xrightarrow{\text{Fast}}$ C$_5$H$_{10}$ + H$_3$O$^+$

This example shows that first-order kinetics are not restricted to reactions with just one starting material, such as the decomposition of nitrogen dioxide or the isomerization of *cis*-2-butene.

The rate constant for the alkyl bromide reaction is equal to the slope of the line. The best way to determine a slope is by doing a linear curve fit using a spreadsheet or graphing calculator. Somewhat less accurately, any two points on the line determine the slope:

$$\text{Rate constant} = k = \text{Slope} = \frac{\Delta y}{\Delta x} = \frac{1.16 - 0}{3.00 \text{ hr} - 0 \text{ hr}} = \frac{1.16}{3.00 \text{ hr}} = 0.387 \text{ hr}^{-1}$$

Equation 15-3 is used to test whether a reaction is first order overall. The concentration of reactant A is monitored as a function of time. If a plot of $\ln\left(\dfrac{[A]_0}{[A]}\right)$ vs. t is linear, the reaction is first order in A. If the plot is not linear, the reaction is not first order. In Example 15-4, we apply this test to the data for the decomposition reaction of NO$_2$ presented in Figure 15-6.

Example 15-4

According to proposed Mechanism I, the decomposition of NO_2 should follow first-order kinetics. Do the experimental data of Figure 15-6 support this mechanism?

First-Order Kinetic Analysis

Strategy: According to the concentration–time form of first-order rate laws, a plot of $\ln\left(\dfrac{[A]_0}{[A]}\right)$ vs. t must be a straight line if Mechanism I is correct.

Solution: Convert the concentration data into ratios, take the logarithms, and then prepare a logarithmic plot to determine whether the reaction is first order:

Time(s)	0	5	10	15	20	30	50	80	120	240
$[NO_2]$ (10^{-2} M)	4.1	3.1	2.5	2.1	1.8	1.4	1.0	0.70	0.50	0.30
$\dfrac{[NO_2]_0}{[NO_2]}$	1.00	1.32	1.64	1.95	2.28	2.93	4.10	5.86	8.20	13.7
$\ln\left(\dfrac{[NO_2]_0}{[NO_2]}\right)$	0.00	0.28	0.50	0.67	0.82	1.08	1.41	1.77	2.10	2.62

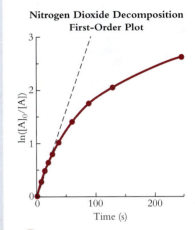

Nitrogen Dioxide Decomposition First-Order Plot

The graph is not linear, so we conclude that the decomposition of NO_2 does not follow first-order kinetics. Consequently, Mechanism I, which predicts first-order behavior, cannot be correct.

Does the Result Make Sense? Remember that reaction order always must be determined by analyzing experiments. The decomposition of NO_2 does not follow first-order kinetics, so a mechanism that predicts first-order behavior cannot be correct, no matter how "reasonable" that mechanism might appear on paper.

Here are experimental data for the reaction $SO_2Cl_2\,(g) \rightarrow SO_2\,(g) + Cl_2\,(g)$

Time (10^3 s)	0	2.5	5.0	7.5	10.0	15.0	20.0	30.0	50.0
$[SO_2Cl_2]$ (torr)	760	720	680	645	610	545	490	395	255

Construct the appropriate graph to determine if these data are consistent with first-order kinetics.

Data must be collected over a relatively long period to determine whether a first-order plot is linear. Notice that the first five points on the plot in Example 15-4 fall reasonably close to the dashed straight line. Only after more than 50% of the reactant has been consumed does this plot deviate substantially from linearity.

Another characteristic of first-order reactions is that the time it takes for half the reactant to disappear is the same, no matter what the concentration. This time is called the **half-life** ($t_{1/2}$). Applying Equation 15-3 to a time interval equal to the half-life results in an equation for $t_{1/2}$. When half the original concentration has been consumed, $[A] = 0.5[A]_0$. The $[A]_0$ terms cancel:

$$kt_{1/2} = \ln\left(\frac{[A]_0}{0.5[A]_0}\right) = \ln 2$$

Rearranging gives an equation for $t_{1/2}$:

$$t_{1/2} = \frac{\ln 2}{k} \tag{15-4}$$

Equation 15-4 does not contain the concentration of A, so the half-life of a first-order reaction is a constant that is independent of how much A is present. The decomposition reactions of radioactive isotopes provide excellent examples of first-order processes, as Example 15-5 illustrates. We describe the use of radioactive isotopes and their half-lives to determine the age of an object in more detail in Chapter 22.

Example 15-5

Half-Lives

The ages of ancient pots can be determined by ^{14}C dating of charcoal found with the pots.

Carbon-14 (^{14}C) is a radioactive isotope with a half-life of 5.73×10^3 years. The fractional amount of ^{14}C present in an object can be used to determine its age. Calculate the rate constant for decay of ^{14}C and determine how long it takes for 90% of the ^{14}C in a sample to decompose.

Strategy: We are asked to calculate two quantities, so this is a quantitative problem to which we can apply the seven-step approach.

1. We are asked to determine a rate constant and the time required for material to be consumed.

2. Visualize the process: atoms of ^{14}C decompose as time passes.

Solution:

3. The only piece of data is the half-life: $t_{1/2} = 5.73 \times 10^3$ years.

4. Equation 15–3 links time and concentration through the rate constant, and Equation 15–4 links the rate constant to the half-life. Knowing a half-life, we can calculate the rate constant using Equation 15–4. Then the value of the rate constant and Equation 15–3 can be used to determine the time required to reach a certain concentration.

5. We need to break the problem into its two parts. To find the value of the rate constant, rearrange Equation 15–4 to isolate k:

$$t_{1/2} = \frac{\ln 2}{k} \qquad so \qquad k = \frac{\ln 2}{t_{1/2}}$$

To find the value for time, rearrange Equation 15–3 to isolate t:

$$\ln\left(\frac{[A]_0}{[A]}\right) = kt \qquad so \qquad t = \frac{\ln\left(\frac{[A]_0}{[A]}\right)}{k}$$

6. Now substitute and calculate the two quantities:

$$k = \frac{0.6931}{5.73 \times 10^3 \text{ yr}} = 1.21 \times 10^{-4} \text{ yr}^{-1}$$

When 90% has decayed, 10% remains, or $[A] = 0.10[A]_0$:

$$\ln\left(\frac{[A]_0}{[A]}\right) = \ln\left(\frac{[A]_0}{0.10[A]_0}\right) = \ln(10)$$

$$t = \frac{\ln(10)}{k} = \frac{2.303}{1.21 \times 10^{-4} \text{ yr}^{-1}} = 1.90 \times 10^4 \text{ yr}$$

7. **Do the Results Make Sense?** It requires 19,000 years for 90% of a sample of ^{14}C to decompose. This is a long time, but it is reasonable because it takes 5730 years for half of the isotope to decompose.

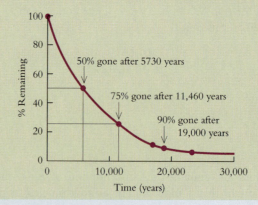

Extra Practice Exercise 15.5

Determine the half-life for the decomposition reaction for which data are given in Extra Practice Exercise 15-4.

Second-Order Reactions

Mechanism II for the decomposition of NO_2 predicts that the reaction should have a second-order rate law:

$$\text{Rate} = -\frac{\Delta[NO_2]}{\Delta t} = k[NO_2]^2$$

The general form of the second-order rate expression is $\text{Rate} = k[A]^2$. Mathematical treatment converts this general form into a time-dependent rate law in which the reciprocal of concentration varies linearly with time:

$$\frac{1}{[A]} - \frac{1}{[A]_0} = kt \qquad (15\text{-}5)$$

As before, [A] is the concentration of reactant A at time t, and $[A]_0$ is the concentration of reactant A at the beginning of the reaction ($t = 0$).

The most important feature of Equation 15-5 is that a graph of $\left(\dfrac{1}{[A]} - \dfrac{1}{[A]_0}\right)$ vs. t gives a straight line whose slope is k. Consequently, experimental data can be tested for second-order behavior by preparing such a graph and observing whether or not it is linear. Figure 15-10 shows the decomposition data of Figure 15-6 plotted in this way. The graph is linear, showing that the reaction is second order in NO_2. Thus, the kinetic behavior of the decomposition of NO_2 matches the predicted rate law for Mechanism II (see earlier discussion). Chemists have accepted Mechanism II as the correct mechanism for this decomposition because it predicts the correct rate law and consists of reasonable elementary reactions.

Example 15-6 provides another example of the analysis of rate data.

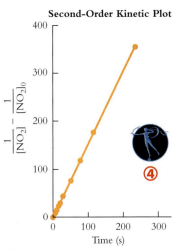

Second-Order Kinetic Plot

Figure 15-10
The decomposition data for NO_2 plotted as the reciprocal of concentration vs. time. This graph is linear, with a slope equal to the second-order rate constant.

Students who have taken calculus will recognize that Equation 15-5 results from integration of the second-order rate law.

Example 15-6

Analysis of Rate Data

The Diels−Alder reaction, in which two alkenes combine to give a new product, is one of the most frequently used reactions for the synthesis of organic compounds. Thousands of examples are found in the chemical literature. The reaction of butadiene is a simple example.

$$\underbrace{\text{⬡} + \text{⬡}}_{\substack{2\,C_4H_6 \\ \text{(Butadiene)}}} \longrightarrow \underset{C_8H_{12}}{\text{⬡}}$$

Use the data in the table to determine the rate law and the rate constant for the Diels−Alder reaction of butadiene:

Time (min)	0	4.0	8.0	12.0	16.0	20.0	30.0
$[C_4H_6]$ (M)	0.130	0.0872	0.0650	0.0535	0.0453	0.0370	0.0281

Strategy: When asked to determine a rate law and rate constant from concentration−time data, we must determine the order of the reaction. The rate law for this reaction has some order x with respect to butadiene (C_4H_6 = A), which we must determine:

$$\text{Rate} = k[A]^x$$

We can use the data provided in the problem to test graphically whether the reaction is first or second order in butadiene. If the reaction is first order, a plot of $\ln\left(\dfrac{[A]_0}{[A]}\right)$ vs. t is linear. If the reaction is second order, a plot of $\left(\dfrac{1}{[A]} - \dfrac{1}{[A]_0}\right)$ vs. t is linear. If neither plot is a straight line, the reaction is something other than first or second order.

Example 15-6

Analysis of Rate Data
(*continued*)

Solution: Begin by calculating the values for the logarithms and the reciprocals of the concentrations:

Time (min)	0	4.0	8.0	12.0	16.0	20.0	30.0
$\ln\left(\dfrac{[A]_0}{[A]}\right)$	0	0.400	0.693	0.888	1.05	1.26	1.53
$\dfrac{1}{[A]} - \dfrac{1}{[A]_0}$	0	3.8	7.7	11.0	14.4	19.3	27.9

Next, prepare first- and second-order plots to see whether either gives a straight line. The graphs are shown below.

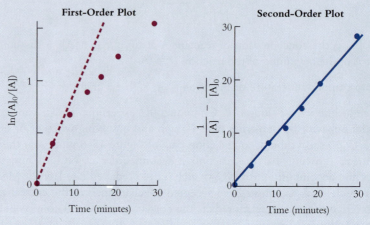

The first-order plot is not linear, so the reaction cannot be first order. The straight line in the $\left(\dfrac{1}{[A]} - \dfrac{1}{[A]_0}\right)$ vs. t plot shows that the reaction is second order in butadiene. This is the rate law for this Diels–Alder reaction:

$$\text{Rate} = k[C_4H_6]^2$$

We could determine the slope using a spreadsheet or graphing calculator. Here, we use the first and last data points to evaluate the slope and rate constant:

$$k = \text{Slope} = \frac{\Delta y}{\Delta x} = \frac{27.9\ \text{M}^{-1} - 0\ \text{M}^{-1}}{30.0\ \text{min} - 0\ \text{min}} = \frac{27.9\ \text{M}^{-1}}{30.0\ \text{min}} = 0.930\ \text{M}^{-1}\ \text{min}^{-1}$$

 Does the Result Make Sense? A second-order dependence is consistent with a mechanism in which the first, rate-determining step is the collisional interaction of two molecules of butadiene.

**Extra Practice
Exercise 15.6**

Construct the appropriate graph to determine if the data in Extra Practice Exercise 15-4 are consistent with a second-order reaction.

Equations 15-3 and 15-5 describe how concentration changes with time when only a *single* reactant is involved. However, most reactions involve concentration changes for more than one species. Although it is possible to develop equations relating concentration and time for such reactions, such equations are more complicated and more difficult to interpret than the equations that involve just one reactant. Fortunately, it is often possible to simplify the experimental behavior of a reaction. We describe two experimental methods that accomplish this, the "isolation" method and the method of initial rates.

"Isolation" Experiments

One way to simplify the behavior of a reaction is to adjust the conditions so that the initial concentration of one starting material is much smaller than the initial concentrations of the others. This establishes experimental conditions under which the concentration of only one of the starting materials changes significantly during the reaction. This concentration is then said to be *isolated*. We use another reaction that is important in atmospheric chemistry to illustrate this **isolation method** in detail.

Trees and shrubs contain a group of fragrant compounds called *terpenes*. The simplest terpene is isoprene. All other terpenes are built around carbon skeletons constructed from one or more isoprene units. Plants emit terpenes into the atmosphere, as anyone who has walked in a pine or eucalyptus forest will have noticed. The possible effect of terpenes on the concentration of ozone in the troposphere has been the subject of much debate and has led to careful measurements of rates of reaction with ozone.

The reactions of terpenes with ozone lead to a complicated array of products, but the rate behavior for ozone reacting with a terpene can be studied by measuring the concentration of the terpene as a function of time:

$$O_3 + \text{terpene} \longrightarrow \text{products} \qquad \text{Rate} = k[O_3]^y[\text{terpene}]^z$$

We expect the reaction rate to depend on two concentrations rather than one, but we can isolate one concentration variable by making the initial concentration of one reactant much smaller than the initial concentration of the other. Data collected under these conditions can then be analyzed using Equations 15-3 and 15-5, which relate concentration to time. For example, an experiment could be performed on the reaction of ozone with isoprene with the following initial concentrations:

$$[O_3]_0 = 5.40 \times 10^{-4}\ M \qquad [\text{isoprene}]_0 = 3.0 \times 10^{-6}\ M$$

Under these initial conditions, the isoprene is consumed entirely before there is an appreciable change in the concentration of ozone, as Figure 15-11 illustrates. For 1:1 reaction stoichiometry and an initial ozone concentration 200 times larger than the initial isoprene concentration, the ozone concentration changes by less than 1%:

$$(5.40 \times 10^{-4}\ M) - (3.0 \times 10^{-6}\ M) = 5.37 \times 10^{-4}\ M$$

| Initial ozone concentration | Concentration decrease from reaction with isoprene | Final ozone concentration |

In other words, the reaction is flooded with a large excess of ozone. Because the concentration of ozone does not change appreciably, the near-constant concentration of ozone can be grouped with the rate constant:

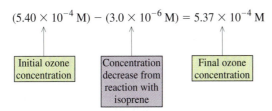

This quantity remains constant over the course of the reaction because $[O_3] \gg [\text{isoprene}]$.

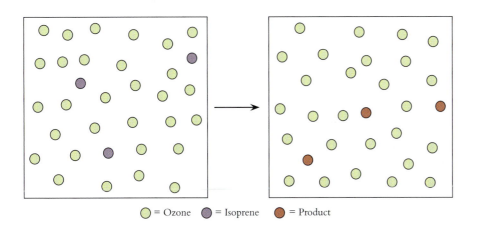

○ = Ozone   ● = Isoprene   ● = Product

Isoprene
C_5H_8

The familiar odors of a pine forest are caused by terpenes, the simplest of which is isoprene.

Figure 15-11
When the initial concentration of ozone (*green circles*) is much greater than that of isoprene (*lavender circles*), isoprene reacts completely to give products (*orange circles*) without changing the concentration of ozone appreciably.

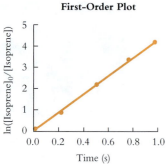

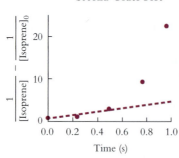

Time (s)	[Isoprene] $(10^{-6}$ M)	$\ln\left(\dfrac{\text{[Isoprene]}_0}{\text{[Isoprene]}}\right)$	$\dfrac{1}{\text{[Isoprene]}} - \dfrac{1}{\text{[Isoprene]}_0}$ $(10^6$ M$^{-1})$
0.000	3.00	0.00	0.00
0.215	1.33	0.813	0.419
0.495	0.364	2.109	2.41
0.765	0.104	3.362	9.28
0.955	0.0443	4.215	22.2

Figure 15-12

Data for the reaction of isoprene with ozone under conditions when the concentration of isoprene is isolated. The experimental behavior is first order under these conditions.

We can identify this combination of constant quantities as a new constant, k_{obs}:

$$k[O_3]^\gamma = k_{obs}$$

Here, "obs" is an abbreviation for "observed." Using k_{obs}, we can simplify the rate law to rate $= k_{obs}[\text{isoprene}]^z$. It is important to remember that k_{obs} is an observed, *experimental* rate constant, but it is not the *true* rate constant (k) that appears in the general rate law.

Isolating isoprene as the only reactant whose concentration changes simplifies the general rate law to a form that can be tested against the single concentration rate laws:

$$\text{First-order case:} \quad \ln\left(\frac{[\text{isoprene}]_0}{[\text{isoprene}]}\right) = k_{obs}t$$

$$\text{Second-order case:} \quad \frac{1}{[\text{isoprene}]} - \frac{1}{[\text{isoprene}]_0} = k_{obs}t$$

Experimental data collected under these conditions are plotted in Figure 15-12. Notice that the first-order plot is linear but the second-order plot is not, demonstrating that the reaction is first order in isoprene. The slope of the first-order plot gives the observed rate constant:

$$\text{Slope} = \frac{4.216 - 0.00}{0.955\text{ s} - 0.00\text{ s}} = 4.4\text{ s}^{-1} = k_{obs} = k[O_3]^\gamma$$

Before we can calculate the true rate constant (k), we must determine γ, the order with respect to ozone. One way to determine γ is to repeat the isolation experiment with a different initial concentration of ozone. For example, the experiment could be repeated with the same initial isoprene concentration but an initial ozone concentration of 2.70×10^{-4} M, half as large as the concentration in the first experiment. When all the isoprene has been consumed, the ozone concentration is 2.67×10^{-4} M, a decrease of about 1%. When this experiment is done, the result is $k_{obs} = 2.2\text{ s}^{-1}$, half as large as k_{obs} in the first experiment:

$$k_{obs,\text{ I}} = k[O_3]_\text{I}^\gamma = 4.4\text{ s}^{-1} \quad k_{obs,\text{ II}} = k[O_3]_\text{II}^\gamma = 2.2\text{ s}^{-1}$$

We need to determine two unknowns, k and γ, but we can eliminate k by taking the ratio of the two expressions for k_{obs}:

$$\frac{k_{obs,\text{ I}}}{k_{obs,\text{ II}}} = \frac{k[O_3]_\text{I}^\gamma}{k[O_3]_\text{II}^\gamma} = \frac{[O_3]_\text{I}^\gamma}{[O_3]_\text{II}^\gamma}$$

⑤ Now substitute the appropriate values to obtain a numerical expression for γ:

$$\frac{k_{obs,\text{ I}}}{k_{obs,\text{ II}}} = \frac{4.4\text{ s}^{-1}}{2.2\text{ s}^{-1}} = 2.0$$

$$\frac{[O_3]_\text{I}^\gamma}{[O_3]_\text{II}^\gamma} = \left(\frac{5.4 \times 10^{-4}\text{ M}}{2.7 \times 10^{-4}\text{ M}}\right)^\gamma = (2.0)^\gamma$$

The ratio of experimental rate constants is 2.0, and the ratio of the ozone concentration terms is $(2.0)^\gamma$. To evaluate γ, we set these ratios equal to each other and take the logarithm of each side of the equality:

$$\ln(2.0) = \gamma \ln(2.0)$$

Solving for y gives $y = \ln(2.0)/\ln(2.0) = 1$, so the reaction is first order in ozone. Thus, the complete experimental rate law for the reaction is

$$\text{Rate} = k[O_3][\text{isoprene}]$$

Finally, the true rate constant can be calculated from k_{obs} using the data from either experiment:

$$k_{obs} = k[O_3] \qquad so \qquad k = \frac{k_{obs}}{[O_3]} = \frac{4.4 \text{ s}^{-1}}{5.4 \times 10^{-4} \text{ M}} = 8.1 \times 10^3 \text{ M}^{-1} \text{ s}^{-1}$$

Example 15-7 provides another illustration of the isolation technique.

Example 15-7

Rate Law from Isolation Experiments

The reaction of hydrogen and bromine produces hydrogen bromide, a highly corrosive gas:

$$H_2(g) + Br_2(g) \longrightarrow 2 \, HBr(g)$$

To determine the rate law for this reaction, a chemist performed two isolation experiments using different initial concentrations. Both experiments gave linear graphs of $\ln\left(\dfrac{[H_2]_0}{[H_2]}\right)$ vs. t, but with different slopes. Here are the details:

$[Br_2]_0$	$[H_2]_0$	Slope of graph
3.50×10^{-5} M	2.50×10^{-7} M	8.87×10^{-4} s^{-1}
2.00×10^{-5} M	2.50×10^{-7} M	6.71×10^{-4} s^{-1}

Determine the rate law and the rate constant for the reaction.

Strategy: When asked to determine a rate law and rate constant, we must determine the order of the reaction. The rate law for this reaction can be expected to contain the concentrations of H_2 and Br_2 raised to powers y and z that must be determined:

$$\text{Rate} = k[H_2]^y[Br_2]^z$$

Because we expect the rate law to contain the concentrations of more than one species, isolation experiments are required to determine the orders of reaction. In the experiments whose data are shown, the initial concentrations of Br_2 are about 100 times the initial concentrations of H_2, so the Br_2 concentration remains essentially constant over the course of the reaction. Thus, the rate law can be rewritten with $[Br_2]$ included as part of the observed rate constant:

$$\text{Rate} = k_{obs}[H_2]^y \qquad where \qquad k_{obs} = k[Br_2]^z$$

Solution: In each experiment a first-order graph gives a straight line, which means that the reaction is first order in hydrogen ($y = 1$). For a first-order reaction, the slope gives k_{obs}:

$$k_{obs, \, I} = \text{Slope}_I = 8.87 \times 10^{-4} \text{ s}^{-1} \qquad and \qquad k_{obs, \, II} = \text{Slope}_{II} = 6.71 \times 10^{-4} \text{ s}^{-1}$$

To find the order of reaction with respect to bromine, we must compare the ratio of the observed rate constants to the ratio of the initial concentrations of Br_2:

$$\frac{k_{obs, \, I}}{k_{obs, \, II}} = \frac{k[Br_2]_I^z}{k[Br_2]_{II}^z}$$

We calculate the ratio of observed rate constants from the slopes of the graphs:

$$\frac{k_{obs, \, I}}{k_{obs, \, II}} = \frac{8.87 \times 10^{-4} \text{ s}^{-1}}{6.71 \times 10^{-4} \text{ s}^{-1}} = 1.32$$

We calculate the ratio of concentrations from the different initial concentrations of Br_2:

$$\frac{[Br_2]_I}{[Br_2]_{II}} = \frac{3.50 \times 10^{-5} \text{ M}}{2.00 \times 10^{-5} \text{ M}} = 1.75$$

Example 15-7

Rate Low from Isolation Experiments *(continued)*

Putting these two together yields an equality where z is the only unknown:

$$1.32 = \frac{k_{obs,\,I}}{k_{obs,\,II}} = \frac{[Br_2]_I^z}{[Br_2]_{II}^z} = (1.75)^z$$

The exponent z can be isolated by taking the logarithm of both sides:

$$\ln(1.32) = z \ln(1.75)$$

$$0.278 = z(0.560) \quad \textit{from which} \quad z = 0.496$$

Reaction orders must be integers or simple fractions, so we round y to 0.5. The reaction is half-order in bromine, and the rate law is as follows:

$$\text{Rate} = k[H_2][Br_2]^{1/2}$$

To calculate the true rate constant (k), use the relationship between the observed rate constant and k, substituting data from either of the two experiments:

$$k_{obs} = k[Br_2]^{1/2} \quad \textit{so} \quad k = \frac{k_{obs}}{[Br_2]^{1/2}}$$

$$k = \frac{8.87 \times 10^{-4}\,s^{-1}}{(3.50 \times 10^{-5}\,M)^{1/2}} = \frac{8.87 \times 10^{-4}\,s^{-1}}{5.92 \times 10^{-3}\,M^{1/2}} = 0.150\,M^{-1/2}\,s^{-1}$$

Does the Result Make Sense? A reaction order of one-half seems strange, but repeated experiments have confirmed this behavior. In Section 15.5, we describe a mechanism that explains the half-order dependence on bromine concentration.

Extra Practice Exercise 15.7

When the reaction of methyl bromide with aqueous base was studied under isolation conditions, the following data were collected:

$[CH_3Br]_0$	$[OH^-]_0$	Experimental order	Slope of graph
5.0×10^{-4} M	0.1 M	First order	$17.2\,s^{-1}$
5.0×10^{-4} M	0.025 M	First order	$4.3\,s^{-1}$

Determine the order with respect to each reactant and find the value of k.

Initial Rates

A second way to simplify the behavior of a reaction is the **method of initial rates.** In this method, we measure the rate at the very beginning of the reaction for different concentrations. A set of experiments is done, changing only one initial concentration each time. Instead of measuring the concentration at many different times during the reaction, we make just one measurement for each set of concentrations. The reaction orders can be evaluated from the relationships between the changes in concentration and the changes in initial rates. We illustrate how this works using a gas-phase reaction of H_2 with NO:

$$2\,H_2\,(g) + 2\,NO\,(g) \longrightarrow N_2\,(g) + 2\,H_2O\,(g)$$

At an elevated temperature of 1000 K, a sequence of initial rate experiments gives the following information:

Experiment	$[H_2\,(g)]$	$[NO\,(g)]$	Initial rate
1	0.0010 M	0.0020 M	1.2×10^{-4} M/s
2	0.0010 M	0.0030 M	1.8×10^{-4} M/s
3	0.0020 M	0.0020 M	4.8×10^{-4} M/s

For each of these experiments, we can write the same general equation for the initial rate:

$$\text{Initial rate} = k\,[H_2]_{initial}^x\,[NO]_{initial}^y$$

This equation contains three unknowns that we have to evaluate: x, y, and k. The essential feature of the initial rate method is that we can take ratios of initial rates under different conditions. First, apply this technique to Experiments 1 and 2, which have the same initial concentration of H_2:

$$\text{Initial rate}_1 = 1.2 \times 10^{-4} \text{ M/s} = k(0.0010 \text{ M})^x(0.0020 \text{ M})^y$$
$$\text{Initial rate}_2 = 1.8 \times 10^{-4} \text{ M/s} = k(0.0010 \text{ M})^x(0.0030 \text{ M})^y$$

When we take the ratio of the second initial rate to the first, the rate constant and the initial concentration term for H_2 cancel:

$$\frac{\text{Initial rate}_2}{\text{Initial rate}_1} = \frac{1.8 \times 10^{-4} \text{ M/s}}{1.2 \times 10^{-4} \text{ M/s}} = \frac{\cancel{k(0.0010 \text{ M})^x}(0.0030 \text{ M})^y}{\cancel{k(0.0010 \text{ M})^x}(0.0020 \text{ M})^y} = \frac{(0.0030 \text{ M})^y}{(0.0020 \text{ M})^y}$$

Evaluating the ratios, we find $1.5 = (1.5)^y$, from which $y = 1$.

Now repeat this analysis for the third experiment and the first experiment, for which the initial concentrations of NO are the same:

$$\frac{\text{Initial rate}_3}{\text{Initial rate}_1} = \frac{4.8 \times 10^{-4} \text{ M/s}}{1.2 \times 10^{-4} \text{ M/s}} = \frac{k(0.0020 \text{ M})^x\cancel{(0.0020 \text{ M})^y}}{k(0.0010 \text{ M})^x\cancel{(0.0020 \text{ M})^y}} = \frac{(0.0020 \text{ M})^x}{(0.0010 \text{ M})^x}$$

This gives $4 = (2)^x$, from which $x = 2$. Thus, the rate law is Rate $= k[H_2][NO]^2$.

Once we know x and y, we can use any of the experiments to calculate the rate constant k:

$$1.2 \times 10^{-4} \text{ M/s} = k(0.0010 \text{ M})^2(0.0020 \text{ M})$$

$$k = \frac{1.2 \times 10^{-4} \text{M/s}}{(0.0010 \text{ M})^2(0.0020 \text{ M})} = 6.0 \times 10^4 \text{ M}^{-2}\text{ s}^{-1}$$

Example 15-8 provides practice in applying the method of initial rates.

Example 15-8

Rate Law from Initial Rates

The reaction of nitrogen dioxide with fluorine generates nitryl fluoride:

$$2 \text{ NO}_2(g) + \text{F}_2(g) \longrightarrow 2 \text{ NO}_2\text{F}(g)$$

To determine the rate law for this reaction, a chemist performed several initial rate experiments using different initial concentrations. Here are the details:

Experiment	$[NO_2(g)]$	$[F_2(g)]$	$[NO_2F(g)]$	Initial rate
1	1.0 mM	5.0 mM	0.10 mM	2.0×10^{-4} M/s
2	2.0 mM	5.0 mM	0.10 mM	4.0×10^{-4} M/s
3	2.0 mM	5.0 mM	1.0 mM	4.0×10^{-4} M/s
4	2.0 mM	7.5 mM	0.10 mM	6.0×10^{-4} M/s

Determine the rate law and the rate constant for the reaction.

Strategy: When asked to determine a rate law and rate constant, we must determine the order of the reaction. The rate law for this reaction may contain the concentrations of NO_2, F_2, and NO_2F raised to powers x, y, and z that must be determined:

$$\text{Rate} = k[NO_2]^x [F_2]^y [NO_2F]^z$$

Because the rate law contains more than one species, we need to use either isolation or initial rates to determine the orders of reaction. In the experiments whose data are shown, initial rate data are obtained for various combinations of initial concentrations. We apply the ratios of these initial rates to evaluate the orders.

Solution: First determine which pairs of initial rates to compare. Only $[NO_2(g)]$ changes from Experiment 1 to Experiment 2, so we can use that ratio to evaluate x. Only $[F_2(g)]$ changes from Experiment 2 to Experiment 4, so we can use that ratio to evaluate y. Finally, Experiments 2 and 3 are identical except for $[NO_2F(g)]$, so we can use that pair to evaluate z.

Example 15-8 **Rate Law from Initial Rates (continued)**	To find x: $$\frac{\text{Initial rate}_2}{\text{Initial rate}_1} = \frac{4.0 \times 10^{-4}\ \text{M/s}}{2.0 \times 10^{-4}\ \text{M/s}} = \frac{k(2.0\ \text{mM})^x(5.0\ \text{mM})^y(0.10\ \text{mM})^z}{k(1.0\ \text{mM})^x(5.0\ \text{mM})^y(0.10\ \text{mM})^z} = \frac{(2.0\ \text{mM})^x}{(1.0\ \text{mM})^x}$$ Simplifying, $2.0 = (2.0)^x$, so $x = 1$. The reaction is first order in NO_2. To find y: $$\frac{\text{Initial rate}_4}{\text{Initial rate}_2} = \frac{6.0 \times 10^{-4}\ \text{M/s}}{4.0 \times 10^{-4}\ \text{M/s}} = \frac{k(2.0\ \text{mM})^x(7.5\ \text{mM})^y(0.10\ \text{mM})^z}{k(2.0\ \text{mM})^x(5.0\ \text{mM})^y(0.10\ \text{mM})^z} = \frac{(7.5\ \text{mM})^y}{(5.0\ \text{mM})^y}$$ Simplifying, $1.5 = (1.5)^y$, so $y = 1$. The reaction is first order in F_2. To find z: $$\frac{\text{Initial rate}_3}{\text{Initial rate}_2} = \frac{4.0 \times 10^{-4}\ \text{M/s}}{4.0 \times 10^{-4}\ \text{M/s}} = \frac{k(2.0\ \text{mM})^x(5.0\ \text{mM})^y(1.0\ \text{mM})^z}{k(2.0\ \text{mM})^x(5.0\ \text{mM})^y(0.10\ \text{mM})^z} = \frac{(1.0\ \text{mM})^z}{(0.10\ \text{mM})^z}$$ Simplifying, $1.0 = (10)^z$; how can this be? We can treat this mathematically, but examining the data also gives the result. Changing the concentration of NO_2F has no effect on the initial rate, so the rate is independent of $[NO_2F]$. This means that $[NO_2F]$ does not appear in the rate law. Mathematically, we set $z = 0$. The reaction is zero order in NO_2F, and the rate law is as follows: $$\text{Rate} = k[NO_2][F_2]$$ To calculate the true rate constant (k), use the expression for the initial rate, substituting data from any of the experiments: $$\text{Initial rate} = k[NO_2][F_2] \qquad so \qquad k = \frac{\text{Initial rate}}{[NO_2][F_2]}$$ $$k = \frac{2.0 \times 10^{-4}\ \text{M s}^{-1}}{(1.0\ \text{mM})(5.0\ \text{mM})(10^{-3}\ \text{M/mM})^2} = 40.\ \text{M}^{-1}\ s^{-1}$$
	Does the Result Make Sense? A rate law that is first order in each reactant is consistent with a mechanism whose first step is a collision between reactant molecules that results in a slow reaction. Both reactants are stable substances, so it makes sense that although they can collide and react, the reaction is slow.
Extra Practice Exercise 15.8	Carbon dioxide can undergo a reaction with itself to produce carbon monoxide: $$2\ CO_2\,(g) \longrightarrow 2\ CO\,(g) + O_2\,(g)$$ Neither CO nor O_2 affects the rate. Initial rate studies on this reaction give the following data: $[CO_2]_{\text{initial}} = 0.025\ \text{M}$, initial rate $= 4.7 \times 10^{-3}\ \text{M/s}$ $[CO_2]_{\text{initial}} = 0.037\ \text{M}$, initial rate $= 7.0 \times 10^{-3}\ \text{M/s}$ Determine the rate law for the reaction and find the value of k.

This is the first example where we have tested to see if the concentration of a product affects the rate of a reaction. It may seem intuitive that products should not be involved in rate laws, but as we show in Section 15.5, a product may influence the rate of a reaction. In careful rate studies, this possibility must be considered. It is common for some reagents in a chemical system to have no effect on the rate of chemical reaction. Although it is unusual for *none* of the reagents to affect the rate, there are some reactions that are zero order overall. Such reactions have a particularly simple rate law: Rate $= k$.

Table 15-1 summarizes the parameters for the experimental kinetics of the three simplest types of rate laws.

Table 15-1   Rate Parameters for Simple Reaction Orders

Order	Rate Law	Concentration Dependence	Linear Plot
0	Rate $= k$	$[A]_o - [A] = kt$	$[A]_o - [A]$ vs. t
1	Rate $= k\,[A]$	$\ln\left(\dfrac{[A]_o}{[A]}\right) = kt$	$\ln\left(\dfrac{[A]_o}{[A]}\right)$ vs. t
2	Rate $= k\,[A]^2$	$\dfrac{1}{[A]} - \dfrac{1}{[A]_o} = kt$	$\dfrac{1}{[A]} - \dfrac{1}{[A]_o}$ vs. t

Section Exercises

15.4.1 N_2O_5 can decompose to nitrogen dioxide and oxygen:

$$2\ N_2O_5\,(g) \longrightarrow 4\ NO_2\,(g) + O_2\,(g).$$

Here are some data for this decomposition reaction:

Time (min)	0	20.0	40.0	60.0	80.0
$[N_2O_5]$ (10^{-2} M)	0.92	0.50	0.28	0.15	0.08

Determine the order and the rate constant by constructing appropriate graphs using these data.

15.4.2 AB_2 reacts with CB as follows:

$$AB_2\,(g) + CB\,(g) \longrightarrow AB\,(g) + CB_2\,(g)$$

Use the following information to determine the rate law: (1) When an experiment is performed with $[AB_2]_0 = 1.75 \times 10^{-3}$ M and $[CB]_0 = 0.15$ M, a plot of

$$\left(\frac{1}{[AB_2]} - \frac{1}{[AB_2]_0}\right)$$ vs. t gives a straight

line. (2) When this experiment is repeated with $[AB_2]_0 = 1.75 \times 10^{-3}$ M and $[CB]_0 = 0.25$ M, the slope of a plot of

$$\left(\frac{1}{[AB_2]} - \frac{1}{[AB_2]_0}\right)$$ vs. t is identical to the

slope of the plot for Experiment (1).

15.4.3 The ammonium ion can react with the nitrite ion in aqueous solution as follows:

$$NH_4^+\,(aq) + NO_2^-\,(aq) \longrightarrow N_2\,(g) + 2H_2O\,(l)$$

Use the following information about initial rates to determine the rate law and evaluate the rate constant for this reaction:

$[NH_4^+\,(aq)]$	$[NO_2^-\,(aq)]$	Initial rate
0.040 M	0.200 M	2.15×10^{-6} M/s
0.060 M	0.200 M	3.23×10^{-6} M/s
0.060 M	0.100 M	1.66×10^{-6} M/s

15.5 LINKING MECHANISMS AND RATE LAWS

The mechanism of any chemical reaction has the following characteristics:

1. The mechanism is one or more elementary reactions describing how the chemical reaction occurs. These elementary reactions may be unimolecular, bimolecular, or (rarely) termolecular.

2. The sum of the individual steps in the mechanism must give the overall balanced chemical equation. Sometimes a step may occur more than once in the mechanism.

3. The reaction mechanism must be consistent with the experimental rate law.

   In this section we show that every mechanism predicts a rate law. If the rate law predicted by a proposed mechanism matches the experimental rate law, the mechanism is a possible description of how the reaction proceeds. On the other hand, if the rate law predicted by the proposed mechanism differs from experimental rate law, the proposed mechanism must be wrong.

Remember that a bimolecular elementary reaction is a collision between two molecules; its rate law contains the concentrations of both reactants.

KEY CONCEPT   *Bimolecular elementary reaction:* $A + B \rightarrow products$
                                                        Elementary rate $= k[A][B]$

A unimolecular elementary reaction is a fragmentation or rearrangement of one chemical species, so its rate law contains the concentration of only that species:

KEY CONCEPT   *Unimolecular elementary reaction:* $C \rightarrow products$
                                                          Elementary rate $= k[C]$

These rate laws for elementary steps are related to the experimental rate law for the overall reaction in a way that depends on which step in the mechanism is rate-determining.

Rate-Determining First Step

Each example we have used to introduce the concepts of chemical mechanisms has a first step that is rate-determining. These mechanisms and their rate laws are summarized in Table 15-2.

In each of these mechanisms, the first elementary reaction is the rate-determining step. Because the overall reaction can go no faster than its rate-determining step, no elementary reaction that occurs after the rate-determining step affects the overall rate of reaction.

KEY CONCEPT   *When the first step of a mechanism is rate-determining, the predicted rate law for the overall reaction is the rate law for that first step.*

The predicted rate law is first order for a reaction whose first step is unimolecular and rate-determining. The predicted rate law is second order overall for a reaction whose first step is bimolecular and rate-determining. For example, the first step of the mechanism for the $C_5H_{11}Br$ reaction is unimolecular and slow, so the rate law predicted by this mechanism is first order: Rate $= k[C_5H_{11}Br]$. Example 15-9 treats the rate law of another mechanism whose first step is rate-determining.

Table 15-2   Some Mechanisms and Rate Laws*

Stoichiometry	Mechanism	Rate Law
$2\,NO_2 \rightarrow N_2O_4$	$NO_2 + NO_2 \rightarrow N_2O_4$	Rate $= k[NO_2]^2$
$2\,NO_2 \rightarrow 2\,NO + O_2$	$2\,NO_2 \rightarrow NO_3 + NO$ (slow) $NO_3 \rightarrow NO + O_2$	Rate $= k[NO_2]^2$
$NO + O_3 \rightarrow NO_2 + O_2$	$NO + O_3 \rightarrow NO_2 + O_2$	Rate $= k[NO][O_3]$

cis-2-butene $\rightarrow$
      trans-2-butene
	cis-2-butene $\rightarrow$
	      trans-2-butene
		Rate $=$
		 $k[$*cis*-2-butene$]$

$C_5H_{11}Br + H_2O \rightarrow$ $C_5H_{10} + H_3O^+ + Br^-$	$C_5H_{11}Br \rightarrow C_5H_{11}^+ + Br^-$ (slow) $C_5H_{11}^+ + H_2O \rightarrow C_5H_{10} + H_3O^+$	Rate $= k[C_5H_{11}Br]$

*Examples from Sections 15.1 through 15.4.

At elevated temperature, NO_2 reacts with CO to produce CO_2 and NO:

$$CO + NO_2 \longrightarrow CO_2 + NO$$

Example 15-9

Predicted Rate Laws

In one possible mechanism for this reaction, products form directly in a one-step bimolecular collision that transfers an oxygen atom from NO_2 to CO:

In another possible mechanism, two NO_2 molecules collide in the rate-determining step to form NO and NO_3. In a second and faster step, the highly reactive NO_3 intermediate transfers an oxygen atom to CO in a bimolecular collision:

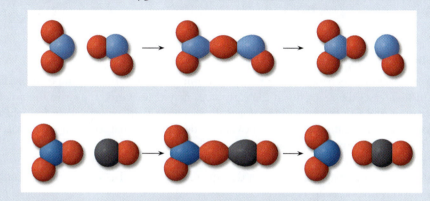

What is the predicted rate law for each of these mechanisms?

Strategy: It is best to work with each mechanism separately. When the first step is rate-determining, the predicted rate law matches the rate expression for that first elementary step.

Solution: First, analyze the one-step mechanism:

$$CO + NO_2 \longrightarrow CO_2 + NO$$

This process is analogous to the reaction of NO and O_3 discussed in Section 15.3. In a simple one-step atom transfer, the reaction is first order in each of the starting materials and second order overall:

One-step mechanism: Rate $= k[CO][NO_2]$

The second mechanism occurs in two steps:

$$2\,NO_2 \longrightarrow NO_3 + NO \qquad \text{(slow, rate-determining)}$$

$$NO_3 + CO \longrightarrow NO_2 + CO_2 \qquad \text{(fast)}$$

In this mechanism the first step is slow and rate-determining. Because the collision partners are identical, this reaction is second order in NO_2 and second order overall:

Two-step mechanism: Rate $= k[NO_2]^2$

Do the Results Make Sense? Carbon monoxide does not appear in the second rate law because it participates in the mechanism *after* the rate-determining step. Remember, any reaction after the rate-determining step does not affect the overall rate of reaction.

Experiments show that this reaction is second order in NO_2, as predicted by the second proposed mechanism. The one-step mechanism can be ruled out because it is not consistent with the experimental rate law. Agreement with the rate law does not

Example 15-9

Predicted Rate Laws
(*continued*)

**Extra Practice
Exercise 15.9**

prove that the second mechanism is the correct one, however, because other mechanisms may predict the same rate law. It is one strong piece of evidence that supports this particular two-step process.

Here is a proposed mechanism for the Diels–Alder reaction:

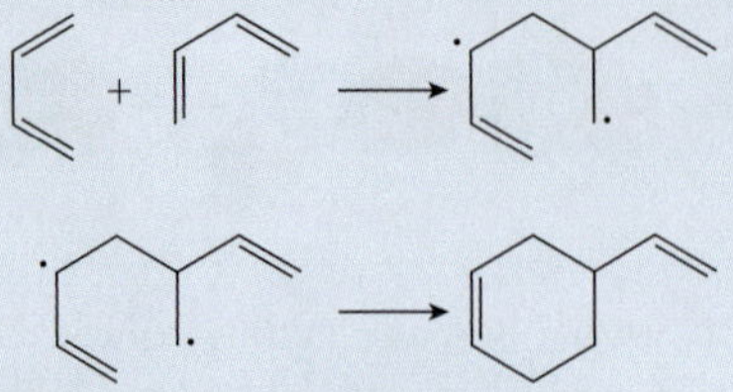

What rate law does this mechanism predict?

Rate-Determining Later Step

For many reaction mechanisms, the rate-determining step occurs after one or more faster steps. In such cases the reactants in the early steps may or may not appear in the rate law. Furthermore, the rate law is likely to depart from simple first- or second-order behavior. Fractional orders, negative orders, and overall orders greater than two, all are signals that a fast first step is followed by a slow subsequent step.

Example 15-7 treats a reaction of this kind. The experimental rate law for the reaction of H_2 gas with Br_2 gas depends on the square root of the Br_2 concentration, and the reaction also is first order in H_2:

$$H_2 + Br_2 \longrightarrow 2\,HBr \qquad \text{Rate} = k[H_2][Br_2]^{1/2}$$

Despite the simple 1:1 stoichiometry of the overall reaction, this experimental rate law cannot be explained by a simple mechanism. For the first step of the mechanism for this reaction to be rate-determining, it would have to include a half-molecule of Br_2. There is no such species, so the first step cannot be rate-determining. Instead, some later step in the mechanism must be rate-determining.

Figure 15-13 uses molecular pictures to show the accepted mechanism for this reaction. The reaction begins with the dissociation of Br_2 molecules into Br atoms:

$$Br_2 \xrightarrow{\ k_1\ } 2\,Br$$

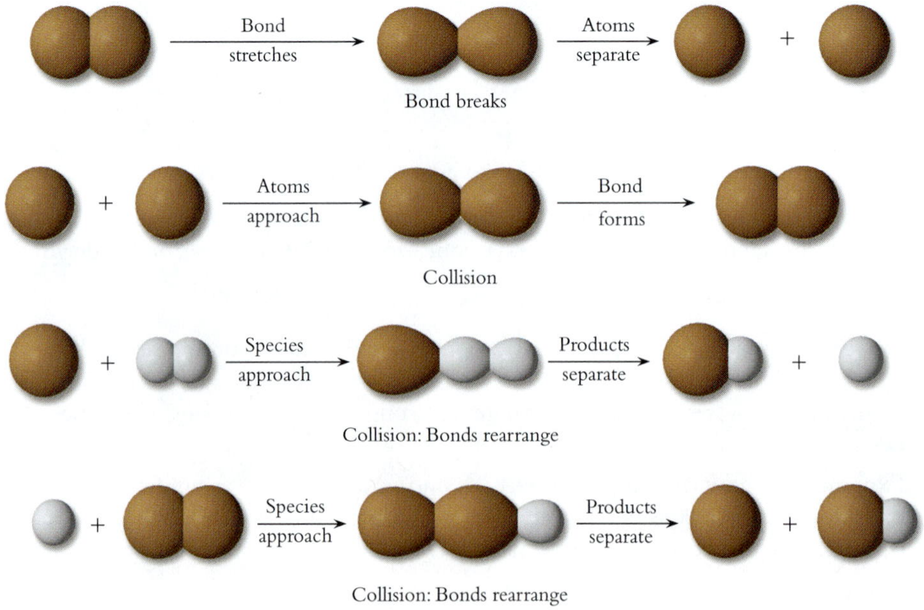

**Figure 15-13
A molecular view of the accepted mechanism for the reaction of H_2 and Br_2 to produce 2 HBr.**

Almost all the Br atoms produced in the first step simply recombine to regenerate Br_2 molecules:

$$2\,Br \xrightarrow{k_{-1}} Br_2$$

This is an example of a **reversible reaction,** one that occurs rapidly in both directions. For simplicity, we combine these two elementary reactions in a single expression that uses double arrows:

$$Br_2 \underset{k_{-1}}{\overset{k_1}{\rightleftharpoons}} 2\,Br \text{ (fast, reversible)}$$

Most bromine atoms produced in the first step recombine, but occasionally a Br atom reacts with a molecule of H_2 to form one molecule of HBr and a hydrogen atom. This step is slow and rate-determining:

$$Br + H_2 \xrightarrow{k_2} HBr + H \text{ (slow, rate-determining)}$$

To complete the mechanism, we need a step that consumes H atoms. The hydrogen atom produced in the rate-determining step undergoes a third elementary reaction, a bimolecular collision with a Br_2 molecule. The H atom rapidly forms a bond with one Br atom to form one molecule of HBr and a Br atom:

$$H + Br_2 \xrightarrow{k_3} HBr + Br \text{ (fast)}$$

Is this mechanism satisfactory? A satisfactory mechanism must meet three criteria. It must be made up entirely of reasonable elementary steps, it must give the correct stoichiometry of the reaction, and it must predict the experimental rate law.

This mechanism has four steps. The first step is the unimolecular decomposition of Br_2. The remaining steps all are simple bimolecular collisions. We can readily visualize each step as a reaction that the molecules can undergo, so the mechanism meets the first criterion.

The easiest way to determine whether the steps lead to the correct stoichiometry is to write all four reactions in the same direction. Then cancel those species that appear both on the left and on the right:

$$\begin{aligned}
\cancel{Br_2} &\longrightarrow 2\,\cancel{Br} \\
2\,\cancel{Br} &\longrightarrow \cancel{Br_2} \\
\cancel{Br} + H_2 &\longrightarrow HBr + \cancel{H} \\
\cancel{H} + Br_2 &\longrightarrow HBr + \cancel{Br} \\
\hline
\text{Net reaction: } H_2 + Br_2 &\longrightarrow 2\,HBr
\end{aligned}$$

Combining the steps of the mechanism leads to the balanced equation for the overall reaction, so the second criterion also is satisfied.

To see if the proposed mechanism predicts the correct rate law, we start with the rate-determining step. The second step in this mechanism is rate-determining, so the overall rate of the reaction is governed by the rate of this step:

$$\text{Rate} = k_2[Br][H_2]$$

This rate law describes the rate behavior predicted by the proposed mechanism accurately, but the law cannot be tested against experiments because it contains the concentration of Br atoms, which are intermediates in the reaction. As mentioned earlier, an intermediate has a short lifetime and is hard to detect, so it is difficult to make accurate measurements of its concentration. Furthermore, it is not possible to adjust the experimental conditions in a way that changes the concentration of an intermediate by a known amount. Therefore, if this proposed rate law is to be tested against experimental behavior, the concentration of the intermediate must be expressed in terms of the concentrations of reactants and products.

Equality of Rates

One way to relate the concentration of an intermediate to other concentrations is by assuming that earlier reversible steps have equal forward and reverse rates. The proposed

Typically, elementary reactions in a mechanism are written with their rate constants above or below the reaction arrow. A reaction that is the reverse of another is labeled accordingly.

mechanism begins with the decomposition of Br_2 molecules into Br atoms, most of which recombine rapidly to give Br_2 molecules. At first, Br_2 molecules decompose faster than Br atoms recombine, but the Br atom concentration quickly becomes large enough for recombination to occur at the same rate as decomposition. When the two rates are equal, so are their rate expressions:

$$\text{Rate of decomposition} = \text{Rate of recombination}$$

$$k_1[Br_2] = k_{-1}[Br]^2$$

This equality can be used to express the concentration of the intermediate in terms of concentrations of the reactants. First solve for [Br] by rearranging the rate expressions and taking the square root of both sides:

$$[Br]^2 = \frac{k_1}{k_{-1}}[Br_2] \quad \textit{from which} \quad [Br] = \left(\frac{k_1}{k_{-1}}\right)^{1/2}[Br_2]^{1/2}$$

Now substitute this equation for [Br] into the expression for the rate-determining step:

$$\text{Rate} = k_2[Br][H_2] = k_2\left(\frac{k_1}{k_{-1}}\right)^{1/2}[H_2][Br_2]^{1/2} = k[H_2][Br_2]^{1/2}$$

In this expression, k, the rate constant for the overall reaction, is related to the elementary rate constants:

$$k = k_2\left(\frac{k_1}{k_{-1}}\right)^{1/2}$$

The predicted rate law is first order in hydrogen and one-half order in bromine, in agreement with the experimental rate law determined in Example 15-7. Thus, the proposed mechanism satisfies the three requirements for a satisfactory mechanism.

As the mechanism for the reaction between H_2 and Br_2 shows, the observation of an order other than first or second indicates that some step beyond the first one in the mechanism is rate-determining. The one-half-order rate dependence for Br_2 comes about because of the initial rapid cleavage reaction in which each Br_2 molecule produces two Br atoms.

Our Chemistry and the Environment Box discusses reactions in the stratosphere, and the connection between mechanisms and rate laws is illustrated further by Example 15-10.

Example 15-10
Reaction Between NO and O_3

Nitrogen oxide converts ozone into molecular oxygen, as follows:

$$O_3 + NO \longrightarrow O_2 + NO_2$$

The experimental rate law is rate = $k[O_3][NO]$. Which of the following mechanisms are consistent with the experimental rate law?

Mechanism I

$$O_3 + NO \xrightarrow{k_1} O + NO_3 \quad \text{(slow)}$$

$$O + O_3 \xrightarrow{k_2} 2\,O_2 \quad \text{(fast)}$$

$$NO_3 + NO \xrightarrow{k_3} 2\,NO_2 \quad \text{(fast)}$$

Mechanism II

$$O_3 \underset{k_{-1}}{\overset{k_1}{\rightleftarrows}} O_2 + O \quad \text{(fast, reversible)}$$

$$NO + O \xrightarrow{k_2} NO_2 \quad \text{(slow)}$$

Mechanism III

$$O_3 + NO \xrightarrow{k_1} O_2 + NO_2 \quad \text{(slow)}$$

Example 15-10

Reaction Between NO and O_3 (*continued*)

Strategy: We need to determine whether the rate law predicted by each mechanism matches the experimental rate law by calculating the rate law predicted by each.

Solution: Mechanism I is a three-step process in which the first step is rate-determining. When the first step of a mechanism is rate-determining, the predicted rate law is the same as the rate expression for that first step. Here, the rate-determining step is a bimolecular collision. The rate expression for a bimolecular collision is first order in each collision partner:

$$\text{Rate} = k_1[O_3][NO]$$

Mechanism I is consistent with the experimental rate law. If we add the elementary reactions, we find that it also gives the correct overall stoichiometry, so this mechanism meets all the requirements for a satisfactory one.

Mechanism II begins with fast reversible ozone decomposition followed by a rate-determining bimolecular collision of an oxygen atom with a molecule of NO. The rate of the slow step is as follows:

$$\text{Rate} = k_2[NO][O]$$

This rate expression contains the concentration of an intermediate, atomic oxygen. To convert the rate expression into a form that can be compared with the experimental rate law, assume that the rate of the first step is equal to the rate of its reverse process. Then solve the equality for the concentration of the intermediate:

$$k_1[O_3] = k_{-1}[O][O_2] \quad \textit{from which} \quad [O] = \frac{k_1[O_3]}{k_{-1}[O_2]}$$

$$\text{Rate} = \frac{k_1 k_2[O_3][NO]}{k_{-1}[O_2]}$$

Mechanism II is inconsistent with the experimental rate law. Notice that this mechanism predicts a rate that is inversely proportional to the concentration of molecular oxygen. That means that oxygen would slow down the reaction: as the concentration of oxygen *increased,* the rate of the reaction would *decrease.*

Mechanism III is a simple one-step bimolecular collision. Its predicted rate law is as follows:

$$\text{Rate} = k_1[O_3][NO]$$

Mechanism III is consistent with the experimental rate law.

Therefore, the O_3 + NO reaction might go by Mechanism I or Mechanism III, but not by Mechanism II.

Do the Results Make Sense? The fact that two different mechanisms give the same predicted rate law illustrates one feature of mechanisms. Chemists can imagine more than one possible mechanism for almost any chemical reaction. We have to use additional chemical knowledge to distinguish among possible mechanisms. Here, Mechanism I is less likely than Mechanism III because the products of the first step, O and NO_3, are less stable than the O_2 and NO_2 formed in Mechanism III. Other factors being equal, molecules generally follow the path of lower energy, which is Mechanism III in this case.

Here is a proposed mechanism for the formation of N_2O_5 from ozone:

$$NO_2(g) + O_3(g) \underset{k_{-1}}{\overset{k_1}{\rightleftharpoons}} NO_3(g) + O_2(g) \qquad \text{(fast, reversible)}$$

$$NO_3(g) + NO_2(g) \xrightarrow{k_2} N_2O_5 \qquad \text{(slow)}$$

What rate law does this mechanism predict?

Extra Practice Exercise 15.10

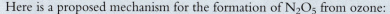

Box 15-1   Chemistry and the Environment: Reactions of Ozone

The ozone layer is found $30-45$ km above the surface of the Earth. Nowhere else in the atmosphere is ozone found in significant amounts. As described in Section 7.7, ozone is limited to this narrow band because only in the ozone layer can O_3 form through sunlight interacting with O_2.

Ozone is formed in the stratosphere by the action of ultraviolet light on oxygen molecules:

$$3\,O_2 + h\nu_{(\lambda<240\ nm)} \longrightarrow 2\,O_3$$

An oxygen molecule absorbs a photon ($h\nu$) of high-energy ultraviolet light ($\lambda < 240$ nm). The energy of the photon breaks O_2 into two O atoms:

$$O_2 + h\nu_{(\lambda<240\ nm)} \longrightarrow O + O \quad \text{Step 1}$$

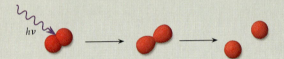

Oxygen atoms are reactive and add to O_2 to form O_3:

$$O + O_2 \longrightarrow O_3 \qquad \text{Step 2}$$

This ozone molecule contains excess energy and can decompose again. The stratosphere contains enough N_2, however, that an ozone molecule usually collides with a nitrogen molecule and gives up its excess energy before it can break apart. Step 1 of the mechanism generates two oxygen atoms, each of which reacts with an O_2 molecule. Thus, each time the first step occurs, the second step occurs twice:

$$O_2 \;+\; h\nu_{(\lambda<240\ nm)} \longrightarrow \text{O} + \text{O}$$
$$\text{O} + O_2 \longrightarrow O_3$$
$$\text{O} + O_2 \longrightarrow O_3$$
$$\overline{\text{Net reaction: } 3\,O_2 + h\nu_{(\lambda<240\ nm)} \longrightarrow 2\,O_3}$$

The first step is the rate-determining step because as soon as an oxygen atom forms, it is snapped up by the nearest available O_2 molecule.

The formation of O_3 goes on continuously, but O_3 is also decomposed by ultraviolet light between 240 and 340 nm:

$$O_3 + h\nu_{(\lambda=240-340\ nm)} \longrightarrow O + O_2$$

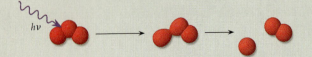

The O atom can react with a second O_3 molecule:

$$O + O_3 \longrightarrow 2\,O_2$$

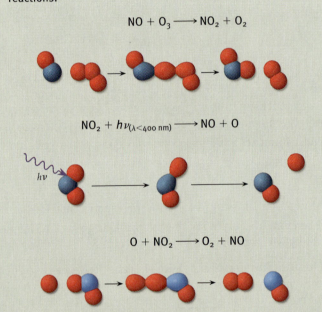

The result is a reduction in the amount of ozone:

$$\text{Net: } 2\,O_3 + h\nu_{(\lambda=240-340\ nm)} \longrightarrow 3\,O_2$$

Short-wavelength UV light forms O_3, whereas long-wavelength UV light decomposes O_3. The reactions create a delicate balance in which the rate of O_3 decomposition matches the rate of O_3 production. The resulting ozone layer absorbs nearly all the solar photons in the 240- to 340-nm range. Photons in this wavelength range have enough energy to damage and even destroy living cells, so the ozone layer protects life on Earth.

Unfortunately, chemical species produced by humans can react in ways that reduce the O_3 concentration. For example, NO changes the oxygen balance through the following reactions:

$$NO + O_3 \longrightarrow NO_2 + O_2$$

$$NO_2 + h\nu_{(\lambda<400\ nm)} \longrightarrow NO + O$$

$$O + NO_2 \longrightarrow O_2 + NO$$

The first step in this process occurs twice to generate the two molecules of NO_2 needed for the second and third steps. The result of this sequence is that NO in the stratosphere increases the rate of O_3 decomposition:

$$\text{Net: } 2\,O_3 + h\nu_{(\lambda<400\ nm)} \longrightarrow 3\,O_2$$

As mentioned in the introduction to this chapter, CFCs also have potentially devastating effects on the ozone layer. In Section 15.7, we describe how CFCs destroy stratospheric ozone.

Section Exercises

■ **15.5.1** A student proposes the following mechanism for the gas-phase decomposition of N_2O_5 (all substances are gases):

Overall reaction: $2 N_2O_5 \longrightarrow 4 NO_2 + O_2$
Step 1: $N_2O_5 \longrightarrow N_2O_4 + O$
Step 2: $O + N_2O_5 \longrightarrow N_2O_4 + O_2$

(a) Propose a third step to complete this mechanism.

(b) Experiments show that this decomposition reaction is first order. Which step in the mechanism must be rate-determining?

■ **15.5.2** What order for N_2O_5 does the mechanism in Section Exercise 15.5.1 predict if the first step is fast and reversible and the second step is rate-determining?

■ **15.5.3** The oxidation of NO by O_2 is an example of a third-order reaction:

$$2 NO(g) + O_2(g) \longrightarrow 2 NO_2(g)$$
$$\text{Rate} = k[NO]^2[O_2]$$

Although a single elementary termolecular mechanism is consistent with the rate law, three-body collisions are rare. It is more likely that the first step is fast and reversible:

$$NO + NO \underset{k_{-1}}{\overset{k_1}{\rightleftharpoons}} N_2O_2$$

(a) Propose a slow second step to complete this mechanism.

(b) Show that this mechanism is consistent with the observed rate law.

15.6 REACTION RATES AND TEMPERATURE

Common experience tells us that chemical reactions proceed faster at higher temperature. Food is stored in refrigerators because food spoils more quickly at high temperature than at low temperature. Wood does not burn at room temperature, but it burns vigorously at high temperature. At the macroscopic level, higher temperature means faster reactions, but how does temperature affect the rate of a reaction at the molecular level? Because temperature is a measure of the energy of motion of molecules, we need to explore the relationship between reaction rates and molecular energy to answer this question.

Energy Changes in a Unimolecular Reaction

In a unimolecular reaction, a molecule fragments into two pieces or rearranges to a different isomer. In either case, a chemical bond breaks. For example, in the fragmentation of bromine molecules, breaking a σ bond gives a pair of bromine atoms:

$$Br_2 \longrightarrow 2 Br$$

Recall that this unimolecular process is the first step of the reaction between molecular hydrogen and molecular bromine to give HBr.

Figure 15-14a shows the energy changes that occur during the fragmentation of a bromine molecule. The reaction begins with a stretching of the Br—Br bond. As the bond becomes longer, the system moves to higher energy. This decreased stability results

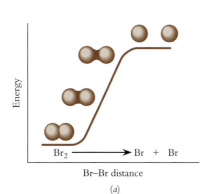

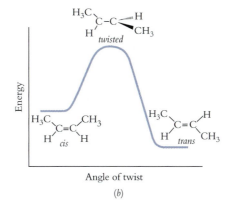

Figure 15-14
Energy profiles for two unimolecular processes: (a) the unimolecular decomposition of a bromine molecule; (b) the unimolecular isomerization of *cis*-2-butene.

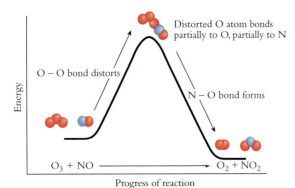

Figure 15-15
A schematic representation of the energy profile for the bimolecular reaction between O_3 and NO molecules.

from the loss of orbital overlap as the bromine atoms move apart. If the bond is stretched far enough, the molecule fragments into two bromine atoms. When a bond breaks, the fragment products are always at higher energy than the starting materials.

Like fragmentations, unimolecular rearrangements are always "uphill" at the beginning of the process, because a bond breaks. Unlike fragmentations, rearrangements are "downhill" at the end as a new bond forms. An example of this kind of energy profile for the isomerization reaction of *cis*-2-butene appears in Figure 15-14*b*.

Energy Changes during Bimolecular Reactions

In a bimolecular reaction, some bonds break and other bonds form. For example, in the reaction between NO and O_3, an O atom is transferred from O_3 to NO:

$$O_3 + NO \longrightarrow O_2 + NO_2$$

During the collision, the ozone molecule distorts as the O—O bond stretches and weakens, and NO must also undergo distortion as it bonds with the incoming oxygen atom from ozone. Distorting a molecule from its most favored configuration always requires the addition of energy. Thus, at the outset of the reaction, the system is destabilized by the molecular distortions required for atom transfer. At the end of the process, on the other hand, the system becomes stabilized as the products adopt their lowest-energy configuration. Figure 15-15 shows a schematic representation of the energy profile for this reaction.

Activation Energy

The energy profiles for these elementary reactions share several common features, one of which is particularly important for a discussion of reaction rates: In each example the system must move "uphill" from the energy of the reactants to an energy maximum. In almost all chemical reactions, the molecules must overcome an energy barrier before starting materials can become products. This energy barrier, the minimum energy that must be supplied before reaction can take place, is the **activation energy (E_a)** of the chemical reaction. Activation energies arise because chemical bonds in reactant molecules must distort or break before new bonds can form in product molecules. The activation energy for any reaction mechanism is independent of both *reactant concentrations* and *temperature*.

Figures 15-14 and 15-15 are **activation energy diagrams** for elementary reactions. They are plots of the change in energy as reactants are transformed into products. Figure 15-16 shows a generalized activation energy diagram labeled with its characteristics. The *x* axis represents the course of the reaction as starting materials are transformed into products. Because these transformations may involve complicated combinations of rotations, vibrations, and atom transfers, accepted practice is to label this axis as the **reaction coordinate** without specifying further details about what changes occur.

The *y* axis of an activation energy diagram is the energy of the molecular system. The graph shows how chemical energy changes during the course of the reaction. Low values of energy represent high molecular stability, and high values represent low molecular stability. The reaction shown in the activation energy diagram in Figure 15-16 releases energy, so its products are lower in energy than its reactants are. For a reaction that absorbs energy, in contrast, the activation energy diagram shows the energy of the products to be higher than the energy of the reactants. Whether the net reaction releases or absorbs energy, however, there always is a positive activation energy because of the requirement to break or distort one chemical bond before another can form.

The molecular arrangement at the point of highest energy along the diagram is known as the **activated complex.** The difference in energy between the reactants and this activated complex is the activa-

Figure 15-16
A generalized activation energy diagram for an elementary reaction. The figure shows a reaction that releases energy, resulting in a negative overall energy change (ΔE).

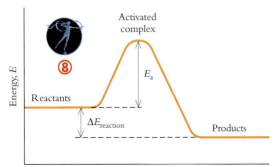

tion energy and is generally labeled E_a. An activated complex is *not* considered to be an intermediate. An intermediate is stable for a short time, perhaps milliseconds, and may be observed with highly sensitive instruments. An activated complex forms and disappears in the time required for a molecular collision, which is about 10^{-13} second. For this reason, an activated complex is also known as a *transition state*. In other words, the activated complex represents the instantaneous configuration of atoms, at the top of the activation energy barrier, during the conversion from reactants with one set of bonds into products with a different set of bonds.

The magnitude of an activation energy depends on the details of the bonding changes that occur during the formation of the activated complex. As described in Chapter 6, the energy change for a reaction (ΔE) can be estimated from average bond energies. In contrast, the activation energy (E_a) is not so easily estimated. Activation energies must be determined by measuring how rate constants vary with temperature.

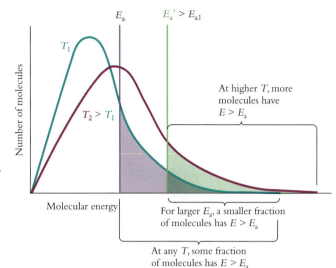

Figure 15-17
The distribution of molecular energies, shown at two different temperatures. Notice that the energy needed for reaction to occur is a fixed value that is independent of T.

Activation Energy and the Rate Constant

A chemical reaction cannot occur unless the starting materials have enough energy to overcome the activation energy barrier. Where do molecules obtain this energy? Recall from Section 6.1 that every molecule has thermal energy associated with translation, rotation, and vibration. During a chemical reaction, some of this thermal energy is transformed into chemical energy as a molecule distorts or as two molecules collide.

Any collection of molecules includes molecules with a wide range of energies. Some molecules have much energy, but others have very little. This distribution of molecular energies, described in Section 6.1, depends on the temperature of the sample, as shown in Figure 15-17. Heating the system increases the translational, rotational, and vibrational energy of the molecules. Thus as the temperature increases, the distribution broadens and shifts toward higher energy.

The energy distributions in Figure 15-17 show that only some of the molecules in any reaction system have energies greater than that required for reaction. How large this fraction is depends on E_a and temperature:

1. At any T, a larger fraction of molecules can react if E_a is small than if E_a is large.
2. For any reaction, a larger fraction of molecules can react if T is high than if T is low.

The larger the fraction of molecules that can react, the faster a reaction will proceed. Thus, a reaction with a low activation energy is faster than one with a high activation energy, and the rate of any reaction increases with increasing temperature.

The rate of a reaction increases with temperature.   **KEY CONCEPT**

The Arrhenius Equation

The value of the rate constant for a particular reaction depends on the activation energy of the reaction, the temperature of the system, and how often a collision occurs in which the atoms are in the required orientation. All these factors can be summarized in a single equation, called the **Arrhenius equation:**

$$k = Ae^{-E_a/RT} \tag{15-6}$$

In this equation, A is the value that the rate constant would have if all molecules had enough energy to react. The exponential term contains the dependence on activation energy (E_a in J/mol) and temperature (T in K). The gas constant (R in J mol^{-1} K^{-1})

serves as a unit conversion factor. The negative sign of the exponential means that the larger the value of the exponent, the smaller the rate constant. Qualitatively, this means that as E_a gets larger, the rate constant gets smaller, and that as T gets larger, the rate constant gets larger.

Equation 15-6 can be converted from exponential form to a form that is easier to treat graphically by taking the natural logarithm of both sides and making use of the fact that the natural logarithm of an exponential is the exponent:

$$\ln k = \ln (Ae^{-E_a/RT}) = \ln A + \ln (e^{-E_a/RT})$$

$$\ln k = \ln A - \frac{E_a}{RT} \tag{15-7}$$

Equation 15-7 leads to a linear graph: plotting $\ln k$ along the y axis vs. $1/T$ along the x axis gives a straight line with a slope of $-E_a/R$ and an intercept of $\ln A$. Thus, A and E_a can be evaluated after k has been measured at several different temperatures. Example 15-11 shows how this is done.

Example 15-11

Graphing to Determine E_a

At high temperature cyclopropane isomerizes to propene:

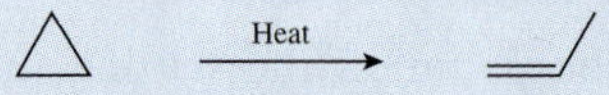

Cyclopropane $\xrightarrow{\text{Heat}}$ Propene
(C_3H_6) (C_3H_6)

When this reaction is studied at different temperatures, the following rate constants are obtained:

T (°C)	477	523	577	623
k (s^{-1})	1.8×10^{-4}	2.0×10^{-3}	2.7×10^{-2}	2.3×10^{-1}

What is the activation energy for the isomerization of cyclopropane?

Strategy: The variation of a rate constant with temperature is described by the Arrhenius equation. According to its logarithmic form (Equation 15-7), a plot of $\ln k$ vs. $1/T$, with temperature expressed in kelvins, should be a straight line.

Solution: We need to convert the data into the appropriate form and then prepare a graph. Here are the values that should be graphed:

T (K)	750.	796	850.	896
$1/T$ (10^{-3} K)	1.33	1.26	1.18	1.12
$\ln k$	-8.62	-6.21	-3.61	-1.43

A graph of these data has the expected linear form, with the following slope:

$$\text{Slope} = -3.27 \times 10^4 \text{ K}$$

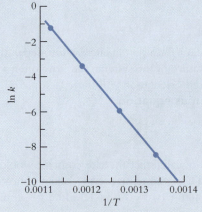

According to Equation 15-7,

$$\text{Slope} = -\frac{E_a}{R} \quad \textit{from which} \quad E_a = -R\,(\text{Slope})$$

$$E_a = -(8.314\ \text{J/mol K})(-3.27 \times 10^4\ \text{K})(10^{-3}\ \text{kJ/J}) = 2.7 \times 10^2\ \text{kJ/mol}$$

Does the Result Make Sense? A C—C bond must break for isomerization to occur, and typical C—C bond energies are around 350 kJ/mol (Table 6-2). Although 270 kJ/mol is considerably smaller than this, it is a reasonable value for the activation energy because the new π bond begins to form before the C—C bond has broken completely. The value is rounded to two significant figures to match the precision of the k values.

Here are some data for the rate constant of the reaction $O_3 + NO \longrightarrow O_2 + NO_2$:

T (°C)	125	175	225	275
k ($10^7\ \text{M}^{-1}\,\text{s}^{-1}$)	3.46	5.07	6.90	8.87

Prepare the appropriate graph and determine the activation energy for this reaction.

Example 15-11

Graphing to Determine E_a
(*continued*)

**Extra Practice
Exercise 15.11**

Graphing, using rate constants measured at several temperatures, is the most accurate method to find the value of E_a. However, a good estimate of E_a can be calculated from rate constants measured at just two temperatures:

$$\ln k_1 = \ln A = \frac{E_a}{RT_1} \quad \textit{and} \quad \ln k_2 = \ln A - \frac{E_a}{RT_2}$$

The value of A changes slowly enough with T that A usually can be treated as a constant for temperature changes of 50 K or less. Thus, the A term can be eliminated by subtracting one equation from the other:

$$\ln k_2 - \ln k_1 = \left(\ln A - \frac{E_a}{RT_2}\right) - \left(\ln A - \frac{E_a}{RT_1}\right) = \frac{E_a}{R}\left(\frac{1}{T_1} - \frac{1}{T_2}\right)$$

$$\ln\left(\frac{k_2}{k_1}\right) = \frac{E_a}{R}\left(\frac{1}{T_1} - \frac{1}{T_2}\right) \tag{15-8}$$

Example 15-12 shows how to use Equation 15-8.

The reactions of NO_2 have been studied as a function of temperature. For the following decomposition reaction, the rate constant is $2.7 \times 10^{-2}\ \text{M}^{-1}\,\text{s}^{-1}$ at 227 °C and $2.4 \times 10^{-1}\ \text{M}^{-1}\,\text{s}^{-1}$ at 277 °C:

$$2\,NO_2 \longrightarrow 2\,NO + O_2$$

Studies of the conversion of NO_2 to N_2O_4 give $k = 5.2 \times 10^8\ \text{M}^{-1}\,\text{s}^{-1}$ at both 25 °C and 87 °C:

$$2\,NO_2 \longrightarrow N_2O_4$$

Calculate the activation energies of these two reactions.

Strategy: The graphical treatment of the variation of rate constants with temperature is appropriate when a set of values is available. When only two values of the rate constant are available, E_a is calculated using Equation 15-8, which can be rearranged to solve for E_a:

$$E_a = \frac{R\ln\left(\dfrac{k_2}{k_1}\right)}{\left(\dfrac{1}{T_1} - \dfrac{1}{T_2}\right)}$$

Example 15-12

**Calculating
an Activation Energy**

Example 15-12	Remember that the units must be consistent, so T must be in K and R in J/mol K.
Calculating an Activation Energy (*continued*)	**Solution:** For the decomposition of NO_2: Convert the temperatures to kelvins:

$$227\,°C + 273 = 500.\ K$$

$$277\,°C + 273 = 550.\ K$$

$$E_a = \frac{(8.314\ J\ mol^{-1}\ K^{-1})\ \ln\left(\dfrac{2.4 \times 10^{-1}\ M^{-1}\ s^{-1}}{2.7 \times 10^{-2}\ M^{-1}\ s^{-1}}\right)}{\left(\dfrac{1}{500.\ K} - \dfrac{1}{550.\ K}\right)}$$

$$E_a = \frac{(8.314\ J\ mol^{-1}\ K^{-1})\ \ln(8.89)}{(2.000 \times 10^{-3}\ K^{-1} - 1.818 \times 10^{-3}\ K^{-1})} = \frac{(8.314\ J\ mol^{-1}\ K^{-1})(2.18)}{1.82 \times 10^{-4}\ K^{-1}}$$

As usual, we carry one additional significant figure until we round the final answer to two significant figures:

$$E_a = (9.96 \times 10^4\ J/mol)(10^{-3}\ kJ/J) = 1.0 \times 10^2\ kJ/mol$$

For the formation of N_2O_4:

$$k_2 = k_1 \qquad so \qquad \ln\left(\frac{k_2}{k_1}\right) = \ln 1 = 0 \qquad giving \qquad E_a = 0\ kJ/mol$$

Do the Results Make Sense? Because bonds must be broken in most reactions, we expect activation energies to be comparable to but smaller than bond energies, so an activation energy of 100 kJ/mol is reasonable. An activation energy of 0 kJ/mol indicates that no bonds need to be distorted or broken in this reaction. We discuss this in more detail shortly.

Extra Practice Exercise 15.12	Cyclobutane decomposes to form ethylene: $C_4H_8 \rightarrow 2\ C_2H_4$. Experimental rate constants for this reaction are $k = 6.1 \times 10^{-8}\ s^{-1}$ at 327 °C and $k = 1.5 \times 10^{-11}\ s^{-1}$ at 245 °C. Determine the activation energy for the reaction.

After the activation energy of a reaction has been determined, we can use the Arrhenius equation to estimate values of the rate constant for the reaction at temperatures where experiments have not been carried out. This is particularly useful for temperatures at which a reaction is too slow or too fast to be studied conveniently. Example 15-13 illustrates this application.

Example 15-13 **Calculating k from E_a** 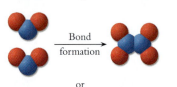 Bond formation or 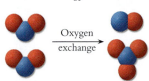 Oxygen exchange	When two NO_2 molecules collide and react, they can form a bond or exchange an oxygen atom. For the oxygen exchange, $E_a = 1.0 \times 10^2$ kJ/mol, and the rate constant at 250 °C = $8.6 \times 10^{-2}\ M^{-1}\ s^{-1}$. Estimate the rate constant for oxygen exchange at room temperature (25 °C). **Strategy:** We can calculate a rate constant at any temperature using the Arrhenius equation provided that we know E_a and the rate constant for the reaction at some other temperature. **Solution:** Equation 15-8 relates rate constants at different temperatures to the activation energy:

$$\ln\left(\frac{k_2}{k_1}\right) = \frac{E_a}{R}\left(\frac{1}{T_1} - \frac{1}{T_2}\right)$$

The values that we need to substitute into this equation are given in the problem:

$$k_1 = 8.6 \times 10^{-2} \text{ M}^{-1} \text{ s}^{-1} \qquad T_1 = 250 \text{ °C} = 523 \text{ K} \qquad T_2 = 25 \text{ °C} = 298 \text{ K}$$

$$E_a = 1.0 \times 10^2 \text{ kJ/mol}$$

Example 15-13

Calculating k from E_a (*continued*)

After substituting, we solve for k_2, the rate constant of the decomposition reaction at 298 K:

$$\ln\left(\frac{k_2}{8.2 \times 10^{-2} \text{ M}^{-1} \text{ s}^{-1}}\right) = \frac{(1.0 \times 10^2 \text{ kJ/mol})(10^3 \text{ J/kJ})}{8.314 \text{ J/mol K}}\left(\frac{1}{523 \text{ K}} - \frac{1}{298 \text{ K}}\right)$$

$$= -17.4$$

$$\frac{k_2}{8.6 \times 10^{-2} \text{ M}^{-1} \text{ s}^{-1}} = e^{-17.4} = 2.8 \times 10^{-8}$$

$$k_2 = 2.4 \times 10^{-9} \text{ M}^{-1} \text{ s}^{-1}$$

Does the Result Make Sense?   Notice that at room temperature, the rate constant for exchange of oxygen atoms, 2×10^{-9} M^{-1} s^{-1}, is 17 powers of ten smaller than the rate constant for bond formation between two N atoms, 5.2×10^8 M^{-1} s^{-1}. This immense difference reflects the fact that exchange of atoms involves a large energy input to break an N—O bond, whereas bond formation between two NO$_2$ molecules requires no energy input.

Estimate the rate constant for the decomposition of cyclobutane at 25 °C (see Extra Practice Exercise 15-12).

Extra Practice Exercise 15.13

Values of Activation Energy

Most reactions between stable molecules have activation energies of 100 kJ/mol or greater, even when the overall reaction is exothermic. As an example, consider the reaction of hydrogen and oxygen:

$$2 \text{ H}_2\,(g) + \text{O}_2\,(g) \longrightarrow 2 \text{ H}_2\text{O}\,(g) \qquad \Delta H° = -484 \text{ kJ mol}^{-1} \quad and \quad \Delta G° = -457 \text{ kJ mol}^{-1}$$

This combustion reaction is so spontaneous and exothermic that it is used to drive the main engines of the space shuttle. Nevertheless, mixtures of hydrogen and oxygen are stable indefinitely at room temperature. This reaction does not occur at room temperature because strong bonds must be broken to transform H$_2$ and O$_2$ into water molecules. In other words, the bonds in the reactant molecules generate a large activation energy for the reaction. As Figure 15-18 shows, a violent explosion occurs when a spark is applied to a mixture of H$_2$ and O$_2$. The spark gives some of the molecules enough energy to

Figure 15-18
A balloon filled with a mixture of hydrogen and oxygen is stable until a source of energy, such as a spark or flame, initiates the reaction. After the reaction begins, the formation of water releases enough energy to cause an explosion.

overcome the activation barrier, and after the reaction starts, the energy released in the formation of water molecules raises the temperature sufficiently that the reaction goes to completion very quickly.

The combustion reactions of hydrocarbons such as natural gas and gasoline, which also release large amounts of energy, have negligible rates at room temperature because they have large activation energy barriers, typically 140–200 kJ/mol. For a hydrocarbon molecule to combine with oxygen to form CO_2 and H_2O, one of its C—H bonds first must be weakened substantially. This accounts for the large activation energy barrier for combustion.

Many elementary reactions have large activation energies. For example, the isomerization of *cis*-2-butene to *trans*-2-butene is a unimolecular rotation whose activation energy is 284 kJ/mol. This high value arises because a C—C π bond must be broken during the course of the isomerization process (see Figure 15-3).

Example 15-12 shows that no energy barrier exists for the combination of two NO_2 molecules to form N_2O_4. The activation energy for this reaction is zero because NO_2 is an odd-electron molecule with a lone electron readily available for bond formation:

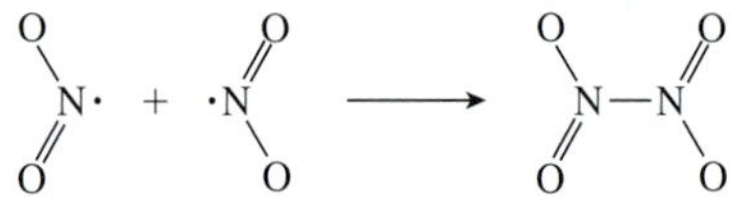

The recombination of two bromine atoms to form a Br_2 molecule has zero activation energy for the same reason.

Section Exercises

15.6.1 Use bonding arguments to predict whether the ammonia synthesis reaction has a high or a low activation energy:

$$N_2(g) + 3\,H_2(g) \longrightarrow 2\,NH_3(g) \qquad \Delta H° = -91.8 \text{ kJ}$$

15.6.2 The reaction of H_2 with Br_2 discussed in Section 15.5 has been studied extensively as a function of temperature. Here are some experimental results:

$T\ (10^2\ \text{K})$	5.00	5.25	5.50	5.75
$k\ (\text{M}^{-1/2}\,\text{s}^{-1})$	8.1×10^{-7}	6.5×10^{-6}	4.0×10^{-5}	2.1×10^{-4}

(a) Determine E_a by graphical analysis, and (b) estimate k at 425 K.

15.6.3 Use standard thermodynamic data to determine ΔE and construct the activation energy diagram for the following reaction. (*Hint:* $\Delta(PV) = 0$ for this reaction, so enthalpies and energies can be used interchangeably.)

$$CO + NO_2 \longrightarrow CO_2 + NO \qquad E_a = 133 \text{ kJ}$$

15.7 CATALYSIS

Many potentially useful chemical reactions proceed slowly at room temperature. To take advantage of these reactions, we must find ways to make them go faster. According to the discussion in Section 15.6, one way to make a reaction go faster is to run it at higher temperature. At higher temperature, however, undesirable reactions also go faster, so increasing the temperature may lead to lower yields because of competing reactions. In addition, the extra energy required to run a reaction at higher temperature adds to production costs. Furthermore, high-temperature reactions introduce safety concerns, and many chemical species are not stable at high temperature. Thus, although some industrial reactions are carried out at high temperature to maximize the amount of product that can be synthesized in a given time, the chemical industry prefers to find other ways to increase reaction rates.

Another way to make a reaction go faster is to add a substance called a **catalyst.** A catalyst functions by changing the mechanism of a reaction in a manner that lowers activation energy barriers. Although the catalyst changes the mechanism of a reaction, it is not part of the overall stoichiometry of the reaction. A catalyst always participates in an early step of a reaction mechanism, but when the reaction is over, the catalyst has been regenerated. When we write a net reaction that is influenced by a catalyst, we write the formula of the catalyst above or below the reaction arrow.

It is important to recognize the difference between a *catalyst* and an *intermediate*. A catalyst is a reactant in an early step of a mechanism and a product of a later step. The opposite is true of an intermediate. An intermediate is a transient species that is a product in an early step of a mechanism but is consumed in a later step:

$$\text{Catalyst} + \text{Reactant}_1 \longrightarrow \text{Intermediate}$$

$$\text{Intermediate} + \text{Reactant}_2 \longrightarrow \text{Products} + \text{Catalyst}$$

Catalysts are vital in the chemical industry. The market for catalysts in the United States exceeds $2.0 billion, including more than $600 million for petroleum refining and more than $750 million for chemical production. Although these are large sums of money, the products made available by catalysts are far more valuable than the catalysts themselves. The total value of fuels and chemicals produced by catalysts exceeds $900 billion.

Catalysis and the Ozone Problem

Catalysts are immensely beneficial in industry, but accidental catalysis in the atmosphere can be disastrous. Recall from Box 15-1 that the chemistry of ozone in the stratosphere involves a delicate balance of reactions that maintain a stable concentration of ozone. Chlorofluorocarbons (CFCs) shift that balance by acting as catalysts for the destruction of O_3 molecules.

Because CFCs are highly resistant to chemical attack, they are stable in the lower atmosphere, where they can exist for up to 100 years. This stability gives CFCs time to diffuse up through the troposphere and into the stratosphere. There, CFCs absorb short-wavelength ultraviolet light from the sun that breaks carbon–chlorine bonds and produces chlorine atoms:

$$CF_2Cl_2 \xrightarrow{h\nu} CF_2Cl + Cl$$

Chlorine atoms react with O_3 molecules to produce O_2 and ClO, as shown by the molecular pictures in Figure 15-19. This is a catalytic process because chlorine monoxide reacts with an oxygen atom to produce a second O_2 molecule and *regenerate the chlorine atom:*

$$Cl + O_3 \longrightarrow ClO + O_2$$
$$\underline{ClO + O \longrightarrow Cl + O_2}$$
$$\text{Net: } O_3 + O \xrightarrow{Cl} 2\,O_2$$

The net reaction for this two-step mechanism is the conversion of an O_3 molecule and an oxygen atom into two O_2 molecules. In this mechanism, chlorine atoms catalyze

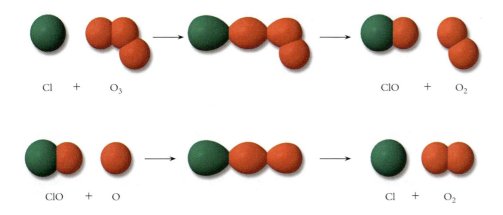

Cl + O$_3$ ClO + O$_2$

ClO + O   Cl + O$_2$

Figure 15-19
Chlorine atoms catalyze the reaction of ozone with oxygen atoms by forming an unstable ClO intermediate that readily reacts with an O atom, forming another O_2 molecule and regenerating the Cl atom.

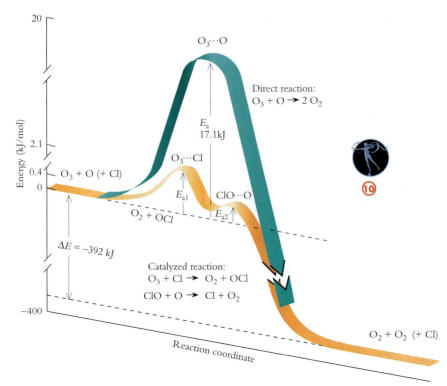

Figure 15-20
The activation energy diagram for the reaction between O_3 and O. Notice the break in the vertical scale: the overall reaction is exothermic by -392 kJ.

ozone decomposition. They participate in the mechanism, but they do not appear in the overall stoichiometry. Although chlorine atoms are consumed in the first step, they are regenerated in the second. The cyclical nature of this process means that each chlorine atom can catalyze the destruction of many O_3 molecules. It has been estimated that each chlorine atom produced by a CFC molecule in the upper stratosphere destroys about 100,000 molecules of ozone before it is removed by other reactions such as recombination:

$$CF_2Cl + Cl \longrightarrow CF_2Cl_2$$

Figure 15-20, an activation energy diagram for the ozone–oxygen atom reaction, illustrates how a catalyst modifies the energy profile of a reaction. The figure shows that the direct reaction between O_3 and O has a substantially higher activation energy than the chlorine-catalyzed reaction sequence. In other words, the activated complex between Cl atoms and O_3 molecules lies at a lower energy than the activated complex between O atoms and O_3 molecules, making chlorine atoms much more effective than oxygen atoms at destroying O_3 molecules.

Notice that both steps of the chlorine-catalyzed reaction appear on the activation energy diagram. Each step of a mechanism has its own activation energy, so the diagram has two activation barriers. Experimental data indicate that the first barrier is higher than the second. Because the uncatalyzed reaction involves just one step, it has only one activation barrier. At the temperature of the ozone layer, 220 K, the rate data for the two pathways are as follows:

$$O_3 + O \longrightarrow 2 O_2 \qquad E_a = 17.1 \text{ kJ/mol} \qquad k = 4.1 \times 10^5 \text{ M}^{-1}\text{ s}^{-1}$$
$$O_3 + Cl \longrightarrow O_2 + OCl \qquad E_{a1} = 2.1 \text{ kJ/mol} \qquad k_1 = 5.2 \times 10^9 \text{ M}^{-1}\text{ s}^{-1}$$
$$ClO + O \longrightarrow Cl + O_2 \qquad E_{a2} = 0.4 \text{ kJ/mol} \qquad k_2 = 2.2 \times 10^{10} \text{ M}^{-1}\text{ s}^{-1}$$

We see that chlorine atoms provide an alternative mechanism for the reaction of ozone with oxygen atoms. The lower-energy pathway breaks down ozone in the stratosphere at a significantly faster rate than in the absence of the catalyst. This disturbs the delicate balance among ozone, oxygen atoms, and oxygen molecules in a way that poses a serious threat to the life-protecting ozone layer.

Homogeneous and Heterogeneous Catalysts

The catalysis of ozone decomposition by chlorine atoms occurs entirely in the gas phase. Chlorine is classified as a **homogeneous catalyst** because the catalyst and the reactants are present in the *same phase,* in this case the gas phase. A **heterogeneous catalyst,** on the other hand, is in a *different phase* than the one where the reaction occurs. A heterogeneous catalyst is usually a solid, and the reactants are gases or are dissolved in a liquid solvent.

The development of the ozone hole over Antarctica is accelerated by heterogeneous catalysis on microcrystals of ice. These microcrystals form in abundance in the Antarctic spring, which is when the ozone hole appears. Ice microcrystals are less common in the Arctic atmosphere, so ozone depletion has not been as extensive in the Northern Hemisphere.

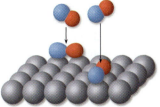

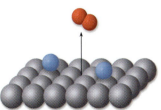

Step 1: Adsorption and bond weakening

Step 2: Migration

Step 3: Bond reorganization

Step 4: Desorption

Figure 15-21
Schematic molecular pictures of the four steps for the conversion of NO to N_2 and O_2 on a platinum metal surface.

Heterogeneous catalysts are the active ingredients in automobile catalytic converters. When combustion occurs in an automobile engine, side reactions generate small amounts of undesired products. Some carbon atoms end up as poisonous CO rather than CO_2. Another reaction that takes place at the high temperatures and pressures in automobile engines is the conversion of N_2 to NO. Furthermore, the combustion process fails to burn all the hydrocarbons. Hydrocarbons, CO, and NO all are undesirable pollutants that can be removed from exhaust gases by the action of heterogeneous catalysts:

$$2\,CO + O_2 \xrightarrow{\text{Catalyst}} 2\,CO_2$$

$$\text{Hydrocarbons} + O_2 \xrightarrow{\text{Catalyst}} CO_2 + H_2O$$

$$2\,NO \xrightarrow{\text{Catalyst}} N_2 + O_2$$

A mixture of catalysts is needed to catalyze the variety of reactions that must be carried out in the converter. The most important catalyst is platinum metal, but palladium and rhodium are used as well, as are transition metal oxides such as CuO and Cr_2O_3.

Heterogeneous reactions are complicated because the reacting species must be transferred from one phase to another before the reaction can occur. Despite much research, chemists still have limited knowledge about the mechanisms of reactions that involve heterogeneous catalysts. However, it is known that heterogeneous catalysis generally proceeds in four steps, as illustrated in Figure 15-21 for the conversion of NO into N_2 and O_2.

1. The starting materials bind to the surface of the catalyst. This process is known as **adsorption.** When a substance is adsorbed, its internal bonds are weakened or broken in favor of bonds to the catalyst.

2. Bound materials migrate over the surface of the catalyst.

3. Bound substances react to form products.

4. Products escape from the surface of the catalyst. This step is **desorption.**

Catalysis in Industry

The business of the chemical industry is to transform inexpensive substances into more valuable ones. In many cases, catalysts play important roles in these processes. Here, we describe the roles of catalysts in some important industrial reactions. Other catalyzed industrial reactions are considered in Chapter 16, where we describe properties of chemical equilibria.

Petroleum refining. Petroleum is the source not only of fuels but also of carbon compounds used for synthesis in the chemical industry. Crude petroleum is a complex mixture containing thousands of compounds, most of which are alkanes, alkenes, and derivatives of benzene. Petroleum refining converts this viscous black liquid into a variety of useful products, from gasoline to lubricants (see Figure 15-22 and Table 9.1). Refining begins with distillation, which separates the various components into fractions with different boiling point ranges. Large, high-boiling hydrocarbons are less versatile than smaller ones, so a further step in refining is the catalytic cracking of large hydrocarbons into smaller ones. Cracking is performed at high temperature in the presence of a heterogeneous catalyst composed of silicon dioxide and aluminum oxide:

$$C_{12} \text{ and higher} \xrightarrow[\text{300 °C}]{SiO_2/Al_2O_3} C_5 \text{ to } C_{10}$$

Kerosene $\qquad\qquad$ Alkanes, alkenes

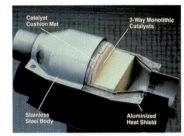

Hexadecane

Dodecane

Figure 15-22
Crude petroleum is refined to produce a variety of useful products. Shown are motor oil and lubricant jelly.

Chemicals from coal. Heterogeneous catalysts are used to convert solid coal into gasoline and other chemicals. Solid coal is not easily transformed into hydrocarbon chains, so the conversion requires two general steps: gasification followed by catalytic hydrocarbon-forming reactions. Coal is first gasified by reaction with steam:

$$C\ (coal) + H_2O\ (g) \longrightarrow CO\ (g) + H_2\ (g) \qquad \Delta H = 1131\ \text{kJ/mol}$$

This reaction is significantly endothermic, so it must be driven by added energy. Burning some of the coal provides the energy required to drive the gasification and produce the steam:

$$C\ (coal) + O_2\ (g) \longrightarrow CO_2\ (g) \qquad \Delta H = -407\ \text{kJ/mol}$$

The reaction of steam with coal generates a 1:1 gas mixture of H_2 and CO. Hydrocarbon formation requires a 2:1 mixture, so additional H_2 is produced by the reaction of steam with some of the CO:

$$CO\ (g) + H_2O\ (g) \longrightarrow CO_2\ (g) + H_2\ (g) \qquad \Delta H = -42\ \text{kJ/mol}$$

A variety of solid metallic catalysts containing oxides of iron, chromium, or copper can be used for hydrocarbon formation. High gas pressure (20 atm) and elevated temperature (250 to 350 °C) are also required. Several hydrocarbon-forming reactions occur; the most important is the stepwise production of alkanes. Here are two simple examples:

$$CO + 3\ H_2 \xrightarrow[\text{High } P,T]{\text{Catalyst}} CH_4 + H_2O$$

$$CO + 2\ H_2 + CH_4 \xrightarrow[\text{High } P,T]{\text{Catalyst}} C_2H_6 + H_2O$$

Another important chemical produced from CO and H_2 is methanol, CH_3OH. The synthesis is carried out at high temperature (200 to 300 °C) and pressure (50 to 100 atm) in the presence of copper(II) oxide or some other metal oxide catalyst:

$$CO + 2\ H_2 \xrightarrow[\text{High } P,T]{\text{Catalyst}} CH_3OH$$

Approximately 5 billion kilograms of methanol are produced annually in the United States.

Mixtures of CO and H_2 have immense potential as a feedstock for the chemical industry. The variety of products available from CO and H_2 is virtually unlimited, and the Earth has huge coal reserves. Unfortunately, the direct conversion of coal is energy-intensive and highly polluting. For all of these reasons, coal conversion has been studied intensively. Despite considerable research, however, coal conversion is currently not economical compared with petroleum refining. The catalytic process is not the main stumbling block. Instead, the problem is the high cost of gasification, both in energy and in capital investment. The development of a catalyst that is effective at low temperature would solve this problem.

Acetic acid production. Although most industrial catalysts are heterogeneous, a growing number of industrial reactions use homogeneous catalysts. One example is the production of acetic acid. Most of the 2.1 billion kilograms of acetic acid produced annually is used in the polymer industry. The reaction of methanol and carbon monoxide to form acetic acid is catalyzed by a rhodium compound that dissolves in methanol:

$$CH_3OH + CO \xrightarrow[\text{173 °C, 700 atm}]{[Rh(CO)_2I_3]} CH_3CO_2H$$

These are just a few examples of catalytic reactions used industrially. Catalyst research is one of the most active areas of chemistry and chemical engineering. Our Chemistry and Technology Box describes one direction such research is taking.

Most catalysts used in the chemical industry are solid metal halides, metal oxides, or pure metals.

Biocatalysis: Enzymes

Living organisms carry out an astonishing variety of chemical processes. Organisms organize small molecules into complex biopolymers such as deoxyribonucleic acid (DNA) and proteins (see Chapter 13). Conversely, organisms break down large, energy-rich molecules in many steps to extract chemical energy in small portions to drive their many

Box 15-2	Chemistry and Technology: Ionic Liquids

Many important industrial reactions rely on heterogeneous catalysts, yet there are inherent drawbacks to heterogeneous reactions. Such reactions occur only when the reactants contact the solid surface of the catalyst. Catalysis can be much more efficient when the catalyst is dissolved in the solvent where the reaction occurs. Unfortunately, the main catalysts used by industry, metals and metal oxides, are not soluble in traditional solvents.

Recently, a new class of solvents has been developed that may revolutionize industrial chemistry. These solvents are ionic liquids. An ionic liquid resembles salts such as sodium chloride in that it consists of cations and anions. There, however, the resemblance ends. Solid salts contain relatively small, spherical or near-spherical ions that pack tightly together, generating substantial lattice energies that give them high melting points. Ionic liquids, in contrast, contain large, oddly shaped ions that cannot pack tightly together. Consequently, although electrical interactions hold these substances in condensed phases, ion–ion interactions are not strong enough to lock the ions into fixed lattice positions. Hence, these substances are ionic but liquid.

The cation in an ionic liquid typically is an organic nitrogen-based ion such as alkyl ammonium, alkyl pyridinium, or dialkylimidazolium, examples of which appear below.

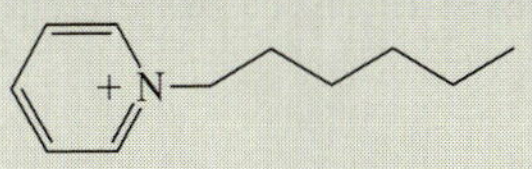

Alkyl pyridinium 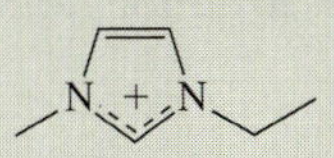   Dialkyl imidazolium

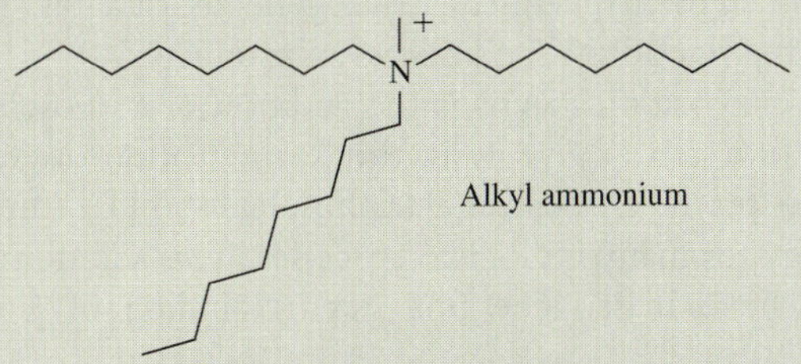

Alkyl ammonium

All these cations are bulky and asymmetric. In addition, their alkyl groups make it possible to modify them almost endlessly. The length of the alkyl chains can be varied, they can be straight or branched, and functional groups such as —OH can be substituted. This feature makes it possible to vary the characteristics of ionic liquids to suit a particular application.

The anion in an ionic liquid can be varied, too. Many ionic liquids contain relatively simple inorganic anions, such as nitrate (NO_3^-), tetrafluoroborate (BF_4^-), or hexafluorophosphate (PF_6^-). Anions that are more exotic also are possible, such as the two shown below. Varying the anion provides another way of tuning the properties of an ionic liquid to match a desired application.

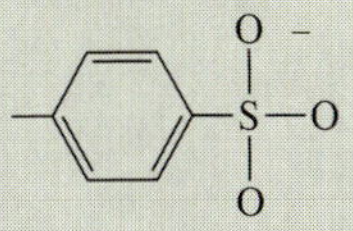

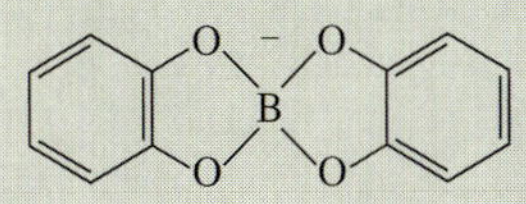

Coupling one of these cations with one of these anions gives an ionic molecule that has high electron density at the anionic end. The computer image at top right shows this for butyl methyl imidazolium hexafluorophosphate.

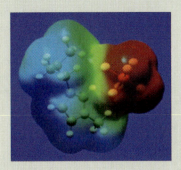

Ionic liquids have great promise as nonpolluting, "green" solvents. Although their chemical structures look menacing, their ionic character gives them two properties that are highly desirable from an environmental perspective. They are highly stable, meaning that they do not decompose to give toxic byproducts. They are nonvolatile, meaning that they do not escape into the atmosphere. In these regards, they are much more environmentally friendly than volatile industrial solvents, such as toluene, hexane, and the chlorohydrocarbons.

How might ionic liquids be used industrially? One example is the selective hydrogenation of double bonds, an important reaction in the food, plastics, and drug industries. Typically, this reaction requires high temperatures and pressures and often gives a range of products. Running the reaction in an ionic liquid solvent can eliminate these shortcomings. One example is the synthesis of (S)-naproxen, a nonsteroidal anti-inflammatory painkiller. The final step in making this drug is hydrogenation of a double bond. This reaction is accomplished cleanly in an imidazolium tetrafluoroborate solvent containing a ruthenium catalyst, as shown below.

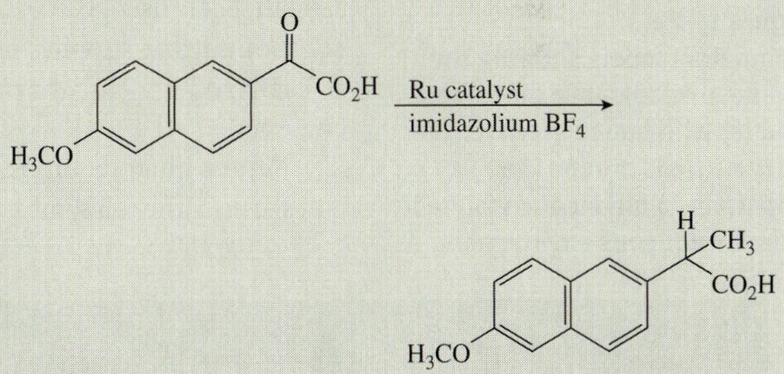

In addition to replacing volatile organic solvents and making some chemical synthesis easier, ionic liquids are being explored for a variety of other chemical uses. For example, the appropriate ionic liquid can dissolve the rubber from discarded tires, after which the rubber can be treated for recycling. In another application that uses the ability of ionic liquids to dissolve otherwise insoluble substances, an ionic liquid with a properly designed cation can form complexes with metal contaminants, including mercury, cadmium, and uranium. These liquids are immiscible with water but can remove these heavy-metal contaminants, making it possible to clean up contaminated water supplies. Ionic liquids also hold promise as components of batteries and other electrical devices such as capacitors. The range of applications for these versatile new materials may be limited only by chemists' imaginations.

activities. Further, organisms produce antibodies, medium-sized molecules that combat bacterial invaders. In many of these processes, organisms break chemical bonds selectively without resorting to high temperature.

A biochemical catalyst is called an **enzyme.** Enzymes are specialized proteins that catalyze specific biochemical reactions. Some enzymes are found in extracellular fluids such as saliva and gastric juices, but most are found inside cells. Each type of cell has a different array of enzymes that act together to determine what role the cell plays in the overall biochemistry of the organism. Enzymes are complicated molecules. Biochemists have determined the molecular structures of some enzymes, but the structures of many enzymes are not yet known.

Despite the wide diversity of enzyme structures, most enzyme activity follows a general mechanism that has several reversible steps. In the first step, a reactant molecule known as a *substrate* (S) binds to a specific location on the enzyme (E), usually a groove or a pocket on the surface of the protein:

$$E + S \rightleftharpoons ES$$

The substrate binds to the active site through intermolecular interactions that usually include significant amounts of hydrogen bonding.

Binding causes the shape of the enzyme to change, and this change distorts the structure of the substrate:

$$ES \rightleftharpoons E \cdots S$$

Distortion allows the substrate to react more easily with another reactant (R) to form the desired product (P). Once the product forms, it no longer binds strongly to the enzyme:

$$E \cdots S + R \longrightarrow E \cdots P \longrightarrow E + P$$

The enzyme is regenerated at the end of this sequence, making it available to bind another substrate molecule. Note that the steps in this enzyme-catalyzed biochemical mechanism are similar to the steps in chemical heterogeneous catalysis: binding with bond weakening, reaction at the bound site, and release of products.

A specific illustration of enzymatic catalysis is the addition of a phosphate group to glucose. This reaction is the first step in glycolysis, the fragmentation of glucose. Virtually all cells use glycolysis to produce adenosine triphosphate (ATP), the energy-rich compound that supplies energy for many biological processes. This addition of phosphate is catalyzed by an enzyme, hexokinase. The first step is binding of glucose by the enzyme.

When glucose binds to hexokinase, chemical interactions between the two molecules cause the enzyme–glucose complex to change its shape, as Figure 15-23 shows.

The names of enzymes usually end in *-ase*, for example hexokinase and nitrogenase.

Figure 15-23
Computer models showing the shape of hexokinase (*a*) without and (*b*) with bound glucose. The enzyme folds around the substrate to bind it and isolate it from its aqueous environment.

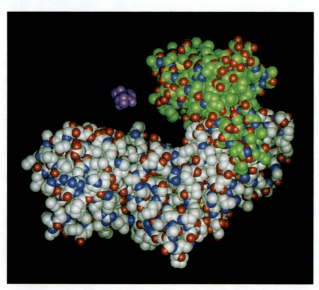
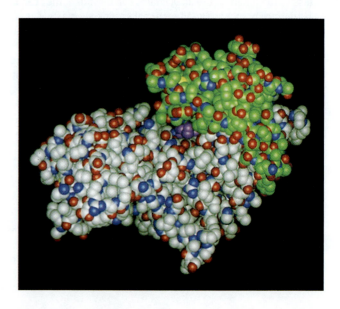

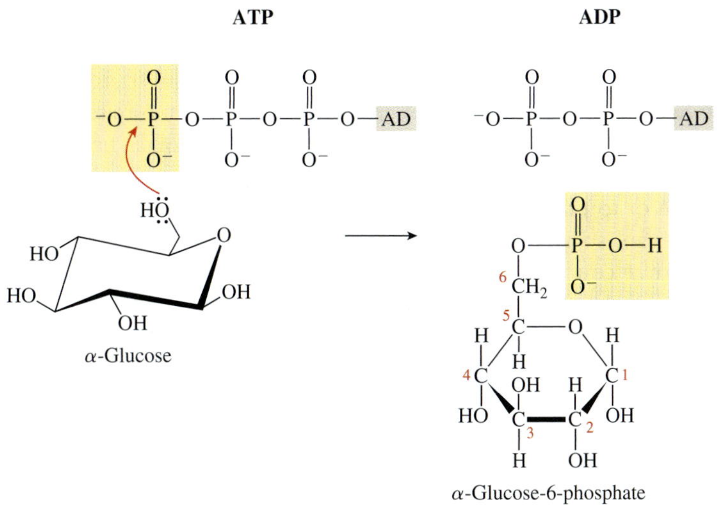

The enzyme folds in on the substrate, isolating the bound glucose molecule from its aqueous environment. This allows an ATP molecule to transfer a phosphate group to the hydroxyl group in position 6 of glucose, as shown in Figure 15-24. The product, glucose-6-phosphate, is too big to bind tightly to hexokinase. Consequently, the enzyme opens up, releases the product molecule, and is ready to bind another glucose molecule.

Figure 15-24
Hexokinase catalyzes the phosphorylation of α-glucose by adenosine triphosphate (ATP). "AD" represents the adenosine portion of ATP and ADP. The complete structure of ATP appears in Figure 14-19.

Although the molecular details of enzyme mechanisms are complex, the kinetic behavior of many enzymatic processes is first order in both the substrate and the enzyme. Example 15-14 shows that the mechanism just outlined is consistent with this kinetic behavior.

Example 15-14

Enzyme Kinetics

Derive the predicted rate law for the general mechanism for enzyme catalysis, assuming that the distortion step is rate-determining:

$$E + S \underset{k_{-1}}{\overset{k_1}{\rightleftharpoons}} ES \xrightarrow{k_2, \text{ slow}} E \cdots S$$

$$E \cdots S + R \xrightarrow{k_3} E + P$$

Strategy: The rate law predicted by a mechanism can be derived by setting the overall rate of reaction equal to the rate of the slowest step. As described in Section 15.5, any earlier steps are assumed to have equal forward and reverse rates.

Solution: The mechanism includes a rate-determining distortion step, for which the rate expression is

$$\text{Rate} = k_2[ES]$$

This rate equation contains the concentration of an intermediate, so use the forward and reverse rates of the first step to derive an expression for [ES] in terms of [E] and [S]:

$$k_1[E][S] = k_{-1}[ES] \quad so \quad [ES] = \frac{k_1}{k_{-1}}[E][S]$$

Now substitute into the original rate expression:

$$\text{Rate} = \frac{k_1 k_2}{k_{-1}}[E][S] = k_{obs}[E][S] \quad where \quad k_{obs} = \frac{k_1 k_2}{k_{-1}}$$

Does the Result Make Sense? The mechanism predicts that the rate of this reaction depends on the concentrations of enzyme (E) and the reactant that binds to it (S) but does not depend on additional reactants (R). This makes sense from a mechanistic point of view provided there is enough of R present that the third step proceeds rapidly once distortion is complete. Experimentally, many enzyme-catalyzed reactions obey this form of rate law.

If the third reaction of this sequence is rate-determining and the distorted complex is at equilibrium with undistorted complex, what is the resulting rate law?

Extra Practice Exercise 15.14

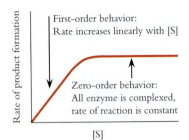

First-order behavior:
Rate increases linearly with [S]

Zero-order behavior:
All enzyme is complexed,
rate of reaction is constant

[S]

Figure 15-25
The typical profile of how the rate of an enzyme-catalyzed reaction varies with substrate concentration.

Enzyme reactions are almost always first order in enzyme, as one would expect for a reaction in which one molecule of enzyme participates in the rate-determining step. As Figure 15-25 illustrates, the rate dependence on substrate concentration generally shows two distinct regions of kinetic behavior. At low substrate concentration, the binding or distortion step is rate-determining, and the reaction is first order in substrate. However, when substrate concentration is high, all the enzyme molecules are bound either to substrate or to product. Under these conditions, the enzyme is catalyzing the reaction as fast as it can, so adding more substrate cannot make the enzyme work any faster. At high substrate concentrations, the reaction is zero order in substrate.

This has been a brief overview of a rich field. Details of enzyme structure and catalytic activity are studied in laboratories worldwide. Moreover, genetic engineering makes it possible to "manufacture" key enzymes in large quantities, so enzymes may become industrial catalysts that accomplish reactions rapidly and selectively.

Section Exercises

15.7.1 When metals such as platinum, palladium, and rhodium are used as catalysts, they are usually deposited as thin layers over a highly porous material such as charcoal. Explain why these materials are less expensive and more effective catalysts than pieces of pure metal.

15.7.2 Which of the following statements are true? Correct the untrue statements so that they are true. (a) The concentration of a homogeneous catalyst appears in the rate law.

(b) A catalyst changes an endothermic reaction into an exothermic reaction. (c) A catalyst lowers the activation energy of the rate-determining step. (d) A catalyst is a reactant in an early step of a mechanism and a product of a later step.

15.7.3 Why would it be advantageous to develop a catalyst for the following reaction?

$$4\,CO + 2\,NO_2 \longrightarrow N_2 + 4\,CO_2$$

CHAPTER 15 VISUAL SUMMARY

Important equations

Essential terms

Key concepts

SKILLS TO MASTER

① Visualizing elementary reactions
② Predicting rate laws from mechanisms
③ Analyzing first order rate data
④ Analyzing second order rate data
⑤ Interpreting isolation experiments

⑥ Interpreting initial rate data
⑦ Testing the consistency of a mechanism
⑧ Interpreting activation energy diagrams
⑨ Applying the Arrhenius equation
⑩ Explaining the action of a catalyst

15.1 WHAT IS A REACTION MECHANISM?

A mechanism is the exact molecular pathway that starting materials follow on their way to becoming products.

Reaction mechanism   ①

Rate determining step   Elementary reaction

Bimolecular reaction

Unimolecular reaction

Intermediate   Rate   Termolecular reaction

15.2 RATES OF CHEMICAL REACTIONS

②   Kinetics

$$a\,A \longrightarrow d\,D + e\,E$$

$$-\left(\frac{1}{a}\right)\left(\frac{\Delta[A]}{\Delta t}\right) = \left(\frac{1}{d}\right)\left(\frac{\Delta[D]}{\Delta t}\right) = \left(\frac{1}{e}\right)\left(\frac{\Delta[E]}{\Delta t}\right)$$

15.3 CONCENTRATION AND REACTION RATES

The order of a reaction must be determined by experiments.

Rate law   Rate $= k[A]^y[B]^z$ Rate constant (k)

Rate $= k[NO_2]$ Rate $= k[NO_2]^2$

First order   Second order

15.4 EXPERIMENTAL KINETICS

First order   Half-life ($t_{1/2}$)   Second order

$$\ln\frac{[A]_0}{[A]} = kt$$ $$t_{1/2} = \frac{\ln 2}{k}$$ $$\frac{1}{[A]} - \frac{1}{[A]_0} = kt$$

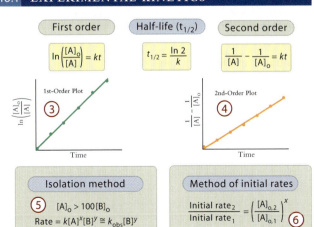

1st-Order Plot   ③

2nd-Order Plot   ④

Isolation method

⑤ $[A]_0 > 100\,[B]_0$
Rate $= k[A]^x[B]^y \cong k_{obs}[B]^y$

Method of initial rates

$$\frac{\text{Initial rate}_2}{\text{Initial rate}_1} = \left(\frac{[A]_{0,2}}{[A]_{0,1}}\right)^x$$ ⑥

15.5 LINKING MECHANISMS AND RATE LAWS

When the 1st step is rate-determining, the predicted rate law for the overall reaction is the rate expression for that 1st step.

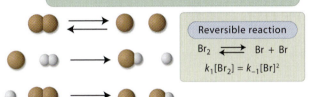

Reversible reaction

$$Br_2 \rightleftharpoons Br + Br$$
$$k_1[Br_2] = k_{-1}[Br]^2$$

Bimolecular elementary reaction: $A + B \longrightarrow products$
Elementary rate $= k[A][B]$

Unimolecular elementary reaction: $C \longrightarrow products$
Elementary rate $= k[C]$

Characteristics of a valid mechanism ⑦

1. The mechanism consists entirely of elementary reactions.
2. The sum of the elementary steps gives the overall balanced equation.
3. The mechanism is consistent with the experimental rate law.

15.6 REACTION RATES AND TEMPERATURE

$$k = Ae^{-E_a/RT}$$ ⑨

Arrhenius equation

$$\ln k = \ln A - \frac{E_a}{RT}$$

$$\ln \frac{k_2}{k_1} = \frac{E_a}{R}\left(\frac{1}{T_1} - \frac{1}{T_2}\right)$$

Activation energy diagram

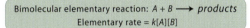

Activated complex   ⑧

Activation energy (E_a)

Reactants

$\Delta E_{reaction}$ Products

Reaction coordinate

The rate of a reaction increases with temperature.

15.7 CATALYSIS

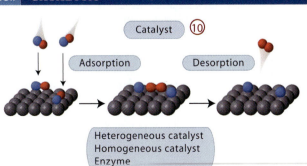

Catalyst ⑩

Adsorption   Desorption

Heterogeneous catalyst
Homogeneous catalyst
Enzyme

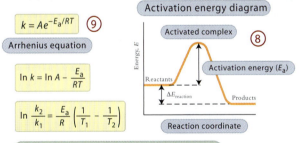
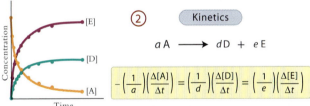
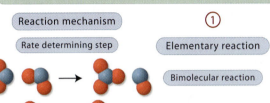

Learning Exercises

15.1 List the requirements of a satisfactory mechanism, and describe the characteristics of the steps of a mechanism.

15.2 Write one-sentence explanations for the following kinetic terms: (a) first order; (b) second order; (c) isolation method; (d) elementary step; (e) catalyst; (f) intermediate; and (g) enzyme.

15.3 Describe the differences among the rate, the rate law, and the rate constant for a chemical reaction.

15.4 Write a paragraph that describes what the activation energy is and how it affects the kinetic behavior of a reaction.

15.5 Update your list of memory bank equations. Include information about how to apply each new equation.

15.6 Make a list of terms new to you that appear in Chapter 15. In your own words, write a one-sentence definition of each. Consult the Glossary if you need help.

 Problems <u>ilw</u> = interactive learning ware problem. Visit the website at www.wiley.com/college/olmsted

What Is a Reaction Mechanism?

15.1 What is the rate-determining step in each of the following processes? (a) A line of people get coffee from a large coffee urn. (b) You go through the express line (10 items, no checks) at a supermarket. (c) A squad of parachutists makes a "jump" from an airplane cargo door.

15.2 What is the rate-determining step in each of the following processes? (a) People buy their lunches at a cafeteria. (b) Music enthusiasts enter an amphitheater for a concert. (c) You leave a pay-as-you-exit parking lot.

15.3 The reaction of H_2 with I_2 is $H_2 + I_2 \longrightarrow 2$ HI

Use reactant molecules to write appropriate elementary reactions that satisfy the following criteria: (a) a unimolecular decomposition that generates I; (b) a bimolecular collision that forms a square H_2I_2 complex; and (c) a bimolecular collision in which a hydrogen atom is transferred between reactants.

15.4 The reaction of NO with Cl_2 is

$$2 \text{ NO} + Cl_2 \longrightarrow 2 \text{ NOCl}$$

Use reactant molecules to write appropriate elementary reactions that satisfy the following criteria: (a) a unimolecular decomposition that generates Cl; (b) a bimolecular collision in which a Cl atom is transferred between reactants; and (c) a termolecular collision leading to the observed products.

15.5 Draw molecular pictures illustrating each part of Problem 15.3.

15.6 Draw molecular pictures illustrating each part of Problem 15.4.

15.7 Write three different satisfactory mechanisms for the reaction in Problem 15.3, one having your elementary reaction (a) as its first step, one having your elementary reaction (b) as its first step, and one having your elementary reaction (c) as its first step.

15.8 Write three different satisfactory mechanisms for the reaction in Problem 15.4, one having your elementary reaction (a) as its first step, one having your elementary reaction (b) as its first step, and one having your elementary reaction (c) as its first step.

Rates of Chemical Reactions

15.9 For the reaction $2 \text{ NO} + Cl_2 \longrightarrow 2 \text{ NOCl}$, do the following: (a) Express the rate in terms of the disappearance of Cl_2. (b) Relate the rate of NOCl formation to the rate of Cl_2 disappearance. (c) If Cl_2 reacts at a rate of 47 M s^{-1}, state how fast NOCl will form.

15.10 Do the following for the ozone decomposition mechanism: (a) Express the rate in terms of O_2 formation. (b) Relate the rate of O_3 consumption to the rate of O_2 production. (c) If O_2 forms at a rate of 2.7×10^{-6} M s^{-1}, state how fast ozone disappears.

15.11 For the reaction $2 \text{ NO} + Cl_2 \longrightarrow 2 \text{ NOCl}$, do the following: (a) Draw a molecular picture showing a sample that contains 12 NO molecules and 5 Cl_2 molecules. (b) Redraw the picture to show the result after four molecules of NO have reacted. (c) Redraw the picture to show the result when one reactant has been consumed completely.

15.12 For the reaction $O_3 + NO \longrightarrow O_2 + NO_2$, do the following: (a) Draw a molecular picture showing a sample that contains 6 NO molecules and 8 O_3 molecules. (b) Redraw the picture to show the result after three molecules of NO have reacted. (c) Redraw the picture to show the result when one reactant has been consumed completely.

15.13 Calcium oxide, an important ingredient in cement, is produced by decomposing calcium carbonate at high temperature:

$$CaCO_3 (s) \longrightarrow CaO (s) + CO_2 (g)$$

In one reaction, 3.5 kg of calcium carbonate is heated at 550 °C in a 5.0-L vessel. The pressure of CO_2 is 0.15 atm after 5.0 minutes. (a) What is the average rate of CO_2 production in mol/min during the 5-minute interval? (b) How many moles of $CaCO_3$ decompose in the 5-minute interval?

15.14 A chemist is studying the rate of the Haber synthesis:

$$N_2 + 3 H_2 \longrightarrow 2 NH_3$$

Starting with a closed reactor containing 1.25 mol/L of N_2 and 0.50 mol/L of H_2, the chemist finds that the H_2 concentration has fallen to 0.25 mol/L after 30 seconds. (a) What is the average rate of reaction over this time? (b) What is the average rate of NH_3 production? (c) What is the N_2 concentration after 30 seconds?

15.15 At high temperature, cyclopropane isomerizes to propene:

Cyclopropane (C_3H_6) → (Heat) → Propene (C_3H_6)

In a small sample containing 10 molecules of cyclopropane, the reaction proceeds at an average rate of 0.25 molecules/min for 20 minutes. (a) Use line drawings to illustrate the small sample before the reaction begins. (b) Redraw the picture after 20 minutes of reaction.

15.16 Under the appropriate conditions, *cis*-dichloroethylene isomerizes to *trans*-dichloroethylene:

cis-Dichloroethylene ($C_2H_2Cl_2$) → *trans*-Dichloroethylene ($C_2H_2Cl_2$)

In a small sample of gas containing 12 molecules of *cis*-dichloroethylene, the reaction proceeds at an average rate of 1.5 molecules/s for 2.0 seconds. (a) Use line drawings to illustrate the small sample before the reaction begins. (b) Redraw the picture after 2.0 seconds of reaction.

Concentration and Reaction Rates

15.17 Radioactive isotopes decay according to first-order kinetics. For one particular isotope, 1.00 nmol registers 1.2×10^6 decays in 1.00 min. (a) How many decays will occur in 1.00 min if 5.00 nmol of this isotope are present? (b) What fraction of the isotope decays per minute in each case? (c) Explain the relationship between your answers to (a) and (b).

15.18 Popcorn kernels pop independently (that is, "unimolecularly"). At constant temperature, 6 kernels pop in 5 seconds when 150 kernels are present. (a) After 50 kernels have popped, how many kernels pop in 5 seconds? (b) Is there a change in the fraction of kernels popping per second? If so, by how much? (c) Explain the relationship between your answers to (a) and (b).

15.19 For the net reaction $2 AB + 2 C \longrightarrow A_2 + 2 BC$, the following slow first step has been proposed:

$$C + AB \longrightarrow BC + A$$

(a) What rate law is predicted by this step? (b) What units are associated with the rate constant for this rate law? (c) Write additional steps that complete the mechanism.

15.20 For the net reaction in Problem 15.19, another possible slow first step is as follows:

$$2 AB \longrightarrow A_2 + 2 B$$

(a) What rate law is predicted by this step? (b) What units are associated with the rate constant for this rate law? (c) Write additional steps that complete the mechanism.

15.21 For the net reaction in Problem 15.19, another possible slow first step is as follows:

$$2 C + AB \longrightarrow AC + BC$$

(a) What rate law is predicted by this step? (b) What units are associated with the rate constant for this rate law? (c) Write additional steps that complete the mechanism.

15.22 For the net reaction in Problem 15.19, another possible slow first step is as follows:

$$AB \longrightarrow A + B$$

(a) What rate law is predicted by this step? (b) What units are associated with the rate constant for this rate law? (c) Write additional steps that complete the mechanism.

15.23 The two pictures shown below represent starting conditions for the following reaction:

$$O_3 + NO \longrightarrow O_2 + NO_2 \qquad Rate = k[O_3][NO]$$

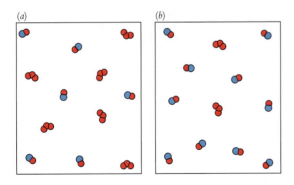

Will Flask *b* react faster or slower or at the same rate as Flask *a*? By how much? Explain your reasoning in terms of molecular collisions.

15.24 The two pictures shown below represent starting conditions for the following reaction:

$$O_2 + NO_2 \longrightarrow O_3 + NO \qquad Rate = k[O_2][NO_2]$$

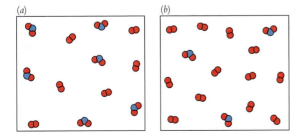

Will Flask *b* react faster or slower or at the same rate as Flask *a*? By how much? Explain your reasoning in terms of molecular collisions.

Experimental Kinetics

ilw 15.25 The following data are obtained for the decomposition of N_2O_5 at 45 °C:

$$2 N_2O_5 (g) \longrightarrow 4 NO_2 (g) + O_2 (g)$$

t (10^2 s)	0	2.00	4.00	6.00	8.00
$[N_2O_5]$ (atm)	2.50	2.22	1.96	1.73	1.53

(a) Use these data to determine the order of the decomposition process. Show your reasoning. (b) Determine the rate constant, including appropriate units. (c) What is the pressure of N_2O_5 after 1600 s? (d) How long will it take for the pressure to decrease to 0.500 atm?

15.26 Radioactive iodine (^{131}I) is used frequently in biological studies. A radiation biologist studies the rate of decomposition of this substance and obtains the following data:

Time (days)	0	4.0	8.0	12.0	16.0
Mass ^{131}I (μg)	12.0	8.48	6.0	4.24	3.0

(a) Use these data to determine the order of the decomposition process. Show your reasoning. (b) Determine the rate constant, including appropriate units. (c) How many micrograms will be left after 32 days? (d) How many days will it take for the sample size to decrease to 1.2 μg?

ilw 15.27 The decomposition reaction of NOBr is second order in NOBr, with a rate constant at 20 °C of $25 \, M^{-1}min^{-1}$. If the initial concentration of NOBr is 0.025 M, find (a) the time at which the concentration will be 0.010 M; and (b) the concentration after 125 min of reaction.

15.28 The condensation reaction of butadiene, C_4H_6, is second order in C_4H_6, with a rate constant of $0.93 \, M^{-1}min^{-1}$ (see Example 15-6). If the initial concentration of C_4H_6 is 0.240 M, find (a) the time at which the concentration will be 0.100 M; and (b) the concentration after 25 min of reaction.

15.29 The conversion of $C_5H_{11}Br$ into C_5H_{10} follows first-order kinetics, with a rate constant of $0.385 \, hr^{-1}$. If the initial concentration of $C_5H_{11}Br$ is 0.125 M, find (a) the time at which the concentration will be 1.25×10^{-3} M; and (b) the concentration after 3.5 hr of reaction.

15.30 The light-emitting decay of excited mercury atoms is first order, with a rate constant of $1.65 \times 10^6 \, s^{-1}$. A sample contains 4.5×10^{-6} M of excited mercury atoms. Find (a) the time at which the concentration will be 4.5×10^{-7} M; and (b) the concentration after 2.5×10^{-6} s.

15.31 The reaction of N_2O_5 with gaseous H_2O is a potentially important reaction in the chemistry of acid rain:

$$N_2O_5 (g) + H_2O (g) \longrightarrow 2 HNO_3 (g)$$

It is possible to monitor the concentration of N_2O_5 with time. Describe a set of isolation experiments from which the rate law and rate constant for this reaction could be determined.

15.32 The reaction of NO_2 with CO is a reaction that may occur in automobile exhaust:

$$NO_2 (g) + CO (g) \longrightarrow NO (g) + CO_2 (g)$$

Because of its color, it is possible to monitor the concentration of NO_2 with time. Describe a set of isolation experiments from which the rate law and rate constant for this reaction could be determined.

15.33 The following initial rate information was collected at 25 °C for this aqueous reaction:

$$2 I^- (aq) + S_2O_8^{2-} (aq) \longrightarrow I_2 (aq) + 2 SO_4^{2-} (aq)$$

$[I^-]_0$, M	$[S_2O_8^{2-}]_0$, M	Initial rate, M/min
0.125	0.150	4.4×10^{-2}
0.375	0.150	1.3×10^{-1}
0.125	0.050	1.5×10^{-2}

Determine the rate law and evaluate the rate constant for this reaction.

15.34 The following initial rate information was collected at 25 °C for this aqueous reaction:

$$2\,ClO_2\,(aq) + 2\,OH^-\,(aq) \longrightarrow ClO_3^-\,(aq) + ClO_2^-\,(aq) + 2\,H_2O$$

$[ClO_2]_0$, M	$[OH^-]_0$, M	Initial rate, M/s
0.050	0.050	2.9×10^{-2}
0.075	0.050	6.5×10^{-2}
0.075	0.075	9.8×10^{-2}

Determine the rate law and evaluate the rate constant for this reaction.

Linking Mechanisms and Rate Laws

15.35 Problem 15.19 asks about a mechanism for the hypothetical reaction of AB with C. Assume that the first step described in Problem 15.19 is fast and reversible and your second step is rate-determining. Derive the rate law under these assumptions.

15.36 Problem 15.8 asks for mechanisms for the reaction of NO with Cl_2. For parts (a) and (b) of that problem, assume that the first step in your mechanism is fast and reversible and the second step is rate-determining. Derive the rate laws under these assumptions.

ilw 15.37 A student proposes the following mechanism for the atmospheric decomposition of ozone to molecular oxygen:

$$O_3 + O_2 \rightleftharpoons O_5 \qquad \text{(fast, reversible)}$$
$$O_5 \longrightarrow 2\,O_2 + O \qquad \text{(slow, rate-determining)}$$

(a) Propose a third step that completes this mechanism. (b) Determine the rate law predicted by this mechanism. (c) Atmospheric chemists would consider this mechanism to be molecularly unreasonable. Explain why.

15.38 The reaction of CO with Cl_2 gives phosgene ($COCl_2$), a nerve gas used in World War I. Even though the stoichiometry is simple, the mechanism has several steps:

$$Cl_2 \rightleftharpoons 2\,Cl \qquad \text{(fast, reversible)}$$
$$Cl + CO \longrightarrow COCl \qquad \text{(slow, rate-determining)}$$
$$COCl + Cl_2 \longrightarrow COCl_2 + Cl \qquad \text{(fast)}$$

(a) Show that this mechanism gives the correct overall stoichiometry. (b) What rate law does this mechanism predict? (c) Identify any reactive intermediates in the mechanism.

15.39 The reaction of NO with O_2 to give NO_2 is an important process in the formation of smog in Los Angeles:

$$2\,NO + O_2 \longrightarrow 2\,NO_2$$

Experiments show that this reaction is third order overall. The following mechanism has been proposed:

$$NO + NO \underset{k_{-1}}{\overset{k_1}{\rightleftharpoons}} N_2O_2$$
$$N_2O_2 + O_2 \overset{k_2}{\longrightarrow} NO_2 + NO_2$$

(a) If the second step is rate-determining, what is the rate law? (b) Is this rate law consistent with the overall third-order behavior? Explain. (c) Draw molecular pictures that show different ways the intermediate species might bind together, and identify the one that is most reasonable with respect to the second step of the mechanism.

15.40 Gaseous N_2O_5 decomposes according to the following equation:

$$2\,N_2O_5 \longrightarrow 4\,NO_2 + O_2$$

Much evidence suggests that the mechanism is as follows:

$$N_2O_5 \rightleftharpoons NO_2 + NO_3 \qquad \text{(fast decomposition)}$$
$$NO_2 + NO_3 \longrightarrow NO + NO_2 + O_2 \qquad \text{(slow)}$$
$$NO + NO_3 \longrightarrow 2\,NO_2 \qquad \text{(fast)}$$

(a) Show that this mechanism gives the correct overall stoichiometry. (b) Determine the rate law predicted by this mechanism. (c) This decomposition is first order experimentally. Does this information prove that Step 2, rather than Step 1, is rate-determining? Explain.

Reaction Rates and Temperature

15.41 If a reaction has an activation energy of zero, how will its rate constant change with temperature? Explain in molecular terms what $E_a = 0$ means.

15.42 If a reaction has an activation energy of zero, how is ΔE for the forward reaction related to E_a for the reverse reaction? Draw an activation energy diagram illustrating your answer.

15.43 Consider the exothermic reaction $AC + B \rightarrow AB + C$. (a) Draw an activation energy diagram for this reaction. (b) Label the energies of reactants and products. (c) Show $\Delta E_{reaction}$ by a double-headed arrow. (d) Show E_a for the forward reaction by a single-headed arrow. (e) Label and draw a molecular picture of the activated complex.

15.44 Consider the endothermic reaction $AB + C \rightarrow AC + B$. (a) Draw an activation energy diagram for this reaction. (b) Label the energies of reactants and products. (c) Show $\Delta E_{reaction}$ by a double-headed arrow. (d) Show E_a for the forward reaction by a single-headed arrow. (e) Label and draw a molecular picture of the activated complex.

15.45 For the isomerization that converts cyclopropane to propene, $\Delta H_{reaction} = -33$ kJ and $E_a = 272$ kJ/mol. (a) Draw a molecular picture of cyclopropane converting into propene. (b) Draw the activation energy diagram for the reaction, and label it completely.

15.46 Nitrogen dioxide in smog can combine in an elementary reaction to form N_2O_4 molecules. The combination reaction is exothermic by 57 kJ/mol. For the reverse reaction, the dissociation of N_2O_4, $E_a = 70$ kJ/mol. (a) Draw a molecular picture of NO_2 combining to form N_2O_4. (b) Draw an activation energy diagram that shows the energy relationships as quantitatively as possible. Label the diagram completely.

ilw 15.47 Fireflies flash at a rate that depends on the temperature. At 29 °C, the average rate is 3.3 flashes every 10 seconds, whereas at 23 °C, the average rate falls to 2.7 flashes every 10 seconds. Calculate the "energy of activation" for the flashing process.

15.48 The rate at which tree crickets chirp is 190/min at 28 °C but only 39.6/min at 5 °C. From these data, calculate the "energy of activation" for the chirping process.

Catalysis

15.49 The industrial process for forming methanol involves a catalyst:

$$CO + 2\,H_2 \xrightarrow[\text{High } P,T]{\text{Catalyst}} CH_3OH$$

Draw molecular pictures showing the bond breakage and formation that must occur in the course of this reaction. Suggest how the catalyst might make it easier for these reactions to occur.

15.50 The industrial process for forming acetic acid involves a catalyst:

$$CH_3OH + CO \xrightarrow[\text{173 °C, 700 atm}]{[Rh(CO)_2I_3]} CH_3CO_2H$$

Draw molecular pictures showing the bond breakage and formation that must occur in the course of this reaction. Suggest how the catalyst might make it easier for these reactions to occur.

15.51 The addition of molecular hydrogen to ethylene is extremely slow unless Pd metal is present. In the presence of the metal, however, H_2 is adsorbed on the metal surface as H atoms, which then add to C_2H_4 when it strikes the surface:

(a) Draw an activation energy diagram for this mechanism that illustrates the effect of the catalyst. (b) Identify any catalysts or intermediates in the mechanism. Explain. (c) Draw a molecular picture that illustrates how this addition takes place.

15.52 NO is an atmospheric pollutant that destroys ozone in the stratosphere. Here is the accepted mechanism:

$$O_3 + NO \longrightarrow NO_2 + O_2 \quad \text{(slow)}$$

$$NO_2 + O \longrightarrow NO + O_2 \quad \text{(fast)}$$

$$O_3 + O \longrightarrow 2\ O_2$$

Additional Paired Problems

15.53 Oxygen reacts with NO to form NO_2 as the only product. Here is a proposed first step of the mechanism:

$$O_2 + NO \longrightarrow NO_2 + O$$

(a) What second step is required to give a satisfactory mechanism? (b) Draw molecular pictures to illustrate each step.

15.54 Oxygen reacts with CO to form CO_2 as the only product. Here is a proposed first step of the mechanism:

$$O_2 + CO \longrightarrow CO_2 + O$$

(a) What second step is required to give a satisfactory mechanism? (b) Draw molecular pictures to illustrate each step.

15.55 With appropriate catalysts, it is possible to make benzene from acetylene. The reaction has simple overall stoichiometry and can be studied in kinetics experiments:

$$3\ C_2H_2\ (g) \longrightarrow C_6H_6\ (g)$$

(a) Write the rate *expression* (see Equation 15-1) for this reaction in terms of the disappearance of C_2H_2 and in terms of the appearance of C_6H_6. (b) If enough information is given to find the rate law, find it. If not, describe the data that you would need and how you would analyze the data to determine the rate law.

15.56 The first step in the Ostwald process for the synthesis of nitric acid is the combustion of ammonia:

$$4\ NH_3 + 5\ O_2 \xrightarrow{\text{Pt gauze}} 4\ NO + 6\ H_2O$$

In a catalytic experiment, the rate of reaction is found to be

$$\text{Rate} = \frac{\Delta[NO]}{\Delta t} = 1.5 \times 10^{-3}\ \text{M s}^{-1}$$

What is the rate expressed in terms of NH_3, O_2, and H_2O?

15.57 Use the following information to construct an activation energy diagram for the reaction, $S\ (s) + O_2\ (g) \longrightarrow SO_2\ (g)$

$$\Delta H_f^\circ(SO_2) = -296.1\ \text{kJ/mol},\ E_a = 150\ \text{kJ/mol}$$

15.58 The reaction of ethylene (C_2H_4) with H_2 to form ethane (C_2H_6) (see Problem 15.51) is exothermic by 137 kJ/mol; has $E_a = 200$ kJ/mol; and is catalyzed by Pd metal, which reduces E_a by 60 kJ/mol. Draw an activation energy diagram for this reaction, and label it clearly. Be sure all the given facts are illustrated in the diagram.

15.59 What happens to the rate of a reaction involving H_2 if the concentration of H_2 is tripled and the reaction is (a) second order in H_2; (b) zero order in H_2; and (c) $\frac{3}{2}$ order in H_2?

15.60 What happens to the rate of a reaction involving CO if the concentration is doubled and the reaction is (a) first order in CO; (b) half order in CO; and (c) inverse first order in CO?

15.61 Azomethane decomposes into nitrogen and ethane at high temperature:

$$H_3C-N{=}N-CH_3 \longrightarrow N_2 + C_2H_6$$

A chemist studying this reaction at 300 °C obtains the following data:

Time (10^2 s)	0	1.00	1.50	2.00	2.50	3.00
[Azomethane] (mM)	7.94	6.15	5.40	4.75	4.20	3.69

Prepare first- and second-order graphic plots of these data, determine the order of the reaction, and calculate its rate constant.

(a) Draw an activation energy diagram for this mechanism. Include as many details as possible. (b) Identify any catalysts or intermediates in the mechanism. Explain. (c) The activation energy for decomposition of ozone promoted by NO is 11.9 kJ/mol. The activation energy for decomposition of ozone promoted by Cl is 2.1 kJ/mol. Which pollutant is a more serious threat to the ozone layer, Cl or NO? Explain.

15.62 Gas A decomposes by the reaction $3\ A \rightarrow B + C$ at 45 °C, and its concentration changes as follows:

Time (10^2 s)	0	2.00	4.00	6.00	8.00
[A] (M)	2.50	2.22	1.96	1.73	1.53

Find the rate law and determine the rate constant.

15.63 The hypothetical reaction $X_2 + 2\ Y \rightarrow 2\ XY$ goes by the following mechanism:

$$X_2 \longrightarrow 2\ X \quad \text{(slow)}$$

$$X + Y \longrightarrow XY \quad \text{(fast)}$$

The pictures that follow represent starting conditions for the reaction. What would the rate of reaction be for Flask *b* compared with that for Flask *a*? Explain.

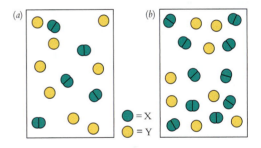

15.64 Here is another set of starting conditions for the reaction described in Problem 15.63. What would the rate of reaction be for Flask *b* compared with that for Flask *a*? Explain.

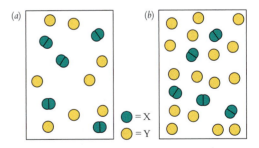

15.65 A chemist carried out kinetic studies on the reaction described in Problem 15.31, obtaining the following data:

Experiment A: $[N_2O_5]_0 = 1.5 \times 10^{-4}$ M, $[H_2O]_0 = 0.025$ M

Time (s)	0	60	120	180	240
[N_2O_5] (10^{-5} M)	15	12	9.6	7.6	6.1

Experiment B: $[N_2O_5]_0 = 1.5 \times 10^{-4}$ M, $[H_2O]_0 = 0.050$ M

Time (s)	0	60	120	180	240
[N_2O_5] (10^{-5} M)	15	9.6	6.1	3.9	2.5

Use graphs to determine the rate law for this reaction, and evaluate the rate constant.

15.66 A chemist carried out kinetic studies on the reaction described in Problem 15.32, obtaining the following data:

Experiment A: $[NO_2]_0 = 3.5 \times 10^{-3}$ M, $[CO]_0 = 0.40$ M

Time (s)	0	240	480	720	960
$[NO_2]$ (10^{-3} M)	3.5	2.5	1.9	1.6	1.3

Experiment B: $[NO_2]_0 = 3.5 \times 10^{-3}$ M, $[CO]_0 = 0.60$ M

Time (s)	0	240	480	720	960
$[NO_2]$ (10^{-3} M)	3.5	2.5	1.9	1.6	1.3

Use graphs to determine the rate law for this reaction, and evaluate the rate constant.

15.67 Cyclopropane converts to propene by first-order kinetics with a rate constant at 500 °C of 5.5×10^{-4} s^{-1}. Calculate how long it takes (a) for 10.0% of a sample to decompose; (b) for 50.0% to decompose; and (c) for 99.9% to decompose.

15.68 The reaction of Cl_2 with H_2S in aqueous solution is first order in each reactant, with $k = 3.5 \times 10^{-2}$ M^{-1} s^{-1} at 28 °C. A solution has $[Cl_2] = 0.035$ M and $[H_2S] = 5.0 \times 10^{-5}$ M. (a) Find the H_2S concentration after 225 s of reaction. (b) Find the time at which the H_2S concentration has fallen to 1.0×10^{-5} M.

15.69 Phosphine (PH_3) decomposes into phosphorus and molecular hydrogen:

$$4\ PH_3\,(g) \longrightarrow P_4\,(g) + 6\ H_2\,(g)$$

Experiments show that this is a first-order reaction, with a rate constant of 1.73×10^{-2} s^{-1} at 650 °C. Which of the following statements are true? (a) The overall reaction is elementary. (b) The data allow E_a to be evaluated. (c) The rate constant will be smaller than 1.73×10^{-2} s^{-1} at 500 °C. (d) The reaction $PH_3 \rightarrow PH_2 + H$ might be the first, rate-determining step in the mechanism of this reaction.

15.70 The Haber reaction for the manufacture of ammonia is

$$N_2 + 3\ H_2 \longrightarrow 2\ NH_3$$

Without doing any experiments, which of the following can you say *must* be true? (a) Reaction rate $= -\Delta[N_2]/\Delta t$. (b) The reaction is first order in N_2. (c) The rate of disappearance of H_2 is three times the rate of disappearance of N_2. (d) The rate of disappearance of N_2 is three times the rate of disappearance of H_2. (e) $\Delta[H_2]/\Delta t$ has a positive value. (f) The reaction is not an elementary reaction. (g) The activation energy is positive.

15.71 For each of the following reactions, what is the order with respect to each starting material and what is the overall order of the reaction?

(a) $2\ N_2O_5 \longrightarrow 4\ NO_2 + O_2$ Rate $= k[N_2O_5]$

(b) $2\ NO + 2\ H_2 \longrightarrow N_2 + 2\ H_2O$ Rate $= k[NO]^2[H_2]$

(c) Glucose + ATP $\xrightarrow{\text{Enzyme}}$ Glucose phosphate + ADP
$$\text{Rate} = k[\text{enzyme}]$$

15.72 For each of the following reactions, what is the order with respect to each starting material, and what is the overall order of the reaction?

(a) $N_2 + 3\ H_2 \xrightarrow{\text{Catalyst}} 2\ NH_3$ Rate $= k$

(b) Sucrose + $H_2O \longrightarrow$ Glucose + Fructose
$$\text{Rate} = k[\text{Sucrose}][H_2O][H_3O^+]$$

(c) $CHCl_3 + Cl_2 \longrightarrow CCl_4 + HCl$ Rate $= k[CHCl_3][Cl]^{1/2}$

15.73 What is the difference between speed and spontaneity of a chemical reaction?

15.74 Can the assumption that the forward and reverse rates are equal ever be exact for a reacting system? Explain your answer.

15.75 Write chemical equations that show how CF_2Cl_2 contributes to the destruction of ozone in the stratosphere.

15.76 Write the chemical equations for the reactions that maintain the balance among O_3, O_2, and O in the unpolluted stratosphere.

15.77 One possible mechanism for the reaction between H_2 and NO follows:

$2\ NO \rightleftharpoons N_2O_2$	(fast)
$N_2O_2 + H_2 \longrightarrow N_2O + H_2O$	(slow)
$N_2O + H_2 \longrightarrow N_2 + H_2O$	(fast)

(a) What is the overall stoichiometry of this reaction? (b) What rate law is predicted by this mechanism?

15.78 A chemical reaction is thought to proceed by the following mechanism:

$$A + B \underset{k_{-1}}{\overset{k_1}{\rightleftharpoons}} C$$
$$C + D \xrightarrow{k_2} A + E$$
$$E + B \xrightarrow{k_3} 2\ F$$

(a) What is the net stoichiometry of the reaction? (b) If the second step is rate-determining, what is the rate law? (c) Identify any catalysts and intermediates.

15.79 Use molecular arguments to explain why each of the following factors speeds up a chemical reaction: (a) a catalyst; (b) an increase in temperature; and (c) an increase in concentration.

15.80 Explain in molecular terms why the Haber synthesis cannot proceed in a single-step elementary reaction.

15.81 Molecule A decomposes to give B and C. The following figures represent two experiments conducted to study this decomposition reaction.

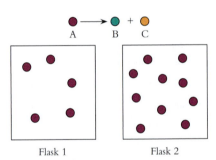

Flask 1   Flask 2

Flask 2 is found to react four times faster than Flask 1. What is the rate law for the decomposition of A? Explain your reasoning in terms of molecular collisions.

15.82 Phosphorus-32 is a radioactive isotope that decomposes in a unimolecular first-order process. The following figures represent portions of two flasks, each containing six atoms of ^{32}P.

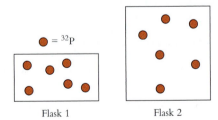

Flask 1   Flask 2

Will the time required for the six atoms in Flask 2 to decompose to three atoms be faster, slower, or the same as that in Flask 1? Explain.

More Challenging Problems

15.83 At least three possible reaction mechanisms exist for the reaction of hydrogen with halogens, $H_2 + X_2 \rightarrow 2\,HX$. Determine the rate law predicted by each of them:

(a) $H_2 + X_2 \longrightarrow 2\,HX$ (bimolecular, direct)
(b) $X_2 \longrightarrow X + X$ (slow)
    $X + H_2 \longrightarrow HX + H$ (fast)
    $H + X \longrightarrow HX$ (fast)
(c) $X_2 \rightleftharpoons X + X$ (fast, reversible)
    $X + H_2 \longrightarrow HX + H$ (slow)
    $H + X \longrightarrow HX$ (fast)

15.84 Consider the following hypothetical reaction, which is $\frac{3}{2}$ order overall and half order in  :

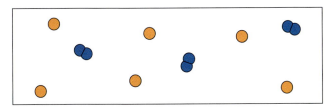

$$2\,\bigcirc + \bigcirc\!\bigcirc \longrightarrow 2\,\bigcirc\!\bigcirc$$

The following figure represents one set of initial conditions for the reaction. Draw a similar figure that represents a set of initial conditions for which the rate would be twice as fast as in the first case:

15.85 The following mechanism has been proposed for the gas-phase reaction of $CHCl_3$ and Cl_2:

$$Cl_2 \underset{k_{-1}}{\overset{k_1}{\rightleftharpoons}} 2\,Cl \qquad \text{(fast, reversible)}$$

$$Cl + CHCl_3 \xrightarrow{k_2} HCl + CCl_3 \qquad \text{(slow)}$$

$$CCl_3 + Cl \xrightarrow{k_3} CCl_4 \qquad \text{(fast)}$$

(a) Write the net balanced equation. (b) What are the intermediates in the reaction? (c) What rate law is predicted by this mechanism?

15.86 Ammonium cyanate (NH_4NCO) rearranges in water to urea (NH_2CONH_2). The following data are obtained at 50. °C:

Time (hr)	0	1.0	2.0	3.0	5.0	7.0	9.0
Conc. (M)	0.500	0.375	0.300	0.250	0.188	0.150	0.125

At 25 °C, the concentration falls from 0.500 M to 0.300 M in 6.0 hours. (a) Determine the rate law. (b) Determine the rate constant at 50. °C. (c) Determine the activation energy.

15.87 Photographers use a rule of thumb, that a 10° C rise in temperature cuts development time in half, to determine how to modify film development time as the temperature varies. (a) Calculate the activation energy for the chemistry of film developing, assuming that "normal" temperature is 20 °C. (b) If a certain film takes 10 minutes to develop at 20 °C, how long will it take at 25 °C?

15.88 If chlorofluorocarbons catalyze O_3 decomposition, which is exothermic by 392 kJ/mol, they must also catalyze O_3 production from O_2. Using the energies shown in Figure 15-20, explain why catalysis results in net O_3 destruction even though reactions in both directions are accelerated.

15.89 Acetaldehyde decomposes to methane and carbon monoxide:

$$\underset{H_3C}{\overset{O}{\underset{}{\overset{\|}{C}}}}\!\!-\!\!H \longrightarrow CH_4 + CO$$

Kinetic data for the decomposition of acetaldehyde follow:

Time (10^3 s)	0	1.000	2.000	3.000	4.000
$[CH_3CHO]$ (mol L^{-1})	0.250	0.118	0.0770	0.0572	0.0455

(a) Determine the rate law for the decomposition of acetaldehyde. (b) What is the value of the rate constant? (c) How long does it take for 75% of the acetaldehyde to decompose?

15.90 A chemist interested in a reaction having overall stoichiometry

$$3A + 2B \longrightarrow \text{Products}$$

carries out two sets of experiments, both at 25 °C. In an experiment in which [B] is 1.00 M and [A] is 0.050 M at the start of the reaction, the chemist obtains these data:

Time (10^2 s)	0	1.0	2.0	3.0	4.0	5.0
[A](M)	0.050	0.040	0.032	0.025	0.020	0.016

Then the chemist performs another experiment with [B] = 1.50 M and [A] = 0.050 M. Again, it takes 3.0×10^2 seconds for the concentration of A to fall to 0.025 M. (a) What is the rate law of the reaction? Show your reasoning. (b) Calculate the rate constant.

15.91 The enzyme urease catalyzes the conversion of urea to ammonia and carbon dioxide:

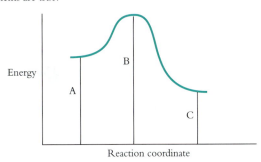

The uncatalyzed reaction has an activation energy of 125 kJ/mol. The enzyme catalyzes a mechanism that has $E_a = 46$ kJ/mol. By what factor does urease increase the rate of urea conversion at 21 °C? (*Hint:* Use the Arrhenius equation.)

15.92 An atmospheric scientist interested in how NO is converted to NO_2 in urban atmospheres carries out two experiments to measure the rate of this reaction. The data are tabulated below. Find the rate law and rate constant for this reaction.

A: $[NO]_0 = 9.63 \times 10^{-3}$ M, $[O_2]_0 = 4.1 \times 10^{-4}$ M:

t (s)	0	3.0	6.0	9.0	12.0
$[O_2](10^{-4}$ M)	4.1	2.05	1.02	0.51	0.25

B: $[NO]_0 = 4.1 \times 10^{-4}$ M, $[O_2]_0 = 9.75 \times 10^{-3}$ M:

t (10^2 s)	0	1.00	2.00	3.00	4.00
$[NO](10^{-4}$ M)	4.1	2.05	1.43	1.02	0.82

15.93 How can you identify intermediates and catalysts in a proposed mechanism for a reaction?

15.94 For the reaction $NO + O_3 \rightarrow NO_2 + O_2$, the experimental rate law is Rate = $k[NO][O_3]$. Which of the following sets of conditions will give the fastest rate? Explain your choice. (a) 0.5 mol of NO and 0.5 mol of O_3 in a 2.0-L vessel; (b) 0.5 mol of NO and 0.5 mol of O_3 in a 1.0-L vessel; and (c) 2.0 mol of NO and 0.1 mol of O_3 in a 1.0-L vessel.

15.95 For the activation energy diagram shown, which of the following statements are true?

(a) $\Delta E_{\text{reaction}} = $ A $-$ C; (b) $\Delta E_{\text{reaction}} = $ B $-$ C;
(c) $E_a(\text{forward}) = E_a$ (reverse); (d) A represents the energy of the starting materials; (e) E_a (forward) = B $-$ C; and
(f) $E_a(\text{forward}) < E_a(\text{reverse})$.

15.96 Below are some data for the decomposition of molecule A. Determine the rate law and rate constant.

t (s)	0	10.0	20.0	30.0
[A](M)	0.64	0.52	0.40	0.28

Group Study Problems

15.97 According to the induced-fit model of enzyme activity, binding a reactant to the enzyme causes a distortion in the conformation of the reactant itself. This distortion decreases the activation energy of the catalyzed reaction. For the hypothetical unimolecular reaction A → B + C, draw a reaction coordinate diagram for the uncatalyzed reaction and for the same reaction catalyzed by an enzyme. Use the enzyme model to explain the differences in the two diagrams.

15.98 In the preparation of cobalt complexes, one reaction involves displacement of H_2O by Cl^-:

$$[Co(NH_3)_5H_2O]^{3+} + Cl^- \longrightarrow [Co(NH_3)_5Cl]^{2+} + H_2O$$

A proposed mechanism for this displacement starts with rapid reversible dissociation of water from the complex:

$$[Co(NH_3)_5H_2O]^{3+} \rightleftharpoons [Co(NH_3)_5]^{3+} + H_2O$$

(a) Propose a slow second step that completes the mechanism and gives the correct overall stoichiometry. (b) Derive the rate law that this mechanism predicts. (c) When the rate is studied in 1 M aqueous HCl solution that is 1 mM in $[Co(NH_3)_5H_2O]^{3+}$, first-order experimental kinetics are observed. Is this observation consistent with the proposed mechanism? State your reasoning clearly and in detail.

15.99 In your own words, describe the induced-fit model of enzyme specificity. Illustrate with diagrams, using a hypothetical enzyme that catalyzes the decomposition of a square but cannot catalyze the decomposition of a triangle:

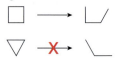

15.100 Red and white Ping-Pong balls with small Velcro patches are placed in an air-blowing machine like the ones used to scramble numbered balls in lottery drawings on TV. If red and white balls collide at the Velcro points, they stick together. When 20 red and 10 white balls are put in the machine, 4 pairs form after 2 minutes of blowing. (a) If 10 balls of each color had been placed in the machine, how many pairs would form under the same conditions? (b) What if there were 20 red and 15 white balls? (c) What if there were 10 red and 20 white balls? (d) What if the machine had been crammed with 40 red and 20 white balls? (e) Explain why the rates change with the numbers of balls.

15.101 Bromide ions catalyze the decomposition of aqueous H_2O_2:

$$2 H_2O_2 (aq) \longrightarrow 2 H_2O (l) + O_2 (g)$$

The first step is formation of BrO^-:

$$H_2O_2 (aq) + Br^- (aq) \longrightarrow H_2O (l) + BrO^- (aq)$$

The BrO^- ion then quickly reacts with another H_2O_2 molecule:

$$BrO^- (aq) + H_2O_2 (aq) \longrightarrow H_2O (l) + A + B$$

(a) Identify A and B. (b) Construct an activation energy diagram for this reaction. Consult thermodynamic tables to get an estimate of ΔE. Clearly show the effect of bromide ion catalysis and locate on your diagram the energy of the BrO^- intermediate.

15.102 The reaction between ozone and nitrogen dioxide follows:

$$2 NO_2 (g) + O_3 (g) \longrightarrow N_2O_5 (g) + O_2 (g) \quad \Delta H° = -200 \text{ kJ}$$

The reaction proceeds according to the experimental rate law, Rate = $k[NO_2][O_3]$, and NO_3 has been identified as an intermediate in the reaction. The activation energy is 50 kJ/mol. (a) Devise a two-step mechanism for the reaction that is consistent with the experimental observations. Identify the rate-determining step. (b) Draw a molecular picture that illustrates your rate-determining step. (c) Sketch the activation energy diagram for this reaction and label it as completely as you can.

15.103 Hydroxide ion has an effect on the rate of the redox reaction between I^- and OCl^-, as the following data show:

$$OCl^- (aq) + I^- (aq) \longrightarrow Cl^- (aq) + IO^- (aq)$$

Experiment	$[OCl^-]_0$ (M)	$[I^-]_0$ (M)	$[OH^-]_0$ (M)	Initial rate (10^{-2} M/s)
1	0.025	0.0025	0.10	3.8
2	0.025	0.0060	0.10	9.1
3	0.037	0.0025	0.10	5.6
4	0.037	0.0025	0.20	2.8

(a) Determine the rate law. (b) Evaluate the rate constant. (c) Devise a mechanism that is consistent with the rate law. The first step should be a rapid equilibrium involving H_2O. (d) Design additional experiments that you could use to test your mechanism.

Answers to Section Exercises

15.1.1 Second product is CO: NO + CO_2 → NO_2 + CO.

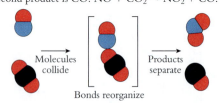

Molecules collide → Bonds reorganize → Products separate

15.1.2 Your description should include highway speeds and the rate at which toll-takers take tolls.

15.1.3 Chemical equation is 2 NO + 2 H_2 → 2 H_2O + N_2, and the intermediates are N_2O_2 and N_2O.

15.2.1

15.2.2 Hydrogen

15.2.3

$$\text{Relative rate} = -\frac{\Delta[N_2]}{\Delta t} = -\frac{1}{3}\frac{\Delta[H_2]}{\Delta t} = +\frac{1}{2}\frac{\Delta[NH_3]}{\Delta t}$$

15.3.1

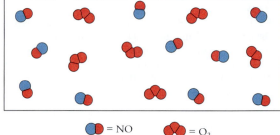

= NO   = O_3

15.3.2 The rate law is Rate = k[Butene], and the rate constant has units of time^{-1}.

15.3.3 The rate law is Rate = $k[H_2][Br_2]^{1/2}$, the overall order is $\frac{3}{2}$, and the rate constant has units of $M^{1/2}$ time^{-1}.

15.4.1 A plot of ln ($[N_2O_5]_0/[N_2O_5]$) vs. t is linear, so the reaction is first order; $k = 3.04 \times 10^{-2}$ min^{-1}.

15.4.2 The rate law is rate = $k[AB_2]^2$.

15.4.3 Rate $= k[NH_4^+][NO_2^-]$; $k = 2.7 \times 10^{-4}$ M^{-1} s^{-1}

15.5.1 (a) $N_2O_4 \rightarrow 2\ NO_2$ (occurs twice); and (b) the first step must be rate-determining.

15.5.2 If the second step is rate-determining, the rate law is second order in N_2O_5.

15.5.3 (a) $N_2O_2 + O_2 \rightarrow 2\ NO_2$; and (b) rate $= k_2[N_2O_2][O_2]$, but using equality of rates, $[N_2O_2] = k_1[NO]^2/k_{-1}$, resulting in Rate $= k[NO]^2[O_2]$.

15.6.1 The N_2 triple bond is extremely strong. The H_2 bond, 435 kJ/mol, is a strong single bond. From these bond strengths, we predict a high activation energy for the reaction.

15.6.2 (a) Graph ln k vs. $1/T$:

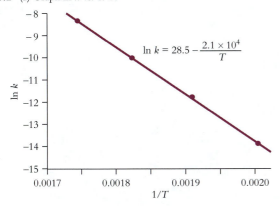

$E_a = 1.8 \times 10^2$ kJ/mol; and (b) 4.4×10^{-10} s^{-1}

15.6.3

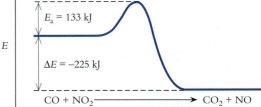

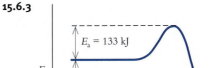

15.7.1 They are less expensive because only a thin layer of the expensive metal, rather than solid pieces, is required. They are more effective because a layer on porous charcoal has a larger surface area (where the catalyzed reaction occurs) than the relatively smooth surface of a pure piece of metal.

15.7.2 (a) True; (b) a catalyst does not change the overall energy change of a reaction; (c) a catalyst lowers the activation energy of an overall reaction by changing its mechanism; and (d) true.

15.7.3 The reaction converts pollutants (CO and NO_2) into benign constituents of the atmosphere (CO_2 and N_2), so catalyzing this reaction could reduce atmospheric pollution levels.

Answers to Extra Practice Exercises

15.1 Reaction rate $= \left(-\dfrac{1}{2}\right)\left(\dfrac{\Delta[NO_2]}{\Delta t}\right) = \left(\dfrac{\Delta[N_2O_4]}{\Delta t}\right)$

15.2 The rate will be the same as the rate for Figure 15-8a, so it will be half as fast as the rate for Figure 15-8b and three times as fast as the rate for the view shown in the example.

15.3 The rate law is first order in each reactant and second order overall, so the rate constant has units of M^{-1} s^{-1}.

15.4 A plot of ln ($[SO_2Cl_2]_0/[SO_2Cl_2]$) will be linear if the reaction is first order:

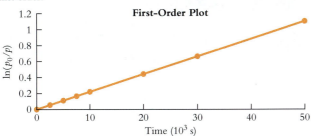

The plot is linear, so the reaction is first order.

15.5 $t_{1/2} = 3.18 \times 10^4$ s

15.6 A plot of $\left(\dfrac{1}{[SO_2Cl_2]} - \dfrac{1}{[SO_2Cl_2]_0}\right)$ will be linear if the reaction is second order:

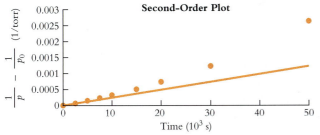

The plot is not linear, so the data are not consistent with a second-order reaction.

15.7 Rate $= k[CH_3Br][OH^-]$ and $k = 172$ M^{-1} s^{-1}

15.8 Rate $= k[CO_2]$ and $k = 0.19$ s^{-1}

15.9 Rate $= k_1[\text{Butadiene}]^2$

15.10 Rate $= \dfrac{k_2 k_1[O_3][NO_2]^2}{k_{-1}[O_2]}$

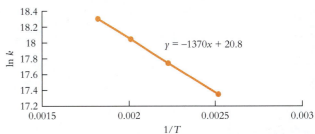

15.11 $E_a = 11.4$ kJ/mol

15.12 $E_a = 262$ kJ/mol

15.13 At 25 °C (298 K), $k = 4.7 \times 10^{-31}$ s^{-1}

15.14 Rate $= \dfrac{k_1 k_2 k_3}{k_{-1} k_{-2}}$ [E][S][R] $= k_{obs}$[E][S][R], with

$k_{obs} = \dfrac{k_1 k_2 k_3}{k_{-1} k_{-2}}$

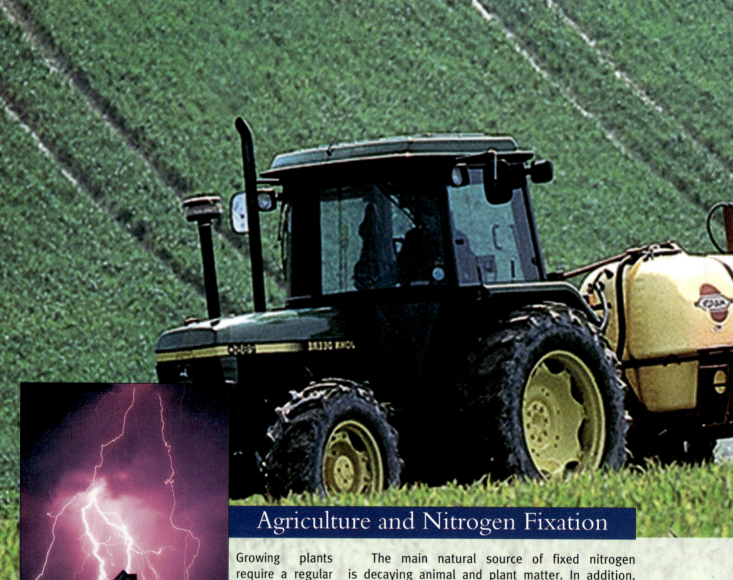

16

Principles of Chemical Equilibrium

Agriculture and Nitrogen Fixation

Growing plants require a regular supply of carbon, hydrogen, oxygen, and nitrogen to synthesize living tissue. Water, atmospheric carbon dioxide, and molecular oxygen supply carbon, hydrogen, and oxygen, but nitrogen is not readily available. Even though molecular nitrogen is the main component of the atmosphere, the nitrogen atoms in N_2 are held together by an exceptionally strong triple bond. Consequently, few organisms make direct use of N_2. Instead, plants extract nitrogen from nitrate and ammonium ions in the soil. The availability of usable nitrogen, known as "fixed" nitrogen, often limits the growth rate of vegetation.

The main natural source of fixed nitrogen is decaying animal and plant matter. In addition, lightning strikes convert atmospheric nitrogen into compounds containing fixed nitrogen, and a few types of bacteria are capable of using N_2 directly. These natural sources fall well short of providing enough nitrogen for intensive agriculture. Farmers must supply nitrogen-containing fertilizers, as our opening photo illustrates. Before World War I, the main source of fertilizer was natural deposits of sodium nitrate, but these supplies were quite limited. In 1898, the English chemist Sir William Ramsay predicted that the limited supply of fixed nitrogen would lead to world famine by the middle of the twentieth century.

Learning Resources

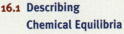 KEY CONCEPTS

 CRITICAL THINKING

 SKILLS TO MASTER

 ADDITIONAL HELP
www.wiley.com/college/olmsted

● TUTORIALS
● ANIMATIONS

In the absence of a new source of fixed nitrogen, Ramsay's prediction might have come true, but just before the start of World War I, German chemists developed the Haber process for producing ammonia from molecular nitrogen and hydrogen:

$$N_2 + 3\,H_2 \longrightarrow 2\,NH_3$$

The Haber process thoroughly dominates the modern fertilizer industry, because it provides a plentiful and relatively inexpensive industrial source of fixed nitrogen.

As an indispensable source of fertilizer, the Haber process is one of the most important reactions in industrial chemistry. Nevertheless, even under optimal conditions the yield of the ammonia synthesis in industrial reactors is only about 13%. This is because the Haber process does not go to completion; the net rate of producing ammonia reaches zero when substantial amounts of N_2 and H_2 are still present. At balance, the concentrations no longer change even though some of each starting material is still present. This balance point represents dynamic chemical equilibrium.

In this chapter, we present basic features of chemical equilibrium. We explain why reactions such as the Haber process cannot go to completion.

We also show why using catalysts and elevated temperatures can accelerate the rate of this reaction but cannot shift its equilibrium position in favor of ammonia and why elevated temperature shifts the equilibrium in the wrong direction. In Chapters 17 and 18, we turn our attention specifically to applications of equilibria, including acid–base chemistry.

16.1 DESCRIBING CHEMICAL EQUILIBRIA

The detailed chemistry describing the synthesis of ammonia is complex, so we introduce the principles of equilibrium using the chemistry of nitrogen dioxide. Molecules in a sample of nitrogen dioxide are always colliding with one another. As described in Chapter 15, a collision in the correct orientation can result in bond formation, producing an N_2O_4 molecule:

$$2\,NO_2 \longrightarrow N_2O_4$$

In a vessel that contains only NO_2 molecules, the production of N_2O_4 is the only reaction that takes place. However, once N_2O_4 molecules are present, the reverse reaction also can occur. An N_2O_4 molecule can fragment after collisions give it sufficient energy to break the N—N bond. These fragmentations regenerate NO_2:

$$N_2O_4 \longrightarrow 2\,NO_2$$

Figure 16-1 depicts these two processes from the molecular perspective.

Dynamic Equilibrium

Collisions between NO_2 molecules produce N_2O_4 and consume NO_2. At the same time, fragmentation of N_2O_4 produces NO_2 and consumes N_2O_4. When the concentration of N_2O_4 is very low, the first reaction occurs more often than the second. As the N_2O_4 concentration increases, however, the rate of fragmentation increases. Eventually, the rate of N_2O_4 production equals the rate of its decomposition. Even though individual molecules continue to combine and decompose, the rate of one reaction exactly balances the rate of the other. This is a **dynamic equilibrium.** At dynamic equilibrium, the rates of the forward and reverse reactions are equal. The system is *dynamic* because individual molecules react continuously. It is *at equilibrium* because there is no *net* change in the system.

The reactions that occur in a mixture of NO_2 and N_2O_4 lead to changes in concentrations. The concentration of NO_2 can be determined experimentally, because NO_2 is orange and N_2O_4 is colorless. The color intensity of the gas mixture is proportional to the

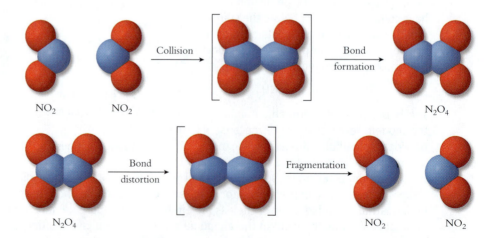

Figure 16-1
When two NO₂ molecules collide in the proper orientation, an N—N bond forms to produce a molecule of N₂O₄. When a molecule of N₂O₄ acquires sufficient energy through molecular collisions, its N—N bond distorts and eventually fragments to produce two molecules of NO₂.

concentration of NO_2. Figure 16-2 summarizes the results of quantitative experiments on the NO_2/N_2O_4 system. The data show that the concentrations of both gases level off to constant values as the reaction reaches equilibrium. If the reaction to form N_2O_4 went to completion, the concentration of NO_2 would drop to zero. If the reaction to decompose N_2O_4 went to completion, the concentration of N_2O_4 would drop to zero. Instead, these reactions reach equilibrium when substantial amounts of both gases are present.

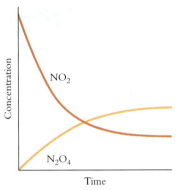

Figure 16-2
Changes in the concentrations of NO_2 and N_2O_4 with time. The system reaches equilibrium before either substance is fully consumed.

The Equilibrium Constant

When a mixture contains both NO_2 and N_2O_4, both the production reaction and the decomposition reaction occur. Each reaction has a characteristic rate constant, k_p for the formation reaction and k_d for the decomposition reaction:

$$\text{Production:} \qquad NO_2 + NO_2 \xrightarrow{k_p} N_2O_4$$

$$\text{Decomposition:} \qquad N_2O_4 \xrightarrow{k_d} NO_2 + NO_2$$

Decomposition is the reverse of production, so we can combine the two reactions in a single equation with a double arrow to show that the reaction proceeds in both directions:

$$NO_2 + NO_2 \underset{k_d}{\overset{k_p}{\rightleftharpoons}} N_2O_4$$

As this designation suggests, the reactions continue at equilibrium, even though the concentrations of NO_2 and N_2O_4 no longer change. That is, the rate of decomposition of N_2O_4 exactly balances the rate of its production, so no net change occurs. The rates of these elementary reactions can be expressed as described in Chapter 15:

$$\text{Rate of production of } N_2O_4 = k_p [NO_2]^2$$

$$\text{Rate of decomposition of } N_2O_4 = k_d [N_2O_4]$$

At equilibrium, the rate at which N_2O_4 decomposes equals the rate at which N_2O_4 forms. We designate equilibrium concentrations with subscript eq:

$$k_p [NO_2]^2_{eq} = k_d [N_2O_4]_{eq}$$

We rearrange this equality to group the rate constants on one side and the concentrations on the other, with product concentrations in the numerator. The concentration ratio is the **equilibrium constant (K_{eq})** for the reaction:

$$K_{eq} = \frac{k_p}{k_d} = \frac{[N_2O_4]_{eq}}{[NO_2]^2_{eq}}$$

The equilibrium constant expresses the relationship between the concentrations of products and reactants at equilibrium.

Reactions in aqueous solution behave in a similar way to those in the gas phase. As the concentration of products increases, the products react to regenerate reactants, and eventually the reaction reaches equilibrium. Example 16-1 treats an equilibrium in aqueous solution.

	Example 16-1

Molecular iodine dissolves in a solution containing iodide anions to form triiodide anions:

$$I_2 (aq) + I^- (aq) \longrightarrow I_3^- (aq)$$

Aqueous Equilibrium

Describe the additional reaction that occurs as this system approaches equilibrium, draw molecular pictures illustrating the reactions that occur at equilibrium, and write the equilibrium constant expression for this reaction.

Strategy: The problem asks for a qualitative analysis of a chemical equilibrium. We must visualize what takes place at the molecular level, describe the system in words, draw pictures that summarize the reactions, and then use the ideas developed for the NO_2/N_2O_4 reaction to write an expression for the equilibrium constant.

Example 16-1

Aqueous Equilibrium
(*continued*)

Solution: Initially, the forward reaction to form triiodide anions is the only process that occurs. As the concentration of the product increases, however, the reverse reaction also takes place:

$$I_3^-(aq) \longrightarrow I_2(aq) + I^-(aq)$$

We need two molecular pictures to illustrate the chemistry. One shows an I_2 molecule colliding with an I^- anion to form I_3^-. A second picture illustrates the reverse process:

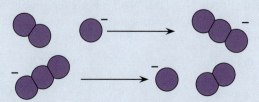

The equilibrium constant expression is determined from the rates for the forward and reverse reactions:

$$\text{Rate of production of } I_3^- = k_p[I_2][I^-]$$

$$\text{Rate of decomposition of } I_3^- = k_d[I_3^-]$$

Setting these rates equal to each other gives the equilibrium condition, which can be solved for the equilibrium constant expression:

$$I_2(aq) + I^-(aq) \rightleftharpoons I_3^-(aq)$$

$$k_p[I_2]_{eq}[I^-]_{eq} = k_d[I_3^-]_{eq} \qquad so \qquad K_{eq} = \frac{k_p}{k_d} = \frac{[I_3^-]_{eq}}{[I_2]_{eq}[I^-]_{eq}}$$

Do the Results Make Sense? From a molecular perspective, just as I_2 and I^- can combine to form I_3^-, it is reasonable that triiodide can break apart into the substances from which it originally formed.

**Extra Practice
Exercise 16.1**

In the gas phase, formic acid molecules (HCO_2H) form molecular pairs, called dimers, through hydrogen bonding. Describe the resulting equilibrium, including the reactions, a molecular picture, and the equilibrium constant expression.

Reversibility

In the NO_2/N_2O_4 system, molecules of NO_2 combine to give N_2O_4 molecules, and N_2O_4 molecules decompose to give NO_2 molecules. This is an example of the **reversibility** of molecular reactions. Look again at Figure 16-1: If two NO_2 molecules can form a bond when they collide, then that bond also can break apart when an N_2O_4 molecule distorts. The concept of reversibility is a general principle that applies to all molecular processes. Every *elementary* reaction that goes in the forward direction can also go in the reverse direction. As a consequence of reversibility, we can write each step in a chemical mechanism using a double arrow to describe what happens at chemical equilibrium.

To illustrate the generality of reversibility and the equilibrium expression, we extend our kinetic analysis to a chemical reaction that has a two-step mechanism. At elevated temperature NO_2 decomposes into NO and O_2 instead of forming N_2O_4. The mechanism for the decomposition reaction, which appears in Chapter 15, contains two steps:

$$NO_2 + NO_2 \xrightarrow{k_1} NO + NO_3$$

$$NO_3 \xrightarrow{k_2} NO + O_2$$

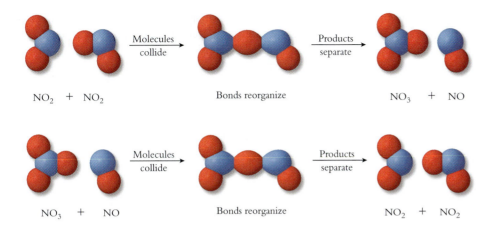

Figure 16-3
If two molecules of NO$_2$ can exchange an O atom when they collide, forming NO$_3$ and NO, then a collision of NO$_3$ and NO can also result in exchange of an O atom, forming two molecules of NO$_2$.

Just as the decomposition of N$_2$O$_4$ is reversible, the decomposition of NO$_3$ is also reversible. As Figure 16-3 shows, NO$_3$ and NO can collide and exchange an O atom. Furthermore, if an O$_2$ molecule and an NO molecule collide in the proper orientation, they can "stick together" to form NO$_3$:

$$NO + O_2 \xrightarrow{k_{-2}} NO_3$$

These reactions play negligible roles early in the decomposition of NO$_2$, when the concentration of NO is low. As the products accumulate, however, collisions involving NO become more likely. Eventually, product concentrations become large enough that the rates of the reverse reactions match the rates of the forward reactions. At this time the system has reached equilibrium, and each step in the mechanism has a forward rate that equals its reverse rate:

$$NO_2 + NO_2 \underset{k_{-1}}{\overset{k_1}{\rightleftharpoons}} NO + NO_3$$

$$NO_3 \underset{k_{-2}}{\overset{k_2}{\rightleftharpoons}} NO + O_2$$

$$k_1[NO_2]_{eq}^2 = k_{-1}[NO]_{eq}[NO_3]_{eq} \quad and \quad k_2[NO_3]_{eq} = k_{-2}[NO]_{eq}[O_2]_{eq}$$

Rearranging each of these rate equalities gives equilibrium expressions for both steps of the mechanism:

$$\frac{k_1}{k_{-1}} = \frac{[NO]_{eq}[NO_3]_{eq}}{[NO_2]_{eq}^2} \quad and \quad \frac{k_2}{k_{-2}} = \frac{[NO]_{eq}[O_2]_{eq}}{[NO_3]_{eq}}$$

If we multiply one of these equations by the other, the concentration of the intermediate, $[NO_3]_{eq}$, cancels to produce an equilibrium expression entirely in terms of concentrations of reactants and products:

$$\frac{k_1}{k_{-1}} \frac{k_2}{k_{-2}} = \frac{[NO]_{eq}\,\cancel{[NO_3]_{eq}}}{[NO_2]_{eq}^2}\frac{[NO]_{eq}[O_2]_{eq}}{\cancel{[NO_3]_{eq}}}$$

The two concentration terms containing NO can be combined, and the ratio of rate constants can be replaced by a single constant, the equilibrium constant for the reaction:

$$K_{eq} = \frac{[NO]_{eq}^2[O_2]_{eq}}{[NO_2]_{eq}^2}$$

Each equilibrium expression described so far contains a ratio of concentrations of products and reactants. Moreover, each concentration is raised to a power equal to its stoichiometric coefficient in the balanced equation for the overall reaction. Concentration ratios always have products in the numerator and reactants in the denominator:

$$2\,NO_2 \rightleftharpoons 2\,NO + O_2 \qquad 2\,NO_2 \rightleftharpoons N_2O_4$$

$$K_{eq} = \frac{[NO]_{eq}^2[O_2]_{eq}}{[NO_2]_{eq}^2} \qquad K_{eq} = \frac{[N_2O_4]_{eq}}{[NO_2]_{eq}^2}$$

It is possible to carry out this type of kinetic analysis whether a mechanism is simple or elaborate. That is, we can always derive the equilibrium expression for a reaction by applying reversibility and setting forward and reverse rates equal to one another at equilibrium. It is unnecessary to go through this procedure for every chemical equilibrium. As our two examples suggest, inspection of the overall stoichiometry *always* gives the correct expression for the equilibrium constant. That is, a reaction of the form

$$aA + bB \rightleftharpoons dD + eE$$

has an equilibrium constant expression given by Equation 16-1:

$$K_{eq} = \frac{[D]_{eq}^d \, [E]_{eq}^e}{[A]_{eq}^a \, [B]_{eq}^b} \qquad (16\text{-}1)$$

Equation 16-1 expresses the fact that the equilibrium expression for any reaction can be written from its overall stoichiometry. Example 16-2 applies Equation 16-1 to the Haber process.

Example 16-2	What is the equilibrium expression for the Haber synthesis of ammonia?
Equilibrium Constant Expression	$$N_2(g) + 3\,H_2(g) \rightleftharpoons 2\,NH_3(g)$$

Strategy: If the mechanism for this equilibrium were given, we could derive an equilibrium expression using the principle of reversibility, but Equation 16-1 shows that we can write the equilibrium expression directly without knowing details of the mechanism.

Solution: The product concentration, raised to a power equal to its stoichiometric coefficient, appears in the numerator. The concentrations of the two starting materials, each raised to a power equal to its stoichiometric coefficient, appear in the denominator:

$$K_{eq} = \frac{[NH_3]_{eq}^2}{[N_2]_{eq} \, [H_2]_{eq}^3}$$

Does the Result Make Sense? Compare the equilibrium constant expression with the equilibrium reaction to verify that products and reactants are in the right places, with the correct powers.

Extra Practice Exercise 16.2	Write the equilibrium constant expression for the conversion of molecular oxygen into ozone, $O_2 \rightarrow O_3$ (unbalanced).

Equilibrium constant expressions have some similarities with the rate laws described in Chapter 15. For example, compare the equilibrium constant expression and the rate law for the NO_2/NO reaction:

$$2\,NO_2 \longrightarrow 2\,NO + O_2 \qquad K_{eq} = \frac{[NO]_{eq}^2 \, [O_2]_{eq}}{[NO_2]_{eq}^2} \qquad \text{Rate} = k[NO_2]^2$$

Both relationships include a constant and both involve concentrations raised to exponential powers. However, a rate law and an equilibrium expression describe fundamentally different aspects of a chemical reaction. A *rate law* describes how the *rate of a reaction changes* with concentration. As we describe in this chapter, an *equilibrium expression* describes the concentrations of reactants and products when *the net rate of the reaction is zero*.

Section Exercises

■ **16.1.1** One possible mechanism for the O_2/O_3 equilibrium in the stratosphere is relatively simple:

$$O_3 \underset{k_{-1}}{\overset{k_1}{\rightleftarrows}} O + O_2$$

$$O + O_3 \underset{k_{-2}}{\overset{k_2}{\rightleftarrows}} 2\,O_2$$

Use this mechanism to derive the expression for the equilibrium constant.

■ **16.1.2** Draw molecular pictures like the ones shown in Figure 16-1 to illustrate all the reactions that occur when the O_3/O_2 system is at equilibrium.

■ **16.1.3** Write the equilibrium expression for the combustion of NH_3 during the synthesis of nitric acid:

$$4\,NH_3(g) + 5\,O_2(g) \rightleftarrows 4\,NO(g) + 6\,H_2O(g)$$

16.2 PROPERTIES OF EQUILIBRIUM CONSTANTS

The equilibrium constant expression given by Equation 16-1 is completely general and can be applied to any chemical reaction. Three features of equilibrium constants are especially important:

---

1. **K_{eq} is related to the stoichiometry of the balanced net reaction.**   **KEY CONCEPT**

The numerator contains only concentrations of products, and the denominator contains only concentrations of reactants. Each concentration is raised to a power equal to its stoichiometric coefficient.

---

2. **K_{eq} applies only at equilibrium.**   **KEY CONCEPT**

We use subscripts (eq) to emphasize that the concentrations of reactants and products used in the ratio must be concentrations at equilibrium.

---

3. **K_{eq} is independent of initial conditions.**   **KEY CONCEPT**

The equilibrium constant is a constant for any particular reaction at a given temperature. Whether initial conditions include pure reactants, pure products, or any composition in between, the system reaches a state of equilibrium that is determined by the value of K_{eq}.

Concentration Units and K_{eq}

Equilibrium constants are dimensionless numbers, yet the concentrations used in an equilibrium constant expression have units. To understand this, we need to explore the reaction quotient Q, introduced in Chapter 14. In Section 16.3 we explore in detail the link between Q and K_{eq}. Here we use Q to address the issue of concentration units and the equilibrium constant.

Recall that Equation 14-12 defines Q:

$$\left[\frac{(c_D)^d (c_E)^e}{(c_A)^a (c_B)^b} \right] = Q \qquad \text{(14-12)}$$

In Equation 14-12, concentrations are represented as c_X, whereas in Equation 16-1, concentrations are expressed as [X]. Either usage describes concentrations, so comparing Equations 16-1 and 14-12, we can see that both contain the same ratio. By either notation and in either equation, concentrations refer to molarity for species in aqueous solution and to bar for gases. Henceforth, to remind you of these conventions, we represent aqueous concentrations (molarities) as [X] and gas concentrations (bar) as p_X.

Recall that from the perspective of thermodynamics, every concentration is measured *relative to a defined standard concentration*. For any gas, the defined standard is 1 bar pressure; for any solute in aqueous solution, the defined standard is 1 M. We treat the concentration terms in equilibrium constant expressions in this same way.

Although not stated explicitly, each concentration in a reaction quotient and in an equilibrium constant expression has been divided by standard concentration (1 bar for gases, 1 M for solutes) to make the equilibrium constant dimensionless. For example,

$$(p_{N_2O_4})_{eq} = \frac{(p_{N_2O_4, \text{ units of bar}})_{eq}}{1 \text{ bar}} \quad \text{and} \quad (p_{NO_2})_{eq} = \frac{(p_{NO_2, \text{ units of bar}})_{eq}}{1 \text{ bar}}$$

When we include these references to standard concentrations, the equilibrium constant becomes a dimensionless number, as demonstrated for the N_2O_4/NO_2 equilibrium, for which experiments show that $K_{eq} = 3.10$ at 25 °C:

$$K_{eq} = \frac{\left(\dfrac{p_{N_2O_4}}{1 \text{ bar}}\right)_{eq}}{\left(\dfrac{p_{NO_2}}{1 \text{ bar}}\right)^2_{eq}} = 3.10$$

An equilibrium constant expression contains concentrations, each of which has been divided by the reference concentration. For convenience, we omit these reference concentrations when we write the expressions for Q or for K_{eq}. You should remember, however, that the implicit presence of reference concentrations means that K values are dimensionless but that concentrations must be in bar for gases and molarity for solutes.

Direction of a Reaction at Equilibrium

For a reaction at equilibrium, the rate of the forward reaction is balanced exactly by the rate of the reverse reaction. For this reason, any equilibrium reaction can be written in either direction. The equilibrium constant for the Haber synthesis of ammonia, for example, can be expressed in two ways:

$$N_2(g) + 3\,H_2(g) \rightleftharpoons 2\,NH_3(g) \qquad K_{eq,\,f} = \frac{(p_{NH_3})^2_{eq}}{(p_{N_2})_{eq}\,(p_{H_2})^3_{eq}}$$

$$2\,NH_3(g) \rightleftharpoons N_2(g) + 3\,H_2(g) \qquad K_{eq,\,r} = \frac{(p_{N_2})_{eq}\,(p_{H_2})^3_{eq}}{(p_{NH_3})^2_{eq}}$$

The two equilibrium constants are reciprocals of each other:

$$K_{eq,\,f} = \frac{1}{K_{eq,\,r}} \tag{16-2}$$

Equilibrium constant   Equilibrium constant
for forward reaction   for reverse reaction

The direction chosen for the equilibrium reaction is determined by convenience. A scientist interested in producing ammonia from N_2 and H_2 would use $K_{eq,f}$. On the other hand, someone studying the decomposition of ammonia on a metal surface would use $K_{eq,r}$. Either choice works as long as the *products* of the net reaction appear in the *numerator* of the equilibrium constant expression and the *reactants* appear in the *denominator*. Example 16-3 applies this reasoning to the iodine–triiodide reaction.

Example 16-3	Write the equilibrium constant expression for the decomposition equilibrium for the triiodide anion, $I_3^-(aq) \rightleftharpoons I_2(aq) + I^-(aq)$ and relate this expression to the equilibrium constant expression for production of triiodide anions.
Formation and Decomposition	**Strategy:** Equilibrium constant expressions are always written with the concentrations of products in the numerator and the concentrations of reactants in the denominator, and when a reaction is reversed, its equilibrium constant expression is inverted.

Solution: We can write the equilibrium constant expression by inspection of the decomposition equilibrium:

$$K_{eq}\text{ (decomposition)} = \frac{[I_2]_{eq}\,[I^-]_{eq}}{[I_3^-]_{eq}}$$

Inverting this expression gives the equilibrium constant expression for the production reaction (Example 16-1):

$$I_2(aq) + I^-(aq) \rightleftharpoons I_3^-(aq)$$

$$K_{eq}\text{ (production)} = \frac{1}{K_{eq}\text{ (decomposition)}} = \frac{[I_3^-]_{eq}}{[I_2]_{eq}\,[I^-]_{eq}}$$

Does the Result Make Sense? When we reverse the direction of a reaction, products become reactants and vice versa. Thus, the concentrations that were in the numerator of Q now are in the denominator, and vice versa.

At 25 °C, experiments give $K_{eq} = 3.10$ for the N_2O_4/NO_2 equilibrium. Write the equilibrium constant expression for the NO_2/N_2O_4 equilibrium and determine the value of this equilibrium constant at 25 °C.

Example 16-3

Formation and Decomposition (*continued*)

Extra Practice Exercise 16.3

Pure Liquids, Pure Solids, and Solvents

Chemical equilibria often involve pure liquids and solids in addition to gases and solutes. The concentration of a pure liquid or solid does not vary significantly. Figure 16-4 shows that although the amount of a solid or liquid can vary, the number of moles per unit volume remains fixed. In other words, the concentrations of pure liquids or solids are always equal to their standard concentrations. Thus, division by standard concentration results in a value of 1 for any pure liquid or solid. This allows us to omit pure liquids and solids from equilibrium constant expressions. For a general reaction:

$$aA + bB \rightleftharpoons dD + sS$$

where S is a pure solid or liquid:

$$K_{eq} = \frac{[D]_{eq}^d}{[A]_{eq}^a\,[B]_{eq}^b}$$

Water often is a reagent in an aqueous equilibrium. For example, when carbon dioxide dissolves in water, it reacts with a water molecule to form carbonic acid:

$$CO_2(g) + H_2O(l) \rightleftharpoons H_2CO_3(aq)$$

$$K_{eq} = \frac{[H_2CO_3]_{eq}}{(p_{CO_2})_{eq}\,[H_2O]_{eq}}$$

Here, carbonic acid concentration is expressed in molarity and carbon dioxide concentration in atmospheres. What units are appropriate for the concentration of water, which is neither a solute nor a pure liquid? The thermodynamic convention is to express the concentration of solvent as its mole fraction, X.

$$\text{Mole fraction of solvent} = X_{solvent} = \frac{n_{solvent}}{n_{total}}$$

The mole fraction of water varies slightly as solutes are added, but in an aqueous solution, the concentration of water is much greater than the

Figure 16-4
Although the amount of a liquid or solid can vary substantially, the concentration of substance in moles/L does not vary, as the highlighted regions illustrate.

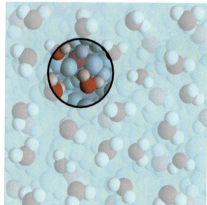
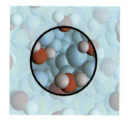
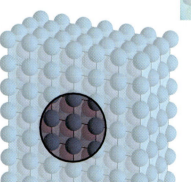

concentration of any solute. For instance, when CO_2 is bubbled through water at room temperature, the equilibrium concentration of H_2CO_3 is only 3.4×10^{-2} M. Water has a density of 1.000 g/mL, so 1 L of liquid contains 55.5 mol of H_2O. Compared to 55.5 mol, the 3.4×10^{-2} mol of carbonic acid in 1 L of this solution is negligible, $n_{total} \cong n_{solvent}$, and the mole fraction of water is negligibly different from 1.00.

Even in relatively concentrated solutions, the mole fraction of water remains close to 1.00. At a solute concentration of 0.50 M, for example, $X_{H_2O} = 0.99$, only 1% different from 1.00. Equilibrium calculations are seldom accurate to better than 5%, so this small deviation from 1.00 can be neglected. Consequently, we treat solvent water just like a pure substance: its concentration is essentially invariant, so it is omitted from the equilibrium constant expression.

Example 16-4 provides practice in manipulating equilibrium constant expressions.

Example 16-4 **Writing an Equilibrium Constant Expression**	Write the equilibrium constant expression for the reaction of iron metal with strong aqueous acid, and indicate the concentration units for each reagent: $$2\, Fe\,(s) + 6\, H_3O^+\,(aq) \rightleftharpoons 2\, Fe^{3+}\,(aq) + 6\, H_2O\,(l) + 3\, H_2\,(g)$$ **Strategy:** The stoichiometry of the reaction determines the form of the equilibrium constant expression. Pure solids, liquids, or solvents do not appear in the expression, since their concentrations are constant. **Solution:** In this reaction, Fe is a pure solid and H_2O is the solvent, so they do not appear in the equilibrium expression. The other reagents have exponents equal to their stoichiometric coefficients: $$K_{eq} = \frac{[Fe^{3+}]_{eq}^2\,(p_{H_2})_{eq}^3}{[H_3O^+]_{eq}^6}$$ The concentrations of the two ionic solutes are expressed as molarities, and the concentration of H_2 is expressed as a partial pressure in bar.

Does the Result Make Sense? The products whose concentrations can vary appear in the numerator, and the reactant whose concentration can vary appears in the denominator, consistent with Equation 16-1.

Extra Practice Exercise 16.4	Write the equilibrium constant expression for HCl gas dissolving in water to produce hydronium ions, and indicate the appropriate concentration units for each: $$HCl\,(g) + H_2O\,(l) \rightleftharpoons H_3O^+\,(aq) + Cl^-\,(aq)$$

Magnitudes of Equilibrium Constants

The magnitudes of equilibrium constants vary over a tremendous range and depend on the nature of the reaction as well as on the temperature of the system. Many reactions have very large equilibrium constants. For example, the reaction between H_2 and Br_2 to form HBr has a huge equilibrium constant:

$$H_2\,(g) + Br_2\,(g) \rightleftharpoons 2\, HBr\,(g) \qquad K_{eq} = \frac{(p_{HBr})_{eq}^2}{(p_{H_2})_{eq}\,(p_{Br_2})_{eq}} = 5.4 \times 10^{18}$$

The large value for this equilibrium constant indicates that the reaction goes virtually to completion. If the initial pressures are 1 bar each for H_2 and Br_2, then the pressure of HBr will be 2 bar when the system reaches equilibrium. The partial pressures of H_2 and Br_2 at equilibrium will be about 10^{-9} bar, which is negligible compared with the initial pressures.

Other reactions have extremely small equilibrium constants. For example, elemental fluorine, a diatomic molecule under standard conditions, is nevertheless at equilibrium with fluorine atoms:

$$F_2\,(g) \rightleftharpoons 2\, F\,(g) \qquad K_{eq} = \frac{(p_F)_{eq}^2}{(p_{F_2})_{eq}} = 2.1 \times 10^{-22}$$

The tiny value of this equilibrium constant indicates that a sample of fluorine at 25 °C consists almost entirely of F_2 molecules. If the partial pressure of F_2 is 1.0 bar at equilibrium, the partial pressure of fluorine atoms is 2.5×10^{-11} bar, which is negligible compared with 1.0 bar. Nevertheless, the equilibrium constant is not zero, indicating that some fluorine atoms are present in the gas.

Some equilibrium constants are neither large nor small. As already mentioned, the equilibrium constant for the formation of N_2O_4 from NO_2 at 25 °C is 3.10. This moderate value indicates that an equilibrium mixture of NO_2 and N_2O_4 at 25 °C contains measurable amounts of each molecule.

Section Exercises

16.2.1 Write expressions for the equilibrium constants for the following reactions:
(a) $PCl_5(s) \rightleftharpoons PCl_3(l) + Cl_2(g)$
(b) $CaCO_3(s) \rightleftharpoons CaO(s) + CO_2(g)$
(c) $Ca_3(PO_4)_2(s) \rightleftharpoons$
$\qquad 3\ Ca^{2+}(aq) + 2\ PO_4^{3-}(aq)$
(d) $HCN(g) + H_2O(l) \rightleftharpoons$
$\qquad H_3O^+(aq) + CN^-(aq)$
(e) $PCl_3(l) + Cl_2(g) \rightleftharpoons PCl_5(s)$
(f) $3\ Ca^{2+}(aq) + 2\ PO_4^{3-}(aq) \rightleftharpoons$
$\qquad Ca_3(PO_4)_2(s)$

16.2.2 Draw a molecular picture illustrating the equilibrium in Exercise 16.2.1(d).

16.2.3 Write the equilibrium constant expression and determine the value of K_{eq} for each of these reactions, which are related to reactions described in this section:

$$2\ F(g) \rightleftharpoons F_2(g)$$

$$2\ HBr(g) \rightleftharpoons H_2(g) + Br_2(g)$$

16.3 THERMODYNAMICS AND EQUILIBRIUM

Why do some reactions go virtually to completion, whereas others reach equilibrium when hardly any of the starting materials have been consumed? At the molecular level, bond energies and molecular organization are the determining factors. These features correlate with the thermodynamic state functions of enthalpy and entropy. As discussed in Chapter 14, free energy (G) is the state function that combines these properties. This section establishes the connection between thermodynamics and equilibrium.

Free Energy and the Equilibrium Constant

Recall from Chapter 14 that the free energy change for a chemical process, ΔG, is a signpost for spontaneity. Equation 14-13 relates ΔG to concentrations through the reaction quotient Q:

$$\Delta G_{reaction} = \Delta G^\circ_{reaction} + RT \ln Q \qquad \text{(14-13)}$$

A negative value for ΔG indicates that a process is spontaneous in the direction written, provided the system remains at constant temperature and pressure.

Consider as an example the reaction of NO_2 to produce N_2O_4:

$$2\ NO_2 \longrightarrow N_2O_4 \qquad Q_{production} = \frac{p_{N_2O_4}}{(p_{NO_2})^2}$$

In a flask that contains only NO_2, $p_{N_2O_4} = 0$ and $Q_{production} = 0$. Because $\ln 0 = -\infty$, under these conditions the reaction that produces N_2O_4 has a negative value for ΔG. In other words, the spontaneous reaction is the production of N_2O_4:

$$2\ NO_2 \longrightarrow N_2O_4 \qquad \Delta G_{production} < 0 \qquad when \qquad p_{N_2O_4} = 0$$

Now consider what happens to Q and ΔG as the reaction proceeds. An increase in the amount of N_2O_4 causes an increase in the numerator of Q, while a decrease in the amount of NO_2 causes a decrease in the denominator of Q. Both of these changes

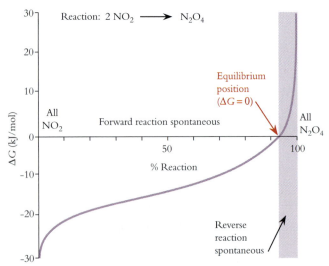

Figure 16-5
Variation in ΔG with Q. At high NO$_2$ concentration (*left*), the spontaneous reaction is 2 NO$_2$ → N$_2$O$_4$; at high N$_2$O$_4$ concentration (*right*), the spontaneous reaction is N$_2$O$_4$ → 2 NO$_2$. As the spontaneous reaction proceeds, Q and ΔG change until the equilibrium condition is reached and ΔG = 0.

increase the overall value of Q, making ΔG progressively less negative. Eventually, Q becomes large enough to make $\Delta G_{\text{production}} = 0$. At this point the reaction is at equilibrium, and $Q = K_{\text{eq}}$, because each concentration in the expression for Q is an equilibrium concentration.

Starting from the other end of the reaction gives an analogous result. If Q is large and positive, so is $\ln Q$, and ΔG has a positive value. In molecular terms, the reaction that forms N$_2$O$_4$ cannot proceed when there is no NO$_2$ present. Instead, the decomposition reaction of N$_2$O$_4$ has a negative value for ΔG, so the decomposition reaction proceeds to form NO$_2$:

$$\text{N}_2\text{O}_4 \longrightarrow 2\,\text{NO}_2 \quad Q_{\text{decomposition}} = \frac{(p_{\text{NO}_2})^2}{p_{\text{N}_2\text{O}_4}}$$

$$\Delta G_{\text{decomposition}} < 0 \quad when \quad p_{\text{NO}_2} = 0$$

Decomposition of N$_2$O$_4$ causes Q to increase; ΔG becomes less negative, and eventually the reaction quotient reaches a value that makes $\Delta G = 0$.

Figure 16-5 shows schematically how ΔG for the NO$_2$/N$_2$O$_4$ system varies with Q. The formation of N$_2$O$_4$ is spontaneous, and the reaction proceeds to the right, if the relative amount of NO$_2$ is large. This causes $\Delta G_{\text{production}}$ to become less negative until $\Delta G_{\text{production}} = 0$ and the equilibrium position is reached. Decomposition of N$_2$O$_4$ is spontaneous, and reaction proceeds to the left, if the relative amount of N$_2$O$_4$ is large. This causes $\Delta G_{\text{decomposition}}$ to become less negative until $\Delta G_{\text{decomposition}} = 0$ when the equilibrium position is reached. To summarize, regardless of the initial concentrations in a reaction system, the system reacts in the direction for which ΔG is negative. This makes ΔG less negative, and eventually $\Delta G = 0$ when the system has reached chemical equilibrium.

As this example indicates, the direction in which a reaction proceeds depends on the relationship between the magnitudes of Q and K_{eq}.

KEY CONCEPTS

When $Q < K_{\text{eq}}$, the reaction goes to the right to make products.

When $Q = K_{\text{eq}}$, the reaction is at equilibrium and there is no net change.

When $Q > K_{\text{eq}}$, the reaction goes to the left to make reactants.

Notice that if the concentration of any *product* is zero, $Q = 0$, and the direction of the reaction must be to the *right,* toward products. Conversely, if any *reactant* concentration is zero, $Q = \infty$, and the direction of the reaction must be to the *left,* toward reactants.

The relationship between Q and K_{eq} signals the direction of a chemical reaction. The free energy change, ΔG, also signals the direction of a chemical reaction. These two criteria can be compared:

Reaction Goes Right when	Equilibrium when	Reaction Goes Left when
$Q < K_{\text{eq}}$	$Q = K_{\text{eq}}$	$Q > K_{\text{eq}}$
$\Delta G_{\text{reaction}} < 0$	$\Delta G_{\text{reaction}} = 0$	$\Delta G_{\text{reaction}} > 0$

Calculating K_{eq} from $\Delta G°$

The relationship between the equilibrium constant and free energy provides a connection between thermodynamics and equilibrium constants. At equilibrium, $\Delta G = 0$ and $Q = K_{\text{eq}}$. We can substitute these equalities into Equation 14-13:

$$\Delta G = \Delta G° + RT \ln Q$$

$$0 = \Delta G° + RT \ln K_{\text{eq}}$$

Rearrange to relate K_{eq} to the standard free energy change:

$$\Delta G° = - RT \ln K_{eq} \qquad (16\text{-}3)$$

Equation 16-3 is extremely important because it links thermodynamic data with equilibrium constants. Recall that Equation 14-9 makes it possible to use tabulated values for $\Delta G_f°$ to calculate the value for $\Delta G°$ for many reactions:

③

$$\Delta G°_{reaction} = \Sigma \text{ coeff}_{products} \Delta G°_{f \, products} - \Sigma \text{ coeff}_{reactants} \Delta G°_{f \, reactants} \qquad (14\text{-}9)$$

Then K_{eq} can be calculated from Equation 16-3. Example 16-5 does this for the Haber process.

Use standard thermodynamic data to calculate the value of K_{eq} at 298 K for the Haber reaction:

Example 16-5

K_{eq} **from Thermodynamics**

$$N_2(g) + 3 H_2(g) \rightleftharpoons 2 NH_3(g)$$

Strategy: To show the logic of this kind of calculation, we apply the seven-step approach to problem solving. We are asked to calculate K_{eq} at 298 K for the Haber reaction. We visualize the process by remembering that there is a connection between free energy, $\Delta G°_{reaction}$, and equilibrium, K_{eq}. The problem provides only the balanced equation and the temperature. Any other necessary data will be found in tables and appendices.

Solution: Appendix D contains the appropriate values for $\Delta G_f°$, which we need for the calculation of $\Delta G°_{reaction}$:

$$\Delta G_f°(kJ/mol): \qquad N_2(g): 0 \qquad H_2(g): 0 \qquad NH_3(g): -16.4 \text{ kJ/mol}$$

We need two equations. We calculate the free energy change for the reaction using Equation 14-9, and Equation 16-3 links free energy with the equilibrium constant.

$$\Delta G°_{reaction} = \Sigma \text{ coeff}_{products} \Delta G°_{f \, products} - \Sigma \text{ coeff}_{reactants} \Delta G°_{f \, reactants} \qquad \Delta G° = -RT \ln K_{eq}$$

Rearrange Equation 16-3 to isolate $\ln K_{eq}$:

$$\ln K_{eq} = \frac{-\Delta G°_{reaction}}{RT}$$

Now substitute and calculate $\Delta G°_{reaction}$ and K_{eq}:

$$\Delta G°_{reaction} = (2 \text{ mol } NH_3)(-16.4 \text{ kJ/mol } NH_3) - 3(0) - 1(0) = -32.8 \text{ kJ}$$

$$\ln K_{eq} = \frac{-(-32.8 \text{ kJ})(10^3 \text{ J/kJ})}{(8.314 \text{ J/K})(298 \text{ K})} = 13.24 \qquad K_{eq} = e^{13.24} = 5.6 \times 10^5$$

Does the Result Make Sense? When evaluating exponents of e, the number of significant figures generally equals the number of decimal places in the ln value. In this case, whereas $e^{13.24} = 5.62 \times 10^5$, $e^{13.25} = 5.68 \times 10^5$ and $e^{13.23} = 5.57 \times 10^5$. The value of K_{eq} is relatively large but reasonable, indicating that at room temperature, the equilibrium position for the Haber reaction favors the product, NH_3.

Use data from Appendix D to determine the equilibrium constant for formation of ozone at 298 K.

Extra Practice Exercise 16.5

Equilibrium Constants and Temperature

Experimental studies on how temperature affects equilibria reveal a consistent pattern. The equilibrium constant of any *exothermic* reaction *decreases* with increasing temperature, whereas the equilibrium constant of any *endothermic* reaction *increases* with increasing temperature. We can use two equations for $\Delta G°$, Equations 16-3 and 14-10, to provide a thermodynamic explanation for this behavior:

$$\Delta G° = -RT \ln K_{eq}$$

$$\Delta G° = \Delta H° - T\Delta S° \qquad (14\text{-}10)$$

We set these two expressions for $\Delta G°$ equal to each other:

$$\Delta H° - T\Delta S° = -RT \ln K_{eq}$$

Solving for $\ln K_{eq}$ gives an equation relating K_{eq} to standard enthalpy and entropy changes:

$$\ln K_{eq} = -\frac{\Delta H°}{RT} + \frac{\Delta S°}{R} \qquad (16\text{-}4)$$

An exothermic reaction has a negative $\Delta H°$, making the first term on the right in Equation 16-4 positive. As T increases, this term decreases, causing K_{eq} to decrease. An endothermic reaction, in contrast, has a positive $\Delta H°$, making the first term on the right in Equation 16-4 negative. As T increases, this term becomes less negative, causing K_{eq} to increase. These variations in K_{eq} with temperature, which often are substantial, can be estimated using Equation 16-4. Example 16-6 applies Equation 16-4 to the Haber synthesis.

Example 16-6
K_{eq} and Temperature

Use tabulated thermodynamic data (see Appendix D) to estimate K_{eq} for the Haber reaction at 500 °C.

Strategy: Again, we follow the seven-step procedure. This example is similar to Example 16-5, but note we are asked to find K_{eq} at a temperature that is different from 298 K. Because the temperature is not 298 K, we cannot use the value of $\Delta G°_{reaction}$ at 298 K. Instead, the process is as follows:

a. Calculate values for $\Delta H°_{reaction}$ and $\Delta S°_{reaction}$ at 298 K using tabulated thermodynamic values and equations from earlier chapters.

$$\Delta H°_{reaction} = \Sigma\ coeff_p\ \Delta H°_f(products) - \Sigma\ coeff_r\ \Delta H°_f(reactants) \qquad (6\text{-}12)$$

$$\Delta S°_{reaction} = \Sigma\ coeff_p\ S°(products) - \Sigma\ coeff_r\ S°(reactants) \qquad (14\text{-}6)$$

b. Use Equation 16-4 to determine the value of the equilibrium constant at 500 °C.

$$\ln K_{eq} = -\frac{\Delta H°}{RT} + \frac{\Delta S°}{R}$$

Solution: Appendix D contains the appropriate values:

Substance	$N_2(g)$	$H_2(g)$	$NH_3(g)$
$\Delta H°_f$ (kJ/mol)	0	0	−45.9
$S°$ (J/mol K)	191.61	130.680	192.8

First, determine $\Delta H°_{reaction}$ and $\Delta S°_{reaction}$:

$$\Delta H°_{reaction} = (2)(-45.9\ kJ/mol) - 3(0) - 0 = -91.8\ kJ/mol$$

$$\Delta S°_{reaction} = (2)(192.8\ J/mol\ K) - (3)(130.680\ J/mol\ K) - 191.61\ J/mol\ K$$

$$\Delta S°_{reaction} = -198.1\ J/mol\ K$$

Now calculate K_{eq} at 500 °C:

$$\ln K_{eq} = -\frac{(-91.8\ kJ/mol)(10^3\ J/kJ)}{(8.314\ J/mol\ K)(500 + 273\ K)} + \frac{-198.1\ J/mol\ K}{8.314\ J/mol\ K}$$

$$\ln K_{eq} = 14.28 - 23.83 = -9.55$$

Example 16-6

Taking the antilogarithm, or e^x, of -9.55 gives the estimated equilibrium constant at 500 °C:

$$K_{eq} = 7.1 \times 10^{-5}$$

K_{eq} **and Temperature**
(*continued*)

Does the Result Make Sense? As in Example 16-5, the result has only two significant figures because of the sensitivity of powers of e to small variations. We see that at this temperature, the equilibrium constant has a small value, indicating that the reactants are favored. This is consistent with the observation that the Haber reaction has only a 13% yield at elevated temperature.

Estimate the equilibrium constant for the ozone formation reaction at 825 °C.

**Extra Practice
Exercise 16.6**

Examples 16-5 and 16-6 underscore a dilemma faced by industrial chemists and engineers. Example 16-5 shows that at 298 K the equilibrium position of the Haber reaction strongly favors the formation of ammonia. Why then is the Haber synthesis not carried out at 298 K? The reason is that even with a catalyst, the reaction is much too slow to be useful at this temperature. Thermodynamics favors the product, but the kinetics of the reaction are unfavorable. For the reaction to proceed at a practical rate, the temperature must be increased. This improves the kinetics of the process but, unfortunately, the equilibrium constant falls dramatically as temperature increases. Example 16-6 shows that at 773 K, a realistic temperature for the Haber reaction, the equilibrium position does not favor NH_3:

$$N_2 + 3\,H_2 \rightleftharpoons 2\,NH_3$$

$$K_{eq,\,298\,K} = 5.6 \times 10^5 \qquad K_{eq,\,773\,K} = 7.1 \times 10^{-5}$$

As we describe in the Chemistry and Life Box in Chapter 14, chemists are searching for a catalyst that will allow this essential reaction to be carried out close to room temperature.

The Haber reaction is a practical example of the effect of temperature on an *exothermic* reaction. A practical example of an *endothermic* reaction is the use of methane and steam to produce the molecular hydrogen needed for the Haber reaction. The production of hydrogen in this way is highly endothermic, so the reaction is carried out at temperatures much greater than 1000 K to force the equilibrium toward the products:

Ammonia for fertilizer is produced in reaction towers at elevated temperature and pressure.

$$CH_4\,(g) + H_2O\,(g) \rightleftharpoons CO\,(g) + 3\,H_2\,(g)$$

$$\Delta H° = +206 \text{ kJ/mol} \qquad K_{eq,\,298\,K} = 1.3 \times 10^{-25} \qquad K_{eq,\,1500\,K} = 1.1 \times 10^4$$

You should be able to do the calculations to verify the value of the equilibrium constant at 1500 K.

Our Chemistry and Technology Box on the next page gives more information about the role of equilibrium in the chemical industry.

Section Exercises

16.3.1 Use thermodynamic data (see Appendix D) to determine K_{eq} for the oxidation of ClO at 298 K:

$$2\,ClO\,(g) + O_2\,(g) \rightleftharpoons 2\,ClO_2\,(g)$$

16.3.2 Does K_{eq} for the reaction described in Section Exercise 16.3.1 increase or decrease when the temperature is raised above room temperature? State your reasoning.

16.3.3 Use thermodynamic data to estimate K_{eq} at 1000 °C for the reaction in Section Exercise 16.3.1.

Box 16-1	Chemistry and Technology: Industrial Equilibria

Reactions that reach equilibrium well before completion are not very attractive to the industrial chemist, because unreacted starting materials must be separated from the products, an expensive process. Nevertheless, gas-phase equilibria are involved in the industrial syntheses of H_2SO_4, NH_3, and HNO_3.

Sulfuric acid is produced from elemental sulfur, which is burned in air to give SO_2. This gas reacts with additional O_2 to produce SO_3. The reaction requires a catalyst such as V_2O_5 and a temperature around 700 K. Even so, this reaction reaches equilibrium well before completion.

$$2\,SO_2(g) + O_2(g) \xrightleftharpoons{V_2O_5,\,700\,K} 2\,SO_3(g)$$

To form sulfuric acid, SO_3 is added to H_2SO_4, and the resulting solution is treated with water:

$$SO_3(g) + H_2SO_4(l) \longrightarrow H_2S_2O_7(l)$$

$$H_2S_2O_7(l) + H_2O(l) \longrightarrow 2\,H_2SO_4(l)$$

The Haber synthesis for the conversion of N_2 into NH_3 is outlined below. The process uses N_2 from the atmosphere, but H_2 must be generated from natural gas (methane) and steam:

$$CH_4(g) + H_2O(g) \xrightleftharpoons{Ni,\,800\,K} CO(g) + 3\,H_2(g)$$

The equilibrium constant for this reaction is small at 800 K, but the CO is used to make more hydrogen:

$$CO(g) + H_2O(g) \xrightleftharpoons{Fe_2O_3,\,Cr_2O_3,\,525\,K} CO_2(g) + H_2(g)$$

This reaction is run at 525 K, a temperature at which it has a large equilibrium constant.

The Haber reaction must be catalyzed, and even then it does not approach completion:

$$N_2(g) + 3\,H_2(g) \xrightleftharpoons{Fe,\,K_2O,\,Al_2O_3,\,725\,K} 2\,NH_3(g)$$

Recycling increases the yield. The product gas is chilled below $-10\ °C$, causing ammonia to liquefy. The remaining gases are recycled back into the reactor to make more ammonia.

Much of the ammonia synthesized by the Haber process is used to make HNO_3. Ammonia and air are heated and passed over Pt gauze:

$$4\,NH_3(g) + 5\,O_2(g) \xrightarrow{Pt\ gauze,\,1200\,K} 4\,NO(g) + 6\,H_2O(g)$$

The combustion reaction goes nearly to completion, but the further reaction of NO and O_2 does not. As it leaves the reactor, NO is at equilibrium with NO_2:

$$2\,NO(g) + O_2(g) \xrightleftharpoons{} 2\,NO_2(g)$$

Because this reaction is exothermic, lowering the temperature favors nitrogen dioxide, so as the products are cooled, more NO is converted into NO_2.

Finally, NO_2 is bubbled through water to give a concentrated aqueous solution of nitric acid, HNO_3:

$$3\,NO_2(g) + H_2O(l) \longrightarrow 2\,HNO_3(aq) + NO(g)$$

Because of the expense that accompanies inefficient syntheses, most industrial processes have been designed using reactions that go virtually to completion.

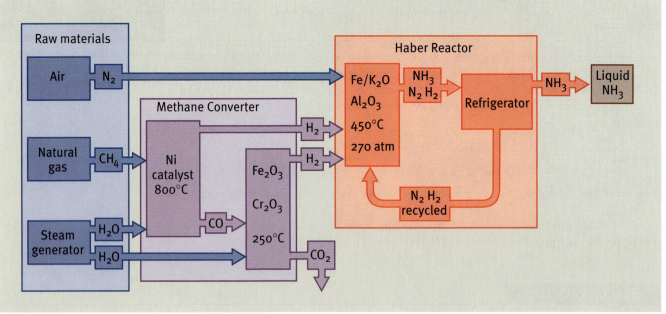

16.4 SHIFTS IN EQUILIBRIUM

What happens when the conditions change for a system that is already at equilibrium? Suppose additional amounts of one or more substances are introduced or the temperature of the system changes. How does a system that had been at equilibrium respond to these changes?

Le Châtelier's Principle

A quick *qualitative* indication of how a system at equilibrium responds to a change in conditions can be obtained using **Le Châtelier's principle,** which was first formulated in 1884 by Henri-Louis Le Châtelier, a French industrial chemist.

> **KEY CONCEPT**
>
> *When a change is imposed on a system at equilibrium, the system will react in the direction that reduces the amount of change.*

Le Châtelier's principle is a compact summary of how different factors influence equilibrium. Introducing a reagent causes a reaction to proceed in the direction that consumes the reagent. Reducing the temperature removes heat from the system and causes the reaction to produce heat by proceeding in the exothermic direction.

Changes in Amounts of Reagents

A change in the amount of any substance that appears in the reaction quotient displaces the system from its equilibrium position. As an example, consider an industrial reactor containing a mixture of methane, hydrogen, steam, and carbon monoxide at equilibrium:

$$CH_4(g) + H_2O(g) \rightleftharpoons CO(g) + 3\,H_2(g) \qquad Q = \frac{(p_{CO})(p_{H_2})^3}{(p_{CH_4})(p_{H_2O})}$$

How does this system respond if more steam is injected into the reactor? Adding one of the starting materials decreases the value of Q, making $Q < K_{eq}$. Because the system responds in a way that restores equilibrium, adding steam causes the reaction to proceed to the right, consuming CH_4 and H_2O and producing CO and H_2 until equilibrium is restored. Figure 16-6 provides a graphical illustration of how concentrations change when steam is added.

Removing a product from a system at equilibrium also makes $Q < K_{eq}$ and leads to the formation of additional products. This behavior is commonly used to advantage in chemical synthesis. For example, calcium oxide (lime), an important material in the construction industry, is made by heating calcium carbonate in a furnace to about 1100 K:

$$CaCO_3(s) \rightleftharpoons CaO(s) + CO_2(g) \qquad K_{eq} = (p_{CO_2})_{eq} = 1.0 \text{ at } 1100 \text{ K}$$

If the pressure of CO_2 in the furnace were to reach 1.0 bar, the system would attain equilibrium, and no additional products would form. If the CO_2 is allowed to escape from the reactor as it forms, on the other hand, the continuous removal of CO_2 drives the reaction to completion. Figure 16-7 illustrates this from a molecular perspective.

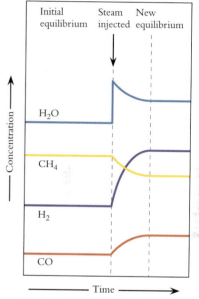

Figure 16-6
Adding some steam to a reactor that is at chemical equilibrium changes the value of Q, so the reaction is no longer at equilibrium. The reaction proceeds in the direction that consumes some of the added reagent in order to reestablish equilibrium.

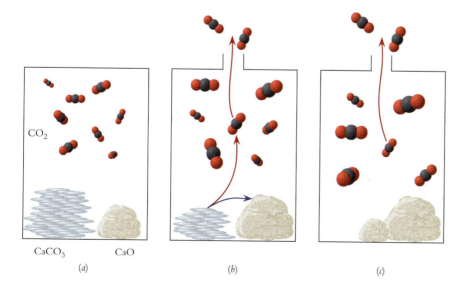

Figure 16-7
(a) If CO_2 cannot escape, the decomposition reaction of $CaCO_3$ quickly establishes equilibrium. (b) Removing the CO_2 gas from a limestone kiln maintains Q at a value below the equilibrium value, allowing the reaction to continue. (c) As long as CO_2 gas can escape, the reaction will proceed to completion.

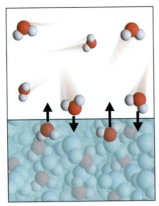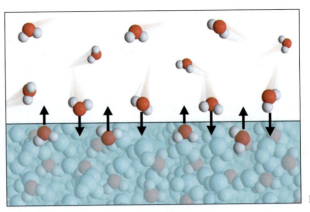

Gas phase

Liquid phase

Small puddle

Large pond

Figure 16-8
The equilibrium pressure (vapor pressure) above a liquid is independent of the amount of liquid, because the ratio of evaporating to condensing molecules is independent of amount.

Continuous removal of a product maintains the pressure of CO_2 below 1.0 bar, so Q has a smaller value than K_{eq}, and the reaction continues until all the $CaCO_3$ has been converted to CaO.

A change in the amount of a chemical species that has no impact on the reaction quotient will not disturb a chemical equilibrium. Thus, adding or removing air has no effect on the equilibrium conditions of the $CaO/CaCO_3/CO_2$ system, because the equilibrium constant depends only on the partial pressure of CO_2, not on the total pressure of gas in the system. Likewise, adding more $CaCO_3$ to a limestone reactor does not result in a higher equilibrium pressure of CO_2, because the concentration of pure solid $CaCO_3$ has a fixed value that is independent of how much solid is present.

Vapor pressure provides a simple illustration of why adding a pure liquid or solid does not change equilibrium concentrations. Recall from Chapter 11 that any liquid establishes a dynamic equilibrium with its vapor, and the partial pressure of the vapor at equilibrium is the vapor pressure. The vapor pressure is independent of the amount of liquid present. Figure 16-8 illustrates that the vapor pressure of water above a small puddle is the same as the vapor pressure above a large pond at the same temperature. More molecules escape from the larger surface of the pond, but more molecules are captured, too. The balance between captures and escapes is the same for both puddle and pond.

The effects of changes in amounts on a system at equilibrium can be summarized in accordance with Le Châtelier's principle:

1. Any change in conditions that *increases* the value of Q causes the reaction to consume products and produce reactants until equilibrium is reestablished.

2. Any change in conditions that *decreases* the value of Q causes the reaction to form products and consume reactants until equilibrium is reestablished.

3. Any change in amounts that has no effect on the value of Q has no effect on the equilibrium position.

Example 16-7 provides practice in applying Le Châtelier's principle at the molecular level.

Example 16-7 **Effects of Concentration Changes**	The left-hand view in the accompanying figure is a molecular picture showing a very small portion of an iodine–triiodide solution at equilibrium. The right-hand view shows this same solution, no longer at equilibrium because additional I_2 has been dissolved in the solution. Redraw this molecular picture to show qualitatively how the system responds to this change.

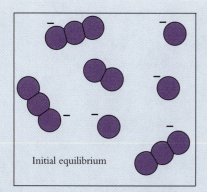

Initial equilibrium

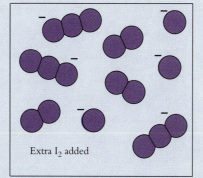

Extra I$_2$ added

Example 16-7

**Effects
of Concentration Changes**
(*continued*)

Strategy: Because this problem asks for a qualitative answer, we do not need to do calculations. It is sufficient to apply Le Châtelier's principle to determine the direction of change and draw the new picture that shows the result of the change.

Solution: The change that has been imposed is addition of a reactant, I$_2$. The system responds by moving in the direction that reduces the concentration of this added reactant. We show this by combining one I$_2$ molecule with an I$^-$ anion to make one more I$_3{}^-$ anion.

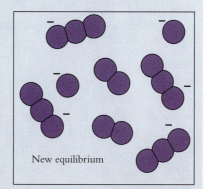

New equilibrium

Does the Result Make Sense? Because this is a qualitative question, we need only make sure that the changes are stoichiometrically correct and in the right direction, consuming added I$_2$. The new view contains one less I$_2$ molecule, one less I$^-$ anion, and one more molecule of the I$_3{}^-$ product.

Suppose that the right-hand view represents equilibrium and the left-hand view represents a displacement from equilibrium. Draw a molecular picture showing how the left-hand view would react to re-establish equilibrium.

**Extra Practice
Exercise 16.7**

Effect of Catalysts

Catalysts *do not* affect the equilibrium constant. Figure 16-9 provides a reminder that a catalyst changes the mechanism of the reaction in a way that reduces the net activation energy barrier, but it does not alter the thermodynamic changes that accompany the reaction. In other words, a catalyst reduces the forward activation energy and the reverse activation energy, so the rate of reaction is increased in *both* directions. However, the catalyst does not affect the nature of the reactants or the products, so the standard free energy change, $\Delta G°$, does not change, and neither does K_{eq}. A catalyst changes the kinetics of a reaction but not its thermodynamics. Catalysis allows a reaction to reach equilibrium *more rapidly,* but it does not alter the equilibrium *position.*

**Figure 16-9
A catalyst changes the
mechanism of a reaction and
lowers the net activation energy
barrier in both directions. It has
no effect on the overall $\Delta G°$ and
K_{eq} for the reaction.**

Effect of Temperature

As shown in Section 16.3, K_{eq} varies with temperature in a way that can be understood using the principles of thermodynamics. Temperature is the *only* variable that causes a change in the value of K_{eq}. The effect of temperature on K_{eq} depends on the enthalpy change of the reaction, ΔH. An *increase* in temperature always shifts the equilibrium position in the *endothermic* direction, and a *decrease* in temperature always shifts the equilibrium position in the *exothermic* direction.

Example 16-8 provides practice in analyzing changes that may lead to shifts in equilibrium.

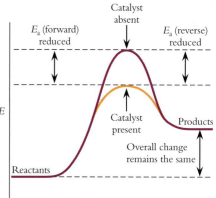

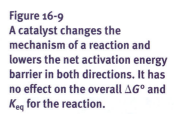

Example 16-8

Shifts in Equilibrium

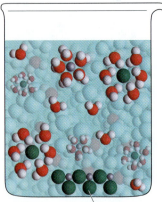

Solid CaCl₂

Consider a saturated solution of $CaCl_2$ at equilibrium with excess $CaCl_2(s)$:

$$CaCl_2(s) \rightleftharpoons Ca^{2+}(aq) + 2\,Cl^-(aq) \qquad \Delta H^\circ = +585 \text{ kJ}$$

How do the following changes affect the amount of dissolved $CaCl_2$?

a. More $CaCl_2(s)$ is added.

b. Some NaCl is dissolved in the solution.

c. Some $NaNO_3$ is dissolved in the solution.

d. Some pure water is added.

e. The solution is heated.

Strategy: According to Le Châtelier's principle, the system will respond in the direction that reduces the amount of change. It will do so, however, only if an appropriate response exists. Changes in quantities of substances must be analyzed for their effect, if any, on the value of Q.

Solution:

a. More $CaCl_2(s)$ is added. Because $Q = [Ca^{2+}][Cl^-]^2$, adding $CaCl_2(s)$ does *not* change Q. There is no effect on the amount of dissolved $CaCl_2$.

b. Some NaCl is dissolved in the solution. This change increases $[Cl^-]$, making $Q > K_{eq}$. The reaction proceeds to the left, and some $CaCl_2$ precipitates.

c. Some $NaNO_3$ is dissolved in the solution. There is no effect on Q, so there is no effect on the amount of dissolved $CaCl_2$.

d. Some pure water is added. Adding water dilutes the solution and lowers $[Ca^{2+}]$ and $[Cl^-]$. Because Q is now less than K_{eq}, the reaction proceeds to the right. More $CaCl_2$ dissolves.

e. The solution is heated. Because the solubility reaction is endothermic (ΔH° is positive), K_{eq} increases as T increases. More $CaCl_2$ dissolves.

Do the Results Make Sense? The reasoning behind each part of the answer matches le Châtelier's principle.

Extra Practice Exercise 16.8

The formation of triiodide from iodide and molecular iodine (Example 16-1) is exothermic. Prepare a list of changes that will lead to an increase in the concentration of triiodide present at equilibrium.

Section Exercises

■ **16.4.1** The following solubility reaction is exothermic by about 70 kJ/mol:

$$Ba(OH)_2(s) \rightleftharpoons Ba^{2+}(aq) + 2\,OH^-(aq)$$

Will more $Ba(OH)_2$ dissolve, or will some precipitate, or will no change occur, when each of the following changes is made on a saturated solution of $Ba(OH)_2$? (a) More $Ba(OH)_2(s)$ is added to the solution; (b) more water is added to the solution; (c) some HCl is added to the solution; and (d) the solution is cooled on an ice bath.

■ **16.4.2** Refer to Examples 16-5 and 16-6. List four changes in conditions that might be used to increase the yield of ammonia in the Haber process.

■ **16.4.3** Chemists are optimistic that a catalyst will be found for the production of ammonia from hydrogen and nitrogen under standard conditions. In contrast, no hope exists of developing a catalyst for the production of hydrogen from methane and steam under standard conditions. Explain.

16.5 WORKING WITH EQUILIBRIA

The quantitative treatment of a reaction equilibrium usually involves one of two things. Either the equilibrium constant must be computed from a knowledge of concentrations, or equilibrium concentrations must be determined from a knowledge of initial conditions and K_{eq}. In this section, we describe the basic reasoning and techniques needed to solve equilibrium problems. Stoichiometry plays a major role in equilibrium calculations, so you may want to review the techniques described in Chapter 4, particularly Section 4.4 on limiting reactants.

Chemistry of Equilibria

Equilibrium conditions are determined by the chemical reactions that occur in a system. Consequently, it is necessary to analyze the chemistry of the system before doing *any* calculations. After the chemistry is known, a mathematical solution to the problem can be developed. We can modify the seven-step approach to problem solving so that it applies specifically to equilibrium problems, proceeding from the chemistry to the equilibrium constant expression to the mathematical solution.

> **SOLVING EQUILIBRIUM PROBLEMS**
>
> **Step 1** Determine what is asked for.
>
> **Step 2** Identify the major chemical species.
>
> **Step 3** Determine what chemical equilibria exist.
>
> **Step 4** Write the K_{eq} expressions.
>
> **Step 5** Organize the data and unknowns.
>
> **Step 6** Carry out the calculations.
>
> **Step 7** Does the result make sense?

We illustrate this approach using the equilibrium shown in Figure 16-10. When solid LiF is added to water, a small amount of the salt dissolves, leading to equilibrium between the solid and a solution. Chemical analysis reveals that the equilibrium concentration of F^- ions in the solution is 6.16×10^{-2} M. We want to determine the equilibrium constant for this process.

1. Determine what is asked for.
This is the usual first step. We need to know our destination before we can map out a route. We are asked for the value of the equilibrium constant for a solid dissolving to form an aqueous solution.

2. Identify the major chemical species.
This step is important because chemical equilibria are dynamic interactions among *molecules and ions*. The key to success in working with equilibria is to "think molecules" (and ions). When solid LiF is added to water (Figure 16-10a), the species present before any reaction occurs are LiF (*s*) and H_2O (*l*).

3. Determine what chemical equilibria exist.
Sometimes the net chemical reaction is provided, but in other cases we have to examine the species present and determine what reactions can occur

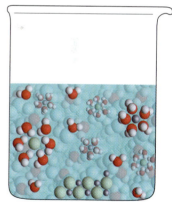

Figure 16-10
(*a*) When solid LiF is added to water, initially there are no ions in solution. (*b*) At equilibrium, Li^+ cations and F^- anions are present.

among them. The statement of the problem indicates that fluoride anions are present in solution when LiF dissolves in water. To maintain electrical neutrality, Li^+ ions must also be present in equal number. Here is the net reaction:

$$LiF (s) \rightleftharpoons Li^+ (aq) + F^- (aq)$$

4. Write the K_{eq} expressions.
This is the equation (or equations) that we must use to complete the calculations. The equilibrium constant expression for LiF dissolving in water is

$$K_{eq} = [Li^+]_{eq} [F^-]_{eq}$$

5. Organize the data and unknowns.
In some problems, concentrations at equilibrium are provided. In other problems concentrations at equilibrium must be calculated, usually by using amounts tables (see Chapter 4). In this example, we are told that a solution of LiF at chemical

equilibrium has $[F^-]_{eq} = 6.16 \times 10^{-2}$ M. The stoichiometric ratio of LiF is 1:1, so an equal amount of Li^+ dissolves: $[Li^+]_{eq} = 6.16 \times 10^{-2}$ M.

6. Substitute and calculate.

Refer to Step 1 to find out what must be calculated, the equilibrium constant or the equilibrium concentrations. In this example, we are after an equilibrium constant, which we calculate using the equilibrium concentrations:

$$K_{eq} = [Li^+]_{eq}\,[F^-]_{eq} = (6.16 \times 10^{-2})(6.16 \times 10^{-2}) = 3.79 \times 10^{-3}$$

7. Does the result make sense?

The concentration of ions in the saturated solution is in the range of 10^{-2} M, so an equilibrium constant in the 10^{-3} range is reasonable.

As the LiF example illustrates, the most direct way to determine the value of an equilibrium constant is to mix substances that can undergo a chemical reaction, wait until the system reaches equilibrium, and measure the concentrations of the species present once equilibrium is established. Although the calculation of an equilibrium constant requires knowledge of the equilibrium concentrations of *all* species whose concentrations appear in the equilibrium constant expression, stoichiometric analysis often can be used to deduce the concentration of one species from the known concentration of another species. Example 16-9 shows how to approach an equilibrium constant problem from a molecular perspective.

Example 16-9 K_{eq} **from a Molecular View**	The figure represents a molecular view of a gas-phase reaction that has reached equilibrium. Assuming that each molecule in the molecular view represents a partial pressure of 1.0 bar, determine K_{eq} for this reaction.

Strategy: Our seven-step approach to equilibrium problems will lead to the correct result. The first four steps are part of the strategy:

2 AB $\longrightarrow$ AB$_2$ + A

View at equilibrium

1. **Determine what is asked for.** The problem asks us to calculate an equilibrium constant, K_{eq}.

2. **Identify the major chemical species.** The chemical species present are the hypothetical gas-phase molecules, AB_2, A, and AB.

3. **Determine what chemical equilibria exist.** The forward reaction is provided. At equilibrium the reverse reaction proceeds at an equal rate:

$$2\,AB \rightleftharpoons AB_2 + A$$

4. **Write the equilibrium constant expression.** We use the chemical reaction to determine the equilibrium constant expression:

$$K_{eq} = \frac{(p_{AB_2})_{eq}(p_A)_{eq}}{(p_{AB})^2_{eq}}$$

Solution:

5. **Organize the data and unknowns.** The problem states that each molecule represents a partial pressure of 1.0 bar, so we can determine the equilibrium concentrations of each reagent by counting molecules in the molecular picture:

$$(p_{AB_2})_{eq} = 4.0\text{ bar} \qquad (p_A)_{eq} = 1.0\text{ bar} \qquad (p_{AB})_{eq} = 4.0\text{ bar}$$

6. **Substitute and calculate.** We substitute each concentration into the equilibrium constant expression and evaluate:

$$K_{eq} = \frac{(p_{AB_2})_{eq}\,(p_A)_{eq}}{(p_{AB})^2_{eq}} = \frac{(4.0\text{ bar})(1.0\text{ bar})}{(4.0\text{ bar})^2} = 0.25$$

7. Does the Result Make Sense? There are comparable numbers of molecules of reactants and products present in this system at equilibrium, so an equilibrium constant close to 1 is reasonable.

Refer to the equilibrium view of the triiodide reaction shown in Example 16-7. If each circle represents a concentration of 1.0 mM, what is the value of the equilibrium constant?

Extra Practice Exercise 16.9

Initial Conditions and Concentration Tables

When we do equilibrium calculations, we are usually interested in the concentrations of species present *at equilibrium*. In many cases, however, we have information about what we call **initial concentrations,** *before* any net change has occurred. Initial concentrations are the concentrations that would be present if it were possible to mix all the reactants but block the reactions that lead to equilibrium. These concentrations are easy to calculate from the initial conditions, but they seldom exist in reality because substances begin to react as soon as they are mixed.

A chemical system reacts, often rapidly, from initial conditions to equilibrium. As this occurs, concentrations of starting materials decrease, and concentrations of products increase. These concentration changes are related in two ways, as described in Section 4.4. To review, the concentration of each reagent at equilibrium is its initial concentration plus the change that has occurred:

$$\left[\begin{array}{c} \text{Equilibrium} \\ \text{concentration} \end{array} \right] = \left[\begin{array}{c} \text{Initial} \\ \text{concentration} \end{array} \right] + \left[\begin{array}{c} \text{Change in concentration} \\ \text{during reaction} \end{array} \right]$$

$$[A]_{eq} = [A]_i + \Delta[A] \qquad (16\text{-}5)$$

Furthermore, stoichiometric ratios link the changes in concentration of the various reagents:

[Change in concentration of B] =

[Stoichiometric ratio of B to A][Change in concentration of A]

$$\Delta[B] = \frac{\text{Coeff B}}{\text{Coeff A}} \Delta[A] \qquad (16\text{-}6)$$

As we show in upcoming examples, some changes in concentration are positive while others are negative, depending on whether a reagent is produced or consumed.

These relationships provide complete stoichiometric information regarding the equilibrium. Just as amounts tables are useful in doing stoichiometric calculations, a **concentration table** that provides initial concentrations, changes in concentrations, and equilibrium concentrations is an excellent way to organize Step 5 of the problem-solving procedure for equilibrium problems. Figure 16-11 diagrams how to complete a concentration table: Equation 16-5 applies to every column, but Equation 16-6 applies only across the "change" row.

Example 16-10 shows how to use stoichiometric reasoning and a concentration table to calculate an equilibrium constant.

Figure 16-11
Schematic view of a concentration table. The arrows indicate how the relationships expressed in Equations 16-5 and 16-6 are used to complete the table.

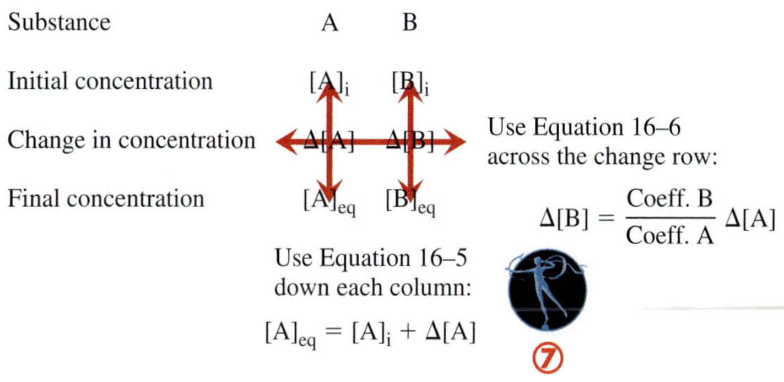

Substance	A	B
Initial concentration	$[A]_i$	$[B]_i$
Change in concentration	$\Delta[A]$	$\Delta[B]$
Final concentration	$[A]_{eq}$	$[B]_{eq}$

Use Equation 16-6 across the change row:

$$\Delta[B] = \frac{\text{Coeff. B}}{\text{Coeff. A}} \Delta[A]$$

Use Equation 16-5 down each column:

$$[A]_{eq} = [A]_i + \Delta[A]$$

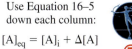

Example 16-10

**Calculating
an Equilibrium Constant**

Benzoic acid is a weak carboxylic acid that undergoes proton transfer with water. When a 0.125 M aqueous solution of benzoic acid ($C_6H_5CO_2H$) reaches equilibrium, $[H_3O^+] = 0.0028$ M. What is K_{eq} for the proton transfer reaction?

Strategy: Again, the seven-step approach to equilibrium problems will lead to the correct result. This is a more complicated example than Example 16-9, so a concentration table as part of Step 5 helps keep track of the stoichiometric relationships.

1. The problem asks us to calculate an equilibrium constant, K_{eq}.
2. The chemical species present initially are benzoic acid and water, $C_6H_5CO_2H$ and H_2O. The H_3O^+ mentioned in the problem is formed as the solution comes to equilibrium.
3. The equilibrium concentration of hydronium ions is provided, which suggests that one of the products is H_3O^+. Benzoic acid is a weak acid. Thus the correct chemical reaction is proton transfer from benzoic acid to water:

$$C_6H_5CO_2H\,(aq) + H_2O\,(l) \rightleftharpoons C_6H_5CO_2^-\,(aq) + H_3O^+\,(aq)$$

Benzoic acid
($C_6H_5CO_2H$)

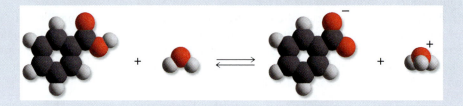

4. Use the chemical reaction to determine the equilibrium constant expression. Water is the solvent, so it is omitted from the equilibrium expression:

$$K_{eq} = \frac{[H_3O^+]_{eq}\,[C_6H_5CO_2^-]_{eq}}{[C_6H_5CO_2H]_{eq}}$$

Solution:
5. Construct a table of initial concentrations, changes in concentration, and equilibrium concentrations for each species that appears in the equilibrium constant expression. The equilibrium concentrations from the last row of the table are needed to find K_{eq}. Start by entering the data given in the problem. The initial concentration of benzoic acid is 0.125 M. Pure water contains no benzoate ions and a negligible concentration of hydronium ions. The problem also states the equilibrium concentration of hydronium ions, 0.0028 M.

Reaction	$H_2O\,(l)$ +	$C_6H_5CO_2H\,(aq) \rightleftharpoons$	$C_6H_5CO_2^-\,(aq)$ +	$H_3O^+\,(aq)$
Initial concentration (M)		0.125	0	0
Change in concentration (M)				
Equilibrium concentration (M)		*Need to find*	*Need to find*	0.0028

We see in the last column of the table that the initial and equilibrium concentrations of H_3O^+ are known, so we can use Equation 16-5 to determine the change in concentration for this species:

$$[H_3O^+]_{eq} = [H_3O^+]_i + \Delta[H_3O^+]$$

$$0.0028\text{ M} = 0\text{ M} + \Delta[H_3O^+] \quad so \quad \Delta[H_3O^+] = 0.0028\text{ M}$$

Once we know one change, Equation 16-6 allows us to complete the change row by calculating the changes in the other concentrations:

$$\Delta[B] = \frac{\text{Coeff B}}{\text{Coeff A}} \Delta[A]$$

Example 16-10

Calculating
an Equilibrium Constant
(*continued*)

All of the stoichiometric coefficients are equal to 1, so all three changes have the same magnitude, but they have different *signs*. As the reaction comes to equilibrium, concentrations of products *increase*, so the sign for changes in products is *positive*. The concentrations of starting materials *decrease* in the reaction, so the sign for the change in benzoic acid is *negative*.

Reaction	$H_2O\,(l)$ +	$C_6H_5CO_2H\,(aq) \rightleftharpoons$	$C_6H_5CO_2^-\,(aq)$ +	$H_3O^+\,(aq)$
Initial concentration (M)		0.125	0	0
Change in concentration (M)		−0.0028	+0.0028	+0.0028
Equilibrium concentration (M)		*Need to find*	*Need to find*	0.0028

Now we know two of the three entries in each of the other columns. To complete the table, we apply Equation 16-5 to each column, obtaining the concentrations at equilibrium for these species:

$$[C_6H_5CO_2^-]_{eq} = 0 + 0.0028\ \text{M} = 0.0028\ \text{M}$$

$$[C_6H_5CO_2H]_{eq} = 0.125\ \text{M} - 0.0028\ \text{M} = 0.122\ \text{M}$$

6. Substitute the values from the last row of the concentration table into the equilibrium constant expression and evaluate the result:

$$K_{eq} = \frac{(2.8 \times 10^{-3})(2.8 \times 10^{-3})}{0.122} = 6.4 \times 10^{-5}$$

7. Does the Result Make Sense? The concentration of hydronium ion at equilibrium is in the 10^{-3} range, so a value of K_{eq} in the 10^{-5} range is reasonable.

A 0.335 M aqueous solution of NH_3 (a weak base) has $[OH^-]_{eq} = 2.46 \times 10^{-3}$ M. Construct an amounts table for this equilibrium and use it to determine the value of K_{eq}.

Calculating Equilibrium Concentrations

The second main type of equilibrium problem asks for values of equilibrium concentrations. We also use concentration tables for this type of problem, with one additional feature. In such problems, we need to assign a variable x to one unknown concentration, and then we use the equilibrium constant to find the value of x by standard algebraic techniques. Examples 16-11 and 16-12 illustrate this use and manipulation of unknowns.

The molecular view represents a set of initial conditions for the reaction described in Example 16-9. Each molecule represents a partial pressure of 1.0 bar. Determine the equilibrium conditions and redraw the picture to illustrate those conditions.

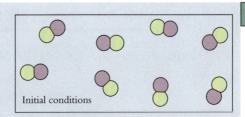

Initial conditions

Example 16-11

Gas Equilibrium
Concentrations

Example 16-11

Gas Equilibrium Concentrations (*continued*)

Strategy: Apply the seven-step strategy.

1. The problem asks for the equilibrium pressures and a molecular view illustrating the equilibrium conditions.

2. The only species present initially is the diatomic gas AB. From Example 16-9, we know that this gas reacts to form A and AB_2.

3. The chemical equilibrium is the same one that we identified in Example 16-9.

4. The equilibrium constant expression is the same as in Example 16-9:

$$2\,AB \rightleftharpoons AB_2 + A \qquad K_{eq} = \frac{(p_{AB_2})_{eq}\,(p_A)_{eq}}{(p_{AB})^2_{eq}} = 0.25$$

Solution:

5. From the molecular picture, we can determine the initial pressure of AB:

$$p(\text{initial, AB}) = 8.0 \text{ bar}$$

To relate this to equilibrium pressures, we need a concentration table. The only information available to us is the initial pressures:

Reaction	$2\,AB \rightleftharpoons AB_2 + A$		
Initial pressure (bar)	8.0	0	0
Change in pressure (bar)			
Equilibrium pressure (bar)			

We assign the unknown x to represent the *change* in pressure of A during the reaction. Then we can find the changes for the other reagents by applying Equation 16-6 across the change row. The stoichiometric ratio indicates that two molecules of AB must be consumed for each molecule of A produced. Because AB molecules are *consumed* as the reaction comes to equilibrium, the change in pressure of AB has a *negative* value. Thus, if the change in pressure of A is $+x$ bar, the change in pressure of AB is $-2x$ bar. By stoichiometry, the change for AB_2 is also $+x$:

Reaction	$2\,AB \rightleftharpoons AB_2 + A$		
Initial pressure (bar)	8.0	0	0
Change in pressure (bar)	$-2x$	$+x$	$+x$
Equilibrium pressure (bar)			

Now we apply Equation 16-5 to obtain expressions for the equilibrium pressures:

Reaction	$2\,AB \rightleftharpoons AB_2 + A$		
Initial pressure (bar)	8.0	0	0
Change in pressure (bar)	$-2x$	$+x$	$+x$
Equilibrium pressure (bar)	$8.0 - 2x$	x	x

6. To evaluate x, we substitute the equilibrium pressures into the equilibrium constant expression:

$$0.25 = \frac{(p_{AB_2})_{eq}\,(p_A)_{eq}}{(p_{AB})^2_{eq}} = \frac{x^2}{(8.0 - 2x)^2}$$

In general, we would have to solve for x using the quadratic equation, but this particular expression can be simplified by taking the square root of each side:

$$0.50 = \frac{x}{8.0 - 2x}$$

Clear the fraction and then multiply through:

$$0.50(8.0 - 2x) = x \quad so \quad 4.0 - x = x$$

Combine terms in x and solve for x:

$$4.0 = 2x \quad and \quad x = 2.0$$

This lets us calculate the equilibrium pressures:

$$(p_{AB})_{eq} = 8.0 - 2(2.0) = 4.0 \text{ bar} \quad (p_{AB_2})_{eq} = (p_A)_{eq} = 2.0 \text{ bar}$$

To complete the problem, we draw a molecular view that shows four molecules of AB, two molecules of AB_2, and two atoms of A:

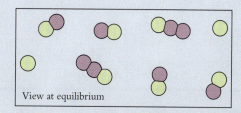

View at equilibrium

7. Does the Result Make Sense? There is less of the starting material and more of the products than were present at the beginning, and atoms of A and B have been conserved. This is a reasonable result.

If the initial conditions for the reaction described in this example are adjusted so that $p_{AB_2} = p_A = 2.0$ bar and $p_{AB} = 0$ bar, determine the equilibrium pressures and draw a molecular picture showing the equilibrium.

xample 16-11

Gas Equilibrium
Concentrations *(continued)*

**Extra Practice
Exercise 16.11**

The equilibrium constant for the proton transfer reaction of benzoic acid, determined in Example 16-10, is 6.4×10^{-5}. Calculate the equilibrium concentration of benzoic acid and benzoate anions in a 5.0×10^{-2} M aqueous solution of the acid.

Strategy: Apply the seven-step procedure.
1. This problem asks about equilibrium concentrations and provides the value of an equilibrium constant. Under these circumstances, we expect to have to define an appropriate unknown to represent the change in concentrations of one reagent.

2. Initially, benzoic acid and water are the only species present.

3. The proton transfer reaction is the same as in Example 16-10:

$$C_6H_5CO_2H\,(aq) + H_2O\,(l) \rightleftharpoons C_6H_5CO_2^-\,(aq) + H_3O^+\,(aq)$$

4. The equilibrium constant expression is the same as in Example 16-10:

$$K_{eq} = \frac{[H_3O^+]_{eq}\,[C_6H_5CO_2^-]_{eq}}{[C_6H_5CO_2H]_{eq}} = 6.4 \times 10^{-5}$$

Solution:
5. Construct a concentration table. The only quantitative information provided in the problem is the value for the initial concentration of benzoic acid (5.0×10^{-2} mol/L).

Example 16-12

Equilibrium Concentrations

Example 16-12

Equilibrium Concentrations (*continued*)

Reaction	$H_2O\,(l)$ +	$C_6H_5CO_2H\,(aq)$ ⇌	$C_6H_5CO_2^-\,(aq)$ +	$H_3O^+\,(aq)$
Initial concentration (M)		0.050	0	0
Change in concentration (M)				
Equilibrium concentration (M)		*Need to find*	*Need to find*	*Need to find*

Completing the table requires an appropriate unknown. If we let x represent the *change* in concentration of H_3O^+ during the reaction, the changes in the other concentrations are $-x$ for benzoic acid and $+x$ for benzoate anion. Then we apply Equation 16-5 to each column and complete the table:

Reaction	$H_2O\,(l)$ +	$C_6H_5CO_2H\,(aq)$ ⇌	$C_6H_5CO_2^-\,(aq)$ +	$H_3O^+\,(aq)$
Initial concentration (M)		0.050	0	0
Change in concentration (M)		$-x$	$+x$	$+x$
Equilibrium concentration (M)		$0.050 - x$	x	x

6. Next we substitute the values from the last row of the concentration table into the equilibrium constant expression:

$$K_{eq} = \frac{[H_3O^+]_{eq}\,[C_6H_5CO_2^-]_{eq}}{[C_6H_5CO_2H]_{eq}} = 6.4 \times 10^{-5} = \frac{x^2}{0.050 - x}$$

Multiplying through to clear the fraction gives

$$x^2 = (3.2 \times 10^{-6}) - (6.4 \times 10^{-5})x \quad or \quad x^2 + (6.4 \times 10^{-5})x - (3.2 \times 10^{-6}) = 0$$

This is a quadratic equation in the form $ax^2 + bx + c = 0$. In this case,

$$a = 1 \qquad b = 6.4 \times 10^{-5} \qquad and \qquad c = -3.2 \times 10^{-6}$$

Use the quadratic formula to find the value of x:

$$x = \frac{-b \pm \sqrt{b^2 - 4ac}}{2a} = \frac{-(6.4 \times 10^{-5}) \pm \sqrt{(6.4 \times 10^{-5})^2 - 4(1)(-3.2 \times 10^{-6})}}{2(1)}$$

$$x = \frac{-(6.4 \times 10^{-5}) \pm \sqrt{(4.096 \times 10^{-9}) + (1.28 \times 10^{-5})}}{2(1)}$$

There are two solutions to this equation, but one gives a negative value for x. We know that x must be positive, because it represents the increase in hydronium ion concentration. Hence, we evaluate the positive solution:

$$x = \frac{-(6.4 \times 10^{-5}) + (3.58 \times 10^{-3})}{2} = 1.76 \times 10^{-3}$$

Now we can solve for the concentrations of the ions at equilibrium:

$$[C_6H_5CO_2^-]_{eq} = x\,M = 1.76 \times 10^{-3}\,M$$

$$[C_6H_5CO_2H]_{eq} = (5.0 \times 10^{-2}\,M - 1.76 \times 10^{-3}\,M) = 4.8 \times 10^{-2}\,M$$

7. Do the Results Make Sense? Both concentrations are reasonable values: they are positive and smaller than the initial concentration of benzoic acid.

Extra Practice Exercise 16.12

Calculate the concentrations of the species present in an aqueous solution of ammonia (see Extra Practice Exercise 16-10) that is 0.0125 M NH_3.

Examples 16-11 and 16-12 involve equilibrium constants with moderate values. However, many chemical reactions have equilibrium constants that are either very small ($K_{eq} < 10^{-8}$) or very large ($K_{eq} > 10^{8}$). For these equilibria we usually can use approximations to simplify the calculations.

A *very small* value of K_{eq} means that concentrations of *products* at equilibrium are *very low* compared with reactant concentrations. As Figure 16-12a shows, when K_{eq} is small, reactions barely get started before they reach equilibrium. One example is the equilibrium that results when solid AgBr is placed in water:

$$AgBr\,(s) \rightleftharpoons Ag^{+}(aq) + Br^{-}(aq)$$
$$K_{eq} = 5.35 \times 10^{-13}$$

The equilibrium constant for this reaction is small, indicating that only a small amount of AgBr will dissolve in water.

At the other extreme, a *very large* value of K_{eq} means that concentrations of *products* at equilibrium are *very high* compared with reactant concentrations. As Figure 16-12b shows, when K_{eq} is large, the reaction proceeds almost to completion before equilibrium is reached. One example is the equilibrium that results from the mixing of aqueous solutions of $AgNO_3$ and KBr:

$$Ag^{+}(aq) + Br^{-}(aq) \rightleftharpoons AgBr\,(s) \qquad K_{eq} = 1/(5.35 \times 10^{-13}) = 1.87 \times 10^{12}$$

The equilibrium constant for this reaction is large, indicating that nearly all of the aqueous ions end up as solid precipitate.

(a)

Equilibrium position

$$AgBr\,(s) \longrightarrow Ag^{+}(aq) + Br^{-}(aq)$$
$$K_{eq} \ll 1$$

Extent of reaction →

Initial conditions:
AgBr (s) only

(b)

Equilibrium position

$$Ag^{+}(aq) + Br^{-}(aq) \longrightarrow AgBr\,(s)$$
$$K_{eq} \gg 1$$

◄ Extent of reaction ►

Initial conditions: Ag⁺ (aq) and Br⁻ (aq) only   Completion: AgBr (s) only

Figure 16-12
Schematic representation of the relationship between K_{eq} and the extent of reaction. (a) When $K_{eq} \ll 1$, the difference from initial conditions, x, is small. (b) When $K_{eq} \gg 1$, the difference from completion, y, is small.

Working with Small Equilibrium Constants

When K_{eq} is small, the changes in concentration are likely to be small compared with initial concentrations. Example 16-13 illustrates the use of approximations to simplify the mathematical calculations in such cases.

Sodium benzoate is a common food preservative. This salt dissolves in water to produce benzoate anions, which accept protons from water:

$$H_2O\,(l) + C_6H_5CO_2^{-}(aq) \rightleftharpoons OH^{-}(aq) + C_6H_5CO_2H\,(aq)$$

$$K_{eq} = \frac{[OH^{-}]_{eq}[C_6H_5CO_2H]_{eq}}{[C_6H_5CO_2^{-}]_{eq}} = 1.5 \times 10^{-10}$$

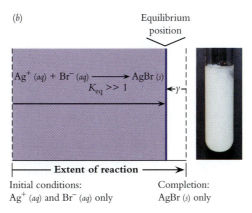

What is the equilibrium concentration of hydroxide ions in a 0.135 M solution of sodium benzoate?

Example 16-13

Small Equilibrium Constant

Benzoate anion
($C_6H_5CO_2^{-}$)

Example 16-13
Small Equilibrium Constant (*continued*)

Strategy: We follow the step-by-step procedure for working with equilibria.

1. The problem asks for the equilibrium concentration of OH^-.

2. Under initial conditions, three chemical species are present: benzoate anions, Na^+, and H_2O.

3. The net reaction is given. A proton is transferred from a water molecule to a benzoate anion. A small amount of benzoic acid forms as the reaction reaches equilibrium. Notice that Na^+ ions do not participate in the net reaction.

4. The equilibrium constant expression is given in the problem.

Solution:

5. Set up a concentration table. The concentration of water is not required because it does not appear in the equilibrium constant expression. A convenient choice for the unknown x is the amount of benzoic acid formed in the reaction. This choice makes several other entries in the table equal to x.

Reaction	H_2O +	$C_6H_5CO_2^-$	$\rightleftharpoons$	$C_6H_5CO_2H$ +	OH^-
Initial concentration (M)		0.135		0	0
Change to equilibrium (M)	$-x$			$+x$	$+x$
Equilibrium concentration (M)		$0.135 - x$		x	x

6. Substitute the equilibrium concentrations in the expression for K_{eq} and solve for the unknown:

$$K_{eq} = \frac{[OH^-]_{eq}\,[C_6H_5CO_2H]_{eq}}{[C_6H_5CO_2^-]_{eq}} = 1.5 \times 10^{-10} = \frac{(x)(x)}{0.135 - x}$$

The equilibrium constant has a small value (10^{-10}). This suggests that we can make an approximation by assuming that the change required to reach equilibrium (x) is very small. This situation is like that shown in Figure 16-12a. As shown in the concentration table, the concentration of benzoate ion is $(0.135 - x)$ M at equilibrium. Knowing that x is small, we make the approximation that x can be neglected relative to 0.135:

$$x \ll 0.135 \qquad so \qquad 0.135 - x \cong 0.135$$

This simplifies the equilibrium expression:

$$1.5 \times 10^{-10} = \frac{x^2}{0.135 - x} \cong \frac{x^2}{0.135}$$

$$x^2 \cong (0.135)(1.5 \times 10^{-10}) = 0.203 \times 10^{-10}$$

$$x \cong 0.45 \times 10^{-5} = 4.5 \times 10^{-6}$$

We use this value of x to calculate the concentrations at equilibrium:

$$[C_6H_5CO_2^-] = (0.135 - x) \text{ M} = (0.135 - 4.5 \times 10^{-6}) \text{ M} = 0.135 \text{ M}$$

$$[C_6H_5CO_2H] = [OH^-] = x \text{ M} = 4.5 \times 10^{-6} \text{ M}$$

7. Do the Results Make Sense? Check that the calculated result is consistent with the approximation. The approximation in this example is $x \ll 0.135$ M. We find $x = 4.5 \times 10^{-6}$ M, which is more than four orders of magnitude less than 0.135, so the approximation is a good one.

Extra Practice Exercise 16.13

Use the approximation method to calculate the concentrations of the species present in concentrated ammonia, which is 14.8 M (see Extra Practice Exercise 16.10 for *K*).

Whenever we make an approximation, we must verify that it is valid by comparing the value calculated using the approximation with the approximation itself. Most equilibrium constants are uncertain by about 5%, so x can be neglected whenever its value is two or more orders of magnitude smaller than the value from which it is subtracted or added.

Notice, however, that we do not neglect the lone x in determining the equilibrium concentrations of hydroxide ions and benzoic acid. We can neglect x *only* when it appears in a sum or difference and *never when it stands alone.*

Usually, x in a sum or difference is small enough to neglect if the equilibrium constant is smaller than 10^{-4} and the initial concentrations of starting materials are equal to or greater than 0.1 M. Nevertheless, the validity of an approximation must always be verified by comparing the result with the approximation. In other words, after calculating the value of x using an approximation such as $(A - x) \cong A$, check to see whether the calculated value for x indeed is less than 5% of A.

Working with Large Equilibrium Constants

When K_{eq} is very large, the situation is that represented by Figure 16-12b. In such a situation, the reaction proceeds nearly to completion, so the difference between completion and equilibrium, represented by y in the figure, is very small. We choose y as the variable rather than x to highlight the fact that the system is coming back to equilibrium from completion as opposed to working forward from initial conditions. Because y is small, working from completion *backward* to equilibrium allows us to neglect y in many cases. When K_{eq} is large, we could never neglect x when working from initial concentrations *forward* to equilibrium.

When an equilibrium constant is very large, we use stoichiometry to determine what the concentrations would be at completion. Then we define y to be the amount of back-reaction leading from completion to equilibrium. Solving for y leads to the true equilibrium concentrations. The equilibrium position of the system lies very near the completion point of the reaction, so y will be very small, allowing us to make approximations.

This is a two-step process, so for a reaction with a large K_{eq} we can construct two concentration tables. The first table helps us determine the concentrations *at completion,* and the second table expresses the concentrations *at equilibrium.* Example 16-14 illustrates an equilibrium position close to completion.

One of the steps in the industrial production of nitric acid is the combustion of NO at room temperature:

$$2\,NO\,(g) + O_2\,(g) \rightleftharpoons 2\,NO_2\,(g) \qquad K_{eq} = 4.2 \times 10^{12}$$

Suppose a reactor is charged with 10.0 bar each of NO and O_2. Find the partial pressure of each gas at equilibrium.

Example 16-14

Large Equilibrium Constant

Strategy: Use the step-by-step approach for an equilibrium problem.

1. The problem asks us to calculate equilibrium pressures of all reagents.

2. The species present before reaction are NO gas and O_2 gas.

3. The reaction is given in the problem:

$$2\,NO\,(g) + O_2\,(g) \rightleftharpoons 2\,NO_2\,(g)$$

4. We obtain the equilibrium constant expression from the reaction:

$$K_{eq} = \frac{(p_{NO_2})^2_{eq}}{(p_{NO})^2_{eq}\,(p_{O_2})_{eq}} = 4.2 \times 10^{12}$$

Because the magnitude of the equilibrium constant is very large, *almost all* of the starting materials will be converted to products. Because of this, the problem is easier to solve by taking the reaction to completion and then returning to equilibrium.

Example 16-14

Large Equilibrium Constant (*continued*)

Solution:

5. Start the solution by setting up a concentration table. Because K_{eq} is very large, we first take the reaction to completion. Finding the concentrations at completion is a limiting reactant problem. Because we have equal amounts of the two reactants and the coefficient for NO is 2, NO would be used up first, so it is the limiting reactant, and its partial pressure would be zero at completion. This requires a change in pressure of -10.0 bar for NO. By stoichiometry, the change of pressure for NO_2 is $+10.0$ bar and the change for O_2 is half the change for NO, -5.0 bar.

Reaction	$2\,NO\,(g)$	$+$	$O_2\,(g)$	$\rightleftharpoons$	$2\,NO_2\,(g)$
Initial pressure (bar)	10.0		10.0		0
Change to completion (bar)	-10.0		-5.0		$+10.0$
Pressure at completion (bar)	0		5.0		10.0

Now we work from completion to equilibrium. The equilibrium constant for this reaction is very large, but the partial pressure of NO cannot be zero at equilibrium. We define y to be the change in NO pressure on going from completion to equilibrium. Then the stoichiometric coefficients and Equation 16-6 give $+0.5y$ for the change in pressure of O_2 and $-y$ for the change in NO_2.

Reaction	$2\,NO\,(g)$	$+$	$O_2\,(g)$	$\rightleftharpoons$	$2\,NO_2\,(g)$
Pressure at completion (bar)	0		5.0		10.0
Change to equilibrium (bar)	$+y$		$+0.5y$		$-y$
Equilibrium pressure (bar)	y		$5.0 + 0.5y$		$10.0 - y$

6. Having obtained expressions for the equilibrium pressures, we substitute these into the expression for K_{eq}:

$$K_{eq} = 4.2 \times 10^{12} = \frac{(10.0 - y)^2}{(y)^2(5.0 + 0.5y)}$$

Because K_{eq} is very large, the amount of the limiting reactant present at equilibrium, y, will be very small compared with the amounts of materials present at completion. That is the point of taking the reaction to completion and then working back toward equilibrium. This leads to two approximations:

$$10.0 - y \cong 10.0 \quad and \quad 5.0 + 0.5y \cong 5.0$$

Now solve for y:

$$4.2 \times 10^{12} \cong \frac{(10.0)^2}{(y)^2\,(5.0)} \quad giving \quad y^2 \cong \frac{(10.0)^2}{(5.0)(4.2 \times 10^{12})}$$

$$y^2 \cong 4.76 \times 10^{-12} \quad from\ which \quad y \cong 2.2 \times 10^{-6}$$

Use the value of y to determine the equilibrium pressures:

$$(p_{NO})_{eq} = y\ bar = 2.2 \times 10^{-6}\ bar$$

$$(p_{O_2})_{eq} = (5.0 + 0.5y)\ bar = 5.0\ bar \qquad (p_{NO_2})_{eq} = (10.0 - y)\ bar = 10.0\ bar$$

Do the Results Make Sense? The calculations show that y is more than five orders of magnitude smaller than 5.0 and 10.0 bar, so the approximations are valid.

Extra Practice Exercise 16.14

Figure 16-12b shows the precipitation reaction to form AgBr. Calculate the amount of AgBr solid that forms and the concentrations of Ag^+ and Br^- ions remaining in solution if 0.500 L each of 1.00 M solutions of $AgNO_3$ and KBr are mixed together.

The examples of this section illustrate the general approach to equilibrium problems. Notice that these examples include gas-phase, precipitation, and acid–base chemistry. We use a variety of equilibrium examples to emphasize that the general strategy for working with equilibria is always the same, no matter what type of equilibrium is involved. In Chapters 17 and 18 we apply these ideas in more detail to important types of equilibria.

Section Exercises

■ **16.5.1** Barium sulfate is a relatively insoluble salt used for medical radiographs of the gastrointestinal tract:

$$BaSO_4\,(s) \rightleftharpoons Ba^{2+}(aq) + SO_4^{\,2-}(aq)$$

The concentration of Ba^{2+} ions in a saturated aqueous solution is 1.05×10^{-5} M. Determine K_{eq} for barium sulfate dissolving in water.

■ **16.5.2** A solution of acetic acid in water is described by the following equilibrium:

$$\underset{\text{Acetic acid}}{CH_3CO_2H\,(aq)} + H_2O\,(l) \rightleftharpoons \underset{\text{Acetate}}{CH_3CO_2^{\,-}(aq)} + H_3O^+(aq)$$

$$K_{eq} = 1.8 \times 10^{-5}$$

Calculate the equilibrium concentrations of acetic acid, acetate ion, and hydronium ion in a 2.5 M solution of acetic acid.

■ **16.5.3** Silver bromide, a solid used in photographic film, can be prepared by mixing solutions of silver nitrate and potassium bromide:

$$Ag^+(aq) + Br^-(aq) \rightleftharpoons AgBr\,(s)$$
$$K_{eq} = 1.87 \times 10^{12}$$

Calculate the concentrations of Ag^+ and Br^- ions remaining in solution after mixing 2.50 L of 0.100 M $AgNO_3$ solution with 2.50 L of 0.500 M KBr solution.

16.6 EQUILIBRIA IN AQUEOUS SOLUTIONS

Equilibria that occur in aqueous solution are of particular interest, because water is the medium of life and a major influence on the geography of our planet. Many substances dissolve in water, and the solutes in an aqueous solution may participate in a number of different types of equilibria. Solubility itself is one important type of equilibrium, as we describe in Chapter 18. Acid–base reactions, considered in detail in Chapter 17, are another. To conclude this chapter, we describe how to determine which equilibria are most important in any particular aqueous solution.

Species in Solution

The species in an aqueous solution can be categorized broadly into two groups present at different relative concentrations. We designate those present in relatively high concentrations as **major species.** We refer to those present in relatively low concentrations as **minor species.** Minor species in aqueous solutions generally have concentrations at least three orders of magnitude lower than the concentrations of the major solute species. In most aqueous solutions, one equilibrium plays the most important role. We call this the **dominant equilibrium.** The dominant equilibrium always has major species as its reactants, and the products of that equilibrium often are minor species. Thus, we focus our attention on major species because they determine which equilibria are most important in the solution.

Tutorial

The first step in analyzing an aqueous equilibrium is to identify the major species. Pure water contains H_2O molecules at a concentration of 55.5 M, so an aqueous solution always contains H_2O as a major species. In pure water, H_2O is the only major species, but an aqueous solution contains two or more major species: H_2O and those solute species that are present at highest concentration.

As discussed in Section 3.6, whenever an ionic solid dissolves in water, the salt breaks apart to give a solution of cations and anions. Thus, in any aqueous salt solution the major species are water molecules and the cations and anions generated by the salt. For example, a solution of potassium chloride contains K^+ and Cl^- ions and H_2O molecules as major species. Likewise, the major species in a solution of ammonium nitrate are NH_4^+, $NO_3^{\,-}$, and H_2O.

Figure 16-13
The major species present in an aqueous solution of nitric acid are water molecules, hydronium ions, and nitrate anions. The concentration of HNO₃ molecules is negligible.

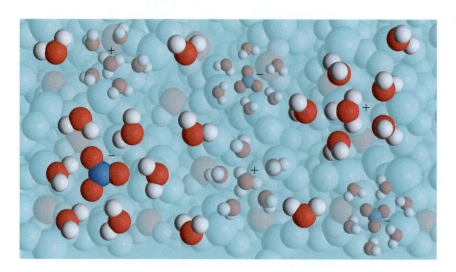

As described in Section 4.6, a strong acid is a substance that reacts virtually completely with water to produce hydronium ions. For example, when nitric acid (HNO₃) dissolves in water, its molecules react with water to produce hydronium ions:

$$H_2O + HNO_3 \longrightarrow H_3O^+(aq) + NO_3^-(aq)$$

This reaction is written with a single arrow to designate that it goes virtually to completion: the equilibrium constant for this proton transfer reaction is so large that the concentration of unreacted acid is negligible. The major species in an aqueous solution of any strong acid are H_2O, H_3O^+, and the anion produced by removing a proton from the acid. This situation is illustrated in Figure 16-13, which shows a molecular picture of aqueous nitric acid. Table 16-1 lists the most important strong acids.

A strong base generates hydroxide ions when it dissolves in water. The most common examples of strong bases (see Table 16-1) are soluble metal hydroxides such as NaOH and KOH. These ionic substances separate into ions when they dissolve in water:

$$KOH(s) \longrightarrow K^+(aq) + OH^-(aq)$$

The major species in any aqueous solution of a strong base are H_2O, OH^-, and the cation generated by the base.

In addition to salts, strong acids, and strong bases, many molecular substances dissolve in water. As Figure 16-14 shows, these solutes

Table 16-1   Common Strong Acids and Bases

Acids of Industrial Importance

H₂SO₄	Sulfuric acid
HNO₃	Nitric acid
HCl	Hydrochloric acid

Acids with Laboratory Applications

HClO₄	Perchloric acid
HBr	Hydrobromic acid
HI	Hydriodic acid

Bases of Industrial Importance

NaOH	Sodium hydroxide
Ca(OH)₂	Calcium hydroxide
KOH	Potassium hydroxide

Bases with Laboratory Applications

LiOH	Lithium hydroxide
Ba(OH)₂	Barium hydroxide

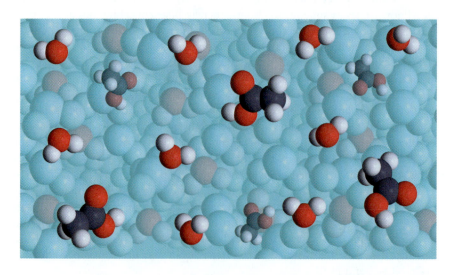

Figure 16-14
A molecular view of the major species present in an aqueous solution of acetic acid (vinegar).

retain their structures, dissolving in aqueous solution to give intact molecules. The figure shows a molecular picture of vinegar, which is a dilute aqueous solution of acetic acid (CH_3CO_2H). The major species present in vinegar are H_2O and CH_3CO_2H.

To identify the major species in any aqueous solution, first categorize the solutes. A soluble salt, strong acid, or strong base generates the appropriate cations and anions as major species. Every other solute generates its molecular species in solution. In addition, H_2O is *always* a major species in aqueous solutions. Example 16-15 provides practice in identifying the major species in solution.

Identify the major species in each of the following aqueous solutions: (a) $NaCH_3CO_2$ (sodium acetate); (b) $HClO_4$ (perchloric acid); (c) $C_6H_{12}O_6$ (glucose, used for intravenous feeding); and (d) NH_3 (ammonia, used for household cleaning).

Example 16-15

Major Species in Solution

Strategy: Identify the nature of each solute. If it is a salt, strong acid, or strong base, it generates ions in aqueous solution. All other solutes give aqueous solutions that contain molecules as the major species.

Solution:

a. From its formula, you should recognize that $NaCH_3CO_2$ is a salt. It contains the alkali metal cation Na^+ and the polyatomic acetate anion, $CH_3CO_2^-$. According to the solubility guidelines from Chapter 4, all sodium salts are soluble. Thus, this salt dissolves, generating an aqueous solution in which the major species are H_2O, Na^+, and $CH_3CO_2^-$:

b. $HClO_4$ is a strong acid, so it transfers a proton to water quantitatively. (See the list of strong acids in Table 16-1.) The major species in aqueous perchloric acid are H_2O, H_3O^+, and ClO_4^-:

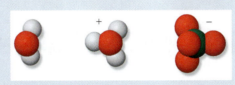

c. Glucose is not a strong acid, a strong base, or a salt. It dissolves in water without reacting, so the major species in intravenous feeding solutions are molecules of $C_6H_{12}O_6$ and H_2O:

Glucose ($C_6H_{12}O_6$)

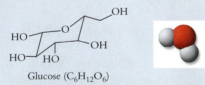

d. NH_3 is not a strong acid, a strong base, or a salt. The major species in household ammonia are molecules of NH_3 and H_2O:

Do the Results Make Sense? Remember that major species are molecules unless the solute is a salt, strong acid, or strong base.

Identify the major species present in aqueous solutions of hydrochloric acid, sodium chlorate ($NaClO_3$), and chloroform ($CHCl_3$).

Extra Practice Exercise 16.15

Types of Aqueous Equilibria

Most aqueous equilibria fall into three broad categories: proton transfer, solubility, or complexation. The nature of the major species in the solution determines which category of equilibrium we need to consider.

One of the most important types of aqueous equilibrium involves proton transfer from an acid to a base. In aqueous solutions, water can act as an acid or a base. In the presence of an acid, symbolized H*A*, water acts as a base by accepting a proton. The equilibrium constant for transfer of a proton from an acid to a water molecule is called the **acid ionization constant (K_a):**

$$HA(aq) + H_2O(l) \rightleftharpoons A^-(aq) + H_3O^+(aq) \qquad K_{eq} = \frac{[A^-]_{eq} [H_3O^+]_{eq}}{[HA]_{eq}} = K_a$$

For instance, when acetic acid, an example of a weak acid, dissolves in water, some acetic acid molecules transfer protons to water molecules to produce acetate anions and hydronium ions:

$$CH_3CO_2H(aq) + H_2O(l) \rightleftharpoons CH_3CO_2^-(aq) + H_3O^+(aq)$$

$$K_a = \frac{[CH_3CO_2^-]_{eq} [H_3O^+]_{eq}}{[CH_3CO_2H]_{eq}}$$

In the presence of a base B, water acts as an acid by donating a proton, and the equilibrium constant for the transfer of a proton from water to a base is called the **base ionization constant (K_b):**

$$B(aq) + H_2O(l) \rightleftharpoons BH^+(aq) + OH^-(aq) \qquad K_{eq} = \frac{[BH^+]_{eq} [OH^-]_{eq}}{[B]_{eq}} = K_b$$

When ammonia, an example of a weak base, dissolves in water, some ammonia molecules accept protons from water to produce ammonium cations and hydroxide ions:

$$NH_3(aq) + H_2O(l) \rightleftharpoons NH_4^+(aq) + OH^-(aq) \qquad K_b = \frac{[NH_4^+]_{eq} [OH^-]_{eq}}{[NH_3]_{eq}}$$

Notice that the expressions for K_a and K_b do not include the water molecules that act as starting materials for the proton transfer reactions. Water, as the solvent, is always present in huge excess. Thus, as described in Section 16.2, the concentration of water does not change significantly during an acid–base reaction and is omitted from K_a and K_b.

Many substances that participate in aqueous reactions are soluble salts. These ionic solids dissolve in water to give solutions of cations and anions. For almost all salts, there is an upper limit to the amount that will dissolve in water. A salt solution is *saturated* when the amount dissolved has reached this upper limit of solubility. Any additional salt added to a saturated solution remains undissolved at the bottom of the vessel. When excess solid salt and a saturated solution are present, a solubility equilibrium exists. At equilibrium, ions dissolve continually, but other ions precipitate out at exactly the same rate. This equilibrium is illustrated for sodium chloride by the molecular picture in Figure 16-15.

By convention, a solubility equilibrium is written in the direction of a solid dissolving to give aqueous ions, and the equilibrium constant for this reaction is called the **solubility product (K_{sp}).** Here, for example, is the reaction describing the solubility equilibrium of copper(II) chloride:

$$CuCl_2(s) \rightleftharpoons Cu^{2+}(aq) + 2\,Cl^-(aq) \qquad K_{eq} = [Cu^{2+}]_{eq} [Cl^-]_{eq}^2 = K_{sp}$$

The color of an aqueous solution of copper(II) chloride shows that copper(II) cations are present in the aqueous phase at equilibrium with the solid salt.

The evaporation of water from a saturated solution leaves a solution in which the ion concentrations exceed the solubility limit. To return to equilibrium, the salt must precipitate from the solution. Evaporation is used to "mine" sodium chloride and other

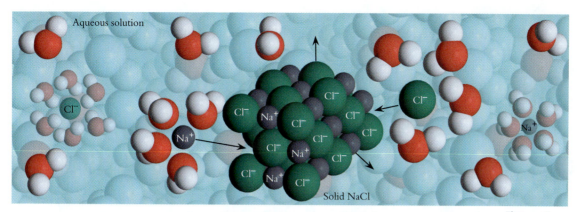

Figure 16-15
A molecular view of the solubility equilibrium for a solution of sodium chloride in water. At equilibrium, ions dissolve from the crystal surface at the same rate they are captured, so the concentration of ions in the solution remains constant.

salts from the highly salty waters of inland seas such as Great Salt Lake in Utah and Israel's Dead Sea.

Metal cations in aqueous solution often form chemical bonds to anions or neutral molecules that have lone pairs of electrons. A silver cation, for example, can associate with two ammonia molecules to form a silver–ammonia complex:

$$H\!-\!\overset{\displaystyle H}{\underset{\displaystyle H}{N}}\!: + \text{Ag}^+ + :\overset{\displaystyle H}{\underset{\displaystyle H}{N}}\!-\!H \rightleftharpoons H\!-\!\overset{\displaystyle H}{\underset{\displaystyle H}{N}}\!-\!\text{Ag}^+\!-\!\overset{\displaystyle H}{\underset{\displaystyle H}{N}}\!-\!H$$

The resulting species is called a **complex ion.** The equilibrium constant for the formation of a complex ion is called its **formation constant (K_f).** Tabulated values of K_f always refer to the equilibrium constant for the complex forming from the metal cation. Here, for example, is the reaction describing the complexation equilibrium between Ag^+ and NH_3:

$$\text{Ag}^+(aq) + 2\,\text{NH}_3(aq) \rightleftharpoons [\text{Ag}(\text{NH}_3)_2]^+(aq) \qquad K_f = \frac{[\text{Ag}(\text{NH}_3)_2^+]_{eq}}{[\text{Ag}^+]_{eq}\,[\text{NH}_3]_{eq}^2}$$

Notice that square brackets are used in a new way for the silver–ammonia complex in the chemical reaction. Chemists use square brackets to identify a complex ion such as $[\text{Ag}(\text{NH}_3)_2]^+$, because the species involved in complexation, NH_3 in this instance, are set off by parentheses. When square brackets appear in an equilibrium constant expression, they always designate concentration; when they appear in a chemical reaction, they designate a complex. We consider the formation of metal complexes in more depth in Chapter 20.

Identifying Types of Equilibria

The major species in an aqueous solution determine which categories of equilibria are important for that solution. Each major species present in the solution must be examined in light of these general categories. Are any of the major species weak acids or weak bases? Are there ions present that combine to form an insoluble salt? Do any of the major species participate in more than one equilibrium? Any chemical reaction can approach equilibrium from either direction. Consequently, there are six different types of aqueous equilibria in which major species are reactants:

1. An acid may donate a proton to water or some other base.
2. A base may accept a proton from water or some other acid.
3. A salt may dissolve in water.
4. Cations and anions in a solution may precipitate to form a solid.
5. A cation may associate with electron-donating species to form a complex.
6. A complex may dissociate.

We consider each of these in more detail in subsequent chapters, but being able to identify types of equilibria helps greatly in solving equilibrium problems. The equilibrium constants for many of these characteristic types of equilibria have been measured and tabulated. Representative K_a, K_b, and K_{sp} values appear in Appendix E, and tables that are more extensive can be found in the *CRC Handbook of Chemistry and Physics*. Example 16-16 provides practice in identifying equilibria.

Example 16-16

Types of Aqueous Equilibria

Write expressions for the equilibrium constants and decide which general category applies to each of these equilibria:

a. $H_2CO_3(aq) + H_2O(l) \rightleftharpoons HCO_3^-(aq) + H_3O^+(aq)$

b. $Fe^{3+}(aq) + 3\,OH^-(aq) \rightleftharpoons Fe(OH)_3(s)$

c. $NH_4^+(aq) + OH^-(aq) \rightleftharpoons NH_3(aq) + H_2O(l)$

Strategy: Examine each reaction closely, and look for the characteristic features of the general equilibria. The equilibrium constant expression can be written by inspecting the overall stoichiometry.

Solution:

a. In this reaction, water accepts a proton from H_2CO_3. The equilibrium constant for a proton transfer reaction that consumes H_2O and produces H_3O^+ is called K_a:

$$\frac{[H_3O^+]_{eq}\,[HCO_3^-]_{eq}}{[H_2CO_3]_{eq}} = K_a$$

b. This is a precipitation reaction. Solid $Fe(OH)_3$ forms in the forward direction. In the reverse direction, solid $Fe(OH)_3$ dissolves. This reverse reaction is a solubility reaction for which $K_{eq} = K_{sp}$. According to Equation 16-2, K_{eq} must be $1/K_{sp}$ for the reaction in the direction written:

$$\frac{1}{[Fe^{3+}]_{eq}\,[OH^-]_{eq}^3} = \frac{1}{K_{sp}}$$

c. This reaction involves proton transfer. In the *forward* direction, a hydroxide ion removes a proton from an ammonium ion. In the *reverse* direction, a proton is transferred from water to ammonia. This reverse reaction is included among the general categories; it is the transfer of a proton from a water molecule to a base. The equilibrium constant for this reaction is K_b. In the direction the reaction is written, the equilibrium constant is $1/K_b$:

$$\frac{[NH_3]_{eq}}{[NH_4^+]_{eq}\,[OH^-]_{eq}} = \frac{1}{K_b}$$

Do the Results Make Sense? First, verify that each concentration quotient has products in the numerator and reactants in the denominator. Then examine the concentration ratio and relate it to the general types of equilibrium constants.

Extra Practice Exercise 16.16

Relate each of the following equilibrium reactions to equilibrium constants of standard types:

a. $HCO_3^-(aq) + H_3O^+(aq) \rightleftharpoons H_2CO_3(aq) + H_2O(l)$

b. $Al(OH)_3(s) \rightleftharpoons Al^{3+}(aq) + 3\,OH^-(aq)$

c. $Co^{3+}(aq) + 6\,NH_3(aq) \rightleftharpoons [Co(NH_3)_6]^{3+}(aq)$

d. $CN^-(aq) + H_2O(l) \rightleftharpoons HCN(aq) + OH^-(aq)$

Spectator Ions

With all the possible equilibria in aqueous systems, most species might be expected to participate in at least one equilibrium. Nonetheless, in many solutions some of the ionic species undergo no significant reactions. These species are classified as **spectator ions.**

Recall from Section 4.5 what happens when a solution of potassium iodide is mixed with a solution of lead(II) nitrate. The major species present are $K^+(aq)$, $I^-(aq)$, $Pb^{2+}(aq)$, $NO_3^-(aq)$, and H_2O. As shown in Figure 16-16, a bright yellow solid precipitates from the mixture. This solid is insoluble PbI_2. The Pb^{2+} and I^- ions that remain in solution are at equilibrium with the precipitate:

$$PbI_2(s) \rightleftharpoons Pb^{2+}(aq) + 2\,I^-(aq) \qquad K_{sp} = 9.8 \times 10^{-9}$$

Two of the major species, K^+ and NO_3^-, are not involved in this equilibrium. They are spectator ions. Example 16-17 provides practice in identifying spectator ions.

Figure 16-16
Mixing aqueous solutions of KI and Pb(NO$_3$)$_2$ results in the formation of a yellow solid, PbI$_2$.

When equal volumes of 0.100 M solutions of sodium bromide and silver nitrate are mixed, a white solid precipitates from the solution. Identify the precipitate, write the net ionic reaction for the solubility equilibrium, and identify any spectator ions.

Example 16-17

Spectator Ions

Strategy: Analyze the problem from the molecular perspective. Begin by identifying the major species present. Then identify the equilibria in which these ions participate.

Solution: Sodium bromide and silver nitrate are both soluble ionic compounds, so the major species are Na^+, Ag^+, Br^-, NO_3^-, and H_2O molecules. The solubility guidelines presented in Section 4.5 identify the precipitate as silver bromide:

$$AgBr(s) \rightleftharpoons Ag^+(aq) + Br^-(aq) \qquad K_{sp} = 5.35 \times 10^{-13}$$

Major ionic species that do not participate in aqueous equilibria are classified as spectator ions. In this case the spectator ions are Na^+ and NO_3^-.

Do the Results Make Sense? All sodium salts are soluble, and so are all nitrate salts, so it makes sense that neither of these ions participates in a solubility equilibrium. Furthermore, nitrate and sodium cations are neither acidic nor basic, so it makes sense that neither participates in an acid–base equilibrium.

Identify the spectator ions present when a solution of sodium hydroxide is mixed with a solution of nitric acid.

Extra Practice Exercise 16.17

The types of aqueous equilibria described in this section have been given special names, and it is essential that you be able to recognize them. Keep in mind, however, that the principles described in the previous sections apply to all chemical equilibria. Chemists categorize equilibria for convenience, but they treat all equilibria the same way. Our Chemistry and the Environment Box explores the roles of these equilibria in a spectacular natural process, the formation of limestone caverns.

Section Exercises

16.6.1 What are the major species present in each of the following solutions? (a) 1.00 M perchloric acid; (b) 0.25 M ammonia; (c) 0.50 M potassium hydrogen carbonate; and (d) 0.010 M hypochlorous acid, HClO

16.6.2 A green precipitate forms when solutions of sodium hydroxide and nickel(II) sulfate are mixed. Identify the precipitate, write the net ionic equation for the solubility equilibrium, and identify the spectator species.

16.6.3 Relate each of the following equilibrium reactions to equilibrium constants of standard types: K_{sp}, K_a, and K_b:
(a) $Al^{3+}(aq) + 3\,OH^-(aq) \rightleftharpoons Al(OH)_3(s)$
(b) $HClO(aq) + H_2O(l) \rightleftharpoons ClO^-(aq) + H_3O^+(aq)$
(c) $NH_3(aq) + H_2O(l) \rightleftharpoons NH_4^+(aq) + OH^-(aq)$

Box 16-2   Chemistry and the Environment: Limestone Caverns

Limestone caverns are among nature's most spectacular displays. These caves occur in many parts of the world. Examples are Carlsbad Caverns in New Mexico, Jeita Caves in Lebanon, the Blue Grotto in Italy, and the Jenolan Caves in Australia. Wherever they occur, the chemistry of their formation involves the aqueous equilibria of limestone, which is calcium carbonate. Three such equilibria, linked to one another by Le Châtelier's principle, play essential roles in cave dynamics.

A spectacular gallery of Carlsbad Caverns.

The starting point for the aqueous chemistry of calcium carbonate is its solubility product:

$$CaCO_3(s) \rightleftharpoons Ca^{2+}(aq) + CO_3^{2-}(aq) \qquad K_{sp} = 3.36 \times 10^{-9}$$

As the small value of K_{sp} indicates, this substance is insoluble. The material from which caverns form was deposited over 250 million years ago, during an age when inland seas covered many regions of the continents. In these seas, corals, mollusks, and other marine organisms flourished, incorporating calcium ions dissolved in seawater into solid calcium carbonate shells and skeletons. As these sea creatures died, their remains accumulated on the ocean floor because of the insolubility of calcium carbonate. Over many eons, the shells and skeletons were compressed into limestone rock, eventually forming deposits hundreds of miles long.

When the climate became drier, the ancient seas evaporated, and these limestone deposits were buried beneath beds of silt. In time, uplift and erosion exposed the limestone deposits to nature's elements—primarily the action of running water. Water would not dissolve limestone were it not for additional equilibria involving carbon dioxide. First is the solubility equilibrium between gaseous and dissolved carbon dioxide:

$$CO_2(g) \rightleftharpoons CO_2(aq)$$

Dissolved CO_2 reacts with one molecule of water to form carbonic acid, which undergoes proton transfer with a second water molecule. These two reactions establish a second equilibrium that makes rainwater slightly acidic:

$$CO_2(aq) + 2 H_2O(l) \rightleftharpoons HCO_3^-(aq) + H_3O^+(aq)$$

When exposed to acidic water, limestone slowly dissolves by forming hydrogen carbonate anions:

$$CaCO_3(s) + H_3O^+(aq) \rightleftharpoons Ca^{2+}(aq) + HCO_3^-(aq) + H_2O(l)$$

Over thousands of years, water percolating through limestone dissolved some of the rock, forming underground chambers.

The story does not end with cavern formation. Many limestone caves, such as Carlsbad Caverns, contain spectacular formations that include stalagmites, stalactites, and limestone columns. We need to examine the equilibria more closely to understand how these structures form.

Inside newly formed caverns, calcium carbonate precipitates from water dripping from the ceilings of the chambers. This happens when water that is saturated with carbon dioxide and calcium hydrogen carbonate comes into contact with air. Some of the dissolved CO_2 escapes into the gas phase. This shifts the two equilibria to the left, and solid calcium carbonate precipitates:

$$CaCO_3(s) + CO_2(g) + H_2O(l) \xleftarrow{\text{Shifts to left}} Ca^{2+}(aq) + 2 HCO_3^-(aq)$$

The amount of calcium carbonate precipitating from any particular drop is imperceptibly small. Nevertheless, over the years, these deposits grow into translucent hollow tubes of $CaCO_3$ called "soda straws" (see photo inset). Soda straws lengthen as water drops fall from their tips. These delicate structures can reach lengths of several feet. In time, water flowing over the outside of the tube adds width to the growing formation, and the soda straw matures into the familiar stalactite.

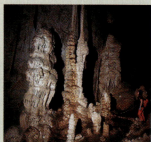

Stalactites (left), "soda straws" (inset), and stalagmites.

Over the eons, the flow and evaporation of water inside a cavern creates a stunning array of rock sculptures. Stalagmites grow upward from the floor, sometimes joining stalactites to form massive columns. Limestone dams create beautiful pools of water. Limestone draperies fall like curtains from water flowing around overhanging rock. Delicate mineral flowers sprout from the walls. All these features result from the aqueous solubility equilibrium of calcium carbonate.

As water slowly dissolves the limestone roof of a cave, the roof becomes weak and may eventually collapse. The result is a sinkhole. When a source of water remains, the sinkhole fills to form a lake that can be very scenic.

Montezuma's Well, a sinkhole in Arizona.

CHAPTER 16 VISUAL SUMMARY

SKILLS TO MASTER

1. Visualizing molecular equilibria
2. Writing K expressions
3. Calculating K from thermodynamic properties
4. Predicting changes in K with T
5. Predicting how changes affect equilibria
6. Calculating K from equilibrium concentrations
7. Completing concentration tables
8. Calculating equilibrium concentrations
9. Using approximations in calculations
10. Identifying major species
11. Recognizing types of equilibria
12. Relating K expressions to tabulated values

16.1 DESCRIBING CHEMICAL EQUILIBRIA

1. Dynamic equilibrium

Reversibility

$$K_{eq} = \frac{[I_3^-]_{eq}}{[I_2]_{eq}[I^-]_{eq}}$$

Equilibrium constant

$$a\text{A} + b\text{B} \rightleftharpoons c\text{D} + e\text{E}$$

$$K_{eq} = \frac{[\text{D}]_{eq}^d [\text{E}]_{eq}^e}{[\text{A}]_{eq}^a [\text{B}]_{eq}^b}$$ 2

16.2 PROPERTIES OF EQUILIBRIUM CONSTANTS

Properties of equilibrium constants

K_{eq} is related to the stoichiometry of the net reaction.

K_{eq} applies only at equilibrium.

K_{eq} is independent of initial conditions.

Pure solids, liquids, or solvents do not appear in the expression for K_{eq}.

$$K_{eq,\,fwd} = \frac{1}{K_{eq,\,rev}}$$

Magnitudes of K_{eq} vary over a wide range.

16.3 THERMODYNAMICS AND EQUILIBRIUM

Reaction goes left: $Q > K_{eq}$ & $\Delta G_{reaction} > 0$	Equilibrium when: $Q = K_{eq}$ & $\Delta G_{reaction} = 0$	Reaction goes right: $Q < K_{eq}$ & $\Delta G_{reaction} < 0$

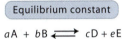

$2\,NO_2 \rightleftharpoons N_2O_4$

$K_{eq} > Q$: NO_2 formation favored

% Reaction

$K_{eq} < Q$: N_2O_4 formation favored

$K_{eq} = Q$: No net change

$$\ln K_{eq} = -\frac{\Delta H^o}{RT} + \frac{\Delta S^o}{R}$$

$$\Delta G^o = RT \ln K_{eq}$$ 3   4

16.4 SHIFTS IN EQUILIBRIUM

LeChâtelier's principle

A system at equilibrium responds to offset changes.

No catalyst

Catalysts do not affect the value of K_{eq}. 5

Catalyst

Products

Overall change remains the same

Reactants

E

Reaction coordinate

Initial equilibrium | Steam injected | New equilibrium

H_2O

CH_4

H_2

CO

Concentration

Time

16.5 WORKING WITH EQUILIBRIA

Substance	B	A
Initial concentrations 7	$[\text{B}]_i$	$[\text{A}]_i$
Changes in concentration	$\Delta[\text{B}]$	$\Delta[\text{A}]$
Final concentrations	$[\text{B}]_{eq}$	$[\text{A}]_{eq}$

$$\Delta[\text{B}] = \left(\frac{\text{Coeff. B}}{\text{Coeff. A}}\right)\Delta[\text{A}]$$

Concentration table

8 $[\text{A}]_{eq} = [\text{A}]_i + \Delta[\text{A}]$ 6

Equilibrium position

$AgBr\,(s) \rightleftharpoons Ag^+\,(aq) + Br^-\,(aq)$
$K_{eq} \ll 1$ 9

Extent of reaction

$[\text{A}]_i - x \cong [\text{A}]_i$

SOLVING EQUILIBRIUM PROBLEMS

1. Determine what is asked for.
2. Identify the major species.
3. Determine what equilibria exist.
4. Write the expression for K_{eq}.
5. Organize the data and unknowns.
6. Carry out the calculations.
7. Check for reasonableness.

16.5 EQUILIBRIA IN AQUEOUS SOLUTION

Major species
Minor species 10
Spectator ions
Dominant equilibrium

Water   Ions   Molecules

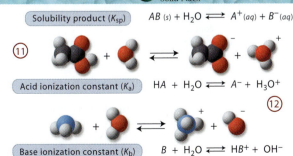

Aqueous solution

Solid NaCl

Solubility product (K_{sp}) $AB\,(s) + H_2O \rightleftharpoons A^+\,(aq) + B^-\,(aq)$

11

Acid ionization constant (K_a) $HA + H_2O \rightleftharpoons A^- + H_3O^+$

12

Base ionization constant (K_b) $B + H_2O \rightleftharpoons HB^+ + OH^-$

2 Formation constant (K_f) $M^{n+} + x\,L \rightleftharpoons [M^{n+}L_x]$

Learning Exercises

16.1 List the types of chemical equilibrium introduced in this chapter and give a specific example of each.

16.2 Draw a molecular picture that illustrates each of the examples that you gave in Learning Exercise 16-1.

16.3 Describe in your own words how to set up and solve a problem that asks for concentrations at equilibrium.

16.4 Describe how to calculate K_{eq} from equilibrium concentrations and from thermodynamic functions.

16.5 Write a paragraph that describes the logic, process, and justification for using approximations in solving equilibrium problems.

16.6 Use your own words to define (a) initial concentrations, (b) change to equilibrium, (c) change to completion, and (d) equilibrium concentrations.

16.7 Summarize in writing the connections between equilibrium constants and thermodynamic functions.

16.8 Update your list of memory bank equations and their uses.

16.9 List all terms new to you that appear in Chapter 16. Write a one-sentence definition for each in your own words. Consult the Glossary if you need help.

 Problems <u>ilw</u> = interactive learning ware problem. Visit the website at www.wiley.com/college/olmsted

---

Describing Chemical Equilibria

16.1 The alkene 2-butene has two isomers, *cis*-butene and *trans*-butene, which interconvert at high temperature or in the presence of a catalyst:

cis-Butene (g)   *trans*-Butene (g)

Suppose that the equilibrium constant for this reaction is 3.0. Draw a qualitative graph that shows how the pressure of each gas changes with time if the system initially contains pure *cis*-butene at a pressure of 1.0 bar.

16.2 Redraw the graph for the reaction in Problem 16.1, showing what happens if the system initially contains pure *trans*-butene at a pressure of 1.0 bar.

16.3 Aqueous solutions of sodium hypochlorite undergo decomposition according to the following two-step mechanism:

$$ClO^-(aq) + ClO^-(aq) \xrightarrow{k_1} Cl^-(aq) + ClO_2^-(aq)$$

$$ClO^-(aq) + ClO_2^-(aq) \xrightarrow{k_2} Cl^-(aq) + ClO_3^-(aq)$$

(a) What additional reactions will be important as the reaction approaches equilibrium?
(b) What is the equilibrium constant expression for the overall reaction?
(c) Express the equilibrium constant in terms of the rate constants of the elementary reactions.

16.4 The reaction of H_2 gas with CO gas to give formaldehyde has been described by this mechanism (all species in the gas phase):

$$H_2 \underset{k_{-1}}{\overset{k_1}{\rightleftharpoons}} 2\,H$$

$$H + CO \xrightarrow{k_2} HCO$$

$$HCO + H \xrightarrow{k_3} H_2CO$$

(a) What reactions describe the decomposition of H_2CO under these same conditions?
(b) What is the equilibrium constant expression for this reaction?
(c) Express the equilibrium constant in terms of rate constants for the elementary reactions.

16.5 Draw molecular pictures that illustrate the reversibility of the reactions involved in the hypochlorite decomposition given in Problem 16.3. (See Figure 16-1 for examples.)

16.6 Draw molecular pictures that illustrate the reversibility of the reactions involved in the reaction of H_2 with CO to form H_2CO given in Problem 16.4. (See Figure 16-1 for examples.)

16.7 Describe experiments that could be done to demonstrate that the decomposition reaction of ClO^- anions (Problem 16.3) is reversible.

16.8 Describe experiments that could be done to demonstrate that the reaction of H_2 with CO (Problem 16.4) is reversible.

Properties of Equilibrium Constants

16.9 Write the equilibrium constant expression for each of the following reactions:

    (a) $I_2(s) + 5\,F_2(g) \rightleftharpoons 2\,IF_5(g)$
    (b) $P_4(s) + 5\,O_2(g) \rightleftharpoons P_4O_{10}(s)$
    (c) $BaCO_3(s) + C(s) \rightleftharpoons BaO(s) + 2\,CO(g)$
    (d) $CO(g) + 2\,H_2(g) \rightleftharpoons CH_3OH(l)$
    (e) $H_3PO_4(aq) + 3\,H_2O(l) \rightleftharpoons PO_4^{3-}(aq) + 3\,H_3O^+(aq)$

16.10 Write the equilibrium constant expression for each of the following reactions:

    (a) $2\,H_2S(g) + 3\,O_2(g) \rightleftharpoons 2\,H_2O(l) + 2\,SO_2(g)$
    (b) $Fe_2O_3(s) + 3\,CO(g) \rightleftharpoons 2\,Fe(s) + 3\,CO_2(g)$
    (c) $Cl_2(g) + 3\,F_2(g) \rightleftharpoons 2\,ClF_3(g)$
    (d) $NH_3(g) + H_3O^+(aq) \rightleftharpoons NH_4^+(aq) + H_2O(l)$
    (e) $SnO_2(s) + 2\,H_2(g) \rightleftharpoons Sn(s) + 2\,H_2O(l)$

16.11 Write each of the chemical reactions of Problem 16.9 in the opposite direction and determine the equilibrium constant expressions for these reactions.

16.12 Write each of the chemical reactions of Problem 16.10 in the opposite direction and determine the equilibrium constant expressions for these reactions.

16.13 State the standard (reference) concentration for each substance appearing in the equilibria of Problem 16.9.

16.14 State the standard (reference) concentration for each substance appearing in the equilibria of Problem 16.10.

16.15 Using copper as an example, draw a molecular picture illustrating that the concentration of atoms in a pure metal is independent of the volume of the sample.

16.16 Using mercury as an example, draw a molecular picture illustrating that the concentration of atoms in a pure liquid is independent of the volume of the sample.

Thermodynamics and Equilibrium

<u>ilw</u>**16.17** Using standard thermodynamic data from Appendix D, calculate K_{sp} at 298 K for LiCl.

16.18 Using standard thermodynamic data from Appendix D, calculate K_{sp} at 298 K for $MgCl_2$.

16.19 Using standard thermodynamic data from Appendix D, calculate the equilibrium constant at 298 K for each of the following chemical equilibria:

    (a) $CO(g) + H_2O(l) \rightleftharpoons CO_2(g) + H_2(g)$
    (b) $2\,CO(g) + O_2(g) \rightleftharpoons 2\,CO_2(g)$
    (c) $BaCO_3(s) + C(s, graphite) \rightleftharpoons BaO(s) + 2\,CO(g)$
    (d) $3\,CO(g) + 6\,H_2(g) \rightleftharpoons C_3H_6(g) + 3\,H_2O(l)$

16.20 Using standard thermodynamic data from Appendix D, calculate the equilibrium constant at 298 K for each of the following chemical equilibria:

    (a) $CH_4(g) + H_2O(l) \rightleftharpoons CO(g) + 3\,H_2(g)$
    (b) $4\,NH_3(g) + 5\,O_2(g) \rightleftharpoons 4\,NO(g) + 6\,H_2O(l)$
    (c) $SnO_2(s) + 2\,H_2(g) \rightleftharpoons Sn(s, white) + 2\,H_2O(l)$
    (d) $3\,Fe(s) + 4\,H_2O(l) \rightleftharpoons Fe_3O_4(s) + 4\,H_2(g)$

16.21 Estimate K_{eq} for the equilibria in Problem 16.19 b and c at 250 K.

16.22 Estimate K_{eq} for the equilibria in Problem 16.20 c and d at 350 K.

16.23 Determine K_{eq} for the equilibria in Problem 16.19 a and d at 395 K if H_2O is present as a gas rather than as a liquid.

16.24 Estimate K_{eq} for the equilibria in Problem 16.20 c and d at 425 K if H_2O is present as a gas rather than as a liquid.

Shifts in Equilibrium

16.25 For each equilibrium in Problem 16.19, predict the effect of injecting additional CO (g) into the system.

16.26 For each equilibrium in Problem 16.20, predict the effect of injecting additional H_2O (l) into the system.

16.27 Consider the following reaction at equilibrium in water:

$$PbCl_2\,(s) \rightleftharpoons Pb^{2+}\,(aq) + 2\,Cl^-\,(aq)$$

Predict whether $PbCl_2$ will dissolve, precipitate, or do neither after each of the following changes: (a) more $PbCl_2$ (s) is added; (b) more H_2O is added; (c) solid NaCl is added; and (d) solid KNO_3 is added.

16.28 The following exothermic gas-phase reaction is at equilibrium:

$$2\,SO_2\,(g) + O_2\,(g) \rightleftharpoons 2\,SO_3\,(g)$$

Predict what happens to the amount of SO_3 in the system when each of the following changes is made: (a) the temperature is raised; (b) more O_2 is added; and (c) some Ar gas is introduced.

16.29 Consider the following gas-phase reaction:

$$SO_2\,(g) + Cl_2\,(g) \rightleftharpoons SO_2Cl_2\,(g) \qquad \text{(exothermic)}$$

Describe four changes that would drive the equilibrium to the left.

16.30 Consider the following gas-phase reaction:

$$PCl_5\,(g) \rightleftharpoons PCl_3\,(g) + Cl_2\,(g) \qquad \text{(endothermic)}$$

Describe four changes that would drive the equilibrium to the left.

Working with Equilibria

ilw 16.31 An industrial chemist puts 1.00 mol each of H_2 (g) and CO_2 (g) in a 1.00-L container at constant temperature of 800 °C. This reaction occurs:

$$H_2\,(g) + CO_2\,(g) \rightleftharpoons H_2O\,(g) + CO\,(g)$$

When equilibrium is reached, 0.49 mol of CO (g) is in the container. Find the value of K_{eq} for the reaction.

16.32 At high temperature, HCl and O_2 react to give Cl_2 gas:

$$4\,HCl\,(g) + O_2\,(g) \rightleftharpoons 2\,Cl_2\,(g) + 2\,H_2O\,(g)$$

If HCl at 2.30 bar and O_2 at 1.00 bar react at 750 K, the equilibrium pressure of Cl_2 is measured to be 0.93 bar. Determine the value of K_{eq} at 750 K.

16.33 When 0.0500 mol of propanoic acid ($C_2H_5CO_2H$) is dissolved in 0.500 L of water, proton transfer occurs:

$$C_2H_5CO_2H + H_2O \rightleftharpoons C_2H_5CO_2^- + H_3O^+$$

The equilibrium concentration of H_3O^+ ions is 1.15×10^{-3} M. Evaluate K_{eq}.

16.34 Cyanic acid, HCNO, is a weak acid:

$$HCNO + H_2O \rightleftharpoons CNO^- + H_3O^+$$

In a 0.20 M aqueous solution of HCNO, the concentration of H_3O^+ cations is 6.5×10^{-3} M. Evaluate K_{eq}.

ilw 16.35 The equilibrium constant for the following reaction is 1.6×10^5 at 1024 K:

$$H_2\,(g) + Br_2\,(g) \rightleftharpoons 2\,HBr\,(g)$$

Find the equilibrium pressures of all gases if 10.0 bar of HBr is introduced into a sealed container at 1024 K.

16.36 At 100 °C, $K_{eq} = 1.5 \times 10^8$ for the following reaction:

$$CO\,(g) + Cl_2\,(g) \rightleftharpoons COCl_2\,(g)$$

Using appropriate approximations, calculate the partial pressure of CO at 100 °C at equilibrium in a chamber that initially contains $COCl_2$ at a pressure of 0.250 bar.

16.37 At 1000 °C, $K_{eq} = 0.403$ for the following reaction:

$$FeO\,(s) + CO\,(g) \rightleftharpoons Fe\,(s) + CO_2\,(g)$$

If CO gas at 5.0 bar is injected into a container at 1000 °C that contains excess FeO, what are the partial pressures of all gases present at equilibrium?

16.38 K_{eq} for the Haber reaction is 2.81×10^{-5} at 472 °C. If a reaction starts with 3.0 bar of H_2 and 5.0 bar of N_2 at 472 °C, what is the equilibrium pressure of NH_3?

Equilibria in Aqueous Solutions

16.39 Identify the major species present in an aqueous solution of each of the following substances: (a) CH_3CO_2H; (b) NH_4Cl; (c) KCl; (d) $NaCH_3CO_2$; and (e) NaOH

16.40 Identify the major species present in an aqueous solution of each of the following substances: (a) HClO; (b) $CaBr_2$; (c) KClO; (d) HNO_3; and (e) HCN

16.41 Identify the acid–base equilibria, if any, for the major species in Problem 16.39.

16.42 Identify the acid–base equilibria, if any, for the major species in Problem 16.40.

16.43 When the following substances dissolve in water, what major species are present? (a) acetone; (b) potassium bromide; (c) lithium hydroxide; and (d) sulfuric acid

16.44 When the following substances dissolve in water, what major species are present? (a) sodium hydrogen carbonate; (b) methanol; (c) hydrogen bromide; and (d) benzoic acid

16.45 Write the equilibrium constant expression for each of the following reactions and relate it to equilibrium constants of standard types (K_{sp}, K_a, and K_b):
(a) $HClO_2\,(aq) + H_2O\,(l) \rightleftharpoons ClO_2^-\,(aq) + H_3O^+\,(aq)$
(b) $Fe^{3+}\,(aq) + 3\,OH^-\,(aq) \rightleftharpoons Fe(OH)_3\,(s)$
(c) $CN^-\,(aq) + H_3O^+\,(aq) \rightleftharpoons HCN\,(aq) + H_2O\,(l)$

16.46 Write the equilibrium constant expression for each of the following reactions and relate it to equilibrium constants of standard types (K_{sp}, K_a, and K_b):
(a) $Ag_2SO_4\,(s) \rightleftharpoons 2\,Ag^+\,(aq) + SO_4^{2-}\,(aq)$
(b) $H_3PO_4\,(aq) + H_2O\,(l) \rightleftharpoons H_2PO_4^-\,(aq) + H_3O^+\,(aq)$
(c) $HCO_2^-\,(aq) + H_3O^+\,(aq) \rightleftharpoons HCO_2H\,(aq) + H_2O\,(l)$

16.47 Identify the spectator ions present when the following are mixed:
(a) a solution of CH_3CO_2H and a solution of NaOH
(b) a solution of $CaCl_2$ and a solution of K_3PO_4
(c) a solution of HNO_3 and a solution of KOH

16.48 Identify the spectator ions present when the following are mixed:
(a) a solution of $FeCl_3$ and a solution of NaOH
(b) a solution of NH_3 and a solution of HNO_3
(c) a solution of KCH_3CO_2 and a solution of $HClO_4$

Additional Paired Problems

16.49 Write the equilibrium constant expressions for the following:
(a) $2\,NaHCO_3\,(s) \rightleftharpoons Na_2CO_3\,(s) + CO_2\,(g) + H_2O\,(g)$
(b) $2\,N_2\,(g) + 6\,H_2O\,(l) \rightleftharpoons 4\,NH_3\,(g) + 3\,O_2\,(g)$
(c) $2\,C_2H_4\,(g) + O_2\,(g) \rightleftharpoons 2\,CH_3CHO\,(g)$
(d) $Ag_2SO_4\,(s) \rightleftharpoons 2\,Ag^+\,(aq) + SO_4^{2-}\,(aq)$
(e) $NH_4HS\,(s) \rightleftharpoons H_2S\,(g) + NH_3\,(g)$

16.50 Write the equilibrium constant expressions for the following:
(a) $4\,NH_3\,(g) + 5\,O_2\,(g) \rightleftharpoons 4\,NO\,(aq) + 6\,H_2O\,(l)$
(b) $HCN\,(aq) + H_2O\,(l) \rightleftharpoons CN^-\,(aq) + H_3O^+\,(aq)$
(c) $2\,NH_4^+\,(aq) + SO_4^{2-}\,(aq) \rightleftharpoons (NH_4)_2SO_4\,(s)$
(d) $SO_4^{2-}\,(aq) + H_2O\,(l) \rightleftharpoons OH^-\,(aq) + HSO_4^-\,(aq)$
(e) $2\,O_3\,(g) \rightleftharpoons 3\,O_2\,(g)$

16.51 Phosgene is a deadly gas whose use in warfare has been outlawed. The gas decomposes at elevated temperature:

$$COCl_2 (g) \rightleftharpoons CO (g) + Cl_2 (g) \qquad K_{eq} = 8.3 \times 10^{-4} \text{ at } 352 \text{ °C}$$

What are the pressures at equilibrium if a 2.55-L metal container is charged with 3.00 bar of phosgene at 25 °C and then heated to 352 °C?

16.52 The equilibrium constant for the dissociation of Cl_2 into atomic chlorine at 1200 K is 2.5×10^{-5}. If Cl_2 gas at 298 K, 0.57 bar is placed in a sealed container that is then heated to 1200 K, what is the equilibrium pressure of Cl atoms?

16.53 The figure represents two chambers containing gases that react as follows:

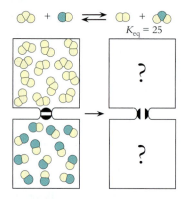

Draw a molecular picture that shows what the system looks like if the stopcock is opened and the reaction proceeds to equilibrium.

16.54 The figure represents two chambers containing gases that react as follows:

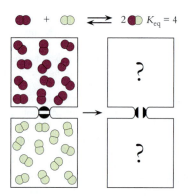

Draw a molecular picture that shows what the system looks like if the stopcock is opened and the reaction proceeds to equilibrium.

16.55 A saturated solution of chloroform ($CHCl_3$) in water contains one drop of excess chloroform. If more liquid chloroform is added to this mixture, does the concentration of chloroform in the aqueous solution change? Explain your answer in molecular terms.

16.56 A student mixes sodium chloride and water until the solid and the aqueous solution are at equilibrium. If more solid sodium chloride is added to this mixture, does the concentration of sodium chloride in the aqueous solution change? Explain your answer in molecular terms.

16.57 At 350 K, K_{eq} for the following reaction is 322:

$$Br_2 (g) + I_2 (g) \rightleftharpoons 2 \text{ IBr} (g)$$

Suppose the equilibrium partial pressure of bromine is 0.512 bar and that of iodine is 0.327 bar. What is the pressure of IBr?

16.58 Coal (solid carbon plus a collection of impurities) can be converted to a mixture of carbon dioxide and hydrogen gas by the following process:

$$C (s) + 2 H_2O (g) \rightleftharpoons CO_2 (g) + 2 H_2 (g) \qquad K_{eq} = 0.38 \text{ at } 1300 \text{ K}$$

At an industrial plant, a mixture of coal and water is placed inside a steel reaction vessel. When the mixture is heated to 1300 K, the equilibrium partial pressure of water is 2.80×10^2 bar. Calculate the partial pressures of H_2 and CO_2.

16.59 Using tabulated standard thermodynamic data from Appendix D, calculate K_{eq} for the reaction of NO_2 to form N_2O_4 at 298 K and at 525 K:

$$2 NO_2 (g) \rightleftharpoons N_2O_4 (g)$$

16.60 Using tabulated standard thermodynamic data from Appendix D, calculate K_{eq} for the following reaction at 298 K and at 825 K:

$$2 N_2O (g) + O_2 (g) \rightleftharpoons 4 NO (g)$$

16.61 Write the equilibrium reaction and equilibrium constant expression for each of the following processes: (a) Trimethylamine, $(CH_3)_3N$, a weak base, is added to water. (b) Hydrofluoric acid, HF, a weak acid, is added to water. (c) Solid calcium sulfate, $CaSO_4$, a sparingly soluble salt, is added to water.

16.62 Write the equilibrium reaction and equilibrium constant expression for each of the following processes: (a) Trichloroacetic acid, Cl_3CCO_2H, is added to water. (b) Aniline, $(C_6H_5)NH_2$, a weak base, is dissolved in water. (c) Solid calcium hydroxide, $Ca(OH)_2$, a sparingly soluble salt, is added to water.

16.63 Predict the effect of each of the following changes on the equilibrium position of the reaction in Example 16-14: (a) The partial pressure of NO_2 is cut in half. (b) The volume of the reactor is doubled. (c) A total of 10.0 bar of Ar gas is added to the reactor.

16.64 Predict the effect each of the following changes will have on the equilibrium position of the reaction in Example 16-12: (a) Some sodium benzoate is dissolved in the solution. (b) An additional 1.0 L of water is added to the solution. (c) Some NaCl is dissolved in the solution.

16.65 Use the appropriate data from Appendix D to calculate K_{sp} for LiCl at 100 °C. Is LiCl more or less soluble at high temperature than at low temperature? (See Problem 16.17.)

16.66 Use the appropriate data from Appendix D to calculate K_{sp} for $MgCl_2$ at 100 °C. Is $MgCl_2$ more or less soluble at high temperature than at low temperature? (See Problem 16.18.)

More Challenging Problems

16.67 At elevated temperature, carbon tetrachloride decomposes to its elements:

$$CCl_4 (g) \rightleftharpoons C (s) + 2 Cl_2 (g)$$

At 700 K, if the initial pressure of CCl_4 is 1.00 bar, the total pressure at equilibrium is 1.35 bar. Use these pressures to calculate K_{eq} at 700 K.

16.68 Suppose that the equilibrium system described in Problem 16.67 is expanded to twice its initial volume. Find the new equilibrium pressure.

16.69 Using tabulated thermodynamic data from Appendix D, compute (a) K_{eq} at 298 K; (b) the temperature at which the equilibrium pressure is 1.00 bar; and (c) K_{eq} at 1050 K for the following reaction:

$$Hg (g) + HgCl_2 (s) \rightleftharpoons Hg_2Cl_2 (s)$$

16.70 For the following reaction, K_{eq} is 1.83×10^{-3} at 390. K:

$$PCl_5 (g) \rightleftharpoons PCl_3 (g) + Cl_2 (g)$$

If 2.00 g of PCl_5 is placed in a 3.00-L bulb at 390 K, what is the equilibrium pressure of Cl_2? (1 atm = 1.013 bar)

16.71 Consider the following reaction:

$$CO_2(g) + 2\,OH^-(aq) \rightleftharpoons CO_3{}^{2-}(aq) + H_2O(l)$$

(a) Write the equilibrium constant expression for this reaction. (b) What will happen to the pressure of CO_2 in this equilibrium system if some Na_2CO_3 solid is dissolved in the solution? (c) What will happen to the pressure of CO_2 in this equilibrium system if some HCl gas is bubbled through the solution?

16.72 Benzene can be sulfonated by concentrated sulfuric acid:

$$C_6H_6 + H_2SO_4 \rightleftharpoons C_6H_5SO_3{}^- + H_3O^+$$

This reaction occurs by the following three-step mechanism:

$$H_2SO_4 \rightleftharpoons SO_3 + H_2O$$

$$SO_3 + C_6H_6 \rightleftharpoons C_6H_6SO_3$$

$$C_6H_6SO_3 + H_2O \rightleftharpoons C_6H_5SO_3{}^- + H_3O^+$$

Determine the relationship between the equilibrium constant for the net reaction and the rate constants for the various elementary steps.

16.73 At 1020 °C, K_{eq} for the conversion of $CO_2(g)$ to $CO(g)$ by solid graphite, C (s), is 167.5. A 1.00-L, high-pressure chamber containing excess graphite powder is charged with 0.494 mol each of CO_2 and CO and then is heated to 1020 °C. What is the equilibrium total pressure? (1 atm = 1.013 bar)

16.74 A chemist claims to have discovered a new gaseous element, effluvium (Ef), which reacts with atmospheric nitrogen to form effluvium nitride:

$$2\,Ef(g) + 3\,N_2(g) \rightleftharpoons 2\,EfN_3(g)$$

In a container initially containing N_2 at 1.00 bar pressure and Ef gas at 0.75 bar, the total gas pressure at equilibrium, according to the chemist's measurements, is 0.85 bar. Compute K_{eq} for this reaction.

16.75 $K_{eq} = 1.1 \times 10^{-33}$ for the following reaction at 298 K:

$$Sn(s) + 2\,H_2(g) \rightleftharpoons SnH_4(g)$$

Find the equilibrium pressure of $SnH_4(g)$ in a container at 298 K containing 10.0 g of Sn (s) and 2.00×10^2 bar of H_2. What volume of container is expected to contain a single molecule of SnH_4? 1(1 atm = 1.013 bar)

16.76 In the gas phase, acetic acid is at equilibrium with a dimer held together by a pair of hydrogen bonds. (a) If the total pressure of acetic acid gas in a glass bulb is 0.75 bar, what is the partial pressure of the

dimer? (b) Is the equilibrium constant for this reaction higher or lower at 200 °C? (*Hint:* Hydrogen bonds must be broken for the dimer to decompose.)

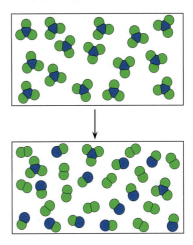

Dimer

16.77 The following figure represents a system coming to equilibrium:

(a) Use molecular pictures similar to those in Problem 16.55 to write a balanced equation for the equilibrium reaction, and (b) What is K_{eq} for the reaction?

16.78 Hydrogen fluoride is a highly reactive gas. It has many industrial uses, but the most familiar property of HF is its ability to react with glass. As a result, HF is used to etch glass and frost the inner surfaces of light bulbs. Hydrogen fluoride gas must be stored in stainless steel containers, and aqueous solutions must be stored in plastic bottles. Hydrogen fluoride can be produced from H_2 and F_2:

$$H_2(g) + F_2(g) \rightleftharpoons 2\,HF(g) \qquad K_{eq} = 115$$

In a particular experiment, 3.00 bar each of H_2 and F_2 are added to a 1.50-L flask. Calculate the equilibrium partial pressures of all species.

Group Study Problems

16.79 (a) A container contains CO_2 at $P = 0.464$ bar. When graphite is added to the container, some CO_2 is converted to CO, and at equilibrium the total pressure is 0.746 bar. Compute K_{eq} for the reaction:

$$CO_2(g) + C(s) \rightleftharpoons 2\,CO(g)$$

(b) Suppose that the equilibrium system is compressed to one-third its initial volume. Find the new equilibrium pressure.

16.80 When 1.00 mol of gaseous HI is sealed in a 1.00-L flask at 225 °C, it decomposes until the equilibrium amount of I_2 present is 0.182 moles:

$$2\,HI(g) \rightleftharpoons H_2(g) + I_2(g)$$

(a) Use these data to calculate K_{eq} for this reaction at 225 °C.
(b) Using standard thermodynamic data from Appendix D, estimate K_{eq} at 625 °C.
(c) Using the result of (b), calculate the equilibrium partial pressures of all reagents under these conditions. (1 atm = 1.013 bar)

16.81 One reaction that plays a role in photochemical smog is the following:

$$O_3(g) + NO(g) \rightleftharpoons O_2(g) + NO_2(g) \qquad K_{eq} = 6.0 \times 10^{34}$$

Automobile engines produce both O_3 and NO. If the morning rush hour results in an atmosphere containing 1.5 ppm NO and 6.5 ppb O_3, calculate the equilibrium partial pressures of the three pollutants (O_2 is 21% of the atmosphere).

16.82 The key step in the synthesis of sulfuric acid is the combustion of SO_2 to give sulfur trioxide:

$$2\,SO_2(g) + O_2(g) \rightleftharpoons 2\,SO_3(g) \qquad K_{eq} = 5.60 \times 10^4 \text{ at } 350 \text{ °C}$$

(a) Sulfur dioxide and oxygen are mixed initially at 0.350 and 0.762 bar, respectively, at 350 °C. What are the partial pressures of the three gases when the mixture reaches equilibrium?
(b) Use thermodynamic data (see Appendix D) to calculate K_{eq} for the sulfur trioxide reaction at 298 K. Based on this result, should this industrial synthesis be run at a higher or lower temperature than 350 °C to improve the yield?
(c) Suggest reasons why this reaction is run industrially at 700 K.

16.83 The combustion of sulfur dioxide is one of the steps in the synthesis of sulfuric acid:

$$2\,SO_2 + O_2 \rightleftharpoons 2\,SO_3$$

The following diagram represents a mixture of SO_2, O_2, and SO_3 at equilibrium at $T = 1100$ K:

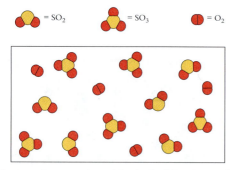

(a) Some gas is removed, resulting in the following conditions.

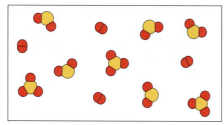

What happens to the position of the equilibrium? Explain.

(b) When the temperature of the system is increased to 1300 K, reaction occurs, leading to the following equilibrium conditions:

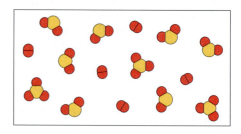

Is the reaction exothermic or endothermic? Use equilibrium arguments to explain your answer.

(c) Some gas is added to the original equilibrium mixture (at 1100 K), giving the following conditions:

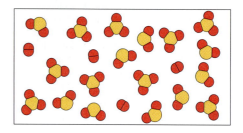

What happens to the position of the equilibrium? Explain.

Answers to Section Exercises

16.1.1 $K_{eq} = \dfrac{[O_2]^3}{[O_3]^2}$

16.1.2

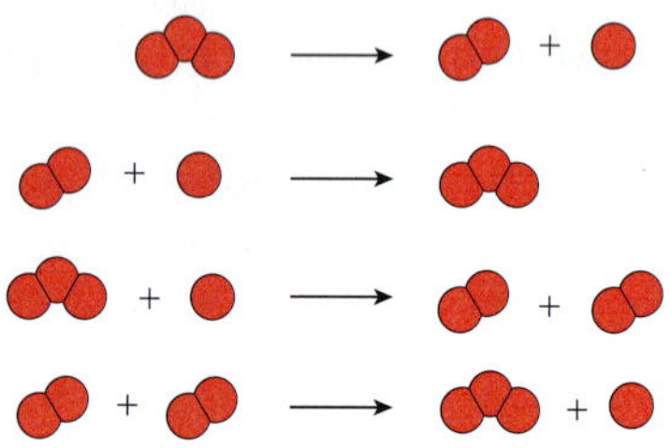

16.1.3 $K_{eq} = \dfrac{[H_2O]^6\,[NO]^4}{[NH_3]^4\,[O_2]^5}$

16.2.1 (a) $K_{eq} = (p_{Cl_2})_{eq}$; (b) $K_{eq} = (p_{CO_2})_{eq}$;

(c) $K_{eq} = [PO_4^{3-}]_{eq}^2[Ca^{2+}]_{eq}^3$;

(d) $K_{eq} = \dfrac{[H_3O^+][CN^-]}{(p_{HCN})_{eq}}$; (e) $K_{eq} = \dfrac{1}{(p_{Cl_2})_{eq}}$; and

(f) $K_{eq} = \dfrac{1}{[PO_4^{3-}]_{eq}^2\,[Ca^{2+}]_{eq}^3}$

16.2.2

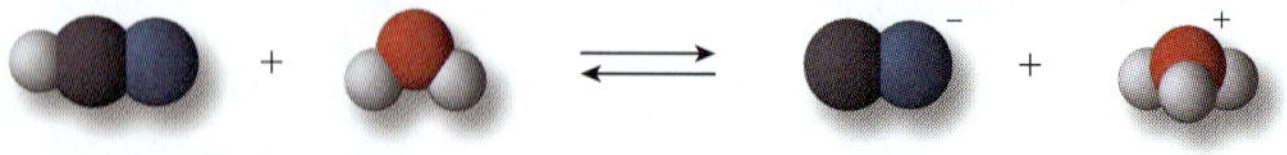

16.2.3 $K_{eq} = \dfrac{(p_{F_2})_{eq}}{(p_F)_{eq}^2} = 4.8 \times 10^{21}$;

$K_{eq} = \dfrac{(p_{H_2})_{eq}\,(p_{Br_2})_{eq}}{(p_{HBr})_{eq}^2} = 1.9 \times 10^{-19}$

16.3.1 $K_{eq} = 1.4 \times 10^{-8}$

16.3.2 Increases because the reaction is endothermic

16.3.3 $K_{eq} = 2.4 \times 10^{-8}$

16.4.1 (a) No change in Q, thus no change; (b), (c), and (d) more $Ba(OH)_2$ dissolves

16.4.2 Lower T, remove NH_3, and increase pressures of H_2 and/or N_2

16.4.3 The H_2–N_2 reaction has a large K_{eq} at room temperature, whereas the CH_4–H_2O reaction has a small K_{eq} at room temperature.

16.5.1 $K_{eq} = 1.10 \times 10^{-10}$

16.5.2 $[CH_3CO_2^-]_{eq} = [H_3O^+]_{eq} = 6.7 \times 10^{-3}$ M and $[CH_3CO_2H]_{eq} = 2.5$ M

16.5.3 $[Ag^+]_{eq} = 2.7 \times 10^{-12}$ M; $[Br^-]_{eq} = 0.200$ M

16.6.1 (a) H_3O^+, ClO_4^-, and H_2O; (b) NH_3 and H_2O; (c) K^+, HCO_3^-, and H_2O; and (d) $HClO$ and H_2O

16.6.2 $Ni(OH)_2$; net reaction is $Ni^{2+}(aq) + 2\,OH^-(aq) \rightarrow Ni(OH)_2\,(s)$; spectators are SO_4^{2-} and Na^+.

16.6.3 (a) $1/K_{sp}$; (b) K_a; and (c) K_b

Answers to Extra Practice Exercises

16.1 $2\,HCO_2H\,(g) \rightleftharpoons (HCO_2H)_2\,(g)$; $K_{eq} = \dfrac{[(HCO_2H)_2]_{eq}}{[HCO_2H]^2_{eq}}$

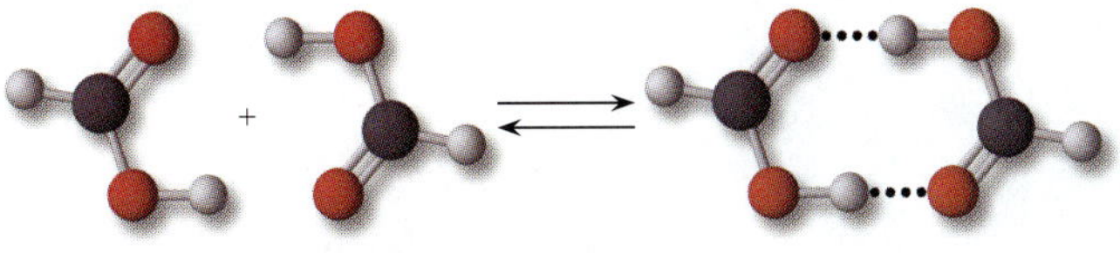

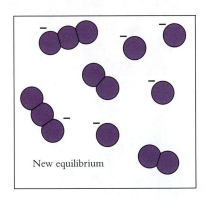

16.2 $3\,O_2\,(g) \rightleftharpoons 2\,O_3\,(g)$; $K_{eq} = \dfrac{[O_3]^2_{eq}}{[O_2]^3_{eq}}$

16.3 $K_{eq} = \dfrac{(p_{NO_2})^2_{eq}}{(p_{N_2O_4})_{eq}} = 1/3.10 = 0.323$

16.4 $K_{eq} = \dfrac{[Cl^-]_{eq}\,[H_3O^+]_{eq}}{(p_{HCl})_{eq}}$; pressure of HCl should be in bar, concentrations of the two ions in molarity

16.5 $K_{eq} = 6.1 \times 10^{-58}$

16.6 $K_{eq} = 1.7 \times 10^{-21}$

16.7 The shift is removal of some I_2, so your new picture should show one less triiodide and one more each of I_2 and iodide.

New equilibrium

16.8 Add I_2, add a soluble iodide salt such as KI, cool the solution, boil off some solvent.

16.9 $K_{eq} = 1.0 \times 10^3$

16.10

Reaction	$H_2O\,(l)$ +	$NH_3\,(aq)$	$\rightleftharpoons$ $OH^-\,(aq)$ +	$NH_4^+\,(aq)$
Initial concentration (M)		0.335	0	0
Change in concentration (M)		−0.00246	+0.00246	+0.00246
Equilibrium concentration (M)		0.333	0.00246	0.00246

$K_{eq} = 1.8 \times 10^{-5}$

16.11 At equilibrium, $p(AB_2) = p(A) = 1.0$ bar and $p(AB) = 2$ bar

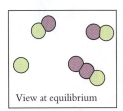

View at equilibrium

16.12 $[NH_4^+] = [OH^-] = 4.7 \times 10^{-4}$ M; $[NH_3] = 1.20 \times 10^{-2}$ M

16.13 $[NH_4^+] = [OH^-] = 1.6 \times 10^{-2}$ M; $[NH_3] = 14.8$ M

16.14 0.50 mol AgBr forms; $[Ag^+] = [Br^-] = 7.3 \times 10^{-7}$ M

16.15 Major species in aqueous hydrochloric acid are H_3O^+, Cl^-, and H_2O; in aqueous sodium chlorate are Na^+, ClO_3^-, and H_2O; and in aqueous chloroform are $CHCl_3$ and H_2O

16.16 (a) $1/K_a$; (b) K_{sp}; (c) K_f; and (d) K_b

16.17 Na^+ and NO_3^-

Acids and Taste

The tangy flavor of a fresh orange, the sour bite of lemon juice, and the tart pleasure of a Granny Smith apple are familiar flavors. Tangy, sour, and tart are taste sensations created by carboxylic acids, which contain the —CO_2H functional group. Although it is not yet known exactly how tastes arise, a reasonable speculation is that the sensing process involves proton transfer reactions from this functional group to protein molecules in our taste buds.

Our molecular inset shows citric acid, which has three carboxylic acid functional groups. Citric acid is present in all citrus fruits as well as many other tart-tasting foods, including berries, pineapples, pears, and tomatoes. Lemons are acidic because they contain as much as 3% citric acid by mass.

The grape is another fruit containing carboxylic acids that contribute to its taste. Grapes contain several carboxylic acids as well as sugar molecules. When grapes are fermented to make wine, most of their sugar is converted into ethanol, but the carboxylic acids remain unchanged. A wine is "sweet" if its residual sugar content masks the sour taste of these residual acids and "dry" if its residual sugar content is less than about 1%.

Vegetables also contain carboxylic acids that contribute to their flavors. One example is oxalic acid, prevalent in spinach and rhubarb. Raw rhubarb leaves are mildly poisonous, and folklore holds that the toxic substance is oxalic acid. However, raw

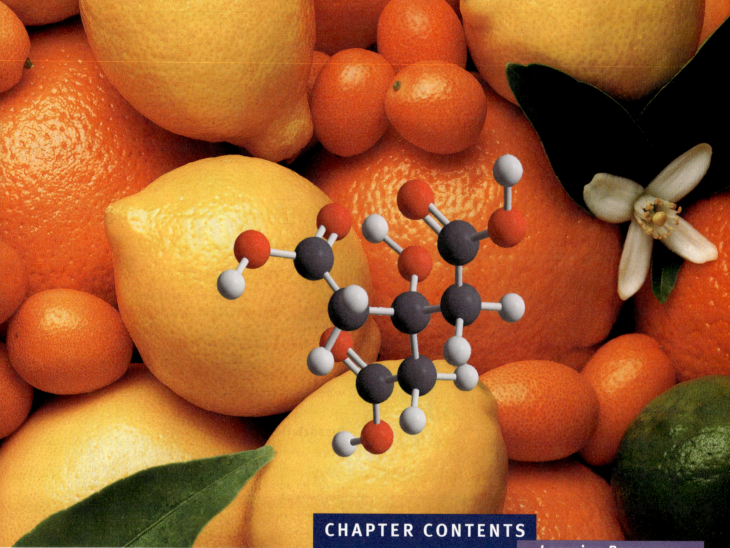

Aqueous Acid–Base Equilibria

17

Learning Resources

 KEY CONCEPTS

 CRITICAL THINKING

 SKILLS TO MASTER

 ADDITIONAL HELP
www.wiley.com/college/olmsted
- TUTORIALS
- ANIMATIONS

spinach can be eaten safely despite its equally high content of this acid.

Bacteria in milk feed on lactose (milk sugar) to produce lactic acid, another carboxylic acid. Under controlled conditions, this process gives yogurt. Similarly, the proper combination of yeast and bacteria acts on flour to form just enough lactic acid and acetic acid (CH_3CO_2H) to give sourdough bread its distinctive tang.

In addition to conveying acidity, the carboxylic acid group is highly polar, so smaller carboxylic acids such as acetic acid are water-soluble. Vinegar is a solution of acetic acid in water, typically about 5% by mass of the acid. Acetic acid can be produced from sugar-containing foods by a combination of fermentation and oxidation, which breaks down glucose molecules into ethanol and then converts ethanol to acetic acid. The word *vinegar* comes from the French term for "sour wine"; indeed, poor wine tastes sour because some of its ethanol has become acetic acid.

This chapter describes the details of acid–base chemistry. We begin with a molecular view and a way of measuring acidity. Then we look at acid–base equilibrium calculations, making use of the general approaches to equilibrium developed in Chapter 16. We describe the important applications of buffers and titrations in Chapter 18.

17.1 PROTON TRANSFERS IN WATER

In an acid–base reaction, a proton (H^+) is transferred from one chemical species to another. A species that donates a proton is an acid, and a species that accepts a proton is a base. This identification of acids and bases is the **Brønsted–Lowry definition** of acid–base reactions. From this perspective, every acid–base reaction has two reactants, an acid and a base. Every acid–base reaction also forms two products:

Acid		**Base**				
$HCl\,(g)$	+	$H_2O\,(l)$	$\longrightarrow$	$Cl^-\,(aq)$	+	$H_3O^+\,(aq)$
$H_2O\,(l)$	+	$PO_4^{3-}\,(aq)$	$\longrightarrow$	$OH^-\,(aq)$	+	$HPO_4^{2-}\,(aq)$
$H_3O^+\,(aq)$	+	$OH^-\,(aq)$	$\longrightarrow$	$H_2O\,(l)$	+	$H_2O\,(l)$
$CH_3CO_2H\,(aq)$	+	$OH^-\,(aq)$	$\longrightarrow$	$CH_3CO_2^-\,(aq)$	+	$H_2O\,(l)$

Although these reactions look different, they all involve the same chemical process: an acid collides with a base and transfers H^+. Figure 17-1 illustrates this process at the molecular level.

Notice that water molecules appear in all of these examples. Aqueous acid–base chemistry usually includes water as either a starting material or a product.

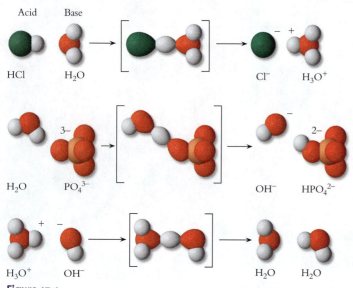

Acid   Base

HCl   H₂O   Cl⁻   H₃O⁺

H₂O   PO₄³⁻   OH⁻   HPO₄²⁻

H₃O⁺   OH⁻   H₂O   H₂O

Figure 17-1
Every acid–base reaction involves transfer of a proton from an acid to a base.

Dissociation of Water

These examples of acid–base reactions show that water can act as either an acid or a base: Water accepts a proton from an HCl molecule, but it donates a proton to a PO_4^{3-} anion. As an acid, water donates a proton to a base and becomes a hydroxide anion. As a base, water accepts a proton from an acid and becomes a hydronium cation. A chemical species that can both donate and accept protons is said to be **amphiprotic.** Water is an amphiprotic molecule.

This ability to serve as both a proton donor and a proton acceptor suggests that a proton transfer equilibrium exists for pure water, as Figure 17-2 illustrates.

$$2\,H_2O\,(l) \rightleftharpoons OH^-\,(aq) + H_3O^+\,(aq) \qquad K_{eq} = K_w = [H_3O^+]_{eq}\,[OH^-]_{eq}$$

The equilibrium constant for this reaction is the **water equilibrium constant (K_w).** Its value is 1.0×10^{-14} at 25 °C.

Figure 17-2
The water equilibrium illustrates the amphiprotic nature of H_2O. In this reaction, one water molecule acts as a proton donor (acid), and another acts as a proton acceptor (base).

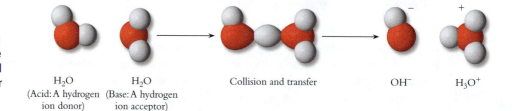

H₂O
(Acid: A hydrogen ion donor)

H₂O
(Base: A hydrogen ion acceptor)

Collision and transfer

OH⁻

H₃O⁺

In pure water, the hydronium and hydroxide ion concentrations are equal: We find these concentrations by taking the square root of K_w:

$$[H_3O^+]_{eq} = [OH^-]_{eq} = 1.0 \times 10^{-7} \text{ M} \qquad \text{(pure water at 25 °C)}$$

Equal concentrations of these two ions means that pure water is neither acidic nor basic.

The water equilibrium describes an inverse relationship between $[H_3O^+]_{eq}$ and $[OH^-]_{eq}$. When an acid dissolves in water, the hydronium ion concentration increases, so the hydroxide ion concentration must decrease to maintain the product of the concentrations at 1.0×10^{-14}. Similarly, the hydroxide ion concentration increases when a base dissolves in water, so the hydronium ion concentration must decrease.

Strong Acids

As noted in Chapter 4, an acid that quantitatively donates protons to water molecules is called a **strong acid**. Table 4-2 lists the six most common strong acids: HNO_3, $HClO_4$, H_2SO_4, HCl, HBr, and HI. In an aqueous solution of a strong acid, the hydronium ion concentration is equal to the concentration of the acid solution. The concentration of hydroxide ions in a solution of a strong acid can be calculated from the concentration of H_3O^+ and K_w. Example 17-1 provides an illustration.

Example 17-1

Ion Concentrations in a Solution of Strong Acid

Determine the ion concentrations in a 5.0×10^{-2} M aqueous solution of $HClO_4$, a strong acid.

Strategy: The seven-step method described in Chapter 16 provides the strategy for working any quantitative problem that involves equilibria. As a reminder, we detail each step in this first example of acid–base equilibria.

1. **Determine what is asked for.** This problem asks for concentrations of all ions in an acid solution.

2. **Identify the major chemical species.** Because $HClO_4$ is a strong acid, it transfers protons to water molecules quantitatively:

$$HClO_4 + H_2O \longrightarrow H_3O^+(aq) + ClO_4^-(aq)$$

   Thus, the major species in solution are H_3O^+, ClO_4^-, and H_2O.

3. **Determine what chemical equilibria exist.** The water equilibrium always exists in aqueous solution:

$$2 H_2O(l) \rightleftharpoons OH^-(aq) + H_3O^+(aq)$$

4. **Write the K_{eq} expressions.** The strong acid dissociates completely, so the water equilibrium is the only one that applies to this problem:

$$K_{eq} = K_w = [H_3O^+]_{eq}[OH^-]_{eq} = 1.0 \times 10^{-14}$$

Solution:

5. **Organize the data and unknowns.** This aqueous solution is 5.0×10^{-2} M $HClO_4$, so $[H_3O^+] = [ClO_4^-] = 5.0 \times 10^{-2}$ M. The final concentrations are found using an equilibrium analysis. Set up a concentration table for the water dissociation equilibrium, and define the change in hydronium ion concentration as x:

Reaction	$2 H_2O(l) \rightleftharpoons$	$H_3O^+(aq)$	$+$ $OH^-(aq)$
Initial concentration (M)		5.0×10^{-2}	0
Change in concentration (M)		$+x$	$+x$
Equilibrium concentration (M)		$5.0 \times 10^{-2} + x$	x

Example 17-1

Ion Concentrations in a Solution of Strong Acid (*continued*)

6. **Substitute and calculate.** The equilibrium constant is much smaller than the initial concentration, 5.0×10^{-2}, so it is reasonable to assume that $x \ll 5.0 \times 10^{-2}$. This leads to an approximation:

$$(5.0 \times 10^{-2} + x) \cong 5.0 \times 10^{-2}$$

Solving for x gives the equilibrium concentration of hydroxide ions:

$$K_w = [H_3O^+]_{eq}[OH^-]_{eq} = (5.0 \times 10^{-2})\,x = 1.0 \times 10^{-14}$$

$$x = \frac{1.0 \times 10^{-14}}{5.0 \times 10^{-2}} = 2.0 \times 10^{-13} \quad so \quad [OH^-]_{eq} = 2.0 \times 10^{-13}\ M$$

7. Does the Result Make Sense? Each of the ion concentrations is equal to or smaller than the starting concentration of the acid, which is reasonable. Also, the approximation $x \ll 5.0 \times 10^{-2}$ is valid, because $x = 2.0 \times 10^{-13}$.

Extra Practice Exercise 17.1

Determine the ion concentrations in a solution of HCl whose concentration is 6.5 mM.

In any solution of an acid, the total hydronium and hydroxide ion concentrations include the 10^{-7} M contribution from the water reaction. This example illustrates, however, that the change in hydronium ion concentration due specifically to the water equilibrium is negligibly small in an aqueous solution of a strong acid. This is true for any strong acid whose concentration is greater than 10^{-5} M. Consequently, the hydronium ion concentration equals the concentration of the strong acid, and

$$[OH^-]_{eq} = \frac{1.0 \times 10^{-14}}{[H_3O^+]_{eq}}\ M$$

Strong Bases

A substance that generates hydroxide ions quantitatively in aqueous solution is a **strong base.** The most common strong bases are the soluble metal hydroxides, among which NaOH perennially ranks among the top ten industrial chemicals. When a soluble metal hydroxide dissolves in water, it generates metal cations and hydroxide anions:

$$NaOH\,(s) \xrightarrow{H_2O} Na^+(aq) + OH^-(aq)$$

For any aqueous strong base, the hydroxide ion concentration can be calculated directly from the overall solution molarity. As is the case for aqueous strong acids, the hydronium and hydroxide ion concentrations are linked through the water equilibrium, as shown by Example 17-2.

Example 17-2

Ion Concentrations in a Strong Base Solution

What are the ion concentrations in 0.500 L of an aqueous solution that contains 5.0 g of NaOH?

Strategy: For this example, we summarize the first four steps of the method: The problem asks for the concentration of ions. Sodium hydroxide is a strong base that dissolves in water to generate Na^+ cations and OH^- anions quantitatively. The concentration of hydroxide ion equals the concentration of the base. The water equilibrium links the concentrations of OH^- and H_3O^+, so an equilibrium calculation is required to determine the concentration of hydronium ion. What remains is to organize the data, carry out the calculations, and check for reasonableness.

Solution: The overall molarity of NaOH can be calculated from mass, molar mass, and volume:

Example 17-2

**Ion Concentrations in a
Strong Base Solution**
(*continued*)

$$MM_{NaOH} = 40.00 \text{ g/mol}$$

$$n_{NaOH} = \frac{m}{MM} = \frac{5.0 \text{ g}}{40.00 \text{ g/mol}} = 0.125 \text{ mol}$$

$$M_{NaOH} = \frac{n}{V} = \frac{0.125 \text{ mol}}{0.500 \text{ L}} = 0.25 \text{ mol/L}$$

The solution is 0.25 M in NaOH, so the initial concentrations of the two ions are $[Na^+] = [OH^-] = 0.25$ M.

Use the water equilibrium to determine the concentration of H_3O^+.

$$2 H_2O (l) \rightleftharpoons OH^- (aq) + H_3O^+ (aq) \qquad K_w = [H_3O^+]_{eq} [OH^-]_{eq} = 1.0 \times 10^{-14}$$

Set up a concentration table, defining the change in hydronium ion concentration as x:

Reaction	$2 H_2O (l)$	$\rightleftharpoons$	$H_3O^+ (aq)$	+	$OH^- (aq)$
Initial concentration (M)			0		0.25
Change in concentration (M)			$+x$		$+x$
Equilibrium concentration (M)			x		$0.25 + x$

The equilibrium constant is much smaller than 0.25, so it is reasonable to assume that $x \ll 0.25$ and make the approximation that $0.25 + x \cong 0.25$. Solving for x gives the equilibrium concentration of hydronium ions:

$$K_w = 1.0 \times 10^{-14} = x (0.25 + x) \cong 0.25 x$$

$$x \cong \frac{1.0 \times 10^{-14}}{0.25} = 4.0 \times 10^{-14} \quad so$$

$$[H_3O^+]_{eq} = 4.0 \times 10^{-14} \text{ M} \quad and \quad [OH^-]_{eq} = 0.25 \text{ M}$$

Does the Result Make Sense? 5.0 g of NaOH is around 0.1 mol, and the solution volume is 0.5 L, so concentrations less than 1 M are reasonable. Note that x is indeed much smaller than 0.250.

Determine the ion concentrations in a solution prepared by dissolving 68 mg of LiOH in enough water to make 125 mL of solution.

**Extra Practice
Exercise 17.2**

Just as for solutions of strong acids, the water equilibrium contributes negligibly to the total hydroxide ion concentration in any solution of strong base whose concentration is greater than 10^{-5} M. Consequently, the hydroxide ion concentration equals the concentration of the strong base, and

$$[H_3O^+]_{eq} = \frac{1.0 \times 10^{-14}}{[OH^-]_{eq}} \text{ M}$$

Section Exercises

17.1.1 Write the net proton transfer reaction and draw a molecular picture to illustrate the reaction between H_3O^+ and SO_4^{2-}.

17.1.2 What is $[OH^-]$ if a sample of acid rain has $[H_3O^+] = 2.5 \times 10^{-4}$ M?

17.1.3 A student prepares 500.0 mL of aqueous solution containing 5.61 g of KOH. Determine the concentrations of all ions in this solution.

17.2 THE pH SCALE

The hydronium ion concentration in aqueous solution ranges from extremely high to extremely low. Here are three examples:

$$6.0 \text{ M HCl, } [H_3O^+] = 6.0 \text{ M}$$

$$\text{pure } H_2O, [H_3O^+] = 1.0 \times 10^{-7} \text{ M}$$

$$0.25 \text{ M NaOH, } [H_3O^+] = 4.0 \times 10^{-14} \text{ M}$$

Chemists use logarithmic notation to express this huge range of concentrations. As one example, $\log (4.0 \times 10^{-14}) = \log (4.0) + \log (10^{-14}) = 0.60 + (-14) = -13.40$. Notice that a logarithm has two parts. The numbers preceding the decimal point determine the power of 10. The numbers after the decimal point determine the numerical value. Therefore, when two digits follow the decimal place of the logarithm, the corresponding numerical value has two significant figures.

Because hydronium concentrations usually involve negative powers of ten, chemists use a negative log scale in expressing these concentrations. Equation 17-1 defines the **pH** scale of acid concentration:

$$pH = -\log [H_3O^+] \tag{17-1}$$

The pH of a solution is obtained by taking the logarithm of the hydronium ion concentration and then changing the sign. For example, the pH of pure water is

$$pH = -\log [H_3O^+] = -\log (1.0 \times 10^{-7}) = -(-7.00) = 7.00$$

The reverse conversion from pH to $[H_3O^+]$ uses powers of 10. For example, lemon juice has a pH of about 2.0:

$$[H_3O^+] = 10^{-pH} = 10^{-2.0} = 1 \times 10^{-2} \text{ M}$$

A logarithm carries no units, but the result of this calculation is a concentration in mol/L (M).

Features of the pH Scale

1. pH = 7.00 defines a "neutral" solution, $[H_3O^+] = [OH^-] = 10^{-7}$ M, neither acidic nor basic. Acidic solutions have pH < 7, and basic solutions have pH > 7.

2. The *more acidic* the solution, the *lower* its pH: Lemon juice, at pH ~ 2, is much more acidic than acid rain, pH ~ 4.

3. A change in pH of one unit reflects a tenfold change in hydronium ion concentration: Normal rainfall has pH ~ 5, but acid rain has ten times larger hydronium ion concentration, pH ~ 4.

4. The immense range of concentrations from >1 M to <10^{-14} M is compressed into a more convenient range, from ~ −1 to ~ +15.

Figure 17-3 shows the range of pH and hydronium ion concentrations. The measurement of pH is a routine operation in most laboratories. Litmus paper, which turns red when dipped in acidic solution and blue when dipped in basic solution, gives a quick, qualitative indication of acidity. As Figure 17-4 shows, approximate measures of pH can be done using pH paper. Universal pH paper displays a range of colors in response to different pH values and is accurate to about 0.5 pH unit. For quantitative pH determinations, scientists use pH meters.

A pH meter uses a probe to make electrical measurements between a reference solution containing acid at known concentration and the sample being tested. The pH meter is calibrated by dipping the probe in a solution of known pH and adjusting the meter to read the correct value. Then the probe is dipped into the sample solution, and the pH appears on a digital display. Example 17-3 illustrates how to convert from pH measurements to ion concentrations.

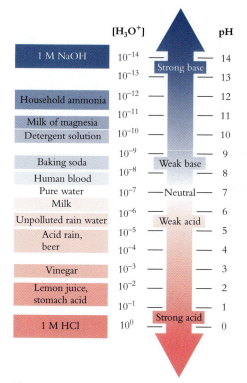

[H₃O⁺]   **pH**

1 M NaOH 10^{-14} —— Strong base —— 14
            10^{-13} —— —— 13
Household ammonia 10^{-12} —— —— 12
Milk of magnesia 10^{-11} —— —— 11
Detergent solution 10^{-10} —— —— 10
            10^{-9} —— —— 9
Baking soda   —— Weak base ——
Human blood 10^{-8} —— —— 8
Pure water 10^{-7} —— Neutral —— 7
Milk
Unpolluted rain water 10^{-6} —— —— 6
            —— Weak acid ——
Acid rain, 10^{-5} —— —— 5
beer 10^{-4} —— —— 4
Vinegar 10^{-3} —— —— 3
Lemon juice, 10^{-2} —— —— 2
stomach acid 10^{-1} —— —— 1
            —— Strong acid ——
1 M HCl 10^{0} —— —— 0

Figure 17-3
The pH values of commonly encountered aqueous solutions range from 0 to 14.

Figure 17-4
pH measurements can be made using universal pH paper or a pH meter.

What are the concentrations of hydronium and hydroxide ions in a beverage whose pH = 3.05?

Strategy: To convert from pH to ion concentrations, first apply Equation 17-1 to calculate $[H_3O^+]$.

$$pH = -\log [H_3O^+]$$

Then make use of the water equilibrium to calculate $[OH^-]$.

Solution: We must rearrange Equation 17-1, solving for the concentration:

$$\log [H_3O^+] = -pH \quad and \quad [H_3O^+] = 10^{-pH}$$

Now substitute and evaluate:

$$[H_3O^+] = 10^{-3.05} = 8.9 \times 10^{-4} \text{ M}$$

To determine the hydroxide ion concentration, we need to rearrange the water equilibrium expression:

$$K_w = 1.0 \times 10^{-14} = [H_3O^+]_{eq} [OH^-]_{eq} \quad and \quad [OH^-]_{eq} = \frac{1.0 \times 10^{-14}}{[H_3O^+]_{eq}}$$

Now substitute and evaluate:

$$[OH^-]_{eq} = \frac{1.0 \times 10^{-14}}{8.9 \times 10^{-4}} = 1.1 \times 10^{-11} \text{ M}$$

Does the Result Make Sense? A pH around 3 represents an acidic solution, so we expect the hydronium ion concentration to be much larger than the hydroxide ion concentration. Remember that although pH and K_w are dimensionless, concentrations of solutes always are expressed in mol/L. Our results have two significant figures because the logarithm has two decimal places.

Example 17-3

pH and Ion Concentrations

Extra Practice Exercise 17.3

According to Figure 17-3, milk of magnesia, an aqueous suspension of $Mg(OH)_2$, has a pH around 10.5. Determine the corresponding concentrations of hydronium ions and hydroxide ions.

Other 'p' Scales

A logarithmic scale is useful not only for expressing hydronium ion concentrations, but also for expressing hydroxide ion concentrations and equilibrium constants. That is, the pH definition can be generalized to other quantities:

$$pOH = -\log [OH^-] \qquad pK_a = -\log K_a \qquad pK_b = -\log K_b$$

Use of these definitions and the properties of logarithms leads to a statement of the water equilibrium constant in terms of pH:

$$K_w = [H_3O^+] [OH^-] = 1.0 \times 10^{-14}$$

$$\log K_w = \log (1.0 \times 10^{-14}) = -14.00$$

$$\log \{[H_3O^+] [OH^-]\} = \log [H_3O^+] + \log [OH^-] = -14.00$$

$$so \qquad -\log [H_3O^+] - \log [OH^-] = 14.00$$

Substituting $pH = -\log [H_3O^+]$ and $pOH = -\log [OH^-]$ gives Equation 17-2, which connects pH and pOH for any aqueous solution:

$$pH + pOH = 14.00 \qquad\qquad (17\text{-}2)$$

② Example 17-4 demonstrates the usefulness of this equation.

Example 17-4

pH and pOH

What is the pH of a 0.25 M solution of NaOH?

Strategy: To find the pH of a solution, first compute $[H_3O^+]$ or $[OH^-]$ and then apply Equation 17-1 or 17-2. Sodium hydroxide is a strong base, and the water equilibrium provides the link between hydroxide and hydronium ion concentrations.

Solution: In this solution of strong base, $[OH^-]$ is equal to the concentration of sodium hydroxide:

$$[OH^-] = 0.25 \text{ M}$$

$$pOH = -\log (0.25) = -(-0.60) = 0.60$$

$$pH = 14.00 - pOH = 14.00 - 0.60 = 13.40$$

Notice how much easier this calculation is than the analysis in Example 17-2. The detailed reasoning of the concentration table, K_w, and an approximation are compressed into two steps.

Does the Result Make Sense? Verify that this result is the same as the one obtained in Example 17-2 by converting from pH to $[H_3O^+]$:

$$13.40 = pH = -\log [H_3O^+]$$

$$\log [H_3O^+] = -13.40 \quad so \quad [H_3O^+] = 10^{-13.40} = 4.0 \times 10^{-14} \text{ M}$$

Extra Practice Exercise 17.4

Determine the pH of the solution of LiOH that appears in Extra Practice Exercise 17.2.

Section Exercises

■ **17.2.1** Determine the pH values of 1.5×10^{-3} M HCl and 2.5×10^{-2} M NaOH.

■ **17.2.2** The most acidic rainfall ever recorded in the United States had a pH of 1.80. What were the hydronium and hydroxide ion concentrations in this rainfall?

17.3 WEAK ACIDS AND BASES

Most acids and bases are weak. A solution of a weak acid contains the acid and water as major species, and a solution of a weak base contains the base and water as major species. Proton-transfer equilibria determine the concentrations of hydronium ions and hydroxide ions in these solutions. To determine the concentrations at equilibrium, we must apply the general equilibrium strategy to these types of solutions.

Weak Acids: Proton Transfer to Water

Measurements of the pH of solutions of acids show that, except for strong acids, the hydronium ion concentration is smaller than would be expected if proton transfer were quantitative. An acid that reaches equilibrium when only a small fraction of its molecules has transferred protons to water is called a **weak acid.** One example is benzoic acid, treated in Example 16-10. A 0.125 M solution of this acid has $[H_3O^+] = 0.0028$ M (pH = 2.55). As another example, the pH of a 0.25 M solution of HF is 1.92, giving $[H_3O^+] = 1.2 \times 10^{-2}$ M. This is substantially lower than 0.25 M (pH = 0.60), the expected result from quantitative proton transfer. In other words, when HF dissolves in water, equilibrium is established when only a fraction of the HF molecules have transferred protons to water molecules:

$$H_2O\,(l) + HF\,(aq) \rightleftharpoons H_3O^+(aq) + F^-(aq)$$

In a solution of a weak acid, the *major* species are water molecules and the acid, H*A*. The products of the proton transfer reaction, H_3O^+ and A^-, are present in smaller concentrations as *minor* species. Figure 17-5 provides a molecular view.

The strength of a weak acid is measured by its acid ionization constant, K_a. This equilibrium constant can be calculated from the measured pH of the solution, as illustrated in Example 17-5.

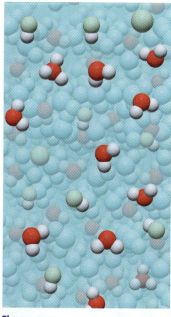

Figure 17-5
In a solution of the weak acid HF, H_2O and HF are present as major species, and H_3O^+ and F^- are present in much lower concentrations.

The pH of a 0.25 M aqueous HF solution is 1.92. Calculate K_a for this weak acid.

Strategy: Use the seven-step method. We are asked to evaluate an equilibrium constant. Visualize the molecular situation: What are the species involved in this equilibrium? After identifying the chemistry, set up and complete the appropriate concentration table and solve for the desired quantity, in this case K_a.

Solution: The problem states that hydrogen fluoride is a weak acid, so the major species in solution are H_2O and HF molecules. In aqueous solution, HF transfers protons to H_2O:

$$H_2O\,(l) + HF\,(aq) \rightleftharpoons H_3O^+(aq) + F^-(aq)$$

From the pH, calculate the concentration of hydronium ions at equilibrium:

$$[H_3O^+] = 10^{-pH} = 10^{-1.92} = 1.2 \times 10^{-2} \text{ M}$$

Now construct a concentration table. The key is to use the initial and equilibrium concentrations of H_3O^+ to complete the "change" row. Shading indicates the concentrations that provide the starting points.

Example 17-5

Calculating K_a

③

	Example 17-5
	Calculating K_a (continued)

Reaction	H_2O	+	HF (aq)	$\rightleftharpoons$	$H_3O^+(aq)$	+	$F^-(aq)$
Initial concentration (M)			0.25		0		0
Change in concentration (M)			-1.2×10^{-2}		$+1.2 \times 10^{-2}$		$+1.2 \times 10^{-2}$
Equilibrium concentration (M)			0.24		1.2×10^{-2}		1.2×10^{-2}

Now substitute numerical values of the concentrations and solve for K_a:

$$K_a = \frac{(1.2 \times 10^{-2})^2}{0.24} = 6.0 \times 10^{-4}$$

Does the Result Make Sense? The equilibrium concentration of hydronium ions is in the 10^{-2} M range, so this value for K_a appears reasonable. Notice that the problem identified this solution as 0.25 M HF, but at equilibrium, [HF] = 0.24 M. Conventionally, a solution's concentration is stated as its initial concentration, even though the equilibrium concentrations may differ slightly from this initial concentration.

Extra Practice Exercise 17.5	An aqueous solution of propanoic acid ($C_2H_5CO_2H$) whose concentration is 85 mM has pH = 2.97. Calculate K_a for this weak acid.

Aqueous hydrofluoric acid is used to etch glass.

Figure 17-6
The concentrations of molecular and ionic species are dramatically different in a solution of a strong acid (HCl) and a solution of a weak acid (HF).

Although it is not strongly reactive as an acid, HF has other unique reactive properties. It is almost the only common substance that reacts with glass. For this reason, artists use aqueous solutions of HF to etch glass, and it is used industrially to frost light bulbs. Because it destroys glass bottles, aqueous HF must be stored in plastic containers. Besides attacking glass, HF has a devastating effect on human nerve tissues. It is rapidly absorbed through the skin, so people who work with HF must always wear strong plastic gloves.

Although proton transfer is never complete for a weak acid dissolved in water, some proton transfer always occurs, so the hydronium ion concentration in any solution of a weak acid is greater than that in pure water. Consequently, the pH of an aqueous solution of a weak acid is always less than 7.00.

Aqueous solutions of strong acids and weak acids both have pH < 7, but their molecular compositions are quite different, as Figure 17-6 displays. Both solutions are acidic, because both contain hydronium ion concentrations greater than 10^{-7} M. The strong acid transfers protons quantitatively, generating a high concentration of hydronium ions and leaving a negligible number of undissociated acid molecules. In contrast, the weak acid undergoes only a small amount of proton transfer, generating a much lower concentration of hydronium ions and remaining almost entirely undissociated.

As Figure 17-6 indicates, only a fraction of weak acid molecules undergoes proton transfer to form hydronium ions. Equation 17-3 defines the percent ionization of any acid solution to be the percentage of the initial concentration of acid that has generated hydronium ions:

$$\% HA \text{ ionized} = 100\% \frac{[H_3O^+]_{eq}}{[HA]_{initial}} \tag{17-3}$$

Notice that the concentration ratio in Equation 17-3 is the *equilibrium* concentration of hydronium ions divided by the *initial* concentration of the acid. The reason for this is that we are interested in how much of the starting material has undergone proton transfer.

When we know the pH and initial concentration of a weak acid solution, we can calculate the percent ionization quickly. For the solution of HF in Example 17-5, we have pH = 1.92, from which $[H_3O^+]_{eq} = 1.2 \times 10^{-2}$ M; and we have $[HF]_{initial} = 0.25$ M. Substituting, we find

$$\% \ HA \ ionized = 100\% \ \frac{[H_3O^+]_{eq}}{[HA]_{initial}} = 100\% \left(\frac{1.2 \times 10^{-2} \ M}{0.25 \ M} \right) = 4.8\%$$

Example 17-6 shows how to determine percent ionization from initial concentration and an acid dissociation constant, K_a.

Determine the percent ionization for an aqueous solution of HF that is 25 mM.	**Example 17-6**

Determining Percent Ionization

Strategy: To determine percent ionization, we need to know the equilibrium concentration of hydronium ions. This requires an equilibrium calculation, for which we follow the seven-step method. We need to set up the appropriate equilibrium expression and solve for $[H_3O^+]_{eq}$, after which we can use Equation 17-3 to obtain the final result.

Solution: As in Example 17-5, the major species in solution are H_2O and HF molecules. In aqueous solution, HF transfers protons to H_2O:

$$H_2O \ (l) + HF \ (aq) \rightleftharpoons H_3O^+ (aq) + F^- (aq)$$

The result of Example 17-5 provides the value for the equilibrium constant:

$$K_a = \frac{[H_3O^+]_{eq} \ [F^-]_{eq}}{[HF]_{eq}} = 6.0 \times 10^{-4}$$

Now construct a concentration table. All we know is that the initial concentration of HF is 25 mM (0.025 M):

Reaction	H_2O +	HF (aq) $\rightleftharpoons$	H_3O^+(aq) +	F^-(aq)
Initial concentration (M)		0.025	0	0
Change in concentration (M)		$-x$	$+x$	$+x$
Equilibrium concentration (M)		$0.025-x$	x	x

We substitute these values into the equilibrium constant expression, then solve for x:

$$K_a = \frac{[H_3O^+]_{eq} \ [F^-]_{eq}}{[HF]_{eq}} = \frac{x^2}{0.025 - x} = 6.0 \times 10^{-4}$$

We need to decide if we can make the approximation that $0.025 \ M - x \cong 0.025$ M. Recall from Chapter 16 that the difference will be small enough to neglect if $K_{eq} < 10^{-4}$ and initial concentrations are greater than 0.10 M. In this case, K_a is marginally larger than 10^{-4}, but the initial HF concentration is *less than* 0.10 M. It is not clear that x will be much smaller than 0.025, so we solve for x using the quadratic formula:

$$x^2 = (0.025 - x)(6.0 \times 10^{-4}) \quad or \quad x^2 + (6.0 \times 10^{-4})x - 1.5 \times 10^{-5} = 0$$

$$x = \frac{-b \pm \sqrt{b^2 - 4ac}}{2a} = \frac{-(6.0 \times 10^{-4}) \pm \sqrt{(6.0 \times 10^{-4})^2 - 4(1)(-1.5 \times 10^{-5})}}{2(1)}$$

Example 17-6

Determining Percent Ionization (*continued*)

$$x = \frac{-(6.0 \times 10^{-4}) \pm \sqrt{(3.6 \times 10^{-7}) + (6.0 \times 10^{-5})}}{2(1)}$$

There are two solutions to this equation, but one gives a negative value for x. We know that x must be positive, because it represents the increase in hydronium ion concentration. Hence, we solve for the positive solution:

$$x = \frac{-(6.0 \times 10^{-4}) + (7.7 \times 10^{-3})}{2} = 3.6 \times 10^{-3}$$

The value of x provides the concentration of hydronium ions at equilibrium:

$$[H_3O^+]_{eq} = x\ M = 3.6 \times 10^{-3}\ M$$

Now substitute this value and the initial concentration of HF into Equation 17-3 to determine the percent ionization:

$$\%\ HA\ \text{ionized} = 100\% \frac{[H_3O^+]_{eq}}{[HA]_{initial}} = 100\% \left(\frac{3.6 \times 10^{-3}\ M}{0.025\ M} \right) = 14\%$$

Does the Result Make Sense? The equilibrium concentration of hydronium ions is smaller than the starting concentration of HF, as it must be. We use two significant figures to match the precision in the K_a value.

Extra Practice Exercise 17.6

Determine the percent ionization for a solution of HF that is 1.0 M.

The results of Example 17-6 and Extra Practice Exercise 17-6 illustrate an important feature of percent ionization. As the concentration of a weak acid increases, the percent ionization decreases. The concentration of hydronium ion decreases as a weak acid is diluted, but it decreases by a smaller fraction than the dilution factor. This behavior of weak acids played an important role in the development of the ionic view of aqueous solutions, as described in our Chemical Milestones Box.

Weak Bases: Proton Transfer from Water

When ammonia dissolves in water, the resulting solution has a pH greater than 7.00, indicating that the solution is basic (Figure 17-3). In other words, the concentration of hydroxide ions in aqueous ammonia is greater than that in pure water. Ammonia molecules accept protons from water molecules, generating ammonium cations and hydroxide anions:

$$NH_3\,(aq) + H_2O\,(aq) \rightleftharpoons NH_4^+\,(aq) + OH^-\,(aq)$$

Ammonia is an example of a **weak base.** A weak base generates hydroxide ions by accepting protons from water but reaches equilibrium when only a fraction of its molecules have done so. The equilibrium constant for this type of equilibrium is designated K_b:

$$K_b = \frac{[NH_4^+]_{eq}\,[OH^-]_{eq}}{[NH_3]_{eq}}$$

Example 17-7 explores the ammonia equilibrium in more detail.

Box 17-1   Chemical Milestones: Arrhenius and the Ionic Theory

The existence of ions in aqueous solutions was first proposed by Svante Arrhenius, a young Swedish chemist, during the 1880s, well before the electronic structure of atoms had been discovered. This insight came while Arrhenius was pursuing his PhD in chemistry, exploring why aqueous solutions conduct electricity.

The types of conductivity measurements carried out by Arrhenius were much like the demonstrations of electrical conductivity used today. Using a voltage source and a current meter, one can determine how much current a solution carries, and this is proportional to its conductivity. A light bulb can serve as a qualitative current meter. As the photos show, solutes fall into three categories. A solution of sugar (*left*) does not conduct, a solution of NaCl (*right*) conducts extremely well, and a solution of acetic acid (*center*) conducts only a little. When Arrhenius began his work, ions were called "electrolytic molecules," so the categories were termed nonelectrolytes, strong electrolytes, and weak electrolytes.

While Arrhenius was studying conductivity, others were characterizing colligative properties of solutions. The Dutch chemist J. T. van't Hoff studied osmotic pressure and derived the law of osmotic pressure, $\pi V = nRT$. Van't Hoff noted the parallel between this law and the ideal gas equation, and he proposed that solute molecules in solution act independently of one another. Van't Hoff's law worked for solutions of nonelectrolytes and weak electrolytes, but for strong electrolytes, van't Hoff had to multiply n by a coefficient, i. For HCl and NaCl the value of i was close to 2, and for $CaCl_2$, i was close to 3. For this reason, strong electrolytes were considered to be exceptions to van't Hoff's law.

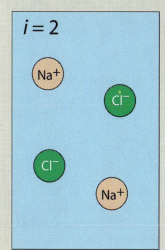
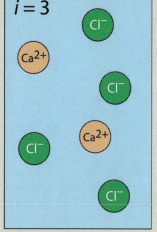

From his conductivity measurements on solutions, Arrhenius concluded that strong electrolytes are not exceptions. Instead they dissociate into ions. When $i = 2$, it meant that each solute species dissociated to give two ions. A compound with $i = 3$ dissociated to give three ions. Moreover, interpreting the results of his experiments at varying levels of concentration, Arrhenius concluded that at sufficiently high dilution, every electrolyte becomes fully dissociated.

Arrhenius devised demonstrations to support his ideas. The diagrams illustrate two of these. Layers of acetic acid solution and pure water, separated by a thin layer of concentrated sugar solution, show low conductivity. When the layers are mixed, the conductivity *increases*. To Arrhenius, this demonstrated that diluting acetic acid leads to more dissociation into ions. Layers of solutions of silver sulfate and barium chloride conduct well. When these layers are mixed, precipitates of silver chloride and barium sulfate form, and the light goes out. To Arrhenius, this showed that ions were present in solutions of salts, and that these ions were removed by precipitation.

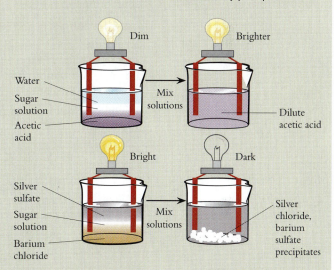

Arrhenius knew that these ideas were revolutionary. He avoided mentioning "ions" in his thesis, for fear his examiners would reject the thesis. Even so, his committee gave him the lowest possible passing grade. In desperation, Arrhenius sent copies of his work to famous professors in several countries. Wilhelm Ostwald, a brilliant chemist, recognized the genius of the ion hypothesis and traveled to Sweden to meet Arrhenius.

Still, the chemical establishment remained opposed to the notion of ions in solution. In an attempt to convince "the wild army of Ionians" of how wrong their ideas were, the British Association scheduled a discussion titled "Theories of Solution," and invited van't Hoff, Arrhenius, and Ostwald to present their views. The rest of the discussion was packed with conservative older chemists, the idea being that reason would prevail and the Ionians would give up their views. Instead, most of the younger chemists sought out the Ionians for spirited exchanges, while the old chemists delivered their lectures to nearly empty rooms.

By 1895, nearly all chemists accepted Arrhenius's views, and Arrhenius finally was awarded a professorship in Stockholm. In 1903, the triumph of Arrhenius became complete, when he was awarded the Nobel Prize for chemistry.

Example 17-7

pH of a Weak Base

Ammonia has ($K_b = 1.8 \times 10^{-5}$). What is the pH of 0.25 M aqueous ammonia?

Strategy: Use the seven-step procedure. We want to determine pH, for which we need to know the equilibrium concentration of either H_3O^+ or OH^-. The major species present in aqueous ammonia are molecules of NH_3 and H_2O. Both of these compounds produce hydroxide ions as minor species in solution:

$$NH_3\,(aq) + H_2O\,(l) \rightleftharpoons NH_4^+(aq) + OH^-(aq) \qquad K_b = 1.8 \times 10^{-5}$$

$$2\,H_2O\,(l) \rightleftharpoons H_3O^+(aq) + OH^-(aq) \qquad K_w = 1.0 \times 10^{-14}$$

These two equilibria are linked by the fact that the solution can have only one hydroxide ion concentration. Both equilibria must be satisfied, but which reaction should we use to calculate $[OH^-]_{eq}$? Notice that both equilibrium constants are much less than 1, but K_b for ammonia is more than a billion times larger than K_w. This indicates that almost all the hydroxide ions in solution come from the ammonia equilibrium. Thus, the ammonia reaction is the appropriate choice for the calculation.

Solution: Determine $[OH^-]_{eq}$ by setting up a concentration table, solving the equilibrium expression for the unknown, and finding $[OH^-]_{eq}$. We know the initial concentrations but must identify x as the change in concentration needed to reach equilibrium.

Reaction	H_2O +	$NH_3\,(aq)$	$\rightleftharpoons$	$NH_4^+(aq)$ +	$OH^-(aq)$
Initial concentration (M)		0.25		0	0
Change in concentration (M)		$-x$		$+x$	$+x$
Equilibrium concentration (M)		$0.25 - x$		x	x

$$K_b = 1.8 \times 10^{-5} = \frac{x^2}{0.25 - x} \cong \frac{x^2}{0.25}$$

$$x^2 = 4.5 \times 10^{-6} \quad so \quad x = 2.1 \times 10^{-3}$$

The table shows that $[OH^-]_{eq} = x$, so $[OH^-]_{eq} = 2.1 \times 10^{-3}$ M.

Complete the problem using Equation 17-2 to calculate the pH of the solution:

$$pOH = -\log[OH^-]_{eq} = -\log(2.1 \times 10^{-3}) = 2.68$$

$$pH + pOH = 14.00 \quad so \quad pH = 14.00 - 2.68 = 11.32$$

Does the Result Make Sense? Note that $x \ll 0.25$, validating the approximation $(0.25 - x) \cong 0.25$. A pH greater than 7 but less than 14 indicates a basic solution, so this is a reasonable result.

Extra Practice Exercise 17.7

Determine the pH of 85 mM aqueous ammonia.

The results of Examples 17-4 and 17-7 illustrate the differences between strong and weak bases:

Strong base
0.25 M NaOH (aq)
$[OH^-]_{eq} = 0.25$ M
pH = 13.40

Weak base
0.25 M NH_3 (aq)
$[OH^-]_{eq} = 2.1 \times 10^{-3}$ M
pH = 11.32

The strong base is a soluble hydroxide that ionizes completely in water, so the concentration of OH^- matches the 0.25 M concentration of the base. For the weak base, in contrast, the equilibrium concentration of OH^- is substantially smaller than the 0.25 M concentration of the base. At any instant, only 0.8% of the ammonia molecules have accepted protons from water molecules, producing a much less basic solution in which OH^- is a minor species. The equilibrium concentration of unprotonated ammonia is nearly equal to the initial concentration. Figure 17-7 summarizes these differences.

Examples 17-5 through 17-7 illustrate the two main types of equilibrium calculations as they apply to solutions of acids and bases. Notice that the techniques are the same as those introduced in Chapter 16 and applied to weak acids in Examples 16-10 and 16-12. We can calculate values of equilibrium constants from a knowledge of concentrations at equilibrium (Examples 16-10 and 17-5), and we can calculate equilibrium concentrations from a knowledge of equilibrium constants and initial concentrations (Examples 16-12, 17-6, and 17-7).

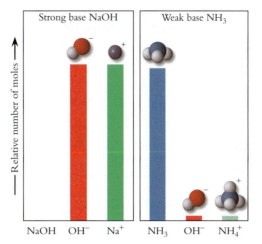

Figure 17-7
The concentrations of molecular and ionic species are dramatically different in a solution of a strong base (NaOH) and a solution of a weak base (NH₃).

⑤

Section Exercises

■ **17.3.1** A 4.8×10^{-2} M solution of hypochlorous acid (HClO) has a pH of 4.36. Compute the percent ionization and K_a for this acid.

■ **17.3.2** Draw a molecular picture showing the proton transfer reaction that takes place in an aqueous solution of HClO.

■ **17.3.3** Construct a table comparing the concentrations of species in solutions of HCl and HClO, each of which is 4.8×10^{-2} M.

17.4 RECOGNIZING ACIDS AND BASES

Learning to recognize the properties of a substance by examining a chemical formula is an important part of mastering chemistry. Fortunately, common acids and bases fall into a small number of structural categories. In this section we describe how to recognize acids and bases. In Section 17.6 we explain how chemical structure influences acidity.

Oxoacids

One general type of acid contains an inner atom bonded to a variable number of oxygen atoms and acidic OH groups. Five common examples of these **oxoacids** are shown in Figure 17-8. An oxoacid usually can be recognized by its formula. The general formula is H_xEO_y, where $x = 1-3$ and $y = 1-4$. The inner atom, E, can be a number of different elements, including B, C, N, P, S, and Cl. Hydrogen atoms of an oxoacid are listed first in the formula to emphasize that these compounds are acids. In almost all cases, the acidic hydrogen atoms are bonded to oxygen atoms, not to the inner atom, E. Three of

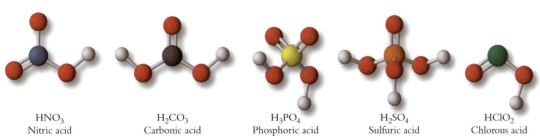

| HNO_3 | H_2CO_3 | H_3PO_4 | H_2SO_4 | $HClO_2$ |
| Nitric acid | Carbonic acid | Phosphoric acid | Sulfuric acid | Chlorous acid |

Figure 17-8
Oxoacids have the general formula H_xEO_y ($x = 1-3$, $y = 1-4$).

Table 17-1 Some Weak Oxoacids

Name	Formula	K_a	Lewis Structure
Chlorous acid	$HClO_2$	1.1×10^{-2}	$:\ddot{O}=\ddot{Cl}-\ddot{O}-H$
Nitrous acid	HNO_2	5.6×10^{-4}	$:\ddot{O}=\ddot{N}-\ddot{O}-H$
Hypochlorous acid	$HClO$	4.0×10^{-8}	$:\ddot{Cl}-\ddot{O}-H$
Hypoiodous acid	HIO	3.2×10^{-11}	$:\ddot{I}-\ddot{O}-H$

the strong acids, HNO_3, H_2SO_4, and $HClO_4$, are oxoacids. Most other oxoacids are weak, as shown by the examples in Table 17–1.

KEY CONCEPT *An oxoacid can be recognized by the general formula H_xEO_y.*

Rules for Naming Oxoacids

1. The name of each oxoacid is based on the name of the polyatomic anion from which it forms, followed by the word "acid." Review Table 3-4 for the names of common polyatomic anions.

2. An oxoacid that forms from a polyatomic anion whose name ends in *-ate* has a name ending in *-ic*. For example, HNO_3 forms by adding a proton to the nitrate polyatomic anion, so HNO_3 is nitric acid. Likewise, $HClO_4$ is perchloric acid from the perchlorate anion.

3. An acid that forms from a polyatomic anion whose name ends in *-ite* has a name ending in *-ous*. For example, HNO_2 forms by adding a proton to the nitrite polyatomic anion, so HNO_2 is nitrous acid. Likewise, $HClO$ is hypochlorous acid from the hypochlorite anion.

Carboxylic Acids

As described in Chapter 13, a **carboxylic acid** can be represented as RCO_2H, in which the symbol R can be as simple as a hydrogen atom or a methyl group (CH_3) or as elaborate as a structure containing dozens of atoms. The R group of a carboxylic acid almost always contains C—H bonds, but hydrogen atoms bonded to carbon are *not* acidic. These atoms do not participate in aqueous proton transfer reactions.

KEY CONCEPT *A carboxylic acid can be recognized by the general formula RCO_2H.*

All carboxylic acids are weak. In an aqueous solution at equilibrium, a small fraction of the carboxylic acid molecules have undergone proton transfer to water molecules, generating hydronium ions and anions that contain the $—CO_2^-$ *carboxylate* group:

$$CH_3CO_2H\,(aq) + H_2O\,(l) \rightleftharpoons CH_3CO_2^-\,(aq) + H_3O^+\,(aq)$$

Acetic acid +   ⇌   Acetate anion +

The name of a carboxylic acid always ends in *-ic*, and the name of the anion ends in *-ate*. Figure 17-9 shows three simple carboxylic acids: formic acid ($K_a = 1.8 \times 10^{-4}$), acetic acid ($K_a = 1.8 \times 10^{-5}$), and benzoic acid ($K_a = 6.4 \times 10^{-5}$). HCO_2^- is formate, $CH_3CO_2^-$ is acetate, and $C_6H_5CO_2^-$ is benzoate. Examples 16-10 and 16-12 treat benzoic acid quantitatively.

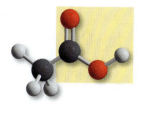
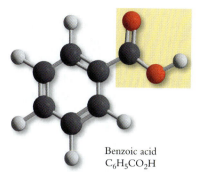

Figure 17-9
Ball-and-stick models and Lewis structures of three common carboxylic acids, with acidic carboxyl groups highlighted.

Formic acid
HCO_2H

Acetic acid
CH_3CO_2H

Benzoic acid
$C_6H_5CO_2H$

Polyprotic Acids

An acid that contains more than one acidic hydrogen atom is called a **polyprotic acid.** Figure 17-8 shows the structures of two polyprotic oxoacids, sulfuric acid (H_2SO_4, diprotic) and phosphoric acid (H_3PO_4, triprotic). Many carboxylic acids are polyprotic, too. Table 17-2 lists several examples of polyprotic acids.

Each successive proton transfer reaction of a polyprotic acid has its own value for K_a, and each successive value is approximately 10^5 times smaller than its predecessor. We explain the reasons for this trend in Section 17.6 and show how to treat calculations involving polyprotic acids in Section 17.7.

Other Acids

A number of small molecules that are acids do not fall into the categories mentioned above. These acids have no clear patterns in their structures, so it is best simply to learn their names and structures. Three hydrogen halides, HCl, HBr, and HI, are strong acids, but the fourth, HF, is a weak acid. Hydrogen sulfide, H_2S, and hydrocyanic acid, HCN, are weak acids that also happen to be quite poisonous. Table 17-3 summarizes the acid–base properties of these and other relatively small weak acids, and Example 17-8 provides some practice in identifying acids.

Table 17-2   Representative Polyprotic Acids

Name	Formula	K_{a1}	Monoanion	K_{a2}
Sulfuric acid	H_2SO_4	Strong	HSO_4^-	1.0×10^{-2}
Oxalic acid	HO_2CCO_2H	5.6×10^{-2}	$HO_2CCO_2^-$	1.5×10^{-4}
Sulfurous acid	H_2SO_3	1.4×10^{-2}	HSO_3^-	6.3×10^{-8}
Phosphoric acid	H_3PO_4	6.9×10^{-3}	$H_2PO_4^-$	6.3×10^{-8}
($K_{a3} = 4.8 \times 10^{-13}$)				
Phthalic acid	$HO_2CC_6H_4CO_2H$	1.1×10^{-3}	$HO_2CC_6H_4CO_2^-$	3.7×10^{-6}
Carbonic acid	H_2CO_3	4.5×10^{-7}	HCO_3^-	4.7×10^{-11}

Table 17-3   Some Weak Acids

Name	Formula	K_a	Lewis Structure
Hydrofluoric acid	HF	6.3×10^{-4}	$:\ddot{F}\!-\!H$
Hydrazoic acid	HN_3	2.5×10^{-5}	$:\ddot{N}\!=\!N\!=\!\ddot{N}\!-\!H$
Hydrogen sulfide	H_2S	8.9×10^{-8}	$H\!-\!\ddot{S}\!-\!H$
Hydrocyanic acid	HCN	6.2×10^{-10}	$:N\!\equiv\!C\!-\!H$
Hydrogen peroxide	H_2O_2	2.4×10^{-12}	$H\!-\!\ddot{O}\!-\!\ddot{O}\!-\!H$

Example 17-8 **Identifying Acids**	Examine the following formulas. Decide if each represents a strong acid, a weak acid, or neither. Justify your conclusions. (a) Cl_3CCO_2H; (b) $CH_3CH_2CH_2OH$; (c) HCN; and (d) $HClO_4$. **Strategy:** We identify acids by recognizing the formulas of oxoacids, carboxylic acids, or one of the few other acids listed in Table 17-3. **Solution:** a. The formula ends with a carboxyl group (—CO_2H). The Cl_3C fragment fills the role of the *R* group in the general formula of a carboxylic acid, RCO_2H. All carboxylic acids are weak. b. There is an —OH fragment at the end of the formula, but it is not part of a carboxyl group. This compound is an alcohol, not a carboxylic acid. c. The hydrogen atom listed first in the formula suggests this compound is an acid. This is hydrocyanic acid, listed in Table 17-3. Since it is not one of the six common strong acids, we can conclude that HCN is a weak acid. d. $HClO_4$ matches the general formula of an oxoacid, H_xEO_y, where $x = 1$, $y = 4$ and $E =$ Cl. This compound is perchloric acid, one of the six common strong acids.

Do the Results Make Sense? The Lewis structures of these molecules confirm the assignments. The acidic hydrogen atoms are highlighted.

Extra Practice Exercise 17.8	Decide if each of the following formulas represents a strong acid, a weak acid, or neither. Justify your conclusions. (a) HF; (b) HCO_2H; (c) $(CH_3)_2O$; and (d) H_2CO_3.

Weak Bases

Besides water, the most common weak base is ammonia, NH_3, whose proton transfer equilibrium with water appears in Section 16.3. Many other weak bases are derivatives of ammonia called **amines.** In these organic compounds, one, two, or three of the N—H bonds in ammonia have been replaced with N—C bonds. The nitrogen atom in an amine, like its counterpart in ammonia, has a lone pair of electrons that can form a bond to a proton. Water does not protonate an amine to an appreciable extent, so all amines are weak bases. Table 17-4 lists several examples of bases derived from ammonia.

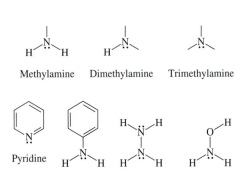

Table 17-4   Representative Organic Bases

Name	Formula	K_b
Methylamine	CH_3NH_2	4.6×10^{-4}
Dimethylamine	$(CH_3)_2NH$	5.4×10^{-4}
Trimethylamine	$(CH_3)_3N$	6.5×10^{-5}
Hydrazine	H_2NNH_2	1.3×10^{-6}
Hydroxylamine	$HONH_2$	8.7×10^{-9}
Pyridine	C_5H_5N	1.7×10^{-9}
Aniline	$C_6H_5NH_2$	7.4×10^{-10}

Example 17-9 shows how the acid–base properties of a molecule can be determined from formulas and structures.

Example 17-9

Acidic and Basic Properties of Amino Acids

Lysine and aspartic acids are two of the amino acids found in proteins. Describe the acid–base characteristics of these compounds.

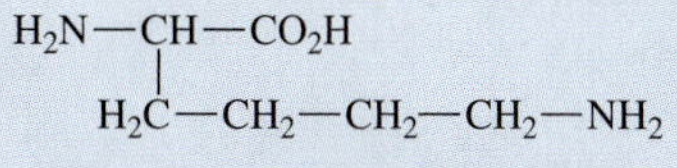

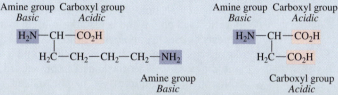

Lysine

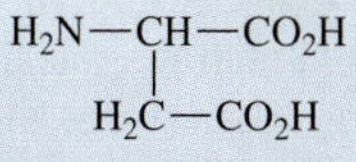

Aspartic acid

Strategy: The structures of the amino acids reveal any functional groups that result in acidic or basic properties. To determine the acid–base properties of these compounds, look for carboxyl groups and amine groups.

Solution: Being amino acids, lysine and aspartic acid each has a carboxyl group, CO_2H, at one end and an amine group, NH_2, at the other. Lysine has a second amine group in the side chain and aspartic acid has a second carboxyl group.

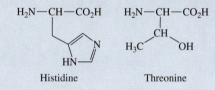

Do the Results Make Sense? Remember that each functional group retains its characteristics, even if other functional groups are present in a molecule. The amine groups of these molecules can accept protons, and the carboxyl groups of the molecules can donate protons.

Describe the acidic and basic characteristics of the amino acids histidine and threonine.

Extra Practice Exercise 17.9

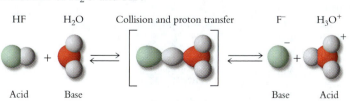

Histidine

Threonine

The amino acids, basic building blocks of proteins, all share this dual acid–base character. See Chapter 13 for a description of the amino acids and their biological chemistry. Organic bases also have a long and varied history as painkillers and narcotics, as our Chemistry and Life Box on the next page describes.

Conjugate Acid–Base Pairs

In a solution of hydrofluoric acid, HF molecules donate protons to water molecules, generating H_3O^+ and F^- ions. As discussed in Chapter 16, the forward reaction and the reverse reaction must occur at equal rates in a solution that is at equilibrium. In an aqueous solution of HF that is at equilibrium, hydronium ions donate protons to fluoride ions to generate molecules of H_2O and HF:

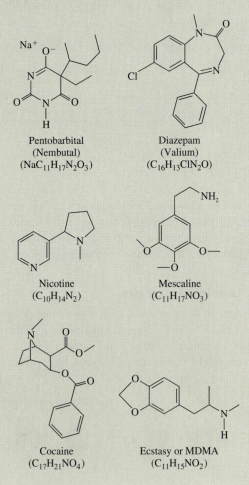

Box 17-2   Chemistry and Life: Drugs and the Brain

F or thousands of years, humans have used plants to make medicines and drugs. Many of the biologically active ingredients in plant extracts contain at least one nitrogen atom with a lone pair of electrons, making them weak bases. Such naturally occurring weak bases are classified as alkaloids. Modern pharmaceutical chemists have built upon nature's chemistry by synthesizing new compounds that are important medicines.

The profound physiological effects of alkaloids have been known for centuries. For example, Socrates was put to death with an extract of hemlock, which contains a poisonous alkaloid, coniine. Other alkaloids have long been valued for their beneficial medical effects. Examples include morphine (a painkiller), quinine (used to treat malaria), and atropine (used to treat Parkinson's disease and in eye drops that dilate the pupils).

Coniine
($C_8H_{17}N$)

Morphine
($C_{17}H_{19}NO_3$)

Quinine
($C_{20}H_{24}N_2O_2$)

Atropine
($C_{17}H_{23}NO_3$)

Because of their basic properties, alkaloids were among the first natural substances that early chemists extracted and purified. Morphine was isolated from poppies in 1805 and was the first alkaloid to be characterized. When treated with aqueous strong acid, alkaloids accept protons to produce water-soluble cations. The protonated alkaloids dissolve, leaving the rest of the plant materials behind. Adding strong base to the aqueous extract reverses the proton-transfer reaction, converts the alkaloid back to its neutral base form, and causes pure alkaloid to precipitate from the solution:

$$\text{Alkaloid}(s) + \text{H}_3\text{O}^+(aq) \rightleftharpoons \text{AlkaloidH}^+(aq) + \text{H}_2\text{O}(l)$$

$$\text{AlkaloidH}^+(aq) + \text{OH}^-(aq) \rightleftharpoons \text{Alkaloid}(s) + \text{H}_2\text{O}(l)$$

Most of us have had to endure the discomfort of dental work, be it repairing cavities or more elaborate oral surgery. Dentists reduce the discomfort immensely by the use of a local painkiller, Novocain, to block the nerves in the mouth. Novocain is a weak base containing two nitrogen atoms. A related analgesic, lidocaine (also known as Xylocaine), is sufficiently strong that applying it to the skin causes the nerves in the immediate area to shut down temporarily.

Procaine
(Novocain)
($C_{13}H_{20}N_2O_2$)

A variety of drugs have been developed that act as sedatives, antidepressants, or stimulants; some of these are effective in treating psychiatric disorders. Many of these drugs are weak bases. Examples are barbiturates such as phenobarbital, tranquilizers like diazepam (Valium), and amphetamines derived from phenylethylamine.

Although nitrogen-containing drugs can be highly beneficial, others—both natural and synthetic—are notorious narcotics. Among alkaloids, morphine is responsible for the narcotic effects of opium and is also the starting material for the synthesis of codeine and heroin. Three other well-known narcotic alkaloids are nicotine, mescaline, and cocaine. Synthetic drugs as well as alkaloids can be abused. Overuse of barbiturates, tranquilizers, or amphetamines can lead to addiction or even death. The darkest side of these substances may be the so-called "designer drugs" such as Ecstasy, which alter brain chemistry in unpredictable and sometimes devastating ways.

Pentobarbital
(Nembutal)
($NaC_{11}H_{17}N_2O_3$)

Diazepam
(Valium)
($C_{16}H_{13}ClN_2O$)

Nicotine
($C_{10}H_{14}N_2$)

Mescaline
($C_{11}H_{17}NO_3$)

Cocaine
($C_{17}H_{21}NO_4$)

Ecstasy or MDMA
($C_{11}H_{15}NO_2$)

The reaction involves proton transfer in both directions, so there are two proton donors, HF and H_3O^+. There are also two proton acceptors, H_2O and F^-. In other words, HF is an acid, and F^- is a base. When an H_2O molecule accepts a proton, the resulting cation is a proton donor. In other words, H_3O^+ is an acid, and H_2O is a base. This is a general feature of proton transfer equilibria. Any proton donor and the species generated by removing one of its protons are called a **conjugate acid–base pair.** The hydronium ion is the **conjugate acid** of H_2O, and the fluoride ion is the **conjugate base** of HF. Thus, HF and F^- are a conjugate acid–base pair; H_3O^+ and H_2O are another conjugate acid–base pair.

Water can act as an acid or a base, so there are two conjugate acid–base pairs for water: H_3O^+ and H_2O are a conjugate acid–base pair, and H_2O and OH^- are another conjugate acid–base pair. Example 17-10 reinforces the structural relationships of conjugate acid–base pairs.

Example 17-10

Conjugate Acid–Base Pairs

Write the chemical formula and the Lewis structure and draw a molecular picture of each of the following: (a) the conjugate acid of NH_3; (b) the conjugate base of HCO_2H; and (c) the conjugate acid of HSO_4^-.

Strategy: Conjugate acid–base pairs are related through proton transfer. Removing a proton from an acid generates its conjugate base. Adding a proton to a base generates its conjugate acid. We use the procedures of Chapter 9 to construct the Lewis structures.

Solution:

a. Add H^+ to NH_3 to obtain its conjugate acid, NH_4^+. With a steric number of 4, the ammonium ion has tetrahedral geometry:

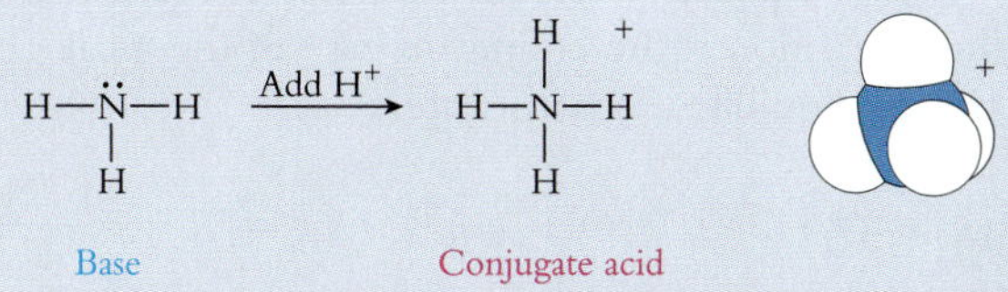

b. Remove H^+ from HCO_2H to obtain its conjugate base, HCO_2^-:

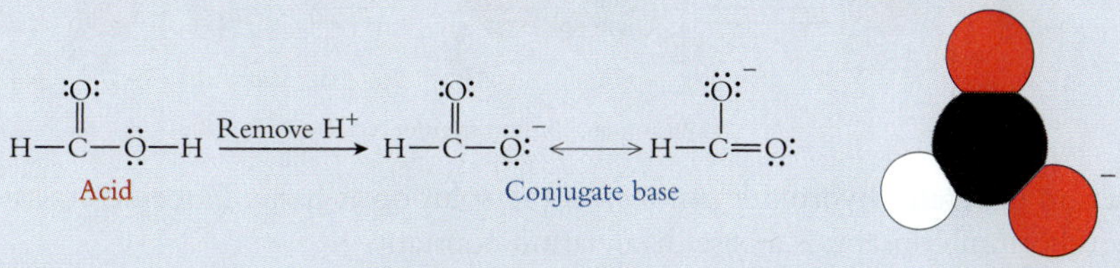

c. Add H^+ to HSO_4^- to obtain its conjugate acid, H_2SO_4:

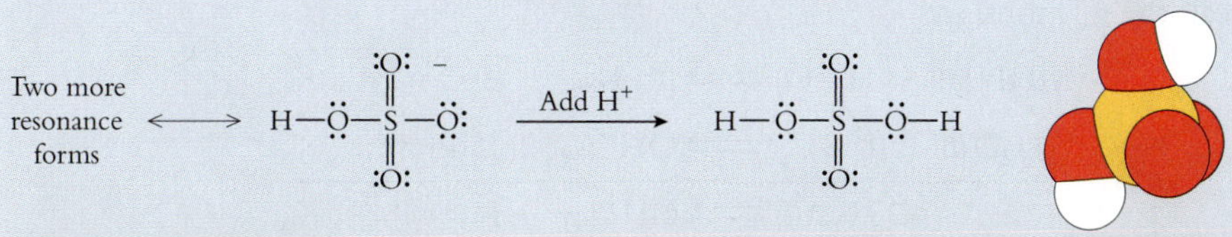

Do the Results Make Sense? An acid always differs from its conjugate base partner by the addition of H^+, and a base always differs from its conjugate acid partner by the removal of H^+. Notice that HSO_4^-, like H_2O, has a conjugate base (SO_4^{2-}) as well as a conjugate acid (H_2SO_4).

Identify the conjugate acid and the conjugate base of the dihydrogen phosphate anion, $H_2PO_4^-$, and the hydrogen phosphate anion, HPO_4^{2-}.

**Extra Practice
Exercise 17.10**

The introduction of conjugate acid–base pairs completes our inventory of acids and bases. In addition to strong bases, ammonia, and amines, the anions of weak acids act as bases.

Section Exercises

■ **17.4.1** Classify each of the following substances as a weak acid, strong acid, weak base, strong base, or neither an acid nor a base: (a) $HClO_4$; (b) $NaOH$; (c) CH_3OH; (d) $C_2H_5CO_2H$; and (e) $C_2H_5NH_2$.

■ **17.4.2** Draw ball-and-stick models for the conjugate acid–base pair of each weak acid and weak base in Section Exercise 17.4.1.

■ **17.4.3** Identify the acid–base properties of nicotine and atropine, whose structures appear below:

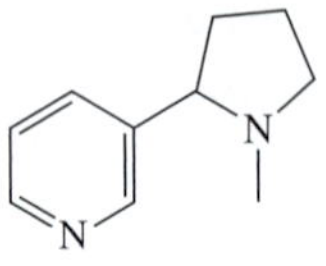

Nicotine
$(C_{10}H_{14}N_2)$

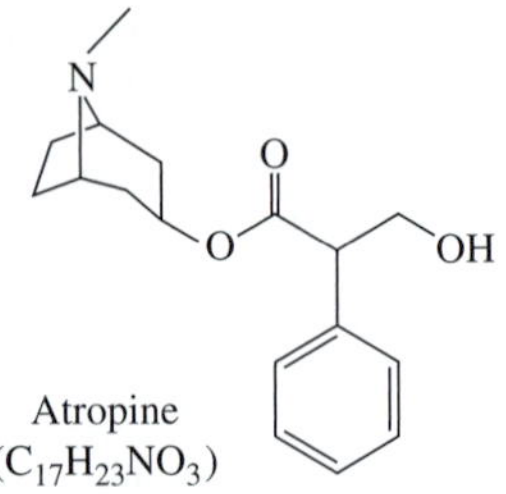

Atropine
$(C_{17}H_{23}NO_3)$

17.5 ACIDIC AND BASIC SALTS

Some compounds that are not immediately recognizable as acids and bases nevertheless display acid–base properties. In this section we describe the acid–base chemistry of salts.

Salts of Weak Acids

An aqueous solution of a soluble salt contains cations and anions. These ions often have acid–base properties. Anions that are conjugate bases of weak acids make a solution basic. For example, sodium fluoride dissolves in water to give Na^+, F^-, and H_2O as major species. The fluoride anion is the conjugate base of the weak acid HF. This anion establishes a proton transfer equilibrium with water:

$$H_2O\,(l) + F^-\,(aq) \rightleftharpoons HF\,(aq) + OH^-\,(aq) \qquad K_{eq} = K_b = \frac{[OH^-]_{eq}\,[HF]_{eq}}{[F^-]_{eq}}$$

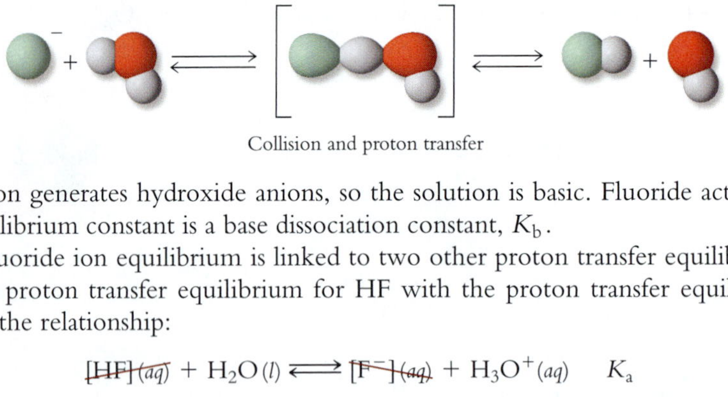

Collision and proton transfer

The reaction generates hydroxide anions, so the solution is basic. Fluoride acts as a base, so the equilibrium constant is a base dissociation constant, K_b.

The fluoride ion equilibrium is linked to two other proton transfer equilibria. Combining the proton transfer equilibrium for HF with the proton transfer equilibrium for F^- reveals the relationship:

$$\text{HF}\,(aq) + H_2O\,(l) \rightleftharpoons \text{F}^-\,(aq) + H_3O^+\,(aq) \qquad K_a$$
$$H_2O\,(l) + \text{F}^-\,(aq) \rightleftharpoons OH^-\,(aq) + \text{HF}\,(aq) \qquad K_b$$
$$\overline{2\,H_2O\,(l) \rightleftharpoons OH^-\,(aq) + H_3O^+\,(aq) \qquad K_w}$$

Combining these two equilibria leads to cancellation of HF and F^-, so the sum of the two is the water equilibrium. How are the equilibrium constants for these three equilibria related?

$$K_a = \frac{[H_3O^+]_{eq}\,[F^-]_{eq}}{[HF]_{eq}} \qquad K_b = \frac{[OH^-]_{eq}\,[HF]_{eq}}{[F^-]_{eq}} \qquad K_w = [H_3O^+]_{eq}\,[OH^-]_{eq}$$

Multiplying the expressions for K_a and K_b leads to cancellation of the concentrations of $[F^-]$ and $[HF]$, just as concentrations cancel when the individual reactions are added.

Whenever two or more equilibria are added, the equilibrium constant for the net reaction is the product of the individual equilibrium constants of the summed reactions:

$$\underbrace{\frac{[H_3O^-]_{eq}\,\cancel{[F^-]_{eq}}}{\cancel{[HF]_{eq}}}}_{K_a}\;\underbrace{\frac{[OH^-]_{eq}\,\cancel{[HF]_{eq}}}{\cancel{[F^-]_{eq}}}}_{K_b} = \underbrace{[H_3O^+]_{eq}\,[OH^-]_{eq}}_{K_w}$$

$$K_a K_b = K_w \qquad\qquad (17\text{-}4)$$

Equation 17-4 applies to any acid and its conjugate base. The equation can also be expressed in logarithmic form using pK notation and $pK_w = 14.00$:

$$pK_a + pK_b = 14.00 \qquad\qquad (17\text{-}5)$$

Because K_a for HF and K_w are known, Equation 17-4 can be used to determine K_b for the fluoride anion. From Example 17-5, $K_a = 6.0 \times 10^{-4}$ for HF, and as always, $K_w = 1.0 \times 10^{-14}$. Thus:

$$K_b = \frac{K_w}{K_a} = \frac{1.0 \times 10^{-14}}{6.0 \times 10^{-4}} = 1.7 \times 10^{-11}$$

Notice that both K_b and K_a are much smaller than 1. This reflects the fact that the conjugate base of a weak acid is itself weak: HF is a weak acid, and F^- is a weak base.

Example 17-11 shows how to deal quantitatively with the basicity of a salt solution.

Example 17-11

Salt of a Weak Acid

Sodium hypochlorite (NaOCl) is the active ingredient in laundry bleach. Typically, bleach contains 5.0% of this salt by mass, which is a 0.67 M solution. Determine the concentrations of all species and compute the pH of laundry bleach.

Strategy: This is a quantitative acid–base equilibrium problem, so we use the seven-step method.

1. We are asked to determine pH, which requires that we find the hydronium ion concentration.

2. Sodium hypochlorite is a salt, so its aqueous solution contains the ionic species Na^+ and OCl^-. Thus, the major species in solution are Na^+, OCl^-, and H_2O.

3. Water is a proton donor and a proton acceptor. The sodium ion is neither an acid nor a base, so it is a spectator ion in this solution. Hypochlorite is the conjugate base of the weak oxoacid, HOCl. The OCl^- anion accepts a proton from a water molecule:

$$H_2O + OCl^- \rightleftharpoons HOCl + OH^- \qquad K_{eq} = K_b$$

Solution:

4. Determine K_b for OCl^- from the information given in Table 17-1:

$$K_a \,(HOCl) = 4.0 \times 10^{-8}$$

$$K_b \,(OCl^-) = \frac{K_w}{K_a} = \frac{1.0 \times 10^{-14}}{4.0 \times 10^{-8}} = 2.5 \times 10^{-7}$$

5. Set up a concentration table for this equilibrium. The initial concentration of OCl^- is given, and x can represent the change in concentration of OCl^-:

Reaction	H_2O +	OCl^-	$\rightleftharpoons$	HOCl +	OH^-
Initial concentration (M)		0.67		0	0
Change in concentration (M)		$-x$		$+x$	$+x$
Equilibrium concentration (M)		$0.67 - x$		x	x

Example 17-11

Salt of a Weak Acid
(continued)

6. Solve the equilibrium constant expression to find the desired concentrations.

$$K_b = \frac{[\text{HOCl}]_{eq}\,[\text{OH}^-]_{eq}}{[\text{OCl}^-]_{eq}} = \frac{(x)(x)}{0.67 - x} = 2.5 \times 10^{-7}$$

Now make the approximation that $x \ll 0.67$:

$$0.67 - x \cong 0.67 \qquad and \qquad 2.5 \times 10^{-7} \cong \frac{x^2}{0.67}$$

$$x^2 \cong 1.68 \times 10^{-7} \qquad from\ which \qquad x \cong 4.1 \times 10^{-4}$$

The value of x provides three of the required concentrations:

$$[\text{OCl}^-] = (0.67 - x)\ \text{M} = (0.67 - 4.1 \times 10^{-4})\ \text{M} = 0.67\ \text{M}$$

$$[\text{OH}^-] = [\text{HOCl}] = x\ \text{M} = 4.1 \times 10^{-4}\ \text{M}$$

The sodium ion concentration is the initial concentration of the salt:

$$[\text{Na}^+] = 0.67\ \text{M}$$

To find $[\text{H}_3\text{O}^+]$, rearrange the expression for K_w:

$$[\text{H}_3\text{O}^+] = \frac{K_w}{[\text{OH}^-]} = \frac{1.0 \times 10^{-14}}{4.1 \times 10^{-4}} = 2.4 \times 10^{-11}\ \text{M}$$

Finally, use $[\text{H}_3\text{O}^+]$ to determine the pH of the bleach solution:

$$\text{pH} = -\log[\text{H}_3\text{O}^+] = -\log(2.4 \times 10^{-11}) = 10.62$$

7. Does the Result Make Sense? The value of x is three orders of magnitude smaller than 0.67, so the approximation is valid. We know that the hypochlorite anion is the conjugate base of hypochlorous acid, so a pH > 7 is a reasonable result.

**Extra Practice
Exercise 17.11**

Potassium nitrite is used as a food preservative. Determine the pH and concentrations of all species in an aqueous solution of KNO_2 whose concentration is 0.22 M (see K_a values in Table 17-1).

Salts of Weak Bases

Salts that contain cations of weak bases are acidic. For example, the ammonium cation is the conjugate *acid* of ammonia. When ammonium salts dissolve in water, NH_4^+ ions transfer protons to H_2O molecules, generating H_3O^+ and making the solution slightly acidic:

$$NH_4^+\,(aq) + H_2O\,(l) \rightleftharpoons NH_3\,(aq) + H_3O^+\,(aq)$$

The equilibrium constant for this reaction can be calculated from Equation 17-4 and K_b for ammonia (Example 17-7):

$$K_a = \frac{K_w}{K_b} = \frac{1.0 \times 10^{-14}}{1.8 \times 10^{-5}} = 5.6 \times 10^{-10}$$

The salt of a weak base, pyridinium chloride, is the subject of Example 17-12.

Example 17-12

Salt of a Weak Base

What are the important acid–base equilibria in an aqueous solution of pyridinium chloride (C_5H_5NHCl)? What are the values of their equilibrium constants?

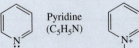 Pyridine
(C_5H_5N)

 Pyridinium ion
$(C_5H_5NH^+)$

Strategy: This is a qualitative problem. First identify the major species, and then list their equilibria. Values for equilibrium constants can be found in tables in this chapter and in Appendix E.

Solution: Pyridinium chloride is a salt that generates ions in solution. The major species are the pyridinium cation ($C_5H_5NH^+$), Cl^-, and H_2O. The formula identifies the pyridinium cation as the conjugate acid of the weak base, pyridine (see Table 17-4). There are two acid−base equilibria with major species as reactants:

$$H_2O + C_5H_5NH^+ \rightleftharpoons H_3O^+ + C_5H_5N \qquad K_a$$

$$H_2O + H_2O \rightleftharpoons H_3O^+ + OH^- \qquad K_w = 1.0 \times 10^{-14}$$

K_b for pyridine, listed in Table 17-4, is used to calculate K_a:

$$K_b \,(C_5H_5N) = 1.7 \times 10^{-9}$$

$$K_a = \frac{K_w}{K_b} = \frac{1.0 \times 10^{-14}}{1.7 \times 10^{-9}} = 5.9 \times 10^{-6}$$

This K_a value is much greater than the value of K_w, so the weak acid determines the pH of a pyridinium chloride solution.

Does the Result Make Sense? The water equilibrium seldom is the dominant one when a solution contains any solute with acid−base properties, so it is reasonable that the pH is determined by K_a.

Identify the major species and the dominant equilibrium for an aqueous solution of anilinium chloride (see Table 17-4).

Example 17-12

Salt of a Weak Base
(continued)

Extra Practice Exercise 17.12

In contrast to the acid−base character of ions arising from weak acids or bases, cations and anions resulting from strong acids or bases do not affect solution pH. The sodium ions in a solution of NaF do not participate in proton transfer reactions, nor do the chloride ions in a solution of pyridinium chloride. Consequently, the water equilibrium determines the pH of a solution of a salt containing neither weak acid nor weak base components. Consider as an example an aqueous slution of sodium chloride. The major ionic species present are Na^+ and Cl^-. Neither of these ions contains a hydrogen atom, so neither is a proton donor. The sodium cation has no lone pairs of electrons, so it is not a base. The chloride anion does have lone pairs, but Cl^- is the conjugate base of a *strong* acid, HCl. Remember that a strong acid donates its protons to water quantitatively. This means that the chloride anion has a negligible tendency to accept protons from hydronium ions in aqueous solution. Because water is a much weaker acid than the hydronium ion, Cl^- will not accept protons from water, either. Both Na^+ and Cl^- are spectator ions, and the only important proton transfer equilibrium in an aqueous solution of NaCl is the water equilibrium.

Summarizing Acids and Bases

In the last two sections we have described several different categories of strong and weak acids and bases. Table 17-5 summarizes the characteristics of these species.

Section Exercises

17.5.1 Determine the concentrations of all species present in a 0.35 M solution of KCN. (Potassium cyanide is a deadly poison.)

17.5.2 Write the acid−base equilibrium that determines the pH of aqueous solutions of each of the following salts, and state whether the resulting solution is acidic, basic, or neither: (a) NH_4I; (b) $NaClO_4$; and (c) $NaCH_3CO_2$.

17.5.3 Calculate the pH of 0.75 M solution of NH_4NO_3 (K_b for $NH_3 = 1.8 \times 10^{-5}$).

Table 17-5   Recognizing Acids and Bases

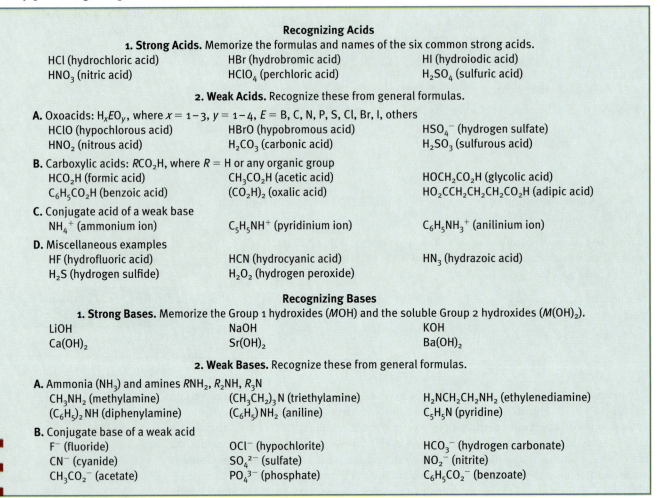

Recognizing Acids

1. Strong Acids. Memorize the formulas and names of the six common strong acids.

HCl (hydrochloric acid)	HBr (hydrobromic acid)	HI (hydroiodic acid)
HNO_3 (nitric acid)	$HClO_4$ (perchloric acid)	H_2SO_4 (sulfuric acid)

2. Weak Acids. Recognize these from general formulas.

A. Oxoacids: H_xEO_y, where $x = 1$–3, $y = 1$–4, $E = $ B, C, N, P, S, Cl, Br, I, others

HClO (hypochlorous acid)	HBrO (hypobromous acid)	HSO_4^- (hydrogen sulfate)
HNO_2 (nitrous acid)	H_2CO_3 (carbonic acid)	H_2SO_3 (sulfurous acid)

B. Carboxylic acids: RCO_2H, where $R = $ H or any organic group

HCO_2H (formic acid)	CH_3CO_2H (acetic acid)	$HOCH_2CO_2H$ (glycolic acid)
$C_6H_5CO_2H$ (benzoic acid)	$(CO_2H)_2$ (oxalic acid)	$HO_2CCH_2CH_2CH_2CO_2H$ (adipic acid)

C. Conjugate acid of a weak base

NH_4^+ (ammonium ion)	$C_5H_5NH^+$ (pyridinium ion)	$C_6H_5NH_3^+$ (anilinium ion)

D. Miscellaneous examples

HF (hydrofluoric acid)	HCN (hydrocyanic acid)	HN_3 (hydrazoic acid)
H_2S (hydrogen sulfide)	H_2O_2 (hydrogen peroxide)	

Recognizing Bases

1. Strong Bases. Memorize the Group 1 hydroxides (MOH) and the soluble Group 2 hydroxides ($M(OH)_2$).

LiOH	NaOH	KOH
$Ca(OH)_2$	$Sr(OH)_2$	$Ba(OH)_2$

2. Weak Bases. Recognize these from general formulas.

A. Ammonia (NH_3) and amines RNH_2, R_2NH, R_3N

CH_3NH_2 (methylamine)	$(CH_3CH_2)_3N$ (triethylamine)	$H_2NCH_2CH_2NH_2$ (ethylenediamine)
$(C_6H_5)_2NH$ (diphenylamine)	$(C_6H_5)NH_2$ (aniline)	C_5H_5N (pyridine)

B. Conjugate base of a weak acid

F^- (fluoride)	OCl^- (hypochlorite)	HCO_3^- (hydrogen carbonate)
CN^- (cyanide)	SO_4^{2-} (sulfate)	NO_2^- (nitrite)
$CH_3CO_2^-$ (acetate)	PO_4^{3-} (phosphate)	$C_6H_5CO_2^-$ (benzoate)

17.6   FACTORS AFFECTING ACID STRENGTH

Why is NaOH a strong base, HNO_3 a strong acid, and CH_3OH neither? Each of these compounds contains O—H bonds, yet their proton transfer properties are strikingly different. In this section we examine the effect of molecular structure on acid strength.

Effect of Charge

The charge on a species has a major effect on its ability to donate or accept protons. Remember that opposite electrical charges attract, and like charges repel. An anion is both a better proton acceptor and a poorer proton donor than is a neutral molecule. Likewise, a cation is a poorer proton acceptor and a better proton donor.

The effect of charge is shown clearly by the acid–base properties of water and its ions (see Figure 17-10). The water molecule is both a weak acid and a weak base.

Figure 17-10
Hydronium cations, neutral water molecules, and hydroxide anions illustrate the effect of charge on acid–base behavior.

Strong acid

Positive charge prohibits proton attachment — H⁺

Positive charge assists proton removal

:Base

Weak acid and weak base

Proton attachment possible — H⁺

Proton removal possible

:Base

Strong base

Negative charge assists proton attachment — H⁺

Negative charge prohibits proton removal

:Base

Hydroxide anion is a strong base because its negative charge strongly attracts protons. Despite having an O—H bond, the hydroxide ion is not acidic because the negative charge of OH^- makes it exceedingly difficult to remove the remaining proton. The hydronium ion, H_3O^+, is a strong acid because its positive charge enhances the removal of one of its protons. The hydronium ion shows negligible basicity even though it has a lone pair of electrons on its oxygen atom. The positive charge of H_3O^+ makes it exceedingly difficult to bind another proton.

Polyprotic acids also show clearly the effect of charge on acidity. As mentioned earlier, the successive K_a values for a polyprotic oxoacid decrease by approximately five orders of magnitude. The neutral parent acid always is a stronger acid than is the anion produced by removing one proton.

Structural Factors

A proton transfer reaction involves breaking a covalent bond. For an acid, an H—X bond breaks as the acid transfers a proton to the base, and the bonding electrons are converted to a lone pair on X. Breaking the H—X bond becomes easier to accomplish as the bond energy becomes weaker and as the bonding electrons become more polarized toward X. Bond strengths and bond polarities help explain trends in acidity among neutral molecules.

The acidities of the binary hydrides increase from left to right across a row of the periodic table. This trend follows bond polarity: as the electronegativity of X increases, the H—X bond polarity increases and so does acidity. The C—H bonds of methane are essentially nonpolar, so the hydrogen atoms are not at all acidic. On the other hand, with an electronegativity difference of 1.9, the electrons of the H—F bond are highly polarized toward the fluorine atom. The resulting partial positive charge on the hydrogen atom assists proton transfer.

Acidity increases moving down a column of the periodic table. The difference between HF (weak acid) and the other hydrogen halides (strong acids) exemplifies the vertical trend. Moving down a column of the periodic table, the principal quantum number of the valence orbitals of the halogen increases, orbital overlap in bonding decreases, and bond strength decreases. Thus, the vertical trend follows bond strength. The H—F bond is substantially stronger than the H—Cl bond, so even though the H—F bond is more polar, HF is a weaker acid than HCl.

Among oxoacids, the strengths of O—H bonds depend on the amount of electron density in the bond. This, in turn, depends on how strongly the rest of the atoms in the molecule attract electrons. The halogen oxoacids show this effect in two ways, as shown in Figure 17-11. The acidity of the hypohalous acids increases as the electronegativity of the halogen increases. This is because the more electronegative the atom, the more it attracts electrons. The acidity of the chlorine oxoacids increases as the number of oxygen atoms increases. This is because each oxygen atom attracts electron density from the rest of the molecule, so the more oxygen atoms there are, the less electron density resides in the O—H bond.

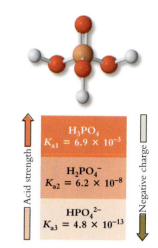

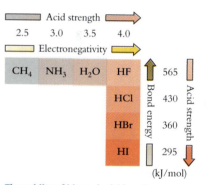

Acid strength decreases as negative charge increases.

The acidity of binary hydrides shows periodic variations.

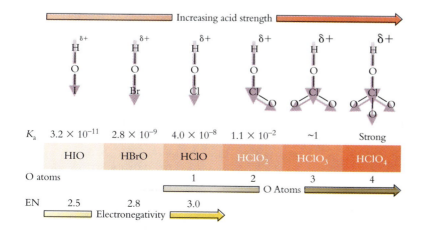

Figure 17-11
The strength of an oxoacid increases as more and more electron density is withdrawn from the O—H bond.

The presence of a second oxygen atom also explains why alcohols like methanol (CH_3OH) are not acidic in aqueous solution, while carboxylic acids like formic acid (HCO_2H) are acidic:

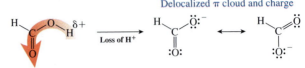

The second oxygen atom in a carboxyl group withdraws electron density from the O—H bond, weakening the bond and enhancing its acidity. The nonpolar C—H bonds in an alcohol do not have the same effect. At least equally important, the carboxylate conjugate base is stabilized by delocalization of two types: both the π electron cloud and the negative charge are spread over both oxygen atoms rather than being localized on one atom.

Example 17-13 explores another feature of relative acid strengths.

Example 17-13

Acidities of Simple Carboxylic Acids

Oxalic acid, $HO_2C—CO_2H$, has $K_{a1} = 1.3 \times 10^{-2}$ and $K_{a2} = 1.5 \times 10^{-4}$. Formic acid, HCO_2H, has $K_a = 1.8 \times 10^{-4}$. Explain why the first proton of oxalic acid is substantially more acidic than the proton of formic acid, but the second proton is less acidic.

Strategy: To explain features of acid strength, we need to look at the effects of bond strength, polarity, and charge.

Solution: Examine the structures of these three species to gain insight into their relative acid strengths:

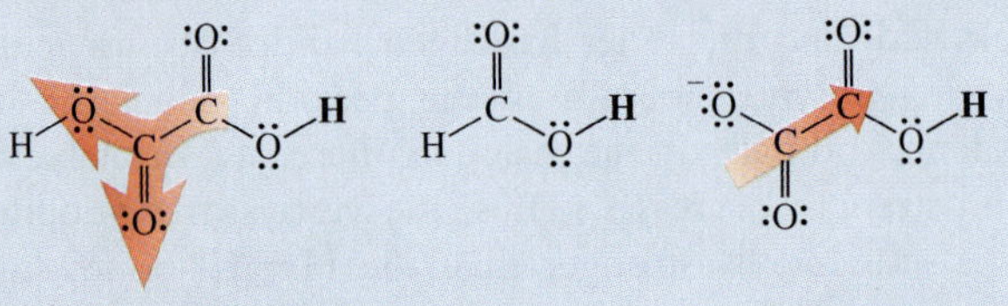

As the left-hand structure indicates, the second —CO_2H group in oxalic acid is better at withdrawing electron density than is the —H in formic acid. This makes the O—H bonds in oxalic acid more polar than the O—H bond in formic acid, so the first proton of oxalic acid is more easily transferred, making oxalic acid stronger than formic acid. The resulting anion has a negative charge that hinders removal of the second proton and makes hydrogen oxalate a weaker acid than formic acid.

Does the Result Make Sense? Notice that the trends and electron shifts are similar to those shown in Figure 17-11.

Extra Practice Exercise 17.13

Iodic acid, HIO_3, has $K_a = 0.17$. Explain why this acid is stronger than four of the acids shown in Figure 17-11 but weaker than the other two acids in that figure.

Section Exercises

■ **17.6.1** Sodium metal reacts with ammonia to form an ionic salt, $NaNH_2$, that contains NH_2^- anions. Is this salt acidic or basic, and why? What reaction occurs if this salt is placed in water?

■ **17.6.2** Among the following pairs of acids, which is stronger and why? (a) H_3PO_4 and H_3PO_3; (b) H_2S and H_2Se; (c) H_2SO_3 and HSO_3^-.

■ **17.6.3** Rank the following acids from weakest to strongest: HBr, HBrO, HClO, and $HClO_2$.

17.7 MULTIPLE EQUILIBRIA

Most of the examples in previous sections appear to involve a single acid–base equilibrium. A closer look reveals that nearly any solution that displays acid–base properties has at least two acid–base equilibria. Look again at Example 17-7, where we list two equilibria as potentially important:

$$NH_3\,(aq) + H_2O\,(l) \rightleftharpoons NH_4^+\,(aq) + OH^-\,(aq) \qquad K_b = 1.8 \times 10^{-5}$$

$$2\,H_2O\,(l) \rightleftharpoons H_3O^+\,(aq) + OH^-\,(aq) \qquad K_w = 1.0 \times 10^{-14}$$

The water equilibrium always exists in aqueous solution. In general, we can focus our initial attention on the equilibria involving other major species (NH_3 in this example). Nevertheless, the water equilibrium does exert its effect on the concentrations of OH^- and H_3O^+. In this example, the concentration of hydroxide anion is established by the ammonia equilibrium, but the concentration of hydronium cations must be found by applying the water equilibrium. We use this feature in several of our examples in this chapter.

Because many species exhibit acid–base properties, it is often possible to write several proton-transfer equilibrium expressions for an aqueous solution. Each such expression is valid if the reactants are species that are actually present in the solution. We have already seen how to focus on the dominant equilibrium: consider only those expressions that have major species as reactants, and look for the one with the largest equilibrium constant.

After completing our analysis of the effects of the dominant equilibrium, we may need to consider the effects of other equilibria. The calculation of $[H_3O^+]$ in a solution of weak base illustrates circumstances where this secondary consideration is necessary. Here, the dominant equilibrium does not include the species, H_3O^+, whose concentration we wish to know. In such cases, we must turn to an equilibrium expression that has the species of interest as a product. The reactants should be species that are involved in the dominant equilibrium, because the concentrations of these species are determined by the dominant equilibrium. We can use these concentrations as the initial concentrations for our calculations based on secondary equilibria. Look again at Example 17-11 for another application of this idea. In that example, the dominant equilibrium is the reaction between hypochlorite anions and water molecules:

$$H_2O\,(l) + OCl^-\,(aq) \rightleftharpoons HOCl\,(aq) + OH^-\,(aq)$$

Working with this equilibrium, we can determine the concentrations of OCl^-, $HOCl$, and OH^-. To find the concentration of hydronium ions, however, we must invoke a second equilibrium, the water equilibrium:

$$2\,H_2O\,(l) \rightleftharpoons H_3O^+\,(aq) + OH^-\,(aq)$$

Polyprotic Acids

Other than the water equilibrium, polyprotic acids provide the most common examples of secondary equilibria that play roles in determining concentrations of minor species in aqueous solution. As noted earlier, the K_a value for the second proton transfer is smaller than K_a for the first. When K_{a1} is three or more orders of magnitude larger than K_{a2}, we can base our initial analysis completely on K_{a1}. To find the concentrations of minor species, however, we must carry out more than one calculation, using more than one concentration table. When multiple equilibria must be taken into account, it is particularly important to identify the dominant equilibrium, which will have major species as reactants. After solving for the concentrations of ions resulting from the dominant equilibrium, we then consider how other equilibria affect the concentrations of the minor species. Example 17-14 shows how this works.

Tutorial

Example 17-14

Ion Concentrations in a Polyprotic Acid Solution

Carbonated water contains carbonic acid, a diprotic acid that forms when carbon dioxide dissolves in water:

$$CO_2\,(g) + H_2O\,(l) \rightleftharpoons H_2CO_3\,(aq)$$

A typical carbonated beverage contains 0.050 M H_2CO_3. Determine the concentrations of the ions present in this solution.

Strategy: The problem describes a weak diprotic acid and asks for ion concentrations. To determine concentrations of all ions, we need to consider more than one equilibrium. This is done in stages, starting with the dominant equilibrium. We apply the seven-step strategy. The problem asks us for the concentrations of the ions in carbonated water, in which the major species are H_2CO_3 and H_2O.

There are two chemical equilibria involving these major species. The acid undergoes proton transfer with equilibrium constant K_{a1}:

$$H_2O\,(l) + H_2CO_3\,(aq) \rightleftharpoons H_3O^+\,(aq) + HCO_3^-\,(aq) \qquad K_{a1} = 4.5 \times 10^{-7}$$

As always, water undergoes proton transfer:

$$H_2O\,(l) + H_2O\,(l) \rightleftharpoons H_3O^+\,(aq) + OH^-\,(aq) \qquad K_w = 1.0 \times 10^{-14}$$

Because carbonic acid is diprotic, a second proton transfer equilibrium has an effect on the ion concentrations:

$$H_2O\,(l) + HCO_3^-\,(aq) \rightleftharpoons H_3O^+\,(aq) + CO_3^{2-}\,(aq) \qquad K_{a2} = 4.7 \times 10^{-11}$$

Solution: Because K_{a1} is several orders of magnitude larger than K_{a2} or K_w, we identify K_{a1} as dominant. Notice, however, that the water equilibrium generates some hydroxide ions in the solution, so this equilibrium must be used to find the concentration of hydroxide ions. The third reaction involves a *minor* species, HCO_3^-, as a reactant, so it cannot be the dominant equilibrium. However, just as the water equilibrium generates some hydroxide ions, the hydrogen carbonate equilibrium generates some carbonate anions, whose concentration must be determined.

We can write an equilibrium constant expression for each equilibrium. Because the initial calculations involve the dominant equilibrium, we start with the expression for K_{a1}:

$$K_{a1} = \frac{[H_3O^+]_{eq}\,[HCO_3^-]_{eq}}{[H_2CO_3]_{eq}}$$

Now we are ready to organize the data and the unknowns and do the calculations. There are multiple equilibria affecting ion concentrations, so we must work with more than one concentration table, starting with the dominant equilibrium. Set up a concentration table to determine concentrations of the ions generated by this reaction:

Reaction	$H_2O\,(l)$	$+$	$H_2CO_3\,(aq)$	$\rightleftharpoons$	$H_3O^+\,(aq)$	$+$	$HCO_3^-\,(aq)$
Initial concentration (M)			0.050		0		0
Change in concentration (M)			$-x$		$+x$		$+x$
Equilibrium concentration (M)			$0.050 - x$		x		x

Substitute the equilibrium concentrations into the equilibrium constant expression and solve for x, making the approximation that $x \ll 0.050$:

$$K_{a1} = 4.5 \times 10^{-7} = \frac{(x)(x)}{0.050 - x} = \frac{x^2}{0.050 - x} \cong \frac{x^2}{0.050}$$

$$x^2 = (0.050)(4.5 \times 10^{-7}) = 2.25 \times 10^{-8} \qquad \textit{from which} \qquad x = 1.5 \times 10^{-4}$$

We round to two significant figures because the K value has two significant figures. The concentrations are

Example 17-14

Ion Concentrations in a Polyprotic Acid Solution (*continued*)

⑧

$$[H_3O^+] = [HCO_3^-] = 1.5 \times 10^{-4} \text{ M}$$

$$[H_2CO_3] = 0.050 - 1.5 \times 10^{-4} = 0.050 - 0.00015 = 0.050 \text{ M}$$

Note that x is about 0.3% of 0.050, so the approximation is valid.

Next, we take into account the proton-transfer equilibrium involving hydrogen carbonate anion. To do this, we complete a second concentration table, using as "initial" concentrations those calculated for the first equilibrium:

Reaction	$H_2O\,(l)$ +	$HCO_3^-\,(aq)$	$\rightleftharpoons$	$H_3O^+\,(aq)$ +	$CO_3^{2-}\,(aq)$
Initial concentration (M)		1.5×10^{-4}		1.5×10^{-4}	0
Change in concentration (M)		$-x$		$+x$	$+x$
Equilibrium concentration (M)		$1.5 \times 10^{-4} - x$		$1.5 \times 10^{-4} + x$	x

Substitute equilibrium concentrations into the equilibrium constant expression and solve for x, making the approximation that $x \ll 1.5 \times 10^{-4}$:

$$K_{a2} = 4.7 \times 10^{-11} = \frac{[H_3O^+]_{eq}\,[CO_3^{2-}]_{eq}}{[HCO_3^-]_{eq}}$$

$$4.7 \times 10^{-11} = \frac{(x)(1.5 \times 10^{-4} + x)}{1.5 \times 10^{-4} - x} \cong \frac{x(1.5 \times 10^{-4})}{1.5 \times 10^{-4}} = x$$

This value is too small to cause a measurable change in the concentrations already calculated, but it does tell us the concentration of carbonate anions in the solution:

$$[CO_3^{2-}] = 4.7 \times 10^{-11} \text{ M}$$

One more ion remains, OH^-, generated from the water equilibrium. It is possible to set up a concentration table, but it is easier to apply the water equilibrium expression directly:

$$K_w = 1.0 \times 10^{-14} = [H_3O^+][OH^-] \quad so$$

$$[OH^-] = \frac{1.0 \times 10^{-14}}{[H_3O^+]} = \frac{1.0 \times 10^{-14}}{1.5 \times 10^{-4}} = 6.7 \times 10^{-11} \text{ M}$$

Do the Results Make Sense? Compare the product concentrations with the concentration of the acid that is responsible for the various ions in the solution, $[H_2CO_3] = 0.050$ M. The ions generated when this acid undergoes proton transfer with water have concentrations that are two orders of magnitude smaller than the concentration of the parent acid.

Products of dominant equilibria: $[H_3O^+] = [HCO_3^-] = 1.5 \times 10^{-4}$ M

The products generated by other equilibria have concentrations that are more than six orders of magnitude smaller than the concentrations of ions produced in the reaction corresponding to the dominant equilibrium.

Products of other equilibria: $[CO_3^{2-}] = 4.7 \times 10^{-11}$ M *and* $[OH^-] = 6.7 \times 10^{-11}$ M

The major species have the highest concentrations; the species generated directly by reactions of major species have the next highest concentrations; and species generated by secondary reactions have very low concentrations. This is a reasonable outcome.

Determine the concentrations of the ionic species present in a 1.0 M solution of phosphoric acid, H_3PO_4 (see Table 17-2 for K_a values, and ignore the third K_a value).

Extra Practice Exercise 17.14

There are several acidic species present in any solution of a polyprotic acid. The solution of carbonic acid in Example 17-14 contains the following concentrations of acidic species:

$$[H_2CO_3] = 0.050 \text{ M} \qquad [H_3O^+] = 1.5 \times 10^{-4} \text{ M} \qquad [HCO_3^-] = 1.5 \times 10^{-4} \text{ M}$$

What happens if we add a base to this solution? The strongest acid donates protons preferentially, but the concentration of hydronium ions is so small that this ion is rapidly consumed. Then the next strongest acid, H_2CO_3, reacts with added base. Generalizing, when a base is added to a solution that contains both a polyprotic acid and its anion, the base accepts protons preferentially from the neutral acid. Only after the neutral acid has been consumed completely does the anion participate significantly in proton transfer. Example 17-15 provides molecular pictures of this feature.

Example 17-15

Molecular View of a Polyprotic Acid

Tutorial

The drawing shows a molecular view of a very small region of an aqueous solution of oxalic acid. For clarity, water molecules are not shown. Redraw this molecular picture to show the solution (a) after two hydroxide ions react with these molecules and (b) after four hydroxide ions react with these molecules. Include in your drawings the water molecules that form as products.

Strategy: To draw molecular pictures illustrating a proton transfer process, we must visualize the chemical reactions that occur, see what products result, then draw the resulting solution. When a strong base is added to a weak acid, hydroxide ions remove protons from the molecules of weak acid. When more than one acidic species is present, the stronger acid loses protons preferentially.

Solution: The oxalic acid solution contains water molecules and $H_2C_2O_4$ molecules as major species. Added hydroxide ions remove protons from the strongest acid, $H_2C_2O_4$:

$$H_2C_2O_4(aq) + OH^-(aq) \rightleftharpoons HC_2O_4^-(aq) + H_2O(l)$$

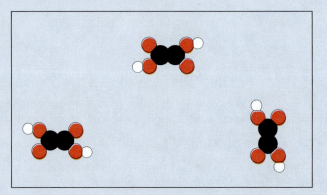

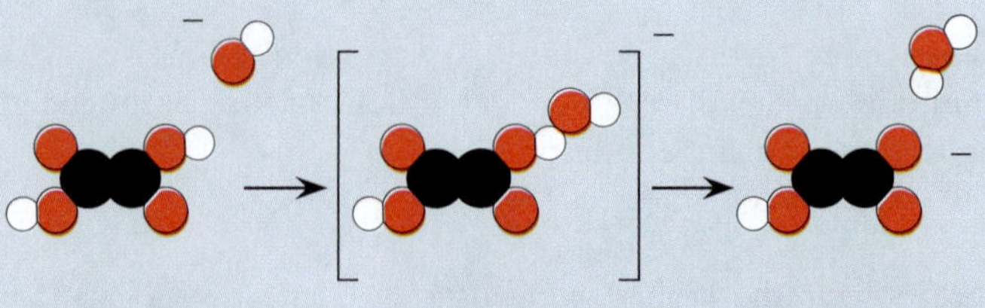

Collision and transfer

Each proton-transfer reaction creates $HC_2O_4^-$, which is itself a weak acid. The K_a values show that $H_2C_2O_4$ is substantially more acidic than $HC_2O_4^-$:

$$H_2C_2O_4 \quad K_{a1} = 5.6 \times 10^{-2} \qquad HC_2O_4^- \quad K_{a2} = 1.5 \times 10^{-4}$$

Example 17-15

**Molecular View
of a Polyprotic Acid**
(*continued*)

Consequently, hydroxide ions react preferentially with $H_2C_2O_4$. Any added OH^- ions react with $H_2C_2O_4$ until every oxalic acid molecule has lost one proton.

a. The first two hydroxide ions convert two $H_2C_2O_4$ molecules into $HC_2O_4^-$ ions. Thus, the molecular picture should contain two $HC_2O_4^-$ anions, two water molecules, and one unreacted $H_2C_2O_4$ molecule:

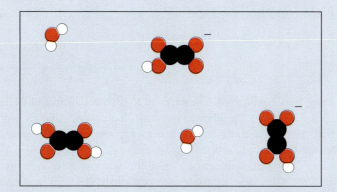

b. When four OH^- ions are added, the first three consume all the $H_2C_2O_4$ molecules. The remaining OH^- ion reacts with the weaker acid, $HC_2O_4^-$:

$$HC_2O_4^-\,(aq) + OH^-\,(aq) \rightleftharpoons C_2O_4^{2-}\,(aq) + H_2O\,(l)$$

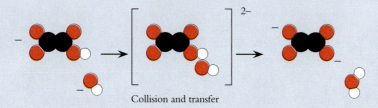

Collision and transfer

This reaction goes virtually to completion, so at equilibrium this system contains two $HC_2O_4^-$, one $C_2O_4^{2-}$, and four water molecules.

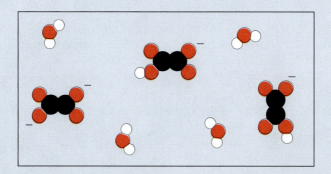

Does the Result Make Sense? Keep in mind that added base accepts protons from acids that are present but leaves the rest of the acid molecule unchanged. Notice that each of the molecular pictures includes the same number of oxalate species. In the first view, each oxalate has two protons attached to it. In the second view, two of them have released one proton each to the added base. In the third view, all three have lost one proton; additionally, one has lost a second proton.

Draw a molecular picture of the carbonic acid solution of Example 17-14, with one molecule for each 0.01 M unit of concentration, showing how the solution looks after the addition of three hydroxide ions.

**Extra Practice
Exercise 17.15**

Salts of Polyprotic Acids

Any anion of a weak acid, including the anions of polyprotic acids, is a weak base. The acid–base properties of monoanions of polyprotic acids are complicated, however, because the monoanion is simultaneously the conjugate base of the parent acid and an acid in its own right. For example, hydrogen carbonate anions undergo two proton-transfer reactions with water:

$$H_2O\,(l) + HCO_3^-\,(aq) \rightleftharpoons H_3O^+\,(aq) + CO_3^{2-}\,(aq) \qquad K_{a2} = 4.7 \times 10^{-11}$$

$$HCO_3^-\,(aq) + H_2O\,(l) \rightleftharpoons H_2CO_3\,(aq) + OH^-\,(aq)$$

$$K_{b1} = \frac{K_w}{K_{a1}} = \frac{1.0 \times 10^{-14}}{4.5 \times 10^{-7}} = 2.2 \times 10^{-8}$$

The larger of these two equilibrium constants is K_{b1}, so a solution of $NaHCO_3$ will be basic. Nevertheless, these two equilibrium constants have values that are sufficiently close together that both equilibria must be taken into account in careful quantitative work. Such calculations are beyond the scope of this text.

In contrast to the intermediate anions, an anion without hydrogen atoms no longer has acidic possibilities. Common examples are phosphate (PO_4^{3-}), sulfate (SO_4^{2-}), carbonate (CO_3^{2-}), and sulfite (SO_3^{2-}). Example 17-16 shows how to treat a solution of one of these salts.

Example 17-16

Ion Concentrations in a Polyprotic Anion Solution

Wines are preserved using sulfites.

Potassium sulfite is commonly used as a food preservative, because the sulfite anion undergoes reactions that release sulfur dioxide, an effective preservative. Determine the concentrations of the ionic species present in a solution of potassium sulfite that is 0.075 M.

Strategy: The problem describes a salt of a polyprotic acid and asks for ion concentrations. To determine concentrations of all ions, we need to consider more than one equilibrium. This is done in stages, starting with the dominant equilibrium. We apply the seven-step strategy. The problem asks for the concentrations of the ions in aqueous potassium sulfite, in which the major species are K^+, SO_3^{2-}, and H_2O. The potassium ion is a spectator ion.

The sulfite anion undergoes proton transfer with equilibrium constant K_{b2}:

$$H_2O\,(l) + SO_3^{2-}\,(aq) \rightleftharpoons OH^-\,(aq) + HSO_3^-\,(aq) \qquad K_{b2} = \frac{[OH^-]_{eq}\,[HSO_3^-]_{eq}}{[SO_3^{2-}]_{eq}}$$

Table 17-2 provides the value of the acid equilibrium constant, $K_{a2} = 6.3 \times 10^{-8}$:

$$K_{b2} = \frac{K_w}{K_{a2}} = \frac{1.0 \times 10^{-14}}{6.3 \times 10^{-8}} = 1.6 \times 10^{-7}$$

This is much larger than K_w, so this is the dominant equilibrium, which we use to begin our calculations.

Because sulfurous acid is diprotic, a second proton transfer equilibrium has an effect on the ion concentrations, and the water equilibrium also plays a secondary role:

$$H_2O\,(l) + HSO_3^-\,(aq) \rightleftharpoons OH^-\,(aq) + H_2SO_3\,(aq) \qquad K_{b1}$$

$$H_2O\,(l) + H_2O\,(l) \rightleftharpoons H_3O^+\,(aq) + OH^-\,(aq) \qquad K_w$$

Solution: Now we are ready to organize the data and the unknowns and do the calculations. The spectator ion is easiest to deal with:

$$[K^+] = 2\,[SO_3^{2-}] = 2(0.075\ M) = 0.15\ M$$

Example 17-16

Ion Concentrations in a Polyprotic Anion Solution *(continued)*

There are multiple equilibria affecting ion concentrations, so we must work with more than one concentration table, starting with the dominant equilibrium. Set up a concentration table to determine concentrations of the ions generated by this reaction:

Reaction	$H_2O\,(l)$ +	$SO_3^{2-}(aq)$ ⇌	$OH^-(aq)$ +	$HSO_3^-(aq)$
Initial concentration (M)		0.075	0	0
Change in concentration (M)		$-x$	$+x$	$+x$
Equilibrium concentration (M)		$0.075 - x$	x	x

Substitute the equilibrium concentrations into the equilibrium constant expression and solve for x, making the approximation that $x \ll 0.075$:

$$K_{b2} = 1.6 \times 10^{-7} = \frac{(x)(x)}{0.75 - x} \cong \frac{x^2}{0.075}$$

$$x^2 = (0.075)(1.6 \times 10^{-7}) = 1.2 \times 10^{-8} \quad \textit{from which} \quad x = 1.1 \times 10^{-4}$$

We round to two significant figures because the K value has two significant figures. The concentrations are

$$[OH^-] = [HSO_3^-] = 1.1 \times 10^{-4}\ \text{M}$$

$$[SO_3^{2-}] = 0.075 - 1.1 \times 10^{-4} = 0.075 - 0.00011 = 0.075\ \text{M}$$

Note that x is about 0.1% of 0.075, so the approximation is valid.

Next, we take into account the proton transfer equilibrium involving hydrogen sulfite anion. To do this, we complete a second concentration table, using as "initial" concentrations those calculated for the first equilibrium:

Reaction	$H_2O\,(l)$ +	$HSO_3^-(aq)$ ⇌	$OH^-(aq)$ +	$H_2SO_3\,(aq)$
Initial concentration (M)		1.1×10^{-4}	1.1×10^{-4}	0
Change in concentration (M)		$-x$	$+x$	$+x$
Equilibrium concentration (M)		$1.1 \times 10^{-4} - x$	$1.1 \times 10^{-4} + x$	x

Substitute equilibrium concentrations into the equilibrium constant expression and solve for x, making the approximation that $x \ll 1.1 \times 10^{-4}$:

$$K_{b1} = \frac{K_w}{K_{a1}} = \frac{1.0 \times 10^{-14}}{1.4 \times 10^{-2}} = 7.1 \times 10^{-13} = \frac{[OH^-]_{eq}[H_2SO_3]_{eq}}{[HSO_3^-]_{eq}}$$

$$7.1 \times 10^{-13} = \frac{(x)(1.1 \times 10^{-4} + x)}{1.1 \times 10^{-4} - x} \cong \frac{x(1.1 \times 10^{-4})}{1.1 \times 10^{-4}} \cong x$$

This value is too small to cause a measurable change in the concentrations already calculated, but it does tell us the concentration of sulfurous acid in the solution:

$$[H_2SO_3] = 7.1 \times 10^{-13}\ \text{M}$$

One more ion remains, H_3O^+, generated from the water equilibrium. It is possible to set up a concentration table, but it is easier to apply the water equilibrium expression directly:

$$K_w = 1.0 \times 10^{-14} = [H_3O^+][OH^-] \quad \textit{so}$$

$$[H_3O^+] = \frac{1.0 \times 10^{-14}}{[OH^-]} = \frac{1.0 \times 10^{-14}}{1.1 \times 10^{-4}} = 9.1 \times 10^{-11}\ \text{M}$$

Example 17-16	**Do the Results Make Sense?** Compare the product concentrations with the concentration of the salt in the solution, 0.075 M. The ions generated when the anion undergoes proton transfer with water have concentrations that are over two orders of magnitude smaller than the concentration of the original solution.

Ion Concentrations in a Polyprotic Anion Solution (*continued*)

Products of dominant equilibria: $[OH^-] = [HSO_3^-] = 1.1 \times 10^{-4}$ M

The products generated by other equilibria have concentrations that are more than six orders of magnitude smaller than the concentrations of ions produced in the reaction corresponding to the dominant equilibrium:

Products of other equilibria: $[H_2SO_3] = 7.1 \times 10^{-13}$ M   *and* $[H_3O^+] = 9.1 \times 10^{-11}$ M

The major species have the highest concentrations; the species generated directly by reactions of major species have the next highest concentrations, and species generated by secondary reactions have very low concentrations. This is a reasonable outcome.

Extra Practice Exercise 17.16	Determine the concentrations of the ionic species present in a 0.36 M solution of sodium carbonate, Na_2CO_3 (see Table 17-2 for K_a values).

Review the quantitative examples in this chapter and compare the techniques used to solve them. You should recognize a common approach. Although the species differ depending on the substances present, identifying the dominant equilibrium is a key step. Once this is done, the approach always is the same: Set up and complete concentration tables, use their results to write algebraic expressions linking concentrations to equilibrium constants, and do the algebra to get the results.

Section Exercises

17.7.1 Calculate the hydroxide ion concentration in a 0.333 M aqueous solution of NH_4Cl.

17.7.2 Write all the acid–base equilibrium reactions that have major species as reactants for a solution of sodium dihydrogen phosphate, NaH_2PO_4.

17.7.3 Determine which of the equilibria in Section Exercise 17.7.2 will be the most important and predict whether this solution is acidic or basic.

CHAPTER 17 VISUAL SUMMARY

Important equations

Essential terms

Key concepts

SKILLS TO MASTER

1. Calculating concentrations of strong acids and bases
2. Converting between concentration, pH, and pOH
3. Calculating K_{eq} from pH
4. Calculating % ionization
5. Calculating concentrations of weak acids and bases
6. Recognizing acids and bases
7. Understanding factors that affect acid strength
8. Working with multiple acid-base equilibria

17.1 PROTON TRANSFERS IN WATER

Strong acid + Base → +

Strong acid   Base

NaOH
Strong base   (1)

$$HA + B \rightleftharpoons A^- + HB$$
Acid   Base

Brønsted-Lowry definition

Acid   Base

Amphiprotic

$K_w = [H_3O^+][OH^-]$

Water equilibrium constant (K_w)

17.2 THE pH SCALE

pH   (2)

$pH = -\log[H_3O]^+$

$pH + pOH = 14.00$

$[H_3O^+]$	pH	
10^{-14}	14	1 M NaOH
10^{-13}	13	
10^{-12}	12	Household ammonia
10^{-11}	11	
10^{-10}	10	Milk of magnesia
10^{-9}	9	Detergent solution
10^{-8}	8	Baking soda
10^{-7}	7	Human blood / Pure water
10^{-6}	6	Milk
10^{-5}	5	Unpolluted rain water
10^{-4}	4	Acid rain, beer
10^{-3}	3	Vinegar
10^{-2}	2	Lemon juice, stomach acid
10^{-1}	1	
10^{0}	0	1 M HCl

Strong base — Weak base — Neutral — Weak acid — Strong acid

17.3 WEAK ACIDS AND BASES

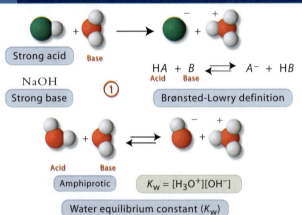

Strong acid HCl   Weak acid HF

Relative number of moles

HCl H_3O^+ Cl⁻   HF H_3O^+ F⁻

Concentration table   (5)

Weak acid

$$K_a = \frac{[H_3O^+]_{eq}[A^-]_{eq}}{[HA]_{eq}}$$

Reaction	$H_2O + HA$	$\rightleftharpoons$	$H_3O^+ + A^-$	
Initial concentration	$[HA]_0$		0	0
Changes in concentration	$-x$		$+x$	$+x$
Final concentrations	$[HA]_0 - x$		x	x

$$K_b = \frac{[HB^+]_{eq}[OH^-]_{eq}}{[B]_{eq}}$$

Weak base

(4) % HA ionized = $100\% \dfrac{[H_3O^+]_{eq}}{[HA]_{initial}}$

17.4 RECOGNIZING ACIDS AND BASES

Recognizing acids

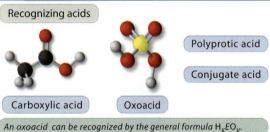

Polyprotic acid

Conjugate acid

Carboxylic acid   Oxoacid

An oxoacid can be recognized by the general formula H_xEO_y.

A carboxylic acid can be recognized by the general formula RCO_2H.   (6)

Recognizing bases

Amine   Conjugate base

Conjugate acid-base pair

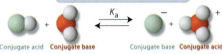

K_a

Conjugate acid   Conjugate base   Conjugate base   Conjugate acid

17.5 ACIDIC AND BASIC SALTS AND OXIDES

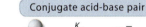

K_b

Conjugate base   Conjugate acid   Conjugate acid   Conjugate base

$K_a K_b = K_w$ $pK_a + pK_b = 14.00$

17.6 FACTORS AFFECTING ACID STRENGTH

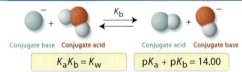

Increasing acid strength →

(7)

K_a	3.2×10^{-11}	2.8×10^{-9}	4.0×10^{-8}	1.1×10^{-2}	~1	Strong
	HIO	HBrO	HClO	$HClO_2$	$HClO_3$	$HClO_4$

O Atoms   1   2   3   4

Electronegativity   2.5   2.8   3.0

17.7 MULTIPLE EQUILIBRIA

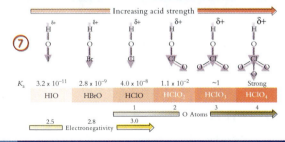

Dominant equilibrium

+ ⇌ + $K_{a,1}$

(8)

Secondary equilibrium

+ ⇌ + $K_{a,2}$

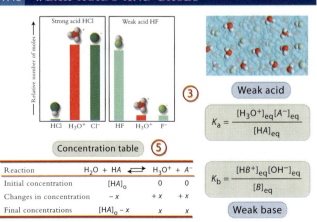
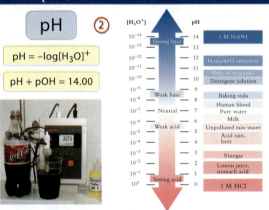
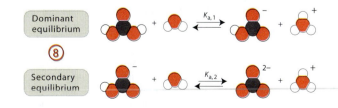

Learning Exercises

17.1 Outline the procedure for working an equilibrium problem for a weak acid–base system.

17.2 Prepare a table listing the various types of acids and bases and the identifying features of each.

17.3 Write a paragraph describing the conjugate acid–base pair and explaining how each interacts with water.

17.4 Update your list of memory bank equations.

17.5 Prepare a list of all terms in Chapter 17 that are new to you. In your own words, write a one-sentence definition of each. Consult the Glossary if you need help.

 Problems **ilw** = interactive learning ware problem. Visit the website at www.wiley.com/college/olmsted

Proton Transfers in Water

17.1 Draw a set of molecular pictures that show the proton transfer reaction occurring when HBr dissolves in water.

17.2 Draw a set of molecular pictures that show the proton transfer reaction occurring when HNO_3 dissolves in water.

17.3 Determine the concentrations of hydronium and hydroxide ions in 1.25×10^{-3} M aqueous perchloric acid.

17.4 Determine the concentrations of hydronium and hydroxide ions in 3.65×10^{-2} M aqueous sodium hydroxide.

ilw 17.5 Concentrated aqueous HCl has a concentration of 12.1 M. Calculate the concentrations of all ions present in a solution prepared by pipetting 1.00 mL of concentrated HCl into a 100-mL volumetric flask and filling to the mark.

17.6 Concentrated aqueous $HClO_4$ has a concentration of 14.8 M. Calculate the concentrations of all ions present in a solution prepared by pipetting 5.00 mL of concentrated HCl into a 1000-mL volumetric flask and filling to the mark.

17.7 Calculate the concentrations of hydronium and hydroxide ions in a solution prepared by dissolving 0.488 g of HCl gas in enough water to make 325 mL of solution.

17.8 Calculate the concentrations of hydronium and hydroxide ions in a solution prepared by dissolving 0.345 g of solid NaOH in enough water to make 225 mL of solution.

The pH Scale

17.9 Convert each of the following H_3O^+ concentrations to pH: (a) 4.0 M; (b) 3.75×10^{-6} M; (c) 4.8 mM; and (d) 0.000 255 M.

17.10 Convert each of the following H_3O^+ concentrations to pH: (a) 1.25 M; (b) 2.95 μM; (c) 0.0366 M; and (d) 7.45×10^{-4} M.

17.11 Convert each of the following OH^- concentrations to pH: (a) 4.0 M; (b) 3.75×10^{-6} M; (c) 4.8 mM; and (d) 0.000 255 M.

17.12 Convert each of the following OH^- concentrations to pH: (a) 1.25 M; (b) 2.95 μM; (c) 0.0366 M; and (d) 7.45×10^{-4} M.

17.13 Convert each of the following pH values into a hydronium ion concentration: (a) 0.66; (b) 7.85; (c) 3.68; and (d) 14.33.

17.14 Convert each of the following pH values into a hydronium ion concentration: (a) 1.56; (b) 3.85; (c) 9.75; and (d) 11.22.

17.15 Convert each of the following pH values into a hydroxide ion concentration: (a) 0.66; (b) 7.85; (c) 3.68; and (d) 14.33.

17.16 Convert each of the following pH values into a hydroxide ion concentration: (a) 1.56; (b) 3.85; (c) 9.75; and (d) 11.22.

Weak Acids and Bases

17.17 Calculate the pH of a 1.5 M solution of each of the following compounds (see Appendix E for K values): (a) C_5H_5N; (b) $HONH_2$; and (c) HCO_2H.

17.18 Calculate the pH of a 2.5×10^{-2} M solution of the following compounds (see Appendix E for K values): (a) NH_3; (b) HClO; and (c) HCN.

17.19 Draw molecular pictures illustrating the proton transfer reactions that determine the pH of the solutions in Problem 17.17 (b) and (c).

17.20 Draw molecular pictures illustrating the proton transfer reactions that determine the pH of the solutions in Problem 17.18 (a) and (c).

ilw 17.21 For a 1.50 M aqueous solution of hydrazoic acid, HN_3, do the following: (a) Identify the major and minor species. (b) Compute concentrations of all species present. (c) Find the pH. (d) Draw a molecular picture illustrating the equilibrium reaction that determines the pH.

17.22 For a 2.75×10^{-2} M aqueous solution of cyanic acid, HCNO, do the following: (a) Identify the major and minor species. (b) Compute concentrations of all species present. (c) Find the pH. (d) Draw a molecular picture illustrating the equilibrium reaction that determines the pH.

17.23 Determine the percent ionization of the solution of hydrazoic acid in Problem 17.21.

17.24 Determine the percent ionization of the solution of cyanic acid in Problem 17.22.

17.25 For a 0.350 M aqueous solution of trimethylamine, $N(CH_3)_3$, do the following: (a) Identify major and minor species. (b) Compute concentrations of all species. (c) Find the pH. (d) Draw a molecular picture illustrating the equilibrium reaction that determines the pH.

17.26 For a 1.85×10^{-3} M aqueous solution of aniline, $C_6H_5NH_2$, do the following: (a) Identify major and minor species. (b) Compute concentrations of all species. (c) Find the pH. (d) Draw a molecular picture illustrating the equilibrium reaction that determines the pH.

17.27 Determine the percent ionization of a solution of acetic acid that is 0.75 M (see Appendix E for K values).

17.28 Determine the percent ionization of a solution of boric acid that is 75 mM (see Appendix E for K values).

Recognizing Acids and Bases

17.29 Identify each of the following substances as a weak acid, strong acid, weak base, strong base, both weak acid and weak base, or neither an acid nor a base: (a) NH_3; (b) HCNO; (c) HClO; and (d) $Ba(OH)_2$.

17.30 Identify each of the following substances as a weak acid, strong acid, weak base, strong base, both weak acid and weak base, or neither an acid nor a base: (a) HNO_3; (b) CH_3CO_2H; (c) HOH; (d) HOCl; and (e) NH_2OH.

17.31 For each molecular picture shown in the figure, determine the chemical formula, identify it as an acid or base, and draw a picture of its conjugate partner.

17.32 For each molecular picture shown in the figure, determine the chemical formula, identify it as an acid or base, and draw a picture of its conjugate partner.

17.33 For each weak acid in Problem 17.17, identify the conjugate base. For each weak base, identify the conjugate acid.

17.34 For each weak acid in Problem 17.18, identify the conjugate base. For each weak base, identify the conjugate acid.

17.35 Write all the conjugate acid–base equilibrium expressions that apply to an aqueous solution of each of the substances in Problem 17.29, and identify each conjugate acid and base.

17.36 Write all the conjugate acid–base equilibrium expressions that apply to an aqueous solution of each of the substances in Problem 17.30, and identify each conjugate acid and base.

Acidic and Basic Salts

ilw 17.37 For a 0.45 M solution of Na_2SO_3, do the following: (a) Identify the major species. (b) Identify the equilibrium that determines the pH. (c) Compute the pH.

17.38 For a 6.75×10^{-3} M solution of sodium benzoate, $NaC_6H_5CO_2$, do the following: (a) Identify the major species. (b) Identify the equilibrium that determines the pH. (c) Compute the pH.

17.39 For a solution that is 0.0100 M in NH_4NO_3, do the following: (a) Identify the major species. (b) Identify the equilibrium that determines the pH. (c) Compute the pH.

17.40 For a solution that is 4.75×10^{-2} M in NH_4Br, do the following: (a) Identify the major species. (b) Identify the equilibrium that determines the pH. (c) Compute the pH.

17.41 Decide if aqueous solutions of the following salts are acidic, basic, or neutral. For each, write the balanced equation that determines the pH of the solution. (a) NaHS; (b) NaOI; (c) $LiClO_4$; and (d) HC_5H_5NCl (pyridinium chloride). (See Appendix E for K values.)

17.42 Decide if aqueous solutions of the following salts are acidic, basic, or neutral. For each, write the balanced equation that determines the pH of the solution. (a) $Mg(NO_3)_2$; (b) Li_2CO_3; (c) $(CH_3)_3NHBr$; and (d) $NaCH_3CO_2$. (See Appendix E for K values.)

17.43 List the following sodium salts in order of increasing pH of their 0.25 M solutions: NaI, NaOH, NaF, Na_3PO_4, and $NaC_6H_5CO_2$. (See Appendix E for K values.)

17.44 List the following substances in order of increasing pH of their 0.25 M solutions: NH_4Cl, $MgCl_2$, HCl, NaOCl, $C_6H_5NH_3Cl$, and $NaClO_2$. (See Appendix E for K values.)

Factors Affecting Acid Strength

17.45 Among the following pairs of acids, which is stronger and why? (a) H_2SO_4 and HSO_4^-; (b) HClO and HIO; and (c) HClO and $HClO_2$.

17.46 Among the following pairs of acids, which is stronger and why? (a) $HBrO_3$ and $HBrO_2$; (b) H_2S and H_2O; and (c) H_2S and HS^-.

17.47 Draw Lewis structures of the acids in Problem 17.45 (b) and (c), and use arrows to show electron density shifts that account for their different acid strengths.

17.48 Draw Lewis structures of the acids in Problem 17.46 (a) and (b), and use arrows to show electron density shifts that account for their different acid strengths.

17.49 Alcohols generally are not acidic, yet phenol (C_6H_5OH) is a weak acid. Draw the Lewis structure of the phenolate anion (the conjugate base of phenol), with a $C=O$ double bond, and use it to explain the acidity of the —OH group in phenol relative to the —OH group in ethanol.

17.50 Acetic acid (CH_3CO_2H) is weaker than chloroacetic acid ($ClCH_2CO_2H$). Draw Lewis structures of these two acids. Draw an arrow indicating the effect of the Cl atom on the electron density in the rest of the molecule, and use this drawing to explain why chloroacetic acid is the stronger acid.

Multiple Equilibria

ilw 17.51 Use two concentration tables to calculate the concentrations of all species present in a 0.250 M solution of sodium acetate, $NaCH_3CO_2$. (See Appendix E for K values.)

17.52 Use two concentration tables to calculate the concentrations of all species present in a 3.45×10^{-2} M solution of KBrO. (See Appendix E for K values.)

17.53 Determine the concentrations of the ionic species present in a 1.55×10^{-2} M solution of the diprotic acid H_2CO_3. (See Table 17-2 for K values.)

17.54 Deermine the concentrations of the ionic species present in a 0.355 M solution of the diprotic acid H_2SO_3. (See Table 17-2 for K values.)

17.55 Determine the concentrations of the ionic species present in a 55 mM solution of sodium carbonate, Na_2CO_3. (See Table 17-2 for K values.)

17.56 Determine the concentrations of the ionic species present in a 0.35 M solution of sodium sulfite, Na_2SO_3. (See Table 17-2 for K values.)

Additional Paired Problems

17.57 Boric acid, H_3BO_3, $K_a = 5.4 \times 10^{-10}$, is frequently used as an eyewash. (a) Use Lewis structures to illustrate the equilibrium reaction of K_a. (b) Calculate the pH of 0.050 M boric acid solution.

17.58 Hydrazine, N_2H_4, has $K_b = 1.3 \times 10^{-6}$. (a) Use Lewis structures to illustrate the equilibrium reaction of K_b. (b) Calculate the pH of a 2.00×10^{-1} M solution of N_2H_4.

17.59 In the mid-1930s a substance was isolated from a fungus that is a parasite of ryes and other grasses. This alkaloid, lysergic acid, has been of great interest to chemists because of its strange, dramatic action on the human mind. Many derivatives of lysergic acid are known, some with medicinal applications. Perhaps the best known derivative of lysergic acid is the potent hallucinogen lysergic acid diethylamide (LSD):

LSD ($C_{20}H_{25}N_3O$)

Like other alkaloids, LSD is a weak base, $K_b = 7.6 \times 10^{-7}$. What is the pH of a 0.55 M solution of LSD?

17.60 The addictive painkiller morphine, $C_{17}H_{19}NO_3$, is the principal molecule in the milky juice that exudes from unripe poppy seed capsules.

Calculate the pH of a 0.015 M solution of morphine, given that $K_b = 7.9 \times 10^{-7}$.

17.61 The pH of a 0.0100 M solution of the sodium salt of a weak acid is 11.00. What is the K_a of the weak acid?

17.62 The pH of a 0.060 M solution of a weak acid is 2.72. Calculate K_a and identify the acid from among those listed in Appendix E.

Morphine ($C_{17}H_{19}NO_3$)

17.63 Determine the percent ionization of the weak acid solution in Problem 17.57.

17.64 Determine the percent ionization of the weak acid solution in Problem 17.62.

17.65 Putrescine, a substance with a vile odor, has two basic amino functional groups, as shown in the line structure above. Use Lewis structures to show the two proton transfer reactions of putrescine and water.

Putrescine ($C_4H_{12}N_2$)

17.66 Draw Lewis structure sketches showing the aqueous-phase reaction between acetic acid and ammonia.

17.67 Calculate the pH of a 0.250 M solution of aqueous NaF.

17.68 Calculate the pH of a 0.025 M aqueous solution of NH_4Cl.

17.69 Determine the concentrations of all species present in 2.00 M H_2SO_4.

17.70 Determine the concentrations of all species present in 0.200 M aqueous $NaNO_2$.

17.71 For each of the following compounds, identify the major species in solution and write a balanced equation for the equilibrium reaction that determines the pH: (a) H_2SO_4; (b) Na_2SO_4; (c) $NaHSO_4$; and (d) NH_4Cl.

17.72 Identify major species and proton transfer equilibria for aqueous solutions of the following: (a) NH_4NO_3; (b) KH_2PO_4; (c) H_3PO_4; and (d) HCO_2H.

17.73 Use tabulated K_{eq} values to find the value of K_{eq} for the following reaction:

$$HPO_4^{2-}(aq) + OH^-(aq) \rightleftharpoons PO_4^{3-}(aq) + H_2O(l)$$

17.74 Use tabulated K_{eq} values to find the value of K_{eq} for the following reaction:

$$HPO_4^{2-}(aq) + H_3O^+(aq) \rightleftharpoons H_2PO_4^-(aq) + H_2O(l)$$

More Challenging Problems

17.75 Oxalic acid, a diprotic carboxylic acid found in many plants, including rhubarb, is an effective stain remover. Consider the following diagram to be a small section of an aqueous solution of oxalic acid:

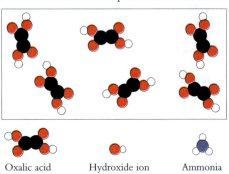

Oxalic acid   Hydroxide ion   Ammonia

(a) Draw a new picture that shows the appearance of the solution at equilibrium after four hydroxide ions enter the region. (b) Draw a new picture that shows the appearance of the solution at equilibrium after eight hydroxide ions enter the region. (c) Draw a new picture that shows the appearance of the solution at equilibrium after four ammonia molecules enter the region ($K_b = 1.8 \times 10^{-5}$).

17.76 By coincidence, the K_a of CH_3CO_2H and the K_b of NH_3 both are 1.8×10^{-5}. Write the appropriate acid–base equilibrium reactions and use them to determine whether each of the following 1.00 M solutions is acidic, basic, or neutral: (a) CH_3CO_2H; (b) NH_3; (c) NH_4Cl; (d) $NaCH_3CO_2$; and (e) $NH_4CH_3CO_2$.

17.77 A solution is prepared by dissolving 3.5 g of P_4O_{10} in 1.50 L of water. The oxide reacts with water quantitatively to form phosphoric

acid. (a) Identify the major species in the solution. (b) Identify the minor species in the solution and rank them in order of concentration, highest first. (c) Calculate the pH of the solution.

17.78 Aqueous solutions of Na_2SO_3 and CH_3CO_2H are mixed. (a) List the major species in each solution. (b) Write the net ionic reaction that occurs on mixing. (c) Identify the acid, base, conjugate acid, and conjugate base.

17.79 In aqueous solution, amino acids exist as zwitterions (German for "double ions"), compounds in which internal proton transfer gives a molecule with two charged functional groups. Use Lewis structures to illustrate the proton transfer equilibrium between the uncharged form of glycine ($NH_2CH_2CO_2H$) and its zwitterion form.

17.80 Vinegar is a dilute aqueous solution of acetic acid. A sample of vinegar has a pH of 2.39 and a density of 1.07 g/mL. What is the mass percentage of acetic acid in the vinegar?

17.81 For each of the following reactions, write a balanced net ionic equation. Use different sizes of arrows to indicate whether the reaction goes nearly to completion or proceeds to only a small extent. (*Hint:* You may need to compare K_a values.)
   - (a) $NaOH(aq) + C_6H_5CO_2H(s) \rightleftharpoons ?$
   - (b) $(CH_3)_3N(aq) + HNO_3(aq) \rightleftharpoons ?$
   - (c) $Na_2SO_4(aq) + CH_3CO_2H(aq) \rightleftharpoons ?$
   - (d) $NH_4Cl(aq) + Ca(OH)_2(aq) \rightleftharpoons ?$
   - (e) $K_2HPO_4(aq) + NH_3(aq) \rightleftharpoons ?$

17.82 Identify each of the following as an acid, base, both, or neither and then write balanced reactions that illustrate the acidic or basic properties of the compound: (a) H_2CO_3; (b) $KHCO_3$; (c) NH_3; (d) $NaCl$; (e) Na_2SO_4; (f) NH_4HSO_4; and (g) NaO_2CCO_2H.

Group Study Problems

17.83 According to its label, each tablet of Alka-Seltzer contains 1.916 g (0.0228 mol) of sodium hydrogen carbonate. When an Alka-Seltzer tablet dissolves in 150 mL of water: (a) What equilibrium determines the pH? (b) What is the pH of the solution?

17.84 When ammonium acetate dissolves in water, both the resulting ions undergo proton transfer reactions with water, but the net reaction can be written without using water:

$$NH_4^+(aq) + CH_3CO_2^-(aq) \rightleftharpoons NH_3(aq) + CH_3CO_2H(aq)$$

(a) Combine proton transfer equilibrium reactions involving water so that the net result is the above reaction, and use this sequence of reactions to derive an expression for K_{eq} for this reaction in terms of K_a, K_b, and K_w. (b) Make a list of all the proton-transfer reactions that occur in this solution. (c) Use K_a and K_b to calculate the concentrations of H_3O^+ and OH^- that result from the reactions of NH_4^+ and $CH_3CO_2^-$ with water in a 0.25 M solution of ammonium acetate. (d) Based on the results of part (c), what pH would you expect to find for this solution?

17.85 In each of the following situations, one reaction goes essentially to completion. In each case, identify the major species in solution under initial conditions and write a balanced net ionic equation for the reaction that goes essentially to completion. (a) Gaseous HBr is bubbled through a 0.015 M solution of $Ca(OH)_2$. (b) 4.0 g of NaOH is added to 0.75 L of 0.055 M $NaHSO_4$. (c) 25.0 mL of saturated NH_4I is mixed with 50.0 mL of 0.95 M $Pb(NO_3)_2$.

17.86 For each of the following reactions, write a balanced net ionic equation. Use different sizes of arrows to indicate whether the reaction goes nearly to completion or proceeds to only a small extent. (*Hint:* You may need to compare K_a values.)
   - (a) $H_2S(aq) + NH_3(aq) \rightleftharpoons ?$
   - (b) $C_2H_5NH_2(aq) + KHSO_4(aq) \rightleftharpoons ?$
   - (c) $C_5H_5N(aq) + HCN(aq) \rightleftharpoons ?$
   - (d) $NH_4Br(aq) + Na_3PO_4(aq) \rightleftharpoons ?$
   - (e) $HClO(aq) + HONH_2(aq) \rightleftharpoons ?$

17.87 Pure sulfuric acid (H_2SO_4) is a viscous liquid that causes severe burns when it contacts the skin. Like water, sulfuric acid is amphiprotic, so a proton transfer equilibrium exists in pure sulfuric acid. (a) Write this proton transfer equilibrium reaction. (*Hint:* H_2O is NOT involved.) (b) Construct the Lewis structure of sulfuric acid and identify the features that allow this compound to function as a base. (c) Perchloric acid ($HClO_4$) is a stronger acid than sulfuric acid. Write the proton transfer reaction that takes place when perchloric acid dissolves in pure sulfuric acid.

17.88 Prepare a graph of the percent ionization of aqueous acetic acid as a function of the logarithm of its concentration. To do this, calculate the percent ionization at a variety of molarities ranging from 1.0 M to 10^{-7} M. Repeat for aqueous hypoiodous acid. What similarities are there between the two graphs? What differences are there? (*Note:* This problem is suitable for spreadsheet analysis.)

Answers to Section Exercises

17.1.1 $H_3O^+(aq) + SO_4^{2-}(aq) \rightleftharpoons H_2O(l) + HSO_4^-(aq)$

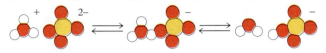

17.1.2 4.0×10^{-11} M
17.1.3 $[K^+] = [OH^-] = 0.200$ M; $[H_3O^+] = 5.0 \times 10^{-14}$ M
17.2.1 1.5×10^{-3} M HCl, pH = 2.82; 2.5×10^{-2} M NaOH, pH = 12.40
17.2.2 $[H_3O^+] = 1.6 \times 10^{-2}$ M; $[OH^-] = 6.3 \times 10^{-13}$ M
17.3.1 % Ionization = 0.091%; $K_a = 4.0 \times 10^{-8}$
17.3.2

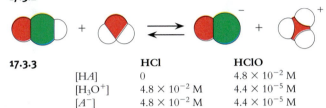

17.3.3

	HCl	HClO
[HA]	0	4.8×10^{-2} M
$[H_3O^+]$	4.8×10^{-2} M	4.4×10^{-5} M
$[A^-]$	4.8×10^{-2} M	4.4×10^{-5} M

17.4.1 (a) strong acid; (b) strong base; (c) neither; (d) weak acid; and (e) weak base

17.4.2

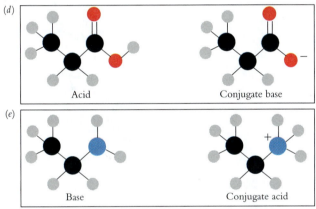

(d)

Acid   Conjugate base

(e)

Base   Conjugate acid

17.4.3 Nicotine is a weak base, having two N atoms with lone pairs of electrons. Atropine also is a weak base, having one N atom with a lone pair of electrons. Atropine also contains O atoms, but it lacks the CO_2H group, so atropine is not acidic.

17.5.1 $[HCN] = [OH^-] = 2.4 \times 10^{-3}$ M; $[CN^-] = [K^+] = 0.35$ M; $[H_3O^+] = 4.2 \times 10^{-12}$ M
17.5.2 (a) $NH_4^+(aq) + H_2O(l) \rightleftharpoons NH_3(aq) + H_3O^+(aq)$, acidic
(b) $2 H_2O(l) \rightleftharpoons OH^-(aq) + H_3O^+(aq)$, neither
(c) $CH_3CO_2^-(aq) + H_2O(aq) \rightleftharpoons CH_3CO_2H(aq) + OH^-(aq)$, basic

17.5.3 pH = 4.69
17.6.1 The salt is basic, because the anion of NH_3 is more basic than NH_3. The anion accepts a proton from water:

$$NH_2^- + H_2O \rightarrow NH_3 + OH^-$$

17.6.2 (a) H_3PO_4 is stronger, more O atoms to withdraw electrons; (b) H_2Se is stronger, higher principal quantum number means weaker H—X bond; and (c) H_2SO_3 is stronger, neutral acid is stronger than anion.
17.6.3 HBrO is weakest, then HClO, $HClO_2$, and HBr (a strong acid) is strongest.
17.7.1 $[OH^-] = 7.4 \times 10^{-10}$ M
17.7.2 The major species are $H_2PO_4^-$, H_2O, and Na^+. There are three acid–base equilibria with these species as reactants:

$$H_2PO_4^-(aq) + H_2O(aq) \rightleftharpoons HPO_4^{2-}(aq) + H_3O^+(aq)$$

$$H_2PO_4^-(aq) + H_2O(aq) \rightleftharpoons H_3PO_4(aq) + OH^-(aq)$$

$$2 H_2O(aq) \rightleftharpoons H_3O^+(aq) + OH^-(aq)$$

17.7.3 The most important equilibrium is the one with the largest K_{eq}. Here are the values:

$$H_2PO_4^-(aq) + H_2O(aq) \rightleftharpoons HPO_4^{2-}(aq) + H_3O^+(aq)$$

$$K_{a2} = 6.3 \times 10^{-8}$$

$$H_2PO_4^-(aq) + H_2O(aq) \rightleftharpoons H_3PO_4(aq) + OH^-(aq)$$

$$K_b = K_w/K_{a1} = 1.4 \times 10^{-12}$$

$$2 H_2O(aq) \rightleftharpoons H_3O^+(aq) + OH^-(aq)$$

$$K_w = 1.0 \times 10^{-14}$$

The value of K_{a2} is larger than the others, so this reaction is the most important. The solution is acidic, since this reaction produces H_3O^+ ions.

Answers to Extra Practice Exercises

17.1 $[H_3O^+] = [Cl^-] = 6.5 \times 10^{-3}$ M; $[OH^-] = 1.5 \times 10^{-12}$ M
17.2 $[Li^+] = [OH^-] = 0.023$ M; $[H_3O^+] = 4.4 \times 10^{-13}$ M
17.3 $[H_3O^+] = 3 \times 10^{-11}$ M; $[OH^-] = 3 \times 10^{-4}$ M
17.4 $pOH^- = 1.64$, pH = 12.36
17.5 $K_a = 1.4 \times 10^{-5}$
17.6 2.4% ionized
17.7 pH = 11.09
17.8 (a), (b), and (d) weak acid; (c) neither
17.9 Both histidine and threonine have a basic amino NH_2 group and an acidic carboxyl group. The histidine side chain has two amine nitrogen atoms in the ring. The side chain of threonine has no significant acid–base properties under normal conditions.
17.10 $H_2PO_4^-$: conjugate acid is H_3PO_4, conjugate base is HPO_4^{2-}; HPO_4^{2-}: conjugate acid is $H_2PO_4^-$, conjugate base is PO_4^{3-}.
17.11 $[K^+] = [NO_2^-] = 0.22$ M; $[OH^-] = [HNO_2] = 2.0 \times 10^{-6}$ M; pH = 8.30
17.12 Major species: $C_6H_5NH_3^+$, Cl^-, and H_2O; dominant equilibrium: $H_2O + C_6H_5NH_3^+ \rightleftharpoons H_3O^+ + C_6H_5NH_2$
17.13 HIO_3 is stronger than HIO, HBrO, HClO, and $HClO_2$ because it has more O atoms to withdraw more electron density from the O—H bond. HIO_3 is weaker than $HClO_3$ because I, being less electronegative than Cl, is less effective at withdrawing electron density. HIO_3 is weaker than $HClO_4$ both because of electronegativity differences and because four O atoms withdraw more electron density than three O atoms do.
17.14 $[H_3PO_4] = 0.92$ M; $[H_3O^+] = [H_2PO_4^-] = 0.080$ M; $[HPO_4^{2-}] = 6.3 \times 10^{-8}$ M; $[OH^-] = 1.3 \times 10^{-13}$ M
17.15 Your picture should contain two carbonic acid molecules, three hydrogen carbonate anions and three "new" water molecules.

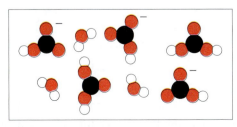

17.16 $[CO_3^{2-}] = 0.35$ M; $[HCO_3^-] = [OH^-] = 1.5 \times 10^{-2}$ M; $[H_2CO_3] = 2.2 \times 10^{-8}$ M; $[H_3O^+] = 6.8 \times 10^{-13}$ M; $[Na^+] = 0.72$ M

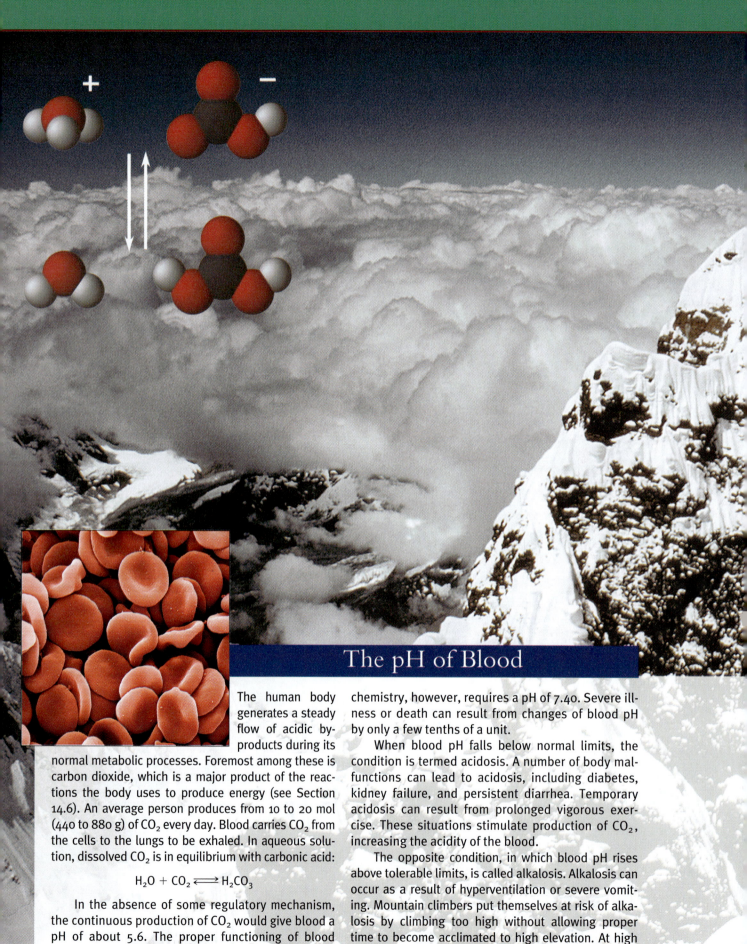

The pH of Blood

The human body generates a steady flow of acidic by-products during its normal metabolic processes. Foremost among these is carbon dioxide, which is a major product of the reactions the body uses to produce energy (see Section 14.6). An average person produces from 10 to 20 mol (440 to 880 g) of CO_2 every day. Blood carries CO_2 from the cells to the lungs to be exhaled. In aqueous solution, dissolved CO_2 is in equilibrium with carbonic acid:

$$H_2O + CO_2 \rightleftharpoons H_2CO_3$$

In the absence of some regulatory mechanism, the continuous production of CO_2 would give blood a pH of about 5.6. The proper functioning of blood chemistry, however, requires a pH of 7.40. Severe illness or death can result from changes of blood pH by only a few tenths of a unit.

When blood pH falls below normal limits, the condition is termed acidosis. A number of body malfunctions can lead to acidosis, including diabetes, kidney failure, and persistent diarrhea. Temporary acidosis can result from prolonged vigorous exercise. These situations stimulate production of CO_2, increasing the acidity of the blood.

The opposite condition, in which blood pH rises above tolerable limits, is called alkalosis. Alkalosis can occur as a result of hyperventilation or severe vomiting. Mountain climbers put themselves at risk of alkalosis by climbing too high without allowing proper time to become acclimated to high elevation. At high

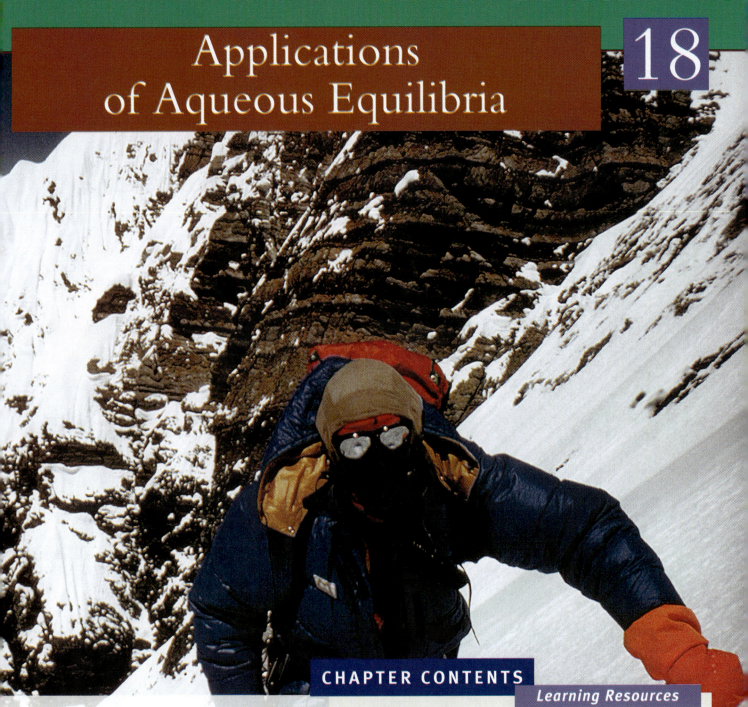

Applications of Aqueous Equilibria

CHAPTER CONTENTS

18.1 Buffer Solutions

18.2 Capacity and Preparation of Buffer Solutions

18.3 Acid–Base Titrations

18.4 Solubility Equilibria

18.5 Complexation Equilibria

Learning Resources

 KEY CONCEPTS

 CRITICAL THINKING

 SKILLS TO MASTER

 ADDITIONAL HELP
www.wiley.com/college/olmsted
- TUTORIALS
- ANIMATIONS

altitude there is less oxygen in the air, and climbers must compensate by breathing faster. Under these conditions, the body may lose CO_2 too rapidly, causing a decrease in blood acidity and a corresponding rise in pH. The result can be severe altitude sickness, which results in death if the climber is not given supplemental oxygen or moved quickly to lower elevation.

Healthy humans do not suffer these ailments, because blood pH is tightly regulated by a pair of reactions that involve H_2CO_3 and HCO_3^-. Hydronium ions that enter the bloodstream react with hydrogen carbonate, and hydroxide anions react with carbonic acid. These reactions work together to maintain the pH of blood at 7.40 ± 0.05:

$$HCO_3^- + H_3O^+ \rightleftharpoons H_2CO_3 + H_2O$$
$$H_2CO_3 + OH^- \rightleftharpoons HCO_3^- + H_2O$$

This process for the chemical control of pH in an aqueous solution is known as buffering.

Despite buffering, blood pH would quickly become acidic were it not for the rapid elimination of CO_2 from the body. In the lungs, CO_2 is transferred from blood cells to the gas phase and then exhaled from our bodies.

This chapter describes several important applications of aqueous equilibria. We begin with a discussion of buffer chemistry, followed by a description of acid and base titration reactions. Then we change our focus to examine the solubility equilibria of inorganic salts. The chapter concludes with a discussion of the equilibria of complex ions.

18.1 BUFFER SOLUTIONS

Human blood contains a variety of acids and bases that maintain the pH very close to 7.4 at all times. Close control of blood pH is critical because death results if the pH of human blood drops below 7.0 or rises above 7.8. This narrow pH range corresponds to only a fivefold change in the concentration of hydronium ions. Chemical equilibria work in the blood to hold the pH within this narrow window. Close control of pH is achieved by a **buffer solution,** so called because it protects, or buffers, the solution against pH variations.

The Composition of Buffer Solutions

To protect a solution against pH variations, a major species in the solution must react with added hydronium ions, and another major species must react with added hydroxide ions. The conjugate base of a weak acid will react readily with hydronium ions, and the weak acid itself will react readily with hydroxide ions. This means that a buffer solution can be defined in terms of its composition.

 KEY CONCEPT *A buffer solution contains both a weak acid and its conjugate base as major species in solution.*

As one example, dissolving sodium acetate in a solution of acetic acid produces a buffer solution in which both acetic acid and acetate anions are major species. Example 18-1 describes an acetic acid–acetate buffer solution.

Example 18-1	A solution contains 0.125 mol of solid sodium acetate dissolved in 1.00 L of 0.250 M acetic acid. Determine the concentrations of hydronium ions, acetate ions, and acetic acid.
Concentrations in a Buffer Solution	

Strategy: The seven-step procedure described in Chapter 16 can be applied to quantitative equilibrium problems.

1. *Determine what is asked for.* This problem asks about concentrations of species in an aqueous solution.

2. *Identify the major chemical species.* The original solution contains water and acetic acid molecules. Adding sodium acetate introduces two new major species, acetate anions and sodium cations. Thus, the resulting buffer solution has four major species: H_2O, Na^+, $CH_3CO_2^-$, and CH_3CO_2H.

3. *Determine what chemical equilibria exist.* Acetic acid is a weak acid, acetate anion is a weak base, water can act as an acid or a base, and Na^+ is a spectator ion. These species are reactants in three acid–base equilibria:

$$CH_3CO_2H\,(aq) + H_2O\,(l) \rightleftharpoons CH_3CO_2^-\,(aq) + H_3O^+\,(aq) \qquad K_a = 1.8 \times 10^{-5}$$

$$CH_3CO_2^-\,(aq) + H_2O\,(l) \rightleftharpoons CH_3CO_2H\,(aq) + OH^-\,(aq) \qquad K_b = \frac{K_w}{K_a} = 5.6 \times 10^{-10}$$

$$H_2O\,(l) + H_2O\,(l) \rightleftharpoons H_3O^+\,(aq) + OH^-\,(aq) \qquad K_w = 1.0 \times 10^{-14}$$

Example 18-1

Optimizing a Provisional Structure (*continued*)

4. *Write the K_{eq} expressions.* Among these reactions, the first one has the largest equilibrium constant, so the acetic acid equilibrium will generate the largest changes from initial concentrations:

$$K_a = 1.8 \times 10^{-5} = \frac{[H_3O^+]_{eq}\,[CH_3CO_2^-]_{eq}}{[CH_3CO_2H]_{eq}}$$

Solution:

5. *Organize the data and unknowns.* Set up a concentration table for this equilibrium:

Reaction	H_2O + CH_3CO_2H	$\rightleftharpoons$	$CH_3CO_2^-$ +	H_3O^+
Initial concentration (M)	0.250		0.125	0
Change in concentration (M)	$-x$		$+x$	$+x$
Equilibrium concentration (M)	$0.250 - x$		$0.125 + x$	x

6. *Substitute and calculate.* Because the initial concentrations are much larger than K_{eq}, we make the approximation that x is negligible compared with the initial concentrations:

$$0.250 - x \cong 0.250 \qquad 0.125 + x \cong 0.125$$

To determine the concentration of hydronium ions, rearrange the expression for K_a and substitute these concentrations:

$$K_a = \frac{[H_3O^+]_{eq}\,[CH_3CO_2^-]_{eq}}{[CH_3CO_2H]_{eq}} \qquad so \qquad [H_3O^+]_{eq} = \frac{K_a[CH_3CO_2H]_{eq}}{[CH_3CO_2^-]_{eq}}$$

$$x = \frac{(1.8 \times 10^{-5})(0.250)}{0.125} = 3.6 \times 10^{-5}$$

$$[H_3O^+]_{eq} = 3.6 \times 10^{-5}\,M$$

$$[CH_3CO_2H]_{eq} = 0.250\,M \qquad [CH_3CO_2^-]_{eq} = 0.125\,M$$

7. Do the Results Make Sense? Notice that the H_3O^+ concentration is much smaller than the concentrations of acetic acid and acetate, so the approximation that x is negligibly small is valid. The concentrations of acetic acid and acetate are reasonable, being comparable to the starting concentrations.

A solution contains 0.360 mol of ammonia dissolved in 1.5 L of 0.125 M ammonium chloride. Determine the concentrations of hydroxide ions, ammonium ions, and ammonia molecules.

The analysis carried out in Example 18-1 reveals one of the key features of buffer solutions: The equilibrium concentrations of both the weak acid and its conjugate base are essentially the same as their initial concentrations.

In the laboratory, chemists prepare buffer solutions in three different ways. Each results in a solution containing a weak acid and its conjugate base as major species. The most straightforward way to produce a buffer solution is by dissolving a salt of a weak acid in a solution of the same weak acid, as described in Example 18-1.

A second way to prepare a buffer is by adding strong base to a solution of a weak acid. This produces a buffer solution if the number of moles of strong base is about half the number of moles of weak acid. As a simple example, if 1 L of 0.5 M NaOH is mixed with 1 L of 1.0 M CH_3CO_2H, hydroxide anions react quantitatively with acetic acid molecules:

$$OH^- + CH_3CO_2H \rightleftharpoons H_2O + CH_3CO_2^- \qquad K_{eq} = \frac{1}{K_b} = 1.8 \times 10^9$$

The 0.5 mol of added hydroxide converts 0.5 mol of acetic acid into acetate, producing a buffer solution containing 0.5 mol of acetate and 0.5 mol of acetic acid.

A third approach to buffer solutions is to add strong acid to a solution of a weak base. This produces a buffer solution if the amount of strong acid is about half the amount of weak base. Continuing with our examples of acetic acid–acetate buffers, if a solution of hydrochloric acid is added to a solution of sodium acetate, then hydronium ions react quantitatively with acetate anions:

$$H_3O^+ + CH_3CO_2^- \rightleftharpoons H_2O + CH_3CO_2H \qquad K_{eq} = \frac{1}{K_a} = 5.6 \times 10^4$$

Example 18-2 treats a buffer solution made in this way.

Example 18-2 **Strong Acid–Weak Base Buffer**	Determine if a solution prepared by mixing 5.0 mL of concentrated HCl (12 M) with 75 mL of 1.0 M sodium acetate is a buffer solution. **Strategy:** The question asks if this is a buffer solution. A buffer solution contains both a weak acid and its conjugate base as major species. Thus, to answer the question, we must calculate the concentrations of acetate anions and acetic acid in the solution. We use a compressed version of the seven-step method to obtain these concentrations. **Solution:** Begin by identifying the major species in the two starting solutions: HCl: H_3O^+ and Cl^- \qquad Sodium acetate: $CH_3CO_2^-$ and Na^+ Because the hydronium ion is a strong acid and the acetate anion is a weak base, mixing the solutions results in near-quantitative reaction between these two ions: $$H_3O^+ + CH_3CO_2^- \rightleftharpoons H_2O + CH_3CO_2H \qquad K_{eq} = \frac{1}{K_a} = 5.6 \times 10^4$$ We calculate the initial amounts of the species from the volumes and concentrations, using $n = MV$: $$n\,(H_3O^+) = (12\ M)(0.0050\ L) = 0.060\ mol$$ $$n\,(CH_3CO_2^-) = (1.0\ M)(0.075\ L) = 0.075\ mol$$ When the solutions are mixed, the reaction between H_3O^+ and $CH_3CO_2^-$ consumes all the hydronium ions and generates acetic acid molecules. After mixing the starting solutions, we have the following amounts: $$n\,(CH_3CO_2^-) = 0.075\ mol - 0.060\ mol = 0.015\ mol$$ $$n\,(CH_3CO_2H) = 0.060\ mol$$ Both the weak acid and its conjugate base are major species in the mixture, so this is indeed a buffer solution.
	Does the Result Make Sense? The sum of the moles of the products is equal to the number of initial moles of acetate, and the final amounts of acid and base are comparable. Thus, the calculations support the conclusion that the final mixture is a buffer solution.
Extra Practice Exercise 18.2	Calculate the concentrations of the major species present in a buffer solution prepared by adding 1.40 g NaOH to 140. mL of 0.750 M NH_4Cl.

Molecular View of a Buffer Solution

The purpose of a buffer solution is to maintain the pH within a narrow range. The reactions that occur when H_3O^+ or OH^- is added to a buffer solution show how this is accomplished. Consider what happens when a small amount of hydroxide ion is added to the acetic acid–acetate buffer solution described in Example 18-1. Hydroxide ion is a

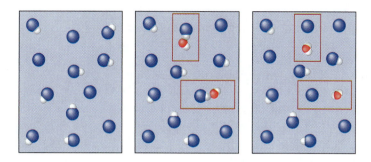

Figure 18-1
When a strong base is added to a buffer solution, the weak acid H*A* donates protons to hydroxide ions to form the conjugate base *A*$^-$, preventing a large increase in hydroxide ion concentration. (All water molecules except those produced in the proton transfer process are omitted for clarity.)

strong base and acetic acid is a weak acid, so proton transfer from CH_3CO_2H to OH^- goes essentially to completion:

Collision and proton transfer

$$OH^- + CH_3CO_2H \rightleftharpoons H_2O + CH_3CO_2^- \qquad K_{eq} = \frac{1}{K_b} = 1.8 \times 10^9$$

Tutorial

As long as the buffer solution contains acetic acid as a major species, a small amount of hydroxide ion added to the solution will be neutralized completely. Figure 18-1 shows two hydroxide ions added to a portion of a buffer solution. When a hydroxide ion collides with a molecule of weak acid, proton transfer forms a water molecule and the conjugate base of the weak acid. As long as there are more weak acid molecules in the solution than the number of added hydroxide ions, the proton transfer reaction goes virtually to completion. Weak acid molecules change into conjugate base anions as they "mop up" added hydroxide.

The same molecular reasoning shows that a buffer solution can absorb added hydronium ions. Consider what happens when some hydronium ions are added to the acetic acid–acetate buffer solution described in Example 18-1. The hydronium ion is a strong acid, and the acetate anion is a weak base, so proton transfer from $CH_3CO_2^-$ to H_3O^+ goes essentially to completion:

Collision and proton transfer

$$H_3O^+ + CH_3CO_2^- \rightleftharpoons H_2O + CH_3CO_2H \qquad K_{eq} = \frac{1}{K_a} = 5.6 \times 10^4$$

As long as the buffer solution contains a weak base as a major species, a small amount of hydronium ion added to the solution will be neutralized completely. Figure 18-2 shows two hydronium ions added to a portion of a buffer solution. When a hydronium ion collides with a weak base ion, proton transfer occurs, forming a water molecule and the conjugate acid of the weak base. Example 18-3 examines another buffer solution at the molecular level.

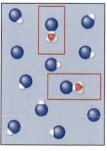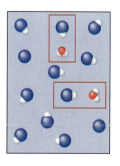

Figure 18-2
When a strong acid is added to a buffer solution, the conjugate base *A*$^-$ accepts protons from hydronium ions to form the weak acid H*A*, preventing a large increase in hydronium ion concentration. (All water molecules except those produced in the proton transfer process are omitted for clarity.)

Example 18-3

Molecular View of a Buffer Solution

The molecular picture at right represents a small portion of a buffer system. Solvent water molecules are omitted for clarity.

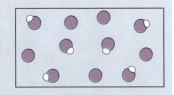

Redraw the original figure to show the equilibrium condition that is established when (a) three hydroxide ions enter the region, and (b) seven hydronium ions enter the region. Include any water molecules that are part of the buffer chemistry.

Strategy: The problem asks for a drawing that represents equilibrium conditions. We need a stoichiometric analysis of the reaction components. A table of amounts helps organize the information. The problem has two parts, and it best to treat them individually.

Solution: For Part a, hydroxide ions are neutralized by the weak acid component of the buffer. Set up a table of amounts, using a molecular picture to represent the balanced equation.

Reaction					
Start (molecules/ions)	6	3		5	"0"
Change	−3	−3		+3	+3
Final	3	0		2	3

Of course, the true amount of water is not "0". Water is the solvent, so there is an abundance of water molecules in the solution. This column represents only the water molecules produced by the buffer chemistry. We now redraw the system to show the molecular species present at the end of the reaction:

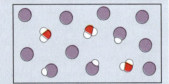

In Part b, the weak base of the buffer system reacts with hydronium ions:

For this example, the weak base is the limiting reactant, leaving excess H_3O^+ in the final solution. The buffer has been destroyed. We leave it to you to construct a table of amounts that leads to the final drawing, which appears at right:

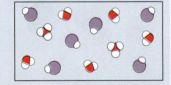

Do the Results Make Sense? Inventories of the components of the figures match the stoichiometry of the reactions, so the final pictures make sense.

Extra Practice Exercise 18.3

Redraw the original figure of Example 18-3 to show the final condition that is established when (a) nine OH^- ions enter the region, and (b) four H_3O^+ ions enter the region.

The Buffer Equation

The pH of a buffer solution depends on the weak acid equilibrium constant and the concentrations of the weak acid and its conjugate base. To show this, we begin by taking the logarithm of the acid equilibrium constant:

$$K_a = \frac{[H_3O^+]_{eq}\,[A^-]_{eq}}{[HA]_{eq}} \qquad so \qquad \log K_a = \log\left(\frac{[H_3O^+]_{eq}\,[A^-]_{eq}}{[HA]_{eq}}\right)$$

The logarithm of the concentration quotient can be separated by using the fact that $\log xy = \log x + \log y$:

$$\log K_a = \log [H_3O^+]_{eq} + \log\left(\frac{[A^-]_{eq}}{[HA]_{eq}}\right)$$

Multiplying both sides of the equation by -1 allows the use of pK_a and pH instead of $\log K_a$ and $\log [H_3O^+]_{eq}$:

$$-\log K_a = -\log [H_3O^+]_{eq} - \log\left(\frac{[A^-]_{eq}}{[HA]_{eq}}\right) \quad and \quad pK_a = pH - \log\left(\frac{[A^-]_{eq}}{[HA]_{eq}}\right)$$

Now we rearrange this equation to solve for pH:

$$pH = pK_a + \log\left(\frac{[A^-]_{eq}}{[HA]_{eq}}\right)$$

This equation is exact, but it can be simplified by applying one of the key features of buffer solutions. Any buffer solution contains both members of a conjugate acid–base pair as major species. In other words, both the weak acid and its conjugate base are present in relatively large amounts. As a result, the change to equilibrium, x, is small relative to each initial concentration, and the equilibrium concentrations are virtually the same as the initial concentrations:

$$[A^-]_{eq} = [A^-]_{initial} - x \cong [A^-]_{initial}$$

$$[HA]_{eq} = [HA]_{initial} + x \cong [HA]_{initial}$$

Thus, instead of using equilibrium concentrations in this equation, we substitute the initial concentrations to give the **buffer equation:**

$$pH = pK_a + \log\left(\frac{[A^-]_{initial}}{[HA]_{initial}}\right) \qquad (18\text{-}1)$$

The buffer equation, which is often called the Henderson–Hasselbalch equation, is used to calculate the equilibrium pH of a buffer solution directly from initial concentrations. The approximation is valid as long as the difference between initial concentrations and equilibrium concentrations is negligibly small. As a rule of thumb, the buffer equation can be applied when initial concentrations of HA and A^- differ by less than a factor of 10. Example 18-4 provides an illustration of the use of the buffer equation.

Example 18-4

The Buffer Equation

Buffer solutions with pH values around 10 are prepared using sodium carbonate (Na_2CO_3) and sodium hydrogen carbonate ($NaHCO_3$). What is the pH of a solution prepared by dissolving 10.0 g each of these two salts in enough water to make 0.250 L of solution?

Strategy: We use the seven-step strategy for equilibrium problems, except that we identify this as a buffer solution. This allows us to use the buffer equation in place of an equilibrium constant expression.

1. We are asked to calculate the pH of a buffer solution.

2. Both compounds are salts that dissolve in water to give their constituent ions, so the major species in this buffer solution are H_2O, Na^+, HCO_3^-, and CO_3^{2-}.

3. A buffer solution must contain a weak acid and its conjugate base as major species. In this solution, HCO_3^- is the weak acid, and CO_3^{2-} is the conjugate base:

$$H_2O\,(l) + HCO_3^-\,(aq) \rightleftharpoons CO_3^{2-}\,(aq) + H_3O^+\,(aq)$$

4. This proton transfer reaction involves the second acidic hydrogen atom of carbonic acid, so the appropriate equilibrium constant is K_{a2}, whose pK is found in

Example 18-4

The Buffer Equation
(continued)

Appendix E: $pK_{a2} = 10.33$. Because this is a buffer solution, we apply the buffer equation:

$$pH = pK_a + \log\left(\frac{[A^-]_{initial}}{[HA]_{initial}}\right)$$

Solution:

5. We find the initial concentrations from the masses of the salts and the volume of the solution:

$$[HCO_3^-]_{initial} = \frac{m}{(MM)V} = \frac{10.0\ g}{(84.01\ g/mol)(0.250\ L)} = 0.476\ M$$

$$[CO_3^{2-}]_{initial} = \frac{m}{(MM)V} = \frac{10.0\ g}{(106.0\ g/mol)(0.250\ L)} = 0.377\ M$$

6. Now substitute the appropriate values into the buffer equation and evaluate:

$$pH = pK_a + \log\left(\frac{[A^-]_{initial}}{[HA]_{initial}}\right) = 10.33 + \log\left(\frac{0.377\ M}{0.476\ M}\right)$$

$$pH = 10.33 + (-0.101) = 10.23$$

7. Does the Result Make Sense? The pH is close to the pK_a of the conjugate acid–base pair, so this is a reasonable result.

Extra Practice Exercise 18.4

An acidic buffer solution can be prepared from phosphoric acid and dihydrogen phosphate. What is the pH of solution prepared by mixing 23.5 g NaH_2PO_4 and 15.0 mL concentrated phosphoric acid (14.7 M) in enough water to give 1.25 L of solution?

Buffer Action

When a strong acid is added to a buffer solution, hydronium ions react with the conjugate base A^-, lowering the concentration of A^- and increasing the concentration of HA. This reaction lowers the pH of the buffer solution, but the increase in H_3O^+ concentration is much smaller than would be generated by the same amount of strong acid in an unbuffered solution. Similarly, when a strong base is added to a buffer solution, hydroxide ions react with the acid HA, lowering the concentration of HA and increasing the concentration of A^-. In this case the pH of the buffer solution rises, but the increase in hydroxide ion concentration is much smaller than would be generated by the same amount of strong base added to an unbuffered solution. Example 18-5 shows a quantitative calculation.

Example 18-5

Change in Buffer pH

By how much does the pH of the buffer solution of Example 18-4 change on the addition of 3.50 mL of 6.0 M HCl?

Strategy: The first four steps of the seven-step strategy are identical to the ones in Example 18-4. In this example, addition of a strong acid or base modifies the concentrations that go into the buffer equation. We need to determine the new concentrations (Step 5) and then apply the buffer equation (Step 6). In dealing with changes in amounts of acid and base, it is often convenient to work with moles rather than molarities. The units cancel in the concentration term of the buffer equation, so the ratio of concentrations can be expressed as a ratio of moles as well as a ratio of molarities:

$$\frac{[A^-]}{[HA]} = \frac{n_{A^-}/V}{n_{HA}/V} = \frac{n_{A^-}}{n_{HA}}$$

Example 18-5

Change in Buffer pH
(*continued*)

Solution: We do this problem using moles. First, determine the amounts present in the solution before addition of the HCl:

$$n_{HA} = \frac{10.0\ \text{g}}{84.01\ \text{g/mol}} = 1.19 \times 10^{-1}\ \text{mol}$$

$$n_{A^-} = \frac{10.0\ \text{g}}{106.0\ \text{g/mol}} = 9.43 \times 10^{-2}\ \text{mol}$$

Next, calculate the amount of hydronium ions added:

$$n_{H_3O^+} = MV = (6.0\ \text{mol/L})(3.5\ \text{mL})(10^{-3}\ \text{L/mL}) = 2.1 \times 10^{-2}\ \text{mol}$$

The hydronium ions react completely with conjugate base anions, increasing the amount of HCO_3^- and reducing the amount of CO_3^{2-}:

$$n_{HA} = 1.19 \times 10^{-1}\ \text{mol} + 2.1 \times 10^{-2}\ \text{mol} = 1.40 \times 10^{-1}\ \text{mol}$$

$$n_{A^-} = 9.43 \times 10^{-2}\ \text{mol} - 2.1 \times 10^{-2}\ \text{mol} = 7.3 \times 10^{-2}\ \text{mol}$$

Finally, substitute these new amounts into the buffer equation to compute the new pH:

$$\text{pH} = \text{p}K_a + \log\left(\frac{n_{A^-}}{n_{HA}}\right) = 10.33 + \log\left(\frac{0.073\ \text{mol}}{0.140\ \text{mol}}\right) = 10.33 - 0.28 = 10.05$$

Does the Result Make Sense? From Example 18-4, the pH of the buffer before adding the acid was 10.23, so the addition of 2.1×10^{-2} mol of acid reduces the pH by 0.18 pH units. This small change is a reasonable outcome for a buffer solution.

By how much does the pH of the buffer solution of Example 18-4 change on the addition of 0.92 g of NaOH?

The change in pH experienced by the buffer solution in Example 18-5 is tiny compared to the change that would result from adding 2.1×10^{-2} mol of acid to 250 mL of water. The pH of water would change from 7.00 to 1.08, a reduction of 5.92 pH units. Figure 18-3 illustrates and contrasts the two changes.

Figure 18-3
Addition of strong acid to a buffer solution changes the pH by much less than does addition of strong acid to water.

Buffer solution $\xrightarrow[\text{6.0 M HCl}]{\text{Add 3.50 mL}}$ $\Delta(\text{pH}) = 0.18$ Pure water $\xrightarrow[\text{6.0 M HCl}]{\text{Add 3.50 mL}}$ $\Delta(\text{pH}) = -6$

Section Exercises

18.1.1 Which of the following sets of chemicals can be used to prepare buffer solutions? For each one that can, specify the weak acid and its conjugate base that are major species in the buffer solution: (a) HCl and KCl; (b) HCl and KNO$_2$; (c) HCl and NH$_4$Cl; (d) NaOH and Na$_2$HPO$_4$; and (e) NaCl and NaC$_2$H$_3$O$_2$.

18.1.2 A buffer solution made from NH$_4$Cl and NH$_3$ is used to control pH from pH = 8 to pH = 10. Write balanced equations that show how this buffer system neutralizes H$_3$O$^+$ and OH$^-$.

18.1.3 A 1.50-L buffer solution is prepared from 0.200 mol NH$_4$Cl and 0.112 mol NH$_3$. (a) What is the pH of the buffer solution? (b) What would be the pH after adding 1.70 g of NaOH to the buffer described in Part a? (c) What would be the pH of a solution made by adding 1.70 g NaOH to 1.50 L of water?

18.2 CAPACITY AND PREPARATION OF BUFFER SOLUTIONS

Buffer solutions are practical and commonplace. In fact, many chemists and biologists use buffer solutions on a daily basis. Thus, it is important to know the limits of a buffer solution's capacity to control pH, as well as how to make buffer solutions.

Buffer Capacity

When small amounts of hydronium or hydroxide ions are added to a buffer solution, the pH changes are very small. There is a limit, however, to the amount of protection that a buffer solution can provide. After either buffering agent is consumed, the solution loses its ability to maintain near-constant pH. The **buffer capacity** of a solution is the amount of added H$_3$O$^+$ or OH$^-$ that the buffer solution can tolerate without exceeding a specified pH range.

Buffer capacity is determined by the amounts of weak acid and conjugate base present in the solution. If enough H$_3$O$^+$ is added to react completely with the conjugate base, the buffer is destroyed. Likewise, the buffer is destroyed if enough OH$^-$ is added to consume all of the weak acid. Consequently, buffer capacity depends on the overall concentration as well as the volume of the buffer solution. A buffer solution whose overall concentration is 0.50 M has five times the capacity as an equal volume of a buffer solution whose overall concentration is 0.10 M. Two liters of 0.10 M buffer solution has twice the capacity as one liter of the same buffer solution. Example 18-6 includes a calculation involving buffer capacity.

Example 18-6	Biochemists and molecular biologists use phosphate buffers to match physiological conditions. A buffer solution that contains H$_2$PO$_4^-$ as the weak acid and HPO$_4^{2-}$ as the weak base has a pH value very close to 7.0. A biochemist prepares 0.250 L of a buffer solution that contains 0.225 M HPO$_4^{2-}$ and 0.330 M H$_2$PO$_4^-$. What is the pH of this buffer solution? Is the buffering action of this solution destroyed by addition of 0.40 g NaOH?
Buffer Capacity	

Strategy: Use the seven-step strategy to calculate the pH of the buffer solution using the buffer equation. Then compare the amount of acid in the solution with the amount of added base. Buffer action is destroyed if the amount of added base is sufficient to react with all the acid. The buffering action of this solution is created by the weak acid H$_2$PO$_4^-$ and its conjugate base HPO$_4^{2-}$. The equilibrium constant for this pair is K_{a2} of phosphoric acid:

$$H_2O \, (l) + H_2PO_4^- \, (aq) \rightleftharpoons HPO_4^{2-} \, (aq) + H_3O^+ \, (aq) \qquad K_{eq} = K_{a2}$$

Solution: Substitute the tabulated value of pK_{a2} and the given concentrations into the buffer equation to calculate pH:

$$pH = pK_a + \log \left(\frac{[A^-]_{initial}}{[HA]_{initial}} \right) = 7.21 + \log \left(\frac{0.225 \text{ M}}{0.330 \text{ M}} \right) = 7.04$$

The question asks whether addition of base destroys the buffering capacity of the solution, so compare the amount of acid present with the amount of base added. As solid NaOH is added, each hydroxide ion that enters the buffer solution consumes one $H_2PO_4^-$ ion and produces one ion of HPO_4^{2-} and one water molecule:

$$OH^-(aq) + H_2PO_4^-(aq) \longrightarrow H_2O(l) + HPO_4^{2-}(aq)$$

$$n\ OH^-\ added = m/MM = 0.40\ g/40.00\ g/mol = 0.010\ mol$$

$$n\ H_2PO_4^-\ present = MV = (0.330\ M)(0.250\ L) = 0.0825\ mol$$

The amount of base added is considerably less than the amount of acid initially present in the buffer solution, so the solution will still act as a buffer. The added base raises the pH, however, because reaction of OH^- with the buffer increases the amount of HPO_4^{2-} and decreases the amount of $H_2PO_4^-$. We leave it to you to use the procedure illustrated in Example 18-5 to show that the pH after addition of this solid is 7.17, an increase of 0.13 pH units.

Do the Results Make Sense?   The pH of the buffer solution, both before and after adding the solid NaOH, is close to the pK_a of the conjugate acid–base pair. Moreover, the pH increases when NaOH is added. The solution becomes more basic as a consequence of the added hydroxide anions.

Addition of 5.25 g of NaOH to the buffer solution described in Example 18-4 would exceed the capacity of the buffer. Calculate the concentration of excess hydroxide ion and the pH of the solution.

Example 18-6

Buffer Capacity (*continued*)

**Extra Practice
Exercise 18.6**

Buffer Preparation

The buffer equation indicates that the pH of a buffer solution is close to the pK_a of the acid used to prepare the buffer:

$$pH = pK_a + \log\left(\frac{[A^-]_{initial}}{[HA]_{initial}}\right)$$

Every weak acid has a specific pK_a that determines the pH range over which it can serve as a buffering agent. Remember that a buffer solution must contain a weak acid and its conjugate weak base as major species. This condition is met when the ratio of weak base to weak acid is between 0.1 and 10. The buffer equation translates this restriction into a pH range over which the acid and its conjugate base can serve as an effective buffer:

$$pH_{low} = pK_a + \log 0.1 = pK_a - 1 \qquad pH_{high} = pK_a + \log 10 = pK_a + 1$$

$$pH\ range = pK_a \pm 1$$

With a given weak acid, a buffer solution can be prepared at any pH within about one unit of its pK_a value. Suppose, for example, that a biochemist needs a buffer system to maintain the pH of a solution close to 5.0. What reagents should be used? According to the previous analysis, the weak acid can have a pK_a between 4.0 and 6.0. As the pK_a deviates from the desired pH, however, the solution has a reduced buffer capacity. Thus, a buffer has maximum capacity when its acid has its pK_a as close as possible to the target pH. Table 18-1 lists some acid–base pairs often used as buffer solutions. For a pH = 5.0 buffer, acetic acid ($pK_a = 4.75$) and its conjugate base, acetate, would be a good choice.

A buffer solution must contain both the acid and its conjugate base, so at least two reagents must be added to water to prepare a buffer solution. An acetate buffer can be prepared, for example, from pure water, concentrated acetic acid, and an acetate salt. The cation contained in the salt should not have acid–base properties of its own, so sodium acetate would be an appropriate choice, but ammonium acetate would not.

②

Table 18-1 Common Buffer Systems

Acid	Conjugate Base	pK_a	pH Range
H_3PO_4	$H_2PO_4^-$	2.16	1–3
HCO_2H	HCO_2^-	3.75	3–5
CH_3CO_2H	$CH_3CO_2^-$	4.75	4–6
$H_2PO_4^-$	HPO_4^{2-}	7.21	6–8
NH_4^+	NH_3	9.25	8–10
HCO_3^-	CO_3^{2-}	10.33	9–11
HPO_4^{2-}	PO_4^{3-}	12.32	11–13

The chemist must prepare a solution that contains acetic acid and acetate in amounts that generate a buffer with pH = 5.00. The buffer equation is used to calculate the desired molar ratio:

$$5.00 = 4.75 + \log \frac{[\text{Acetate}]}{[\text{Acetic acid}]} \qquad so \qquad \log \frac{[\text{Acetate}]}{[\text{Acetic acid}]} = 0.25$$

$$\frac{[\text{Acetate}]}{[\text{Acetic acid}]} = 10^{0.25} = 1.8$$

For the preparation of a buffer solution of pH = 5.00, sodium acetate and acetic acid should be added to pure water in a molar ratio of 1.8:1.0. The exact amounts must be calculated using the volume and concentration of the solution, as Example 18-7 illustrates.

Example 18-7

Buffer Preparation

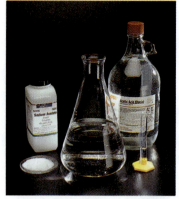

Ingredients of an acetate buffer

What mass of sodium acetate ($NaCH_3CO_2$, $MM = 82.04$ g/mol) and what volume of concentrated acetic acid (17.45 M) should be used to prepare 1.5 L of a buffer solution at pH = 5.00 that is 0.150 M overall?

Strategy: Because we know we are dealing with a buffer solution made from a specific conjugate acid–base pair, we can work directly with the buffer equation. We need to calculate the ratio of concentrations of conjugate base and acid that will produce a buffer solution of the desired pH. Then we use mole–mass–volume relationships to translate the ratio into actual quantities.

Solution: The problem specifies a solution volume of 1.5 L with a total molarity of 0.150 mol/L. The total molarity is the combined concentration of the two buffer components:

$$M_{\text{acetate}} + M_{\text{acetic acid}} = 0.150 \text{ M}$$

Use the total volume of the solution, 1.5 L, to determine the total number of moles in the system:

$$n = MV = (0.150 \text{ mol/L})(1.5 \text{ L}) = 0.225 \text{ mol}$$

$$n_{\text{acetate}} + n_{\text{acetic acid}} = 0.225 \text{ mol}$$

The calculation involving the buffer equation appears in the text: A buffer solution with pH = 5.00 requires an acetate–acetic acid molar ratio of 1.8. This ratio can be rewritten as a molar equality:

$$\frac{n_{\text{acetate}}}{n_{\text{acetic acid}}} = 1.8 \qquad so \qquad n_{\text{acetate}} = 1.8 \, n_{\text{acetic acid}}$$

Now substitute and calculate the required moles of acetic acid:

$$1.8 \, n_{\text{acetic acid}} + n_{\text{acetic acid}} = 0.225 \text{ mol}$$

$$n_{\text{acetic acid}} = \frac{0.225 \text{ mol}}{2.8} = 8.04 \times 10^{-2} \text{ mol}$$

The rest of the 0.225 mol must be acetate:

$$n_{\text{acetate}} = 0.225 \text{ mol} - 8.04 \times 10^{-2} \text{ mol} = 0.145 \text{ mol}$$

Finally, use molarity and molar mass to convert from moles to measurable amounts:

$$m_{\text{sodium acetate}} = (0.145 \text{ mol})(82.04 \text{ g/mol}) = 12 \text{ g}$$

$$V_{\text{acetic acid}} = \frac{(8.04 \times 10^{-2} \text{ mol})(10^3 \text{ mL/L})}{17.45 \text{ mol/L}} = 4.6 \text{ mL}$$

The final values are rounded to two significant figures to match the precision of the mole ratio and the total volume of the solution.

Does the Result Make Sense? The units are correct and the quantities are appropriate for approximately tenth-molar concentrations, so the answer appears to be reasonable.

What masses of Na_2HPO_4 and NaH_2PO_4 are required to prepare 2.0 L of a buffer solution that is at pH = 7.40 and a total concentration of 0.20 M?

Example 18-7

Buffer Preparation (*continued*)

Extra Practice Exercise 18.7

As described in Section 18.1, there are two other ways to prepare the buffer solution described in Example 18-7. Concentrated acetic acid could be added to pure water, followed by enough sodium hydroxide to generate the required 1.8:1.0 ratio of acetate to acetic acid:

$$CH_3CO_2H(aq) + OH^-(aq) \longrightarrow CH_3CO_2^-(aq) + H_2O(l)$$

Alternatively, a solution of sodium acetate could be prepared and a strong acid added to reach the proper acetate–acetic acid ratio:

$$CH_3CO_2^-(aq) + H_3O^+(aq) \longrightarrow CH_3CO_2H(aq) + H_2O(l)$$

Buffer preparation requires detailed, step-by-step calculations. Example 18-8 illustrates the complete procedure.

A technician wants to prepare a buffer solution at pH = 9.00 with an overall concentration of 0.125 mol/L. The technician has solutions of 1.00 M HCl and NaOH and bottles of all common salts. What reagents should be used, and in what quantities, to prepare 1.00 L of a suitable buffer?

Strategy: A practical problem in solution preparation usually requires a different strategy than our standard seven-step procedure. The technician must first identify a suitable conjugate acid–base pair and decide what reagents to use. Then the concentrations must be calculated, using pH and total concentration. Finally, the technician must determine the amounts of starting materials. The technician needs a buffer at pH = 9.00. Of the buffer systems listed in Table 18-1, the combination of NH_3 and NH_4^+ has the proper pH range for the required buffer solution.

Apparently, no bottles of aqueous ammonia are present in the laboratory, so the components of the buffer solution must come from the salts. The technician needs an ammonium salt with a counter anion that has no acid–base properties. Ammonium chloride (NH_4Cl) would be an appropriate choice. This salt contains the conjugate acid, NH_4^+, and the technician can generate NH_3 by adding strong base to the ammonium chloride solution:

$$NH_4^+(aq) + OH^-(aq) \longrightarrow NH_3(aq) + H_2O(l)$$

Example 18-8

Preparing a Buffer

Example 18-8

Preparing a Buffer
(*continued*)

Solution: What concentrations of NH_4^+ and NH_3 are required? First use the buffer equation to find the proper ratio of base to acid, then use the total molarity of the solution to determine the concentrations of NH_3 and NH_4^+. The ratio of ammonia to ammonium ion must produce a pH of 9.00:

$$pH = pK_a + \log \frac{[Base]}{[Acid]} \qquad 9.00 = 9.25 + \log \frac{[NH_3]}{[NH_4^+]}$$

$$\log \frac{[NH_3]}{[NH_4^+]} = -0.25 \quad \textit{from which} \quad \frac{[NH_3]}{[NH_4^+]} = 0.56 \quad \textit{and} \quad [NH_3] = 0.56\,[NH_4^+]$$

Use the total molarity of the buffer solution to find the concentrations:

$$[NH_3] + [NH_4^+] = 0.125 \text{ M}$$

Substitute for $[NH_3]$: $\qquad 0.56\,[NH_4^+] + [NH_4^+] = 0.125 \text{ M}$

$$[NH_4^+] = \frac{0.125 \text{ M}}{1.56} = 0.0801 \text{ M}$$

$$[NH_3] = 0.125 \text{ M} - 0.0801 \text{ M} = 0.045 \text{ M}$$

Finally, use stoichiometry to calculate actual amounts. The buffer solution contains both ammonia and ammonium ions, but both species are derived from ammonium chloride. For the preparation of the solution, some of the NH_4^+ ions in an aqueous solution of ammonium chloride must be converted into NH_3 molecules. Thus, the amount of salt required is found from the total molarity of the buffer solution:

$$n_{NH_4Cl} = MV = (0.125 \text{ mol/L})(1.00 \text{ L}) = 0.125 \text{ mol}$$

$$m_{NH_4Cl} = n\,MM = (0.125 \text{ mol})(53.49 \text{ g/mol}) = 6.69 \text{ g}$$

Ammonia is generated from NH_4^+ by adding sodium hydroxide solution:

$$n_{OH^-} = n_{NH_3} = (0.045 \text{ M})(1.00 \text{ L}) = 0.045 \text{ mol}$$

$$\text{Volume of } 1.00 \text{ mol NaOH} (aq) = \frac{0.045 \text{ mol}}{1.00 \text{ mol/L}} = 0.045 \text{ L}$$

To make the buffer, the technician should mix together 6.69 g of NH_4Cl, 45 mL of 1.00 M NaOH, and enough water to make 1.00 L of solution.

Does the Result Make Sense? The units are correct and the quantities seem reasonable for the target volume and concentrations, so the answer does make sense.

Extra Practice
Exercise 18.8

A chemist needs a buffer solution at pH = 5.25 with an overall concentration of 0.175 M. The chemist has bottles of all common salts and solutions of 1.50 M HCl and NaOH. What reagents should be used, and in what quantities, to prepare 2.50 L of a suitable buffer solution?

Section Exercises

18.2.1 A student adds 30. mL of 5.00 M HCl to the buffer solution described in Section Exercise 18.1.3. Is the buffering capacity of the solution destroyed? What is the final pH of the solution?

18.2.2 Determine the mass of solid sodium formate ($NaHCO_2$) and the volume of 0.500 M HCl solution required to generate 250 mL of

buffer solution with pH = 3.50 and a total concentration (conjugate acid plus base) of 0.225 M.

18.2.3 Describe how you would prepare 2.5 L of a pH = 10.50 buffer solution with a total concentration of 0.15 M, using an appropriate salt and solid NaOH. Use Table 18-1 to choose the buffer system.

18.3 ACID–BASE TITRATIONS

As described in Chapter 4, acid–base reactions that go to completion can be exploited in chemical analysis using the method of titration. Titrations can be understood in greater detail from the perspective of acid–base equilibria. Protonation of a weak base by a strong acid is a reaction that goes virtually to completion because of its large equilibrium constant:

$$B + H_3O^+ \longrightarrow BH^+ + H_2O \qquad K_{eq} \gg 1$$

Any base, including OH^-

Likewise, when strong base is added to a solution containing an acid, the reaction goes essentially to completion, again because of a large equilibrium constant:

$$OH^- + HA \longrightarrow H_2O + A^- \qquad K_{eq} \gg 1$$

Any acid, including H_3O^+

Even though these reactions go essentially to completion, equilibria determine the pH of the solution before, during, and at the stoichiometric point of the titration.

Titrations are treated like any other equilibrium analysis, but we must pay special attention to the major species present in the solution, because these change during the titration. The most common titrations are analysis of a weak acid using a solution of strong base and analysis of a weak base using a solution of strong acid.

Titration of a Weak Acid by OH⁻ Ions

The titration of a solution of a weak acid with a solution of a strong base is shown graphically in Figure 18-4. This titration curve has four distinct regions, each characterized by different major species:

1. Near the beginning of the titration, HA and H_2O are the only major species. The pH of the solution can be calculated using K_a and the initial concentration of HA.

2. During most of the titration, both HA and its conjugate base, A^-, are major species. As a result, the solution is buffered, and the pH changes relatively slowly with added hydroxide. In this region, the pH of the solution can be calculated using the buffer equation.

3. When nearly all of the HA molecules have reacted with added OH^-, the only major species are A^- and H_2O. Here the pH is determined by K_b and the concentration of A^-. Addition of OH^- ions causes a sharp increase in pH.

4. After all the HA molecules have reacted, the solution contains excess A^- and OH^- ions as the major species in solution. The pH is determined by the excess concentration of hydroxide ion.

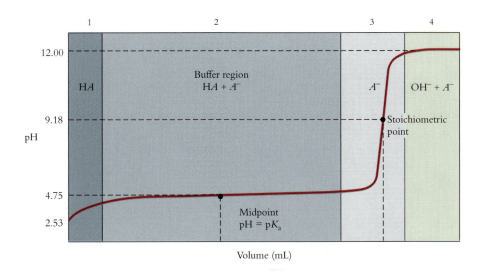

Figure 18-4
Schematic profile of the titration curve for a weak acid HA titrated with hydroxide ions. The titration can be divided into four regions that differ in the major species present in solution. The pH values are those for titration of 0.500 M acetic acid.

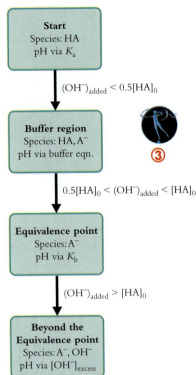

Figure 18-5
A flow chart summarizes the major species in solution and the pH calculations for the four key regions of a weak acid titration curve.

These features of the weak acid titration curve are summarized in the flow chart shown in Figure 18-5.

As an example, consider the titration of 0.150 L of 0.500 M acetic acid solution with 2.50 M potassium hydroxide. Very early in the titration, the major species are water and acetic acid, and the equilibrium that determines solution pH is proton transfer between these major species:

$$H_2O\,(l) + CH_3CO_2H\,(aq) \rightleftharpoons H_3O^+(aq) + CH_3CO_2^-(aq) \qquad K_a = 1.8 \times 10^{-5}$$

Using the standard method for solving equilibrium problems, you should be able to show that the initial solution contains 0.0750 mol of acetic acid, has a hydronium ion concentration of 3.0×10^{-3} M, and has a pH of 2.52.

As the titration proceeds, the hydroxide ions in each added volume of titrant convert acetic acid molecules to acetate ions:

As a result, both acetate ions and acetic acid molecules are present as major species in solution. The presence of an acid and its conjugate base means that in this region of the titration, the solution is buffered, so the pH changes slowly as hydroxide ions are added to the solution.

This buffer region contains the **midpoint** of the titration, the point at which the amount of added OH^- is equal to exactly half the weak acid originally present. In the current example, the solution at the midpoint contains 0.0375 mol each of acetic acid and acetate. Applying the buffer equation reveals the key feature of the midpoint:

$$pH = pK_a + \log \frac{n_{acetate}}{n_{acetic\ acid}} = 4.75 + \log 1 = 4.75$$

KEY CONCEPT *At the midpoint of a titration of a weak acid by a strong base, the pH of the solution equals the pK$_a$ of the weak acid.*

Beyond the buffer region, when nearly all of the acetic acid has been consumed, the pH increases sharply with each added drop of hydroxide solution. The titration curve passes through an almost vertical region before leveling off again. Recall from Chapter 4 that the **stoichiometric point** of an acid titration (also called the equivalence point) is the point at which the number of moles of added base is exactly equal to the number of moles of acid present in the original solution. At the stoichiometric point of a weak acid titration, the conjugate base is a major species in solution, but the weak acid is not.

Example 18-9 illustrates that the pH at the stoichiometric point in a titration of a weak acid differs from 7.0.

Example 18-9 **pH** **at the Stoichiometric Point**	What is the pH at the stoichiometric point of the titration of 0.150 L of 0.500 M acetic acid with 2.50 M KOH solution? **Strategy:** This is another equilibrium calculation to which the standard seven-step procedure applies. Special attention must be given, however, to analyzing the initial conditions *at the stoichiometric point,* bearing in mind that the reaction between hydroxide ions and a weak acid goes essentially to completion. 1. We are asked to calculate the pH at the stoichiometric point. 2. At the stoichiometric point, the amount of added hydroxide ions equals the amount of acetic acid that was originally present: mol OH^- added = mol CH_3CO_2H originally present

In the flow chart (Figure 18-5):

Start
Species: HA
pH via K_a

$(OH^-)_{added} < 0.5[HA]_0$

Buffer region
Species: HA, A$^-$
pH via buffer eqn.

③

$0.5[HA]_0 < (OH^-)_{added} < [HA]_0$

Equivalence point
Species: A$^-$
pH via K_b

$(OH^-)_{added} > [HA]_0$

Beyond the Equivalence point
Species: A$^-$, OH$^-$
pH via $[OH^-]_{excess}$

Example 18-9

pH
at the Stoichiometric Point
(*continued*)

This added hydroxide reacts essentially to completion with acetic acid:

$$CH_3CO_2H\,(aq) + OH^-\,(aq) \longrightarrow CH_3CO_2^-\,(aq) + H_2O\,(l)$$

Thus, the major acid–base species present at the stoichiometric point are $CH_3CO_2^-$ and H_2O.

3. The pH is determined by the proton transfer from water to acetate ions:

$$CH_3CO_2^-\,(aq) + H_2O\,(l) \rightleftharpoons CH_3CO_2H\,(aq) + OH^-\,(aq)$$

4. The equilibrium constant expression for this equilibrium is related to K_a for acetic acid:

$$K_{eq} = K_{b,\text{ acetate}} = \frac{K_w}{K_{a,\text{ acetic acid}}} = 5.6 \times 10^{-10}$$

Solution:

5. We need to calculate "initial" conditions for the stoichiometric point. Recall that the added base has reacted to form acetate anions:

$$n\ OH^-\text{ added} = M_{acid}V_{acid} = (0.500\text{ mol/L})(0.150\text{ L}) = 0.0750\text{ mol}$$

$$n\,(CH_3CO_2H) = 0,\ n\,(OH^-) = 0,\text{ and } n\,(CH_3CO_2^-) = 0.0750\text{ mol}$$

Before constructing a concentration table, convert moles of acetate to molarity. The volume at the stoichiometric point is the original volume plus the volume of added titrant:

$$V_{initial} = 0.150\text{ L}$$

$$V_{titrant} = \frac{n}{M} = \frac{0.0750\text{ mol}}{2.50\text{ mol/L}} = 0.0300\text{ L}$$

$$V_{total} = 0.150\text{ L} + 0.0300\text{ L} = 0.180\text{ L}$$

$$[CH_3CO_2^-]_{initial} = \frac{0.0750\text{ mol}}{0.180\text{ L}} = 0.417\text{ M}$$

Reaction	$H_2O\,(l)$ +	$CH_3CO_2^-\,(aq)$	$\rightleftharpoons$	$CH_3CO_2H\,(aq)$ +	$OH^-\,(aq)$
Initial conc. (M)		0.417		0	0
Change in conc. (M)		$-x$		$+x$	$+x$
Equilibrium conc. (M)		$0.417 - x$		x	x

6. We have done this type of calculation many times, so you should be able to show that the pH of the solution at the stoichiometric point is 9.18.

Does the Result Make Sense? It may surprise you that the pH at the stoichiometric point is not 7.0, but this is a reasonable outcome. The major species present at the stoichiometric point are water and a conjugate base, so the resulting solution is basic in nature.

What is the pH at the stoichiometric point of the titration of 0.250 L of 0.0350 M hypochlorous acid (HClO) with 1.00 M NaOH?

Although the exact pH at the stoichiometric point depends on what weak acid is being titrated, the qualitative result of Example 18-9 is reproduced for every titration of a weak acid with a strong base. At the stoichiometric point of a weak acid titration, the exact value of the pH is determined by K_b for the conjugate base, and it is always *greater than* 7.0.

Figure 18-6
Schematic profile of the titration curve for a weak base B titrated with hydronium ions. The pH values are those for titration of ephedrine, a weak base that is the active ingredient in many decongestants.

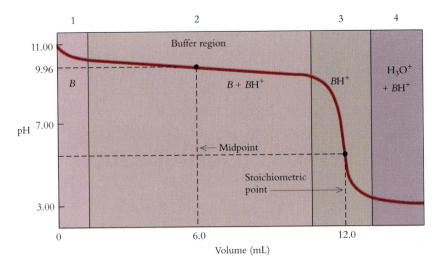

Beyond the stoichiometric point, in the final region of the titration curve, the concentration of acetic acid is very close to zero. There are no acid molecules to react with any further hydroxide ions, so excess hydroxide ions are present in solution as a major species. Beyond the stoichiometric point, the pH of the solution is determined by the amount of excess hydroxide ion.

Titration of a Weak Base with H_3O^+ Ions

The principles that describe the titration of a weak acid also describe the titration of a weak base with hydronium ions. The titration curve for a weak base is shown in Figure 18-6.

Notice that the titration curve for a weak base has the same four regions seen in the titration curve of a weak acid:

1. At the beginning of the titration, the base B and water are the only major species.
2. In the long, flat buffer region, both B and its conjugate acid BH^+ are major species. The midpoint of the titration occurs in the buffer region. At this point, the pH of the solution is equal to the pK_a of the conjugate acid of the base.
3. When nearly all the B molecules have been protonated, the solution is no longer buffered. Further addition of hydronium ions causes a sharp drop in pH.
4. Beyond the stoichiometric point the solution contains excess H_3O^+ ions, which determine the pH.

Example 18-10 demonstrates that at the stoichiometric point in any weak base titration, the pH is less than 7.0.

Example 18-10	Ephedrine, a weak base, is the active ingredient in many commercial decongestants. To analyze a sample of ephedrine dissolved in 0.200 L of water, a chemist carries out a titration with 0.900 M HCl, monitoring the pH continuously. The data obtained in this titration are shown in Figure 18-6. Calculate K_b for ephedrine and determine the pH of the solution at the stoichiometric point.
Titration of a Weak Base	

Strategy: This problem asks for two different quantities, requiring two separate analyses. Thus, the problem should be solved in two stages. The reaction that takes place during the titration is proton transfer from hydronium ions (added from the buret) to ephedrine molecules (in the solution). For simplicity, we designate ephedrine as Ep and its conjugate acid as EpH^+. The problem provides a titration curve (Figure 18-6) and asks about two different equilibrium results, K_b and pH at the stoichiometric point. Ephedrine is not listed in Appendix E, so K_b must be determined by analyzing the titration curve. The pH calculation requires the standard procedure using K_b.

Example 18-10

Titration of a Weak Base
(*continued*)

Solution:

Determination of K_b. The titration curve does not give K_b directly. However, at the midpoint of the titration the concentration of Ep and EpH$^+$ are identical. Use this information in the buffer equation to show that at the midpoint of the titration, the pH of the solution equals the pK_a for EpH$^+$:

$$EpH^+ (aq) + H_2O (l) \rightleftharpoons Ep (aq) + H_3O^+ (aq)$$

$$pH = pK_a + \log \frac{[\text{Base}]}{[\text{Acid}]} = pK_a + \log (1) = pK_a$$

Ephedrine
$C_{10}H_{15}NO$

We are asked to determine K_b, which is related to pK_a through Equation 17-5:

$$pK_a + pK_b = 14.00$$

To locate the midpoint of the titration, use the volume needed to reach the stoichiometric point. At the stoichiometric point, all the ephedrine has been converted to EpH$^+$ ions, whereas at the midpoint, exactly half the ephedrine has been converted to EpH$^+$ ions. Thus, the volume of the HCl solution at the midpoint of the titration is exactly half the volume needed to reach the stoichiometric point. According to the titration curve, the volume needed to reach the stoichiometric point is 12.0 mL. The midpoint, therefore, corresponds to the addition of 6.0 mL, and the pH at the midpoint is the pH reading when 6.0 mL of solution has been added. Reading from Figure 18-6:

$$pH_{\text{midpoint}} = pK_a = 9.96$$

Now apply Equation 17-5 to find K_b for ephedrine:

$$pK_a + pK_b = 14.00$$

$$pK_b = 14.00 - pK_a = 14.00 - 9.96 = 4.04$$

$$K_b = 10^{-4.04} = 9.1 \times 10^{-5}$$

pH at the Stoichiometric Point. At the stoichiometric point, virtually all the ephedrine molecules have been converted to EpH$^+$ ions by proton transfer from H$_3$O$^+$:

$$Ep (aq) + H_3O^+ (aq) \longrightarrow EpH^+ (aq) + H_2O (l)$$

At this point, the solution contains EpH$^+$, H$_2$O, and Cl$^-$ as major species. The chloride ion is the conjugate base of a *strong* acid, HCl, so the pH of the solution is determined by the acid–base equilibrium of water and the acid, EpH$^+$:

$$EpH^+ (aq) + H_2O (l) \rightleftharpoons Ep (aq) + H_3O^+ (aq) \qquad pK_a = 9.96$$

$$K_a = \frac{[Ep]_{eq} [H_3O^+]_{eq}}{[EpH^+]_{eq}} = 10^{-9.96} = 1.1 \times 10^{-10}$$

The amount of weak acid present at the stoichiometric point is equal to the amount of added hydronium ion:

$$n(EpH^+) = n(H_3O^+) \text{ needed to reach the stoichiometric point}$$

$$n(H_3O^+) \text{ needed} = (0.0120 \text{ L})(0.900 \text{ mol/L}) = 0.0108 \text{ mol}$$

$$n(EpH^+) = 0.0108 \text{ mol}$$

Example 18-10

Titration of a Weak Base
(*continued*)

The initial concentration of EpH^+ is found by dividing the number of moles by the total volume of the original solution plus the volume of added HCl solution:

$$[EpH^+]_{initial} = \frac{0.0108 \text{ mol}}{0.200 \text{ L} + 0.0120 \text{ L}} = 0.0509 \text{ M}$$

You should be able to set up a concentration table and use the K_a expression to solve for $[H_3O^+]_{eq}$. The pH of the solution at the stoichiometric point is 5.63.

Do the Results Make Sense? Ephedrine is a weak base, so $K_b = 9.1 \times 10^{-5}$ is a reasonable value. At the stoichiometric point, the only major species present are the conjugate acid and water, so the acidic pH of 5.63 is also reasonable.

Extra Practice Exercise 18.10

Triethylamine, $N(CH_3)_3$, ($pK_b = 4.19$) is a weak base with an odor of fish. A chemist analyzed a 1.36-g sample of $N(CH_3)_3$ dissolved in 0.250 L of water. Titration with 0.974 M HCl required 23.6 mL to reach the stoichiometric point. Determine (a) the pH at the beginning of the titration, (b) the pH at the midpoint, (c) the pH at the stoichiometric point, and (d) the pH after 27.5 mL of the HCl solution had been added.

You may wonder why we did not use the titration curve to determine the pH at the stoichiometric point of the ephedrine titration, as we did to find the pH at the midpoint. Notice from Figure 18-6 that near the stoichiometric point, the pH changes very rapidly with added H_3O^+. At the stoichiometric point, the curve is nearly vertical. Thus, there is much uncertainty in reading a graph to determine the pH at the stoichiometric point. In contrast, a titration curve is nearly flat in the vicinity of the midpoint, minimizing uncertainty caused by errors in graph reading.

The key to understanding the titrations of weak acids and bases is to be familiar with the species in solution and the dominant equilibrium at each point along the titration curve. Example 18-11 reinforces these qualitative features.

Example 18-11

Qualitative Features of a Titration

The figures below represent a small portion of a weak acid solution and a titration curve for the acid. Redraw the molecular picture to show how the figure should look for each of the points A–C along the titration curve. Include in your drawings any water molecules formed as part of the titration process.

Strategy: Identify the major species in solution by assigning each of the points to one of the four characteristic regions of the titration curve. In the titration reaction, hydroxide ions react with molecules of weak acid:

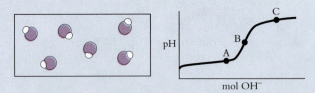

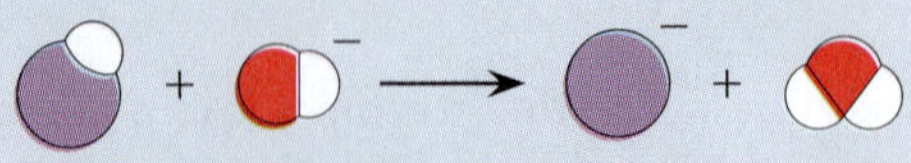

The figure shows six molecules of H*A* at the start of the titration. Apply stoichiometric reasoning to determine the numbers of H*A*, *A*⁻, and H_2O molecules to include in the figures.

Solution: Point A lies along the section of the titration curve known as the buffer region. Buffering action comes from the presence of a weak acid and its conjugate base as major species in solution. Moreover, Point A lies beyond the midpoint of the titration, which tells us that more than half of the weak acid has been consumed. We represent this solution with two molecules of HA, four ions of A^-, and four H_2O molecules:

Example 18-11

Qualitative Features of a Titration (*continued*)

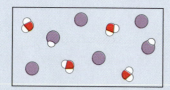

Point B represents the stoichiometric point of the titration. All of the weak acid has been converted to the conjugate base, so the figure should show six A^- ions and the six H_2O molecules formed in the titration:

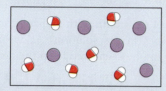

Beyond the equivalence point there is no weak acid present to neutralize added hydroxide ions. The figure for Point C should show six A^- ions, six molecules of H_2O, and several OH^- ions:

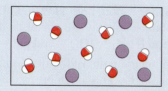

Do the Results Make Sense? The sequence of figures, from the initial state to Point C, shows a progressive loss of HA with matching increases in A^- and H_2O, as we would expect for a weak acid titration. The figures make sense.

The molecular view below represents a small portion of a solution that matches Point A on the corresponding titration curve. Redraw the molecular picture to show how the figure should look for each of the points B–D along the titration curve. Your drawings should show any water molecules formed as part of the titration process.

Extra Practice Exercise 18.11

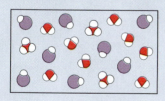 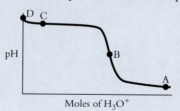

Moles of H_3O^+

Titration of Polyprotic Acids

In the titration of a polyprotic acid, the added base reacts first with the more acidic hydrogen atoms of the neutral acid. For example, the titration of maleic acid takes place in two steps. For removal of one acidic hydrogen atom of maleic acid, $pK_a = 1.82$ ($K_{a1} = 1.5 \times 10^{-2}$):

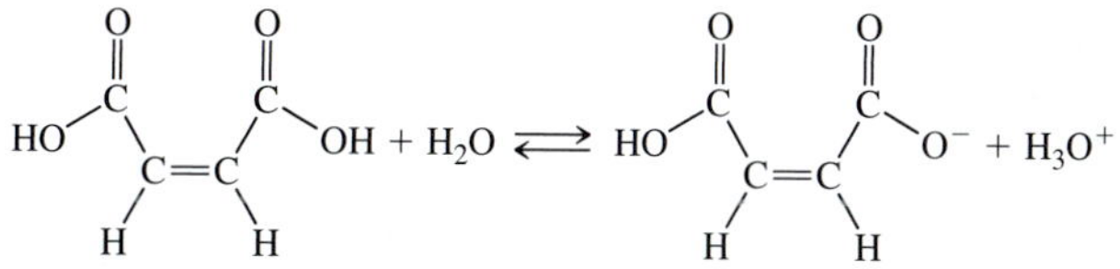

The second acidic hydrogen atom, with a pK_a of 6.59 ($K_{a2} = 2.6 \times 10^{-7}$), is dramatically less acidic than the first. Consequently, the second proton transfer reaction does not occur until all of the neutral diprotic acid has been consumed:

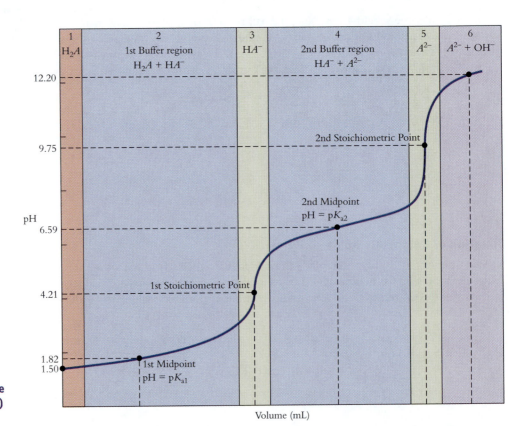

$$pK_{a2} = 6.59$$

The titration curve for maleic acid, which appears in Figure 18-7, shows this two-step process clearly. These data represent the titration of 50.0 mL of 0.10 M maleic acid with 1.0 M NaOH. The titration curve has six distinct regions.

Region 1. At the beginning of the titration, the diprotic acid (represented by H_2A) and H_2O are the only major species in the solution. As we describe in Chapter 17, the hydronium ion concentration can be calculated from the familiar equilibrium expression:

$$K_{a1} = 1.5 \times 10^{-2} = \frac{[HA^-]_{eq}\,[H_3O^+]_{eq}}{[H_2A]_{eq}} = \frac{x^2}{0.10 - x}$$

Solving the quadratic equation gives $x = [H_3O^+]_{eq} = 3.19 \times 10^{-2}$ M and an initial pH of 1.50.

Region 2. Within the first buffer region, both H_2A and HA^- are major species in solution, and we can apply the buffer equation to calculate the pH. Halfway to the first equivalence point of the titration $[H_2A] = [HA^-]$ and pH = pK_{a1} = 1.82.

Region 3. At the first stoichiometric point of the titration, all the diprotic acid has been converted to its conjugate base, HA^-. This amphiprotic anion can react with itself, analogous to the self-ionization of water:

$$2\,HA^-\,(aq) \rightleftharpoons H_2A\,(aq) + A^{2-}\,(aq) \qquad K_{eq} = \frac{[H_2A]_{eq}\,[A^{2-}]_{eq}}{[HA^-]_{eq}^2}$$

Figure 18-7
The titration of a diprotic acid, H_2A, can be divided into six regions that differ in the major species present in solution. Shown here is the titration curve for maleic acid (50.0 mL, 0.10 M) titrated with 1.0 M NaOH.

We can relate this equilibrium constant to K_{a1} and K_{a2}:

$$H_2O\,(l) + HA^-(aq) \rightleftharpoons H_3O^+(aq) + A^{2-}(aq) \qquad K_{eq} = K_{a2}$$

$$H_3O^+(aq) + HA^-(aq) \rightleftharpoons H_2O\,(l) + H_2A\,(aq) \qquad K_{eq} = 1/K_{a1}$$

Adding these two equations gives the reaction of the anion with itself, and as we show in Chapter 16, when equilibria are added their equilibrium constants are multiplied. Thus:

$$K_{eq} = \frac{[H_2A]_{eq}\,[A^{2-}]_{eq}}{[HA^-]_{eq}^2} = \frac{K_{a2}}{K_{a1}} = \frac{2.6 \times 10^{-7}}{1.5 \times 10^{-2}} = 1.7 \times 10^{-5}$$

This value is larger than the equilibrium constants for the other acid–base equilibria involving major species, so we can treat the self-reaction of HA^- as the dominant equilibrium at the first stoichiometric point. The equilibrium constant expression does not contain $[H_3O^+]$, however, so we need to do further manipulations to relate the dominant equilibrium to $[H_3O^+]$.

At the first stoichiometric point $[H_2A]_{eq} = [A^{2-}]_{eq}$, which allows us to write

$$\frac{K_{a2}}{K_{a1}} = \frac{[H_2A]_{eq}^2}{[HA^-]_{eq}^2}$$

Next, we rearrange the expression for K_{a1}:

$$K_{a1} = \frac{[HA^-]_{eq}\,[H_3O^+]_{eq}}{[H_2A]_{eq}} \qquad giving \qquad \frac{[H_2A]_{eq}}{[HA^-]_{eq}} = \frac{[H_3O^+]_{eq}}{K_{a1}}$$

Substitute this expression into the ratio of equilibrium constants and solve for $[H_3O^+]_{eq}$:

$$\frac{K_{a2}}{K_{a1}} = \frac{[H_3O^+]_{eq}^2}{K_{a1}^{\,2}} \qquad from\ which \qquad [H_3O^+]_{eq} = \sqrt{K_{a1}K_{a2}}$$

Now evaluate $[H_3O^+]_{eq}$:

$$[H_3O^+]_{eq} = \sqrt{(2.6 \times 10^{-7})(1.5 \times 10^{-2})} = 6.2 \times 10^{-5}\ M$$

$$pH = -\log (6.2 \times 10^{-5}) = 4.21$$

The pH at the first stoichiometric point is 4.21. This derivation is completely general, so Equation 18-2 can be applied at the first stoichiometric point of any polyprotic acid titration:

$$[H_3O^+]_{eq,\ 1st\ stoichiometric\ point} = \sqrt{K_{a1}K_{a2}} \qquad\qquad (18\text{-}2)$$

Region 4. Moving beyond the first stoichiometric point, the titration enters the second buffer region. Here, the major species are HA^- and its conjugate base, A^{2-}. The pH in this region is given by the buffer equation, using the pK_a of HA^-. At the second midpoint, $[HA^-] = [A^{2-}]$, and $pH = pK_{a2}$. For the titration of maleic acid, $pH = 6.59$ at the second midpoint.

Region 5. At the second stoichiometric point, all of the diprotic acid has been converted to A^{2-}. The dominant equilibrium for maleic acid is proton transfer from water:

$$A^{2-}(aq) + H_2O\,(l) \rightleftharpoons HA^-(aq) + OH^-(aq) \qquad K_b = \frac{K_w}{K_{a2}}$$

The titration of maleic acid requires 10.0 mL to reach the second equivalence point, which gives a concentration of 0.833 M for A^{2-}. The equilibrium constant expression is straightforward:

$$\frac{[HA^-]_{eq}\,[OH^-]_{eq}}{[A^{2-}]_{eq}} = \frac{x^2}{0.833 - x} \cong \frac{x^2}{0.833} = \frac{1.00 \times 10^{-14}}{2.6 \times 10^{-7}} = 3.85 \times 10^{-8}$$

Solving for x gives $[OH^-] = 5.66 \times 10^{-5}$, a value that justifies the usual assumption. The pH at the second stoichiometric point is 9.75.

Region 6. Finally, beyond the second stoichiometric point, the major species are H_2O, A^{2-}, and OH^-. The pH of the solution is determined by the concentration of excess hydroxide. For example, 11.0 mL of NaOH solution added in the maleic acid titration gives 0.016 M OH^- and pH = 12.20. You should be able to verify these calculations.

By recognizing species in solution and their dominant equilibrium, we can construct titration curves for other diprotic acids. Example 18-12 shows how this is done for sulfurous acid.

Example 18-12

Drawing a Titration Curve

Sulfurous acid, H_2SO_3, has two acidic hydrogen atoms, with pK_a values of 1.85 ($K_{a1} = 1.40 \times 10^{-2}$) and 7.20 ($K_{a2} = 6.3 \times 10^{-8}$). Construct a titration curve for the titration of 125 mL of 0.150 M sulfurous acid with 0.800 M NaOH.

Strategy: Calculate a pH value for each of the six regions of the titration, and then extrapolate to complete the titration curve.

Solution:
Region 1: The starting pH of the solution is calculated using K_{a1} and the initial molarity of the diprotic acid.

We use the standard approach to a weak acid equilibrium:

$$\frac{[HSO_3^-]_{eq}\,[H_3O^+]_{eq}}{[H_2SO_3]_{eq}} = \frac{x^2}{0.150 - x} = 1.4 \times 10^{-2}$$

Using the quadratic equation lets use determine that $x = 0.0664$:

$$[H_3O^+]_{eq} = 6.64 \times 10^{-2}\,M \quad giving \quad pH = 1.18$$

Region 2: At the first midpoint of the titration, the pH of the solution equals pK_{a1}:

$$pK_{a1} = 1.85$$

Region 3: We solve Equation 18-2 to determine the concentration of $[H_3O^+]_{eq}$ at the first stoichiometric point:

$$[H_3O^+]_{eq} = \sqrt{K_{a1}K_{a2}} = \sqrt{(1.4 \times 10^{-2})(6.3 \times 10^{-8})} = 2.75 \times 10^{-5}\,M$$

$$pH = 4.56$$

Region 4: The pH at the second midpoint is given by pK_{a2}:

$$pK_{a2} = 7.20$$

Region 5: At the second stoichiometric point, all of the sulfurous acid has been converted to sulfite, and the pH is determined using K_{b2}. For this calculation, we need the concentration of SO_3^{2-}, for which we need to know the volume of NaOH solution required to reach the second stoichiometric point. Start by finding the number of moles of sulfurous acid:

$$n = MV = (0.150\,mol/L)(0.125\,L) = 0.01875\,mol\,H_2SO_3$$

To calculate the required volume of NaOH solution, remember that the acid has two acidic hydrogen atoms, so two moles of OH^- are required for each mole of H_2SO_3:

$$V_{base} = 0.01875\,mol\,H_2SO_3\left(\frac{2\,mol\,NaOH}{1\,mol\,H_2SO_3}\right)\left(\frac{1\,L}{0.800\,mol\,NaOH}\right) = 0.0469\,L$$

Now calculate the sulfite concentration using the total volume of the solution:

$$M_{SO_3^{2-}} = \frac{n_{SO_3^{2-}}}{V_{total}} = \frac{0.01875\,mol}{0.125\,L + 0.0469\,L} = 0.1091\,M$$

Next we need K_{b2} for the sulfite anion:

$$K_{b2} = \frac{K_w}{K_{a2}} = \frac{1.00 \times 10^{-14}}{6.3 \times 10^{-8}} = 1.59 \times 10^{-7}$$

We are now ready to set up the base equilibrium expression for SO_3^{2-}:

$$K_{b2} = \frac{[HSO_3^-]_{eq}\,[OH^-]_{eq}}{[SO_3^{2-}]_{eq}} = \frac{x^2}{0.1091 - x} \cong \frac{x^2}{0.1091} = 1.59 \times 10^{-7}$$

Example 18-12

Drawing a Titration Curve (*continued*)

This gives $x = 1.32 \times 10^{-4}$, from which we find $[OH^-]_{eq}$:

$$[OH^-]_{eq} = 1.32 \times 10^{-4}\,M \quad giving \quad pH = 10.12$$

Region 6: Finally, beyond the second stoichiometric point the pH is determined by the amount of excess hydroxide ion. Calculate the amount of hydroxide ion in 1.00 mL of the NaOH solution:

$$n = MV = (0.800\ mol/L)(0.00100) = 8.00 \times 10^{-4}\ mol$$

The total volume is now $0.125\ L + 0.0469\ L + 0.00100\ L = 0.173\ L$:

$$[OH^-]_{eq} = \frac{8.00 \times 10^{-4}\ mol}{0.173\ L} = 4.62 \times 10^{-3}\ M \quad giving \quad pH = 11.67$$

We now have six pH values for key regions along the titration curve. A table helps to organize these pH values. You should be able to verify that the volumes are correct:

Region	1	2	3	4	5	6
pH	1.18	1.85	4.56	7.20	10.12	11.67
Volume (mL)	0	11.7	23.4	35.1	46.9	47.9

The titration curve plots pH against volume:

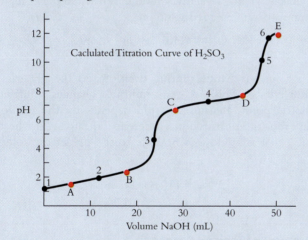

Caclulated Titration Curve of H_2SO_3

When plotted on a graph of pH vs. volume of NaOH solution, these six points reveal the gross features of the titration curve. Adding additional calculated points helps define the pH curve. On the curve shown here, the red points A–D were calculated using the buffer equation with base/acid ratios of 1/3 and 3/1. Point E was generated from excess hydroxide ion concentration, 2.00 mL beyond the second stoichiometric point. You should verify these additional five calculations.

Does the Result Make Sense? The calculated curve shows the general features of the pH titration curve for a diprotic acid. The pH of the solution is acidic at the first stoichiometric point (major species = weak acid HA^-) and basic at the second (major species = conjugate base A^{2-}). These features tell us the answer does make sense.

Amino acids form diprotic cations. For example, the diprotic form of alanine is shown here:

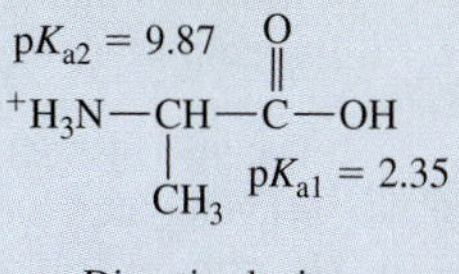

Estimate the titration curve for 0.200 L of 0.120 M diprotic alanine with 3.00 M NaOH.

$pK_{a2} = 9.87$

$^+H_3N-CH-C-OH$ (with O double-bonded above C)

$\underset{\displaystyle CH_3}{|}$ $pK_{a1} = 2.35$

Diprotic alanine

Extra Practice Exercise 18.12

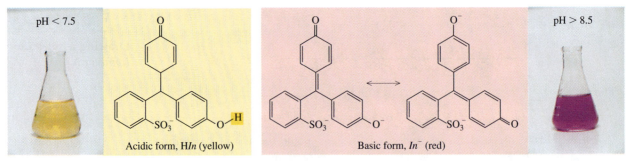

Figure 18-8
The line structures of phenol red in its acidic and basic forms.

Indicators

An acid–base titration can be performed without measuring the pH of the solution, if an **indicator** is present that changes color as the titration passes the stoichiometric point. An indicator is a weak organic acid that contains a highly delocalized π bonding network. We use pK_{In} to designate the pK_a of an indicator. At $pH < pK_{In}$, the indicator is in its protonated form, which we designate H*In*. At $pH > pK_{In}$, the indicator gives up its proton and is converted to its conjugate base, designated *In*$^-$.

Deprotonation of H*In* changes the structure of the indicator, which alters the pattern of electron delocalization in the π system. This change in the delocalized π system causes a change in the color of the indicator. Figure 18-8 shows these changes for one acid–base indicator, phenol red. The pK_{In} of phenol red is 7.9. When the pH is less than 7.9, the acid dominates, giving a yellow solution. When the pH is higher than 7.9, the conjugate base dominates, giving a red solution:

$$\underset{\text{Yellow}}{\text{H}In} + \text{H}_2\text{O} \rightleftharpoons \text{H}_3\text{O}^+ + \underset{\text{Red}}{In^-} \qquad pK_{In} = 7.9$$

The pH of a solution changes during titration. As the pH passes the value of pK_{In}, the indicator changes color. A good indicator changes color very near the pH of the stoichiometric point. Because this pH depends on the pK_a of the substance being titrated, different titrations require different indicators. The best indicator for a titration has a pK_{In} that is as close as possible to the pH of the solution at the stoichiometric point:

$$pK_{In} \cong pH_{\text{stoichiometric point}}$$

Table 18-2   Indicators and Their pH Ranges

Indicator	pK_{In}*	pH Range	Colors Acid	Base
Thymol blue**	1.75	1.2–2.8	Red	Yellow
Methyl orange	3.40	3.1–4.4	Red	Yellow
Bromocresol green	4.68	4.0–5.6	Yellow	Blue
Methyl red	4.95	4.4–6.2	Red	Yellow
Bromocresol purple	6.3	5.2–6.8	Yellow	Purple
Phenol red	7.9	6.4–8.0	Yellow	Red
Thymol blue**	8.9	8.0–9.6	Yellow	Blue
Phenolphthalein	9.4	8.0–10.0	Colorless	Red
Thymolphthalein	10.0	9.4–10.6	Colorless	Blue

* The acid dissociation constant of an indicator is designated K_{In}.
**Thymol blue has two acidic hydrogen atoms, so it can be used as an indicator for two different pH regions.

Because the pH changes rapidly near the stoichiometric point, an indicator is suitable for a titration if the pH at the stoichiometric point is within one unit of pK_{In}:

$$pH_{\text{stoichiometric point}} = pK_{In} \pm 1 \qquad (18\text{-}3)$$

Table 18-2 lists a selection of acid–base indicators, and Example 18-13 shows how to select an appropriate indicator.

Example 18-13

Selecting an Indicator

A student wants to titrate a solution of ammonia whose approximate concentration is 10^{-2} M. What indicator would be appropriate?

Strategy: The student needs an indicator whose color changes as close to the stoichiometric point as possible:

$$pH_{\text{stoichiometric point}} = pK_{In} \pm 1$$

Thus, the student must calculate the pH at the stoichiometric point and then consult Table 18-2.

Solution: The titration reaction is

$$NH_3\,(aq) + H_3O^+(aq) \longrightarrow NH_4^+(aq) + H_2O\,(l)$$

At the stoichiometric point, all the ammonia molecules have been converted to ammonium ions, so the major species present are NH_4^+ and H_2O. The pH of the solution is thus determined by the acid–base equilibrium of NH_4^+ ions:

$$NH_4^+(aq) + H_2O\,(l) \rightleftharpoons NH_3\,(aq) + H_3O^+(aq)$$

$$K_{eq} = \frac{[NH_3]_{eq}\,[H_3O^+]_{eq}}{[NH_4^+]_{eq}} = \frac{K_w}{K_b} = 5.6 \times 10^{-10}$$

Set up a concentration table at the stoichiometric point:

Reaction	H$_2$O	+	NH$_4^+$	$\rightleftharpoons$	NH$_3$	+	H$_3$O$^+$
Initial concentration (M)			10^{-2}		0		0
Change in concentration (M)			$-x$		$+x$		$+x$
Equilibrium concentration (M)			$10^{-2} - x$		x		x

Now substitute and solve:

$$5.6 \times 10^{-10} = \frac{(x)(x)}{10^{-2} - x} \cong \frac{x^2}{10^{-2}}$$

$$x^2 = 5.6 \times 10^{-12} \quad giving \quad x = 2.4 \times 10^{-6}$$

$$pH = 5.62$$

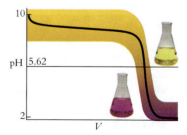

Table 18-2 shows that bromocresol purple ($pK_{In} = 6.3$) or methyl red ($pK_{In} = 4.95$) would be suitable indicators for the titration of ammonia. Methyl red is typically used for titrations of weak bases by strong acids. The figure shows its color behavior.

Does the Result Make Sense? We see that x is small relative to 10^{-2}, and an acidic pH at the stoichiometric point is what we expect for this titration of weak base by strong acid.

A chemist needs to titrate a solution of hydrofluoric acid that is approximately 0.15 M. What indicator would be appropriate for the titration?

Extra Practice Exercise 8.13

The objective of an acid–base titration is to determine the amount of an acid or base in a solution. Because an indicator is itself a weak acid, it may appear that adding an indicator would alter equilibrium concentrations and influence the titration. However, a useful indicator gives a noticeable color to a solution at a concentration of 10^{-6} M. This is negligible compared with the concentration of the solution being titrated, which is usually in the range of 10^{-2} to 10^{-3} M.

The low concentration of an indicator also explains why the presence of this weak acid does not change the pH of the solution. Indicators are always present as minor species in solution, never as major species. Thus, the dominant equilibrium that determines the pH of a solution never involves the indicator. The K_a of the substance being titrated establishes the equilibrium concentration of hydronium ions. This, in turn, establishes whether the indicator is in its acidic or basic form.

Section Exercises

■ **18.3.1** Glycolic acid ($HOCH_2CO_2H$), a constituent of sugar cane juice, has a pK_a of 3.9. Sketch the titration curve for the titration of 60.0 mL of 0.010 M glycolic acid with 0.050 M KOH. Indicate the stoichiometric point, the buffer region, and the point of the titration where pH = pK_a. Sketch the curve qualitatively without doing any quantitative calculations.

■ **18.3.2** What is the pH of the glycolic acid solution at the stoichiometric point? What is the pH of the solution when 3.0 mL of the KOH solution has been added?

■ **18.3.3** Choose an indicator suitable for the titration of glycolic acid. What is the color of the solution at the stoichiometric point?

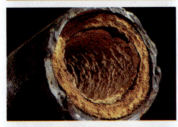

18.4 SOLUBILITY EQUILIBRIA

The limestone deposits that decorate Carlsbad and other caverns are the result of the solubility equilibrium of calcium carbonate in groundwater, as described in Chapter 16:

$$CaCO_3\,(s) \rightleftharpoons Ca^{2+}(aq) + CO_3^{2-}(aq) \qquad K_{sp} = [Ca^{2+}]_{eq}\,[CO_3^{2-}]_{eq}$$

Solubility equilibria are important in the natural environment and in our daily lives. Coral reefs are built up over many years through deposits of calcium carbonate from tiny sea organisms. Thermal vents on the ocean's floor spew out insoluble metal sulfides and other minerals. Deposits from hard water can wreak havoc with plumbing.

In an aqueous solubility reaction, a salt dissolves to yield ions in solution. The amount of a salt that dissolves in water varies over a large range, as the following examples show:

$$AgI\,(s) \rightleftharpoons Ag^+(aq) + I^-(aq) \qquad K_{sp} = 8.5 \times 10^{-17}$$
$$Cu(OH)_2\,(s) \rightleftharpoons Cu^{2+}(aq) + 2\,OH^-(aq) \qquad K_{sp} = 1.1 \times 10^{-15}$$
$$CaCO_3\,(s) \rightleftharpoons Ca^{2+}(aq) + CO_3^{2-}(aq) \qquad K_{sp} = 3.36 \times 10^{-9}$$
$$Ag_2SO_4\,(s) \rightleftharpoons 2\,Ag^+(aq) + SO_4^{2-}(aq) \qquad K_{sp} = 1.2 \times 10^{-5}$$
$$NaCl\,(s) \rightleftharpoons Na^+(aq) + Cl^-(aq) \qquad K_{sp} = 6.2$$

Substances that have $K_{sp} \ll 1$, such as calcium carbonate, copper(II) hydroxide, and silver iodide, are classified as **insoluble.** Salts that have K_{sp} between 10^{-2} and 10^{-5}, such as silver sulfate, are **slightly soluble,** and solids such as NaCl, which have $K_{sp} > 10^{-2}$, are **soluble.** This classification provides a more specific description of soluble and insoluble than the qualitative solubility guidelines presented in Chapter 4.

Solubility equilibria can be treated using our standard approach to equilibrium, as demonstrated in Examples 18-14, 18-15, and 18-16.

Gypsum is a relatively soft rock made of calcium sulfate. Rainwater percolates through gypsum, dissolves some of the rock, and eventually becomes saturated with Ca^{2+} ions and SO_4^{2-} ions. A geochemist takes a sample of groundwater from a cave and finds that it contains 8.4×10^{-3} M SO_4^{2-} and 5.8×10^{-3} M Ca^{2+}. (The ratio is not 1:1 because other sulfate rock contributes some of the SO_4^{2-} ions to the solution.) Use these data to determine the solubility product of calcium sulfate.

Strategy: This problem is straightforward, so a truncated version of the seven-step strategy is appropriate.
We are asked to evaluate K_{sp} of $CaSO_4$:

$$CaSO_4\,(s) \rightleftharpoons Ca^{2+}(aq) + SO_4^{2-}(aq) \qquad K_{sp} = [Ca^{2+}]_{eq}\,[SO_4^{2-}]_{eq}$$

Solution: The problem gives the required equilibrium concentrations, since it is stated that the groundwater sample is saturated (that is, at equilibrium) with respect to these ions.

$$K_{sp} = (5.8 \times 10^{-3})(8.4 \times 10^{-3}) = 4.9 \times 10^{-5}$$

Does the Result Make Sense? The equilibrium constant is small, indicating a slightly soluble salt, as we would expect for a naturally occurring mineral like gypsum.

Calcium hydroxide, known as slaked lime, is an ingredient of mortar and plaster. A saturated solution of $Ca(OH)_2$ has $[Ca^{2+}]_{eq} = 0.011$ M and $[OH^-]_{eq} = 0.0216$ M. Calculate the solubility product of calcium hydroxide.

> **Example 18-14**
>
> **Solubility Products**

> **Extra Practice**
> **Exercise 18.14**

Example 18-15 shows a calculation in which the stoichiometry differs from 1:1.

When solid PbI_2 is added to pure water at 25 °C, the salt dissolves until the concentration of Pb^{2+} reaches 1.35×10^{-3} M. After this concentration is reached, excess solid remains undissolved. What is K_{sp} for this salt?

Strategy: The problem asks for an equilibrium constant, which means we need to find equilibrium concentrations of the species involved in the solubility reaction. Use the seven-step strategy, which we present here without step numbers.

The starting materials are solid PbI_2 and H_2O. These are the only major species present before any solid dissolves. The only equilibrium involving these reactants is the solubility reaction. Solid PbI_2 dissolves in water to produce its ions in solution, Pb^{2+} and I^-:

$$PbI_2\,(s) \rightleftharpoons Pb^{2+}(aq) + 2\,I^-(aq) \qquad K_{eq} = K_{sp} = [Pb^{2+}]_{eq}\,[I^-]_{eq}^2$$

Solution: The equilibrium concentration of Pb^{2+} ions is stated to be 1.35×10^{-3} M, but the equilibrium concentration of iodide ions is not stated:

$$[Pb^{2+}]_{eq} = 1.35 \times 10^{-3}\,M \qquad [I^-]_{eq} = ?$$

The equilibrium concentration of I^- is determined by stoichiometric analysis. Initially, the system contains only pure water and solid lead(II) iodide. Enough PbI_2 dissolves to make $[Pb^{2+}]_{eq} = 1.35 \times 10^{-3}$ M. One formula unit of PbI_2 contains one Pb^{2+} cation and *two* I^- anions. Thus, twice as many iodide ions as lead ions enter the solution. The concentration of I^- at equilibrium is double that of Pb^{2+} cations:

$$[I^-]_{eq} = 2\,[Pb^{2+}]_{eq} = 2.70 \times 10^{-3}\,M$$

> **Example 18-15**
>
> **Calculating K_{sp}**
>
>
>
> ⑤

Example 18-15	Substitute the values of the concentrations at equilibrium into the equilibrium expression and calculate the result:

Calculating K_{sp} (*continued*)

$$K_{sp} = [Pb^{2+}]_{eq}\,[I^-]^2_{eq} = (1.35 \times 10^{-3})(2.70 \times 10^{-3})^2 = 9.84 \times 10^{-9}$$

Does the Result Make Sense? Remember that although equilibrium calculations require concentrations in molarities for solutes, the equilibrium constant expression is dimensionless. The solubility product has a small value, which is reasonable given the small concentrations, 10^{-3} M, of the ions at equilibrium.

Extra Practice Exercise 18.15

A saturated solution of magnesium hydroxide has a pH of 10.35. What is the solubility product of $Mg(OH)_2$?

Example 18-16 deals with the second type of calculation, determining a concentration at equilibrium when the value of the solubility product is known.

Example 18-16

Solubility Calculations

Cadmium is an extremely toxic metal that finds its way into the aqueous environment as a result of some human activities. A major cause of cadmium pollution is zinc mining and processing, because natural deposits of ZnS ores usually also contain CdS. During the processing of these ores, highly insoluble cadmium sulfide ($K_{sp} = 7.9 \times 10^{-27}$) may be converted into considerably less insoluble cadmium hydroxide ($K_{sp} = 7.2 \times 10^{-15}$). What mass of $Cd(OH)_2$ will dissolve in 1.00×10^2 L of an aqueous solution?

Strategy: Determining what mass will dissolve requires that we find equilibrium concentrations, so we apply our standard approach:

Major species present: $Cd(OH)_2\,(s)$ and $H_2O\,(l)$

Equilibrium: $Cd(OH)_2\,(s) \rightleftharpoons Cd^{2+}(aq) + 2\,OH^-(aq)$

$$K_{sp} = [Cd^{2+}]_{eq}\,[OH^-]^2_{eq} = 7.2 \times 10^{-15}$$

Solution: Set up a concentration table:

Reaction	$Cd(OH)_2\,(s)$	$\rightleftharpoons$	$Cd^{2+}(aq)$	+	$2\,OH^-(aq)$
Initial concentration (M)	(Solid)		0		0
Change in concentration (M)	(Solid)		$+x$		$+2x$
Equilibrium concentration (M)	(Solid)		x		$2x$

Evaluation:

$$7.2 \times 10^{-15} = (x)(2x)^2 = 4x^3 \quad so \quad x^3 = 1.8 \times 10^{-15} \quad and \quad x = 1.2 \times 10^{-5}$$

The concentration of Cd^{2+} ions in a saturated solution is 1.2×10^{-5} M. Therefore, in 1.00×10^2 L of solution, there will be

$$(1.2 \times 10^{-5}\text{ mol/L})(1.00 \times 10^2\text{ L}) = 1.2 \times 10^{-3}\text{ mol }Cd^{2+}$$

Cadmium hydroxide generates Cd^{2+} in a 1:1 stoichiometric ratio, so the number of moles of Cd^{2+} in solution equals the number of moles of $Cd(OH)_2$ that dissolves. Find the amount in grams using the molar mass of $Cd(OH)_2$:

$$m = n\,MM = (1.2 \times 10^{-3}\text{ mol})(146\text{ g/mol}) = 0.18\text{ g}$$

In contrast, only 1.3×10^{-9} g of CdS dissolves in the same volume of water.

Does the Result Make Sense? The units are correct, and the overall solubility is low, as we expect for a hydroxide salt.

Extra Practice Exercise 18.16

What mass of lead(II) bromide ($pK_{sp} = 5.18$) will dissolve in 1.25 L of water?

Precipitation Equilibria

The solubility product (K_{sp}) describes the equilibrium of a salt dissolving in water. In the laboratory and in industry, solubility equilibria are often exploited in the opposite direction. Two solutions are mixed to form a new solution in which the solubility product of one substance is exceeded. This salt precipitates and can be collected by filtration. Example 18-17 illustrates how precipitation techniques can be used to remove toxic heavy metals from aqueous solutions.

Example 18-17

Precipitation Reactions

As illustrated in Example 18-16, wastewater resulting from metal processing often contains significant amounts of toxic heavy metal ions that must be removed before the water can be returned to the environment. One method uses sodium hydroxide solution to precipitate insoluble metal hydroxides. Suppose that 1.00×10^2 L of wastewater containing 1.2×10^{-5} M Cd^{2+} is treated with 1.0 L of 6.0 M NaOH solution. What is the residual concentration of Cd^{2+} after treatment, and what mass of $Cd(OH)_2$ precipitates?

Strategy: Again, we use the standard approach to an equilibrium calculation. In this case the reaction is a precipitation, for which the equilibrium constant is quite large. Thus, taking the reaction to completion by applying limiting reactant stoichiometry is likely to be the appropriate approach to solving the problem.

The wastewater contains Cd^{2+}, so an anion must also be present in the solution to balance the charge of the cadmium ions. Other species may exist as well. The problem asks only about the cadmium in the wastewater, so assume that any other ions are spectators. The sodium hydroxide solution contains Na^+ and OH^-, so the major species in the treated wastewater include H_2O, Cd^{2+}, OH^-, and Na^+. The equilibrium constant for the precipitation reaction is the inverse of K_{sp} for $Cd(OH)_2$:

$$Cd^{2+}(aq) + 2\,OH^-(aq) \rightleftharpoons Cd(OH)_2(s)$$

$$K_{eq} = \frac{1}{[Cd^{2+}]_{eq}\,[OH^-]_{eq}^2} = \frac{1}{K_{sp}} = \frac{1}{7.2 \times 10^{-15}} = 1.4 \times 10^{14}$$

Solution: The problem asks for the residual concentration of cadmium ions. In other words, what is the Cd^{2+} ion concentration in the solution after the NaOH is added? Before a concentration table can be completed, initial ion concentrations must be determined. The volume of the mixed solutions is 1.01×10^2 L, which is less than 5% different from 1.00×10^2 L. This means the initial Cd^{2+} concentration is 1.2×10^{-5} M. Do a dilution calculation to find the "initial" concentration (after mixing but before reaction) of hydroxide ions:

$$M_2 = \frac{M_1 V_1}{V_2} = \frac{(6.0 \text{ M})(1.0 \text{ L})}{1.00 \times 10^2\,\text{L}} = 6.0 \times 10^{-2} \text{ M} = [OH^-]_{initial}$$

Now set up a concentration table. The equilibrium constant for precipitation is very large, so imagine the precipitation in two steps (see Example 16-14). First, take the reaction to completion by applying limiting reactant stoichiometry. Then "switch on" the solubility equilibrium:

Reaction	$Cd^{2+}(aq)$	$+$	$2\,OH^-(aq)$	$\rightleftharpoons$	$Cd(OH)_2(s)$
Initial concentration (M)	1.2×10^{-5}		6.0×10^{-2}		(solid)
Change to completion (M)	-1.2×10^{-5}		-2.4×10^{-5}		(solid)
Concentration at completion (M)	0		6.0×10^{-2}		(solid)
Change to equilibrium (M)	$+\gamma$		$+2\gamma$		(solid)
Equilibrium concentration (M)	γ		$6.0 \times 10^{-2} + 2\gamma$		(solid)

A tiny amount of $Cd(OH)_2$ dissolves in this approach to equilibrium, so we use the solubility equilibrium and K_{sp}:

$$K_{sp} = [Cd^{2+}]_{eq}\,[OH^-]_{eq}^2 = 7.2 \times 10^{-15}$$

Example 18-17

Precipitation Reactions
(*continued*)

The variable y represents the change to equilibrium in a reaction with a small equilibrium constant, so make the approximation that y will be small compared with 6.0×10^{-2}:

$$K_{sp} = (y)(6.0 \times 10^{-2} + 2y)^2 \cong (y)(6.0 \times 10^{-2})^2 = 7.2 \times 10^{-15}$$

$$y = \frac{7.2 \times 10^{-15}}{(6.0 \times 10^{-2})^2} = 2.0 \times 10^{-12} \quad and \quad [Cd^{2+}]_{eq} = 2.0 \times 10^{-12} \text{ M}$$

The problem also asks for the mass of $Cd(OH)_2$ that precipitates. The number of moles of precipitate equals the number of moles of Cd^{2+} ions removed by hydroxide treatment. The necessary information is contained in the concentration table in the row that gives the completion conditions. The precipitation equilibrium constant is so large ($> 10^{14}$) that equilibrium considerations can be neglected for this calculation:

$$n_{Cd(OH)_2} = n_{Cd^{2+}} = MV$$

$$= (1.2 \times 10^{-5} \text{ M})(1.00 \times 10^2 \text{ L}) = 1.2 \times 10^{-3} \text{ mol}$$

$$m_{Cd(OH)_2} = n \, MM = (1.2 \times 10^{-3} \text{ mol})(146 \text{ g/mol}) = 0.18 \text{ g}$$

Does the Result Make Sense? Notice that virtually all the mass that dissolved originally (Example 18-16) has been recovered, and the residual concentration of Cd^{2+} is very small. Both these observations are consistent with a water treatment process.

Extra Practice Exercise 18.17

A student mixes 0.175 L of 0.500 M $AgNO_3$ with 0.225 L of 0.900 M Na_2CO_3. A white precipitate of silver carbonate forms. Calculate the mass of the precipitate and the residual concentration of Ag^+ ion in the final solution. (See Appendix E for K_{sp} values.)

The Common-Ion Effect

The concentrations of Cd^{2+} ions in Example 18-16 reveal an important feature of aqueous equilibria. Notice in the calculations that the amount of dissolved Cd^{2+} ions decreased by a factor of 10^{10} upon addition of excess OH^-. This reduction in concentration is an example of the **common-ion effect**.

The solubility product of a salt contains the concentrations of both cations and anions, as shown here for cadmium hydroxide:

$$Cd(OH)_2 \, (s) \rightleftharpoons Cd^{2+}(aq) + 2 \, OH^-(aq) \qquad K_{sp} = [Cd^{2+}]_{eq} [OH^-]^2_{eq} = 7.2 \times 10^{-15}$$

By the common-ion effect, increasing the amount of hydroxide ion in a saturated solution of $Cd(OH)_2$ causes a drop in the concentration of Cd^{2+} ions and a precipitation of solid $Cd(OH)_2$. The common-ion effect is an application of Le Châtelier's principle (see Section 16.4). Added OH^- shifts the equilibrium position in the direction that reduces this added hydroxide concentration. The drop in $[Cd^{2+}]$ is a response to the increase in $[OH^-]$, moving the system to a new equilibrium position.

Figure 18-9 is a molecular illustration of the common-ion effect, and Example 18-18 shows a quantitative application.

Figure 18-9
When additional anions (*yellow circles*) are added to a solution of a salt at equilibrium (*left*), the solubility product is exceeded (*center*), and some solid precipitates, reducing the concentration of cations (*blue circles*) in the solution. At the new equilibrium (*right*), there is a higher concentration of anions, but a lower concentration of cations. Water molecules have been omitted for clarity.

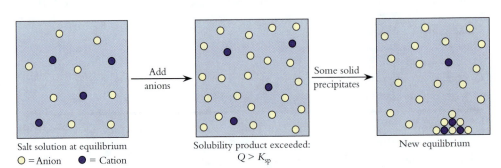

Salt solution at equilibrium
○ = Anion ● = Cation

Solubility product exceeded:
$Q > K_{sp}$

New equilibrium

Example 18-18

The Common-Ion Effect

The concentration of chloride ion in seawater is around 0.55 M. To compare the solubility of Pb^{2+} in freshwater vs. seawater, calculate the solubility in g/L of $PbCl_2$ in pure water and in 0.55 M NaCl.

Strategy: The problem asks for the g/L solubility of $PbCl_2$ in pure H_2O and in 0.55 M NaCl. The same solubility equilibrium applies to each solution.

$$PbCl_2\,(s) \rightleftharpoons Pb^{2+}(aq) + 2\,Cl^-(aq) \qquad K_{sp} = [Pb^{2+}]_{eq}\,[Cl^-]_{eq}^2 = 1.2 \times 10^{-5}$$

The standard approach to equilibrium problems works well for both parts of this Example. Solve the parts independently.

Solution:

Pure water. We have shown this type of calculation several times, so we summarize the results here. Review Example 18-16, if necessary.

$$K_{sp} = [Pb^{2+}]_{eq}\,[Cl^-]_{eq}^2 = 4x^3 = 1.2 \times 10^{-5} \qquad and \qquad x = 0.014$$

$$[Pb^{2+}]_{eq} = 0.014\ M$$

Knowing that $PbCl_2$ produces Pb^{2+} cations in 1:1 ratio tells us that 0.014 mol of $PbCl_2$ dissolves in 1 L of pure water. Convert to grams to finish the calculation. The molar mass of $PbCl_2$ is 278.1 g/mol:

$$(0.014\ mol/L)(278.1\ g/mol) = 3.9\ g/L\ PbCl_2$$

Salt water. Set up a table of amounts that incorporates the starting concentration of chloride ion:

Reaction	$PbCl_2\,(s)$	$\rightleftharpoons$	$Pb^{2+}(aq)$	+	$2\,Cl^-(aq)$
Initial concentration (M)	(solid)		0		0.55
Change in concentration (M)	(solid)		$+x$		$+2x$
Final concentration (M)	(solid)		x		$0.55 + 2x$

Substitute the results in the K_{sp} expression and solve for x:

$$K_{sp} = [Pb^{2+}]_{eq}\,[Cl^-]_{eq}^2 = (x)(0.55 + 2x)^2 \cong 0.303x = 1.2 \times 10^{-5}$$

$$x = 4.0 \times 10^{-5} \qquad and \qquad [Pb^{2+}]_{eq} = 4.0 \times 10^{-5}\ M$$

The value of x is small enough to justify the approximation. Now calculate the mass solubility. Again, the concentration of Pb^{2+} in solution equals the number of moles of solid that dissolves:

$$(4.0 \times 10^{-5}\ mol/L)(278.1\ g/mol) = 0.011\ g/L\ PbCl_2.$$

Do the Results Make Sense? Both solubilities are low, as we would expect for a salt with a small value of K_{sp}. Notice that $PbCl_2$ is about 350 times less soluble in the NaCl solution. This makes sense in terms of the common-ion effect. The excess chloride ion suppresses the solubility of Pb^{2+} by Le Châtelier's principle. The actual concentration of lead in seawater is much less than 4.0×10^{-5} M. This is because other lead salts are much less soluble than lead(II) chloride. The ocean contains carbonate, for example, and K_{sp} for lead(II) carbonate is quite small, 7.4×10^{-14}.

(a) Calculate the hydronium ion concentration in 1.25 M NH_4Cl. (b) Calculate the hydronium ion concentration in a solution that is 1.25 M in NH_4Cl and 0.755 M NH_3. (c) Explain the difference in concentrations.

Extra Practice
Exercise 18.18

Figure 18-10
The pH of an acetic acid solution is attenuated by the presence of added acetate ion. Added acetate ion decreases the concentration of hydronium ion, thereby modifying the pH of the solution.

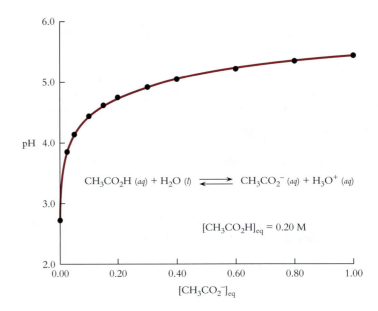

$$CH_3CO_2H \ (aq) + H_2O \ (l) \rightleftharpoons CH_3CO_2^- \ (aq) + H_3O^+ \ (aq)$$

$$[CH_3CO_2H]_{eq} = 0.20 \ M$$

The common-ion effect is quite general. The chemistry of buffer solutions is another important application of this principle. A buffer solution relies on the common-ion effect to suppress the concentration of hydronium ions and maintain a steady pH:

$$CH_3CO_2H \ (aq) + H_2O \ (l) \rightleftharpoons CH_3CO_2^- \ (aq) + H_3O^+ (aq) \qquad K_a = 1.8 \times 10^{-5}$$

The graph of Figure 18-10 shows that the pH of an acetic acid solution is attenuated by added acetate anion. As acetate concentration increases, the equilibrium position shifts in the direction of more acetic acid and less H_3O^+.

Effects of pH

Calcium carbonate is insoluble in pure water but dissolves in weakly acidic water. The role of this solubility phenomenon in the geochemistry of caverns is described in Box 16-2. We can understand this dependence on pH by examining the acid–base properties of the species involved in the solubility equilibrium.

$$CaCO_3 \ (s) \rightleftharpoons Ca^{2+} (aq) + CO_3^{2-} (aq)$$

Carbonate is the conjugate base of hydrogen carbonate, so we can write an acid–base equilibrium for a saturated solution of calcium carbonate:

$$CO_3^{2-} (aq) + H_2O \ (l) \rightleftharpoons HCO_3^- \ (aq) + OH^- (aq)$$

As a result of this equilibrium, a saturated solution of $CaCO_3$ is slightly basic. Now consider what happens if some acid is added to this saturated solution. Hydronium ions react with hydroxide ions to form water:

$$OH^- (aq) + H_3O^+ (aq) \rightleftharpoons 2 \ H_2O \ (l)$$

By Le Châtelier's principle, the system responds in the direction that will reduce the effect of this added acid, meaning that additional carbonate ions accept protons from water molecules:

$$CO_3^{2-} (aq) + H_2O \ (l) \rightleftharpoons HCO_3^- \ (aq) + OH^- (aq)$$

Responding to the reduction in carbonate ion concentration, additional calcium carbonate dissolves:

$$CaCO_3 \ (s) \rightleftharpoons Ca^{2+} (aq) + CO_3^{2-} (aq)$$

Qualitatively, making the solution more acidic increases the amount of calcium carbonate that dissolves in water.

The analysis can be made quantitative by writing the various equilibria and their K_{eq} values. The reactions can be added to obtain the net reaction that occurs when calcium carbonate is exposed to acidic water, and the equilibrium constant for the net reaction is the product of the individual K_{eq} values:

$$CaCO_3 (s) \rightleftharpoons Ca^{2+}(aq) + CO_3^{2-}(aq) \qquad K_{eq} = K_{sp}$$

$$CO_3^{2-}(aq) + H_2O (l) \rightleftharpoons HCO_3^-(aq) + OH^-(aq) \qquad K_{eq} = K_b = \frac{K_w}{K_a}$$

$$OH^-(aq) + H_3O^+(aq) \rightleftharpoons 2 H_2O (l) \qquad K_{eq} = \frac{1}{K_w}$$

Net: $CaCO_3 (s) + H_3O^+(aq) \rightleftharpoons Ca^{2+}(aq) + HCO_3^-(aq) + H_2O (l)$

$$K_{acidic} = K_{sp} \left(\frac{K_w}{K_a} \right) \left(\frac{1}{K_w} \right) = \frac{K_{sp}}{K_a}$$

Substituting the numerical values gives the equilibrium constant in acidic solution:

$$K_{acidic} = \frac{K_{sp}}{K_a} = \frac{3.36 \times 10^{-9}}{4.7 \times 10^{-11}} = 71$$

Taking the ratio of the equilibrium constants shows that calcium carbonate is ten orders of magnitude more soluble in acidic solution than in water:

$$\frac{K_{acidic}}{K_{neutral}} = \frac{71}{K_{sp}} = \frac{71}{3.36 \times 10^{-9}} = 2.1 \times 10^{10}$$

Our Chemistry and the Environment Box on page 799 explores the effects of equilibria on CO_2 in the atmosphere, and Example 18-19 shows that the acidity of the solvent places an upper limit on the amount of salt that will dissolve.

Example 18-19

Salt Dissolving in Acid

The most acid rain on record had pH = 2.00. Calculate the concentration of Ca^{2+} cations in a solution formed when this rain becomes saturated with calcium carbonate.

Strategy: We apply the seven-step method. The key feature is to identify the dominant equilibrium in this acidic solution. We do that by paying close attention to the major species present initially.

1. We are asked to calculate the equilibrium concentration of calcium ions, $[Ca^{2+}]_{eq}$.

2. The major species present initially, other than water, are solid calcium carbonate and aqueous hydronium ions.

3. These major species participate in the equilibrium described in the text:

$$CaCO_3 (s) + H_3O^+(aq) \rightleftharpoons Ca^{2+}(aq) + HCO_3^-(aq) + H_2O (l)$$

4. The equilibrium constant expression is taken from the balanced equation. The value of the equilibrium constant was determined in the text:

$$K_{acidic} = \frac{[Ca^{2+}]_{eq} [HCO_3^-]_{eq}}{[H_3O^+]_{eq}} = 71$$

Solution:

5. Organize the data using a concentration table. The large value for K_{acidic} indicates that the reaction should be taken to completion by applying limiting reactant stoichiometry and then brought back to equilibrium. We can omit H_2O and $CaCO_3$ from the table, because their concentrations do not appear in the equilibrium constant expression.

Example 18-19

Salt Dissolving in Acid
(*continued*)

Reaction	$CaCO_3\,(s) + H_3O^+(aq) \rightleftharpoons Ca^{2+}(aq) + HCO_3^-(aq) + H_2O\,(l)$		
Initial concentration (M)	0.010	0	0
Change to completion (M)	−0.010	+0.010	+0.010
Concentration at completion (M)	0	0.010	0.010
Change to equilibrium (M)	$+\gamma$	$-\gamma$	$-\gamma$
Equilibrium concentration (M)	γ	$0.010 - \gamma$	$0.010 - \gamma$

6. We expect γ to be small, so we make the following approximation:

$$(0.010 - \gamma) \cong 0.010$$

Next substitute and solve for γ:

$$\frac{[Ca^{2+}]_{eq}\,[HCO_3^-]_{eq}}{[H_3O^+]_{eq}} = \frac{(0.010)^2}{\gamma} = 71$$

$$\gamma = \frac{(0.010)^2}{71} = 1.4 \times 10^{-6}$$

This is quite small relative to 0.010, so

$$[Ca^{2+}]_{eq} = (0.010\ M - \gamma) = 0.010\ M$$

7. **Does the Result Make Sense?** To determine if the overall calculation is reasonable, notice that the hydronium concentration in acid rain is five orders of magnitude higher than in pure water. This increased hydronium ion concentration can convert a substantial amount of carbonate anions into hydrogen carbonate anions, promoting the solubility of calcium carbonate. The resulting equilibrium concentration of calcium cations is about five orders of magnitude larger than the concentration in a saturated solution in pure water.

Extra Practice
Exercise 18.19

Calculate the g/L solubility of $CaCO_3$ in pure water and in acidic solutions with pH = 3.00 and pH = 1.00.

The equilibrium constant expression derived for calcium carbonate dissolving in acidic solution is a general one that applies to any salt containing the conjugate base of a weak acid:

$$K_{acidic} = \frac{K_{sp}}{K_a} \qquad\qquad (18\text{-}4)$$

Thus, many salts are more soluble in acidic solution than in pure water. In addition to the carbonates, important examples include the phosphates (PO_4^{3-}) and the sulfides (S^{2-}). Insoluble hydroxides also dissolve in acidic solution, as illustrated by magnesium hydroxide:

$$Mg(OH)_2\,(s) \rightleftharpoons Mg^{2+}(aq) + 2\,OH^-(aq) \qquad K_{sp} = 5.6 \times 19^{-12}$$

$$2\,[OH^-(aq) + H_3O^+(aq) \rightleftharpoons 2\,H_2O\,(l)] \qquad K = \left(\frac{1}{K_w}\right)^2$$

$$\text{Net: } Mg(OH)_2\,(s) + 2\,H_3O^+(aq) \rightleftharpoons Mg^{2+}(aq) + 2\,H_2O\,(l) \qquad K = \frac{K_{sp}}{K_w^{\,2}}$$

Knowing that $K_w = 1.0 \times 10^{-14}$, you should be able to show that the hydroxide salt is 28 orders of magnitude more soluble in acid solution than in pure water.

Box 18-1	Chemistry and the Environment: The Carbon Cycle

Carbon accounts for only 0.08% of our planet's mass. Nevertheless, life on Earth is based on carbon. Living organisms take in carbon, process it through biochemical reactions, and expel it as waste.

Carbon dioxide is especially important in the exchange of carbon between the biosphere and its environment. Green plants store energy from sunlight by converting CO_2 and H_2O into carbohydrates. Most carbohydrates have the general formula $(CH_2O)_n$, so the chemical equation for carbohydrate formation appears simple, even though carbohydrates are complex (as described in Chapter 13):

$$n\, CO_2 + n\, H_2O \xrightarrow{h\nu} (CH_2O)_n + n\, O_2$$

Plants and animals use enzymes to catalyze the conversion of carbohydrates back to CO_2 and H_2O, a process that releases chemical energy and powers life processes:

$$(CH_2O)_n + n\, O_2 \xrightarrow{Enzymes} n\, CO_2 + n\, H_2O + energy$$

These two processes cycle carbon between the atmosphere, where it is found primarily as CO_2, and the biosphere, where it is found primarily as carbohydrates.

The carbon cycle is complicated by several reactions that involve CO_2. These reactions transfer carbon between the atmosphere, the hydrosphere (Earth's surface waters), and the lithosphere (Earth's crustal solids). The processes that move carbon from one sphere to another are illustrated schematically in the figure below.

Carbon dioxide is divided between the atmosphere and the hydrosphere. This division strongly favors the hydrosphere, because the large volume of water in the oceans has an immense capacity to absorb CO_2.

Almost all the Earth's carbon is found in the lithosphere as carbonate sediments that have precipitated from the oceans. Shells of aquatic animals also contribute $CaCO_3$ to the lithosphere. Carbon returns to the hydrosphere as carbonate minerals dissolve in water percolating through the Earth's crust. This process is limited by the solubility products for carbonate salts, so lithospheric carbonates represent a relatively inaccessible storehouse of carbon.

Until recently, the distribution of carbon among the different terrestrial spheres was stable. When humans began burning fossil fuels, however, such burning transferred carbon into the atmosphere as CO_2. This has become a rapidly changing feature of the overall carbon cycle. Over the last quarter century, the atmospheric concentration of CO_2 has grown by more than 10%.

Why should there be concern about an increase in atmospheric CO_2 whose concentration in air is only 350 parts per million? Even a doubling of CO_2 concentration still leaves it well below 0.1% of the atmosphere. Despite its low concentration, CO_2 exerts a major effect on the average temperature of Earth's surface. Recall from Chapter 7 that CO_2 is a "greenhouse" gas that traps outgoing radiation. Increasing CO_2 in the atmosphere may contribute to global warming that could have catastrophic consequences. In a warmer climate, for example, much of the polar ice caps would melt, and the oceans would expand as their temperatures increased. Sea level could rise by several meters and inundate many of the world's most populous regions, such as Bangladesh, the Netherlands, and the East and Gulf Coasts of the United States. In addition, even small temperature fluctuations may alter global weather patterns and perhaps lead to serious droughts.

Our planet is such a complicated, dynamic set of interconnected systems that scientists do not know the result of doubling the concentration of atmospheric CO_2. Scientists are studying the carbon cycle in hopes of learning what lies ahead for the planet. There is some urgency to such studies because recent global weather patterns show some of the characteristics predicted from computer models of global warming: average temperatures higher than those in previous years and an increased incidence of extreme weather such as hurricanes and heavy rainfall.

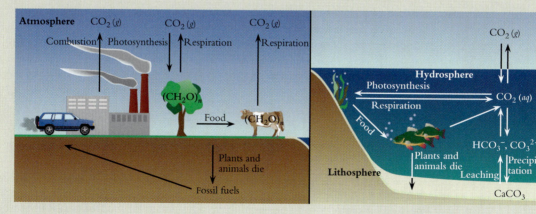

Section Exercises

■ **18.4.1** What is the concentration of Mg^{2+} ions in a saturated aqueous solution of $Mg(OH)_2$?

■ **18.4.2** Determine the concentration of Mg^{2+} ions remaining in solution after 0.150 L of 0.125 M $MgCl_2$ solution is treated with 0.100 L of 0.65 M NaOH solution. Does NaCl also precipitate from this solution?

■ **18.4.3** Determine the net reaction and the value of K_{eq} for $Al(OH)_3$ ($K_{sp} = 4.6 \times 10^{-33}$) dissolving in acid solution.

18.5 COMPLEXATION EQUILIBRIA

In Section 16-6, we describe how metal cations in aqueous solution can form bonds to anions or neutral molecules that have lone pairs of electrons. This leads to formation of complex ions and to chemical equilibria involving complexation. The complexation equilibrium between Ag^+ and NH_3 is an example:

$$Ag^+(aq) + 2\,NH_3(aq) \rightleftharpoons [Ag(NH_3)_2]^+(aq) \qquad K_f = \frac{[Ag(NH_3)_2^+]_{eq}}{[Ag^+]_{eq}\,[NH_3]_{eq}^2}$$

The treatment of complexation equilibria is more complicated than treatments for acid–base or solubility equilibria, both because the stoichiometries of complexes are not obvious and because complex formation and dissociation frequently involve multiple steps.

Stoichiometry of Complexes

A species that bonds to a metal cation to form a complex is known as a **ligand.** Any species that has a lone pair of electrons has the potential to be a ligand, but in this section, we confine our description to a few of the most common ligands: ammonia, compounds derived from ammonia, cyanide, and halides. We describe additional examples in Chapter 20, which addresses the chemistry of the transition metals.

The stoichiometry of a metal complex is described by its chemical formula. For example, each cation of the silver–ammonia complex contains one Ag^+ cation bound to two neutral NH_3 ligands and carries a net charge of $+1$, as shown in Figure 18-11. The formula of a complex ion is enclosed in square brackets, as in $[Ag(NH_3)_2]^+$. The coordination number, which is the number of ligand atoms bonded directly to the metal, provides a shorthand notation for the stoichiometry of a metal complex. Coordination numbers of 6 and 4 are most common, but 5 and 2 are not unusual. There are no simple and reliable guidelines for predicting coordination numbers. Thus, coordination numbers must be determined by experiment. In this section, we provide information about the coordination number for each complex that we describe.

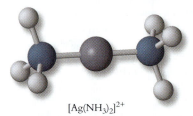

$[Ag(NH_3)_2]^{2+}$

Figure 18-11
Silver forms a linear complex ion with 2 neutral ammonia molecules as ligands.

Complexation Calculations

In principle, the calculation of concentrations of species of a complexation equilibrium is no different from any other calculation involving equilibrium constant expressions. In practice, we have to consider multiple equilibria whenever a complex is present. This is because each ligand associates with the complex in a separate process with its own equilibrium expression. For instance, the silver–ammonia equilibrium is composed of two steps:

$$Ag^+(aq) + NH_3(aq) \rightleftharpoons [Ag(NH_3)]^+(aq) \qquad K_1 = \frac{[Ag(NH_3)^+]_{eq}}{[Ag^+]_{eq}\,[NH_3]_{eq}} = 2.1 \times 10^3$$

$$[Ag(NH_3)]^+(aq) + NH_3(aq) \rightleftharpoons [Ag(NH_3)_2]^+(aq)$$

$$K_2 = \frac{[Ag(NH_3)_2^+]_{eq}}{[Ag(NH_3)^+]_{eq}\,[NH_3]_{eq}} = 8.2 \times 10^3$$

Because the two equilibrium constant expressions have similar magnitudes, a solution of the silver–ammonia complex generally has a significant concentration of each of the species that participate in the equilibria. The details of such calculations are beyond the scope of general chemistry. When the solution contains a large excess of ligand, however, each step in the complexation process proceeds nearly to completion. Under these conditions we can apply the standard seven-step approach to a single expression that describes the formation reaction of the complete complex.

One of the characteristics of complexation equilibria is that, in the presence of excess concentration of a complexing ligand, formation of a complex often reduces the concentration of a free metal cation essentially to zero. This is another application of the common-ion effect discussed in the previous section. Example 18-20 treats a situation of this sort.

The small amounts of gold contained in low-grade ores can be extracted using a combination of oxidation and complexation. Gold is oxidized to Au^+, which forms a very strong complex with cyanide anions:

$$Au^+(aq) + 2\,CN^-(aq) \rightleftharpoons [Au(CN)_2]^-(aq) \qquad K_f = 2 \times 10^{38}$$

Suppose that a sample of ore containing 2.5×10^{-3} mol of gold is extracted with 1.0 L of 4.0×10^{-2} M aqueous KCN solution. Calculate the concentrations of the three species involved in the complexation equilibrium.

Strategy: Determining concentrations requires a quantitative equilibrium calculation, so we apply the seven-step strategy. The complexation equilibrium is given in the problem.

1. The problem asks for the equilibrium concentrations of the three ionic species: $[Au^+]_{eq}$, $[CN^-]_{eq}$, and $[Au(CN)_2^-]_{eq}$.

2. Potassium cyanide is a salt that dissociates completely into ions, and the problem states that gold is oxidized, so the major species present are Au^+, CN^-, K^+, and H_2O.

3. Both CN^- and H_2O have acid–base properties, but the problem asks only about the species involved in the complexation equilibrium, so the important equilibrium is the complexation reaction:

$$Au^+(aq) + 2\,CN^-(aq) \rightleftharpoons [Au(CN)_2]^-(aq)$$

4. Now we write the equilibrium constant expression for formation of the complex ion:

$$K_f = \frac{[Au(CN)_2^-]_{eq}}{[Au^+]_{eq}\,[CN^-]_{eq}^2} = 2 \times 10^{38}$$

Solution:

5. Organize the data using a concentration table. Initial concentrations can be found from the data stated in the problem, and because the equilibrium constant is quite large, we take the reaction to completion and then allow back-reaction to equilibrium:

Reaction	$Au^+(aq)$	$+$	$2\,CN^-(aq)$	$\rightleftharpoons$	$[Au(CN)_2]^-(aq)$
Initial concentration (M)	2.5×10^{-3}		4.0×10^{-2}		0
Change to completion (M)	-2.5×10^{-3}		-5.0×10^{-3}		$+2.5 \times 10^{-3}$
Concentration at completion (M)	0		3.5×10^{-2}		2.5×10^{-3}
Change to equilibrium (M)	$+y$		$+2y$		$-y$
Equilibrium concentration (M)	y		$3.5 \times 10^{-2} + 2y$		$2.5 \times 10^{-3} - y$

6. We expect y to have a small value, so we make two approximations:

$$3.5 \times 10^{-2} + 2y \cong 3.5 \times 10^{-2} \quad and \quad 2.5 \times 10^{-3} - y \cong 2.5 \times 10^{-3}$$

$$K_f = 2 \times 10^{38} = \frac{2.5 \times 10^{-3}}{y(3.5 \times 10^{-2})^2} \quad so \quad y = \frac{2.5 \times 10^{-3}}{K_f (3.5 \times 10^{-2})^2}$$

$$y = \frac{2.5 \times 10^{-3}}{(2 \times 10^{38})(3.5 \times 10^{-2})^2} = 1.0 \times 10^{-38}$$

$$[Au^+]_{eq} = 1 \times 10^{-38}\,M \quad [CN^-]_{eq} = 4 \times 10^{-2}\,M \quad [Au(CN)_2^-]_{eq} = 3 \times 10^{-3}\,M$$

Example 18-20

Formation of a Gold Complex

$[Au(CN)_2]^{2-}$

7. Do the Results Make Sense? Each equilibrium concentration has only one significant figure, matching the single significant figure in the K_f value. The calculated concentration for Au^+ is too small to measure, but this is reasonable in light of the very large value of K_f. The tiny value of γ shows that the approximations are reasonable.

**Extra Practice
Exercise 18.20**

Suppose that 0.275 g of silver nitrate dissolves in 0.85 L of 0.250 M ammonia:

$$Ag^+(aq) + 2\ NH_3(aq) \rightleftharpoons [Ag(NH_3)_2^+](aq) \qquad K_f = 1.6 \times 10^{17}$$

Calculate the equilibrium concentrations of all species involved in the complexation equilibrium.

The Chelate Effect

Ligands that have two or more donor atoms are **chelating** ligands. One of the most common chelating ligands is ethylenediamine (often abbreviated "en"), $H_2NCH_2CH_2NH_2$. Each nitrogen atom has a lone pair of electrons, so both can be donor atoms at the same time. Thus, ethylenediamine is said to be a **bidentate** ligand that can coordinate to a metal at two sites simultaneously. The word *dentate* is derived from the Latin *dentis,* meaning "tooth." A monodentate ligand has "one tooth" for binding to a metal, a bidentate ligand has "two teeth", and so on.

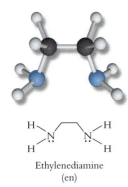

H—N...N—H
H   H

Ethylenediamine
(en)

Ethylene diamine (en)

Chelating ligands bind much more tightly to their metal cations than do ligands that possess only one donor atom. A good example is the ethylenediamine complex with Ni^{2+}. The ethylenediamine complex is much more stable than the analogous ammonia complex:

$$Ni^{2+}(aq) + 6\ NH_3(aq) \rightleftharpoons [Ni(NH_3)_6]^{2+}(aq) \qquad K_f = 2.0 \times 10^8$$

$$Ni^{2+}(aq) + 3\ en(aq) \rightleftharpoons [Ni(en)_3]^{2+}(aq) \qquad K_f = 4.1 \times 10^{17}$$

Even though both ligands form Ni—N bonds of similar strength, ethylenediamine binds to Ni^{2+} many orders of magnitude more strongly than does NH_3. Why is this? Think about complexation at the molecular level. One of the Ni—N bonds in either complex can be broken fairly easily. When this happens to $[Ni(NH_3)_6]^{2+}$, the ammonia molecule drifts away and is replaced by another ligand, typically a water molecule from the solvent:

$$[Ni(NH_3)_6]^{2+} \longrightarrow [Ni(NH_3)_5]^{2+} + NH_3$$

$$[Ni(NH_3)_5]^{2+} + H_2O \longrightarrow [Ni(NH_3)_5H_2O]^{2+}$$

The result is an exchange of ligands.

This simple ligand exchange reaction is less likely for $[Ni(en)_3]^{2+}$. As Figure 18-12 shows, when one Ni—N bond breaks and one NH_2 group of an ethylenediamine ligand dissociates from the metal complex, the ligand is still bound to the metal complex by the second NH_2 group. Consequently, the en ligand is not free to float away in solution, so the first NH_2 group binds again to the metal rather than being replaced by a competing ligand. For an ethylenediamine ligand to be replaced, both ends of the ligand must be

**Figure 18-12
Breaking one bond in an ethylenediamine complex leaves the en ligand dangling but "tethered" to the metal by the second nitrogen atom. (Each carbon atom bonds to two hydrogen atoms that are not shown.)**

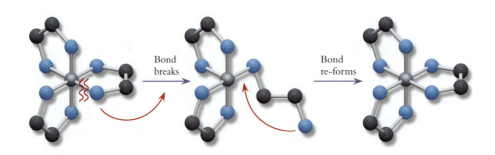

Bond
breaks

Bond
re-forms

released at almost the same time. This stabilization of a metal complex by a ligand with more than one donor atom is known as the **chelate effect.** (The word *chelate* comes from the Greek *chele,* meaning "claw".)

The chelate effect has important practical applications. Heavy metal ions such as Hg^{2+} and Pb^{2+} are extremely poisonous, and their effects are cumulative because the body has no mechanism for excreting heavy metal ions. Fortunately, chelating ligands can be used to treat poisoning by mercury and lead. One important example of a ligand used to treat metal ion poisoning is ethylenediamine tetraacetate ion, abbreviated $[EDTA]^{4-}$ or simply EDTA. With six donor atoms that can bond to a single metal ion, EDTA is a hexadentate ligand.

Dosages of EDTA are delivered as the calcium disodium salt, $Na_2[Ca(EDTA)]$. The calcium complex prevents EDTA from extracting iron from the blood. Unlike iron, heavy metal ions such as Hg^{2+} preferentially displace calcium cations from $[Ca(EDTA)]^{2-}$:

$$[Ca(EDTA)]^{2-}(aq) + Hg^{2+}(aq) \longrightarrow [Hg(EDTA)]^{2-}(aq) + Ca^{2+}(aq)$$

The kidneys are able to pass chelated heavy metal ions into the urine, so treatment with $[Ca(EDTA)]^{2-}$ cleanses mercury from the blood, counteracting its poisonous effects.

$[Ca(EDTA)]^{2-}$, with the hydrogen atoms removed for clarity

Complex Formation and Solubility

The effect of complex formation on the solubility of a solid can be observed in the home. Silver dinnerware eventually becomes discolored by an unsightly black tarnish of Ag_2S, formed from the reaction of the silver surface with small amounts of H_2S present in the atmosphere. Silver sulfide is highly insoluble in water. Commercial silver polishes contain ligands that form strong soluble complexes with Ag^+ ions. If a tarnished serving pan is rubbed with a polish, the black tarnish dissolves, returning the silver to its brilliant shine.

Silver salts demonstrate how complexation enhances solubility. Figure 18-13 shows that silver will precipitate or dissolve in aqueous solution depending on the species that are present. Starting with a solution of silver nitrate, addition of aqueous NaCl causes AgCl to precipitate:

$$AgCl(s) \rightleftharpoons Ag^+(aq) + Cl^-(aq) \qquad K_{sp} = 1.8 \times 10^{-10}$$

If aqueous ammonia is added to the mixture, the precipitate dissolves by forming the silver–ammonia complex. The formation equilibrium combines with the solubility equilibrium to give the overall process:

$$Ag^+(aq) + 2\,NH_3(aq) \rightleftharpoons [Ag(NH_3)_2]^+(aq) \qquad\qquad K_f = 1.6 \times 10^7$$

$$\underline{AgCl(s) \rightleftharpoons Ag^+(aq) + Cl^-(aq) \qquad\qquad\qquad\quad K_{sp} = 1.8 \times 10^{-10}}$$

$$\text{Net: } AgCl(s) + 2\,NH_3(aq) \rightleftharpoons [Ag(NH_3)_2]^+(aq) + Cl^-(aq) \qquad K_{eq} = K_{sp}K_f = 2.9 \times 10^{-3}$$

Figure 18-13
Silver precipitates from solution when a solution is added that contains an anion that forms an insoluble silver compound (*a* and *c*). The precipitates redissolve when a solution is added that contains a species that can form a strong complex with Ag^+ (*b* and *d*).

(a)   AgCl(s), $\xrightarrow{NH_3(aq)}$ (b) $[Ag(NH_3)_2]^+(aq)$ $\xrightarrow{NaBr(aq)}$ (c)   AgBr(s), $\xrightarrow{Na_2S_2O_3(aq)}$ (d) $[Ag(S_2O_3)_2]^{3-}(aq)$

$K_{sp} = 1.8 \times 10^{-10}$ $K_{sp} = 5.35 \times 10^{-13}$

Although still less than 1, this equilibrium constant is large enough to allow solid AgCl to dissolve in strong aqueous ammonia. In fact, if you open a bottle of commercial polish, you will surely notice the sharp odor of ammonia. These polishes contain a variety of ligands, including ammonia.

Addition of a solution of NaBr results in a new precipitate, because AgBr is significantly less soluble than AgCl:

$$[Ag(NH_3)_2]^+(aq) \rightleftharpoons Ag^+(aq) + 2\,NH_3(aq) \qquad \frac{1}{K_f} = \frac{1}{1.6 \times 10^7}$$

$$Ag^+(aq) + Br^-(aq) \rightleftharpoons AgBr(s) \qquad \frac{1}{K_{sp}} = \frac{1}{5.35 \times 10^{-13}}$$

Net: $[Ag(NH_3)_2]^+(aq) + Br^-(aq) \rightleftharpoons AgBr(s) + 2\,NH_3(aq)$

$$K_{eq} = \frac{1}{K_f K_{sp}} = \frac{1}{(1.6 \times 10^7)(5.35 \times 10^{-13})} = 1.2 \times 10^5$$

This large positive value indicates that the equilibrium position lies well to the right, favoring formation of solid AgBr.

Upon addition of sodium thiosulfate solution, the AgBr precipitate dissolves, because $S_2O_3^{2-}$ forms a very strong complex with Ag^+. This ligand is capable of dissolving AgBr:

$$AgBr(s) \rightleftharpoons Ag^+(aq) + Br^-(aq) \qquad K_{sp} = 5.35 \times 10^{-13}$$

$$Ag^+(aq) + 2\,S_2O_3^{2-}(aq) \rightleftharpoons [Ag(S_2O_3)_2]^{3-}(aq) \qquad K_f = 2.0 \times 10^{13}$$

Net: $AgBr(s) + 2\,S_2O_3^{2-}(aq) \rightleftharpoons [Ag(S_2O_3)_2]^{3-}(aq) + Br^-(aq) \qquad K_{eq} = K_{sp}K_f = 10.7$

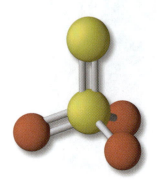

These reactions find practical applications in photography. Most photographic films contain silver halides, primarily silver bromide. Exposure to light causes some Ag^+ cations to gain electrons, forming Ag granules that, after developing, generate the images captured on the film. But what about the unreacted silver bromide? If it were left in the film, further exposure to light would cause further reaction and eventual blackening of the image. To prevent this, film must be "fixed" by removing all unreacted AgBr. This is done using thiosulfate, which forms a complex with othewise insoluble AgBr and allows it to be washed from the film. As Example 18-21 shows, fixer solution has a limited capacity to accept AgBr, so after repeated use, a solution must be discarded and a new solution prepared.

Example 18-21	What mass of AgBr will dissolve in 1.5 L of fixer solution that contains 0.50 M of thiosulfate anions?

**Capacity
of Thiosulfate Fixer**

Thiosulfate, $S_2O_3^{2-}$
(one resonance form)

Strategy: We can calculate the mass that dissolves from the total concentration of Ag^+ ions at equilibrium. We apply the seven-step method.

1. The problem asks for mass of AgBr that will dissolve. To find this, we must calculate the equilibrium concentration of silver-containing ions in the solution.

2. In addition to water, the initial species are $AgBr(s)$, $S_2O_3^{2-}(aq)$, and an unspecified spectator cation.

3. The equilibrium is the one described in the text, by which thiosulfate dissolves AgBr through complexation:

$$AgBr(s) + 2\,S_2O_3^{2-}(aq) \rightleftharpoons [Ag(S_2O_3)_2]^{3-}(aq) + Br^-(aq)$$

4. This reaction is the sum of the solubility reaction of AgBr and the complexation reaction between Ag and thiosulfate, so its K_{eq} is the product of the two equilibrium constants as described in the text:

$$AgBr(s) + 2\,S_2O_3^{2-}(aq) \rightleftharpoons [Ag(S_2O_3)_2]^{3-}(aq) + Br^-(aq) \qquad K_{eq} = 10.7$$

Solution:

5. Set up a concentration table for the aqueous species, using the initial concentration of thiosulfate, 0.50 M:

Reaction	$AgBr\,(s)$	$+$	$2\,S_2O_3{}^{2-}\,(aq)$	$\rightleftharpoons$	$[Ag(S_2O_3)_2]^{3-}\,(aq)$	$+$	$Br^-\,(aq)$
Initial concentration (M)			0.50		0		0
Change to equilibrium (M)			$-2x$		$+x$		$+x$
Equilibrium concentration (M)			$0.50 - 2x$		x		x

Example 18-21

Capacity
of Thiosulfate Fixer
(*continued*)

6. Substitute the equilibrium concentrations into the equilibrium constant expression and solve for x:

$$\frac{[Ag(S_2O_3)_2^{3-}]_{eq}\,[Br^-]_{eq}}{[S_2O_3^{2-}]_{eq}^2} = \frac{x^2}{(0.50 - 2x)^2} = 10.7$$

We could solve by the quadratic equation, but notice that we can simplify this expression by taking the square root of each side:

$$\frac{x}{0.50 - 2x} = \sqrt{10.7} = 3.27 \qquad \textit{from which} \qquad x = 1.64 - 6.54x$$

$$7.54x = 1.64 \qquad \textit{so} \qquad x = 0.217 = 0.22$$

Thus, the equilibrium concentration of silver complex and bromide ion in the solution is 0.22 M. Multiplying by the volume of the solution gives the amounts in moles:

$$n_{Br^-} = MV = (0.22\ \text{M})(1.5\ \text{L}) = 0.33\ \text{mol}$$

The stoichiometric ratio between AgBr and the ions in solution is 1:1, so this amount is also the amount of AgBr that dissolves. Convert to mass by multiplying by the molar mass of AgBr:

$$m = n\,MM = (0.33\ \text{mol})(187.8\ \text{g/mol}) = 62\ \text{g}$$

7. Does the Result Make Sense? Notice that all the variables for the solution are somewhat less than 1 M, and the solution volume is somewhat greater than 1 L, so a result that is somewhat less than the molar mass of AgBr appears reasonable.

Zinc oxalate, $Zn(C_2O_4)$, is sparingly soluble in water ($K_{sp} = 1.4 \times 10^{-8}$). The Zn^{2+} ion forms a tetrahedral-shaped complex with ammonia. The formation constant for the complex is 4.1×10^8. How many moles of zinc oxalate will dissolve in 1.0 L of 0.200 M aqueous ammonia?

**Extra Practice
Exercise 18.21**

$[Zn(NH_3)_4]^{2+}$

Buffer action and complexation occur together in many natural settings. We conclude this chapter with a Chemistry and Life Box that describes one important example, the pH of soil.

Section Exercises

■ **18.5.1** Determine the concentration of the species present at equilibrium if 0.10 mol of ethylenediamine is dissolved in 2.00 L of a solution that contains 6.5×10^{-3} M Ni^{2+}. ($K_f = 4.1 \times 10^{17}$ for $[Ni(en)_3]^{2+}$.)

■ **18.5.2** When Zn^{2+} is added to aqueous ammonia, four ammonia ligands complex in stepwise fashion. Write all the equilibria involving Zn^{2+} and NH_3.

■ **18.5.3** Calculate the mass of AgCl that will dissolve in 250 mL of 1.0 M aqueous ammonia.

| Box 18-2 | Chemistry and Life: The pH of Soil |

I n 1914, two European immigrants, Karen Blixen and her husband, established a coffee plantation outside of Nairobi in what is now Kenya. Conditions seemed perfect for coffee, as both the altitude and latitude closely matched that of Colombia, a highly successful coffee-growing country. Nevertheless, the coffee plants withered and eventually died, because the soil in that part of Kenya is too acidic for coffee plants to flourish. Blixen eventually became famous as the author of *Out of Africa,* which was made into an Oscar-winning movie.

Every type of plant requires soil whose pH falls within a particular range. No crops like strongly acid soils. If soil pH falls much below 5, only grasses grow well. Blixen and her husband might have been successful had they tried cattle ranching rather than growing coffee.

Ironically, coffee does need relatively acidic soil, with pH between 5 and 6. Conifers and shrubs such as azaleas and rhododendrons thrive on soils with this acidity, as do tea, potatoes, rice, and rye. The vast majority of crop plants, including most vegetables, need soils just on the acidic side of neutral, pH between 6 and 7. Only a few crops—barley, sugar beets, cotton, and sugarcane—like soils on the mildly basic side, between pH 7 and 8, and only desert plants can cope with soils whose pH is greater than 8.

Plants are highly sensitive to soil acidity because many equilibria involving plant nutrients are affected by pH. Phosphorus is a primary example. This essential element for plant growth occurs in soils mainly as phosphates, which are subject to phosphate–hydrogen phosphate equilibria. Consequently, phosphorus is most readily available to plants when the soil pH is near neutral, between 6 and 8. Conversely, aluminum, which is toxic to most plants, is soluble in acidic soil but occurs as insoluble hydroxides at pH values above about 5. In highly acidic soils, plants are deprived of phosphorus and subjected to toxic aluminum.

The acidity of soils also strongly affects the availability of metal cations such as K^+, Mg^{2+}, and Ca^{2+}. Most soils contain significant amounts of clay, whose chemical composition is dominated by aluminosilicates. Silicates are anionic, and the anions can be neutralized either by accepting protons or by associating with metal cations. When soil is too acidic, protons replace these metal cations, and the soil becomes depleted in these essential nutrients.

The colors of flowering plants such as hydrangeas are highly sensitive to soil acidity. At pH > 6.5, these showy flowers are deep pink, but at pH < 5, the blossoms are vivid blue. The chemistry of these changes involves complexation of aluminum by pigments that have acidic groups, as the structures show.

Delphinidin-3-glucoside

Chlorogenic acid

Low pH: blue

High pH: pink

Several processes act to increase soil acidity. As organic matter decays, it forms many organic carboxylic acids. Peat moss, for example, is entirely organic and has quite low pH, around 4. Acid rain, the result of industrial pollution, can seriously acidify soils. Chemical fertilizers that contain ammonium cations generate hydronium ions. Intensive agriculture increases soil acidity, because growing crops remove basic metal cations from the soil, replacing them with hydronium ions.

To reduce soil acidity and keep cropland productive, farmers amend the soil by "liming," which is the application of limestone. Limestone is calcium carbonate, and the carbonate anion is the conjugate base of hydrogen carbonate. Liming therefore increases the concentration of hydroxide in the soil, thereby increasing the pH:

$$CO_3^{2-}(aq) + H_2O(l) \rightleftharpoons HCO_3^-(aq) + OH^-(aq)$$

CHAPTER 18 VISUAL SUMMARY

Important equations

Essential terms

Key concepts

SKILLS TO MASTER

① Using the buffer equation
② Preparing buffer solutions
③ Analyzing titration curves
④ Choosing an indicator

⑤ Working solubility equilibrium problems
⑥ Determining the effects of acids and ions on solubility
⑦ Working complexation equilibrium problems
⑧ Determining the effects of complexing agents on solubility

18.1 BUFFER SOLUTIONS

A buffer soution contains both a weak acid and its conjugate base as major species in solution.

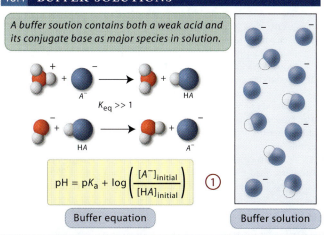

$K_{eq} \gg 1$

$$pH = pK_a + \log\left(\frac{[A^-]_{initial}}{[HA]_{initial}}\right)$$ ①

Buffer equation

Buffer solution

18.2 CAPACITY AND PREPARATION OF BUFFER SOLUTIONS

pH range = $pK_a \pm 1$ Buffer capacity   ②

Table 18-1 Common Buffer Systems

Acid	Base	pK_a	pH Range
H_3PO_4	$H_2PO_4^-$	2.16	1–3
HCO_2H	HCO_2^-	3.75	3–5
CH_3CO_2H	$CH_3CO_2^-$	4.75	4–6
$H_2PO_4^-$	HPO_4^{2-}	7.21	6–8
NH_4^+	NH_3	9.25	8–10
HCO_3^-	CO_3^{2-}	10.33	9–11
HPO_4^{2-}	PO_4^{3-}	12.32	11–13

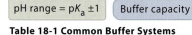

18.3 ACID–BASE TITRATIONS

Buffer region
$HA + A^-$

HA ③ A^- $OH^- + A^-$

Stoichiometric point

At the midpoint of the titration: $pH_{solution} = pK_{a,\ weak\ acid}$.

$pH = pK_a$

Midpoint

V (mL)

Indicator

$$[H_3O^+]_{eq,\ 1st\ stoichiometric\ point} = \sqrt{K_{a1}K_{a2}}$$

$$pH_{stoichiometric\ point} = pK_{In} \pm 1$$ ④

V (mL)

18.4 SOLUBILITY EQUILIBRIA

$$Cd(OH)_2(s) \rightleftharpoons Cd^{2+}(aq) + 2\ OH^-(aq)$$

Insoluble	$K_{sp} < 10^{-5}$
Slightly soluble	$10^{-5} < K_{sp} < 10^{-2}$
Soluble	$K_{sp} > 10^{-2}$

$$K_{acidic} = \frac{K_{sp}}{K_a}$$

Substance		B	A
Initial concentration	⑤	$[B]_i$	$[A]_i$
Changes in concentration		$\Delta[B]$	$\Delta[A]$
Final concentrations		$[B]_{eq}$	$[A]_{eq}$

Common-ion effect   ⑥

Salt solution at equilibrium

Add anions →

Solubility exceeded: $Q > K_{sp}$

Solid precipitates →

New equilibrium

18.5 COMPLEXATION EQUILIBRIA

$$2\ NH_3 + Ag^+ \rightleftharpoons [Ag(NH_3)_2]^+$$

Complex ion   ⑦

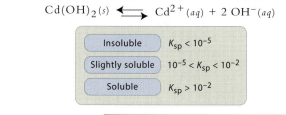

Ligand

Ethylenediamine (en)

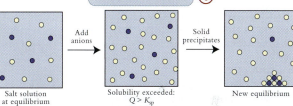

Bidentate ligand

Chelating ligand   EDTA

Chelate effect

(a) $AgCl(s)$, $K_{sp} = 1.8 \times 10^{-10}$

$NH_3(aq)$ → (b) $[Ag(NH_3)_2]^+(aq)$

$NaBr(aq)$ → (c) $AgBr(s)$, $K_{sp} = 5.35 \times 10^{-13}$

$Na_2S_2O_3(aq)$ → (d) $[Ag(S_2O_3)_2]^{3-}(aq)$

⑧

Learning Exercises

18.1 Write a paragraph that explains the chemical principles of buffer action.

18.2 Explain why the pH is not necessarily 7.0 at the stoichiometric point in an acid–base titration.

18.3 List all the types of calculations described in Chapter 18 in which acid–base equilibrium expressions play a role.

18.4 Update your list of memory bank equations and their uses.

18.5 Use your own words to define (a) buffer solution, (b) stoichiometric point, (c) common-ion effect, and (d) chelate.

 Problems **ilw = interactive learning ware problem. Visit the website at www.wiley.com/college/olmsted**

Buffer Solutions

18.1 Which of the following solutions show buffer properties? Compute the pH of each solution that is buffered.
  (a) 0.100 L of 0.25 M NH_4Cl + 0.150 L of 0.25 M NaOH
  (b) 0.100 L of 0.25 M NH_4Cl + 0.050 L of 0.25 M NaOH
  (c) 0.100 L of 0.25 M NH_4Cl + 0.050 L of 0.25 M HCl
  (d) 0.100 L of 0.25 M NH_3 + 0.050 L of 0.25 M HCl

18.2 Which of the following solutions show buffer properties? Compute the pH of each solution that is buffered.
  (a) 0.100 L of 0.25 M $NaCH_3CO_2$ + 0.150 L of 0.25 M HCl
  (b) 0.100 L of 0.25 M $NaCH_3CO_2$ + 0.050 L of 0.25 M HCl
  (c) 0.100 L of 0.25 M $NaCH_3CO_2$ + 0.050 L of 0.25 M NaOH
  (d) 0.100 L of 0.25 M CH_3CO_2H + 0.050 L of 0.25 M NaOH

18.3 Compute the change in pH resulting from the addition of 5.0 mmol of strong acid to each solution in Problem 18.1 that is buffered.

18.4 Compute the change in pH resulting from the addition of 2.5 mmol of strong base to each solution in Problem 18.2 that is buffered.

18.5 Compute how many moles of base it takes to change the pH of each buffered solution in Problem 18.1 by 0.10 pH unit.

18.6 Compute how many moles of acid it takes to change the pH of each buffered solution in Problem 18.2 by 0.20 pH unit.

18.7 A buffer solution made from $NaHCO_3$ and Na_2CO_3 has a pH in the range of 9–11. Write balanced equations that show how this buffer system neutralizes H_3O^+ and OH^-.

18.8 A buffer solution made from formic acid (HCO_2H) and sodium formate ($NaHCO_2$) has a pH in the range of 3–5. Write balanced equations that show how this buffer system neutralizes H_3O^+ and OH^-.

18.9 Which of the following sets of chemicals can be used to prepare buffer solutions? For each one that can, specify the weak acid and the conjugate base that make up the major species of the buffer solution: (a) $NaHCO_3$ + NaOH; (b) NaOH + NH_3; (c) H_3PO_4 + HCl; and (d) HCl + Na_2CO_3.

18.10 Which of the following sets of chemicals can be used to prepare buffer solutions? For each one that can, specify the weak acid and the conjugate base that make up the major species of the buffer solution: (a) Na_2HPO_4 + HCl; (b) NH_4Cl + NH_3; (c) Na_3PO_4 + NaOH; and (d) NH_4Cl + $NaNO_3$.

Capacity and Preparation of Buffer Solutions

18.11 From Table 18-1, select the best conjugate acid–base pairs for buffer solutions at pH 3.50 and 12.60. If you were going to add HCl solution as part of the buffer preparation, what other substance should you use in each case?

18.12 From Table 18-1, select the best conjugate acid–base pairs for buffer solutions at pH 6.85 and 10.80. If you were going to add NaOH solution as part of the buffer preparation, what other substance should you use in each case?

ilw 18.13 What mass of ammonium chloride must be added to 1.25 L of 0.25 M ammonia to make a buffer solution whose pH = 8.90?

18.14 What mass of sodium acetate must be added to 2.50 L of 0.55 M acetic acid to make a buffer solution whose pH = 5.75?

18.15 What is the maximum volume of 2.0 M HCl (aq) that the buffer described in Problem 18.13 can tolerate without showing a pH change greater than 0.25 units?

18.16 What is the maximum volume of 2.5 M NaOH (aq) that the buffer described in Problem 18.14 can tolerate without showing a pH change greater than 0.25 units?

18.17 Determine the mass of Na_2CO_3 and the volume of 0.500 M HCl solution required to make 1.5 L of a buffer solution at pH = 9.85 and at a total concentration of 0.350 M.

18.18 Determine the mass of NaOH and the volume of 3.50 M H_3PO_4 solution required to make 2.5 L of a buffer solution at pH = 2.25 and at a total concentration of 0.175 M.

18.19 How many grams of solid NaOH must be added to the buffer solution described in Problem 18.17 to change the pH by 0.15 units? After this NaOH has dissolved, is the solution still a buffer? Explain.

18.20 How many mL of 2.00 M HCl must be added to the buffer solution described in Problem 18.18 to change the pH by 1.25 units? After this HCl has dissolved, is the solution still a buffer? Explain.

Acid–Base Titrations

18.21 For each of the following, decide whether the pH at the stoichiometric point is greater than, less than, or equal to 7. In each case, identify the equilibrium that determines the pH. (a) NH_3 (aq) titrated with $HClO_4$ (aq); (b) $HClO_4$ (aq) titrated with KOH (aq); and (c) $NaCH_3CO_2$ (aq) titrated with HCl (aq).

18.22 For each of the following, decide whether the pH at the stoichiometric point is greater than, less than, or equal to 7. In each case, identify the equilibrium that determines the pH. (a) NaClO (aq) titrated with HCl (aq); (b) HNO_3 (aq) titrated with KOH (aq); and (c) NH_4Cl (aq) titrated with NaOH (aq).

18.23 A laboratory technician wants to determine the aspirin content of a headache pill by acid–base titration. Aspirin has a K_a of 3.0×10^{-4}. The pill is dissolved in water to give a solution that is about 10^{-2} M and is then titrated with KOH solution. Find the pH at each of the following points, neglecting dilution effects: (a) before titration begins; (b) at the stoichiometric point; and (c) at the midpoint of the titration.

18.24 Sleeping pills often contain barbital, which is weakly acidic ($pK_a = 8.0$). For analysis of the barbital content of a sleeping pill, a titration is carried out with strong base. It takes 12.00 mL of 0.200 M base to reach the stoichiometric point. If the initial acid concentration is 0.0120 M and the solution volume is 200 mL, what is the pH of the solution after adding the following volumes? (a) 0.0 mL; (b) 5.0 mL; (c) 12.00 mL; and (d) 13.0 mL

18.25 Sketch the titration curve for the titration described in Problem 18.23, and choose an appropriate indicator for this titration.

18.26 Sketch the titration curve for the titration described in Problem 18.24, and choose an appropriate indicator for this titration.

ilw 18.27 Calculate the pH at the second stoichiometric point when 150 mL of a 0.015 M solution of phthalic acid ($K_{a1} = 1.1 \times 10^{-3}$, $K_{a2} = 3.9 \times 10^{-6}$) is titrated with 1.00 M NaOH. What is a suitable indicator for this titration?

18.28 Calculate the pH at the second stoichiometric point when 250 mL of a 0.025 M solution of tartaric acid ($K_{a1} = 9.2 \times 10^{-4}$, $K_{a2} = 4.3 \times 10^{-5}$) is titrated with 1.00 M NaOH. What is a suitable indicator for this titration?

18.29 The dihydrogen phosphate ion is a diprotic acid ($pK_{a1} = 7.21$, $pK_{a2} = 12.32$). A titration sample was prepared by dissolving 1.51 g of

NaH_2PO_4 in enough water to make 0.100 L. The solution was titrated with 1.452 M NaOH. Calculate the pH at the following points: (a) before the titration; (b) the 1st midpoint; (c) the 1st stoichiometric point; (d) the 2nd midpoint; and (e) the 2nd stoichiometric point.

18.30 A sample of carbonic acid (0.125 L, 0.120 M, $pK_{a1} = 6.35$, $pK_{a2} = 10.33$) was titrated with 1.504 M NaOH. Calculate the pH at the following points: (a) before the titration; (b) the 1st midpoint; (c) the 1st stoichiometric point; (d) the 2nd midpoint; and (e) the 2nd stoichiometric point.

Solubility Equilibria

18.31 For the following salts, write a balanced equation showing the solubility equilibrium and write the solubility product expression for each: (a) silver chloride; (b) barium sulfate; (c) iron(II) hydroxide; and (d) calcium phosphate.

18.32 For the following salts, write a balanced equation showing the solubility equilibrium and write the solubility product expression for each: (a) lead(II) chloride; (b) magnesium carbonate; (c) nickel(II) hydroxide; and (d) silver acetate.

18.33 For each of the salts in Problem 18.31, determine the mass that dissolves in 475 mL of water at 25 °C. (See Appendix E for K_{sp} values.)

18.34 For each of the salts in Problem 18.32, determine the mass that dissolves in 375 mL of water at 25 °C. (See Appendix E for K_{sp} values; for $AgC_2H_3O_2$, $K_{sp} = 1.94 \times 10^{-3}$.)

18.35 Only 6.1 mg of calcium oxalate (CaC_2O_4) dissolves in 1.0 L of water at 25 °C. What is the solubility product of calcium oxalate?

18.36 The solubility of sodium sulfate (Na_2SO_4) in water is 9.5 g/100 mL. What is K_{sp} for sodium sulfate?

ilw 18.37 How many grams of $PbCl_2$ ($K_{sp} = 1.7 \times 10^{-5}$) will dissolve in 0.750 L of 0.650 M $Pb(NO_3)_2$ solution?

18.38 How many grams of BaF_2 ($K_{sp} = 1.8 \times 10^{-7}$) will dissolve in 0.500 L of 0.100 M NaF solution?

18.39 Determine the value of K_{eq} for $CaSO_3$ ($K_{sp} = 1.0 \times 10^{-4}$) dissolving in acid solution, given that $K_{a2} = 6.3 \times 10^{-8}$ for H_2SO_3.

18.40 Determine the value of K_{eq} for $FeCO_3$ ($K_{sp} = 3.1 \times 10^{-11}$) dissolving in acid solution, given that $K_{a2} = 4.7 \times 10^{-11}$ for H_2CO_3.

18.41 Determine the concentration of Ca^{2+} in a solution obtained by treating solid $CaSO_3$ with 0.125 M HCl.

18.42 Determine the concentration of Fe^{2+} in a solution obtained by treating solid $FeCO_3$ with 0.255 M HNO_3.

Complexation Equilibria

18.43 Write the chemical formulas, including charge, of the complexes that form between the following metal cations and ligands: (a) Fe^{3+} and CN^-, coordination number = 6; (b) Zn^{2+} and NH_3, coordination number = 4; and (c) V^{3+} and ethylenediamine (en), coordination number = 6.

18.44 Write the chemical formulas, including charge, of the complexes that form between the following metal cations and ligands: (a) Co^{2+} and SCN^-, coordination number = 4; (b) Hg^{2+} and NH_3, coordination number = 2; and (c) Mg^{2+} and $EDTA^{4-}$, coordination number = 6.

18.45 Zinc forms a tetrahedral complex ion with four molecules of ammonia ($K_f = 4.1 \times 10^8$). A solution was prepared by dissolving 0.275 g of $ZnCl_2$ in 375 mL of 0.250 M ammonia. Calculate the equilibrium concentrations of Zn^{2+}, $[Zn(NH_3)_4]^{2+}$, and NH_3.

18.46 Copper forms a trigonal planar complex ion with three chloride ions ($K_f = 5.0 \times 10^5$). A solution was prepared by dissolving 0.325 g of $CuCl_2$ in 425 mL of 0.250 M NaCl. Calculate $[Cu^{2+}]_{eq}$, $[CuCl_3^-]_{eq}$, and $[Cl^-]_{eq}$.

ilw 18.47 Determine the concentration of the species present at equilibrium after 0.25 mol of NaCl is dissolved in 1.50 L of a solution that contains 7.5×10^{-3} M Pb^{2+}. ($K_f = 2.5 \times 10^{15}$ for $[PbCl_4]^{2-}$.)

18.48 Determine the concentration of the species present after 0.15 mol of bipyridyl (bipy, a bidentate ligand) is dissolved in 2.50 L of a solution that contains 4.5×10^{-2} M Fe^{2+}. ($K_f = 1.6 \times 10^{17}$ for $[Fe(bipy)_3]^{2+}$.)

18.49 When a solution of ethylenediamine (en) is added to a solution containing Mn^{2+}, three en ligands bond to the metal in stepwise fashion. Write all the equilibria involving Mn^{2+} and en.

18.50 When a solution containing Co^{2+} is mixed with a solution containing oxalate anions ($C_2O_4^{2-}$), three bidentate oxalate ligands bond to the metal in stepwise fashion. Write all the equilibria involving Co^{2+} and $C_2O_4^{2-}$.

18.51 Calculate the mass of $CaSO_3$ that will dissolve in 0.50 L of 0.25 M solution of the tridentate ligand nitrilotriacetate (NTA^{3-}):

$$Ca^{2+} + 2\ NTA^{3-} \rightleftharpoons [Ca(NTA)_2]^{4-} \qquad K_f = 3.2 \times 10^{11}$$

18.52 Calculate the mass of $FeCO_3$ that will dissolve in 0.33 L of a solution that is 0.20 M in the oxalate anion ($C_2O_4^{2-}$):

$$Fe^{2+} + 3\ C_2O_4^{2-} \rightleftharpoons [Fe(C_2O_4)_3]^{4-} \qquad K_f = 3.3 \times 10^{20}$$

Additional Paired Problems

18.53 The pH of a formic acid/formate buffer solution is 4.04. Calculate the acid/conjugate base ratio for this solution. Draw a molecular picture that shows a small region of the buffer solution. (You may omit spectator ions and water molecules.) Use the following symbols:

 = Formic acid   = Formate

18.54 The pH of an NH_4^+/NH_3 buffer solution is 8.77. Calculate the acid/conjugate base ratio for this solution. Draw a molecular picture that shows a small region of the buffer solution. (You may omit spectator ions and water molecules.) Use the following symbols:

= NH_4^+ = NH_3

18.55 The solubility product for $PbCl_2$ is 2×10^{-5}. If 0.50 g of solid $PbCl_2$ is added to 300 mL of water, will all the solid dissolve? (*Hint:* Calculate Q if all the solid dissolves.)

18.56 The solubility product for $MgCO_3$ is 7×10^{-6}. If 1.55 g of solid $MgCO_3$ is added to 255 mL of water, will all the solid dissolve? (*Hint:* Calculate Q if all the solid dissolves.)

18.57 Consider a buffer solution that contains 0.50 M NaH_2PO_4 and 0.20 M Na_2HPO_4. (a) Calculate its pH. (b) Calculate the change in pH if 0.120 g of solid NaOH is added to 150 mL of this solution. (c) If the acceptable buffer range of the solution is ± 0.10 pH units, calculate how many moles of H_3O^+ can be neutralized by 250 mL of the buffer.

18.58 Consider a buffer solution that contains 0.45 M H_3PO_4 and 0.55 M NaH_2PO_4. (a) Calculate its pH. (b) Calculate the change in pH if 0.260 g of solid KOH is added to 250 mL of this buffer solution. (c) If the acceptable buffer range of the solution is ± 0.15 pH units, calculate how many moles of OH^- can be neutralized by 150 mL of the buffer.

18.59 Bromocresol purple is an indicator that changes color from yellow to purple in the pH range from 5.2 to 6.8. Without doing a calculation, determine the color of the indicator in each of the following solutions: (a) 0.15 M HCl; (b) 0.25 M NaOH; (c) 1.0 M KCl; and (d) 0.55 M NH_3.

18.60 A solution has a pH of 8.5. What would be the color of the solution if the following indicators were present in the solution? (a) methyl orange; (b) phenol red; (c) bromocresol green; and (d) thymol blue (See Table 18-2.)

18.61 When a solution of leucine (an amino acid) in water is titrated with strong base, the pH before titration is 1.85, the pH at the midpoint of the titration is 2.36, and the pH at the stoichiometric point is 6.00. Determine the value of K_a for leucine.

18.62 The following figure shows the data obtained in the pH titration of a biochemical substance that is a weak acid. From the information provided, determine the pK_a of this compound.

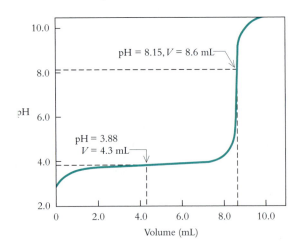

18.63 If 150 mL of 1.50×10^{-2} M MgSO$_4$ solution is mixed with 350 mL of 2.00×10^{-2} M CaCl$_2$ solution, will solid CaSO$_4$ form? ($K_{sp} = 4.9 \times 10^{-5}$.)

18.64 If 200 mL of 2.50×10^{-2} M NaCl solution is mixed with 300 mL of 4.00×10^{-2} M Pb(NO$_3$)$_2$ solution, will solid PbCl$_2$ form? ($K_{sp} = 1.7 \times 10^{-5}$.)

18.65 When lead(II) fluoride dissolves in water at 25 °C, the equilibrium concentration of Pb^{2+} is 2.0×10^{-3} M. Use this information to calculate K_{sp} for PbF$_2$.

18.66 When silver sulfate dissolves in water at 25 °C, the equilibrium concentration of Ag$^+$ is 3.0×10^{-2} mol/L. Use this information to calculate K_{sp} for Ag$_2$SO$_4$.

18.67 Ammonia is a convenient buffer system in the slightly basic range. (a) What is the pH of a buffer solution containing 35.0 g of NH$_4$Cl dissolved in 1.00 L of 1.00 M NH$_3$? (b) How many moles of acid are required to change the pH of this solution by 0.05 pH units? (c) Suppose 5.0 mL of 12.0 M HCl solution is added to 250 mL of the solution of Part (a). Calculate the new pH.

18.68 Acetic acid is a convenient buffer system in the slightly acidic range. (a) How many grams of sodium acetate must be added to 0.750 L of 0.100 M solution of acetic acid to make a buffer of pH = 5.00? (b) How many moles of acid are required to change the pH of this solution by 0.05 pH units? (c) Suppose 5.0 mL of 12.0 M HCl solution is added to 250 mL of the solution of Part (a). Calculate the new pH.

18.69 The solubility of calcium phosphate, Ca$_3$(PO$_4$)$_2$, in water is 3.5×10^{-5} g/L. Use this information to calculate K_{sp} for this salt.

18.70 The solubility of calcium arsenate, Ca$_3$(AsO$_4$)$_2$, in water is 0.036 g/L. Use this information to calculate K_{sp} for this salt.

18.71 Suppose you titrate 0.300 L of a 0.200 M solution of sodium formate with 6.0 M HCl. (K_a for formic acid is 1.8×10^{-4}.) (a) What is the pH of the solution before beginning the titration? (b) What is the pH of the solution halfway through the titration? (c) What is the pH at the stoichiometric point? (d) What is a suitable indicator for this titration?

18.72 Suppose you titrate 0.200 L of a 0.150 M solution of NaNO$_2$ with 6.0 M HCl. (K_a for HNO$_2$ is 5.6×10^{-4}.) (a) What is the pH of the solution before beginning the titration? (b) What is the pH of the solution halfway through the titration? (c) What is the pH at the stoichiometric point? (d) What is a suitable indicator for this titration?

18.73 One way to remove heavy metal ions from water is by treatment with sodium carbonate. Calculate the mass of the precipitate that will form and the concentration of Mn^{2+} remaining in solution after 0.750 L of a solution that contains Mn^{2+} at 2.50×10^{-2} M is mixed with 0.150 L of 0.500 M Na$_2$CO$_3$ solution. (K_{sp} for MnCO$_3$ is 2.2×10^{-11}.)

18.74 Calculate the mass of the precipitate that will form and the concentration of Ag$^+$ remaining in solution after 0.250 L of 0.200 M AgNO$_3$ solution is mixed with 0.350 L of 0.300 M Na$_2$CO$_3$ solution. (K_{sp} for Ag$_2$CO$_3$ is 8.46×10^{-12}.)

18.75 If 0.10 g of Zn(OH)$_2$ is added to 1.00 L of water, what mass will dissolve? (See Appendix E for K_{sp} values.)

18.76 If 0.025 g of Fe(OH)$_3$ is added to 1.50 L of water, what mass will dissolve? (See Appendix E for K_{sp} values.)

18.77 In a biochemistry laboratory, you are asked to prepare a buffer solution to be used as a solvent for isolation of an enzyme. On the shelf are the following solutions, all 1.00 M: formic acid ($K_a = 1.8 \times 10^{-4}$), acetic acid ($K_a = 1.8 \times 10^{-5}$), sodium formate (NaHCO$_2$), and sodium acetate (NaCH$_3$CO$_2$). Describe how you would prepare 1.0 L of a pH = 4.80 buffer solution whose total molarity is 0.35 M.

18.78 You are doing undergraduate research for a biology professor. Your first assignment is to prepare a pH = 7.50 phosphate buffer solution to be used in the isolation of DNA from a cell culture. The buffer must have a total concentration of 0.500 M. On the shelf you find the following chemicals: solid NaOH; concentrated HCl (12.0 M); concentrated H$_3$PO$_4$ (14.7 M); KH$_2$PO$_4$; and K$_2$HPO$_4$. Write a quantitative detailed set of instructions that describe how you would prepare 1.5 L of the buffer solution.

18.79 Zinc forms an octahedral complex ion with three bidentate oxalate ions:

$$Zn^{2+}(aq) + 3\ C_2O_4^{2-}(aq) \rightleftharpoons [Zn(C_2O_4)_3]^{4-}(aq) \qquad K_f = 1.4 \times 10^8$$

A solution was prepared by dissolving 0.275 g of ZnCl$_2$ in 0.450 L of 0.250 M Na$_2$C$_2$O$_4$. Calculate the equilibrium concentrations of Zn^{2+}, $[Zn(C_2O_4)_3]^{4-}$, and C$_2$O$_4^{2-}$.

18.80 Cadmium forms a tetrahedral complex ion with cyanide ions:

$$Cd^{2+}(aq) + 4\ CN^-(aq) \rightleftharpoons [Cd(CN)_4]^{2-}(aq) \qquad K_f = 6.0 \times 10^{18}$$

A solution was prepared by dissolving 0.125 g of CdCl$_2$ in 0.225 L of 0.210 M NaCN. Calculate the equilibrium concentrations of Cd^{2+}, $[Cd(CN)_4]^{2-}$, and CN$^-$.

More Challenging Problems

18.81 At 25 °C, $[Mg^{2+}] = 1.14 \times 10^{-3}$ M in a saturated aqueous solution of MgF$_2$. (a) Write the equilibrium reaction and the equilibrium constant expression. (b) Use the concentration of the saturated solution to compute K_{sp} for MgF$_2$ at 25 °C. (c) Using standard thermodynamic data from Appendix D, calculate K_{sp} at 100.0 °C.

18.82 Using the appropriate K_{sp} values from Appendix E, find the concentrations of all ions in the solution at equilibrium after 0.500 L of 0.300 M aqueous Cu(NO$_3$)$_2$ solution is mixed with 0.500 L of 0.400 M aqueous KOH solution.

18.83 Calculate the concentration of Hg^{2+} cations in a saturated solution of HgS, $K_{sp} = 4.0 \times 10^{-53}$. What volume of saturated solution can be expected to contain a single Hg^{2+} cation?

18.84 A biochemist wants to use X-ray diffraction to determine the structure of a protein. The biochemist must isolate crystals of the protein

from buffered solutions. Write instructions for preparing 1.0 L of a buffer that holds the pH of the protein solution at 5.20. Make the solution 0.50 M in an appropriate acid, and decide what else must be added to complete the preparation.

18.85 Phosphate ions are a major pollutant of water supplies. They can be removed by precipitation using solutions of Ca^{2+} ions because the K_{sp} of calcium phosphate is 2.0×10^{-33}. Suppose that 3.00×10^3 L of wastewater containing PO$_4^{3-}$ at 2.2×10^{-3} M is treated by adding 120 moles of solid CaCl$_2$ (which dissolves completely). (a) What is the concentration of phosphate ions after treatment? (b) What mass of calcium phosphate precipitates?

18.86 The pH of an acetic acid–acetate buffer solution is 4.27. (a) Calculate the HA:A$^-$ concentration ratio. (b) Draw a molecular picture that shows a small region of the buffer solution. (You may omit spectator ions and water molecules.) Use the following symbols:

= Acetic acid   = Acetate

18.87 Strontium iodate, $Sr(IO_3)_2$, is considerably more soluble in hot than in cold water. At 25 °C, 0.030 g of this compound dissolves in 0.100 L of water, but 0.80 g dissolves at 100.0 °C. Calculate K_{sp} and $\Delta G°$ at each temperature.

18.88 The titration of 25.0 mL of a mixture of hydrochloric acid and formic acid with 0.578 M NaOH gave the titration curve below. What are the molarities of HCl and HCO_2H in the solution?

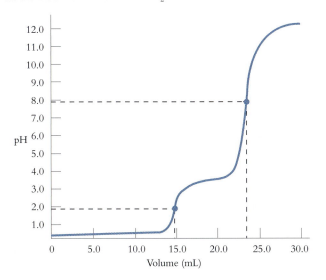

Volume (mL)

18.89 Limestone caverns are formed by the reaction of H_2O and CO_2 with natural deposits of calcium carbonate:

$$CaCO_3(s) + CO_2(g) + H_2O(l) \rightleftharpoons Ca^{2+}(aq) + 2\ HCO_3^-(aq)$$
$$K_{eq} = 1.56 \times 10^{-8}$$

The partial pressure of CO_2 in the atmosphere is 3.2×10^{-4} atm. What is the equilibrium concentration of calcium ions in groundwater?

18.90 The solubility of Ni(II) hydroxide can be increased by adding either $NH_3(aq)$ or $HCl(aq)$. Explain, including balanced equations. Will $NH_3(aq)$ or $HCl(aq)$ increase, decrease, or have no effect on the solubility of Ni(II) chloride? Explain your answer, including balanced equations.

18.91 Divide the silver salts listed in Appendix E into two sets: those that are more soluble in acidic solution than in pure water and those whose solubility is independent of pH.

18.92 One of the most common buffers used in protein chemistry is a weak base called TRIS (the nitrogen atom of the amino group is the basic portion of the molecule):

HO

⎯NH₂   = $(HOCH_2)_3CNH_2$ =   TRIS

HO   OH

(a) Write a balanced equation that shows how TRIS acts as a weak base in water. (b) Buffer solutions are prepared from TRIS by adding enough 12 M HCl to produce concentrations appropriate for the desired pH. Write a balanced equation that shows what happens when 12 M HCl is added to an aqueous solution of TRIS. (c) Write a balanced equation that shows what happens when hydroxide ions are added to a TRIS buffer solution. (d) Write a balanced equation that shows what happens when hydronium ions are added to a TRIS buffer solution.

18.93 A biochemist prepares a buffer solution by adding enough TRIS (see Problem 18.92) and 12 M HCl to give 1.0 L of a buffer solution whose concentrations are [TRIS] = 0.30 M and [TRISH⁺] = 0.60 M. (pK_b of TRIS is 5.91.) (a) Calculate the pH of the buffer solution.

(b) Suppose that 5.0 mL of 12 M HCl is added to 1.0 L of the buffer solution. Calculate the new pH of the solution.

18.94 A technician accidentally pours 35 mL of 12 M HCl into the 1.0 L of buffer solution freshly prepared as described in Problem 18.93. (a) Do a calculation to determine whether the buffer has been ruined. (b) Is it possible to bring the buffer solution back to the original pH calculated in Problem 18.93? If so, what reagent, and how much, must be added to restore the buffer?

18.95 The graph shows the titration curve for a solution of formic acid, which is a principal component in the venom of stinging ants. Identify the major species in the solution at equilibrium, identify the dominant equilibrium, and write the equilibrium constant expression for points A, B, C, and D on the curve.

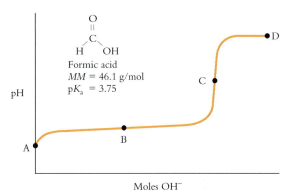

Moles OH^-

18.96 Quinine, an alkaloid derived from a tree that grows in tropical rain forests, is used in the treatment of malaria. Like all alkaloids, quinine is a sparingly soluble weak base: 1.00 g of quinine will dissolve in 1.90×10^2 L of water. (a) What is the pH of a saturated solution of quinine? (b) A 100.0-mL sample of saturated quinine is titrated with 0.0100 M HCl solution. What is the pH at the stoichiometric point of the titration?

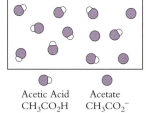

Quinine
$(C_{20}H_{24}N_2O_2)$
$pK_b = 5.1$
$MM = 324$ g/mol

18.97 The following figure represents a small portion of an acetic acid–acetate buffer. Solvent molecules have been omitted for clarity (pK_a of acetic acid = 4.75). (a) What is the pH of the buffer solution? (b) Redraw the original figure to show the equilibrium condition that is established when one hydroxide ion enters the region. (c) Redraw the original figure to show the equilibrium condition that is established when one ion of HSO_4^- enters the region. (pK_a of $HSO_4^- = 2.0$)

Acetic Acid   Acetate
CH_3CO_2H $CH_3CO_2^-$

18.98 Determine the solubility (g/L) of silver carbonate (Ag_2CO_3, $K_{sp} = 8.46 \times 10^{-12}$) in a pH = 10.75 buffer solution containing CO_3^{2-} and HCO_3^- at a total concentration of 0.860 M.

18.99 Nickel(II) cations form a strong complex with cyanide:

$$Ni^{2+}(aq) + 4\ CN^-(aq) \rightleftharpoons [Ni(CN)_4]^{2-}(aq) \qquad K_f = 1 \times 10^{22}$$

The complexation process allows aqueous cyanide solutions to dissolve insoluble nickel hydroxide:

$$Ni(OH)_2(s) + 4\ CN^-(aq) \rightleftharpoons [Ni(CN)_4]^{2-}(aq) + 2\ OH^-(aq)$$

Use appropriate information from Appendix E to determine the equilibrium constant for this reaction, and calculate the mass of $Ni(OH)_2$ that will dissolve in 225 mL of 0.500 M aqueous cyanide solution.

Group Study Problems

18.100 The amine group of an amino acid readily accepts a proton, and the protonated form of an amino acid can be viewed as a diprotic acid. The pK_a values for serine ($H_2NCHRCO_2H$, $R = CH_2OH$) are pK_a (H_3N^+) = 9.2 and pK_a (CO_2H) = 2.2. (a) What is the chemical formula of the species that forms when serine dissolves in pure water? (b) If this species is titrated with strong acid, what reaction occurs? (c) 10.00 mL of 1.00 M HCl is added to 200. mL of 0.0500 M serine solution. This mixture is then titrated with 0.500 M NaOH. Draw the titration curve, indicating the pH at various stages of this titration.

18.101 Virtually all investigations in cell biology and biochemistry must be carried out in buffered aqueous solutions. Imagine that you are studying an enzyme that is active only between pH = 7.1 and 7.4 and that you need to prepare 1.5 L of a phosphate buffer at pH = 7.25 whose total phosphate concentration is 0.085 M. On the laboratory shelves, you find the following reagents: concentrated H_3PO_4 (14.75 M), solid KH_2PO_4, and solid K_2HPO_4. (a) Which of these will you use to prepare the buffer solution? (b) What quantities of each will you use? (c) The enzyme generates H_3O^+ as it functions. If you are running an experiment in 250 mL of the buffer, how many moles of H_3O^+ can the enzyme generate before it loses its activity?

18.102 Seawater is approximately 0.5 M each in Na^+ and Cl^- ions. By evaporation, NaCl (K_{sp} = 6.2) can be precipitated from this solution. If 1.00×10^2 L of seawater is evaporated, at what volume will the first solid NaCl appear?

18.103 The K_{sp} of $Fe(OH)_3$ is 2.8×10^{-39}, and the K_{sp} of $Zn(OH)_2$ is 3×10^{-17}. (a) Can Fe^{3+} be separated from Zn^{2+} by the addition of an NaOH solution to an acidic solution that contains 0.300 M Zn^{2+} and

0.100 M Fe^{3+}? (b) At what hydroxide ion concentration will the second cation precipitate? (c) What is the concentration of the other cation at that hydroxide ion concentration?

18.104 The figure below shows the titration curves for 50.0-mL samples of weak bases A, B, and C. The titrant was 0.10 M HCl. (a) Which is the strongest base? (b) Which base has the largest pK_b? (c) What are the initial concentrations of the three bases? (d) What are the approximate pK_a values of the conjugate acids of the three bases? (e) Which of the bases can be titrated quantitatively using indicators? Explain your answers, and identify an appropriate indicator for each base that can be titrated successfully.

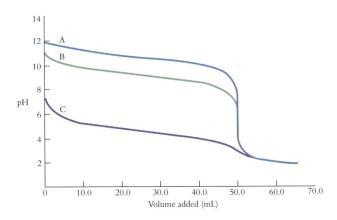

Answers to Section Exercises

18.1.1 (a) not suitable; (b) HNO_2 and NO_2^-; (c) not suitable; (d) HPO_4^{2-} and PO_4^{3-}; and (e) not suitable

18.1.2 $NH_4^+ + OH^- \longrightarrow NH_3 + H_2O$ and $NH_3 + H_3O^+ \longrightarrow NH_4^+ + H_2O$

18.1.3 (a) 9.00; (b) 9.24; and (c) 12.45

18.2.1 Buffering capacity destroyed, final pH = 1.6

18.2.2 3.8 g sodium formate and 72 mL of 0.500 M HCl solution

18.2.3 Dissolve 9.0 g NaOH and 35 g $NaHCO_3$ in enough water to make 2.5 L of solution.

18.3.1

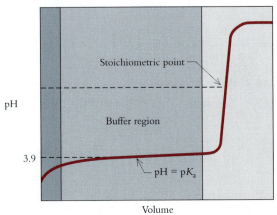

18.3.2 pH = 8.41 at stoichiometric point and 3.4 after 3.0 mL of KOH is added

18.3.3 Phenol red would be suitable; it would be red at the stoichiometric point.

18.4.1 1.7×10^{-4} M

18.4.2 1.5×10^{-9} M; NaCl does not precipitate.

18.4.3 $Al(OH)_3 (s) + 3 H_3O^+(aq) \rightleftharpoons Al^{3+}(aq) + 6 H_2O (l)$; $K_{eq} = 4.6 \times 10^9$

18.5.1 $[Ni^{2+}] = 5.6 \times 10^{-16}$ M; $[en] = 3.0 \times 10^{-2}$ M; $[Ni(en)_3^{2+}] = 6.5 \times 10^{-3}$ M

18.5.2 $Zn^{2+}(aq) + NH_3(aq) \rightleftharpoons [Zn(NH_3)]^{2+}(aq)$; $[Zn(NH_3)]^{2+}(aq) + NH_3(aq) \rightleftharpoons [Zn(NH_3)_2]^{2+}(aq)$; $[Zn(NH_3)_2]^{2+}(aq) + NH_3(aq) \rightleftharpoons [Zn(NH_3)_3]^{2+}(aq)$; $[Zn(NH_3)_3]^{2+}(aq) + NH_3(aq) \rightleftharpoons [Zn(NH_3)_4]^{2+}(aq)$

18.5.3 1.8 g of AgCl dissolves.

Answers to Extra Practice Exercises:

18.1 $[OH^-]_{eq} = 3.36 \times 10^{-5}$ M; $[NH_4^+]_{eq} = 0.125$ M; $[NH_3]_{eq} = 0.240$ M

18.2 $[NH_3]_{eq} = 0.25$ M; $[NH_4^+]_{eq} = 0.50$ M; $[Cl^-]_{eq} = 0.75$ M

18.3 (a)

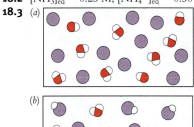

(b)

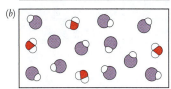

18.4 pH = 2.10

18.5 Increases by 0.18, to pH = 10.41

18.6 $[OH^-] = 0.049$ M, pH = 12.69

18.7 34 g Na_2HPO_4 and 18 g NaH_2PO_4

18.8 The buffer should contain CH_3CO_2H and $CH_3CO_2^-$, made from sodium acetate solid and hydrochloric acid. Use 35.8 g $NaCH_3CO_2$ and 0.700 L HCl(*aq*).

18.9 pH = 9.96

18.10 (a) 11.39; (b) 9.81; (c) 5.44; (d) 1.85

18.11

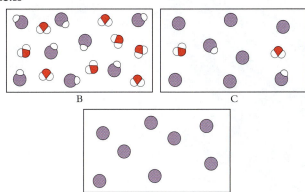

B

C

D

18.12

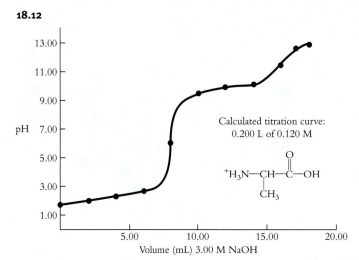

Calculated titration curve:
0.200 L of 0.120 M

18.13 Thymol blue

18.14 $K_{sp} = 5.1 \times 10^{-6}$

18.15 $K_{sp} = 5.6 \times 10^{-12}$

18.16 5.4 g

18.17 $m_{Ag_2CO_3} = 12.1$ g; $[CO_3^{2-}]_{eq} = 0.0688$ M; $[Ag^+]_{eq} = 1.11 \times 10^{-5}$ M

18.18 (a) $[H_3O^+]_{eq} = 2.67 \times 10^{-5}$ M;
(b) $[H_3O^+]_{eq} = 9.45 \times 10^{-10}$ M;
(c) NH_3, being a product of the reaction that generates H_3O^+, drives the equilibrium to the left, reducing $[H_3O^+]_{eq}$.

18.19 H_2O: 5.8×10^{-3} g;
pH = 3.00: 0.10 g;
pH = 1.00: 10. g

18.20 $[Ag^+]_{eq} = 2.0 \times 10^{-19}$ M; $[NH_3]_{eq} = 0.246$ M; $[Ag(NH_3)^{2+}]_{eq} = 0.0019$ M

18.21 0.025 mol

Electron Transfer Reactions

Corrosion

Reactions that involve the transfer of electrons are immensely beneficial in a vast range of chemical processes, from the operation of a battery to the transmission of nerve impulses. However, in the form of corrosion, electron-transfer reactions also can be highly destructive. The rusting of iron objects is a prominent example of corrosion.

Metallic iron is made up of neutral iron atoms held together by shared electrons (see Section 10.7). The formation of rust involves electron-transfer reactions. Iron atoms lose three electrons each, forming Fe^{3+} cations. At the same time, molecular oxygen gains electrons from the metal, each molecule adding four electrons to form a pair of oxide anions. As our inset figure shows, the Fe^{3+} cations combine with O^{2-} anions to form insoluble Fe_2O_3, rust. Over time, the surface of an iron object becomes covered with flaky iron(III) oxide and pitted from loss of iron atoms.

Corrosion of iron has serious economic impacts. Approximately 20% of the iron and steel produced in the United States is used to replace rusted metal, and the cost of replacing corroded materials is several billion dollars per year. Consequently, the prevention of corrosion is a major business.

Corrosion can be prevented by covering the metal surface with a protective coating that blocks oxygen and water from reaching the metal surface. Paint is a liquid mixture that sets to form a solid film

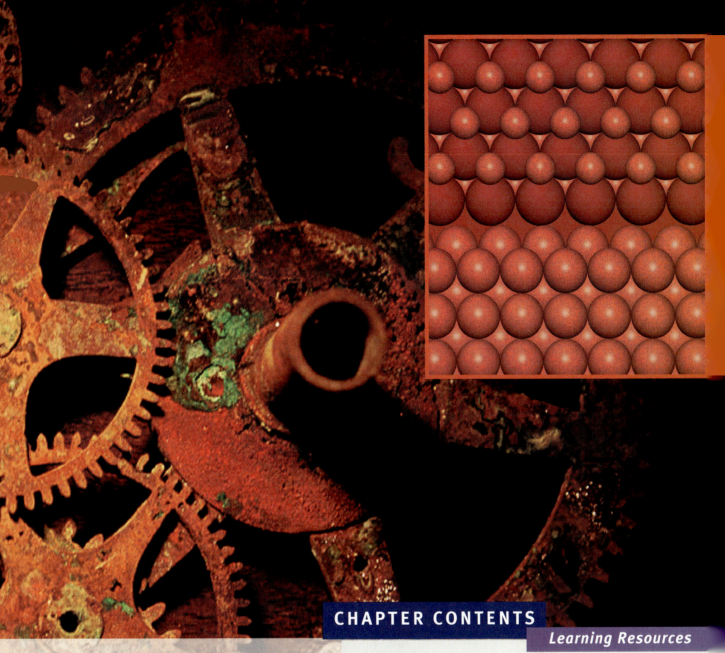

on a surface. Structures such as San Francisco's Golden Gate Bridge are continually being painted, ensuring unbroken layers that protect steel cables and towers from corrosion.

Electron-transfer reactions occur all around us. Objects made of iron become coated with rust when they are exposed to moist air. Animals obtain energy from the reaction of carbohydrates with oxygen to form carbon dioxide and water. Turning on a flashlight generates a current of electricity from a chemical reaction in the batteries. In an aluminum refinery, huge quantities of electricity drive the conversion of aluminum oxide into aluminum metal. These different chemical proces-

ses share one common feature: Each is an oxidation–reduction reaction, commonly called a redox reaction, in which electrons are transferred from one chemical species to another.

CHAPTER CONTENTS

Learning Resources

 KEY CONCEPTS

 CRITICAL THINKING

 SKILLS TO MASTER

 ADDITIONAL HELP
www.wiley.com/college/olmsted
- TUTORIALS
- ANIMATIONS

Redox reactions can proceed by direct transfer of electrons between chemical species. Examples include the rusting of iron and the metabolic breakdown of carbohydrates. Redox processes also can take place by indirect electron transfer from one chemical species to another via an electrical circuit. When a chemical reaction is coupled with electron flow through a circuit, the process is electrochemical. Flashlight batteries and aluminum smelters involve electrochemical processes.

We begin this chapter with a discussion of the principles of redox reactions, including redox stoichiometry. Then we introduce the principles of electrochemistry. Practical examples of redox chemistry, including corrosion, batteries, and metallurgy, appear throughout the chapter.

19.1 RECOGNIZING REDOX REACTIONS

Recall from Section 4.7 that in a redox reaction, one species loses electrons while another species gains electrons. A chemical species that loses electrons is oxidized, and a chemical species that gains electrons is reduced.

Remember also that electrons are conserved in all chemical processes. That is, electrons can be *transferred* from one species to another, but they are neither created nor destroyed. Thus, it is impossible for oxidation to occur without reduction, because when one species gains electrons, another species must lose electrons.

KEY CONCEPTS

Oxidation is the loss of electrons from a substance.
Reduction is the gain of electrons by a substance.
Oxidation and reduction always occur together.

A simple example of a redox reaction is the violent reaction of magnesium metal with molecular oxygen to generate magnesium oxide, shown in Figure 19-1:

$$2\,\text{Mg}\,(s) + \text{O}_2\,(g) \longrightarrow 2\,\text{MgO}\,(s)$$

In magnesium metal, each atom is neutral; the electrons exactly counterbalance the nuclear charge. When a magnesium atom reacts with an oxygen molecule, the metal atom loses its valence electrons to produce a +2 cation:

$$\text{Mg} \longrightarrow \text{Mg}^{2+} + 2e^-$$

Figure 19-1
When set ablaze with a match, magnesium burns with an intense flame. The reaction involves electron transfer from the metal to oxygen.

Magnesium atoms are oxidized in this reaction, so some other species must be reduced. Oxygen molecules accept the electrons lost by the magnesium atoms. Each oxygen *atom* gains two electrons from a magnesium atom, generating two oxide anions:

$$\text{O}_2 + 4e^- \longrightarrow 2\,\text{O}^{2-}$$

Magnesium cations and oxide anions attract each other strongly, forming the ionic solid, MgO. Notice that in the balanced redox reaction there is no net change in the number of electrons; two Mg atoms lose four electrons, and one O_2 molecule gains four electrons.

The species that loses electrons in a redox process causes the reduction of some other species. Consequently, the species that loses electrons is a **reducing agent.** Magnesium metal acts as a reducing agent in the presence of atmospheric oxygen. Similarly, the species that gains electrons causes the oxidation of some other chemical species and is an **oxidizing agent.** Molecular oxygen acts as an oxidizing agent in the presence of magnesium metal. Every redox reaction has both an oxidizing agent and a reducing agent. Figure 19-2 summarizes these features using the magnesium–oxygen reaction as an example.

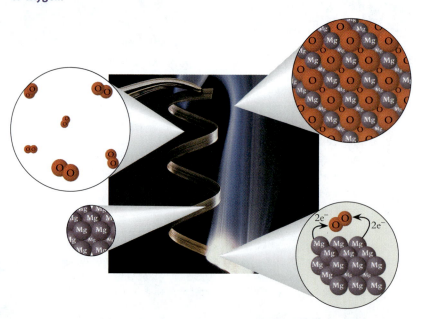

Electron transfer is often difficult to recognize, because electrons normally do not appear in chemical formulas and equations. The formula of a molecule or ion specifies the atomic composition and the net charge but does not account for the electrons. Likewise, electrons never appear in a balanced chemical equation. The electrons are "hidden" in the chemical formulas of the species involved in the reaction. For example, the formulas Mg, O_2, and MgO do not suggest immediately that oxygen is reduced to oxide anions and magnesium oxidized to magnesium cations when magnesium burns. To recognize when electron transfer occurs, we need a method to keep track of electrons in chemical species.

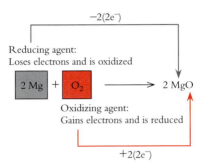

Figure 19-2
Schematic view of the processes involved in redox reactions.

Oxidation Numbers

The difficulty in recognizing redox reactions is illustrated by two of the reactions that occur during the extraction of iron from iron ores, a process that we describe in detail in Chapter 20:

$$FeO\,(s) \;+\; CO\,(g) \xrightarrow{\text{Heat}} Fe\,(l) \;+\; CO_2\,(g)$$

$$FeCO_3\,(s) \xrightarrow{\text{Heat}} FeO\,(s) \;+\; CO_2\,(g)$$

One of these is a redox reaction, but the other is not; how can we determine which one involves electron transfer?

To determine whether electrons are transferred in chemical reactions, chemists use a procedure that assigns an **oxidation number** (also known as an *oxidation state*) to each atom in each chemical species. In a redox reaction, electron transfer causes some of the atoms to change their oxidation numbers. Thus, we can identify redox reactions by noting changes in oxidation numbers.

To determine oxidation numbers, we assign each valence electron to a specific atom in a compound. This means that the oxidation number of an atom is the charge it would have if the compound were composed of ions. Ionic FeO, for example, would contain Fe^{2+} cations and O^{2-} anions. Thus, in FeO we assign iron an oxidation number of $+2$ and oxygen an oxidation number of -2.

Most substances contain covalent bonds rather than ions. Nevertheless, the electrons in a bond between atoms of two different elements, such as FeO or CO, are polarized toward the more electronegative atom (see Figure 19-3 for electronegativities). For oxidation number purposes, we imagine that these electrons are transferred *completely* to the more electronegative atom.

Figure 19-3
Electronegativities of the elements.

The bonding electrons in CO and CO_2 are polarized in the direction of the O atoms, as shown by the electronegativities of C (2.5) and O (3.5). Thus, if these substances were ionic, CO would consist of C^{2+} cations and O^{2-} anions, and CO_2 would contain C^{4+} cations and O^{2-} anions. Accordingly, we assign O an oxidation number of -2 in both these compounds, whereas C has an oxidation number of $+2$ in CO and $+4$ in CO_2.

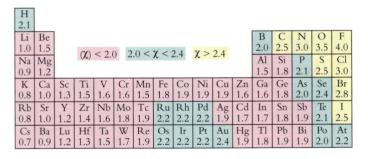

It is always possible to determine oxidation numbers starting from electronegativity differences. A more systematic method for determining oxidation numbers uses the following four guidelines:

1. Treat polyatomic ions separately.

2. The sum of all the oxidation numbers must equal the charge of the species.

3. Hydrogen usually has an oxidation number of $+1$.

4. The most electronegative atom in a species has a negative oxidation number equal to the number of electrons needed to complete its valence octet.

Some examples illustrate these guidelines.

1. In an ionic compound, each ion is an individual chemical species with its own set of oxidation numbers. For example, we treat ammonium nitrate, NH_4NO_3, as NH_4^+ cations and NO_3^- anions.

2. The sum of the oxidation numbers for NO_3^- must be -1, and the sum for NH_4^+ must be $+1$. For a neutral molecule such as NH_3, the sum of the oxidation numbers must be zero. By this guideline the oxidation number of every atom in pure neutral elements is zero. As examples, each atom in H_2, Fe, O_2, O_3, P_4, and C_{60} has an oxidation number of zero. In addition, the oxidation number of any monatomic ion equals its charge: the oxidation number of K^+ is $+1$, Al^{3+} is $+3$, and Cl^- is -1.

3. Hydrogen has a $+1$ oxidation number in most of its compounds. In NH_3, for example, each hydrogen atom is assigned an oxidation number of $+1$, leaving a -3 value for nitrogen. There is one common exception to this guideline: In the metal hydrides such as NaH and CaH_2, hydrogen is assigned an oxidation number of -1. These assignments are consistent with electronegativities:

$$Na\ (0.9)\ and\ Ca\ (1.0) < H\ (2.1) < C\ (2.5)\ and\ Cl\ (3.0)$$

4. Nitrogen, which is in Group 15 of the periodic table, has five valence electrons and needs three more to complete its valence octet. Thus, in the cyanide anion, CN^-, nitrogen is the more electronegative atom, so N is assigned an oxidation number of -3. In contrast, oxygen is the more electronegative atom in the nitrate ion, NO_3^-, so each oxygen atom is assigned an oxidation number of -2 (Group 16, six valence electrons). With this assignment, the nitrogen atom in a nitrate anion must have a $+5$ oxidation number to satisfy Guideline 2. The only important exceptions to Guideline 4 are the peroxides, species that contain an O—O bond: Oxygen has a -1 oxidation number in the peroxides. Examples include hydrogen peroxide, H_2O_2 (H is $+1$ by Guideline 3, so O is -1 by Guideline 2), and sodium peroxide, Na_2O_2 (treated as Na^+ and O_2^{2-} by Guideline 1, making O -1 by Guideline 2).

Oxidation numbers do not indicate how charge is *actually* distributed in a molecule. Remember that most bonds are polar, meaning that the electrons in a bond are skewed toward the more electronegative atom. In contrast, we determine oxidation numbers by *assuming* that all bonds are ionic. Consequently, oxidation numbers are useful for assessing whether electron exchange occurs during a reaction, but they do not represent actual charge distributions. A good example is CN^-, where we assign oxidation numbers of -3 to N and $+2$ to C. In reality, the ion carries -1 net charge, and both atoms have slightly negative charges.

Examples 19-1 and 19-2 illustrate the application of these guidelines.

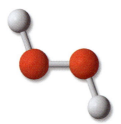

Hydrogen peroxide

Example 19-1	Assign oxidation numbers to all atoms in these two reactions involving iron:

Assigning Oxidation Numbers

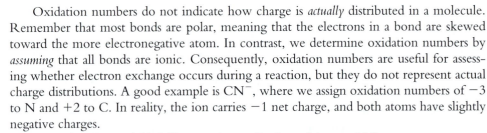

Strategy: Consider each substance separately, and follow the four guidelines.

Solution:

FeO: Oxygen has an oxidation number of -2 (Guideline 4). For the sum of the oxidation numbers to be zero (Guideline 2), iron must be $+2$.

CO: Oxygen has a value of -2 (Guideline 4). For the sum of the oxidation numbers to be zero (Guideline 2), carbon must have an oxidation number of $+2$.

Fe: This is a neutral element, so iron has an oxidation number of zero (Guideline 2).

CO_2: Each oxygen is -2, so carbon must be $+4$.

$FeCO_3$: This ionic compound contains the carbonate polyatomic anion, CO_3^{2-}. The -2 charge on carbonate requires that iron have $+2$ oxidation number (Guideline 2). In the carbonate anion, each oxygen atom is -2, for a total of -6. The oxidation numbers must add up to the net charge on the anion (-2), so the carbon atom must have an oxidation number of $+4$.

Example 19-1

**Assigning
Oxidation Numbers**
(*continued*)

Here are the reactions again, showing all oxidation numbers:

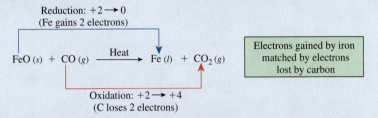

$$\overset{+2}{\text{FeO}}(s) + \overset{+2}{\text{CO}}(g) \xrightarrow{\text{Heat}} \overset{0}{\text{Fe}}(l) + \overset{+4}{\text{CO}_2}(g)$$
$$\underset{-2}{} \underset{-2}{} \underset{-2}{}$$

$$\overset{+2+4}{\text{FeCO}_3}(s) \xrightarrow{\text{Heat}} \overset{+2}{\text{FeO}}(s) + \overset{+4}{\text{CO}_2}(g)$$
$$\underset{-2}{} \underset{-2}{} \underset{-2}{}$$

The first reaction is a redox reaction because the oxidation number of carbon increases (oxidation), and that of iron decreases (reduction).

Reduction: $+2 \rightarrow 0$
(Fe gains 2 electrons)

$$\text{FeO}(s) + \text{CO}(g) \xrightarrow{\text{Heat}} \text{Fe}(l) + \text{CO}_2(g)$$

Electrons gained by iron
matched by electrons
lost by carbon

Oxidation: $+2 \rightarrow +4$
(C loses 2 electrons)

In the second reaction, none of the atoms changes its oxidation number, so this is not a redox process.

Do the Results Make Sense? Although these reactions look similar, the redox reaction includes a pure element. Nearly all reactions involving pure elements are redox, because the oxidation number changes from zero to either a positive or a negative value.

Here are two reactions involved in the metallurgy of nickel. Assign oxidation numbers and determine if either is a redox reaction.

$$\text{NiO}(s) + \text{C}(s) \xrightarrow{\text{Heat}} \text{Ni}(l) + \text{CO}(g)$$

$$\text{Ni}(s) + 4\,\text{CO}(g) \xrightarrow{\text{Heat}} \text{Ni(CO)}_4(g)$$

Assign oxidation numbers to each element in the following substances: H_3PO_4, O_3, NaOCl, ClF_3, and KH.

Example 19-2

**Examples
of Oxidation Numbers**

Strategy: Apply the guidelines for oxidation numbers to each substance.

Solution:

H_3PO_4: Phosphoric acid is a covalent compound with a net charge of zero. Each hydrogen atom has an oxidation number of $+1$ (Guideline 3), and each oxygen has an oxidation number of -2 (Guideline 4). Now add the contributions from these atoms: $3(+1) + 4(-2) = -5$. For the oxidation numbers to sum to zero (Guideline 2), the phosphorus atom of phosphoric acid must have an oxidation number of $+5$.

O_3: Ozone is a neutral form of the element, so the oxidation number of its oxygen atoms is zero (Guideline 2).

NaOCl: Sodium hypochlorite is ionic, containing Na^+ cations and OCl^- anions. The sodium cation has oxidation number equal to its charge, $+1$. In the anion, oxygen is -2 (Guideline 4), so chlorine must be $+1$ for the sum of the oxidation numbers to match the -1 charge of the hypochlorite anion (Guideline 2).

ClF_3: Fluorine is the more electronegative atom, so each fluorine atom has an oxidation number of -1 (Guideline 4). For the sum of the oxidation numbers to be zero (Guideline 2), chlorine must be $+3$.

Example 19-2	KH: In most cases hydrogen has an oxidation number of +1. In compounds with metals, however, the metal has a positive oxidation number and the oxidation number of hydrogen is -1. This requires potassium to be $+1$.

Examples of Oxidation Numbers (*continued*)

Do the Results Make Sense? Use chemical behavior and electronegativities to assess the reasonableness of the assignments. Phosphoric acid can be viewed as the phosphate anion (-3 charge) associated with three H^+ ions. Sodium, an alkali metal, readily forms cations with $+1$ charge. In KH, the assignments are consistent with the electronegativities (χ) of the two elements: $\chi_K = 0.8$, $\chi_H = 2.1$.

Extra Practice Exercise 19.2

Assign oxidation numbers for all atoms in H_2SO_4 and $KClO_3$.

Section Exercises

19.1.1 Determine the oxidation numbers of all atoms in each of the following substances: (a) $KMnO_4$ (potassium permanganate, a powerful oxidizing agent found in many general chemistry laboratories); (b) $SiCl_4$ (an intermediate in the production of ultrapure silicon for computer chips); (c) Na_2SO_3 (sodium sulfite, a food preservative); (d) S_8, the most stable form of sulfur; and (e) $LiAlH_4$ (a reducing agent in organic chemistry. *Hint:* This is an ionic compound with a lithium cation and a polyatomic anion.)

19.1.2 Determine which of the following are redox reactions. For each redox reaction, state which species is oxidized and which is reduced.

(a) $CH_4(g) + 2\,O_2(g) \longrightarrow$
$$CO_2(g) + 2\,H_2O(l)$$
(b) $HCl(g) + NH_3(aq) \longrightarrow NH_4Cl(aq)$
(c) $SiCl_4(l) + 2\,Mg(s) \longrightarrow$
$$Si(s) + 2\,MgCl_2(s)$$
(d) $PbS(s) + 4\,H_2O_2(aq) \longrightarrow$
$$PbSO_4(s) + 4\,H_2O(l)$$
(e) $3\,O_2(g) \longrightarrow 2\,O_3(g)$

19.1.3 Nitrogen can have many different oxidation states, as illustrated by the following examples. Assign oxidation states to the nitrogen atoms in each species. (a) NH_3; (b) N_2; (c) N_2H_4; (d) HNO_3; (e) $NaNO_2$; (f) NO; and (g) NO_2.

19.2 BALANCING REDOX REACTIONS

Some redox reactions have relatively simple stoichiometry and can be balanced by inspection. Others are much more complicated. Because redox reactions involve the transfer of electrons from one species to another, electrical charges must be considered explicitly when balancing complicated redox equations.

The key to balancing complicated redox equations is to balance electrons as well as atoms. Because electrons do not appear in chemical formulas or balanced net reactions, however, the number of electrons transferred in a redox reaction often is not obvious. To balance complicated redox reactions, therefore, we need a procedure that shows the electrons involved in the oxidation and the reduction. One such procedure separates redox reactions into two parts, an oxidation and a reduction. Each part is a **half-reaction** that describes half of the overall redox process.

Half-Reactions

In any redox reaction, some species are oxidized and others are reduced. To reveal the number of electrons transferred, we must separate the oxidation and reduction parts. This is most easily accomplished by dividing the redox reaction into two half-reactions, one describing the oxidation and the other describing the reduction. In half-reactions, electrons appear as reactants or products.

A starting material loses electrons in an oxidation, so electrons appear among the *products* of the oxidation half-reaction. A starting material gains electrons in a reduction, so electrons appear among the *reactants* of the reduction half-reaction. The reaction of magnesium metal with hydronium ions to produce hydrogen gas provides an example:

$$Mg(s) + 2\,H_3O^+(aq) \longrightarrow Mg^{2+}(aq) + H_2(g) + 2\,H_2O(l)$$

Here are the half-reactions for this redox reaction:

Oxidation: $$Mg(s) \longrightarrow Mg^{2+}(aq) + 2\,e^-$$

Reduction: $$2\,H_3O^+(aq) + 2\,e^- \longrightarrow H_2(g) + 2\,H_2O(l)$$

The half-reactions show that two electrons are required to produce each H_2 molecule in the reduction, and that the oxidation of one magnesium atom releases two electrons. In other words, electron transfer is balanced when one magnesium atom reacts for every H_2 molecule that forms. This example reveals an essential feature of redox reactions.

Mg undergoes a redox reaction with strong aqueous acid.

The number of electrons lost in oxidation must equal the number of electrons gained in reduction.

KEY CONCEPT

The first step in balancing a redox reaction is to divide the unbalanced equation into half-reactions. Identify the participants in each half-reaction by noting that each half-reaction must be balanced. That is, each element in each half-reaction must be conserved. Consequently, any element that appears as a reactant in a half-reaction must also appear among the products. Hydrogen and oxygen frequently appear in both half-reactions, but other elements usually appear in just one of the half-reactions. Water, hydronium ions, and hydroxide ions often play roles in the overall stoichiometry of redox reactions occurring in aqueous solution. Chemists frequently omit these species in preliminary descriptions of such redox reactions.

When reactions do not divide clearly into an oxidation and a reduction, changes in oxidation numbers reveal how to divide the starting materials into half-reactions. Example 19-3 illustrates the separation process.

Separate the following unbalanced redox processes into half-reactions. All occur in aqueous solution:

a. $Cr_2O_7^{2-}(aq) + Fe^{2+}(aq) \longrightarrow Cr^{3+}(aq) + Fe^{3+}(aq)$

b. $H_2O_2(aq) + SO_3^{2-}(aq) \longrightarrow SO_4^{2-}(aq)$

c. $BrO_2^-(aq) \longrightarrow BrO_3^-(aq) + Br^-(aq)$

Example 19-3

Identifying Half-Reactions

Strategy: A redox equation usually can be separated by placing elements other than hydrogen and oxygen in separate half-reactions. If this does not work, use oxidation numbers to identify the oxidized and reduced species. Occasionally, an element other than O or H may be involved in both half-reactions.

Solution:

a. This reaction can be divided by inspection. Iron must appear in one half-reaction and chromium in the other. Oxygen, which is present only in $Cr_2O_7^{2-}$, will be balanced in a later step.

$$Fe^{2+} \longrightarrow Fe^{3+} \qquad Cr_2O_7^{2-} \longrightarrow Cr^{3+}$$

b. In this reaction, sulfur is the only element other than hydrogen and oxygen. Thus, sulfur must appear on both sides of one half-reaction. Furthermore, since sulfur appears in only one of the starting materials and in only one of the products, the second half-reaction must contain only hydrogen and oxygen. No product containing hydrogen is given, but water always is present in aqueous solution. The second half-reaction is the conversion of hydrogen peroxide to water:

$$SO_3^{2-} \longrightarrow SO_4^{2-} \qquad H_2O_2 \longrightarrow H_2O$$

It is convenient to omit the phase designation when doing the steps in balancing redox reactions.

Example 19-3

Identifying Half-Reactions
(continued)

c. How can a reaction that contains just one reactant be divided into half-reactions? The only way is for this reactant to act as *both* the oxidizing agent and the reducing agent. Anions of BrO_2^- react with one another; some are reduced and others are oxidized. Therefore, BrO_2^- must appear in *both* half-reactions:

$$BrO_2^- \longrightarrow BrO_3^- \qquad BrO_2^- \longrightarrow Br^-$$

Do the Results Make Sense? The separation in Part a is obvious. Oxidation numbers confirm the other assignments. In Part b, recall that whereas the oxygen atom in H_2O has an oxidation number of -2, the oxygen atoms in H_2O_2 have oxidation number -1. In Part c, the oxidation number of the bromine atoms in BrO_2^- is $+3$. Among the products, bromine is $+5$ in BrO_3^- and -1 in Br^-. Thus, bromite ions are oxidized to bromate ions in one half-reaction and reduced to bromide ions in the other half-reaction.

Extra Practice Exercise 19.3

Divide the following reactions into half-reactions:

a. $Cu(s) + NO_3^-(aq) \longrightarrow Cu^{2+}(aq) + NO(g)$

b. $ClO^-(aq) \longrightarrow ClO_3^-(aq) + Cl^-(aq)$

Balancing Half-Reactions

After a redox reaction is divided into half-reactions, each half-reaction can be balanced independently. This is done in four steps, which *must* be done *in the order listed:*

a. Balance all elements except oxygen and hydrogen by adjusting stoichiometric coefficients.

b. Balance oxygen by adding H_2O to the side that is deficient in oxygen.

c. Balance hydrogen without unbalancing oxygen, as follows: Unless the solution is basic, add H_3O^+ cations to the side that is deficient in hydrogen and an equal number of H_2O molecules to the other side. If the solution is basic, add H_2O to the side that is deficient in hydrogen and an equal number of OH^- anions to the other side.

d. Balance net charge by adding electrons to the side that has the more positive (less negative) net charge.

These four steps result in a balanced half-reaction, in which elements, electrons, and total charge all are conserved. Example 19-4 illustrates the procedure.

Example 19-4

Balancing Half-Reactions

Balance the redox equation presented in Example 19-3a:

$$Cr_2O_7^{2-} + Fe^{2+} \longrightarrow Cr^{3+} + Fe^{3+}$$

Strategy: The half-reactions were identified in Example 19-3. Balance each half-reaction using the four-step procedure.

Solution:
Iron half-reaction: $Fe^{2+} \longrightarrow Fe^{3+}$

Steps a, b, c: Iron is already balanced, and neither oxygen nor hydrogen is present.

Step d: The net charge on the left is $+2$, and on the right it is $+3$. The charge on the right needs to be made less positive: To balance the charges, add one electron on the right:

$$Fe^{2+} \longrightarrow Fe^{3+} + \boxed{1\ e^-}$$

Now for the chromium half-reaction: $Cr_2O_7^{2-} \longrightarrow Cr^{3+}$

Step a: To balance chromium, give Cr^{3+} a coefficient of 2:

$$Cr_2O_7^{2-} \longrightarrow \boxed{2}\ Cr^{3+}$$

Example 19-4

Balancing Half-Reactions
(*continued*)

Step b: There are seven oxygen atoms on the left, so add seven water molecules on the right:

$$Cr_2O_7{}^{2-} \longrightarrow 2\,Cr^{3+} + \boxed{7\,H_2O}$$

Step c: The solution is identified as aqueous in Example 19-3, so it is not basic. There are 14 hydrogen atoms on the right, requiring 14 H_3O^+ ions on the left and 14 additional H_2O molecules on the right:

$$\boxed{14\,H_3O^+} + Cr_2O_7{}^{2-} \longrightarrow 2\,Cr^{3+} + \boxed{21}\,H_2O$$

Step d: The left-hand side has $14(+1) + 1(-2) = +12$ charge. The right-hand side has $2(+3) = +6$ charge. Charge balance requires six electrons on the left:

$$14\,H_3O^+ + Cr_2O_7{}^{2-} + \boxed{6\,e^-} \longrightarrow 2\,Cr^{3+} + 21\,H_2O$$

Do the Results Make Sense? We can verify the electron counts by looking at oxidation numbers. The oxidation number of Fe increases by one, meaning that one electron must be released. The oxidation number of Cr changes from +6 to +3, meaning that each Cr atom must gain three electrons.

Balance the half-reactions from Extra Practice Exercise 19-3.

**Extra Practice
Exercise 19.4**

Balancing Redox Equations

After oxidation and reduction half-reactions are balanced, they can be combined to give the balanced chemical equation for the overall redox process. Although electrons are reactants in reduction half-reactions and products in oxidation half-reactions, they must cancel in the overall redox equation. To accomplish this, multiply each half-reaction by an appropriate integer that makes the number of electrons in the reduction half-reaction equal to the number of electrons in the oxidation half-reaction. The entire half-reaction must be multiplied by the integer to maintain charge balance. Example 19-5 illustrates this procedure.

Example 19-5

Combining Half-Reactions

Complete the balancing of the reaction presented in Example 19-3 a:

$$Cr_2O_7{}^{2-} + Fe^{2+} \longrightarrow Cr^{3+} + Fe^{3+}$$

Strategy: The balanced half-reactions appear in Example 19-4. Balance the overall equation by combining the half-reactions in such a way that electrons cancel.

Solution: The half-reaction for iron contains one electron on the right, whereas the half-reaction for chromium contains six electrons on the left. To combine these half-reactions so that the electrons cancel, multiply the iron half-reaction by 6 and add it to the chromium half-reaction:

$$6\,Fe^{2+} \longrightarrow 6\,Fe^{3+} + 6\,e^-$$
$$\underline{14\,H_3O^+ + Cr_2O_7{}^{2-} + 6\,e^- \longrightarrow 2\,Cr^{3+} + 21\,H_2O}$$
$$14\,H_3O^+ + Cr_2O_7{}^{2-} + 6\,Fe^{2+} \longrightarrow 6\,Fe^{3+} + 2\,Cr^{3+} + 21\,H_2O$$

Does the Result Make Sense? An inventory of the reaction verifies that atoms and charge are balanced:

Reactants	Products
42 H, 2 Cr, 21 O, 6 Fe	42 H, 2 Cr, 21 O, 6 Fe
$14(+1) + (-2) + 6(+2) = +24$ charge	$6(+3) + 2(+3) = +24$ charge

All the elements are balanced, and so is the total charge, so the overall equation is balanced.

Complete the balancing of the reactions from Extra Practice Exercise 19.3.

**Extra Practice
Exercise 19.5**

Our Chemistry and the Environment Box on the next page describes a detoxification problem that was solved by applying redox chemistry.

| Box 19-1 | Chemistry and the Environment: Purifying Groundwater |

I ndustrial operations use large quantities of chemicals that pose contamination threats to groundwater. In the recent past, industrial wastes such as chlorinated hydrocarbons (used as solvents and for cleaning) and chromium-containing solutions (used in metal plating) were routinely dumped into storage tanks from which they eventually leaked, leading to groundwater contamination. Heightened public awareness of the health and environmental hazards of such practices resulted in laws restricting the dumping of chemical waste, but preventing groundwater contamination remains a major problem. Many former dump sites continue to leak hazardous wastes, and industrial processes continue to generate chemical wastes whose safe disposal is costly and difficult.

The ideal disposal method is a chemical treatment that can convert hazardous waste into environmentally benign materials. For example, trichloroethylene ($Cl_2C{=}CHCl$) is highly toxic to aquatic life, but this compound can be made nontoxic by chemical treatment that converts its chlorine atoms into chloride anions. Similarly, the chromium-containing waste from electroplating operations contains highly toxic CrO_4^{2-} anions, but a chemical treatment that converts CrO_4^{2-} into Cr^{3+} causes the chromium to precipitate from the solution as insoluble $Cr(OH)_3$. This removal of chromium detoxifies the water.

Both these conversion processes involve the addition of electrons to the toxic substances. The trichloroethylene molecule is electrically neutral and must gain electrons in reactions that generate negatively charged chloride anions. In addition, water or hydronium ions must supply hydrogen atoms that replace the chlorine atoms in the organic substance. The detailed reaction is complicated, but the net reaction is relatively simple:

$$Cl_2C{=}CHCl + 3\,H_3O^+ + 6\,e^- \longrightarrow$$
$$H_2C{=}CH_2 + 3\,Cl^- + 3\,H_2O$$

The chromium atom in CrO_4^{2-} has an oxidation number of $+6$, so it must gain three electrons to be converted into Cr^{3+}. Again, water or hydronium ions must supply hydrogen atoms that combine with the O atoms in chromate to form water:

$$CrO_4^{2-} + 8\,H_3O^+ + 3\,e^- \longrightarrow Cr^{3+} + 12\,H_2O$$

Because electrons are conserved, chemical detoxification of water containing either of these contaminants requires a substance that easily donates electrons. A suitable electron donor for groundwater cleanup must also be relatively inexpensive and must not itself be a source of toxicity.

These are stringent requirements; yet a very common material does this job spectacularly well: iron filings. Metallic iron contains neutral iron atoms that readily donate electrons to become Fe^{2+} or Fe^{3+} cations, neither of which is significantly toxic.

$$Fe \longrightarrow Fe^{2+} + 2e^-$$

Thus, when iron reacts with trichloroethylene or chromate, all the products are relatively nontoxic.

How does this process work in practice? As the figure indicates, it turns out to be amazingly simple, taking advantage of the same natural water flow that leads to contamination. First, the source of contamination and direction in which groundwater flows are determined. A pit is then dug downstream from the contamination source. This pit is filled with a mixture of sand and iron filings. The contaminant reacts with iron filings as the groundwater flows through the pit, and the water emerging downstream from the pit is contaminant-free. The first full-scale operation of this type treated a trichloroethylene-contaminated site in Sunnyvale, California. At a cost of $1.5 million, a site was recovered that had been predicted to remain too contaminated for human use for up to 30 years.

Another groundwater contaminant that can be removed by reduction reactions is perchlorate. Ammonium perchlorate is a solid rocket fuel, and wastes from its manufacturing have contaminated some irrigation and drinking water sources, among them the lower Colorado River in the American Southwest. Elevated perchlorate levels have been linked to human health effects, and because perchlorate has been found in vegetable crops such as lettuce, there is concern that this contamination could have serious consequences.

A promising technology for removing perchlorate from water uses microorganisms. A number of bacteria thrive on perchlorate, reducing it ultimately to harmless chloride anions in the course of their metabolism.

$$ClO_4^- + 8\,H_3O^+ + 8\,e^- \longrightarrow Cl^- + 12\,H_2O$$

These bacteria use energy-releasing redox reactions in which electrons are transferred from hydrogen-rich substances to perchlorate. Microbiologists and biochemists are studying these bacteria and the enzymes that they contain, in hopes of discovering how to use these life-forms to purify water that is contaminated with perchlorate.

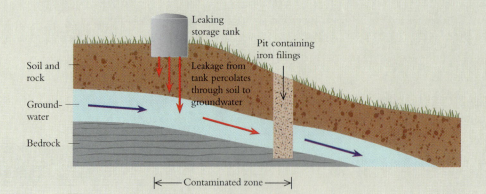

Soil and rock

Leaking storage tank

Pit containing iron filings

Leakage from tank percolates through soil to groundwater

Groundwater

Bedrock

|← Contaminated zone →|

When half-reactions are combined, there is often a duplication of some chemical species, particularly H_2O and H_3O^+ or OH^-. The overall equation is "cleaned up" by combining species that appear twice on the same side. Also, when a species appears on both sides of the balanced equation, equal numbers of the species are subtracted from each side.

To summarize, the balancing of a redox equation occurs in several steps that are carried out in sequence.

BALANCING REDOX REACTIONS

Step 1 Break the unbalanced equation into half-reactions by inspection.
Step 2 Balance each half-reaction, following the four-step procedure.
    a Balance all elements except oxygen and hydrogen by adjusting stoichiometric coefficients.
    b Balance oxygen by adding H_2O to the side that is deficient in oxygen.
    Balance hydrogen:
    c If the solution is neutral or acidic, add H_3O^+ to the side that is deficient in H and H_2O to the other side. If the solution is basic, add H_2O to the side that is deficient in H and OH^- to the other side.
    d Balance net charge by adding electrons to the side that is deficient in negative charge.
Step 3 Multiply the half-reactions by integers that will lead to cancellation of electrons.
Step 4 Recombine the half-reactions, and simplify by combining and canceling duplicated species.

Example 19-6 illustrates the entire balancing process, and Example 19-7 illustrates the balancing procedure in basic solution.

Example 19-6

Balancing a Redox Equation

Concentrated aqueous sulfuric acid, H_2SO_4, is a strong oxidizing agent that can react with elemental carbon:

$$H_2SO_4 + C \longrightarrow CO_2 + SO_2 \quad \text{(unbalanced)}$$

What is the balanced equation for this process?

Strategy: Divide the reaction into half-reactions, balance each using the stepwise procedure, combine the half-reactions, and then clean up the result to eliminate duplicated species.

Solution:

1. The half-reactions can be identified by inspection:

$$C \longrightarrow CO_2 \qquad H_2SO_4 \longrightarrow SO_2$$

2. Each half-reaction is balanced separately using the four-step process. Concentrated sulfuric acid contains some water, so H_3O^+ and H_2O can be used as needed. The new feature in each step is highlighted by a yellow background.

Step a: C is already balanced   S is already balanced

Step b: $C + 2\,H_2O \longrightarrow CO_2$ $H_2SO_4 \longrightarrow SO_2 + 2\,H_2O$

Step c: $C + 6\,H_2O \longrightarrow CO_2 + 4\,H_3O^+$

                    $H_2SO_4 + 2\,H_3O^+ \longrightarrow SO_2 + 4\,H_2O$

Step d: $C + 6\,H_2O \longrightarrow CO_2 + 4\,H_3O^+ + 4\,e^-$

              $H_2SO_4 + 2\,H_3O^+ + 2\,e^- \longrightarrow SO_2 + 4\,H_2O$

Example 19-6 **Balancing a Redox Equation** (*continued*)	3. Multiply the sulfur equation by 2, so the electrons will cancel, and add the half-reactions: $$C + 6\,H_2O + 4\,H_3O^+ + 2\,H_2SO_4 + 4\,e^- \longrightarrow$$ $$CO_2 + 4\,H_3O^+ + 4\,e^- + 2\,SO_2 + 8\,H_2O$$ 4. Cancel common terms: $4\,H_3O^+$ on each side and $6\,H_2O$ on each side: $$C + 2\,H_2SO_4 \longrightarrow CO_2 + 2\,SO_2 + 2\,H_2O$$

Does the Result Make Sense?   An inventory verifies that this equation is balanced:

Reactants	Products
1 C, 4 H, 2 S, 8 O	1 C, 4 H, 2 S, 8 O
0 charge	0 charge

Extra Practice Exercise 19.6	Balance this reaction in aqueous solution: $MnO_4^- + ClO_3^- \longrightarrow MnO_2 + ClO_4^-$.

Example 19-7 **Redox in Basic Solution**  **Never mix household ammonia and chlorine bleach!**	Household ammonia should never be mixed with chlorine bleach, because a redox reaction occurs that generates toxic chlorine gas and hydrazine: $$NH_3 + OCl^- \longrightarrow Cl_2 + N_2H_4 \quad \text{(unbalanced)}$$ Balance this equation. **Strategy:** Follow the step-by-step procedure. **Solution:** 1. Break into half-reactions, one containing Cl and the other containing N: $$NH_3 \longrightarrow N_2H_4 \qquad\qquad OCl^- \longrightarrow Cl_2$$ 2. Balance the half-reactions individually. Step a. Balance N and Cl by inspection: $$2\,NH_3 \longrightarrow N_2H_4 \qquad\qquad 2\,OCl^- \longrightarrow Cl_2$$ Step b. Add H_2O to balance oxygen atoms: $$2\,NH_3 \longrightarrow N_2H_4 \qquad\qquad 2\,OCl^- \longrightarrow Cl_2 + 2\,H_2O$$ Step c. Add H_2O and OH^- to balance hydrogen atoms: $$2\,NH_3 + 2\,OH^- \longrightarrow N_2H_4 + 2\,H_2O$$ $$2\,OCl^- + 4\,H_2O \longrightarrow Cl_2 + 2\,H_2O + 4\,OH^-$$ Two water molecules can be canceled from each side of the second half-reaction: $$2\,OCl^- + 2\,H_2O \longrightarrow Cl_2 + 4\,OH^-$$ Step d. Add electrons to balance the charges. The nitrogen half-reaction needs two electrons on the right, and the chlorine reaction needs two electrons on the left: $$2\,NH_3 + 2\,OH^- \longrightarrow N_2H_4 + 2\,H_2O + 2\,e^-$$ $$2\,OCl^- + 2\,H_2O + 2\,e^- \longrightarrow Cl_2 + 4\,OH^-$$ 3. Because there are two electrons in each half-reaction, adding the two half-reactions leads to cancellation of the electrons.

4. Now recombine the half-reactions and cancel common terms:

$$2\,NH_3 + 2\,OH^- + 2\,H_2O + 2\,OCl^- + 2\,e^- \longrightarrow$$

$$N_2H_4 + 2\,e^- + 2\,H_2O + Cl_2 + 4\,OH^-$$

Example 19-7

Redox in Basic Solution (*continued*)

Cancel two OH^- ions:

$$2\,NH_3 + 2\,OCl^- \longrightarrow N_2H_4 + Cl_2 + 2\,OH^-$$

Does the Result Make Sense? Make sure the equation is balanced by taking an inventory:

Reactants	Products
2 N, 6 H, 2 Cl, 2 O	2 N, 6 H, 2 Cl, 2 O
−2 charge	−2 charge

The equation is balanced. Although we could have balanced this equation by inspection, the detailed procedure reveals its redox nature.

Balance this reaction in basic aqueous solution: $MnO_4^- + Cd \rightarrow MnO_2 + Cd(OH)_2$.

Extra Practice Exercise 19.7

Section Exercises

■ **19.2.1** Balance the following redox reactions from Example 19-3:
(a) $H_2O_2 + SO_3^{2-} \longrightarrow SO_4^{2-}$ (basic)
(b) $BrO_2^- \longrightarrow BrO_3^- + Br^-$ (acidic)

■ **19.2.2** The tarnish that collects on objects made of silver is silver sulfide, a black solid. Tarnish can be removed by heating the tarnished object in an aluminum pan containing mildly basic water. Balance this equation:

$$Al\,(s) + Ag_2S\,(s) \longrightarrow$$
$$Ag\,(s) + Al(OH)_3\,(s) + H_2S\,(g)$$

■ **19.2.3** Copper refining traditionally involves "roasting" sulfide ores with oxygen, which produces large quantities of polluting SO_2. An alternative process uses aqueous nitric acid to convert the CuS into soluble copper(II) hydrogen sulfate without generating SO_2. Balance this redox equation:

$$CuS\,(s) + NO_3^-\,(aq) \longrightarrow$$
$$NO\,(g) + Cu^{2+}\,(aq) + HSO_4^-\,(aq)$$

19.3 GALVANIC CELLS

The reaction of potassium metal with water to generate hydrogen gas, shown in Figure 19-4, is highly spontaneous:

$$2\,K\,(s) + 2\,H_2O\,(l) \xrightarrow{\text{Rapid}} 2\,K^+\,(aq) + 2\,OH^-\,(aq) + H_2\,(g)$$

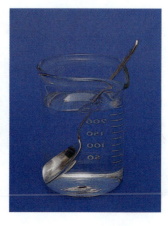

Figure 19-4
Potassium reacts spontaneously and violently with water, but silver does not react with water at all.

In contrast, no reaction occurs when a silver spoon is dipped into water. The balanced redox equation for the possible oxidation of silver by water looks similar to that for potassium:

$$2\,Ag\,(s) + 2\,H_2O\,(l) \xrightarrow{\text{No reaction}} 2\,Ag^+(aq) + 2\,OH^-(aq) + H_2(g)$$

One metal reacts with water so vigorously that a fire or explosion may result, whereas the other does not react with water, even at high temperature. Despite their apparent similarities, the reaction of potassium is spontaneous, but that of silver is not. Thermodynamic calculations verify these observations. Remember from Chapter 14 that spontaneous reactions have negative values for ΔG. We can calculate the standard free energy changes for these reactions using tabulated free energies of formation. The $\Delta G°$ for the reaction of potassium has a large negative value, but the value for the reaction of silver is large and positive:

$$2\,K\,(s) + 2\,H_2O\,(l) \longrightarrow 2\,K^+(aq) + 2\,OH^-(aq) + H_2(g) \qquad \Delta G°_{\text{reaction}} = -407\ kJ$$

$$2\,Ag\,(s) + 2\,H_2O\,(l) \longrightarrow 2\,Ag^+(aq) + 2\,OH^-(aq) + H_2(g) \qquad \Delta G°_{\text{reaction}} = +314\ kJ$$

Experiments and calculations both indicate that electron transfer from potassium to water is spontaneous and rapid, whereas electron transfer from silver to water does not occur. In redox terms, potassium oxidizes easily, but silver resists oxidation. Because oxidation involves the loss of electrons, these differences in reactivity of silver and potassium can be traced to how easily each metal loses electrons to become an aqueous cation. One obvious factor is their first ionization energies, which show that it takes much more energy to remove an electron from silver than from potassium: 731 kJ/mol for Ag and 419 kJ/mol for K. The other alkali metals with low first ionization energies, Na, Rb, Cs, and Fr, all react violently with water.

Direct and Indirect Electron Transfer

The reactivities of potassium and silver with water represent extremes in the spontaneity of electron-transfer reactions. The redox reaction between two other metals illustrates less drastic differences in reactivity. Figure 19-5 shows the reaction that occurs

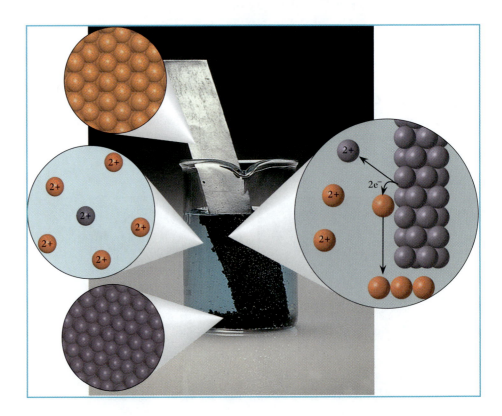

Figure 19-5
When a strip of zinc metal is dipped in a solution of copper(II) sulfate, zinc is oxidized to Zn^{2+} (aq), and Cu^{2+} (aq) is reduced to copper metal. The insoluble metal precipitates from the solution. In the molecular views, water molecules and spectator anions have been omitted for clarity.

between zinc metal and an aqueous solution of copper(II) sulfate: zinc slowly dissolves, and copper metal precipitates. This spontaneous reaction has a negative standard free energy change, as does the reaction of potassium with water:

$$Cu^{2+}(aq) + Zn(s) \longrightarrow Cu(s) + Zn^{2+}(aq) \qquad \Delta G^{\circ}_{reaction} = -213 \text{ kJ}$$

Each zinc atom transfers two electrons to a copper cation. Balanced half-reactions show the electron gains and losses more clearly:

$$\text{Reduction:} \qquad Cu^{2+}(aq) + 2 \text{ e}^- \longrightarrow Cu(s)$$

$$\text{Oxidation:} \qquad Zn(s) \longrightarrow Zn^{2+}(aq) + 2 \text{ e}^-$$

The oxidation of Zn metal by Cu^{2+} ions is an example of *direct* electron transfer. A copper ion accepts two electrons when it collides with the surface of the zinc strip. Direct electron transfer occurs when electrons are transferred during a collision between the species being oxidized and the species being reduced, as shown by the molecular views in Figure 19-5.

Spontaneous redox reactions can also occur by *indirect* electron transfer. In an indirect electron transfer, species involved in the redox chemistry are not allowed to come into direct contact with one another. Instead, the oxidation occurs at one end of a wire and transfers electrons to the wire. Reduction occurs at the other end of the wire and removes electrons from the wire. The wire conducts electrons between the oxidation site and the reduction site.

Figure 19-6 is a schematic view of two beakers set up for indirect electron transfer. The beaker on the left contains an aqueous solution of zinc sulfate and a strip of zinc metal. The beaker on the right contains an aqueous solution of copper(II) sulfate and a strip of copper metal. A wire connects the two metal strips to allow indirect electron transfer. The oxidation half-reaction transfers electrons to the wire and releases Zn^{2+} ions into the solution containing the zinc electrode:

$$Zn(s) \longrightarrow Zn^{2+}(aq) + 2 \text{ e}^-(wire)$$

Because metals conduct charge, electrons can move from the zinc strip through the wire into the copper strip. Copper cations can collect electrons when they collide with the surface of the copper strip:

$$Cu^{2+}(aq) + 2 \text{ e}^-(wire) \longrightarrow Cu(s)$$

The arrangement shown in Figure 19-6 does not generate a sustained flow of electrons, however, because as electrons are transferred, the solutions in both beakers quickly become unbalanced in electrical charge. Oxidation of the zinc strip releases Zn^{2+} ions into the solution, generating excess positive charge. On the other side, reduction of Cu^{2+} ions to copper metal produces a solution that is deficient in positive charge. This charge imbalance generates an electrical force that stops the further flow of electrons.

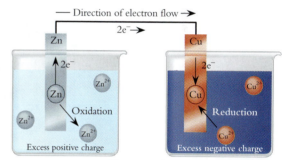

Figure 19-6
If the zinc- and copper-containing portions of this redox system are physically separated, electron transfer can occur only through an external wire.

Ion Transport

Redox by way of indirect electron transfer cannot continue unless there is a way to remove the charge imbalance created by electron flow. Ion movement can remove this charge imbalance, if we provide a way for anions to move from the region of excess negative charge to the region of excess positive charge. (Cations will move in the opposite direction.) Such ion migration must not allow the oxidation–reduction pair to come into direct contact, however, because then direct electron transfer would replace indirect electron transfer.

In our zinc–copper example, electrons flowing through the wire from left to right generate surplus positive charge on the left and surplus negative charge on the right. Metal cations moving from left to right and sulfate anions moving from right to left can remove the surplus charges, but this flow of ions must occur in a way that prevents the two solutions from mixing freely. Figure 19-7 shows an arrangement that meets these needs. A porous glass plate connects the two vessels. This plate allows ions to migrate

Figure 19-7
Schematic view of an arrangement that allows a sustained redox reaction accompanied by an external flow of electrons.

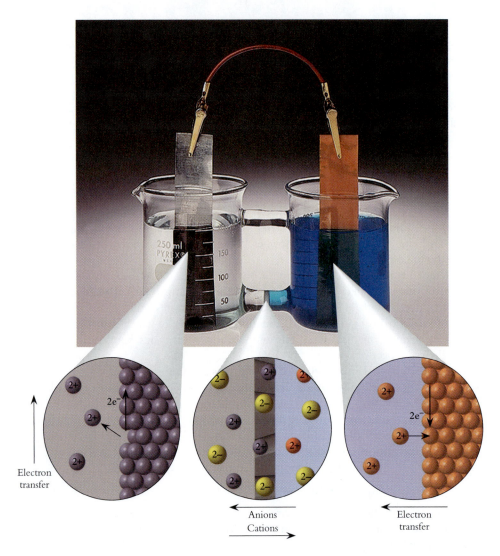

between the two compartments, but the mixing process is too slow to lead to direct electron transfer. Electrons flow through the wire, and ions diffuse through the plate. Adding the porous plate completes a circuit that can be used to generate a sustained flow of electrons through the external wire.

Salt bridges and membranes also can serve as separators that allow ion migration without free mixing of solutions. A porous plate contains many tiny channels that allow water molecules and ions to diffuse slowly from one side to the other. A salt bridge contains gelatin impregnated with cations and anions that are similar in size and mobility, usually K^+ and Cl^- or NH_4^+ and NO_3^-. The gelatin prevents solvent from moving freely but allows the ions to diffuse slowly. A membrane functions like a porous plate but is very thin, often only microns in width.

Electrodes

The spontaneous redox reaction shown in Figure 19-7 takes place at the surfaces of metal plates, where electrons are gained and lost by metal atoms and ions. These metal plates are examples of **electrodes.** At an electrode, redox reactions transfer electrons between the aqueous phase and the external circuit. An oxidation half-reaction releases electrons to the external circuit at one electrode. A reduction half-reaction withdraws electrons from the external circuit at the other electrode. The electrode where oxidation occurs is the **anode,** and the electrode where reduction occurs is the **cathode.**

Some electrodes are made of substances that participate in the redox reactions that transfer electrons. These are **active** electrodes. Other electrodes serve only to supply or accept electrons but are not part of the redox chemistry; these are **passive** electrodes. In Figure 19-7, both metal strips are active electrodes. During the redox reaction, zinc metal dissolves from the anode while copper metal precipitates at the cathode. The reactions that take place at these active electrodes are conversions between the metals contained in the electrodes and their aqueous cations.

At another type of active electrode, found in many batteries, the reaction is the conversion between a metal and an insoluble salt. At the surface of this type of electrode, metal cations combine with anions from the solution to form the salt. One example is the cadmium anode of a rechargeable nickel–cadmium battery, at whose surface cadmium metal loses electrons and forms Cd^{2+} cations. These cations combine immediately with hydroxide ions in solution to form insoluble cadmium hydroxide:

$$Cd(s) + 2\, OH^-(aq) \longrightarrow Cd(OH)_2(s) + 2\, e^-$$

A redox half-reaction at an active electrode also may convert one metal salt into another. For example, the cathode in a nickel–cadmium battery is NiO(OH), which is reduced to nickel(II) hydroxide. The half-reaction reduces Ni(III) to Ni(II):

$$NiO(OH)(s) + H_2O(l) + e^- \rightleftharpoons Ni(OH)_2(s) + OH^-(aq)$$

In contrast with these active electrodes, a passive electrode conducts electrons to and from the external circuit but does not participate chemically in the half-reactions. Figure 19-8 shows a redox setup that contains passive electrodes. One compartment contains an aqueous solution of iron(III) chloride in contact with a platinum electrode. Electron transfer at this electrode reduces $Fe^{3+}(aq)$ to $Fe^{2+}(aq)$:

Reduction: $Fe^{3+}(aq) + e^- \longrightarrow Fe^{2+}(aq)$

In the second compartment, hydrogen gas bubbles over the surface of a platinum electrode. On the surface of this passive electrode, hydrogen molecules are oxidized, giving up two electrons to form pairs of protons, which bond to water molecules to give H_3O^+ ions:

Oxidation: $H_2 + 2\, H_2O \longrightarrow 2\, H_3O^+ + 2\, e^-$

Each half-reaction occurs at a platinum metal electrode, but notice that Pt is not a participant in either half-reaction. Neither electrode takes part in the redox chemistry.

Platinum metal is often used as a passive electrode because platinum is one of the least reactive elements. Platinum has a large ionization energy, so it can act as an "electron shuttle" without participating in redox chemistry.

The combination of hydrogen gas, H_3O^+ ions, and a platinum electrode is referred to as a *hydrogen electrode*. This electrode appears in the right-hand portion of Figure 19-8. When a hydrogen electrode operates under standard conditions, $p_{H_2} = 1.00$ bar and $[H_3O^+] = 1.00$ M, it is a standard hydrogen electrode (SHE). The standard hydrogen electrode is particularly important in electrochemistry, as we describe in Section 19.4.

Figure 19-8
Passive electrodes serve only to conduct electrons to and from the interfaces. They do not take part in the redox reactions.

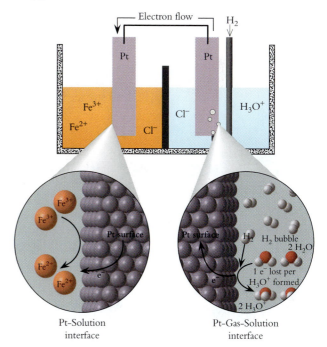

Pt-Solution interface

Pt-Gas-Solution interface

Components of Galvanic Cells

Electrochemistry is the coupling of a chemical redox process with electron flow through a wire. The process represented in Figure 19-7 is electrochemical because the redox reaction releases electrons that flow through an external wire as an electrical current. On the other hand, Figure 19-5 shows a redox process that is not electrochemical, because direct electron transfer cannot generate an electrical current through a wire.

A **galvanic cell** uses redox reactions to generate electrical current. Galvanic cells find wide usage in the form commonly called *batteries*. To give just one example, flashlight batteries make use of electrochemical reactions to generate electrical current. When

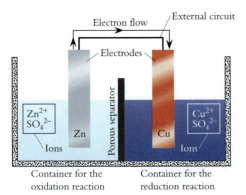

Figure 19-9
The essential components of a galvanic cell.

a flashlight is turned on, chemical redox reactions occur in the battery and generate an electron flow through the light bulb.

The cell shown diagrammatically in Figure 19-9 uses the Cu^{2+}/Zn reaction to illustrate the essential components of a galvanic cell:

1. *A spontaneous redox reaction.* In Figure 19-9, the spontaneous reaction is electron transfer between zinc and copper:

$$Zn\,(s) + Cu^{2+}(aq) \longrightarrow Zn^{2+}(aq) + Cu\,(s)$$

2. *A physical barrier blocking direct electron transfer.* Here, placing the solutions in two separate compartments provides the barrier.

3. *Physical contact between chemical and electrical parts of the cell.* This is the role of the electrodes. The electrodes in Figure 19-9 are strips of copper and zinc.

4. *An external electrical circuit.* This circuit may simply transfer electrons, as in the Zn/Cu example. In useful applications, the flow of electrons in the external circuit provides power to operate a flashlight, radio, or other device.

5. *Some means of completing the circuit.* In Figure 19-9, the porous glass plate allows ions to diffuse slowly between the two solutions, completing the circuit.

6. *Ions present in the solutions.* All parts of the circuit must allow the passage of charge. When an aqueous phase is present, it must contain ions, which are supplied by a soluble salt or strong acid.

To apply the features that characterize galvanic cells, Example 19-8 describes the lead storage battery.

Example 19-8
Describing a Galvanic Cell

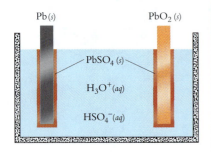

The electrical current needed to start an automobile engine is provided by a lead storage battery. This battery contains aqueous sulfuric acid in contact with two electrodes. One electrode is metallic lead, and the other is solid PbO_2. Each electrode becomes coated with solid $PbSO_4$ as the battery operates. Determine the balanced half-reactions, the overall redox reaction, and the anode and cathode in this galvanic cell.

Strategy: We are asked to identify the redox chemistry occurring in this battery. The problem provides a description of the chemical composition of a galvanic cell. To determine what redox reactions take place, examine the species present at each electrode. Then use the standard procedure to balance the half-reactions.

Solution: The description of the battery states that one electrode is Pb in contact with $PbSO_4$, and the other electrode is PbO_2 in contact with $PbSO_4$. This information identifies the two half-reactions:

$$Pb\,(s) \longrightarrow PbSO_4\,(s) \qquad PbO_2\,(s) \longrightarrow PbSO_4\,(s)$$

Lead is already balanced. To balance sulfur, we need to work with the major species in solution. The aqueous solution contains a strong acid, H_2SO_4, which generates H_3O^+ and HSO_4^-. Use the sulfur-containing species, HSO_4^-, to balance sulfur:

$$Pb + HSO_4^- \longrightarrow PbSO_4 \qquad PbO_2 + HSO_4^- \longrightarrow PbSO_4$$

Now all elements other than hydrogen and oxygen are balanced.

Here are the additional steps for the left half-reaction:

Oxygen is already balanced. Add H_3O^+ and H_2O to balance hydrogen:

$$Pb + HSO_4^- + H_2O \longrightarrow PbSO_4 + H_3O^+$$

Add electrons to balance charge:

$$Pb + HSO_4^- + H_2O \longrightarrow PbSO_4 + H_3O^+ + 2\,e^-$$

Electrons appear as products, so this half-reaction is the oxidation, which takes place at the anode. The lead electrode is an active anode in a lead storage battery.

Now balance the second half-reaction. Add water to balance oxygen:

$$PbO_2 + HSO_4^- \longrightarrow PbSO_4 + 2\,H_2O$$

Add H_3O^+ and H_2O to balance hydrogen:

$$PbO_2 + HSO_4^- + 3\,H_3O^+ \longrightarrow PbSO_4 + 5\,H_2O$$

Add e^- to balance charge:

$$PbO_2 + HSO_4^- + 3\,H_3O^+ + 2\,e^- \longrightarrow PbSO_4 + 5\,H_2O$$

Electrons appear as starting materials, so this half-reaction is the reduction, which takes place at the cathode. Lead(IV) oxide is an active cathode in a lead storage battery.

Because two electrons are transferred in each half-reaction, the overall redox reaction can be obtained by adding the half-reactions and cleaning up:

$$PbO_2\,(s) + Pb\,(s) + 2\,HSO_4^-\,(aq) + 2\,H_3O^+\,(aq) \longrightarrow 2\,PbSO_4\,(s) + 4\,H_2O\,(l)$$

Do the Results Make Sense? You should verify that each of the elements is balanced in this reaction. In addition, the net charge is zero on each side.

The mercury battery, use of which is being discontinued because of the toxicity of mercury, contains HgO and Zn in contact with basic aqueous solution. The redox products are Hg and ZnO. Determine the oxidation and reduction half-reactions and the overall reaction for these batteries.

Example 19-8

Describing a Galvanic Cell (*continued*)

To remember how electrodes are named, notice that *oxidation* and *anode* both begin with vowels; *reduction* and *cathode* both begin with consonants.

Extra Practice Exercise 19.8

Section Exercises

19.3.1 The thermite reaction between aluminum metal and iron oxide is so rapid and exothermic that it generates a fountain of sparks (see photo) and can melt the container in which it takes place. The spontaneity of this reaction suggests the possibility of a galvanic cell involving aluminum and iron:

$$2\,Al + 6\,OH^- \longrightarrow Al_2O_3 + 3\,H_2O + 6\,e^-$$
$$Fe_2O_3 + 3\,H_2O + 6\,e^- \longrightarrow 2\,Fe + 6\,OH^-$$

Draw a schematic diagram, similar to Figure 19-7, showing the operation of this galvanic cell. Include the directions of all charge flows.

19.3.2 Use tabulated thermodynamic data to verify that the galvanic cell of Section Exercise 19.3.1 is spontaneous in the direction written.

19.3.3 Draw molecular pictures similar to the one shown in Figure 19-5 to illustrate the difference between an active electrode and a passive electrode. Use an iron anode in contact with a solution containing Fe^{2+} ions to illustrate an active electrode. Use a platinum cathode immersed in a solution containing Fe^{2+} and Fe^{3+} ions to illustrate a passive electrode.

The thermite reaction

19.4 CELL POTENTIALS

Batteries supply electrical current that consists of electrons flowing from one place to another. In some ways, flowing electrons are like flowing water. Because of gravitational force, water always flows downhill, from higher altitude to lower altitude. Similarly, electrons flow under the influence of electrical force. Electrons flow from regions of more negative electrical potential energy to regions of more positive electrical potential energy.

Water at high altitude has gravitational potential energy that can be converted to other forms. As water flows, the gravitational potential energy that it loses is converted into kinetic energy in a waterfall, mechanical energy in a waterwheel, or electrical energy in a hydroelectric turbine. Similarly, as electrons flow, electrical potential energy can be converted to other forms: light energy in a flashlight, mechanical energy in the starter motor of an automobile, or thermal energy in an electrical heater.

A potential energy difference is referred to simply as a *potential*. Continuing our analogy with water, differences in gravitational potential are measured by height differences. The top and bottom of the spillway of a dam are at different heights and have different gravitational potentials: water falling down the spillway converts gravitational potential energy into kinetic energy of motion. Analogously, one electrode generally is at higher electrical potential than another.

Whereas gravitational potentials are measured in heights, electrical potential differences are measured in volts. For electrons, a point at a more negative electrical potential is "uphill" from a point at a more positive electrical potential. The parallel between gravitational potential and electrical potential is summarized in Figure 19-10.

Electrode Equilibrium

To visualize how electrochemical cells generate electrical potential differences, consider a zinc electrode dipped into a solution of zinc sulfate. From the macroscopic perspective, nothing happens. At the molecular level, however, some of the zinc atoms of the electrode are oxidized to Zn^{2+} ions:

$$Zn\,(s) \longrightarrow Zn^{2+}\,(aq) + 2\,e^-\,(metal)$$

The ions move off into the solution, leaving an excess of electrons in the metal strip. The reverse process also occurs. A Zn^{2+} ion that collides with the metal surface may capture two electrons from the metal and be reduced to a neutral zinc atom:

$$Zn^{2+}\,(aq) + 2\,e^-\,(metal) \longrightarrow Zn\,(s)$$

When a zinc strip is dipped into the solution, the initial rates of these two processes are different. The different rates of reaction lead to a charge imbalance across the metal−solution interface. If the concentration of zinc ions in solution is low enough, the initial rate of oxidation is more rapid than the initial rate of reduction. Under these con-

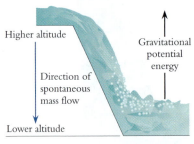

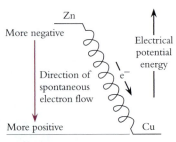

Figure 19-10
Water flows down a spillway from higher to lower gravitational potential. Electrons flow "downhill" through a wire from higher to lower negative electrical potential.

ditions, excess electrons accumulate in the metal, and excess cationic charges accumulate in the solution. As excess charge builds, however, the rates of reaction change until the rate of reduction is balanced by the rate of oxidation. When this balance is reached, the system is at dynamic equilibrium. Oxidation and reduction continue, but the net rate of exchange is zero:

$$Zn(s) \rightleftharpoons Zn^{2+}(aq) + 2\,e^-(metal)$$

Figure 19-11 illustrates the zinc equilibrium at the molecular level. At equilibrium, the charge imbalance in the zinc strip is about one excess electron for every 10^{14} zinc atoms. This is negligible from the macroscopic perspective but significant at the molecular level.

An analogous dynamic equilibrium is established when a strip of copper is dipped in a dilute solution of copper sulfate:

$$Cu(s) \rightleftharpoons Cu^{2+}(aq) + 2\,e^-(metal)$$

Once again, nothing appears to happen at the macroscopic level. At the molecular level, however, some of the copper atoms lose electrons and enter the solution as Cu^{2+} ions, and some of the Cu^{2+} ions capture electrons from the metal and deposit on the metal as Cu atoms. As with zinc, dipping a strip of copper metal in the solution generates a small charge imbalance.

The charge imbalances for copper and zinc have different values, because zinc is easier to oxidize than copper. Consequently, zinc creates a greater charge imbalance than copper. The concentration of excess electrons in the zinc electrode is greater than the concentration of excess electrons in the copper electrode, giving the zinc electrode more excess charge than the copper electrode.

When two electrodes contain different amounts of excess charge, there is a difference in electrical potential between them. Because it has more excess electrons, the zinc electrode is at a higher electrical potential than the copper electrode. In a galvanic cell, the difference in electrical potential causes electrons to flow from a region where the concentration of electrons is higher to a region where the concentration of electrons is lower. In this case, electrons flow from the zinc electrode toward the copper electrode, as shown at the molecular level in Figure 19-12.

The difference in electrical potential between two electrodes is the **cell potential,** designated E and measured in volts (V). The magnitude of E increases as the amount of charge imbalance between the two electrodes increases. For any galvanic cell, the value of E and the direction of electron flow can be determined experimentally by inserting a voltmeter in the external circuit.

Electrons always flow spontaneously "downhill" from higher electrical potential to lower electrical potential. In a galvanic cell, the electrode with the higher potential is

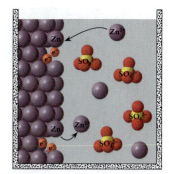

Figure 19-11
A strip of zinc metal immersed in a solution of zinc sulfate reaches a dynamic equilibrium when a few excess charges occur in each phase and the rates of oxidation and reduction are equal.

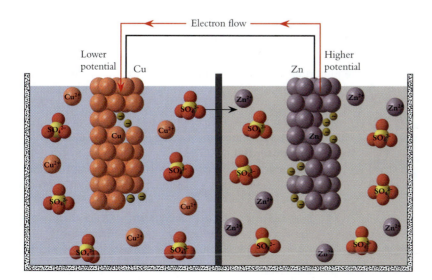

Animation

Figure 19-12
In a galvanic cell, electrons flow "downhill" from a region of higher electrical potential to a region of lower electrical potential.

The cathode of a battery is at lower potential and is labeled with a +, designating that electrons flow to this terminal from the negative (anode) terminal.

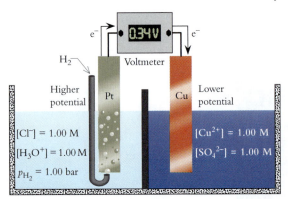

Figure 19-13
Diagram of a copper/zinc electrochemical cell operating under standard conditions.

designated by convention as the *negative electrode*. The electrode with the lower potential is designated as the *positive electrode*.

Standard Electrical Potential

Electrochemical cells can be constructed using an almost limitless combination of electrodes and solutions, and each combination generates a specific potential. Keeping track of the electrical potentials of all cells under all possible situations would be extremely tedious without a set of standard reference conditions. By definition, the standard electrical potential is the potential developed by a cell in which all chemical species are present under standard thermodynamic conditions. Recall that standard conditions for thermodynamic properties include concentrations of 1 M for solutes in solution and pressures of 1 bar for gases. Chemists use the same standard conditions for electrochemical properties. As in thermodynamics, standard conditions are designated with a superscript °. A standard electrical potential is designated $E°$.

The zinc−copper galvanic cell is under standard conditions when the concentration of each ion is 1.00 M, as shown in Figure 19-13. The cell potential under these conditions can be determined by connecting the electrodes to a voltmeter. The measured potential is 1.10 V, with the Zn electrode at the higher (more negative) potential, so Zn gives up electrons and $E°_{cell} = 1.10$ V:

$$Zn(s) + Cu^{2+}(aq, 1\ M) \longrightarrow Zn^{2+}(aq, 1\ M) + Cu(s) \qquad E°_{cell} = 1.10\ V$$

A table giving the cell potentials of all possible redox reactions would be immense. Instead, chemists use the fact that any redox reaction can be broken into two distinct half-reactions, an oxidation and a reduction. They assign a potential to every half-reaction and tabulate $E°$ values for all half-reactions. The standard cell potential for any redox reaction can then be obtained by combining the potentials for its two half-reactions.

To do this, one particular half-reaction has to be selected as a reference reaction with zero potential. Once a reference half-reaction has been selected, all other half-reactions can then be assigned values relative to this reference value of 0 V. This is necessary because an experiment always measures the *difference* between two potentials rather than an absolute potential. The standard potential of 1.10 V for the Zn/Cu cell, for example, is the difference between the $E°$ values of its two half-reactions.

Chemists have chosen the reduction of hydronium ions to hydrogen gas as the reference half-reaction:

$$2\ H_3O^+(aq, 1\ M) + 2\ e^- \rightleftharpoons H_2(g, 1\ bar) + 2\ H_2O(l) \qquad E° = 0\ V\ \text{by definition}$$

This standard hydrogen electrode (SHE) is shown schematically in Figure 19-14.

In addition to defined standard conditions and a reference potential, tabulated half-reactions have a defined reference *direction*. As the double arrow in the previous equation indicates, $E°$ values for half-reactions refer to electrode equilibria. Just as the value of an equilibrium constant depends on the direction in which the equilibrium reaction is written, the values of $E°$ depend on whether electrons are reactants or products. For half-reactions, the conventional *reference* direction is *reduction*, with electrons always appearing as *reactants*. Thus, each tabulated $E°$ value for a half-reaction is a **standard reduction potential.**

It is important to remember that every redox reaction includes both an oxidation and a reduction. Half-reactions are tabulated as reductions by convention, but in any real redox process, one half-reaction occurs as a reduction and another occurs as an oxidation.

Defining a reference value for the SHE makes it possible to determine $E°$ values of all other redox half-reactions. As an example, Figure 19-14 shows a cell in which a standard hydrogen electrode is connected to a copper electrode in contact with a 1.00 M solution of

Figure 19-14
Schematic diagram of a cell for measuring $E°$ of the Cu^{2+}/Cu half-reaction relative to the standard hydrogen electrode (SHE).

Table 19-1   Representative Standard Reduction Potentials

$$2\,H_3O^+ + 2\,e^- \rightleftharpoons H_2\,(g) + 2\,H_2O\,(l) \qquad E^\circ = 0\,V\,(\text{defined})$$

Reaction Positive Values	E° (v)	Reaction Negative Values	E° (v)
$F_2\,(g) + 2e^- \rightleftharpoons 2\,F^-$	2.866	$Fe^{3+} + 3\,e^- \rightleftharpoons Fe$	-0.037
$PbO_2 + SO_4^{2-} + 4\,H_3O^+ + 2\,e^- \rightleftharpoons PbSO_4 + 6\,H_2O\,(l)$	1.6913	$Pb^{2+} + 2\,e^- \rightleftharpoons Pb$	-0.1262
$Au^{3+} + 3\,e^- \rightleftharpoons Au$	1.498	$Sn^{2+} + 2\,e^- \rightleftharpoons Sn$	-0.137
$Cl_2\,(g) + 2e^- \rightleftharpoons 2\,Cl^-$	1.35827	$PbSO_4 + 2\,e^- \rightleftharpoons Pb + SO_4^{2-}$	-0.3588
$Cr_2O_7^{2-} + 14\,H_3O^+ + 6\,e^- \rightleftharpoons 2\,Cr^{3+} + 21\,H_2O$	1.232	$Fe^{2+} + 2\,e^- \rightleftharpoons Fe$	-0.447
$O_2\,(g) + 4\,H_3O^+ + 4\,e^- \rightleftharpoons 6\,H_2O$	1.229	$Ni(OH)_2 + 2\,e^- \rightleftharpoons Ni + 2\,OH^-$	-0.72
$Pt^{2+} + 2\,e^- \rightleftharpoons Pt$	1.18	$Cr^{3+} + 3\,e^- \rightleftharpoons Cr$	-0.744
$Br_2\,(l) + 2\,e^- \rightleftharpoons 2\,Br^-$	1.066	$Zn^{2+} + 2\,e^- \rightleftharpoons Zn$	-0.7618
$Hg^{2+} + 2\,e^- \rightleftharpoons Hg\,(l)$	0.851	$Cd(OH)_2 + 2\,e^- \rightleftharpoons Cd + 2\,OH^-$	-0.860
$Ag^+ + e^- \rightleftharpoons Ag$	0.7996	$Al^{3+} + 3\,e^- \rightleftharpoons Al$	-1.662
$I_2 + 2\,e^- \rightleftharpoons 2\,I^-$	0.5355	$Mg^{2+} + 2\,e^- \rightleftharpoons Mg$	-2.37
$Cu^+ + e^- \rightleftharpoons Cu$	0.521	$Na^+ + e^- \rightleftharpoons Na$	-2.71
$O_2\,(g) + 2\,H_2O\,(l) + 4\,e^- \rightleftharpoons 4\,OH^-$	0.401	$Ca^{2+} + 2\,e^- \rightleftharpoons Ca$	-2.868
$Cu^{2+} + 2\,e^- \rightleftharpoons Cu$	0.3419	$Ca(OH)_2 + 2\,e^- \rightleftharpoons Ca + 2\,OH^-$	-3.02
$AgCl + e^- \rightleftharpoons Ag + Cl^-$	0.22233	$Li^+ + e^- \rightleftharpoons Li$	-3.0401

Note: All ions are aqueous, all neutrals are solid unless noted.

Cu^{2+}. Measurements on this cell show that the SHE is at higher electrical potential than the copper electrode, indicating that electrons flow from the SHE to the Cu electrode. Reduction of Cu^{2+} and oxidation of H_2 occur spontaneously to generate this electron flow:

$$Cu^{2+}(aq,\,1\,M) + H_2\,(g,\,1\,bar) + 2\,H_2O\,(l) \longrightarrow Cu\,(s) + 2\,H_3O^+(aq,\,1\,M)$$

In other words, oxidation takes place at the SHE (anode), and reduction occurs at the copper electrode (cathode).

Measurements show that the Cu/H_2 cell has a voltage of 0.34 V. Because the potential of the SHE is defined to be 0 V, this means that the reduction half-reaction Cu^{2+}/Cu has $E^\circ = +0.34$ V *relative to the SHE*:

$$Cu^{2+}(aq,\,1\,M) + 2\,e^- \rightleftharpoons Cu\,(s) \qquad E^\circ = 0.34\,V\ (\text{relative to SHE})$$

Over the years, chemists have carried out many measurements similar to the one depicted in Figure 19-14. The resulting values for standard reduction potentials are tabulated in reference sources such as the *CRC Handbook of Chemistry and Physics*. Many values appear in Appendix F, and some representative values appear in Table 19-1.

Every standard reduction potential has a specific sign. When a substance is easier to reduce than hydronium ions under standard conditions, its E° is positive. When a substance is more difficult to reduce than hydronium ions under standard conditions, its E° is negative.

Standard Cell Voltages

In any galvanic cell that is under standard conditions, electrons are produced by the half-reaction with the more negative standard reduction potential and consumed by the half-reaction with the more positive standard reduction potential. In other words, the half-reaction with the more negative E° value occurs as the oxidation, and the half-reaction with the more positive E° value occurs as the reduction. Figure 19-15 summarizes the conventions used to describe galvanic cells.

Tabulated standard reduction potentials allow us to determine the potential of any cell under standard conditions. This net standard cell potential is obtained by *subtracting* the more negative standard reduction potential from the more positive standard reduction potential, giving a positive overall potential.

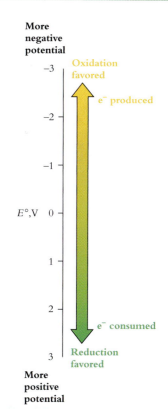

Figure 19-15
The half-reaction with the more negative E° produces electrons at higher electrical potential through oxidation. The half-reaction with the more positive E° consumes electrons at lower electrical potential through reduction.

KEY CONCEPTS

The half-reaction with the more negative reduction potential occurs at the anode as oxidation.

The half-reaction with the more positive reduction potential occurs at the cathode as reduction.

Combining these features gives an equation that summarizes the calculation of the standard potential for a galvanic cell:

④

$$E^\circ_{cell} = E^\circ_{cathode} - E^\circ_{anode} \qquad\qquad (19\text{-}1)$$

Here is this procedure for the copper–hydrogen cell illustrated in Figure 19-14. We write the two half-reactions as reductions:

$$Cu^{2+}(aq,\ 1\ bar) + 2\ e^- \rightleftharpoons Cu\,(s) \qquad\qquad E^\circ = +0.34\ V$$

$$2\ H_3O^+(aq,\ 1\ M) + 2\ e^- \rightleftharpoons H_2\,(g,\ 1\ bar) + 2\ H_2O \qquad E^\circ = 0\ V$$

To obtain the overall reaction, we reverse the half-reaction that has the more negative potential, and we find E°_{cell} by subtracting that E° from the more positive E°:

$$H_2\,(g,\ 1\ bar) + 2\ H_2O + Cu^{2+}(aq,\ 1\ M) \longrightarrow 2\ H_3O^+(aq,\ 1\ M) + Cu\,(s)$$

$$E^\circ_{cell} = (E^\circ_{Cu^{2+}/Cu}) - (E^\circ_{H_3O^+/H_2}) = (0.34\ V) - (0\ V) = 0.34\ V$$

In this spontaneous redox reaction, oxidation of H_2 releases electrons at the platinum anode. Excess electrons in the anode create a negative electrical potential. In the other compartment of the cell, Cu^{2+} ions are reduced to copper metal by capturing electrons from the copper cathode. In the wire connecting the electrodes, electrons flow "downhill" from the region of more negative electrical potential (the anode, $E^\circ = 0$ V) to the region of less negative electrical potential (the cathode, $E^\circ = 0.34$ V).

Consider another galvanic cell, one in which a standard Zn^{2+}/Zn half-reaction combines with the standard hydrogen electrode. According to Table 19-1, E° for the standard Zn^{2+}/Zn half-reaction is -0.76 V:

$$Zn^{2+}(aq,\ 1\ bar) + 2\ e^- \rightleftharpoons Zn\,(s) \qquad\qquad E^\circ = -0.76\ V$$

$$2\ H_3O^+(aq,\ 1\ bar) + 2\ e^- \rightleftharpoons H_2\,(g,\ 1\ bar) + 2\ H_2O \qquad E^\circ = 0\ V$$

The Zn^{2+}/Zn reduction potential is more negative than the H_3O^+/H_2 reduction potential (-0.76 V vs. 0 V), so zinc is the anode in this cell. Zinc is oxidized and hydronium ions are reduced, causing electrons to flow from the more negative zinc electrode to the less negative SHE. Again, we reverse the direction of the half-reaction with the more negative potential and find E°_{cell} by subtracting the half-cell potentials:

$$2\ H_3O^+(aq,\ 1\ bar) + Zn\,(s) \longrightarrow H_2\,(g,\ 1\ bar) + 2\ H_2O + Zn^{2+}(aq,\ 1\ bar)$$

$$E^\circ_{cell} = E^\circ_{H_3O^+/H_2} - E^\circ_{Zn^{2+}/Zn} = (0\ V) - (-0.76\ V) = 0.76\ V$$

The overall voltage generated by a standard galvanic cell is always obtained by *subtracting* one standard reduction potential from the other in the way that gives a *positive* value for E°_{cell}. Example 19-9 applies this reasoning to zinc and iron.

Example 19-9	A galvanic cell can be constructed from a zinc electrode immersed in a solution of zinc sulfate and an iron electrode immersed in a solution of iron(II) sulfate. What is the standard potential of this cell, and what is its spontaneous direction under standard conditions?
Standard Cell Potential	**Strategy:** First, identify the half-reactions. Then look up the standard reduction potentials in a table, and subtract the more negative value from the more positive value.

Solution: Write both half-reactions as reductions, and then look up the $E°$ values in Table 19-1:

$$Zn^{2+}(aq) + 2\ e^- \rightleftharpoons Zn\,(s) \qquad E°_{Zn^{2+}/Zn} = -0.7618\ V$$

$$Fe^{2+}(aq) + 2\ e^- \rightleftharpoons Fe\,(s) \qquad E°_{Fe^{2+}/Fe} = -0.447\ V$$

Now, subtract the more negative value (-0.7618 V) from the more positive value (-0.447 V) to obtain the standard cell potential:

$$E°_{cell} = E°_{Fe^{2+}/Fe} - E°_{Zn^{2+}/Zn} = (-0.447\ V) - (-0.7618\ V) = 0.315\ V$$

Under standard conditions, the iron–zinc cell operates spontaneously in the direction that reduces Fe^{2+} and oxidizes Zn:

$$Fe^{2+}(aq) + Zn\,(s) \longrightarrow Fe\,(s) + Zn^{2+}(aq) \qquad E°_{cell} = 0.315\ V$$

Does the Result Make Sense? The calculation shows that zinc is oxidized preferentially over iron. Later in this chapter we describe the use of zinc as a sacrificial anode to prevent corrosion of iron.

If Cu/Cu^{2+} is used in place of Zn/Zn^{2+}, what is the standard potential and spontaneous direction of reaction under standard conditions?

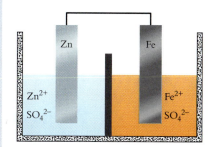

Example 19-9

Standard Cell Potential
(*continued*)

**Extra Practice
Exercise 19.9**

Conventions for Standard Reduction Potentials

Here is a summary of the definitions and conventions for working with electrochemical potentials:

1. Standard conditions for electrochemical half-reactions are the same as those in thermodynamics: 25 °C (298.15 K), 1 M for solutes, and 1 bar for gases.
2. Reduction is the reference direction for electrochemical half-reactions: Electrons are reactants.
3. The standard potential for reducing $H_3O^+(aq)$ to $H_2\,(g)$ is defined to be 0 V.
4. The standard potential for any galvanic cell is determined by subtracting the more negative standard reduction potential from the more positive standard reduction potential. A positive $E°$ indicates spontaneity under standard conditions.
5. In any galvanic cell, the half-reaction with the more negative reduction potential occurs as oxidation at the anode, and the half-reaction with the more positive reduction potential occurs as reduction at the cathode.

Example 19-10 provides another illustration of standard cell potentials.

The following half-reactions occur in the rechargeable nickel–cadmium battery:

$$Cd(OH)_2\,(s) + 2\ e^- \rightleftharpoons Cd\,(s) + 2\ OH^-(aq)$$

$$NiO(OH)\,(s) + H_2O\,(l) + e^- \rightleftharpoons Ni(OH)_2\,(s) + OH^-(aq)$$

This battery has a potential of 1.35 V under standard conditions, with nickel as the cathode. Determine the net reaction, and use the tabulated standard reduction potential for the cadmium half-reaction to find $E°$ for the nickel half-reaction.

Strategy: We need to determine which reaction occurs as reduction and which occurs as oxidation when this battery is operating in its spontaneous direction. Then we can use Equation 19-1 to determine $E°$ for the nickel half-reaction.

Solution: The problem states that the nickel electrode is the cathode. Because reduction takes place at the cathode, we know that the nickel half-reaction occurs as reduction. In the oxidation half-reaction, cadmium is oxidized at the anode.

Example 19-10

Half-Cell Potentials

Example 19-10

Half-Cell Potentials
(continued)

The problem also states that $E_{cell}^{\circ} = 1.35$ V. The E° value for the cadmium reaction appears in Table 19-1, but we must determine the value for the nickel half-reaction:

$$E_{cell}^{\circ} = E_{cathode}^{\circ} - E_{anode}^{\circ}$$

$$E_{cell}^{\circ} = E_{NiO(OH)/Ni(OH)_2}^{\circ} - E_{Cd(OH)_2/Cd}^{\circ}$$

$$1.35 \text{ V} = E_{NiO(OH)/Ni(OH)_2}^{\circ} - (-0.860 \text{ V})$$

$$E_{NiO(OH)/Ni(OH)_2}^{\circ} = (1.35 \text{ V}) + (-0.860 \text{ V}) = +0.49 \text{ V}$$

To balance the net reaction, multiply the nickel reduction half-reaction by 2 to balance the electrons and combine the result with the cadmium oxidation half-reaction. Electrons and hydroxide ions cancel:

$$2 \text{ NiO(OH)}(s) + 2 \text{ H}_2\text{O}(l) + 2 \text{ e}^- \longrightarrow 2 \text{ Ni(OH)}_2(s) + 2 \text{ OH}^-(aq)$$

$$\text{Cd}(s) + 2 \text{ OH}^-(aq) \longrightarrow \text{Cd(OH)}_2(s) + 2 \text{ e}^-$$

$$\overline{\text{Net: } 2 \text{ NiO(OH)}(s) + 2 \text{ H}_2\text{O}(l) + \text{Cd}(s) \longrightarrow 2 \text{ Ni(OH)}_2(s) + \text{Cd(OH)}_2(s)}$$

Does the Result Make Sense? The relatively simple chemistry of this redox reaction is one reason why nickel–cadmium batteries are rechargeable. As we show later in this chapter, applying an external voltage can reverse this reaction.

**Extra Practice
Exercise 19.10**

A cell is constructed with europium metal and a 1.00 M solution of $EuCl_2$ in one compartment and Cu metal in a 1.00 M solution of $CuSO_4$ in the other compartment. The cell has a voltage of 3.15 V, with the copper electrode being the cathode. Determine the net reaction and the standard potential for the Eu/Eu^{2+} half-reaction.

The calculation of E° for this cell illustrates an important feature of cell potentials. A standard cell potential is the difference between two standard reduction potentials. This difference does not change when one half-reaction is multiplied by 2 to cancel electrons in the overall redox reaction.

KEY CONCEPT *When a reaction is multiplied by any integer, its cell potential remains unchanged.*

To understand why this is so, recall that cell potentials are analogous to altitude differences for water. Whether 10,000 or 20,000 L of water flows down a spillway, the altitude difference between the top and bottom of the spillway is the same. In the same way, multiplying a reaction by some integer changes the total number of moles of electrons transferred, but it does not change the potential difference through which the electrons are transferred. We return to this point in Section 19.5.

Section Exercises

■ **19.4.1** Use standard reduction potentials to determine the net reaction and standard cell potential for cells of two compartments, each containing a 1.00 M solution of the indicated cation in contact with an electrode of that neutral metal (that is, cells similar to the one shown in Figure 19-13): (a) Fe^{2+} and Cr^{3+}; (b) Cu^{2+} and Pb^{2+}; and (c) Au^{3+} and Ag^+.

■ **19.4.2** Draw molecular pictures that illustrate the charge transfer process that takes place in the cell in Section Exercise 19.4.1(b).

■ **19.4.3** Using the appropriate values from Table 19-1, calculate E° for one cell of a lead storage battery. (Six of these cells are connected in series in an automobile storage battery.)

$$\text{PbO}_2(s) + \text{Pb}(s) + 2 \text{ HSO}_4^-(aq) + 2 \text{ H}_3\text{O}^+(aq) \longrightarrow$$
$$2 \text{ PbSO}_4(s) + 4 \text{ H}_2\text{O}(l)$$

19.5 FREE ENERGY AND ELECTROCHEMISTRY

Electrochemically, a spontaneous reaction generates a positive cell potential, E_{cell}. Thermodynamically, a spontaneous reaction has a negative change in free energy, ΔG. Thus, a reaction that has a negative change in free energy generates a positive cell potential:

$\Delta G_{reaction} < 0$:   Spontaneous redox reaction

$E_{cell} > 0$:   Spontaneous redox reaction

Cell Potential and Free Energy

The linkage between free energy and cell potentials can be made quantitative. The more negative the value of $\Delta G°$ for a reaction, the more positive its standard cell potential, as the following two examples illustrate:

$Cu^{2+}(aq) + Pb(s) \longrightarrow Cu(s) + Pb^{2+}(aq)$ $\Delta G° = -71.0 \text{ kJ/mol}$ $E°_{cell} = +0.368 \text{ V}$

$Pb^{2+}(aq) + Sn(s) \longrightarrow Pb(s) + Sn^{2+}(aq)$ $\Delta G° = -2.1 \text{ kJ/mol}$ $E°_{cell} = +0.011 \text{ V}$

For further insight into how these two quantities are related, consider again the analogy between gravitational and electrical potentials. When water tumbles down a spillway, the total energy change depends on the height through which the water falls and how much water flows. Similarly, when electrons flow in an electrochemical cell, the change in free energy, ΔG, depends on the potential difference (E) and on the amount of charge that flows through the cell (n, the number of moles of electrons).

To complete the connection between electrical potential and free energy, electrical units must be converted to energy units. The fundamental unit of electrical charge is the *coulomb* (C), defined as the quantity of electricity transferred by a current of 1 ampere in 1 second (the ampere and second are fundamental SI units). The charge on a single electron (e) has been measured to a very high degree of accuracy: $e = 1.602\,176\,46 \times 10^{-19}$ C. The amount of charge provided by one mole of electrons is obtained by multiplying the charge on a single electron by Avogadro's number (N_A). This quantity, the **Faraday constant (F),** is the link between chemical amounts (moles) and electrical amounts (coulombs):

$$F = e\,N_A = (1.602\,176\,46 \times 10^{-19} \text{ C})(6.022\,142 \times 10^{23} \text{ mol}^{-1}) = 96,485.34 \text{ C mol}^{-1}$$

Putting these factors together, we find that the molar free energy change is the product of the electrical potential, the number of electrons transferred, and the Faraday constant:

$$\Delta G = -nFE \qquad\qquad \textbf{(19-2)}$$

Michael Faraday, shown here delivering a popular lecture to the Royal Institution in 1855, developed many of the fundamental principles of electricity.

The units of ΔG are J/mol. On the right side of Equation 19-2, the Faraday constant has units of C/mol. Potential differences are in volts, and $1 \text{ J} = 1 \text{ V C}$, so $1 \text{ V} = 1 \text{ J/C}$ and the product FE has units of J/mol. In this equation, n is dimensionless because it is a *ratio,* the number of electrons transferred per atom reacting. Equation 19-2 has a negative sign because a spontaneous reaction has a negative value for ΔG but a positive value for E.

Remember that the number of electrons transferred is not explicitly stated in a net redox equation. This means that any overall redox reaction must be broken down into its balanced half-reactions to determine n, the ratio between the number of electrons transferred and the stoichiometric coefficients for the chemical reagents.

Equation 19-2 applies under all conditions, including standard conditions. At standard conditions, $\Delta G = \Delta G°$ and $E = E°$:

$$\Delta G° = -nFE° \qquad\qquad \textbf{(19-3)}$$

Equation 19-3 expresses an important link between two standard quantities. The equation lets us calculate standard electrical potentials from tabulated values for standard free energies. Equally important, accurate potential measurements on galvanic cells yield

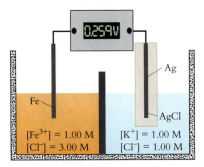

Figure 19-16
A galvanic cell can be constructed from a silver–silver chloride electrode in contact with a solution containing chloride anions and an iron electrode in contact with a solution containing iron(III) cations.

experimental values for standard potentials that can be used to calculate standard free energy changes for reactions.

The cell shown in Figure 19-16 can serve as an example for calculations using Equation 19-3. One cell contains aqueous 1.00 M iron(III) chloride in contact with an iron metal electrode, and the other cell contains 1.00 M KCl in contact with a silver–silver chloride (AgCl/Ag) electrode. The half-reactions for these electrodes follow:

$$AgCl(s) + e^- \rightleftharpoons Ag(s) + Cl^-(aq) \qquad E° = 0.2223 \text{ V}$$
$$Fe^{3+}(aq) + 3\,e^- \rightleftharpoons Fe(s) \qquad E° = -0.037 \text{ V}$$

In the spontaneous redox reaction under standard conditions, AgCl is reduced and Fe is oxidized:

$$Fe(s) + 3\,AgCl(s) \longrightarrow Fe^{3+}(aq) + 3\,Ag(s) + 3\,Cl^-(aq)$$

$$E° = (0.2223 \text{ V}) - (-0.037 \text{ V}) = 0.259 \text{ V}$$

Equation 19-3 can be used to calculate the standard free energy change for this net reaction, applying the appropriate conversion factors:

$$\Delta G° = -nFE° = -(3)(9.6485 \times 10^4 \text{ C mol}^{-1})(0.259 \text{ V})\left(\frac{1 \text{ J}}{1 \text{ V C}}\right)$$

$$\Delta G° = (-7.50 \times 10^4 \text{ J mol}^{-1})\left(\frac{1 \text{ kJ}}{10^3 \text{ J}}\right) = -75.0 \text{ kJ/mol}$$

Cell Potentials and Chemical Equilibrium

Equation 19-3 links $\Delta G°$ for a reaction with $E°$. From Chapter 16, Equation 16-3 links $\Delta G°$ for a reaction to K_{eq}:

$$\Delta G° = -nFE°$$

$$\Delta G° = -RT \ln K_{eq} \tag{16-3}$$

Combining these two equalities and grouping the constants gives a new equation that links standard potentials directly to equilibrium constants:

$$-RT \ln K_{eq} = -nFE°$$

$$E° = \frac{RT}{nF} \ln K_{eq} \tag{19-4}$$

Many calculations using Equation 19-4 refer to standard temperature, 298.15 K. Furthermore, K_{eq} often has a very large or very small value that is expressed using power-of-ten notation. For such values, calculations using the logarithm to base 10 (log) rather than ln are more convenient: $\log x = 2.302\,585 \ln x$. We can substitute these values and the value for F and then evaluate the multiplier for the log term at standard temperature:

$$\frac{2.302\,585\,RT}{F} = \frac{(2.302\,585)(8.31451 \text{ J/mol K})(298.15 \text{ K})(1 \text{ V C}/1 \text{ J})}{96,485.34 \text{ C/mol}} = 0.059\,159\,68 \text{ V}$$

Although this conversion factor is accurate to seven significant figures, experimental equilibrium constants typically are accurate to no more than three significant figures, so we round the value of the constant to three significant figures:

$$E° = \frac{0.0592 \text{ V}}{n} \log K_{eq} \tag{19-5}$$

Equations 19-4 and 19-5 provide a method for calculating equilibrium constants from tables of standard reduction potentials. Example 19-11 illustrates the technique.

Example 19-11

Cell Potentials and Equilibrium

Use tabulated standard reduction potentials to determine K_{eq} for the following redox reaction:

$$Cu(s) + Br_2(l) \rightleftharpoons Cu^{2+}(aq) + 2\, Br^-(aq)$$

Strategy: This is a quantitative calculation, so it is appropriate to use the seven-step problem-solving strategy. We are asked to determine an equilibrium constant from standard reduction potentials. Visualizing the problem involves breaking the redox reaction into its two half-reactions:

$$Br_2(l) + 2\, e^- \rightleftharpoons 2\, Br^-(aq) \quad and \quad Cu(s) \rightleftharpoons Cu^{2+}(aq) + 2\, e^-$$

Then we use tabulated values and Equation 19-5 to calculate K_{eq}.

Solution:
Table 19-1 provides standard reduction potentials for the two reactions:

$$Br_2(l) + 2\, e^- \rightleftharpoons 2\, Br^-(aq) \qquad E^\circ = 1.066\ V$$

$$Cu^{2+}(aq) + 2\, e^- \rightleftharpoons Cu(s) \qquad E^\circ = 0.3419\ V$$

The link between standard reduction potentials and the equilibrium constant is Equation 19-5.

$$E^\circ = \frac{0.0592\ V}{n} \log K_{eq}$$

To obtain the cell potential, subtract the less positive standard reduction potential from the more positive standard reduction potential. The spontaneous direction is reduction of Br_2 and oxidation of Cu.

$$E^\circ = E^\circ_{cathode} - E^\circ_{anode} = (1.066\ V) - (0.3419\ V) = 0.724\ V$$

Now use Equation 19-5 with the appropriate value of n, found from the half-reactions:

$$E^\circ = \frac{0.0592\ V}{n} \log K_{eq}$$

$$\log K_{eq} = \frac{nE^\circ}{0.0592\ V} = \frac{(2)(0.724\ V)}{0.0592\ V} = 24.5$$

$$K_{eq} = 10^{24.5} = 3 \times 10^{24}$$

Does the Result Make Sense? The magnitude of this equilibrium constant indicates that the redox reaction goes essentially to completion. This reflects the fact that bromine is a potent oxidizing agent and copper is relatively easy to oxidize.

Use values in Table 19-1 to calculate the equilibrium constant for the redox reaction $Pb(s) + Sn^{2+}(aq) \rightleftharpoons Pb^{2+}(aq) + Sn(s)$

The Nernst Equation

In most laboratories, electrochemistry is practiced under nonstandard conditions. That is, concentrations of dissolved solutes often are not 1 M, and gases are not necessarily at 1 bar. Recall from Chapter 14 that ΔG changes with concentration and pressure. The equation that links ΔG° with free energy changes under nonstandard conditions is Equation 14-13:

$$\Delta G = \Delta G^\circ + RT \ln Q \qquad (14\text{-}13)$$

Here, Q is the reaction quotient.

We can see how concentration and pressure affect cell potentials by substituting for ΔG and ΔG° using Equations 19-2 and 19-3.

$$\Delta G = -nFE \quad and \quad \Delta G^\circ = -nFE^\circ$$

$$-nFE = -nFE^\circ + RT \ln Q$$

Dividing both sides by $-nF$ gives the **Nernst equation:**

$$E = E° - \frac{RT}{nF} \ln Q \qquad \text{(19-6)}$$

The Nernst equation is used to convert between standard cell potentials and potentials of electrochemical cells operating under nonstandard concentration conditions.

When measurements are made at standard temperature, 298 K, the numerical value calculated earlier can replace RT/F:

$$E = E° - \frac{0.0592 \text{ V}}{n} \log Q \qquad when \qquad T = 298 \text{ K}$$

Recall that the numerator of Q contains concentrations of products, and the denominator contains concentrations of reactants, all raised to powers equal to their stoichiometric coefficients. Solvents and pure solids and liquids do not appear in the concentration quotient. Thus, only solutes and gases appearing in the cell reaction affect its cell potential. Nevertheless, most cell reactions involve solutes, so a typical cell potential differs from $E°$. Example 19-12 provides an illustration.

Example 19-12

Nonstandard Cell Conditions

The permanganate ion is a powerful oxidizing agent that oxidizes water to oxygen under standard conditions. Here are the half-cell reactions:

$$MnO_4^-(aq) + 8 H_3O^+(aq) + 5 e^- \rightleftharpoons Mn^{2+}(aq) + 12 H_2O \qquad E° = 1.507 \text{ V}$$

$$O_2(g) + 4 H_3O^+(aq) + 4 e^- \rightleftharpoons 6 H_2O \qquad E° = 1.229 \text{ V}$$

What is the potential of a permanganate–oxygen cell operating at pH = 7.00, oxygen at $p = 0.200$ bar, 0.100 M MnO_4^-, and 0.100 M Mn^{2+}?

Strategy: This is a quantitative problem, so we follow the standard strategy. The problem asks about an actual potential under nonstandard conditions. Before we determine the potential, we must visualize the electrochemical cell and determine the balanced chemical reaction. The half-reactions are given in the problem. To obtain the balanced equation, reverse the direction of the reduction half-reaction with the less positive standard potential, balance the electrons, and combine the two half-reactions:

$$4 MnO_4^-(aq) + 32 H_3O^+(aq) + 20 e^- \longrightarrow 4 Mn^{2+}(aq) + 48 H_2O$$

$$30 H_2O \longrightarrow 5 O_2(g) + 20 H_3O^+(aq) + 20 e^-$$

$$4 MnO_4^-(aq) + 12 H_3O^+(aq) \longrightarrow 4 Mn^{2+}(aq) + 5 O_2(g) + 18 H_2O$$

Solution:

The data are the $E°$ values for the half-reactions, which are given in the problem. Under standard conditions, permanganate is reduced to Mn^{2+} at the cathode, and water is oxidized at the anode:

$$E°_{cell} = E°_{cathode} - E°_{anode} = 1.507 \text{ V} - 1.229 \text{ V} = 0.278 \text{ V}$$

The cell operates under nonstandard conditions, so use the Nernst equation.

$$E = E° - \frac{0.0592 \text{ V}}{n} \log Q$$

The form of the reaction quotient comes from the balanced overall redox equation.

$$Q = \frac{[p_{O_2}]^5 [Mn^{2+}]^4}{[MnO_4^-]^4 [H_3O^+]^{12}}$$

Use the concentrations to evaluate Q:

$$Q = \frac{[p_{O_2}]^5 [Mn^{2+}]^4}{[MnO_4^-]^4 [H_3O^+]^{12}} = \frac{(0.200)^5 (0.100)^4}{(0.100)^4 (1.00 \times 10^{-7})^{12}} = 3.20 \times 10^{80}$$

$$\log Q = \log (3.20 \times 10^{80}) = 80.505$$

Substitute into the Nernst equation to find the cell potential. The balanced half-reactions show that $n = 20$:

$$E = E° - \frac{0.0592 \text{ V}}{n} \log Q = 0.278 \text{ V} - \left(\frac{0.0592 \text{V}}{20}\right)(80.505) = 0.040 \text{ V}$$

Does the Result Make Sense? The result has three decimal places because the $E°$ value is accurate to three decimal places, and the correction is a subtraction. The low concentration of H_3O^+ in pure water reduces this cell potential nearly to zero, but the reaction is still spontaneous, so permanganate can oxidize water. Solutions of potassium permanganate slowly deteriorate and cannot be stored for long times. The reaction is slow enough, however, that significant oxidation does not occur over days or weeks.

Determine E for the reaction of Pb with Sn^{2+} described in Extra Practice Exercise 19.11, if $[Sn^{2+}] = 0.255$ M and $[Pb^{2+}] = 4.75 \times 10^{-4}$ M.

Example 19-12

Nonstandard Cell Conditions (*continued*)

Extra Practice Exercise 19.12

The pH Meter

As the Nernst equation describes, cell potentials are linked quantitatively to concentrations. One practical consequence of this relationship is that potential measurements can be used to determine concentrations of ions in solution. The most common example is the **pH meter,** which scientists use to measure the acidity of aqueous solutions. Figure 19-17a illustrates this device: A probe is dipped into the sample, and the pH appears on a digital display. A pH meter measures the electrical potential difference between the sample and a reference solution containing acid at known concentration.

Contemporary pH meters use single probes that contain two reference electrodes, shown diagrammatically in Figure 19-17b. One electrode contains a buffer solution of known pH. A glass membrane separates this buffer solution from the solution whose pH is to be measured, so this electrode is called a glass electrode. Because hydronium ions participate in the cell reaction of the glass electrode, the overall cell potential depends on the hydronium ion concentration in the solution whose pH is being measured.

Glass pH electrodes are simple to use and maintain. They respond selectively to hydronium ion concentration and provide accurate measurements of pH values between about 0 and 10. They can be small enough to be implanted into blood vessels or even inserted into individual living cells. In precision work, these electrodes are calibrated before each use, because their characteristics change somewhat with time and exposure to solutions. The electrode is dipped into a buffer solution of known pH, and the meter is electronically adjusted until it reads the correct value.

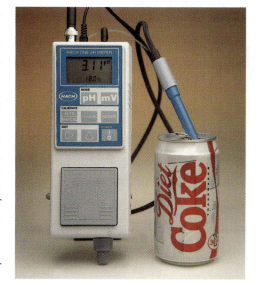

(a)

Electrochemical Stoichiometry

The coefficients of any balanced redox equation describe the stoichiometric ratios between chemical species, just as for other balanced chemical equations. Additionally, in redox reactions we can relate moles of chemical change to moles of electrons. Because electrons always cancel in a balanced redox equation, however, we need to look at half-reactions to determine the stoichiometric coefficients for the electrons. A balanced half-reaction provides the stoichiometric coefficients needed to compute the number of moles of electrons transferred for every mole of reagent.

Electricity is normally measured in units of charge, the coulomb (C), or as rate of electrical current flow, the ampere (A; 1 A = 1 C/s). The total amount of charge is the product of the current flow, symbolized by I, and the time for which this current flows:

$$\text{Charge} = It$$

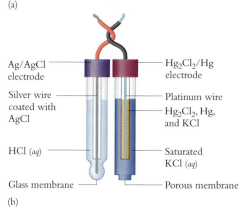

Ag/AgCl electrode

Silver wire coated with AgCl

HCl (*aq*)

Glass membrane

(b)

Hg₂Cl₂/Hg electrode

Platinum wire

Hg₂Cl₂, Hg, and KCl

Saturated KCl (*aq*)

Porous membrane

Figure 19-17
(a) Modern pH meters are compact digital electronic devices. (b) The probe consists of two electrodes whose compositions are shown in the diagram.

Just as molar mass provides the link between mass and moles, the Faraday constant provides the link between charge and moles. The number of moles of electrons transferred in a specific amount of time is the charge in coulombs divided by the charge per mole, F:

$$\text{Moles of electrons} = n = \frac{It}{F} \qquad (19\text{-}7)$$

$I = $ Current (C s^{-1}), $t = $ Time(s), and $F = 96{,}485$ C mol^{-1}

Here n refers to moles of electrons reacting and has units, whereas in Equations 19-3 through 19-6, n is a dimensionless ratio that expresses the coefficient for the electrons involved in a balanced chemical reaction.

Equation 19-7 links the stoichiometry of a redox reaction with the characteristics of an electrochemical cell, and Example 19-13 shows how to apply this equation.

Example 19-13	Automobile headlights typically draw 5.9 A of current. The galvanic cell of a lead storage battery, described in Example 19-8, consumes Pb and PbO_2 as it operates. A typical electrode contains about 250 g of PbO_2. Assuming that the battery can supply 5.9 A of current until all the PbO_2 has been consumed, how long will it take for a battery to run down if the lights are left on after the engine is turned off?

Electron Stoichiometry

Strategy: This is an electrochemical stoichiometry problem, in which an amount of a chemical substance is consumed as electrical current flows. We use the seven-step strategy in summary form. The question asks how long the battery can continue to supply current. Current flows as long as there is lead(IV) oxide present to accept electrons, and the battery dies when all the lead(IV) oxide is consumed. We need to have a balanced half-reaction to provide the stoichiometric relationship between moles of electrons and moles of PbO_2.

Solution: There are 250 g of PbO_2, and the headlights draw 5.9 A of current. Equation 19-7 links current with moles of electrons. Moles of electrons and moles of PbO_2 are related as described by the balanced half-reaction, determined in Example 19-8:

$$PbO_2\,(s) + HSO_4^-\,(aq) + 3\,H_3O^+\,(aq) + 2\,e^- \longrightarrow PbSO_4\,(s) + 5\,H_2O\,(l)$$

This gives us the stoichiometric ratio needed to relate moles of electrons to moles of chemical substance:

$$n_{e^-} = n_{PbO_2}\left(\frac{2 \text{ mol } e^-}{1 \text{ mol } PbO_2}\right)$$

Use the molar mass of PbO_2 to convert mass of PbO_2 to moles:

$$n_{PbO_2} = \frac{m_{PbO_2}}{MM_{PbO_2}} = \frac{250 \text{ g}}{239.2 \text{ g/mol}} = 1.05 \text{ mol } PbO_2$$

$$n_{e^-} = \left(\frac{2 \text{ mol } e^-}{1 \text{ mol } PbO_2}\right)(1.05 \text{ mol } PbO_2) = 2.10 \text{ mol electrons}$$

We rearrange Equation 19-7, which provides the link between moles of electrons and the time it takes for the battery to become discharged:

$$n_{e^-} = \frac{It}{F} \qquad or \qquad t = \frac{n_{e^-}F}{I}$$

$$t = \frac{(2.10 \text{ mol})(96{,}485 \text{ C/mol})}{(5.9 \text{ C/s})(60 \text{ s/min})(60 \text{ min/hr})} = 9.5 \text{ hours}$$

This is the maximum time that the battery could supply current. In practice, the battery runs down before all the PbO_2 is consumed, because the voltage drops as the battery operates.

Does the Result Make Sense? If you leave your automobile headlights on for several hours when the engine is off, you will have a dead battery, as this calculation predicts. When an automobile engine is running, however, the engine drives an electrical alternator that provides the current required for the headlights and other needs.

The reaction in a nickel–cadmium battery is

$$2 \, NiO(OH) + 2 \, H_2O + Cd \longrightarrow 2 \, Ni(OH)_2 + Cd(OH)_2$$

If a battery contains 0.355 g of Cd and excess $NiO(OH)$, how long can it operate if it draws 1.25 mA of current?

Example 19-13

Electron Stoichiometry (*continued*)

Extra Practice Exercise 19.13

Redox reactions may involve solids, solutes, gases, or charge flows. Consequently, you must be prepared for all the various conversions from molar amounts to measurable variables. As a reminder, Table 19-2 lists the four relationships used for mole calculations.

Galvanic cells use redox reactions to generate electrical current. Electrical current can also drive redox reactions, and the same stoichiometric relationships apply to such processes, as we describe in Section 19.7.

Table 19-2   Relationships Used to Calculate Moles

Type of Material	Equation	Section Reference
Pure substance	$n = m/MM$	Sections 2.4 and 3.4
Solutes	$n = MV$	Section 3.6
Gases	$n = PV/RT$	Section 5.6
Electrons	$n = It/F$	Section 19.5

Section Exercises

19.5.1 Using standard reduction potentials, determine $\Delta G°$ and K_{eq} at 25 °C for each of the following reactions:
(a) $2 \, Cu(s) + Hg^{2+}(aq) \rightleftharpoons$
$$2 \, Cu^+(aq) + Hg(l)$$
(b) $4 \, Cr^{3+}(aq) + 21 \, H_2O(l) \rightleftharpoons$
$$2 \, Cr(s) + Cr_2O_7^{2-}(aq) + 14 \, H_3O^+(aq)$$

19.5.2 Use the Nernst equation to determine the potential at 298 K for each of the cells described in Section Exercise 19.4.1, if all cation concentrations are 2.5×10^{-2} M rather than 1.00 M: (a) Fe^{2+} and Cr^{3+}; (b) Cu^{2+} and Pb^{2+}; and (c) Au^{3+} and Ag^+.

19.5.3 In one form of lithium battery, the spontaneous cell reaction is

$$4 \, Li + FeS_2 \longrightarrow Fe + 2 \, Li_2S$$

Suppose that a lithium battery contains 250. mg each of Li and FeS_2. (a) Which reagent will be consumed first? (b) How many coulombs of charge can the battery deliver before this reagent is consumed? (c) What is the lifetime of the battery (in days) if it powers a cell phone requiring 1.50 mA of current?

19.6 REDOX IN ACTION

Electrochemical reactions have many practical applications. Some are spontaneous, and others are driven "uphill" by applying an external potential. In this section, we present practical examples of spontaneous redox processes. We describe externally driven redox reactions in Section 19.7.

Batteries power hearing aids as well as automobiles.

Batteries

A battery is a galvanic cell that generates electrical current to power a practical device. Batteries can be as small as the "buttons" that power cameras and hearing aids or large charge storage banks like those of electric automobiles.

A battery must use cell reactions that generate and maintain a large electrical potential difference. This requires two half-reactions with substantially different standard reduction potentials. The ideal battery would be compact, inexpensive, rechargeable, and environmentally safe. This is a stringent set of requirements. No battery meets all of them, and only a few come close.

A flashlight battery usually is an **alkaline dry cell,** so called because it is basic (alkaline) but contains a minimum amount of water (dry). In these batteries, zinc and manganese dioxide are the working materials (see schematic diagram in Figure 19-18). The dry cell has a zinc anode in contact with a moist paste of MnO_2, KOH, and graphite powder. The paste contacts a passive graphite cathode. Both half-reactions involve multiple steps, but the chemistry can be approximated by single half-reactions:

Cathode: $2\ MnO_2(s) + H_2O(l) + 2\ e^- \longrightarrow Mn_2O_3(s) + 2\ OH^-(paste)$

Anode: $Zn(s) + 2\ OH^-(paste) \longrightarrow Zn(OH)_2(s) + 2\ e^-$

Net: $2\ MnO_2(s) + H_2O(l) + Zn(s) \longrightarrow Mn_2O_3(s) + Zn(OH)_2(s)$

A dry cell generates a potential of about 1.5 V. These cells run irreversibly, and their cell potential slowly decreases with time. Nevertheless, dry cells have many uses because they are compact and made of inexpensive materials with low toxicities. Approximately a billion of these batteries are produced annually in the United States.

The **Zinc–air battery** is more expensive than the dry cell and deteriorates relatively quickly once it is exposed to air. High capacity and a cell potential that does not vary with use offset these disadvantages. Like the dry cell, a zinc–air battery uses zinc for the anode reaction. Uniquely among batteries in common use, this battery relies on molecular oxygen from the atmosphere for its cathode reaction.

Cathode: $O_2(g) + 2\ H_2O(l) + 4\ e^- \longrightarrow 4\ OH^-(aq)$

Anode: $2\ Zn(s) + 4\ OH^-(aq) \longrightarrow 2\ ZnO(s) + 2\ H_2O(l) + 4\ e^-$

Net: $O_2(g) + 2\ Zn(s) \longrightarrow 2\ ZnO(s)$

Figure 19-19 shows the molecular processes that occur in a zinc–air battery. In this cell, none of the components of the redox reaction changes concentration as the battery operates. Both zinc-containing substances are solids, and atmospheric air provides a constant supply of O_2 gas through a small hole in the battery casing. As the balanced redox reaction indicates, there is no net change in concentration of hydroxide ions, which are consumed at the cathode but regenerated at the anode. The aqueous zinc paste allows free diffusion of hydroxide ions from the anode to the cathode. Because none of the concentrations change as the battery functions, this battery delivers a constant voltage of approximately 1.4 V throughout its lifetime. Zinc–air batteries are used widely in devices that require constant voltage, including watches, calculators, hearing aids, and cameras.

The **lead storage battery** provides electrical power in automobiles. It is well suited for this use because it supplies the large current needed to drive starter motors and headlights and can be recharged easily. Figure 19-20 shows the lead storage cell in a schematic view. The half-reactions are the subject of Example 19-8:

$PbO_2(s) + HSO_4^-(aq) + 3\ H_3O^+(aq) + 2\ e^- \longrightarrow PbSO_4(s) + 5\ H_2O(l)$
$$E^\circ = +1.6913\ V$$

$Pb(s) + HSO_4^-(aq) + H_2O(l) \longrightarrow PbSO_4(s) + H_3O^+(aq) + 2\ e^- \quad E^\circ = -0.3588\ V$

Here is the net reaction:

$Pb(s) + 2\ HSO_4^-(aq) + PbO_2(s) + 2\ H_3O^+(aq) \longrightarrow 2\ PbSO_4(s) + 4\ H_2O(l)$
$$E^\circ = (1.6913) - (-0.3588) = 2.0501\ V$$

Figure 19-18
Schematic diagram and examples of alkaline dry cells. These batteries provide electrical current for many portable devices.

+

Graphite cathode

Paste of MnO_2, KOH, and graphite powder

Porous separator

Zinc anode

−

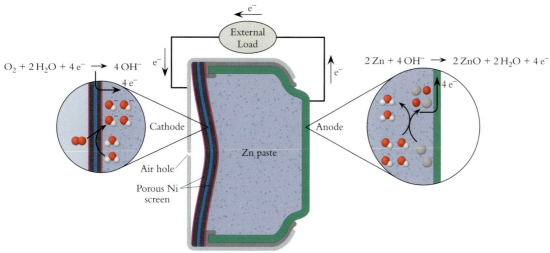

Figure 19-19
In a zinc–air battery, zinc is oxidized and molecular oxygen is reduced, but no net change occurs in the concentrations of any species in solution. The migration of OH^- through the zinc paste carries current and maintains a uniform concentration.

The anodes are lead plates that become coated with $PbSO_4$ as the battery discharges. The cathodes are lead containing PbO_2, which also become coated with $PbSO_4$ as the battery discharges. Both sets of electrodes are immersed in aqueous sulfuric acid that is about 5 M. Although the concentrations are not standard, the working potential of the cell is close to its $E°$ value of about 2 V. Automobile batteries use six such cells, connected in series to generate a total electrical potential of 12 V. This battery does not require a porous separator between its electrodes because the starting materials and products for both half-reactions are insoluble in sulfuric acid. The lead-containing materials never contact each other, so there is no possibility of direct electron transfer.

Any battery has a finite life. When its chemicals are consumed, it no longer generates electricity and the battery is "dead." The lead storage battery can be recharged by using electricity to run its reactions in the opposite direction. Lead sulfate produced by the redox reaction adheres to the electrodes of a lead storage battery, so the reactions are easily reversed. While the auto is running, its motor drives an electrical generator, known as an alternator, that meets the vehicle's electrical needs. The alternator also recharges the battery by supplying the energy needed to drive its redox reactions in the reverse direction (see Section 19.7). When the electrical system is functioning properly,

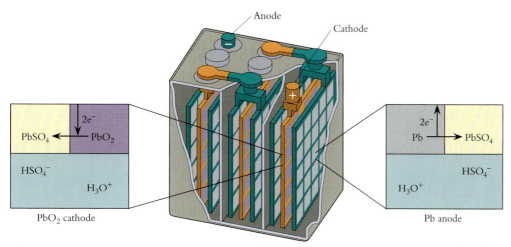

Figure 19-20
Schematic view of a 6-V lead storage battery. Lead is oxidized and lead(IV)oxide reduced during the operation of this battery, which contains three cells connected in series.

the battery is maintained at its optimum charge level, giving these batteries long lifetimes. This process does not work for dry cells or zinc–air batteries, because applying a reverse voltage to these batteries causes other reactions to occur.

The **nickel–cadmium battery** is another common battery that can be recharged. As described in Example 19-10, nickel and cadmium are the working substances in this battery:

Cathode: $NiO(OH)(s) + H_2O(l) + e^- \longrightarrow Ni(OH)_2(s) + OH^-(aq)$

Anode: $Cd(s) + 2\,OH^-(aq) \longrightarrow Cd(OH)_2(s) + 2\,e^-$

Net: $2\,NiO(OH)(s) + Cd(s) + 2\,H_2O(l) \longrightarrow 2\,Ni(OH)_2(s) + Cd(OH)_2(s)$

$$E = E° = 1.35\ V$$

Nickel–cadmium batteries are rechargeable because the nickel and cadmium hydroxides products adhere tightly to the electrodes. More than 1.5 billion Ni–Cd batteries are produced every year. These batteries are not without drawbacks, however. Cadmium is a toxic heavy metal that adds weight to the battery. Moreover, even rechargeable batteries degrade eventually and must be recycled or disposed of in a benign way.

Battery technology continues to advance at a steady pace. Lithium batteries and nickel-metal-hydride batteries are now commonplace. These new rechargeable batteries eliminate the need for toxic cadmium and store more energy per unit mass. The detailed chemistry that underlies the newest advances in battery technology involves principles that are beyond the scope of an introductory course.

Another recent development in batteries is the fuel cell, which is described in our Chemistry and Technology Box.

Corrosion

Corrosion is a natural redox process that returns refined metals to their more stable metal oxides. The oxidizing agent in corrosion chemistry is atmospheric oxygen. Oxygen is a potent oxidizing agent, particularly in the presence of aqueous acids:

$$O_2(g) + 2\,H_2O(l) + 4\,e^- \rightleftharpoons 4\,OH^-(aq) \qquad E° = +0.401\ V$$

$$O_2(g) + 4\,H_3O^+(aq) + 4\,e^- \rightleftharpoons 6\,H_2O(l) \qquad E° = +1.229\ V$$

Atmospheric O_2 has a partial pressure of 0.20 bar, and atmospheric water vapor is saturated with carbon dioxide. This dissolved CO_2 forms carbonic acid, which generates a hydronium ion concentration of about 2.0×10^{-6} M. The Nernst equation allows calculation of the half-cell potential for the reduction of $O_2(g)$ under these conditions:

$$E = E° - \frac{0.0592\ V}{n} \log\left(\frac{1}{p_{O_2}[H_3O^+_{(aq)}]^4}\right)$$

$$E = 1.229\ V - \frac{0.0592\ V}{4} \log\left(\frac{1}{(0.20)(2.0 \times 10^{-6})^4}\right) = 0.88\ V$$

Consequently, common terrestrial conditions promote corrosion.

In damp air, materials with standard reduction potentials less than 0.88 V oxidize spontaneously. Atmospheric O_2 easily oxidizes iron and aluminum, the most important structural metals:

$$4\,Fe(s) + 3\,O_2(g) + 12\,H_3O^+(aq) \longrightarrow 4\,Fe^{3+}(aq) + 18\,H_2O(l)$$

$$E = (0.88\ V) - (-0.037\ V) = +0.92\ V$$

$$4\,Al(s) + 3\,O_2(g) + 12\,H_3O^+(aq) \longrightarrow 4\,Al^{3+}(aq) + 18\,H_2O(l)$$

$$E = (0.88\ V) - (-1.662\ V) = +2.54\ V$$

Box 19-2	**Chemistry and Technology: Fuel Cells**

Imagine an automobile that runs in silence and without polluting emissions. Such an automobile, long a dream of the environmentally conscious, has recently become a reality. The power source is a fuel cell, an electrochemical cell that uses a combustion reaction to produce electricity. Hydrocarbons such as natural gas and propane can be used in fuel cells, but the "cleanest" fuel is molecular hydrogen.

The advantage of a fuel cell over a conventional battery is that the fuel for electrical power can be replenished easily. Just as we pull into a service station to refill the gas tank, owners of automobiles powered by fuel cells will refill their fuel tanks with hydrogen or butane.

Although the principle was first proposed in 1839, making a practical fuel cell eluded scientists for a century and a half. The concept is simple, but the chemistry is difficult. A hydrogen fuel cell must cleanly convert H_2 into H_3O^+ at one electrode and cleanly convert O_2 into OH^- at the other electrode. In addition, the fuel cell must contain a medium that allows these ions to diffuse and combine stoichiometrically.

In a hydrogen fuel cell, oxidation of H_2 at the anode releases electrons into the circuit and produces aqueous H_3O^+. Reduction of O_2 at the cathode consumes electrons and generates OH^-, which combines with H_3O^+ to produce H_2O. The schematic diagram shows these processes.

Anode: $\quad\quad 2\,H_2(g) + 4\,H_2O(l) \longrightarrow 4\,H_3O^+(aq) + 4\,e^-$

Cathode: $\quad\quad O_2(g) + 2\,H_2O(l) + 4\,e^- \longrightarrow 4\,OH^-(aq)$

Net reaction: $\quad\quad 2\,H_2(g) + O_2(g) \longrightarrow 2\,H_2O(l)$

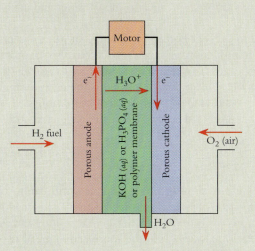

A hydrogen fuel cell is environmentally friendly, but H_2 is much more difficult to store than liquid fuels. The production, distribution, and storage of hydrogen present major difficulties, so researchers are working on fuel cells that use liquid hydrocarbon fuels. One such fuel cell is composed of layers of yttria-stabilized zirconia (YSZ), which is solid ZrO_2 containing around 5% Y_2O_3. This cell uses the combustion of a hydrocarbon as its energy source:

Anode: $\quad\quad 2\,(C_4H_{10} + 13\,O^{2-} \longrightarrow 4\,CO_2 + 5\,H_2O + 26\,e^-)$

Cathode: $\quad\quad 13\,(O_2 + 4\,e^- \longrightarrow 2\,O^{2-})$

Net reaction: $\quad\quad 2\,C_4H_{10} + 13\,O_2 \longrightarrow 8\,CO_2 + 10\,H_2O$

As the reaction shows, a hydrocarbon fuel cell produces carbon dioxide, a "greenhouse" gas. These fuel cells would nevertheless be less polluting than internal-combustion engines, because fuel cells are more efficient and do not generate polluting byproducts such as CO and NO_x.

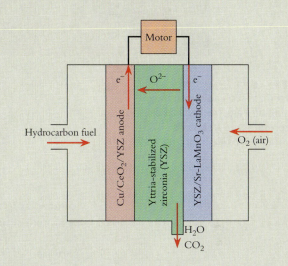

Alkali fuel cells containing KOH and platinum- and gold-coated electrodes were developed for the space program, but these are too expensive for down-to-earth vehicles. In addition, these cells require pure oxygen rather than CO_2-containing air.

Two new technologies have reduced the cost of alkali fuel cells to the point where a European company markets taxis that use them. One is the use of CO_2 "scrubbers" to purify the air supply, making it possible to use atmospheric O_2 rather than purified oxygen. The other is the development of ultrathin films of platinum so that a tiny mass of this expensive metal can provide the catalytic surface area needed for efficient fuel-cell operation.

Automobile manufacturers have invested heavily in fuel cells. Buses powered by fuel cells are on the road, and prototype cars are being tested. The chances are excellent that there is a fuel-cell car in your future.

Oxidation: $Fe \longrightarrow Fe^{2+} + 2\,e^-$

Reduction: $O_2 + 4\,H_3O^+(aq) + 4\,e^- \longrightarrow 6\,H_2O$

Further Redox: $2\,Fe^{2+} + 6\,H_2O + \frac{1}{2}\,O_2 \longrightarrow Fe_2O_3 + 4\,H_3O^+$

Figure 19-21

A water droplet on an iron surface is a miniature electrochemical cell that corrodes iron metal and generates iron oxide (rust).

Both Fe^{3+} and Al^{3+} form insoluble oxides that precipitate from solution. Overall, then, O_2 oxidizes each metal to give its metal oxide:

$$4\,Fe\,(s) + 3\,O_2\,(g) \longrightarrow 2\,Fe_2O_3\,(s)$$

$$4\,Al\,(s) + 3\,O_2\,(g) \longrightarrow 2\,Al_2O_3\,(s)$$

The formation of rust on a steel surface, shown in our chapter opener illustrations, is an obvious manifestation of corrosion. Even more damaging effects of corrosion take place beneath the surface of the metal, where tiny cracks weaken the metal.

The corrosion of iron occurs particularly rapidly when an aqueous solution is present. This is because water that contains ions provides an oxidation pathway with an activation energy that is much lower than the activation energy for the direct reaction of iron with oxygen gas. As illustrated schematically in Figure 19-21, oxidation and reduction occur at different locations on the metal surface. In the absence of dissolved ions to act as charge carriers, a complete electrical circuit is missing, so the redox reaction is slow. In contrast, when dissolved ions are present, such as in salt water and acidic water, corrosion can be quite rapid.

Various strategies are employed to prevent corrosion. The use of paint as a protective coating is described in our chapter introduction. A metal surface can also be protected by coating it with a thin film of a second metal. When the second metal is easier to oxidize than the first, the process is *galvanization*. Objects made of iron, including automobile bodies and steel girders, are dipped in molten zinc to provide sacrificial coatings. If a scratch penetrates the zinc film, the iron is still protected because zinc oxidizes preferentially:

$$Fe^{2+} + 2\,e^- \rightleftharpoons Fe \qquad E° = -0.447 \text{ V}$$

$$Zn^{2+} + 2\,e^- \rightleftharpoons Zn \qquad E° = -0.7618 \text{ V}$$

Many iron objects, such as the hulls of ships or oil drilling platforms, are too large to dip in pots of molten zinc. Instead, these objects are protected from corrosion by connecting them to blocks of some more easily oxidized metal. Magnesium and zinc frequently are used for this purpose. As in galvanization, the more active metal is oxidized preferentially. The sacrificial metal must be replaced as it is consumed, but the cost of replacing a block of zinc or magnesium is minuscule compared with the cost of replacing a sophisticated iron structure.

Another way to protect a metal uses an impervious metal oxide layer. This process is known as *passivation*. In some cases, passivation is a natural process. Aluminum oxidizes readily in air, but the result of oxidation is a thin protective layer of Al_2O_3 through which O_2 cannot readily penetrate. Aluminum oxide adheres to the surface of unoxidized aluminum, protecting the metal from further reaction with O_2. Passivation is not effective for iron, because iron oxide is porous and does not adhere well to the metal. Rust continually flakes off the surface of the metal, exposing fresh iron to the atmosphere. Alloying iron with nickel or chromium, whose oxides adhere well to metal surfaces, can be used to prevent corrosion. For example, stainless steel contains as much as 17% chromium and 10% nickel, whose oxides adhere to the metal surface and prevent corrosion.

Section Exercises

■ **19.6.1** Mercury batteries have the following half-cell reactions:

Cathode:
$$HgO(s) + H_2O(l) + 2 e^- \longrightarrow Hg(l) + 2 OH^-(aq)$$

Anode:
$$Zn(s) + 2 OH^-(aq) \longrightarrow ZnO(s) + H_2O(l) + 2e^-$$

Draw molecular pictures to illustrate these processes.

■ **19.6.2** In Section Exercise 19.3.1, an aluminum−iron galvanic cell was proposed. Using information presented in this section, explain why this cell is unlikely to make a good battery.

■ **19.6.3** "Tin" cans are made of iron, coated with a thin film of tin. After a break occurs in the film, a tin can corrodes much more rapidly than zinc-coated iron. Using standard reduction potentials, explain this behavior.

19.7 ELECTROLYSIS

In a galvanic cell, a spontaneous chemical reaction generates an electrical current. It is also possible to use an electrical current to drive a nonspontaneous chemical reaction. The recharging of a dead battery uses an external electrical current to drive the battery reaction in the reverse, or "uphill," direction.

An electrochemical reaction can be driven in the nonspontaneous direction only if the external electrical power source generates a larger potential than the potential developed by the cell. Compare this relationship with the flow of water under the influence of gravity. The spontaneous direction for the flow of water is "downhill," from higher to lower gravitational potential. Nevertheless, we can pump water uphill—in the "nonspontaneous" direction—if we exert an opposing force that is greater than the force of gravity. Similarly, we can reverse the direction of a galvanic cell by imposing an external potential to drive electrons "uphill." In the presence of an opposing voltage that is greater than its cell potential, the cell reaction is reversed. This is an **electrolytic cell** rather than a galvanic cell. The process of using electrical current to drive redox reactions is **electrolysis.**

Electrolysis is illustrated by the lead storage battery. The electrolytic (recharging) process is shown in Figure 19-22. As a galvanic cell, this battery oxidizes lead metal and reduces lead(IV) oxide spontaneously, generating an electrical potential of 2.05 V. Electrons flow from the Pb electrode to the PbO_2 electrode. The reaction can be driven in the opposite direction by an external potential of *opposite sign and greater than 2.05 V.* Under these conditions, the external potential drives electrons through the external wire from PbO_2 toward Pb. At the $PbSO_4$−PbO_2 interface, lead sulfate is oxidized to lead oxide:

$$PbSO_4(s) + 5 H_2O(l) \longrightarrow$$
$$PbO_2(s) + HSO_4^-(aq) + 3 H_3O^+(aq) + 2 e^-$$

The electrons released in this reaction are pushed into the lead electrode, and at the $PbSO_4$−Pb interface, lead sulfate captures electrons and is reduced to lead:

$$PbSO_4(s) + H_3O^+(aq) + 2 e^- \longrightarrow Pb(s) + HSO_4^-(aq) + H_2O(l)$$

Whether the cell operates galvanically or electrolytically, the electrode where oxidation occurs is always called the anode. Thus, under recharging conditions the PbO_2 electrode is the anode, and the Pb electrode is the cathode.

Electrolysis of Water

As world deposits of petroleum and coal are exhausted, new sources of hydrogen will have to be developed for use as a

Figure 19-22
A lead storage battery is recharged by forcing electrons to flow in the opposite direction from galvanic operation.

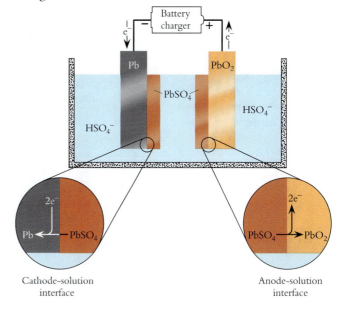

Cathode-solution interface

Anode-solution interface

Figure 19-23
A photograph and a schematic diagram of a cell for the electrolysis of water. The cell generates H_2 gas and O_2 gas in a 2:1 ratio.

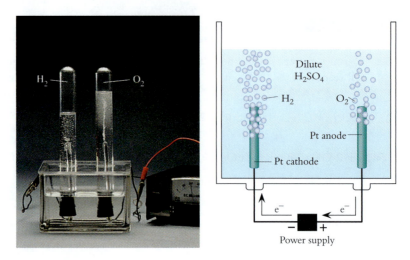

fuel and in the production of ammonia for fertilizer. At present, most hydrogen gas is produced from hydrocarbons, but hydrogen gas can also be generated by the electrolysis of water. Figure 19-23 shows an electrolytic cell set up to decompose water. Two plat-inum electrodes are dipped in a dilute solution of sulfuric acid. The cell requires just one compartment because hydrogen and oxygen escape from the cell much more rapidly than they react with each other.

During the electrolysis of water, hydronium ions capture electrons from the cathode, producing hydrogen gas. Water molecules lose electrons to the anode, producing oxygen gas:

Oxidation at anode: $\qquad$ $6\,H_2O\,(l) \longrightarrow O_2\,(g) + 4\,H_3O^+\,(aq) + 4\,e^-$

Reduction at cathode: $\qquad$ $4\,H_3O^+\,(aq) + 4\,e^- \longrightarrow 2\,H_2\,(g) + 4\,H_2O\,(l)$

Overall: $\qquad$ $2\,H_2O\,(l) \longrightarrow 2\,H_2\,(g) + O_2\,(g)$

The potential supplied to an electrolytic cell determines whether or not electrolysis can occur, but the current flow and the time of electrolysis determine the amount of material electrolyzed. Recall Equation 19-7 from Section 19.5:

$$\text{Moles of electrons} = n = \frac{It}{F}$$

This equation provides the link between electrical measurements and amount of electrons, and the balanced half-reactions for the electrolytic process provide the link between the amount of electrons and amounts of chemical substances. Example 19-14 shows a calculation regarding electrolytic stoichiometry.

Example 19-14
Electrolysis and Stoichiometry

If the electrolytic cell shown in Figure 19-23 draws a current of 0.775 A for 45.0 minutes, calculate the volumes of H_2 and O_2 produced if each gas is collected at 25 °C and $P = 1.00$ atm.

Strategy: "Volumes of H_2 and O_2" identifies this as a stoichiometry problem, for which our seven-step approach, presented here in compact form, is appropriate. The problem asks about amounts of chemicals produced in electrolysis and provides data about current and time. Visualize the electrolysis shown in Figure 19-23. Calculating amounts in electrochemistry requires the use of Equation 19-7 and knowledge of the numbers of electrons transferred in the half-reactions.

Solution: We calculate moles of electrons using Equation 19-7. Recall that 1 A = 1 C/s:

$$n_{\text{electrons}} = \frac{It}{F} = \frac{(0.775\ \text{C/s})(45.0\ \text{min})(60\ \text{s/min})}{96,485\ \text{C/mol}} = 2.169 \times 10^{-2}\ \text{mol}$$

To convert from moles of electrons to moles of chemical species, we need the molar ratios between electrons and chemical species, which we find by examining the half-reactions:

Example 19-14

Electrolysis and Stoichiometry (*continued*)

$$6 H_2O\,(l) \longrightarrow O_2(g) + 4 H_3O^+(aq) + 4\,e^-$$

$$4 H_3O^+(aq) + 4\,e^- \longrightarrow 2 H_2(g) + 4 H_2O\,(l)$$

The transfer of four moles of e^- produces two moles of H_2 and one mole of O_2:

$$n_{H_2} = (2.169 \times 10^{-2}\ \text{mol}\ e^-)\left(\frac{2\ \text{mol}\ H_2}{4\ \text{mol}\ e^-}\right) = 1.084 \times 10^{-2}\ \text{mol}\ H_2$$

$$n_{O_2} = (2.169 \times 10^{-2}\ \text{mol}\ e^-)\left(\frac{1\ \text{mol}\ O_2}{4\ \text{mol}\ e^-}\right) = 0.5423 \times 10^{-2}\ \text{mol}\ O_2$$

To complete the calculation, convert moles of gas into volumes using the ideal gas equation:

$$V = \frac{nRT}{P}$$

We leave this stoichiometric calculation to you. The volume of H_2 is 0.265 L, and that of O_2 is 0.133 L.

Do the Results Make Sense? Notice that the volume of H_2 gas is twice the volume of O_2 gas, as required by the stoichiometry of the overall reaction.

Calculate the mass (in mg) of Li transferred from the cathode to the anode when a lithium ion battery is recharged for 35 minutes using a constant current of 0.155 A.

Extra Practice Exercise 19.14

Competitive Electrolysis

When an electrolytic cell is designed, care must be taken in the selection of the cell components. For example, consider what happens when an aqueous solution of sodium chloride is electrolyzed using platinum electrodes. Platinum is used for passive electrodes, because this metal is resistant to oxidation and does not participate in the redox chemistry of the cell. There are three major species in the solution: H_2O, Na^+, and Cl^-. Chloride ions cannot be reduced further, so there are just two candidates for reduction, water molecules and sodium ions:

$$Na^+(aq) + e^- \rightleftharpoons Na\,(s) \qquad E° = -2.71\ \text{V}$$

$$2 H_2O\,(l) + 2\,e^- \rightleftharpoons H_2(g) + 2\,OH^-(aq) \qquad E° = -0.828\ \text{V}$$

In an electrolytic cell, the most easily oxidized species is oxidized, and the most easily reduced species is reduced. The standard potentials show that water is much easier to reduce than sodium ions, so the electrolysis will produce hydrogen and hydroxide ions rather than sodium metal.

The possibilities for oxidation are more varied in this solution. Water and chloride ions can be oxidized, but sodium ions cannot. The various possibilities for oxidation can be found in Appendix F. Because oxidation is the reverse of reduction, the half-reactions chosen must have *products* that include Cl^- or H_2O. Although several possibilities exist, only the two with the lowest reduction potentials need to be considered:

$$O_2(g) + 4 H_3O^+(aq) + 4\,e^- \rightleftharpoons 6 H_2O\,(l) \qquad E° = 1.229\ \text{V}$$

$$Cl_2(g) + 2\,e^- \rightleftharpoons 2\,Cl^-(aq) \qquad E° = 1.358\ \text{V}$$

In general, we expect that an external potential will drive the pair of reactions whose spontaneous reaction has the least positive cell potential. In this case, combining the reduction of water to hydrogen and hydroxide with the oxidation of water to H_3O^+ and oxygen gas gives the least positive $E°$:

$$2 H_2(g) + O_2(g) \longrightarrow 2 H_2O(l) \qquad E°_{cell} = (1.229 \text{ V}) - (-0.828 \text{ V}) = +2.057 \text{ V}$$

Notice, however, that the cell potential for the reaction involving Cl^- is only slightly larger:

$$H_2(g) + 2 OH^-(aq) + Cl_2(g) \longrightarrow 2 H_2O(l) + 2 Cl^-(aq)$$

$$E°_{cell} = (1.358 \text{ V}) - (-0.828 \text{ V}) = +2.186 \text{ V}$$

Experimentally, when an opposing potential sufficient to cause electrolysis is applied to a cell containing aqueous NaCl, the electrolytic reaction produces chlorine gas rather than oxygen gas at the anode:

$$\text{Cathode:} \qquad 2 H_2O(l) + 2 e^- \longrightarrow H_2(g) + 2 OH^-(aq)$$

$$\text{Anode:} \qquad 2 Cl^-(aq) \longrightarrow Cl_2(g) + 2 e^-$$

The reason is that electrochemical reactions are determined by kinetic considerations as well as thermodynamics. The oxidation of water is favored by thermodynamics, but the oxidation of chloride is a much faster reaction. Notice that the oxidation of water requires four electrons per water molecule, but the oxidation of chloride anions requires only two electrons per chloride ion. When reactions have similar potentials, the one that requires fewer electrons often is favored because it proceeds faster than the one that requires more electrons.

Electroplating

Electrolysis can be used to deposit one metal on top of another, a process known as **electroplating.** Electroplating is done for cosmetic reasons as well as for protection against corrosion and wear. A thin layer of gold or silver is often plated on jewelry and tableware made from inexpensive metals such as iron. Figure 19-24 shows a schematic diagram of an electroplating apparatus. In this example, the spoon acts as a cathode where silver cations are reduced. The anode is a silver rod. An external power source removes electrons from the silver rod, causing oxidation to Ag^+ cations, which enter the solution. Electrons flow into the spoon, where silver ions in solution capture the electrons to become neutral silver atoms. These neutral atoms adhere to the iron surface as a layer of silver metal.

The quantity of metal deposited during a plating process is linked stoichiometrically to the current flow through Equation 19-7, as Example 19-15 illustrates.

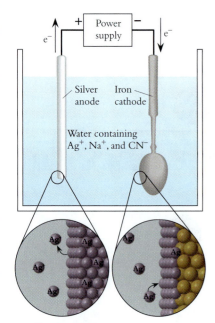

Figure 19-24
The electroplating of silver.

Example 19-15	An electroplating apparatus is used to coat jewelry with gold. What mass of gold can be deposited from a solution that contains $Au(CN)_4^-$ ions if a current of 5.0 A flows for 30.0 min?

Electroplating

Strategy: This, like Example 19-14, is an electrochemical stoichiometry problem. It asks about the amount of gold deposited in an electrolytic reaction. The method is the same as that of Example 19-14.

Solution:

$$n_{electrons} = \frac{It}{F} = \frac{(5.0 \text{ C/s})(30.0 \text{ min})(60 \text{ s/min})}{96{,}485 \text{ C/mol}} = 0.0933 \text{ mol e}^-$$

To determine moles of gold, we need a balanced half-reaction. The reactant is $Au(CN)_4^-$, and the product is Au. Apply the usual procedure for balancing a half-reaction, recognizing the presence of cyanide ions:

$$Au(CN)_4^- + 3 e^- \longrightarrow Au + 4 CN^-$$

The stoichiometric coefficients, 1 and 3, indicate that one mole of gold will be plated for every three moles of electrons:

$$(0.0933 \text{ mol e}^-)\left(\frac{1 \text{ mol Au}}{3 \text{ mol e}^-}\right) = 0.0311 \text{ mol Au}$$

Finally, the molar mass of gold gives the mass of gold in grams:

$$(0.0311 \text{ mol Au})(197 \text{ g/mol}) = 6.1 \text{ g Au}$$

Example 19-15

Electroplating (*continued*)

Does the Result Make Sense? The final result is rounded to two significant figures to match the precision of the current measurement. The amount of gold is significantly less than one mole, consistent with the total charge.

Calculate the mass of silver that will be plated onto a spoon from a plating solution containing $[Ag(NH_3)_2]^+$ if a current of 3.75 A flows for 27.5 min.

**Extra Practice
Exercise 19.15**

Steel objects are often protected from corrosion by electroplating with chromium. The most straightforward process would be to electrolyze a solution of Cr^{3+} cations. This fails because aqueous Cr^{3+} ions are not reduced at a useful rate. Instead, solutions containing chromate anions are used:

$$8 \text{ H}_3\text{O}^+(aq) + \text{CrO}_4{}^{2-}(aq) + 6 \text{ e}^- \longrightarrow \text{Cr}(s) + 12 \text{ H}_2\text{O}(l)$$

This reaction proceeds much faster than the direct reduction of Cr^{3+}. It requires twice as much electrical current, however, since 6 mol of electrons must be supplied to deposit 1 mol of Cr rather than 3 mol for direct electrolysis of Cr^{3+}.

Chromium is plated on the surface of iron to prevent corrosion and to improve appearance.

Section Exercises

■ **19.7.1** Fluorine is manufactured by the electrolysis of hydrogen fluoride dissolved in molten KF:
$2 \text{ HF } (KF \text{ melt}) \rightarrow \text{H}_2(g) + \text{F}_2(g)$
(a) Identify the half-reactions. (The only species present in the molten phase are HF, K^+, and F^-.)
(b) $E° = -0.187$ V for reducing HF to H_2. What is the minimum external potential required to drive the electrolysis of HF at standard concentrations?

■ **19.7.2** If a commercial reactor for the electrolysis of HF operates continuously at a current of 1500 A, how many kilograms of F_2 are produced in 24 hours of operation? What is the volume of the tank required to store the F_2 at $T = 298$ K, $P = 60$ atm?

■ **19.7.3** Explain why neither aqueous KF nor pure liquid HF can be used for the electrolytic production of fluorine, even though both liquids are easier to handle than molten potassium fluoride.

CHAPTER 19 VISUAL SUMMARY

Important equations

Essential terms

Key concepts

SKILLS TO MASTER

① Determining oxidation numbers
② Balancing redox reactions using half-reactions
③ Describing galvanic cells
④ Calculating standard cell potentials
⑤ Relating cell potentials and equilibrium constants

⑥ Doing Nernst equation calculations
⑦ Doing charge–mole conversions
⑧ Describing examples of batteries
⑨ Analyzing electrolytic cells

19.1 RECOGNIZING REDOX REACTIONS

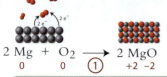

Reducing agent

Oxidizing agent

$$2\ Mg\ +\ O_2\ \longrightarrow\ 2\ MgO$$
$$0\ \ 0\ \ ①\ \ \ +2\ \ -2$$

Oxidation numbers

Oxidation is the loss of electrons from a substance.
Reduction is the gain of electrons by a substance.

Oxidation and reduction always occur together.

19.2 BALANCING REDOX REACTIONS

Electrons lost in oxidation = Electrons gained in reduction

1. Break into half-reactions
2. Balance each half-reaction   Half-reaction
   a. Balance all elements except O and H.
   b. Balance O by adding H_2O.
   c. Balance H by adding H_3O^+ and H_2O (acid), or
      Balance H by adding OH^- and H_2O (base).
   d. Balance charge by adding electrons.
3. Multiply and cancel electrons   ②
4. Recombine and simplify

19.3 GALVANIC CELLS

Galvanic cell

Anode   Cathode   ③

Electrodes

Passive   Active

Voltmeter 0.34V

$[Cl^-]$ = 1.00 M $[Cu^{2+}]$ = 1.00 M
$[H_3O^+]$ = 1.00 M $[SO_4^{2-}]$ = 1.00 M
p_{H_2} = 1.00 atm

19.4 CELL POTENTIAL

The half-reaction with the most negative reduction potential occurs at the anode as oxidation.

The half-reaction with the least negative reduction potential occurs at the cathode as reduction.

$$Zn\ (s)\ +\ Cu^{2+}(aq)\ \rightleftharpoons\ Zn^{2+}(aq)\ +\ Cu\ (s)$$

Anode   Cathode

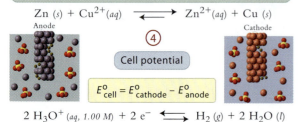

④ Cell potential

$$E^o_{cell} = E^o_{cathode} - E^o_{anode}$$

$$2\ H_3O^+\ (aq,\ 1.00\ M)\ +\ 2\ e^-\ \rightleftharpoons\ H_2\ (g)\ +\ 2\ H_2O\ (l)$$

Standard reduction potential E^o = 0 V (defined)

When a reaction is multiplied by an integer, its cell potential remains unchanged.

19.5 FREE ENERGY AND ELECTROCHEMISRTRY

$$\Delta G = -nFE$$
$$\Delta G^o = -nFE^o$$

$$n_{electrons} = \frac{It}{F}\quad ⑦$$

Faraday constant (F)   Charge-mole equivalence

pH meter

$$E^o = \frac{RT}{nF}\ \ln K_{eq}\quad ⑤$$

$$E^o = \frac{0.0592\ V}{n}\ \log K_{eq}$$

⑥ Nernst equation

$$E = E^o - \frac{RT}{nF}\ \ln Q$$

19.6 REDOX IN ACTION

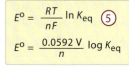

Anode   Cathode

⑧

Alkaline dry cell

Nickel–cadmium battery

Lead storage battery

Zinc–air battery

Corrosion

$$\tfrac{1}{2}O_2\ +\qquad Fe^{2+}$$
$$\qquad\quad Fe^{2+}$$
$$\qquad\qquad\qquad\qquad 6H_2O\ \ 4H_3O^+ + O_2$$

Rust, Fe_2O_3 Fe Fe
Iron surface $2e^-$ $2e^-$ $4e^-$

External Load

e^- e^-

Cathode   Zn paste
Air hole
Porous Ni
screen   Anode

19.7 ELECTROLYSIS

Electroplating   Electrolysis   Electrolytic cell

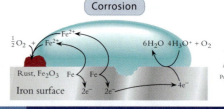

Power supply

e^- e^-

Ag anode
Fe cathode

⑨

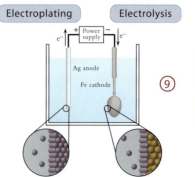

H_2 O_2

Dilute H_2SO_4

Pt anode

Pt cathode

e^- e^-
Power supply

Learning Exercises

19.1 Describe the steps that must be followed in (a) balancing a redox equation; (b) drawing a molecular diagram of a galvanic cell; (c) calculating the potential of a galvanic cell operating under nonstandard conditions; and (d) determining what reactions occur in an electrolytic cell.

19.2 Describe the mechanism for charge movement in each of the following components of an electrochemical cell: (a) the external wire; (b) the electrode–solution interface; (c) the solution; and (d) the porous barrier.

19.3 Write a paragraph explaining the linkages among cell potential, free energy, and the equilibrium constant.

19.4 List at least six practical examples of redox chemistry.

19.5 Update your list of memory bank equations. For each new equation, specify the type of calculations for which it is useful.

19.6 List all terms new to you that appear in Chapter 19, and write a one-sentence definition of each in your own words. Consult the Glossary if you need help.

 Problems **ilw = interactive learning ware problem. Visit the website at www.wiley.com/college/olmsted**

Recognizing Redox Reactions

19.1 Determine the oxidation numbers of all atoms in the following: (a) $Fe(OH)_3$; (b) NF_3; (c) CH_3OH; (d) K_2CO_3; (e) NH_4NO_3; (f) $TiCl_4$; (g) $PbSO_4$; and (h) P_4.

19.2 Determine the oxidation numbers of all atoms in the following: (a) H_2CO; (b) $AlCl_3$; (c) XeF_4; (d) F_2O; (e) $K_2Cr_2O_7$; (f) $NaIO_3$; (g) P_2O_5; and (h) Na_2O_2.

19.3 Which of the following are redox reactions?
(a) $HBr(g) + H_2O(l) \longrightarrow H_3O^+(aq) + Br^-(aq)$
(b) $2\,Fe^{2+}(aq) + H_2O_2(aq) \longrightarrow 2\,Fe^{3+}(aq) + 2\,OH^-(aq)$
(c) $Fe^{2+}(aq) + 2\,OH^-(aq) \longrightarrow Fe(OH)_2(s)$
(d) $2\,CH_3OH(l) + 3\,O_2(g) \longrightarrow 2\,CO_2(g) + 4\,H_2O(l)$
(e) $N_2(g) + 3\,H_2(g) \longrightarrow 2\,NH_3(g)$

19.4 Which of the following are redox reactions?
(a) $CO(g) + H_2O(g) \longrightarrow CO_2(g) + H_2(g)$
(b) $CO_2(g) + 2\,H_2O(l) \longrightarrow H_3O^+(aq) + HCO_3^-(aq)$
(c) $2\,CuS(s) + O_2(g) \longrightarrow 2\,Cu(s) + SO_2(g)$
(d) $2\,AgNO_3(aq) + Cu(s) \longrightarrow Cu(NO_3)_2(aq) + 2\,Ag(s)$
(e) $2\,AgNO_3(aq) + Na_2SO_4(aq) \longrightarrow 2\,NaNO_3(aq) + Ag_2SO_4(s)$

19.5 Chlorine displays a wide range of oxidation numbers. Determine the oxidation number of chlorine in each of the following species: (a) ClF_5; (b) $MgCl_2$; (c) $NaClO_4$; (d) Cl_2; (e) $KClO_2$; and (f) $NaClO$.

19.6 Sulfur displays a wide range of oxidation numbers. Determine the oxidation number of sulfur in each of the following species: (a) H_2S; (b) S_2Cl_2; (c) Li_2SO_4; (d) S_8; (e) Na_2SO_3; and (f) SF_4.

Balancing Redox Reactions

19.7 Determine the half-reactions for the following redox processes: (a) Sodium metal reacts with water to give hydrogen gas. (b) Gold metal dissolves in "aqua regia" (a mixture of HCl and HNO_3) to give $[AuCl_4]^-$ and NO. (c) Acidic potassium permanganate reacts with aqueous $K_2C_2O_4$, producing carbon dioxide and Mn^{2+}. (d) Coal (solid carbon) reacts with steam to produce molecular hydrogen and carbon monoxide.

19.8 What are the half-reactions for these redox processes? (a) Aqueous hydrogen peroxide acts on Co^{2+}, and the products are hydroxide and Co^{3+}, in basic solution. (b) Methane reacts with oxygen gas and produces water and carbon dioxide. (c) To recharge a lead storage battery, lead(II) sulfate is converted to lead metal and to lead(IV) oxide. (d) Zinc metal dissolves in aqueous hydrochloric acid to give Zn^{2+} ions and hydrogen gas.

19.9 Balance the following half-reactions:
(a) $Cu^+ \longrightarrow CuO$ (acidic solution)
(b) $S \longrightarrow H_2S$ (acidic solution)
(c) $AgCl \longrightarrow Ag$ (basic solution)
(d) $I^- \longrightarrow IO_3^-$ (basic solution)
(e) $IO_3^- \longrightarrow IO^-$ (basic solution)
(f) $H_2CO \longrightarrow CO_2$ (acidic solution)

19.10 Balance the following half-reactions:
(a) $SbH_3 \longrightarrow Sb$ (acidic solution)
(b) $AsO_2^- \longrightarrow As$ (basic solution)
(c) $BrO_3^- \longrightarrow Br_2$ (acidic solution)
(d) $Cl^- \longrightarrow ClO_2^-$ (basic solution)

(e) $Sb_2O_5 \longrightarrow Sb_2O_3$ (acidic solution)
(f) $H_2O_2 \longrightarrow O_2$ (basic solution)

19.11 Balance the net redox reaction resulting from combining the following half-reactions in Problem 19.9 (the first listed in each case is the reduction): (a) 9a and 9b; (b) 9d and 9c; (c) 9d and 9e; and (d) 9f and 9b.

19.12 Balance the net redox reaction resulting from combining the following half-reactions in Problem 19.10 (the first listed in each case is the reduction): (a) 10a and 10c; (b) 10a and 10e; (c) 10b and 10d; and (d) 10b and 10f.

19.13 Consider the following redox reaction:

$$2\,MnO_4^- + 10\,Cl^- + 16\,H_3O^+ \longrightarrow 2\,Mn^{2+} + 5\,Cl_2 + 24\,H_2O$$

(a) Which species is oxidized? (b) Which species is reduced? (c) Which species is the oxidizing agent? (d) Which species is the reducing agent? (e) Which species gains electrons? (f) Which species loses electrons?

19.14 Consider the following redox reaction:

$$4\,NO + 3\,O_2 + 6\,H_2O \longrightarrow 4\,NO_3^- + 4\,H_3O^+$$

(a) Which species is oxidized? (b) Which species is reduced? (c) Which species is the oxidizing agent? (d) Which species is the reducing agent? (e) Which species gains electrons? (f) Which species loses electrons?

19.15 Balance the following redox equations:
(a) $CN^- + MnO_4^- \longrightarrow CNO^- + MnO_2$ (basic)
(b) $O_2 + As \longrightarrow HAsO_2 + H_2O$ (acidic)
(c) $Br^- + MnO_4^- \longrightarrow MnO_2 + BrO_3^-$ (basic)
(d) $NO_2 \longrightarrow NO_3^- + NO$ (acidic)
(e) $ClO_4^- + Cl^- \longrightarrow ClO^- + Cl_2$ (acidic)
(f) $AlH_4^- + H_2CO \longrightarrow Al^{3+} + CH_3OH$ (basic)

19.16 Balance the following redox equations:
(a) $H_5IO_6 + Cr \longrightarrow IO_3^- + Cr^{3+}$ (acidic)
(b) $Se + Cr(OH)_3 \longrightarrow Cr + SeO_3^{2-}$ (basic)
(c) $HClO_2 + Co \longrightarrow Cl_2 + Co^{2+}$ (acidic)
(d) $CH_3COH + Cu^{2+} \longrightarrow CH_3CO_2^- + Cu_2O$ (basic)
(e) $NO_3^- + H_2O_2 \longrightarrow NO + O_2$ (basic)
(f) $BrO_3^- + Fe^{2+} \longrightarrow Br^- + Fe^{3+}$ (acidic)

Galvanic Cells

19.17 Use standard thermodynamic values to determine whether or not each of the following redox reactions is spontaneous under standard conditions:
(a) $O_2 + 2\,Cu \longrightarrow 2\,CuO$
(b) $O_2 + 2\,Hg \longrightarrow 2\,HgO$
(c) $CuS + O_2 \longrightarrow Cu + SO_2$
(d) $FeS + O_2 \longrightarrow Fe + SO_2$

19.18 Use standard thermodynamic values to determine whether or not each of the following redox reactions is spontaneous under standard conditions:
(a) $H_2O + CO \longrightarrow CO_2 + H_2$
(b) $2\,Al + 3\,MgO \longrightarrow 3\,Mg + Al_2O_3$
(c) $PbS + Cu \longrightarrow CuS + Pb$
(d) $N_2 + 2\,O_2 \longrightarrow 2\,NO_2$

19.19 Draw a sketch that shows a molecular view of the charge transfer processes that take place at a silver–silver chloride electrode in contact with aqueous HCl, undergoing reduction:

$$AgCl\,(s)\ +\ e^-\ \longrightarrow\ Ag\,(s)\ +\ Cl^-(aq)$$

19.20 Draw a sketch that shows a molecular view of the charge transfer processes that take place at a nickel electrode in contact with aqueous nickel(II) chloride, undergoing oxidation:

$$Ni^{2+}(aq)\ +\ 2e^-\ \longrightarrow\ Ni\,(s)$$

19.21 Draw a sketch of a cell that could be used to study the following redox reaction:

$$H_2\,(g)\ +\ 2\,H_2O\,(l)\ +\ Cl_2\,(g)\ \rightleftharpoons\ 2\,H_3O^+(aq)\ +\ 2\,Cl^-(aq)$$

19.22 Draw a sketch of a cell that could be used to study the following redox reaction:

$$2\,Cu^+(aq)\ \rightleftharpoons\ Cu\,(s)\ +\ Cu^{2+}(aq)$$

19.23 Which of the electrodes in Problems 19.19 and 19.21 are active, and which are passive?

19.24 Which of the electrodes in Problems 19.20 and 19.22 are active, and which are passive?

Cell Potentials

19.25 Use standard reduction potentials in Appendix F to calculate $E°$ for the reactions in Problem 19.15 a, b, and e.

19.26 Use standard reduction potentials in Appendix F to calculate $E°$ for the reactions in Problem 19.16 a, b, and e.

19.27 Balance and calculate the standard potential for the following reaction in acidic solution:

$$NO\ \longrightarrow\ N_2O\ +\ NO_3^-$$

19.28 Balance and calculate the standard potential for the following reaction in acidic solution:

$$ClO_3^-\ \longrightarrow\ ClO_4^-\ +\ Cl^-$$

19.29 Describe a cell that could be used to measure the $E°$ of the F_2/F^- reduction reaction. Include a sketch similar to that shown in Figure 19-8. Which electrode would be the anode?

19.30 A cell is set up with two Cu wire electrodes, one immersed in a 1.0 M solution of $CuNO_3$, the other in a 1.0 M solution of $Cu(NO_3)_2$. Determine $E°$ of this cell, identify the anode, and draw a picture that shows the direction of electron flow at each electrode and in the external circuit.

19.31 Consult Appendix F and list the metals that can reduce Be^{2+} to Be ($E° = -1.97$ V) under standard conditions. What characteristics do these metals have in common?

19.32 Consult Appendix F and list the elements that can oxidize H_2O to O_2 under standard conditions. What characteristics do these elements have in common?

Free Energy and Electrochemistry

19.33 Use standard reduction potentials in Appendix F to calculate $\Delta G°$ for the reactions in Problem 19.15 a, b, and e.

19.34 Use standard reduction potentials in Appendix F to calculate $\Delta G°$ for the reactions in Problem 19.16 a, b, and e.

19.35 Example 19-10 describes the nickel–cadmium battery. What potential does this battery produce if its hydroxide ion concentration is 1.50×10^{-2} M?

19.36 If the cell illustrated in Figure 19-14 contains 1.00×10^{-3} M concentrations of HCl and $CuSO_4$, what potential does it produce?

19.37 If it takes 15 seconds to start your car engine and the battery provides 5.9 A of current to the starter motor, what masses of Pb and PbO_2 are used in each battery cell?

19.38 Suppose that automobiles were equipped with "thermite" batteries as described in Section Exercise 19.3.1. What masses of Al and Fe_2O_3 would be consumed in the process described in Problem 19.37?

19.39 An alternator recharges an automobile battery when the engine is running. If the alternator produces 1.750 A of current and the operation of the engine requires 1.350 A of this current, how long will it take to convert 0.850 g of $PbSO_4$ back into Pb metal?

19.40 A digital watch draws 0.20 mA of current provided by a zinc–air battery, whose net reaction is

$$O_2\,(g)\ +\ 2\,Zn\,(s)\ \longrightarrow\ 2\,ZnO\,(s)$$

If a partially used battery contains 1.00 g of zinc, for how many more hours will the watch run?

19.41 Dichromate ions, $Cr_2O_7^{2-}$, oxidize acetaldehyde, CH_3CHO, to acetic acid, CH_3CO_2H, and are reduced to Cr^{3+}. The reaction takes place in acidic solution. Balance the redox reaction and determine how many moles of electrons are required to oxidize 1.00 g of acetaldehyde. What mass of sodium dichromate would be required to deliver this many electrons?

19.42 Permanganate ions, MnO_4^-, oxidize acetaldehyde, CH_3CHO, to acetic acid, CH_3CO_2H, and are reduced to MnO_2. The reaction takes place in acidic solution. Balance the redox reaction and determine how many moles of electrons are required to oxidize 1.00 g of acetaldehyde. What mass of $KMnO_4$ would be required to deliver this many electrons?

Redox in Action

19.43 Set up the Nernst equation for the standard dry cell. Using this equation, show that the voltage of a dry cell must decrease with use.

19.44 Set up the Nernst equation for the lead storage cell, and use it to show that the voltage of this cell must decrease with use.

19.45 If a chromium-plated steel bicycle handlebar is scratched, breaking the protective film, will the chromium or the steel corrode? Use standard potentials to support your prediction.

19.46 Using standard potentials, explain why the steel propeller of an ocean-going yacht has a zinc collar.

19.47 Explain why zinc–air batteries find extensive use for cameras and pacemakers but are not used to start automobiles.

19.48 Explain why the lead storage battery, despite being the battery of choice for automobiles, is not suitable for space flights.

Electrolysis

19.49 In the electrolysis of aqueous NaCl, what mass of Cl_2 is generated by a current of 4.50 A flowing for 200.0 min?

19.50 Ni–Cd rechargeable batteries power a portable CD player that draws 150 mA of current. Compute the masses of Cd and NiO_2 consumed when a disk is played whose length is 65 minutes.

19.51 The Zn–Cu galvanic cell shown in Figure 19-7 is not rechargeable because of competitive electrolysis. Describe the reactions that take place if an opposing potential sufficient to reverse the reactions is applied to this galvanic cell.

19.52 The alkaline dry cell is not rechargeable. The solid products separate from the electrodes, so the reverse reactions cannot occur. What reactions may take place if an opposing potential sufficient to reverse the reactions is applied to a dry cell?

19.53 After use, a nickel–cadmium battery has 1.55 g of $Cd(OH)_2$ deposited on its anode. It is inserted in a recharger that supplies 125 mA of current at a voltage of 1.45 V. (a) To which electrode, Ni or Cd, should the negative wire from the charger be connected? Write the half-reaction occurring at this electrode during charging. (b) Compute the time in hours needed to convert all 1.55 g of $Cd(OH)_2$ back to Cd metal.

19.54 A "dead" 12-V lead storage battery has 4.80 g of $PbSO_4$ deposited on each of its anodes. It is connected to a "trickle charger" that supplies 0.120 A of current at a voltage of 13 V. (a) To which electrode, Pb or PbO_2, should the negative wire from the charger be connected? Write the half-reaction occurring at this electrode during charging. (b) Compute the time in hours needed to convert all 4.80 g of $PbSO_4$ back to HSO_4^- ions in solution.

19.55 Electrochemistry can be used to measure electrical current in a silver coulometer, in which a silver cathode is immersed in a solution containing Ag^+ ions. The cathode is weighed before and after passage of current. A silver cathode initially has a mass of 10.77 g, and its mass increases to 12.89 g after current has flowed for 15.0 minutes. Compute the quantity of charge in coulombs and the current in amperes.

19.56 When Thomas Edison first sold electricity, he used zinc coulometers to measure charge consumption (see Problem 19.55). If the zinc plate in one of Edison's coulometers increased in mass by 7.55 g, how much charge had been consumed?

Additional Paired Problems

19.59 Here are two standard reduction potentials:

$$O_2(g) + 2 H_2O + 4 e^- \rightleftharpoons 4 OH^-(aq) \qquad E° = 0.401 \text{ V}$$

$$O_2(g) + 4 H_3O^+(aq) + 4 e^- \rightleftharpoons 6 H_2O \qquad E° = 1.229 \text{ V}$$

Are these data sufficient to allow calculation of K_w at 298 K? If so, do the calculation. If not, explain in detail what additional data (thermodynamic or electrochemical) you would need to do the calculation.

19.60 Here are two standard reduction potentials:

$$SO_4^{2-}(aq) + 4 H_3O^+(aq) + 2 e^- \rightleftharpoons SO_2(g) + 6 H_2O(l)$$
$$E° = 0.20 \text{ V}$$

$$SO_4^{2-}(aq) + 4 H_3O^+(aq) + 2 e \rightleftharpoons H_2SO_3(aq) + 5 H_2O(l)$$
$$E° = 0.18 \text{ V}$$

Are these data sufficient to allow calculation of the equilibrium vapor pressure of $SO_2(g)$ over a 1.00 M solution of $H_2SO_3(aq)$ at 298 K? If so, do the calculation. If not, explain in detail what additional data (thermodynamic or electrochemical) you would need to do the calculation.

19.61 Use data in Appendix F to calculate the equilibrium constant of the following reaction:

$$Zn^{2+}(aq) + 4 NH_3(aq) \rightleftharpoons Zn(NH_3)_4^{2+}(aq)$$

19.62 Use data in Appendix F to calculate K_{sp} for AgI.

19.63 For the reduction of $Cr(OH)_3$ by H_2 in basic solution to give Cr and H_2O, do the following: (a) Write the balanced net equation. (b) Compute $E°$. (c) Compute $\Delta G°$.

19.64 Consider the following redox reaction:

$$MnO_2 \longrightarrow MnO_4^- + Mn^{2+} \qquad \text{(unbalanced)}$$

(a) Balance this reaction in acidic solution. (b) Write the expression for Q. (c) Determine n, $E°$, $\Delta G°$, and K_{eq} at 298 K.

19.65 Potassium chromate, K_2CrO_4, dissolves in acidic solution to generate strongly oxidizing $Cr_2O_7^{2-}$ ions:

$$K_2CrO_4(s) + H_3O^+(aq) \longrightarrow 2 K^+(aq) + HCrO_4^-(aq) + H_2O(l)$$

$$2 HCrO_4^-(aq) \longrightarrow Cr_2O_7^{2-}(aq) + H_2O(l)$$

This oxidizing agent is reduced to Cr^{3+} as it oxidizes other substances. (a) Balance the reduction half-reaction. (b) What mass of K_2CrO_4 produces 0.250 mol of electrons?

19.66 Breathalyzers determine the alcohol content in a person's breath by a redox reaction using dichromate ions:

$$C_2H_5OH + Cr_2O_7^{2-} \longrightarrow CH_3CO_2H + Cr^{3+} \qquad \text{(unbalanced)}$$

(a) Balance this reaction. (b) If a breath sample generates a concentration of 4.5×10^{-4} M Cr^{3+} in 50.0 mL of solution, how many milligrams of alcohol were in the sample?

19.67 The figure below shows a schematic sketch of a galvanic cell.

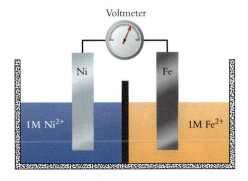

19.57 In a silver coulometer (see Problem 19.55), both electrodes are silver metal. Draw a molecular picture that illustrates the reactions that occur during operation of this coulometer.

19.58 In Edison's zinc coulometer (see Problem 19.56), both electrodes were zinc metal. Draw a molecular picture that illustrates the reactions that occur during operation of this coulometer.

(a) Identify the two half-reactions. (b) Determine the potential of this cell. (c) Identify the anode and cathode. (d) Redraw the sketch to show the direction of electron flow and the molecular processes occurring at each electrode.

19.68 The figure below shows a schematic sketch of a galvanic cell.

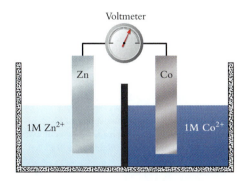

(a) Identify the two half-reactions. (b) Determine the potential of this cell. (c) Identify the anode and cathode. (d) Redraw the sketch to show the direction of electron flow and the molecular processes occurring at each electrode.

19.69 For the galvanic cell in Problem 19.67, which solution concentration would have to be reduced, and to what concentration, to reduce the cell potential to 0.0 V?

19.70 For the galvanic cell in Problem 19.68, which solution concentration would have to be reduced, and to what concentration, to reduce the cell potential to 0.0 V?

19.71 Balance the redox reactions between MnO_4^- and each of the following sulfur-containing species. The final products are Mn^{2+} and HSO_4^- and the solution is acidic: (a) H_2SO_3; (b) SO_2; (c) H_2S; and (d) $H_2S_2O_3$.

19.72 Balance the redox reactions between $Cr_2O_7^{2-}$ and each of the following nitrogen-containing species. The final products are Cr^{3+} and NO_3^- and the solution is acidic: (a) NO; (b) NO_2; (c) HNO_2; and (d) NH_4^+.

19.73 How long would it take to electroplate all the Cu^{2+} in 0.250 L of 0.245 M $CuSO_4$ solution with an applied potential difference of 0.225 V and a current of 2.45 A?

19.74 Calcium metal is obtained by the direct electrolysis of molten $CaCl_2$. If a metallurgical electrolysis apparatus operates at 27.6 A and 1.2 V, what mass of calcium metal will it produce in 24 hours of operation?

19.75 Draw a sketch of the electroplating apparatus that illustrates the process occurring in Problem 19.73. Include arrows showing the direction of electron flow, label the anode and cathode, and draw molecular pictures showing the processes occurring at each electrode.

19.76 Draw a sketch of an electroplating apparatus that illustrates the process occurring in Problem 19.74. Include arrows showing the direction of electron flow, label the anode and cathode, and draw molecular pictures showing the processes occurring at each electrode.

19.77 For each of the following pairs of species, select the one that is the better oxidizing agent under standard conditions (see Appendix F): (a) in acidic solution, $Cr_2O_7^{2-}$ or MnO_4^-; (b) in basic solution, O_2 or NO_3^-; and (c) Fe^{2+} or Sn^{2+}.

19.78 For each of the following pairs of species, select the one that is the better reducing agent under standard conditions (see Appendix F): (a) Cu or Ag; (b) Fe^{2+} or Co^{2+}; and (c) H_2 or I^- (acid solution).

More Challenging Problems

19.79 Determine whether O_2 is capable of oxidizing each of the substances in Problem 19.78, under standard conditions (basic).

19.80 The following are spontaneous reactions. List the species in these reactions in order of oxidizing strength, from strongest oxidizing agent to weakest oxidizing agent:

$$2\ Cr^{2+} + Sn^{2+} \longrightarrow Sn + 2\ Cr^{3+}$$
$$Fe + Sn^{2+} \longrightarrow Sn + Fe^{2+}$$
$$Fe + U^{4+} \longrightarrow No\ reaction$$
$$U^{3+} + Cr^{3+} \longrightarrow U^{4+} + Cr^{2+}$$
$$Fe + 2\ Cr^{3+} \longrightarrow 2\ Cr^{2+} + Fe^{2+}$$

19.81 A cell is set up using two zinc wires and two solutions, one containing 0.250 M $ZnCl_2$ solution and the other containing 1.25 M $Zn(NO_3)_2$ solution. (a) What electrochemical reaction occurs at each electrode? (b) Draw a molecular picture showing spontaneous electron transfer processes at the two zinc electrodes. (c) Compute the potential of this cell.

19.82 Using standard reduction potentials, determine K_{eq} for the decomposition reaction of hydrogen peroxide:

$$2\ H_2O_2 \rightleftharpoons 2\ H_2O + O_2.$$

What does this value tell you about the stability of H_2O_2?

19.83 The first battery to find widespread commercial use was the carbon–zinc dry cell, in which the cathode reaction is

$$2\ MnO_2\,(s) + Zn^{2+}\,(aq) + 2\ e^- \longrightarrow ZnMn_2O_4\,(s)$$

In a flashlight, one of these batteries provides 0.0048 A. If the battery contains 4.0 g of MnO_2 and fails after 90% of its MnO_2 is consumed, calculate the operating life of the flashlight.

19.84 Electrolytic reactions, like other chemical reactions, are not 100% efficient. In a copper purification apparatus depositing Cu from $CuSO_4$ solution, operation for 5.0 hours at constant current of 5.8 A deposits 32 g of Cu metal. What is the efficiency?

19.85 A galvanic cell consists of a Pt electrode immersed in a solution containing Fe^{2+} at 1.00 M and Fe^{3+} at unknown concentration, as well as a Cu electrode immersed in a 1.00 M solution of Cu^{2+}. The cell voltage is 0.00 V. What is the concentration of Fe^{3+} ions?

19.86 A galvanic cell is constructed using a silver wire coated with silver chloride and a nickel wire immersed in a beaker containing 1.50×10^{-2} M $NiCl_2$. (a) Determine the balanced cell reaction. (b) Calculate the potential of the cell. (c) Draw a sketch showing the electron transfer reaction occurring at each electrode.

19.87 The sketch below shows a cell set up to electrolyze molten NaCl.

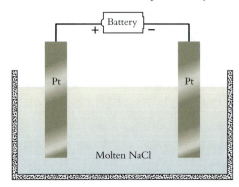

(a) What reactions occur? (b) Identify the anode and cathode. (c) Redraw the sketch showing the direction of electron flow, and include molecular pictures showing the processes at each electrode.

19.88 Given that $E° = -0.34$ V for the reduction of Tl^+ to Tl, find the voltage developed by a cell consisting of Tl metal dipping in an aqueous solution that is 0.050 M in Tl^+, connected by a porous bridge to a 0.50 M aqueous solution of HCl in contact with a Pt electrode over which H_2 gas is bubbling at $p = 0.90$ bar.

19.89 Draw a sketch of the cell in Problem 19.88. Include molecular views of the processes taking place at the electrodes.

19.90 List all the metals that could be used as sacrificial anodes for iron. Which of these could also be sacrificial anodes for aluminum?

19.91 For the reaction between strontium and magnesium, $K_{eq} = 2.7 \times 10^{12}$:

$$Sr\,(s) + Mg^{2+}(aq) \rightleftharpoons Sr^{2+}(aq) + Mg\,(s)$$

Calculate $E°$ for a strontium–magnesium battery.

Group Study Problems

19.92 Use standard reduction potentials from Table 19-1 and Appendix F to determine K_{sp} for as many metal hydroxides as the table allows. Compare your values with those in Appendix E. If there are K_{sp} values for hydroxides in Appendix E that cannot be calculated from standard reduction potentials in Appendix F, use the K_{sp} values to calculate the appropriate standard reduction potentials.

19.93 Consider an electrochemical cell consisting of two vessels connected by a porous separator. One vessel contains 0.500 M HCl solution and an Ag wire electrode coated with AgCl solid. The other vessel contains 1.00 M $MgCl_2$ solution and an Mg wire electrode. (a) Determine the net reaction. (b) Calculate E for the cell (see Appendix F). (c) Draw a molecular picture showing the reactions at each electrode.

19.94 Use data from Appendix F for the following reaction:

$$5\ I^-\,(aq) + IO_3^-\,(aq) + 6\ H_3O^+(aq) \longrightarrow 3\ I_2\,(s) + 9\ H_2O\,(l)$$

(a) Determine the spontaneous direction at pH = 2.00 and

$[I^-] = [IO_3^-] = 0.100$ M. (b) Repeat the calculation at pH = 11.00. (c) At what pH is this redox reaction at equilibrium at these concentrations of I^- and IO_3^-?

19.95 From the standard reduction potentials appearing in Table 19-1 and Appendix F, identify reaction pairs that are candidates for batteries that would produce more than 5 V of electrical potential under standard conditions. Suggest chemical reasons why no such battery has been commercially developed.

19.96 A chemist wanted to determine $E°$ for the Ru^{3+}/Ru reduction reaction. The chemist had all the equipment needed to make potential measurements, but the only chemicals available were $RuCl_3$, a piece of ruthenium wire, $CuSO_4$, copper wire, and water. Describe and sketch a cell that the chemist could set up to determine this $E°$. Show how the measured voltage would be related to $E°$ of the half-reaction. If the cell has a measured voltage of 1.44 V, with the ruthenium wire being negative, determine $E°$ for Ru^{3+}/Ru.

Answers to Section Exercises

19.1.1 (a) K = +1, O = -2, Mn = +7; (b) Cl = -1, Si = +4; (c) Na = +1, O = -2, S = +4; (d) S = 0; and (e) Li = +1, Al = +3, H = -1

19.1.2 (a) redox, C oxidized, O_2 reduced; (b) not redox; (c) redox, Mg oxidized, Si reduced; (d) redox, S oxidized, O reduced; and (e) not redox

19.1.3 (a) -3; (b) 0; (c) -2; (d) $+5$; (e) $+3$; (f) $+2$; and (g) $+4$

19.2.1 (a) $SO_3^{2-} + H_2O_2 \longrightarrow SO_4^{2-} + H_2O$;
(b) $3\,BrO_2^- \longrightarrow Br^- + 2\,BrO_3^-$

19.2.2 $2\,Al + 3\,Ag_2S + 6\,H_2O \longrightarrow 6\,Ag + 2\,Al(OH)_3 + 3\,H_2S$

19.2.3 $3\,CuS + 11\,H_3O^+ + 8\,NO_3^- \longrightarrow$
$$8\,NO + 15\,H_2O + 3\,Cu^{2+} + 3\,HSO_4^-$$

19.3.1

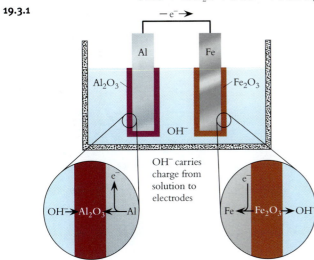

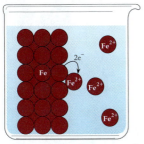 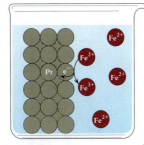

Al loses electrons to become Al_2O_3

Fe_2O_3 gains electrons to become Fe

OH^- carries charge from solution to electrodes

19.3.2 $\Delta G^\circ = -840.1$ kJ; negative value verifies spontaneity.

19.3.3

Active electrode: When an Fe^{2+} ion collides with the surface of the active electrode, Fe^{2+} gains two electrons and sticks to the electrode as an Fe atom.

Passive electrode: When an Fe^{2+} ion collides with the surface of the passive electrode, Fe^{2+} loses one electron and becomes an Fe^{3+} ion, which remains dissoved in solution.

19.4.1 (a) $2\,Cr + 3\,Fe^{2+} \longrightarrow 2\,Cr^{3+} + 3\,Fe$, $E^\circ = 0.297$ V
(b) $Pb + Cu^{2+} \longrightarrow Cu + Pb^{2+}$, $E^\circ = 0.4681$ V
(c) $Au^{3+} + 3\,Ag \longrightarrow Au + 3\,Ag^+$, $E^\circ = 0.698$ V

19.4.2

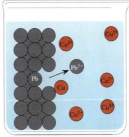

When a Cu^{2+} ion collides with the Pb surface, Cu^{2+} is reduced to Cu (elemental copper), and a Pb atom is oxidized to Pb^{2+}. Cu is insoluble in water, so it sticks to the Pb surface; Pb^{2+} is soluble in water, so it dissolves.

19.4.3 2.0501 V

19.5.1 (a) $\Delta G^\circ = -63.7$ kJ, $K_{eq} = 1.4 \times 10^{11}$
(b) $\Delta G^\circ = 1.14 \times 10^3$ kJ, $K_{eq} = 10^{-200}$

19.5.2 (a) 0.281 V; (b) 0.4681 V; and (c) 0.761 V

19.5.3 (a) FeS_2; (b) 804 C; and (c) 6.21 days

19.6.1

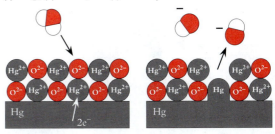

HgO captures 2 electrons to form Hg and O^{2-}

O^{2-} reacts with H_2O to form 2 OH^-

19.6.2 At the aluminum electrode, the deposit of aluminum oxide is impervious to penetration by ions, so it blocks the passage of current.

19.6.3 E° for tin is less negative than that for iron, whereas E° for zinc is more negative than that for iron. Thus, iron oxidizes instead of tin, but zinc oxidizes instead of iron.

19.7.1 (a) $2\,HF + 2\,e^- \longrightarrow H_2 + 2\,F^-$ and $2\,F^- \longrightarrow F_2 + 2\,e^-$; and (b) 3.053 V

19.7.2 25.5 kg, requiring a tank whose volume is 274 L

19.7.3 In aqueous KF, water would be electrolyzed rather than KF, generating O_2 instead of F_2; liquid HF does not conduct electricity and therefore does not support electrolysis.

Answers to Extra Practice Exercises

19.1 NiO, O is -2, Ni is $+2$; C, C is 0; Ni, Ni is 0; CO, O is -2, C is $+2$; redox reaction. Ni, Ni is 0; CO, O is -2, C is $+2$; $Ni(CO)_4$, CO has not changed, so O is -2, C is $+2$, and Ni is 0; not redox

19.2 H_2SO_4, O is -2, H is $+1$, S is $+6$; $KClO_3$, O is -2, K is $+1$, Cl is $+5$

19.3 (a) $Cu \longrightarrow Cu^{2+}$ and $NO_3^- \longrightarrow NO$; (b) $ClO^- \longrightarrow ClO_3^-$ and $ClO^- \longrightarrow Cl^-$

19.4 (a) $Cu \longrightarrow Cu^{2+} + 2\,e^-$ and
$$NO_3^- + 4\,H_3O^+ + 3\,e^- \longrightarrow NO + 6\,H_2O$$
(b) $ClO^- + 6\,H_2O \longrightarrow ClO_3^- + 4\,H_3O^+ + 4\,e^-$ and
$$ClO^- + 2\,H_3O^+ + 2\,e^- \longrightarrow Cl^- + 3\,H_2O$$

19.5 (a) $3\,Cu + 2\,NO_3^- + 8\,H_3O^+ \longrightarrow 3\,Cu^{2+} + 2\,NO + 12\,H_2O$;
(b) $3\,ClO^- \longrightarrow ClO_3^- + 2\,Cl^-$

19.6 $2\,MnO_4^- + 3\,ClO_3^- + 2\,H_3O^+ \longrightarrow$
$$2\,MnO_2 + 3\,ClO_4^- + 3\,H_2O$$

19.7 $2\,MnO_4^- + 3\,Cd + 4\,H_2O \longrightarrow$
$$2\,MnO_2 + 3\,Cd(OH)_2 + 2\,OH^-$$

19.8 Oxidation half-reaction is
$Zn + 2\,OH^- \longrightarrow ZnO + H_2O + 2\,e^-$;
reduction half-reaction is $HgO + H_2O + 2\,e^- \longrightarrow Hg + 2\,OH^-$;
overall reaction is $Zn + HgO \longrightarrow ZnO + Hg$.

19.9 $E^\circ = 0.789$ V; net reaction is
$Cu^{2+}(aq) + Fe(s) \longrightarrow Cu(s) + Fe^{2+}(aq)$.

19.10 $Eu^{2+}(aq) + 2\,e^- \rightleftharpoons Eu(s)$, $E^\circ = -2.81$ V; net reaction is
$Cu^{2+}(aq) + Eu(s) \longrightarrow Cu(s) + Eu^{2+}(aq)$.

19.11 $K_{eq} = 0.43$

19.12 $E = 0.070$ V

19.13 135 hours

19.14 23 mg of Li is transferred.

19.15 6.92 g of Ag plates onto the spoon.

The Transition Metals

Transition Metals

| V | Cr | Mn | Fe | Co | Ni | Cu | Zn |

Mo

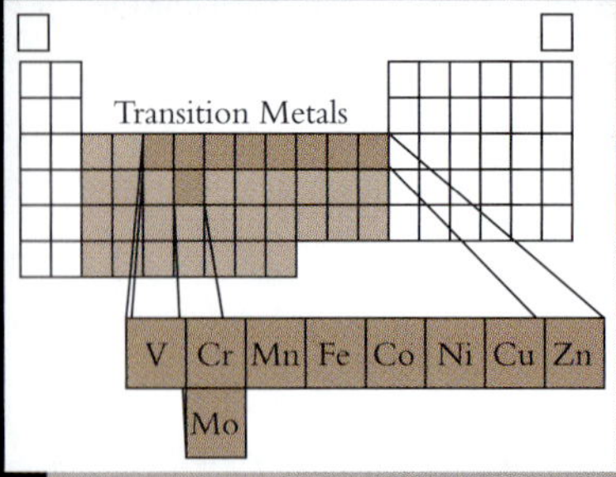

Metals Essential for Life

We tend to think of metals as lustrous solids: copper, tin, gold, silver, iron. We are less likely to think of individual metal atoms in biological "machines." Nevertheless, plants and animals require the presence of tiny amounts of transition metals. Humans require most of the elements in the first transition metal series and at least one element from the second transition metal series, molybdenum.

Among the transition metals, the biochemistry of iron is known in the most detail. Iron is essential for the synthesis of chlorophyll, the green pigment that drives photosynthesis. Our background photograph shows citrus trees suffering from the yellowing of leaves that is characteristic of iron deficiency.

To combat iron deficiency, gardeners use fertilizers that contain iron chelates, octahedral complexes that slowly release their iron into the soil.

Iron is the transition metal present in largest quantities in the human body. The best-known biological iron-containing compound is the protein hemoglobin, the red component of blood that is responsible for the transport of oxygen. The central feature of hemoglobin is the porphyrin structure, a plane of carbon and nitrogen atoms that forms multiple rings with an extensive delocalized π system. In hemoglobin, an Fe^{2+} cation bonds to four nitrogen atoms that are part of the planar porphyrin ring; this structure is known as *heme*. An oxygen molecule can bind to the iron at a site perpendicular to the plane of the porphyrin ring. Hemoglobin exploits

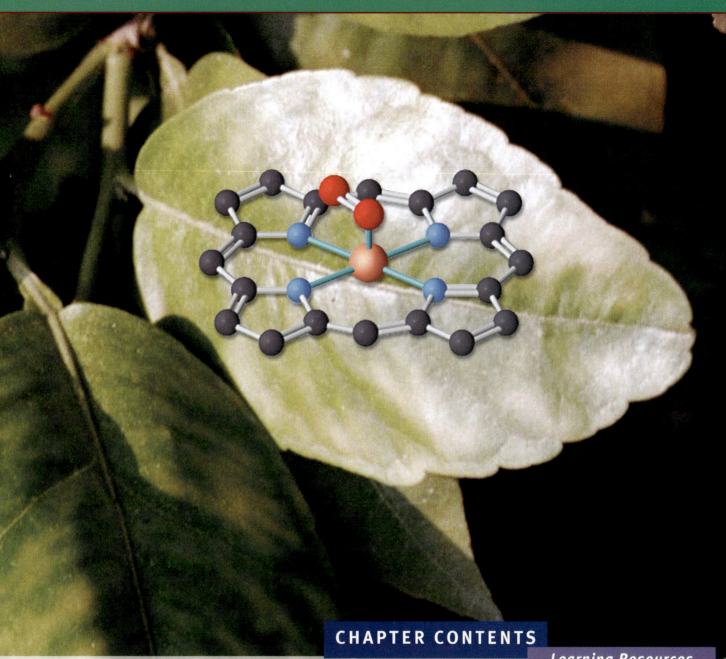

one of the special bonding properties of transition metal ions: their ability to form bonds by accepting electron pairs from atoms, such as nitrogen and oxygen, that contain lone pairs. The structure of heme appears in our molecular inset. Variations of the heme structure are found in all life forms.

Other iron-containing proteins feature clusters of iron and sulfur atoms. The iron atoms in these clusters have tetrahedral bonding geometry. Iron–sulfur proteins are essential for electron transfer processes, in which the clusters of iron and sulfur atoms gain and lose electrons. Here, another special property of transition metal ions is exploited:

their ability to take on several different oxidation states.

Metals display a remarkable range of chemical properties. The individual atoms or cations in biological macromolecules readily form and break bonds to other species as well as gain or lose electrons. Geologically, metals typically occur in minerals, where metal cations are associated with oxygen, sulfur, or polyatomic anions such as carbonate and silicate. In pure metals and alloys, metals participate in delocalized bonding that generates high strength and electrical conductivity.

In this chapter, we survey the diversity of transition metals, beginning with an overview. Then we describe the structure and bonding in transition metal complexes. We describe metallurgy, the processes by which pure metals are extracted from mineral ores. The chapter ends with a presentation of some properties of transition metals and their biological roles.

20.1 OVERVIEW OF THE TRANSITION METALS

The **transition metals** lie in the *d* block, at the center of the periodic table, between the *s*-block metals and the elements in the *p* block, as Figure 20-1 shows. As we describe in Chapter 8, most transition metal atoms in the gas phase have valence electron configurations of $s^2 d^{(x-2)}$, where *x* is the group number of the metal. Titanium, for example, is in Group 4 and has valence configuration $4s^2 3d^2$. Recall, however, that the $(n + 1)s$ orbital and the set of *nd* orbitals have nearly the same energy. As a result, several of the transition metals have valence configurations that differ from the general pattern, as listed in Table 8-2. Remember also that for transition metal *cations*, the *nd* orbital *always* is more stable than the $(n + 1)s$ orbital, so transition metal cations have vacant $(n + 1)s$ orbitals.

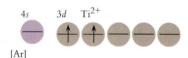

A pure transition metal is best described by the band theory of solids, as introduced in Chapter 10. In this model, the valence *s* and *d* electrons form extended bands of orbitals that are delocalized over the entire network of metal atoms. These valence electrons are easily removed, so most elements in the *d* block react readily to form compounds: oxides such as Fe_2O_3, sulfides such as ZnS, and mineral salts such as zircon, $ZrSiO_4$.

Between barium (Group 2, element 56) and lutetium (Group 3, element 71), the *4f* orbitals fill with electrons, giving rise to the **lanthanides,** a set of 14 metals named for lanthanum, the first member of the series. The lanthanides are also called the *rare earths,* although except for promethium they are not particularly rare. Between radium (Group 2, element 88) and lawrencium (Group 3, element 103), are the 14 **actinides,** named for the first member of the set, actinium. The lanthanides and actinides are also known as the **inner transition metals.**

Physical Properties

As pure elements, almost all the transition metals are solids that conduct heat and electricity and are malleable and ductile. Although they share these general properties, transition metals display variations in other properties that can be traced to their different numbers of valence electrons.

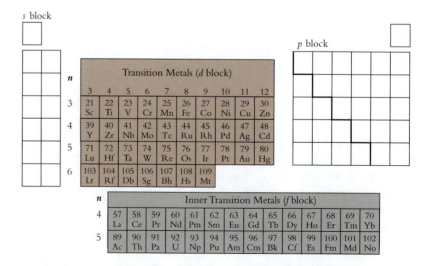

Figure 20-1
The transition metals are the elements in the *d* block of the periodic table. The inner transition metals are those in the *f* block.

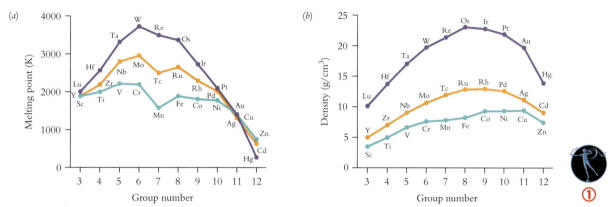

Figure 20-2

The melting points of the transition metals increase from Group 3 to Group 6 and then decrease to the end of the block. The densities of transition metals increase with atomic mass except at the end of the block.

Most pure transition metals have the shiny gray appearance that is termed "silvery" because of the appearance of silver metal. However, some transition metals have other colors—for example, the orange color of copper and the yellow hue of gold.

The melting points of the transition metals show a periodic variation that is displayed in Figure 20-2a. Up to Group 6, the melting point increases with each added valence electron. This is because the first six valence electrons occupy bonding orbital bands that increase the overall bonding strength of the metal. Beyond Group 6, in contrast, the melting point decreases, falling precipitously for the elements of Group 12. This decrease is sufficient to make mercury unique among metals. This element has a melting point of −39 °C (234 K), making Hg a liquid at room temperature. Melting points decrease with added electrons beyond Group 6 because the added electrons occupy antibonding bands, weakening the overall bonding strength of the metal.

The densities of transition metals also display regular periodic trends, as Figure 20-2b shows. Density increases moving down each column of the periodic table and increases smoothly across the first part of each row. Increasing atomic mass accounts for both these trends. The volume occupied by an individual atom in the metallic lattice varies slowly within the d block, so the more massive the nucleus, the greater the density of the metal. Toward the end of each row, density decreases for the same reason that melting point decreases. The added electrons occupy antibonding orbitals, and this leads to a looser array of atoms, larger atomic volume, and decreased density.

Transition metals are good conductors of both electricity and heat. In general, these two properties are linked, because mobile electrons can transport both charge (electricity) and energy (heat) through bulk materials. The metals of Group 11 (copper, silver, and gold) have the highest electrical conductivity of any of the elements. Silver is the best conductor, but copper is more abundant and consequently less expensive, so it is the metal of choice for electrical wiring. The Group 11 elements also are the best conductors of heat among the metals.

Samples of metals from the first transition series. The metals are Cu (wire), Fe (block), Ni (balls), and Zn (sticks).

Redox Behavior

The chemistry of the transition metals is determined in part by their atomic ionization energies. Metals of the 3d and 4d series show a gradual increase in ionization energy with atomic number (Z), whereas the trend for the 5d series is more pronounced (Figure 20-3).

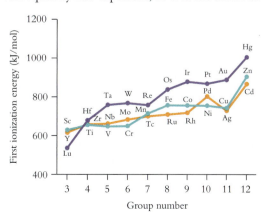

Figure 20-3

The first ionization energies of transition metals show gradual upward trends across each row of the periodic table.

Table 20-1
Oxidation States Displayed by the 3d Transition Metals*

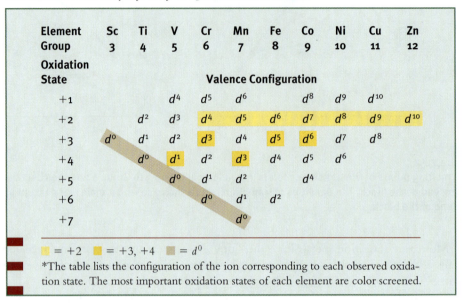

Element	Sc	Ti	V	Cr	Mn	Fe	Co	Ni	Cu	Zn
Group	3	4	5	6	7	8	9	10	11	12
Oxidation State					Valence Configuration					
+1			d^4	d^5	d^6		d^8	d^9	d^{10}	
+2		d^2	d^3	d^4	d^5	d^6	d^7	d^8	d^9	d^{10}
+3	d^0	d^1	d^2	d^3	d^4	d^5	d^6	d^7	d^8	
+4		d^0	d^1	d^2	d^3	d^4	d^5	d^6		
+5			d^0	d^1	d^2		d^4			
+6				d^0	d^1	d^2				
+7					d^0					

☐ = +2   ☐ = +3, +4   ☐ = d^0

*The table lists the configuration of the ion corresponding to each observed oxidation state. The most important oxidation states of each element are color screened.

First ionization energies for transition metals in the 3d and 4d series are between 650 and 750 kJ/mol, somewhat higher than the values for Group 2 alkaline earth metals but lower than the typical values for nonmetals in the p block.

Except for the elements at the ends of the rows, each transition metal can exist in several different oxidation states. The oxidation states displayed by the 3d transition metals are shown in Table 20-1. The most important oxidation states are highlighted in the table. The most common oxidation state for the 3d transition metals is +2, known for all the elements except Sc. Chromium, iron, and cobalt are also stable in the +3 oxidation state, and for vanadium and manganese the +4 oxidation state is stable. Elements from scandium to manganese have a particularly stable oxidation state corresponding to the loss of all the valence electrons ($s^0 d^0$ configuration).

Cations with charges greater than +3 are generally unstable even in an ionic environment, because the energy required to remove an electron from an ion with +3 charge is very large. To convert Ti^{3+} to Ti^{4+}, for instance, requires an energy input of 4175 kJ/mol. Consequently, oxidation states greater than +3 are found in covalently bonded oxides, oxoanions, and halides but not in ionic compounds such as sulfates and phosphates. Table 20-2 lists representative examples of compounds displaying the important oxidation states of the 3d transition metals.

Transition Metal Compounds

A few of the transition metals, including gold, platinum, and iridium, are found in nature as pure elements, but most of the others are found associated with either sulfur or oxygen. Iron, manganese, and the metals of Groups 3 to 6 (except for Mo) are most often found as oxides; less often, they occur as sulfates or carbonates. Molybdenum and the metals of Groups 7 to 12 (except for Mn and Fe) are most often found as sulfides.

Although naturally occurring compounds of transition metals are restricted in scope, a wide variety of compounds can be synthesized in the laboratory. Representative compounds appear in Table 20-2. These compounds fall into three general categories: There are many binary halides and oxides in a range of oxidation numbers. Ionic compounds containing transition metal cations and polyatomic oxoanions also are common; these include nitrates, carbonates, sulfates, phosphates, and perchlorates. Finally, there are numerous ionic compounds in which the transition metal is part of an oxoanion.

Table 20-2   Representative Compounds of 3d Transition Metals in Their Commonly Occurring Oxidation States

Element	Oxidation State	Compound	Oxidation State	Compound	Oxidation State	Compound
Sc	+3	$Sc(NO_3)_3 \cdot 6\,H_2O$				
Ti			+4	$TiBr_4, FeTiO_3$		
V			+4	VCl_4, VO_2	+5	$Pb_2(VO_4)Cl$
Cr	+2	$Cr(CH_3CO_2)_2$	+3	$Cr(OH)_3$	+6	$CrF_6, K_2Cr_2O_7$
Mn	+2	$MnCO_3, MnCl_2$	+4	MnO_2, MnF_4	+7	$KMnO_4$
Fe	+2	$FeSO_4 \cdot H_2O, FeS_2$	+3	$FeCl_3, FePO_4 \cdot 2\,H_2O$		
Co	+2	$CoS, Co(ClO_4)_2$	+3	$Co(OH)_3, CoF_3$		
Ni	+2	$NiSO_4 \cdot 7\,H_2O$				
Cu	+2	$Cu(NO_3)_2 \cdot 6\,H_2O$				
Zn	+2	$ZnS, ZnCO_3$				

The synthesis of transition metal compounds must be tailored to a particular product. The details are complex, and existing patterns of synthesis contain so many exceptions that an overall view is beyond the scope of an introductory course in chemistry.

Section Exercises

20.1.1 Predict which element of the following pairs will have the higher melting point and support your prediction using periodic trends: (a) Cr or W; (b) W or Os; (c) Pd or Ag; and (d) Y or Nb.

20.1.2 Predict which element of the following pairs will have the higher density and support your prediction using periodic trends: (a) Cr or W; (b) W or Os; (c) Pd or Ag; and (d) Y or Nb.

20.1.3 The following compounds can be purchased from chemical companies. Determine the oxidation state of the transition metal in each of them.
(a) $TaCl_5$;
(b) $Fe(NO_3)_3 \cdot 9H_2O$;
(c) Rh_2O_3;
(d) CrO_2Cl_2; and
(e) $Cu_2(OH)PO_4$.

20.2 COORDINATION COMPLEXES

Earlier chapters described how metal salts dissolve in water to give solutions of aqueous ions. Here are three examples:

$$Cu(ClO_4)_2\,(s) \xrightarrow{H_2O} Cu^{2+}\,(aq) + 2\,ClO_4^-\,(aq)$$

$$NiSO_4\,(s) \xrightarrow{H_2O} Ni^{2+}\,(aq) + SO_4^{2-}\,(aq)$$

$$FeCl_3\,(s) \xrightarrow{H_2O} Fe^{3+}\,(aq) + 3\,Cl^-\,(aq)$$

Section 18-5 addresses the formation of complexes between metal ions and ligands. For example, $Ni^{2+}\,(aq)$ forms a complex with ammonia:

$$Ni^{2+}\,(aq) + 6\,NH_3\,(aq) \rightleftharpoons [Ni(NH_3)_6]^{2+}\,(aq)$$

For complexation purposes, a **ligand** is a species that has lone pairs of electrons available to donate to a metal atom or cation. Water molecules possess lone pairs of electrons, so water is a ligand that readily forms complex ions with metal cations. Although solubility and complexation equations show uncomplexed metal ions in solution, dissolved cations actually form chemical bonds to water molecules of the solvent. For example, $Ni^{2+}\,(aq)$ bonds to six water molecules in octahedral geometry, as illustrated in Figure 20-4 on the next page. The aqueous $[Ni(H_2O)_6]^{2+}$ cation contains Ni^{2+} bonded to six water ligands.

Figure 20-4
The Ni^{2+} cations in $[Ni(H_2O)_6]^{2+}$ and $[Ni(NH_3)_6]^{2+}$ bond to six ligands, one at each vertex of an octahedron.

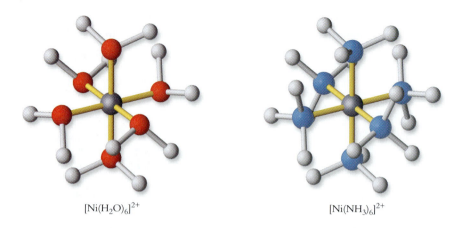

$[Ni(H_2O)_6]^{2+}$ $[Ni(NH_3)_6]^{2+}$

Figure 20-5
Solid nickel(II) sulfate, nearly colorless, dissolves in water to give a green solution containing $[Ni(H_2O)_6]^{2+}$ cations. The addition of ammonia produces a blue solution of $[Ni(NH_3)_6]^{2+}$ cations. Solvent evaporation gives a blue-violet precipitate of $[Ni(NH_3)_6]SO_4$.

Color changes often provide evidence for the interaction of ligands and metal cations, particularly for the transition metals. The Ni^{2+} cation provides an example. Figure 20-5 shows that nickel(II) sulfate, a white crystalline solid, dissolves in water to give a green solution. The green color cannot be due to Ni^{2+} or SO_4^{2-}, which the white solid shows to be colorless. Rather, the color comes from the octahedral complex that forms when each nickel ion binds to six water molecules:

$$NiSO_4\,(s) + 6\;H_2O\,(l) \longrightarrow [Ni(H_2O)_6]^{2+}(aq) + SO_4^{2-}(aq)$$

Replacing the water ligands with other ligands also can result in color changes. Figure 20-5 shows the color change that occurs when ammonia is added to a solution of hydrated Ni^{2+}. Ammonia molecules replace water ligands to give the blue species $[Ni(NH_3)_6]^{2+}$, whose ball-and-stick structure appears in Figure 20-4:

$$\underset{\text{green}}{[Ni(H_2O)_6]^{2+}} + 6\;NH_3 \longrightarrow \underset{\text{blue}}{[Ni(NH_3)_6]^{2+}} + 6\;H_2O$$

Complex ions, also called **coordination complexes,** have well-defined stoichiometries and structural arrangements. Usually, the formula of a coordination complex is enclosed in brackets to show that the metal and all its ligands form a single structural entity. When an ionic coordination complex is isolated from aqueous solution, the product is composed of the complex ion and enough counter-ions to give a neutral salt. In the chemical formula, the counter-ions are shown outside the brackets. Examples include the sulfate salt of $[Ni(NH_3)_6]^{2+}$, $[Ni(NH_3)_6]SO_4$, and the potassium salt of $[Fe(CN)_6]^{3-}$, $K_3[Fe(CN)_6]$.

Nature of Ligands

A ligand must have a lone pair of electrons that it donates to form a bond to the metal. For example, the Ni—N bonds in $[Ni(NH_3)_6]^{2+}$ form by overlap of an empty valence orbital on the metal with the lone pair sp^3 orbital on the nitrogen atom. The water ligands in $[Ni(H_2O)_6]^{2+}$ coordinate to the metal in a similar manner, with the oxygen atoms donating lone pairs to form Ni—O bonds.

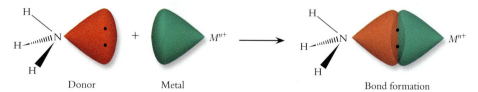

Donor   Metal   Bond formation

Hundreds of different ligands are known. Some of the most common are listed in Table 20-3. The simplest ligands coordinate to the metal through a single donor atom. The most common donor atoms in coordination chemistry are nitrogen, oxygen, and the halogens, but other important ligands have carbon, phosphorus, or sulfur donor atoms. Recall from Chapter 18 that a ligand with one donor atom is called monodentate.

Table 20-3 Common Ligands

	Combining Names	Formula or structure (abbreviation)	Donor atom
Monodentate	Bromo	Br^-	Br
	Chloro	Cl^-	Cl
	Fluoro	F^-	F
	Iodo	I^-	I
	Cyano	CN^-	C
	Carbonyl	CO	C
	Ammine*	NH_3	N
	N-Nitrito	NO_2^-	N
	Pyridine	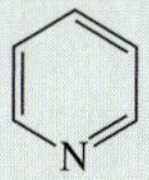 (pyr)	N
	N-thiocyanato	NCS^-	N
	Aqua	H_2O	O
	Oxo	O^{2-}	O
	Hydroxo	OH^-	O
	O-Nitrito	NO_2^-	O
	Carbonato	CO_3^{2-}	O
	Hydrido	H^-	H
	S-Thiocyanato	NCS^-	S
	Trimethylphosphine†	$P(CH_3)_3$	P
Bidentate	Ethylenediamine	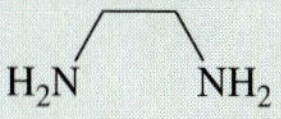 (en) H_2N ... NH_2	N, N
	Bipyridine	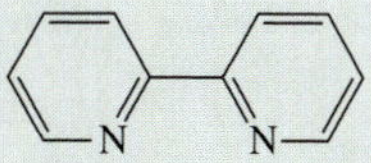 (bipy)	N, N
	Glycinato‡	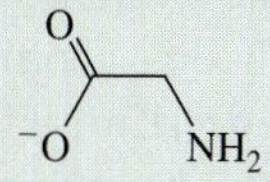	N, O
	Oxalato	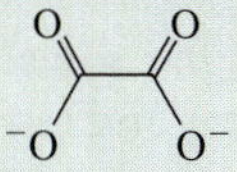 (ox)	O, O
Tetradentate	Porphinato	See Figure 20-25	N, N, N, N
Hexadentate	Ethylenediaminetetraacetato	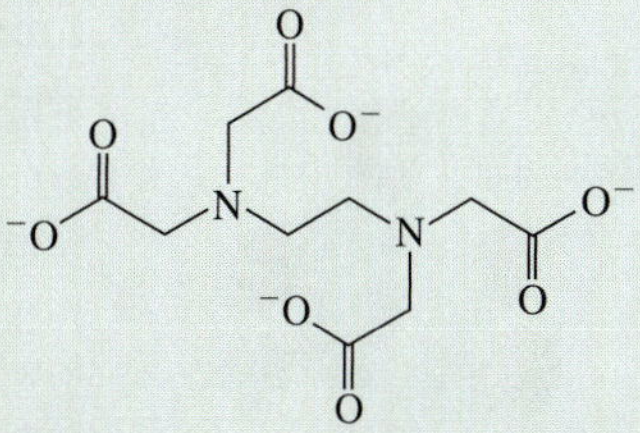 (EDTA)	N, N, O, O, O, O

*Amines (derivatives of NH_3), such as methylamine (CH_3NH_2), also are common ligands.
†This is the simplest of a group of ligands called *phosphines*, which have three organic groups bonded to phosphorus.
‡All of the common amino acids form bidentate ligands.

Ammonia is a monodentate ligand with a nitrogen donor atom. Other examples are water, halide ions, cyanide ion, hydroxide ion, and carbon monoxide.

A chelating ligand contains two or more donor atoms in a structure that allows the ligand to wrap around the metal. Examples featured in Chapter 18 are the bidentate ligand ethylenediamine (en, chemical formula $H_2NCH_2CH_2NH_2$) and the hexadentate ethylenediaminetetraacetate (EDTA).

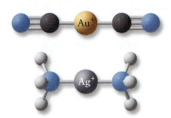

Figure 20-6
Complexes with coordination number 2 always adopt linear geometry about the metal cation.

Structures of Coordination Complexes

Coordination complexes are a remarkably diverse group of molecules that form from virtually all transition metals in a variety of oxidation states. These compounds involve an extensive array of ligands, and they adopt several molecular geometries.

The molecular geometry of a complex depends on the **coordination number,** which is the number of ligand atoms bonded to the metal. The most common coordination number is 6, and almost all metal complexes with coordination number 6 adopt octahedral geometry. This preferred geometry can be traced to the valence shell electron pair repulsion (VSEPR) model introduced in Chapter 9. The ligands space themselves around the metal as far apart as possible, to minimize electron–electron repulsion.

Although 6 is most prevalent, a coordination number of 4 is also common, and several important complexes have a coordination number of 2. In addition, a few complexes display coordination numbers of 3, 5, and 7. Examples of coordination number 2 include the silver–ammonia complex and the gold–cyanide complex, both described in Chapter 18. To minimize ligand–ligand repulsions, a complex with a coordination number of 2 is invariably linear, as Figure 20-6 shows.

Figure 20-7
Four-coordinate complexes adopt either square planar geometry, as in [AuCl$_4$]$^-$, or tetrahedral geometry, as in [NiCl$_4$]$^-$.

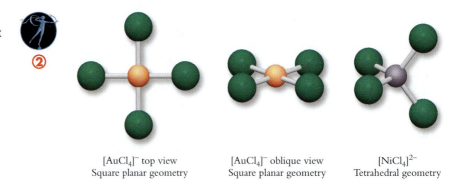

[AuCl$_4$]$^-$ top view
Square planar geometry

[AuCl$_4$]$^-$ oblique view
Square planar geometry

[NiCl$_4$]$^{2-}$
Tetrahedral geometry

Four-coordinate complexes may be either square planar or tetrahedral, as Figure 20-7 shows. Tetrahedral geometry is most common among the first-row transition metals. Examples include [Zn(NH$_3$)$_4$]$^{2+}$, [Cd(en)$_2$]$^{2+}$, [FeCl$_4$]$^-$, and [Ni(CO)$_4$]. Square planar geometry is characteristic of transition metal ions such as palladium(II), gold(III), and others that have eight d electrons in the valence shell. We describe the differences between square planar and tetrahedral geometry in more detail in Section 20.3.

Figure 20-8
There is only one isomer of [Co(NH$_3$)$_5$Cl]$^{2-}$. The top structure is identical to the bottom structure because the two can be superimposed on each other after a 90-degree rotation.

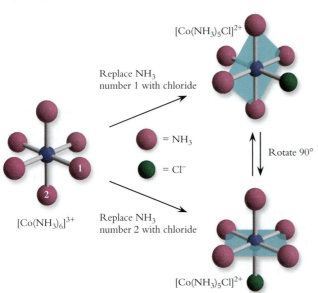

[Co(NH$_3$)$_5$Cl]$^{2+}$

Replace NH$_3$
number 1 with chloride

= NH$_3$

= Cl$^-$

Rotate 90°

[Co(NH$_3$)$_6$]$^{3+}$

Replace NH$_3$
number 2 with chloride

[Co(NH$_3$)$_5$Cl]$^{2+}$

Isomers

Chemical compounds that have the same formula but different structures are called **isomers.** Although they have the same formula, isomers usually have different chemical and physical properties. Isomers may have different colors, different melting or boiling points, different solubilities, and different rates of reactions with a common reagent. Isomerism is one of the most intriguing aspects of chemistry, not only among inorganic compounds but also among organic and biochemical substances.

In [Co(NH$_3$)$_6$]$^{3+}$, one ammonia ligand occupies each apex of an octahedron; this complex has no isomers. Replacement of one of the six NH$_3$ ligands with a chloride ion generates [Co(NH$_3$)$_5$Cl]$^{2+}$. Regardless of which ligand is replaced, the geometry of the resulting complex is the same, because the six positions around an octahedron are equivalent by symmetry, as illustrated in Figure 20-8. Any octahedral coordination complex of the general formula ML_5X has no isomers.

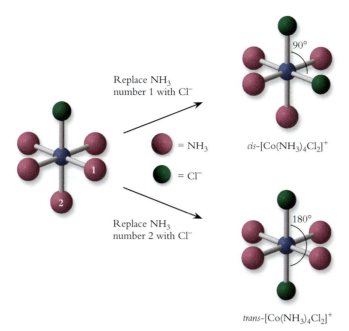

Figure 20-9
There are two isomers of
$[Co(NH_3)_4Cl_2]^+$. The *cis* isomer is
violet, and the *trans* isomer is
green.

Animation

Now suppose a second ammonia ligand is replaced with a chloride ion. If an ammonia ligand *adjacent to* the first chloride ion is replaced, the two chloride ions are separated by a bond angle of 90°. This is called the *cis* isomer (Figure 20-9). If the ammonia ligand *opposite* the chloride ion is replaced, the two chloride ions are separated by a bond angle of 180°. This is called the *trans* isomer. These two isomers of $[Co(NH_3)_4Cl_2]^+$ have the same chemical bonds but display different geometries, so they are **geometric isomers.** Geometric isomers have different properties. For example, whereas *cis*-$[Co(NH_3)_4Cl_2]^+$ is violet, the *trans* isomer is green. Example 20-1 introduces two more geometric isomers of coordination complexes.

Cis means "next to," and *trans* means "across from."

Draw ball-and-stick models of all possible isomers of the octahedral compound $[Cr(NH_3)_3Cl_3]$.

Strategy: It is best to approach this problem by starting with $[Cr(NH_3)_6]^{3+}$ and replacing one ligand at a time, considering all possible structures. The first substitution gives only one compound. Introducing a second chloride ion creates *cis* and *trans* isomers of $[Cr(NH_3)_4Cl_2]^+$. Choose each of these isomers in turn, and consider all possible structures that result from introducing a third chloride ligand. This is best accomplished by drawing pictures.

Solution: For the *trans* isomer, replacing any of the four remaining NH₃ ligands gives the same product, in which three of the same ligand are arranged in a T shape about the metal ion. This shape is designated *mer*:

Example 20-1

Isomers
of Coordination Complexes

In this isomer, three of the same ligands lie on an equatorial circle about the complex. Such a circle is called a *meridian,* and this isomer is the meridianal isomer, abbreviated *mer.*

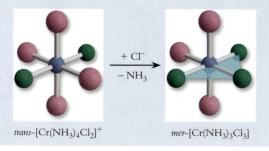

Animation

Example 20-1

**Isomers
of Coordination
Complexes** (*continued*)

Starting with *cis*-[Cr(NH₃)₄Cl₂]⁺, it is possible to make two isomers of the final product. This can be seen most clearly by labeling the four NH₃ ligands. Replacing an NH₃ ligand at site *a* or *b* with Cl⁻ generates the T-shaped arrangement. However, replacement at position *c* or *d* gives a new isomer in which the three chloride ligands occupy the corners of a triangle. This shape is designated *fac*:

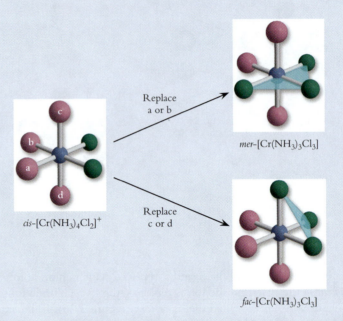

The three like ligands lie in a plane that forms one face of the octahedron, shown in outline in the drawing. This is called the *facial* isomer, abbreviated *fac*.

This exhausts all possible structures of [Cr(NH₃)₃Cl₃]. Octahedral coordination complexes of general formula *ML₃X₃* have two isomers.

Do the Results Make Sense?   No matter how many times the relative positions of NH₃ or Cl are exchanged around the metal center, only two possible structures result: *fac* and *mer*.

**Extra Practice
Exercise 20.1**

Platinum forms a large collection of square planar complexes. Draw ball-and-stick models similar to those in Example 20-1 for the *cis* and *trans* isomers of [Pt(NH₃)₂Cl₂].

Another type of isomerism displayed by coordination complexes is based on the bonding of the ligand. **Linkage isomers** occur when a ligand can bond to a metal using either of two donor atoms. Figure 20-10 shows the two linkage isomers of [Co(NH₃)₅NO₂]²⁺, in which either the nitrogen atom or the oxygen atom of the nitrite ion can bind to cobalt. Linkage isomers are distinct compounds that have different properties. For example, the linkage isomers of [Co(NH₃)₅NO₂]²⁺ have slightly different colors.

Naming Coordination Compounds

Originally, compounds containing coordination complexes were given common names such as Prussian blue (KFe[Fe(CN)₆]), which is deep blue, or Reinecke's salt (NH₄[Cr(NH₃)₂(NCS)₄]), named for its first maker. Eventually, coordination compounds became too numerous for chemists to keep track of all the common names. To solve the nomenclature problem, the International Union of Pure and Applied Chemistry (IUPAC) created a systematic procedure for naming coordination compounds. The following guidelines are used to determine the name of a coordination compound from its formula, or vice versa:

1. As with all salts, name the cation before the anion.

2. Within the complex, first name the ligands in alphabetical order, and then name the metal.

3. If the ligand is an anion, add the suffix –*o* to the stem name: *bromo* (Br⁻), *cyano* (CN⁻), and *hydroxo* (OH⁻). The simplest neutral ligands have special names: *aqua* (H_2O), *ammine* (NH_3), and *carbonyl* (CO). Other neutral ligands retain their usual names. Some common ligands and their names appear in Table 20-3.

4. Use a Greek prefix (*di-, tri-, tetra-, penta-, hexa-*) to indicate the number of identical ligands. If the name of the ligand already incorporates one of these prefixes (as in ethylenediamine), enclose the ligand name in parentheses and use the alternative prefixes *bis-* (two), *tris-* (three), and *tetrakis-* (four). These numerical prefixes are ignored in determining the alphabetical order of the ligands.

5. If the coordination complex is an anion, add the suffix –*ate* to the stem name of the metal.

6. After the name of the metal, give the oxidation number of the metal in parentheses as a Roman numeral or as 0 if the oxidation number is zero.

Example 20-2 provides practice in applying these guidelines.

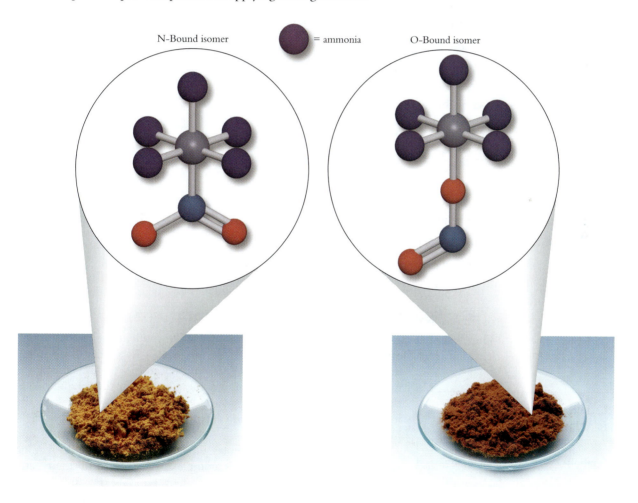

Figure 20-10
Linkage isomers occur when a ligand is capable of bonding to a metal through either of two donors. In $[Co(NH_3)_5NO_2]^{2+}$, the nitro ligand can bond to cobalt through the N atom or the O atom. These isomers have slightly different colors, as the photos show.

Example 20-2

Naming Coordination Compounds

What is the IUPAC name for each of the following coordination compounds? (a) $[Ni(H_2O)_6]SO_4$; (b) $[Cr(en)_2Cl_2]Cl$; and (c) $K_2[CoCl_4]$.

Strategy: Follow the guidelines for naming coordination compounds. Break the complex ion down one piece at a time.

Solution:

a. The cation is the complex ion. Name the ligand and include the appropriate prefix: hexaaqua. The formula lists one sulfate anion outside the brackets. Ions outside the brackets are not part of the complex, so the net charge on the nickel-containing cation is +2. Aqua is neutral, requiring that Ni be in the +2 oxidation state: hexaaquanickel(II) sulfate.

b. The ligands are chloro and ethylenediamine. The coordination compound has a +1 charge and contains two anionic ligands, so chromium must be in the +3 oxidation state. The en ligand contains *di* as part of its formal name, so we use the prefix *bis* to specify two en ligands: dichlorobis(ethylenediamine)-chromium(III) chloride.

c. The cation, potassium, is named first. The complex anion has four negative chloro ligands. The complex is an anion with −2 charge, requiring that Co be in the +2 oxidation state. Since the complex is an anion, the metal is given a suffix of *-ate*: potassium tetrachlorocobaltate(II).

Do the Results Make Sense? Verify that the numerical prefixes give the correct total number of ligands, that the ligands are named alphabetically, and that the overall species is charge-neutral.

Extra Practice Exercise 20.2

Name the following coordination compounds:
(a) $[Co(NH_3)_5CO_3]NO_3$;
(b) $[Mo(CO)_3(NH_3)_3]$; and
(c) $[Al(H_2O)_5OH]Cl_2$.

The following examples illustrate further applications of the IUPAC guidelines:

$Na_2[FeEDTA]$	Sodium ethylenediaminetetraacetatoferrate(II)
$[Zn(NH_3)_4](NO_3)_2$	Tetraamminezinc nitrate
$K[Ag(CN)_2]$	Potassium dicyanoargentate(II)
$[Rh(NH_3)_5Br]Br_2$	Pentaamminebromorhodium(III) bromide

These examples show that when information is not needed to identify the compound, it is omitted from the name. In the first name, for instance, it is not necessary to tell how many sodium ions are present, because we can deduce the number from the name of the complex anion. In the second name, the oxidation state of zinc is omitted because it is always +2. In the fourth name, the single bromo ligand is not preceded by the prefix mono.

The first and third examples illustrate a nuance of the naming rules. Iron and silver in anionic complexes are named by their Latin roots, *ferr-* and *argent-,* from which their symbols (Fe and Ag) are derived. Metals taking their Latin names in anionic coordination complexes are listed in Table 20-4.

Example 20-3 provides more practice in working with the names of coordination compounds. Our Chemical Milestones Box describes the detective work that led to the birth of coordination chemistry.

Table 20-4 Latin Names of Metals in Anionic Complexes

Metal (Symbol)	Latin Name
Iron (Fe)	Ferrate
Copper (Cu)	Cuprate
Silver (Ag)	Argentate
Tin (Sn)	Stannate
Gold (Au)	Aurate
Lead (Pb)	Plumbate

| BOX 20-1 | **Chemical Milestones: The Birth of Coordination Chemistry** |

L ate in the nineteenth century, just as the principles of chemical bonding were being discovered, chemists carried out many studies of the interactions of ammonia with cations such as Cr^{3+}, Co^{3+}, Pt^{4+}, and Pd^{2+}. The most intriguing results were obtained for cobalt(III) chloride. By 1890, several ammonia compounds of $CoCl_3$ had been isolated. These coordination compounds differed in several of their properties, the most striking of which were their beautiful colors. At the time, the formulas of these cobalt complexes were written as follows:

$CoCl_3 \cdot 6\,NH_3$	Yellow-orange
$CoCl_3 \cdot 5\,NH_3$	Rose
$CoCl_3 \cdot 4\,NH_3$	Green
$CoCl_3 \cdot 3\,NH_3$	Green

Chemists were convinced that all the chlorine atoms had to be bonded to the cobalt in some way, and since ammonia is a gas at room temperature they could not understand why the ammonia molecules did not evaporate away. One of the first proposed explanations was the chain theory, in which the ammonia molecules were assumed to form chains between the metal and the chloride. For example, a prominent coordination chemist of the time proposed the following structure for $CoCl_3 \cdot 6\,NH_3$:

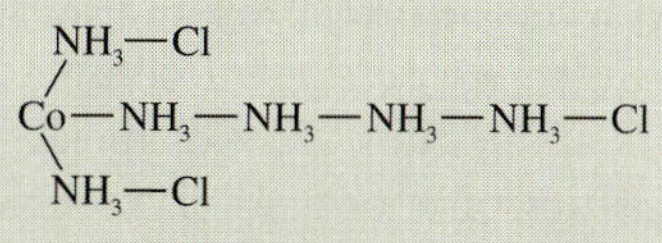

$$Co \begin{matrix} NH_3\!\!-\!\!Cl \\ -NH_3\!\!-\!\!NH_3\!\!-\!\!NH_3\!\!-\!\!NH_3\!\!-\!\!Cl \\ NH_3\!\!-\!\!Cl \end{matrix}$$

During the 1890s, these cobalt complexes attracted the attention of a Swiss chemist, Alfred Werner (1866–1919). Only in his early twenties, Werner had just earned his Ph.D. in organic chemistry. He studied the cobalt complexes in detail and developed the basis for our understanding of coordination chemistry.

Werner found that adding aqueous $AgNO_3$ to solutions of the various cobalt ammine complexes gave different amounts of silver chloride precipitate. For $CoCl_3 \cdot 6\,NH_3$, all three chloride ions precipitated as AgCl. Only two chloride ions precipitated for $CoCl_3 \cdot 5\,NH_3$, just one precipitated for $CoCl_3 \cdot 4\,NH_3$, and there was no precipitate for $CoCl_3 \cdot 3\,NH_3$. Werner explained these results by proposing that the cobalt ion in all four compounds had a coordination number of 6. He reformulated the compounds as follows:

$$CoCl_3 \cdot 6\,NH_3 = [Co(NH_3)_6]Cl_3$$
$$CoCl_3 \cdot 5\,NH_3 = [Co(NH_3)_5Cl]Cl_2$$
$$CoCl_3 \cdot 4\,NH_3 = [Co(NH_3)_4Cl_2]Cl$$
$$CoCl_3 \cdot 3\,NH_3 = [Co(NH_3)_3Cl_3]$$

The chloride ions that appear outside the brackets represent chloride anions that balance the positive charge on the coordination compound. When a coordination compound dissolves in water, the ligands (inside the brackets) remain bound to the metal cation, but the nonligands (outside the brackets) exist as individual ions. These chloride ions precipitate in the presence of silver ions. The chloride ions inside the brackets, which are ligands bonded to the cobalt center, do not precipitate as AgCl.

This elegant insight into chemical structure, based on simple stoichiometric relations, was one reason why Werner was awarded the Nobel Prize in Chemistry in 1913. Since then, a long path of discovery has uncovered many remarkable properties of coordination complexes. To give just one example, there are two isomers of the square planar platinum complex $[Pt(NH_3)_2Cl_2]$. The *cis* isomer, known as cisplatin, is an effective anticancer drug, but the *trans* isomer shows no anticancer activity at all. The mechanism by which the *cis* isomer destroys cancer cells is not fully understood, but research indicates that cisplatin disrupts cell duplication by inserting into the DNA double helix. It appears that DNA binds this platinum complex by replacing the two chloride ions. In a sense, the DNA molecule becomes a huge ligand for the platinum atom. Apparently, the coordination complex can form bonds to DNA only when the chloride ions are adjacent to each other in the *cis* configuration.

cis-$[Pt(NH_3)_2Cl_2]$

Example 20-3	Determine the formulas of the following coordination compounds:
Formula of a Coordination Compound	a. *fac*-Triamminetriiodoruthenium(III)
	b. *cis*-Chlorohydridobis(trimethylphosphine)platinum(II)
	c. Sodium hexacyanoferrate(II)

Strategy: To obtain the formula, break the name down, one piece at a time.

Solution:

a. Determine the number of each type of ligand: triammine = 3 NH_3; triiodo = 3 I^-. Identify the metal (ruthenium) and its oxidation state (III). Retain the italicized prefix to identify a particular isomer: $[fac\text{-}Ru(NH_3)_3I_3]$.

b. Chloro = Cl^-, Hydrido = H^-, and bis(trimethylphosphine) = 2 $P(CH_3)_3$. The metal is Pt^{2+}: $cis\text{-}[PtClH(P(CH_3)_3)_2]$.

c. Sodium is the cation. The metal is Fe^{2+}, and there are six cyano anions. Overall, then, the complex anion has a net charge of -4. Four Na^+ cations are required to balance the charge: $Na_4[Fe(CN)_6]$.

Do the Results Make Sense? Verify that the number of ligands matches the numerical prefixes in the name, that the ligands are listed alphabetically, and that the overall species is charge-neutral.

Extra Practice Exercise 20.3	Determine the formulas of the following coordination compounds: (a) tris(bipyridine)ruthenium(II) chloride; (b) potassium hexachloroplatinate(IV); (c) pentaaquaoxoiron(III) ion.

Section Exercises

20.2.1 For each of the following, determine the charge on the complex ion, the oxidation state of the metal, and the coordination number of the metal: (a) $K_3[Fe(CN)_6]$; (b) $[V(NH_3)_4Cl_2]$; and (c) $[Ni(en)_2]SO_4$.

20.2.2 Draw all of the isomers of $[Cr(en)(NH_3)_2I_2]^+$.

20.2.3 Name the following coordination compounds:
(a) $[Cr(NH_3)_5I]SO_4$;
(b) $K_4[PtCl_6]$;
(c) *cis*-$[Fe(CO)_4Cl_2]$; and
(d) $[Fe(H_2O)_6]Cl_2$.

20.3 BONDING IN COORDINATION COMPLEXES

In the simplest view, a metal−ligand σ bond can be described as the overlap of a filled donor orbital on the ligand with an empty acceptor orbital on the metal. There is one such interaction for each ligand donor atom, so an octahedral complex of the general formula $[ML_6]^{n+}$ has six σ bonds. Although simple, this view of metal−ligand bonding does not explain the colors and other properties of coordination complexes. Molecular orbital theory provides the most complete description of the bonding in coordination complexes, but this approach is beyond the scope of introductory chemistry. Instead, we use a model called **crystal field theory** to explain the colors and magnetic properties of coordination compounds.

Crystal field theory focuses on electrical interactions between a transition metal ion and its ligands. In this view, the complex is held together by attractive forces between the negatively charged electrons of the ligand lone pairs and the positive charge of the metal ion. Consequently, bonding can be described much like the bonding in an ionic crystal. Like all models, crystal field theory has benefits and limitations. The model is easy to apply and can explain several important properties of transition metal complexes. Nevertheless, all metal−ligand bonds have covalent character, which limits the usefulness of an ionic model.

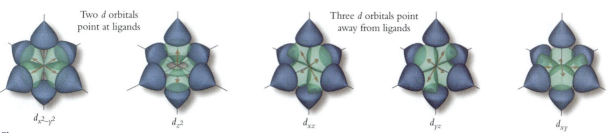

Figure 20-11
The five *d* orbitals and their relationship to an octahedral set of ligands. Whereas two orbitals point directly at the ligands, the other three orbitals point between the ligands.

Orbital Stability in Octahedral Complexes

In a free metal ion without any ligands, all five *d* orbitals have identical energies, but what happens to the *d* orbitals when six ligands are placed around a metal in octahedral geometry? The complex is stabilized by attractions between the positive charge of the metal ion and negative electrons of the ligands. At the same time, the electrons in the metal *d* orbitals repel the electrons on the ligands. Electron–electron repulsion affects some of the *d* orbitals more than others. Figure 20-11 shows that two orbitals, $d_{x^2-y^2}$ and d_{z^2}, point directly toward the ligands. As a consequence, electrons in these orbitals experience greater electron–electron repulsion than do electrons in d_{xz}, d_{yz}, and d_{xy}, which point between the ligands. This means that metal valence electrons are more stable when they occupy the *d* orbitals that point away from the ligands.

Figure 20-12 summarizes the electrical interactions of an octahedral complex ion. The three orbitals that are more stable are called t_{2g} orbitals, and the two less stable orbitals are called e_g orbitals. The difference in energy between the two sets is known as the **crystal field splitting energy,** symbolized by the Greek letter Δ.

Populating the *d* Orbitals

Electron configurations of transition metal complexes are governed by the principles described in Chapter 8. The Pauli exclusion principle states that no two electrons can have identical descriptions, and Hund's rule requires that all unpaired electrons have the same spin orientation. These concepts are used in Chapter 8 for atomic configurations and in Chapters 9 and 10 to describe the electron configurations of molecules. They also determine the electron configurations of transition metal complexes.

The number of electrons in the *d* orbitals depends on the electron configuration of the metal. That configuration can be found from the oxidation state of the metal and its atomic number. As an example, consider $[Cr(NH_3)_6]^{3+}$. The ionic charge on the complex is +3, and because ammonia is a neutral ligand, all the charge must arise from the metal. In other words, chromium has an oxidation state of +3, which means that three of its valence electrons have been removed. Chromium is in Group 6 of the periodic table, and Group 6 atoms have six valence electrons. Thus, Cr^{3+} has three valence electrons, all in 3*d* orbitals: $[Ar] 3d^3$. One *d* electron occupies each of the three t_{2g} orbitals, and all three have the same spin:

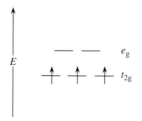

The crystal field electron configuration of $[Cr(NH_3)_6]^{3+}$ can be summarized as $t_{2g}^3 e_g^0$. Example 20-4 describes the electron configuration of another coordination complex.

The names t_{2g} and e_g are derived from symmetry properties of the orbitals that are not important for general chemistry.

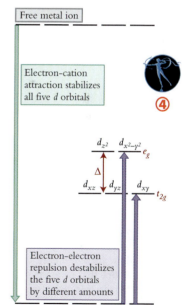

Figure 20-12
The crystal field energy level diagram for octahedral coordination complexes. The energies of the *d* orbitals differ because of differing amounts of electron–electron repulsion. The t_{2g} orbitals point between the ligands, have less electron repulsion, and are more stable than the e_g orbitals, which point directly at the ligands.

| Example 20-4 | Draw an energy level diagram and write the d electron configuration of $[Pt(en)_3]Cl_2$. |

Electron Configurations

Strategy: Identify the ligands and the geometry of the coordination complex, construct the crystal field energy level diagram, count d electrons from the metal and place them according to the Pauli principle and Hund's rule.

Solution: Ethylenediamine is a bidentate ligand, so there are six donor atoms, giving the complex ion octahedral geometry. The two chlorides outside the bracket indicate that the complex ion is a $+2$ cation. Since en is a neutral ligand, Pt is in its $+2$ oxidation state. Platinum (Group 10) has ten valence electrons, two of which are removed to give the $+2$ oxidation state, leaving eight d electrons to be placed in the energy level diagram. The Pauli principle and Hund's rule dictate the result:

The electron configuration is $t_{2g}^{6}e_{g}^{2}$.

Does the Result Make Sense? The d electron count is correct (8), and the filling order follows both the Pauli principle and Hund's rule.

Extra Practice Exercise 20.4

Draw an energy level diagram and write the d electron configuration for $[V(H_2O)_6]^{3+}$.

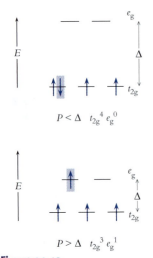

Figure 20-13
The two possible electron configurations for a d^4 metal complex. Notice that whereas Δ can be shown on this diagram, P cannot, because P measures electron–electron repulsion, not an orbital energy.

The examples given so far lead to unambiguous electron configurations, but not all configurations are this straightforward. Consider any octahedral complex containing a metal atom with four d electrons. Following the standard filling procedure, the first three electrons are placed in the t_{2g} orbitals. The fourth electron might also be placed in a t_{2g} orbital, but there is a price to pay in energy: Two electrons in the same orbital are destabilized because two negative charges are confined to the same region of space (electron–electron repulsion). This destabilization is called the **pairing energy (P).** There is no pairing energy if the fourth electron is placed in an e_g orbital, but these orbitals are higher in energy than the t_{2g} orbitals. The energy difference between the two sets is the crystal field splitting energy, Δ.

Whether $t_{2g}^{4}e_g^{0}$ or $t_{2g}^{3}e_g^{1}$ is more stable depends on the relative magnitudes of P and Δ. If the energy required to pair electrons in a t_{2g} orbital is less than the energy required to populate an e_g orbital ($P < \Delta$), the ground state configuration is $t_{2g}^{4}e_g^{0}$. On the other hand, if $P > \Delta$, the most stable configuration is $t_{2g}^{3}e_g^{1}$. Both possibilities are shown in Figure 20-13 for the four-electron, octahedral case.

In conformity with Hund's rule, a metal complex with the configuration $t_{2g}^{3}e_g^{1}$ has four unpaired electrons. This configuration maximizes the number of electrons with unpaired spins, so it is described as **high-spin.** The alternative configuration, $t_{2g}^{4}e_g^{0}$, has just two unpaired electrons and is described as **low-spin.**

Magnetic Properties of Coordination Complexes

What experiments can be done to reveal whether a complex with four d electrons is $t_{2g}^{4}e_g^{0}$ or $t_{2g}^{3}e_g^{1}$? One of the most common ways to determine the electron configuration of a coordination complex is to measure its magnetic properties.

Recall from Chapters 7 and 8 that the spin of an electron generates magnetism. When electrons are paired, their spins point in opposite directions, causing their magnetism to cancel. A molecule with all electrons paired has no magnetism from its electron spins and is diamagnetic, but a molecule that has unpaired electrons is paramagnetic. The amount of magnetism in a paramagnetic molecule depends on the number of unpaired electrons: The more unpaired electrons in the d orbitals of a complex, the greater the magnetism. The magnetic properties of a complex ion can be measured with an instrument known as a *Gouy balance,* as shown schematically in Figure 20-14. The paramagnetic sample is attracted into the magnetic field, creating a downward force on the left pan. Weights are added to the pan on the right until the downward force is balanced. The magnetic strength of the sample is proportional to the mass required to balance the pans. This provides an experimental method for determining the ground-state electron configuration of a coordination complex.

Example 20-5 illustrates the magnetic properties of another configuration.

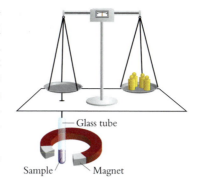

Figure 20-14
The number of unpaired electrons in a paramagnetic complex can be determined with a Gouy balance.

Glass tube

Sample   Magnet

$[Fe(NH_3)_6]^{2+}$ is paramagnetic, but $[Co(NH_3)_6]^{3+}$ is not. Write the electron configuration for each of these metal complexes and draw energy level diagrams showing which has the higher Δ.	**Example 20-5** **High- and Low-Spin Complexes**

Strategy: Use the formula of a complex to determine its geometry, the form of its energy level diagram, and the number of d electrons. The spin information indicates how many of these electrons are paired.

Solution: Both complexes have six monodentate ligands, so both are octahedral.

Because ammine ligands are neutral molecules, the oxidation state of each metal is the same as the charge on the complex. Iron loses two of its eight valence electrons to reach the +2 oxidation state, leaving six electrons for the d orbitals. Likewise, cobalt in its +3 oxidation state has six d electrons.

The iron complex is paramagnetic, which means that it has unpaired electrons. This happens when the crystal field splitting energy is less than the pairing energy, so the energy level diagram should show a relatively small Δ. By Hund's rule, we place one electron in each of the five d orbitals before pairing any. The sixth d electron pairs up with an electron in any one of the t_{2g} orbitals.

The cobalt complex is not paramagnetic, which means that it has all of its valence electrons paired. This happens when Δ is larger than P, so the energy level diagram should show a relatively large Δ. In a low-spin complex, the lower-energy t_{2g} orbitals are filled completely before putting electrons in e_g. Here are the correct diagrams:

$$[Fe(NH_3)_6]^{2+}$$
High-spin
$\Delta < P$
$t_{2g}{}^4 e_g{}^2$

$$[Co(NH_3)_6]^{3+}$$
Low-spin
$\Delta > P$
$t_{2g}{}^6 e_g{}^0$

The paramagnetic configuration is $t_{2g}{}^4 e_g{}^2$, with four unpaired electrons. The other configuration is $t_{2g}{}^6 e_g{}^0$, with no unpaired electrons and zero magnetism.

Does the Result Make Sense? Both configurations are d^6, and both follow Hund's rule, so the answer is reasonable.

$[Mn(H_2O)_6]^{2+}$ is more paramagnetic than $[Re(H_2O)_6]^{2+}$. Write the d electron configuration for each of these metal complexes and draw energy level diagrams showing which has the higher Δ.	**Extra Practice Exercise 20.5**

Contributions to Crystal Field Splitting Energy

Whether a complex is low-spin or high-spin depends on the balance between pairing energy and crystal field splitting energy. Pairing energy changes very little from one coordination complex to the next. Consequently, electron configurations of coordination complexes are governed by the crystal field splitting energy, Δ.

The most important factor that influences the value of Δ is the identity of the ligands. Compare $[Fe(CN)_6]^{4-}$ and $[FeCl_6]^{4-}$, two coordination complexes similar in every respect except for their ligands. Both contain Fe^{2+} ions with six d electrons, both have six ligands arranged in an octahedral geometry, and both have the same net charge. Despite these similarities, the cyano complex is low-spin and diamagnetic, whereas the chloro complex is high-spin and paramagnetic. For $[Fe(CN)_6]^{4-}$, the crystal field splitting energy is larger than the pairing energy, so the molecule is low-spin. In contrast, $P > \Delta$ for $[FeCl_6]^{4-}$. The chloro ligand generates a small energy gap between the t_{2g} and e_g orbitals, giving a high-spin coordination complex.

Studies of many coordination complexes reveal a common pattern in the energetic effects of the various ligands. This pattern is described by the **spectrochemical series,** in which ligands are listed in order of increasing energy level splitting:

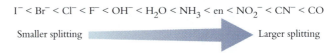

$$I^- < Br^- < Cl^- < F^- < OH^- < H_2O < NH_3 < en < NO_2^- < CN^- < CO$$

Smaller splitting ⟶ Larger splitting

The spectrochemical series was established from experimental measurements. The ranking of ligands cannot be fully rationalized using crystal field theory, and more advanced bonding theories are beyond the scope of general chemistry.

The oxidation state of the metal also contributes to the crystal field splitting energy. For a given ligand, Δ increases as the oxidation state of the metal increases. This trend is easy to interpret using electrical forces. As the charge on the metal increases, so does the attraction between the metal and its ligands. The ligands approach the metal more closely, resulting in stronger orbital interactions and a larger crystal field splitting (Figure 20-15). The complexes of Example 20-5, $[Fe(NH_3)_6]^{2+}$ (high-spin) and $[Co(NH_3)_6]^{3+}$ (low-spin) illustrate this feature. These two species have the same ligands and the same molecular geometry, and both are d^6 metal ions, but Co^{3+} generates a stronger crystal field than Fe^{2+}.

Another influence on the magnitude of the crystal field splitting is the position of the metal in the periodic table. Crystal field splitting energy increases substantially as valence orbitals change from $3d$ to $4d$ to $5d$. Again, orbital shapes explain this trend. Orbital size increases as n increases, and this means that the d orbital set becomes more exposed to orbital interactions with approaching ligands. For example, the value of Δ for

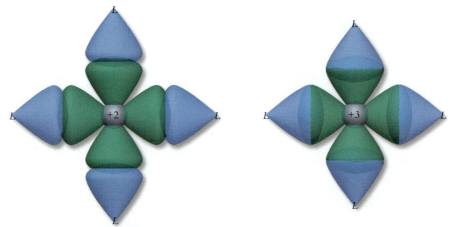

Figure 20-15
As the charge on the metal ion increases, electrical attraction pulls the ligands closer to the cation. This leads to greater repulsive interactions between valence d electrons and ligand electrons.

complexes of rhodium is about 50% greater than that for analogous cobalt complexes, and there is another 25% increase on moving from rhodium to iridium. As a result of this trend, coordination complexes of $4d$ and $5d$ elements are almost always low-spin. In contrast, the $3d$ transition metals have both low- and high-spin complexes. Example 20-6 applies the principles of crystal field splitting to a series of coordination complexes.

Arrange the following complexes in order of increasing crystal field splitting: $[Fe(H_2O)_6]^{2+}$, $[Fe(H_2O)_6]^{3+}$, $[FeCl_6]^{4-}$, and $[Ru(H_2O)_6]^{3+}$.	**Example 20-6** **Crystal Field Splitting Energy**

Strategy: The strength of splitting depends on the ligand, the charge on the metal, and the position of the metal in the periodic table. Examine these factors independently.

Solution:

Ligand: Cl^- is lower in the spectrochemical series than H_2O. Thus, $[FeCl_6]^{4-}$ has low splitting relative to the others.

Cation charge: In $[FeCl_6]^{4-}$ and $[Fe(H_2O)_6]^{2+}$, iron is in its +2 oxidation state, but the metal is +3 in the other two complexes. Complexes with +3 oxidation states have larger splitting:

$$[FeCl_6]^{4-} < [Fe(H_2O)_6]^{2+} < [Fe(H_2O)_6]^{3+} \text{ and } [Ru(H_2O)_6]^{3+}$$

Valence orbitals: Three complexes contain iron, a metal with $3d$ valence electrons, but the fourth contains ruthenium, with $4d$ valence electrons. All other things being equal, the crystal field splitting for a $4d$ metal is about 50% larger than that for a $3d$ metal. Applying all three trends gives an unambiguous arrangement of these complexes:

$$[FeCl_6]^{4-} < [Fe(H_2O)_6]^{2+} < [Fe(H_2O)_6]^{3+} < [Ru(H_2O)_6]^{3+}$$

Lowest splitting ⟶ Highest splitting

Does the Result Make Sense? The ranking is consistent with the three trends for determining Δ: nature of the ligands, oxidation state of the metal, and position of the metal in the periodic table.

Arrange the following complexes in order of increasing crystal field splitting: $[Cr(NH_3)_6]^{3+}$, $[Cr(H_2O)_6]^{2+}$, $[Mo(NH_3)_6]^{3+}$, and $[Mo(H_2O)_6]^{3+}$.	**Extra Practice Exercise 20.6**

Color in Coordination Complexes

Color is a spectacular property of coordination complexes. For example, the hexaaqua cations of $3d$ transition metals display colors ranging from orange through violet (see photo at right). The origin of these colors lies in the d orbital energy differences and can be understood using crystal field theory.

Color is caused by absorption of some of the light from the visible spectrum (Figure 20-16). The wavelengths of light absorbed by a collection of molecules are lost from the rest of the spectrum. Light that is not absorbed is reflected by an opaque sample and is

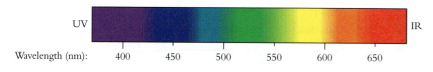

Figure 20-16
The visible spectrum, showing the colors as a function of wavelength.

transmitted through a transparent sample. In either case, an observer sees all wavelengths except those absorbed by the molecules. If a sample absorbs all wavelengths of visible light except blue, that sample will appear blue to the observer. A substance that appears black absorbs all wavelengths of visible light. A substance that absorbs no visible wavelengths appears white if light is reflected by the surface and colorless if the light is transmitted through the sample.

Molecules that absorb a small wavelength region have a color that is different from the color absorbed. For example, a molecule that absorbs only orange light appears blue. Likewise, a molecule that absorbs only blue appears orange. Orange and blue are said to be **complementary colors.** The complementary colors are listed in Table 20-5.

When a molecule absorbs light, it gains energy from the absorbed photons. Energy conservation requires that the energy change for the molecule equals the energy of the absorbed photon:

$$\Delta E_{molecule} = E_{photon} = h\nu$$

This requirement means that when a coordination complex absorbs light, the crystal field splitting energy, Δ, must match the energy of the absorbed light. Thus, the crystal field splitting energy of a complex can be determined from the wavelength of visible light that the complex absorbs:

$$\Delta E_{molecule} = \Delta = h\nu = \frac{hc}{\lambda}$$

Spectroscopic properties of coordination compounds are highlighted by a series of d^3 Cr^{3+} complexes, as shown in Figure 20-17. Each molecule has a ground-state configuration with three unpaired electrons in the t_{2g} orbitals. Absorption of a photon excites one electron into an e_g orbital, giving an excited-state configuration, $t_{2g}{}^2 e_g{}^1$. These three complexes have different values of Δ, so they absorb light of different wavelengths. The cyano ligand is high in the spectrochemical series, generating a large crystal field splitting for $[Cr(CN)_6]^{3-}$. A photon of high-energy violet light is required to promote an electron from the t_{2g} set to the e_g set. Absorption of violet light gives $[Cr(CN)_6]^{3-}$ a yellow color. The fluoro ligand is near the bottom of the spectrochemical series, generating a weak crystal field. The fluoro complex absorbs red light, generating a green color for $[CrF_6]^{3-}$. When H_2O is the ligand, there is an intermediate energy gap. Yellow-green light is absorbed, so $[Cr(H_2O)_6]^{3+}$ is violet.

Table 20-5   Relationships Among Wavelength, Color, and Crystal Field Splitting Energy (Δ)

Wavelength (nm)	Color Absorbed	Complementary Color	Δ (kJ/mol)
>720	Infrared	Colorless	<165
720	Red	Green	166
680	Red-orange	Blue-green	176
610	Orange	Blue	196
580	Yellow	Indigo	206
560	Yellow-green	Violet	214
530	Green	Purple	226
500	Blue-green	Red	239
480	Blue	Orange	249
430	Indigo	Yellow	279
410	Violet	Lemon-yellow	292
<400	Ultraviolet	Colorless	>299

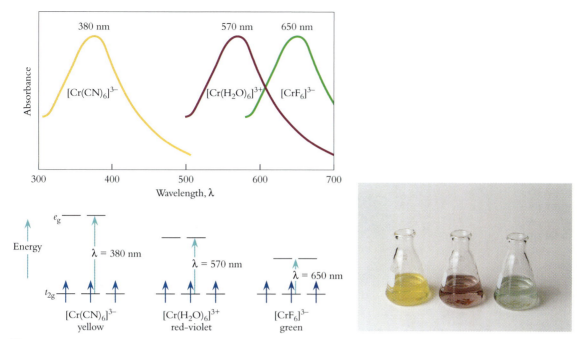

Figure 20-17
The colors of Cr^{3+} coordination complexes depend on the positions of the ligands in the spectrochemical series.

Example 20-7 shows how to determine the value of Δ from an absorption spectrum.

Example 20-7

Determining the Value of Δ

Titanium(III) chloride dissolves in water to give $[Ti(H_2O)_6]^{3+}$. This complex ion has the absorption spectrum shown. From the wavelength at which maximum absorption occurs, predict the color of the solution and calculate Δ in kilojoules per mole.

Strategy: The spectrum shows wavelengths of light absorbed by the metal complex. The wavelength absorbed most strongly corresponds to Δ, but appropriate conversions are needed to find the molar energy associated with this wavelength. The color of the solution is the complementary color of the most strongly absorbed wavelength.

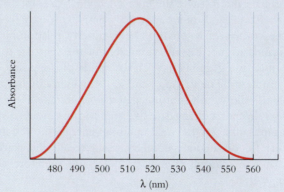

Solution: Interpolate on the graph to find that the maximum intensity occurs at 514 nm. Use this wavelength to find the energy of one photon:

$$E_{photon} = \frac{hc}{\lambda} = \frac{(6.626 \times 10^{-34}\,J\,s)(2.998 \times 10^{8}\,m\,s^{-1})}{(514\,nm)(10^{-9}\,m/nm)} = 3.86 \times 10^{-19}\,J/photon$$

Avogadro's number is used to convert to kilojoules per mole:

$$(3.86 \times 10^{-19}\,J/photon)(10^{-3}\,kJ/J)(6.022 \times 10^{23}\,photons/mol) = 232\,kJ/mol = \Delta$$

The $[Ti(H_2O)_6]^{3+}$ ion absorbs at 514 nm, which is in the green to blue-green region of the visible spectrum. According to Table 20-5, the solution should be reddish purple.

$[Ti(H_2O)_6]^{3+}$

886 / Chapter 20 / The Transition Metals

Does the Result Make Sense? The calculated value of Δ has the right units, and the numerical value matches the values found in Table 20-5. The color is right, as shown by the photo in the margin of page 885.

Extra Practice Exercise 20.7

The complex ion $[\text{Fe(ox)}_3]^{3-}$ has $\Delta = 169$ kJ/mol. Calculate the wavelength absorbed by the complex, and determine the color of the complex.

Figure 20-18
In square planar geometry, the $d_{x^2-y^2}$ orbital is most strongly destabilized by interactions with the ligands. The d_{xy} orbital also lies in the ligand plane, but it points *between* the ligands.

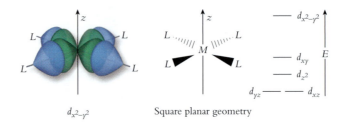

Square planar geometry

Square Planar and Tetrahedral Complexes

In a square planar complex, the four ligands lie along the x- and y-axes. As shown in Figure 20-18, $d_{x^2-y^2}$ is the only d orbital that points directly at the four ligands. Thus, in square planar geometry $d_{x^2-y^2}$ is significantly higher in energy than the other four d orbitals. Although the d_{xy} orbital also lies in the metal–ligand plane, it points *between* the ligands and is subject to less electron–electron repulsion than $d_{x^2-y^2}$. Because the remaining three d orbitals point out of the metal–ligand plane, these orbitals are repelled less by the ligands and are lower in energy than d_{xy}. The d_{z^2} orbital is slightly less stable than d_{xz} and d_{yz} because of the small band of electron density that circles the metal in the xy plane. The energy level diagram shown in Figure 20-18 shows the splitting pattern for a square planar complex. Square planar geometry is most common for d^8 metal ions with large crystal field splitting energies. The d^8 configuration fills the four lower-energy d orbitals, but it leaves the high-energy $d_{x^2-y^2}$ empty.

The ligands of a tetrahedral complex occupy the corners of a tetrahedron rather than the corners of a square. The symmetry relationships between the d orbitals and these ligands are not easy to visualize, but the splitting pattern of the d orbitals can be determined using geometry. The result is the opposite of the pattern found in octahedral complexes: The d_{xz}, d_{xy}, and d_{yz} orbitals are higher in energy than $d_{x^2-y^2}$ and d_{z^2}. Figure 20-19 shows the splitting pattern for tetrahedral coordination. Because there are only four ligands rather than six, and because none of the d orbitals point directly at the ligands, the crystal field splitting energy for tetrahedral geometry is only about half that for octahedral geometry. As a result, the pairing energy is almost always greater than the splitting energy ($P > \Delta$), and therefore almost all tetrahedral complexes are high-spin.

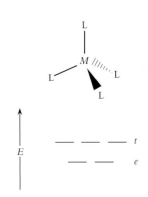

Figure 20-19
The crystal field energy level diagram for tetrahedral complexes. The d orbitals are split into two sets, with three orbitals destabilized relative to the two others.

Section Exercises

■ **20.3.1** Draw crystal field splitting diagrams that show the electron configurations for the following complex ions: (a) $[\text{Cr(H}_2\text{O)}_6]^{2+}$; (b) $[\text{IrCl}_6]^{3-}$; (c) $[\text{V(en)}_3]^{3+}$; and (d) $[\text{NiCl}_4]^{2-}$ (tetrahedral).

■ **20.3.2** Explain why hexacyano complexes of metals in their +2 oxidation state are usually yellow, but the corresponding hexaaqua compounds are often blue or green.

■ **20.3.3** The value of Δ for $[\text{RhCl}_6]^{3-}$ is 243 kJ/mol. What wavelength of light will promote an electron from the t_{2g} set to the e_g set? What color is the complex?

20.4 METALLURGY

Metallurgy is the production and purification of metals from naturally occurring deposits called *ores*. It has an ancient history and may represent the earliest useful application of chemistry. Metallurgical advances have had profound influences on the course of human civilization, so much so that historians speak of the Bronze Age (ca. 3000 to 1000 BC) and the Iron Age (starting ca. 1000 BC). Except for aluminum and tin, the story of metallurgy is primarily about the extraction and purification of transition metals from their ores. In this section, we discuss some of the techniques of metallurgy.

Nearly all transition metals are oxidized readily, so most ores are compounds in which the metals have positive oxidation numbers. Examples include oxides (TiO_2, rutile; Fe_2O_3, hematite; Cu_2O, cuprite), sulfides (ZnS, sphalerite; MoS_2, molybdenite), phosphates ($CePO_4$, monazite; YPO_4, xenotime; both found mixed with other rare earth metal phosphates), and carbonates ($FeCO_3$, siderite). Other minerals contain oxoanions ($MnWO_4$, wolframite) and even more complex structures such as carnotite, $K_2(UO_2)_2(VO_4)_2 \cdot 3\ H_2O$.

Figure 20-20 shows in schematic fashion some of the alternative paths leading from ores to pure metals. These paths include four general processes of which the essential chemical process is **reduction** to yield the neutral metal. First is *separation*. Generally, a metal ore obtained from a mine contains a particular compound of some desired metal mixed with various other materials. The mineral must be separated from these other contaminants. Separation often is followed by *conversion*, in which the mineral is treated chemically to convert it into a form that can be easily reduced. The third step is *reduction*. After a suitable compound has been obtained, it is reduced to free metal by chemical reaction with a reducing agent or by electrolysis. The metal obtained by reduction often contains small amounts of impurities, so the final step is *refining* to purify the metal.

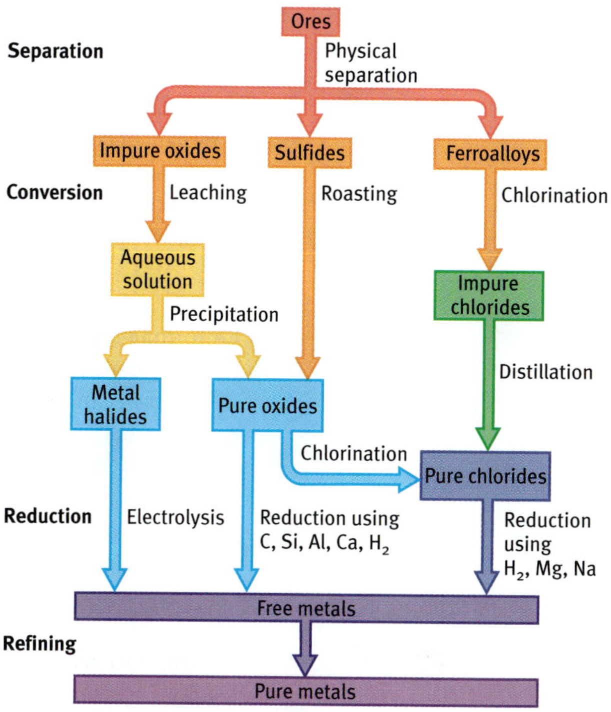

Figure 20-20
Metallurgy includes separation, conversion, reduction, and refining steps. The starting material is an impure ore, and the end product is pure metal.

Overview of Metallurgical Processes

Ore obtained from a mining operation contains a desired mineral contaminated with other components, which may include sand, clay, and organic matter. This economically valueless portion of the ore, which is called gangue (pronounced "gang"), must be removed before the metal can be extracted and refined. Ores can be separated into components by physical or chemical methods.

Flotation is a common physical separation process in which the ore is crushed and mixed with water to form a thick slurry. As shown in Figure 20-21, the slurry is transferred to a flotation vessel and mixed with oil and a surfactant (see Chapter 12). The polar head groups of the surfactant coat the surface of the mineral particles, but the nonpolar tails point outward, making the surfactant-coated mineral particles hydrophobic. Air is blown vigorously through the mixture, carrying the oil and the coated mineral to the surface, where they become trapped in the froth. Because the gangue has a much lower affinity for the surfactant, it absorbs water and sinks to the bottom of the flotation vessel. The froth is removed at the top, and the gangue is removed at the bottom.

A second separation technique is **leaching,** which uses solubility properties to separate the components of an ore. For example, modern gold production depends on the extraction of tiny particles of gold from gold-bearing rock deposits. After the rock is crushed, it is treated with an aerated aqueous basic solution of sodium cyanide. Molecular oxygen oxidizes the metal, which forms a soluble coordination complex with the cyanide anion:

$$4\ Au\ (s) + 8\ CN^-\ (aq) + O_2\ (g) + 2\ H_2O\ (l) \longrightarrow 4\ [Au(CN)_2]^-\ (aq) + 4\ OH^-\ (aq)$$

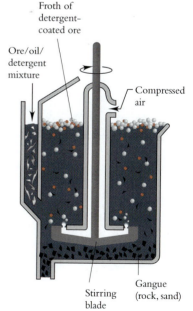

Froth of detergent-coated ore

Ore/oil/detergent mixture

Compressed air

Stirring blade

Gangue (rock, sand)

Figure 20-21
In the flotation process, surfactant-coated mineral particles float to the surface of the mixture, where they become trapped in the froth. The gangue settles to the bottom.

The aqueous gold-containing solution is then treated with zinc dust, which reduces gold back to the free metal:

$$Zn\,(s) + 2\,[Au(CN)_2]^-(aq) \longrightarrow 2\,Au\,(s) + [Zn(CN)_4]^{2-}(aq)$$

Sulfide ores may be treated chemically to convert them to oxides before extraction can occur. The process of oxidizing an ore by heating to a high temperature in the presence of air is known as **roasting.** In the roasting of a sulfide ore, sulfide anions are oxidized and molecular oxygen is reduced. The conversion of galena is a typical example:

$$2\,ZnS\,(s) + 3\,O_2\,(g) \longrightarrow 2\,ZnO\,(s) + 2\,SO_2\,(g)$$

Unfortunately, roasting produces copious amounts of highly polluting SO_2 gas that has seriously damaged the environment around smelters for sulfide ores.

Today, zinc and other metals can be extracted from sulfides by aqueous conversion processes that avoid the generation of SO_2. Aqueous acid reacts with the sulfides to generate free sulfur or sulfate ions rather than SO_2. Here are two examples:

$$2\,ZnS\,(s) + 4\,H_3O^+(aq) + O_2\,(g) \longrightarrow 2\,Zn^{2+}(aq) + 2\,S\,(s) + 6\,H_2O\,(l)$$

$$3\,CuS\,(s) + 8\,NO_3^-(aq) + 8\,H_3O^+(aq) \longrightarrow$$

$$8\,NO\,(g) + 3\,Cu^{2+}(aq) + 3\,SO_4^{2-}(aq) + 12\,H_2O\,(l)$$

Taking into account the cost of SO_2 pollution of the atmosphere, these more elaborate aqueous separation procedures are economically competitive with conversion by roasting.

Once an ore is in suitably pure form, it can be reduced to the free metal. This is accomplished either chemically or electrolytically. Electrolysis is costly because it requires huge amounts of electrical energy. For this reason, chemical reduction is used unless the metal is too reactive for chemical reducing agents to be effective.

Mercury and lead are sufficiently easily reduced that roasting the sulfide ore frees the metal. Sulfide ion is the reducing agent, and both O_2 and the metal ion gain electrons:

$$HgS\,(s) + O_2\,(g) \longrightarrow Hg\,(l) + SO_2\,(g)$$

This reduction produces SO_2, which must be removed from the exhaust gases.

One of the most common chemical reducing agents for metallurgy is coke, a form of carbon made by heating coal at high temperature until all of the volatile impurities have been removed. Metals whose cations have moderately negative reduction potentials— Co, Ni, Fe, and Zn—are reduced by coke. For example, direct reaction with coke in a furnace frees nickel from its oxide:

$$NiO\,(s) + C\,(s) \longrightarrow Ni\,(l) + CO\,(g)$$

Chemical reduction of an ore usually gives metal that is not pure enough for its intended use. Further **refining** of the metal removes undesirable impurities. Several important metals, including Cu, Ni, Zn, and Cr, are refined by electrolysis, either from an aqueous solution of the metal salt or from anodes prepared from the impure metal. To give one example, Zn^{2+} ions, obtained by dissolving ZnS or ZnO in acidic solution, can be reduced while water is oxidized:

$$Zn^{2+}(aq) + 2\,e^- \longrightarrow Zn\,(s)$$

$$\underline{3\,H_2O\,(l) \longrightarrow \tfrac{1}{2}\,O_2\,(g) + 2\,H_3O^+(aq) + 2\,e^-}$$

$$Zn^{2+}(aq) + 3\,H_2O\,(l) \longrightarrow Zn\,(s) + \tfrac{1}{2}\,O_2\,(g) + 2\,H_3O^+(aq)$$

Table 20-6 provides a summary of the chemical species and processes involved in the metallurgy of many transition metals. The following survey of several metals provides further examples of the four phases of metallurgy.

Table 20-6 Metallurgy of Transition Metals

Metal	Ore	Separation Method	Intermediate*	Reducing Agent
Ti	TiO_2	Chlorination	$TiCl_4$	Mg
Cr	$FeCr_2O_4$	Roasting	Cr_2O_3	Al
Mn	MnO_2		Mn_2O_3	Al
Fe	Fe_3O_4	Slag formation	$(CaSiO_3)$	C
Co	CoAsS	Roasting	CoO	C
Ni	Ni_9S_8	Complexation	$Ni(CO)_4$	H_2
Cu	$CuFeS_2$	Leaching	$Cu^{2+}, (SO_4^{2-})$	
Zn	ZnS	Roasting	ZnO	C
Mo	MoS_2	Roasting	MoO_3	H_2
W	$CaWO_4$	Leaching	WO_4^{2-}, WO_3	H_2
Au	Au	Leaching	$[Au(CN)_2]^-$	Zn
Hg	HgS	Roasting	(SO_2)	S^{2-}

*Intermediates that represent impurities are shown in parentheses.

Iron and Steel

Iron has been the dominant structural material of modern times, and despite the growth in importance of aluminum and plastics, iron still ranks first in total use. Worldwide production of steel (iron strengthened by additives) is on the order of 700 million tons per year. The most important iron ores are two oxides, hematite (Fe_2O_3) and magnetite (Fe_3O_4). The production of iron from its ores involves several chemical processes that take place in a blast furnace. As shown in Figure 20-22, this is an enormous chemical reactor where heating, reduction, and purification all occur together.

Figure 20-22
A diagrammatic view showing the chemical reactions occurring within a blast furnace, which operates continuously at fiery temperatures (see photo).

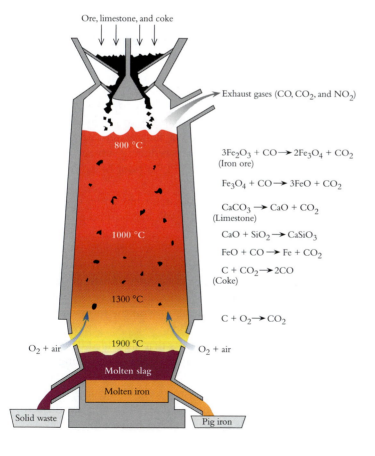

Ore, limestone, and coke

Exhaust gases (CO, CO_2, and NO_2)

800 °C

$$3Fe_2O_3 + CO \longrightarrow 2Fe_3O_4 + CO_2$$
(Iron ore)

$$Fe_3O_4 + CO \longrightarrow 3FeO + CO_2$$

$$CaCO_3 \longrightarrow CaO + CO_2$$
(Limestone)

1000 °C

$$CaO + SiO_2 \longrightarrow CaSiO_3$$

$$FeO + CO \longrightarrow Fe + CO_2$$

$$C + CO_2 \longrightarrow 2CO$$
(Coke)

1300 °C

$$C + O_2 \longrightarrow CO_2$$

O_2 + air   1900 °C O_2 + air

Molten slag

Molten iron

Solid waste

Pig iron

The raw materials placed in a blast furnace include the ore (usually hematite) and coke, which serves as the reducing agent. The ores always contain various amounts of silicon dioxide (SiO_2), which is removed chemically by reaction with limestone ($CaCO_3$). To begin the conversion process, pellets of ore, coke, and limestone are mixed and fed into the top of the furnace, and a blast of hot air is blown in at the bottom. As the starting materials fall through the furnace, the burning coke generates intense heat:

$$2\ C\,(s) + O_2\,(g) \longrightarrow CO\,(g) \qquad \Delta H° = -221\ kJ$$

The result is a temperature gradient ranging from about 800 °C at the top of the furnace to 1900 °C at the bottom.

The reduction of iron oxide takes place in several stages in different temperature zones within the furnace. The reducing agent is CO produced from burning coke. Here are the key reactions:

$$3\ Fe_2O_3\,(s) + CO\,(g) \longrightarrow 2\ Fe_3O_4\,(s) + CO_2\,(g)$$

$$Fe_3O_4\,(s) + CO\,(g) \longrightarrow 3\ FeO\,(s) + CO_2\,(g)$$

$$FeO\,(s) + CO\,(g) \longrightarrow Fe\,(l) + CO_2\,(g)$$

Once liberated from its oxides, the iron melts when the temperature reaches 1500 °C. Molten iron collects in a pool at the bottom of the furnace.

At the same time that heating and reduction occur, limestone decomposes into calcium oxide and CO_2. The CaO then reacts with SiO_2 impurities in the ore to generate calcium silicate:

$$CaCO_3\,(s) \xrightarrow{\text{Heat}} CaO\,(s) + CO_2\,(g)$$

$$CaO\,(s) + SiO_2\,(s) \xrightarrow{\text{Heat}} CaSiO_3\,(l)$$

At blast furnace temperatures, calcium silicate is a liquid, called *slag*. Being less dense than iron, slag pools on the surface of the molten metal. Both products are drained periodically through openings in the bottom of the furnace.

Although this chemistry is complex, the basic process is reduction of iron oxide by carbon in an atmosphere depleted of oxygen. Archaeologists have found ancient smelters in Africa (in what is now Tanzania) that exploited this chemistry to produce iron in prehistoric times. Early African peoples lined a hole with a fuel of termite residues and added iron ore. Charred reeds and charcoal provided the reducing substance. Finally, a chimney of mud was added. When this furnace was "fired," a pool of iron collected in the bottom.

The iron formed in a blast furnace, called *pig iron,* contains impurities that make the metal brittle. These include phosphorus and silicon from silicate and phosphate minerals that contaminated the original ore, as well as carbon and sulfur from the coke. This iron is refined in a converter furnace. Here, a stream of O_2 gas blows through molten impure iron. Oxygen reacts with the nonmetal impurities, converting them to oxides. As in the blast furnace, CaO is added to convert SiO_2 into liquid calcium silicate, in which the other oxides dissolve. The molten iron is analyzed at intervals until its impurities have been reduced to satisfactory levels. Then the liquid metal, now in the form called *steel,* is poured from the converter and allowed to solidify.

Most steels contain various amounts of other elements that are added deliberately to give the metal particular properties. These additives may be introduced during the converter process or when the molten metal is poured off. One of the most important additives is manganese, which adds strength and hardness to steel. Manganese is added to nearly every form of steel in amounts ranging from less than 1% to higher than 10%. More than 80% of manganese production ends up incorporated into steel.

Other Metals

Titanium. The metallurgy of titanium illustrates the purification of one metal by another. The major titanium ores are rutile (TiO_2) and ilmenite ($FeTiO_3$). Either is converted to titanium(IV) chloride by a redox reaction with chlorine gas and coke. For rutile:

$$TiO_2(s) + C(s) + 2\,Cl_2(g) \xrightarrow{500\,°C} TiCl_4(g) + CO_2(g)$$

In this reaction, carbon is oxidized and chlorine is reduced. When the hot gas cools, titanium tetrachloride (bp = 140 °C) condenses to a liquid that is purified by distillation.

Titanium metal is obtained by reduction of $TiCl_4$ with molten magnesium metal at high temperature. The reaction gives solid titanium metal (mp = 1660 °C) and liquid magnesium chloride (mp = 714 °C):

$$TiCl_4(g) + 2\,Mg(l) \xrightarrow{850\,°C} Ti(s) + 2\,MgCl_2(l)$$

Copper. Copper is found mainly in the sulfide ore chalcopyrite ($FeCuS_2$), but chalcocite (Cu_2S), cuprite (Cu_2O), and malachite ($Cu_2CO_3(OH)_2$) are also important. Copper ores often have concentrations of copper less than 1% by mass, so achieving economic viability requires mining operations on a huge scale. The extraction and purification of copper is complicated by the need to remove iron from chalcopyrite. The first step in the process is flotation, which concentrates the ore to around 15% Cu by mass. In the next step, the concentrated ore is roasted to convert $FeCuS_2$ to CuS and FeO. Copper(II) sulfide is unaffected if the temperature is kept below 800 °C:

$$2\,FeCuS_2(s) + 3\,O_2(g) \longrightarrow 2\,CuS(s) + 2\,FeO(s) + 2\,SO_2(g)$$

Copper mining is carried out on a huge scale.

Heating the mixture of CuS and FeO to 1400 °C in the presence of silica (SiO_2) causes the material to melt and separate into two layers. The top layer is molten $FeSiO_3$ formed from the reaction of SiO_2 with FeO. As this takes place, the copper in the bottom layer is reduced from CuS to Cu_2S. This bottom layer consists of molten Cu_2S contaminated with FeS. Reduction of the Cu_2S takes place in a converter furnace following the same principle that converts impure iron into steel. Silica is added, and oxygen gas is blown through the molten mixture. Iron impurities are converted first to FeO and then to $FeSiO_3$, which is a liquid that floats to the surface. At the same time, Cu_2S is converted to Cu_2O, which reacts with more Cu_2S to give copper metal and SO_2:

$$2\,Cu_2S(l) + 3\,O_2(g) \longrightarrow 2\,Cu_2O(l) + 2\,SO_2(g)$$

$$2\,Cu_2O(l) + Cu_2S(l) \longrightarrow 6\,Cu(l) + SO_2(g)$$

Copper metal obtained from the converter furnace must be refined to better than 99.95% purity before it can be used to make electrical wiring. This is accomplished by electrolysis, as illustrated in Figure 20-23. The impure copper is formed into slabs that serve as

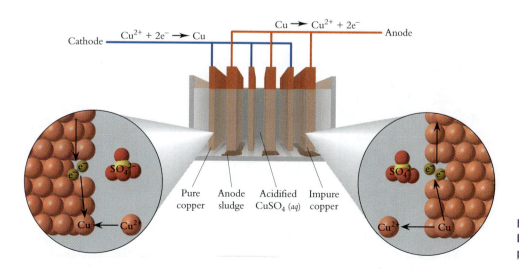

Figure 20-23
Diagram of an electrolytic cell for purification of copper.

Copper is used to make electrical wire.

anodes in electrolysis cells. The cathodes are constructed from thin sheets of very pure copper. These electrodes are immersed in a solution of $CuSO_4$ dissolved in dilute sulfuric acid. Application of a controlled voltage causes oxidation in which copper, along with iron and nickel impurities, is oxidized to its cations. Less reactive metal contaminants, including silver, gold, and platinum, are not oxidized. As the electrolysis proceeds and the anode dissolves, these and other insoluble impurities fall to the bottom of the cell. This sludge is a valuable source of the precious metals, as described in Section 20.5.

The metal cations released from the anode migrate through the solution to the cathode. Because Cu^{2+} is easier to reduce than Fe^{2+} and Ni^{2+}, careful control of the applied voltage makes it possible to reduce Cu^{2+} to Cu metal, leaving Fe^{2+} and Ni^{2+} dissolved in solution.

Section Exercises

■ **20.4.1** Construct a flowchart that summarizes the chemistry that takes place in a blast furnace.

■ **20.4.2** Examine Appendix F and explain why ZnO can be reduced with coke but Cr_2O_3 requires a more reactive metal such as aluminum.

■ **20.4.3** Which of the reactions listed in Figure 20-22 are redox reactions? Identify the reducing agent in each case.

20.5 APPLICATIONS OF TRANSITION METALS

A complete discussion of all the transition metals is beyond the scope of an introductory course in chemistry. Instead, we provide a brief survey of several metals that highlights the diversity and utility of this group of elements.

Titanium

Titanium, the ninth most abundant element in the Earth's crust, is characterized by its high strength, its low density (~57% that of steel), and its stability at very high temperature. When alloyed with small amounts of aluminum or tin, titanium has the highest strength-to-weight ratio of all the engineering metals. Its major use is in the construction of aircraft frames and jet engines. Because titanium is also highly resistant to corrosion, it is used in the construction of pipes, pumps, and vessels for the chemical industry.

Titanium finds uses as diverse as aircraft turbine blades, lightweight bicycle frames, and TiO_2 "smoke" for skywriting.

Because it is difficult to purify and fabricate, titanium is an expensive metal. For example, although titanium bicycles are highly prized by avid riders, they are quite expensive, averaging well over $1000 for just the frame.

The most important compound of titanium is titanium(IV) oxide, TiO_2. More than 2 million tons of TiO_2 are produced every year, much of it by the controlled combustion of $TiCl_4$:

$$TiCl_4(g) + O_2(g) \xrightarrow{1200\,°C} TiO_2(s) + 2\,Cl_2(g)$$

The Cl_2 produced in the combustion reaction is recycled to produce more $TiCl_4$ from rutile ore.

Titanium dioxide is brilliant white, highly opaque, chemically inert, and nontoxic. Consequently, it finds wide uses as a pigment in paints and other coatings, in paper, sunscreens, cosmetics, and toothpaste. Almost all white-colored commercial products contain TiO_2.

Chromium

Chromium makes up just 0.012% of the Earth's crust, yet it is an important industrial metal. The main use of chromium is in metal alloys. Stainless steel, for example, contains as much as 20% chromium. Nichrome, a 60:40 alloy of nickel and chromium, is used to make heat-radiating wires in electrical devices such as toasters and hair dryers. Another important application of chromium metal is as a protective and decorative coating for the surface of metal objects, as described in Chapter 19.

The only important ore of chromium is chromite, $FeCr_2O_4$. Reduction of chromite with coke gives ferrochrome, an iron–chromium compound:

$$FeCr_2O_4(s) + 4\,C(s) \longrightarrow FeCr_2(s) + 4\,CO(g)$$

Chromium is used as a protective and decorative coating.

Ferrochrome is mixed directly with molten iron to form chromium-containing stainless steel.

Chromium compounds of high purity can be produced from chromite ore without reduction to the free metal. The first step is the roasting of chromite ore in the presence of sodium carbonate:

$$4\,FeCr_2O_4 + 8\,Na_2CO_3 + 7\,O_2 \xrightarrow{1100\,°C} 8\,Na_2CrO_4 + 2\,Fe_2O_3 + 8\,CO_2$$

The product is converted to sodium dichromate by reaction with sulfuric acid:

$$2\,Na_2CrO_4 + H_2SO_4 \longrightarrow Na_2Cr_2O_7 + Na_2SO_4 + H_2O$$

When the resulting solution is concentrated by evaporation, $Na_2Cr_2O_7 \cdot 2H_2O$ precipitates from the solution. This compound is the most important source of chromium compounds for the chemical industry. It is the starting material for most other chromium-containing compounds of commercial importance, including $(NH_4)_2Cr_2O_7$ (ammonium dichromate), Cr_2O_3 (chromium(III) oxide), and CrO_3 (chromium(VI) oxide).

Pure chromium metal is made by a two-step reduction sequence. First, sodium dichromate is reduced to chromium(III) oxide by heating in the presence of charcoal:

$$Na_2Cr_2O_7 + C \xrightarrow{Heat} Cr_2O_3 + Na_2CO_3 + CO$$

Dissolving Cr_2O_3 in sulfuric acid gives an aqueous solution of Cr^{3+} cations:

$$Cr_2O_3(s) + 6\,H_3O^+(aq) \longrightarrow 2\,Cr^{3+}(aq) + 9\,H_2O(l)$$

Electrolysis of this solution reduces the cations to pure Cr metal, which forms a hard, durable film on the surface of the object serving as the cathode.

The name "chromium" is derived from the Greek word for color, *chroma,* because this metal forms a wide variety of compounds with beautiful colors. Chromium compounds have been used for many years as pigments in paints and other coatings: $Na_2Cr_2O_7$ is bright orange, Cr_2O_3 is green, and the Zn and Pb salts of CrO_4^{2+} are bright yellow. However, in recent years the use of chromium pigments has diminished because chromium in the $+6$ oxidation state is highly toxic.

Chromium compounds display a striking range of beautiful colors. Shown here are Na_2CrO_4 (yellow), $K_2Cr_2O_7$ (orange), $CrCl_3$ (green), and CrO_3 (dark purple).

Chromium is also important in converting animal hides into leather. In the tanning process, hides are treated with basic solutions of Cr(III) salts, which causes cross-linking of collagen proteins. The hides toughen and become pliable and resistant to biological decay.

Copper, Silver, and Gold

The first three pure metals known to humanity probably were copper, silver, and gold, known as the coinage metals because they found early use as coins. All three are found in nature in their pure elemental form, and all have been highly valued throughout civilization. The oldest known gold coins were used in Egypt around 3400 BC. At about the same time, copper was obtained in the Middle East from charcoal reduction of its ores. The first metallurgy of silver was developed in Asia Minor (Turkey) about 500 years later. All three of these metals are excellent electrical conductors and are highly resistant to corrosion.

These properties, coupled with its relatively low cost, make copper one of the most useful metals in modern society. About half of all copper produced is for electrical wiring, and the metal is also widely used for plumbing pipes. Copper is used to make several important alloys, the most important of which are bronze and brass. Both alloys contain copper mixed with lesser amounts of tin and zinc in various proportions. In bronze, the amount of tin exceeds that of zinc, whereas the opposite is true for brass. The discovery of bronze sometime around 3000 BC launched the advance of civilization known today as the Bronze Age. Because bronze is harder and stronger than other metals known in antiquity, it became a mainstay of the civilizations of India and the Mediterranean, used for tools, cookware, weapons, coins, and objects of art. Today the principal use of bronze is for bearings, fittings, and machine parts.

Copper is resistant to oxidation, but over the course of time the metal acquires a coating of green corrosion called *patina*. The green compound is a mixed salt of Cu^{2+}, hydroxide, sulfate, and carbonate that is formed by air oxidation in the presence of carbon dioxide and small amounts of sulfur dioxide:

$$3\ Cu + 2\ H_2O + SO_2 + 2\ O_2 \longrightarrow Cu_3(OH)_4SO_4$$

$$2\ Cu + H_2O + CO_2 + O_2 \longrightarrow Cu_2(OH)_2CO_3$$

Although trace amounts of copper are essential for all forms of life, the metal is toxic in large amounts. Thus, copper(II) salts, particularly $CuSO_4 \cdot 5\ H_2O$, are used as pesticides and wood preservatives. Wood soaked in solutions of Cu^{2+} or coated with paints containing Cu^{2+} resist degradation resulting from bacteria, algae, and fungi.

Silver is usually found as a minor component of ores of more abundant metals such as copper and zinc. Most commercial silver is produced as a by-product of the production of these common metals. For example, electrolytic refining of copper generates a solid anodic residue that is rich in silver and other precious metals. The silver from this residue can be isolated by oxidizing the metal into nitrate-containing solutions, silver nitrate being one of the few soluble silver salts. The pure metal is then deposited electrolytically. Silver is used for tableware in the form of sterling silver, an alloy containing small amounts of copper to make the metal harder. Silver is also used in jewelry, mirrors, and batteries, but the single most important use of the metal, accounting for about a third of all production, is in photography, described in our Chemistry and Technology Box.

Silver does not form a simple oxide by direct oxidation in air, but the metal does form a black tarnish with oxygen and trace amounts of hydrogen sulfide in the atmosphere:

$$4\ Ag + 2\ H_2S + O_2 \longrightarrow \underset{\text{black}}{2\ Ag_2S} + 2\ H_2O$$

The extraction of gold by leaching is described in the previous section. Gold is used extensively in the manufacture of jewelry. Interestingly, Au(I) compounds are very effective in the treatment of rheumatoid arthritis, and there is recent evidence that certain gold-containing compounds have anticancer properties.

The coffin of Tutankhamun contained over 100 kg of gold. Tutankhamun was only a minor pharaoh, who died at age 18.

The name *copper* and the symbol Cu are derived from the Latin *cuprum*, after the island of Cyprus, where the Romans first obtained copper metal. The symbols Ag and Au for silver and gold come from the Latin names for these elements: *argentum* and *aurum*.

The green coating on copper objects such as the Statue of Liberty is a mixture of copper salts, $Cu_3(OH)_4SO_4$ and $Cu_2(OH)_2CO_3$.

Box 20-2   Chemistry and Technology: How Are Images Captured Chemically?

P hotography has become an almost routine part of life. The front page of every newspaper contains several photos, often in color. Tourists the world over capture memories of their trips on photographic film. All photographic films rely on the redox chemistry of silver to "capture" an image on film.

Photographic film is a transparent plastic coated with a gelatinous film containing a silver salt, usually AgBr. When the film is exposed to light, individual photons absorbed by the Ag^+ ions cause an electron-transfer reaction that produces neutral silver atoms:

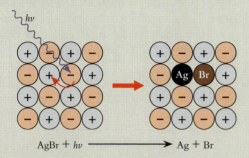

$$AgBr + h\nu \longrightarrow Ag + Br$$

Exposed film is developed with an aqueous reducing agent such as hydroquinone, which can reduce Ag^+ cations when catalyzed by Ag atoms. The reducing agent reacts selectively with those Ag^+ cations located next to already-reduced Ag atoms, generating clusters of Ag:

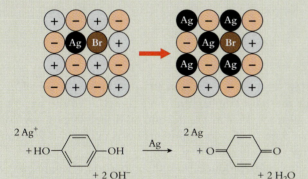

$$2\,Ag^+ + HO\!-\!\!\bigcirc\!\!-\!OH + 2\,OH^- \xrightarrow{Ag} 2\,Ag + O\!=\!\!\bigcirc\!\!=\!O + 2\,H_2O$$

Without intervention, this process would continue until all the Ag^+ had been reduced, giving a completely black film. To prevent this, developing is stopped after an appropriate time, and the film is treated with an aqueous solution containing thiosulfate anions, $S_2O_3^{2-}$. This anion forms several soluble complexes with Ag^+, the simplest of which is $[Ag(S_2O_3)(H_2O)_3]^-$:

$$AgBr(s) + S_2O_3^{2-}(aq) + 3\,H_2O(l) \longrightarrow$$

$$\left[\begin{array}{c} H_2O \\ H_2O\!-\!Ag\!-\!S\!-\!S\!-\!O \\ H_2O \end{array} \right]^- (aq) + Br^-(aq)$$

Rinsing with water removes all the remaining silver cations from the film, leaving islands of black clumps of Ag atoms wherever photons were absorbed.

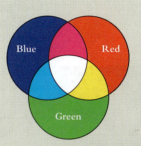

Color photographic films use the photosensitivity of AgBr combined with dyes of different colors. A color image uses various proportions of red, blue, and green to create any other color of the visible spectrum.

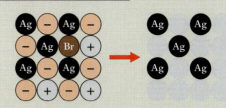

A color film makes use of sensitizer molecules that absorb photons and then reduce silver ions. A color film contains three emulsions overlaid on one another, each emulsion containing a different sensitizer. One sensitizer selectively absorbs red light, one selectively absorbs blue light, and the third selectively absorbs green light. In each layer, absorption of photons of the selected color results in clumps of neutral silver atoms. These clumps occur in different places in each layer, as determined by where photons of each color were absorbed.

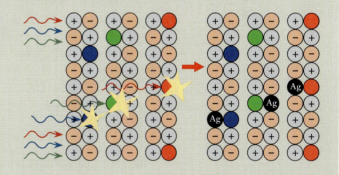

In addition to the sensitizer, each layer contains molecules that combine during the developing process to generate a colored dye. The combining reactions occur only where there are silver cations, so regions of the film that absorbed light and converted the silver cations into silver metal do not become colored. After development, the free silver atoms are removed from the film by treating with an oxidizing agent that converts Ag back to Ag^+ cations. What is left are islands of colored dye that match the colors of the scene that was photographed.

Film photography may soon be replaced by electronic imaging. Digital cameras offer high light sensitivity and ease of editing. Consequently, the physics of photodiodes now competes effectively with the chemistry of film photography as the best way to capture images.

Zinc and Mercury

Zinc and mercury are found in the Earth's crust as sulfide ores, the most common of which are sphalerite (ZnS) and cinnabar (HgS). See the previous section for descriptions of the extraction and purification of these metals.

Most of the world's zinc output is used to prevent the corrosion of steel. Zinc is easier to oxidize than iron, as shown by the more negative reduction potential of Zn^{2+}:

$$Fe^{2+} + 2\,e^- \rightleftharpoons Fe \qquad E° = -0.447\ V$$

$$Zn^{2+} + 2\,e^- \rightleftharpoons Zn \qquad E° = -0.7618\ V$$

Consequently, a zinc coating oxidizes preferentially and protects steel from corrosion. Zinc coatings are applied in several ways: by immersion in molten zinc, by paint containing powdered zinc, or by electroplating.

Zinc is also combined with copper and tin to make brass and bronze. Finally, large amounts of zinc are used to make several types of batteries, as discussed in Chapter 19.

Zinc oxide is the most important zinc compound. The principal industrial use of zinc oxide is as a catalyst to shorten the time of vulcanization in the production of rubber. The compound also is used as a white pigment in paints, cosmetics, and photocopy paper. In everyday life, ZnO is also a common sunscreen.

The use of mercury for extracting silver and gold from their ores has been known for many centuries. Gold and silver form amalgams with liquid mercury, which is then distilled away to leave the pure precious metal. The Romans mined the mineral cinnabar (HgS) from deposits in Spain 2000 years ago, and in the sixteenth century the Spanish shipped mercury obtained from the same ore deposits to the Americas for the extraction of silver. Mercury is an important component of street lamps and fluorescent lights. It is used in thermometers and barometers and in gas-pressure regulators, electrical switches, and electrodes.

Zinc oxide is an effective sunscreen.

The Platinum Metals

Six of the transition metals—Ru, Os, Rh, Ir, Pd, and Pt—are known as the *platinum metals*. The group is named for the most familiar and most abundant of the six. These elements are usually found mingled together in ore deposits, and they share many common features. Although they are rare (total annual production is only about 200 tons), the platinum metals play important roles in modern society.

The platinum metals are valuable by-products from the extraction of common metals such as copper and nickel. The anodic residue that results from copper refining is a particularly important source. The chemistry involved in their purification is too complicated to describe here, except to note that the final reduction step involves reaction of molecular hydrogen with metal halide complexes.

By far the most important use of the platinum metals is for catalysis. The largest single use is in automobile catalytic converters. Platinum is the principal catalyst, but catalytic converters also contain rhodium and palladium. These elements also catalyze a wide variety of reactions in the chemical and petroleum industry. For example, platinum metal is the catalyst for ammonia oxidation in the production of nitric acid, as described in Chapter 16:

$$4\,NH_3 + 5\,O_2 \xrightarrow{\text{Pt gauze, 1200 K}} 4\,NO + 6\,H_2O$$

Palladium is used as a catalyst for hydrogenation reactions in the food industry, and a rhodium catalyst is used in the production of acetic acid:

$$CH_3OH + CO \xrightarrow[\text{175 °C, 1 atm}]{\text{Rh-containing catalyst}} CH_3CO_2H$$

The combustion of ammonia is catalyzed by a gauze of platinum metal.

■ **20.5.1** In recent years, copper has replaced galvanized iron as the material of choice for plumbing pipes. Explain why copper is a better material for pipes than zinc-coated iron. Justify your answer with balanced equations.

■ **20.5.2** The chemistry of chromium and zinc can be used to remove traces of oxygen from bottled gases such as nitrogen and argon. The gas is bubbled through a solution of Cr^{2+} in the presence of zinc metal. Trace oxygen in the gas reacts quickly with Cr^{2+} to give Cr^{3+} and water. The zinc in turn reduces Cr^{3+} back to Cr^{2+}. Write balanced equations for these reactions.

■ **20.5.3** Refer to the standard reduction potentials for various metal cations (see Appendix F) and use these to predict which metals will be found in the anodic residue that results from the electrolytic refining of copper.

20.6 TRANSITION METALS IN BIOLOGY

Almost 90% of the atoms that make up a human body are either hydrogen or oxygen. Most of these are in water, which constitutes around 70% of a human. The organic structures that make up the body, as well as the molecules involved in biosynthesis and energy production, are made almost entirely of C, H, N, and O. These four elements account for 99% of all the atoms present in a human. Another seven elements—Na, K, Ca, Mg, P, S, and Cl—are essential for all known life-forms. These seven add another 0.9% to the total atom count of a human being. The remaining 0.1%, the so-called "trace elements," are required by most biological organisms. Although these elements are present in only minute amounts, they are essential for healthy function. Figure 20-24 shows a periodic table summarizing the elemental composition of living organisms.

Metalloproteins

The trace elements include nine transition metals: all members of the first row from vanadium to zinc, and molybdenum from the second row. Most transition metals in the body are natural constituents of proteins, biological macromolecules made of long chains of amino acids that are described in Chapter 13. These **metalloproteins** play three essential roles in biochemistry. Some act as transport and storage agents, moving small molecules from place to place within an organism. Others are enzymes, catalysts for a diverse group of biochemical reactions. Both transport and catalysis depend on the ability of transition metals to bind and release ligands. The third role of metalloproteins is to serve as redox reagents, adding or removing electrons in many different reactions. Transition metals are ideal for this purpose because of their abilities to shuttle between two or more oxidation states.

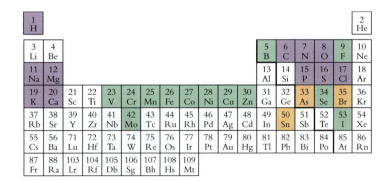

□ = Bulk biological elements
□ = Trace elements believed to be essential for plants or animals
□ = Trace elements that may be essential for plants or animals

Figure 20-24
Elements essential for life.

Metalloproteins typically are macromolecules containing many thousands of atoms. Determining the bonding, structure, and geometries of these macromolecules is a challenging task. Nevertheless, it is known that transition metals are bound to proteins through ligand–metal interactions of the sort described in this chapter. Among the 20 amino acids found in proteins, several have side chains that act as ligands: Histidine, tyrosine, cysteine, glutamic acid, aspartic acid, and methionine bind to metals through lone pairs on nitrogen, oxygen, or sulfur. Structures of the amino acid side chains appear in Figure 13-32.

Transport and Storage Proteins

Organisms extract energy from food by using molecular oxygen to oxidize fats and carbohydrates. For most animals, the movement and storage of O_2 is accomplished by the iron-containing proteins *hemoglobin* and *myoglobin*. The iron atom in hemoglobin binds O_2 and transports this vital molecule from the lungs or the gills to various parts of the body where oxidation takes place. Hemoglobin makes O_2 about 70 times more soluble in blood than in water. Whereas hemoglobin transports O_2, myoglobin stores O_2 in tissues such as muscle that require large amounts of oxygen.

Hemoglobin and myoglobin have closely related structures. Both contain the heme structure shown in the introduction to this chapter. In myoglobin the heme is bound to a polypeptide chain of 153 amino acids arranged in helical arrays. The ribbon structure of myoglobin is shown in Figure 20-25. The polypeptide chain folds in a manner that creates a "pocket" in the protein for a heme group. Hemoglobin is made up of four polypeptide chains, each of which is similar in shape and structure to a myoglobin molecule.

Each heme unit in myoglobin and hemoglobin contains one Fe^{2+} ion bound to four nitrogen donor atoms in a square planar arrangement. This leaves the metal with two axial coordination sites to bind other ligands. One of these sites is bound to a histidine

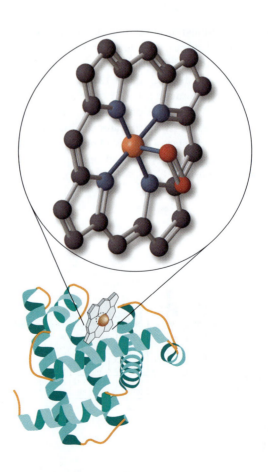

Figure 20-25
Myoglobin is a globular protein with a heme unit embedded in a pocket created by the folding of the protein chain.

side chain that holds the heme in the pocket of the protein. The other axial position is where reversible binding of molecular oxygen takes place.

The binding and release of oxygen by hemoglobin can be represented as a ligand-exchange equilibrium at the sixth coordination site on the Fe^{2+} ion. Each of the four polypeptide chains of hemoglobin contains one heme unit, so a molecule of hemoglobin can bind as many as four oxygen molecules:

$$[\text{Heme}]_4 + 4\,O_2 \rightleftharpoons [\text{Heme}(O_2)]_4$$
$$\underset{\text{Deoxyhemoglobin}}{} \qquad\qquad\qquad \underset{\text{Oxyhemoglobin}}{}$$

In the absence of oxygen, the iron center in each heme remains five-coordinate, with square pyramidal geometry, as shown in Figure 20-26a. In this form of hemoglobin, the d^6 metal ion is in a high-spin environment with four unpaired electrons in a $t_{2g}^4 e_g^2$ arrangement. Deoxyhemoglobin has a bluish color because the energy separation between the valence d orbitals of the iron cations is small, and the molecule absorbs red light at the low-energy end of the visible spectrum. When oxygen moves into the pocket of the protein, it binds to the sixth coordination site on the metal, as shown in Figure 20-26b. One of the effects of the O_2 ligand is to increase the crystal field splitting energy. As a result, in oxyhemoglobin the metal is in a low-spin $t_{2g}^6 e_g^0$ environment and is diamagnetic. In this form, hemoglobin absorbs light at the blue end of the visible spectrum and appears bright red.

In the lungs, hemoglobin "loads" its four oxygen molecules and then moves through the bloodstream. In tissues, the concentration of O_2 is very low, but there is plenty of carbon dioxide, the end product of metabolism. The concentration of CO_2 has an important effect on hemoglobin–oxygen binding. Like oxygen, carbon dioxide can bind to hemoglobin. However, carbon dioxide binds to specific amino acid side chains of the protein, not to the heme group. Binding carbon dioxide to the protein causes the shape of the hemoglobin molecule to change in ways that reduce the equilibrium constant for O_2 binding. The reduced binding constant allows hemoglobin to unload its O_2 molecules in oxygen-deficient, CO_2-rich tissue. The bloodstream carries this deoxygenated hemoglobin back to the lungs, where it releases carbon dioxide and loads four more molecules of oxygen. This CO_2 effect does not operate in myoglobin, which binds and stores the oxygen released by hemoglobin.

Carbon monoxide seriously impedes transport of oxygen. The deadly effect of inhaled CO results from its reaction with hemoglobin. A CO molecule is almost the same size and shape as O_2, so it fits into the binding pocket of the hemoglobin molecule. In addition, the carbon atom of CO forms a stronger bond to Fe^{2+} than does O_2. Under typical conditions in the lungs, hemoglobin binds carbon monoxide 230 times more strongly than it binds O_2. Hemoglobin complexed to CO cannot transport oxygen, so when a significant fraction of hemoglobin contains CO, oxygen "starvation" occurs at the cellular level, leading to loss of consciousness and then to death.

Many deaths occur through accidental carbon monoxide poisoning. Burning fossil fuels generates some CO, particularly in an oxygen-depleted environment. Automobile engines, gas heaters, and charcoal braziers all produce some CO. The presence of colorless, odorless CO goes undetected, and in a poorly ventilated room carbon monoxide may build up to lethal concentrations without the occupants being aware of its presence.

An adult human contains about 4 g of iron, most of it in the form of heme-containing proteins. Yet the daily requirement of iron in the diet is only about 1 mg, indicating that the body recycles iron rather than excreting it. The recycling of iron requires a transport system and a storage mechanism. Transport of iron is accomplished by a protein called *transferrin*. Transferrin collects iron in the spleen and liver, where hemoglobin is degraded, and carries it to the bone marrow where fresh red blood cells are synthesized.

The protein that stores iron in the body is called *ferritin*. A ferritin molecule consists of a protein coat and an iron-containing core. The outer coat is made up of 24 polypeptide chains, each with about 175 amino acids. As Figure 20-27 shows, the polypeptides pack together to form a sphere. The sphere is hollow, and channels through the protein

Deoxyhemoglobin:

Histidine

Fe^{2+} Porphyrin ring

(a)

Oxyhemoglobin:

Fe^{2+}

(b) Molecular oxygen

Figure 20-26
(*a*) In deoxyhemoglobin, iron is five-coordinate in a square pyramidal shape. (*b*) On coordination of molecular oxygen, the metal adopts octahedral geometry.

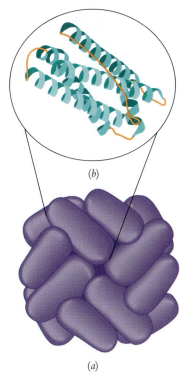

(b)

(a)

Figure 20-27
Schematic representation of ferritin, the iron storage protein. (*a*) The protein contains 24 nearly identical polypeptides. (*b*) A ribbon structure of one of the polypeptide chains.

coat allow movement of iron in and out of the molecule. The core of the protein contains hydrated iron(III) oxide, $Fe_2O_3 \cdot H_2O$. The protein retains its shape whether or not iron is stored on the inside. When filled to capacity, one ferritin molecule holds as many as 4500 iron atoms, but the core is only partially filled under normal conditions. In this way, the protein has the capacity to provide iron as needed for hemoglobin synthesis or to store iron if an excess is absorbed by the body.

Enzymes

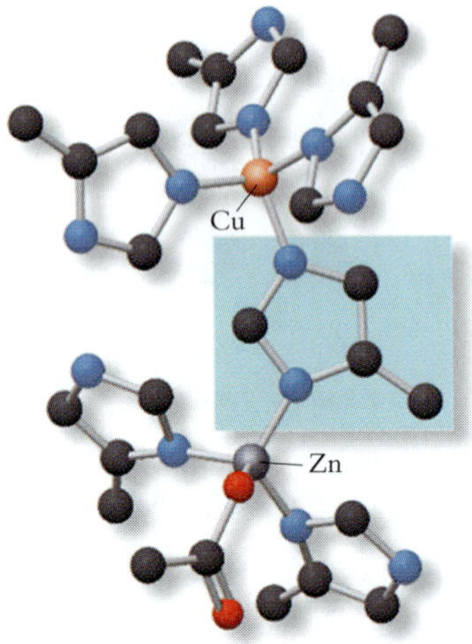

As described in Section 15.7, enzymes are the catalysts of biological reactions. Without enzymes, most of the reactions that occur in a cell would be imperceptibly slow. Cations of transition metals play essential roles in the mechanisms of many enzyme-catalyzed reactions. Here we introduce just one representative example, superoxide dismutase.

Molecular oxygen is essential to life, but reduced oxygen species such as superoxide, O_2^-, damage cells and are thought to play a role in the aging process. Cells contain enzymes that destroy these contaminants. One such enzyme, superoxide dismutase (SOD), is abundant in virtually every type of aerobic organism. The catalytic portion of the SOD enzyme contains one copper atom and one zinc atom, each bound in a tetrahedral arrangement. As Figure 20-28 shows, the two metals are held close together by a histidine ligand that forms a bridge between Cu and Zn.

The role of SOD is to convert superoxide ion to hydrogen peroxide:

$$2\,O_2^- + 2\,H_3O^+ \xrightarrow{\text{SOD}} O_2 + H_2O_2 + 2\,H_2O$$

Figure 20-28
The structure of the enzymatic site of superoxide dismutase. The Zn^{2+} and Cu^{2+} cations lie in close proximity, with a histidine side chain (color screened) acting as a bridge between the metals.

The mechanism is believed to be a two-step process involving reduction and oxidation of the copper center:

$$O_2^- + Cu^{2+}(SOD) \longrightarrow O_2 + Cu^+(SOD)$$

$$O_2^- + Cu^+(SOD) + 2\,H_3O^+ \longrightarrow Cu^{2+}(SOD) + H_2O_2 + 2\,H_2O$$

First, a superoxide ion transfers an electron to Cu^{2+}, giving molecular oxygen and Cu^+. In the second step, another superoxide anion reoxidizes the copper center back to Cu^{2+}. The resulting O_2^{2-} anion is protonated rapidly to give hydrogen peroxide. The source of the two protons is still unclear. It appears that the role of the Zn^{2+} ion is to provide structural stability to the protein, because if zinc is removed, the protein degrades quite easily.

Interest in superoxide dismutase has increased in recent years with the discovery that a mutation in the gene coding for SOD is linked to certain types of the neurodegenerative disease amyotrophic lateral sclerosis (ALS), commonly known as Lou Gehrig's disease. Exactly how mutant forms of SOD are involved in ALS is a subject of intense research.

Electron Transfer Proteins

From biochemical synthesis to bioenergetics, redox reactions are fundamental parts of the life process. Molecular oxygen is the oxidizing agent in most of these redox reactions. It has been estimated that the reduction of oxygen to water accounts for 90% of all the O_2 consumed in the biosphere:

$$O_2 + 4\,H_3O^+ + 4\,e^- \rightleftharpoons 6\,H_2O \qquad E° = 1.23\text{ V}$$

The many redox reactions that take place within a cell make use of metalloproteins with a wide range of electron transfer potentials. To name just a few of their functions, these proteins play key roles in respiration, photosynthesis, and nitrogen fixation. Some of them simply shuttle electrons to or from enzymes that require electron transfer as part of their catalytic activity. In many other cases, a complex enzyme may incorporate its own electron transfer centers. There are three general categories of transition metal redox centers: cytochromes, blue copper proteins, and iron–sulfur proteins.

Cytochromes. A cytochrome is a protein containing a heme with an iron cation bonded to four donor nitrogen atoms in a square planar array. Figure 20-29*a* shows the structure of cytochrome *c*, in which a histidine nitrogen atom and a cysteine sulfur atom occupy the fifth and sixth coordination sites of the octahedral iron center.

The electron transfer properties of the cytochromes involve cycling of the iron between the +2 and +3 oxidation states:

$$(\text{Cytochrome})\text{Fe}^{3+} + \text{e}^- \rightleftharpoons (\text{Cytochrome})\text{Fe}^{2+} \qquad E° = -0.3 \text{ V to } + 0.4 \text{ V}$$

Different cytochromes have different side groups attached to the porphyrin ring. These side groups modify the electron density in the delocalized π system of the porphyrin, which in turn changes the redox potential of the iron cation in the heme.

Blue copper proteins. A typical blue copper redox protein contains a single copper atom in a distorted tetrahedral environment. Copper performs the redox function of the protein by cycling between Cu^+ and Cu^{2+}. Usually the metal binds to two N atoms and two S atoms through a methionine, a cysteine, and two histidines. An example is plasto-cyanin, shown in Figure 20-29*b*. As their name implies, these molecules have a beautiful

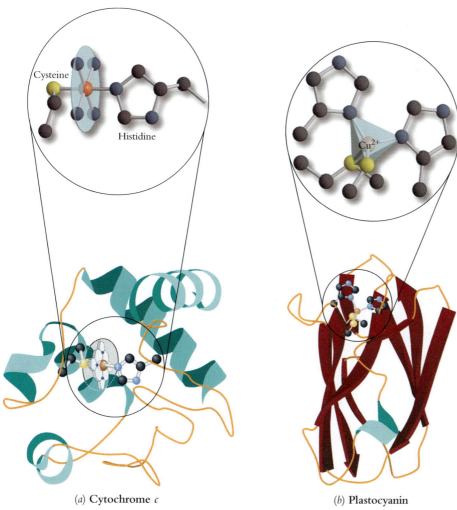

(*a*) **Cytochrome** *c*

(*b*) **Plastocyanin**

Figure 20-29
Ribbon structures of two redox proteins, cytochrome *c* (*a*) and plastocyanin (*b*). The blowups show the active sites where transition metal atoms are located.

Figure 20-30
There are three general types of iron–sulfur redox centers: (*a*) a single iron atom (brown) surrounded by four cysteine sulfur atoms (yellow); (*b*) two iron atoms bound to two cysteines and a pair of bridging sulfide ligands; and (*c*) a cubelike structure consisting of four irons and four sulfurs.

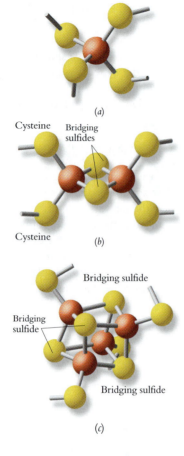

(*a*)

Cysteine   Bridging sulfides

Cysteine   (*b*)

Bridging sulfide

Bridging sulfide

Bridging sulfide

(*c*)

deep blue color that is attributed to photon-induced charge transfer from the sulfur atom of cysteine to the copper cation center.

Iron–sulfur proteins. In an iron–sulfur protein, the metal center is surrounded by a group of sulfur donor atoms in a tetrahedral environment. Box 14-2 describes the roles that iron-sulfur proteins play in nitrogenase, and Figure 20-30 shows the structures about the metal in three different types of iron–sulfur redox centers. One type (Figure 20-30*a*) contains a single iron atom bound to four cysteine ligands. The electron transfer reactions at these centers involve cycling of the metal between Fe^{2+} and Fe^{3+}.

A second type (Figure 20-30*b*) contains two iron atoms, each bound to two cysteine ligands. The metals are connected through a pair of sulfide ligands. The most complicated redox center of the iron–sulfur proteins contains four iron atoms and four sulfur atoms arranged in a distorted cube (Figure 20-30*c*). Each iron is bonded to one cysteine and three bridging sulfides. Individual iron cations in iron–sulfur clusters do not undergo simple changes in oxidation state. Instead, it appears that the electron transfer reactions involve orbitals that are delocalized over the entire cluster.

Section Exercises

20.6.1 Make a table that summarizes the biological chemistry of iron discussed in this section. Your table should include the names of the metalloproteins, their functions, and the coordination environment around the metal.

20.6.2 Most iron-containing proteins have a reddish color, and most copper proteins are blue. In contrast, zinc proteins are colorless. Why are zinc proteins colorless?

CHAPTER 20 VISUAL SUMMARY

Important equations

Essential terms

Key concepts

SKILLS TO MASTER

① Predicting properties of transition metals
② Drawing structures of coordination compounds
③ Naming coordination compounds
④ Comparing orbital stabilities in complexes
⑤ Correlating properties with electron configurations

20.1 OVERVIEW OF THE TRANSITION METALS

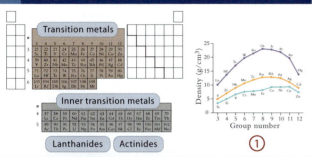

Transition metals

Inner transition metals

Lanthanides   Actinides

Density (g/cm³) vs Group number ①

20.2 COORDINATION COMPLEXES

Coordination complexes   Geometric isomers   Isomers

② Cis   Trans   Mer   Fac

$H_2\ddot{N}$ — en — $\ddot{N}H_2$

③ Ligand   Chelating ligand

Coordination number

Linkage isomers

Nomenclature

1. Name the cation first.
2. Name the ligands, then the metal.
3. Anionic ligands end in -o.
4. Use Greek prefixes for the number of ligands.
5. Anionic complexes end in -ate.
6. List the oxidation state as a Roman numeral.

20.3 BONDING IN COORDINATION COMPLEXES

Crystal field theory   Crystal field splitting energy

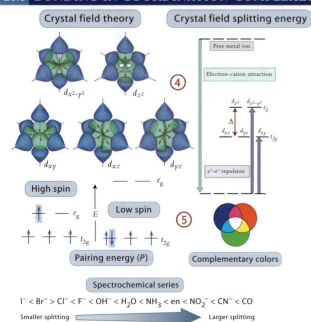

$d_{x^2-y^2}$ d_{z^2}

d_{xy} d_{xz} d_{yz}

Free metal ion

Electron–cation attraction

d_{z^2} $d_{x^2-y^2}$ e_g

Δ

d_{xz} d_{yz} d_{xy} t_{2g}

e^-–e^- repulsion

High spin e_g

Low spin E e_g

t_{2g} ⑤ t_{2g}

Pairing energy (P)   Complementary colors

Spectrochemical series

$I^- < Br^- > Cl^- < F^- < OH^- < H_2O < NH_3 < en < NO_2^- < CN^- < CO$

Smaller splitting → Larger splitting

20.4 METALLURGY

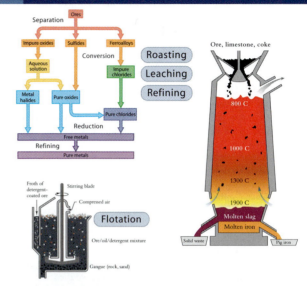

Ores

Separation

Impure oxides   Sulfides   Ferroalloys

Aqueous solution   Conversion   Impure chlorides

Metal halides   Pure oxides   Pure chlorides

Reduction

Refining   Free metals

Pure metals

Roasting
Leaching
Refining

Ore, limestone, coke

800 C
1000 C
1300 C
1900 C
Molten slag
Molten iron
Solid waste   Pig iron

Froth of detergent-coated ore   Stirring blade   Compressed air

Flotation

Ore/oil/detergent mixture

Gangue (rock, sand)

20.5 APPLICATIONS OF TRANSITION METALS

Ti Cr Cu Ag Au Zn Hg Pt metals

20.6 TRANSITION METALS IN BIOLOGY

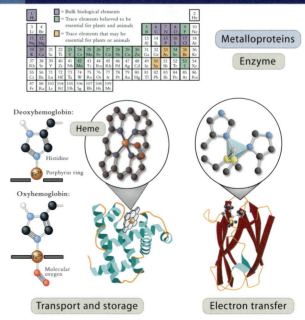

= Bulk biological elements
= Trace elements believed to be essential for plants and animals
= Trace elements that may be essential for plants or animals

Metalloproteins

Enzyme

Deoxyhemoglobin:
Histidine
Porphyrin ring
Fe

Oxyhemoglobin:
Fe
Molecular oxygen

Heme

Transport and storage   Electron transfer

Learning Exercises

20.1 Write a description of the features that transition metals have in common.

20.2 Describe the features of the transition metals that are exploited in biology. Use an example to illustrate each feature.

20.3 Draw ball-and-stick models of all possible isomers of linear, tetrahedral, square planar, and octahedral complexes containing two different ligands.

20.4 Make a list of the transition metals discussed in this chapter and summarize their applications.

20.5 Write a summary of the biological chemistry of iron and copper described in this chapter.

20.6 List all terms new to you that appear in Chapter 20, and write a one-sentence definition of each in your own words. Consult the Glossary if you need help.

 Problems   **ilw = interactive learning ware problem. Visit the website at www.wiley.com/college/olmsted**

Overview of the Transition Metals

20.1 The following compounds can be purchased from chemical supply companies. Determine the oxidation states of the transition metals in each: (a) $MnCO_3$; (b) $MoCl_5$; (c) Na_3VO_4; (d) Au_2O_3; and (e) $Fe_2(SO_4)_3 \cdot 5H_2O$.

20.2 The following compounds can be purchased from chemical supply companies. Determine the oxidation state of the transition metal in each: (a) $NiSO_4$; (b) $KMnO_4$; (c) $(NH_4)_2WO_4$; (d) $PbCrO_4$; and (e) $ZrOCl_2 \cdot 8H_2O$.

20.3 Give the names and symbols for the elements that have the following valence configurations: (a) $4s^1 3d^5$; (b) $5s^2 4d^{10}$; and (c) $4s^1 3d^{10}$.

20.4 Give the names and symbols for the elements that have the following valence configurations: (a) $5s^1 4d^8$; (b) $4s^2 3d^2$; and (c) $5s^1 4d^4$.

20.5 Write valence electron configurations for the following transition metal cations: (a) Mn^{2+}; (b) Ir^{3+}; (c) Ni^{2+}; and (d) Mo^{2+}.

20.6 Write valence electron configurations for the following transition metal cations: (a) Ti^{2+}; (b) Fe^{3+}; (c) Pt^{2+}; and (d) Nb^{3+}.

20.7 In each of the following pairs of transition metals, select the one with the higher value for the indicated property and give the reason: (a) melting points of Pd and Cd; (b) densities of Cu and Au; and (c) first ionization energies of Cr and Co.

20.8 In each of the following pairs of transition metals, select the one with the higher value for the indicated property and give the reason: (a) melting points of Zr and Mo; (b) densities of Ti and Cr; and (c) first ionization energies of Zr and Ag.

Coordination Complexes

20.9 Determine the oxidation states and d electron counts for the metal ions in the following coordination complexes: (a) $[Ru(NH_3)_6]Cl_2$; (b) trans-$[Cr(en)_2I_2]I$; (c) cis-$[PdCl_2(P(CH_3)_3)_2]$; (d) fac-$[Ir(NH_3)_3Cl_3]$; and (e) $[Ni(CO)_4]$.

20.10 Determine the oxidation states and d electron counts for the metal ions in the following coordination complexes: (a) $[Rh(en)_3]Cl_3$; (b) cis-$[Mo(CO)_4Br_2]$; (c) $Na_3[IrCl_6]$; (d) mer-$[Ir(NH_3)_3Cl_3]$; and (e) $[Mn(CO)_5Cl]$.

20.11 Name the compounds in Problem 20.9.

20.12 Name the compounds in Problem 20.10.

20.13 Draw structures for the metal complexes in Problem 20.9.

20.14 Draw structures for the metal complexes in Problem 20.10.

20.15 Write the formulas of the following complex ions: (a) cis-tetraamminechloronitrocobalt(III); (b) amminetrichloroplatinate(II); (c) trans-diaquabis(ethylenediamine)copper(II); and (d) tetrachloroferrate(III).

20.16 Write the formulas of the following compounds: (a) potassium tetrachloroplatinate(II); (b) pentaammineaquachromium(III) iodide; (c) tris(ethylenediamine)manganese(II) chloride; and (d) pentaammineiodocobalt(III) nitrate.

20.17 Draw the structure for each complex ion in Problem 20.15.

20.18 Draw the structure for each complex ion in Problem 20.16.

Bonding in Coordination Complexes

20.19 For an octahedral complex of each of the following metal ions, draw a crystal field energy diagram that shows the electron populations of the various d orbitals. Where appropriate show both the high-spin and low-spin configurations: (a) Ti^{2+}; (b) Cr^{3+}; (c) Mn^{2+}; and (d) Fe^{3+}.

20.20 For an octahedral complex of each of the following metal ions, draw a crystal field energy diagram that shows the electron populations of the various d orbitals. Where appropriate show both the high-spin and low-spin configurations: (a) Zn^{2+}; (b) Cr^{2+}; (c) Co^{2+}; and (d) Rh^{3+}.

20.21 Predict whether each of the following complexes is diamagnetic or paramagnetic. If a complex is paramagnetic, state its number of unpaired electrons. (a) $[Ir(NH_3)_6]^{3+}$; (b) $[Cr(H_2O)_6]^{2+}$; (c) $[PtCl_4]^{2-}$; and (d) $[Pd(P(CH_3)_3)_4]$.

20.22 Predict whether each of the following complexes is diamagnetic or paramagnetic. If a complex is paramagnetic, state its number of unpaired electrons. (a) $[Ru(CN)_6]^{4-}$; (b) $[Co(NH_3)_6]^{3+}$; (c) $[CoBr_4]^{2-}$ (tetrahedral); and (d) $[Pt(en)Cl_2]$.

20.23 Compounds of Zr^{2+} are dark purple, but most Zr^{4+} compounds are colorless. Explain.

20.24 Of the coordination complexes $[Cr(H_2O)_6]^{3+}$ and $[Cr(NH_3)_6]^{3+}$, one is violet, and the other is orange. Decide which is which and explain your reasoning.

20.25 The complex $[Fe(en)_3]Cl_3$ is low-spin. Provide the following information: (a) the coordination number of the metal; (b) the oxidation number and d electron count of the metal; (c) the geometry of the complex; (d) whether the complex is diamagnetic or paramagnetic; and (e) the number of unpaired electrons.

20.26 The complex $[Mn(H_2O)_6]SO_4$ is high-spin. Provide the following information: (a) the coordination number of the metal; (b) the oxidation number and d electron count of the metal; (c) the geometry of the complex; (d) whether the complex is diamagnetic or paramagnetic; and (e) the number of unpaired electrons.

ilw 20.27 An aqueous solution of $Cr^{2+}(aq)$ is paramagnetic. Addition of sodium cyanide makes the solution diamagnetic. Use crystal field energy diagrams to explain why the magnetic properties of the solutions are different.

20.28 An aqueous solution of $CoCl_2$ is paramagnetic. Addition of hydrogen peroxide (an oxidizing agent) causes the solution to become diamagnetic. Use crystal field energy diagrams to explain why the magnetic properties of the solutions are different.

ilw 20.29 The complex $[Cr(NH_3)_6]^{3+}$ has its maximum absorbance at 465 nm. Calculate the crystal field splitting energy for the compound and predict its color.

20.30 The complex $[Fe(H_2O)_6]^{3+}$ has its maximum absorbance at 724 nm. Calculate the crystal field splitting energy for the compound and predict its color.

Metallurgy

20.31 Write balanced chemical equations for the following metallurgical processes: (a) roasting of $CuFeS_2$; (b) removal of silicon from steel in a converter; and (c) reduction of titanium tetrachloride using sodium metal.

20.32 Write balanced chemical equations for these metallurgical processes: (a) heating NiS in air; (b) reducing Co_3O_4 using Al metal; and (c) reduction of MnO_2 by coke.

20.33 A copper ore contains 2.37% Cu_2S by mass. If 5.60×10^4 kg of this ore is heated in air, compute the mass of copper metal that is obtained and the volume of SO_2 gas produced at ambient conditions, 755 torr and 23.5 °C.

20.34 What mass of limestone, in kilograms, should be added for every kilogram of iron ore processed in a blast furnace if the limestone is 95.5% $CaCO_3$ and the iron ore contains 9.75% SiO_2?

20.35 Calculate the standard free energy change at 25 °C for reduction of ZnO to Zn using carbon and using carbon monoxide.

20.36 Determine $\Delta G°$ for each oxidation reaction that occurs in a steel-making converter. Compare your values with $\Delta G°$ for the reaction of iron with O_2 to give Fe_2O_3.

Applications of Transition Metals

20.37 Identify the coinage metals and describe some of their applications.

20.38 Identify the platinum metals and describe some of their applications.

20.39 What features of titanium account for its use as an engineering metal?

20.40 Explain why titanium(IV) oxide is used extensively as a white pigment.

Transition Metals in Biology

20.41 Summarize the differences between hemoglobin and myoglobin.

20.42 What features do myoglobin and the cytochromes have in common?

20.43 Draw a crystal field splitting diagram that illustrates the electron transfer reaction of the simple iron redox protein shown in Figure 20-29a.

20.44 Draw a crystal field splitting diagram that illustrates the electron transfer reaction of a cytochrome.

Additional Paired Problems

20.45 Draw all possible isomers of the following compounds: (a) $[Ir(NH_3)_3Cl_3]$; (b) $[Pd(P(CH_3)_3)_2Cl_2]$; (c) $[Cr(CO)_4Br_2]$; and (d) $[Cr(en)(NH_3)_2I_2]$.

20.46 Draw the structures of all possible isomers of the following coordination compounds: (a) tetraamminedibromocobalt(III) bromide; (b) triamminetrichlorochromium(III); and (c) dicarbonylbis-(trimethylphosphine)platinum(0).

20.47 Several commercial rust removers contain the bidentate ligand oxalate, $(O_2CCO_2)^{2-}$. Explain how these household products remove rust. Include a structural drawing of the species that forms.

20.48 The carbonate ion can be either a monodentate or a bidentate ligand. Make sketches that show the ligand binding to a metal cation in both modes.

20.49 Write electron configurations for the following: (a) Cr, Cr^{2+}, and Cr^{3+}; (b) V, V^{2+}, V^{3+}, V^{4+}, and V^{5+}; and (c) Ti, Ti^{2+}, and Ti^{4+}.

20.50 Write electron configurations for the following: (a) Au, Au^+, and Au^{3+}; (b) Ni, Ni^{2+}, and Ni^{3+}; and (c) Mn, Mn^{2+}, Mn^{4+}, and Mn^{7+}.

20.51 Name the following coordination compounds:

20.52 Name the following coordination compounds:

20.53 Write a balanced equation for the reaction catalyzed by the enzyme superoxide dismutase. Make a sketch of the coordination environment around the two metal atoms.

20.54 In superoxide dismutase it is the Cu^{2+} center that oxidizes O_2^{2-}. Why is copper more suitable than the Zn^{2+} center for the role of the oxidizing agent in SOD?

20.55 The Cu^{2+} ion forms tetrahedral complexes with some anionic ligands. When $CuSO_4 \cdot 5H_2O$ dissolves in water, a blue solution results. The addition of aqueous KF solution results in a green precipitate, but the addition of aqueous KCl results in a bright green solution. Identify each green species and write chemical reactions for these processes.

20.56 The Cu^{2+} ion forms tetrahedral complexes with some anionic ligands. When $CuSO_4 \cdot 5H_2O$ dissolves in water, a blue solution results. The addition of aqueous KCN solution at first results in a white precipitate, but the addition of more KCN causes the precipitate to dissolve. Identify the precipitate and the dissolved species and write chemical reactions for these processes.

20.57 Both vanadium and silver are lustrous silvery metals. Suggest why silver is widely used for jewelry, but vanadium is not.

20.58 Titanium is nearly 100 times more abundant in the Earth's crust than copper; yet copper was exploited as a metal in antiquity, and titanium has found applications only in recent times. Explain.

20.59 Draw a ball-and-stick model of the *mer* isomer of $[NiCl_3F_3]^{4-}$, oriented so that the fluoride ions are in the top, bottom, and left forward positions.

20.60 Draw a ball-and-stick model of the *fac* isomer of $[NiCl_3F_3]^{4-}$, oriented so that the fluoride ions are in the top and two rear positions.

20.61 What is the chemical name and the formula of the black tarnish that accumulates on objects made of silver? Write a balanced equation that shows how this black tarnish forms.

20.62 What is the chemical name and the composition of the green tarnish that accumulates on objects made of copper? Write balanced equations that show how this green tarnish forms.

20.63 As ligands, chloride and cyanide are at opposite ends of the spectrochemical series. Nevertheless, experiments show that $[CrCl_6]^{3-}$ and $[Cr(CN)_6]^{3-}$ have about the same amount of magnetism. Explain how this can be so.

20.64 In octahedral geometry, the amount of magnetism in $d^4 - d^7$ coordination compounds of the 3d transition metals depends on the ligands and the oxidation state of the metal. On the other hand, $d^1 - d^3$ and $d^8 - d^9$ compounds always have approximately the same amount of magnetism. Explain.

20.65 The d_{z^2} and $d_{x^2-y^2}$ orbitals have the same stability in an octahedral complex. However, in a square planar complex, the $d_{x^2-y^2}$ orbital is much less stable than the d_{z^2} orbital. Use orbital sketches to explain the difference.

20.66 Use orbital sketches to explain why the d_{xz} and $d_{x^2-y^2}$ orbitals have different stabilities in octahedral coordination complexes.

More Challenging Problems

20.67 Consider the chromium complexes $[Cr(NH_3)_5L]^{n+}$, where L is Cl^-, H_2O, or NH_3. (a) What is the value of $n+$ for each of them? (b) The compounds absorb at 515 nm, 480 nm, and 465 nm, respectively. Predict the colors of the three compounds, and (c) calculate the crystal field splitting energy in kJ/mol for each of them and explain the trend in values.

20.68 The complex ion $[Ag(NH_3)_2]^+$ has linear geometry. Predict the crystal field splitting diagram for this complex. Place the ligands along the z-axis.

20.69 Tetracarbonylnickel(0) is $[Ni(CO)_4]$, and tetracyanozinc(II) is $[Zn(CN)_4]^{2-}$. Predict the geometry and color of each complex.

20.70 Carbon monoxide and tungsten form an octahedral complex. The molecule is colorless even though $W(0)$ is d^6. Sketch the ligand field diagram for hexacarbonyltungsten(0), and suggest why the complex is colorless.

20.71 Some researchers use the term "the brass enzyme" to describe superoxide dismutase. Can you suggest a reason for this nickname?

20.72 Blue copper proteins are blue when they contain Cu^{2+} but colorless as Cu^+ compounds. The color comes from an interaction in which a photon causes an electron to transfer from a sulfur lone pair on a cysteine ligand to the copper center. Why does this charge transfer interaction occur for Cu^{2+} but not Cu^+?

20.73 The complex $[Ni(CN)_4]^{2-}$ is diamagnetic, but $[NiCl_4]^{2-}$ is paramagnetic. Propose structures for the two complexes and explain why they have different magnetic properties.

20.74 Predict whether each of the following complexes is high-spin or low-spin: (a) $[Fe(CN)_6]^{4-}$; (b) $[MnCl_4]^{2-}$; (c) $[Rh(NH_3)_6]^{3+}$; and (d) $[Co(H_2O)_6]^{2+}$.

20.75 The iron storage protein ferritin usually is neither empty of iron nor filled to capacity. Why is this situation advantageous for an organism?

20.76 Explain how liquid mercury can be used to purify metals such as gold and silver.

20.77 One of the most common approaches to the investigation of metalloproteins is to replace the naturally occurring metal ion with a different one that has a property advantageous for chemical studies. For example, zinc proteins are often studied by visible spectroscopy after Co^{2+} has been substituted for Zn^{2+}. Explain, using crystal field energy diagrams, why Co^{2+} is a better metal than Zn^{2+} for visible spectroscopy.

20.78 Draw a molecular picture of a surfactant coating a mineral particle in a flotation process. (See Chapter 12 for a review of surfactants.)

20.79 In Zn purification by electrochemistry, ZnO is dissolved in sulfuric acid. Before deposition, zinc powder is added to displace less active metals, such as cadmium. Use $E°$ values and balanced equations to show how this is accomplished.

20.80 Determine the Lewis structure and draw a ball-and-stick model showing the geometry of the dichromate anion, which contains one bridging oxygen atom.

Group Study Problems

20.81 Oxyhemoglobin is bright red, but deoxyhemoglobin is blue. In both cases the iron is in the +2 oxidation state. Give a detailed explanation for the difference in color. How would you test your hypothesis? Based on your explanation, what color would you predict for a sample of blood that is saturated with carbon monoxide?

20.82 A portion of the absorption spectrum of a complex ion, $[Cr(H_2O)_4Cl_2]^+$, is represented by the following graph:

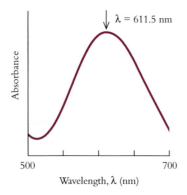

(a) Estimate the crystal field splitting energy Δ (in kilojoules per mole). (b) What color is the complex? (c) Name the complex cation. (d) Draw all possible isomers of the complex. (e) Draw the crystal field energy level diagram and show the electronic transition that gives the complex its color.

20.83 On Earth, two posttransition metals, Al and Pb, are used when low- and high-density metals are desired. Suppose you are transported to a planet elsewhere in our galaxy, where all transition metals are readily available but posttransition metals are rare. Where among the transition metals would you seek a replacement for Al for low-density uses, and where would you seek a replacement for Pb for high-density uses? Explain your reasoning.

20.84 Often, a detailed figure can give an informative summary of a complex concept. The figure in Box 18-1 summarizing the carbon cycle is one example. Design a figure that summarizes the transport and storage of oxygen by hemoglobin and myoglobin.

20.85 In the 1890s, Alfred Werner prepared several platinum complexes that contained both ammonia and chlorine. He determined the formulas of these species by precipitating the chloride ions with Ag^+. The empirical formulas and number of chloride ions that precipitate per formula unit follow:

Empirical formula	# of Cl^- ions
$PtCl_4 \cdot 2\ NH_3$	0
$PtCl_4 \cdot 3\ NH_3$	1
$PtCl_4 \cdot 4\ NH_3$	2
$PtCl_4 \cdot 5\ NH_3$	3
$PtCl_4 \cdot 6\ NH_3$	4

Determine the molecular formulas, name these compounds, and draw the structures of the platinum complexes.

20.86 Design a flowchart that summarizes the metallurgy of copper from chalcopyrite.

Answers to Section Exercises

20.1.1 (a) W, mp increases down a column; (b) W, mp decreases after Column 6; (c) Pd, mp decreases after Column 6; and (d) Nb, mp increases up to Column 6.

20.1.2 (a) W, density increases with Z; (b) Os, density increases with Z; (c) Pd, density decreases at the end of a row; and (d) Nb, density increases with Z.

20.1.3 (a) Ta(V); (b) Fe(III); (c) Rh(III); (d) Cr(VI); and (e) Cu(II)

20.2.1 (a) +3, Fe(III), coordination number = 6; (b) 0, V(II), coordination number = 6; (c) +2, Ni(II), coordination number = 4

20.2.2

[Structural diagrams of three chromium complexes with NH_3 and I ligands and ethylenediamine ($N\ H_2$) groups]

20.2.3 (a) pentaammineiodochromium(III) sulfate; (b) potassium hexachloroplatinate(II); (c) *cis*-tetracarbonyldichloroiron(II); and (d) hexaaquairon(II) chloride

20.3.1 (a)   (b)   (c)

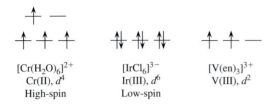

$[Cr(H_2O)_6]^{2+}$ $[IrCl_6]^{3-}$ $[V(en)_3]^{3+}$
Cr(II), d^4 Ir(III), d^6 V(III), d^2
High-spin   Low-spin

(d)

$[NiCl_4]^{2-}$
Ni(II), d^8

20.3.2 Because cyanide is near the top of the spectrochemical series, it generates a relatively large energy gap between the two sets of *d* orbitals. The hexacyano complexes are yellow because they absorb high-energy indigo light. The corresponding aqua complexes have a much smaller crystal field splitting energy. They absorb orange or red light, thus appearing blue or green.

20.3.3 492 nm, orange

20.4.1

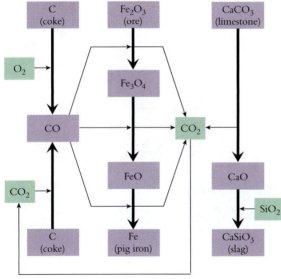

20.4.2 The standard reduction potentials are Cr^{3+}, -0.744 V and Zn^{2+}, -0.7618 V, suggesting that these oxides should be equally easily reduced. However, Cr^{2+} has a potential of -0.913 V, so reduction of chromium(III) oxide using coke stops at Cr^{2+}.

20.4.3

$$3\ Fe_2O_3 + CO \longrightarrow 2\ Fe_3O_4 + CO_2;\ \text{reducing agent is CO}$$
$$Fe_3O_4 + CO \longrightarrow 3\ FeO + CO_2;\ \text{reducing agent is CO}$$
$$FeO + CO \longrightarrow Fe + CO_2;\ \text{reducing agent is CO}$$
$$CO_2 + C \longrightarrow 2\ CO;\ \text{reducing agent is C}$$
$$C + O_2 \longrightarrow CO_2;\ \text{reducing agent is C}$$

20.5.1 Iron pipes are coated with zinc to prevent corrosion, zinc being easier to oxidize than iron. Copper is much more difficult to oxidize than either zinc or iron. Thus, copper makes better piping than galvanized iron because it is much less susceptible to corrosion.

20.5.2 Trace oxygen in the gas reacts with Cr^{2+}:

$$4\ Cr^{2+}(aq) + O_2(g) + 4\ H_3O^+(aq) \longrightarrow 4\ Cr^{3+}(aq) + 6\ H_2O(l)$$

Zinc reduces Cr^{3+} back to Cr^{2+}:

$$Zn(s) + 2\ Cr^{3+}(aq) \longrightarrow Zn^{2+}(aq) + 2\ Cr^{2+}(aq)$$

20.5.3 Metals with large positive standard reduction potentials will not electrolyze and will remain in the anodic residue: Au, Hg, Pt, Ag.

20.6.1

Table: **Biological Roles of Iron**

Protein	Function	Geometry	Coordination
Hemoglobin	O_2 transport	Octahedral and square pyramid	4N (porphyrin), 1N (His), O_2
Myoglobin	O_2 storage	Octahedral and square pyramid	4N (porphyrin), 1N (His), O_2
Transferrin	Fe transport	Octahedral	Not specified in text
Ferritin	Fe storage	Not specified in text	$Fe_2O_3 \cdot H_2O$
Cytochromes	Electron transfer	Octahedral	4N (porphyrin), 2N (His) or 1N (His), 1S (Met)
Fe-S Proteins	Electron transfer	Tetrahedral	4S (Cys) or 2S (Cys), 2S (sulfide)

20.6.2 Colors in coordination compounds come from electronic transitions, either from a lower energy *d* orbital to a higher energy *d* orbital or from a ligand to an empty *d* orbital (charge transfer). In zinc proteins, Zn^{2+} ion has a d^{10} configuration with all the *d* orbitals filled. Thus, electronic transitions in the visible portion of the spectrum are impossible.

Answers to Extra Practice Problems

20.1

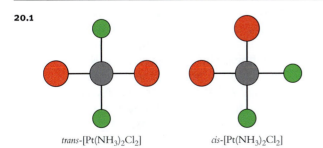

trans-$[Pt(NH_3)_2Cl_2]$ *cis*-$[Pt(NH_3)_2Cl_2]$

20.2 (a) pentaamminecarbonatocobalt(III) nitrate;
(b) triamminetricarbonylmolybdenum(0); and
(c) pentaaquahydroxoaluminum chloride

20.3 (a) $[Ru(bipy)_3]Cl_2$; (b) $K_2[PtCl_6]$; (c) $[Fe(H_2O)_5O]^+$

20.4

e_g
t_{2g}

20.5

e_g

e_g
t_{2g} t_{2g}

$[Mn(H_2O)_6]^{2+}$ $[Re(H_2O)_6]^{2+}$
$\Delta < P$ $\Delta > P$
High-spin   Low-spin
$t_{2g}^3 e_g^2$ $t_{2g}^5 e_g^0$

20.6 $[Cr(H_2O)_6]^{2+} < [Cr(NH_3)_6]^{3+} < [Mo(H_2O)_6]^{3+} < [Mo(NH_3)_6]^{3+}$

20.7 708 nm; green

13	14	15	16	17	18
5 B	6 C	7 N	8 O	9 F	10 Ne
13 Al	14 Si	15 P	16 S	17 Cl	18 Ar
31 Ga	32 Ge	33 As	34 Se	35 Br	36 Kr
49 In	50 Sn	51 Sb	52 Te	53 I	54 Xe
81 Tl	82 Pb	83 Bi	84 Po	85 At	86 Rn

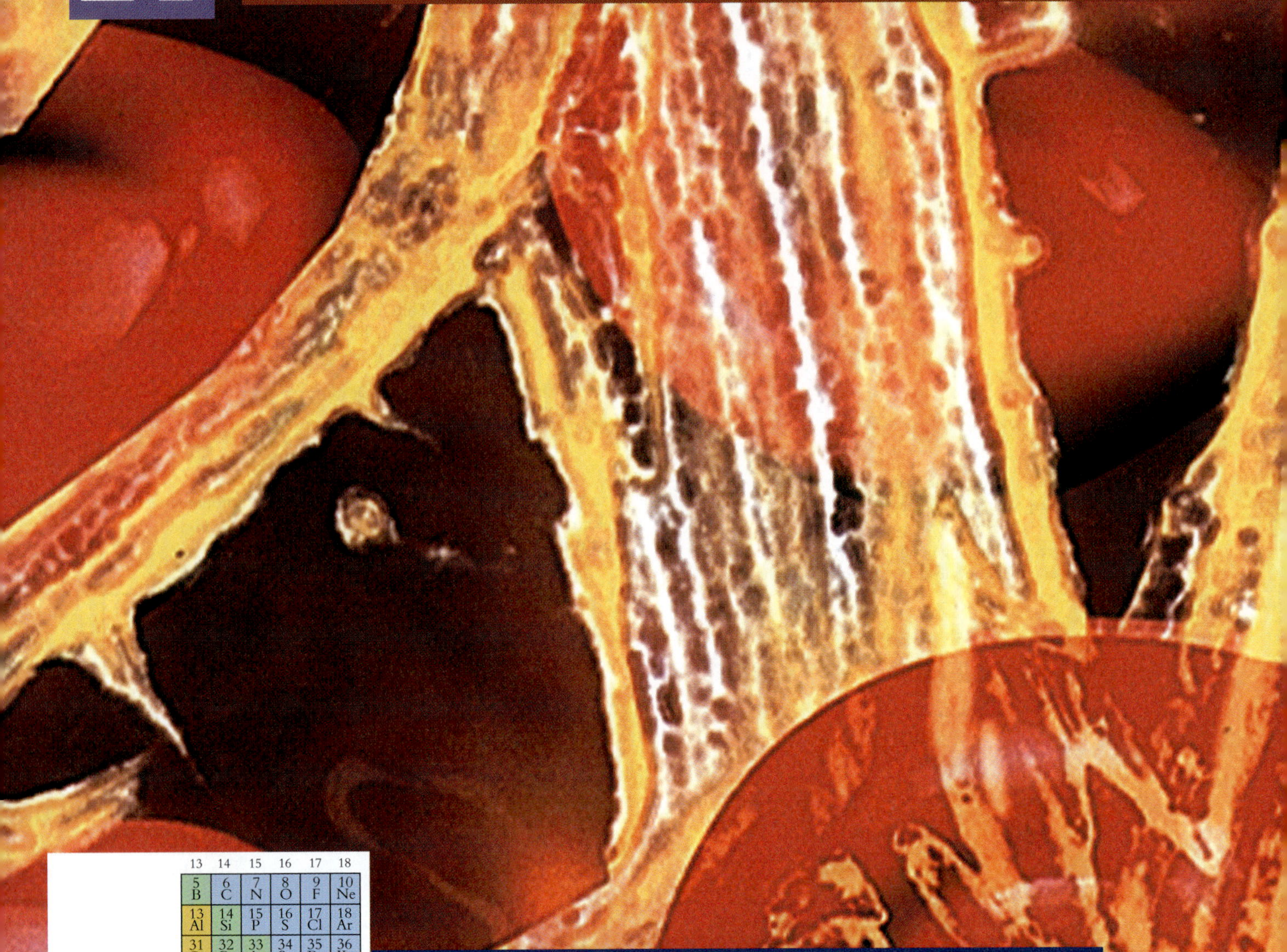

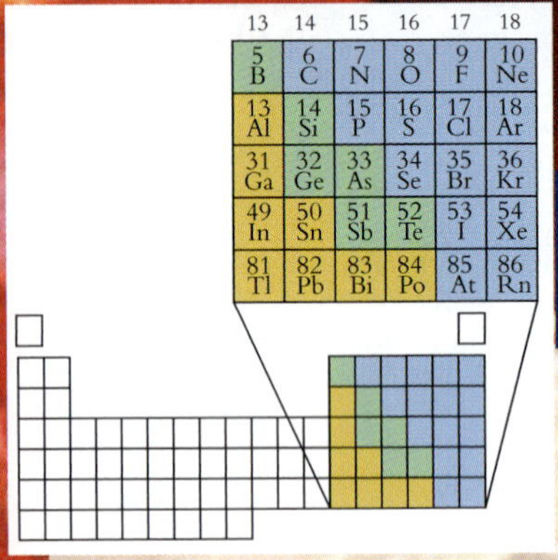

New Discoveries about an Old Compound

One might think that everything had long since been learned about a molecule as simple as nitrogen oxide (NO). After all, Joseph Priestley made the compound for the first time more than 200 years ago. Yet, recent research reveals that NO plays intriguing roles in a number of biochemical processes. Many animals, including humans and fruit flies, barnacles and trout, synthesize NO. This little molecule plays roles in nerve function, regulation of blood pressure, blood clotting, and immune system responses. In many of these processes, NO is produced in one part of an organism and moves to another part, where its chemical properties trigger a biochemical reaction.

Nitrogen oxide helps to adjust the interactions between blood and blood vessels. One example is blood clotting, shown in our background photo. When the wall of a blood vessel is cut, an enzyme in the cell wall sends chemical signals that interact with blood platelets, causing clotting. Studies have shown that NO inhibits this process, an effect exploited by mosquitoes, whose saliva contains NO-generating chemicals that prevent their victims' blood from clotting while the mosquito feeds. Another example is the dilation of blood vessels. Nitroglycerin has been prescribed for more than a century as a treatment of angina because it dilates blood vessels, thereby decreasing blood pressure and increasing blood flow. A century later, it was discovered that NO is involved in the mechanism of this

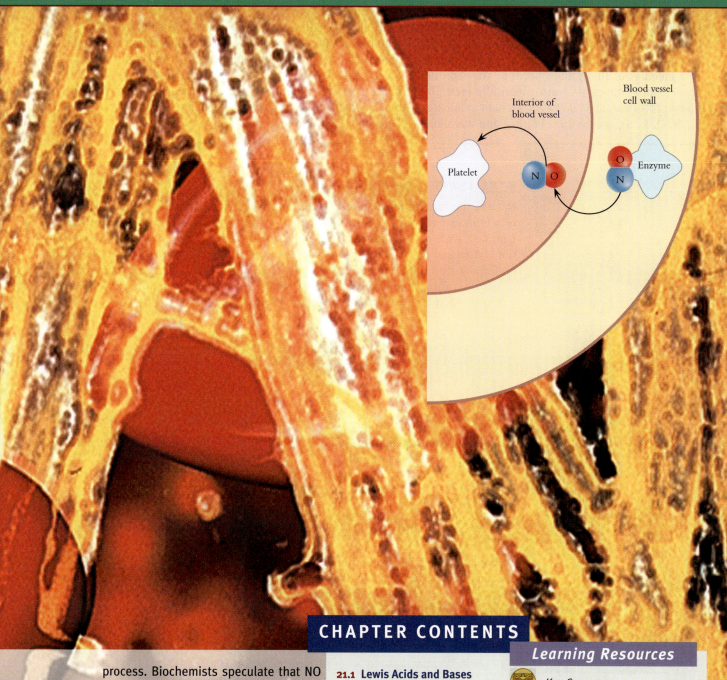

Interior of
blood vessel

Blood vessel
cell wall

Platelet

 N O

 O
Enzyme
 N

process. Biochemists speculate that NO binds to iron in the heme unit of the enzyme that causes dilation, as our molecular inset shows, but detailed knowledge is still lacking.

There is evidence that the neural processes leading to memory include the production and diffusion of NO. An enzyme in the brain produces NO from an amino acid. How does NO interact with more complex molecules as the brain functions? One possibility is that once a neural trigger fires, an adjacent neuron produces NO, which diffuses back to the original neuron and stimulates further activity. This could be how long-term memory develops.

CHAPTER CONTENTS

Learning Resources

KEY CONCEPTS

CRITICAL THINKING

SKILLS TO MASTER

ADDITIONAL HELP
www.wiley.com/college/olmsted
• TUTORIALS
• ANIMATIONS

Chemical reactions display a near-infinite variety, and the story of nitrogen oxide illustrates that there are still exciting discoveries to be made in chemistry. Many of these are likely to involve the elements in the *p* block of the periodic table, because the *p*-block elements are remarkably diverse. Aluminum, one of the most reactive metals, is a member of this group. So, too, are neon and argon, noble gases that are chemically inert. The elements of Groups 14 to 16 include nonmetals (C, N, O, P, S), metals (Sn, Pb, Bi), and metalloids (Si, Ge, Te). In this chapter, we describe selected features of this diverse chemistry. The chapter opens with sections describing a system for organizing chemical reactions, including those of *p*-block elements. Then we survey metals in the *p* block, metalloids, phosphorus, sulfur, and the halogens.

21.1 LEWIS ACIDS AND BASES

Because the breadth of chemical behavior can be bewildering in its complexity, chemists search for general ways to organize chemical reactivity patterns. Two familiar patterns are Brønsted acid—base (proton transfer) and oxidation—reduction (electron transfer) reactions. A related pattern of reactivity can be viewed as the donation of a pair of electrons to form a new bond. One example is the reaction between gaseous ammonia and trimethyl boron, in which the ammonia molecule uses its nonbonding pair of electrons to form a bond between nitrogen and boron:

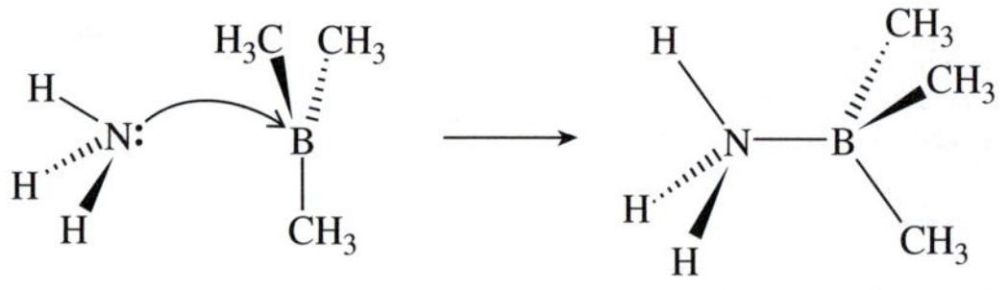

$$NH_3 + B(CH_3)_3 \longrightarrow H_3NB(CH_3)_3$$

Viewed from the perspective of electrons, this reaction is similar to the transfer of a proton to an ammonia molecule:

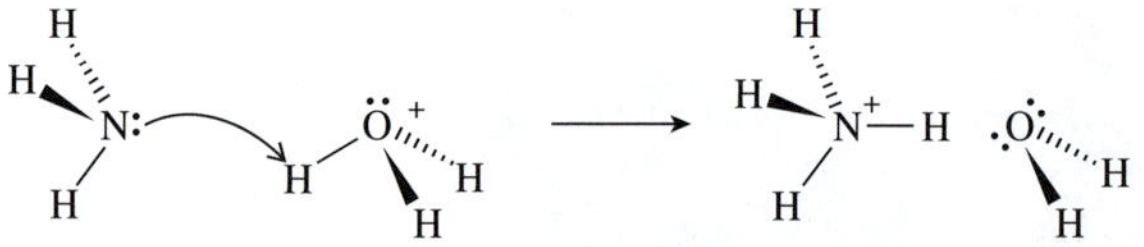

In both cases the nitrogen atom uses its pair of nonbonding electrons to make a new covalent bond. This similarity led G. N. Lewis to classify ammonia as a base in its reaction with $B(CH_3)_3$ as well as in its reaction with H_3O^+. Whereas the Brønsted definition focuses on proton transfer, the Lewis definition of acids and bases focuses on electron pairs.

KEY CONCEPTS *Any chemical species that acts as an electron-pair donor is a Lewis base.*
Any chemical species that acts as an electron-pair acceptor is a Lewis acid.

Ammonia is a prime example of a **Lewis base.** In addition to its three N—H bonds, this molecule has a lone pair of electrons on its nitrogen atom, as Figure 21-1 shows. Although all of the valence orbitals of the nitrogen atom in NH_3 are occupied, the nonbonding pair can form a fourth covalent bond with a bonding partner that has a vacant valence orbital available.

Trimethylboron is an example of one type of **Lewis acid.** This molecule has trigonal planar geometry in which the boron atom is sp^2 hybridized with a vacant $2p$ orbital perpendicular to the plane of the molecule (Figure 21-1). Recall from Chapter 9 that atoms tend to use all their valence *s* and *p* orbitals to form covalent bonds. Second-row elements such as boron and nitrogen are most stable when surrounded by eight valence electrons divided among covalent bonds and lone pairs. The boron atom in $B(CH_3)_3$ can use its vacant $2p$ orbital to form a fourth covalent bond to a new partner, provided that the new partner supplies both electrons. Trimethyl boron is a Lewis acid because it forms an additional bond by accepting a pair of electrons from some other chemical species.

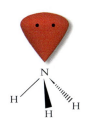

Nonbonding sp^3 pair

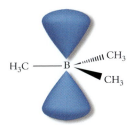

Vacant valence $2p$ orbital

Figure 21-1
Ammonia, which has a pair of nonbonding valence electrons, is a typical Lewis base. Trimethylboron, which has a vacant valence orbital, represents one type of Lewis acid.

Formation of Lewis Acid–Base Adducts

The simplest type of Lewis acid–base reaction is the combination of a Lewis acid and a Lewis base to form a compound called an **adduct.** The reaction of ammonia and trimethyl boron is an example. A new bond forms between boron and nitrogen, with both electrons supplied by the lone pair of ammonia (see Figure 21-2). Forming an adduct with ammonia allows boron to use all of its valence orbitals to form covalent bonds. As this occurs, the geometry about the boron atom changes from trigonal planar to tetrahedral, and the hybrid description of the boron valence orbitals changes from sp^2 to sp^3.

A general equation summarizes the formation of Lewis acid–base adducts:

$$A \; + \; :B \; \longrightarrow \; A\!-\!B$$

Lewis acid   Lewis base   Adduct

Figure 21-2
The N—B bond between NH_3 and $B(CH_3)_3$ can be described by the overlap of sp^3 orbitals on each atom, with the two electrons supplied by the lone pair of the N atom.

Here are two more examples of adduct formation:

$$SiF_4 + :F^- \longrightarrow SiF_5^- \qquad AlCl_3 + :PCl_3 \longrightarrow Cl_3Al\!-\!PCl_3$$

Acid   Base   Adduct   Acid   Base   Adduct

The reaction between two PCl_5 molecules demonstrates two additional features of Lewis acid–base reactions. First, substances such as PCl_5 can act as both Lewis acids and Lewis bases. The chlorine atoms in PCl_5 have electron pairs to donate, imparting Lewis base properties to the molecule. The phosphorus atom in PCl_5, whose bonding can be described using dsp^3 hybrid orbitals, has additional vacant valence d orbitals. One of these orbitals can form an additional bond by accepting an electron pair, resulting in d^2sp^3 hybridization. This gives PCl_5 Lewis acid properties. Second, in addition to formation of a new P—Cl bond, the reaction between two PCl_5 molecules involves P—Cl bond breaking. A more detailed view of the process uses curved arrows to indicate shifts of electrons.

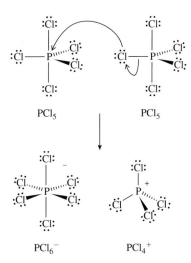

PCl_5 PCl_5

PCl_6^- PCl_4^+

Recognizing Lewis Acids and Bases

A Lewis *base* must have valence electrons available for bond formation. Any molecule whose Lewis structure shows nonbonding electrons can act as a Lewis base. Ammonia, phosphorus trichloride, and dimethyl ether, each of which contains lone pairs, are Lewis bases. Anions can also act as Lewis bases. In the first example of adduct formation above, the fluoride ion, with eight valence electrons in its $2s$ and $2p$ orbitals, acts as a Lewis base.

A Lewis *acid* must be able to accept electrons to form a new bond. Because bond formation can occur in several ways, compounds with several different structural characteristics can act as Lewis acids. Nevertheless, most Lewis acids fall into the following categories:

1. *A molecule that has vacant valence orbitals.* A good example is $B(CH_3)_3$, which uses a vacant $2p$ orbital to form an adduct with ammonia. The elements in the p block beyond the second row of the periodic table have empty valence d orbitals that allow them to act as Lewis acids. The silicon atom in SiF_4 is an example.

2. *A molecule with delocalized π bonds involving oxygen.* Examples are CO_2, SO_2, and SO_3. Each of these molecules can form a σ bond between its central atom and a Lewis base, at the expense of a π bond. For example, the hydroxide anion, a good Lewis base, attacks the carbon atom of CO_2 to form hydrogen carbonate:

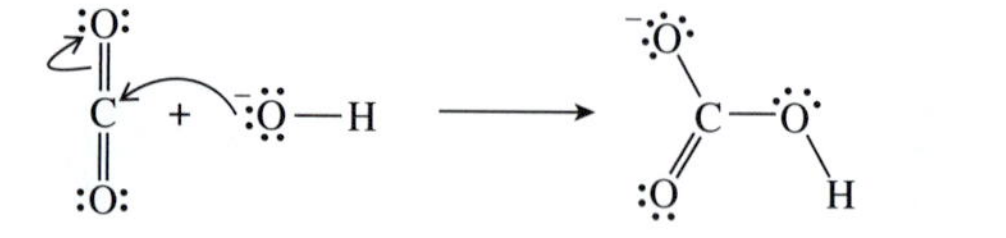

Lewis acid   Lewis base   Hydrogen carbonate adduct

In this reaction, the oxygen atom of the hydroxide ion donates a pair of electrons to make a new C—O bond. Because all the valence orbitals of the carbon atom in CO_2 are involved in bonding to oxygen, one of the C—O π bonds must be broken to make an orbital available to overlap with the occupied orbital of the hydroxide anion.

3. *A metal cation.* Removing electrons from a metal atom always generates vacant valence orbitals. As described in Chapter 20, many transition metal cations form complexes with ligands in aqueous solution. In these complexes, the ligands act as Lewis bases, donating pairs of electrons to form metal–ligand bonds. The metal cation accepts these electrons, so it acts as a Lewis acid. Metal cations from the p block also act as Lewis acids. For example, $Pb^{2+}(aq)$ forms a Lewis acid–base adduct with four CN^- anions, each of which donates a pair of electrons:

$$Pb^{2+}(aq) + 4\ CN^-(aq) \longrightarrow [Pb(CN)_4]^{2-}(aq)$$

Each of the four lead–carbon bonds forms by the overlap of an empty valence orbital on the metal ion with a lone pair on a carbon atom.

Example 21-1 provides practice in recognizing Lewis acids and bases.

Example 21-1	Identify the Lewis acids and bases in each of the following reactions and draw structures of the resulting adducts:
Lewis Acids and Bases	a. $AlCl_3 + Cl^- \rightarrow AlCl_4^-$

b. $Co^{3+} + 6\ NH_3 \rightarrow [Co(NH_3)_6]^{3+}$

c. $SO_2 + OH^- \rightarrow HSO_3^-$

Strategy: Every Lewis base has one or more lone pairs of valence electrons. A Lewis acid can have vacancies in its valence shell, or it can sacrifice a π bond to make a valence orbital available for adduct formation. To decide whether a molecule or ion acts as a Lewis acid or base, examine its Lewis structure for these features. Review Section 9.2 for procedures used to determine Lewis structures.

Solution: On the Lewis structures, we show possible transfers of electron pairs (blue) with curved red arrows.

a. Both species of this pair contain chlorine atoms with lone pairs, so either might act as a Lewis base if a suitable Lewis acid is present. The aluminum atom of $AlCl_3$ has a vacant $3p$ orbital perpendicular to the molecular plane. The empty p orbital accepts a pair of electrons from the Cl^- anion to form the fourth Al—Cl bond. The Lewis acid is $AlCl_3$, and the Lewis base is Cl^-.

Acid   Base   Adduct

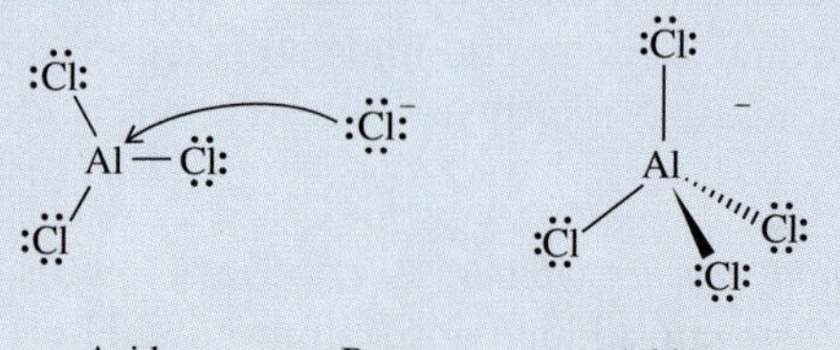

Example 21-1

Lewis Acids and Bases
(*continued*)

b. As already noted, ammonia is a Lewis base because it has a donor pair of electrons on the nitrogen atom. Like other transition metal cations, Co^{3+} is a Lewis acid. The cation uses vacant $3d$ orbitals to form bonds to NH_3:

$$Co^{3+} \quad :N\begin{matrix} H \\ H \\ H \end{matrix} \xrightarrow{\text{6 times}} \quad \begin{matrix} & NH_3 & \\ H_3N & | & NH_3 \\ & Co^{3+} & \\ H_3N & | & NH_3 \\ & NH_3 & \end{matrix}$$

| Acid | Base | Adduct |

c. Sulfur dioxide has delocalized π bonds, indicating Lewis acidity. The sulfur atom of SO_2 has a set of $3d$ orbitals that can be used to form an adduct. In this case, the hydroxide ion acts as a Lewis base. The anion uses one lone pair of electrons to form a new bond to sulfur:

| Acid | Base | | Adduct |

Do the Results Make Sense? The Lewis structures verify that each species identified as a Lewis base possesses lone pairs of electrons and that the Lewis acids have orbitals available to accept electrons.

Identify the Lewis acid and Lewis base, and draw the reaction showing electron transfers, in the formation of the aquo complex $[Ni(H_2O)_6]^{2+}$.

Section Exercises

21.1.1 Draw the Lewis structures of each of the reactants in the following reactions:
(a) $SO_3 + OH^- \rightarrow HSO_4^-$
(b) $SnCl_2 + Cl^- \rightarrow [SnCl_3]^-$
(c) $AsF_3 + SbF_5 \rightarrow [AsF_2]^+ [SbF_6]^-$

21.1.2 Identify the Lewis acid and the Lewis base in each of the reactions that appears in Section Exercise 21.1.1.

21.1.3 Draw structures that show the donation of electrons that takes place in each reaction in Section Exercise 21.1.1 and draw the Lewis structures of the products.

21.2 HARD AND SOFT LEWIS ACIDS AND BASES

Many years ago, geochemists recognized that whereas some metallic elements are found as sulfides in the Earth's crust, others are usually encountered as oxides, chlorides, or carbonates. Copper, lead, and mercury are most often found as sulfide ores; Na and K are found as their chloride salts; Mg and Ca exist as carbonates; and Al, Ti, and Fe are all found as oxides. Today chemists understand the causes of this differentiation among metal compounds. The underlying principle is how tightly an atom binds its valence electrons. The strength with which an atom holds its valence electrons also determines the ability of that atom to act as a Lewis base, so we can use the Lewis acid–base model to describe many affinities that exist among elements. This notion not only explains the natural distribution of minerals, but also can be used to predict patterns of chemical reactivity.

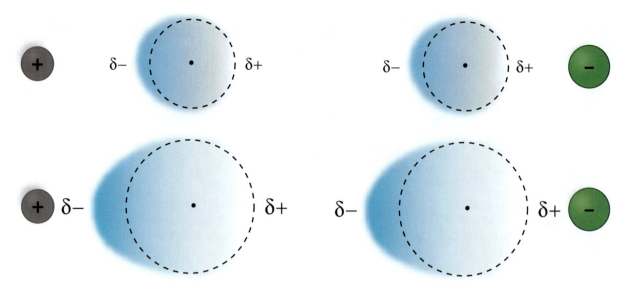

Figure 21-3
The electron cloud of an atom is polarized toward a positive charge and away from a negative charge. A smaller atom binds its valence electrons more tightly and is less polarizable than a larger atom, which binds its electrons loosely.

Polarizability

Polarizability, described in Chapter 11, is a measure of how tightly electrons are bound to an atom or molecule. The polarizability of an atom, ion, or molecule is the ease with which its electron cloud can be distorted by an electrical charge. An electron cloud is polarized toward a positive charge and away from a negative charge, as shown in Figure 21-3. Pushing the electron cloud to one side of an atom causes a polarization of charge. The side with the concentrated electron density builds up a small negative charge; the protons in the nucleus give the opposite side a small positive charge.

Polarizability shows periodic variations that correlate with periodic trends in how tightly valence electrons are bound to the nucleus:

1. Polarizability decreases from left to right in any row of the periodic table. As the effective nuclear charge (Z_{eff}) increases, the nucleus holds the valence electrons more tightly.
2. Polarizability increases from top to bottom in any column of the periodic table. As the principal quantum number (*n*) increases, the valence orbitals become larger. This reduces the net attraction between valence electrons and the nucleus.

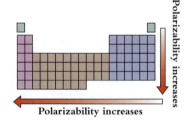

Polarizability increases

The Hard–Soft Concept

Lewis acids and bases can be organized according to their polarizability. If polarizability is low, the species is categorized as "hard." If polarizability is high, the species is "soft."

A **hard Lewis base** has electron pairs of low polarizability. This characteristic correlates with high electronegativity. Fluoride, the anion of the most electronegative element, is the hardest Lewis base because it contains a small, dense sphere of negative charge. Molecules and ions that contain oxygen or nitrogen atoms are also hard bases, although not as hard as fluorine. Examples include H_2O, CH_3OH, OH^-, NH_3, and H_2NCH_3.

A **soft Lewis base** has a large donor atom of high polarizability and low electronegativity. Iodide ion has its valence electrons in large *n* = 5 orbitals, making this anion highly polarizable and a very soft base. Other molecules and polyatomic anions with donor atoms from rows 3 to 6 are also soft bases. To summarize, the donor atom becomes softer from top to bottom of a column of the periodic table.

Group 17

F — Hardest

Cl

Br

I

At — Softest

A **hard Lewis acid** has an acceptor atom with low polarizability. Most metal atoms and ions are hard acids. In general, the smaller the ionic radius and the larger the charge, the harder the acid. The Al^{3+} ion, with an ionic radius of only 67 pm, is a prime example of a hard Lewis acid. The nucleus exerts a strong pull on the compact electron cloud, giving the ion very low polarizability.

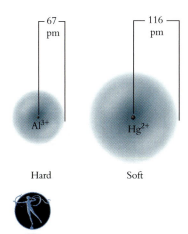

Hard   Soft

The designation of hard acids is not restricted to metal cations. For example, in BF_3 the small boron atom in its $+3$ oxidation state is bonded to three highly electronegative fluorine atoms. All the B—F bonds are polarized away from a boron center that is already electron-deficient. Boron trifluoride is a hard Lewis acid.

A **soft Lewis acid** has a relatively high polarizability. Large atoms and low oxidation states often convey softness. Contrast Al^{3+} with Hg^{2+}, a typical soft acid. The ionic radius of Hg^{2+} is 116 pm, almost twice the size of Al^{3+}, because the valence orbitals of Hg^{2+} have a high principal quantum number, $n = 6$. Consequently, Hg^{2+} is a highly polarizable, very soft Lewis acid. The relatively few soft transition metal ions are located around gold in the periodic table.

The terms *hard* and *soft* are relative, so there is no sharp dividing line between the two, and many Lewis acids and bases are intermediate between hard and soft. Example 21-2 shows how to categorize Lewis acids and bases according to their hard–soft properties.

③

Example 21-2

Ranking Hardness and Softness

5 B		7 N	8 O	
13 Al		15 P	16 S	
31 Ga			34 Se	

Rank the following groups of Lewis acids and bases from softest to hardest: (a) H_2S, H_2O, and H_2Se; (b) Fe^0, Fe^{3+}, and Fe^{2+}; and (c) BCl_3, $GaCl_3$, and $AlCl_3$.

Strategy: The first task is to decide whether the members of a given group are Lewis acids or bases. Then evaluate the relative softness and hardness based on polarizability, taking into account correlations with electronegativity, size, and charge. Refer to the periodic table in assessing the trends.

Solution:

a.

$$\ddot{O} \qquad H—\ddot{N}—H \qquad H—\ddot{P}—H$$
$$H \quad H \qquad\quad H \qquad\qquad H$$

These three molecules have lone pairs, so they are Lewis bases. The central atoms are in the same column of the periodic table, so their polarizability and the softness of the molecules increases moving down the column. Thus, H_2Se is softer than H_2S, which is softer than H_2O.

b. $Fe^0 \quad Fe^{2+} \quad Fe^{3+}$

Metal atoms and cations are Lewis acids. As valence electrons are removed from a metal atom, the remaining electron cloud undergoes an ever-larger pull from the nuclear charge. This decreases the size of the ion as well as its polarizability. Thus, Fe^0 is softer than Fe^{2+}, which is softer than Fe^{3+}.

c.

$$:\ddot{Cl} \quad \ddot{Cl}: \qquad :\ddot{Cl} \quad \ddot{Cl}: \qquad :\ddot{Cl} \quad \ddot{Cl}:$$
$$Ga \qquad\qquad Al \qquad\qquad B$$
$$:\ddot{Cl}: \qquad\qquad :\ddot{Cl}: \qquad\qquad :\ddot{Cl}:$$

These three molecules have trigonal planar geometry with sp^2 hybridized central atoms. Each has a vacant valence p orbital perpendicular to the molecular plane, making the molecules Lewis acids. The size, polarizability, and softness of the central acceptor atom increases going down the column. A gallium atom is larger and more polarizable than an aluminum atom. Thus, $GaCl_3$ is softer than $AlCl_3$, which is softer than BCl_3. We have already noted that aluminum is hard. Thus gallium trichloride is a soft Lewis acid, whereas $AlCl_3$ and BCl_3 are both hard.

Do the Results Make Sense? Hardness and softness are determined by atomic size, and the order of hardness in each of these sets is consistent with the size trends among the species.

Identify H_2O, NH_3, and PH_3 as Lewis acids or Lewis bases, and rank in order of increasing hardness.

The Hard–Soft Acid–Base Principle

The concept of hard and soft acids and bases can be used to interpret many trends in chemical reactivity. These trends are summarized in the hard–soft acid–base principle (HSAB principle), an empirical summary of results collected from many chemical reactions studied through decades of research.

KEY CONCEPTS *Hard Lewis acids tend to combine with hard Lewis bases.*
Soft Lewis acids tend to combine with soft Lewis bases.

The geochemical distribution of metals conforms to the HSAB principle. Metals that form hard acid cations have strong affinities for hard bases such as oxide, fluoride, and chloride. Most metal ions that are hard acids are found bonded to the oxygen atoms of various silicate anions. These elements are concentrated in the Earth's mantle. Hard acid metals also occur in combinations with other hard bases, including oxides or, less often, halides, sulfates, and carbonates. Examples include rutile (TiO_2), limestone ($CaCO_3$), gypsum ($CaSO_4$), and sylvite (KCl).

Metals that are soft acids, such as gold and platinum, have low affinities for hard oxygen atoms, so they are not affected by O_2 in the atmosphere. Consequently, these metals, including Ru, Rh, Pd, Os, Ir, Pt, and Au, are found in the crust of the Earth in their elemental form.

Other soft metals occur in nature as sulfides. The sulfide anion is a soft base, so this category includes some soft acids and many intermediate cases. Soft acids also may occur either in elemental form or as arsenide or telluride minerals. Examples include galena (PbS), cinnabar (HgS), chalcopyrite ($CuFeS_2$), argentite (Ag_2S), calaverite ($AuTe_2$), and sperrylite ($PtAs_2$).

The Group 13 elements illustrate the trend in hardness and softness among Lewis acids. At the top of the group, boron is a hard Lewis acid, and so is aluminum. Moving down the column, the valence orbitals increase in size and polarizability. Thus, gallium is a borderline acid, and indium is soft. The order of reactivity for the trihalides of these elements depends on the Lewis base. For a hard base such as ammonia, the reactivity trend for adduct formation is $BCl_3 > AlCl_3 > GaCl_3 > InCl_3$, reflecting the preference that the hard base has for the harder acid. The order is reversed when the trihalides form adducts with dimethylsulfide, $(CH_3)_2S$, a soft base.

Metathesis Reactions

A **metathesis reaction** is an exchange of bonding partners. Lewis acids and bases often undergo such exchanges, as exemplified by the following reaction:

$$BI_3 + GaF_3 \longrightarrow BF_3 + GaI_3$$

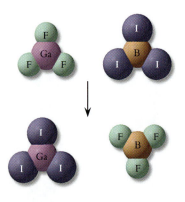

In this reaction, boron and gallium trade their halogen partners. According to the HSAB principle, metathesis reactions proceed in the direction that couples the harder acid with the harder base. Here, the sizes of the atoms reflect the fact that boron is harder than gallium, and fluoride is harder than iodide. Notice that the iodine atoms are significantly crowded in BI_3 but much more easily accommodated in GaI_3.

Metathesis is an important method for making molecules that have boron–carbon and aluminum–carbon bonds. These organoboron and organoaluminum species are valuable reagents in organic chemistry. Such compounds are often synthesized using alkyllithium reagents, powerful Lewis bases that are used widely in chemistry as sources of carbon atoms. Although highly reactive (they ignite spontaneously in air), alkyllithium reagents are available commercially, and they are easy to prepare in the laboratory by reacting alkyl halides with lithium metal:

$$ClCH_2CH_3 + 2\,Li \longrightarrow LiCH_2CH_3 + LiCl$$

Aluminum halides and boron halides undergo metathesis reactions with alkyllithium reagents according to the HSAB principle. Both boron and aluminum are softer acids than lithium, and carbon in these compounds is a very soft base. Metathesis generates lithium chloride and an organoaluminum or organoboron compound. To give a specific example, triethylaluminum, a colorless liquid that burns spontaneously in air, can be prepared by treating one mole of $AlCl_3$ with three moles of ethyllithium, $LiCH_2CH_3$:

$$3 \, LiCH_2CH_3 + AlCl_3 \longrightarrow Al(CH_2CH_3)_3 + 3 \, LiCl$$
Triethylaluminum

Many other alkyl groups can be used in place of $—CH_2CH_3$. Organoaluminum compounds, including triethylaluminum, are used to make catalysts for the polymer industry.

Group 14 metal halides also undergo metathesis reactions. For instance, organotin compounds are prepared on an industrial scale using organoaluminum reagents. These reactions take place because tin is a softer Lewis acid than aluminum, and carbon (in an alkyl group, represented by R) is a softer Lewis base than chlorine:

$$3 \, SnCl_4 + 4 \, AlR_3 \longrightarrow 3 \, SnR_4 + 4 \, AlCl_3$$

Organotin compounds are important industrial chemicals. One major use is as stabilizers for poly(vinyl chloride) (PVC) plastics. These additives, one example of which is dioctyltinmaleate, inhibit degradation of the polymer by heat, light, and oxygen. In the absence of these tin compounds, PVC yellows and becomes brittle.

Organotin compounds are also used extensively in agriculture. More than a third of the world's food crops are lost to fungi, bacteria, insects, or weeds. Organotin compounds such as tributylhydroxytin, $(CH_3CH_2CH_2CH_2)_3SnOH$, inhibit growth of fungi among crops such as potatoes, peanuts, sugar beets, and rice. Marine paints for wooden boats also contain organotin compounds that inhibit the attachment of barnacles. Organotins are added to cellulose and wool to inhibit attack by moths. The tin compounds used in these applications are specific in their toxicity, so they present little danger to mammalian life.

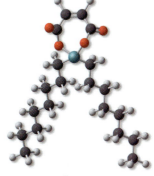

Dioctyltinmaleate

Section Exercises

21.2.1 Rank the following from softest to hardest:
(a) NCl_3, NH_3, and NF_3; (b) Pb^{2+}, Pb^{4+}, and Zn^{2+}; (c) ClO_4^-, ClO_2^-, and ClO_3^-; and (d) PCl_3, $SbCl_3$, and PF_3.

21.2.2 Iron is always found in nature in compounds, often with oxygen. The other members of Column 8 in the periodic table, ruthenium and osmium, occur in elemental form. Explain these observations using the HSAB principle.

21.2.3 Sulfur tetrafluoride fluorinates boron trichloride according to the following unbalanced equation:

$$BCl_3 + SF_4 \longrightarrow BF_3 + SCl_2 + Cl_2$$

This is both a redox reaction and a metathesis reaction. (a) Balance the equation. (b) Identify the elements that change oxidation state. (c) Explain the metathesis portion of the reaction using hard−soft acid−base arguments.

21.3 THE MAIN GROUP METALS

The elements in the lower left portion of the p-block of the periodic table are the **main group metals.** Although the most important metals of technological society are transition metals from the d block, three main group metals, aluminum, lead, and tin, have considerable technological importance.

Production of Aluminum

Aluminum is unique among the main group metals. All other p block metals have filled valence d orbitals. As a consequence, these metals have much in common with their transition metal neighbors. They tend to be soft Lewis bases. Aluminum, on the other hand, lacks a filled d orbital set and is a hard Lewis acid that has more in common with its nearest neighbor, magnesium. Highly reactive, aluminum is found naturally in the +3 oxidation state and is difficult to reduce to the pure metal. Thus, although tin and lead have been known since antiquity, aluminum was not discovered until 1825 and did not become a common commodity until more than 60 years later.

Bauxite, the main aluminum ore, is a mixed oxide−hydroxide, $Al(O)OH$, contaminated with SiO_2, Fe_2O_3, clay, and other hydroxide salts. To isolate the aluminumcontaining material, the ore is treated with a strongly basic solution, whose high

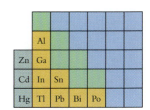

The main group metals are found in the corner of the p block closest to the transition metals.

hydroxide concentration causes the solid to dissolve as a soluble complex ion, $[Al(OH)_4]^-$:

$$Al(O)OH(s) + OH^-(aq) + H_2O \longrightarrow [Al(OH)_4]^-(aq)$$

The impurities in bauxite are not soluble in strong base, so the impurities remain behind when the solution is separated from the undissolved solids. Although $[Al(OH)_4]^-$ is soluble, $Al(OH)_3$ is not, so diluting the strongly basic solution causes solid aluminum hydroxide to precipitate:

$$[Al(OH)_4]^-(aq) \xrightarrow{\text{Added water}} Al(OH)_3(s) + OH^-(aq)$$

After this separation is complete, $Al(OH)_3$ is heated to drive off water, which converts the hydroxide into pure aluminum oxide:

$$2\ Al(OH)_3(s) \xrightarrow{125\,°C} Al_2O_3(s) + 3\ H_2O(g)$$

Aluminum metal is produced from aluminum oxide by electrolysis using the **Hall–Héroult process,** whose story is detailed in our Chemical Milestones Box. The melting point of Al_2O_3 is too high (2015 °C) and its electrical conductivity too low to make direct electrolysis commercially viable. Instead, Al_2O_3 is mixed with cryolite (Na_3AlF_6) containing about 10% CaF_2. This mixture has a melting point of 1000 °C, still a high temperature but not prohibitively so. Aluminum forms several complex ions with fluoride and oxide, so the molten mixture contains a variety of species, including AlF_4^-, AlF_6^{3-}, and $AlOF_3^{2-}$. These and other ions move freely through the molten mixture as electrolysis occurs.

Figure 21-4 shows a schematic representation of an electrolysis cell for aluminum production. An external electrical potential drives electrons into a graphite cathode, where Al^{3+} ions are reduced to Al metal:

$$\text{Cathode:} \quad Al^{3+}(melt) + 3\ e^- \longrightarrow Al(l)$$

The anode, which is also made of graphite, is oxidized during electrolysis. Carbon from the anode combines with oxide ions to form CO_2 gas:

$$\text{Anode:} \quad 2\ O^{2-}(melt) + C(s) \longrightarrow CO_2(g) + 4\ e^-$$

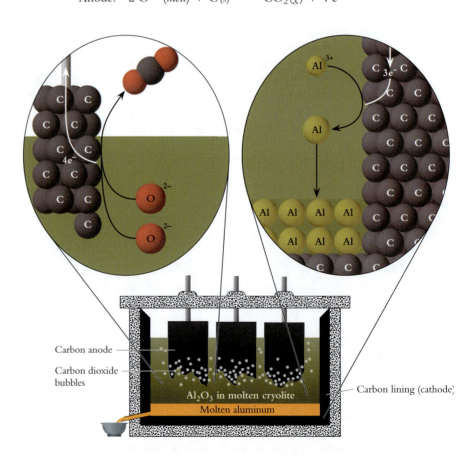

Figure 21-4
Aluminum metal is produced by electrolysis of aluminum oxide dissolved in molten cryolite. Al^{3+} is reduced to Al at the cathode, and C is oxidized to CO_2 at the anode.

Carbon anode

Carbon dioxide bubbles

Al_2O_3 in molten cryolite

Molten aluminum

Carbon lining (cathode)

Box 21-1 — Chemical Milestones: The Story of the Hall–Héroult Process

Aluminum is the third most abundant element in the Earth's crust and the most abundant metal. Nevertheless, aluminum was not discovered until 1825 and was still a precious rarity 60 years later. The reason for this elusiveness is the high stability of Al^{3+}. The reduction of aluminum compounds to the free metal requires stronger reducing power than common chemical reducing agents can provide. The discovery of aluminum had to await the birth of electrochemistry and the development of electrolysis.

Twenty-five years after its discovery, aluminum was a precious metal. Then a French chemist developed procedures for reducing aluminum compounds using sodium metal. The price of the metal dropped 100-fold. Even so, in 1885 aluminum was a semiprecious metal used for esoteric purposes such as a prince's baby rattle and the cap for the Washington Monument.

To convert aluminum from the stuff of princes' toys into recyclable kitchen foil required an inexpensive electrolytic reduction process. Two 22-year-old scientists, the American chemist Charles Hall and the French metallurgist Paul Héroult, discovered the same process independently in 1886. Both became famous as founders of the aluminum industry, Hall in the United States and Héroult in Europe.

Charles Hall was inspired by his chemistry professor at Oberlin College, who observed that whoever perfected an inexpensive way of producing aluminum would become rich and famous. After his graduation, Hall set to work in his home laboratory, trying to electrolyze various compounds of aluminum. He was aided by his sister Julia, who had studied chemistry and shared Charles' interests.

Julia helped to prepare chemicals and witnessed many of the electrolysis experiments. After only eight months of work, Hall had successfully produced globules of the metal. Meanwhile, Héroult was developing the identical process in France.

Charles Hall   Julia Hall

Hall capitalized on his discovery by founding a company for the manufacture of aluminum. That company became immensely successful, eventually growing into Alcoa. It made the Halls very rich.

Successful electrolysis of aluminum requires a liquid medium other than water that can conduct electricity. The key to the Hall–Héroult process is the use of molten cryolite, Na_3AlF_6, as a solvent. Cryolite melts at an accessible temperature, it dissolves Al_2O_3, and it is available in good purity. A second important feature is the choice of graphite to serve as the anode. Graphite provides an easy oxidation process, the oxidation of carbon to CO_2.

How did two persons working independently on two different continents come up with an identical process at the same time? The reasoning that led Hall and Héroult to the identical process was probably similar. Electrolysis was recognized as a powerful reducing method. All attempts to reduce aqueous aluminum cations failed, making it clear that some molten salt would have to be used. Experimenting with various salts, no doubt guided by the principle that "like dissolves like," the two young men eventually tried Na_3AlF_6, a mineral whose constituent elements should not interfere with the reduction of aluminum. Graphite electrodes were already in use, so experimenting with them would have been a natural choice.

The ingredients for this invention may have all been in place, but that does not detract from its brilliance. Hall and Héroult had the courage to explore new procedures in homemade laboratories without the support of research grants. They explored the possibilities systematically to find a process that was a spectacular success. In more than 100 years of growth in the aluminum industry, the only significant change to the Hall–Héroult process has been the addition of CaF_2 to the melt to lower the operating temperature.

Because of the variety of ionic species present in the melt, the reactions that take place at the electrodes are considerably more complex than the simple representations given here. Nevertheless, the net reaction is the one given by these simplified reactions:

$$4\ Al^{3+}(melt) + 6\ O^{2-}(melt) + 3\ C\,(s) \longrightarrow 3\ CO_2(g) + 4\ Al\,(l)$$

The electrolysis apparatus operates well above the melting point of aluminum (660 °C), and liquid aluminum has a higher density than the molten salt mixture, so pure liquid metal settles to the bottom of the reactor. The pure metal is drained through a plug and cast into ingots.

Aluminum refining consumes huge amounts of electricity. Approximately 5% of all electricity consumed in the United States is used to produce aluminum.

Uses of Aluminum

Aluminum is one of society's most important structural metals. It has one of the lowest densities among the metals, yet it is very strong, and it resists corrosion by forming a thin layer of aluminum oxide on the surface of the metal. Alloys of aluminum are used for aircraft bodies, trailers, cooking utensils, highway signs, storage tanks, beverage cans, and many other objects.

Although the major use of aluminum by far is as a metal, some aluminum compounds also are economically important. Foremost among these is aluminum chloride, which is an important industrial catalyst.

The aluminum atom in aluminum chloride behaves as a Lewis acid, and the chlorine atoms are Lewis bases. As a consequence, two $AlCl_3$ molecules form a Lewis acid–base adduct, Al_2Cl_6. As Figure 21-5 shows, the compound has two tetrahedral aluminum atoms held together by a pair of chlorine atoms that bridge the two metal atoms. The 91° Al—Cl—Al bond angle suggests that the chlorine atom uses two valence p orbitals to form the bridge. There is an equilibrium between $AlCl_3$ and Al_2Cl_6 molecules, but except in the gas phase above 750 °C, the formation of Al_2Cl_6 goes essentially to completion.

In the solid state, aluminum chloride exists in a crystalline lattice. Each aluminum atom is surrounded by six chlorine atoms arranged around the metal atoms at the corners of an octahedron. Aluminum bromide and aluminum iodide form Al_2X_6 molecules in all three phases.

The aluminum trihalides are particularly important Lewis acids in the chemical industry. They promote or catalyze a large variety of reactions. One of the

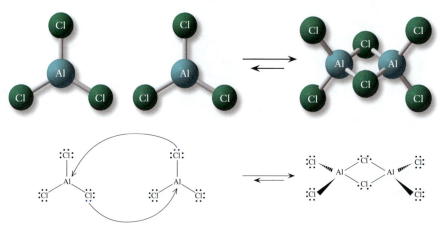

Figure 21-5
The Al atom in $AlCl_3$ has Lewis acid character, whereas the Cl atoms are Lewis bases. The compound forms a self-adduct in which two Cl atoms donate electron pairs to form additional Al—Cl bonds.

most important applications is the Friedel–Crafts reaction, in which two molecules combine, forming a new C—C bond. For example, aluminum chloride or some other Lewis acid catalyzes the reaction between an acid chloride and benzene to form acetophenone:

Benzene   Acetyl chloride   Acetophenone

The Lewis acid activates the acid chloride by forming an adduct with a chlorine atom bridge between carbon and aluminum. This chlorine bridge is similar to the one found in Al_2Cl_6:

Lewis base   Lewis acid   Adduct

The adduct fragments into an $[AlCl_4]^-$ anion and a very reactive cation:

Adduct   Reactive cation

> An acid chloride is a derivative of a carboxylic acid in which the hydroxyl group is replaced by a chlorine atom.

Reaction with a benzene molecule produces acetophenone and HCl and regenerates the $AlCl_3$ catalyst.

Friedel–Crafts chemistry is used in the synthesis of dyes, flavorings, fragrances, surfactants, pesticides, and many other types of organic compounds. Many different Lewis acids, including $AlCl_3$, BF_3, $TiCl_4$, $SnCl_4$, SbF_5, and $ZnCl_2$, serve as catalysts for these reactions.

Aluminum sulfate, $Al_2(SO_4)_3$, is widely used in water purification to remove finely divided particulate matter. When added to water, aluminum sulfate forms a precipitate of aluminum hydroxide that has a very open structure and large surface area. This precipitate, called a *gel*, traps dispersed particulate matter as it settles out of the liquid phase.

Tin and Lead

Tin and lead are neighbors in Group 14 of the periodic table. Each metal displays two common oxidation states, +2 and +4, and each can be reduced to the free metal with relative ease. The major ore of lead is galena, PbS, in which lead is in the +2 state. In contrast, the major ore of tin is cassiterite, SnO_2, where tin is in the +4 state. The Pb^{2+} state is substantially softer than the Sn^{4+} state, not only because of lower charge but also because of a larger value for n. This accounts for lead being found associated with the softer base, S^{2-}, whereas tin combines with the harder base, O^{2-}.

Carbon in the form of charcoal is a sufficiently powerful reducing agent to convert lead and tin ores to the free metals. In the case of galena, the first step is roasting in air to form PbO:

$$2\ PbS + 3\ O_2 \xrightarrow{Heat} 2\ PbO + 2\ SO_2$$

$$PbO + C \xrightarrow{Heat} Pb + CO$$

$$SnO_2 + 2\ C \xrightarrow{Heat} Sn + 2\ CO$$

A bed of hot charcoal can supply both the heat and the reducing agent, and the simplicity of the process accounts for the early metallurgical development of these metals.

In former times, tin was used widely as a constituent of metal alloys, of which bronze, solder, and pewter are common examples. Bronze is an alloy of copper containing approximately 20% tin and smaller amounts of zinc. Pewter is another Cu–Sn alloy

Bronze

Pewter

that contains tin as the major component ($\approx 85\%$), with roughly equal portions of copper, bismuth, and antimony. Solder consists of 67% lead and 33% tin.

Because of its relatively low melting point (232 °C) and good resistance to oxidation, tin is used to provide protective coatings on metals such as iron that oxidize more readily. "Tin cans" are iron cans dipped in molten tin to provide a thin surface film of tin. Traditional metalsmiths use a similar process, coating copperware with a thin film of tin.

Metals that are soft Lewis acids, for example cadmium, mercury, and lead, are extremely hazardous to living organisms. Tin, in contrast, is not. One reason is that tin oxide is highly insoluble, so tin seldom is found at measurable levels in aqueous solution. Perhaps more important, the toxic metals generally act by binding to sulfur in essential enzymes. Tin is a harder Lewis acid than the other heavy metals, so it has a lower affinity for sulfur, a relatively soft Lewis base.

The history of lead is almost as ancient as that of tin. In Roman times, lead was formed into pipes that were used for water supplies (hence our word *plumbing,* derived from the same Latin word, *plumbum,* that gives us the symbol Pb). Lead is a component of pewter and also was used as a glaze on drinking vessels. "White lead," $Pb_3(OH)_2(CO_3)_2$, is an enduring white substance that was used for many years as a paint pigment and even as a component of cosmetics.

More recently, the major use for lead has been in the automobile industry. The lead storage battery, described in Chapter 19, generates a high electrical potential, can deliver large currents, and is easily recharged. Tetraethyllead, $(C_2H_5)_4Pb$, is excellent at preventing the pre-ignition of gasoline in automobile cylinders. Once used widely as an "antiknock" fuel additive, $(C_2H_5)_4Pb$ is no longer added to gasoline because lead is highly toxic.

Unfortunately, lead easily enters the biosphere. Lead poisoning in humans causes learning impairment and behavioral disorders in children, and at higher levels it triggers mood swings, irritability, and loss of coordination. Historians think that the widespread use of lead in piping and drinking vessels caused the Romans to suffer a relatively high incidence of lead poisoning, which may even have contributed to the downfall of their empire. Lead-based paints may have caused a similarly high incidence of lead poisoning among Europeans, and it is known that present-day American children living in low-income housing are at high risk because of the presence of lead-based paints that have not been removed.

Because of these damaging effects, most uses of lead that involve direct exposure for humans are being phased out. Unleaded gasoline and lead-free paints have replaced two former major commercial uses of lead. Lead has proved to be indispensable, however, in the lead storage battery, which now provides the major use of this metal. Although leakage from damaged batteries is still a potential hazard, contemporary batteries are manufactured in such a way that human exposure to battery contents is minimized.

Section Exercises

■ **21.3.1** Draw molecular pictures similar to those in Figure 21-5 that illustrate the electron pair donation that occurs when acetyl chloride (CH_3COCl, formed from acetic acid by replacing —OH with —Cl), forms an adduct with aluminum chloride.

■ **21.3.2** Indium is a relatively soft Lewis acid. Use this fact to predict properties of indium compared with its horizontal neighbor (tin) and its diagonal neighbor (lead) in the periodic table.

■ **21.3.3** Lead poisoning can be treated using the dianion of 2,3-dimercaptopropanol. Two ions bind one Pb^{2+} ion in a soluble tetrahedral complex that can be excreted from the body. Using hard–soft acid–base concepts, draw the expected structure for this complex.

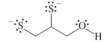

2,3-Dimercaptopropanol dianion

21.4 THE METALLOIDS

Six elements are **metalloids**: B, Si, Ge, As, Sb, and Te. Of these, silicon is by far the most abundant, making up over 27% of the Earth's crust, more than any other element except oxygen. In fact, SiO_2 and silicate minerals account for 80% of the atoms near the Earth's surface. Despite its great abundance, silicon was not discovered until 1824, probably because the strong bonds it forms with oxygen makes silicon difficult to isolate. Two much rarer metalloids, antimony (known to the ancients) and arsenic (discovered ca. 1250 AD) were isolated and identified long before silicon.

Until quite recently, chemical interest in the metalloids consisted mainly of isolated curiosities, such as the poisonous nature of arsenic and the mildly therapeutic value of borax. The development of metalloid semiconductors, which we describe in Chapter 10, focused intense study on these elements.

Boron

As is typical of second-row elements, boron has properties that distinguish it from the other elements in Group 13 as well as from the rest of the metalloids. The unique features of boron chemistry can be attributed to characteristics of its electron configuration. As a second-row element, boron has no valence d orbitals that can participate in bonding. Its ionization energies are considerably higher than those of aluminum and the other Group 13 elements, all of which are metals. However, the three valence electrons of the boron atom are insufficient to form the extensive bonding networks that are characteristic of carbon, its second-row neighbor.

Boric acid (H_3BO_3) and boron trifluoride (BF_3) exemplify the bonding patterns of boron compounds. As Figure 21-6 shows, both these compounds contain delocalized π bonds. Each acts as a Lewis acid, readily adding an additional anion to form $[B(OH)_4]^-$ and $[BF_4]^-$ adducts. Both adducts are tetrahedral species in which the bonding can be described using four σ bonds formed from sp^3 hybrid orbitals on the boron atom.

The changes in bonding that accompany adduct formation are reflected in differences in bond lengths. The B—F bond lengths are 131 pm in BF_3 and 139 pm in $[BF_4]^-$; B—O bond lengths are 137 pm in H_3BO_3 and 147 pm in $[B(OH)_4]^-$. The shorter bonds in the neutral species provide clear evidence for multiple bonding. The unusual strength of the B—F bond in BF_3 also indicates the presence of π bonding: 645 kJ/mol is an energy more consistent with double bonds than single bonds.

Because of their low polarizability, BF_3 and BCl_3 are gases. The larger, more polarizable electron clouds on bromine and iodine make BBr_3 a volatile liquid and BI_3 a solid at room temperature.

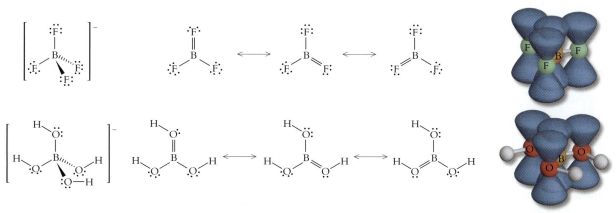

Figure 21-6
The Lewis structures of BF₃ and H₃BO₃ indicate that these species contain delocalized π bonds. The adducts [BF₄]⁻ and [B(OH)₄]⁻, in contrast, contain only single σ bonds.

Boron trihalides are strong Lewis acids that react with a wide collection of Lewis bases. Many adducts form with donor atoms from Group 15 (N, P, As) or Group 16 (O, S). Metal fluorides transfer F^- ion to BF_3 to give tetrafluoroborate salts:

$$LiF + BF_3 \longrightarrow LiBF_4$$

Tetrafluoroborate anion is an important derivative of BF_3 because it is nonreactive. With four σ bonds, $[BF_4]^-$ anion has no tendency to coordinate further ligands. Tetrafluoroborate salts are used in synthesis when a bulky inert anion is necessary.

The observed order of reactivity for the boron halides is

$$BF_3 < BCl_3 < BBr_3 < BI_3$$

From electronegativity considerations, we might expect the opposite trend. The electronegativity difference between boron ($\chi = 2.0$) and fluorine ($\chi = 4.0$) is 2, whereas boron and iodine ($\chi = 2.5$) differ by only 0.5. Thus, fluorine atoms withdraw more electron density from boron than iodine atoms do, resulting in a more positive boron atom that should be a stronger Lewis acid. The observed reactivity trend indicates that some mechanism returns electron density to boron in the lighter boron halides. Such a mechanism is provided by the delocalized π bonding resulting from the overlap between filled halogen p orbitals and the empty p orbital on the boron atom. The π system shifts electron density from the halogen atoms into the bonding region between the atoms. Fluorine, being the smallest halogen, forms the strongest π bond with boron, since the side-by-side overlap needed for π bonding is greater when $n = 2$ than for larger, more diffuse p orbitals.

Silicon

The most abundant compounds of silicon are SiO_2 and the related silicate anions, all of which contain Si—O bonds. See Sections 9.3 and 10.3 for descriptions of the structure and bonding of these compounds, which involve σ bond networks and tetrahedral geometry. As already mentioned, many minerals are combinations of hard silicate anions and hard metal cations.

Even though silicon is extremely abundant, only one silicon-containing compound appears in the list of top 50 industrial chemicals. That is sodium silicate, Na_2SiO_3, used for the manufacture of silica gel and glass. Nevertheless, with the advent of the electronic age silicon has become an extremely important substance that is the primary ingredient of most semiconductors. Because these are microscale devices, the quantity of production of silicon remains small compared with that of fertilizers and construction materials. Although relatively small in quantity, the value of silicon products is quite high.

The main source of silicon for semiconductor chips is silicon dioxide (silica). Silica can be reduced directly to elemental form by intense heating with coke in an electric arc furnace. At the temperature of the furnace, silicon is a gas:

$$SiO_2 + C \xrightarrow{3000\ °C} Si + 2\ CO$$

The product of this process has a purity of around 98%, which is sufficient for uses such as alloy formation but not nearly high enough for semiconductors.

To prepare pure silicon, silica is first converted to $SiCl_4$. A redox reaction between coke and chlorine gas is coupled with a metathesis reaction to give $SiCl_4$, which liquefies on cooling:

$$SiO_2 + 2\ C + 2\ Cl_2 \xrightarrow{High\ T} SiCl_4 + 2\ CO$$

Distillation yields $SiCl_4$ of very high purity, which is then reduced with magnesium:

$$SiCl_4 + 2\ Mg \longrightarrow Si + 2\ MgCl_2$$

After the removal of $MgCl_2$ by washing with water, the silicon is purified still further by the technique of zone refining, shown schematically in Figure 21-7. A rod of impure silicon is melted and resolidified many times by a heating coil that passes back and forth along the rod. During this process, impurities remain preferentially in the liquid phase, yielding solid silicon containing less than 1 part per billion of impurities.

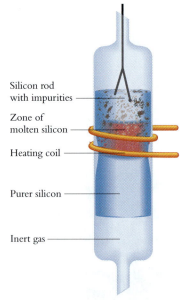

Silicon rod with impurities

Zone of molten silicon

Heating coil

Purer silicon

Inert gas

Figure 21-7
In zone refining. a rod of impure silicon passes along a heating coil. The impurities concentrate in the molten zone, leaving solid material behind that is of higher purity. Repeated passage through the coil moves the impurities to the end of the rod.

Silicones

In addition to its uses for electronic devices, silicon is a major component of silicone polymers. The silicone backbone consists of alternating silicon and oxygen atoms. The synthesis of these polymers begins with an organic chloride such as methyl chloride and an alloy of silicon and copper:

$$2 \ CH_3Cl + Si(Cu) \longrightarrow (CH_3)_2SiCl_2 + Cu$$

Treatment of the resulting chlorosilane with water replaces the chlorine atoms with hydroxyl groups:

$$(CH_3)_2SiCl_2 + 2 \ H_2O \longrightarrow (CH_3)_2Si(OH)_2 + 2 \ HCl$$

Finally, the dihydroxysilane eliminates water in a condensation reaction to give a polymer with an Si—O—Si linkage, as Figure 21-8 shows.

Approximately 70,000 tons of silicone polymers are produced each year in the United States. Silicones are used as greases, caulking, gaskets, biomedical devices, cosmetics, surfactants, antifoaming agents, hydraulic fluids, and water repellents.

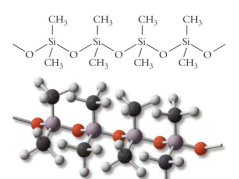

Figure 21-8
Structural formula and ball-and-stick model of a portion of a silicone polymer chain.

Other Metalloids

The remaining metalloids are Ge, As, Sb, and Te. Each has an abundance less than 2 ppm in the Earth's crust. As already mentioned, arsenic is highly poisonous. Besides being the stuff of many murder mysteries, arsenic is used as a pesticide. The major current use of antimony is in lead–acid batteries. Battery electrodes are much less likely to electrolyze water during recharging when they are made from a lead–antimony alloy containing up to 5% Sb. This virtually eliminates the danger of gas buildup and subsequent rupturing of the battery and has allowed the production of sealed batteries, which have much improved lifetimes compared with unsealed batteries that use pure lead electrodes.

A major and growing use of the minor metalloids is in semiconductor fabrication. Germanium, like silicon, exhibits semiconductor properties. Binary compounds between elements of Groups 13 and 15 also act as semiconductors. These 13–15 compounds, such as GaAs and InSb, have the same number of valence electrons as Si or Ge. The energy gap between the valence band and the conduction band of a 13–15 semiconductor can be varied by changing the relative amounts of the two components. This allows the properties of 13–15 semiconductors to be fine-tuned.

Section Exercises

■ **21.4.1** The anion present in aqueous solutions of boric acid is $[B(OH)_4]^-$ rather than $H_2BO_3^-$. This is because H_3BO_3 undergoes Lewis acid–base adduct formation with one water molecule, and the adduct then transfers a proton to a second water molecule to generate $[B(OH)_4]^-$ and a hydronium ion. Write Lewis structural diagrams illustrating these two transfer reactions. Show all formal charges, and include arrows that show the movement of electrons.

■ **21.4.2** Draw band gap diagrams (review Section 10.7 if necessary) illustrating that silicon is a semiconductor but carbon (diamond) is not. What feature of the valence atomic orbitals accounts for this difference?

21.5 PHOSPHORUS

Of all the elements, phosphorus is the only one that was first isolated from a human source. The element was extracted from human urine in 1669 using an unsavory process: After a sample of urine was allowed to stand for several days, the putrefied liquid was boiled until only a paste remained. Further heating of the paste at high temperature produced a gas that condensed to a waxy white solid when the vapor was bubbled into water. It wasn't until 1779 that phosphorus was discovered in mineral form, as a component of phosphate minerals.

Despite its relatively late discovery, phosphorus is the eleventh most abundant element in Earth's crustal rock. It has been estimated that world reserves of "phosphate rock" are sufficient to last for several hundred years. Virtually all phosphorus deposits contain apatite, whose general formula is $Ca_5(PO_4)_3X$, where X = F, OH, or Cl. Fluoroapatite is the least soluble, hence most abundant, of the three apatite minerals. Phosphorus is found in aqueous systems as HPO_4^{2-} and $H_2PO_4^-$ ions. In biological organisms, phosphorus is a component of nucleic acids and energy-shuttling molecules such as ATP.

Elemental Phosphorus

Modern production of elemental phosphorus uses a technique similar to the metallurgical processes described in Chapter 20. Apatite is mixed with silica and coke and then heated strongly in the absence of oxygen. Under these conditions, coke reduces phosphate to elemental phosphorus, the silica forms liquid calcium silicate, and the fluoride ions in apatite dissolve in the liquid calcium silicate. The reactions are not fully understood, but the stoichiometry for the calcium phosphate part of apatite is as follows:

$$2\ Ca_3(PO_4)_2 + 6\ SiO_2 + 10\ C \xrightarrow{1450\ °C} P_4 + 6\ CaSiO_3 + 10\ CO$$

Elemental phosphorus is a gas at the temperature of the reaction, so the product distills off along with carbon monoxide. When cooled to room temperature, P_4 condenses as a waxy white solid.

White phosphorus consists of individual P_4 molecules with the four atoms at the corners of a tetrahedron. Each atom bonds to three others and has one lone pair of electrons, for a steric number of 4. However, the triangular geometry of the faces of the tetrahedron constrains the bond angles in the P_4 tetrahedron to 60°, far from the optimal four-coordinate geometry of 109.5°. As a result, P_4 is highly reactive. As shown in Figure 21-9, samples of white phosphorus are stored under water because P_4 burns spontaneously in the presence of oxygen. The glow that emanates from P_4 as it burns in the dark led to its name, from the Greek: *phos* ("light") and *phoros* ("bringing"). This form of phosphorus is very toxic: As little as 50 mg can cause death. As described later, most of the white phosphorus produced by the chemical industry is used to manufacture phosphoric acid.

Heating white phosphorus in the absence of oxygen causes the discrete P_4 units to link together, giving a chemically distinct elemental form, red phosphorus. As Figure 21-10 shows, one P—P bond of each tetrahedron breaks to allow formation of the bonds that link the P_4 fragments. The bond angles that involve the links are much closer to 109.5°, making red phosphorus less strained and less reactive than white phosphorus. Red phosphorus undergoes the same chemical reactions as the white form, but higher temperatures are required. In addition, red phosphorus is essentially nontoxic, and therefore it is easier and safer to handle.

The most thermodynamically stable form of the element is black phosphorus, which can be prepared by heating red phosphorus under high pressure. The black form contains chains of P_4 units cross-linked by P—P bonds, making this form even more polymerized and less strained than red phosphorus. Example 21-3 explores another difference between the elemental forms of phosphorus.

Figure 21-9
The strained 60° bond angles in tetrahedral P_4 make white phosphorus highly reactive. This form of the element must be kept out of contact with air, in which it spontaneously burns.

Animation

60°

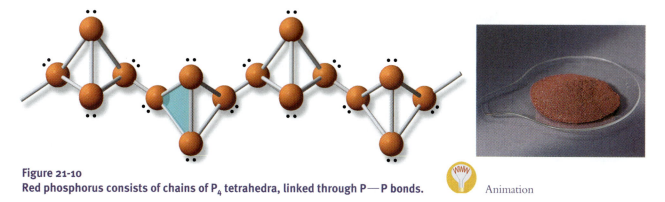

Figure 21-10
Red phosphorus consists of chains of P_4 tetrahedra, linked through P—P bonds.

Animation

The melting point of white phosphorus is 44.1 °C. In contrast, red phosphorus remains a solid up to 400 °C. Account for this large difference in melting point.

Strategy: As described in Chapter 11, melting points of solids may depend on both covalent bonds and intermolecular forces, so we must explain the melting point difference with reference to the bonding differences between the two forms.

Solution: White phosphorus consists of individual P_4 molecules. Because there are no polar bonds, the molecules are held in the solid state only by dispersion forces. This is a molecular solid, and such solids have relatively low melting temperatures. Red phosphorus is made of long chains of P_4 groups, each of which is held together by covalent P—P bonds, as shown in Figure 21-10. Thus, red phosphorus consists of macromolecules held together by dispersion forces, much like polyethylene, as described in Chapter 13. To melt, red phosphorus must break some chemical bonds. Bond breakage requires much more energy than the energy needed to overcome dispersion forces, giving red phosphorus the higher melting temperature.

Does the Result Make Sense? The molecular models in Figures 21-9 and 21-10 show that white phosphorus is a molecular solid, whereas red phosphorus is intermediate between molecular and network. This correlates well with the melting points.

Predict whether black phosphorus melts at a higher or lower temperature than the other two forms, and rank the three forms of elemental phosphorus in order of increasing density, giving a reason for your ranking.

Example 21-3

Melting Points of Phosphorus

Extra Practice Exercise 21.3

Phosphoric Acid

The most important commercial product of phosphorus is phosphoric acid, H_3PO_4. Phosphoric acid consistently ranks among the top ten industrial chemicals in the United States, with a yearly production of over 10,000 tons.

Almost all phosphoric acid is produced directly from apatite. The ore is partially purified, crushed, and then slurried with aqueous sulfuric acid:

$$Ca_5(PO_4)_3F(s) + 5\ H_2SO_4(aq) \longrightarrow 3\ H_3PO_4(aq) + 5\ CaSO_4(s) + HF(aq)$$

The dilute phosphoric acid obtained from this process is concentrated by evaporation. It is usually dark green or brown because of the presence of many metal ion impurities in the phosphate rock. However, this impure acid is suitable for the manufacture of phosphate fertilizers, which consumes almost 90% of phosphoric acid production.

High-purity H_3PO_4 is obtained using a more expensive redox process that starts from the pure element. Controlled combustion of white phosphorus gives phosphorus(V) oxide, P_4O_{10}, whose structure is shown in Figure 21-11:

$$P_4 + 5\ O_2 \longrightarrow P_4O_{10}$$

Animation

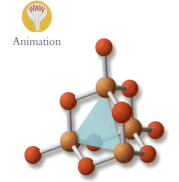

Figure 21-11
Phosphorus(V) oxide, P_4O_{10}, contains the same triangular arrangement of phosphorus atoms as in P_4, but an oxygen atom is inserted into each P—P bond. An additional terminal O atom is double-bonded to each P atom.

Addition of water to P_4O_{10} generates highly pure phosphoric acid:

$$P_4O_{10}(s) + 6\,H_2O\,(l) \longrightarrow 4\,H_3PO_4\,(aq)$$

More than 80% of the elemental phosphorus produced is converted to phosphoric acid. This pure product, which constitutes about 10% of the total industrial output of phosphoric acid, is the starting material for making food additives, pharmaceuticals, and detergents.

The elemental phosphorus that is not converted into phosphoric acid is used mainly to produce phosphorus chlorides (PCl_3 and PCl_5) and phosphorus oxychloride ($POCl_3$). These are important reagents for production of agrochemicals, drugs, and other specialty products.

Phosphorus Fertilizers

Plants typically contain about 0.2% phosphorus by weight, but the element is easily depleted from soils. For this reason, the #1 commercial application of phosphorus is in fertilizers. There are several common phosphorus fertilizers, but the single most important one is ammonium hydrogen phosphate, $(NH_4)_2HPO_4$. This phosphate compound is particularly valuable because it is highly soluble and provides both phosphorus and nitrogen. The compound can be made by treating phosphoric acid with ammonia:

$$H_3PO_4 + 2\,NH_3 \longrightarrow (NH_4)_2HPO_4$$

This is the predominant industrial route to $(NH_4)_2HPO_4$, accounting for the single largest use of phosphoric acid.

Phosphate Condensations

Phosphoric acid can undergo phosphate condensation, a reaction that is similar to the condensation reactions of other species that contain O—H bonds (see Chapter 13):

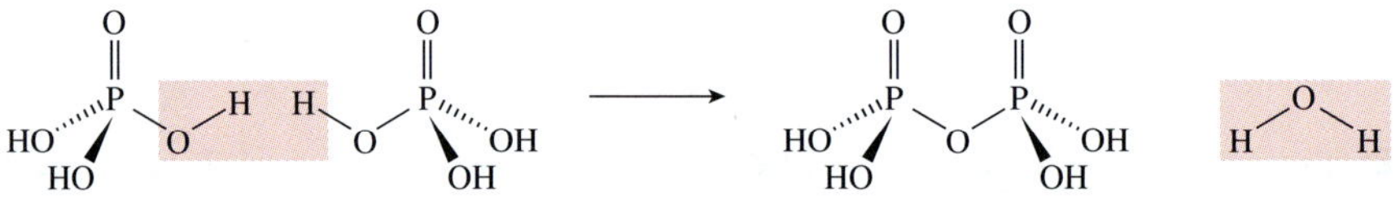

The product, $H_4P_2O_7$, is pyrophosphoric acid. A second condensation leads to triphosphoric acid:

$$H_4P_2O_7 + H_3PO_4 \longrightarrow H_5P_3O_{10} + H_2O$$

The sodium salts of these acids, sodium pyrophosphate ($Na_4P_2O_7$) and sodium triphosphate ($Na_5P_3O_{10}$), have been used widely in detergents. The polyphosphate anions are good additives for cleansing agents because they form complexes with metal ions, including those that make water "hard" (Ca^{2+}, Mg^{2+}) and those that cause color stains (Fe^{3+}, Mn^{2+}). Moreover, polyphosphates are nontoxic, nonflammable, and noncorrosive; they do not attack dyes or fabrics, and they are readily decomposed during wastewater treatment.

These advantages appear to make polyphosphates ideal for use in cleaning agents. Unfortunately, adding phosphates to water leads to an imbalance in aquatic biosystems, particularly in lakes. Too much phosphate leads to runaway algae growth, which can overwhelm the lake with decaying organic matter. Organic decay consumes oxygen, depleting the lake of this life-sustaining substance and leading to the death of fish and other aquatic life-forms. This contributes more organic decay and still more oxygen depletion, until eventually the lake supports no animal life.

Phosphate condensation reactions play an essential role in metabolism. Recall from Section 14.6 that the conversion of adenosine diphosphate (ADP) to adenosine triphosphate (ATP) requires an input of free energy:

$$ADP + H_3PO_4 \longrightarrow ATP + H_2O \qquad \Delta G° = +30.6\ \text{kJ}$$

As also described in that section, ATP serves as a major biochemical energy source, releasing energy in the reverse, hydrolysis, reaction. The ease of interchanging O—H and O—P bonds probably accounts for the fact that nature chose a phosphate condensation/hydrolysis reaction for energy storage and transport.

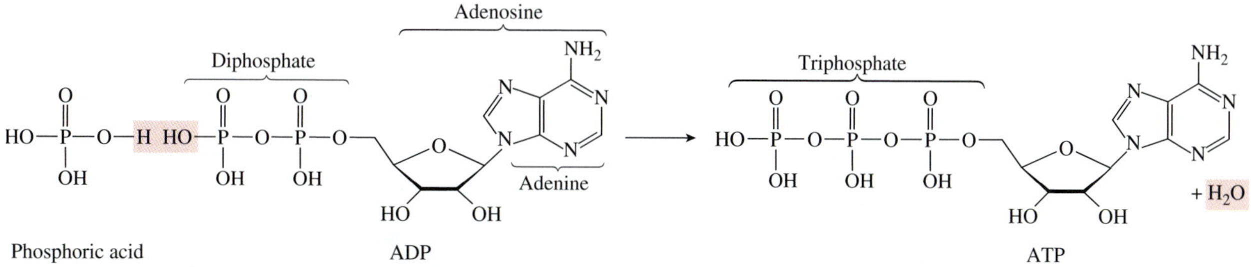

In ADP and ATP, one end of the polyphosphate chain links to adenosine through an O—C bond. These bonds form in condensation reactions between hydrogen phosphate and alcohols. Other biochemical substances contain P—O—C linkages. An example is the thickening agent used in instant puddings and pie fillings, which is produced by the reaction of starch with sodium dihydrogen phosphate:

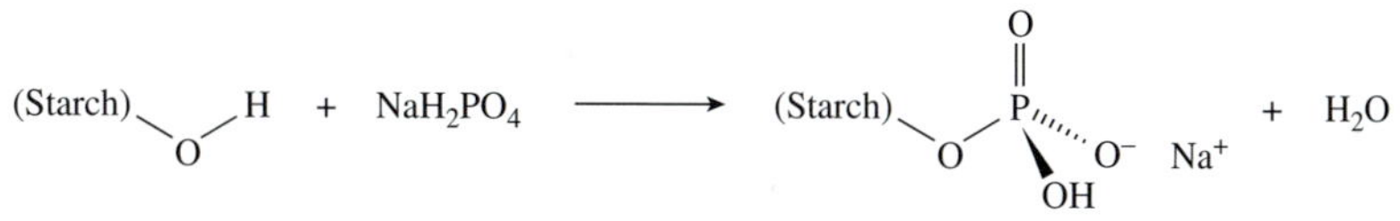

Starch molecules have many exposed O—H bonds, so this phosphorylation reaction results in multiple phosphate groups attached to each starch molecule. The remaining —OH group on each phosphate can condense with an O—H bond on another starch molecule. This cross-linking of starch chains gives the desired thick consistency of puddings and pies.

Organophosphorus Compounds

Strictly speaking, the biochemical substances just mentioned fall into the class of organophosphorus compounds, which are substances containing both phosphorus and carbon. However, the term is usually used more specifically to describe smaller molecules of this type. Organophosphorus compounds have varied uses, ranging from gasoline additives to herbicides.

The phosphate structure plays a central role in many familiar insecticides and herbicides. These compounds often have other elements substituted in place of one or more O atoms in the phosphate group, giving structures that can penetrate living cells and disrupt their normal activities. Figure 21-12 shows the line structures of some representatives of these species. Organophosphorus compounds such as Parathion and Malathion are potent insecticides. They act by phosphorylating an essential enzyme. Herbicides are represented by amiprophos, a compound that has a highly modified phosphate group. One oxygen atom is replaced with sulfur and another with nitrogen.

Not all toxic organophosphorus compounds have uses beneficial to humans. Sarin is an extremely toxic nerve gas that is lethal to humans. In March 1995 this substance was released in a terrorist attack on a Japanese subway, resulting in several deaths and many serious injuries. Sarin and related nerve gases bind an amino acid in the enzyme responsible for muscle action. When this enzyme is deactivated, muscles contract but cannot relax. Even a small dose can be lethal if the nerve gas reaches the muscles of the heart.

Figure 21-12
Line structures of four toxic organophosphorus compounds. Each has a modified phosphate group with atoms of other elements replacing one or more O atoms.

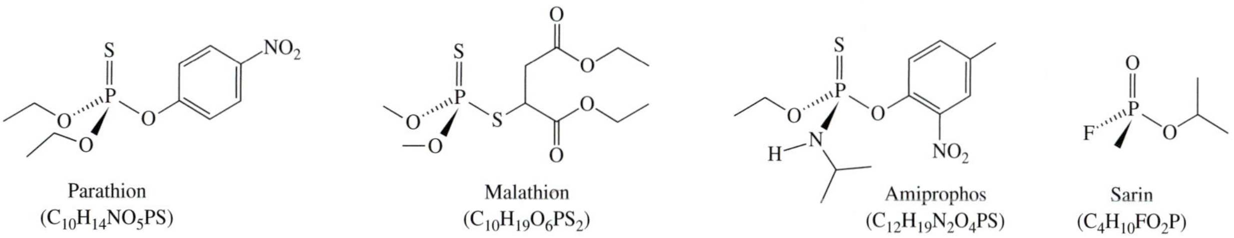

Parathion
($C_{10}H_{14}NO_5PS$)

Malathion
($C_{10}H_{19}O_6PS_2$)

Amiprophos
($C_{12}H_{19}N_2O_4PS$)

Sarin
($C_4H_{10}FO_2P$)

Section Exercises

■ **21.5.1** Figure 21-11 highlights the triangular arrangement of phosphorus atoms in P_4O_{10}. As every P atom bonds to four O atoms, however, the geometry about each P atom is tetrahedral, just as in the phosphate anion. Redraw Figure 21-11 in a way that highlights one of the tetrahedral PO_4 units.

■ **21.5.2** Casein is a polypeptide found in milk, in which the side chains of serine amino acids of the polypeptide have condensed with phosphoric acid to form a phosphoprotein that binds calcium phosphate. Draw the line structure of a serine−phosphate group. (See Section 13.7 for the structures of serine and polypeptides.)

■ **21.5.3** Shown below is a ball-and-stick model of the active ingredient in Roundup, a commercial herbicide. Draw the Lewis structure of Roundup and describe its functional groups.

21.6 OTHER NONMETALS

Among the nonmetals, the second-row elements carbon, nitrogen, and oxygen display a rich variety of chemical properties that provide many of the examples used throughout this textbook. Because the structure and chemistry of compounds of these elements are described in other chapters, we do not present further details here. The noble gases of Group 18 are unreactive, for the most part. See Box 9-2 for a description of their limited chemical behavior. The remaining reactive nonmetals, all members of the p block, are sulfur from the third row of the periodic table and the halogens of Group 17 (F, Cl, Br, and I). Two of these elements are components of leading industrial chemicals: Sulfuric acid perennially ranks #1 by a wide margin, and molecular chlorine ranked in the top ten until very recently. Our survey of sulfur and the halogens emphasizes commercially important uses.

Sulfur

Sulfur displays rich and varied chemical behavior, but from the commercial standpoint, sulfuric acid dominates the chemistry of this element. Sulfuric acid is used in every major chemical-related industry: fertilizers (60% of annual production), chemical manufacture (6%), petroleum refining (5%), metallurgy (5%), detergents, plastics, fibers, paints and pigments, and paper-making. In many of these cases, sulfur itself is not of interest and is eventually discarded as sulfate waste. Instead, H_2SO_4 is exploited for its Brønsted acidity, its oxidizing ability, its affinity for water, and its ability to form sulfate precipitates.

Sulfur is encountered most often in sulfide minerals such as pyrite (FeS_2), molybdenite (MoS_2), chalcocite (Cu_2S), cinnabar (HgS), and galena (PbS). It is also present in huge amounts as H_2S in natural gas and in sulfur-containing organic compounds in crude oil and coal. Moreover, sulfur is found in abundance as the pure element, particularly around hot springs and volcanoes and in capping layers over natural salt deposits. References to sulfur occur throughout recorded history, dating back as far as the sixteenth century BC. As shown in Figure 21-13, elemental sulfur is a yellow crystalline solid that consists of individual S_8 molecules. The eight atoms in each molecule form a ring that puckers in such a way that four atoms lie in one plane and the other four atoms lie in a second plane.

Elemental sulfur can be obtained from underground deposits. A well is drilled into the deposit, and superheated pressurized water (165 °C) is forced into the hole. The hot water melts the sulfur (mp = 119 °C), and the liquid sulfur is forced to the surface when pressurized air is injected into the well. This method supplies about one-third of the world's sulfur. However, the major present-day source of the element is from hydrogen sulfide produced as a by-product of oil and gas refining. Many petroleum and natural gas

Figure 21-13
Under normal conditions, sulfur forms yellow crystals. The crystals consist of individual S_8 molecules, with the eight sulfur atoms of each molecule arranged in a puckered ring.

supplies contain some sulfur—up to 25% in some cases. Besides being undesirable in the final products, sulfur poisons many of the catalysts used in oil refining; hence, it must be removed from crude petroleum as a first step in refining. Sulfur is produced in a gaseous redox reaction between hydrogen sulfide and sulfur dioxide:

$$16 \ H_2S + 8 \ SO_2 \xrightarrow{Fe_2O_3} 3 \ S_8 + 16 \ H_2O$$

If a supply of SO_2 is not already available, this gas is generated by treating H_2S with molecular oxygen:

$$2 \ H_2S + 3 \ O_2 \longrightarrow 2 \ SO_2 + 2 \ H_2O$$

This process and the subsequent oxidation of sulfur to sulfuric acid are relatively inexpensive because the reactions are exothermic. Thus, these reactions produce energy rather than requiring energy expenditure.

Sulfuric acid is manufactured from elemental sulfur by the process described in our Chemistry and Technology Box in Chapter 16. About 90% of world output of sulfur is converted to sulfuric acid, and more than 60% of that sulfuric acid is used to extract phosphoric acid from phosphate minerals. This major use exploits the fact that sulfuric acid is the least expensive strong Brønsted acid. In addition to protonating the phosphate anions, the sulfate group of sulfuric acid sequesters Ca^{2+} as $CaSO_4$, a waste by-product of the process. Another industrial use of sulfuric acid also exploits its acidity: The treatment of calcium fluoride (CaF_2) with sulfuric acid produces hydrogen fluoride (HF):

$$CaF_2 (s) + H_2SO_4 (l) \longrightarrow 2 \ HF (g) + CaSO_4 (s)$$

This source of hydrogen fluoride provides about 70% of the fluorine used industrially.

Sulfate solubility properties are the basis for other industrial uses of sulfuric acid. One example is in the metallurgy of titanium. One of the major ores of titanium is $FeTiO_3$, which is treated with sulfuric acid to separate titanium from iron:

$$FeTiO_3 (s) + 2 \ H_2SO_4 (aq) + 5 \ H_2O (l) \xrightarrow{150-180\,°C} TiOSO_4 (aq) + FeSO_4 \cdot 7H_2O (s)$$

$$TiOSO_4 (aq) + H_2O (l) \xrightarrow{90\,°C} TiO(OH)_2 (aq) + H_2SO_4 (aq)$$

The $TiO(OH)_2$ precipitate is washed with dilute sulfuric acid to remove impurities and then converted to TiO_2 by heating to 1000 °C. As described in Section 20.4, TiO_2 is subsequently converted to $TiCl_4$, purified by distillation, and then reduced to titanium metal.

Another use for sulfuric acid is in water treatment. Aluminum sulfate, which is produced by treating sulfuric acid with aluminum oxide, is a top-50 industrial chemical because of its widespread use as a coagulant. Addition of $Al_2(SO_4)_3$ and $Ca(OH)_2$ generates a gelatinous precipitate of aluminum hydroxide mixed with calcium sulfate:

$$Al_2(SO_4)_3 (aq) + Ca(OH)_2 (aq) \longrightarrow Al(OH)_3 (s) + CaSO_4 (s)$$

The gel traps finely suspended particulate material as it settles slowly out of the solution. This process is used in the primary stage of wastewater treatment, which is the removal of materials that make the water cloudy.

Although not as important industrially as processes involving sulfate, sulfur displays a rich chemistry with the halogens. For example, sulfur forms seven different binary fluorides. One of these, SF_4, has significant Lewis acid–base properties. Sulfur tetrafluoride can act as either an electron pair donor or an electron pair acceptor. With a steric number of 5, SF_4 adopts a trigonal bipyramidal arrangement of valence orbitals with the lone pair of electrons in one of the equatorial positions. The molecule acts as a Lewis base but not as a sulfur lone-pair donor. Instead, SF_4 is a fluoride donor, as shown in the following reactions. The donor atom in SF_4 is fluoride, which is a harder base than sulfur. After forming an adduct with the Lewis acid, the S—F bond breaks to give the ionic products:

$$SF_4 + BF_3 \longrightarrow [SF_3]^+[BF_4]^-$$

$$SF_4 + PF_5 \longrightarrow [SF_3]^+[PF_6]^-$$

Sulfur tetrafluoride is an important industrial fluorinating agent for both organic and inorganic compounds.

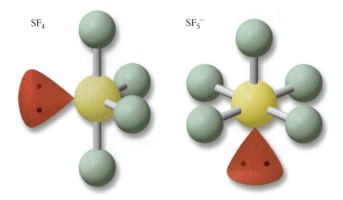

SF$_4$ SF$_5^-$

With its empty $3d$ orbitals, SF$_4$ also acts as a Lewis *acid*. This molecule forms adducts with pyridine and with fluoride ion. Each adduct has square pyramidal geometry, with a lone pair of electrons completing an octahedral arrangement of valence orbitals:

$$SF_4 + Pyr \longrightarrow SF_4 \cdot Pyr$$

$$SF_4 + CsF \longrightarrow Cs^+ + [SF_5]^-$$

Another key feature of sulfur chemistry is the Lewis acidity of sulfur dioxide. Sulfur dioxide is a common atmospheric pollutant that results from burning coal to produce electricity. Most coal reserves in North America include significant amounts of sulfur-containing impurities. When coal is burned, sulfur combines with O$_2$ to form SO$_2$, a hard Lewis acid.

In SO$_2$, the two S=O bonds are polarized toward the electronegative oxygen atoms, creating a partial positive charge on sulfur, as shown in Figure 21-14. The sulfur atom can use its empty d orbitals to accept a pair of electrons from a water molecule to form an adduct. A second water molecule subsequently acts as a Brønsted base and transfers a proton from one oxygen atom to another to give H$_2$SO$_3$:

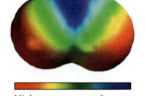

Highest electron density   Lowest electron density

Figure 21-14
An electron density model of sulfur dioxide shows low electron density (*blue*) around the sulfur atom and high electron density (*red*) on the oxygen atoms.

Oxidation by molecular oxygen or NO$_2$ converts this weak acid into H$_2$SO$_4$, a major contributor to acid rain.

Chlorine

As mentioned earlier, molecular chlorine has long been one of the leading industrial chemicals. Table 21-1 provides a summary of the industrial importance of chlorine.

Table 21-1   Major Uses of Chlorine

Reactant	Intermediate	Final Products
Organic Reagents		
Alkenes + SO$_2$	Sulfonated alkenes	Surfactants, detergents
Benzene	Chlorobenzenes	Plastics, dyestuffs
Butadiene	Chloroprene	Neoprene (chlorinated rubber)
Ethylene	Chloroethenes	Plastics (PVC), solvents
Methane	Chloromethanes	Silicones
Inorganic reagents		
CO	COCl$_2$	Plastics (polycarbonate)
NaOH	NaOCl	Bleaches, disinfectants
Phosphorus	PCl$_3$, POCl$_3$	Pesticides, flame retardants
Rutile (TiO$_2$ ore)	TiCl$_4$	Paint pigment (pure TiO$_2$)

The starting material for all industrial chlorine chemistry is sodium chloride, obtained primarily by evaporation of seawater. The chloride ion is highly stable and must be oxidized electrolytically to produce chlorine gas. This is carried out on an industrial scale using the chlor−alkali process, which is shown schematically in Figure 21-15. The electrochemistry involved in the chlor−alkali process is discussed in Section 19.7. As with all electrolytic processes, the energy costs are very high, but the process is economically feasible because it generates three commercially valuable products: H_2 gas, aqueous NaOH, and Cl_2 gas.

Although chlorine displays a range of oxidation states from −1 to +7, the most stable by far is Cl^-, as the positive standard reduction potentials in Table 21-2 indicate. Thus, Cl_2 and all oxochloro anions are good oxidizing agents. This property accounts for two of the major uses of chlorine compounds. Molecular chlorine is used extensively as a purifying agent for water supplies because it destroys harmful bacteria, producing harmless Cl^- in the process. Sodium hypochlorite, NaOCl, is the active ingredient in most bleaches. Hypochlorite ions oxidize many organic materials that are colored, breaking them into smaller, colorless substances that are easily removed by detergents.

A second major use for chlorine derives from its reactivity with organic materials, in particular with hydrocarbons. Two chlorinated hydrocarbons, ethylene dichloride (IUPAC name: 1,2-dichloroethane) and vinyl chloride (IUPAC name: chloroethene), rank among the leading industrial chemicals. Both substances are manufactured by reactions of Cl_2 with ethylene.

The first step in this process is a direct reaction carried out in liquid 1,2-dichloroethane with a metal chloride catalyst:

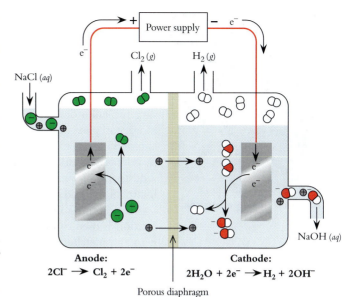

Figure 21-15
Schematic view of the electrolytic chlor−alkali process showing the molecular species involved in the redox reactions.

Anode: $2Cl^- \rightarrow Cl_2 + 2e^-$

Cathode: $2H_2O + 2e^- \rightarrow H_2 + 2OH^-$

$$H_2C=CH_2 + Cl_2 \xrightarrow{FeCl_3} CH_2Cl-CH_2Cl$$

Thermal decomposition of this product yields vinyl chloride and HCl:

$$CH_2Cl-CH_2Cl \xrightarrow{Heat} CHCl=CH_2 + HCl$$

The HCl produced in this reaction is used in an oxychlorination reaction to chlorinate additional ethylene:

$$2\,H_2C=CH_2 + 2\,HCl + O_2 \longrightarrow 2\,CHCl=CH_2 + 2\,H_2O$$

More than 95% of vinyl chloride is converted to PVC polymer.

Table 21-2   Standard Reduction Potentials for Cl Species

Reduction Half-reaction	E° (V)
$Cl_2 + 2\,e^- \rightleftharpoons 2\,Cl^-$	+1.35827
$HClO + H_3O^+ + 2\,e^- \rightleftharpoons Cl^- + 2\,H_2O$	+1.482
$HClO_2 + 3\,H_3O^+ + 4\,e^- \rightleftharpoons Cl^- + 5\,H_2O$	+1.570
$ClO_3^- + 6\,H_3O^+ + 6\,e^- \rightleftharpoons Cl^- + 9\,H_2O$	+1.451
$ClO_4^- + 8\,H_3O^+ + 8\,e^- \rightleftharpoons Cl^- + 12\,H_2O$	+1.389

PVC is used for a variety of common products.

The reaction of Cl_2 with ethylene to form vinyl chloride is just one example of a general process for the chlorination of hydrocarbons. One chlorine atom reacts with a hydrogen atom from the hydrocarbon to form HCl while a second chlorine atom replaces the hydrogen atom in the hydrocarbon. The generalized reaction is

$$RH + Cl_2 \longrightarrow RCl + HCl$$

Organic chlorides produced in this way are key starting materials for the synthesis of more complex organic compounds. The HCl by-product in these reactions may be exploited in an oxychlorination reaction as previously described, or it may be collected in water and marketed as hydrochloric acid. Sufficient HCl is produced through various industrial chlorination reactions to account for over 90% of the production of hydrochloric acid.

In addition to making organic chlorine compounds, a significant fraction of Cl_2 production is used to make inorganic halides. One important use, described in Chapter 20, is in the metallurgy of titanium, in which molecular chlorine is used to convert TiO_2 into $TiCl_4$, which is easy to purify by distillation.

$$TiO_2 + C + 2\ Cl_2 \longrightarrow TiCl_4 + CO_2$$

In similar fashion that is described earlier in this chapter, easily-purified $SiCl_4$ is produced from molecular chlorine and SiO_2 in the presence of coke:

$$SiO_2 + 2\ C + 2\ Cl_2 \longrightarrow SiCl_4 + 2\ CO$$

Another use of Cl_2 is the halogenation of phosphorus as an early step in the synthesis of organophosphorus compounds:

$$P_4 + 6\ Cl_2 \longrightarrow 4\ PCl_3$$

As Table 21-1 suggests, molecular chlorine is a tremendously versatile industrial chemical. This element is a leading industrial chemical because of this versatility rather than any single application, although polymers account for about one third of its uses. In recent years, however, the industrial use of chlorine has come under strong attack from many environmentally conscious groups. One major reason is that dioxins, one class of by-products of chlorine reactions, have a very detrimental effect on biosystems. The controversy over industrial chlorine is described in our Chemistry and the Environment Box on page 936.

Other Halogens

No compound containing a halogen other than chlorine appears among the top 50 industrial chemicals, but fluorine nevertheless has considerable commercial value. Fluorine occurs as the mineral fluorite (CaF_2) and is prevalent in phosphate-bearing rock. As already mentioned, HF produced from sulfuric acid treatment of fluorite supplies some 70% of industrial HF; the remainder comes as a by-product of phosphoric acid production.

One major use of HF is in the manufacture of fluorinated hydrocarbons. Fluorinated ethylene is used for several specialty polymers. For example, the Teflon coating on nonstick cookware is made from polytetrafluoroethylene (PTFE). This polymer is made from tetrafluoroethylene.

The starting material here is chloroform, $CHCl_3$, which is treated with HF to form the chlorofluorocarbon $CHClF_2$:

$$\underset{\overset{|}{Cl}}{\overset{\overset{Cl}{|}}{H-C-Cl}} + 2\ HF \longrightarrow \underset{\overset{|}{F}}{\overset{\overset{Cl}{|}}{H-C-F}} + 2\ HCl$$

Subsequent heating in the presence of a platinum catalyst results in partial conversion to tetrafluoroethylene:

$$2\ \underset{\overset{|}{F}}{\overset{\overset{Cl}{|}}{H-C-F}} \xrightarrow[700\ °C]{Pt} \underset{F\ \ \ \ \ F}{\overset{F\ \ \ \ \ F}{C=C}} + 2\ HCl$$

A coating of polytetrafluoroethylene gives Teflon cookware its nonstick surface.

Until recently, chlorofluorocarbons (CFCs) for refrigeration were major end products of HF chemistry, but these compounds are being phased out in accord with the Montreal Protocols because of their effect on the ozone layer (see Chapter 15).

Some industrial uses of fluorine require molecular fluorine, F_2, which is produced by electrolysis of HF. As Figure 21-16 shows, the cell uses liquid HF to which KF is added as an electrolyte. The redox chemistry is straightforward:

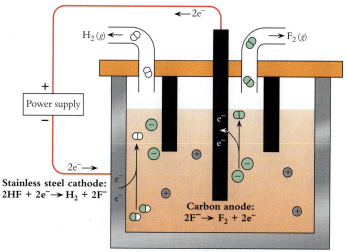

Anode:	$2\,F^- \longrightarrow F_2 + 2\,e^-$	
Cathode:	$2\,HF + 2\,e^- \longrightarrow H_2 + 2\,F^-$	
Net:	$2\,HF \longrightarrow F_2 + H_2$	

Figure 21-16
Schematic view of an electrolytic cell used for the production of molecular fluorine, showing the molecular species involved in the redox reactions.

Molecular fluorine reacts readily with a variety of organic and inorganic substances. It oxidizes metals to give metal fluorides. One useful metal fluoride is NaF, added to toothpaste to inhibit tooth decay. Uranium hexafluoride, UF_6, is used to purify uranium for nuclear power plants. Molecular fluorine also combines with nonmetals to give binary fluorides. The most important of these is sulfur hexafluoride, an inert and nontoxic gas that is used in electrical transformers:

$$S_8 + 24\,F_2 \longrightarrow 8\,SF_6$$

In addition, F_2 reacts with the other halogens to form interhalogen compounds in which an atom of the second halogen is bonded to an odd number of F atoms. Examples are ClF_3, BrF_5, and IF_7.

The remaining halogens, bromine and iodine, are isolated from brines by treatment with Cl_2, which oxidizes the halide ions to their elemental forms:

$$2\,Br^-(aq) + Cl_2(g) \longrightarrow 2\,Cl^-(aq) + Br_2(l)$$

$$2\,I^-(aq) + Cl_2(g) \longrightarrow 2\,Cl^-(aq) + I_2(s)$$

Iodine is an essential element for humans because it is present in thyroxine, a hormone that regulates the rate of cellular use of oxygen:

People whose diets are sparse in seafoods are susceptible to iodine-deficiency diseases, which are easily prevented by providing iodine in the diet. Iodized salt, which contains 0.1% KI, performs this function.

Molecular bromine is highly toxic, as is methyl bromide (CH_3Br), a dense gas used as an insecticide. Methyl bromide is produced by bromination of methane:

$$CH_4 + Br_2 \longrightarrow CH_3Br + HBr$$

Molecular bromine is also used in the synthesis of dyes and pharmaceuticals.

The most valuable bromine compound probably is silver bromide, which is the light-absorbing species on which most film photography is based (see Box 20.2, Chemistry and Technology, on page 895). Bromine compounds are also used extensively as fire retardants, particularly in carpets, rugs, and clothing

| Box 21-2 | Chemistry and the Environment: The Case Against the Industrial Use of Chlorine |

Molecular chlorine has been a leading industrial chemical because it is the precursor for an immense variety of useful products—some 15,000 in all—that contain chlorine. Even a short list of the uses of chlorine indicates its versatility and commercial value: sugar refining, flame retardants, photography, deodorants, leather finishing, automobile bumpers, magnetic tape, and cosmetics. Nevertheless, groups ranging from environmental activists to government agencies have in recent years called for the reduction or even total elimination of chlorine as an industrial chemical.

The chloride anion is a major species in the oceans and plays an essential role in biochemistry. Compounds containing carbon–chlorine bonds occur much less frequently in nature. Volcanos emit some halocarbons, and marine algae generate chloromethane. Other marine species produce toxic organohalogen molecules that protect them from predators. Nevertheless, organic chlorine compounds are uncommon, and consequently there are few mechanisms that degrade them.

The ecological problems caused by chlorine-containing compounds were first recognized years ago and called to public attention by Rachel Carson's best-selling book, *Silent Spring*. The chlorinated insecticide DDT accumulates in the tissues of animals. Birds of prey such as the peregrine falcon produced fragile, thin-shelled eggs that broke during incubation, and these species became endangered. Fortunately, the cause was determined and the use of DDT was discontinued in time to allow gradual recovery of bird populations.

DDT

More recently, another class of organic chlorine compounds has emerged as an environmental hazard. These are the dioxins, which, like DDT, contain ring compounds with chlorine substituents. A relatively simple example is 2,3,7,8-tetrachlorodibenzo-*p*-dioxin:

Dioxin

This structure looks nothing like the structures of chlorine-containing compounds used in industrial processes. In fact, no dioxin is deliberately manufactured anywhere in the diverse chlorine industry. Nevertheless, dioxins are of concern for two

reasons: First, dioxins appear to be inevitable trace by-products of some reactions involving chlorine, particularly combustion; and second, dioxins accumulate in the biosphere, where they have highly deleterious effects.

Dioxins have effects similar to and potentially even more far-reaching than those of DDT, because they apparently affect a wide variety of species. Predatory birds are especially susceptible, and there is growing evidence that humans may be at risk. Tests have shown that when the concentration of dioxins in the blood of laboratory animals reaches a critical level, reproductive and immune-system defects result. Moreover, recent data indicate that the concentration of dioxins in the blood of the average U.S. resident has nearly reached that level. A major reason is that dioxins are hydrophobic, so they accumulate in fatty tissue rather than being readily processed and excreted from the body.

Unfortunately, there is no easy way to eliminate production of dioxins without severely curtailing the use of all chlorinated compounds. This is because dioxins are formed when useful chlorine-containing compounds degrade, as for example when industrial wastes are incinerated. Consequently, several groups, including the American Public Health Association and the International Joint Commission overseeing the quality of the Great Lakes as well as environmental groups such as the Natural Resources Defense Council and the Sierra Club, have issued calls for the complete phase-out of the use of industrial chlorine.

Complete phase-out of chlorinated compounds is being resisted not only by chlorine producers, but also by the many industries that use chlorine compounds in the manufacture of products from paper to pharmaceuticals. Meanwhile, chemists seek ways to degrade dioxins to nontoxic substances.

Section Exercises

21.6.1 Determine the changes in oxidation states for the reactions that convert H_2S into S_8.

$$2 H_2S + 3 O_2 \longrightarrow 2 SO_2 + 2 H_2O$$

$$16 H_2S + 8 SO_2 \xrightarrow{Fe_2O_3} 3 S_8 + 16 H_2O$$

21.6.2 Determine the Lewis structures and describe the bonding and geometry of the following

compounds that appear in this section: (a) SF_4, (b) BrF_5, and (c) $F_2C{=}CF_2$.

21.6.3 Although Cl_2 can be manufactured by electrolysis of aqueous brine, the analogous reaction cannot be used to make F_2. Explain why anhydrous HF must be used for production of F_2.

SKILLS TO MASTER

Important equations

Essential terms

Key concepts

① Identifying Lewis acids and bases

② Depicting Lewis acid–base reactions

③ Identifying hard and soft Lewis acids and bases

④ Recognizing Lewis acid–base aspects of main group chemistry

21.1 LEWIS ACIDS AND BASES

A Lewis base acts as an electron pair donor.

A Lewis acid acts as an electron pair acceptor.

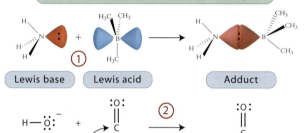

Lewis base	Lewis acid	Adduct

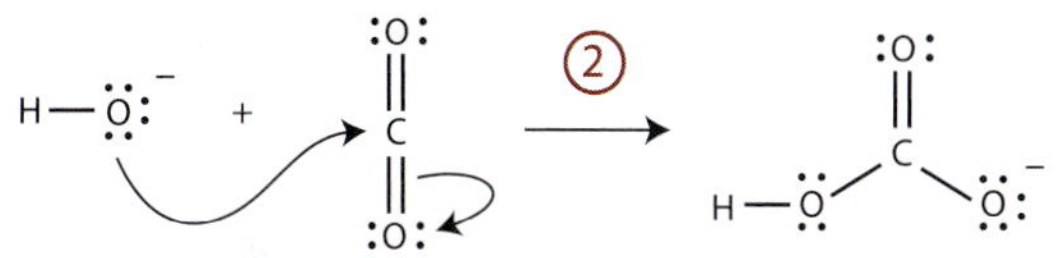

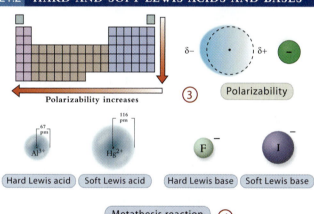

21.2 HARD AND SOFT LEWIS ACIDS AND BASES

Polarizability increases

③ Polarizability

Hard Lewis acid	Soft Lewis acid	Hard Lewis base	Soft Lewis base
Al^{3+} (67 pm)	Hg^{2+} (116 pm)	F^-	I^-

Metathesis reaction ④

Soft Lewis acids tend to combine with soft Lewis acids.

Hard Lewis acids tend to combine with hard Lewis bases.

21.3 THE MAIN GROUP METALS

Hall-Héroult process

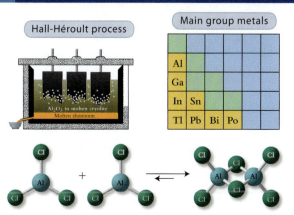

Main group metals

Al			
Ga			
In	Sn		
Tl	Pb	Bi	Po

Al_2O_3 in molten cryolite
Molten aluminum

21.4 THE METALLOIDS

Metalloids

B			
Si			
Ge	As		
	Sb	Te	

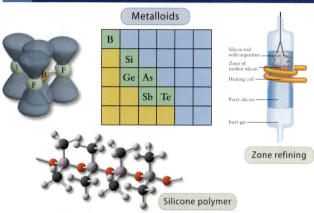

Silicon rod with impurities
Zone of molten silicon
Heating coil
Purer silicon
Inert gas

Zone refining

Silicone polymer

21.5 PHOSPHORUS

Red phosphorus

H_3PO_4

White phosphorus

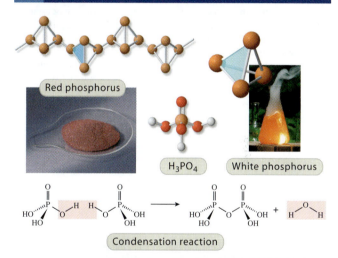

Condensation reaction

21.6 OTHER NONMETALS

Chlor–alkali process

C	N	O	F	Ne
	P	S	Cl	Ar
		Se	Br	Kr
			I	Xe
		At	Rn	

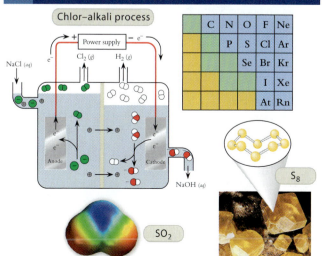

Power supply

e^- $+$ $-$ e^-

NaCl (aq) Cl_2 (g) H_2 (g)

e^- e^-

Anode   Cathode

NaOH (aq)

S_8

SO_2

Learning Exercises

21.1 There are no new "memory bank" equations in this chapter, but several concepts from earlier chapters play major roles in the chemistry of main group elements. List the key concepts from earlier chapters that you are asked to apply in this chapter.

21.2 Draw molecular pictures showing a typical Lewis acid–base reaction and a typical Brønsted acid–base reaction. Describe in words the differences and similarities of these two reactions.

21.3 State the hard–soft acid–base (HSAB) principle. Define and give examples of hard and soft acids and bases.

21.4 List the major industrial compounds containing main group elements and write a one-sentence description of the industrial importance of each.

21.5 Write the structural formula for an example of each of the following: (a) Lewis acid–base adduct; (b) polyphosphate; (c) fluorine-containing polymer; and (d) alkyllithium reagent.

 Problems __ilw__ = interactive learning ware problem. Visit the website at www.wiley.com/college/olmsted

Lewis Acids and Bases

21.1 In each of the following reactions, identify the Lewis acid and the Lewis base:
(a) $Ni + 4\,CO \longrightarrow [Ni(CO)_4]$
(b) $SbCl_3 + 2\,Cl^- \longrightarrow [SbCl_5]^{2-}$
(c) $(CH_3)_3P + AlBr_3 \longrightarrow (CH_3)_3P{-}AlBr_3$
(d) $BF_3 + ClF_3 \longrightarrow [ClF_2]^+[BF_4]^-$

21.2 In each of the following reactions, identify the Lewis acid and the Lewis base:
(a) $AlCl_3 + CH_3Cl \longrightarrow CH_3^+ + [AlCl_4]^-$
(b) $Zn^{2+} + 4\,CN^- \longrightarrow [Zn(CN)_4]^{2-}$
(c) $I_2 + I^- \longrightarrow I_3^-$
(d) $CaO + CO_2 \longrightarrow CaCO_3$

21.3 Make a sketch that shows the three-dimensional structure of each of the following Lewis acids. Identify the specific orbital that acts as the acceptor during adduct formation: (a) AlF_3; (b) SbF_5; and (c) SO_2.

21.4 Make a sketch that shows the three-dimensional structure of each of the following Lewis acids. Identify the specific orbital that acts as the acceptor during adduct formation: (a) $SnCl_4$; (b) PF_5; and (c) CO_2.

__ilw__ **21.5** The reaction between CO_2 and H_2O to form carbonic acid (H_2CO_3) can be described in two steps: formation of a Lewis acid–base adduct followed by Brønsted proton transfer. Draw Lewis structures illustrating these two steps, showing electron and proton movement by curved arrows.

21.6 The reaction between SO_3 and H_2O to form sulfuric acid (H_2SO_4) can be described in two steps: formation of a Lewis acid–base adduct followed by Brønsted proton transfer. Draw Lewis structures illustrating these two steps, showing electron and proton movement by curved arrows.

Hard and Soft Lewis Acids and Bases

21.7 Rank the following ions in order of increasing polarizability, and explain your reasoning: Fe^{3+}, Fe^{2+}, Pb^{2+}, and V^{3+}.

21.8 Rank the following ions in order of increasing polarizability, and explain your reasoning: SO_3^{2-}, NO_3^-, CO_3^{2-}, and ClO_3^-.

21.9 Rank the following Lewis acids from hardest to softest and explain your reasoning: (a) BCl_3, BF_3, and $AlCl_3$; (b) Al^{3+}, Tl^{3+}, and Tl^+; and (c) $AlCl_3$, AlI_3, and $AlBr_3$.

21.10 Rank the following Lewis bases from hardest to softest and explain your rankings: (a) NH_3, SbH_3, and PH_3; (b) PO_4^{3-}, ClO_4^-, and SO_4^{2-}; and (c) O^{2-}, Se^{2-}, and S^{2-}.

21.11 Explain why SO_3 is a harder Lewis acid than SO_2.

21.12 Explain why iodide is a soft base but chloride is a hard base.

21.13 For each of the following pairs of substances, determine whether metathesis will occur and if so, identify the products: (a) $AlI_3 + NaCl$; (b) $TiCl_4 + TiI_4$; (c) $CaO + H_2S$; (d) $CH_3Li + PCl_3$; and (e) $AgI + SiCl_4$.

21.14 For each of the following, state whether a reaction occurs, and write a balanced equation for any reaction:
(a) $NBr_3 + GaCl_3 \longrightarrow ?$
(b) $Al(CH_3)_3 + LiCH_3 \longrightarrow ?$
(c) $SiF_4 + LiF \longrightarrow ?$
(d) $LiCH_2CH_2CH_2CH_3 + SnCl_4 \longrightarrow ?$

The Main Group Metals

21.15 Describe in detail the bonding in Al_2Cl_6. Explain the formation of the molecule in terms of Lewis acid–base chemistry.

21.16 Write Lewis structures and describe the bonding in these three species found in the solution that is electrolyzed to form aluminum metal: $[AlF_4]^-$, $[AlF_6]^{3-}$, and $[OAlF_3]^{2-}$.

21.17 From its position in the periodic table, predict the properties of thallium (Element 81).

21.18 From its position in the periodic table, predict the properties of gallium (Element 31).

21.19 Determine the Lewis structure of $SnCl_4$ and explain how it functions as a Lewis acid.

21.20 Determine the Lewis structure of SbF_5 and explain how it functions as a Lewis acid.

The Metalloids

21.21 Borazine ($B_3N_3H_6$) is a planar molecule analogous to benzene (C_6H_6). Write the Lewis structure and describe the bonding of borazine.

21.22 Boron nitride (BN) is a planar covalent solid analogous to graphite. Write a portion of the Lewis structure and describe the bonding of boron nitride, which has alternating B and N atoms.

21.23 Reasoning from periodic trends, determine whether Ge or Si has a larger band gap. Use orbital overlap arguments to support your choice.

21.24 Explain why, among the 13–15 semiconductors, Ga pairs with As whereas In pairs with Sb. Predict which of these has the smaller band gap.

__ilw__ **21.25** Write the reactions that generate a silicone polymer starting from ethyl chloride (C_2H_5Cl), and draw a portion of the structure of the resulting polymer. Show at least four repeat units of the polymer.

21.26 Write the reactions that generate a silicone polymer starting from a 1:1 mixture of $(F_3CCH_2CH_2)(CH_3)Si(OH)_2$ and $(CH_3)_2Si(OH)_2$, and draw a portion of the structure of the resulting polymer. Show at least four repeat units of the polymer.

Phosphorus

21.27 Pyrophosphate ($P_2O_7^{4-}$) and triphosphate ($P_3O_{10}^{5-}$) are the first two polyphosphate anions. What is the chemical formula of the next largest polyphosphate anion? Draw a ball-and-stick model of this anion.

21.28 Three phosphate anions can condense to form a ring whose chemical formula is $P_3O_9^{3-}$. Determine the Lewis structure and draw a ball-and-stick model of this anion.

21.29 Describe the reactions that convert apatite to phosphoric acid. Identify Brønsted acid–base reactions and redox reactions, if any.

21.30 Describe the reactions that convert calcium phosphate to phosphoric acid. Identify Brønsted acid–base reactions and redox reactions, if any.

21.31 Write structural formulas showing the conversion of white phosphorus into red phosphorus. Use curved arrows to show how electrons move during this conversion.

21.32 Write structural formulas showing the reaction of ATP with water to form ADP. What is the other product?

Other Nonmetals

21.33 Prepare a list of the industrial reactions described in Section 21.6 that exploit the Brønsted acid character of sulfuric acid.

21.34 Prepare a list of the industrial reactions described in Section 21.6 that use sulfuric acid because of sulfate solubility characteristics.

21.35 Draw a portion of the repeating structure of polyvinylchloride.

21.36 Draw a portion of the repeating structure of polytetrafluoroethylene.

21.37 Identify the oxidizing agent, reducing agent, and changes of oxidation state that occur in the reaction forming $TiCl_4$ from TiO_2.

21.38 Identify the oxidizing agent, reducing agent, and changes of oxidation state that occur in the reaction forming $SiCl_4$ from SiO_2.

Additional Paired Problems

21.39 What mass of bauxite rock, $Al(O)OH$, must be processed to produce 2500 kg of pure aluminum if the bauxite rock is 85% pure and the processing steps have a net yield of 75%?

21.40 What mass of calcium phosphate rock, $Ca_3(PO_4)_2$, must be processed to produce 1500 kg of pure P_4 if the phosphate rock is 87% pure and the processing steps have a net yield of 68%?

21.41 Explain why SF_6 forms and is quite stable, whereas neither OF_6 nor SBr_6 is known.

21.42 Explain why the known forms of elemental carbon include forms with single, double, and triple bonds, whereas the known forms of elemental phosphorus all have single bonds.

21.43 Arsenic trichloride can act both as a Lewis acid and as a Lewis base. Explain why this is so and write a balanced equation for each using BF_3 and Cl^- as reaction partners.

21.44 Boron trifluoride is a very strong Lewis acid, but trimethylboron, $B(CH_3)_3$, is a mild Lewis acid. Given the π bonding character in BF_3, does this order of reactivity surprise you? Why or why not? Use bonding arguments to explain the trend in reactivity.

21.45 In basic aqueous solution, Al acts as a strong reducing agent, being oxidized to AlO_2^-. Balance this half-reaction, and determine balanced net reactions for Al reduction of the following: (a) NO_3^- to NH_3; (b) H_2O to H_2; and (c) SnO_3^{2-} to Sn.

21.46 In acidic aqueous solution, $SnCl_2$ acts as a mild reducing agent, tin being oxidized to Sn^{4+} in the process. Balance this half-reaction, and determine balanced net reactions for this reagent causing the following reductions: (a) MnO_4^- to Mn^{2+}; (b) $Cr_2O_7^{2-}$ to Cr^{3+}; and (c) Hg_2^{2+} to Hg.

21.47 Briefly describe the changes in chemical properties that take place from top to bottom of Group 15 of the periodic table.

21.48 Briefly describe the changes in chemical properties that take place from top to bottom of Group 13 of the periodic table.

21.49 Describe the chemical reactions by which Al metal is obtained from its ore.

21.50 Describe the chemical reactions by which Pb metal is obtained from its ore.

21.51 Describe the industrial preparation of pure Si, starting from impure SiO_2. Include balanced chemical reactions.

21.52 Describe the industrial preparation of pure H_3PO_4 starting from phosphate rock. Include balanced chemical reactions.

21.53 The only important ore of mercury is cinnabar, HgS. In contrast, zinc is found in several ores, including sulfides, carbonates, silicates, and oxides. Explain these observations in terms of hard and soft acids and bases.

21.54 The only important ore of lead is galena, PbS. In contrast, tin is found primarily as cassiterite, SnO_2. Explain this difference in terms of hard and soft acids and bases.

More Challenging Problems

21.55 The pressure of gaseous Al_2Cl_6 increases more rapidly with temperature than would be predicted by the ideal gas equation. Explain this behavior.

21.56 Water in thermal hot springs often is unpalatable due to dissolved H_2S. Treatment with Cl_2 oxidizes the sulfur to S_8, which precipitates (the other product is HCl). Balance this reaction and calculate the mass of Cl_2 required to purify 6.0×10^3 L of water (the average amount used daily by one person in the United States) containing 25 ppm (by mass) dissolved H_2S.

21.57 Calcium dihydrogen phosphate is a common phosphorus fertilizer that is made by treating fluoroapatite with phosphoric acid. Hydrogen fluoride is a by-product of the synthesis. Write a balanced equation for the production of this fertilizer and calculate the mass percent of phosphorus in the fertilizer.

21.58 The black tarnish that forms on pure silver metal is the sulfide Ag_2S, formed by reaction with H_2S in the atmosphere:

$$4\,Ag + 2\,H_2S + O_2 \longrightarrow 2\,Ag_2S + 2\,H_2O$$

The sulfide forms in preference to Ag_2O, even though the atmosphere is 20% O_2 with just a slight trace of H_2S. Use Lewis acid–base arguments to explain this behavior.

21.59 Thionyl chloride, $SOCl_2$, is used to remove water of hydration from metal halide hydrates. Besides the anhydrous metal halide, the products are SO_2 and HCl. (a) Draw the Lewis structure of $SOCl_2$. (b) Balance the reaction of iron(III) chloride hexahydrate with $SOCl_2$.

21.60 Ammonium dihydrogen phosphate and ammonium hydrogen phosphate are common fertilizers that provide both nitrogen and phosphorus to growing plants. In contrast, ammonium phosphate is rarely used as a fertilizer because this compound has a high vapor pressure of toxic ammonia gas. Write balanced equations showing how each of these solid phosphates could generate ammonia gas. Why does ammonium phosphate have the highest vapor pressure of NH_3 among these three compounds?

21.61 Boron trichloride is a gas, boron tribromide is a liquid, and boron triiodide is a solid. Explain this trend in terms of intermolecular forces and polarizability.

21.62 Aluminum refining requires large amounts of electricity. Calculate the masses of Al and Na that are produced per mole of charge by electrolytic refining of Al_2O_3 and NaCl.

21.63 Trisodium phosphate forms strongly basic solutions that are used as cleansers. Write balanced equations that show why Na_3PO_4 solutions are strongly basic. Include pK_a values to support your equations.

21.64 The fluorides BF_3, AlF_3, SiF_4, and PF_5 are Lewis acids. They all form very stable fluoroanions when treated with lithium fluoride. In contrast, three other fluorides, CF_4, NF_3, and SF_6, do not react with lithium fluoride. Use Lewis acid–base concepts to explain this behavior.

21.65 The first commercially successful method for the production of aluminum metal was developed in 1854 by H. Deville. The process relied on earlier work by the Danish scientist H. Oersted, who discovered that aluminum chloride is produced when chlorine gas is passed over hot aluminum oxide. Deville found that aluminum chloride reacts with sodium metal to give aluminum metal. Write balanced equations for these two reactions.

21.66 Metal oxides and sulfide ores are usually contaminated with silica, SiO_2. This impurity must be removed when the ore is reduced to the pure element. Silica can be removed by adding calcium oxide to the reactor. Silica reacts with CaO to give $CaSiO_3$. Write a balanced equation for this reaction, and describe the reaction in terms of Lewis acids and bases.

21.67 Complete the following reactions:
  (a) $AlCl_3 + LiCH_3 \longrightarrow ?$
  (b) $SO_3 + Excess\ H_2O \longrightarrow ?$
  (c) $SbF_5 + LiF \longrightarrow ?$
  (d) $SF_4 + AsCl_5 \longrightarrow ?$

21.68 Some pure liquid interhalogen compounds are good electrical conductors, indicating that they contain cations and anions. Show a Lewis acid–base reaction between two bromine trifluoride molecules that would generate ionic species.

21.69 Phosphorus(V) oxide has a very strong affinity for water; hence, it is often used as a drying agent in laboratory desiccators. One mole of P_4O_{10} reacts with six moles of water. Based on this stoichiometry, identify the product of the reaction and balance the equation.

21.70 A company that manufactures photographic film generates 2550 L/day of aqueous waste containing 0.125 g/L of Br^- ions. To recover the bromine in the form of Br_2, the company bubbles Cl_2 gas through this waste. Calculate the volume of gas that is consumed daily if the gas is delivered at 1.05 atm and 21 °C.

Group Study Problems

21.71 Summarize the arguments for and against using phosphate-based detergents. Do you think these detergents should be used? Explain your position.

21.72 Construct a table of bond lengths that supports the existence of π bonding in boron trifluoride and boric acid. The relevant data can be found in the text. Label your table thoroughly.

21.73 Describe the ways in which the structures of DDT and dioxin (see Chemistry and the Environment Box) are similar. Describe the ways in which these structures differ.

21.74 Summarize the arguments for and against using industrial chlorine. Do you think the industrial use of chlorine should be phased out? Explain your position.

21.75 One of the factors that controls the interaction of a Lewis base with a Lewis acid is the size of the molecules involved. For example, boron trifluoride forms a stronger adduct with tetrahydrofuran than with dimethyltetrahydrofuran. Use this example to explain size effects in adduct formation. (*Hint:* Think about electron–electron repulsion.)

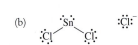

Tetrahydrofuran   Dimethyltetrahydrofuran
    (C_4H_8O) $(C_6H_{12}O)$

Answers to Section Exercises

21.1.1

(a) [Lewis structures of SO_3 and $O-H$] (b) [Lewis structures of $SnCl_2$ and Cl^-]

(c) [Lewis structures of AsF_5 and SbF_5]

21.1.2 (a) Lewis acid: SO_3, Lewis base: OH^-; (b) Lewis acid: $SnCl_2$, Lewis base: Cl^-; and (c) Lewis acid: SbF_5, Lewis base: AsF_3

21.1.3

(a) [reaction scheme] (One of three resonance structures)

(b) [reaction scheme]

(c) [reaction scheme]

21.2.1 (a) NH_3, NCl_3, NF_3; (b) Pb^{2+}, Pb^{4+}, Zn^{2+}; (c) ClO_4^-, ClO_3^-, ClO_2^-; and (d) $SbCl_3$; PCl_3; PF_3

21.2.2 Only the softest metals, those clustered around gold in the periodic table, exist in nature in elemental form. Ru and Os are in this category. Iron forms harder cations with an affinity for the hard anions abundant in the environment, including carbonate, oxide, and the silicates.

21.2.3 (a) $4 BCl_3 + 3 SF_4 \rightarrow 4 BF_3 + 3 SCl_2 + 3 Cl_2$;
　　　 (b) Chlorine is oxidized, and sulfur is reduced; and
　　　 (c) Sulfur is softer than boron, and chlorine is softer than fluorine. According to the HSAB principle, sulfur prefers to bond with the softer chlorine, and boron prefers to bond with the harder fluorine.

21.3.1

[reaction scheme with Al and Cl atoms; second reaction scheme below]

21.3.2 Indium, being relatively soft, has properties more like lead, which is softer than tin. Thus, indium is toxic, occurs naturally as a sulfide, and so on.

21.3.3 Bonding is between the relatively soft Lewis-base atoms, S, of 2,3-dimercaptopropanol and the soft Pb^{2+} cation:

[structure of Pb complex]

21.4.1

[reaction schemes with B and O atoms]

21.4.2 The more diffuse **n** = 3 valence orbitals of Si do not overlap with one another as effectively as the compact **n** = 2 valence orbitals of C, so the difference in energy between bonding and antibonding orbitals (valence and conduction bands) is smaller for Si than for C:

Carbon: Large band gap   Silicon: Small band gap

21.5.1

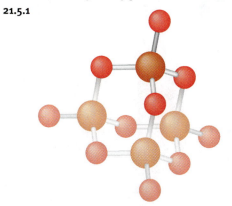

21.5.2

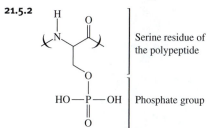

Serine residue of the polypeptide

Phosphate group

21.5.3 The structure includes a modified phosphate (P—C bond in place of one P—O bond), a carboxylic acid, and an amine:

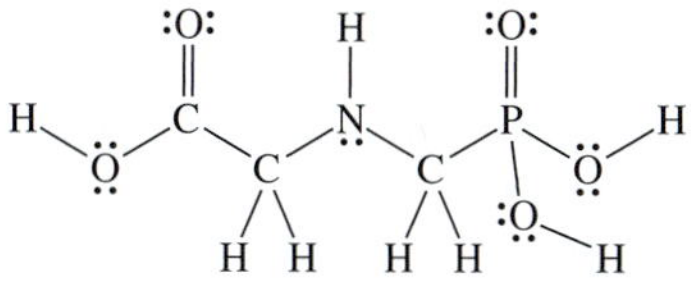

21.6.1 Oxidation states are shown beneath the atoms:

$$2\ H_2S + 3\ O_2 \longrightarrow 2\ SO_2 + 2\ H_2O$$
$$-2 \qquad 0 \qquad +4 \qquad -2$$

Sulfur is oxidized and oxygen is reduced.

$$16\ H_2S + 8\ SO_2 \longrightarrow 3\ S_8 + 16\ H_2O$$
$$-2 \qquad +4 \qquad 0$$

The sulfur in H_2S is oxidized (loses electrons), and the sulfur in SO_2 is reduced (gains electrons).

21.6.2 (a)

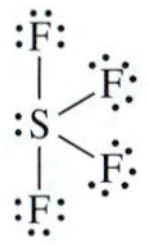

The S atom has SN = 5 and trigonal bipyramidal electron pair geometry. The lone pair is equatorial, giving a seesaw shape. Bonds can be described as formed from $2p$ atomic orbitals on F overlapping with sp^3d hybrids on S.

(b)

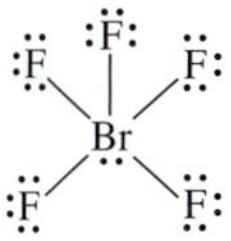

The Br atom has SN = 6 and octahedral electron geometry. The one lone pair results in a molecular shape that is square pyramidal. Bonds can be described as formed from $2p$ atomic orbitals on F overlapping with sp^3d^2 hybrids on Br.

(c)

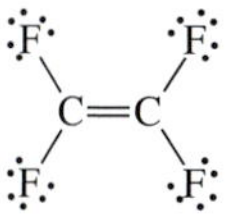

Each C atom has SN = 3, giving trigonal planar geometry and a planar molecule. There is a σ bond network that can be described using sp^2 hybrids from C and $2p$ atomic orbitals from F. Side-by-side overlap of $2p$ orbitals from C gives a π bond.

21.6.3 The oxidation reactions and their standard potentials are as follows:

$$6\ H_2O \longrightarrow O_2 + 4\ H_3O^+(aq) + 4e^- \qquad -1.229\ V$$

$$2\ Cl^- \longrightarrow Cl_2 + 2e^- \qquad -1.358\ V$$

$$2\ F^- \longrightarrow F_2 + 2e^- \qquad -2.866\ V$$

As described in Chapter 19, even though Cl^- is thermodynamically harder to oxidize than H_2O, the reaction kinetically is much faster, so Cl^- is preferentially oxidized in aqueous solution containing Cl^-. Oxidation of F^- is thermodynamically so much more difficult that water is preferentially oxidized in aqueous solutions containing F^-.

Answers to Extra Practice Exercises

21.1 Ni^{2+} is the Lewis acid, H_2O is the Lewis base:

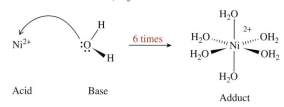

Acid   Base   Adduct

21.2 Lewis bases: PH_3 is softer than NH_3, which is softer than H_2O.

21.3 Because of its covalent cross-linking, black phosphorus melts at a higher temperature than either red or white phosphorus. White phosphorus is least dense, red is intermediate in density, and black is most dense, because added bonds decrease the interatomic distances.

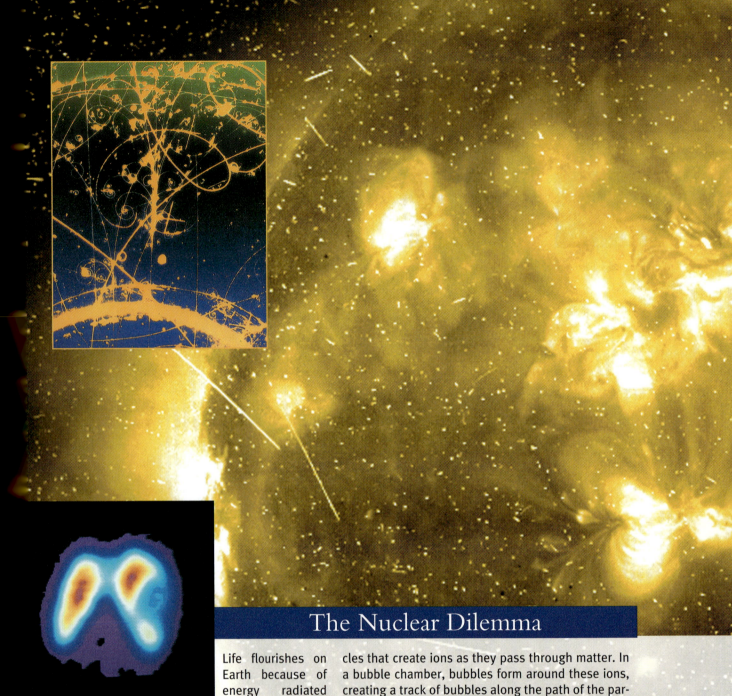

The Nuclear Dilemma

Life flourishes on Earth because of energy radiated by the sun. That energy comes from transformations of atomic nuclei. At the extreme temperature of the sun, small nuclei fuse together to form larger nuclei. The vast amounts of energy released by these solar fusion reactions continuously churn the matter making up the sun, generating the hot spots and flares that appear in the X-ray image in our opening photo.

To unlock the energy secrets of the sun, it was necessary to study matter at the subatomic level. Our inset figure shows results from an experiment carried out in an instrument called a bubble chamber. Nuclear reactions generate high-energy parti-cles that create ions as they pass through matter. In a bubble chamber, bubbles form around these ions, creating a track of bubbles along the path of the particle. Each type of particle creates a track with a different shape and length. The tracks tell scientists about the nature of nuclear reactions and have helped us understand what fuels the stars.

Unlocking the secrets of the nucleus was a mixed blessing, for in addition to our understanding of the sun, we also acquired nuclear weapons of immense destructive potential. The bombing of Hiroshima and Nagasaki with nuclear weapons was one of the last acts of the Second World War but the beginning of the nuclear dilemma: More than 50 years later, controversies still rage over how society should use the fruits of nuclear science.

Nuclear Chemistry and Radiochemistry

CHAPTER CONTENTS

The nuclear dilemma results partly from the devastating power of nuclear weapons. Equally troublesome are the health hazards associated with the radioactive products of nuclear reactions. These hazards include genetic effects, cancer, and other illnesses that can be fatal. Unfortunately, there is no known way to make radioactive elements non-lethal. Additionally, once radioactive substances are created, some of them will last for tens of thousands of years.

Learning Resources

 KEY CONCEPTS

 CRITICAL THINKING

 SKILLS TO MASTER

 ADDITIONAL HELP
www.wiley.com/college/olmsted
- TUTORIALS
- ANIMATIONS

The benefits of nuclear reactions lie primarily in power generation and medical diagnosis and treatment. At present, society is divided as to whether these benefits outweigh their accompanying hazards.

Nuclear medicine is firmly established. Many hospitals use radioactive materials for diagnosis and treatment; for example, to generate the images of diseased thyroid glands shown in our inset photo.

Still, society has not solved the problem of what to do with the resulting radioactive waste.

One hundred years after the discovery of radioactivity and fifty years after the dawn of the "nuclear age," society continues to debate the benefits and costs of nuclear technology. Understanding nuclear transformations and the properties of radio-

activity is necessary for intelligent discussions of the nuclear dilemma. In this chapter, we explore the nucleus and the nuclear processes that it undergoes. We describe the factors that make nuclei stable or unstable, the various types of nuclear reactions that can occur, and the effects and applications of radioactivity.

22.1 NUCLEAR STABILITY

In earlier chapters we treated the nucleus as a structure that never changes. Although this is true for normal chemical processes, under the right conditions nuclei undergo transformations that change their structures. These processes depend strongly on energy as well as nuclear structure. Our discussion of nuclear chemistry begins with the structure and energetics of the nucleus.

Nuclear Composition

As described in Section 2.2, nuclei are unimaginably small. The radius of a nucleus is about 10^{-14} m, ten thousand times smaller than the radius of an atom. An atom the size of a football stadium would have a nucleus the size of a pea, but the density of nuclei is so great that such a pea would have a mass of more than 250 million tons.

Nuclei are composed of two different fundamental particles, protons and neutrons, which are called **nucleons.** A proton has a charge of $+1.60218 \times 10^{-19}$ C and a mass of 1.672622×10^{-27} kg. A neutron has almost the same mass, 1.674927×10^{-27} kg, but is electrically neutral. Ionized H atoms, which form from H_2 in electrical discharges, are free protons. A free proton does not last long, because it quickly captures an electron to become a hydrogen atom. Free neutrons are often generated in the course of nuclear reactions, but they do not last long either, because they are easily captured again when they collide with nuclei. Table 22-1 summarizes the properties of these nuclear building blocks. The other fundamental atomic particle, the electron, is included for comparison.

Each particular type of nucleus is called a **nuclide.** Nuclides are characterized by the number of protons (Z) and neutrons (N) that they possess. The number of protons in a nuclide is always the same as the atomic number of the element, Z. Recall from Chapter 2, however, that the number of neutrons can vary, and that isotopes are nuclides with the same number of protons but different numbers of neutrons. Copper, for example, has two stable isotopes. Each isotope has 29 protons, but one isotope has 34 neutrons, whereas the other has 36 neutrons.

As described in Chapter 2, the mass number of a nuclide, A, is its total number of protons and neutrons: $A = Z + N$. Because protons and neutrons each have molar mass near 1 g/mol, A is always close to the numerical value of the molar mass of that isotope. For example, fluorine has a molar mass of 18.998 g/mol and $A = 19$. A particular nuclide can be described by its elemental symbol, X, preceded by the value of A as a superscript and that of Z as a subscript: Examples include $^{63}_{29}Cu$, ^4_2He, and $^{12}_6C$, whose nuclei are shown schematically in Figure 22-1. The elemental symbol identifies Z, so the

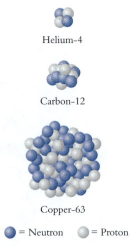

Helium-4

Carbon-12

Copper-63

● = Neutron ○ = Proton

Figure 22-1
Each nucleus contains Z protons and N neutrons. Shown here in schematic fashion are the nuclei of ^4_2He, $^{12}_6C$, and $^{63}_{29}Cu$.

Table 22-1 Properties of Fundamental Particles

Particle	Symbol	Charge (10^{-19} C)	Mass (10^{-27} kg)	Molar Mass (g/mol)
Proton	p	+1.60218	1.672 622	1.007 276
Neutron	n	0	1.674 927	1.008 665
Electron	e	−1.60218	0.000 911	5.486×10^{-4}

subscript is often omitted (for example, ^{63}Cu). Alternatively, the name of the element is followed by its mass number, as in copper-63. Example 22-1 provides some practice in writing the symbols of nuclides.

Example 22-1

Nuclide Symbols

Write the nuclear symbols for the following nuclides: (a) the one that contains 92 protons and 143 neutrons; and (b) the carbon isotope that has 8 neutrons.

Strategy: When determining symbols for nuclides, the key is to remember that the atomic number and number of protons are the same and that the mass number is the sum of the number of protons plus the number of neutrons.

Solution:
a. The nuclide with 92 protons and 143 neutrons has an atomic number of 92 and a mass number that is the sum of the number of protons and neutrons: $A = 92 + 143 = 235$. Atomic number 92 corresponds to uranium, so the symbol for this nuclide is $^{235}_{92}U$.

b. All carbon isotopes have the same atomic number, $Z = 6$. From the definition of A, $A = Z + N = 8 + 6 = 14$. The symbol for this nuclide is $^{14}_{6}C$.

Do the Results Make Sense? The symbol of an element must match its atomic number. Verify this for uranium and carbon by consulting a periodic table. The superscript must be larger than the subscript, since the superscript is the sum of the number of protons (the subscript) and the number of neutrons.

Write the nuclear symbols for the following nuclides: (a) the nuclide that contains 15 protons and 16 neutrons; (b) neon with the same number of neutrons as protons.

**Extra Practice
Exercise 22.1**

Nuclear Binding Energy

As described in Chapter 2, nuclei with more than one nucleon are held together by the *strong nuclear force*. Energy must be provided to overcome this force and remove a nucleon from a nucleus. This energy is called the **nuclear binding energy.**

Every nucleus contains Z protons and $(A - Z)$ neutrons, so we can visualize a nuclear "formation reaction" for any nuclide, in which protons and neutrons combine to form the product nucleus:

$$Zp + (A - Z)n \longrightarrow {}^{A}_{Z}X$$

Protons and neutrons are symbolized formally as $^{1}_{1}p$ and $^{1}_{1}n$, but they are often simplified to p and n.

Figure 22-2 illustrates this process schematically for fluorine:

$$9p + 10n \longrightarrow {}^{19}_{9}F$$

Because any stable nucleus is more stable than its separated nucleons, nuclear formation reactions of all stable nuclides are exothermic.

Formation reactions for nuclides are easy to visualize but impossible to carry out, so scientists must measure energies of nuclide formation indirectly. Accurate mass measurements are the best way to do this. Einstein recognized that mass is a form of stored energy. His famous equation links energy to mass and the speed of light, c:

$$E = mc^2 \qquad (22\text{-}1)$$

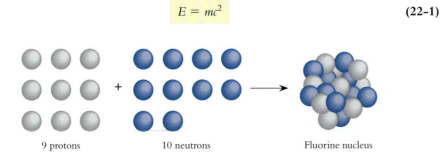

9 protons   10 neutrons   Fluorine nucleus

Figure 22-2
The nuclear formation reaction for
$^{19}_{9}F$: $Z = 9, A - Z = 10.$

According to this equation, formation of a nucleus releases an amount of energy that is related to the change in mass accompanying the formation reaction:

$$\Delta E = (\Delta m)c^2 \tag{22-2}$$

Equation 22-2 is the fundamental equation of nuclear energetics. It allows us to calculate the change in energy that accompanies any nuclear reaction from the masses of the particles before and after the transformation. When mass decreases, the energy change is negative. Recall that a negative energy change for a system means that energy is released to the surroundings. In other words, nuclear reactions in which mass decreases are exothermic.

Nuclide masses are ordinarily tabulated using units of grams per mole. Therefore, to compute ΔE, it is useful to rewrite Equation 22-2 with energy in units of kilojoules per mole and the change in mass in grams per mole:

$c = 2.998 \times 10^8$ m/s
$1 \text{ J} = 1 \text{ kg m}^2/\text{s}^2$

$$\Delta E = (\Delta m)c^2 = \Delta m (2.998 \times 10^8 \text{ m/s})^2 (10^{-3} \text{ kg/g})(10^{-3} \text{ kJ/J})$$

$$\Delta E = (\Delta m)(8.988 \times 10^{10}) \left(\frac{\text{kg m}^2 \text{ kJ}}{\text{s}^2 \text{ J g}} \right)$$

$$\Delta E = (\Delta m)(8.988 \times 10^{10} \text{ kJ/g}) \tag{22-3}$$

② The large positive exponent in Equation 22-3 indicates that a tiny change in mass results in a huge change in energy.

The change in mass that accompanies the formation of a nucleus can be found from the difference between the mass of the product nucleus and the masses of its component nucleons:

$$\Delta m = m_{\text{nucleus}} - [Z m_{\text{proton}} + (Z - A) m_{\text{neutron}}]$$

Once this change in mass is known, Equation 22-3 can be used to calculate the binding energy of any particular nuclide.

The tabulated molar mass of an element divided by Avogadro's number is the *average* mass per atom of that element, but it is not the *exact* mass of an individual nucleus. There are two reasons for this. First, molar masses refer to neutral atoms. The tabulated molar mass of an element includes the mass of its electrons in addition to the mass of its nucleus. Consequently, the mass of Z electrons must be subtracted from the isotopic molar mass in computing the energy of formation of a nuclide. Second, molar masses of the elements are weighted averages of all naturally occurring isotopes of that element. As an example, the most abundant isotope of hydrogen, ^{1}H, has an isotopic molar mass of 1.007 825 g/mol, but the presence of small amounts of the isotope ^{2}H makes the elemental molar mass of naturally occurring hydrogen slightly larger, 1.007 94 g/mol. In making mass–energy conversions, we must work with isotopic molar masses rather than elemental molar masses.

The isotopic molar masses of all stable and many unstable isotopes have been determined using mass spectrometry, as described in Section 2.3. These masses can be found in standard data tables. We provide values as needed for calculations. Example 22-2 illustrates the calculation of nuclear binding energies from isotopic molar masses.

Example 22-2	The most abundant isotope of helium has two neutrons and an isotopic molar mass of 4.002 60 g/mol. Compute the nuclear binding energy of this nuclide.
Nuclear Binding Energies	**Strategy:** A particular nuclide is made from the combination of Z protons and $(A - Z)$ neutrons. Thus, a neutral atom of a specific isotope contains Z protons, Z electrons, and $(A - Z)$ neutrons. When these particles are brought together, a small amount of mass is converted to energy. To calculate that energy, first count protons, neutrons, and electrons, and then do a mass–energy calculation using Equation 22-3.

Example 22-2

Nuclear Binding Energies
(*continued*)

Solution: The helium isotope has two neutrons, two electrons, and two protons:

Protons:	$2(1.007\,276$ g mol$^{-1})$	$= 2.014\,552$ g mol^{-1}
Neutrons:	$2(1.008\,665$ g mol$^{-1})$	$= 2.017\,330$ g mol^{-1}
Electrons:	$2(0.000\,5486$ g mol$^{-1})$	$= 0.001\,0972$ g mol^{-1}

Total molar mass of components $= 4.032\,979$ g mol^{-1}

The actual mass of the helium isotope is slightly smaller than the mass of the component particles. The difference represents the amount of mass that is converted to energy when the nuclide forms:

$$\Delta m = (4.00260 \text{ g mol}^{-1}) - (4.032\,979 \text{ g mol}^{-1}) = 0.03038 \text{ g mol}^{-1}$$

Equation 22-3 lets us calculate an energy change from a mass change:

$$\Delta E = (-0.03038 \text{ g/mol}) (8.988 \times 10^{10} \text{ kJ/g}) = -2.731 \times 10^{9} \text{ kJ/mol}$$

Notice that we carry all the significant figures for the masses. It is essential to do this, because a very small mass difference translates into an extremely large amount of energy. Even though the mass difference in this example is less than 1%, the energy difference is more than 10^9 kJ/mol.

Does the Result Make Sense? The energy change is negative, indicating that this helium nuclide is more stable than its separate component particles. We expect this for any stable nuclide.

The most abundant isotope of uranium is ^{238}U, with an isotopic molar mass of 238.0508 g/mol. Compute the nuclear binding energy of this nuclide.

Calculations show that as the mass number increases, so does the total binding energy of the nuclide. More important for the overall stability of the nucleus, however, is the binding energy *per nucleon*. This quantity is the total binding energy divided by the mass number A. It describes how tightly each nucleon is bound to the nucleus. As the binding energy per nucleon becomes more negative, nuclides become more stable. Figure 22-3 shows the binding energy per nucleon plotted vs. mass number. Notice the broad minimum around $A = 60$, $Z = 26$, at a binding energy of -8.3×10^8 kJ/mol of nucleons. The most stable nuclide of all is $^{56}_{26}$Fe.

The variations in binding energy shown in Figure 22-3 indicate that there are two types of nuclear reactions that release energy. When heavy nuclides fragment in a process called **fission,** energy is released. When light nuclides combine in a process called **fusion,** energy is also released. We describe these two energy-releasing processes in Sections 22.4 and 22.5.

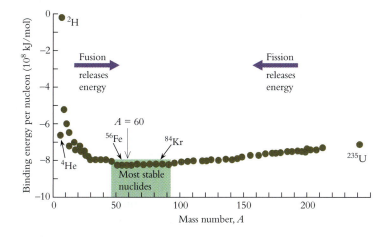

Figure 22-3
Plot of the binding energy per nucleon vs. mass number A. The most stable nuclides lie in the region around $A = 60$.

When the energy change accompanying a process is known, Equation 22-3 can be used to calculate how much the mass changes during the process. Example 22-3 shows how to do this for a chemical reaction.

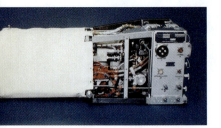

Fuel cell used in the space shuttle

Example 22-3
Mass Equivalence of Chemical Energy

Hydrogen, a fuel that releases a large amount of chemical energy when it burns, is used as an energy source in fuel cells. Use standard enthalpies of formation to calculate the change in mass that occurs when 1.00 mol of H_2 is burned.

Strategy: Whereas Example 22-2 addresses the energy change that accompanies a specific change in mass, Example 22-3 asks about the change in mass when a specific amount of energy is released in an exothermic chemical reaction. First determine ΔE for the chemical reaction and then use Equation 22-3.

Solution: The combustion reaction per mole of hydrogen is as follows:

$$H_2(g) + \tfrac{1}{2} O_2(g) \longrightarrow H_2O(g)$$

This reaction also is the formation reaction of gaseous H_2O. Its ΔH_f° is listed in Appendix D. Enthalpy changes and energy changes are nearly the same, so $\Delta E \cong \Delta H$:

$$\Delta E \cong \Delta H = \Delta H_f^\circ = -242 \text{ kJ/mol}$$

To obtain the mass change, substitute this value into Equation 22-3:

$$\Delta E = (8.988 \times 10^{10} \text{ kJ/g}) \, \Delta m$$

$$-242 \text{ kJ/mol} = (8.988 \times 10^{10} \text{ kJ g}^{-1}) \, \Delta m$$

$$\Delta m = -2.69 \times 10^{-9} \text{ g/mol}$$

Do the Results Make Sense?   When 1.00 mole of hydrogen reacts with oxygen, a few nanograms are converted to energy. This amount, which is typical of the mass consumed in conventional chemical reactions, is too small to detect. Thus, within the precision of measurements, mass is conserved in ordinary chemical reactions.

Extra Practice Exercise 22.3

Use standard enthalpies of formation to calculate the change in mass that occurs when 1.00 mol of methane (CH_4) is burned to form gaseous products.

The binding energy of a typical nuclide is about 8×10^8 kJ/mol of nucleons. To obtain a better feel for just how immense this quantity is, compare the energy released in a nuclear reaction with the energy released in a chemical reaction. As noted in Example 22-3, the chemical reaction of H_2 with O_2 is highly exothermic. When 1.00 mol of H_2 reacts chemically with O_2, 242 kJ of energy is released. If that same 1.00 mol of H_2 could be fused with neutrons to form He nuclei, the calculation of Example 22-2 indicates that 2.73×10^9 kJ of energy would be released. There is a difference of seven orders of magnitude between these two quantities; in other words, *fusion* of hydrogen nuclei releases 10,000,000 times as much energy as *combustion* of molecular hydrogen.

Energy Barriers

If fission of large nuclides and fusion of small nuclides are vastly exothermic, why have these processes not converted all elements into the most stable one, iron-56? The reason is that nuclear reactions have huge energy barriers. These barriers are analogous to the activation energy barriers for conventional combustion reactions. The combustion of gasoline, for example, is highly exothermic ($\Delta H > -5000$ kJ/mol), but the rate of the reaction is negligible at room temperature because an activation energy barrier prevents combustion from occurring without an external boost.

Massive nuclides do not fragment into lighter, more stable nuclides because an immense amount of energy must be supplied to overcome strong attractions among nucleons and pull a nucleus apart. Light nuclides do not fuse to give more massive, more stable nuclides because an immense amount of energy must be supplied to overcome electrical repulsion and bring nuclei together. The following analysis shows the magnitudes of typical activation energies for nuclear reactions.

Nuclei cannot fuse without first overcoming the repulsive electrical forces between them. Recall that this repulsion prevents the nuclei in molecules from approaching closer than the lengths of chemical bonds, which are about 100 pm. For nuclei to fuse, they must be brought within about 10^{-3} pm of each other. Equation 6-1, which describes electrical energy, allows calculation of the magnitude of this barrier.

$$E_{electrical} = k\frac{q_1 q_2}{r} \qquad (6\text{-}1)$$

For a pair of nuclei, the charges q_1 and q_2 are equal to the nuclear charges Z_1 and Z_2, and the value of k is 1.389×10^5 kJ pm/mol when charges are expressed in units of nuclear charge:

$$E_{electrical} = \frac{(1.389 \times 10^5 \text{ kJ pm/mol})(Z_1)(Z_2)}{d}$$

Because isotopes of hydrogen have the smallest possible nuclear charge ($+1$), they also have the minimal energy barrier to fusion. Consider, for example, the fusion of two deuterium nuclei:

$$^2_1H + {}^2_1H \longrightarrow {}^4_2He$$

For fusion of deuterium nuclei, Z_1 and Z_2 are each $+1$, and d is the sum of the two nuclear radii, 2.8×10^{-3} pm. Substituting these values gives $E = 5.0 \times 10^7$ kJ/mol. This is small relative to the energy released by fusion but very large relative to typical chemical activation energies of about 100 kJ/mol. The graph in Figure 22-4 is a schematic representation of the energy barriers for fusion of deuterium.

Activation energy barriers for nuclear reactions are more than a million times larger than the activation energies of conventional chemical reactions, so nuclear processes require immense energy inputs before they can occur. As one example, deuterium nuclei cannot fuse unless they first acquire 5×10^7 kJ/mol of energy. At room temperature, the average kinetic energy of deuterium nuclei is only 3.7 kJ/mol. A sample of deuterium needs only the energy boost from a spark or flame to react chemically with oxygen, but the temperature must be raised to 5×10^6 K before deuterium nuclei acquire enough kinetic energy to overcome the barrier to fusion.

The energy profile in Figure 22-4 also indicates that nuclei cannot eject nucleons without overcoming the strong nuclear forces that hold nucleons together. Nuclear binding energies are on the order of 10^8 kJ/mol. To summarize, electrical repulsion keeps nuclei from fusing, and the strong nuclear attractive force keeps nuclei from disintegrating.

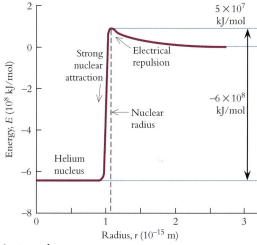

Figure 22-4
The energy profile for nuclear fusion of two deuterium nuclei.

Stable Nuclides

Although nuclides with mass numbers around 60 are the most stable, the balance of electrical repulsion and strong nuclear attraction makes many combinations of protons and neutrons stable for indefinite times. Nevertheless, many other combinations decompose spontaneously. For example, all hydrogen nuclides with $A > 2$ are so unstable that only one of them—tritium, ^3_1H—exists even briefly. If other nuclides of hydrogen could be made, they would decompose rapidly by expelling neutrons. Hydrogen has two stable nuclides, and fluorine has just one stable nuclide, $^{19}_9F$. In contrast, tin has ten, with mass numbers 112, 114, 115, 116, 117, 118, 119, 120, 122, and 124. All stable nuclides, as well as a few that are unstable, are found in the Earth's crust and atmosphere. Those unstable nuclides that occur naturally have natural lifetimes that are longer than the age of the Earth or are produced in naturally occurring nuclear reactions.

As described in Chapter 2 (see Figure 2-20), stable nuclides fall within a "belt of stability" with roughly equal numbers of neutrons and protons. Lighter nuclides lie along

Table 22-2   Distribution of Stable Nuclides

Protons	Neutrons	Stable Nuclides	%
Even	Even	154	58.8
Even	Odd	53	20.2
Odd	Even	50	19.1
Odd	Odd	5	1.9

the $N = Z$ line, but as the mass of the nuclide increases, the $N{:}Z$ ratio rises slowly until it reaches 1.54. The trend is illustrated by the $N{:}Z$ ratios of four representative nuclides: $^{19}_{9}F$, 1.11; $^{93}_{41}Nb$, 1.27; $^{159}_{65}Tb$, 1.45; and $^{209}_{83}Bi$, 1.54. Any nuclide whose ratio of neutrons to protons falls outside the belt of stability is unstable and decomposes spontaneously. Chapter 2 explains that the nuclides within the belt of stability contain just the right number of neutrons. These neutrons generate enough strong nuclear force to offset the electrical repulsions among protons. When there are too many protons, this becomes impossible: no nuclide with $Z > 83$ is stable.

Within the belt of stability, Table 22-2 shows that nuclides with even numbers of protons and neutrons are more prevalent than those with odd numbers of protons or neutrons. Almost 60% of all stable nuclides have even numbers of both protons and neutrons, whereas fewer than 2% have odd numbers of both. Moreover, of the five stable odd–odd nuclides, four are the lightest odd-Z elements: $^{2}_{1}H$, $^{6}_{3}Li$, $^{10}_{5}B$, $^{14}_{7}N$. Above $Z = 7$, there are 152 stable even–even nuclides and just one, $^{158}_{57}La$, that is odd–odd.

Section Exercises

22.1.1 Write the symbols and determine Z, N, and A for the following nuclides: (a) an oxygen nucleus with the same number of neutrons and protons; (b) element 43 with 55 neutrons; and (c) the unstable nuclide of hydrogen with the lowest mass.

22.1.2 Predict whether each of the following is stable or unstable. If you predict that it is unstable, give your reason: (a) the nuclide with 94 protons and 150 neutrons; (b) the iodine nuclide with 73 neutrons; (c) $^{154}_{64}Gd$; and (d) the oxygen nuclide with 6 neutrons.

22.1.3 Fluorine has only one stable isotope, $^{19}_{9}F$. Compute the total binding energy and the binding energy per nucleon for this nuclide.

22.2   NUCLEAR DECAY

Unstable nuclides decompose spontaneously into other, more stable nuclides. These decompositions are called **nuclear decay**, and unstable nuclides are called **radioactive.** Three features characterize nuclear decays: the products, the energy released, and the rate of decay. The products of nuclear decomposition include electrons, helium nuclei, and high-energy photons. Nuclear decay releases energy because the decay products are more stable than the starting materials. Furthermore, each unstable nuclide decomposes at some particular rate, which can take anywhere from less than a second to many millions of years.

Nuclear transformations always obey two fundamental conservation laws:

KEY CONCEPTS   *Mass number is conserved.*

Electrical charge is conserved.

That is, the sum of the number of protons and neutrons is the same after transformation as before, and the sum of the charges also is the same after transformation as before. Any description of a nuclear reaction must take these conservation requirements into account.

As an example, consider the decay of free neutrons. A neutron has $A = 1$, so its decay products must also have $A = 1$. The stable particle with $A = 1$ is a proton, and neutron decay results in a proton. The neutron has zero charge, so the sum of the charges of its decay products must also be zero. Because the proton carries a +1 charge, another particle with a −1 charge is required. This particle must have $A = 0$ to ensure that the mass number is conserved. The only particle with these properties is the electron. Thus, a neutron decays into a proton and an electron:

$$\, ^1_0 n \longrightarrow \, ^1_1 p + \, ^0_{-1} e$$

Although mass *number* is conserved, total mass is *not* conserved in nuclear reactions. Instead, some mass is converted into energy, or some energy is converted into mass. The notion of conservation of mass, introduced in Chapter 2, is valid within experimental error for all chemical transformations, as Example 22-3 illustrates. For nuclear processes, however, the amounts of energy produced or consumed are large enough to generate measurable mass changes. In nuclear decay, some mass is always converted into energy. We can use Equation 22-3, the mass of the decaying nucleus, and the masses of its products to calculate how much energy is released.

This reasoning is easily applied to the decay of a free neutron. The masses of the three participants are given in Table 22-1:

Neutron: $1.008\,665$ g/mol   Proton: $1.007\,276$ g/mol   Electron: $0.000\,5486$ g/mol

$$\Delta m = 1.007\,276 + 0.000\,5486 - 1.008\,665 = -0.000\,840 \text{ g/mol}$$

$$\Delta E = (-0.000\,840 \text{ g/mol})(8.988 \times 10^{10} \text{ kJ/g}) = -7.55 \times 10^{10} \text{ kJ/mol}$$

To summarize, the equation for a nuclear reaction is balanced when the total *charge* and total *mass number* of the products equals the total charge and total mass number of the reactants. This conservation requirement is one reason why the symbol for any nuclide includes its charge number (Z) as a subscript and its mass number as a superscript. These features provide a convenient way to keep track of charge and mass balances. Notice that in the equation for neutron decay, the sum of the subscripts for reactants equals the sum of the subscripts for products. Likewise, the sum of the superscripts for reactants equals the sum of the superscripts for products. We demonstrate how to balance equations for other reactions as they are introduced.

Decay Processes

There are five fundamental types of nuclear decay process, as listed in Table 22-3. Figure 22-5 on the next page diagrams how nuclear decays affect N and Z. As the figure suggests, the decay process of any particular unstable nuclide depends on the reason for its instability.

Many nuclides that are too massive to be stable lose mass by emitting energetic helium nuclei, reducing A by 4 units and Z by 2 units. Energetic helium nuclei are called α *particles,* so these nuclides undergo $\boldsymbol{\alpha}$ **emission.** Here are some examples:

$$\, ^{146}_{62} \text{Sm} \longrightarrow \, ^{142}_{60} \text{Nd} + \, ^4_2 \alpha \qquad \text{(see Figure 22-5)}$$

$$\, ^{222}_{86} \text{Rn} \longrightarrow \, ^{218}_{84} \text{Po} + \, ^4_2 \alpha$$

$$\, ^{218}_{84} \text{Po} \longrightarrow \, ^{214}_{82} \text{Pb} + \, ^4_2 \alpha$$

Table 22-3   Types of Nuclear Decay

Type of Decay	Cause of Instability	Emission	ΔZ of Emitter	ΔA of Emitter
α emission	Too massive ($Z > 83$)	$\alpha = \, ^4_2 \text{He}$	−2	−4
β emission	N/Z too large	$\beta = \, ^0_{-1} e$	+1	0
β^+ emission	N/Z too small	$\beta^+ = \, ^0_{+1} e$	−1	0
Electron capture	N/Z too small	X-ray photon	−1	0
γ emission	Excited nucleus	γ-ray photon	0	0

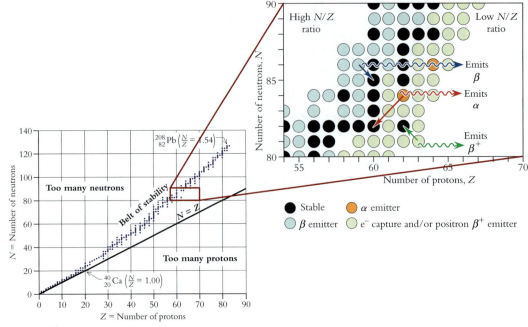

Figure 22-5
A detailed view of one portion of the *N* vs. *Z* plot of nuclides, illustrating the modes of nuclear decay for nuclides on either side of the belt of stability.

Ernest Rutherford observed that the paths taken by energetic particles emitted by radioactive uranium and thorium responded in three ways to magnetic fields: slightly bent, strongly bent, and unaffected. He gave them the designations $\overline{\alpha}$, β, and γ. Even though scientists soon identified the particles, they still use these names to emphasize that they are nuclear decay products.

Nuclides that lie *above* the belt of stability have ratios of neutrons to protons that are too high. To become stable, these nuclides need to increase their nuclear charges. Such a nuclide converts a neutron into a proton plus an energetic electron. The nucleus expels the electron, thereby increasing its charge by 1 unit while leaving its mass number unchanged. Energetic electrons are called β *particles,* so these nuclides undergo **β emission.** Here are some examples:

$$^{3}_{1}H \longrightarrow ^{3}_{2}He + ^{0}_{-1}\beta$$

$$^{145}_{59}Pr \longrightarrow ^{145}_{60}Nd + ^{0}_{-1}\beta \qquad \text{(see Figure 22-5)}$$

$$^{208}_{79}Au \longrightarrow ^{208}_{80}Hg + ^{0}_{-1}\beta$$

Nuclides that lie *below* the belt of stability have low neutron–proton ratios and must reduce their nuclear charges to become stable. These nuclides can convert protons into neutrons by **positron emission.** Positrons (symbolized β^{+}) are particles with the same mass as electrons but with a charge of $+1$ instead of -1. Here are three examples:

$$^{18}_{9}F \longrightarrow ^{18}_{8}O + ^{0}_{+1}\beta^{+}$$

$$^{52}_{26}Fe \longrightarrow ^{52}_{25}Mn + ^{0}_{+1}\beta^{+}$$

$$^{144}_{63}Nd \longrightarrow ^{144}_{62}Eu + ^{0}_{+1}\beta^{+} \qquad \text{(see Figure 22-5)}$$

Positrons cannot be observed directly because, as Figure 22-6*a* illustrates, when a positron encounters an electron, the two particles annihilate each other, converting their entire mass into a pair of photons. The occurrence of positron emission can be inferred from the observation of such a pair of photons. Each photon produced in this process has a specific energy: $E_{\text{photon}} = 9.87 \times 10^{7}$ kJ/mol. Photons with such high energy are called **γ rays.**

A nuclide with a low neutron–proton ratio can also reduce its nuclear charge by capturing one of its 1*s* orbital electrons in a process called **electron capture.** The captured electron combines with a proton to give a neutron, so *Z* drops by one unit, while *A* remains fixed. For example:

$$^{26}_{13}Al + ^{0}_{-1}e \longrightarrow ^{26}_{12}Mg$$

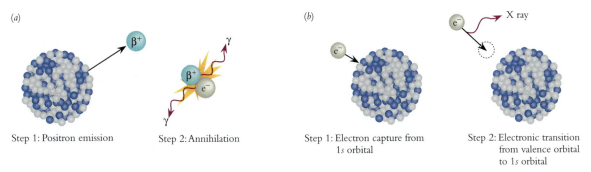

(a) Step 1: Positron emission   Step 2: Annihilation   (b) Step 1: Electron capture from 1s orbital   Step 2: Electronic transition from valence orbital to 1s orbital

Figure 22-6
Positron emission (a) and electron capture (b) are observed indirectly from the high-energy radiation that each produces.

Like positron emission, electron capture is never observed directly. However, after electron capture, the product atom is missing one of its 1s electrons, as shown schematically in Figure 22-6b. When an electron from an outer orbital occupies this vacancy in the 1s orbital, a photon is emitted whose energy falls in the X-ray region of the spectrum ($E \cong 10^6$ kJ/mol).

Nuclides that are unstable because they have odd–odd composition can be converted into stable nuclides with even–even composition through any of three processes: electron emission, positron emission, or electron capture. Each process changes Z and N by one unit. The $^{64}_{29}$Cu nuclide provides a convenient illustration of all three processes because it decays by all three modes:

$$^{64}_{29}\text{Cu} \longrightarrow {}^{64}_{30}\text{Zn} + {}^{0}_{-1}\beta$$

$$^{64}_{29}\text{Cu} \longrightarrow {}^{64}_{28}\text{Ni} + {}^{0}_{1}\beta^+$$

$$^{64}_{29}\text{Cu} + {}^{0}_{-1}\text{e} \longrightarrow {}^{64}_{28}\text{Ni}$$

> The electron captured by a nucleus during electron capture is symbolized e, not β, because β is reserved for electrons ejected during nuclear decay.

It is unusual for a nuclide to decay by three different paths. Generally, an odd–odd nuclide decays preferentially by only one mode. However, predicting which mode of decay predominates is beyond the scope of this book.

In addition to the γ rays that result from the annihilation of a positron and an electron, γ rays are also emitted directly by many nuclides. The emission of a γ ray from an unstable nuclide changes neither Z nor A, because γ rays are photons with zero mass and zero charge. Many nuclear reactions generate γ rays because of the release of energy that accompanies these reactions. Although most of this energy appears as kinetic energy of the products, some energy may be retained by the product nuclide as excitation energy. The excited nuclide can get rid of this excess energy by emitting a γ ray, just as an excited atom can lose excess energy by emitting a photon. Excited nuclides (also called metastable nuclides) are identified with a superscript m after the value of A. For example, α emission from radium-226 gives an excited radon nuclide, $^{222\text{m}}$Rn, which loses its excess energy by emitting a γ ray:

$$^{226}\text{Ra} \longrightarrow {}^{222\text{m}}\text{Rn} + \alpha$$

$$^{222\text{m}}\text{Rn} \longrightarrow {}^{222}\text{Rn} + \gamma$$

This decay sequence illustrates a common feature of nuclear decays of very heavy elements: several decays often occur in sequence, because the product of the initial decay is also unstable. As another example, ^{222}Rn is unstable, undergoing α decay:

$$^{222}\text{Rn} \longrightarrow {}^{218\text{m}}\text{Po} + \alpha$$

Example 22-4 traces this example of nuclear decay through its entire sequence of nuclear decompositions.

Example 22-4

Decay Sequences

Radon-222 is an unstable nuclide that has been detected in the air of some homes. Its presence is a concern because of high health hazards associated with exposure to its radioactivity. Gaseous radon easily enters the lungs, and once it decays, the products are solids that remain embedded in lung tissue. Radon-222 transmutes to a stable nuclide by emitting α and β particles. The first four steps are α, α, β, β. Write this sequence of nuclear reactions and identify each product.

Strategy: Charge number and mass number must be conserved in each reaction. Thus, each α particle decreases the nuclear charge by two units and the mass number by four units. Similarly, each β emission increases the nuclear charge by one unit but leaves the mass number unchanged. Consult a periodic table to identify the elemental symbol of each product nuclide.

Solution: The starting nuclide is ^{222}Rn, which, according to the decay sequence, emits an α particle:

$$^{222}_{86}\text{Rn} \longrightarrow {}^{A}_{Z}\text{X} + {}^{4}_{2}\alpha$$

The product has $A = 222 - 4 = 218$ and $Z = 86 - 2 = 84$. Element 84 is Po, so radon-222 decomposes to give an α particle and polonium-218:

$$^{222}_{86}\text{Rn} \longrightarrow {}^{218}_{84}\text{Po} + {}^{4}_{2}\alpha$$

The polonium nucleus is unstable, and the sequence indicates that it decays via α emission:

$$^{218}_{84}\text{Po} \longrightarrow {}^{214}_{82}\text{Pb} + {}^{4}_{2}\alpha$$

The decay sequence indicates that lead-214 is unstable as well. This nuclide converts a neutron to a proton by β emission:

$$^{214}_{82}\text{Pb} \longrightarrow {}^{A}_{Z}\text{X} + {}^{0}_{-1}\beta$$

For this product, $A = 214 - 0 = 214$ and $Z = 82 - (-1) = 83$. The product nucleus is element 83, bismuth:

$$^{214}_{82}\text{Pb} \longrightarrow {}^{214}_{83}\text{Bi} + {}^{0}_{-1}\beta$$

The fourth decay in the sequence is also β emission, which produces another isotope of element 84, polonium-214:

$$^{214}_{83}\text{Bi} \longrightarrow {}^{214}_{84}\text{Po} + {}^{0}_{-1}\beta$$

Do the Results Make Sense: You can confirm that both A and Z are conserved in these processes. Extra-heavy nuclides like ^{222}Rn frequently decay through sequences of α and β decays, because to reach the belt of stability they must not only shed mass (α decays) but also decrease their $N{:}Z$ ratios (β decays).

Extra Practice Exercise 22.4

This decay sequence continues with four more steps: α, β, β, and α. Determine the correct equations for these reactions.

Rates of Nuclear Decay

A sample of any unstable nuclide undergoes nuclear decay continuously as its individual nuclei undergo reaction. All nuclear decays obey the first-order rate law: Rate $= kc$. This rate law can be treated mathematically to give Equation 15-3, which relates concentration, c, to time, t, for a first-order process (c_0 is the concentration present at $t = 0$):

$$\ln\left(\frac{c_0}{c}\right) = kt \tag{15-3}$$

Recall from our discussion of kinetics in Chapter 15 that elementary unimolecular reactions are first order. The observed first-order kinetics of nuclear decays indicates that each nuclear decay mode is elementary and "uninuclear."

Table 22-4 Half-Lives of Representative Nuclides

Nuclide	Mode	Half-Life
^{214}Po	α	1.6×10^{-4} s
^{210}Tl	β	1.32 min
^{239}U	β	24 min
^{60}Co	β, γ	5.26 yr
^{3}H	β	12.3 yr
^{90}Sr	β	28.1 yr
^{14}C	β	5.73×10^3 yr
^{235}U	α	7.0×10^8 yr
^{40}K	β	1.28×10^9 yr
^{238}U	α	4.51×10^9 yr

Recall also from Chapter 15 that for first-order reactions, the time required for exactly half of the substance to react is independent of how much material is present. This constant time interval is the half-life, $t_{1/2}$. Equation 15-4 relates the half-life to the reaction rate constant:

$$t_{1/2} = \frac{\ln 2}{k} \qquad (15\text{-}4)$$

By convention, nuclear decay rates are expressed using half-lives rather than rate constants. Every unstable nuclide has its own characteristic half-life. Nuclide half-lives range from shorter than a second to longer than a billion years. The half-lives of some representative nuclides appear in Table 22-4.

The particles emitted during nuclear decay are so energetic that it is possible to count them individually. For this reason, the rate equations for nuclear decay are often expressed using the number of nuclei present, N, rather than molar concentrations:

$$\text{Rate} = \frac{\Delta N}{\Delta t} = \frac{-N \ln 2}{t_{1/2}} \qquad (22\text{-}4)$$

$$\ln\left(\frac{N_0}{N}\right) = \frac{t \ln 2}{t_{1/2}} \qquad (22\text{-}5)$$

⑤

Equations 22-4 and 22-5 can be derived from the first-order kinetic equations presented in Chapter 15 by making appropriate algebraic substitutions.

In these equations, N is the number of nuclei present at time t, and N_0 is the number of nuclei present at $t = 0$.

As shown in Example 22-5, Equation 22-4 is used to find a nuclear half-life from measurements of nuclear decays. Equation 22-5 is used to find how much of a radioactive substance will remain after a certain time, or how long it will take for the amount of substance to fall by a given amount. Example 22-6 provides an illustration of this type of calculation. In Section 22.7, we show that Equation 22-5 also provides a way to determine the age of a material that contains radioactive nuclides.

Example 22-5

Nuclear Half-Life

Plutonium is a synthetic element that is used in nuclear weapons. It is also proposed for use as a nuclear fuel. A sample of ^{239}Pu whose mass is 1.00 mg decays at a rate of 2.3×10^6 counts/s. (Each count corresponds to the decay of one nucleus.) What is the half-life of this isotope?

Strategy: This is a quantitative problem, so our standard seven-step approach is appropriate. We are asked to determine the half-life of an isotope from the information

Example 22-5

Nuclear Half-Life
(continued)

about its decay rate. A simple block diagram helps us visualize the process and summarize the data:

The emission rate is $-\Delta N/\Delta t$, so rearrange Equation 22-4 to calculate $t_{1/2}$:

$$\frac{\Delta N}{\Delta t} = \frac{-N \ln 2}{t_{1/2}} \quad or \quad t_{1/2} = \frac{N \ln 2}{-\Delta N/\Delta t}$$

Solution: The number of nuclei can be calculated from the mass of the sample and the molar mass of the isotope. Because the molar mass of an isotope is nearly equal to its mass number, we can use the mass number without introducing significant error:

$$N = \text{Number of nuclei} = \frac{(\text{Mass})(N_A)}{\text{Isotopic molar mass}}$$

$$N = \frac{(1.00 \times 10^{-3}\ \text{g})(6.022 \times 10^{23}\ \text{nuclei/mol})}{239\ \text{g/mol}} = 2.52 \times 10^{18}\ \text{nuclei}$$

Now we can substitute to determine the half-life:

$$t_{1/2} = \frac{(2.52 \times 10^{18}\ \text{nuclei})(0.693)}{2.3 \times 10^6\ \text{nuclei/s}} = 7.6 \times 10^{11}\ \text{s}$$

This half-life is easier to interpret if expressed in years rather than seconds:

$$t_{1/2} = (7.6 \times 10^{11}\ \text{s})\left(\frac{1\ \text{min}}{60\ \text{s}}\right)\left(\frac{1\ \text{hr}}{60\ \text{min}}\right)\left(\frac{1\ \text{day}}{24\ \text{hr}}\right)\left(\frac{1\ \text{year}}{365\ \text{day}}\right) = 2.4 \times 10^4\ \text{years}$$

Does the Result Make Sense? A half-life of many years is reasonable, given that isotopic half-lives vary from less than a second to millions of years. It takes 24,000 years for half of the plutonium nuclei in any sample of ^{239}Pu to decay.

Extra Practice
Exercise 22.5

There are no stable isotopes of technetium, but several radioactive isotopes can be prepared. A sample of ^{99}Tc whose mass is 2.35 μg emits 6.65×10^6 β particles in a time interval of 75 min. What is the half-life (in years) of this nuclide?

Example 22-6

Radioactive Decay

One of the problems with radioactive nuclides such as ^{239}Pu is that their decay cannot be stopped, so any sample of the nuclide continues to emit dangerous amounts of radiation until most of it has decayed. Use the result of Example 22-5 and assume that the radiation level from ^{239}Pu is no longer a hazard when 99% of it has decayed. How long will this take?

Strategy: The question asks for the time it takes for 99% of a sample of plutonium to decay. The half-life is known from the previous Example. Equation 22-5 relates the ratio N_0/N to time and the half-life for decay. This equation can be solved for t, the time at which the ratio reaches the desired value:

$$\ln\left(\frac{N_0}{N}\right) = \frac{t \ln 2}{t_{1/2}} \quad or \quad t = \frac{t_{1/2} \ln(N_0/N)}{\ln 2}$$

Solution: When 99% of the Pu has decayed, 1% of the original amount, in other words $0.01N_0$, remains. Use this information to evaluate the ratio in the ln term:

$$\frac{N_0}{N} = \frac{N_0}{0.01N_0} = 100$$

Example 22-6

Now substitute values and calculate t:

$$t = \frac{(2.4 \times 10^4 \text{ years})(\ln 100)}{\ln 2} = 1.6 \times 10^5 \text{ years}$$

Radioactive Decay (*continued*)

Does the Result Make Sense?   160,000 years is a very long time, but the half-life of this nuclide is 24,000 years, and it makes sense that it will take many half-lives before the amount of Pu is 1% of its original amount. It will be a very long time before the plutonium that is stockpiled for use as a nuclear fuel decays sufficiently that it no longer is a potential hazard.

How long will it take for two-thirds of the ^{99}Tc in Extra Practice Exercise 22-5 to decay?

Extra Practice Exercise 22.6

Section Exercises

22.2.1 Write the modes of decay that describe the following reactions: (a) ^{213}Bi to ^{213}Po; (b) ^{213}Bi to ^{209}Tl; (c) ^{213m}Bi to ^{213}Bi; and (d) ^{207}Bi to ^{207}Pb.

22.2.2 Identify the nuclides that decay in the following manner: (a) A nuclide undergoes β and γ decay to give $Z = 58$ and $A = 140$; (b) A nuclide undergoes α decay to give polonium-218; and (c) A nuclide captures an orbital electron to give tellurium with 73 neutrons.

22.2.3 Radioisotopes are used in many research applications. Because they decay continuously, the shelf life of a radioisotope is limited. An isotope used for bone marrow scanning is ^{111}In, $t_{1/2} = 2.8$ days. Within what time must this isotope be used if it is effective down to 5% of its initial activity?

22.3 INDUCED NUCLEAR REACTIONS

Stable nuclides remain the same indefinitely, whereas unstable nuclides disintegrate continuously. In time, therefore, every element should be composed entirely of stable isotopes. On Earth, most elements have no naturally occurring unstable isotopes. With two exceptions, the unstable nuclides that are present either have half-lives longer than the age of the Earth (U, Ra, Th) or are products of the decays of these long-lived nuclides (Rn, Po). The exceptions are ^{3}H ($t_{1/2} = 12.3$ years) and ^{14}C ($t_{1/2} = 5730$ years). These half-lives are very short compared with the age of the Earth, so any ^{3}H and ^{14}C present during the Earth's formation decayed long ago. Thus, the ^{3}H and ^{14}C on Earth today must come from ongoing nuclear reactions. These nuclides are not formed in any nuclear-decay schemes. Instead, each forms through a binuclear reaction in which a nuclear projectile collides and reacts with another nucleus. Such reactions, which are called **induced nuclear reactions,** are categorized according to the nature of the nuclear projectile that induces the reaction.

Neutron-Capture Reactions

One way to create unstable nuclides is by neutron capture. Because neutrons have no electrical charge, they readily penetrate any nucleus and may be captured as they pass through a nucleus. The sun emits neutrons, so a continuous stream of solar neutrons bathes the Earth's atmosphere. The most abundant nuclide in the atmosphere, nitrogen-14, can capture a neutron to form the unstable nuclide ^{15m}N. This nucleus rapidly ejects a proton, producing carbon-14:

$$^{14}_{7}\text{N} + ^{1}_{0}\text{n} \longrightarrow (^{15m}_{7}\text{N}) \longrightarrow ^{14}_{6}\text{C} + ^{1}_{1}\text{p}$$

Although carbon-14 decays via β emission with $t_{1/2} = 5730$ years, it is replenished continuously by this neutron-capture reaction. We show in Section 22.7 how this isotope is used to estimate the age of carbon-containing artifacts.

The other short-lived nuclide in the atmosphere, ^{3}H, is also produced by disintegration of metastable ^{15}N nuclei. When nitrogen-14 captures a very energetic neutron, the metastable nucleus has sufficient excess energy that it fragments into ^{12}C and ^{3}H rather than ^{14}C and a proton:

$$^{14}_{7}\text{N} + {}^{1}_{0}\text{n} \longrightarrow ({}^{15\text{m}}_{7}\text{N}) \longrightarrow {}^{12}_{6}\text{C} + {}^{3}_{1}\text{H}$$

Almost every nuclide undergoes neutron capture if a source of neutrons is available. Unstable nuclides used in radiochemical applications are manufactured by neutron bombardment. A sample containing a suitable target nucleus is exposed to neutrons coming from a nuclear reactor (see Section 22.4). When a target nucleus captures a neutron, its mass number increases by one:

$$^{A}_{Z}X + {}^{1}_{0}\text{n} \longrightarrow {}^{(A+1)\text{m}}_{Z}X$$

Neutron capture always is exothermic, because the neutron is attracted to the nucleus by the strong nuclear force. Consequently, neutron capture generates a product nuclide in a metastable, excited state. These excited nuclei typically lose energy by emitting either γ rays or protons:

$$^{(A+1)\text{m}}_{Z}X \longrightarrow {}^{(A+1)}_{Z}X + \gamma \qquad {}^{(A+1)\text{m}}_{Z}X \longrightarrow {}^{A}_{(Z-1)}Y + {}^{1}_{1}\text{p}$$

The following are specific examples of each type of decay:

$$^{98}_{42}\text{Mo} + {}^{1}_{0}\text{n} \longrightarrow {}^{99}_{42}\text{Mo} + \gamma \qquad {}^{14}_{7}\text{N} + {}^{1}_{0}\text{n} \longrightarrow {}^{14}_{6}\text{C} + {}^{1}_{1}\text{p}$$

$$^{207}_{79}\text{Au} + {}^{1}_{0}\text{n} \longrightarrow {}^{208}_{79}\text{Au} + \gamma \qquad {}^{3}_{2}\text{He} + {}^{1}_{0}\text{n} \longrightarrow {}^{3}_{1}\text{H} + {}^{1}_{1}\text{p}$$

One of the most important isotopes in nuclear medicine, $^{99\text{m}}$Tc, is produced by bombarding molybdenum with neutrons. The initial product, ^{99}Mo, has a high ratio of neutrons to protons, so it decomposes by releasing a β particle. The product is an excited technetium nuclide that emits a γ ray:

$$^{98}_{42}\text{Mo} + {}^{1}_{0}\text{n} \longrightarrow {}^{99}_{42}\text{Mo} + \gamma$$

$$^{99}_{42}\text{Mo} \longrightarrow {}^{99\text{m}}_{43}\text{Tc} + {}^{0}_{-1}\beta$$

$$^{99\text{m}}_{43}\text{Tc} \longrightarrow {}^{99}_{43}\text{Tc} + \gamma$$

The product of this three-step nuclear process, ^{99}Tc, is also unstable but lasts much longer than its predecessor. It decays by β emission with $t_{1/2} = 2.12 \times 10^5$ years.

Other Binuclear Reactions

Neutrons readily induce nuclear reactions, but they always produce nuclides on the *high* neutron−proton side of the belt of stability. Protons must be added to the nucleus to produce an unstable nuclide with a *low* neutron−proton ratio. Because protons have positive charges, this means that the bombarding particle must have a positive charge. Nuclear reactions with positively charged particles require projectile particles that possess enough kinetic energy to overcome the electrical repulsion between two positive particles.

Ernest Rutherford was the first person to observe a binuclear reaction. In 1919, he exposed a sample of nitrogen to α particles from a naturally radioactive source. He observed the production of protons and deduced from the requirements of charge and mass balance that the other product was oxygen-17:

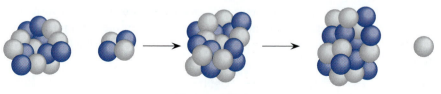

Compound Nucleus

The immediate product of a reaction between two nuclei is a **compound nucleus.** It has a charge equal to the sum of the charges of the reactants and a mass number equal to the sum of the mass numbers of the reactants. Every compound nucleus has excess energy that must be released after the two reactants bind together. Compound nuclei lose this excess energy by emitting one or more neutrons, protons, α particles, or deuterons (^{2}H nuclei). As Example 22-7 shows, conservation of charge and mass number permit identification of the participants in binuclear reactions.

⑥

Identify X and Y in each of the following nuclear reactions:

a. $^3\text{He} + \text{n} \rightarrow (^m X) \rightarrow Y + \text{p}$

b. $^{112}\text{Sn} + X \rightarrow (^{113\text{m}}\text{Sn}) \rightarrow {}^{113}\text{Sn} + Y$

c. $X + \alpha \rightarrow (^{113\text{m}}\text{In}) \rightarrow Y + 2\,\text{n}$

Example 22-7

Balancing Binuclear Reactions

Strategy: Charge number and mass number are conserved in nuclear reactions, so the missing components can be identified from the atomic numbers of the elements and the charge and mass numbers of elementary particles.

Solution:

a. Both reactants are given: ^{3}He and n. Remember that a neutron has $Z = 0$ and $A = 1$. Therefore, the compound nucleus must have $A = 4$ and $Z = 2$ (He). It then loses a proton, leaving $A = 3$ and $Z = 1$ (H). To emphasize the conservation of charge and mass number, we write the reaction showing all charge and mass numbers:

$$\,^3_2\text{He} + \,^1_0\text{n} \longrightarrow (^{4\text{m}}_2\text{He}) \longrightarrow \,^3_1\text{H} + \,^1_1\text{p}$$

b. Because all the nuclides are isotopes of tin, $Z = 50$ for all of them. Therefore, X and Y each must have a charge number of zero. The possibilities are a neutron and a γ ray. To balance mass numbers, X must have a mass number of one (neutron), and Y must have a mass number of zero (γ ray):

$$\,^{112}_{50}\text{Sn} + \,^1_0\text{n} \longrightarrow (^{113\text{m}}_{50}\text{Sn}) \longrightarrow \,^{113}_{50}\text{Sn} + \gamma$$

c. The compound nucleus is $^{113\text{m}}$In, so X must have this configuration minus an α particle, $^4_2\alpha$. Thus, X must have $A = 113 - 4 = 109$ and $Z = 49 - 2 = 47$, and X is $^{109}_{47}\text{Ag}$. Similarly, Y must have the configuration $^{113\text{m}}_{49}\text{In}$ minus two neutrons. Therefore, Z does not change, but $A = 113 - 2 = 111$, so Y is $^{111}_{49}\text{In}$:

$$\,^{109}_{47}\text{Ag} + \,^4_2\alpha \longrightarrow (^{113\text{m}}_{49}\text{In}) \longrightarrow \,^{111}_{49}\text{In} + 2\,^1_0\text{n}$$

Do the Results Make Sense?   Confirm that each nuclear reaction is correctly balanced by verifying that both A and Z are conserved.

Write the complete reaction for two ^{12}C nuclei fusing and emitting an α particle.

Extra Practice Exercise 22.7

Making Synthetic Elements

Elements 43 (technetium), 61 (promethium), 85 (astatine), and all elements with $Z > 92$ do not exist naturally on the Earth, because no isotopes of these elements are stable. After the discovery of nuclear reactions early in the twentieth century, scientists set out to make these "missing" elements. Between 1937 and 1945, the gaps were filled and three actinides, neptunium ($Z = 93$), plutonium ($Z = 94$), and americium ($Z = 95$) also were made.

The production of synthetic elements requires binuclear reactions between two positive nuclei that must be forced together against the force of electrical repulsion. This necessitates the use of nuclear accelerators to give extremely high kinetic energies to positive projectile nuclei. The first instrument applied to this task was the **cyclotron,** a particle accelerator developed by E. O. Lawrence at the University of California, Berkeley.

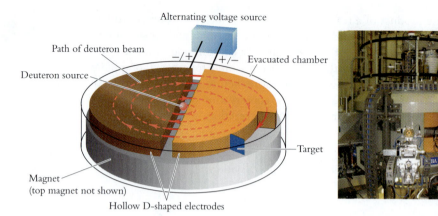

Alternating voltage source

Path of deuteron beam

Deuteron source

Evacuated chamber

Target

Magnet
(top magnet not shown)

Hollow D-shaped electrodes

Figure 22-7
A cyclotron accelerates particles in spiral paths. The photograph shows a modern cyclotron. The magnets (orange) are partially visible.

The operation of the cyclotron is shown schematically in Figure 22-7. A combination of pulsed electrical and magnetic fields is used to accelerate a beam of particles to high kinetic energy. Lawrence used high-energy deuterons to increase the atomic number of target nuclei by one unit:

$$^{98}_{42}\text{Mo} + ^{2}_{1}\text{H} \longrightarrow ^{99}_{43}\text{Tc} + ^{1}_{0}\text{n}$$

$$^{238}_{92}\text{U} + ^{2}_{1}\text{H} \longrightarrow ^{238}_{93}\text{Np} + 2\,^{1}_{0}\text{n}$$

The most efficient way to make elements 93 and 94 uses neutrons produced during fission in nuclear reactors instead of accelerated positive nuclei. Neutron capture by ^{238}U followed by β emission give isotopes with mass number of 239:

$$^{238}_{92}\text{U} + \text{n} \longrightarrow ^{239}_{92}\text{U} + \gamma$$

$$^{239}_{92}\text{U} \longrightarrow ^{239}_{93}\text{Np} + \beta$$

$$^{239}_{93}\text{Np} \longrightarrow ^{239}_{94}\text{Pu} + \beta$$

As Z increases, the overall yield of the target element falls sharply, because many steps are required. Plutonium (element 94) has been produced in ton quantities by neutron bombardment of uranium-238. Up to curium (element 96), production in kilogram quantities is possible, but the yields fall by about one order of magnitude for each successive element beyond $Z = 96$.

Beyond $Z = 100$, synthesis by neutron bombardment of uranium is no longer effective. Instead, nuclides in the $Z = 95$ to 99 range are bombarded with beams of light nuclei. For example, mendelevium ($Z = 101$) was first made in 1955 by a team of Berkeley chemists led by Glenn Seaborg. They bombarded element 99 with helium nuclei that had been accelerated in the cyclotron:

$$^{253}_{99}\text{Es} + \alpha \longrightarrow ^{256}_{101}\text{Md} + \text{n}$$

Beyond element 101, increasingly heavier nuclei must be used as the projectiles, as the following examples illustrate:

$$^{246}_{96}\text{Cm} + ^{12}_{6}\text{C} \longrightarrow ^{(258-x)}_{102}\text{No} + x\,\text{n}$$

$$^{252}_{98}\text{Cf} + ^{11}_{5}\text{B} \longrightarrow ^{(263-x)}_{103}\text{Lw} + x\,\text{n}$$

$$^{249}_{98}\text{Cf} + ^{16}_{8}\text{O} \longrightarrow ^{263}_{106}\text{Sg} + 2\,\text{n}$$

This requires a different type of accelerator that is linear rather than circular. The schematic diagram in Figure 22-8 shows how a heavy-ion **linear accelerator** works. A packet of ions is accelerated down the center of the accelerator by a series of electrically charged cylindrical tubes. The electrical potentials are varied so that the tube just ahead of the ion packet is always negatively charged, and this pulls the packet ahead at ever-increasing kinetic energy.

As the charge of the projectile increases, it takes ever greater acceleration for the projectile to penetrate the target nucleus. Even when successful, bombardment generates

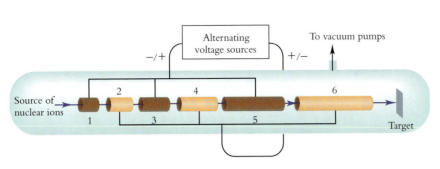

Figure 22-8
The beam in a linear accelerator follows a straight line. The photograph shows the interior of the Fermilab linear accelerator.

only a few nuclei of a new element, and these are so unstable that they decay in a matter of seconds or less. Nevertheless, nuclear scientists continue to work at extending the list of known elements, in part because theory predicts that the element with $Z = 114$ will be more stable than its lighter neighbors. Recently, formation of the $Z = 110$ element has been confirmed, and this element has been named darmstadtium (Ds). Nuclides with Z ranging from 111 to 118 have been reported, but more experiments are required before these are confirmed and given official names.

Section Exercises

22.3.1 The only stable isotope of cobalt is ^{59}Co. What induced reaction would generate each of the following nuclides from ^{59}Co? Identify the compound nucleus in each case.
(a) ^{59}Fe; (b) ^{60}Co; (c) ^{62}Ni; and (d) ^{58}Ni.

22.3.2 The two best-characterized isotopes of promethium ($Z = 61$) are ^{145}Pm and ^{147}Pm. The elements next to Pm in the periodic table have many naturally occurring isotopes. Neodymium ($Z = 60$) has $A = 142$, 143, 144, 145, 146, 148, and 150; and samarium ($Z = 62$) has $A = 144$, 147, 148, 149, 150,

152, and 154. Write nuclear reactions describing neutron capture by naturally occurring isotopes of Nd and Sm, followed by decay to produce ^{145}Pm or ^{147}Pm.

22.3.3 The following partial nuclear reactions show one reactant, the compound nucleus, and one product. Identify the other reactant and any additional products:
(a) ^{112}In $\longrightarrow$ $(^{113m}$In$)$ $\longrightarrow$ ^{113}In
(b) α $\longrightarrow$ $(^{58m}$Ni$)$ $\longrightarrow$ n
(c) ^{60}Ni $\longrightarrow$ $(^{61m}$Cu$)$ $\longrightarrow$ ^{57}Co

22.4 NUCLEAR FISSION

The graph of binding energy in Figure 22-3 shows that large amounts of energy are released when heavy nuclei split into lighter ones (fission) and when light nuclei combine into heavier ones (fusion). Thus, both fission reactions and fusion reactions can serve as sources of energy. Fission is the subject of this section, and fusion is the subject of Section 22.5.

Characteristics of Fission

Fission splits a nucleus into two fragments, each with a much lower Z value. Several free neutrons are released during each fission event. For example, uranium-235 undergoes fission after it captures a neutron:

$$^{235}_{92}\text{U} + {}^{1}_{0}\text{n} \longrightarrow (^{236m}_{92}\text{U}) \begin{cases} \nearrow {}^{81}_{32}\text{Ge} + {}^{152}_{60}\text{Nd} + 3\,{}^{1}_{0}\text{n} \\ \searrow {}^{103}_{42}\text{Mo} + {}^{131}_{50}\text{Sn} + 2\,{}^{1}_{0}\text{n} \end{cases}$$

Figure 22-9
Schematic view of fission. Neutron capture produces a highly unstable nucleus that distorts and then splits into smaller nuclei and a few free neutrons.

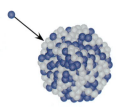
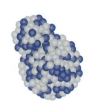
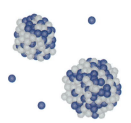

1. Neutron capture destabilizes the nucleus

2. The unstable nucleus distorts

3. The nucleus splits into smaller fragment nuclei and free neutrons

The compound nucleus formed on neutron capture, ^{236m}U, is highly unstable. As Figure 22-9 shows, it quickly splits into fragment nuclei plus several neutrons. Fission results in a wide range of product nuclides, of which we show two examples. There are many modes of fragmentation, but free neutrons are always generated, and charge and mass number are always conserved.

Among naturally occurring nuclides, only ^{235}U undergoes fission, but neutron capture followed by β decay convert two other naturally occurring nuclides, ^{238}U and ^{232}Th, into nuclides, ^{239}Pu and ^{233}U, that undergo fission.

The characteristics of nuclear fission can be summarized as follows:

1. Fission follows neutron capture by a small number of the heaviest nuclides, notably ^{235}U, ^{239}Pu, and ^{233}U.

2. Fission gives a range of product nuclides. Neutron-induced fission of ^{235}U yields the distribution shown in Figure 22-10. This distribution includes nuclides from $A = 77$ to $A = 157$. The most likely products are $A = 95$ and $A = 138$, but no single nuclide makes up more than 7% of the product fragments.

3. Each fission reaction generates from one to four free neutrons, with two or three being most common.

4. The process of fission releases large amounts of energy. The energy released for one set of fission fragments is computed in Example 22-8. Although each set of fission fragments has a slightly different total mass, the average mass loss is about 0.2 g/mol. This translates into an average energy released in ^{235}U fission of 1.8×10^{10} kJ/mol.

5. Fission products often are radioactive. This is because the fissioning nucleus has an N:Z ratio of 1.54, so its products have a similar N:Z ratio. In contrast, stable nuclides in the $A = 77$ to 157 range have ratios of around 1.3, so the products of fission have excess neutrons, making them unstable.

Figure 22-10
The isotopic "signature" of the nuclear fission of ^{235}U. Different mass numbers are produced in widely different percentages.

Example 22-8	
Fission Energy	

Use isotopic molar masses (given below) to compute the energy released per mole, per nucleus, and per gram when ^{235}U undergoes fission in the following manner:

$$^{235}\text{U} + \text{n} \rightarrow {}^{138}\text{Xe} + {}^{95}\text{Sr} + 3\,\text{n}$$

Nuclide	n	^{235}U	^{138}Xe	^{95}Sr
MM, g/mol	1.0087	235.0439	137.908	94.913

Strategy: As this is a quantitative problem, the seven-step strategy is appropriate. The problem asks for energy released during fission and provides molar masses. The nuclear reaction, which is provided, lets us visualize what is occurring. Equation 22-3 relates the energy released in a nuclear transformation to the "mass defect," Δm, which is the loss of mass per mole of reaction. The usual mole−mass conversion factors give energy released per nucleus and per gram.

Example 22-8

Fission Energy (*continued*)

Solution: The mass defect is the difference between the total mass of all products and the total mass of all reactants:

$$\Delta m = [(1\ ^{138}\text{Xe})(137.908\ \text{g/mol}) + (1\ ^{95}\text{Sr})(94.913\ \text{g/mol}) + (3\ \text{n})(1.0087\ \text{g/mol})]$$

$$-[(1\ ^{235}\text{U})(235.0439\ \text{g/mol}) + (1\ \text{n})(1.0087\ \text{g/mol})] = -0.2055\ \text{g/mol}$$

Substitute this mass change into Equation 22-3 to find the energy released:

$$\Delta E = (\Delta m)(8.988 \times 10^{10}\ \text{kJ/g})$$

$$\Delta E = (8.988 \times 10^{10}\ \text{kJ/g})(-0.2055\ \text{g/mol}) = -1.847 \times 10^{10}\ \text{kJ/mol}$$

Finally, use Avogadro's number and the molar mass to convert to energy per nucleus and energy per gram:

$$\Delta E_{\text{per nucleus}} = \frac{-1.847 \times 10^{10}\ \text{kJ/mol}}{6.022 \times 10^{23}\ \text{nuclei/mol}} = -3.07 \times 10^{-14}\ \text{kJ/nucleus}$$

$$\Delta E_{\text{per gram}} = \frac{-1.847 \times 10^{10}\ \text{kJ/mol}}{235.0439\ \text{g/mol}} = -7.86 \times 10^{7}\ \text{kJ/g}$$

Do the Results Make Sense? The isotopic molar masses are precise to five or more significant figures, so we are tempted to express the result with five significant figures. The mass defect is determined by addition and subtraction, however, and two of the isotopic molar masses are known to just three decimal places, so the mass defect is precise to three decimal places, and the results are precise to only three significant figures. Fission of one gram of ^{235}U releases enough energy to raise the temperature of about 250 million liters of water (66 million gallons) from 25 to 100 °C. For comparison, about 1.65 million grams of octane must be burned to release the same amount of energy. This immense energy explains why nuclear bombs are so devastating and why nuclear energy has been developed for power generation.

Calculate the energy that is released per gram of ^{235}U when induced fission gives ^{144}Ba ($MM = 143.92294$ g/mol), ^{90}Kr ($MM = 89.91953$ g/mol), and two neutrons.

Every fission reaction releases some neutrons, and these neutrons can be recaptured by other nuclei, causing more fission reactions. When the amount of fissionable material is small, most neutrons escape from the sample, and only a few neutrons are recaptured. Increasing the amount of material increases the likelihood that neutrons will be recaptured and cause additional fission reactions. The **critical mass** is defined as the amount of material that is just large enough to recapture one neutron, on average, for every fission reaction.

As long as the amount of fissionable material is less than the critical mass, the rate of fission events does not grow, and the rate of energy release remains low. In contrast, a sample behaves quite differently when the amount of fissionable material is larger than the critical mass. Above the critical mass, more than one neutron, on average, is recaptured for every fission that occurs. Now the number of fission reactions grows rapidly. As an illustration, consider what happens when two neutrons are recaptured from each fission reaction. As shown in Figure 22-11 on the next page, the neutrons produced by the first fission reaction trigger fragmentation of two more nuclei. Neutrons from these two fission events are recaptured by four additional nuclei, and this fission cascade goes on, doubling in each successive round. The result is a "chain" reaction that grows quickly to explosive proportions.

The recapture ratio does not have to be 2 for this effect to occur. Any recapture value larger than 1.0 results in explosive growth of the fission chain. The critical mass is called "critical" because any mass greater than this value sustains a chain reaction and may explode.

Figure 22-11
Schematic view of the start of a fission chain reaction. The chain continues to grow if more than one neutron, on average, is captured for every fission event.

Animation

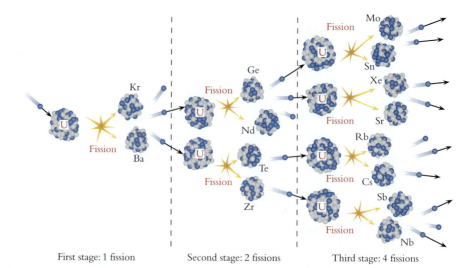

First stage: 1 fission   Second stage: 2 fissions   Third stage: 4 fissions

Nuclear Weapons

The potential of nuclear fission was first realized in the atomic bomb. In 1945, the United States dropped two bombs of unprecedented power, one on Hiroshima and the other on Nagasaki, Japan. Both were fission weapons. The Hiroshima bomb contained ^{235}U, and the Nagasaki bomb contained ^{239}Pu. These tremendously destructive weapons had been developed in a secret wartime project that involved an international team of outstanding physicists and chemists.

The central feature of a fission explosion is a growing chain of fission reactions. There are three requirements:

1. The fissionable nuclide must be concentrated enough to become critical.
2. Subcritical portions of this fissionable nuclide must be combined into a critical mass.
3. The critical mass must be held together long enough for the chain to multiply to immense size.

The second and third requirements posed engineering problems that were met by a carefully designed detonation of a chemical explosive. This explosion propelled subcritical masses of fissionable material together and confined them while the fission chain multiplied.

The first requirement, on the other hand, posed formidable chemical problems. Three nuclides can serve as nuclear fuels, ^{239}Pu, ^{233}U, and ^{235}U. Neither plutonium-239 nor uranium-233 is a naturally occurring nuclide, so a bomb could not be made from these materials without first finding a means of synthesizing these nuclides. Naturally occurring uranium contains only 0.72% ^{235}U. At this concentration, nearly all the neutrons produced by fission of ^{235}U are captured by ^{238}U, which does not undergo fission. Thus, no mass of natural uranium is enough to be critical. Therefore, before a bomb could be constructed from ^{235}U, a way had to be found to increase the percentage of ^{235}U in natural uranium.

The slight mass difference between ^{235}U and ^{238}U was exploited to increase the content of ^{235}U. Uranium was reacted with F_2 gas to produce UF_6, which is a gas at moderate temperatures. This gas was diffused through porous filters, whereupon the lighter ^{235}U-containing molecules moved slightly faster than those containing ^{238}U. The first fraction emerging from the filters was collected, and the hexafluoride was reduced to regenerate pure uranium metal enriched in ^{235}U.

Plutonium-239 was produced by the neutron bombardment of ^{238}U. The first nuclear reactor, constructed at the University of Chicago in 1942, was designed and used for this purpose. Significant amounts of plutonium form when natural uranium is placed in a reactor and bombarded with neutrons. The resulting mixture can be treated by appropriate chemical methods to recover the plutonium in pure form.

The mushroom cloud from the Nagasaki atomic bomb

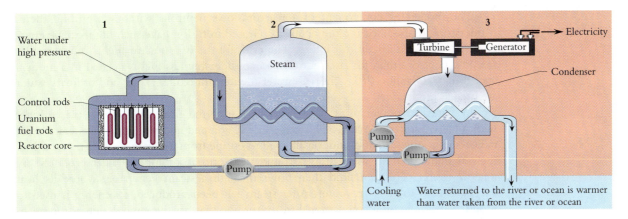

1. **Nuclear reactor:** Water under high pressure carries heat generated in the nuclear reactor core to the steam generator.

2. **Steam generator:** Heat from the reactor vaporizes water in the steam generator, creating steam.

3. **Turbine and condenser:** Steam from the steam generator powers a turbine, producing useable electricity. The condenser uses cooling water from a river or ocean to recondense the steam from the turbine.

Nuclear Reactors

A nuclear bomb is a terrifying example of the enormous amount of energy released by nuclear fission. A bomb, however, is not the only way to extract the energy produced by nuclear fission. Instead, nuclear fission can be used to generate electrical power if the rate of fission is controlled by adjusting the number of recaptured neutrons. The fission of just one gram of ^{235}U releases as much energy as the combustion of 600 gallons of gasoline or 6 tons of coal. Consequently, nuclear power is attractive, especially for countries that lack supplies of petroleum or coal.

Nuclear power plants, such as the one diagrammed in Figure 22-12, use the energy released in fission to generate electricity. The fission reaction takes place in a core that is heated by the released energy. A circulating fluid transfers this thermal energy to a heat exchanger, where the energy is used to convert water into steam. The steam drives a turbine connected to an electrical generator. Cooling water must be supplied to recondense the steam.

The turbine and generator components of a nuclear power plant have exact counterparts in power plants fueled by fossil fuels. The uniqueness of the nuclear power plant lies in its core. The core is a nuclear reactor where fission takes place under conditions that keep the reactor operating just below the critical level. The core contains three parts: fuel rods, moderators, and control rods. These components act on the flow of neutrons within the core, as shown in Figure 22-13. The fate of neutrons must be controlled carefully. Fission must be sustained at a steady rate that produces sufficient energy to run a generator, but the rate must not be allowed to increase and destroy the reactor.

A nuclear reactor runs on its nuclear fuel rods, which contain fissionable material such as uranium that has been enriched in ^{235}U. The fuel rods contain more than the critical mass of fuel, but the rate of fission is kept under control by movable control rods. These rods contain ^{112}Cd, which absorbs neutrons but does not undergo fission. The control rods capture neutrons that would otherwise trigger additional fission reactions. When pushed into the reactor, the control rods capture more neutrons, and the rate of fission decreases. When pulled out, the control rods capture fewer neutrons, and the rate of fission increases.

For the start-up of a reactor, its core is allowed to heat up by withdrawal of the control rods until the recapture ratio is slightly greater than 1.0. When the optimum operating temperature is reached, the control rods are inserted until the capture ratio is exactly 1, and the reaction proceeds at a steady rate. For the shutdown of a reactor, its control rods are fully inserted, reducing the recapture ratio to nearly zero.

Figure 22-12
Schematic view of a nuclear power plant. The energy source is the core, in which a fission reaction occurs. The rest of the plant is designed to transfer the energy released during fission and convert it into electricity.

Figure 22-13
The neutron processes occurring in the core of a fission reactor. Fission of ^{235}U gives from one to four fast neutrons, which are slowed down by the moderator. Some neutrons are captured by other nuclides.

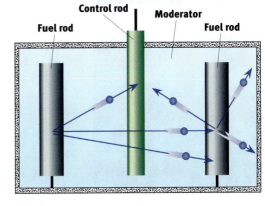

The moderator component of a reactor slows neutrons without capturing them. Moderators are used because the neutrons released in fission have such high kinetic energies that they are difficult to capture. The critical mass of a nuclear fuel is much smaller for slow neutrons than for fast neutrons, so considerably less fuel is needed in a moderated reactor. Graphite (^{12}C), "heavy" water (deuterium oxide, D_2O), and ordinary water are used as moderators.

In a nuclear power plant, heat must be transferred from the core to the turbines without any transfer of matter. This is because fission and neutron capture generate lethal radioactive products that cannot be allowed to escape from the core. A heat-transfer fluid such as liquid sodium metal flows around the core, absorbing the heat produced by nuclear fission. This hot fluid then flows through a steam generator, where its heat energy is used to vaporize water. After steam has been produced, its energy is used to produce electricity in a conventional steam turbine that has the same design as the steam turbines in fossil fuel power plants.

The radioactive materials that are an unavoidable by-product of nuclear fission are life-threatening. To operate safely, therefore, a nuclear power plant must confine all its radioactive products until they can be disposed of safely.

The main danger in the operation of a nuclear power plant is potential loss of control over the nuclear reaction. If the core overheats, it may either explode or "melt down." In either event, radioactive materials escape from the reactor to contaminate the environment. Designers attempt to make nuclear reactors "fail-safe" by providing mechanisms that automatically shut the core down on overheating. One way this has been done is to design the control rods to fall into the core if their control mechanism fails.

Despite such safeguards, nuclear accidents have occurred. The worst occurred in 1986 at Chernobyl in Ukraine, shown in Figure 22-14a. During a test of the reactor, most of the control rods were removed. The reactor surged out of control, probably because the safety systems had been disabled in the course of the test. The excess heat generated steam at high pressure, which blew off the roof of the reactor facility, spewing substantial amounts of radioactive material into the atmosphere. A fire released more radioactive material. Contamination was most severe within 30 km of the plant, where radioactivity levels became so high that the entire population had to be relocated. Significant contamination was also detected 1000 km away in Germany, and increased levels of radioactivity attributable to Chernobyl appeared throughout the Northern Hemisphere.

Nuclear power plants in the United States are supposed to be designed well enough to prevent accidents as serious as the one at Chernobyl. Nevertheless, the Three Mile Island plant in Pennsylvania, an aerial view of which is shown in Figure 22-14b, experienced a partial meltdown in 1979. This accident was caused by a malfunctioning coolant system. A small amount of radioactivity was released into the environment, but because there was no explosion, the extent of contamination was minimal.

A few nations rely heavily on nuclear power despite the possibility of accidents. In France and Japan, fission power from nuclear reactors provides two thirds or more of overall energy needs. A French plant appears in Figure 22-14c.

Even when operated safely, nuclear power plants produce long-lived radioactive wastes, which must be sequestered from the biosphere until their radioactivity diminishes to acceptable levels. Plutonium-239, formed in nuclear fuel rods when ^{238}U captures neutrons and decays, is a particularly dangerous nuclear waste that is extremely toxic and has a half-life of nearly 25,000 years. In Europe, nuclear waste is placed far underground in abandoned salt mines. Deep-sea burial and ejection into outer space have also been proposed. Antinuclear activists contend that no current technology is acceptable, because there is no proof that radioactive material can be contained for the thousands of years necessary for these dangerous materials to decompose.

Modern nuclear reactors are highly technological, carefully engineered creations of advanced human societies, so it may seem impossible that a nuclear reactor could result from natural conditions. Our Chemistry and the Environment Box describes evidence indicating that such a natural nuclear reactor did exist.

Figure 22-14
Aerial views of three nuclear power plants. (*a*) The Chernobyl nuclear power plant, site of a major nuclear accident in 1986. (*b*) The Three Mile Island power plant, site of a minor nuclear accident in 1979. (*c*) A plant in France, which has operated nuclear power plants safely for nearly 30 years.

(a)

(b)

(c)

Box 22-1   **Chemistry and the Environment: A Natural Nuclear Reactor?**

Naturally occurring uranium mined today contains 99.28% ^{238}U and 0.72% ^{235}U. This is too low a percentage to sustain a fission reaction. In the past, however, the percentage of ^{235}U was higher than it is today, because ^{235}U has a shorter half-life than ^{238}U (7.0×10^8 yr compared with 4.5×10^9 yr). Calculations using these half-lives show that 1.8 billion years ago, ^{235}U made up 3% of natural uranium. This is comparable to the enrichment levels used in nuclear fuels. Long ago, then, the composition of uranium ores was such that a self-sustaining nuclear reaction could have occurred.

But did this ever happen? A sustained reaction requires a neutron recapture ratio of at least one. Besides enrichment in ^{235}U, there must be a high degree of purity that minimizes the fraction of neutrons captured by other nuclides. A sustained reaction also requires a moderator to slow the neutrons before they escape from the uranium mass. Because water is a relatively good moderator, the need for a moderator could be met by groundwater seeping through the deposit. These unique conditions are illustrated in our molecular view.

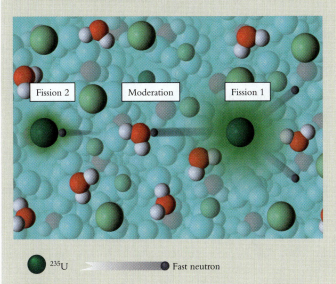

Fission 2   Moderation   Fission 1

^{235}U   →   Fast neutron

^{238}U   →   Slow neutron

It may seem unlikely that all these conditions could have been met, but at least one deposit of uranium ore has characteristics indicating that, long ago, it operated as a natural nuclear reactor. At Oklo in the Gabon Republic near the western coast of equatorial Africa (see photo), there are uranium deposits of high purity that are about 1.8 billion years old.

Thus, this uranium originally contained the 3% abundance of ^{235}U needed for a sustained reaction. Currently, the deposits in this mine differ from other uranium deposits in two ways that indicate that Oklo experienced sustained fission at one time. First, the ^{235}U content is slightly depleted, and second, there are unusual amounts of elements with mass numbers between 80 and 150.

The usual percentage of ^{235}U is 0.7207%, but the Oklo deposits contain 0.7071% of this isotope. This difference is small, but it is significant because it indicates that, at some time in the past, ^{235}U was consumed faster than its spontaneous decay rate. The question is whether that faster rate was caused by sustained fission.

What convinces scientists that sustained fission once occurred at Oklo is the presence of characteristic fission products in the ore. Elements of mass numbers between 75 and 160 occur in the ore in larger amounts than elsewhere. Furthermore, mass analysis of the elements in Oklo ore shows that they are distributed in the characteristic pattern shown in Figure 22-12. This isotopic "signature," which is not found in any other naturally occurring materials, is so characteristic that it has convinced most scientists that the ore deposits at Oklo once formed a huge nuclear reactor.

From the size of the ore deposit and the extent of ^{235}U depletion, scientists estimate that the Oklo deposits functioned as a natural nuclear reactor for about 100,000 years, generating 5×10^{14} kJ of energy. This is an average power output of about 150 kilowatts per year, which would be about enough to meet the needs of ten present-day Americans.

Section Exercises

■ **22.4.1** Fission products have molar masses that average 0.09 g/mol less than their mass number: For ^{106}Ru, $MM = 105.91$ g/mol. Using this value, calculate mass losses and energy releases for ^{235}U fissions that release 1 neutron, 2 neutrons, and 3 neutrons.

■ **22.4.2** Calculate the isotopic abundances of naturally occurring uranium 10^8 years ago. (*Hint:* Start with a convenient amount of uranium of present-day abundances, and use half-lives to calculate the amounts that were present 10^8 years ago.)

22.5 NUCLEAR FUSION

Fusion is a more attractive energy source than fission. Whereas fission occurs for only a few rare, extremely heavy nuclides, fusion is possible for abundant light nuclides such as ^{1}H. Moreover, some fusion reactions release more energy per unit mass than fission reactions do. For example, the fusion of two hydrogen isotopes, deuterium and tritium, releases 3.4×10^8 kJ/g, compared to 7.9×10^7 kJ/g released in the fission of ^{235}U. Another attractive feature of fusion reactions is that the product nuclides are usually stable, so smaller amounts of radioactive by-products result from fusion than from fission.

The Threshold for Fusion

The major impediment to fusion reactions is that the reacting nuclei must have very high kinetic energies to overcome the electrical repulsion between positive particles. The fusion of two hydrogen nuclei has the lowest possible repulsion barrier because it involves two $Z = 1$ nuclei. Even so, this reaction requires kinetic energies equivalent to temperatures of 10^7 K or greater. As Z increases, so does this energy requirement: The fusion of two ^{12}C nuclei requires kinetic energies equivalent to a temperature of 10^9 K.

There are several ways to produce nuclei with enough kinetic energy to fuse. One method uses particle accelerators to generate small quantities of fast-moving nuclei. The characteristics of fusion reactions are studied with such accelerators, but the scale of fusion events in these experiments is too small to release useful amounts of energy. A second way to induce fusion uses the energy released in gravitational attraction to generate a temperature hot enough to start the reaction. The sun and other stars operate in this way. A third way is the fusion bomb, which uses a fission bomb to generate a temperature high enough for fusion. None of these provides a method for harnessing fusion as a practical source of power on Earth. However, nuclear scientists have proposed that radiation might be able to initiate a useful fusion reaction by heating a gaseous plasma confined in a magnetic field. If successful, this method may harness fusion as a source of power.

> A plasma is a high-temperature ionized gas consisting of free electrons and nuclei.

Fusion Bombs

A "hydrogen bomb," which uses nuclear fusion for its destructive power, is three bombs in one. A conventional explosive charge triggers a fission bomb, which in turn triggers a fusion reaction. Such bombs can be considerably more powerful than fission bombs because they can incorporate larger masses of nuclear fuel. In a fission bomb, no component of fissionable material can exceed the critical mass. In fusion, there is no critical mass because fusion begins at a threshold *temperature* and is independent of the *amount* of nuclear fuel present. Thus, there is no theoretical limit on how much nuclear fuel can be squeezed into a fusion bomb.

Hydrogen bombs contain ^{2}H, ^{3}H, and ^{6}Li. The energy released in the fission explosion heats the two hydrogen isotopes hot enough to fuse:

$$^2_1\text{H} + {^3_1\text{H}} \longrightarrow (^5_2\text{He}) \longrightarrow {^4_2\text{He}} + {^1_0\text{n}} \qquad \Delta E = -1.7 \times 10^9 \text{ kJ/mol}$$

Lithium captures neutrons from this reaction in another energy-releasing process:

$$^6_3\text{Li} + {^1_0\text{n}} \longrightarrow (^7_3\text{Li}) \longrightarrow {^4_2\text{He}} + {^3_1\text{H}} \qquad \Delta E = -4.6 \times 10^8 \text{ kJ/mol}$$

Tritium nuclei produced in this reaction can fuse with additional deuterons, thus beginning the process again. All these processes occur in an extremely short time to release such an immense amount of energy that the bomb is blown apart.

Controlled Fusion

A long-standing goal of nuclear science has been to harness the energy of nuclear fusion in a sustained process. The technological problems are immense, however, because the fusing material must be confined to a volume small enough to generate high nuclear densities and many nuclear collisions. Because all materials vaporize at temperatures well below that required to sustain fusion reactions, containers that would withstand sustained fusion cannot be built.

The first fusion bomb completely vaporized a small islet of the Eniwetok atoll.

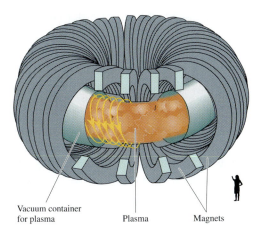

Vacuum container for plasma   Plasma   Magnets

Figure 22-15
A version of a fusion reactor that has been the subject of intense development efforts is the tokamak, whose design and an experimental prototype are shown here.

⑧

The approach most often taken to address these problems is to use a hot, ionized gaseous plasma that contains cations of ^2H, ^3H, and ^6Li, along with enough electrons to maintain charge neutrality. This plasma is heated intensely, usually by powerful laser beams. To keep the plasma from flying apart, it is confined by a donut-shaped magnetic field. Figure 22-15 shows a diagram of such a magnetic field and a photo of a prototype that is in operation in Princeton, New Jersey.

Fusion cannot become a practical source of energy unless the energy released by fusion exceeds the energy used to heat and confine the plasma. Unfortunately, the energy input requirement is enormous, even in small-scale experimental studies. The best experiments conducted so far have managed to achieve fusion but not sustain it long enough to achieve any net production of energy.

Assuming that a sustained fusion reaction is achieved, some way must be found to harness its energy output. Because the field confining the plasma prevents the escape of charged particles, most of the energy output of the reactor is expected to be photons, ranging in energy from infrared rays to γ rays. These photons will have to be absorbed by a suitable material that converts their energy, directly or indirectly, into electricity.

Stellar Nuclear Reactions

The sun and all other stars produce energy at a huge rate from sustained nuclear fusion. Over time, stars evolve through several stages, including stellar explosions. The products of a stellar explosion can form stars of more complex composition. Three distinct generations of stars have been identified, each fueled by a different set of fusion reactions.

The formation of a star begins with a cloud of matter. Gravitational attraction pulls the matter together, and during this collapse, gravitational potential energy is converted into kinetic energy of motion: Atoms move faster and faster, and the temperature rises. The temperature attained in the collapsing cloud increases with the total mass of the cloud, and if the mass is larger than about 2×10^{29} kg, the temperature rises high enough to initiate the fusion of hydrogen nuclei.

First-Generation Stars

Astrophysicists believe that the early universe was composed mostly of hydrogen. As a cloud of hydrogen collapses, heating breaks its hydrogen atoms into a plasma of protons and electrons. A **first-generation star** forms when the interior temperature of this plasma reaches 4×10^7 K. Above this temperature, protons combine in a reaction sequence that yields helium nuclei:

$$^1_1H + {}^1_1H \longrightarrow ({}^2_2He) \longrightarrow {}^2_1H + {}^0_1\beta^+$$

$$^2_1H + {}^1_1H \longrightarrow {}^3_2He + \gamma$$

$$^0_1\beta^+ + {}^0_{-1}e \longrightarrow 2\,\gamma$$

$$^3_2He + {}^3_2He \longrightarrow ({}^6_4Be) \longrightarrow {}^4_2He + 2\,{}^1_1H$$

The color of a hydrogen-burning star depends on its mass; the higher the mass, the higher the temperature of the star and the more blue it appears.

The net reaction converts protons and electrons into helium nuclei and radiation, releasing about 2.5×10^9 kJ/mol of energy:

$$4\,_1^1\text{H} + 2\,_{-1}^0\text{e} \longrightarrow \,_2^4\text{He} + 6\,\gamma \qquad \Delta E \cong -2.5 \times 10^9 \text{ kJ/mol}$$

The pressure exerted by the radiation escaping from inner portions of the star counteracts the force of gravity, and the balance of the two opposing forces keeps the volume of the star constant as long as hydrogen fusion continues.

In time (about 10^{10} years for a star the size of our sun), a star consumes most of its protons. Then the preceding sequence of nuclear reactions ceases, gravitational forces take over once more, and the star collapses until its temperature reaches about 10^8 K. At this temperature, a new fusion reaction begins in which helium nuclei fuse to form beryllium-8, which can fuse with a third helium nucleus:

$$_2^4\text{He} + \,_2^4\text{He} \rightleftharpoons (_4^8\text{Be})$$

$$(_4^8\text{Be}) + \,_2^4\text{He} \longrightarrow \,_6^{12}\text{C} + \gamma$$

$$\text{Net: } 3\,_2^4\text{He} \longrightarrow \,_6^{12}\text{C} + \gamma \qquad \Delta E \cong -7.7 \times 10^8 \text{ kJ/mol}$$

The ^{12}C that is produced also reacts with ^4He:

$$_6^{12}\text{C} + \,_2^4\text{He} \longrightarrow \,_8^{16}\text{O} + \gamma$$

A star undergoing helium fusion has a dense, hot core and a large, cooler outer mantle. These stars appear large and red, so they are called red giants.

About 10^8 years after fusion of helium begins, a star runs out of ^4He fuel. When this happens, the star enters a new stage of gravitational collapse until the temperature increases to about 10^9 K. This triggers an entirely new set of nuclear reactions between nuclides of carbon and oxygen. Here is one example:

$$_6^{12}\text{C} + \,_8^{16}\text{O} \longrightarrow \,_{12}^{24}\text{Mg} + \,_2^4\text{He}$$

As a result of these and other fusion reactions, the star eventually contains nuclides with Z and A values all the way up to iron-56, which is the most stable of all nuclides.

At this stage of their evolution, some stars generate energy more rapidly than it can be dissipated. These stars explode like giant hydrogen bombs, ejecting their various nuclides, ranging from hydrogen to iron, into space. The remaining core collapses on itself. Such explosions, called **supernovae,** have been observed by astronomers over the centuries, and the debris from supernovae explosions has been detected at numerous locations in the heavens. Figure 22-16 contains before-and-after photographs of a supernova that appeared in 1987.

The drawing in Figure 22-17 shows the evolutionary stages of a first-generation star.

Second-Generation Stars

Explosions of first-generation stars spew nuclides from $Z = 1$ to $Z = 26$ into interstellar space. There, this matter mixes with interstellar hydrogen, and eventually enough matter

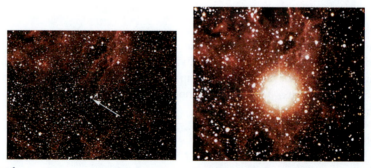

Figure 22-16
Before-and-after photos of a supernova that appeared in 1987. The arrow in the left photo indicates the star that exploded. The picture on the right shows the aftermath of the explosion.

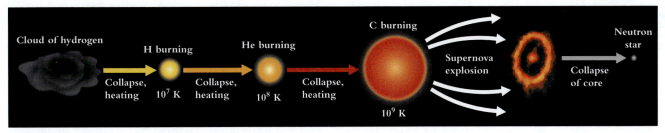

Figure 22-17
Schematic depiction of the evolutionary stages of a first-generation star.

clumps together under the force of gravity to collapse and form a **second-generation star.** Young second-generation stars, like young first-generation stars, contain large amounts of hydrogen. Unlike first-generation stars, however, they also contain higher-Z nuclides. One of these, ^{12}C, catalyzes the fusion of H to form He. As shown schematically in Figure 22-18, protons add to ^{12}C in a sequence that ultimately produces ^{4}He. Here are the reactions:

$$^{1}_{1}H + ^{12}_{6}C \longrightarrow (^{13}_{7}N) \longrightarrow ^{13}_{6}C + ^{0}_{1}\beta^{+}$$

$$^{1}_{1}H + ^{13}_{6}C \longrightarrow ^{14}_{7}N$$

$$^{1}_{1}H + ^{14}_{7}N \longrightarrow (^{15}_{8}O) \longrightarrow ^{15}_{7}N + ^{0}_{1}\beta^{+}$$

$$^{1}_{1}H + ^{15}_{7}N \longrightarrow ^{12}_{6}C + ^{4}_{2}He$$

$$2[^{0}_{1}\beta^{+} + ^{0}_{-1}e \longrightarrow 2\ \gamma]$$

Net: $4\ ^{1}_{1}H + 2\ ^{0}_{-1}e \longrightarrow ^{4}_{2}He + 4\ \gamma$ $\Delta E \cong -2.5 \times 10^{9}$ kJ/mol

When most of the hydrogen has been consumed, a second-generation star collapses until it is hot enough for helium fusion to occur. Now a larger range of reactions takes place because nuclides such as ^{13}C generate neutrons when they fuse with ^{4}He:

$$^{13}_{6}C + ^{4}_{2}He \longrightarrow ^{16}_{8}O + ^{1}_{0}n$$

These neutrons are captured by iron to yield Co after β emission:

$$^{56}_{26}Fe + ^{1}_{0}n \longrightarrow ^{57}_{27}Co + ^{0}_{-1}\beta$$

Neutron capture and β emission forms nuclei of ever higher atomic number. Neutron capture and β emission by Co ($Z = 27$) produces Ni ($Z = 28$), Ni produces Cu ($Z = 29$), and so on up the atomic-number ladder: Neutron capture and β emission form all possible stable nuclides during the lifetime of a second-generation star.

Like first-generation stars, second-generation stars often become unstable when they reach the carbon-burning stage. If they explode in supernovae, their nuclear debris includes their content of heavier nuclides. In addition, supernova explosions generate large numbers of neutrons that can be captured by lighter nuclides in the exploding star to form still more heavy nuclides.

Our planet Earth contains significant amounts of elements all the way up to $Z = 92$. This indicates that our solar system resulted from the gravitational collapse of a cloud of matter that included debris from second-generation stellar supernovae. Thus, our sun most likely is a **third-generation star.** The composition of a third-generation star includes high-Z nuclides, but the nuclear reactions are the same as those in a second-generation star.

The nuclear reactions described in this section can account for the elements found on Earth, but much about the composition of the universe remains to be discovered.

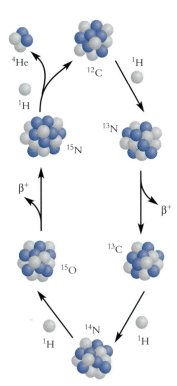

Figure 22-18
In a second-generation star, fusion of protons to produce helium occurs in a catalytic cycle by the sequential addition of protons to ^{12}C.

Section Exercises

■ **22.5.1** One proposal for controlled fusion involves using an accelerator to propel deuterons into a lithium target, inducing the following reactions:

$$^{7}_{3}\text{Li} + ^{2}_{1}\text{H} \rightarrow ^{8}_{3}\text{Li} + ^{1}_{1}\text{H} \qquad ^{8}_{3}\text{Li} \rightarrow ^{0}_{-1}\beta + 2\,^{4}_{2}\text{He}$$

(a) The first reaction can be viewed as the transfer of a neutron. Draw a nuclear picture illustrating this reaction.

(b) The reaction has a barrier that is 50% of the calculated barrier due to electrical force. What minimum kinetic energy must the deuterons have, if the radius of ^{2}H is 1.4×10^{-3} pm and that of ^{7}Li is 2.4×10^{-3} pm? (Refer to Section 22.1 to calculate the electrical energy barrier.)

(c) Use the following isotopic masses, all in grams per mole, to calculate the energy change for each step of this fusion reaction:

^{7}Li	^{8}Li	^{1}H	^{2}H	^{4}He
7.016005	8.022488	1.007822	2.0141022	4.0026036

(d) If 5.0 kg of ^{2}H is completely consumed in the reaction, how much energy can be produced?

■ **22.5.2** To show how zinc could form in a second-generation star, write a sequence of nuclear reactions that starts at ^{56}Fe, ends at ^{66}Zn, and is composed entirely of neutron capture and β emission. Your sequence should stay within the belt of stability depicted in Figure 22-5 (see also Figure 2-20).

■ **22.5.3** Describe how the compositions of planets around a second-generation star differ from that of the Earth. Could life as we know it exist on such a planet? Why or why not?

22.6 EFFECTS OF RADIATION

The nuclear explosions that devastated Hiroshima and Nagasaki killed 100,000 to 200,000 people instantaneously. Probably an equal number died later, victims of the radiation released in those explosions. Millions of people were exposed to the radioactivity released by the accident at the Chernobyl nuclear power plant. The full health effects of that accident may never be known, but 31 people died of radiation sickness within a few weeks of the accident, and more than 2000 people have developed thyroid cancer through exposure to radioactive iodine released in the accident. Even low levels of radiation can cause health problems. For this reason, workers in facilities that use radioisotopes monitor their exposure to radiation continually, and they must be rotated to other duties if their total exposure exceeds prescribed levels.

Radiation Damage

Nuclear radiation causes damage because of its high energy content. Radiation passing through matter transfers energy to atoms and molecules in its path. The major result of this energy transfer is ionization. Electrons are torn from molecules, creating positive ions. A single nuclear emission has enough energy to generate many positive ions and free electrons. A typical α particle, for example, carries an energy of about 10^{-12} J, whereas ionization energies are typically 2×10^{-18} J. Only about one third of the energy of an α particle goes into generating ions, the rest being converted into heat. Still, one α particle generates around 150,000 cations. Beta and gamma rays have similar energies, and they also generate ions as they pass through matter.

As a high-energy particle passes through matter, it creates an ionization "track" that contains positive ions. These ions are chemically reactive because their bonds are weakened by the loss of bonding electrons. Even though each cation eventually recaptures an electron to return to electrical neutrality, many ions first undergo chemical reactions that are the source of the damage done by nuclear radiation.

Immediate Health Effects

Living cells are delicately balanced chemical machines. The ionization track generated by a nuclear particle upsets this balance, almost always destroying the cell in the process.

Table 22-5   Health Effects of Radiation Doses

Dose (rem)	Effect
0–25	Increased susceptibility to cancer; possible DNA damage
25–50	Reduced amounts of white blood cells
50–100	Fatigue and nausea in half of persons exposed
100–200	Nausea and vomiting; hair loss
200–400	Damage to bone marrow and spleen; 50% fatality rate
>600	Usually death, even with treatment

Although the body has a remarkable ability to repair and replace damaged cells, exposure to radiation can overload these control mechanisms, causing weakness, illness, and even death.

Because nuclear radiation varies considerably in energy, the potential to cause damage cannot be assessed simply by counting the number of emissions. The energy of emissions must also be taken into account. Furthermore, the three different types of nuclear radiation affect human cells to different extents. When the amount, energy content, and type of radiation are taken into account, the result is a measure of the effect of radiation on the human body. This is expressed using a unit called the **rem.**

Different types of body cells show different sensitivities to nuclear radiation. Cells that divide most rapidly tend to be most easily damaged. These include bone marrow, white blood cells, blood platelets, the lining of the gastrointestinal tract, and cells in the gonads. Consequently, the symptoms of radiation sickness include loss of blood functions and gastrointestinal distress.

Individual ability to tolerate radiation damage varies, so a statistical variation exists in the relationship between dose level and health effects. Also, there are effective treatments, such as blood transfusions, for some radiation effects. The statistical patterns of human response to radiation are summarized in Table 22-5. Doses of over 600 rem are almost always fatal.

Long-Term Effects

Exposure to low doses of radiation causes no short-term damage but makes the body more susceptible to cancers. In particular, people who have been exposed to increased radiation levels have a much higher incidence of leukemia than the general population has. Marie Curie, the discoverer of radium, eventually died of leukemia brought on by exposure to radiation in the course of her experiments. Medical researchers estimate that about 10% of all cancers are caused by exposure to high-energy radiation.

Another cumulative effect of radiation can be an irreversible alteration of DNA sequences. If part of a DNA molecule is ionized, its molecular chain may be broken. Chain breaks are repaired in the body, but after a serious rupture, the repaired unit may have a different sequence. This type of changed sequence is a genetic mutation. Altered DNA sequences in the reproductive organs are transmitted faithfully, thus passing on the genetic mutations to future generations. Because these effects are cumulative, individuals of childbearing age need to be especially careful about radiation exposure.

All humans are exposed to some level of radiation. Cosmic radiation continually bathes the Earth. This radiation and the naturally occurring radioactivity from nuclides such as ^{14}C, ^{40}K, and ^{222}Rn expose the average individual to about 0.1 rem per year. This exposure is increased by radiation from human activities. Medical procedures, notably the use of X rays for imaging teeth and bones, contribute the largest amount. A typical medical X-ray procedure exposes the patient to 0.2 rem. Although this is well below the level at which immediate effects occur, the possibility of cumulative effects makes it prudent to have X rays taken no more often than necessary for maintaining good health.

Total exposures vary considerably with human activities as well. "Frequent flyers," for example, receive higher doses of radiation because the intensity of cosmic radiation is significantly greater at high altitude than it is at ground level. Residents in locations such as Montana and Idaho, where there are uranium deposits, receive higher doses of radiation from radon, one of the radioactive decay products of uranium.

Figure 22-19
Different types of radiation penetrate matter to different degrees and produce damage at different rates along their paths.

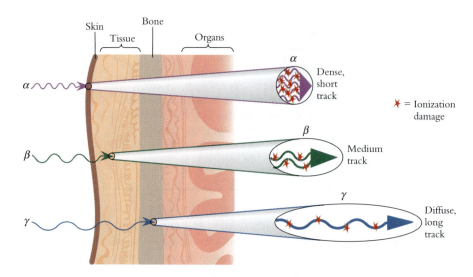

Radiation Shielding

Radiation exposure can be reduced by placing the radiation source or the potential target behind a shield that captures the radiation. During exposure to X rays for dental imaging, the patient wears a lead-lined pad, because X rays are absorbed more effectively by lead than by any other material. A lead shield a few millimeters thick is sufficient to stop X rays.

Radiation of different types penetrates matter to different extents, as Figure 22-19 illustrates. Even though α particles are the most damaging radiation, they are the most easily stopped by shielding. In fact, α particles travel less than 1 mm in solid materials before losing all their excess energy. Beta particles travel much farther, requiring shields 10 to 100 mm thick. Plastic shields several centimeters thick provide excellent shielding from both types of radiation.

The γ rays that accompany many nuclear decay processes can be more dangerous than α or β particles, because γ rays travel long distances before losing their energy. The radiation from ^{222}Rn, which is responsible for as much as 40% of the background radiation to which humans are exposed, is typical. This nuclide decays by α emission accompanied by γ rays. The α particle is stopped within 10 cm in air or about 1 mm in a solid. The accompanying γ ray, on the other hand, may travel many meters and penetrate walls before finally losing its destructive power. Consequently, radioactive materials that emit γ rays must be shielded using lead blocks that are many centimeters thick.

Section Exercises

■ **22.6.1** Potassium-40 is a radioactive nuclide that occurs naturally. This nuclide emits β particles with the following properties:

$$E = 1.32 \times 10^8 \text{ kJ/mol}$$

$$t_{1/2} = 1.28 \times 10^9 \text{ yr}$$

Our bodies contain about 2.4 mg of potassium per kilogram of body weight, and ^{40}K makes up 0.0118% of the element. To how many β particles, with how much total energy, is a person weighing 70 kg exposed daily from this nuclide?

■ **22.6.2** Although X-ray exposures pose little threat to patients, medical technicians must be careful to avoid exposure to the beam. Suppose that an X-ray machine is "leaking" radiation that exposes the technician to 1% of the dose received by a patient. After how many exposures would the technician begin to show reduced numbers of white blood cells? If the technician administers 40 X-ray studies daily and works 250 days per year, would this damage show up?

22.7 **APPLICATIONS OF RADIOACTIVITY**

Considering the frighteningly destructive potential of nuclear weapons and the health risks posed by exposure to too much nuclear radiation, radioactivity may seem like a curse. Nevertheless, scientists and medical practitioners use radioactivity routinely in many beneficial ways. Our Introduction to Chapter 2 features positron emission tomography, a medical imaging technique that uses radioactive positron emitters. Section 2.3 elaborates on medical uses of radioactivity for cancer treatment as well as imaging. In this section we describe two techniques that have broad scientific applications, radioactive dating and radioactive tracing. Dating techniques, which rely on the constant half-life of radioactive decay, allow scientists to estimate the ages of human artifacts and of the Earth itself. Tracing techniques exploit the high sensitivity of radiation detectors to study complicated processes such as the mechanisms of chemical reactions and the percolation of water through geological formations.

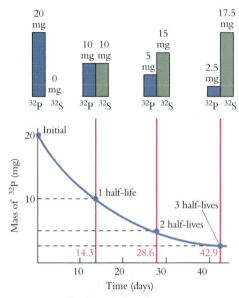

Figure 22-20
Decay of ^{32}P, a radioactive nuclide with a half-life of 14.3 days. The amount of the nuclide present is cut in half every 14.3 days.

Dating Using Radioactivity

Each radioactive nuclide has a characteristic, constant half-life. This means that it acts as a clock, "ticking" (decaying) at a constant rate. Suppose that a nuclear reactor generates 20 mg of ^{32}P, an isotope with a half-life of 14.3 days. As Figure 22-20 illustrates, after 14.3 days, 10 mg will remain, and there will be 10 mg of the decay product, ^{32}S. After 28.6 days, 5 mg of ^{32}P will remain, and there will be 15 mg of ^{32}S. A chemist who analyzes a sample of ^{32}P and finds that it contains 15 mg of ^{32}S and only 5 mg of ^{32}P can calculate that the sample was generated 28.6 days ago.

Radioactive dating is most valuable in estimating the age of materials that predate human records, such as the Earth itself. Calculating the age of a sample from the amounts of its radioactive nuclides and their decay products forms the basis for radioactive dating techniques. Equation 22-5 is used for such calculations:

$$\ln\left(\frac{N_0}{N}\right) = \frac{t \ln 2}{t_{1/2}}$$

In this equation, N_0 and N are amounts of the nuclide present initially and at time t.

The Earth's age can be estimated by using the half-lives of unstable nuclides that are still present and the half-lives of those that are missing. Among those listed in Table 22-4, the shortest-lived naturally occurring nuclide is ^{235}U ($t_{1/2} = 7.0 \times 10^8$ yr). Nuclides with half-lives shorter than 10^8 years are missing. Examples are ^{146}Sm ($t_{1/2} = 7 \times 10^7$ yr), ^{205}Pb ($t_{1/2} = 3 \times 10^7$ yr), and ^{129}I ($t_{1/2} = 1.7 \times 10^7$ yr). If the Earth were less than 10^9 years old, its crust would contain some of these nuclides. If it were older than 10^{10} years, ^{235}U would no longer be present. Thus isotopic abundances indicate that the Earth is older than 10^9 years but younger than 10^{10} years.

More precise estimates come from accurate measurements of isotope ratios. Three pairs of radioactive isotopes and their products are abundant enough for such ratios to be measured:

$$^{238}\text{U} \longrightarrow \text{(Decay chain)} \longrightarrow {}^{206}\text{Pb} \qquad t_{1/2} = 4.5 \times 10^9 \text{ yr}$$
$$^{40}\text{K} \longrightarrow {}^{40}\text{Ar} + \beta \qquad t_{1/2} = 1.28 \times 10^9 \text{ yr}$$
$$^{87}\text{Rb} \longrightarrow {}^{87}\text{Sr} + \beta \qquad t_{1/2} = 4.9 \times 10^{11} \text{ yr}$$

If the amount of a radioactive nuclide in a rock sample is N, the sum of this amount plus the amount of its product nuclide is N_0. For argon dating, N_0 is the sum of potassium-40 and argon-40 present in a sample of rock. Assuming that Ar gas escapes from molten rock but is trapped when the rock cools and solidifies, the lifetime obtained by substituting these values into Equation 22-5 is the time since the rock solidified. Such analyses show that the oldest rock samples on Earth are 3.8×10^9 years old.

Analyses of this type are correct only if all of the product nuclide comes from radioactive decay. This is not known with certainty, but when age estimates using different pairs of nuclides give the same age and samples from different locations also agree, the age estimate is likely to be accurate. Note also that 3.8×10^9 years agrees with the qualitative limits derived from naturally occurring radioactive nuclides.

Samples that are 4.6×10^9 years old have been found in meteorites. This is the best present estimate for the age of the solar system. Example 22-9 illustrates this type of calculation for rock from the Earth's moon.

Example 22-9 **Argon Dating**  A sample of rock from the moon	When a sample of moon rock was analyzed by mass spectroscopy, the ratio of ^{40}K to ^{40}Ar was found to be 0.1295. Based on this ratio, how old is the moon? **Strategy:** This is a question about radioactive dating. Convert the isotopic ratio of the radioactive nuclide and its product into an amount ratio, and then use Equation 22-5. **Solution:** Equation 22-5 requires a value for N_0/N, the ratio of potassium-40 present when the moon was formed to the amount present today. We can determine N_0/N from the isotopic ratio N_K/N_{Ar} given in the problem. First note that $N = N_K$. Then relate N_0 to N_K/N_{Ar}. Each ^{40}K nuclide that decays generates a ^{40}Ar nuclide, so the initial amount, N_0, is the sum of the amounts currently present: $$N_0 = N_K + N_{Ar}$$ Divide both sides of this equality by N_K:

$$\frac{N_K}{N_{Ar}} = 0.1295 \qquad so \qquad \frac{N_{Ar}}{N_K} = \frac{1}{0.1295} = 7.722$$

$$\frac{N_0}{N_K} = 1 + 7.722 = 8.722$$

Now rearrange Equation 22-5 to solve for t:

$$t = \frac{t_{1/2}}{\ln 2} \ln\left(\frac{N_0}{N_K}\right)$$

Find the half-life of ^{40}K from Table 22-4: $t_{1/2} = 1.28 \times 10^9$ yr:

$$t = \frac{1.28 \times 10^9 \text{ yr}}{0.693} \ln(8.722) = 4.00 \times 10^9 \text{ yr}$$

According to this analysis, the moon solidified about 4 billion years ago.

 Does the Result Make Sense? The age estimate for the moon is similar to age estimates for the Earth based on several methods of radioactive dating. Thus, this is a sensible result.

Extra Practice Exercise 22.9	A meteor that is believed to have come from Mars has a ratio of ^{40}K to ^{40}Ar of 0.1660. What age does this indicate?

Radioactivity serves as a useful clock only for times that are the same order of magnitude as the decay half-life. At times much longer than $t_{1/2}$, the amount of radioactive nuclide is too small to measure accurately. At times much shorter than $t_{1/2}$, the amount of the product nuclide is too small to measure. By an accident of nature, one naturally occurring radioisotope, ^{14}C, has a half-life that is close to the age of human civilization. As a result, this isotope is used to determine the age of human artifacts.

The logic underlying carbon-14 dating differs from that for potassium–argon dating. Recall that bombardment of the upper atmosphere by cosmic rays generates small but measurable levels of ^{14}C. This rate of production of ^{14}C by cosmic rays equals its rate of decay by β emission, so the percentage of ^{14}C in the atmosphere remains nearly constant. Living plants incorporate CO_2 from the atmosphere into carbohydrates, and animals use these carbohydrates as food. As a result, the percentage of carbon-14 in the tissues of living organisms is the same as that in the atmosphere. On death, however, the uptake of carbon stops. The carbon-14 in an object that was once alive slowly disappears with $t_{1/2} = 5.73 \times 10^3$ years. This makes it possible to date objects such as bone, wood, and cloth by assaying their levels of ^{14}C.

The product of ^{14}C decay is ^{14}N, but there is so much ^{14}N already present in the atmosphere that this isotope is not an accurate measure of how much ^{14}C has decayed. Instead, the ratio of ^{14}C to ^{12}C is used for radiocarbon dating. In the atmosphere and in living objects, the fraction of ^{14}C is 1.33×10^{-12}. That is, there is 1 atom of ^{14}C for every 7.54×10^{11} atoms of ^{12}C. Although this is a tiny amount, it nevertheless results in 15.3 decays of ^{14}C per gram of carbon per minute. Using modern mass spectrometers, the fractional abundance of ^{14}C can also be measured directly.

After a living organism dies, its ^{14}C content decreases according to the first-order decay law, whereas its ^{12}C content remains fixed. After 5.73×10^3 years, there is half as much ^{14}C present, for a fractional content of 6.63×10^{-13} or one atom for every 1.508×10^{12} atoms of ^{12}C. The decay rate from a sample that is 5700 years old will be 7.65 decays g^{-1} min^{-1}.

The usual procedure for radiocarbon dating is to burn a tiny sample of the object to be dated, collect the CO_2 that is produced, and compare its rate of radioactive decay with that of a fresh CO_2 sample. The ratio of counts gives N_0/N, which can then be substituted into Equation 22-5 to calculate t. Mass spectroscopic isotope analysis can also be used to obtain the N_0/N value, as Example 22-10 illustrates.

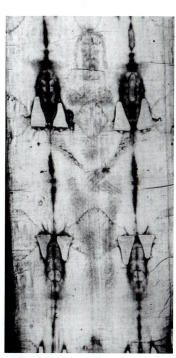

Example 22-10

Isotopic Carbon Dating

In 1988, the Shroud of Turin, claimed by some to have been used to bury Christ, was age-dated by isotopic analysis of its carbon content. The fractional content of ^{14}C from a sample from the shroud was 1.22×10^{-12}. How old is the shroud? If it were 2000 years old, what fractional content would it have?

Strategy: Calculation of an age requires the value of N_0/N, which is used in Equation 22-5 to calculate age, t. The ratio can be found from the isotopic analyses of a fresh carbon sample and the old sample.

Solution: The data include the fractional content of ^{14}C in the sample from the shroud, 1.22×10^{-12}. The text states that the fractional content for a fresh sample is 1.33×10^{-12}. To obtain N_0/N, divide the ratio for a fresh sample by the ratio for the old sample:

$$\frac{N_0}{N} = \frac{1.33 \times 10^{-12}}{1.22 \times 10^{-12}} = 1.09$$

$$t = \frac{t_{1/2}}{\ln 2} \ln\left(\frac{N_0}{N}\right)$$

$$t = \frac{5730 \text{ yr}}{0.693} \ln(1.09) = 714 \text{ yr}$$

Thus, the cotton or flax from which the shroud was woven was harvested around 1300 AD. The shroud might be younger than this, but it cannot be older.

Use the same equation to compute the isotopic ratio expected for an artifact that is 2000 years old:

$$\ln\left(\frac{N_0}{N}\right) = \frac{(2000 \text{ yr})(0.693)}{5730 \text{ yr}} = 0.242 \qquad \frac{N_0}{N} = e^{0.242} = 1.274$$

$$N = \frac{N_0}{1.274} = \frac{1.33 \times 10^{-12}}{1.274} = 1.04 \times 10^{-12}$$

The Shroud of Turin is imprinted with an image that resembles a male human, thought by some to be a representation of Christ.

Does the Result Make Sense? The ratio N_0/N must be a number greater than one, because the amount of radioactive carbon always decreases with time. The fraction for the shroud, 1.09, is significantly smaller than the ratio expected for a 2000-year-old object, 1.274. The smaller ratio means that less radioactive carbon has disintegrated, indicating that the object is not as old as 2000 years.

Mass spectrometric analysis of charcoal from an archaeological dig gave a carbon-14 fractional abundance of 6.5×10^{-13}. What is the age of this dig?

Extra Practice Exercise 22.10

Calculations of ages, as illustrated in Example 22-10, make the assumption that the level of ^{14}C in the atmosphere at the time of death was the same as it is today. To check the validity of ^{14}C dating, its age estimates have been compared with those obtained by tree-ring dating, which involves fewer assumptions than radiocarbon dating. Wood from a Roman shipwreck, for example, was dated at 40 BC ± 3 years using tree-ring dating and 80 BC ± 200 years using radiocarbon dating. A large number of such comparisons show small but consistent differences, indicating that the fractional amount of ^{14}C in the atmosphere has changed slowly but steadily with time. Archaeologists use a small multiplying factor to correct for this small drift when they use radiocarbon dating on objects that do not retain tree rings, such as charcoal and cloth.

Radioactive Tracers

Radioactive isotopes are chemically identical to their natural, nonradioactive counterparts, but their high-energy decays allow them to be detected even though they may compose only a tiny fraction of the overall isotopic composition. As already described, radiocarbon dating can detect ^{14}C at a concentration of one part in 10^{12} in samples that contain carbon. If an isotopically enriched sample is introduced into some dynamic process, the course of that process can be followed by tracing the whereabouts of the radioactive isotope.

Tracer techniques are applied in fields that range from geology to medicine. For example, a holding tank at an industrial plant might be suspected of leaking contaminated water into wells in its vicinity, as shown diagrammatically in Figure 22-21. The path of water draining from the holding tank could be traced by adding water enriched in radioactive tritium to the holding tank. The water from the wells could then be sampled for radioactivity at regular intervals. If, after a few days, water in some wells showed significant radioactivity while the remaining wells were free of contamination even after several months, that would show which of the wells were receiving water from the holding tank.

Tracer techniques are used widely in biology. Botanists, for example, work to develop new plant hybrids that grow more rapidly. One common way to determine how fast plants grow is to measure how quickly they take up elemental phosphorus from the soil. New hybrids can be planted in a plot and fertilized with phosphorus enriched in radioactive ^{32}P. When leaves from the plants are assayed for ^{32}P at regular intervals, the rate of increase in radioactivity measures the rate of phosphorus uptake.

One of the first chemical applications of radioactive tracers was a set of elegant experiments on photosynthesis performed in the 1950s by Melvin Calvin. His goal was to determine the set of reactions used by plants to transform atmospheric CO_2 into carbohydrates. Calvin supplied growing plant cells with CO_2 enriched with ^{14}C. By harvesting cells and using radioactive counting to determine which molecules gained enhanced carbon-14 radioactivity and in what sequence, Calvin was able to unravel the

Figure 22-21
Radioactive tracers can show flow pathways that cannot be detected using other methods of detection. Tritiated water added to a holding tank results in radioactivity in the water from wells that draw from the ground water supply into which the holding tank drains.

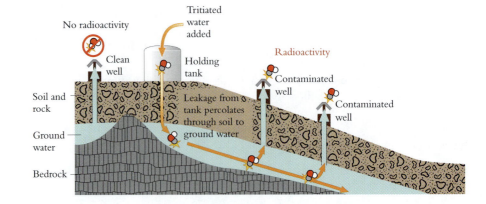

BOX 22-2	Chemistry and Life: Preserving Food with Radiation

I n a society where canned and frozen foods are so prevalent, we tend to take food preservation for granted. It is easy to forget that prevention of spoilage is a major undertaking when food is harvested at one time and place but eaten at another. A contemporary example is the space station, where the astronauts must be supplied with edible food that is preserved without refrigeration for long times.

Spoilage is a biological process. Molds, bacteria, and vermin eat foodstuffs, rendering them unfit for human consumption. To stop spoilage, processors treat food to kill microorganisms, chill it to slow the metabolism of destruction, and keep it sealed to ward off pests. Even in fully developed societies, these procedures are only partially successful, and in economically developing countries, up to 50% of crops may be lost to spoilage.

The bactericidal effects of ionizing radiation have been known for a century, and for over 80 years it has been known that ionizing radiation kills the *Trichinella spiralis* parasite, which infects raw pork. By the mid-1970s, international experts had concluded that irradiating foods preserves them without creating any toxicological hazards. Only in the last decade, however, has preservation by irradiation been applied commercially.

Irradiation of foods can be used for a variety of purposes. Low doses help preserve potatoes, onions, and garlic by inhibiting sprouting. Such doses also kill the parasite in pork that results in trichinosis, and they inactivate insects that feed on stored grain. Medium doses inhibit the growth of mold on fruits such as strawberries. Medium doses also help to preserve meat, poultry, and fish by killing microorganisms that cause spoilage and reducing the levels of pathogens such as salmonella. High doses of radiation are sufficiently lethal to microorganisms to provide long-term protection for meat as well as for spices and seasonings.

Irradiation is done inside irradiation chambers.

Although γ-irradiation of food is used widely in Europe, usage has been more restricted in the United States. The U.S. Food and Drug Administration has been slow to approve the process because of concern that irradiation may generate unhealthy and perhaps even carcinogenic by-products. Nevertheless, irradiation has been approved for numerous food products. Wheat and wheat products are irradiated to prevent insect infestation, fresh and frozen poultry and red meat to control pathogens, and fresh eggs to kill salmonella bacteria. At Thanksgiving, the astronauts aboard the space station have dined on turkey that had been prevented from spoiling by irradiation.

Astronauts have dined on irradiated turkey meat.

Opposition to irradiated foods arises from two main concerns. In the process of killing microorganisms, irradiation might also destroy important nutrients such as vitamins. Secondly, irradiation produces ions and free radicals such as OH that may react with foodstuffs to generate harmful compounds such as carcinogens. While it is difficult to prove beyond a doubt that these effects do not occur, all the scientific evidence gathered thus far indicates that properly irradiated food is both nutritious and safe.

Furthermore, it is important to keep in mind that conventional methods of preservation also carry risks. For example, ethylene dibromide, which was used for years to kill pests in fruit, has been banned because it is implicated as a carcinogen. Nitrites, which are used widely in preserving bacon, bologna, and other processed meats, are now known to be unhealthy at high exposures. Inadequate treatment of foodstuffs, particularly raw meat and poultry, has led to several outbreaks of illness in the United States, some of which have resulted in deaths from pathogens such as *E. coli*. Irradiation is highly effective in eliminating the possibility of trichinosis from undercooked pork, salmonella from raw eggs, or gastrointestinal distress from medium-rare hamburgers. Those who like their beef rare may increasingly join the American astronauts in feasting on irradiated meats.

Irradiation preserves strawberries and other fruit.

Irradiation is a relatively simple process. The food to be irradiated is placed inside a chamber where it is exposed to γ rays from a radioactive source such as ^{60}Co or ^{137}Cs. Alternatively, an accelerated electron beam can be used, mimicking β radiation.

elaborate chain of reactions that is now known as the Calvin cycle. Dr. Calvin received the Nobel Prize in chemistry in 1961 in recognition of this work.

Our Chemistry and Life Box discusses another use for radiation, the controversial area of food preservation.

Section Exercises

■ **22.7.1** An archaeologist discovers a new site where small charcoal bits are mixed with human remains. Burning a small sample of this charcoal gives gaseous CO_2 that, when placed in a counter, registers 1.75 decays per second. When an equal mass of CO_2 from fresh charcoal is placed in this counter, the count rate is 3.85 decays per second. How old is the archaeological site?

■ **22.7.2** Describe how radioactive iodine (^{131}I) could be used to determine how rapidly each of the following proposed chemical exchange reactions occurs: (a) $I_2(s)$ exchanging with I^- in aqueous solution; and (b) $I_2(s)$ exchanging with $I_2(g)$ in a closed container.

CHAPTER 22 VISUAL SUMMARY

Important equations

Essential terms

Key concepts

SKILLS TO MASTER

1. Writing and interpreting nuclide symbols
2. Calculating energies of nuclear reactions
3. Predicting nuclide stability
4. Describing nuclear decay processes
5. Working with nuclear half-lives
6. Balancing binuclear reactions
7. Describing details of fission and fusion
8. Describing stellar evolution and nuclear synthesis
9. Describing medical and health aspects of radiation
10. Dating ancient objects

22.1 NUCLEAR STABILITY

Nucleons → Nuclide

⑦ ①

9 protons   10 neutrons

$$^A_Z X$$

Fluorine–19, $^{19}_9 F$

Nuclear binding energy

$E = mc^2$ ② $\Delta E = (\Delta m)c^2$

$\Delta E = (\Delta m)(8.988 \times 10^{10} \text{ kJ/g})$

③

Fission   Fusion

Binding energy (10^8 kJ/mol) vs Mass number, A

^2H
Fusion releases energy → ← Fission releases energy
^{56}Fe $A = 60$ ^{84}Kr
^4He Most stable nuclides ^{235}U

22.2 NUCLEAR DECAY

Radioactive   Nuclear decay

④

α emission → $^4_2 \alpha$

Positron emission → β^+, γ rays, γ

β emission → $^0_{-1}\beta$

Electron capture → X ray

Electrical charge is conserved.   *Mass number is conserved.*

$\text{Rate} = \dfrac{\Delta N}{\Delta t} = \dfrac{-N \ln 2}{t_{1/2}}$ ⑤ $\ln\left(\dfrac{N_0}{N}\right) = \dfrac{t \ln 2}{t_{1/2}}$

22.3 INDUCED NUCLEAR REACTIONS

⑥ Induced nuclear reaction

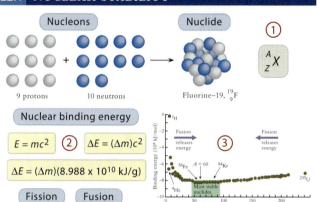

$^{14}_7 N$ + $^4_2 He$ → $[^{18m}_9 F]$ → $^{17}_8 O$ + $^1_1 H$

Compound nucleus

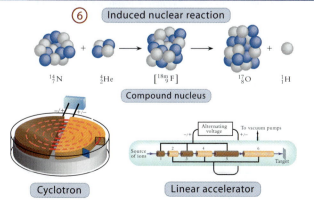

Cyclotron   Linear accelerator

22.4 NUCLEAR FISSION

Critical mass   Fission

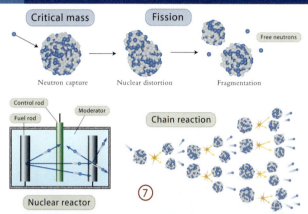

Neutron capture   Nuclear distortion   Fragmentation

Free neutrons

Control rod   Moderator
Fuel rod

Chain reaction

Nuclear reactor   ⑦

22.5 NUCLEAR FUSION

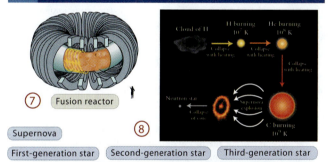

⑦ Fusion reactor

Cloud of H   H burning 10^7 K   He burning 10^8 K
Collapse with heating
Collapse with heating
Neutron star   Supernova explosion   Collapse of core
C burning 10^9 K
⑧

Supernova

First-generation star   Second-generation star   Third-generation star

22.6 EFFECTS OF RADIATION

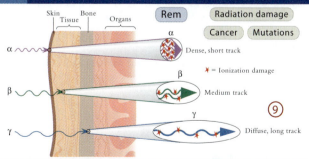

Skin Bone Organs
Tissue

Rem   Radiation damage

Cancer   Mutations

α   Dense, short track

✳ = Ionization damage

β   Medium track

γ   Diffuse, long track   ⑨

22.7 APPLICATIONS OF RADIOACTIVITY

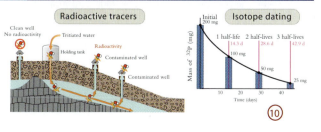

Radioactive tracers   Isotope dating

Clean well No radioactivity   Tritiated water   Radioactivity   Holding tank   Contaminated well   Contaminated well

Initial 200 mg
1 half-life 14.3 d 100 mg
2 half-lives 28.6 d 50 mg
3 half-lives 42.9 d 25 mg
Mass of ^{32}P (mg) vs Time (days)

⑩

Learning Exercises

22.1 Write a paragraph summarizing the important features of each of the following topics: (a) nuclear stability; (b) nuclear decay; (c) fission; (d) fusion; and (e) binding energy.

22.2 Outline the requirements for a balanced nuclear equation, and explain how they differ from those for a balanced chemical equation.

22.3 Prepare a list of the beneficial and detrimental effects that have resulted from the discovery of nuclear reactions.

22.4 Make a list of all terms introduced in Chapter 22 that are new to you. Write a one-sentence definition for each, using your own words. If you need help, consult the Glossary.

22.5 Update your list of memory bank equations to include those that apply to nuclear chemistry and radiochemistry.

 Problems **ilw = interactive learning ware problem. Visit the website at www.wiley.com/college/olmsted**

Nuclear Stability

22.1 Determine Z, A, and N for each of the following nuclides: (a) the nuclide of neon that contains the same number of protons and neutrons; (b) the nuclide of lead that contains 1.5 times as many neutrons as protons; and (c) the nuclide of zirconium whose neutron–proton ratio is 1.25.

22.2 Determine Z, A, and N for each of the following nuclides: (a) the helium nuclide with one less neutron than proton; (b) the nuclide of barium whose neutron–proton ratio is 1.25; and (c) the nuclide of thorium that contains 1.5 times as many neutrons as protons.

22.3 Manganese has only one stable isotope. Calculate the N:Z ratio for each of the following four nuclides and determine which is the stable one: ^{53}Mn, ^{54}Mn, ^{55}Mn, and ^{56}Mn.

22.4 Rhodium has only one stable isotope. Calculate the N:Z ratio for each of the following four nuclides and determine which is the stable one: ^{101}Rh, ^{102}Rh, ^{103}Rh, and ^{104}Rh.

22.5 Compute the energy released in kilojoules when the sun converts 1.00 metric ton of matter into energy.

22.6 Compute the energy released in joules per event and in kilojoules per mole when antiprotons (the antimatter corresponding to protons) annihilate with protons.

ilw 22.7 Use atomic masses to compute the total binding energy and the binding energy per nucleon for elemental cesium, which has just one stable nuclide.

22.8 Use atomic masses to compute the total binding energy and the binding energy per nucleon for elemental cobalt, which has just one stable nuclide.

22.9 Is the repulsive barrier for fusion of ^{1}H and ^{6}Li larger, smaller, or about the same as that for the fusion of two ^{4}He nuclei? Which gives a more stable product? Explain.

22.10 Is the repulsive barrier for fusion of two protons larger, smaller, or about the same as that for the fusion of two deuterium nuclei? Which of the resulting helium nuclei is less stable? Explain.

Nuclear Decay

22.11 Write the correct symbol and give the name for each of the following products of nuclear decay: (a) a photon; (b) a positive particle with mass number 4; and (c) a positive particle with the same mass as an electron.

22.12 What are the charge number, mass number, and symbol for each of the following? (a) an alpha particle; (b) a beta particle; and (c) a neutron

22.13 Identify the product of each of the following decay processes: (a) tellurium with 73 neutrons emits a γ ray; (b) tellurium with 71 neutrons captures an electron; and (c) tellurium with 75 neutrons undergoes β and γ decay.

22.14 Identify the product of each of the following decay processes: (a) ^{52}Fe emits a positron and a γ ray; (b) ^{55}Fe captures an electron; and (c) ^{59}Fe undergoes β and γ decay.

22.15 Preidct the most likely mode or modes of decay for each of the isotopes of Mn listed in Problem 22.3.

22.16 Predict the most likely mode or modes of decay for each of the isotopes of Rh listed in Problem 22.4.

ilw 22.17 The iron isotope with 33 neutrons is used in medical applications. It is a β emitter. A sample of iron containing 1.33 picograms (pico $= 10^{-12}$) of this isotope registers 242 decays per second. Calculate the half-life of the iron isotope.

22.18 A radioactive nuclide of mass number 94 has been prepared by neutron bombardment. If 4.7 μg of this nuclide registers 20 counts per minute on a radioactivity counter, what is the half-life of this nuclide?

22.19 Thorium-232 decays by the following sequence of emissions: α, β, β, α, α, α, β, β, α. Identify the final product and all of the unstable intermediates.

22.20 Radon-222 decays by the following sequence of emissions: α, α, β, β, α, β, β, α. Identify the final product and all of the unstable intermediates.

Induced Nuclear Reactions

22.21 Identify the compound nucleus and final product resulting from each of the following nuclear reactions: (a) carbon-12 captures a neutron and then emits a proton; (b) the nuclide with eight protons and eight neutrons captures an α particle and emits a γ ray; and (c) curium-247 is bombarded with boron-11, and the product loses three neutrons.

22.22 Identify the compound nucleus and final product resulting from each of the following nuclear reactions: (a) lithium-7 captures a proton, and the product emits a neutron; (b) the nuclide with six protons and six neutrons captures an α particle and emits a γ ray; and (c) plutonium-244 is bombarded with carbon-12, and the product loses two neutrons.

22.23 Draw a nuclear picture (see Figure 22-1 for pictures of nuclei) that illustrates the capture of an α particle by ^{14}N, followed by the emission of a positron.

22.24 Draw a nuclear picture (see Figure 22-1 for pictures of nuclei) that illustrates the nuclear reaction responsible for the production of carbon-14. Include a picture of the compound nucleus.

$$^{14}\text{N} + \text{n} \longrightarrow {}^{14}\text{C} + \text{p}$$

Nuclear Fission

22.25 When uranium-235 undergoes fission, the product with higher mass is most likely to have a mass number between 132 and 142. Use Figures 22-5 and 2-20 to identify which elements are most likely to result.

22.26 When uranium-235 undergoes fission, the product with lower mass is most likely to have a mass number between 88 and 100. Use Figures 22-5 and 2-20 to identify which elements are most likely to result.

ilw 22.27 Compute the energy released in the following fission reaction:

$$^{235}\text{U} + \text{n} \longrightarrow {}^{152}\text{Nd} + {}^{81}\text{Ge} + 3\,\text{n}$$

The nuclide masses in grams per mole are U, 235.0439; Nd, 151.9233; and Ge, 80.9199. Compare your result with the calculation in Section Exercise 22.4.1.

22.28 Compute the energy released in the following fission reaction:

$$^{235}\text{U} + \text{n} \longrightarrow {}^{100}\text{Mo} + {}^{134}\text{Sn} + 2\,\text{n}$$

The nuclide masses in grams per mole are U, 235.0439; Mo, 99.9076; and Sn, 133.9125. Compare your result with the calculation in Section Exercise 22.4.1.

22.29 A coal-fired power plant requires just one heat exchanger, in which water is converted to the steam needed to drive the generator. A nuclear power plant requires two. Describe the features of nuclear reactions that require this second heat exchanger.

22.30 Describe the features of radioactivity that make an accident in a nuclear power plant more devastating than an accident in a coal-burning power plant.

Nuclear Fusion

22.31 How much energy is released if 2.50 g of deuterons fuse with tritium and 50% of the neutrons are captured by ^{6}Li? (Consult reactions and energies given in the text.)

22.32 How much energy will be released if 1.50 g of deuterons and 1.50 g of lithium-6 undergo fusion in a hydrogen bomb? (Consult reactions and energies given in the text.)

ilw 22.33 Compute the speed of a tritium nucleus with enough kinetic energy to fuse with a deuteron. Calculate nuclear radii using $r = 1.2A^{1/3}$ fm (1 fm = 10^{-15} m).

22.34 Compute the speed of an alpha particle with enough kinetic energy to fuse with lithium-6. Calculate nuclear radii using $r = 1.2A^{1/3}$ fm.

22.35 Describe the features of fusion that have prevented fusion power from being commercialized as fission power has been.

22.36 Describe the features of fusion that make it more attractive as a power source than fission.

22.37 Tabulate the differences in temperature and composition of a first-generation star at the hydrogen-, helium-, and carbon-burning stages of its evolution.

22.38 Repeat the tabulation in Problem 22.37 for a second-generation star. List the differences between the two tabulations.

22.39 Using the characteristics of nuclear reactions, explain why no elements with $Z > 26$ form during the evolution and eventual explosion of first-generation stars.

22.40 Second- and third-generation stars may contain significant amounts of carbon; yet carbon burning does not begin until the third stage of their evolution. Explain this.

Effects of Radiation

ilw 22.41 The isotope ^{131}I has a half-life of 8.07 days and undergoes β decay with an energy of 9.36×10^7 kJ/mol. If a patient with a thyroid disorder is given this isotope and 7.45 pg is incorporated into the thyroid gland, how much energy will the gland receive in one day?

22.42 A patient weighing 65 kg is given an intravenous dose of ^{99m}Tc, a radioactive isotope that decays by γ emission with energy of 1.35×10^7 kJ/mol and with a 6.0-hour half-life. If 15.0 mL of a

2.50-nanomolar solution is administered and all of the radioactive nuclide decays within the patient's body, what total energy does the patient receive?

22.43 How many β particles are emitted per second by the dose of ^{131}I described in Problem 22.41?

22.44 How many γ rays are emitted per second by the dose of ^{99m}Tc described in Problem 22.42?

22.45 Why does exposure to radiation in cancer therapy result in nausea and reduced resistance to infection?

22.46 Why is it more important to protect the reproductive organs from radiation than other organs?

Applications of Radioactivity

ilw 22.47 Extremely old rock samples can be dated using the ratio of ^{87}Sr to ^{87}Rb because $t_{1/2}$ for ^{87}Rb is 4.9×10^{11} yr. What is the approximate age of a meteorite that has a ^{87}Sr to ^{87}Rb mass ratio of 0.0050?

22.48 Uranium deposits are dated by determining the ratio of ^{238}U to its final decay product, ^{206}Pb. The half-life of ^{238}U is 4.5×10^9 yr. If the ratio of ^{238}U to ^{206}Pb in an ore sample is 1.21, what is the age of the ore?

22.49 A river contains a high concentration of iron cations, and environmental activists contend that an industrial manufacturing plant is the source. The manufacturer says that although the plant generates aqueous iron waste, it is processed on site and does not contaminate the river. How might this disagreement be resolved using radioactive tracers?

22.50 Many places have underground water, called aquifers, that can be tapped using wells. Depending on the types of geological formations above these aquifers, water from the surface may or may not reach these underground reservoirs. How could radioactive tracers be used to determine if rainwater falling on a particular location makes its way down into an aquifer beneath that location?

22.51 In the hydrolysis of an ester, one oxygen atom in the products comes from the ester and the other comes from a water molecule:

$$RCO-O-R' + H_2O \longrightarrow RCO-OH + HO-R'$$

Describe experiments that would determine if the oxygen atom in the alcohol, HO—R', comes from water or from the ester.

22.52 In aqueous solution, the hydrogen atom in the —CO$_2$H group of a weak organic acid rapidly exchanges with a hydrogen atom in a water molecule through proton transfer:

$$RCO_2H + H_2O \rightleftharpoons RCO_2^- + H_3O^+$$

In contrast, the hydrogen atom in the —OH group of an alcohol is not thought to undergo this reaction. A research chemist claims that this is incorrect, and that rapid exchange of H atoms occurs between alcoholic —OH groups and water molecules. Describe experiments that would test this claim.

Additional Paired Problems

22.53 Naturally occurring bismuth contains only one isotope, ^{209}Bi. Compute the total molar binding energy and molar binding energy per nucleon of this element.

22.54 Naturally occurring gold contains only one isotope, ^{197}Au. Compute the total molar binding energy and molar binding energy per nucleon of this element.

22.55 Polonium-210 has a half-life of 138.4 days. If a sample contains 5.0 mg of ^{210}Po, how many milligrams will remain after 365 days? How many emissions per second will this residue emit?

22.56 What fraction of ^{209}Po, which has a half-life of 103 years, will remain intact after 365 days? If a sample of polonium contains equimolar amounts of ^{210}Po and ^{209}Po, what will the isotope ratio be after 10 years (see Problem 22.55)?

22.57 Neutrons decay into protons. What is the other product of this decay? If all of the decay energy is converted into kinetic energy of this other product, how much kinetic energy does it have?

22.58 A positron has the same mass as an electron. When a positron and an electron annihilate, both masses are converted entirely into the energy of a pair of γ rays. Calculate the energy per γ ray and the energy of 1 mol of γ rays.

22.59 The long-lived isotope of radium, ^{226}Ra, decays by α emission with a half-life of 1622 years. Calculate how long it will take for 1% of this nuclide to disappear and how long until 1% of it remains.

22.60 Phosphorus-30, which has a half-life of 150 seconds, decays by positron emission. How long will it take for (a) 2% and (b) 99.5% of this nuclide to decay?

22.61 Two isotopes used in positron-emission imaging are ^{11}C and ^{15}O. On which side of the belt of stability are these nuclides located? Write the nuclear reactions for their disintegrations.

22.62 Neutron bombardment is an elegant way of "doping" silicon with phosphorus to convert it into an *n*-type semiconductor. Bombardment generates stable ^{31}P from ^{30}Si. Write the nuclear reactions for this transformation. (Doped semiconductors are described in Chapter 10.)

22.63 Sodium-24, $t_{1/2} = 15.0$ hr, can be used to study the sodium balance in animals. If a saline solution containing 25 μg of ^{24}Na is injected into an animal, how long can the study continue before only 1 μg is left?

22.64 What is the shelf life of a radioactive nuclide with a half-life of 14.6 days if it loses its usefulness when less than 15% of the radioactivity remains?

22.65 Complete the following nuclear reactions: (a) alpha emission from ^{238}U; (b) n + $^{60}Ni \rightarrow$? + p; (c) ^{239}Np + $^{12}C \rightarrow$? + 3 n; (d) p + $^{35}Cl \rightarrow \alpha$ + ?; and (e) β emission from ^{60}Co.

22.66 Complete the following nuclear reactions: (a) positron emission from ^{26}Si; (b) electron capture by ^{82}Sr; (c) $^{210}Po \rightarrow \alpha$ + ?; (d) α + $^9Be \rightarrow$ n + ?; and (e) $^{99m}Tc \rightarrow \gamma$ + ?

22.67 The heaviest transuranium elements are formed by bombardment with relatively heavy nuclides such as ^{48}Ti. What nuclide could be formed by bombarding lead-208 with this nuclide?

22.68 The heaviest transuranium elements are formed by bombardment with relatively heavy nuclides such as ^{58}Fe. What nuclide could be formed by bombarding the stable bismuth nuclide with this nuclide?

22.69 Gadolinium-152 ($MM = 151.9205$ g/mol) emits an α particle, $E_{kinetic} = 3.59 \times 10^{-13}$ J, to form ^{148}Sm ($MM = 147.9146$ g/mol). How much energy (in J/nucleus) is released in this transformation? What fraction of the energy is carried off by the α particle?

22.70 Samarium-146 ($MM = 145.9129$ g/mol) emits an α particle, $E_{kinetic} = 3.94 \times 10^{-13}$ J, and becomes ^{142}Nd ($MM = 141.9075$ g/mol). How much energy (in J/nucleus) is released in this transformation? What fraction of the energy is carried off by the α particle?

22.71 State why each of the following nuclides is unstable: (a) carbon-14; (b) plutonium-244; (c) ^{56}Mn; and (d) a lithium nucleus with five neutrons.

22.72 State why each of the following nuclides is unstable: (a) tritium (one proton, two neutrons); (b) uranium-238; (c) ^{40}K; and (d) a beryllium nucleus with four neutrons.

22.73 Describe the roles of the moderator and the control rods in the operation of a nuclear power plant.

22.74 Describe how irradiation protects foods against spoilage, and describe how irradiation of foods might also produce undesirable by-products.

More Challenging Problems

22.75 Free neutrons are unstable. Since they cannot be collected and weighed, it is difficult to measure their half-life accurately. Early estimates gave $t_{1/2} = 1.1 \times 10^3$ s, but more refined experiments give $t_{1/2} = 876 \pm 21$ s. Suppose a neutron source generates 10^5 neutrons per second for 30 seconds. Assuming no neutrons are captured by nuclei, how many will be left after 5 hours, according to each of these half-lives?

22.76 Radioactive ^{64}Cu ($MM = 63.92976$ g/mol) decays either by emission of β particles with 9.3×10^{-14} J of energy or by emission of positrons with 1.04×10^{-13} J of energy. (a) Write the two decay reactions. (b) Calculate the molar masses of the two elemental products using mass–energy equivalence.

22.77 Nuclear power plants often use boron control rods. Boron-10 absorbs neutrons efficiently, emitting α particles as a consequence. Write the nuclear reaction. Does this reaction pose a significant health hazard? Explain your reasoning.

22.78 The radius of a ^{12}C nucleus is 3.0×10^{-15} m. Compute the energy barrier in J and in kJ/mol for the fusion of two ^{12}C nuclei.

22.79 A 250-mg sample of CO_2 collected from a small piece of wood at an archaeological site gave 1020 counts over a 24-hour period. In the same counting apparatus, 1.00 g of CO_2 from freshly cut wood gave 18,400 counts in 20 hours. What age does this give for the site?

22.80 One isotope of nitrogen and one isotope of fluorine are positron emitters with relatively long half-lives. Identify them and write their decay reactions.

22.81 The amount of radioactive carbon in any once-living sample eventually drops too low for accurate dating. This detection limit is about 0.03/g min, whereas fresh samples exhibit a count rate of 15.3/g min. What is the upper limit for age determinations using carbon dating?

22.82 Carbon-14 dating gives 3250 years as the age of a charcoal sample, assuming a constant level of cosmic radiation. If the cosmic ray level in the atmosphere was 20% higher at the time the tree grew, what is the correct age of the sample?

22.83 Strontium-90 is a dangerous fission product because it passes into human bodies and lodges in bones. It decays in two steps to give ^{90}Zr. The molar masses of these two nuclides are 89.9073 and 89.9043 g/mol. (a) Write the decay reactions that lead from ^{90}Sr to ^{90}Zr. (b) Which nuclide has the higher molar mass? (c) Calculate the total energy released in the decay scheme. (*Reminder:* Include the masses of the other decay products.)

22.84 Suppose that fusion of tritium and deuterium involves two nuclei with equal kinetic energies totaling 75% of the electrical repulsion barrier. Calculate their speeds.

22.85 Iodine-123, $t_{1/2} = 13.2$ hr, is used in medical imaging of the thyroid gland. If 0.5 mg of this isotope is injected into a patient's bloodstream and 45% of it binds to the thyroid gland, how long will it take for the amount of iodine in the thyroid gland to fall to less than 0.1 μg?

22.86 In 1934, Irene Curie and Frederic Joliot produced the first artificial radioisotope by bombarding ^{27}Al with α particles. The resulting compound nucleus decayed by neutron emission. Write the reaction and predict whether or not the final product is stable. If it is not stable, predict its mode of decay.

22.87 Technetium has no stable isotopes. Which nuclide of this element would you predict to be nearly stable? (Make use of Figures 22-5 and 2-20.)

22.88 Calculate the mass of ^{235}U that reacts in a 20-kiloton bomb, given that one nucleus releases 2.9×10^{-11} J. The designation "20-kiloton" means the same energy as 20×10^3 metric tons of TNT, which releases 2500 kJ of energy per kilogram.

22.89 As discussed in Chapter 13, a condensation reaction between a carboxylic acid and an alcohol produces an ester with the elimination of water:

Explain how the unstable isotope ^{18}O can be used to show whether the oxygen atom in the water molecule comes from the carboxylic acid or from the alcohol.

22.90 A radioactivity counter gave a reading of 350 min^{-1} for a 12.5-mg sample of cobalt(II) chloride partially enriched with cobalt-60 ($t_{1/2} = 5.2$ yr). What percentage of the cobalt atoms is cobalt-60?

22.91 A small amount of NaBr containing the radioactive isotope sodium-24 is dissolved in a hot solution of sodium nitrate containing the naturally occurring nonradioactive isotope sodium-23. The solution is cooled, and sodium nitrate precipitates from the solution. Will the precipitate be radioactive? Explain your answer.

Group Study Problems

22.92 From the mass number distribution given in Figure 22-10 and the belt of stability shown in Figures 22-5 and 2-20, predict which elements should be found in greatest abundance in the remains of a fission event like that of the Oklo reactor (Box 22-1).

22.93 The Earth captures 3.4×10^{17} J/s of radiant energy, which is one part in 4.5×10^{-10} of the sun's total energy output. (a) By how much does the mass of the sun change every second? (b) The sun is in the hydrogen-burning stage of its evolution. Calculate how many moles of ^{1}H are converted into ^{4}He per second.

22.94 For how many years could ^{235}U supply the world's energy needs, given that uranium contains 0.72% of ^{235}U, the world reserves of uranium are 1.0×10^7 metric tons, world energy consumption is 2.0×10^{17} kJ/yr, and each ^{235}U nucleus releases 2.9×10^{-11} J?

22.95 Smoke detectors contain small quantities of ^{241}Am, which decays by emitting an α particle with 5.44×10^8 kJ/mol energy and a 458-year half-life. It also emits γ rays. (a) What is the neutron–proton ratio for ^{241}Am? (b) If a smoke detector operates at 5.0 decays per second, what mass of ^{241}Am does it contain? (c) If this smoke detector malfunctions when its decay rate falls below 3.5 decays per second, for how many years could it be used? (d) If your bed were directly under

this smoke detector, would you be exposed to its nuclear radiation? Explain.

22.96 Unlike most other elements, different samples of lead have different molar masses. This is because lead is the stable end product of two different radioactive decay schemes, one starting from ^{238}U and the other from ^{232}Th. Both decay schemes consist entirely of α, β, and γ emissions. The most abundant stable lead isotopes have mass numbers 206, 207, and 208. Will a sample of lead, some of which came from ^{238}U, have a higher or lower molar mass than another sample of lead, some of which came from ^{232}Th? Draw a decay scheme that illustrates your reasoning.

22.97 List the advantages and disadvantages of fission power and fusion power. Based on your list, do you think that the United States should continue to develop fission power plants? What about fusion power plants?

22.98 Uranium-238 undergoes eight consecutive α emissions to give stable lead. In a uranium ore sample, all the α particles are quickly stopped, becoming trapped ^{4}He atoms. If analysis of a rock sample shows that it contains 6.0×10^{-5} cm^3 of helium-4 (1 atm, 298 K) and 1.3×10^{-7} g of ^{238}U per gram of rock, estimate the age of the rock.

Answers to Section Exercises

22.1.1 (a) $^{16}_{8}$O, $Z = 8$, $N = 8$, $A = 16$; (b) $^{98}_{43}$Tc, $Z = 43$, $N = 55$, $A = 98$; and (c) $^3_1$H, $Z = 1$, $N = 2$, $A = 3$

22.1.2 (a) unstable, $Z > 83$; (b) unstable, odd–odd; (c) stable; and (d) unstable, $N < Z$

22.1.3 Total binding energy is -1.43×10^{10} kJ/mol; binding energy per nucleon is -7.52×10^8 kJ/mol.

22.2.1 (a) β decay; (b) α decay; (c) γ emission; and (d) positron emission or electron capture

22.2.2 (a) $^{140}_{57}$La; (b) $^{222}_{86}$Rn; and (c) $^{125}_{53}$I

22.2.3 12 days

22.3.1 (a) $^{59}_{27}$Co + $^1_0$n → $(^{60m}_{27}$Co$)$ → $^{59}_{26}$Fe + $^1_1$p;
(b) $^{59}_{27}$Co + $^1_0$n → $(^{60m}_{27}$Co$)$ → $^{60}_{27}$Co + γ;
(c) $^{59}_{27}$Co + $^4_2\alpha$ → $(^{63m}_{29}$Cu$)$ → $^{62}_{28}$Ni + $^1_1$p; and
(d) $^{59}_{27}$Co + $^1_1$p → $(^{60m}_{28}$Ni$)$ → $^{58}_{28}$Ni + 2 $^1_0$n

22.3.2 There are several possibilities. Here are two:
$^{144}_{60}$Nd + $^1_0$n → $(^{145m}_{60}$Nd$)$ → $^{145}_{61}$Pm + $^0_{-1}\beta$
$^{147}_{62}$Sm + $^1_0$n → $(^{148m}_{52}$Sm$)$ → $^{147}_{61}$Pm + $^1_1$p

22.3.3 (Answers are other reactant, then other product)
(a) neutron, γ ray; (b) ^{54}Fe, ^{57}Ni; and (c) proton, α particle

22.4.1 One neutron: -0.22 g/mol, -2.0×10^{10} kJ/mol
Two neutrons: -0.22 g/mol, -1.9×10^{10} kJ/mol
Three neutrons: -0.21 g/mol, -1.9×10^{10} kJ/mol

22.4.2 99.22% ^{238}U and 0.78% ^{235}U

22.5.1 (a)

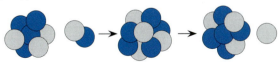

(b) 5.5×10^7 kJ/mol; (c) the first step is endothermic by 1.82×10^7 kJ/mol (but this energy could be provided by the kinetic energy of the deuterons); the second step is exothermic by -1.50×10^9 kJ/mol; and (d) 4×10^{12} kJ

22.5.2 More than one sequence is possible. Here is one:

$$^{56}\text{Fe} \xrightarrow{+n} {}^{57}\text{Fe} \xrightarrow{+n} {}^{58}\text{Fe} \xrightarrow{+n} {}^{59}\text{Fe} \xrightarrow{-\beta} {}^{59}\text{Co}$$

$$^{59}\text{Co} \xrightarrow{+n} {}^{60}\text{Co} \xrightarrow{-\beta} {}^{60}\text{Ni}$$

$$^{60}\text{Ni} \xrightarrow{+n} {}^{61}\text{Ni} \xrightarrow{+n} {}^{62}\text{Ni} \xrightarrow{+n} {}^{63}\text{Ni} \xrightarrow{-\beta} {}^{63}\text{Cu}$$

$$^{63}\text{Cu} \xrightarrow{+n} {}^{64}\text{Cu} \xrightarrow{+n} {}^{65}\text{Cu} \xrightarrow{+n} {}^{66}\text{Cu} \xrightarrow{-\beta} {}^{66}\text{Zn}$$

22.5.3 Planets around a second-generation star contain no elements with $Z > 26$. Simple life as we know it requires no elements with $Z > 26$, but mammalian life depends on enzymes containing copper, zinc, and cobalt.

22.6.1 4.4×10^5 β particles, which deposit 9.6×10^{-8} J of energy.

22.6.2 12,500 X rays; the damage would not show up in a year.

22.7.1 6.5×10^3 years

22.7.2 Prepare a sample of solid I_2 containing some of the radioactive isotope. (a) Put some of this solid in an aqueous solution containing a salt such as KI. Withdraw samples at various times, precipitate the I^- as AgI, and measure the radioactivity of the solid. (b) Put some of the solid in a closed container equipped with a gas sampling device. Collect gas samples at various times and measure the radioactivity of the gas.

Answers to Extra Practice Exercises

22.1 (a) $^{31}_{15}$P and (b) $^{20}_{10}$Ne

22.2 $\Delta E = -1.738 \times 10^{11}$ kJ/mol

22.3 $\Delta m = -8.93 \times 10^{-9}$ g/mol

22.4 $^{214}_{84}$Po → $^4_2\alpha$ + $^{210}_{82}$Pb; $^{210}_{82}$Pb → $^{210}_{83}$Bi + $^0_{-1}\beta$;
$^{210}_{83}$Bi → $^{210}_{84}$Po + $^0_{-1}\beta$; and $^{210}_{84}$Po → $^4_2\alpha$ + $^{206}_{82}$Pb

22.5 $t_{1/2} = 2.1 \times 10^5$ years

22.6 $t = 3.4 \times 10^5$ years

22.7 $^{12}_{6}$C + $^{12}_{6}$C → $(^{24}_{12}$Mg$)$ → $^4_2\alpha$ + $^{20}_{10}$Ne

22.8 $\Delta E_{\text{per gram}} = -7.37 \times 10^7$ kJ/g

22.9 3.6 billion years

22.10 5.9×10^3 years

Appendix A: Scientific Notation

S pecial notations are useful for expressing the precisions of very large and very small quantities. The diameter of a helium atom, for instance, is 0.000 000 000 24 meters, and the average distance from the Earth to its moon is 384,000,000 meters. Writing numbers with this many zeros is cumbersome, and it is easy to make a mistake. To shorten the writing of small and large numbers, scientists commonly use **scientific notation.**

Scientific notation is based on the fact that any number can be expressed as a number between 1 and 10 multiplied or divided by ten an appropriate number of times. For example, 384,000,000 meters can be written as follows:

$$(3.84)(10)(10)(10)(10)(10)(10)(10)(10) = 3.84 \times 10^8 \text{ meters}$$

Similarly, 0.000 000 000 24 meters is written as follows:

$$2.4 \div (10)(10)(10)(10)(10)(10)(10)(10)(10)(10)$$
$$= 2.4 \times 10^{-10} \text{ meters}$$

The numeral "8" in 10^8 and the numeral "-10" in 10^{-10} are examples of **exponents** or **powers.** The scientific notation, 3.84×10^8 m, contains an exponent (or power of ten) of eight.

CONVERTING TO SCIENTIFIC NOTATION

When a number is *divided* by ten, the decimal point moves one place to the *left;* when a number is *multiplied* by ten, the decimal point moves one place to the *right:*

$$\frac{384}{10} = 38.4 \quad and \quad (0.0024)(10) = 0.024$$

Multiplying *or* dividing a number by ten changes its value, but multiplying *and* dividing a number by ten leaves its value unchanged. Thus, to convert a number larger than ten to scientific notation, first divide by ten the number of times that gives a number between one and ten, then multiply by ten that same number of times. Here, for example, is the conversion of the number of seconds in a day into scientific notation:

$$\frac{86,400 \text{ s/day}}{(10)(10)(10)(10)} \times (10)(10)(10)(10)$$

$$\downarrow \qquad\qquad \downarrow$$

$$8.6400 \qquad 10^4 \qquad = 8.6400 \times 10^4 \text{ s/day}$$

The four divisions by ten move the decimal point four places to the left, and the four multiplications by ten are written as 10^4.

A number larger than ten can be quickly converted into scientific notation by moving the decimal point to the left enough places to give a number between one and ten, and multiplying by ten raised to the positive power that equals the number of places moved. For example:

$$96{,}485 = 9.6485 \times 10^4$$
4 places

To convert a number smaller than one into scientific notation, first multiply by ten as many times as needed to give a number between one and ten, then divide by ten that same number of times. Here, for example, is the conversion into scientific notation of the fraction of a day represented by one second:

$$0.000011574 \text{ day/s } (10)(10)(10)(10)(10) \times \frac{1}{(10)(10)(10)(10)(10)}$$

$$\downarrow \qquad\qquad\qquad \downarrow$$

$$1.1574 \qquad\qquad 10^{-5}$$
$$= 1.1574 \times 10^{-5} \text{ day/s}$$

To convert a number smaller than ten into scientific notation, move the decimal point to the right enough places to give a number between one and ten and multiply by ten raised to the negative power equal to the number of places moved.

With practice, these conversions can be carried out quite quickly. The key is to count the number of places the decimal point must be moved. Here are some additional examples:

$$5260 \text{ ft/mi} = 5.260 \times 10^3 \text{ ft/mi}$$
Move left three places

$$0.08206 \text{ L atm/mol K} = 8.206 \times 10^{-2} \text{ L atm/mol K}$$
Move right two places

$$384{,}000{,}000 \text{ mi} = 3.84 \times 10^8 \text{ mi}$$
Move left eight places

$$0.000 000 000 24 \text{ m} = 2.4 \times 10^{-10} \text{ m}$$
Move right ten places

WORKING WITH POWERS OF TEN

Chemistry students must be able to add, subtract, multiply, and divide numbers that are expressed in scientific notation. This is easily accomplished on a calculator by entering the numbers using power-of-ten notation

$(10^x$ function key). It is also useful to understand how these operations are carried out. Multiplication and division are handled in one fashion, but addition and subtraction are handled in another fashion.

To *multiply* two numbers expressed in scientific notation, multiply the values and *add* the exponents:

$$(2.450 \times 10^2)(1.680 \times 10^3) = (2.450)(1.680) \times 10^{(2+3)}$$
$$= 4.116 \times 10^5$$

Retain signs when adding exponents:

$$(2.450 \times 10^2)(1.680 \times 10^{-3}) = (2.450)(1.680) \times 10^{(2-3)}$$
$$= 4.116 \times 10^{-1}$$

To *divide* numbers expressed in power of ten notation, divide the values and *subtract* the exponent of the divisor from that of the number divided:

$$\frac{2.450 \times 10^2}{1.680 \times 10^3} = \frac{2.450}{1.680} \times 10^{(2-3)} = 1.458 \times 10^{-1}$$

As with multiplication, retain signs when subtracting exponents:

$$\frac{2.450 \times 10^2}{1.680 \times 10^{-3}} = \frac{2.450}{1.680} \times 10^{[2-(-3)]} = 1.458 \times 10^5$$

Addition and subtraction require a different approach. Consider, for example, the addition of these two numbers:

$$2.450 \times 10^2 + 1.680 \times 10^3 = ?$$

Using standard notation, this addition is as follows:

$$\begin{array}{r} 1680 \\ \underline{245.0} \\ 1925 \end{array}$$

To obtain this result in scientific notation, express both numbers using the *same* power of ten, most conveniently the largest such power:

$$2.450 \times 10^2 = 0.2450 \times 10^3$$

$$0.2450 \times 10^3 + 1.680 \times 10^3 = 1.925 \times 10^3$$

To add or subtract numbers expressed in power of ten notation, first express all numbers using the *same power of ten*. Then add or subtract the values and retain the same power of ten. Calculators automatically take care of these power-of-ten manipulations.

Subtraction may give a result that is smaller than one. When this happens, the convention is to change the exponent to make the numerical value between one and ten. Here, for example, is the difference between the atomic radii of oxygen atoms and fluorine atoms:

O atom   F atom   Difference
$$1.40 \times 10^{-10}\text{m} - 1.35 \times 10^{-10}\text{m} = 0.05 \times 10^{-10}\text{m}$$

Proper notation
$$= 5 \times 10^{-12}\text{m}$$

To express this result using a number between one and ten, we move the decimal point two places to the right and increase the negative power of ten by two.

This last example helps to show how the use of scientific notation simplifies the operations with the very small and very large numbers encountered in chemistry. Carried out in standard notation, this subtraction requires a whole lot of zeros:

$$\begin{array}{r} 0.000\,000\,000\,140 \text{ m} \\ -\underline{0.000\,000\,000\,135 \text{ m}} \\ 0.000\,000\,000\,005 \text{ m} \end{array}$$

Appendix B: Quantitative Observations

Chemists usually want to know not only *what* is happening but also to what extent (*how much*). Quantitative measurements have three equally important parts: a *numerical value,* appropriate *units,* and a *precision.* Any experimentally measured quantity *always includes all these parts.* Numerical value refers to the numbers. Units are what allow us to scale a numerical value appropriately. For example, 20 *miles* is a very different length than 20 *centimeters.* Precision requires further description.

PRECISION AND SIGNIFICANT FIGURES

The precision of a quantitative value is the degree of certainty with which it is known. For example, "about 20 miles" is a less precise statement than "21.5 miles." The basic rule for precision is that the number of digits in the numerical value expresses the precision of the measurement.

As an example of this rule, consider measuring the length of a table. After a quick measurement, you might estimate, "This table is 1.8 meters long." You have expressed the length as a two-digit number, meaning that you know the table to be longer than 1.7 meters but shorter than 1.9 meters. A more careful measurement with a tape measure might yield a result containing four digits, 1.826 meters. This means that the table is longer than 1.825 meters but shorter than 1.827 meters.

The number of digits in a numerical result is called its number of **significant figures.** Unless otherwise stated, the measurement is precise to within one unit in the last significant figure. Thus, a result reported as 1.826 meters has four significant figures. Unless additional information is given, the last digit (6) is uncertain by one unit: the length is $1.826 \pm .001$ meters. That is, the length falls somewhere between 1.825 meters and 1.827 meters.

When quantities are expressed in scientific notation, extra zeros always indicate extra precision. For instance, 3.840×10^8 meters means "not less than 3.839×10^8, nor more than 3.841×10^8 meters." On the other hand, 3.84×10^8 meters means "not less than 3.83×10^8, nor more than 3.85×10^8 meters."

The precision of a measurement depends on the quality of the measuring device used to obtain the measurement. Whereas very sensitive instruments yield measurements of high precision, less sensitive instruments yield results of lower precision. An electronic balance can determine the mass of 24 pennies to the nearest 0.0001 g, giving a measurement with six significant figures: 63.5465 g. A pan balance, on the other hand, can determine this same mass only to the nearest 0.01 g, giving a measurement with four significant figures: 63.55 g.

When a quantity fluctuates in value, the precision in its measurement is also limited by these fluctuations. For example, the distance between the Earth and its moon fluctuates with time, because the moon's orbit is elliptical rather than perfectly circular. The average distance is 3.844×10^5 km, but the actual distance varies between 3.564 and 4.067×10^5 km. Thus, the instantaneous distance between the Earth and its moon varies by 4.97×10^4 kilometers in the course of a month.

RELATIVE AND ABSOLUTE PRECISION

The degree of precision in a given experimental value can be expressed either as an **absolute** precision or as a **relative** precision. Absolute precision is the numerical uncertainty in the experimental value, and relative precision is the absolute precision divided by the experimental value. Absolute precisions have units, but relative precisions are always dimensionless ratios. Each of these ways of looking at precision has useful applications.

To illustrate these concepts, return to the example of table length. If we measure a table to be 1.826 meters long, the absolute precision of the measurement is ± 0.001 meters. The relative precision is the absolute precision divided by the length, $(0.001 \text{ m})/(1.826 \text{ m})$, or 5×10^{-4}. This relative precision can be expressed as $1/1826$, or 5×10^{-4}, or 0.05%, or 5 parts per 10,000. Each of these expressions is a ratio of two values that have the same units, so each ratio is dimensionless. Notice that because the absolute precision contains only one significant figure, this relative precision also contains only one significant figure.

When measurements are added or subtracted to compute a result, the *absolute* precision is more useful. When measurements are multiplied or divided to compute a result, the *relative* precision is more useful. As illustrations, return to the table example and consider the area and perimeter of the table top. Suppose that measurements show its width to be 3.20×10^{-1} meters. The absolute precision of this measurement is 0.001 meters. The relative precision of the width measurement is the appropriate ratio: $(0.001 \text{ m})/(3.20 \times 10^{-1} \text{ m})$, which is $1/320$, or 3×10^{-3}.

Now combine these measurements to determine the perimeter P and the area A. Perimeter is $P = 2L + 2W$: $P = 2(1.826) + 2(3.20 \times 10^{-1}) = 4.292$ m, and area is $A = (L)(W) = (1.826)(3.20 \times 10^{-1}) = 5.84 \times 10^{-1}$ m^2. How precisely is each of these computed values known? We find the precision of a computed result from the precisions of individual measurements by considering the largest possible variation in each individual measurement.

First, consider the table's perimeter. According to the precision of the measurements, the length might be as large as 1.827 m or as small as 1.825 m. The width might be as large as 0.321 m or as small as 0.319 m. Thus, the perimeter could be as large as $2(1.827) + 2(.321) = 4.296$ m, and it might be as small as $2(1.825) + 2(.319) = 4.288$ m. The precision in the perimeter is ± 0.004 m. This is the *sum of the absolute precisions* of the individual values: $0.001 + 0.001 + 0.001 + 0.001 = 0.004$.

Applying the same logic to the area reveals that it might be as large as $1.827 \times 0.321 = 5.87 \times 10^{-1}$ m^2 or as small as $1.825 \times 0.319 = 5.82 \times 10^{-1}$ m^2. The area is $5.84 \times 10^{-1} \pm 0.025 \times 10^{-1}$ m^2, giving an imprecision of $(0.025 \times 10^{-1})/(5.84 \times 10^{-1}) = 4 \times 10^{-3}$. This is the *sum of the relative precisions* of length (5×10^{-4}) and width (3×10^{-3}).

RULES FOR DETERMINING PRECISION

The precision of a composite result can always be determined by the type of analysis described above, but this procedure is tedious. Fortunately the outcome is always the same. For *addition and subtraction,* the *absolute* precision of the result is the sum of the *absolute* precisions of the individual values. For *multiplication and division,* the *relative* precision of the result is the sum of the *relative* precisions of the individual values.

Another example illustrates the application of these rules. To determine the density of a liquid, a student filled a 25-mL graduated cylinder until it contained 25.0 mL of liquid. The mass of the full cylinder was 47.5764 g. The student removed liquid until the cylinder contained 20.0 mL of liquid, whereupon its mass was 43.0464 g. Table B-1 on the next page summarizes the measurements, computations, and precisions involved in this example.

The absolute precision of each *measured quantity* is found directly from the measurement. The absolute precision of the mass of the cylinder is the precision of the balance, 0.0001 g. Each relative precision is the absolute precision divided by the measured value; thus, the relative precision of the mass of the cylinder is $1/475,000$. The precision of each *computed quantity* is determined by adding the precisions of quantities used in the calculation. The values in bold face are found directly. The volume transferred (5.0 mL) is obtained by *subtracting* two volumes, so

Table B-1
Measured and Derived Quantities and Precisions

Quantity	Value	Absolute Precision	Relative Precision
Measured Quantities			
Initial cylinder volume	25.0 mL	0.1 mL	1/250
Initial cylinder mass	47.5764 g	0.0001 g	1/475,000
Final cylinder volume	20.0 mL	0.1 mL	1/200
Final cylinder mass	43.0464 g	0.0001 g	1/430,000
Computed Quantities			
Volume transferred	5.0 mL	**0.2 mL**	2/50 = 1/25
Mass transferred	4.5300 g	**0.0002 g**	1/20,000
Liquid density	0.9060 g/mL	0.04	**1/25**

its *absolute* precision is the sum of the *absolute* precisions of the volumes (0.1 + 0.1 = 0.2 mL). Similar reasoning applies to the mass transferred. The density (0.9060 g/mL) is obtained by *dividing* mass by volume, so its *relative* precision is the sum of the *relative* precisions of transferred mass and transferred volume (1/25 + 1/20,000 = 1/25). Once the relative precision of the density has been calculated, it can be used to compute the absolute precision of the density: (1/25)(0.9060 g/mL) = ± 0.04 g/mL.

ROUNDING OFF

In the above example, the density might be as large as 0.94 g/mL or as small as 0.87 g/mL. It would be incorrect to write it as 0.9060 g/mL, because the use of four significant figures implies that the value is known to ±0.0001 g/mL. All nonsignificant figures must be eliminated, a process that is called **rounding off.** Because the uncertainty in the density is in the *second* decimal place, this result is rounded off to two decimals: 0.91 g/mL. This still overstates the precision, because 0.91 g/mL implies a precision of 0.01 g/mL, but this result is actually known only to ±0.04 g/mL. However, rounding off one more place to give 0.9 g/mL would imply an uncertainty of 0.1 g/mL, which is larger than the actual uncertainty. The convention is to round off until dropping one more digit would result in an uncertainty larger than the actual uncertainty.

When the digit following the last significant digit is 5 or greater, the remaining digit is increased by 1 unit. For example, 0.9060 becomes 0.91. If the digit following the last significant digit is less than 5, the remaining digit remains unchanged. For example, 0.9045 rounded to 2 significant figures is 0.90.

SHORTCUTS TO PRECISION

You can spend much effort figuring out the appropriate number of significant figures in a result. Every time a scientist makes a measurement, precision and significant figures are a concern, but scientists are more interested in what experiments reveal than they are in significant figures. Precision may be extremely important, for example when determining the level of a particular carcinogen in a sample of ground water. Because the amount of carcinogen may be critical to human health, the precision of this measurement would have to be carefully stated. Precision is not so important to a chemist whose goal is to prepare new compounds. The chemist wishes to achieve a high yield but is not concerned about whether a yield is 90% or 90.05%.

The following simple guidelines are sufficient to determine the appropriate number of significant figures in General Chemistry:

1. To determine the number of significant figures in an individual measurement, count the number of digits, from left to right, beginning with the first one that is non-zero.

305	3 significant figures
1.00	3 significant figures
0.020	2 significant figures

2. When adding or subtracting, set the number of *decimal places* in the answer equal to the number of *decimal places* in the number with the fewest places. The number of significant figures is irrelevant:

	0.1|2	2 decimal places
	1.6|	**1 decimal place**
	11.4|90	3 decimal places
Sum:	13.2|	**1 decimal place**

3. When multiplying or dividing, set the number of *significant figures* in the answer equal to that of the quantity with the fewest *significant figures*. The number of decimal places is irrelevant:

$$1.365 \times 2.63 \times 0.33 = 1.2$$

Significant figures:	4	3	**2**	**2**

The last example is one where the simplified procedure gives a different result than the more elaborate one. One number (0.33) has a relative precision of 1/33, so the result, 1.1847, could be rounded to 1.18 (1 part in 118) rather than to 1.2 (1 part in 12). If the importance of the result is its *quantitative value*, 1.18 would be the more appropriate number to report; if its importance is in its *qualitative* significance, report 1.2. (But don't spend a lot of time worrying about it: either way of reporting is legitimate! What is *not* legitimate is to report this result as **1.1847.**)

4. Calculators do not necessarily give results with the correct number of significant figures. They automatically drop trailing zeroes even when they are significant (try multiplying or adding 3.00 and 5.00 on your calculator) and they carry extra decimal places even when they are insignificant (try dividing 5.00 by 3.00 on your calculator). *Never believe the number of significant figures on your calculator!*

Appendix C: Ionization Energies and Electron Affinities of the First 36 Elements

Z	Symbol	EA	IE_1	IE_2	IE_3	$n^\dagger$
1	H	−72.8	1312	—	—	
2	He	>0*	2372	5250	—	1
3	Li	−59.7	520.2	7298	11,815	
4	Be	>0	899.4	1757	14,848	
5	B	−26.8	800.6	2427	3660	
6	C	−121.9	1086	2353	4620	
7	N	>0	1402	2856	4578	
8	O	−141.1	1314	3388	5300	
9	F	−328.0	1681	3374	6050	2
10	Ne	>0	2081	3952	6122	
11	Na	−52.9	495.6	4562	6912	
12	Mg	>0	737.7	1451	7733	
13	Al	−42.7	577.6	1817	2745	
14	Si	−133.6	786.4	1577	3232	
15	P	−72.0	1012	1908	2912	
16	S	−200.4	999.6	2251	3357	3
17	Cl	−348.8	1251	2297	3822	
18	Ar	>0	1520	2666	3931	
19	K	−48.4	418.8	3051	4420	
20	Ca	−2.4	589.8	1145	4912	
21	Sc	−18.2	633	1235	2389	
22	Ti	−7.7	658	1310	2653	
23	V	−50.8	650	1414	2828	
24	Cr	−64.4	652.8	1591	2987	
25	Mn	>0	717.4	1509	3248	
26	Fe	−14.6	763	1561	2957	
27	Co	−63.9	758	1646	3232	
28	Ni	−111.6	736.7	1753	3396	4
29	Cu	−119.2	745.4	1958	3554	
30	Zn	>0	906.4	1733	3833	
31	Ga	−28.9	578.8	1979	2963	
32	Ge	−119	762.1	1537	3302	
33	As	−78	947	1798	2736	
34	Se	−195.0	940.9	2045	2974	
35	Br	−324.6	1140	2103	3500	
36	Kr	>0	1351	2350	3565	

All values are in kJ/mol.

* A value >0 means that the anion is unstable, so its electron affinity cannot be experimentally determined.

† *n* is the principal quantum number of the electron whose ionization energy is listed.

Source: *CRC Handbook of Chemistry and Physics, 82nd Edition*, David R. Lide, editor-in-chief, CRC Press LLC, Boca Raton, FL (2001), pp. **10**−47, 48, 175.

Appendix D:
Standard Thermodynamic Functions

$T = 298.15$ K; $P = 1.000$ bar; Aqueous species at 1.000 M

Substance	ΔH_f^o (kJ/mol)	ΔG_f^o (kJ/mol)	S^o (J/mol K)	Substance	ΔH_f^o (kJ/mol)	ΔG_f^o (kJ/mol)	S^o (J/mol K)
Aluminum				**Carbon**			
Al (s)	0	0	28.3	C (s, graphite)	0	0	5.7
AlCl$_3$ (s)	−704.2	−628.8	109.3	C (s, diamond)	1.9	2.9	2.4
Al$_2$O$_3$ (s)	−1675.7	−1582.3	50.9	CH$_4$ (g)	−74.6	−50.5	186.3
Antimony				C$_2$H$_2$ (g)	227.4	209.9	200.9
Sb (s)	0	0	45.7	C$_2$H$_4$ (g)	52.4	68.4	219.3
Sb$_4$O$_6$ (s)	−1417.1	−1253.0	246.0	C$_2$H$_6$ (g)	−84.0	−32.0	229.2
Argon				C$_3$H$_6$ (g)	20.0	74.62	226.9
Ar (g)	0	0	154.843	C$_3$H$_8$ (g)	−103.8	−23.4	270.3
Arsenic				C$_6$H$_6$ (l)	49.1	124.5	173.4
As (s)	0	0	35.1	CO (g)	−110.5	−137.2	197.7
As$_2$O$_5$ (s)	−924.9	−782.3	105.4	CO$_2$ (g)	−393.5	−394.4	213.8
AsCl$_3$ (l)	−305.0	−259.4	216.3	HCN (g)	135.1	124.7	201.8
Barium				CS$_2$ (l)	89.0	64.6	151.3
Ba (s)	0	0	62.5	CCl$_4$ (l)	−128.2	−65.21	216.40
BaO (s)	−548.0	−520.3	72.1	CHCl$_3$ (l)	−134.1	−73.7	201.7
BaCO$_3$ (s)	−1213.0	−1134.4	112.1	CH$_3$CHO (g)	−166.2	−127.6	263.8
BaCl$_2$ (s)	−855.0	−806.7	123.7	CH$_3$CO$_2$H (l)	−484.3	−389.9	159.8
BaSO$_4$ (s)	−1473.2	−1362.2	132.2	CH$_3$OH (l)	−239.2	−166.6	126.8
Beryllium				CH$_3$CH$_2$OH (l)	−277.6	−174.8	160.7
Be (s)	0	0	9.5	CH$_3$CN (l)	40.6	86.5	149.6
Be(OH)$_2$ (s)	−902.5	−815.0	51.9	**Chlorine**			
Boron				Cl (g)	121.3	105.3	165.2
B (s)	0	0	5.9	Cl$_2$ (g)	0	0	223.1
B$_2$O$_3$ (s)	−1273.5	−1194.3	54.0	Cl$^-$ (aq)	−167.1	−131.0	56.5
H$_3$BO$_3$ (s)	−1094.3	−968.9	90.0	HCl (g)	−92.3	−95.3	186.9
BCl$_3$ (g)	−403.8	−388.7	290.1	ClO$^-$ (aq)	−107.1	−36.8	42
Bromine				ClO$_2^-$ (aq)	−67	17	101
Br (g)	111.9	82.4	175.0	ClO$_3^-$ (aq)	−104	−3	162
Br$_2$ (l)	0	0	152.2	ClO$_4^-$ (aq)	−128.1	−8.52	184.0
Br$_2$ (g)	30.9	3.1	245.5	ClO (g)	101.8	98.1	226.6
Br$^-$ (aq)	−121.4	−104.0	82.4	ClO$_2$ (g)	102.5	120.5	256.8
BrO$^-$ (aq)	−94.1	−33.4	42	**Chromium**			
HBr (g)	−36.3	−53.4	198.7	Cr (s)	0	0	23.8
Cadmium				Cr$_2$O$_3$ (s)	−1139.7	−1058.1	81.2
Cd (s)	0	0	51.8	CrCl$_3$ (s)	−556.5	−486.1	123.0
CdO (s)	−258.4	−228.7	54.8	**Cobalt**			
CdCl$_2$ (s)	−391.5	−343.9	115.3	Co (s)	0	0	30.0
Calcium				CoO (s)	−237.9	−214.2	53.0
Ca (s)	0	0	41.6	CoCl$_2$ (s)	−312.5	−269.8	109.2
Ca^{2+} (aq)	−543.0	−553.6	−56.2	**Copper**			
CaO (s)	−634.9	−603.3	38.1	Cu (s)	0	0	33.2
Ca(OH)$_2$ (s)	−985.2	−897.5	83.4	CuO (s)	−157.3	−129.7	42.6
CaCO$_3$ (s)	−1207.6	−1129.1	91.7	CuS (s)	−53.1	−53.6	66.5
CaCl$_2$ (s)	−795.4	−748.8	108.4	CuCl$_2$ (s)	−220.1	−175.7	108.1
CaF$_2$ (s)	−1228	−1176	68.5	CuCl (s)	−137.2	−119.9	86.2
CaH$_2$ (s)	−185.1	−142.5	41.4	CuBr (s)	−104.6	−100.8	96.1
CaS (s)	−482.4	−477.4	56.5	CuI (s)	−67.8	−69.5	96.7
CaSO$_4$ (s)	−1434.5	−1322.0	106.5	CuSO$_4$ (s)	−771.4	−662.2	109.2

Substance	ΔH_f^o (kJ/mol)	ΔG_f^o (kJ/mol)	S^o (J/mol K)	Substance	ΔH_f^o (kJ/mol)	ΔG_f^o (kJ/mol)	S^o (J/mol K)
Fluorine				**Manganese**			
F (g)	79.4	62.3	158.8	Mn (s)	0	0	32.0
F_2 (g)	0	0	202.8	MnO (s)	−385.2	−362.9	59.7
F^- (aq)	−335.4	−278.79	−13.8	MnO_2 (s)	−520.0	−465.1	53.1
HF (g)	−273.3	−275.4	173.8	$MnCl_2$ (s)	−481.3	−440.5	118.2
Germanium				$MnCO_3$ (s)	−894.1	−816.7	85.8
Ge (s)	0	0	31.1	$MnSO_4$ (s)	−1065.25	−957.36	112.1
$GeCl_4$ (g)	−495.8	−457.3	347.7	**Mercury**			
GeO_2 (s)	−580.0	−521.4	39.7	Hg (l)	0	0	75.9
Gold				Hg (g)	61.4	31.8	175.0
Au (s)	0	0	47.4	HgO (s)	−90.8	−58.5	70.3
Helium				$HgCl_2$ (s)	−224.3	−178.6	146.0
He (g)	0	0	126.153	Hg_2Cl_2 (s)	−265.4	−210.7	191.6
Hydrogen				HgS (s)	−58.2	−50.6	82.4
H_2 (g)	0	0	130.680	**Neon**			
H_2O (l)	−285.83	−237.1	69.95	Ne (g)	0	0	146.328
H_2O (g)	−241.83	−228.72	188.835	**Nickel**			
H_2O_2 (l)	−187.8	−120.4	109.6	Ni (s)	0	0	29.9
H^+ (aq)	0	0	0	NiO (s)	−239.7	−211.7	37.99
H_3O^+ (aq)	−285.83	−237.1	69.95	$NiCl_2$ (s)	−305.3	−259.0	97.7
Iodine				**Nitrogen**			
I_2 (s)	0	0	116.1	N_2 (g)	0	0	191.61
I_2 (g)	62.4	19.3	260.7	NH_3 (g)	−45.9	−16.4	192.8
I^- (aq)	−56.78	−51.57	106.5	NH_3 (aq)	−80.29	−26.50	111.3
HI (g)	26.5	1.7	206.6	NH_4^+ (aq)	−133.3	−79.31	111.2
Iron				N_2H_4 (l)	50.6	149.3	121.2
Fe (s)	0	0	27.3	NO (g)	91.3	87.6	210.8
Fe_2O_3 (s)	−824.2	−742.2	87.4	NO_2 (g)	33.2	51.3	240.1
FeO (s)	−272.0	−251.4	60.8	N_2O (g)	81.6	103.7	220.0
Fe_3O_4 (s)	−1118.4	−1015.4	146.4	N_2O_4 (g)	11.1	99.8	304.4
$Fe(OH)_3$ (s)	−823.0	−696.5	106.7	N_2O_5 (g)	13.3	117.1	355.7
$FeCl_2$ (s)	−341.8	−302.3	118.0	NH_4Cl (s)	−314.4	−202.9	94.6
$FeCl_3$ (s)	−399.5	−334.0	142.3	NH_4NO_3 (s)	−365.6	−183.9	151.1
$FeSO_4$ (s)	−928.4	−820.8	107.5	$(NH_2)_2CO$ (s)	−333.1	−198	105
FeS (s)	−100.0	−100.4	60.3	HNO_3 (g)	−133.9	−73.5	266.9
FeS_2 (s)	−178.2	−166.9	52.9	**Oxygen**			
Krypton				O_2 (g)	0	0	205.152
Kr (g)	0	0	164.085	O_3 (g)	142.7	163.2	238.9
Lead				OH^- (aq)	−230.0	−157.244	−10.9
Pb (s)	0	0	64.8	**Phosphorus**			
$PbCl_2$ (s)	−359.4	−314.1	136.0	P (s, white)	0	0	41.1
PbO (s)	−217.3	−187.9	68.7	P (s, red)	−17.6	−12.1	22.8
PbO_2 (s)	−277.4	−217.3	68.6	P_4 (g)	58.9	24.4	280.0
PbS (s)	−100.4	−98.7	91.2	PH_3 (g)	5.4	13.4	210.2
Lithium				P_4O_{10} (s)	−2984.0	−2697.7	228.86
Li (s)	0	0	29.1	H_3PO_4	−1271.7	−1123.6	150.8
Li^+ (aq)	−278.47	−293.31	12.2	PCl_5 (g)	−374.9	−305.0	364.6
Li_2O (s)	−597.9	−561.2	37.6	PCl_3 (g)	−287.0	−267.8	311.8
LiOH (s)	−484.9	−439.0	42.8	**Potassium**			
LiCl (s)	−408.6	−384.4	59.3	K (s)	0	0	64.7
Magnesium				K^+ (aq)	−252.1	−283.27	101.2
Mg (s)	0	0	32.67	KO_2 (s)	−284.9	−239.4	116.7
Mg^{2+} (aq)	−467.0	−454.8	−137	KOH (s)	−424.6	−378.9	78.9
MgO (s)	−601.6	−569.3	27.0	KCl (s)	−436.5	−408.5	82.6
$Mg(OH)_2$ (s)	−924.5	−833.5	63.2	$KClO_3$ (s)	−397.7	−296.3	143.1
$MgCl_2$ (s)	−641.3	−591.8	89.6	KI (s)	−327.9	−324.9	106.3
MgF_2 (s)	−1124.2	−1071.1	57.2	**Selenium**			
$MgCO_3$ (s)	−1095.8	−1012.1	65.7	Se (s)	0	0	42.4
$MgSO_4$ (s)	−1284.9	−1170.6	91.6	H_2Se (g)	29.7	15.9	219.0

Substance	ΔH_f° (kJ/mol)	ΔG_f° (kJ/mol)	S° (J/mol K)	Substance	ΔH_f° (kJ/mol)	ΔG_f° (kJ/mol)	S° (J/mol K)
Silicon				**Sulfur (cont.)**			
Si (s)	0	0	18.8	SO_3 (g)	−395.7	−371.1	256.8
SiH_4 (g)	34.3	56.9	204.6	H_2SO_4 (l)	−814	−690.0	156.9
SiO_2 (s)	−910.7	−856.3	41.5	SO_4^{2-} (aq)	−909.3	−774.53	18.5
$SiCl_4$ (l)	−687.0	−619.8	239.7	SF_6 (g)	−1220.5	−1116.5	291.5
SiC (s)	−65.3	−62.8	16.6	**Tellurium**			
Silver				Te (s)	0	0	49.7
Ag (s)	0	0	42.6	TeO_2 (s)	−322.6	−270.3	79.5
Ag^+ (aq)	105.8	77.107	73.4	**Tin**			
Ag_2O (s)	−31.1	−11.2	121.3	Sn(s, white)	0	0	51.2
AgCl (s)	−127.0	−109.8	96.3	Sn(s, gray)	−2.1	0.13	44.1
AgBr (s)	−100.4	−96.9	107.1	SnO (s)	−280.7	−251.9	57.2
$AgNO_3$ (s)	−124.4	−33.4	140.9	SnO_2 (s)	−577.6	−515.8	49.0
Sodium				$SnCl_4$ (s)	−511.3	−440.1	258.6
Na (s)	0	0	51.3	**Titanium**			
Na^+ (aq)	−240.3	−261.905	58.5	Ti (s)	0	0	30.7
Na_2O (s)	−414.2	−375.5	75.1	$TiCl_4$ (l)	−804.2	−737.2	252.3
NaOH (s)	−425.6	−379.5	64.5	$TiCl_4$ (g)	−763.2	−726.3	353.2
NaF (s)	−576.6	−546.3	51.1	TiO_2 (s)	−944.0	−888.8	50.6
NaCl (s)	−411.2	−384.1	72.1	**Uranium**			
NaBr (s)	−361.1	−349.0	86.8	U (s)	0	0	50.2
NaI (s)	−287.8	−286.1	98.5	UO_2 (s)	−1085.0	−1031.8	77.0
Na_2CO_3 (s)	−1130.7	−1044.4	135.0	UF_6 (g)	−2147.4	−2063.7	377.9
Strontium				**Xenon**			
Sr (s)	0	0	55.0	Xe (g)	0	0	169.685
SrO (s)	−592.0	−561.9	54.4	XeF_4 (s)	−261.5	−138	316
$SrCl_2$ (s)	−828.9	−781.1	114.9	**Zinc**			
Sulfur				Zn (s)	0	0	41.63
S_8 (s)	0	0	31.80	ZnO (s)	−350.5	−320.5	43.7
S_8 (g)	102.30	49.63	430.23	$ZnCl_2$ (s)	−415.1	−369.4	111.5
H_2S (g)	−20.6	−33.4	205.8	ZnS (s)	−206.0	−201.3	57.7
SO_2 (g)	−296.8	−300.1	248.2				

Main Source: *CRC Handbook of Chemistry and Physics, 82nd Edition,* David R. Lide, editor-in-chief, CRC Press LLC, Boca Raton, FL (2001), pp. **5**-1 through **5**-60.

Appendix E: Equilibrium Constants

K_b VALUES

Name	Formula	pK_b	K_b
Ammonia	NH_3	4.75	1.8×10^{-5}
Aniline	$C_6H_5NH_2$	9.13	7.4×10^{-10}
Diethylamine	$(C_2H_5)_2NH$	3.16	6.9×10^{-4}
Dimethylamine	$(CH_3)_2NH$	3.27	5.4×10^{-4}
Ethylamine	$C_2H_5NH_2$	3.35	4.5×10^{-4}
Ethylenediamine	$(CH_2NH_2)_2$	4.08	8.3×10^{-5}
Hydrazine	H_2NNH_2	5.89	1.3×10^{-6}
Hydroxylamine	$HONH_2$	8.06	8.7×10^{-9}
Methylamine	CH_3NH_2	3.34	4.6×10^{-4}
Pyridine	C_5H_5N	8.77	1.7×10^{-9}
Triethylamine	$(C_2H_5)_3N$	3.25	5.6×10^{-4}
Trimethylamine	$(CH_3)_3N$	4.19	6.5×10^{-5}
Urea	H_2NCONH_2	13.82	1.5×10^{-14}

K_a VALUES

Name	Formula	pK_a	K_a
Oxoacids			
Arsenic	H_3AsO_4	2.26	5.5×10^{-3}
Boric	H_3BO_3	9.27	5.4×10^{-10}
Carbonic	H_2CO_3	6.35	4.5×10^{-7}
K_{a2}	HCO_3^-	10.33	4.7×10^{-11}
Chlorous	$HClO_2$	1.94	1.1×10^{-2}
Chromic	H_2CrO_4	0.74	1.8×10^{-1}
K_{a2}	$HCrO_4^-$	6.49	3.2×10^{-7}
Hypobromous	$HBrO$	8.55	2.8×10^{-9}
Hypochlorous	$HClO$	7.40	4.0×10^{-8}
Hypoiodous	HIO	10.5	3.2×10^{-11}
Iodic	HIO_3	0.78	1.7×10^{-1}
Nitrous	HNO_2	3.25	5.6×10^{-4}
Paraperiodic	H_5IO_6	1.55	2.8×10^{-2}
Periodic	HIO_4	1.64	7.3×10^{-2}
Phosphoric	H_3PO_4	2.16	6.9×10^{-3}
K_{a2}	$H_2PO_4^-$	7.21	6.2×10^{-8}
K_{a3}	HPO_4^{2-}	12.32	4.8×10^{-13}
Phosphorous	H_3PO_3	1.3	5.0×10^{-2}
K_{a2}	$H_2PO_3^-$	6.70	2.0×10^{-7}
Sulfuric	H_2SO_4	strong	
K_{a2}	HSO_4^-	1.99	1.0×10^{-2}
Sulfurous	H_2SO_3	1.85	1.4×10^{-2}
K_{a2}	HSO_3^-	7.20	6.3×10^{-8}

Name	Formula	pK_a	K_a
Carboxylic acids			
Acetic	CH_3CO_2H	4.75	1.8×10^{-5}
Benzoic	$C_6H_5CO_2H$	4.20	6.3×10^{-5}
Chloroacetic	$ClCH_2CO_2H$	2.87	1.4×10^{-3}
Formic	HCO_2H	3.75	1.8×10^{-4}
Oxalic	$(CO_2H)_2$	1.25	5.6×10^{-2}
K_{a2}	$HO_2CCO_2^-$	3.81	1.5×10^{-4}
Propanoic	$C_2H_5CO_2H$	4.87	1.3×10^{-5}
Trichloroacetic	Cl_3CCO_2H	0.66	2.2×10^{-1}
Other acids			
Cyanic	$HCNO$	3.46	3.5×10^{-4}
Hydrazoic	HN_3	4.6	2.5×10^{-5}
Hydrocyanic	HCN	9.21	6.2×10^{-10}
Hydrofluoric	HF	3.20	6.3×10^{-4}
Hydrogen peroxide	H_2O_2	11.62	2.4×10^{-12}
Hydrogen sulfide	H_2S	7.05	8.9×10^{-8}
K_{a2}	HS^-	19	1×10^{-19}
Phenol	C_6H_5OH	9.99	1.0×10^{-10}

K_{sp} VALUES, SALTS AT 25 °C

Formula	pK_{sp}	K_{sp}
Bromides		
AgBr	12.27	5.35×10^{-13}
CuBr	8.20	6.3×10^{-9}
Carbonates		
Ag_2CO_3	11.07	8.46×10^{-12}
$BaCO_3$	8.59	2.6×10^{-9}
$CaCO_3$	8.47	3.36×10^{-9}
$CdCO_3$	12.00	1.0×10^{-12}
$CoCO_3$	12.85	1.4×10^{-13}
$CuCO_3$	9.85	1.4×10^{-10}
$FeCO_3$	10.51	3.1×10^{-11}
$MgCO_3$	5.17	6.8×10^{-6}
$MnCO_3$	10.66	2.2×10^{-11}
$NiCO_3$	6.84	1.4×10^{-7}
$PbCO_3$	13.13	7.4×10^{-14}
$SrCO_3$	9.25	5.6×10^{-10}
$ZnCO_3$	9.82	1.5×10^{-10}
Chlorides		
AgCl	9.74	1.8×10^{-10}
CuCl	6.76	1.7×10^{-7}
$PbCl_2$	4.77	1.7×10^{-5}
Fluorides		
BaF_2	6.74	1.8×10^{-7}
CaF_2	10.46	3.45×10^{-11}
PbF_2	7.48	3.3×10^{-8}
Hydroxides		
$Al(OH)_3$	32.34	4.6×10^{-33}
$Ba(OH)_2$	3.59	2.55×10^{-4}
$Ca(OH)_2$	5.30	5.0×10^{-6}
$Cd(OH)_2$	14.14	7.2×10^{-15}
$Co(OH)_2$	14.23	5.9×10^{-15}

Formula	pK_{sp}	K_{sp}
Hydroxides (cont.)		
$Co(OH)_3$	43.8	1.6×10^{-44}
$Cu(OH)_2$	14.96	1.1×10^{-15}
$Fe(OH)_2$	16.32	4.8×10^{-17}
$Fe(OH)_3$	38.55	2.8×10^{-39}
$Mg(OH)_2$	11.25	5.6×10^{-12}
$Ni(OH)_2$	15.26	5.5×10^{-16}
$Pb(OH)_2$	19.85	1.4×10^{-20}
$Sn(OH)_2$	26.26	5.5×10^{-27}
$Zn(OH)_2$	16.52	3×10^{-17}
Phosphates		
$Ba_3(PO_4)_2$	22.47	3.4×10^{-23}
$Ca_3(PO_4)_2$	32.70	2.0×10^{-33}
Sulfates		
Ag_2SO_4	4.92	1.2×10^{-5}
$BaSO_4$	9.96	1.1×10^{-10}
$CaSO_4$	4.31	4.9×10^{-5}
$PbSO_4$	7.60	2.5×10^{-8}
Sulfides		
Ag_2S	49.20	6.3×10^{-50}
CoS	20.40	4.0×10^{-21}
Cu_2S	47.60	2.5×10^{-48}
CuS	25.20	6.3×10^{-26}
FeS	18.80	1.6×10^{-19}
Hg_2S	47.00	1.0×10^{-47}
HgS	52.40	4.0×10^{-53}
MnS	13.34	4.6×10^{-14}
NiS	20.97	1.1×10^{-21}
PbS	28.05	8.9×10^{-29}
SnS	27.49	3.2×10^{-28}
ZnS	24.53	3.0×10^{-25}

Main Source: *CRC Handbook of Chemistry and Physics, 82nd Edition*, David R. Lide, editor-in-chief, CRC Press LLC, Boca Raton, FL (2001), pp. **8**-44 through **8**-56 and **8**-117 through **8**-120.

Appendix F:
Standard Reduction Potentials, $E°$

$T = 298.15$ K, and $P = 1.00$ atm.
Listed alphabetically by element; all values are in volts.

All ionic species are aqueous. H_2O is liquid.
All other neutrals are solids unless otherwise specified.

Aluminum

$Al^{3+} + 3\,e^- \rightleftharpoons Al$	-1.662
$Al(OH)_3 + 3\,e^- \rightleftharpoons Al + 3\,OH^-$	-2.31

Arsenic

$HAsO_2\,(aq) + 3\,H_3O^+ + 3\,e^- \rightleftharpoons As + 5\,H_2O$	$+0.248$

Barium

$Ba^{2+} + 2\,e^- \rightleftharpoons Ba$	-2.912
$Ba(OH)_2 + 2\,e^- \rightleftharpoons Ba + 2\,OH^-$	-2.99

Boron

$H_3BO_3\,(aq) + 3\,H_3O^+ + 3\,e^- \rightleftharpoons B + 6\,H_2O$	-0.8698

Bromine

$Br_2\,(aq) + 2\,e^- \rightleftharpoons 2\,Br^-$	$+1.0873$
$Br_2\,(l) + 2\,e^- \rightleftharpoons 2\,Br^-$	$+1.066$
$HBrO\,(aq) + H_3O^+ + 2\,e^- \rightleftharpoons Br^- + 2\,H_2O$	$+1.331$
$BrO_3^- + 6\,H_3O^+ + 5\,e^- \rightleftharpoons Br_2\,(l) + 3\,H_2O$	$+1.423$

Cadmium

$Cd^{2+} + 2\,e^- \rightleftharpoons Cd$	-0.403
$Cd(OH)_2 + 2\,e^- \rightleftharpoons Cd + 2\,OH^-$	-0.809

Calcium

$Ca^{2+} + 2\,e^- \rightleftharpoons Ca$	-2.868
$Ca(OH)_2 + 2\,e^- \rightleftharpoons Ca + 2\,OH^-$	-3.02

Carbon

$CO_2\,(g) + 2\,H_3O^+ + 2\,e^- \rightleftharpoons HCO_2H + 2\,H_2O$	-0.199
$2\,CO_2\,(g) + 2\,H_3O^+ + 2\,e^- \rightleftharpoons$	
$\qquad\qquad H_2C_2O_4 + 2\,H_2O$	-0.49
$CNO^- + H_2O + 2\,e^- \rightleftharpoons CN^- + 2\,OH^-$	-0.970
$p\text{–Benzoquinone} + H_3O^+ + 2\,e^- \rightleftharpoons$	
$\qquad\qquad \text{Hydroquinone} + H_2O$	$+0.6992$

Cesium

$Cs^+ + e^- \rightleftharpoons Cs$	-3.026

Chlorine

$Cl_2\,(g) + 2\,e^- \rightleftharpoons 2\,Cl^-$	$+1.35827$
$2\,HClO\,(aq) + 2\,H_3O^+ + 2\,e^- \rightleftharpoons Cl_2\,(g) + 2\,H_2O$	$+1.611$
$ClO^- + H_2O + 2\,e^- \rightleftharpoons Cl^- + 2\,OH^-$	$+0.81$
$ClO_3^- + 6\,H_3O^+ + 6\,e^- \rightleftharpoons Cl^- + 9\,H_2O$	$+1.451$
$ClO_4^- + 2\,H_3O^+ + 2\,e^- \rightleftharpoons ClO_3^- + 3\,H_2O$	$+1.189$
$2\,ClO_4^- + 16\,H_3O^+ + 14\,e^- \rightleftharpoons$	
$\qquad\qquad Cl_2\,(g) + 24\,H_2O$	$+1.39$
$ClO_4^- + 6\,H_3O^+ + 6\,e^- \rightleftharpoons ClO^- + 9\,H_2O$	$+1.36$
$ClO_4^- + 8\,H_3O^+ + 8\,e^- \rightleftharpoons Cl^- + 12\,H_2O$	$+1.389$

Chromium

$Cr^{3+} + 3\,e^- \rightleftharpoons Cr$	-0.744
$Cr^{3+} + e^- \rightleftharpoons Cr^{2+}$	-0.407
$Cr^{2+} + 2\,e^- \rightleftharpoons Cr$	-0.913
$Cr_2O_7^{2-} + 14\,H_3O^+ + 6\,e^- \rightleftharpoons 2\,Cr^{3+} + 21\,H_2O$	$+1.232$
$Cr(OH)_3 + 3\,e^- \rightleftharpoons Cr + 3\,OH^-$	-1.48

Cobalt

$Co^{3+} + e^- \rightleftharpoons Co^{2+}$	$+1.92$
$Co^{2+} + 2\,e^- \rightleftharpoons Co$	-0.28
$[Co(NH_3)_6]^{3+} + e^- \rightleftharpoons [Co(NH_3)_6]^{2+}$	-0.108

Copper

$Cu^{2+} + 2\,e^- \rightleftharpoons Cu$	$+0.3419$
$Cu^{2+} + e^- \rightleftharpoons Cu^+$	$+0.153$
$Cu^+ + e^- \rightleftharpoons Cu$	$+0.521$
$Cu(OH)_2 + 2\,e^- \rightleftharpoons Cu + 2\,OH^-$	-0.222

Fluorine

$F_2\,(g) + 2\,e^- \rightleftharpoons 2\,F^-$	$+2.866$

Gold

$Au^{3+} + 3\,e^- \rightleftharpoons Au$	$+1.498$
$Au(CN)_2^- + e^- \rightleftharpoons Au + 2\,CN^-$	-0.60

Hydrogen

$2\,H_3O^+ + 2\,e^- \rightleftharpoons H_2\,(g) + 2\,H_2O$	0 (by definition)
$2\,H_2O + 2\,e^- \rightleftharpoons H_2\,(g) + 2\,OH^-$	-0.828
$H_2O_2\,(aq) + 2\,H_3O^+ + 2\,e^- \rightleftharpoons 2\,H_2O$	$+1.776$

Iodine

$I_2 + 2\,e^- \rightleftharpoons 2\,I^-$	$+0.5355$
$I_3^- + 2\,e^- \rightleftharpoons 3\,I^-$	$+0.536$
$2\,IO_3^- + 12\,H_3O^+ + 10\,e^- \rightleftharpoons I_2 + 18\,H_2O$	$+1.195$
$H_5IO_6 + H_3O^+ + 2\,e^- \rightleftharpoons IO_3^- + 4\,H_2O$	$+1.601$

Iron

$Fe^{3+} + 3\,e^- \rightleftharpoons Fe$	-0.037
$Fe^{2+} + 2\,e^- \rightleftharpoons Fe$	-0.447
$Fe^{3+} + e^- \rightleftharpoons Fe^{2+}$	$+0.771$
$Fe(CN)_6^{3-} + e^- \rightleftharpoons Fe(CN)_6^{4-}$	$+0.358$

Lead

$Pb^{2+} + 2\,e^- \rightleftharpoons Pb$	-0.1262
$PbO_2 + SO_4^{2-} + 4\,H_3O^+ + 2\,e^- \rightleftharpoons$	
$\qquad\qquad PbSO_4 + 6\,H_2O$	$+1.6913$
$PbO_2 + 4\,H_3O^+ + 2\,e^- \rightleftharpoons Pb^{2+} + 6\,H_2O$	$+1.455$
$PbSO_4 + 2\,e^- \rightleftharpoons Pb + SO_4^{2-}$	-0.3588

Lithium

$Li^+ + e^- \rightleftharpoons Li$	-3.0401

Magnesium

$Mg^{2+} + 2\,e^- \rightleftharpoons Mg$	-2.37
$Mg(OH)_2 + 2\,e^- \rightleftharpoons Mg + 2\,OH^-$	-2.69

Manganese

$Mn^{2+} + 2\,e^- \rightleftharpoons Mn$	-1.185
$MnO_4^- + 4\,H_3O^+ + 3\,e^- \rightleftharpoons MnO_2 + 6\,H_2O$	$+1.679$
$MnO_4^- + 2\,H_2O + 3\,e^- \rightleftharpoons MnO_2 + 4\,OH^-$	$+0.595$
$MnO_4^- + 8\,H_3O^+ + 5\,e^- \rightleftharpoons Mn^{2+} + 12\,H_2O$	$+1.507$
$MnO_2 + 4\,H_3O^+ + 2\,e^- \rightleftharpoons Mn^{2+} + 6\,H_2O$	$+1.224$

Mercury

$Hg^{2+} + 2\,e^- \rightleftharpoons Hg\,(l)$	$+0.851$
$Hg_2Cl_2 + 2\,e^- \rightleftharpoons 2\,Hg\,(l) + 2\,Cl^-$	$+0.26808$
$2\,Hg^{2+} + 2\,e^- \rightleftharpoons Hg_2^{2+}$	$+0.920$
$Hg_2^{2+} + 2\,e^- \rightleftharpoons 2\,Hg\,(l)$	$+0.7973$

Nickel

$Ni^{2+} + 2\,e^- \rightleftharpoons Ni$	-0.257
$Ni(OH)_2 + 2\,e^- \rightleftharpoons Ni + 2\,OH^-$	$+0.72$
$NiO_2 + 4\,H_3O^+ + 2\,e^- \rightleftharpoons Ni^{2+} + 6\,H_2O$	$+1.678$

Nitrogen

$N_2(g) + 8 H_3O^+ + 6 e^- \rightleftharpoons 2 NH_4^+ + 8 H_2O$	$+0.092$
$NO_2(g) + 2 H_3O^+ + 2 e^- \rightleftharpoons NO(g) + 3 H_2O$	$+1.03$
$N_2O(g) + 2 H_3O^+ + 2 e^- \rightleftharpoons N_2(g) + 3 H_2O(l)$	$+1.766$
$NO_3^- + 4 H_3O^+ + 3 e^- \rightleftharpoons NO(g) + 6 H_2O$	$+0.957$
$2 NO(g) + 2 H_3O^+ + 2 e^- \rightleftharpoons N_2O(g) + 3 H_2O$	$+1.591$
$NO_2^-(g) + H_2O + 3 e^- \rightleftharpoons NO(g) + 2 OH^-$	-0.46
$NO_3^- + 2 H_2O + 3 e^- \rightleftharpoons NO(g) + 4 OH^-$	$+0.109$

Oxygen

$O_2(g) + 2 H_3O^+ + 2 e^- \rightleftharpoons H_2O_2(l) + 2 H_2O$	$+0.695$
$O_2(g) + 2 H_2O + 2 e^- \rightleftharpoons H_2O_2(l) + 2 OH^-$	-0.146
$O_2(g) + 4 H_3O^+ + 4 e^- \rightleftharpoons 6 H_2O$	$+1.229$
$O_2(g) + 2 H_2O + 4 e^- \rightleftharpoons 4 OH^-$	$+0.401$
$O_3(g) + 2 H_3O^+ + 2 e^- \rightleftharpoons O_2(g) + 3 H_2O$	$+2.076$

Platinum

$Pt^{2+} + 2 e^- \rightleftharpoons Pt$	$+1.18$

Potassium

$K^+ + e^- \rightleftharpoons K$	-2.931

Selenium

$SeO_3^{2-} + 3 H_2O + 4 e^- \rightleftharpoons Se + 6 OH^-$	-0.366

Silver

$Ag^+ + e^- \rightleftharpoons Ag$	$+0.7996$
$AgCl + e^- \rightleftharpoons Ag + Cl^-$	$+0.22233$
$AgBr + e^- \rightleftharpoons Ag + Br^-$	$+0.07133$
$AgI + e^- \rightleftharpoons Ag + I^-$	-0.15224

Sodium

$Na^+ + e^- \rightleftharpoons Na$	-2.71

Sulfur

$S + 2 H_3O^+ + 2 e^- \rightleftharpoons H_2S + 2 H_2O$	$+0.14$
$S_4O_6^{2-} + 2 e^- \rightleftharpoons 2 S_2O_3^{2-}$	$+0.08$

Tin

$Sn^{2+} + 2 e^- \rightleftharpoons Sn$	-0.137
$Sn^{4+} + 2 e^- \rightleftharpoons Sn^{2+}$	$+0.151$

Titanium

$Ti^{2+} + 2 e^- \rightleftharpoons Ti$	-1.630
$Ti^{3+} + e^- \rightleftharpoons Ti^{2+}$	-0.85
$TiO_2 + 4 H_3O^+ + 2 e^- \rightleftharpoons Ti^{2+} + 6 H_2O$	-0.502

Zinc

$Zn^{2+} + 2 e^- \rightleftharpoons Zn$	-0.7618
$[Zn(NH_3)_4]^{2+} + 2 e^- \rightleftharpoons Zn + 4 NH_3(aq)$	-1.04

Main Source: *CRC Handbook of Chemistry and Physics, 82nd Edition,* David R. Lide, editor-in-chief, CRC Press LLC, Boca Raton, FL (2001), pp. **8**-21 through **8**-26.

Solutions to Odd-Numbered Problems

Chapter 1

1.1 Examples: What are appropriate standards for "clean" air and water? What should be the regulations for testing new drugs? How can toxic waste sites be cleaned up? Should chlorofluorocarbons be banned?

1.3 Examples: weighing, volume measurements and unit conversions; knowing chemical compatibility of different drugs; identifying similar or chemically equivalent drugs; and protecting drugs from degradation

1.5 The chemist should redo the experiment to determine whether or not the results are correct. If the results consistently differ from what theory predicts, then the chemist should revise the theory to accommodate the results.

1.7 (a) H; (b) He; (c) Hf; (d) N; (e) Ne; and (f) Nb

1.9 (a) arsenic; (b) argon; (c) aluminum; (d) americium; (e) silver; (f) gold; (g) astatine; and (h) actinium

1.11 (a) Br_2; (b) HCl; (c) C_2H_5I; and (d) C_3H_6O

1.13 (a) CCl_4; (b) H_2O_2; (c) P_4O_{10}; and (d) Fe_2S_3

1.15 Cesium

1.17 Vertical (similar chemical properties): oxygen (O) and selenium (Se); horizontal: phosphorus (P) and chlorine (Cl)

1.19 Metals from Group 1: Li, Na, K, Rb, and Cs; other metals: Cu, Ag, and Au

1.21 Lithium, Li; beryllium, Be; boron, B; carbon, C; oxygen, O; fluorine, F; and neon, Ne

1.23 (a) pure substance; (b) and (d) solution; (c), (e), and (f) heterogeneous mixture

1.25 (a) liquid; (b) solid; (c) solid; and (d) gas

1.27 (a) physical; (b) physical; and (c) chemical

1.29 (a), (c), (e), and (f) mixture; (b) compound; and (d) element

1.31 (a) 1.00000×10^5; (b) 1.0×10^4; (c) 4.00×10^{-4}; (d) 3×10^{-4}; and (e) 2.753×10^2

1.33 (a) 4.32×10^2 kg; (b) 6.24×10^{-10} s; (c) 1.024×10^{-9} kg; (d) 9.300×10^7 m; (e) 8.6400×10^4 s; and (f) 1.08×10^{-3} m

1.35 7.10×10^{-3} kg

1.37 9.46×10^2 g

1.39 0.79 g/cm³

1.41 5.70 cm³

1.43 9.98×10^2 kg/m³

1.45 Silver cube

1.47 11.8 mL

1.49 8: Ti (titanium), Tc (technetium), Te (tellurium), Ta (tantalum), Tl (thallium), Tb (terbium), Tm (thulium), and Th (thorium)

1.51

1.53 Lead (Pb)

1.55 11.12 s

1.57 Silver and copper

1.59

1.61 Intensive: (a), (c), and (e); extensive: (b) and (d)

1.63 261.7 K

1.65 (a) 4.52×10^{34}; and (b) 1×10^{-4}

1.67 (b) matches (d); (c) matches (e); (a) matches (f)

1.69 (a) $CHClF_2$; (b) CH_2O_2; (c) BrF_3; and (d) C_4H_{10}

1.71 (a) 7.44×10^{-3} m³; (b) 16 m s⁻¹; (c) 2.1 m; and (d) 1.03×10^2 kg

1.73 (a) Sr; (b) In; (c) Br and O; (d) Co; (e) Ne; and (f) Pu

1.75

HCN H_2O CO N_2O

1.77 393 minutes

1.79 Physical properties: appearance, melting point, softness, density; chemical properties: reaction with chlorine and reaction with water

1.81 (a) fluorine (F), chlorine (Cl), bromine (Br), iodine (I); (b) beryllium (Be), magnesium (Mg), calcium (Ca), strontium (Sr), barium (Ba); (c) examples include actinium (Ac), uranium (U), and berkelium (Bk); and (d) examples include helium (He), neon (Ne), argon (Ar), and krypton (Kr)

1.83 9.4604×10^{12} km

1.85 (a) copper, Cu; sulfur, S; and oxygen, O; and (b) $CuSO_4$

1.87 1.3×10^8 atoms

1.89 3.6×10^8 ft; 1.1×10^5 km

1.91 Your list should include questions about which you have a special interest.

1.93 (a) vinegar; (b) table salt; and (c) iron

1.95 3.1557×10^9 s

Chapter 2

2.1 (a) gaseous helium (b) solid tungsten and (c) liquid gallium

2.3 (1) All starting materials and products are made from atoms of carbon and oxygen. (2) All oxygen atoms are in diatomic molecules in the starting materials and CO molecules in the products; likewise, C atoms behave all in the same way. (3) C and O combine in 1:1 atomic ratio to give CO molecules. (4) Atoms of each type are conserved: 4 O atoms and 13 C atoms.

2.5 When magnesium burns in air, magnesium atoms combine with oxygen from the air to form magnesium oxide. The mass of the solid increases, but the mass of solid plus gas remains constant.

2.7   Solid   Liquid   Gaseous

2.9   The molecules that give roses their aroma evaporate from the surface of the flower. Once in the gas phase, the molecules move slowly away from the rose until, when they reach a nose, they are sensed by the olfactory sensors.

2.11   Iodine molecules sublime from the crystals at the bottom of the flask into the gas phase, where their presence imparts a pale violet color to the gas. Some of these molecules then condense on the surfaces of the flask, forming crystals. At any time, there is a constant number of molecules in the gas phase, but some molecules are subliming into the gas from the crystals while equal numbers of molecules are condensing onto the crystals.

2.13   (a) 6.24×10^{12} electrons; and (b) 5.68×10^{-18} kg

2.15   Positively charged particles would be attracted upward. The proper electrical field would cause some of the positively charged particles to be suspended in space. Because each of these particles had lost one or more electrons, the charge on the electron could be determined from these observations.

2.17 9.0×10^{23} protons; 1.4×10^{5} C

2.19   (a) 8 p, 8 n, 10 e; (b) 5 p, 6 n, 5 e; (c) 25 p, 30 n, 22 e; (d) 17 p, 18 n, 18 e; and (e) 17 p, 20 n, 16 e

2.21   (a) $^{56}_{26}Fe$; (b) $^{236}_{92}U$; (c) $^{38}_{18}Ar$; and (d) $^{19}_{9}F$

2.23

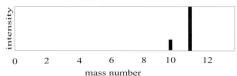

2.25

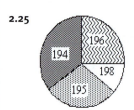

2.27   (a) unstable, too few neutrons; (b) unstable, $Z > 83$; and (c) stable

2.29

Name	Symbol	Z
Tungsten	W	74
Cobalt	Co	27
Mercury	Hg	80

2.31   39.948 g/mol

2.33   (a) 0.141 mol; (b) 5.45×10^{-6} mol; (c) 1.67×10^{-4} mol; and (d) 5.41×10^{4} mol

2.35   (a) 3.92×10^{20} atoms; (b) 1.14×10^{20} atoms; (c) 3.87×10^{19} atoms; and (d) 1.48×10^{19} atoms

2.37   Cation: Cr^{3+}; $Cr \rightarrow Cr^{3+} + 3\ e^{-}$; anion: Cl^{-}, $Cl + e^{-} \rightarrow Cl^{-}$; and neutral: C

2.39   (a) Cl^{-}; (b) Na^{+}; and (c) O^{2-}

2.41   (a) Rb^{+}; (b) F^{-}; and (c) Ba^{2+}

2.43   RbF and BaF_2

2.45 Al_2O_3; AlF_3

2.47 1.3×10^{25} atoms

2.49

Part:	(a)	(b)	(c)	(d)	(e)	(f)
Z	3	20	92	52	10	82
A	6	43	238	130	20	205
N	3	23	146	78	10	123

2.51   (a) $^{4}_{2}He$; (b) $^{184}_{74}W$; (c) $^{60}_{28}Ni$; and (d) $^{26}_{12}Mg$

2.53 7.696×10^{13} m

2.55

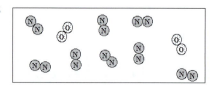

2.57

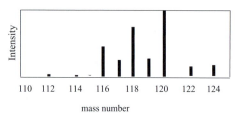

2.59   (a) 43 p, 56 n, 43 e; (b) 26 p, 26 n, 26 e; (c) 54 p, 79 n, 54 e; and (d) 53 p, 78 n, 53 e

2.61   0.205 g

2.63   (a) Ca^{2+}, Cl^{-}; (b) $CaCl_2$; and (c)

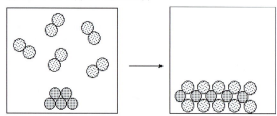

2.65 $^{40}_{17}Cl$, $^{40}_{18}Ar$, $^{40}_{19}K$, $^{40}_{20}Ca$, and $^{40}_{21}Sc$

2.67

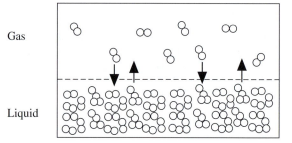

2.69   (a) 11 p, 10 e; (b) 7 p, 10 e; (c) 22 p, 18 e; and (d) 53 p, 54 e

2.71

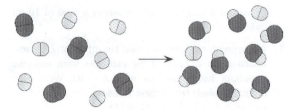

2.73 $^{45}_{20}Ca$, $^{54}_{24}Cr$, $^{63}_{28}Ni$, $^{72}_{32}Ge$, $^{81}_{36}Kr$, and $^{90}_{40}Zr$

2.75

Solid   Liquid

2.77   Not correct;

Chapter 3

3.1 (a) CH_4; (b) C_2H_4; (c) C_2H_6O; (d) HBr; (e) PCl_3; (f) CH_4N_2O; and (g) C_2H_5I

3.3

(a) a methane structure: H—C—H with H above and below

(e) a PCl_3 structure with P and three Cl

(b) ethylene structure $H_2C=CH_2$

(f) urea structure with O, C, N, H

(c) dimethyl ether structure H_3C—O—CH_3

(g) structure H—C—C—H with I

(d) H—Br

3.5

(a) (skeletal structure)

(d) (styrene skeletal structure)

(b) (piperidine ring with N—H)

(e) (skeletal structure with O—H)

(c) (ketone skeletal structure with O)

(f) (cyclopentane with two O—CH₃ groups)

3.7

(a) through (i) various structural formulas

3.9 (a) CO_2, carbon dioxide; (b) HCl, hydrogen chloride; and (c) CCl_4, carbon tetrachloride

3.11 (a) H_2S, hydrogen sulfide; (b) SF_4, sulfur tetrafluoride; and (c) HF, hydrogen fluoride

3.13 (a) CH_4; (b) HI; (c) CaH_2; (d) PCl_3; (e) N_2O_5; (f) SF_6; and (g) BF_3

3.15 (a) disulfur dichloride; (b) iodine heptafluoride; (c) hydrogen bromide; (d) dinitrogen trioxide; (e) silicon carbide; and (f) methanol

3.17 (a) not ionic, HF; (b) ionic, CaF_2; (c) ionic, $Al_2(SO_4)_3$; (d) ionic, $(NH_4)_2S$; (e) not ionic, SO_2; and (f) not ionic, CCl_4

3.19 (a) not ionic, dichloromethane; (b) not ionic, carbon dioxide; (c) ionic, calcium oxide; (d) ionic, potassium carbonate; (e) not ionic, phosphorus tribromide; (f) not ionic, hydrogen bromide; and (g) ionic, sodium hydrogen phosphate

3.21 (a) Na_2SO_4; (b) K_2S; (c) KH_2PO_4; (d) $CoF_2 \cdot 4H_2O$; (e) PbO_2; (f) $NaHCO_3$; and (g) $LiBrO_4$

3.23 (a) calcium chloride hexahydrate; (b) iron(II) ammonium sulfate; (c) potassium carbonate; (d) tin(II) chloride dihydrate; (e) sodium hypochlorite; (f) silver sulfate; (g) copper(II) sulfate; (h) potassium dihydrogen phosphate; (i) sodium nitrate; (j) calcium sulfite; and (k) potassium permanganate

3.25 (a) 153.81 g/mol; (b) 110.27 g/mol; (c) 48.00 g/mol; (d) 86.84 g/mol; (e) 144.64 g/mol; and (f) 169.88 g/mol

3.27 (a) $C_9H_{11}NO_3$; 181.19 g/mol; (b) $C_{11}H_{12}N_2O_2$, 204.23 g/mol; (c) $C_5H_9NO_4$, 147.13 g/mol; and (d) $C_6H_{14}N_2O_2$, 146.19 g/mol

3.29 (a) 2.20×10^{20} molecules; (b) 2.57×10^{19} molecules; (c) 7.66×10^{19} molecules; and (d) 1.00×10^{19} molecules

3.31 (a) 9.99×10^{-18} g; (b) 7.6×10^{-13} g; and (c) 1.484×10^{-21} g

3.33 2.6×10^{-6} mol; 1.6×10^{18} molecules; $\#_H = 4.8 \times 10^{19}$ atoms; 8.0×10^{-2} mg

3.35 CaO: O, 28.53%, Ca, 71.47%; SiO_2: O, 53.25%, Si, 46.75%; Al_2O_3: O, 47.08%, Al, 52.92%; Fe_2O_3: O, 30.06%, Fe, 69.94%

3.37 $C_{10}H_{14}N_2$

3.39 HgS

3.41 % C = 64.4%; % H = 5.41%; % Fe = 30.2%; $C_{10}H_{10}Fe$

3.43 Empirical formula is C_4H_6O, molecular formula is $C_{12}H_{24}O_3$

3.45 (a) $M_{Mg^{2+}} = 0.328$ M; $M_{Cl^-} = 0.656$ M; and (b)

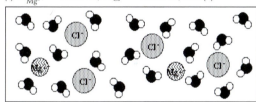

3.47 (a) $[K^+] = [OH^-] = 0.308$ M; (b) $[K^+] = [OH^-] = 0.077$ M; and (c):

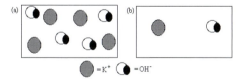

$\bigcirc$ = K⁺ $\bigcirc$ = OH⁻

3.49 5.17 mL

3.51 (a) $[CO_3^{2-}] = 0.175$ M; $[Na^+] = 0.350$ M; (b) $[NH_4^+] = [Cl^-] = 0.0331$ M; and (c) $[SO_4^{2-}] = 1.84 \times 10^{-3}$ M; $[K^+] = 3.69 \times 10^{-3}$ M

3.53 (a) carbon dioxide; (b) potassium nitrate; (c) sodium chloride; (d) sodium hydrogen carbonate; (e) sodium carbonate; (f) sodium hydroxide; (g) calcium oxide; and (h) magnesium hydroxide

3.55 (a) Fe = 54.81%; Si = 13.78%; O = 31.40%; (b) Na = 8.77%; Al = 10.29%; Si = 32.13%; O = 48.81%; (c) Al = 20.90%; Si = 21.76%; O = 55.78%; H = 1.56%; and (d) Mg = 5.62%; Si = 25.97%; O = 66.55%; H = 1.86%

3.57 2.86 kg; 4.72 kg; 2.14 kg; and 4.71 kg

3.59 (a) $C_{15}H_{20}O_4$, $MM = 264.31$ g/mol; (b) $C_{10}H_9NO_2$, $MM = 175.18$ g/mol; and (c) $C_{10}H_{13}N_5O$, $MM = 219.25$ g/mol

3.61 (a) H_2O, NH_4^+, SO_4^{2-}; (b) H_2O, CO_2; (c) H_2O, Na^+, F^-; (d) H_2O, K^+, CO_3^{2-}; (e) H_2O, Na^+, HSO_4^-; and (f) H_2O, Cl_2

3.63 (a) $NaNO_2$ and $NaNO_3$; (b) K_2CO_3 and $KHCO_3$; (c) FeO and Fe_2O_3; and (d) I_2 and I^-

3.65 (a) 342.17 g/mol; (b) 7.31×10^{-2} mol; (c) Al: 15.77%, S: 28.12%, O: 56.11%; and (d) 28.5 g

3.67 2.6×10^{-4} mol/L

3.69 (a) ammonium chloride; (b) xenon tetrafluoride; (c) iron(III) oxide; (d) sulfur dioxide; and (e) potassium perchlorate

3.71 (a) 454.59 g/mol; (b) $2.640 = 10^{-4}$ mol; and (c) 3.179×10^{20} atoms N

3.73 (a) 14 M; and (b) 1.1×10^2 mL

3.75 (a) $n(Mg^{2+}) = 6.95 \times 10^{-2}$ mol, $n(NO_3^-) = 1.39 \times 10^{-1}$ mol, # $(Mg^{2+}) = 4.19 \times 10^{22}$ ions, # $(NO_3^-) = 8.37 \times 10^{22}$ ions; (b) $n(K^+) = 3.00 \times 10^{-2}$ mol, $n(SO_4^{2-}) = 1.50 \times 10^{-2}$ mol, # $(K^+) = 1.81 \times 10^{22}$ ions, # $(SO_4^{2-}) = 9.03 \times 10^{21}$ ions; (c) $n(PO_4^{3-}) = 2.92 \times 10^{-4}$ mol, $n(Na^+) = 8.77 \times 10^{-4}$ mol, # $(PO_4^{3-}) = 1.76 \times 10^{20}$ ions, # $(Na^+) = 5.28 \times 10^{20}$ ions

3.77 55.5 mol/L

3.79 78.1% F and 21.9% S; SF_6

3.81 (a) hydrogen peroxide; (b) carbon dioxide; (c) methanol

3.83 (a) 0.49 mol; (b) 0.72 g; (c) 3.7×10^{-5} mol; 2.2×10^{19} molecules; and (d) 7.4×10^{-6} mol

3.85 1.212×10^9 mol H_3PO_4; 1.8×10^8 mol P; 5.6×10^6 kg P

3.87 To determine a molecular formula, mass percent composition and an approximate molar mass must be known. The mass of compound that was burned must be known. Unless the percentages of C and H together total 100%, further information about the other elements present in the compound is needed.

3.89 (a) 1.44 g Al; (b) 2.96×10^{19} phosphate ions; and (c) +2

3.91 (a) 23.682% C, 3.180% H, 13.809% N, 41.008% O, 18.321% P; (b) 6.23×10^{15} atoms; (c) 1.5 ng; and (d) 11.7 mg

3.93 1.36×10^3 g/mol

3.95 (a) 2.4×10^{22} atoms; (b) 5.17 g; (c) 1.5 M; (d) 3.8×10^{-6} mol

3.97 To prepare the solution, the worker should mix 0.090 L of sodium acetate solution, 0.045 L of acetic acid solution, and enough water to make the final volume 1.5 L.

3.99 0.783 mol $CuSO_4$ and 196 g $CuSO_4 \cdot 5H_2O$

Chapter 4

4.1 (a) $NH_4NO_3 \rightarrow N_2O + 2\,H_2O$; (b) $P_4O_{10} + 6\,H_2O \rightarrow 4\,H_3PO_4$; (c) $2\,HIO_3 \rightarrow I_2O_5 + H_2O$; and (d) $2\,As + 5\,Cl_2 \rightarrow 2\,AsCl_5$

4.3

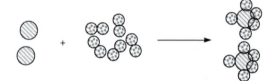

4.5 (a) $2\,H_2 + CO \rightarrow CH_3OH$; (b) $CaO + 3\,C \rightarrow CO + CaC_2$; and (c) $2\,C_2H_4 + O_2 + 4\,HCl \rightarrow 2\,C_2H_4Cl_2 + 2\,H_2O$

4.7

4.9 (a) $3\,Ca(OH)_2 + 2\,H_3PO_4 \rightarrow 6\,H_2O + Ca_3(PO_4)_2$; (b) $Na_2O_2 + 2\,H_2O \rightarrow 2\,NaOH + H_2O_2$; (c) $BF_3 + 3\,H_2O \rightarrow 3\,HF + H_3BO_3$; and (d) $2\,NH_3 + 3\,CuO \rightarrow 3\,Cu + N_2 + 3\,H_2O$

4.11 (a) 34.7 g CO; (b) 3.21 g C; and (c) 2.85 g O_2

4.13 1.77×10^3 g

4.15 511 g

4.17 45.5 kg HF; 138 kg CCl_2F_2; and 83.0 kg HCl

4.19 68.6%

4.21 28 kg

4.23 76.3%; 258 kg CCl_4; and 67.2 kg HF

4.25 lettuce; 66

4.27 91.3 kg

4.29 (a) Limiting reactant = CO, 1.14 metric ton; (b) Limiting reactant = CaO, 0.499 metric ton CO and 1.14 metric ton CaC_2; and (c) Limiting reactant = HCl, 1.36 metric ton $C_2H_4Cl_2$ and 0.247 metric ton H_2O

4.31 8.60 g; 1.7 g O_2

4.33 (a) NH_4^+, Cl^-, H_2O; (b) Fe^{2+}, ClO_4^-, H_2O; (c) Na^+, SO_4^{2-}, H_2O; (d) Br_2, H_2O; and (e) K^+, Br^-, H_2O

4.35 (a) K^+, HPO_4^-, H_2O; (b) CH_3CO_2H, H_2O; (c) Na^+, $CH_3CO_2^-$, H_2O; (d) NH_3, H_2O; and (e) NH_4^+, Cl^-, H_2O

4.37 (a) AgCl precipitates with a; no precipitate with b, c, or d; AgBr precipitates with e; (b) no precipitate with a, c, d, or e; $FeCO_3$ precipitates with b; and (c) no precipitate with a, d, or e; $Fe(OH)_2$ precipitates with b; $BaSO_4$ precipitates with c.

4.39 (a) $Ag^+(aq) + OH^-(aq) \rightarrow AgOH(s)$, spectator ions: NO_3^-, K^+; (b) $2\,Fe^{3+}(aq) + 3\,C_2O_4^{2-}(aq) \rightarrow Fe_2(C_2O_4)_3(s)$, spectator ions: ClO_4^-, NH_4^+; (c) $Pb^{2+}(aq) + 2\,Br^-(aq) \rightarrow PbBr_2(s)$, spectator ions: NO_3^-, Na^+; and (d) $Ni^{2+}(aq) + 2\,OH^-(aq) \rightarrow Ni(OH)_2(s)$, spectator ions: K^+, SO_4^{2-}

4.41 0.381 g; CO_3^{2-}, NO_3^-, and K^+

4.43 (a) $H_3O^+(aq) + OH^-(aq) \rightarrow 2\,H_2O(l)$, spectators: Ca^{2+}, Cl^-; (b) $H_3PO_4(aq) + 3\,OH^-(aq) \rightarrow 3\,H_2O(l) + PO_4^{3-}(aq)$, spectator ions: Li^+; (c) $NH_3(aq) + H_3O^+(aq) \rightarrow H_2O(l) + NH_4^+(aq)$, spectator ions: NO_3^-; and (d) $CH_3CO_2H(aq) + OH^-(aq) \rightarrow H_2O(l) + CH_3CO_2^-(aq)$, spectator ions: K^+

4.45 0.705 g

4.47 $[OH^-] = 4.00 \times 10^{-3}$ M; $[Ba^{2+}] = 1.20 \times 10^{-2}$ M; $[Cl^-] = 2.00 \times 10^{-2}$ M; $[H_3O^+] = 0$

4.49 0.660 M

4.51 8.886×10^{-2} M

4.53 (a) and (b) no reaction; (c) $Cu(s) + 2\,Ag^+(aq) \rightarrow Cu^{2+}(aq) + 2\,Ag(s)$; and (d) $2\,K(s) + 2\,H_2O(l) \rightarrow 2\,K^+(aq) + H_2(g) + 2\,OH^-(aq)$

4.55 (a) $2\,Sr + O_2 \rightarrow 2\,SrO$; (b) $4\,Cr + 3\,O_2 \rightarrow 2\,Cr_2O_3$; and (c) $Sn + O_2 \rightarrow SnO_2$.

4.57 3.98×10^{-2} g

4.59 (a) $Na^+ + HCO_3^- \rightarrow NaHCO_3(s)$; (b) NH_4^+ and Cl^- are spectator ions; (c) 55.6%; and (d) $[NH_4^+] = 0.750$ M; $[Cl^-] = 3.00$ M; $[Na^+] = 2.58$ M; $[HCO_3^-] = 0.333$ M.

4.61 50.6%; 2.47 g

4.63 54 metric tons

4.65 (a) $4\,Al + 3\,O_2 \rightarrow 2\,Al_2O_3$ (redox); (b) $C_3H_8(g) + 5\,O_2(g) \rightarrow 3\,CO_2(g) + 4\,H_2O(g)$ (redox); (c) $Mg(s) + 2\,H_3O^+(aq) \rightarrow Mg^{2+}(aq) + H_2(g) + 2\,H_2O(l)$ (redox); (d) $OH^-(aq) + H_3O^+(aq) \rightarrow 2\,H_2O(l)$ (acid−base); (e) $Pb^{2+}(aq) + CO_3^{2-}(aq) \rightarrow PbCO_3(s)$ (precipitation); and (f) $OH^-(aq) + H_3O^+(aq) \rightarrow 2\,H_2O(aq)$ (acid−base)

4.67 $4\,Ru + 3\,O_2 \rightarrow 2\,Ru_2O_3$, 3 electrons; $Ru + O_2 \rightarrow RuO_2$, 4 electrons; $2\,Ru + 3\,O_2 \rightarrow 2\,RuO_3$, 6 electrons; $Ru + 2\,O_2 \rightarrow RuO_4$, 8 electrons

4.69 $Y + 2\,X \rightarrow YX_2$

4.71 (a) $5 H_2 + 2 NO \rightarrow 2 NH_3 + 2 H_2O$;
(b) $2 CO + 2 NO \rightarrow N_2 + 2 CO_2$;
(c) $2 NH_3 + 2 O_2 \rightarrow N_2O + 3 H_2O$;
(d) $6 NO + 4 NH_3 \rightarrow 5 N_2 + 6 H_2O$;
(e) $6 H_2O + 4 NO \rightarrow 5 O_2 + 4 NH_3$; and
(f) $2 H_2 + O_2 \rightarrow 2 H_2O$

4.73 (a) Mix solutions of Na_3PO_4 and $FeCl_3$. The net reaction is
$Fe^{3+}(aq) + PO_4^{3-}(aq) \rightarrow FePO_4(aq)$; 2.69 kg $FeCl_3$ and 2.72 kg
Na_3PO_4; (b) Mix solutions of $NaOH$ and $ZnCl_2$. The net reaction is $Zn^{2+}(aq) + 2 OH^-(aq) \rightarrow Zn(OH)_2(s)$; 3.42 kg of $ZnCl_2$
and 2.01 kg $NaOH$; and (c) Mix solutions of Na_2CO_3 and
$NiCl_2$. The net reaction is $Ni^{2+}(aq) + CO_3^{2-}(aq) \rightarrow NiCO_3(s)$;
2.73 kg $NiCl_2$ and 2.24 kg Na_2CO_3.

4.75 (a) H_2;
(b)

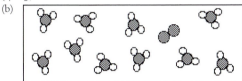

and (c) 170. g NH_3 produced and 28 g N_2 remaining

4.77 (a) 0.30 mol; (b) 1.8×10^{23} molecules; and (c) 5.4 g

4.79 (a) Species present are NH_3, H_2O, H_3O^+ and Cl^-;
$NH_3 + H_3O^+ \rightarrow NH_4^+ + H_2O$; (b) H_2O, Ca^{2+}, Cl^-, Na^+,
and SO_4^{2-}; $Ca^{2+} + SO_4^{2-} \rightarrow CaSO_4(s)$; (c) H_2O, K^+, OH^-,
H_3O^+, and Br^-; $H_3O^+ + OH^- \rightarrow 2 H_2O$; and (d) HNO_2,
H_2O, K^+, and OH^-; $HNO_2 + OH^- \rightarrow NO_2^- + H_2O$

4.81 (a) 48 kg; (b) 35 kg

4.83 $H_3O^+ + OH^- \rightarrow 2 H_2O$

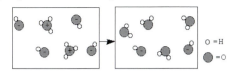

4.85

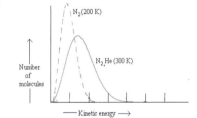

4.87 (a) $CaCO_3 + 2 H_3O^+ \rightarrow Ca^{2+} + CO_2 + 3 H_2O$; (b) 1.1 g;
and (c) $[Cl^-] = 0.10$ M and $[Ca^{2+}] = 0.050$ M

4.89 (a) $Ca + 2 H_3O^+ \rightarrow Ca^{2+} + H_2 + 2 H_2O$ (redox);
(b) $4 Li + O_2 \rightarrow 2 Li_2O$ (redox);
(c) $NH_3 + H_3O^+ \rightarrow NH_4^+ + 2 H_2O$ (acid-base); and
(d) $C_3H_8 + 5 O_2 \rightarrow 3 CO_2 + 4 H_2O$ (redox)

4.91 $4 C_3H_5N_3O_9(l) \rightarrow 6 N_2(g) + 12 CO_2(g) + 10 H_2O(g) + O_2(g)$

4.93 Theoretical yield = 102 kg; % yield = 67%; 28 kg N_2 and
12 kg H_2

4.95 1:3.03

4.97 10.0% impurities

4.99 (a) $2 H_2SO_4 + Ca_3(PO_4)_2 \rightarrow 2 CaSO_4 + Ca(H_2PO_4)_2$;
(b) $Ca(H_2PO_4)_2$: 30.6 kg; H_2SO_4: 19.4 kg; and (c) 198 mol

4.101 $2 C_5H_{12} + 15 O_2 \rightarrow 12 H_2O + 10 CO_2$; 0.132 mol

4.103 Cu: 13.0%; Ag: 87.0%

Chapter 5

5.1 A pinhole in the top of the tube would let air leak into the
space until the internal pressure matched the external,
atmospheric pressure. The barometer would then indicate zero
pressure.

5.3 When air is pumped into a flat tire, the tire expands and resists
compression. Both these observations are the result of the pressure exerted by the gas molecules making up air.

5.5 (a) 6.07×10^4 Pa, 0.607 bar; (b) 2.48×10^5 Pa, 2.48 bar;
(c) 61 Pa, 6.1×10^{-4} bar; and (d) 1.35×10^2 Pa,
1.35×10^{-3} bar

5.7 4.09 mol; 1.00×10^2 L

5.9 (a) $P_i/T_i = P_f/T_f$; (b) $V = nRT/P$; and (c) $P_iV_i = P_fV_f$

5.11 0.221 L

5.13 (a) equation is valid; (b), (c), and (d) not valid

5.15

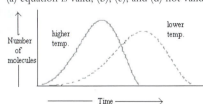

5.17

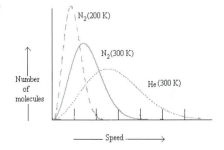

5.19

5.21 (a) 1.93×10^3 m/s, 11.2 kJ/mol; (b) 394 m/s, 3.74 kJ/mol; and
(c) 319 m/s, 11.2 kJ/mol

5.23 (a) At very high pressure, molecules are very close together, so
their volumes are significant compared to the volume of their
container; the first condition is not met and the gas is not ideal.
(b) At very low temperature, molecules move very slowly, so
the forces between molecules, even though small, are sufficient
to influence molecular motion; the second condition is not met
and the gas is not ideal.

5.25 (a) 1 atm, generated by gas molecules colliding with the face of
the piston; and (b) increase in average molecular speed causes an
increase in pressure, so the piston will move outward, leading to
a lower frequency of collisions with the wall, reducing the pressure until the internal pressure is once again 1 atm.

5.27 As a gas cools, its molecules move more slowly, so they impart
smaller impulses on the walls of their container. This reduces
the internal pressure, so the balloon collapses until the increase
in gas density inside the balloon brings the internal pressure
back up to 1.000 atm.

5.29 121 g/mol; CF_2Cl_2

5.31 5.89 g/L

5.33 226 m/s

5.35 CH_4 diffuses faster, because it has the smaller *MM*.

5.37 7.8×10^{-7} atm

5.39 $p(N_2) = 593.4$ torr; $p(O_2) = 159.2$ torr; $p(Ar) = 7.10$ torr; and $p(CO_2) = 2.47 \times 10^{-1}$ torr

5.41 (a) He; (b) He; and (c) 0.67

5.43 $p(CH_4) = 2.05$ atm; $p(C_2H_6) = 0.28$ atm; and $p(C_3H_8) = 9.5 \times 10^{-3}$ atm

5.45 3.94 L

5.47 0.438 atm

5.49 2.28×10^5 g

5.51 3.0×10^4 g

5.53 + 0.0048

5.55 20.8 °C

5.57 6.7×10^2 L

5.59 6.18 g; 4.44×10^{22} atoms

5.61 (a) Chamber B, because it contains the most atoms; (b) 0.50 atm, because the pressures in the chambers are proportional to the number of atoms; (c) 4.5 atm, because the pressure will increase in proportion to the number of atoms; and (d) When the valves are opened, the number of atoms in each chamber equalize: 27/3 = 9, giving 0.38 atm.

5.63 359 m/s

5.65 2×10^{12} molecules/L

5.67 (a) 30.9 °C; (b) 4.6 °C; and (c) 6.1 °C

5.69 Pump out a bulb of known volume, weigh it, fill with oxygen at a measured pressure, and weigh again. If gaseous oxygen were monatomic, this experiment would give $MM = 16.00$ g/mol, while the diatomic gas gives $MM = 32.00$ g/mol.

5.71 $m(C_3H_6) = 0.692$ g; $m(O_2) = 2.10$ g

5.73 (a) H_2; (b) O_2; (c) O_2; and (d) H_2

5.75 (a) 4×10^{-21} J; and (b) 4×10^2 m/s

5.77 Sulfur in coal combines with oxygen from the air to form SO_2, and SO_2 reacts with water to form H_2SO_4: $S + O_2 \rightarrow SO_2$; $2\,SO_2 + O_2 \rightarrow 2\,SO_3$; $SO_3 + H_2O \rightarrow H_2SO_4$.

5.79 (a) 2.09 L; (b) 788 torr; and (c) 621 torr

5.81 (a) Your drawing should show the same atomic density as Figure 5-16 but with longer "tails" on the atoms, indicating that they are moving at higher speeds. (b) Your drawing should show the same lengths of "tails" as Figure 5-16, but there should be 1/3 fewer atoms. (c) Your drawing should be the same as Figure 5-16 except that half the atoms should be replaced with diatomic molecules.

5.83 (a) *P* is proportional to number of moles of gas; (b) true; and (c) *P/T* is constant

5.85 0.911 L

5.87 1.0×10^{-9} atm; 2.5×10^{10} molecules/cm³

5.89 10.2 atm

5.91 (a) 1.18 g/L; and (b) 1.17 g/L

5.93 6.32×10^5 L

5.95 64.1%

5.97

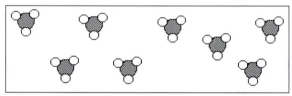

5.99 $Ni(CO)_4$

5.101 At higher elevations, Mount Everest, the atmospheric pressure, 250 mm, is much lower than at ground level, 760 mm. A lower pressure corresponds to a smaller molecular density. "The air is thin" refers to the fact that there is less air at higher elevations.

5.103 $p(H_2O) = 1.6 \times 10^{-2}$ atm; $p(\text{dry air}) = 0.991$ atm; $p(N_2) = 0.774$ atm; $p(O_2) = 0.208$ atm; $p(Ar) = 9.26 \times 10^{-3}$ atm; $p(CO_2) = 3.22 \times 10^{-4}$ atm; $p(Ne) = 1.80 \times 10^{-5}$ atm; $p(He) = 5.19 \times 10^{-6}$ atm; $p(CH_4) = 1.4 \times 10^{-6}$ atm

5.105 $C_6H_{15}N$

Chapter 6

6.1 (a) On the tree, an apple possesses some gravitational potential energy. As an apple falls, that potential energy is converted into kinetic energy of motion. (b) When an apple hits the ground, the impact transfers energy to molecules in the earth and in the apple. As a result, there is a slight increase in temperature; the kinetic energy of the apple has been converted into thermal energy.

6.3 9.43×10^{-20} J

6.5 7.22×10^{-19} J

6.7 (a) Radiant energy is consumed and thermal energy is produced. (b) Chemical potential energy is consumed and kinetic energy is produced. (c) Chemical potential energy is consumed and thermal energy is produced.

6.9 2.12×10^2 m/s

6.11 (a) 17.8 °C; (b) 30.6 °C; (c) 33.8 °C; and (d) 22.2 °C

6.13 water: 8.29×10^5 J; kettle: 4.37×10^4 J

6.15 23.6 °C

6.17 131 kJ

6.19 14 km

6.21 (a) $2\,C_7H_6O_2 + 15\,O_2 \rightarrow 14\,CO_2 + 6\,H_2O$; (b) -3.221×10^3 kJ/mol; and (c) -4.295×10^2 kJ/mol

6.23 Moving down the column, the bond energy decreases. We predict that the H—Se bond has an energy between H—S and H—Te, and the H—Po bond is weaker than H—Te.

6.25 165 kJ

6.27

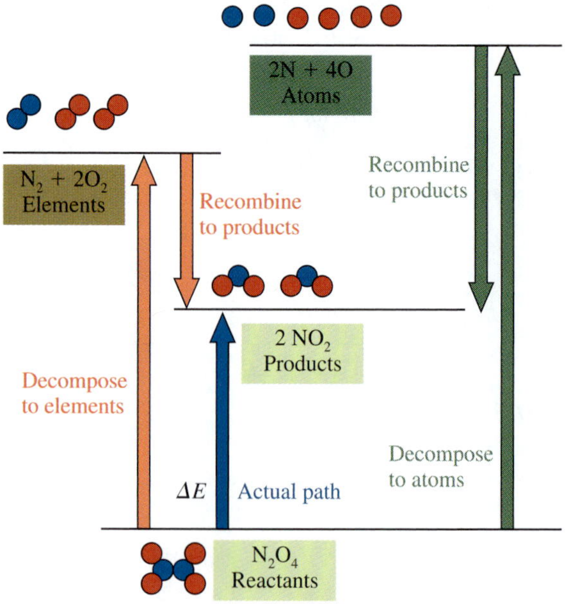

(diagram with labels)
- 2N + 4O Atoms
- Recombine to products
- $N_2 + 2O_2$ Elements
- Recombine to products
- 2 NO₂ Products
- Decompose to elements
- Decompose to atoms
- ΔE | Actual path
- N_2O_4 Reactants

6.29 H_2O, -475 kJ; NH_3, -90 kJ; and $CH_4 + H_2O$, -205 kJ

6.31 -1.4×10^3 J

6.33 14.4 kJ/°C
6.35 34.46 kJ; -4.96×10^3 kJ/mol
6.37 -2.5×10^2 J
6.39 (a) -1411.1 kJ; (b) 91.8 kJ; (c) -1597.0 kJ; and (d) -21.2 kJ
6.41 (a) -1406.1 kJ; (b) 86.8 kJ; (c) -1597.0 kJ; and (d) -11.3 kJ
6.43 (a) $3\,K\,(s) + P\,(s) + 2\,O_2\,(g) \rightarrow K_3PO_4\,(s)$;
(b) $2\,C\,(graphite) + 2\,H_2\,(g) + O_2\,(g) \rightarrow CH_3CO_2H\,(l)$;
(c) $3\,C\,(graphite) + \frac{9}{2}\,H_2\,(g) + \frac{1}{2}\,N_2\,(g) \rightarrow (CH_3)_3N\,(g)$; and
(d) $2\,Al\,(s) + \frac{3}{2}\,O_2\,(g) \rightarrow Al_2O_3\,(s)$
6.45 (a) -1166.2 kJ; and (b) -1531.4 kJ
6.47 1×10^{17} J
6.49 Total is 7.89×10^{19} kJ

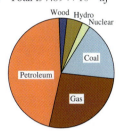

6.51 Under idling conditions
6.53 4.96×10^3 J
6.55 2.93 g
6.57 (a) $C_6H_{12}O_6 + 6\,O_2 \rightarrow 6\,CO_2 + 6\,H_2O$;
(b) -2.83×10^3 kJ/mol; and (c) -1.25×10^3 kJ/mol
6.59 Ethanol is more stable by 30 kJ/mol.
6.61 Conversion of stored chemical energy of the airplane's fuel into heat, kinetic energy, and gravitational potential energy of the airplane

6.63 20.1 J/mol K
6.65 $\Delta E_{ethane} = -46.3$ kJ/g, $\Delta E_{ethylene} = -45.6$ kJ/g, and $\Delta E_{acetylene} = -46.7$ kJ/g. Acetylene releases slightly more energy per unit mass than ethane or ethylene.
6.67 3.37×10^5 J
6.69 (a) -197.8 kJ; (b) -55.3 kJ; and (c) -851.5 kJ
6.71 -45 kJ for HCl reaction; -150 kJ for Cl_2 reaction
6.73 -197 kJ
6.75 $C = 0.382$ J g^{-1} °C^{-1}, Cu
6.77 (a) -539.8 kJ/mol; and (b) -534.3 kJ/mol
6.79 (a) wood; (b) natural gas production peaked in about 1972, at 23×10^{18} kJ; (c) 91%
6.81 (a) negative; (b) positive; and (c) 0
6.83 25.2 J mol^{-1} °C^{-1}
6.85 133 g cereal
6.87 81.3 °C; Al spoon more effective
6.89 (a) $\Delta E \neq q$ when volume changes; (b) $\Delta H \neq q$ when pressure changes; (c) $q_v \neq q$ when volume changes; (d) $q_p \neq q$ when pressure changes; (e) always true
6.91 1.9×10^2 s
6.93 5.85 °C
6.95

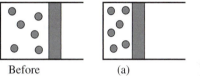

Before   (a)   (b)

6.97 $\Delta H_1° = 163.2$ kJ; $\Delta H_2° = 85.7$ kJ; $2\,NO_2 + N_2O_4$
6.99 (a) 37 kJ; (b) -692 kJ/mol

Chapter 7

7.1 (a) $V_{Ag} = 1.706 \times 10^{-29}$ m^3/atom, $V_{Pb} = 3.034 \times 10^{-29}$ m^3/atom;
(b) $d_{Ag} = 2.57 \times 10^{-10}$ m, $d_{Pb} = 3.12 \times 10^{-10}$ m; and
(c) thickness$_{Ag} = 1.7 \times 10^{-3}$ m, thickness$_{Pb} = 2.0 \times 10^{-3}$ m
7.3 Gas pressure results from transfer of momentum between atoms and container walls, indicating that atoms have momentum and mass; mass spectrometers sort atomic and molecular ions according to their masses; all matter is made up of atoms and has mass.
7.5

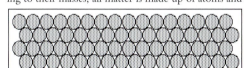

7.7 (a) 6.92×10^{16} Hz; (b) 1.28×10^{18} Hz; (c) 4.08×10^8 Hz; and (d) 6.56×10^{13} Hz
7.9 (a) 6.29×10^{-2} m; (b) 1.04×10^6 cm; (c) 5.0×10^9 mm; and (d) 1.04×10^8 μm
7.11 (a) 4.049×10^{-19} J/photon, 2.438×10^2 kJ/mol;
(b) 7.79×10^{-18} J/photon, 4.69×10^3 kJ/mol; and
(c) 1.6855×10^{-23} J/photon, 1.0150×10^{-2} kJ/mol
7.13 1.7×10^{16} photons
7.15 (a) 161 nm, 1.86×10^{15} Hz; and (b) 560 nm, 5.35×10^{14} Hz
7.17 (a) 231 nm; (b) 3.4×10^{-19} J; and (c) 580 nm
7.19

7.21 (a) 1 to 1000 m; (b) yellow; and (c) radar region
7.23 745 kJ/mol

7.25 $\lambda_{8-1} = 92.57$ nm, $\lambda_{9-1} = 92.27$ nm; both are in the ultraviolet region
7.27 Under normal conditions, absorptions occur between the lowest state and any higher-energy state. Emission originates from any higher state, to any lower state.
7.29 $5.485\,7990 \times 10^{-4}$ g/mol
7.31 (a) 1.45 nm; (b) 6.38 nm; and (c) 4.41 μm
7.33 (a) 1.71×10^{-20} J; (b) 1.11×10^{-38} J; and (c) 3.08×10^{-33} J
7.35

n	6	6	6	6	6	6
l	1	1	1	1	1	1
m_l	$+1$	$+1$	0	0	-1	-1
m_s	$+\frac{1}{2}$	$-\frac{1}{2}$	$+\frac{1}{2}$	$-\frac{1}{2}$	$+\frac{1}{2}$	$-\frac{1}{2}$

7.37 $l = 0$, $m_l = 0$; $l = 1$, $m_l = 1, 0$, or -1; and $l = 2$, $m_l = +2$, $+1, 0, -1$, or -2; in all cases, $m_s = +\frac{1}{2}$ or $-\frac{1}{2}$
7.39 (a) nonexistent: m_s must be $+\frac{1}{2}$ or $-\frac{1}{2}$; (b) actual; (c) nonexistent: l must be less than n; and (d) actual
7.41 3, 2, 2, $+\frac{1}{2}$
7.43

7.45 x and z axis y axis

7.47 (a) $2p$, $n = 2$, $l = 1$; (b) and (c) $3d$, $n = 3$, $l = 2$

7.49 Electron density plots fail to show whether or not an orbital has spherical symmetry, and they do not directly show electron densities.

7.51

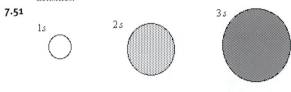

7.53

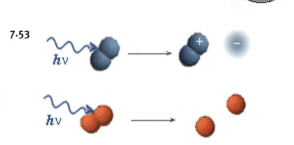

7.55 $E = 352$ kJ/mol; $E_{kinetic} = 2.10 \times 10^{-19}$ J

7.57 (a) O_2, N_2, and O_3 (thermosphere); and (b) O_3 (stratosphere)

7.59 (a) ~50 km; (b) upper stratosphere; and (c) formation of ozone

7.61 (a) 6.7×10^{-6} m; (b) 3.0×10^{-20} J; and (c) 1.53 s

7.63 492 nm; yes

7.65 Six:

n	4	4	4	4	4	4
l	1	1	1	1	1	1
m_l	+1	+1	0	0	−1	−1
m_s	$+\frac{1}{2}$	$-\frac{1}{2}$	$+\frac{1}{2}$	$-\frac{1}{2}$	$+\frac{1}{2}$	$-\frac{1}{2}$

7.67

Original   2x frequency

7.69 552.7 nm

7.71 (a) 3.30×10^{-19} J; (b) 3.31×10^{-7} m; and (c) 9.45×10^{-10} m

7.73 11.0 m; 1.81×10^{-26} J

7.75 (a) 2×10^{-4} atm; (b) N O_2, O_3; (c) ~35 km; and (d) stratosphere

7.77

λ (nm)	487	514	543	553	578
ν (10^{14} s^{-1})	6.16	5.83	5.52	5.42	5.19
E (kJ/mol)	246	233	220	216	207

7.79 95.0 nm; UV

7.81 No

7.83 $V_{atom} = 4 \times 10^{-30}$ m^3; $V_{nucleus} = 4 \times 10^{-45}$ m^3; $1/10^{15}$

7.85 6 photons

7.87 Electron: 4.853×10^{-11} m; proton: 2.642×10^{-14} m

7.89 $Photon_1 = 565$ nm (d to c); $Photon_2 = 121$ nm (c to b); $Photon_3 = 152$ nm (b to a). The energies relative to a are: b, 1.31×10^{-18} J; c, 2.95×10^{-18} J; d, 3.30×10^{-18} J.

7.91 (a) $E_{488} = 4.07 \times 10^{-19}$ J; $E_{514} = 3.86 \times 10^{-19}$ J;

(b)

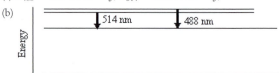

and (c) 4.17×10^{15} s^{-1} and 71.9 nm

7.93 (a) $E_{589.6} = 202.8$ kJ/mol; $E_{590.0} = 202.7$ kJ/mol;

(b)

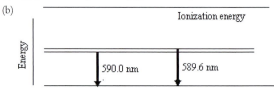

and (c) 423 nm

7.95 $E_A = 2.7 \times 10^{-19}$ J, $E_B = 4.3 \times 10^{-19}$ J; Metal B has the higher binding energy. (b) 1.59×10^{-18} J; and (c) 460 nm to 750 nm

Chapter 8

8.1 (a) He $1s$, lower value of n; (b) Kr $5s$, lower value of l; and (c) He$^+$ $2s$, less screened by $1s$ electrons

8.3 H atom, no screening effect, so orbital energy depends only on n and Z, and all $n = 3$ orbitals have identical energy. He atom, the second electron screens the first, and the amount of screening decreases as l increases.

8.5 (a) IE of the He $2p$ orbital is not much larger than that of the H $2p$ orbital. (b) IE of the He$^+$ $2p$ orbital is four times larger than that of the H $2p$ orbital.

8.7

8.9 Atomic number = 114; valence orbital = $7p$

8.11 Row 7, Group 11 (directly below gold)

8.13 O, 6 valence electrons; V, 5; Rb, 1; Sn, 4; Cd, 2

8.15 (a) Be, 4 electrons, $1s^2\, 2s^2$

n	2	2
l	0	0
m_l	0	0
m_s	$+\frac{1}{2}$	$-\frac{1}{2}$

(b) O, 8, $1s^1\, 2s^2\, 2p^4$;

n	2	2	2	2	2	2
l	0	0	1	1	1	1
m_l	0	0	1	0	−1	1
m_s	$+\frac{1}{2}$	$-\frac{1}{2}$	$+\frac{1}{2}$	$+\frac{1}{2}$	$+\frac{1}{2}$	$-\frac{1}{2}$

(c) Ne, 10 electrons, $1s^1\, 2s^2\, 2p^6$; first six valence electrons the same as for O; remaining two:

n	2	2
l	1	1
m_l	0	−1
m_s	$-\frac{1}{2}$	$-\frac{1}{2}$

(d) P, 15 electrons, $1s^2\, 2s^2\, 2p^6\, 3s^2\, 3p^3$

n	3	3	3	3	3
l	0	0	1	1	1
m_l	0	0	−1	1	0
m_s	$+\frac{1}{2}$	$-\frac{1}{2}$	$+\frac{1}{2}$	$+\frac{1}{2}$	$+\frac{1}{2}$

8.17

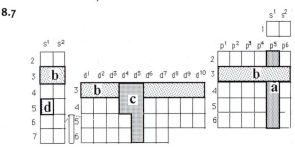

8.19 (a) Pauli-forbidden; (b) Pauli-forbidden; (c) excited-state configuration; (d) excited-state configuration; (e) ground-state configuration; (f) non-existent orbital; and (g) non-existent orbital

8.21

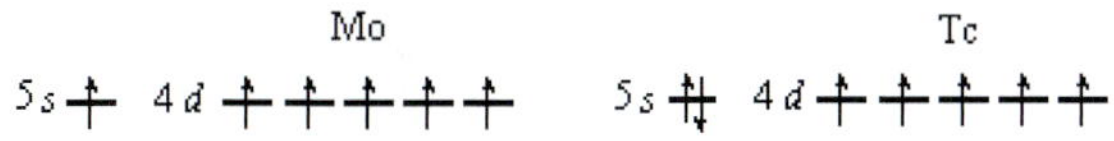

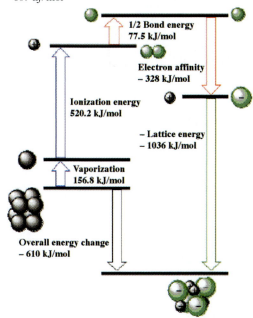

Mo   Tc

$5s$ ⇅ $4d$ ↑ ↑ ↑ ↑ ↑ $5s$ ⇅ $4d$ ↑ ↑ ↑ ↑ ↑

8.23 C: [He] $2s^2 2p^2$; Cr: [Ar] $4s^1 3d^5$; Sb: [Kr] $5s^2 4d^{10} 5p^3$; Br: [Ar] $4s^2 3d^{10} 4p^5$

8.25 Seven: $1s^1 2s^2 2p^4$; $2s^2 2p^5$; $1s^1 2s^1 2p^5$; $1s^2 2s^1 2p^4$; $1s^2 2p^5$; $2s^1 2p^6$; $1s^1 2p^6$

8.27 Ar > Cl > K > Cs

8.29 Column 1 (Cs)

8.31 N, electron adds to an already-occupied orbital; electron–electron repulsion makes this a disfavored process. Mg, electron adds to the next higher orbital, which is significantly higher in energy. Zn, electron adds to the next higher orbital, which is significantly higher in energy.

8.33 Y^{3+} < Sr^{2+} < Rb^+ < Br^- < Se^{2-} < As^{3-}

8.35 Stable anion: Cl; stable cations: Ca, Cu, Cs, Cr

8.37 -270 kJ/mol

8.39 (a) $Ba^{3+} O^{3-}$; (b) $Ba^+ O^-$; (c) $Ba^{2+} O^{2-}$. The gain in lattice energy more than offsets the additional energy required to form the ions, but the energy required to remove the third electron from Ba is prohibitive.

8.41 -607 kJ/mol

8.43

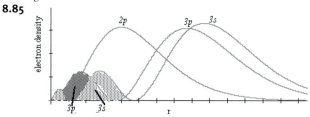

1/2 Bond energy
77.5 kJ/mol

Electron affinity
– 328 kJ/mol

Ionization energy
520.2 kJ/mol

– Lattice energy
– 1036 kJ/mol

Vaporization
156.8 kJ/mol

Overall energy change
– 610 kJ/mol

8.45 The decrease from Rb_2O to Cs_2O will be about the same as from K_2O to Rb_2O, 75 kJ/mol. The predicted value is $2163 - 75 = 2088$ kJ/mol.

8.47 Metal: Po; nonmetal: O, S, Se; metalloid: Te

8.49 Metal: Ca, Cu, Cs, Cr; nonmetal: C, Cl

8.51 Al, Ga, Sn, Bi

8.53 (a) Cr, $4s^1 3d^5$; and Cu, $4s^1 3d^{10}$. (b) In column 6, Cr and Mo have $s^1 d^5$ configurations, while W and Sg have $s^2 d^4$ configurations. (c) The $6d$ and $5f$ orbitals fill in the second f block, where Pa, U, Np, and Cm have both these orbitals partly filled, indicating nearly equal energies.

8.55 Mn^{2+}: $1s^2 2s^2 2p^6 3s^2 3p^6 3d^5$;

n	3	3	3	3	3
l	2	2	2	2	2
m_l	2	1	0	-1	-2
m_s	$+\frac{1}{2}$	$+\frac{1}{2}$	$+\frac{1}{2}$	$+\frac{1}{2}$	$+\frac{1}{2}$

8.57 P: spin $= \frac{3}{2}$; Br^-: spin $= 0$; Cu^+: spin $= 0$

8.59 Na < O < N < Ne < Na^+

8.61 $Br^- > Cl^- > Cl > K^+$

8.63 $Z = 9$ is F: [He] $2s^2 2p^5$; $Z = 20$ is Ca: [Ar] $4s^2$; and $Z = 33$ is As: [Ar] $4s^2 3d^{10} 4p^3$

8.65 Average difference between alkali fluorides and alkali chlorides is 125 kJ/mol; average difference between alkali bromides and alkali iodides is 42.6 kJ/mol. Values decrease with the size of the alkali cation, because a larger cation cannot get as close to the anion as can a smaller cation.

8.67

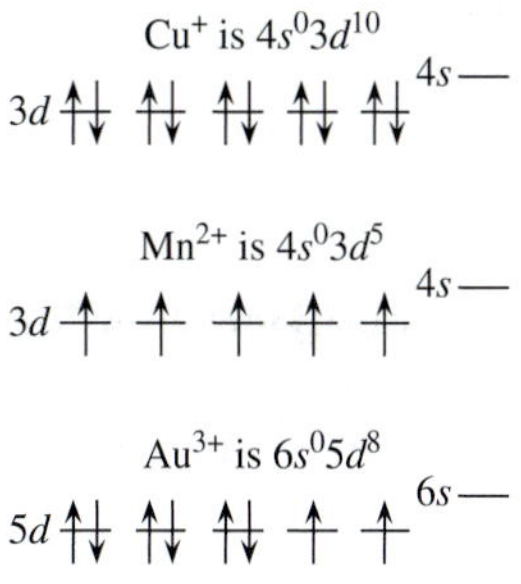

Cu^+ is $4s^0 3d^{10}$

$3d$ ⇅ ⇅ ⇅ ⇅ ⇅ $4s$ —

Mn^{2+} is $4s^0 3d^5$

$3d$ ↑ ↑ ↑ ↑ ↑ $4s$ —

Au^{3+} is $6s^0 5d^8$

$5d$ ⇅ ⇅ ⇅ ↑ ↑ $6s$ —

8.69 (a) In all one-electron atoms, orbital energy depends only on n and Z, both of which are the same for the hydrogen atom $2s$ and $2p$ orbitals. (b) In multi-electron atoms, n orbitals with different l values are screened to different extents. The orbital with the lower l value is less screened, hence more stable.

8.71 (a) from element 80 (Hg) to 81 (Tl); (b) Cs, element 55; (c) $Z = 56$ to 71, $Z = 39$ to 46, and $Z = 20$ to 29; and (d) elements 15 (P), 16 (S), 33 (As), 34 (Se), 53 (I), 80 (Hg), 85 (At), and 86 (Rn)

8.73 S^+

8.75 Ground state is (c), (a) and (b) are excited states, but (d) is non-existent.

8.77 (a) MgS; (b) KF; and (c) $BeCl_2$

8.79 (a) b is most stable (smallest n value), and a is more stable than c; (b) $n = 3$, $l = 0$, $m_l = 0$, $m_s = +\frac{1}{2}$; $n = 3$, $l = 0$, $m_l = 0$, $m_s = -\frac{1}{2}$; (c) Any element from $Z = 12$ to $Z = 20$; (d) Several $3d$ transition metal cations, for example, Fe^{3+}; (e) eight; and (f) smaller

8.81 Be: [He] $2s^1 2p^1$; O^{2-}: [He] $2s^2 2p^5 3s^1$; Br^-: [Ar] $4s^2 3d^{10} 4p^5 5s^1$; Ca^{2+}: [Ne] $3s^2 3p^5 4s^1$; Sb^{3+}: [Kr] $4d^{10} 5s^1 5p^1$

8.83 The first ionization energies involve the $2s$ orbital for both atoms, so Be, with larger Z, has a larger IE. The second electron removed from Li is a core electron, so the IE is much greater than the second IE for Be.

8.85

[graph: electron density vs r, showing curves labeled $2p$, $3p$, $3s$ and shaded regions labeled $3p$ and $3s$]

8.87 Francium is an alkali metal (Column 1) whose properties should closely resemble those of cesium: low IE, highly reactive, forms $+1$ cation, soft metal, melts at a low temperature.

8.89 Removing electrons from any atom or ion reduces screening and electron–electron repulsion and therefore stabilizes the orbitals. The Li^{2+} cation has a more stable $2s$ orbital.

8.91 At what distance from the nucleus has the electron density decreased sufficiently that we consider it to be zero? That question has no unambiguous answer.

Chapter 9

9.1 (a) O: $1s^2\,2s^2\,2p^4$, six $n = 2$ electrons; (b) P: $1s^2\,2s^2\,2p^6\,3s^2\,3p^3$, five $n = 3$ electrons; (c) B: $1s^2\,2s^2\,2p^1$, three $n = 2$ electrons; and (d) Br: $1s^2\,2s^2\,2p^6\,3s^2\,3p^6\,3d^{10}\,4s^2\,4p^5$, seven $n = 4$ electrons

9.3 One valence $5p$ orbital of the iodine atom overlaps with the $1s$ orbital of the hydrogen atom to form a bond:

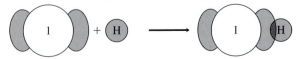

9.5 (a) Al, Group 13, three valence electrons; (b) As, Group 15, five valence electrons; (c) F, Group 17, seven valence electrons; and (d) Sn, Group 14, four valence electrons.

9.7 The $2s$ orbital of a lithium atom can overlap with the $1s$ orbital of a hydrogen atom to produce an orbital with high electron density between the nuclei:

9.9 (a) N (3.0) will attract electrons more than C (2.5); (b) S (2.5) will attract electrons more than H (2.1); (c) I (2.5) will attract electrons more than Zn (1.6); and (d) S (2.5) will attract electrons more than As (2.0).

9.11 (a) δ^+ Si—O δ^-; (b) δ^- N—C δ^+; (c) δ^+ Cl—F δ^-; and (d) δ^- Br—C δ^+.

9.13 $PH_3 < H_2S < NH_3 < H_2O$

9.15 (a) 32; (b) 90; (c) 24; and (d) 32

9.17

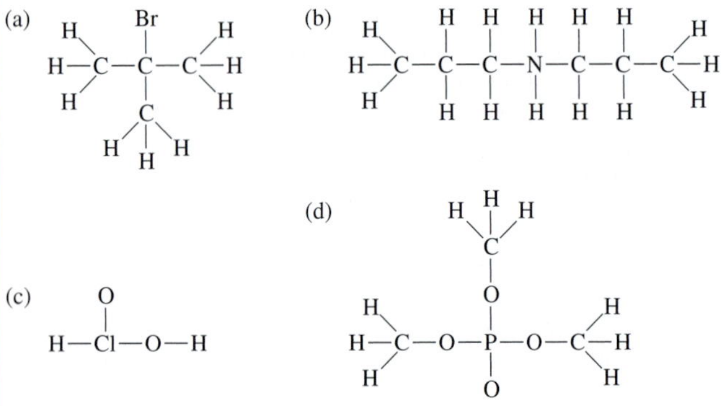

(a), (b), (c), (d) structures

9.19 structures (a), (b), (c)

9.21 structures (a), (b), (c)

9.23 (a), (b), (c) structures

9.25 (a), (b), (c), (d) structures

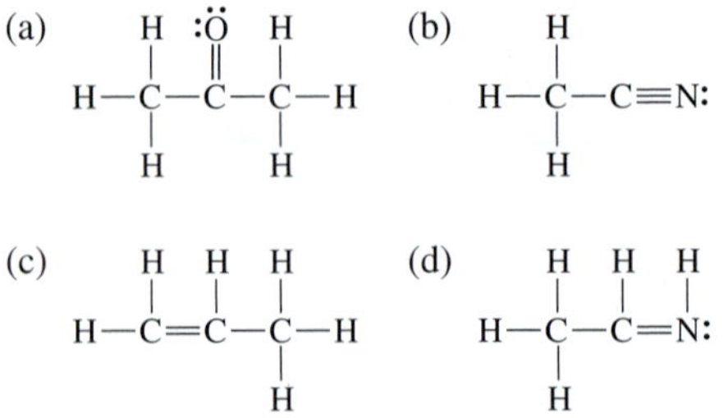

9.27

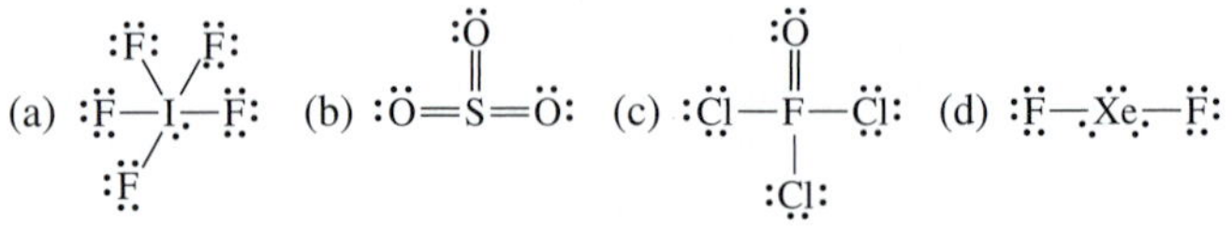

(a), (b), (c), (d) structures

9.29 (a), (b), (c), (d) structures

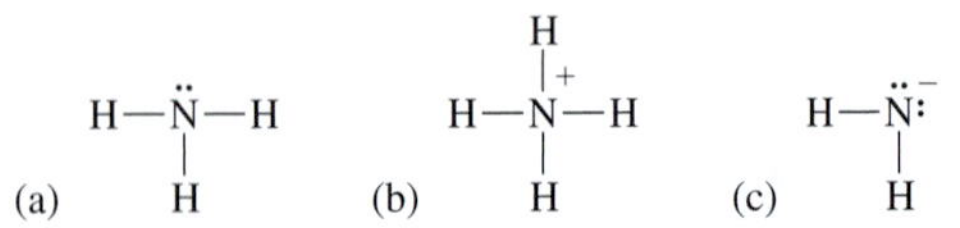

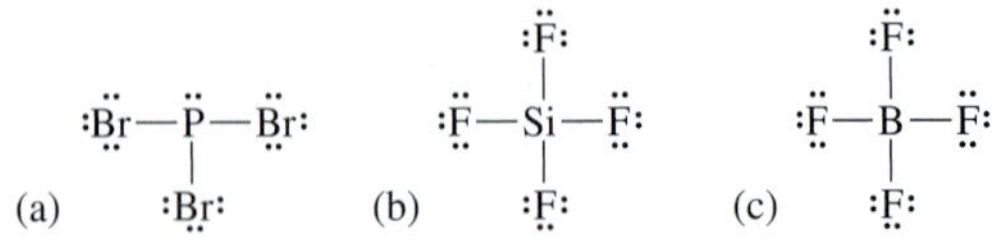

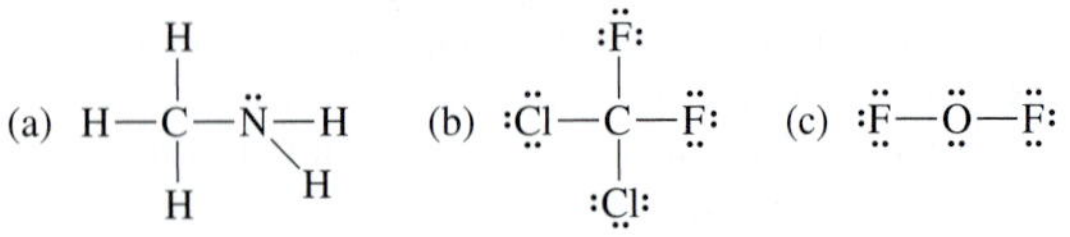

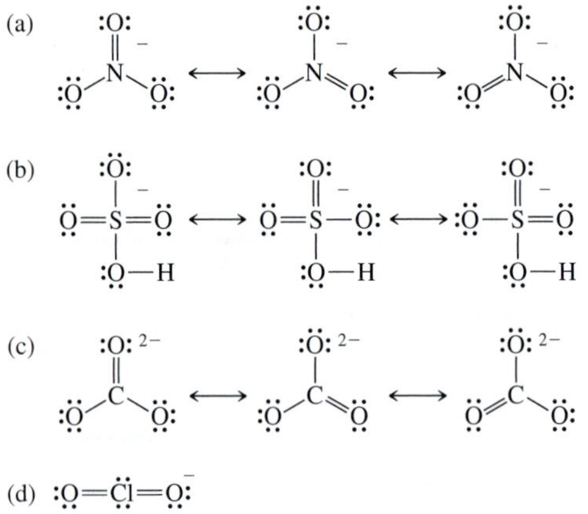

9.31 (a) tetrahedral; (b) tetrahedral; and (c) trigonal pyramidal

(a), (b), (c) structures

9.33

structure

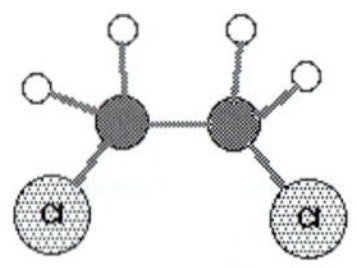

9.35 "Straight chain": C—C—C—C—C—C

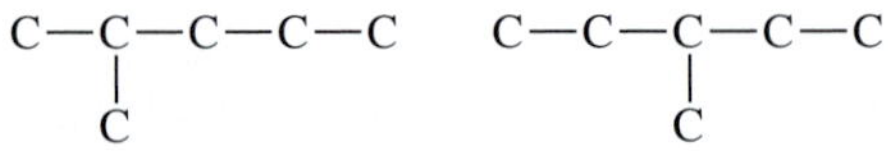

structures. Putting the branched carbon in Position 4 gives the same isomer as putting it in Position 2

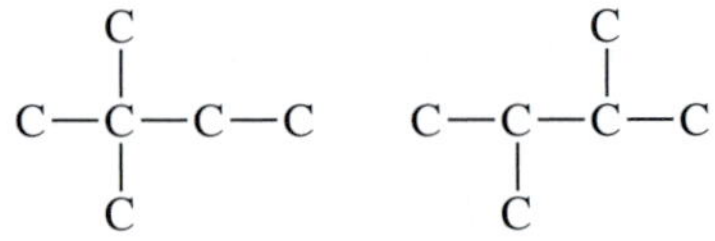

9.37 Shape about the N atom is trigonal pyramidal; shape about C atoms is tetrahedral

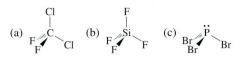
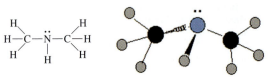

9.39

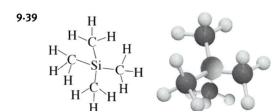

Hybridization = sp^3; geometry = tetrahedral

9.41 (a) T-shaped; (b) seesaw; (c) trigonal planar; and (d) square pyramidal

9.43 GeF$_4$, tetrahedron, sp^3; SeF$_4$, seesaw, dsp^3; XeF$_4$, square plane, d^2sp^3

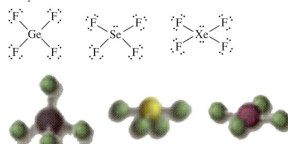

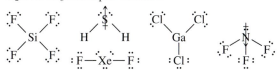

9.45 (a) bent, 120°; (b) trigonal bipyramidal, 90° and 120°; (c) seesaw, 90° and 120°; and (d) square planar, 90°

9.47 H$_2$S and NF$_3$ have dipole moments.

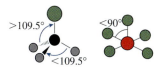

9.49 CO$_2$ has no lone pairs on the inner C atom, giving SN = 2, linear shape, and bond polarities that cancel. SO$_2$ has a lone pair on the inner S atom, bent shape, and polar bonds whose moments do not cancel.

9.51 (a) ideal; (b) and (c) deviate

9.53 H—N < C≡N < N≡N < N—N < Cl—N

9.55 C—C < H—N < C≡C < C=O < N≡N

9.57

(a)
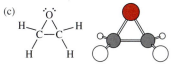

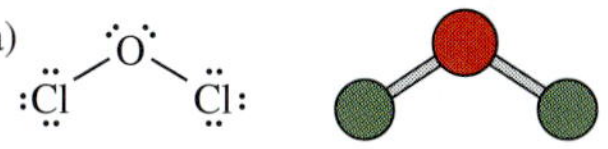

(b)

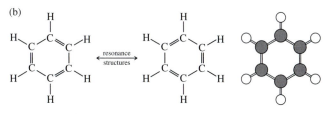

(c)

9.59 SiCl$_4$ and CI$_4$, tetrahedral; SeF$_4$, seesaw

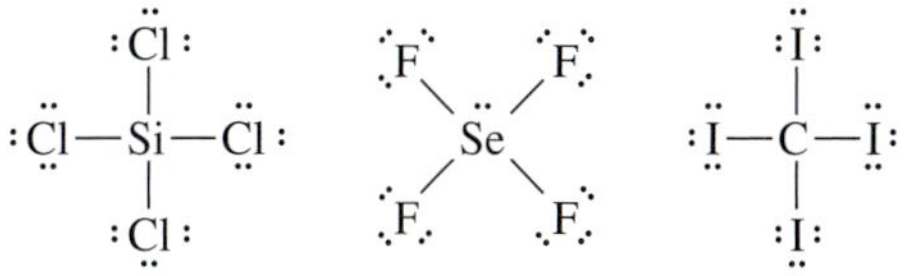

9.61 (a) C—N is shorter but weaker than C—C; (b) C—N is longer but stronger than N—N.

9.63 Two isomers

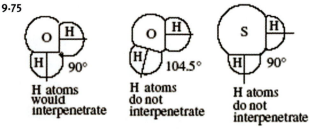

9.65 Empirical formula is SiO$_4$$^{4-}$. Orthosilicate networks are ionic, containing SiO$_4$$^{4-}$ anions and metal cations. The Si—O bonds form by overlap of oxygen $2p$ orbitals with silicon sp^3 orbitals. Anions are tetrahedral.

9.67 (a) FC$_{Br}$ = 0

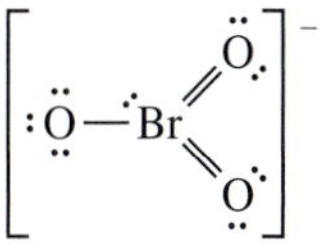

$$\left[:\overset{..}{O}—\overset{..}{Br}\begin{smallmatrix}\\ =\overset{..}{O}\\ \backslash\\ \overset{..}{O} \end{smallmatrix} \right]^{-} + \text{two others}$$

(b) FC$_N$ = 0

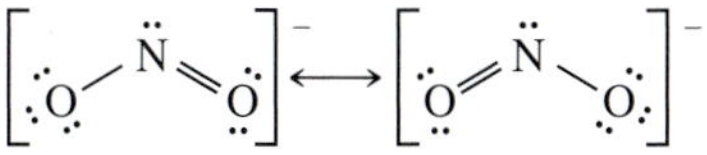

$$\left[:\overset{..}{O}—N=\overset{..}{O}: \right]^{-} \longleftrightarrow \left[:\overset{..}{O}=N—\overset{..}{O}: \right]^{-}$$

(c) FC$_P$ = 0

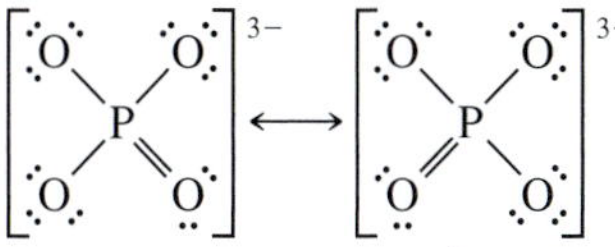

$$\left[\begin{smallmatrix} :\overset{..}{O}: & :\overset{..}{O}:\\ \backslash & /\\ & P\\ / & \backslash\\ :\overset{..}{O}: & :\overset{..}{O}: \end{smallmatrix} \right]^{3-} \longleftrightarrow \left[\begin{smallmatrix} :\overset{..}{O}: & :\overset{..}{O}\\ \backslash & /\\ & P\\ / & \backslash\\ :\overset{..}{O}: & :\overset{..}{O}: \end{smallmatrix} \right]^{3-} + \text{two others}$$

(d) FC$_C$ = 0; FC$_O$ = 0

$$\left[\begin{smallmatrix} :\overset{..}{O}:\\ \backslash\\ C—\overset{..}{O}—H\\ /\\ :\overset{..}{O}: \end{smallmatrix} \right]^{-} \longleftrightarrow \left[\begin{smallmatrix} :\overset{..}{O}\\ \backslash\\ C—\overset{..}{O}—H\\ /\\ :\overset{..}{O}: \end{smallmatrix} \right]^{-}$$

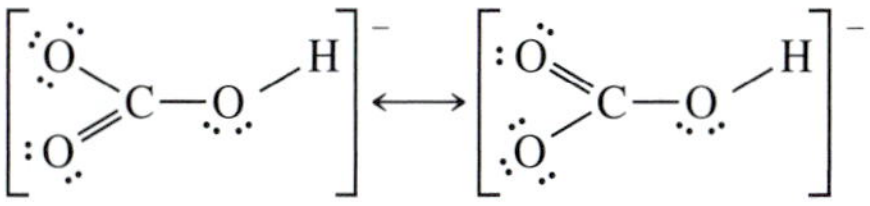

9.69 (a) seesaw, 90° and 120°, one lone pair, SF$_4$; (b) square planar, 90°, two lone pairs, XeF$_4$; and (c) tetrahedron, 109.5°, no lone pairs, CH$_4$

9.71 H—O < H—C < C≡N < C=O

9.73 (a)

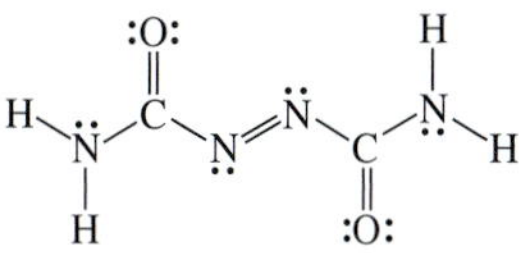

(b) end N atoms, SN = 4, with one lone pair, trigonal pyramidal; C atoms and the center N atoms, SN = 3, trigonal planar

9.75

H atoms would interpenetrate   H atoms do not interpenetrate   H atoms do not interpenetrate

9.77 Si—H < C—H < N—H < O—H < F—H

9.79 They are the same compound, because they can be superimposed after rotating one by 109.5°.

9.81 The six-membered ring of benzyne requires bond angles of 120°. The C atoms in the triple bond have SN = 2, for which the optimum VSEPR angle is 180°. Thus, these atoms react readily.

9.83 N_2O must have the structure N—N—O if it is to have a dipole moment.

$$\ddot{\text{N}}{=}\text{N}{=}\ddot{\text{O}} \qquad \ddot{\text{N}}{=}\text{O}{=}\ddot{\text{N}}$$

9.85 The isomer with Cl atoms at opposite ends is nonpolar, and the isomer with Cl atoms adjacent to each other has the largest dipole moment.

9.87 Compounds with formula XF_5 have five electron pairs associated with the inner atom. This is possible for phosphorus, a third-row element that has *d* orbitals available for bonding. It is not possible for nitrogen, a second-row element that lacks valence *d* orbitals.

9.89 (a) TeF_2; (b) TeF_3^-; (c) TeF_5^-; (d) TeF_5^+; (e) TeF_6; and (f) TeF_4

9.91 (a)
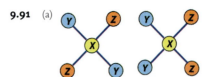

(b) Like atoms opposite each other, no dipole moment; like atoms adjacent, dipole moment

9.93 IF_7

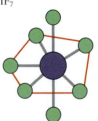

Chapter 10

10.1 The bond forms by overlap of two 4*p* orbitals.

10.3 The bond in LiH forms by overlap of the hydrogen 1*s* orbital and the lithium 2*s* orbital.

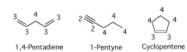

10.5 Each bond results from overlap between a 5*p* orbital from Sb and a 2*p* orbital from F. There are three identical bonds that point at near-right angles to one another.

10.7 (a) sp^3; (b) sp^3; (c) sp^2; and (d) sp^3d^2

10.9 (a) sp^3; (b) sp; and (c) sp^3d^2

10.11 (a) sp^3; (b) sp^3d; (c) sp^2; and (d) sp

10.13 (a) sp^2; (b) sp^3d^2; (c) sp^3d; and (d) sp^3

10.15 C has SN = 4, is tetrahedral, and uses sp^3 hybrids to form three C—Cl bonds and one C—H bond. Cl uses a 3*p* orbital for bond formation and has lone pairs in the 3*s* and two other 3*p* orbitals. H atom uses its 1s orbital to form the C—H bond.

C—H bond   C—Cl bond

10.17 4 sp^3 (N) − 1*s* (H) σ bonds, 1 sp^3 (N) − sp^3 (N) σ bond, 2 lone pairs in sp^3 hybrids

N—N bond   N—H bond

10.19 Acetone has three inner C atoms. Two have only single bonds (one C—C and three C—H); these have SN = 4. The atom bonded to O has one π bond and SN = 3.

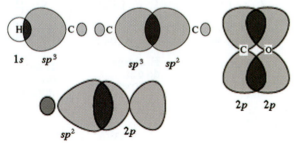

1s sp^3 sp^3 sp^2

2p   2p

sp^2 2p

10.21 Some of the inner atoms have all single bonds, SN = 4, sp^3 hybridization, and tetrahedral geometry; others have one double bond, SN = 3, sp^2 hybridization, and trigonal planar geometry. The SN values for the 20 carbon atoms are shown in the line structure at right.

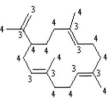

10.23

1,4-Pentadiene   1-Pentyne   Cyclopentene

10.25 (a) σ bond; (b) no bond; (c) σ bond; and (d) π bond

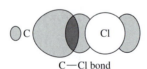

$2p_z$ $2p_z$ $2p_3$ $2p_z$
(a)   (b) $2p_x$ $2p_y$ (c)   (d) $2p_y$ $2p_y$

10.27 From weakest to strongest bond, the order is H_2^{2-}, H_2^-, H_2

10.29 (a) CO; (b) N_2; (c) CN^-

10.31

(a)
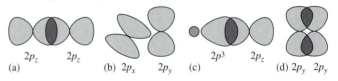
(b)

σ σ
σ^* σ^*

10.33 There are eight bonding electrons and six antibonding electrons, for a net of two bonding electrons and a single bond for this ionic species.

σ^*
π^*
π
σ
σ^*
σ

10.35 C atom has *sp* hybrids that overlap with 3*p* orbitals from the S atoms to form two σ bonds. There are two delocalized π systems at right angles to each other, each made up of a 2*p* orbital on the C atoms overlapping side-by-side with a 3*p* orbital on each S atom. As in CO_2, eight electrons occupy the delocalized π orbitals.

10.37 Isoelectric with CO_2; Lewis structure is

$$:\ddot{N}=C=\ddot{N}:^{2-}$$

C atom has SN = 2, sp hybridization, and linear geometry. There are two sets of three-center delocalized π orbitals.

10.39 N bonded to H is bent, sp^2 hybridized, angle between 109–120°; outer N is not hybridized, middle N atom is linear, sp hybridized.

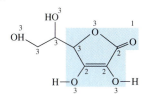

10.41 Xanthin has a delocalized π system in which all 26 double-bonded C atoms contribute.

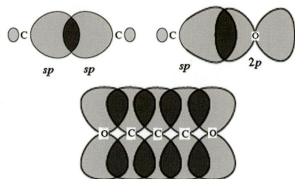

10.43 (a) Hybridization is as shown on the line drawing ($3 = sp^3$, $2 = sp^2$).
(b–c) 2 π bonds, on adjacent atoms, four electrons in the extended π system; and (d) eight atoms screened in blue

10.45 Lewis structure:

$$:\ddot{O}=C=C=C=\ddot{O}:$$

Linear, σ bond network can be described using sp hybrids; two sets of delocalized π orbitals, each extending over all five atoms:

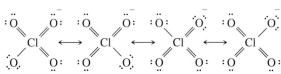

10.47 Tetrahedral geometry, sp^3 hybrids on Cl overlap with $2p$ orbitals from O; π system extends over all five atoms, incorporating d orbitals from the inner Cl atom; three bonding π orbitals occupied by six electrons.

10.49 (a) undoped; (b) n-type; and (c) undoped

10.51 Each K atom contributes one valence electron to the valence (bonding) band of the solid, each Fe atom contributes eight electrons. Thus, iron has much greater bonding, making it harder and giving it a higher melting point than potassium.

10.53 p-type semiconductor

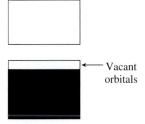

← Vacant orbitals

10.55 The S atom in each compound has SN = 3, so sp^2 hybrids overlapping with oxygen $2p$ orbitals describe the σ bonds: two in SO_2 and three in SO_3. SO_2 is bent, and SO_3 is trigonal planar. There are two π bonds in SO_2 and three π bonds in SO_3. All the π orbitals extend over the entire molecule.

10.57 $\nu = 8.70 \times 10^{13}$/s, far infrared

10.59 4 sp^2-1s C—H σ bonds, 2 sp^2-sp^2 C—C σ bonds, 1 sp^2-$2p$ C—O σ bond, delocalized π network extending over four atoms; two bonding π orbitals; one $2s$ and one $2p$ lone pair on the outer O atom.

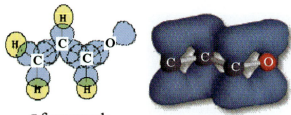

σ framework π system

10.61 (a)

Cinnamic acid
$C_9H_8O_2$

(b, c) A: sp^2 and p, 120°; B: $2p$, no angle (outer); C: sp^3, 109.5°; D: sp^2 and p, 120°; E: atomic $1s$, no angle (outer); and (d) five π bonds

10.63 (a) and (d), diamagnetic; (b) and (c), paramagnetic

10.65

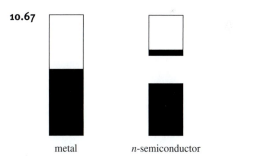

All C's use sp^2 C and N use sp^3

10.67

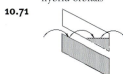

metal n-semiconductor

10.69 Central C, SN = 3, trigonal planar geometry, sp^2 hybrid orbitals; methyl C atoms, SN = 4, tetrahedral geometry, sp^3 hybrid orbitals

10.71

10.73 O_2: $(\sigma_s)^2(\sigma_s^*)^2(\sigma_p)^2(\pi_x)^2(\pi_y)^2(\pi_x^*)^1(\pi_y^*)^1$ Bond order = 2

O_2^-: $(\sigma_s)^2(\sigma_s^*)^2(\sigma_p)^2(\pi_x)^2(\pi_y)^2(\pi_x^*)^2(\pi_y^*)^1$ Bond order = 1.5

O_2^{2-}: $(\sigma_s)^2(\sigma_s^*)^2(\sigma_p)^2(\pi_x)^2(\pi_y)^2(\pi_x^*)^2(\pi_y^*)^2$ Bond order = 1

O_2 has the strongest, shortest bond and O_2^{2-} the longest, weakest bond. O_2 and O_2^- are both magnetic, with O_2 showing the largest magnetism.

10.75 Cl uses sp^3 hybrids that overlap with $2p$ orbitals from O. The π system of the chlorate ion is extended over all four atoms, and there are two bonding π orbitals occupied by four electrons. The π system of the chlorite ion is extended over all three atoms, with one bonding π orbital occupied by two electrons.

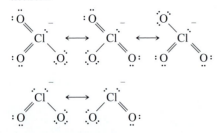

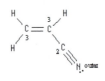

10.77 Higher bond order should be accompanied by increased bond strength and decreased bond length. Bond length and bond strength can be measured.

10.79 (a) The carbon atoms with SN = 3 have σ bonding that can be described using sp^2 hybrid orbitals, and the σ bonding of the carbon atom with SN = 2 can be described using sp hybrids.

(b)

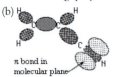

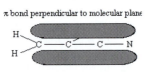

(c) There is an extended π network that includes the three C atoms and the N atom.

10.81 There are 7 bonding electrons and 2 antibonding electrons, for a net of 5, or bond order 2.5. The least stable occupied orbital is a σ_p orbital:

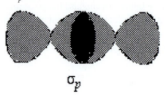

σ_p

10.83 Its geometry is tetrahedral, the σ bonds can be described using sp^3 hybrids from Cl, and the molecule has a bond angle near 109°. There is an extended π system formed from d orbitals on Cl overlapping side-by-side with $2p$ orbitals from the two O atoms. It is unusual because it has an odd number of electrons.

10.85 MO diagram shows a bond order of zero for Be_2, which cannot be prepared.

10.87

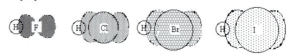

10.89 Multiple bonds involving π overlap are strong for N and O, but π overlap is very slight for P and S.

10.91

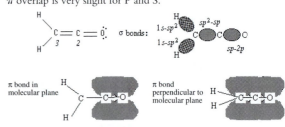

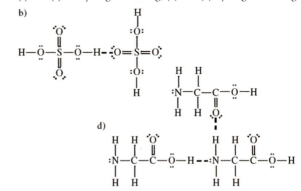

Chapter 11

11.1

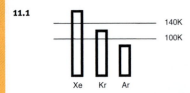

11.3 (a) less significant; (b) more significant; and (c) less significant

11.5 $Ag_{(s)}$ $Ar_{(g)}$ $Hg_{(l)}$

11.7 -7.9%

11.9 21.5 atm; -25% deviation from ideal gas behavior

11.11 CH_4 (hardest to liquefy) $< CF_4 < CCl_4$ (easiest to liquefy)

11.13 CH_4 CCl_4

11.15 Propane (lowest bp) $< n$-pentane $<$ ethanol (highest bp)

11.17 (a) and (c) no hydrogen bonding; (b) and (d) hydrogen bonding

b)

d)

11.19 (a)

(b)

11.21 Pentane (C_5) $<$ gasoline (C_8) $<$ fuel oil (C_{12})

11.23 Water will be concave, mercury will be convex.

11.25 Paper towels contain cellulose, which forms many hydrogen bonds with water. There are no strong intermolecular interactions between an oil and the fibers of paper towels.

11.27 (a) benzene, no dipole moment; (b) hexane, no H-bonding; and (c) heptane, smaller dispersion forces

11.29 Sn: metallic solid; S_8: molecular solid; Se: network solid; SiO_2: network solid; Na_2SO_4: ionic solid

11.31 (a) The bonding in metals comes from extended networks of delocalized electrons, while the bonding in network solids includes many individual covalent bonds. (b) Metals conduct electricity, are malleable and ductile, and are shiny in appearance. Network solids are nonconductors or semiconductors, are brittle, and often have dull appearances.

11.33 (a) molecular solid; (b) ionic solid; (c) metallic solid; (d) network solid; and (e) molecular solid

11.35 The density of a solid depends not only on the nature of the elements it contains but also on the tightness of bonding. Graphite has a more open bonding pattern than diamond.

11.37 $CaTiO_3$

11.39

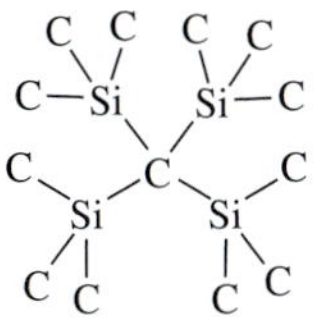

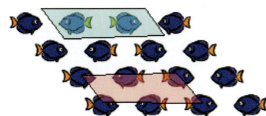

11.41 8

11.43

11.45 (a) ethane, larger dispersion forces; (b) ethanol, hydrogen bonding; and (c) argon, higher polarizability

11.47

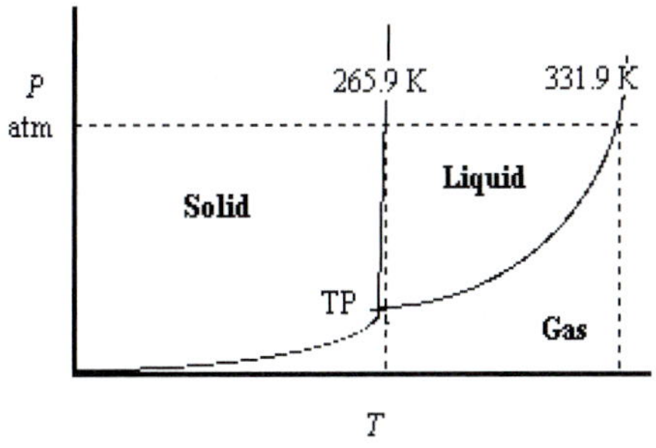

11.49 (a) Gas cools, liquefies at 331.9 K, and solidifies at 265.9 K. (b) Compressing the gas causes it to liquefy at about $P = 6 \times 10^{-2}$ atm and solidify at about 0.5 atm. (c) Solid sublimes to the vapor at about 265 K.

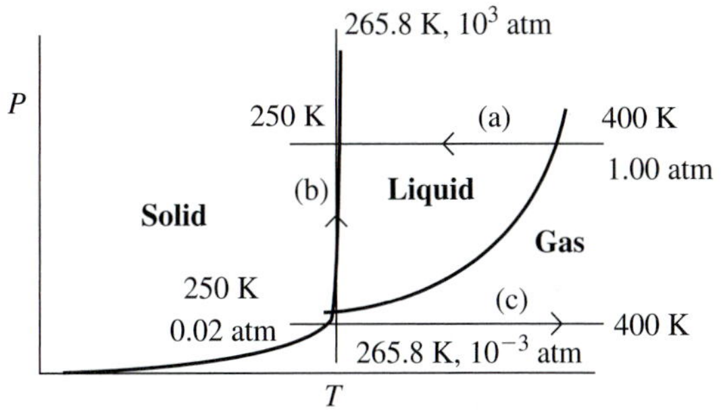

11.51 980. J

11.53 (a) dispersion interactions and hydrogen bonds; (b) dispersion interactions and polarity; (c) dispersion interactions; and (d) dispersion interactions

11.55 Face-centered cubic structure contains more atoms in a given volume and is denser.

11.57 (a) CH_3OH (hydrogen bond); (b) SiO_2 (covalent bonds); (c) HF (hydrogen bond); and (d) I_2 (larger dispersion forces)

11.59 (a) At 70 K, 1 atm, nitrogen is liquid. When the pressure falls below about 0.5 atm, the liquid boils and forms a gas. (b) The sample remains a gas. (c) The gas liquefies at 77 K, and the liquid solidifies at 63 K.

11.61 *trans*, no dipole moment, 47 °C; *cis*, dipole moment, 60 °C

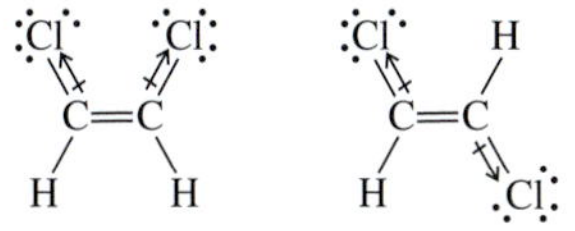

cis
dipole moment
bp = 60 °C

trans
no dipole moment
bp = 47 °C

11.63 Cl_2, greater dispersion forces

11.65 At ~ 600 °C, α quartz converts to β quartz. At ~ 850 °C, the solid becomes tridymite, and just below 1500 °C, this converts to cristobalite. At ~ 1700 °C, cristobalite melts to give liquid silica.

11.67 More closely packed is stable at high pressure: body-centered cubic

11.69 (a) network; (b) metal; and (c) ionic

11.71 (a) $T > 153$ °C, $P > 1420$ atm; (b) T, P combinations ranging from 95.3 °C, 5.1×10^{-6} atm to 153 °C, 1420 atm; (c) $P < 5.1 \times 10^{-6}$ atm

11.73 (a) simple cubic, and (b) 1 Rh, 3 O, RhO_3

11.75 Forces are smaller.

11.77 (a) At any pressure between 3.2×10^{-5} atm and 1420 atm, there is a temperature at which each phase is stable. Below 95.3 °C, the rhombic phase is stable. At a temperature between 95.3 and 153 °C, the monoclinic phase becomes stable. At a temperature between 115.2 and 153 °C, monoclinic sulfur melts and the liquid becomes stable. At a still higher temperature that depends strongly on the pressure, the vapor phase is stable. (b) At temperatures between 115.2 and 153 °C, there is a pressure at which each phase is stable. Below 3.2×10^{-5} atm, the vapor is stable. Above this pressure, the liquid is stable up to a pressure that depends on the exact temperature, but at higher pressure, the monoclinic solid phase is stable. At extremely high pressure, the rhombic phase becomes stable.

11.79 The magnitudes of dispersion forces depend on how extended the valence electrons are. In a molecule, valence electrons are spread over a larger volume because they are shared between atoms.

11.81 Each ion has six nearest neighbors of opposite charge:

11.83 (a) mp > 900 °C; (b) sillimanite; (c) yes, kyanite; and (d) the stable phase at highest pressure, kyanite, is densest. The stable phase at lowest pressure, andalusite, is the least dense.

11.85

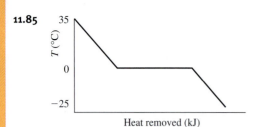

11.87 (a)

$$H-C-C \overset{\ddot{O}:\cdots H-\ddot{O}}{\underset{\ddot{O}-H\cdots:\ddot{O}}{}} C-C-H$$

(b) Fraction decreases as H-bonds break.

Chapter 12

12.1 (a) solvent, liquid water; primary solute, carbon dioxide (normally a gas); (b) solvent, liquid water; primary solute, ethanol (normally a liquid); (c) solvent, gaseous N_2; primary solutes, O_2 (normally a gas) and water (normally a liquid)

12.3 $X_{methanol} = 0.0747$, $X_{water} = 0.925$

12.5 $c_m = 1.88 \times 10^{-1}$ mol/kg; $X = 1.08 \times 10^{-2}$; % = 2.14 %

12.7 $X_{hydrogen\ peroxide} = 0.18$, $X_{water} = 0.82$; $M = 9.8$ M; $c_m = 13$ mol/kg

12.9 $X_{ammonia} = 0.042$; $X_{water} = 0.958$; Mass fraction ammonia = 0.0399; $c_m = -2.44$ mol/kg

12.11 (a) 120. g; (b) 6.50 mol/kg

12.13 The hydrocarbon molecules in salad oil have dispersion-type intermolecular forces but low polarity and no hydrogen-bonding capability. A solution of acetic acid in water has high polarity and a large hydrogen-bonding capability. Hydrogen bonds would have to be broken for the two liquids to mix, making mixing energetically unfavorable.

12.15 Ionic salts and polar organic materials such as alcohols

12.17 H_2O has strong hydrogen bonding, and so does CH_3OH, so this pair is miscible; C_8H_{18} and CCl_4 both have only dispersion forces, so this pair is miscible.

12.19 CCl_4 is most like I_2; NaCl is a salt, so it only dissolves in water; Au, a metal, requires another metal, Hg; and paraffin, a hydrocarbon, is best matched by *n*-octane, another hydrocarbon.

12.21

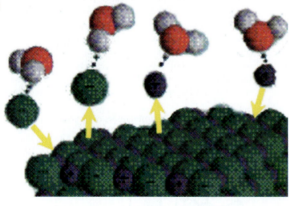

12.23 Methane does not form H-bonds, so life is unlikely.

12.25

12.27 Energy must be supplied to the molecules that escape.

12.29 48%

12.31 $K_H = 0.0735$; $p = 34$ torr

12.33 28.7 torr

12.35 -5.5 °C

12.37 Not enough information, because ethanol is volatile.

12.39 0.25 M NH_3 (highest) > 0.30 M NaCl > 0.25 M $MgCl_2$ > 0.75 M NH_3 (lowest)

12.41 (a) 287 g/mol; (b) 132 g/mol

12.43 37.6 atm

12.45 Similarities: colloidal suspensions in a gas, aerosols; difference: fog particles are liquid, smoke particles are mostly solids.

12.47 Best surfactant: sodium lauryl sulfate; worst surfactant: propanoic acid

(a)

(b)

(c)

12.49 Nonpolar, because the nonpolar "tails" of the surfactant are attracted by the liquid

12.51 $c_m = 58$ mol/kg; $X = 0.51$; $M = 15$ M

12.53 Hydrophobic; it will concentrate in fatty tissues; the effects of DDT are cumulative, especially for organisms high on the "food chain" such as predator birds.

12.55 11 mol/kg

12.57 Benzene (lowest solubility < pentanol < erythritol (highest solubility)

12.59 Iodine is a nonpolar solid, water is highly polar and CCl_4 is nonpolar, so these two liquids are immiscible and I_2 preferentially dissolves in CCl_4.

12.61 Nonpolar substances (oils and greases) and network solids (metal oxides)

12.63 (a) 2.34 °C kg/mol; (b) 2.6×10^2 g/mol

12.65 Brine has much higher osmotic pressure than fresh water, so when fish is placed in brine, water leaches from the cells, dehydrating them. Bacteria are not able to survive and multiply, so the fish does not spoil.

12.67 5.4 mol/kg

12.69 Poor choice, because gasoline floats on water, so the fire will spread.

12.71 4 atm

12.73 (a) 0.37 g; (b) 0.48 atm

Chapter 13

13.1

(a) Amine

(b) Alcohol

(c) Ketone

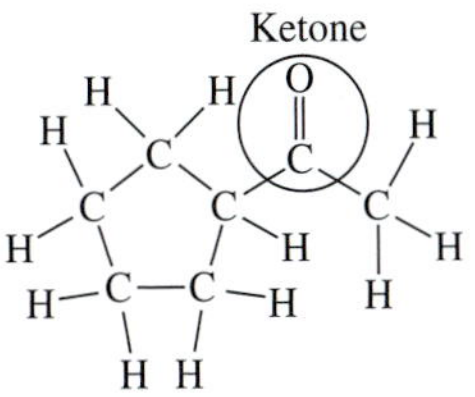

(d) Alkene
Carboxylic acid

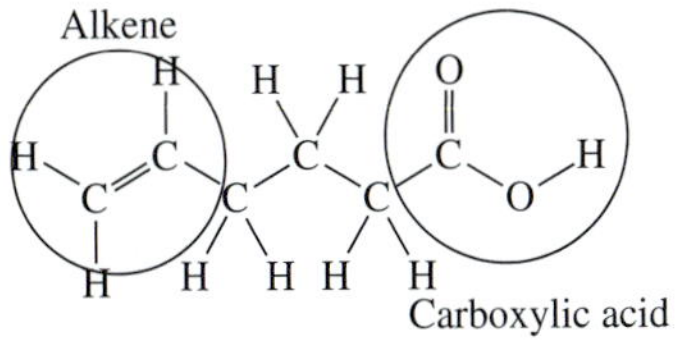

13.3

(a) Amine
Phenyl

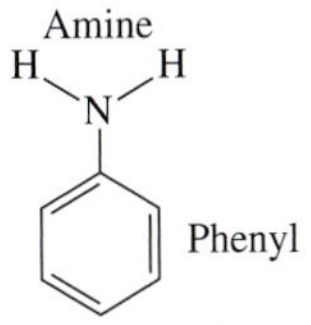

(b) Carboxylic acid
Carboxylic acid

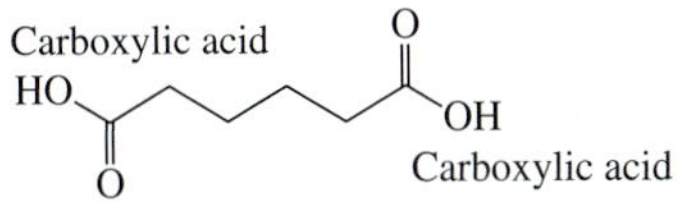

(c) Thiol   **(d)** Alcohol

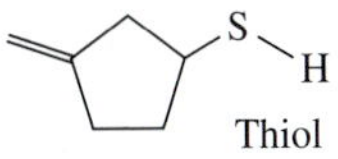

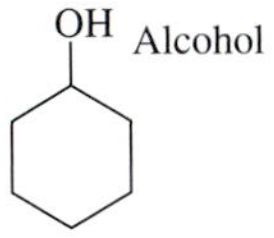

13.5

(a) Amide

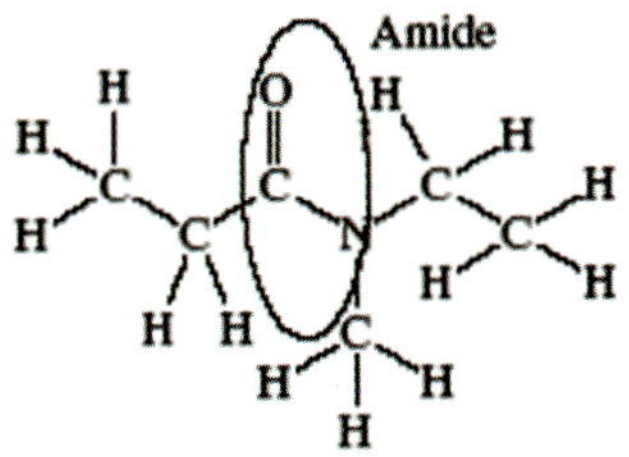

(b) Ester

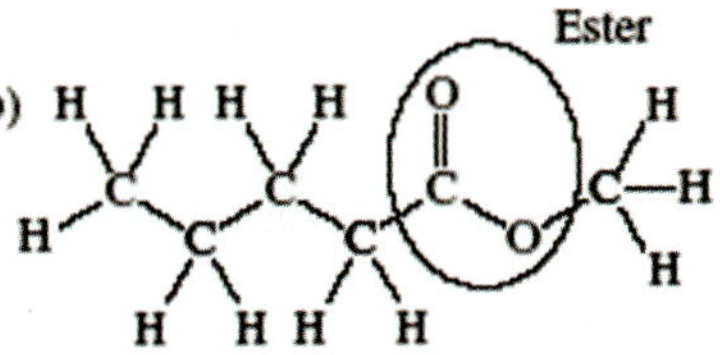

(c) Ether

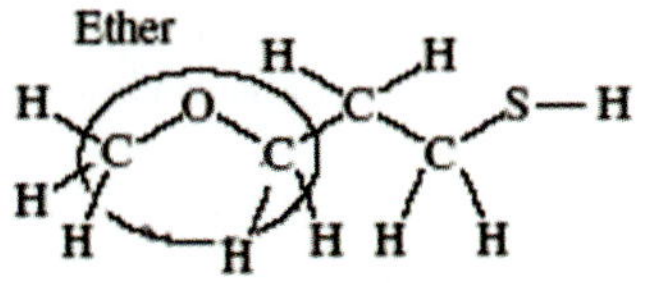

(d) Ether

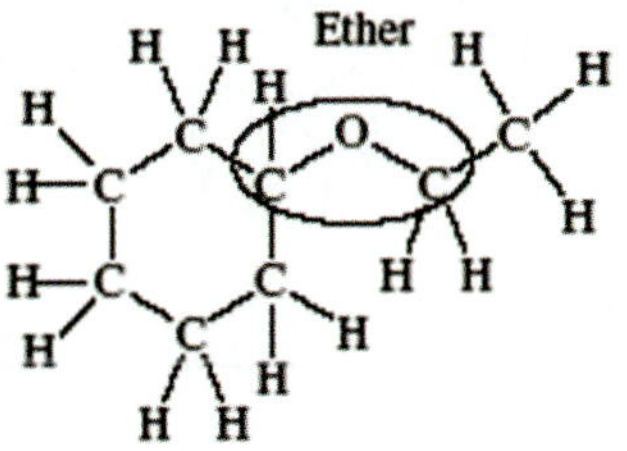

13.7

(a) Other isomers are also possible.

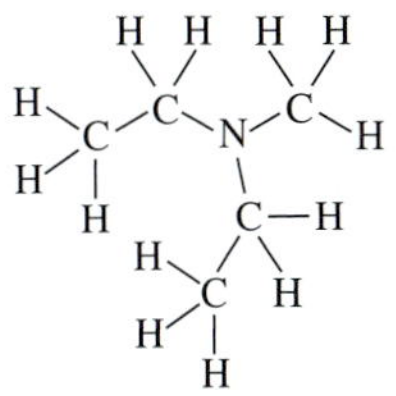

(b) Other isomers are also possible.

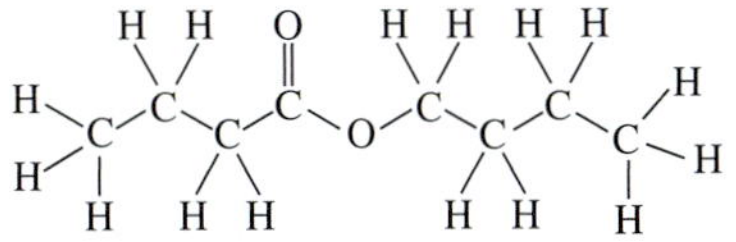

(c) Others are also possible.

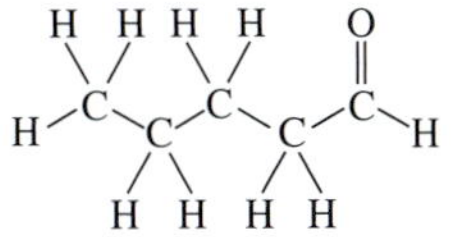

(d) Others are also possible.

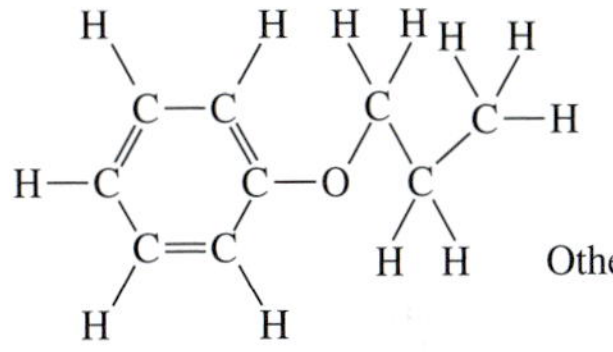

13.9

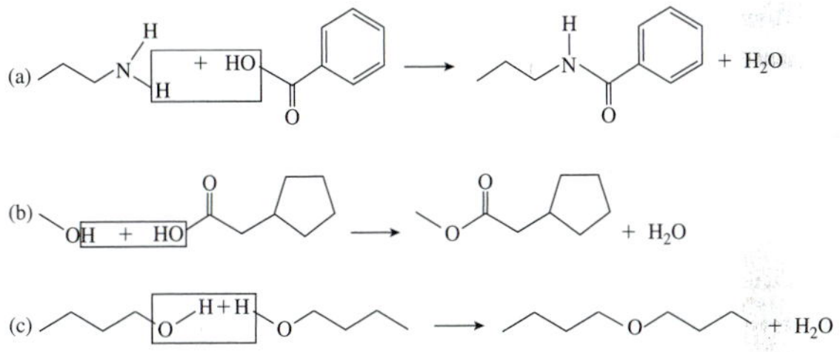

(a) + H₂O

(b) + H₂O

(c) + H₂O

13.11

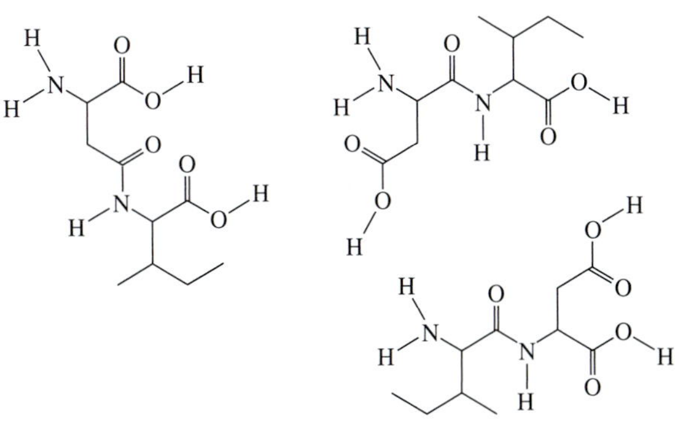

13.13

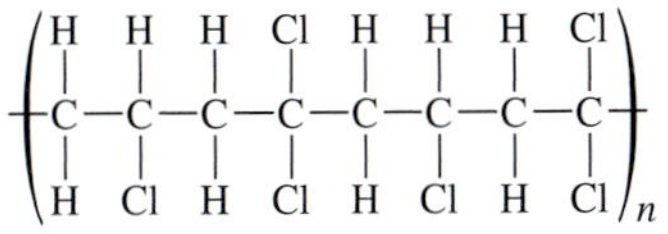

13.15 (a)   (b)

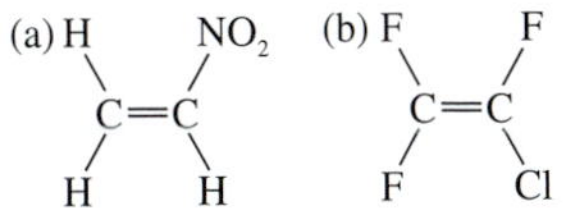

13.17

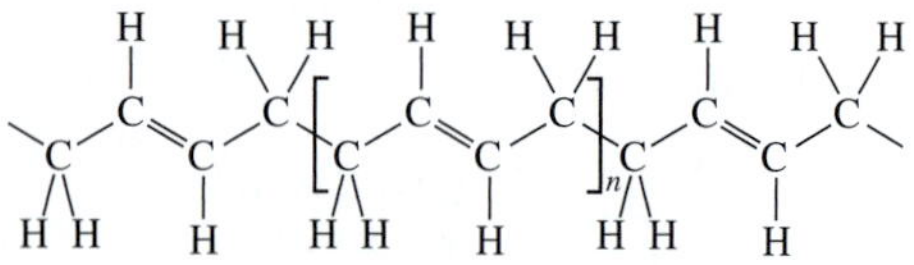

Polybutadiene contains double bonds, whereas polyethylene does not.

13.19

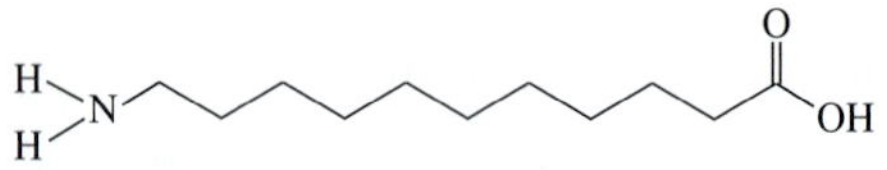

13.21

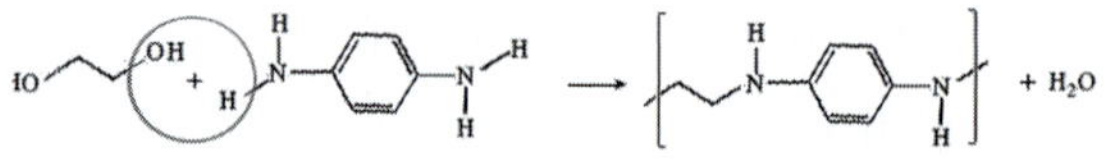

13.23 Automobile tire

13.25 (a) elastomers; (b) fibers; and (c) plastics

13.27 Dioctylphthalate reduces the amount of cross-linking and adds a fluid component to the polymer, so the polymer becomes more flexible.

13.29

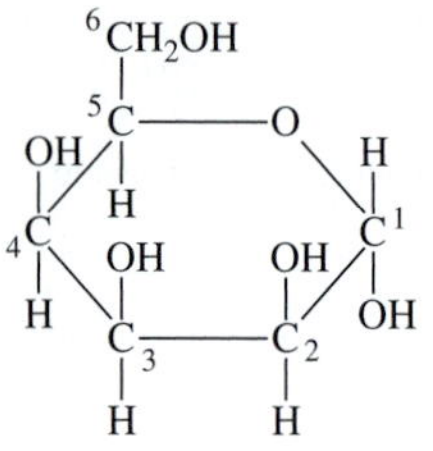

α-talose

13.31

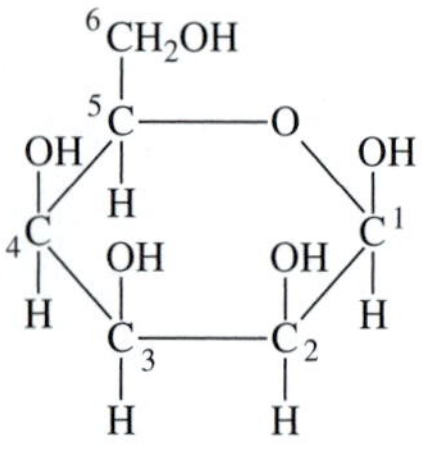

β-talose

13.33 Polymers of α-glucose (e.g., glycogen) coil on themselves; polymers of β-glucose (e.g., cellulose) form planar sheets.

13.35 T-T-A-C-G-T-G-A-C

13.37

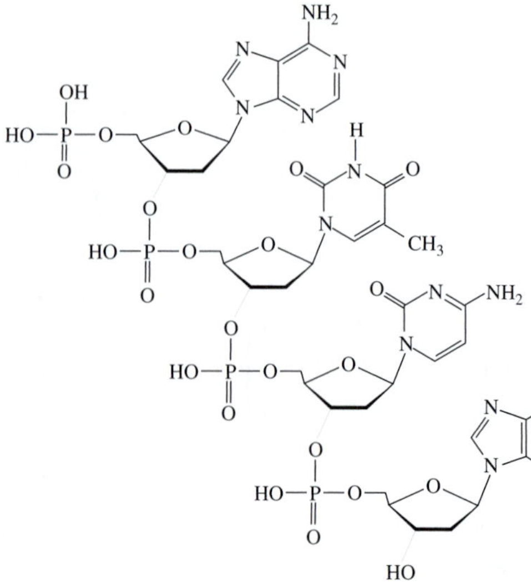

13.39 The complementary sequence is T-A-G.

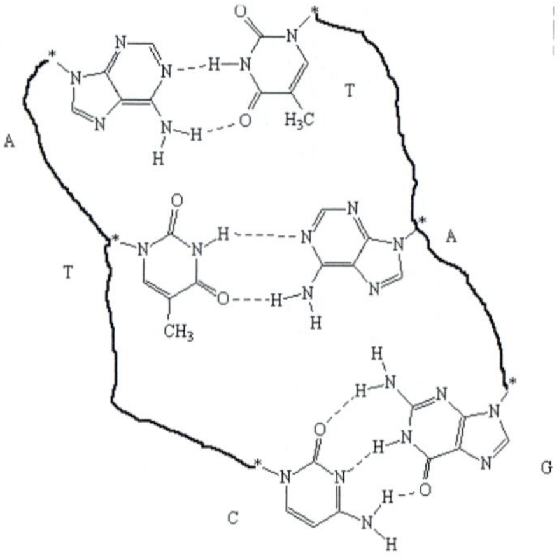

13.41

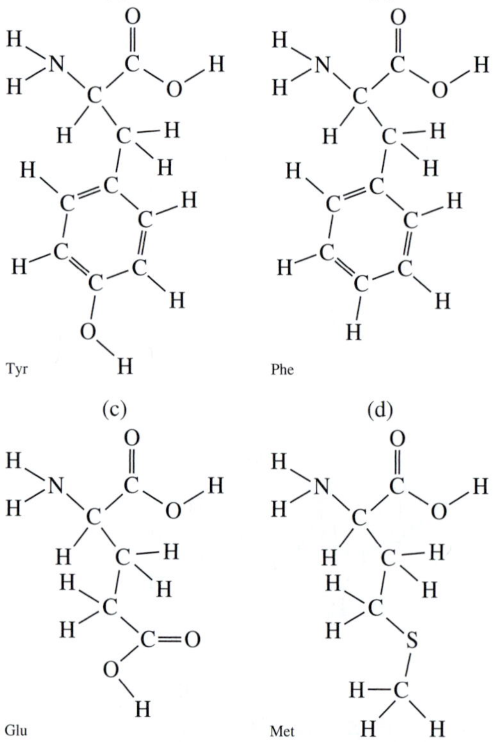

(a)

Tyr

(b)

Phe

(c)

Glu

(d)

Met

13.43 Hydrophilic side chains: Tyr and Glu; hydrophobic side chains: Met and Phe

13.45 Glycine, hydrophobic side chain; (b) serine, hydrophilic; and (c) cysteine, hydrophilic

13.47 Six combinations: Met-Ala, Ala-Met, Ala-Glu, Glu-Ala, Met-Glu, and Glu-Met. Only Ala-Glu is shown here:

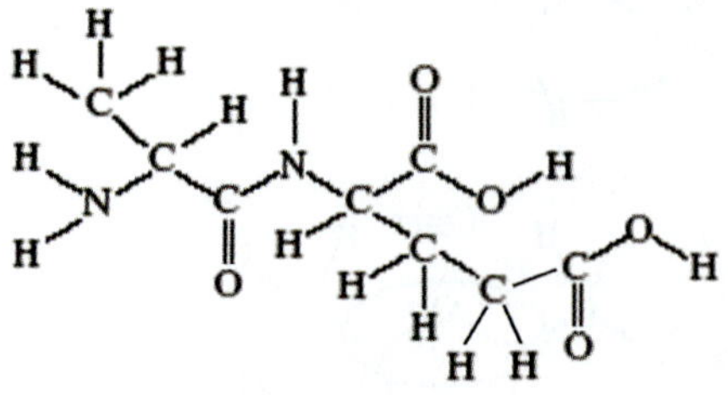

13.49 2.09×10^4 g/mol

13.51 16,777,216

13.53 (a) random; (b)

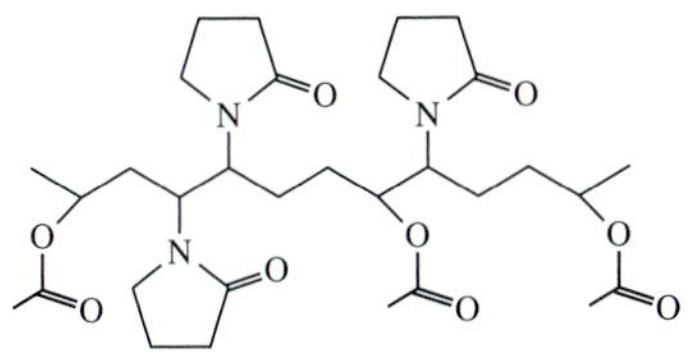

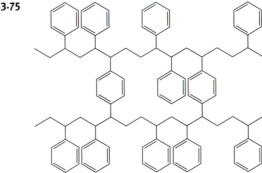

and (c) the backbone of this polymer is hydrophobic, but its side chains contain highly polar C=O groups that interact with one another and with polar and H-bonding groups on hair.

13.55 C - G

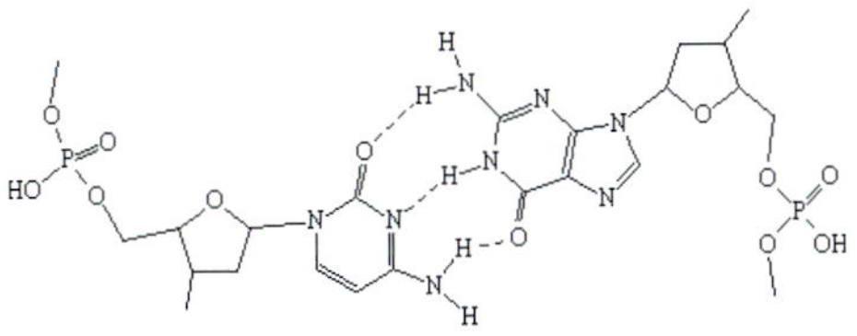

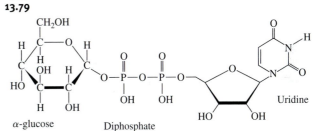

13.57 Initiation

Propagation

Termination:

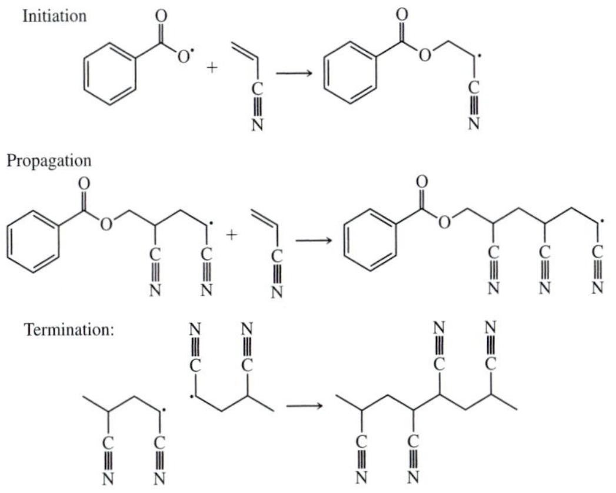

13.59

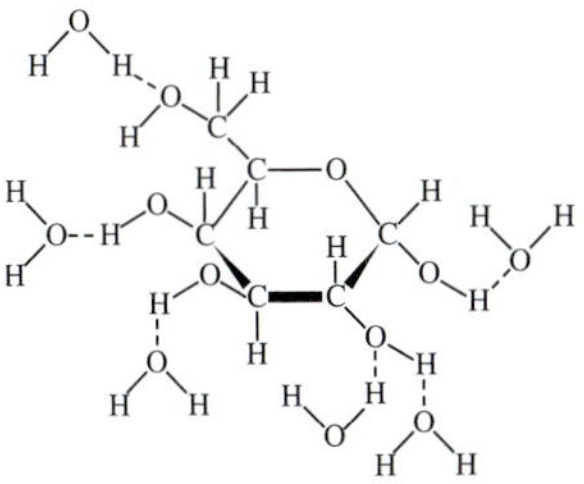

13.61 Both contain amide linkage groups. Nylons contain one or at most two different monomers, each of which typically contains several carbon atoms that form part of the backbone of the polymer. Proteins contain backbones that are absolutely regular repetitions of amide—C—amide bonding, but the carbon atoms in the backbone have a variety of substituent groups attached to them.

13.63 (a) terephthalic acid and phenylenediamine; (b) PET is made from ethylene glycol and terephthalic acid; and (c) styrene

13.65 High-density polyethylene is all straight chains that nest together readily; low-density polyethylene has many side chains that cannot nest easily, so low-density polyethylene has more open space.

13.67 Complementary base pairing requires that the molar ratios of A to T and G to C be 1.0. Chargoff's observations indicate that this relationship holds even though DNA from different sources has different sequences, some AT-rich and others GC-rich.

13.69 1 = cytosine; 2 = uracil; 3 = adenine; 4 = guanine

13.71 Hydrophilic amino acids are most stable when in contact with aqueous solvent, while hydrophobic amino acids are most stable when in contact with the protein backbone. The tertiary structure of a protein is the manner in which the individual amino acids orient themselves, so solvent interactions are the primary determinants of tertiary structure.

13.73

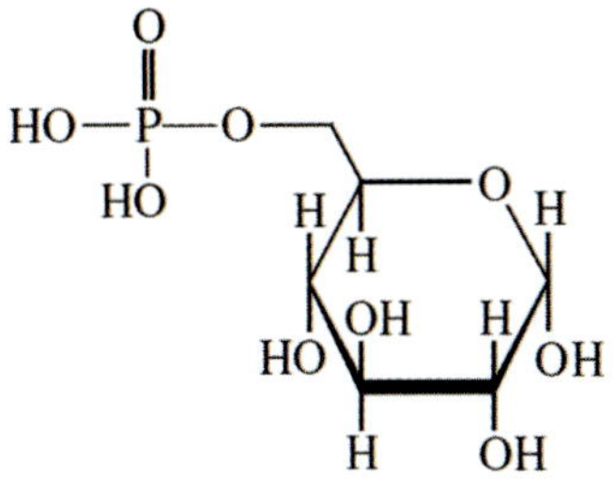

13.75

13.77 4

13.79

α-glucose   Diphosphate   Uridine

13.81 sp^2, allowing the lone pair to be part of a three-center π system

13.83

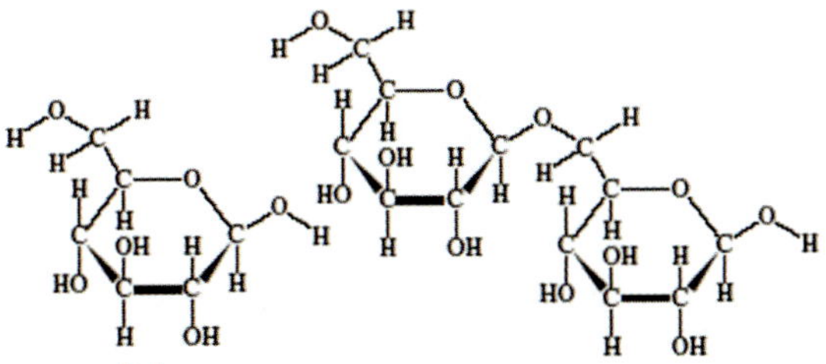

β-glucose   Gentobiose

13.85

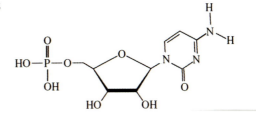

Chapter 14

14.1 (a) A sand castle is organized in a particular way. Waves disperse the sand. (b) Mixing liquids disperses their molecules. (c) Sticks in a bundle are aligned in one direction. When dropped, the sticks lose their alignment and become more dispersed. (d) Water molecules in the gas phase are more dispersed than those in the liquid phase.

14.3 Initially the ink is concentrated in a droplet, but the molecules gradually become dispersed throughout the liquid.

14.5 (a) The air molecules in a tire are relatively constrained. A puncture allows gas molecules to escape from the tire and become more dispersed. (b) The fragrant molecules in a perfume bottle are relatively constrained. When the bottle is open, molecules escape from the confined volume and become more dispersed.

14.7 (a) Energy dispersal requires that heat flows from high to low temperature, never in the opposite direction. (b) Matter dispersal requires that the spontaneous direction is toward greater dispersal, and a "sand castle" represents constrained matter.

14.9 72

14.11 (a) 16.8 J/K; (b) −15.3 J/K; and (c) 1.5 J/K

14.13 (a) $\Delta S_{sys} > 0$, $\Delta S_{surr} < 0$; (b) $\Delta S_{sys} > 0$, $\Delta S_{surr} < 0$; and (c) $\Delta S_{sys} > 0$, $\Delta S_{surr} < 0$. $\Delta S_{total} > 0$ always.

14.15 −94.0 J/K; $\Delta S_{surr} > 94.0$ J/K

14.17 (a) 16 J/mol K; (b) 9.3 J/mol K; (c) 48.8 J/mol K; and (d) 1.00×10^2 J/mol K

14.19 (a) $MgCl_2$, because it produces more moles of ions per mole of substance; (b) HgS, because it has a higher molar mass; and (c) $Br_2(l)$, because it is a liquid

14.21 Ozone has three atoms per molecule, whereas O_2 and H_2 have only two. H_2 has lower molar mass than O_2.

14.23 He: $S° = 126.153$ J/mol K; H_2: $S° = 65.340$ J/mol K; CH_4: $S° = 37.26$ J/mol K; C_3H_6: $S° = 25.21$ J/mol K. Entropy per mole of atoms decreases as the number of atoms in a molecule increase, because tying together atoms into a molecule increases the amount of order among those atoms.

14.25 (a) 416 J/K; (b) 1.7×10^2 J/K; and (c) 9.0×10^1 J/K

14.27 (a) −198.1 J/K; (b) −137.7 J/K; (c) 12.3 J/K; and (d) −267.3 J/K.

14.29 (a) reduction in number of moles of gaseous substances; (b) reduction in number of moles of gaseous substances; (c) all reagents are relatively ordered solid substances; and (d) reduction in number of moles of gaseous substances

14.31 (a) positive; (b) negative; and (c) negative

14.33 (a) $\Delta G_{system} < 0$ for any spontaneous process at constant T and P; (b) the free energy of a system decreases in any process at constant T and P; and (c) $\Delta G = \Delta H - T\Delta S$ at constant T

14.35 (a) −32.8 kJ; (b) 326.4 kJ; (c) −206.1 kJ; and (d) −1331.4 kJ

14.37 (14.27b) 311.3 kJ; and (14.27c) −204.3 kJ

14.39 (a) −46.2 kJ; and (b) 319.7 kJ

14.41 $\Delta H°_{reaction} = -559.6$ kJ; $\Delta S°_{reaction} = -574.9$ J/K; $\Delta G°_{reaction} = -388.2$ kJ

14.43 (a) no temperature; and (b) zero concentration of products

14.45 5.2×10^{-5} bar

14.47 The first "reaction" would be child 1 starting on the ground and rising into the air. The second "reaction" would be child 2 starting in the air and falling to the ground. In this case child 2 would undergo the spontaneous reaction. The motion of child 1 would be non-spontaneous. The common intermediate is the seesaw board, which acts as a lever transferring energy from one child to the other.

14.49 ATP + $C_6H_{12}O_6$ (fructose) + $C_6H_{12}O_6$ (glucose) → ADP + $C_{12}H_{22}O_{11}$ + H_3PO_4; $\Delta G° = -7.6$ kJ

14.51 (a) not spontaneous; (b) absorbs heat; and (c) more ordered

14.53 (a) −83 J/K; (b) 328.7 J/K; and (c) 365.3 J/K

14.55 (a) −40 kJ; (b) −148.4 kJ; and (c) −1997.7 kJ

14.57 (a) −42 kJ; (b) −168.1 kJ; and (c) −2020.7 J/K

14.59 (a) $w_{sys} = 0$ (no volume change); (b) $q_{sys} > 0$ (energy must be supplied to break chemical bonds); and (c) $\Delta S_{surr} < 0$ ($q_{sys} > 0$, $q_{surr} < 0$)

14.61 (a) and (b) not thermodynamically feasible; and (c) thermodynamically feasible

14.63 When a cell needs energy, it "spends" ATP; when fat or carbohydrates ("capital") are consumed, the energy is stored by converting ADP to ATP ("buying" ATP).

14.65 (a) spontaneous; (b) absorbs energy; and (c) yes

14.67 (a) −189 J/K; (b) 15 J/K; and (c) yes

14.69 Ammonia is a toxic gas, while urea and ammonium nitrate both are relatively nontoxic solids. Thus, the transport and application of ammonia entail significant risks to humans. Even in aqueous solution, ammonia is highly irritating, as anyone knows who has used strong ammonia as a cleanser.

14.71 Glucose, −15.9 kJ/g; palmitic acid, −38.2 kJ/g

14.73 (a) negative; (b) negative; and (c) positive

14.75 (a) $\Delta E_{total} = 0$; (b) $\Delta E_{teaspoon} > 0$; (c) $\Delta S_{total} > 0$; (d) $\Delta S_{water} < 0$; and (e) $q_{teaspoon} > 0$

14.77 (0.5 mol, l, 298 K) < (1 mol, l, 298 K) < (1 mol, l, 373 K) < (1 mol, g, 1 bar, 373 K) < (1 mol, g, 0.1 bar, 373 K)

14.79 $\Delta H° = 182.6$ kJ; $\Delta S° = 24.83$ J/K; $\Delta G° = 175.2$ kJ; spontaneous above 7350 K; reaction occurs because of low pressures of products.

14.81 (a) $\Delta S°_{vap} = 87.1$ J/K; $\Delta G°_{vap} = 0$; (b) 5.16 kJ/mol; and (c) 0.14 bar

14.83 (a)

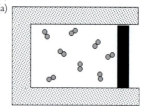

(b) $\Delta S_{total} = 0$

14.85 Attractive forces are larger for CaO, restricting dispersion by way of vibration.

14.87 (a) $q_v = \Delta E$; (b) $q_p = \Delta H$; and (c) $q_T = T\Delta S$

14.89

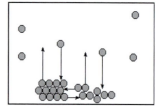

$\Delta G = 0$ for all three processes; when matter becomes more constrained in the system, energy is dispersed into the surroundings, and vice versa.

14.91 (a) ΔH is negative, ΔS is negative; (b) ΔH is positive, ΔS is positive; and (c) ΔH is negative, ΔS is small

14.93 Temperature is 37 °C (310 K) and no solutes are present at 1 M. In particular, hydronium ion concentration is around 10^{-7} M.

14.95 40.6 %; 7.8 g

14.97 Both processes release energy (consume ATP, generate light), have positive ΔS (increase in number of molecules) and have negative ΔG (spontaneous reactions).

14.99 The HCl molecule is heteronuclear, while H_2 and Cl_2 are homonuclear. The homonuclear molecules are more symmetrical than the heteronuclear, so they have fewer possible rotational orientations in space. This leads to less rotational disorder and accounts for the difference.

Chapter 15

15.1 (a) pouring the coffee from the urn into the cup; (b) entering the items on the cash register (if the market has a good laser scanner, paying and receiving change may be rate-determining); and (c) preparing for the jump and passing through the door

15.3 (a) $I_2 \longrightarrow I + I$; (b) $H_2 + I_2 \longrightarrow H_2I_2$; and (c) $H_2 + I_2 \longrightarrow H + HI_2$

15.5 (a)

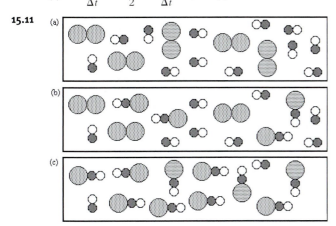

(b)

(c)

15.7 (a) $I_2 \longrightarrow I + I$
$I + H_2 \longrightarrow HI + H$
$H + I \longrightarrow HI$
(b) $H_2 + I_2 \longrightarrow H_2I_2$
$H_2I_2 \longrightarrow HI + HI$
(c) $H_2 + I_2 \longrightarrow H + HI_2$
$H + HI_2 \longrightarrow HI + HI$

15.9 (a) Rate $= -\dfrac{\Delta[Cl_2]}{\Delta t}$
(b) $-\dfrac{\Delta[Cl_2]}{\Delta t} = \dfrac{1}{2}\dfrac{\Delta[NOCl]}{\Delta t}$; and (c) 94 M s^{-1}

15.11 (a)

(b)

(c)

15.13 (a) 2.2×10^{-3} mol/min; and (b) 1.1×10^{-2} mol

15.15

15.17 (a) 6.0×10^6; (b) 2.0×10^{-9}; and (c) in first-order reactions, the fraction reacting is independent of concentration.

15.19 (a) Rate $= k[C][AB]$; (b) units $=$ (conc.)$^{-1}$ (time)$^{-1}$; and
(c) $C + AB \longrightarrow BC + A$
$A + AB \longrightarrow B + A_2$
$B + C \longrightarrow BC$
Net: $2 C + 2 AB \longrightarrow 2 BC + A_2$

15.21 (a) Rate $= k[C]^2[AB]$; (b) units $=$ (conc.)$^{-2}$ (time)$^{-1}$; and
(c) $2 C + AB \longrightarrow BC + AC$
$AC + AB \longrightarrow BC + A_2$
Net: $2 C + 2 AB \longrightarrow 2 BC + A_2$

15.23 Slower by a factor of 0.56

15.25 (a) This reaction is first-order; (b) $k = 6.14 \times 10^{-4}$ s^{-1}; (c) 0.936 atm; and (d) 2.62×10^3 s

15.27 (a) 2.4 min; and (b) 3.2×10^{-4} M

15.29 (a) 12.0 hr; and (b) 0.0324 M

15.31 Do two experiments; in each $[H_2O]_0 > 100$ $[N_2O_5]_0$ but with two different values for $[H_2O]_0$. Plot $\ln\left(\dfrac{[N_2O_5]_0}{[N_2O_5]}\right)$ and $\dfrac{1}{[N_2O_5]} - \dfrac{1}{[N_2O_5]_0}$ vs. t to determine order with respect to N_2O_5, and use the ratio of slope values and H_2O concentrations to determine order with respect to H_2O.

15.33 Rate $= k[S_2O_8^{2-}][I^-]$; $k = 2.3$ M^{-1} min^{-1}

15.35 Rate $= k_2\left(\dfrac{k_1}{k_{-1}}\right)\dfrac{[C][AB]^2}{[BC]}$

15.37 (a) $O_3 + O \to 2 O_2$; (b) Rate $= k_2\dfrac{k_1}{k_{-1}}[O_3][O_2]$; and (c) it is improbable to have two simultaneous fragmentations of O_5.

15.39 (a) Rate $= k_2(\dfrac{k_1}{k_{-1}})[NO]^2[O_2]$; (b) yes; and (c) the intermediate with an N—N bond is the most likely.

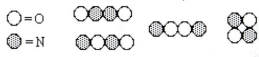

15.41 When $E_a = 0$, k is independent of temperature. This happens when a reaction can occur without first breaking any chemical bonds.

15.43

Activated complex:

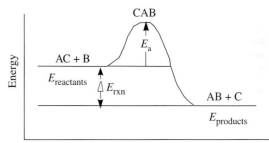

Reaction coordinate

15.45

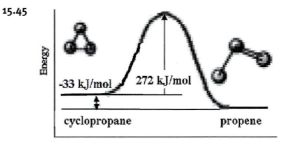

Reaction coordinate

15.47 25 kJ/mol

15.49 Bond breakage (wavy lines):   Bond formation (dotted lines):

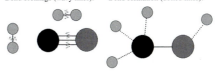

Hydrogen gas could adsorb on the catalyst's surface as H atoms which will then react readily with CO molecules.

15.51 (a)

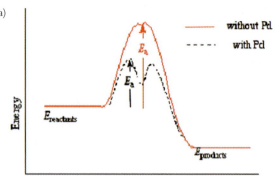

(b) catalyst: Pd metal; intermediate: H atoms on the Pd metal surface; and (c)

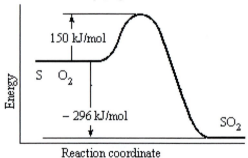

H₂ adding to Pd metal and dissociating   C taking one H   2nd C taking other H   Ethane molecule

15.53 (a) $O + NO \longrightarrow NO_2$; and (b)

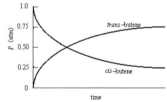

15.55 (a) Rate $\dfrac{\Delta[C_6H_6]}{\Delta t} = -\dfrac{1}{3}\left(\dfrac{\Delta[C_2H_2]}{\Delta t}\right)$

(b) Experiments would have to be carried out measuring the rate as a function of $[C_2H_2]$.

15.57

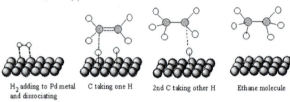
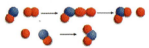

15.59 (a) nine-fold increase; (b) no change; and (c) increases by 5.2 times

15.61 First order, with $k = 2.55 \times 10^{-3}$ s^{-1}

15.63 The rate for b will be twice that for a, because the concentration of X_2 is twice as great.

15.65 First-order plot is linear, with slope = $k_{obs} = 0.0038$ s^{-1}; Rate = $k[N_2O_5][H_2O]$, $k = 0.15$ M^{-1} s^{-1}

15.67 (a) 1.9×10^2 s; (b) $1.3 \times ^3$ s; and (c) 1.3×10^4 s

15.69 (a) false; (b) false; (c) true; and (d) true

15.71 (a) first order in N_2O_5 and first order overall; (b) second order in NO, first order in H_2, and third order overall; and (c) first order in enzyme and first order overall.

15.73 The speed of a chemical reaction refers to how fast it proceeds. The spontaneity of a chemical reaction refers to whether or not the reaction can go in the direction written without outside intervention.

15.75 $CF_2Cl_2 \xrightarrow{h\nu} CF_2Cl + Cl$

$O_3 \xrightarrow{h\nu} O_2 + O$

$Cl + O_3 \longrightarrow ClO + O_2$

$ClO + O \longrightarrow Cl + O_2$

15.77 (a) $2\,NO + 2\,H_2 \longrightarrow N_2 + 2\,H_2O$; and

(b) Rate $= k_2 \dfrac{k_1}{k_{-1}}[NO]^2[H_2]$

15.79 (a) A catalyst binds to reactant species in a way that weakens chemical bonds. (b) The average kinetic energy of the molecules increases, so a larger fraction of molecular collisions leads to reaction. (c) There are more molecules to react and a higher rate of molecular collisions.

15.81 Rate $= k[A]^2$

15.83 (a) Rate $= k[H_2][X_2]$; (b) Rate $= k[X_2]$; and

(c) Rate $= k_2\left(\dfrac{k_1}{k_{-1}}\right)^{1/2}[H_2][X_2]^{1/2}$

15.85 (a) $Cl_2 + CHCl_3 \longrightarrow HCl + CCl_4$; (b) Cl and CCl_3; and

(c) Rate $= k_2(\dfrac{k_1}{k_{-1}})^{1/2}[CHCl_3][Cl_2]^{1/2}$

15.87 (a) 51 kJ/mol; (b) 7.0 min

15.89 (a) Rate $= k[CH_3CHO]^2$; (b) 4.5×10^{-3} M^{-1} s^{-1}; and (c) 2.7×10^3 s

15.91 1.1×10^{14}

15.93 Neither appears in the overall stoichiometry. Catalysts are reactants in early steps but are products of later steps, while intermediates are products of early steps and are consumed in later steps.

15.95 Statements (d) and (f) are true.

Chapter 16

16.1

16.5 Your molecular pictures of elementary reactions should show the reactants, the products, and (if applicable) the intermediate collision complex:

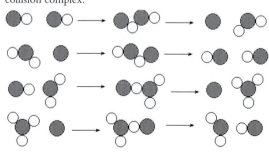

16.3 (a) $Cl^-(aq) + ClO_2^-(aq) \xrightarrow{k_{-1}} ClO^-(aq) + ClO^-(aq)$

$Cl^-(aq) + ClO_3^-(aq) \xrightarrow{k_{-2}} ClO^-(aq) + ClO_2^-(aq)$

(b) $K_{eq} = \dfrac{[ClO_3^-]_{eq}[Cl^-]_{eq}^2}{[ClO^-]_{eq}^3}$; and (c) $K_{eq} = \dfrac{k_1k_2}{k_{-1}k_{-2}}$

16.7 To test the reversibility of a reaction, set up a system containing the products and observe whether or not reactants form. Here, a solution containing Cl^- and ClO_3^- ions should react to form some ClO^- ions.

16.9 (a) $K_{eq} = \dfrac{(p_{IF_5})^2_{eq}}{(p_{F_2})^5_{eq}}$; (b) $K_{eq} = \dfrac{1}{(p_{O_2})^5_{eq}}$;

(c) $K_{eq} = (p_{CO})^2_{eq}$; (d) $K_{eq} = \dfrac{1}{(p_{CO})_{eq}(p_{H_2})^2_{eq}}$; and

(e) $K_{eq} = \dfrac{[H_3O^+]^3_{eq}[PO_4^{3-}]_{eq}}{[H_3PO_4]_{eq}}$

16.11 (a) $2\,IF_5\,(g) \rightleftharpoons I_2\,(s) + 5\,F_2\,(g)$, $K_{eq} = \dfrac{(p_{F_2})^5_{eq}}{(p_{IF_5})^2_{eq}}$;

(b) $P_4O_{10}\,(s) \rightleftharpoons P_4\,(s) + 5\,O_2\,(g)$, $K_{eq} = (p_{O_2})^5_{eq}$;

(c) $BaO\,(s) + 2\,CO\,(g) \rightleftharpoons BaCO_3\,(s) + C\,(s)$, $K_{eq} = \dfrac{1}{(p_{CO})^2_{eq}}$;

(d) $CH_3OH\,(l) \rightleftharpoons CO\,(g) + 2\,H_2\,(g)$, $K_{eq} = (p_{CO})_{eq}(p_{H_2})^2_{eq}$; and

(e) $PO_4^{3-}\,(aq) + 3\,H_3O^+\,(aq) \rightleftharpoons H_3PO_4\,(aq) + 3\,H_2O\,(l)$,

$$K_{eq} = \dfrac{[H_3PO_4]_{eq}}{[H_3O^+]^3_{eq}\,[PO_4^{3-}]_{eq}}$$

16.13 (a) $p = 1$ bar for F_2 and IF_5, $X = 1$ for I_2;
(b) $p = 1$ bar for O_2, $X = 1$ for P_4 and P_4O_{10};
(c) $p = 1$ bar for CO, $X = 1$ for others;
(d) $p = 1$ bar for CO and H_2, $X = 1$ for CH_3OH; and
(e) $c = 1$ M for H_3PO_4, H_3O^+, and PO_4^{3-}, $X = 1$ for H_2O

16.15 Your views should be like those in Figure 16-4, showing two different-sized samples with equal-sized portions of each highlighted to show that equal volumes contain equal numbers of atoms:

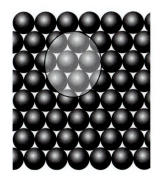

16.17 9.9×10^6

16.19 (a) 3.3×10^3; (b) 1×10^{90}; (c) 3×10^{-60}; and (d) 2.9×10^{39}

16.21 2×10^{109}; 3×10^{-75}

16.23 1.8×10^3; 9.2×10^{18}

16.25 (a), (b), and (d), adding CO causes the reaction to go to the right, forming products; (c), adding CO causes the reaction to proceed to the left, forming reactants.

16.27 (a) no effect; (b) more solid will dissolve; (c) some solid will precipitate; and (d) no effect

16.29 This reaction can be driven to the left by removing SO_2, removing Cl_2, adding SO_2Cl_2, or increasing the temperature.

16.31 0.92

16.33 1.3×10^{-5}

16.35 $(p_{H_2})_{eq} = (p_{Br_2})_{eq} = 2.5 \times 10^{-2}$ bar; $(p_{HBr})_{eq} = 10.0$ bar

16.37 $(p_{CO_2})_{eq} = 1.4$ bar and $(p_{CO})_{eq} = 3.6$ bar

16.39 (a) H_2O and CH_3CO_2H; (b) H_2O, NH_4^+, and Cl^-;
(c) H_2O, K^+, and Cl^-; (d) H_2O, Na^+, and $CH_3CO_2^-$; and
(e) H_2O, Na^+, and OH^-

16.41 (a) $CH_3CO_2H\,(aq)\ H_2O\,(l) \rightleftharpoons CH_3CO^-\,(aq) + H_3O^+\,(aq)$;
(b) $NH_4^+\,(aq) + H_2O\,(l) \rightleftharpoons NH_3\,(aq) + H_3O^+\,(aq)$;
(c) no equilibria other than the water equilibrium;
(d) $CH_3CO_2^-\,(aq) + H_2O\,(l) \rightleftharpoons CH_3CO_2H\,(aq) + OH^-\,(aq)$; and
(e) no equilibria other than the water equilibrium

16.43 (a) $(CH_3)_2CO$ (acetone) and H_2O; (b) H_2O, K^+, and Br^-;
(c) H_2O, Li^+, and OH^-; and (d) H_2O, H_3O^+, and HSO_4^-

16.45 (a) $K_{eq} = \dfrac{[ClO_2^-][H_3O^+]}{HClO_2} = K_a$;

(b) $K_{eq} = \dfrac{1}{[Fe^{3+}][OH^-]^3} = \dfrac{1}{K_{sp}}$; and

(c) $K_{eq} = \dfrac{[HCN]}{[CN^-][H_3O^+]} = \dfrac{1}{K_a}$

16.47 (a) Na^+; (b) Cl^- and K^+; and (c) K^+ and NO_3^-

16.49 (a) $K_{eq} = (p_{CO_2})_{eq}\,(p_{H_2O})_{eq}$; (b) $K_{eq} = \dfrac{(p_{NH_3})^4_{eq}\,(p_{O_2})^3_{eq}}{(p_{N_2})^2_{eq}}$;

(c) $K_{eq} = \dfrac{(p_{CH_3CHO})^2_{eq}}{(p_{C_2H_4})^2_{eq}\,(p_{O_2})}$; (d) $K_{eq} = [Ag^+]^2_{eq}\,[SO_4^{2-}]_{eq}$; and

(e) $K_{eq} = (p_{H_2S})_{eq}\,(p_{NH_3})_{eq}$

16.51 $(p_{CO})_{eq} = 7.2 \times 10^{-2}$ bar; $(p_{COCl_2})_{eq} = 6.2$ bar

16.53 The molecular picture should show 2 each of the reactants and 10 each of products:

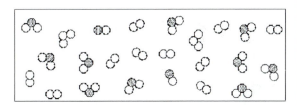

16.55 Adding more solute to a solution that is saturated increases the total surface area, so the overall rate at which molecules enter the solution becomes greater. However, the rate at which molecules are captured also becomes greater. The rate per unit surface area is unchanged, so the concentration remains the same.

16.57 7.34 bar

16.59 3.0 at 298 K; 2.1×10^{-4} at 525 K

16.61 (a) $(CH_3)_3N + H_2O \rightleftharpoons (CH_3)_3NH^+ + OH^-$,

$$K_b = \dfrac{[(CH_3)_3NH^+]_{eq}\,[OH^-]_{eq}}{[(CH_3)_3N]_{eq}}$$

(b) $HF + H_2O \rightleftharpoons F^- + H_3O^+$, $K_a = \dfrac{[F^-]_{eq}\,[H_3O^+]_{eq}}{[HF]_{eq}}$;

(c) $CaSO_4\,(s) \rightleftharpoons Ca^{2+}\,(aq) + SO_4^{2-}\,(aq)$,

$$K_{sp} = [Ca^{2+}]_{eq}\,[SO_4^{2+}]_{eq}$$

16.63 (a) goes to right; (b) goes to left; and (c) no effect

16.65 4.7×10^5; less soluble at higher T

16.67 0.75

16.69 (a) 1.6×10^{11}; (b) 792 K; and (c) 2.2×10^{-2}

16.71 (a) $K_{eq} = \dfrac{[CO_3^{2-}]_{eq}}{(p_{CO_2})_{eq}\,[OH^-]^2_{eq}}$;

(b) increases; and (c) increases

16.73 1.20×10^2 bar

16.75 4.4×10^{-29} bar; 9.4×10^5 L

16.77 (a)

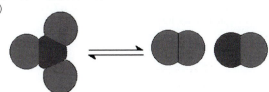

(b) 48

Chapter 17

17.1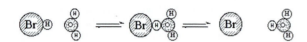

17.3 $[H_3O^+] = 1.25 \times 10^{-3}$ M; $[OH^-] = 8.00 \times 10^{-12}$ M
17.5 $[H_3O^+] = [Cl^-] = 0.121$ M; $[OH^-] = 8.26 \times 10^{-14}$ M
17.7 $[H_3O^+] = 4.12 \times 10^{-2}$ M; $[OH^-] = 2.43 \times 10^{-13}$ M
17.9 (a) -0.60; (b) 5.426; (c) 2.32; and (d) 3.593
17.11 (a) 14.60; (b) 8.574; (c) 11.68; and (d) 10.407
17.13 (a) 0.22 M; (b) 1.4×10^{-8} M; (c) 2.1×10^{-4} M; and
(d) 4.7×10^{-15} M
17.15 (a) 4.6×10^{-14} M; (b) 7.1×10^{-7} M; (c) 4.8×10^{-11} M; and
(d) 2.1 M
17.17 (a) 9.70; (b) 10.04; and (c) 1.80
17.19

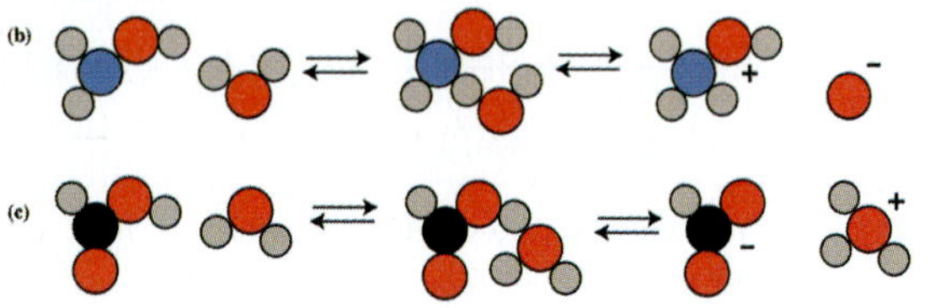

17.21 (a) major species: HN_3 and H_2O, minor species: N_3^- and
H_3O^+; (b) $[H_3O^+] = [N_3^-] = 6.1 \times 10^{-3}$ M and
$[HN_3] = 1.50$ M; (c) 2.21; and (d)

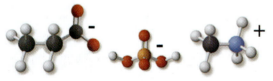

17.23 0.41 %
17.25 (a) major species: $N(CH_3)_3$ and H_2O, minor species:
$HN(CH_3)_3^+$ and OH^-; (b) $[OH^-] = [HN(CH_3)_3^+] =$
4.8×10^{-3} M, $[N(CH_3)_3] = 0.345$ M; (c) 11.68; and (d)

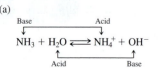

17.27 0.49 %
17.29 (a) weak base; (b) weak acid; (c) weak acid; and (d) strong base
17.31

Propanoic   Phosphoric acid   Methylamine
(acid)   (acid)   (base)

17.33 C_5H_5N conjugate acid is $C_5H_5NH^+$; $HONH_2$ conjugate acid is
$HONH_3^+$; HCO_2H conjugate base is HCO_2^-; NH_3 conjugate
acid is NH_4^+; HCNO conjugate base is CNO^-; HClO conjugate base is ClO^-.
17.35

(a)
Base   Acid
$NH_3 + H_2O \rightleftharpoons NH_4^+ + OH^-$
Acid   Base

(b)
Base   Acid
$H_2O + HCNO \rightleftharpoons H_3O^+ + CNO^-$
Acid   Base

(c)
Base   Acid
$H_2O + HClO \rightleftharpoons H_3O^+ + ClO^-$
Acid   Base

(d)
Base   Acid
$OH^- + H_3O^+ \rightleftharpoons H_2O + H_2O$
Acid   Base

17.37 (a) major species are Na^+, SO_3^{2-}, and H_2O;
(b) $H_2O + SO_3^{2-} \rightleftharpoons HSO_3^- + OH^-$; and (c) 10.43
17.39 (a) major species are NH_4^+, NO_3^-, and H_2O;
(b) $H_2O + NH_4^+ \rightleftharpoons NH_3 + H_3O^+$; and (c) 5.63
17.41 (a) basic, $H_2O + HS^- \rightleftharpoons OH^- + H_2S$;
(b) basic, $H_2O + CO_3^{2-} \rightleftharpoons OH^- + HCO_3^-$;
(c) neutral, $H_2O\,(l) + H_2O\,(l) \rightleftharpoons OH^-(aq) + H_3O^+(aq)$; and
(d) acidic,
$H_2O\,(l) + HC_5H_5N^+(aq) \rightleftharpoons H_3O^+(aq) + C_5H_5N\,(aq)$
17.43 $NaI < NaF < NaC_6H_5CO_2 < Na_3PO_4 < NaOH$
17.45 (a) H_2SO_4 (anions are poorer proton donors than neutrals);
(b) HClO (Cl is more electronegative); and (c) $HClO_2$ (more
O atoms)
17.47

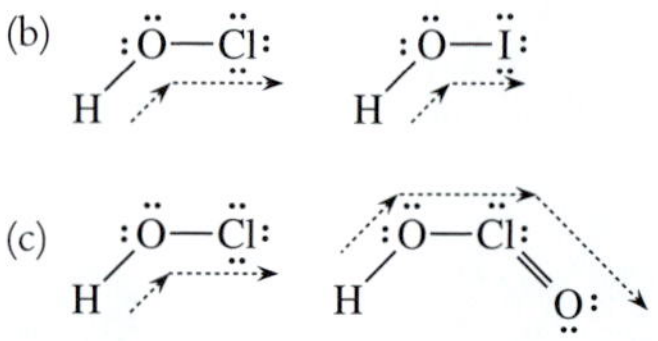

(b)   :Ö—Cl: :Ö—Ï:
       H   H

(c)   :Ö—Cl: :Ö—Cl:
       H   H   Ö:

17.49

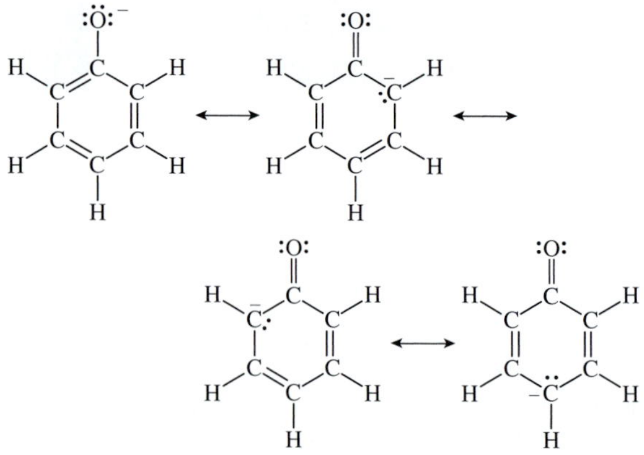

Charge in the phenolate anion is distributed around the ring,
stabilizing the anion.
17.51 $[Na^+] = [C_2H_3O_2^-] = 0.250$ M; $[OH^-] = 1.2 \times 10^{-5}$ M;
$[H_3O^+] = 8.3 \times 10^{-10}$ M
17.53 $[H_3O^+] = [HCO_3^-] = 8.4 \times 10^{-5}$ M;
$[CO_3^{2-}] = 4.7 \times 10^{-11}$ M; $[OH^-] = 1.2 \times 10^{-10}$ M
17.55 $[Na^+] = 0.11$ M; $[OH^-] = [HCO_3^-] = 3.2 \times 10^{-3}$ M;
$[CO_3^{2-}] = 0.052$ M; $[H_2CO_3] = 2.2 \times 10^{-9}$ M;
$[H_3O^+] = 3.1 \times 10^{-12}$ M

17.57 (a)

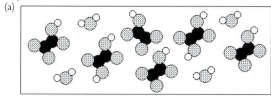

(b) 5.28
17.59 10.81
17.61 9.1×10^{-11}
17.63 0.010 %
17.65

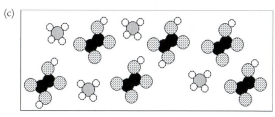

17.67 8.30
17.69 $[H_3O^+] = 2.01$ M; $[SO_4^{2-}] = 1.0 \times 10^{-2}$ M;
$[HSO_4^-] = 1.99$ M

17.71 (a) major species: H_2O, HSO_4^-, and H_3O^+;
$H_2O + HSO_4^- \rightleftharpoons SO_4^{2-} + H_3O^+$
(b) major species: H_2O, SO_4^{2-}, and Na^+;
$SO_4^{2-} + H_2O \rightleftharpoons HSO_4^- + OH^-$
(c) major species: H_2O, HSO_4^-, and H_3O^+;
$H_2O + HSO_4^- \rightleftharpoons SO_4^{2-} + H_3O^+$
(d) major species: H_2O, Cl^-, and NH_4^+;
$H_2O + NH_4^+ \rightleftharpoons NH_3 + H_3O^+$
17.73 $K_{eq} = 48$
17.75 (a)

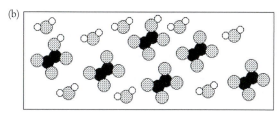

(b)

(c)

17.77 (a) H_2O and H_3PO_4; (b) $H_2PO_4^-$, H_3O^+, HPO_4^{2-}, OH^-, and PO_4^{3-}; and (c) pH = 1.92
17.79

17.81 (a) $OH^- + C_6H_5CO_2H \rightleftharpoons H_2O + C_6H_5CO_2^-$
(b) $H_3O^+ + (CH_3)_3N \rightleftharpoons H_2O + (CH_3)_3NH^+$
(c) $SO_4^{2-} + CH_3CO_2H \rightleftharpoons HSO_4^- + CH_3CO_2^-$
(d) $OH^- + NH_4^+ \rightleftharpoons H_2O + NH_3$
(e) $HPO_4^{2-} + NH_3 \rightleftharpoons PO_4^{3-} + NH_4^+$

Chapter 18

18.1 (a) no buffer properties; (b) buffer, 9.25; (c) no buffer properties; and (d) buffer, 9.25
18.3 8.88
18.5 1.4×10^{-3} mol
18.7 $H_3O^+(aq) + CO_3^{2-}(aq) \rightleftharpoons H_2O(l) + HCO_3^-(aq)$;
$OH^-(aq) + HCO_3^-(aq) \rightleftharpoons H_2O(l) + CO_3^{2-}(aq)$
18.9 (a) buffer containing HCO_3^- and CO_3^{2-}; (b), (c) no buffer; and (d) buffer containing HCO_3^- and CO_3^{2-}
18.11 For pH = 3.50, the HCO_2H/HCO_2^- system should be used. Sodium formate could be used along with HCl solution. For pH = 12.60, the HPO_4^{2-}/PO_4^{3-} system should be used. Potassium phosphate could be used along with HCl solution.
18.13 37 g
18.15 55 mL

18.17 55.7 g sodium carbonate, 0.789 L HCl solution
18.19 1.5 g, still a buffer
18.21 (a) $NH_4^+(aq) + H_2O(l) \rightleftharpoons NH_3(aq) + H_3O^+(aq)$, pH < 7;
(b) $H_2O(l) + H_2O(l) \rightleftharpoons OH^-(aq) + H_3O^+(aq)$; pH = 7; and
(c) $CH_3CO_2H(aq) + H_2O(l) \rightleftharpoons CH_3CO_2^-(aq) + H_3O^+(aq)$, pH < 7
18.23 (a) 2.8; (b) 7.8; and (c) 3.5
18.25 Phenol red

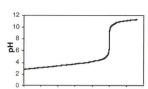

18.27 8.79; thymol blue

18.29 (a) 4.05; (b) 7.21; (c) 9.76; (d) 12.32; and (e) 12.62

18.31 (a) $AgCl(s) \rightleftharpoons Ag^+(aq) + Cl^-(aq)$, $K_{sp} = [Ag^+]_{eq} [Cl^-]_{eq}$;
(b) $BaSO_4(s) \rightleftharpoons Ba^{2+}(aq) + SO_4{}^{2-}(aq)$,
$$K_{sp} = [Ba^{2+}]_{eq} [SO_4{}^{2-}]_{eq};$$
(c) $Fe(OH)_2(s) \rightleftharpoons Fe^{2+}(aq) + 2\,OH^-(aq)$,
$$K_{sp} = [Fe^{2+}]_{eq} [OH^-]_{eq}{}^2; \text{ and}$$
(d) $Ca_3(PO_4)_2(s) \rightleftharpoons 3\,Ca^{2+}(aq) + 2\,PO_4{}^{3-}(aq)$,
$$K_{sp} = [Ca^{2+}]_{eq}{}^3 [PO_4{}^{3-}]_{eq}{}^2$$

18.33 (a) 9.1×10^{-4} g; (b) 1.2×10^{-3} g; (c) 9.8×10^{-5} g; and (d) 1.6×10^{-5} g

18.35 2.3×10^{-9}

18.37 0.53 g

18.39 1.6×10^3

18.41 0.125 M

18.43 (a) $[Fe(CN)_6]^{3-}$; (b) $[Zn(NH_3)_4]^{2+}$; and (c) $[V(en)_3]^{3+}$

18.45 $[Zn^{2+}] = 4.9 \times 10^{-9}$ M; $[Zn(NH_3)_4{}^{2+}] = 5.4 \times 10^{-3}$ M; $[NH_3] = 0.228$ M

18.47 $[Cl^-] = 0.14$ M; $[Pb^{2+}] = 8.5 \times 10^{-15}$ M; $[Na^+] = 0.17$ M; $[PbCl_4{}^{2-}] = 0.0075$ M

18.49 $Mn^{2+} + en \rightleftharpoons [Mn(en)]^{2+}$
$[Mn(en)]^{2+} + en \rightleftharpoons [Mn(en)_2]^{2+}$
$[Mn(en)_2]^{2+} + en \rightleftharpoons [Mn(en)_3]^{2+}$

18.51 7.5 g

18.53 2.0;

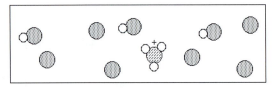

18.55 All the salt dissolves.

18.57 (a) 6.81; (b) 0.06; and (c) 7.5×10^{-3} mol

18.59 4.4×10^{-3}

18.61 (a) yellow; (b) purple; (c) purple; and (d) purple

18.63 Precipitate forms.

18.65 3.2×10^{-8}

18.67 (a) 9.43; (b) 0.050 mol; and (c) 9.32

18.69 2.0×10^{-33}

18.71 (a) 8.52; (b) 3.74; (c) 2.23; and (d) thymol blue

18.73 2.16 g; $[Mn^{2+}] = 3.5 \times 10^{-10}$ M

18.75 2.0×10^{-4} g

18.77 Mix together 15 g sodium acetate, 0.17 L of 1.0 M acetic acid, and enough water to make 1.0 L.

18.79 $[C_2O_4{}^{2-}] = 0.237$ M; $[Zn^{2+}] = 2.4 \times 10^{-9}$ M; $[[Zn(C_2O_4)_3]^{4-}] = 4.5 \times 10^{-3}$ M

18.81 (a) $MgF_2(s) \rightleftharpoons Mg^{2+}(aq) + 2F^-(aq)$, $K_{sp} = [Mg^{2+}]_{eq} [F^-]_{eq}{}^2$; (b) 5.93×10^{-9}; and (c) 2.1×10^{-10}

18.83 2.6×10^2 L

18.85 (a) 6.2×10^{-13} M; and (b) 1.0×10^3 g

18.87 At 25 °C, $K_{sp} = 1.29 \times 10^{-9}$, $\Delta G° = 50.7$ kJ/mol; at 100 °C, $K_{sp} = 2.5 \times 10^{-5}$, $\Delta G° = 33$ kJ/mol

18.89 1.1×10^{-4} M

18.91 More soluble in acidic solution: Ag_2CO_3, Ag_2SO_4, and Ag_2S; independent of pH: AgBr and AgCl

18.93 (a) 7.79; and (b) 7.65

18.95 Point A: Major species are H_2O and HCO_2H, dominant equilibrium is
$$H_2O(l) + HCO_2H(aq) \rightleftharpoons HCO_2{}^-(aq) + H_3O^+(aq),$$
$$K = \frac{[HCO_2{}^-][H_3O^+]}{[HCO_2H]};$$
Point B: major species are H_2O, HCO_2H, and $HCO_2{}^-$, dominant equilibrium is
$$H_2O(l) + HCO_2H(aq) \rightleftharpoons HCO_2{}^-(aq) + H_3O^+(aq),$$
$$K = \frac{[HCO_2{}^-][H_3O^+]}{[HCO_2H]};$$
Point C: major species are H_2O and $HCO_2{}^-$, dominant equilibrium is
$$H_2O(l) + HCO_2{}^-(aq) \rightleftharpoons HCO_2H(aq) + OH^-(aq),$$
$$K = \frac{[HCO_2H][OH^-]}{[HCO_2{}^-]};$$
Point D: major species are H_2O, OH^-, and $HCO_2{}^-$, dominant equilibrium is
$$H_2O(l) + HCO_2{}^-(aq) \rightleftharpoons HCO_2H(aq) + OH^-(aq),$$
$$K = \frac{[HCO_2H][OH^-]}{[HCO_2{}^-]}.$$
The cation of the strong base will also be present as a major species at points B, C, and D.

18.97 (a) 4.27; (b)

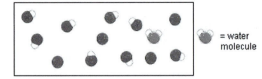

and (c)

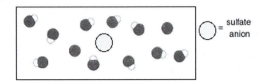

18.99 $K_{eq} = 6 \times 10^6$; 3 g

Chapter 19

19.1 (a) O is −2, H is +1, and Fe is +3; (b) F is −1, N is +3; (c) O is −2, H is +1, C is −2; (d) K^+ is +1, O is −2, and C is +4; (e) in $NH_4{}^+$, H is +1 and N is −3; in $NO_3{}^-$, O is −2 and N is +5; (f) Cl is −1, Ti is +4; (g) Pb is +2; in sulfate, O is −2, S is +6; and (h) P is 0

19.3 (a) not redox; (b) redox; (c) not redox; (d) redox; and (e) redox

19.5 (a) +5; (b) −1; (c) +7; (d) 0; (e) +3; and (f) +1

19.7 (a) $Na \rightarrow Na^+ + e^-$ and $2\,H_3O^+ + 2\,e^- \rightarrow H_2 + 2\,H_2O$;
(b) $Au + 4\,Cl^- \rightarrow [AuCl_4]^- + 3\,e^-$ and $NO_3{}^- + 4\,H_3O^+ + 3\,e^- \rightarrow NO + 6\,H_2O$;
(c) $MnO_4{}^- + 8\,H_3O^+ + 5\,e^- \rightarrow Mn^{2+} + 12\,H_2O$ and $C_2O_4{}^{2-} \rightarrow 2\,CO_2 + 2\,e^-$; and

(d) $C + 3\,H_2O \rightarrow CO + 2\,H_3O^+ + 2\,e^-$ and $2\,H_3O^+ + 2\,e^- \rightarrow H_2 + 2\,H_2O$

19.9 (a) $Cu^+ + 3\,H_2O \rightarrow CuO + 2\,H_3O^+ + e^-$;
(b) $S + 2\,H_3O^+ + 2\,e^- \rightarrow H_2S + 2\,H_2O$;
(c) $Ag_2O + H_2O + 2\,e^- \rightarrow 2\,Ag + 2\,OH^-$;
(d) $I^- + 6\,OH^- \rightarrow IO_3{}^- + 3\,H_2O + 6\,e^-$;
(e) $IO_3{}^- + 2\,H_2O + 4\,e^- \rightarrow IO^- + 4\,OH^-$; and
(f) $H_2CO + 5\,H_2O \rightarrow CO_2 + 4\,H_3O^+ + 4\,e^-$

19.11 (a) $2\,Cu^+ + 4\,H_2O + S \rightarrow 2\,CuO + 2\,H_3O^+ + H_2S$;
(b) $3\,Ag_2O + I^- \rightarrow 6\,Ag + IO_3{}^-$; (c) $2\,I^- + IO_3{}^- \rightarrow 3\,IO^-$; and (d) $2\,S + H_2CO + H_2O \rightarrow 2\,H_2S + CO_2$

19.13 (a) Cl^-; (b) $MnO_4{}^-$; (c) $MnO_4{}^-$; (d) Cl^-; (e) $MnO_4{}^-$; and (f) Cl^-

19.15 (a) $3 CN^- + 2 MnO_4^- + H_2O \rightarrow 3 CNO^- + 2 MnO_2 + 2 OH^-$; (b) $4 As + 3 O_2 + 2 H_2O \rightarrow 4 HAsO_2$; (c) $Br^- + 2 MnO_4^- + H_2O \rightarrow BrO_3^- + 2 MnO_2 + 2 OH^-$; (d) $3 NO_2 + 3 H_2O \rightarrow 2 NO_3^- + NO + 2 H_3O^+$; (e) $ClO_4^- + 6 Cl^- + 6 H_3O^+ \rightarrow ClO^- + 9 H_2O + 3 Cl_2$; and (f) $AlH_4^- + 4 H_2CO + 4 H_2O \rightarrow Al^{3+} + 4 CH_3OH + 4 OH^-$

19.17 (a), (b), (c) and (d) spontaneous

19.19

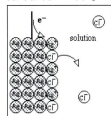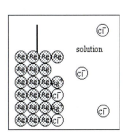

19.21

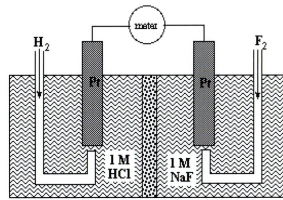

19.23 The silver–silver chloride electrode (19.19) is active, whereas the platinum electrodes (19.21) are passive.

19.25 (a) 1.565 V; (b) 0.981 V; and (e) 0.002 V

19.27 $8 NO + 3 H_2O \rightarrow 3 N_2O + 2 NO_3^- + 2 H_3O^+$; $E°_{cell} = 0.634$ V

19.29 Set up a standard hydrogen electrode on one side and a Pt electrode immersed in 1 M solution of NaF with F_2 bubbling over the electrode. The H_2/H_3O^+ electrode will be the anode.

19.31 Ba, Ca, Cs, Li, Mg, K, and Na. All these metals lie in the s block of the periodic table and easily lose 1 or 2 electrons.

19.33 (a) -906.0 kJ; (b) -1.14×10^3 kJ; and (e) -1 kJ

19.35 $E = 1.35$ V

19.37 $m_{Pb} = 9.5 \times 10^{-2}$ g; $m_{PbO_2} = 0.11$ g

19.39 1.35×10^3 s

19.41 1.98 g

19.43 $Q = \dfrac{[OH^-]_{cathode}}{[OH^-]_{anode}}$; as the battery operates, the anode concentration decreases while the cathode concentration increases, $Q > 1$ and $\ln Q > 0$, and the potential of the battery decreases with use.

19.45 Cr corrodes.

19.47 Zinc–air batteries are characterized by a stable potential and compact size but limited current capacity, making them well suited for use where large currents are not needed, such as pacemakers and cameras. They cannot supply the large current needed to start an automobile engine; moreover, they are irreversible and would have to be replaced very frequently.

19.49 19.9 g

19.51 The oxidation reaction is $Cu \rightarrow Cu^{2+} + 2 e^-$. The reduction of H_3O^+ occurs: $2 H_3O^+ + 2 e^- \rightarrow 2 H_2O + H_2$ ($E° = 0$ V)

19.53 (a) Attach the negative wire to the Cd electrode; $Cd (s) + 2 OH^- (aq) \rightleftarrows Cd(OH)_2 (s) + 2 e^-$; (b) 4.56 hr

19.55 1.90×10^3 C; 2.10 A

19.57

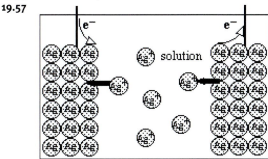

19.59 $K_{eq} = 1.0 \times 10^{-14}$

19.61 3.0×10^9

19.63 (a) $3 H_2 + 2 Cr(OH)_3 \rightarrow 6 H_2O + 2 Cr$; (b) -0.65 V; and (c) 3.8×10^2 kJ

19.65 (a) $Cr_2O_7^{2-} + 14 H_3O^+ + 6 e^- \rightarrow 2 Cr^{3+} + 21 H_2O$; and (b) 16.2 g

19.67 (a) on the left, $Ni^{2+} + 2e^- \rightarrow Ni$, $E° = -0.257$ V; on the right, $Fe^{2+} + 2e^- \rightarrow Fe$, $E° = -0.447$ V; (b) $E° = 0.190$ V; (c) Fe electrode is the anode and Ni is the cathode; and (d)

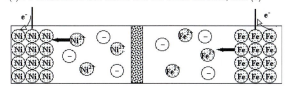

19.69 $[Ni^{2+}] = 3.8 \times 10^{-7}$ M gives $E = 0.0$ V.

19.71 (a) $2 MnO_4^- + 5 H_2SO_3 + H_3O^+ \rightarrow 2 Mn^{2+} + 5 HSO_4^- + 4 H_2O$; (b) $2 MnO_4^- + 5 SO_2 + H_2O + H_3O^+ \rightarrow 2 Mn^{2+} + 5 HSO_4^-$; (c) $8 MnO_4^- + 5 H_2S + 19 H_3O^+ \rightarrow 8 Mn^{2+} + 5 HSO_4^- + 31 H_2O$; and (d) $8 MnO_4^- + 5 H_2S_2O_3 + 14 H_3O^+ \rightarrow 8 Mn^{2+} + 10 HSO_4^- + 21 H_2O$

19.73 80.3 min

19.75

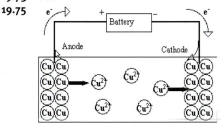

19.77 (a) MnO_4^-; (b) O_2; and (c) Sn^{2+}

19.79 O_2 can oxidize Cu, Ag, Fe^{2+}, H_2, and I^-.

19.81 (a) $Zn(NO_3)_2$ soln (cathode): $Zn^{2+} + 2\,e^- \rightarrow Zn$; $ZnCl_2$ soln (anode): $Zn \rightarrow Zn^{2+} + 2\,e^-$; (c) 0.0207 V; and
(b)

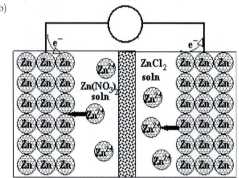

19.83 2.3×10^2 hr

19.85 5.6×10^{-8} M

19.87 (a) $Na^+ + e^- \rightarrow Na$ and $2\,Cl^- \rightarrow Cl_2 + 2\,e^-$; (b) cathode is the Pt electrode connected to the negative pole, and anode is the Pt electrode connected to the positive pole of the battery;

(c)

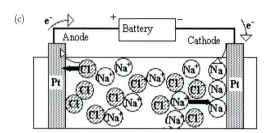

19.89

19.91 0.368 V

Chapter 20

20.1 (a) Mn is $+2$; (b) Mo is $+5$; (c) V is $+5$; (d) Au is $+3$; and (e) Fe is $+3$

20.3 (a) chromium, Cr; (b) cadmium, Cd; and (c) copper, Cu

20.5 (a) $3d^5$; (b) $5d^6$; (c) $3d^8$; and (d) $4d^4$

20.7 (a) Pd; (b) Au; and (c) Co

20.9 (a) Ru has oxidation state $+2$, d^6; (b) Cr has oxidation state $+3$, d^3; (c) Pd has oxidation state $+2$, d^8; (d) Ir has oxidation state $+3$, d^6; and (e) Ni has oxidation state 0, $s^2 d^8$

20.11 (a) hexaammineruthenium(II) chloride; (b) *trans*-(ethylenediamine)diiodochromium(III) iodide; (c) *cis*-dichlorobis(trimethylphosphine)palladium(II); (d) *fac*-triamminetrichloroiridium(III); and (e) tetracarbonylnickel(0)

20.13 (a) Six NH_3 in an octahedron around the central Ru:

(b) Two I at opposite ends of one axis, two en in a square plane around the central Cr:

(c) Square planar arrangement about Pd, with two Cl adjacent each other:

(d) Octahedral arrangement around Ir, with three Cl in a triangular face:

(e) Tetrahedral arrangement of $C\equiv O$ about a central Ni:

20.15 (a) *cis*-$[Co(NH_3)_4ClNO_2]^+$; (b) $[PtNH_3Cl_3]^-$; (c) *trans*-$[Cu(en)_2(H_2O)_2]^{2+}$; and (d) $[FeCl_4]^-$

20.17 (a) (b) (c) (d)

20.19 (a) and (b)
d^2 d^3

(c) and (d)

d^5
High-spin   Low-spin

20.21 (a) diamagnetic; (b) paramagnetic with four unpaired electrons; (c) paramagnetic with two unpaired electrons; and (d) diamagnetic

20.23 The colors of transition-metal complexes are generally determined by d–d transitions, but Zr^{4+} has no valence electron that can undergo a transition that would absorb visible light.

20.25 (a) 6; (b) $+3$, $3d^5$; (c) octahedral; (d) paramagnetic; and (e) one

20.27

High-spin (H_2O)   Low-spin (CN^-)

20.29 257 kJ/mol, orange

20.31 (a) $CuFeS_2 + 2 O_2 \rightarrow CuO + FeO + 2 SO_2$;
(b) $Si + O_2 + CaO \rightarrow CaSiO_3$; and
(c) $TiCl_4 + 4 Na \rightarrow 4 NaCl + Ti$

20.33 1.06×10^6 g; 2.05×10^5 L

20.35 183.3 kJ; 63.3 kJ

20.37 The coinage metals are copper, silver, and gold. They are used for money, electrical wire, and decorative objects such as jewelry.

20.39 Titanium is used as an engineering metal because of its relatively low density, high bond strength, resistance to corrosion, and ability to withstand high temperatures, all of which make it a favored structural material.

20.41 Hemoglobin has four subunits, whereas myoglobin has just one. As a consequence, hemoglobin has a more complex cooperative chemical behavior than myoglobin.

20.43 Remove 1 electron

20.45 (a)

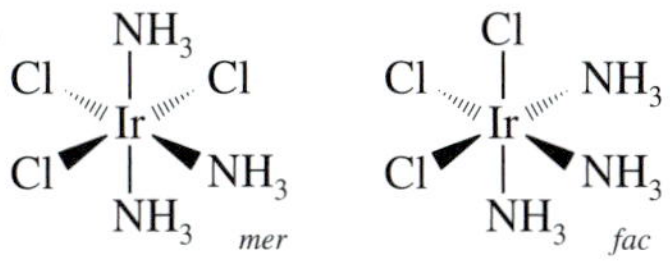

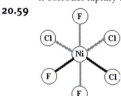

(b)

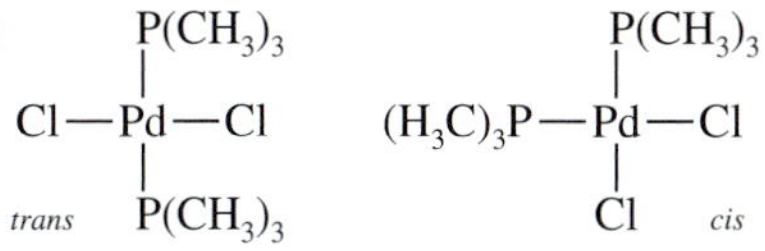

$$\text{Cl—Pd—Cl} \qquad (H_3C)_3P\text{—Pd—Cl}$$

(c)

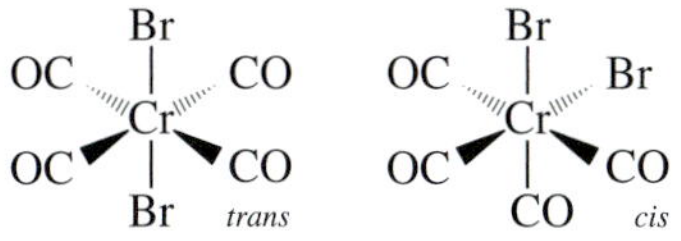

(d)

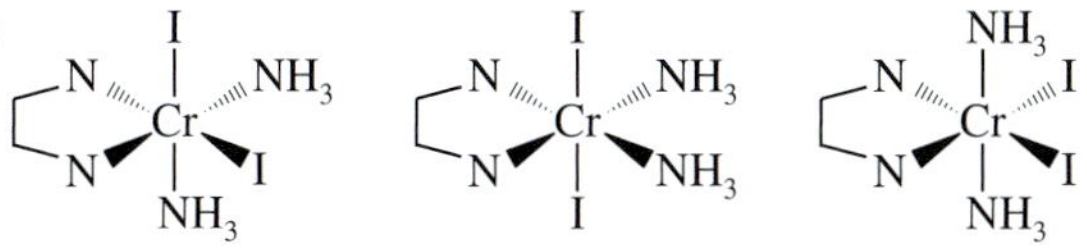

20.47 Bidentate ligands form complexes with two links. Each Fe ion forms an octahedral complex with three oxalate anions:

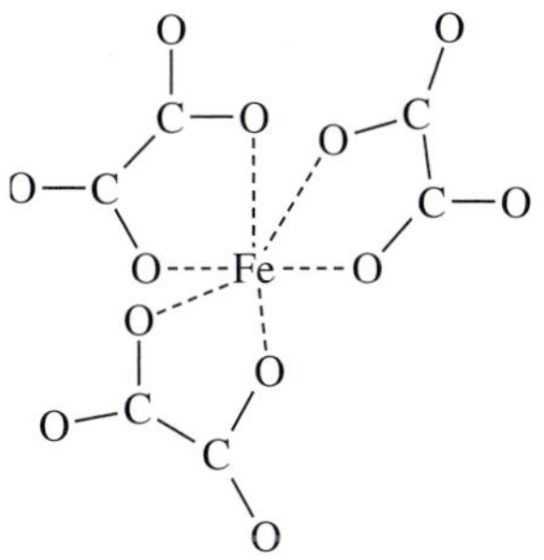

20.49 (a) Cr is $[Ar]4s^1 3d^5$; Cr^{2+} is $[Ar]3d^4$; Cr^{3+} is $[Ar]3d^3$;
(b) V is $[Ar]4s^2 3d^3$; V^{2+} is $[Ar]3d^3$; V^{3+} is $[Ar]3d^2$;
V^{4+} is $[Ar]3d^1$; V^{5+} is $[Ar]$; and
(c) Ti is $[Ar]4s^2 3d^2$; Ti^{2+} is $[Ar]3d^2$; Ti^{4+} is $[Ar]$

20.51 (a) *cis*-tetraaquadichlorochromium(III) chloride;
(b) bromopentacarbonylmanganese(I); and
(c) *cis*-diamminedichloroplatinum(II)

20.53 $2 O_2^- + 2 H_3O^+ \xrightarrow{SOD} O_2 + H_2O_2 + 2 H_2O$

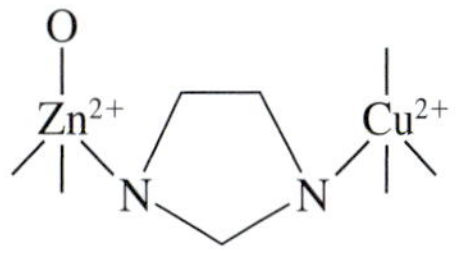

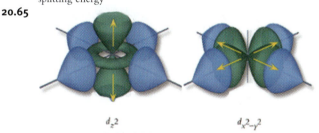

20.55 Cu(II) in water forms an aqua complex. Addition of fluoride produces the insoluble green salt CuF_2, while addition of chloride produces the tetrachlorocopper(II) complex $[CuCl_4]^{2-}$:

$$Cu^{2+}(aq) + 2 F^-(aq) \longrightarrow CuF_2(s)$$
$$Cu^{2+}(aq) + 4 Cl^-(aq) \longrightarrow [CuCl_4]^{2-}(aq)$$

20.57 Silver is difficult to oxidize, so it has good resistance to corrosion and is suitable for jewelry. Vanadium is readily oxidized, so it corrodes rapidly and is unsuited to jewelry.

20.59

20.61 Silver sulfide;
$4 Ag (s) + 2 H_2S (g) + O_2 (g) \rightarrow 2 Ag_2S (s) + 2 H_2O (l)$

20.63 Cr^{3+} is d^3, each electron in a different orbital regardless of the splitting energy

20.65

d_{z^2} $d_{x^2-y^2}$

20.67 (a) 2+, 3+, 3+; (b) red-violet, orange, orange; and
(c) 232 kJ/mol, 249 kJ/mol, 257 kJ/mol

20.69 $[Ni(CO)_4]$ has tetrahedral geometry, colorless; $[Zn(CN)_4]^{2-}$ has tetrahedral geometry, colorless

20.71 Brass is an alloy of zinc and copper, and superoxide dismutase contains zinc and copper ions in its reaction center.

20.73 $[Ni(CN)_4]^{2-}$ is square planar and $[NiCl_4]^{2-}$ is tetrahedral.

20.75 When ferritin is neither empty nor filled to capacity, it can provide iron as needed for hemoglobin synthesis, or store iron if an excess is absorbed by the body.

20.77 Because Zn^{2+} has d^{10} configuration, its d orbitals are completely filled and the lowest unoccupied orbital is quite high in energy. Co^{2+} has d^7 configuration, giving this cation unfilled d orbitals. Consequently, metalloproteins that contain Co^{2+} absorb visible light, making it possible to study them with visible spectroscopy.

20.79 Zn displaces any less active metal cation from the solution. Cd is one example:

$$Zn (s) + Cd^{2+}(aq) \longrightarrow Cd (s) + Zn^{2+}(aq) \quad E^0 > 0$$

Chapter 21

21.1   (a) Lewis acid is Ni, Lewis base is CO; (b) Lewis acid is $SbCl_3$, Lewis base is Cl^-; (c) Lewis acid is $AlBr_3$, Lewis base is $(CH_3)_3P$; and (d) Lewis acid is BF_3, Lewis base is ClF_3.

21.3   (a) acceptor: 3p orbital   (b) acceptor: 3d orbital

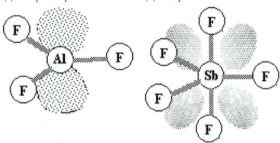

(c) acceptor: delocalized π orbital

21.5   In the first step, an electron pair from the O atom of H_2O displaces a π bond in Lewis acid-base adduct formation. Then a proton from H_2O migrates to a C—O oxygen atom:

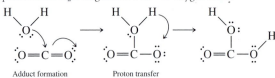

Adduct formation   Proton transfer

21.7 V^{3+} (lowest Z, high charge) < Fe^{3+} (high charge) < Fe^{2+} < Pb^{2+} (large n)

21.9   (a) hardest is BF_3, then BCl_3, and $AlCl_3$ is softest; (b) the hardest acid is Al^{3+} (row 3), then Tl^{3+} (row 6), and Tl^+ (low charge) is softest; and (c) polarizability increases with n, so hardest is $AlCl_3$, then $AlBr_3$, and AlI_3 is softest.

21.11 When an electronegative O atom bonds to a less negative S atom, it withdraws electron density from S, decreasing the polarizability about S and increasing the hardness of the base.

21.13 (a) metathesis occurs, giving $AlCl_3$ and NaI; (b) no reaction occurs; (c) metathesis occurs, giving H_2O and CaS; (d) metathesis occurs, giving $(CH_3)_3P$ and $LiCl$; and (e) no reaction occurs.

21.15 Al_2Cl_6 contains two "bridging" chlorine atoms. Standard procedures would predict tetrahedral geometry about all inner atoms, but the Al—Cl—Al bond angles of 91° indicate that Cl uses p orbitals. Each Al atom can be described as using sp^3 hybrids to form four σ bonds to four different Cl atoms. In Lewis acid-base terms, the bridged molecule forms from two $AlCl_3$ units linking together in double-adduct formation between the Al Lewis-acid atoms and two Cl Lewis-base atoms.

21.17 Thallium lies below indium and gallium, so its properties should be similar to those metals: valence of 3, soft Lewis acid. Like its neighbor Pb, it is toxic.

21.19 $SnCl_4$ can function as a Lewis acid because Sn has empty d orbitals from which it can accept electrons to form more bonds:

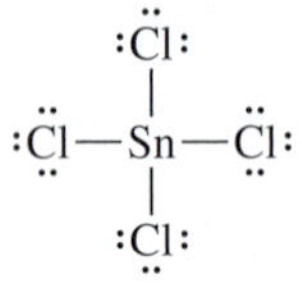

$$:\overset{\displaystyle :\overset{..}{Cl}:}{\underset{\displaystyle :\overset{..}{Cl}:}{\overset{|}{Cl}-Sn-\overset{..}{Cl}:}}$$

21.21 Lewis structure:

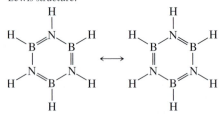

The B and N atoms have bonding and geometry that can be described using sp^2 hybrid orbitals, resulting in three σ bonds around each ring atom. In addition, there is a delocalized π bonding network encompassing all six ring atoms and containing six electrons.

21.23 Si has a larger band gap.

21.25 $2\ C_2H_5Cl + Si(Cu) \longrightarrow (C_2H_5)_2SiCl_2 + Cu$
$(C_2H_5)_2SiCl_2 + 2\ H_2O \longrightarrow (C_2H_5)_2Si(OH)_2 + 2\ HCl$
Condensation will eliminate water to give the polymer:

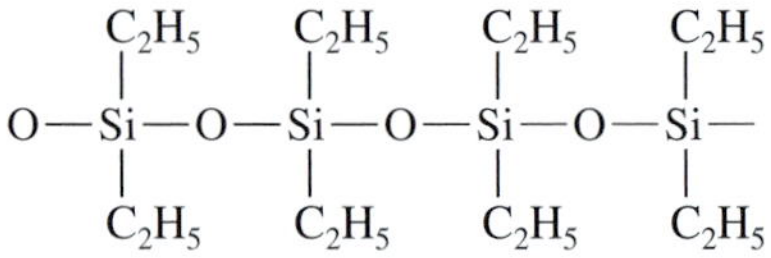

$$\underset{\displaystyle C_2H_5}{O-\overset{\displaystyle C_2H_5}{\overset{|}{Si}}-O-\overset{\displaystyle C_2H_5}{\underset{\displaystyle C_2H_5}{\overset{|}{Si}}}-O-\overset{\displaystyle C_2H_5}{\underset{\displaystyle C_2H_5}{\overset{|}{Si}}}-O-\overset{\displaystyle C_2H_5}{\underset{\displaystyle C_2H_5}{\overset{|}{Si}}}-}$$

21.27 $P_4O_{13}{}^{6-}$:

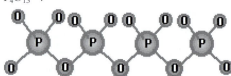

21.29 Brønsted acid–base reaction:
$Ca_5(PO_4)_3F\,(s) + 5\ H_2SO_4\,(aq) \longrightarrow$
$\qquad\qquad 3\ H_3PO_4\,(aq) + 5\ CaSO_4\,(s) + HF\,(aq)$

21.31

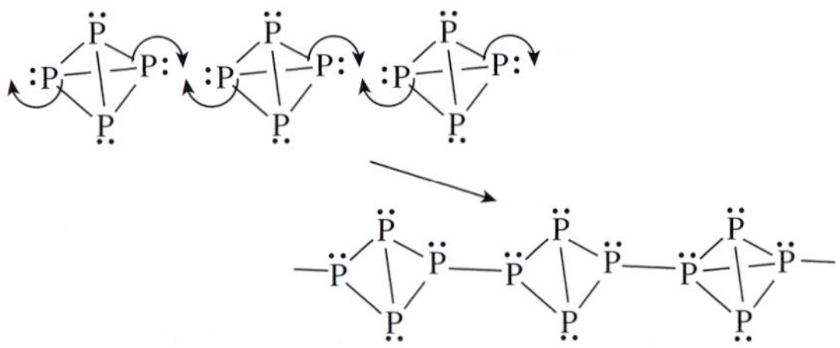

21.33 There are two industrial reactions in Section 21.6 in which sulfuric acid acts as a Brønsted acid:

$CaF_2\,(s) + H_2SO_4\,(l) \longrightarrow 2\ HF\,(g) + CaSO_4\,(s)$
$Ca_5(PO_4)_3F\,(s) + 5\ H_2SO_4\,(aq) \longrightarrow$
$\qquad\qquad 3\ H_3PO_4\,(aq) + 5\ CaSO_4\,(s) + HF\,(aq)$

21.35

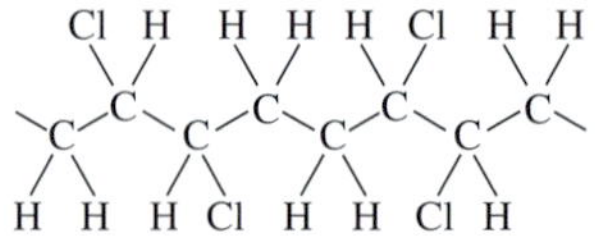

21.37 $TiO_2\,(s) + C\,(s) + 2\ Cl_2\,(g) \longrightarrow TiCl_4\,(l) + CO_2\,(g)$
C is oxidized from 0 to +4 and is the reducing agent; Cl is reduced from −1 to 0 and is the oxidizing agent.

21.39 8.7×10^3 kg bauxite rock

21.41 Sulfur has d orbitals available for bond formation, allowing the formation of SF_6, in which the bonding can be described using d^2sp^3 hybrid orbitals on the S atom. The Br atom is too large for six Br atoms to be accommodated around a central S atom. Oxygen has no valence d orbitals available.

21.43 The As atom in $AsCl_3$ has a lone pair that it donates, giving this compound Lewis-base character. The As atoms also can accommodate additional electron pairs by using valence d orbitals, giving this compound Lewis-acid character.

$$AsCl_3 + BF_3 \longrightarrow Cl_3As{-}BF_3 \text{ (As acts as Lewis base.)}$$
$$AsCl_3 + Cl^- \longrightarrow AsCl_4^- \text{ (As acts as Lewis acid.)}$$

21.45 $Al + 2 H_2O \longrightarrow AlO_2^- + 4 H^+ + 3 e^-$;
(a) $8 Al + 5 OH^- + 3 NO_3^- + 2 H_2O \longrightarrow 3 NH_3 + 8 AlO_2^-$;
(b) $2 Al + 2 OH^- + 2 H_2O \longrightarrow 3 H_2 + 2 AlO_2^-$; and
(c) $4 Al + 3 SnO_3^{2-} + H_2O \longrightarrow 3 Sn + 2 OH^- + 4 AlO_2^-$

21.47 Nitrogen is a nonmetal that forms polar bonds most readily with other nonmetals (B, C, O, the halogens). Phosphorus has a similar pattern of reactivity but can also involve d orbitals in its bonding. Arsenic and antimony are metalloids with useful semiconductor properties, and bismuth is metallic.

21.49 $Al(O)OH (s) + OH^- (aq) + H_2O \longrightarrow [Al(OH)_4]^- (aq)$
$[Al(OH)_4]^- (aq) \longrightarrow Al(OH)_3 (s) + OH^- (aq)$
$2 Al(OH)_3 (s) \xrightarrow{1250\,°C} Al_2O_3 (s) + 3 H_2O (g)$
$2 Al_2O_3 (s) + 3 C (s) \longrightarrow 3 CO_2 (g) + 4 Al (l)$

21.51 Silicon dioxide is first converted into silicon tetrachloride by reaction with molecular chlorine:

$$SiO_2 (s) + 2 C (s) + 2 Cl_2 (g) \xrightarrow{\text{High } T} SiCl_4 (g) + 2 CO (g)$$

The $SiCl_4$ is purified by distillation and then reduced by reaction with Mg metal:
$SiCl_4 + 2 Mg \longrightarrow Si + 2 MgCl_2$

21.53 Mercury is a very soft Lewis acid, while zinc is intermediate in hard/soft character. Thus mercury preferentially associates with sulfur, a soft Lewis base, while zinc forms stable bonds with both soft (S-containing) and hard (O-containing) Lewis bases.

21.55 Gaseous Al_2Cl_6 is in equilibrium with gaseous $AlCl_3$:

$$Al_2Cl_6 \rightleftharpoons 2 AlCl_3$$

Le Châtelier's principle predicts that because the forward reaction is endothermic (bonds must be broken), an increase in temperature shifts the position of the equilibrium to the right. Thus, as temperature increases, the number of moles of gaseous substance increases, so the pressure increases faster than would be predicted by the ideal gas equation.

21.57 $Ca_5(PO_4)_3F (s) + 7 H_3PO_4 (aq) \rightarrow 5 Ca(H_2PO_4)_2 (s) + HF (aq)$; 26.47%

21.59 (a)

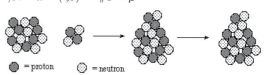

(b) $6 SOCl_2 (l) + FeCl_3 \cdot 6H_2O (s) \longrightarrow$
$\qquad\qquad\qquad 6 SO_2 (g) + 12 HCl (g) + FeCl_3 (s)$

21.61 As the size of an atom increases, so does its polarizability, and greater polarizability leads to larger dispersion forces. Thus, the larger the atoms in a molecule, the larger the intermolecular forces and the easier it is to condense the substance. In the sequence BCl_3, BBr_3, BI_3, the halogens are increasing in size, accounting for the different stable phases at room temperature.

21.63 The phosphate anion is the conjugate base of a weak acid, HPO_4^{2-}:
$PO_4^{3-} + H_2O \longrightarrow HPO_4^{2-} + OH^-$;
$pK_b = 14.00 - pK_a = 14.00 - 12.32 = 1.68$

21.65 $2 Al_2O_3 + 6 Cl_2 \longrightarrow 4 AlCl_3 + 3 O_2$;
$AlCl_3 + 3 Na \longrightarrow Al + 3 NaCl$

21.67 (a) $AlCl_3 + 3 LiCH_3 \longrightarrow Al(CH_3)_3 + 3 LiCl$;
(b) $SO_3 + H_2O \longrightarrow H_2SO_4 (aq)$;
(c) $SbF_5 + LiF \longrightarrow Li(SbF_6)$; and
(d) no reaction

21.69 $P_4O_{10} + 6 H_2O \longrightarrow 4 H_3PO_4$

Chapter 22

22.1
Part:	(a)	(b)	(c)
Z	10	82	40
A	20	205	90
N	10	123	50

22.3 ^{53}Mn: unstable, too few neutrons; ^{54}Mn: unstable, odd−odd; ^{55}Mn: stable; and ^{56}Mn: unstable, too many neutrons and odd−odd

22.5 $\Delta E = 8.99 \times 10^{16}$ kJ

22.7 $\Delta E = -1.079 \times 10^{11}$ kJ/mol;
ΔE (per nucleon) $= -8.11 \times 10^8$ kJ/mol nucleon

22.9 The repulsive barrier is greater for $^1H + {^6Li} \rightarrow {^7Be}$ than for $^4He + {^4He} \rightarrow {^8Be}$; ^7Be is more stable.

22.11
Description	Symbol	Name
(a) high-energy photon	γ	gamma ray
(b) mass number 4	α	alpha particle
(c) positive particle with m_e	β^+	positron

22.13 (a) $^{125}_{52}Te$; (b) $^{123}_{51}Sb$; and (c) $^{127}_{53}I$

22.15 ^{53}Mn, β^+ or EC; ^{54}Mn, β, β^+ or EC; ^{55}Mn is stable; ^{56}Mn, β

22.17 $t_{1/2} = 1.2$ yr

22.19 $^{232}_{90}Th \rightarrow \alpha + {^{228}_{88}Ra} \rightarrow \beta + {^{228}_{89}Ac} \rightarrow \beta + {^{228}_{90}Th} \rightarrow \alpha + {^{224}_{88}Ra}$
$^{224}_{88}Ra \rightarrow \alpha + {^{220}_{86}Rn} \rightarrow \alpha + {^{216}_{84}Po} \rightarrow \beta + {^{216}_{85}At} \rightarrow \alpha + {^{212}_{83}Bi}$
$^{212}_{83}Bi \rightarrow \beta + {^{212}_{84}Po} \rightarrow \alpha + {^{208}_{82}Pb}$

22.21 (a) $^{12}_6C + n \rightarrow ({^{13}_6C}) \rightarrow {^{12}_5B} + p$;
(b) $^{16}_8O + \alpha \rightarrow ({^{20m}_{10}Ne}) \rightarrow {^{20}_{10}Ne} + \gamma$; and
(c) $^{247}_{96}Cm + {^{11}_5B} \rightarrow ({^{258}_{101}Md}) \rightarrow {^{255}_{101}Md} + 3 n$

22.23 The reaction in this problem is
$$^{14}_7N + \alpha \rightarrow ({^{18}_9F}) \rightarrow {^{18}_8O} + \beta^+$$

○ = proton   ◎ = neutron

22.25 Cs, Ba, La, Ce

22.27 $\Delta E = -1.648 \times 10^{10}$ kJ/mol, somewhat less than Section Exercise 22.4.1 result

22.29 The core of a nuclear reactor generates radiation that converts any material in its vicinity into radioactive substances. Thus, the heat exchanger in immediate contact with the core becomes radioactive and must be separated from the turbine that generates electricity. The primary heat exchanger transfers energy to a secondary heat exchanger, which does not become radioactive and can safely drive the turbine.

22.31 $\Delta E_{total} = -2.4 \times 10^9$ kJ

22.33 $u = 5.3 \times 10^6$ m/s

22.35 Your description should include the extremely high energies required to initiate fusion and the difficulties in containing the fusion components at the temperature required for nuclei to have these high energies.

22.37

Stage	Temperature	Composition
H-burning	4×10^7 K	C, N, H, He, e^-
He-burning	10^8 K	He, Be, C, and O
C-burning	10^9 K	All nuclides from $Z = 6$ (C) up to $Z = 26$ (Fe)

22.39 Your description should include the fact that elements beyond $Z = 26$ are less stable than Fe, so they cannot be generated by fusion of lighter elements.

22.41 $E = 4.4 \times 10^{-4}$ J

22.43 3.41×10^4 β/s

22.45 Exposure to radiation results first in damage to those cells that reproduce most quickly, including the white blood cells that are responsible for fighting infection and the mucous membrane lining of the intestinal tract. Thus the early symptoms of radiation exposure include reduced resistance to infection and nausea due to disruption of the digestive tract.

22.47 3.5×10^9 yr

22.49 Add a radioactive iron isotope to the manufacturer's iron waste stream before treatment, and monitor the iron content of the river downstream. If radioactivity appears in the river water, the source is the manufacturer; if not, the iron in the river comes from some other source.

22.51 Run the hydrolysis in water that is enriched with a radioactive isotope of oxygen. Isolate the two products and analyze them for radioactivity. Whichever product shows radioactivity is the one whose added oxygen atom comes from water.

22.53 $\Delta E = -1.583 \times 10^{11}$ kJ/mol; ΔE (per nucleon) $= -7.574 \times 10^8$ kJ/mol nucleon

22.55 $N = 0.80$ mg; $\dfrac{\Delta N}{\Delta t} = 1.3 \times 10^{11}$/s

22.57 The other product is an electron; E_{kinetic} (electron) $= 1.25 \times 10^{-13}$ J

22.59 1% decays in 24 yr; 1% remains after 1.1×10^4 yr

22.61 Both isotopes are neutron-deficient and lie below (to the right of) the belt of stability: $^{11}\text{C} \rightarrow \beta^+ + ^{11}\text{B}$ and $^{15}\text{O} \rightarrow \beta^+ + ^{15}\text{N}$

22.63 70 hr

22.65 (a) $^{238}_{92}\text{U} \rightarrow \alpha + ^{234}_{90}\text{Th}$; (b) $^{60}_{28}\text{Ni} + \text{n} \rightarrow \text{p} + ^{60}_{27}\text{Co}$; (c) $^{239}_{93}\text{Np} + ^{12}_{6}\text{C} \rightarrow 3\,\text{n} + ^{248}_{99}\text{Es}$; (d) $^{35}_{17}\text{Cl} + \text{p} \rightarrow ^{16}_{32}\text{S} + \alpha$; and (e) $^{60}_{27}\text{Co} \rightarrow \beta + ^{60}_{28}\text{Ni}$

22.67 $^{256}_{104}\text{Rf}$

22.69 ΔE (per event) $= -4.9 \times 10^{-13}$ J; fraction carried off by the α particle $= 0.73$

22.71 (a) too many neutrons; (b) $Z > 83$; (c) odd–odd; and (d) too many neutrons

22.73 In a nuclear reactor, the moderator serves to slow down fast neutrons so they are more efficiently captured by the nuclear fuel. This reduces the amount of fuel required to sustain the reaction. The control rods serve to absorb some of the neutrons, allowing the reactor to operate just below its critical point and generate large quantities of heat without heating up beyond control.

22.75 If $t_{1/2} = 1100$ s, $N = 36$; if $t_{1/2} = 876$ s, $N = 2$

22.77 $^{10}_{5}\text{B} + \text{n} \rightarrow ^{11}_{5}\text{B} \rightarrow ^{7}_{3}\text{Li} + \alpha$; the reaction does not pose a significant health hazard.

22.79 1.4×10^4 yr

22.81 5.2×10^4 yr

22.83 (a) $^{90}_{38}\text{Sr} \rightarrow ^{90}_{39}\text{Y} + \beta$; $^{90}_{39}\text{Y} + ^{90}_{40}\text{Zr} + \beta$; (b) Sr has the larger mass; and (c) $\Delta E = -1.7 \times 10^8$ kJ/mol

22.85 1.5×10^2 hr

22.87 $^{99}_{43}\text{Tc}$

22.89 Prepare a sample of the alcohol that is enriched in radioactive ^{18}O and run the reaction using this sample. Separate the products and measure the radioactivity of the ester and the water. If the C—OH bond in the alcohol breaks during the condensation, the ^{18}O will appear in the water, while if the C—OH bond in the carboxylic acid breaks during the condensation, the ^{18}O will appear in the ester.

22.91 The precipitate will contain radioactive Na.

Photo Credits

Chapter 1

Pages 0–1: Reuters/Corbis Images Page 0 *(center left):* Courtesy NASA. Page 4 *(top left):* Courtesy NASA. Page 4 *(top right):* Courtesy NASA. Page 4 *(bottom left):* Courtesy Jet Propulsion Lab/NASA. Page 4 *(bottom right):* Dudley Foster//Woods Hole Oceanographic Institution. Page 5 *(left):* Arthur S. Aubry/Photo Disc. Page 5 *(right):* Courtesy of The Hendrix Group, Inc., Houston, Texas. Page 6: Charles Lenars/Corbis Images. Page 9 *(top left):* Martyn F. Chillmaid/Photo Researchers. Page 9 *(top center left):* Paul Silverman/Fundamental Photographs. Page 9 *(top center):* David Wrobel/Visuals Unlimited. Page 9 *(top center right):* Phillip Hayson/Photo Researchers. Page 9 *(top right):* Richard Treptow/Photo Researchers. Page 9 *((bottom left):* Klaus Guldbrandsen/Photo Researchers. Page 9 *(bottom center left):* Charles D. Winters/ Photo Researchers. Page 9 *(bottom center):* OPC, Inc. Page 9 *(bottom center right):* Courtesy Atlantic Metals and Alloys, Inc. Page 9 *(bottom right):* OPC, Inc. Page 11 *(left):* Layne Kennedy/Corbis Images. Page 11 *(right):* John Olmsted. Page 13: Pat O'Hara/Corbis Images. Page 14 *(top left & bottom photos):* Courtesy of IBM Almaden Research Center. Page 14 *(top right):* Courtesy of Digital Instruments. Page 15 *(top right):* Michael Watson. Page 15 *(bottom left):* Spencer Grant/PhotoEdit. Page 15 *(bottom right):* Dr. Helen Aspinall, University of Liverpool. Page 17: Don Mason/Corbis Stock Market. Page 18: Tom McHugh/Photo Researchers. Page 21: Andy Washnik. Page 22: Steve Taylor/Stone/ Getty Images. Page 23: John Olmsted. Page 25: Courtesy International Silver Plating, Inc. Page 30 *(top left):* OPC, Inc. Page 30 *(top center):* Peter Lerman. Page 30 *(top right):* Stephen Frisch. Page 31 *(bottom right):* Stephen Frisch.

Chapter 2

Pages 32–33: Roger Ressmeyer/Corbis Corp. Page 32 inset: Collection CNRI/Phototake. Page 34: Richard Megna/Fundamental Photographs. Page 35: National Renewable Energy Laboratory. Page 36: Patrick Watson. Page 37 *(top right):* Courtesy IBM Almaden Research Center. Page 37 *(bottom left):* "Molecular Dynamics Simulation of Carbon-Nanotube Based Gears." Jie Han, Al Globus, Richard Jaffe and Glen Deardorff, Numerical Aerospace Simulation (NAS) Systems Division at NASA Ames Research Center. Nanotechnology, vol. 8, #3, Sept. 3, 1997, pages 95–102. Page 39: Patrick Watson. Page 40: Jump Run Productions/The Image Bank/Getty Images. Page 41: Patrick Watson. Page 46 *(inset):* Courtesy of Sachtleben Chemie GmbH. Page 46: M. Freeman/ PhotoLink/Photo Disc. Page 50 *(left):* Photo Researchers. Page 50 *(center):* Charing Cross Hospital/Photo Researchers. Page 50 *(bottom):* ©ISM/Phototake. Page 52: Andy Washnik . Page 55: Courtesy Michael P. Doukas, USGS. Page 57 *(left):* Ken Karp. Page 57 *(center):* A. Fenn/ Time Life Books. Page 57 *(right):* Charles Falco/Photo Researchers. Page 60 *(top):* Peter Lerman. Page 60 *(bottom):* John Olmsted. Page 61: John Olmsted. Page 62 *(top):* Michael Scott/Stone/Getty Images. Page 62 *(bottom):* Richard Megna/Fundamental Photographs.

Chapter 3

Pages 64–65: Dennis Kunkel/Phototake. Page 64 *(inset):* Andy Washnik. Page 72 *(top):* Courtesy of Carbon Nanotechnologies, Inc. Page 72 *(bottom):* Lee Snyder/The Image Works. Page 81: Courtesy John Olmsted. Page 82: L.S. Stepanowicz/Visuals Unlimited. Page 91: Richard

Megna/Fundamental Photographs. Page 97: Stephen Frisch. Page 99: Ken Karp. Page 101 *(top):* John Olmsted. Page 101 *(bottom)* & 103: Tony Freeman/PhotoEdit.

Chapter 4

Pages 112–113: Corbis Images. Page 112 *(inset):* Santokh Kochar/PhotoDisc, Inc. Page 119: John Olmsted. Page 121: Richard Megna/Fundamental Photographs. Page 122: John Olmsted. Page 124: Courtesy Cargill, Inc. Page 132: Stephen Frisch. Page 134: Michael Dalton/Fundamental Photographs. Page 135: Patrick Watson. Page 138 *(top):* CNRI/ Science Photo Library/Photo Researchers. Page 138 *(bottom):* Courtesy Newport Corporation. Page 145: Stephen Frisch. Page 148: Richard Megna/Fundamental Photographs. Page 149 *(top):* Royal Ontario Museum/Corbis Images. Page 149 *(bottom):* Ken Whitmore/Stone/ Getty Images. Page 150: Patrick Watson. Page 153 *(top):* Stephen Frisch. Page 153 *(bottom):* Kevin R. Morris/Corbis. Page 157: John Olmsted. Page 158 *(top):* Richard Megna/Fundamental Photographs. Page 158 *(bottom):* Patrick Watson.

Chapter 5

Pages 164–165: Daniel Biggs/SUPERSTOCK. Page 164 *(inset):* Richard Dobson/Getty Images. Page 181: Joseph Nettis/Photo Researchers. Page 182: Courtesy John Olmsted. Page 184 *(top):* Courtesy Thermionics Laboratory, Inc., San Leandro, California. Page 184 *(bottom):* Courtesy Stanford Linear Accelerator Center, US Department of Energy. Page 185: Patrick Watson. Page 188 *(top):* Jonathan Blair/Corbis Images. Page 188 *(bottom):* UlfE. Wallin/The Image Bank/Getty Images. Page 192: Courtesy Armfield Limited. Page 193: Richard Megna/Fundamental Photographs. Page 195: NASA. Page 197 *(top):* Hans Pfletschinger/Peter Arnold, Inc. Page 197 *(bottom):* Corbis Images. Page 199 *(left):* Jim Zuckerman/Corbis Images. Page 199 *(right):* Bernard Edmaier/Photo Researchers. Page 200: Vanessa Miles Photography.

Chapter 6

Pages 208–209: European Space Agency/Science Photo Library/Photo Researchers. Page 208 *(inset):* Simon Fraser/Photo Researchers. Page 210 *(top):* Tony Demin/Image State. Page 210 *(bottom):* Jim Cummins/Corbis Corp. Page 212: Graham French/Masterfile. Page 214: Chris Sorenson/Corbis Stock Market. Page 218 *(top):* Danny Lehman/Corbis Images. Page 218 *(bottom):* Monika Graff/The Image Works. Page 219 *(top):* Jeff Greenberg/Photo Researchers. Page 219 *(bottom):* Picture Press/Corbis Images. Page 227: Courtesy NASA. Page 232 *(top):* Patrick Watson. Page 232 *(bottom):* OPC, Inc. Page 233: Courtesy Parr Instrument Company. Page 234: Culver Pictures, Inc. Page 242: Stephen Frisch. Page 248: PixFolio/Alamy Images. Page 249 *(top):* Jim Wark/Index Stock. Page 249 *(bottom):* Daryl Benson/Masterfile. Page 250: Peter Bowater/Photo Researchers. Page 251: Photo by Dan Aiello/Courtesy of Arizona Solar Center.

Chapter 7

Pages 258–259: Roger Ressmeyer/Corbis Images. Page 258 *(inset):* Yoav Levy/Phototake. Page 259 *(inset):* Courtesy Electron Physics

Group, National Institute of Standards and Technology. Page 260: John Olmsted. Page 262: Pal Hermansen/Stone/Getty Images. Page 264: Courtesy Bausch & Lomb. Page 266: Time Life Pictures/Getty Images. Page 268: Michael Dalton/Fundamental Photographs. Page 270: Courtesy Bausch & Lomb. Page 274: Tom Tracy/Corbis Stock Market. Page 275 (top left): William Sterne, Jr., Sterne Photography. Page 275 (top right): Courtesy David Malin, Anglo-Australian Telescope Board. Page 275 (bottom left): Andy Washnik. Page 275 (bottom center): Andy Washnik. Page 275 (bottom right): Andy Washnik. Page 277 (top left): Richard Megna/Fundamental Photographs. Page 277 (top center): VU-NIH/Visuals Unlimited. Page 277 (top right): Courtesy IBM Research Division. Page 281: Andy Washnik. Page 287: Sanford/Agliolo/Corbis Stock Market. Page 290: Paul Silverman/Fundamental Photographs.

Chapter 8

Pages 298–299: Masterfile. Page 305: Science Photo Library/Photo Researchers. Page 310 (left): Courtesy Bente Lebech, Materials Research Department, Risoe National Laboratory. Page 310 (right): ©Index Stock. Page 317: Paul Silverman/Fundamental Photographs. Page 324: Michael Watson. Page 329: C. Van Der Lende/The Image Bank. Page 330: Walt Anderson/Visuals Unlimited.

Chapter 9

Pages 338–339 & inset: Courtesy of James Gimzewski, IBM Research, Zurich Research Laboratory. Page 344: Roger Ressmeyer/Corbis Images. Page 351: Courtesy SGL Carbon. Page 360: Rich LaSalle/Stone/Getty Images. Page 363: Geoff Topkinson/Science Photo Library/Photo Researchers. Page 365 (left): John Veevaert/Trinity Mineral Company. 365 (center left): Jeffrey Scovil. Page 365 (center right): Paul Silverman/Fundamental Photographs. Page 365 (right): Charles D. Winters/Photo Researchers. Page 370: Photo by Richard L. Battaglia, Department of Microelectronic Engineering, Rochester Institute of Technology, Rochester, NY.

Chapter 10

Pages 386–387: Mark Tomalty/Masterfile. Page 386 (inset): Newcomb & Wergin/Stone/Getty Images. Page 404 (left): Joseph Van Os/The Image Bank/Getty Images. Page 404 (right): Tom Mareschal/The Image Bank/Getty Images. Page 412: Richard Megna/Fundamental Photographs. Page 420: Courtesy the Lucite Store and Home Gallery, Naples, Fl. (www.TheLuciteStore.com). Page 423: John Olmsted. Page 425 (top): John Olmsted. Page 425 (bottom): Andy Washnik. Page 426: John Olmsted.

Chapter 11

Pages 438–439: Alfred Pasieka/Photo Researchers. Page 438 (inset): Gail Shumway/Getty Images. Page 439 (inset): S. John, G.A. Ozin from Nature 405, 437, 2000. Photo courtesy Sajeev John. Page 440 (top): Yoav Levy/Phototake. Page 440 (center & bottom): Courtesy John Olmsted. Page 441: Lester V. Bergman/Corbis. Page 443: ©SUPERSTOCK. Page 451: John Olmsted. Page 452 (top): Richard Megna/Fundamental Photographs. Page 452 (bottom): John Olmsted. Page 455: Stephen Frisch. Page 457 (top): Richard Megna/Fundamental Photographs. Page 457 (center): John Olmsted. Page 457 (bottom): Adam Woolfitt/Corbis Images. Page 459: Courtesy American Superconductor. Reproduced with permission. Page 460 (top): Courtesy John Olmsted. Page 460 (bottom left): Sinclair Stammers/Science Photo Library/ Photo Researchers. Page 460 (bottom center): Geoff Tompkinson/Science Photo Library/Photo Researchers. Page 460 (bottom right): National Museum of Natural History ©2004 Smithsonian Institution. Photo by Dane Penland. Page 461: SYMMETRY DRAWING E18 by M.C. Escher. © 2001 Cordon Art-Baarn-Holland. All Rights Reserved. Page 464: Andy Washnik. Page 466 (top): Richard Hutchings/PhotoEdit. Page 466 (bot-

tom): Richard Megna/Fundamental Photographs. Page 467 (left): Dr. Gopal Murti/Visuals Unlimited. Page 467 (right): Reichelt/Photo Researchers. Page 468: Vaughan Fleming/Science Photo Library/Photo Researchers. Page 469: Patrick Watson. Page 470: Photex/Corbis Images.

Chapter 12

Pages 482–483: Bob Daemmrich/The Image Works. Page 482 (inset): Andy Washnik. Page 484: The Image Bank/Getty Images. Pages 486–488: John Olmsted. Page 491: L.S. Stepanowicz/Visuals Unlimited. Page 492: John Olmsted. Page 494: Ed Reschke/Peter Arnold, Inc. Page 496 (top): Richard Megna/Fundamental Photographs. Page 496 (bottom): Stephen Frisch. Page 497: Chris Sorenson/Corbis Stock Market. Page 498 (left): Jon Feingersh/Tom Stack & Associates. Page 498 (right): David Hall/The Image Works. Page 501: Amy Etra/PhotoEdit. Page 503: Diane Hirsch/Fundamental Photographs. Page 505: David M. Phillips/Visuals Unlimited. Page 506 (top): Courtesy City of Melbourne, Florida. Page 507: John Olmsted. Page 508 (top): Michael Rutherford/SUPERSTOCK. Page 508 (center): Seawif Image/ORBIMAGES. Page 508 (bottom): Adam Jones/Photo Researchers. Page 511: Dr. Patricia J. Shulz/Peter Arnold, Inc.

Chapter 13

Pages 518–519: Michael Orton/Image State. Page 518 (left): ©Nik Wheeler. Page 518 (right): Hans Pfletschinger/Peter Arnold, Inc. Page 521: Tom J. Ulrich/Visuals Unlimited. Page 526: Felicia Martinez/PhotoEdit. Page 528 (left): Owen Franken/Corbis. Page 528 (right): Laurence Fordyce; Eye Ubiquitous/Corbis Images. Page 530: Andy Washnik. Page 531 (left): Bernard Annebicque/Corbis Images. Page 531 (right): ©Telegraph Colour Library/Taxi/Getty Images. Page 532: Tom Pantages. Page 533: ©VU/SIU/Visuals Unlimited. Page 534 (left): SUPERSTOCK. Page 534 (right): ©Science Vu/EP-AS/Visuals Unlimited. Page 535 (top): SUPERSTOCK. Page 535 (bottom): M. Greenlar/The Image Works. Page 536: Arthur Tilley/Stone/Getty Images. Page 537 (top): Richard Megna/Fundamental Photographs. Page 537 (bottom): Diane Hirsch/Fundamental Photographs. Page 540: Digital Vision/Getty Images. Page 541: Alvin E. Staffan/Photo Researchers. Page 542: Andrew Syred/Photo Researchers. Page 543 (top): James Bell/Photo Researchers. Page 543 (bottom): Michael J. Doolittle/The Image Works. Page 547 (top): Courtesy Oesper Collection in the History of Chemistry, University of Cincinnati. Page 547 (bottom): Ken Eward/Biografix/Photo Researchers. Page 554: Patrick Watson.

Chapter 14

Pages 566–567: Martin Jones/Corbis Images. Page 566 (inset): M. Antman/The Image Works. Page 569 (top): Patrick Watson. Page 569 (bottom): Dennis M. Gottlieb/Corbis Stock Market. Page 570: Andy Washnik. Page 580: Courtesy J.F. Allen, St. Andrews University. Page 584: Sonda Dawes/The Image Works. Page 588 (top left): Stephen Frisch. Page 588 (top center): Courtesy John Olmsted. Page 588 (top right): Martin Bond/Science Photo Library/Photo Researchers. Page 595: Courtesy John Olmsted. Page 598 (top): Rick Raymond/Index Stock. Page 599 (bottom): Pat O'Hara/Corbis Images. Page 600: Lester V. Bergman/Corbis. Page 601: W. Cody/Corbis Images. Page 602: Dr. R. Clark & M. Goff/Science Photo Library/Photo Researchers.

Chapter 15

Pages 613–614: Courtesy GSFC/NASA. Page 615: Richard Megna/Fundamental Photographs. Page 617 (top): Kevin Fleming/Corbis Images. Page 617 (bottom): The Image Bank/Getty Images. Page 630: Stuart Cohen/The Image Works. Page 633: Layne Kennedy/Corbis Images. Page 653: Stephen Frisch. Page 657 (top): Courtesy Walker Manufacturing. Page 657 (center): Telegraph Colour Library/Taxi/Getty Images.

Page 657 *(bottom):* Peticolas/Megna/Fundamental Photographs. Page 658: Courtesy Englehard Corporation. Page 659: Timothy I. Morrow and Edward J. Maginn, Dept. of Chemical and Biomolecular Engineering, University of Notre Dame, Notre Dame, In. Page 660: Courtesy Thomas Steitz, Yale University.

Chapter 16

Pages 672–673: Nigel Cattlin/Holt Studios International/Photo Researchers. Page 672 *(inset):* Craig Aurness/Corbis Images. Page 701: Courtesy John Olmsted. Page 707: Courtesy Cargill, Inc. Page 708: Courtesy John Olmsted. Page 711: Lawrence Migdale/Science Source/Photo Researchers. Page 712 *(top left):* David Muench/Corbis Images. Page 712 *(center left):* Luis Veiga/The Image Bank/Getty Images. Page 712 *(center right):* Dave Bunnell. Page 712 *(bottom):* Almay Images.

Chapter 17

Pages 720–721: Foodpix/PictureArts Corp. Page 720 *(inset):* Burke/Triolo Productions/Foodpix/PictureArts Corp. Page 727 *(left & right):* John Olmsted. Page 727 *(center):* Yoav Levy/Phototake. Page 730: Richard Megna/Fundamental Photographs. Page 733 *(top left):* ©1970 FP/Fundamental Photographs. Page 733 *(bottom center):* ©1970 FP/Fundamental Photographs. Page 733 *(bottom right):* ©1970 FP/Fundamental Photographs. Page 743: OPC, Inc. Page 750: Andy Washnik. Page 754: Courtesy John Olmsted.

Chapter 18

Pages 762–763: Robert Holmes/Corbis. Page 762 *(inset):* SPL/Photo Researchers. Page 771: Courtesy John Olmsted. Page 774: Stephen Frisch. Pages 781 & 788: Patrick Watson. Page 790 *(top):* Paul Chesley/Stone/Getty Images. Page 790 *(top center):* Ralph White/Corbis Images. Page 790 *(bottom left):* Corbis Images. Page 790 *(bottom):* Courtesy Betz Company. Page 803 *(top):* Courtesy John Olmsted. Page 803 *(bottom):* Charles D. Winters/Photo Researchers. Page 804: Creatas/Fotosearch, LLC. Page 806 *(top left):* Inga Spence/Visuals Unlimited. Page 806 *(top right):* Henry Beeker/Age Fotostock America, Inc. Page 806 *(bottom right):* Ron Sanford/Corbis Images.

Chapter 19

Pages 814–815 : Todd Gipstein/Corbis Images. Page 814 *(inset):* Roger Ressmeyer/Corbis Images. Page 816: Richard Megna/Fundamental Photographs. Page 821: Richard Megna/Fundamental Photographs. Page 826: Courtesy John Olmsted. Page 827 *(left):* Stephen Frisch. Page 827 *(right):* Tom Pantages. Page: Alamy Images. Pages 828, 830, 833: Richard Megna/Fundamental Photographs. Page 834: Thomas Del Brase. Page 836: Andy Washnik. Page 839: Tom Pantages. Page 841: Alexander Blaikley, Michael Faraday Lecturing in the Theatre of the Royal Institution, c. 1856. Colored lithograph. The Royal Insitution, UK/Bridgeman Art Library/NY. Page: Courtesy Nissan North America.

Page 845: Charles D. Winters/Photo Researchers. Page 846: Tony Freeman/PhotoEdit. Page 848 *(top):* Patrick Watson. Page 848 *(bottom):* Tony Freeman/PhotoEdit. Page 851: Courtesy Nissan North America. Page 852: Alamy Images. Page 854: Stephen Frisch. Page 857: Patrick Ward/Corbis Images.

Chapter 20

Pages 865–866: Courtesy John Olmsted. Page 870: Stephen Frisch. Page 875: Richard Megna/Fundamental Photographs. Page 877: Courtesy John Olmsted. Page 883: Stephen Frisch. Page 885: Patrick Watson. Page 890: Phil Degginger/Stone/Getty Images. Page 891: James L. Amos/Corbis. Page 892 *(top):* Charles D. Winters/Photo Researchers. Page 892 *(bottom left):* Will & Deni McIntyre/Corbis Images. Page 892 *(bottom center):* Colorstock/Taxi/Getty Images. Page 892 *(bottom right):* Erich Schrempp Photography. Page 893 *(top):* Bill Bachmann/PhotoEdit. Page 893 *(bottom):* Stephen Frisch. Page 894 *(top):* George Holton/Photo Researchers. Page 894 *(bottom):* Andy Levin/Photo Researchers. Page 896 *(top):* Peter Cade/Stone/Getty Images. Page 896 *(bottom):* James L. Amos/Corbis.

Chapter 21

Pages 908–909: Alfred Pasieka/Photo Researchers. Page 919 *(top):* Courtesy Alcoa. Page 919 *(center):* James L. Amos/Corbis. Page 919 *(bottom):* Tony Freeman/PhotoEdit. Page 920: Charles D. Winters/Photo Researchers. Page 922 *(top):* Gianni Dagli Orti/Corbis Images. Page 922 *(bottom):* David J. & Janice L. Frent Collection/Corbis Images. Page 926 *(left):* Richard Megna/Fundamental Photographs. Page 926 *(right):* Andy Washnik. Page 927: Stephen Frisch. Page 930: Gary Retherford/Photo Researchers. Page 934 *(top):* Charles D. Winters/Photo Researchers. Page 934 *(bottom):* Courtesy Regal Ware, Inc. Page 935: Thomas D. Mangelsen/Peter Arnold, Inc.

Chapter 22

Pages 942–943: NASA/GSFC. Page 942 *(inset):* Science Photo Library/Photo Researchers. Page 948: Courtesy NASA. Page 960: CourtesyTechnical University of Eindhoven, Accelerator Lab. Page 961: Courtesy Fermilab. Page 964: Courtesy U.S. Department of Defense. Page 966 *(left):* Agence France Presse/Corbis Images. Page 966 *(center):* George Lepp/Corbis Images. Page 966 *(right):* AP/Wide World Photos. Page 967: Courtesy Hirsohi Hidaka, Hiroshima University. Page 969 *(top):* Courtesy U. S. Dept. of Energy. Page 969 *(bottom):* Courtesy Princeton University Plasma Physics Lab. Page 970 *(top):* Courtesy NASA. Page 970 *(center):* Courtesy David Malin, Anglo Australian Telescope Board. Page 970 *(bottom):* Courtesy NASA and Space Telescope Science Institute. Page 976: Courtesy NASA. Page 977: Agence France Presse/Corbis Images. Page 979 *(top):* Courtesy of Texas Engineering Experiment Station, Texas A & M University. Page 979 *(center):* Courtesy NASA. Page 979 *(bottom):* Courtesy MDS Nordion.

Glossary

Absolute entropy. (14.3) The amount of disorder contained in a chemical substance.

Absorption spectrum. (7.3) The distribution of wavelengths of light absorbed by a species.

Accuracy. (1.5) How close a measurement is to the true value.

Acid. (Brønsted definition) (4.6) A substance that acts as a proton donor.

Acid ionization constant (K_a). (16.6) The equilibrium constant for proton transfer between an acid and water

$$K_a = \frac{[H_3O^+]_{eq}\,[A^-]_{eq}}{[HA]_{eq}}.$$

Acid rain. (5.7) Rain whose pH is lower than 5.6, which is the pH of water saturated with carbon dioxide.

Actinide element. (1.3) Any of the elements in the $5f$ block of the periodic table, between $Z = 89$ and $Z = 102$.

Activated complex. (15.6) The unstable molecular species through which an elementary reaction proceeds as it evolves from reactant(s) to product(s).

Activation energy (E_a). (15.6) The minimum amount of energy reactants must possess to react to form products.

Activation energy diagram. (15.6) A graph showing how the energy of a reacting set of molecules varies with the course of the reaction.

Active electrode. (19.3) An electrode whose chemical constituents take part in the redox reaction that occurs at its surface.

Activity series. (4.7) A list of elements in order of how easily they are oxidized in aqueous solution.

Actual yield. (4.3) The amount of chemical product formed in a chemical reaction.

Adduct. (21.1) A compound formed by bond formation between a Lewis base and a Lewis acid.

Adenine. (13.6) A two-ring organic base found in DNA and RNA.

Adenosine triphosphate (ATP). (14.6) A biochemical molecule containing adenine and three phosphate groups instrumental in biochemical energetics.

Adhesive force. (11.3) A force of attraction between molecules in one phase and different molecules in another phase.

Adsorption. (15.7) The physical attachment of molecules to a surface.

Alcohol. (13.1) An organic compound that contains the hydroxyl group, —OH.

Aldehyde. (13.1) An organic compound that contains the carbonyl group (C=O) attached to a hydrogen atom, —CHO.

Alkali metal. (1.3) Any of the Group 1 elements, all of which are reactive metals with s^1 valence configurations.

Alkaline dry cell. (19.6) A galvanic cell whose working components are Zn and MnO_2.

Alkaline earth metal. (1.3) Any of the Group 2 elements, all of which are reactive metals with s^2 valence configurations.

Alkane. (3.2) A compound containing only carbon and hydrogen, with general formula $C_nH_{(2n+2)}$, in which all carbon atoms possess four σ bonds.

Alkene. (3.2) A compound containing only carbon and hydrogen in which there is at least one C=C bond.

Alkyne. (3.2) A compound containing only carbon and hydrogen in which there is at least one C≡C bond.

Alloy. (12.2) A metallic solution or mixture of two or more metals.

Alpha emission. (22.2) Nuclear decay in which a helium nucleus is emitted by the decaying nucleus.

Alpha particle (α). (2.2) An energetic helium nucleus emitted by a radioactive nuclide.

Amalgam. (12.3) A solution of a metal in mercury.

Amide. (13.1) An organic compound that contains the —C(=O)—N(H)— linkage.

Amine. (13.1) A compound whose molecules can be viewed as ammonia with one or more N—H bonds replaced by N—C bonds.

Amino acid. (13.7) A compound that contains both an amine (—NH_2) and a carboxylic acid (—CO_2H) group.

Amorphous. (11.5) Without any organized regular repeating pattern.

Ampere (A). (1.5) The base unit of electric current in the SI system.

Amphiprotic. (17.1) Able to act as both an acid and a base.

Amplitude. (7.2) The height of a wave.

Analytical balance. (1.5) An accurate mass-measuring instrument.

Anion. (2.5) An ion that possesses negative charge.

Anode. (19.3) The electrode at which oxidation occurs.

Antibonding molecular orbital. (10.4) A molecular orbital that has electron density concentrated outside the bonding region, making it less stable than the atomic orbitals from which it forms.

Aqueous. (3.6) A solution with water as the solvent.

Area. (1.5) The total surface of something.

Arrhenius equation. (15.6) The equation describing how rate constants depend on temperature and activation energy, $k = Ae^{-E_a/RT}$.

Atmosphere (atm). (5.1) A unit of pressure based on the normal pressure exerted by the Earth's atmosphere at sea level. 1 atm = 101.325 kPa.

Atom. (1.2) The basic unit of chemical matter, consisting of a nucleus and enough electrons to convey electrical neutrality.

Atomic number (Z). (2.3) The number of protons in an atomic nucleus.

Atomic orbital. (7.5) A description of an atomic electron that provides the distribution of electron density about the nucleus.

Atomic radius. (8.4) The distance from the nucleus of an atom at which electron-electron repulsion prevents closer approach of another atom.

Atomic symbol. (2.3) The letter designation for an element.

Atomic theory. (2.1) The description of matter as composed of atoms that retain their identities during all physical and chemical processes.

Atomic view. (1.2) How matter looks when viewed at the level of atoms.

Aufbau principle. (8.2) The statement that the most stable arrangement of electrons in an atom results from placing each successive electron in the most stable available atomic orbital.

Avogadro's number (N_A). (2.4) The number of particles contained in 1 mol, 6.022142×10^{23} particles/mol.

Axial position. (9.4) A position along the z axis in a trigonal bipyramid.

Azimuthal quantum number (l). (7.5) The quantum number, restricted to integers from 0 to $n - 1$, that indexes the shape of an atomic orbital.

Balanced chemical equation. (4.1) A description of a chemical reaction using chemical formulas in which the coefficients describe the ratios of molecules of each species that react.

Ball-and-stick model. (3.1) A representation of a molecule that shows the atoms as small balls and the chemical bonds as sticks.

Band gap. (10.7) The difference in energy between the highest filled orbital and the lowest vacant orbital in a solid.

Band theory. (10.7) The description of bonding in solids using delocalized orbitals.

Bar. (5.1) A pressure unit equal to 10^5 pascals.

Barometer. (5.1) An instrument for measuring atmospheric pressure.

Base (Brønsted definition). (4.6) A substance that acts as a proton acceptor.

Base ionization constant (K_b). (16.6) The equilibrium constant for the proton transfer from water to a base,

$$K_b = \frac{[BH^+]_{eq}[OH^-]_{eq}}{[B]_{eq}}.$$

Base pairing. (13.6) The formation of hydrogen bonds between strands of DNA or RNA.

Battery. (19.3) One or more galvanic cells that serve as a source of electric current or voltage.

Belt of stability. (2.3) On a graph of N vs. Z, the region that contains the stable nuclides.

Bent shape. (9.3) The arrangement of a triatomic molecule having a bond angle less than 180°.

Beta emission. (22.2) Nuclear decay in which an energetic electron is emitted by the decaying nucleus.

Beta particle (β). (22.3) An energetic electron emitted by a radioactive nuclide.

Bidentate. (18.5) A ligand that can bind to a metal through two sites.

Bilayer. (12.5) A double layer of atoms or molecules.

Bimolecular reaction. (15.1) An elementary step in which two molecules collide and undergo a chemical reaction.

Binary compound. (3.1) A chemical substance that contains only two different elements.

Body centered cubic (BCC). (11.5) The crystal structure whose unit cell is a cube with identical atoms at its corners and at its center.

Boiling point. (11.1) The temperature at which the vapor pressure of a liquid matches the external pressure.

Boiling point elevation. (12.4) The increase in boiling point of a liquid caused by the presence of nonvolatile solutes in solution.

Bond. (3.1) A strong attractive force generated by sharing of electrons between atoms or electrical attraction between ions in a substance.

Bond angle. (9.5) The angle between two atoms that are bonded to a third atom.

Bond energy. (6.3) The energy (usually expressed per mole) required to break one particular chemical bond in a gaseous substance.

Bond length. (9.1) The average distance between the nuclei of two bonded atoms in a molecule.

Bond order. (10.4) The net number of pairs of bonding electrons in a chemical bond.

Bonding molecular orbital. (10.4) An orbital formed from atomic orbitals that conveys bonding by placing electron density between the bonded atoms.

Bonding orbital. (10.1) An orbital with high electron density between atoms, formed by constructive interaction between atomic orbitals.

Boundary. (6.2) The imaginary line separating a system from its surroundings.

Brønsted-Lowry. (17.1) The proton-transfer definition of acids and bases.

Buffer capacity. (18.2) The amount of added hydronium or hydroxide ions that can be added to a buffer solution without exceeding a specified pH range.

Buffer equation. (18.1) The equation linking the pH of a buffer solution with the pK_a and concentrations of the acid-base pair.

Buffer solution. (18.1) A solution containing both a weak acid and its conjugate base, whose pH changes slowly on addition of small amounts of an acid or base.

Buret. (4.6) A volume-delivering device that is calibrated to allow measurement of the amount of fluid delivered.

Calorimeter. (6.4) An apparatus for measuring the amount of heat absorbed or emitted during some process.

Capillary action. (11.3) The upward movement of a liquid in a narrow tube against the force of gravity.

Carbohydrate. (13.5) A biochemical substance whose empirical formula is $(CH_2O)_n$.

Carbonyl group. (13.1) The $C{=}O$ functional group.

Carboxylic acid. (4.6) An acid that contains the $-CO_2H$ functional group.

Catalyst. (15.7) A substance that increases the rate of a chemical reaction without being consumed or produced in the reaction.

Cathode. (19.3) The electrode at which reduction occurs.

Cation. (2.5) A positively charged ion.

Cell potential. (19.4) The difference in electrical potential (voltage) between the two electrodes in a galvanic cell.

Cellulose. (13.5) The polymer, composed of sugar monomers, that makes up the woody tissue of plants.

Celsius scale. (1.5) The temperature scale defined by 0 °C as the normal freezing point and 100 °C as the normal boiling point of water.

Change of state. (6.2) Any process in which there is a change in one or more of the variables (P, V, T, n) describing a system.

Chelate effect. (18.5) The stabilization of a metal complex by a multidentate ligand.

Chelating. (18.5) A ligand containing two or more donor atoms.

Chemical energy. (6.1) Energy stored in bonds between atoms or ions.

Chemical equation. (4.1) An equation using chemical formulas that describes the identities and relative amounts of reactants and products in a chemical reaction.

Chemical formula. (1.2) An expression using chemical symbols that describes the identities and relative amounts of elements in a substance.

Chemical transformation. (1.4) A process in which chemical substances react to form other chemical substances.

Chemistry. (1.1) The study of matter and its interactions.

Close packed. (11.5) The most efficient arrangement for packing atoms, molecules, or ions in a regular crystal.

Cohesive force. (11.3) A force of attraction between like molecules in the same phase.

Colligative property. (12.4) A property of a solution that is proportional to the concentration of solute species.

Colloid. (12.5) A particle whose dimensions are between 1 nm and 1 μm.

Colloidal suspension. (12.5) A colloid−liquid mixture that does not readily separate into its components.

Combustion. (3.5) A reaction of a substance with oxygen that releases chemical energy.

Combustion analysis. (3.5) The determination of chemical composition by burning a sample.

Common ion effect. (18.4) Reduction in solubility caused by the presence of one of the ions contained in a salt.

Complementary base pair. (13.6) Two nitrogen-containing organic bases that hydrogen bonds in DNA or RNA.

Complementary colors. (20.3) Colors that are related in that if light of one color is absorbed by a substance, the substance displays the complementary color.

Complex ion. (16.6) A species formed by bonding of species containing lone pairs of electrons to a metal ion.

Compound. (1.2) A substance formed of atoms of two or more elements chemically bonded in fixed proportions.

Compound nucleus. (22.3) The transient nucleus that forms from the collision of a nuclear projectile with another nucleus.

Concentration. (3.6) The amount of a solute contained in a standard amount of a solution.

Concentration table. (16.5) A table of amounts for a chemical reaction.

Condensation. (11.1) Conversion of a vapor into its liquid.

Condensation reaction. (13.1) A reaction that joins two molecules, accompanied by the elimination of a small molecule such as water.

Conduction band. (10.6) The set of delocalized orbitals in a solid that, when partially occupied, contains mobile electrons and conveys high electrical conductivity.

Conductor. (10.6) A substance that is a good conductor of electrical current.

Conjugate acid-base pair. (17.4) A Brønsted acid and the base that results when the acid loses one proton.

Conjugate base. (17.4) The chemical species formed by removal of a proton from an acid.

Conjugated π system. (10.6) A set of π bonds that is spread over more than two atoms.

Conservation law. (2.1) A law stating that a quantity is conserved, in other words has a constant amount.

Conserved. (2.1) Unchanging in amount.

Conversion ratio. (1.5) A ratio equal to 1 that converts a quantity from one unit to another.

Coordination complex. (20.2) A combination of a metal atom or cation with two or more species (ligands) that bind covalently to the metal.

Coordination number. (20.2) The number of atoms bonded to an inner atom in a molecule or metal complex.

Copolymerization. (13.2) The formation of a polymer from two or more different monomers.

Core electrons. (8.2) The inner atomic electrons with principal quantum number less than that of the valence electrons.

Corrosion. (19.6) The oxidation of a metal in contact with its environment.

Coupled reactions. (14.6) A pair of reactions, one of which is driven in what is otherwise its nonspontaneous direction by the influence of the second, highly spontaneous reaction.

Covalent bond. (9.1) A bond resulting from the sharing of electrons between atoms.

Critical mass. (22.4) The minimum mass of fissionable material required to generate a self-sustaining nuclear fission reaction.

Critical point. (11.6) The temperature and pressure above which the distinction between the liquid and vapor phases disappears.

Cross-linking. (13.2) The formation of additional chemical bonds between the chains of a polymer.

Crystal field splitting energy. (20.3) The difference in energy atomic d orbitals created by the effects of neighboring ions in a crystal.

Crystal field theory. (20.3) A model describing the electronic structure of transition metal complexes in terms of the interactions between metal d orbitals and the electric field of the ligands.

Crystalline. (11.5) Containing a regular array of atoms, molecules, or ions.

Crystalline defect. (11.5) An imperfection in a regular solid.

Cubic close packing. (11.5) The close packing arrangement in which hexagonal layers are stacked in an ABCABC . . . pattern.

Cyclotron. (22.3) A particle accelerator used to generate high energy beams of nuclear particles.

Cytosine. (13.6) A one-ring organic base found in DNA and RNA.

Dalton's law of partial pressures. (5.5) The law stating that in a gas mixture, each gas exerts a pressure equal to the pressure that it would exert if present by itself under otherwise identical conditions.

de Broglie equation. (7.4) The equation describing the wave nature of particles, $\lambda = h/mu$.

Delocalized. (7.4) Spread out over space.

Delocalized bond. (10.1) A bond in which electron density is distributed over more than two atoms.

Delocalized π orbital. (10.5) An orbital, generally formed by side-to-side overlap of atomic p orbitals, that spreads over more than two atoms.

Density (ρ). (1.6) The ratio of an object's mass to its volume, $\rho = m/V$.

Deoxyribonucleic acid (DNA). (13.6) A polymer of nucleotide units that stores genetic information in the chromosomes.

Desorption. (15.7) Detachment of an adsorbed molecule from a surface.

Deuterium (D). (2.3) An atom (an isotope of hydrogen) containing one proton and one neutron in its nucleus.

Diamagnetic. (8.3) The quality of being repelled by a magnetic field. Substances with no unpaired electrons are diamagnetic.

Diatomic molecule. (1.2) A molecule that contains exactly two atoms.

Diffusion. (2.1) The gradual mixing of a solute in a solution by molecular motion.

Dilution. (3.6) Production of a less-concentrated solution by addition of solvent.

Dipolar force. (11.2) The attractive force between polar molecules that results from the negative end of one molecule aligning with the positive end of its neighbor.

Dipole moment. (9.5) The net electrical character arising from an asymmetric charge distribution.

Diprotic acid. (17.4) An acid each of whose molecules contains two acidic hydrogen atoms.

Dispersion force. (11.2) The net attractive force between molecules generated by temporary dipole moments arising from polarization of their electron clouds.

Disulfide bridge. (13.7) An S—S bond linking two portions of a protein molecule.

Dominant equilibrium. (16.6) The equilibrium that is most important in determining equilibrium concentrations of major species in a solution.

Donor atom. (20.2) An atom that possesses one or more lone pairs of electrons that it can contribute to form a covalent bond.

Doped semiconductor. (10.6) A metalloid to which a small amount of another element has been added, usually to convey higher conductivity.

Double bond. (9.2) A chemical bond containing two pairs of bonding electrons.

Double helix. (13.6) The pair of intertwined spirals that is the secondary structure of DNA.

Ductile. (1.3) Able to be drawn into tubes or wires.

Dynamic. (2.1) Containing objects that are in continual motion.

Dynamic equilibrium. (2.1) The condition in which a forward and a reverse process occur at equal rates, so the system undergoes no net change.

Effective nuclear charge. (8.1) The net positive charge, equal to the nuclear charge minus the effects of screening, that an electron in an atomic orbital experiences.

Effusion. (5.4) The escape of a gas through a pinhole from a container into a vacuum.

Elastomer. (13.4) A flexible polymer.

Electrical Energy. (6.1) Energy of an object or system created by the action of electrical force.

Electrical Force. (2.2) The attraction or repulsion between objects caused by their electrical charge.

Electrode. (19.3) An electrical conductor that establishes an interface between an external circuit and an electrochemical cell.

Electrolysis. (19.7) The use of electrical energy to drive a nonspontaneous chemical reaction.

Electrolytic cell. (19.7) An electrochemical cell whose redox reaction is driven by an externally applied electrical potential.

Electromagnetic radiation. (7.2) Wave phenomena having both electrical and magnetic components; visible light is one type.

Electron. (2.2) The smallest unit of negative electrical charge, whose mass is 9.11×10^{-31} kg and whose charge is -1.602×10^{-19} C.

Electron affinity. (8.4) The energy change accompanying the attachment of an electron to an atom or anion.

Electron capture. (22.2) The mode of nuclear decay in which an unstable nucleus captures a core electron and converts a proton into a neutron.

Electron configuration. (8.3) The distribution of electrons among the various orbitals of an atom, molecule, or ion.

Electron contour drawing. (7.6) A depiction of an orbital that shows a con-

tour surface within which most of the electron density is located.

Electron density. (7.6) The distribution of electron probability in space around an atom or molecule.

Electron group. (9.3) A set of from one to six valence electrons that occupies a particular region around an atom.

Electron group geometry. (9.3) The shape created by the electron groups around an atom.

Electronegativity. (9.1) A measure of the ability of an atom in a molecule to attract the shared electrons in a chemical bond.

Electroplating. (19.7) The electrolytic deposition of a thin metal film on the surface of an object.

Element. (1.2) A substance composed of atoms all of whose nuclei have the same amount of positive charge.

Elemental analysis. (3.5) Determination of the percent by mass of each element in a compound.

Elemental symbol. (2.3) A letter or letters designating a particular chemical element.

Elementary reaction. (15.1) A chemical reaction that describes a process as it occurs at the molecular level.

Emission spectrum. (7.3) The distribution of wavelengths of light given off by a species in an excited state.

Empirical formula. (3.5) The chemical formula of a substance expressed using the smallest possible integers.

Endothermic. (6.4) Accompanied by the absorption of energy.

Energy. (6.1) The capacity to do work or move matter.

Energy level. (7.3) A specific energy value for a species.

Energy level diagram. (7.3) A diagram of the specific energy levels of a species, showing increasing energy on the y axis.

Enthalpy (H). (6.5) The energy-related thermodynamic state function defined by the equation $H = E + PV$.

Enthalpy of vaporization. (6.5) The enthalpy change accompanying the conversion of a liquid into a gas.

Entropy (S). (14.2) The thermodynamic state function that measures the degree of dispersal of matter and energy.

Enzyme. (13.7) A biological catalyst.

Equatorial position. (8.6) A position within the trigonal plane of a trigonal bipyramid.

Equilibrium. (2.1) The state in which a system shows no change in properties over time.

Equilibrium constant (K). (16.1) The value of the ratio of equilibrium concentrations of products to equililbrium concentrations of reactants, each raised to the power equal to its stoichiometric coefficient.

Ester. (13.1) An organic compound that contains a $-\overset{\overset{\text{O}}{\|}}{\text{C}}-\text{O}-$ linkage.

Ether. (13.1) An organic compound that contains a C—O—C linkage.

Evaporation. (5.7) The escape of molecules from the liquid to the gas phase.

Excited state. (7.3) An orbital description of a species that is not the most stable description because one or more electrons possesses more energy than in the most stable description.

Exothermic. (6.4) Accompanied by the release of energy.

Extensive property. (1.6) A property that increases in proportion to the amount of material in the sample.

Face-centered cubic. (11.5) A crystalline structure whose unit cell has atoms at each corner of a cube plus additional atoms in the center of each face of the cube.

Faraday constant (F). (19.5) The electrical charge of 1 mol of electrons, 96,485 C.

Fiber. (13.4) A long thin strand of polymeric material.

Fibrous protein. (13.7) A protein that forms long linear strands or coils that are water insoluble.

First-generation star. (22.5). A star formed from a collapsing cloud that contains only hydrogen and electrons.

First law of thermodynamics. (6.2) The statement that energy is conserved, often expressed as an equation for change in energy, $\Delta E = q + w$.

First order. (15.3) Dependent on concentration to the first power.

Fission. (22.1) A nuclear reaction in which one nucleus fragments into two smaller ones.

Flotation. (20.4) The metallurgical process in which water is used to separate components of an ore.

Formal charge. (9.3) The charge that an atom in a molecule would have if each of its bonding electrons were equally shared with its bonding partner.

Formation constant. (16.6) The equilibrium constant for the formation of a complex ion.

Formation reaction. (6.5) A reaction in which 1 mol of a compound forms from chemical elements in their standard states.

Free energy (G). (14.4) The energy-related thermodynamic state function defined by the equation, $G = H - TS$.

Free radical. (13.2) A highly reactive chemical species containing an unpaired lone electron.

Free radical polymerization. (13.2) Polymer formation that proceeds by a mechanism that involves free radicals.

Freezing point. (11.1) The temperature at which a liquid crystallizes to form a solid.

Freezing point depression. (1.4) The reduction in freezing point caused by the presence of solutes in a solution.

Frequency (v). (7.2) The number of wave crests of a wave that pass a fixed point in unit time.

Functional group. (13.1) An atom or small group of atoms in a molecule that gives the molecule characteristic chemical properties.

Fusion. (22.1) A nuclear reaction in which two nuclei combine (fuse) into one more massive nucleus.

Galvanic cell. (19.3) An electrochemical cell operating spontaneously to produce a voltage from a chemical reaction.

Gamma ray (γ). (22.2) A high-energy photon emitted in the course of nuclear decay.

Gas. (1.4) A fluid in which intermolecular forces are so small that the substance expands, contracts, and flows freely to take on the size and shape of its container.

Gas constant (R). (5.2) The fundamental constant linking temperature with molar energy units. $R = 8.314$ J/mol K $= 0.08206$ L atm/mol K.

Geometric isomer. (20.2) One of two or more isomers that have identical chemical bonds distributed differently in space.

Glass. (11.5) An amorphous solid that has cooled to a solid state without crystallizing.

Globular protein. (13.7) A protein with a compact, roughly spherical tertiary structure.

Gravitational energy. (6.1) Energy that is created by gravitational force.

Gravitational force. (2.2) The attraction between objects caused by their mass.

Greenhouse effect. (7.7) The heating of the Earth's surface caused by gases in the atmosphere that absorb outgoing infrared radiation.

Greenhouse gases. (7.7) Gases, primarily carbon dioxide and methane, that trap radiation and lead to heating of the Earth's surface.

Ground state. (7.3) An orbital description of a species in which the electrons are in the most stable description possible.

Ground state configuration. (8.2) The most stable orbital arrangement of electrons in an atom or molecule.

Group. (1.3) The elements in a column of the periodic table.

Guanine. (13.6) A two-ring organic base found in DNA and RNA.

Haber process. (4.1) The industrial process for preparing ammonia from molecular nitrogen and hydrogen.

Half-life ($t_{1/2}$). (15.4) The time required to consume half the initial concentration of a reactant, $t_{1/2} = \dfrac{\ln 2}{k}$.

Half-reaction. (4.7) A reaction that isolates a reduction or oxidation and explicitly shows the electrons involved.

Hall-Héroult process. (21.3) The industrial process for producing aluminum electrolytically.

Halogen. (1.3) Any of the elements in Group 17 of the periodic table.

Hard acid. (21.2) A Lewis acid whose acceptor atom has a low polarizability.

Hard base. (21.2) A Lewis base whose donor electron pairs are tightly bound, resulting in low polarizability.

Heat (q). (6.2) A transfer of thermal energy between a system and its surroundings.

Heat capacity. (6.2) The amount of energy required to raise the temperature of a given quantity of a substance by 1 K (or 1 °C).

Heat of dilution. (12.3) The enthalpy change accompanying a dilution process.

Heat of solution. (12.3) The enthalpy change accompanying the dissolving of a solute.

Helix. (13.7) A regular, spirally twisted shape taken on by nucleic acids and proteins.

Henry's law. (12.3) The statement that the concentration of a gas in solution is directly proportional to the partial pressure of that gas above the solution.

Henry's law constant. (12.3) The proportionality constant relating the solubility of a gas to its partial pressure.

Hess' law. (6.5) The statement that the enthalpy change for any process is equal to the sum of the enthalpy changes for any set of steps that leads from the initial to final conditions.

Heterogeneous. (1.4) Containing more than one physically distinct component and therefore not uniform in composition.

Heterogeneous catalyst. (15.7) A catalyst that exists in a different phase than the one where the chemical reaction of interest is taking place.

Hexagonal close packing. (11.5) The close packing arrangement in which hexagonal layers are stacked in an ABAB . . . pattern.

High-spin complex. (20.3) A complex in which the maximum number of electrons is unpaired.

Homogeneous. (1.4) Uniform in composition.

Homogeneous catalyst. (15.7) A catalyst that is contained in the phase where the chemical reaction of interest is taking place.

Hund's rule. (8.3) The observation that the most stable arrangement of electrons among orbitals of equal energy is the one that maximizes the number of unpaired electron spins.

Hybrid orbital. (10.2) An atomic orbital obtained by combining two or more valence orbitals on the same atom.

Hybridization. (10.2) The formation of a set of hybrid orbitals with favorable directional characteristics by mixing together two or more valence orbitals of the same atom.

Hydrate. (3.3) A solid compound or complex ion that has water molecules incorporated into its structure.

Hydrocarbon. (9.3) A compound that contains only hydrogen and carbon.

Hydrogen bond. (11.2) A moderately strong intermolecular attraction caused by the partial sharing of electrons between a highly electronegative atom of F, O, or N and the polar hydrogen atom in a F—H, O—H, or N—H bond.

Hydrophilic. (12.5) Compatible with water.

Hydrophobic. (12.5) Incompatible with water.

Hydroxyl group. (13.1) The —OH group.

Hypothesis. (1.1) An explanatory proposition that is not yet established.

Ideal gas. (5.2) A gas in which molecular volumes and intermolecular forces both are negligible.

Ideal gas equation. (5.2) The equation describing the behavior of an ideal gas, $PV = nRT$.

Indicator. (4.6) A substance that can be used to identify the stoichiometric point of an acid–base titration because its color is sensitive to pH changes.

Induced nuclear reaction. (22.3) A reaction that occurs as a result of a nuclear projectile colliding with an atomic nucleus.

Initial concentration. (16.5) The concentration of a reagent immediately after a system is prepared but before any chemical reaction takes place.

Initiation. (13.2) A chemical reaction that begins the process of polymerization.

Initiator. (13.2) A substance that starts a polymerization chain reaction.

Inner atom. (9.3) Any atom in a molecule that is bonded to more than one other atom.

Inner transition metal. (20.1) The lanthanide and actinide elements.

Insoluble. (4.5) Unable to dissolve in a liquid. Salts with $K_{sp} \ll 1$ are classified as insoluble in water.

Intensity. (7.2) The brightness (number of photons) of light.

Intensive property. (1.6) A property whose magnitude is independent of the amount of the substance present.

Intermediate. (15.1) A species that is not part of the overall stoichiometry but is produced in an early step of a reaction mechanism and consumed in a later step.

Intermolecular forces. (11.1) Forces that exist between molecules.

International System of Units (SI). (1.5) The set of metric units that have been chosen by international agreement for expressing scientific quantities.

Ion. (2.5) A charged species resulting from the gain or loss of electrons from a neutral atom or molecule.

Ionic compound. (2.5) A neutral compound composed of cations and anions.

Ionic solid. (11.4) A solid containing cations and anions that are attracted to each other by electrical interactions rather than covalent bonds.

Ionization energy. (8.1) The energy required to remove an electron from an isolated species.

Isoelectronic. (8.3) Having the same number of electrons.

Isolation method. (15.4) An experimental technique in chemical kinetics that sets the initial concentration of one reactant substantially lower than those of all others.

Isomers. (20.2) Molecules that have the same chemical formulas but different molecular structures.

Isotonic. (12.4) Having identical total molarity of solutes.

Isotopes. (2.3) Atoms of the same element whose nuclei contain different numbers of neutrons.

Joule (J). (6.1) The SI energy unit, defined to be $1 \ J = 1 \ kg \ m^2 \ s^{-2}$.

Kelvin (K). (1.5) The SI temperature unit, based on absolute zero, with a unit size equal to $1/100$ the difference between the normal freezing point and normal boiling point of water.

Ketone. (13.1) An organic compound that contains the carbonyl group (C=O) linked to two carbon atoms.

Kilogram (kg). (1.5) The SI base unit for mass.

Kinetic energy. (6.1) The energy of motion of an object, $E_{kinetic} = \frac{1}{2}mu^2$.

Kinetics. (15.2) The study of chemical reaction rates.

Lanthanide element. (1.3) Any of the elements in the $4f$ block of the periodic table, between $Z = 57$ and $Z = 70$.

Lattice. (8.5) An organization of ions in a regular, alternating array.

Lattice energy. (8.5) The energy released when individual gaseous cations and anions condense to form a regular three-dimensional array.

Law of conservation of energy. (6.2) The statement that energy cannot be created or destroyed.

Law of conservation of mass. (2.1) The statement that mass cannot be created or destroyed.

Leaching. (20.4) The metallurgical process in which one component of an ore mixture is separated by dissolving.

Le Châtelier's principle. (16.4) The observation that when a change of conditions is applied to a system at equilibrium, the system responds in the manner that reduces the amount of change.

Lead battery. (19.6) A voltaic cell, widely used for automobile storage batteries, in which both half-reactions involve reactions of lead.

Length. (1.5) The distance that something extends.

Lewis acid. (21.1) An electron-pair acceptor.

Lewis base. (21.1) An electron-pair donor.

Lewis structure. (9.3) A representation of covalent bonding that uses symbols for the elements, dots for nonbonding valence electrons, and lines for pairs of bonding valence electrons.

Ligand. (9.3) An atom or group of atoms that is bonded to an inner atom.

Light. (7.2) Electromagnetic radiation in the visible portion of the spectrum, between $\lambda = 400$ nm and $\lambda = 700$ nm.

Limiting reactant. (4.4) The reactant whose amount falls to zero in a chemical reaction.

Linear. (9.4) Lying along a straight line.

Linear accelerator. (22.3) An apparatus that accelerates atomic nuclei to high energy along a linear trajectory.

Line structure. (3.1) A compact representation of a carbon-containing molecule that shows the molecular structure in a simplified fashion.

Linkage group. (13.1) A chemical group that joins two monomers in a polymer molecule.

Linkage isomers. (20.2) Metal complexes with identical chemical formulas that differ in the way in which one or more ligands binds to the metal.

Liquid. (1.4) The phase of matter in which intermolecular forces are large enough that molecules cannot escape into space but small enough that molecules move freely past one another. A liquid has a distinct volume but no distinct shape.

Liter (L). (1.5) A unit of volume equal to $1000 \ cm^3 \ (10^{-3} \ m^3)$.

Localized bond. (10.1) A chemical bond that involves only two atoms.

Lone pair (nonbonding pair). (9.3) A pair of valence electrons that is localized on an atom rather than involved in bonding.

Low-spin complex. (20.3) A complex in which the maximum number of electrons is paired.

Macromolecule. (13.1) A molecule that contains a large number of atoms (typically, more than several hundred).

Macroscopic. (1.2) Visible to the naked eye.

Magnet. (2.2) An object that attracts or repels a moving electrical charge but carries no net electric charge.

Magnetic force. (2.2) The attraction or repulsion between objects caused by their magnetism.

Magnetic quantum number (m_l). (7.5) The quantum number, restricted to integers between $+l$ and $-l$, that indexes the orientation in space of an atomic orbital.

Magnitude. (1.5) The size of a number.

Main group element. (1.3) Any of the elements in the *p* block of the periodic table.

Main group metal. (21.3) Any element displaying metallic properties and located in the *p* block of the periodic table.

Major species. (16.6) The species (cations, anions, molecules) present in relatively high concentration in aqueous solution.

Malleable. (1.3) Able to be formed into various shapes, including thin sheets.

Manometer. (5.1) A bent tube, containing a liquid and open at both ends, used to measure differences in pressure.

Mass. (1.5) The quantity of matter in a substance.

Mass number (*A*). (2.3) The total number of protons and neutrons in a nucleus.

Mass percent composition. (3.5) The makeup of a chemical substance expressed in parts per hundred by mass of each component.

Mass spectrometer. (2.3) An instrument that separates the atomic/molecular components of a sample according to their masses.

Matter. (1.1) Anything that occupies space and possesses mass.

Mechanism. (15.1) The detailed molecular processes by which a chemical reaction proceeds.

Meniscus. (11.3) The curved surface of a liquid contained in a narrow tube.

Mesosphere. (7.7) The region of the Earth's atmosphere between 50 km and 85 km above its surface.

Metabolism. (14.6) The process of breaking down organic molecules in biochemical cells.

Metal. (1.3) An element that is lustrous, conducts heat and electricity well, and tends to lose electrons to form cations.

Metal displacement reaction. (4.7) An oxidation−reduction process in which one metal dissolves and another precipitates.

Metallic solid. (11.4) An elemental solid whose atoms are held together by valence electrons occupying delocalized orbitals spread across the entire solid.

Metalloid. (1.3) An element with properties intermediate between those of metals and nonmetals.

Metalloprotein. (20.6) A biological macromolecule made of long chains of amino acids and containing at least one metal atom.

Metallurgy. (20.4) The science of extracting, purifying, and forming useful objects from metals.

Metathesis reaction. (21.2) A chemical reaction in which there is an exchange of bonding partners between two atoms or ions.

Meter (m). (1.5) The SI unit of length.

Micelle. (12.5) A cluster of molecules in aqueous solution organized with their hydrophobic ends pointing inward and their hydrophilic ends pointing outward.

Microscopic. (1.2) Requiring magnification to be seen.

Midpoint. (18.3) The point in a titration where the amount of added titrant is exactly half the amount required for complete reaction.

Minor species. (16.6) The species (ions, molecules) present in relatively low concentration in aqueous solution.

Miscible. (12.2) Soluble in all proportions.

Mixture. (1.4) A material containing two or more substances.

Moderator. (22.4) A substance used to slow down neutrons in a nuclear reactor.

Molality. (12.1) The concentration measure defined by moles of solute divided by mass of solvent in kilograms.

Molar heat capacity (*C*). (6.2) The amount of energy required to raise the temperature of one mole of a substance by 1 °C (1 K).

Molar heat of solution. (12.3) The net energy change that occurs as a substance dissolves.

Molar mass (*MM*). (2.4) The mass in grams of 1 mol of atoms or molecules.

Molarity (*M*). (3.6) The amount of moles of solute dissolved in 1 L of solution.

Mole. (2.4) The amount of a substance that contains the same number of units as the number of atoms in exactly 12 g of carbon-12, 6.022×10^{23}.

Mole fraction. (5.5) The ratio of the number of moles of one component to the total number of moles of all components.

Molecular density. (5.3) The number of molecules per unit volume.

Molecular orbital. (10.4) A three-dimensional wave that encompasses an entire molecule and describes a bound electron.

Molecular orbital diagram. (10.4) An energy level diagram displaying the molecular orbitals of a particular molecule.

Molecular shape. (9.3) The overall shape of a molecule, arising from the relative positions of its atomic nuclei.

Molecular solid. (11.4) A solid containing discrete molecules that do not have chemical bonds between them.

Molecule. (1.2) A group of atoms linked together by chemical bonds.

Monodentate. (20.2) A ligand with one donor atom.

Monolayer. (12.5) A single two-dimensional layer of atoms or molecules.

Monomer. (13.1) A molecule from which a polymer is synthesized.

Monosaccharide. (13.5) A carbohydrate (sugar) molecule containing between three and six carbon atoms.

***n*-type semiconductor.** (10.6) A metalloid that contains a dopant that gives it excess valence electrons.

Nernst equation. (19.5) The equation that relates the potential of an electrochemical cell to standard potentials and concentrations, $E = E^\circ - \dfrac{RT}{nF} \ln Q$.

Net ionic equation. (4.5) A chemical equation showing the actual participants in a reaction of ions in solution.

Network solid. (1.4) A solid made up of atoms held together in a crystalline array by covalent bonds.

Neutral atom. (2.3) An atom containing the same number of electrons as it has protons in its nucleus.

Neutralization reaction. (4.6) Proton transfer between an acid and a base to generate a pair of neutral molecules.

Neutron. (2.2) A fundamental nuclear particle that possesses mass but zero electric charge.

Newton. (5.1) The SI unit for force.

Nickel−cadmium battery. (19.6) A galvanic cell whose working components are $NiO(OH)$ and Cd.

Nitrogen fixation. (14.5) The conversion of molecular nitrogen into nitrogen-containing species such as ammonia or nitrate.

Noble gas. (1.3) Any of the elements of Group 18 of the periodic table.

Noble gas configuration. (8.3) The portion of the distribution of electrons among the orbitals of an atom that matches that of one of the noble gases.

Node. (7.6) A point, line, or surface where the electron density of an orbital is zero.

Nomenclature. (3.2) A systematic procedure for naming chemical compounds.

Nonbonding electron. (9.2) A valence electron that does not participate in bond formation.

Nonbonding molecular orbital (π_{nb}). (10.5) A delocalized orbital whose occupancy by electrons neither stabilizes nor destabilizes the molecule.

Nonmetal. (1.3) An element that lacks the properties of metals, in particular the tendency to form cations.

Normal boiling point. (11.1) The boiling point of a substance under one atmosphere pressure.

Nuclear binding energy. (22.1) The amount of energy per nucleon that binds an atomic nucleus together.

Nuclear decay. (22.2) The spontaneous decomposition of an unstable nucleus.

Nucleic acid. (13.6) A biochemical macromolecule containing nucleotide units.

Nucleon. (22.1) One of the protons or neutrons in a nucleus.

Nucleotide. (13.6) The repeating unit in DNA and RNA, containing a base, a five-carbon sugar, and a phosphate group.

Nucleus. (2.2) The central core of an atom, where nearly all the mass is concentrated.

Nuclide. (22.1) One particular nucleus, characterized by its charge number (Z) and mass number (A).

Octahedral. (9.4) Having the shape of a regular octahedron, with eight triangular faces and six vertices.

Octet. (9.2) A set of eight valence electrons associated with one atom.

Orbital. (7.5) A three-dimensional wave describing a bound electron.

Orbital density picture. (7.6) A two-dimensional dot drawing representing the distribution of electron density in an orbital.

Orbital mixing. (10.4) Interactions among s and p atomic orbitals that change the relative energies of the molecular orbitals resulting from these atomic orbitals.

Orbital overlap. (10.1) The extent to which two orbitals on different atoms interact.

Order of reaction. (15.3) The exponent to which a concentration is raised in a rate law.

Organic chemistry. (3.2) The chemistry of carbon and its compounds.

Osmosis. (12.4) The net movement of solvent molecules through a semipermeable membrane.

Osmotic pressure (Π). (12.4) The pressure difference that must be applied to a solution to prevent osmosis from pure solvent, $\Pi = MRT$.

Outer atom. (9.2) Any atom in a molecule that bonds to only one other atom.

Overall order. (15.3) The sum of the reaction orders of all species in a rate law.

Oxidation. (4.7) Loss of electrons by a chemical species.

Oxidation number. (19.1) The charge that an atom would have if each of its bonding electrons were assigned to the more electronegative atom involved in the bond.

Oxidation-reduction reactions. (4.7) A class of chemical reactions that involves transfers of electrons between chemical species.

Oxidizing agent. (19.1) A chemical species that can gain electrons from another substance.

Oxoacid. (4.6) An acid containing an inner atom bonded to OH groups and O atoms.

Oxoanion. (3.3) An anion of general formula XO_m, containing a central atom bonded to two or more oxygen atoms.

Ozone layer. (7.7) The region of the Earth's atmosphere that contains ozone (O_3) and absorbs potentially lethal ultraviolet radiation.

p-type semiconductor. (10.6) A metalloid containing a dopant that gives it a deficiency of valence electrons.

Pairing energy. (20.3) The electron-electron repulsion energy arising from placement of two electrons in the same orbital.

Paramagnetic. (8.3) Attracted by a magnetic field, as a consequence of having unpaired electron spins.

Partial pressure. (5.5) The pressure exerted by one component of a gaseous mixture.

Parts per billion. (5.5) How many of one particular component are present in one billion objects.

Parts per million. (5.5) How many of one particular component are present in one million objects.

Pascal (Pa). (5.1) The SI unit of pressure, one newton per square meter.

Passive electrode. (19.3) An electrode that serves only to conduct electrons between a wire and a solution; its chemical constituents do not take part in the redox reaction that occurs at its surface.

Path function. (6.2) A quantity whose change depends on the path along which a change takes place.

Pauli exclusion principle. (8.2) The requirement that no two electrons in a chemical species can be described by the same wave function.

Peptide. (13.7) A small polymer of amino acids.

Percent yield. (4.3) The actual yield of a reaction divided by its theoretical yield and multiplied by 100%.

Periodic table. (1.3) The table of the chemical elements arranged in rows of increasing atomic number and columns of similar chemical behavior.

pH. (17.2) The negative logarithm of the hydronium ion concentration.

pH meter. (19.5) A device for measuring solution pH.

Phase. (1.4) Any of the three states of matter: gas, liquid, or solid.

Phase change. (1.4) Transformation from one state of matter to another.

Phase diagram. (11.6) A pressure-temperature graph showing the conditions under which a substance exists as solid, liquid, and gas.

Phospholipid. (12.5) A biochemical surfactant molecule that is one component of cell membranes.

Photochemical smog. (5.7) The mixture of pollutants that results from the action of sunlight on the emissions from internal combustion engines.

Photoelectric effect. (7.2) The ejection of electrons from a metal surface by light.

Photoelectron spectroscopy. (8.1) The technique that analyzes the kinetic energies of electrons that are ejected from gaseous atoms or molecules by photons of a specific frequency.

Photon. (7.2) A particle of light, characterized by energy $E = h\nu$.

Pi (π) bond. (10.3) A chemical bond formed by side-by-side orbital overlap so that electron density is concentrated above and below the bond axis.

Physical transformation. (1.4) A change in properties that is not accompanied by chemical rearrangements.

Physical property. (1.5) A property that can be observed without causing any chemical rearrangements.

Planck's constant (h). (7.5) The physical constant, 6.63×10^{-34} J s, that relates the energy of a photon to its frequency.

Plastic. (13.4) A polymer that exists as blocks or sheets.

Plasticizer. (13.4) A substance added to a plastic to make it more flexible.

Pleated sheet. (13.7) The protein secondary structure formed like a corrugated plane.

Polar covalent bond. (9.2) A bond that possesses an asymmetric distribution of electrons.

Polarizability. (11.2) The ease with which the electron density about an atom or molecule can be distorted.

Polyamide. (13.3) A polymer containing the amide linkage group.

Polyatomic. (3.3) Containing many bonded atoms.

Polyatomic ion. (3.3) A charged chemical species containing more than one atom.

Polyester. (13.3) A polymer containing the ester linkage group.

Polyethylene. (13.2) The macromolecule formed by the combination of many ethylene molecules.

Polymer. (13.1) A molecule that contains a large number of identical individual units (monomers) linked together.

Polypeptide. (13.7) A protein molecule.

Polyprotic acid. (17.4) A molecule that contains more than one acidic hydrogen atom.

Polysaccharide. (13.5) A macromolecule formed by the condensation of many monosaccharide units.

Positron. (2.3) A subatomic particle with the mass of an electron but a positive unit charge.

Positron emission. (22.2) Nuclear decay by emission of a positron, which decreases the atomic number by one unit without changing the mass number.

Potential energy. (6.1) The energy an object has by virtue of some force (for example, gravitational or electrical) acting on it.

Precipitate. (4.5) An insoluble solid that separates from a solution.

Precipitation. (4.5) Formation of an insoluble solid that separates from a solution.

Precision. (1.5) How reproducible an experimental quantity is.

Pressure. (5.1) Force per unit area.

Primary structure. (13.6) The sequence of monomer units making up DNA or a protein.

Principal quantum number (n). (7.5) The quantum number, restricted to positive integers, that indexes the energy and size of an atomic orbital.

Product. (4.1) A substance that is formed in a chemical reaction.

Propagation. (13.2) The continuation of a polymerization chain reaction.

Protein. (13.7) A biochemical polymer composed of amino acids.

Proton. (2.2) The subatomic constituent of nuclei that possesses unit positive charge, 1.602×10^{-19} C, and a mass of 1.673×10^{-27} kg.

Proton transfer. (4.6) The transfer of H^+ from one chemical species to another.

Pure substance. (1.4) A material that contains only one chemical compound.

Quantized. (7.3) Having discrete allowed values.

Quantum number. (7.5) An integer or half-integer describing the allowed values of some quantized property.

Radiant energy. (6.1) The energy possessed by electromagnetic radiation (photons).

Radiation therapy. (2.3) The use of radioactive nuclides to treat cancers.

Radioactivity. (2.3) The spontaneous breakdown of a nucleus by giving off energetic particles.

Raoult's law. (12.4) The equation that relates vapor pressure to the mole fraction of a solution.

Rate. (15.1) Change per unit time.

Rate constant (k). (15.3) The constant of proportionality linking a reaction rate with concentrations of reagents.

Rate law. (15.3) An expression relating the rate of a reaction to the concentrations of reagents.

Rate-determining step. (15.1) The slowest step in a reaction mechanism.

Reactant. (4.1) A species that is consumed in a chemical reaction.

Reaction coordinate. (15.6) The course of a reaction as reactants are converted to products.

Reaction quotient (Q). (14.4) The ratio of concentrations of products to concentrations of reactants, each raised to its stoichiometric coefficient.

Reagent. (4.1) Any reactant or product of a chemical reaction.

Redox reaction. (4.7) A reaction in which electrons are transferred between species.

Reducing agent. (19.1) A substance that loses electrons during a reaction.

Reduction. (4.7) Gain of electrons.

Refining. (20.4) The metallurgical process by which an impure metal is purified.

Rem. (22.6) The unit of measure for the effect on humans of the energetic emissions from radioactive elements.

Resonance. (9.2) The use of two or more equivalent Lewis structures to describe a substance that contains delocalized electrons.

Resonance structure. (9.2) One of two or more Lewis structures that are equivalent to one another.

Reverse osmosis. (12.4) The forced passage of solvent out of a solution through a semi-permeable membrane.

Reversible reaction. (15.5) A molecular process, which can proceed readily in both directions.

Reversibility. (16.1) The ability of any elementary chemical reaction to proceed in either direction.

Ribonucleic acid (RNA). (13.6) A polymer of nucleotide units that transmits genetic information in the cell.

Roasting. (20.4) The metallurgical process by which a metal ore is oxidized by heating in air.

Root-mean-square speed. (5.4) Average speed obtained by taking the square root of the mean value of the squares of the individual speeds.

Rounding. (1.6) The procedure for expressing a numerical result with the correct precision (number of significant figures).

Saturated. (12.2) Containing the maximum possible concentration of a solute.

Schrödinger equation. (7.5) The equation describing the behavior of electrons in atoms and molecules.

Scientific notation. (1.5) Expression of a number as a value between 1 and 10 multiplied by the appropriate power of ten.

Screening. (8.1) The reduction in effective nuclear charge caused by electrons in orbitals.

Second (s). (1.5) The SI base unit of time.

Second law of thermodynamics. (14.2) The assertion that entropy (dispersal) always increases.

Second-generation star. (22.5) A star formed from the collapse of interstellar matter, including (besides hydrogen and electrons) elements with $Z \leq 26$ that are the debris from supernovae explosions of first-generation stars.

Second-order reaction. (15.3) A reaction whose rate law has an overall order equal to two.

Secondary structure. (13.6) The structural arrangement of DNA or a string of amino acids.

Seesaw shape. (9.4) The molecular shape that resembles a seesaw.

Semiconductor. (10.7) A substance that is intermediate in electrical conductivity between metals (good conductors) and nonmetals (poor conductors).

Semipermeable membrane. (12.4) A thin sheet that allows the passage of some types of molecules (typically, solvent) but prevents the passage of others (typically, solutes).

SI (Système International). (1.5) The system of units that has been adopted by scientists for general use.

Sigma (σ) bond. (10.3) A bond formed by end-on overlap of atomic orbitals, giving electron density that is concentrated along the bond axis.

Significant figure. (1.5) A digit in a numerical value that is known with certainty or has an uncertainty of one unit.

Silicate. (9.3) A compound containing one or more metal cations and a network of Si—O bonds.

Simple cubic. (11.5) The crystal form whose unit cell contains one atom at each corner of a cube.

Single bond. (9.2) A chemical bond formed by one pair of electrons shared between two atoms.

Slightly soluble. (18.4) Able to dissolve to a modest extent. Salts whose K_{sp} lies between 10^{-2} and 10^{-5} are classified as slightly soluble.

Soft acid. (21.2) A Lewis acid whose acceptor atom has a high polarizability.

Soft base. (21.2) A Lewis base whose donor electron pairs are loosely bound, resulting in high polarizability.

Solid. (1.4) The state of matter characterized by a defined volume and shape.

Solubility. (12.2) The amount of a solute that will dissolve in a given amount of solution.

Solubility product (K_{sp}). (16.5) The equilibrium constant for the solubility equilibrium of an ionic compound in water.

Soluble. (4.5) Able to dissolve in a liquid. Salts with $K_{sp} > 10^{-2}$ are classified as soluble in water.

Solute. (3.6) A substance that dissolves in a solvent to form a solution.

Solution. (1.4) A homogeneous mixture of two or more substances.

Solvent. (3.6) The component of a solution that defines its phase. Generally, the solvent is the component present in largest amount.

sp hybrid orbitals. (10.2) Two atomic orbitals constructed by the interaction between an s orbital and a p orbital on the same atom.

sp^2 hybrid orbitals. (10.2) Three atomic orbitals constructed by the interactions among an s orbital and two p orbitals on the same atom.

sp^3 hybrid orbitals. (10.2) Four atomic orbitals constructed by the interactions among an s orbital and three p orbitals on the same atom.

sp^3d hybrid orbitals. (10.2) Five atomic orbitals constructed by the interactions among an s orbital, three p orbitals, and one d orbital on the same atom.

sp^3d^2 hybrid orbitals. (10.2) Six atomic orbitals constructed by the interactions among an s orbital, three p orbitals, and two d orbitals on the same atom.

Space-filling model. (3.1) A representation of a molecule that shows the space occupied by its electron cloud.

Spectator ions. (4.5) Ions that are present in a solution but do not participate in a chemical reaction.

Spectrochemical series. (20.3) The listing of ligands in order of increasing energy-level splitting.

Spectrum. (7.2) A graph of the intensity of light as a function of either frequency or wavelength.

Spin. (7.4) The intrinsic angular momentum of electrons and protons that gives them magnetism.

Spin orientation quantum number (m_s). (7.5) The quantum number, restricted to either $+\frac{1}{2}$ or $-\frac{1}{2}$, that indexes the orientation of electron spin.

Spontaneous. (14.1) Able to occur without outside intervention.

Square planar. (9.4) Having the shape of a square.

Square pyramid. (9.4) A pyramid with a square base.

Standard conditions. (6.5) Unit concentrations (1 M for solutes, 1 bar for gases). Unless otherwise specified, standard conditions also means 298 K.

Standard enthalpy of formation (ΔH_f°). (6.5) The enthalpy change accompanying a formation reaction.

Standard hydrogen electrode (SHE). (19.4) The reference standard for standard reduction potentials, with a defined value of exactly 0 V. The electrode is a platinum wire immersed in an acid solution that is 1.00 M in hydronium ion, over which hydrogen gas passes at a pressure of 1.00 bar.

Standard molar free energy of formation (ΔG_f°). (14.4) The change in thermodynamic free energy accompanying the reaction in which one mole of a substance forms from elements in their standard states.

Standard reduction potential (E°). (19.4) The electrode potential for reduction under standard conditions.

Standard solution. (4.6) A solution whose concentration is accurately known.

Standard state. (6.5) The most stable phase of a substance under standard conditions.

Standardization. (4.6) Accurate determination of the concentration of a solution.

Starch. (13.5) The carbohydrate that plants use to store chemical energy.

State function. (6.2) A property that depends only on the present state but not on the previous history of the system.

Steric number. (9.3) The sum of the coordination number and lone pairs for an inner atom.

Stoichiometric coefficient. (4.1) An integer giving the relative number of molecules of a species that react in a chemical reaction.

Stoichiometric point. (4.6) The point in a titration at which the amount of added titrant is exactly enough to react completely with the species being titrated.

Stoichiometric ratio. (4.2) The ratio of coefficients of two reagents in a balanced chemical reaction.

Stoichiometry. (4.2) The amount relationships among chemical substances undergoing reactions.

Stratosphere. (7.7) The region of the Earth's atmosphere between 10 km and 50 km above its surface.

Strong acid. (4.6) An acid that generates virtually stoichiometric amounts of hydronium ions in water.

Strong base. (4.6) A base that generates virtually stoichiometric amounts of hydroxide ions in water.

Structural formula. (3.1) A molecular formula that shows how atoms are bonded together.

Structural isomers. (9.3) Compounds that have identical chemical formulas but different molecular structures.

Sublimation. (11.6) The phase change between solid and vapor.

Supernova. (22.5) An exploding star, which produces unusually bright emission.

Surface tension. (11.3) The resistance of a liquid to an increase in its surface area.

Surfactant. (12.5) A molecule containing hydrophilic and hydrophobic parts that is used to modify the behavior of aqueous solutions.

Surroundings. (6.2) All of the universe outside of a system.

System. (6.2) Any specific, well-defined part of the universe that is of interest.

Temperature. (1.5) The property of an object that measures the amount of random energy of motion of its molecules and determines the direction of spontaneous heat flow.

Termination. (13.2) The completion of a polymerization chain reaction.

Termolecular reaction. (15.1) An elementary reaction in which three molecular species collide and react.

Tertiary structure. (13.7) The overall shape of a protein molecule.

Tetrahedron. (9.3) Pyramid with four identical faces that are equilateral triangles.

Theoretical yield. (4.3) The amount of a product that would be formed if a reaction proceeded to completion without any losses.

Theory. (1.1) A unifying principle that explains a collection of facts.

Thermal energy. (6.1) Energy associated with the random motion of atoms and molecules.

Thermal pollution. (14.5) Heating of the environment as a byproduct of industrial operations.

Thermodynamics. (6.2) The scientific study of the relationships among heat and other forms of energy.

Thermosphere. (7.7) The region of the Earth's atmosphere more than 90 km above its surface.

Thiol. (11.1) An organic compound that contains the —SH group.

Third law of thermodynamics. (14.3) The statement that the entropy of any pure, perfect crystalline substance is zero at 0 K.

Third-generation star. (22.5) A star formed from the collapse of interstellar matter, including (besides hydrogen and electrons) elements with all Z values that are the debris from supernovae explosions of second-generation stars.

Thymine. (13.6) A one-ring organic base found in DNA.

Time. (1.5) The continuum of experience in which events occur.

Titrant. (4.6) The liquid solution added during a titration.

Titration. (4.6) The gradual addition of measured amounts of one solution to another until a chemical reaction between them is complete.

Torr. (5.1) The pressure unit that is 1/760 of a standard atmosphere.

Transition metal. (1.3) Any of the elements in the d block of the periodic table.

Trigonal bipyramid. (9.4) A double pyramid with a triangular base and two apices along a linear axis perpendicular to the base.

Trigonal planar. (9.4) The molecular shape in which a central atom is bonded to three other atoms lying in a plane at 120° angles to one another.

Trigonal pyramid. (9.3) A pyramid with a base that is an equilateral triangle and sides that are isosceles triangles.

Triple bond. (9.2) A bond between two atoms consisting of three pairs of bonding electrons.

Triple point. (11.6) The temperature and pressure at which solid, liquid, and vapor can coexist at equilibrium.

Troposphere. (5.7) The region of the Earth's atmosphere between its surface and an altitude of 10 km.

T-shaped. (9.4) The molecular shape that resembles the letter T.

Uncertainty principle. (7.4) The assertion that position and momentum cannot both be exactly known.

Unimolecular reaction. (15.1) An elementary reaction in which there is only one reactant molecule.

Unit. (1.5) A standard reference value for a quantity.

Unit cell. (11.5) The simplest repeating unit of a regular pattern, such as an atomic or molecular crystal.

Uracil. (13.6) A one-ring organic base found in RNA.

Valence electrons. (8.2) The electrons of an atom that occupy orbitals of highest principal quantum number and incompletely filled orbitals.

Valence shell electron pair repulsion (VSEPR). (9.3) The principle of minimizing electron-electron repulsion by placing electron pairs as far apart as possible.

van der Waals equation. (11.1) An equation that corrects the ideal gas equation for the effects of molecular size and intermolecular forces.

Vapor. (5.7) The gas phase of a substance that exists as a liquid or solid under normal conditions.

Vapor pressure. (5.7) The partial pressure of a vapor at equilibrium with a condensed phase.

Vaporization. (6.2) The conversion of a liquid into a gas.

Vesicle. (12.5) An enclosed bilayer made up of surfactant molecules.

Viscosity. (11.3) The resistance to flow of a fluid.

Visible light. (7.2) Photons in the wavelength range between 400 and 700 nm, to which the human eye is sensitive.

Volt (V). (19.4) The SI unit for electrical potential.

Volume. (1.5) Three-dimensional space occupied by something.

Volumetric flask. (3.6) A vessel calibrated to hold a specified volume of liquid.

Water equilibrium constant (K_w). (17.1) The equilibrium constant for proton transfer between two water molecules, $K_w = [H_3O^+]_{eq} [OH^-]_{eq}$.

Wave. (7.2) A periodic variation that can be described by amplitude, wavelength, and frequency.

Wave properties. (7.4) Properties characteristic of a wave, such as amplitude, frequency, and wavelength.

Wavelength. (7.2) The distance between successive crests in a wave.

Weak acid. (4.6) An acid that undergoes incomplete proton-transfer in water.

Weak base. (4.6) A base that undergoes incomplete proton-transfer in water.

Work (w). (6.2) Energy transfer that is described by the product of a force times a displacement, $w = fd$.

Yield. (4.3) The amount of a product obtained from a chemical reaction.

Zinc–air battery. (19.6) A galvanic cell whose working components are Zn and O_2 from the air.

Index of Equations

Bold entries in parentheses are equation numbers used in the textbook.

A
Arrhenius equation,
$$k = Ae^{-E_a/RT}, \textbf{(15-6)}, 649$$

B
Bond order,
Bond order $= \frac{1}{2}$ (Number of bonding electrons − Number of antibonding electrons), **(10-1)**, 409

Buffer equation,
$$pH = pK_a + \log\left(\frac{[A^-]_{initial}}{[HA]_{initial}}\right), \textbf{(18-1)}, 769$$

C
Concentration, change in,
$$\Delta[B] = \left(\frac{\text{Coeff. B}}{\text{Coeff. A}}\right)\Delta[A], \textbf{(16-6)}, 695$$

Concentration, equilibrium,
$$[A]_{eq} = [A]_i + \Delta[A], \textbf{(16-5)}, 695$$

D
Density,
$$\rho = \frac{m}{V}, \textbf{(1-2)}, 21$$

E
Electrical potential,
$$E = E° - \frac{RT}{nF}\ln Q, \textbf{(19-6)}, 844$$

Electrical potential, standard,
$$E°_{cell} = E°_{cathode} - E°_{anode}, \textbf{(19-1)}, 838$$
$$E° = \frac{RT}{nF}\ln K_{eq}, \textbf{(19-4)}, 842$$
$$E° = \left(\frac{0.0592\ \text{V}}{n}\right)\log K_{eq}, \textbf{(19-5)}, 842$$

Energy,
$$E = mc^2, \textbf{(22-1)}, 945$$

Energy, atomic,
$$\Delta E_{atom} = \pm\, h\nu_{photon}, \textbf{(7-4)}, 268$$

Energy, average kinetic,
$$\bar{E}_{kinetic} = \frac{3RT}{2N_A}, \textbf{(5-3)}, 176$$

Energy, change in,
$$\Delta E = q + w\ \textbf{(6-3)}, 218$$
$$\Delta E_{sys} = -\Delta E_{surr}\ \textbf{(6-4)}, 218$$

Energy-mass equivalence,
$$\Delta E = (\Delta m)c^2, \textbf{(22-2)}, 946$$
$$\Delta E = (\Delta m)(8.988 \times 10^{10}\ \text{kJ/g}), \textbf{(22-3)}, 946$$

Energy, change in molar,
$$\Delta E_{molar} = \frac{\Delta E}{n}, \textbf{(6-7)}, 235$$

Energy, electrical,
$$E_{electrical} = k\frac{q_1 q_2}{r}, \textbf{(6-1)}, 211$$

Energy, kinetic,
$$E_{kinetic} = \frac{1}{2}mu^2, \textbf{(5-2)}, 175$$

Energy, of photon,
$$E_{photon} = h\nu_{photon}, \textbf{(7-2)}, 264$$

Energy, of photon and atom,
$$E_{photon} = |\Delta E_{atom}|, \textbf{(7-6)}, 271$$

Energy, of reaction,
$$\Delta E_{reaction} = \sum BE_{bonds\ broken} - \sum BE_{bonds\ formed}, \textbf{(6-5)}, 227$$

Enthalpy,
$$H = E + PV, \textbf{(6-9)}, 238$$

Enthalpy, change in,
$$\Delta H = q_p, \textbf{(6-10)}, 238$$

Enthalpy, of reaction,
$$\Delta H_{reaction} \cong \Delta E_{reaction} + RT\Delta n_{gases}, \textbf{(6-11)}, 239$$
$$\Delta H_{reaction} = \sum coeff_{products}\, \Delta H°_{f\ products} - \sum coeff_{reactants}\, \Delta H°_{f\ reactants}, \textbf{(6-12)}, 243$$

Entropy,
$$S = k\ln W, \textbf{(14-1)}, 572$$
$$S_{(pure,\ perfect\ crystal;\ T = 0\ K)} = 0, \textbf{(14-4)}, 578$$
$$S_{(p \neq 1\ atm)} = S° - R\ln p\ and$$
$$S_{(c \neq 1\ M)} = S° - R\ln c, \textbf{(14-5)}, 584$$

Entropy, change in,
$$\Delta S = \frac{q_T}{T}, \textbf{(14-2)}, 573$$

Entropy, change in total,
$$\Delta S_{total} = \Delta S_{system} + \Delta S_{surroundings}, \textbf{(14-3)}, 575$$

Entropy, of reaction,
$$\Delta S°_{reaction} = \sum coeff_{products}\, S°_{products} - \sum coeff_{reactants}\, S°_{reactants}, \textbf{(14-6)}, 585$$

P

pH,
$$\text{pH} = -\log [H_3O^+], \textbf{(17-1)}, 726$$
$$\text{pH} + \text{pOH} = 14.00, \textbf{(17-2)}, 728$$

pK, of indicator,
$$\text{pH}_{\text{stoichiometric point}} = \text{p}K_{\text{In}} \pm 1, \textbf{(18-3)}, 789$$

Percent ionization,
$$\% \text{ H}A \text{ ionized} = 100\% \frac{[H_3O^+]_{\text{eq}}}{[HA]_{\text{initial}}}, \textbf{(17-3)}, 730$$

Photoelectric effect,
$$E_{\text{kinetic}} \text{ (electron)} = h\nu - h\nu_0, \textbf{(7-3)}, 265$$

Pressure, osmotic,
$$\Pi = MRT, \textbf{(12-6)}, 504$$

Pressure, partial,
$$p_A = X_A P_{\text{total}}, \textbf{(5-5)}, 188$$

Pressure, vapor,
$$\ln vp = -\frac{\Delta H_{\text{vap}}}{RT} + \frac{\Delta S^{\circ}_{\text{vap}}}{R}, \textbf{(14-14)}, 599$$

R

Rate constant,
$$\ln k = \ln A - \frac{E_a}{RT}, \textbf{(15-7)}, 650$$
$$\ln \left(\frac{k_2}{k_1} \right) = \frac{E_a}{R} \left(\frac{1}{T_1} - \frac{1}{T_2} \right), \textbf{(15-8)}, 651$$

Rate law,
$$\text{Rate} = k[A]^y[B]^z, \textbf{(15-2)}, 624$$

Raoult's law,
$$vp_{\text{solution}} = X_{\text{solvent}} \, vp_{\text{pure solvent}}, \textbf{(12-3)}, 501$$

Reaction quotient,
$$Q = \left[\frac{(c_D)^d (c_E)^e}{(c_A)^a (c_B)^b} \right], \textbf{(14-12)}, 592$$

Reaction rate,
$$\text{Rate} = \left(-\frac{1}{a} \right)\left(\frac{\Delta[A]}{t} \right) = \left(-\frac{1}{b} \right)\left(\frac{\Delta[B]}{t} \right)$$
$$= \left(\frac{1}{d} \right)\left(\frac{\Delta[D]}{\Delta t} \right) = \left(\frac{1}{e} \right)\left(\frac{\Delta[E]}{\Delta t} \right), \textbf{(15-1)}, 621$$

Reaction rate, first-order,
$$\ln \left(\frac{[A]_0}{[A]} \right) = kt, \textbf{(15-3)}, 627$$

Reaction rate, second-order,
$$\frac{1}{[A]} - \frac{1}{[A]_0} = kt, \textbf{(15-5)}, 631$$

S

Speed,
$$\bar{u} = \left(\frac{3RT}{MM} \right)^{1/2}, \textbf{(5-4)}, 183$$

T

Temperature, conversion of,
$$T \text{ (K)} = T \text{ (°C)} + 273.15 \text{ and}$$
$$T \text{ (°C)} = T \text{ (K)} - 273.15, \textbf{(1-1)}, 19$$

Temperature, depression of freezing,
$$\Delta T_f = K_f c_m, \textbf{(12-4)}, 502$$

Temperature, elevation of boiling,
$$\Delta T_b = K_b c_m, \textbf{(12-5)}, 502$$

V

van der Waals equation,
$$\left(P + \frac{n^2 a}{V^2} \right)(V - nb) = nRT, \textbf{(12-1)}, 442$$

W

Wavelength, particle,
$$\lambda_{\text{particle}} = \frac{h}{mu}, \textbf{(7-7)}, 277$$

Work, expansion,
$$w_{\text{sys}} = -P\Delta V_{\text{sys}}, \textbf{(6-8)}, 237$$

Y

Yield, percent,
$$\text{Percent yield} = 100\% \left(\frac{\text{Actual amount}}{\text{Theoretical amount}} \right),$$
$$\textbf{(4-2)}, 123$$

Index

Note: The following codes are used after page numbers in this index: A b denotes a Box; f denotes a figure; t denotes a table.

TABLE OF THE ELEMENTS, THEIR ATOMIC SYMBOLS, AND THEIR MOLAR MASSES*

Element	Symbol	Atomic Number	Molar Mass	Element	Symbol	Atomic Number	Molar Mass
Actinium	Ac	89	(227)	Meitnerium	Mt	109	(268)
Aluminum	Al	13	26.98	Mendelevium	Md	101	(258)
Americium	Am	95	(243)	Mercury	Hg	80	200.6
Antimony	Sb	51	121.8	Molybdenum	Mo	42	95.94
Argon	Ar	18	39.95	Neodymium	Nd	60	144.2
Arsenic	As	33	74.92	Neon	Ne	10	20.18
Astatine	At	85	(210)	Neptunium	Np	93	(237)
Barium	Ba	56	137.3	Nickel	Ni	28	58.69
Berkelium	Bk	97	(247)	Niobium	Nb	41	92.91
Beryllium	Be	4	9.012	Nitrogen	N	7	14.01
Bismuth	Bi	83	209.0	Nobelium	No	102	(259)
Bohrium	Bh	107	(264)	Osmium	Os	76	190.2
Boron	B	5	10.81	Oxygen	O	8	16.00
Bromine	Br	35	79.90	Palladium	Pd	46	106.4
Cadmium	Cd	48	112.4	Phosphorus	P	15	30.97
Calcium	Ca	20	40.08	Platinum	Pt	78	195.1
Californium	Cf	98	(251)	Plutonium	Pu	94	(244)
Carbon	C	6	12.01	Polonium	Po	84	(209)
Cerium	Ce	58	140.1	Potassium	K	19	39.10
Cesium	Cs	55	132.9	Praseodymium	Pr	59	140.9
Chlorine	Cl	17	35.45	Protactinium	Pa	91	231.0
Chromium	Cr	24	52.00	Promethium	Pm	61	(145)
Cobalt	Co	27	58.93	Radium	Ra	88	(226)
Copper	Cu	29	63.55	Radon	Rn	86	(222)
Curium	Cm	96	(247)	Rhenium	Re	75	186.2
Darmstadtium	Ds	110	(271)	Rhodium	Rh	45	102.9
Dubnium	Db	105	(262)	Rubidium	Rb	37	85.47
Dysprosium	Dy	66	162.5	Ruthenium	Ru	44	101.1
Einsteinium	Es	99	(252)	Rutherfordium	Rf	104	(261)
Erbium	Er	68	167.3	Samarium	Sm	62	150.4
Europium	Eu	63	152.0	Scandium	Sc	21	44.96
Fermium	Fm	100	(257)	Seaborgium	Sg	106	(263)
Fluorine	F	9	19.00	Selenium	Se	34	78.96
Francium	Fr	87	(223)	Silicon	Si	14	28.09
Gadolinium	Gd	64	157.3	Silver	Ag	47	107.9
Gallium	Ga	31	69.72	Sodium	Na	11	22.99
Germanium	Ge	32	72.61	Strontium	Sr	38	87.62
Gold	Au	79	197.0	Sulfur	S	16	32.07
Hafnium	Hf	72	178.5	Tantalum	Ta	73	180.9
Hassium	Hs	108	(265)	Technetium	Tc	43	(98)
Helium	He	2	4.003	Tellurium	Te	52	127.6
Holmium	Ho	67	164.9	Terbium	Tb	65	158.9
Hydrogen	H	1	1.008	Thallium	Tl	81	204.4
Indium	In	49	114.8	Thorium	Th	90	232.0
Iodine	I	53	126.9	Thulium	Tm	69	168.9
Iridium	Ir	77	192.2	Tin	Sn	50	118.7
Iron	Fe	26	55.85	Titanium	Ti	22	47.87
Krypton	Kr	36	83.80	Tungsten	W	74	183.8
Lanthanum	La	57	138.9	Uranium	U	92	238.0
Lawrencium	Lr	103	(262)	Vanadium	V	23	50.94
Lead	Pb	82	207.2	Xenon	Xe	54	131.3
Lithium	Li	3	6.941	Ytterbium	Yb	70	173.0
Lutecium	Lu	71	175.0	Yttrium	Y	39	88.91
Magnesium	Mg	12	24.31	Zinc	Zn	30	65.39
Manganese	Mn	25	54.94	Zirconium	Zr	40	91.22

*All molar masses are in g/mol and have been rounded to four significant figures.
Values in parentheses represent the molar mass of the isotope with the longest half-life.

Values are from Atomic Weights of the Elements 1997, IUPAC Commission on Atomic Weights and Isotopic Abundances, http://www.chem.qmw.ac.uk/iupac/AtWt/